THERMAL RADIATIVE PROPERTIES

Metallic Elements and Alloys

THERMOPHYSICAL PROPERTIES OF MATTER

The TPRC Data Series

A Comprehensive Compilation of Data by the
Thermophysical Properties Research Center (TPRC), Purdue University

Y. S. Touloukian, Series Editor
C. Y. Ho, Series Technical Editor

Volume 1. Thermal Conductivity–Metallic Elements and Alloys
Volume 2. Thermal Conductivity–Nonmetallic Solids
Volume 3. Thermal Conductivity–Nonmetallic Liquids and Gases
Volume 4. Specific Heat–Metallic Elements and Alloys
Volume 5. Specific Heat–Nonmetallic Solids
Volume 6. Specific Heat–Nonmetallic Liquids and Gases
Volume 7. Thermal Radiative Properties–Metallic Elements and Alloys
Volume 8. Thermal Radiative Properties–Nonmetallic Solids
Volume 9. Thermal Radiative Properties–Coatings
Volume 10. Thermal Diffusivity
Volume 11. Viscosity
Volume 12. Thermal Expansion–Metallic Elements and Alloys
Volume 13. Thermal Expansion–Nonmetallic Solids

New data on thermophysical properties are being constantly accumulated at TPRC. Contact TPRC and use its interim updating services for the most current information.

THERMOPHYSICAL PROPERTIES OF MATTER
VOLUME 7

THERMAL RADIATIVE PROPERTIES

Metallic Elements and Alloys

Y. S. Touloukian
Director
Thermophysical Properties Research Center
and
Distinguished Atkins Professor of Engineering
School of Mechanical Engineering
Purdue University
and
Visiting Professor of Mechanical Engineering
Auburn University

D. P. DeWitt
Deputy Director and
Associate Senior Researcher
Thermophysical Properties Research Center
and
Associate Professor of Mechanical Engineering
Purdue University

IFI/PLENUM • NEW YORK-WASHINGTON • 1970

Library of Congress Catalog Card Number 73-129616

SBN (13-Volume Set) 306-67020-8

SBN (Volume 7) 306-67027-5

IFI/Plenum Data Corporation is a subsidiary of
Plenum Publishing Corporation
227 West 17th Street, New York, N.Y. 10011

Distributed in Europe by Heyden & Son, Ltd.
Spectrum House, Alderton Crescent
London N.W. 4, England

Printed in the United States of America

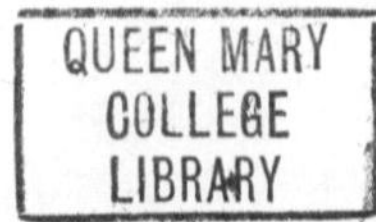

"In this work, when it shall be found that much is omitted, let it not be forgotten that much likewise is performed..."

SAMUEL JOHNSON, A.M.

From last paragraph of Preface to his two-volume *Dictionary of the English Language,* Vol. I, page 5, 1755, London, Printed by Strahan.

Foreword

In 1957, the Thermophysical Properties Research Center (TPRC) of Purdue University, under the leadership of its founder, Professor Y. S. Touloukian, began to develop a coordinated experimental, theoretical, and literature review program covering a set of properties of great importance to science and technology. Over the years, this program has grown steadily, producing bibliographies, data compilations and recommendations, experimental measurements, and other output. The series of volumes for which these remarks constitute a foreword is one of these many important products. These volumes are a monumental accomplishment in themselves, requiring for their production the combined knowledge and skills of dozens of dedicated specialists. The Thermophysical Properties Research Center deserves the gratitude of every scientist and engineer who uses these compiled data.

The individual nontechnical citizen of the United States has a stake in this work also, for much of the science and technology that contributes to his well-being relies on the use of these data. Indeed, recognition of this importance is indicated by a mere reading of the list of the financial sponsors of the Thermophysical Properties Research Center; leaders of the technical industry of the United States and agencies of the Federal Government are well represented.

Experimental measurements made in a laboratory have many potential applications. They might be used, for example, to check a theory, or to help design a chemical manufacturing plant, or to compute the characteristics of a heat exchanger in a nuclear power plant. The progress of science and technology demands that results be published in the open literature so that others may use them. Fortunately for progress, the useful data in any single field are not scattered throughout the tens of thousands of technical journals published throughout the world. In most fields, fifty percent of the useful work appears in no more than thirty or forty journals. However, in the case of TPRC, its field is so broad that about 100 journals are required to yield fifty percent. But that other fifty percent! It is scattered through more than 3500 journals and other documents, often items not readily identifiable or obtainable. Nearly 50,000 references are now in the files.

Thus, the man who wants to use existing data, rather than make new measurements himself, faces a long and costly task if he wants to assure himself that he has found all the relevant results. More often than not, a search for data stops after one or two results are found—or after the searcher decides he has spent enough time looking. Now with the appearance of these volumes, the scientist or engineer who needs these kinds of data can consider himself very fortunate. He has a single source to turn to; thousands of hours of search time will be saved, innumerable repetitions of measurements will be avoided, and several billions of dollars of investment in research work will have been preserved.

However, the task is not ended with the generation of these volumes. A critical evaluation of much of the data is still needed. Why are discrepant results obtained by different experimentalists? What undetected sources of systematic error may affect some or even all measurements? What value can be derived as a "recommended" figure from the various conflicting values that may be reported? These questions are difficult to answer, requiring the most sophisticated judgment of a specialist in the field. While a number of the volumes in this series do contain critically evaluated and recommended data, these are still in the minority. The data are now being more intensively evaluated by the staff of TPRC as an integral part of the effort of the National Standard Reference Data System (NSRDS). The task of the National Standard Reference Data System is to organize and operate a comprehensive program to prepare compilations of critically evaluated data on the properties of substances. The NSRDS is administered by the National Bureau of Standards under a directive from the Federal Council for Science and Technology, augmented by special legislation of the Congress of the United States. TPRC is one of the national resources participating in the National Standard Reference Data System in a united effort to satisfy the needs of the technical community for readily accessible, critically evaluated data.

As a representative of the NBS Office of Standard Reference Data, I want to congratulate Professor Touloukian and his colleagues on the accomplishments represented by this Series of reference data books. Scientists and engineers the world over are indebted to them. The task ahead is still an awesome one and I urge the nation's private industries and all concerned Federal agencies to participate in fulfilling this national need of assuring the availability of standard numerical reference data for science and technology.

EDWARD L. BRADY
Associate Director for Information Programs
National Bureau of Standards

Preface

Thermophysical Properties of Matter, the TPRC Data Series, is the culmination of twelve years of pioneering effort in the generation of tables of numerical data for science and technology. It constitutes the restructuring, accompanied by extensive revision and expansion of coverage, of the original *TPRC Data Book*, first released in 1960 in loose-leaf format, 11″ × 17″ in size, and issued in June and December annually in the form of supplements. The original loose-leaf *Data Book* was organized in three volumes: (1) metallic elements and alloys, (2) nonmetallic elements, compounds, and mixtures which are solid at N.T.P., and (3) nonmetallic elements, compounds, and mixtures which are liquid or gaseous at N.T.P. Within each volume, each property constituted a chapter.

Because of the vast proportions the *Data Book* began to assume over the years of its growth and the greatly increased effort necessary in its maintenance by the user, it was decided in 1967 to change from the loose-leaf format to a conventional publication. Thus, the December 1966 supplement of the original *Data Book* was the last supplement disseminated by TPRC.

While the manifold physical, logistic, and economic advantages of the bound volume over the loose-leaf oversize format are obvious and welcome to all who have used the unwieldy original volumes, the assumption that this work will no longer be kept on a current basis because of its bound format would not be correct. Fully recognizing the need of many important research and development programs which require the latest available information, TPRC has instituted a *Data Update Plan* enabling the subscriber to inquire, by telephone if necessary, for specific information and receive, in many instances, same-day response on any new data processed or revision of published data since the latest edition. In this context, the TPRC Data Series departs drastically from the conventional handbook and giant multivolume classical works, which are no longer adequate media for the dissemination of numerical data of science and technology without a continuing activity on contemporary coverage. The loose-leaf arrangements of many works fully recognize this fact and attempt to develop a combination of bound volumes and loose-leaf supplement arrangements as the work becomes increasingly large. TPRC's *Data Update Plan* is indeed unique in this sense since it maintains the contents of the TPRC Data Series current and live on a day-to-day basis between editions. In this spirit, I strongly urge all purchasers of these volumes to complete in detail and return the *Volume Registration Certificate* which accompanies each volume in order to assure themselves of the continuous receipt of annual listing of corrigenda during the life of the edition.

The TPRC Data Series consists initially of 13 independent volumes. The initial ten volumes will be published in 1970, and the remaining three by 1972. It is also contemplated that subsequent to the first edition, each volume will be revised, updated, and reissued in a new edition approximately every fifth year. The organization of the TPRC Data Series makes each volume a self-contained entity available individually without the need to purchase the entire Series.

The coverage of the specific thermophysical properties represented by this Series constitutes the most comprehensive and authoritative collection of numerical data of its kind for science and technology.

Whenever possible, a uniform format has been used in all volumes, except when variations in presentation were necessitated by the nature of the property or the physical state concerned. In spite of the wealth of data reported in these volumes, it should be recognized that all volumes are not of the same degree of completeness. However, as additional data are processed at TPRC on a continuing basis, subsequent editions will become increasingly more complete and up to date. Each volume in the Series basically comprises three sections, consisting of a text, the body of numerical data with source references, and a material index.

The aim of the textual material is to provide a complementary or supporting role to the body of numerical data rather than to present a treatise on the subject of the property. The user will find a basic theoretical treatment, a comprehensive presentation of selected works which constitute reviews, or compendia of empirical relations useful in estimation of the property when there exists a paucity of data or when data are completely lacking. Established major experimental techniques are also briefly reviewed.

The body of data is the core of each volume and is presented in both graphical and tabular format for convenience of the user. Every single point of numerical data is fully referenced as to its original source and no secondary sources of information are used in data extraction. In general, it has not been possible to critically scrutinize all the original data presented in these volumes, except to eliminate perpetuation of gross errors. However, in a significant number of cases, such as for the properties of liquids and gases and the thermal conductivity of all the elements, the task of full evaluation, synthesis, and correlation has been completed. It is hoped that in subsequent editions of this continuing work, not only new information will be reported but the critical evaluation will be extended to increasingly broader classes of materials and properties.

The third and final major section of each volume is the material index. This is the key to the volume, enabling the user to exercise full freedom of access to its contents by any choice of substance name or detailed alloy and mixture composition, trade name, synonym, etc. Of particular interest here is the fact that in the case of those properties which are reported in separate companion volumes, the material index in each of the volumes also reports the contents of the other companion volumes.* The sets of companion volumes are as follows:

Thermal conductivity:	Volumes 1, 2, 3
Specific heat:	Volumes 4, 5, 6
Radiative properties:	Volumes 7, 8, 9
Thermal expansion:	Volumes 12, 13

The ultimate aims and functions of TPRC's Data Tables Division are to extract, evaluate, reconcile, correlate, and synthesize all available data for the thermophysical properties of materials with the result of obtaining internally consistent sets of property values, termed the "recommended reference values." In such work, gaps in the data often occur, for ranges of temperature, composition, etc. Whenever feasible, various techniques are used to fill in such missing information, ranging from empirical procedures to detailed theoretical calculations. Such studies are resulting in valuable new estimation methods being developed which have made it possible to estimate values for substances and/or physical conditions presently unmeasured or not amenable to laboratory investigation. Depending on the available information for a particular property and substance, the end product may vary from simple tabulations of isolated values to detailed tabulations with generating equations, plots showing the concordance of the different values, and, in some cases, over a range of parameters presently unexplored in the laboratory.

The TPRC Data Series constitutes a permanent and valuable contribution to science and technology. These constantly growing volumes are invaluable sources of data to engineers and scientists, sources in which a wealth of information heretofore unknown or not readily available has been made accessible. We look forward to continued improvement of both format and contents so that TPRC may serve the scientific and technological community with ever-increasing excellence in the years to come. In this connection, the staff of TPRC is most anxious to receive comments, suggestions, and criticisms from all users of these volumes. An increasing number of colleagues are making available at the earliest possible moment reprints of their papers and reports as well as pertinent information on the more obscure publications. I wish to renew my earnest request that this procedure become a universal practice since it will prove to be most helpful in making TPRC's continuing effort more complete and up to date.

It is indeed a pleasure to acknowledge with gratitude the multisource financial assistance received from over fifty of TPRC's sponsors which has made the continued generation of these tables possible. In particular, I wish to single out the sustained major support being received from the Air Force Materials Laboratory–Air Force Systems Command, the Office of Standard Reference Data–National Bureau of Standards, and the Office of Advanced Research and Technology–National Aeronautics and Space Administration. TPRC is indeed proud to have been designated as a National Information Analysis Center for the Department of Defense as well as a component of the National

*For the first edition of the Series, this arrangement was not feasible for Volume 7 due to the sequence and the schedule of its publication. This situation will be resolved in subsequent editions.

Standard Reference Data System under the cognizance of the National Bureau of Standards.

While the preparation and continued maintenance of this work is the responsibility of TPRC's Data Tables Division, it would not have been possible without the direct input of TPRC's Scientific Documentation Division and, to a lesser degree, the Theoretical and Experimental Research Divisions. The authors of the various volumes are the senior staff members in responsible charge of the work. It should be clearly understood, however, that many have contributed over the years and their contributions are specifically acknowledged in each volume. I wish to take this opportunity to personally thank those members of the staff, research assistants, graduate research assistants, and supporting graphics and technical typing personnel without whose diligent and painstaking efforts this work could not have materialized.

Y. S. Touloukian

Director
Thermophysical Properties Research Center
Distinguished Atkins Professor of Engineering

Purdue University
Lafayette, Indiana
July 1969

Introduction to Volume 7

This volume of *Thermophysical Properties of Matter*, the TPRC Data Series, was initiated in recent years and follows the general format of the Center's earlier works on other properties except where departures and innovations were found necessary for the effective presentation of thermal radiative properties.

The volume consists of three major sections: the front text material together with its bibliography, the main body of numerical data and its references, and the material index.

The text material is intended to assume a role complementary to the main body of numerical data, the presentation of which is the primary purpose of this volume. It is felt that a moderately detailed discussion of the theoretical nature of the properties under consideration together with a review of predictive procedures and recognized experimental techniques will be appropriate in a major reference work of this kind. The extensive reference citations given in the text should lead the interested reader to sufficient literature for a detailed study. It is hoped, however, that enough detail is presented for this volume to be self-contained for the practical user.

The main body of the volume consists of the presentation of numerical data compiled over the years in a most comprehensive and meticulous manner. The scope of coverage includes the metallic elements and metallic alloys of engineering importance. The extraction of all data directly from their original sources ensures freedom from errors of transcription. Furthermore, some gross errors appearing in the original source documents have been corrected. The organization and presentation of the data together with other pertinent information in the use of the tables and figures are discussed in detail in the text of the section entitled *Numerical Data*.

In addition to the original data presentation in the *Numerical Data* section, the Analyzed Data Graphs will give the user an evaluative review of the data. This analysis work is first a filtering process; it identifies data which are felt to be reliably or typically identified with the materials and gives the user a good feeling of "relief" from the "spaghetti" type of presentation shown in the original or archival data graphs. The analyzed curves are based, in some instances, on experiences of the research team as well as the original data sources. This analysis work is an innovative feature of the radiative properties volumes and should not be considered as recommended reference values identified in other volumes of this Series; the work is intended to make the best reliable data available to the thermal designer.

As stated earlier, all data have been obtained from their original sources and each data set is so referenced. TPRC has in its files all documents cited in this volume. Those that cannot readily be obtained elsewhere are available from TPRC in microfiche form.

This volume has grown out of activities supported principally by the National Aeronautics and Space Administration, Office of Advanced Research, under the monitorship of Mr. Conrad Mook. We wish to acknowledge the benefit of extensive discussions with Mr. D. W. Gates, Space Sciences Laboratory, NASA–MSC, and Mr. W. F. Carroll, Materials Section, Jet Propulsion Laboratory, who have followed the progress of work from the outset.

Over the past four years, many graduate students, research assistants, and technical staff have contributed to the preparation of this volume for varying periods under the authors' supervision. In chronological order of their association with TPRC, we wish to acknowledge the contributions of I. M. Yeyinmen, R. S. Hernicz, B. Compani-Tabrizi, M. C. Muinzer, P. Sioshansi, J. J. Hsia, C. K. Hsieh, and R. L. Jones. Special mention should be made of two of these contributors still at TPRC: Mr. Muinzer was responsible for the final organization of the tables, figures, and text, and for the demanding task of checking details. Mr. Hernicz participated in nearly all phases of the program since its outset and made valuable contributions to the data analysis activity in particular.

The authors acknowledge the assistance of Mr. Joseph C. Richmond, National Bureau of Standards, who provided valuable suggestions relating to nomenclature and subproperty classification; many of the ideas in the text portion of this volume are a result of his direct effort.

Inherent in the character of this work is the fact that in the preparation of this volume we have drawn most heavily upon the scientific literature and feel a debt of gratitude to the authors of the referenced articles. While their often discordant results have caused us much difficulty in reconciling their findings, we consider this to be our challenge and our contribution to negative entropy of information, as an effort is made to create from the randomly distributed data a condensed, more orderly state.

While this volume is primarily intended as a reference work for the designer, researcher, experimentalist, and theoretician, the teacher at the graduate level may also use it as a teaching tool to point out to his students the topography of the state of knowledge on the thermal radiative properties of metals. We believe there is also much food for reflection by the specialist and the academician concerning the meaning of "original" investigation and its "information content."

The authors and their contributing associates are keenly aware of the possibility of many weaknesses in a work of this scope. We hope that we will not be judged too harshly and that we will receive the benefit of suggestions regarding references omitted, additional material groups needing more detailed treatment, improvements in presentation or in recommended values, and, most important, any inadvertent errors. If the *Volume Registration Certificate* accompanying this volume is returned, the reader will assure himself of receiving annually a list of corrigenda as possible errors come to our attention.

Lafayette, Indiana
July 1969

Y. S. Touloukian
D. P. DeWitt

Contents

Numerical Data

Material Index

GROUPING OF MATERIALS AND LIST OF FIGURES AND TABLES

1. ELEMENTS

Note: Figure number with "A" indicates analyzed data graph.
*No figure

1. ELEMENTS (continued)

Note: Figure number with "A" indicates analyzed data graph.
*No figure

1. ELEMENTS (continued)

Note: Figure number with "A" indicates analyzed data graph.
*No figure

1. ELEMENTS (continued)

Note: Figure number with "A" indicates analyzed data graph.
*No figure

1. ELEMENTS (continued)

Note: Figure number with "A" indicates analyzed data graph.
*No figure

1. ELEMENTS (continued)

Note: Figure number with "A" indicates analyzed data graph.
*No figure

1. ELEMENTS (continued)

Note: Figure number with "A" indicates analyzed data graph.
*No figure

1. ELEMENTS (continued)

2. BINARY ALLOYS

Note: Figure number with "A" indicates analyzed data graph.
*No figure

2. BINARY ALLOYS (continued)

3. MULTIPLE ALLOYS

Note: Figure number with "A" indicates analyzed data graph.
*No figure

3. MULTIPLE ALLOYS (continued)

Note: Figure number with "A" indicates analyzed data graph.
*No figure

3. MULTIPLE ALLOYS (continued)

Note: Figure number with "A" indicates analyzed data graph.
*No figure

3. MULTIPLE ALLOYS (continued)

Theory, Estimation, and Measurement

Notation

c	Speed of light in a medium
c_0	Speed of light in a vacuum
c_1	Planck's first radiation constant, $2\pi c_0{}^2 h$
c_2	Planck's second radiation constant, $c_0 h/k$
C_p	Specific heat (constant pressure)
dA	Elemental area on radiating surface
e	Base of natural logarithms
E	Irradiance
E	Electric field strength
g	Solid angle of cone
h	Planck constant
H	Magnetic field strength
I	Radiant intensity
j	Imaginary unit
k	Index of absorption
k	Boltzmann constant
K^*	Complex dielectric constant
K'	Relative permittivity or real dielectric constant, real part of K^*
K''	Relative loss factor, imaginary part of K^*
l	Mean free path
L	Radiance
M	Sample mass
M	Radiant exitance
m	Electronic mass
m	RMS slope (surface roughness parameter)
n	Index of refraction
N	Number density of free electrons
P	Degree of polarization
q	Electronic charge
Q	Radiant energy
r	Electrical resistivity
R	Reflectance factor
s	Solar (wavelength distribution)
t	Time
t	Total (blackbody wavelength distribution)
T	Temperature (absolute)
v	Volume
W	Radiant density
x	Distance
Z^*	Intrinsic impedance

α	Absorptance; Absorptivity
α_s	Solar absorptance
α	Plane half angle of right circular cone
α	Attenuation factor, real part of γ
β	Radiance factor
β	Phase factor, imaginary part of γ
γ	Complex propagation factor
ϵ	Emittance; Emissivity
ϵ_0	Permittivity of free space
ϵ^*	Complex permittivity
ϵ'	Real permittivity, real part of ϵ^*
ϵ''	Loss factor, imaginary part of ϵ^*
θ	Angle between surface normal and direction of incident flux, zenith angle, or co-latitude
θ'	Angle between surface normal and direction of reflected or emitted flux
λ	Wavelength
$\bar{\lambda}$	Integrated (wavelength distribution)
μ_0	Permeability of free space
μ^*	Complex permeability
ν	Frequency
π	Constant, 3.14 . . .
ρ	Reflectance; Reflectivity
σ	Electrical conductivity (Roberts' model only)
σ	Stefan–Boltzmann constant
σ	RMS roughness (surface roughness parameter)
τ	Transmittance
ϕ	Azimuthal angle of incident flux
ϕ'	Azimuthal angle of reflected or emitted flux
Φ	Radiant flux
ω	Solid angle of incident beam
ω'	Solid angle of reflected or emitted beam
ω	$2\pi \times$ frequency

Subscripts

s	Perpendicular polarized component
p	Parallel polarized component
b	Blackbody conditions
λ	Spectral concentration
t	Total (blackbody wavelength distribution)
s	Solar (spectral distribution)

Thermal Radiative Properties of Metals

1. INTRODUCTION

Radiation is one of the three fundamental modes of heat transfer, the others being conduction and convection. Radiation differs from the other modes in two important respects; first, no medium is required for transport of energy by radiation, and second, the rate of heat dissipation by radiation varies approximately as the fourth power of the absolute temperature, while that by the other modes varies approximately as the first power of temperature. For these reasons radiation becomes the dominant mode of heat transfer at high temperatures and in the absence of an atmosphere.

There has been a marked increase of interest in radiant heat transfer and thermal radiative properties since about 1957, which is reflected in the scientific literature. In large measure this is due to the advent of the space age, because in the vacuum of space, radiation is the only mode of heat transfer to or from a satellite or space vehicle that does not also involve mass transfer.

The thermal radiative properties of the opaque metallic materials are strongly influenced by surface effects arising from methods of preparation, surface finish, thermal history, and environmental interaction. Oxide films in particular may significantly affect the thermal radiative properties and may change in thickness and in character as a result of heating unless very careful precautions are taken to prevent formation of such films, both prior to and during testing.

The difficulties of characterizing data—unambiguously relating the measured property data to the conditions of the specimen—and of understanding environmental influences have frequently required the designer to measure the desired property of the actual surface as it will be used in the environment of the application. Much of such effort can be reduced through proper use of the extensive compendia presented in this volume; also through the availability of such a bulk of data, it is likely that attention to the characterization of materials (surfaces) will increase.

The thermal radiative properties are descriptive of a radiant energy–matter interaction; this interaction can be phenomenologically described by other properties as well, such as the optical constants, complex dielectric constant, or propagation factor, each of which is especially convenient for studying various aspects of the interaction. If the designer is to make effective use of radiative properties data, it is helpful that he be aware of the basic mechanisms of the radiant energy–matter interaction which models the radiative transport process.

The purpose of the text material following is to expose the user to some of the pitfalls and limitations in the use of existing data. The exposition begins with Section 2 briefly defining the terms—processes, things, quantities, properties, and modifiers—used in discussing thermal radiation phenomena. Following this, Section 3 presents a more rigorous definition of the properties and the notion of their dependency upon wavelength, temperature, polarization, and geometric directions. Section 4 reviews the physics of thermal radiation; then Section 5 discusses the interaction of radiant energy with materials. The interrelations between the properties—as described by geometric and wavelength modifiers—need to be understood in order for the user to synthesize fragments of data as available; Section 7 treats the more important concepts.

Theoretical models, Section 8, are presented which can be used in some instances to predict data values or to check the consistency of measurements. Such tools are useful to only ideally smooth materials; engineering materials as a rule do not fall in this category and detail regarding the surface must be understood if the theoretical tools are to be of any value. Section 9 discusses surface characterization and, where possible, quantitative information concerning influences is shown. This section serves to focus attention on the pitfalls in use of data and to justify the detail that must be presented to characterize radiative properties data. Section 10 deals with measurement techniques and apparatus, to inform the reader regarding the more conventional approaches used by the experimentalist. It is not the aim here to evaluate and recommend techniques, but rather to briefly review the limitations and capability of the various techniques.

The depth of presentation has been aimed at the reader with a background in the physics of thermal radiation; special attention has been given to providing general references, usually standard texts or extensive review articles, for the reader's benefit. The development of thermal radiation studies over the past ten years can be followed through the conference proceedings starting as an informal one in 1958 to the annual society meetings in most recent years [3, 17, 21, 29, 71, 72, 82, 114]. The data compilations of references [65] and [143] are most useful supplements for property coverage on materials not contained in the data section of this volume; the *TPRC Retrieval Guide* [142] permits rapid access to the literature on the thermal radiative properties—and twelve other thermophysical properties—of all classes of materials. Through these references, the reader will have access to a very vital portion of knowledge created in the field of thermal radiation and a full appreciation of the technical problems involved in application of the basic principles to real situations.

2. DEFINITION OF TERMS [4]

A. Processes

Radiation. The process by which radiant energy is emitted by a body, also the process by which energy is transferred in the form of radiant energy.

Reflection. The process by which radiant energy incident on a surface or medium leaves that surface or medium from the incident side.

Transmission. The process by which radiant energy incident on a surface or medium leaves that surface or medium on a side other than the incident side.

Absorption. The process by which radiant energy is converted into another form of energy.

Refraction. The process by which a beam of radiant energy, on transmission through the interface between two media of different index of refraction, is deviated toward the normal to the interface in the medium of higher index of refraction.

Propagation. The process or processes by which radiant energy is transferred from one point to another in space.

B. Things

Radiator. A source of radiant energy.

Thermal Radiator. A radiator that emits thermal radiant energy, as a consequence of its temperature only.

Blackbody. A body or surface that absorbs all of the radiant energy incident upon it, and emits the maximum possible amount of thermal radiant energy at each frequency for a body at its temperature.

Reflector. A body that reflects incident radiant energy.

Transmitter. A body that transmits incident radiant energy.

Absorber. A body that absorbs incident radiant energy.

Retroreflector. A reflector that reflects incident radiant energy in directions close to the direction of incidence.

C. Quantities

Radiant energy, Q. Energy in the form of electromagnetic waves or photons. Joules, ergs, or kilowatt-hours.

Thermal Radiant Energy, Q. Radiant energy that is emitted by a thermal radiator.

Radiant Density, W. $W = dQ/dv$. Radiant energy per unit volume. Joule per cubic meter, erg per cubic centimeter.

Radiant Flux, Φ. $\Phi = dQ/dt$. Time rate of flow of radiant energy. Erg per second, watt.

Radiant Intensity, I. $I = d\Phi/d\omega$. Flux per unit solid angle from a source. Watt per steradian.

Radiance, L. $L = d^2\Phi/d\omega dA \cos\theta$. Flux propagated in a given direction, per unit solid angle about that direction and per unit area projected normal to the direction.

Exitance, M. $M = d\Phi/dA$. Flux per unit area leaving a surface.

Irradiance, E. $E = d\Phi/dA$. Flux per unit area incident on a surface.

D. Properties*

Reflectance, ρ. The ratio of reflected flux to incident flux.

**Note:* Properties ending in "ance" are properties of real specimens, regardless of thickness or surface condition. Properties ending in "ivity" are instrinsic properties of the material of which the specimen is composed, and can only be approached by values measured on real specimens that have clean optically smooth surfaces and are opaque.

Absorptance, α. The ratio of absorbed flux to incident flux.

Transmittance, τ. The ratio of transmitted flux to incident flux.

Emittance, ϵ. The ratio of the radiant exitance of a body at a given temperature to that of a blackbody radiator at the same temperature.

Reflectivity, ρ, ρ_{∞}. The reflectance of a specimen that has an optically smooth surface and is thick enough to be opaque.

Absorptivity, α, α_{∞}. The absorptance of a specimen that has an optically smooth surface and is thick enough to be opaque.

Emissivity, ϵ, ϵ_{∞}. The emittance of a specimen that has an optically smooth surface and is thick enough to be opaque.

Reflectance Factor, R. The ratio of the flux reflected by a specimen under specified conditions of irradiation and viewing to that reflected by the ideal completely reflecting, perfectly diffusing surface, identically irradiated and viewed.

Radiance Factor, β. The ratio of the reflected radiance of a specimen in a given direction under specified conditions of irradiation to that of the ideal completely reflecting, perfectly diffusing surface, identically irradiated.

E. Modifiers

Spectral. For a property, at a given wavelength, designated by (λ) following the symbol for the property. For a quantity, spectral concentration, designated by the subscript λ, such as $\Phi_{\lambda} = d\Phi/d\lambda$.

Total.* Refers to blackbody wavelength distribution. For a quantity, the spectral quantity is integrated and designated by the subscript t. For the property emittance, which is a ratio, the numerator and denominator are integrated separately and designated by the symbol (t) following the property symbol.

Solar. Having the spectral distribution of solar energy, or integrated over the solar spectrum. It is designated by the symbol (s) following the property symbol.

Integrated.† Having a spectral distribution prescribed by integration over some specified portion of the spectrum; designated by the symbol ($\bar{\lambda}$) following the symbol for the property and necessarily some comments regarding the spectrum must be given.

Directional. In a given direction. The direction is completely specified by two angles, θ and ϕ; θ is the angle between the specified direction and the normal to the surface, and ϕ is the azimuth of the specified direction measured from some fiducial mark on the specimen. For quantities, direction is denoted by the subscripts θ, ϕ (ϕ is only required when the surface structure has lay). For properties, the symbols θ, ϕ indicating direction are enclosed in parentheses following the symbol for the property as ($\theta, \phi; \theta', \phi'$); those indicating the incident direction first, followed by a semicolon, then the primed symbols indicating the direction of the reflected or transmitted rays.

Normal. A special case of directional where $\theta = 0°$; in the context of this volume, this modifier includes conditions where $\theta < 15°$, see Section 6.B.a for further detail.

Angular. A more general case of directional where $\theta > 0°$, that is, for cases other than normal; in the context of this volume, this modifier includes $\theta \geq 15°$, see Section 6.B.a for further detail.

Conical.* Over a finite solid angle smaller than a hemisphere. Both the size and direction of the solid angle must be specified. If the angle is a right circular cone, the direction is the axis of the cone and is designated by the symbols θ, ϕ as above, and the size is designated by the plane half angle of the cone, α. If the solid angle is not a right circular cone, it must be described in detail in the text and is designated by the symbol g. As above, primed symbols are used to indicate transmitted or reflected beams.

Hemispherical.* Over a complete hemisphere, designated by the symbol 2π replacing the θ, ϕ, g or θ' ϕ', g' in parentheses following the symbol for the property or quantity.

Specular. In the direction of mirror reflection. Under these conditions, $\theta' = \theta$ and $\phi' = \phi + 180°$.

Diffuse. Applied to a reflector or transmitter; reflecting or transmitting in all directions over a hemisphere. Applied to incident radiant energy, incident from angles over a hemisphere.

*Frequently the modifier *total* is used to include any wavelength distribution including blackbody; in this volume it refers only to blackbody conditions and as such is applicable only to the property emittance.

†This modifier is not in widespread usage but is most convenient for sub-property classification purposes in this volume.

**Note:* Unless otherwise indicated, it is assumed that the incident radiance is uniform over the specified solid angle for conical or hemispherical irradiation. No such assumption is made for emitted, reflected or transmitted radiance.

Perfectly Diffuse. With equal radiance in all directions from a surface.

3. THERMAL RADIATIVE PROPERTIES*

A. Interrelationships of Properties

All matter is continually emitting radiant energy as a result of the themral vibration of the particles (electrons, ions, atoms, and molecules) of which it is composed. This process is called thermal radiation, and the radiant energy so emitted is called thermal radiant energy.

Each solid body is not only continually emitting thermal radiant energy, but it is also continually being bombarded by radiant energy from its surroundings, some of which is absorbed. The net rate of heat transfer by radiation to or from the body is equal to the difference in the rates of emission and absorption. Hence, the properties of the body that influence these rates are called thermal radiative properties.

When a body is irradiated, part of the incident radiant energy is reflected, part is absorbed, and the rest is transmitted. Nothing else can happen to it. The incident flux Φ_i is equal to the sum of the reflected flux Φ_r, the absorbed flux Φ_a, and the transmitted flux Φ_t:

$$\Phi_i = \Phi_r + \Phi_a + \Phi_t \qquad (1)$$

The reflectance ρ is the ratio of reflected flux to incident flux; the absorptance α is the ratio of absorbed flux to incident flux; and the transmittance τ is the ratio of transmitted flux to incident flux. Dividing both sides of equation (1) by Φ_i gives

$$1 = \rho + \alpha + \tau \qquad (2)$$

For opaque materials, $\tau = 0$, hence for such materials

$$\rho + \alpha = 1 \qquad (\tau = 0) \qquad (3)$$

Kirchhoff's law states that the absorptance is equal to the emittance

$$\alpha = \epsilon \qquad (4)$$

Thus, for an opaque material

$$\rho + \epsilon = 1 \qquad (5)$$

and the thermal radiative properties of an opaque body are fully described by either the reflectance or the emittance. However, there are certain restrictions that apply to equations (2) through (5) which will be discussed later.

*References for general background review are [67, 127, 133, 134, 137, 140, 145].

B. Blackbody Radiation

A blackbody radiator absorbs all radiant energy incident upon it and emits the maximum possible amount of flux per unit area at any given wavelength or wavelength interval for any body at its temperature.

The only true blackbody radiator that exists is a completely enclosed cavity with opaque walls at a uniform temperature. All real materials reflect part of the radiant energy incident upon them and emit less radiant energy than a blackbody radiator at the same temperature. Nevertheless, the concept of a blackbody radiator is indispensable to a discussion of thermal radiation processes. The radiant exitance M (the flux emitted per unit area) of a blackbody radiator is given by the Stefan–Boltzmann equation

$$M = \sigma T^4 \qquad (6)$$

in units of watts per square meter, where σ is the Stefan–Boltzmann constant, $5.6697 \times 10^{-8}\ \mathrm{W \cdot m^{-2} \cdot K^{-4}}$ and T is temperature in kelvins. The spectral, or wavelength, distribution of this flux is given by the Planck equation

$$M_\lambda = c_1 \lambda^{-5} [e^{c_2/\lambda T} - 1]^{-1} \qquad (7)$$

in which M_λ is the spectral exitance in watts per square meter and meter wavelength interval, c_1 is the first radiation constant, $3.7415 \times 10^{-16}\ \mathrm{W \cdot m^2}$, c_2 is the second radiation constant, $1.43879 \times 10^{-2}\ \mathrm{m \cdot K}$, and e is the base of natural logarithms. The geometric distribution of this radiant exitance is given by Lambert's cosine law

$$I_\theta = I_0 \cos \theta \qquad (8)$$

in which I_θ is the directional intensity of a plane source in the direction θ (measured from the normal to the surface) and I_0 is the intensity of the source in a direction normal to its surface.

While the radiation laws expressed in equations (6), (7), and (8) apply only to blackbody radiators,* they can be applied to real surfaces by using the emittance as a proportionality factor. For instance,

*There is a further restriction. These equations apply rigorously only to the case where the blackbody is emitting into a vacuum. Whem emitting into a medium of index of refraction greater than unity, the emitted flux is increased by n^2, where n is the index of refraction of the medium. The increase in emitted energy when radiating into air ($n = 1.0003$ approx.) is too small to be detected in ordinary measurements.

the exitance M of a real specimen is given by

$$M = \epsilon(2\pi; t)\sigma T^4 \tag{9}$$

where $\epsilon(2\pi; t)$ is the hemispherical total emittance of the specimen at temperature T, and the spectral exitance M_λ is given by

$$M_\lambda = \epsilon(2\pi; \lambda)c_1\lambda^{-5}[e^{c_2/\lambda T} - 1]^{-1} \tag{10}$$

where $\epsilon(2\pi; \lambda)$ is the hemispherical spectral emittance of the specimen at wavelength λ and temperature T. The directional radiance L_θ, where θ is the angle from the given direction to the normal to the surface, is given by

$$L_\theta = \epsilon(\theta; t)\sigma\pi^{-1}T^4 \tag{11}$$

where $\epsilon(\theta; t)$ is the total directional emittance in direction θ at temperature T. The directional spectral radiance, $L_{\theta,\lambda}$, of a specimen is given by

$$L_{\theta,\lambda} = \epsilon(\theta; \lambda)c_1\pi^{-1}\lambda^{-5}[e^{c_2/\lambda T} - 1]^{-1} \tag{12}$$

where $\epsilon(\theta, \lambda)$ is the directional spectral emittance of the specimen at wavelength λ and temperature T in the direction θ.

Equations (9), (10), (11), and (12) suggest that the emittance of a specimen may change with wavelength, angle of incidence, and temperature. This is indeed the case. All thermal radiative properties vary with wavelength, direction (measured from the normal to the surface) of the incident or exitent radiant energy, temperature of the specimen, the degree of polarization, and, for polarized incident or exitent flux, with the angle between the plane of polarization and the plane of incidence. The plane of incidence is the plane defined by the direction of the incident ray and the normal to the surface. The modifiers defined in the Definitions of Terms are used to indicate the conditions under which the properties or quantities were evaluated.

The variation of the thermal radiative properties with temperature, wavelength, and geometric conditions (including polarization) of irradiation and viewing pose certain restrictions on equations (2), (3), (4), and (5). For equations (2) and (3) to be valid, α, ρ, and τ must be evaluated under the same conditions, which means that the temperature of the specimen must be the same, and the spectral composition, direction, solid angle, and degree and direction of polarization of the incident radiant energy must be identical, and all of the reflected and transmitted radiant energy must be measured.

C. Kirchhoff's Law

Kirchhoff's law, equation (4), is derived for the condition that the specimen is irradiated in a blackbody cavity with walls at the same temperature as the specimen, which means that the specimen is uniformly irradiated over a hemisphere with unpolarized radiant energy having the spectral distribution of that of a blackbody radiator at the temperature of the specimen. However, it can be proved that equation (4) is also valid for the two conditions: (1) any solid angle less than a hemisphere if the direction and solid angle of the incident beam for the absorption evaluation is identical to the direction and solid angle (but opposite in sense) of the emitted beam for the emittance evaluation, and (2) for plane-polarized radiant energy with the plane of polarization at any given angle to the plane of measurement, provided that it is the same for the incident radiant energy for the absorption evaluation and the emitted radiant energy for the emittance evaluation. Even with these modifications, equation (4) applies strictly only provided the spectral composition of the incident radiant energy for the absorptance is that of blackbody radiant energy at the temperature of the specimen. This would appear to impose a severe restriction on the general applicability of equation (5). However, it can also be shown that equation (4) applies to any small wavelength band, as well as to total blackbody radiant energy. The properties of reflectance, absorptance, and transmittance do not vary with the amount of incident radiant energy until very high flux densities are reached. Within the narrow wavelength band used in measuring spectral thermal radiative properties the spectral distribution of radiant energy from almost any thermal source is approximately the same as that from a blackbody radiator at the temperature of the specimen. Also, polarization effects are completely absent for normally incident radiant energy and are negligible at angles near the normal. Hence equations (4) and (5) can be considered valid for normal spectral properties and can be used to convert normal hemispherical reflectance to normal emittance with but little error.

4. PHYSICS OF THERMAL RADIATION*

A. The Nature of Radiant Energy

Radiant energy can be treated in two ways, as electromagnetic waves, or as photons. In both forms it travels in straight lines at the speed of light, which,

*General references are [22, 147].

in vacuum, is a fundamental constant of nature c_0, with a value of 2.997925×10^{10} cm sec^{-1}.

Photons are particles of zero rest mass, each of which contains or consists of a fixed amount of energy, called a quantum. Quantum mechanical treatment of photons is the most convenient way of studying the generation and interaction of radiant energy with matter on a micro scale, but interactions on a macro scale are handled more readily by wave mechanics, in which radiant energy is considered as waves.

Electromagnetic waves are characterized by frequency or wavelength and amplitude. Frequency ν and wavelength λ are related by

$$c = \nu\lambda \tag{13}$$

The energy content of a wave is the square of the amplitude.

The frequency ν is constant, regardless of the medium through which the wave is propagated, but the speed and wavelength change with the index of refraction n of the medium

$$n = c_0/c_n = \lambda_0/\lambda_n \tag{14}$$

where λ_n and c_n are the wavelength and speed in the medium of index of refraction n, and λ_0 and c_0 are the corresponding values in vacuum.

Electromagnetic waves consist of magnetic (H) and electric (E) field vectors which successively oscillate in directions perpendicular to the direction of travel and to each other; this is a characteristic of transverse waves as described by the Maxwell equations governing electromagnetic wave phenomena. The plane of polarization of the wave is defined as the plane containing the electric field vector. If the waves are oriented with their planes of polarization parallel, the beam is *plane polarized.* If the planes are randomly oriented, the wave is termed *unpolarized.* The terms *circularly* or *elliptically* polarized identify other conditions of amplitude and oscillation direction of the electric field. Most practical radiant energy sources and beams within optical systems—particularly after reflectance from a plane surface—are partially polarized; that is, there is a partial, but not complete, preferred orientation of the planes of polarization of the waves making up the beam.

The plane of polarization of such a partially polarized beam is taken as the plane of polarization for which the intensity of the beam is maximum. The degree of polarization P of such a beam is measured as

$$P = \frac{I_p - I_s}{I_p + I_s} \tag{15}$$

where I_p is the intensity of the beam when polarized parallel to its plane of polarization, and I_s is the intensity when polarized normal to its plane of polarization.

The frequency, and hence the wavelength, of electromagnetic radiation theoretically can vary from zero to infinity. Thermal radiant energy, however, is generally restricted to the wavelength range of 0.1 μm to 1 mm. This overall range has been broken down into seven subranges [4], as follows:

Range	*Designation*	*Wavelength Range,* μm
Ultraviolet	UVC	0.1–0.28
	UVB	0.28–0.315
	UVA	0.315–0.400
Visible	VIS	0.380–0.780
Infrared	IRA	0.78–1.4
	IRB	1.4–3
	IRC	3–1000

B. Basic Laws

The Planck equation, (7), is the basis for all thermal radiation measurements. All of the other important relationships can be derived from this simple equation:

$$M_\lambda = c_1\lambda^{-5}[e^{c_2/\lambda T} - 1]^{-1} \tag{7}$$

This equation was derived by quantum mechanics. The radiation constants c_1 and c_2 are related to more fundamental constants as follows:

$$c_1 = 2\pi c_0{}^2 h$$

$$c_2 = c_0 h/k$$

where c_0 is the speed of light in vacuum, h is the Planck constant, 6.6256×10^{-34} J·sec, and k is the Boltzmann constant, 1.38054×10^{-23} J·K^{-1}.

Since for a blackbody radiator, $L = M/\pi$, equation (7) can be rewritten to give spectral radiance:

$$L_\lambda = c_1\pi^{-1}\lambda^{-5}[e^{c_2/\lambda T} - 1]^{-1} \tag{16}$$

An inspection of equation (16) shows that the right-hand side can be reduced to a single variable, λT, by multiplying both sides of the equation by T^{-5}:

$$L_\lambda T^{-5} = c_1\pi^{-1}(\lambda T)^{-5}[e^{c_2/\lambda T} - 1]^{-1} \tag{17}$$

Equation (17) shows that the shape of the spectral distribution of the radiance of a blackbody radiator

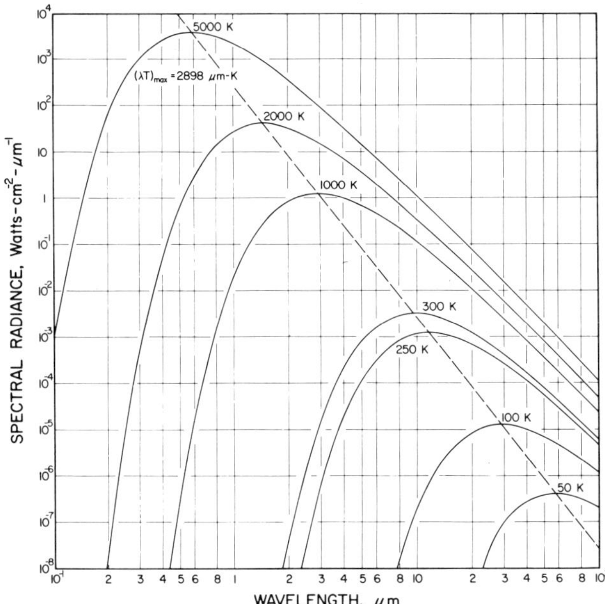

Fig. 1. The Planck distribution law, spectral radiance of blackbody radiation as a function of temperature and wavelength.

is a function of λT and not of λ and T separately. This is one form of Wien's displacement law, which states that the peak of the spectral distribution curve occurs at a wavelength that is inversely proportional to the absolute temperature. This is shown in Fig. 1. Setting $dL/d\lambda$ equal to zero and solving for $(\lambda T)_{\max}$ gives

$$(\lambda T)_{\max} = 2898\ \mu\text{m}\cdot\text{K} \tag{18}$$

which is the Wien displacement equation.

Two other equations were developed prior to the Planck equation and are important not only for their historical significance but also because they are useful approximations in certain wavelength ranges. Both were developed by classical wave mechanics. The first is known as Wien's law, and in its original form was expressed as

$$L_\lambda = F(\lambda) e^{-f(\lambda)/T} \tag{19}$$

in which $F(\lambda)$ and $f(\lambda)$ were unknown functions of λ. It was later found that the true equation is

$$L_\lambda = c_1 \pi^{-1} \lambda^{-5} e^{-c_2/\lambda T} \tag{20}$$

Note that it is identical to the Planck equation except for the minus one in the denominator. This equation is a useful approximation at short wavelengths, and is much used in optical pyrometry. It gives values of L_λ that are too low at long wavelengths. The error is less than 1 percent when λT is less than 2898 μm·K. The second important relation is the Rayleigh–Jeans law, and is expressed as

$$L_\lambda = \frac{c_1 T}{c_2 \pi \lambda^4} \tag{21}$$

Equation (21) is valid at very long wavelengths where $\lambda T > 10{,}000\ \mu$m·K with but small error.

If equation (7) is integrated over all wavelengths,

$$M = \int_0^\infty \frac{c_1 \lambda^{-5}\, d\lambda}{e^{c_2/\lambda T} - 1} = \frac{\pi^4 c_1}{15 c_2^4} T^4 = \sigma T^4 \tag{22}$$

which is equation (6). Thus, σ, the Stefan–Boltzmann constant, is expressed in terms of other constants as

$$\sigma = \frac{\pi^4 c_1}{15 c_2^4} = \frac{2k^4 \pi^5}{15 c_0^2 h^3} \tag{23}$$

The fraction of the total exitance of a blackbody at temperature T that is emitted within the wavelength interval 0 to λ, may be obtained as follows. The exitance in the wavelength interval will be designated as $M_{(0-\lambda)T}$, and is given by

$$\begin{aligned} M_{(0-\lambda)T} &= \int_0^\lambda M_\lambda\, d\lambda = \int_0^\lambda \frac{c_1 \lambda^{-5}\, d\lambda}{e^{c_2/\lambda T} - 1} \\ &= \int_0^\lambda \frac{T^5 c_1 (\lambda T)^{-5}\, d(\lambda T)}{e^{c_2/\lambda T} - 1} \\ &= \int_0^{\lambda T} \frac{T^4 c_1 (\lambda T)^{-5}\, d(\lambda T)}{e^{c_2/\lambda T} - 1} \end{aligned}$$

The total exitance is $M_t = \sigma T^4$, thus

$$\begin{aligned} M_{(0-\lambda)T}/M_t &= \int_0^{\lambda T} \frac{T^4 c_1 (\lambda T)^{-5}\, d(\lambda T)}{e^{c_2/\lambda T} - 1} \Big/ \sigma T^4 \\ &= \frac{1}{\sigma} \int_0^{\lambda T} \frac{c_1 (\lambda T)^{-5}\, d(\lambda T)}{e^{c_2/\lambda T} - 1} \end{aligned} \tag{24}$$

Substituting M_λ within the integral for its equivalent, $c_1 \lambda^{-5}/(e^{c_2/\lambda T} - 1)$, gives

$$M_{(0-\lambda)T}/M_t = \frac{1}{\sigma T^5} \int_0^{\lambda T} M_{\lambda T}\, d(\lambda T) \tag{25}$$

Since for blackbody $M = \pi L$, then equation (25) is equally valid if L is substituted for M wherever it occurs.

The ratio $L_{(0-\lambda)T}/L_t$ or $M_{(0-\lambda)T}/M_t$ is given in Table I for various values of λT. The fraction of the total exitance or radiance of a blackbody at temperature T, occurring in any wavelength interval λ_1 to λ_2 ($\lambda_2 > \lambda_1$) may be obtained by subtracting the value from the table for $\lambda_1 T$ from that for $\lambda_2 T$. The

Table I. Blackbody Radiation Functions

λT, μm·K	$L_{b,\lambda}/\sigma T^5$, $m^{-1}\cdot K^{-1}\cdot sr^{-1}$		$\pi L_b\ (0-\lambda T)/\sigma T^4$		$L_{b,\lambda}/(L_{b,\lambda})_{\lambda T=2898}$	
	p	q	p	q	p	q
200	0.375195	−21	0.341796	−26	0.519451	−23
400	0.490424	−7	0.186468	−11	0.678981	−9
600	0.104056	−2	0.929299	−7	0.144064	−4
800	0.991183	−1	0.164351	−4	0.137277	−2
1000	0.118508	+1	0.320780	−3	0.164072	−1
1200	0.523935	+1	0.213431	−2	0.725376	−1
1400	0.134411	+2	0.779084	−2	0.186089	0
1600	0.249128	+2	0.197204	−1	0.344913	0
1800	0.375563	+2	0.393449	−1	0.519959	0
2000	0.493422	+2	0.667347	−1	0.683133	0
2200	0.589636	+2	0.100897	0	0.816338	0
2400	0.658848	+2	0.140268	0	0.912161	0
2600	0.701271	+2	0.183135	0	0.970894	0
2800	0.720216	+2	0.227908	0	0.997123	0
2898	0.722294	+2	0.250126	0	0.100000	+1
3000	0.720229	+2	0.273252	0	0.997142	0
3200	0.705948	+2	0.318124	0	0.977369	0
3400	0.681517	+2	0.361760	0	0.943546	0
3600	0.650369	+2	0.403633	0	0.900422	0
3800	0.615199	+2	0.443411	0	0.851730	0
4000	0.578040	+2	0.480907	0	0.800283	0
4200	0.540370	+2	0.516046	0	0.748131	0
4400	0.503231	+2	0.548830	0	0.696712	0
4600	0.467321	+2	0.579316	0	0.646996	0
4800	0.433089	+2	0.607597	0	0.599602	0
5000	0.400794	+2	0.633786	0	0.554890	0
5200	0.370562	+2	0.658011	0	0.513035	0
5400	0.342428	+2	0.680402	0	0.474084	0
5600	0.316361	+2	0.701090	0	0.437994	0
5800	0.292287	+2	0.720203	0	0.404664	0
6000	0.270108	+2	0.737864	0	0.373958	0
6200	0.249710	+2	0.754187	0	0.345718	0
6400	0.230973	+2	0.769282	0	0.319777	0
6600	0.213775	+2	0.783248	0	0.295967	0
6800	0.197997	+2	0.796180	0	0.274123	0
7000	0.183524	+2	0.808160	0	0.254085	0
7200	0.170247	+2	0.819270	0	0.235703	0
7400	0.158065	+2	0.829580	0	0.218837	0
7600	0.146883	+2	0.839157	0	0.203356	0
7800	0.136614	+2	0.848060	0	0.189139	0
8000	0.127177	+2	0.856344	0	0.176075	0
8500	0.106766	+2	0.874666	0	0.147816	0
9000	0.901414	+1	0.890090	0	0.124798	0
9500	0.765296	+1	0.903147	0	0.105953	0
10000	0.653243	+1	0.914263	0	0.904400	−1
10500	0.560490	+1	0.923775	0	0.775987	−1
11000	0.483294	+1	0.931956	0	0.669110	−1
11500	0.418701	+1	0.939027	0	0.579683	−1
12000	0.364373	+1	0.945167	0	0.504466	−1
13000	0.279441	+1	0.955210	0	0.386880	−1

Table I (continued)

λT, μm·K	$L_{b,\lambda}/\sigma T^5$, $m^{-1} \cdot K^{-1} \cdot sr^{-1}$		$\pi L_b\,(0-\lambda T)/\sigma T^4$		$L_{b,\lambda}/(L_{b,\lambda})_{\lambda T=2898}$	
	p	q	p	q	p	q
14000	0.217628	+1	0.962970	0	0.301301	−1
15000	0.171855	+1	0.969056	0	0.237930	−1
16000	0.137421	+1	0.973890	0	0.190257	−1
18000	0.908187	0	0.980939	0	0.125736	−1
20000	0.623273	0	0.985683	0	0.862908	−2
25000	0.276458	0	0.992299	0	0.382750	−2
30000	0.140461	0	0.995427	0	0.194465	−2
40000	0.473862	−1	0.998057	0	0.656052	−3
50000	0.201592	−1	0.999045	0	0.279100	−3
75000	0.418572	−2	0.999807	0	0.579503	−4
100000	0.135744	−2	1.000000	0	0.187934	−4

Note: Value in each case is $p \times 10^q$.

value for the exitance $M_{(\lambda_1-\lambda_2)T}$ may be obtained by multiplying the resulting value by σT^5, and the radiance $L_{(\lambda_1-\lambda_2)T}$, by multiplying the value by $\sigma T^5/\pi$.

Figure 2A is a plot of the ratio $L_{(0-\lambda)T}/\sigma T^4$, dotted line, and $L_\lambda/\sigma T^5$, solid line, both plotted as a function of λT.

Another useful relationship is the fractional change in L_λ or M_λ produced by a fractional change in the temperature T of a blackbody radiator. Differentiating equation (16) with respect to T gives

$$\frac{dL_\lambda}{L_\lambda} \bigg/ \frac{dT}{T} = (c_2/\lambda T)\frac{e^{c_2/\lambda T}}{e^{c_2/\lambda T} - 1} \qquad (26)$$

This ratio is plotted as a function of λT in Fig. 2B, dotted line, and shows that the change in spectral radiance with temperature increases with decrease in λT, particularly at values of λT below 2898 μm·K. The term on the right of equation (26) can be simplified to $c_2/\lambda T$ for use at $\lambda T < 2898$ μm·K with an error of considerably less than 1 percent. Again the equation applies to either spectral radiance L_λ or spectral exitance M_λ.

The value of the right side of equation (26) can be used for the exponent x in the relationship

$$L_\lambda \sim T^x \qquad (27)$$

which is a good approximation when the change in temperature is a small fraction of T. Again M_λ can be substituted for L_λ in equation (27). The value of x is 5.0 at 2898 μm·K, and varies as $1/\lambda T$, hence is about 10 at 1500 μm·K, 15 at 1000 μm·K, and 2.5 at 6000 μm·K.

The relationship shown in equation (26) and

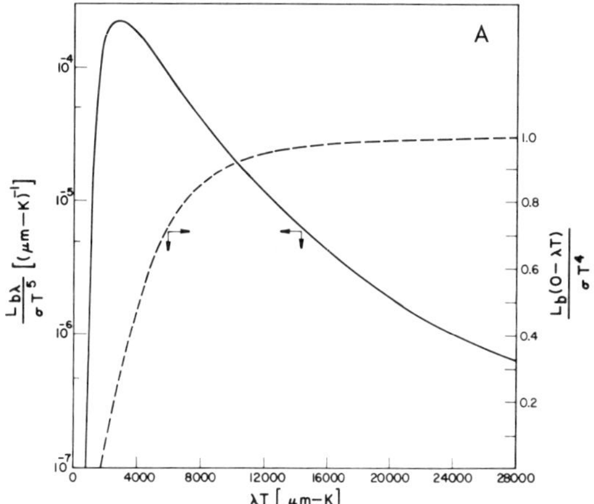

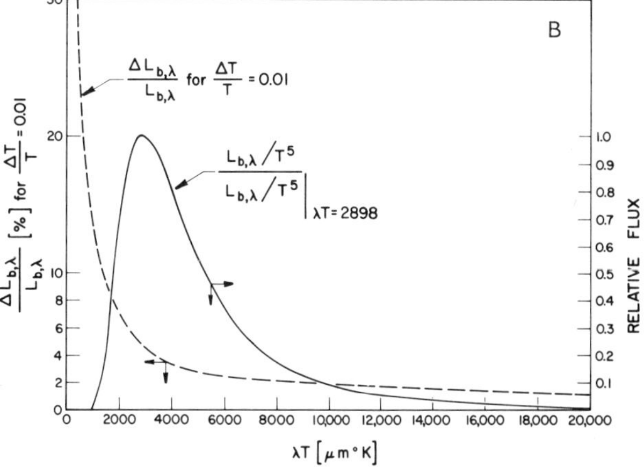

Fig. 2. (A) Relative spectral radiance of blackbody radiation as a function of λT, solid curve; and fraction of spectral radiance in the wavelength interval 0–λT, dashed curve. (B) Relative spectral radiance of blackbody radiation as a function of λT, solid curve; and percentage increase in spectral radiance produced by a 1 percent increase in absolute temperature as a function of λT, dashed curve.

plotted in Fig. 2B is extremely important to remember in making measurements of L_λ by direct comparison of the radiance of a hot specimen to that of a blackbody radiator at the same temperature; it shows the effect of a small temperature difference on the accuracy of the results obtained.

5. INTERACTION OF RADIANT ENERGY WITH MATTER

A. Wave Behavior [22, 147]

When an electromagnetic wave in vacuum is incident on the plane surface of an optically homogeneous specimen, interaction of the wave with the material of the specimen will occur. The electrical and magnetic properties of the specimen will be different from those of the vacuum, and, as a result, there may be a change in the direction of propagation of the wave, its velocity, amplitude, wavelength, and phase, and it may be separated into two portions, one reflected and one transmitted. The transmitted portion will be partially or totally absorbed. The only property of the wave that never changes is its frequency.

Similar changes in the wave will occur whenever it is incident on an interface between two media of different properties. The changes can be computed from the properties of the material, or the differences in properties on the two sides of the interface, and from the direction of propagation of the wave relative to the interface and the direction of its plane of polarization relative to the plane containing the direction of incidence and the normal to the interface at the point of incidence.

B. Optical Properties [59]

The Maxwell equations describe the change in a wave as it crosses an interface and propagates into the material:

$$E = E_0 \exp(j\omega t - \gamma x) \tag{28}$$

where

$$\gamma = \alpha + j\beta \tag{29}$$

Equation (28) indicates that the electric field vector E_0 at the interface where $x = 0$ (x is the distance from the interface in the material) is attenuated on penetrating into the material by an amount α and that a phase change β has occurred in crossing the interface. Thus, two parameters (or one complex one) of the material are needed to define the changes in amplitude and phase which occur in crossing the interface.

Many different sets of parameters, called optical properties in this discussion, can be found in the literature, each having some merit in interpreting the interaction of a wave with a material. Table II presents several of the more commonly used optical properties, together with a brief description and summary of their interrelationship.

C. Thermal Radiative Properties

The optical properties describe the interaction of an electromagnetic wave with matter in terms of phase and amplitude, while the thermal radiative

Table II. Parameters and Relations Descriptive of Electromagnetic Wave–Material Interaction

Complex permittivity	$\epsilon^* = \epsilon' - j\epsilon''$	ϵ'—Real permittivity ϵ''—Loss factor
Complex permeability	$\mu^* = \mu_0$	Valid assumption for almost all applications.
Complex propagation factor	$\gamma = \alpha + j\beta$	α—Attenuation factor β—Phase factor
Complex dielectric constant	$K^* = \dfrac{\epsilon^*}{\epsilon_0} = K' - jK''$	K' — Relative permittivity or real dielectric constant K'' — Relative loss factor
Complex index of refraction	$n^* = n - jk$	n—Index of refraction, ratio of phase velocities k—Index of absorption
Intrinsic impedance	$Z^* = \sqrt{\dfrac{\mu^*}{\epsilon^*}}$	Z^*—Ratio of electric to magnetic field vectors
Normal spectral reflectivity	$\rho(0; \lambda) = \dfrac{(n-1)^2 + k^2}{(n+1)^2 + k^2}$	Ratio of squares of the electric field vector of reflected and incident waves
$\gamma = j\omega(\epsilon^*\mu^*)^{1/2}$	$n^* = \sqrt{K^*}$	$\epsilon^* = \epsilon_0\left(1 - j\dfrac{1}{\omega r}\right)$

properties describe the energy transfer during the interaction. It is obvious that the two types of properties, optical and thermal radiative, are related. In some cases the relationships are simple.

One situation in which the relation is not simple is that for the general case of a wave incident on an interface. By solving the Maxwell equations for the boundary conditions, the Fresnel relations for specular reflection can be derived. The specular reflectance at the interface (fraction of incident flux reflected in the direction of mirror reflectance) is given as

$$\rho_s(\theta) = \frac{a^2 + b^2 - 2a\cos\theta + \cos^2\theta}{a^2 + b^2 + 2a\cos\theta + \cos^2\theta} \tag{30}$$

$$\rho_p(\theta) = \rho_s(\theta)\frac{a^2 + b^2 - 2a\sin\theta\tan\theta + \sin^2\theta\tan^2\theta}{a^2 + b^2 + 2a\sin\theta\tan\theta + \sin^2\theta\tan^2\theta} \tag{31}$$

where

$$2a^2 = [(n^2 - k^2 - \sin^2\theta)^2 + 4n^2k^2]^{1/2} + (n^2 - k^2 - \sin^2\theta) \tag{32}$$

$$2b^2 = [(n^2 - k^2 - \sin^2\theta)^2 + 4n^2k^2]^{1/2} - (n^2 - k^2 - \sin^2\theta) \tag{33}$$

The angle θ is the angle between the incident ray and the normal to the interface, ρ_s is the reflectance for plane-polarized incident radiant energy with its plane of polarization normal to the plane of incidence (the plane containing the incident ray and the normal to the interface at the point of incidence), and ρ_p is the reflectance for plane-polarized incident radiant energy with its plane of polarization parallel to the plane of incidence.

If the incident radiant energy is completely unpolarized, it can be shown that

$$\rho(\theta) = \tfrac{1}{2}[\rho_s(\theta) + \rho_p(\theta)] \tag{34}$$

The Fresnel equations, (30) and (31), have been expressed in terms of n and k, but the relations are found in various forms in the literature. The simplest case occurs for normal incidence ($\theta = 0$), where the equations reduce to

$$\rho_p(0) = \rho_s(0) \tag{35}$$

and

$$a = n \qquad b = k \tag{36}$$

Hence, for radiant energy incident from vacuum or a medium of index of refraction of 1,

$$\rho(0) = \frac{(n-1)^2 + k^2}{(n+1)^2 + k^2} \tag{37}$$

The above relations express the energy transfer in terms of electromagnetic wave theory, and no specification of the material, other than its optical properties n and k, is necessary. As indicated previously, the reflected, absorbed, and transmitted flux sum to the incident flux, hence

$$\rho + \tau + \alpha = 1 \tag{2}$$

6. DETAILED DISCUSSION OF THERMAL RADIATIVE PROPERTIES

A. Primary Properties

The primary properties—emissivity, reflectivity, absorptivity and transmissivity—have been previously introduced. The suffix *ivity* denotes the property of the ideal material—optically smooth and homogeneous. For metallic materials that are neither smooth nor homogeneous, the radiative parameters are not unique properties of the bulk material, but rather are properties of the surface. In these more frequent situations the properties are denoted by the suffix *ance* and hereafter the properties are referred to and defined as follows:

- ϵ, emittance —Ratio of the radiant exitance of the specimen to that emitted by a blackbody radiator at the same temperature and under the same geometric and wavelength conditions.
- ρ, reflectance —Ratio of the reflected radiant flux to the incident flux.
- α, absorptance —Ratio of absorbed radiant flux to the incident flux.
- τ, transmittance—Ratio of some specified portion of the transmitted flux to the incident radiant flux.

For each of these primary properties, it is necessary to specify the geometric and wavelength conditions to which the particular property corresponds. Unfortunately, there is no generally accepted convention on the choice of symbols or terminology to describe these conditions. After due consideration and consultation, a nomenclature system suitable to classification needs in the organization of this compendium was adopted. To assist the reader, the system used in this volume will be fully explained and then related when appropriate to the various terminologies found in the literature [4, 6, 81, 98].

B. Subproperty Descriptors

a. Geometric Descriptors

Figure 3 shows the general case of reflection at a

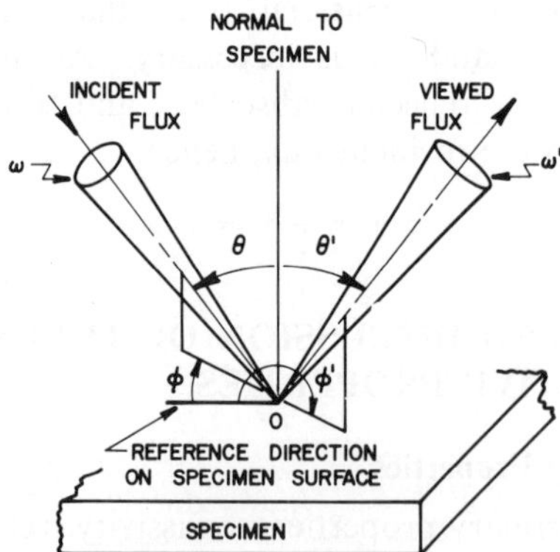

Fig. 3. Geometric parameters descriptive of reflection from a surface. θ is the zenith angle, or co-latitude, in degrees; ϕ is the azimuthal angle, or longitude, in degrees; ω is the beam solid angle, in steradians; and the symbol $'$ refers to viewing conditions.

surface and indicates the necessary geometric parameters required to fully describe the incident and reflected fluxes. The beams representing the incident or viewed flux are described by the zenith angles for θ or θ' and by the beam solid angles ω or ω'. The longitudinal angles ϕ and ϕ' relate the axes of the beams to each other and some reference line on the specimen; as a practical matter very few measurements so specify this angular descriptor. It is the convention in this volume to distinguish three sets of conditions as follows:

Normal —Conditions for incidence and/or viewing through a solid angle ω or ω', normal to the specimen; that is θ or $\theta' < 15°$

Angular —Conditions for incidence and/or viewing through a solid angle ω or ω' at some direction specified by θ or $\theta' \geq 15°$

Hemispherical—Conditions for incidence and/or viewing of flux over a hemispherical region; that is ω or $\omega' = 2\pi$

The descriptors normal and angular do not fully describe the geometric conditions; ω and/or ω' and θ and/or θ' must be provided to fully specify the geometry.

It has been suggested that other descriptors be used to indicate the two extreme conditions for ω or ω'. If the incident or viewed beam is parallel, then ω or $\omega' = 0$; this condition can be approximated in practice where ω or ω' is so small that slight increases have no influence. In this case, the term *directional* is used and when ω or ω' is not negligible, the term *conical* is applied. As a practical matter for categorization of subproperties, only the terms *normal*, *angular*, and *hemispherical* are separately distinguished in the subsequent data tables. Whenever information is given on ω or ω', it is reported in an appropriate column. To a great many thermophysicists details on ω or ω' may not be essential, but they should be fully aware that such information is available.

For the subproperties of emittance and absorptance, only one geometric descriptor is required to designate the conditions of viewing and incidence respectively. For the subproperties of reflectance and transmittance, two geometric descriptors are required since both incidence and viewing conditions need to be specified. While the three descriptors selected for categorization are not fully descriptive, they give good practical resolution from a classification and retrieval viewpoint.

b. Wavelength Descriptors

These descriptors indicate spectral or wavelength conditions for which the subproperties are specified. Each of the four prime properties needs to be characterized by the wavelength conditions of the radiant flux—emitted, incident, reflected, or transmitted. The following terminology is used in this volume to refer to the common conditions:

Spectral—For a very narrow band of wavelengths, also referred to as monochromatic; no maximum band width criteria has been established by convention; symbolically denoted by λ.

Total —Refers to blackbody wavelength distribution; this descriptor is applicable only to emittance; symbolically denoted by *t*. The temperature of the blackbody should be given, since spectral distribution varies with temperature.

The properties, being dimensionless ratios, cannot be integrated. The term *integrated*, when applied to such ratios, means that the numerator and denominator are integrated separately, then divided. An "integrated" ratio can also be thought of as a weighted average, where the weighting factor is the spectral distribution of the source, or blackbody in the case of emittance.

Integrated—Relative to some specified wavelength distribution of the irradiating source or some specified broad band; for such subproperties it is necessary that details of the spectral characteristics of the source be fully presented; symbolically denoted by $\bar{\lambda}$.

Solar —Having the wavelength distribution of the sun; the solar distribution can be either natural or artificial—lamps, arcs, etc., in which case the nature of the source needs to be specified; symbolically denoted by s.

In a later section, interrelationships between the various wavelength conditions will be discussed with a view to exposing the reader to methods of computing the subproperties from fragments of other subproperty data.

c. Symbolic Representation

In subsequent discussion the various subproperties are expressed according to the following convention. The symbols for the four primary properties—ϵ, ρ, α, and τ—have already been presented. The geometric (incidence and viewing condition) and wavelength descriptors, in that same order, are symbolically represented within the parentheses being separated by semicolons. The most general case would be (using reflectance as an example):

$$\rho(\theta, \phi, \omega; \theta', \phi', \omega'; \lambda)$$

where the wavelength descriptors used in this text are λ, t, $\bar{\lambda}$, and s indicating spectral conditions previously defined.

As a practical matter not all the designations are always needed and many are omitted for convenience sake; usually ϕ and ϕ' are not used and, of course, for emittance and absorptance, the incidence and viewing geometry symbols, respectively, are not applicable.

C. Reflectance Factor

The *reflectance factor R* is defined as the ratio of the flux reflected by a specimen under specified geometric conditions of irradiation and viewing to that reflected by the ideal completely reflecting, perfectly diffusing surface, under identical conditions of irradiation and viewing [81]. A perfectly diffusing surface is defined as a surface whose reflected radiance is independent of direction of viewing and hence obeys Lambert's law.

Reflectance factor is relatively easy to measure because comparatively simple equipment can be used. Integrating sphere reflectometers are frequently used for these measurements. While the ideal completely reflecting, perfectly diffusing surface does not exist, several materials can be used that are reasonably close approximations to such a surface. Examples are freshly smoked magnesium oxide, magnesium carbonate, high-purity barium sulphate, sodium chloride, flowers of sulphur, and some good white paints [2, 64, 89]. For those conditions where the specimen is diffusely irradiated over a hemisphere, a high-quality first-surface mirror can be used as the reference standard. When integrating sphere reflectometers are used, the sphere coating itself may be used as the reference.

Integrating sphere reflectometers are generally restricted to operation over the solar range (0.25 to 2.5 μm) where conventional sphere coatings such as smoked magnesium oxide and barium sulphate have high reflectance. Operation has been extended farther into the infrared by use of flowers of sulfur and sodium chloride sphere coatings.

Reflectometers such as the hohlraum reflectometer, where the specimen is irradiated over a hemisphere, actually measure reflectance factor. This also applies to the ellipsoidal mirror, paraboloid mirror, and Coblentz hemisphere reflectometers when used in the inverse mode, so that the specimen is irradiated over a hemisphere.

The primary distinction between true reflectance and reflectance factor measurements is that in true reflectance measurements the incident flux is measured directly, then the reflected flux is measured directly, and the reported reflectance is the ratio of the second measured value to the first. In reflectance factor measurements either (1) the incident flux is measured indirectly by measuring the flux reflected by a diffusely reflecting standard, or (2) the radiance of the flux incident on the specimen is measured over a small solid angle, either by viewing the source directly as in a hohlraum, or by viewing it indirectly by use of a mirror standard, as in the ellipsoidal mirror, paraboloid mirror, or Coblentz hemisphere instruments.

Reflectance factor data are useful to the thermophysicist, because for a diffuse reflector the reflectance factor measured under any given geometric conditions of irradiation and viewing is a good estimate of the reflectance of the specimen measured under the same conditions of irradiation, but with hemispherical viewing. The error in the estimate is zero if the specimen is a perfectly diffuse reflector, and increases with both the departure of the specimen from a perfectly diffuse reflector and with the variation of the solid angle of viewing from a hemisphere.

The fundamental relationship by the use of which the various types of reflectance and reflectance factor can be related to each other is [51]

$$\frac{\rho(\theta_1, \phi_1, \omega_1; \theta_2, \phi_2, \omega_2)}{\cos \theta_2 \omega_2} = \frac{\rho(\theta_2, \phi_2, \omega_2; \theta_1, \phi_1, \omega_1)}{\cos \theta_1 \omega_1} \quad (38)$$

where the bidirectional reflectance, $\rho(\theta, \phi, d\omega; \theta', \phi', d\omega')$ is defined as

$$\rho(\theta, \phi, d\omega; \theta', \phi', d\omega') = \frac{L'(\theta', \phi') \cos \theta' \, d\omega'}{L(\theta, \phi) \cos \theta \, d\omega} \quad (39)$$

where L is the radiance of the beam incident on the specimen over the elementary solid angle $d\omega$ from direction θ, ϕ, and L' is the reflected radiance of the specimen over the elementary solid angle $d\omega'$ in the direction θ', ϕ'. The geometry for equations (38) and (39) is shown in Fig. 3.

The only restriction on the general application of equation (38) to solid angles of any size is that L, the incident radiance, must not vary with angle over the solid angle of incidence in either case.

By the use of equation (38) it can be shown that the following relationships hold:

$$R(2\pi; \theta', \omega') = R(\theta, \omega; 2\pi) \quad (40)$$

where $\theta = \theta'$ and $\omega = \omega'$.

$$\rho(2\pi; 2\pi) = R(2\pi; 2\pi) \quad (41)$$

$$\rho(\theta, \omega; 2\pi) = R(\theta, \omega; 2\pi) \quad (42)$$

where θ and ω are the same for ρ and R.

$$R(\theta_1, \omega_1; \theta_2, \omega_2) = R(\theta_2, \omega_2; \theta_1, \omega_1) \quad (43)$$

$$\rho(\theta, \omega; \theta', \omega') = \frac{\omega' \cos \theta'}{\pi} R(\theta, \omega; \theta', \omega') \quad (44)$$

where for large solid angles, $\omega' \cos \theta' = \int_{\omega'} \cos \theta' \cdot \sin \theta' \, d\theta' \, d\phi'$.

7. INTERRELATIONSHIPS BETWEEN PROPERTIES

A. Geometric Relations

In many cases the data required for a specific application will not be available in the literature. In some of these cases, it is possible to obtain the desired data from other data that may be available for the same or a similar material by judicious use of the relationships previously mentioned and realistic estimates of the geometric distribution of the reflected or emitted radiant energy.

The geometric distribution of the radiant energy emitted or reflected from a metal surface is greatly influenced by the surface roughness, or rather by the ratio of the surface roughness, σ, evaluated in one dimension, to the wavelength of the radiant energy involved, or σ/λ. An optically smooth metal surface, $\sigma/\lambda < 0.05$, reflects specularly. A very rough surface, $\sigma/\lambda > 10$, such as might be produced by severe sandblasting, reflects diffusely, although a rough metal surface is not a close approximation to a perfectly diffusing reflector. A smooth metal surface emits diffusely in the sense that it emits radiant energy in all directions according to the Fresnel relations, but it is far from a Lambertian emitter. The radiance of a smooth metal surface increases with angle from the normal to a near-grazing angle ($\theta' > 85°$) then drops to zero at grazing ($\theta' = 90°$). The angle at which the radiance is a maximum is a function of both the electronic structure (resistivity, in the most simple case) of the metal and the wavelength of the emitted radiant energy. The polarization of the emitted radiant energy also increases with angle from the normal. As the surface becomes rougher, it approaches a Lambertian emitter, and the degree of polarization decreases.

The two extremes for geometric distribution of reflected flux are thus specular and diffuse reflection. These are useful limits that are only approximately attained by real metal surfaces. The equations given below will be found quite useful in converting data from one parameter to another. Additional detail is given in references [137] and [74]; reference [81] treats reflectance and reflectance factor subproperties.

From Kirchhoff's law [equations (4) and (5)]

$$\epsilon = \alpha = 1 - \rho \quad (45)$$

Equation (45) is restricted by the geometric and wavelength distribution of the reflected and emitted radiant energy. Considering the geometric distribution only, for opaque specimens

$$\alpha(\theta, \omega) = 1 - \rho(\theta, \omega; 2\pi) \quad (46)$$

where θ, ω are the same for α and ρ. and

$$\epsilon(\theta', \omega') = \alpha(\theta, \omega) \quad (47)$$

where $\theta = \theta'$ and $\omega = \omega'$. Equation (46) was derived on the basis of conservation of energy. Incident radiant energy that is not reflected must be

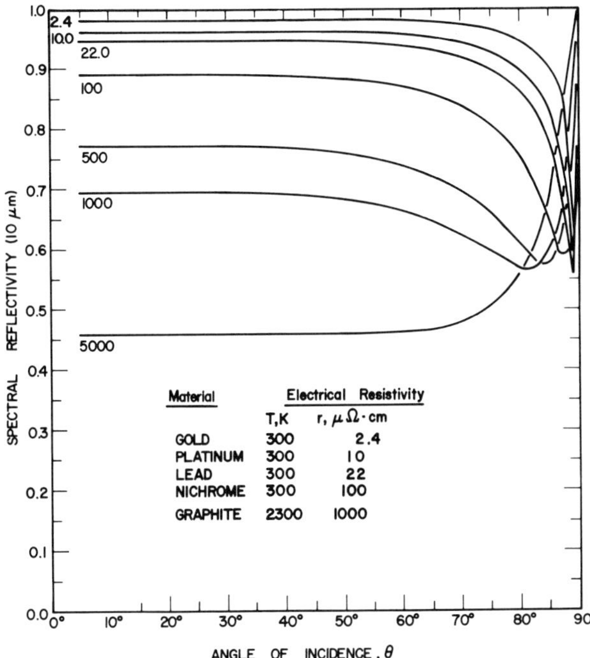

Fig. 4. Reflectance of polished metals, at 10 μm, computed from the Schmidt–Eckert equations.

absorbed and equation (47) is a statement of Kirchhoff's law. Equations (46) and (47) can be used to convert one type of data (subproperty) to another. If normal emittance data are not available, for instance, normal absorptance or normal hemispherical reflectance can be used to compute the desired values.

In the classification system used in the data section of this volume, reflectance and transmittance subproperties are grouped (geometrically) by common incidence conditions. The relations which can be used to compute one subproperty of reflectance from another for different geometric conditions are as follows:

For perfectly diffuse reflecting surfaces

$$\rho(\theta, \omega; 2\pi) = \text{constant, for all } \theta \tag{48}$$

$$\rho(\theta, \omega; 2\pi) = \frac{\pi}{\omega' \cos\theta'} \rho(\theta, \omega; \theta', \omega') \tag{49}$$

where θ and ω are the same on both sides of the equation.

$$R(\theta_1, \omega_1; \theta_1', \omega_1') = R(\theta_2, \omega_2; \theta_2', \omega_2') \tag{50}$$

where θ_1 and θ_2, θ_1' and θ_2', ω_1 and ω_2, ω_1' and ω_2' can have any values.

For perfectly specular reflecting surfaces

$$\rho(\theta, \phi, \omega; 2\pi) = \rho(\theta, \phi, \omega; \theta', \phi', \omega') \tag{51}$$

if and only if $\theta' = \theta, \phi' = \phi + \pi$ and $\omega' \geq \omega$.

The geometric distribution of reflected radiant energy for perfectly diffuse reflectors is described by Lambert's cosine law and that for perfectly specular reflectors by the Fresnel equations. Figure 4 shows how specular spectral reflectance at 10 μm changes with angle of incidence for surfaces with different electrical resistivities, computed from the Schmidt–Eckert relation discussed later in Section 8 and Table V. The curve for r = 5000 is typical of dielectric materials in general. The reflectance is essentially constant out to angles of about 50° from the normal, then decreases to a minimum which occurs somewhere between 80° and 90° from the normal. It should be emphasized, however, that the curves are for unpolarized incident radiant energy. When the incident radiant energy is polarized in the plane of incidence, the reflectance increases with angle of incidence to a value of 1.0 at 90°, with no minimum. When the incident radiant energy is polarized normal to the plane of incidence, the reflectance decreases with increasing angle of incidence to a minimum then increases to 1.0 at 90°. The curves for the two polarized components at 10 μm are shown in Fig. 5 for a metal with r = 100. The reflectance for unpolarized incident radiant energy is the average of the two reflectances for polarized incident radiant energy.

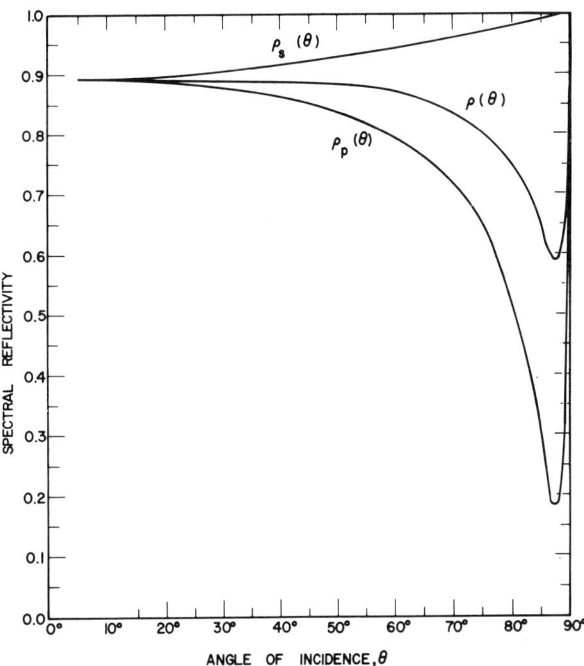

Fig. 5. Reflectance of a polished metal (10μm, r = 100) for incident radiant energy polarized normal $\rho_s(\theta)$ and parallel $\rho_p(\theta)$ to the plane of incidence and for unpolarized incident monochromatic radiant energy $\rho(\theta)$.

The different reflectances for incident radiant flux polarized in different directions means that unpolarized incident radiant energy is partially polarized after reflection. The degree of such polarization, defined earlier as equation (15), is shown in Fig. 6, for four different metals at 10 μm. The degree of polarization is small at angles out to about 70° from the normal, especially for metals with high reflectance (low r). The degree of polarization is inversely related to the reflectance.

The radiant energy emitted by a polished metal is also partially polarized. Figure 7 is a plot of the degree of polarization at 10 μm for two metals having r values of 2.4 and 500, respectively. The degree of polarization increases with angle of viewing in a smooth S-curve, and is not a strong function of the resistivity of the metal, and hence of the reflectance. Polarization of emitted radiant energy begins to become significant at angles of viewing greater than about 25° to the normal.

To compute one subproperty of emittance from another, for different geometric conditions, the following equation may be used.

$$\epsilon(2\pi) = \frac{\int_{\theta=0}^{\pi/2} \int_{\phi=0}^{2\pi} L_b \epsilon(\theta', \phi', \omega') \cos\theta' \sin\theta' \, d\theta' \, d\phi'}{\pi L_b} \quad (52)$$

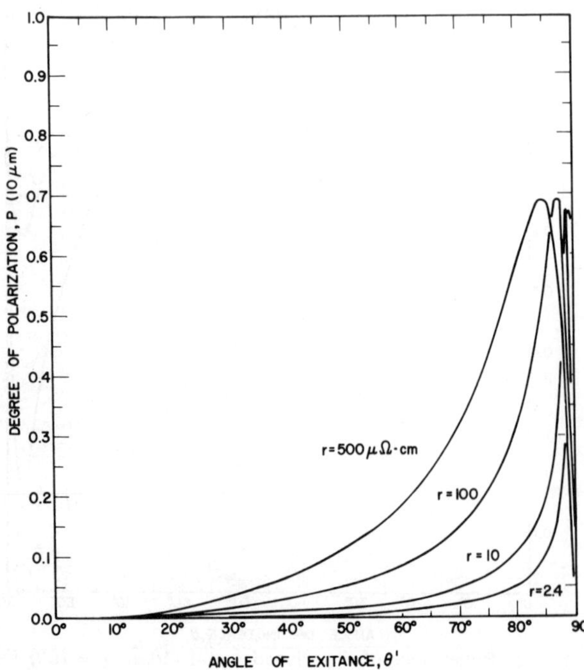

Fig. 6. Degree of polarization (P) of reflected radiant energy (10 μm) from polished metals as a function of their electrical resistivity (r).

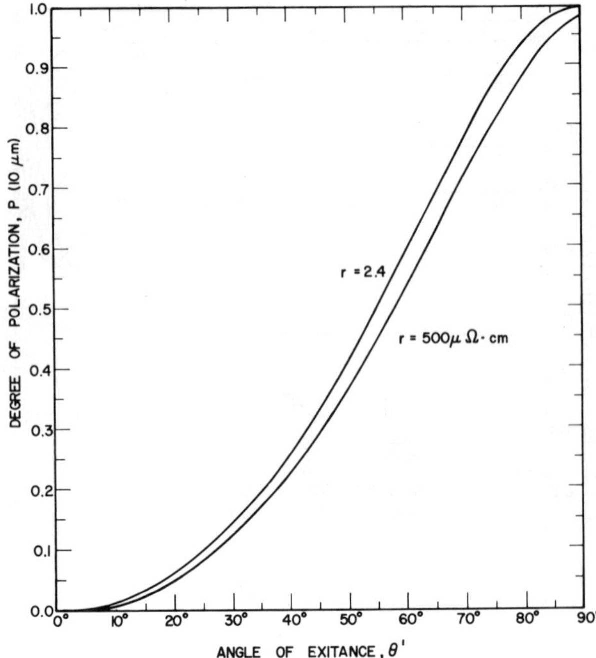

Fig. 7. Degree of polarization of emitted radiant energy (10 μm) from polished metals as a function of their electrical resistivity (r).

This equation is sometimes written

$$\epsilon(2\pi) = \frac{1}{\pi} \int_{\Omega} \epsilon(\theta', \omega') \cos\theta' \, d\omega' \quad (53)$$

For a perfectly diffuse emitter, which may be approximated by a roughened metal surface,

$$\epsilon(2\pi) = \epsilon(\theta', \omega') = \epsilon(0, \omega') \quad (54)$$

The geometric distribution of radiant energy emitted by a polished metal surface is given by the Fresnel relations, and by the Schmidt–Eckert relation that will be discussed in a following section. The ratio of the hemispherical to normal emittance is given by another Schmidt–Eckert relation, which shows $\epsilon(2\pi)/\epsilon(0)$ to vary from 1.33 to 1.05 as a function of $\epsilon(0)$.

As a matter of convenience and practicality in much of the development for theoretical relations, little attention has been given to specification of the solid angles of illumination or viewing (ω or ω'). Two cases appear in the literature distinguishing the importance of the relative size of the solid angle. The term *conical* is applied when the solid angle is sufficiently large that there is a variation of radiance across the front of the solid angle. In the limit when the solid angle is 2π, the term *hemispherical* has been used. The term *directional* is applied when the solid angle is so small that there is insignificant variation of the radiance within this solid angle. This concept is carried through the literature for bidirectional reflectance [$\rho(\theta, \omega; \theta', \omega')$ where $\omega = \omega' \cong 0$] and directional emittance [$\epsilon(\theta', \omega')$ where $\omega' \cong 0$].

More detailed considerations of angular or geometric relations for the reflectance (and transmittance) properties utilize the concept of a distribution function [94]. This function permits computation of any of the various geometric subproperties of reflectance. There are not too many instances where such a function is available in the literature and normally there are insufficient fragments of subproperty data available to permit one to formulate it.

B. Wavelength Relations

The four specific wavelength conditions identified in the data section of this volume have been defined earlier as: spectral (λ), total (t), integrated ($\bar{\lambda}$), and solar (s). Straightforward methods can be used to obtain one subproperty from another, although in some instances the calculations can be laborious.

The interrelation between *spectral* and *total* emittance which follows directly from their definitions is

$$\epsilon(t) = \frac{\int_0^\infty \epsilon(\lambda) M_b(\lambda, T)\, d\lambda}{\int_0^\infty M_b(\lambda, T)\, d\lambda} \tag{55}$$

If the function $\epsilon(\lambda)$ can be simply prescribed, the integration can be performed by a numerical technique on a high-speed computer. In practice, it is convenient to redefine the limits of integration to correspond to the wavelength limits between which some specific fraction of the total energy is found.

The Planck blackbody function indicates the distribution of energy throughout the spectrum. From such considerations, Fig. 8 is derived with the cross-hatched region indicating the wavelength band

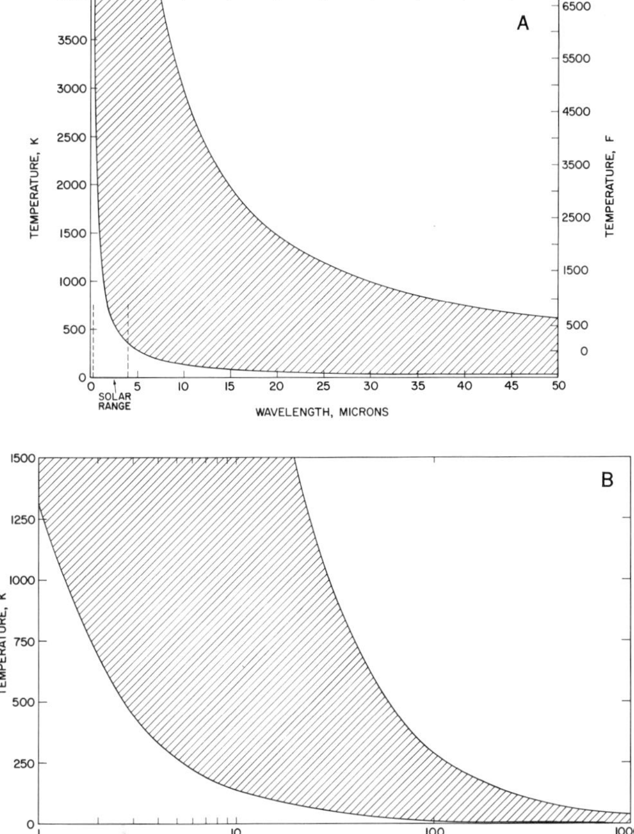

Fig. 8. Wavelength band containing 99% of the flux emitted by a blackbody at the indicated temperature: (A) for high temperatures and (B) for low temperatures.

encompassing 99 percent of the flux emitted by a blackbody source at the specified temperature. For example, 99 percent of the energy emitted by a blackbody at 1000 K is in the wavelength band 2 to 28 μm. Hence, for this temperature, the integration limits of equation (55) above can be simplified with small error in the result.

In the process of computing total emittance data from spectral values, it is essential to identify the wavelength band over which most of the radiant energy is emitted. To do so permits one to assess the validity of a very common assumption used in heat-transfer calculations; namely, that the surface in question is a grey body which has no spectral variation in emittance. Another situation arises when it is necessary to evaluate the possible effect extrapolation of limited spectral data may have. An example of this arises when considering metallic conductors in which the spectral emittance in the visible and near infrared (say to 1 μm) is quite variable, but in the longer wavelength region is monotonic of a form approximated by the Hagen–Rubens relation.

The computations by integration of equation (55) can be done by finite wavelength intervals

$$\epsilon(t) = \frac{\sum_{\lambda_1}^{\lambda_2} \epsilon(\lambda) M_b(\lambda, T)\Delta\lambda}{\sum_{\lambda_1}^{\lambda_2} M_b(\lambda, T)\Delta\lambda} \tag{56}$$

where the limits λ_1 and λ_2 are selected to include substantially all of the flux emitted by a blackbody radiator at the temperature T.

In the Weighted Ordinate Method, uniform values of $\Delta\lambda$ are used, and each value of $\epsilon(\lambda)$ must be weighted by a factor proportional to $L_b(\lambda, T)$, the flux emitted per unit area, unit solid angle, and, within the wavelength interval λ and $(\lambda + \Delta\lambda)$, by a blackbody at the test temperature T. In the Selected Ordinate Method, the area under the blackbody spectral distribution curve is divided vertically into X units of equal area. The X selected ordinates are then the X median wavelengths of the X areas. In this case, $\Delta\lambda$ varies, but the product $L(\lambda, T)\Delta\lambda$ is held constant and for $X = 100$ this is 0.01. Under these conditions then

$$\epsilon(t) \cong 0.01 \sum_{\lambda_1}^{\lambda_{100}} \epsilon(\lambda) \tag{57}$$

A high degree of precision in the computation of $\epsilon(t)$ can be attained using either the Weighted or Selected Ordinate Method by taking a large enough number of ordinates. With any given number of ordinates, the computation error will be smaller by the Selected Ordinate Method than by the Weighted Ordinate Method. Since the spectral emittance curves for solids do not normally have sharp peaks or valleys, the 100-Selected Ordinate Method gives values of adequate precision for most applications.

The wavelengths representing the 100 selected ordinates for the temperature range 600 to 1400 K are presented in Table III [117]. This table along with equation (55) should be most useful for com-computing total emittance from spectral emittance data.

The spectral subproperties of reflectance, absorptance, and transmittance are used to find the corresponding *integrated* values according to the relation (shown for absorptance)

$$\alpha(\bar{\lambda}) = \frac{\int_a^b \alpha(\lambda) E(\lambda)\, d\lambda}{\int_a^b E(\lambda)\, d\lambda} \tag{58}$$

where $E(\lambda)$ is the spectral irradiance of the illuminating source. The integration is performed over the wavelength interval or band denoted by the limits of a and b.

Should the source be a blackbody at some temperature T, the absorptance can be evaluated by use of an equation similar to equation (55). A special case of the integrated subproperty is where the illuminating source is that of the solar disc, natural or artificial. The solar absorptance is normally specified for the geometric condition of normal or angular incidence and can be determined from spectral data in the following manner.

$$\alpha(\theta; s) = \frac{\int_0^\infty \alpha(\theta; \lambda) E_s(\lambda)\, d\lambda}{\int_0^\infty E_s(\lambda)\, d\lambda} \tag{59}$$

where $E_s(\lambda)$ denotes the solar irradiance [80, 116]. Frequently, the solar absorptance is determined from room temperature reflectance data

$$\alpha(\theta; \lambda) = 1 - \rho(\theta; 2\pi; \lambda) \tag{60}$$

using Kirchhoff's law.

Table III. Wavelengths for Computation of Total Emittance from Spectral Data by the 100 Selected Ordinate Method [117]

		λ, μm, at various values of *T*								
%*	λ*T*, μK	600 K	700 K	800 K	900 K	1000 K	1100 K	1200 K	1300 K	1400 K
0.5	1322	2.203	1.889	1.652	1.469	1.322	1.202	1.102	1.017	0.944
1.5	1534	2.557	2.191	1.918	1.704	1.534	1.395	1.278	1.180	1.096
2.5	1662	2.770	2.374	2.078	1.847	1.622	1.511	1.385	1.278	1.187
3.5	1762	2.937	2.517	2.202	1.958	1.762	1.602	1.468	1.355	1.258
4.5	1846	3.077	2.637	2.308	2.051	1.846	1.678	1.538	1.420	1.318
5.5	1920	3.200	2.743	2.400	2.133	1.920	1.745	1.600	1.477	1.371
6.5	1989	3.315	2.841	2.486	2.210	1.989	1.808	1.657	1.530	1.424
7.5	2052	3.420	2.931	2.565	2.280	2.052	1.865	1.710	1.578	1.466
8.5	2111	3.518	3.016	2.639	2.346	2.111	1.919	1.759	1.624	1.508
9.5	2168	3.613	3.097	2.710	2.409	2.168	1.971	1.806	1.668	1.549
10.5	2222	3.703	3.174	2.778	2.469	2.222	2.020	1.852	1.709	1.587
11.5	2274	3.790	3.249	2.842	2.527	2.274	2.067	1.895	1.749	1.624
12.5	2325	3.875	3.321	2.906	2.583	2.325	2.114	1.937	1.788	1.661
13.5	2374	3.957	3.391	2.968	2.638	2.374	2.158	1.978	1.826	1.696
14.5	2423	4.038	3.461	3.029	2.692	2.423	2.203	2.019	1.864	1.731
15.5	2470	4.117	3.529	3.088	2.744	2.470	2.245	2.058	1.900	1.764
16.5	2517	4.195	3.596	3.146	2.797	2.517	2.288	2.097	1.936	1.798
17.5	2563	4.271	3.662	3.204	2.848	2.563	2.330	2.136	1.972	1.831
18.5	2609	4.348	3.727	3.261	2.899	2.609	2.372	2.174	2.007	1.864
19.5	2654	4.423	3.792	3.318	2.949	2.654	2.413	2.212	2.042	1.896
20.5	2698	4.496	3.854	3.372	2.998	2.698	2.453	2.248	2.075	1.927
21.5	2743	4.571	3.919	3.429	3.048	2.743	2.494	2.286	2.110	1.959
22.5	2887	4.645	3.982	3.484	3.097	2.787	2.534	2.322	2.144	1.991
23.5	2831	4.718	4.044	3.539	3.146	2.831	2.574	2.359	2.178	2.022
24.5	2876	4.793	4.109	3.595	3.196	2.876	2.615	2.397	2.212	2.054
25.5	2920	4.866	4.172	2.650	3.244	2.920	2.655	2.433	2.246	2.086
26.5	2964	4.940	4.234	3.705	3.293	2.964	2.695	2.470	2.280	2.117
27.5	3008	5.013	4.297	3.760	3.342	3.008	2.735	2.507	2.314	2.149
28.5	3052	5.086	4.360	3.815	3.391	3.052	2.775	2.543	2.348	2.180
29.5	3097	5.161	4.424	3.871	3.441	3.097	2.815	2.581	2.382	2.212
30.5	3141	5.235	4.487	3.926	3.490	3.141	2.855	2.617	2.416	2.244
31.5	3186	5.310	4.552	3.982	3.540	3.186	2.896	2.655	2.451	2.275
32.5	3231	5.385	4.616	4.039	3.590	3.231	2.937	2.692	2.485	2.308
33.5	3277	5.461	4.682	4.096	3.641	3.277	2.979	2.731	2.521	2.341
34.5	3323	5.538	4.747	4.154	3.692	3.323	3.021	2.769	2.556	2.374
35.5	3369	5.615	4.813	4.211	3.743	3.369	3.063	2.807	2.592	2.406
36.5	3415	5.691	4.879	4.269	3.794	3.415	3.105	2.846	2.627	2.439
37.5	3462	5.770	4.946	4.328	3.847	3.462	3.147	2.885	2.663	2.473
38.5	3510	5.850	5.014	4.388	3.900	3.510	3.191	2.925	2.700	2.507
39.5	3558	5.930	5.083	4.448	3.953	3.558	3.235	2.965	2.737	2.542
40.5	3607	6.011	5.152	4.509	4.008	3.607	3.279	3.006	2.775	2.576
41.5	3656	6.093	5.223	4.570	4.062	3.656	3.324	3.047	2.812	2.611
42.5	3706	6.176	5.294	4.632	4.118	3.706	3.369	3.088	2.851	2.647
43.5	3757	6.261	5.367	4.696	4.174	3.757	3.415	3.131	2.890	2.684
44.5	3809	6.348	5.441	4.761	4.232	3.809	3.463	3.174	2.930	2.721
45.5	3861	6.435	5.516	4.826	4.290	3.861	3.510	3.217	2.970	2.758
46.5	3914	6.523	5.592	4.892	4.349	3.914	3.558	3.262	3.011	2.796
47.5	3968	6.613	5.669	4.960	4.409	3.968	3.607	3.307	3.052	2.834
48.5	4023	6.705	5.747	5.029	4.470	4.023	3.657	3.352	3.095	2.874
49.5	4079	6.798	5.827	5.099	4.532	4.079	3.708	3.399	3.138	2.914
50.5	4136	6.893	5.909	5.170	4.596	4.136	3.750	3.447	3.182	2.954
51.5	4194	6.990	5.992	5.242	4.660	4.194	3.813	3.495	3.226	2.996
52.5	4254	7.090	6.077	5.318	4.727	4.254	3.867	3.545	3.272	3.039
53.5	4314	7.190	6.163	5.392	4.793	4.314	3.922	3.595	3.318	3.081
54.5	4377	7.295	6.253	5.471	4.863	4.377	3.979	3.647	3.367	3.126

Table III (continued)

%*	λT, μK	λ, μm, at various values of T								
		600 K	700 K	800 K	900 K	1000 K	1100 K	1200 K	1300 K	1400 K
55.5	4440	7.400	6.343	5.550	4.933	4.440	4.036	3.700	3.415	3.171
56.5	4505	7.508	6.436	5.631	5.006	4.505	4.095	3.754	3.465	3.218
57.5	4572	7.620	6.532	5.715	5.080	4.572	4.156	3.810	3.517	3.266
58.5	4640	7.733	6.629	5.800	5.156	4.640	4.218	3.867	3.569	3.314
59.5	4710	7.856	6.729	5.888	5.233	4.710	4.282	3.925	3.623	3.364
60.5	4782	7.970	6.832	5.978	5.313	4.782	4.347	3.985	3.678	3.416
61.5	4856	8.093	6.937	6.070	5.396	4.856	4.415	4.047	3.735	3.469
62.5	4932	8.220	7.046	6.165	5.480	4.932	4.484	4.110	3.794	3.523
63.5	5010	8.350	7.157	6.262	5.567	5.010	4.555	4.175	3.854	3.579
64.5	5091	8.485	7.273	6.364	5.657	5.091	4.628	4.242	3.916	3.636
65.5	5175	8.625	7.393	6.469	5.750	5.175	4.705	4.312	3.981	3.696
66.5	5262	8.770	7.517	6.578	5.847	5.262	4.784	4.385	4.048	3.759
67.5	5351	8.918	7.644	6.689	5.945	5.351	4.865	4.459	4.116	3.822
68.5	5444	9.073	7.777	6.805	6.049	5.444	4.945	4.536	4.188	3.889
69.5	5541	9.235	7.916	6.926	6.157	5.541	5.037	4.617	4.262	3.958
70.5	5641	9.401	8.059	7.051	6.268	5.641	5.128	4.701	4.339	4.029
71.5	5745	9.575	8.207	7.181	6.383	5.745	5.223	4.787	4.419	4.105
72.5	5854	9.756	8.363	7.318	6.504	5.854	5.322	4.878	4.503	4.181
73.5	5968	9.946	8.526	7.460	6.631	5.968	5.425	4.923	4.591	4.263
74.5	6087	10.145	8.696	7.609	6.783	6.087	5.533	5.072	4.682	4.348
75.5	6212	10.353	8.874	7.765	6.902	6.212	5.647	5.176	4.778	4.437
76.5	6343	10.571	9.062	7.929	7.048	6.343	5.766	5.286	4.879	4.531
77.5	6482	10.803	9.260	8.102	7.202	6.482	5.893	5.401	4.986	4.630
78.5	6628	11.046	9.469	8.285	7.364	6.628	6.025	5.523	5.098	4.734
79.5	6783	11.304	9.690	8.479	7.537	6.783	6.166	5.652	5.218	4.845
80.5	6948	11.580	9.926	8.685	7.720	6.948	6.316	5.790	5.345	4.963
81.5	7123	11.871	10.176	8.904	7.914	7.123	6.475	5.936	5.479	5.088
82.5	7311	12.185	10.444	9.139	8.123	7.311	6.646	6.092	5.624	5.222
83.5	7514	12.523	10.735	9.392	8.349	7.514	6.831	6.261	5.780	5.367
84.5	7732	12.886	11.046	9.665	8.591	7.732	7.029	6.443	5.948	5.523
85.5	7969	13.281	11.385	9.961	8.854	7.969	7.245	6.641	6.130	5.692
86.5	8228	13.712	11.755	10.285	9.142	8.228	7.480	6.856	6.329	5.877
87.5	8513	14.188	12.162	10.641	9.459	8.513	7.739	7.094	6.548	6.081
88.5	8829	14.714	12.613	11.036	9.810	8.829	8.026	7.357	6.691	6.306
89.5	9183	15.304	13.119	11.479	10.203	9.183	8.348	7.652	7.064	6.559
90.5	9583	15.971	13.690	11.979	10.648	9.583	8.712	7.986	7.371	6.845
91.5	10042	16.746	14.346	12.552	11.158	10.042	9.129	8.386	7.726	7.173
92.5	10577	17.628	15.110	13.221	11.752	10.577	9.615	8.813	8.136	7.555
93.5	11215	18.524	16.022	14.019	12.461	11.215	10.195	9.345	8.627	8.011
94.5	11996	19.993	17.137	14.995	13.329	11.996	10.878	9.996	9.228	8.569
95.5	12990	21.649	18.558	16.238	14.433	12.990	11.809	10.825	9.992	9.278
96.5	14327	23.877	20.468	17.909	15.919	14.327	13.025	11.939	10.021	10.233
97.5	16295	27.157	23.279	20.369	18.105	16.295	14.814	13.579	12.535	11.639
98.5	19724	32.872	28.178	24.655	21.915	19.724	17.931	16.436	15.172	14.088
99.5	29372	48.951	41.961	36.715	32.635	29.372	26.702	24.476	22.594	20.980

*The wavelength λ is chosen so that the indicated percentage of blackbody radiation occurs at wavelengths shorter than the indicated wavelength.

8. THEORETICAL MODELS—IDEAL SURFACES

A. Classical Free-Electron Theory

The earliest attempts to predict the optical properties of metals were made by Lorentz, Drude [40], Kronig [88], and Mott and Zener [103], who assumed the metal to contain electrons which were essentially free to move under the influence of the electric field induced by the incident electromagnetic wave. These free electrons are the valence electrons in the outer shell of the atoms constituting the metal.

Table IV. Classical Models for the Optical Properties of Metals (MKS Units)

Drude Free Electron. Assumes the metal contains free electrons which are subjected to an oscillating electric field and a viscous damping force proportional to the velocity of the electrons arising from collisions between accelerated electrons and the atomic lattice.	$K^* = 1 - \left(\frac{\lambda}{\lambda_0}\right)^2 \frac{1 + j(\lambda/\lambda_f)}{1 + (\lambda/\lambda_f)^2}$ $\lambda_f = 2\pi c\tau$ $\lambda_0 = \left[\frac{\pi m c^2 \epsilon_0}{q^2 N}\right]^{1/2}$
Simplified Drude Free Electron. Drude theory valid for long wavelengths where currents in the metal are in phase with electric field.	$K^* = -j\frac{\lambda}{c\epsilon_0 r}$
Extension of Drude Theory—Roberts' Model. Drude theory extended to include several types of free and bound electrons subjected to viscous damping and elastic restoring forces.	$K^* = 1 + \sum_m \frac{K_{om}\lambda^2}{\lambda^2 - \lambda^2_{sm} + j\delta_m\lambda_{sm}\lambda} - \frac{\lambda^2}{2\pi c\epsilon_0}\sum_n \frac{\sigma_n}{\lambda_{rn} - j\lambda}$

When the wave is incident upon its surface, an oscillating electric field parallel to the surface is induced in the metal and the free electrons will oscillate under the influence of this field at the frequency of the incident wave. There is a phase difference between the oscillation of the electrons and that of the field, caused by a viscous damping force arising from collisions between accelerated electrons and the atomic lattice. To describe the optical behavior of the material requires two parameters: the number density of free electrons, N, being excited by the induced field, and the average time (relaxation time τ) between collisions of the electron with the atomic lattice. These two parameters can be estimated from the number of valence electrons per unit volume, the electrical conductivity and the assumption of a spherical Fermi surface. This is called the *Drude Free Electron* model, and is shown in Table IV expressing the complex dielectric constant as a function of the two parameters N and τ.

If the phase change arising from electronic collisions can be neglected, the model describing the optical behavior of the material is greatly simplified. This situation occurs when the relaxation time is zero or when the time between electronic collisions is much less than the period of the induced electric field. For this condition, the optical behavior can be completely described by one material parameter—the dc electrical conductivity. Table IV presents the resulting model for the complex dielectric constant, labeled the *Simplified Drude Free Electron* model.

This simplified model for the optical constants serves as the basis for relations used to compute the thermal radiative properties of materials from knowledge of the electrical conductivity (or resistivity) as a function of temperature. If the appropriate relation between the complex dielectric constant, K^*, and $\epsilon(0; \lambda)$ is used with the simplified Drude model, the normal spectral emissivity can be expressed as a function of the electrical resistivity, r, in the series form

$$\epsilon(0; \lambda) = 0.365\,(r/\lambda)^{1/2} - 0.0464\,(r/\lambda) + \ldots \quad (61)$$

where the units are r(ohm-m) and λ(m). This celebrated relation is frequently referred to as the Hagen–Rubens relation.

From the above discussions, the assumptions used to derive this basic model limit the Hagen–Rubens relation to long wavelengths (usually beyond 10 μm) and high temperatures for metals in which the electronic structure can be approximated by one class of free electrons as the current carriers. This relationship has found extensive use in engineering applications.

From the basic theory yielding the Hagen–Rubens relation, various models have been developed for expressing total and angular property values. Table V shows these relations, expressed in terms of the emissivity, which have been developed over the years [117]. Aschkinass [8] obtained an expression for the normal total emissivity, $\epsilon(0; t)$, by normalization of the first term of the Hagen–Rubens relation with the Planck blackbody distribution curve.

$$\epsilon(0; t) = \frac{\int_0^\infty \epsilon(0; \lambda) L_b(\lambda, T)\, d\lambda}{\int_0^\infty L_b(\lambda, T)\, d\lambda} \quad (62)$$

The result of the integration gives two terms

$$\epsilon(0; t) = 0.5736(rT)^{1/2} - 0.1769(rT) \quad (63)$$

For many metals, particularly at higher temperatures and for relatively high purity, the electrical resistivity is proportional to temperature; through integration of equation (55) it can be shown that the normal total emissivity is proportional to the first power of the temperature. For a blackbody radiator, the radiant exitance, M, is proportional to the fourth power of temperature, that is

$$M = \sigma T^4 \tag{64}$$

and for a nonblackbody

$$M = \epsilon(0; t)\sigma T^4 \tag{65}$$

The Aschkinass relation, coupled with the linearity of electrical resistivity and temperature, shows that the radiant exitance of a metallic surface is to a first approximation proportional to the *fifth* power of the temperature.

Foote [49] improved the utility of the Aschkinass relation by including the second term of the Hagen–Rubens relation in the integration. This gives rise to a third term in the $\epsilon(0; t)$ relation as shown in the Table V.

The relations so far have considered only the case of emission normal to the surface. The classical electromagnetic wave theory can prescribe the angular variation in the optical behavior of the material. Equations (30) and (31) relate the two polarization components of reflectivity to the angle as measured from the surface normal in terms of the optical constants n and k. In the simple Drude theory, n and k are determined by the electrical resistivity. Davisson and Weeks [33] were the first to treat this problem of determining the hemispherical total emissivity using the Fresnel relation for angular dependency.

$$\epsilon(2\pi; t) = 0.751\,(rT)^{1/2} - 0.632\,(rT) + \dots \quad \text{for } rT < 0.1 \tag{66}$$

The limitation in the product rT is usually not more restrictive than the assumptions of the basic model itself. Schmidt and Eckert [130] removed the product of rT limitation and also developed the relations for hemispherical spectral emissivity as shown in Table V. These four relations, along with the Hagen–Rubens relation, constitute the conventional and basic equations used to estimate the radiative properties of metals.

The Schmidt–Eckert relations also yield the ratio of hemispherical to normal emissivity of metallic surfaces. Table VI presents the results showing that for metals the hemispherical emissivity ranges from 1.33 to 1.05 times the normal emissivity; the values are also shown graphically in Fig. 9 and compared to the theory for nonconductors.

Useful relations for predicting the angular variation of spectral emissivity for the two degrees of polarization are:

Table V. Classical Models for the Emissivity of Metals (MKS Units)

Model	Relation	Range
Hagen and Rubens [66]. Also called the Drude–Zener relations; applicable at long wavelengths and/or high temperature; derived from classical electromagnetic theory assuming negligible phase change of electrical field within the metal.	$\epsilon(0; \lambda) = 0.365\,(r/\lambda)^{1/2} - 0.0464\,(r/\lambda) + \dots$	
Aschkinass [8]. Obtained by integration of Hagen–Rubens' first term with Planck distribution curve; when combined with r vs T linearity relation then radiant energy is proportional to the fifth power of the temperature.	$\epsilon(0; t) = 0.5736\,(rT)^{1/2} - 0.1769\,(rT)$	
Foote [49]. Obtained by integration of Hagen–Rubens' first and second terms with Planck distribution curve.	$\epsilon(0; t) = 0.578\,(rT)^{1/2} - 0.178\,(rT) + 0.0584\,(rT)^{3/2}$	
Davisson and Weeks [33]. Using the Fresnel relation for angular dependency, hemispherical emissivity is obtained.	$\epsilon(2\pi; t) = 0.751\,(rT)^{1/2} - 0.632\,(rT) + \dots$	$rT < 0.1$
Schmidt and Eckert [130]. With graphical integration techniques, the limit of applicability of the Davisson–Weeks model is extended.	$\epsilon(2\pi; \lambda) = 0.476\,(r\lambda)^{1/2} - 0.148\,(r/\lambda)$	$0 < r/\lambda < 0.5$
	$\epsilon(2\pi; \lambda) = 0.442\,(r/\lambda)^{1/2} - 0.0995\,(r/\lambda)$	$0.5 < r/\lambda < 2.5$
	$\epsilon(2\pi; t) = 0.751\,(rT)^{1/2} - 0.396\,(rT)$	$0 < rT < 0.2$
	$\epsilon(2\pi; t) = 0.698\,(rT)^{1/2} - 0.266\,(rT)$	$0.2 < rT < 0.5$
Parker and Abbott [109]. Assumed the Drude theory with finite relaxation time; expressions for $\epsilon(0; t)$, $\epsilon(2\pi; \lambda)$ and $\epsilon(2\pi; t)$ are also developed.	$\epsilon(0; \lambda) = 2\sum_{n=1}^{\infty} (-1)^{n+1} j^n \left(\frac{r}{30\lambda}\right)^{n/2}$; where $j = [(1 + y^2)^{1/2} - y]^{1/2}$, $y = \frac{2\pi cm}{\lambda r N q^2}$, $r = \frac{m}{Nq^2\tau}$	$rT \ll \frac{7.8}{1 + y^2}$

$$\epsilon(\theta;\lambda)_s = \frac{4(30\lambda/r)^{1/2}\cos\theta}{60\lambda/r + 2(30\lambda/r)^{1/2}\cos\theta + \cos^2\theta} \tag{67}$$

$$\epsilon(\theta;\lambda)_p = \frac{4(30\lambda/r)^{1/2}}{60(\lambda/r)\cos^2\theta + 2(30\lambda/r)^{1/2}\cos\theta + 1} \tag{68}$$

The wavelength dependence is based upon the Hagen–Rubens relation; the angular dependence and polarization are prescribed by the Fresnel relation as a direct consequence of electromagnetic wave theory and the interaction process with polished metals [77].

The models prescribed above all assume that the phase difference between the induced electric field and the electron current carriers is negligible. That is, the relaxation time of the electron carriers is small compared to the period of the induced electric field. Parker and Abbott [109] have developed a relation for the emissivity which accounts for a finite relaxation time. The resulting equation for the normal spectral emissivity is presented in Table V. To use this equation, one needs to know two parameters of the material: the relaxation time, τ, and the free electron density, N. This model develops relations for other subproperties of emissivity but does not find widespread use because of the difficulty in evaluating τ and N as functions of temperature. As a result, this model would find more utility in evaluation of measured data rather than in estimation of property values.

Figures 10 and 11 compare the classical theory with selected data on the normal total and hemispherical total emissivity of several metallic elements. The dashed line—denoted as theoretical—refers to negligible relaxation time; thus the appropriate equations in Table V are those of Schmidt–Eckert and of Parker–Abbott.

Figure 12 shows the results of a recent study by Bennett and Bennett [14] regarding the validity of the Drude theory; the infrared reflectance of carefully

Table VI. Ratio of Hemispherical to Normal Emissivity of Metallic Conductors [77]

$\epsilon(2\pi; t)$	$\epsilon(2\pi; t)/\epsilon(0; t)$
0	1.33
0.05	1.27
0.10	1.225
0.15	1.185
0.20	1.145
0.25	1.105
0.30	1.075
0.35	1.055

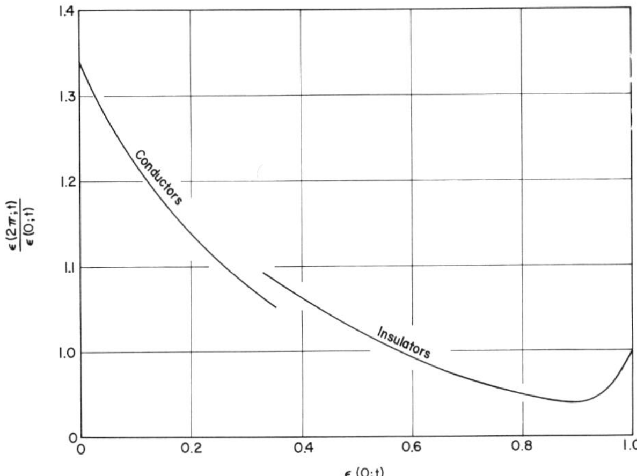

Fig. 9. Ratio of hemispherical to normal total emissivity of radiating surfaces [77].

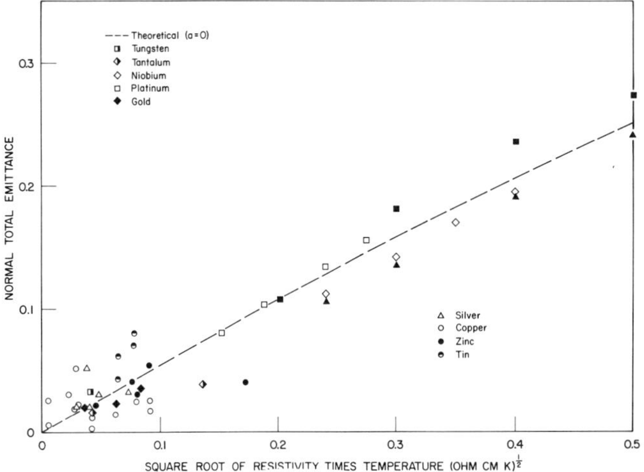

Fig. 10. Normal total emittance of various metallic elements [109].

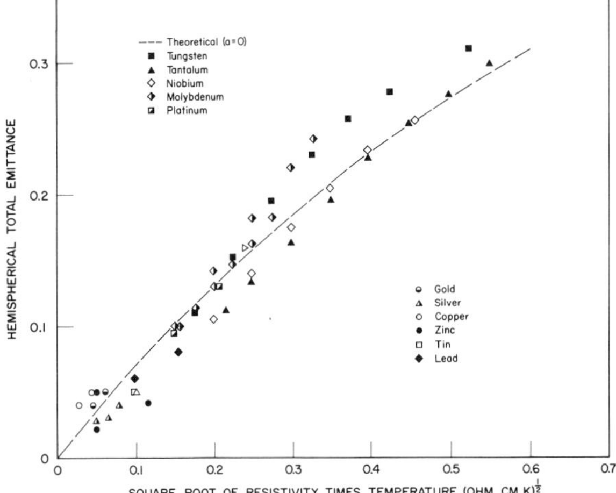

Fig. 11. Hemispherical total emittance of various metallic elements [109].

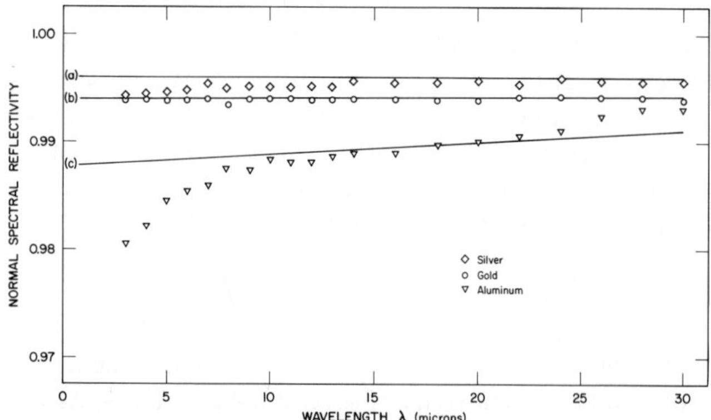

Fig. 12. The infrared reflectance of silver, gold, and aluminum high-vacuum evaporated films compared with the Simple Drude model [14].

prepared ultrahigh-vacuum-deposited silver, gold, and aluminum films are shown to be in excellent agreement with the theoretical predictions. Their work points out the need for proper specimen preparation if the appropriate theoretical models are to be in agreement.

B. Other Models

The failure of the free-electron models to yield computed values that agree with experimental data in the wavelength regions of interest should not be interpreted as a failure of the classical approach, but rather should be ascribed to the over-simplification of the mechanism of the absorption process. Quantum theory states that both free and bound electrons exist in a metal and that both types can exist in different energy state.

The basic classical model of a metal consisting of one type of free electron which is responsible for the absorption process is too simple to explain the behavior of many metals. There may be several types of free electrons each characterized by a different relaxation time and density function. Contributions to the absorption process arising from the influence of the induced electric field can come from electrons which are bound to the lattice. It is normally assumed that the electrons are elastically bound in their equilibrium positions and each type of bound electron is characterized by particular values of the three parameters—an elastic restoring force, a viscous damping force, and the number density of the electron type.

The equations for the optical constants for the most general case of a metal with several types of free and bound electrons are presented in Table IV. Roberts [121, 122, 123] has shown that his form of the relation is successful in describing "humps" and inflection points observed in the optical constants data. The relation is primarily useful for correlation of measured data, and in general this approach does not permit prediction or estimation of the optical properties. Despite the large number of parameters, the correlation is not an arbitrary curve-fitting operation, as there are certain physical constraints that the relation parameters must satisfy.

The influence of bound-electron contributions on optical behavior can only be ignored in the long-wavelength region of the spectrum. The resonance wavelength of the bound electrons in metallic materials is usually found in the visible or very near infrared (less than 1 μm), but the effect or contribution to the absorption process is not confined to the immediate wavelength region about the resonance wavelength. Using the terminology of Roberts, the influence of individual terms of the multi-free bound electron model on emissivity is presented in Fig. 13 for typical metals.

Seban [131] studied the Roberts model for the emissivity of the transition metals iron, nickel, and platinum in the spectral region 2 to 15 μm. Several types of free-electron carriers and one resonance term are required to correlate the data. The study deals with the possibility of predicting high-temperature property values from room-temperature measurements and knowledge of the temperature dependence of the various model constants, especially the conductivity terms.

Edwards and deVolo [47] studied the influence of free-electron carriers for some twenty metals for the purpose of developing useful approximate

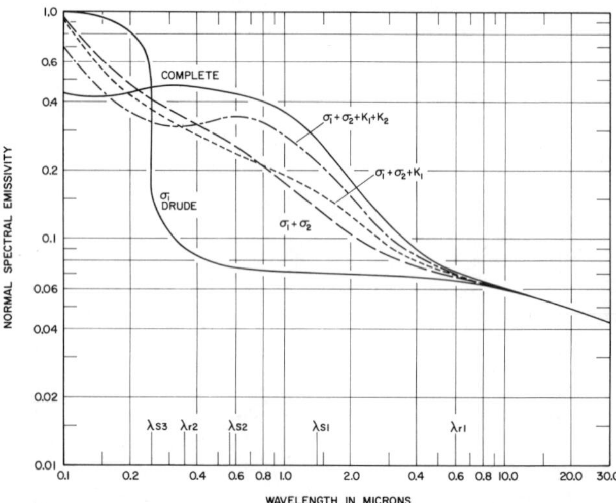

Fig. 13. The effect of individual terms of the multi-free/bound electron (Roberts') model on normal spectral emissivity of a representative metal.

expressions for correlating experimental data. No simple techniques have been developed for estimating or predicting the Roberts or multi-free/bound model parameters, consequently this more complicated but realistic model of the absorption process finds limited applications, primarily in research studies.

The classical free-electron theory resulting in the several models discussed previously all assume that the relaxation time is small compared to the period of the electrical field induced by the absorbed radiant energy. At low temperature and short wavelengths (high frequency) it is no longer permissible to make this simplifying assumption. The strength of the electric field changes appreciably within the mean free path of the electrons, and hence the current density (free-electron flow) at a given point within the metal is determined only by the value of the electric field at that point. The theory to account for this effect is called the anomalous skin effect and is applicable to low-temperature regions for the alkali and noble metals [38, 135]. Relations have not been developed to treat metals which are other than one band type, and even then the comparison of theory and data is only just fair.

Recently a more generalized quantum theory for the optical constants has been developed [132]. The theory deals with normal dispersion (in contrast to anomalous dispersion) so that the frequency spectrum involved must be sufficiently removed from the anomalous region in which photoelectric resonances of bound electrons become important. As a result of theory, a simple expression for the normal spectral emissivity is

$$\epsilon(0; \lambda) = 2(\omega r/2\pi)^{1/2}[b(\omega, T)]^{1/2} \tag{69}$$

The term $b(\omega, T)$ is a quantum correction factor, which in the limit of the wavelength spectrum approaches unity. In this sense, the formula above may be referred to as the generalized Hagen–Rubens relation.

The quantum "b-factor" is determined from the overall electronic damping coefficient which is the sum of contributions by electron–phonon collisions, electron–electron collisions, and impurity scattering. The electron–phonon process is of primary importance for pure metals while the electron–electron process may be significant either at very low temperatures or in the high-frequency limit of the spectrum. By comparison of the theory with optical constants data, it is possible to determine the damping coefficients along with other lattice parameters. Then with these values at one temperature, the theory provides for determining the parameters at any other temperature. While the theory has not been rigorously tested, there is sufficient evidence to believe that the approach could eventually be utilized for engineering applications.

9. SURFACE CHARACTERIZATION

A. Real Surface Effects [37, 54, 124]

It has been understood for many years that the surface condition of metallic specimens plays a dominant role in the magnitude of the radiative properties. The literature abounds with examples of test surfaces shown to be very sensitive to methods of preparation, thermal history, and environmental conditions. Despite this awareness, descriptions of test surfaces are generally inadequate because of our modest understanding of the important mechanisms of real surface effects and how to properly characterize a surface.

From consideration of wave theory, it is apparent that the reflection process is influenced by the surface condition. The depth of penetration of an incident electric field can range from very small to very large values depending upon the absorption and scattering properties of the material. For metals, the absorption property dominates the process, since the extinction coefficient is quite large in the thermal spectrum and radiant energy will not travel more than a few hundred angstroms before being totally absorbed. This is not the case for nonmetallic or dielectric materials, which are shown to be less sensitive to surface conditions [118]. In the emission process for metals, internally generated radiant energy reaching the surface originates from atomic and molecular oscillators characteristic of the chemical composition and physical state in a very thin surface layer from 0 to perhaps 1000 Å deep.

Topographical, chemical, and physical (structural) characteristics all influence the properties of the metallic surface. The topographical characteristics describe the profile or geometry of the surface—the boundary between the material and the surrounding medium. The chemical characteristics describe the composition of the surface layer including such features as inhomogeneities and contaminants. The physical characteristics describe the structure of the surface such as crystal lattice orientation, particle size, strain, and other features which might affect the radiant energy exchange process.

To isolate the individual surface characteristics as outlined is a difficult task. For most materials it is not practical to alter one characteristic without causing an influence on another. The control of the many variables required to study surface characterization in a logical manner is a complex problem. As a result only the simplest of surface profiles or compositional effects have been studied or are understood. In the succeeding sections studies on characterization are discussed.

The most important influences on the radiative properties of metals arise from surface roughness and films (oxide growth). The effect is most pronounced on the spectral radiative properties when the characteristic profile variation or film thickness is on the same order as the wavelength of interest. For some situations a thin dielectric film has a more significant influence on emittance properties than does surface roughness of the same dimension. These changes in spectral properties are also apparent as changes in angular distribution of reflected or emitted energy.

The influences of surface characteristics—topographical, chemical, physical—can be considerably dependent upon the energy spectrum of importance to the radiative property of interest. For example, the description of a surface for use as a room temperature absorber ($5 < \lambda < 40\ \mu$m) will be quite different than that for a solar absorber ($0.25 < \lambda < 4\ \mu$m). Also the techniques required to study each will be quite different.

B. Topographical Characterization

The profiles of real metal surfaces are always shown as irregular patterns of peaks and valleys. Various parameters are in common use to describe the topography of a surface including RMS (root mean square) height, CLA (center line average) height, lay, average slope, height distribution, etc. [1, 83, 136]. Such parameters are obtained primarily from stylus-type profilometers and to some extent from interferometry techniques.

The effect of surface roughness on the optical properties of materials was first studied by Lord Rayleigh, but only recently has this problem been of intense interest. If the size of the irregularities is of the order of the wavelength or larger, the interaction can be described by geometrical optics [141]. In this case, the facets of the surfaces reflect in various directions, and the properties/orientation of the facets must be described by some statistical process in order to explain the optical behavior of the surface. If, however, the surface irregularities are much smaller than the wavelength, the optical behavior can be explained by diffraction phenomena.

The diffraction problem was originally studied by Rice [111] and Davies [32] and their work was extended and experimentally verified by Bennett and Porteus [16]. Their expression for the relative reflectance ratio of the rough, ρ, to smooth, ρ_0, surface at normal incidence is given as

$$\frac{\rho}{\rho_0} = \exp[-(4\pi\sigma/\lambda)^2] + 32\pi^4(\sigma/\lambda)^4(\Delta\theta)^2/m \quad (70)$$

where σ is the RMS roughness, m is the RMS slope, and $\Delta\theta$ is the half angle of acceptance of the optical system. The first term represents the coherently or specularly reflected fraction and the second term the incoherent or diffusely reflected term. The second term is shown proportional to $(\sigma/\lambda)^4$, and hence for longer wavelengths and smoother surfaces the first term predominates.

Porteus [110] showed that when the second term is important, that is for larger values of σ/λ still within the diffraction range, an increasingly exact statistical description of the surface is required. This necessarily follows when taken to the limiting case of the geometrical optics region. For a surface of Gaussian height distribution, approximated by aluminized ground glass, the theory is in good agreement with the data [12].

This theory, adequate for the range of σ/λ within the diffraction region, assumes that there is no change in the hemispherical reflectance of the roughened surface when compared to the smooth one, but rather a change in the ratio of specular to diffuse reflected fractions. It would then follow directly from Kirchhoff's law that no change in total emittance can occur in this region because of surface roughness; a value of $\sigma/\lambda = 0.15$ is an upper limit for the diffraction range. For room temperature the maximum energy occurs at 10 μm and hence, for a surface of roughness of 1.5 μm or 60 μin. (a rather rough finished metal surface) there is a negligible effect on the energy radiated.

For very rough surfaces and short wavelengths, geometrical optical effects predominate. Zipin [149] studied the directional reflectance of a series of regular V-shaped specimens of RMS heights (σ) 2, 5, 12, and 24 μin. in the visible region. The combination of σ/λ is such that the two limiting cases of geometrical and diffraction effects could be studied conveniently. In the range of geometrical effects, only regular-shaped asperities can be handled without resorting to elaborate statistical tools [10, 13]. Little progress has been made in the correlation of rough surface characteristics in this region [19, 125, 126].

C. Chemical Characterization

The real metal surface unavoidably has a surface film of one type or another. In engineering applications, these films may be greases or other deposits, but normally they are oxides of the base metal. Even if only the natural oxide layer is considered, characterization is difficult. The layer is a mixture of several different chemical species (metal atoms, oxygen ions, and one or more metal ions); the interface between oxide and base metal is not smooth; the rate of oxide growth is dependent upon the base metal surface topography, structure, temperature, and atmospheric conditions. Numerous reviews and studies have been reported on oxide-layer growth, but the basic mechanism is not completely understood for some metals, for there are many conflicting theories and opinions. It is evident that considerable work is necessary to further our understanding of how to chemically characterize a surface.

The coatings or films formed by oxidation can have very significant effects on the thermal performance of a surface. Generally speaking, the use of coatings—combinations of layers or films—for tailoring thermal performance is very much a topic of current interest. The behavior of the material ranges from that of a pure base metal surface to the condition where the oxide (or other film) dominates behavior and the surface has the characteristics of a nonmetallic material. Oxidized metallic surfaces can be stabilized and are useful for many purposes; oxidized Kanthal and sand-blasted/oxidized Inconel are two such materials and are in current use as reference standards of emittance [115].

D. Physical (Structural) Characterization

The surface characteristics just discussed—topographical and chemical—are easily understood to have important influences on optical behavior. However, the effects caused by physical structure beneath the surface are not so apparent [39]. In the case of a perfectly smooth surface free from surface films, a surface layer several hundred angstroms deep (skin depth of penetration of radiation) is responsible for the optical behavior of the metal. Structural features of this layer, such as adsorbed gas atoms, lattice imperfections, and crystallinity variations, can be expected to have an influence on the optical behavior.

It has been shown that some mechanical polishing processes cause structural changes in layers near the surface [128]. There is evidence of a layer of supercooled fluid metal that fills the scratches caused by the abrasive [11]. This layer (called the Beilby layer) has no clearly defined interface with the underlying metal. The outermost regions of this layer appear to be completely noncrystalline, and the structure becomes more crystalline as the depth from the outer surface increases. In some cases, the layer will contain oxide. It has also been suggested that the polished layer is unstable and will revert, in time, to the ordinary crystalline state.

Heat-treating, cold-working, various surface preparation treatments, and many other processes can give rise to variations in physical and chemical characteristics between the bulk and surface layers. However, concrete evidence is lacking that these variations can in all cases cause significant changes in the optical behavior of metals.

10. METHODS OF MEASUREMENT

A. Basic Techniques

Many experimental techniques for measuring the thermal radiative properties of materials have been described in the literature. It is obviously impossible to describe all of the methods in detail here. Instead, the general principles involved will be discussed briefly, and a few general methods will be outlined. The reader is referred to the original references for detailed descriptions of the methods.

In selecting a method for evaluating a given property, it is well to keep in mind the equations previously discussed, which can be used to compute one property from another. It frequently happens that the desired property data can be obtained more easily by computation from data obtained by direct measurement of another property, than by direct measurement of the desired property.

Methods of evaluating radiant flux fall into two general categories, calorimetric and radiometric. In calorimetric techniques the radiant flux absorbed or emitted by a sample is evaluated in terms of the heat lost or gained by the sample. In radiometric techniques, the radiant flux is measured directly. In general, calorimetric techniques tend to be relatively simple and free from systematic error, but of relatively low precision. Radiometric techniques, on the other hand, tend to require more elaborate equipment, and, while capable of relatively high precision, are subject to systematic errors that may be difficult to evaluate.

Calorimetric techniques are used to measure absorptance and emittance; they are not suitable for direct measurement of reflectance. In essence, a sample is placed in an environment where all heat transfer to or from it is by radiation, or is in a form such that it can be evaluated directly in power units, such as power input to a direct electrically heated sample, or heat loss in terms of the rate of boiling of a liquid such as water or nitrogen. If a steady-state condition is achieved, the heat input can be equated to the heat output, and the net radiant heat transfer can be related to the temperature of the sample. If the temperature of the sample is changing, the net heat transfer to or from the sample can be related to the temperature and rate of temperature change. The desired thermal radiative properties can be computed from the measured rate of radiant heat transfer, the geometry of the system, and the temperature of the sample and its surroundings.

In radiometric techniques the emitted incident and/or reflected radiant flux is measured directly and the desired property is computed as the ratio of measured fluxes. Rather elaborate optical systems may be required to collect the radiant flux from the source and focus it onto the sample for reflectance measurements and to collect the flux reflected or emitted by the specimen and focus it onto the detector for reflectance and emittance measurements. If spectral measurements are made, a monochromator must be included in the optical path. The major

Table VII. Primary Reflectance Techniques

Type of reflectometer	Property	Wavelength range, μm	Temperature range, K	References	Remarks
Heated Cavity (Hohlraum)	$\rho(0; 2\pi)$	2–35	RT–900	58	A versatile infrared instrument; in widespread use.
Integrating Spheres					
Substitution/comparative type	$R(0; 2\pi)$	0.3–2.6	RT	61, 76	Sample mounted at sphere surface; technique used in many commercial spectrophotometers; may be sensitive to sample texture.
Absolute	$\rho(\theta; 2\pi)$	0.2–2.6	RT	48	Sample mounted at sphere center; suitable for sample of arbitrary reflection distribution function.
Laser source	$R(0; 2\pi)$	0.63, 1.15	RT–2500	85	Laser used as conventional source.
Integrating Mirrors					
Coblentz hemisphere	$\rho(\theta; 2\pi)$	1–15	RT–600	30, 78	Sample and detector or source are located at conjugate focal points of a hemispherical mirror; aberrations are a serious source of error.
Paraboloidal	$\rho(\theta; 2\pi)$	2–100	RT	42, 104	Suitable for samples of arbitrary reflection distribution function.
Ellipsoidal	$\rho(\theta; 2\pi)$	1–15	RT–400	43	Aberrations are reduced by using an ellipsoidal mirror with true focal points instead of a hemispherical mirror.
Multiple Reflection					
Strong technique	$\rho(0; 0)$	0.3–35	RT	50, 56, 139	For specular reflectors only.
Bennett–Koehler modification	$\rho(0; 0)$	0.2–35	RT	15	Errors minimized by unique optical design, accuracy state of the art ± 0.001 unit.

errors in radiometric determinations arise from flux losses in the optical system.

As mentioned in Section 9, the thermal radiative properties of metallic surfaces are as much related to the character of the surface as to the material of which it is composed. Any report of thermal radiative property measurements should therefore include a complete description of the surface, giving all available information on its physical and chemical characteristics.

B. Reflectance Techniques [44]

The more important techniques for measuring reflectance are summarized in Table VII. Only methods that measure essentially all of the reflected radiant energy are included in the table, since methods that measure only a portion of the reflected radiant energy have only limited application. Goniometric reflectance methods, which measure the geometric distribution of the reflected radiant energy are described in a separate section.

Five separate techniques are described in the following sections; they are: (a) heated cavity reflectometers, (b) integrating sphere reflectometers, (c) integrating mirror reflectometers, (d) specular reflectometers, and (e) gonioreflectometers.

a. Heated Cavity Reflectometers

The Gier–Dunkle reflectometer [58] is sketched in Figs. 14A and B. The sample forms part of the wall of a blackbody cavity and is irradiated over a hemisphere by blackbody radiation from the hot cavity walls. Images of the sample and a spot on the cavity wall are alternately focused by the optical system onto the entrance slit of the monochromator. The output of the monochromator is the ratio of the radiances in the two beams, which is a measure of the hemispherical directional reflectance factor $R(2\pi;\, \theta',\, \omega')$ of the specimen.

The sample is usually viewed at an angle of about 4° from the normal, so that the specular component is included in the measured value. If desired, the sample can be viewed normally, so that the specular component is not included in the measured value.

The double-beam monochromator frequently used with the hohlraum reflectometer is a prism instrument with a thermocouple detector that covers the wavelength range of 1 to 35 μm. The wavelength range is set primarily by the energy available for measurement and the sensitivity of the detector.

Because $R(2\pi;\, \theta',\, \omega') = \rho(\theta,\, \omega;\, 2\pi)$, the results are frequently reported as the directional hemispherical reflectance and are reported as such in the data compilation of this volume, with a footnote to the effect that $R(2\pi;\, \theta',\, \omega')$ was actually measured.

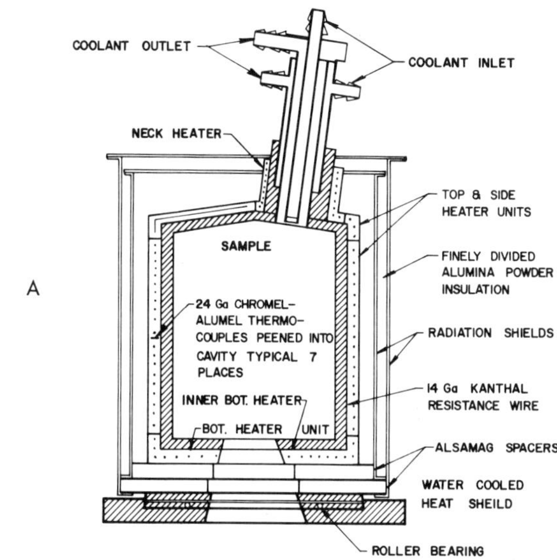

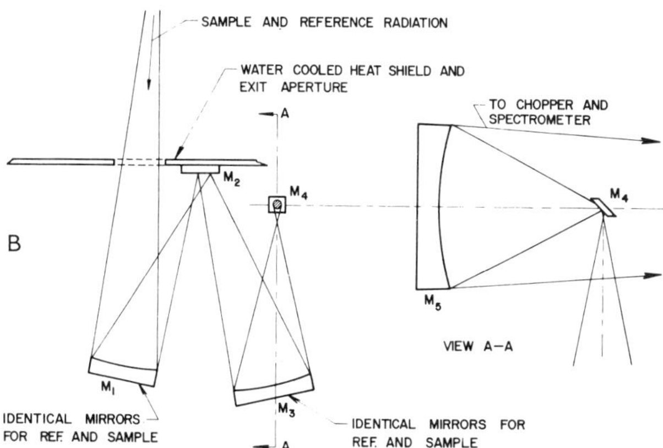

Fig. 14. Schematic of the heated cavity (Hohlraum) reflectometer—the cavity and specimen arrangement. (A) Cavity and sample arrangement. (B) Auxiliary optics—sample and reference beam viewing.

There are many sources of error in heated cavity reflectance measurements which have been discussed in detail in the literature [42, 73, 138]. The principal sources of error arise due to nonuniform irradiance of the sample over the hemisphere as a result of the aperture and thermal gradients in the cavity, emission from the sample due to heating, and failure of the radiance of the spot viewed on the cavity wall to represent the average radiance of the cavity walls, again due to thermal gradients in the cavity.

The hohlraum reflectometer irradiates the sample over nearly a complete hemisphere with nearly uniform radiance, hence the measured values are nearly independent of the geometric distribution of flux reflected by the sample. The instrument can be used to measure both specularly and diffusely reflecting samples. It is not recommended that measurements be made on heated samples since it is not possible to separate the radiant energy reflected and emitted by the sample.

Several versions of the heated cavity reflectometer have been built and the reader is referred to the literature for details. One major advance [41] has been to place the water-cooled specimen at the center of the cavity, away from the walls, so that the angle of viewing can be varied from normal to near grazing. This modification has the further advantage that cooling of the cavity walls by the cooling water for the specimen is virtually eliminated, hence thermal gradients within the cavity are usually much smaller than for the type of cavity shown in Fig. 14A.

b. Integrating Sphere Reflectometers

Integrating sphere reflectometers are used in greater numbers than any other type, and many different commercial instruments of this type are available. The reflectometer uses an integrating sphere to collect the radiant energy reflected by a sample. The reflected flux is distributed uniformly over the surface of the sphere, where it can be sampled by a detector. The sphere has an inside surface that has a high uniform reflectance and is a near-perfect diffuse reflector. There are several apertures in the sphere, one for the incident beam, one for the detector, and one for the sample. There may also be one for the comparison standard.

The theory of the integrating sphere is based on two fundamental laws of radiation: (1) the flux received by an elemental area from a point source is inversely proportional to the square of the distance from the source and directly proportional to the cosine of the angle between the normal to the surface and the direction of incidence, and (2) the flux reflected by a perfect diffuser follows the cosine distribution law, which means that the flux per unit solid angle reflected from a unit surface area in a given direction is proportional to the cosine of the angle between the normal to the surface and the given direction. Thus, for a sphere having an inner surface of uniform perfectly diffuse reflectance, the flux reflected by an area on the sphere wall is uniformly distributed over the surface of the sphere.

Integrating sphere instruments can be operated in the direct or indirect mode. In the direct mode the sample is irradiated directly, and the detector views an area on the sphere wall. A similar reading is then taken on a comparison standard of known reflectance under the same conditions. The measured directional-hemispherical reflectance factor $R(\theta, \omega; 2\pi)$ is then the ratio of the sample reading to the standard reading times the known reflectance of the standard.

In the indirect mode the beam from the source is first incident on the sphere wall, and the sample is irradiated uniformly over the hemisphere. The detector views the sample directly. A similar reading is then taken on a comparison standard of known reflectance under the same conditions. The measured hemispherical-directional reflectance factor $R(2\pi; \theta', \omega')$ is then the ratio of the sample reading to the standard reading times the known reflectance of the standard. The measured reflectance factors $R(\theta, \omega; 2\pi)$ and $R(2\pi; \theta', \omega')$ are each equal to the directional hemispherical reflectance $\rho(\theta, \omega; 2\pi)$ and are frequently reported as such.

The sample may be removed and replaced by the standard (substitution method), or there may be separate sample and standard apertures which are alternately irradiated or viewed (comparison method). Practically all double-beam recording instruments use the comparison method. In this method the sample and standard apertures should be, and usually are, located symmetrically with respect to the entrance aperture, the detector aperture, and the area of sphere wall viewed by the detector or irradiated by the source.

Errors which may be significant for some specimens can be avoided if there is an internal shield in the sphere which prevents the radiant energy reflected by this specimen, in direct-mode operation, from falling directly onto the area of sphere wall viewed by the detector, or, in indirect-mode operation, that prevents the radiant energy reflected by the sphere wall on the first reflection from falling on the specimen.

Integrating sphere reflectometers frequently include a monochromator, which usually is located in the incident beam between the source and the sphere in instruments operated in the direct mode, and between the sphere and the detector in instruments operated in the indirect mode.

The wavelength range over which an integrating sphere reflectometer can be used is determined by the reflectance characteristics of the sphere walls. For the most commonly used wall coatings, magnesium

actual geometric distribution of reflected radiant energy from the sample being measured, and the final error can be reduced to less than one percent. A high-quality mirror of known reflectance is used as the reference standard. The useful range of the instrument used in this manner is about 0.4 to 8.0 μm with the long-wavelength limit being imposed by the low efficiency of the averaging sphere. Detector scanning devices [119] can be used to extend the wavelength range to 15 μm.

An improved version of this instrument [119] employs a large area source surrounding the second focal point, which irradiates a sample at the first focal point uniformly over a hemisphere. The specimen is viewed through a small hole in the mirror, and the reflected beam is passed through a monochromator in front of the detector. When used in this mode, satisfactory operation was attained at wavelengths out to beyond 30 μm.

The ellipsoidal mirror reflectometer measures directional-hemispherical reflectance factor $R(\theta,\omega;2\pi)$ when used in the direct mode (detector at second focal point) and hemispherical directional reflectance factor $R(2\pi;\theta',\omega')$ when used in the inverse mode (source at the second focal point).

d. Specular Reflectometers

For high-quality mirrors the multiple-reflection technique [50, 56, 139] is simple and accurate. Figure 21 shows a schematic of the optical path. The

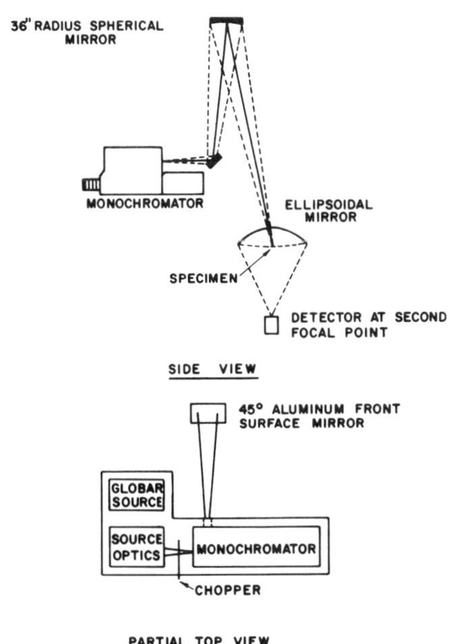

Fig. 20. Ellipsoidal mirror reflectometer.

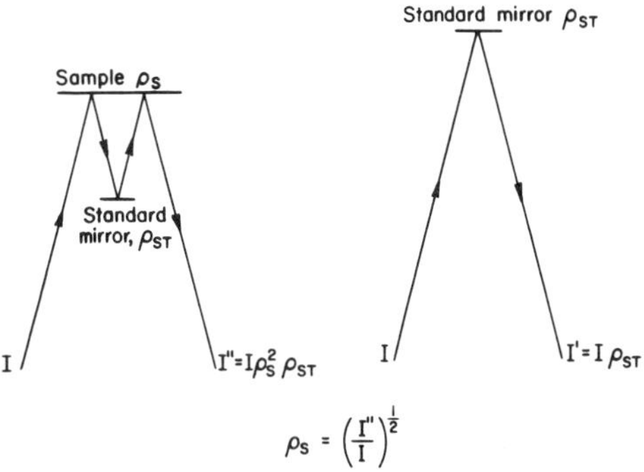

Fig. 21. The Strong [139] multiple-reflection technique.

reflectance is computed as the square root of the ratio of two measured fluxes, hence the uncertainty in the reflectance is about half that in the measured ratio. Very accurate alignment of the mirrors, and optically flat mirrors, are required to maintain the same beam geometry for the two configurations. The major source of error is due to slight displacement of the reflected beam on the detector, since for most detectors the output varies with the position of the incident beam on the detector. Flux-averaging devices in front of the detector can help to minimize this error [15].

Bennett and Koehler [19] have improved the optical system, as illustrated in Fig. 22, to reduce the errors due to slight shifts in the alignment of the sample mirror, or slight deviations of the sample from optical flatness. They report accuracies of ± 0.001 reflectance units over the wavelength range of 0.45 to 22.5 μm.

e. Gonioreflectometers

A gonioreflectometer measures the bidirectional reflectance as a function of angle of reflectance for any given angle of incidence. Several instruments [25, 62, 85] have been constructed to measure bidirectional reflectance in the plane of incidence. More elaborate instruments [24, 31, 101] have been developed to measure bidirectional reflectance over the entire hemisphere. The latter instruments permit scanning over nearly the entire hemisphere for any given direction of incidence, and the angle of incidence can be varied from normal to near grazing. In addition, the measurements can be made with monochromatic incident radiant energy over the wavelength range of about 0.5 to 1.0 μm.

C. Radiometric Emittance Techniques

Thermal emittance may be measured directly by measuring the ratio of the radiance of a sample to

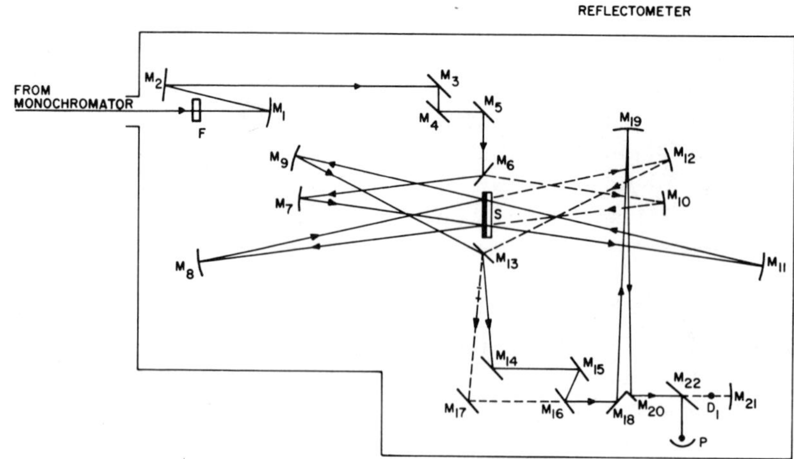

Fig. 22. The Bennett–Koehler [15] high-precision reflectometer.

that of a blackbody radiator at the same temperature, and under the same geometric and wavelength conditions of viewing. The measured property is then the directional emittance of the sample, either total, if all wavelengths are measured at once with a nonselective detector, or spectral, if the measurement is confined to a narrow spectral band.

The comparison blackbody may be either an integral blackbody cavity, whose walls are formed by the sample [9, 36, 75, 102, 120] or a separate blackbody controlled at the temperature of the sample [63, 113]. The integral blackbody is preferred when working at very high temperatures, where temperature measurement and control may be difficult, or at very short wavelengths, where extremely close temperature control is required for accurate measurements. This technique is particularly suitable when measuring samples that are heated in a vacuum or controlled atmospheres, where they must be viewed through a window. The transmittance of the window may be changed as a result of condensation on its inner surface of material volatilized from the hot specimen. Such change in transmittance can cause serious errors, which are compensated when both the sample and comparison blackbody are viewed through the same window. The separate blackbody technique is most accurate at temperatures below 1800 K, where temperature measurement and control by use of conventional thermocouples present no serious problems.

Spectral emittance is usually desired at all wavelengths where significant amounts of radiant energy are emitted. This range was shown as a function of temperature in Fig. 8. As a practical matter, most measurements are made in the 1 to 15 μm range, with a few being extended out to 30 or 35 μm. Measurements at wavelengths below about 1 μm are difficult because of temperature control problems; measurements at long wavelengths are difficult because of the low energy available for measurement.

The change in temperature required to change the spectral radiance of a blackbody radiator by 0.5 percent is given in Table VIII for several wavelengths and temperatures. These data indicate the degree of temperature control that is required for accurate emittance measurements, particularly by the separate blackbody technique.

An example of the separate blackbody technique, described in reference [113], is shown in Fig. 23. The sample, in the form of a flat strip, is heated by passing a current through it and is enclosed in a water-cooled shield to reduce thermal gradients due to convective cooling. Two identical laboratory blackbodies are used, designed to have an effective emittance of 0.998 or better. One blackbody always is used as the source for the comparison beam of the spectrometer; either the hot sample or the reference blackbody can be used as the source for the sample beam of the spectrometer.

In operation, the comparison blackbody is brought to the desired temperature by manual adjustment of the power input. The reference blackbody is placed in position to serve as the source for the sample beam of the spectrometer, and its temperature is controlled to be the same as that of the comparison blackbody by means of a differential thermocouple, one junction of which is in each blackbody. The ratio of the signals from the two beams, plotted as a function of wavelength, is referred to as the "100 percent line." The reference

Table VIII. Allowable Temperature Variation for a $\pm 1/2\%$ Variation of Blackbody Radiance

	Temperature differences, ΔT (K)					
T (K)	$\lambda = 0.4\ \mu m$	$\lambda = 0.5\ \mu m$	$\lambda = 0.75\ \mu m$	$\lambda = 1.0\ \mu m$	$\lambda = 5\ \mu m$	$\lambda = 15\ \mu m$
500	0.035	0.043	0.065	0.087	0.433	1.112
1000	0.139	0.174	0.261	0.348	1.640	3.215
1500	0.313	0.391	0.586	0.782	3.335	5.538
2000	0.556	0.695	1.042	1.389	5.302	7.949
2500	0.869	1.086	1.628	2.164	7.425	10.389
3000	1.251	1.564	2.343	3.102	9.639	12.848

blackbody is then replaced by the specimen furnace, and the sample is controlled at the same temperature as the comparison blackbody, again by means of a differential thermocouple, one junction of which is in the blackbody cavity, and the other being formed by separate wire welds to the sample. The ratio of the signals from the two beams, plotted as a function of wavelength, is referred to as the "sample line." The specimen beam is then blocked near the specimen furnace, and the ratio of the signals from the two beams is referred to as the "zero line." The spectral emittance at any wavelength, $\epsilon(0; \lambda)$ is then computed as

$$\epsilon(0; \lambda) = \frac{S - Z}{H - Z} \tag{71}$$

where S is the height of the sample line, H the height of the 100 percent line, and Z the height of the zero line at wavelength λ. Standards of emittance [117] have been calibrated by this technique, and are available from the Office of Standard Reference Materials, National Bureau of Standards, Washington, D.C. 20234.

The integral blackbody approach is illustrated in Fig. 24. A thin-walled tubular specimen is heated by passing a current through it. The small hole in the wall of the tube approximates a blackbody at the temperature of the wall if the tube walls are thin and of uniform thickness and the hole diameter is small compared to the diameter of the tube [36]. Thermal gradients along the tube are virtually eliminated by proper adjustment of the power input to the end heaters. Larrabee [91, 92] has generated very-high-precision emittance measurements on tungsten with this approach although Kunz [90] has raised a question regarding manner of viewing to obtain the "zero" reading.

Total normal emittance can be measured by the

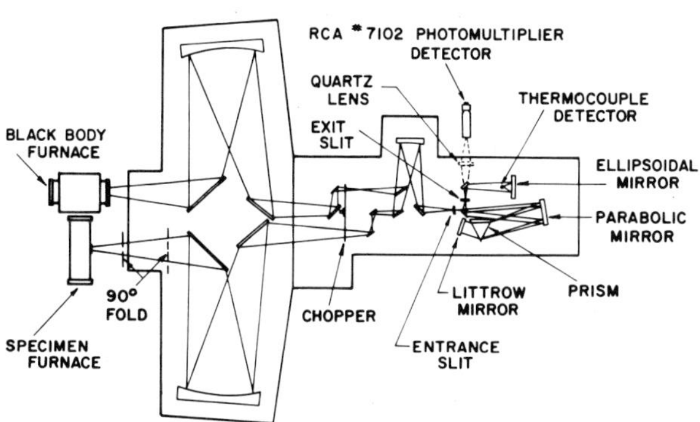

Fig. 23. Schematic of apparatus for measurement of normal spectral emittance [113].

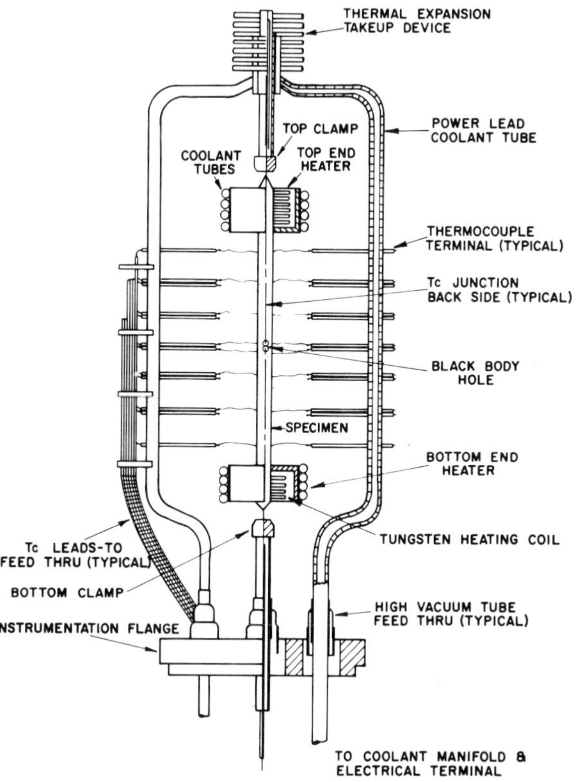

Fig. 24. Typical arrangement of blackbody technique for normal spectral emittance measurements.

techniques described above if a detector is used that has essentially the same spectral response over the wave-length range of emitted radiant energy, or if the sample is a graybody emitter [93]. This technique has been used [7] to study the variation of total directional emittance with angle of viewing. Normal total emittance data are frequently computed from normal spectral data, as has been discussed previously.

D. Calorimetric Techniques

In calorimetric techniques the emittance or absorptance is measured in terms of the heat lost or gained by the specimen through radiant heat transfer. The ratio of the observed rate of heat transfer to the computed rate of radiant heat transfer for a black-body surface under the same conditions gives the emittance or absorptance. The technique is only suitable for total emittance, and usually hemispherical total emittance, but can be used for absorptance for incident radiant energy of any desired spectral distribution, including narrow spectral bands.

One common factor in all calorimetric methods is that the sample is thermally isolated from its surroundings, so that essentially all heat transfer to or from it is by radiation. The rate of heat transfer is evaluated in terms of the rate of temperature change of the sample, or the power input or output to or from the specimen is in a form such that it can be accurately evaluated. This assures that when the sample reaches an equilibrium temperature, the radiant heat transfer can be equated to the heat transfer by other means.

a. Measurements at Temperatures above 500 K

The simplest calorimetric technique for measuring hemispherical total emittance is the hot-filament method [148]. The sample, in the form of a long filament, is heated in vacuum by passing a current through it. If the wire is of uniform cross section and resistance, a section at the center will quickly come to a uniform equilibrium temperature. A small length of this central portion is instrumented with thermocouples, which can also serve as potential leads. After the filament has come to equilibrium, its temperature, the current flowing through it, and the potential drop across the central section are measured. The total hemispherical emittance is computed from the following equation

$$\epsilon(2\pi; t) = VI/A\sigma(T_1^4 - T_2^4) \qquad (72)$$

where $\epsilon(2\pi; t)$ is the hemispherical total emittance, V is the voltage drop across the central section, I is the current flowing in the filament, A is the surface area of the center section, σ is the Stefan–Boltzmann constant, T_1 is the temperature of the central section, and T_2 is the temperature of the surroundings. Several simplifying assumptions were made in deriving equation (72), which are discussed in the literature. The same general method has been used for metal strip samples [9, 112] with the addition of a water-cooled chamber, and provision for expansion of the heated specimen, and even guard heaters for the ends of the sample. Figure 25 is a drawing of one such apparatus. The overall accuracy is dependent upon the magnitude of $T_1 - T_2$, and the relative magnitude of the heat losses by conduction in the atmosphere of the chamber, through thermocouple leads and the ends of the sample, and the assumptions made in deriving equation (72). With careful work, accuracy on the order of ± 2 percent is easily attained.

Several error analyses of the method [95, 105] have been published. Measurements to 1800 K are relatively simple and reliable, with the use of conven-

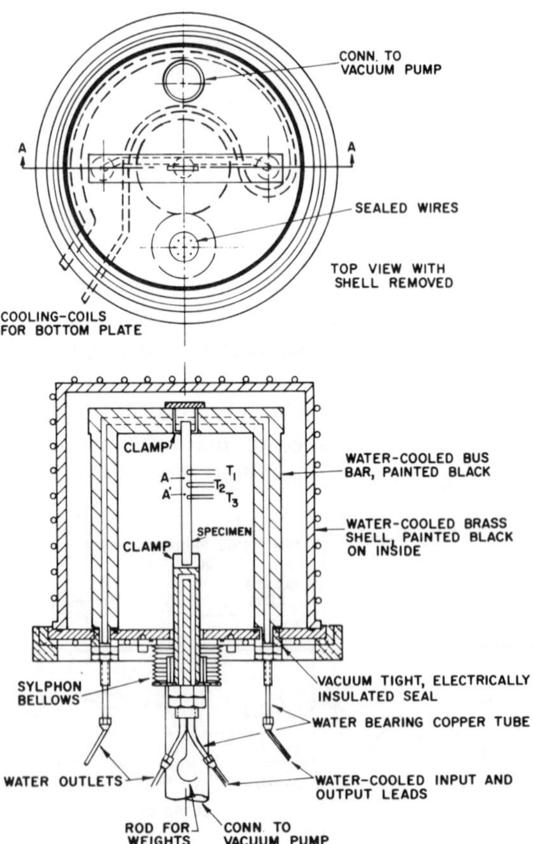

Fig. 25. Typical arrangement of calorimetric technique for hemispherical total emittance measurements, moderate-to-high-temperature region.

tional noble metal thermocouples. Measurements have been made to 3000 K with the use of refractory metal thermocouples, with a significant loss in accuracy due to the instability of the thermocouples.

b. Measurements in the Range 270 *to* 500 K

The hot-filament method can be used at lower temperatures by reducing T_2. This is commonly done by liquid nitrogen cooling and extends the useful temperature range downwards to about 270 K [108]. Heat losses by conduction become a much larger fraction of the total heat loss at these temperatures, and more stringent precautions must be taken to reduce them. Also longer times are required to reach thermal equilibrium.

In this temperature range the heat input is frequently by radiation [55, 57]. The sample may be heated by a beam from a solar simulator, and the $\alpha(s)/\epsilon(2\pi; t)$ ratio computed from the equilibrium temperature attained by use of the equation [26]

$$\alpha(s)/\epsilon(2\pi; t) = A_2\sigma(T_1^4 - T_2^4)/EA_1 \tag{73}$$

or during heating by use of the equation

$$\alpha(s) = [\epsilon(2\pi; t)A_2\sigma(T_1^4 - T_2^4) + MC_p\, dT_1/dt]/EA_1 \tag{74}$$

where $\alpha(s)$ is the solar absorptance, $\epsilon(2\pi; t)$ is the hemispherical total emittance, A_2 is the total surface area of the sample, σ is the Stefan–Boltzmann constant, T_1 is the temperature of the sample, T_2 is the temperature of the chamber walls, E is the irradiance on the sample, A_1 is the irradiated area of the sample, M is the mass of the sample, C_p is the specific heat of the sample, and dT_1/dt is the rate of temperature change of the sample. The source is then turned off, and the sample is allowed to cool. The hemispherical total emittance can then be computed by the equation

$$\epsilon(2\pi; t) = \frac{MC_p\, dT_1/dt}{\sigma A_2(T_1^4 - T_2^4)} \tag{75}$$

c. Measurements below 270 K

For measurements at temperatures below 270 K the chamber walls may be cooled by liquid helium [26, 79] and the $\alpha(s)/\epsilon(2\pi; t)$ ratio measured by the procedures outlined above. For steady-state measurements, the heat input may be electrical [27, 52, 53]. Figure 26 shows a helium-cooled chamber used for measurements at temperatures in the 20 to 300 K range, in which electrical heating is employed. The total hemispherical emittance can be computed from the net heat-transfer rate between two parallel plates held at different temperatures. Figure 27 is one such type of apparatus [68, 69] in which two samples of the same material are used. The lower sample, 8 in. in diameter, is electrically heated and maintained at about 290 K, and the upper sample, 4 in. in diameter, is cooled by the liquid nitrogen to about 80 K. The net heat-transfer rate is measured by the rate of boil-off of the liquid nitrogen from the inner dewar. The emittance, ϵ, is computed from

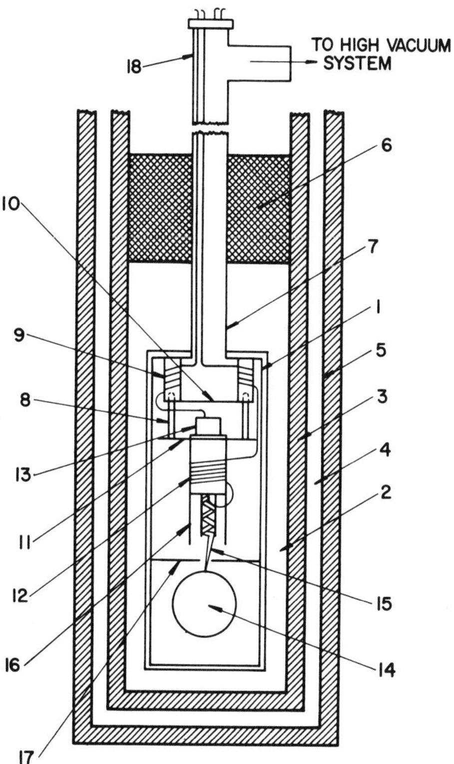

Fig. 26. Concentric sphere technique for low-temperature hemispherical total emittance [27]: 1) Sample container, 2) liquid helium space, 3) super-insulated dewar, 4) liquid nitrogen space, 5) dewar, 6) heat exchanger, 7) support tube, 8) nylon studs, 9) copper posts, 10) copper plate, 11) copper plate, 12) copper block, 13) germanium resistance thermometer, 14) sample, 15) differential thermocouple, 16) radiation shield, 17) radiation shield, 18) hermetic seal.

$$\Phi = \sigma\frac{\epsilon}{2 - \epsilon}A(T_1^4 - T_2^4) \tag{76}$$

where Φ is the measured heat-transfer rate, A is the area of the cold (upper) sample, T_1 is the temperature of the hot sample, and T_2 is the temperature of the cold sample. The measured emittance, ϵ, is an average total hemispherical emittance for the sample material at the two temperatures. The factor $\epsilon/(2 - \epsilon)$

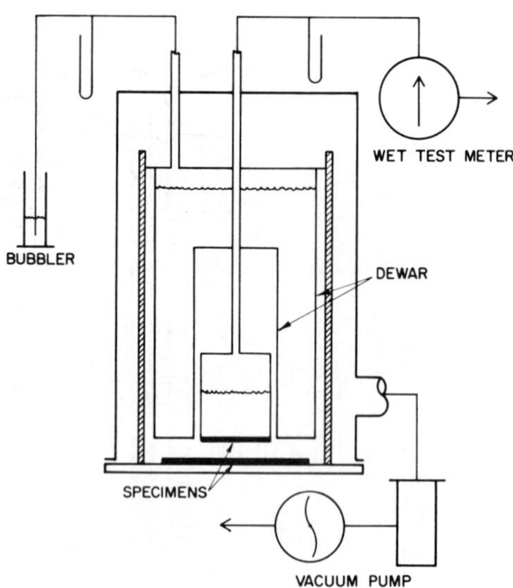

Fig. 27. Flat-plate technique for averaged hemispherical total emittance, low-temperature region [69].

in the equation arises from the geometrical view factor between infinite parallel plates. This condition is closely approximated by the small separation between the plates, $\frac{1}{4}$ in., and the larger size of the lower plate. Caren [28] describes an instrument for the range 10 to 300 K of a similar nature.

Biondi [18] measured the spectral absorptance and reflectance of specularly reflecting samples at 4.2 K. The sample and a black absorber are thermally bonded to copper stages of appreciable thermal mass, which are in turn thermally connected to the liquid helium sink through heat leads of the proper thermal conductivity to give the stage a thermal time constant of about 10. The black absorber is positioned to receive the radiant energy specularly reflected by the sample. The stages can also be electrically heated. The sample is irradiated with a beam of monochromatic radiant energy of known irradiance and allowed to come to thermal equilibrium. The temperature of the copper stages is then measured, as well as that of the liquid helium sink. The source is then turned off, and the copper stages are heated electrically to the same temperature observed during irradiation. The radiant flux absorbed by the sample and black absorber is then equated to the electrical power required to maintain each stage at its respective temperature, and the reflectance and absorptance of the specimen is computed.

References to Text

1. American Standards Association, "Surface Roughness, Waviness and Lay," ASA B46-1, ASME., 1955.
2. Preparation and Reference White Reflectance Standards, ASTM Description E259-66, ASTM Standards, General Test Methods, Part 30, 803–5, May 1967.
3. Fifth Sagamore Ordnance Materials Research Conference—Materials in Space Environment, Syracuse University Research Institute, September 1958.
4. Glossary, Electromagnetic Radiation Definitions, from *Progress in Astronautics and Aeronautics Vol. 20* (G. B. Heller, Editor), Academic Press, 947–61, 1967.
5. Standards for Checking the Calibration of Spectrophotometers (200 to 1000 μm), Nat. Bur. Std. (US) Letter Circular 1017.
6. USA Standard—Nomenclature and Definitions for Illuminating Engineering, PP-16-USAS Z7. 1–1967, Revision of Z7. 1–1942 UBC 653. 014, 8:62132.
7. Abbott, G. L., "Total Normal and Total Hemispherical Emittance of Polished Metals," *Measurement of Thermal Radiation Properties of Solids* (J. C. Richmond, Editor), NASA SP-31, 293–306, 1963.
8. Aschkinass, E., "Die Wärmestrahlung der Metalle," *Ann. Physik* **17**, 960–76, 1905.
9. Askwyth, W. H., Yahes, R. J., House, R. D., and Mikk, G., "Determination of Emissivity of Materials," Vol. I, NASA-CR-56496, 1962; Vol. II, NASA-CR-56497, 1963; Vol. III, NASA-CR-56498, 1964.
10. Beckmann, P. and Spizzichino, A., *The Scattering of Electromagnetic Waves from Rough Surfaces*, Pergamon Press and MacMillan, 1963.
11. Beilby, G., *Aggregation and Flow of Solids*, MacMillan and Co., 1921.
12. Bennett, H. E., "Specular Reflectance of Aluminized Ground Glass and the Height Distribution of Surface Irregularities," *J. Opt. Soc. Am.* **53**, 1389–1894, 1963.
13. Bennett, H. E., "Influence of Surface Roughness, Surface Damage, and Oxide Films on Emittance," *Symposium on Thermal Radiation of Solids* (S. Katzoff, Editor), NASA SP-55, 145–52, 1965.
14. Bennett, H. E. and Bennett, J. M., "Validity of the Drude Theory," *Optical Properties and Electronic Structure of Metals and Alloys* (F. Abeles, Editor), North Holland Publishing Co., 173–88, 1966.
15. Bennett, H. E. and Koehler, W. F., "Precision Measurement of Absolute Specular Reflectance with Minimized Systematic Errors," *J. Opt. Soc. Am.* **50**, 1–6, 1960.
16. Bennett, H. E. and Porteus, J. O., "Relation between Surface Roughness and Specular Reflectance at Normal Incidence," *J. Opt. Soc. Am.* **51**, 123, 1961.
17. Bevans, J. T. (Editor), "Thermal Design Principles of Spacecraft and Entry Bodies," *Progress in Astronautics and Aeronautics, Vol. 21*, Academic Press, 1969.
18. Biondi, A., "Optical Absorption of Copper and Silver at 4.2 K," *Phys. Rev.*, **102**, 964–7, 1956.
19. Birkebak, R. C. and Eckert, E. R. G., "Effects of Roughness of Metal Surfaces on Angular Distribution of Monochromatic Reflected Radiation," *Trans. ASME* **87C**, 85–94, 1965.
20. Birkebak, R. C. and Hartnett, J. P., "Measurements of the Total Absorptivity for Solar Radiation of Several Engineering Materials," *Trans. ASME* **80**, 373–78, 1958.
21. Blau, H. and Fischer, H. (Editors), *Radiative Transfer from Solid Materials*, MacMillan and Co., 257 pp., 1962.
22. Born, W. and Wolf, E., *Principles of Optics*, Pergamon Press, 3rd Ed., 1965.
23. Brandenberg, W. M., "Focusing Properties of Hemispherical and Ellipsoidal Mirror Reflectometers," *J. Opt. Soc. Am.* **54**, 1235–7, 1964.
24. Brandenberg, W. M. and Neu, J. T., "Unidirectional Reflectance of Imperfectly Diffuse Surfaces," *J. Opt. Soc. Am.* **56** (1), 97–103, 1966.
25. Brookshier, R. K., presented at the Optical Society of America Meeting, 1966.
26. Butler, C. P. and Jenkins, R. J., "Space Chamber Emittance Measurements," *Measurement of Thermal Radiation Properties of Solids* (J. C. Richmond, Editor), NASA SP-31, 39–43, 1963.
27. Caren, R. P., "Cryogenic Emittance Measurements," *Measurement of Thermal Radiation Properties of Solids* (J. C. Richmond, Editor), NASA SP-31, 45–7, 1963.
28. Caren, R. P., "Low-Temperature Emittance Determinations," *Progress in Astronautics and Aeronautics, Vol. 18*, AIAA (G. B. Heller, Editor), Academic Press, 61–73, 1966.
29. Clauss, F. J. (Editor), *First Symposium—Surface Effects on Spacecraft Materials*, Wiley and Sons, 404 pp., 1960.
30. Coblentz, W. W., "The Diffuse Reflecting Power of Various Substances," *Bull. Nat. Bur. Std. (US)* **9**, 283–325, 1913.
31. Comstock, D. F., Jr., A. D. Little Report to Jet Propulsion Laboratory, Contract No. 950867, Subcontract Under NAS 7-100, March 1966. Presented at Optical Society of America Meeting.
32. Davies, H., "Reflection of Electromagnetic Waves from Rough Surfaces," *Proc. Inst. Elec. Engrs.* **101**, 209–14, 1954.
33. Davisson, D. and Weeks, J. R., "The Relationship between the Total Thermal Emissive Power of a Metal and Its Electrical Resistivity," *J. Opt. Soc. Am.* **8**, 581–605, 1929.

34. Derksen, W. L., Monahan, T. I., and Lawes, A. J., "Automatic Recording Reflectometer for Measuring Diffuse Reflectance in the Visible and Infrared Regions," *J. Opt. Soc. Am.* **47**, 995–9, 1957.

35. Derksen, W. L. and Monahan, T. I., "Automatic Recording Reflectometer for Measuring Diffuse Reflectance in the Visible and Infrared Regions," *J. Opt. Soc. Am.* **42**, 263–5, 1952.

36. DeVos, J. S., "A New Determination of the Emissivity of Tungsten Ribbon," *Physica* **20**, 690–714, 1954.

37. DeWitt, D. P., "The Surface Characterization of Real Metals," *Symposium on Thermal Radiation of Solids* (S. Katzoff, Editor), NASA SP-55, 141–4, 1965.

38. Dingle, R. B., "The Anomalous Skin Effect and the Reflectivity of Metals, I. Theory," *Physica* **19**, 311–47, 1953, and "II. Comparison between Theoretical and Experimental Optical Properties," *Physica* **19**, 348–64, 1953.

39. Donovan, T. M., Ashley, E. J., and Bennett, H. E., "Effect of Surface Damage on the Reflectance of Germanium in the 2650–10,000 Å Region," *J. Opt. Soc. Am.* **53**, 1403–9, 1963.

40. Drude, P., "Bestimmung der optischen Constanten der Metalle," *Ann. Physik* **64**, 159, 1898; available as AFCRL (Hanscom Field) translation "Determination of the Optical Constants of Metals," March 1963. [AD 430 983]

41. Dunkle, R. V., Edwards, D. K., Gier, J. T., Nelson, K. E., and Roddick, R. D., "Heated Cavity Reflectometer for Angular Reflectance Measurement," *Progress in International Research on Thermodynamic and Transport Properties*, ASME, 100–6, 1962.

42. Dunkle, R. V., "Spectral Reflectance Measurements," *Surface Effects on Spacecraft Materials* (F. J. Clauss, Editor), Wiley and Sons, 117–37, 1960.

43. Dunn, S. T., Richmond, J. C., and Wiebelt, J. A., "Ellipsoidal Mirror Reflectometer," *J. Res. Nat. Bur. Std.* **70C** (2), 75–88, 1966.

44. Dunn, S. T., Richmond, J. C., and Parmer, J. F., "Survey of Infrared Techniques and Computational Methods in Radiant Heat Transfer," *J. Spacecraft Rockets* **3**, 961–75, 1966.

45. Dunn, S. T., "Application of Sulphur Coatings to Integrating Spheres," *Appl. Opt.* **4** (4), 377, 1965.

46. Dunn, S. T., "Flux Averaging Devices for the Infrared," Nat. Bur. Std. (US) Technical Note 279, December 1965.

47. Edwards, D. K. and de Volo, B. N., "Useful Approximations for the Spectral and Total Emissivity of Smooth Bare Metals," *Advances in Thermophysical Properties at Extreme Temperatures and Pressures* (S. Gratch, Editor), ASME, 174–88, 1965.

48. Edwards, D. K., Gier, J. T., Nelson, E. K., and Roddick, R. D., "Integrating Sphere for Imperfectly Diffuse Samples," *Appl. Opt.* **51**, 1279–88, 1961.

49. Foote, P. D., "The Emissivity of Metals and Oxides," *Bull. Nat. Bur. Std.* (US), **11**, 607–12, 1914–15.

50. Fowler, P., "Far Infrared Absorptance of Gold," 57 pp., January 1960. [AD 418 456]

51. Fragstein, C. V., "On the Formulation of Kirchhoff's Law and Its Use for a Suitable Definition of Diffuse Reflection Factors," *Optik* **12**, 60–70, 1955.

52. Fulk, M. M., Reynolds, M. M., and Park, O. E., "Thermal Radiation Absorption by Metals," *Proceedings of the 1954 Cryogenic Engineering Conference*, Nat. Bur. Std. Report 3517, 1955. [AD 125 047]

53. Fulk, M. M., Reynolds, M. M., and Park, O. E., "Thermal Radiation Absorption by Metals," *Advances in Cryogenic Engineering, Vol. 1* (K. D. Timmerhaus, Editor), Plenum Press, 224–9, 1960.

54. Funai, A. I. and Rolling, R. E., "Inspection Techniques for the Characterization of Smooth, Rough, and Oxide Surfaces," *Progress in Aeronautics and Astronautics*, AIAA (G. B. Heller, Editor), 41–64, 1967.

55. Fussell, W. B., Triolo, J. J., and Henninger, J. H., "A Dynamic Thermal Vacuum Technique for Measuring the Solar Absorption and Thermal Emittance of Spacecraft Coatings," *Measurement of Thermal Radiation Properties of Solids* (J. C. Richmond, Editor), NASA SP-31, 83–101, 1963.

56. Gates, D. M., Shaw, C. C., and Beaumont, D., "Infrared Reflectance of Evaporated Metal Films," *J. Opt. Soc. Am.* **48** (2), 88–9, 1958.

57. Gaumer, R. E. and Stewart, J. V., "Calorimetric Determination of Infrared Emittance and the α_s/ϵ Ratio," *Measurement of Thermal Radiation Properties of Solids* (J. C. Richmond, Editor), NASA SP-31, 127–33, 1963.

58. Gier, J. T., Dunkle, R. V., and Bevans, J. T., "Measurement of Absolute Spectral Reflectivity from 1.0 to 15 Microns," *J. Opt. Soc. Am.* **44**, 558–62, 1954.

59. Givens, M. P., "Optical Properties of Metals," *Solid State Physics, Advances in Research and Applications, Vol. 6* (F. Seitz and D. Turnbull, Editors), Academic Press, 313–52, 1958.

60. Goebel, D. G., Caldwell, B. P., and Hammond, H. K., III, "Use of an Auxiliary Sphere with a Spectroreflectometer to Obtain Absolute Reflectance," *J. Opt. Soc. Am.* **56**, 783–8, 1966.

61. Goebel, D. G., "Generalized Integrating Sphere Theory," *Appl. Opt.* **6**, 125–8, 1967.

62. Goebel, D. G., Personal communication, to be published in *Appl. Opt.*

63. Gravina, A., Bastian, R., and Dyer, J., "Instrumentation for Emittance Measurements in the 400 to 1800 F Temperature Range," *Measurement of Thermal Radiation Properties of Solids* (J. C. Richmond, Editor), NASA SP-31, 329–36, 1963.

64. Grum, F. and Luckey, G. W., "Optical Sphere Paint and a Working Standard of Reflectance," *Appl. Opt.* **7** (11), 2289–94, 1968.

65. Gubareff, G. G., Janssen, J. E., and Torborg, R. H., *Thermal Radiation Properties Survey*, Minneapolis-Honeywell Regulator Co., 1960.

66. Hagen, E. and Rubens, H., "Emissionsvermögen und electrische Leitfähigkeit der Metallegierungen," *Verhandl. Deutsch. Phys. Ges.* **6**, 128, 1904.

67. Harrison, T. R., *Radiation Pyrometry and its Underlying Principles of Radiant Heat Transfer*, Wiley and Sons, 1960.

68. Haury, G. L., "An Apparatus for the Measurement of Total Hemispherical Emissivity and Thermal Conductivity between Ambient and Liquid Nitrogen Temperatures," ASD-TDR-63-146, 16 pp., 1960. [AD 411 140]

69. Haury, G. L., "An Apparatus for Measuring Total Hemispherical Emittance between Ambient and Liquid

Nitrogen Temperatures," *Measurement of Thermal Radiation Properties of Solids* (J. C. Richmond, Editor), NASA SP-31, 51–4, 1963.
70. Heinisch, R. P., Personal communication, to be published in *Appl. Opt.*
71. Heller, G. B. (Editor), "Thermophysics and Temperature Control of Spacecraft and Entry Vehicles," *Progress in Astronautics and Aeronautics, Vol. 18*, AIAA, Academic Press, 867 pp, 1966.
72. Heller, G. B. (Editor), "Thermophysics of Spacecraft and Planetary Bodies, Radiation Properties of Solids and the Electromagnetic Radiation Environment in Space," *Progress in Astronautics and Aeronautics, Vol. 20*, AIAA, Academic Press, 975 pp., 1967.
73. Hembach, R. J., Hemmerdinger, L., and Katz, A. J., "Heated Cavity Reflectometer Modifications," *Measurement of Thermal Radiation Properties of Solids* (J. C. Richmond, Editor), NASA SP-31, 153–67, 1963.
74. Hottel, H. C. and Sarofim, A. F., *Radiative Transfer*, McGraw-Hill, 1968.
75. House, R. D., Lyons, G. J., and Askwyth, W. H., "Measurement of Spectral Normal Emittance of Materials under Simulated Spacecraft Powerplant Operating Conditions," *Measurement of Thermal Radiation Properties of Solids* (J. C. Richmond, Editor), NASA SP-31, 343–55, 1963.
76. Jacquez, J. A. and Kuppenheim, H. F., "Theory of the Integrating Sphere," *J. Opt. Soc. Am.* **45**, 460–70, 1954.
77. Jakob, M., *Heat Transfer*, Academic Press, 51–2, 1949.
78. Janssen, J. E. and Torborg, R. H., "Measurement of Spectral Reflectance Using an Integrating Hemisphere," *Measurement of Thermal Radiation Properties of Solids* (J. C. Richmond, Editor), NASA SP-31, 169–82, 1963.
79. Jenkins, R. J., Butler, C. P., and Parker, W. J., "Total Hemispherical Emittance Measurements over the Temperature Range 77 to 300 K," USNRDL-TR-663, 57 pp., 1963. [AD 419 067]
80. Johnson, F. S., "The Solar Constant," *J. Meteorol.* **11**, 431, 1954.
81. Judd, D. B., "Terms, Definitions and Symbols in Reflectometry," *J. Opt. Soc. Am.* **57** (4), 445–52, 1967.
82. Katzoff, S. (Editor), *Symposium on Thermal Radiation of Solids*, NASA SP-55, U.S. Government Printing Office, 1965.
83. Keegan, H. J., Schleter, J. C., and Wiedner, V. R., "Effect of Surface Texture on Diffuse Spectral Reflectance, Part A: Diffuse Spectral Reflectance of Metal Surfaces," *Symposium on Thermal Radiation of Solids* (S. Katzoff, Editor), NASA SP-55, 165–9, 1965.
84. Keegan, H. J. and Weidner, V. R., "Infrared Spectral Reflectance of Frost," *J. Opt. Soc. Am.* **56** (4), 523–4, April 1966.
85. Kneissl, G. J. and Richmond, J. C., "A Laser Source Integrating Sphere Reflectometer," Nat. Bur. Std. (US) Tech. Note 439, 1968.
86. Kneissl, G. J., Richmond, J. C. and Wiebelt, J. A., "A Laser Source Integrating Sphere for the Measurement of Directional Hemispherical Reflectance at High Temperature," *Progress in Aeronautics and Astronautics, Vol. 20*, AIAA (G. B. Heller, Editor), Academic Press, 177–202, 1967.
87. Kozyrev, B. P. and Vershinin, O. E., "Determination of Spectral Coefficients of Diffuse Reflection of Infrared Radiation from Blackened Surfaces," *Opt. Spectry.* **6**, 345–50, 1959.
88. Kronig, R. D. L., "The Quantum Theory of Dispersion in Metallic Conductors," *Proc. Royal Soc.* **133A**, 255, 1931.
89. Kronstein, M., Krauschaar, R. J., and Deacle, R. E., "Sulphur as a Standard of Reflectance in the Infrared," *J. Opt. Soc. Am.* **53**, 458, 1963.
90. Kunz, H., "Prüfen technischer Strahlungspyromtere," *VDI-Berichte* **112**, 37–46, 1966.
91. Larrabee, R. D., "The Spectral Emissivity and Optical Properties of Tungsten," Technical Report 328, Res. Lab. of Electronics, MIT, April 1957. [AD 156 602]
92. Larrabee, R. D., "Spectral Emissivity of Tungsten," *J. Opt. Soc. Am.* **49**, 619–25, 1959.
93. Limperis, T., Szeles, D. M., and Wolfe, W. L., "The Measurement of Total Normal Emittance of Three Nuclear Reactor Materials," *Measurement of Thermal Radiation Properties of Solids* (J. C. Richmond, Editor), NASA SP-31, 357–64, 1963.
94. Love, T. J. and Francis, R. E., "Experimental Determination of Reflectance Function for Type 302 Stainless Steel," *Progress in Astronautics and Aeronautics, Vol. 20*, AIAA (G. B. Heller, Editor), Academic Press, 115–35, 1967.
95. McElroy, D. L. and Kollie, T. G., "The Total Hemispherical Emittance of Platinum, Columbium-1% Zirconium, and Polished and Oxidized INOR-8 in the Range 100 to 1200 C," *Measurement of Thermal Radiation Properties of Solids* (J. C. Richmond, Editor), NASA SP-31, 365–79, 1963.
96. McNicholas, H. J.," Absolute Methods in Reflectometry," *J. Res. Nat. Bur. Std. (US)* **1**, 29–73, 1928.
97. Martin, W. E., "Hemispherical Spectral Reflectance of Solids," *Measurement of Thermal Radiation Properties of Solids* (J. C. Richmond, Editor), NASA SP-31, 183–92, 1963.
98. Meyer-Arendt, Jurgen R., "Radiometry and Photometry; Units and Conversion Factors," *J. Appl. Opt.* **7** (10), 2081–4, 1968.
99. Middleton, W. E. K. and Sanders, C. L., "The Absolute Spectral Diffuse Reflectance of Magnesium Oxide," *J. Opt. Soc. Am.* **41**, 419–24, 1951.
100. Middleton, W. E. K. and Sanders, C. L., "An Improved Sphere Paint," *Illum. Engr.* **48**, 254–6, 1953.
101. Miller, E. R. and VunKannon, R. S., "Development and Use of a Bidirectional Spectroreflectometer," *Progress in Aeronautics and Astronautics, Vol. 20* (G. B. Heller, Editor), Academic Press, 219–33, 1967.
102. Moore, D. G., "Investigation of Shallow Reference Cavities for High Temperature Emittance Measurements," *Measurement of Thermal Radiation Properties of Solids* (J. C. Richmond, Editor), NASA SP-31, 515–25, 1963.
103. Mott, N. F. and Zener, C., *Proc. Cambridge Phil. Soc.* **30**, 249, 1934.
104. Neher, R. T. and Edwards, D. K., "Far Infrared Reflectometer for Imperfectly Diffuse Specimens," *Appl. Opt.* **4**, 775–80, 1965.
105. Nelson, K. E. and Bevans, J. T., "Errors of the Calorimetric Method of Total Emittance Measurements,"

Measurement of Thermal Radiation Properties of Solids (J. C. Richmond, Editor), NASA SP-31, 55–65, 1963.

106. Neu, John T., Personal communication, to be published in *Appl. Opt.*
107. Null, M. R. and Lozier, W. W., "Measurement of Reflectance and Emittance at High Temperatures with a Carbon Arc Image Furnace," *Measurement of Thermal Radiation Properties of Solids* (J. C. Richmond, Editor), NASA SP-31, 535–51, 1963.
108. Nyland, T. W., "Apparatus for the Measurement of Hemispherical Emittance from 270 to 650 K," *Measurement of Thermal Radiation Properties of Solids* (J. C. Richmond, Editor), NASA SP-31, 393–401, 1963.
109. Parker, W. J. and Abbott, G. C., "Total Emittance of Metals," *Symposium on Thermal Radiation of Solids* (S. Katzoff, Editor), NASA SP-55, 11–28, 1965.
110. Porteus, J. P., "Relation between the Height Distribution of a Rough Surface and the Reflectance at Normal Incidence," *J. Opt. Soc. Am.* **53**, 1394–1402, 1963.
111. Rice, S. O., "Reflection of Electromagnetic Waves from Slightly Rough Surfaces," *Commun. Pure Appl. Math.* **4**, 351–78, 1951.
112. Richmond, J. C. and Harrison, W. N., "Equipment and Procedures for Evaluation of Total Hemispherical Emittance," *Am. Ceram. Soc. Bull.* **39**, 668–73, 1960.
113. Richmond, J. C., Harrison, W. N., and Shorten, F. J., "An Approach to Thermal Emittance Standards," *Measurement of Thermal Radiation Properties of Solids* (J. C. Richmond, Editor), NASA SP-31, 403–23, 1963.
114. Richmond, J. C. (Editor), *Measurement of Thermal Radiation Properties of Solids*, NASA SP-31, U.S. Government Printing Office, 1963.
115. Richmond, J. C., Dunn, S. T., DeWitt, D. P., and Hayes, W. D., Jr., "Procedures for Precise Determination of Thermal Radiative Properties," Nat. Bur. Std. (US) Tech. Note 252, November 1964.
116. "Solar Electromagnetic Radiation," NASA SP-8005, June 1965.
117. Richmond, J. C., Dunn, S. T., DeWitt,D. P., and Hayes, W. D., Jr., "Procedures for Precise Determination of Thermal Radiation Properties," Nat. Bur. Std. (US) Tech. Note 267, December 1965.
118. Richmond, J. C., "Effect of Surface Roughness on Emittance of Nonmetals," *J. Opt. Soc. Am.* **56**, 253–4, 1966.
119. Richmond, J. C. and Geist, Jon C., "Infrared Reflectance Measurements," Nat. Bur. Std. (US) Report 10071.
120. Riethof, T. R. and DeSantis, V. J., "Techniques of Measuring Normal Spectral Emissivity of Conductive Refractory Compounds at High Temperatures," *Measurement of Thermal Radiation Properties of Solids* (J. C. Richmond, Editor), NASA SP-31, 565–84, 1963.
121. Roberts, S., "Optical Properties of Copper," *Phys. Rev.* **118**, 1509, 1960.
122. Roberts, S., "Optical Properties of Nickel and Tungsten and Their Interrelation According to Drude's Formula," *Phys. Rev.* **114**, 104, 1959.
123. Roberts, S., "Interpretation of the Optical Properties of Metal Surfaces," *Phys. Rev.* **100**, 1667, 1955.
124. Rolling, R. E., "The Effect of Slight Surface Roughness on Emittance," *Symposium on Thermal Radiation of Solids* (S. Katzoff, Editor), NASA SP-55, 153–5, 1965.
125. Rolling, R. E., Funai, A. I., and Grammer, J. R., "Investigation of the Effect of Surface Condition on the Radiant Properties of Metals," AFML-TR-64-363, November 1964.
126. Rolling, R. E. and Funai, A. I., "Investigation of the Effect of Surface Conditions on the Radiant Properties of Metals, Part II, Measurements on Roughened Platinum and Oxidized Stainless Steel," AFML-TR-64-363, Part II, April 1967.
127. Rutgers, G. A. W., "Temperature Radiation of Solids," *Handbuch der Physik* **26**, 9, 1958.
128. Samuels, L. E., "Modern Ideas on the Mechanical Polishing of Metals," *Res. Appl. Ind.* **13**, 344, 1960.
129. Sanderson, J. A., "The Diffuse Spectral Reflectance of Paints in the Near Infrared," *J. Opt. Soc. Am.* **37**, 771–7, 1947.
130. Schmidt, E. and Eckert, E., "Ueber die Richtungsverteilung der Wärmestrahlung von Oberflächen," *Forsch. Gebiete Ingenievrw.* **6**, 175–83, 1935.
131. Seban, R. A., "The Emissivity of Transition Metals in the Infrared," *J. Heat Transfer* **C87**, 173–6, 1965.
132. Schocken, K. (Editor), "Optical Properties of Satellite Materials—The Theory of Optical and Infrared Properties of Metals," Research Projects Division, GMSFC, NASA-TN D-1523, March 1963.
133. Siegel, R. and Howell, J. R., "Thermal Radiation Heat Transfer: Blackbody, Electromagnetic Theory and Materials Properties," NASA SP-164, **1**, 1968.
134. Snyder, N. W., "Radiation in Metals," *Trans. ASME* **76**, 541–8, 1954.
135. Sokolov, A. V., *Optical Properties of Metals*, American Elsevier, 472 pp., 1967.
136. Spangenberg, D. B., Strang, A. G., and Chamberlin, J. L., "Effect of Surface Texture on Diffuse Spectral Reflectance, Part B: Surface Texture Measurements of Metal Surfaces," *Symposium on Thermal Radiation of Solids* (S. Katzoff, Editor), NASA SP-55, 169–77, 1965.
137. Sparrow, E. M. and Cess, R. D., *Radiation Heat Transfer*, Brooks/Cole Publishing Co., 1966.
138. Streed, E. R., McKellar, L. A., Rollings, R., Jr., and Smith, C. A., "Errors Associated with Hohlraum Radiation Characteristics Determinations," *Measurement of Thermal Radiation Properties of Solids* (J. C. Richmond, Editor), NASA SP-31, 237–52, 1963.
139. Strong, J., *Procedures in Experimental Physics*, Prentice-Hall, 376 pp., 1938.
140. Svet, D. Ya., *Thermal Radiation* (*Metals, Semiconductors, Ceramics, Partly Transparent Bodies, and Films*), New York, Consultants Bureau, 98 pp., 1965.
141. Torrance, K. E. and Sparrow, E. M., "Off-Specular Peaks in the Directional Distribution of Reflected Thermal Radiation," *J. Heat Transfer* **C88**, 223–30, 1966.
142. Touloukian, Y. S. and Gerritsen, J. K. (Editors), *Thermophysical Properties Research Literature Guide*, Plenum Press, 1967.
143. Touloukian, Y. S. (Editor), *Thermophysical Properties of High Temperature Solid Materials*, MacMillan Co., New York, 1967.
144. White, J. U., "New Method for Measuring Diffuse Reflectance in the Infrared," *J. Opt. Soc. Am.* **54**, 1332–7, 1964.

145. Wiebelt, J. A., *Engineering Radiation Heat Transfer*, Holt, Rinehart and Winston Publishing Co., 1966.
146. Wilson, R. G., "Hemispherical Spectral Emittance of Ablation Chars, Carbon, and Zirconia (to 3700 K)," *Symposium on Thermal Radiation of Solids* (S. Katzoff, Editor), NASA SP-55, 259–75, 1965.
147. Wood, R. W., *Physical Optics*, Macmillan Co., 3rd Ed., 1934.
148. Worthing, A. G., "Temperature Radiation Emissivities and Emittances," *Temperature, Its Measurement and Control in Science and Industry*, Reinholt Publishing Corp., 1164–87, 1941.
149. Zipin, R. B., "A Preliminary Investigation of the Bidirectional Spectral Reflectance of V-grooved Surfaces," *Appl. Opt.* **5**, 1954–7, 1966.

Numerical Data

Data Presentation and Related General Information

1. SCOPE OF COVERAGE

Included in this volume are data of 43 metallic elements, 37 major classes of binary alloys, and 37 major classes of multiple alloys. These data were obtained by systematically searching over 1000 documents on radiative properties (journals, reports, theses, etc.) dated anywhere from 1895 to 1967 and with a few selected 1968 papers; 371 contained usable data. These 1000 documents were collected, organized, and coded by TPRC's Scientific Documentation Division, which scans the world literature through abstracting journals and certain primary technical and scientific journals.

Since the major area of interest in thermal radiative properties is on materials in their solid state, the data presented cover only the temperature range from near 0 K to the melting point of the materials presented.

The four primary properties covered in this work, as defined in an earlier section, are emittance, reflectance, absorptance, and transmittance. The suffix *ance* is used to indicate that the radiative parameters are not unique properties of the bulk material, but rather depend strongly on the character of the surface. This contrasts with the use of the suffix *ivity*, which denotes the property of the ideal optically smooth, and homogeneous material. Application of the geometric and wavelength descriptors to these four primary properties results in the 33 subproperties listed in Table IX.

The thermal radiation spectrum of engineering importance includes a relatively small portion of the total electromagnetic wave spectrum. Typically, the interval runs from 0.3 to 50 microns. However, there are many applications where there exist significant amounts of energy outside these conventional limits. For this reason, spectral data in this volume covers the expanded wavelength range 0.05 to 1000 microns.

Table IX. Subproperty Designation

Emittance

Hemispherical total emittance
Normal total emittance
Angular total emittance

Hemispherical spectral emittance
Normal spectral emittance
Angular spectral emittance

*Reflectance**

Hemispherical integrated reflectance
Normal integrated reflectance
Angular integrated reflectance

Hemispherical spectral reflectance
Normal spectral reflectance
Angular spectral reflectance

Hemispherical solar reflectance
Normal solar reflectance
Angular solar reflectance

Absorptance

Hemispherical integrated absorptance
Normal integrated absorptance
Angular integrated absorptance

Hemispherical spectral absorptance
Normal spectral absorptance
Angular spectral absorptance

Hemispherical solar absorptance
Normal solar absorptance
Angular solar absorptance

*Transmittance**

Hemispherical integrated transmittance
Normal integrated transmittance
Angular integrated transmittance

Hemispherical spectral transmittance
Normal spectral transmittance
Angular spectral transmittance

Hemispherical solar transmittance
Normal solar transmittance
Angular solar transmittance

*The geometry descriptors refer to the conditions of the incident radiant flux.

2. PRESENTATION OF DATA

For each material, subproperty data are separately presented in graphical and tabular form, accompanied by a table presenting details of the test conditions and specimen preparation and characterization.

The format for the presentation of the thermal radiative properties is designed specifically to supply the reader with the aspects of the properties in a comprehensive yet concise form. Each presentation consists of four sections*; Original Data Plot, Analyzed Data Graph, Specification Table, and Data Table, in that same order.

The Original Data Plot is a graphical representation which presents most of the tabulated data. In overcrowded figures some of the data which are repetitive in nature are omitted. Occasionally comments on the test conditions are included to add clarity to the presentation.

The Analyzed Data Graph presents a new and powerful approach to increasing the effectiveness of literature data. It is an evaluative review identifying and "recommending" reliable or typical data for various surface or environmental conditions. The study to generate these figures considers the interrelationships between the subproperties to give a consistent set of data. Where the data are well characterized or highly reliable, it is represented by a solid curve; where there exists some speculation, the data are represented by dashed lines or a shaded band.

The Specification Table gives the most important information concerning a set of data: the curve number of correlating the information on the Specification Table with that of the figures and Data Table, the reference number corresponding to the number given in the listed references, the year of the publication from which the data were extracted, independent variable range (temperature or wavelength), geometric descriptors (θ, θ', ω, ω'), and the error (%) reported by the author.

The "Composition (weight percent), Specification, and Remarks" column of the Specification Table provides all the available pertinent information about the specimen and test conditions. The presentation is standardized in the following order:

(1) trade name of material if applicable
(2) composition (weight percent)
(3) film or foil thickness
(4) specimen preparation processes
(5) surface conditions (roughness, etc.)
(6) environment
(7) form in which data was presented in original source document (smooth curve, tabular, raw data, etc.)
(8) reference standard used in data observation or reduction
(9) other pertinent information
(10) author's sample designation

Following the Specification Table is the Data Table, a tabular presentation of all the property values shown or not shown on the figure and described in the Specification Table.

Table IX lists the grouping of the various subproperties that are presented in this volume. It shows the 33 subproperties which are classified for organizational and retrieval purposes. The amount of existing data for some of these subproperties is quite small, but there are good reasons to present the data using this generalized scheme. First, the clarity of presentation is better by not grouping together data which logically are unrelated. Also, this scheme lends itself especially well to the systematic updating and expansion contemplated in this continuing work.

3. CLASSIFICATION OF MATERIALS

The classification scheme for materials in Volume 7, shown in Table X, is based upon bulk composition rather than surface chemical composition which varies due to weathering, oxidation, etc.

4. SYMBOLS AND ABBREVIATIONS USED IN THE FIGURES AND TABLES

Emittance	ϵ
Reflectance	ρ
Absorptance	α
Transmittance	τ
Wavelength, microns (μ)	λ
Temperature, K	T
Melting point	M.P.
Zenith angle, degrees	θ, θ'
Azimuthal angle, degrees	ϕ, ϕ'
Solid angle, steradians	ω, ω'
Viewing conditions	′ (prime)
Greater than	>
Less than	<
Approximately	~

*In certain cases, where there exists only a small amount of data, the *Original Data Plot* and/or the *Analyzed Data Graph* may be omitted.

Table X. Classification of Materials

		X_1, %	$X_1 + X_2$, %	X_2, %	X_3, %
Metallic elements		>99.5	—	<0.2	<0.2
Alloys	Binary alloys	—	≥99.5	≥0.2	≤0.2
	Multiple alloys	—	≥99.5	>0.2	>0.2
	Multiple alloys	—	<99.5	≥0.2	≤0.2
	Multiple alloys	—	<99.5	>0.2	>0.2
	Multiple alloys	≤99.5	—	<0.2	<0.2

*Nomenclature: X_1 is the major constituent, X_2 is the second highest constituent, X_3 is the third highest constituent, and % is the weight percent.

Curve number ③
Single data point [4]

5. CONVENTION FOR BIBLIOGRAPHIC CITATIONS

Periodicals

a. Author(s)—last name first, followed by initials
b. Name of the journal—standard literature abbreviations as used in *Chemical Abstracts*
c. Series, volume, and number
d. Pages
e. Year

Reports

a. Author(s)
b. Name of the responsible organization (not the name of the organization that executed the research, if they are different)
c. Report or bulletin, circular, etc.
d. Number
e. Part
f. Pages
g. Year
h. ASTIA's AD number—given in brackets whenever available

Books

a. Author(s)
b. Title
c. Volume
d. Edition
e. Publisher
f. Pages
g. Year

Theses

a. Author
b. Title
c. Degree—Thesis (MS, ScD, PhD)
d. Pages
e. Year
f. Department
g. University
h. City and State

6. CRYSTAL STRUCTURES, TRANSITION TEMPERATURES, AND OTHER PERTINENT PHYSICAL CONSTANTS OF THE ELEMENTS

Table XI contains information on the crystal structures, transition temperatures, and certain pertinent physical constants of the elements. No attempt has been made to critically evaluate the temperatures and constants given in Table XI and they should not be considered recommended values. This table has an independent series of numbered references, and these references immediately follow the table.

7. CONVERSION FACTORS

Wavelength

1 micron (μm) $= 10^{-6}$ meter (m)
$= 10^{-4}$ centimeter (cm)
$= 10^{-3}$ millimeter (mm)
$= 10^{3}$ nanometers (nm)
$= 10^{3}$ millimicrons (mμ)
$= 10^{4}$ angstroms (Å)
$= 3.937 \times 10^{-5}$ inch
$= 10^{4}$/wave number (cm^{-1})
$= 1.24$/photon energy (eV)
$= 2.998 \times 10^{14}$/frequency (cps)

Temperature

$C = K - 273.2$
$F = (9/5)(K - 273.2) + 32$
$R = (9/5)K$

C = °Centigrade, F = °Fahrenheit, R = °Rankine, K = °Kelvin

Table XI. Crystal Structures, Transition Temperatures, and Other Pertinent Physical Constants of the Elements

Name	Atomic Number	Atomic[a] Weight	Density[b], kg m^{-3}·10^{-3}	Crystal Structure	Phase Transition Temp., K	Superconducting Transition Temp., K	Curie Temp., K	Néel Temp., K	Debye Temperature at 0 K, K	Debye Temperature at 298 K, K	Melting Point, K	Boiling Point, K	Critical Temp., K
Actinium	89	(227)	10.07[1c]	f.c.c.[2]					124[3]	100[4] (at~50 K)	1323[5]	3200±300[6]	
Aluminum	13	26.9815	2.702[5]	f.c.c.[7]		1.196[5] 1.17[8] 1.18[9]			423±5[3]	390[3]	933.2[3,10]	2723[29]	8650[11] 7740[109]
Americium	95	(243)	11.7[5]	Double c.p.h.[2]							1473[29]	2880[108]	
Antimony	51	121.75	6.684[29]	r.[2] (?) ? (?) ? (?)	367.8[13] (?-?) 690[13] (?-?)	2.6[8] (Sb II, high-pressure modification)			150[3]	200[14]	903.7[13] 903.65[23]	1907±10[3]	2989[15]
Argon	18	39.948	0.0017824[29] (at 273.2 K and 1 atm)	f.c.c.[16]						90[4] (at~45 K)	83.8[17]	87.29[13]	151[15]
Arsenic	33	74.9216	5.73[29] (gray, at 287.2 K) 4.7[29] (black) 2.0[29] (yellow)	r.[7] (gray) c.[5] (yellow)					236[3]	275[18]	1090[13] (35.8 atm) subl. 886[5]	1090[13] (35.8 atm)	
Astatine	85	(210)									573.2[19]	650[20]	
Barium	56	137.34	3.5[29]	b.c.c.[2] (α) ? (β)[13]	648[13,21] (α-β)				110.5±1.8[22]	116[23]	998.2[5]	1910[3]	3663[15] 3920[109]
Berkelium	97	(249)											
Beryllium	4	9.0122	1.85[29]	c.p.h.[2] (α) b.c.c.[2] (β)	1533[24] (α-β)	~6[108] ~8.4[108]			1160[25]	1031[3]	1550[26]	3142±100[3]	6153[15]
Bismuth	83	208.980	9.78[29]	r.[2]		3.9[8] (Bi II, at 25 kbar) 7.2[8] (Bi III, at 27 kbar)			119±2[3]	116±5[3]	544.525[3,111]	1824±8[3]	4620[27]
Boron	5	10.811	2.50[42]	Simple r.[2] (α) r.[2] (β)	1473[2] (α-β)				1315[53]	1362[3]	2573[5]	4050±100[30]	
Bromine	35	79.909	3.119[29]	orthorh.[16]							266.0[17]	331.93[29]	584[15]

[a] Atomic weights are based on $^{12}C = 12$ as adopted by the International Union of Pure and Applied Chemistry in 1961; those in parentheses are the mass numbers of the isotopes of longest known half-life.

[b] Density values are given at 293.2 K unless otherwise noted.

[c] Superscript numbers designate references listed at the end of the table.

Name	Atomic Number	Atomic[a] Weight	Density,[b] $kg\,m^{-3}\cdot 10^{-3}$	Crystal Structure	Phase Transition Temp., K	Superconducting Transition Temp., K	Curie Temp., K	Néel Temp., K	Debye Temperature at 0 K, K	Debye Temperature at 298 K, K	Melting Point, K	Boiling Point, K	Critical Temp., K
Cadmium	48	112.40	8.65[29]	c.p.h.[2] b.c.c.[4](?)		0.56[5] 0.52[9]			252 ± 48[3]	221[3] 170[4](b.c.c., at ~85 K)	594.18[3,10] Subl. 594.1[13] (at 0.11 mm Hg)	1038[3]	1903[15] 3560[109]
Calcium	20	40.08	1.55[29]	f.c.c.[7](α) b.c.c.[7](β)	737[62](α–β)				234 ± 5[3]	230[3]	1123[19] Subl. 1123[13] (at 0.35 mm Hg)	1765[3]	3267[15]
Californium	98	(251)											
Carbon (amorphous)	6	12.01115	1.8~2.1[29]								Subl. 3925–3970[5]	4473[5]	
Carbon (diamond)	6	12.01115	3.51[29]	d.[16]					2240 ± 5[31]	1874[3]	> 3823[5]	5100[5]	
Carbon (graphite)	6	12.01115	2.26[29](α)	h.[2](α) r.[2](β)					402 ± 11[3]	1550[3]	Subl. 3925–3970[5]	4473[5]	
Cerium	58	140.12	6.90[29]	f.c.c.(α)[32] Double c.p.h.?[8](β) f.c.c.(γ)[32] b.c.c.(δ)[32]	103 ± 5[33](α–β) 263 ± 5[33](β–γ) 1003[32](γ–δ)			13[32]	146[3]	138[34]	1077[26]	3972[3]	10400[109]
Cesium	55	132.905	1.873[29]	b.c.c.[2]					40 ± 5[3]	43[23]	301.9[29] Subl. 301.9[13] (at 1.2 μHg)	939[35]	2060[113,114,115] 1900[109]
Chlorine	17	35.453	0.003214[29] (at 273.2 K)	t.[16]						115[4,36] (at ~58 K)	172.2[26]	239.10[13]	417[15]
Chromium	24	51.996	7.16[42]	c.p.h.[17,d](α) b.c.c.[7](β)	~299[17](α–β)[d]			311[37]	598 ± 32[3]	424[3]	2118[38]	2918 ± 35[3]	
Cobalt	27	58.9332	8.862[42]	c.p.h.[7](α) f.c.c.[17](β)	690[39](α–β)		1400[40]		452 ± 17[3]	386[3]	1765[3,10]	3229[3]	
Copper	29	63.54	8.933[29]	f.c.c.[2]					342 ± 2[3]	310[3]	1356[3,10]	2811 ± 20[41]	8500[11] 8280[109]
Curium	96	(247)	7[42]	Double c.p.h.[8]									
Dysprosium	66	162.50	8.556[42]	c.p.h.[2](α) b.c.c.[2](β)	Near m.p.[2](α–β)			174[43] 83.5[43] (ferro–antiferromag.)	172 ± 35[3]	158[44]	1773[12]	3011[44]	7640[109]

[d] Close-packed hexagonal crystalline modification of chromium may be formed by electrodeposition below 293 K under special conditions of deposition process. This c.p.h. form is unstable and will irreversibly transform into b.c.c. form on heating.

Name	Atomic Number	Atomic Weight[a]	Density,[b] kg m^{-3} · 10^{-3}	Crystal Structure	Phase Transition Temp., K	Superconducting Transition Temp., K	Curie Temp., K	Neel Temp., K	Debye Temperature at 0 K, K	Debye Temperature at 298 K, K	Melting Point, K	Boiling Point, K	Critical Temp., K
Einsteinium	99	(254)											
Erbium	68	167.26	9.06[42]	c.p.h.[2] (α) b.c.c.[2] (β)	1643[2] (α–β)		19[4]	80[4]	134 ± 10[45]	163[44]	1770[26]	3000[3]	7250[109]
Europium	63	151.96	5.245[28]	b.c.c.[7]				~90[4]	127[3]		1099[5]	1971[46]	4600[109]
Fermium	100	(253)											
Fluorine	9	18.9984	0.001695[29] (at 273.2 K and 1 atm)	c.[108] (β-F_2)							53.58[5]	85.24[13]	144[15]
Francium	87	(223)							39[3]		300.2[19]	879[108]	
Gadolinium	64	157.25	7.87[42]	c.p.h.[2] (α) b.c.c.[2] (β)	1535[32] (α–β)		292[40]		170[3]	155 ± 3[3]	1579[19]	3540[3]	8670[109]
Gallium	31	69.72	5.91[29]	orthorh.[4] (α) t.[4] (β)	275.6[13] (α–β) (at 8.86 x 10^6 mm Hg)	1.091[5] 7.2[38] (Ga II, high-pressure modification)			317[3]	240[14] 125[4] (tetra at ~63 K)	302.93[5] 275.6[13] (at 8.86 x 10^6 mm Hg)	2510[3]	7620[27]
Germanium	32	72.59	5.36[29]	d.[7]		5.5[47] (at ~118 kbar) 8.4[108]			378 ± 22[3]	403[3]	1210.6[5]	3100[3]	5642[15]
Gold	79	196.967	19.3[42]	f.c.c.[7]					165 ± 1[3]	178 ± 8[3]	1336.2[3,10] 1336.15[23]	3240[3]	9500[11] 8060[109]
Hafnium	72	178.49	13.28[42]	c.p.h.[48] (α) b.c.c.[48] (β)	2023 ± 20[48] (α–β)	0.16[9] 0.35[108]			256 ± 5[3]	213[23]	2495[19]	4575 ± 150[49]	
Helium	2	4.0026	0.0001785[29] (at 273.2 K and 1 atm)	c.p.h.[16]						30[4] (at ~15 K)	3.45[29] 1.8 ± 0.2[17] (at 30 atm)	4.216[13] 4.22[23]	5.3[15]
Holmium	67	164.930	8.80[29]	c.p.h.[2] (α) b.c.c.[2] (β)	Near m.p.[50] (α–β)		20[4]	132[4]	114 ± 7[45]	161[44]	1734[19]	3228[51]	
Hydrogen	1	1.00797	0.00008987[29] (at 273.2 K and 1 atm)	c.p.h.[16]						116[36] (para., at ~58 K) 105[36] (ortho., at ~53 K)	13.8 ± 0.1[17]	20.39[13] 20.37[23]	33.3[15]
Indium	49	114.82	7.3[29]	f.c.t.[7]		3.4035[5]			108.8 ± 0.3[3]	129[14]	429.76[3,110]	2279 ± 6[3]	4377[15] 7050[109]
Iodine	53	126.9044	4.93[29]	orthorh.[16]						105[4] (at ~53 K)	386.8[29] subl. 298.16[13] (at 0.31 mm Hg)	457.50[29]	785[15]
Iridium	77	192.2	22.5[42]	f.c.c.[7]		0.14[5,9]			425 ± 5[3]	228[3]	2716[3,10]	4820 ± 30[3]	

Name	Atomic Number	Atomic Weight[a]	Density,[b] kg m⁻³ . 10⁻³	Crystal Structure	Phase Transition Temp., K	Superconducting Transition Temp., K	Curie Temp., K	Néel Temp., K	Debye Temperature at 0 K, K	Debye Temperature at 298 K, K	Melting Point, K	Boiling Point, K	Critical Temp., K
Iron	26	55.847	7.87[28]	b.c.c.-ferromag.[7] (α) b.c.c.-paramag.[7] (β) f.c.c.[7] (γ) b.c.c.[7] (δ)	1183[2] (β-γ) 1673[13] (γ-δ)		1043[40]		457 ±12[3]	373[3]	1810[19]	3160[20]	~~10350~~[27] 6750[123] 9400[109]
Krypton	36	83.80	0.003708[29] (at 273.2 K and 1 atm)	f.c.c.[16]						60[4] (at~30 K)	116.6[5]	119.93[13]	209.4[15]
Lanthanum	57	138.91	6.18[42]	Double c.p.h.[8] (α) f.c.c.[2] (β) b.c.c.[2] (γ)	583[32] (α-β) 1141[32] (β-γ)	4.9[8] (α) 6.3[8] (β)			142 ±3[52]	135 ±5[44]	1193[5]	3713 ±70[3]	10500[109]
Lawrencium	103	(257)											
Lead	82	207.19	11.34[29]	f.c.c.[2]		7.193[5]			102 ±5[3]	87 ±1[3]	600.576[3,111]	2022 ±10[41]	5400[27] 4760[109]
Lithium	3	6.939	0.534[29]	b.c.c.[7]	Martensitic transformation at low temp.[56]				352 ±17[3]	448[3]	453.7[19]	1599[13]	4150[11] 3720[109]
Lutetium	71	174.97	9.85[29]	c.p.h.[2] (α) b.c.c.[2] (β)	Near m.p.[50] (α-β)				210[54]	116[3]	1923[19]	4140[3]	
Magnesium	12	24.312	1.74[29]	c.p.h.[7]					396 ±54[3]	330[3]	923[55]	1385[3]	3530[109]
Manganese	25	54.9380	7.43(α)[28] 7.29(β)[28] 7.18(γ)[28]	~~c.[7] (α)~~ b.c.c.[93] (α) c.[7] (β) ~~f.c.t.[7] (γ)~~ f.c.c.[93] (γ) b.c.c.[7] (δ)	1000[13] (α-β) 1374[13] (β-γ) 1410[13] (γ-δ)			95[5]	418 ±32[3]	363[3]	1517 ±3[5]	2360[13]	6050[109]
Mendelevium	101	(256)											
Mercury	80	200.59	13.546[29] 14.19[29] (at 234.25 K)	r.[7] (α) b.c.t.-pressure induced structure (β)	Martensitic transformation at low temp.[56]	4.153[5] (α) 3.949[5] (β)			~ 75[58]	92 ±8[3]	234.28[3,10]	629.73[3,10]	1733[27] 1705[109]
Molybdenum	42	95.94	10.24[42]	b.c.c.[2]		0.92[5,9]			459 ±11[3]	377[3]	2883[13]	5785 ±175[3]	17000[11] 16800[109]
Neodymium	60	144.24	7.007[29]	Double c.p.h.[8] (α) b.c.c.[32] (β)	1135[32] (α-β)			8[4] (ordinary) 19[4] (special)	159[3]	148 ±8[3]	1292[19]	2956[60]	7900[109]
Neon	10	20.183	0.0009002[29] (at 273.2 K and 1 atm)	f.c.c.[16]						60[4] (at~30 K)	24.48[5]	27.23[5] 27.06[23]	44.5[15]

Name	Atomic Number	Atomic Weight[a]	Density,[b] kg m^{-3} · 10^{-3}	Crystal Structure	Phase Transition Temp., K	Superconducting Transition Temp., K	Curie Temp., K	Néel Temp., K	Debye Temperature at 0 K, K	Debye Temperature at 298 K, K	Melting Point, K	Boiling Point, K	Critical Temp., K
Neptunium	93	(237)	20.46[42]	orthorh.[2] (α) t.[2] (β) b.c.c.[2] (γ)	551[2] (α–β) 813[2] (β–γ)				121[3]	163[3]	913.2[5]	4150[3]	
Nickel	28	58.71	8.90[42]	f.c.c.[7]			631[40]		427 ±14[3]	345[3]	1726[3,10] 1726 ±4[61]	3055[63]	6294[15] 11750[109]
Niobium	41	92.906	8.57[42]	b.c.c.[7]		9.13[5] 9.09[8] 9.1[9]			241 ±13[3]	260[64]	2741 ±27[3] 2688[65]	4813[66]	19000[109]
Nitrogen	7	14.0067	0.0012506[29]	c.[16] (α) h.[107] (β)	35.62[13] (α–β)					70[4] (at∼35 K)	63.29[5]	77.34[13,23]	126.2[15]
Nobelium	102	(254)											
Osmium	76	190.2	22.48[29]	c.p.h.[2]		0.655[5] 0.65[8]			500[67]	400[68]	3283 ±10[69]	5300 ±100[70]	
Oxygen	8	15.9994	0.001429[29] (at 273.2 K and 1 atm)	b.c. orthorh.[7] (α) r.[7] (β) c.[7] (γ)	23.876 ± 0.01[112] (α–β) 43.818 ± 0.01[112] (β–γ)					250[4] (at∼125 K) 500[36] (at∼250 K)	54.8[5]	90.19[13] 90.18[23]	154.8[15]
Palladium	46	106.4	12.02[28]	f.c.c.[2]					283 ±16[3]	275[14]	1825[3,10]	3200[3]	
Phosphorus	15	30.9738	1.82[29] (β) 2.22[29] (γ) 2.69[29] (δ)	h. ?[7] (α) b.c.c.[7] (β) c.[7] (γ) f.c. orthorh.[17] (δ)	196[71] (α–β) 298.16[13] (β–γ) 298.16[13] (β–δ)				193[3] (white) 325[3] (red)	576[3] (white) 800[3] (red)	317.3[5] (white) 1300[72] (black)	553[13]	993.8[15]
Platinum	78	195.09	21.45[29]	f.c.c.[2]					234 ±1[3]	225 ±5[3]	2042[3,10]	4100[3]	8280[15]
Plutonium	94	(242)	19.737[29] (at 298.2 K)	Simple monocl.[2] (α) b.c. monocl.[2] (β) f.c. orthorh.[2] (γ) f.c.c.[2] (δ) b.c.t.[2] (δ') b.c.c.[2] (ϵ)	396.7[73] (α–β) 475[73] (β–γ) 591.4[73] (γ–δ) 729[73] (δ–δ') 757 ±3[73] (δ'–ϵ)				171[74]	176[74]	912.7[5]	3727[75]	
Polonium	84	(210)	9.3[29] (α) 9.5[29] (β)	Simple c.[7] (α) r.[7] (β)	327 ±1.5[76] (α–β)				81[3]		527.2[5]	1235[20]	2281[15]
Potassium	19	39.102	0.86[29]	b.c.c.[7]					89.4 ±0.5[3]	100[3]	336.8[5]	1027[35]	2450[11] 2140[109]
Praseodymium	59	140.907	6.769[29]	Double c.p.h.[8] (α) b.c.c.[2] (β)	1071[32] (α–β)			25[77]	85 ±1[45]	138[78]	1192 ±2[79]	3616[80]	8900[109]

Name	Atomic Number	Atomic Weight[a]	Density,[b] kg m^{-3} · 10^{-3}	Crystal Structure	Phase Transition Temp., K	Superconducting Transition Temp., K	Curie Temp., K	Néel Temp., K	Debye Temperature at 0 K, K	Debye Temperature at 298 K, K	Melting Point, K	Boiling Point, K	Critical Temp., K
Promethium	61	(145)		h.[7] (α) b.c.c.[120] (β)	1185[120] (α–β)			6[120]			1353 ± 10[81]	2730[3]	
Protactinium	91	(231)	15.37[42]	b.c.t.[2]		1.4[9]			159[3]	262[3]	1503[5]	4680[3]	
Radium	88	(226)	5[29]						89[3]		973.2[5]	1900[3]	
Radon	86	(222)	0.00973[29] (at 273.2 K and 1 atm)	f.c.c.[7]						400[4] (at ~200 K)	202.2[5]	211[13]	377.16[15]
Rhenium	75	186.2	21.1[42]	c.p.h.[2]		1.698[26]			429 ± 22[3]	275[23]	3453[5]	6035 ± 135[3]	20000[11]
Rhodium	45	102.905	12.45[42]	f.c.c.[7]	possible transformation at 1373–1473 K[57]				480 ± 32[3]	350[3]	2233[3,10,82]	3960 ± 60[3]	
Rubidium	37	85.47	1.53[29]	b.c.c.[2]					54 ± 4[3]	59[23]	312.04[5]	959[35]	2100[113,115,116] 2030[109]
Ruthenium	44	101.07	12.2[29]	c.p.h.[7] (α) ? (β) ? (γ) ? (δ)	1308[13,121] (α–β) 1473[13,121] (β–γ) 1773[13,121] (γ–δ)	0.49[5,9]			600[67]	415[3]	2523 ± 10[69]	4325 ± 25[3]	
Samarium	62	150.35	7.54[29]	r.[32] (α) b.c.c.[32] (β)	1190[32] (α–β)		14[8]	106[8]	116[45]	184 ± 4[3]	1345.2[83]	2140[3]	5400[109]
Scandium	21	44.956	3.00[42]	c.p.h.[2] (α) b.c.c.[2] (β)	1607[2] (α–β)				470 ± 80[52]	476[3]	1812[5]	3537 ± 30[3]	
Selenium	34	78.96	4.50[29] (α) 4.80[29] (β)	monocl.[7] (α) h.[7] (β) amorphous[7]	304[84,117] (vitrification) 398[13] (vit.–β) 423[13] (α–β)	7.3[85] (at ~118 kbar)			151.7 ± 0.4[86]	89[36] (at ~45 K) 150[4] (at ~75 K)	490.2[5]	1009[13] (Se_6) 958.0[13] ($Se_{4.37}$) 1027[13] (Se_2)	1757[15]
Silicon	14	28.086	2.33[42]	d.[7]		7.5[47] (at 118–128 kbar)			647 ± 11[3]	692[3]	1685 ± 2[87]	2753[28]	5159[15]
Silver	47	107.870	10.5[29]	f.c.c.[2]					228 ± 3[3]	221[3]	1234.0[3,13]	2468 ± 15[41]	7460[11]
Sodium	11	22.9898	0.9712[29]	b.c.c.[2]	Martensitic transformation at low temp.[56]				157 ± 1[3]	155 ± 5[3]	371.0[13]	1154[35]	2800[11] 2400[109]
Strontium	38	87.62	2.60[28]	f.c.c.[88] (α) c.p.h.[7] (β) b.c.c.[7] (γ)	488[88] (α–β) 878[88] (β–γ)				147 ± 1[22]	148[23]	1042[5]	1645[3]	3059[15] 3810[109]
Sulfur	16	32.064	2.07[29] (α) 1.96[29] (β)	r.[7] (α) monocl.[7] (β)	368.6[13] (α–β)				200[3] (β)	527[89] (α) 250[89] (α, at 40 K)	386.0[5] (α) 392.2[5] (β) Subl. 368.6[13] (at 0.0047 mm Hg)	717.75[3,10]	1313[15]
Tantalum	73	180.948	16.6[42]	b.c.c.[2]		4.483[5] 4.48[9]			247 ± 13[3]	225[14]	3269[5]	5760 ± 60[3]	22000[11]

Name	Atomic Number	Atomic Weight[a]	Density,[b] kg m^{-3} · 10^{-3}	Crystal Structure	Phase Transition Temp., K	Superconducting Transition Temp., K	Curie Temp., K	Neel Temp., K	Debye Temperature at 0 K, K	Debye Temperature at 298 K, K	Melting Point, K	Boiling Point, K	Critical Temp., K
Technetium	43	(99)	11.50[29]	c.p.h.[2]		8.22[5] 11.2[9]			351[3]	422[3]	2473 ± 50[5]	5300[3]	
Tellurium	52	127.60	6.24[29] (α) 6.00[5] (amorph.)	h.[7] (α) ? (β)[7] amorph.[5]	621[13] (α–β)	3.3[8] (Te II, at 56 kbar)			141 ± 12[3]		722.7[5]	1163 ± 1[3]	2329[15]
Terbium	65	158.924	8.25[29]	c.p.h.[2, 32] (α) b.c.c.[2] (β)	Near m.p.[2] (α–β)		219[90]	230[90]	150[91]	158[44]	1629[19]	3810[3]	
Thallium	81	204.37	11.85[29]	c.p.h.[2] (α) b.c.c.[2] (β)	508.3[5] (α–β)	2.39[5] 2.38[8] 2.37[9]			88 ± 1[3]	96[14]	576.2[19]	1939[92]	3219[15]
Thorium	90	232.038	11.7[42]	f.c.c.[2] (α) b.c.c.[2] (β)	1673 ± 25[93] (α–β)	1.368[5] 1.37[9]			170[94]	100[14]	2023[19]	4500[20]	14550[109]
Thulium	69	168.934	9.32[29]	c.p.h.[2] (α) b.c.c.[2] (β)	Near m.p.[50] (α–β)		22[95] (ferro.-antiferro.)	53[96]	127 ± 1[45]	167[44]	1818[5]	2266[97]	6430[109]
Tin	50	118.69	5.750[29] (α) 7.31[29] (β)	f.c.c.[7] (α) b.c.t.[7] (β) r.[29] (?)	286.2 ± 3[98] (α–β)	3.722[5] (β)			236 ± 24[3] (gray) 196 ± 9[3] (white)	254[3] (gray) 170[14] (white)	505.06[3, 10]	2766 ± 14[3]	8000[11] 9300[109]
Titanium	22	47.90	4.5[29]	c.p.h.[7] (α) b.c.c.[7] (β)	1155[13] (α–β)	0.39[5, 9]			426 ± 5[3]	380[14]	1953[99]	3586[100]	
Tungsten	74	183.85	19.3[29]	b.c.c.[2]		0.011[122]			388 ± 17[3]	312 ± 3[3]	3653[3, 10, 13]	6000 ± 200[3]	23000[11]
Uranium	92	238.03	19.07[28]	orthorh.[7] (α) t.[7] (β) b.c.c.[7] (γ)	37 ± 2[118] (α_0–α) 938[13] (α–β) 1049[13] (β–γ)	0.68[5] (α) 1.80[5] (γ)			200[94]	300[3]	1405.6 ± 0.6[101]	3950 ± 250[102]	12500[27] 12000[109]
Vanadium	23	50.942	6.1[28]	b.c.c.[2]		5.3[5] 5.03[9]			326 ± 54[3]	390[14]	2192 ± 2[61]	3582 ± 42[3]	11200[109]
Xenon	54	131.30	0.005851[29] (at 273.2 K and 1 atm)	f.c.c.[16]							161.2[26]	165.1[13]	289.75[15]
Ytterbium	70	173.04	7.02[42]	f.c.c.[32] (α) b.c.c.[32] (β)	1071[2, 5] (α–β)				118[103]		1097[12]	1970[3]	4420[109]
Yttrium	39	88.905	4.47[29]	c.p.h.[32] (α) b.c.c.[32] (β)	1753[119] (α–β)				268 ± 32[3]	214[104]	1798[119]	3670[105]	8950[109]
Zinc	30	65.37	7.140[29]	c.p.h.[2]		0.875[5] 0.85[9]			316 ± 20[3]	237 ± 3[3]	692.655[3, 110]	1175[106]	2169[15] 2910[109]
Zirconium	40	91.22	6.57[59]	c.p.h.[7] (α) b.c.c.[7] (β)	1135[13] (α–β)	0.546[5] 0.55[9]			289 ± 24[3]	250[14]	2125[19]	4650[20]	12300[109]

REFERENCES

(Crystal Structures, Transition Temperatures, and Other Pertinent Physical Constants of the Elements)

1. Farr, J.D., Giorgi, A.L., and Bowman, M.G., USAEC Rept. LA-1545, 1-13, 1953.
2. Elliott, R.P., Constitution of Binary Alloys, 1st Suppl., McGraw-Hill, 1965.
3. Gschneider, K.A, Jr., Solid State Physics (Sietz, F. and Turnbull, D., Editors), 16, 275-426, 1964.
4. Gopal, E.S.R., Specific Heat at Low Temperatures, Plenum Press, 1966.
5. Weast, R.C. (Editor), Handbook of Chemistry and Physics, 47th Ed., The Chemical Rubber Co., 1966-67.
6. Foster, K.W. and Fauble, L.G., J. Phys. Chem., 64, 958-60, 1960.
7. The Institution of Metallurgists, Annual Yearbook, pp. 68-73, 1960-61.
8. Meaden, G.T., Electrical Resistance of Metals, Plenum Press, 1965.
9. Matthias, B.T., Geballe, T.H., and Compton, V.B., Rev. Mod. Phys., 35, 1-22, 1963.
10. Stimson, H.F., J. Res. NBS, 42, 209, 1949.
11. Grosse, A.V., Rev. Hautes Tempér. et Réfract., 3, 115-46, 1966.
12. Spedding, F.H. and Daane, A.H., J. Metals, 6 (5), 504-10, 1954.
13. Rossini, F.D., Wagman, D.D., Evans, W.H., Levine, S., and Jaffe, I., NBS Circ. 500, 537-822, 1952.
14. deLaunay, J., Solid State Physics, 2, 219-303, 1956.
15. Gates, D.S. and Thodos, G., AIChE J., 6 (1), 50-4, 1960.
16. Gray, D.E. (Coordinating Editor), American Institute of Physics Handbook, McGraw-Hill, 1957.
17. Sasaki, K. and Sekito, S., Trans. Electrochem. Soc., 59, 437-60, 1931.
18. Anderson, C.T., J. Am. Chem. Soc., 52, 2296-300, 1930.
19. Trombe, F., Bull. Soc. Chim. (France), 20, 1010-2, 1953.
20. Stull, D.R. and Sinke, G.C., Thermodynamic Properties of the Elements in Their Standard State, American Chemical Soc., 1956.
21. Rinck, E., Ann. Chim. (Paris), 18 (10), 455-531, 1932.
22. Roberts, L.M., Proc. Phys. Soc. (London), B70, 738-43, 1957.
23. Zemansky, M.W., Heat and Thermodynamics, 4th Ed., McGraw-Hill, 1957.
24. Martin, A.J. and Moore, A., J. Less-Common Metals, 1, 85, 1959.
25. Hill, R.W. and Smith, P.L., Phil. Mag., 44 (7), 636-44, 1953.
26. Moffatt, W.G., Pearsall, G.W., and Wulff, J., The Structure and Properties of Materials, Vol. I, pp. 205-7, 1964.
27. Grosse, A.V., Temple Univ. Research Institute Rept., 1-40, 1960.
28. Lyman, T. (Editor), Metals Handbook, Vol. 1, 8th Ed., American Soc. for Metals, 1961.
29. Lange, N.A. (Editor), Handbook of Chemistry, Revised 10th Edition, McGraw-Hill, 1967.
30. Paule, R.C., Dissertation Abstr., 22, 4200, 1962.
31. Burk, D.L. and Friedberg, S.A., Phys. Rev., 111 (5), 1275-82, 1958.
32. Spedding, F.H. and Daane, A.H. (Editors), The Rare Earths, John Wiley, 1961.
33. McHargue, C.J., Yakel, H.L., and Letter, C.K., ACTA Cryst., 10, 832-33, 1957.
34. Arajs, S. and Colvin, R.V., J. Less-Common Metals, 4, 159-68, 1962.
35. Bonilla, C.F., Sawhney, D.L., and Makansi, M.M., Trans. Am. Soc. Metals, 55, 877, 1962.
36. Rosenberg, H.M., Low Temperature Solid State Physics, Oxford at Clarendon Press, 1965.
37. Arajs, S., J. Less-Common Metals, 4, 46-51, 1962.
38. Edwards, A.R. and Johnstone, S.T.M., J. Inst. Metals, 84 (8), 313-7, 1956.
39. Lagneborg, R. and Kaplow, R., ACTA Metallurgica, 15 (1), 13-24, 1967.
40. Kittel, C., Introduction to Solid State Physics, 3rd Ed., John Wiley, 1967.
41. Kirshenbaum, A.D. and Cahill, J.A., J. Inorg. and Nucl. Chem., 25 (2), 232-34, 1963.
42. Touloukian, Y.S. (Ed.), Thermophysical Properties of High Temperature Solid Materials, MacMillan, Vol. 1, 1967.
43. Griffel, M., Skochdopole, R.E., and Spedding, F.H., J. Chem. Phys., 25 (1), 75-9, 1956.
44. Gschneidner, K.A., Jr., Rare Earth Alloys, Van Nostrand, 1961.
45. Dreyfus, B., Goodman, B.B., Lacaze, A., and Trolliet, G., Compt. Rend., 253, 1764-6, 1961.

46. Spedding, F.H., Hanak, J.J., and Daane, A.H., Trans. AIME, 212, 379, 1958.
47. Buckel, W. and Wittig, J., Phys. Lett. (Netherland), 17 (3), 187-8, 1965.
48. Deardorff, D.K. and Kata, H., Trans. AIME, 215, 876-7, 1959.
49. Panish, M.B. and Reif, L., J. Chem. Phys., 38 (1), 253-6, 1963.
50. Miller, A.E. and Daane, A.H., Trans. AIME, 230, 568-72, 1964.
51. Spedding, F.H. and Daane, A.H., USAEC Rept. IS-350, 22-4, 1961.
52. Montgomery, H. and Pells, G.P., Proc. Phys. Soc. (London), 78, 622-5, 1961.
53. Kaufman, L. and Clougherty, E.V., ManLabs, Inc., Semi-Annual Rept. No. 2, 1963.
54. Lounasmaa, O.V., Proc. 3rd Rare Earth Conf., 1963, Gordon and Breach, New York, 1964.
55. Baker, H., WADC TR 57-194, 1-24, 1957.
56. Reed, R.P. and Breedis, J.F., ASTM STP 387, pp. 60-132, 1966.
57. Hansen, M., Constitution of Binary Alloys, 2nd Edition, McGraw-Hill, p. 1268, 1958.
58. Smith, P.L., Conf. Phys. Basses Temp., Inst. Intern. du Froid, Paris, 281, 1956.
59. Powell, R.W. and Tye, R.P., J. Less-Common Metals, 3, 202-15, 1961.
60. Yamamoto, A.S., Lundin, C.E., and Nachman, J.F., Denver Res. Inst. Rept., NP-11023, 1961.
61. Oriena, R.A. and Jones, T.S., Rev. Sci. Instr., 25, 248-51, 1954.
62. Smith, J.F., Carlson, O.N., and Vest, R.W., J. Electrochem. Soc., 103, 409-13, 1956.
63. Edwards, J.W. and Marshal, A.L., J. Am. Chem. Soc., 62, 1382, 1940.
64. Morin, F.J. and Maita, J.P., Phys. Rev., 129 (3), 1115-20, 1963.
65. Pendleton, W.N., ASD-TDR-63-164, 1963.
66. Woerner, P.F. and Wakefield, G.F., Rev. Sci. Instr., 33 (12), 1456-7, 1962.
67. Walcott, N.M., Conf. Phys. Basses Temp., Inst. Intern. du Froid, Paris, 286, 1956.
68. White, G.K. and Woods, S.B., Phil. Trans. Roy. Soc. (London), A251 (995), 273-302, 1959.
69. Douglass, R.W. and Adkins, E.F., Trans. AIME, 221, 248-9, 1961.
70. Panish, M.B. and Reif, L., J. Chem. Phys., 37 (1), 128-31, 1962.
71. Bridgman, P.W., J. Am. Chem. Soc., 36 (7), 1344-63, 1914.
72. Slack, G.A., Phys. Rev., A139 (2), 507-15, 1965.
73. Sandenaw, T.A. and Gibney, R.B., J. Phys. Chem. Solids, 6 (1), 81-8, 1958.
74. Sandenaw, T.A., Olsen, C.E., and Gibney, R.B., Plutonium 1960, Proc. 2nd Intern. Conf. (Grison, E., Lord, W.B.H., and Fowler, R.D., Editors), 66-79, 1961.
75. Mulford, R.N.R., USAEC Rept. LA-2813, 1-11, 1963.
76. Goode, J.M., J. Chem. Phys., 26 (5), 1269-71, 1957.
77. Cable, J.W., Moon, R.M., Koehler, W.C., and Wollan, E.O., Phys. Rev. Letters, 12 (20), 553-5, 1964.
78. Murao, T., Progr. Theoret. Phys. (Kyoto), 20 (3), 277-86, 1958.
79. Grigor'ev, A.T., Sokolovskaya, E.M., Budennaya, L.D., Iyutina, I.A., and Maksimona, M.V., Zhur. Neorg. Khim., 1, 1052-63, 1956.
80. Daane, A.H., USAEC AECD-3209, 1950.
81. Weigel, F., Angew. Chem., 75, 451, 1963.
82. Nassau, K. and Broyer, A.M., J. Am. Ceram. Soc., 45 (10), 474-8, 1962.
83. McKeown, J.J., State Univ. of Iowa, Ph.D. Dissertation, 1-113, 1958.
84. Abdullaev, G.B., Mekhtiyeva, S.I., Abdinov, D.Sh., and Aliev, G.M., Phys. Letters, 23 (3), 215-6, 1966.
85. Wittig, J., Phys. Rev. Letters, 15 (4), 159, 1965.
86. Fukuroi, T. and Muto, Y., Tohoku Univ. Res. Inst. Sci. Rept., A8, 213-22, 1956.
87. Olette, M., Compt. Rend., 244, 1033-6, 1957.
88. Sheldon, E.A., and King, A.J., ACTA Cryst., 6, 100, 1953.
89. Eastman, E.D. and McGavock, W.C., J. Am. Chem. Soc., 59, 145-51, 1937.
90. Arajs, S. and Colvin, R.V., Phys. Rev., A136 (2), 439-41, 1964.
91. Roach, P.R. and Lounasmaa, O.V., Bull. Am. Phys. Soc., 7, 408, 1962.
92. Shchukarev, S.A., Semenov, G.A., and Rat'kovskii, I.A., Zh. Neorgan. Khim., 7, 469, 1962.
93. Pearson, W.B., A Handbook of Lattice Spacings and Structures of Metals and Alloys, Pergamon Press, 1958.

94. Smith, P.L. and Walcott, N.M., Conf. Phys. Basses Temp., Inst. Intern. du Froid, 283, 1956.

95. Davis, D.D. and Bozorth, R.M., Phys. Rev., 118 (6), 1543-5, 1960.

96. Aliev, N.G. and Volkenstein, N.V., Soviet Physics - JETP, 22 (5), 997-8, 1966.

97. Spedding, F.H., Barton, R.J., and Daane, A.H., J. Am. Chem. Soc., 79, 5160, 1957.

98. Raynor, G.V. and Smith, R.W., Proc. Roy. Soc. (London), A244, 101-9, 1958.

99. Savitskii, E.M. and Burhkanov, G.S., Zhur. Neorg. Khim., 2, 2609-16, 1957.

100. Argent, B.B. and Milne, J.G.C., Niobium, Tantalum, Molybdenum and Tungsten, Elsevier Publ. Co. (Quarrell, A.G., Editor), pp. 160-8, 1961.

101. Argonne National Laboratory, USAEC Rept. ANL-5717, 1-67, 1957.

102. Holden, A.N., Physical Metallurgy of Uranium, Addison-Wesley, 1958.

103. Lounasmaa, O.V., Phys. Rev., 129, 2460-4, 1963.

104. Jennings, L.D., Miller, R.E., and Spedding, F.H., J. Chem. Phys., 33 (6), 1849-52, 1960.

105. Ackerman, R.J. and Rauh, E.G., J. Chem. Phys., 36 (2), 448-52, 1962.

106. Rosenblatt, G.M. and Birchenall, C.E., J. Chem. Phys., 35 (3), 788-94, 1961.

107. Streib, W.E., Jordan, T.H., and Lipscomb, W.N., J. Chem. Phys., 37 (12), 2962-5, 1962.

108. Samsonov, G.V. (Editor), Handbook of the Physicochemical Properties of the Elements, Plenum Press, 1968.

109. Kopp, I.Z., Russ. J. Phys. Chem., 41 (6), 782-3, 1967.

110. Stimson, H.F., in Temperature, Its Measurement and Control in Science and Industry (Herzfeld, C.M., Ed.), Vol. 3, Part 1, Reinhold, New York, pp. 59-66, 1962.

111. McLaren, E.H., in Temperature, Its Measurement and Control in Science and Industry (Herzfeld, C.M., Ed.), Vol. 3, Part 1, Reinhold, New York, pp. 185-98, 1962.

112. Orlova, M.P., in Temperature, Its Measurement and Control in Science and Industry (Herzfeld, C.M., Ed.), Vol. 3, Part 1, Reinhold, New York, pp. 179-83, 1962.

113. Grosse, A.V., J. Inorg. Nucl. Chem., 28, 2125-9, 1966.

114. Hochman, J.M. and Bonilla, C.F., in Advances in Thermophysical Properties at Extreme Temperatures and Pressures (Gratch, S., Ed.), ASME 3rd Symposium on Thermophysical Properties, Purdue University, March 22-25, 1965, ASME, pp. 122-30, 1965.

115. Dillon, I.G., Illinois Institute of Technology, Ph.D. Thesis, June 1965.

116. Hochman, J.M., Silver, I.L., and Bonilla, C.F., USAEC Rept. CU-2660-13, 1964.

117. Abdullaev, G.B., Mekhtieva, S.I., Abdinov, D.Sh., Aliev, G.M., and Alieva, S.G., Phys. Status Solidi, 13(2), 315-23, 1966.

118. Fisher, E.S. and Dever, D., Phys. Rev., 2, 170(3), 607-13, 1968.

119. Beaudry, B.J., J. Less-Common Metals, 14(3), 370-2, 1968.

120. Williams, R.K. and McElroy, D.L., USAEC Rept. ORNL-TM 1424, 1-32, 1966.

121. Jaeger, F.M. and Rosenbaum, E., Proc. Nederland Akademie van Wetenschappen, 44, 144-52, 1941.

122. Gibson, J.W. and Hein, R.A., Phys. Letters, 12(25), 688-90, 1964.

123. Grosse, A.V., Research Instit. of Temple Univ., Report on USAEC CONTRACT No. AT(30-1)-2082, 1-71, 1965.

1. ELEMENTS

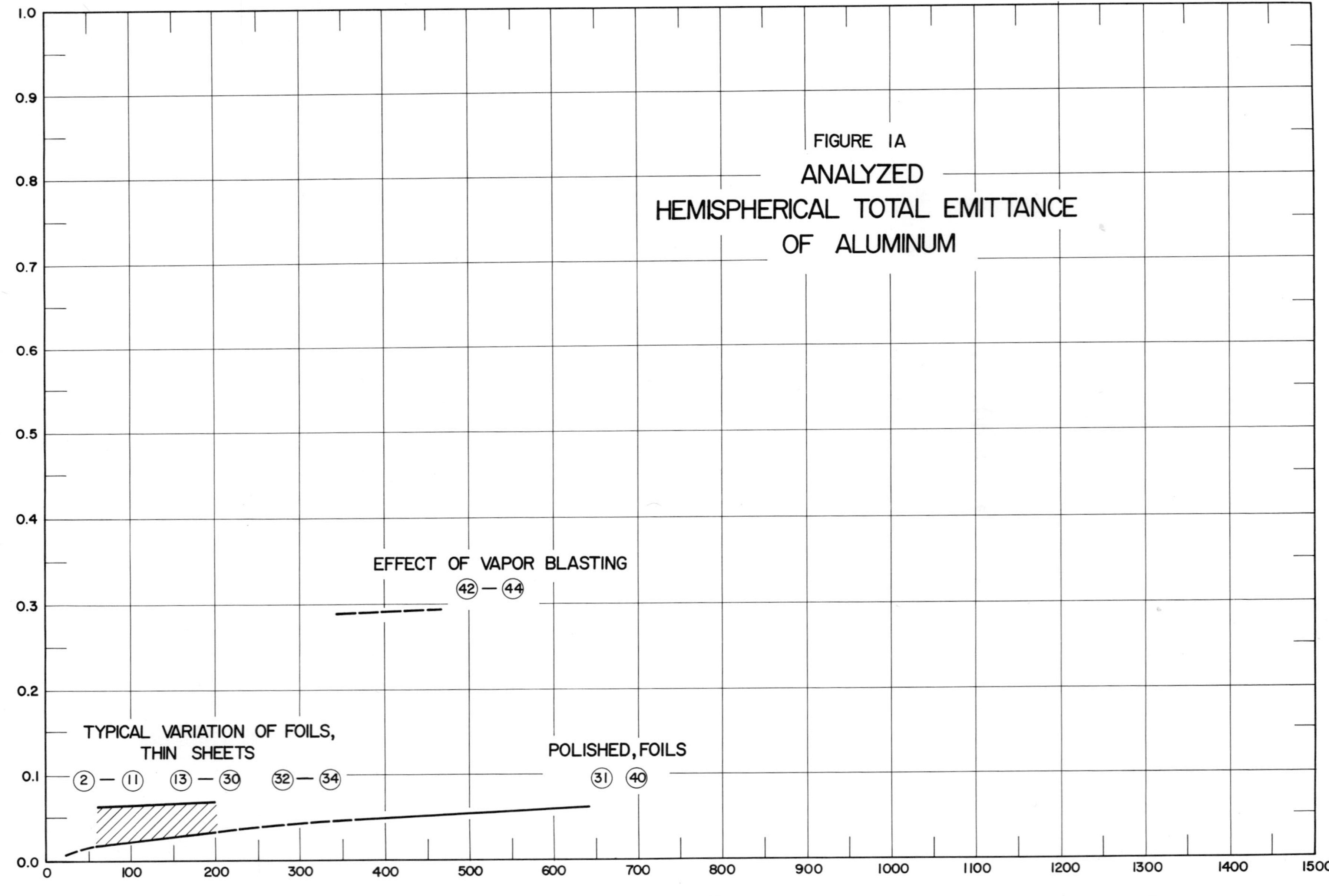
FIGURE 1A
ANALYZED
HEMISPHERICAL TOTAL EMITTANCE
OF ALUMINUM
EFFECT OF VAPOR BLASTING
(42) — (44)
TYPICAL VARIATION OF FOILS,
THIN SHEETS
(2) — (11) (13) — (30) (32) — (34)
POLISHED, FOILS
(31) (40)
HEMISPHERICAL TOTAL EMITTANCE
1.0
0.9
0.8
0.7
0.6
0.5
0.4
0.3
0.2
0.1
0.0
0
100
200
300
400
500
600
700
800
900
1000
1100
1200
1300
1400
1500
TEMPERATURE, K

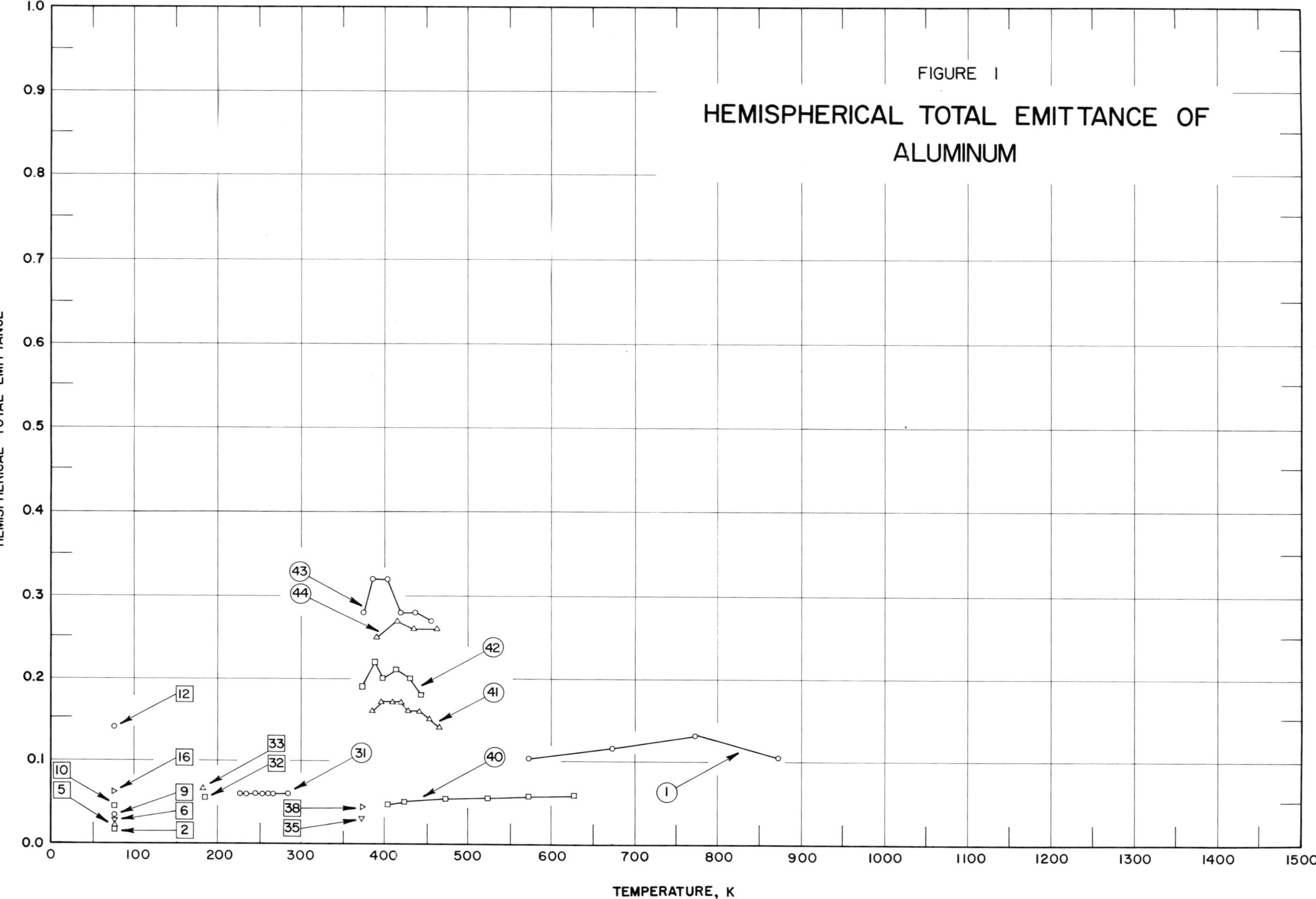
FIGURE I
HEMISPHERICAL TOTAL EMITTANCE OF ALUMINUM
HEMISPHERICAL TOTAL EMITTANCE
TEMPERATURE, K
0.0
0.1
0.2
0.3
0.4
0.5
0.6
0.7
0.8
0.9
1.0
0
100
200
300
400
500
600
700
800
900
1000
1100
1200
1300
1400
1500
1
2
5
6
9
10
12
16
31
32
33
35
38
40
41
42
43
44

SPECIFICATION TABLE NO. 1 HEMISPHERICAL TOTAL EMITTANCE OF ALUMINUM

Curve No.	Ref. No.	Year	Temperature Range, K	Reported Error, %	Composition (weight percent), Specifications and Remarks
1	1	1955	573-873		Calculated from spectral data.
2	3	1955	76	5	Kaiser foil (0.001 in. thick); unannealed; measured in vacuum (10^{-6} to 10^{-7} mm Hg); emittance for 300 K black body incident radiation; authors assumed $\alpha = \epsilon$.
3	3	1955	76	5	Cockron home foil (0.0015 in. thick); measured in vacuum (10^{-6} to 10^{-7} mm Hg); emittance for 300 K black body incident radiation; authors assumed $\alpha = \epsilon$.
4	3	1955	76	5	Hurwich home foil (0.0015 in. thick); measured on mat side in vacuum (10^{-6} to 10^{-7} mm Hg); emittance for 300 K black body incident radiation; authors assumed $\alpha = \epsilon$.
5	3	1955	76	5	Hurwich home foil (0.0015 in. thick); measured on bright side in vacuum (10^{-6} to 10^{-7} mm Hg); emittance for 300 K black body incident radiation; authors assumed $\alpha = \epsilon$.
6	3	1955	76	5	Sheet (0.020 in. thick); cold acid cleaned; measured in vacuum (10^{-6} to 10^{-7} mm Hg); emittance for 300 K black body incident radiation; authors assumed $\alpha = \epsilon$.
7	3	1955	76	5	Alcoa No. 2 reflector plate (0.020 in. thick); measured in vacuum (10^{-6} to 10^{-7} mm Hg); emittance for 300 K black body incident radiation; authors assumed $\alpha = \epsilon$.
8	3	1955	76	5	Alcoa No. 2 reflector plate (0.020 in. thick); sanded with fine emery; measured in vacuum (10^{-6} to 10^{-7} mm Hg); emittance for 300 K black body incident radiation; authors assumed $\alpha = \epsilon$.
9	3	1955	76	5	Alcoa No. 2 reflector plate (0.020 in. thick); cleaned with alkali; measured in vacuum ($10^{-6}$to 10^{-7} mm Hg); emittance for 300 K black body incident radiation; authors assumed $\alpha = \epsilon$.
10	3	1955	76	5	Sheet (0.020 in. thick); cleaned with wire brush, emery paper, steel wool and cold acid; measured in vacuum (10^{-6} to 10^{-7} mm Hg); emittance for 300 K black body incident radiation; authors assumed $\alpha = \epsilon$.
11	3	1955	76	5	Sheet (0.020 in. thick); wire brush cleaned; measured in vacuum (10^{-6} to 10^{-7} mm Hg); emittance for 300 K black body incident radiation; authors assumed $\alpha = \epsilon$.
12	3	1955	76	5	Sheet (0.020 in. thick); liquid honed; measured in vacuum ($10^{-6}$to 10^{-7} mm Hg); emittance for 300 K black body incident radiation; authors assumed $\alpha = \epsilon$.
13	3	1955	76	5	Sheet (0.020 in. thick); hot acid cleaned (Alcoa process); measured in vacuum (10^{-6} to 10^{-7} mm Hg); emittance for 300 K black body incident radiation; authors assumed $\alpha = \epsilon$.
14	4	1953	76		Cockron foil (0.0015 in. thick); measured in vacuum ($< 10^{-6}$ mm Hg); emittance for 294 K black body radiation; authors assumed $\alpha = \epsilon$.
15	4	1953	76		Cockron foil (0.0015 in. thick); measured in vacuum ($< 10^{-6}$ mm Hg); emittance for 294 K black body radiation; authors assumed $\alpha = \epsilon$.
16	4	1953	76		Sheet (0.020 in. thick); cleaned with wire brush; measured in vacuum ($< 10^{-6}$ mm Hg); emittance for 294 K black body radiation; authors assumed $\alpha = \epsilon$.
17	4	1953	76		Sheet (0.020 in. thick); cleaned with wire brush, emery paper, and steel wool; measured in vacuum ($< 10^{-6}$ mm Hg); emittance for 294 K black body radiation; authors assumed $\alpha = \epsilon$.
18	4	1953	76		Sheet (0.020 in. thick); cleaned with wire brush, emery paper, steel wool, and cold acid; measured in vacuum ($< 10^{-6}$ mm Hg); emittance for 294 K black body radiation; authors assumed $\alpha = \epsilon$.

SPECIFICATION TABLE NO. 1 (continued)

Curve No.	Ref. No.	Year	Temperature Range, K	Reported Error, %	Composition (weight percent), Specifications and Remarks
19	4	1953	76		Sheet (0.020 in. thick); alkali cleaned; measured in vacuum ($< 10^{-6}$ mm Hg); emittance for 294 K black body radiation; authors assumed $\alpha = \epsilon$.
20	4	1953	76		Alcoa No. 2 reflector plate (0.020 in. thick, $\epsilon = 0.30$ for visible light at 298 K); measured in vacuum ($< 10^{-6}$ mm Hg); emittance for 294 K black body radiation; authors assumed $\alpha = \epsilon$.
21	4	1953	76		Sheet (0.020 in. thick); cold acid cleaned; measured in vacuum ($< 10^{-6}$ mm Hg); emittance for 294 K black body radiation; authors assumed $\alpha = \epsilon$.
22	4	1953	76		Sheet (0.020 in. thick); cold acid cleaned; measured in vacuum ($< 10^{-6}$ mm Hg); emittance for 294 K black body radiation; authors assumed $\alpha = \epsilon$.
23	4	1953	76		Hurwich home foil (0.0015 in. thick); measured on bright side in vacuum ($< 10^{-6}$ mm Hg); emittance for 294 K black body radiation; authors assumed $\alpha = \epsilon$.
24	4	1953	76		Hurwich home foil (0.0015 in. thick); measured on mat side in vacuum ($< 10^{-6}$ mm Hg); emittance for 294 K black body radiation; authors assumed $\alpha = \epsilon$.
25	4	1953	76		Cockron home foil (0.0015 in. thick); measured in vacuum ($< 10^{-6}$ mm Hg); emittance for 294 K black body radiation; authors assumed $\alpha = \epsilon$.
26	4	1953	76		Cockron home foil (0.0015 in. thick); measured in vacuum ($< 10^{-6}$ mm Hg); emittance for 294 K black body radiation; authors assumed $\alpha = \epsilon$.
27	4	1953	76		Kaiser foil (0.001 in. thick); unannealed; measured in vacuum ($< 10^{-6}$ mm Hg); emittance for 294 K black body radiation; authors assumed $\alpha = \epsilon$.
28	4	1953	76		Kaiser foil (0.001 in. thick); unannealed; measured in vacuum ($< 10^{-6}$ mm Hg); emittance for 294 K black body radiation; authors assumed $\alpha = \epsilon$.
29	4	1953	76		Kaiser home foil (0.00075 in. thick); measured in vacuum ($< 10^{-6}$ mm Hg); emittance for 294 K black body radiation; authors assumed $\alpha = \epsilon$.
30	4	1953	76		Foil (0.020 in. thick); hot acid cleaned (Alcoa process); measured in vacuum ($< 10^{-6}$ mm Hg); emittance for 294 K black body incident radiation; authors assumed $\alpha = \epsilon$.
31	6	1963	227-282	± 3	Plate (0.2 in. thick); hand polished; measured in vacuum (10^{-3} mm Hg).
32	7	1963	185	± 2.5	Foil (0.0005 in. thick); bright side; measured in vacuum ($\le 0.2 \times 10^{-4}$ mm Hg).
33	7	1963	182	± 2.5	Foil (0.00024 in. thick); bright side; measured in vacuum ($\le 0.2 \times 10^{-4}$ mm Hg).
34	7	1963	181	± 2.5	Foil (0.001 in. thick); bright side; measured in vacuum ($\le 0.2 \times 10^{-4}$ mm Hg).
35	8	1963	373		99.99 pure; prefinished with 600 grit aluminum oxide powder on felt and then electropolished; measured in vacuum (10^{-5} mm Hg).
36	8	1963	373		Above specimen and conditions except bombarded with hydrogen ions (3.275×10^{20} ions cm^{-2}).
37	8	1963	373		Above specimen and conditions except bombarded with hydrogen ions (6.60×10^{20} ions cm^{-2}).
38	8	1963	373		Above specimen and conditions except bombarded with hydrogen ions (9.840×10^{20} ions cm^{-2}).

SPECIFICATION TABLE NO. 1 (continued)

Curve No.	Ref. No.	Year	Temperature Range, K	Reported Error, %	Composition (weight percent), Specifications and Remarks
39	9	1960	77.4		Household type foil wrapped loosely around G. E. No. 7031 surface; cleaned with acetone; measured in vacuum (10^{-5} mm Hg).
40	10	1949	403–628	± 5	Hollow sphere; polished to approx. 5 μ then rinsed with distilled water and alcohol, dried in a stream of nitrogen; measured in vacuum; data extracted from smooth curve.
41	11	1962	384 -465	± 3	Hand polished; measured in vacuum (10^{-3} mm Hg).
42	11	1962	373 -441	± 3	Above specimen and conditions except vapor blasted for 20 sec.
43	11	1962	375 -454	± 3	Above specimen and conditions except vapor blasted for 2 min.
44	11	1962	390 -462	± 3	Above specimen and conditions except vapor blasted for 10 min.

DATA TABLE NO. 1 HEMISPHERICAL TOTAL EMITTANCE OF ALUMINUM

[Temperature, T, K; Emittance, ϵ]

T	ϵ
CURVE 1	
573	0.102
673	0.115
773	0.130
773	0.130*
873	0.113
873	0.113*
873	0.113*
CURVE 2	
76	0.018
CURVE 3*	
76	0.018
CURVE 4*	
76	0.021
CURVE 5	
76	0.022
CURVE 6	
76	0.028
CURVE 7*	
76	0.026
CURVE 8*	
76	0.032
CURVE 9	
76	0.035
CURVE 10	
76	0.045
CURVE 11*	
76	0.06
CURVE 12	
76	0.14
CURVE 13*	
76	0.029
CURVE 14*	
76	0.0204
CURVE 15*	
76	0.020
CURVE 16	
76	0.0615
CURVE 17*	
76	0.0615
CURVE 18*	
76	0.0452
CURVE 19*	
76	0.0356
CURVE 20*	
76	0.0327
CURVE 21*	
76	0.0317
CURVE 22*	
76	0.0283
CURVE 23*	
76	0.0217
CURVE 24*	
76	0.0212
CURVE 25*	
76	0.0204
CURVE 26*	
76	0.0186
CURVE 27*	
76	0.0213
CURVE 28*	
76	0.0184
CURVE 29*	
76	0.0186
CURVE 30*	
76	0.0294
CURVE 31	
227	0.06
234	0.06
244	0.06
252	0.06
260	0.06
266	0.06
282	0.06
CURVE 32	
185	0.055
CURVE 33	
182	0.065
CURVE 34*	
181	0.060
CURVE 35	
373	0.0300
CURVE 36*	
373	0.0413
CURVE 37*	
373	0.0480
CURVE 38	
373	0.0443
CURVE 39*	
77.4	0.043
CURVE 40	
403	0.049
423	0.050
473	0.052
523	0.054
573	0.057
628	0.060
CURVE 41	
384	0.16
395	0.17
409	0.17
419	0.17
428	0.16
440	0.16
452	0.15
465	0.14
CURVE 42	
373	0.19
388	0.22
398	0.20
413	0.21
430	0.20
441	0.18
CURVE 43	
375	0.28
385	0.32
403	0.32
419	0.28
436	0.28
454	0.27
CURVE 44	
390	0.25
415	0.27
433	0.26
462	0.26

*Not shown on plot

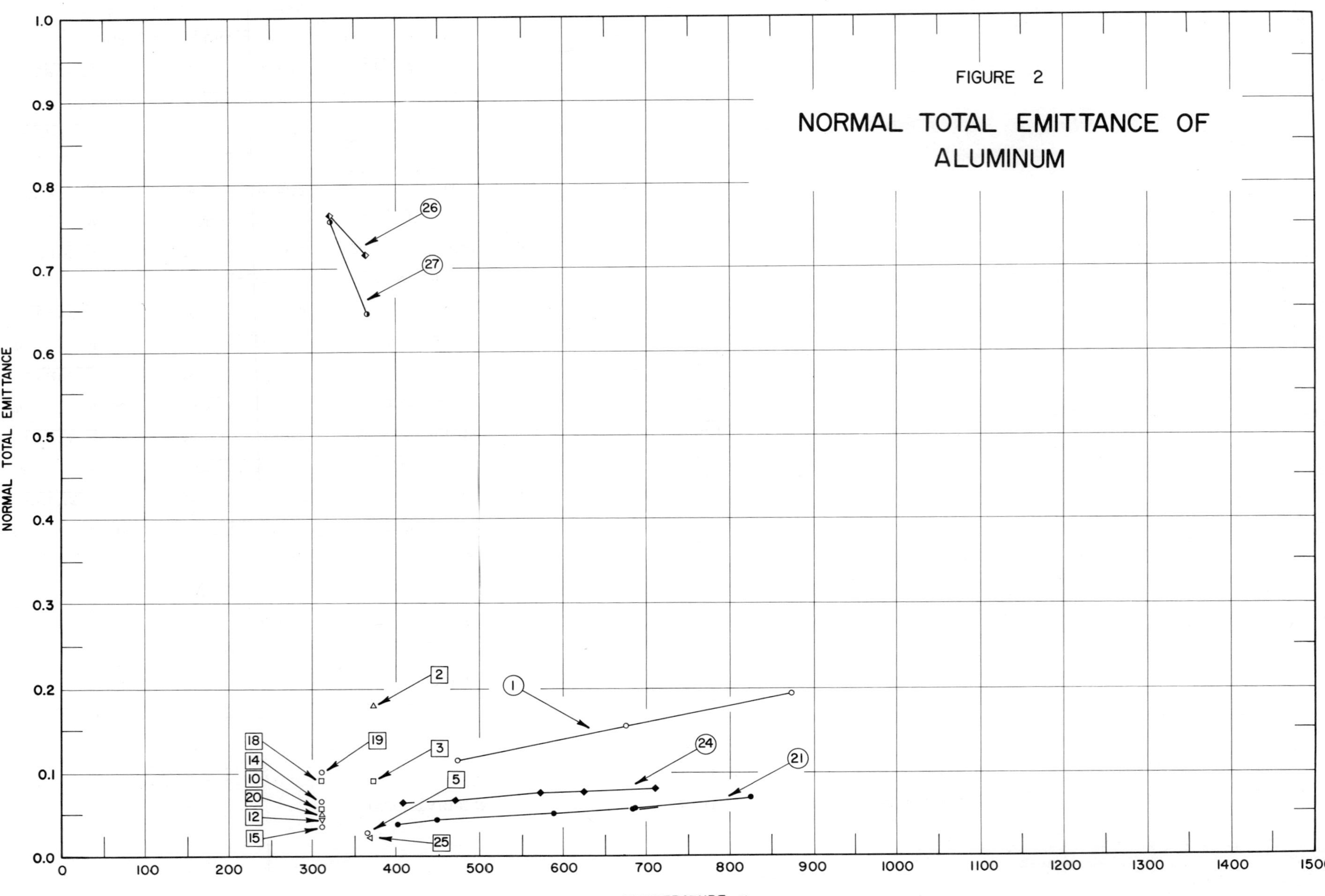
FIGURE 2
NORMAL TOTAL EMITTANCE OF
ALUMINUM
NORMAL TOTAL EMITTANCE
TEMPERATURE, K
0.0
0.1
0.2
0.3
0.4
0.5
0.6
0.7
0.8
0.9
1.0
0
100
200
300
400
500
600
700
800
900
1000
1100
1200
1300
1400
1500
1
2
3
5
10
12
14
15
18
19
20
21
24
25
26
27

SPECIFICATION TABLE NO. 2 NORMAL TOTAL EMITTANCE OF ALUMINUM

Curve No.	Ref. No.	Year	Temperature Range, K	Geometry θ'	Reported Error, %	Composition (weight percent), Specifications and Remarks
1	14	1913	473-873	$\sim 0^\circ$		Cleaned, polished, and oxidized.
2	15	1947	373	$\sim 0^\circ$		Rough polished.
3	15	1947	373	$\sim 0^\circ$		Commercial sheet.
4	15	1947	373	$\sim 0^\circ$		Sheet; polished.
5	16	1937	366	$\sim 0^\circ$	± 1.1	Foil.
6	17	1938	311	$\sim 0^\circ$		0.0020 Si, 0.0001 Fe, 0.0010 Cu, 0.0003 Na,, 0.0003 Ca, 0.0003 Mg; electrolytically brightened; density 2.6978 g cm^{-3} at 298 K; aluminum foil reference (ϵ = 0.05).
7	13	1939	311	$\sim 0^\circ$		Foil; embossed; [Author's designation: Pattern No. 1].
8	13	1939	311	$\sim 0^\circ$		Foil; embossed; [Author's designation: Pattern No. 2].
9	13	1939	311	$\sim 0^\circ$		Foil; embossed; [Author's designation: Pattern No. 3].
10	13	1939	311	$\sim 0^\circ$		Foil; embossed; [Author's designation: Pattern No. 4].
11	13	1939	311	$\sim 0^\circ$		Foil; embossed; [Author's designation: Pattern No. 5].
12	13	1939	311	$\sim 0^\circ$		Foil; bright; roughened with abrasive cloth (No. 120 Aloxite cloth).
13	13	1939	311	$\sim 0^\circ$		Foil; bright; roughened with abrasive cloth (No. 120 Aloxite cloth).
14	13	1939	311	$\sim 0^\circ$		Foil; bright; roughened with abrasive cloth (No. 120 Aloxite cloth).
15	13	1939	311	$\sim 0^\circ$		Foil; bright; etched in hot sodium hydroxide solution for times of 0.5 to 2 min. and dipped in nitric acid.
16	13	1939	311	$\sim 0^\circ$		Foil; 2 years exposure to salt and moisture at the seashore; [Author's designation: Sample No. 1].
17	13	1939	311	$\sim 0^\circ$		Foil; surface film formed by means of exposure to corrosive attack; foil-covered card-board taken from foil-insulated dry-ice cabinet exposed to weather on beach at Coney Island from Sept 1934 to Apr 1935; [Author's designation: Sample No. 2].
18	13	1939	311	$\sim 0^\circ$		Foil; chemically oxidized by treating with hot solution of sodium carbonate and chromate; [Author's designation: Sample No. 3].
19	13	1939	311	$\sim 0^\circ$		Foil; lacquer coated; heated to partially decompose lacquer and color it brown; [Author's designation: Sample No. 4].
20	13	1939	311	$\sim 0^\circ$		Foil; suspended vertically in the laboratory for 3 years and measured with the accumulated dust and fume; [Author's designation: Sample No. 5].
21	217	1930	401-824	$\sim 0^\circ$	2	Pure; highly polished; [Author's designation: Specimen 1].
22	217	1930	414.0-733	$\sim 0^\circ$	2	Pure; highly polished; [Author's designation: Specimen 2].
23	217	1930	425.4-780	$\sim 0^\circ$	2	Pure; rolled; polished; [Author's designation: Specimen 7].
24	217	1930	408.8-711	$\sim 0^\circ$	2	Pure; rolled; as received; [Author's designation: Specimen 13].

SPECIFICATION TABLE NO. 2 (continued)

Curve No.	Ref. No.	Year	Temperature Range, K	Geometry θ'	Reported Error, %	Composition (weight percent), Specifications and Remarks
25	218	1963	370	$\sim 0^\circ$		Aluminum 1100; nominal composition: > 99.00 Al; polished with MgO and water; measured in vacuum (1.6×10^{-4} mm Hg).
26	218	1963	320-363	$\sim 0^\circ$		Paint, white; aluminized with leafing 3.0-mil Al; measured in vacuum (1.6×10^{-4} mm Hg).
27	218	1963	321-367	$\sim 0^\circ$		Paint, white; aluminized with non-leafing 3.0 mil Al; measured in vacuum (1.6×10^{-4} mm Hg).

DATA TABLE NO. 2 NORMAL TOTAL EMITTANCE OF ALUMINUM

[Temperature, T, K; Emittance, ϵ]

T	ϵ
CURVE 1	
473	0.113
673	0.153
873	0.192
CURVE 2	
373	0.18
CURVE 3	
373	0.09
CURVE 4*	
373	0.095
CURVE 5	
366	0.028
CURVE 6*	
311	0.045
CURVE 7*	
311	0.045
CURVE 8*	
311	0.056
CURVE 9*	
311	0.047
CURVE 10	
311	0.056
CURVE 11*	
311	0.061
CURVE 12	
311	0.044
CURVE 13*	
311	0.063
CURVE 14	
311	0.066
CURVE 15	
311	0.035
311	0.040*
311	0.050*
CURVE 16*	
311	0.10
CURVE 17*	
311	0.04
CURVE 18	
311	0.09
CURVE 19	
311	0.10
CURVE 20	
311	0.05
CURVE 21	
401	0.0397
449	0.0444
589	0.0510
685	0.058
824	0.0685
683	0.057
CURVE 22*	
414.0	0.0486
512	0.0549
610	0.0576
695	0.0641
733	0.065
CURVE 23*	
425.4	0.0370
535	0.0419
647	0.0450
704	0.0468
780	0.0497
CURVE 24	
408.8	0.0643
471.5	0.0685
573	0.0750
625	0.0785
711	0.081
CURVE 25	
370	0.022
CURVE 26	
320	0.763
363	0.716
CURVE 27	
321	0.757
367	0.645

*Not shown on plot

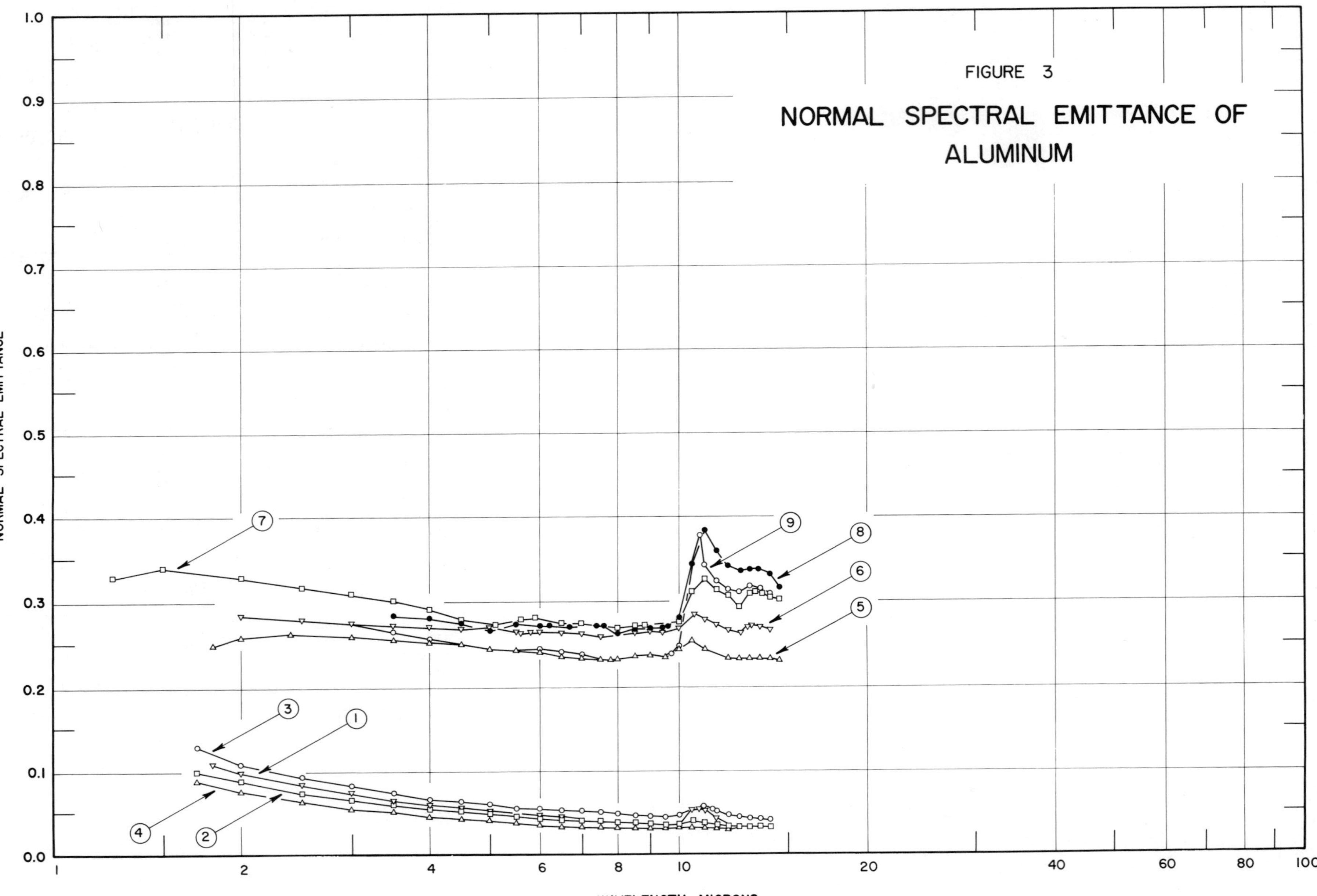
FIGURE 3
NORMAL SPECTRAL EMITTANCE OF
ALUMINUM
NORMAL SPECTRAL EMITTANCE
1.0
0.9
0.8
0.7
0.6
0.5
0.4
0.3
0.2
0.1
0.0
1
2
4
6
8
10
20
40
60
80
100
WAVELENGTH, MICRONS
7
9
8
6
5
3
1
4
2

SPECIFICATION TABLE NO. 3 NORMAL SPECTRAL EMITTANCE OF ALUMINUM

Curve No.	Ref. No.	Year	Temperature K	Wavelength Range, μ	Geometry θ'	Reported Error, %	Composition (weight percent), Specifications and Remarks
1	2	1961	599	1.8-12.5	$\sim 0^{\circ}$	± 20	99.7 Al, 0.11 Fe, 0.11 Si, 0.01 Cu, 0.01 Mg, < 0.01 Mn, Ni and Zn; cylindrical tube; heated at 467 K for 15 hrs; polished with Carnu on Selvyt cloth; surface roughness 3 microinches (center line average); data extracted from smooth curve; error given in the wavelength range 2 to 10μ.
2	2	1961	697	1.7-14.0	$\sim 0^{\circ}$	± 20	Above specimen and conditions except heated at 697 K for 20 hrs before measurement.
3	2	1961	805	1.7-14.0	$\sim 0^{\circ}$	± 20	Above specimen and conditions except heated at 805 K for 15 hrs before measurement.
4	2	1961	599	1.7-12.5	$\sim 0^{\circ}$	± 20	Above specimen and conditions.
5	2	1961	462	1.8-14.5	$\sim 0^{\circ}$	± 10	99.7 Al, 0.11 Fe, 0.11 Si, 0.01 Cu, 0.01 Mg, < 0.01 Mn, Ni and Zn; tube; heated for 25 hrs at 462 K; roughened and knurled with grade 180 silicon carbide paper; surface roughness 115 microinches (center line average); data extracted from a smooth curve; error given over the wavelength range 2 to 10μ.
6	2	1961	599	2.0-14.0	$\sim 0^{\circ}$	± 10	Above specimen and conditions except heated at 598 K for 22 hrs before measurements.
7	2	1961	715	1.25-14.5	$\sim 0^{\circ}$	± 10	Above specimen and conditions except heated at 715 K for 27 hrs before measurement.
8	2	1961	803	3.5-14.5	$\sim 0^{\circ}$	± 10	Above specimen and conditions except heated at 787 K for 17 hrs before measurement.
9	2	1961	461	3.0-14.0	$\sim 0^{\circ}$	± 10	Above specimen and conditions.

DATA TABLE NO. 3 NORMAL SPECTRAL EMITTANCE OF ALUMINUM

[Wavelength, λ, μ; Emittance, ϵ; Temperature, T, K]

CURVE 1 T = 599

λ	ϵ
1.8	0.110
2.0	0.100
2.5	0.086
3.0	0.076
3.5	0.068
4.0	0.062
4.5	0.057
5.0	0.053
5.5	0.050*
6.0	0.048
6.5	0.046
7.0	0.044*
7.5	0.042*
8.0	0.041*
8.5	0.039*
9.0	0.038*
9.5	0.037*
10.0	0.040*
10.5	0.054
10.7	0.056
11.0	0.054
11.5	0.044
12.0	0.039*
12.5	0.036*

CURVE 2 T = 697

λ	ϵ
1.7	0.100
2.0	0.090
2.5	0.075
3.0	0.067
3.5	0.061
4.0	0.057
4.5	0.053
5.0	0.050
5.5	0.048
6.0	0.046
6.5	0.044
7.0	0.043
7.5	0.042
8.0	0.041
8.5	0.040
9.0	0.039

CURVE 2 (cont.)

λ	ϵ
9.5	0.038
10.0	0.039
10.6	0.041
11.0	0.040
11.5	0.038
12.0	0.037
12.5	0.036
13.0	0.035
13.5	0.035
14.0	0.034

CURVE 3 T = 805

λ	ϵ
1.7	0.130
2.0	0.110
2.5	0.094
3.0	0.083
3.5	0.075
4.0	0.068
4.5	0.065
5.0	0.061
5.5	0.058
6.0	0.056
6.5	0.054
7.0	0.052
7.5	0.051
8.0	0.050
8.5	0.049
9.0	0.048
9.5	0.047
10.0	0.049
10.5	0.057*
10.9	0.059
11.3	0.055
11.5	0.051
12.0	0.049
12.5	0.047
13.0	0.045
13.5	0.043
14.0	0.042

CURVE 4 T = 599

λ	ϵ
1.7	0.0900
2.0	0.0780
2.5	0.0650
3.0	0.0568
3.5	0.0515
4.0	0.0470
4.5	0.0446
5.0	0.0420
5.5	0.0400
6.0	0.0386
6.5	0.0372
7.0	0.0365
7.5	0.0360
8.0	0.0355
8.5	0.0348
9.0	0.0347
9.5	0.0346
10.0	0.0345
10.5	0.0344
11.0	0.0343
11.5	0.0342
12.0	0.0341
12.5	0.0341*

CURVE 5 T = 462

λ	ϵ
1.8	0.250
2.0	0.260
2.4	0.264
3.0	0.261*
3.5	0.258
4.0	0.254
4.5	0.251
5.0	0.247
5.5	0.244
6.0	0.241
6.5	0.238
7.0	0.235
7.5	0.233
7.8	0.232
8.0	0.233
8.5	0.238
9.0	0.239

CURVE 5 (cont.)

λ	ϵ
9.5	0.237
10.0	0.246
10.5	0.256
11.0	0.248
12.0	0.234
12.5	0.233
13.0	0.233
13.5	0.233
14.0	0.233
14.5	0.231

CURVE 6 T = 599

λ	ϵ
2.0	0.284
2.5	0.280
3.0	0.277
3.5	0.273
4.0	0.271
4.5	0.270
5.0	0.270
5.5	0.266
5.6	0.265
5.8	0.266
6.0	0.266
6.5	0.265
7.0	0.264
7.5	0.263
8.0	0.262*
8.5	0.266
9.0	0.268
9.4	0.267
10.0	0.270
10.6	0.288
11.0	0.280
11.5	0.273
12.0	0.269
12.6	0.266
12.9	0.271
13.1	0.273
13.5	0.271
14.0	0.269

CURVE 7 T = 715

λ	ϵ
1.25	0.330
1.5	0.340
2.0	0.330
2.5	0.319
3.0	0.310
3.5	0.301
4.0	0.292
4.5	0.280
5.1	0.274
5.6	0.280
5.9	0.281
6.5	0.276
7.0	0.273
7.5	0.271*
8.0	0.270
8.5	0.272
8.8	0.272
9.4	0.272*
10.0	0.279
10.5	0.312
11.0	0.328
11.5	0.314
12.0	0.308
12.5	0.294
13.0	0.310
13.3	0.312
13.6	0.310
14.0	0.304
14.5	0.301

CURVE 8 T = 803

λ	ϵ
3.5	0.284
4.0	0.282
4.5	0.277
5.0	0.268
5.5	0.274
6.0	0.272
6.2	0.272
6.7	0.271
7.4	0.272
7.6	0.272
8.0	0.263

CURVE 8 (cont.)

λ	ϵ
8.5	0.268
9.0	0.270
9.4	0.270
9.6	0.272
10.0	0.282
10.5	0.345
11.0	0.384
11.5	0.360
12.0	0.343
12.6	0.337
13.0	0.339
13.4	0.339
14.0	0.333
14.5	0.328

CURVE 9 T = 465

λ	ϵ
3.0	0.275*
3.5	0.267
4.0	0.259
4.5	0.252*
5.0	0.246*
5.5	0.243*
6.0	0.246
6.5	0.243
7.0	0.239
7.5	0.236*
8.0	0.235*
8.5	0.240*
9.0	0.239*
9.4	0.237*
9.7	0.240
10.0	0.249
10.5	0.345*
10.8	0.377
11.0	0.345
11.5	0.326
12.0	0.315
12.5	0.312
13.0	0.320
13.5	0.314
14.0	0.310

*Not shown on plot

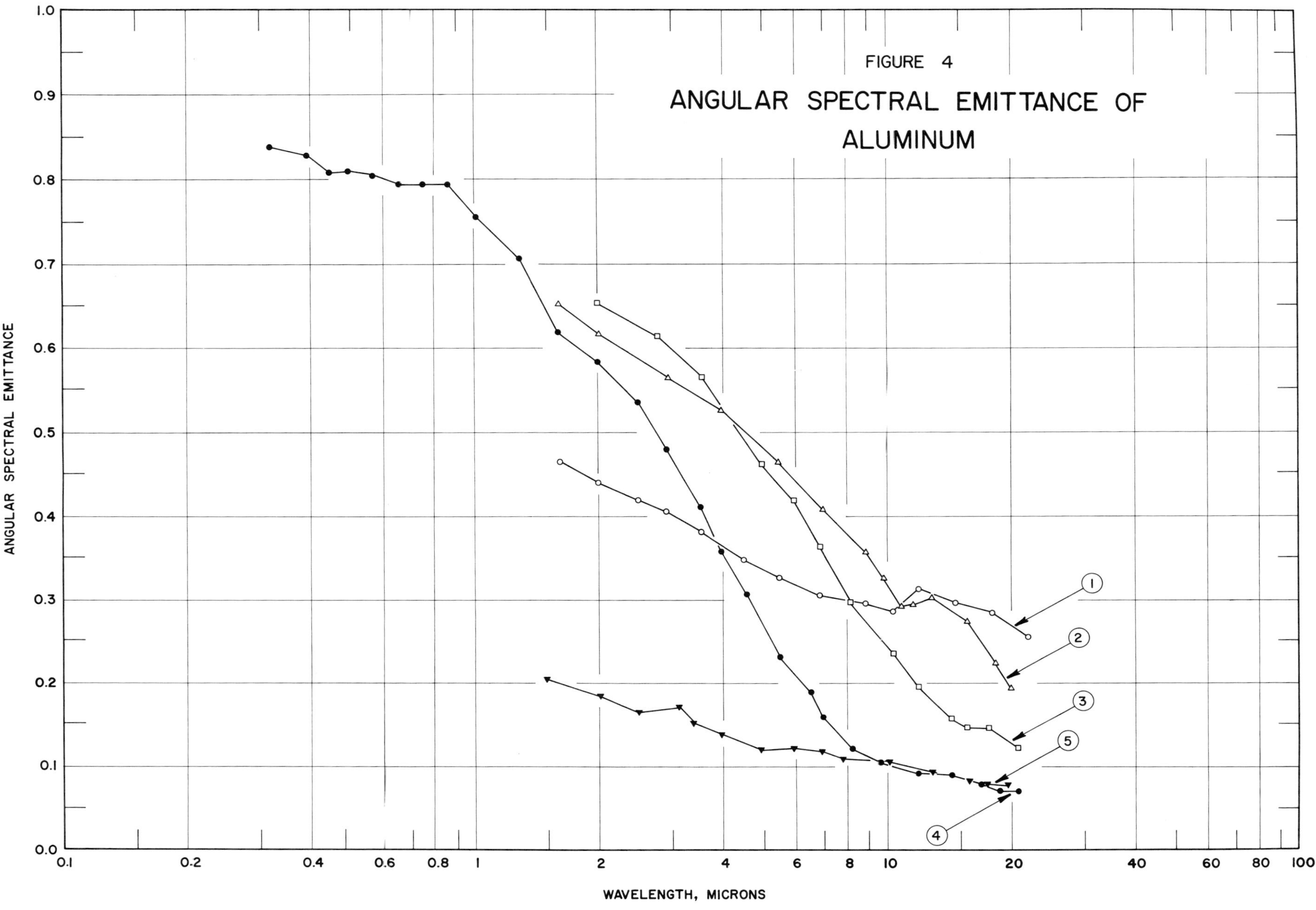

FIGURE 4
ANGULAR SPECTRAL EMITTANCE OF ALUMINUM

SPECIFICATION TABLE NO. 4 ANGULAR SPECTRAL EMITTANCE OF ALUMINUM

Curve No.	Ref. No.	Year	Temperature K	Wavelength Range, μ	Geometry θ'	Reported Error, %	Composition (weight percent), Specifications and Remarks
1	219	1965	306	1.61-22.00	25°		Aluminum 1100; nominal composition: 99.00 min Al; sandblasted with 120 mesh alumina (mesh opening 125μ); authors assumed $\epsilon = \alpha = 1-\rho(25°, 2\pi)$.
2	219	1965	306	1.60-20.00	25°		Aluminum 1100; nominal composition: 99.00 min Al; sandblasted with 320 mesh alumina (mesh opening 46μ); authors assumed $\epsilon = \alpha = 1-\rho(25°, 2\pi)$.
3	219	1965	306	1.99-20.80	25°		Aluminum 1100; nominal composition: 99.00 min Al; sandblasted with 600 mesh alumina (mesh opening 24μ); authors assumed $\epsilon = \alpha = 1-\rho(25°, 2\pi)$.
4	219	1965	306	0.320-20.70	25°		Aluminum 1100; nominal composition: 99.00 min Al; sandblasted with 1000 mesh alumina (mesh opening 14μ); authors assumed $\epsilon = \alpha = 1-\rho(25°, 2\pi)$.
5	219	1965	306	1.51-19.80	25°		Aluminum 1100; nominal composition: 99.00 min Al; sanded with 280 mesh silicon carbide paper; authors assumed $\epsilon = \alpha = 1-\rho(25°, 2\pi)$.

DATA TABLE NO. 4 ANGULAR SPECTRAL EMITTANCE OF ALUMINUM

[Wavelength, λ, μ; Emittance, ∈; Temperature, T, K]

λ	∈
CURVE 1	
T = 306	
1.61	0.463
2.00	0.439
2.49	0.419
2.92	0.405
3.54	0.381
4.53	0.348
5.51	0.327
6.92	0.306
8.93	0.297
10.40	0.286
11.90	0.313
11.90	0.313
14.80	0.298
18.00	0.284
22.00	0.255
CURVE 2	
T = 306	
1.60	0.653
2.00	0.618
2.95	0.564
3.95	0.525
5.46	0.463
7.00	0.409
8.93	0.358
9.82	0.327
10.80	0.293
11.50	0.296
12.80	0.303
15.60	0.274
18.30	0.224
20.00	0.193
CURVE 3	
T = 306	
1.99	0.653
2.79	0.614
3.57	0.564
5.00	0.461
5.98	0.419
6.89	0.364
8.15	0.298
CURVE 3 (cont.)	
10.30	0.236
11.90	0.194
14.30	0.157
15.70	0.147
17.70	0.147
20.80	0.122
CURVE 4	
T = 306	
0.320	0.840
0.393	0.830
0.440	0.809
0.498	0.811
0.570	0.805
0.655	0.796
0.750	0.796
0.865	0.796
1.02	0.757
1.28	0.706
1.60	0.619
2.00	0.582
2.50	0.535
2.94	0.479
3.54	0.411
3.98	0.357
4.55	0.306
5.50	0.231
6.28	0.188
7.00	0.158
8.22	0.121
9.68	0.105
11.80	0.092
14.30	0.089
16.90	0.079
18.70	0.071
20.70	0.071
CURVE 5	
T = 306	
1.51	0.205
2.01	0.185
2.50	0.165
3.14	0.170
CURVE 5 (cont.)	
3.39	0.150
3.99	0.137
4.97	0.120
5.98	0.122
6.97	0.118
7.82	0.109
10.10	0.105
12.90	0.094
15.70	0.084
17.50	0.080
19.80	0.079

SPECIFICATION TABLE NO. 5 ANGULAR INTEGRATED REFLECTANCE OF ALUMINUM

Curve No.	Ref. No.	Year	Temperature K	Angular Range, °	Geometry θ	θ'	ω'	Reported Error, %	Composition (weight percent), Specifications and Remarks
1	221	1964	298	18.4-71.6	θ	θ'			Angle of viewing within 2.5° of angle of specular reflection.
2	220	1965	298	0-80	45°	θ'			Sand blasted; $\phi = 0°$, $\phi' = 0°$; tungsten filament lamp source.
3	220	1965	298	0-20	45°	θ'			Above specimen and conditions except $\phi' = 180°$.
4	220	1965	298	0-80	45°	θ'			Different sample, same as curve 2 specimen and conditions.
5	220	1965	298	0-20	45°	θ'			Above specimen and conditions except $\phi' = 180°$.
6	220	1965	298	0-80	45°	θ'			Curve 4 specimen and conditions except green filter used.
7	220	1965	298	0-20	45°	θ'			Above specimen and conditions except $\phi' = 180°$.
8	220	1965	298	0-80	45°	θ'			Curve 4 specimen and conditions except amber filter used.
9	220	1965	298	0-20	45°	θ'			Above specimen and conditions except $\phi' = 180°$.
10	220	1965	298	0-80	45°	θ'			Curve 4 specimen and conditions except blue filter used.
11	220	1965	298	0-20	45°	θ'			Above specimen and conditions except $\phi' = 180°$.
12	220	1965	298	0-80	65°	θ'			Different sample, same as curve 2 specimen and conditions.
13	220	1965	298	0-40	65°	θ'			Above specimen and conditions except $\phi' = 180°$.

DATA TABLE NO. 5 ANGULAR INTEGRATED REFLECTANCE OF ALUMINUM

[Angle, °; Reflectance, ρ; Temperature, T,K]

θ'	ρ
CURVE 1*	
T = 298	
18.4	0.518
33.1	0.550
45.0	0.585
56.9	0.618
71.6	0.671
CURVE 2*	
T = 298	
0	0.17
5	0.18
10	0.18
15	0.19
20	0.20
25	0.21
30	0.22
35	0.24
40	0.26
45	0.28
50	0.24
55	0.21
60	0.18
65	0.15
70	0.11
75	0.09
80	0.04
CURVE 3*	
T = 298	
0	0.17
5	0.17
10	0.16
15	0.15
20	0.15
CURVE 4*	
T = 298	
0	0.18
10	0.20
20	0.25
30	0.28
40	0.31
CURVE 4*(cont.)	
45	0.31
50	0.30
60	0.23
70	0.16
80	0.13
CURVE 5*	
T = 298	
0	0.18
10	0.14
20	0.12
CURVE 6*	
T = 298	
0	0.018
10	0.020
20	0.025
30	0.029
40	0.033
45	0.033
50	0.030
60	0.025
70	0.017
80	0.011
CURVE 7*	
T = 298	
0	0.018
10	0.015
20	0.011
CURVE 8*	
T = 298	
0	0.018
10	0.020
20	0.026
30	0.029
40	0.031
45	0.031
50	0.030
60	0.024
70	0.017
80	0.011
CURVE 9*	
T = 298	
0	0.018
10	0.017
20	0.012
CURVE 10*	
T = 298	
0	0.007
10	0.008
20	0.010
30	0.011
40	0.012
45	0.012
50	0.012
60	0.010
70	0.007
80	0.004
CURVE 11*	
T = 298	
0	0.007
10	0.006
20	0.005
CURVE 12*	
T = 298	
0	0.08
10	0.09
20	0.10
30	0.13
40	0.16
50	0.20
60	0.21
70	0.18
80	0.12
CURVE 13*	
T = 298	
0	0.08
10	0.07
20	0.06
30	0.05
40	0.04

* Not shown on plot

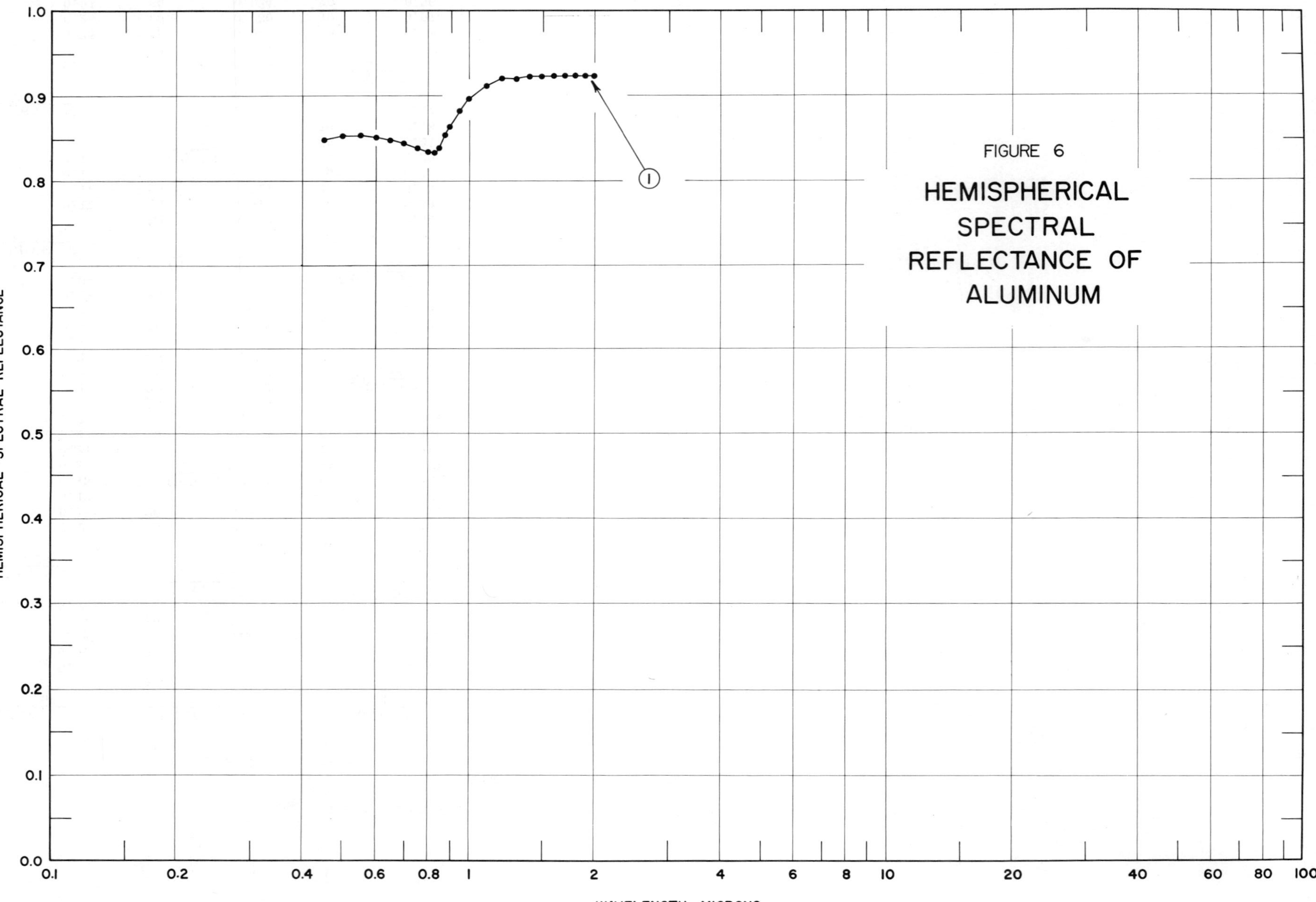

FIGURE 6
HEMISPHERICAL SPECTRAL REFLECTANCE OF ALUMINUM

SPECIFICATION TABLE NO. 6 HEMISPHERICAL SPECTRAL REFLECTANCE OF ALUMINUM

Curve No.	Ref. No.	Year	Temperature K	Wavelength Range, μ	Geometry ω	θ'	ω'	Reported Error, %	Composition (weight percent), Specifications and Remarks
1	282	1966	298	0.45-2.0	2π	$\sim 0°$			Evaporated film on mechanically polished and electropolished stainless steel.

DATA TABLE NO. 6 HEMISPHERICAL SPECTRAL REFELCTANCE OF ALUMINUM

[Wavelength, λ, μ; Reflectance, ρ; Temperature, T, K]

λ	ρ
CURVE 1	
T = 298	
0.45	0.850
0.50	0.853
0.55	0.854
0.60	0.852
0.65	0.848
0.70	0.846
0.75	0.840
0.80	0.836
0.825	0.835
0.850	0.840
0.875	0.855
0.90	0.865
0.95	0.883
1.00	0.897
1.10	0.912
1.20	0.920
1.30	0.920
1.40	0.922
1.50	0.923
1.60	0.923
1.70	0.923
1.80	0.923
1.90	0.923
2.0	0.923

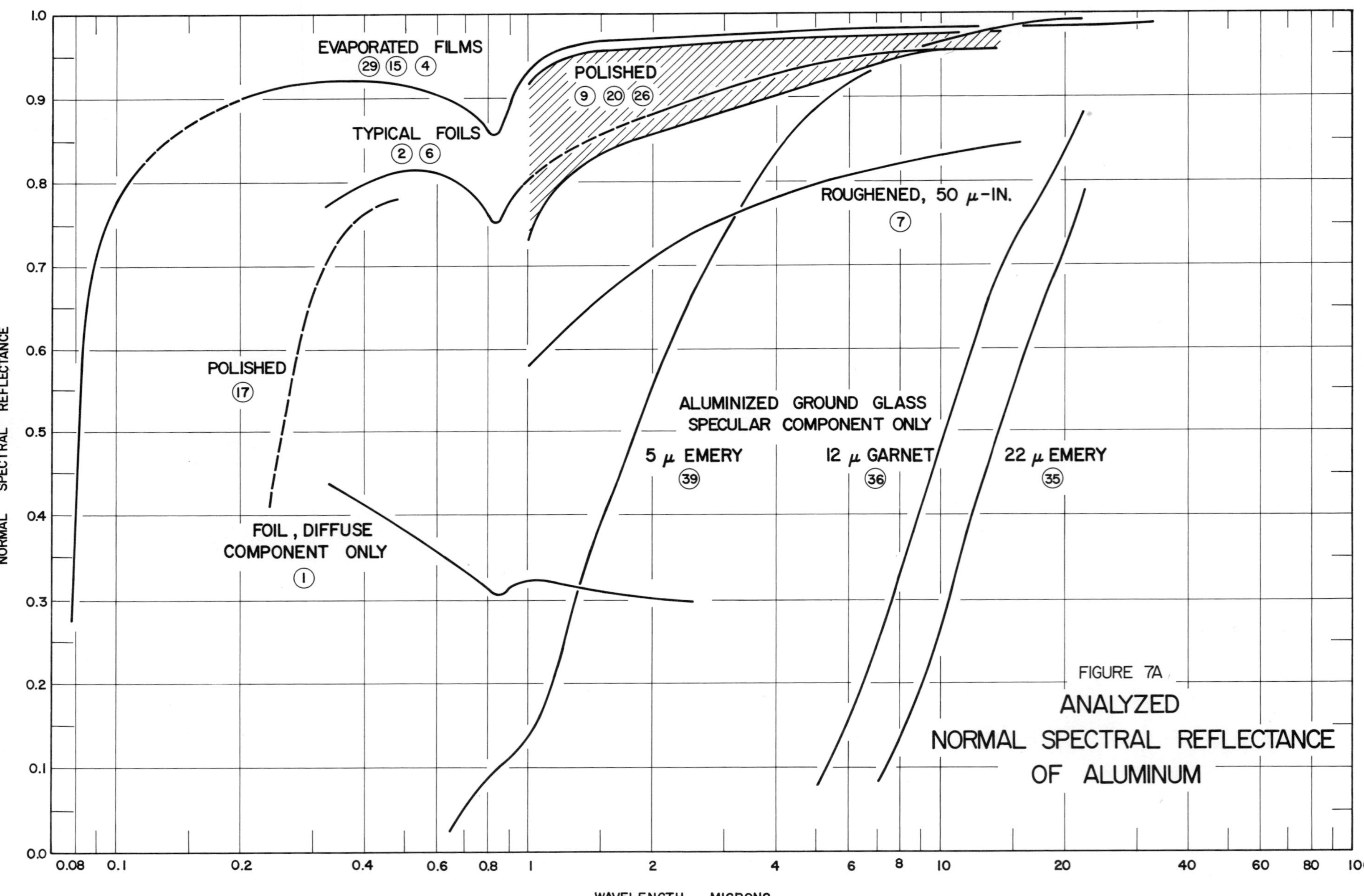

FIGURE 7A
ANALYZED
NORMAL SPECTRAL REFLECTANCE
OF ALUMINUM

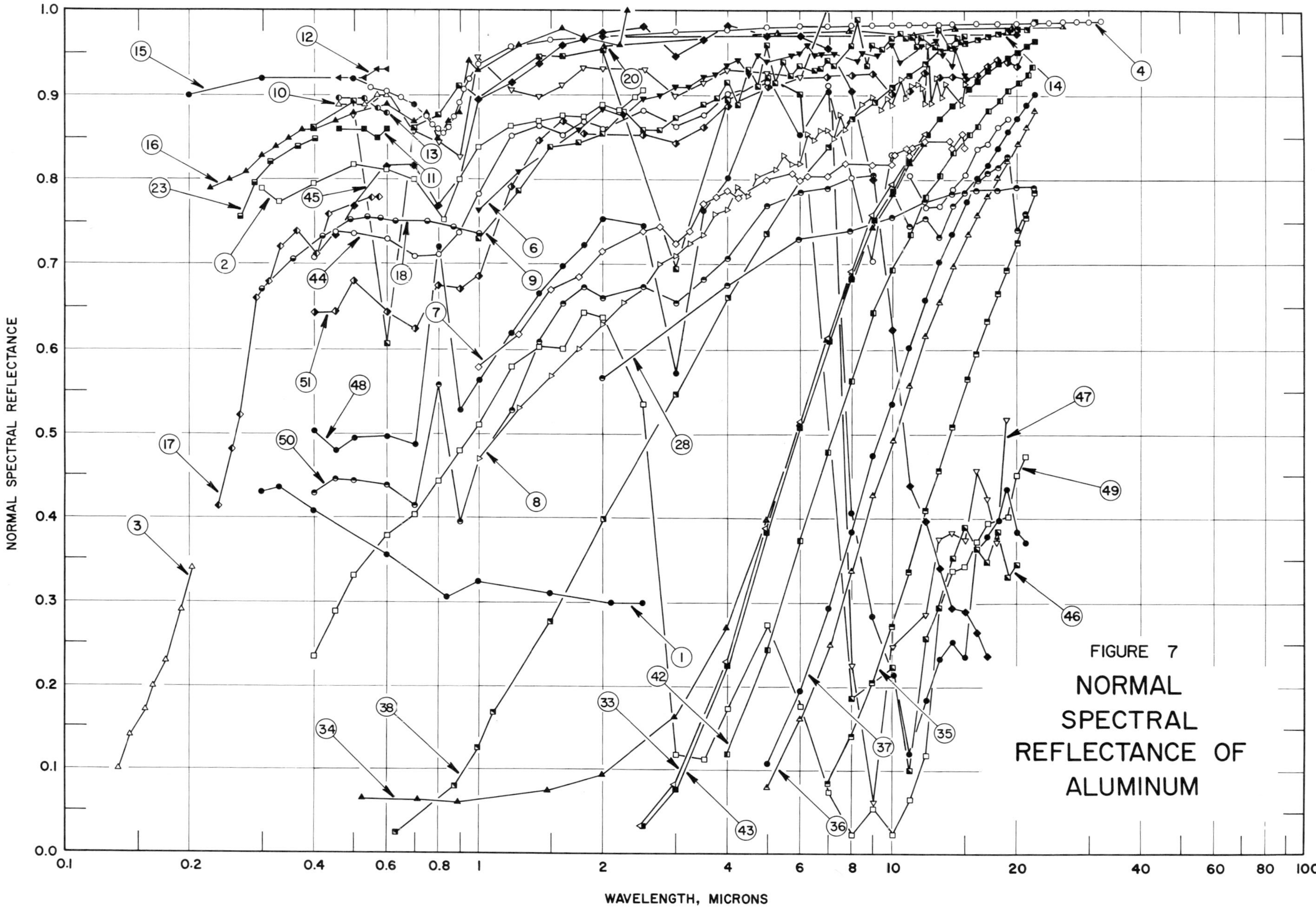

FIGURE 7
NORMAL SPECTRAL REFLECTANCE OF ALUMINUM

SPECIFICATION TABLE NO. 7 NORMAL SPECTRAL REFLECTANCE OF ALUMINUM

Curve No.	Ref. No.	Year	Temperature K	Wavelength Range, μ	Geometry θ	θ' ω'	Reported Error, %	Composition (weight percent), Specifications and Remarks
1	123	1960	298	0.30-2.50	$\sim 0°$	2π		Foil; MgO reference; diffuse reflectance.
2	123	1960	298	0.30-2.50	$\sim 0°$	2π		Foil; cemented on fiberglass laminate; MgO reference.
3	124	1941	298	0.1347-0.2026	$\sim 0°$	$\sim 0°$		An opaque film on glass deposited by the evaporation process; measured in vacuum (0.001 mm Hg).
4	125	1962	298	0.550-32	5°	5°	± 0.1	99.998 pure; Al film (0.065 to 0.11 μ thick), evaporated at 1 x 10^{-5} mm Hg, supersmooth fused quartz optical flats as substrate, no watermarks or other blemishes on the substrate surface, no shadows or streaks in the evaporated Al film; freshly prepared; measured in dry nitrogen.
5	125	1962	298	0.550-32	5°	5°	± 0.1	99.998 pure; Al film (0.065 to 0.11 μ thick), evaporated at 1 x 10^{-5} mm Hg, supersmooth fused quartz optical flats as substrate, no watermarks or other blemishes on the substrate surfaces, no shadows or streaks in the evaporated Al film; aged in air for several weeks; measured in dry nitrogen.
6	126	1953	300	1.00-15.00	5°	2π	± 2.6	Foil (0.001 in. thick); data extracted from smooth curve; converted from R (2π, 5°).
7	126	1953	300	1.00-15.00	5°	2π	± 2.6	Disc (0.032 in. thick); polished, roughened (roughness approximately 50 microinches); data extracted from smooth curve; converted from R(2π, 5°).
8	126	1953	300	1.00-15.00	5°	2π	± 4.3	Disc; commercial finish; data extracted from smooth curve; converted from R (2π, 5°).
9	126	1953	300	1.00-15.00	5°	2π	± 2.7	Disc; polished; data extracted from smooth curve; converted from R(2π, 5°).
10	127	1955	298	0.46-0.60	10°	2π	± 0.5	99 pure; vacuum deposited on glass; measured immediately after removed from vacuum chamber; calculated by authors from $\rho = 1 - \alpha$ using an incandescent tungsten lamp as source.
11	127	1955	298	0.46-0.60	10°	2π	± 0.5	Above specimen and conditions except exposed to the atmosphere for 8 days.
12	127	1955	298	0.46-0.60	10°	2π		99.99 pure; vacuum deposited on glass; measured immediately after removed from vacuum chamber; calculated by authors from $\rho = 1 - \alpha$ using an incandescent tungsten lamp as source.
13	127	1955	298	0.46-0.60	10°	2π		Above specimen and conditions except except exposed to atmosphere for 8 days.
14	128	1962	298	2.00-20.00	$\sim 0°$	2π		Evaporated Al on mylar substrate (0.20 μ thick); illumination solid angle is cone of 0.034 steradians; converted from R (2π, 0).
15	129	1964	298	0.20-0.70	$\sim 0°$	2π		Evaporated aluminum; data extracted from smooth curve.
16	130	1934	298	0.225-2.3	$\sim 0°$	$\sim 0°$		Deposited on a mirror by evaporation.
17	133	1934	298	0.235-0.578	$\sim 0°$	$\sim 0°$	2	Disc; cold worked, annealed, etch tested, polished, stored in a solution of NaOH + NaF, washed and dried.
18	220	1965	298	0.300-1.000	$\sim 0°$	2π		Sand blasted.
19	222	1960	298	0.450-0.600	$\sim 7°$	$\sim 7°$	< 0.16	Measured in air.
20	223	1962	298	2.01-25.96	$\sim 0°$	2π		Polished; converted from R(2π, 0°).

SPECIFICATION TABLE NO. 7 (continued)

Curve No.	Ref. No.	Year	Temperature K	Wavelength Range, μ	Geometry θ	θ'	ω'	Reported Error, %	Composition (weight percent), Specifications and Remarks
21	223	1962	298	1.57-25.94	~0°		2π		Above specimen and conditions except after particle impact.
22	223	1962	77	1.91-26.00	~0°		2π		Above specimen and conditions.
23	224	1931	298	0.2653-0.4038	~5°		2π		Acid-etched.
24	216	1949	~298	1.01-15.00	0°		2π	5	Foil; data extracted from smooth curve.
25	285	1962	298	1.97-13.05	~5°	~5°			No details given.
26	336	1964	298	2.00-23.99	~0°	~0°			Polished.
27	336	1964	298	2.00-23.99	~0°	~0°			Above specimen and conditions except cratered with spherical particles (100 μ dia) of Zircalloy at 1.5 km sec^{-1}; average crater dia 123 μ; average crater depth 289 μ; Knoop hardness 22 (100 g load).
28	336	1964	298	2.00-22.00	~0°	~0°			Different sample, same as above specimen and conditions except cratered with spherical particles (100 μ dia) of tungsten at 7 km sec^{-1}; average crater depth 54 μ; average crater depth 183 μ; Knoop hardness 22 (100 g load).
29	341	1967	298	0.079-0.1175	~0°	~0°			Evaporated film; 99.999 pure; evaporated on microscope slide at 3 x 10^{-8} mm Hg; measured in vacuum (3 x 10^{-8} mm Hg) 4 min after evaporation.
30	341	1967	298	0.079-0.1175	~0°	~0°			Different sample, same as above specimen and conditions except measured 8 min after evaporation.
31	341	1967	298	0.079-0.1175	~0°	~0°			Different sample, same as above specimen and conditions except measured 12 min after evaporation.
32	341	1967	298	0.079-0.1175	~0°	~0°			Different sample, same as above specimen and conditions except measured 16 min after evaporation.
33	344	1963	298	2.47-12.08	~5°	~5°			Aluminized ground glass; aluminum mirror reference; ω' = 0.03 Sr.
34	344	1963	298	0.52-12.07	~5°	~5°			Aluminized ground steel; aluminum mirror reference; ω' = 0.03 Sr.
35	344	1963	298	7.07-22.10	~5°	~5°			Aluminized ground glass; glass ground with M302 grinding powder (Al_2O_3 emery) with average particle size of 22 μ; aluminum mirror reference; ω' = 0.03 Sr.
36	344	1963	298	5.08-22.19	~5°	~5°			Aluminized ground glass; glass ground with W6 grinding powder ($Fe_3Al_2(SiO_4)_3$ garnet) with average particle size of 12 μ; aluminum mirror reference; ω' = 0.03 Sr.
37	344	1963	298	5.07-22.12	~5°	~5°			Aluminized ground glass; glass ground with M 303.5 grinding powder (Al_2O_3 emery) with average particle size of 11 μ; aluminum mirror reference; ω' = 0.03 Sr.
38	344	1963	298	0.63-22.11	~5°	~5°			Aluminized ground glass; glass ground with W10 grinding powder ($Fe_3Al_2(SiO_4)_3$ garnet) with average particle size of 5 μ; aluminum mirror reference; ω' = 0.03 Sr.
39	344	1963	298	0.64-22.14	~5°	~5°			Aluminized ground glass; glass ground with M 305 grinding powder (Al_2O_3 emery) with average particle size of 5 μ; aluminum mirror reference; ω' = 0.03 Sr.
40	344	1963	298	3.93-22.23	~5°	~5°			Aluminized dense flint; flint ground with M303.5 grinding powder (Al_2O_3 emery) with average particle size of 11 μ; aluminum mirror reference; ω' = 0.03 Sr.

SPECIFICATION TABLE NO. 7 (continued)

Curve No.	Ref. No.	Year	Temperature K	Wavelength Range, μ	Geometry θ	θ'	ω'	Reported Error, %	Composition (weight percent), Specifications and Remarks
41	344	1963	298	3.88-22.25	~5°	~5°			Aluminized plate glass; glass ground with M303.5 grinding powder (Al_2O_3 emery) with average particle size of 11 μ; aluminum mirror reference; ω' = 0.03 Sr.
42	344	1963	298	4.06-22.23	~5°	~5°			Aluminized Pyrex; Pyrex ground with M303.5 grinding powder (Al_2O_3 emery) with average particle size of 11 μ; aluminum mirror reference; ω' = 0.03 Sr.
43	344	1963	298	2.49-22.21	~5°	~5°			Aluminized fused quartz; quartz ground with M303.5 grinding powder (Al_2O_3 emery) with average particle size of 11 μ; aluminum mirror reference; ω' = 0.03 Sr.
44	122	1961	298	0.4-20	~0°		2π		1075 Aluminum; mechanically polished, electropolished; measured in vacuum (~10^{-6} Hg); tungsten lamp source (0.4 to 1.0μ), globar source (1.0 to 22μ); converted from $\beta(2\pi, 0°)$; [Authors' designation: Specimen No. 18].
45	122	1961	298	0.45-17	~0°		2π		Different sample, same as curve 44 specimen and conditions except heated in air at 700 K for 30 min; [Authors' designation: Specimen No. 56].
46	122	1961	298	0.4-20	~0°		2π		Different sample, same as curve 44 specimen and conditions; [Authors' designation: Specimen No. 71].
47	122	1961	422	0.4-19	~0°		2π		Above specimen and conditions.
48	122	1961	298	0.4-21	~0°		2π		Different sample, same as above specimen and conditions except mechanically polished only; [Authors' designation: Specimen No. 72].
49	122	1961	298	0.4-21	~0°		2π		Different sample, same as curve 44 specimen and conditions except mill finish; electro-polished; [Authors' designation: Specimen No. 500].
50	122	1961	298	0.4-21	~0°		2π		Different sample, same as curve 44 specimen and conditions except mechanically polished only; [Authors' designation: Specimen No. 79]
51	122	1961	298	0.4-21	~0°		2π		Different sample, same as curve 44 specimen and conditions except mill finish; electro-polished; [Authors' designation: Specimen No. 100].
52	362	1965	298	1.5-7.0	7°		2π		Opaque film (99.999 pure) vacuum evaporated on glass.
53	362	1965	298	1.5-7.0	7°		2π		Above specimen and conditions.
54	362	1965	298	1.5-7.0	7°		2π		Above specimen and conditions.
55	362	1965	298	1.5-7.0	7°		2π		Above specimen and conditions.
56	362	1965	298	1.5-7.0	7°		2π		Above specimen and conditions.
57	362	1965	298	1.5-7.0	7°		2π		Above specimen and conditions.
58	121	1962	298	0.295-2.291	~0°	~0°			Plate; sandblasted; data extracted from smooth curve.
59	359	1965	298	3.0-30.1	~5°	~5°			>99.999 Al; opaque ultrahigh vacuum film; vacuum evaporated (10^{-9} mm Hg) on super-smooth fused quartz optical flats; measured in nitrogen.

DATA TABLE NO. 7 NORMAL SPECTRAL REFLECTANCE OF ALUMINUM

[Wavelength λ, μ; Reflectance ρ; Temperature T, K]

λ	ρ
CURVE 1 T = 298	
0.30	0.430
0.33	0.435
0.40	0.408
0.60	0.355
0.84	0.305
1.00	0.325
1.50	0.311
2.10	0.300
2.50	0.300
CURVE 2 T = 298	
0.30	0.790
0.33	0.775
0.40	0.796
0.50	0.819
0.60	0.812
0.70	0.800
0.825	0.752
0.90	0.800
1.00	0.840
1.20	0.865
1.40	0.870
1.60	0.877
1.80	0.877
2.00	0.890
2.20	0.884
2.40	0.899
2.50	0.905
CURVE 3 T = 298	
0.1347	0.10
0.1438	0.14
0.1570	0.17
0.1640	0.20
0.1857	0.23
0.1901	0.29
0.2026	0.34
CURVE 4 T = 298	
0.550	0.9094
0.600	0.9048
0.650	0.8989
0.700	0.8900
0.750	0.8761
0.775	0.8678
0.800	0.8604
0.825	0.8569
0.850	0.8622
0.875	0.8759
0.900	0.8920
0.925	0.9072
0.950	0.9192
1.000	0.9360
1.200	0.9596
1.500	0.9676
2.000	0.9718
3.000	0.9765
4.000	0.9795
5	0.9812
6	0.9823
7	0.9831
8	0.9837
9	0.9841
10	0.9845
11	0.9849
12	0.9854
13	0.9857
14	0.9861
16	0.9868
18	0.9873
20	0.9878
22	0.9883
24	0.9887
26	0.9890
28	0.9893
30	0.9896
32	0.9898
CURVE 5* T = 298	
0.550	0.9049
0.600	0.9021
CURVE 5 (cont.)* T = 298	
0.650	0.8976
0.700	0.8886
0.750	0.8761
0.775	0.8678
0.800	0.8596
0.825	0.8556
0.850	0.8596
0.875	0.8730
0.900	0.8894
0.925	0.9030
0.950	0.9154
1.000	0.9224
1.200	0.9585
1.500	0.9658
2.000	0.9699
3.000	0.9736
4.000	0.9758
5	0.9772
6	0.9784
7	0.9794
8	0.9801
9	0.9807
10	0.9812
11	0.9816
12	0.9821
13	0.9826
14	0.9830
16	0.9838
18	0.9845
20	0.9852
22	0.9856
24	0.9861
26	0.9864
28	0.9867
30	0.9870
32	0.9872
CURVE 6 T = 300	
1.00	0.765
1.25	0.810
1.50	0.840
1.75	0.860
CURVE 6 (cont.) T = 300	
2.00	0.885
2.25	0.878
2.50	0.895
2.75	0.900
3.00	0.910
3.25	0.910
3.50	0.923
3.75	0.925
4.00	0.935
4.25	0.941
4.50	0.925
4.75	0.948
5.00	0.940
5.25	0.939
5.50	0.940
5.75	0.948
6.00	0.950
6.25	0.960
6.50	0.948
6.75	0.950
7.25	0.950
7.50	0.945
7.75	0.949
8.00	0.945
8.25	0.942
8.50	0.951
9.25	0.949
10.00	0.960
10.50	0.940
10.75	0.955
11.00	0.949
11.25	0.960
11.50	0.951
11.75	0.960
12.50	0.960
12.75	0.965
13.00	0.960
13.25	0.950
13.75	0.950
14.00	0.935
14.25	0.955
14.75	0.936
15.00	0.925
CURVE 7 T = 300	
1.00	0.579
1.25	0.618
1.50	0.670
1.75	0.685
2.00	0.715
2.50	0.740
2.75	0.745
3.00	0.725
3.25	0.740
3.50	0.772
3.75	0.780
4.00	0.788
4.25	0.780
4.50	0.793
4.75	0.800
5.00	0.800
5.25	0.802
5.50	0.805
5.75	0.809
6.00	0.800
6.25	0.805
7.00	0.805
7.75	0.821
8.00	0.819
8.25	0.821
8.50	0.820
8.75	0.819
9.00	0.819
9.25	0.821
9.50	0.818
9.75	0.819
10.00	0.820
10.25	0.831
10.50	0.831
10.75	0.839
11.00	0.831
11.25	0.841
11.50	0.841
11.75	0.840
12.00	0.848
12.25	0.845
12.50	0.849
12.75	0.845
13.25	0.846
13.50	0.842
CURVE 7 (cont.) T = 300	
13.75	0.848
14.00	0.840
14.25	0.841
14.50	0.843
14.75	0.855
15.00	0.840
CURVE 8 T = 300	
1.00	0.470
1.25	0.530
1.50	0.569
1.75	0.601
2.00	0.630
2.25	0.655
2.50	0.670
2.75	0.700
3.00	0.710
3.25	0.725
3.50	0.735
3.75	0.760
4.00	0.765
4.25	0.792
4.50	0.782
4.75	0.805
5.00	0.810
5.25	0.813
5.50	0.830
5.75	0.820
6.00	0.820
6.25	0.855
6.50	0.849
6.75	0.860
7.00	0.860
7.25	0.850
7.50	0.879
7.75	0.862
8.00	0.870
8.25	0.889
8.50	0.891
8.75	0.888
9.00	0.899
9.50	0.885
9.75	0.898
CURVE 8 (cont.) T = 300	
10.00	0.890
10.25	0.898
10.50	0.920
10.75	0.898
11.00	0.900
11.25	0.912
11.50	0.913
11.75	0.910
12.00	0.892
12.25	0.960
12.50	0.892
12.75	0.922
13.00	0.915
13.25	0.899
13.50	0.915
13.75	0.919
14.00	0.915
14.25	0.900
14.50	0.895
14.75	0.891
15.00	0.915
CURVE 9 T = 300	
1.00	0.731
1.25	0.788
1.50	0.840
1.75	0.845
2.00	0.855
2.25	0.881
2.50	0.860
2.75	0.860
3.00	0.875
3.25	0.879
3.50	0.891
3.75	0.898
4.00	0.915
4.25	0.890
4.50	0.931
4.75	0.910
5.00	0.960
5.25	0.915
5.50	0.940
5.75	0.925

*Not shown on plot

DATA TABLE NO. 7 (continued)

λ	ρ
CURVE 9 (cont.) T = 300	
6.00	0.936
6.50	0.930
6.75	0.932
7.00	0.942
7.25	0.935
8.00	0.960
8.25	0.990
8.75	0.935
9.00	0.960
9.25	0.960
9.50	0.955
9.75	0.969
10.00	0.969
10.50	0.975
10.75	0.970
11.00	0.972
11.25	0.972
11.50	0.959
11.75	0.972
12.00	0.960
12.25	0.960
12.50	0.963
12.75	0.959
13.00	0.980
13.25	0.969
13.50	0.964
13.75	0.960
14.00	0.965
14.25	0.951
14.50	0.969
14.75	0.965
15.00	0.972
CURVE 10 T = 298	
0.46	0.89
0.53	0.89
0.57	0.90
0.60	0.90
CURVE 11 T = 298	
0.46	0.86
0.53	0.86
0.57	0.85
0.60	0.86
CURVE 12 T = 298	
0.46	0.92
0.53	0.92
0.57	0.93
0.60	0.93
CURVE 13 T = 298	
0.46	0.895
0.53	0.895
0.57	0.885
0.60	0.880
CURVE 14 T = 298	
2.00	0.970
10.00	0.970*
20.00	0.975
CURVE 15 T = 298	
0.20	0.90
0.30	0.92
0.50	0.92
0.70	0.89
CURVE 16 T = 298	
0.225	0.79
0.250	0.80
0.275	0.81
0.300	0.83
0.325	0.84
0.350	0.85
0.375	0.86
0.400	0.86
0.450	0.87
0.500	0.88
0.600	0.89
0.70	0.87
0.75	0.88
0.8	0.85
0.85	0.87
0.9	0.88
0.95	0.94
CURVE 16 (cont.) T = 298	
1.00	0.93
1.25	0.96
1.6	0.98
1.8	0.97
2.2	0.96
2.3	1.00
CURVE 17 T = 298	
0.235	0.413
0.254	0.482
0.265	0.523
0.293	0.662
0.312	0.679
0.334	0.722
0.366	0.740
0.406	0.712
0.435	0.759
0.546	0.780
0.578	0.781
CURVE 18 T = 298	
0.300	0.671
0.357	0.706
0.419	0.733
0.488	0.753
0.538	0.756
0.579	0.754
0.627	0.751
0.752	0.751
0.871	0.747
1.000	0.738
CURVE 19* T = 298	
0.450	0.9083
0.500	0.9082
0.550	0.9064
0.600	0.9026
CURVE 20 T = 298	
2.01	0.960
5.27	0.973
7.94	0.979
11.36	0.981
14.07	0.982
19.27	0.982
25.96	0.984
CURVE 21* T = 298	
1.57	0.890
4.34	0.929
7.17	0.946
8.53	0.946
11.73	0.953
14.46	0.954
17.38	0.961
19.33	0.961
22.54	0.957
25.94	0.952
CURVE 22* T = 77	
1.91	0.908
3.95	0.939
5.94	0.937
7.90	0.945
9.94	0.955
11.97	0.952
13.88	0.953
15.89	0.959
17.92	0.962
19.94	0.960
21.86	0.957
23.93	0.949
26.00	0.949
CURVE 23 T = 298	
0.2653	0.756
0.2890	0.796
0.3133	0.822
0.3664	0.839
0.4038	0.849
CURVE 24* T = ~298	
1.01	0.929
1.26	0.928
1.50	0.944
1.79	0.949
1.98	0.955
2.25	0.957
2.50	0.959
2.75	0.955
2.98	0.956
3.27	0.964
3.50	0.955
3.75	0.963
3.96	0.974
4.24	0.940
4.48	0.968
4.76	0.969
4.98	0.969
5.24	0.972
5.50	0.969
5.73	0.965
5.98	0.977
6.26	0.975
6.49	0.975
6.74	0.971
6.98	0.969
7.26	0.969
7.51	0.970
7.76	0.976
7.98	0.974
8.26	0.974
8.53	0.978
8.78	0.968
9.00	0.967
9.29	0.972
9.55	0.971
9.79	0.974
10.01	0.973
10.27	0.973
10.53	0.973
10.77	0.973
11.04	0.973
11.30	0.974
11.52	0.976
11.78	0.976
12.02	0.976
12.27	0.977
12.51	0.974
12.79	0.976
CURVE 24 (cont.)* T = ~298	
13.01	0.976
13.28	0.976
13.50	0.977
13.79	0.976
14.01	0.976
14.28	0.977
14.53	0.973
14.75	0.966
15.00	0.971
CURVE 25* T = 298	
1.97	0.9670
2.97	0.9712
3.95	0.9747
4.98	0.9765
6.00	0.9774
7.03	0.9778
8.03	0.9783
9.00	0.9790
10.00	0.9791
11.09	0.9796
12.04	0.9798
13.05	0.9805
CURVE 26* T = 298	
2.00	0.964
4.00	0.970
5.99	0.973
8.00	0.977
10.00	0.980
12.00	0.981
14.00	0.980
16.00	0.980
18.00	0.980
20.00	0.979
22.00	0.979
23.99	0.978
CURVE 27* T = 298	
2.00	0.897
4.00	0.918
5.99	0.933
8.00	0.945
10.00	0.954
12.00	0.952
14.00	0.954
16.00	0.958
18.00	0.958
20.00	0.955
22.00	0.946
23.99	0.946
CURVE 28 T = 298	
2.00	0.566
4.00	0.676
5.99	0.731
8.00	0.741
10.00	0.757
12.00	0.777
14.00	0.786
16.00	0.789
18.00	0.790
20.00	0.792
22.00	0.792
CURVE 29* T = 298	
0.0790	0.325
0.0833	0.624
0.0920	0.767
0.1175	0.815
CURVE 30* T = 298	
0.0790	0.308
0.0833	0.588
0.0920	0.728
0.1175	0.811

* Not shown on plot

DATA TABLE NO. 7 (continued)

CURVE 31* T = 298

λ	ρ
0.0790	0.292
0.0833	0.559
0.0920	0.689
0.1175	0.811

CURVE 32* T = 298

λ	ρ
0.0790	0.276
0.0833	0.528
0.0920	0.653
0.1175	0.811

CURVE 33 T = 298

λ	ρ
2.47	0.033
2.99	0.082
3.97	0.230
4.98	0.388
6.00	0.515
7.00	0.615
8.02	0.692
9.00	0.760
10.04	0.797
11.04	0.831*
12.08	0.857

CURVE 34 T = 298

λ	ρ
0.52	0.065
0.71	0.063
0.89	0.060
1.47	0.075
2.00	0.093
2.99	0.163
3.99	0.270
4.99	0.398
6.00	0.513
7.00	0.612
8.01	0.688
9.02	0.746
10.03	0.784
11.06	0.822
12.07	0.849

CURVE 35 T = 298

λ	ρ
7.07	0.083
8.04	0.140
9.06	0.206
10.04	0.272
11.04	0.337
12.01	0.409
13.00	0.456
14.06	0.509
15.25	0.565
16.07	0.597
17.07	0.634
18.05	0.668
19.07	0.694
20.12	0.728
21.12	0.755
22.10	0.788

CURVE 36 T = 298

λ	ρ
5.08	0.079
6.07	0.161
7.08	0.250
8.06	0.339
9.03	0.428
10.08	0.492
11.04	0.559
12.08	0.617
13.06	0.656
14.10	0.699
15.34	0.738
16.12	0.759
17.03	0.782
18.04	0.802
19.07	0.823
20.13	0.846
21.16	0.864
22.19	0.883

CURVE 37 T = 298

λ	ρ
5.07	0.108
6.02	0.197
7.05	0.294
8.05	0.384

CURVE 37 (cont.)

λ	ρ
9.06	0.474
10.03	0.537
11.07	0.602
12.08	0.660
13.08	0.703
14.07	0.738
15.26	0.777
16.12	0.803
17.10	0.820
18.10	0.840
19.10	0.859
20.13	0.877
21.14	0.891
22.12	0.902

CURVE 38 T = 298

λ	ρ
0.63	0.025
0.87	0.079
0.99	0.126
1.08	0.168
1.49	0.277
2.09	0.398
3.06	0.548
4.06	0.661
5.07	0.738
6.07	0.801*
7.04	0.841
8.06	0.873
9.10	0.896
10.01	0.911
11.06	0.926
12.06	0.939
13.12	0.953
14.12	0.958
15.08	0.963
16.10	0.969
17.08	0.972
18.09	0.974
19.13	0.978
20.18	0.981
21.12	0.981
22.11	0.989

CURVE 39* T = 298

λ	ρ
0.64	0.027
0.77	0.077
0.97	0.126
1.07	0.172
1.56	0.412
2.08	0.575
3.11	0.752
4.08	0.845
5.07	0.892
6.05	0.919
7.10	0.942
8.10	0.956
9.09	0.965
10.09	0.970
11.08	0.974
12.09	0.980
13.11	0.983
14.12	0.986
15.08	0.989
16.15	0.993
17.13	0.993
18.13	0.993
19.11	0.994
20.16	0.994
21.15	0.994
22.14	0.998

CURVE 40* T = 298

λ	ρ
3.93	0.005
5.05	0.095
6.03	0.190
7.01	0.284
8.02	0.373
9.02	0.452
10.03	0.524
11.02	0.588
12.09	0.649
13.10	0.684
14.10	0.722
15.28	0.762
16.15	0.783
17.11	0.806
18.17	0.829
19.20	0.845

CURVE 40 (cont.)*

λ	ρ
20.23	0.862
21.23	0.876
22.23	0.892

CURVE 41* T = 298

λ	ρ
3.88	0.014
5.03	0.107
6.01	0.197
7.01	0.292
8.03	0.380
9.03	0.467
10.03	0.531
11.07	0.597
12.12	0.655
13.07	0.694
14.12	0.733
15.30	0.769
16.13	0.795
17.12	0.814
18.16	0.832
19.17	0.853
20.20	0.872
21.25	0.886
22.25	0.898

CURVE 42 T = 298

λ	ρ
4.06	0.119
5.03	0.243
6.05	0.372
7.05	0.479
8.02	0.564
9.07	0.646
10.05	0.694
11.10	0.738
12.09	0.781
13.08	0.807
14.14	0.833
15.32	0.856
16.13	0.867
17.17	0.883
18.24	0.897
19.16	0.908
20.22	0.918
21.22	0.927
22.23	0.939

CURVE 43 T = 298

λ	ρ
2.49	0.033
3.09	0.077
4.07	0.225
5.06	0.382
6.08	0.509
7.08	0.611
8.07	0.683
9.13	0.753
10.11	0.789
11.09	0.824
12.10	0.855
13.07	0.874
14.10	0.890
15.35	0.910
16.16	0.918
17.13	0.930
18.18	0.939
19.17	0.946
20.25	0.952
21.23	0.960
22.21	0.967

CURVE 44 T = 298

λ	ρ
0.4	0.708
0.45	0.740
0.5	0.737
0.6	0.731
0.7	0.710
0.8	0.711
0.9	0.740
1.0	0.783
1.2	0.852
1.4	0.865
1.6	0.853
1.8	0.870
2.0	0.861
2.5	0.883
3.0	0.865
3.5	0.879
4	0.902
5	0.917
6	0.982*
7	0.912

CURVE 44 (cont.)

λ	ρ
8	0.803
9	0.703
10	0.831
11	0.807
12	0.769
13	0.770
14	0.790
15	0.808
16	0.839
17	0.844
18	0.863
19	0.874
20	0.876*

CURVE 45 T = 298

λ	ρ
0.45	0.736
0.5	0.770
0.6	0.817
0.7	0.819
0.8	0.770
0.9	0.802*
1.0	0.896
1.2	0.916
1.4	0.938
1.6	0.959
1.8	0.968
2.0	0.974
2.5	0.981
3.0	0.947
3.5	0.966
4	0.981
5	0.971
6	0.971
7	0.956
8	0.907
9	0.801
10	0.625
11	0.439
12	0.398
13	0.342
14	0.293
15	0.290
16	0.266
17	0.238

* Not shown on plot

DATA TABLE NO. 7 (continued)

λ	ρ
CURVE 46	
T = 298	
0.4	0.862
0.5	0.896
0.6	0.608
0.7	0.862
0.8	0.878
0.9	0.911
1.0	0.898*
1.4	0.944
1.6	0.947
1.8	0.969*
2.0	0.954
3.0	0.696
4	0.897
5	0.921
6	0.902
8	0.187
10	0.224
11	0.100
12	0.258
13	0.294
14	0.352
15	0.390
16	0.363
17	0.349
18	0.383
19	0.431
20	0.345
CURVE 47	
T = 422	
0.4	0.861*
0.5	0.879
0.6	0.903
0.7	0.859*
0.8	0.845*
0.9	0.828
1.0	0.945
1.2	0.907
1.4	0.899
1.6	0.911
1.8	0.932
2.0	0.931
2.5	0.932
3.0	0.900

λ	ρ
CURVE 47 (cont.)	
3.5	0.916
4	0.930
5	0.926
6	0.921
7	1.011*
8	0.227
9	0.062
10	0.249
11	0.108
12	0.287
13	0.375
14	0.382
15	0.375
16	0.458
17	0.422
18	0.372
19	0.523
CURVE 48	
T = 298	
0.4	0.503
0.45	0.480
0.5	0.497
0.6	0.499
0.7	0.488
0.8	0.721
0.9	0.529
1.0	0.564
1.2	0.620
1.4	0.667
1.6	0.699
1.8	0.723
2.0	0.754
2.5	0.746
3.0	0.572
3.5	0.766
4	0.803
5	0.920*
6	0.855
7	0.904
8	0.408
9	0.283
10	0.217
11	0.121
12	0.186

λ	ρ
CURVE 48 (cont.)	
13	0.232
14	0.304
15	0.336
16	0.363*
17	0.379
18	0.398
19	0.435
20	0.384
21	0.371
CURVE 49	
T = 298	
0.4	0.236
0.45	0.289
0.5	0.332
0.6	0.379
0.7	0.404
0.8	0.443
0.9	0.480
1.0	0.511
1.2	0.580
1.4	0.603
1.6	0.601
1.8	0.644
2.0	0.639
2.5	0.535
3.0	0.119
3.5	0.113
4	0.174
5	0.274
6	0.177
7	0.074
8	0.023
9	0.056
10	0.026
11	0.067
12	0.219
13	0.295*
14	0.339
15	0.342
16	0.372
17	0.394
18	0.398*
19	0.402
20	0.451
21	0.475

λ	ρ
CURVE 50	
T = 298	
0.4	0.429
0.45	0.446
0.5	0.444
0.6	0.439
0.7	0.415
0.8	0.558
0.9	0.396
1.0	0.470*
1.2	0.528
1.4	0.610
1.6	0.654
1.8	0.674
2.0	0.661
2.5	0.673
3.0	0.655
3.5	0.682
4	0.708
5	0.770
6	0.788
7	0.791
8	0.805*
9	0.807
10	0.788*
11	0.746
12	0.756
13	0.733
14	0.770
15	0.787
16	0.801
17	0.810
18	0.817
19	0.829
20	0.741
21	0.760
CURVE 51	
T = 298	
0.4	0.644
0.45	0.645
0.5	0.681
0.6	0.644
0.7	0.627
0.8	0.676
0.9	0.672

λ	ρ
CURVE 51 (cont.)	
1.0	0.687
1.2	0.793
1.4	0.848
1.6	0.870
1.8	0.856
2.0	0.854*
2.5	0.856
3.0	0.843
3.5	0.863
4	0.890
5	0.911
6	0.923*
7	0.922
8	0.925
9	0.926
10	0.903
11	0.907
12	0.929
13	0.922
14	0.923
15	0.921
16	0.924
17	0.935
18	0.943
19	0.941
20	0.935
21	0.960*
CURVE 52*	
T = 298	
1.5	0.961
2.0	0.975
2.5	0.977
3.5	0.985
4.5	0.985
5.5	0.985
6.5	0.985
7.0	0.980
CURVE 53*	
T = 298	
1.5	0.961
2.0	0.975
2.5	0.976

λ	ρ
CURVE 53 (cont.)*	
3.5	0.983
4.5	0.985
5.5	0.986
6.5	0.985
7.0	0.988
CURVE 54*	
T = 298	
1.5	0.962
2.0	0.972
2.5	0.974
3.5	0.984
4.5	0.982
5.5	0.983
6.5	0.983
7.0	0.984
CURVE 55*	
T = 298	
1.5	0.960
2.0	0.975
2.5	0.976
3.5	0.981
4.5	0.984
5.5	0.985
6.5	0.985
7.0	0.985
CURVE 56*	
T = 298	
1.5	0.962
2.0	0.976
2.5	0.975
3.5	0.982
4.5	0.984
5.5	0.986
6.5	0.986
7.0	0.986
CURVE 57*	
T = 298	
1.5	0.959
2.0	0.972

λ	ρ
CURVE 57 (cont.)*	
2.5	0.976
3.5	0.982
4.5	0.984
5.5	0.986
6.5	0.987
7.0	0.987
CURVE 58*	
T = 298	
0.295	0.337
0.368	0.358
0.449	0.393
0.582	0.411
0.711	0.393
0.796	0.383
0.908	0.449
0.997	0.482
1.095	0.500
1.265	0.529
1.393	0.546
1.672	0.572
1.787	0.577
1.865	0.601
2.104	0.608
2.196	0.617
2.291	0.633
CURVE 59*	
T = 298	
3.0	0.9805
4.0	0.9822
5.0	0.9844
6.0	0.9853
7.0	0.9858
8.0	0.9873
9.0	0.9872
10.0	0.9881
11.1	0.9881
12.0	0.9881
13.0	0.9886
14.1	0.9888
16.0	0.9888
18.0	0.9896
20.0	0.9898
22.1	0.9905

*Not shown on plot

DATA TABLE NO. 7 (continued)

λ	ρ
CURVE 59 (cont.)*	
24.1	0.9911
26.1	0.9923
28.1	0.9932
30.1	0.9932

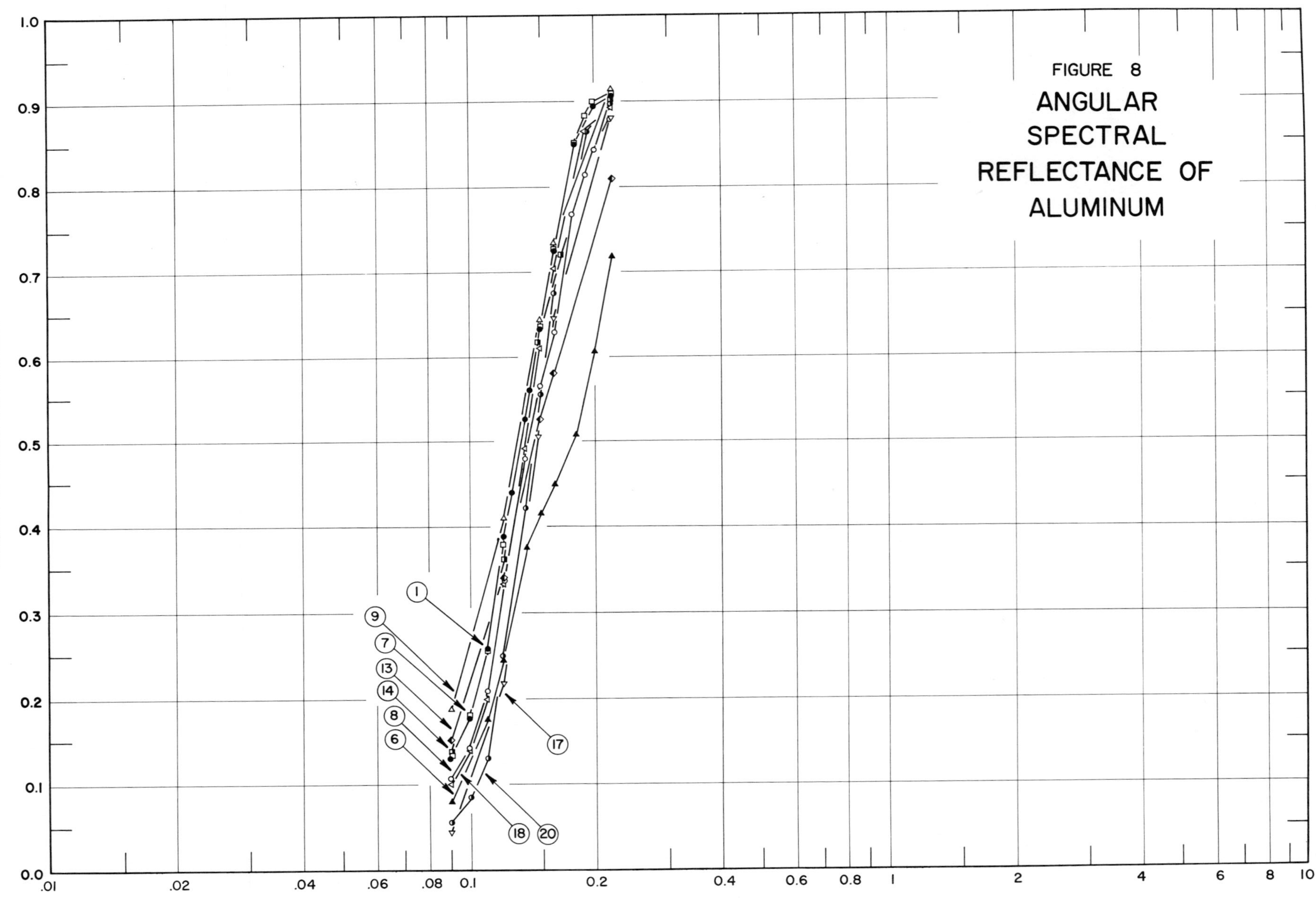

FIGURE 8
ANGULAR SPECTRAL REFLECTANCE OF ALUMINUM

SPECIFICATION TABLE NO. 8 ANGULAR SPECTRAL REFLECTANCE OF ALUMINUM

Curve No.	Ref. No.	Year	Temperature K	Wavelength Range, μ	Geometry ϑ	ϑ'	ω'	Reported Error, %	Composition (weight percent), Specifications and Remarks
1	306	1959	~298	0.0898-0.2200	~18°	~18°			Evaporated film (800 Å thick); substrate cleaned by a high-voltage dc glow discharge; deposited film showed no trace of tungsten spectrographically; evaporated at 10^{-5} mm Hg for 2 sec; exposed to air for 24 hrs.
2	306	1959	~298	0.0899-0.2199	~18°	~18°			Different sample, same as above specimen and conditions except evaporation time 55 sec.
3	306	1959	~298	0.0899-0.2200	~18°	~18°			Different sample, same as above specimen and conditions except evaporation time 130 sec.
4	306	1959	~298	0.0903-0.2201	~18°	~18°			Different sample, same as curve 1 specimen and conditions except evaporated at 10^{-4} mm Hg for 1-2 sec.
5	306	1959	~298	0.0904-0.2201	~18°	~18°			Different sample, same as above specimen and conditions except evaporation time 55 sec.
6	306	1959	~298	0.0901-0.2200	~18°	~18°			Different sample, same as above specimen and conditions except evaporation time 120 sec.
7	306	1959	~298	0.0900-0.2199	~18°	~18°			Evaporated film (900 Å thick); substrate cleaned by a high-voltage dc glow discharge; evaporated at 10^{-5} mm Hg for 2 sec using 99.99 pure Al; film was 24 hrs old.
8	306	1959	~298	0.0898-0.2200	~18°	~18°			Different sample, same as above specimen and conditions except 99.5 pure Al used for evaporation.
9	306	1959	~298	0.0900-0.2200	~18°	~18°			Evaporated film (800 Å thick); substrate cleaned by a high-voltage dc glow discharge; substrate temperature 308 K at time of deposition; film 2-3 hrs old when measurement was made.
10	306	1959	~298	0.0900-0.2200	~18°	~18°			Different sample, same as above specimen and conditions except substrate at 323 K at time of deposition.
11	306	1959	~298	0.0900-0.2200	~18°	~18°			Different sample, same as above specimen and conditions except substrate at 373 K at time of deposition.
12	306	1959	~298	0.0900-0.2200	~18°	~18°			Different sample, same as above specimen and conditions except substrate at 423 K at time of deposition.
13	306	1959	~298	0.0900-0.2200	~18°	~18°			Different sample, same as above specimen and conditions except substrate at 673 K at time of deposition.
14	306	1959	~298	0.0900-0.2200	~18°	~18°			Evaporated film; substrate cleaned by a high-voltage dc glow discharge; aged 10 hrs before measurement; data extracted from smooth curve.
15	306	1959	~298	0.0900-0.2000	~18°	~18°			Different sample, same as above specimen and conditions except aged 100 hrs.
16	306	1959	~298	0.0900-0.2000	~18°	~18°			Different sample, same as above specimen and conditions except aged 1000 hrs.
17	306	1959	~298	0.0900-0.2200	~18°	~18°			Different sample, same as above specimen and conditions except aged 10,000 hrs.
18	306	1959	~298	0.0901-0.2202	~18°	~18°			Evaporated film; substrate cleaned by a high-voltage dc glow discharge; stored in dry air for 63 days before measurement.
19	306	1959	~298	0.0900-0.2201	~18°	~18°			Different sample, same as above specimen and conditions except stored in normal air (30 to 50% humidity) for 63 days before measurement.

SPECIFICATION TABLE NO. 8 (continued)

Curve No.	Ref. No.	Year	Temperature K	Wavelength Range, μ	Geometry θ	θ'	ω'	Reported Error, %	Composition (weight percent), Specifications and Remarks
20	306	1959	~298	0.0901-0.2198	~18°	~18°			Evaporated film; substrate cleaned by a high-voltage dc glow discharge; prepared under optimum conditions; placed 8 in. from a 435-W quartz mercury burner and irradiated for 20 hrs in desiccated air; surface exposed to irradiation immediately following deposition.
21	306	1959	~298	0.0901-0.2199	~18°	~18°			Different sample, same as above specimen and conditions except irradiated for 20 hrs in normal air (30% humidity).
22	306	1959	~298	0.0901-0.2203	~18°	~18°			Different sample, same as above specimen and conditions except irradiated for 20 hrs in moist air (> 90% humidity).

DATA TABLE NO. 8 ANGULAR SPECTRAL REFLECTANCE OF ALUMINUM

[Wavelength, λ, μ; Reflectance, ρ; Temperature, T, K]

λ	ρ
CURVE 1	
T = ~298	
0.0898	0.129
0.0999	0.175
0.1097	0.258
0.1211	0.388
0.1261	0.439
0.1358	0.526
0.1398	0.560
0.1486	0.633
0.1605	0.727
0.1798	0.850
0.2003	0.894
0.2200	0.907
CURVE 2*	
T = ~298	
0.0899	0.110
0.0999	0.155
0.1103	0.236
0.1213	0.351
0.1260	0.406
0.1356	0.494
0.1397	0.521
0.1480	0.579
0.1603	0.640
0.1803	0.752
0.2006	0.856
0.2199	0.886
CURVE 3*	
T = ~298	
0.0899	0.093
0.0998	0.130
0.1099	0.195
0.1210	0.302
0.1260	0.345
0.1359	0.407
0.1399	0.421
0.1486	0.461
0.1610	0.505
0.1806	0.584
0.2000	0.673
0.2200	0.775
CURVE 4*	
T = ~298	
0.0903	0.127
0.1004	0.164
0.1102	0.238
0.1212	0.360
0.1492	0.618
0.1609	0.720
0.1800	0.847
0.2000	0.897
0.2201	0.908
CURVE 5*	
T = ~298	
0.0904	0.111
0.1004	0.147
0.1102	0.227
0.1211	0.331
0.1484	0.502
0.1602	0.553
0.1795	0.655
0.1997	0.773
0.2201	0.849
CURVE 6	
T = ~298	
0.0901	0.079
0.1107	0.173
0.1212	0.243
0.1363	0.375
0.1487	0.413
0.1607	0.449
0.1807	0.508
0.2000	0.606
0.2200	0.720
CURVE 7	
T = ~298	
0.0900	0.134
0.0997	0.181
0.1103	0.258
0.1212	0.378
0.1358	0.529
CURVE 7 (cont.)	
0.1480	0.635
0.1600	0.729
0.1799	0.851
0.1901	0.883
0.1999	0.900
0.2199	0.908
CURVE 8	
T = ~298	
0.0898	0.108
0.0999	0.143
0.1099	0.210
0.1213	0.339
0.1357	0.479
0.1480	0.564
0.1603	0.630
0.1797	0.769
0.1901	0.814
0.2002	0.845
0.2200	0.882
CURVE 9	
T = ~298	
0.0900	0.188
0.1216	0.410
0.1486	0.644
0.1606	0.737
0.2200	0.915
CURVE 10*	
T = ~298	
0.0900	0.185
0.1216	0.407
0.1486	0.644
0.1606	0.731
0.2200	0.912
CURVE 11*	
T = ~298	
0.0900	0.177
0.1216	0.395
CURVE 11 (cont.)*	
0.1486	0.626
0.1606	0.702
0.2200	0.900
CURVE 12*	
T = ~298	
0.0900	0.170
0.1216	0.375
0.1486	0.585
0.1606	0.655
0.2200	0.871
CURVE 13	
T = ~298	
0.0900	0.151
0.1216	0.341
0.1486	0.526
0.1606	0.580
0.2200	0.810
CURVE 14	
T = ~298	
0.0900	0.137
0.1216	0.361
0.1466	0.620
0.1606	0.722
0.2200	0.901
CURVE 15*	
T = ~298	
0.0900	0.102
0.1216	0.314
0.1466	0.580
0.1606	0.694
0.2000	0.899
CURVE 16*	
T = ~298	
0.0900	0.075
0.1216	0.264
CURVE 16 (cont.)*	
0.1466	0.543
0.1606	0.672
0.2000	0.896
CURVE 17	
T = ~298	
0.0900	0.045
0.1216	0.218
0.1466	0.504
0.1606	0.646
0.2200	0.881
CURVE 18	
T = ~298	
0.0901	0.100
0.1003	0.140
0.1106	0.203
0.1213	0.334
0.1360	0.491
0.1484	0.611
0.1605	0.704
0.1902	0.867
0.2202	0.893
CURVE 19*	
T = ~298	
0.0900	0.058
0.1002	0.095
0.1106	0.152
0.1213	0.281
0.1363	0.435
0.1486	0.549
0.1608	0.663
0.1905	0.846
0.2201	0.889
CURVE 20	
T = ~298	
0.0901	0.056
0.1001	0.084
0.1104	0.131
CURVE 20 (cont.)	
0.1213	0.250
0.1361	0.421
0.1486	0.556
0.1605	0.675
0.1901	0.865
0.2198	0.895
CURVE 21*	
T = ~298	
0.0901	0.035
0.1001	0.060
0.1101	0.101
0.1215	0.203
0.1358	0.363
0.1484	0.529
0.1607	0.650
0.1900	0.842
0.2199	0.894
CURVE 22*	
T = ~298	
0.0901	0.016
0.1001	0.028
0.1104	0.054
0.1212	0.141
0.1359	0.295
0.1486	0.443
0.1605	0.566
0.1899	0.814
0.2203	0.876

* Not shown on plot

SPECIFICATION TABLE NO. 9 ANGULAR SPECTRAL REFLECTANCE OF ALUMINUM

Curve No.	Ref. No.	Year	Temperature K	Wavelength, μ	Angular Range°	Geometry θ	θ'	ω'	Reported Error, %	Composition (weight percent), Specifications and Remarks
1	306	1959	~298	0.0585	6.80-89.5	θ	θ'			Evaporated film (2100 Å thick); glass substrate cleaned by a high-voltage dc glow discharge; $\theta = \theta'$; data extracted from smooth curve.
2	306	1959	~298	0.0735	4.80-88.3	θ	θ'			Above specimen and conditions.

DATA TABLE NO. 9 ANGULAR SPECTRAL REFLECTANCE OF ALUMINUM

[Angle, θ, °; Reflectance, ρ; Temperature, T, K; Wavelength, λ, μ]

θ	ρ
CURVE 1*	
T = ~298	
λ = 0.0585	
6.8	0.0149
15.5	0.0398
19.4	0.0247
23.1	0.0404
29.0	0.0373
32.4	0.0138
36.2	0.0334
39.0	0.0199
41.1	0.0352
43.0	0.0334
48.3	0.1191
50.8	0.1923
57.5	0.2911
73.4	0.5598
89.5	0.9638
CURVE 2*	
T = ~298	
λ = 0.0735	
4.8	0.0043
10.6	0.0070
13.5	0.0019
15.7	0.0057
18.5	0.0090
21.4	0.0076
24.5	0.0135
29.7	0.0714
52.1	0.2004
88.3	0.9506

*Not shown on plot

SPECIFICATION TABLE NO. 10 NORMAL SOLAR REFLECTANCE OF ALUMINUM

Curve No.	Ref. No.	Year	Temperature Range, K	Geometry θ	θ'	ω'	Reported Error, %	Composition (weight percent), Specifications and Remarks
1	131	1964	298	$\sim 0^{\circ}$		2π		Pure Al; 0.3 μ thick opaque layer on glass; freshly prepared; MgO reference; computed from spectral data.
2	131	1964	298	$\sim 0^{\circ}$		2π		Above specimen and conditions; diffuse component only.
3	131	1964	298	$\sim 0^{\circ}$		2π		Different sample, same as curve 1 specimen and conditions.
4	131	1964	298	$\sim 0^{\circ}$		2π		Above specimen and conditions; diffuse component only.
5	40	1962	298	$\sim 0^{\circ}$		2π		Aluminum foil Reynolds Wrap - heavy duty; measured in air.
6	40	1962	298	$\sim 0^{\circ}$		2π		Pure aluminum on optical flat (8×10^{-6} in. thick); measured in air.
7	40	1962	298	$\sim 0^{\circ}$		2π		Aluminum film on 1/4 mil thick Mylar; measured in air.
8	40	1962	298	$\sim 0^{\circ}$		2π		Reynolds Wrap - heavy duty foil; measured in air.
9	40	1962	298	$\sim 0^{\circ}$		2π		Pure aluminum on optical flat (8×10^{-6} in. thick); measured in air.
10	40	1962	298	$\sim 0^{\circ}$		2π		Aluminum film on 1/4 mil thick Mylar; measured in air.

DATA TABLE NO. 10 NORMAL SOLAR REFLECTANCE OF ALUMINUM

[Temperature, T, K; Reflectance, ρ]

T	ρ
CURVE 1*	
298	0.90
CURVE 2*	
298	0.01
CURVE 3*	
298	0.91
CURVE 4*	
298	0.01
CURVE 5*	
298	0.88
CURVE 6*	
298	0.90
CURVE 7*	
298	0.89
CURVE 8*	
298	0.88
CURVE 9*	
298	0.90
CURVE 10*	
298	0.88

*Not shown on plot

SPECIFICATION TABLE NO. 11 HEMISPHERICAL INTEGRATED ABSORPTANCE OF ALUMINUM

Curve No.	Ref. No.	Year	Temperature Range, K	Reported Error, %	Composition (weight percent), Specifications and Remarks
1	3	1955	76	5	Kaiser foil (0.001 in. thick); unannealed; measured in vacuum (10^{-6} to 10^{-7} mm Hg); absorptance for 300 K black body incident radiation.
2	3	1955	76	5	Cockron home foil (0.0015 in. thick); measured in vacuum (10^{-6} to 10^{-7} mm Hg); absorptance for 300 K black body incident radiation.
3	3	1955	76	5	Hurwich home foil (0.0015 in. thick); measured on mat side in vacuum (10^{-6} to 10^{-7} mm Hg); absorptance for 300 K black body incident radiation.
4	3	1955	76	5	Hurwich home foil (0.0015 in. thick); measured on bright side in vacuum (10^{-6} to 10^{-7} mm Hg); absorptance for 300 K black body incident radiation.
5	3	1955	76	5	Sheet (0.020 in. thick); cold acid cleaned; measured in vacuum (10^{-6} to 10^{-7} mm Hg); absorptance for 300 K black body incident radiation.
6	3	1955	76	5	Alcoa No. 2 reflector plate (0.020 in. thick); measured in vacuum (10^{-6} to 10^{-7} mm Hg); absorptance for 300 K black body incident radiation.
7	3	1955	76	5	Alcoa No. 2 reflector plate (0.020 in. thick); sanded with fine emery; measured in vacuum (10^{-6} to 10^{-7} mm Hg); absorptance for 300 K black body incident radiation.
8	3	1955	76	5	Alcoa No. 2 reflector plate (0.020 in. thick); cleaned with alkali; measured in vacuum (10^{-6} to 10^{-7} mm Hg); absorptance for 300 K black body incident radiation.
9	3	1955	76	5	Sheet (0.020 in. thick); cleaned with wire brush, emery paper, steel wool and cold acid; measured in vacuum (10^{-6} to 10^{-7} mm Hg); absorptance for 300 K black body incident radiation.
10	3	1955	76	5	Sheet (0.020 in. thick); wire brush cleaned; measured in vacuum (10^{-6} to 10^{-7} mm Hg); absorptance for 300 K black body incident radiation.
11	3	1955	76	5	Sheet (0.020 in. thick); liquid honed; measured in vacuum (10^{-6} to 10^{-7} mm Hg); absorptance for 300 K black body incident radiation.
12	3	1955	76	5	Sheet (0.020 in. thick); hot acid cleaned (Alcoa process); measured in vacuum (10^{-6} to 10^{-7} mm Hg); abosrptance for 300 K black body incident radiation.
13	4	1953	76		Cockron foil (0.0015 in. thick); measured in vacuum ($< 10^{-6}$ mm Hg); absorptance for 294 K black body incident radiation.
14	4	1953	76		Cockron foil (0.0015 in. thick); measured in vacuum ($< 10^{-6}$ mm Hg); absorptance for 294 K black body incident radiation.
15	4	1953	76		Sheet (0.020 in. thick); cleaned with wire brush; measured in vacuum ($< 10^{-6}$ mm Hg); absorptance for 294 K black body incident radiation.
16	4	1953	76		Sheet (0.020 in. thick); cleaned with wire brush, emery paper, and steel wool; measured in vacuum ($< 10^{-6}$ mm Hg); absorptance for 294 K black body incident radiation.
17	4	1953	76		Sheet (0.020 in. thick); cleaned with wire brush, emery paper, steel wool, and cold acid; measured in vacuum ($< 10^{-6}$ mm Hg); absorptance for 294 K black body incident radiation.
18	4	1953	76		Sheet (0.020 in. thick); alkali cleaned; measured in vacuum ($< 10^{-6}$ mm Hg); absorptance for 294 K black body incident radiation.

SPECIFICATION TABLE NO. 11 (continued)

Curve No.	Ref. No.	Year	Temperature Range, K	Reported Error, %	Composition (weight percent), Specifications and Remarks
19	4	1953	76		Alcoa No. 2 reflector plate (0.020 in. thick, $\epsilon = 0.30$ for visible light at 298 K); measured in vacuum ($< 10^{-6}$ mm Hg); absorptance for 294 K black body incident radiation.
20	4	1953	76		Sheet (0.020 in. thick); cold acid cleaned; measured in vacuum ($< 10^{-6}$ mm Hg); absorptance for 294 K black body incident radiation.
21	4	1953	76		Sheet (0.020 in. thick); cold acid cleaned; measured in vacuum ($< 10^{-6}$ mm Hg); absorptance for 294 K black body incident radiation.
22	4	1953	76		Hurwich home foil (0.0015 in. thick); measured on bright side in vacuum ($< 10^{-6}$ mm Hg); absorptance for 294 K black body incident radiation.
23	4	1953	76		Hurwich home foil (0.0015 in. thick); measured on mat side in vacuum ($< 10^{-6}$ mm Hg); absorptance for 294 K black body incident radiation.
24	4	1953	76		Cockron home foil (0.0015 in. thick); measured in vacuum ($< 10^{-6}$ mm Hg); absorptance for 294 K black body incident radiation.
25	4	1953	76		Cockron home foil (0.0015 in. thick); measured in vacuum ($<10^{-6}$ mm Hg); absorptance for 294 K black body incident radiation.
26	4	1953	76		Kaiser foil (0.001 in. thick); unannealed; measured in vacuum ($< 10^{-6}$ mm Hg); absorptance for 294 K black body incident radiation.
27	4	1953	76		Kaiser foil (0.001 in. thick); unannealed; measured in vacuum ($< 10^{-6}$ mm Hg); abosrptance for 294 K black body incident radiation.
28	4	1953	76		Kaiser home foil (0.00075 in. thick); measured in vacuum ($< 10^{-6}$ mm Hg); absorptance for 294 K black body incident radiation.
29	4	1953	76		Foil (0.020 in. thick); hot acid cleaned (Alcoa process); measured in vacuum ($< 10^{-6}$ mm Hg); absorptance for 294 K black body incident radiation.

DATA TABLE NO. 11 HEMISPHERICAL INTEGRATED ABSORPTANCE OF ALUMINUM

[Temperature, T, K; Absorptance, α]

T	α	T	α	T	α
CURVE 1*		CURVE 13*		CURVE 25*	
76	0.018	76	0.0204	76	0.0186
CURVE 2*		CURVE 14*		CURVE 26*	
76	0.018	76	0.0200	76	0.0213
CURVE 3*		CURVE 15*		CURVE 27*	
76	0.021	76	0.0615	76	0.0184
CURVE 4*		CURVE 16*		CURVE 28*	
76	0.022	76	0.0615	76	0.0186
CURVE 5*		CURVE 17*		CURVE 29*	
76	0.028	76	0.0452	76	0.0294
CURVE 6*		CURVE 18*			
76	0.026	76	0.0356		
CURVE 7*		CURVE 19*			
76	0.032	76	0.0327		
CURVE 8*		CURVE 20*			
76	0.035	76	0.0317		
CURVE 9*		CURVE 21*			
76	0.045	76	0.0283		
CURVE 10*		CURVE 22*			
76	0.060	76	0.0217		
CURVE 11*		CURVE 23			
76	0.140	76	0.0212		
CURVE 12*		CURVE 24*			
76	0.029	76	0.0204		

*Not shown on plot

SPECIFICATION TABLE NO. 12 NORMAL INTEGRATED ABSORPTANCE OF ALUMINUM

Curve No.	Ref. No.	Year	Temperature Range, K	Geometry θ	Reported Error, %	Composition (weight percent), Specifications and Remarks
1	134	1952	2	$\sim 0^{\circ}$	1	Electropolished; absorptance for 298 K black body incident radiation.

DATA TABLE NO. 12 NORMAL INTEGRATED ABSORPTANCE OF ALUMINUM

[Temperature, T, K; Absorptance, α]

T	α
CURVE 1*	
2	0.0111

*Not shown on plot

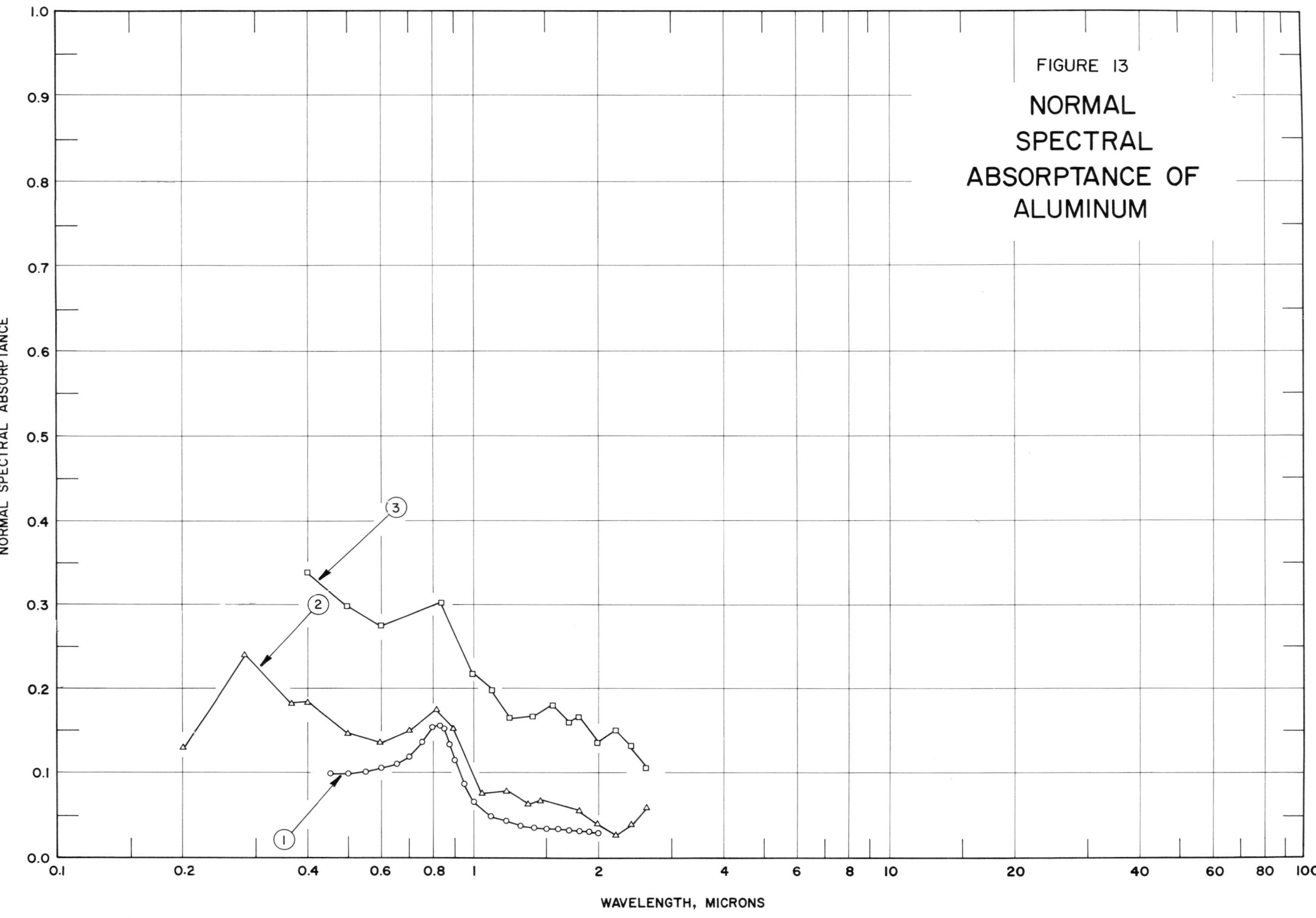
FIGURE 13
NORMAL
SPECTRAL
ABSORPTANCE OF
ALUMINUM
NORMAL SPECTRAL ABSORPTANCE
1.0
0.9
0.8
0.7
0.6
0.5
0.4
0.3
0.2
0.1
0.0
0.1
0.2
0.4
0.6
0.8
1
2
4
6
8
10
20
40
60
80
100
WAVELENGTH, MICRONS
1
2
3

SPECIFICATION TABLE NO. 13 NORMAL SPECTRAL ABSORPTANCE OF ALUMINUM

Curve No.	Ref. No.	Year	Temperature K	Wavelength Range, μ	Geometry θ	Reported Error, %	Composition (weight percent), Specifications and Remarks
1	282	1966	298	0.45-2.00	~10°	±1.4	Evaporated film; evaporation rate 300 Å sec^{-1} at 2 x 10^{-5} mm Hg; measured in vacuum; aged 8 days before measurement.
2	307	1954	~298	0.204-2.600	~ 0°		Data extracted from smooth curve.
3	307	1954	~298	0.402-2.600	~ 0°		Polished; data extracted from smooth curve.

DATA TABLE NO. 13 NORMAL SPECTRAL ABSORPTANCE OF ALUMINUM

[Wavelength, λ, μ; Absorption, α; Temperature, T, K]

λ	α	λ	α
CURVE 1 T = 298		CURVE 3 T = ~298	
0.45	0.0982	0.402	0.338
0.50	0.0987	0.498	0.298
0.55	0.1001	0.607	0.273
0.60	0.1037	0.841	0.301
0.65	0.1096	0.997	0.217
0.70	0.1189	1.119	0.198
0.75	0.1331	1.220	0.163
0.80	0.1511	1.397	0.167
0.825	0.1533	1.553	0.179
0.850	0.1501	1.704	0.159
0.875	0.1311	1.800	0.165
0.90	0.1142	2.000	0.134
0.95	0.0846	2.200	0.149
1.00	0.0663	2.400	0.131
1.10	0.0498	2.600	0.105
1.20	0.0420		
1.30	0.0387		
1.40	0.0357		
1.50	0.0341		
1.60	0.0331		
1.70	0.0322		
1.80	0.0319		
1.90	0.0313		
2.00	0.0308		
CURVE 2 T = ~298			
0.204	0.128		
0.281	0.239		
0.367	0.181		
0.401	0.182		
0.506	0.145		
0.597	0.134		
0.709	0.149		
0.815	0.173		
0.895	0.151		
1.048	0.075		
1.200	0.078		
1.352	0.062		
1.440	0.067		
1.798	0.055		
2.000	0.040		
2.200	0.028		
2.400	0.039		
2.600	0.059		

SPECIFICATION TABLE NO. 14 ANGULAR SPECTRAL ABSORPTANCE OF ALUMINUM

Curve No.	Ref. No.	Year	Temperature K	Wavelength Range, μ	Geometry θ	Reported Error, %	Composition (weight percent), Specifications and Remarks
1	225	1965	306	1.52-21.20	25°		Aluminum foil; measured in dry nitrogen; heated cavity at approx 1056 K with platinum reference; authors assumed $\alpha = 1 - R(2\pi, 25°)$.

DATA TABLE NO. 14 ANGULAR SPECTRAL ABSORPTANCE OF ALUMINUM

[Wavelength, λ, μ; Absorptance, α; Temperature, T, K]

λ	α
CURVE 1*	
T = 306	
1.52	0.037
2.05	0.035
3.14	0.029
4.25	0.024
6.98	0.020
9.16	0.019
12.20	0.017
14.80	0.015
18.00	0.015
21.20	0.014

*Not shown on plot

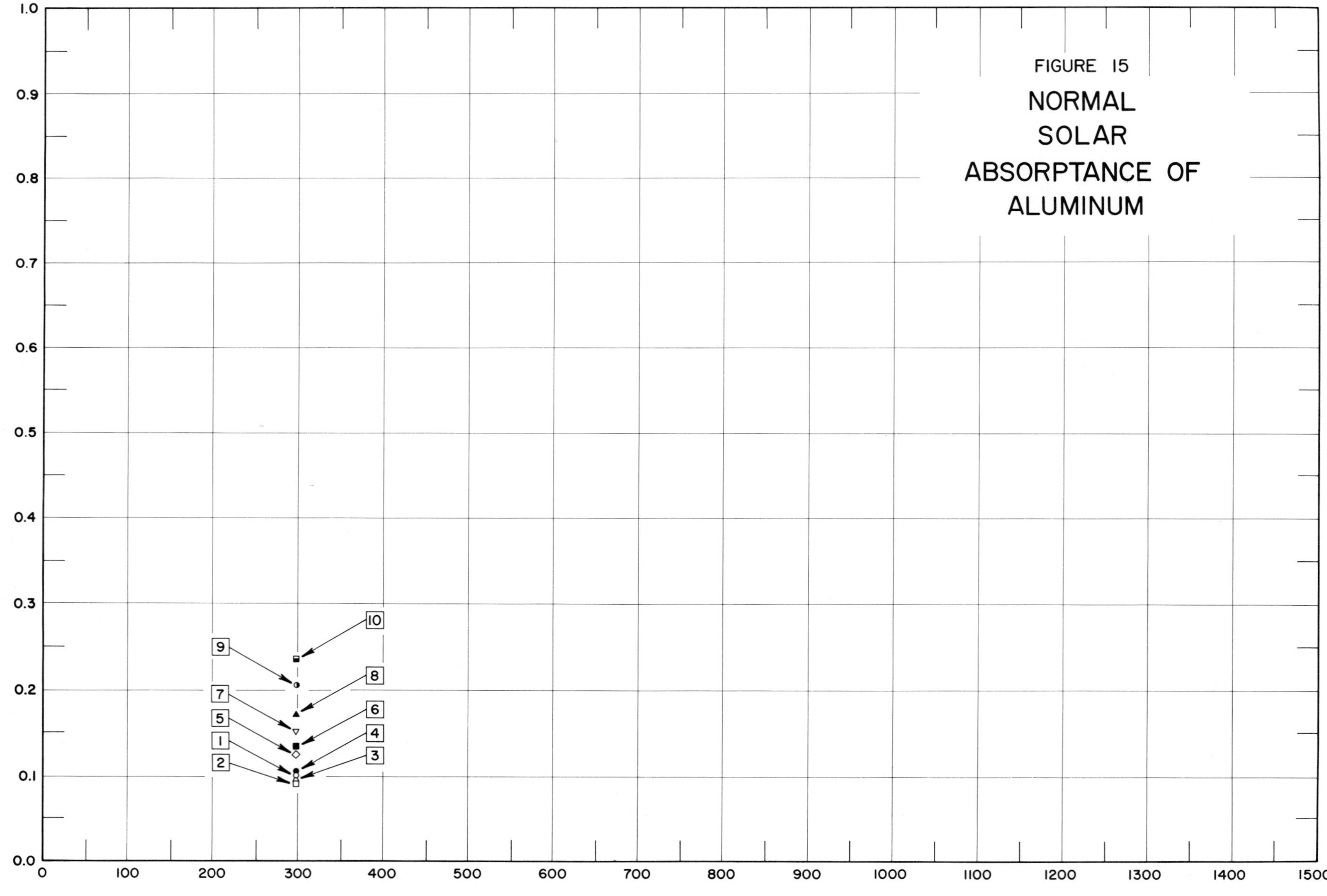
FIGURE 15
NORMAL
SOLAR
ABSORPTANCE OF
ALUMINUM
NORMAL SOLAR ABSORPTANCE
TEMPERATURE, K
1.0
0.9
0.8
0.7
0.6
0.5
0.4
0.3
0.2
0.1
0.0
0
100
200
300
400
500
600
700
800
900
1000
1100
1200
1300
1400
1500
10
9
8
7
6
5
4
1
3
2

SPECIFICATION TABLE NO. 15 NORMAL SOLAR ABSORPTANCE OF ALUMINUM

Curve No.	Ref. No.	Year	Temperature Range, K	Geometry θ	Reported Error, %	Composition (weight percent), Specifications and Remarks
1	131	1964	298	$\sim 0^\circ$		Pure Al; 0.3 μ thick opaque layer on glass; freshly prepared; calculated from $(1-\rho)$.
2	131	1964	298	$\sim 0^\circ$		Different sample, same as curve 1 specimen and conditions.
3	352	1963	298	$\sim 0^\circ$		99.99 pure; electropolished; computed from spectral reflectance.
4	352	1963	298	$\sim 0^\circ$		Curve 3 specimen and conditions except hydrogen ion bombarded (0.25×10^{20} ions-cm^{-2}).
5	352	1963	298	$\sim 0^\circ$		Curve 3 specimen and conditions except hydrogen ion bombarded (0.84×10^{20} ions-cm^{-2}).
6	352	1963	298	$\sim 0^\circ$		Curve 3 specimen and conditions except hydrogen ion bombarded (1.67×10^{20} ions-cm^{-2}).
7	352	1963	298	$\sim 0^\circ$		Curve 3 specimen and conditions except hydrogen ion bombarded (3.20×10^{20} ions-cm^{-2}).
8	352	1963	298	$\sim 0^\circ$		Curve 3 specimen and conditions except hydrogen ion bombarded (4.92×10^{20} ions-cm^{-2}).
9	352	1963	298	$\sim 0^\circ$		Curve 3 specimen and conditions except hydrogen ion bombarded (7.45×10^{20} ions-cm^{-2}).
10	352	1963	298	$\sim 0^\circ$		Curve 3 specimen and conditions except hydrogen ion bombarded (9.86×10^{20} ions-cm^{-2}).

DATA TABLE NO. 15 NORMAL SOLAR ABSORPTANCE OF ALUMINUM

[Temperature, T, K; Absorptance, α]

T	α
CURVE 1	
298	0.10
CURVE 2	
298	0.09
CURVE 3	
298	0.096
CURVE 4	
298	0.104
CURVE 5	
298	0.123
CURVE 6	
298	0.133
CURVE 7	
298	0.151
CURVE 8	
298	0.170
CURVE 9	
298	0.204
CURVE 10	
298	0.236

SPECIFICATION TABLE NO. 16 ANGULAR SOLAR ABSORPTANCE OF ALUMINUM

Curve No.	Ref. No.	Year	Temperature Range, K	Geometry θ	Reported Error, %	Composition (weight percent), Specifications and Remarks
1	154	1960	298	15°		Foil (Reynolds Wrap); computed from spectral reflectance data for above atmosphere conditions; [Authors' designation: Sample 8].
2	154	1960	298	15°		Different sample, same as above specimen and conditions except exposed to 700 Mev proton irradiation, 1.9×10^{13} protons cm^{-2}; [Authors' designation: Sample 9].

DATA TABLE NO. 16 ANGULAR SOLAR ABSORPTANCE OF ALUMINUM

[Temperature, T, K; Absorptance, α]

T	α
CURVE 1*	
298	0.10
CURVE 2*	
298	0.12

*Not shown on plot

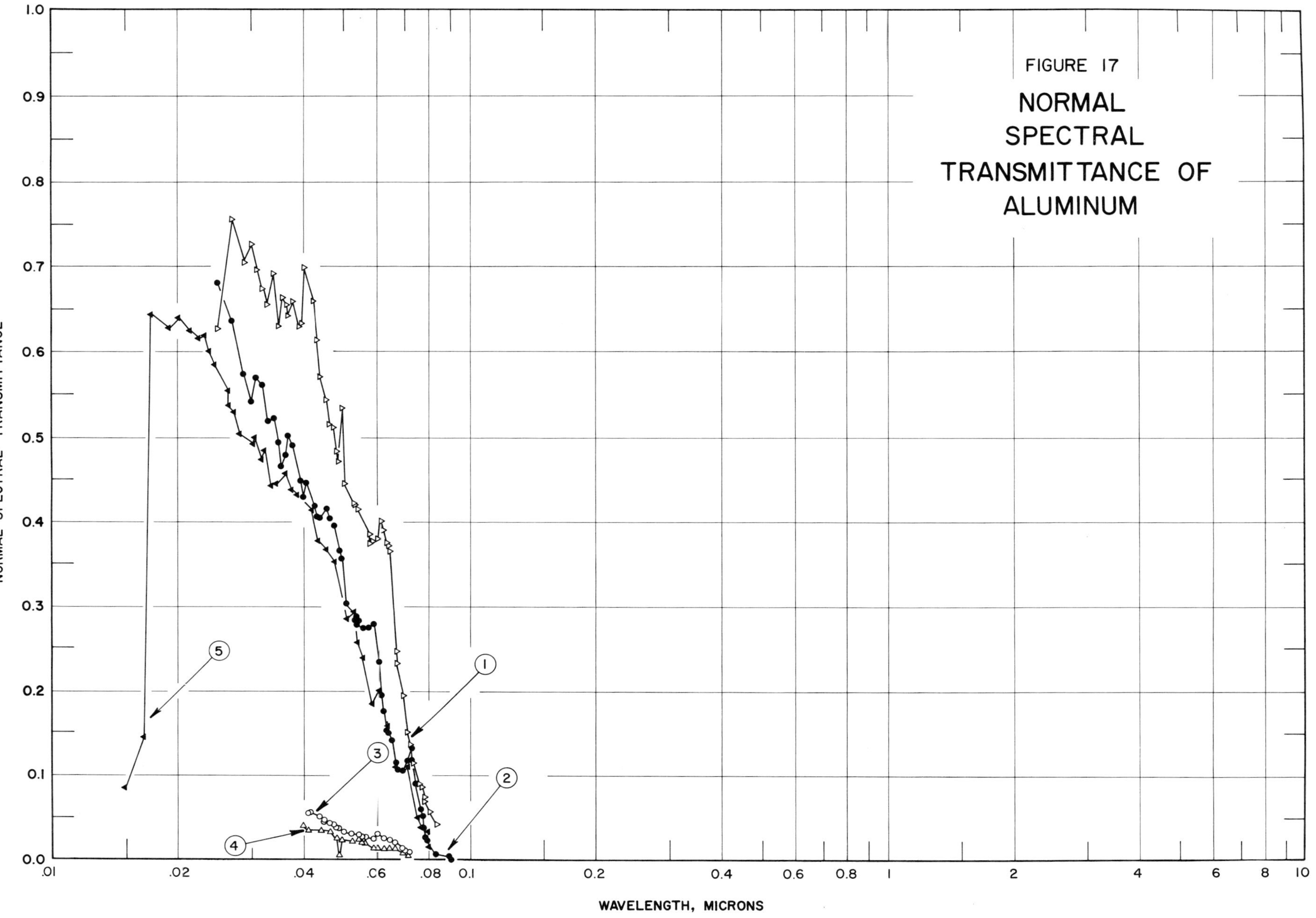
FIGURE 17
NORMAL SPECTRAL TRANSMITTANCE OF ALUMINUM
NORMAL SPECTRAL TRANSMITTANCE
WAVELENGTH, MICRONS
1.0
0.9
0.8
0.7
0.6
0.5
0.4
0.3
0.2
0.1
0.0
.01
.02
.04
.06
.08
0.1
0.2
0.4
0.6
0.8
1
2
4
6
8
10
1
2
3
4
5

SPECIFICATION TABLE NO. 17 NORMAL SPECTRAL TRANSMITTANCE OF ALUMINUM

Curve No.	Ref. No.	Year	Temperature K	Wavelength Range, μ	Geometry θ	θ'	ω'	Reported Error, %	Composition (weight percent), Specifications and Remarks
1	308	1965	298	0.0249-0.0832	~0°	~0°			Evaporated film (600 ± 50 Å thick); glass slide substrate at room temperature at evaporation; evaporated at 2 x 10^{-5} mm Hg in 10-15 sec; error ~10% at $\lambda > 0.035\ \mu$; error < 25% at $\lambda < 0.035\ \mu$.
2	308	1965	298	0.0250-0.0899	~0°	~0°			Different sample, same as above specimen and conditions except film thickness 1000 ± 100 Å.
3	337	1967	298	0.0411-0.0718	0°	0°			Evaporated film (470 Å thick); unsupported; exposed to air at atmospheric pressure.
4	337	1967	298	0.0406-0.0717	0°	0°			Different sample, same as above specimen and conditions except 1380 Å thick.
5	290	1965	298	0.0150-0.0800	~0°	~0°			Unbacked film (~800 Å thick); evaporated at several hundred Å sec^{-1}; measured in vacuum (< 10^{-5} mm Hg); slight oxide layer.

DATA TABLE NO. 17 NORMAL SPECTRAL TRANSMITTANCE OF ALUMINUM

[Wavelength, λ, μ; Transmittance, τ; Temperature, T, K]

λ	τ
CURVE 1	
T = 298	
0.0249	0.629
0.0270	0.758
0.0288	0.706
0.0302	0.729
0.0308	0.697
0.0320	0.674
0.0328	0.657
0.0340	0.692
0.0349	0.631
0.0356	0.664
0.0363	0.654
0.0368	0.643
0.0378	0.660
0.0391	0.631
0.0397	0.635
0.0401	0.700
0.0423	0.660
0.0429	0.614
0.0437	0.570
0.0453	0.543
0.0464	0.516
0.0473	0.512
0.0481	0.484
0.0484	0.471
0.0494	0.534
0.0506	0.445
0.0525	0.422
0.0530	0.411
0.0539	0.416
0.0574	0.386
0.0574	0.373
0.0585	0.377
0.0605	0.380
0.0614	0.402
0.0617	0.390
0.0629	0.375
0.0633	0.373
0.0642	0.316
0.0667	0.248
0.0674	0.232
0.0689	0.195
0.0705	0.150
0.0717	0.137
0.0725	0.113
0.0752	0.089

λ	τ
CURVE 1 (cont.)	
0.0765	0.086
0.0770	0.073
0.0775	0.068
0.0800	0.056
0.0832	0.042
CURVE 2	
T = 298	
0.0250	0.682
0.0270	0.638
0.0288	0.575
0.0302	0.542
0.0308	0.570
0.0320	0.561
0.0329	0.520
0.0340	0.522
0.0350	0.496
0.0356	0.467
0.0364	0.479
0.0368	0.502
0.0378	0.491
0.0394	0.450
0.0400	0.430
0.0403	0.448
0.0425	0.420
0.0432	0.408
0.0437	0.406
0.0454	0.416
0.0464	0.404
0.0473	0.397
0.0486	0.366
0.0496	0.357
0.0508	0.303
0.0530	0.283
0.0534	0.278
0.0532	0.289
0.0541	0.282
0.0556	0.275
0.0577	0.275
0.0585	0.278
0.0605	0.233
0.0614	0.195
0.0620	0.176
0.0633	0.152

λ	τ
CURVE 2 (cont.)	
0.0639	0.149
0.0653	0.141
0.0660	0.116
0.0670	0.107
0.0693	0.106
0.0705	0.117
0.0725	0.118
0.0725	0.132
0.0756	0.089
0.0761	0.060
0.0770	0.052
0.0775	0.039
0.0780	0.028
0.0790	0.024
0.0832	0.007
0.0886	0.005
0.0899	0.000
CURVE 3	
T = 298	
0.0411	0.0540
0.0415	0.0556
0.0437	0.0510
0.0442	0.0484
0.0451	0.0463*
0.0457	0.0444
0.0461	0.0436
0.0471	0.0415
0.0476	0.0404*
0.0478	0.0389
0.0487	0.0384
0.0492	0.0373*
0.0501	0.0344
0.0509	0.0331*
0.0521	0.0317
0.0526	0.0295*
0.0541	0.0302
0.0545	0.0290*
0.0549	0.0283
0.0554	0.0279*
0.0556	0.0282*
0.0561	0.0287
0.0564	0.0305*
0.0586	0.0264

λ	τ
CURVE 3 (cont.)	
0.0590	0.0266*
0.0595	0.0275*
0.0600	0.0316
0.0612	0.0278*
0.0620	0.0260
0.0632	0.0235*
0.0635	0.0235*
0.0641	0.0225
0.0642	0.0237*
0.0646	0.0220*
0.0650	0.0228*
0.0659	0.0204
0.0687	0.0145
0.0700	0.0111
0.0710	0.0099*
0.0716	0.0094*
0.0718	0.0090
CURVE 4	
T = 298	
0.0406	0.0407
0.0412	0.0359
0.0416	0.0384*
0.0426	0.0367*
0.0441	0.0366
0.0449	0.0351*
0.0457	0.0336*
0.0462	0.0333
0.0466	0.0330*
0.0472	0.0302*
0.0476	0.0286*
0.0480	0.0277
0.0487	0.0057
0.0494	0.0246
0.0501	0.0233*
0.0509	0.0229*
0.0520	0.0230
0.0526	0.0236*
0.0540	0.0241
0.0547	0.0232*
0.0550	0.0214*
0.0553	0.0220
0.0558	0.0209*
0.0562	0.0213

λ	τ
CURVE 4 (cont.)	
0.0565	0.0197*
0.0573	0.0187*
0.0587	0.0135
0.0591	0.0141*
0.0597	0.0134*
0.0597	0.0138*
0.0601	0.0140
0.0611	0.0131*
0.0621	0.0122
0.0631	0.0125*
0.0635	0.0125*
0.0639	0.0132
0.0643	0.0131*
0.0645	0.0137*
0.0652	0.0133*
0.0661	0.0136
0.0687	0.0080
0.0717	0.0056
CURVE 5	
T = 298	
0.0150	0.0853
0.0166	0.144
0.0173	0.643
0.0192	0.628
0.0202	0.640
0.0215	0.627
0.0226	0.617
0.0232	0.621
0.0239	0.601
0.0246	0.586
0.0266	0.556
0.0266	0.538
0.0274	0.531
0.0283	0.504
0.0304	0.492
0.0307	0.501
0.0320	0.474
0.0324	0.485
0.0336	0.442
0.0345	0.445
0.0364	0.458
0.0376	0.438
0.0388	0.432

λ	τ
CURVE 5 (cont.)	
0.0421	0.416
0.0436	0.378
0.0453	0.368
0.0474	0.352
0.0508	0.286
0.0526	0.294
0.0539	0.257
0.0556	0.239
0.0584	0.183
0.0608	0.201
0.0637	0.158
0.0661	0.109
0.0685	0.105*
0.0713	0.111
0.0750	0.0501
0.0778	0.0402
0.0791	0.0337
0.0800	0.0152

* Not shown on plot

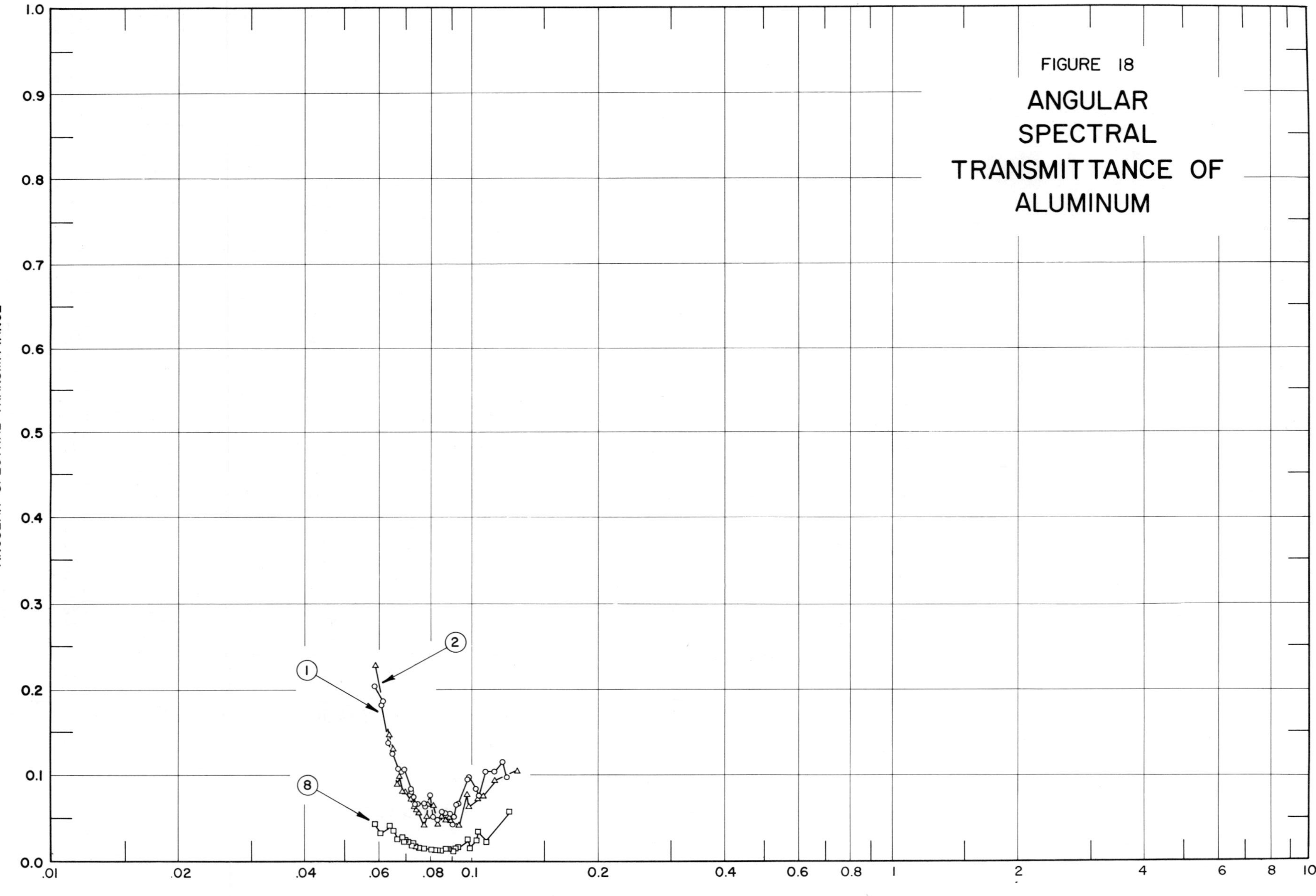
FIGURE 18
ANGULAR SPECTRAL TRANSMITTANCE OF ALUMINUM
ANGULAR SPECTRAL TRANSMITTANCE
WAVELENGTH, MICRONS
1.0
0.9
0.8
0.7
0.6
0.5
0.4
0.3
0.2
0.1
0.0
.01
.02
.04
.06
.08
0.1
0.2
0.4
0.6
0.8
1
2
4
6
8
10
1
2
8

SPECIFICATION TABLE NO. 18 ANGULAR SPECTRAL TRANSMITTANCE OF ALUMINUM

Curve No.	Ref. No.	Year	Temperature K	Wavelength Range, μ	Geometry θ ϑ' ω	Reported Error, %	Composition (weight percent), Specifications and Remarks
1	348	1965	298	0.0584-0.1211	0° 0°		Unbacked foil (300 Å thick); evaporated in vacuum (3×10^{-6} mm Hg); exposed to air; condensed spark discharge in argon light source; p-polarization dominant; data uncorrected for partial polarization effects.
2	348	1965	298	0.0587-0.1208	0° 0°		Above specimen and conditions except s-polarization is dominant.
3	348	1965	298	0.0578-0.1207	$20^\circ 20^\circ$		Curve 1 specimen and conditions.
4	348	1965	298	0.0579-0.1201	$20^\circ 20^\circ$		Above specimen and conditions except s-polarization is dominant.
5	348	1965	298	0.0576-0.1203	$30^\circ 30^\circ$		Curve 1 specimen and conditions.
6	348	1965	298	0.0576-0.1204	$30^\circ 30^\circ$		Above specimen and conditions except s-polarization is dominant.
7	348	1965	298	0.0592-0.1220	$45^\circ 45^\circ$		Curve 1 specimen and conditions.
8	348	1965	298	0.0594-0.1220	$45^\circ 45^\circ$		Above specimen and conditions except s-polarization is dominant.

DATA TABLE NO. 18 ANGULAR SPECTRAL TRANSMITTANCE OF ALUMINUM

[Wavelength, λ, μ; Transmittance, τ; Temperature, T, K]

CURVE 1
T = 298

λ	τ
0.0584	0.2028
0.0605	0.1820
0.0608	0.1871
0.0639	0.1381
0.0645	0.1250
0.0666	0.1077
0.0683	0.1023
0.0692	0.1028
0.0719	0.0813
0.0720	0.0793
0.0729	0.0740
0.0738	0.0658
0.0743	0.0672
0.0769	0.0672
0.0775	0.0630
0.0800	0.0776
0.0809	0.0506
0.0836	0.0490
0.0849	0.0581
0.0868	0.0573
0.0878	0.0552
0.0900	0.0411
0.0904	0.0518
0.0920	0.0641
0.0929	0.0566
0.0981	0.0955
0.0987	0.0989
0.1024	0.0824
0.1031	0.0760
0.1081	0.1028
0.1132	0.1028
0.1197	0.1151
0.1211	0.0971

CURVE 2
T = 298

λ	τ
0.0587	0.2296
0.0638	0.1473
0.0650	0.1306
0.0670	0.0895
0.0676	0.0998
0.0688	0.0804
0.0694	0.0804

CURVE 2 (cont.)

λ	τ
0.0719	0.0718
0.0732	0.0641
0.0739	0.0607
0.0749	0.0565
0.0771	0.0541
0.0781	0.0511
0.0798	0.0664
0.0812	0.0641
0.0837	0.0434
0.0852	0.0522
0.0872	0.0492
0.0881	0.0512
0.0885	0.0482
0.0935	0.0424
0.0979	0.0780
0.0989	0.0635
0.1033	0.0716
0.1081	0.0766
0.1131	0.0929
0.1208	0.1047

CURVE 3*
T = 298

λ	τ
0.0578	0.2009
0.0607	0.1950
0.0631	0.1637
0.0640	0.1542
0.0661	0.1170
0.0673	0.1045
0.0676	0.1143
0.0685	0.1312
0.0697	0.0946
0.0715	0.1202
0.0720	0.0971
0.0726	0.0914
0.0737	0.1002
0.0741	0.0912
0.0765	0.1052
0.0795	0.0538
0.0807	0.0424
0.0828	0.0353
0.0845	0.0360
0.0862	0.0448
0.0872	0.0474

CURVE 3 (cont.)*

λ	τ
0.0875	0.0493
0.0895	0.0447
0.0913	0.0494
0.0928	0.0535
0.0973	0.1026
0.0987	0.0975
0.1016	0.0791
0.1027	0.0843
0.1076	0.1143
0.1129	0.1143
0.1190	0.1479
0.1207	0.1183

CURVE 4*
T = 298

λ	τ
0.0579	0.2255
0.0612	0.1986
0.0631	0.1473
0.0640	0.1524
0.0669	0.0871
0.0687	0.0840
0.0714	0.0916
0.0715	0.0759
0.0725	0.0824
0.0738	0.0802
0.0767	0.0891
0.0796	0.0673
0.0808	0.0647
0.0831	0.0486
0.0846	0.0513
0.0864	0.0513
0.0872	0.0498
0.0885	0.0570
0.0893	0.0488
0.0914	0.0426
0.0927	0.0515
0.0974	0.0867
0.0983	0.0745
0.1016	0.0695
0.1028	0.1040
0.1076	0.1265
0.1201	0.1033

CURVE 5*
T = 298

λ	τ
0.0576	0.1442
0.0600	0.1507
0.0629	0.1312
0.0636	0.1312
0.0659	0.1033
0.0670	0.1047
0.0676	0.1047
0.0688	0.1189
0.0692	0.1135
0.0710	0.1107
0.0719	0.0968
0.0725	0.0923
0.0729	0.0895
0.0739	0.0738
0.0764	0.0650
0.0794	0.0269
0.0804	0.0178
0.0829	0.0154
0.0842	0.0175
0.0864	0.0234
0.0872	0.0256
0.0893	0.0249
0.0911	0.0279
0.0925	0.0309
0.0973	0.0605
0.0981	0.0411
0.1015	0.0507
0.1026	0.0507
0.1076	0.0735
0.1203	0.0813

CURVE 6*
T = 298

λ	τ
0.0576	0.1191
0.0604	0.1148
0.0635	0.0697
0.0638	0.0908
0.0664	0.0724
0.0668	0.0644
0.0677	0.0611
0.0686	0.0762
0.0696	0.0634
0.0712	0.0624
0.0715	0.0568
0.0726	0.0540

CURVE 6 (cont.)*

λ	τ
0.0734	0.0524
0.0743	0.0497
0.0764	0.0500
0.0796	0.0334
0.0806	0.0316
0.0828	0.0301
0.0844	0.0313
0.0862	0.0309
0.0875	0.0301
0.0885	0.0333
0.0895	0.0267
0.0912	0.0305
0.0925	0.0309
0.0972	0.0564
0.0981	0.0314
0.1015	0.0525
0.1025	0.0650
0.1074	0.0847
0.1123	0.0728
0.1204	0.0735

CURVE 7*
T = 298

λ	τ
0.0592	0.0655
0.0609	0.0628
0.0647	0.0594
0.0650	0.0520
0.0670	0.0491
0.0680	0.0474
0.0685	0.0500
0.0694	0.0458
0.0706	0.0496
0.0718	0.0353
0.0722	0.0274
0.0729	0.0225
0.0735	0.0179
0.0748	0.0152
0.0771	0.0115
0.0800	0.0079
0.0810	0.0055
0.0832	0.0060
0.0850	0.0062
0.0871	0.0077
0.0882	0.0082

CURVE 7 (cont.)*

λ	τ
0.0888	0.0090
0.0904	0.0087
0.0924	0.0104
0.0937	0.0109
0.0985	0.0229
0.0996	0.0144
0.1029	0.0254
0.1036	0.0247
0.1093	0.0325
0.1139	0.0408
0.1220	0.0408

CURVE 8
T = 298

λ	τ
0.0594	0.0446
0.0607	0.0334
0.0647	0.0407
0.0653	0.0379
0.0667	0.0270
0.0685	0.0286
0.0693	0.0253
0.0705	0.0266
0.0718	0.0247
0.0725	0.0196
0.0729	0.0213
0.0735	0.0180
0.0756	0.0171
0.0772	0.0175
0.0804	0.0167
0.0815	0.0160
0.0834	0.0148
0.0855	0.0148
0.0872	0.0151
0.0882	0.0153
0.0893	0.0159
0.0905	0.0116
0.0925	0.0152
0.0937	0.0162
0.0985	0.0276
0.0996	0.0160
0.1030	0.0251
0.1043	0.0344
0.1093	0.0238
0.1220	0.0508

* Not shown on plot

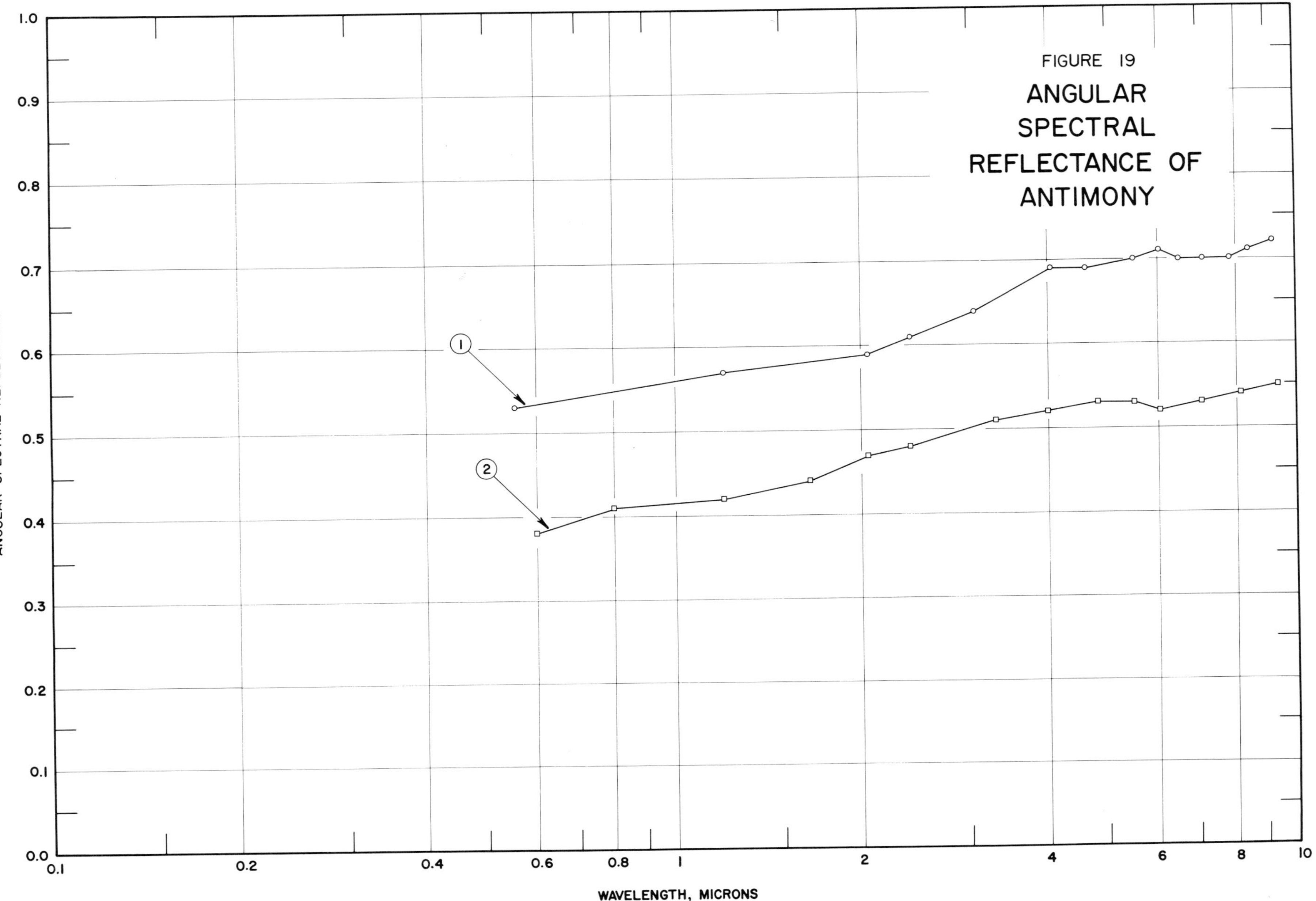

FIGURE 19
ANGULAR SPECTRAL REFLECTANCE OF ANTIMONY
ANGULAR SPECTRAL REFLECTANCE
1.0
0.9
0.8
0.7
0.6
0.5
0.4
0.3
0.2
0.1
0.0
0.1
0.2
0.4
0.6
0.8
1
2
4
6
8
10
WAVELENGTH, MICRONS
1
2

SPECIFICATION TABLE NO. 19 ANGULAR SPECTRAL REFLECTANCE OF ANTIMONY

Curve No.	Ref. No.	Year	Temperature K	Wavelength Range, μ	Geometry θ	θ'	ω'	Reported Error, %	Composition (weight percent), Specifications and Remarks
1	132	1911	298	0.55-9.20	15°	15°		≤ 3	Formed in vacuo by cathode disintegration; silvered glass mirror reference.
2	132	1911	298	0.60-9.35	15°	15°		≤ 3	Polished upon cleavage plane of a large crystal of the metal; silvered glass mirror reference.

DATA TABLE NO. 19 ANGULAR SPECTRAL REFLECTANCE OF ANTIMONY

[Wavelength λ, μ; Reflectance ρ; Temperature T, K]

λ	ρ
CURVE 1	
T = 298	
0.55	0.53
1.20	0.57
2.05	0.59
2.40	0.61
3.05	0.64
4.05	0.69
4.60	0.69
5.50	0.70
6.05	0.71
6.50	0.70
7.10	0.70
7.85	0.70
8.40	0.71
9.20	0.72
CURVE 2	
T = 298	
0.60	0.38
0.80	0.41
1.20	0.42
1.65	0.44
2.05	0.47
2.40	0.48
3.30	0.51
4.00	0.52
4.80	0.53
5.50	0.53
6.05	0.52
7.05	0.53
8.15	0.54
9.35	0.55

SPECIFICATION TABLE NO. 20 NORMAL SPECTRAL TRANSMITTANCE OF ANTIMONY

Curve No.	Ref. No.	Year	Temperature K	Wavelength Range, μ	Geometry θ θ' ω'	Reported Error, %	Composition (weight percent), Specifications and Remarks
1	308	1965	298	0.0249-0.0685	~0° ~0°		Al backed Sb film (1000 ± 100 Å thick); Al evaporated on glass slide, then Sb evaporated over that; evaporations at 2 x 10^{-5} mm Hg in 10-15 sec; error ~10% for $\lambda > 0.035$ μ and < 25% for $\lambda < 0.035$ μ.

DATA TABLE NO. 20 NORMAL SPECTRAL TRANSMITTANCE OF ANTIMONY

[Wavelength, λ, μ; Transmittance, τ; Temperature, T, K]

λ	τ
CURVE 1* T = 298	
0.0249	0.000
0.0270	0.000
0.0288	0.000
0.0302	0.003
0.0308	0.040
0.0320	0.037
0.0328	0.021
0.0340	0.000
0.0350	0.000
0.0356	0.000
0.0363	0.000
0.0367	0.000
0.0380	0.014
0.0392	0.142
0.0396	0.262
0.0401	0.254
0.0423	0.244
0.0432	0.243
0.0435	0.240
0.0453	0.228
0.0464	0.224
0.0471	0.203
0.0481	0.203
0.0486	0.196
0.0494	0.165
0.0506	0.169
0.0523	0.136
0.0530	0.131
0.0537	0.135
0.0556	0.107
0.0571	0.080
0.0582	0.075
0.0602	0.055
0.0611	0.042
0.0617	0.034
0.0626	0.028
0.0636	0.027
0.0649	0.024
0.0685	0.005

*Not shown on plot

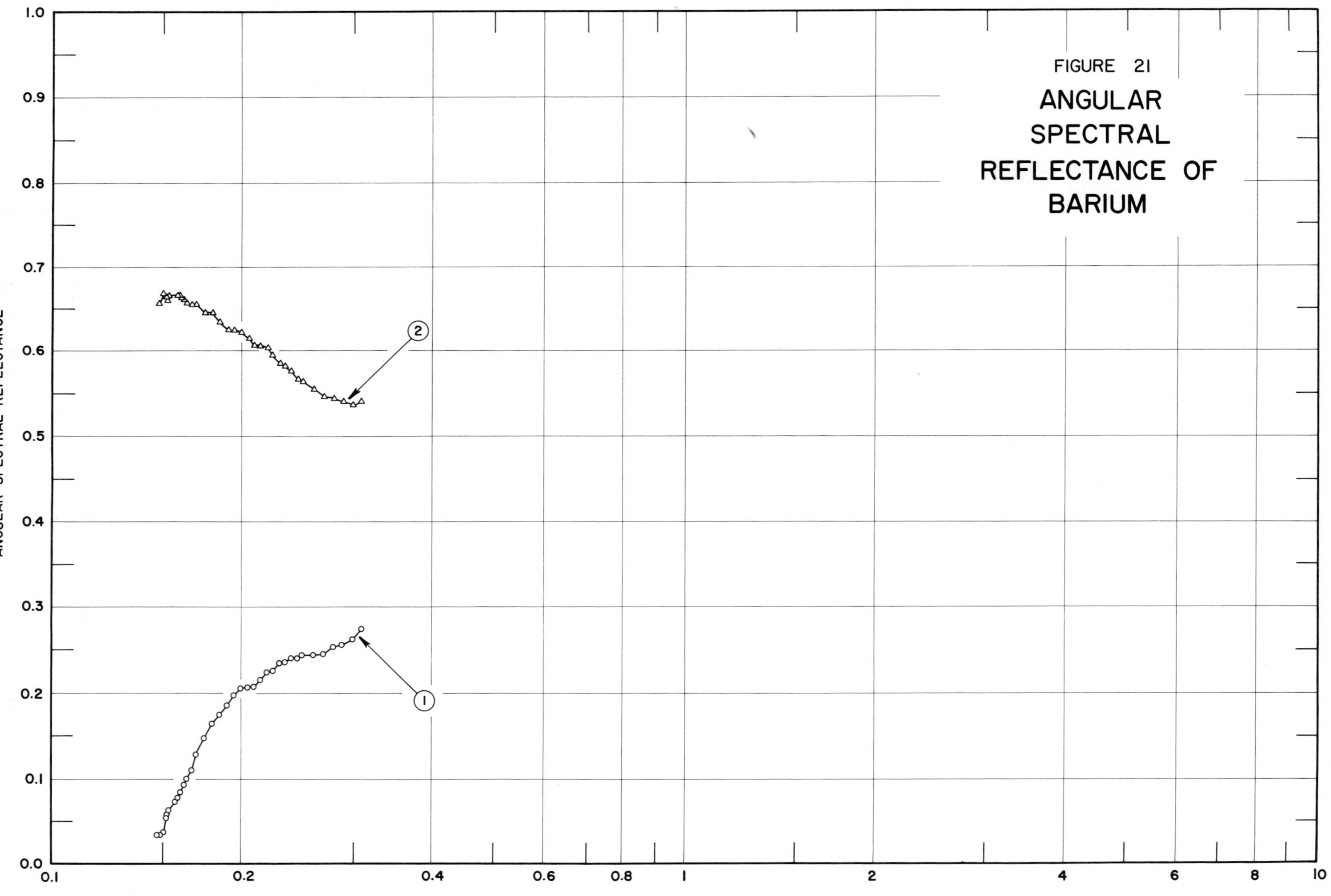

FIGURE 21
ANGULAR SPECTRAL REFLECTANCE OF BARIUM
ANGULAR SPECTRAL REFLECTANCE
WAVELENGTH, MICRONS
0.0
0.1
0.2
0.3
0.4
0.5
0.6
0.7
0.8
0.9
1.0
0.1
0.2
0.4
0.6
0.8
1
2
4
6
8
10
1
2

SPECIFICATION TABLE NO. 21 ANGULAR SPECTRAL REFLECTANCE OF BARIUM

Curve No.	Ref. No.	Year	Temperature T	Wavelength Range, μ	Geometry θ	θ'	ω'	Reported Error, %	Composition (weight perceint), Specifications and Remarks
1	309	1966	298	0.1483-0.3100	17.5°	17.5°			Evaporated film (2000 to 3000 Å thick); evaporation time < 5 sec; evaporated and measured at 5×10^{-10} mm Hg.
2	309	1966	298	0.1483-0.3092	72.5°	72.5°			Different sample, same as above specimen and conditions.

DATA TABLE NO. 21 ANGULAR SPECTRAL REFLECTANCE OF BARIUM

[Wavelength, λ, μ; Reflectance, ρ; Temperature, T, K]

λ	ρ
CURVE 1 T = 298	
0.1483	0.034
0.1490	0.034
0.1501	0.038
0.1520	0.054
0.1527	0.059
0.1546	0.063
0.1576	0.073
0.1590	0.078
0.1610	0.085
0.1621	0.094
0.1640	0.100
0.1671	0.111
0.1706	0.129
0.1754	0.148
0.1800	0.164
0.1854	0.174
0.1908	0.186
0.1956	0.198
0.2000	0.204
0.2056	0.206
0.2102	0.207
0.2149	0.216
0.2206	0.225
0.2242	0.227
0.2313	0.233
0.2348	0.235
0.2403	0.240
0.2455	0.240
0.2495	0.244
0.2594	0.244
0.2719	0.246
0.2799	0.254
0.2897	0.257
0.3002	0.262
0.3100	0.274
CURVE 2 T = 298	
0.1483	0.657
0.1494	0.657*
0.1501	0.669
0.1520	0.664
0.1531	0.660
0.1548	0.667
0.1582	0.667
CURVE 2 (cont.) T = 298	
0.1596	0.667
0.1612	0.664
0.1627	0.663
0.1642	0.657
0.1673	0.655
0.1703	0.655
0.1751	0.647
0.1802	0.647
0.1848	0.635
0.1911	0.627
0.1956	0.627
0.2003	0.623
0.2060	0.617
0.2102	0.606
0.2149	0.606
0.2206	0.602
0.2242	0.595
0.2313	0.587
0.2348	0.582
0.2394	0.577
0.2451	0.568
0.2500	0.563
0.2600	0.555
0.2713	0.548
0.2812	0.545
0.2904	0.541
0.3002	0.538
0.3092	0.541

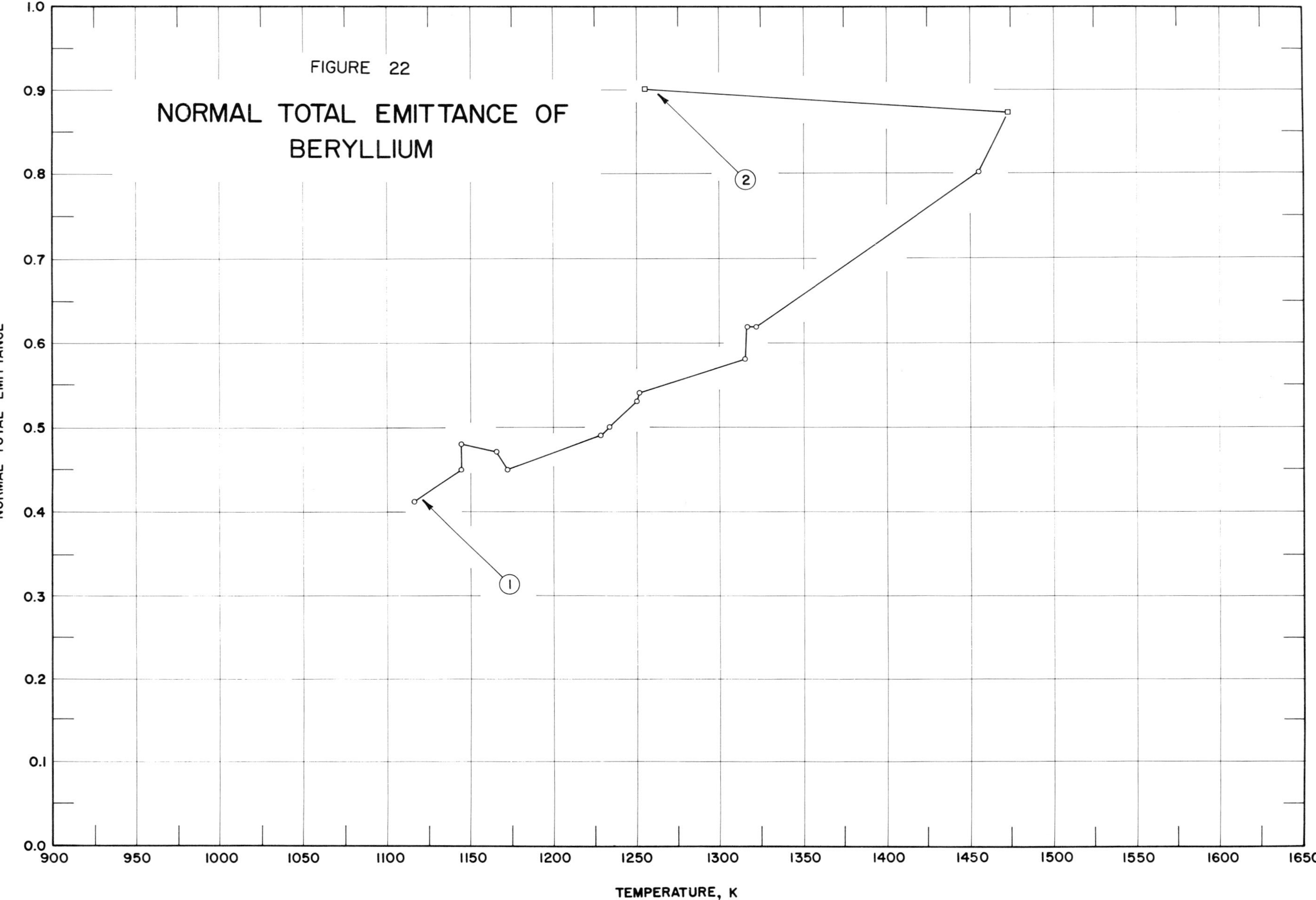

FIGURE 22
NORMAL TOTAL EMITTANCE OF BERYLLIUM
1
2
NORMAL TOTAL EMITTANCE
0.0
0.1
0.2
0.3
0.4
0.5
0.6
0.7
0.8
0.9
1.0
TEMPERATURE, K
900
950
1000
1050
1100
1150
1200
1250
1300
1350
1400
1450
1500
1550
1600
1650

SPECIFICATION TABLE NO. 22 — NORMAL TOTAL EMITTANCE OF BERYLLIUM

Curve No.	Ref. No.	Year	Temperature Range, K	Geometry θ'	Reported Error, %	Composition (weight percent), Specifications and Remarks
1	18	1963	1116-1473	$\sim 0^\circ$		Heating cycle.
2	18	1963	1473-1255	$\sim 0^\circ$		Cooling cycle for above specimen.

DATA TABLE NO. 22 NORMAL TOTAL EMITTANCE OF BERYLLIUM

[Temperature, T, K; Emittance, ϵ]

T	ϵ
CURVE 1	
1116	0.412
1144	0.450
1144	0.480
1166	0.470
1172	0.450
1228	0.490
1233	0.500
1250	0.530
1253	0.540
1314	0.580
1316	0.620
1322	0.620
1455	0.800
1473	0.870*
CURVE 2	
1473	0.870
1255	0.900

*Not shown on plot

SPECIFICATION TABLE NO. 23 NORMAL SPECTRAL EMITTANCE OF BERYLLIUM

Curve No.	Ref. No.	Year	Wavelength μ	Temperature Range, K	Geometry θ'	Reported Error, %	Composition (weight percent), Specifications and Remarks
1	19	1914	0.65	1743	$\sim 0°$	1	Film; tungsten substrate; melted in hydrogen then oxidized in air by heating; measured in air; Pt reference ($\epsilon = 0.33$ for $\lambda = 0.650\ \mu$ at all temp).

DATA TABLE NO. 23 NORMAL SPECTRAL EMITTANCE OF BERYLLIUM

[Temperature, T, K; Emittance, ϵ; Wavelength, λ, μ]

T	ϵ
CURVE 1*	
$\lambda = 0.65\mu$	
1743	0.37

* Not shown on plot

SPECIFICATION TABLE NO. 24 NORMAL SPECTRAL EMITTANCE OF BERYLLIUM

Curve No.	Ref. No.	Year	Temperature K	Wavelength Range, μ	Geometry θ'	Reported Error, %	Composition (weight percent), Specifications and Remarks
1	19	1914	< M. P	0.55-0.65	$\sim 0^\circ$	1	Film; tungsten substrate; measured in hydrogen; platinum reference ($\epsilon = 0.33$ for $\lambda = 0.65\ \mu$ and $\epsilon = 0.38$ for $\lambda = 0.547\ \mu$ at all temp).

DATA TABLE NO. 24 NORMAL SPECTRAL EMITTANCE OF BERYLLIUM

[Wavelength, λ, μ; Emittance, ϵ; Temperature, T, K]

λ	ϵ
CURVE 1*	
T = <M. P.	
0.55	0.61
0.65	0.61

*Not shown on plot

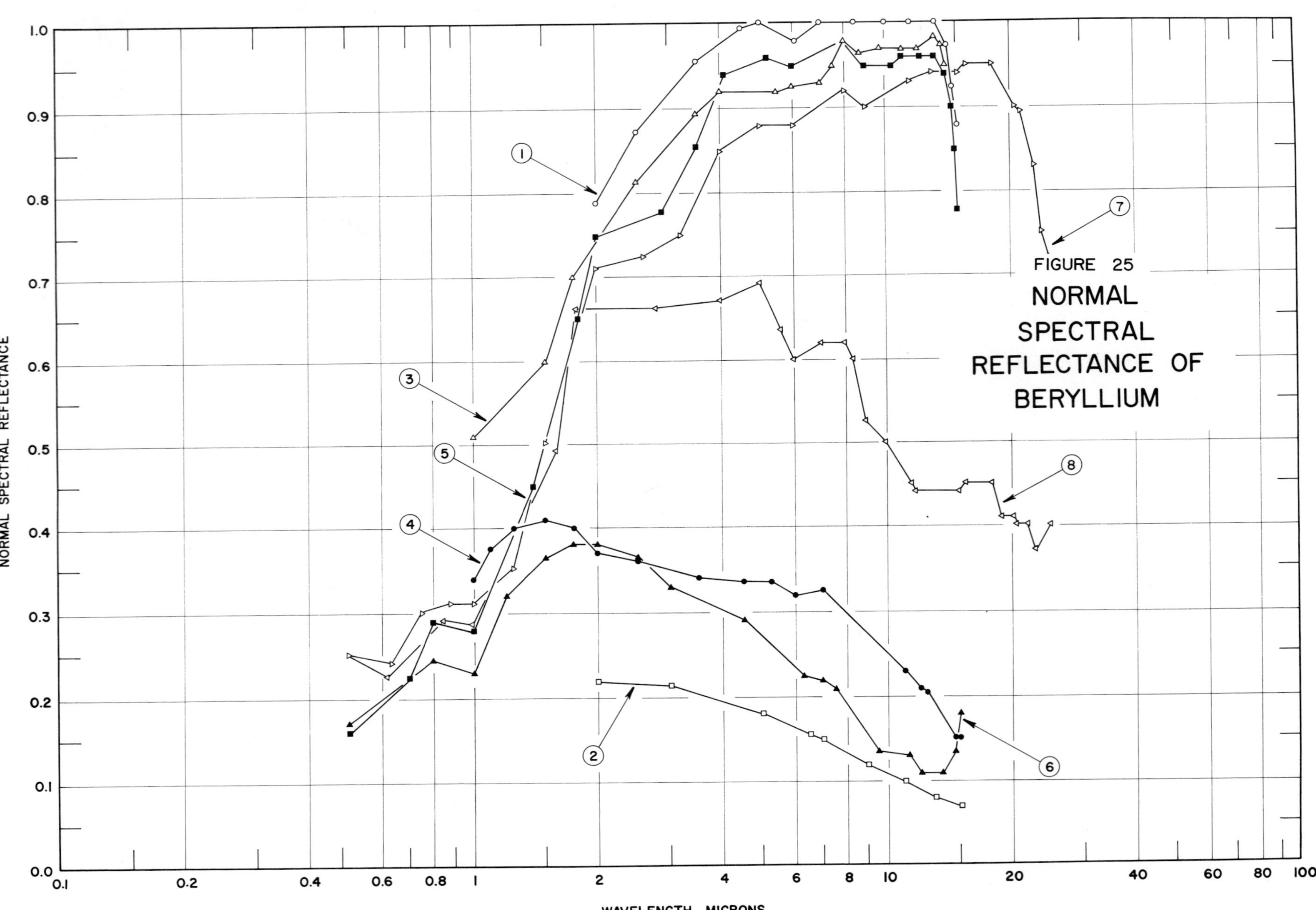
FIGURE 25
NORMAL
SPECTRAL
REFLECTANCE OF
BERYLLIUM
NORMAL SPECTRAL REFLECTANCE
WAVELENGTH, MICRONS
1.0
0.9
0.8
0.7
0.6
0.5
0.4
0.3
0.2
0.1
0.0
0.1
0.2
0.4
0.6
0.8
1
2
4
6
8
10
20
40
60
80
100
1
2
3
4
5
6
7
8

SPECIFICATION TABLE NO. 25 NORMAL SPECTRAL REFLECTANCE OF BERYLLIUM

Curve No.	Ref. No.	Year	Temperature K	Wavelength Range, μ	Geometry θ	θ'	ω'	Reported Error, %	Composition (weight percent), Specifications and Remarks
1	86	1961	~ 322	2.00-15.00	~0°		2π	< 2	As received; data extracted from smooth curve; hohlraum at 523 K; converted from R (2π, 0°).
2	86	1961	~ 322	2.00-15.00	~0°		2π	< 2	Above specimen and conditions; diffuse component only.
3	86	1961	~ 322	1.00-14.00	~0°		2π	< 2	Different sample, same as Curve 1 specimen and conditions except hohlraum at 773 K.
4	86	1961	~ 322	1.00-15.00	~0°		2π	< 2	Above specimen and conditions; diffuse component only.
5	86	1961	~ 322	0.50-15.00	~0°		2π	< 2	Different sample, same as Curve 1 specimen and conditions except hohlraum at 1273 K.
6	86	1961	~ 322	0.50-15.00	~0°		2π	< 2	Above specimen and conditions; diffuse component only.
7	153	1962	322	0.50-25.00	~0°		2π	< 2	Beryllium Extrusion #30, Brush S - 100B; as received; cleaned; data extracted from smooth curve; hohlraum at 1273 K; converted from R (2π, 0°).
8	153	1962	322	0.50-25.00	~0°		2π	< 2	Above specimen and conditions; diffuse component only.
9	153	1962	322	0.50-25.00	~0°		2π	< 2	Different sample, same as Curve 7 specimen and conditions except grit blasted using 60 grit silicon carbide with air pressure of 110 to 120 psi for 30 to 45 sec.
10	153	1962	322	0.50-25.00	~0°		2π	< 2	Above specimen and conditions; diffuse component only.
11	153	1962	322	0.50-25.00	~0°		2π	< 2	Different sample, same as Curve 7 specimen and conditions except exposed to vacuum (< 4 x 10^{-8} mm Hg) for 24 hrs.
12	153	1962	322	0.50-25.00	~0°		2π	< 2	Above specimen and conditions; diffuse component only.
13	153	1962	322	0.50-25.00	~0°		2π	<2	Different sample, same as Curve 7 specimen and conditions except X-ray exposed in vacuum (4 x 10^{-8} mm Hg) for 24 hrs.
14	153	1962	322	0.50-25.00	~0°		2π	< 2	Above specimen and conditions; diffuse component only.

DATA TABLE NO. 25 NORMAL SPECTRAL REFLECTANCE OF BERYLLIUM

[Wavelength, λ, μ; Reflectance, ρ; Temperature, T, K]

λ	ρ
CURVE 1 T = ~322	
2.00	0.790
2.50	0.875
3.50	0.955
4.50	0.995
5.00	1.000
6.10	0.980
7.00	1.000
8.50	1.000
10.00	1.000
11.50	1.000
13.25	1.000
14.10	0.975
14.65	0.925
15.00	0.880
CURVE 2 T = ~322	
2.00	0.220
3.00	0.215
5.00	0.180
6.50	0.155
7.00	0.150
9.00	0.120
11.00	0.100
13.00	0.080
15.00	0.070
CURVE 3 T = ~322	
1.00	0.510
1.50	0.600
1.75	0.700
2.50	0.815
3.50	0.895
4.00	0.920
5.50	0.920
6.00	0.925
7.00	0.930
7.50	0.950
8.00	0.980
8.75	0.965
CURVE 3 (cont.)	
9.75	0.970
11.00	0.970
12.00	0.970
13.25	0.985
13.75	0.975
14.00	0.950
CURVE 4 T = ~322	
1.00	0.340
1.10	0.375
1.25	0.400
1.50	0.410
1.75	0.400
2.00	0.370
2.50	0.360
3.50	0.340
4.50	0.335
5.25	0.335
6.00	0.320
7.00	0.325
11.00	0.230
12.00	0.210
12.50	0.205
14.50	0.150
15.00	0.150
CURVE 5 T = ~322	
0.50	0.160
0.70	0.225
0.80	0.290
1.00	0.280
1.40	0.450
1.80	0.650
2.00	0.750
2.90	0.780
3.50	0.855
4.10	0.940
5.20	0.960
6.00	0.950
8.00	0.980*
CURVE 5 (cont.)	
9.00	0.950
10.40	0.950
11.00	0.960
12.20	0.960
13.25	0.960
14.00	0.940
14.50	0.900
14.85	0.850
15.00	0.780
CURVE 6 T = ~322	
0.50	0.170
0.80	0.245
1.00	0.230
1.20	0.320
1.50	0.365
1.75	0.380
2.00	0.380
2.50	0.365
3.00	0.330
4.50	0.290
6.25	0.225
7.00	0.220
7.50	0.210
9.50	0.135
11.25	0.130
12.00	0.110
13.50	0.110
14.50	0.135
14.75	0.150*
15.00	0.180
CURVE 7 T = 322	
0.50	0.250
0.63	0.240
0.75	0.300
0.88	0.310
1.00	0.310
1.25	0.350
1.50	0.500
CURVE 7 (cont.)	
2.00	0.710
2.60	0.725
3.20	0.750
4.00	0.850
5.00	0.880
6.00	0.880
8.00	0.920
9.00	0.900
11.50	0.930
13.00	0.940
15.00	0.940
15.70	0.950
18.10	0.950
20.50	0.900
21.20	0.895
23.00	0.830
24.00	0.750
25.00	0.720
CURVE 8 T = 322	
0.50	0.250*
0.62	0.225
0.85	0.290
1.00	0.285
1.60	0.490
2.80	0.660
4.00	0.670
5.00	0.690
5.60	0.635
6.00	0.600
7.00	0.620
8.00	0.620
8.40	0.600
9.00	0.525
10.00	0.500
11.50	0.450
11.70	0.440
15.00	0.440
15.50	0.450
18.00	0.450
19.00	0.410
20.30	0.410
CURVE 8 (cont.)	
20.60	0.400
22.00	0.400
23.00	0.370
25.00	0.400
CURVE 9* T = 322	
0.50	0.140
0.70	0.170
0.80	0.200
0.88	0.240
1.00	0.250
1.50	0.400
2.00	0.525
4.00	0.700
5.10	0.720
8.00	0.850
9.50	0.850
12.00	0.900
14.00	0.900
15.20	0.880
16.00	0.930
17.00	0.950
19.10	0.950
21.10	0.930
23.10	0.850
24.00	0.760
25.00	0.700
CURVE 10* T = 322	
0.50	0.180
0.60	0.200
0.75	0.230
1.05	0.230
1.60	0.390
1.90	0.475
2.20	0.510
4.00	0.610
5.00	0.630
6.10	0.620
7.00	0.630
CURVE 10* (cont.)	
8.00	0.620
9.10	0.575
11.00	0.550
13.00	0.520
14.00	0.520
15.00	0.540
15.80	0.520
17.00	0.520
17.50	0.510
18.50	0.475
19.00	0.450
20.00	0.440
21.00	0.430
21.70	0.430
23.00	0.410
24.00	0.410
25.00	0.440
CURVE 11* T = 322	
0.50	0.260
0.78	0.300
0.88	0.315
1.00	0.325
1.35	0.450
1.60	0.550
2.00	0.640
2.40	0.550
2.50	0.400
2.60	0.250
3.00	0.200
3.20	0.210
2.40	0.400
3.60	0.550
4.20	0.650
5.00	0.700
5.50	0.650
5.60	0.550
6.00	0.470
6.60	0.550
7.00	0.650
8.00	0.700
8.50	0.600
8.60	0.400
CURVE 11* (cont.)	
9.00	0.370
10.00	0.450
11.00	0.500
12.70	0.470
15.00	0.530
16.00	0.620
17.10	0.620
18.00	0.650
19.30	0.650
21.00	0.700
22.00	0.680
23.00	0.700
23.90	0.635
25.00	0.630
CURVE 12* T = 322	
0.50	0.215
0.80	0.260
0.85	0.285
1.00	0.285
1.50	0.450
2.00	0.590
2.10	0.540
2.20	0.400
2.30	0.240
3.00	0.170
3.70	0.400
4.00	0.550
5.00	0.650
6.00	0.450
7.00	0.550
8.00	0.510
8.50	0.300
9.00	0.270
10.00	0.270
11.00	0.310
12.20	0.265
13.20	0.265
13.50	0.270
16.00	0.440
17.00	0.450
18.00	0.460
CURVE 12* (cont.)	
19.00	0.440
19.80	0.460
22.00	0.460
23.00	0.470
24.00	0.460
25.00	0.480
CURVE 13* T = 322	
0.50	0.175
0.68	0.250
0.80	0.265
0.85	0.370
1.00	0.380
1.25	0.440
1.50	0.600
1.75	0.760
2.50	0.810
3.10	0.825
4.00	0.930
5.00	0.940
6.00	0.900
7.00	0.910
8.00	0.950
9.00	0.930
10.00	0.940
12.00	0.940
12.60	0.950
15.20	0.950
16.00	0.975
20.00	0.975
25.00	0.975
CURVE 14* T = 322	
0.50	0.175
0.70	0.235
0.80	0.240
0.80	0.310
1.00	0.320
1.40	0.450
1.50	0.550

*Not shown on plot

DATA TABLE NO. 25 (continued)

λ	ρ
CURVE 14*(cont.)	
2.00	0.710
3.30	0.715
3.50	0.775
4.00	0.800
5.10	0.800
6.00	0.700
7.00	0.750
7.50	0.750
9.30	0.705
10.01	0.695
12.60	0.665
14.00	0.650
15.00	0.670
18.10	0.670
19.00	0.610
20.00	0.615
21.10	0.590
22.00	0.610
23.10	0.580
25.00	0.615

*Not shown on plot

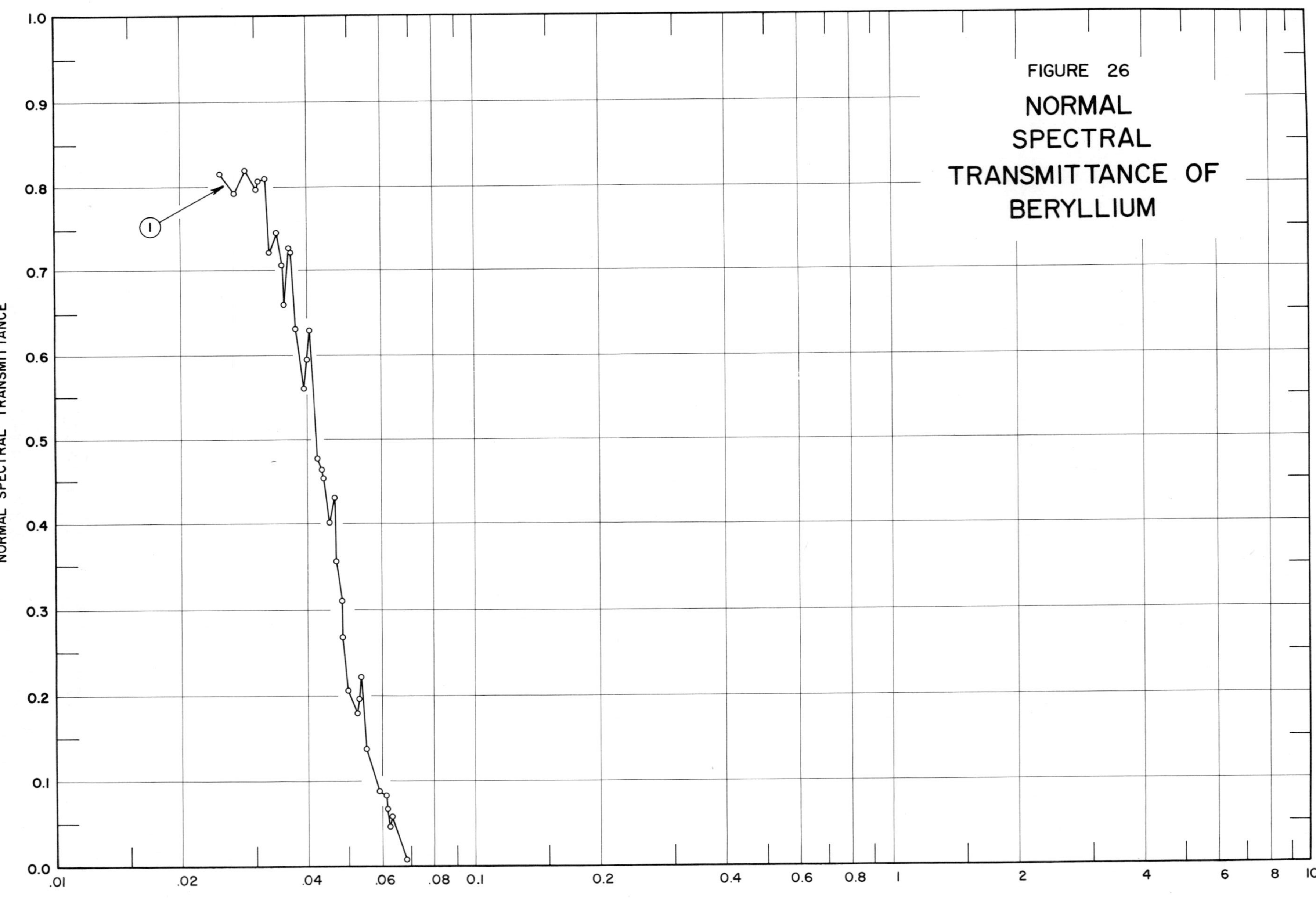
FIGURE 26
NORMAL SPECTRAL TRANSMITTANCE OF BERYLLIUM
NORMAL SPECTRAL TRANSMITTANCE
WAVELENGTH, MICRONS
1.0
0.9
0.8
0.7
0.6
0.5
0.4
0.3
0.2
0.1
0.0
.01
.02
.04
.06
.08
0.1
0.2
0.4
0.6
0.8
1
2
4
6
8
10
1

SPECIFICATION TABLE NO. 26 NORMAL SPECTRAL TRANSMITTANCE OF BERYLLIUM

Curve No.	Ref. No.	Year	Temperature K	Wavelength Range, μ	Geometry θ	ϑ'	ω'	Reported Error, %	Composition (weight percent), Specifications and Remarks
1	308	1965	298	0.0250-0.0681	$\sim 0^\circ$	$\sim 0^\circ$			Al backed Be film (875 ± 100 Å thick); Al evaporated on glass slide, then Be evaporated over that; evaporations at 2×10^{-5} mm Hg in 10-15 sec; error ~10% for $\lambda > 0.035\ \mu$ and < 25% for $\lambda < 0.035\ \mu$; data corrected for transmittance of Al and glass.

DATA TABLE NO. 26 NORMAL SPECTRAL TRANSMITTANCE OF BERYLLIUM

[Wavelength, λ, μ; Transmittance, τ; Temperature, T, K]

λ	τ
CURVE 1	
T = 298	
0.0250	0.817
0.0270	0.792
0.0288	0.820
0.0303	0.798
0.0308	0.806
0.0320	0.810
0.0328	0.721
0.0340	0.748
0.0351	0.706
0.0355	0.660
0.0364	0.729
0.0368	0.722
0.0379	0.631
0.0394	0.560
0.0400	0.595
0.0404	0.630
0.0423	0.477
0.0432	0.462
0.0437	0.452
0.0454	0.404
0.0463	0.430
0.0471	0.355
0.0481	0.310
0.0484	0.267
0.0500	0.206
0.0523	0.179
0.0528	0.196
0.0537	0.221
0.0551	0.137
0.0596	0.087
0.0611	0.081
0.0614	0.066
0.0623	0.046
0.0633	0.057
0.0681	0.008

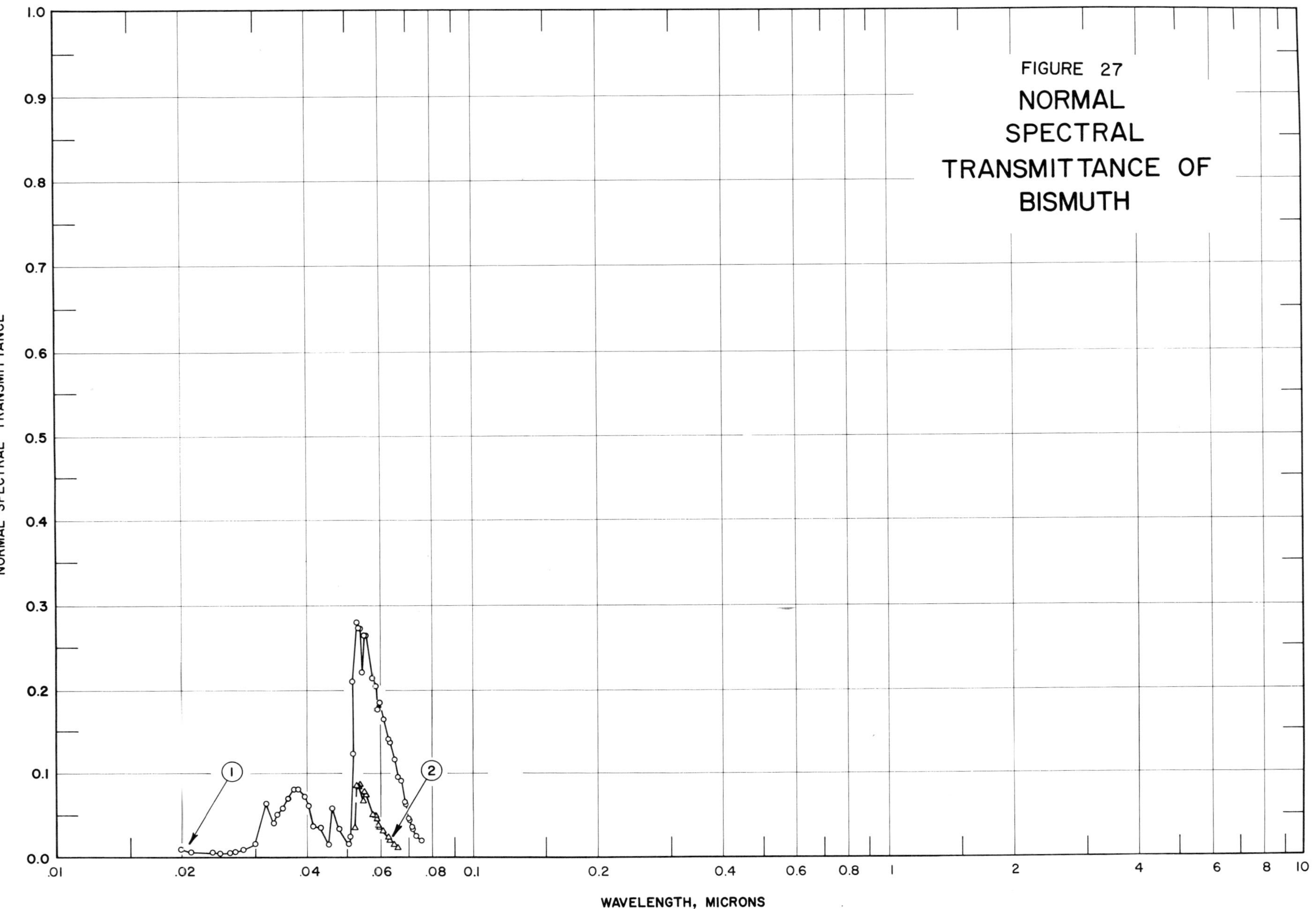

FIGURE 27 NORMAL SPECTRAL TRANSMITTANCE OF BISMUTH

SPECIFICATION TABLE NO. 27 NORMAL SPECTRAL TRANSMITTANCE OF BISMUTH

Curve No.	Ref. No.	Year	Temperature K	Wavelength Range, μ	Geometry θ θ' ω'	Reported Error, %	Composition (weight percent), Specifications and Remarks
1	290	1965	298	0.0200-0.0748	$\sim 0^\circ \sim 0^\circ$		Unbacked film (950 Å thick); evaporated at several hundred Å sec^{-1}; measured in vacuum ($< 10^{-5}$ mm Hg); slight oxide layer.
2	290	1965	298	0.0520-0.0659	$\sim 0^\circ \sim 0^\circ$		Different sample, same as above specimen and conditions except 1990 Å thick.

DATA TABLE NO. 27 NORMAL SPECTRAL TRANSMITTANCE OF BISMUTH

[Wavelength, λ, μ; Transmittance, τ; Temperature, T, K]

λ	τ
CURVE 1 T = 298	
0.0200	0.0100
0.0211	0.0083
0.0236	0.0076
0.0246	0.0062
0.0262	0.0071
0.0269	0.0085
0.0281	0.0100
0.0301	0.0187
0.0319	0.0634
0.0332	0.0412
0.0340	0.0520
0.0350	0.0597
0.0360	0.0700
0.0371	0.0800
0.0380	0.0807
0.0396	0.0718
0.0406	0.0610
0.0416	0.0387
0.0430	0.0371
0.0449	0.0171
0.0459	0.0589
0.0475	0.0347
0.0504	0.0173
0.0506	0.0249
0.0516	0.123
0.0516	0.210
0.0523	0.280
0.0527	0.272
0.0535	0.275
0.0541	0.220
0.0547	0.264
0.0552	0.264
0.0574	0.214
0.0583	0.206
0.0593	0.177
0.0598	0.184
0.0608	0.165
0.0622	0.141
0.0629	0.137
0.0643	0.117
0.0659	0.0957
0.0671	0.0804
0.0681	0.0667
0.0685	0.0637
CURVE 1 (cont.)	
0.0699	0.0481
0.0703	0.0457
0.0714	0.0371
0.0718	0.0351
0.0728	0.0271
0.0748	0.0209
CURVE 2 T = 298	
0.0520	0.0374
0.0523	0.0857
0.0531	0.0847
0.0537	0.0861
0.0543	0.0687
0.0547	0.0783
0.0554	0.0750
0.0574	0.0513
0.0581	0.0513
0.0586	0.0481
0.0593	0.0394
0.0598	0.0381
0.0609	0.0324
0.0625	0.0251
0.0630	0.0217
0.0646	0.0160
0.0659	0.0121

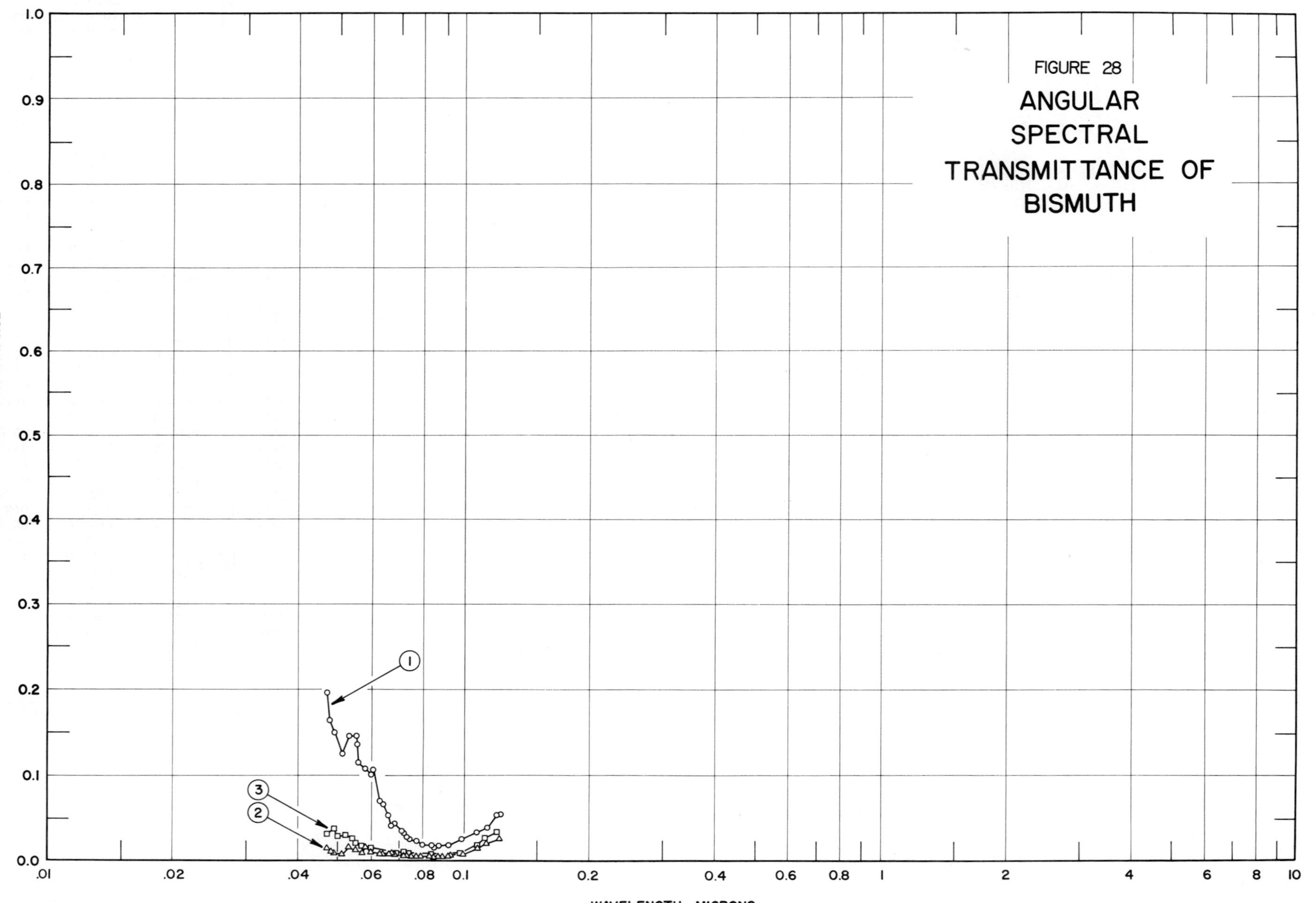
FIGURE 28
ANGULAR
SPECTRAL
TRANSMITTANCE OF
BISMUTH
ANGULAR SPECTRAL TRANSMITTANCE
1.0
0.9
0.8
0.7
0.6
0.5
0.4
0.3
0.2
0.1
0.0
.01
.02
.04
.06
.08
0.1
0.2
0.4
0.6
0.8
1
2
4
6
8
10
WAVELENGTH, MICRONS
1
2
3

SPECIFICATION TABLE NO. 28 ANGULAR SPECTRAL TRANSMITTANCE OF BISMUTH

Curve No.	Ref. No.	Year	Temperature K	Wavelength Range, μ	Geometry θ	θ'	ω'	Reported Error, %	Composition (weight percent), Specifications and Remarks
1	348	1965	298	0.0474-0.1214	0°	0°			Unbacked foil (425 Å thick); evaporated in vacuum (3×10^{-6} mm Hg); exposed to air; condensed spark discharge in argon light source.
2	348	1965	298	0.0476-0.1217	60°	60°			Above specimen and conditions except p-polarization is dominant; data uncorrected for partial polarization effects.
3	348	1965	298	0.0470-0.1218	60°	60°			Above specimen and conditions except s-polarization is dominant.

DATA TABLE NO. 28 ANGULAR SPECTRAL TRANSMITTANCE OF BISMUTH

[Wavelength, λ, μ; Transmittance, τ; Temperature, T, K]

CURVE 1
T = 298

λ	τ
0.0474	0.1982
0.0479	0.1641
0.0491	0.1500
0.0513	0.1259
0.0532	0.1466
0.0541	0.1466
0.0556	0.1352
0.0574	0.1169
0.0581	0.1094
0.0606	0.1009
0.0617	0.1062
0.0635	0.0700
0.0644	0.0641
0.0666	0.0515
0.0675	0.0415
0.0680	0.0440
0.0688	0.0434*
0.0698	0.0414*
0.0719	0.0352
0.0726	0.0323
0.0732	0.0296
0.0740	0.0285
0.0746	0.0285*
0.0770	0.0239
0.0775	0.0187*
0.0795	0.0186*
0.0800	0.0190
0.0801	0.0198*
0.0809	0.0185*
0.0829	0.0181
0.0852	0.0160
0.0871	0.0176
0.0879	0.0179*
0.0888	0.0177*
0.0922	0.0183
0.0933	0.0186*
0.0990	0.0207
0.1085	0.0333
0.1138	0.0399
0.1202	0.0525
0.1214	0.0524

CURVE 2
T = 298

λ	τ
0.0476	0.0136
0.0484	0.0101
0.0490	0.0090
0.0510	0.0087
0.0530	0.0156
0.0538	0.0153*
0.0553	0.0117
0.0577	0.0096
0.0583	0.0106
0.0607	0.0100
0.0637	0.0074
0.0643	0.0078
0.0667	0.0073
0.0674	0.0071
0.0685	0.0068
0.0721	0.0055
0.0726	0.0056*
0.0731	0.0055
0.0739	0.0042*
0.0745	0.0036*
0.0752	0.0036
0.0770	0.0034
0.0799	0.0030
0.0836	0.0030
0.0850	0.0026
0.0870	0.0035
0.0879	0.0033*
0.0886	0.0034
0.0921	0.0039
0.0930	0.0042
0.0991	0.0073
0.1084	0.0135
0.1133	0.0202
0.1217	0.0262

CURVE 3
T = 298

λ	τ
0.0470	0.0319
0.0477	0.0288*
0.0490	0.0308

CURVE 3 (cont.)

λ	τ
0.0509	0.0296
0.0529	0.0333
0.0541	0.0273
0.0556	0.0201*
0.0575	0.0174
0.0581	0.0183*
0.0605	0.0141
0.0638	0.0099*
0.0643	0.0096*
0.0667	0.0083*
0.0681	0.0079*
0.0691	0.0077
0.0721	0.0063
0.0731	0.0063*
0.0740	0.0061
0.0743	0.0057*
0.0774	0.0051*
0.0796	0.0050*
0.0836	0.0051*
0.0849	0.0047
0.0872	0.0054
0.0885	0.0055*
0.0887	0.0054*
0.0921	0.0065*
0.0932	0.0070*
0.0989	0.0081
0.1087	0.0185
0.1136	0.0271
0.1218	0.0323

*Not shown on plot

SPECIFICATION TABLE NO. 29 HEMISPHERICAL TOTAL EMITTANCE OF CADMIUM

Curve No.	Ref. No.	Year	Temperature Range, K	Reported Error, %	Composition (weight percent), Specifications and Remarks
1	3	1955	76	5	A very mossy and smeared-looking plated surface; measured in vacuum (10^{-6} to 10^{-7} mm Hg); emittance for 300 K black body incident radiation; authors assumed $\alpha = \epsilon$.

DATA TABLE NO. 29 HEMISPHERICAL TOTAL EMITTANCE OF CADMIUM

[Temperature, T, K; Emittance, ∈]

T	∈
CURVE 1*	
76	0.03

*Not shown on plot

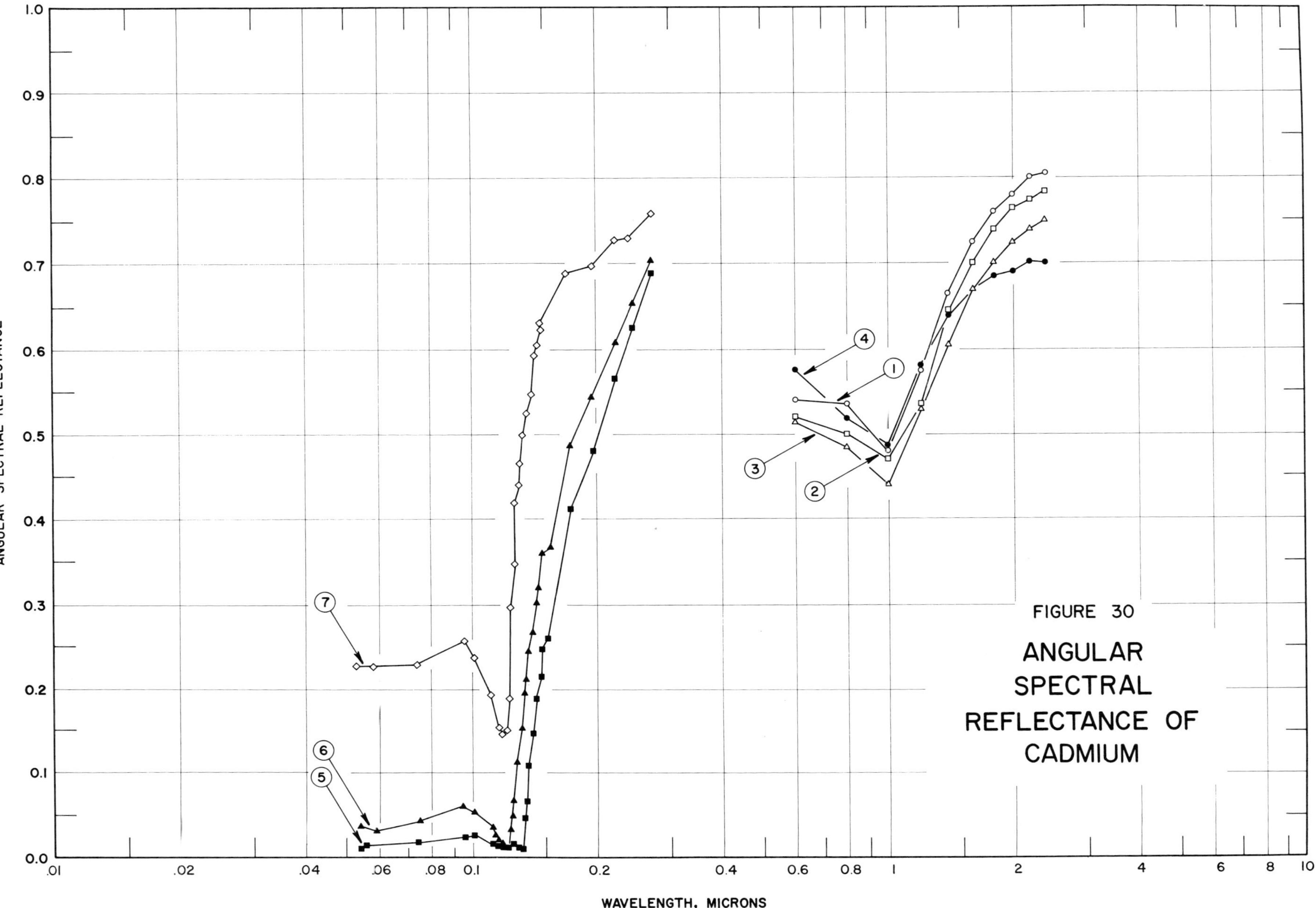

FIGURE 30
ANGULAR SPECTRAL REFLECTANCE OF CADMIUM

SPECIFICATION TABLE NO. 30 ANGULAR SPECTRAL REFLECTANCE OF CADMIUM

Curve No.	Ref. No.	Year	Temperature K	Wavelength Range, μ	Geometry θ	θ'	ω'	Reported Error, %	Composition (weight percent), Specifications and Remarks
1	141	1948	298	0.6-2.4	50°	50°			Mirror surface prepared by vacuum evaporation; data extracted from smooth curve, incident beam polarized parallel to the plane of incidence.
2	141	1948	298	0.6-2.4	60°	60°			Above specimen and conditions, incident beam polarized parallel to the plane of incidence.
3	141	1948	298	0.6-2.4	70°	70°			Above specimen and conditions, incident beam polarized parallel to the plane of incidence.
4	141	1948	298	0.6-2.4	80°	80°			Above specimen and conditions, incident beam polarized parallel to the plane of incidence.
5	310	1966	298	0.054-0.271	20°	20°			Evaporated film (1200 Å thick); 99.8 pure; evaporated onto a Sn coated glass substrate at a distance of 6 cm; exposed to air for 5 to 15 min.
6	310	1966	298	0.054-0.271	45°	45°			Different sample, same as above specimen and conditions.
7	310	1966	298	0.053-0.272	70°	70°			Different sample, same as above specimen and conditions.

DATA TABLE NO. 30 ANGULAR SPECTRAL REFLECTANCE OF CADMIUM

[Wavelength, λ, μ; Reflectance, ρ; Temperature, T, K]

λ	ρ
CURVE 1 T = 298	
0.6	0.540
0.8	0.535
1.0	0.480
1.2	0.575
1.4	0.665
1.6	0.725
1.8	0.760
2.0	0.780
2.2	0.800
2.4	0.805
CURVE 2 T = 298	
0.6	0.520
0.8	0.500
1.0	0.470
1.2	0.535
1.4	0.645
1.6	0.700
1.8	0.740
2.0	0.765
2.2	0.775
2.4	0.785
CURVE 3 T = 298	
0.6	0.515
0.8	0.485
1.0	0.440
1.2	0.530
1.4	0.605
1.6	0.670
1.8	0.700
2.0	0.725
2.2	0.740
2.4	0.750
CURVE 4 T = 298	
0.6	0.575
0.8	0.518
1.0	0.485
1.2	0.580
1.4	0.640
1.6	0.670*
1.8	0.685
2.0	0.690
2.2	0.700
2.4	0.700
CURVE 5 T = 298	
0.054	0.010
0.059	0.013
0.074	0.018
0.096	0.023
0.101	0.026
0.112	0.016
0.115	0.013
0.118	0.011
0.122	0.011
0.125	0.016
0.128	0.012
0.131	0.010
0.134	0.048
0.135	0.066
0.137	0.108
0.140	0.146
0.143	0.188
0.147	0.215
0.149	0.248
0.152	0.260
0.174	0.411
0.198	0.480
0.222	0.565
0.247	0.626
0.271	0.688
CURVE 6 T = 298	
0.054	0.037
0.059	0.031
0.075	0.042
0.095	0.060
0.102	0.051
0.112	0.036
0.113	0.027
0.115	0.020
0.117	0.018
0.123	0.014*
0.123	0.033
0.124	0.049
0.125	0.066
0.128	0.111
0.131	0.151
0.133	0.195
0.134	0.211
0.137	0.244
0.140	0.268
0.143	0.301
0.145	0.320
0.149	0.360
0.155	0.369
0.174	0.486
0.197	0.541
0.222	0.608
0.246	0.652
0.271	0.701
CURVE 7 T = 298	
0.053	0.227
0.058	0.227
0.074	0.229
0.096	0.256
0.101	0.236
0.111	0.191
0.116	0.151
0.117	0.144
0.121	0.148
0.122	0.186
0.123	0.297
CURVE 7 (cont.)	
0.126	0.348
0.127	0.418
0.130	0.439
0.131	0.465
0.134	0.499
0.137	0.522
0.140	0.547
0.143	0.592
0.145	0.605
0.149	0.622
0.148	0.631
0.173	0.689
0.198	0.696
0.222	0.728
0.247	0.730
0.272	0.758

*Not shown on plot

SPECIFICATION TABLE NO. 31 HEMISPHERICAL INTEGRATED ABSORPTANCE OF CADMIUM

Curve No.	Ref. No.	Year	Temperature Range, K	Reported Error, %	Composition (weight percent), Specifications and Remarks
1	3	1955	76	5	A very mossy and smeared-looking plated surface; measured in vacuum (10^{-6} to 10^{-7} mm Hg); absorptance for 300 K black body incident radiation.

DATA TABLE NO. 31 HEMISPHERICAL INTEGRATED ABSORPTANCE OF CADMIUM

[Temperature, T, K; Absorptance, α]

T	α
CURVE 1*	
76	0.03

*Not shown on plot

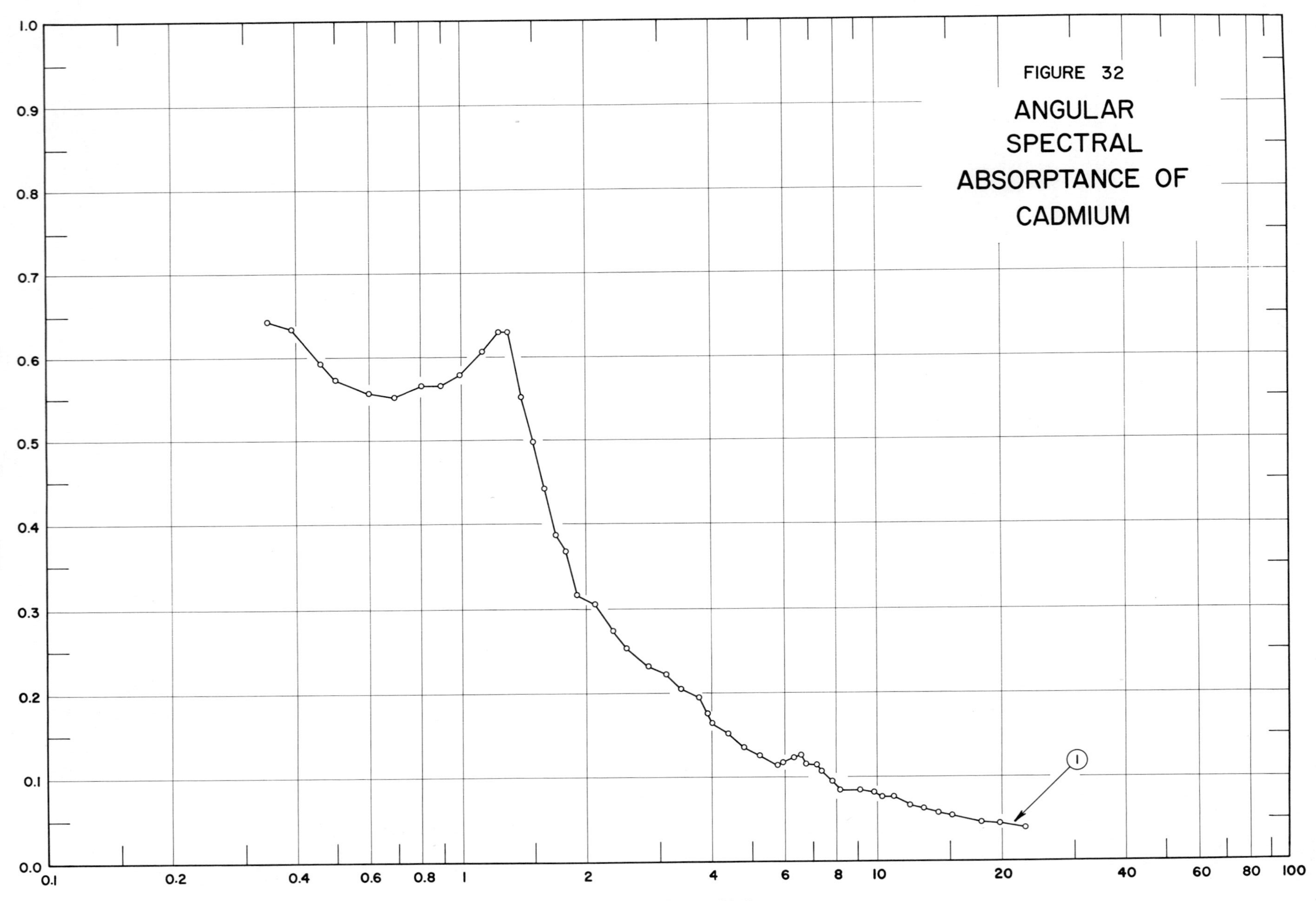

FIGURE 32
ANGULAR SPECTRAL ABSORPTANCE OF CADMIUM

SPECIFICATION TABLE NO. 32 ANGULAR SPECTRAL ABSORPTANCE OF CADMIUM

Curve No.	Ref. No.	Year	Temperature K	Wavelength Range, μ	Geometry θ	Reported Error, %	Composition (weight percent), Specifications and Remarks
1	225	1965	306	0.343-22.9	25°		Rolled plate cadmium from Belmont Smelting and Refining Works; measured in dry nitrogen; heated cavity at approx 1056 K with platinum reference; authors assumed $\alpha = 1 - R(2\pi, 25°)$.

DATA TABLE NO. 32 ANGULAR SPECTRAL ABSORPTANCE OF CADMIUM

[Wavelength, λ, μ; Absorptance, α; Temperature, T, K]

λ	α
CURVE 1 T = 306	
0.343	0.644
0.391	0.635
0.460	0.592
0.502	0.572
0.600	0.556
0.693	0.551
0.809	0.565
0.897	0.565
0.998	0.578
1.14	0.607
1.23	0.631
1.30	0.631
1.40	0.551
1.50	0.498
1.59	0.440
1.69	0.386
1.79	0.366
1.91	0.316
2.10	0.304
2.32	0.272
2.50	0.252
2.82	0.231
3.11	0.222
3.38	0.205
3.72	0.193
3.90	0.174
4.04	0.163
4.40	0.150
4.79	0.134
5.12	0.125
5.77	0.113
5.98	0.118
6.30	0.123
6.59	0.126
6.75	0.115
7.15	0.114
7.35	0.107
7.82	0.094
8.17	0.084
9.10	0.085
9.84	0.081
10.3	0.075
11.0	0.076
12.0	0.067
CURVE 1 (cont.)	
13.0	0.062
14.1	0.058
15.2	0.054
17.9	0.047
19.9	0.045
22.9	0.040

SPECIFICATION TABLE NO. 33 HEMISPHERICAL TOTAL EMITTANCE OF CHROMIUM

Curve No.	Ref. No.	Year	Temperature Range, K	Reported Error, %	Composition (weight percent), Specifications and Remarks
1	9	1960	77.4		Plated on monel; measured in vacuum (10^{-5} mm Hg).

DATA TABLE NO. 33 HEMISPHERICAL TOTAL EMITTANCE OF CHROMIUM

[Temperature, T, K; Emittance, ϵ]

T	ϵ
CURVE 1*	
77.4	0.084

*Not shown on plot

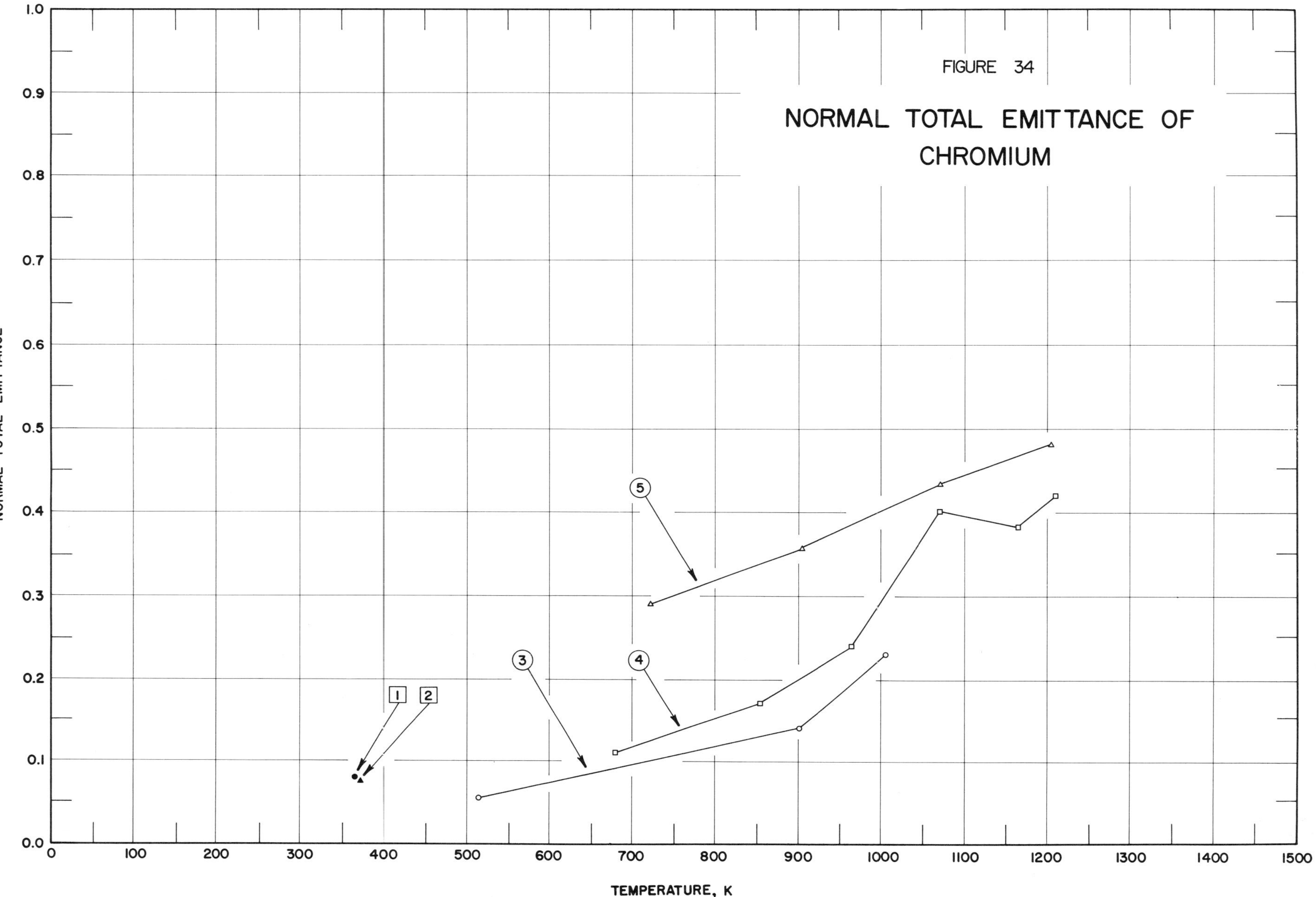

FIGURE 34

NORMAL TOTAL EMITTANCE OF CHROMIUM

SPECIFICATION TABLE NO. 34 NORMAL TOTAL EMITTANCE OF CHROMIUM

Curve No.	Ref. No.	Year	Temperature Range, K	Geometry θ'	Reported Error, %	Composition (weight percent), Specifications and Remarks
1	16	1937	367	$\sim 0°$	± 1.1	Plated on iron; polished.
2	15	1947	373	$\sim 0°$		Polished.
3	34	1957	516-1005	$\sim 0°$	±10	Pure; strip (0.005 in. thick); same results obtained for 4 different surface treatments (a. as received, b. cleaned with liquid detergent, c. polished, d. oxidized in air at red heat for 30 min.); measured in air (5×10^{-4} mm Hg); increasing temp, Cycle 1.
4	34	1957	680-1216	$\sim 0°$	±10	Above specimen and conditions, Cycle 2.
5	34	1957	722-1205	$\sim 0°$	±10	Above specimen and conditions, Cycle 3.

DATA TABLE NO. 34 NORMAL TOTAL EMITTANCE OF CHROMIUM

[Temperature, T, K; Emittance, ϵ]

T	ϵ
CURVE 1	
367	0.08
CURVE 2	
373	0.075
CURVE 3	
516	0.055
903	0.141
1005	0.230
CURVE 4	
680	0.110
855	0.171
966	0.240
1072	0.405
1166	0.382
1216	0.420
CURVE 5	
722	0.290
905	0.355
1072	0.435
1205	0.480

SPECIFICATION TABLE NO. 35 NORMAL SPECTRAL EMITTANCE OF CHROMIUM

Curve No.	Ref. No.	Year	Wavelength μ	Temperature Range, K	Geometry θ'	Reported Error, %	Composition (weight percent), Specifications and Remarks
1	19	1914	0.65	1703	$\sim 0°$	1	Film; tungsten substrate; melted in hydrogen then oxidized in air by heating; measured in air; Pt reference ($\epsilon = 0.33$ for $\lambda = 0.650\ \mu$ at all temp).
2	39	1948	0.669	1550	$\sim 0°$		Heated in hydrogen for one week at 1493° K; measured in hydrogen; ϵ independent of temp up to 1550 °K.

DATA TABLE NO. 35 NORMAL SPECTRAL EMITTANCE OF CHROMIUM

[Temperature, T, K; Emittance, ϵ; Wavelength, λ, μ]

T	ϵ
CURVE 1*	
$\lambda = 0.65$	
1703	0.60
CURVE 2*	
$\lambda = 0.669$	
1550	0.334

*Not shown on plot

SPECIFICATION TABLE NO. 36 NORMAL SPECTRAL EMITTANCE OF CHROMIUM

Curve No.	Ref. No.	Year	Temperature K	Wavelength Range, μ	Geometry θ'	Reported Error, %	Composition (weight percent), Specifications and Remarks
1	19	1914	1733	0.55-0.65	$\sim 0°$		Film; tungsten substrate; measured in hydrogen; Pt reference ($\epsilon = 0.33$ for $\lambda = 0.650\ \mu$ and $\epsilon = 0.38$ for $\lambda = 0.547\ \mu$ at all temp).

DATA TABLE NO. 36 NORMAL SPECTRAL EMITTANCE OF CHROMIUM

[Wavelength, λ, μ; Emittance, ϵ; Temperature, T, K]

λ	ϵ
CURVE 1*	
T = 1733	
0.55	0.53
0.65	0.39

*Not shown on plot

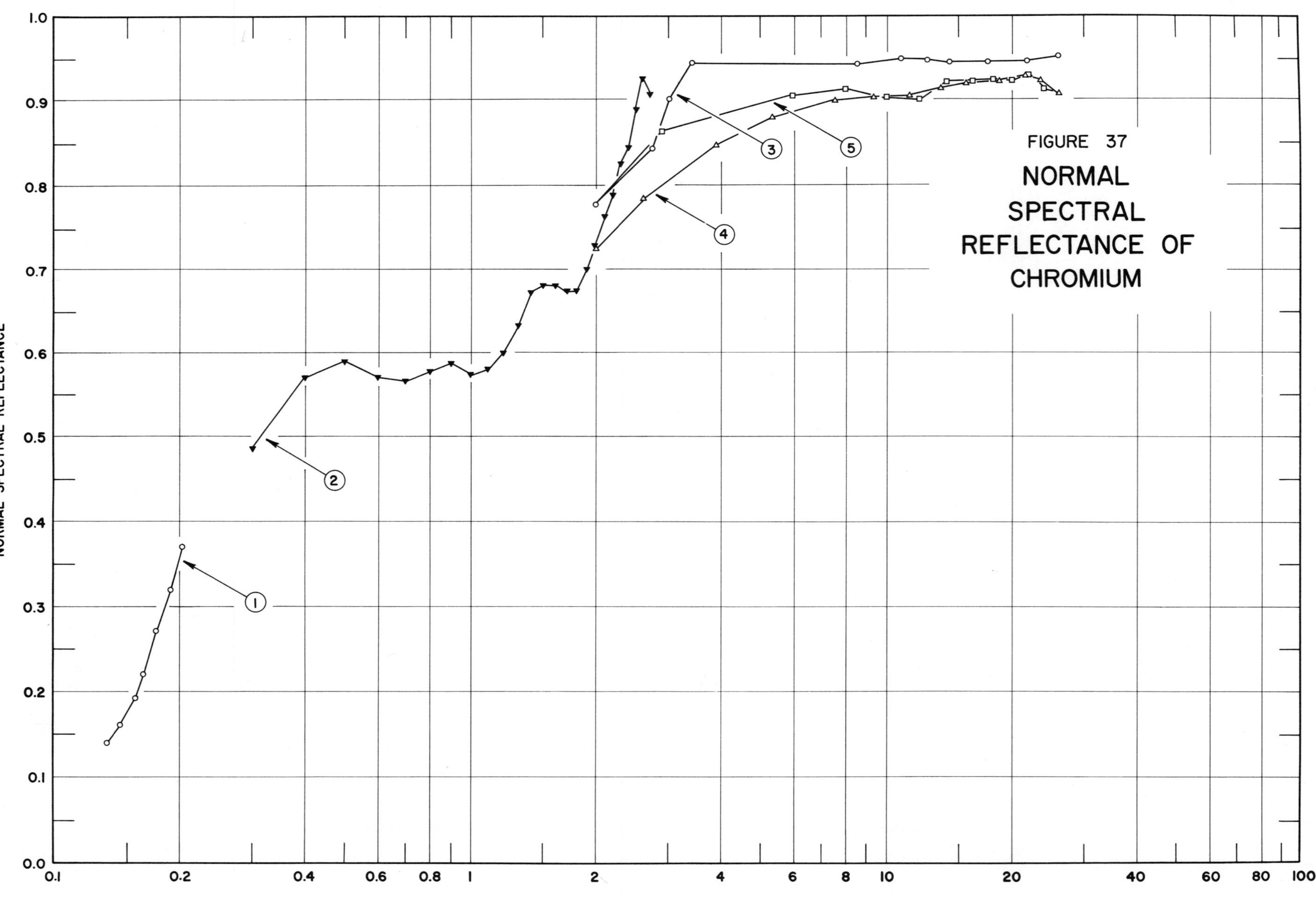

FIGURE 37
NORMAL SPECTRAL REFLECTANCE OF CHROMIUM

SPECIFICATION TABLE NO. 37 NORMAL SPECTRAL REFLECTANCE OF CHROMIUM

Curve No.	Ref. No.	Year	Temperature K	Wavelength Range, μ	Geometry θ	θ'	ω'	Reported Error, %	Composition (weight percent), Specifications and Remarks
1	124	1941	298	0.1347-0.2026	$\sim 0°$	$\sim 0°$			Polished; measured in vacuum ($\sim 10^{-3}$ mm Hg).
2	146	1958	298	0.3-2.7	9°		2π		Electroplated; measured in vacuum; data extracted from smooth curve; $MgCO_3$ reference.
3	223	1962	298	2.00-25.99	$\sim 0°$		2π		Polished; converted from R(2π, 0°).
4	223	1962	298	2.00-26.00	$\sim 0°$		2π		Above specimen and conditions except after particle impact.
5	223	1962	77	2.00-26.00	$\sim 0°$		2π		Above specimen and conditions.
6	235	1967	298	2.5-30.0	$\sim 0°$	$\sim 0°$			5N pure chromium; mechanically polished with 0.25 μ diamond grit; electro-polished; annealed in a high vacuum.

DATA TABLE NO. 37 NORMAL SPECTRAL REFLECTANCE OF CHROMIUM

[Wavelength, λ, μ; Reflectance, ρ; Temperature, T, K]

λ	ρ
CURVE 1	
T = 298	
0.1347	0.14
0.1438	0.16
0.1570	0.19
0.1640	0.22
0.1757	0.27
0.1901	0.32
0.2026	0.37
CURVE 2	
T = 298	
0.3	0.485
0.4	0.570
0.5	0.589
0.6	0.570
0.7	0.565
0.8	0.577
0.9	0.587
1.0	0.572
1.1	0.580
1.2	0.600
1.3	0.633
1.4	0.672
1.5	0.681
1.6	0.681
1.7	0.673
1.8	0.674
1.9	0.700
2.0	0.730
2.1	0.764
2.2	0.790
2.3	0.826
2.4	0.845
2.5	0.890
2.6	0.927
2.7	0.905
CURVE 3	
T = 298	
2.00	0.780
2.73	0.844
4.07	0.902
6.41	0.943
8.50	0.942
10.87	0.950
12.51	0.949
14.18	0.946
17.60	0.946
21.94	0.948
25.99	0.953
CURVE 4	
T = 298	
2.00	0.728
2.61	0.788
3.89	0.849
5.31	0.881
7.49	0.900
9.31	0.904
11.33	0.906
13.49	0.915
15.57	0.921
18.79	0.923
21.53	0.930
23.57	0.925
26.00	0.909
CURVE 5	
T = 77	
2.00	0.727*
3.89	0.864
5.98	0.906
7.99	0.913
9.94	0.903
11.99	0.901
13.98	0.922
16.02	0.922
18.02	0.925
20.05	0.924
21.90	0.930
23.98	0.913
26.00	0.909*
CURVE 6*	
T = 298	
2.5	0.709
3.0	0.778
4.0	0.848
5.0	0.879
6.0	0.890
6.4	0.900
7.0	0.915
8.0	0.927
9.0	0.938
10.0	0.946
11.0	0.955
12.0	0.960
13.0	0.964
14.0	0.969
15.0	0.972
20.0	0.975
25.0	0.976
30.0	0.977

*Not shown on plot

SPECIFICATION TABLE NO. 38 ANGULAR SPECTRAL REFLECTANCE OF CHROMIUM

Curve No.	Ref. No.	Year	Temperature K	Wavelength Range, μ	Geometry θ	Geometry θ'	Geometry ω'	Reported Error, %	Composition (weight percent), Specifications and Remarks
1	132	1911	298	0.55-9.40	15°	15°		≤ 3	Polished; silvered glass mirror reference.

DATA TABLE NO. 38 ANGULAR SPECTRAL REFLECTANCE OF CHROMIUM

[Wavelength, λ, μ; Reflectance, τ; Temperature, T, K]

λ	ρ
CURVE 1*	
T = 298	
0.55	0.550
1 17	0.575
1.55	0.605
2.05	0.623
2.45	0.650
3.05	0.700
4.05	0.765
4.80	0.795
5.45	0.825
6.10	0.855
6.60	0.860
7.10	0 870
7.50	0.875
8.10	0.885
8.45	0.900
8.80	0.903
9.40	0.920

*Not shown on plot

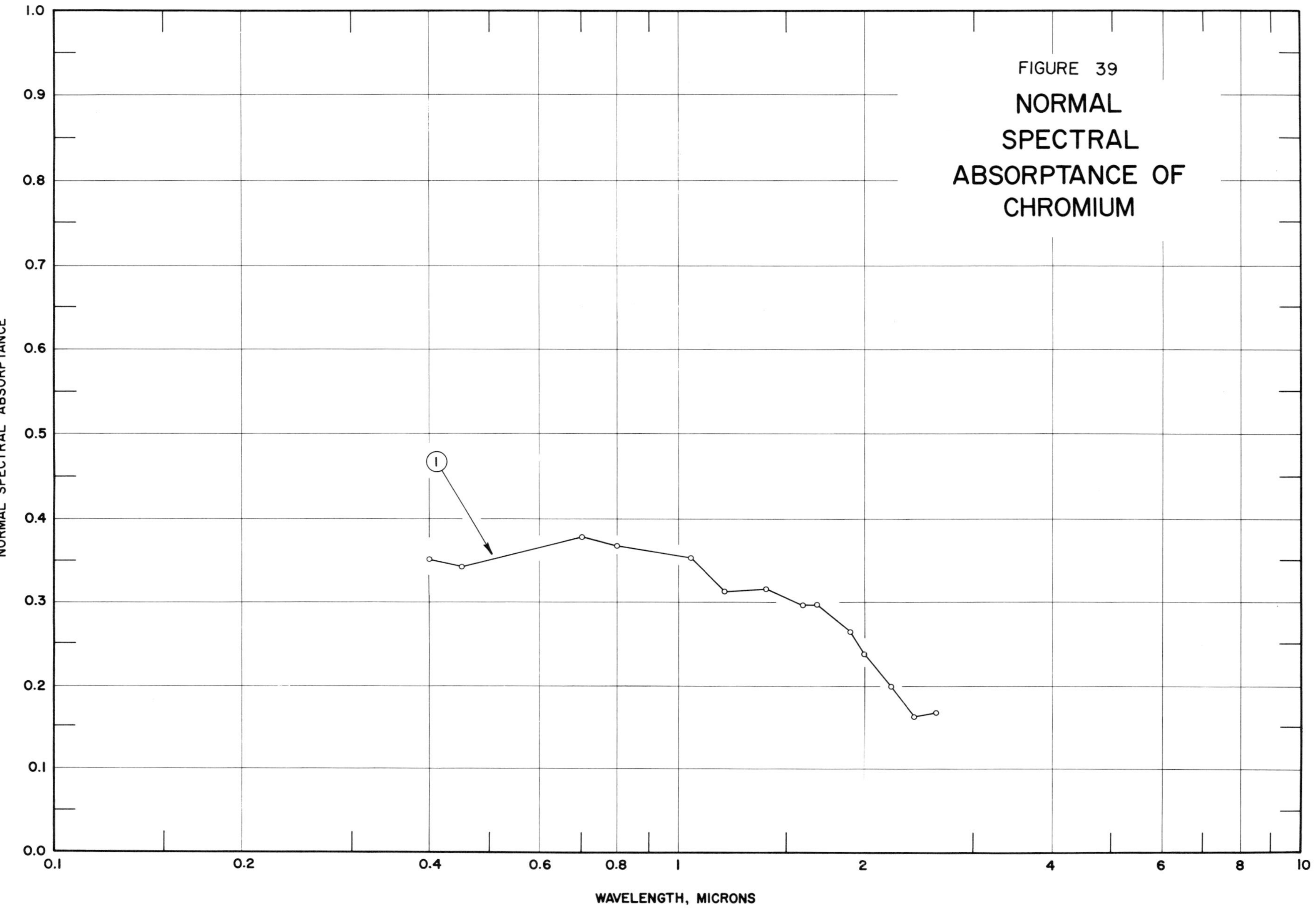

FIGURE 39
NORMAL SPECTRAL ABSORPTANCE OF CHROMIUM

SPECIFICATION TABLE NO. 39 NORMAL SPECTRAL ABSORPTANCE OF CHROMIUM

Curve No.	Ref. No.	Year	Temperature K	Wavelength Range, μ	Geometry ϑ	Reported Error, %	Composition (weight percent), Specifications and Remarks
1	307	1954	~298	0.400-2.600	~0°		Polished; data extracted from smooth curve.

DATA TABLE NO. 39 NORMAL SPECTRAL ABSORPTANCE OF CHROMIUM

[Wavelength, λ, μ; Absorptance, α; Temperature, T, K]

λ	α
CURVE 1	
T = ~298	
0.400	0.351
0.451	0.343
0.708	0.379
0.801	0.368
1.057	0.352
1.193	0.313
1.481	0.317
1.596	0.298
1.684	0.298
1.800	0.265
2.000	0.238
2.200	0.200
2.400	0.163
2.600	0.169

SPECIFICATION TABLE NO. 40 NORMAL SOLAR ABSORPTANCE OF CHROMIUM

Curve No.	Ref. No.	Year	Temperature Range, K	Geometry θ	Reported Error, %	Composition (weight percent), Specifications and Remarks
1	146	1958	298	9°		Electroplated; computed from spectral reflectivity for sea level conditions.
2	146	1958	298	9°		Electroplated; computed from spectral reflectivity for above atmosphere conditions.

DATA TABLE NO. 40 NORMAL SOLAR ABSORPTANCE OF CHROMIUM

[Temperature, T, K; Absorptance, α]

T	α
CURVE 1*	
298	0.415
CURVE 2*	
298	0.397

*Not shown on plot

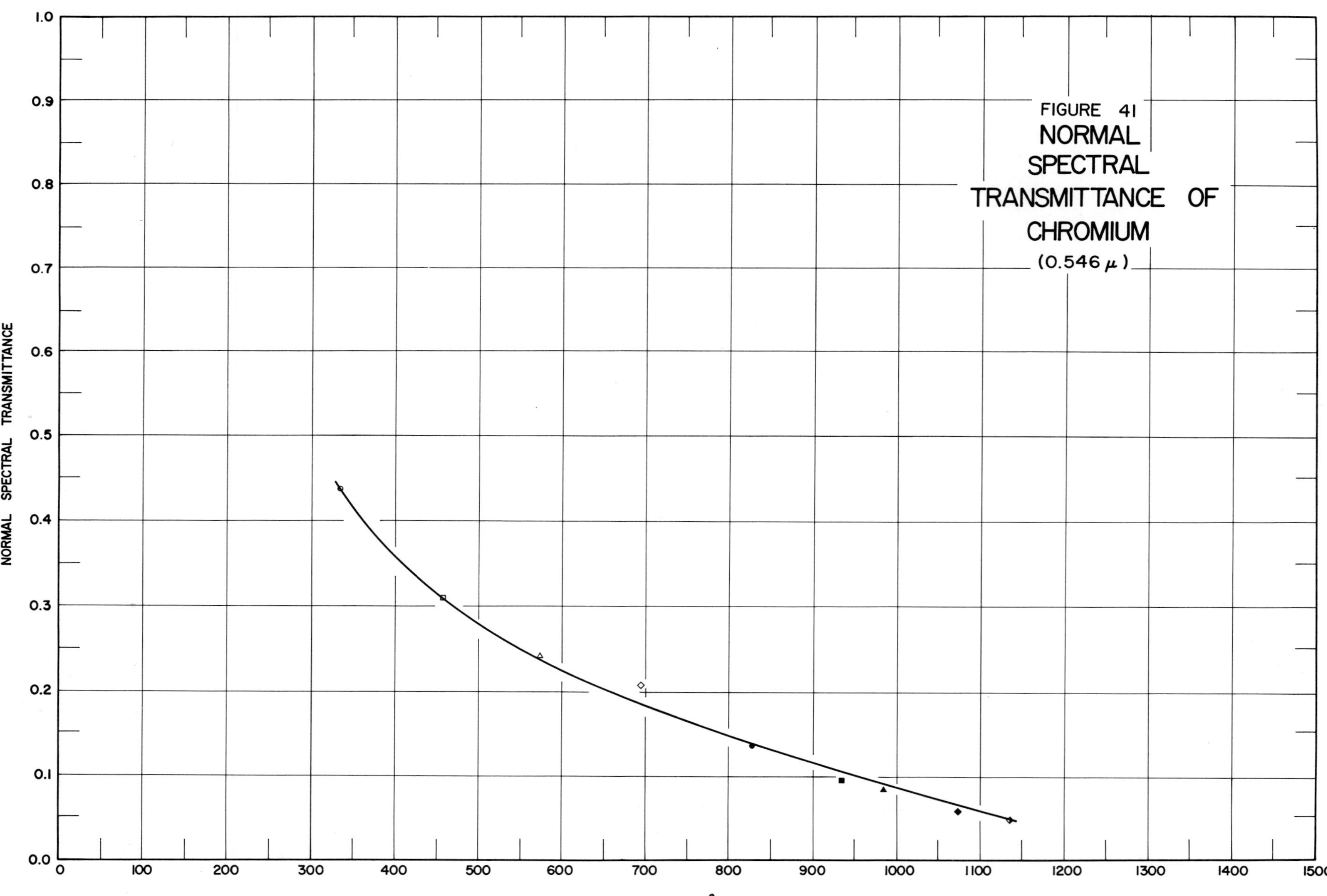

FIGURE 41
NORMAL SPECTRAL TRANSMITTANCE OF CHROMIUM
(0.546 μ)

SPECIFICATION TABLE NO. 41 NORMAL SPECTRAL TRANSMITTANCE OF CHROMIUM

Curve No.	Ref. No.	Year	Temperature K	Wavelength Range, μ	Geometry θ	θ'	ω'	Reported Error, %	Composition (weight percent), Specifications and Remarks
1	311	1966	298	0.546	0°	0°		± 1	Evaporated film (optical thickness 334 Å); evaporated onto glass microscope slide (at 298 K) in vacuum (2×10^{-5} mm Hg); aged in desiccator at 298 K for 10 days.
2	311	1966	298	0.546	0°	0°		± 1	Different sample, same as above specimen and conditions except optical thickness 457 Å.
3	311	1966	298	0.546	0°	0°		± 1	Different sample, same as above specimen and conditions except optical thickness 573 Å.
4	311	1966	298	0.546	0°	0°		± 1	Different sample, same as above specimen and conditions except optical thickness 695 Å.
5	311	1966	298	0.546	0°	0°		± 1	Different sample, same as above specimen and conditions except optical thickness 829 Å.
6	311	1966	298	0.546	0°	0°		± 1	Different sample, same as above specimen and conditions except optical thickness 935 Å.
7	311	1966	298	0.546	0°	0°		± 1	Different sample, same as above specimen and conditions except optical thickness 983 Å.
8	311	1966	298	0.546	0°	0°		± 1	Different sample, same as above specimen and conditions except optical thickness 1072 Å.
9	311	1966	298	0.546	0°	0°		± 1	Different sample, same as above specimen and conditions except optical thickness 1134 Å.

DATA TABLE NO. 41 NORMAL SPECTRAL TRANSMITTANCE OF CHROMIUM

[Wavelength, λ, μ; Transmittance, τ; Temperature, T, K]

λ	τ
CURVE 1 T = 298	
0.546	0.438
CURVE 2 T = 298	
0.546	0.309
CURVE 3 T = 298	
0.546	0.240
CURVE 4 T = 298	
0.546	0.206
CURVE 5 T = 298	
0.546	0.134
CURVE 6 T = 298	
0.546	0.092
CURVE 7 T = 298	
0.546	0.082
CURVE 8 T = 298	
0.546	0.057
CURVE 9 T = 298	
0.546	0.048

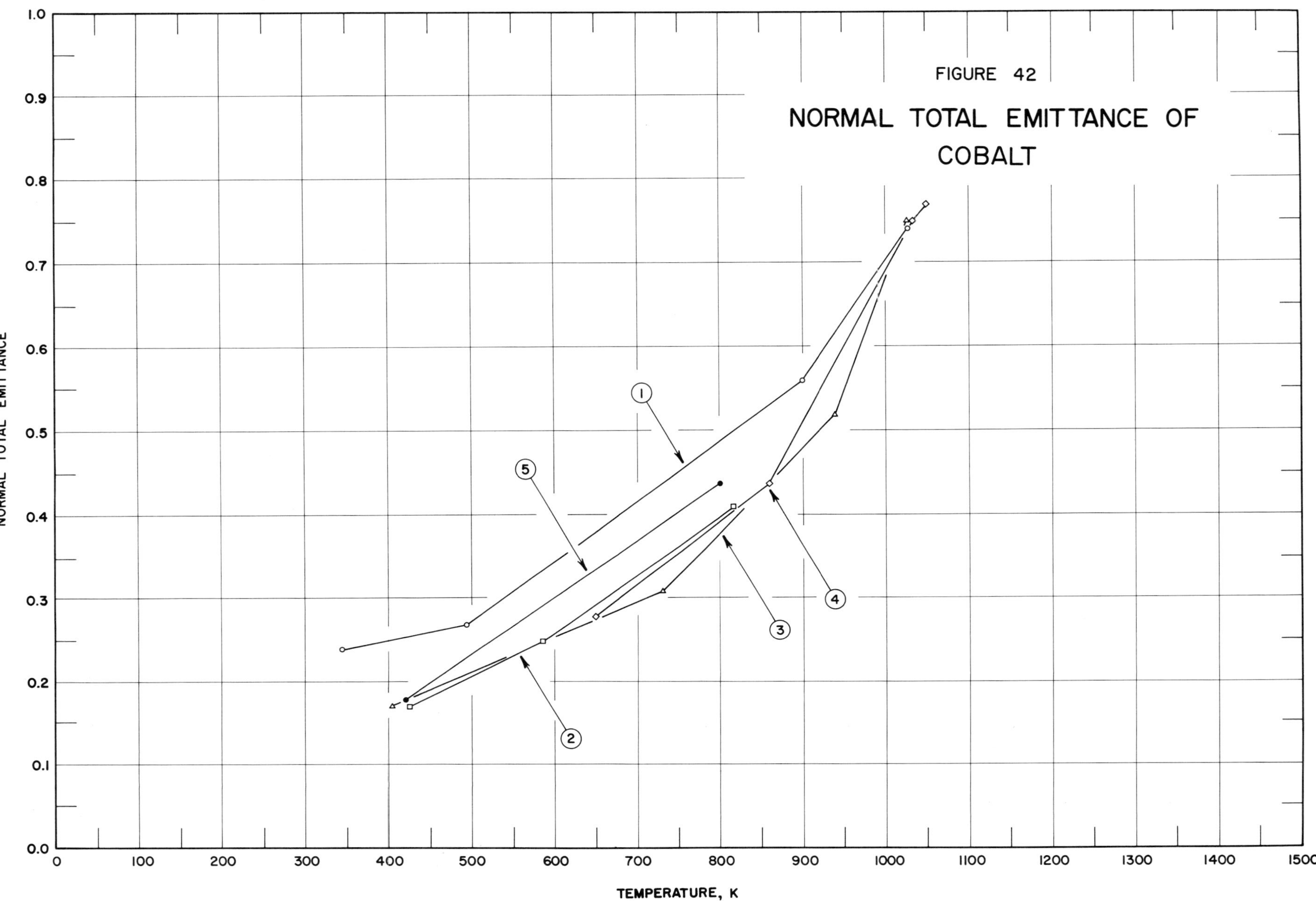

FIGURE 42
NORMAL TOTAL EMITTANCE OF COBALT

SPECIFICATION TABLE NO. 42 NORMAL TOTAL EMITTANCE OF COBALT

Curve No.	Ref. No.	Year	Temperature Range, K	Geometry θ'	Reported Error, %	Composition (weight percent), Specifications and Remarks
1	40	1962	347-1027	$\sim 0^\circ$		Cobalt film on platinum clad Carpenter No. 20 stainless steel; Cycle 1.
2	40	1962	427-816	$\sim 0^\circ$		Above specimen and conditions, increasing temp, Cycle 2.
3	40	1962	1027-405	$\sim 0^\circ$		Above specimen and conditions, decreasing temp, Cycle 2.
4	40	1962	650-1050	$\sim 0^\circ$		Above specimen and conditions, increasing temp, Cycle 3.
5	40	1962	800-422	$\sim 0^\circ$		Above specimen and conditions, decreasing temp, Cycle 3.

DATA TABLE NO. 42 NORMAL TOTAL EMITTANCE OF COBALT

[Temperature, T, K; Emittance, ϵ]

T	ϵ
CURVE 1	
347	0.24
497	0.27
900	0.56
1027	0.74
CURVE 2	
427	0.17
586	0.25
816	0.41
CURVE 3	
1027	0.75
939	0.52
730	0.31
405	0.17
CURVE 4	
650	0.28
861	0.44
1033	0.75
1050	0.77
CURVE 5	
800	0.44
422	0.18

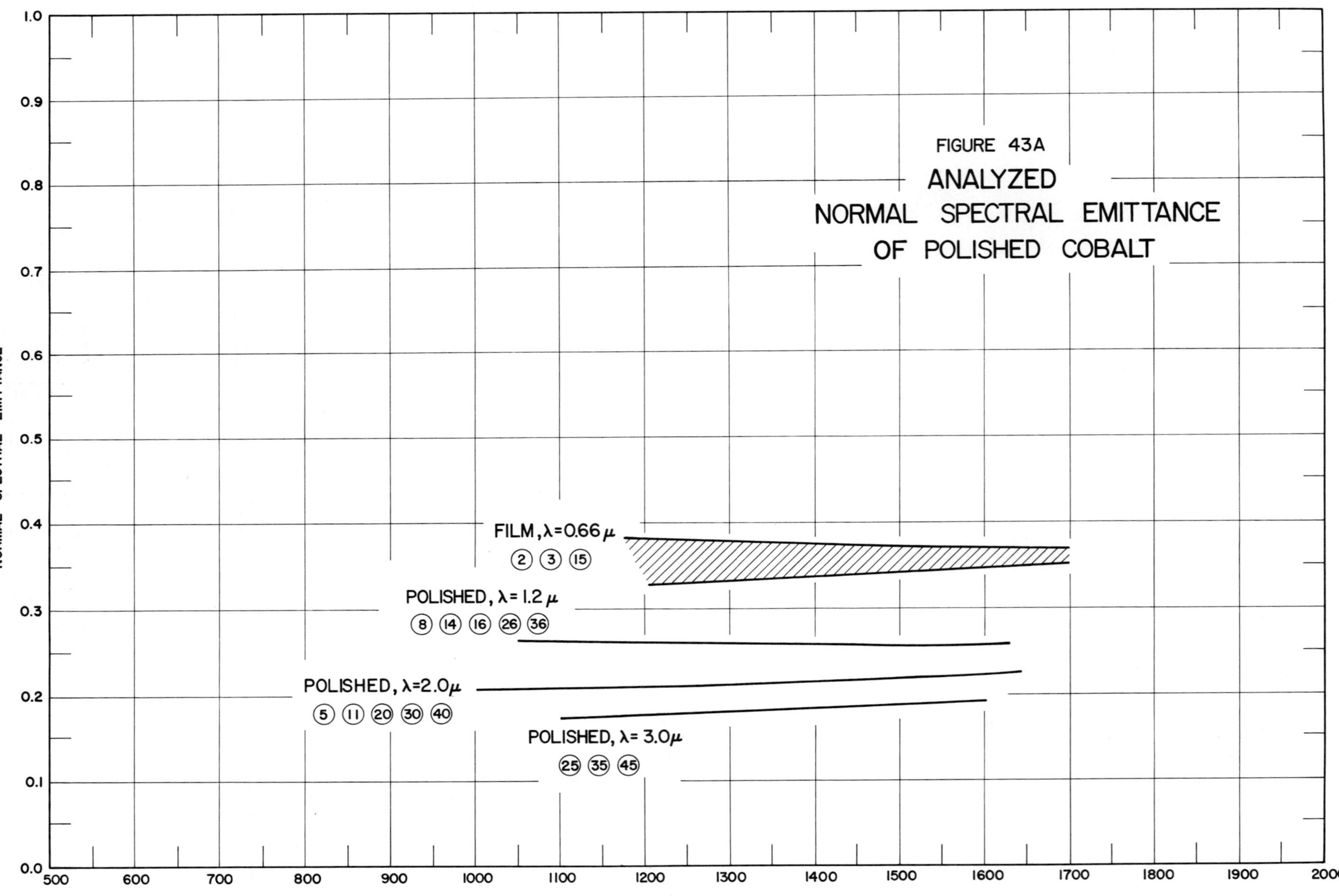
FIGURE 43A
ANALYZED
NORMAL SPECTRAL EMITTANCE
OF POLISHED COBALT
FILM, λ=0.66 μ
② ③ ⑮
POLISHED, λ= 1.2 μ
⑧ ⑭ ⑯ ㉖ ㊱
POLISHED, λ=2.0 μ
⑤ ⑪ ⑳ ㉚ ㊵
POLISHED, λ= 3.0 μ
㉕ ㉟ ㊺
NORMAL SPECTRAL EMITTANCE
0.0
0.1
0.2
0.3
0.4
0.5
0.6
0.7
0.8
0.9
1.0
TEMPERATURE, K
500
600
700
800
900
1000
1100
1200
1300
1400
1500
1600
1700
1800
1900
2000

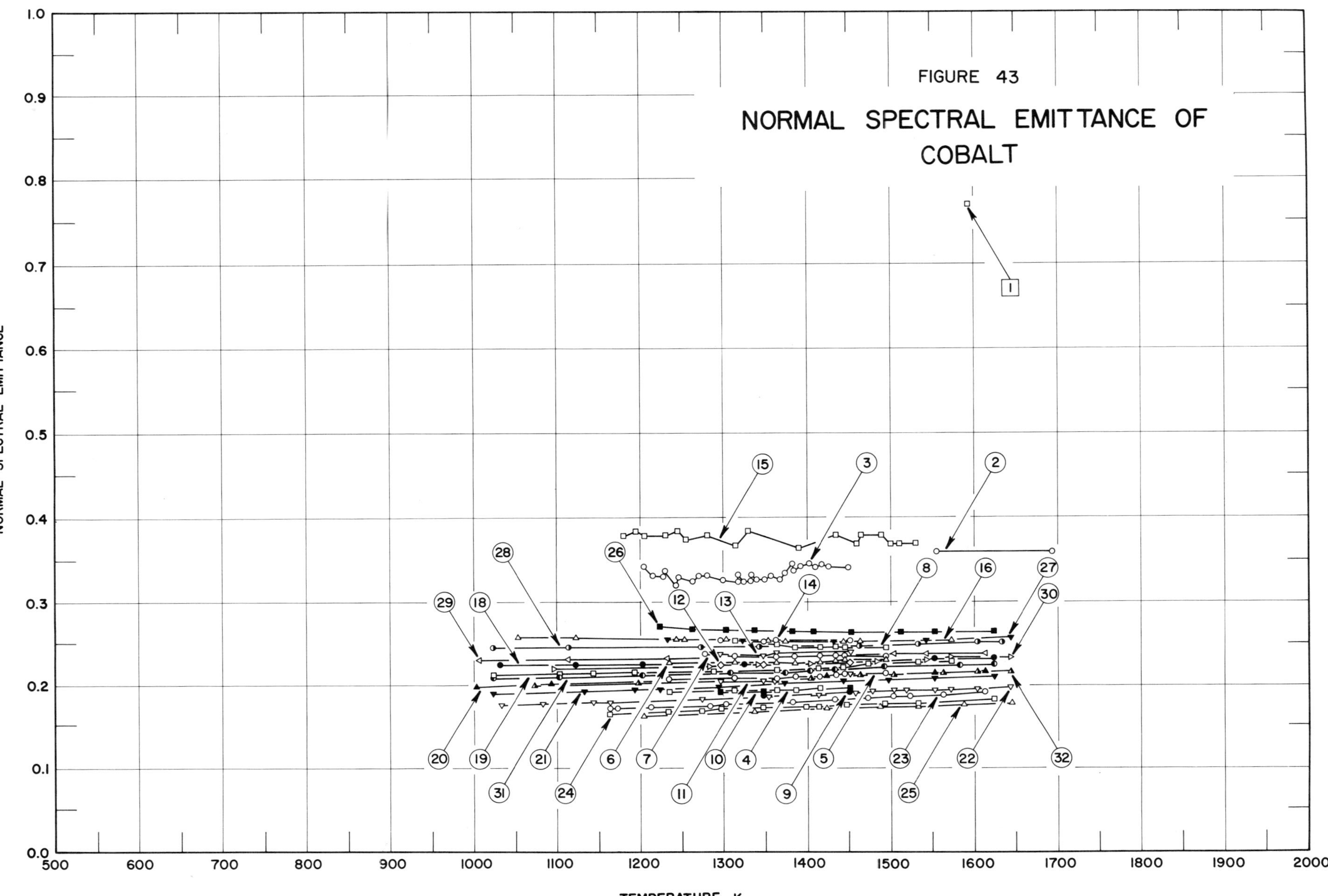

FIGURE 43

NORMAL SPECTRAL EMITTANCE OF COBALT

SPECIFICATION TABLE NO. 43 NORMAL SPECTRAL EMITTANCE OF COBALT

Curve No.	Ref. No.	Year	Wavelength μ	Temperature Range, K	Geometry θ'	Reported Error, %	Composition (weight percent), Specifications and Remarks
1	19	1914	0.65	1593	$\sim 0^\circ$	1	Film; tungsten substrate; melted in hydrogen then oxidized in air by heating; measured in air; Pt reference ($\epsilon = 0.33$ for $\lambda = 0.650$ μ at all temp).
2	19	1914	0.65	1552-1693	$\sim 0^\circ$	1	Film; tungsten substrate; measured in hydrogen; Pt reference ($\epsilon = 0.33$ for $\lambda = 0.650\,\mu$ at all temp).
3	44	1948	0.667	1204-1450	$\sim 0^\circ$		Prepared electrolytically from a chloride bath; measured in vacuum.
4	43	1952	2.4	1234-1415	$\sim 0^\circ$		Polished; washed with ether; measured in vacuum; [Author's designation: Specimen B].
5	43	1952	2.0	1234-1494	$\sim 0^\circ$		Above specimen and conditions.
6	43	1952	1.6	1234-1494	$\sim 0^\circ$		Above specimen and conditions.
7	43	1952	1.4	1278-1494	$\sim 0^\circ$		Above specimen and conditions.
8	43	1952	1.2	1313-1494	$\sim 0^\circ$		Above specimen and conditions.
9	43	1952	2.6	1362-1451	$\sim 0^\circ$		Polished; washed with ether; measured in vacuum; [Author's designation: Specimen A].
10	43	1952	2.4	1297-1451	$\sim 0^\circ$		Above specimen and conditions.
11	43	1952	2.0	1297-1451	$\sim 0^\circ$		Above specimen and conditions.
12	43	1952	1.6	1297-1451	$\sim 0^\circ$		Above specimen and conditions.
13	43	1952	1.4	1297-1451	$\sim 0^\circ$		Above specimen and conditions.
14	43	1952	1.2	1297-1451	$\sim 0^\circ$		Above specimen and conditions.
15	42	1942	0.667	1180-1530	$\sim 0^\circ$		Baked at 623 K for 200 hrs; measured in vacuum (4×10^{-8} mm Hg).
16	41	1956	1.2	1053-1573	$\sim 0^\circ$	8	Polished with emery paper and lapped with jeweler's rouge; measured in vacuum; [Author's designation: Specimen A].
17	41	1956	1.4	1063-1633	$\sim 0^\circ$	8	Above specimen and conditions.
18	41	1956	1.6	1033-1623	$\sim 0^\circ$	8	Above specimen and conditions.
19	41	1956	1.8	1023-1573	$\sim 0^\circ$	8	Above specimen and conditions.
20	41	1956	2.0	1003-1613	$\sim 0^\circ$	8	Above specimen and conditions.
21	41	1956	2.2	1023-1623	$\sim 0^\circ$	8	Above specimen and conditions.
22	41	1956	2.4	1033-1643	$\sim 0^\circ$	8	Above specimen and conditions.
23	41	1956	2.6	1163-1613	$\sim 0^\circ$	8	Above specimen and conditions.
24	41	1956	2.8	1163-1623	$\sim 0^\circ$	8	Above specimen and conditions.
25	41	1956	3.0	1203-1643	$\sim 0^\circ$	8	Above specimen and conditions.
26	41	1956	1.2	1223-1623	$\sim 0^\circ$	8	Polished with emery paper and lapped with jeweler's rouge; measured in vacuum; [Author's designation: Specimen B].

SPECIFICATION TABLE NO. 43 (continued)

Curve No.	Ref. No.	Year	Wavelength μ	Temperature Range, K	Geometry θ'	Reported Error, %	Composition (weight percent), Specifications and Remarks
27	41	1956	1.4	1233-1643	$\sim 0^\circ$	8	Above specimen and conditions.
28	41	1956	1.6	1023-1633	$\sim 0^\circ$	8	Above specimen and conditions.
29	41	1956	1.8	1008-1613	$\sim 0^\circ$	8	Above specimen and conditions.
30	41	1956	2.0	1098-1643	$\sim 0^\circ$	8	Above specimen and conditions.
31	41	1956	2.2	1023-1623	$\sim 0^\circ$	8	Above specimen and conditions.
32	41	1956	2.4	1073-1643	$\sim 0^\circ$	8	Above specimen and conditions.
33	41	1956	2.6	1093-1613	$\sim 0^\circ$	8	Above specimen and conditions.
34	41	1956	2.8	1133-1588	$\sim 0^\circ$	8	Above specimen and conditions.
35	41	1956	3.0	1163-1553	$\sim 0^\circ$	8	Above specimen and conditions.
36	41	1956	1.2	1113-1633	$\sim 0^\circ$	8	Polished with emery paper and lapped with jeweler's rouge; measured in vacuum; [Author's designation: Specimen C].
37	41	1956	1.4	1063-1643	$\sim 0^\circ$	8	Above specimen and conditions.
38	41	1956	1.6	1033-1643	$\sim 0^\circ$	8	Above specimen and conditions.
39	41	1956	1.8	1103-1623	$\sim 0^\circ$	8	Above specimen and conditions.
40	41	1956	2.0	1023-1553	$\sim 0^\circ$	8	Above specimen and conditions.
41	41	1956	2.2	1023-1583	$\sim 0^\circ$		Above specimen and conditions.
42	41	1956	2.4	1033-1573	$\sim 0^\circ$	8	Above specimen and conditions.
43	41	1956	2.6	1033-1633	$\sim 0^\circ$	8	Above specimen and conditions.
44	41	1956	2.8	1033-1558	$\sim 0^\circ$	8	Above specimen and conditions.
45	41	1956	3.0	1103-1598	$\sim 0^\circ$	8	Above specimen and conditions.

DATA TABLE NO. 43 NORMAL SPECTRAL EMITTANCE OF COBALT

[Temperature, T, K; Emittance, ϵ Wavelength, λ, μ]

CURVE 1, λ = 0.65

T	ϵ
1593	0.77

CURVE 2, λ = 0.65

T	ϵ
1552	0.36
1693	0.36

CURVE 3, λ = 0.667

T	ϵ
1204	0.342
1215	0.332
1228	0.331
1230	0.338
1242	0.320
1247	0.330
1262	0.325
1270	0.331
1281	0.331
1300	0.328
1319	0.325
1319	0.333
1320	0.322*
1323	0.325
1327	0.327*
1332	0.326
1334	0.332
1340	0.328
1341	0.326*
1350	0.328
1354	0.327*
1358	0.331
1368	0.329
1371	0.327*
1372	0.330*
1376	0.336
1379	0.341*
1382	0.347
1384	0.339
1388	0.344*
1392	0.343
1393	0.341*
1396	0.347*

CURVE 3 (cont.), λ = 0.667

T	ϵ
1402	0.346
1406	0.344*
1410	0.342
1412	0.343*
1418	0.343*
1418	0.347
1420	0.341*
1421	0.346*
1426	0.343
1450	0.342

CURVE 4, λ = 2.4

T	ϵ
1234	0.192
1313	0.194
1363	0.194
1386	0.194
1415	0.198

CURVE 5, λ = 2.0

T	ϵ
1234	0.208
1313	0.209
1363	0.209
1386	0.210
1415	0.210
1434	0.211
1459	0.212
1494	0.212

CURVE 6, λ = 1.6

T	ϵ
1234	0.227
1313	0.227
1363	0.227
1386	0.226
1415	0.228
1434	0.229
1459	0.230
1494	0.230

CURVE 7, λ = 1.4

T	ϵ
1278	0.238
1313	0.234
1363	0.233
1386	0.233
1415	0.234
1434	0.234
1459	0.234
1494	0.234

CURVE 8, λ = 1.2

T	ϵ
1313	0.253
1363	0.250
1386	0.247
1415	0.247
1434	0.248
1459	0.247
1494	0.247

CURVE 9, λ = 2.6

T	ϵ
1362	0.192*
1349	0.189
1451	0.193

CURVE 10, λ = 2.4

T	ϵ
1297	0.193
1362	0.195*
1349	0.193
1451	0.197

CURVE 11, λ = 2.0

T	ϵ
1297	0.206
1362	0.208
1349	0.206
1451	0.212

CURVE 12, λ = 1.6

T	ϵ
1297	0.225
1362	0.227*
1349	0.224
1451	0.229

CURVE 13, λ = 1.4

T	ϵ
1297	0.237
1362	0.239
1349	0.235
1451	0.240

CURVE 14, λ = 1.2

T	ϵ
1297	0.255
1362	0.253
1349	0.252
1451	0.253

CURVE 15, λ = 0.667

T	ϵ
1180	0.380
1195	0.385
1205	0.380
1230	0.280
1245	0.285
1255	0.375
1280	0.380
1315	0.367
1330	0.385
1390	0.365
1435	0.380
1460	0.370
1465	0.380
1490	0.380
1500	0.370
1510	0.370
1530	0.370

CURVE 16, λ = 1.2

T	ϵ
1053	0.259
1123	0.258
1243	0.256
1253	0.255
1303	0.255
1353	0.253
1373	0.252
1443	0.252
1463	0.253
1573	0.254

CURVE 17*, λ = 1.4

T	ϵ
1063	0.237
1163	0.237
1203	0.236
1263	0.236
1353	0.236
1393	0.235
1443	0.237
1483	0.238
1503	0.238
1553	0.239
1633	0.241

CURVE 18, λ = 1.6

T	ϵ
1033	0.224
1123	0.224
1203	0.224
1323	0.225
1383	0.226*
1453	0.228*
1553	0.231
1623	0.232

CURVE 19, λ = 1.8

T	ϵ
1023	0.212
1103	0.213
1143	0.214

CURVE 19 (cont.), λ = 1.8

T	ϵ
1193	0.216
1288	0.217
1363	0.218
1413	0.220
1443	0.221
1473	0.226
1533	0.227
1573	0.228

CURVE 20, λ = 2.0

T	ϵ
1003	0.199
1093	0.201
1233	0.204*
1348	0.206*
1383	0.210*
1423	0.211
1493	0.213*
1553	0.214
1613	0.216

CURVE 21, λ = 2.2

T	ϵ
1023	0.190
1133	0.192
1193	0.195
1223	0.196
1293	0.198
1373	0.201
1443	0.204
1498	0.206
1553	0.208
1623	0.210

CURVE 22, λ = 2.4

T	ϵ
1033	0.176
1083	0.178
1143	0.180
1163	0.180
1253	0.184*

CURVE 22 (cont.), λ = 2.4

T	ϵ
1273	0.184
1353	0.187
1413	0.189
1458	0.191
1478	0.193
1503	0.194
1518	0.194
1553	0.195
1573	0.196
1603	0.197
1643	0.199

CURVE 23, λ = 2.6

T	ϵ
1163	0.171
1173	0.171
1213	0.172
1283	0.176
1303	0.178
1383	0.180
1433	0.182
1473	0.185
1503	0.187
1523	0.187
1563	0.189
1613	0.191

CURVE 24, λ = 2.8

T	ϵ
1163	0.165
1233	0.167
1273	0.169
1298	0.171
1348	0.172
1398	0.174
1413	0.173
1448	0.177
1493	0.178
1533	0.179
1623	0.183

CURVE 25, λ = 3.0

T	ϵ
1203	0.162
1338	0.167
1423	0.171
1488	0.173
1533	0.175
1588	0.177
1643	0.179

CURVE 26, λ = 1.2

T	ϵ
1223	0.271
1263	0.269
1303	0.268
1338	0.268
1383	0.266
1408	0.265
1463	0.264
1513	0.264
1563	0.264
1623	0.265

CURVE 27, λ = 1.4

T	ϵ
1233	0.255
1323	0.253
1363	0.252*
1433	0.252
1543	0.255
1643	0.257

CURVE 28, λ = 1.6

T	ϵ
1023	0.245
1113	0.246
1273	0.247
1343	0.247
1463	0.248
1493	0.249*
1533	0.250
1603	0.252
1633	0.251

* Not shown on plot

DATA TABLE NO. 43 (continued)

CURVE 29, $\lambda = 1.8$

T	∈
1008	0.230
1113	0.231
1233	0.232
1313	0.233*
1363	0.233*
1433	0.235*
1453	0.235*
1503	0.237
1543	0.237
1613	0.238

CURVE 30, $\lambda = 2.0$

T	∈
1098	0.220
1203	0.222*
1283	0.223
1348	0.224
1383	0.225*
1403	0.226
1453	0.227*
1483	0.229
1543	0.230
1573	0.231
1643	0.231

CURVE 31, $\lambda = 2.2$

T	∈
1023	0.209
1103	0.210
1203	0.212
1308	0.214
1373	0.215
1403	0.217
1433	0.219
1463	0.220*
1493	0.221
1583	0.223
1623	0.223

CURVE 32, $\lambda = 2.4$

T	∈
1073	0.200
1198	0.203
1298	0.205*
1363	0.208*
1403	0.208
1463	0.211
1503	0.212
1563	0.214
1603	0.215
1643	0.216

CURVE 33*, $\lambda = 2.6$

T	∈
1093	0.187
1223	0.192
1308	0.192
1353	0.195
1413	0.196
1478	0.198
1523	0.200
1528	0.200
1563	0.201
1613	0.203

CURVE 34*, $\lambda = 2.8$

T	∈
1133	0.185
1203	0.186
1298	0.190
1373	0.193
1433	0.195
1488	0.196
1538	0.200
1588	0.201

CURVE 35*, $\lambda = 3.0$

T	∈
1163	0.181
1373	0.189
1433	0.192
1503	0.195
1553	0.197

CURVE 36*, $\lambda = 1.2$

T	∈
1113	0.269
1173	0.266
1263	0.265
1343	0.263
1363	0.263
1433	0.261
1468	0.260
1523	0.260
1633	0.261

CURVE 37*, $\lambda = 1.4$

T	∈
1063	0.249
1193	0.243
1303	0.247
1378	0.246
1463	0.246
1553	0.248
1643	0.250

CURVE 38*, $\lambda = 1.6$

T	∈
1033	0.238
1163	0.239
1268	0.240
1313	0.241
1393	0.241
1423	0.241
1493	0.243
1548	0.245
1573	0.245
1643	0.245

CURVE 39*, $\lambda = 1.8$

T	∈
1103	0.226
1148	0.226
1223	0.227
1293	0.228
1373	0.228
1403	0.228
1433	0.229
1503	0.230
1623	0.232

CURVE 40*, $\lambda = 2.0$

T	∈
1023	0.212
1143	0.213
1173	0.214
1248	0.214
1283	0.215
1353	0.216
1413	0.219
1453	0.221
1483	0.221
1553	0.223

CURVE 41*, $\lambda = 2.2$

T	∈
1023	0.202
1123	0.205
1183	0.206
1233	0.208
1313	0.209
1373	0.211
1423	0.212
1478	0.214
1523	0.216
1588	0.218

CURVE 42*, $\lambda = 2.4$

T	∈
1033	0.194
1133	0.197
1158	0.197
1208	0.199
1248	0.200
1313	0.202
1393	0.204
1463	0.206
1473	0.206
1493	0.209
1518	0.210
1573	0.212

CURVE 43*, $\lambda = 2.6$

T	∈
1033	0.186
1063	0.188
1193	0.192
1283	0.195
1353	0.197
1443	0.201
1473	0.204
1533	0.207
1573	0.208
1633	0.210

CURVE 44*, $\lambda = 2.8$

T	∈
1033	0.177
1093	0.179
1163	0.182
1233	0.185
1293	0.187
1363	0.190
1443	0.194
1513	0.197
1558	0.198

CURVE 45*, $\lambda = 3.0$

T	∈
1103	0.176
1163	0.179
1233	0.182
1283	0.184
1358	0.187
1423	0.191
1488	0.193
1598	0.197

*Not shown on plot

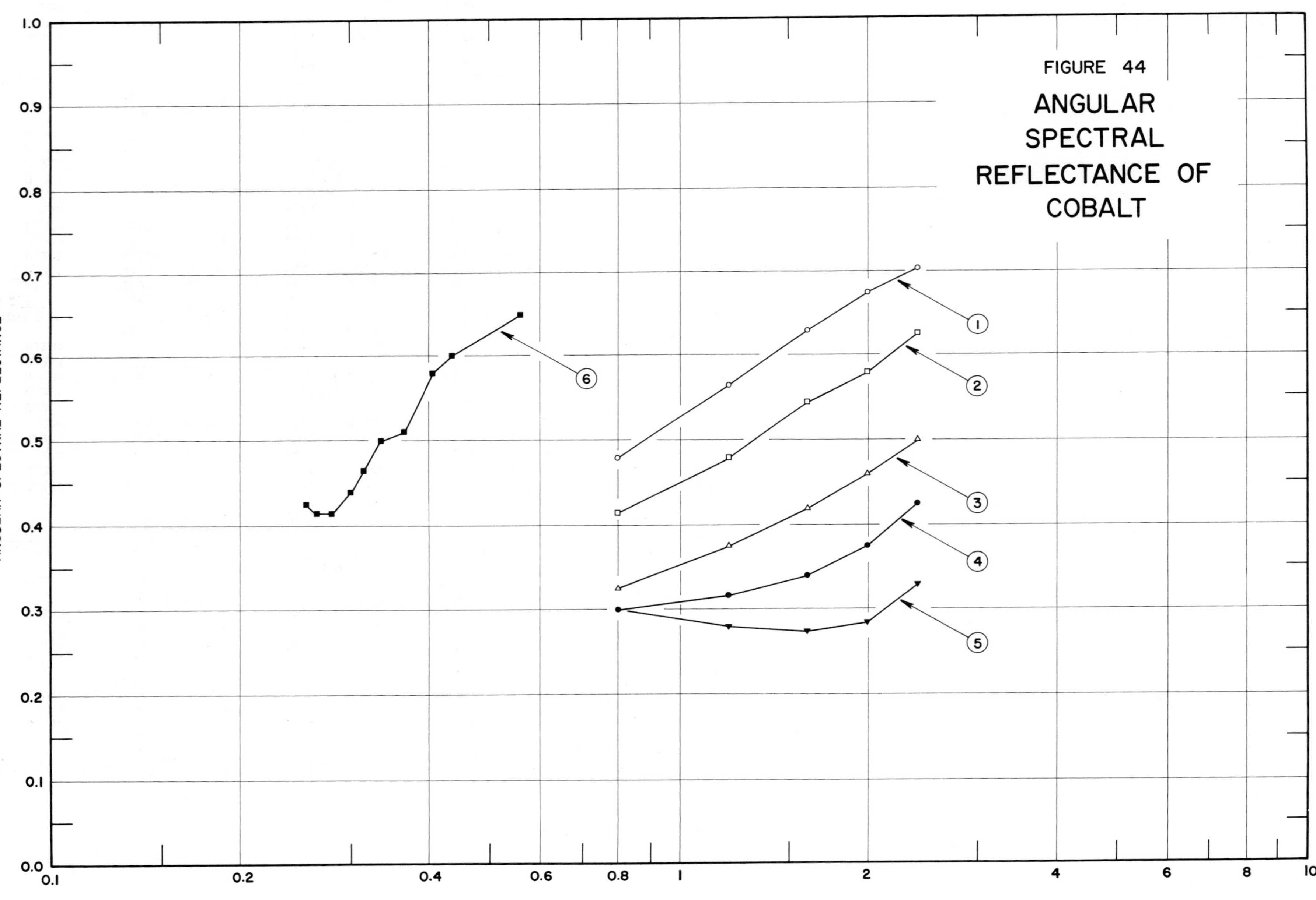
FIGURE 44
ANGULAR SPECTRAL REFLECTANCE OF COBALT
1
2
3
4
5
6
ANGULAR SPECTRAL REFLECTANCE
1.0
0.9
0.8
0.7
0.6
0.5
0.4
0.3
0.2
0.1
0.0
WAVELENGTH, MICRONS
0.1
0.2
0.4
0.6
0.8
1
2
4
6
8
10

SPECIFICATION TABLE NO. 44 ANGULAR SPECTRAL REFLECTANCE OF COBALT

Curve No.	Ref. No.	Year	Temperature K	Wavelength Range, μ	Geometry θ	θ'	ω'	Reported Error, %	Composition (weight percent), Specifications and Remarks
1	141	1948	298	0.8-2.4	50°	50°			Mirror surface prepared by vacuum evaporation; data extracted from smooth curve, incident beam polarized parallel to the plane of incidence.
2	141	1948	298	0.8-2.4	60°	60°			Above specimen and conditions, incident beam polarized parallel to the plane of incidence.
3	141	1948	298	0.8-2.4	70°	70°			Above specimen and conditions, incident beam polarized parallel to the plane of incidence.
4	141	1948	298	0.8-2.4	79°	79°			Above specimen and conditions, incident beam polarized parallel to the plane of incidence.
5	141	1948	298	0.8-2.4	80°	80°			Above specimen and conditions, incident beam polarized parallel to the plane of incidence.
6	143	1929	298	0.255-0.560	45°	45°			Highly polished.

DATA TABLE NO. 44 ANGULAR SPECTRAL REFLECTANCE OF COBALT

[Wavelength, λ, μ; Reflectance, ρ; Temperature, T, K]

λ	ρ	λ	ρ
CURVE 1 T = 298		**CURVE 6** T = 298	
0.8	0.480	0.255	0.425
1.2	0.565	0.265	0.415
1.6	0.630	0.280	0.415
2.0	0.675	0.300	0.440
2.4	0.702	0.315	0.465
		0.335	0.500
CURVE 2 T = 298		0.365	0.510
		0.405	0.580
		0.435	0.600
0.8	0.415	0.560	0.650
1.2	0.480		
1.6	0.545		
2.0	0.580		
2.4	0.625		
CURVE 3 T = 298			
0.8	0.325		
1.2	0.375		
1.6	0.420		
2.0	0.460		
2.4	0.500		
CURVE 4 T = 298			
0.8	0.300		
1.2	0.318		
1.6	0.340		
2.0	0.375		
2.4	0.425		
CURVE 5 T = 298			
0.8	0.300*		
1.2	0.280		
1.6	0.275		
2.0	0.285		
2.4	0.330		

* Not shown on plot

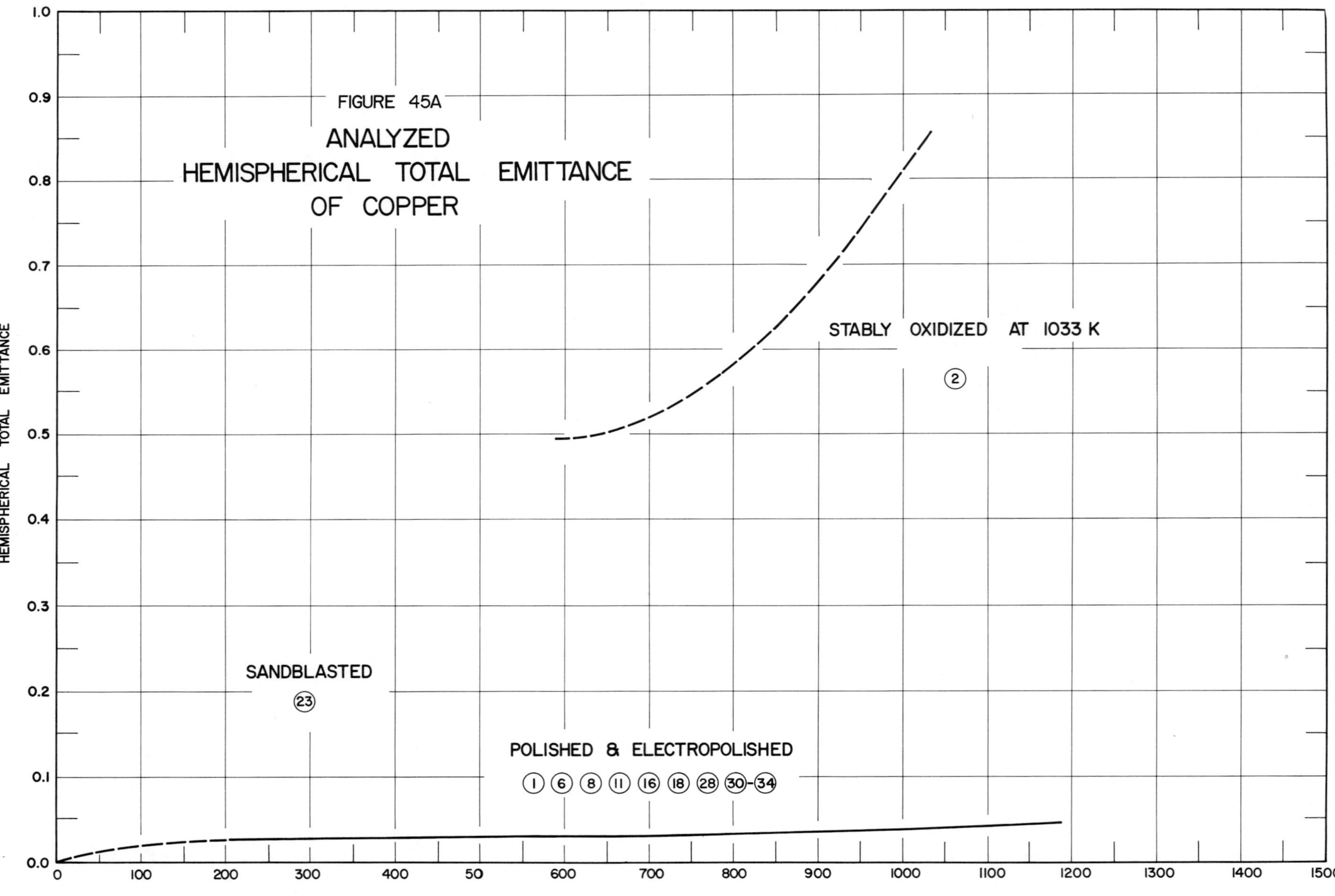
FIGURE 45A
ANALYZED
HEMISPHERICAL TOTAL EMITTANCE
OF COPPER
STABLY OXIDIZED AT 1033 K
(2)
SANDBLASTED
(23)
POLISHED & ELECTROPOLISHED
(1) (6) (8) (11) (16) (18) (28) (30)-(34)
HEMISPHERICAL TOTAL EMITTANCE
1.0
0.9
0.8
0.7
0.6
0.5
0.4
0.3
0.2
0.1
0.0
0
100
200
300
400
50
600
700
800
900
1000
1100
1200
1300
1400
1500
TEMPERATURE, K

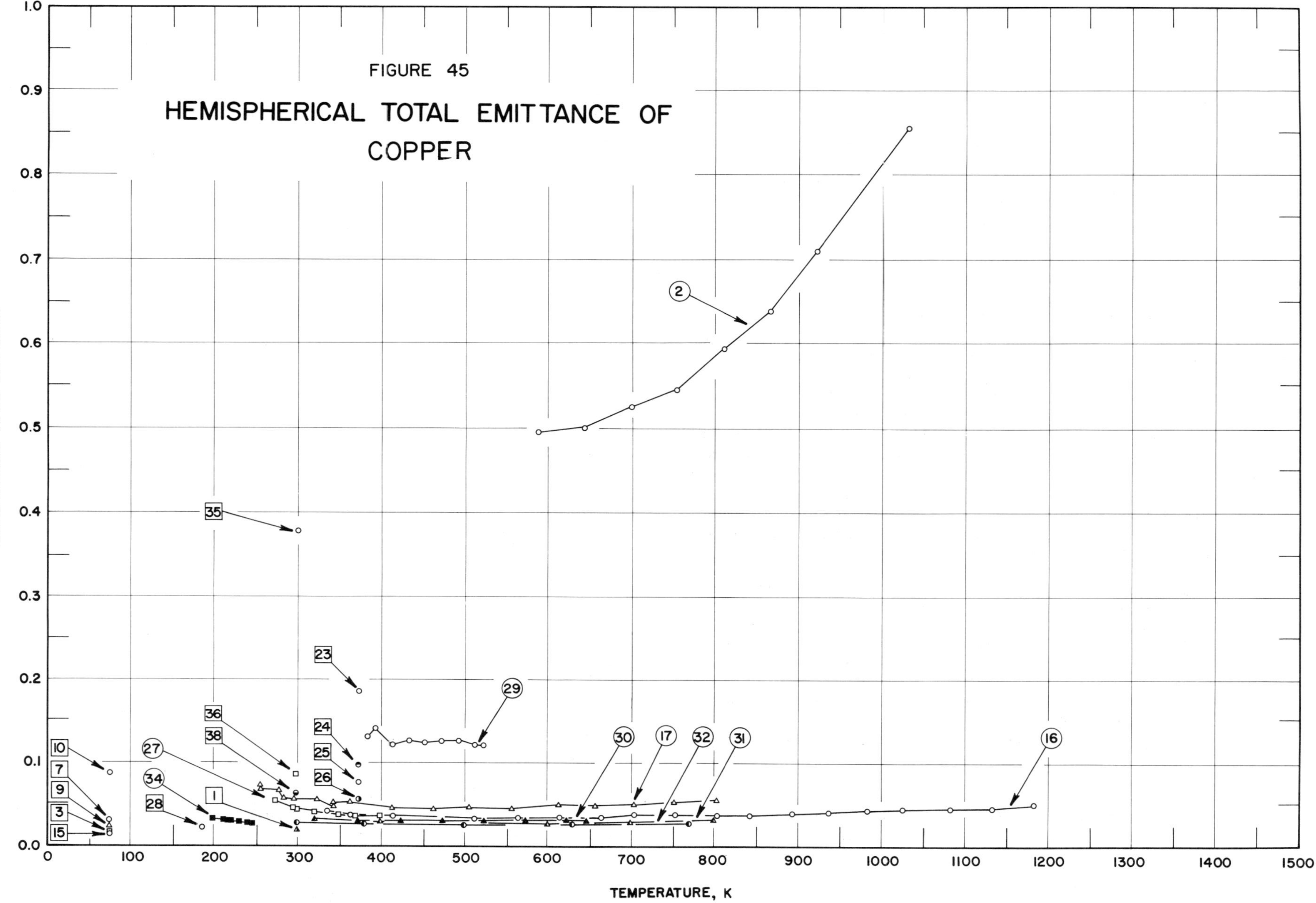

FIGURE 45
HEMISPHERICAL TOTAL EMITTANCE OF COPPER

SPECIFICATION TABLE NO. 45 HEMISPHERICAL TOTAL EMITTANCE OF COPPER

Curve No.	Ref. No.	Year	Temperature Range, K	Reported Error, %	Composition (weight percent), Specifications and Remarks
1	23	1955	300		Polished; measured in vacuum ($< 3 \times 10^{-6}$ mm Hg).
2	22	1958	589-1033	≤ 2	Stably oxidized at 1033 K in quiescent air, calculated from Normal Total Emittance.
3	4	1953	76		Foil (0.005 in. thick); treated by dilute chromic acid dip; measured in vacuum ($< 10^{-6}$ mm Hg); emittance for 294 K black body radiation; authors assumed $\alpha = \epsilon$.
4	4	1953	76		Foil (0.005 in. thick); wet polished with Dr. Lyons tooth powder; measured in vacuum ($< 10^{-6}$ mm Hg); emittance for 294 K black body radiation; authors assumed $\alpha = \epsilon$.
5	4	1953	76		Foil (0.005 in. thick); dry polished with plastic polishing wax-abrasive; measured in vacuum ($< 10^{-6}$ mm Hg); emittance for 294 K black body radiation; authors assumed $\alpha = \epsilon$.
6	4	1953	76		Foil (0.005 in. thick); electrolytic cleaned; measured in vacuum ($<10^{-6}$ mm Hg); emittance for 294 K black body radiation; authors assumed $\alpha = \epsilon$; emittance value may be high because trace of frost was observed on foil.
7	3	1955	76	5	Commercial copper sphere; Oakite No. 33 cleaned; measured in vacuum (10^{-6} to 10^{-7} mm Hg); emittance for 300 K black body incident radiation; authors assumed $\alpha = \epsilon$.
8	3	1955	76	5	Commercial copper sphere; polished; measured in vacuum (10^{-6} to 10^{-7} mm Hg); emittance for 300 K black body incident radiation; authors assumed $\alpha = \epsilon$.
9	3	1955	76	5	Sheet (0.005 in. thick); polished with fine emery; measured in vacuum (10^{-6} to 10^{-7} mm Hg); emittance for 300 K black body incident radiation; authors assumed $\alpha = \epsilon$.
10	3	1955	76	5	Sheet (0.020 in. thick); liquid honed; measured in vacuum (10^{-6} to 10^{-7} mm Hg); emittance for 300 K black body incident radiation; authors assumed $\alpha = \epsilon$.
11	3	1955	76	5	Sheet (0.005 in. thick); electrolytically cleaned; measured in vacuum (10^{-6} to 10^{-7} mm Hg); emittance for 300 K black body incident radiation; authors assumed $\alpha = \epsilon$.
12	3	1955	76	5	Sheet (0.005 in. thick); dry polished with plastic polishing wax abrasive; measured in vacuum (10^{-6} to 10^{-7} mm Hg); emittance for 300 K black body incident radiation; authors assumed $\alpha = \epsilon$.
13	3	1955	76	5	Sheet (0.005 in. thick); wet polished with Dr. Lyons tooth powder; measured in vacuum (10^{-6} to 10^{-7} mm Hg); emittance for 300 K black body incident radiation; authors assumed $\alpha = \epsilon$.
14	3	1955	76	5	Sheet (0.005 in. thick); cleaned in dilute chromic acid dip; measured in vacuum (10^{-6} to 10^{-7} mm Hg); emittance for 300 K black body incident radiation; authors assumed $\alpha = \epsilon$.
15	3	1955	76	5	Mill run sheet (0.005 in. thick); annealed and cleaned; measured in vacuum (10^{-6} to 10^{-7} mm Hg); emittance for 300 K black body incident radiation; authors assumed $\alpha = \epsilon$.
16	27	1959	338-1183		99.985 Cu (OFHC); polished on a cloth lap saturated with water and alumina; cycled to 1263 K in vacuum several times; measured in vacuum (5×10^{-5} mm Hg).
17	27	1959	258-803		99.985 Cu (OFHC); polished on a cloth lap saturated with water and alumina; cycled to 823 K in vacuum several times; measured in vacuum (5×10^{-5} mm Hg).
18	8	1963	373		Prefinished with crocus cloth and then electropolished; measured in vacuum (10^{-5} mm Hg).
19	8	1963	373		Above specimen and conditions except bombarded with hydrogen ions (0.8×10^{20} ions cm^{-2}).

SPECIFICATION TABLE NO. 45 (continued)

Curve No.	Ref. No.	Year	Temperature Range, K	Reported Error, %	Composition (weight percent), Specifications and Remarks
20	8	1963	373		Above specimen and conditions except bombarded with hydrogen ions (1.63×10^{20} ions cm^{-2}) .
21	8	1963	373		Above specimen and conditions except bombarded with hydrogen ions (6.60×10^{20} ions cm^{-2}) .
22	8	1963	373		Above specimen and conditions except bombarded with hydrogen ions (9.85×10^{20} ions cm^{-2}) .
23	8	1963	373		Prefinished with 600 grid silicon carbide paper and then sandblasted; measured in vacuum (10^{-5} mm Hg).
24	8	1963	373		Above specimen and conditions except bombarded with hydrogen ions (3.30×10^{20} ions cm^{-2}).
25	8	1963	373		Above specimen and conditions except bombarded with hydrogen ions (6.60×10^{20} ions cm^{-2}).
26	8	1963	373		Above specimen and conditions except bombarded with hydrogen ions (9.85×10^{20} ions cm^{-2}).
27	26	1962	275-400	7	Cleaned in sodium-dichromate and dilute nitric acid solutions, buffed on felt buffing wheel, cleaned with CCl_4 and acetone; measured in vacuum (10^{-6} mm Hg) ; data extracted from smooth curve.
28	7	1963	187	±2.5	Mechanically polished; measured in vacuum ($\leq 0.2 \times 10^{-4}$ mm Hg) ; temp stated is mean of hot and cold surfaces.
29	11	1962	384-523	±3	Polished by hand using fine abrasive papers; measured in vacuum (10^{-3} mm Hg) ; data extracted from smooth curve.
30	10	1949	373-648	<5	Polished; rinsed with distilled water and alcohol, dried in a stream of nitrogen; surface roughness 5 μ ; measured in vacuum; data extracted from smooth curve.
31	25	1962	320-800		Pure; electropolished; measured in vacuum (10^{-5} mm Hg) using equilibrium method; surface roughness 16 microinch (rms) .
32	25	1962	200-770		Curve 31 specimen and conditions except transient method used.
33	25	1962	250-850		Curve 32 specimen and conditions.
34	25	1962	200-247		Pure; electropolished; measured in vacuum (10^{-5} mm Hg) ; calculated from α/ϵ
35	24	1948	300		Oxidized and corroded; measured in air.
36	24	1948	299		Partially cleaned surface; measured in air.
37	24	1948	301		Fair polished; some pits; measured in air.
38	24	1948	299		Some spots and pits on surface; measured in air.

DATA TABLE NO. 45 HEMISPHERICAL TOTAL EMITTANCE OF COPPER

[Temperature, T, K; Emittance, ∈]

T	∈
CURVE 1	
300	0.020
CURVE 2	
589	0.495
644	0.500
700	0.525
755	0.545
811	0.595
867	0.640
922	0.710
1033	0.855
CURVE 3	
76	0.0176
CURVE 4*	
76	0.0185
CURVE 5*	
76	0.0192
CURVE 6*	
76	0.0178
CURVE 7	
76	0.03
CURVE 8*	
76	0.03
CURVE 9	
76	0.023
CURVE 10	
76	0.088
CURVE 11*	
76	0.017
CURVE 12*	
76	0.019
CURVE 13*	
76	0.018
CURVE 14*	
76	0.017
CURVE 15	
76	0.015
CURVE 16	
338	0.041
363	0.038
413	0.036
463	0.036
513	0.033
563	0.034
613	0.035
663	0.034
703	0.039
753	0.039
803	0.039
843	0.039
893	0.040
938	0.041
983	0.043
1028	0.045
1083	0.046
1133	0.047
1183	0.050
CURVE 17	
258	0.071
258	0.067
278	0.066
283	0.057
298	0.056
323	0.055
343	0.048
CURVE 17 (cont.)	
343	0.050
363	0.051
413	0.047
463	0.045
508	0.047
558	0.045
613	0.050
658	0.049
703	0.050
753	0.052
803	0.055
CURVE 18*	
373	0.0208
CURVE 19*	
373	0.0280
CURVE 20*	
373	0.0272
CURVE 21*	
373	0.0280
CURVE 22*	
373	0.0280
CURVE 23	
373	0.1876
CURVE 24	
373	0.0980
CURVE 25	
373	0.0760
CURVE 26	
373	0.0576
CURVE 27	
275	0.053
295	0.046
300	0.045
320	0.041
350	0.038
370	0.037
400	0.036
CURVE 28	
187	0.022
CURVE 29	
384	0.130
393	0.140
413	0.120
433	0.125
453	0.122
473	0.125
493	0.125
513	0.120
523	0.120
CURVE 30	
373	0.0300
423	0.0302
473	0.0305
523	0.0308
573	0.0310
623	0.0314
648	0.0316
CURVE 31	
320	0.0325
400	0.0290
600	0.0275
700	0.0290
800	0.0315
CURVE 32	
200	0.0310*
300	0.0280
380	0.0260
500	0.0255
CURVE 32 (cont.)	
630	0.0265
770	0.0285
CURVE 33*	
250	0.0360
300	0.0330
455	0.0300
550	0.0300
650	0.0300
850	0.0350
CURVE 34	
200	0.0346
211	0.0329
216	0.0315
221	0.0304
231	0.0290
240	0.0280
247	0.0270
CURVE 35	
300	0.38
CURVE 36	
299	0.085
CURVE 37*	
301	0.042
CURVE 38	
299	0.062

*Not shown on plot

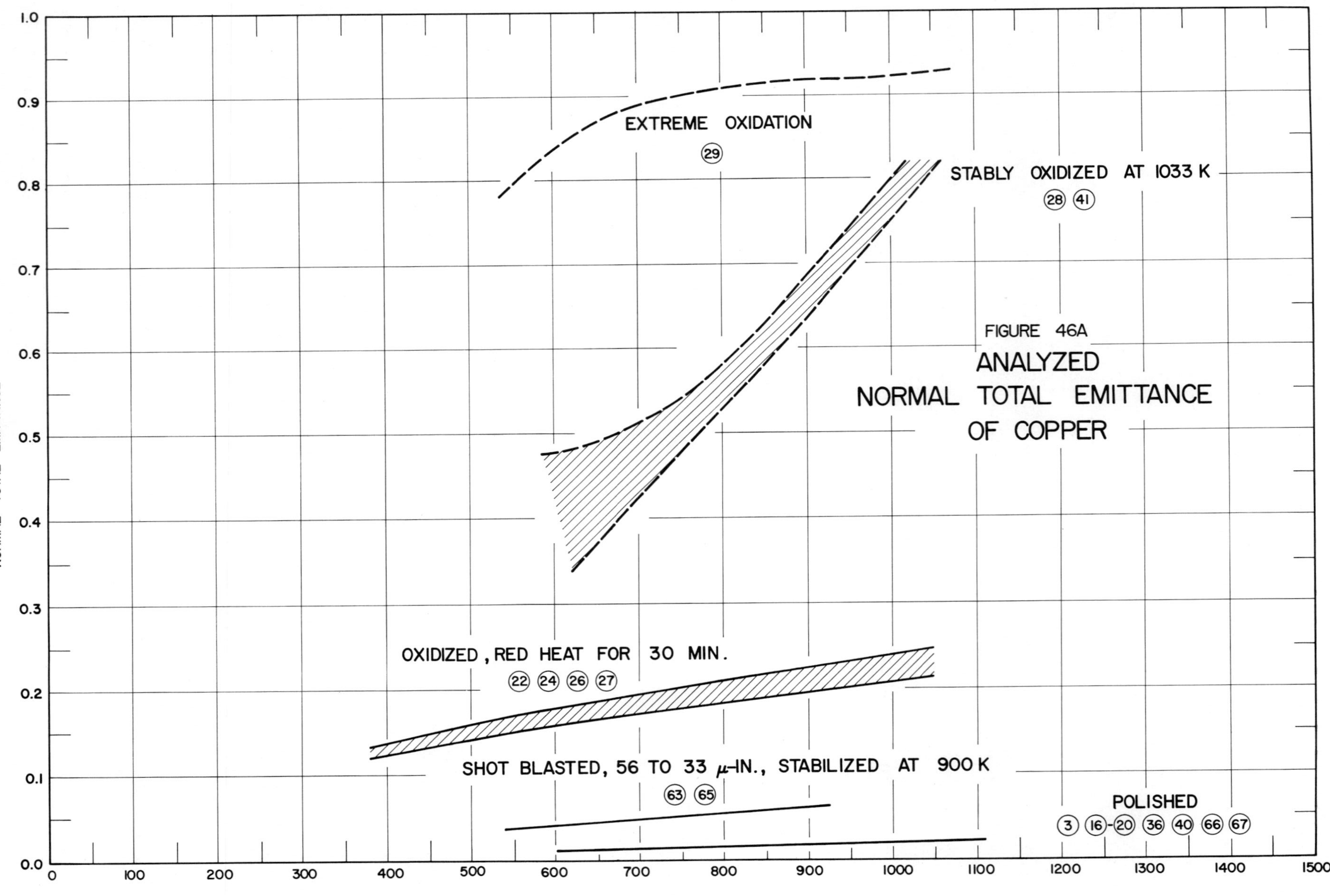
EXTREME OXIDATION
29
STABLY OXIDIZED AT 1033 K
28 41
FIGURE 46A
ANALYZED
NORMAL TOTAL EMITTANCE
OF COPPER
OXIDIZED, RED HEAT FOR 30 MIN.
22 24 26 27
SHOT BLASTED, 56 TO 33 μ-IN., STABILIZED AT 900 K
63 65
POLISHED
3 16-20 36 40 66 67
NORMAL TOTAL EMITTANCE
1.0
0.9
0.8
0.7
0.6
0.5
0.4
0.3
0.2
0.1
0.0
TEMPERATURE, K
0
100
200
300
400
500
600
700
800
900
1000
1100
1200
1300
1400
1500

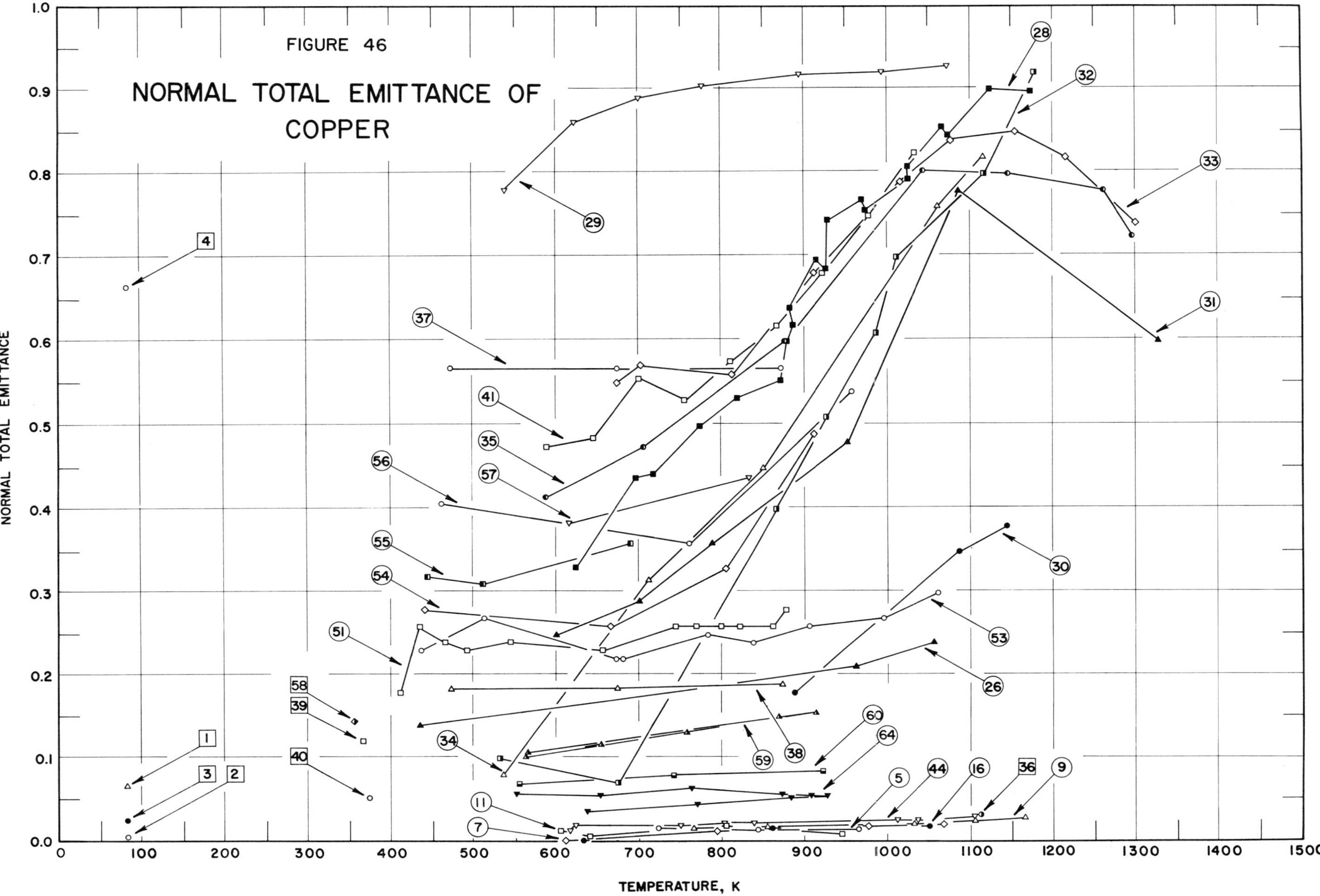
FIGURE 46
NORMAL TOTAL EMITTANCE OF COPPER
NORMAL TOTAL EMITTANCE
TEMPERATURE, K
1.0
0.9
0.8
0.7
0.6
0.5
0.4
0.3
0.2
0.1
0.0
0
100
200
300
400
500
600
700
800
900
1000
1100
1200
1300
1400
1500
1
2
3
4
5
7
9
11
16
26
28
29
30
31
32
33
34
35
36
37
38
39
40
41
44
51
53
54
55
56
57
58
59
60
64

SPECIFICATION TABLE NO. 46 NORMAL TOTAL EMITTANCE OF COPPER

Curve No.	Ref. No.	Year	Temperature Range, K	Geometry θ'	Reported Error, %	Composition (weight percent), Specifications and Remarks
1	34	1957	83.2	$\sim 0°$	± 10	Electrolytic, tough pitch; federal specification QQ-C-576 or QQ-C-502; nominal composition: 99.90 Cu, 0.04 O; strip (0.005 in. thick); as received; measured in air (5×10^{-4} mm Hg).
2	34	1957	83.2	$\sim 0°$	± 10	Different sample, same as Curve 1 specimen and conditions except cleaned with liquid detergent.
3	34	1957	83.2	$\sim 0°$	± 10	Different sample, same as Curve 1 specimen and conditions except polished.
4	34	1957	83.2	$\sim 0°$	± 10	Different sample, same as Curve 1 specimen and conditions except oxidized in air at red heat for 30 min.
5	34	1957	722-966	$\sim 0°$	± 10	Electrolytic, tough pitch; federal specification QQ-C-576 or QQ-C-502; nominal composition: 99.90 Cu, 0.04 O; strip (0.005 in. thick); as received; measured in air (5×10^{-4} mm Hg); increasing temp, Cycle 1.
6	34	1957	1014-741	$\sim 0°$	± 10	Above specimen and conditions; decreasing temp, Cycle 1.
7	34	1957	611-1069	$\sim 0°$	± 10	Above specimen and conditions; increasing temp, Cycle 2.
8	34	1957	989-603	$\sim 0°$	± 10	Above specimen and conditions; decreasing temp, Cycle 2.
9	34	1957	766-1166	$\sim 0°$	± 10	Above specimen and conditions; increasing temp, Cycle 3.
10	34	1957	1044-633	$\sim 0°$	± 10	Above specimen and conditions; decreasing temp, Cycle 3.
11	34	1957	605-944	$\sim 0°$	± 10	Different sample, same as Curve 5 specimen and conditions except cleaned with liquid detergent; increasing temp, Cycle 1.
12	34	1957	855-697	$\sim 0°$	± 10	Above specimen and conditions; decreasing temp, Cycle 1.
13	34	1957	583-950	$\sim 0°$	± 10	Above specimen and conditions; increasing temp, Cycle 2.
14	34	1957	666-916	$\sim 0°$	± 10	Above specimen and conditions; increasing temp, Cycle 3.
15	34	1957	794-711	$\sim 0°$	± 10	Above specimen and conditions; decreasing temp, Cycle 3.
16	34	1957	633-1050	$\sim 0°$	± 10	Different sample, same as Curve 5 specimen and conditions except polished; increasing temp, Cycle 1.
17	34	1957	842-622	$\sim 0°$	± 10	Above specimen and conditions; decreasing temp, Cycle 1.
18	34	1957	783-1089	$\sim 0°$	± 10	Above specimen and conditions; increasing temp, Cycle 2.
19	34	1957	883-600	$\sim 0°$	± 10	Above specimen and conditions; decreasing temp, Cycle 2.
20	34	1957	811-1078	$\sim 0°$	± 10	Above specimen and conditions; increasing temp, Cycle 3.
21	34	1957	672	$\sim 0°$	± 10	Above specimen and conditions; decreasing temp, Cycle 3.
22	34	1957	383-855	$\sim 0°$	± 10	Different sample, same as Curve 5 specimen and conditions except oxidized in air at red heat for 30 min.; increasing temp, Cycle 1.
23	34	1957	816-422	$\sim 0°$	± 10	Above specimen and conditions; decreasing temp, Cycle 1.

SPECIFICATION TABLE NO. 46 (continued)

Curve No.	Ref. No.	Year	Temperature Range, K	Geometry θ'	Reported Error, %	Composition (weight percent), Specifications and Remarks
24	34	1957	400-878	$\sim 0^\circ$	± 10	Above specimen and conditions; increasing temp, Cycle 2.
25	34	1957	739-439	$\sim 0^\circ$	± 10	Above specimen and conditions; decreasing temp, Cycle 2.
26	34	1957	433-1055	$\sim 0^\circ$	± 10	Above specimen and conditions; increasing temp, Cycle 3.
27	34	1957	839-683	$\sim 0^\circ$	± 10	Above specimen and conditions; decreasing temp, Cycle 3.
28	33	1958	622-1173	$\sim 0^\circ$		Polished and oxidized.
29	28	1958	539-1072	$\sim 0^\circ$		Oxidized.
30	31	1963	888-1144	$\sim 0^\circ$	± 10	Smooth; measured in dry air; run 1; [Author's designation: Specimen No. 1].
31	31	1963	600-1327	$\sim 0^\circ$	± 10	Above specimen and conditions; run 2.
32	31	1963	531-1177	$\sim 0^\circ$	± 10	Smooth; measured in dry air; run 1; [Author's designation: Specimen No. 3].
33	31	1963	673-1300	$\sim 0^\circ$	± 10	Above specimen and conditions; run 2.
34	31	1963	536-1116	$\sim 0^\circ$	± 10	Smooth; measured in dry air; run 1; [Author's designation: Specimen No. 4].
35	31	1963	587-1297	$\sim 0^\circ$	± 10	Above specimen and conditions; run 2.
36	30	1963	1111	$\sim 0^\circ$		Mechanically polished; heated at 450 K in air for 3 hrs and at 922 K in air for 3 hrs and then kept at 1242 K in vacuum for 45 hrs; data calculated from normal spectral emittance and corrected to 1111 K.
37	14	1913	473-873	$\sim 0^\circ$		Cleaned, polished and then oxidized.
38	14	1913	473-873	$\sim 0^\circ$		Calorized; surface impregnated with aluminum, cleaned, polished and then oxidized.
39	16	1937	367	$\sim 0^\circ$	± 1.1	Buffed.
40	15	1947	373	$\sim 0^\circ$		Polished.
41	22	1958	589-1033	$\sim 0^\circ$	≤ 2	Stably oxidized at 1033 K in quiescent air.
42	29	1956	721-1093	$\sim 0^\circ$	± 11	Electrolytic tough pitch copper (Federal specifications QQ-C-576 and QQ-C-502 from Central Steel and Wire Co.); as received; measured in vacuum (10^{-5} to 10^{-4} mm Hg); Cycle 1, heating.
43	29	1956	611-1078	$\sim 0^\circ$	± 11	Above specimen and conditions; Cycle 2, heating.
44	29	1956	616-1103	$\sim 0^\circ$	± 11	Above specimen and conditions; Cycle 3, heating.
45	29	1956	616-947	$\sim 0^\circ$	± 11	Different sample, same as Curve 42 specimen and conditions except cleaned with detergent soap and smoothed by pressing between steel plates; Cycle 1, heating.
46	29	1956	658-922	$\sim 0^\circ$	± 11	Above specimen and conditions; Cycle 2, heating.
47	29	1956	589-933	$\sim 0^\circ$	± 11	Above specimen and conditions; Cycle 3, heating.
48	29	1956	633-1047	$\sim 0^\circ$	± 11	Different sample, same as Curve 42 specimen and conditions except smoothed, cleaned, and polished; Cycle 1.

SPECIFICATION TABLE NO. 46 (continued)

Curve No.	Ref. No.	Year	Temperature Range, K	Geometry θ'	Reported Error, %	Composition (weight percent), Specifications and Remarks
49	29	1956	622-1089	$\sim 0^\circ$	± 11	Above specimen and conditions; Cycle 2.
50	29	1956	661-1022	$\sim 0^\circ$	± 11	Above specimen and conditions; Cycle 3.
51	29	1956	411-878	$\sim 0^\circ$	± 11	Different sample, same as Curve 42 specimen and conditions except oxidized in air 30 min. at 922 K; Cycle 1.
52	29	1956	400-811	$\sim 0^\circ$	± 11	Above specimen and conditions; Cycle 2.
53	29	1956	436-1061	$\sim 0^\circ$	± 11	Above specimen and conditions; Cycle 3.
54	29	1956	440-911	$\sim 0^\circ$	± 11	Different sample, same as Curve 51 specimen and conditions except measured in air; Cycle 1, heating.
55	29	1956	689-444	$\sim 0^\circ$	± 11	Above specimen and conditions; Cycle 1, cooling.
56	29	1956	461-958	$\sim 0^\circ$	± 11	Above specimen and conditions; Cycle 2, heating.
57	29	1956	833-616	$\sim 0^\circ$	± 11	Above specimen and conditions; Cycle 2, cooling.
58	218	1963	356	$\sim 0^\circ$		Sand blasted; measured in vacuum (1.6×10^{-4} mm Hg).
59	227	1964	562-913	$\sim 0^\circ$		99.9 +% pure; steel shot blasted; preheated in air at 800 K for 2 minutes; surface roughness before emittance test 348 microinches (rms), after emittance test 204 microinches (rms); measured in vacuum (6×10^{-5} mm Hg); $\omega' = 3.4 \times 10^{-4}$ sr.; first temperature cycle; [Authors' designation: Sample 8].
60	227	1964	555-921	$\sim 0^\circ$		99.9 +% pure; Glas-Shot blasted; preheated in air at 800 K for 3 minutes; surface roughness before emittance test 176 microinches (rms); measured in vacuum (5×10^{-5} mm Hg); $\omega' = 3.4 \times 10^{-4}$ sr.; [Authors' designation: Sample 7].
61	227	1964	923-561	$\sim 0^\circ$		Above specimen and conditions except second cycle after stabilizing at 923 K.
62	227	1964	552-866	$\sim 0^\circ$		99.9 + % pure; Glas-Shot blasted; preheated in vacuum at 800 K; surface roughness before emittance test 121 microinches (rms), after emittance test 56 microinches (rms); measured in vacuum (5×10^{-5} mm Hg); $\omega' = 3.4 \times 10^{-4}$ sr.; first temperature cycle; [Authors' designation: Sample 5].
63	227	1964	913-546	$\sim 0^\circ$		Above specimen and conditions except second cycle after stabilizing at 913 K.
64	227	1964	551-928	$\sim 0^\circ$		99.9 + % pure; Glas-Shot blasted; preheated in vacuum at 800 K for 0.5 hr; surface roughness before emittance test 76 microinches (rms), after emittance test 33 microinches (rms); measured in vacuum (4×10^{-5} mm Hg); $\omega' = 3.4 \times 10^{-4}$ sr.; first temperature cycle; [Authors' designation: Sample 3].
65	227	1964	901-545	$\sim 0^\circ$		Above specimen and conditions except second temperature cycle after stabilizing at 900 K.
66	227	1964	545-1035	$\sim 0^\circ$		99.9 + % pure; precleaned, electro-polished, hot and cold water rinsed, dipped in HBF_4 (1 : 2), cold water and alcohol rinsed; heated in vacuum to 550 K for 0.5 hr; surface roughness before emittance test 5.2 microinches (rms), after emittance test 5.7 microinches (rms); measured in vacuum (5×10^{-6} mm Hg); $\omega' = 3.4 \times 10^{-4}$ sr.; [Authors' designation: Sample 1].
67	227	1964	755-865	$\sim 0^\circ$		Above specimen and conditions except second cycle after recrystallization.

DATA TABLE NO. 46 NORMAL TOTAL EMITTANCE OF COPPER

[Temperature, T, K; Emittance, ∈]

T	∈
CURVE 1	
83.2	0.066
CURVE 2	
83.2	0.066
CURVE 3	
83.2	0.025
CURVE 4	
83.2	0.665
CURVE 5	
722	0.016
844	0.014
966	0.014
CURVE 6*	
1014	0.021
922	0.018
741	0.016
CURVE 7	
611	0.001
794	0.013
978	0.019
1069	0.021
CURVE 8*	
989	0.017
878	0.018
744	0.012
678	0.010
603	0.008

T	∈
CURVE 9	
766	0.016
855	0.018
1033	0.021
1105	0.025
1166	0.029
CURVE 10*	
1044	0.015
955	0.016
633	0.012
CURVE 11	
605	0.013
639	0.007
805	0.019
944	0.009
CURVE 12*	
855	0.010
697	0.011
CURVE 13*	
583	0.010
628	0.011
755	0.013
872	0.015
950	0.016
CURVE 14*	
666	0.012
728	0.012
816	0.012
839	0.015
916	0.015
CURVE 15*	
794	0.015
711	0.014

T	∈
CURVE 16	
633	0.002
861	0.016
1050	0.019
CURVE 17*	
842	0.015
622	0.006
CURVE 18*	
783	0.015
911	0.017
1011	0.020
1089	0.021
CURVE 19*	
883	0.016
755	0.014
600	0.006
CURVE 20*	
811	0.016
892	0.017
1078	0.018
CURVE 21*	
672	0.012
CURVE 22*	
383	0.12
483	0.15
644	0.18
722	2.00
794	2.00
855	2.05
CURVE 23*	
816	0.20

T	∈
CURVE 23* (cont.)	
766	0.20
666	0.18
544	0.175
466	0.175
422	0.185
CURVE 24*	
400	0.120
566	0.155
666	0.150
700	0.180
878	0.210
CURVE 25*	
739	0.195
561	0.175
439	0.170
CURVE 26	
433	0.14
961	0.21
1055	0.24
CURVE 27*	
839	0.19
683	0.17
CURVE 28	
622	0.332
697	0.440
717	0.443
773	0.500
819	0.532
871	0.552
880	0.600
886	0.620
883	0.640
915	0.697
927	0.686

T	∈
CURVE 28 (cont.)	
929	0.743
970	0.769
973	0.757
1025	0.792
1025	0.809
1068	0.855
1074	0.846
1123	0.900
1173	0.897
CURVE 29	
539	0.780
622	0.864
700	0.890
778	0.905
894	0.919
994	0.922
1072	0.929
CURVE 30	
888	0.18
1088	0.35
1144	0.38
CURVE 31	
600	0.25
700	0.29
788	0.36
951	0.48
1086	0.78
1327	0.60
CURVE 32	
531	0.10
673	0.07
866	0.40
927	0.51
986	0.61
1011	0.70
1116	0.80
1177	0.92

T	∈
CURVE 33	
673	0.55
701	0.57
812	0.56
911	0.68
1016	0.79
1077	0.84
1155	0.85
1216	0.82
1300	0.74
CURVE 34	
536	0.080
711	0.315
850	0.450
1061	0.760
1116	0.820
CURVE 35	
587	0.415
705	0.475
877	0.600
1044	0.803
1147	0.800
1261	0.780
1297	0.725
CURVE 36	
1111	0.032
CURVE 37	
473	0.568
673	0.568
873	0.568
CURVE 38	
473	0.185
673	0.185
873	0.190

T	∈
CURVE 39	
367	0.12
CURVE 40	
373	0.052
CURVE 41	
589	0.475
644	0.485
700	0.505
755	0.530
811	0.575
867	0.620
922	0.680
978	0.750
1033	0.825
CURVE 42*	
721	0.0222
743	0.0200
858	0.0320
928	0.0260
976	0.0340
1022	0.0262
1093	0.0290
CURVE 43*	
611	0.016
627	0.012
683	0.017
743	0.020
789	0.020
886	0.024
983	0.026
990	0.025
1078	0.028
CURVE 44	
616	0.013
622	0.020
750	0.020

T	∈
CURVE 44 (cont.)	
883	0.023
889	0.023
1011	0.026
1036	0.026
1103	0.030
CURVE 45*	
616	0.020
644	0.015
694	0.021
811	0.023
861	0.021
947	0.019
CURVE 46*	
658	0.020
716	0.018
722	0.019
803	0.020
816	0.018
922	0.017
CURVE 47*	
589	0.020
616	0.030
722	0.017
750	0.020
811	0.020
839	0.022
861	0.020
933	0.020
CURVE 48*	
633	0.013
639	0.007
850	0.023
866	0.022
1047	0.027

*Not shown on plot

DATA TABLE NO. 46 (continued)

T	∈
CURVE 49*	
622	0.011
766	0.020
789	0.019
894	0.020
922	0.020
1022	0.025
1089	0.028
CURVE 50*	
661	0.015
783	0.019
786	0.017
880	0.021
916	0.025
1022	0.026
CURVE 51	
411	0.18
433	0.26
466	0.24
491	0.23
544	0.24
655	0.23
744	0.26
769	0.26
800	0.26
822	0.26
861	0.26
878	0.28
CURVE 52*	
400	0.18
500	0.24
561	0.24
566	0.21
666	0.25
672	0.20
741	0.25
761	0.24
811	0.26

T	∈
CURVE 53	
436	0.23
514	0.27
672	0.22
680	0.22
783	0.25
839	0.24
905	0.26
997	0.27
1061	0.30
CURVE 54	
440	0.280
666	0.260
805	0.330
911	0.490
CURVE 55	
689	0.360
511	0.310
444	0.320
CURVE 56	
461	0.407
761	0.360
958	0.540
CURVE 57	
833	0.440
616	0.385
CURVE 58	
356	0.147
CURVE 59	
562	0.101
757	0.130
867	0.149
913	0.155
913	0.156*
565	0.107
652	0.117

T	∈
CURVE 60	
555	0.068
741	0.079
921	0.084
CURVE 61*	
923	0.082
891	0.081
868	0.075
750	0.067
645	0.061
561	0.054
923	0.080
CURVE 62*	
552	0.072
660	0.086
748	0.093
858	0.075
866	0.057
574	0.046
CURVE 63*	
913	0.063
865	0.059
749	0.050
546	0.032
660	0.044
902	0.062
CURVE 64	
551	0.057
653	0.055
763	0.063
873	0.057
907	0.056
928	0.056
883	0.052
770	0.045
636	0.036

T	∈
CURVE 65*	
901	0.058
857	0.056
748	0.049
656	0.041
545	0.033
CURVE 66*	
545	0.004
755	0.015
865	0.020
915	0.023
1035	0.025
CURVE 67*	
755	0.016
865	0.019

*Not shown on plot

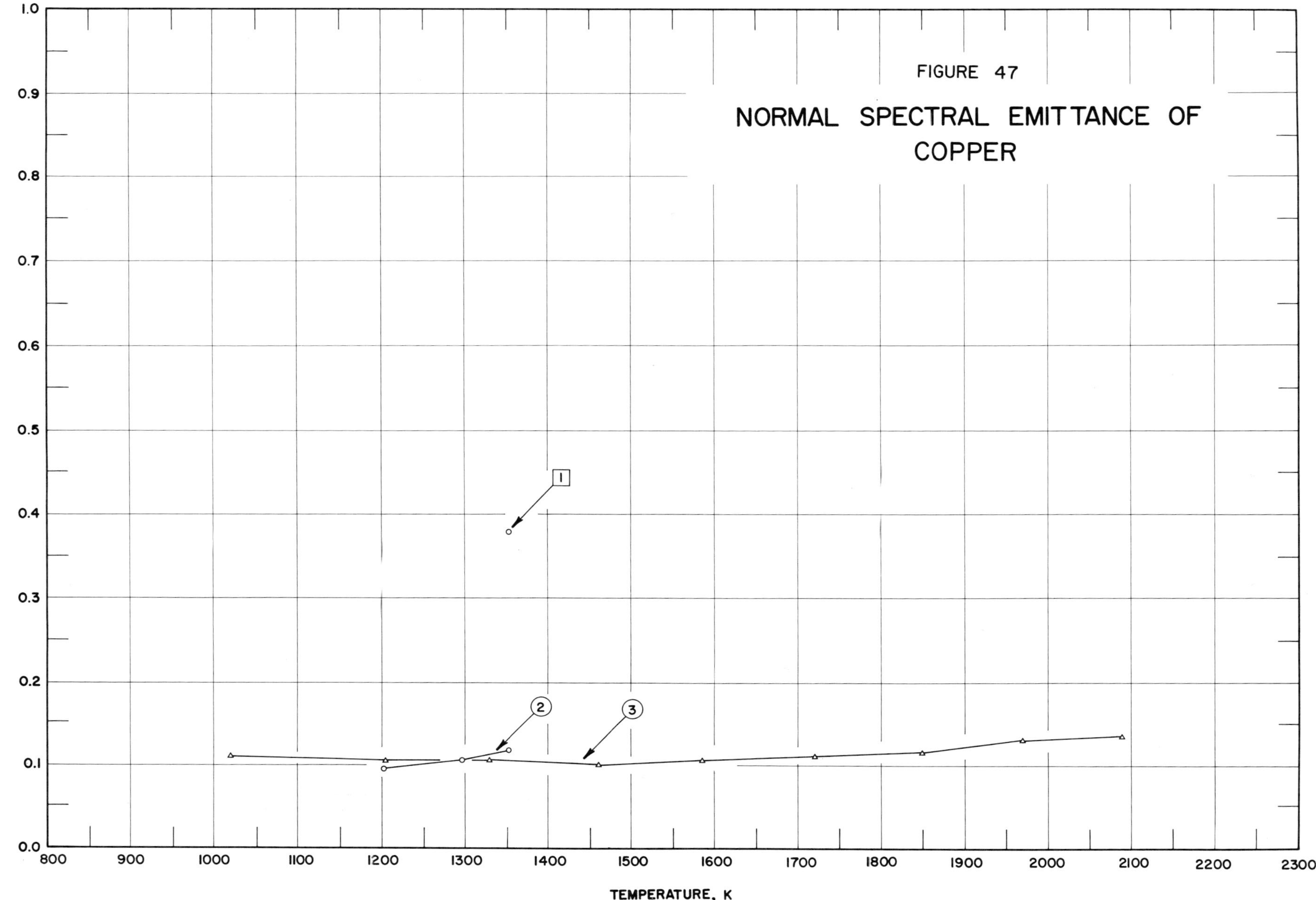
FIGURE 47
NORMAL SPECTRAL EMITTANCE OF
COPPER
1.0
0.9
0.8
0.7
0.6
0.5
0.4
0.3
0.2
0.1
0.0
NORMAL SPECTRAL EMITTANCE
800
900
1000
1100
1200
1300
1400
1500
1600
1700
1800
1900
2000
2100
2200
2300
TEMPERATURE, K
1
2
3

SPECIFICATION TABLE NO. 47 NORMAL SPECTRAL EMITTANCE OF COPPER

Curve No.	Ref. No.	Year	Wavelength μ	Temperature Range, K	Geometry θ'	Reported Error, %	Composition (weight percent), Specifications and Remarks
1	19	1914	0.55	1353	$\sim 0^\circ$	1	Film; tungsten substrate; measured in hydrogen; Pt reference ($\epsilon = 0.38$ for $\lambda = 0.547\ \mu$ at all temp).
2	19	1914	0.65	1203-1353	$\sim 0^\circ$	1	Film; tungsten substrate; measured in hydrogen; Pt reference ($\epsilon = 0.33$ for $\lambda = 0.650\ \mu$ at all temp).
3	35	1914	0.66	1020-2090	$\sim 0^\circ$		Solid and liquid states; measured in hydrogen.

DATA TABLE NO. 47 NORMAL SPECTRAL EMITTANCE OF COPPER

[Temperature, T, K; Emittance, ϵ; Wavelength, λ, μ]

T	ϵ
CURVE 1 $\lambda = 0.55$	
1353	0.38
CURVE 2 $\lambda = 0.65$	
1203	0.096
1298	0.105
1353	0.117
CURVE 3 $\lambda = 0.66$	
1020	0.110
1205	0.105
1330	0.105
1460	0.100
1585	0.105
1720	0.110
1850	0.115
1970	0.130
2090	0.135

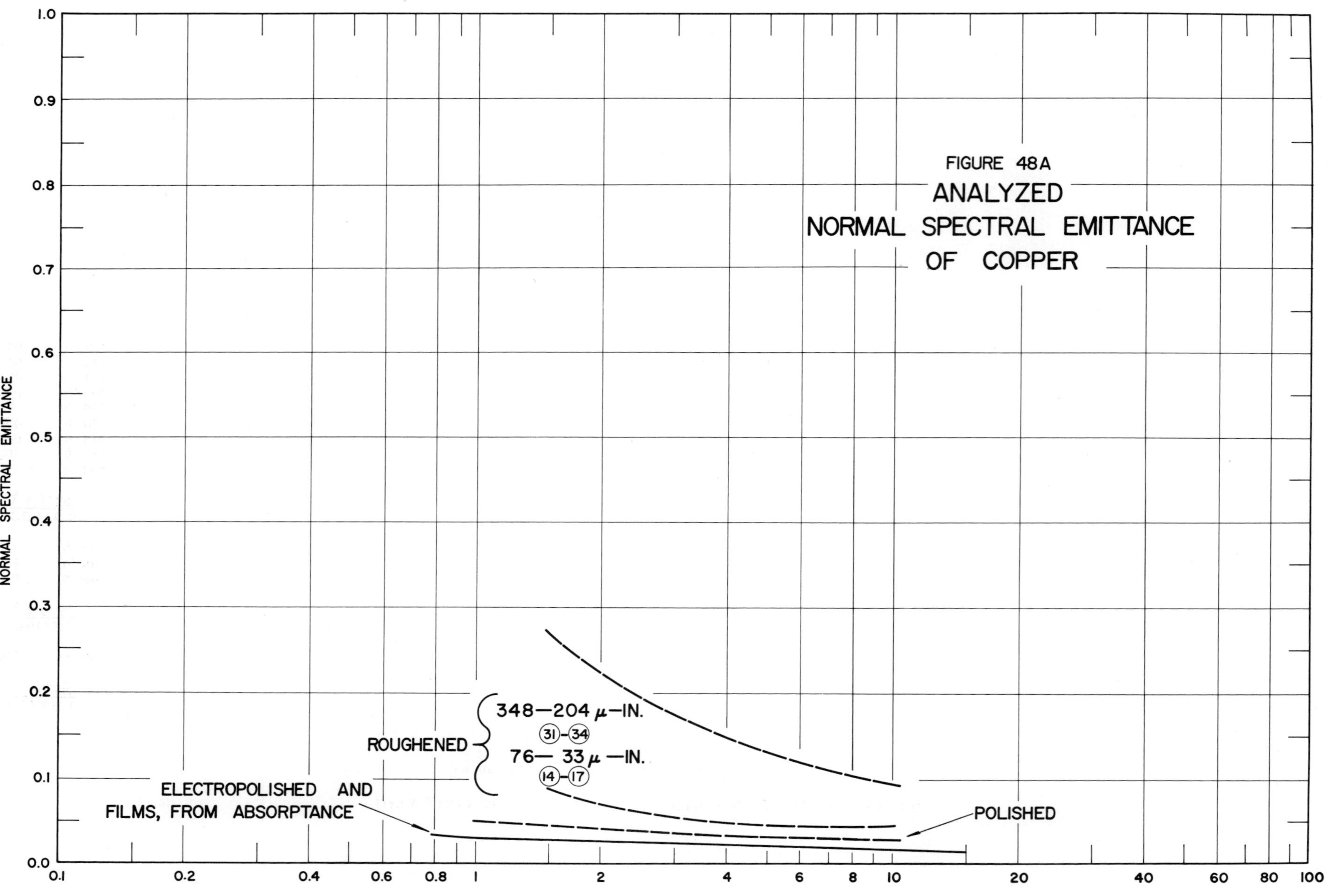
FIGURE 48A
ANALYZED
NORMAL SPECTRAL EMITTANCE
OF COPPER
1.0
0.9
0.8
0.7
0.6
0.5
0.4
0.3
0.2
0.1
0.0
NORMAL SPECTRAL EMITTANCE
348—204 μ—IN.
(31)-(34)
ROUGHENED
76— 33 μ —IN.
(14)-(17)
ELECTROPOLISHED AND
FILMS, FROM ABSORPTANCE
POLISHED
0.1
0.2
0.4
0.6
0.8
1
2
4
6
8
10
20
40
60
80
100
WAVELENGTH, MICRONS

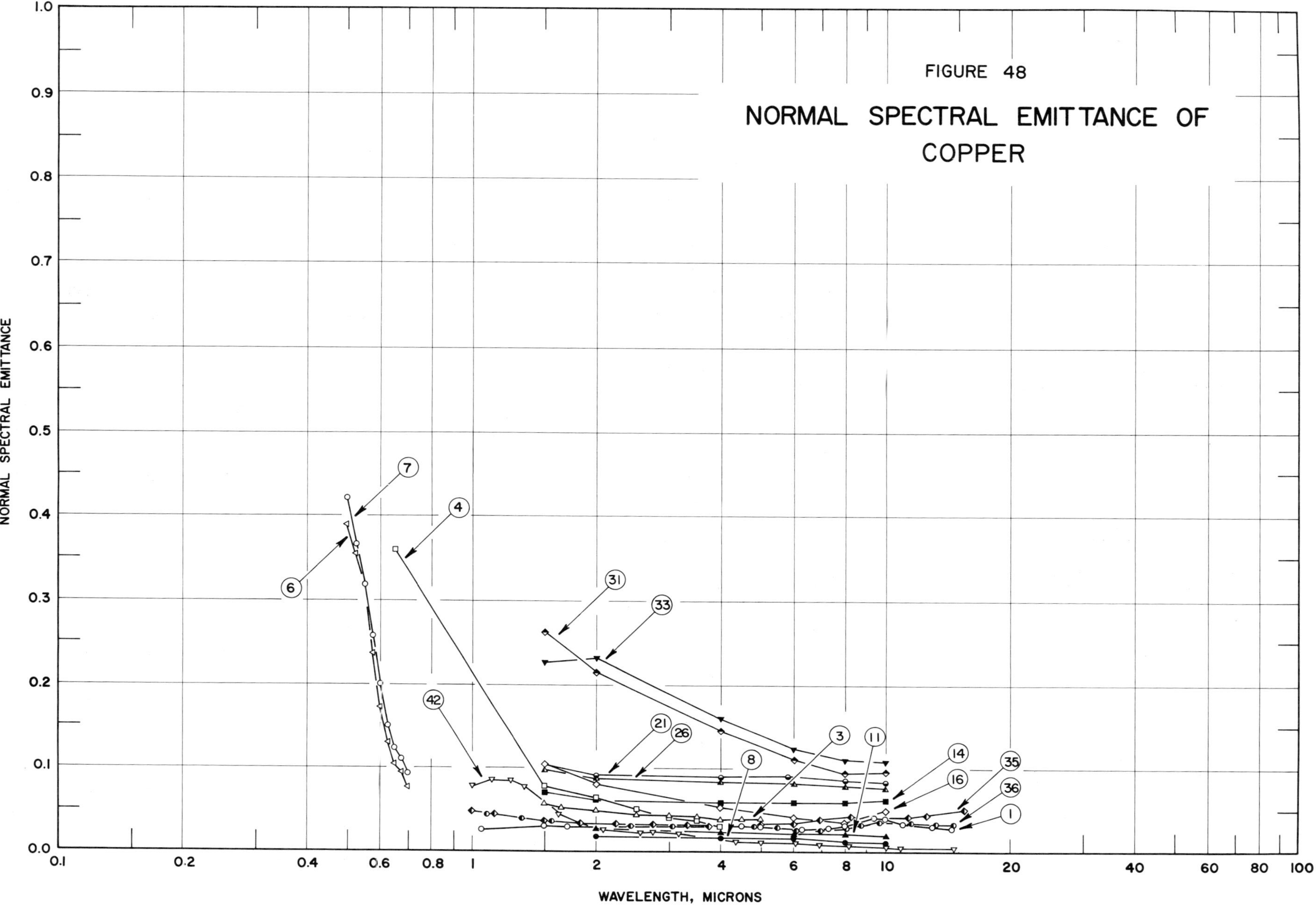
FIGURE 48
NORMAL SPECTRAL EMITTANCE OF COPPER
NORMAL SPECTRAL EMITTANCE
WAVELENGTH, MICRONS
0.0
0.1
0.2
0.3
0.4
0.5
0.6
0.7
0.8
0.9
1.0
0.1
0.2
0.4
0.6
0.8
1
2
4
6
8
10
20
40
60
80
100
7
4
6
31
33
42
21
26
8
3
11
14
16
35
36
1

SPECIFICATION TABLE NO. 48 NORMAL SPECTRAL EMITTANCE OF COPPER

Curve No.	Ref. No.	Year	Temperature K	Wavelength Range, μ	Geometry θ'	Reported Error, %	Composition (weight percent), Specifications and Remarks
1	30	1963	1242	1.06-14.50	$\sim 0^\circ$		Polished; heated at 450 K for 3 hrs and at 922 K for 3 hrs; surface roughness at this stage 0.05 μ peak to peak and 10 μ lateral; again heated at 1222 K for 45 hrs in vacuum; measured in vacuum (5×10^{-6} mm Hg); data extracted from smooth curve; [Author's designation: Sample 3].
2	38	1933	973	1.5-5.0	$<4^\circ$		Washed and polished; measured in vacuum.
3	38	1933	1123	1.5-5.0	$<4^\circ$		Above specimen and conditions.
4	37	1947	1174	0.65-4.0	$\sim 0^\circ$	5-10	Measured in vacuum.
5	36	1913	1262	0.500-0.700	$\sim 0^\circ$		Prepared electrolytically; polished and washed; measured in reducing atmosphere.
6	36	1913	1264	0.500-0.700	$\sim 0^\circ$		Above specimen and conditions.
7	36	1913	1326	0.500-0.700	$\sim 0^\circ$		Above specimen and conditions.
8	227	1964	755	2-10	$\sim 0^\circ$		>99.9 Cu; precleaned, electropolished, hot and cold water rinsed, dipped in HBF_4 (1:2), cold water and alcohol rinsed; heated in vacuum to 550 K for 0.5 hr; surface roughness before emittance test 5.2 microinches rms, after emittance test 5.7 microinches rms; measured in vacuum (5×10^{-6} mm Hg); $\omega' = 3.4 \times 10^{-4}$ sr; [Author's designation: Sample 1].
9	227	1964	865	1.5-10	$\sim 0^\circ$		Above specimen and conditions.
10	227	1964	915	1.5-10	$\sim 0^\circ$		Above specimen and conditions.
11	227	1964	1035	1.5-10	$\sim 0^\circ$		Above specimen and conditions.
12	227	1964	755	1.5-10	$\sim 0^\circ$		Curve 8 specimen and conditions; second cycle after recystallization.
13	227	1964	865	1.5-10	$\sim 0^\circ$		Above specimen and conditions.
14	227	1964	763	1.5-10	$\sim 0^\circ$		>99.9 Cu; Glas-Shot blasted; preheated in vacuum at 800 K for 0.5 hr; surface roughness before emittance test 76 microinches rms, after emittance test 33 microinches rms; measured in vacuum (4×10^{-5} mm Hg); $\omega' = 3.4 \times 10^{-4}$ sr; first temperature cycle; [Author's designation: Sample 3].
15	227	1964	873	1.5-10	$\sim 0^\circ$		Above specimen and conditions.
16	227	1964	907	1.5-10	$\sim 0^\circ$		Above specimen and conditions.
17	227	1964	636	2-8	$\sim 0^\circ$		Above specimen and conditions.
18	227	1964	901	1.5-10	$\sim 0^\circ$		Curve 14 specimen and conditions; second temperature cycle after stabilizing at 900 K.
19	227	1964	857	1.5-8	$\sim 0^\circ$		Above specimen and conditions.
20	227	1964	748	1.5-8	$\sim 0^\circ$		Above specimen and conditions.
21	227	1964	748	1.5-10	$\sim 0^\circ$		>99.9 Cu; Glas-Shot blasted; preheated in vacuum at 800 K; surface roughness before emittance test 121 microinches rms, after emittance test 56 microinches rms; measured in vacuum (5×10^{-5} mm Hg); $\omega' = 3.4 \times 10^{-4}$ sr; first temperature cycle; [Author's designation: Sample 5].

SPECIFICATION TABLE NO. 48 (continued)

Curve No.	Ref. No.	Year	Temperature K	Wavelength Range, μ	Geometry θ'	Reported Error, %	Composition (weight percent), Specifications and Remarks
22	227	1964	858	1.5-10	~0°		Above specimen and conditions.
23	227	1964	913	1.5-10	~0°		Curve 21 specimen and conditions; second cycle after stabilizing at 913 K.
24	227	1964	865	1.5-10	~0°		Above specimen and conditions.
25	227	1964	749	1.5-10	~0°		Above specimen and conditions.
26	227	1964	741	1.5-10	~0°		>99.9 Cu; Glas-Shot blasted; preheated in air at 800 K for 3 min; surface roughness before emittance test 176 microinches rms; measured in vacuum (5×10^{-5} mm Hg); $\omega' = 3.4 \times 10^{-4}$ sr.; [Author's designation: Sample 7].
27	227	1964	891	1.5-10	~0°		Above specimen and conditions; second cycle after stabilizing at 923 K.
28	227	1964	868	1.5-10	~0°		Above specimen and conditions.
29	227	1964	750	1.5-10	~0°		Above specimen and conditions.
30	227	1964	923	1.5-10	~0°		Above specimen and conditions.
31	227	1964	757	1.5-10	~0°		>99.9 Cu; Steel-shot blasted; preheated in air at 800 K for 2 min; surface roughness before emittance test 348 microinches rms, after emittance test 204 microinches rms; measured in vacuum (6×10^{-5} mm Hg); $\omega' = 3.4 \times 10^{-4}$ sr.; [Author's designation: Sample 8].
32	227	1964	867	1.5-10	~0°		Above specimen and conditions.
33	227	1964	913	1.5-10	~0°		Above specimen and conditions.
34	227	1964	652	2-10	~0°		Above specimen and conditions.
35	317	1962	1244	1.000-15.49	0°		Sanded to a 4/0 finish with emery paper; buffed to a mirror finish with an aluminum oxide solution on a buffing wheel; annealed 3 hrs at 449 K and 3 hrs at 922 K; measured in vacuum.
36	317	1962	1258	1.000-14.72	0°		Above specimen and conditions except annealed 14 hrs at 1255 K.
37	317	1962	1241	1.000-15.14	0°		Above specimen and conditions except annealed 36 hrs at 1255 K.
38	317	1962	1235	1.000-14.35	0°		Above specimen and conditions except annealed 42 hrs at 1255 K.
39	317	1962	1202	1.028-13.43	0°		Above specimen and conditions except annealed 62 hrs at 1255 K.
40	317	1962	294	1.000-14.89	~0°		Above specimen and conditions except annealed 62 hrs at 1255 K; calculated from reflectance data.
41	317	1962	294	1.000-14.55	~0°		Sanded to a 4/0 finish with emery paper; buffed to a mirror finish with an aluminum oxide solution on a buffing wheel; measured in vacuum; calculated from reflectance data; [Author's designation: Specimen No. 3].
42	317	1962	294	1.000-14.76	~0°		Above specimen and conditions except annealed for 3 hrs at 449 K and 3 hrs at 922 K.

DATA TABLE NO. 48 NORMAL SPECTRAL EMITTANCE OF COPPER

[Wavelength, λ, μ; Emittance, ∈; Temperature, T, K]

λ	∈
CURVE 1	**T = 1242**
1.06	0.028
1.50	0.032
1.70	0.031
4.50	0.032
5.00	0.031
6.30	0.029
7.30	0.030
8.00	0.033
9.40	0.042
10.00	0.041
11.00	0.036
13.00	0.032
14.50	0.030
CURVE 2*	**T = 973**
1.5	0.061
1.75	0.056
2.0	0.0505
2.5	0.045
3.0	0.042
3.5	0.039
4.0	0.038
4.5	0.036
5.0	0.0355
CURVE 3	**T = 1123**
1.5	0.058
1.75	0.053
2.0	0.050
2.5	0.0465
3.0	0.045
3.5	0.043
4.0	0.039
4.5	0.040
5.0	0.040
CURVE 4	**T = 1174**
0.65	0.360
1.5	0.079
2.0	0.065
CURVE 4 (cont.)	**T = 1174**
2.5	0.052
3.0	0.043
3.5	0.038
4.0	0.032
CURVE 5*	**T = 1262**
0.500	0.415
0.525	0.352
0.550	0.317
0.575	0.246
0.600	0.193
0.625	0.147
0.650	0.116
0.675	0.107
0.700	0.101
CURVE 6	**T = 1264**
0.500	0.389
0.525	0.355
0.550	0.308*
0.575	0.237
0.600	0.173
0.625	0.130
0.650	0.104
0.675	0.095
0.700	0.077
CURVE 7	**T = 1326**
0.500	0.421
0.525	0.367
0.550	0.319
0.575	0.257
0.600	0.201
0.625	0.150
0.650	0.124
0.675	0.111
0.700	0.094
CURVE 8	**T = 755**
2	0.019
4	0.018
6	0.018
8	0.013
10	0.012
CURVE 9*	**T = 865**
1.5	0.024
2	0.022
4	0.022
6	0.020
8	0.016
10	0.015
CURVE 10*	**T = 915**
1.5	0.026
2	0.026
4	0.024
6	0.023
8	0.019
10	0.017
CURVE 11	**T = 1035**
1.5	0.031
2	0.029
4	0.025
6	0.023
8	0.023
10	0.021
CURVE 12*	**T = 755**
1.5	0.027
2	0.020
4	0.019
6	0.016
8	0.014
10	0.016
CURVE 13*	**T = 865**
1.5	0.023
2	0.023
4	0.022
6	0.019
8	0.017
10	0.013
CURVE 14	**T = 763**
1.5	0.070
2	0.063
4	0.059
6	0.060
8	0.060
10	0.063
CURVE 15*	**T = 873**
1.5	0.104
2	0.076
4	0.053
6	0.047
8	0.040
10	0.047
CURVE 16	**T = 907**
1.5	0.104
2	0.081
4	0.053
6	0.042
8	0.037
10	0.051
CURVE 17*	**T = 636**
2	0.063
4	0.039
6	0.031
8	0.026
CURVE 18*	**T = 901**
1.5	0.093
2	0.075
4	0.057
6	0.054
10	0.059
CURVE 19*	**T = 857**
1.5	0.092
2	0.071
4	0.051
6	0.044
8	0.050
CURVE 20*	**T = 748**
1.5	0.097
2	0.063
4	0.050
6	0.042
8	0.051
CURVE 21	**T = 748**
1.5	0.104*
2	0.091
4	0.090
6	0.091
8	0.087
10	0.084
CURVE 22*	**T = 858**
1.5	0.109
2	0.093
4	0.085
6	0.077
8	0.076
10	0.082
CURVE 23*	**T = 913**
1.5	0.109
2	0.085
4	0.059
6	0.050
8	0.046
10	0.058
CURVE 24*	**T = 865**
1.5	0.087
2	0.074
4	0.052
6	0.047
8	0.046
10	0.058
CURVE 25*	**T = 749**
1.5	0.092
2	0.073
4	0.049
6	0.046
8	0.041
10	0.058
CURVE 26	**T = 741**
1.5	0.097
2	0.089
4	0.084
6	0.082
8	0.079
10	0.077
CURVE 27*	**T = 861**
1.5	0.112
2	0.094
4	0.083
6	0.072
8	0.070
10	0.077
CURVE 28*	**T = 868**
1.5	0.126
2	0.102
4	0.088
6	0.083
8	0.068
10	0.077
CURVE 29*	**T = 750**
1.5	0.097
2	0.082
4	0.070
6	0.069
8	0.069
10	0.070
CURVE 30*	**T = 923**
1.5	0.106
2	0.104
4	0.080
6	0.073
8	0.075
10	0.082
CURVE 31	**T = 757**
1.5	0.262
2	0.214
4	0.146
6	0.111
8	0.095
10	0.097
CURVE 32*	**T = 867**
1.5	0.280
2	0.229
4	0.154
6	0.119
8	0.106
10	0.108
CURVE 33	**T = 913**
1.5	0.227
2	0.231
4	0.160
6	0.124
8	0.111
10	0.109
CURVE 34*	**T = 652**
2	0.228
4	0.145
6	0.102
8	0.088
10	0.080
CURVE 35	**T = 1244**
1.000	0.050
1.138	0.045
1.500	0.037
1.824	0.034
2.223	0.033
2.735	0.033
3.319	0.032
3.741	0.032
5.047	0.033
6.012	0.036
6.982	0.040
8.337	0.043
11.46	0.044
12.79	0.045
15.49	0.051
CURVE 36	**T = 1258**
1.000	0.049*
1.099	0.046
1.318	0.040
1.552	0.037
1.837	0.034*
2.410	0.032
3.034	0.031

*Not shown on plot

DATA TABLE NO. 48 (continued)

λ	ε
CURVE 36 (cont.)	
4.018	0.030*
4.819	0.031
5.495	0.030
6.053	0.028
6.966	0.028
8.054	0.030
8.770	0.034
9.683	0.039
11.53	0.036
13.18	0.035
14.72	0.036
CURVE 37*	
T = 1241	
1.000	0.027
1.052	0.027
1.205	0.028
1.510	0.030
1.762	0.029
1.995	0.028
2.805	0.027
3.999	0.027
5.000	0.029
6.039	0.031
7.063	0.036
8.375	0.041
8.892	0.046
9.419	0.051
10.76	0.049
12.97	0.042
15.14	0.038
CURVE 38*	
T = 1235	
1.000	0.045
1.274	0.039
1.611	0.034
2.004	0.032
2.449	0.031
2.748	0.030
3.639	0.029
3.972	0.029
4.955	0.029
5.781	0.029
CURVE 38 (cont.)*	
6.531	0.029
7.568	0.030
8.356	0.032
8.831	0.037
9.226	0.039
9.863	0.040
11.89	0.033
14.35	0.030
CURVE 39*	
T = 1202	
1.028	0.033
1.291	0.029
1.507	0.027
1.742	0.028
2.307	0.031
2.805	0.032
3.715	0.036
4.624	0.040
6.026	0.044
7.031	0.045
8.054	0.046
9.506	0.049
11.12	0.058
13.43	0.071
CURVE 40*	
T = 294	
1.000	0.071
1.104	0.069
1.361	0.072
1.552	0.068
1.828	0.060
2.042	0.063
2.500	0.057
2.858	0.065
3.428	0.059
4.074	0.050
4.498	0.052
5.508	0.047
6.138	0.051
8.810	0.044
10.33	0.042
11.94	0.037
14.13	0.037
14.89	0.033
CURVE 41*	
T = 294	
1.000	0.063
1.161	0.061
1.390	0.052
1.622	0.045
2.000	0.038
2.317	0.035
2.951	0.034
3.483	0.031
4.977	0.024
6.281	0.021
7.621	0.018
9.795	0.016
12.65	0.015
14.55	0.015
CURVE 42	
T = 294	
1.000	0.079
1.125	0.085
1.245	0.084
1.387	0.077
1.500	0.059*
1.626	0.047
1.849	0.036*
2.084	0.029
2.588	0.024
2.735	0.025
3.199	0.022
4.385	0.014
5.000	0.012
6.081	0.012
6.998	0.011
8.166	0.010
10.00	0.008
10.96	0.007
14.76	0.006

*Not shown on plot

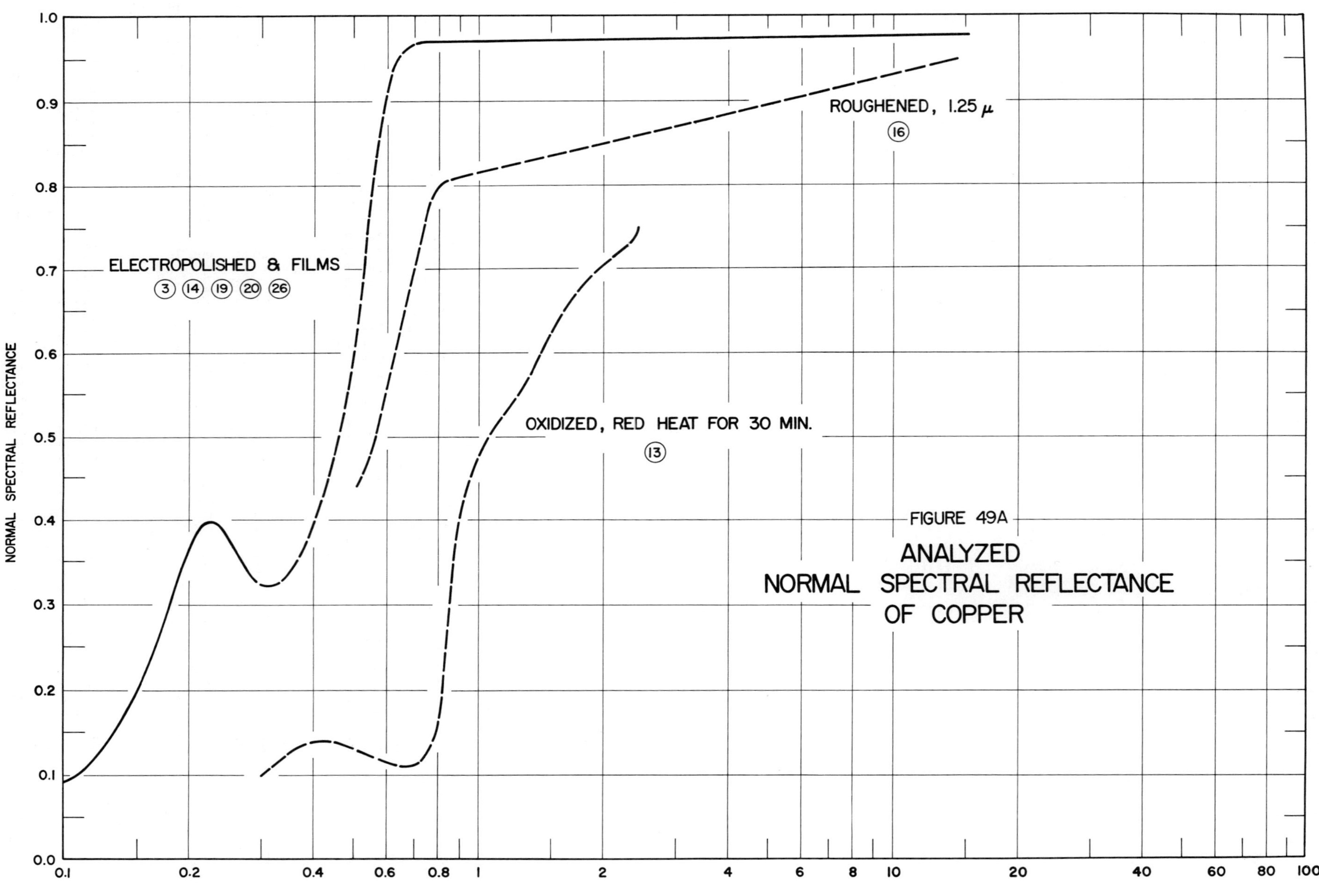

FIGURE 49A
ANALYZED
NORMAL SPECTRAL REFLECTANCE
OF COPPER

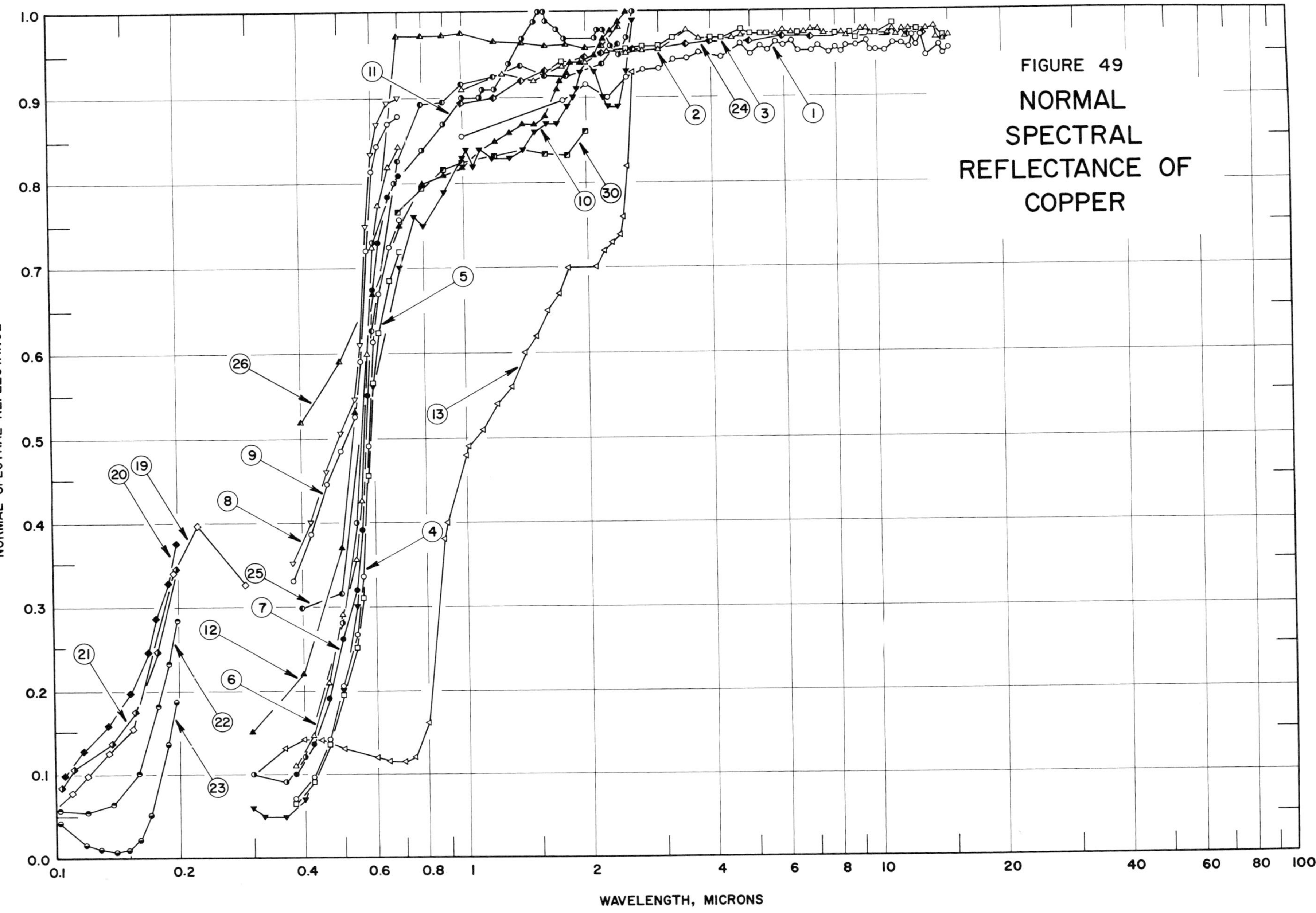
FIGURE 49
NORMAL SPECTRAL REFLECTANCE OF COPPER
NORMAL SPECTRAL REFLECTANCE
WAVELENGTH, MICRONS
0.0 0.1 0.2 0.3 0.4 0.5 0.6 0.7 0.8 0.9 1.0
0.1 0.2 0.4 0.6 0.8 1 2 4 6 8 10 20 40 60 80 100
1 2 3 4 5 6 7 8 9 10 11 12 13 19 20 21 22 23 24 25 26 30

SPECIFICATION TABLE NO. 49 NORMAL SPECTRAL REFLECTANCE OF COPPER

Curve No.	Ref. No.	Year	Temperature K	Wavelength Range, μ	Geometry θ	Geometry θ'	Geometry ω'	Reported Error, %	Composition (weight percent), Specifications and Remarks
1	126	1953	~298	1.00-15.00	5°		2π	± 2.3	Buffed; data extracted from smooth curve; calculated from R(2π, 5°).
2	126	1953	298	1.00-15.00	5°		2π	± 2.2	Buffed and alcohol washed; data extracted from smooth curve; calculated from R (2π, 5°).
3	136	1954	294	1.00-15.00	5°		2π	< ± 2	Washed in separate baths of acetone and ethyl alcohol; polished with a soft rag; MgO reference; hohlraum at approx 1089 K; converted from R (2π, 5°).
4	29	1956	298	0.38-0.70	9°		2π		Electrolytic, tough pitch; federal specification QQ-C-576 and QQ-C-502; as received; data extracted from smooth curve; MgO reference.
5	29	1956	298	0.38-0.70	9°		2π		Above specimen and conditions except rotated 90° in its own plane.
6	29	1956	298	0.38-0.70	9°		2π		Different sample, same as Curve 4 specimen and conditions except smoothed and cleaned.
7	29	1956	298	0.38-0.70	9°		2π		Above specimen and conditions except rotated 90° in its own plane.
8	29	1956	298	0.38-0.70	9°		2π		Different sample, same as Curve 4 specimen and conditions except polished.
9	29	1956	298	0.38-0.70	9°		2π		Above specimen and conditions except rotated 90° in its own plane.
10	34	1957	298	0.30-2.60	9°		2π	± 4	Electrolytic, tough pitch; federal specification QQ-C-576 or QQ-C-502; nominal composition: 99.90 Cu, 0.04 O; as received; data extracted from a smooth curve; Magnesium Carbonate reference.
11	34	1957	298	0.30-2.60	9°		2π	± 4	Different sample, same as above specimen and conditions except cleaned with liquid detergent.
12	34	1957	298	0.30-2.60	9°		2π	± 4	Different sample, same as Curve 10 specimen and conditions except polished.
13	34	1957	298	0.30-2.60	9°		2π	± 4	Different sample, same as Curve 10 specimen and conditions except oxidized in air at red heat for 30 min.
14	65	1962	295	0.40-2.00	~0°		2π		Electropolished.
15	65	1962	295	0.41-14.40	~0°		2π		Mechanically polished.
16	65	1962	295	0.41-14.20	~0°		2π		Above specimen and conditions except roughened with sand paper; surface roughness 1.25 μ.
17	228	1900	298	0.45-0.70	~0°	~0°			Pure; mirror-like surface; platinum filament lamp source.
18	216	1949	~298	1.00-15.00	0°		2π	5	Polished; data extracted from smooth curve.
19	313	1962	300	0.050-0.291	~0°	~0°			Electrolytically polished; data extracted from smooth curve.
20	315	1965	295	0.1034-0.2000	~4°	~4°			Evaporated film (800-1000 Å thick); > 99.99 pure; evaporated at 10^{-6} mm Hg; measured in vacuum; data extracted from smooth curve.
21	315	1965	295	0.1030-0.1999	~4°	~4°			Above specimen and conditions except aged 1 day in air.

SPECIFICATION TABLE NO. 49 (continued)

Curve No.	Ref. No.	Year	Temperature K	Wavelength Range, μ	Geometry θ	θ'	ω'	Reported Error, %	Composition (weight percent), Specifications and Remarks
22	315	1965	295	0.1027-0.2000	~4°	~4°			Above specimen and conditions except aged 5 weeks in air.
23	315	1965	295	0.1027-0.1999	~4°	~4°			Above specimen and conditions except aged 6 months in air.
24	318	1961	298	1.00-14.00	0°		2π		Commercial copper; sanded with increasingly finer grades of emery paper; buffed to a high polish with a buffing wheel using relevigated alumina as the abrasive compound; rinsed in water and alcohol successively and immediately dried; fresh sample; [Author's designation: C0].
25	318	1961	298	0.40-2.40	0°		2π		Above specimen and conditions except MgO reference.
26	318	1961	298	0.40-2.40	0°		2π		Above specimen and conditions except electropolished.
27	318	1961	298	1.00-14.00	0°		2π		Commercial copper; sanded with 0000 emery paper; sanded in one direction to produce a fine-grooved appearance; peak to peak height of asperity as measured by profilometer 0.25 μ; mean groove spacing 9 μ; mean groove angle 3°; fresh sample; [Author's designation: C1].
28	318	1961	298	0.60-1.80	0°		2π		Above specimen and conditions except MgO reference.
29	318	1961	298	1.30-14.00	0°		2π		Commercial copper; sanded with 00 emery paper; sanded in one direction to produce a fine-grooved appearance; peak to peak height of asperity as measured by profilometer 1.25 μ; mean groove spacing 9 μ; mean groove angle 15°; fresh sample; [Author's designation: C2].
30	318	1961	298	0.70-2.00	0°		2π		Above specimen and conditions except MgO reference.

DATA TABLE NO. 49 NORMAL SPECTRAL REFLECTANCE OF COPPER

[Wavelength, λ, μ; Reflectance, ρ; Temperature, T, K]

CURVE 1, T = 298

λ	ρ
1.00	0.855
1.75	0.899
2.00	0.915
2.25	0.900
2.50	0.923
2.75	0.931
3.00	0.933
3.25	0.945
3.50	0.946
3.75	0.951
4.25	0.948
4.75	0.961
5.00	0.950
5.25	0.960
5.50	0.955
5.75	0.964
6.00	0.960
6.25	0.967
6.50	0.953
7.00	0.953
7.25	0.960
7.75	0.951
8.00	0.959
8.25	0.952
8.50	0.960
9.00	0.960
9.25	0.965*
9.50	0.965
9.75	0.955
10.00	0.955
10.25	0.951*
10.50	0.953
10.75	0.951*
11.00	0.962
11.50	0.962
11.75	0.959*
12.00	0.959
12.25	0.971
12.50	0.960
12.75	0.971
13.25	0.949
13.50	0.949*
13.75	0.958*

CURVE 1 (cont.)

λ	ρ
14.25	0.960
14.50	0.950
14.75	0.955*
15.00	0.955

CURVE 2, T = 298

λ	ρ
1.00	0.910
1.25	0.929
1.50	0.920
1.75	0.936
2.00	0.940
2.25	0.954
2.50	0.951
2.75	0.955
3.00	0.956
3.50	0.980
3.75	0.970
4.50	0.970
4.75	0.973
5.75	0.975
6.00	0.980
6.25	0.977
6.75	0.977
7.00	0.980
7.28	0.980
7.50	0.975
8.25	0.975
8.50	0.971
8.75	0.973*
9.25	0.980
9.50	0.980
9.75	0.975
10.75	0.975
11.00	0.980
11.25	0.973
11.50	0.981
11.75	0.971
12.00	0.980
13.25	0.980
13.50	0.982*
13.75	0.982
14.00	0.970

CURVE 2 (cont.)

λ	ρ
14.50	0.969
14.75	0.969*
15.00	0.971

CURVE 3, T = 294

λ	ρ
1.00	0.910*
1.25	0.930*
1.50	0.920*
1.75	0.940
2.00	0.940*
2.25	0.955*
2.50	0.955
2.75	0.960
3.00	0.960
3.25	0.970
3.50	0.980*
3.75	0.970*
4.00	0.970
4.25	0.970
4.50	0.970*
4.75	0.980
5.00	0.975
5.25	0.975
5.50	0.975
5.75	0.975*
6.00	0.980*
6.25	0.975*
6.50	0.975
6.75	0.975*
7.00	0.980*
7.25	0.980*
7.50	0.975*
7.75	0.980*
8.00	0.975
8.25	0.980*
8.50	0.970*
8.75	0.975
9.00	0.970
9.25	0.980*
9.50	0.980*
9.75	0.975*
10.00	0.975

CURVE 3 (cont.)

λ	ρ
10.25	0.975*
10.50	0.975*
10.75	0.975*
11.00	0.985
11.25	0.975*
11.50	0.980*
11.75	0.970*
12.00	0.980*
12.25	0.980*
12.50	0.980
12.75	0.980*
13.00	0.980*
13.25	0.980*
13.50	0.980*
13.75	0.980*
14.00	0.975*
14.25	0.975*
14.50	0.970
14.75	0.970*
15.00	0.970*

CURVE 4, T = 298

λ	ρ
0.38	0.070
0.42	0.095
0.46	0.140
0.50	0.205
0.54	0.265
0.56	0.335
0.58	0.490
0.60	0.615
0.62	0.670
0.66	0.725
0.70	0.755

CURVE 5, T = 298

λ	ρ
0.38	0.065
0.42	0.090
0.46	0.135
0.50	0.195
0.54	0.250

CURVE 5 (cont.)

λ	ρ
0.56	0.310
0.58	0.455
0.60	0.565
0.62	0.625
0.66	0.685
0.70	0.720

CURVE 6, T = 298

λ	ρ
0.38	0.110
0.42	0.145
0.46	0.210
0.50	0.290
0.54	0.355
0.56	0.425
0.58	0.600
0.60	0.725
0.62	0.775
0.66	0.820
0.70	0.845

CURVE 7, T = 298

λ	ρ
0.38	0.100
0.42	0.135
0.46	0.190
0.50	0.260
0.54	0.320
0.56	0.390
0.58	0.550
0.60	0.675
0.62	0.730
0.66	0.785
0.70	0.810

CURVE 8, T = 298

λ	ρ
0.38	0.350
0.42	0.400
0.46	0.460
0.50	0.505

CURVE 8 (cont.)

λ	ρ
0.54	0.545
0.56	0.610
0.58	0.750
0.60	0.835
0.62	0.870
0.66	0.895
0.70	0.900

CURVE 9, T = 298

λ	ρ
0.38	0.330
0.42	0.385
0.46	0.445
0.50	0.485
0.54	0.525
0.56	0.590
0.58	0.720
0.60	0.815
0.62	0.845
0.66	0.870
0.70	0.880

CURVE 10, T = 298

λ	ρ
0.30	0.06
0.32	0.05
0.36	0.05
0.40	0.07
0.50	0.20
0.54	0.30
0.60	0.56
0.70	0.70
0.76	0.76
0.80	0.75
0.90	0.79
1.00	0.83
1.02	0.84
1.06	0.82
1.10	0.84
1.18	0.83
1.20	0.83*
1.30	0.83

CURVE 10 (cont.)

λ	ρ
1.40	0.84
1.50	0.86
1.60	0.87
1.70	0.87
1.80	0.89
1.86	0.90
1.90	0.91
1.94	0.93
2.00	0.94*
2.10	0.93
2.20	0.90
2.26	0.89
2.30	0.89*
2.38	0.89
2.40	0.89*
2.50	0.93
2.60	0.99

CURVE 11, T = 298

λ	ρ
0.30	0.10
0.36	0.09
0.40	0.12
0.50	0.28
0.54	0.40
0.60	0.73
0.68	0.80
0.70	0.81*
0.80	0.84
0.90	0.87
1.00	0.90
1.10	0.90
1.12	0.91
1.20	0.91
1.30	0.94
1.40	0.97
1.50	0.99
1.54	1.00
1.58	1.00
1.60	0.99
1.70	0.98
1.78	0.97
1.80	0.97*

CURVE 11 (cont.)

λ	ρ
1.90	0.97*
2.00	0.97*
2.10	0.97
2.14	0.98
2.20	0.98
2.30	0.96
2.40	0.95
2.44	0.95*
2.50	0.97
2.60	1.00

CURVE 12, T = 298

λ	ρ
0.30	0.15
0.40	0.22
0.50	0.37
0.54	0.53
0.60	0.62
0.70	0.75
0.80	0.80
0.90	0.81
1.00	0.82
1.10	0.84*
1.20	0.84
1.30	0.86
1.40	0.87
1.50	0.87
1.58	0.88
1.60	0.88*
1.70	0.91
1.72	0.92
1.80	0.93
1.84	0.94
1.90	0.94*
1.96	0.94
2.00	0.94*
2.10	0.95
2.20	0.97
2.30	0.98
2.40	0.99
2.50	1.00
2.60	1.00*

*Not shown on plot

DATA TABLE NO. 49 (continued)

λ	ρ
CURVE 13 T = 298	
0.30	0.10*
0.36	0.13
0.40	0.14
0.44	0.14
0.50	0.13
0.60	0.12
0.64	0.11
0.70	0.11
0.74	0.12
0.80	0.16
0.88	0.38
0.90	0.40
1.00	0.48
1.02	0.49
1.10	0.51
1.20	0.54
1.30	0.56
1.40	0.60
1.50	0.62
1.60	0.65
1.70	0.67
1.80	0.70
1.90	0.70*
2.00	0.70*
2.08	0.70*
2.10	0.70
2.20	0.72
2.26	0.73*
2.30	0.73
2.40	0.74
2.44	0.76
2.50	0.82
2.60	0.93
CURVE 14* T = 295	
0.40	0.525
0.50	0.590
0.54	0.640
0.70	0.969
0.80	0.969
0.92	0.969
CURVE 14* (cont.)	
1.00	0.970
2.00	0.969
CURVE 15* T = 295	
0.41	0.27
0.50	0.33
0.60	0.63
0.70	0.83
0.80	0.88
0.90	0.90
1.00	0.91
1.20	0.92
1.40	0.93
2.00	0.94
2.40	0.95
3.00	0.95
4.00	0.96
5.00	0.97
7.00	0.97
9.00	0.98
11.00	0.97
13.00	0.97
14.40	0.97
CURVE 16* T = 295	
0.41	0.42
0.51	0.44
0.55	0.48
0.80	0.80
0.90	0.81
1.00	0.82
1.20	0.83
1.44	0.84
1.80	0.84
2.70	0.86
3.50	0.88
4.05	0.89
4.55	0.89
5.00	0.89
6.00	0.90
CURVE 16* (cont.)	
7.00	0.91
8.00	0.92
9.00	0.93
10.00	0.93
11.20	0.94
12.00	0.94
13.00	0.94
14.20	0.95
CURVE 17* T = 298	
0.45	0.488
0.50	0.533
0.55	0.595
0.60	0.835
0.65	0.890
0.70	0.907
CURVE 18* T = ~298	
1.00	0.923
1.28	0.918
1.52	0.939
1.78	0.936
2.01	0.951
2.27	0.959
2.52	0.959
2.78	0.963
3.01	0.968
3.28	0.974
3.51	0.973
3.76	0.977
4.01	0.979
4.27	0.965
4.52	0.985
5.00	0.980
5.26	0.989
5.50	0.981
5.75	0.990
6.00	0.986
6.25	0.989
6.50	0.978
CURVE 18* (cont.)	
6.73	0.984
6.99	0.984
7.26	0.984
7.50	0.985
7.75	0.990
7.99	0.987
8.29	0.983
8.53	0.984
8.76	0.986
9.01	0.980
9.28	0.981
9.51	0.984
9.78	0.987
10.02	0.985
10.27	0.979
10.52	0.987
10.76	0.984
11.00	0.990
11.28	0.989
11.53	0.981
11.76	0.967
12.04	0.996
12.16	0.994
12.26	0.986
12.51	0.989
12.76	0.990
13.00	0.994
13.26	0.984
13.52	0.987
13.74	0.992
13.99	0.988
14.27	0.987
14.50	0.979
14.71	0.988
15.00	0.987
CURVE 19 T = 300	
0.050	0.140*
0.052	0.118*
0.054	0.090*
0.057	0.070*
0.060	0.077*
CURVE 19 (cont.)	
0.072	0.113*
0.078	0.118*
0.090	0.108*
0.098	0.090*
0.110	0.079
0.120	0.099
0.135	0.126
0.156	0.153
0.195	0.340
0.223	0.398
0.291	0.327
CURVE 20 T = 295	
0.1034	0.099
0.1198	0.129
0.1354	0.158
0.1539	0.199
0.1700	0.246
0.1799	0.286
0.1900	0.329
0.2000	0.376
CURVE 21 T = 295	
0.1030	0.085
0.1199	0.107
0.1399	0.136
0.1598	0.172
0.1799	0.247
0.1999	0.345
CURVE 22 T = 298	
0.1027	0.058
0.1200	0.055
0.1398	0.065
0.1600	0.101
0.1801	0.181
0.1900	0.233
0.2000	0.284
CURVE 23 T = 295	
0.1027	0.042
0.1198	0.018
0.1298	0.011
0.1401	0.009
0.1501	0.011
0.1601	0.023
0.1701	0.053
0.1899	0.136
0.1999	0.188
CURVE 24 T = 298	
1.00	0.896
1.20	0.900
1.40	0.920
1.60	0.931
1.80	0.936
2.00	0.946
2.20	0.951
2.40	0.951
2.60	0.956
3.00	0.956*
3.50	0.961
4.00	0.966
5.00	0.966
7.00	0.971
11.00	0.971
13.00	0.975
14.00	0.972
CURVE 25 T = 298	
0.40	0.299
0.50	0.315
0.60	0.627
0.70	0.828
0.80	0.882
0.90	0.897
1.00	0.915
1.20	0.923
1.40	0.937
CURVE 25 (cont.)	
1.60	0.927
1.80	0.927
2.00	0.942
2.20	0.939
2.40	0.961
CURVE 26 T = 298	
0.40	0.519
0.50	0.590
0.60	0.677*
0.70	0.973
0.80	0.973
0.90	0.973
1.00	0.976
1.20	0.967
1.40	0.967
1.60	0.960
1.80	0.963
2.00	0.959
2.20	0.961
2.40	0.986
CURVE 27* T = 298	
1.00	0.871
1.20	0.876
1.40	0.890
1.60	0.903
1.80	0.905
2.00	0.910
2.20	0.923
2.40	0.925
2.60	0.931
3.00	0.941
4.00	0.953
5.00	0.962
6.00	0.967
7.00	0.967
8.00	0.970
9.00	0.972
10.00	0.974

* Not shown on plot

DATA TABLE NO. 49 (continued)

λ	ρ
CURVE 27 (cont.)*	
11.00	0.974
12.00	0.974
13.00	0.979
14.00	0.979
CURVE 28* T = 298	
0.60	0.790
0.70	0.862
0.80	0.887
1.00	0.909
1.20	0.918
1.60	0.915
1.80	0.914
CURVE 29* T = 298	
1.30	0.804
1.40	0.812
1.80	0.837
2.00	0.845
2.40	0.855
2.80	0.862
3.50	0.880
4.00	0.885
4.50	0.892
5.00	0.892
6.00	0.895
7.00	0.909
8.00	0.916
9.00	0.925
10.00	0.930
11.00	0.935
12.00	0.940
13.00	0.943
14.00	0.947
CURVE 30 T = 298	
0.70	0.766
0.80	0.796
0.90	0.816
1.00	0.826

λ	ρ
CURVE 30 (cont.)	
1.20	0.833
1.40	0.840*
1.60	0.834
1.80	0.833
2.00	0.860

* Not shown on plot

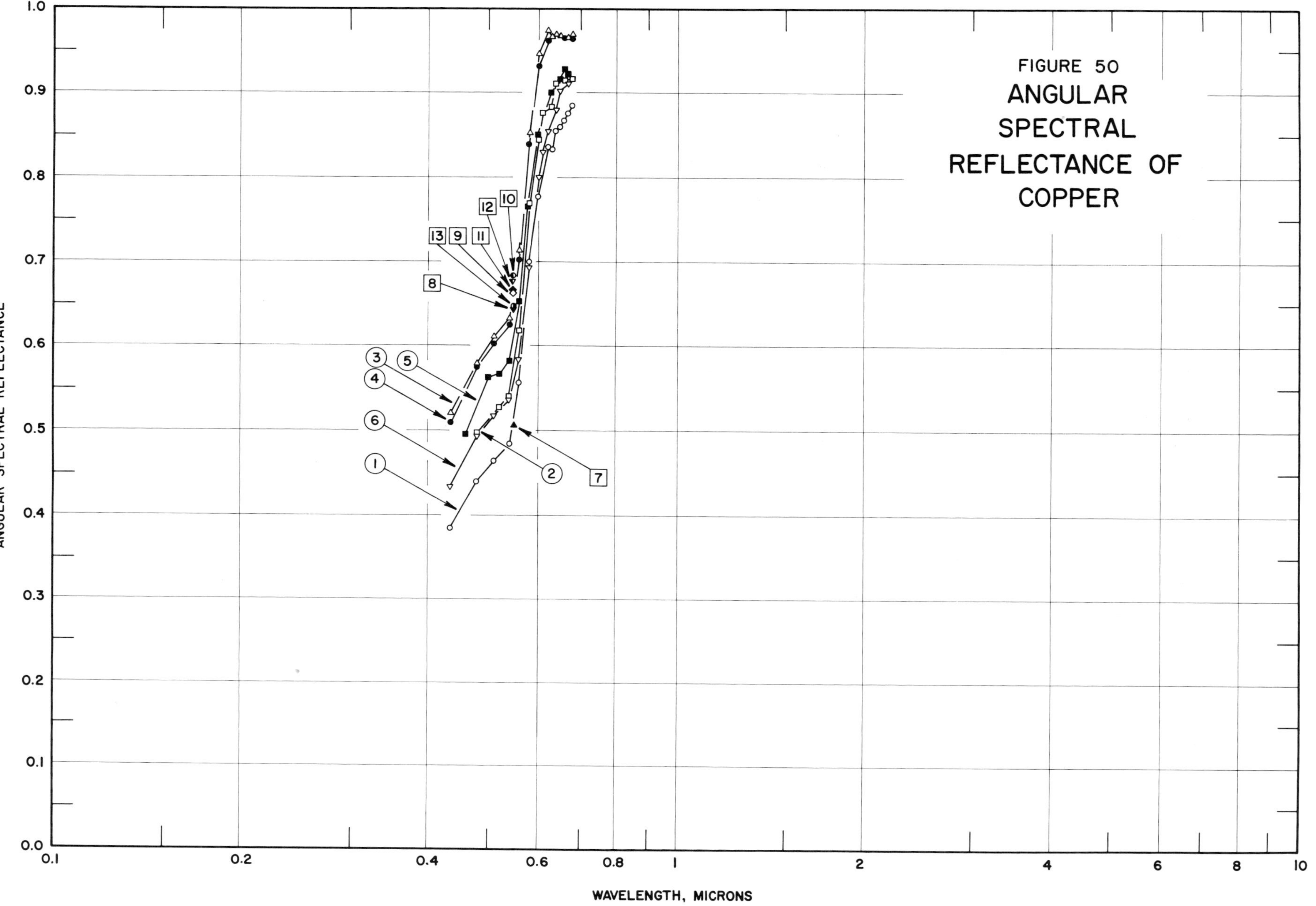

FIGURE 50 ANGULAR SPECTRAL REFLECTANCE OF COPPER

SPECIFICATION TABLE NO. 50 ANGULAR SPECTRAL REFLECTANCE OF COPPER

Curve No.	Ref. No.	Year	Temperature K	Wavelength Range, μ	Geometry θ	θ'	ω'	Reported Error, %	Composition (weight percent), Specifications and Remarks
1	312	1936	298	0.4358-0.6800	70°	70°			Cast copper; filed, ground, annealed in vacuo for 1 hr at 973 K, rubbed, and washed; mercury arc and carbon arc sources; computed from optical constants measured for band width of approx 20 Å; [Authors' designation: Specimen A (M. P. A.)].
2	312	1936	298	0.4800-0.6800	70°	70°			Different sample, same as above specimen and conditions; [Authors' Specimen B (M. P. B.)].
3	312	1936	298	0.4358-0.6800	70°	70°			Curve 1 specimen and conditions except electrolytically polished; [Authors' designation: Specimen A (E. P. A.)].
4	312	1936	298	0.4358-0.6800	70°	70°			Curve 2 specimen and conditions except electrolytically polished; [Authors' designation: Specimen B (E. P. B.)].
5	312	1936	298	0.4600-0.6700	70°	70°			Opaque film evaporated on glass in vacuum; mercury arc and carbon arc sources; computed from optical constants measured for band width of approx 20 Å; [Authors' designation: Specimen I (Ev. 1)].
6	312	1936	298	0.4358-0.6700	70°	70°			Different sample, same as above specimen and conditions; [Authors' designation: Specimen II (Ev. 2)].
7	312	1936	298	0.5461	70°	70°			Cast copper; filed, ground, annealed in vacuo for 1 hr at 973 K, rubbed, and washed; mercury arc source; computed from optical constants measured for band width of approx 20 Å; [Authors' designation: Specimen B].
8	312	1936	298	0.5461	70°	70°			Above specimen and conditions except electrolytically polished for 6 min.
9	312	1936	298	0.5461	70°	70°			Above specimen and conditions except electrolytically polished for 9 additional min.
10	312	1936	298	0.5461	70°	70°			Above specimen and conditions except electrolytically polished for 7 additional min.
11	312	1936	298	0.5461	70°	70°			Above specimen and conditions except electrolytically polished for 5 additional min.
12	312	1936	298	0.5461	70°	70°			Above specimen and conditions except electrolytically polished for 6 additional min.
13	312	1936	298	0.5461	70°	70°			Above specimen and conditions except electrolytically polished for 10 additional min.
14	312	1936	298	0.5461	70°	70°			Above specimen and conditions except electrolytically polished for 15 additional min.
15	137	1962	298	0.5770	70°	70°			Specimen prepared in an induction furnace, in air, from electrolytic copper; specimen fairly free from blow-holes; lightly polished; azimuth angle of polarization referenced to reflection plane 78° 37'; [Author's designation: Specimen A].
16	137	1932	298	0.5770	71.5°	71.5°			Specimen was a piece of cold-drawn high-conductivity copper rod; obtained from the Broughton Copper Company, Manchester; specimen entirely free from blow-holes; lightly polished; azimuth angle of polarization referenced to reflection plane 80° 26'; [Author's designation: Specimen B].
17	137	1932	298	0.5770	70°	70°			High-conductivity copper from Broughton Copper Company; cold drawn; specimen entirely free from blow-holes; surface slightly scratched; azimuth angle of polarization referenced to reflection plane 80° 7'; etched for 4 sec; [Author's designation: Specimen B].

SPECIFICATION TABLE NO. 50 (continued)

Curve No.	Ref. No.	Year	Temperature K	Wavelength Range, μ	Geometry θ θ' ω'	Reported Error, %	Composition (weight percent), Specifications and Remarks
18	137	1932	298	0.5770	70° 70°		High-conductivity copper from Broughton Copper Company; cold drawn; specimen entirely free from blow-holes; light burnish surface; azimuth angle of polarization referenced to reflection plane 79° 2'; etched for 25 sec; [Author's designation: Specimen B].
19	137	1932	298	0.5770	70° 70°		High-conductivity copper from Broughton Copper Company; cold drawn; specimen entirely free from blow-holes; heavy burnish surface; azimuth angle of polarization referenced to reflection plane 77° 14'; etched for several minutes; [Author's designation: Specimen B].

DATA TABLE NO. 50 ANGULAR SPECTRAL REFLECTANCE OF COPPER

[Wavelength, λ, μ; Reflectance, ρ; Temperature, T, K]

λ	ρ
CURVE 1 T = 298	
0.4358	0.385
0.4800	0.440
0.5100	0.465
0.5400	0.485
0.5600	0.557
0.5800	0.700
0.6000	0.777
0.6200	0.836
0.6300	0.835
0.6400	0.857
0.6500	0.860
0.6600	0.868
0.6700	0.877
0.6800	0.887
CURVE 2 T = 298	
0.4800	0.497
0.5200	0.528
0.5400	0.540
0.5600	0.621
0.5800	0.770
0.6000	0.845
0.6100	0.877
0.6300	0.884
0.6400	0.912
0.6600	0.916
0.6800	0.918
CURVE 3 T = 298	
0.4358	0.520
0.4800	0.581
0.5100	0.613
0.5400	0.633
0.5600	0.713
0.5800	0.854
0.6000	0.949
0.6200	0.975
0.6300	0.968
0.6400	0.971
0.6500	0.969
0.6700	0.968
0.6800	0.971
CURVE 4 T = 298	
0.4358	0.510
0.4800	0.577
0.5100	0.601
0.5400	0.627
0.5600	0.702
0.5800	0.840
0.6000	0.933
0.6200	0.962
0.6400	0.970*
0.6600	0.968
0.6800	0.967
CURVE 5 T = 298	
0.4600	0.497
0.5000	0.564
0.5200	0.569
0.5400	0.582
0.5600	0.652
0.5800	0.764
0.6000	0.850
0.6300	0.900
0.6500	0.918
0.6600	0.930
0.6700	0.925
CURVE 6 T = 298	
0.4358	0.433
0.4800	0.494
0.5100	0.518
0.5400	0.538
0.5600	0.583
0.5800	0.692
0.6000	0.800
0.6100	0.833
0.6200	0.855
0.6300	0.880
0.6500	0.901
0.6700	0.912
CURVE 7 T = 298	
0.5461	0.507
CURVE 8 T = 298	
0.5461	0.645
CURVE 9 T = 298	
0.5461	0.663
CURVE 10 T = 298	
0.5461	0.682
CURVE 11 T = 298	
0.5461	0.666
CURVE 12 T = 298	
0.5461	0.678
CURVE 13 T = 298	
0.5461	0.649
CURVE 14* T = 298	
0.5461	0.635
CURVE 15* T = 298	
0.5770	0.737
CURVE 16* T = 298	
0.5770	0.798
CURVE 17* T = 298	
0.5770	0.746
CURVE 18* T = 298	
0.5770	0.701
CURVE 19* T = 298	
0.5770	0.637

*Not shown on plot

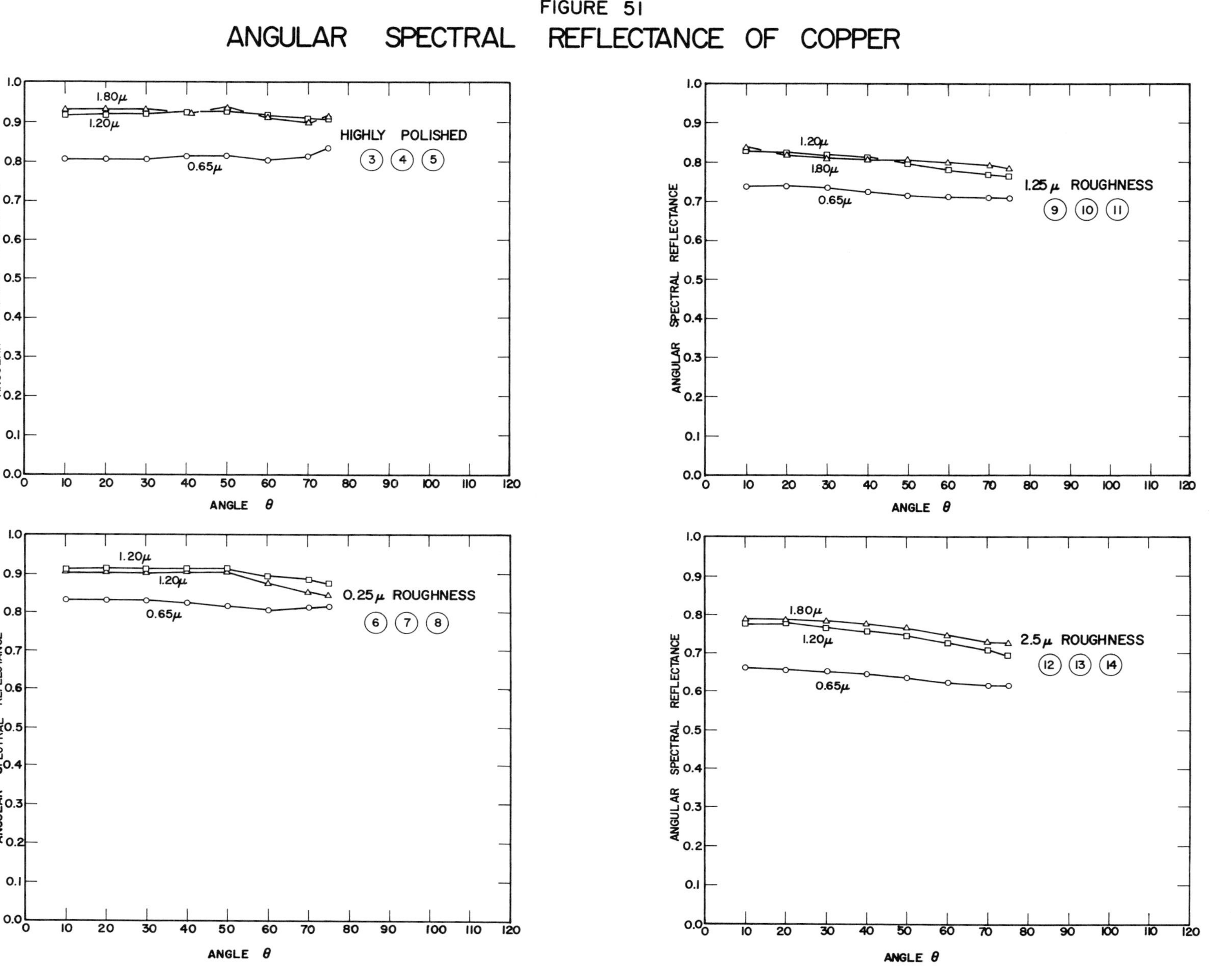
FIGURE 51
ANGULAR SPECTRAL REFLECTANCE OF COPPER
1.80μ
1.20μ
0.65μ
HIGHLY POLISHED
3 4 5
1.20μ
1.80μ
0.65μ
1.25μ ROUGHNESS
9 10 11
1.20μ
1.20μ
0.65μ
0.25μ ROUGHNESS
6 7 8
1.80μ
1.20μ
0.65μ
2.5μ ROUGHNESS
12 13 14
ANGULAR SPECTRAL REFLECTANCE
ANGLE θ

SPECIFICATION TABLE NO. 51 ANGULAR SPECTRAL REFLECTANCE OF COPPER

Curve No.	Ref. No.	Year	Temperature K	Wavelength, μ	Angular Range,°	Geometry θ θ' ω'	Reported Error, %	Composition (weight percent), Specifications and Remarks
1	138	1960	298	1.2	0-70	θ 2π		Polished; MgO reference.
2	138	1960	298	1.2	0-80	θ 2π		Roughened by a grid of grooves of 45° opening angle, 0.005 in. (127 μ) deep; surface created is an array of pyramids; MgO reference.
3	318	1961	298	0.65	10-75	θ 2π		Commercial copper; sanded with increasingly finer grades of emery paper; buffed to a high polish with a buffing wheel using relevigated alumina as the abrasive compound; rinsed in water and alcohol successively and immediately dried; fresh sample; MgO reference; incident beam polarized in the direction perpendicular to the incident plane of the sample; [Authors's designation: C0].
4	318	1961	298	1.20	10-75	θ 2π		Above specimen and conditions.
5	318	1961	298	1.80	10-75	θ 2π		Above specimen and conditions.
6	318	1961	298	0.65	10-75	θ 2π		Commercial copper; sanded with 0000 emery paper; sanded in one direction to produce a fine-grooved appearance; peak to peak height of asperity as measured by profilometer 0.25 μ; mean groove spacing 9 μ; mean groove angle 3°; fresh sample; MgO reference; incident beam polarized in the direction perpendicular to the incident plane of the sample; [Author's designation: C1].
7	318	1961	298	1.20	10-75	θ 2π		Above specimen and conditions.
8	318	1961	298	1.80	10-75	θ 2π		Above specimen and conditions.
9	318	1961	298	0.65	10-75	θ 2π		Commercial copper; sanded with 00 emery paper; sanded in one direction to produce a fine-grooved appearance; peak to peak height of asperity as measured by profilometer 1.25 μ; mean groove spacing 9 μ; mean groove angle 15°; fresh sample; MgO reference; incident beam polarized in the direction perpendicular to the incident plane of the sample; [Author's designation: C2].
10	318	1961	298	1.20	10-75	θ 2π		Above specimen and conditions.
11	318	1961	298	1.80	10-75	θ 2π		Above specimen and conditions.
12	318	1961	298	0.65	10-75	θ 2π		Commercial copper; sanded with grade 1 emery paper; sanded in one direction to produce a fine-grooved appearance; peak to peak height of asperity as measured by profilometer 2.5 μ; mean groove spacing 10 μ; mean groove angle 27°; fresh sample; MgO reference; incident beam polarized in a direction perpendicular to the incident plane of the sample; [Author's designation: C3].
13	318	1961	298	1.20	10-75	θ 2π		Above specimen and conditions.
14	318	1961	298	1.80	10-75	θ 2π		Above specimen and conditions.

DATA TABLE NO. 51 ANGULAR SPECTRAL REFLECTANCE OF COPPER

[Angle, °, θ; Reflectance, ρ; Temperature, T, K, Wavelength, λ, μ]

θ	ρ
CURVE 1*	
T = 298	
λ = 1.2	
0	0.915
10	0.960
20	0.980
30	0.953
40	0.912
50	0.890
60	0.885
70	0.870
CURVE 2*	
T = 298	
λ = 1.2	
0	0.820
10	0.816
20	0.825
30	0.830
40	0.835
50	0.835
60	0.830
70	0.830
80	0.790
CURVE 3*	
T = 298	
λ = 0.65	
10	0.808
20	0.808
30	0.808
40	0.813
50	0.816
60	0.807
70	0.815
75	0.836
CURVE 4*	
T = 298	
λ = 1.20	
10	0.920
20	0.920
30	0.920
40	0.925
CURVE 4 (cont.)*	
50	0.930
60	0.920
70	0.910
75	0.910
CURVE 5*	
T = 298	
λ = 1.80	
10	0.933
20	0.933
30	0.933
41	0.923
50	0.934
60	0.915
70	0.904
75	0.914
CURVE 6*	
T = 298	
λ = 0.65	
10	0.831
20	0.831
30	0.830
40	0.822
50	0.816
60	0.804
70	0.811
75	0.815
CURVE 7*	
T = 298	
λ = 1.20	
10	0.916
20	0.915
30	0.915
40	0.915
50	0.915
60	0.896
70	0.886
75	0.875
CURVE 8*	
T = 298	
λ = 1.80	
10	0.912
20	0.908
30	0.904
40	0.911
50	0.903
60	0.875
70	0.851
75	0.844
CURVE 9*	
T = 298	
λ = 0.65	
10	0.739
20	0.739
30	0.734
40	0.725
50	0.715
60	0.711
70	0.709
75	0.709
CURVE 10*	
T = 298	
λ = 1.20	
10	0.830
20	0.824
30	0.820
40	0.811
50	0.799
60	0.781
70	0.770
75	0.765
CURVE 11*	
T = 298	
λ = 1.80	
10	0.833
20	0.822
30	0.809
40	0.808
50	0.808
CURVE 11 (cont.)*	
60	0.799
70	0.794
75	0.787
CURVE 12*	
T = 298	
λ = 0.65	
10	0.661
20	0.656
30	0.653
40	0.647
50	0.636
60	0.623
70	0.617
75	0.617
CURVE 13*	
T = 298	
λ = 1.20	
10	0.775
20	0.775
30	0.765
40	0.756
50	0.743
60	0.726
70	0.706
75	0.691
CURVE 14*	
T = 298	
λ = 1.80	
10	0.789
20	0.784
30	0.780
41	0.774
50	0.765
60	0.745
70	0.727
75	0.726

* Not shown on plot

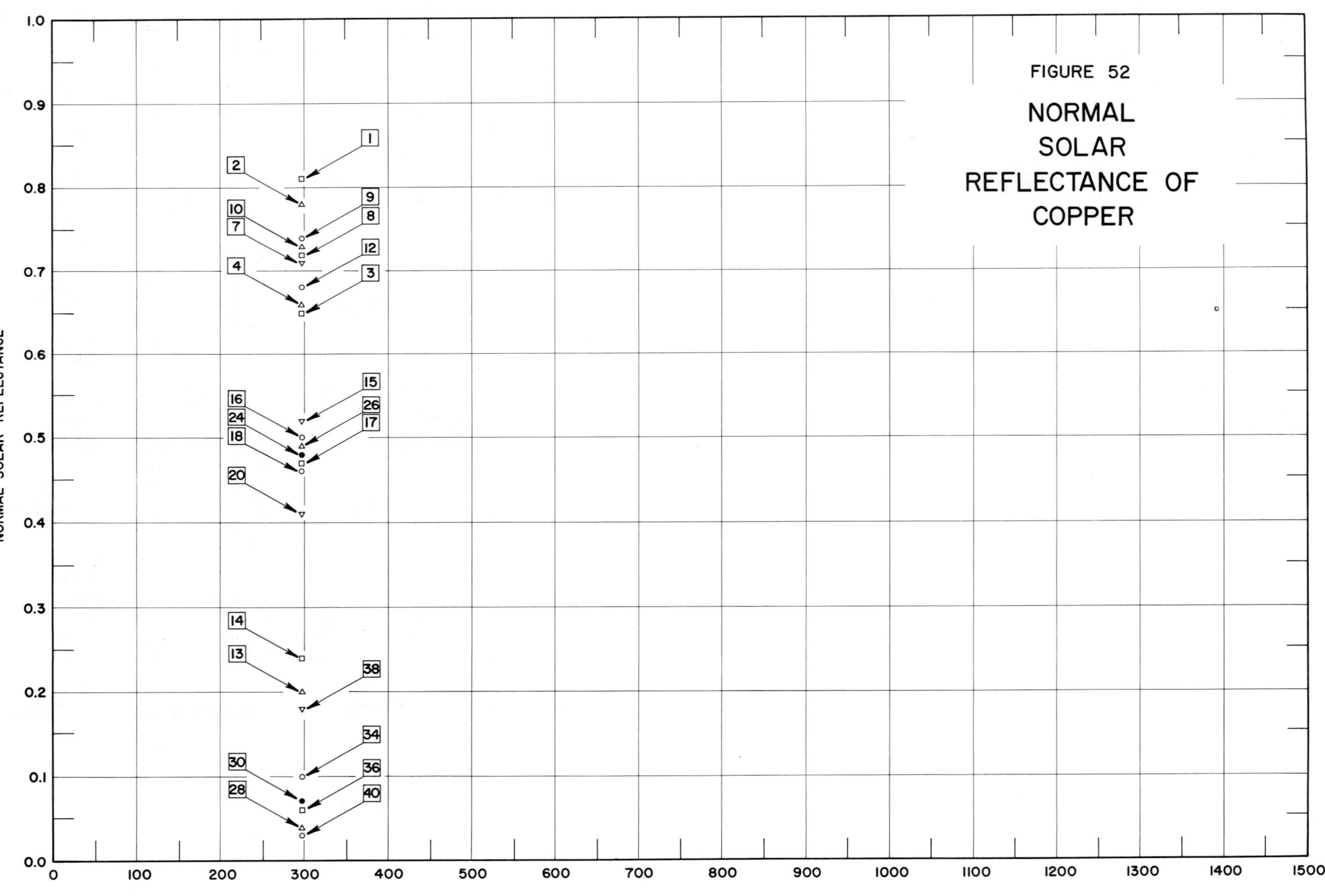

FIGURE 52 NORMAL SOLAR REFLECTANCE OF COPPER

SPECIFICATION TABLE NO. 52 NORMAL SOLAR REFLECTANCE OF COPPER

Curve No.	Ref. No.	Year	Temperature Range, K	Geometry θ θ'	ω'	Reported Error, %	Composition (weight percent), Specifications and Remarks
1	40	1962	298	~0°	2π		Copper film on 1/4 mil thick Mylar.
2	40	1962	298	~0°	2π		Copper film on 1/4 mil thick Mylar.
3	131	1964	298	~0°	2π		B.S. 1433; 99.9 Cu; turned; measured immediately after preparation.
4	131	1964	298	~0°	2π		Different sample, same as above specimen and conditions.
5	131	1964	298	~0°	2π		B.S. 1433; 99.9 Cu; turned; exposed to indoor atmosphere for 1 month.
6	131	1964	298	~0°	2π		Different sample, same as above specimen and conditions.
7	131	1964	298	~0°	2π		B.S. 1433; 99.9 Cu; polished; measured immediately after preparation.
8	131	1964	298	~0°	2π		Different sample, same as above specimen and conditions.
9	131	1964	298	~0°	2π		Different sample, same as above specimen and conditions.
10	131	1964	298	~0°	2π		Different sample, same as above specimen and conditions.
11	131	1964	298	~0°	2π		B.S. 1433; 99.9 Cu; polished; exposed to indoor atmosphere for 1 month.
12	131	1964	298	~0°	2π		Different sample, same as above specimen and conditions.
13	131	1964	298	~0°	2π		B.S. 1433; 99.9 Cu; polished; exposed to outdoor atmosphere for 1 month.
14	131	1964	298	~0°	2π		Different sample, same as above specimen and conditions.
15	131	1964	298	~0°	2π		B.S. 1433; 99.9 Cu; vapor blasted (400 mesh alumina and a nozzle pressure of 70 psi); measured immediately after preparation.
16	131	1964	298	~0°	2π		Different sample, same as above specimen and conditions.
17	131	1964	298	~0°	2π		B.S. 1433; 99.9 Cu; vapor blasted (400 mesh alumina and a nozzle pressure of 70 psi); exposed to indoor atmosphere for 1 month.
18	131	1964	298	~0°	2π		Different sample, same as above specimen and conditions.
19	131	1964	298	~0°	2π		B.S. 1433 (British Standard Specification); nominal composition: 99.90 min Cu, 0.005 max Pb, 0.0010 max Bi, 0.03 max others; turned; freshly prepared; MgO reference; computed from spectral data.
20	131	1964	298	~0°	2π		Above specimen and conditions; diffuse component only.
21	131	1964	298	~0°	2π		Curve 19 specimen and conditions except exposed within clean laboratory area for 1 month.
22	131	1964	298	~0°	2π		Above specimen and conditions; diffuse component only.
23	131	1964	298	~0°	2π		Different sample, same as Curve 19 specimen and conditions.
24	131	1964	298	~0°	2π		Above specimen and conditions; diffuse component only.
25	131	1964	298	~0°	2π		Curve 23 specimen and conditions except exposed within clean laboratory area for 1 month.

SPECIFICATION TABLE NO. 52 (continued)

Curve No.	Ref. No.	Year	Temperature Range, K	Geometry θ	θ'	ω'	Reported Error, %	Composition (weight percent), Specifications and Remarks
26	131	1964	298	~0°		2π		Above specimen and conditions; diffuse component only.
27	131	1964	298	~0°		2π		B.S. 1433 (British Standard Specification); nominal composition: 99.90 min Cu, 0.005 max Pb, 0.0010 max Bi, 0.03 max others; machine polished; freshly prepared; MgO reference; computed from spectral data.
28	131	1964	298	~0°		2π		Above specimen and conditions; diffuse component only.
29	131	1964	298	~0°		2π		Curve 27 specimen and conditions except exposed within clean laboratory area for 1 month.
30	131	1964	298	~0°		2π		Above specimen and conditions; diffuse component only.
31	131	1964	298	~0°		2π		Different sample, same as Curve 27 specimen and conditions.
32	131	1964	298	~0°		2π		Above specimen and conditions; diffuse component only.
33	131	1964	298	~0°		2π		Curve 31 specimen and conditions except specimen exposed within clean laboratory area for 1 month.
34	131	1964	298	~0°		2π		Above specimen and conditions; diffuse component only.
35	131	1964	298	~0°		2π		Different sample, same as Curve 27 specimen and conditions.
36	131	1964	298	~0°		2π		Above specimen and conditions; diffuse component only.
37	131	1964	298	~0°		2π		Curve 35 specimen and conditions except exposed to outside atmosphere for 1 month.
38	131	1964	298	~0°		2π		Above specimen and conditions; diffuse component only.
39	131	1964	298	~0°		2π		Different sample, same as Curve 27 specimen and conditions.
40	131	1964	298	~0°		2π		Above specimen and conditions; diffuse component only.
41	131	1964	298	~0°		2π		Curve 39 specimen and conditions except exposed to outside atmosphere for 1 month.
42	131	1964	298	~0°		2π		Above specimen and conditions; diffuse component only.
43	131	1964	298	~0°		2π		B.S. 1433 (British Standard Specification); nominal composition: 99.90 min Cu, 0.005 max Pb, 0.0010 max Bi, 0.03 max others; vapor blasted with 400 mesh alumina, 70 psi nozzle pressure; freshly prepared; MgO reference; computed from spectral data.
44	131	1964	298	~0°		2π		Above specimen and conditions; diffuse component only.
45	131	1964	298	~0°		2π		Curve 43 specimen and conditions except exposed within clean laboratory area for 1 month.
46	131	1964	298	~0°		2π		Above specimen and conditions; diffuse component only.
47	131	1964	298	~0°		2π		Different sample, same as Curve 43 specimen and conditions.
48	131	1964	298	~0°		2π		Above specimen and conditions; diffuse component only.

SPECIFICATION TABLE NO. 52 (continued)

Curve No.	Ref. No.	Year	Temperature Range, K	Geometry θ	θ' ω'	Reported Error, %	Composition (weight percent), Specifications and Remarks
49	131	1964	298	~0°	2π		Curve47 specimen and conditions except exposed within clean laboratory area for 1 month.
50	131	1964	298	~0°	2π		Above specimen and conditions; diffuse component only.

DATA TABLE NO. 52 NORMAL SOLAR REFLECTANCE OF COPPER

[Temperature, T,K; Reflectance, ρ]

T	ρ	T	ρ	T	ρ	T	ρ	T	ρ
CURVE 1		CURVE 13		CURVE 25*		CURVE 37*		CURVE 49*	
298	0.81	298	0.20	298	0.65	298	0.20	298	0.46
CURVE 2		CURVE 14		CURVE 26		CURVE 38		CURVE 50*	
298	0.78	298	0.24	298	0.49	298	0.18	298	0.46
CURVE 3		CURVE 15		CURVE 27*		CURVE 39*			
298	0.65	298	0.52	298	0.71	298	0.73		
CURVE 4		CURVE 16		CURVE 28		CURVE 40			
298	0.66	298	0.50	298	0.04	298	0.03		
CURVE 5*		CURVE 17		CURVE 29*		CURVE 41*			
298	0.65	298	0.47	298	0.66	298	0.24		
CURVE 6*		CURVE 18		CURVE 30		CURVE 42*			
298	0.65	298	0.46	298	0.07	298	0.20		
CURVE 7		CURVE 19*		CURVE 31*		CURVE 43*			
298	0.71	298	0.65	298	0.72	298	0.52		
CURVE 8		CURVE 20		CURVE 32*		CURVE 44*			
298	0.72	298	0.41	298	0.07	298	0.52		
CURVE 9		CURVE 21*		CURVE 33*		CURVE 45*			
298	0.74	298	0.65	298	0.68	298	0.47		
CURVE 10		CURVE 22*		CURVE 34		CURVE 46*			
298	0.73	298	0.41	298	0.10	298	0.47		
CURVE 11*		CURVE 23*		CURVE 35*		CURVE 47*			
298	0.66	298	0.66	298	0.74	298	0.50		
CURVE 12		CURVE 24		CURVE 36		CURVE 48*			
298	0.68	298	0.48	298	0.06	298	0.50		

*Not shown on plot

SPECIFICATION TABLE NO. 53 HEMISPHERICAL INTEGRATED ABSORPTANCE OF COPPER

Curve No.	Ref. No.	Year	Temperature Range, K	Reported Error, %	Composition (weight percent), Specifications and Remarks
1	4	1953	76		Foil (0.005 in. thick); electrolytic cleaned; measured in vacuum (< 10^{-6} mm Hg) ; absorptance for 294 K black body incident radiation.
2	4	1953	76		Foil (0.005 in. thick) ; dry polished with polishing wax-abrasive; measured in vacuum (< 10^{-6} mm Hg) ; absorptance for 294 K black body incident radiation.
3	4	1953	76		Foil (0.005 in. thick) ; wet polished with Dr. Lyons tooth powder; measured in vacuum (< 10^{-6} mm Hg) ; absorptance for 294 K black body incident radiation.
4	4	1953	76		Foil (0.005 in. thick) ; treated by dilute chromic acid dip; measured in vacuum (< 10^{-6} mm Hg) ; absorptance for 294 K black body incident radiation.
5	3	1955	76	5	Mill run sheet (0.005 in. thick) ; annealed and cleaned; measured in vacuum (10^{-6} to 10^{-7} mm Hg) ; absorptance for 300 K black body incident radiation.
6	3	1955	76	5	Sheet (0.005 in. thick) ; cleaned in dilute chromic acid dip; measured in vacuum (10^{-6} to 10^{-7} mm Hg) ; absorptance for 300 K black body incident radiation.
7	3	1955	76	5	Sheet (0.005 in. thick) ; wet polished with Dr. Lyons Tooth Powder; measured in vacuum (10^{-6} to 10^{-7} mm Hg) ; absorptance for 300 K black body incident radiation.
8	3	1955	76	5	Sheet (0.005 in. thick) ; dry polished with plastic polishing wax abrasive; measured in vacuum (10^{-6} to 10^{-7} mm Hg) ; absorptance for 300 K black body incident radiation.
9	3	1955	76	5	Sheet (0.005 in. thick) ; electrolytically cleaned; measured in vacuum (10^{-6} to 10^{-7} mm Hg); absorptance for 300 K black body incident radiation.
10	3	1955	76	5	Sheet (0.020 in. thick) ; liquid honed; measured in vacuum (10^{-6} to 10^{-7} mm Hg) ; absorptance for 300 K black body incident radiation.
11	3	1955	76	5	Sheet (0.005 in. thick) ; polished with fine emery; measured in vacuum (10^{-6} to 10^{-7} mm Hg); absorptance for 300 K black body incident radiation.
12	3	1955	76	5	Commercial copper sphere; polished; measured in vacuum (10^{-6} to 10^{-7} mm Hg) ; absorptance for 300 K black body incident radiation.
13	3	1955	76	5	Commercial copper sphere; Oakite No. 33 cleaned; measured in vacuum (10^{-6} to 10^{-7} mm Hg); absorptance for 300 K black body incident radiation.
14	23	1955	77.3	± 10	Plating (0.0003 in. thick) ; polished; measured in vacuum (<3 x 10^{-6} mm Hg) ; absorptance for incident radiation from a 300 K black body.

DATA TABLE NO. 53 HEMISPHERICAL INTEGRATED ABSORPTANCE OF COPPER

[Temperature, T, K; Absorptance, α]

T	α	T	α
CURVE 1*		CURVE 13*	
76	0.0178	76	0.03
CURVE 2*		CURVE 14*	
76	0.0192	77.3	0.015
CURVE 3*			
76	0.0185		
CURVE 4*			
76	0.0176		
CURVE 5*			
76	0.015		
CURVE 6*			
76	0.017		
CURVE 7*			
76	0.018		
CURVE 8*			
76	0.019		
CURVE 9*			
76	0.017		
CURVE 10*			
76	0.088		
CURVE 11*			
76	0.023		
CURVE 12*			
76	0.03		

*Not shown on plot

SPECIFICATION TABLE NO. 54 NORMAL INTEGRATED ABSORPTANCE OF COPPER

Curve No.	Ref. No.	Year	Temperature Range, K	Geometry θ	Reported Error, %	Composition (weight percent), Specifications and Remarks
1	139	1959	5-305	$\sim 0^{\circ}$		Annealed and electropolished; wavelength range 1.8 to 4 μ.
2	134	1952	2	$\sim 0^{\circ}$	1.0	Mechanically polished; irradiation from a 300 K black body.
3	134	1952	2	$\sim 0^{\circ}$	1.0	Electropolished surface; irradiation from a 300 K black body.

DATA TABLE NO. 54 NORMAL INTEGRATED ABSORPTANCE OF COPPER

[Temperature, T, K; Absorptance, α]

T	α
CURVE 1*	
5	0.0045
80	0.0043
90	0.0045
120	0.0046
130	0.0047
160	0.0048
165	0.0048
180	0.0052
203	0 0056
205	0 0060
225	0.0061
245	0.0067
268	0.0070
270	0.0069
303	0.0079
305	0 0080
CURVE 2*	
2	0.00147
CURVE 3*	
2	0.0062

*Not shown on plot

SPECIFICATION TABLE NO. 55 NORMAL SPECTRAL ABSORPTANCE OF COPPER

Curve No.	Ref. No.	Year	Wavelength μ	Temperature Range, K	Geometry θ	Reported Error, %	Composition (weight percent), Specifications and Remarks
1	35	1914	0.66	873-2073	$\sim 0^{\circ}$		Measured in burning hydrogen; author assumes $\alpha = \epsilon$.

DATA TABLE NO. 55 NORMAL SPECTRAL ABSORPTANCE OF COPPER

[Temperature, T, K; Absorptance, α; Wavelength, λ, μ]

T	α
CURVE 1*	
$\lambda = 0.66\ \mu$	
873	0.105
2073	0.105

*Not shown on plot

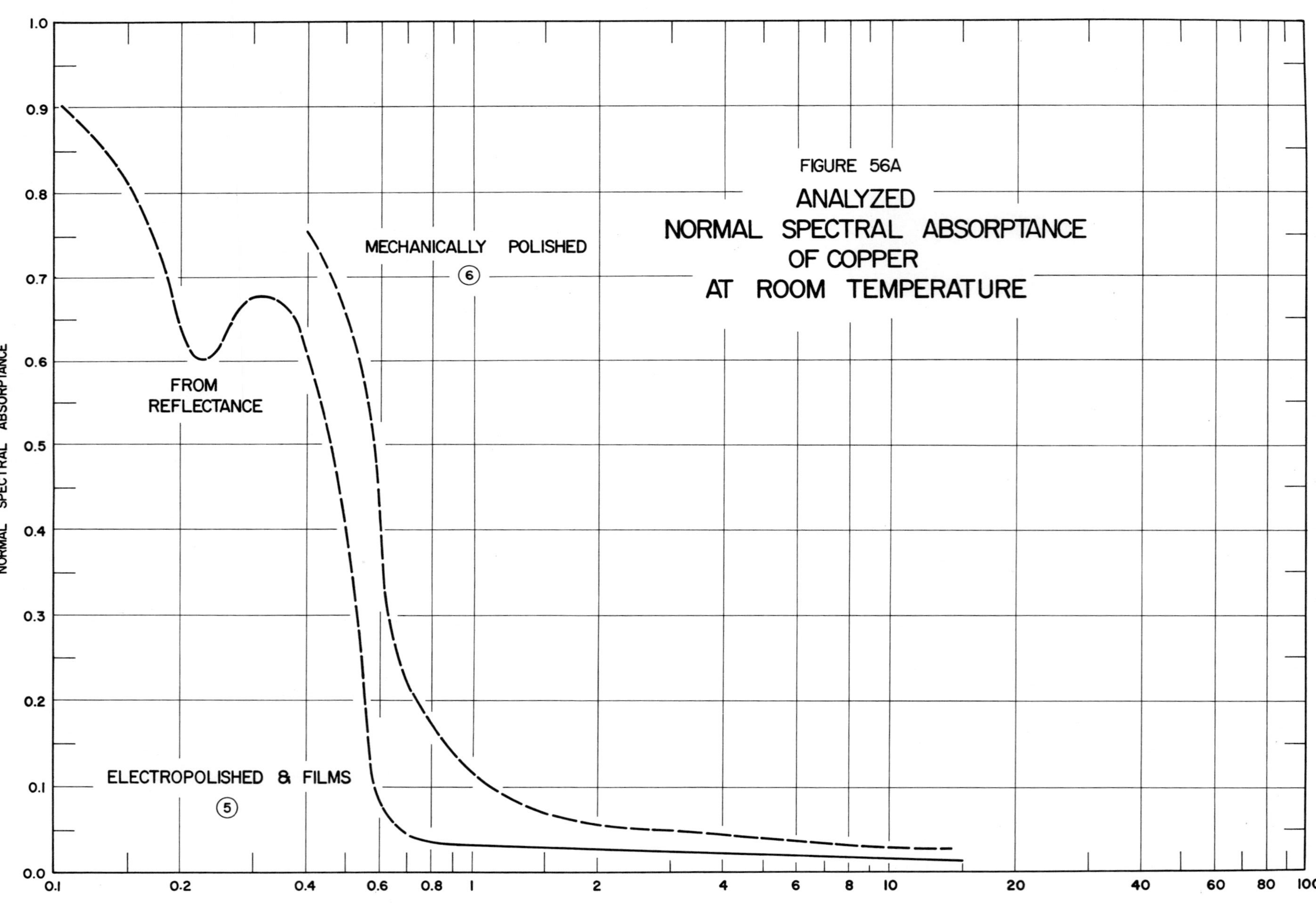

FIGURE 56A
ANALYZED
NORMAL SPECTRAL ABSORPTANCE
OF COPPER
AT ROOM TEMPERATURE
MECHANICALLY POLISHED
6
FROM
REFLECTANCE
ELECTROPOLISHED & FILMS
5
NORMAL SPECTRAL ABSORPTANCE
1.0
0.9
0.8
0.7
0.6
0.5
0.4
0.3
0.2
0.1
0.0
0.1
0.2
0.4
0.6
0.8
1
2
4
6
8
10
20
40
60
80
100
WAVELENGTH, MICRONS

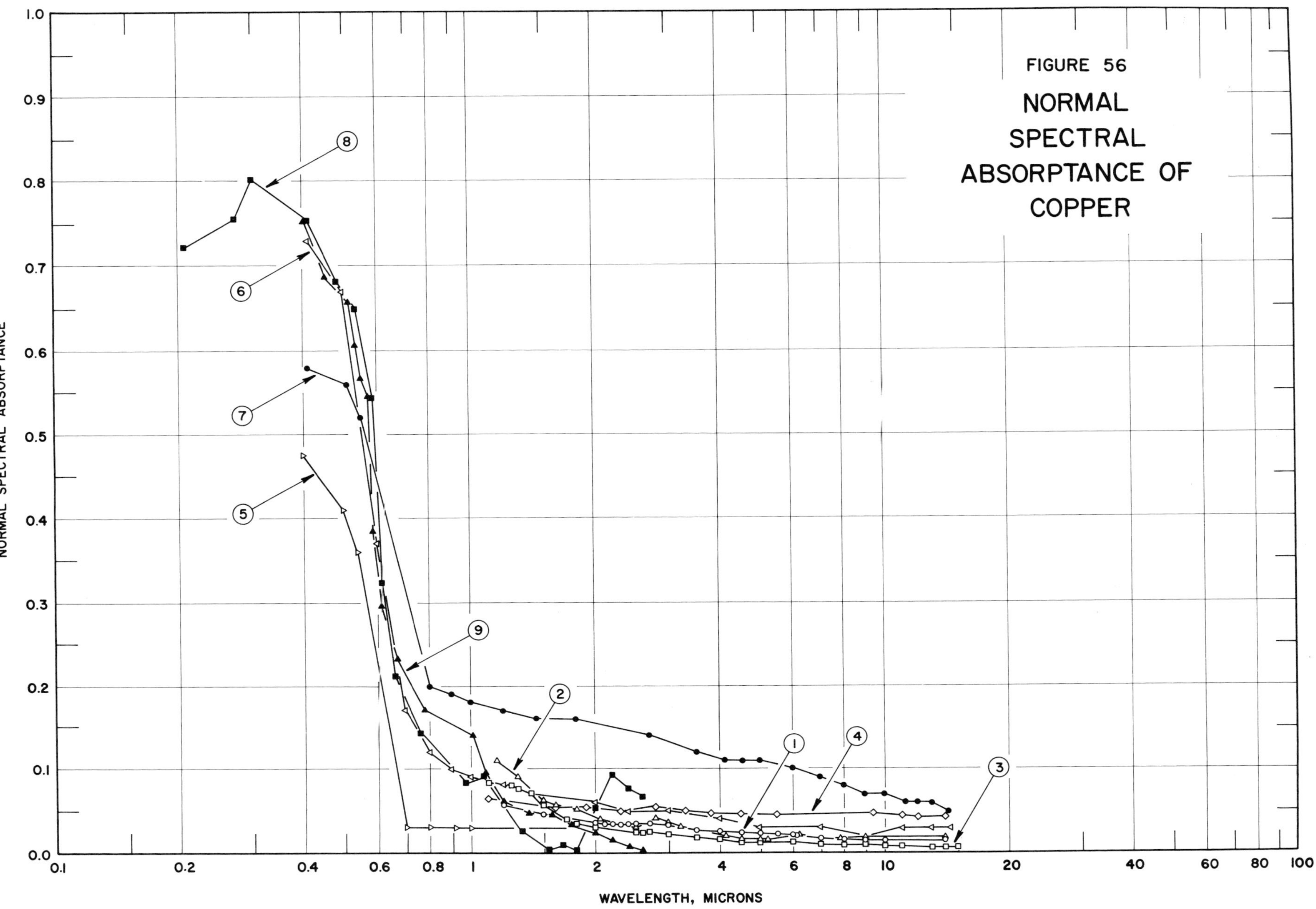
FIGURE 56
NORMAL
SPECTRAL
ABSORPTANCE OF
COPPER
NORMAL SPECTRAL ABSORPTANCE
WAVELENGTH, MICRONS
1.0
0.9
0.8
0.7
0.6
0.5
0.4
0.3
0.2
0.1
0.0
0.1
0.2
0.4
0.6
0.8
1
2
4
6
8
10
20
40
60
80
100
8
6
7
5
9
2
1
4
3

SPECIFICATION TABLE NO. 56 NORMAL SPECTRAL ABSORPTANCE OF COPPER

Curve No.	Ref. No.	Year	Temperature K	Wavelength Range, μ	Geometry θ	Reported Error, %	Composition (weight percent), Specifications and Remarks
1	30	1963	294	1.20-14.00	$\sim 0^\circ$		Mechanically polished (surface roughness 0.02μpeak to peak and 5μ lateral); measured in air; data extracted from smooth curve; [Author's designation: Sample 3].
2	30	1963	294	1.15-14.00	$\sim 0^\circ$		Above specimen and conditions except heated at 450 K for 3 hrs; surface oxidation possible.
3	30	1963	294	1.10-15.00	$\sim 0^\circ$		Above specimen and conditions except heated at 922 K for 3 hrs.
4	30	1963	294	1.10-14.00	$\sim 0^\circ$		Above specimen and conditions except heated at 1222 K for 102 hrs.
5	65	1962	294	0.40-2.00	$\sim 0^\circ$		Electropolished; calculated from $(1-\rho)$.
6	65	1962	294	0.41-14.40	$\sim 0^\circ$		Mechanically polished; calculated from $(1-\rho)$.
7	65	1962	294	0.41-14.20	$\sim 0^\circ$		Above specimen and conditions except roughened with sand paper; surface roughness 1.25μ.
8	307	1954	~298	0.207-2.600	$\sim 0^\circ$		Data extracted from smooth curve.
9	307	1954	~298	0.401-2.600	$\sim 0^\circ$		Polished; data extracted from smooth curve.

DATA TABLE NO. 56 NORMAL SPECTRAL ABSORPTANCE OF COPPER

[Wavelength, λ, μ; Absorptance, α; Temperature, T, K]

λ	α
CURVE 1	
T = 294	
1.20	0.057
1.50	0.046
2.00	0.036
2.10	0.034
2.20	0.034
2.30	0.033
2.40	0.034
2.50	0.035
2.70	0.035
3.00	0.033
3.50	0.028
4.00	0.027
4.50	0.025
5.00	0.023
5.50	0.022
6.00	0.021
6.10	0.021
7.00	0.018
8.00	0.017
9.00	0.015
10.00	0.015
14.00	0.015
CURVE 2	
T = 294	
1.15	0.110
1.30	0.090
1.50	0.062
1.60	0.057
1.80	0.051
2.05	0.040
2.30	0.032*
2.50	0.031
2.80	0.041
3.00	0.035
3.20	0.031
3.50	0.028*
4.10	0.020
4.50	0.017
5.00	0.016
5.20	0.016
6.00	0.020*
CURVE 2 (cont.)	
6.20	0.021
7.00	0.019*
7.80	0.018
9.00	0.018
14.00	0.019
CURVE 3	
T = 294	
1.10	0.082
1.25	0.080
1.30	0.076
1.40	0.070
1.50	0.057
1.70	0.040
1.80	0.035
2.00	0.031
2.50	0.024
2.60	0.024
2.70	0.025
3.00	0.022
3.50	0.019
4.00	0.016
4.50	0.013
5.00	0.012
6.00	0.012
7.00	0.011
8.00	0.010
9.00	0.009
10.00	0.008
11.00	0.008
13.00	0.007
14.00	0.007
15.00	0.007
CURVE 4	
T = 294	
1.10	0.064
1.50	0.055*
1.60	0.054
1.90	0.054
2.30	0.050
2.80	0.055
CURVE 4 (cont.)	
3.30	0.050
3.80	0.048
4.50	0.047
5.50	0.046
9.40	0.048
11.00	0.044
12.00	0.043
14.00	0.043
CURVE 5	
T = 294	
0.40	0.475
0.50	0.410
0.54	0.360
0.70	0.031
0.80	0.031
0.92	0.031
1.00	0.030
2.00	0.031*
CURVE 6	
T = 294	
0.41	0.73
0.50	0.67
0.60	0.37
0.70	0.17
0.80	0.12
0.90	0.10
1.00	0.09
1.20	0.08
1.40	0.07*
2.00	0.06
2.40	0.05
3.00	0.05
4.00	0.04
5.00	0.03
7.00	0.03
9.00	0.02*
11.00	0.03
13.00	0.03
14.40	0.03
CURVE 7	
T = 294	
0.41	0.58
0.51	0.56
0.55	0.52
0.80	0.20
0.90	0.19
1.00	0.18
1.20	0.17
1.44	0.16
1.80	0.16
2.70	0.14
3.50	0.12
4.05	0.11
4.55	0.11
5.00	0.11
6.00	0.10
7.00	0.09
8.00	0.08
9.00	0.07
10.00	0.07
11.20	0.06
12.00	0.06
13.00	0.06
14.20	0.05
CURVE 8	
T = ~298	
0.207	0.724
0.275	0.755
0.301	0.802
0.415	0.752
0.484	0.681
0.535	0.649
0.583	0.542
0.615	0.324
0.660	0.212
0.759	0.141
0.971	0.083
1.075	0.090
1.340	0.026
1.555	0.004
1.688	0.010
1.801	0.003
CURVE 8 (cont.)	
2.000	0.053
2.200	0.091
2.400	0.075
2.600	0.067
CURVE 9	
T = ~298	
0.401	0.751
0.457	0.686
0.515	0.658
0.539	0.607
0.549	0.567
0.579	0.544
0.583	0.389
0.612	0.296
0.664	0.231
0.779	0.170
1.001	0.104
1.096	0.095
1.200	0.060
1.394	0.048
1.594	0.047
1.798	0.034
2.000	0.023
2.200	0.014
2.400	0.008
2.600	0.023

*Not shown on plot

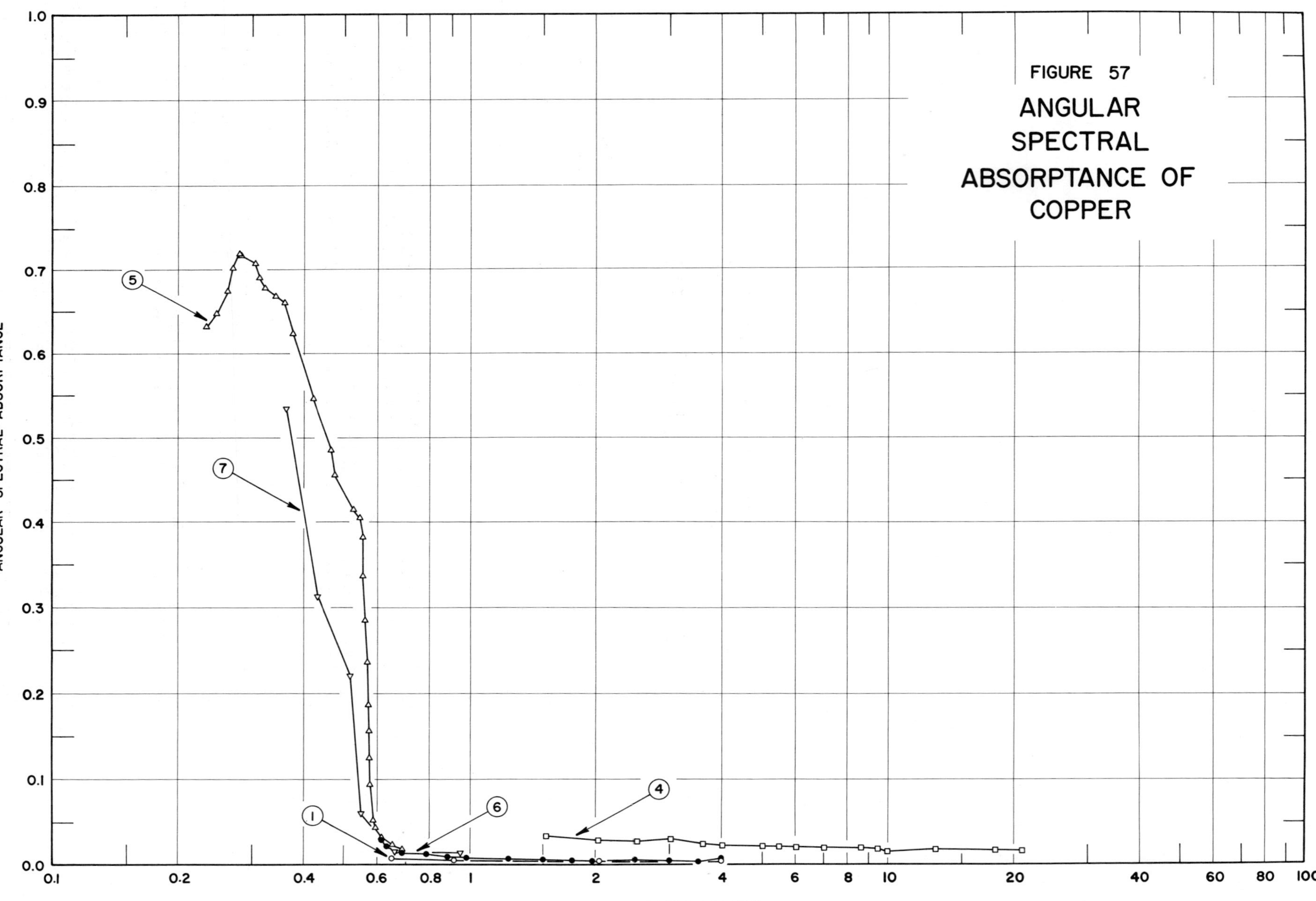
FIGURE 57
ANGULAR SPECTRAL ABSORPTANCE OF COPPER
ANGULAR SPECTRAL ABSORPTANCE
1.0
0.9
0.8
0.7
0.6
0.5
0.4
0.3
0.2
0.1
0.0
0.1
0.2
0.4
0.6
0.8
1
2
4
6
8
10
20
40
60
80
100
WAVELENGTH, MICRONS
5
7
1
6
4

SPECIFICATION TABLE NO. 57 ANGULAR SPECTRAL ABSORPTANCE OF COPPER

Curve No.	Ref. No.	Year	Temperature K	Wavelength Range, μ	Geometry θ	Reported Error, %	Composition (weight percent), Specifications and Remarks
1	140	1954	4.2	0.65-4.00	15°		Single crystal; vacuum annealed at 1313 K for 45 hrs; electropolished; measured in vacuum; run 1.
2	140	1954	4.2	0.65-4.00	15°		Above specimen and conditions; run 2.
3	140	1954	4.2	0.65-4.00	15°		Polycrystal; hydrogen annealed at 1293 K for 6 hrs ; electropolished; measured in vacuum.
4	225	1965	306	1.53-21.0	25°		99.99 Cu from Belmont Smelting and Refining Works; dry sanded and polished on metallurgical felt wheels; measured in dry nitrogen; heated cavity at approx. 1056 K with platinum reference; authors assumed $\alpha = 1 - R(2\pi, 25^{\circ})$.
5	286	1959	4.2	0.2325-0.6997	15°		High-purity copper; cold worked; maintained at 1073 K for 24 hrs; annealed in helium; electropolished and washed; measured in vacuum.
6	286	1959	4.2	0.616-4.004	15°		Above specimen and conditions.
7	287	1956	4.2	0.362-0.951	15°		OFHC Copper; faced with a diamond cutting tool; mechanically polished with metallographic polishing paper and electropolished to remove abrasives in the surface layer; annealed for ~10 hrs either in vacuum or in a hydrogen atmosphere at ~303 K below melting point; electropolished; rinsed in absolute alcohol and boiled distilled water; mounted in the apparatus under a helium atmosphere to prevent oxidation; measured in vacuum.

DATA TABLE NO. 57 ANGULAR SPECTRAL ABSORPTANCE OF COPPER

[Wavelength, λ, μ; Absorptance, α; Temperature, T, K]

λ	α
CURVE 1 T = 4.2	
0.65	0.0069
0.92	0.0051
2.05	0.0038
4.00	0.0042
CURVE 2* T = 4.2	
0.65	0.0064
0.92	0.0055
2.10	0.0038
4.00	0.0040
CURVE 3* T = 4.2	
0.65	0.0061
0.92	0.0050
2.10	0.0039
4.00	0.0039
CURVE 4 T = 306	
1.53	0.032
2.03	0.029
2.51	0.028
3.03	0.030
3.62	0.025
4.02	0.022
5.01	0.021
5.56	0.021
6.05	0.020
7.02	0.020
8.63	0.020
9.48	0.019
10.0	0.018
13.0	0.018
18.2	0.018
21.0	0.018
CURVE 5 T = 4.2	
0.2325	0.632
0.2474	0.647
0.2621	0.673
0.2717	0.702
0.2804	0.720
0.2836	0.719
0.3006	0.708
0.3147	0.690
0.3226	0.678
0.3410	0.669
0.3602	0.660
0.3796	0.624
0.4203	0.547
0.4616	0.485
0.4790	0.456
0.5208	0.414
0.5401	0.405
0.5534	0.381
0.5592	0.337
0.5633	0.285
0.5656	0.236
0.5699	0.187
0.5721	0.156
0.5752	0.125
0.5781	0.093
0.5886	0.051
0.5996	0.043
0.6196	0.032
0.6510	0.023
0.6997	0.019
CURVE 6 T = 4.2	
0.616	0.0300
0.635	0.0208
0.687	0.0168
0.789	0.0113
0.880	0.0097
0.985	0.0083
CURVE 6 (cont.)	
1.232	0.0061
1.502	0.0052
1.764	0.0049
1.986	0.0048
2.491	0.0050
3.003	0.0045
3.511	0.0043
4.004	0.0044
CURVE 7 T = 4.2	
0.362	0.534
0.430	0.312
0.519	0.220
0.550	0.059
0.649	0.014
0.951	0.013

* Not shown on plot

SPECIFICATION TABLE NO. 58 HEMISPHERICAL SOLAR ABSORPTANCE OF COPPER

Curve No.	Ref. No.	Year	Temperature Range, K	Reported Error, %	Composition (weight percent), Specifications and Remarks
1	26	1962	200-400	5	Cleaned in both sodium dichromate and dilute nitric acid solutions, buffed on a felt buffing wheel, cleaned with CCl_4 and acetone; measured in vacuum (10^{-6} mm Hg); data extracted from smooth curve.

DATA TABLE NO. 58 HEMISPHERICAL SOLAR ABSORPTANCE OF COPPER

[Temperature, T, K; Absorptance, α]

T	α
CURVE 1*	
200	0.520
275	0.515
285	0.518
325	0.505
375	0.492
400	0.488

*Not shown on plot

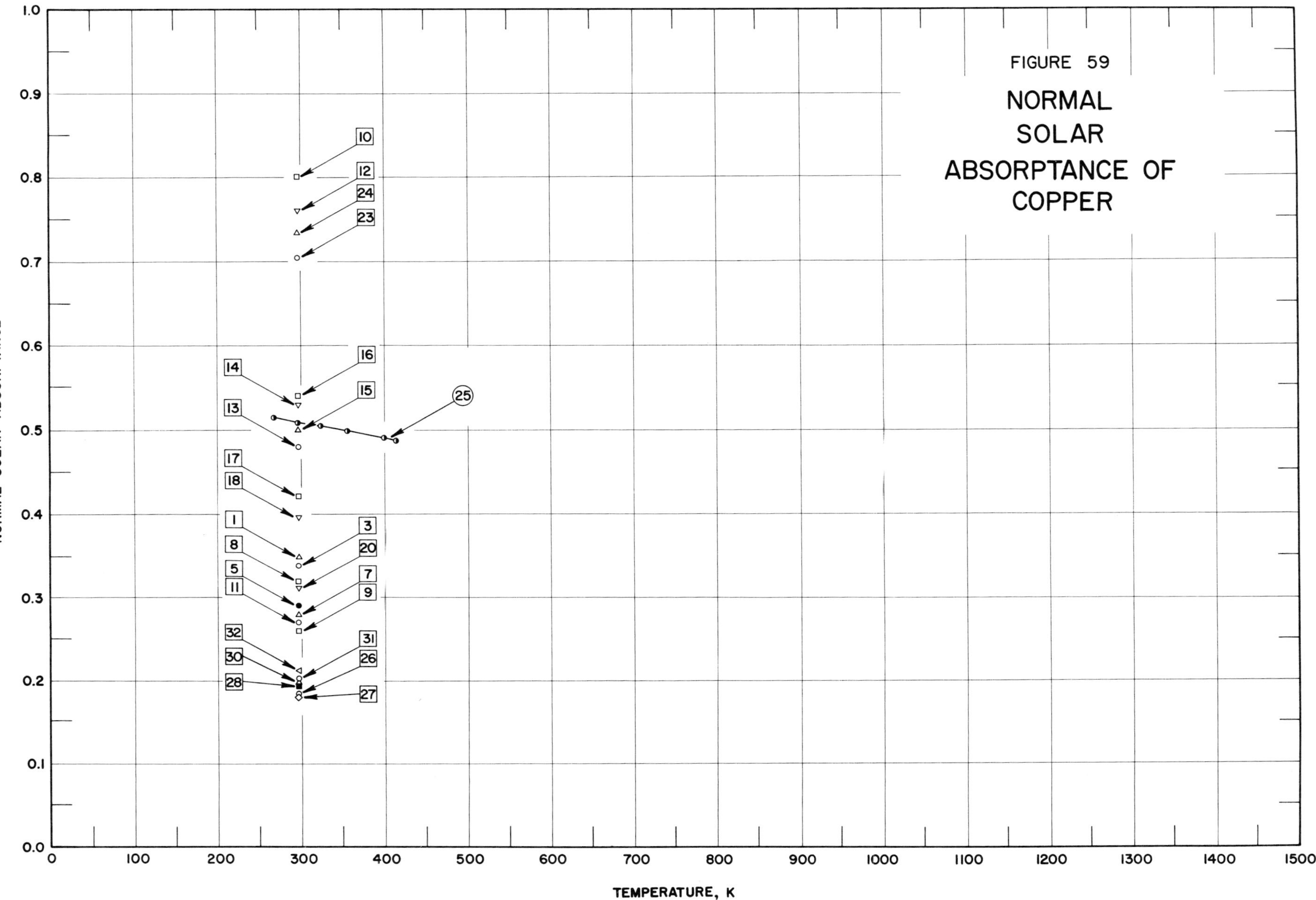

FIGURE 59
NORMAL SOLAR ABSORPTANCE OF COPPER
NORMAL SOLAR ABSORPTANCE
TEMPERATURE, K
0.0 0.1 0.2 0.3 0.4 0.5 0.6 0.7 0.8 0.9 1.0
0 100 200 300 400 500 600 700 800 900 1000 1100 1200 1300 1400 1500
10 12 24 23 14 16 15 13 25 17 18 1 3 8 20 5 7 11 9 32 31 30 26 28 27

SPECIFICATION TABLE NO. 59 NORMAL SOLAR ABSORPTANCE OF COPPER

Curve No.	Ref. No.	Year	Temperature Range, K	Geometry θ	Reported Error, %	Composition (weight percent), Specifications and Remarks
1	131	1964	298	~0°		B.S. 1433 (British Standard Specification); nominal composition: 99.90 min Cu, 0.005 max Pb, 0.0010 max Bi, 0.03 max others; turned; freshly prepared; calculated from $(1 - \rho)$.
2	131	1964	298	~0°		Curve 1 specimen and conditions except exposed within clean laboratory area for 1 month.
3	131	1964	298	~0°		Above specimen and conditions; diffuse component only.
4	131	1964	298	~0°		Curve 3 specimen and conditions except exposed within clean laboratory area for 1 month.
5	131	1964	298	~0°		B.S. 1433 (British Standard Specification); nominal composition: 99.90 min Cu, 0.005 max Pb, 0.0010 max Bi, 0.03 max others; machine polished; freshly prepared; calculated from $(1 - \rho)$.
6	131	1964	298	~0°		Curve 5 specimen and conditions except exposed within clean laboratory area for 1 month.
7	131	1964	298	~0°		Different sample, same as Curve 5 specimen and conditions.
8	131	1964	298	~0°		Curve 7 specimen and conditions except exposed within clean laboratory area for 1 month.
9	131	1964	298	~0°		Different sample, same as Curve 5 specimen and conditions.
10	131	1964	298	~0°		Curve 9 specimen and conditions except exposed to outside atmosphere for 1 month.
11	131	1964	298	~0°		Different sample, same as Curve 5 specimen and conditions.
12	131	1964	298	~0°		Curve 11 specimen and conditions except exposed to outside atmosphere for 1 month.
13	131	1964	298	~0°		B.S. 1433 (British Standard Specification); nominal composition: 99.90 min Cu, 0.005 max Pb, 0.0010 max Bi, 0.03 max others; vapor blasted with 400 mesh alumina, 70 psi nozzle pressure; freshly prepared; calculated from $(1 - \rho)$.
14	131	1964	298	~0°		Curve 13 specimen and conditions except exposed within clean laboratory area for 1 month.
15	131	1964	298	~0°		Different sample, same as Curve 13 specimen and conditions.
16	131	1964	298	~0°		Curve 15 specimen and conditions except exposed within clean laboratory area for 1 month.
17	34	1957	298	9°		Electrolytic, tough pitch; federal specification QQ-C-576 or QQ-C-502; nominal composition: 99.9 Cu, 0.04 O; as received; data calculated from spectral absorptance data measured above atmosphere.
18	34	1957	298	9°		Above specimen and conditions except measured at sea level.

SPECIFICATION TABLE NO. 59 (continued)

Curve No.	Ref. No.	Year	Temperature Range, K	Geometry θ	Reported Error, %	Composition (weight percent), Specifications and Remarks
19	34	1957	298	9°		Different sample, same as Curve 17 specimen except cleaned with liquid detergent.
20	34	1957	298	9°		Above specimen and conditions except measured at sea level.
21	34	1957	298	9°		Different sample, same as Curve 17 specimen and conditions except polished with a fine polishing compound on a buffing wheel.
22	34	1957	298	9°		Above specimen and conditions except measured at sea level.
23	34	1957	298	9°		Different sample, same as Curve 17 specimen and conditions except oxidized in air at red heat for 30 min.
24	34	1957	298	9°		Above specimen and conditions except measured at sea level.
25	26	1964	269-414	~0°		Commerically pure; cleaned, buffed, cleaned with carbon tetrachloride and acetone; measured in vacuum (1×10^{-6} mm Hg); data extracted from smooth curve.
26	352	1963	298	~0°		Electrolytic copper; electropolished; computed from spectral reflectance.
27	352	1963	298	~0°		Above specimen and conditions except hydrogen ion bombarded (0.12×10^{20} ions - cm^{-2}).
28	352	1963	298	~0°		Above specimen and conditions except hydrogen ion bombarded (0.72×10^{20} ions - cm^{-2}).
29	352	1963	298	~0°		Above specimen and conditions except hydrogen ion bombarded (1.62×10^{20} ions - cm^{-2}).
30	352	1963	298	~0°		Above specimen and conditions except hydrogen ion bombarded (3.41×10^{20} ions - cm^{-2}).
31	352	1963	298	~0°		Above specimen and conditions except hydrogen ion bombarded (6.52×10^{20} ions - cm^{-2}).
32	352	1963	298	~0°		Above specimen and conditions except hydrogen ion bombarded (9.83×10^{20} ions - cm^{-2}).

DATA TABLE NO. 59 NORMAL SOLAR ABSORPTANCE OF COPPER

[Temperature, T, K; Absorptance, α]

T	α
CURVE 1	
298	0.35
CURVE 2*	
298	0.35
CURVE 3	
298	0.34
CURVE 4*	
298	0.35
CURVE 5	
298	0.29
CURVE 6*	
298	0.34
CURVE 7	
298	0.28
CURVE 8	
298	0.32
CURVE 9	
298	0.26
CURVE 10	
298	0.80
CURVE 11	
298	0.27
CURVE 12	
298	0.76

T	α
CURVE 13	
298	0.48
CURVE 14	
298	0.53
CURVE 15	
298	0.50
CURVE 16	
298	0.54
CURVE 17	
298	0.421
CURVE 18	
298	0.397
CURVE 19*	
298	0.339
CURVE 20	
298	0.311
CURVE 21*	
298	0.346
CURVE 22*	
298	0.328
CURVE 23	
298	0.704
CURVE 24	
298	0.733

T	α
CURVE 25	
269	0.517
297	0.509
324	0.505
356	0.499
400	0.491
414	0.488
CURVE 26	
298	0.184
CURVE 27	
298	0.182
CURVE 28	
298	0.195
CURVE 29*	
298	0.196
CURVE 30	
298	0.197
CURVE 31	
298	0.202
CURVE 32	
298	0.212

*Not shown on plot

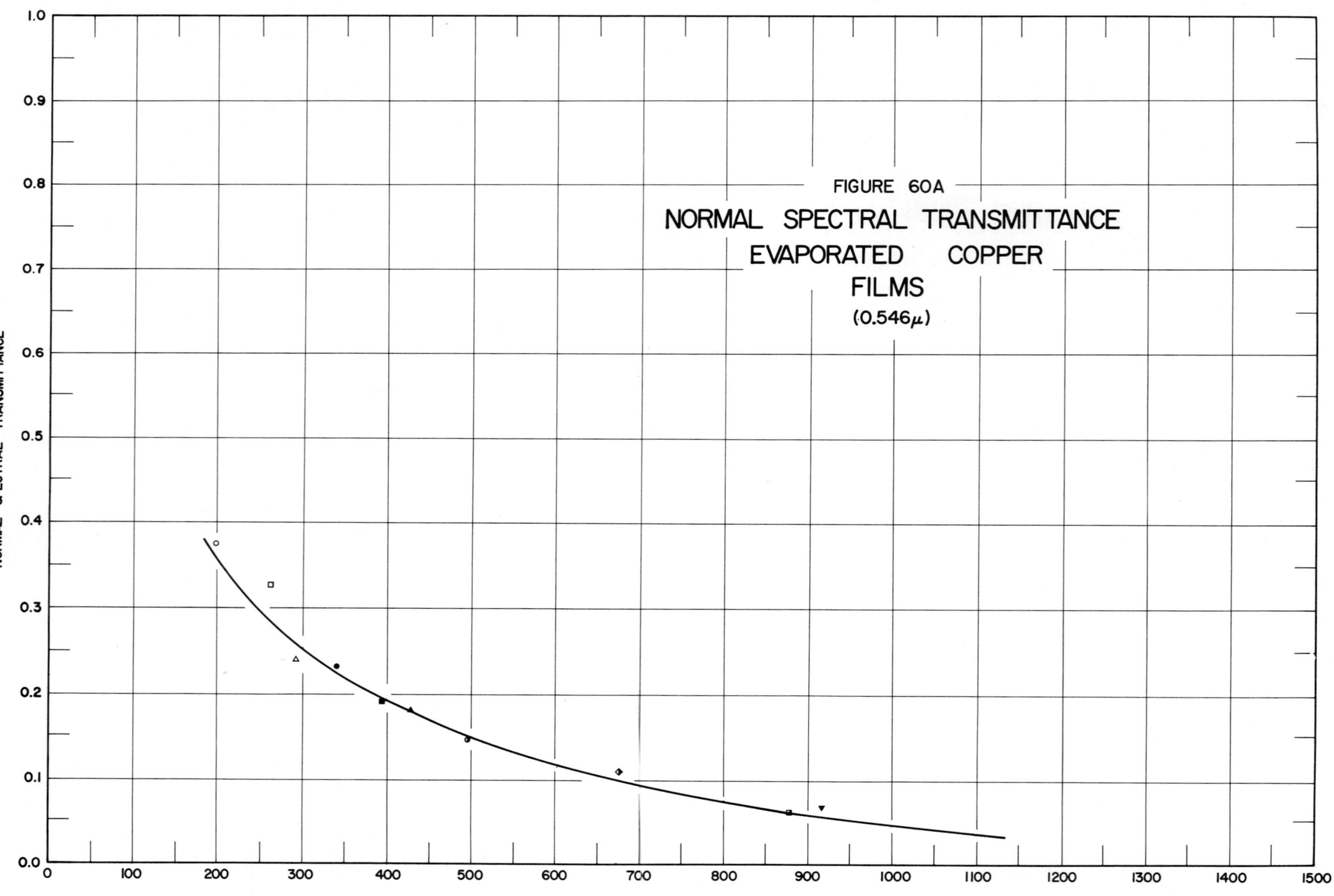

FIGURE 60A

NORMAL SPECTRAL TRANSMITTANCE
EVAPORATED COPPER
FILMS
(0.546μ)

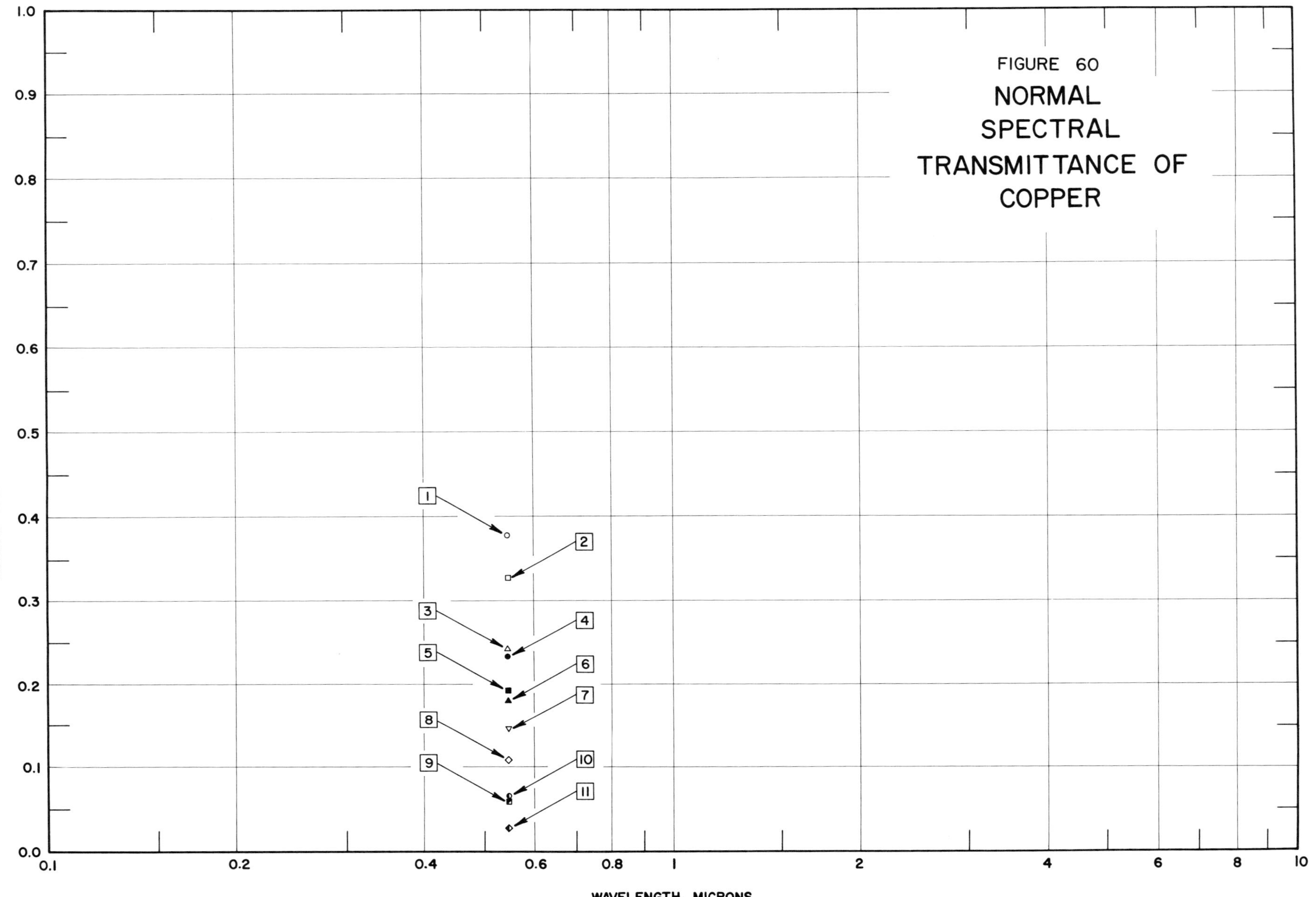

FIGURE 60
NORMAL SPECTRAL TRANSMITTANCE OF COPPER

SPECIFICATION TABLE NO. 60 NORMAL SPECTRAL TRANSMITTANCE OF COPPER

Curve No.	Ref. No.	Year	Temperature K	Wavelength Range, μ	Geometry θ	θ'	ω'	Reported Error, %	Composition (weight percent), Specifications and Remarks
1	311	1966	298	0.546	0°	0°		± 1	Evaporated film; glass microscope slide substrate; evaporated at 2 x 10^{-5} mm Hg; optical thickness 196 Å; aged in desiccator jar at room temperature for 10 days.
2	311	1966	298	0.546	0°	0°		± 1	Different sample, same as above specimen and conditions except optical thickness 261 Å.
3	311	1966	298	0.546	0°	0°		± 1	Different sample, same as above specimen and conditions except optical thickness 290 Å.
4	311	1966	298	0.546	0°	0°		± 1	Different sample, same as above specimen and conditions except optical thickness 340 Å.
5	311	1966	298	0.546	0°	0°		± 1	Different sample, same as above specimen and conditions except optical thickness 392 Å.
6	311	1966	298	0.546	0°	0°		± 1	Different sample, same as above specimen and conditions except optical thickness 428 Å.
7	311	1966	298	0.546	0°	0°		± 1	Different sample, same as above specimen and conditions except optical thickness 492 Å.
8	311	1966	298	0.546	0°	0°		± 1	Different sample, same as above specimen and conditions except optical thickness 672 Å.
9	311	1966	298	0.546	0°	0°		± 1	Different sample, same as above specimen and conditions except optical thickness 876 Å.
10	311	1966	298	0.546	0°	0°		± 1	Different sample, same as above specimen and conditions except optical thickness 915 Å.
11	311	1966	298	0.546	0°	0°		± 1	Different sample, same as above specimen and conditions except optical thickness 1106 Å.

DATA TABLE NO. 60 NORMAL SPECTRAL TRANSMITTANCE OF COPPER

[Wavelength, λ, μ; Transmittance, τ; Temperature, T, K]

λ	τ	λ	τ
CURVE 1 T = 298		CURVE 10 T = 298	
0.546	0.378	0.546	0.065
CURVE 2 T = 298		CURVE 11 T = 298	
0.546	0.328	0.546	0.028
CURVE 3 T = 298			
0.546	0.241		
CURVE 4 T = 298			
0.546	0.232		
CURVE 5 T = 298			
0.546	0.192		
CURVE 6 T = 298			
0.546	0.180		
CURVE 7 T = 298			
0.546	0.146		
CURVE 8 T = 298			
0.546	0.109		
CURVE 9 T = 298			
0.546	0.060		

SPECIFICATION TABLE NO. 61 NORMAL SPECTRAL EMITTANCE OF ERBIUM

Curve No.	Ref. No.	Year	Wavelength μ	Temperature Range, K	Geometry θ'	Reported Error, %	Composition (weight percent), Specifications and Remarks
1	19	1914	0.65	< M.P.	$\sim 0^{\circ}$	1	Film; tungsten substrate; measured in hydrogen; Pt reference ($\epsilon = 0.33$ for $\lambda = 0.650$ μ at all temp).

DATA TABLE NO. 61 NORMAL SPECTRAL EMITTANCE OF ERBIUM

[Temperature, T, K; Emittance, ϵ; Wavelength, λ, μ]

T	ϵ
CURVE 1* $\lambda = 0.65\mu$	
<M.P.	0.55

*Not shown on plot

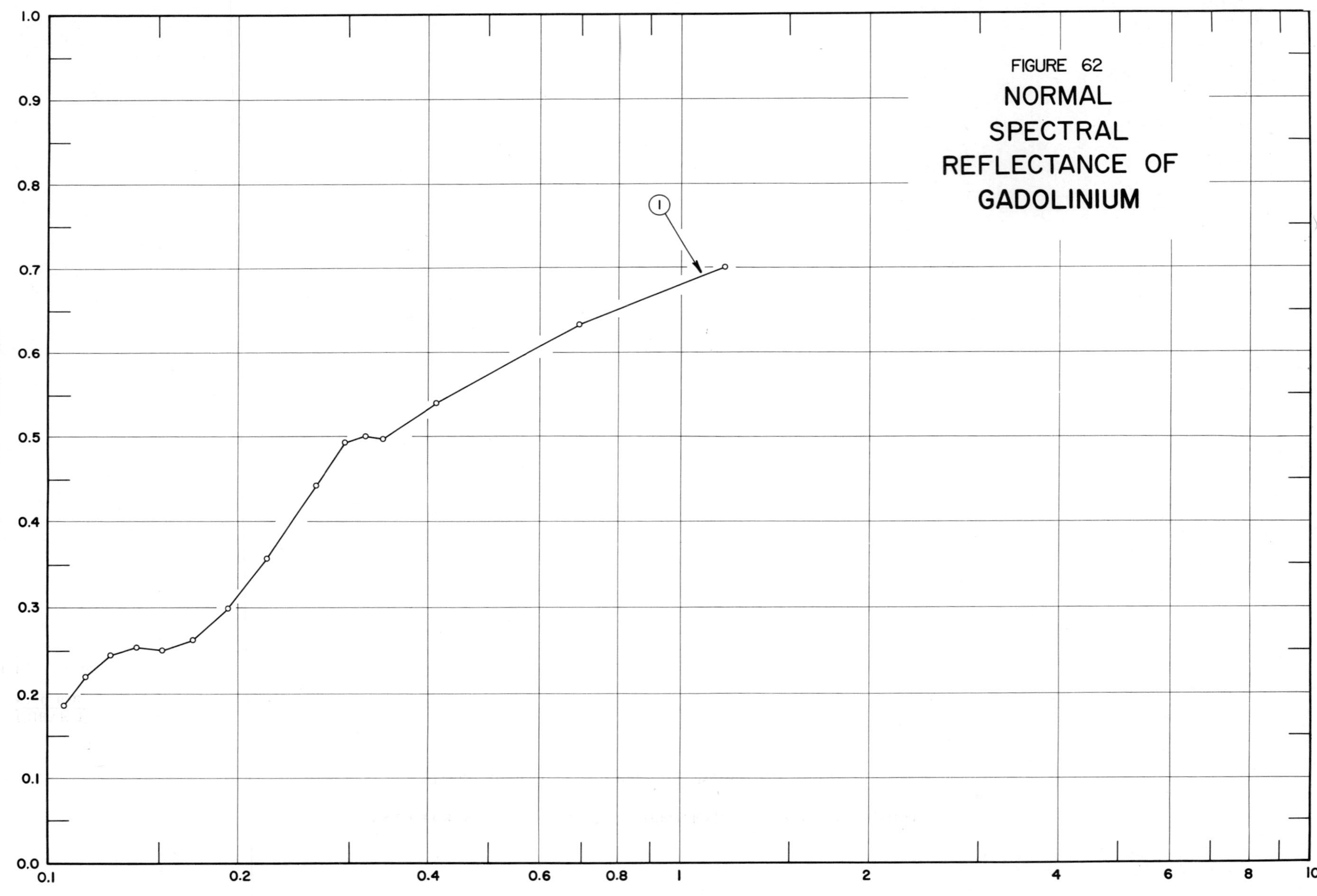

FIGURE 62
NORMAL SPECTRAL REFLECTANCE OF GADOLINIUM
NORMAL SPECTRAL REFLECTANCE
WAVELENGTH, MICRONS
1.0
0.9
0.8
0.7
0.6
0.5
0.4
0.3
0.2
0.1
0.0
0.1
0.2
0.4
0.6
0.8
1
2
4
6
8
10
1

SPECIFICATION TABLE NO. 62 NORMAL SPECTRAL REFLECTANCE OF GADOLINIUM

Curve No.	Ref. No.	Year	Temperature K	Wavelength Range, μ	Geometry θ	θ'	ω'	Reported Error, %	Composition (weight percent), Specifications and Remarks
1	347	1965	298	0.106-1.170	$\sim 0^\circ$	$\sim 0^\circ$			Film; evaporated in vacuum ($< 10^{-9}$ mm Hg); data extracted from smooth curve.

DATA TABLE NO. 62 NORMAL SPECTRAL REFLECTANCE OF GADOLINIUM

[Wavelength, λ, μ; Reflectance, ρ; Temperature, T, K]

λ	ρ
CURVE 1 T = 298	
0.106	0.1879
0.115	0.2193
0.126	0.2443
0.138	0.2529
0.152	0.2506
0.170	0.2624
0.193	0.2999
0.222	0.3573
0.266	0.4426
0.296	0.4943
0.319	0.5000
0.339	0.4989
0.412	0.5408
0.693	0.6339
1.170	0.7063

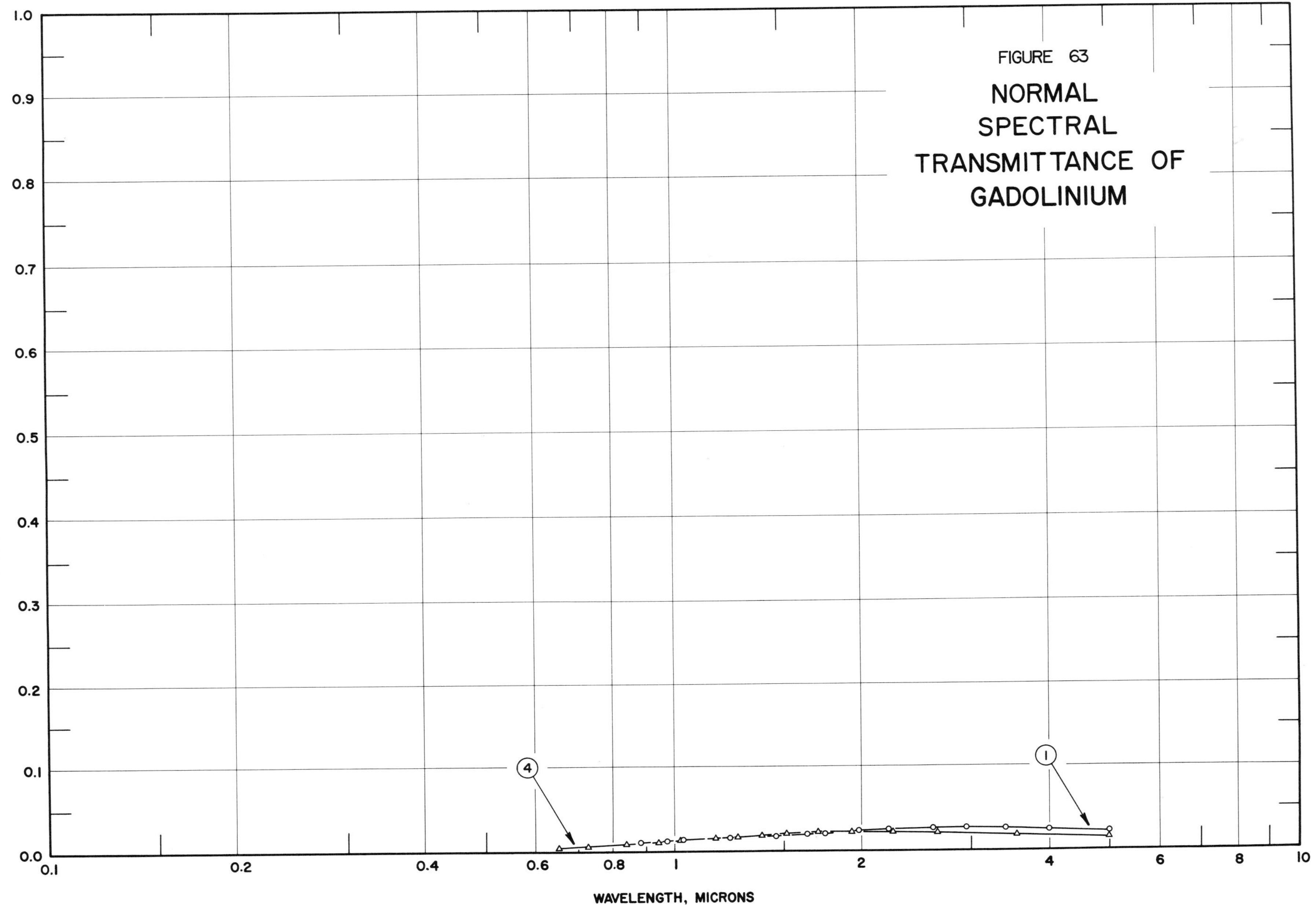

FIGURE 63
NORMAL
SPECTRAL
TRANSMITTANCE OF
GADOLINIUM
1
4
NORMAL SPECTRAL TRANSMITTANCE
1.0
0.9
0.8
0.7
0.6
0.5
0.4
0.3
0.2
0.1
0.0
0.1
0.2
0.4
0.6
0.8
1
2
4
6
8
10
WAVELENGTH, MICRONS

SPECIFICATION TABLE NO. 63 NORMAL SPECTRAL TRANSMITTANCE OF GADOLINIUM

Curve No.	Ref. No.	Year	Temperature, K	Wavelength Range, μ	Geometry θ	θ'	ω'	Reported Error, %	Composition (weight percent), Specifications and Remarks
1	316	1965	80	0.886-4.980	~0°	~0°			Film (925 Å thick); deposited in ultrahigh vacuum on sapphire; measured in vacuum; data extracted from smooth curve.
2	316	1965	160	0.886-4.980	~0°	~0°			Above specimen and conditions.
3	316	1965	220	0.886-4.940	~0°	~0°			Above specimen and conditions.
4	316	1965	293	0.653-4.980	~0°	~0°			Above specimen and conditions.

DATA TABLE NO. 63 NORMAL SPECTRAL TRANSMITTANCE OF GADOLINIUM

[Wavelength, λ, μ; Transmittance, τ; Temperature, T, K]

λ	τ
CURVE 1	
T = 80	
0.886	0.0109
0.973	0.0123
1.033	0.0132
1.223	0.0160
1.459	0.0185
1.634	0.0200
1.751	0.0208
1.987	0.0231
2.210	0.0248
2.616	0.0264
2.945	0.0268
3.416	0.0264
4.000	0.0251
4.980	0.0220
CURVE 2*	
T = 160	
0.886	0.0107
0.946	0.0117
1.167	0.0154
1.375	0.0181
1.527	0.0197
1.706	0.0209
1.879	0.0222
2.172	0.0238
2.446	0.0243
2.638	0.0244
3.085	0.0241
3.636	0.0230
4.276	0.0215
4.980	0.0200
CURVE 3*	
T = 220	
0.886	0.0104
0.966	0.0119
1.122	0.0149
1.263	0.0170
1.401	0.0191
1.494	0.0201
1.564	0.0206
1.710	0.0212
CURVE 3 (cont.)*	
1.959	0.0225
2.105	0.0230
2.305	0.0232
2.756	0.0227
3.360	0.0215
3.974	0.0203
4.940	0.0187
CURVE 4	
T = 293	
0.653	0.0048
0.727	0.0064
0.836	0.0090
0.940	0.0112
1.033	0.0132
1.161	0.0159
1.269	0.0180
1.381	0.0196
1.518	0.0210
1.701	0.0220
1.925	0.0224
2.230	0.0220
2.638	0.0209
3.543	0.0189
4.980	0.0168

*Not shown on plot

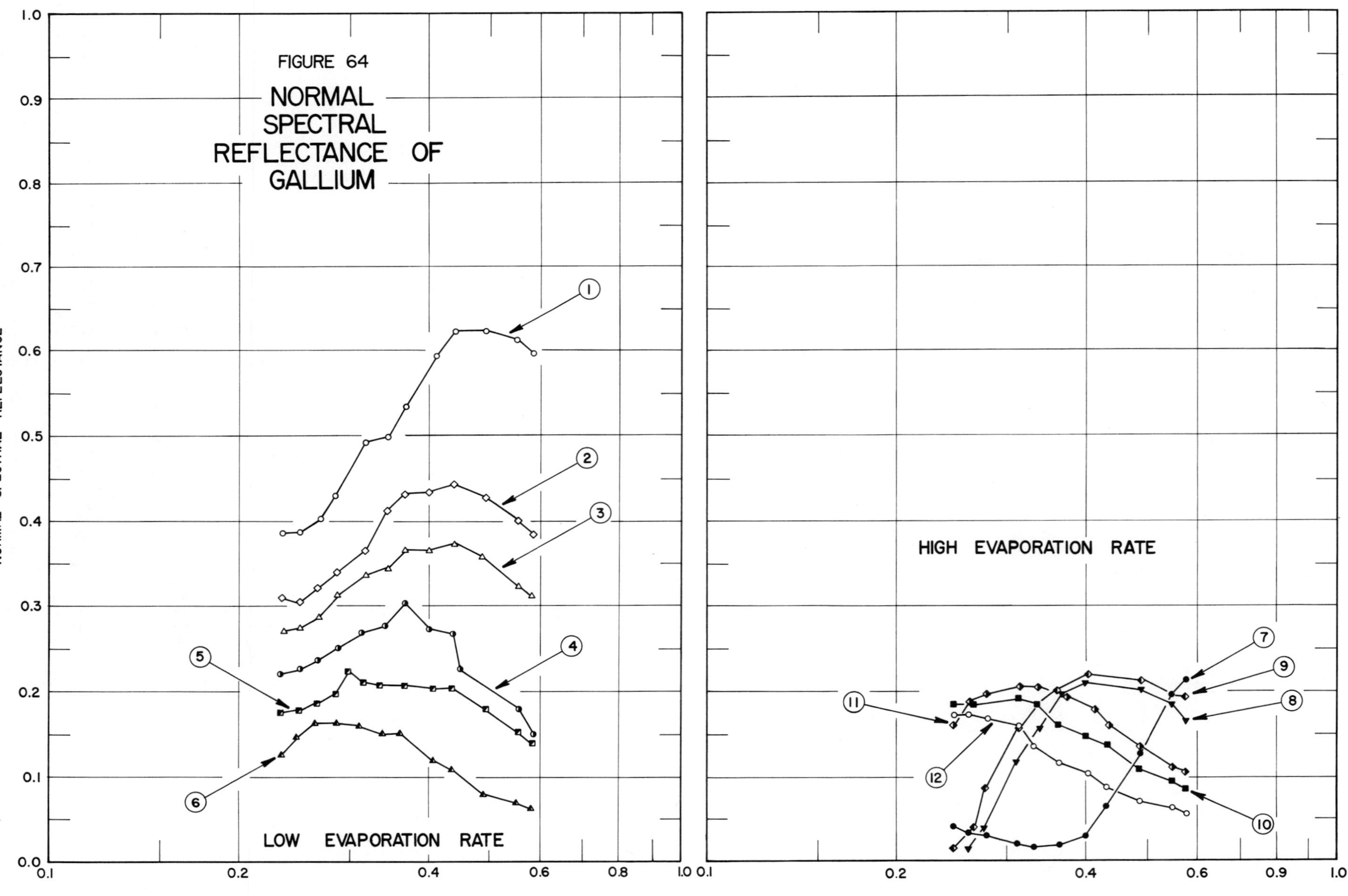
FIGURE 64
NORMAL SPECTRAL REFLECTANCE OF GALLIUM
NORMAL SPECTRAL REFLECTANCE
LOW EVAPORATION RATE
HIGH EVAPORATION RATE
WAVELENGTH , MICRONS
1.0
0.9
0.8
0.7
0.6
0.5
0.4
0.3
0.2
0.1
0.0
0.1
0.2
0.4
0.6
0.8
1.0
0.1
0.2
0.4
0.6
0.9
1.0
1
2
3
4
5
6
7
8
9
10
11
12

SPECIFICATION TABLE NO. 64 NORMAL SPECTRAL REFLECTANCE OF GALLIUM

Curve No.	Ref. No.	Year	Temperature K	Wavelength Range, μ	Geometry θ θ' ω'	Reported Error, %	Composition (weight percent), Specifications and Remarks
1	229	1963	298	0.235-0.584	~0° ~0°		Vacuum deposited thin film of gallium (19 mμ thick); measured in vacuum; spectral Philips Zn, Cd, and Hg Lamp sources; 3.5 Å min^{-1} evaporation rate.
2	229	1963	298	0.233-0.584	~0° ~0°		Different sample, same as above specimen and conditions except 11 mμ thick; 2 Å min^{-1} evaporation rate.
3	229	1963	298	0.236-0.582	~0° ~0°		Different sample, same as above specimen and conditions except 9.5 mμ thick.
4	229	1963	298	0.232-0.583	~0° ~0°		Different sample, same as above specimen and conditions except 5 mμ thick; 2 Å min^{-1} evaporation rate.
5	229	1963	298	0.232-0.582	~0° ~0°		Different sample, same as above specimen and conditions except 4.5 mμ thick.
6	229	1963	298	0.234-0.579	~0° ~0°		Different sample, same as above specimen and conditions except 2.5 mμ thick; 2 Å min^{-1} evaporation rate.
7	229	1963	298	0.246-0.577	~0° ~0°		Different sample, same as above specimen and conditions except 42 mμ thick; 300 Å evaporation rate; supercooled liquid suspected in evaporated film.
8	229	1963	298	0.260-0.577	~0° ~0°		Different sample, same as curve 7 specimen and conditions except 28 mμ thick; 11 Å min^{-1} evaporation rate.
9	229	1963	298	0.247-0.575	~0° ~0°		Different sample, same as curve 7 specimen and conditions except 21 mμ thick.
10	229	1963	298	0.247-0.575	~0° ~0°		Different sample, same as curve 7 specimen and conditions except 14 mμ thick; 11 Å min^{-1} evaporation rate.
11	229	1963	298	0.247-0.575	~0° ~0°		Different sample, same as curve 7 specimen and conditions except 10 mμ thick.
12	229	1963	298	0.247-0.577	~0° ~0°		Different sample, same as curve 7 specimen and conditions except 7 mμ thick; 11 Å min^{-1} evaporation rate.

DATA TABLE NO. 64 NORMAL SPECTRAL REFLECTANCE OF GALLIUM

[Temperature, T, K; Reflectance, ρ; Wavelength, λ, μ]

λ	ρ
CURVE 1	
T = 298	
0.235	0.386
0.251	0.388
0.268	0.402
0.284	0.430
0.318	0.493
0.344	0.499
0.368	0.534
0.410	0.596
0.441	0.623
0.494	0.623
0.552	0.612
0.584	0.597
CURVE 2	
T = 298	
0.233	0.309
0.251	0.303
0.266	0.321
0.286	0.339
0.317	0.365
0.342	0.412
0.366	0.432
0.409	0.434
0.439	0.442
0.493	0.427
0.554	0.400
0.584	0.383
CURVE 3	
T = 298	
0.236	0.270
0.251	0.272
0.267	0.288
0.286	0.312
0.318	0.336
0.343	0.345
0.367	0.366
0.406	0.366
0.439	0.372
0.487	0.358
0.554	0.322
0.582	0.311
CURVE 4	
T = 298	
0.232	0.220
0.250	0.227
0.266	0.236
0.287	0.251
0.312	0.268
0.340	0.277
0.367	0.302
0.409	0.271
0.437	0.266
0.490	0.227
0.552	0.179
0.583	0.150
CURVE 5	
T = 298	
0.232	0.175
0.249	0.178
0.265	0.186
0.285	0.197
0.299	0.222
0.315	0.210
0.334	0.207
0.367	0.207
0.408	0.202
0.436	0.202
0.492	0.179
0.553	0.152
0.582	0.139
CURVE 6	
T = 298	
0.234	0.126
0.247	0.147
0.263	0.163
0.285	0.163
0.310	0.159
0.338	0.151
0.361	0.151
0.405	0.120
0.434	0.109
0.489	0.080
0.551	0.071
0.579	0.063
CURVE 7	
T = 298	
0.246	0.041
0.260	0.032
0.278	0.029
0.311	0.021
0.330	0.016
0.363	0.019
0.402	0.029
0.431	0.065
0.489	0.127
0.547	0.195
0.577	0.212
CURVE 8	
T = 298	
0.260	0.014
0.275	0.039
0.311	0.116
0.337	0.156
0.368	0.197
0.403	0.209
0.490	0.200
0.548	0.183
0.577	0.164
CURVE 9	
T = 298	
0.247	0.016
0.263	0.039
0.276	0.085
0.312	0.157
0.333	0.183*
0.360	0.200
0.404	0.218
0.490	0.211
0.549	0.195*
0.575	0.192
CURVE 10	
T = 298	
0.247	0.182
0.265	0.185
0.312	0.191
CURVE 10 (cont.)	
0.312	0.191
0.334	0.183
0.360	0.160
0.400	0.147
0.431	0.136
0.486	0.108
0.548	0.094
0.575	0.085
CURVE 11	
T = 298	
0.247	0.160
0.263	0.186
0.278	0.197
0.314	0.206
0.335	0.206
0.373	0.192
0.412	0.178
0.433	0.159
0.488	0.134
0.548	0.111
0.575	0.104
CURVE 12	
T = 298	
0.247	0.172
0.261	0.172
0.279	0.168
0.311	0.158
0.330	0.135
0.362	0.116
0.403	0.102
0.430	0.088
0.487	0.072
0.548	0.064
0.577	0.057

*Not shown on plot

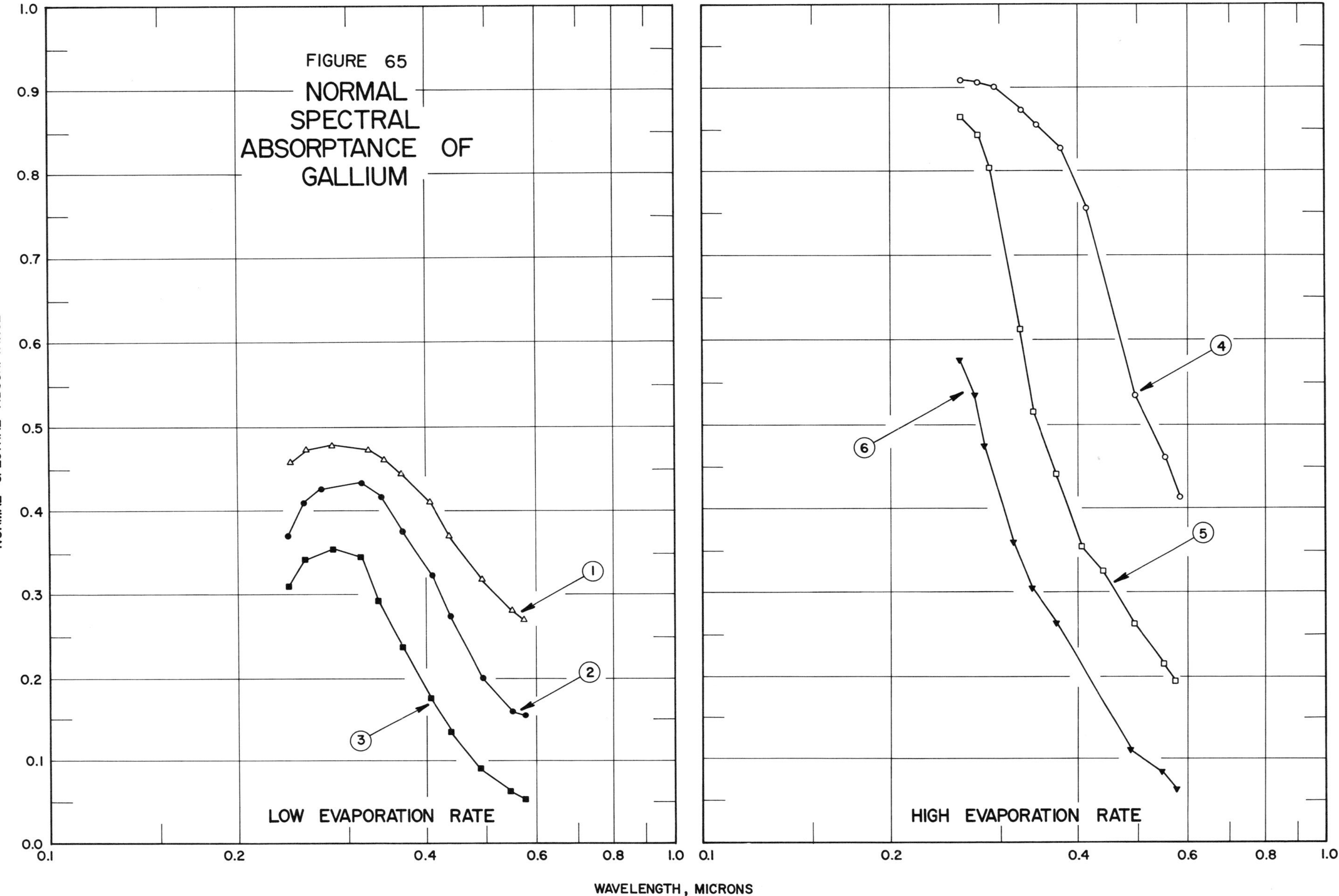

FIGURE 65
NORMAL SPECTRAL ABSORPTANCE OF GALLIUM

SPECIFICATION TABLE NO. 65 NORMAL SPECTRAL ABSORPTANCE OF GALLIUM

Curve No.	Ref. No.	Year	Temperature K	Wavelength Range, μ	Geometry θ	θ'	ω'	Reported Error, %	Composition (weight percent), Specifications and Remarks
1	229	1963	298	0.243-0.573	~0°	~0°			Vacuum deposited thin film of gallium (11 mμ thick); measured in vacuum; spectral Philips Zn, Cd, and Hg Lamp sources; 2 Å min^{-1} evaporation rate; computed from $(1-\rho-\tau)$.
2	229	1963	298	0.242-0.576	~0°	~0°			Different sample, same as above specimen and conditions except 5 mμ thick.
3	229	1963	298	0.242-0.575	~0°	~0°			Different sample, same as above specimen and conditions except 2.5 mμ thick.
4	229	1963	298	0.260-0.583	~0°	~0°			Different sample, same as above specimen and conditions except 42 mμ thick and 300 Å min^{-1} evaporation rate; supercooled liquid suspected in evaporated film.
5	229	1963	298	0.261-0.573	~0°	~0°			Different sample, same as curve 4 specimen and conditions except 21 mμ thick.
6	229	1963	298	0.259-0.578	~0°	~0°			Different sample, same as curve 4 specimen and conditions except 10 mμ thick.

DATA TABLE NO. 65 NORMAL SPECTRAL ABSORPTANCE OF GALLIUM

[Temperature, T, K; Absorptance, α; Wavelength, λ, μ]

λ	α	λ	α
CURVE 1 T = 298		CURVE 4 T = 298	
0.243	0.458	0.260	0.910
0.257	0.474	0.277	0.906
0.282	0.479	0.295	0.901
0.324	0.472	0.325	0.873
0.342	0.461	0.344	0.857
0.367	0.444	0.376	0.829
0.408	0.411	0.414	0.755
0.436	0.370	0.496	0.533
0.492	0.318	0.554	0.460
0.550	0.281	0.583	0.413
0.573	0.270		
CURVE 2 T = 298		CURVE 5 T = 298	
0.242	0.370	0.261	0.865
0.255	0.410	0.277	0.843
0.272	0.426	0.290	0.802
0.316	0.434	0.323	0.612
0.340	0.417	0.341	0.515
0.368	0.375	0.370	0.440
0.411	0.322	0.407	0.354
0.438	0.275	0.440	0.325
0.494	0.200	0.493	0.262
0.550	0.161	0.550	0.214
0.576	0.155	0.573	0.194
CURVE 3 T = 298		CURVE 6 T = 298	
0.242	0.309	0.259	0.577
0.256	0.342	0.273	0.534
0.284	0.355	0.283	0.474
0.315	0.345	0.315	0.360
0.336	0.292	0.338	0.302
0.367	0.237	0.370	0.263
0.408	0.177	0.488	0.111
0.439	0.134	0.546	0.084
0.490	0.090	0.578	0.063
0.546	0.063		
0.575	0.052		

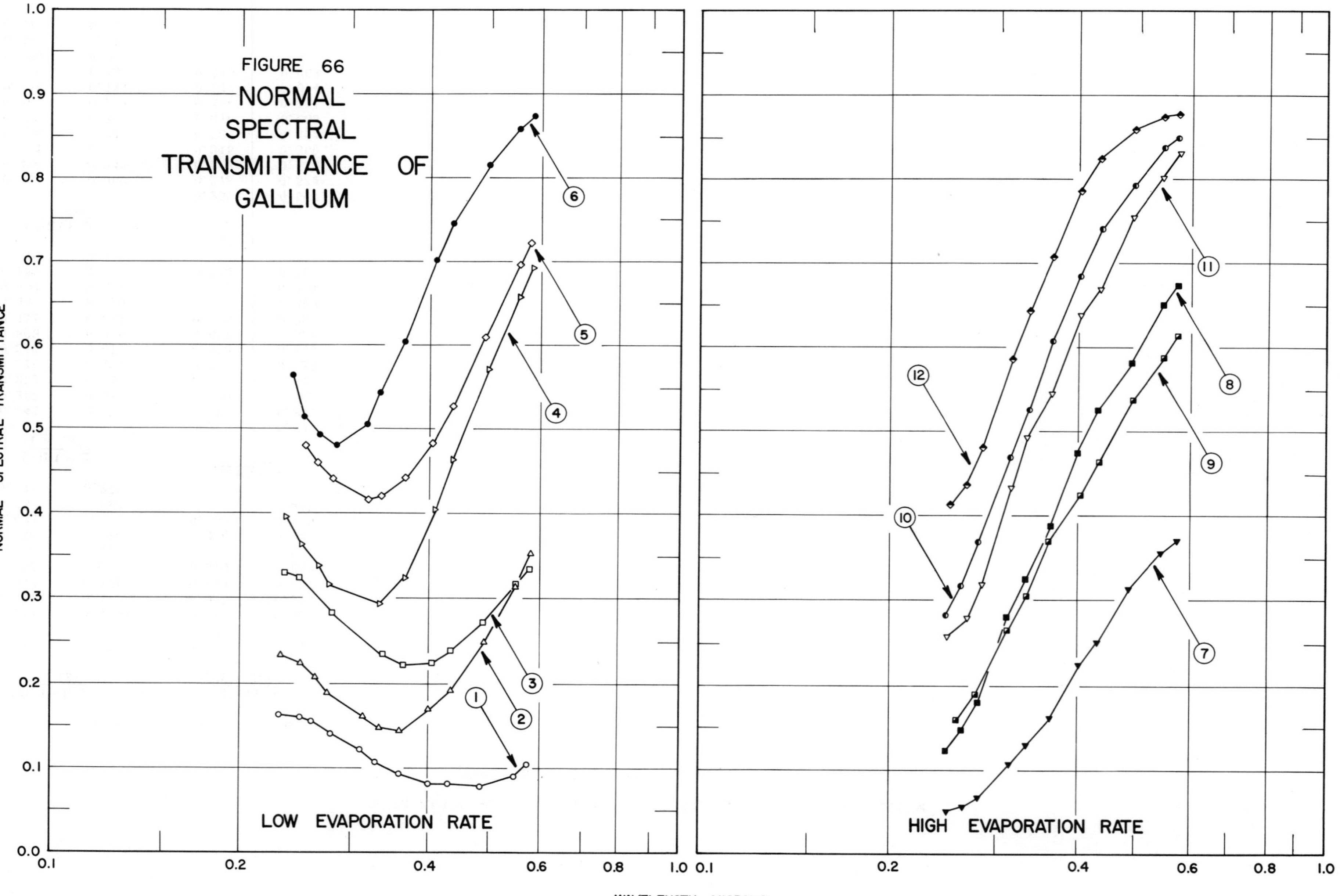

FIGURE 66 NORMAL SPECTRAL TRANSMITTANCE OF GALLIUM

SPECIFICATION TABLE NO. 66 NORMAL SPECTRAL TRANSMITTANCE OF GALLIUM

Curve No.	Ref. No.	Year	Temperature K	Wavelength Range, μ	Geometry θ	θ'	ω'	Reported Error, %	Composition (weight percent), Specifications and Remarks
1	229	1963	298	0.231-0.574	~0°	~0°			Vacuum deposited thin film of gallium (19 mμ thick); measured in vacuum; spectral Philips Zn, Cd and Hg Lamp sources; 3.5 Å min^{-1} evaporation rate.
2	229	1963	298	0.233-0.579	~0°	~0°			Different sample, same as above specimen and conditions except 11 mμ thick; 2 Å min^{-1} evaporation rate.
3	229	1963	298	0.236-0.578	~0°	~0°			Different sample, same as above specimen and conditions except 9.5 mμ thick.
4	229	1963	298	0.236-0.581	~0°	~0°			Different sample, same as above specimen and conditions except 5 mμ thick; 2 Å min^{-1} evaporation rate.
5	229	1963	298	0.242-0.577	~0°	~0°			Different sample, same as above specimen and conditions except 4.5 mμ thick.
6	229	1963	298	0.242-0.584	~0°	~0°			Different sample, same as above specimen and conditions except 2.5 mμ thick; 2 Å min^{-1} evaporation rate.
7	229	1963	298	0.250-0.579	~0°	~0°			Different sample, same as above specimen and conditions except 42 mμ thick; 300 Å min^{-1} evaporation rate; supercooled liquid suspected in the evaporated film.
8	229	1963	298	0.248-0.578	~0°	~0°			Different sample, same as curve 7 specimen and conditions except 28 mμ thick; 11 Å min^{-1} evaporation rate.
9	229	1963	298	0.258-0.579	~0°	~0°			Different sample, same as curve 7 specimen and conditions except 21 mμ thick.
10	229	1963	298	0.247-0.579	~0°	~0°			Different sample, same as curve 7 specimen and conditions except 14 mμ thick; 11 Å min^{-1} evaporation rate.
11	229	1963	298	0.249-0.581	~0°	~0°			Different sample, same as curve 7 specimen and conditions except 10 mμ thick.
12	229	1963	298	0.252-0.581	~0°	~0°			Different sample, same as curve 7 specimen and conditions except 7 mμ thick; 11 Å min^{-1} evaporation rate.

DATA TABLE NO. 66 NORMAL SPECTRAL TRANSMITTANCE OF GALLIUM

[Temperature, T, K; Transmittance, τ; Wavelength, λ, μ]

λ	τ
CURVE 1 T = 298	
0.231	0.162
0.249	0.160
0.261	0.155
0.278	0.140
0.311	0.122
0.329	0.107
0.359	0.094
0.401	0.081
0.430	0.081
0.484	0.078
0.547	0.091
0.574	0.105
CURVE 2 T = 298	
0.233	0.233
0.250	0.224
0.263	0.207
0.276	0.189
0.314	0.161
0.333	0.148
0.360	0.144
0.402	0.170
0.434	0.193
0.490	0.250
0.549	0.313
0.579	0.352
CURVE 3 T = 298	
0.236	0.330
0.248	0.324
0.280	0.282
0.337	0.234
0.364	0.222
0.405	0.225
0.433	0.239
0.489	0.272
0.549	0.316
0.578	0.335
CURVE 4 T = 298	
0.236	0.396
0.250	0.362
0.267	0.338
0.277	0.317
0.333	0.294
0.365	0.324
0.408	0.404
0.435	0.463
0.495	0.572
0.551	0.658
0.581	0.693
CURVE 5 T = 298	
0.242	0.504
0.253	0.481
0.265	0.459
0.279	0.441
0.319	0.414
0.336	0.420
0.366	0.442
0.404	0.483
0.436	0.527
0.490	0.610
0.557	0.697
0.577	0.722
CURVE 6 T = 298	
0.242	0.564
0.253	0.515
0.267	0.493
0.284	0.481
0.318	0.505
0.334	0.544
0.364	0.604
0.409	0.701
0.435	0.747
0.497	0.818
0.554	0.860
0.584	0.875
CURVE 7 T = 298	
0.250	0.050
0.265	0.057
0.280	0.068
0.313	0.107
0.333	0.130
0.364	0.161
0.405	0.224
0.432	0.251
0.487	0.314
0.548	0.356
0.579	0.370
CURVE 8 T = 298	
0.248	0.122
0.262	0.147
0.279	0.180
0.311	0.280
0.332	0.325
0.364	0.388
0.402	0.473
0.433	0.529
0.490	0.582
0.549	0.651
0.578	0.673
CURVE 9 T = 298	
0.258	0.158
0.275	0.190
0.311	0.264
0.333	0.303
0.362	0.370
0.406	0.424
0.434	0.462
0.491	0.538
0.550	0.589
0.579	0.614
CURVE 10 T = 298	
0.247	0.282
0.262	0.318
0.278	0.369
0.314	0.468
0.336	0.525
0.367	0.607
0.405	0.684
0.438	0.741
0.494	0.794
0.550	0.838
0.579	0.851
CURVE 11 T = 298	
0.249	0.257
0.268	0.279
0.283	0.318
0.315	0.432
0.334	0.492
0.365	0.544
0.406	0.639
0.436	0.679
0.493	0.755
0.549	0.802
0.581	0.833
CURVE 12 T = 298	
0.252	0.412
0.268	0.437
0.284	0.482
0.317	0.587
0.336	0.644
0.366	0.708
0.407	0.787
0.436	0.825
0.496	0.860
0.551	0.875
0.581	0.879

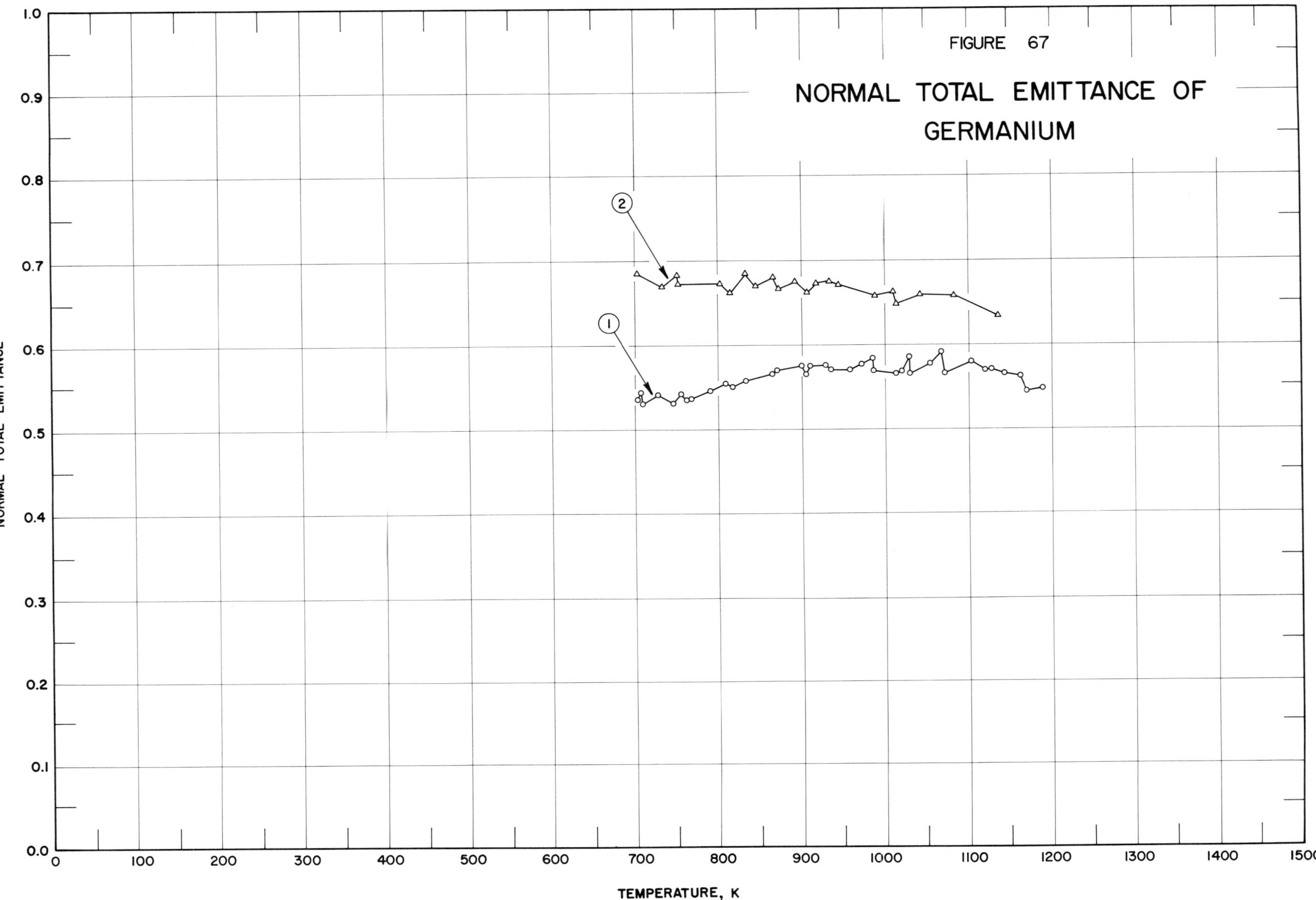

FIGURE 67
NORMAL TOTAL EMITTANCE OF GERMANIUM

SPECIFICATION TABLE NO. 67 NORMAL TOTAL EMITTANCE OF GERMANIUM

Curve No.	Ref. No.	Year	Temperature Range, K	Geometry θ'	Reported Error, %	Composition (weight percent), Specifications and Remarks
1	230	1964	703-1191	$\sim 0^\circ$		Polished; measured in vacuum (4 x 10^{-4} mm Hg).
2	230	1964	704-1136	$\sim 0^\circ$		Buffed; measured in vacuum (4 x 10^{-4} mm Hg).

DATA TABLE NO. 67 NORMAL TOTAL EMITTANCE OF GERMANIUM

[Temperature, T,K; Emittance, ϵ]

T	ϵ	T	ϵ
CURVE 1		CURVE 2	
703	0.535	704	0.685
707	0.543	732	0.670
710	0.530	751	0.682
727	0.541	753	0.672
746	0.531	803	0.673
755	0.542	815	0.663
762	0.535	832	0.686
768	0.537	847	0.670
790	0.547	866	0.668
810	0.554	873	0.668
818	0.550	892	0.676
833	0.558	908	0.664
865	0.566	918	0.672
871	0.570	934	0.675
901	0.574	946	0.671
905	0.566	989	0.659
910	0.577	1011	0.663
929	0.577	1013	0.649
935	0.571	1042	0.660
959	0.571	1083	0.659
972	0.578	1136	0.634
986	0.583		
988	0.569		
1014	0.566		
1021	0.569		
1030	0.587		
1031	0.565		
1055	0.578		
1069	0.591		
1073	0.566		
1103	0.580		
1107	0.566		
1121	0.570		
1129	0.570		
1143	0.566		
1163	0.562		
1171	0.543		
1191	0.548		

SPECIFICATION TABLE NO. 68 NORMAL SPECTRAL EMITTANCE OF GERMANIUM

Curve No.	Ref. No.	Year	Wavelength μ	Temperature Range, K	Geometry θ'	Reported Error, %	Composition (weight percent), Specifications and Remarks
1	20	1957	0.65	1000-1200	$\sim 0^{\circ}$	± 10	Etched; measured in vacuum (10^{-7} - 10^{-9} mm Hg).

DATA TABLE NO. 68 NORMAL SPECTRAL EMITTANCE OF GERMANIUM

[Temperature, T, K; Emittance, ϵ; Wavelength, λ, μ]

T	ϵ
CURVE 1*	
$\lambda = 0.65$	
1000	0.56
1100	0.55
1200	0.53

*Not shown on plot

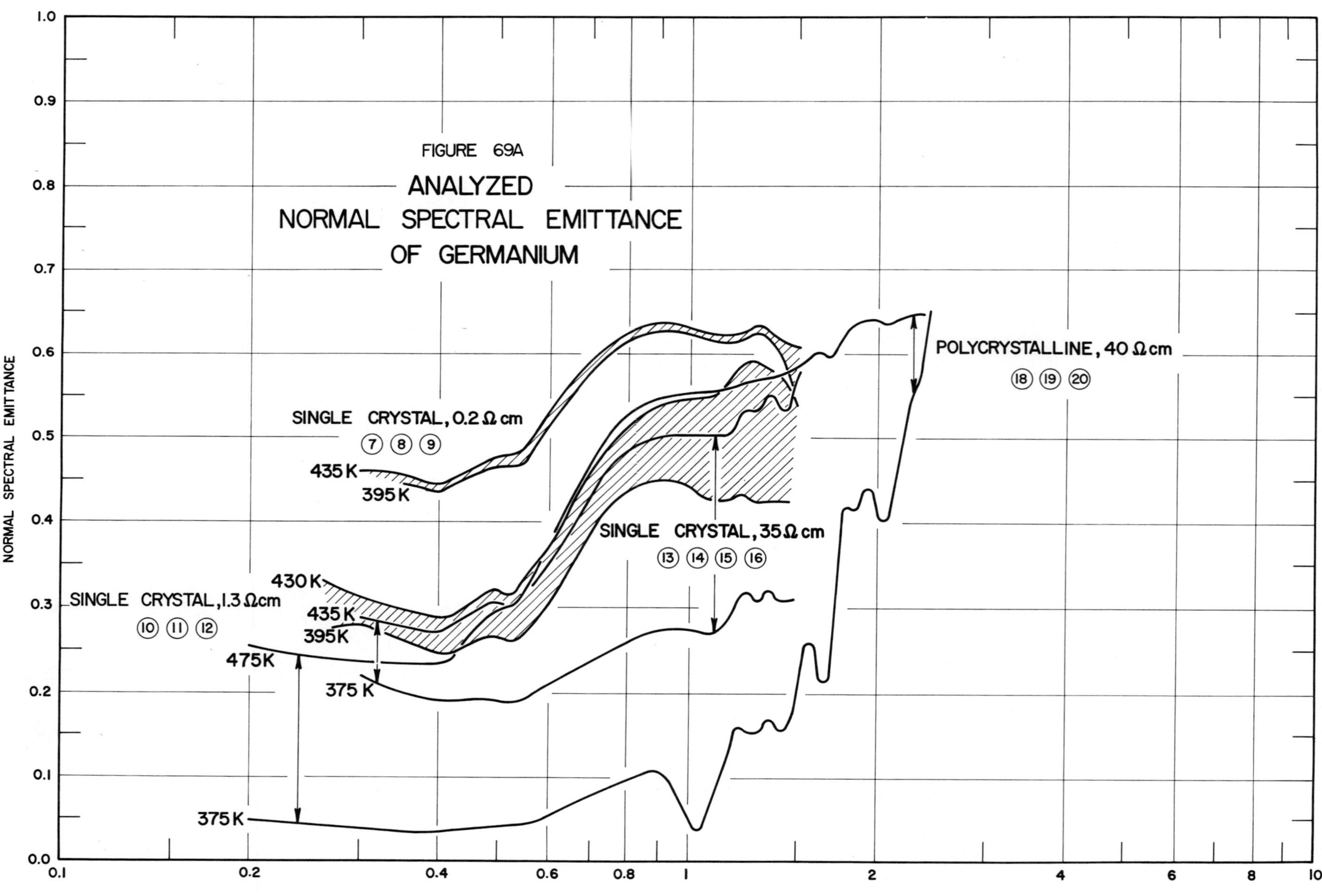
FIGURE 69A
ANALYZED
NORMAL SPECTRAL EMITTANCE
OF GERMANIUM
SINGLE CRYSTAL, 0.2 Ω cm
⑦ ⑧ ⑨
435 K
395 K
SINGLE CRYSTAL, 35 Ω cm
⑬ ⑭ ⑮ ⑯
POLYCRYSTALLINE, 40 Ω cm
⑱ ⑲ ⑳
430 K
SINGLE CRYSTAL, 1.3 Ω cm
⑩ ⑪ ⑫
435 K
395 K
475 K
375 K
375 K
NORMAL SPECTRAL EMITTANCE
1.0
0.9
0.8
0.7
0.6
0.5
0.4
0.3
0.2
0.1
0.0
0.1
0.2
0.4
0.6
0.8
1
2
4
6
8
10
WAVELENGTH, MICRONS

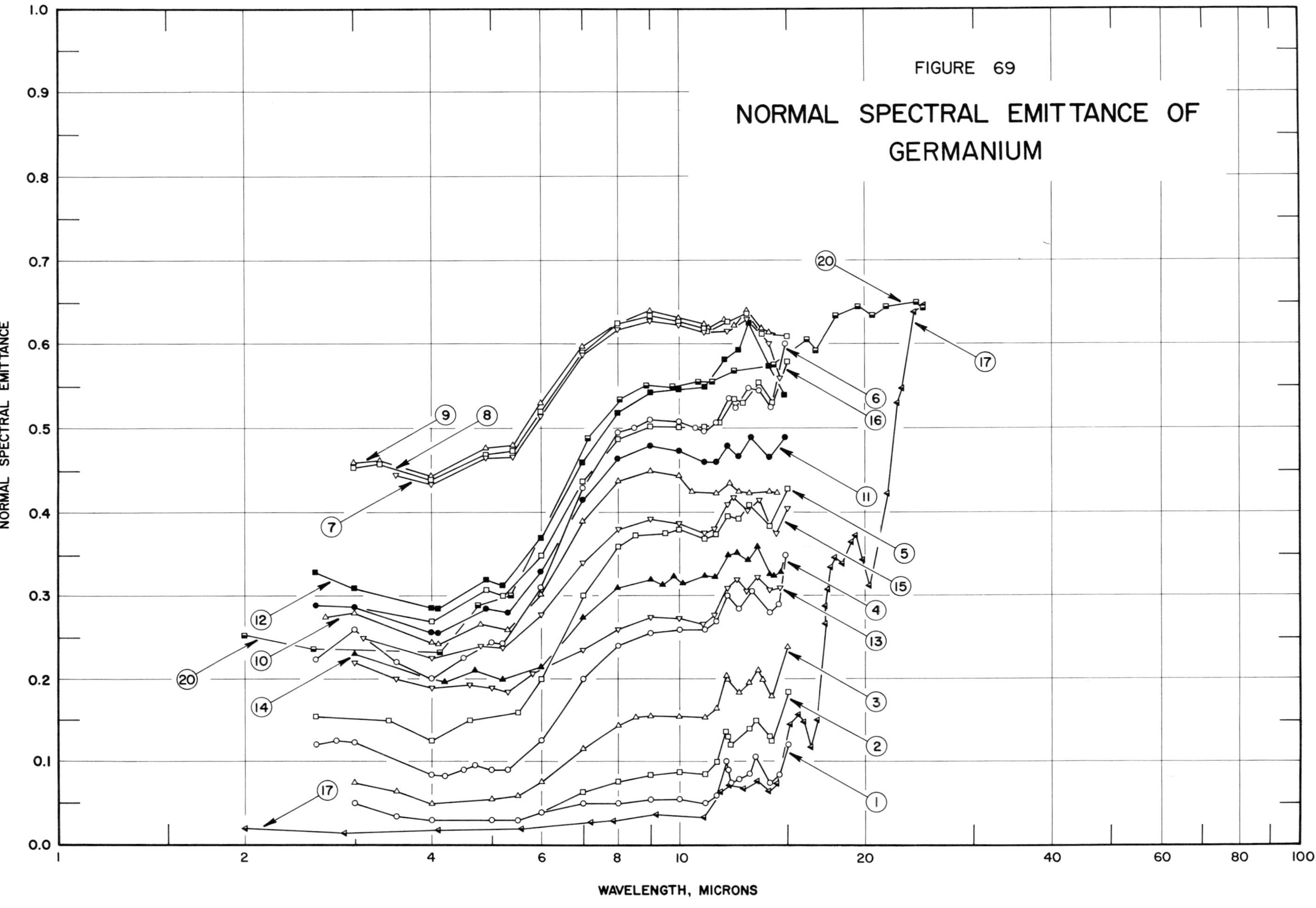
FIGURE 69
NORMAL SPECTRAL EMITTANCE OF GERMANIUM
NORMAL SPECTRAL EMITTANCE
WAVELENGTH, MICRONS
0.0
0.1
0.2
0.3
0.4
0.5
0.6
0.7
0.8
0.9
1.0
1
2
4
6
8
10
20
40
60
80
100
9
8
7
12
10
20
14
17
20
17
6
16
11
5
15
4
13
3
2
1

SPECIFICATION TABLE NO. 69 NORMAL SPECTRAL EMITTANCE OF GERMANIUM

Curve No.	Ref. No.	Year	Temperature K	Wavelength Range, μ	Geometry θ'	Reported Error, %	Composition (weight percent), Specifications and Remarks
1	21	1961	333	3.0-15.0	$\sim 0^{\circ}$	1	Polycrystalline; ground and polished; resistivity 40 ohm cm; measured in vacuum; data extracted from smooth curve.
2	21	1961	353	3.0-15.0	$\sim 0^{\circ}$		Above specimen and conditions.
3	21	1961	373	3.0-15.0	$\sim 0^{\circ}$		Above specimen and conditions.
4	21	1961	393	2.6-14.9	$\sim 0^{\circ}$		Above specimen and conditions.
5	21	1961	413	2.6-15.0	$\sim 0^{\circ}$		Above specimen and conditions.
6	21	1961	433	2.6-14.9	$\sim 0^{\circ}$		Above specimen and conditions.
7	21	1961	393	3.5-14.6	$\sim 0^{\circ}$		Single crystal; zinc doped; ground and polished; resistivity 0.2 ohm cm; measured in vacuum; data extracted from smooth curve.
8	21	1961	413	3.0-15.0	$\sim 0^{\circ}$		Above specimen and conditions.
9	21	1961	433	3.0-15.0	$\sim 0^{\circ}$		Above specimen and conditions.
10	21	1961	393	2.7-14.4	$\sim 0^{\circ}$		Single crystal; gold-doped; ground and polished; resistivity 1.3 ohm cm; measured in vacuum; data extracted from smooth curve.
11	21	1961	413	2.6-14.9	$\sim 0^{\circ}$		Above specimen and conditions.
12	21	1961	430	2.6-14.8	$\sim 0^{\circ}$		Above specimen and conditions.
13	21	1961	373	3.0-14.6	$\sim 0^{\circ}$	< 1	Single crystal; ground and polished; resistivity 35 ohm cm; measured in vacuum; data extracted from smooth curve.
14	21	1961	393	3.0-14.9	$\sim 0^{\circ}$	< 1	Above specimen and conditions.
15	21	1961	413	3.1-15.0	$\sim 0^{\circ}$	< 1	Above specimen and conditions.
16	21	1961	433	3.0-15.0	$\sim 0^{\circ}$	< 1	Above specimen and conditions.
17	214	1962	323	2.00-25.02	$\sim 0^{\circ}$		Polycrystalline germanium; resistivity 40 ohm cm.
18	214	1962	373	2.00-25.00	$\sim 0^{\circ}$		Above specimen and conditions.
19	214	1962	423	2.00-25.00	$\sim 0^{\circ}$		Above specimen and conditions.
20	214	1962	473	2.00-24.99	$\sim 0^{\circ}$		Above specimen and conditions.

DATA TABLE NO. 69 NORMAL SPECTRAL EMITTANCE OF GERMANIUM

[Wavelength, λ; Emittance, ∈; Temperature, T, K]

λ	∈
CURVE 1	**T = 333**
3.0	0.050
3.5	0.035
4.0	0.030
5.0	0.030
5.5	0.030
6.0	0.040
7.0	0.050
8.0	0.050
9.0	0.055
10.0	0.055
11.0	0.050
11.5	0.060
11.9	0.100
12.0	0.090
12.2	0.075
12.5	0.080
13.0	0.085
13.3	0.106
14.0	0.075
14.1	0.073*
14.5	0.085
15.0	0.120
CURVE 2	**T = 353**
3.0	0.050*
3.5	0.035*
4.0	0.030*
5.0	0.030*
5.5	0.030*
6.0	0.040*
7.0	0.064
8.0	0.077
9.0	0.084
10.0	0.088
11.0	0.085
11.5	0.100
11.9	0.138
12.0	0.130
12.2	0.120
CURVE 2 (cont.)	
13.0	0.140
13.3	0.150
14.0	0.130
14.1	0.125
15.0	0.185
CURVE 3	**T = 373**
3.0	0.075
3.5	0.065
4.0	0.050
5.0	0.056
5.5	0.060
6.0	0.075
7.0	0.116
8.0	0.144
8.5	0.153
9.0	0.155
10.0	0.155
11.0	0.154
11.5	0.165
11.9	0.202
12.0	0.200
12.5	0.185
13.0	0.197
13.4	0.210
13.7	0.200
14.1	0.180
15.0	0.240
CURVE 4	**T = 393**
2.6	0.120
2.8	0.123
3.0	0.122
4.0	0.084
4.2	0.082
4.5	0.090
4.7	0.095
5.0	0.090
5.3	0.090
CURVE 4 (cont.)	
6.0	0.125
7.0	0.200
8.0	0.240
9.0	0.257
10.0	0.260
11.0	0.260
11.5	0.270
12.0	0.300
12.5	0.285
13.2	0.305
14.0	0.280
14.2	0.278*
14.5	0.290
14.9	0.350
CURVE 5	**T = 413**
2.6	0.153
3.4	0.150
4.0	0.125
4.6	0.150
5.5	0.160
6.0	0.200
7.0	0.300
8.0	0.360
8.5	0.373
9.5	0.375
10.0	0.380
11.0	0.370
11.5	0.374
12.0	0.397
12.5	0.393
13.0	0.410
14.0	0.385
14.2	0.383*
15.0	0.430
CURVE 6	**T = 433**
2.6	0.223
CURVE 6 (cont.)	
3.0	0.260
3.5	0.220
4.0	0.200
4.5	0.225
5.0	0.245
5.2	0.242
6.0	0.310
7.0	0.430
8.0	0.496
8.5	0.500
9.0	0.510
10.0	0.509
10.7	0.500
11.0	0.498
11.5	0.506
12.1	0.535
12.4	0.525
13.0	0.548
13.5	0.545
14.1	0.525
14.9	0.600
CURVE 7	**T = 393**
3.5	0.445
4.0	0.435
4.9	0.465
5.4	0.466
6.0	0.514
7.0	0.588
8.0	0.618
9.0	0.629
10.0	0.623
11.0	0.614
12.0	0.616
12.9	0.630
14.0	0.600
14.6	0.560
CURVE 8	**T = 413**
3.0	0.453
3.3	0.458
4.0	0.440
4.9	0.470
5.4	0.473
6.0	0.520
7.0	0.590
8.0	0.625
9.0	0.634
10.0	0.628
11.0	0.620
11.2	0.615
12.0	0.628
12.9	0.637
13.7	0.612
15.0	0.610
CURVE 9	**T = 433**
3.0	0.460
3.3	0.460
4.0	0.442
4.9	0.478
5.4	0.480
6.0	0.530
7.0	0.598
8.0	0.628*
9.0	0.640
10.0	0.632
11.0	0.624
11.2	0.620
11.8	0.630
12.3	0.623
12.9	0.640
13.6	0.620
14.0	0.616
15.0	0.610*
CURVE 10	**T = 393**
2.7	0.275
3.0	0.280
4.0	0.245
4.1	0.242
4.8	0.267
5.3	0.260
6.0	0.301
7.0	0.390
8.0	0.438
9.0	0.450
10.0	0.445
10.5	0.427
11.5	0.425
12.1	0.435
12.5	0.427
13.0	0.425
14.0	0.427
14.4	0.425
CURVE 11	**T = 413**
2.6	0.290
3.0	0.288
4.0	0.257
4.1	0.256
4.9	0.285
5.3	0.280
6.0	0.330
7.0	0.416
8.0	0.465
9.0	0.480
10.0	0.473
11.0	0.460
11.5	0.460
12.1	0.480
12.5	0.467
13.2	0.490
14.0	0.468
14.9	0.490
CURVE 12	**T = 430**
2.6	0.330
3.0	0.310
4.0	0.286
4.1	0.285
4.9	0.320
5.2	0.312
6.0	0.370
7.0	0.460
8.0	0.519
9.0	0.542
10.0	0.548
11.0	0.550
11.8	0.582
12.5	0.593
13.0	0.629
13.1	0.630*
14.0	0.575
14.8	0.540
CURVE 13	**T = 373**
3.0	0.220
3.5	0.200
4.0	0.190
4.6	0.193
5.0	0.190
5.3	0.185
5.8	0.206
7.0	0.235
8.0	0.260
9.0	0.276
10.0	0.274
10.8	0.268
11.4	0.280
12.0	0.310
12.4	0.320
12.8	0.305
13.4	0.323
14.0	0.308
14.6	0.310
CURVE 14	**T = 393**
3.0	0.230
4.0	0.200*
4.2	0.198
4.7	0.210
5.2	0.200
6.0	0.215
7.0	0.276
8.0	0.310
9.0	0.320
9.4	0.316
9.8	0.325
10.2	0.318
11.0	0.325
11.4	0.323
12.0	0.350
12.4	0.352
12.9	0.345
13.4	0.360
14.0	0.328
14.2	0.325
14.6	0.330
14.9	0.350*
CURVE 15	**T = 413**
3.1	0.250
4.0	0.225
4.8	0.240
5.2	0.238
6.0	0.279
7.0	0.340
8.0	0.380
9.0	0.392
10.0	0.388
11.0	0.375
11.4	0.380
12.0	0.410
12.3	0.419
12.9	0.402
13.5	0.415

*Not shown on plot

DATA TABLE NO. 69 (continued)

λ	ε
CURVE 15 (cont.) T = 413	
14.0	0.382*
14.4	0.375
15.0	0.405
CURVE 16 T = 433	
3.0	0.288*
4.0	0.270
4.9	0.308
5.2	0.300
6.0	0.350
7.0	0.438
8.0	0.488
9.0	0.501
10.0	0.501
11.0	0.501
11.6	0.506
12.3	0.534
12.7	0.530
13.5	0.555
14.2	0.530
15.0	0.580
CURVE 17 T = 323	
2.00	0.020
2.89	0.015
4.09	0.019
5.57	0.020
6.72	0.028
7.84	0.029
9.14	0.037
10.91	0.032
11.63	0.064
12.11	0.073
12.73	0.068
13.33	0.078
13.93	0.066
14.42	0.075
15.15	0.146
15.58	0.158
15.83	0.149
CURVE 17 (cont.)	
16.34	0.118
16.73	0.152
17.23	0.268
17.23	0.289
17.48	0.309
17.59	0.336
17.85	0.348
18.38	0.340
19.02	0.367
19.39	0.372
19.94	0.344
20.43	0.312
20.98	0.331
21.72	0.423
22.65	0.531
23.06	0.549
24.14	0.639
25.02	0.648
CURVE 18* T = 373	
2.00	0.049
3.74	0.034
5.61	0.044
7.53	0.090
9.07	0.112
10.30	0.034
11.57	0.127
11.94	0.161
12.31	0.152
13.08	0.158
13.43	0.171
14.03	0.153
14.60	0.172
15.32	0.250
15.70	0.261
16.13	0.221
16.46	0.208
16.78	0.232
17.77	0.421
18.44	0.413
19.24	0.444
19.83	0.427
CURVE 18* (cont.)	
20.42	0.394
21.26	0.424
22.50	0.543
23.19	0.568
24.18	0.651
25.00	0.649
CURVE 19* T = 423	
2.00	0.135
4.13	0.101
4.88	0.130
5.23	0.121
5.87	0.161
7.71	0.302
8.86	0.344
11.15	0.352
11.57	0.351
12.03	0.375
12.54	0.369
13.49	0.377
14.21	0.353
15.11	0.405
15.92	0.431
16.48	0.411
17.22	0.461
18.11	0.549
18.50	0.544
19.24	0.566
20.57	0.539
21.52	0.563
22.60	0.614
23.68	0.639
25.00	0.646
CURVE 20 T = 473	
2.00	0.253
2.58	0.238
4.13	0.232
4.76	0.290
5.37	0.301
CURVE 20 (cont.)	
7.15	0.489
8.08	0.534
8.91	0.551
9.81	0.550
10.75	0.557
11.35	0.557
12.34	0.570
14.31	0.576
16.03	0.605
16.68	0.592
18.00	0.636
19.54	0.646
20.73	0.636
21.71	0.646
24.28	0.650
24.99	0.646

*Not shown on plot

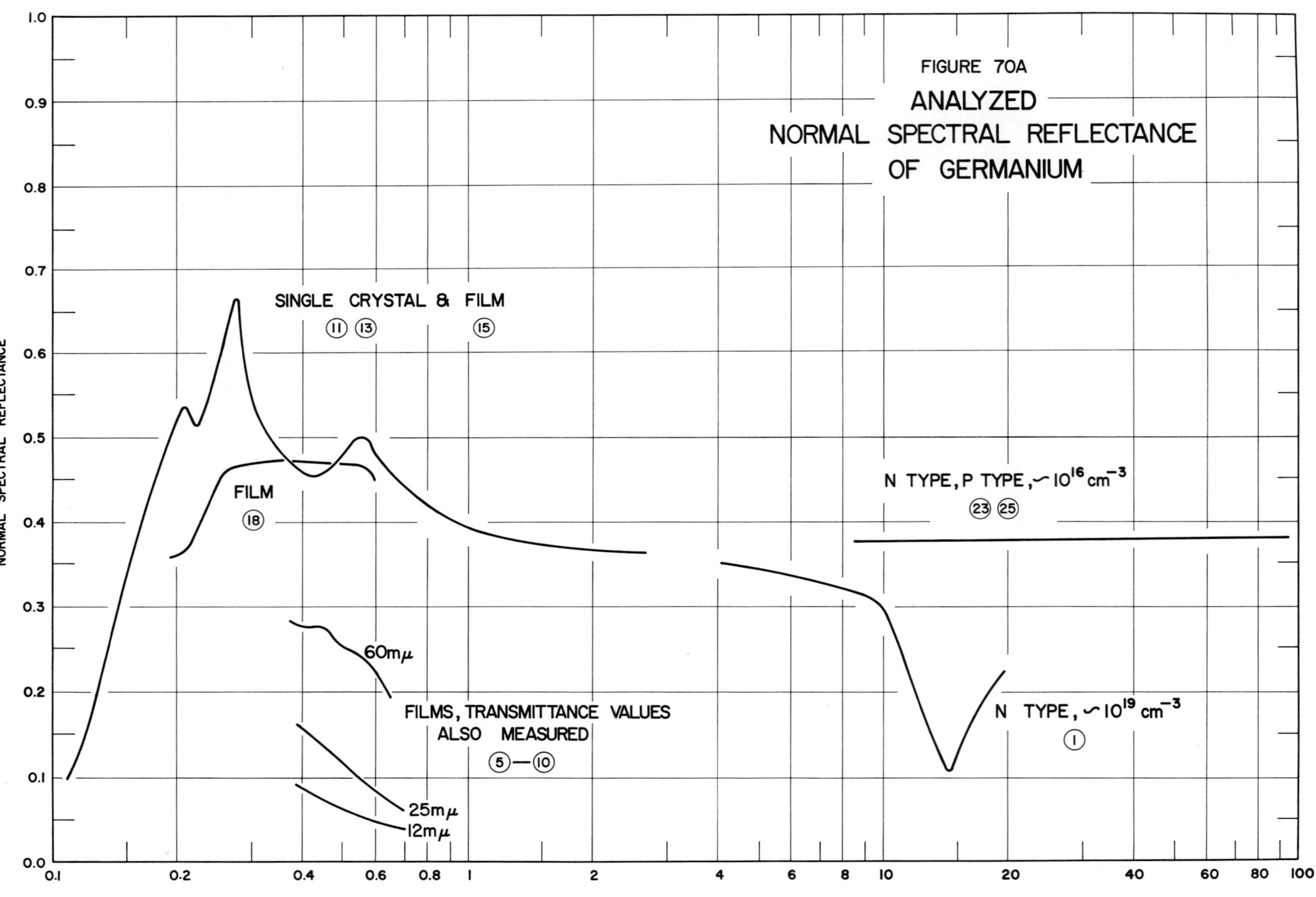
FIGURE 70A
ANALYZED
NORMAL SPECTRAL REFLECTANCE
OF GERMANIUM
SINGLE CRYSTAL & FILM
⑪ ⑬ ⑮
FILM
⑱
60mμ
FILMS, TRANSMITTANCE VALUES
ALSO MEASURED
⑤—⑩
25mμ
12mμ
N TYPE, P TYPE, ~10^16 cm^-3
㉓ ㉕
N TYPE, ~10^19 cm^-3
①
NORMAL SPECTRAL REFLECTANCE
1.0
0.9
0.8
0.7
0.6
0.5
0.4
0.3
0.2
0.1
0.0
0.1
0.2
0.4
0.6
0.8
1
2
4
6
8
10
20
40
60
80
100
WAVELENGTH, MICRONS

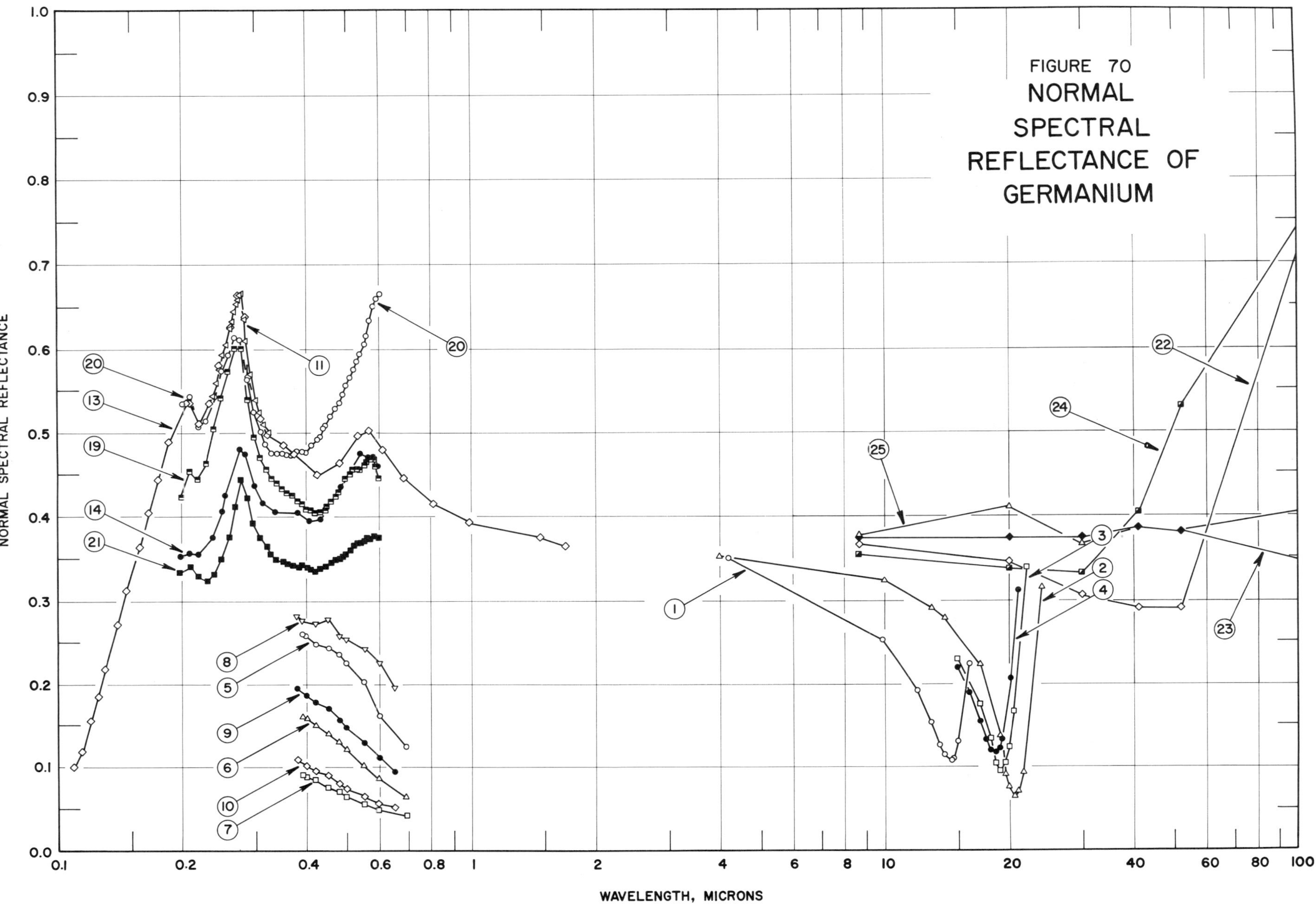
FIGURE 70
NORMAL
SPECTRAL
REFLECTANCE OF
GERMANIUM
NORMAL SPECTRAL REFLECTANCE
WAVELENGTH, MICRONS
0.0
0.1
0.2
0.3
0.4
0.5
0.6
0.7
0.8
0.9
1.0
0.1
0.2
0.4
0.6
0.8
1
2
4
6
8
10
20
40
60
80
100
1
2
3
4
5
6
7
8
9
10
11
13
14
19
20
21
22
23
24
25

SPECIFICATION TABLE NO. 70 NORMAL SPECTRAL REFLECTANCE OF GERMANIUM

Curve No.	Ref. No.	Year	Temperature K	Wavelength Range, μ	Geometry θ	θ'	ω'	Reported Error, %	Composition (weight percent), Specifications and Remarks
1	231	1964	298	4.1-16.1	~0°	~0°			Antimony doped, (N = 1.1 x 10^{19} cm^{-3}); polished; measured in vacuum (10^{-6} mm Hg).
2	231	1964	298	4.0-24.1	~0°	~0°			Above specimen and conditions except heat treated for 1 hr at 1140 K degrees.
3	231	1964	298	15.1-22.1	~0°	~0°			Above specimen and conditions except etched in 20:1 mixture of HNO_3 to HF for 5 sec.
4	231	1964	298	15.1-21.1	~0°	~0°			Above specimen and conditions except etched in 20:1 mixture of HNO_3 to HF for 10 sec.
5	232	1963	298	0.390-0.698	~0°	~0°			Film (50 mμ thick); measured in air.
6	232	1963	298	0.390-0.698	~0°	~0°			Different sample, same as above specimen and conditions except 25 mμ thick.
7	232	1963	298	0.390-0.699	~0°	~0°			Different sample, same as above specimen and conditions except 12 mμ thick.
8	232	1963	298	0.380-0.649	~0°	~0°			Different sample, same as above specimen and conditions except 60 mμ thick.
9	232	1963	298	0.381-0.649	~0°	~0°			Different sample, same as above specimen and conditions except 30 mμ thick.
10	232	1963	298	0.381-0.649	~0°	~0°			Different sample, same as above specimen and conditions except 15 mμ thick.
11	319	1959	300	0.238-0.326	~0°	~0°			Single crystal; etched.
12	319	1959	80	0.238-0.326	~0°	~0°			Single crystal; etched.
13	319	1959	300	0.109-2.695	~0°	~0°			Single crystal; etched.
14	339	1966	298	0.199-0.599	7°	7°			Evaporated film (1020 Å thick); fused quartz substrate; substrate at 1053 K during evaporation at 1-3 x 10^{-6} mm Hg at ~3000 Å min^{-1}.
15	339	1966	298	0.200-0.598	7°	7°			Different sample, same as above specimen and conditions except deposition rate 216 Å min^{-1}; film thickness 432 Å; substrate temperature 873 K.
16	339	1966	298	0.198-0.599	7°	7°			Different sample, same as above specimen and conditions except deposition rate 122 Å min^{-1}; film thickness 245 Å; substrate temperature 723 K.
17	339	1966	298	0.196-0.597	7°	7°			Different sample, same as above specimen and conditions except deposition rate 152 Å min^{-1}; film thickness 300 Å; substrate temperature 573 K.
18	339	1966	298	0.196-0.599	7°	7°			Different sample, same as above specimen and conditions except deposition rate ~150Å min^{-1}; film thickness 910 Å; substrate temperature 298 K.
19	339	1966	298	0.200-0.599	7°	7°			185 mμ epitaxial film; [111] growth direction; CaF_2 substrate at 873 K during 925 Å min^{-1} evaporation at 1-3 x 10^{-6} mm Hg.
20	339	1966	298	0.201-0.602	7°	7°			25 mμ epitaxial film; [111] growth direction; CaF_2 substrate at 873 K during 750 Å min^{-1} evaporation at 1-3 x 10^{-6} mm Hg.
21	339	1966	298	0.198-0.601	7°	7°			13.5 mμ epitaxial film; [111] growth direction; CaF_2 substrate at 893 K during 810 Å min^{-1} evaporation at 1-3 x 10^{-6} mm Hg.

SPECIFICATION TABLE NO. 70 (continued)

Curve No.	Ref. No.	Year	Temperature K	Wavelength Range, μ	Geometry θ θ' ω'	Reported Error, %	Composition (weight percent), Specifications and Remarks
22	262	1949	~293	8.7-152	~0° ~0°		n-type germanium; resistivity at room temperature 0.006 Ωcm; N~5 x 10^{18} cm^{-3}; optically polished.
23	262	1949	~293	8.7-152	~0° ~0°		n-type germanium; resistivity at room temperature ~3 Ωcm; N~10^{16} cm^{-3}; optically polished.
24	262	1949	~293	8.7-152	~0° ~0°		p-type germanium; resistivity at room temperature 0.005 Ωcm; N~5 x 10^{18} cm^{-3}; optically polished.
25	262	1949	~293	8.7-152	~0° ~0°		p-type germanium; resistivity at room temperature 0.5 Ωcm; N~10^{16} cm^{-3}; optically polished.

DATA TABLE NO. 70 NORMAL SPECTRAL REFLECTANCE OF GERMANIUM

[Wavelength, λ, μ ; Reflectance, ρ ; Temperature, T, K]

λ	ρ
CURVE 1 T = 298	
4.1	0.350
9.9	0.252
12.0	0.191
12.9	0.152
13.6	0.127
13.9	0.113
14.4	0.108
14.6	0.110
15.1	0.130
16.1	0.224
CURVE 2 T = 298	
4.0	0.352
10.0	0.325
13.0	0.291
14.0	0.279
17.0	0.222
19.0	0.137
19.6	0.099
20.0	0.076
20.6	0.065
21.1	0.071
21.6	0.093
24.1	0.316
CURVE 3 T = 298	
15.1	0.230
17.0	0.174
18.1	0.132
18.5	0.102
19.0	0.095
19.6	0.103
20.0	0.123
20.5	0.167
22.1	0.341

λ	ρ
CURVE 4 T = 298	
15.1	0.219
16.0	0.188
17.1	0.153
17.5	0.132
18.0	0.119
18.5	0.117
19.0	0.121
19.2	0.132
20.2	0.206
21.1	0.311
CURVE 5 T = 298	
0.390	0.260
0.399	0.259
0.420	0.249
0.450	0.244
0.480	0.237
0.500	0.226
0.548	0.202
0.600	0.161
0.698	0.124
CURVE 6 T = 298	
0.390	0.160
0.400	0.158
0.421	0.150
0.451	0.140
0.479	0.130
0.500	0.121
0.549	0.101
0.599	0.086
0.698	0.066
CURVE 7 T = 298	
0.390	0.090
0.400	0.088
0.419	0.083
0.450	0.075

λ	ρ
CURVE 7 (cont.)	
0.480	0.070
0.500	0.065
0.550	0.056
0.599	0.050
0.699	0.042
CURVE 8 T = 298	
0.380	0.281
0.391	0.277
0.421	0.273
0.450	0.278
0.481	0.258
0.501	0.252
0.551	0.242
0.600	0.224
0.649	0.195
CURVE 9 T = 298	
0.381	0.194
0.400	0.185
0.422	0.178
0.450	0.170
0.480	0.156
0.501	0.148
0.549	0.128
0.600	0.111
0.649	0.094
CURVE 10 T = 298	
0.381	0.108
0.400	0.101
0.420	0.095
0.450	0.089
0.480	0.080
0.500	0.074
0.550	0.065
0.599	0.057
0.649	0.052

λ	ρ
CURVE 11 T = 300	
0.238	0.544
0.243	0.559
0.248	0.575
0.253	0.593
0.258	0.607
0.263	0.625
0.267	0.633
0.269	0.645
0.273	0.653
0.276	0.659
0.278	0.663
0.280	0.666
0.285	0.636
0.288	0.610
0.295	0.569
0.303	0.539
0.310	0.523
0.317	0.510
0.326	0.501
CURVE 12* T = 80	
0.238	0.553
0.243	0.571
0.248	0.593
0.253	0.614
0.258	0.640
0.263	0.664
0.266	0.670
0.267	0.681
0.270	0.689
0.271	0.701
0.273	0.712
0.275	0.720
0.278	0.704
0.279	0.685
0.285	0.615
0.289	0.582
0.295	0.547
0.303	0.526
0.310	0.515
0.317	0.507
0.326	0.501

λ	ρ
CURVE 13 T = 300	
0.109	0.100
0.114	0.119
0.121	0.156
0.126	0.186
0.131	0.219
0.140	0.272
0.147	0.312
0.158	0.365
0.166	0.406
0.176	0.444
0.187	0.489
0.206	0.535
0.209	0.535
0.221	0.511
0.233	0.537
0.248	0.581
0.264	0.628
0.276	0.664
0.279	0.663*
0.285	0.635
0.293	0.569
0.311	0.518
0.325	0.499
0.354	0.486
0.375	0.475
0.424	0.450
0.480	0.463
0.534	0.498
0.568	0.502
0.613	0.479
0.688	0.446
0.813	0.415
0.992	0.392
1.476	0.374
2.695	0.364
CURVE 14 T = 298	
0.199	0.352
0.208	0.357
0.220	0.355
0.238	0.376
0.251	0.407

λ	ρ
CURVE 14 (cont.)	
0.254	0.425
0.278	0.481
0.285	0.473
0.300	0.436
0.314	0.416
0.338	0.406
0.381	0.404
0.409	0.393
0.433	0.398
0.486	0.434
0.541	0.474
0.563	0.469
0.580	0.470
0.559	0.458
CURVE 15* T = 298	
0.200	0.551
0.206	0.556
0.219	0.550
0.233	0.561
0.243	0.598
0.269	0.636
0.281	0.626
0.291	0.583
0.306	0.535
0.327	0.505
0.373	0.470
0.410	0.450
0.439	0.460
0.492	0.491
0.533	0.501
0.551	0.500
0.583	0.512
0.598	0.524
CURVE 16* T = 298	
0.198	0.542
0.208	0.544
0.220	0.530
0.236	0.542
0.254	0.580

λ	ρ
CURVE 16 (cont.)*	
0.274	0.598
0.289	0.578
0.307	0.541
0.333	0.522
0.372	0.525
0.401	0.523
0.424	0.532
0.500	0.582
0.570	0.642
0.585	0.651
0.599	0.653
CURVE 17* T = 298	
0.196	0.557
0.206	0.555
0.221	0.568
0.251	0.586
0.261	0.581
0.286	0.575
0.319	0.558
0.335	0.558
0.353	0.557
0.391	0.562
0.411	0.560
0.494	0.603
0.540	0.626
0.562	0.630
0.579	0.640
0.597	0.643
CURVE 18* T = 298	
0.196	0.357
0.213	0.369
0.256	0.456
0.278	0.466
0.357	0.472
0.401	0.469
0.426	0.466
0.481	0.461
0.543	0.456
0.575	0.463

λ	ρ
CURVE 18 (cont.)*	
0.589	0.460
0.599	0.452
CURVE 19 T = 298	
0.200	0.423
0.209	0.453
0.219	0.443
0.229	0.462
0.239	0.506
0.249	0.542
0.259	0.572
0.270	0.601
0.280	0.601
0.289	0.539
0.300	0.495
0.310	0.469
0.321	0.456
0.330	0.444
0.340	0.439
0.350	0.432
0.360	0.428
0.370	0.425
0.380	0.418
0.390	0.416
0.400	0.409
0.410	0.408
0.420	0.405
0.429	0.405
0.440	0.407
0.450	0.412
0.459	0.418
0.470	0.423
0.479	0.429
0.488	0.436*
0.499	0.445
0.509	0.450
0.519	0.457
0.531	0.457
0.540	0.457
0.550	0.461
0.560	0.465
0.569	0.468
0.580	0.468

* Not shown on plot

DATA TABLE NO. 70 (continued)

λ	ρ
CURVE 19 (cont.)	
0.590	0.459*
0.599	0.445
CURVE 20	
T = 298	
0.201	0.533
0.211	0.543
0.220	0.509
0.229	0.513
0.241	0.545
0.251	0.572
0.261	0.593
0.269	0.615
0.278	0.611
0.290	0.563
0.300	0.524
0.311	0.501
0.321	0.487
0.331	0.476
0.340	0.475
0.351	0.475
0.360	0.473
0.370	0.472
0.380	0.478
0.390	0.478
0.401	0.477
0.412	0.485
0.426	0.492
0.430	0.497
0.440	0.505
0.450	0.508
0.459	0.519
0.470	0.528
0.481	0.536
0.491	0.547
0.500	0.557
0.511	0.567
0.521	0.577
0.530	0.586
0.540	0.595
0.551	0.606
0.561	0.617
0.571	0.633
0.580	0.651
0.591	0.659
0.602	0.665

λ	ρ
CURVE 21	
T = 298	
0.198	0.334
0.211	0.341
0.220	0.330
0.231	0.324
0.240	0.332
0.250	0.350
0.261	0.375
0.269	0.412
0.279	0.444
0.289	0.422
0.299	0.391
0.310	0.375
0.322	0.363
0.329	0.355
0.399	0.349
0.352	0.348
0.361	0.344
0.370	0.343
0.382	0.341
0.391	0.342
0.401	0.340
0.411	0.338
0.420	0.337
0.430	0.339
0.441	0.341
0.451	0.341*
0.460	0.345
0.470	0.349
0.482	0.350
0.491	0.352
0.500	0.356
0.512	0.360
0.521	0.366
0.532	0.369
0.542	0.369
0.552	0.371
0.561	0.374
0.571	0.373
0.582	0.377
0.591	0.377
0.601	0.377

λ	ρ
CURVE 22	
T = ~293	
8.7	0.365
20.0	0.345
30.0	0.306
41.0	0.290
52.0	0.290
117.0	0.810*
152.0	0.500*
CURVE 23	
T = ~293	
8.7	0.372
20.0	0.372
30.0	0.372
41.0	0.385
52.0	0.380
117.0	0.340*
152.0	0.390*
CURVE 24	
T = ~293	
8.7	0.354
20.0	0.338
30.0	0.332
41.0	0.403
52.0	0.530
117.0	0.790*
152.0	0.460*
CURVE 25	
T = ~293	
8.7	0.373
20.0	0.410
30.0	0.367
41.0	0.385*
52.0	0.380*
117.0	0.410*
152.0	0.360*

*Not shown on plot

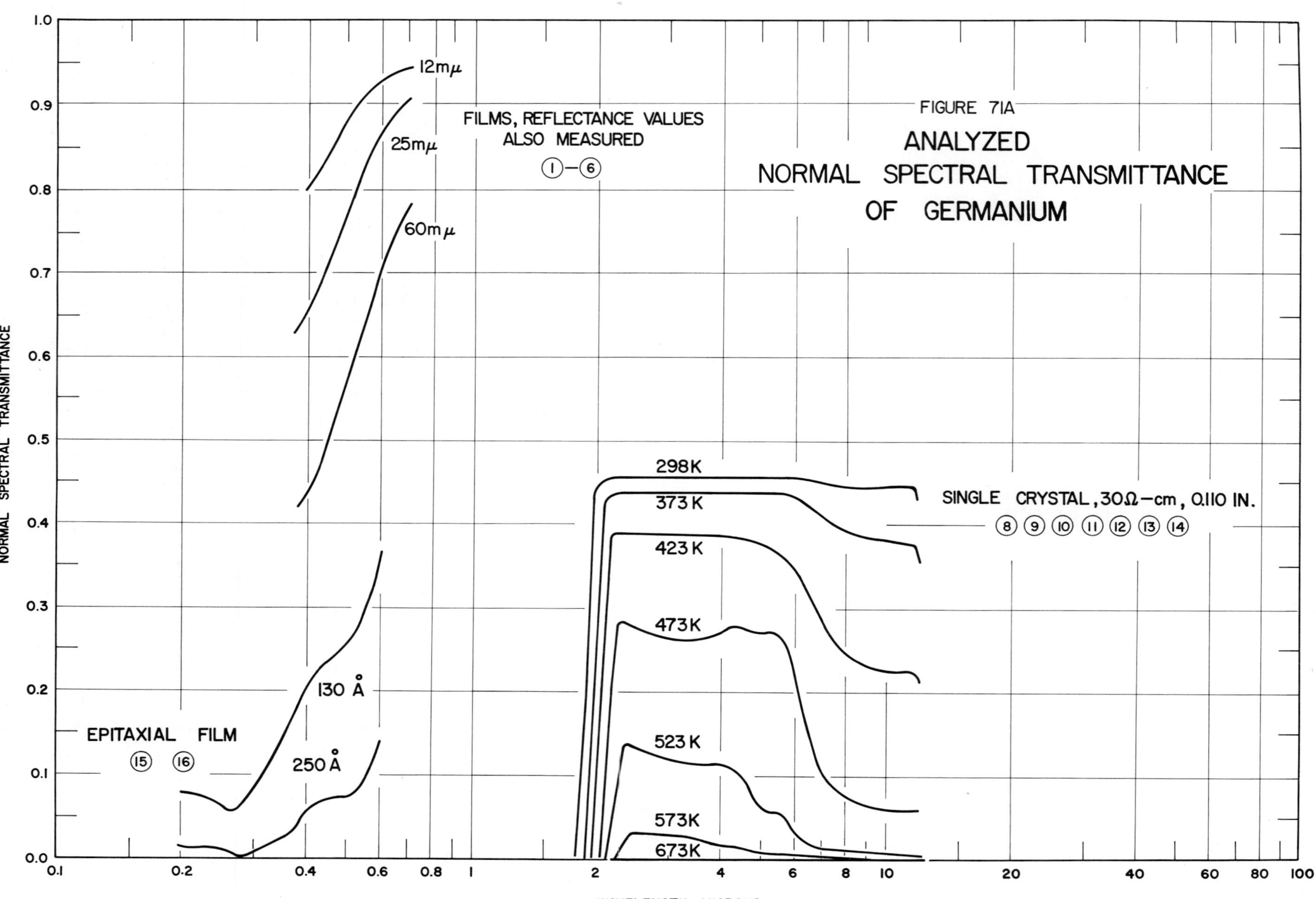
FIGURE 71A
ANALYZED
NORMAL SPECTRAL TRANSMITTANCE
OF GERMANIUM
FILMS, REFLECTANCE VALUES
ALSO MEASURED
① – ⑥
12mμ
25mμ
60mμ
130 Å
250 Å
EPITAXIAL FILM
⑮ ⑯
298K
373 K
423 K
473K
523 K
573K
673K
SINGLE CRYSTAL, 30Ω–cm, 0.110 IN.
⑧ ⑨ ⑩ ⑪ ⑫ ⑬ ⑭
NORMAL SPECTRAL TRANSMITTANCE
1.0
0.9
0.8
0.7
0.6
0.5
0.4
0.3
0.2
0.1
0.0
0.1
0.2
0.4
0.6
0.8
1
2
4
6
8
10
20
40
60
80
100
WAVELENGTH, MICRONS

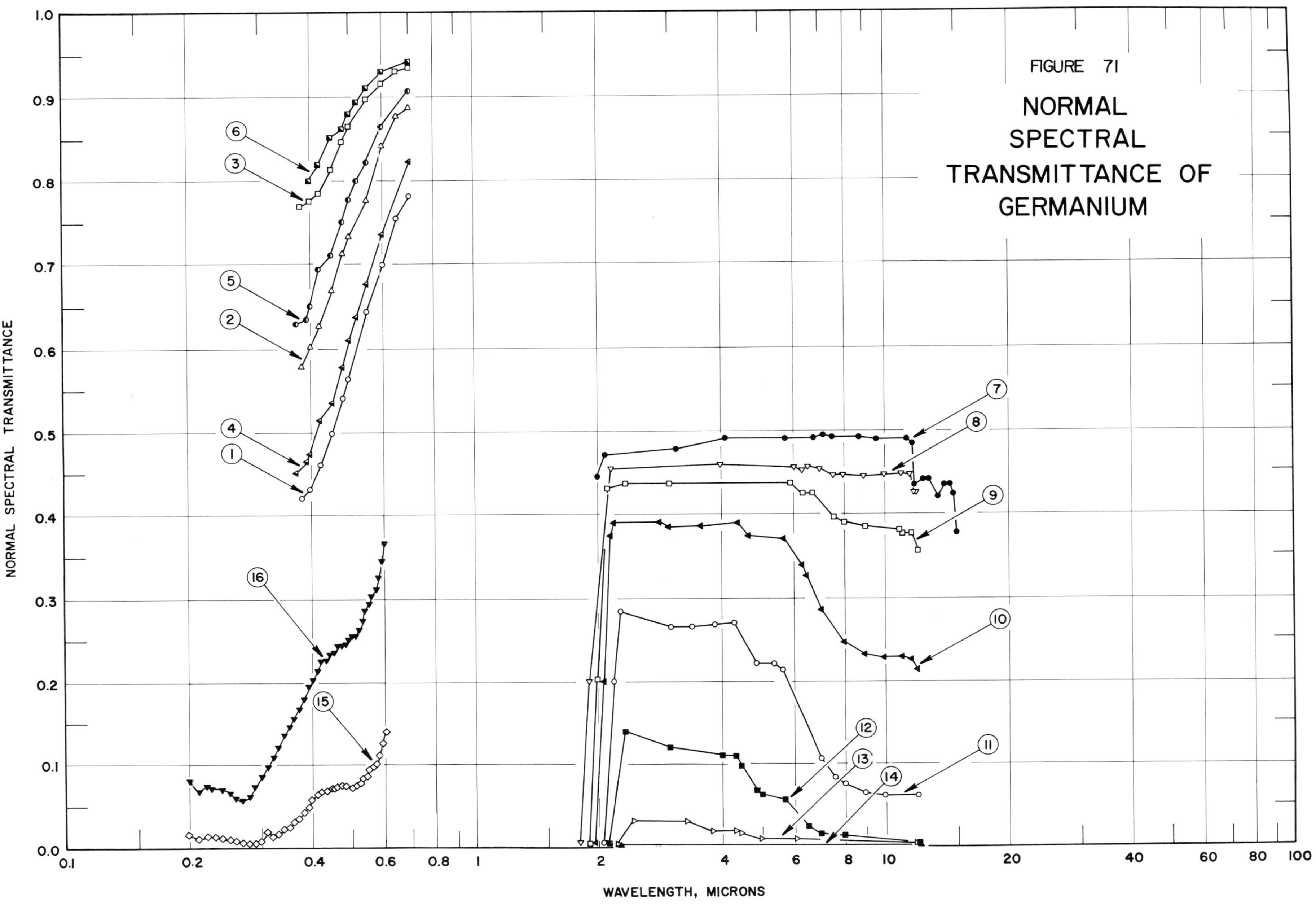

FIGURE 71
NORMAL SPECTRAL TRANSMITTANCE OF GERMANIUM
NORMAL SPECTRAL TRANSMITTANCE
WAVELENGTH, MICRONS
0.0
0.1
0.2
0.3
0.4
0.5
0.6
0.7
0.8
0.9
1.0
0.1
0.2
0.4
0.6
0.8
1
2
4
6
8
10
20
40
60
80
100
1
2
3
4
5
6
7
8
9
10
11
12
13
14
15
16

SPECIFICATION TABLE NO. 71 NORMAL SPECTRAL TRANSMITTANCE OF GERMANIUM

Curve No.	Ref. No.	Year	Temperature K	Wavelength Range, μ	Geometry θ	θ'	ω'	Reported Error, %	Composition (weight percent), Specifications and Remarks
1	232	1963	298	0.379-0.703	~0°	~0°			Film (60 mμ thick); measured in air.
2	232	1963	298	0.381-0.703	~0°	~0°			Different sample, same as above specimen and conditions except 30 mμ thick.
3	232	1963	298	0.380-0.702	~0°	~0°			Different sample, same as above specimen and conditions except 15 mμ thick.
4	232	1963	298	0.369-0.703	~0°	~0°			Different sample, same as above specimen and conditions except 50 mμ thick.
5	232	1963	298	0.370-0.702	~0°	~0°			Different sample, same as above specimen and conditions except 25 mμ thick.
6	232	1963	298	0.400-0.702	~0°	~0°			Different sample, same as above specimen and conditions except 12 mμ thick.
7	226	1963	298	2.00-15.00	~0°	~0°			1 ppb total impurities; from Knapic Electro-physics, Inc.; both surfaces polished optically flat to within 5 green mercury fringes; refractive index 4.108; data extracted from smooth curve.
8	288	1964	298	1.79-11.99	~0°	~0°			1 ppb p-type germanium; from Knapic Electro-physics, Inc.; single crystal; electrical resistivity 30 ohm cm; polished; Nernst glower source; data extracted from smooth curve; 0.110 in. thick.
9	288	1964	373	1.89-12.00	~0°	~0°			Above specimen and conditions.
10	288	1964	423	1.95-12.00	~0°	~0°			Above specimen and conditions.
11	288	1964	473	2.04-11.99	~0°	~0°			Above specimen and conditions.
12	288	1964	523	2.11-12.02	~0°	~0°			Above specimen and conditions.
13	288	1964	573	2.21-11.99	~0°	~0°			Above specimen and conditions.
14	288	1964	673	2.22-12.01	~0°	~0°			Above specimen and conditions.
15	339	1966	298	0.199-0.600	7°	7°			25 mμ epitaxial film; [111] growth direction; evaporated at 1-3 x 10^{-6} mm Hg; CaF_2 substrate at 873 K, evaporation rate 750 Å min^{-1}.
16	339	1966	298	0.200-0.601	7°	7°			13.5 mμ epitaxial film; [111] growth direction; evaporated at 1-3 x 10^{-6} mm Hg; CaF_2 substrate at 893 K, evaporation rate 810 Å min^{-1}.

DATA TABLE NO. 71 NORMAL SPECTRAL TRANSMITTANCE OF GERMANIUM

[Wavelength, λ, μ; Transmittance, τ; Temperature, T, K]

λ	τ
CURVE 1	**T = 298**
0.379	0.420
0.399	0.431
0.421	0.460
0.450	0.499
0.480	0.540
0.499	0.563
0.550	0.643
0.601	0.699
0.651	0.754
0.703	0.781
CURVE 2	**T = 298**
0.381	0.579
0.401	0.602
0.421	0.628
0.450	0.669
0.481	0.712
0.501	0.734
0.550	0.777
0.601	0.842
0.652	0.878
0.703	0.889
CURVE 3	**T = 298**
0.380	0.770
0.401	0.776
0.420	0.787
0.451	0.813
0.481	0.848
0.501	0.865
0.550	0.898
0.601	0.916
0.652	0.930
0.702	0.934
CURVE 4	**T = 298**
0.369	0.451
0.390	0.466
0.399	0.474
CURVE 4 (cont.)	
0.419	0.515
0.449	0.535
0.479	0.578
0.500	0.610
0.519	0.638
0.550	0.677
0.600	0.736
0.703	0.823
CURVE 5	**T = 298**
0.370	0.630
0.392	0.636
0.400	0.651
0.420	0.694
0.450	0.711
0.480	0.751
0.501	0.778
0.520	0.800
0.551	0.822
0.600	0.866
0.702	0.907
CURVE 6	**T = 298**
0.400	0.800
0.421	0.820
0.450	0.852
0.480	0.862
0.500	0.880
0.520	0.895
0.549	0.911
0.601	0.930
0.702	0.942
CURVE 7	**T = 298**
2.00	0.444
2.08	0.470
3.10	0.478
4.04	0.491
5.87	0.491
6.76	0.491
CURVE 7 (cont.)	
7.05	0.496
7.42	0.492
8.71	0.492
9.59	0.490
11.23	0.490
11.61	0.484
11.89	0.433
12.40	0.441
12.97	0.441
13.49	0.420
14.05	0.435
14.48	0.435
14.70	0.422
15.00	0.377
CURVE 8	**T = 298**
1.79	0.005
1.89	0.200
1.99	0.441
2.17	0.454
4.00	0.459
6.02	0.455
6.34	0.451
6.49	0.456
6.98	0.452
7.50	0.445
7.97	0.445
8.99	0.444
10.00	0.447
11.00	0.448
11.63	0.447
11.86	0.426
11.99	0.426
CURVE 9	**T = 373**
1.89	0.003
1.99	0.202
2.11	0.431
2.33	0.437
2.99	0.437
5.90	0.437
6.32	0.424
CURVE 9 (cont.)	
6.64	0.424
7.52	0.396
8.00	0.389
9.00	0.383
10.82	0.380
11.00	0.376
11.67	0.376
12.00	0.355
CURVE 10	**T = 423**
1.95	0.004
2.06	0.200
2.14	0.373
2.19	0.389
2.83	0.389
2.99	0.383
3.56	0.384
4.40	0.388
4.63	0.374
5.65	0.369
6.23	0.330
6.43	0.323
7.00	0.284
7.98	0.246
8.99	0.231
9.99	0.228
10.99	0.228
11.56	0.225
12.00	0.213
CURVE 11	**T = 473**
2.04	0.004
2.18	0.200
2.26	0.283
3.00	0.265
3.38	0.265
3.82	0.268
4.29	0.270
4.84	0.221
5.37	0.221
5.61	0.213
6.99	0.105
CURVE 11 (cont.)	
7.54	0.082
7.99	0.075
8.98	0.064
9.99	0.061
11.99	0.061
CURVE 12	**T = 523**
2.11	0.003
2.32	0.138
2.99	0.119
4.01	0.111
4.34	0.109
4.50	0.096
4.83	0.067
5.00	0.062
5.69	0.056
6.46	0.024
6.99	0.016
7.98	0.013
12.02	0.004
CURVE 13	**T = 573**
2.21	0.002
2.42	0.032
3.27	0.031
3.78	0.019
4.28	0.019
4.41	0.016
4.99	0.009
6.02	0.009
11.99	0.003
CURVE 14	**T = 673**
2.22	0.001
12.01	0.001
CURVE 15	**T = 298**
0.199	0.0157
0.208	0.0123
0.219	0.0144
0.231	0.0145
0.241	0.0132
0.251	0.0106
0.260	0.0079
0.269	0.0055
0.279	0.0038
0.289	0.0047
0.299	0.0074
0.310	0.0109
0.319	0.0146
0.329	0.0187
0.339	0.0234
0.350	0.0278
0.359	0.0324
0.369	0.0374
0.380	0.0438
0.390	0.0498
0.398	0.0560
0.410	0.0624
0.419	0.0662
0.430	0.0678
0.440	0.0705
0.449	0.0705
0.459	0.0716
0.470	0.0728
0.480	0.0726
0.489	0.0726
0.500	0.0726
0.510	0.0748
0.520	0.0783
0.529	0.0830
0.541	0.0865
0.548	0.0927
0.561	0.0966
0.570	0.1045
0.580	0.1132
0.590	0.1259
0.600	0.1396
CURVE 16	**T = 298**
0.200	0.0791
0.210	0.0679
0.220	0.0736
0.228	0.0706
0.238	0.0693
0.249	0.0643
0.259	0.0596
0.268	0.0574
0.279	0.0608
0.288	0.0719
0.299	0.0841
0.308	0.0975
0.319	0.1094
0.330	0.1208
0.341	0.1340
0.350	0.1446
0.359	0.1552
0.370	0.1671
0.380	0.1795
0.389	0.1928
0.400	0.2047
0.410	0.2128
0.420	0.2234
0.430	0.2260
0.439	0.2328
0.450	0.2372
0.460	0.2432
0.470	0.2449
0.480	0.2478
0.490	0.2512
0.498	0.2565
0.510	0.2565
0.519	0.2649
0.529	0.2742
0.539	0.2851
0.548	0.2958
0.559	0.3027
0.570	0.3112
0.580	0.3274
0.590	0.3452
0.601	0.3656

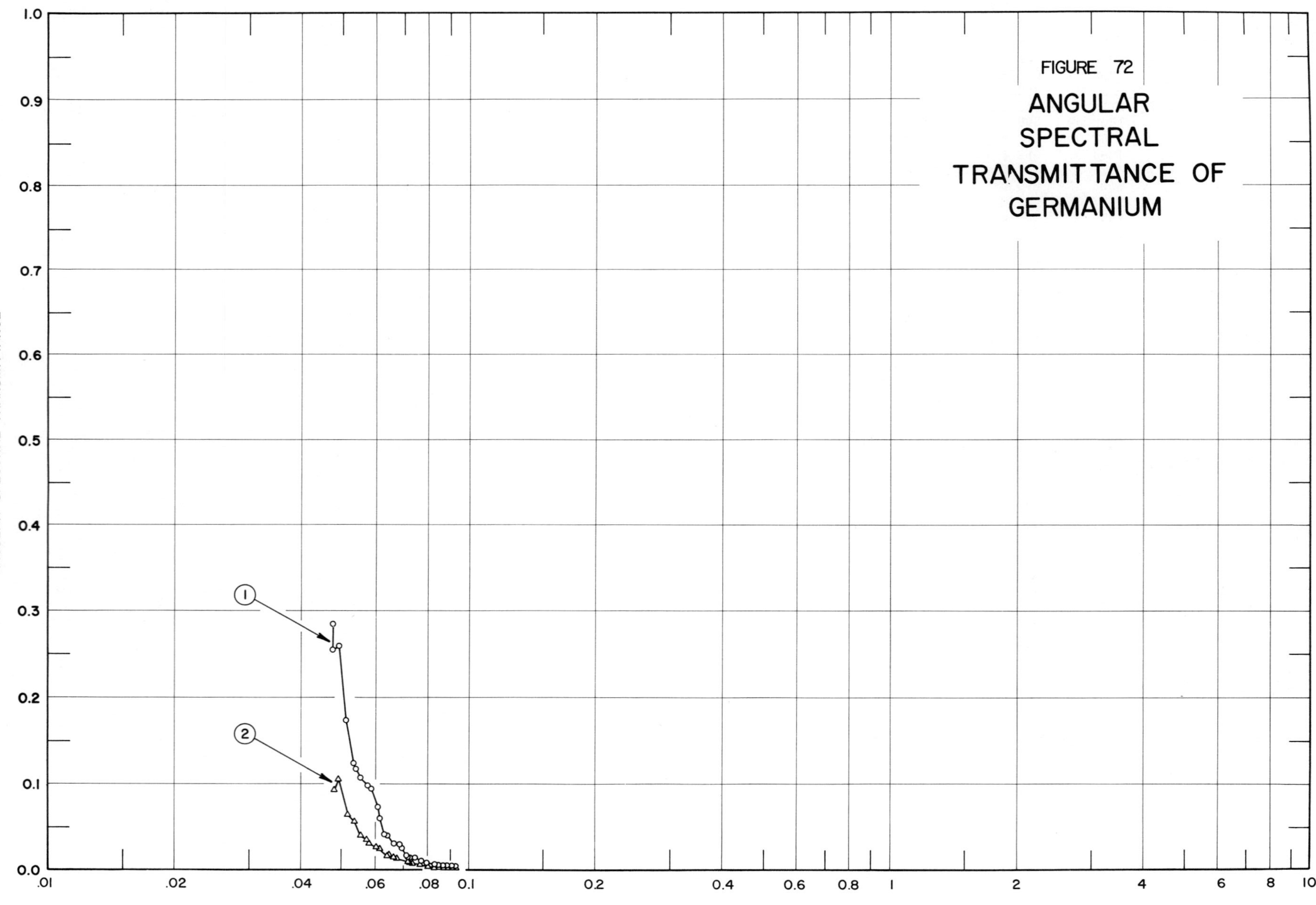
FIGURE 72
ANGULAR SPECTRAL TRANSMITTANCE OF GERMANIUM
ANGULAR SPECTRAL TRANSMITTANCE
1.0
0.9
0.8
0.7
0.6
0.5
0.4
0.3
0.2
0.1
0.0
.01
.02
.04
.06
.08
0.1
0.2
0.4
0.6
0.8
1
2
4
6
8
10
WAVELENGTH, MICRONS
1
2

SPECIFICATION TABLE NO. 72 ANGULAR SPECTRAL TRANSMITTANCE OF GERMANIUM

Curve No.	Ref. No.	Year	Temperature K	Wavelength Range, μ	Geometry θ	θ'	ω'	Reported Error, %	Composition (weight percent), Specification and Remarks
1	348	1965	298	0.0475-0.0930	0°	0°			Unbacked foil (530 Å thick); evaporated in vacuum (3 x 10^{-6} mm Hg); exposed to air; condensed spark discharge in argon light source.
2	348	1965	298	0.0481-0.0931	30°	30°			Above specimen and conditions except p-polarization is dominant; data uncorrected for partial polarization effects.
3	348	1965	298	0.0479-0.0930	30°	30°			Above specimen and conditions except s-polarization is dominant.

DATA TABLE NO. 72 ANGULAR SPECTRAL TRANSMITTANCE OF GERMANIUM

[Wavelength, λ, μ; Transmittance, τ; Temperature, T, K]

λ	τ
CURVE 1	
T = 298	
0.0475	0.2831
0.0476	0.2541
0.0493	0.2582
0.0514	0.1722
0.0533	0.1236
0.0540	0.1178
0.0557	0.1074
0.0576	0.0984
0.0585	0.0931
0.0609	0.0719
0.0615	0.0594
0.0631	0.0405
0.0642	0.0394
0.0666	0.0300
0.0689	0.0296
0.0698	0.0251
0.0715	0.0173
0.0721	0.0152
0.0726	0.0166*
0.0732	0.0156
0.0738	0.0145*
0.0743	0.0130
0.0771	0.0098
0.0802	0.0075
0.0836	0.0043
0.0850	0.0038
0.0867	0.0030
0.0879	0.0031*
0.0888	0.0030
0.0917	0.0029
0.0930	0.0028
CURVE 2	
T = 298	
0.0481	0.0914
0.0492	0.1052
0.0515	0.0632
0.0535	0.0562
0.0557	0.0396
0.0574	0.0352
0.0580	0.0308
0.0605	0.0263
CURVE 2 (cont.)	
0.0616	0.0237
0.0640	0.0165
0.0645	0.0176
0.0667	0.0136
0.0677	0.0122
0.0721	0.0092
0.0726	0.0086
0.0731	0.0076*
0.0741	0.0062
0.0768	0.0049
0.0778	0.0047*
0.0799	0.0047*
0.0807	0.0022
0.0835	0.0018
0.0851	0.0020
0.0869	0.0018
0.0877	0.0016*
0.0889	0.0016
0.0921	0.0013
0.0931	0.0013
CURVE 3*	
T = 298	
0.0479	0.1062
0.0490	0.0995
0.0513	0.0713
0.0534	0.0533
0.0559	0.0444
0.0576	0.0378
0.0583	0.0358
0.0607	0.0210
0.0618	0.0170
0.0639	0.0139
0.0646	0.0142
0.0665	0.0115
0.0672	0.0107
0.0679	0.0105
0.0729	0.0091
0.0731	0.0083
0.0741	0.0079
0.0772	0.0069
0.0802	0.0058
0.0809	0.0052
CURVE 3 (cont.)*	
0.0836	0.0037
0.0873	0.0022
0.0881	0.0020
0.0889	0.0023
0.0920	0.0017
0.0930	0.0018

*Not shown on plot

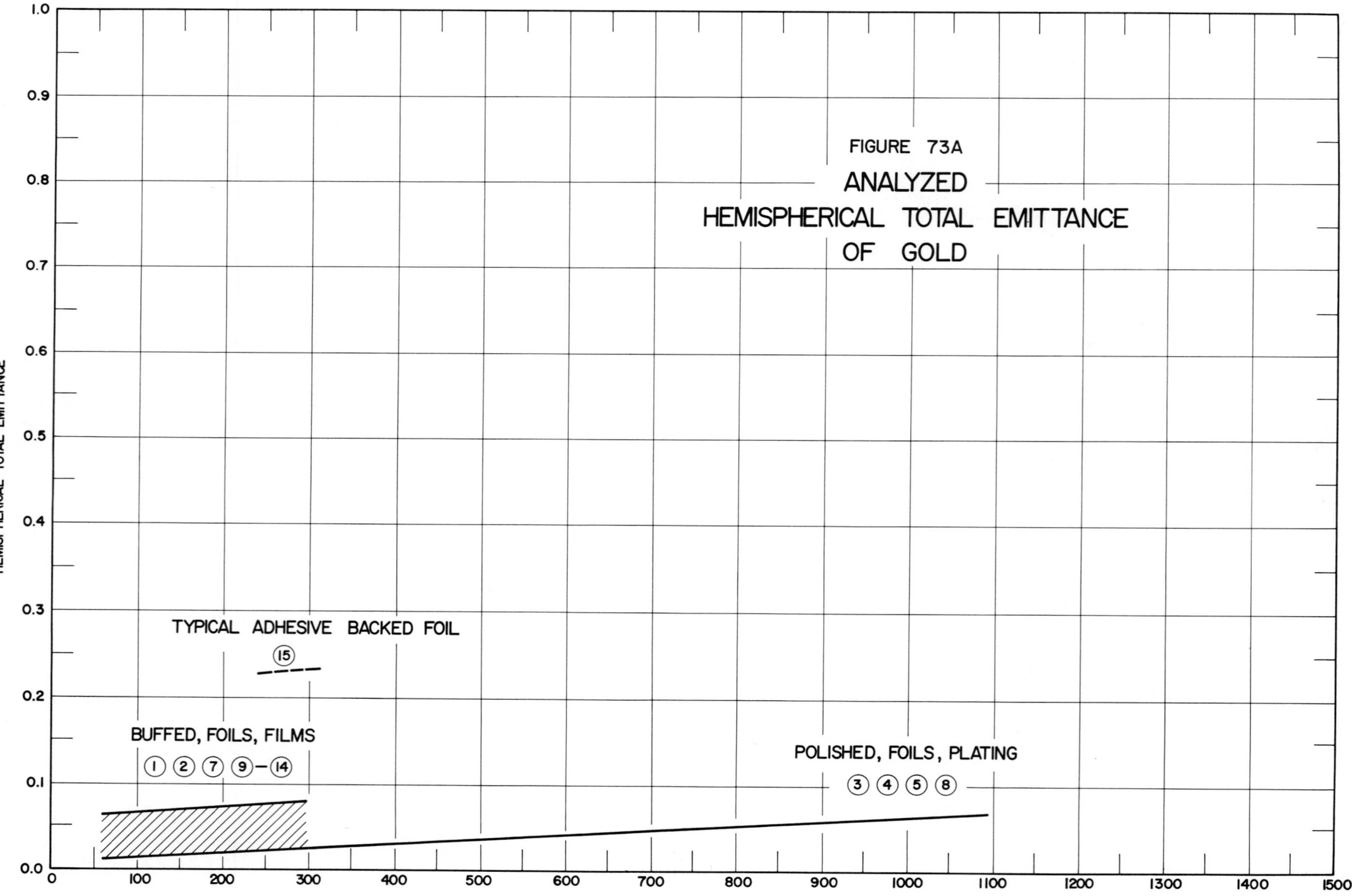

FIGURE 73A
ANALYZED
HEMISPHERICAL TOTAL EMITTANCE
OF GOLD
TYPICAL ADHESIVE BACKED FOIL
⑮
BUFFED, FOILS, FILMS
① ② ⑦ ⑨ – ⑭
POLISHED, FOILS, PLATING
③ ④ ⑤ ⑧
HEMISPHERICAL TOTAL EMITTANCE
1.0
0.9
0.8
0.7
0.6
0.5
0.4
0.3
0.2
0.1
0.0
0
100
200
300
400
500
600
700
800
900
1000
1100
1200
1300
1400
1500
TEMPERATURE, K

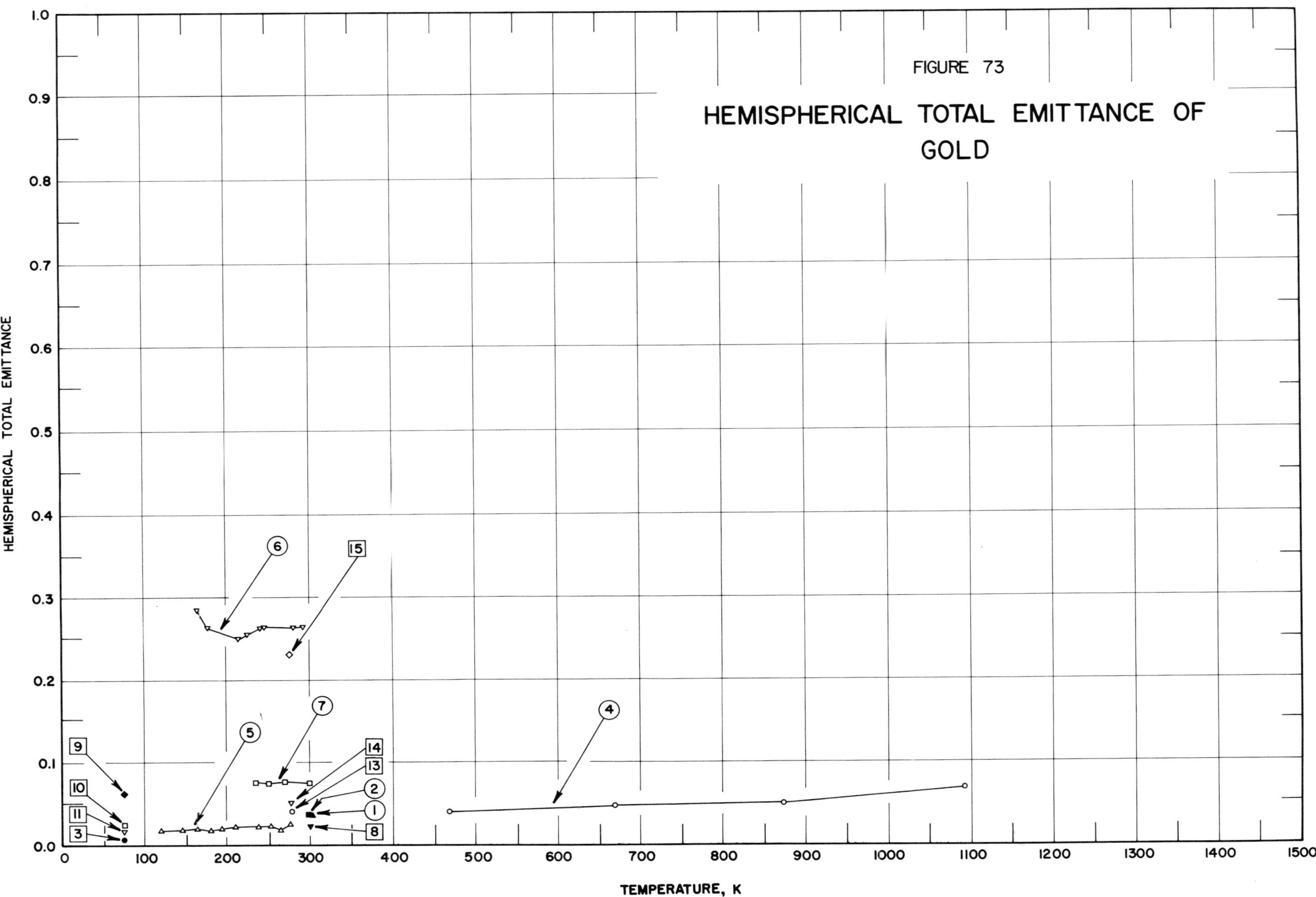
FIGURE 73
HEMISPHERICAL TOTAL EMITTANCE OF GOLD
HEMISPHERICAL TOTAL EMITTANCE
TEMPERATURE, K
0.0
0.1
0.2
0.3
0.4
0.5
0.6
0.7
0.8
0.9
1.0
0
100
200
300
400
500
600
700
800
900
1000
1100
1200
1300
1400
1500
1
2
3
4
5
6
7
8
9
10
11
13
14
15

SPECIFICATION TABLE NO. 73 HEMISPHERICAL TOTAL EMITTANCE OF GOLD

Curve No.	Ref. No.	Year	Temperature Range, K	Reported Error, %	Composition (weight percent), Specifications and Remarks
1	48	1964	301.1-301.9	3.3-7.7	Plated on aluminum; buffed; measured in vacuum (2 x 10^{-7} mm Hg).
2	48	1964	299.6-301.4	3.3-7.7	Plated on aluminum; buffed; measured in vacuum (2 x 10^{-7} mm Hg).
3	4	1953	76		Foil (0.0015 in. thick); solvent cleaned; measured in vacuum ($<10^{-6}$ mm Hg); emittance for 294 K black body radiation; authors assumed $\alpha = \epsilon$.
4	47	1961	468-1093	≤ 10	Ground with 600 grit carborundum and polished on a wet cloth lap; measured in vacuum (10^{-5} mm Hg).
5	5	1963	126-275	± 10	Polished; measured in vacuum (5 x 10^{-5} mm Hg).
6	46	1962	165-290	< 5	Foil (0.0015 in. thick); measured in vacuum (10^{-6} mm Hg);contamination suspected.
7	26	1962	235-300	3	Commercial foil (0.010 in. thick); cleaned in both sodium dichromate and dilute nitric acid solutions, buffed on a felt buffing wheel, cleaned with CCl_4 and acetone; measured in vacuum (10^{-6}mm Hg); data extracted from smooth curve.
8	23	1955	300	± 20	Gold plating (0.0003 in. thick); polished; kerosene buff; measured in vacuum (3 x 10^{-6} mm Hg).
9	3	1955	76	5	Leaf (1 x 10^{-5} in. thick); measured in vacuum (10^{-6} to 10^{-7} mm Hg); emittance for 300 K black body incident radiation; authors assumed $\alpha = \epsilon$.
10	3	1955	76	5	Foil (4.0 x 10^{-5} in. thick); solvent cleaned; measured in vacuum (10^{-6} to 10^{-7} mm Hg); emittance for 300 K black body incident radiation; authors assumed $\alpha = \epsilon$.
11	3	1955	76	5	Foil (0.0005 in. thick); solvent cleaned; measured in vacuum (10^{-6} to 10^{-7} mm Hg); emittance for 300 K black body incident radiation; authors assumed $\alpha = \epsilon$.
12	3	1955	76	5	Foil (0.0015 in. thick); solvent cleaned; measured in vacuum (10^{-6} to 10^{-7} mm Hg); emittance for 300 K black body incident radiation; authors assumed $\alpha = \epsilon$.
13	45	1961	278	10	Vacuum deposited on aluminum.
14	45	1961	278	10	Vacuum deposited on buffed titanium.
15	45	1961	278	10	Bright foil; typical of adhesive backed metals.

DATA TABLE NO. 73 HEMISPHERICAL TOTAL EMITTANCE OF GOLD

[Temperature, T, K; Emittance, ∈]

T	∈
CURVE 1	
301.1	0.0386
301.9	0.0378*
CURVE 2	
299.6	0.0381
301.4	0.0389*
CURVE 3	
76	0.0099
CURVE 4	
468	0.040
668	0.048
873	0.050
1093	0.068
CURVE 5	
126	0.0190
146	0.0195
164	0.0200
180	0.0190
195	0.0205
210	0.0220
223	0.0220
237	0.0220
251	0.0220
264	0.0190
275	0.0250
CURVE 6	
165	0.285
178	0.263
214	0.250
224	0.255
240	0.263
846	0.265
280	0.263
290	0.264
CURVE 7	
235	0.0750
250	0.0745
CURVE 7 (cont.)	
270	0.0750
300	0.0755
CURVE 8	
300	0.0225
CURVE 9	
76	0.062
CURVE 10	
76	0.023
CURVE 11	
76	0.016
CURVE 12*	
76	0.010
CURVE 13	
278	0.04
CURVE 14	
278	0.05
CURVE 15	
278	0.23

*Not shown on plot

SPECIFICATION TABLE NO. 74 NORMAL TOTAL EMITTANCE OF GOLD

Curve No.	Ref. No.	Year	Temperature Range, K	Geometry θ'	Reported Error, %	Composition (weight percent), Specifications and Remarks
1	218	1963	360	$\sim 0^\circ$		Electrolytically polished; measured in vacuum (1.6×10^{-4} mm Hg).
2	215	1962	600-1000	0°		Pure gold strip; measurement covers 2.8 to 10 μ bandwidth.

DATA TABLE NO. 74 NORMAL TOTAL EMITTANCE OF GOLD

[Temperature, T, K; Emittance, ∈]

T	∈
CURVE 1*	
360	0.021
CURVE 2*	
600	0.0146
650	0.0146
700	0.0167
801	0.0191
900	0.0221
949	0.0235
1000	0.0248

*Not shown on plot

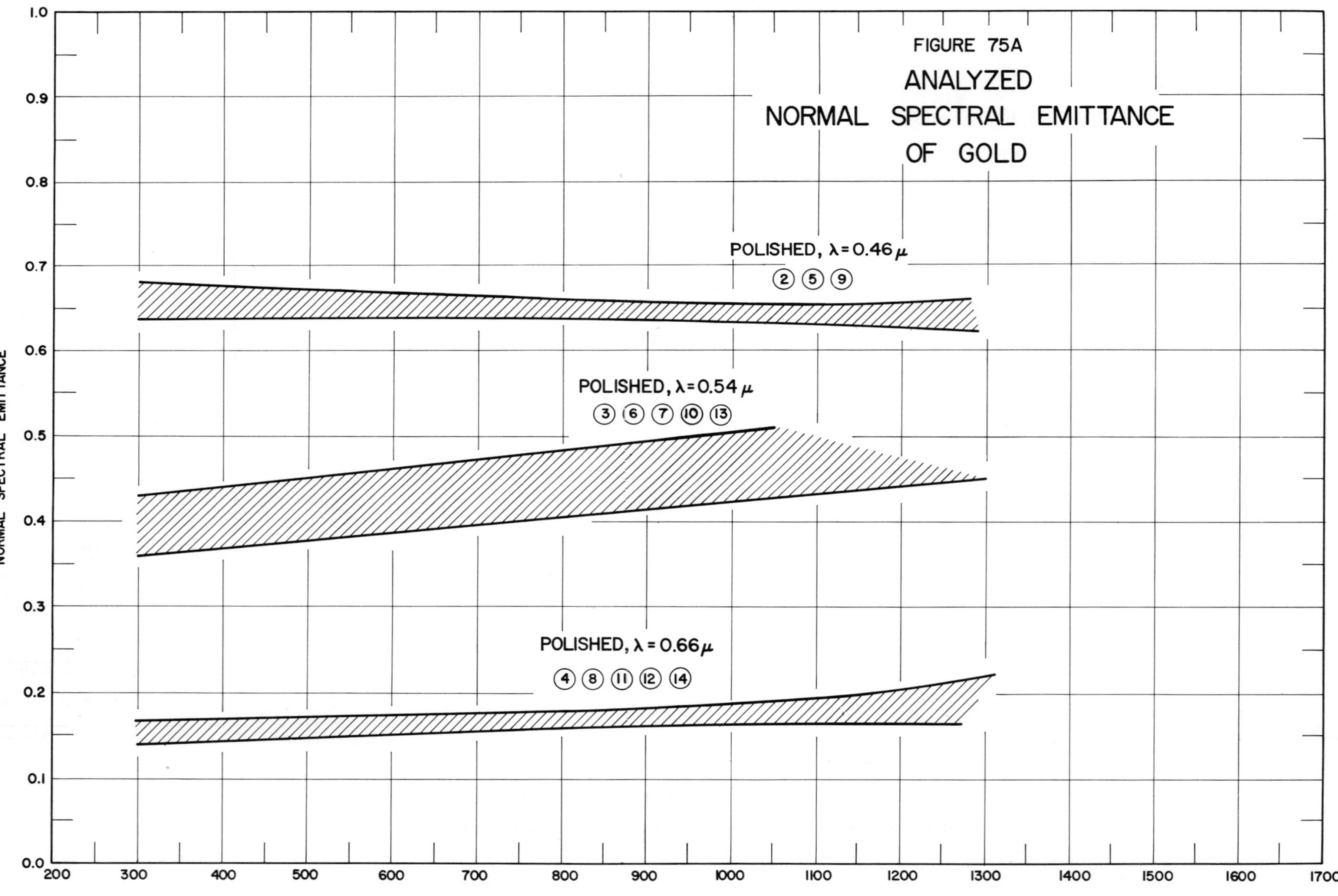
FIGURE 75A
ANALYZED
NORMAL SPECTRAL EMITTANCE
OF GOLD
POLISHED, λ = 0.46 μ
2 5 9
POLISHED, λ = 0.54 μ
3 6 7 10 13
POLISHED, λ = 0.66 μ
4 8 11 12 14
NORMAL SPECTRAL EMITTANCE
1.0
0.9
0.8
0.7
0.6
0.5
0.4
0.3
0.2
0.1
0.0
200
300
400
500
600
700
800
900
1000
1100
1200
1300
1400
1500
1600
1700
TEMPERATURE, K

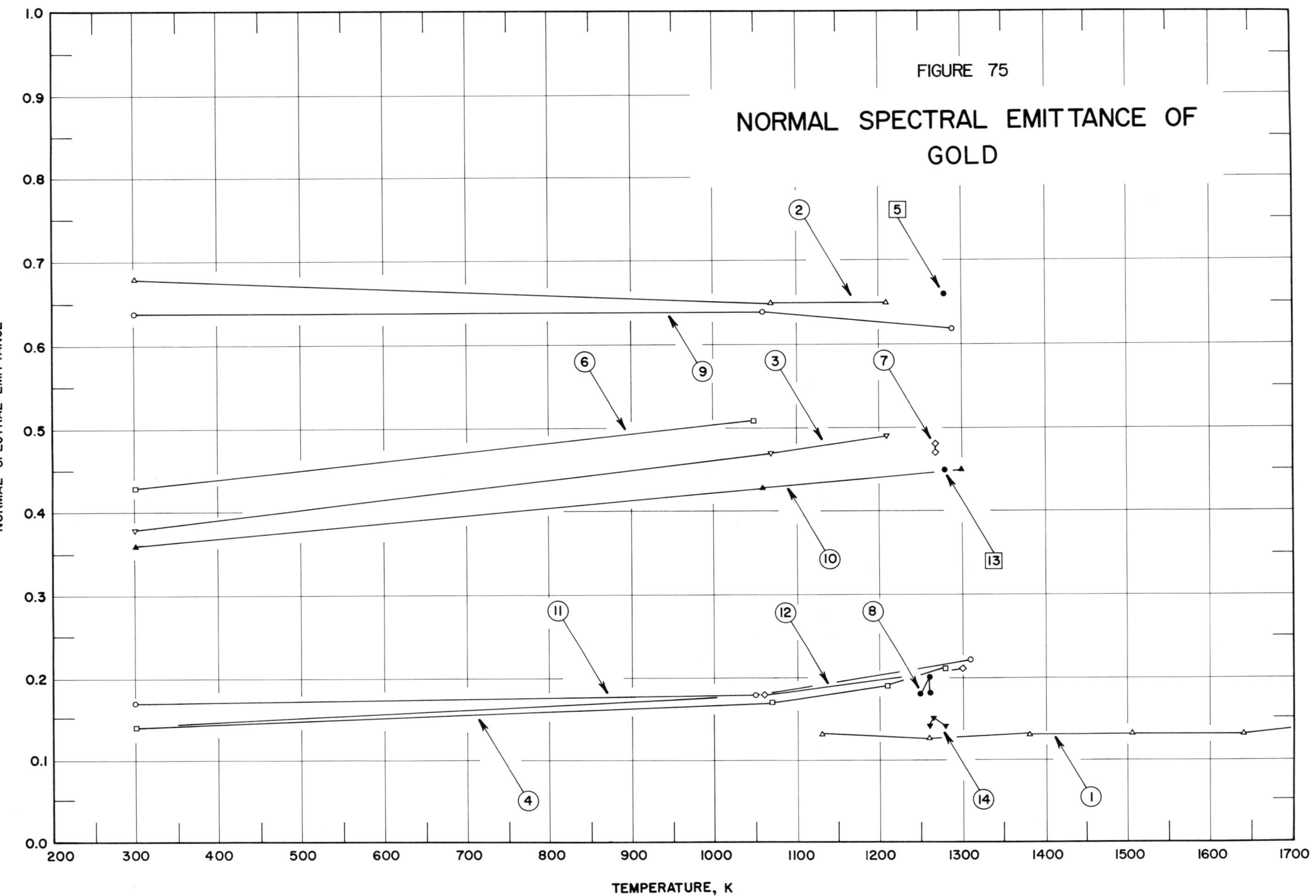
FIGURE 75
NORMAL SPECTRAL EMITTANCE OF GOLD
NORMAL SPECTRAL EMITTANCE
TEMPERATURE, K
0.0
0.1
0.2
0.3
0.4
0.5
0.6
0.7
0.8
0.9
1.0
200
300
400
500
600
700
800
900
1000
1100
1200
1300
1400
1500
1600
1700
2
5
9
6
3
7
10
13
11
12
8
4
14
1

SPECIFICATION TABLE NO. 75 NORMAL SPECTRAL EMITTANCE OF GOLD

Curve No.	Ref. No.	Year	Wavelength μ	Temperature Range, K	Geometry θ'	Reported Error, %	Composition (weight percent), Specifications and Remarks
1	35	1914	0.66	1130-2010	$\sim 0^\circ$		Measured in burning hydrogen.
2	51	1926	0.460	300-1210	$\sim 0^\circ$		"Very pure"; polished; surface showed etching due to heating; author assumed $\epsilon = 1-\rho$.
3	51	1926	0.535	300-1210	$\sim 0^\circ$		Above specimen and conditions.
4	51	1926	0.665	300-1280	$\sim 0^\circ$		Above specimen and conditions.
5	51	1926	0.460	1280	$\sim 0^\circ$		Different sample, same as curve 2 specimen and conditions.
6	51	1926	0.535	300-1050	$\sim 0^\circ$		Above specimen and conditions.
7	51	1926	0.535	1270	$\sim 0^\circ$		Above specimen and conditions.
8	51	1926	0.665	1250-1260	$\sim 0^\circ$		Above specimen and conditions.
9	51	1926	0.460	300-1290	$\sim 0^\circ$		Different sample, same as above specimen and conditions.
10	51	1926	0.535	300-1300	$\sim 0^\circ$		Above specimen and conditions.
11	51	1926	0.665	300-1310	$\sim 0^\circ$		Above specimen and conditions.
12	51	1926	0.665	300-1300	$\sim 0^\circ$		Above specimen and conditions.
13	51	1926	0.535	1280	$\sim 0^\circ$		"Very pure"; measured in argon.
14	51	1926	0.665	1260-1280	$\sim 0^\circ$		Above specimen and conditions.

DATA TABLE NO. 75 NORMAL SPECTRAL EMITTANCE OF GOLD

[Temperature, T, K; Emittance, ϵ; Wavelength, λ, μ]

T	ϵ
CURVE 1 $\lambda = 0.66$	
1130	0.130
1260	0.125
1380	0.130
1505	0.130
1640	0.130
1760	0.140*
1880	0.140*
2010	0.150*
CURVE 2 $\lambda = 0.460$	
300	0.68
1070	0.65
1210	0.65
CURVE 3 $\lambda = 0.535$	
300	0.38
1070	0.47
1210	0.49
CURVE 4 $\lambda = 0.665$	
300	0.14
1070	0.17
1210	0.19
1280	0.21
CURVE 5 $\lambda = 0.460$	
1280	0.66
CURVE 6 $\lambda = 0.535$	
300	0.43
1050	0.51
CURVE 7 $\lambda = 0.535$	
1270	0.48
1270	0.47
CURVE 8 $\lambda = 0.665$	
1250	0.18
1260	0.20
1260	0.18
CURVE 9 $\lambda = 0.460$	
300	0.64
1060	0.64
1290	0.62
CURVE 10 $\lambda = 0.535$	
300	0.36
1060	0.43
1300	0.45
CURVE 11 $\lambda = 0.665$	
300	0.17
1050	0.18
1310	0.22
CURVE 12 $\lambda = 0.665$	
300	0.14*
1060	0.18
1300	0.21
CURVE 13 $\lambda = 0.535$	
1280	0.45
CURVE 14 $\lambda = 0.665$	
1260	0.14
1265	0.15
1280	0.14

*Not shown on plot

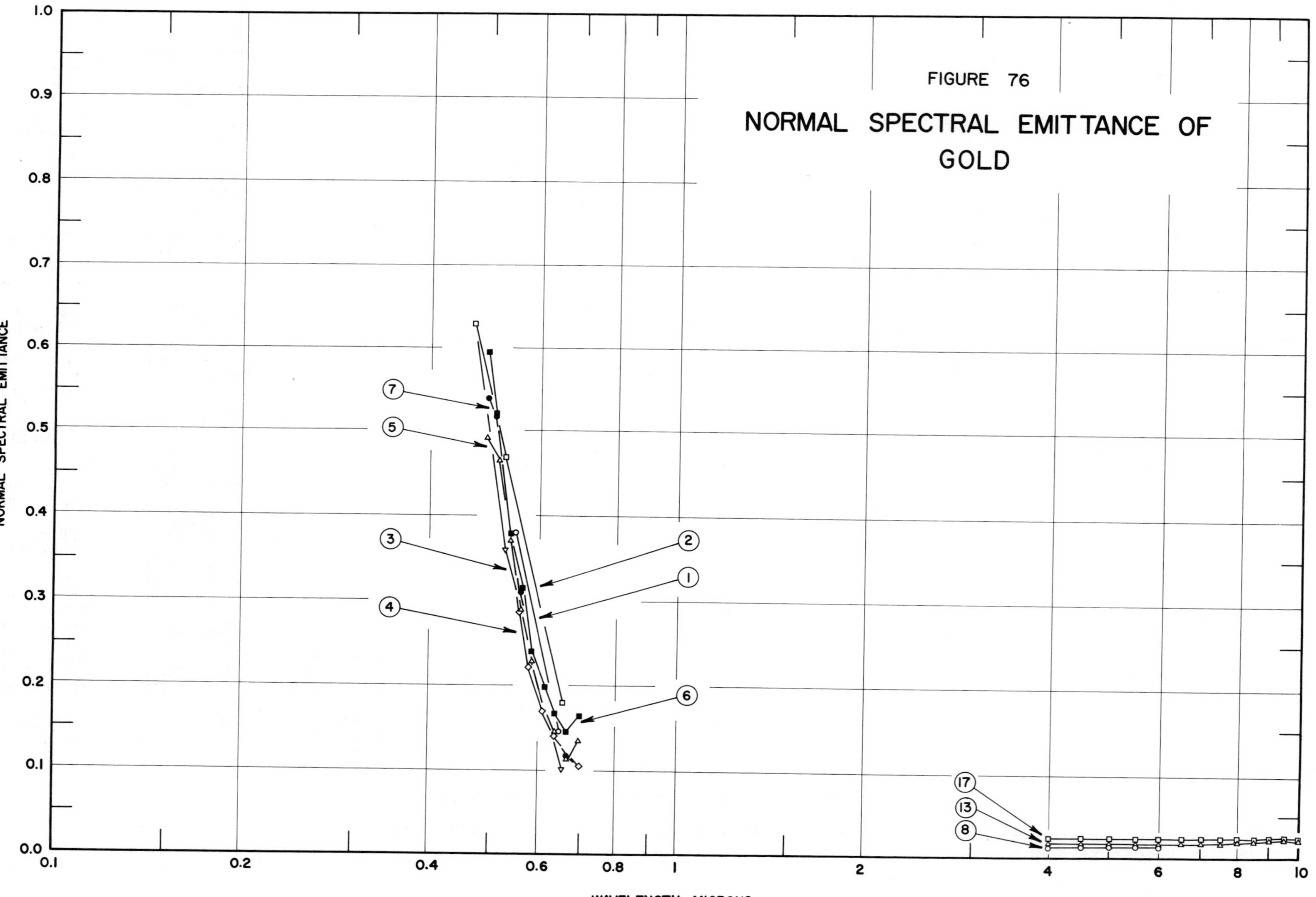

FIGURE 76

NORMAL SPECTRAL EMITTANCE OF GOLD

SPECIFICATION TABLE NO. 76 NORMAL SPECTRAL EMITTANCE OF GOLD

Curve No.	Ref. No.	Year	Temperature K	Wavelength Range, μ	Geometry θ'	Reported Error, %	Composition (weight percent), Specifications and Remarks
1	19	1914	1273	0.55-0.65	$\sim 0^\circ$	1	Film; tungsten substrate; measured in hydrogen; Pt reference (ϵ = 0.33 for λ = 0.650 μ at all temp).
2	50	1921	1300	0.47-0.66	$\sim 0^\circ$		Annealed at 1250 K for several min ; polished.
3	50	1921	298	0.47-0.66	$\sim 0^\circ$		Annealed at 1250 K for several min ; polished.
4	49	1912	1334	0.7014-0.5418	$\sim 0^\circ$	3	99.95 - 99.99 pure; polished and repeatedly treated with borax.
5	49	1912	1313	0.7014-0.4961	$\sim 0^\circ$	3	Above specimen and conditions.
6	49	1912	1222	0.7014-0.4961	$\sim 0^\circ$	3	Above specimen and conditions.
7	49	1912	1319	0.7014-0.4961	$\sim 0^\circ$	3	99.95 - 99.99 pure; clear surface naturally crystallized from the liquid.
8	119	1962	550	4.0-6.0	$\le 5^\circ$		High purity gold ribbon; mirror smooth surface; washed with distilled water and acetone and heated to 800 K for several hrs; samples maintained above 373 K until measurements were completed; smoothed values.
9	119	1962	600	4.0-7.5	$\le 5^\circ$		Different sample, same as above specimen and conditions.
10	119	1962	650	4.0-8.5	$\le 5^\circ$		Different sample, same as above specimen and conditions.
11	119	1962	700	4.0-8.5	$\le 5^\circ$		Different sample, same as above specimen and conditions.
12	119	1962	750	4.0-9.0	$\le 5^\circ$		Different sample, same as above specimen and conditions.
13	119	1962	800	4.0-10.0	$\le 5^\circ$		Different sample, same as above specimen and conditions.
14	119	1962	850	4.0-11.0	$\le 5^\circ$		Different sample, same as above specimen and conditions.
15	119	1962	900	4.0-11.0	$\le 5^\circ$		Different sample, same as above specimen and conditions.
16	119	1962	950	4.0-12.0	$\le 5^\circ$		Different sample, same as above specimen and conditions.
17	119	1962	1000	4.0-13.0	$\le 5^\circ$		Different sample, same as above specimen and conditions.

DATA TABLE NO. 76 NORMAL SPECTRAL EMITTANCE OF GOLD

[Wavelength, λ, μ; Emittance, ε; Temperature, T, K]

λ	ε
CURVE 1	**T = 1273**
0.55	0.38
0.65	0.145
CURVE 2	**T = 1300**
0.47	0.63
0.53	0.47
0.66	0.18
CURVE 3	**T = 298**
0.47	0.63*
0.53	0.36
0.66	0.10
CURVE 4	**T = 1334**
0.7014	0.103
0.6712	0.114*
0.6409	0.143
0.6149	0.171
0.5895	0.221
0.5649	0.289
0.5418	0.366*
CURVE 5	**T = 1313**
0.7014	0.134
0.6712	0.114
0.6409	0.144
0.6149	0.172*
0.5895	0.229
0.5649	0.291
0.5418	0.371
0.5186	0.465
0.4961	0.492
CURVE 6	**T = 1222**
0.7014	0.164
0.6712	0.146
0.6409	0.169
0.6149	0.198
0.5895	0.241
0.5649	0.315
0.5418	0.379
0.5186	0.520
0.4961	0.595
CURVE 7	**T = 1319**
0.7014	0.103*
0.6712	0.116
0.6149	0.178*
0.5649	0.316
0.5186	0.516
0.4961	0.541
CURVE 8	**T = 550**
4.0	0.014
4.5	0.0142
5.0	0.0145
5.5	0.0145
6.0	0.0142
CURVE 9*	**T = 600**
4.0	0.0147
4.5	0.0151
5.0	0.0155
5.5	0.0155
6.0	0.0152
6.5	0.0150
7.0	0.015
7.5	0.016
CURVE 10*	**T = 650**
4.0	0.0158
4.5	0.0163
5.0	0.0165
5.5	0.0165
6.0	0.0162
6.5	0.0160
7.0	0.0162
7.5	0.0169
8.0	0.018
8.5	0.019
CURVE 11*	**T = 700**
4.0	0.0169
4.5	0.0174
5.0	0.0176
5.5	0.0176
6.0	0.0173
6.5	0.0170
7.0	0.0172
7.5	0.0178
8.0	0.019
8.5	0.020
CURVE 12*	**T = 750**
4.0	0.0181
4.5	0.0184
5.0	0.0187
5.5	0.0187
6.0	0.0183
6.5	0.0181
7.0	0.0183
7.5	0.0187
8.0	0.0197
8.5	0.021
9.0	0.022
CURVE 13	**T = 800**
4.0	0.0193
4.5	0.0196
5.0	0.0198
5.5	0.0198
6.0	0.0195
6.5	0.0193
7.0	0.0194
7.5	0.0198
8.0	0.0206
8.5	0.0218
9.0	0.0228
9.5	0.023
10.0	0.022
CURVE 14*	**T = 850**
4.0	0.0206
4.5	0.0208
5.0	0.0210
5.5	0.0210
6.0	0.0208
6.5	0.0206
7.0	0.0207
7.5	0.0209
8.0	0.0215
8.5	0.0225
9.0	0.0235
9.5	0.0234
10.0	0.023
11.0	0.023
CURVE 15*	**T = 900**
4.0	0.0219
4.5	0.0221
5.0	0.0223
5.5	0.0223
6.0	0.0221
CURVE 15* (cont.)	
6.5	0.0220
7.0	0.0220
7.5	0.0221
8.0	0.0225
8.5	0.0233
9.0	0.0242
9.5	0.0241
10.0	0.0238
11.0	0.023
CURVE 16*	**T = 950**
4.0	0.0232
4.5	0.0234
5.0	0.0236
5.5	0.0236
6.0	0.0235
6.5	0.0234
7.0	0.0234
7.5	0.0234
8.0	0.0235
8.5	0.0240
9.0	0.0249
9.5	0.0248
10.0	0.0245
11.0	0.024
12.0	0.023
CURVE 17	**T = 1000**
4.0	0.0246
4.5	0.0248
5.0	0.0249
5.5	0.0249
6.0	0.0249
6.5	0.0249
7.0	0.0248
7.5	0.0246
CURVE 17 (cont.)	
8.0	0.0244
8.5	0.0247
9.0	0.0256
9.5	0.0255
10.0	0.0252
11.0	0.024*
12.0	0.024*
13.0	0.023*

* Not shown on plot

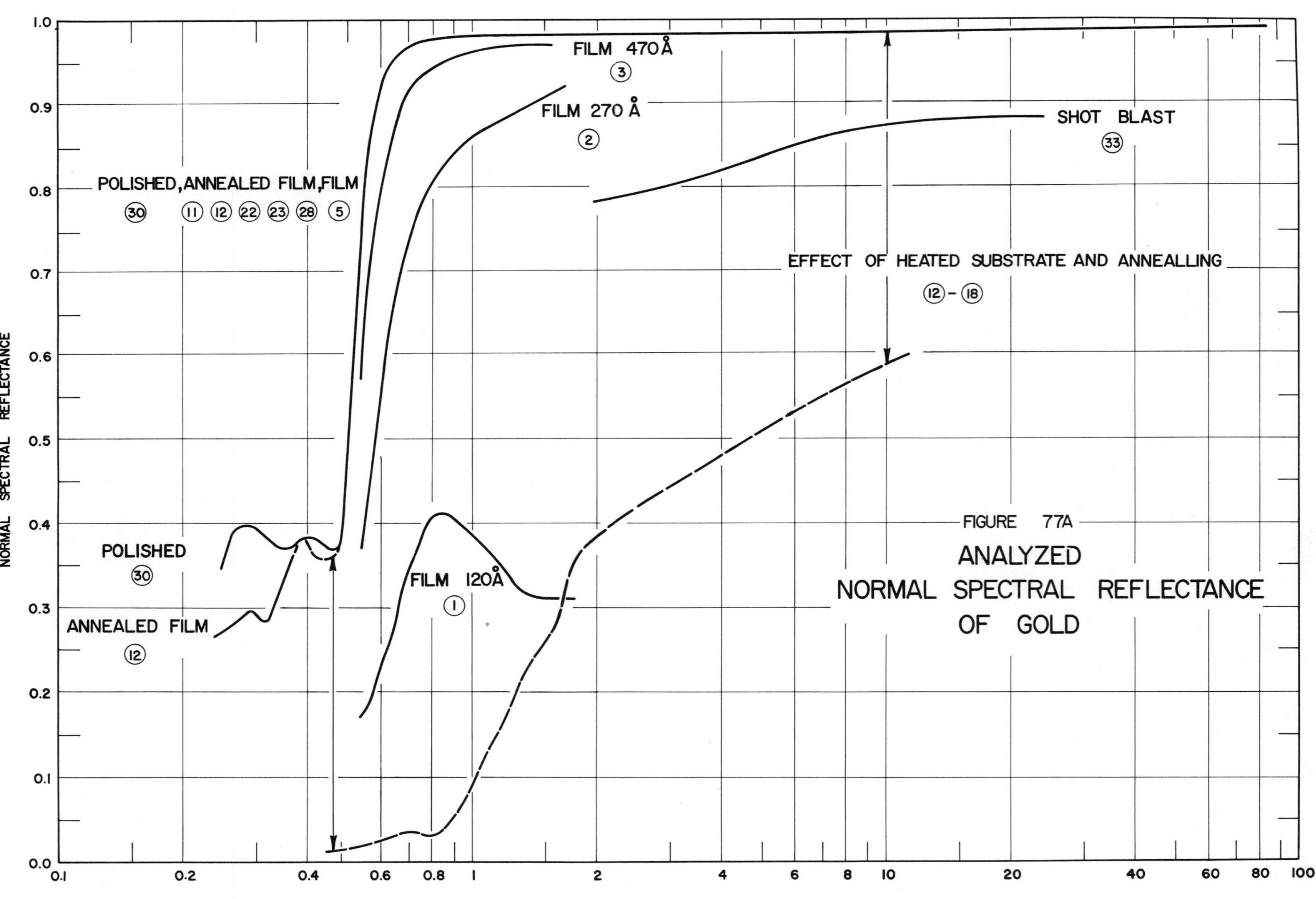

FIGURE 77A
ANALYZED
NORMAL SPECTRAL REFLECTANCE
OF GOLD

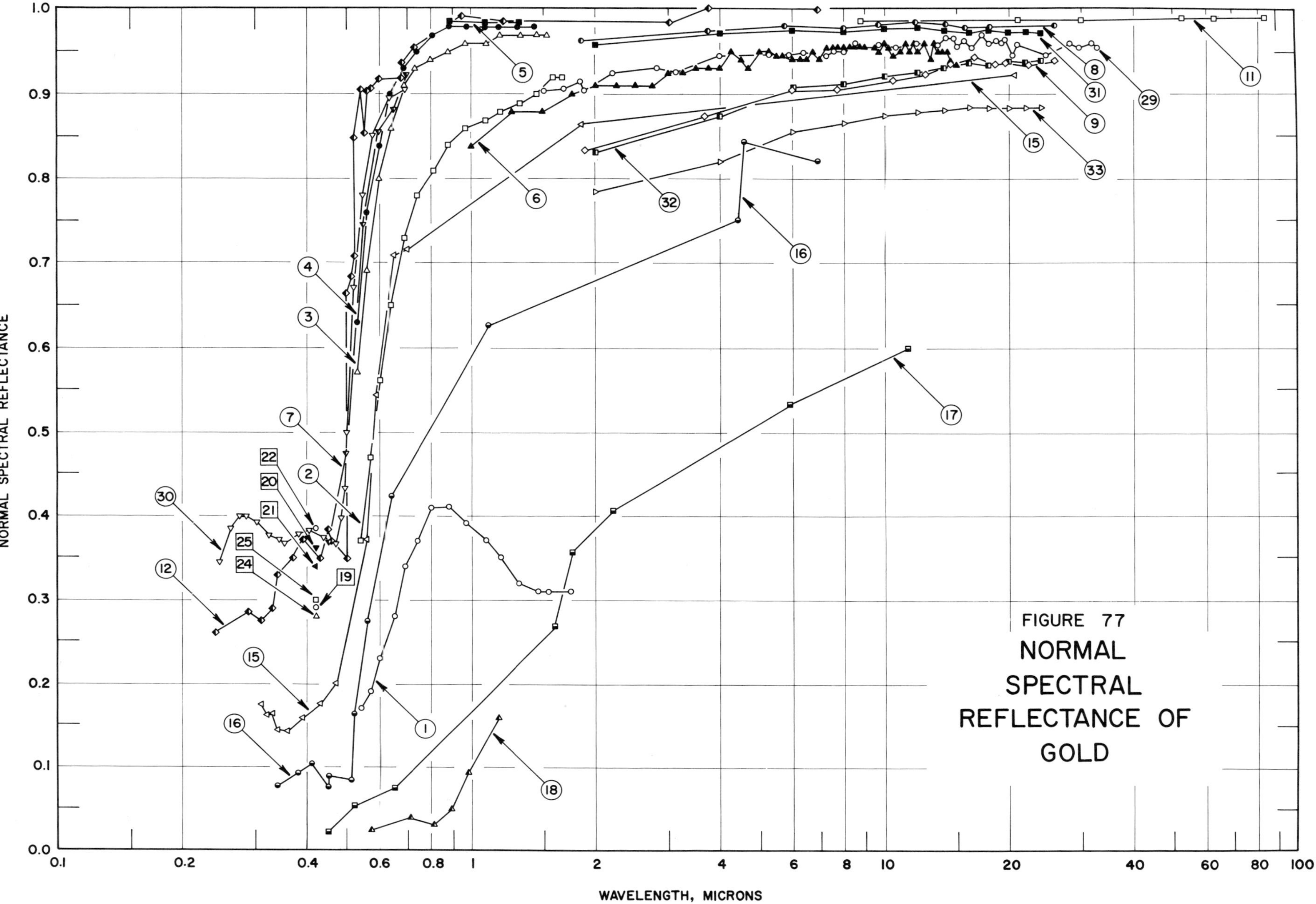

FIGURE 77 NORMAL SPECTRAL REFLECTANCE OF GOLD

SPECIFICATION TABLE NO. 77 NORMAL SPECTRAL REFLECTANCE OF GOLD

Curve No.	Ref. No.	Year	Temperature K	Wavelength Range, μ	Geometry θ θ' ω'	Reported Error, %	Composition (weight percent), Specifications and Remarks
1	144	1962	298	0.54-1.74	$\sim 0° \sim 0°$		99.99 Au; 0.012μ film evaporated on a glass slide; silver mirror as reference.
2	144	1962	298	0.54-1.66	$\sim 0° \sim 0°$		Different sample, same as above specimen and conditions except 0.027 μ thick.
3	144	1962	298	0.53-1.53	$\sim 0° \sim 0°$		Different sample, same as above specimen and conditions except 0.047 μ thick.
4	144	1962	298	0.53-1.44	$\sim 0° \sim 0°$		Different sample, same as above specimen and conditions except 0.063 μ thick.
5	144	1962	298	0.89-1.31	$\sim 0° \sim 0°$		Different sample, same as above specimen and conditions except 0.099 μ thick.
6	136	1954	294	1.00-15.00	5° 2π		Prepared electrolytically; polished with soft rag; washed in separate baths of acetone and ethyl alcohol; hohlraum at approx 1089 K; converted from R (2π, 5°).
7	228	1900	298	0.45-0.70	$\sim 0° \sim 0°$		Pure; mirror like surface; platinum filament lamp source.
8	223	1962	298	1.86-25.92	$\sim 0°$ 2π		Polished; converted from R(2π, 0°); data extracted from smooth curve.
9	223	1962	298	1.89-25.91	$\sim 0°$ 2π		Above specimen and conditions except after damaged by particle impact.
10	223	1962	77	1.86-25.92	$\sim 0°$ 2π		Above specimen and conditions.
11	145	1960	298	8.8-83.7	$\sim 0° \sim 0°$		99.999 Au; vacuum deposited on microscope slides; annealed in vacuum at 543 K for one hr; averaged values.
12	240	1965	300	0.24-6.89	$\sim 0° \sim 0°$		Gold film; deposited on quartz at 308K and annealed for 2 hrs in vacuum at 673K.
13	240	1965	300	0.25-10.33	$\sim 0° \sim 0°$		Gold film; vacuum deposited on quartz held at 513 K; unannealed.
14	240	1965	300	0.31-17.71	$\sim 0° \sim 0°$		Gold film; vacuum deposited on quartz held at 308K; unannealed.
15	240	1965	300	0.31-20.67	$\sim 0° \sim 0°$		Gold film; vacuum deposited on quartz held at 623 K; unannealed.
16	240	1965	300	0.34-6.89	$\sim 0° \sim 0°$		Gold film; vacuum deposited on quartz held at 673 K; unannealed.
17	240	1965	300	0.45-11.27	$\sim 0° \sim 0°$		Gold film; vacuum deposited on quartz held at 773 K; unannealed.
18	240	1965	300	0.57-1.16	$\sim 0° \sim 0°$		Gold film; vacuum deposited on quartz held at 1023K; unannealed.
19	240	1965	300	0.413	$\sim 0° \sim 0°$		Gold film (2700 Å thick); vacuum deposited on quartz held at 313 K; electrical resistivity at 300 K 4.97 x 10^{-6} ohm - cm; unannealed; [Authors' designation: Sample A].
20	240	1965	300	0.413	$\sim 0° \sim 0°$		Above specimen and conditions except annealed 1 hr at 703 K; electrical resistivity at 300 K 3.56 x 10^{-6} ohm - cm.
21	240	1965	300	0.413	$\sim 0° \sim 0°$		Curve 19 specimen and conditions except annealed 2.2 hrs at 703 K; electrical resistivity at 300 K 3.00 x 10^{-6} ohm - cm.
22	240	1965	300	0.413	$\sim 0° \sim 0°$		Curve 19 specimen and conditions except annealed 5.4 hrs at 703 K; electrical resistivity at 300 K 3.00 x 10^{-6} ohm - cm.
23	240	1965	300	0.413	$\sim 0° \sim 0°$		Curve 19 specimen and conditions except annealed 23.7 hrs at 703 K; electrical resistivity at 300 K 2.83 x 10^{-6} ohm - cm.

SPECIFICATION TABLE NO. 77 (continued)

Curve No.	Ref. No.	Year	Temperature K	Wavelength Range, μ	Geometry θ	θ'	ω'	Reported Error, %	Composition (weight percent), Specifications and Remarks
24	240	1965	300	0.413	~0°	~0°			Gold film (1800 Å thick); vacuum deposited on quartz held at 278 K; unannealed; electrical resistivity at 300 K 6.80×10^{-6} ohm-cm; [Authors' designation: Sample B].
25	240	1965	300	0.413	~0°	~0°			Above specimen and conditions except annealed 1.0 hr at 703 K; electrical resistivity at 300 K 4.55×10^{-6} ohm-cm.
26	240	1965	300	0.413	~0°	~0°			Curve 24 specimen and conditions except annealed 2.2 hrs at 703 K; electrical resistivity at 300 K 5.93×10^{-6} ohm-cm.
27	240	1965	300	0.413	~0°	~0°			Curve 24 specimen and conditions except annealed 5.4 hrs at 703 K; electrical resistivity at 300 K 4.02×10^{-6} ohm-cm.
28	240	1965	300	0.413	~0°	~0°			Curve 24 specimen and conditions except annealed 23.7 hrs at 703 K; electrical resistivity at 300 K 4.10×10^{-6} ohm-cm.
29	244	1963	311	1.50-32.8	~0°		2π		Vacuum-evaporated 24 K gold on fiberglass; converted from R(2π, 0°).
30	297	1963	298	0.248-0.689	~10°	~10°			Hand lapped with silicon carbide papers, lapped on a metallographic polishing wheel, and electrolytically slide polished; tungsten filament source for 0.365 to 0.689 μ and a hydrogen lamp source at lower wavelengths.
31	336	1964	298	2.00-24.00	~0°	~0°			Polished.
32	336	1964	298	2.00-24.00	~0°	~0°			Above specimen except cratered with spherical particles (100 μ dia) of Zircalloy at 1.5 Km sec^{-1}; average crater dia 213 μ; average crater depth 241 μ; 40 Knoop hardness (100 g load).
33	336	1964	298	2.00-24.00	~0°	~0°			Different sample, same as above specimen and conditions except cratered with spherical particles (100 μ dia) of tungsten at 5.2 Km sec^{-1}; average crater dia 63 μ; average crater depth 61 μ; 40 Knoop hardness (100 g load).
34	362	1965	298	1.5-7.0	7°		2π		Evaporated film; chromium film on glass, substrate.
35	362	1965	298	1.5-7.0	7°		2π		Above specimen and conditions.
36	362	1965	298	1.5-7.0	7°		2π		Above specimen and conditions.
37	362	1965	298	1.5-7.0	7°		2π		Above specimen and conditions.
38	362	1965	298	1.5-7.0	7°		2π		Above specimen and conditions.
39	362	1965	298	1.5-7.0	7°		2π		Above specimen and conditions.
40	362	1965	298	1.5-5.5	7°		2π		Above specimen and conditions.
41	359	1965	298	3.0-30.1	~5°	~5°			>99.999 Au; opaque ultrahigh vacuum film; vacuum evaporated (10^{-9} mm Hg) on super-smooth fused quartz optical flats; measured in nitrogen.
42	359	1965	298	0.798-31.8	~5°	~5°			Different sample, same as above specimen and conditions.

DATA TABLE NO. 77 NORMAL SPECTRAL REFLECTANCE OF GOLD

[Wavelength, λ, μ; Reflectance, ρ; Temperature, T, K]

λ	ρ
CURVE 1	
T = 298	
0.54	0.17
0.57	0.19
0.60	0.23
0.65	0.28
0.69	0.34
0.74	0.37
0.80	0.41
0.88	0.41
0.97	0.39
1.08	0.37
1.17	0.35
1.30	0.32
1.45	0.31
1.54	0.31
1.74	0.31
CURVE 2	
T = 298	
0.54	0.37
0.57	0.47
0.60	0.56
0.64	0.65
0.69	0.73
0.74	0.78
0.81	0.81
0.88	0.84
0.97	0.86
1.08	0.87
1.17	0.88
1.31	0.89
1.44	0.90
1.58	0.92
1.66	0.92
CURVE 3	
T = 298	
0.53	0.57
0.56	0.69
0.60	0.80
0.64	0.86
0.69	0.91
CURVE 3 (cont.)	
0.73	0.93
0.80	0.94
0.88	0.95
0.97	0.96
1.08	0.96
1.17	0.97
1.31	0.97
1.44	0.97
1.53	0.97
CURVE 4	
T = 298	
0.53	0.63
0.56	0.76
0.60	0.84
0.64	0.90
0.69	0.93
0.74	0.95
0.81	0.97
0.89	0.98
0.98	0.98
1.08	0.98
1.17	0.98
1.30	0.98
1.44	0.98
CURVE 5	
T = 298	
0.89	0.985
1.08	0.985
1.31	0.985
CURVE 6	
T = 294	
1.00	0.840
1.25	0.880
1.50	0.880
1.75	0.900
2.00	0.910
2.25	0.910
2.50	0.910
CURVE 6 (cont.)	
2.75	0.910
3.00	0.925
3.25	0.925
3.50	0.930
3.75	0.930
4.00	0.930
4.25	0.950
4.50	0.940
4.75	0.930
5.00	0.950
5.25	0.950
5.50	0.945
5.75	0.945
6.00	0.940
6.25	0.940
6.50	0.940
6.75	0.950
7.00	0.940
7.25	0.955
7.50	0.955
7.75	0.955
8.00	0.955
8.25	0.955
8.50	0.955
8.75	0.955
9.00	0.955
9.25	0.955*
9.50	0.955
9.75	0.950
10.00	0.960
10.25	0.955
10.50	0.945
10.75	0.950*
11.00	0.950
11.25	0.960
11.50	0.950
11.75	0.960
12.00	0.950
12.25	0.950*
12.50	0.960
12.75	0.960*
13.00	0.940
13.25	0.960
13.50	0.950
CURVE 6 (cont.)	
13.75	0.950
14.00	0.950*
14.25	0.950
14.50	0.940
14.75	0.940*
15.00	0.935
CURVE 7	
T = 298	
0.45	0.368
0.50	0.473
0.55	0.747
0.60	0.856
0.65	0.882
0.70	0.923
CURVE 8	
T = 298	
1.86	0.963
3.74	0.976
5.87	0.981
8.00	0.979
9.71	0.982
11.95	0.985
14.10	0.982
15.75	0.980
17.79	0.981
25.92	0.982
CURVE 9	
T = 298	
1.89	0.835
3.67	0.876
6.00	0.904
7.69	0.906
10.38	0.917
12.67	0.924
14.61	0.936
16.39	0.944
18.68	0.936
19.86	0.938
22.30	0.936
25.91	0.940
CURVE 10*	
T = 77	
1.86	0.822
4.80	0.901
7.87	0.901
10.87	0.930
13.94	0.948
15.88	0.947
17.93	0.944
19.92	0.944
21.94	0.940
23.98	0.941
25.92	0.940
CURVE 11	
T = 298	
8.8	0.988
21.1	0.989
30.2	0.989
52.7	0.991
63.2	0.991
83.7	0.991
CURVE 12	
T = 300	
0.24	0.261
0.29	0.286
0.31	0.275
0.33	0.289
0.34	0.330
0.37	0.351
0.39	0.372
0.43	0.349
0.45	0.383
0.46	0.369
0.50	0.349
0.50	0.664
0.51	0.684
0.52	0.708
0.52	0.849
0.54	0.905
0.55	0.853
0.56	0.922
CURVE 12 (cont.)	
0.57	0.906
0.60	0.918
0.68	0.920
0.68	0.937
0.73	0.954
0.95	0.992
1.20	0.987
3.02	0.984
3.65	1.000
6.89	1.000
CURVE 13*	
T = 300	
0.25	0.274
0.26	0.279
0.29	0.301
0.32	0.273
0.32	0.302
0.34	0.325
0.34	0.298
0.35	0.337
0.39	0.393
0.39	0.373
0.41	0.336
0.41	0.383
0.42	0.399
0.42	0.349
0.44	0.357
0.44	0.338
0.46	0.371
0.46	0.326
0.48	0.326
0.50	0.336
0.51	0.433
0.51	0.458
0.53	0.553
0.53	0.679
0.55	0.696
0.55	0.883
0.58	0.876
0.59	0.914
0.72	0.989
0.83	0.995
CURVE 13* (cont.)	
1.01	0.996
1.16	0.996
1.33	1.000
1.53	0.998
2.00	0.999
2.64	0.999
3.76	0.999
10.33	0.992
CURVE 14*	
T = 300	
0.31	0.301
0.33	0.301
0.35	0.296
0.37	0.286
0.37	0.308
0.40	0.313
0.41	0.299
0.44	0.300
0.47	0.313
0.50	0.433
0.51	0.404
0.54	0.660
0.55	0.752
0.66	0.867
0.68	0.886
0.74	0.879
0.83	0.907
1.88	0.918
4.00	0.918
17.71	0.909
CURVE 15	
T = 300	
0.31	0.176
0.32	0.162
0.33	0.164
0.34	0.144
0.36	0.142
0.39	0.159
0.43	0.176
0.47	0.201
CURVE 15 (cont.)	
0.56	0.371
0.59	0.544
0.65	0.709
0.70	0.717
1.85	0.866
20.67	0.923
CURVE 16	
T = 300	
0.34	0.077
0.38	0.092
0.41	0.104
0.45	0.077
0.45	0.089
0.51	0.183
0.52	0.164
0.56	0.274
0.57	0.301
0.60	0.369
0.64	0.424
1.10	0.628
4.43	0.751
4.59	0.847
6.89	0.822
CURVE 17	
T = 300	
0.45	0.023
0.52	0.053
0.65	0.075
1.59	0.269
1.75	0.358
2.18	0.408
5.90	0.533
11.27	0.600
CURVE 18	
T = 300	
0.57	0.025
0.71	0.040
0.81	0.032
0.89	0.050
0.98	0.143
1.16	0.160

*Not shown on plot

DATA TABLE NO. 77 (continued)

λ	ρ
CURVE 19 T = 300	
0.413	0.29
CURVE 20 T = 300	
0.413	0.36
CURVE 21 T = 300	
0.413	0.34
CURVE 22 T = 300	
0.413	0.383
CURVE 23* T = 300	
0.413	0.383
CURVE 24 T = 300	
0.413	0.28
CURVE 25 T = 300	
0.413	0.30
CURVE 26* T = 300	
0.413	0.34
CURVE 27* T = 300	
0.413	0.36
CURVE 28* T = 300	
0.413	0.383
CURVE 29 T = 311	
1.50	0.902
1.67	0.906
1.34	0.915
1.38	0.913
2.20	0.924
2.82	0.930
3.15	0.926
3.99	0.946
5.32	0.948
5.86	0.945
6.35	0.949
7.21	0.945
7.56	0.950
8.01	0.950
8.49	0.959
8.99	0.955*
9.72	0.957
10.6	0.954
11.8	0.959
13.3	0.958
14.1	0.968
14.7	0.967
15.0	0.957
15.5	0.962
16.2	0.954
17.2	0.971
17.9	0.960
18.7	0.963
19.2	0.961
19.6	0.964
20.4	0.946
20.9	0.959
24.7	0.947
28.1	0.960
29.7	0.954
31.8	0.960
32.8	0.954
CURVE 30 T = 298	
0.248	0.347
0.262	0.384
0.275	0.399
0.285	0.399
0.304	0.391
CURVE 30 (cont.)	
0.324	0.377
0.344	0.371
0.356	0.368
0.382	0.378
0.406	0.382
0.438	0.373
0.468	0.366
0.486	0.399
0.500	0.500
0.521	0.670
0.544	0.781
0.579	0.852
0.633	0.897
0.689	0.908
CURVE 31 T = 298	
2.00	0.959
4.00	0.972
6.00	0.977
8.00	0.976
10.00	0.979
12.00	0.980
13.99	0.977
16.00	0.976
17.98	0.978
20.00	0.975
22.00	0.975
24.00	0.974
CURVE 32 T = 298	
2.00	0.832
4.00	0.876
6.00	0.907
8.00	0.912
10.00	0.921
12.00	0.926
13.99	0.931
16.00	0.937
17.98	0.935
20.00	0.938
22.00	0.937
24.00	0.940
CURVE 33 T = 298	
2.00	0.787
4.00	0.821
6.00	0.855
8.00	0.868
10.00	0.875
12.00	0.880
13.99	0.882
16.00	0.882
17.98	0.885
20.00	0.886
22.00	0.886
24.00	0.886
CURVE 34* T = 298	
1.5	0.980
2.0	0.984
2.5	0.984
3.5	0.987
4.5	0.986
5.5	0.987
6.5	0.988
7.0	0.988
CURVE 35* T = 298	
1.5	0.981
2.0	0.984
2.5	0.986
3.5	0.987
4.5	0.988
5.5	0.989
6.5	0.987
7.0	0.987
CURVE 36* T = 298	
1.5	0.980
2.0	0.984
2.5	0.986
3.5	0.988
4.5	0.987
5.5	0.988
6.5	0.988
7.0	0.988
CURVE 37* T = 298	
1.5	0.981
2.0	0.983
2.5	0.982
3.5	0.987
4.5	0.988
5.5	0.986
6.5	0.987
7.0	0.992
CURVE 38* T = 298	
1.5	0.982
2.0	0.983
2.5	0.984
3.5	0.986
4.5	0.987
5.5	0.985
6.5	0.988
7.0	0.989
CURVE 39* T = 298	
1.5	0.983
2.0	0.983
2.5	0.984
3.5	0.987
4.5	0.986
5.5	0.986
6.5	0.988
7.0	0.989
CURVE 40* T = 298	
1.5	0.979
2.0	0.981
2.5	0.984
3.5	0.988
4.5	0.987
5.5	0.988
CURVE 41* T = 298	
3.0	0.9941
4.0	0.9940
5.0	0.9938
5.9	0.9941
7.0	0.9935
8.0	0.9939
9.0	0.9940
10.0	0.9942
11.0	0.9940
12.0	0.9939
13.0	0.9939
14.0	0.9939
16.0	0.9939
18.0	0.9937
20.0	0.9939
22.0	0.9943
24.0	0.9943
26.1	0.9943
28.0	0.9943
30.1	0.9938
CURVE 42* T = 298	
0.798	0.9791
0.902	0.9860
1.00	0.9865
1.24	0.9884
1.50	0.9885
1.75	0.9901
1.98	0.9925
2.98	0.9940
4.00	0.9939
4.97	0.9939
6.00	0.9938
7.01	0.9940
7.96	0.9935
8.93	0.9938
9.93	0.9940
12.0	0.9938
14.0	0.9938
15.8	0.9939
18.0	0.9937
20.0	0.9940
22.0	0.9942
24.0	0.9941
CURVE 42 (cont.)*	
26.2	0.9943
28.0	0.9943
30.0	0.9938
31.8	0.9944

*Not shown on plot

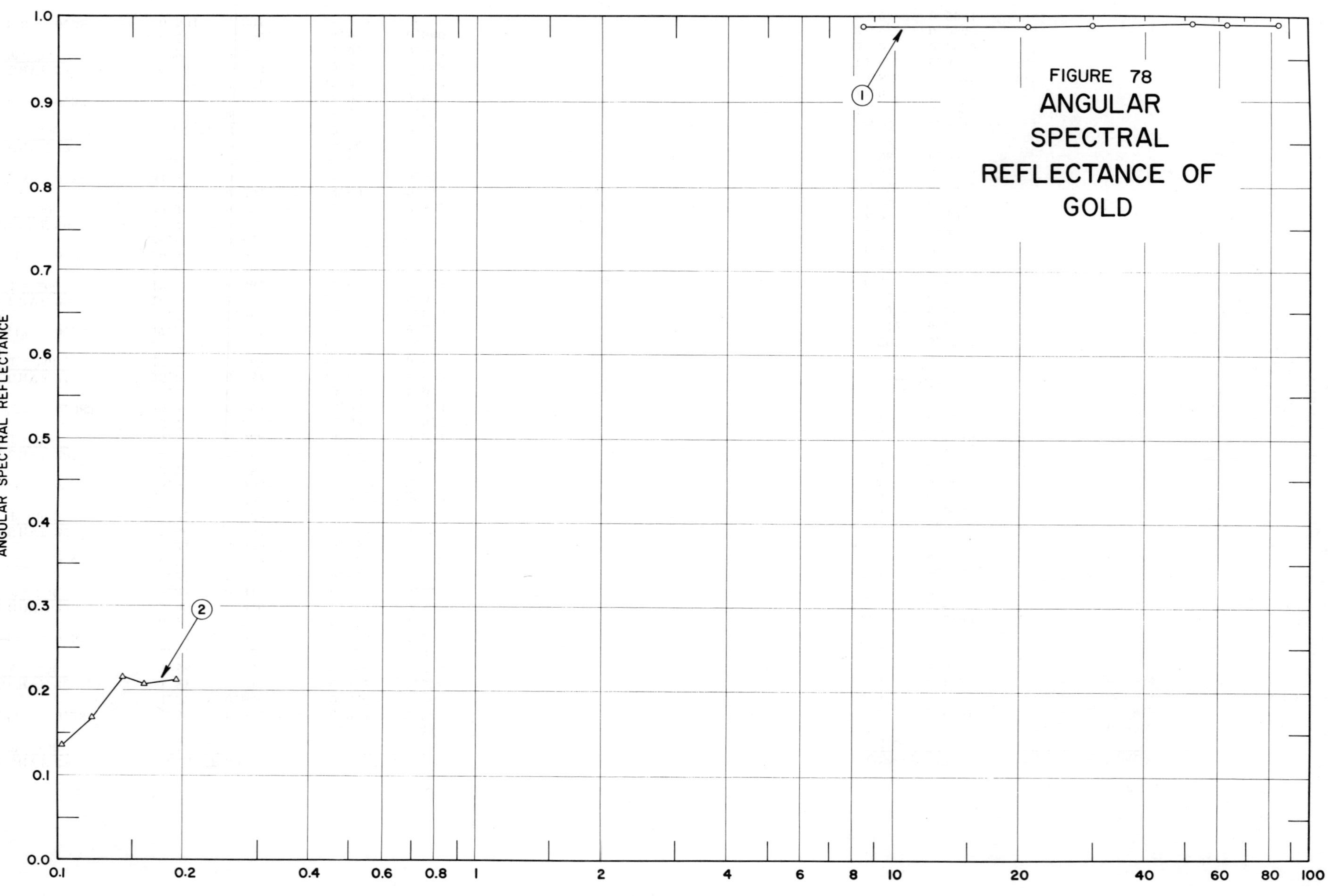

FIGURE 78 ANGULAR SPECTRAL REFLECTANCE OF GOLD

SPECIFICATION TABLE NO. 78 ANGULAR SPECTRAL REFLECTANCE OF GOLD

Curve No.	Ref. No.	Year	Temperature K	Wavelength Range, μ	Geometry θ	θ'	ω'	Reported Error, %	Composition (weight percent), Specifications and Remarks
1	145	1961	294	8.5-84	23°	23°		0.2	99.999 pure (0.23 - 0.60 μ thick); vacuum deposition on carefully polished, cleaned microscope slides; baked before deposition, annealed after deposition; microchemical analysis showed <0.003 Mo; resistivity 2.67 to 3.05 μohm-cm at 294 K.
2	243	1965	~298	0.0584-0.1930	45°	45°			Opaque; measured in vacuum.

DATA TABLE NO. 78 ANGULAR SPECTRAL REFLECTANCE OF GOLD

[Wavelength, λ, μ; Reflectance, ρ; Temperature, T, K]

λ	ρ
CURVE 1 T = 294	
8.5	0.9890
21	0.9895
30	0.9900
52	0.9920
63	0.9917
84	0.9916
CURVE 2 T = ~ 298	
0.0584	0.150*
0.0736	0.117*
0.1026	0.136
0.1216	0.168
0.1436	0.216
0.1606	0.208
0.1930	0.213

*Not shown on plot

SPECIFICATION TABLE NO. 79 NORMAL SOLAR REFLECTANCE OF GOLD

Curve No.	Ref. No.	Year	Temperature Range, K	Geometry θ θ' ω'	Reported Error, %	Composition (weight percent), Specifications and Remarks
1	40	1962	298	$\sim 0^{\circ}$ 2π		Gold film on 1/4 mil thick Mylar.
2	40	1962	298	$\sim 0^{\circ}$ 2π		Gold film on 1/4 mil thick Mylar.

DATA TABLE NO. 79 NORMAL SOLAR REFLECTANCE OF GOLD

[Temperature, T, K; Reflectance, ρ]

T	ρ
CURVE 1*	
298	0.81
CURVE 2*	
298	0.82

*Not shown on plot

SPECIFICATION TABLE NO. 80 HEMISPHERICAL INTEGRATED ABSORPTANCE OF GOLD

Curve No.	Ref. No.	Year	Temperature Range, K	Reported Error, %	Composition (weight percent), Specifications and Remarks
1	3	1955	76	5	Foil (0.0005 in. thick) ; solvent cleaned; measured in vacuum (10^{-6} to 10^{-7} mm Hg) ; absorptance for 300 K black body incident radiation.
2	3	1955	76	5	Foil (0.0015 in. thick) ; solvent cleaned; measured in vacuum (10^{-6} to 10^{-7} mm Hg) ; absorptance for 300 K black body incident radiation
3	3	1955	76	5	Foil (4.0 x 10^{-5} in. thick); solvent cleaned; measured in vacuum (10^{-6} to 10^{-7} mm Hg); absorptance for 300 K black body incident radiation.
4	3	1955	76	5	Leaf (1 x 10^{-5} in. thick); measured in vacuum (10^{-6} to 10^{-7} mm Hg) ; absorptance for 300 K black body incident radiation.
5	23	1955	77.3	±20	Plating (0.0003 in. thick) ; polished; kerosene buff; measured in vacuum (3 x 10^{-6} mm Hg) ; absorptance for radiation from a 300 K surface.
6	23	1955	77.3	±20	Plating (0.0003 in. thick) ; dry buffed; measured in vacuum (3 x 10^{-6} mm Hg); absorptance for radiation from a 300 K surface.
7	23	1955	77.3	±20	Gold wash; measured in vacuum (3 x 10^{-6} mm Hg) ; absorptance for radiation from a 300 K surface.
8	23	1955	77.3	±20	Plating (0.0003 in. thick) ; matte surface; measured in vacuum (3 x 10^{-6} mm Hg) ; absorptance for radiation from a 300 K surface.
9	4	1953	76		Foil (0.0015 in. thick) ; solvent cleaned; measured in vacuum ($< 10^{-6}$ mm Hg) ; absorptance for 294 K black body incident radiation.

DATA TABLE NO. 80 HEMISPHERICAL INTEGRATED ABSORPTANCE OF GOLD

[Temperature, T, K; Absorptance, α]

T	α
CURVE 1*	
76	0.016
CURVE 2*	
76	0.010
CURVE 3*	
76	0.023
CURVE 4*	
76	0.062
CURVE 5*	
77.3	0.018
CURVE 6*	
77.3	0.016
CURVE 7*	
77.3	0.015
CURVE 8*	
77.3	0.014
CURVE 9*	
76	0.0099

*Not shown on plot

SPECIFICATION TABLE NO. 81 NORMAL SPECTRAL ABSORPTANCE OF GOLD

Curve No.	Ref. No.	Year	Wavelength μ	Temperature Range, K	Geometry θ	Reported Error, %	Composition (weight percent), Specifications and Remarks
1	35	1914	0.66	873-2073	$\sim 0^\circ$		Measured in burning hydrogen; author assumes $\alpha = \epsilon$.

DATA TABLE NO. 81 NORMAL SPECTRAL ABSORPTANCE OF GOLD

[Temperature, T, K; Absorptance, α; Wavelength, λ, μ]

T	α
CURVE 1*	
$\lambda = 0.66$	
873	0.125
2073	0.125

*Not shown on plot

SPECIFICATION TABLE NO. 82 NORMAL SPECTRAL ABSORPTANCE OF GOLD

Curve No.	Ref. No.	Year	Temperature K	Wavelength Range, μ	Geometry θ	Reported Error, %	Composition (weight percent), Specifications and Remarks
1	307	1954	~298	0.229-2.600	~0°		Data extracted from smooth curve.

DATA TABLE NO. 82 NORMAL SPECTRAL ABSORPTANCE OF GOLD

[Wavelength, λ, μ; Absorptance, α; Temperature, T, K]

λ	α
CURVE 1*	
T = ~298	
0.229	0.831
0.286	0.784
0.345	0.770
0.420	0.741
0.467	0.742
0.492	0.713
0.506	0.551
0.529	0.422
0.564	0.262
0.621	0.174
0.724	0.120
0.804	0.120
0.992	0.098
1.128	0.126
1.280	0.135
1.707	0.100
1.800	0.111
2.000	0.152
2.200	0.181
2.400	0.189
2.600	0.173

* Not shown on plot

SPECIFICATION TABLE NO. 83 ANGULAR SPECTRAL ABSORPTANCE OF GOLD

Curve No.	Ref. No.	Year	Temperature K	Wavelength Range, μ	Geometry θ	Reported Error, %	Composition (weight percent), Specifications and Remarks
1	225	1965	306	1.00-20.10	25°		99.99 Au from Wilkinson Co.; dry sanded and polished on metallurgical felt wheels; measured in dry nitrogen; heated cavity at approx 1056 K with platinum reference; authors assumed $\alpha = 1-R(2\pi, 25^{\circ})$.

DATA TABLE NO. 83 ANGULAR SPECTRAL ABSORPTANCE OF GOLD

[Wavelength, λ, μ; Absorptance, α; Temperature, T, K]

λ	α
CURVE 1*	
T = 306	
1.00	0.043
2.01	0.028
3.57	0.024
5.09	0.022
7.02	0.022
10.40	0.020
15.30	0.020
18.10	0.020
20.10	0.020

*Not shown on plot

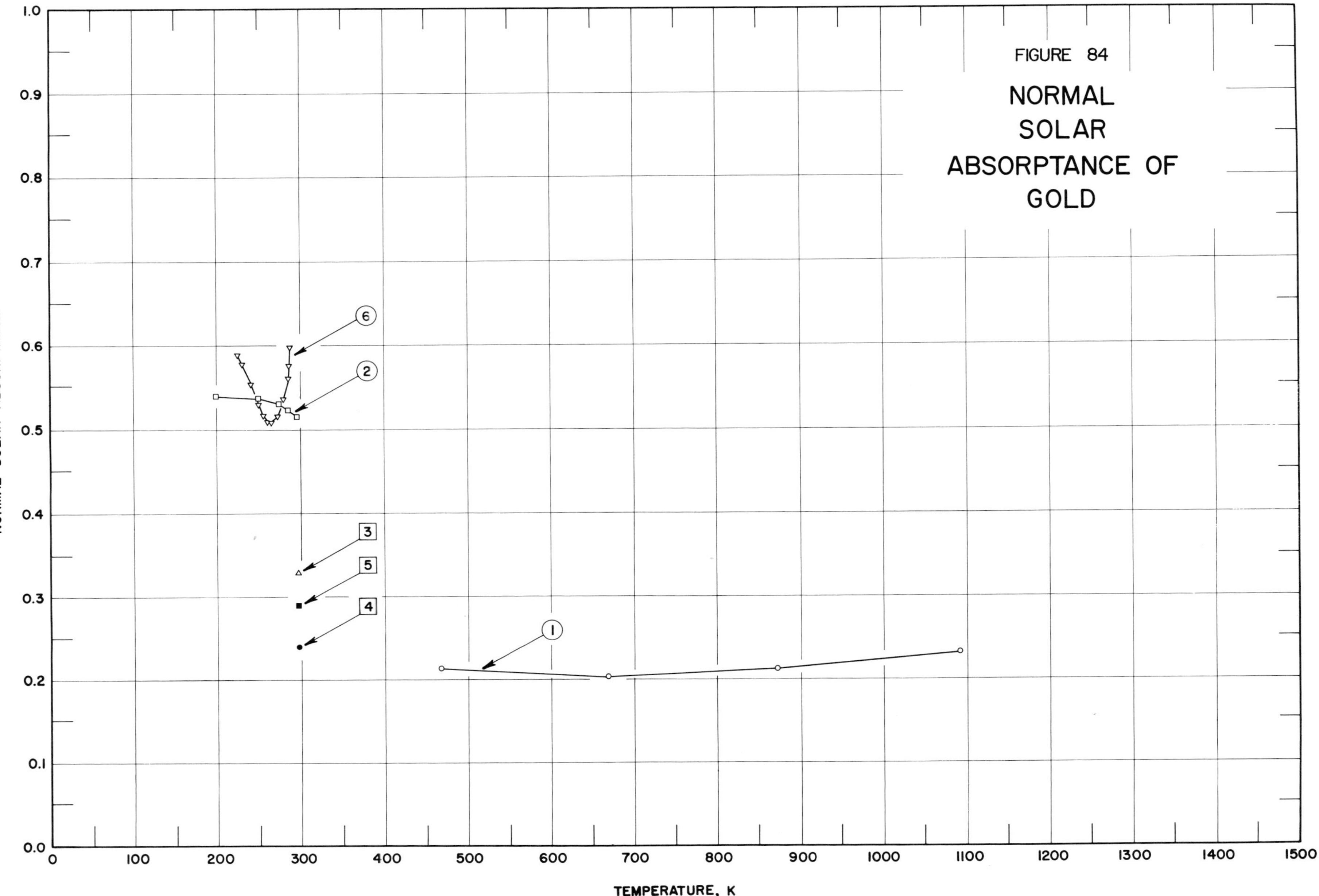
FIGURE 84
NORMAL SOLAR ABSORPTANCE OF GOLD
NORMAL SOLAR ABSORPTANCE
TEMPERATURE, K
1.0
0.9
0.8
0.7
0.6
0.5
0.4
0.3
0.2
0.1
0.0
0
100
200
300
400
500
600
700
800
900
1000
1100
1200
1300
1400
1500
6
2
3
5
4
1

SPECIFICATION TABLE NO. 84 NORMAL SOLAR ABSORPTANCE OF GOLD

Curve No.	Ref. No.	Year	Temperature Range, K	Geometry θ	Reported Error, %	Composition (weight percent), Specifications and Remarks
1	47	1961	468-1093	$\sim 0^\circ$	≤ 10	99.95 Au; ground with 600 grit Carborundum and polished on a wet cloth lap with either Linde Alumina Type B-5125 or unlevigated jewelers rouge; measured in vacuum (10^{-5}mm Hg).
2	26	1962	200-295	$\sim 0^\circ$	4	Cleaned in both sodiumdichromate and dilute nitric acid solutions, buffed on a felt buffing wheel, cleaned with CCl_4 and acetone; measured in vacuum (10^{-6}mm Hg); data extracted from smooth curve.
3	45	1961	298	$\sim 0^\circ$	10	Vacuum deposit on buffed titanium.
4	45	1961	298	$\sim 0^\circ$	10	Vacuum deposit on aluminum.
5	45	1961	298	$\sim 0^\circ$	10	Bright-foil.
6	26	1964	226-287	$\sim 0^\circ$		Commercially pure; cleaned, buffed, cleaned with carbon tetrachloride and acetone; measured in vacuum (1×10^{-6} mm Hg); data extracted from smooth curve.

DATA TABLE NO. 84 NORMAL SOLAR ABSORPTANCE OF GOLD

[Temperature, T, K; Absorptance, α]

T	α
CURVE 1	
468	0.212
668	0.203
873	0.212
1093	0.232
CURVE 2	
200	0.540
250	0.538
275	0.530
285	0.525
295	0.517
CURVE 3	
298	0.33
CURVE 4	
298	0.24
CURVE 5	
298	0.29
CURVE 6	
226	0.588
231	0.577
241	0.551
250	0.530
256	0.516
261	0.510
266	0.508
274	0.516
281	0.536
285	0.561
286	0.575
287	0.596

SPECIFICATION TABLE NO. 85 HEMISPHERICAL TOTAL EMITTANCE OF HAFNIUM

Curve No.	Ref. No.	Year	Temperature Range, K	Reported Error, %	Composition (weight percent), Specifications and Remarks
1	346	1966	1350-1991	±7	Iodide Hafnium; density 13.06 g cm^{-3}; surface roughness ~3 μ; aged at 1750 K; measured in vacuum (5×10^{-5} mm Hg).

DATA TABLE NO. 85 HEMISPHERICAL TOTAL EMITTANCE OF HAFNIUM

[Temperature, T, K; Emittance, ∈]

T	∈
CURVE 1*	
1350	0.306
1360	0.287
1468	0.311
1476	0.315
1557	0.330
1585	0.309
1587	0.317
1736	0.329
1743	0.322
1747	0.327
1750	0.318
1757	0.306
1795	0.327
1799	0.317
1826	0.305
1853	0.325
1864	0.320
1883	0.315
1909	0.321
1945	0.328
1945	0.333
1953	0.334
1964	0.322
1986	0.338
1991	0.327

*Not shown on plot

SPECIFICATION TABLE NO. 86 NORMAL SPECTRAL EMITTANCE OF HAFNIUM

Curve No.	Ref. No.	Year	Wavelength μ	Temperature Range, K	Geometry θ'	Reported Error, %	Composition (weight percent), Specifications and Remarks
1	236	1966	0.65	1630-1790	$\sim 0^{\circ}$	± 2	Measured in vacuum (1×10^{-6} to 2×10^{-6} mm Hg); peak wavelength of transmission band of red filter of pyrometer was 0.65 μ; referenced to emittance of tungsten.

DATA TABLE NO. 86 NORMAL SPECTRAL EMITTANCE OF HAFNIUM

[Temperature, T, K; Emittance, ϵ; Wavelength, λ, μ]

T	ϵ
CURVE 1*	
$\lambda = 0.65$	
1630	0.45
1790	0.45

*Not shown on plot

SPECIFICATION TABLE NO. 87 NORMAL SPECTRAL EMITTANCE OF HAFNIUM

Curve No.	Ref. No.	Year	Temperature K	Wavelength Range, μ	Geometry θ'	Reported Error, %	Composition (weight percent), Specifications and Remarks
1	237	1965	1510	0.4665-0.6980	$\sim 0°$		Chemically pure; measured in vacuum; authors assumed $\epsilon = 1 - \rho$; ρ calculated from optical constants.
2	237	1965	1735	0.4665-0.6980	$\sim 0°$		Above specimen and conditions.

DATA TABLE NO. 87 NORMAL SPECTRAL EMITTANCE OF HAFNIUM

[Wavelength, λ, μ; Emittance, ϵ; Temperature, T, K]

λ	ϵ
CURVE 1*	
T = 1510	
0.4665	0.475
0.5745	0.462
0.5740	0.456
0.6540	0.446
0.6980	0.438
CURVE 2*	
T = 1735	
0.4665	0.487
0.5475	0.472
0.5740	0.472
0.6540	0.462
0.6980	0.451

*Not shown on plot

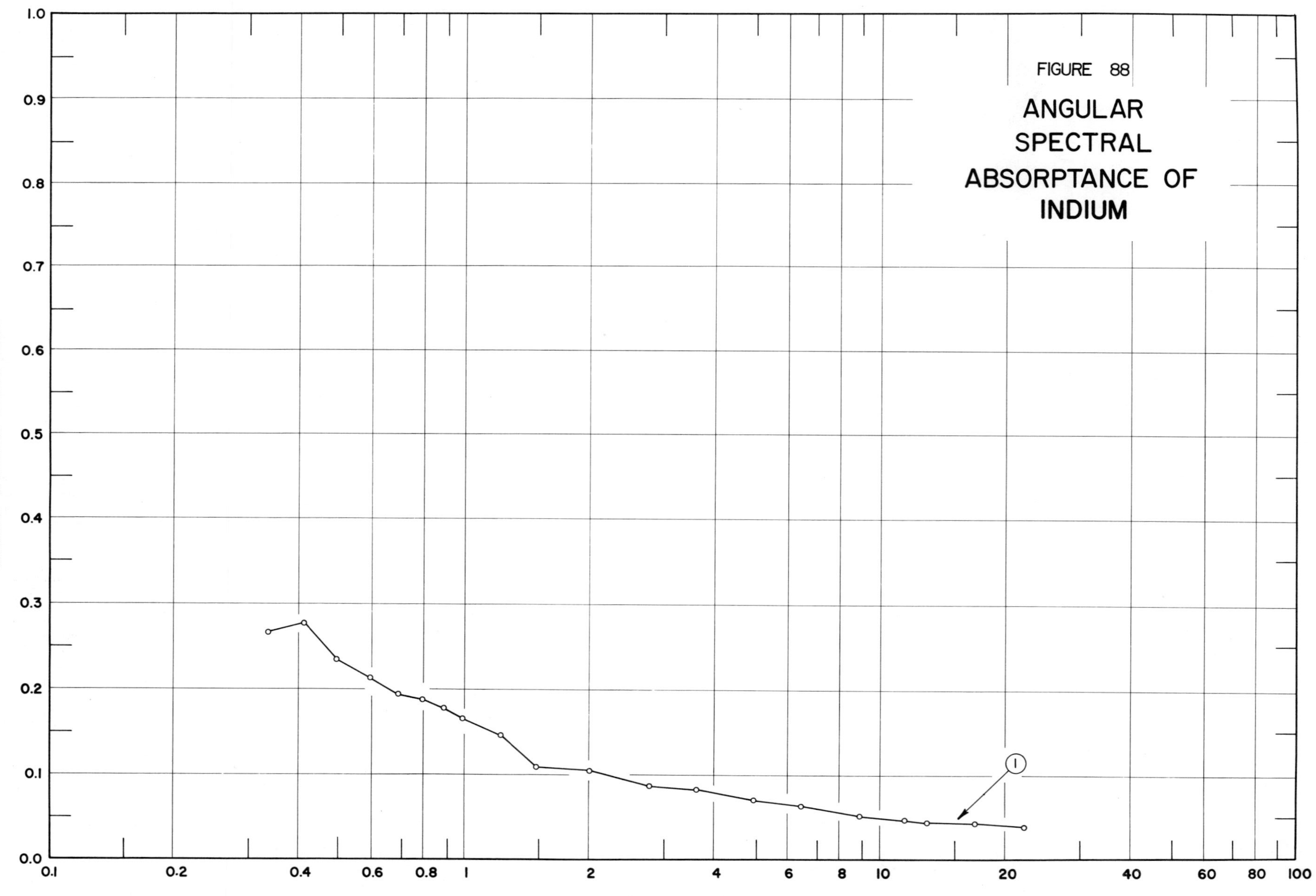

FIGURE 88
ANGULAR SPECTRAL ABSORPTANCE OF INDIUM

SPECIFICATION TABLE NO. 88 ANGULAR SPECTRAL ABSORPTANCE OF INDIUM

Curve No.	Ref. No.	Year	Temperature K	Wavelength Range, μ	Geometry θ'	Reported Error, %	Composition (weight percent), Specifications and Remarks
1	225	1965	306	0.338-22.3	25°		Indium from Belmont Smelting and Refining Works; scraped; measured in dry nitrogen; heated cavity at approx 1056 K with platinum reference; authors assumed $\alpha = 1 - R(2\pi, 25°)$.

DATA TABLE NO. 88 ANGULAR SPECTRAL ABSORPTANCE OF INDIUM

[Wavelength, λ, μ ; Absorptance, α ; Temperature, T, K]

λ	α
CURVE 1	
T = 306	
0.338	0.266
0.410	0.277
0.497	0.235
0.596	0.214
0.697	0.195
0.798	0.188
0.895	0.177
0.989	0.165
1.22	0.147
1.50	0.119
2.01	0.106
2.81	0.087
3.64	0.082
5.00	0.070
6.49	0.064
9.00	0.052
11.5	0.048
13.1	0.046
16.9	0.043
22.3	0.040

SPECIFICATION TABLE NO. 89 NORMAL SPECTRAL EMITTANCE OF IRIDIUM

Curve No.	Ref. No.	Year	Wavelength μ	Temperature Range, K	Geometry θ'	Reported Error, %	Composition (weight percent), Specifications and Remarks
1	19	1914	0.65	2023	~0°	1	Film; tungsten substrate; measured in hydrogen; Pt reference (ϵ = 0.33 for λ = 0.650 μ at all temp).
2	289	1966	0.65	1200-1595	~0°		Iridium; 0.012 Si, 0.0053 Fe, and 0.000048 Cu; highly polished.

DATA TABLE NO. 89 NORMAL SPECTRAL EMITTANCE OF IRIDIUM

[Temperature, T, K; Emittance, ϵ; Wavelength, λ, μ]

T	ϵ
CURVE 1* λ = 0.65	
2023	0.298
CURVE 2* λ = 0.65	
1200	0.366
1442	0.341
1587	0.329
1595	0.321

* Not shown on plot

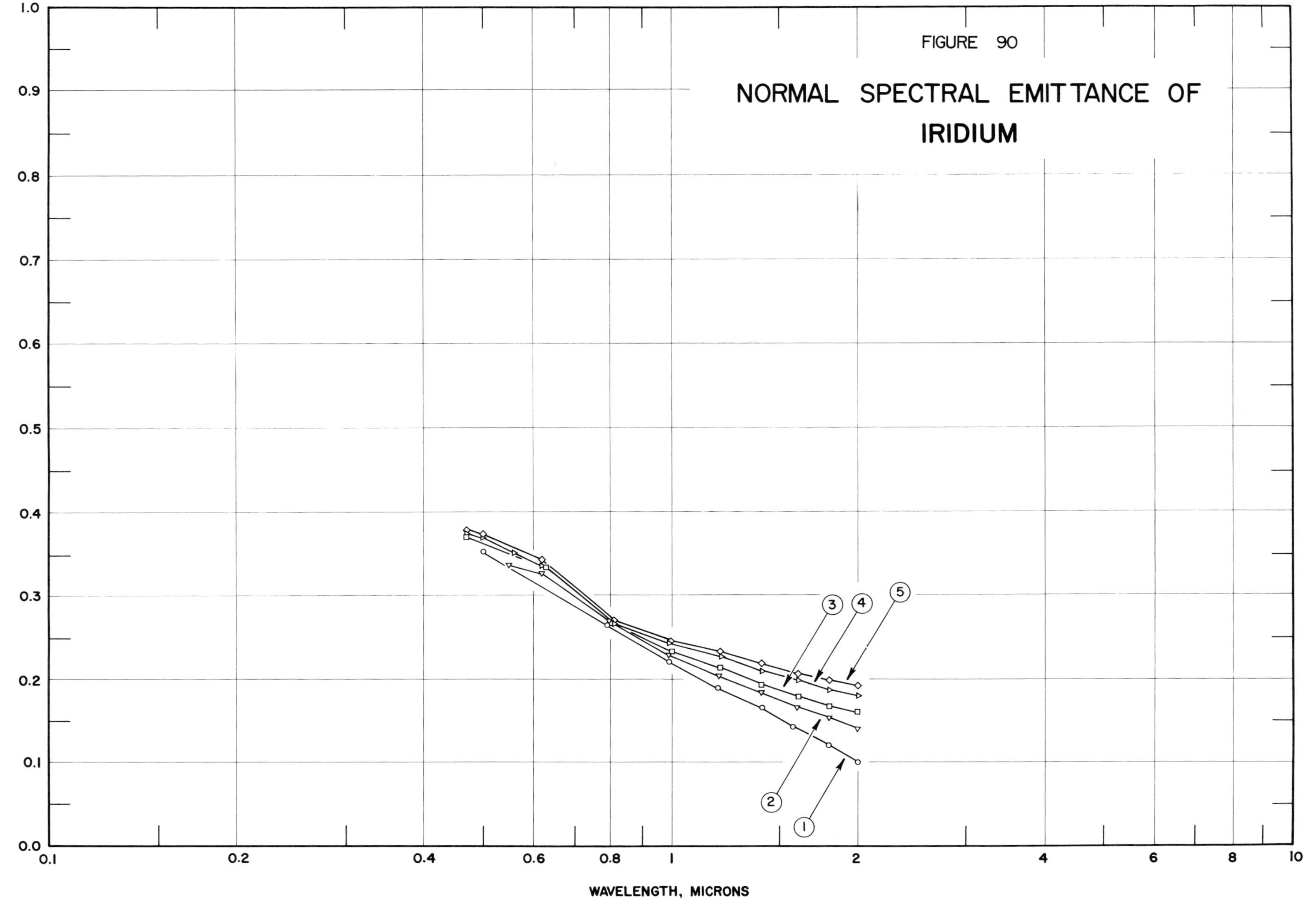
FIGURE 90
NORMAL SPECTRAL EMITTANCE OF
IRIDIUM
NORMAL SPECTRAL EMITTANCE
WAVELENGTH, MICRONS
1.0
0.9
0.8
0.7
0.6
0.5
0.4
0.3
0.2
0.1
0.0
0.1
0.2
0.4
0.6
0.8
1
2
4
6
8
10
1
2
3
4
5

SPECIFICATION TABLE NO. 90 NORMAL SPECTRAL EMITTANCE OF IRIDIUM

Curve No.	Ref. No.	Year	Temperature K	Wavelength Range, μ	Geometry θ'	Reported Error, %	Composition (weight percent), Specifications and Remarks
1	323	1966	300	0.50-2.00	0°		Filament (0.25-0.32 mm in dia); baked for 1 hr at 798 K in vacuum, cooled, heated for 5-10 min in vacuum, and cooled; measured in argon (600 mm Hg); data calculated from optical constants.
2	323	1966	1100	0.55-2.01	0°		Above specimen and conditions.
3	323	1966	1500	0.47-2.01	0°		Above specimen and conditions.
4	323	1966	2000	0.47-2.01	0°		Above specimen and conditions.
5	323	1966	2400	0.47-2.01	0°		Above specimen and conditions.

DATA TABLE NO. 90 NORMAL SPECTRAL EMITTANCE OF IRIDIUM

[Wavelength, λ, μ; Emittance, ϵ; Temperature, T, K]

λ	ϵ
CURVE 1 T = 300	
0.50	0.353
0.79	0.266
0.99	0.221
1.19	0.190
1.40	0.167
1.59	0.142
1.79	0.120
2.00	0.100
CURVE 2 T = 1100	
0.55	0.338
0.62	0.328
0.80	0.270
0.99	0.229
1.19	0.203
1.39	0.185
1.59	0.167
1.80	0.152
2.01	0.140
CURVE 3 T = 1500	
0.47	0.371
0.63	0.335
0.80	0.269
1.00	0.233
1.20	0.213
1.40	0.194
1.60	0.180
1.80	0.169
2.01	0.160
CURVE 4 T = 2000	
0.47	0.376
0.50	0.371
0.56	0.352
0.62	0.338
0.81	0.268
0.99	0.243
1.21	0.228

λ	ϵ
CURVE 4 (cont.)	
1.40	0.210
1.60	0.199
1.80	0.188
2.01	0.180
CURVE 5 T = 2400	
0.47	0.378
0.50	0.372
0.62	0.343
0.81	0.271
1.00	0.247
1.20	0.233
1.40	0.219
1.60	0.207
1.81	0.199
2.01	0.192

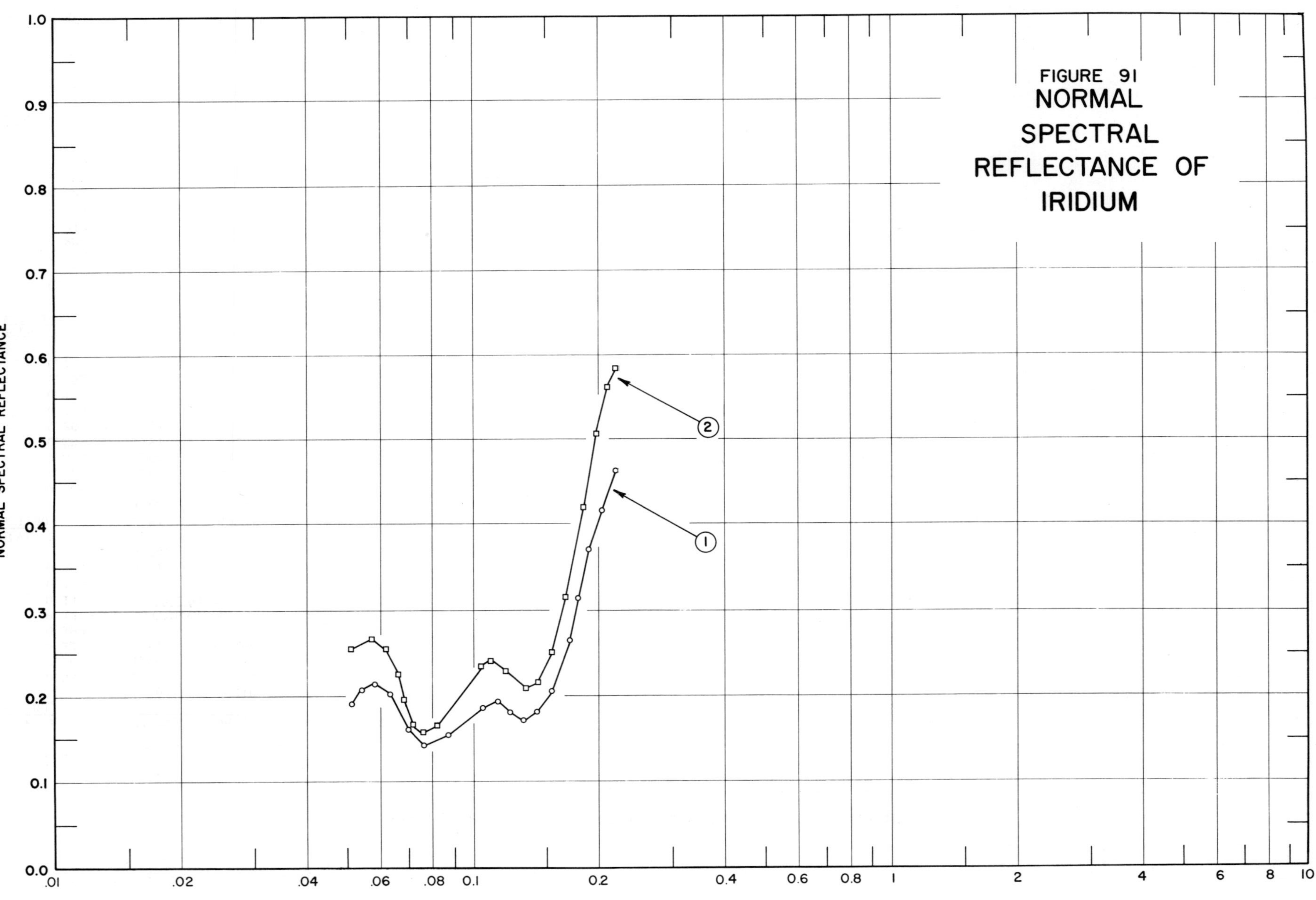
FIGURE 91
NORMAL
SPECTRAL
REFLECTANCE OF
IRIDIUM
NORMAL SPECTRAL REFLECTANCE
1.0
0.9
0.8
0.7
0.6
0.5
0.4
0.3
0.2
0.1
0.0
.01
.02
.04
.06
.08
0.1
0.2
0.4
0.6
0.8
1
2
4
6
8
10
WAVELENGTH, MICRONS
2
1

SPECIFICATION TABLE NO. 91 NORMAL SPECTRAL REFLECTANCE OF IRIDIUM

Curve No.	Ref. No.	Year	Temperature K	Wavelength Range, μ	Geometry θ	θ'	ω'	Reported Error, %	Composition (weight percent), Specifications and Remarks
1	322	1967	298	0.051-0.220	~0°	~0°			Opaque film, 99.98 pure; evaporated at 30 Å sec^{-1} in vacuum (1×10^{-5} mm Hg) on glass at at 313 K.
2	322	1967	298	0.051-0.220	~0°	~0°			Opaque film, 99.98 pure; evaporated at 30 Å sec^{-1} in vacuum (1×10^{-5} mm Hg) on glass at 573 K.

DATA TABLE NO. 91 NORMAL SPECTRAL REFLECTANCE OF IRIDIUM

[Wavelength, λ, μ; Reflectance, ρ; Temperature, T, K]

λ	ρ
CURVE 1 T = 298	
0.051	0.190
0.054	0.208
0.058	0.214
0.063	0.202
0.070	0.160
0.076	0.141
0.087	0.153
0.105	0.185
0.114	0.192
0.123	0.180
0.132	0.171
0.142	0.181
0.155	0.206
0.170	0.264
0.178	0.315
0.189	0.370
0.203	0.417
0.220	0.462
CURVE 2 T = 298	
0.051	0.254
0.057	0.267
0.062	0.254
0.066	0.226
0.068	0.196
0.072	0.167
0.076	0.158
0.082	0.164
0.104	0.234
0.110	0.240
0.120	0.229
0.133	0.210
0.143	0.217
0.155	0.251
0.167	0.317
0.183	0.420
0.198	0.506
0.211	0.561
0.220	0.583

SPECIFICATION TABLE NO. 92 ANGULAR SPECTRAL REFLECTANCE OF IRIDIUM

Curve No.	Ref. No.	Year	Temperature K	Wavelength Range, μ	Geometry θ θ' ω'	Reported Error, %	Composition (weight percent), Specifications and Remarks
1	132	1911	298	1.03-9.30	15° 15°	≤ 3	Polished; silvered glass mirror reference.

DATA TABLE NO. 92 ANGULAR SPECTRAL REFLECTANCE OF IRIDIUM

[Wavelength, λ, μ; Reflectance, ρ; Temperature, T, K]

λ	ρ
CURVE 1* T = 298	
1.03	0 795
1.80	0.845
3.03	0.907
3.97	0.927
5.23	0.947
6.42	0.950
7.10	0.950
9.30	0.955

*Not shown on plot

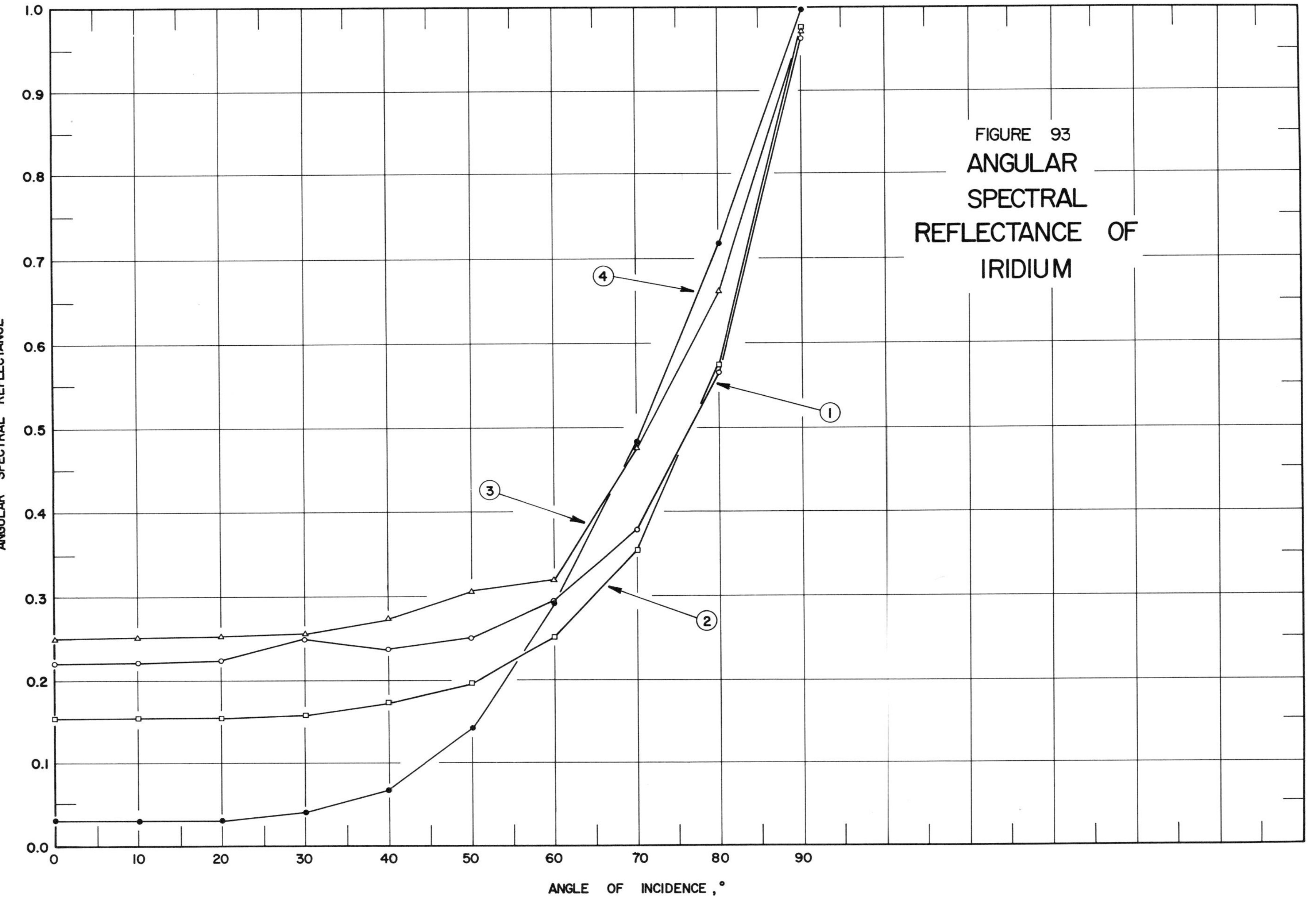

FIGURE 93 ANGULAR SPECTRAL REFLECTANCE OF IRIDIUM

SPECIFICATION TABLE NO. 93 ANGULAR SPECTRAL REFLECTANCE OF IRIDIUM

Curve No.	Ref. No.	Year	Temperature K	Wavelength, μ	Angular Range	Geometry θ θ' ω'	Reported Error, %	Composition (weight percent), Specifications and Remarks
1	322	1967	298	0.1216	0-90	θ θ'		Opaque film, 99.98 pure; evaporated at 30 Å sec^{-1} in vacuum (1 x 10^{-5} mm Hg) on glass at 573 K; specular reflection, $\theta = \theta'$; calculated from optical constants.
2	322	1967	298	0.0736	0-90	θ θ'		Above specimen and conditions.
3	322	1967	298	0.0584	0-90	θ θ'		Above specimen and conditions.
4	322	1967	298	0.0304	0-90	θ θ'		Above specimen and conditions.

DATA TABLE NO. 93 ANGULAR SPECTRAL REFLECTANCE OF IRIDIUM

[Angle, θ, °; Reflectance, ρ; Temperature, T, K; Wavelength, λ, μ]

θ	ρ	θ	ρ
CURVE 1*		CURVE 4*	
$\lambda = 0.1216$		$\lambda = 0.0304$	
T = 298		T = 298	
0	0.220	0	0.030
10	0.220	10	0.030
20	0.222	20	0.030
30	0.225	30	0.040
40	0.236	40	0.065
50	0.250	50	0.141
60	0.293	60	0.291
70	0.378	70	0.482
80	0.565	80	0.718
90	0.962	90	0.996
CURVE 2*			
$\lambda = 0.0736$			
T = 298			
0	0.152		
10	0.152		
20	0.152		
30	0.158		
40	0.171		
50	0.196		
60	0.251		
70	0.355		
80	0.574		
90	0.975		
CURVE 3*			
$\lambda = 0.0584$			
T = 298			
0	0.250		
10	0.251		
20	0.252		
30	0.257		
40	0.274		
50	0.305		
60	0.369		
70	0.476		
80	0.661		
90	0.971		

*Not shown on plot

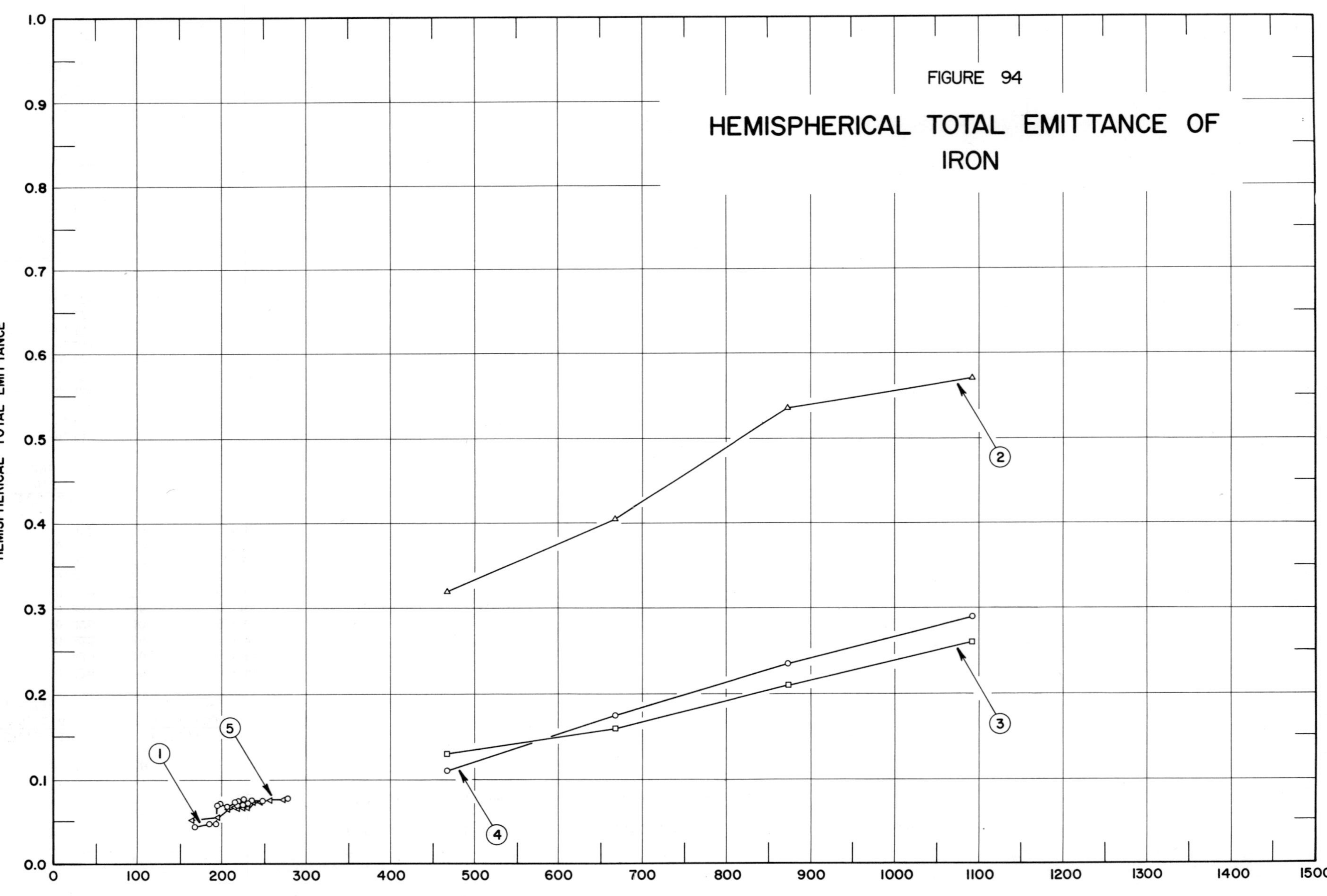
FIGURE 94
HEMISPHERICAL TOTAL EMITTANCE OF
IRON
HEMISPHERICAL TOTAL EMITTANCE
1.0
0.9
0.8
0.7
0.6
0.5
0.4
0.3
0.2
0.1
0.0
0
100
200
300
400
500
600
700
800
900
1000
1100
1200
1300
1400
1500
TEMPERATURE, K
1
2
3
4
5

SPECIFICATION TABLE NO. 94 HEMISPHERICAL TOTAL EMITTANCE OF IRON

Curve No.	Ref. No.	Year	Temperature Range, K	Reported Error, %	Composition (weight percent), Specifications and Remarks
1	40	1962	168-278		Pure; polished; measured in air.
2	47	1961	468-1093	< 10	Armco ingot iron; 0.11 Cu, 0.032 Mn, 0.017 S, 0.014 C, 0.003 P, 0.003 Si; disc (0.04 in. thick); polished; etched by sandblasting with a jet of SS White Abrasive No. 2; measured in vacuum (10^{-5} mm Hg).
3	47	1961	468-1093	< 10	Different sample, same as curve 2 specimen and conditions except ground with 600 grit Carborundum and polished on a wet cloth lap with unlevigated jewelers rouge.
4	47	1961	468-1093	< 10	Different sample, same as curve 2 specimen and conditions except blued by immersing in a niter bath.
5	324	1960	169-272	7	Sample wiped with "00" sandpaper, polished with crocus cloth and jeweler's rouge, then cleaned with cotton saturated with acetone; measured in vacuum (10^{-6} mm Hg); computed from cylindrical fin temperature distribution.

DATA TABLE NO. 94 HEMISPHERICAL TOTAL EMITTANCE OF IRON

[Temperature, T, K; Emittance, ϵ]

T	ϵ	T	ϵ
CURVE 1		**CURVE 5**	
168	0.045	169	0.051
185	0.048	195	0.054
194	0.048	207	0.066
195	0.069	214	0.068
199	0.071	219	0.066
204	0.067*	225	0.067
207	0.068	231	0.068
208	0.069*	236	0.073
212	0.070*	237	0.073*
214	0.070*	245	0.073
215	0.074	255	0.075
215	0.074*	272	0.077
219	0.069		
220	0.075		
225	0.070		
226	0.077		
231	0.072		
235	0.075		
236	0.074*		
236	0.076		
244	0.075*		
249	0.075		
278	0.078		
CURVE 2			
468	0.320		
668	0.405		
873	0.535		
1093	0.570		
CURVE 3			
468	0.13		
668	0.16		
873	0.21		
1093	0.26		
CURVE 4			
468	0.110		
668	0.175		
873	0.235		
1093	0.290		

*Not shown on plot

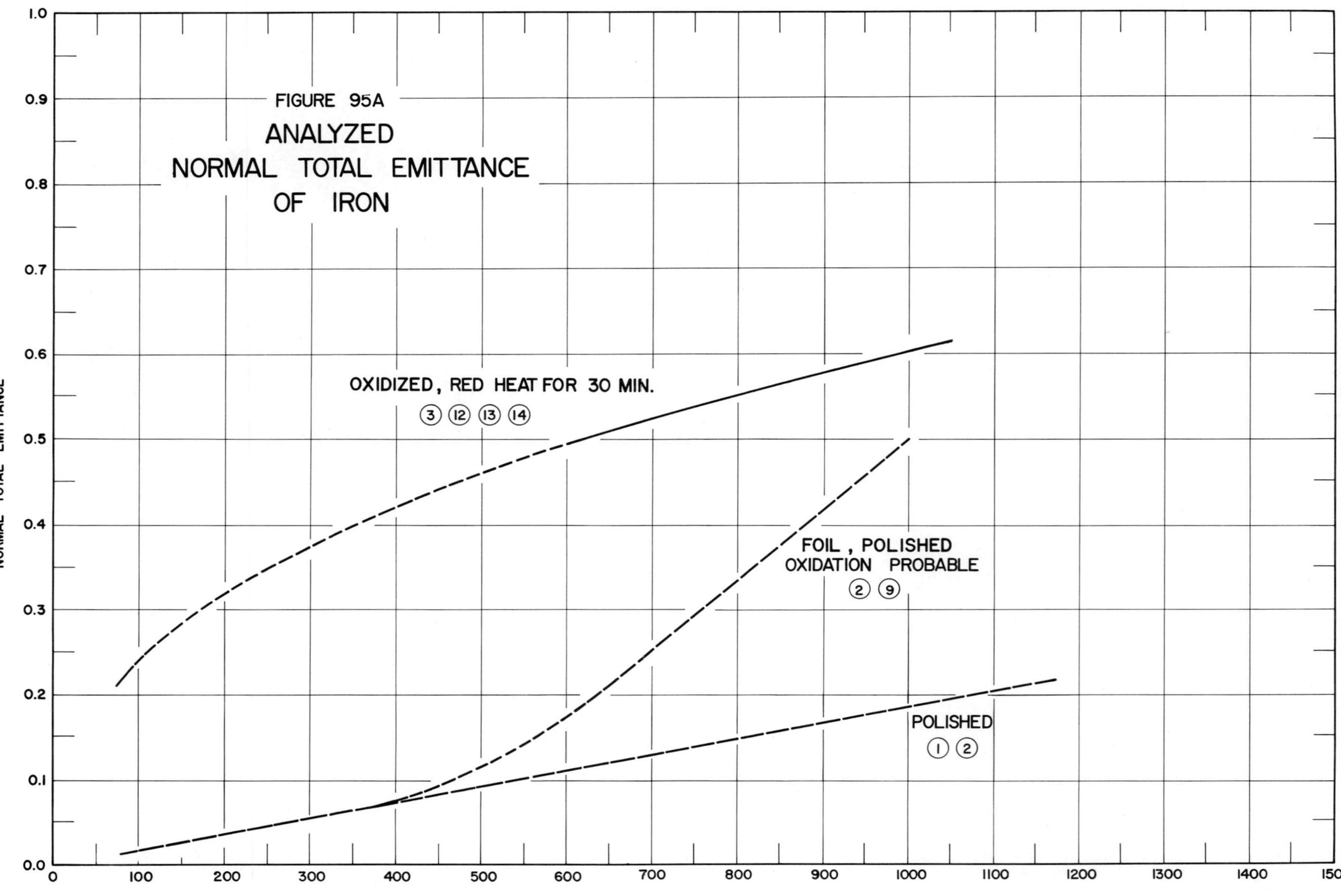
FIGURE 95A
ANALYZED
NORMAL TOTAL EMITTANCE
OF IRON
OXIDIZED, RED HEAT FOR 30 MIN.
③ ⑫ ⑬ ⑭
FOIL, POLISHED
OXIDATION PROBABLE
② ⑨
POLISHED
① ②
NORMAL TOTAL EMITTANCE
TEMPERATURE, K
1.0
0.9
0.8
0.7
0.6
0.5
0.4
0.3
0.2
0.1
0.0
0
100
200
300
400
500
600
700
800
900
1000
1100
1200
1300
1400
1500

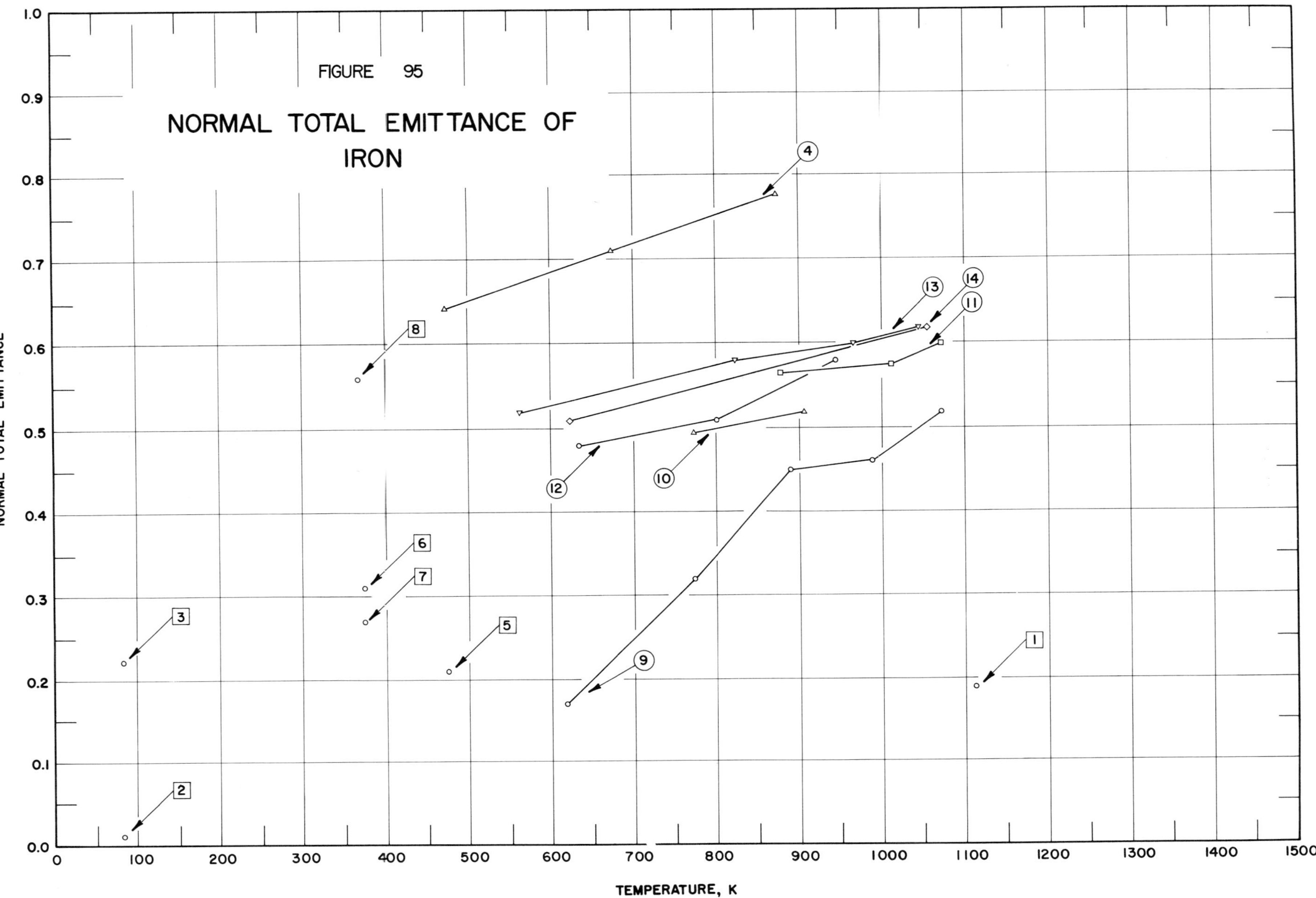

FIGURE 95
NORMAL TOTAL EMITTANCE OF IRON
NORMAL TOTAL EMITTANCE
TEMPERATURE, K
0.0
0.1
0.2
0.3
0.4
0.5
0.6
0.7
0.8
0.9
1.0
0
100
200
300
400
500
600
700
800
900
1000
1100
1200
1300
1400
1500
1
2
3
4
5
6
7
8
9
10
11
12
13
14

SPECIFICATION TABLE NO. 95 NORMAL TOTAL EMITTANCE OF IRON

Curve No.	Ref. No.	Year	Temperature Range, K	Geometry θ'	Reported Error, %	Composition (weight percent), Specifications and Remarks
1	30	1963	1111	$\sim 0^\circ$		Mechanically polished with aluminum oxide and cleaned with water; heated at 1058 K for 2 1/2 hrs; measured in argon; computed from spectral emittance.
2	34	1957	83.2	$\sim 0^\circ$	±10	Armco Ingot iron; nominal composition: 0.03 max C, Fe balance; strip (0.005 in. thick); finish having an RMS of about 2 microinches; measured in air (5×10^{-4} mm Hg).
3	34	1957	83.2	$\sim 0^\circ$	±10	Different sample, same as curve 2 specimen; same conditions except oxidized in air at red heat for 30 min.
4	14	1913	473-873	$\sim 0^\circ$		Cast iron; cleaned, polished and oxidized.
5	14	1913	473	$\sim 0^\circ$		Cast iron; cleaned and polished.
6	15	1947	373	$\sim 0^\circ$		Dark gray surface.
7	15	1947	373	$\sim 0^\circ$		Roughly polished.
8	16	1937	367	$\sim 0^\circ$	± 1.1	Black.
9	34	1957	616-1072	$\sim 0^\circ$	±10	Armco Ingot iron; nominal composition: 0.03 max C, Fe balance; strip (0.005 in. thick); finish having an RMS of about 2 microinches; measured in air (5×10^{-4} mm Hg); increasing temp, cycle 1.
10	34	1957	772-905	$\sim 0^\circ$	±10	Above specimen and conditions; cycle 2.
11	34	1957	878-1072	$\sim 0^\circ$	±10	Above specimen and conditions; cycle 3.
12	34	1957	633-944	$\sim 0^\circ$	±10	Armco Ingot iron; nominal composition: 0.03 max C, Fe balance; strip (0.005 in. thick); finish having an RMS of about 2 microinches; oxidized in air at red heat for 30 min; measured in air (5×10^{-4} mmHg); increasing temp, cycle 1.
13	34	1957	561-1044	$\sim 0^\circ$	±10	Above specimen and conditions; cycle 2.
14	34	1957	622-1055	$\sim 0^\circ$	±10	Above specimen and conditions; cycle 3.

DATA TABLE NO. 95 NORMAL TOTAL EMITTANCE OF IRON

[Temperature, T, K; Emittance ∈]

T	∈
CURVE 1	
1111	0.19
CURVE 2	
83.2	0.01
CURVE 3	
83.2	0.222
CURVE 4	
473	0.643
673	0.710
873	0.777
CURVE 5	
473	0.210
CURVE 6	
373	0.31
CURVE 7	
373	0.27
CURVE 8	
367	0.56
CURVE 9	
616	0.17
772	0.32
889	0.45
989	0.46
1072	0.52
CURVE 10	
772	0.495
905	0.52
CURVE 11	
878	0.565
1011	0.575
1072	0.60
CURVE 12	
633	0.48
800	0.51
944	0.58
CURVE 13	
561	0.52
822	0.58
966	0.60
1044	0.62
CURVE 14	
622	0.51
1055	0.62

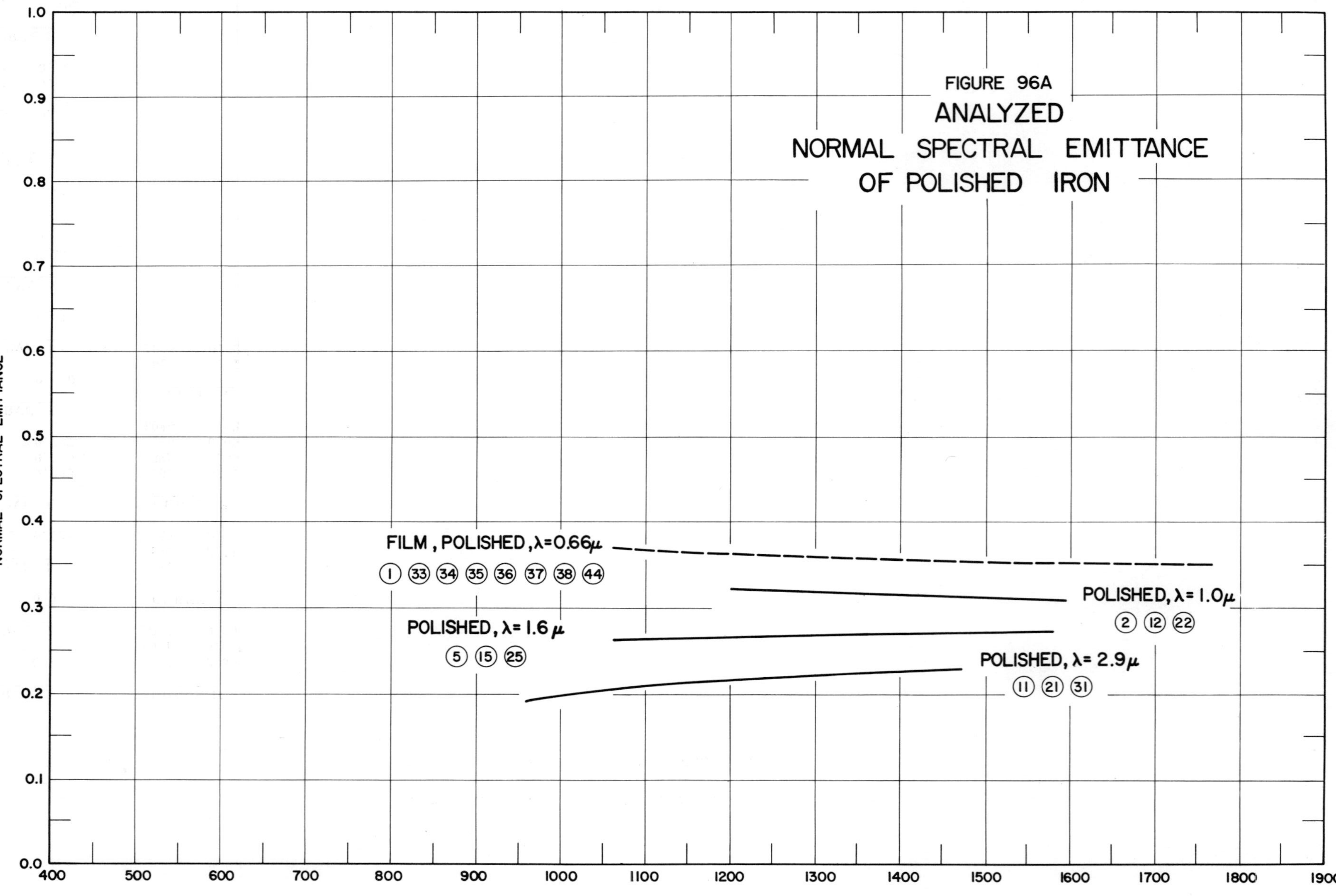

FIGURE 96A
ANALYZED
NORMAL SPECTRAL EMITTANCE
OF POLISHED IRON

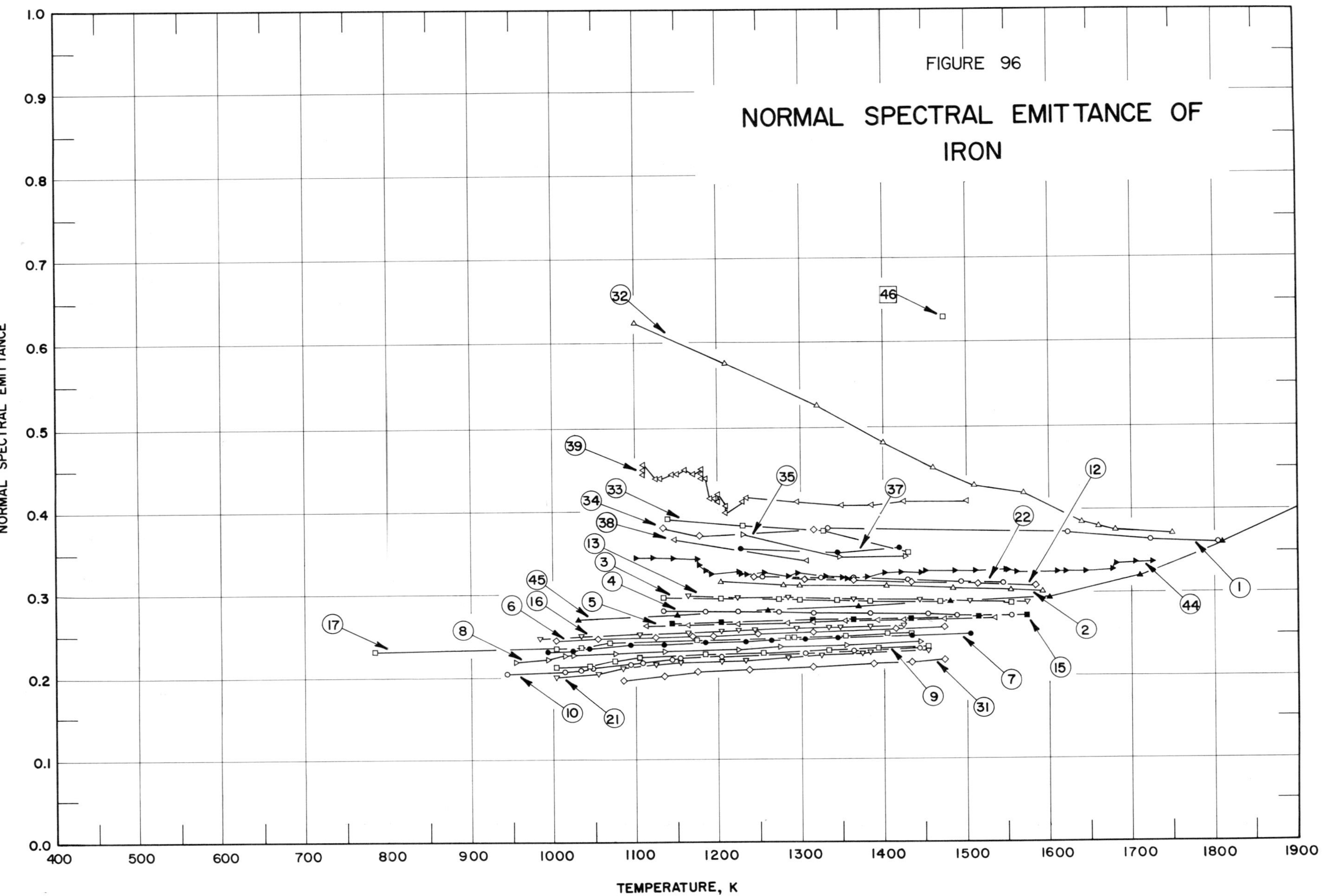

FIGURE 96
NORMAL SPECTRAL EMITTANCE OF IRON

SPECIFICATION TABLE NO. 96 NORMAL SPECTRAL EMITTANCE OF IRON

Curve No.	Ref. No.	Year	Wavelength μ	Temperature Range, K	Geometry θ'	Reported Error, %	Composition (weight percent), Specifications and Remarks
1	19	1914	0.65	1333-1803	$\sim 0^\circ$	1	Film; tungsten substrate; measured in hydrogen; Pt reference (ϵ = 0.33 for λ = 0.650 μ at all temp).
2	41	1956	1.0	1203-1593	$\sim 0^\circ$	8	Polished with emery paper and lapped with jeweler's rouge; measured in vacuum; Specimen A.
3	41	1956	1.2	1133-1553	$\sim 0^\circ$	8	Above specimen and conditions.
4	41	1956	1.4	1133-1553	$\sim 0^\circ$	8	Above specimen and conditions.
5	41	1956	1.6	1113-1533	$\sim 0^\circ$	8	Above specimen and conditions.
6	41	1956	1.8	1003-1473	$\sim 0^\circ$	8	Above specimen and conditions.
7	41	1956	2.0	993-1503	$\sim 0^\circ$	8	Above specimen and conditions.
8	41	1956	2.2	953-1443	$\sim 0^\circ$	8	Above specimen and conditions.
9	41	1956	2.4	1003-1453	$\sim 0^\circ$	8	Above specimen and conditions.
10	41	1956	2.7	943-1443	$\sim 0^\circ$	8	Above specimen and conditions.
11	41	1956	2.9	943-1403	$\sim 0^\circ$	8	Above specimen and conditions.
12	41	1956	1.0	1243-1583	$\sim 0^\circ$	8	Different sample, same as curve 2 specimen and conditions.
13	41	1956	1.2	1163-1573	$\sim 0^\circ$	8	Above specimen and conditions.
14	41	1956	1.4	1133-1483	$\sim 0^\circ$	8	Above specimen and conditions.
15	41	1956	1.6	1143-1573	$\sim 0^\circ$	8	Above specimen and conditions.
16	41	1956	1.8	983-1423	$\sim 0^\circ$	8	Above specimen and conditions.
17	41	1956	2.0	783-1433	$\sim 0^\circ$	8	Above specimen and conditions.
18	41	1956	2.2	923-1423	$\sim 0^\circ$	8	Above specimen and conditions.
19	41	1956	2.4	973-1463	$\sim 0^\circ$	8	Above specimen and conditions.
20	41	1956	2.7	943-1513	$\sim 0^\circ$	8	Above specimen and conditions.
21	41	1956	2.9	1003-1453	$\sim 0^\circ$	8	Above specimen and conditions.
22	41	1956	1.0	1253-1543	$\sim 0^\circ$	8	Different sample, same as curve 2 specimen and conditions.
23	41	1956	1.2	1133-1543	$\sim 0^\circ$	8	Above specimen and conditions.
24	41	1956	1.4	1143-1513	$\sim 0^\circ$	8	Above specimen and conditions.
25	41	1956	1.6	1068-1538	$\sim 0^\circ$	8	Above specimen and conditions.
26	41	1956	1.8	973-1453	$\sim 0^\circ$	8	Above specimen and conditions.
27	41	1956	2.0	1003-1503	$\sim 0^\circ$	8	Above specimen and conditions.

SPECIFICATION TABLE NO. 96 (continued)

Curve No.	Ref. No.	Year	Wavelength μ	Temperature Range, K	Geometry θ'	Reported Error, %	Composition (weight percent), Specifications and Remarks
28	41	1956	2.2	973-1483	$\sim 0^\circ$	8	Above specimen and conditions.
29	41	1956	2.4	983-1473	$\sim 0^\circ$	8	Above specimen and conditions.
30	41	1956	2.65	1023-1483	$\sim 0^\circ$	8	Above specimen and conditions.
31	41	1956	2.9	1083-1473	$\sim 0^\circ$	8	Above specimen and conditions.
32	53	1913	0.65	1100-1750	$\sim 0^\circ$		Measured in nitrogen and hydrogen.
33	34	1957	0.665	1139-1430	$\sim 0^\circ$		Nominal composition: 0.03 max C, Fe balance; finish having an RMS of 15 or 2 micro-inches; measured in air (5×10^{-4} mm Hg); increasing temp, cycle 1.
34	34	1957	0.665	1316-1133	$\sim 0^\circ$		Above specimen and conditions; decreasing temp, cycle 1.
35	34	1957	0.665	1230-1425	$\sim 0^\circ$		Above specimen and conditions; increasing temp, cycle 2.
36	34	1957	0.665	1277-1139	$\sim 0^\circ$		Above specimen and conditions; decreasing temp, cycle 2.
37	34	1957	0.665	1227-1419	$\sim 0^\circ$		Above specimen and conditions; increasing temp, cycle 3.
38	34	1957	0.665	1307-1147	$\sim 0^\circ$		Above specimen and conditions; decreasing temp, cycle 3.
39	42	1942	0.667	1110-1500	$\sim 0^\circ$		Baked at 623 K for 200 hrs; measured in vacuum (4×10^{-8} mm Hg).
40	43	1952	1.4	1273-1533	$\sim 0^\circ$		Polished; washed with ether; measured in vacuum.
41	43	1952	1.6	1233-1533	$\sim 0^\circ$		Above specimen and conditions.
42	43	1952	2.0	1263-1523	$\sim 0^\circ$		Above specimen and conditions.
43	43	1952	2.4	1273-1528	$\sim 0^\circ$		Above specimen and conditions.
44	44	1948	0.667	1100-1725	$\sim 0^\circ$		Electrolytically deposited on a stainless steel strip from a ferrous chloride bath; annealed in hydrogen; rolled with frequent annealings; measured in hydrogen (15 cm).
45	35	1914	0.66	1030-2090	$\sim 0^\circ$		Measured in burning hydrogen.
46	19	1914	0.65	1473	$\sim 0^\circ$	1	Film; tungsten substrate; melted in hydrogen, then oxidized in air by heating; measured in air; Pt reference (ϵ = 0.33 for λ = 0.650 μ at all temp).

DATA TABLE NO. 96 NORMAL SPECTRAL EMITTANCE OF IRON

[Temperature, T, K; Emittance, ∈; Wavelength, λ, μ]

T	∈
CURVE 1	
λ = 0.65	
1333	0.379
1623	0.372
1723	0.363
1803	0.360
CURVE 2	
λ = 1.0	
1203	0.314
1278	0.312
1298	0.309
1403	0.308
1483	0.305
1553	0.303
1593	0.301
CURVE 3	
λ = 1.2	
1133	0.295
1203	0.294
1273	0.293
1298	0.292
1343	0.291
1383	0.290
1469	0.289
1553	0.287
CURVE 4	
λ = 1.4	
1133	0.280
1183	0.279
1223	0.279
1273	0.278
1313	0.277
1383	0.276
1453	0.275
1488	0.274
1553	0.273
CURVE 5	
λ = 1.6	
1113	0.262
1163	0.263
1223	0.264
1283	0.265
1313	0.266
1353	0.267
1393	0.268
1423	0.268
1473	0.269
1533	0.270
CURVE 6	
λ = 1.8	
1003	0.244
1053	0.247
1123	0.248
1168	0.250
1193	0.250
1298	0.252
1313	0.254
1413	0.257
1473	0.259
CURVE 7	
λ = 2.0	
993	0.230
1023	0.231
1043	0.234
1093	0.238
1133	0.238
1183	0.241
1233	0.242
1263	0.243
1303	0.245
1383	0.247
1433	0.249
1503	0.251
CURVE 8	
λ = 2.2	
953	0.218
993	0.221
1013	0.225
1023	0.226
1073	0.228
1133	0.230
1223	0.233
1273	0.235
1353	0.238
1443	0.241
CURVE 9	
λ = 2.4	
1003	0.212
1043	0.213
1073	0.219
1103	0.223
1183	0.226
1253	0.227
1333	0.231
1393	0.234
1453	0.235
CURVE 10	
λ = 2.7	
943	0.203
1013	0.206
1033	0.208
1048	0.210
1093	0.215
1108	0.217
1143	0.220
1153	0.222
1203	0.224
1303	0.227
1363	0.230
1443	0.233
CURVE 11*	
λ = 2.9	
943	0.198
943	0.200
983	0.202
1013	0.207
1043	0.210
1073	0.214
1123	0.218
1183	0.220
1253	0.223
1333	0.226
1403	0.229
CURVE 12	
λ = 1.0	
1243	0.320
1303	0.317
1363	0.315
1433	0.313
1513	0.311
1583	0.308
CURVE 13	
λ = 1.2	
1163	0.298
1223	0.295
1283	0.295
1363	0.293
1443	0.292
1503	0.290
1573	0.288
CURVE 14*	
λ = 1.4	
1133	0.283
1223	0.282
1313	0.279
1383	0.278
1483	0.276
CURVE 15	
λ = 1.6	
1143	0.265
1203	0.266
1313	0.268
1363	0.269
1443	0.270
1513	0.272
1573	0.273
CURVE 16	
λ = 1.8	
983	0.247
1033	0.250
1103	0.252
1163	0.254
1203	0.255
1223	0.256
1243	0.256
1293	0.258
1333	0.259
1348	0.259
1383	0.260
1423	0.261
CURVE 17	
λ = 2.0	
783	0.231
1003	0.234
1033	0.235
1068	0.241
1173	0.244
1283	0.247
1290	0.248
1353	0.249
1403	0.251
1433	0.252
CURVE 18*	
λ = 2.2	
923	0.227
1033	0.232
1083	0.235
1133	0.237
1178	0.238
1233	0.239
1273	0.241
1313	0.242
1373	0.244
1423	0.245
CURVE 19*	
λ = 2.4	
973	0.216
1033	0.221
1093	0.224
1133	0.226
1248	0.231
1293	0.233
1353	0.236
1433	0.239
1463	0.241
CURVE 20*	
λ = 2.7	
943	0.202
973	0.203
993	0.204
1033	0.208
1078	0.214
1123	0.216
1173	0.218
1203	0.219
1253	0.221
1313	0.225
1318	0.224
1393	0.228
1463	0.231
1513	0.233
CURVE 21	
λ = 2.9	
1003	0.200
1053	0.204
1083	0.210
1123	0.215
1153	0.216
1203	0.218
1233	0.219
1283	0.222
1323	0.224
1373	0.226
1383	0.227
1453	0.230
CURVE 22	
λ = 1.0	
1253	0.321
1323	0.319
1363	0.318
1428	0.315
1493	0.313
1543	0.311
CURVE 23*	
λ = 1.2	
1133	0.300
1223	0.298
1273	0.297
1328	0.295
1403	0.294
1483	0.292
1543	0.291
CURVE 24*	
λ = 1.4	
1143	0.282
1243	0.281
1298	0.279
1398	0.278
CURVE 24 (cont.)*	
1443	0.277
1513	0.276
CURVE 25*	
λ = 1.6	
1068	0.266
1143	0.267
1223	0.268
1293	0.269
1363	0.270
1433	0.271
1493	0.272
1538	0.273
CURVE 26*	
λ = 1.8	
973	0.248
1033	0.251
1113	0.255
1173	0.256
1233	0.258
1273	0.259
1353	0.261
1403	0.262
1453	0.264
CURVE 27*	
λ = 2.0	
1003	0.228
1058	0.232
1103	0.241
1143	0.244
1248	0.247
1323	0.250
1373	0.251
1448	0.253
1503	0.254

*Not shown on plot

DATA TABLE NO. 96 (continued)

T	∈
CURVE 28*	
λ = 2.2	
973	0.218
1033	0.224
1073	0.230
1113	0.233
1133	0.234
1243	0.237
1243	0.237
1313	0.240
1363	0.242
1403	0.243
1433	0.244
1483	0.246
CURVE 29*	
λ = 2.4	
983	0.211
1023	0.213
1073	0.221
1103	0.224
1133	0.225
1253	0.230
1303	0.232
1303	0.231
1373	0.234
1423	0.236
1473	0.239
CURVE 30*	
λ = 2.65	
1023	0.215
1103	0.218
1158	0.221
1178	0.222
1243	0.224
1283	0.226
1353	0.229
1413	0.231
1483	0.234
CURVE 31	
λ = 2.9	
1083	0.196
1133	0.201
1173	0.206
1238	0.208
1313	0.211
1388	0.214
1433	0.217
1473	0.219
CURVE 32	
λ = 0.65	
1100	0.625
1210	0.575
1320	0.525
1400	0.480
1460	0.450
1510	0.430
1570	0.420
1640	0.385
1660	0.380
1680	0.375
1750	0.370
CURVE 33	
λ = 0.665	
1139	0.390
1230	0.382
1327	0.375
1430	0.350
CURVE 34	
λ = 0.665	
1316	0.376
1177	0.370
1133	0.380
CURVE 35	
λ = 0.665	
1230	0.372
CURVE 35 (cont.)	
1347	0.345
1425	0.345
CURVE 36*	
λ = 0.665	
1277	0.356
1139	0.360
CURVE 37	
λ = 0.665	
1227	0.355
1344	0.350
1419	0.355
CURVE 38	
λ = 0.665	
1307	0.340
1147	0.366
CURVE 39	
λ = 0.667	
1110	0.445
1110	0.450
1110	0.455
1125	0.440
1130	0.440
1145	0.445
1150	0.445
1160	0.450
1170	0.445
1180	0.450
1180	0.445
1180	0.440
1185	0.440
1190	0.415
1195	0.415
1200	0.410
1200	0.420
1210	0.405
CURVE 39 (cont.)	
1210	0.397
1230	0.410
1235	0.415
1295	0.410
1350	0.405
1385	0.405
1425	0.410
1500	0.410
CURVE 40*	
λ = 1.4	
1273	0.257
1338	0.256
1378	0.255
1448	0.255
1533	0.254
CURVE 41*	
λ = 1.6	
1233	0.245
1343	0.245
1373	0.246
1433	0.246
1533	0.247
CURVE 42*	
λ = 2.0	
1263	0.229
1373	0.233
1443	0.233
1523	0.235
CURVE 43*	
λ = 2.4	
1273	0.220
1333	0.221
1353	0.223
1433	0.225
1528	0.226
CURVE 44	
λ = 0.667	
1100	0.345
1120	0.345
1145	0.344
1173	0.343
1175	0.336
1184	0.329
1190	0.324
1225	0.326
1228	0.324
1234	0.324
1255	0.325
1284	0.321
1295	0.325
1330	0.321
1350	0.319
1355	0.317
1380	0.319
1400	0.324
1420	0.325
1446	0.324
1450	0.327
1485	0.327
1510	0.327
1548	0.328
1550	0.327
1560	0.325
1610	0.326
1620	0.326
1645	0.326
1677	0.328
1680	0.336
1705	0.337
1725	0.338
CURVE 45	
λ = 0.66	
1030	0.270
1150	0.275
1260	0.280
1370	0.285
1480	0.290
1600	0.295
CURVE 45 (cont.)	
1710	0.320
1810	0.360
1905	0.410*
2000	0.480*
2090	0.530*
CURVE 46	
λ = 0.65	
1473	0.63

*Not shown on plot

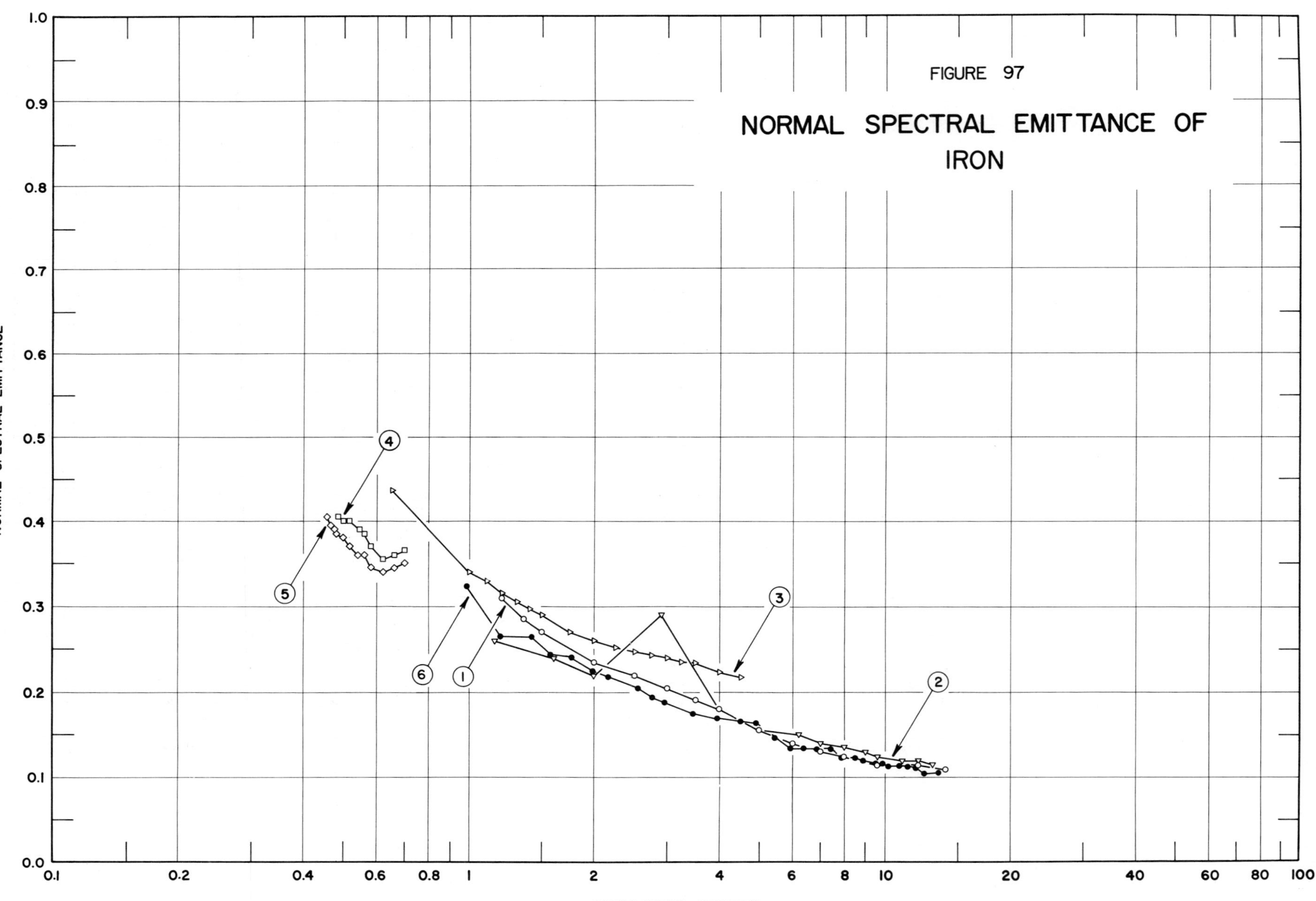
FIGURE 97
NORMAL SPECTRAL EMITTANCE OF IRON
NORMAL SPECTRAL EMITTANCE
1.0
0.9
0.8
0.7
0.6
0.5
0.4
0.3
0.2
0.1
0.0
0.1
0.2
0.4
0.6
0.8
1
2
4
6
8
10
20
40
60
80
100
WAVELENGTH, MICRONS
1
2
3
4
5
6

SPECIFICATION TABLE NO. 97 NORMAL SPECTRAL EMITTANCE OF IRON

Curve No.	Ref. No.	Year	Temperature K	Wavelength Range, μ	Geometry θ'	Reported Error, %	Composition (weight percent), Specifications and Remarks
1	30	1963	1078	1.20 -14.00	$\sim 0^\circ$		Mechanically polished with aluminum oxide and cleaned with water; heated at 1058 K for 2 1/2 hrs; measured in argon; data extracted from smooth curve; oxidization of surface was avoided.
2	30	1963	1321	1.15 -13.00	$\sim 0^\circ$		Above specimen and conditions except heated at 1316 K for 5 hrs.
3	37	1947	1518	0.65 - 4.5	$\sim 0^\circ$		99.96 pure; heated in hydrogen at 1473 K for 12 hrs; measured in vacuum.
4	52	1952	1147	0.700 - 0.485	$\sim 0^\circ$		99.42 pure; sintered 48 hrs in hydrogen at 1273 K; cold rolled and annealed; measured in dry hydrogen; alpha phase.
5	52	1952	1288	0.700 - 0.457	$\sim 0^\circ$		Different sample, same as curve 4 specimen; same conditions except gamma phase.
6	245	1965	1122	0.986-13.4	$\sim 0^\circ$	5	Measured in argon.

DATA TABLE NO. 97 NORMAL SPECTRAL EMITTANCE OF IRON

[Wavelength, λ, μ; Emittance, ∈; Temperature, T, K]

λ	∈	λ	∈	λ	∈
CURVE 1 T = 1078		CURVE 3 (cont.)		CURVE 6 T=1122	
		1.4	0.298		
1.20	0.310	1.5	0.290	0.986	0.324
1.35	0.285	1.75	0.270	1.18	0.264
1.50	0.270	2.0	0.260	1.41	0.264
2.00	0.235	2.25	0.252	1.57	0.244
2.50	0.220	2.5	0.248	1.77	0.240
3.00	0.205	2.75	0.244	1.99	0.226
3.50	0.190	3.0	0.240	2.16	0.218
4.00	0.180	3.25	0.237	2.55	0.205
5.00	0.155	3.5	0.235	2.76	0.194
6.00	0.140	4.0	0.225	2.96	0.188
7.00	0.130	4.5	0.218	3.45	0.175
8.00	0.125			3.97	0.169
9.60	0.115	CURVE 4 T = 1147		4.48	0.166
12.00	0.115			4.92	0.164
14.00	0.110			5.42	0.147
		0.700	0.365	5.97	0.134
CURVE 2 T = 1321		0.660	0.360	6.38	0.135
		0.620	0.355	6.87	0.134
		0.580	0.370	7.41	0.135
1.15	0.260	0.560	0.385	7.91	0.126
1.60	0.240	0.545	0.390	8.45	0.123
2.00	0.220	0.515	0.400	8.89	0.122
2.90	0.290	0.500	0.400	9.51	0.117
4.00	0.180	0.485	0.405	9.84	0.117
5,00	0.155			10.2	0.115
6.20	0.150	CURVE 5 T = 1288		10.9	0.115
7.00	0.140			11.3	0.114
8.00	0.135			11.8	0.112
9.00	0.130	0.700	0.350	12.4	0.105
9.60	0.125	0.660	0.345	13.4	0.107
11.00	0.120	0.620	0.340		
12.00	0.120	0.580	0.345		
13.00	0.115	0.560	0.360		
		0.540	0.360		
CURVE 3 T = 1518		0.515	0.370		
		0.500	0.380		
		0.480	0.385		
0.65	0.437	0.477	0.390		
1.0	0.340	0.467	0.395		
1.1	0.330	0.457	0.405		
1.2	0.316				
1.3	0.306				

SPECIFICATION TABLE NO. 98 NORMAL SPECTRAL REFLECTANCE OF IRON

Curve No.	Ref. No.	Year	Wavelength μ	Temperature Range, K	Geometry θ	θ'	ω'	Reported Error, %	Composition (weight percent), Specifications and Remarks
1	238	1956	0.2537	1128-1252	$\sim 0^{\circ}$		2π		Electrolytic iron; cold rolled and thermally etched; measured in vacuum (10^{-5} mm Hg).

DATA TABLE NO. 98 NORMAL SPECTRAL REFLECTANCE OF IRON

[Temperature, T, K; Reflectance, ρ; Wavelength, λ, μ]

T	ρ
CURVE 1* $\lambda = 0.2537$	
1128	0.428
1145	0.438
1160	0.439
1172	0.439
1184	0.456
1187	0.466
1194	0.492
1200	0.504
1202	0.508
1215	0.508
1252	0.510

*Not shown on plot

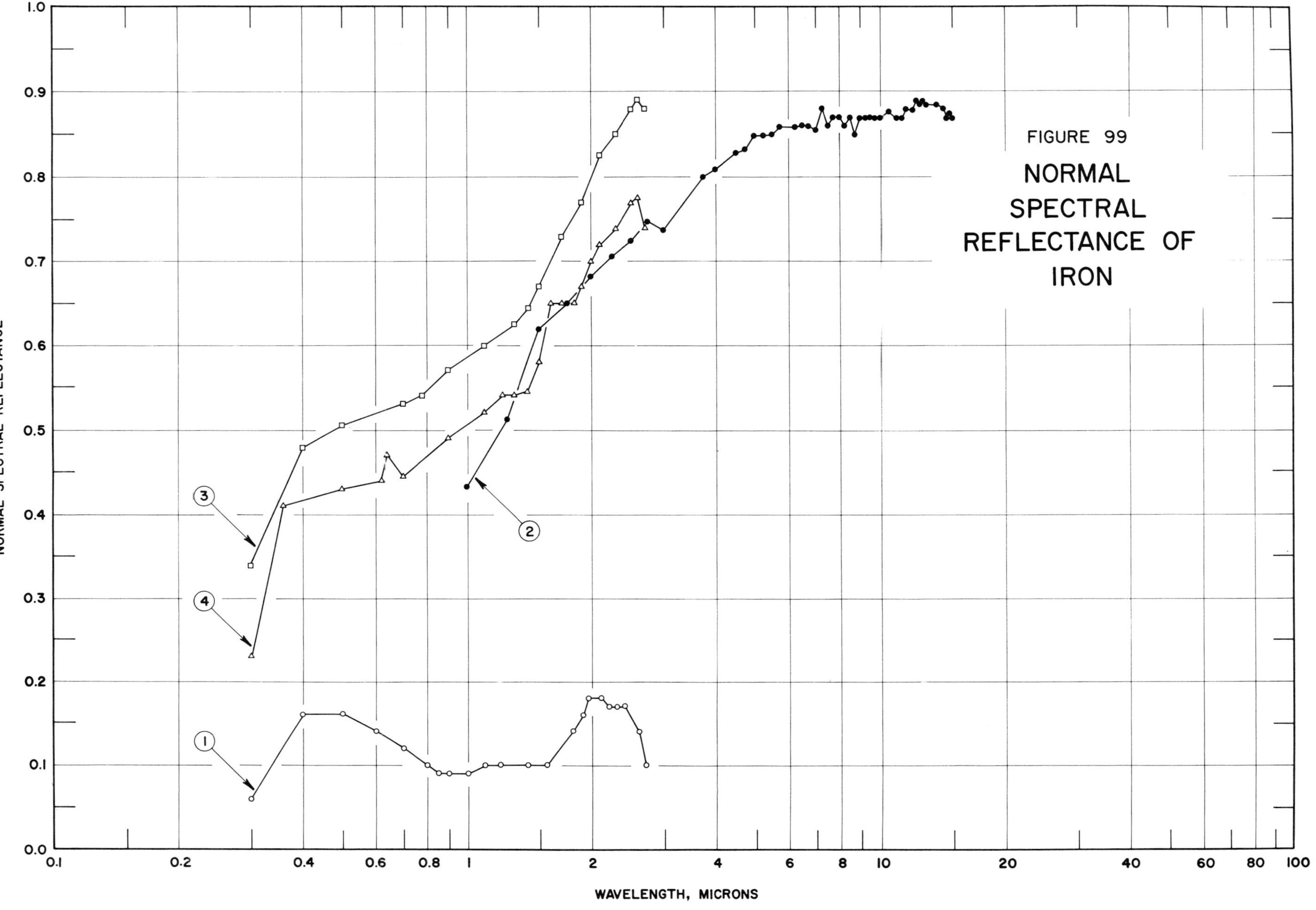
FIGURE 99
NORMAL SPECTRAL REFLECTANCE OF IRON
NORMAL SPECTRAL REFLECTANCE
1.0
0.9
0.8
0.7
0.6
0.5
0.4
0.3
0.2
0.1
0.0
WAVELENGTH, MICRONS
0.1
0.2
0.4
0.6
0.8
1
2
4
6
8
10
20
40
60
80
100
1
2
3
4

SPECIFICATION TABLE NO. 99 NORMAL SPECTRAL REFLECTANCE OF IRON

Curve No.	Ref. No.	Year	Temperature K	Wavelength Range, μ	Geometry θ	Geometry θ' ω'	Reported Error, %	Composition (weight percent), Specifications and Remarks
1	146	1958	298	0.30-2.70	9°	2π		Armco Ingot Iron; nominal composition: 0.03 max C, Fe balance.
2	126	1953	298	1.00-15.00	5°	2π	± 2	Galvanized iron; plate (0.022 in. thick); commercial finish; data extracted from smooth curve; converted from $R(2\pi,5^\circ)$.
3	34	1957	298	0.30-2.70	9°	2π	± 4	Armco Ingot Iron; nominal composition: 0.03 max C, Fe balance; finish having an RMS of about 2 microinches; data extracted from a smooth curve; magnesium carbonate reference.
4	34	1957	298	0.30-2.70	9°	2π	± 4	Armco Ingot Iron; nominal composition: 0.03 max C, Fe balance; finish having an RMS of about 15 microinches; data extracted from a smooth curve; magnesium carbonate reference.

DATA TABLE NO. 99 NORMAL SPECTRAL REFLECTANCE OF IRON

[Wavelength, λ, μ; Reflectance, ρ; Temperature, T, K]

λ	ρ
CURVE 1	
T = 298	
0.30	0.06
0.40	0.16
0.50	0.16
0.60	0.14
0.70	0.12
0.80	0.10
0.85	0.09
0.90	0.09
1.00	0.09
1.10	0.10
1.20	0.10
1.40	0.10
1.55	0.10
1.80	0.14
1.90	0.16
1.95	0.18
2.10	0.18
2.20	0.17
2.30	0.17
2.40	0.17
2.60	0.14
2.70	0.10
CURVE 2	
T = 298	
1.00	0.432
1.25	0.512
1.50	0.620
1.75	0.650
2.00	0.682
2.25	0.705
2.50	0.725
2.75	0.748
3.00	0.738
3.75	0.800
4.00	0.810
4.50	0.829
4.75	0.832
5.00	0.848
5.25	0.848
5.50	0.850
5.75	0.859
6.25	0.859
CURVE 2 (cont.)	
6.50	0.861
6.75	0.860
7.00	0.855
7.25	0.881
7.50	0.860
8.00	0.872
8.25	0.860
8.50	0.870
8.75	0.850
9.00	0.870
9.25	0.868
9.50	0.870
9.75	0.870
10.00	0.870
10.25	0.868*
10.50	0.877
10.75	0.871*
11.00	0.870
11.25	0.870
11.50	0.880
11.75	0.880*
12.00	0.880
12.25	0.890
12.50	0.886
12.75	0.890
13.00	0.885
13.75	0.885
14.00	0.881*
14.25	0.881
14.50	0.870
14.75	0.875
15.00	0.870
CURVE 3	
T = 298	
0.30	0.340
0.40	0.480
0.50	0.505
0.70	0.530
0.78	0.540
0.90	0.570
1.10	0.600
1.30	0.625
1.42	0.645
CURVE 3 (cont.)	
1.50	0.670
1.70	0.730
1.90	0.770
2.10	0.825
2.30	0.850
2.50	0.880
2.60	0.890
2.70	0.880
CURVE 4	
T = 298	
0.30	0.230
0.36	0.410
0.50	0.430
0.62	0.440
0.70	0.445
0.64	0.470
0.90	0.490
0.92	0.495*
1.10	0.520
1.22	0.540
1.30	0.540
1.40	0.545
1.50	0.580
1.60	0.650
1.70	0.650
1.82	0.650
1.90	0.670
2.00	0.700
2.10	0.720
2.30	0.740
2.50	0.770
2.60	0.775
2.70	0.740

*Not shown on plot

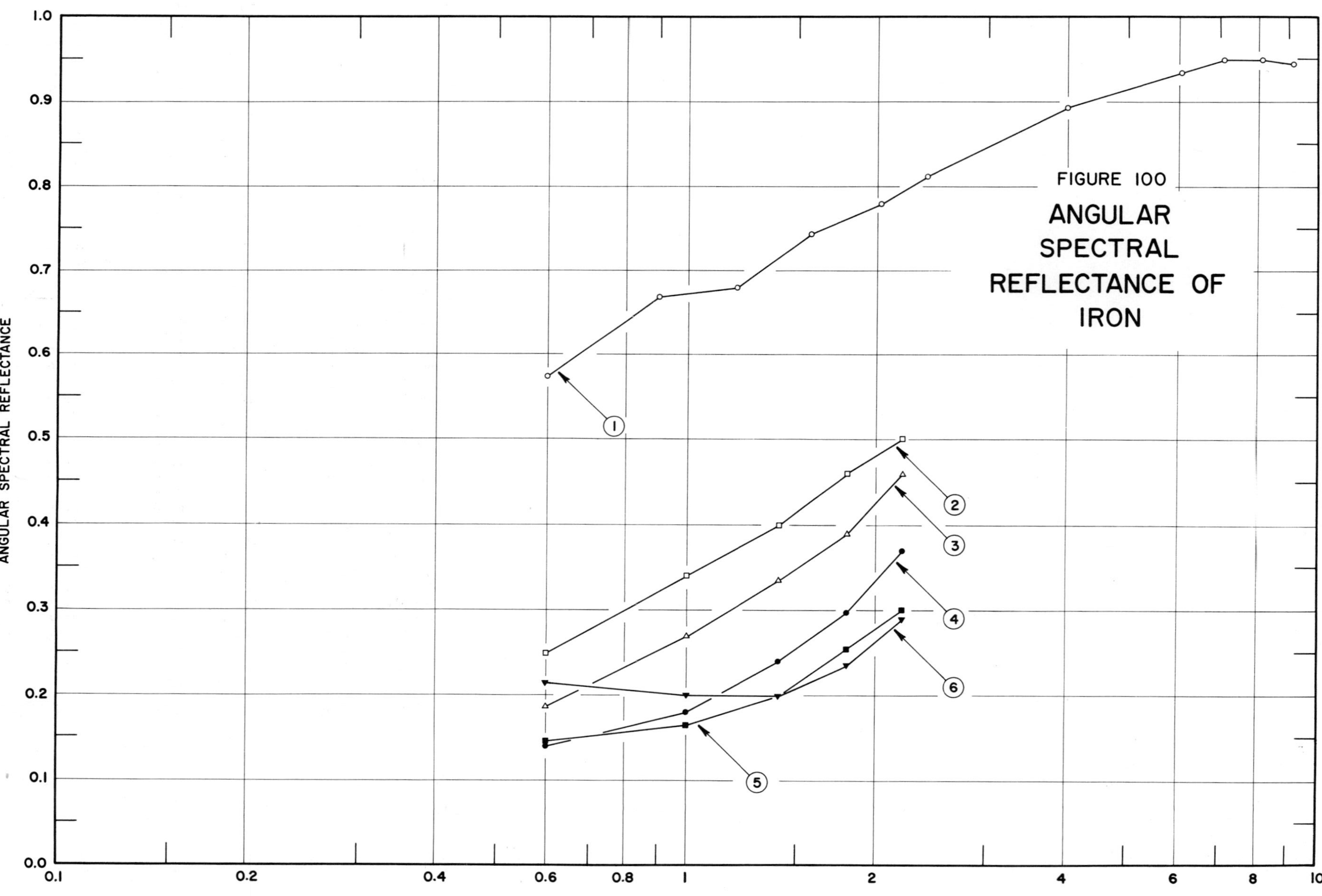
FIGURE 100
ANGULAR SPECTRAL REFLECTANCE OF IRON
ANGULAR SPECTRAL REFLECTANCE
WAVELENGTH, MICRONS
0.0
0.1
0.2
0.3
0.4
0.5
0.6
0.7
0.8
0.9
1.0
0.1
0.2
0.4
0.6
0.8
1
2
4
6
8
10
1
2
3
4
5
6

SPECIFICATION TABLE NO. 100 ANGULAR SPECTRAL REFLECTANCE OF IRON

Curve No.	Ref. No.	Year	Temperature K	Wavelength Range, μ	Geometry θ	θ'	ω'	Reported Error, %	Composition (weight percent), Specifications and Remarks
1	132	1911	298	0.60-9.13	15°	15°		≤3	0.15 Cu, 0.02 Mn, 0.02 C, Fe balance; polished; silvered glass mirror reference.
2	141	1948	298	0.6-2.2	50°	50°			Evaporated mirror surface; data extracted from smooth curve, incident beam polarized parallel to the plane of incidence.
3	141	1948	298	0.6-2.2	60°	60°			Above specimen and conditions, incident beam polarized parallel to the plane of incidence.
4	141	1948	298	0.6-2.2	70°	70°			Above specimen and conditions, incident beam polarized parallel to the plane of incidence.
5	141	1948	298	0.6-2.2	79°	79°			Above specimen and conditions, incident beam polarized parallel to the plane of incidence.
6	141	1958	298	0.6-2.2	80°	80°			Above specimen and conditions, incident beam polarized parallel to the plane of incidence.

DATA TABLE NO. 100 ANGULAR SPECTRAL REFLECTANCE OF IRON

[Wavelength, λ, μ; Reflectance, ρ; Temperature, T, K]

λ	ρ	λ	ρ
CURVE 1 T = 298		CURVE 5 T = 298	
0.60	0.575	0.6	0.145
0.90	0.620	1.0	0.165
1.20	0.680	1.4	0.200*
1.57	0.743	1.8	0.255
2.03	0.780	2.2	0.300
2.40	0.813		
4.00	0.895	CURVE 6 T = 298	
6.07	0.935		
7.07	0.945		
8.13	0.950	0.6	0.215
9.13	0.947	1.0	0.200
		1.4	0.200
CURVE 2 T = 298		1.8	0.235
		2.2	0.290
0.6	0.250		
1.0	0.340		
1.4	0.400		
1.8	0.460		
2.2	0.500		
CURVE 3 T = 298			
0.6	0.188		
1.0	0.270		
1.4	0.335		
1.8	0.390		
2.2	0.460		
CURVE 4 T = 298			
0.6	0.140		
1.0	0.180		
1.4	0.240		
1.8	0.298		
2.2	0.370		

* Not shown on plot

SPECIFICATION TABLE NO. 101 NORMAL SPECTRAL ABSORPTANCE OF IRON

Curve No.	Ref. No.	Year	Wavelength μ	Temperature Range, K	Geometry θ	Reported Error, %	Composition (weight percent), Specifications and Remarks
1	35	1914	0.66	873-2073	$\sim 0^{\circ}$		Measured in burning hydrogen; author assumes $\alpha = \epsilon$.

DATA TABLE NO. 101 NORMAL SPECTRAL ABSORPTANCE OF IRON

[Temperature, T, K; Absorptance, α; Wavelength, λ, μ]

T	α
CURVE 1*	
$\lambda = 0.66$	
873	0.27
2073	0.29

*Not shown on plot

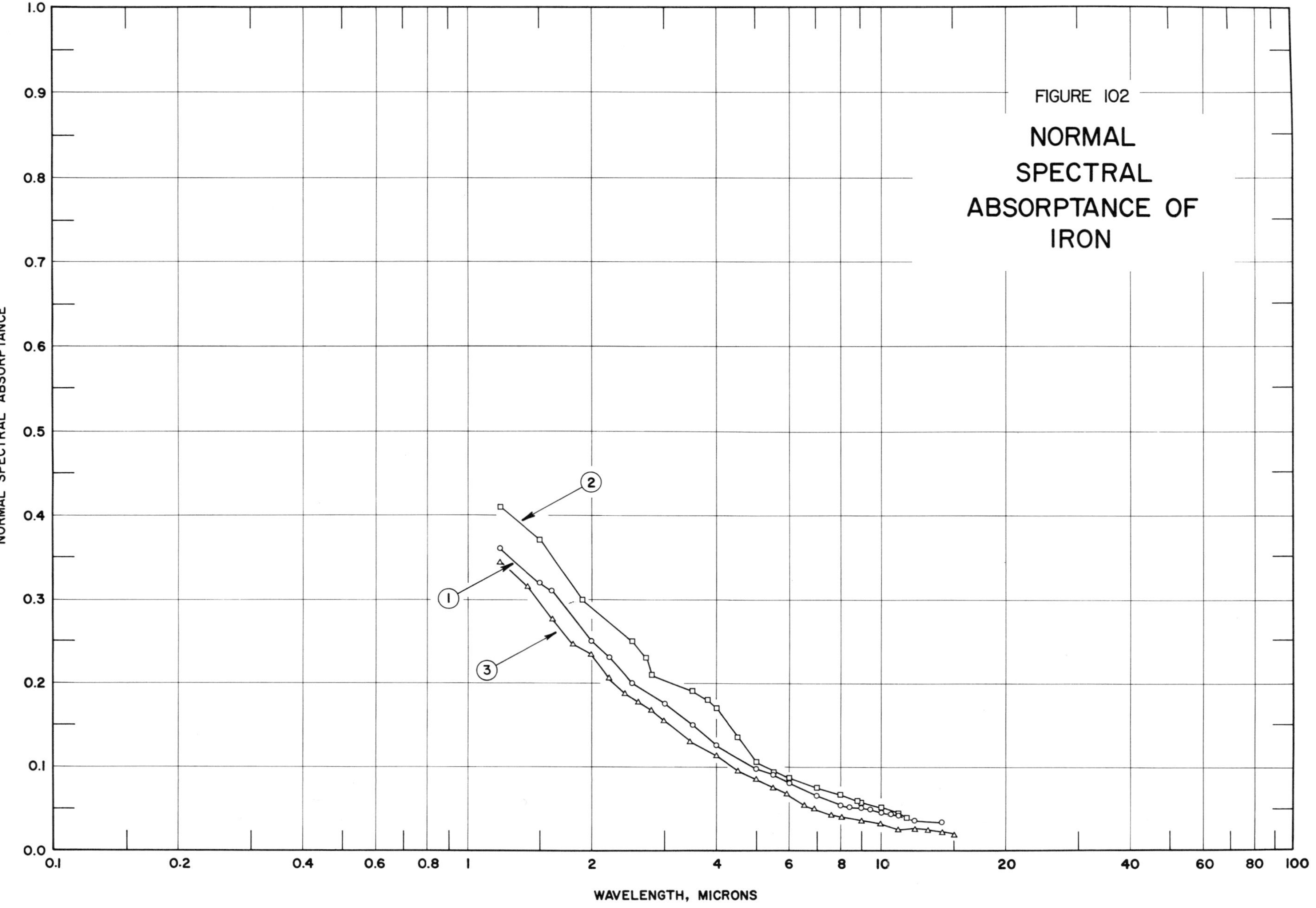
FIGURE 102
NORMAL SPECTRAL ABSORPTANCE OF IRON
NORMAL SPECTRAL ABSORPTANCE
1.0
0.9
0.8
0.7
0.6
0.5
0.4
0.3
0.2
0.1
0.0
0.1
0.2
0.4
0.6
0.8
1
2
4
6
8
10
20
40
60
80
100
WAVELENGTH, MICRONS
1
2
3

SPECIFICATION TABLE NO. 102 NORMAL SPECTRAL ABSORPTANCE OF IRON

Curve No.	Ref. No.	Year	Temperature K	Wavelength Range, μ	Geometry θ	Reported Error, %	Composition (weight percent), Specifications and Remarks
1	30	1963	294	1.2-14.0	$\sim 0°$		Mechanically polished with aluminum oxide and cleaned with water; measured in air; data extracted from smooth curve; measured immediately after polishing.
2	30	1963	294	1.2-11.5	$\sim 0°$		Mechanically polished with aluminum oxide and cleaned with water; heated at 1058 K for 2 1/2 hrs and at 1316 K for 5 hrs; measured in air; data extracted from smooth curve; surface roughness (measured after experimentation) 0.5μ peak to peak and 50μ lateral; oxidization of surface was avoided.
3	245	1965	294	1.20-15.0	$\sim 0°$	<5	Measured after emissivity determination; measured in hohlraum.

DATA TABLE NO. 102 NORMAL SPECTRAL ABSORPTANCE OF IRON

[Wavelength, λ, μ; Absorptance, α; Temperature, T, K]

λ	α
CURVE 1	
T = 294	
1.2	0.360
1.5	0.320
1.6	0.310
2.0	0.250
2.2	0.230
2.5	0.200
3.0	0.175
3.5	0.150
4.0	0.125
5.0	0.098
5.5	0.090
6.0	0.080
7.0	0.066
8.0	0.055
8.4	0.053
9.0	0.052
9.4	0.050
10.0	0.047
10.5	0.045
11.0	0.042
12.0	0.038
14.0	0.036
CURVE 2	
T = 294	
1.2	0.410
1.5	0.370
1.9	0.300
2.5	0.250
2.7	0.230
2.8	0.210
3.5	0.190
3.8	0.180
4.0	0.170
4.5	0.135
5.0	0.105
5.5	0.094
6.0	0.088
7.0	0.076
8.0	0.068
CURVE 2 (cont.)	
8.8	0.060
9.0	0.058
10.0	0.052
11.0	0.045
11.5	0.040
CURVE 3	
T = 294	
1.20	0.342
1.40	0.313
1.61	0.275
1.81	0.246
1.99	0.232
2.20	0.205
2.40	0.186
2.59	0.177
2.80	0.168
2.99	0.155
3.48	0.129
4.01	0.112
4.49	0.0936
5.01	0.0843
5.46	0.0748
5.93	0.0686
6.50	0.0544
6.93	0.0508
7.57	0.0438
8.09	0.0406
9.02	0.0371
9.98	0.0329
11.0	0.0279
12.1	0.0280
12.9	0.0266
14.1	0.0219
15.0	0.0203

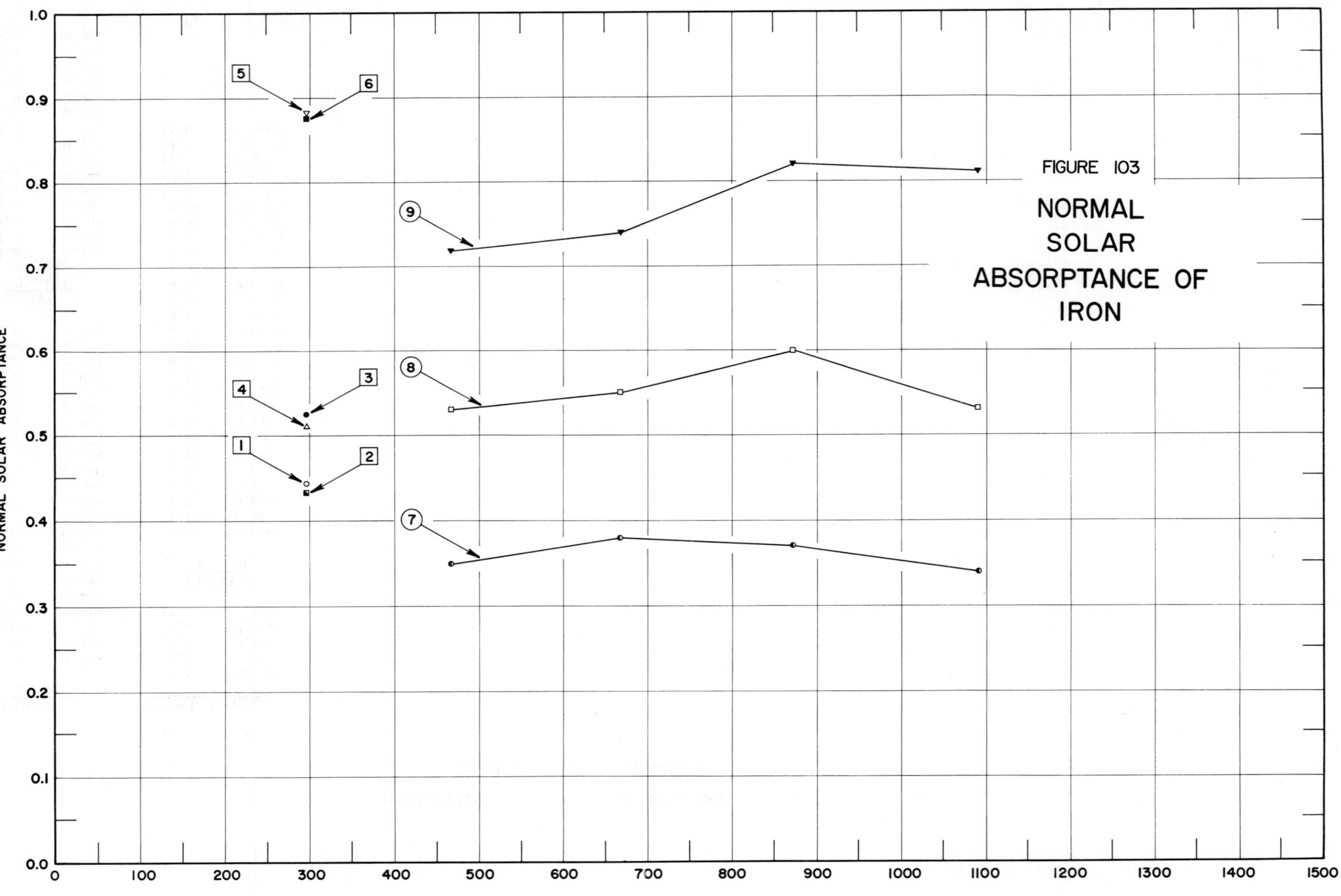

FIGURE 103 NORMAL SOLAR ABSORPTANCE OF IRON

SPECIFICATION TABLE NO. 103 NORMAL SOLAR ABSORPTANCE OF IRON

Curve No.	Ref. No.	Year	Temperature Range, K	Geometry θ	Reported Error, %	Composition (weight percent), Specifications and Remarks
1	34	1957	298	9°		Armco Ingot iron; nominal composition: 0.03 max C, Fe balance ; finish of 2 microinch RMS; computed from spectral reflectance data for sea level conditions.
2	34	1957	298	9°		Above specimen and conditions except computed for above atmosphere conditions.
3	34	1957	298	9°		Armco Ingot iron; nominal composition: 0.03 max C, Fe balance ; finish of 15 microinch RMS; computed from spectral reflectance data for sea level conditions.
4	34	1957	298	9°		Above specimen and conditions except computed for above atmosphere conditions.
5	146	1958	298	9°		Armco Ingot iron: nominal composition: 0.03 max C, Fe balance; oxidized; computed from spectral reflectivity for sea level conditions.
6	146	1958	298	9°		Above specimen and conditions except computed for above atmosphere conditions.
7	47	1961	468-1093	~0°		Armco Ingot iron; 0.11 Cu, 0.032 Mn, 0.017 S, 0.014 C, 0.003 P, 0.003 Si; disc (0.04 in.thick); ground with 600 grit carborundum and polished on a wet cloth lap with unlevigated jewelers rouge; measured in vacuum (10^{-5} mm Hg).
8	47	1961	468-1093	~0°		Armco Ingot iron; 0.11 Cu, 0.032 Mn, 0.017 S, 0.014 C, 0.003 P, 0.003 Si; disc (0.04 in. thick); blued by immersing in a niter bath; measured in vacuum (10^{-5} mm Hg).
9	47	1961	468-1093	~0°		Armco Ingot iron; 0.11 Cu, 0.032 Mn, 0.017 S, 0.014 C, 0.003 P, 0.003 Si; disc (0.04 in. thick); etched by sandblasting with a jet of SS White Abrasive No. 2; measured in vacuum (10^{-5} mm Hg).

DATA TABLE NO. 103 NORMAL SOLAR ABSORPTANCE OF IRON

[Temperature, T, K; Absorptance, α]

T	α
CURVE 1	
298	0.445
CURVE 2	
298	0.433
CURVE 3	
298	0.525
CURVE 4	
298	0.510
CURVE 5	
298	0.882
CURVE 6	
298	0.877
CURVE 7	
468	0.35
668	0.38
873	0.37
1093	0.34
CURVE 8	
468	0.53
668	0.55
873	0.60
1093	0.53
CURVE 9	
468	0.72
668	0.74
873	0.82
1093	0.81

SPECIFICATION TABLE NO. 104 HEMISPHERICAL TOTAL EMITTANCE OF LEAD

Curve No.	Ref. No.	Year	Temperature Range, K	Reported Error, %	Composition (weight percent), Specifications and Remarks
1	3	1955	76	5	Commercial sheet foil (0.004 in. thick); cleaned; measured in vacuum (10^{-6} to 10^{-7} mm Hg); emittance for 300 K black body incident radiation; authors assumed $\alpha = \epsilon$.

DATA TABLE NO. 104 HEMISPHERICAL TOTAL EMITTANCE OF LEAD

[Temperature, T, K; Emittance, ∈]

T	∈
CURVE 1*	
76	0.036

* Not shown on plot

SPECIFICATION TABLE NO. 105 NORMAL TOTAL EMITTANCE OF LEAD

Curve No.	Ref. No.	Year	Temperature Range, K	Geometry θ'	Reported Error, %	Composition (weight percent), Specifications and Remarks
1	14	1913	473	$\sim 0^\circ$		Disc; cleaned and polished.

DATA TABLE NO. 105 NORMAL TOTAL EMITTANCE OF LEAD

[Temperature, T, K; Emittance, ∈]

T	∈
CURVE 1*	
473	0.631

* Not shown on plot

SPECIFICATION TABLE NO. 106 HEMISPHERICAL INTEGRATED ABSORPTANCE OF LEAD

Curve No.	Ref. No.	Year	Temperature Range, K	Reported Error, %	Composition (weight percent), Specifications and Remarks
1	3	1955	76	5	Commercial sheet foil (0.004 in. thick) ; cleaned; measured in vacuum (10^{-6} to 10^{-7} mm Hg); absorptance for 300 K black body incident radiation; authors assumed $\alpha = \epsilon$.

DATA TABLE NO. 106 HEMISPHERICAL INTEGRATED ABSORPTANCE OF LEAD

[Temperature, T, K; Absorptance, α]

T	α
CURVE 1*	
76	0.036

* Not shown on plot

SPECIFICATION TABLE NO. 107 NORMAL INTEGRATED ABSORPTANCE OF LEAD

Curve No.	Ref. No.	Year	Temperature Range, K	Geometry θ	Reported Error, %	Composition (weight percent), Specifications and Remarks
1	134	1952	2	$\sim 0^{\circ}$		Electropolished; absorptance for 298 K black body incident radiation.

DATA TABLE NO. 107 NORMAL INTEGRATED ABSORPTANCE OF LEAD

[Temperature, T, K; Absorptance, α]

T	α
CURVE 1*	
2	0.0115

* Not shown on plot

SPECIFICATION TABLE NO. 108 NORMAL SPECTRAL ABSORPTANCE OF LEAD

Curve No.	Ref. No.	Year	Temperature K	Wavelength Range, μ	Geometry θ	Reported Error, %	Composition (weight percent), Specifications and Remarks
1	307	1954	~298	0.226-2.600	~0°		Lead; 5 mils thick; data extracted from smooth curve.

DATA TABLE NO. 108 NORMAL SPECTRAL ABSORPTANCE OF LEAD

[Wavelength, λ, μ; Absorptance, α; Temperature, T, K]

λ	α
CURVE 1*	
T = ~298	
0.226	0.583
0.325	0.496
0.362	0.429
0.466	0.377
0.527	0.387
0.670	0.365
0.913	0.224
1.076	0.175
1.189	0.170
1.399	0.139
1.515	0.124
1.614	0.137
1.696	0.126
1.801	0.143
2.000	0.167
2.200	0.220
2.400	0.210
2.600	0.206

* Not shown on plot

SPECIFICATION TABLE NO. 109 ANGULAR SPECTRAL ABSORPTANCE OF LEAD

Curve No.	Ref. No.	Year	Temperature K	Wavelength Range, μ	Geometry θ	Reported Error, %	Composition (weight percent), Specifications and Remarks
1	225	1965	306	0.343-22.8	25°		Lead from Belmont Smelting and Refining Works; scraped; measured in dry nitrogen; heated cavity at approx 1056 K with platinum reference; authors assumed $\alpha = 1 - R(2\pi, 25^{\circ})$.

DATA TABLE NO. 109 ANGULAR SPECTRAL ABSORPTANCE OF LEAD

[Wavelength, λ, μ; Absorptance, α; Temperature, T, K]

λ	α
CURVE 1*	
T = 306	
0.343	0.524
0.411	0.518
0.498	0.541
0.611	0.538
0.700	0.494
0.809	0.463
0.922	0.404
1.01	0.351
1.26	0.284
1.51	0.268
2.03	0.250
2.83	0.225
3.70	0.172
5.12	0.153
6.61	0.144
9.20	0.130
13.1	0.112
17.1	0.095
22.8	0.084

* Not shown on plot

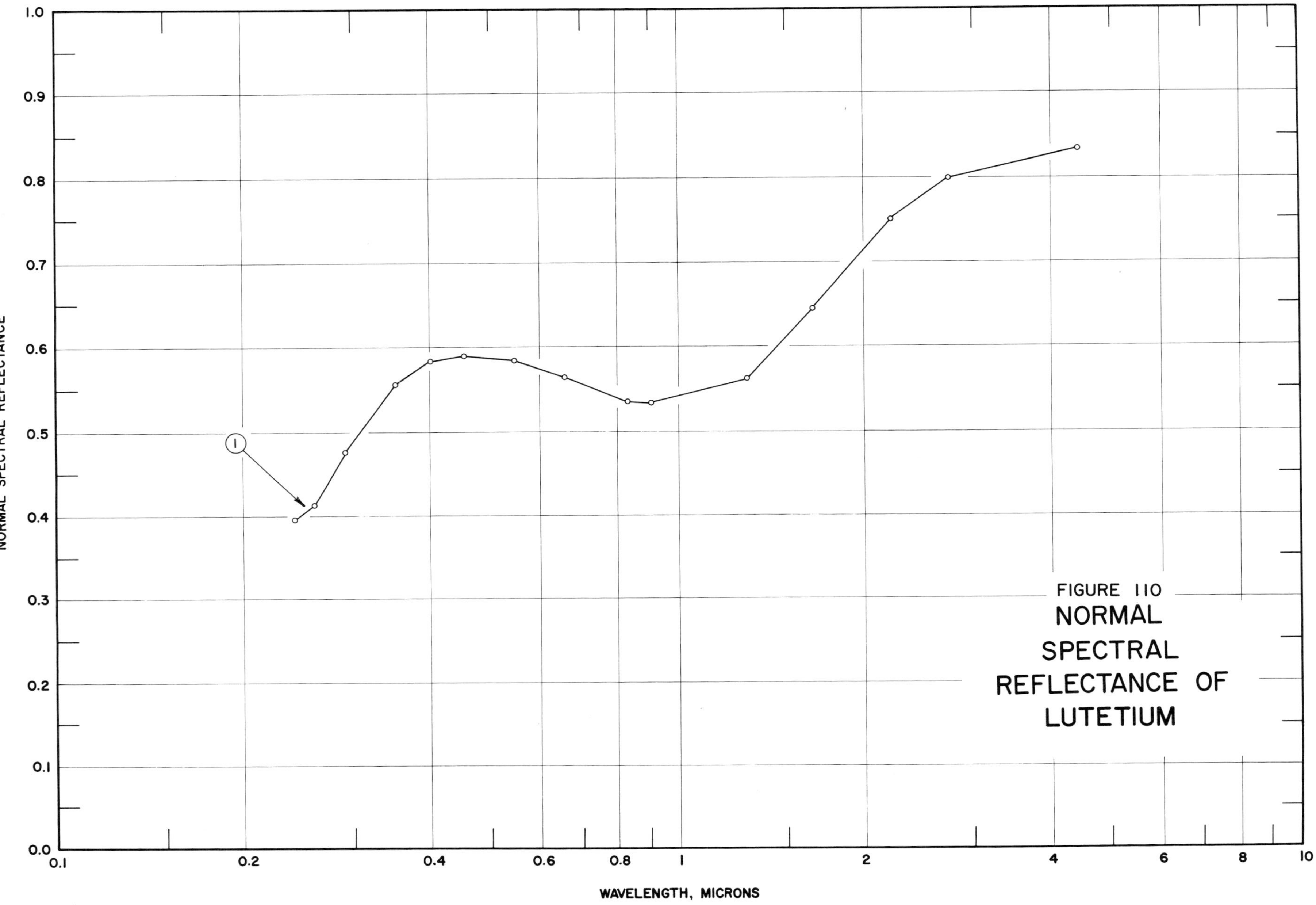

FIGURE 110
NORMAL SPECTRAL REFLECTANCE OF LUTETIUM

SPECIFICATION TABLE NO. 110 NORMAL SPECTRAL REFLECTANCE OF LUTETIUM

Curve No.	Ref. No.	Year	Temperature K	Wavelength Range, μ	Geometry θ θ' ω'	Reported Error, %	Composition (weight percent), Specifications and Remarks
1	316	1965	298	0.243-4.429	~0° ~0°		Film; deposited in ultrahigh vacuum on sapphire; measured in vacuum; data extracted from smooth curve.

DATA TABLE NO. 110 NORMAL SPECTRAL REFLECTANCE OF LUTETIUM

[Wavelength, λ, μ; Reflectance, ρ; Temperature, T, K]

λ	ρ
CURVE 1	
T = 298	
0.243	0.395
0.260	0.412
0.292	0.478
0.352	0.558
0.401	0.584
0.454	0.591
0.546	0.585
0.660	0.565
0.838	0.537
0.905	0.535
1.292	0.569
1.653	0.646
2.214	0.751
2.756	0.800
4.429	0.836

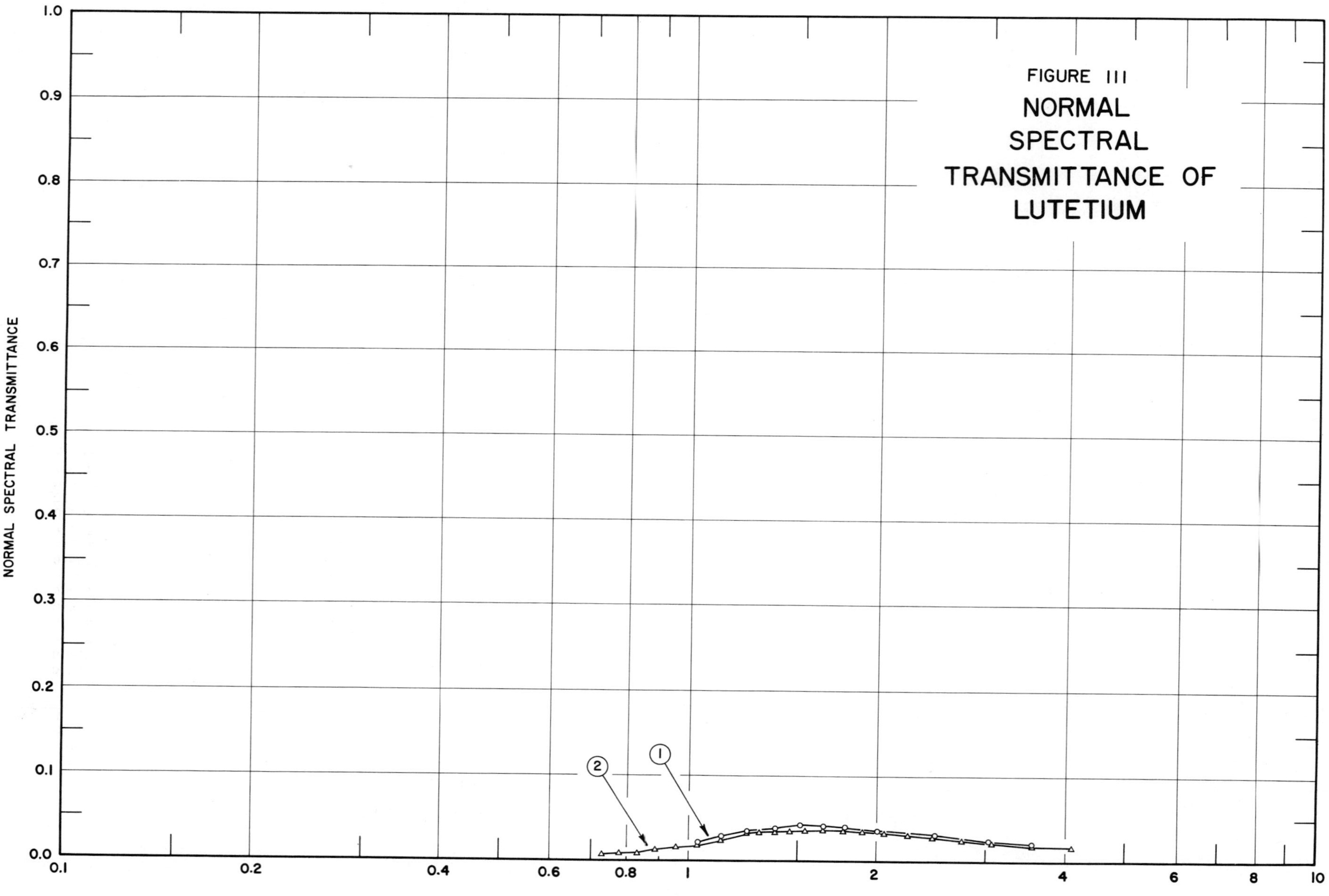
FIGURE 111
NORMAL
SPECTRAL
TRANSMITTANCE OF
LUTETIUM
NORMAL SPECTRAL TRANSMITTANCE
1.0
0.9
0.8
0.7
0.6
0.5
0.4
0.3
0.2
0.1
0.0
1
2
0.1
0.2
0.4
0.6
0.8
1
2
4
6
8
10
WAVELENGTH, MICRONS

SPECIFICATION TABLE NO. 111 NORMAL SPECTRAL TRANSMITTANCE OF LUTETIUM

Curve No.	Ref. No.	Year	Temperature K	Wavelength Range, μ	Geometry θ θ' ω'	Reported Error, %	Composition (weight percent), Specifications and Remarks
1	316	1965	80	1.033-3.543	~0° ~0°		Film (1200 Å thick); deposited in ultrahigh vacuum on sapphire; measured in vacuum.
2	316	1965	293	0.729-4.092	~0° ~0°		Above specimen and conditions.

DATA TABLE NO. 111 NORMAL SPECTRAL TRANSMITTANCE OF LUTETIUM

[Wavelength, λ, μ; Transmittance, τ; Temperature, T, K]

λ	τ
CURVE 1	
T = 80	
1.033	0.0218
1.128	0.0288
1.240	0.0344
1.376	0.0385
1.546	0.0422
1.647	0.0415
1.771	0.0405
2.056	0.0359
2.480	0.0303
3.077	0.0234
3.543	0.0210
CURVE 2	
T = 293	
0.729	0.0060
0.775	0.0075
0.827	0.0097
0.885	0.0122
0.953	0.0154
1.032	0.0189
1.126	0.0246
1.241	0.0303
1.304	0.0329
1.372	0.0331
1.455	0.0347
1.546	0.0359
1.642	0.0371
1.769	0.0359
1.899	0.0344
2.056	0.0323
2.242	0.0308
2.470	0.0281
2.737	0.0246
3.069	0.0218
3.523	0.0197
4.092	0.0174

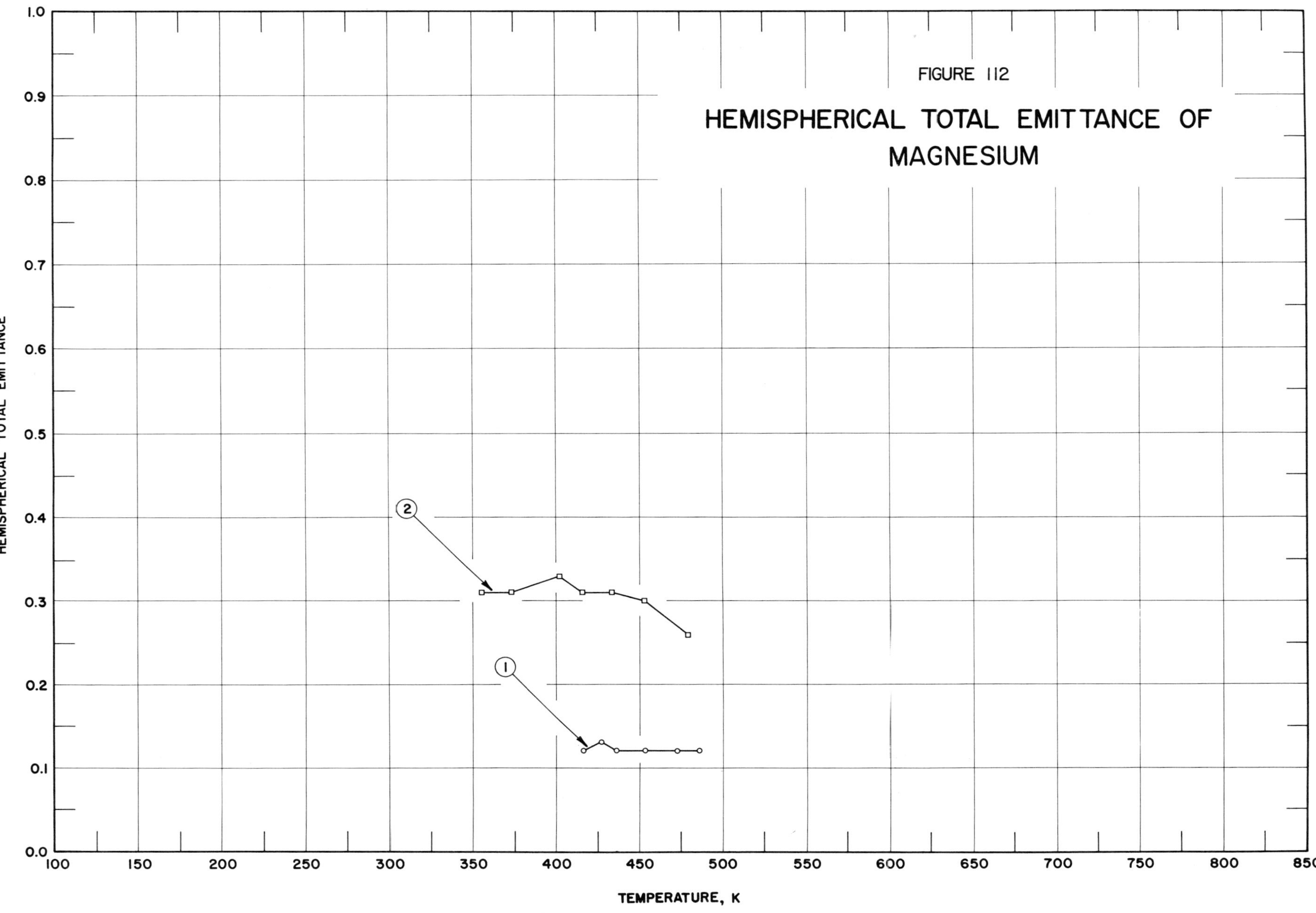

FIGURE 112

HEMISPHERICAL TOTAL EMITTANCE OF MAGNESIUM

SPECIFICATION TABLE NO. 112 HEMISPHERICAL TOTAL EMITTANCE OF MAGNESIUM

Curve No.	Ref. No.	Year	Temperature Range, K	Reported Error, %	Composition (weight percent), Specifications and Remarks
1	11	1962	416-485	± 3	Hand polished using fine abrasive papers; measured in vacuum (10^{-3} mm Hg); data extracted from smooth curve.
2	11	1962	355-479	± 3	Different sample, same as curve 1 specimen and conditions except vapor blasted.

DATA TABLE NO. 112 HEMISPHERICAL TOTAL EMITTANCE OF MAGNESIUM

[Temperature, T, K; Emittance, ∈]

T	∈
CURVE 1	
416	0.12
427	0.13
436	0.12
453	0.12
473	0.12
485	0.12
CURVE 2	
355	0.31
373	0.31
402	0.33
415	0.31
433	0.31
453	0.30
479	0.26

SPECIFICATION TABLE NO. 113 NORMAL SPECTRAL REFLECTANCE OF MAGNESIUM

Curve No.	Ref. No.	Year	Temperature K	Wavelength Range, μ	Geometry θ	θ'	ω'	Reported Error, %	Composition (weight percent), Specifications and Remarks
1	133	1934	298	0.2350-0.5780	$\sim 0^{\circ}$	$\sim 0^{\circ}$			Cold worked; annealed in an inert gas for 6-24 hrs ; polished; stored in dilute solutions of NaOH, NaOH + NaF, and HNO_3.

DATA TABLE NO. 113 NORMAL SPECTRAL REFLECTANCE OF MAGNESIUM

[Wavelength, λ, μ; Reflectance, ρ; Temperature, T, K]

λ	ρ
CURVE 1*	
T = 298	
0.2350	0.232
0.2537	0.328
0.2650	0.350
0.2970	0.386
0.3125	0.401
0.3340	0.432
0.3660	0.521
0.4355	0.643
0.5460	0.725
0.5780	0.726

* Not shown on plot

SPECIFICATION TABLE NO. 114 ANGULAR SPECTRAL REFLECTANCE OF MAGNESIUM

Curve No.	Ref. No.	Year	Temperature K	Wavelength Range, μ	Geometry θ	θ'	ω'	Reported Error, %	Composition (weight percent), Specifications and Remarks
1	132	1911	298	0.55-8.77	15°	15°		≤ 3	Wet-ground on fine emery paper using tin oxide; polished on chamois skin; silvered glass mirror reference.

DATA TABLE NO. 114 ANGULAR SPECTRAL REFLECTANCE OF MAGNESIUM

[Wavelength, λ, μ; Reflectance, ρ; Temperature, T, K]

λ	ρ
CURVE 1*	
T = 298	
0.55	0.717
0.90	0.735
1.20	0.747
1.60	0.757
2.40	0.780
4.00	0.837
6.10	0.883
7.10	0.917
8.00	0.933
8.77	0.937

* Not shown on plot

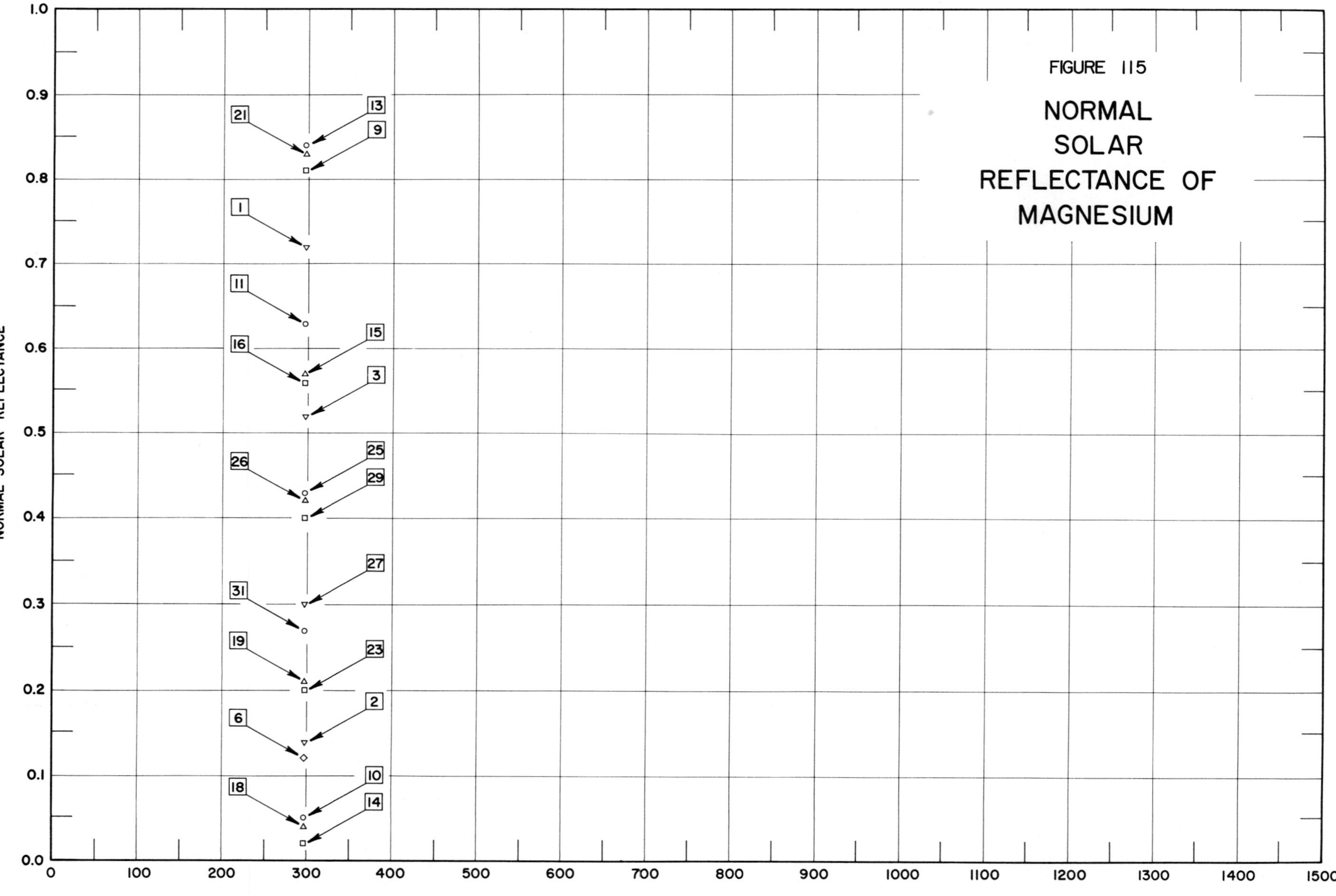
FIGURE 115
NORMAL SOLAR REFLECTANCE OF MAGNESIUM
NORMAL SOLAR REFLECTANCE
TEMPERATURE, K
1.0
0.9
0.8
0.7
0.6
0.5
0.4
0.3
0.2
0.1
0.0
0
100
200
300
400
500
600
700
800
900
1000
1100
1200
1300
1400
1500
21
13
9
1
11
16
15
3
26
25
29
27
31
19
23
2
6
18
10
14

SPECIFICATION TABLE NO. 115 NORMAL SOLAR REFLECTANCE OF MAGNESIUM

Curve No.	Ref. No.	Year	Temperature Range, K	Geometry θ	θ'	ω'	Reported Error, %	Composition (weight percent), Specifications and Remarks
1	131	1964	298	$\sim 0^\circ$		2π		L120 (British Aircraft Material Specification); nominal composition: 99.8 min Mg, 0.05 Mn, 0.04 Fe, 0.03 Si, 0.03 Zn, 0.02 Al, 0.005 Cu, 0.002 Ni; turned; freshly prepared; MgO reference; computed from spectral data.
2	131	1964	298	$\sim 0^\circ$		2π		Above specimen and conditions; diffuse component only.
3	131	1964	298	$\sim 0^\circ$		2π		Curve 1 specimen and conditions except exposed within clean laboratory area for one month.
4	131	1964	298	$\sim 0^\circ$		2π		Above specimen and conditions; diffuse component only.
5	131	1964	298	$\sim 0^\circ$		2π		Different sample, same as Curve 1 specimen and conditions.
6	131	1964	298	$\sim 0^\circ$		2π		Above specimen and conditions; diffuse component only.
7	131	1964	298	$\sim 0^\circ$		2π		Curve 5 specimen and conditions except exposed within clean laboratory area for one month.
8	131	1964	298	$\sim 0^\circ$		2π		Above specimen and conditions; diffuse component only.
9	131	1964	298	$\sim 0^\circ$		2π		L120 (British Aircraft Material Specification); nominal composition: 99.8 min Mg, 0.05 Mn, 0.04 Fe, 0.03 Si, 0.03 Zn, 0.02 Al, 0.005 Cu, 0.002 Ni; machine polished; freshly prepared; MgO reference; computed from spectral data.
10	131	1964	298	$\sim 0^\circ$		2π		Above specimen and conditions; diffuse component only.
11	131	1964	298	$\sim 0^\circ$		2π		Curve 9 specimen and conditions except exposed within clean laboratory area for one month.
12	131	1964	298	$\sim 0^\circ$		2π		Above specimen and conditions; diffuse component only.
13	131	1964	298	$\sim 0^\circ$		2π		Different sample, same as Curve 9 specimen and conditions.
14	131	1964	298	$\sim 0^\circ$		2π		Above specimen and conditions; diffuse component only.
15	131	1964	298	$\sim 0^\circ$		2π		Curve 13 specimen and conditions except exposed within clean laboratory area for one month.
16	131	1964	298	$\sim 0^\circ$		2π		Above specimen and conditions; diffuse component only.
17	131	1964	298	$\sim 0^\circ$		2π		Different sample, same as Curve 9 specimen and conditions.
18	131	1964	298	$\sim 0^\circ$		2π		Above specimen and conditions; diffuse component only.
19	131	1964	298	$\sim 0^\circ$		2π		Curve 17 specimen and conditions except exposed to outside atmosphere for one month.
20	131	1964	298	$\sim 0^\circ$		2π		Above specimen and conditions; diffuse component only.
21	131	1964	298	$\sim 0^\circ$		2π		Different sample, same as Curve 9 specimen and conditions.
22	131	1964	298	$\sim 0^\circ$		2π		Above specimen and conditions; diffuse component only.

SPECIFICATION TABLE NO. 115 (continued)

Curve No.	Ref. No.	Year	Temperature Range, K	Geometry θ θ'	ω'	Reported Error, %	Composition (weight percent), Specifications and Remarks
23	131	1964	298	~0°	2π		Curve 21 specimen and conditions except exposed to outside atmosphere for one month.
24	131	1964	298	~0°	2π		Above specimen and conditions; diffuse component only.
25	131	1964	298	~0°	2π		L120 (British Aircraft Material Specification); nominal composition: 99.8 min Mg, 0.05 Mn, 0.04 Fe, 0.03 Si, 0.03 Zn, 0.02 Al, 0.005 Cu, 0.002 Ni; vapor blasted with 400 mesh alumina; 70 psi nozzle pressure; freshly prepared; MgO reference; computed from spectral data.
26	131	1964	298	~0°	2π		Above specimen and conditions; diffuse component only.
27	131	1964	298	~0°	2π		Curve 25 specimen and conditions except exposed within clean laboratory area for one month.
28	131	1964	298	~0°	2π		Above specimen and conditions; diffuse component only.
29	131	1964	298	~0°	2π		Different sample, same as Curve 25 specimen and conditions.
30	131	1964	298	~0°	2π		Above specimen and conditions; diffuse component only.
31	131	1964	298	~0°	2π		Curve 29 specimen and conditions except exposed within clean laboratory area for one month.
32	131	1964	298	~0°	2π		Above specimen and conditions; diffuse component only.

DATA TABLE NO. 115 NORMAL SOLAR REFLECTANCE OF MAGNESIUM

[Temperature, T, K; Reflectance, ρ]

T	ρ	T	ρ	T	ρ
CURVE 1		CURVE 13		CURVE 25	
298	0.72	298	0.84	298	0.43
CURVE 2		CURVE 14		CURVE 26	
298	0.14	298	0.02	298	0.42
CURVE 3		CURVE 15		CURVE 27	
298	0.52	298	0.57	298	0.30
CURVE 4*		CURVE 16		CURVE 28*	
298	0.52	298	0.56	298	0.30
CURVE 5*		CURVE 17*		CURVE 29	
298	0.72	298	0.84	298	0.40
CURVE 6		CURVE 18		CURVE 30*	
298	0.12	298	0.04	298	0.40
CURVE 7*		CURVE 19		CURVE 31	
298	0.52	298	0.21	298	0.27
CURVE 8*		CURVE 20*		CURVE 32*	
298	0.52	298	0.21	298	0.27
CURVE 9		CURVE 21			
298	0.81	298	0.83		
CURVE 10		CURVE 22*			
298	0.05	298	0.05		
CURVE 11		CURVE 23			
298	0.63	298	0.20		
CURVE 12*		CURVE 24*			
298	0.63	298	0.20		

* Not shown on plot

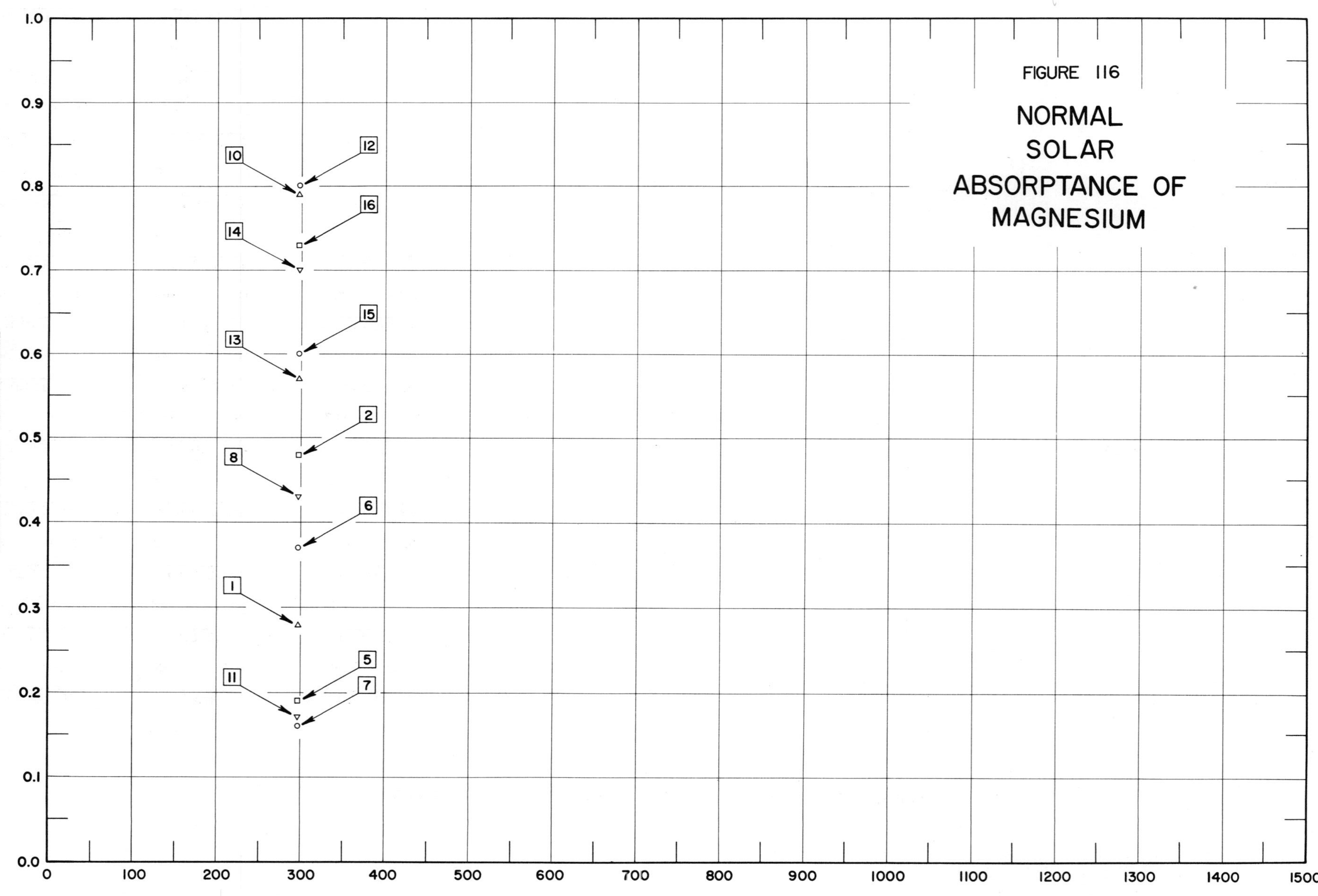

FIGURE 116 NORMAL SOLAR ABSORPTANCE OF MAGNESIUM

SPECIFICATION TABLE NO. 116 NORMAL SOLAR ABSORPTANCE OF MAGNESIUM

Curve No.	Ref. No.	Year	Temperature Range, K	Geometry θ	Reported Error, %	Composition (weight percent), Specifications and Remarks
1	131	1964	298	$\sim 0°$		L120 (British Aircraft Material Specification); nominal composition: 99.8 min Mg, 0.05 Mn, 0.04 Fe, 0.03 Si, 0.03 Zn, 0.02 Al, 0.005 Cu, 0.002 Ni; turned; freshly prepared; calculated from $(1 - \rho)$.
2	131	1964	298	$\sim 0°$		Curve 1 specimen and conditions except exposed within clean laboratory area for one month.
3	131	1964	298	$\sim 0°$		Different sample, same as Curve 1 specimen and conditions.
4	131	1964	298	$\sim 0°$		Curve 3 specimen and conditions except exposed within clean laboratory area for one month.
5	131	1964	298	$\sim 0°$		L120 (British Aircraft Material Specification); nominal composition: 99.8 min Mg, 0.05 Mn, 0.04 Fe, 0.03 Si, 0.03 Zn, 0.02 Al, 0.005 Cu, 0.002 Ni; machine polished; freshly prepared; calculated from $(1-\rho)$.
6	131	1964	298	$\sim 0°$		Curve 5 specimen and conditions except exposed within clean laboratory area for one month.
7	131	1964	298	$\sim 0°$		Different sample, same as Curve 5 specimen and conditions.
8	131	1964	298	$\sim 0°$		Curve 7 specimen and conditions except exposed within clean laboratory area for one month.
9	131	1964	298	$\sim 0°$		Different sample, same as Curve 5 specimen and conditions.
10	131	1964	298	$\sim 0°$		Curve 9 specimen and conditions except exposed to outside atmosphere for one month.
11	131	1964	298	$\sim 0°$		Different sample, same as Curve 5 specimen and conditions.
12	131	1964	298	$\sim 0°$		Curve 11 specimen and conditions except exposed to outside atmosphere for one month.
13	131	1964	298	$\sim 0°$		L120 (British Aircraft Material Specification); nominal composition: 99.8 min Mg, 0.05 Mn, 0.04 Fe, 0.03 Si, 0.03 Zn, 0.02 Al, 0.005 Cu, 0.002 Ni; vapor blasted with 400 mesh alumina; 70 psi nozzle pressure; calculated from $(1 - \rho)$.
14	131	1964	298	$\sim 0°$		Curve 13 specimen and conditions except exposed within clean laboratory area for one month.
15	131	1964	298	$\sim 0°$		Different sample, same as Curve 13 specimen and conditions.
16	131	1964	298	$\sim 0°$		Curve 15 specimen and conditions except exposed within clean laboratory area for one month.

DATA TABLE NO. 116 NORMAL SOLAR ABSORPTANCE OF MAGNESIUM

[Temperature, T, K; Absorptance, α]

T	α	T	α
CURVE 1		CURVE 13	
298	0.28	298	0.57
CURVE 2		CURVE 14	
298	0.48	298	0.70
CURVE 3*		CURVE 15	
298	0.28	298	0.60
CURVE 4*		CURVE 16	
298	0.48	298	0.73
CURVE 5			
298	0.19		
CURVE 6			
298	0.37		
CURVE 7			
298	0.16		
CURVE 8			
298	0.43		
CURVE 9*			
298	0.16		
CURVE 10			
298	0.79		
CURVE 11			
298	0.17		
CURVE 12			
298	0.80		

* Not shown on plot

SPECIFICATION TABLE NO. 117 NORMAL SPECTRAL TRANSMITTANCE OF MAGNESIUM

Curve No.	Ref. No.	Year	Temperature K	Wavelength Range,	Geometry θ θ' ω'	Reported Error, %	Composition (weight percent), Specifications and Remarks
1	290	1965	298	0.0216-0.0510	~0°~0°		Unbacked film (few thousand Å thick); evaporated at several hundred Å sec^{-1}; measured in vacuum (< 10^{-5} mm Hg); slight oxide layer.

DATA TABLE NO. 117 NORMAL SPECTRAL TRANSMITTANCE OF MAGNESIUM

[Wavelength, λ, μ; Transmittance, τ; Temperature, T, K]

λ	τ
CURVE 1*	
T = 298	
0.0216	0.148
0.0231	0.094
0.0244	0.682
0.0251	0.608
0.0258	0.743
0.0265	0.603
0.0290	0.421
0.0305	0.494
0.0330	0.422
0.0345	0.436
0.0360	0.391
0.0369	0.393
0.0378	0.386
0.0402	0.348
0.0418	0.387
0.0436	0.336
0.0445	0.319
0.0465	0.322
0.0490	0.273
0.0510	0.241

* Not shown on plot

SPECIFICATION TABLE NO. 118 NORMAL SPECTRAL EMITTANCE OF MANGANESE

Curve No.	Ref. No.	Year	Wavelength μ	Temperature Range, K	Geometry θ'	Reported Error, %	Composition (weight percent), Specifications and Remarks
1	19	1914	0.65	1473	$\sim 0^\circ$	1	Film; tungsten substrate; measured in hydrogen; Pt reference (ϵ=0.33 for λ 0.650 μ at all temp).

DATA TABLE NO. 118 NORMAL SPECTRAL EMITTANCE OF MANGANESE

[Temperature, T, K; Emittance, ϵ; Wavelength, λ, μ]

T	ϵ
CURVE 1*	
$\lambda = 0.65$	
1473	0.59

* Not shown on plot

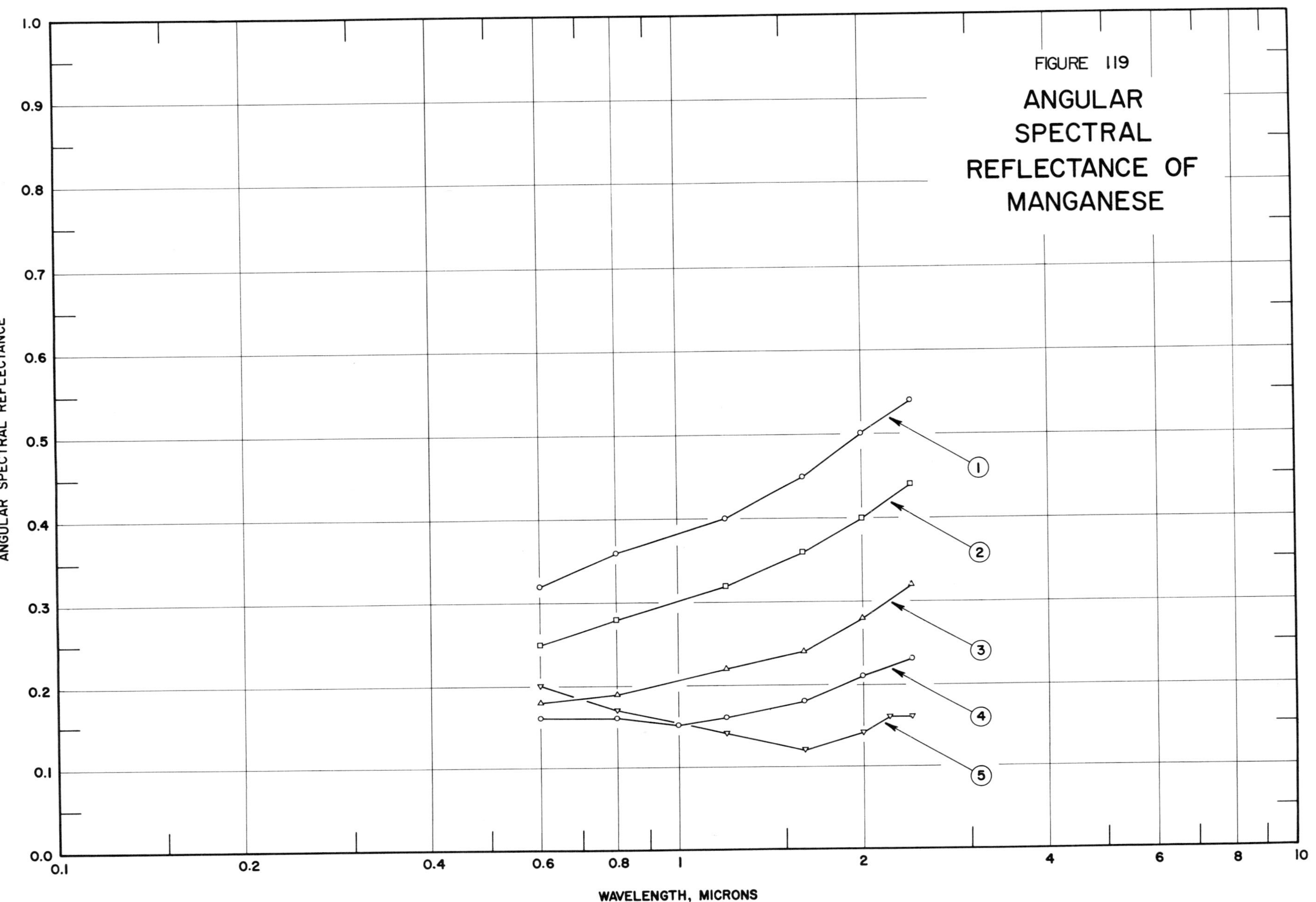

FIGURE 119
ANGULAR SPECTRAL REFLECTANCE OF MANGANESE
1
2
3
4
5
ANGULAR SPECTRAL REFLECTANCE
1.0
0.9
0.8
0.7
0.6
0.5
0.4
0.3
0.2
0.1
0.0
WAVELENGTH, MICRONS
0.1
0.2
0.4
0.6
0.8
1
2
4
6
8
10

SPECIFICATION TABLE NO. 119 ANGULAR SPECTRAL REFLECTANCE OF MANGANESE

Curve No.	Ref. No.	Year	Temperature K	Wavelength Range, μ	Geometry θ θ' ω'	Reported Error, %	Composition (weight percent), Specifications and Remarks
1	141	1948	298	0.6-2.4	$50^\circ 50^\circ$		Evaporated mirror surface; data extracted from smooth curve, incident beam polarized parallel to the plane of incidence.
2	141	1948	298	0.6-2.4	$60^\circ 60^\circ$		Above specimen and conditions, incident beam polarized parallel to the plane of incidence.
3	141	1948	298	0.6-2.4	$70^\circ 70^\circ$		Above specimen and conditions, incident beam polarized parallel to the plane of incidence.
4	141	1948	298	0.6-2.4	$79^\circ 79^\circ$		Above specimen and conditions, incident beam polarized parallel to the plane of incidence.
5	141	1948	298	0.6-2.4	$80^\circ 80^\circ$		Above specimen and conditions, incident beam polarized parallel to the plane of incidence.

DATA TABLE NO. 119 ANGULAR SPECTRAL REFLECTANCE OF MANGANESE

[Wavelength, λ, μ; Reflectance, ρ; Temperature, T, K]

λ	ρ	λ	ρ
CURVE 1		CURVE 5	
T = 298		T = 298	
0.6	0.32	0.6	0.20
0.8	0.36	0.8	0.17
1.2	0.40	1.2	0.14
1.6	0.45	1.6	0.13
2.0	0.50	2.0	0.14
2.4	0.54	2.2	0.16
		2.4	0.16
CURVE 2			
T = 298			
0.6	0.25		
0.8	0.28		
1.2	0.32		
1.6	0.36		
2.0	0.40		
2.4	0.44		
CURVE 3			
T = 298			
0.6	0.18		
0.8	0.19		
1.2	0.22		
1.6	0.24		
2.0	0.28		
2.4	0.32		
CURVE 4			
T = 298			
0.6	0.16		
0.8	0.16		
1.0	0.15		
1.2	0.16		
1.6	0.18		
2.0	0.21		
2.4	0.23		

SPECIFICATION TABLE NO. 120 ANGULAR SPECTRAL ABSORPTANCE OF MANGANESE

Curve No.	Ref. No.	Year	Temperature K	Wavelength Range, μ	Geometry θ	Reported Error, %	Composition (weight percent), Specifications and Remarks
1	225	1965	306	2.02-22.9	25°		Manganese (electrolytic plate) from Belmont Smelting and Refining Works; measured in dry nitrogen; heated cavity at approx 1056 K with platinum reference; authors assumed $\alpha = 1-R(2\pi, 25^{\circ})$.

DATA TABLE NO. 120 ANGULAR SPECTRAL ABSORPTANCE OF MANGANESE

[Wavelength, λ, μ; Absorptance, α; Temperature, T, K]

λ	α
CURVE 1*	
T = 306	
2.02	0.488
2.83	0.433
3.76	0.373
5.05	0.315
6.59	0.281
9.18	0.231
11.7	0.194
13.2	0.176
17.3	0.157
22.9	0.136

* Not shown on plot

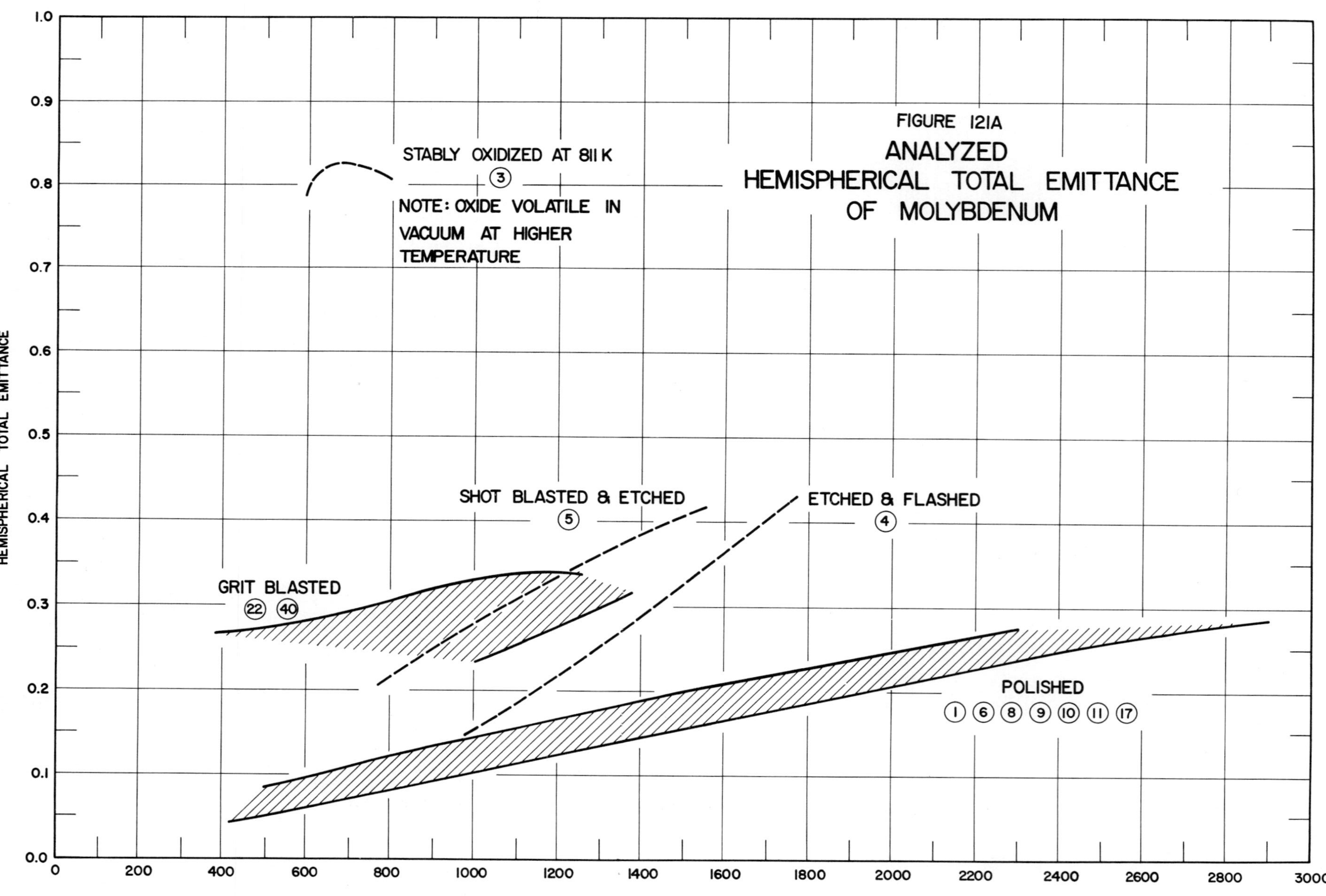
FIGURE 121A
ANALYZED
HEMISPHERICAL TOTAL EMITTANCE
OF MOLYBDENUM
STABLY OXIDIZED AT 811 K
3
NOTE: OXIDE VOLATILE IN
VACUUM AT HIGHER
TEMPERATURE
SHOT BLASTED & ETCHED
5
ETCHED & FLASHED
4
GRIT BLASTED
22 40
POLISHED
1 6 8 9 10 11 17
HEMISPHERICAL TOTAL EMITTANCE
0.0
0.1
0.2
0.3
0.4
0.5
0.6
0.7
0.8
0.9
1.0
0
200
400
600
800
1000
1200
1400
1600
1800
2000
2200
2400
2600
2800
3000
TEMPERATURE, K

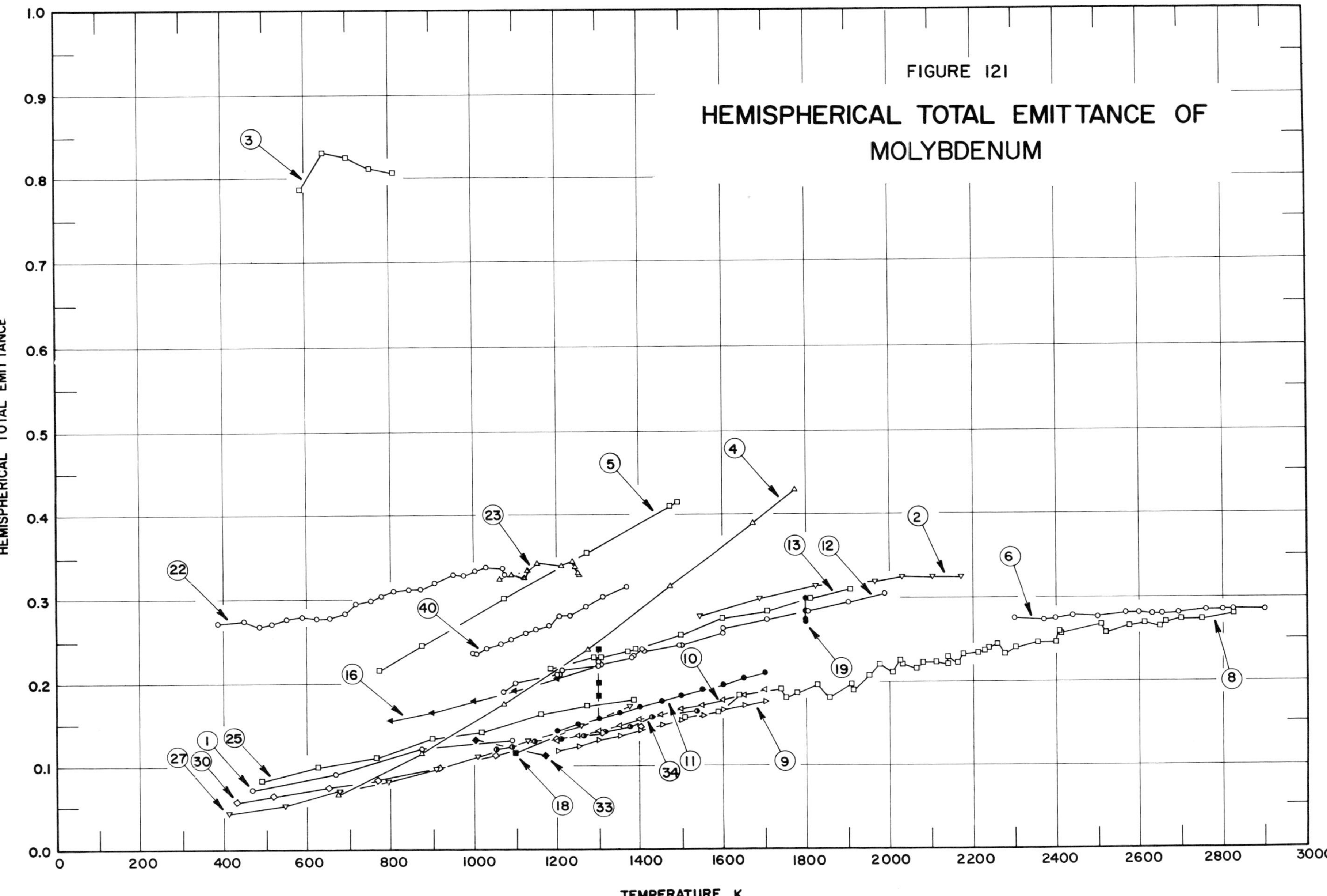

FIGURE 121

HEMISPHERICAL TOTAL EMITTANCE OF MOLYBDENUM

SPECIFICATION TABLE NO. 121 HEMISPHERICAL TOTAL EMITTANCE OF MOLYBDENUM

Curve No.	Ref. No.	Year	Temperature Range, K	Reported Error, %	Composition (weight percent), Specifications and Remarks
1	47	1961	468-1093	<10	Vacuum arc cast, machined, extruded, recrystallized, rolled; disc (0.04 in. thick); ground with 600 grit carborundum and polished on a wet cloth lap with unlevigated jewelers rouge; measured in vacuum (10^{-5} mm Hg).
2	69	1960	1544-2172	±10	Measured in vacuum.
3	22	1958	589-811	≤ 2	Stably oxidized at 811 K in quiescent air.
4	58	1961	673-1773	± 2.5	Lightly etched and flashed in vacuum at 2073 K for 10 min; measured in vacuum (<5 x 10^{-6} mm Hg); data extracted from smooth curve.
5	58	1961	773-1493	± 2.5	Shot-blasted and pickled in hydrochloric acid to remove iron; measured in vacuum (<5 x 10^{-6} mm Hg); data extracted from smooth curve.
6	70	1960	2300-2900		0.18 Fe, 0.073 Si, 0.04 C, 0.036 Mn, 0.005 O_2, 0.01 others, Mo balance; cast under inert gas; hot rolled; successively polished with No. 1-, 0-, 00-, 000-, and 0000- abrasive papers; measured in argon.
7	71	1962	1540-2180	±10	Measured in vacuum (<10^{-5} mm Hg).
8	72	1963	1506-2825		0.07 - 0.09 Fe, 0.04 - 0.06 Nb, 0.001 - 0.003 Mn, 0.001 - 0.003 Si, 0.0004 - 0.0006 Cu, 0.0001 - 0.0005 Mg, Mo balance; thin walled tube; polished using felt with a GOI paste; annealed; measured in vacuum.
9	73	1964	1200-1700	< 2.3	99.96 Mo, 0.004 SiO_2, 0.004 CaO and MgO, 0.026 sesquioxides; prepared by rubbing with abrasive paper; surface roughness 0.063 - 0.050 μ RMS; measured in vacuum (10^{-3} to 10^{-4} mm Hg); [Author's designation: Specimen 1].
10	73	1964	1200-1700	< 2.3	Different sample, same as curve 9 specimen and conditions; [Author's designation: Specimen 2].
11	73	1964	1200-1700	< 2.3	Different sample, same as curve 9 specimen; same conditions except surface roughness 1.25 - 1.00 μ RMS; [Author's designation: Specimen 3].
12	54	1962	1070-1990	± 4	Degreased with acetone, cleaned with a rubber eraser, wiped with acetone; measured in vacuum (10^{-4} to 10^{-6} mm Hg); same data reported for both samples; [Author's designation: Sample No. 1 and Sample No. 2].
13	54	1962	1185-1905	± 4	Degreased with acetone, cleaned with a rubber eraser, wiped with acetone; aged for 1 hr at 1773 K; measured in vacuum (10^{-4} to 10^{-6} mm Hg); [Author's designation: Sample 2].
14	54	1962	1100-1800	± 4	Polished using rouge in wax on a buffing wheel; measured in vacuum (10^{-4} to 10^{-6} mm Hg); [Author's designation: Sample No. 3].
15	54	1962	800-1300	± 4	Polished using fine aluminum oxide powder on a circular rotatable drum with a rotating lap; measured in vacuum (10^{-4} to 10^{-5} mm Hg); cycle 1; [Author's designation: Sample No. 5].
16	54	1962	800-1300	± 4	Above specimen and conditions; cycle 2.
17	54	1962	1400-1800	± 4	Different sample, same as curve 15 specimen and conditions; cycle 1; [Author's designation: Sample No. 6].
18	54	1962	1100-1300	± 4	Above specimen and conditions; cycle 2.

SPECIFICATION TABLE NO. 121 (continued)

Curve No.	Ref. No.	Year	Temperature Range, K	Reported Error, %	Composition (weight percent), Specifications and Remarks
19	54	1962	1800	± 4	Above specimen and conditions; cycle 3.
20	54	1962	1300-1500	± 4	Above specimen and conditions; cycle 4.
21	54	1962	1000-2000	± 4	Above specimen and conditions; cycle 5.
22	12	1962	385.6-1075.1	± 2.7	Grit blasted with aluminum oxide No. 90 (PMC-3043A); measured in vacuum (<2.9 x 10^{-6} mm Hg); Run No. 1.
23	12	1962	1061.1-1251.5	± 2.7	Above specimen and conditions; Run No. 2A.
24	12	1962	1255.1-1235.9	± 2.7	Above specimen and conditions; Run No. 2B.
25	12	1962	491.2-1385.2	± 2.7	Vapor-blasted with Techline Liquabrasive, PMC-3067, grit No. 325; measured in vacuum (<5 x 10^{-6} mm Hg); Run No. 1.
26	12	1962	545.2-1375.7	± 2.7	Above specimen and conditions; Run No. 2.
27	12	1962	412-1373	± 2.7	Above specimen and conditions; Run No. 3.
28	12	1962	410-1244	± 2.7	Above specimen and conditions; Run No. 4A.
29	12	1962	1378-1273	± 2.7	Above specimen and conditions; Run No. 4B.
30	12	1962	429.2-1401.2	± 2.7	Chemically cleaned; measured in vacuum (<5 x 10^{-6} mm Hg); Run No. 1.
31	12	1962	449.7-1405.2	± 2.7	Above specimen and conditions; Run No. 2.
32	12	1962	407.2-1374.2	± 2.7	Above specimen and conditions; Run No. 3A.
33	12	1962	1002.4-1169.2	± 2.7	Above specimen and conditions; Run No. 3B.
34	12	1962	1054-1538	± 2.3	As received; measured in vacuum (<2 x 10^{-6} mm Hg); Run No. 1A.
35	12	1962	1539-1100	± 2.3	Above specimen and conditions; Run No. 1B.
36	12	1962	1242-1540	± 2.3	Above specimen and conditions; Run No. 2.
37	12	1962	1045-1539	± 2.7	As received; measured in vacuum (<2 x 10^{-6} mm Hg); Run No. 1A.
38	12	1962	1535-1097	± 2.7	Above specimen and conditions; Run No. 1B.
39	12	1962	1239-1541	± 2.7	Above specimen and conditions; Run No. 2.
40	12	1962	1368-998	± 2.7	Grit blasted with aluminum oxide No. 90 (PMC-3043A); measured in vacuum (<5.1 x 10^{-6} mm Hg); Run No. 1.

DATA TABLE NO. 121 HEMISPHERICAL TOTAL EMITTANCE OF MOLYBDENUM

[Temperature, T, K; Emittance, ∈]

T	∈
CURVE 1	
468	0.07
668	0.09
873	0.12
1093	0.13
CURVE 2	
1544	0.280
1686	0.300
1822	0.315
1967	0.320
2033	0.325
2103	0.325
2172	0.325
CURVE 3	
589	0.785
644	0.830
700	0.825
755	0.810
811	0.805
CURVE 4	
673	0.065
873	0.115
1073	0.175
1273	0.240
1473	0.315
1673	0.390
1773	0.430
CURVE 5	
773	0.215
873	0.245
1073	0.300
1273	0.355
1473	0.410
1493	0.415
CURVE 6	
2300	0.276
2370	0.274
2400	0.276
CURVE 6 (cont.)	
2440	0.279
2500	0.278
2565	0.282
2600	0.282
2630	0.281
2655	0.281
2695	0.282
2760	0.285
2800	0.285
2825	0.285
2900	0.285
CURVE 7*	
1540	0.275
1690	0.300
1830	0.310
1970	0.320
2040	0.325
2110	0.326
2180	0.326
CURVE 8	
1506	0.1580
1588	0.1650
1638	0.1850
1738	0.1925
1750	0.1825
1775	0.1875
1825	0.1975
1856	0.1825
1906	0.1987
1912	0.1912
1950	0.2075
1975	0.2200
2006	0.2125
2025	0.2250
2031	0.2200
2062	0.2150
2081	0.2225
2112	0.2225
2137	0.2200
2137	0.2275
2162	0.2225
CURVE 8 (cont.)	
2175	0.2325
2212	0.2337
2225	0.2375
2218	0.2400
2237	0.2450
2275	0.2325
2300	0.2400
2356	0.2475
2400	0.2475
2406	0.2600
2412	0.2575
2518	0.2575
2506	0.2687
2575	0.2675
2612	0.2700
2662	0.2712
2650	0.2662
2700	0.2750
2750	0.2750
2825	0.2812
CURVE 9	
1200	0.117
1250	0.123
1300	0.130
1350	0.136
1400	0.143
1450	0.149
1500	0.155
1550	0.161
1600	0.167
1650	0.173
1700	0.178
CURVE 10	
1200	0.124
1250	0.136
1300	0.142
1350	0.149
1400	0.156
1450	0.162
1500	0.168
1550	0.174
1600	0.180
CURVE 10 (cont.)	
1650	0.186
1700	0.192
CURVE 11	
1200	0.143
1250	0.150
1300	0.157
1350	0.164
1400	0.171
1450	0.178
1500	0.185
1550	0.192
1600	0.198
1650	0.205
1700	0.211
CURVE 12	
1070	0.190
1100	0.200
1200	0.210
1205	0.210
1210	0.215
1300	0.220
1300	0.225
1305	0.223
1380	0.230
1385	0.233
1405	0.240
1410	0.237
1495	0.245
1500	0.245
1500	0.245*
1600	0.260
1600	0.263*
1600	0.265
1705	0.275
1705	0.275*
1805	0.285
1900	0.245
1990	0.305
CURVE 13	
1185	0.217
1200	0.217
CURVE 13 (cont.)	
1290	0.230
1305	0.230
1370	0.237
1390	0.240
1500	0.257
1600	0.277
1705	0.285
1810	0.300
1905	0.310
CURVE 14*	
1100	0.195
1195	0.207
1400	0.235
1600	0.255
1800	0.280
CURVE 15*	
800	0.085
900	0.090
1000	0.100
1100	0.113
1200	0.132
1300	0.157
1310	0.170
1310	0.177
1310	0.185
1300	0.210
1300	0.217
1300	0.222
CURVE 16	
800	0.155
900	0.165
1000	0.180
1100	0.192
1200	0.205
1300	0.222*
CURVE 17*	
1400	0.160
1600	0.190
1800	0.217
CURVE 18	
1100	0.115
1200	0.135*
1300	0.157*
1300	0.185
1300	0.200
1300	0.240
CURVE 19	
1800	0.300
1800	0.275
1800	0.273
1800	0.285
CURVE 20*	
1300	0.215
1300	0.230
1500	0.262
CURVE 21*	
1000	0.185
1100	0.200
1200	0.215
1300	0.230
1400	0.247
1600	0.280
1700	0.295
1800	0.305
1900	0.310
2000	0.312
CURVE 22	
385.6	0.272
446.7	0.276
485.6	0.268
514.6	0.272
554.2	0.277
587.8	0.280
621.3	0.277
655.0	0.278
688.9	0.284
716.6	0.295
750.4	0.298
751.8	0.297*
CURVE 22 (cont.)	
778.2	0.303
808.5	0.310
845.3	0.312
870.6	0.312
904.0	0.320
952.2	0.330
975.7	0.329
1001.7	0.334
1030.2	0.339
1071.5	0.337
1075.1	0.330
CURVE 23	
1061.1	0.323
1092.2	0.329
1123.6	0.326
1134.1	0.335
1156.4	0.344
1210.2	0.341
1234.8	0.346
1241.6	0.339
1247.8	0.332
1251.5	0.328
CURVE 24*	
1255.1	0.324
1235.9	0.320
CURVE 25	
491.2	0.082
625.2	0.099
762.2	0.109
900.2	0.133
1018.2	0.141
1161.2	0.163
1270.2	0.173
1385.2	0.180
CURVE 26*	
545.2	0.055
685.7	0.065
813.2	0.082
936.2	0.101

* Not shown on plot

DATA TABLE NO. 121 (continued)

T	∈
CURVE 26 (cont.)	
545.2	0.055
685.7	0.065
813.2	0.082
936.2	0.101
1084.2	0.126
1191.7	0.143
1308.7	0.163
1375.7	0.173
CURVE 27	
412	0.043
545	0.052
675	0.068
797	0.081
909	0.096
1007	0.111
1131	0.129
1255	0.148
1373	0.172
CURVE 28*	
410	0.038
553	0.050
646	0.061
768	0.073
892	0.088
990	0.100
1114	0.121
1244	0.139
CURVE 29*	
1378	0.163
1273	0.144
CURVE 30	
429.2	0.056
520.2	0.063
648.2	0.073
763.2	0.082
919.2	0.097
1054.2	0.112
1198.2	0.128
1307.7	0.139
1401.2	0.147

T	∈
CURVE 31*	
449.7	0.043
594.2	0.054
696.2	0.062
784.2	0.069
913.7	0.081
1052.9	0.098
1177.2	0.112
1318.2	0.135
1405.2	0.146
CURVE 32*	
407.2	0.034
502.8	0.041
590.2	0.055
767.0	0.066
889.2	0.077
1159.2	0.113
1283.2	0.130
1374.2	0.142
CURVE 33	
1002.4	0.130
1169.2	0.112
CURVE 34	
1054	0.118
1089	0.123
1145	0.128
1209	0.132
1263	0.136
1315	0.142
1374	0.147
1426	0.158
1538	0.165
CURVE 35*	
1539	0.165
1481	0.156
1431	0.149
1387	0.143
1335	0.135
1290	0.128
1243	0.122
1188	0.115

T	∈
CURVE 35 (cont.)	
1135	0.109
1100	0.104
CURVE 36*	
1242	0.122
1290	0.128
1337	0.134
1387	0.143
1431	0.149
1481	0.156
1540	0.165
CURVE 37*	
1045	0.122
1083	0.125
1140	0.131
1201	0.135
1258	0.139
1311	0.143
1366	0.150
1475	0.158
1539	0.165
CURVE 38*	
1535	0.167
1480	0.156
1429	0.150
1385	0.143
1333	0.136
1284	0.131
1237	0.124
1184	0.117
1131	0.110
1097	0.105
CURVE 39*	
1239	0.124
1286	0.130
1333	0.136
1385	0.144
1429	0.150
1480	0.156
1541	0.164

T	∈
CURVE 40	
1368	0.314
1308	0.302
1266	0.290
1233	0.281
1204	0.280
1180	0.269
1149	0.264
1125	0.260
1093	0.253
1063	0.248
1034	0.242
1004	0.235
998	0.236

* Not shown on plot

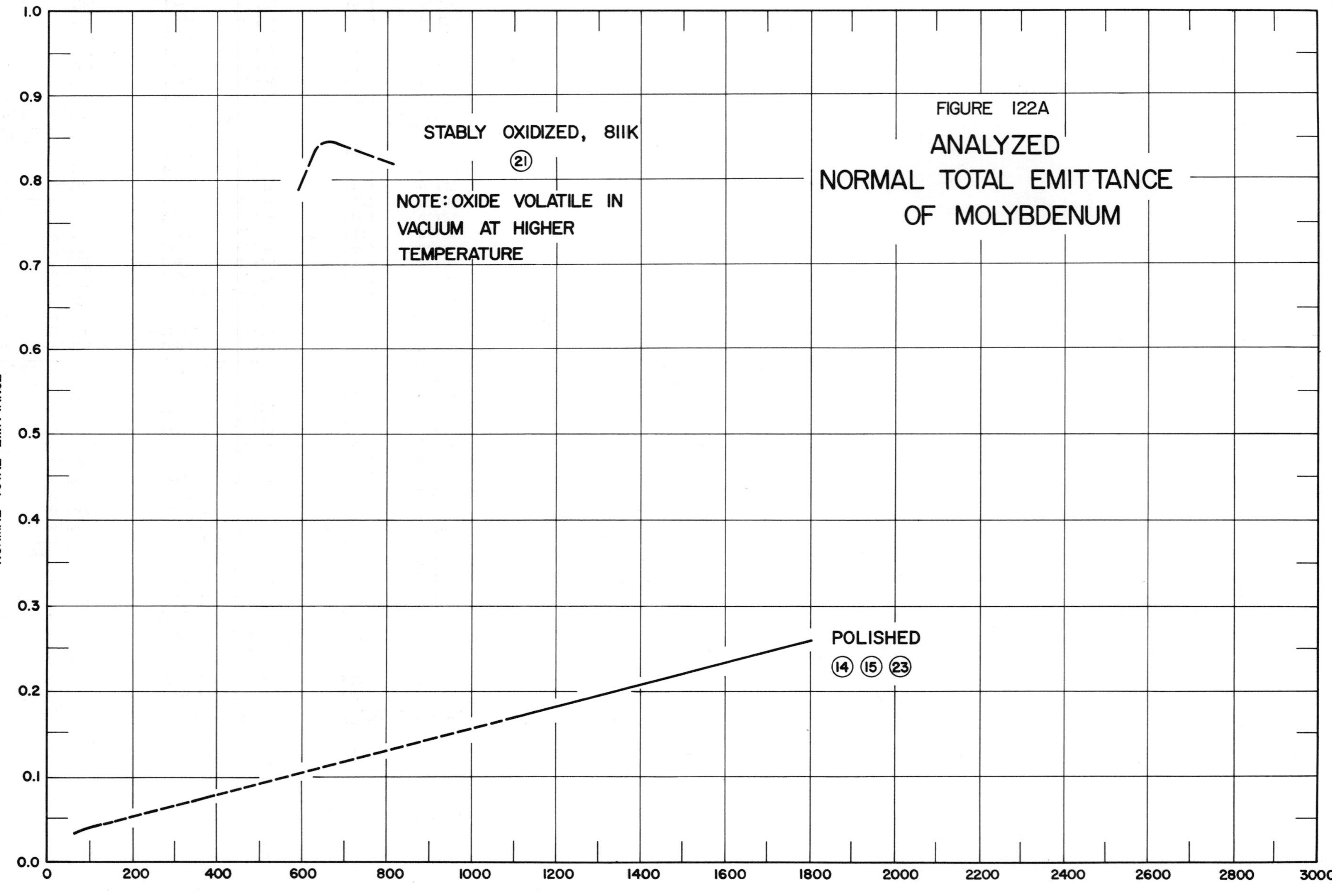
FIGURE 122A
ANALYZED
NORMAL TOTAL EMITTANCE
OF MOLYBDENUM
STABLY OXIDIZED, 811K
㉑
NOTE: OXIDE VOLATILE IN
VACUUM AT HIGHER
TEMPERATURE
POLISHED
⑭ ⑮ ㉓
NORMAL TOTAL EMITTANCE
1.0
0.9
0.8
0.7
0.6
0.5
0.4
0.3
0.2
0.1
0.0
0
200
400
600
800
1000
1200
1400
1600
1800
2000
2200
2400
2600
2800
3000
TEMPERATURE, K

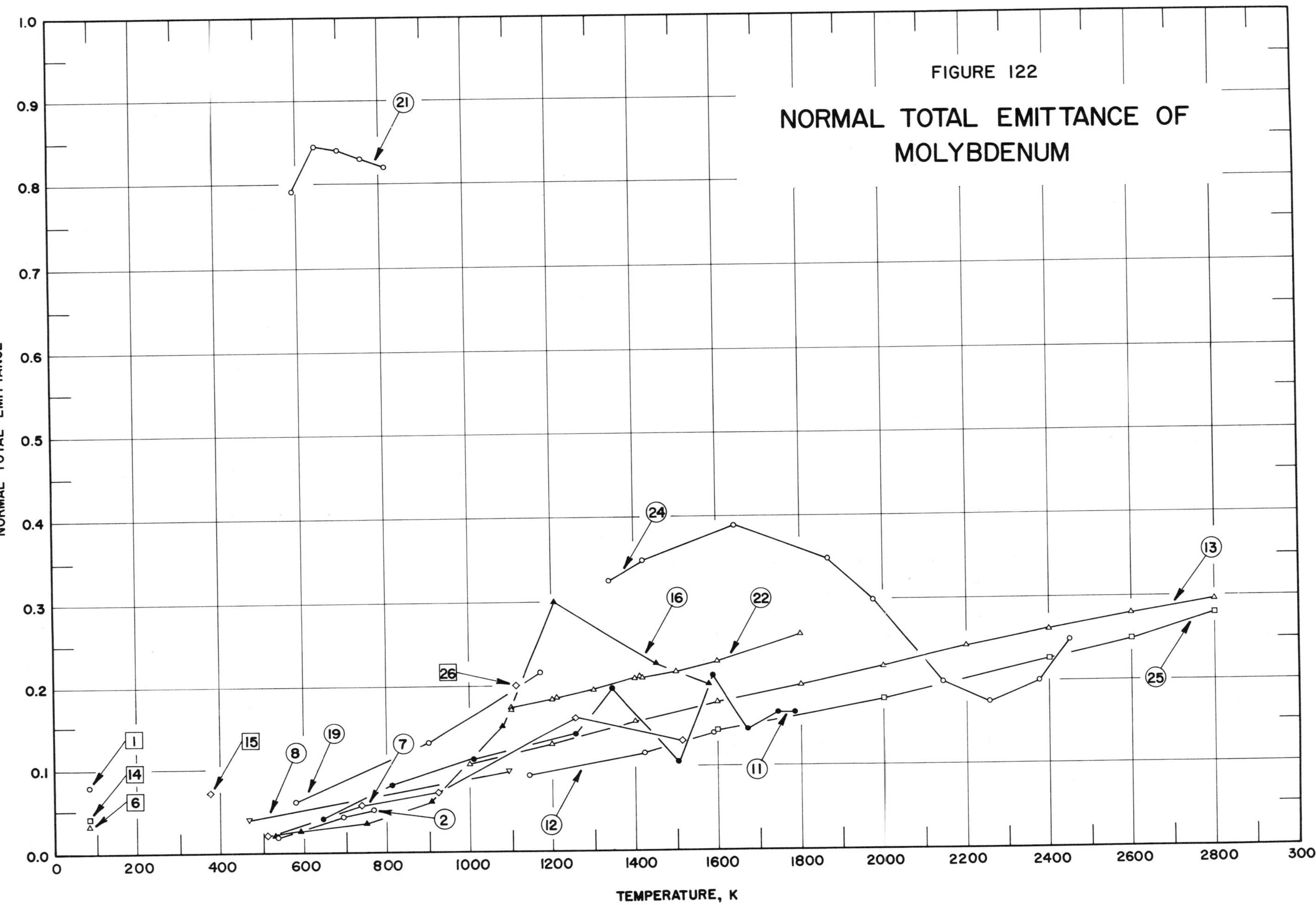
FIGURE 122
NORMAL TOTAL EMITTANCE OF MOLYBDENUM
NORMAL TOTAL EMITTANCE
TEMPERATURE, K
0.0
0.1
0.2
0.3
0.4
0.5
0.6
0.7
0.8
0.9
1.0
0
200
400
600
800
1000
1200
1400
1600
1800
2000
2200
2400
2600
2800
3000
1
14
6
15
8
19
7
2
26
12
16
22
24
11
21
13
25

SPECIFICATION TABLE NO. 122 NORMAL TOTAL EMITTANCE OF MOLYBDENUM

Curve No.	Ref. No.	Year	Temperature Range, K	Geometry θ'	Reported Error, %	Composition (weight percent), Specifications and Remarks
1	34	1957	83.2	~0°	±10	Arc melted, unalloyed (Climax Molybdenum Co.); strip (0.005 in. thick); oxidized in air at red heat for 30 min; measured in air (5×10^{-4} mm Hg).
2	34	1957	536-766	~0°	±10	Arc melted, unalloyed (Climax Molybdenum Co.); strip (0.005 in. thick); oxidized in air at red heat for 30 min; measured in air (5×10^{-4} mm Hg); increasing temp; cycle 1.
3	34	1957	744-519	~0°	±10	Above specimen and conditions; decreasing temp, cycle 1.
4	34	1957	544-736	~0°	±10	Above specimen and conditions; cycle 2.
5	34	1957	780	~0°	±10	Above specimen and conditions; cycle 3.
6	34	1957	83.2	~0°	±10	Arc melted, unalloyed (Climax Molybdenum Co.); strip (0.005 in. thick); as received or cleaned with liquid detergent; measured in air (5×10^{-4} mm Hg).
7	34	1957	516-1511	~0°	±10	Arc melted, unalloyed (Climax Molybdenum Co.); strip (0.005 in. thick); cleaned with liquid detergent; measured in air (5×10^{-4} mm Hg); increasing temp, cycle 1.
8	34	1957	1094-466	~0°	±10	Above specimen and conditions; decreasing temp, cycle 1.
9	34	1957	683-1328	~0°	±10	Above specimen and conditions; cycle 2.
10	34	1957	1100-578	~0°	±10	Above specimen and conditions; decreasing temp, cycle 2.
11	34	1957	644-1783	~0°	±10	Above specimen and conditions; cycle 3.
12	34	1957	1589-1144	~0°	±10	Above specimen and conditions; decreasing temp, cycle 3.
13	74	1955	1000-2800	~0°		Heated electrically at 2100 K for 10 hrs; measured in vacuum (10^{-7} mm Hg); rolled 2 mil sheet.
14	34	1957	83.2	~0°	±10	Arc melted unalloyed (Climax Molybdenum Co.); strip (0.005 in. thick); polished; measured in air (5×10^{-4} mm Hg).
15	15	1947	373	~0°		Polished.
16	34	1957	533-1578	~0°	±10	Arc melted, unalloyed (Climax Molybdenum Co.); strip (0.005 in. thick); as received; measured in air (5×10^{-4} mm Hg); increasing temp, cycle 1.
17	34	1957	1378-700	~0°	±10	Above specimen and conditions; decreasing temp, cycle 1.
18	34	1957	516-1428	~0°	±10	Above specimen and conditions; cycle 2.
19	34	1957	1172-578	~0°	±10	Above specimen and conditions; decreasing temp, cycle 2.
20	34	1957	655-1544	~0°	±10	Above specimen and conditions; cycle 3.
21	22	1958	589-811	~0°	≤ 2	Stably oxidized at 811 K in quiescent air.
22	54	1962	1100-1800	~0°	± 4	Degreased with acetone, cleaned with a rubber eraser, wiped with acetone; measured in vacuum (10^{-4}-10^{-6} mm Hg); [Author's designation: Sample No. 1].
23	54	1962	1100-1800	~0°	± 4	Polished, using rouge in wax, on a buffing wheel; measured in vacuum (10^{-4}-10^{-6} mm Hg); [Author's designation: Sample No. 3].

SPECIFICATION TABLE NO. 122 (continued)

Curve No.	Ref. No.	Year	Temperature Range, K	Geometry θ'	Reported Error, %	Composition (weight percent), Specifications and Remarks
24	75	1962	1339-2450	$\sim 0°$		Ground to a smooth finish; measured in dried argon or helium; data extracted from smooth curve.
25	76	1962	1600-2800	$\sim 0°$		Prepared from micronized powder; hot pressed at >2273 K; sintered, polished, etched, then degassed by heating to ~973 K; measured in argon; data extracted from smooth curve; computed from spectral data.
26	30	1963	1111	$\sim 0°$		Mechanically polished with aluminum oxide and cleaned with water; heated at 1144 K for 1 hr; measured in vacuum (5×10^{-6} mm Hg); computed from spectral emittance.

DATA TABLE NO. 122 NORMAL TOTAL EMITTANCE OF MOLYBDENUM

[Temperature, T, K; Emittance, ∈]

T	∈
CURVE 1	
83.2	0.079
CURVE 2	
536	0.018
694	0.043
766	0.050
CURVE 3*	
744	0.050
700	0.046
519	0.018
CURVE 4*	
544	0.017
736	0.043
CURVE 5*	
780	0.051
CURVE 6	
83.2	0.033
CURVE 7	
516	0.020
739	0.055
922	0.070
1255	0.160
1511	0.130
CURVE 8	
1094	0.095
466	0.040
CURVE 9*	
683	0.035
961	0.080
1144	0.110
1328	0.165
CURVE 10*	
1100	0.120
578	0.035
CURVE 11	
644	0.040
811	0.080
1011	0.110
1255	0.140
1344	0.195
1505	0.105
1589	0.210
1672	0.145
1744	0.165
1783	0.165
CURVE 12	
1589	0.140
1422	0.115
1144	0.090
CURVE 13	
1000	0.104
1200	0.127
1400	0.154
1600	0.178
1800	0.199
2000	0.219
2200	0.243
2400	0.262
2600	0.281
2800	0.298
CURVE 14	
83.2	0.041
CURVE 15	
373	0.071
CURVE 16	
533	0.020
589	0.025
750	0.035
905	0.060
1078	0.150
1211	0.300
1450	0.225
1578	0.200
CURVE 17*	
1378	0.185
1117	0.140
700	0.050
CURVE 18*	
516	0.025
1061	0.155
1428	0.180
CURVE 19	
1172	0.215
900	0.130
578	0.060
CURVE 20*	
655	0.075
1116	0.220
1266	0.305
1544	0.200
CURVE 21	
589	0.790
644	0.845
700	0.840
755	0.830
811	0.820
CURVE 22	
1100	0.170
1100	0.173
1200	0.183
1210	0.185
1300	0.195
1400	0.207
1410	0.210
1420	0.207
1500	0.215
1600	0.227
1800	0.260
CURVE 23*	
1100	0.170
1100	0.173
1200	0.183
1210	0.185
1300	0.195
1400	0.207
1410	0.210
1420	0.207
1500	0.215
1600	0.227
1800	0.260
CURVE 24	
1339	0.325
1422	0.350
1644	0.390
1867	0.350
1978	0.300
2144	0.200
CURVE 24 (cont.)	
2255	0.175
2378	0.200
2450	0.250
CURVE 25	
1600	0.144
2000	0.180
2400	0.225
2600	0.250
2800	0.280
CURVE 26	
1111	0.20

* Not shown on plot

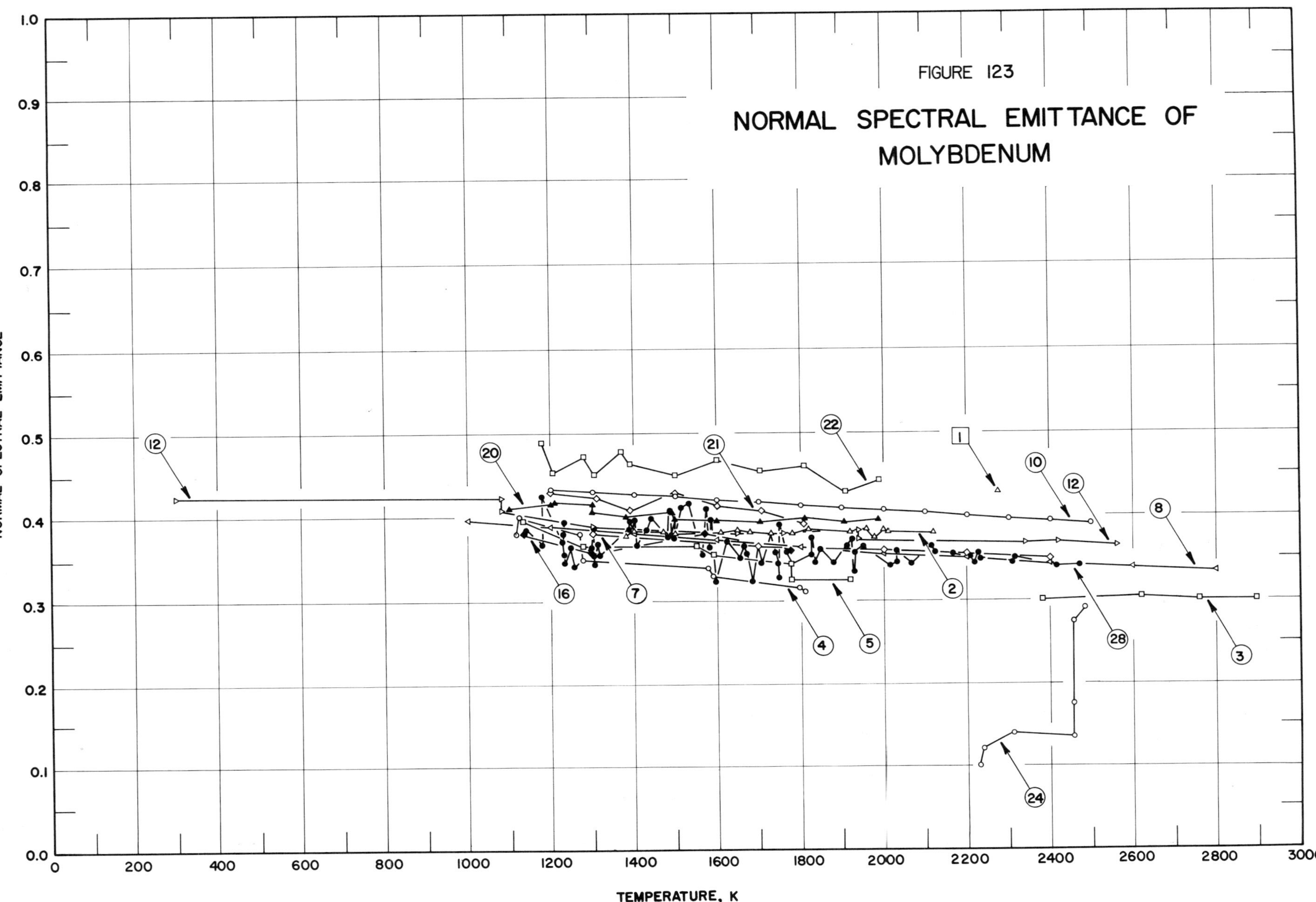
FIGURE 123
NORMAL SPECTRAL EMITTANCE OF MOLYBDENUM
NORMAL SPECTRAL EMITTANCE
TEMPERATURE, K
0.0
0.1
0.2
0.3
0.4
0.5
0.6
0.7
0.8
0.9
1.0
0
200
400
600
800
1000
1200
1400
1600
1800
2000
2200
2400
2600
2800
3000
1
2
3
4
5
7
8
10
12
12
16
20
21
22
24
28

SPECIFICATION TABLE NO. 123 NORMAL SPECTRAL EMITTANCE OF MOLYBDENUM

Curve No.	Ref. No.	Year	Wavelength μ	Temperature Range, K	Geometry θ'	Reported Error, %	Composition (weight percent), Specifications and Remarks
1	19	1914	0.65	2273	~0°	1	Film; tungsten substrate; measured in hydrogen; Pt reference (ϵ = 0.33 for λ = 0.650 μ at all temp).
2	55	1935	0.667	1380-2120	~0°		Baked for 100 hrs, outgassed for 400 hrs; measured in vacuum (4 x 10^{-8} mm Hg).
3	70	1960	0.65	2385-2895	~0°		0.18 Fe, 0.073 Si, 0.04 C, 0.036 Mn, 0.01 others, 0.005 O_2, Mo balance; cast under inert gas in water cooled copper mold using consumable electrode; hot worked and hot rolled; successively polished with No. 1-, 0-, 00-, 000-, and 0000-abrasive papers; measured in argon.
4	34	1957	0.665	1114-1811	~0°		Arc melted,unalloyed (Climax Molybdenum Co.); same data obtained for 3 different surface treatments: (a. as received, b. cleaned with liquid detergent, c. polished); measured in air (5 x 10^{-4} mm Hg); increasing temp, cycle 1.
5	34	1957	0.665	1133-1922	~0°		Above specimen and conditions, cycle 2.
6	34	1957	0.665	1102-1800	~0°		Above specimen and conditions, cycle 3.
7	77	1936	0.660	1300-2400	~0°	± 2	Extruded from paste; sintered in hydrogen; polished electrolytically in KOH bath; polished with 00, 000, and 0000 polishing papers; measured in vacuum (10^{-5} mm Hg).
8	74	1955	0.642	1000-2800	~0°		Heated electrically at 2100 K for 10 hrs; measured in vacuum (10^{-7} mm Hg).
9	78	1927	0.652	1200-2500	~0°		Measured in vacuum.
10	78	1927	0.541	1200-2500	~0°		Measured in vacuum.
11	74	1955	0.665	1000-2800	~0°	< 5	Heated electrically at 2100 K for 10 hrs; measured in vacuum (10^{-7} mm Hg).
12	79	1925	0.475	300-2560	~0°		High purity.
13	79	1925	0.665	300-2550	~0°		High purity.
14	80	1956	0.665	1117-1815	~0°	< 1.1	As received; measured in vacuum (1.0-3.0 x 10^{-4} mm Hg).
15	80	1956	0.665	1114-1984	~0°	< 0.8	As received; measured in vacuum (2.0-9.0 x 10^{-4} mm Hg).
16	80	1956	0.665	1133-1776	~0°	< 0.9	As received; measured in vacuum (1.5-6.0 x 10^{-4} mm Hg).
17	80	1956	0.665	1118-1803	~0°	< 1.0	Polished; measured in vacuum (1.0-6.0 x 10^{-4} mm Hg).
18	80	1956	0.665	1132-1775	~0°	< 1.2	Polished; measured in vacuum (2.0-9.0 x 10^{-4} mm Hg).
19	80	1956	0.665	1103-1802	~0°	< 0.8	Polished; measured in vacuum (1.0-8.0 x 10^{-4} mm Hg).
20	54	1962	0.65	1100-1990	~0°	± 10-20	Degreased with acetone, cleaned with a rubber eraser, wiped with acetone; measured in vacuum (10^{-4} - 10^{-6} mm Hg); same data reported for both samples; [Author's designation: Sample No. 1 and Sample No. 2]; initial heating.
21	54	1962	0.65	1810-1200	~0°	± 10-20	Degreased with acetone; cleaned with a rubber eraser, wiped with acetone; measured in vacuum (10^{-4} - 10^{-6} mm Hg); [Author's designation: Sample No. 2]; cooling.
22	54	1962	0.65	1180-1990	~0°	± 10-20	Degreased with acetone, cleaned with a rubber eraser, wiped with acetone; aged for 1 hr at 1773 K; measured in vacuum (10^{-4} - 10^{-6} mm Hg); [Author's designation: Sample No. 2]; heating and cooling cycle.

SPECIFICATION TABLE NO. 123 (continued)

Curve No.	Ref. No.	Year	Wavelength μ	Temperature Range, K	Geometry θ'	Reported Error, %	Composition (weight percent), Specifications and Remarks
23	54	1962	0.65	1100-1800	~0°	±20	Polished, using rouge in wax, on a buffing wheel; measured in vacuum (10^{-4}-10^{-6} mm Hg); [Author's designation: Sample No. 3]; heating cycle.
24	75	1962	0.69	2228-2483	~0°		Ground to a smooth finish; measured in dried argon.
25	177	1933	0.660	1300-2000	~0°		Electrolytically polished in KOH and then by using 00, 000, 0000 polishing papers; measured in vacuum.
26	325	1963	0.65	1166-2258	~0°		Polished; mean square deviation of roughness 0.5 μ; measured in vacuum.
27	325	1963	0.65	1168-2369	~0°		Specimen from the cutter; mean square deviation of roughness 1.5 μ; measured in vacuum.
28	325	1963	0.65	1171-2470	~0°		Foil; mean square deviation of roughness 0.28 μ; measured in vacuum.

DATA TABLE NO 123 NORMAL SPECTRAL EMITTANCE OF MOLYBDENUM

[Temperature, T, K; Emittance, ∈; Wavelength, λ, μ]

T	∈
CURVE 1 $\lambda = 0.65$	
2273	0.43
CURVE 2 $\lambda = 0.667$	
1380	0.378
1400	0.384
1470	0.383
1500	0.380
1610	0.382
1650	0.384
1680	0.382
1730	0.380
1730	0.382*
1780	0.380
1820	0.385
1920	0.382
1960	0.385
1980	0.375
2000	0.384
2010	0.382
2120	0.381
CURVE 3 $\lambda = 0.65$	
2385	0.300
2620	0.303
2760	0.300
2895	0.300
CURVE 4 $\lambda = 0.665$	
1114	0.38
1122	0.40
1269	0.38
1277	0.35
1577	0.34
1589	0.33
1797	0.315
1811	0.31
CURVE 5 $\lambda = 0.665$	
1133	0.395
1277	0.365
1547	0.365
1589	0.355
1777	0.345
1777	0.325
1922	0.325
CURVE 6* $\lambda = 0.665$	
1102	0.390
1147	0.380
1325	0.365
1566	0.335
1577	0.375
1777	0.355
1800	0.340
CURVE 7 $\lambda = 0.660$	
1300	0.380
1700	0.365
2000	0.360
2200	0.355
2400	0.350
CURVE 8 $\lambda = 0.642$	
1000	0.396
1200	0.388
1400	0.380
1600	0.372
1800	0.364
2000	0.356
2200	0.350
2400	0.344
2600	0.339
2800	0.334
CURVE 9* $\lambda = 0.652$	
1200	0.386
1300	0.385
1400	0.383
1500	0.381
1600	0.380
1700	0.378
1800	0.376
1900	0.375
2000	0.373
2100	0.371
2200	0.370
2300	0.368
2400	0.366
2500	0.365
CURVE 10 $\lambda = 0.541$	
1200	0.434
1300	0.431
1400	0.427
1500	0.424
1600	0.420
1700	0.417
1800	0.413
1900	0.410
2000	0.407
2100	0.404
2200	0.400
2300	0.397
2400	0.394
2500	0.390
CURVE 11* $\lambda = 0.665$	
1000	0.391
1200	0.383
1400	0.376
1600	0.368
1800	0.361
2000	0.354
CURVE 11 (cont.)*	
2200	0.348
2400	0.342
2600	0.337
2800	0.332
CURVE 12 $\lambda = 0.475$	
300	0.424
1080	0.422
1080	0.407
1300	0.389
1650	0.386*
1760	0.380
1940	0.384
1940	0.373
2340	0.368
2420	0.370
2560	0.364
CURVE 13* $\lambda = 0.665$	
300	0.420
760	0.400
1080	0.390
1080	0.393
1280	0.377
1280	0.382
1280	0.384
1420	0.372
1520	0.366
1550	0.362
1550	0.372
1620	0.369
1650	0.363
1760	0.358
1860	0.345
1860	0.347
1860	0.349
1860	0.354
1860	0.357
1930	0.349
1930	0.356
1940	0.352
CURVE 13 (cont.)*	
2000	0.344
2060	0.345
2060	0.353
2120	0.346
2220	0.345
2240	0.347
2260	0.339
2300	0.345
2340	0.343
2420	0.334
2550	0.335
2550	0.337
CURVE 14* $\lambda = 0.665$	
1117	0.400
1277	0.380
1586	0.337
1815	0.312
CURVE 15* $\lambda = 0.665$	
1114	0.398
1276	0.365
1546	0.369
1768	0.340
1984	0.328
CURVE 16 $\lambda = 0.665$	
1133	0.380
1296	0.358
1569	0.380
1776	0.359
CURVE 17* $\lambda = 0.665$	
1118	0.38
1282	0.35
CURVE 17 (cont.)*	
1574	0.34
1803	0.31
CURVE 18* $\lambda = 0.665$	
1132	0.39
1292	0.35
1593	0.36
1775	0.32
CURVE 19* $\lambda = 0.665$	
1103	0.390
1271	0.358
1577	0.340
1802	0.330
CURVE 20 $\lambda = 0.65$	
1100	0.410
1200	0.414
1210	0.407
1300	0.415
1300	0.405
1380	0.400
1490	0.404
1500	0.397
1600	0.395
1705	0.393
1810	0.398
1905	0.393
1990	0.396
CURVE 21 $\lambda = 0.65$	
1810	0.392
1710	0.407
1600	0.413
1500	0.430
CURVE 21 (cont.)	
1390	0.408
1310	0.424
1200	0.430
CURVE 22 $\lambda = 0.65$	
1180	0.488
1205	0.453
1280	0.472
1305	0.451
1370	0.478
1390	0.463
1500	0.450
1600	0.467
1705	0.454
1810	0.460
1910	0.430
1990	0.444
CURVE 23* $\lambda = 0.65$	
1100	0.357
1200	0.360
1400	0.371
1600	0.371*
1800	0.350
CURVE 24 $\lambda = 0.69$	
2228	0.100
2236	0.120
2311	0.140
2455	0.135
2455	0.175
2455	0.275
2483	0.290
CURVE 25* $\lambda = 0.660$	
1300	0.378
1700	0.364
2000	0.360
CURVE 26* $\lambda = 0.65$	
1166	0.393
1235	0.380
1260	0.355
1279	0.362
1298	0.367
1363	0.374
1394	0.374
1475	0.378
1519	0.386
1531	0.380
1552	0.376
1574	0.365
1601	0.370
1625	0.377
1669	0.356
1672	0.369
1717	0.357
1738	0.367
1769	0.368
1859	0.361
1916	0.370
1979	0.359
2032	0.361
2059	0.370
2081	0.362
2168	0.358
2198	0.361
2258	0.356
CURVE 27* $\lambda = 0.65$	
1168	0.401
1173	0.429
1201	0.415
1216	0.396
1225	0.409
1243	0.365
1250	0.423
1254	0.386
1285	0.392
1298	0.414
1307	0.373
1372	0.403
1375	0.435

* Not shown on plot

DATA TABLE NO. 123 (continued)

T	∈	T	∈	T	∈
CURVE 27 (cont.)*		CURVE 28 λ = 0.65		CURVE 28 (cont.)	
1395	0.434	1171	0.383	1747	0.328
1399	0.395	1180	0.367	1753	0.390
1473	0.450	1180	0.424	1764	0.357
1477	0.436	1227	0.371	1774	0.344*
1481	0.412	1230	0.355	1824	0.354
1494	0.389	1230	0.380	1824	0.373
1558	0.386	1233	0.395	1837	0.347
1558	0.401	1233	0.346	1844	0.361
1567	0.456	1243	0.363	1879	0.347
1572	0.385	1257	0.341	1912	0.363
1594	0.451	1293	0.362	1925	0.372
1663	0.392	1300	0.386	1931	0.358
1669	0.383	1301	0.355	1931	0.335
1674	0.446	1304	0.344	1954	0.364
1707	0.424	1311	0.368	2015	0.342
1726	0.432	1314	0.354	2033	0.344
1731	0.374	1386	0.384	2033	0.357
1773	0.426	1389	0.395	2066	0.345
1785	0.435	1394	0.381*	2112	0.363
1810	0.433	1397	0.395	2122	0.356
1814	0.394	1403	0.366	2163	0.354
1820	0.382	1432	0.385	2205	0.353
1831	0.394	1440	0.400	2219	0.345
1914	0.424	1481	0.376	2223	0.357
1914	0.373	1487	0.408	2234	0.350
1917	0.415	1490	0.379	2307	0.345
1917	0.396	1492	0.397*	2313	0.351
1940	0.388	1497	0.374	2417	0.340
1993	0.386	1511	0.410	2470	0.342
1993	0.372	1533	0.416		
2012	0.421	1562	0.355		
2020	0.387	1574	0.409		
2036	0.417	1577	0.382		
2095	0.425	1581	0.362		
2120	0.383	1584	0.397		
2149	0.382	1598	0.322		
2191	0.392	1623	0.371		
2204	0.380	1658	0.350		
2230	0.370	1663	0.363		
2253	0.362	1672	0.355		
2273	0.381	1685	0.322		
2295	0.373	1695	0.363*		
2369	0.370	1708	0.345		
		1729	0.381*		
		1740	0.357		
		1743	0.345		

* Not shown on plot

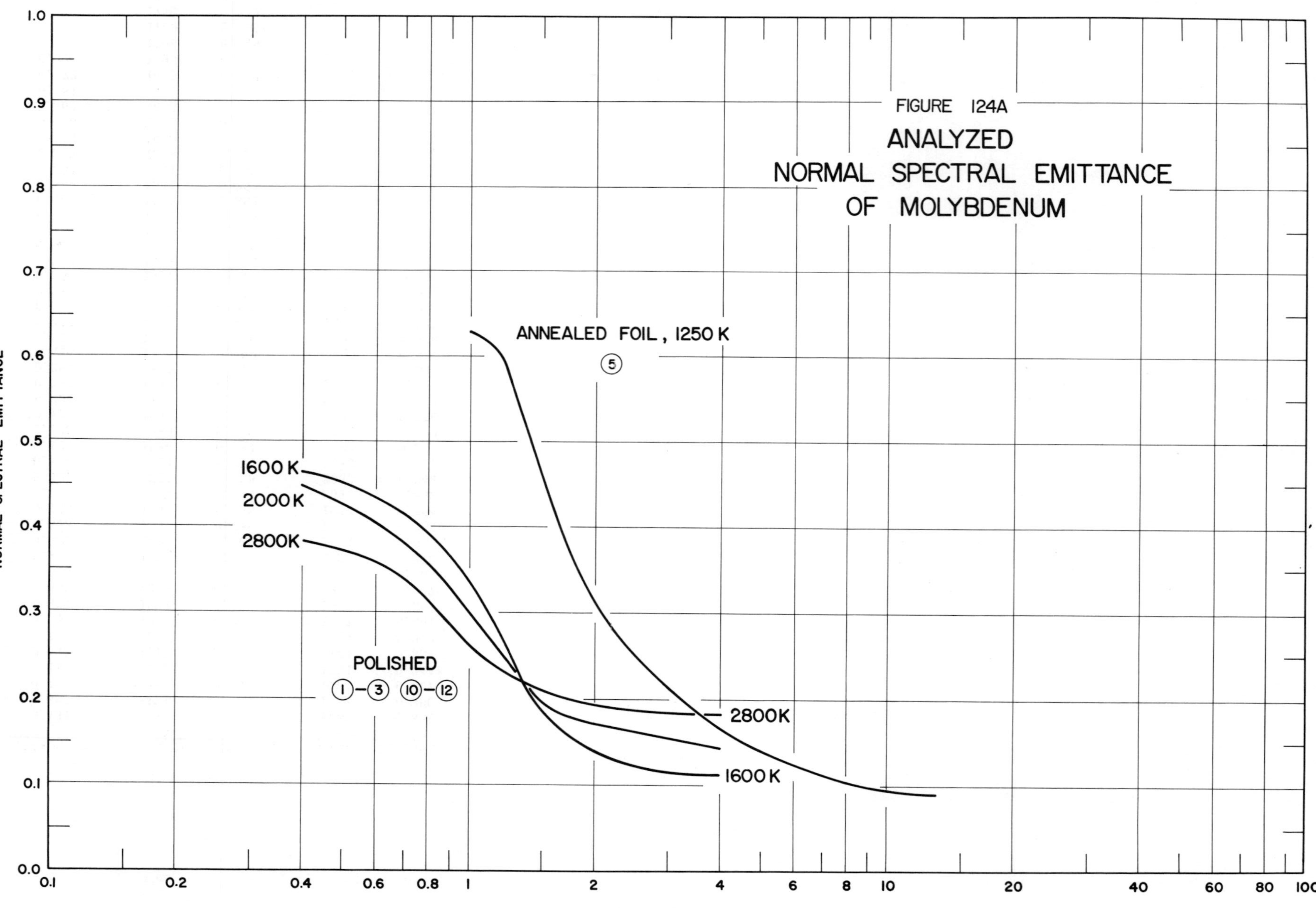

FIGURE 124A
ANALYZED
NORMAL SPECTRAL EMITTANCE
OF MOLYBDENUM

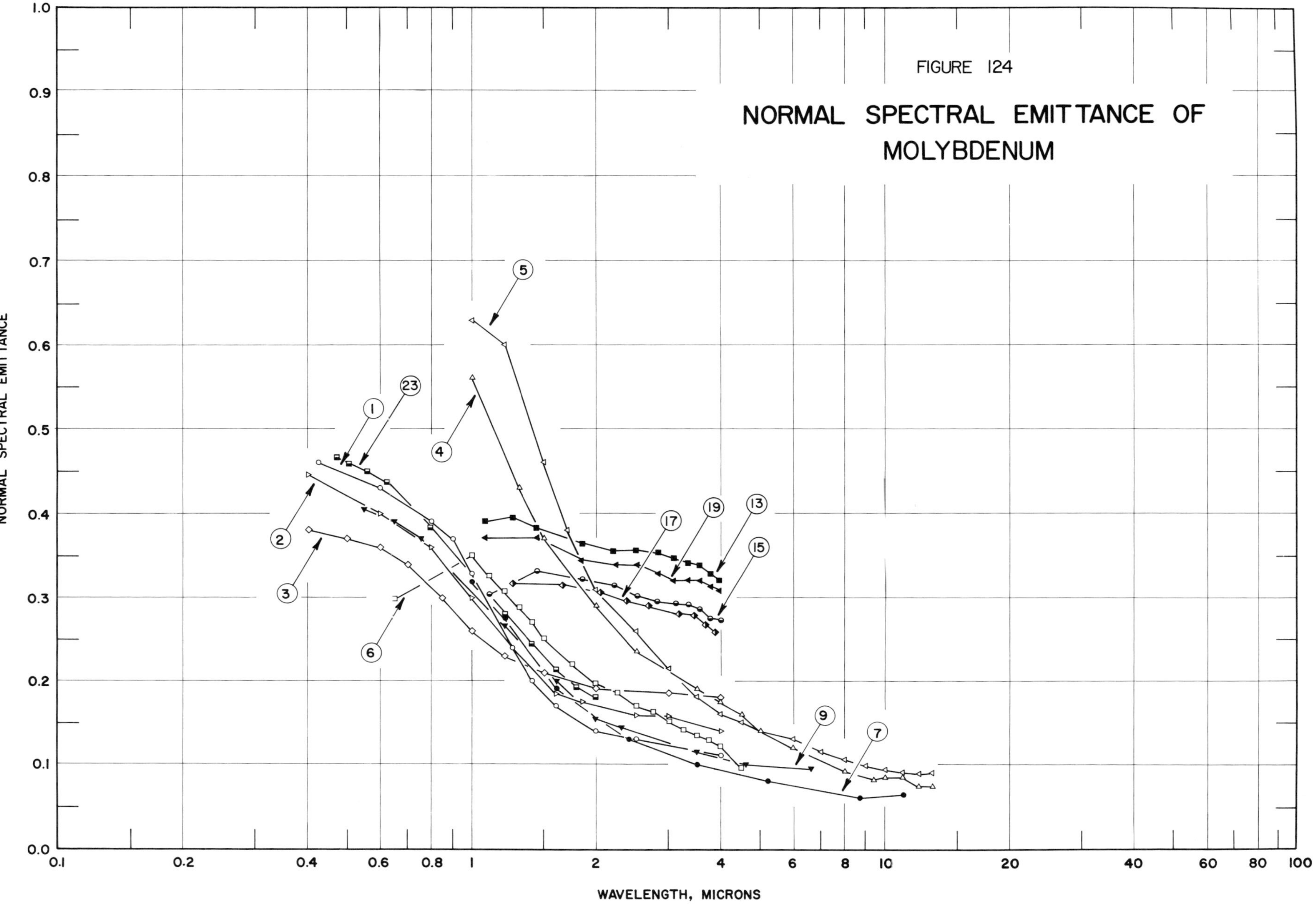
FIGURE 124
NORMAL SPECTRAL EMITTANCE OF MOLYBDENUM
NORMAL SPECTRAL EMITTANCE
WAVELENGTH, MICRONS
0.0
0.1
0.2
0.3
0.4
0.5
0.6
0.7
0.8
0.9
1.0
0.1
0.2
0.4
0.6
0.8
1
2
4
6
8
10
20
40
60
80
100
1
2
3
4
5
6
7
9
13
15
17
19
23

SPECIFICATION TABLE NO. 124 NORMAL SPECTRAL EMITTANCE OF MOLYBDENUM

Curve No.	Ref. No.	Year	Temperature K	Wavelength Range, μ	Geometry θ'	Reported Error, %	Composition (weight percent), Specifications and Remarks
1	76	1962	1600	0.425-4.000	$\sim0°$		Prepared from micronized powder; hot pressed at >2273 K; sintered, polished, etched, then degassed by heating to ~973 K; measured in argon; data extracted from smooth curve.
2	76	1962	2000	0.403-4.000	$\sim0°$		Above specimen and conditions.
3	76	1962	2800	0.403-4.000	$\sim0°$		Above specimen and conditions.
4	30	1963	1111	1.0-13.0	$\sim0°$		Foil (0.005 in. thick); cleaned with acetone; heated at T >1144 K for 1 hr; measured in argon; data extracted from smooth curve.
5	30	1963	1244	1.0-13.0	$\sim0°$		Above specimen and conditions except heated at T >1144 K for 8 additional hrs.
6	37	1947	1499	0.65-4.50	$\sim0°$	2	Polished; heated in static hydrogen.
7	12	1962	1255	1.0-11.1	$\sim0°$		As received; measured in vacuum ($<10^{-7}$ mm Hg).
8	12	1962	1366	2.0-11.7	$\sim0°$		Above specimen and conditions.
9	12	1962	1478	0.55-6.6	$\sim0°$		Above specimen and conditions.
10	95	1963	1600	0.40-3.99	$\sim0°$		99.9 Mo from General Electric Co.; polished with 240, 400, and 600 grit carbide paper, silk cloth, and felt cloth; washed in acetone, then alcohol, and dried with dry nitrogen; data extracted from smooth curve.
11	95	1963	2000	0.40-3.98	$\sim0°$		Above specimen and conditions.
12	95	1963	2800	0.40-3.98	$\sim0°$		Above specimen and conditions.
13	239	1959	1386	1.085-3.980	$\sim0°$		Polished; measured in argon.
14	239	1959	1349	1.092-3.995	$\sim0°$		Different sample, same as above specimen and conditions.
15	239	1959	1320	1.102-4.000	$\sim0°$		Different sample, same as above specimen and conditions.
16	239	1959	1284	1.250-3.899	$\sim0°$		Different sample, same as above specimen and conditions.
17	113	1961	1284	1.255-3.896	$\sim0°$		Trace of surface oxidation observed; measured in vacuum.
18	113	1961	1320	1.104-3.998	$\sim0°$		Above specimen and conditions.
19	113	1961	1349	1.086-3.995	$\sim0°$		Above specimen and conditions except measured in argon (760 mm Hg).
20	113	1961	1386	1.082-3.981	$\sim0°$		Above specimen and conditions.
21	241	1963	2210	0.9-1.7	$0°$		Single crystal; oriented so that surface of interest coincided with closed packed plane; optically polished; heated at several hundred degrees above temperature of interest in 90 Ar + 10 H atm; computed from optical constants.
22	241	1963	2310	0.9-1.1	$0°$		Above specimen and conditions.

SPECIFICATION TABLE NO. 124 (continued)

Curve No.	Ref. No.	Year	Temperature K	Wavelength Range, μ	Geometry θ'	Reported Error, %	Composition (weight percent), Specifications and Remarks
23	323	1966	1500	0.47-2.01	0°		Filament (0.25-0.32 mm in dia); baked for 1 hr at 798 K in vacuum, cooled, heated for 5-10 min in vacuum, and cooled; measured in argon (600 mm Hg); data calculated from optical constants.
24	323	1966	2000	0.47-2.01	0°		Above specimen and conditions.
25	323	1966	2400	0.46-2.01	0°		Above specimen and conditions.

DATA TABLE NO. 124 NORMAL SPECTRAL EMITTANCE OF MOLYBDENUM

[Wavelength, λ, μ; Emittance, ∈; Temperature, T, K]

λ	∈
CURVE 1 T = 1600	
0.425	0.460
0.600	0.430
0.800	0.390
0.900	0.370
1.000	0.330
1.250	0.240
1.400	0.200
1.600	0.170
2.000	0.140
2.500	0.130
4.000	0.112
CURVE 2 T = 2000	
0.403	0.445
0.600	0.400
0.800	0.360
1.000	0.300*
1.250	0.240
1.600	0.185
1.850	0.175
2.500	0.160
3.000	0.159
4.000	0.140
CURVE 3 T = 2800	
0.403	0.380
0.500	0.370
0.600	0.360
0.700	0.340
0.850	0.300
1.000	0.260
1.200	0.230
1.500	0.210
2.000	0.190
3.000	0.185
4.000	0.180
CURVE 4 T = 1111	
1.0	0.560
CURVE 4 (cont.)	
1.3	0.430
1.5	0.370
2.0	0.290
2.5	0.235
3.5	0.190
4.0	0.175
4.5	0.160
5.0	0.140
6.0	0.120
8.0	0.092
9.4	0.082
10.0	0.080
11.0	0.080
12.0	0.074
13.0	0.074
CURVE 5 T = 1244	
1.0	0.630
1.2	0.600
1.5	0.460
1.7	0.380
2.0	0.310
2.5	0.260
3.0	0.215
3.5	0.180
4.0	0.160
4.5	0.150*
5.0	0.140
6.0	0.130
7.0	0.115
8.0	0.105
9.0	0.098
10.0	0.094
11.0	0.090
12.0	0.089
13.0	0.090
CURVE 6 T = 1499	
0.65	0.299
1.00	0.350
1.10	0.327
CURVE 6 (cont.)	
1.20	0.308
1.30	0.288
1.40	0.270
1.50	0.251
1.75	0.220
2.00	0.197
2.25	0.185
2.50	0.170
2.75	0.163
3.00	0.151
3.25	0.142
3.50	0.135
3.75	0.129
4.00	0.122
4.50	0.095
CURVE 7 T = 1255	
1.0	0.320
1.2	0.275
1.6	0.190
2.4	0.130
3.5	0.100
5.2	0.080
8.7	0.062
11.1	0.065
CURVE 8* T = 1366	
2.0	0.155
2.7	0.118
3.5	0.100
4.6	0.095
5.2	0.090
6.6	0.082
7.7	0.075
8.7	0.075
9.6	0.070
10.4	0.075
11.7	0.075
CURVE 9 T = 1478	
0.55	0.405
0.65	0.390
0.75	0.370
1.2	0.265
1.6	0.200
2.3	0.145
3.5	0.115
4.6	0.100
6.6	0.095
CURVE 10* T = 1600	
0.40	0.467
0.48	0.454
0.57	0.436
0.68	0.417
0.78	0.396
0.89	0.370
0.98	0.336
1.09	0.301
1.20	0.261
1.33	0.216
1.47	0.185
1.63	0.163
1.87	0.146
2.19	0.131
2.52	0.124
3.01	0.117
3.50	0.113
3.99	0.111
CURVE 11* T = 2000	
0.40	0.442
0.48	0.422
0.59	0.401
0.68	0.381
0.78	0.358
0.89	0.330
0.99	0.304
1.10	0.273
1.12	0.246
1.35	0.220
CURVE 11*(cont.)	
1.47	0.201
1.63	0.185
1.84	0.173
2.00	0.169
2.39	0.165
2.85	0.157
3.28	0.151
3.98	0.140
CURVE 12* T = 2800	
0.40	0.382
0.48	0.375
0.58	0.361
0.68	0.339
0.79	0.315
0.89	0.283
0.98	0.259
1.08	0.240
1.12	0.223
1.39	0.212
1.70	0.197
1.99	0.190
2.30	0.186
2.77	0.183
3.30	0.179
3.98	0.177
CURVE 13 T = 1386	
1.085	0.391
1.249	0.396
1.444	0.383
1.845	0.364
2.201	0.356
2.504	0.358
2.823	0.354
3.082	0.348
3.328	0.342
3.571	0.340
3.791	0.331
3.980	0.322
CURVE 14* T = 1349	
1.092	0.375
1.440	0.375
1.851	0.349
2.225	0.345
2.516	0.345
2.831	0.335
3.090	0.326
3.337	0.326
3.573	0.326
3.793	0.318
3.995	0.313
CURVE 15 T = 1320	
1.102	0.305
1.446	0.333
1.847	0.322
2.206	0.316
2.516	0.302
2.809	0.298
3.103	0.294
3.343	0.292
3.580	0.287
3.782	0.276
4.000	0.274
CURVE 16* T = 1284	
1.250	0.321
1.655	0.319
2.044	0.310
2.369	0.301
2.686	0.293
3.199	0.284
3.455	0.281
3.692	0.271
3.899	0.261
CURVE 17 T = 1284	
1.255	0.318
1.661	0.317
CURVE 17 (cont.)	
2.044	0.307
2.375	0.298
2.688	0.290
3.198	0.281
3.449	0.279
3.689	0.268
3.896	0.259
CURVE 18* T = 1320	
1.104	0.303
1.449	0.331
1.850	0.319
2.210	0.313
2.519	0.299
2.815	0.295
3.104	0.291
3.341	0.290
3.580	0.285
3.786	0.274
3.998	0.272
CURVE 19 T = 1349	
1.086	0.371
1.441	0.371
1.856	0.346
2.228	0.341
2.516	0.341
2.832	0.331
3.092	0.322
3.334	0.322
3.575	0.322
3.793	0.314
3.995	0.310
CURVE 20* T = 1386	
1.082	0.387
1.247	0.391
1.445	0.379
1.848	0.360
CURVE 20*(cont.)	
2.205	0.352
2.503	0.353
2.825	0.349
3.085	0.343
3.328	0.337
3.571	0.335
3.793	0.326
3.981	0.318
CURVE 21* T = 2210	
0.9	0.317
1.0	0.291
1.1	0.274
1.3	0.249
1.4	0.236
1.6	0.210
1.7	0.210
CURVE 22* T = 2310	
0.9	0.305
1.1	0.275
CURVE 23 T = 1500	
0.47	0.467
0.50	0.460
0.56	0.450
0.62	0.436
0.80	0.383
1.00	0.329*
1.20	0.280
1.40	0.244
1.61	0.214
1.80	0.192
2.01	0.180

* Not shown on plot

DATA TABLE NO. 124 (continued)

λ	∈
CURVE 24* T = 2000	
0.47	0.434
0.50	0.428
0.56	0.418
0.62	0.410
0.80	0.357
1.00	0.316
1.21	0.279
1.40	0.252
1.60	0.225
1.79	0.212
2.01	0.200
CURVE 25* T = 2400	
0.46	0.400
0.51	0.400
0.56	0.396
0.62	0.385
0.80	0.337
1.00	0.300
1.20	0.278
1.39	0.256
1.60	0.235
1.81	0.220
2.01	0.205

* Not shown on plot

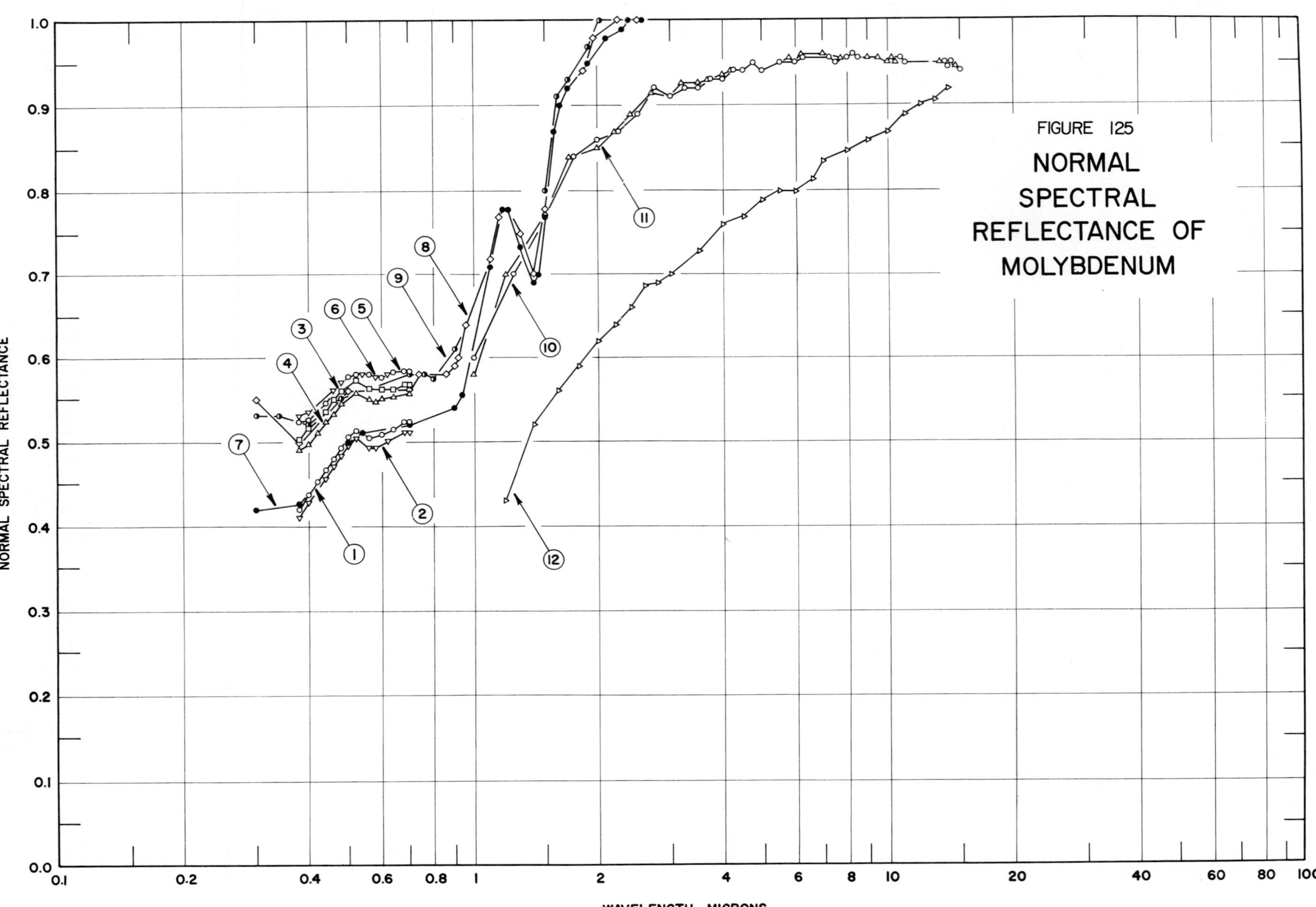

FIGURE 125
NORMAL SPECTRAL REFLECTANCE OF MOLYBDENUM

SPECIFICATION TABLE NO. 125 NORMAL SPECTRAL REFLECTANCE OF MOLYBDENUM

Curve No.	Ref. No.	Year	Temperature K	Wavelength Range, μ	Geometry θ	Geometry θ' ω'	Reported Error, %	Composition (weight percent), Specifications and Remarks
1	29	1956	298	0.38-0.70	9°	2π		Powder, unalloyed; as received; data extracted from smooth curve; MgO reference.
2	29	1956	298	0.38-0.70	9°	2π		Above specimen and conditions except rotated 90° in its own plane.
3	29	1956	298	0.38-0.70	9°	2π		Powder, unalloyed; smoothed and cleaned; data extracted from smooth curve; MgO reference.
4	29	1956	298	0.38-0.70	9°	2π		Above specimen and conditions except rotated 90° in its own plane.
5	29	1956	298	0.38-0.70	9°	2π		Powder, unalloyed; polished; data extracted from smooth curve; MgO reference.
6	29	1956	298	0.38-0.70	9°	2π		Above specimen and conditions except rotated 90° in its own plane.
7	34	1957	298	0.30-2.58	9°	2π	± 4	Arc melted, unalloyed (Climax Molybdenum Co.); as received; data extracted from a smooth curve; magnesium carbonate reference.
8	34	1957	298	0.30-2.60	9°	2π	± 4	Arc melted, unalloyed (Climax Molybdenum Co.); cleaned with liquid detergent; data extracted from a smooth curve; magnesium carbonate reference.
9	34	1957	298	0.30-2.60	9°	2π		Arc melted, unalloyed (Climax Molybdenum Co.); polished; data extracted from a smooth curve; magnesium carbonate reference.
10	136	1954	294	1.00-15.00	5°	2π		Washed in separate baths of acetone and ethyl alcohol; polished with a soft rag; hohlraum at approx 1089 K; converted from $R(2\pi, 5^\circ)$.
11	126	1953	298	1.0 -15.0	5°	2π	± 2	Polished; data extracted from smooth curve; converted from $R(2\pi, 5^\circ)$.
12	65	1962	294	1.20-14.00	$\sim 0^\circ$	2π		Machined smooth.
13	65	1962	294	0.40-21.00	$\sim 0^\circ$	2π		Polished.

DATA TABLE NO. 125 NORMAL SPECTRAL REFLECTANCE OF MOLYBDENUM

[Wavelength, λ, μ; Reflectance, ρ; Temperature, T, K]

CURVE 1
T = 298

λ	ρ
0.38	0.420
0.40	0.437
0.42	0.453
0.44	0.467
0.46	0.480
0.48	0.493
0.50	0.505
0.52	0.512
0.54	0.508*
0.56	0.504
0.58	0.505*
0.60	0.508
0.62	0.511*
0.64	0.513
0.66	0.519*
0.68	0.522
0.70	0.522

CURVE 2
T = 298

λ	ρ
0.38	0.410
0.40	0.427
0.42	0.443*
0.44	0.456
0.46	0.470
0.48	0.483
0.50	0.495
0.52	0.502
0.54	0.497*
0.56	0.493
0.58	0.493
0.60	0.496*
0.62	0.500
0.64	0.503*
0.66	0.508*
0.68	0.510
0.70	0.510

CURVE 3
T = 298

λ	ρ
0.38	0.502
0.40	0.515
0.42	0.526*
0.44	0.535
0.46	0.550
0.48	0.560
0.50	0.567*
0.52	0.572
0.54	0.567*
0.56	0.580
0.58	0.578*
0.60	0.562
0.62	0.562*
0.64	0.563
0.66	0.567*
0.68	0.569
0.70	0.569

CURVE 4
T = 298

λ	ρ
0.38	0.490
0.40	0.498
0.42	0.510
0.44	0.522
0.46	0.533
0.48	0.545
0.50	0.555*
0.52	0.558
0.54	0.554*
0.56	0.550
0.58	0.548
0.60	0.550
0.62	0.550*
0.64	0.553
0.66	0.553*
0.68	0.556*
0.70	0.557

CURVE 5
T = 298

λ	ρ
0.38	0.522
0.40	0.525
0.42	0.536*
0.44	0.545
0.46	0.557*
0.48	0.567*
0.50	0.577
0.52	0.580
0.54	0.580*
0.56	0.580
0.58	0.578*
0.60	0.577
0.62	0.580*
0.64	0.582
0.66	0.582*
0.68	0.583
0.70	0.583

CURVE 6
T = 298

λ	ρ
0.38	0.530
0.40	0.533
0.42	0.540*
0.44	0.548*
0.46	0.560
0.48	0.570
0.50	0.577*
0.52	0.580*
0.54	0.580
0.56	0.580*
0.58	0.578
0.60	0.577*
0.62	0.580
0.64	0.582*
0.66	0.583*
0.68	0.583*
0.70	0.584*

CURVE 7
T = 298

λ	ρ
0.30	0.420
0.38	0.425
0.40	0.430*
0.48	0.490*
0.50	0.500
0.54	0.510
0.70	0.520
0.90	0.540
0.94	0.550
1.10	0.710
1.18	0.780
1.22	0.780
1.30	0.735
1.40	0.690
1.44	0.700
1.50	0.770
1.58	0.870
1.62	0.900
1.70	0.920
1.90	0.950
2.10	0.980
2.30	0.990
2.38	1.000
2.50	1.000*
2.58	1.000

CURVE 8
T = 298

λ	ρ
0.30	0.55
0.38	0.50
0.50	0.56
0.70	0.56
0.74	0.58
0.86	0.58
0.90	0.59
0.92	0.60
0.96	0.64
1.10	0.72
1.16	0.77
1.18	0.78*
1.22	0.78*
1.30	0.75
1.40	0.70
1.50	0.78
1.58	0.87*
1.62	0.90*
1.70	0.92*
1.86	0.94
1.90	0.95*
1.96	0.98
2.22	0.99*
2.24	1.00
2.30	1.00*
2.50	1.00
2.60	1.00*

CURVE 9
T = 298

λ	ρ
0.30	0.53
0.34	0.53
0.40	0.52
0.48	0.55
0.50	0.56*
0.70	0.58
0.76	0.58
0.80	0.575
0.90	0.61
0.96	0.64*
1.10	0.72*
1.16	0.77*
1.18	0.78*
1.22	0.78*
1.30	0.75*
1.40	0.70*
1.50	0.80
1.60	0.91
1.70	0.93
1.90	0.97
2.02	1.00
2.10	1.00*
2.30	1.00*
2.50	1.00*
2.60	1.00*

CURVE 10
T = 294

λ	ρ
1.00	0.600
1.25	0.700
1.50	0.770*
1.75	0.840
2.00	0.860
2.25	0.870
2.50	0.890
2.75	0.920
3.00	0.910
3.25	0.920
3.50	0.920
3.75	0.930
4.00	0.930
4.25	0.490
4.50	0.940
4.75	0.950
5.00	0.940
5.25	0.945*
5.50	0.950
5.75	0.950*
6.00	0.950
6.25	0.955
6.50	0.955*
6.75	0.955*
7.00	0.955*
7.25	0.955
7.50	0.950
7.75	0.955*
8.00	0.955
8.25	0.960
8.50	0.955
8.75	0.955*
9.00	0.955*
9.25	0.955*
9.50	0.955*
9.75	0.955*
10.00	0.955*
10.25	0.955*
10.50	0.955*
10.75	0.955
11.00	0.950
11.25	0.950*
11.50	0.950*
11.75	0.950*
12.00	0.950*
12.25	0.950*
12.50	0.950*
12.75	0.950*
13.00	0.950*
13.25	0.950*
13.50	0.950*
13.75	0.950
14.00	0.945
14.25	0.950
14.50	0.945*
14.75	0.945*
15.00	0.940

CURVE 11
T = 298

λ	ρ
1.0	0.580
1.2	0.700
1.5	0.775
1.7	0.840
2.0	0.860*
2.2	0.870
2.4	0.890
2.7	0.915
3.0	0.910*
3.2	0.925
3.5	0.925
3.7	0.930
4.0	0.935
4.2	0.940
4.5	0.940*
4.7	0.950*
5.0	0.940*
5.8	0.955
6.0	0.950*
6.2	0.960
7.0	0.960
7.5	0.950*
7.7	0.955
8.2	0.960*
8.4	0.955*
9.0	0.955
9.5	0.955
10.0	0.950
10.2	0.955
10.4	0.950
11.0	0.950*
11.5	0.950*
12.0	0.950*
12.5	0.950*
13.0	0.950*
13.5	0.950
14.0	0.950*
14.2	0.950*
14.4	0.945
14.7	0.945
15.0	0.940

CURVE 12
T = 294

λ	ρ
1.20	0.430
1.40	0.520
1.60	0.560
1.80	0.590
2.00	0.620
2.20	0.640
2.40	0.660
2.60	0.685
2.80	0.690
3.00	0.700
3.50	0.730
4.00	0.760
4.50	0.770
5.00	0.790
5.50	0.800
6.00	0.800
6.60	0.815
7.00	0.836
8.00	0.848
9.00	0.860
10.00	0.870
11.00	0.890
12.00	0.900
13.00	0.905
14.00	0.919

CURVE 13*
T = 294

λ	ρ
0.40	0.610
0.50	0.560
0.60	0.560
0.70	0.560
0.80	0.550
0.90	0.560
1.00	0.615
1.00	0.755
1.20	0.695
1.20	0.670
1.40	0.814
1.60	0.884
1.60	0.868
1.80	0.865
2.00	0.880

* Not shown on plot

DATA TABLE NO. 125 (continued)

λ	ρ
CURVE 13 (cont.)*	
2.20	0.920
2.40	0.930
2.60	0.930
2.80	0.890
3.00	0.910
3.50	0.940
4.00	0.929
4.50	0.950
5.00	0.960
5.50	0.960
6.00	0.930
6.50	0.945
7.00	0.950
7.60	0.950
8.00	0.960
9.00	0.960
9.60	0.950
10.00	0.960
10.00	0.970
11.50	0.970
13.00	0.970
15.00	0.970
15.50	0.975
17.50	0.975
18.00	0.970
21.00	0.970

* Not shown on plot

SPECIFICATION TABLE NO. 126 ANGULAR SPECTRAL REFLECTANCE OF MOLYBDENUM

Curve No.	Ref. No.	Year	Temperature K	Wavelength Range, μ	Geometry θ	θ'	ω'	Reported Error, %	Composition (weight percent), Specifications and Remarks
1	132	1911	298	0.65-10.05	15°	15°		≤ 3	Polished; silvered glass mirror reference.

DATA TABLE NO. 126 ANGULAR SPECTRAL REFLECTANCE OF MOLYBDENUM

[Wavelength, λ, μ; Reflectance, ρ; Temperature, T, K]

λ	ρ
CURVE 1*	
T = 298	
0.65	0.455
0.65	0.480
0.75	0.490
0.80	0.510
0.95	0.525
1.10	0.575
1.25	0.630
1.65	0.745
2.10	0.815
2.55	0.845
3.10	0.870
3.65	0.890
4.10	0.900
4.70	0.920
5.55	0.925
6.10	0.930
6.90	0.930
7.90	0.935
8.80	0.940
9.00	0.935
9.35	0.940
10.05	0.940

* Not shown on plot

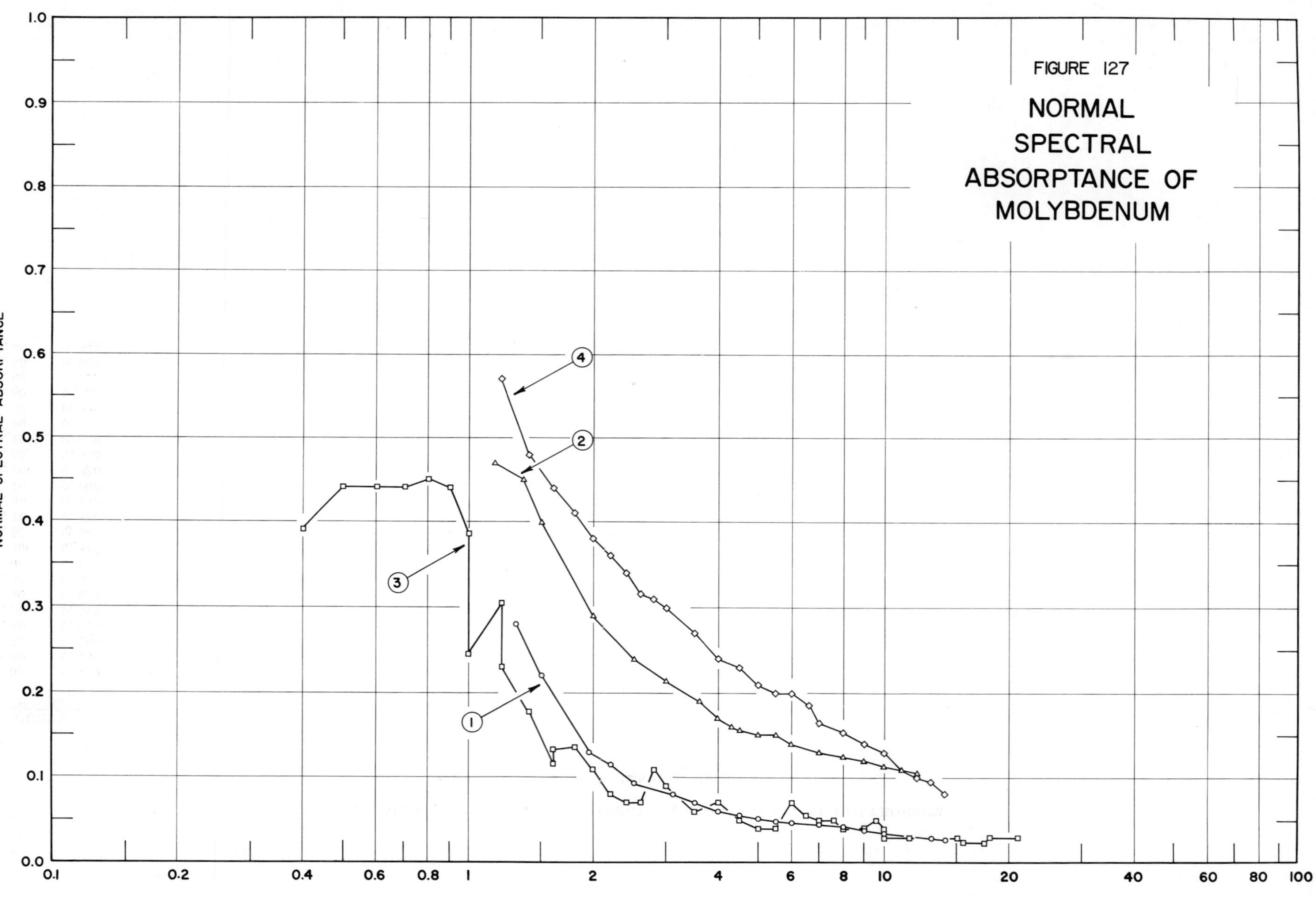
FIGURE 127
NORMAL
SPECTRAL
ABSORPTANCE OF
MOLYBDENUM
NORMAL SPECTRAL ABSORPTANCE
WAVELENGTH, MICRONS
0.0
0.1
0.2
0.3
0.4
0.5
0.6
0.7
0.8
0.9
1.0
0.1
0.2
0.4
0.6
0.8
1
2
4
6
8
10
20
40
60
80
100
1
2
3
4

SPECIFICATION TABLE NO. 127 NORMAL SPECTRAL ABSORPTANCE OF MOLYBDENUM

Curve No.	Ref. No.	Year	Temperature K	Wavelength Range, μ	Geometry θ	Reported Error, %	Composition (weight percent), Specifications and Remarks
1	30	1963	294	1.30-14.00	$\sim 0^\circ$		Foil (0.005 in. thick); cleaned with acetone; measured in air; data extracted from smooth curve.
2	30	1963	294	1.15-12.00	$\sim 0^\circ$		Foil (0.005 in. thick); cleaned with acetone; heated in vacuum at 1273 K for 1 hr at 10^{-5} mm Hg; measured in air; data extracted from smooth curve.
3	65	1962	294	0.40-21.00	$\sim 0^\circ$		Polished; calculated from $(1 - \rho)$.
4	65	1962	294	1.20-14.00	$\sim 0^\circ$		Machined smooth; calculated from $(1 - \rho)$.

DATA TABLE NO. 127 NORMAL SPECTRAL ABSORPTANCE OF MOLYBDENUM

[Wavelength, λ, μ; Absorptance, α; Temperature, T, K]

CURVE 1
T = 294

λ	α
1.30	0.280
1.50	0.220
1.95	0.130
2.20	0.115
2.50	0.092
3.10	0.080
3.50	0.070
4.00	0.060
4.50	0.055
5.00	0.051
5.50	0.049
5.70	0.048*
6.00	0.048
7.00	0.045
8.00	0.042
9.00	0.038
10.00	0.035
13.00	0.030
14.00	0.029

CURVE 2
T = 294

λ	α
1.15	0.470
1.35	0.450
1.50	0.400
2.00	0.290
2.50	0.240
3.00	0.215
3.60	0.190
4.00	0.170
4.30	0.160
4.50	0.155
5.00	0.150
5.50	0.150
6.00	0.140
7.00	0.130
8.00	0.125
9.00	0.120
10.00	0.115
11.00	0.110
12.00	0.105

CURVE 3
T = 294

λ	α
0.40	0.390
0.50	0.440
0.60	0.440
0.70	0.440
0.80	0.450
0.90	0.440
1.00	0.385
1.00	0.245
1.20	0.305
1.20	0.230
1.40	0.176
1.60	0.116
1.60	0.132
1.80	0.135
2.00	0.110
2.20	0.080
2.40	0.070
2.60	0.070
2.80	0.110
3.00	0.090
3.50	0.060
4.00	0.071
4.50	0.050
5.00	0.040
5.50	0.040
6.00	0.070
6.50	0.055
7.00	0.050
7.60	0.050
8.00	0.040
9.00	0.040
9.60	0.050
10.00	0.040
10.00	0.030
11.50	0.030
13.00	0.030*
15.00	0.030
15.50	0.025
17.50	0.025
18.00	0.030
21.00	0.030

CURVE 4
T = 294

λ	α
1.20	0.570
1.40	0.480
1.60	0.440
1.80	0.410
2.00	0.380
2.20	0.360
2.40	0.340
2.60	0.315
2.80	0.310
3.00	0.300
3.50	0.270
4.00	0.240
4.50	0.230
5.00	0.210
5.50	0.200
6.00	0.200
6.60	0.185
7.00	0.164
8.00	0.152
9.00	0.140
10.00	0.130
11.00	0.110*
12.00	0.100
13.00	0.095
14.00	0.081

* Not shown on plot

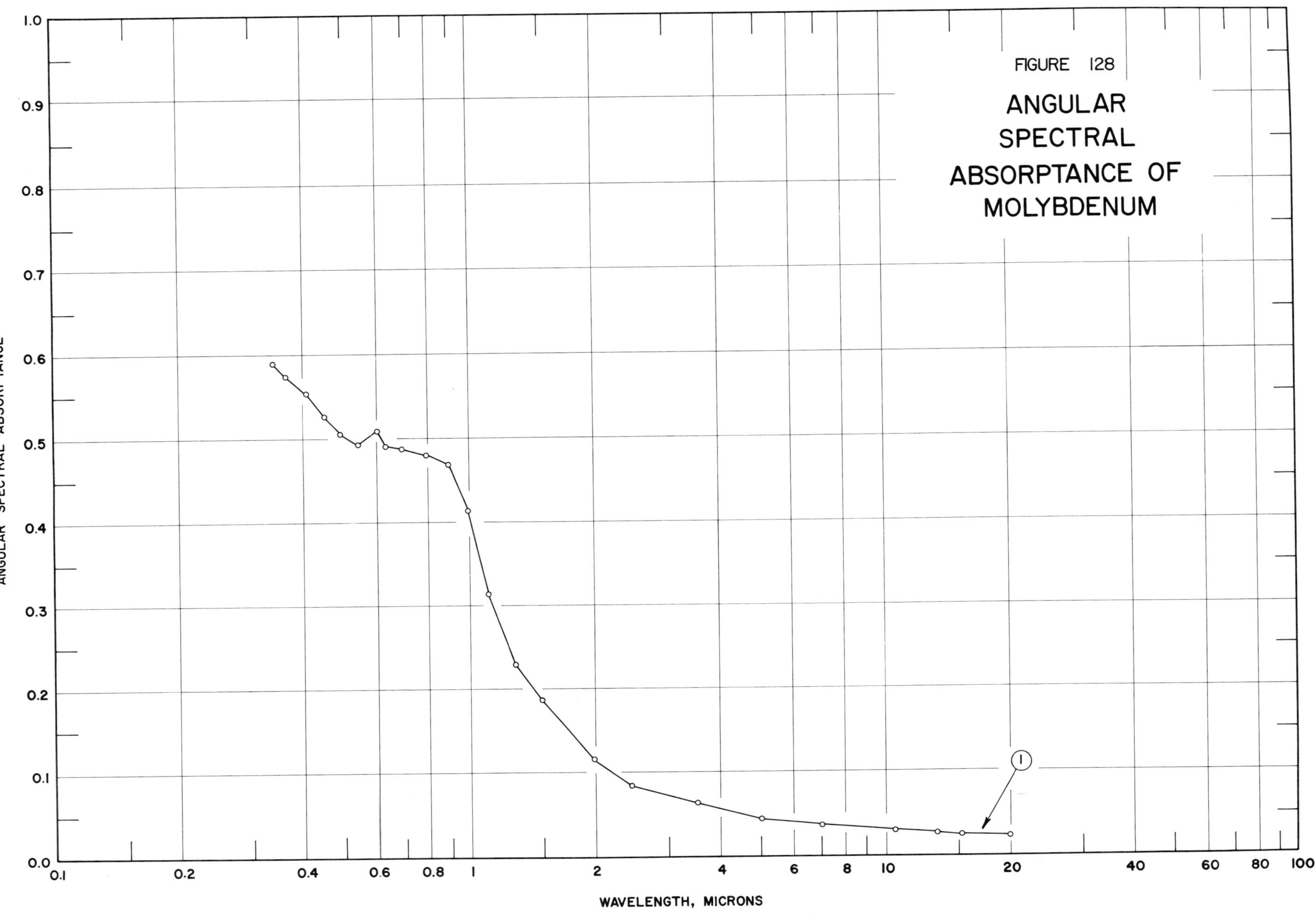
FIGURE 128
ANGULAR
SPECTRAL
ABSORPTANCE OF
MOLYBDENUM
1
ANGULAR SPECTRAL ABSORPTANCE
1.0
0.9
0.8
0.7
0.6
0.5
0.4
0.3
0.2
0.1
0.0
0.1
0.2
0.4
0.6
0.8
1
2
4
6
8
10
20
40
60
80
100
WAVELENGTH, MICRONS

SPECIFICATION TABLE NO. 128 ANGULAR SPECTRAL ABSORPTANCE OF MOLYBDENUM

Curve No.	Ref. No.	Year	Temperature K	Wavelength Range, μ	Geometry θ	Reported Error, %	Composition (weight percent), Specifications and Remarks
1	225	1965	306	0.339-20.1	25°		Molybdenum from Belmont Smelting and Refining Works; measured in dry nitrogen; heated cavity at approx 1056 K with platinum reference; authors assumed $\alpha = 1 - R(2\pi, 25^\circ)$.

DATA TABLE NO. 128 ANGULAR SPECTRAL ABSORPTANCE OF MOLYBDENUM

[Wavelength, λ, μ; Abosrptance, α; Temperature, T, K]

λ	α
CURVE 1 T = 306	
0.339	0.589
0.364	0.572
0.409	0.552
0.456	0.527
0.494	0.506
0.541	0.493
0.601	0.509
0.632	0.491
0.687	0.488
0.791	0.481
0.889	0.469
0.986	0.413
1.13	0.313
1.28	0.229
1.49	0.187
1.98	0.116
2.43	0.083
3.51	0.063
5.02	0.044
7.02	0.038
10.5	0.031
13.2	0.028
15.2	0.026
20.1	0.024

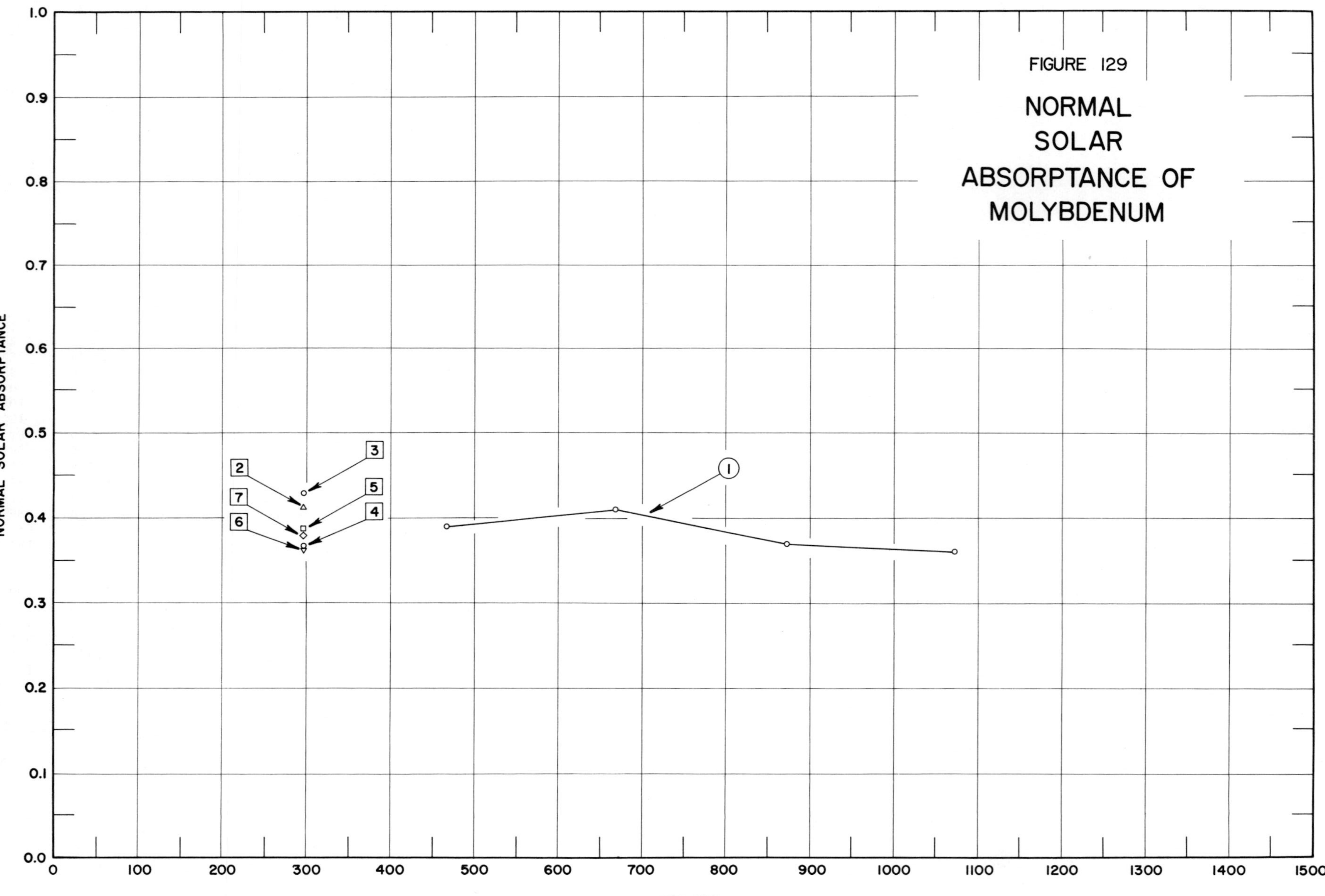

FIGURE 129 NORMAL SOLAR ABSORPTANCE OF MOLYBDENUM

SPECIFICATION TABLE NO. 129 NORMAL SOLAR ABSORPTANCE OF MOLYBDENUM

Curve No.	Ref. No.	Year	Temperature Range, K	Geometry θ	Reported Error, %	Composition (weight percent), Specifications and Remarks
1	47	1961	468–1073	~0°		Vacuum arc cast, machined, extruded, recrystallized, rolled; disc (0.04 in. thick); ground with 600 grit carborundum and polished on a wet cloth lap with unlevigated jewelers rouge; measured in vacuum (10^{-5} mm Hg).
2	34	1957	298	9°		Arc melted, unalloyed (Climax Molybdenum Co.); as received; computed from spectral reflectance data for above atmosphere conditions.
3	34	1957	298	9°		Above specimen and conditions except calculated for sea level conditions.
4	34	1957	298	9°		Different sample, same as Curve 2 specimen and conditions except cleaned with liquid detergent.
5	34	1957	298	9°		Above specimen and conditions except calculated for sea level conditions.
6	34	1957	298	9°		Different sample, same as Curve 2 specimen and conditions except polished.
7	34	1957	298	9°		Above specimen and conditions except calculated for sea level conditions.

DATA TABLE NO. 129 NORMAL SOLAR ABSORPTANCE OF MOLYBDENUM

[Temperature, T,K; Absorptance, α]

T	α
CURVE 1	
468	0.39
668	0.41
873	0.37
1073	0.36
CURVE 2	
298	0.411
CURVE 3	
298	0.430
CURVE 4	
298	0.368
CURVE 5	
298	0.388
CURVE 6	
298	0.361
CURVE 7	
298	0.380

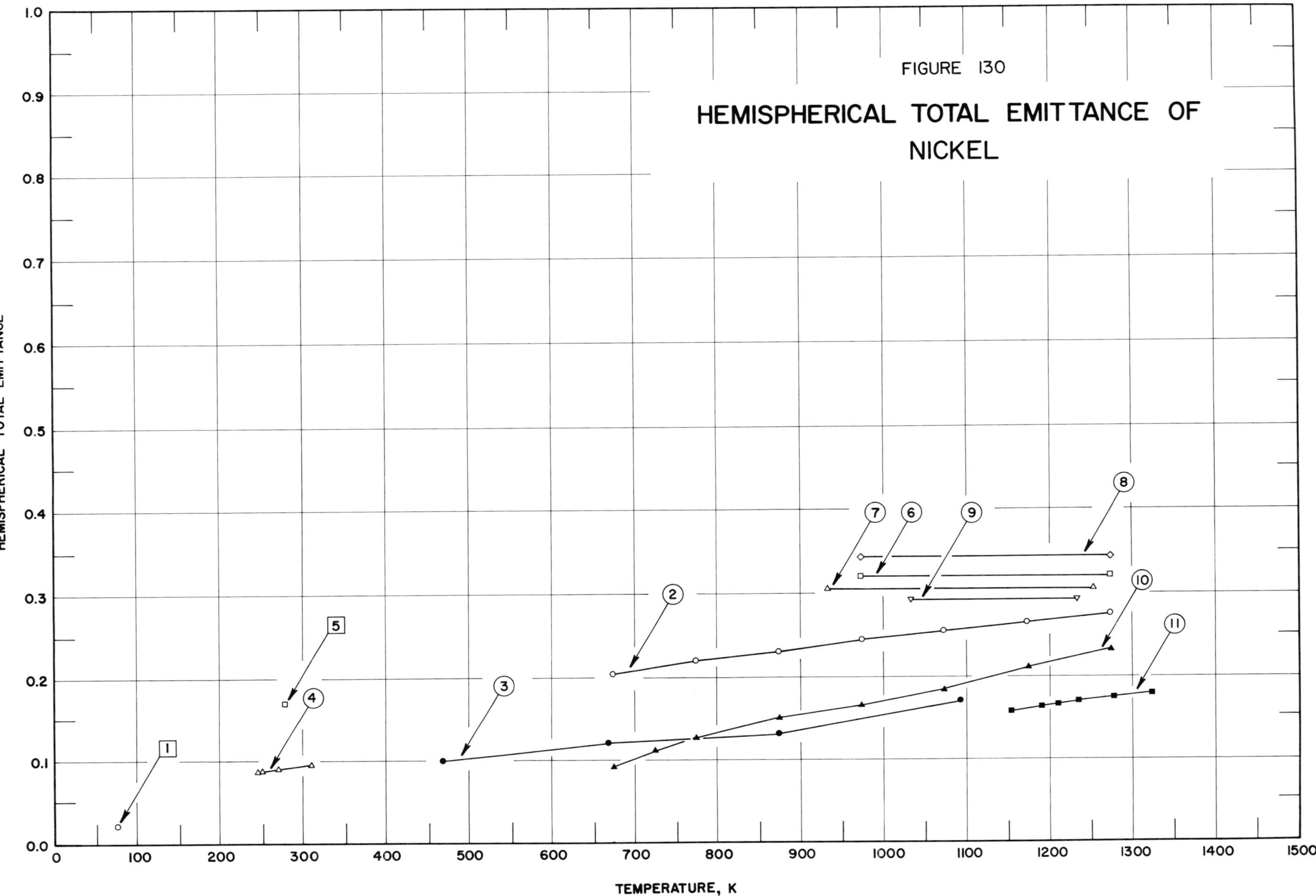

FIGURE 130

HEMISPHERICAL TOTAL EMITTANCE OF NICKEL

SPECIFICATION TABLE NO. 130 HEMISPHERICAL TOTAL EMITTANCE OF NICKEL

Curve No.	Ref. No.	Year	Temperature Range, K	Reported Error, %	Composition (weight percent), Specifications and Remarks
1	3	1955	76	5	Foil (0.004 in. thick); solvent cleaned; measured in vacuum (10^{-6} to 10^{-7} mm Hg); emittance for 300 K black body incident radiation; authors assumed $\alpha = \epsilon$.
2	58	1961	673-1273	± 2.5	Sandblasted; measured in vacuum ($<5 \times 10^{-6}$ mm Hg); data extracted from smooth curve.
3	47	1961	468-1093	< 10	Commercially pure; ground with 600 grit carborundum and polished on a wet cloth lap with unlevigated jewelers rouge; measured in vacuum (10^{-5} mm Hg).
4	26	1962	245-310	3	Commercial sheet; cleaned with both sodium dichromate and dilute nitric acid solutions, buffed on a felt buffing wheel and cleaned with carbon tetrachloride and acetone; measured in vacuum (1×10^{-6} mm Hg); data extracted from smooth curve.
5	45	1961	278	10	Electroless nickel.
6	81	1941	973-1273	< 1	Filament used in the radio tube type DL24; coated with oxide; heated in hydrogen at 1098 K for 15 min.
7	81	1941	933-1253	< 1	Filament used in the radio tube type DL25; coated with oxide; heated in hydrogen at 1098 K for 15 min.
8	81	1941	973-1273	< 1	Filament used in the radio tube type DL27; coated with oxide; heated in hydrogen at 1098 K for 15 min.
9	81	1941	1033-1233	< 1	Filament used in the radio tube type DB217; coated with oxide; heated in hydrogen at 1098 K for 15 min.
10	58	1961	673-1273	± 2.5	Commercially pure; vacuum cooked for 15 min at 1473 K; measured in vacuum ($<5 \times 10^{-6}$ mm Hg); data extracted from smooth curve; bright commercial surface.
11	246	1967	1153-1322		99.95 Ni.

DATA TABLE NO. 130 HEMISPHERICAL TOTAL EMITTANCE OF NICKEL

[Temperature, T, K; Emittance, ϵ]

T	ϵ
CURVE 1	
76	0.022
CURVE 2	
673	0.205
773	0.220
873	0.230
973	0.245
1073	0.255
1173	0.265
1273	0.275
CURVE 3	
468	0.10
668	0.12
873	0.13
1093	0.17
CURVE 4	
245	0.087
250	0.088
270	0.090
310	0.095
CURVE 5	
278	0.17
CURVE 6	
973	0.320
1273	0.320
CURVE 7	
933	0.305
1253	0.305
CURVE 8	
973	0.343
1273	0.343
CURVE 9	
1033	0.292
1233	0.292

T	ϵ
CURVE 10	
673	0.090
723	0.110
773	0.125
873	0.150
973	0.165
1073	0.185
1173	0.210
1273	0.230
CURVE 11	
1153	0.157
1190	0.163
1210	0.166
1236	0.169
1277	0.173
1322	0.179

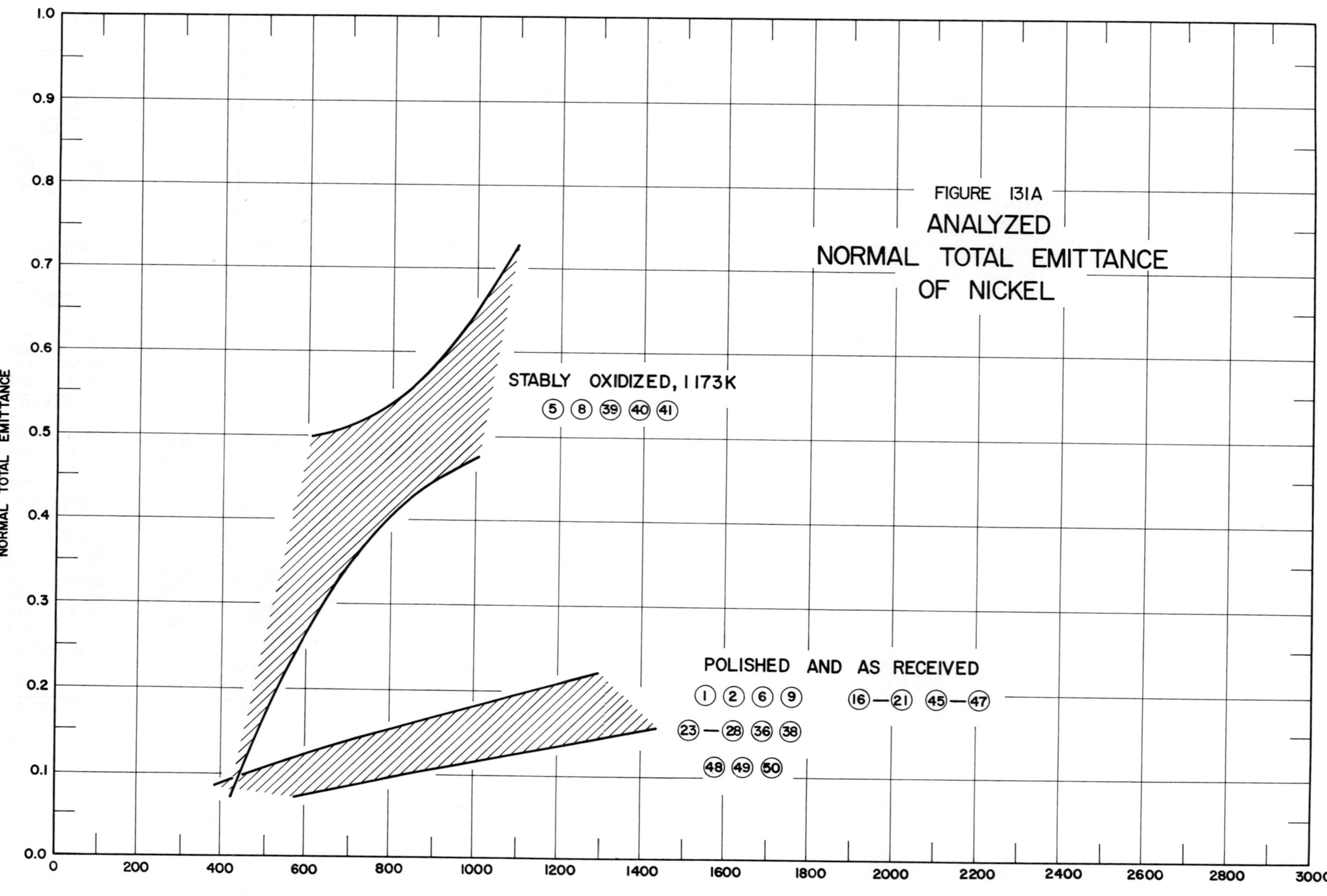

FIGURE 131A
ANALYZED
NORMAL TOTAL EMITTANCE
OF NICKEL
STABLY OXIDIZED, 1173K
5 8 39 40 41
POLISHED AND AS RECEIVED
1 2 6 9 16—21 45—47
23—28 36 38
48 49 50
NORMAL TOTAL EMITTANCE
1.0
0.9
0.8
0.7
0.6
0.5
0.4
0.3
0.2
0.1
0.0
0
200
400
600
800
1000
1200
1400
1600
1800
2000
2200
2400
2600
2800
3000
TEMPERATURE, K

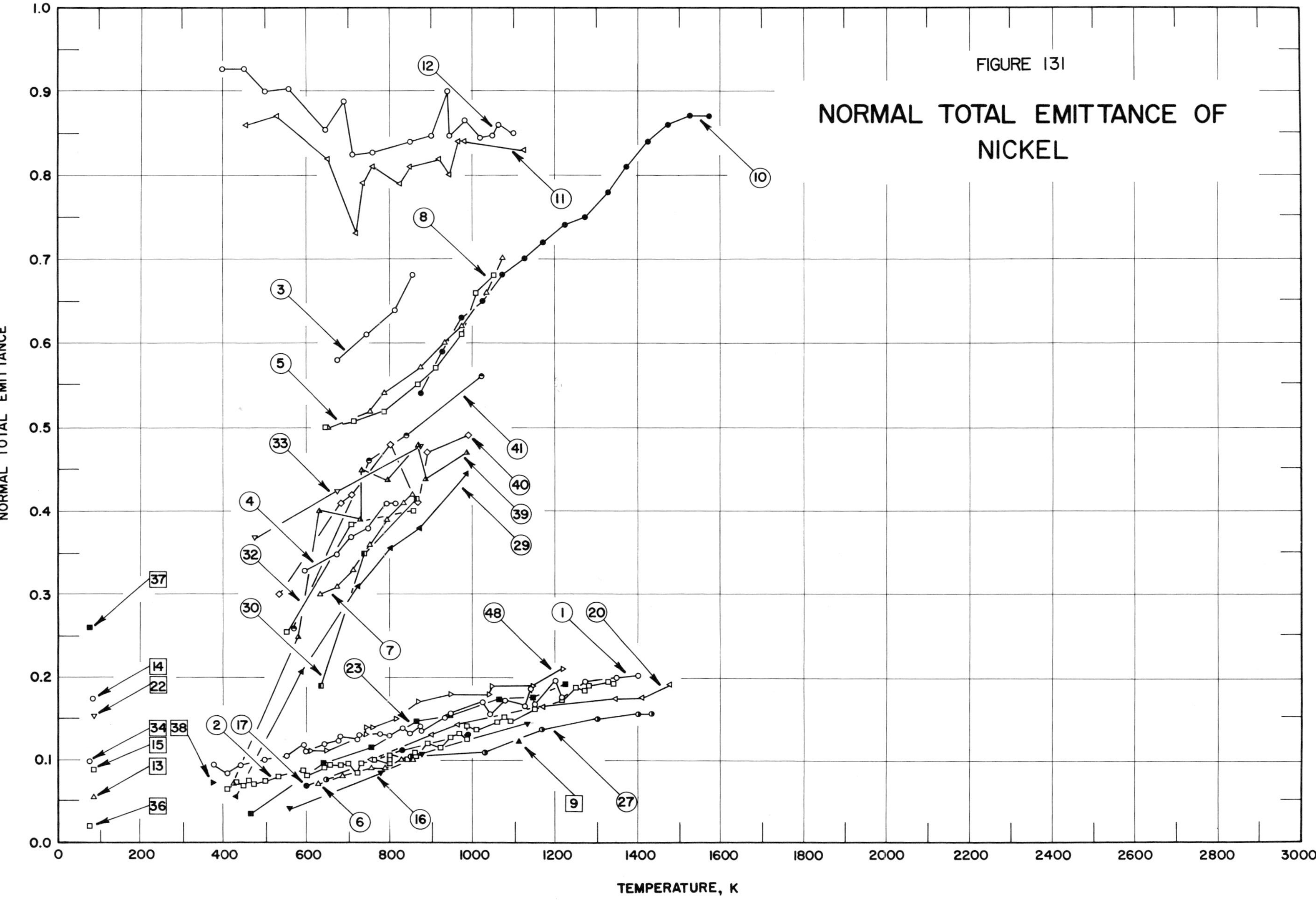

FIGURE 131
NORMAL TOTAL EMITTANCE OF
NICKEL
NORMAL TOTAL EMITTANCE
1.0
0.9
0.8
0.7
0.6
0.5
0.4
0.3
0.2
0.1
0.0
0
200
400
600
800
1000
1200
1400
1600
1800
2000
2200
2400
2600
2800
3000
TEMPERATURE, K

SPECIFICATION TABLE NO. 131 NORMAL TOTAL EMITTANCE OF NICKEL

Curve No.	Ref. No.	Year	Temperature Range, K	Geometry θ'	Reported Error, %	Composition (weight percent), Specifications and Remarks
1	82	1929	375-1400	$\sim 0^\circ$	1-5	Polished; measured in vacuum; increasing temp.
2	82	1929	410-1340	$\sim 0^\circ$		Above specimen and conditions; decreasing temp.
3	68	1952	673-853	$\sim 0^\circ$		Pure; blasted with fused alumina; oxidized at 873 K until steady state reached; cleaned with CCl_4; measured in air.
4	68	1952	593-813	$\sim 0^\circ$		Pure; blasted with fused alumina; cleaned with CCl_4; measured in air.
5	68	1952	653-1073	$\sim 0^\circ$		Pure; blasted with fused alumina; oxidized at 1173 K until steady state reached; cleaned with CCl_4; measured in air.
6	68	1952	623 -853	$\sim 0^\circ$		Pure; buffed, cleaned with CCl_4; measured in air.
7	68	1952	633-853	$\sim 0^\circ$		Pure; buffed; oxidized at 873 K until steady state reached; cleaned with CCl_4; measured in air.
8	68	1952	643-1053	$\sim 0^\circ$		Pure; buffed; oxidized at 1173 K until steady state reached; cleaned with CCl_4; measured in air.
9	30	1963	1111	$\sim 0^\circ$		Mechanically polished with aluminum oxide and cleaned with water; heated in vacuum for 1 hr at 1256 K; measured in vacuum (5×10^{-6} mm Hg); computed from spectral emittance.
10	83	1914	873-1573	$\sim 0^\circ$		High purity nickel with traces of Fe; oxidized in air; average of several runs on several specimens.
11	82	1929	455-1125	$\sim 0^\circ$	4-8	Soot covered nickel; measured in vacuum; cycle 1.
12	82	1929	400-1100	$\sim 0^\circ$		Above specimen and conditions, cycle 2.
13	34	1957	83.2	$\sim 0^\circ$	± 10	Commercial grade A; strip (0.005 in. thick); polished; measured in air (5×10^{-4} mm Hg).
14	34	1957	83.2	$\sim 0^\circ$	± 10	Commercial grade A; strip (0.005 in. thick); oxidized in air at red heat for 30 min; measured in air (5×10^{-4} mm Hg).
15	34	1957	83.2		± 10	Commercial grade A; strip (0.005 in. thick); cleaned with liquid detergent; measured in air (5×10^{-4} mm Hg).
16	34	1957	555-1133	$\sim 0^\circ$	± 10	Commercial grade A; strip (0.005 in. thick); cleaned with liquid detergent or as received; measured in air (5×10^{-4} mm Hg); increasing temp, cycle 1.
17	34	1957	600-989	$\sim 0^\circ$	± 10	Above specimen and conditions; decreasing temp, cycle 1.
18	34	1957	778-1211	$\sim 0^\circ$	± 10	Above specimen and conditions; cycle 2.
19	34	1957	605-1044	$\sim 0^\circ$	± 10	Above specimen and conditions; dereasing temp, cycle 2.
20	34	1957	755-1477	$\sim 0^\circ$	± 10	Above specimen and conditions; cycle 3.
21	34	1957	661-1300	$\sim 0^\circ$	± 10	Above specimen and conditions; decreasing temp, cycle 3.

SPECIFICATION TABLE NO. 131 (continued)

Curve No.	Ref. No.	Year	Temperature Range, K	Geometry θ'	Reported Error, %	Composition (weight percent), Specifications and Remarks
22	34	1957	83.1	$\sim 0^\circ$	± 10	Commercial grade A; strip (0.005 in. thick); as received; measured in air (5×10^{-4} mm Hg).
23	34	1957	466-1222	$\sim 0^\circ$	± 10	Commercial grade A; strip (0.005 in. thick); polished; measured in air (5×10^{-4} mm Hg); increasing temp, cycle 1.
24	34	1957	605-1028	$\sim 0^\circ$	± 10	Above specimen and conditions; decreasing temp, cycle 1.
25	34	1957	566-1172	$\sim 0^\circ$	± 10	Above specimen and conditions; cycle 2.
26	34	1957	733-872	$\sim 0^\circ$	± 10	Above specimen and conditions; decreasing temp, cycle 2.
27	34	1957	644-1433	$\sim 0^\circ$	± 10	Above specimen and conditions; cycle 3.
28	34	1957	683-1266	$\sim 0^\circ$	± 10	Above specimen and conditions; decreasing temp, cycle 3.
29	34	1957	428-983	$\sim 0^\circ$	± 10	Commercial grade A; strip (0.005 in. thick); oxidized in air at red heat for 30 min; measured in air (5×10^{-4} mm Hg); increasing temp, cycle 1.
30	34	1957	633-861	$\sim 0^\circ$	± 10	Above specimen and conditions; decreasing temp, cycle 1.
31	34	1957	683-994	$\sim 0^\circ$	± 10	Above specimen and conditions; cycle 2.
32	34	1957	550-855	$\sim 0^\circ$	± 10	Above specimen and conditions; cycle 3.
33	14	1913	473-873	$\sim 0^\circ$		Cleaned, polished,and oxidized.
34	80	1956	77	$\sim 0^\circ$		As received.
35	80	1956	77	$\sim 0^\circ$		Smoothed and cleaned.
36	80	1956	77	$\sim 0^\circ$		Polished.
37	80	1956	77	$\sim 0^\circ$		Oxidized in air at 1273 K for 30 min.
38	15	1947	373	$\sim 0^\circ$		Polished.
39	29	1956	422-983	$\sim 0^\circ$	± 11	Oxidized in air for 30 min at 1089 K; measured in vacuum ($< 1 \times 10^{-3}$ mm Hg); cycle 1.
40	29	1956	533-989	$\sim 0^\circ$	± 11	Above specimen and conditions; cycle 2.
41	29	1956	566-1022	$\sim 0^\circ$	± 11	Above specimen and conditions; cycle 3.
42	29	1956	647-1105	$\sim 0^\circ$	± 11	Grade A pure; measured in vacuum ($< 1 \times 10^{-3}$ mm Hg); first cycle.
43	29	1956	553-1128	$\sim 0^\circ$	± 11	Above specimen and conditions; cycle 2.
44	29	1956	719-1239	$\sim 0^\circ$	± 11	Above specimen and conditions; cycle 3.
45	29	1956	566-1144	$\sim 0^\circ$	± 11	Grade A pure; cleaned and smoothed; measured in vacuum ($< 1 \times 10^{-3}$ mm Hg); cycle 1.
46	29	1956	611-1211	$\sim 0^\circ$	± 11	Above specimen and conditions; cycle 2.

SPECIFICATION TABLE NO. 131 (continued)

Curve No.	Ref. No.	Year	Temperature Range, K	Geometry θ'	Reported Error, %	Composition (weight percent), Specifications and Remarks
47	29	1956	650-1294	$\sim 0^{\circ}$	± 11	Above specimen and conditions; cycle 3.
48	29	1956	611-1216	$\sim 0^{\circ}$	± 11	Grade A pure; cleaned, smoothed, and polished; measured in vacuum ($< 1 \times 10^{-3}$ mm Hg); cycle 1.
49	29	1956	550-1172	$\sim 0^{\circ}$	± 11	Above specimen and conditions; cycle 2.
50	29	1956	655-1255	$\sim 0^{\circ}$	± 11	Above specimen and conditions; cycle 3.

DATA TABLE NO. 131 NORMAL TOTAL EMITTANCE OF NICKEL

[Temperature, T, K; Emittance, ∈]

T	∈
CURVE 1	
375	0.094
410	0.084
440	0.094
488	0.100
550	0.105
590	0.118
597	0.110
640	0.118
675	0.123
680	0.128
720	0.124
725	0.130
775	0.131
800	0.129
830	0.139
845	0.132
870	0.142
875	0.135
930	0.152
945	0.157
1022	0.170
1040	0.157
1075	0.172
1125	0.166
1140	0.187
1150	0.168
1200	0.196
1215	0.176
1270	0.196
1345	0.200
1400	0.202
CURVE 2	
410	0.064
430	0.072
445	0.068
460	0.074
470	0.070
500	0.073
530	0.080
593	0.087
600	0.081
CURVE 2 (cont.)	
640	0.090
655	0.094
680	0.093
700	0.095
720	0.084
730	0.096
760	0.100
800	0.095
800	0.099
800	0.105
840	0.100
858	0.108
860	0.103
890	0.119
920	0.114
945	0.127
965	0.131
985	0.123
985	0.142
1010	0.137
1060	0.147
1075	0.152
1090	0.146
1150	0.162
1215	0.172
1250	0.188
1270	0.184
1280	0.190
1315	0.194
1340	0.192
CURVE 3	
673	0.58
743	0.61
813	0.64
853	0.68
CURVE 4	
593	0.33
673	0.35
CURVE 4 (cont.)	
703	0.37
743	0.38
793	0.41
813	0.41
CURVE 5	
653	0.50
753	0.52
793	0.54
873	0.57
933	0.60
973	0.62
1033	0.66
1073	0.70
CURVE 6	
623	0.07
683	0.08
753	0.09
783	0.09
823	0.10
853	0.10
CURVE 7	
633	0.30
673	0.31
713	0.33
753	0.36
793	0.39
833	0.41
853	0.42
CURVE 8	
643	0.50
713	0.51
783	0.52
863	0.55
913	0.57
973	0.61
CURVE 8 (cont.)	
1003	0.66
1053	0.68
CURVE 9	
1111	0.121
CURVE 10	
873	0.54
923	0.59
973	0.63
1023	0.65
1073	0.68
1123	0.70
1173	0.72
1223	0.74
1273	0.75
1323	0.78
1373	0.81
1423	0.84
1473	0.86
1523	0.87
1573	0.87
CURVE 11	
455	0.86
530	0.87
650	0.82
720	0.73
738	0.79
760	0.81
815	0.79
850	0.81
920	0.82
945	0.80
965	0.84
980	0.84
1125	0.83
CURVE 12	
400	0.927
450	0.927
500	0.900
555	0.903
645	0.855
690	0.838
710	0.825
760	0.828
850	0.840
900	0.847
940	0.900
945	0.847
980	0.865
1020	0.845
1050	0.848
1065	0.860
1100	0.850
CURVE 13	
83.2	0.055
CURVE 14	
83.2	0.174
CURVE 15	
83.2	0.088
CURVE 16	
555	0.041
778	0.083
878	0.107
1133	0.144
CURVE 17	
600	0.068
827	0.111
989	0.130
CURVE 18*	
778	0.097
889	0.113
1050	0.153
1211	0.153
CURVE 19*	
605	0.079
716	0.100
872	0.125
1044	0.148
CURVE 20	
755	0.100
900	0.130
961	0.143
1172	0.165
1344	0.173
1411	0.175
1477	0.192
CURVE 21*	
661	0.090
811	0.100
1183	0.134
1300	0.166
CURVE 22	
83.1	0.153
CURVE 23	
466	0.035
639	0.096
755	0.114
861	0.147
944	0.154
1061	0.173
1144	0.176
1222	0.192
CURVE 24*	
605	0.094
822	0.111
1028	0.162
CURVE 25*	
566	0.055
728	0.100
911	0.146
1044	0.160
1172	0.186
CURVE 26*	
733	0.107
872	0.135
CURVE 27	
644	0.076
850	0.103
1033	0.108
1166	0.137
1300	0.150
1400	0.156
1433	0.156
CURVE 28*	
683	0.082
1005	0.106
1266	0.150
CURVE 29	
428	0.055
583	0.205
722	0.310
800	0.355
872	0.380
983	0.445
CURVE 30	
633	0.190
739	0.350
861	0.415
CURVE 31*	
683	0.360
800	0.385
894	0.430
994	0.445
CURVE 32	
550	0.255
705	0.335
855	0.400
CURVE 33	
473	0.369
673	0.424
873	0.478
CURVE 34	
77	0.099
CURVE 35*	
77	0.099
CURVE 36	
77	0.020
CURVE 37	
77	0.261
CURVE 38	
373	0.072

* Not shown on plot

DATA TABLE NO. 131 (continued)

T	∈
CURVE 39	
422	0.07
578	0.25
628	0.40
728	0.39
730	0.45
794	0.44
866	0.48
872	0.44
983	0.47
CURVE 40	
533	0.30
683	0.41
705	0.42
800	0.48
866	0.41
891	0.47
989	0.49
CURVE 41	
566	0.26
750	0.46
839	0.49
1022	0.56
CURVE 42*	
647	0.120
669	0.090
761	0.120
769	0.120
933	0.140
989	0.120
1033	0.160
1105	0.160
CURVE 43*	
553	0.070
700	0.110
758	0.110
816	0.140
833	0.130
CURVE 43* (cont.)	
966	0.140
1000	0.150
1128	0.160
CURVE 44*	
719	0.100
750	0.150
855	0.130
944	0.140
994	0.140
1122	0.170
1211	0.180
1239	0.180
CURVE 45*	
566	0.06
589	0.09
778	0.10
822	0.13
883	0.13
983	0.14
1144	0.16
CURVE 46*	
611	0.08
728	0.12
783	0.12
878	0.15
900	0.14
1050	0.16
1072	0.17
1211	0.15
CURVE 47*	
650	0.11
750	0.12
811	0.12
950	0.16
1116	0.14
1155	0.19
1294	0.17
CURVE 48	
611	0.11
644	0.11
741	0.13
741	0.14
758	0.14
816	0.15
865	0.17
944	0.18
1039	0.18
1044	0.19
1144	0.19
1216	0.21
CURVE 49*	
550	0.07
872	0.16
922	0.15
1055	0.19
1172	0.20
CURVE 50*	
655	0.113
666	0.100
847	0.121
1005	0.171
1028	0.142
1144	0.152
1255	0.161

* Not shown on plot

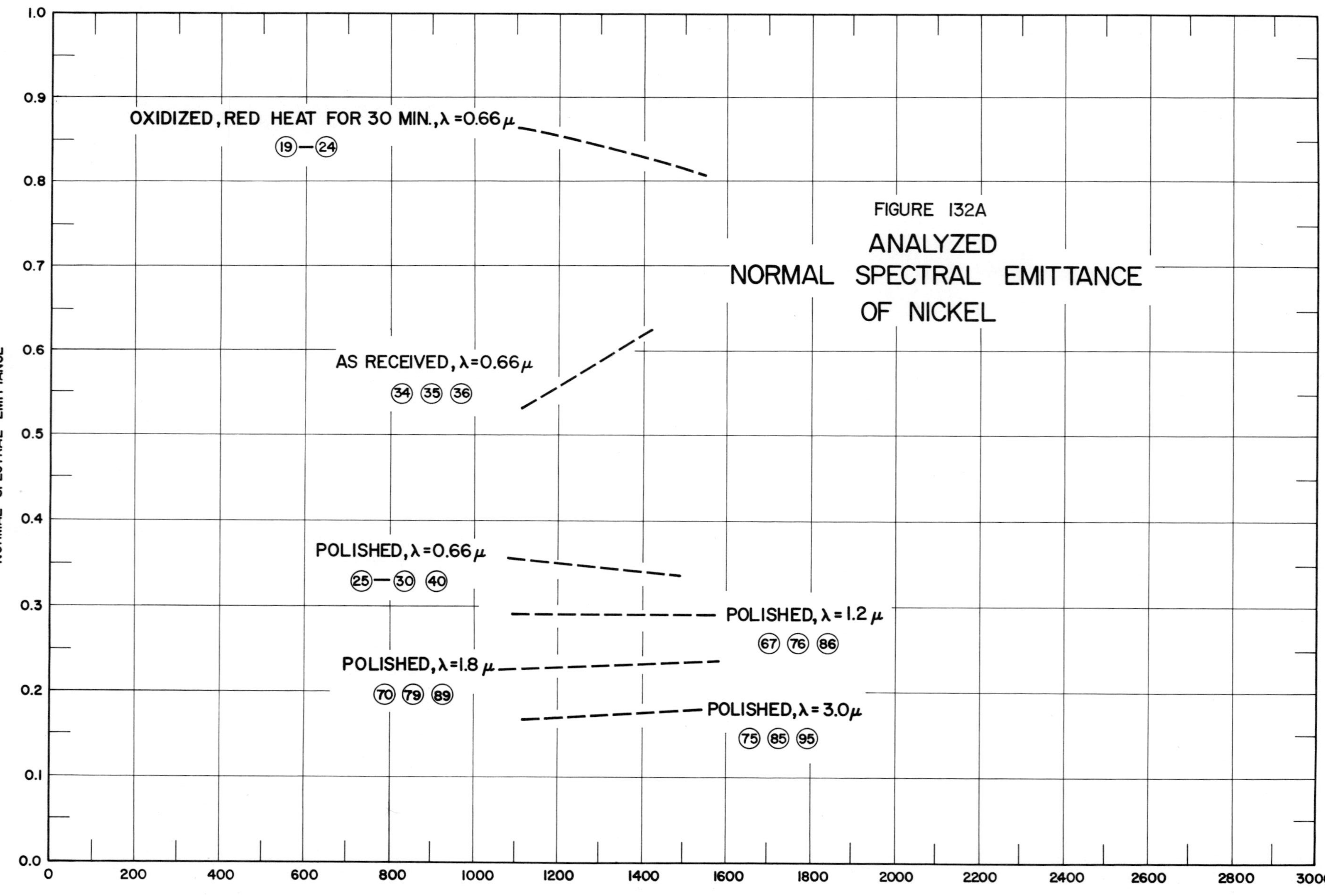

FIGURE 132A
ANALYZED
NORMAL SPECTRAL EMITTANCE
OF NICKEL

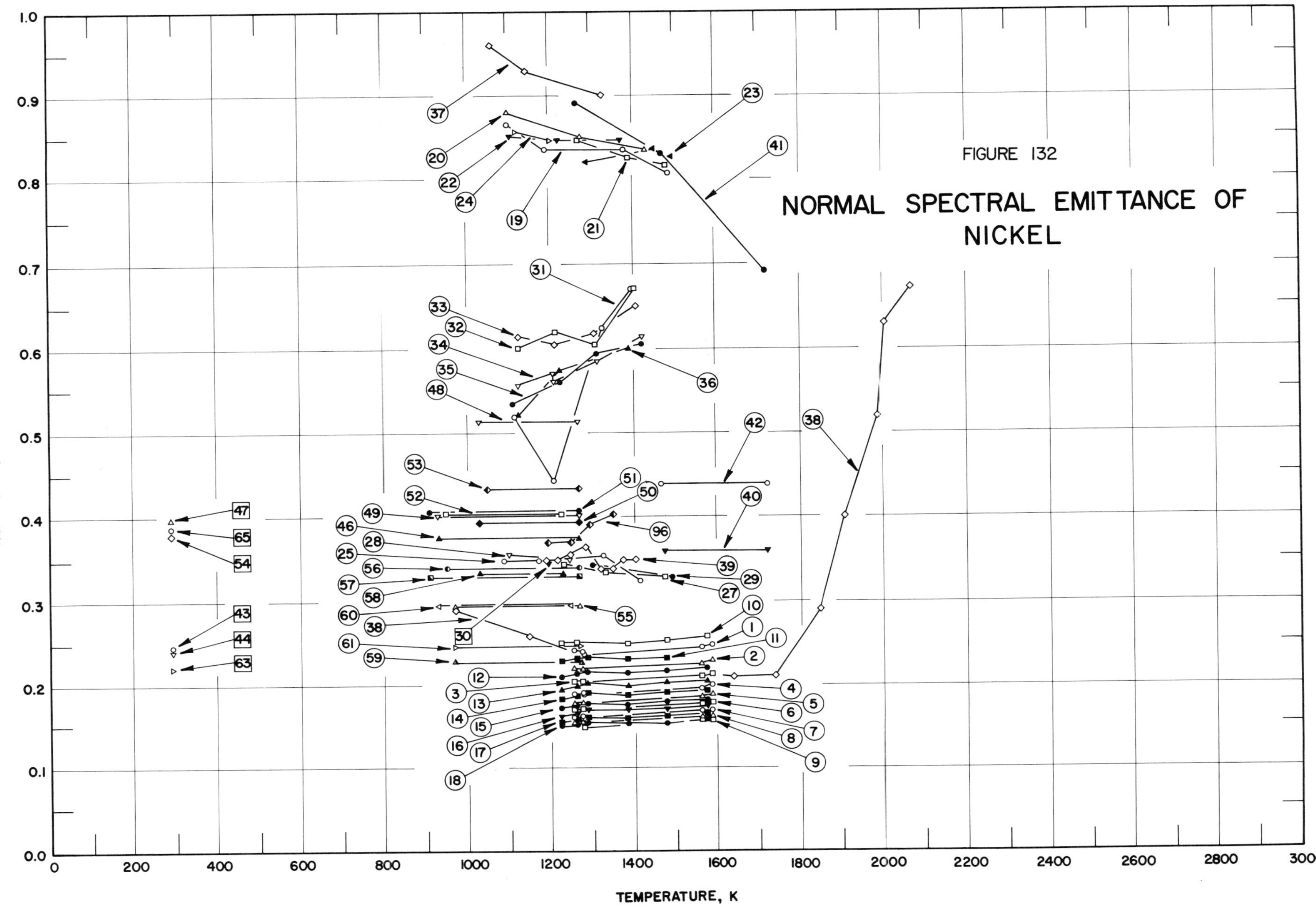

FIGURE 132
NORMAL SPECTRAL EMITTANCE OF NICKEL

SPECIFICATION TABLE NO. 132 NORMAL SPECTRAL EMITTANCE OF NICKEL

Curve No.	Ref. No.	Year	Wavelength μ	Temperature Range, K	Geometry θ'	Reported Error, %	Composition (weight percent), Specifications and Remarks
1	43	1952	1.0	1255-1587	$\sim 0^{\circ}$		Polished; washed with ether; measured in vacuum; [Author's designation; Specimen D].
2	43	1952	1.2	1255-1587	$\sim 0^{\circ}$		Above specimen and conditions.
3	43	1952	1.4	1255-1587	$\sim 0^{\circ}$		Above specimen and conditions.
4	43	1952	1.6	1255-1587	$\sim 0^{\circ}$		Above specimen and conditions.
5	43	1952	1.8	1255-1587	$\sim 0^{\circ}$		Above specimen and conditions.
6	43	1952	2.0	1255-1587	$\sim 0^{\circ}$		Above specimen and conditions.
7	43	1952	2.2	1255-1587	$\sim 0^{\circ}$		Above specimen and conditions.
8	43	1952	2.4	1255-1587	$\sim 0^{\circ}$		Above specimen and conditions.
9	43	1952	2.6	1279-1587	$\sim 0^{\circ}$		Above specimen and conditions.
10	43	1952	1.0	1226-1574	$\sim 0^{\circ}$		Polished; washed with ether; measured in vacuum; [Author's designation; Specimen A].
11	43	1952	1.2	1226-1478	$\sim 0^{\circ}$		Above specimen and conditions.
12	43	1952	1.4	1226-1574	$\sim 0^{\circ}$		Above specimen and conditions.
13	43	1952	1.6	1226-1574	$\sim 0^{\circ}$		Above specimen and conditions.
14	43	1952	1.8	1226-1574	$\sim 0^{\circ}$		Above specimen and conditions.
15	43	1952	2.0	1226-1574	$\sim 0^{\circ}$		Above specimen and conditions.
16	43	1952	2.2	1226-1574	$\sim 0^{\circ}$		Above specimen and conditions.
17	43	1952	2.4	1226-1574	$\sim 0^{\circ}$		Above specimen and conditions.
18	43	1952	2.6	1226-1574	$\sim 0^{\circ}$		Above specimen and conditions.
19	34	1957	0.665	1108-1491	$\sim 0^{\circ}$		Commercial grade A; oxidized in air at red heat for 30 min; measured in air (5×10^{-4} mm Hg); increasing temp, cycle 1.
20	34	1957	0.665	1436-1105	$\sim 0^{\circ}$		Above specimen and conditions; decreasing temp, cycle 1.
21	34	1957	0.665	1275-1486	$\sim 0^{\circ}$		Above specimen and conditions; cycle 2.
22	34	1957	0.665	1375-1111	$\sim 0^{\circ}$		Above specimen and conditions; decreasing temp, cycle 2.
23	34	1957	0.665	1294-1500	$\sim 0^{\circ}$		Above specimen and conditions; cycle 3.
24	34	1957	0.665	1208-1122	$\sim 0^{\circ}$		Above specimen and conditions; decreasing temp, cycle 3.
25	34	1957	0.665	1089-1416	$\sim 0^{\circ}$		Commercial grade A; polished; measured in air (5×10^{-4} mm Hg); increasing temp, cycle 1.

SPECIFICATION TABLE NO. 132 (continued)

Curve No.	Ref. No.	Year	Wavelength μ	Temperature Range, K	Geometry θ'	Reported Error, %	Composition (weight percent), Specifications and Remarks
26	34	1957	0.665	1475-1089	$\sim 0^{\circ}$		Above specimen and conditions; decreasing temp, cycle 1.
27	34	1957	0.665	1233-1477	$\sim 0^{\circ}$		Above specimen and conditions; cycle 2.
28	34	1957	0.665	1244-1100	$\sim 0^{\circ}$		Above specimen and conditions; decreasing temp, cycle 2.
29	34	1957	0.665	1302-1497	$\sim 0^{\circ}$		Above specimen and conditions; cycle 3.
30	34	1957	0.665	1197	$\sim 0^{\circ}$		Above specimen and conditions; decreasing temp, cycle 3.
31	34	1957	0.665	1119-1400	$\sim 0^{\circ}$		Commercial grade A; as received; measured in air (5×10^{-4} mm Hg); increasing temp, cycle 1.
32	34	1957	0.665	1127-1408	$\sim 0^{\circ}$		Above specimen and conditions; cycle 2.
33	34	1957	0.665	1133-1414	$\sim 0^{\circ}$		Above specimen and conditions; cycle 3.
34	34	1957	0.665	1125-1422	$\sim 0^{\circ}$		Commercial grade A; cleaned with liquid detergent; measured in air (5×10^{-4} mm Hg); increasing temp, cycle 1.
35	34	1957	0.665	1116-1422	$\sim 0^{\circ}$		Above specimen and conditions; cycle 2.
36	34	1957	0.665	1122-1397	$\sim 0^{\circ}$		Above specimen and conditions; cycle 3.
37	83	1914	0.65	1073-1336	$\sim 0^{\circ}$		Ni strip; oxidized in air; authors assumed $\alpha = \epsilon$.
38	35	1914	0.66	970-2070	$\sim 0^{\circ}$		Measured in burning hydrogen.
39	42	1942	0.667	1195-1405	$\sim 0^{\circ}$		Baked at 623 K for 200 hrs; measured in vacuum (4×10^{-8} mm Hg).
40	19	1914	0.65	1473-1723	$\sim 0^{\circ}$	1	Film; tungsten substrate; measured in hydrogen; Pt reference ($\epsilon = 0.33$ for $\lambda = 0.65\mu$ at all temp).
41	19	1914	0.65	1273-1723	$\sim 0^{\circ}$	1	Film; tungsten substrate; melted in hydrogen then oxidized in air by heating; measured in air; Pt reference ($\epsilon = 0.33$ for $\lambda = 0.650\mu$ at all temp).
42	19	1914	0.55	1473-1723	$\sim 0^{\circ}$	1	Film; tungsten substrate; measured in hydrogen; Pt reference ($\epsilon = 0.33$ for $\lambda = 0.65\mu$ at all temp).
43	81	1941	0.65	298	$\sim 0^{\circ}$		Filament used in the radio tube type DL23; coated with oxide; heated in hydrogen at 1098 K for 15 min.
44	81	1941	0.65	298	$\sim 0^{\circ}$		Filament used in the radio tube type DL24; coated with oxide; heated at 1098 K in hydrogen for 15 min.
45	81	1941	0.65	298	$\sim 0^{\circ}$		Filament used in the radio tube type DL27; coated with oxide; heated at 1098 K in hydrogen for 15 min.
46	81	1941	0.65	933-1273	$\sim 0^{\circ}$	< 1	Grade A Nickel; 99+ percent pure with Co, Fe, Mn, Cu, and C impurities (filament used in the radio tube type DM212); heated at 1098 K in hydrogen for 15 min; authors assumed $\alpha = \epsilon$.

SPECIFICATION TABLE NO. 132 (continued)

Curve No.	Ref. No.	Year	Wavelength μ	Temperature Range, K	Geometry θ'	Reported Error, %	Composition (weight percent), Specifications and Remarks
47	81	1941	0.65	298	~0°		Grade A Nickel; 99+ percent pure with Co, Fe, Mn, Cu, and C impurities (filament used in the radio tube type DM214); heated at 1098 K in hydrogen for 15 min; authors assumed $\alpha = \epsilon$.
48	81	1941	0.65	1013-1273	~0°	< 1	Grade A Nickel; 99+ percent pure with Co, Fe, Mn, Cu,and C impurities (filament used in the radio tube type DB216); heated in hydrogen at 1098 K for 15 min; authors assumed $\alpha = \epsilon$.
49	81	1941	0.65	933-1273	~0°	< 1	Grade A Nickel; 99+ percent pure with Co, Fe, Mn, Cu,and C impurities (filament used in the radio tube type DM215); heated in hydrogen at 1098 K for 15 min; authors assumed $\alpha = \epsilon$.
50	81	1941	0.65	1033-1273	~0°	< 1	Grade A Nickel; 99+ percent pure with Co, Fe, Mn, Cu,and C impurities (filament used in the radio tube type DM214); heated at 1098 K in hydrogen for 15 min; authors assumed $\alpha = \epsilon$.
51	81	1941	0.65	913-1273	~0°	< 1	Grade A Nickel; 99+ percent pure with Co, Fe, Mn, Cu,and C impurities (filament used in the radio tube type DM213); heated at 1098 K in hydrogen for 15 min; authors assumed $\alpha = \epsilon$.
52	81	1941	0.65	953-1233	~0°	< 1	Grade A Nickel; 99+ percent pure with Co, Fe, Mn, Cu,and C impurities (filament used in the radio tube type DB215); heated in hydrogen at 1098 K for 15 min; authors assumed $\alpha = \epsilon$.
53	81	1941	0.65	1053-1273	~0°	< 1	Grade A Nickel; 99+ percent pure with Co, Fe, Mn, Cu,and C impurities (filament used in the radio tube type DB214); heated in hydrogen at 1098 K for 15 min; authors assumed $\alpha = \epsilon$.
54	81	1941	0.65	298	~0°		Grade A Nickel; 99+ percent pure with Co, Fe, Mn, Cu,and C impurities (filament used in the radio tube type DM212); heated at 1098 K in hydrogen for 15 min; authors assumed $\alpha = \epsilon$.
55	81	1941	0.65	973-1273	~0°	< 1	Filament used in the radio tube type DL23; coated with oxide; authors assumed $\alpha = \epsilon$.
56	81	1941	0.65	953-1273	~0°	< 1	Filament used in the radio tube type DL22; coated with oxide; authors assumed $\alpha = \epsilon$.
57	81	1941	0.65	913-1273	~0°	< 1	Filament used in the radio tube type DB218; coated with oxide; heated in hydrogen at 1098 K for 15 min; authors assumed $\alpha = \epsilon$.
58	81	1941	0.65	1033-1233	~0°	< 1	Filament used in the radio tube type DB217; coated with oxide; heated in hydrogen at 1098 K for 15 min; authors assumed $\alpha = \epsilon$.
59	81	1941	0.65	973-1273	~0°	< 1	Filament used in the radio tube type DL27; coated with oxide; heated in hydrogen at 1098 K for 15 min; authors assumed $\alpha = \epsilon$.

SPECIFICATION TABLE NO. 132 (continued)

Curve No.	Ref. No.	Year	Wavelength μ	Temperature Range, K	Geometry θ'	Reported Error, %	Composition (weight percent), Specifications and Remarks
60	81	1941	0.65	933-1253	$\sim 0^\circ$	< 1	Filament used in the radio tube type DL 25; coated with oxide; heated in hydrogen at 1098 K for 15 min; authors assumed $\alpha = \epsilon$.
61	81	1941	0.65	973-1273	$\sim 0^\circ$	< 1	Filament used in the radio tube type DL 24; coated with oxide; heated hydrogen at 1098 K for 15 min; authors assumed $\alpha = \epsilon$.
62	81	1941	0.65	298	$\sim 0^\circ$		Filament used in the radio tube type DL 24; coated with oxide; heated in hydrogen at 1098 K for 15 min; authors assumed $\alpha = \epsilon$.
63	81	1941	0.65	298	$\sim 0^\circ$		Filament used in the radio tube type DL 27; coated with oxide; heated in hydrogen at 1098 K for 15 min; authors assumed $\alpha = \epsilon$.
64	81	1941	0.65	298	$\sim 0^\circ$		Filament used in the radio tube type DL 23; coated with oxide; authors assumed $\alpha = \epsilon$.
65	81	1941	0.65	298	$\sim 0^\circ$		Grade A Nickel; 99+ percent pure with Co, Fe, Mn, Cu, and C impurities (Filament used in the radio tube type DM 214); heated in hydrogen at 1098 K for 15 min.
66	81	1941	0.65	298	$\sim 0^\circ$		Grade A Nickel; 99+ percent pure with Co, Fe, Mn, Cu, and C impurities (Filament used in the radio tube type DM 212); heated in hydrogen at 1098 K for 15 min.
67	41	1956	1.2	1183-1563	$\sim 0^\circ$	1/2	0.24 Fe, 0.11 Mg, 0.05 C, Ni balance; polished with emery paper and lapped with jeweler's rouge; measured in vacuum; [Author's designation: Specimen A].
68	41	1956	1.4	1033-1583	$\sim 0^\circ$	1/2	Above specimen and conditions.
69	41	1956	1.6	1043-1523	$\sim 0^\circ$	1/2	Above specimen and conditions.
70	41	1956	1.8	1143-1583	$\sim 0^\circ$	1/2	Above specimen and conditions.
71	41	1956	2.2	973-1553	$\sim 0^\circ$	1/2	Above specimen and conditions.
72	41	1956	2.4	1013-1553	$\sim 0^\circ$	1/2	Above specimen and conditions.
73	41	1956	2.6	1003-1593	$\sim 0^\circ$	1/2	Above specimen and conditions.
74	41	1956	2.75	1023-1553	$\sim 0^\circ$	1/2	Above specimen and conditions.
75	41	1956	3.0	1173-1533	$\sim 0^\circ$	1/2	Above specimen and conditions.
76	41	1956	1.2	1093-1523	$\sim 0^\circ$	1/2	Different sample, same as above specimen and conditions; [Author's designation: Specimen B].
77	41	1956	1.4	1193-1573	$\sim 0^\circ$	1/2	Above specimen and conditions.
78	41	1956	1.6	1023-1593	$\sim 0^\circ$	1/2	Above specimen and conditions.
79	41	1956	1.8	1103-1573	$\sim 0^\circ$	1/2	Above specimen and conditions.
80	41	1956	2.0	1033-1573	$\sim 0^\circ$	1/2	Above specimen and conditions.

SPECIFICATION TABLE NO. 132 (continued)

Curve No.	Ref. No.	Year	Wavelength μ	Temperature Range, K	Geometry θ'	Reported Error, %	Composition (weight percent), Specifications and Remarks
81	41	1956	2.2	1013-1553	$\sim 0^\circ$	1/2	Above specimen and conditions.
82	41	1956	2.4	993-1548	$\sim 0^\circ$	1/2	Above specimen and conditions.
83	41	1956	2.6	993-1573	$\sim 0^\circ$	1/2	Above specimen and conditions.
84	41	1956	2.75	1023-1553	$\sim 0^\circ$	1/2	Above specimen and conditions.
85	41	1956	3.0	1163-1528	$\sim 0^\circ$	1/2	Above specimen and conditions.
86	41	1956	1.2	1103-1593	$\sim 0^\circ$	1/2	Different sample, same as above specimen and conditions; [Author's designation: Specimen D].
87	41	1956	1.4	993-1573	$\sim 0^\circ$	1/2	Above specimen and conditions.
88	41	1956	1.6	1088-1543	$\sim 0^\circ$	1/2	Above specimen and conditions.
89	41	1956	1.8	1043-1583	$\sim 0^\circ$	1/2	Above specimen and conditions.
90	41	1956	2.0	1103-1573	$\sim 0^\circ$	1/2	Above specimen and conditions.
91	41	1956	2.2	1003-1538	$\sim 0^\circ$	1/2	Above specimen and conditions.
92	41	1956	2.4	1073-1573	$\sim 0^\circ$	1/2	Above specimen and conditions.
93	41	1956	2.6	1043-1563	$\sim 0^\circ$	1/2	Above specimen and conditions.
94	41	1956	2.8	1033-1558	$\sim 0^\circ$	1/2	Above specimen and conditions.
95	41	1956	3.0	1123-1498	$\sim 0^\circ$	1/2	Above specimen and conditions.
96	246	1967	0.65	1199-1349	$\sim 0^\circ$		99.95 Ni.

DATA TABLE NO. 132 NORMAL SPECTRAL EMITTANCE OF NICKEL

[Temperature, T, K; Emittance, ϵ; Wavelength, λ, μ]

T	ϵ
CURVE 1	λ = 1.0
1255	0.242
1275	0.240
1279	0.237
1561	0.246
1587	0.249
CURVE 2	λ = 1.2
1255	0.220
1275	0.219
1279	0.218
1561	0.225
1587	0.229
CURVE 3	λ = 1.4
1255	0.204
1275	0.204
1279	0.203
1561	0.210
1587	0.213
CURVE 4	λ = 1.6
1255	0.189
1275	0.191
1279	0.189
1561	0.196
1587	0.200
CURVE 5	λ = 1.8
1255	0.178
1275	0.179
1279	0.178
1561	0.186
1587	0.188
CURVE 6	λ = 2.0
1255	0.169
1275	0.170
1279	0.169
1561	0.176
1587	0.178
CURVE 7	λ = 2.2
1255	0.161
1275	0.162
1279	0.161
1561	0.168
1587	0.169
CURVE 8	λ = 2.4
1255	0.154
1275	0.155
1279	0.154
1561	0.161
1587	0.162
CURVE 9	λ = 2.6
1279	0.148
1561	0.156
1587	0.156
CURVE 10	λ = 1.0
1226	0.251
1263	0.252
1383	0.250
1478	0.254
1574	0.258
CURVE 11	λ = 1.2
1226	0.228
1263	0.232
1288	0.233
1383	0.231
1478	0.233
CURVE 12	λ = 1.4
1226	0.209
1263	0.213
1288	0.215
1383	0.214
1478	0.216
1574	0.220
CURVE 13	λ = 1.6
1226	0.194
1263	0.198
1288	0.202
1383	0.199
1478	0.202
1574	0.204
CURVE 14	λ = 1.8
1226	0.183
1263	0.186
1288	0.190
1383	0.187
1478	0.191
1574	0.193
CURVE 15	λ = 2.0
1226	0.172
1263	0.174
CURVE 15 (cont.)	
1288	0.178
1383	0.176
1478	0.180
1574	0.182
CURVE 16	λ = 2.2
1226	0.162
1263	0.165
1288	0.169
1383	0.168
1478	0.170
1574	0.173
CURVE 17	λ = 2.4
1226	0.155
1263	0.158
1288	0.161
1383	0.160
1478	0.163
1574	0.165
CURVE 18	λ = 2.6
1226	0.150
1263	0.151
1288	0.155
1383	0.153
1478	0.153
1574	0.159
CURVE 19	λ = 0.665
1108	0.865
1197	0.835
1386	0.835
1491	0.805
CURVE 20	λ = 0.665
1436	0.835
1283	0.850
1105	0.880
CURVE 21	λ = 0.665
1275	0.845
1394	0.825
1486	0.815
CURVE 22	λ = 0.665
1375	0.845
1227	0.845
1111	0.850
CURVE 23	λ = 0.665
1294	0.820
1455	0.835
1500	0.825
CURVE 24	λ = 0.665
1208	0.845
1122	0.855
CURVE 25	λ = 0.665
1089	0.350
1172	0.350
1327	0.355
1416	0.325
CURVE 26*	λ = 0.665
1475	0.310
1319	0.335
1177	0.360
1089	0.360
CURVE 27	λ = 0.665
1233	0.345
1330	0.335
1477	0.330
CURVE 28	λ = 0.665
1244	0.350
1100	0.355
CURVE 29	λ = 0.665
1302	0.345
1497	0.330
CURVE 30	λ = 0.665
1197	0.345
CURVE 31	λ = 0.665
1119	0.520
1216	0.440
1330	0.625
1400	0.670
CURVE 32	λ = 0.665
1127	0.600
CURVE 32 (cont.)	
1219	0.620
1316	0.605
1408	0.670
CURVE 33	λ = 0.665
1133	0.615
1219	0.605
1316	0.620
1414	0.650
CURVE 34	λ = 0.665
1125	0.555
1216	0.570
1219	0.560
1319	0.585
1422	0.615
CURVE 35	λ = 0.665
1116	0.535
1222	0.560
1319	0.595
1422	0.605
CURVE 36	λ = 0.665
1122	0.525
1325	0.575
1397	0.600
CURVE 37	λ = 0.65
1073	0.96
1157	0.93
1336	0.90
CURVE 38	λ = 0.66
970	0.29
1150	0.26
1270	0.24*
1395	0.23*
1640	0.21
1740	0.21
1850	0.29
1910	0.40
1990	0.52
2050	0.63
2070	0.67
CURVE 39	λ = 0.667
1195	0.350
1220	0.350
1255	0.355
1285	0.365
1320	0.340
1350	0.340
1375	0.350
1405	0.350
CURVE 40	λ = 0.65
1473	0.36
1723	0.36
CURVE 41	λ = 0.65
1273	0.89
1473	0.83
1723	0.69
CURVE 42	λ = 0.55
1473	0.44
1723	0.44

* Not shown on plot

DATA TABLE NO. 132 (continued)

T	ϵ
CURVE 43	
$\lambda = 0.65$	
298	0.248
CURVE 44	
$\lambda = 0.65$	
298	0.240
CURVE 45*	
$\lambda = 0.246$	
298	0.246
CURVE 46	
$\lambda = 0.65$	
933	0.376
1273	0.376
CURVE 47	
$\lambda = 0.65$	
298	0.399
CURVE 48	
$\lambda = 0.65$	
1013	0.513
1273	0.513
CURVE 49	
$\lambda = 0.65$	
933	0.40
1273	0.40
CURVE 50	
$\lambda = 0.65$	
1033	0.393
1273	0.393
CURVE 51	
$\lambda = 0.65$	
913	0.406
1273	0.406
CURVE 52	
$\lambda = 0.65$	
953	0.402
1233	0.402
CURVE 53	
$\lambda = 0.65$	
1053	0.438
1273	0.438
CURVE 54	
$\lambda = 0.65$	
298	0.380
CURVE 55	
$\lambda = 0.65$	
973	0.294
1273	0.294
CURVE 56	
$\lambda = 0.65$	
953	0.341
1273	0.341
CURVE 57	
$\lambda = 0.65$	
913	0.330
1273	0.330
CURVE 58	
$\lambda = 0.65$	
1033	0.334
1233	0.334
CURVE 59	
$\lambda = 0.65$	
973	0.229
1273	0.229
CURVE 60	
$\lambda = 0.65$	
933	0.295
1253	0.295
CURVE 61	
$\lambda = 0.65$	
973	0.248
1273	0.248
CURVE 62*	
$\lambda = 0.65$	
298	0.245
CURVE 63	
$\lambda = 0.65$	
298	0.220
CURVE 64*	
$\lambda = 0.65$	
298	0.245
CURVE 65	
$\lambda = 0.65$	
298	0.389
CURVE 66*	
$\lambda = 0.65$	
298	0.381
CURVE 67*	
$\lambda = 1.2$	
1183	0.283
1278	0.283
1283	0.282
1393	0.281
1488	0.281
1563	0.280
CURVE 68*	
$\lambda = 1.4$	
1033	0.258
1133	0.259
1143	0.259
1213	0.259
1303	0.260
1393	0.261
1463	0.261
1583	0.262
CURVE 69*	
$\lambda = 1.6$	
1043	0.238
1183	0.241
1253	0.242
1303	0.242
1303	0.243
1373	0.244
1523	0.246
CURVE 70*	
$\lambda = 1.8$	
1143	0.221
1173	0.221
1283	0.223
1363	0.224
1413	0.225
1463	0.226
1533	0.227
1583	0.228
CURVE 71*	
$\lambda = 2.2$	
973	0.191
1023	0.191
1113	0.193
1203	0.194
1293	0.195
1333	0.196
1413	0.197
1503	0.198
1553	0.199
CURVE 72*	
$\lambda = 2.4$	
1013	0.176
1128	0.178
1183	0.179
1243	0.181
1333	0.183
1403	0.184
1493	0.185
1553	0.187
CURVE 73*	
$\lambda = 2.6$	
1003	0.169
1053	0.170
1133	0.172
1193	0.173
1273	0.175
1363	0.177
1443	0.179
1523	0.180
1593	0.182
CURVE 74*	
$\lambda = 2.75$	
1023	0.165
1143	0.168
1223	0.171
1333	0.174
1453	0.177
1553	0.180
CURVE 75*	
$\lambda = 3.0$	
1173	0.164
1243	0.166
1333	0.169
1413	0.171
1453	0.172
1533	0.175
CURVE 76*	
$\lambda = 1.2$	
1093	0.294
1198	0.294
1263	0.293
1298	0.294
1333	0.291
1343	0.291
1403	0.291
1523	0.289
CURVE 77*	
$\lambda = 1.4$	
1193	0.270
1203	0.269
1273	0.270
1328	0.270
1353	0.270
1373	0.271
1473	0.271
1573	0.271
CURVE 78*	
$\lambda = 1.6$	
1023	0.250
1163	0.251
1273	0.252
1383	0.253
1453	0.254
1523	0.254
1593	0.255
CURVE 79*	
$\lambda = 1.8$	
1103	0.224
1223	0.226
1353	0.227
1388	0.227
1443	0.229
1513	0.229
1573	0.231
CURVE 80*	
$\lambda = 2.0$	
1033	0.214
1143	0.216
1173	0.216
1248	0.217
1298	0.218
1343	0.219
1373	0.219
1423	0.220
1483	0.220
1523	0.221
1573	0.222
CURVE 81*	
$\lambda = 2.2$	
1013	0.198
1123	0.200
1213	0.202
1293	0.203
1343	0.205
1383	0.205
1443	0.206
1498	0.207
1553	0.208
CURVE 82*	
$\lambda = 2.4$	
993	0.186
1063	0.187
1143	0.188
1233	0.190
1293	0.192
1363	0.194
1438	0.194
1508	0.195
1548	0.196
CURVE 83*	
$\lambda = 2.6$	
993	0.172
1093	0.174
1173	0.177
1263	0.178
CURVE 83*(cont.)	
1353	0.180
1448	0.183
1453	0.183
1493	0.184
1573	0.186
CURVE 84*	
$\lambda = 2.75$	
1023	0.169
1123	0.171
1193	0.173
1273	0.175
1373	0.178
1393	0.178
1443	0.181
1553	0.181
CURVE 85*	
$\lambda = 3.0$	
1163	0.168
1303	0.172
1403	0.175
1493	0.178
1528	0.180
CURVE 86*	
$\lambda = 1.2$	
1103	0.302
1233	0.301
1283	0.300
1393	0.298
1503	0.296
1593	0.295
CURVE 87*	
$\lambda = 1.4$	
993	0.272
1103	0.274
1193	0.272
1293	0.274
1373	0.275

* Not shown on plot

DATA TABLE NO. 132 (continued)

T	∈
CURVE 87*(cont.)	
1423	0.276
1523	0.276
1573	0.277
CURVE 88*	
λ = 1.6	
1088	0.257
1243	0.258
1373	0.259
1468	0.260
1543	0.259
CURVE 89*	
λ = 1.8	
1043	0.234
1123	0.236
1173	0.237
1253	0.238
1273	0.239
1323	0.240
1473	0.242
1583	0.245
CURVE 90*	
λ = 2.0	
1103	0.222
1283	0.226
1473	0.229
1523	0.230
1573	0.231
CURVE 91*	
λ = 2.2	
1003	0.202
1138	0.205
1283	0.208
1363	0.210
1383	0.211
1473	0.212
1498	0.213
1538	0.215

T	∈
CURVE 92*	
λ = 2.4	
1073	0.191
1133	0.192
1163	0.193
1253	0.195
1273	0.195
1373	0.196
1463	0.200
1503	0.199
1543	0.201
1573	0.200
CURVE 93*	
λ = 2.6	
1043	0.177
1178	0.180
1193	0.181
1263	0.183
1363	0.185
1463	0.187
1473	0.188
1563	0.190
CURVE 94*	
λ = 2.8	
1033	0.172
1133	0.174
1233	0.177
1353	0.182
1463	0.185
1558	0.188
CURVE 95*	
λ = 3.0	
1123	0.170
1173	0.171
1293	0.176
1413	0.179
1498	0.182

T	∈
CURVE 96	
λ = 0.65	
1199	0.370
1250	0.371
1299	0.390
1349	0.401

* Not shown on plot

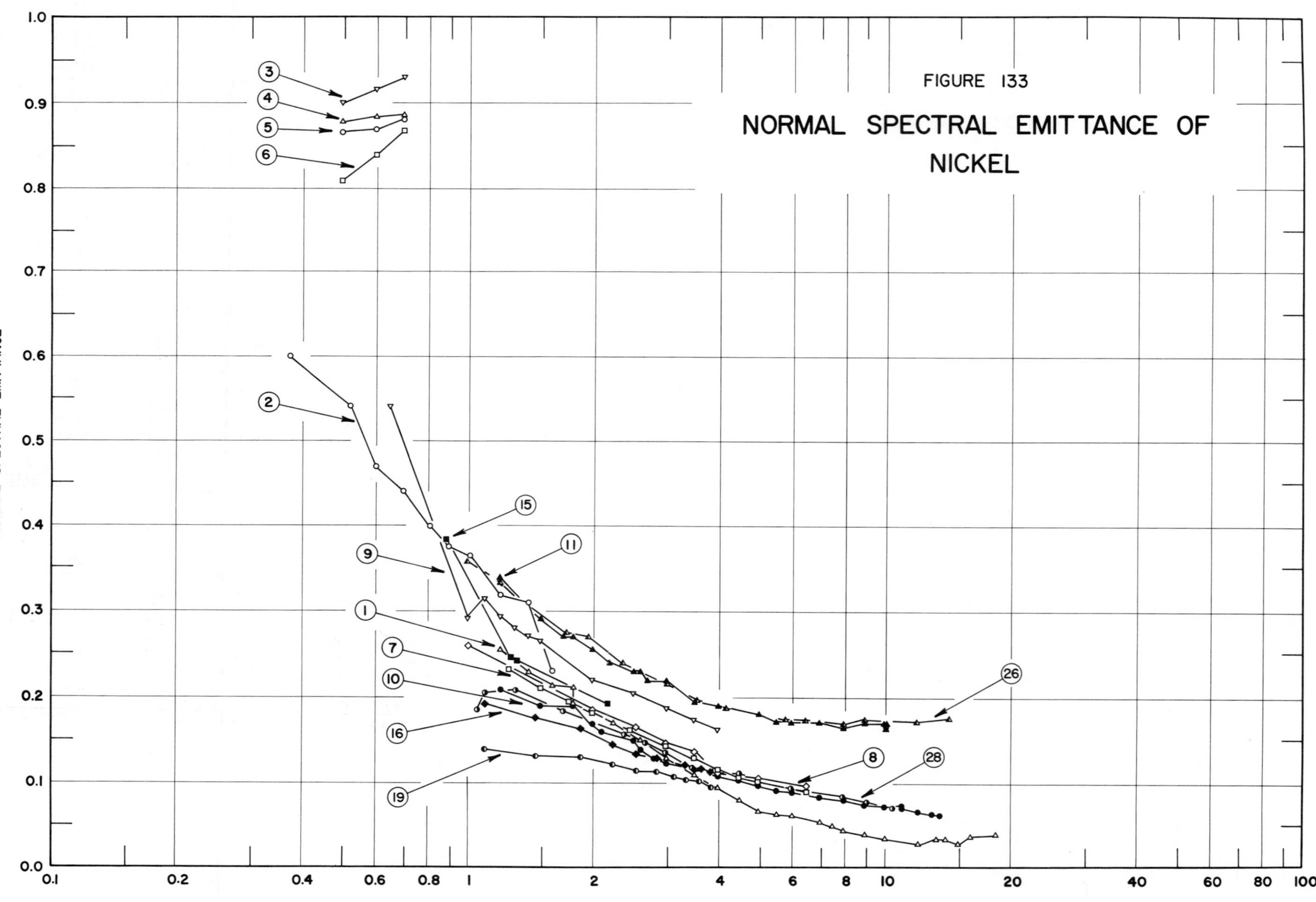

FIGURE 133
NORMAL SPECTRAL EMITTANCE OF NICKEL
NORMAL SPECTRAL EMITTANCE
WAVELENGTH, MICRONS
1.0
0.9
0.8
0.7
0.6
0.5
0.4
0.3
0.2
0.1
0.0
0.1
0.2
0.4
0.6
0.8
1
2
4
6
8
10
20
40
60
80
100
3
4
5
6
2
15
9
11
1
7
10
16
19
26
8
28

SPECIFICATION TABLE NO. 133 NORMAL SPECTRAL EMITTANCE OF NICKEL

Curve No.	Ref. No.	Year	Temperature K	Wavelength Range, μ	Geometry θ'	Reported Error, %	Composition (weight percent), Specifications and Remarks
1	65	1962	294	1.20-18.50	~0°		Polished; measured in air; calculated from (1 - ρ).
2	65	1962	294	0.375-1.600	~0°		Different sample, same as curve 1 specimen and conditions.
3	83	1914	1209	0.5-0.7	~0°		High purity Nickel with minute traces of Fe; oxidized in air for several min.
4	83	1914	1331	0.5-0.7	~0°		Above specimen and conditions.
5	83	1914	1432	0.5-0.7	~0°		Above specimen and conditions.
6	83	1914	1528	0.5-0.7	~0°		Above specimen and conditions.
7	38	1933	1123	1.2-6.5	~0°		99.5 pure; measured in vacuum.
8	38	1933	1273	1.0-6.5	~0°		99.5 pure; measured in vacuum.
9	37	1947	1383	0.6-4.0	~0°		99.97 pure; measured in vacuum.
10	30	1963	1267	1.2-13.5	~0°		Mechanically polished with aluminum oxide and cleaned with water; heated in vacuum for 3 hrs at 588 K, 3 hrs at 1089 K, and 4 hrs at 1267 K; measured in vacuum (5×10^{-6} mm Hg); data extracted from smooth curve; [Authors designation: Sample 7].
11	30	1963	1428	1.2-10.1	~0°		Above specimen and conditions except heated in vacuum at T > 1256 K for 48 hrs.
12	30	1963	1267	1.2-13.5	~0°		Mechanically polished with aluminum oxide and cleaned with water; heated in vacuum at 1372 K for 4 hrs and at 1331 K for 1 hr; data extracted from smooth curve; [Author's designation: Sample 10].
13	245	1965	294	1.00-11.7	~0°	5	Measured in argon.
14	248	1965	1100	0.823-1.660	~0°	<±10	Grade NP-3 nickel.
15	248	1965	1100	0.898-2.196	~0°	<±10	Grade NP-3 nickel.
16	239	1959	1396	1.104-4.055	~0°	5	Polished; measured in argon.
17	239	1959	1350	1.097-4.041	~0°	5	Different sample, same as above specimen and conditions.
18	239	1959	1318	1.090-4.050	~0°	5	Different sample, same as above specimen and conditions.
19	239	1961	1297	1.100-4.040	~0°	5	Different sample, same as above specimen and conditions.
20	113	1961	1297	1.082-4.043	~0°		Trace of surface oxidation observed; measured in vacuum.
21	113	1961	1318	1.069-4.049	~0°		Above specimen and conditions.
22	113	1961	1350	1.073-4.045	~0°		Above specimen and conditions except thermally etched and measured in argon (760 mm Hg).
23	113	1961	1396	1.074-4.051	~0°		Above specimen and conditions.
24	30	1963	298	1.20-12.50	~0°		Mechanically polished with aluminum oxide and cleaned with water (surface roughness 0.025 μ peak to peak and 25 μ lateral); heated in vacuum for 3 hrs at 588 K and 3 hrs at 1089 K; measured in air; data extracted from smooth curve; authors assumed $\alpha = \epsilon$; [Author's designation: Sample 7].

SPECIFICATION TABLE NO. 133 (continued)

Curve No.	Ref. No.	Year	Temperature K	Wavelength Range, μ	Geometry θ'	Reported Error, %	Composition (weight percent), Specifications and Remarks
25	30	1963	298	1.20-12.50	~0°		Mechanically polished with aluminum oxide and cleaned with water; heated in vacuum at 1372 K for 4 hrs; measured in air; data extracted from smooth curve; authors assumed $\alpha = \epsilon$;[Author's designation: Sample 10].
26	317	1962	1428	1.000-14.32	0°		Sanded to a 4/0 finish with emery paper and buffed to mirror finish with aluminum oxide; aged 16 hrs at 1144 K and 10 hrs at 1422 K; measured in vacuum; surface roughness after measurements 7 μ.
27	317	1962	1430	1.000-14.79	0°		Sanded to a 4/0 finish with emery paper and buffed to mirror finish with aluminum oxide; aged 12 hrs at 1255 K and 4 hrs at 1422 K; measured in vacuum; surface roughness after measurements 7 μ.
28	317	1962	1269	1.000-10.05	0°		Sanded to a 4/0 finish with emery paper and buffed to mirror finish with aluminum oxide; aged 6 hrs at 1255 K; measured in vacuum; surface roughness after measurements 7 μ.
29	317	1962	298	1.000-14.96	~0°		Sanded to a 4/0 finish with emery paper and buffed to mirror finish with aluminum oxide; aged 25 hrs at 1255 K and 25 hrs at 1422 K; computed from reflectance.

DATA TABLE NO. 133 NORMAL SPECTRAL EMITTANCE OF NICKEL

[Wavelength, λ, μ; Emittance, ∈; Temperature, T, K]

λ	∈
CURVE 1	
T = 294	
1.20	0.255
1.40	0.230
1.60	0.215
1.80	0.212
1.80	0.193
2.25	0.170
2.60	0.150
3.00	0.130
3.50	0.110
4.00	0.095
4.50	0.080
5.00	0.067
5.50	0.064
6.00	0.061
7.00	0.055
7.50	0.050
8.00	0.045
9.00	0.040
10.00	0.035
12.00	0.030
13.20	0.035
14.00	0.034
15.00	0.030
16.00	0.039
18.50	0.040
CURVE 2	
T = 294	
0.375	0.600
0.510	0.540
0.600	0.470
0.700	0.440
0.810	0.400
0.900	0.375
1.010	0.365
1.200	0.320
1.400	0.310
1.600	0.230
CURVE 3	
T = 1209	
0.5	0.900
0.6	0.915
0.7	0.930
CURVE 4	
T = 1331	
0.5	0.880
0.6	0.885
0.7	0.890
CURVE 5	
T = 1432	
0.5	0.868
0.6	0.870
0.7	0.883
CURVE 6	
T = 1528	
0.5	0.810
0.6	0.840
0.7	0.870
CURVE 7	
T = 1123	
1.2	0.234
1.5	0.211
1.8	0.195
2.0	0.182
2.5	0.162
3.0	0.145
3.5	0.130
4.0	0.118
4.5	0.108
5.0	0.102
6.5	0.090
CURVE 8	
T = 1273	
1.0	0.260
1.2	0.231*
1.5	0.210*
1.8	0.196*
2.0	0.185
2.5	0.166
3.0	0.149
3.5	0.135
4.0	0.123
4.5	0.114*
5.0	0.107
6.5	0.097
CURVE 9	
T = 1383	
0.6	0.540
1.0	0.292
1.1	0.314
1.2	0.292
1.3	0.280
1.4	0.270
1.5	0.250
2.0	0.220
2.5	0.205
3.0	0.187
3.5	0.174
4.0	0.162
CURVE 10	
T = 1267	
1.2	0.210
1.5	0.190
1.8	0.190
2.0	0.170
2.1	0.160
2.5	0.150
2.6	0.140
CURVE 10 (cont.)	
2.8	0.130
3.0	0.125
3.5	0.115
4.0	0.110
4.5	0.105
5.0	0.098
5.5	0.092
6.0	0.090
7.0	0.084
8.0	0.080
9.0	0.076
10.0	0.074
11.0	0.074
11.0	0.072
12.0	0.068
13.0	0.065
13.5	0.064
CURVE 11	
T = 1428	
1.2	0.340
1.5	0.290
1.7	0.270
1.8	0.270
2.0	0.255
2.2	0.240
2.5	0.230
2.6	0.230
2.7	0.220
3.0	0.220
3.5	0.195
4.0	0.190
5.0	0.180
5.5	0.170
6.0	0.170
7.0	0.170
8.0	0.165
9.0	0.170
10.0	0.170
CURVE 11 (cont.)	
10.1	0.165
10.1	0.165*
10.1	0.170
CURVE 12*	
T = 1267	
1.2	0.210
1.5	0.190
1.8	0.190
2.0	0.170
2.1	0.160
2.5	0.150
2.6	0.140
2.8	0.130
3.0	0.125
3.5	0.115
4.0	0.110
4.5	0.105
5.0	0.098
5.5	0.092
6.0	0.090
7.0	0.084
8.0	0.080
9.0	0.076
10.0	0.074
11.0	0.072
12.0	0.068
13.0	0.065
13.5	0.064
CURVE 13*	
T = 294	
1.00	0.259
1.18	0.251
1.36	0.228
1.56	0.205
1.78	0.198
1.96	0.190
2.17	0.177
2.37	0.167
CURVE 13 (cont.)*	
2.56	0.156
2.75	0.148
2.97	0.144
3.45	0.126
3.94	0.119
4.40	0.114
4.98	0.108
5.92	0.0998
6.89	0.0912
7.85	0.0847
8.75	0.0822
9.84	0.0787
10.6	0.0752
11.7	0.0741
CURVE 14*	
T = 1100	
0.823	0.458
0.998	0.349
1.231	0.299
1.660	0.232
CURVE 15	
T = 1100	
0.898	0.385
1.277	0.248
1.325	0.242
2.196	0.194
CURVE 16	
T = 1396	
1.104	0.1921
1.471	0.1761
1.888	0.1604
2.254	0.1468
2.563	0.1348
2.876	0.1301
3.132	0.1268
3.374	0.1223
CURVE 16 (cont.)	
3.622	0.1180
3.852	0.1135
4.055	0.1135*
CURVE 17*	
T = 1350	
1.097	0.1701
1.468	0.1722
1.864	0.1557
2.243	0.1419
2.562	0.1309
2.868	0.1266
3.123	0.1241
3.363	0.1190
3.606	0.1158
3.830	0.1099
4.041	0.1024
CURVE 18*	
T = 1318	
1.090	0.1450
1.469	0.1386
1.872	0.1366
2.253	0.1282
2.569	0.1211
2.879	0.1222
3.128	0.1174
3.385	0.1108
3.617	0.1071
3.858	0.1018
4.050	0.1051
CURVE 19	
T = 1297	
1.100	0.1401
1.474	0.1339
1.871	0.1311
2.233	0.1225
2.556	0.1168
CURVE 19 (cont.)	
2.864	0.1161
3.139	0.1094
3.376	0.1066
3.613	0.1027
3.842	0.0968
4.040	0.0940*
CURVE 20*	
T = 1297	
1.082	0.1399
1.459	0.1340
1.856	0.1312
2.220	0.1228
2.545	0.1170
2.852	0.1163
3.129	0.1097
3.371	0.1066
3.615	0.1032
3.846	0.0971
4.043	0.0941
CURVE 21*	
T = 1318	
1.069	0.1448
1.454	0.1387
1.860	0.1370
2.242	0.1288
2.557	0.1217
2.869	0.1225
3.124	0.1173
3.383	0.1112
3.618	0.1075
3.859	0.1020
4.049	0.1053
CURVE 22*	
T = 1350	
1.073	0.1703
1.444	0.1723

* Not shown on plot

DATA TABLE NO. 133 (continued)

λ	∈
CURVE 22 (cont.)*	
1.847	0.1563
2.225	0.1426
2.545	0.1312
2.859	0.1271
3.113	0.1246
3.356	0.1194
3.608	0.1162
3.827	0.1101
4.045	0.1028
CURVE 23*	
T = 1396	
1.074	0.1926
1.448	0.1765
1.868	0.1608
2.236	0.1473
2.550	0.1352
2.866	0.1304
3.125	0.1273
3.369	0.1226
3.620	0.1183
3.853	0.1138
4.051	0.1138
CURVE 24*	
T = 298	
1.20	0.240
1.50	0.200
2.00	0.160
2.50	0.130
3.00	0.110
3.50	0.094
4.00	0.080
4.50	0.070
5.00	0.060
5.50	0.055
6.00	0.051
7.00	0.043
7.40	0.040
8.00	0.038
9.00	0.034
10.00	0.031
11.00	0.029
12.00	0.027
12.50	0.026

λ	∈
CURVE 25*	
T = 298	
1.20	0.240
1.50	0.200
2.00	0.160
2.50	0.130
3.00	0.110
3.50	0.094
4.00	0.080
4.50	0.070
5.00	0.060
5.50	0.055
6.00	0.051
7.00	0.043
7.40	0.040
8.00	0.038
9.00	0.034
10.00	0.031
11.00	0.029
12.00	0.027
12.50	0.026
CURVE 26	
T = 1428	
1.000	0.3598
1.202	0.3342
1.493	0.2931
1.722	0.2761
1.963	0.2704
2.360	0.2405
3.034	0.2193
3.491	0.1995
4.178	0.1875
5.808	0.1742
6.486	0.1742
7.980	0.1679
8.974	0.1730
11.94	0.1710
14.32	0.1766

λ	∈
CURVE 27*	
T = 1430	
1.000	0.3112
1.114	0.3126
1.297	0.3020
1.710	0.2661
2.183	0.2372
2.710	0.2099
3.342	0.1849
3.981	0.1714
4.764	0.1622
5.984	0.1493
7.925	0.1456
8.954	0.1406
10.94	0.1439
12.25	0.1426
14.79	0.1656
CURVE 28	
T = 1269	
1.000	0.2051
1.104	0.2128
1.306	0.2099
1.726	0.1875
2.399	0.1578
2.673	0.1486
2.992	0.1362
3.508	0.1194
3.999	0.1122*
4.487	0.1102
5.998	0.09441
7.980	0.08299
9.036	0.07944
10.05	0.07113
CURVE 29*	
T = 298	
1.000	0.3148
1.130	0.3436
1.219	0.3476
1.521	0.3206
1.897	0.2858
2.333	0.2512
2.748	0.2312
2.992	0.2344
3.459	0.2042

λ	∈
CURVE 29 (cont.)*	
4.477	0.1648
5.702	0.1377
7.980	0.1135
11.32	0.09683
12.53	0.09376
13.80	0.09529
14.96	0.08975

* Not shown on plot

FIGURE 134A

NORMAL INTEGRATED REFLECTANCE OF NICKEL Vs. SURFACE ROUGHNESS

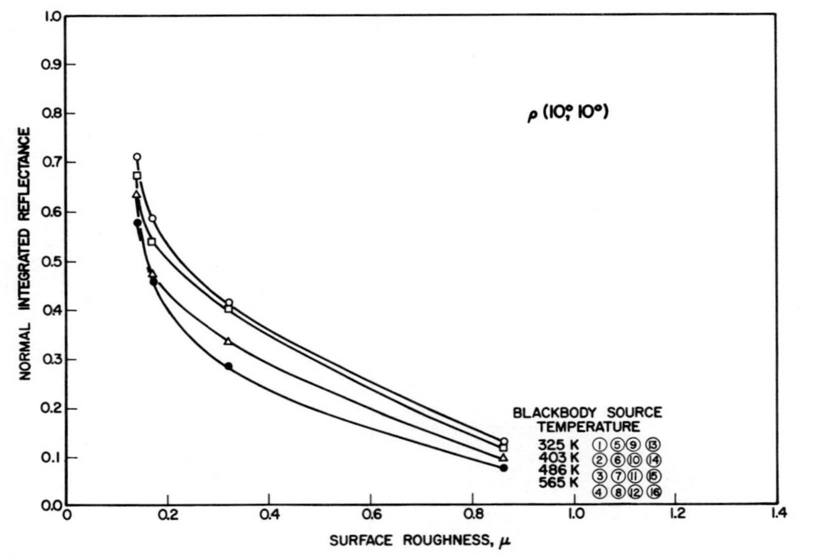

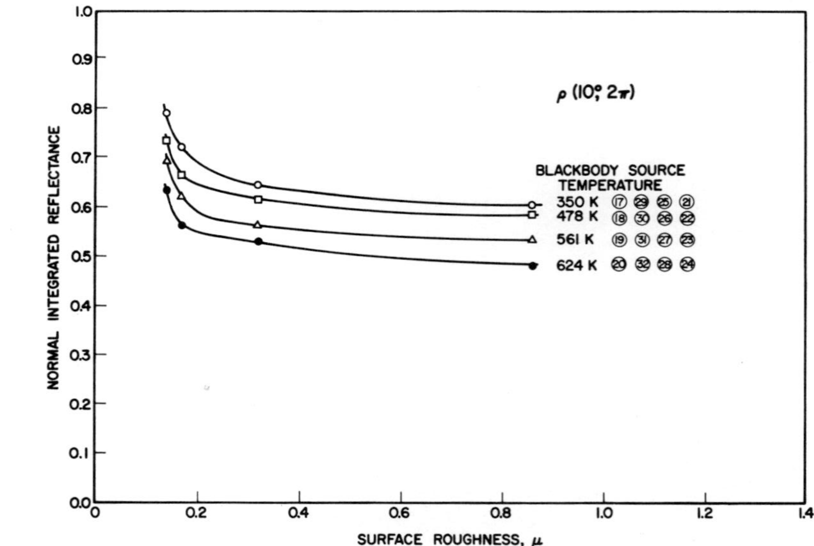

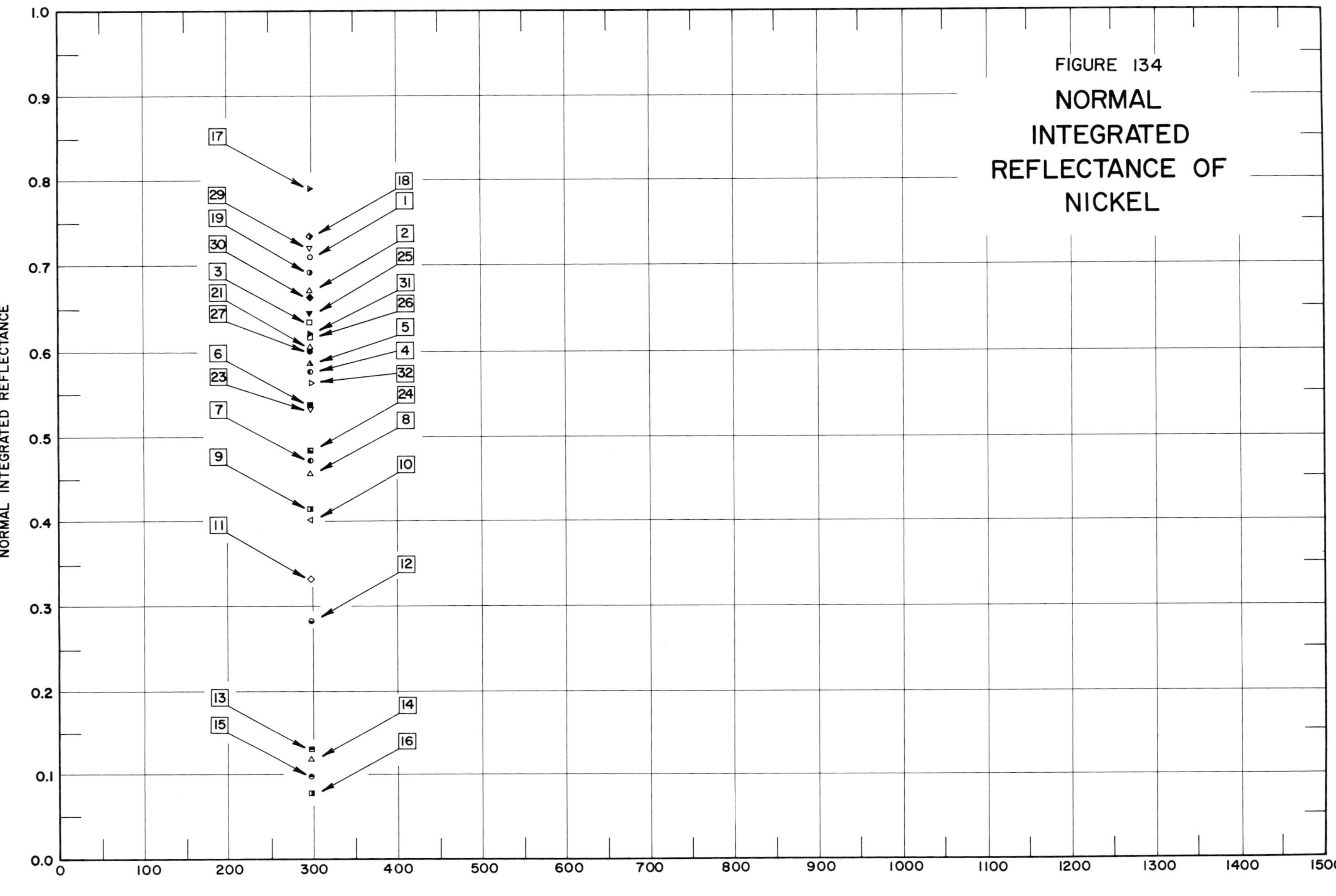
FIGURE 134
NORMAL
INTEGRATED
REFLECTANCE OF
NICKEL
NORMAL INTEGRATED REFLECTANCE
1.0
0.9
0.8
0.7
0.6
0.5
0.4
0.3
0.2
0.1
0.0
TEMPERATURE, K
0
100
200
300
400
500
600
700
800
900
1000
1100
1200
1300
1400
1500
17
29
19
30
3
21
27
6
23
7
9
11
13
15
18
1
2
25
31
26
5
4
32
24
8
10
12
14
16

SPECIFICATION TABLE NO. 134 NORMAL INTEGRATED REFLECTANCE OF NICKEL

Curve No.	Ref. No.	Year	Temperature Range, K	Geometry θ	θ'	ω'	Reported Error, %	Composition (weight percent), Specifications and Remarks
1	247	1964	298	10°	10°			Ground; surface roughness 0.14 μ RMS; polished nickel reference; black body source at 325 K.
2	247	1964	298	10°	10°			Above specimen and conditions except source at 403 K.
3	247	1964	298	10°	10°			Above specimen and conditions except source at 486 K.
4	247	1964	298	10°	10°			Above specimen and conditions except source at 565 K.
5	247	1964	298	10°	10°			Ground; surface roughness 0.17 μ RMS; polished nickel reference; black body source at 325 K.
6	247	1964	298	10°	10°			Above specimen and conditions except source at 403 K.
7	247	1964	298	10°	10°			Above specimen and conditions except source at 486 K.
8	247	1964	298	10°	10°			Above specimen and conditions except source at 565 K.
9	247	1964	298	10°	10°			Ground; surface roughness 0.32 μ RMS; polished nickel reference; black body source at 325 K.
10	247	1964	298	10°	10°			Above specimen and conditions except source at 403 K.
11	247	1964	298	10°	10°			Above specimen and conditions except source at 486 K.
12	247	1964	298	10°	10°			Above specimen and conditions except source at 565 K.
13	247	1964	298	10°	10°			Ground; surface roughness 0.86 μ RMS; polished nickel reference; black body source at 325 K.
14	247	1964	298	10°	10°			Above specimen and conditions except source at 403 K.
15	247	1964	298	10°	10°			Above specimen and conditions except source at 486 K.
16	247	1964	298	10°	10°			Above specimen and conditions except source at 565 K.
17	247	1964	298	10°		2π		Ground; surface roughness 0.14 μ RMS; polished nickel reference; black body source at 350 K.
18	247	1964	298	10°		2π		Above specimen and conditions except source at 478 K.
19	247	1964	298	10°		2π		Above specimen and conditions except source at 561 K.
20	247	1964	298	10°		2π		Above specimen and conditions except source at 624 K.
21	247	1964	298	10°		2π		Ground; surface roughness 0.86 μ RMS; polished nickel reference; black body source at 350 K.
22	247	1964	298	10°		2π		Above specimen and conditions except source at 478 K.
23	247	1964	298	10°		2π		Above specimen and conditions except source at 561 K.
24	247	1964	298	10°		2π		Above specimen and conditions except source at 624 K.
25	247	1964	298	10°		2π		Ground; surface roughness 0.32 μ RMS; polished nickel reference; black body source at 350 K.

SPECIFICATION TABLE NO. 134 (continued)

Curve No.	Ref. No.	Year	Temperature Range, K	Geometry θ	θ'	ω'	Reported Error, %	Composition (weight percent), Specifications and Remarks
26	247	1964	298	10°		2π		Above specimen and conditions except source at 478 K.
27	247	1964	298	10°		2π		Above specimen and conditions except source at 561 K.
28	247	1964	298	10°		2π		Above specimen and conditions except source at 624 K.
29	247	1964	298	10°		2π		Ground; surface roughness 0.17 μ RMS; polished nickel reference; black body source at 350 K.
30	247	1964	298	10°		2π		Above specimen and conditions except source at 478 K.
31	247	1964	298	10°		2π		Above specimen and conditions except source at 561 K.
32	247	1964	298	10°		2π		Above specimen and conditions except source at 624 K.

DATA TABLE NO. 134 NORMAL INTEGRATED REFLECTANCE OF NICKEL

[Temperature, T, K; Reflectance, ρ]

T	ρ	T	ρ	T	ρ
CURVE 1		CURVE 13		CURVE 25	
298	0.710	298	0.129	298	0.645
CURVE 2		CURVE 14		CURVE 26	
298	0.671	298	0.117	298	0.618
CURVE 3		CURVE 15		CURVE 27	
298	0.635	298	0.098	298	0.600
CURVE 4		CURVE 16		CURVE 28*	
298	0.578	298	0.076	298	0.531
CURVE 5		CURVE 17		CURVE 29	
298	0.587	298	0.790	298	0.721
CURVE 6		CURVE 18		CURVE 30	
298	0.538	298	0.733	298	0.663
CURVE 7		CURVE 19		CURVE 31	
298	0.472	298	0.692	298	0.621
CURVE 8		CURVE 20*		CURVE 32	
298	0.457	298	0.634	298	0.563
CURVE 9		CURVE 21			
298	0.414	298	0.606		
CURVE 10		CURVE 22*			
298	0.401	298	0.589		
CURVE 11		CURVE 23			
298	0.334	298	0.533		
CURVE 12		CURVE 24			
298	0.283	298	0.483		

* Not shown on plot

FIGURE 135A(I)

ANGULAR INTEGRATED REFLECTANCE OF NICKEL vs. SURFACE ROUGHNESS

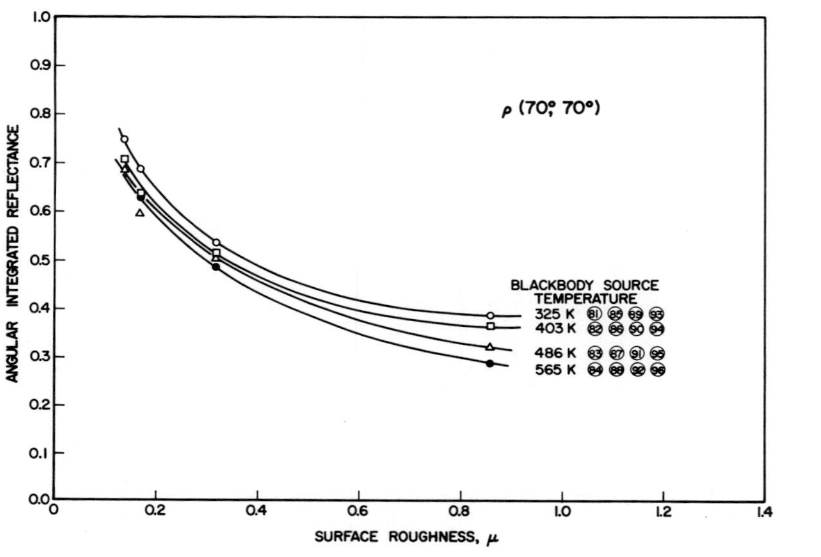

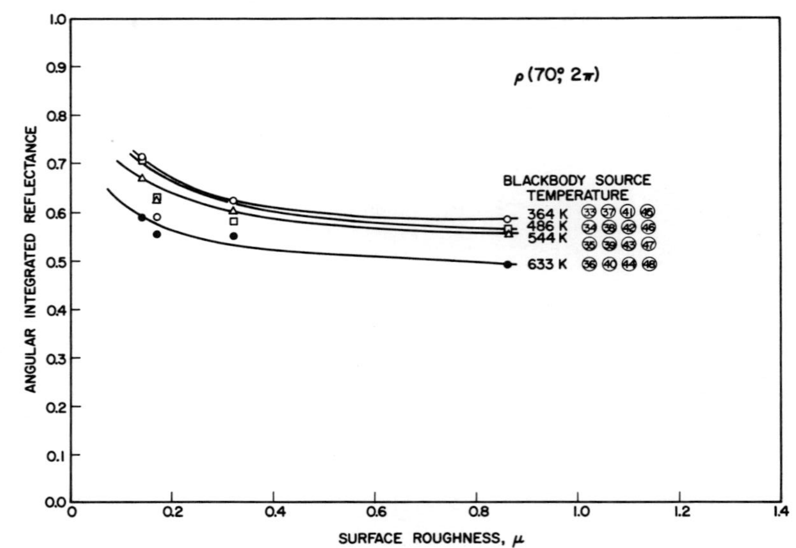

FIGURE 135A(2)

ANGULAR INTEGRATED REFLECTANCE OF NICKEL vs. SURFACE ROUGHNESS

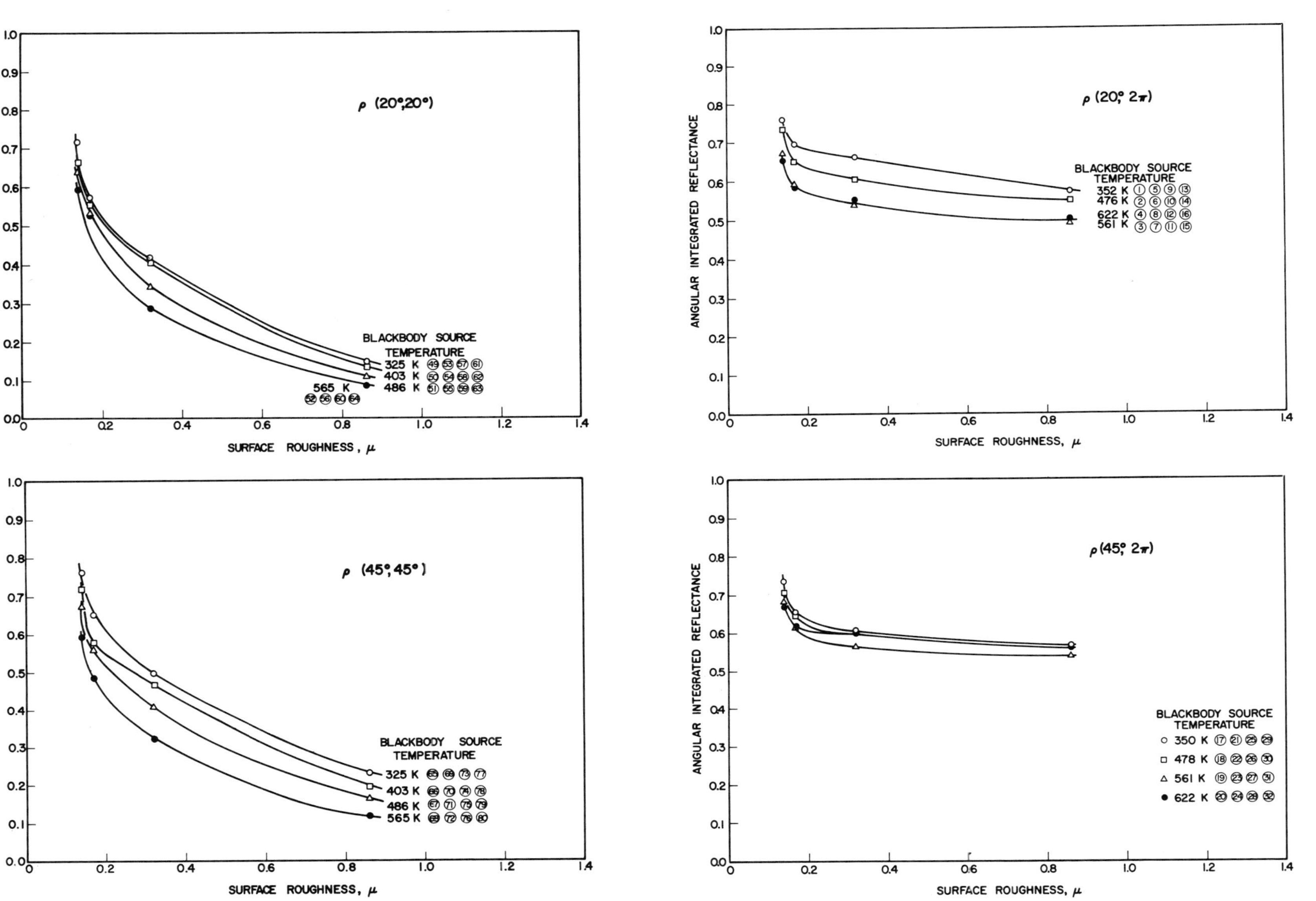

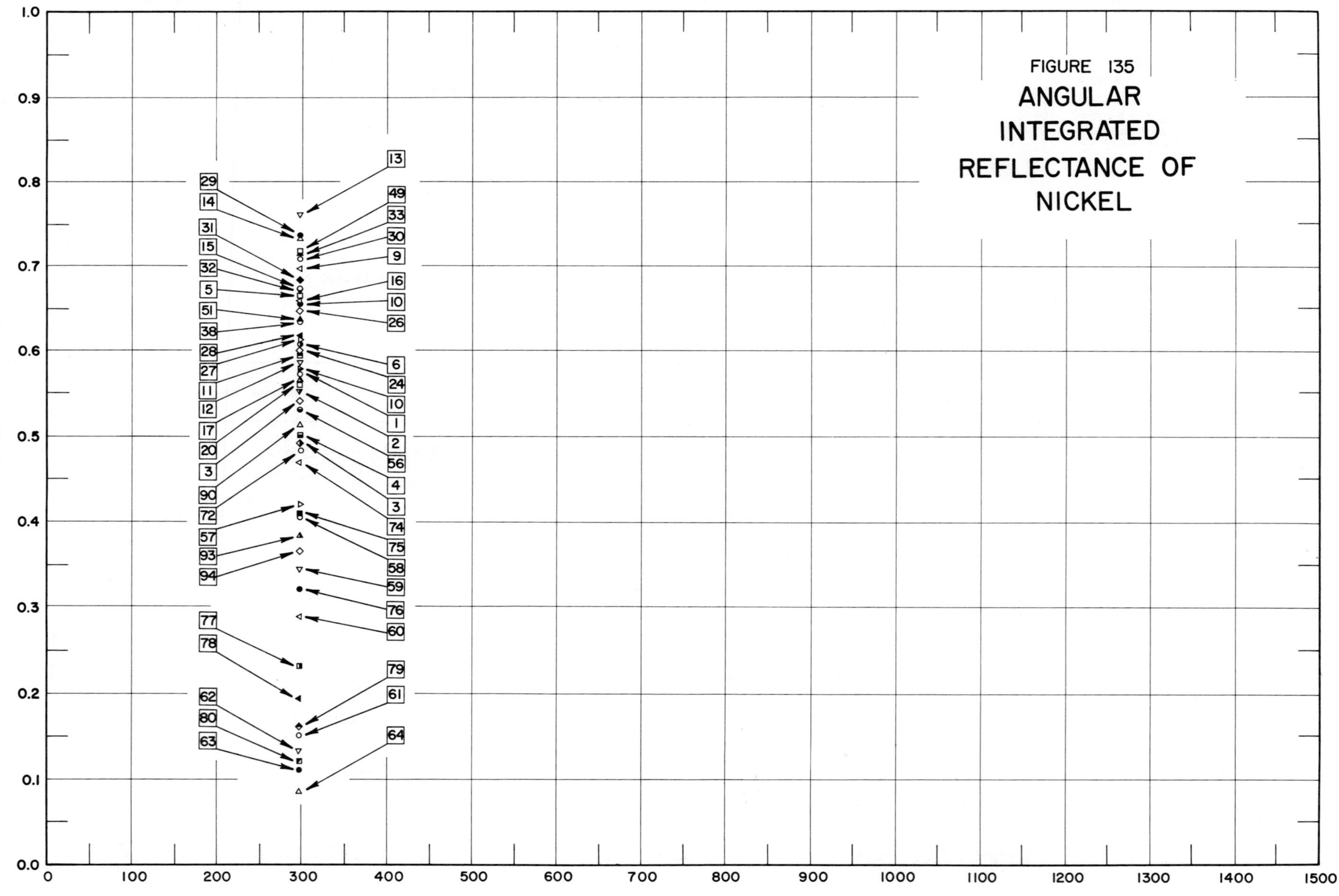
FIGURE 135
ANGULAR
INTEGRATED
REFLECTANCE OF
NICKEL
ANGULAR INTEGRATED REFLECTANCE
TEMPERATURE, K

SPECIFICATION TABLE NO. 135 ANGULAR INTEGRATED REFLECTANCE OF NICKEL

Curve No.	Ref. No.	Year	Temperature Range, K	Geometry θ	θ'	ω'	Reported Error, %	Composition (weight percent), Specifications and Remarks
1	247	1964	298	20°		2π		Ground; surface roughness 0.86 μ RMS; polished nickel reference; black body source at 352 K.
2	247	1964	298	20°		2π		Above specimen and conditions except source at 476 K.
3	247	1964	298	20°		2π		Above specimen and conditions except source at 565 K.
4	247	1964	298	20°		2π		Above specimen and conditions except source at 622 K.
5	247	1964	298	20°		2π		Ground; surface roughness 0.32 μ RMS; polished nickel reference; black body source at 352 K.
6	247	1964	298	20°		2π		Above specimen and conditions except source at 476 K.
7	247	1964	298	20°		2π		Above specimen and conditions except source at 561 K.
8	247	1964	298	20°		2π		Above specimen and conditions except source at 622 K.
9	247	1964	298	20°		2π		Ground; surface roughness 0.17 μ RMS; polished nickel reference; black body source at 350 K.
10	247	1964	298	20°		2π		Above specimen and conditions except source at 478 K.
11	247	1964	298	20°		2π		Above specimen and conditions except source at 561 K.
12	247	1964	298	20°		2π		Above specimen and conditions except source at 622 K.
13	247	1964	298	20°		2π		Ground; surface roughness 0.14 μ RMS; polished nickel reference; black body source at 350 K.
14	247	1964	298	20°		2π		Above specimen and conditions except source at 478 K.
15	247	1964	298	20°		2π		Above specimen and conditions except source at 561 K.
16	247	1964	298	20°		2π		Above specimen and conditions except source at 622 K.
17	247	1964	298	45°		2π		Ground; surface roughness 0.86 μ RMS; polished nickel reference; black body source at 350 K.
18	247	1964	298	45°		2π		Above specimen and conditions except source at 478 K.
19	247	1964	298	45°		2π		Above specimen and conditions except source at 561 K.
20	247	1964	298	45°		2π		Above specimen and conditions except source at 622 K.
21	247	1964	298	45°		2π		Ground; surface roughness 0.32 μ RMS; polished nickel reference; black body source at 350 K.
22	247	1964	298	45°		2π		Above specimen and conditions except source at 478 K.
23	247	1964	298	45°		2π		Above specimen and conditions except source at 561 K.
24	247	1964	298	45°		2π		Above specimen and conditions except source at 622 K.
25	247	1964	298	45°		2π		Ground; surface roughness 0.17 μ RMS; polished nickel reference; black body source at 350 K.

SPECIFICATION TABLE NO. 135 (continued)

Curve No.	Ref. No.	Year	Temperature Range, K	Geometry θ	Geometry θ'	Geometry ω'	Reported Error, %	Composition (weight percent), Specifications and Remarks
26	247	1964	298	45°		2π		Above specimen and conditions except source at 478 K.
27	247	1964	298	45°		2π		Above specimen and conditions except source at 561 K.
28	247	1964	298	45°		2π		Above specimen and conditions except source at 622 K.
29	247	1964	298	45°		2π		Ground; surface roughness 0.14 μ RMS; polished nickel reference; black body source at 350 K.
30	247	1964	298	45°		2π		Above specimen and conditions except source at 478 K.
31	247	1964	298	45°		2π		Above specimen and conditions except source at 561 K.
32	247	1964	298	45°		2π		Above specimen and conditions except source at 622 K.
33	247	1964	298	70°		2π		Ground; surface roughness 0.14 μ RMS; polished nickel reference; black body source at 364 K.
34	247	1964	298	70°		2π		Above specimen and conditions except source at 486 K.
35	247	1964	298	70°		2π		Above specimen and conditions except source at 544 K.
36	247	1964	298	70°		2π		Above specimen and conditions except source at 633 K.
37	247	1964	298	70°		2π		Ground; surface roughness 0.17 μ RMS; polished nickel reference; black body source at 364 K.
38	247	1964	298	70°		2π		Above specimen and conditions except source at 486 K.
39	247	1964	298	70°		2π		Above specimen and conditions except source at 544 K.
40	247	1964	298	70°		2π		Above specimen and conditions except source at 633 K.
41	247	1964	298	70°		2π		Ground; surface roughness 0.32 μ RMS; polished nickel reference; black body source at 364 K.
42	247	1964	298	70°		2π		Above specimen and conditions except source at 486 K.
43	247	1964	298	70°		2π		Above specimen and conditions except source at 544 K.
44	247	1964	298	70°		2π		Above specimen and conditions except source at 633 K.
45	247	1964	298	70°		2π		Ground; surface roughness 0.86 μ RMS; polished nickel reference; black body source at 364 K.
46	247	1964	298	70°		2π		Above specimen and conditions except source at 486 K.
47	247	1964	298	70°		2π		Above specimen and conditions except source at 544 K.
48	247	1964	298	70°		2π		Above specimen and conditions except source at 633 K.
49	247	1964	298	20°	20°			Ground; surface roughness 0.14 μ RMS; polished nickel reference; black body source at 325 K.
50	247	1964	298	20°	20°			Above specimen and conditions except source at 403 K.

SPECIFICATION TABLE NO. 135 (continued)

Curve No.	Ref. No.	Year	Temperature Range, K	Geometry θ	θ'	ω'	Reported Error, %	Composition (weight percent), Specifications and Remarks
51	247	1964	298	20°	20°			Above specimen and conditions except source at 486 K.
52	247	1964	298	20°	20°			Above specimen and conditions except source at 565 K.
53	247	1964	298	20°	20°			Ground; surface roughness 0.17 μ RMS; polished nickel reference; black body source at 325 K.
54	247	1964	298	20°	20°			Above specimen and conditions except source at 403 K.
55	247	1964	298	20°	20°			Above specimen and conditions except source at 486 K.
56	247	1964	298	20°	20°			Above specimen and conditions except source at 565 K.
57	247	1964	298	20°	20°			Ground; surface roughness 0.32 μ RMS; polished nickel reference; black body source at 325 K.
58	247	1964	298	20°	20°			Above specimen and conditions except source at 403 K.
59	247	1964	298	20°	20°			Above specimen and conditions except source at 486 K.
60	247	1964	298	20°	20°			Above specimen and conditions except source at 565 K.
61	247	1964	298	20°	20°			Ground; surface roughness 0.86 μ RMS; polished nickel reference; black body source at 325 K.
62	247	1964	298	20°	20°			Above specimen and conditions except source at 403 K.
63	247	1964	298	20°	20°			Above specimen and conditions except source at 486 K.
64	247	1964	298	20°	20°			Above specimen and conditions except source at 565 K.
65	247	1964	298	45°	45°			Ground; surface roughness 0.14 μ RMS; polished nickel reference; black body source at 325 K.
66	247	1964	298	45°	45°			Above specimen and conditions except source at 403 K.
67	247	1964	298	45°	45°			Above specimen and conditions except source at 486 K.
68	247	1964	298	45°	45°			Above specimen and conditions except source at 565 K.
69	247	1964	298	45°	45°			Ground; surface roughness 0.17 μ RMS; polished nickel reference; black body source at 325 K.
70	247	1964	298	45°	45°			Above specimen and conditions except source at 403 K.
71	247	1964	298	45°	45°			Above specimen and conditions except source at 486 K.
72	247	1964	298	45°	45°			Above specimen and conditions except source at 565 K.
73	247	1964	298	45°	45°			Ground; surface roughness 0.32 μ RMS; polished nickel reference; black body source at 325 K.
74	247	1964	298	45°	45°			Above specimen and conditions except source at 403 K.
75	247	1964	298	45°	45°			Above specimen and conditions except source at 486 K.

SPECIFICATION TABLE NO. 135 (continued)

Curve No.	Ref. No.	Year	Temperature Range, K	Geometry θ	θ'	ω'	Reported Error, %	Composition (weight percent), Specifications and Remarks
76	247	1964	298	45°	45°			Above specimen and conditions except source at 565 K.
77	247	1964	298	45°	45°			Ground; surface roughness 0.86 μ RMS; polished nickel reference; black body source at 325 K.
78	247	1964	298	45°	45°			Above specimen and conditions except source at 403 K.
79	247	1964	298	45°	45°			Above specimen and conditions except source at 486 K.
80	247	1964	298	45°	45°			Above specimen and conditions except source at 565 K.
81	247	1964	298	70°	70°			Ground; surface roughness 0.14 μ RMS; polished nickel reference; black body source at 325 K.
82	247	1964	298	70°	70°			Above specimen and conditions except source at 403 K.
83	247	1964	298	70°	70°			Above specimen and conditions except source at 486 K.
84	247	1964	298	70°	70°			Above specimen and conditions except source at 565 K.
85	247	1964	298	70°	70°			Ground; surface roughness 0.17 μ RMS; polished nickel reference; black body source at 325 K.
86	247	1964	298	70°	70°			Above specimen and conditions except source at 403 K.
87	247	1964	298	70°	70°			Above specimen and conditions except source at 486 K.
88	247	1964	298	70°	70°			Above specimen and conditions except source at 565 K.
89	247	1964	298	70°	70°			Ground; surface roughness 0.32 μ RMS; polished nickel reference; black body source at 325 K.
90	247	1964	298	70°	70°			Above specimen and conditions except source at 403 K.
91	247	1964	298	70°	70°			Above specimen and conditions except source at 486 K.
92	247	1964	298	70°	70°			Above specimen and conditions except source at 575 K.
93	247	1964	298	70°	70°			Ground; surface roughness 0.86 μ RMS; polished nickel reference; black body source at 325 K.
94	247	1964	298	70°	70°			Above specimen and conditions except source at 403 K.
95	247	1964	298	70°	70°			Above specimen and conditions except source at 486 K.
96	247	1964	298	70°	70°			Above specimen and conditions except source at 565 K.

DATA TABLE NO. 135 ANGULAR INTEGRATED REFLECTANCE OF NICKEL

[Temperature, T, K; Reflectance, ρ]

T	ρ	T	ρ	T	ρ	T	ρ	T	ρ	T	ρ	T	ρ	T	ρ
CURVE 1		CURVE 13		CURVE 25*		CURVE 37*		CURVE 49		CURVE 61		CURVE 73*		CURVE 85*	
298	0.572	298	0.760	298	0.654	298	0.592	298	0.718	298	0.150	298	0.499	298	0.684
CURVE 2		CURVE 14		CURVE 26		CURVE 38		CURVE 50*		CURVE 62		CURVE 74		CURVE 86*	
298	0.551	298	0.733	298	0.647	298	0.633	298	0.664	298	0.133	298	0.469	298	0.637
CURVE 3		CURVE 15		CURVE 27		CURVE 39*		CURVE 51		CURVE 63		CURVE 75		CURVE 87*	
298	0.491	298	0.672	298	0.614	298	0.629	298	0.638	298	0.110	298	0.409	298	0.595
CURVE 4		CURVE 16		CURVE 28		CURVE 40*		CURVE 52*		CURVE 64		CURVE 76		CURVE 88*	
298	0.501	298	0.657	298	0.618	298	0.558	298	0.592	298	0.086	298	0.321	298	0.631
CURVE 5		CURVE 17		CURVE 29		CURVE 41*		CURVE 53*		CURVE 65*		CURVE 77		CURVE 89*	
298	0.664	298	0.566	298	0.736	298	0.633	298	0.574	298	0.761	298	0.230	298	0.534
CURVE 6		CURVE 18*		CURVE 30		CURVE 42*		CURVE 54*		CURVE 66*		CURVE 78		CURVE 90	
298	0.607	298	0.565	298	0.708	298	0.584	298	0.556	298	0.720	298	0.194	298	0.513
CURVE 7*		CURVE 19*		CURVE 31		CURVE 43*		CURVE 55*		CURVE 67*		CURVE 79		CURVE 91*	
298	0.541	298	0.539	298	0.682	298	0.604	298	0.532	298	0.675	298	0.162	298	0.502
CURVE 8*		CURVE 20		CURVE 32		CURVE 44*		CURVE 56		CURVE 68*		CURVE 80		CURVE 92*	
298	0.553	298	0.561	298	0.670	298	0.553	298	0.530	298	0.591	298	0.120	298	0.485
CURVE 9		CURVE 21*		CURVE 33		CURVE 45*		CURVE 57		CURVE 69*		CURVE 81*		CURVE 93	
298	0.697	298	0.608	298	0.715	298	0.582	298	0.420	298	0.652	298	0.748	298	0.384
CURVE 10		CURVE 22*		CURVE 34*		CURVE 46*		CURVE 58		CURVE 70*		CURVE 82*		CURVE 94	
298	0.655	298	0.602	298	0.709	298	0.568	298	0.404	298	0.579	298	0.704	298	0.366
CURVE 11		CURVE 23*		CURVE 35*		CURVE 47*		CURVE 59		CURVE 71*		CURVE 83*		CURVE 95*	
298	0.593	298	0.566	298	0.671	298	0.558	298	0.345	298	0.560	298	0.684	298	0.320
CURVE 12		CURVE 24		CURVE 36*		CURVE 48*		CURVE 60		CURVE 72		CURVE 84*		CURVE 96*	
298	0.586	298	0.600	298	0.592	298	0.491	298	0.288	298	0.484	298	0.684	298	0.287

* Not shown on plot

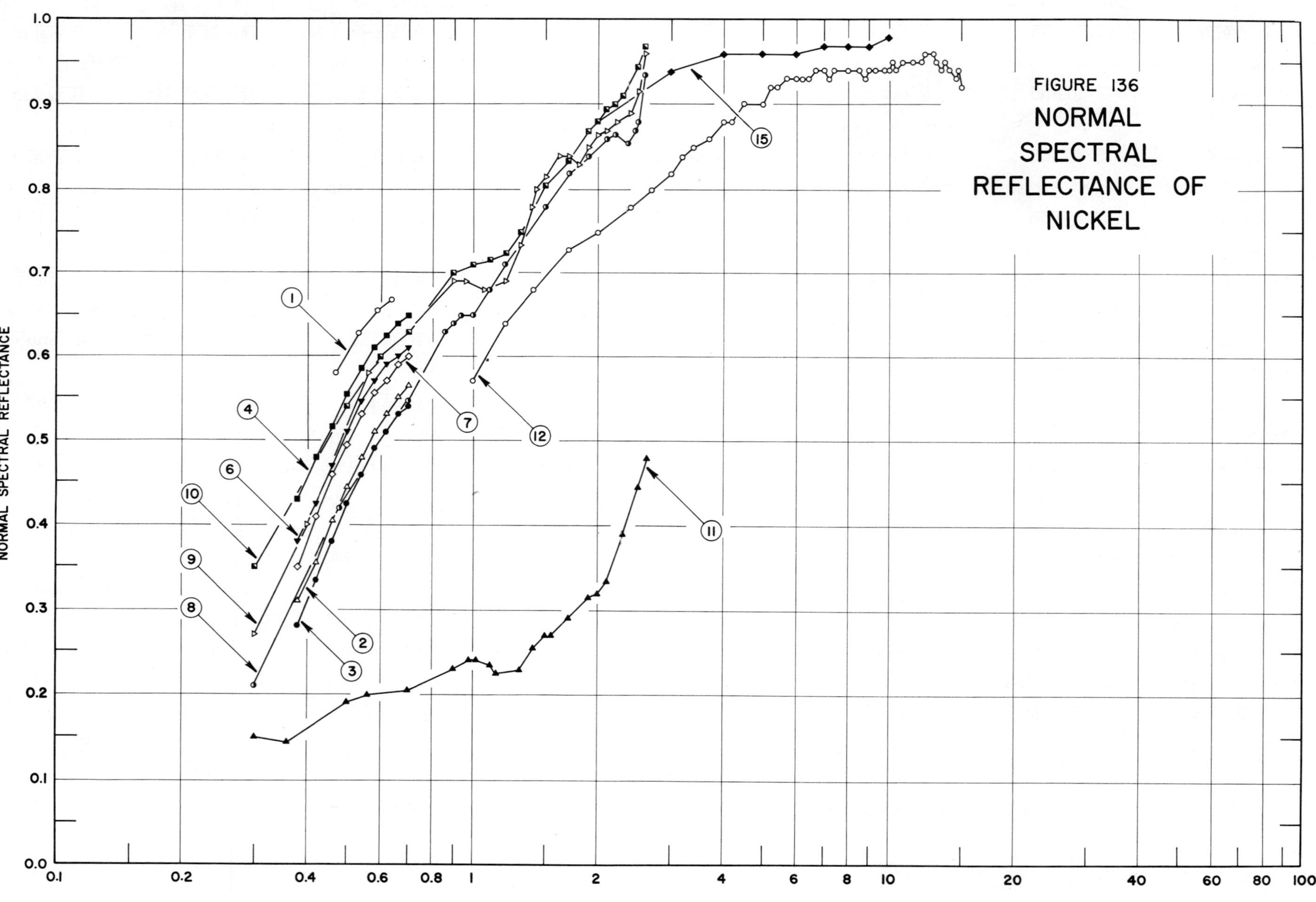
FIGURE 136
NORMAL SPECTRAL REFLECTANCE OF NICKEL
NORMAL SPECTRAL REFLECTANCE
0.0
0.1
0.2
0.3
0.4
0.5
0.6
0.7
0.8
0.9
1.0
WAVELENGTH, MICRONS
0.1
0.2
0.4
0.6
0.8
1
2
4
6
8
10
20
40
60
80
100
1
2
3
4
6
7
8
9
10
11
12
15

SPECIFICATION TABLE NO. 136 NORMAL SPECTRAL REFLECTANCE OF NICKEL

Curve No.	Ref. No.	Year	Temperature K	Wavelength Range, μ	Geometry θ	θ'	ω'	Reported Error, %	Composition (weight percent), Specifications and Remarks
1	142	1947	298	0.47-0.64	<15°	<15°		5-9	Cold rolled; polished by using a pitch surface.
2	29	1956	298	0.38-0.70	9°		2π		Grade A pure; as received; data extracted from smooth curve; MgO reference.
3	29	1956	298	0.38-0.70	9°		2π		Above specimen and conditions except rotated 90° in its own plane.
4	29	1956	298	0.38-0.70	9°		2π		Grade A pure; polished; data extracted from smooth curve; MgO reference.
5	29	1956	298	0.38-0.70	9°		2π		Above specimen and conditions except rotated 90° in its own plane.
6	29	1956	298	0.38-0.70	9°		2π		Grade A pure; smoothed and cleaned; data extracted from smooth curve; MgO reference.
7	29	1956	298	0.38-0.70	9°		2π		Above specimen and conditions except rotated 90° in its own plane.
8	34	1957	298	0.30-2.60	9°		2π	±4	Commercial grade A; as received; data extracted from a smooth curve; magnesium carbonate reference.
9	34	1957	298	0.30-2.60	9°		2π	±4	Commercial grade A; cleaned with liquid detergent; data extracted from a smooth curve; magnesium carbonate reference.
10	34	1957	298	0.30-2.60	9°		2π	±4	Commercial grade A; polished; data extracted from a smooth curve; magnesium carbonate reference.
11	34	1957	298	0.30-2.62	9°		2π	±4	Commercial grade A; oxidized in air at red heat for 30 min; data extracted from a smooth curve; magnesium carbonate reference.
12	126	1953	298	1.0-15.0	5°		2π	±2	Plate (0.019 in. thick); polished; data extracted from smooth curve; converted from R (2π, 5°).
13	65	1962	294	0.375-1.600	~0°		2π		Polished; measured in air.
14	65	1962	294	1.20-18.50	~0°		2π		Polished; measured in air.
15	249	1960	344	2.0-10.0	~5°		2π		Buffed with tripoli, hand rubbed, and washed in acetone and bensol; data extracted from smooth curve; pure silver mirror reference standard; converted from R(2π, 5°).
16	228	1900	298	0.45-0.70	~0°	~0°			Pure; mirror like surface; platinum filament lamp source.
17	253	1963	298	0.107-1.042	~0°	~0°			Electrolytically etched; data extracted from smooth curve.

DATA TABLE NO. 136 NORMAL SPECTRAL REFLECTANCE OF NICKEL

[Wavelength, λ, μ; Reflectance, ρ; Temperature, T, K]

CURVE 1
T = 298

λ	ρ
0.47	0.580
0.53	0.628
0.59	0.655
0.64	0.668

CURVE 2
T = 298

λ	ρ
0.38	0.310
0.42	0.355
0.46	0.405
0.50	0.445
0.54	0.480
0.58	0.510
0.62	0.530
0.66	0.550
0.70	0.565

CURVE 3
T = 298

λ	ρ
0.38	0.280
0.42	0.335
0.46	0.380
0.50	0.425
0.54	0.460
0.58	0.490
0.62	0.510
0.66	0.530
0.70	0.540

CURVE 4
T = 298

λ	ρ
0.38	0.430
0.42	0.480
0.46	0.515
0.50	0.555
0.54	0.585
0.58	0.610
0.62	0.625
0.66	0.640
0.70	0.650

CURVE 5*
T = 298

λ	ρ
0.38	0.430
0.42	0.470
0.46	0.515
0.50	0.555
0.54	0.580
0.58	0.605
0.62	0.620
0.66	0.635
0.70	0.645

CURVE 6
T = 298

λ	ρ
0.38	0.380
0.42	0.425
0.46	0.470
0.50	0.510
0.54	0.545
0.58	0.570
0.62	0.590
0.66	0.600
0.70	0.610

CURVE 7
T = 298

λ	ρ
0.38	0.350
0.42	0.410
0.46	0.460
0.50	0.495
0.54	0.530
0.58	0.555
0.62	0.570
0.66	0.590
0.70	0.600

CURVE 8
T = 298

λ	ρ
0.30	0.210
0.48	0.420
0.50	0.430*
0.70	0.545

CURVE 8 (cont.)

λ	ρ
0.86	0.630
0.90	0.640
0.94	0.650
1.00	0.650
1.10	0.680
1.20	0.710
1.30	0.735*
1.50	0.780
1.70	0.820
1.78	0.830*
1.84	0.835*
1.90	0.840
2.10	0.860
2.22	0.865
2.26	0.865*
2.30	0.860*
2.36	0.855
2.40	0.860*
2.46	0.870
2.50	0.880
2.60	0.935

CURVE 9
T = 298

λ	ρ
0.30	0.270
0.40	0.400
0.50	0.500*
0.56	0.580
0.70	0.610*
0.90	0.690
0.96	0.690
1.06	0.680
1.10	0.680*
1.20	0.690
1.30	0.735
1.38	0.780
1.42	0.800
1.50	0.815
1.62	0.840
1.70	0.840
1.74	0.840*
1.80	0.830
1.84	0.840*

CURVE 9 (cont.)

λ	ρ
1.90	0.850
2.00	0.865
2.10	0.870
2.16	0.880*
2.22	0.880
2.30	0.885*
2.36	0.890*
2.40	0.890
2.50	0.915
2.60	0.960

CURVE 10
T = 298

λ	ρ
0.30	0.350
0.50	0.540
0.60	0.600
0.70	0.630
0.90	0.700
1.00	0.710
1.10	0.715
1.20	0.725
1.30	0.750
1.38	0.780*
1.42	0.800*
1.50	0.805
1.70	0.835
1.90	0.870
2.00	0.880
2.10	0.895
2.14	0.900*
2.20	0.900
2.30	0.910
2.50	0.945
2.60	0.970

CURVE 11
T = 298

λ	ρ
0.30	0.150
0.36	0.145
0.50	0.190
0.56	0.200
0.70	0.205

CURVE 11 (cont.)

λ	ρ
0.90	0.230
0.98	0.240
1.02	0.240
1.10	0.235
1.24	0.225
1.30	0.230
1.40	0.255
1.50	0.270
1.54	0.270
1.70	0.290
1.90	0.315
2.00	0.320
2.10	0.335
2.30	0.390
2.50	0.445
2.62	0.480

CURVE 12
T = 298

λ	ρ
1.0	0.57
1.2	0.64
1.4	0.68
1.7	0.73
2.0	0.75
2.4	0.78
2.7	0.80
3.0	0.82
3.2	0.84
3.4	0.85
3.7	0.86
4.0	0.88
4.2	0.88
4.5	0.90
4.7	0.90*
4.9	0.90*
5.0	0.90
5.2	0.92
5.4	0.92
5.7	0.93
6.0	0.93
6.2	0.93
6.4	0.93
6.7	0.94

CURVE 12 (cont.)

λ	ρ
7.0	0.94
7.2	0.93
7.4	0.94
8.0	0.94
8.5	0.94
8.8	0.93
9.0	0.94
9.3	0.94
9.8	0.94
10.0	0.94
10.2	0.95
10.4	0.94
10.7	0.95
11.0	0.95*
11.3	0.95*
11.5	0.95
12.0	0.95
12.2	0.96
12.8	0.96
13.0	0.95
13.2	0.95*
13.4	0.94
13.7	0.95
14.0	0.94
14.2	0.94*
14.5	0.93
14.7	0.94
15.0	0.92

CURVE 13*
T = 294

λ	ρ
0.375	0.400
0.510	0.460
0.600	0.530
0.700	0.560
0.810	0.600
0.900	0.625
1.010	0.635
1.200	0.680
1.400	0.690
1.600	0.770

CURVE 14*
T = 294

λ	ρ
1.20	0.745
1.40	0.770
1.60	0.785
1.80	0.788
1.80	0.807
2.25	0.830
2.60	0.850
3.00	0.870
3.50	0.890
4.00	0.905
4.50	0.920
5.00	0.933
5.50	0.936
6.00	0.939
7.00	0.945
7.50	0.950
8.00	0.955
9.00	0.960
10.00	0.965
12.00	0.970
13.20	0.965
14.00	0.966
15.00	0.970
16.00	0.961
18.50	0.960

CURVE 15
T = 344

λ	ρ
2.0	0.88*
3.0	0.94
4.0	0.96
5.0	0.96
6.0	0.96
7.0	0.97
8.0	0.97
9.0	0.97
10.0	0.98

CURVE 16*
T = 298

λ	ρ
0.45	0.585

CURVE 16 (cont.)*

λ	ρ
0.50	0.608
0.55	0.626
0.60	0.649
0.65	0.659
0.70	0.698

CURVE 17*
T = 298

λ	ρ
0.107	0.108
0.109	0.106
0.114	0.113
0.124	0.142
0.138	0.180
0.155	0.230
0.177	0.277
0.207	0.404
0.228	0.462
0.310	0.408
0.388	0.512
0.459	0.604
1.042	0.753

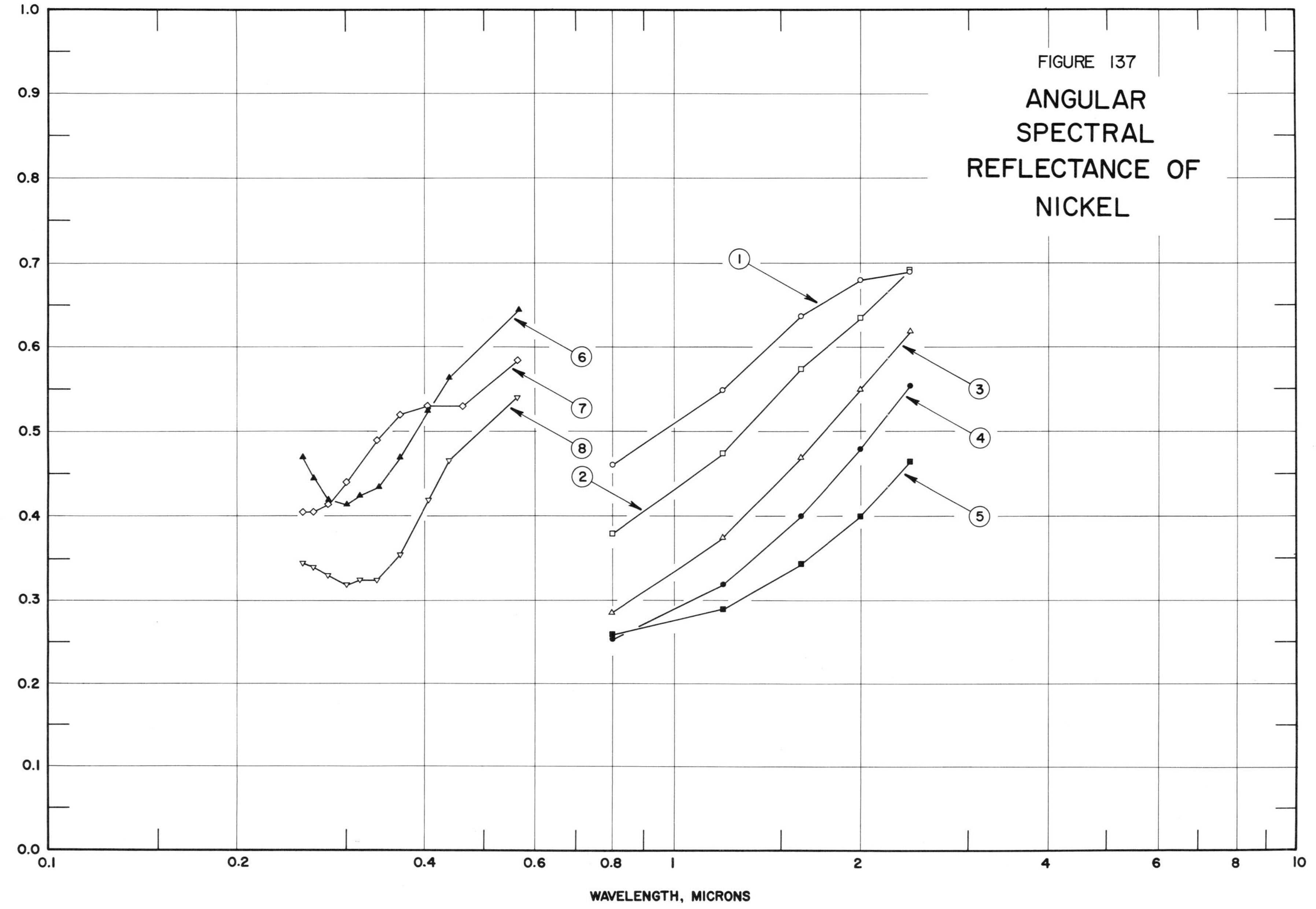

FIGURE 137
ANGULAR SPECTRAL REFLECTANCE OF NICKEL

SPECIFICATION TABLE NO. 137 ANGULAR SPECTRAL REFLECTANCE OF NICKEL

Curve No.	Ref. No.	Year	Temperature K	Wavelength Range, μ	Geometry θ	θ'	ω'	Reported Error, %	Composition (weight percent), Specifications and Remarks
1	141	1948	298	0.8-2.4	50°	50°			Mirror surface prepared by vacuum evaporation; data extracted from smooth curve, incident beam polarized parallel to the plane of incidence.
2	141	1948	298	0.8-2.4	60°	60°			Above specimen and conditions, incident beam polarized parallel to the plane of incidence.
3	141	1941	298	0.8-2.4	70°	70°			Above specimen and conditions, incident beam polarized parallel to the plane of incidence.
4	141	1948	298	0.8-2.4	79°	79°			Above specimen and conditions, incident beam polarized parallel to the plane of incidence.
5	141	1948	298	0.8-2.4	80°	80°			Above specimen and conditions, incident beam polarized parallel to the plane of incidence.
6	143	1928	298	0.255-0.565	45°	45°			Fused and solidified in vacuum; highly polished.
7	143	1928	298	0.255-0.563	45°	45°			Electroplated on steel plate; highly polished.
8	143	1928	298	0.255-0.563	45°	45°			A cross section sawed from a rolled rod of refined nickel.

DATA TABLE NO. 137 ANGULAR SPECTRAL REFLECTANCE OF NICKEL

[Wavelength, λ, μ; Reflectance, ρ; Temperature, T, K]

λ	ρ
CURVE 1	**T = 298**
0.8	0.460
1.2	0.550
1.6	0.638
2.0	0.680
2.4	0.690
CURVE 2	**T = 298**
0.8	0.380
1.2	0.475
1.6	0.575
2.0	0.635
2.4	0.692
CURVE 3	**T = 298**
0.8	0.285
1.2	0.375
1.6	0.470
2.0	0.550
2.4	0.620
CURVE 4	**T = 298**
0.8	0.255
1.2	0.320
1.6	0.400
2.0	0.480
2.4	0.555
CURVE 5	**T = 298**
0.8	0.260
1.2	0.290
1.6	0.345
2.0	0.400
2.4	0.465

λ	ρ
CURVE 6	**T = 298**
0.255	0.470
0.265	0.445
0.280	0.420
0.300	0.415
0.315	0.425
0.338	0.435
0.365	0.470
0.405	0.525
0.438	0.565
0.565	0.645
CURVE 7	**T = 298**
0.255	0.405
0.265	0.405
0.280	0.415
0.300	0.440
0.335	0.490
0.365	0.520
0.405	0.530
0.460	0.530
0.563	0.585
CURVE 8	**T = 298**
0.255	0.345
0.265	0.340
0.280	0.330
0.300	0.320
0.315	0.325
0.335	0.325
0.365	0.355
0.405	0.420
0.438	0.465
0.563	0.540

SPECIFICATION TABLE NO. 138 HEMISPHERICAL INTEGRATED ABSORPTANCE OF NICKEL

Curve No.	Ref. No.	Year	Temperature Range, K	Reported Error, %	Composition (weight percent), Specifications and Remarks
1	3	1955	76	5	Foil (0.004 in. thick); solvent cleaned; measured in vacuum (10^{-6} to 10^{-7} mm Hg); absorptance for 300 K black body incident radiation.

DATA TABLE NO. 138 HEMISPHERICAL INTEGRATED ABSORPTANCE OF NICKEL

[Temperature, T, K; Absorptance, α]

T	α
CURVE 1*	
76	0.022

* Not shown on plot

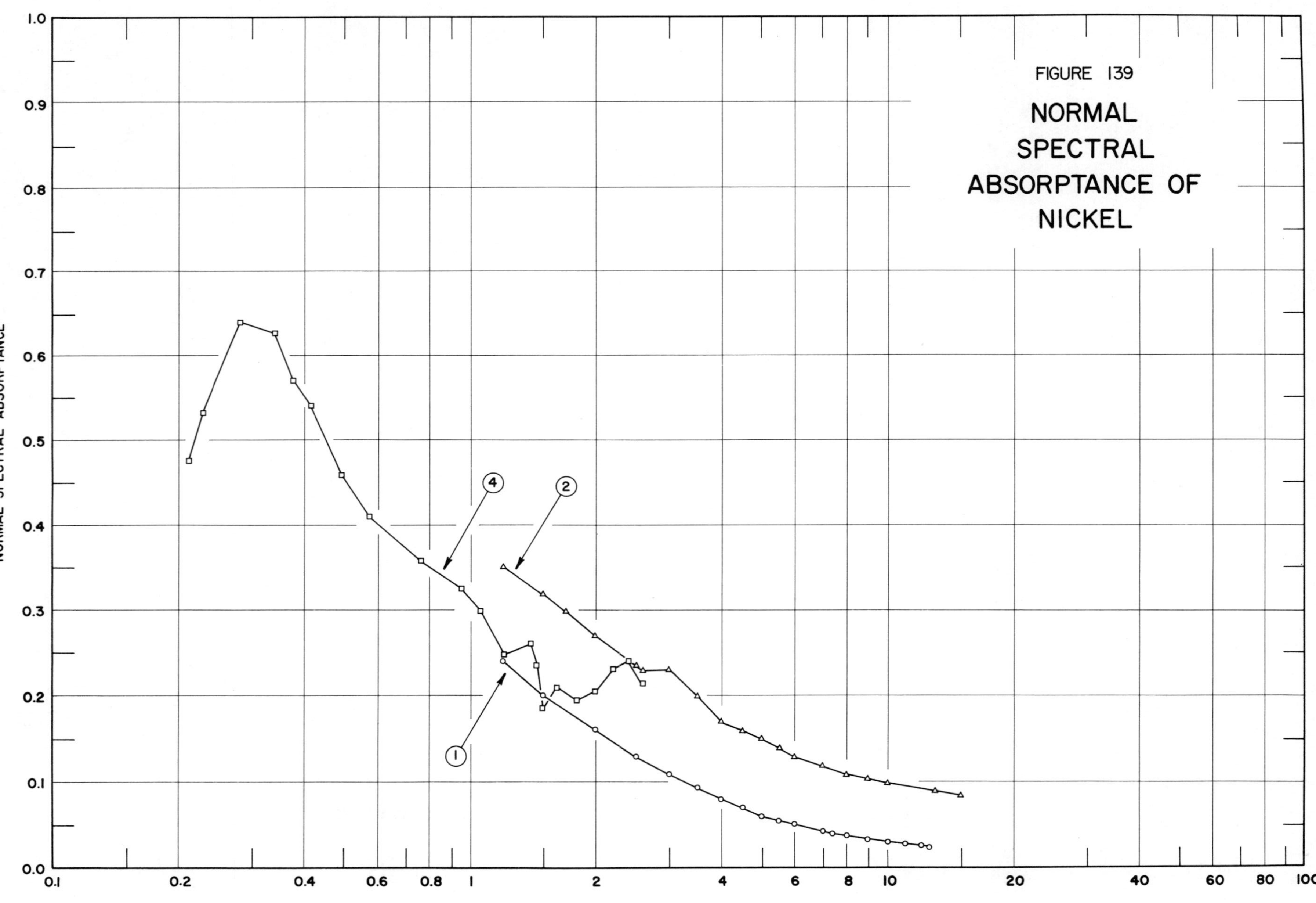
FIGURE 139
NORMAL SPECTRAL ABSORPTANCE OF NICKEL
NORMAL SPECTRAL ABSORPTANCE
WAVELENGTH, MICRONS
0.0
0.1
0.2
0.3
0.4
0.5
0.6
0.7
0.8
0.9
1.0
0.1
0.2
0.4
0.6
0.8
1
2
4
6
8
10
20
40
60
80
100
1
2
4

SPECIFICATION TABLE NO. 139 NORMAL SPECTRAL ABSORPTANCE OF NICKEL

Curve No.	Ref. No.	Year	Temperature K	Wavelength Range, μ	Geometry θ	Reported Error, %	Composition (weight percent), Specifications and Remarks
1	30	1963	298	1.20-12.50	~ 0°		Mechanically polished with aluminum oxide and cleaned with water (surface roughness 0.025 μ peak to peak and 25 μ lateral); measured in air; data extracted from smooth curve; [Authors designation: Sample 7].
2	30	1963	298	1.20-15.00	~ 0°		Mechanically polished with aluminum oxide and cleaned with water; heated in vacuum for 3 hrs at 588 K, 3 hrs at 1089 K, 4 hrs at 1267 K, and 48 hrs at T > 1256 K; data extracted from smooth curve; surface roughness (measured after 4 months) 7.5 μ peak to peak and 25 μ lateral; [Author's designation: Sample 7].
3	245	1965	294	1.18-12.8	~0°	< 5	Measured in hohlraum.
4	307	1954	~298	0.213-2.600	~0°		Nickel 3 mils thick; data extracted from smooth curve.
5	35	1914	873	0.66	~0°		Measured in burning hydrogen; author assumed $\alpha = \epsilon$.

DATA TABLE NO. 139 NORMAL SPECTRAL ABSORPTANCE OF NICKEL

[Wavelength, λ, μ; Absorptance, α; Temperature, T, K]

λ	α
CURVE 1 T = 298	
1.20	0.240
1.50	0.200
2.00	0.160
2.50	0.130
3.00	0.110
3.50	0.094
4.00	0.080
4.50	0.070
5.00	0.060
5.50	0.055
6.00	0.051
6.00	0.043
7.40	0.040
8.00	0.038
9.00	0.034
10.00	0.031
11.00	0.039
12.00	0.027
12.50	0.026
CURVE 2 T = 298	
1.20	0.350
1.50	0.320
1.70	0.300
2.00	0.270
2.50	0.235
2.60	0.230
3.00	0.230
3.50	0.200
4.00	0.170
4.50	0.160
5.00	0.150
5.50	0.140
6.00	0.130
7.00	0.120
8.00	0.110
9.00	0.105
10.00	0.100
13.00	0.090
15.00	0.086
CURVE 3* T = 294	
1.18	0.259*
1.39	0.238
1.58	0.214
1.76	0.187
1.98	0.179
2.14	0.160
2.37	0.146
2.57	0.137
2.76	0.130
2.96	0.119
3.42	0.101
3.93	0.0832
4.46	0.0702
4.94	0.0589
5.42	0.0526
5.96	0.0506
6.43	0.0436
6.98	0.0394
7.50	0.0345
7.94	0.0313
8.91	0.0298
9.89	0.0265
10.7	0.0254
11.9	0.0243
12.8	0.0231
CURVE 4 T = ~298	
0.213	0.473
0.229	0.531
0.280	0.640
0.341	0.627
0.377	0.569
0.416	0.540
0.493	0.459
0.574	0.410
0.763	0.358
0.954	0.326
1.054	0.299
1.206	0.248
1.391	0.260
CURVE 4 (cont.)	
1.438	0.234
1.497	0.184
1.613	0.210
1.801	0.193
2.000	0.205
2.200	0.231
2.400	0.241
2.600	0.213
CURVE 5* T = 873	
0.66	0.355

* Not shown on plot

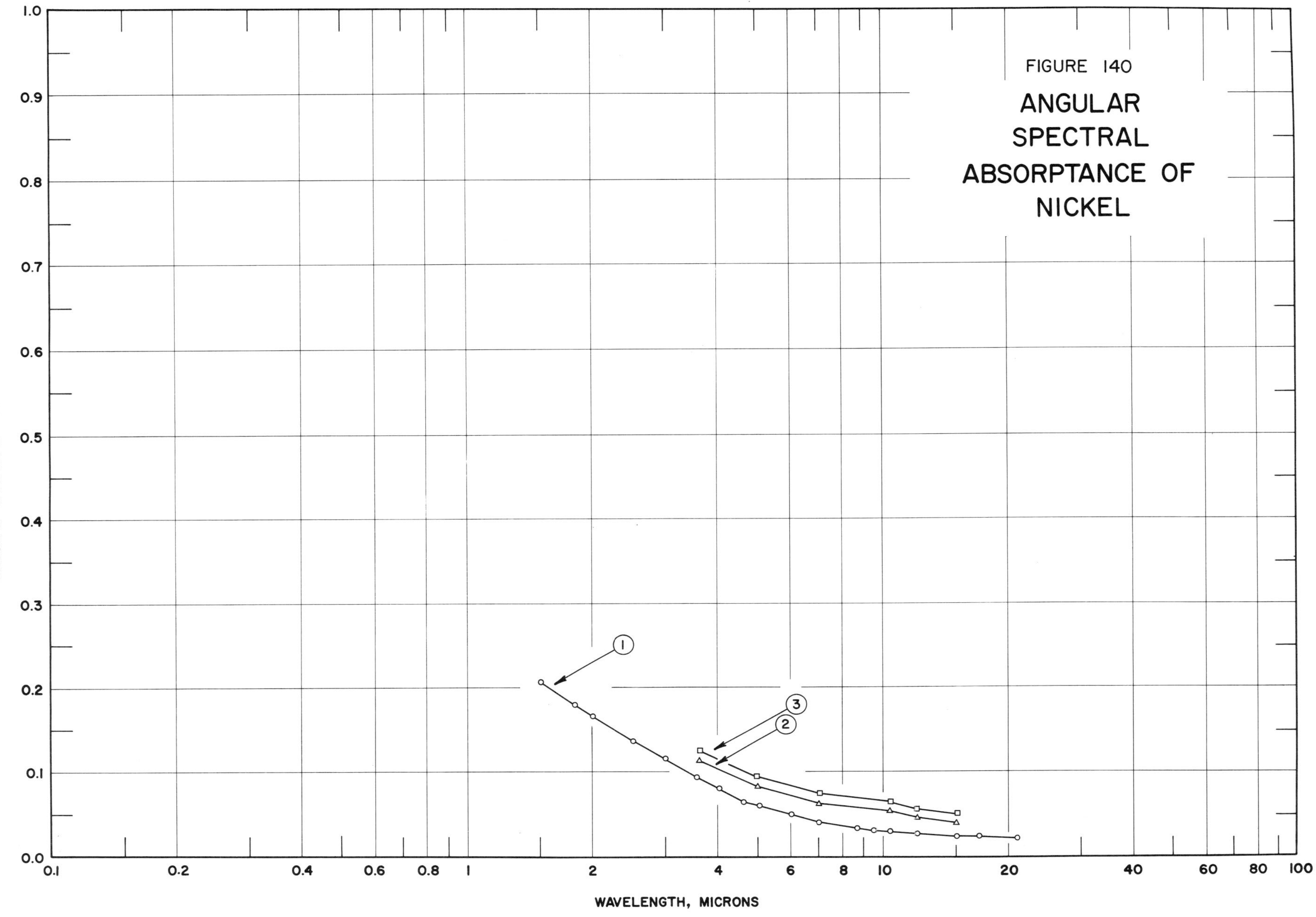

FIGURE 140
ANGULAR SPECTRAL ABSORPTANCE OF NICKEL

SPECIFICATION TABLE NO. 140 ANGULAR SPECTRAL ABSORPTANCE OF NICKEL

Curve No.	Ref. No.	Year	Temperature K	Wavelength Range, μ	Geometry θ	Reported Error, %	Composition (weight percent), Specifications and Remarks
1	225	1965	306	1.52-21.1	25°		99.99 Ni from Belmont Smelting and Refining Works; dry sanded and polished on metallurgical felt wheels; measured in dry nitrogen; heated cavity at approx. 1056 K with platinum reference; authors assumed $\alpha = 1 - R(2\pi, 25°)$.
2	255	1965	583	3.61-15.1	25°		Above specimen and conditions.
3	255	1965	722	3.63-15.1	25°		Above specimen and conditions.

DATA TABLE NO. 140 ANGULAR SPECTRAL ABSORPTANCE OF NICKEL

[Wavelength, λ, μ; Absorptance, α; Temperature, T, K]

λ	α
CURVE 1 T = 306	
1.52	0.207
1.82	0.180
2.01	0.167
2.51	0.136
3.00	0.117
3.57	0.094
4.02	0.080
4.62	0.065
5.05	0.060
6.01	0.050
7.02	0.040
8.71	0.034
9.53	0.031
10.4	0.030
12.1	0.027
15.1	0.024
17.0	0.024
21.1	0.022
CURVE 2 T = 583	
3.61	0.113
5.00	0.082
7.03	0.063
10.4	0.053
12.1	0.047
15.1	0.040
CURVE 3 T = 722	
3.63	0.125
5.00	0.095
7.02	0.074
10.4	0.065
12.1	0.057
15.1	0.051

SPECIFICATION TABLE NO. 141 HEMISPHERICAL SOLAR ABSORPTANCE OF NICKEL

Curve No.	Ref. No.	Year	Temperature Range, K	Reported Error, %	Composition (weight percent), Specifications and Remarks
1	26	1962	230-400	5	Commercial sheet; cleaned with both sodium dichromate and dilute nitric acid solutions; buffed on a felt buffing wheel and cleaned with CCl_4 and acetone; measured in vacuum (1×10^{-6} mm Hg) ; data extracted from smooth curve.

DATA TABLE NO. 141 HEMISPHERICAL SOLAR ABSORPTANCE OF NICKEL

[Temperature, T, K; Absorptance, α]

T	α
CURVE 1*	
230	0.485
250	0.495
275	0.515
280	0.520
300	0.515
325	0.505
350	0.497
375	0.490
400	0.485

* Not shown on plot

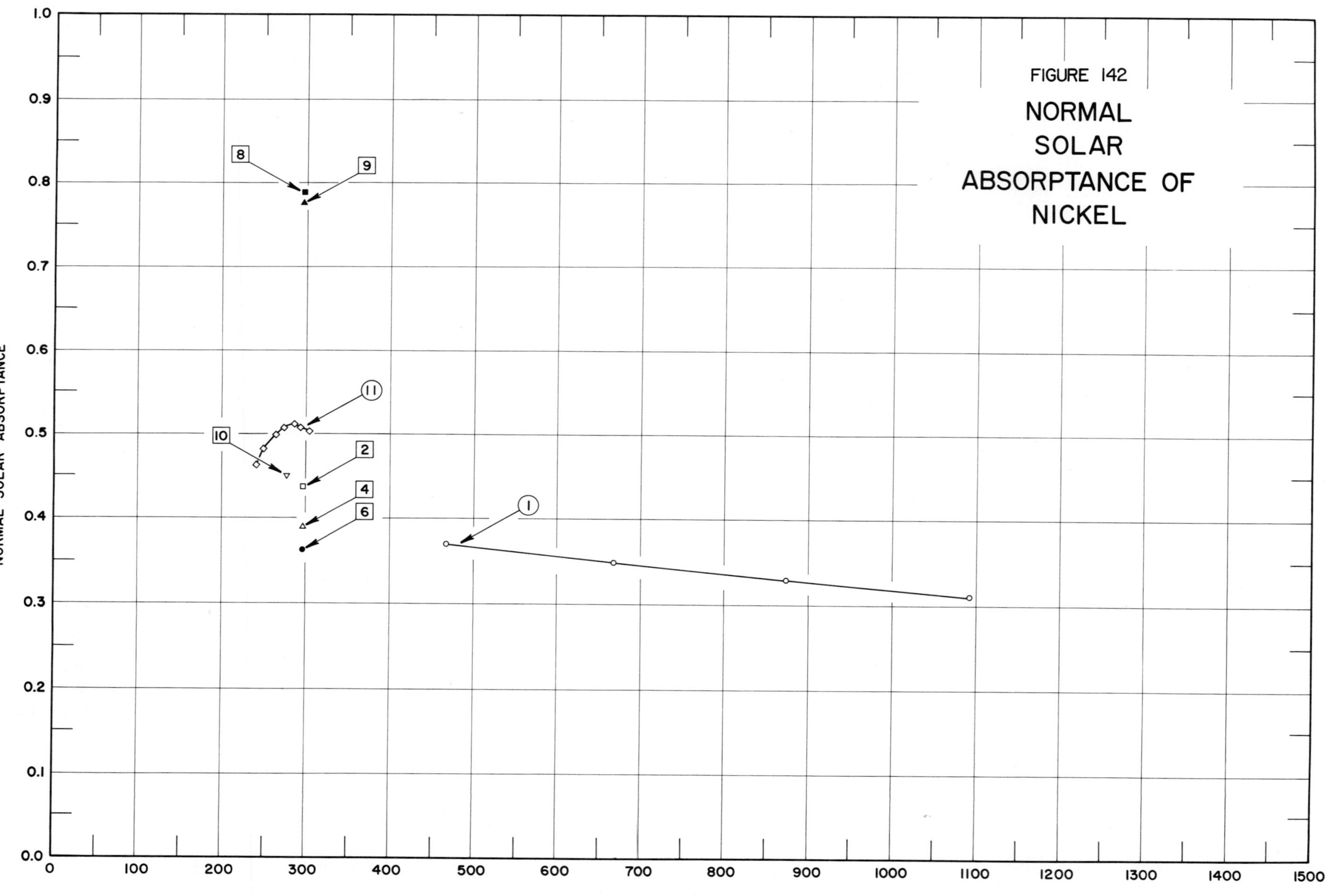

FIGURE 142 NORMAL SOLAR ABSORPTANCE OF NICKEL

SPECIFICATION TABLE NO. 142 NORMAL SOLAR ABSORPTANCE OF NICKEL

Curve No.	Ref. No.	Year	Temperature Range, K	Geometry θ	Reported Error, %	Composition (weight percent), Specifications and Remarks
1	47	1961	468-1093	$\sim 0^\circ$		Commercially pure; ground with 600 grit carborundum and polished on a wet cloth lap with unlevigated jewelers rouge; measured in vacuum (10^{-5} mm Hg).
2	34	1957	298	9°		Commercial grade A; as received; computed from spectral reflectance data for sea level conditions.
3	34	1957	298	9°		Above specimen and conditions except computed for above atmosphere conditions.
4	34	1957	298	9°		Different sample, same as Curve 2 specimen and conditions except cleaned with liquid detergent.
5	34	1957	298	9°		Above specimen and conditions except computed for above atmosphere conditions.
6	34	1957	298	9°		Different sample, same as Curve 2 specimen and conditions except polished.
7	34	1957	298	9°		Above specimen and conditions except computed for above atmosphere conditions.
8	34	1957	298	9°		Different sample, same as Curve 2 specimen and conditions except oxidized in air at red heat for 30 min.
9	34	1957	298	9°		Above specimen and conditions except computed for above atmosphere conditions.
10	45	1961	278	$\sim 0^\circ$	10	Electroless nickel; extraterrestrial.
11	26	1964	241-303	$\sim 0^\circ$		Commercially pure; cleaned, buffed, cleaned with carbon tetrachloride and acetone; measured in vacuum (1×10^{-6} mm Hg); data extracted from smooth curve.

DATA TABLE NO. 142 NORMAL SOLAR ABSORPTANCE OF NICKEL

[Temperature, T,K; Absorptance, α]

T	α	T	α
CURVE 1		**CURVE 11**	
468	0.37	241	0.464
668	0.35	249	0.481
873	0.33	263	0.498
1093	0.31	274	0.507
CURVE 2		285	0.511
298	0.439	292	0.507
CURVE 3*		303	0.501
298	0.437		
CURVE 4			
298	0.390		
CURVE 5*			
298	0.390		
CURVE 6			
298	0.363		
CURVE 7*			
298	0.360		
CURVE 8			
298	0.789		
CURVE 9			
298	0.774		
CURVE 10			
278	0.45		

* Not shown on plot

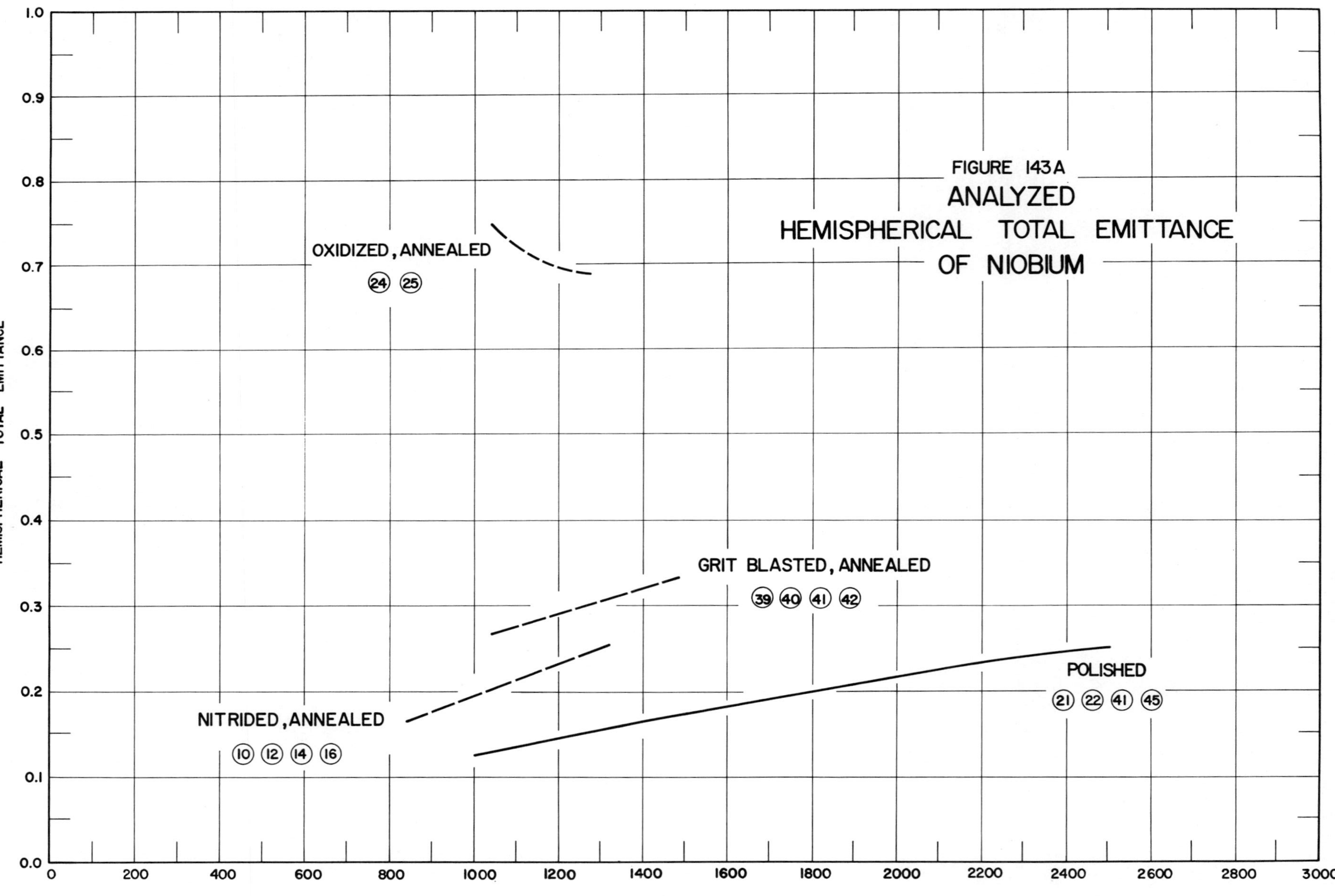

FIGURE 143A
ANALYZED
HEMISPHERICAL TOTAL EMITTANCE
OF NIOBIUM

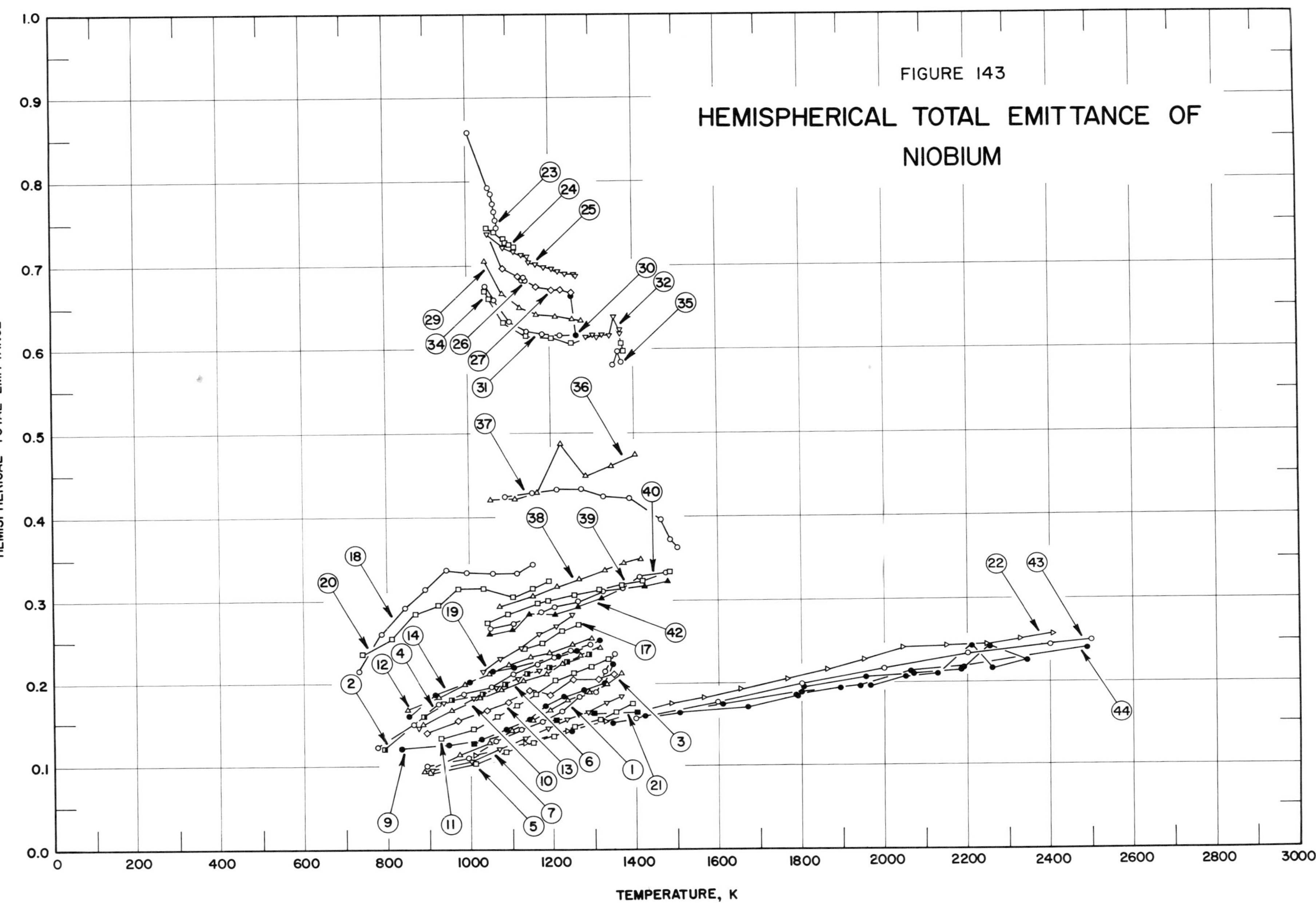

FIGURE 143
HEMISPHERICAL TOTAL EMITTANCE OF NIOBIUM

SPECIFICATION TABLE NO. 143 HEMISPHERICAL TOTAL EMITTANCE OF NIOBIUM

Curve No.	Ref. No.	Year	Temperature Range, K	Reported Error, %	Composition (weight percent), Specifications and Remarks
1	87	1962	896-1348	$<$ 5	As received; measured in vacuum (10^{-3} mm Hg); initial run; [Author's designation: Tube 1A].
2	87	1962	797-1288	$<$ 5	Above specimen and conditions; final run.
3	87	1962	888-1366	$<$ 5	Different sample, same as curve 1 specimen and conditions; initial run; [Author's designation: Tube 1B].
4	87	1962	778-1293	$<$ 5	Above specimen and conditions; final run.
5	87	1962	903-1396	$<$ 5	Deoxidized by pickling in an acid mixture; measured in vacuum (10^{-3} mm Hg); initial run; [Author's designation: Tube 2A].
6	87	1962	885-1313	$<$ 5	Above specimen and conditions; final run.
7	87	1962	908-1366	$<$ 5	Different sample, same as curve 5 specimen and conditions; initial run; [Author's designation: Tube 2B].
8	87	1962	878-1301	$<$ 5	Above specimen and conditions; final run.
9	87	1962	835-1343	$<$ 5	Nitrided from 473 K to 973 K linearly in 1 hr; measured in vacuum (10^{-3} mm Hg); initial run; [Author's designation: Tube 3A].
10	87	1962	873-1316	$<$ 5	Above specimen and conditions; final run.
11	87	1962	930-1337	$<$ 5	Different sample, same as curve 9 specimen and conditions; initial run; [Author's designation: Tube 3B].
12	87	1962	848-1298	$<$ 5	Above specimen and conditions; final run.
13	87	1962	895-1349	$<$ 5	Nitrided from 473 K to 973 K linearly in 1 hr and from 973 K to 1173 K linearly in 1 hr; measured in vacuum (10^{-3} mm Hg); initial run; [Author's designation: Tube 4A].
14	87	1962	856-1313	$<$ 5	Above specimen and conditions; final run.
15	87	1962	873-1313	$<$ 5	Different sample, same as curve 13 specimen and conditions; initial run; [Author's designation: Tube 4B].
16	87	1962	857-1301	$<$ 5	Above specimen and conditions; final run.
17	87	1962	875-1266	$<$ 5	Nitrided from 473 K to 973 K linearly in 1 hr, from 973 K to 1173 K linearly in 1 hr, from 1173 K to 1273 K linearly in 0.5 hr, and from 1273 K to 1373 K linearly in 1.5 hr; measured in vacuum (10^{-3} mm Hg); initial run; [Author's designation: Tube 5A].
18	87	1962	733-1157	$<$ 5	Above specimen and conditions; final run.
19	87	1962	923-1250	$<$ 5	Different sample, same as curve 17 specimen and conditions; initial run; [Author's designation: Tube 5B].
20	87	1962	743-1196	$<$ 5	Above specimen and conditions; final run.
21	85	1963	1008-1404	± 5	Optically smooth, opaque, contamination-free ribbon (0.005 in. thick); measured in vacuum (3×10^{-7} to 1×10^{-5} mm Hg).

SPECIFICATION TABLE NO. 143 (continued)

Curve No.	Ref. No.	Year	Temperature Range, K	Reported Error, %	Composition (weight percent), Specifications and Remarks
22	85	1963	1012-2404	± 5	Aged, optically smooth, opaque, contamination-free ribbon (0.005 in. thick); measured in vacuum (3×10^{-7} to 1×10^{-5} mm Hg).
23	12	1962	1005-1078	± 2.3	Oxidized in air at 922 K for 5 min, then brushed with fine steel wool; measured in vacuum ($< 5 \times 10^{-5}$ mm Hg).
24	12	1962	1051-1119	± 2.3	Above specimen and conditions except measured in vacuum ($< 10^{-7}$ mm Hg); run 1.
25	12	1962	1055-1264	± 2.3	Above specimen and conditions; run 2.
26	12	1962	1145-1138	± 2.3	Above specimen and conditions; measurements taken at intervals during 95.28 hr endurance run.
27	12	1962	1044-1257	± 2.3	Above specimen and conditions; run 3, heating.
28	12	1962	1146-1142	± 2.3	Above specimen and conditions; run 3, cooling.
29	12	1962	1279-1046	± 2.3	Above specimen and conditions; run 4.
30	12	1962	1258-1266	± 2.3	Above specimen and conditions; measurements taken at intervals during 46.05 hr endurance run.
31	12	1962	1266-1049	± 2.3	Above specimen and conditions; run 5, cooling.
32	12	1962	1266-1368	± 2.3	Above specimen and conditions; run 5, heating.
33	12	1962	1271-1058	± 2.3	Above specimen and conditions; run 5, cooling.
34	12	1962	1045-1379	± 2.3	Above specimen and conditions; run 6, heating.
35	12	1962	1376-1354	± 2.3	Above specimen and conditions; run 6, cooling.
36	12	1962	1405-1055	± 2.3	Above specimen and conditions; run 7.
37	12	1962	1089-1504	± 2.3	Grit blasted with No. G-25 steel grit; measured in vacuum ($< 10^{-7}$ mm Hg); run no. 1, heating.
38	12	1962	1416-1073	± 2.3	Above specimen and conditions; run no. 1, cooling.
39	12	1962	1043-1481	± 2.3	Above specimen and conditions; run no. 2, heating.
40	12	1962	1479-1051	± 2.3	Above specimen and conditions; run no. 2, cooling.
41	12	1962	1478	± 2.3	Above specimen and conditions; measurements taken at intervals during 71.88 hr endurance run.
42	12	1962	1479-1053	± 2.3	Above specimen and conditions; run no. 3.
43	250	1966	1400-2500	8	Niobium; 99.7 Nb + Ta, 0.17 Ta, 0.06 Si, 0.03 Fe, and 0.025 Ti; finished to "eighth-class" surface; density 8.56 g cm^{-3}; measured in vacuum (5×10^{-5} mm Hg); smoothed values.
44	291	1963	1803-2494		0.4-0.6 Ta, 0.01-0.03 Mg, 0.007-0.009 Ti, 0.004-0.006 Fe, 0.001-0.003 Mn, 0.0004-0.0006 Cu, Nb balance.
45	291	1963	1200-2500		0.4-0.6 Ta, 0.01-0.03 Mg, 0.007-0.009 Ti, 0.004-0.006 Fe, 0.001-0.003 Mn, 0.0004-0.0006 Cu, Nb balance; averaged values.

DATA TABLE NO. 143 HEMISPHERICAL TOTAL EMITTANCE OF NIOBIUM

[Temperature, T, K; Emittance, ∈]

CURVE 1

T	∈
896	0.100
998	0.110
1061	0.130
1121	0.144
1173	0.152
1221	0.166
1263	0.184
1306	0.190
1323	0.216
1348	0.236

CURVE 2

T	∈
797	0.120
888	0.160
957	0.180
1030	0.187
1083	0.200
1140	0.212
1193	0.220
1238	0.228
1288	0.236

CURVE 3

T	∈
888	0.094
973	0.114
1048	0.128
1100	0.143
1153	0.154
1196	0.169
1238	0.180
1288	0.190
1330	0.200
1366	0.212

CURVE 4

T	∈
778	0.122
865	0.150
926	0.176
993	0.188
1053	0.198
1103	0.212
1163	0.224
1206	0.232
1246	0.240
1293	0.248

CURVE 5

T	∈
903	0.092
1017	0.103
1086	0.118
1153	0.128
1203	0.136
1253	0.148
1313	0.156
1353	0.164
1396	0.176

CURVE 6

T	∈
885	0.150
953	0.168
1023	0.184
1076	0.194
1130	0.204
1183	0.212
1226	0.224
1268	0.236
1313	0.244

CURVE 7

T	∈
908	0.095
1003	0.106
1073	0.120
1133	0.132
1190	0.144
1236	0.154
1286	0.164
1330	0.176
1366	0.184

CURVE 8*

T	∈
878	0.152
955	0.165
1021	0.180
1078	0.192
1178	0.212
1218	0.224
1255	0.240
1301	0.246

CURVE 9

T	∈
835	0.120

CURVE 9 (cont.)

T	∈
948	0.124
1028	0.132
1088	0.144
1143	0.156
1183	0.172
1228	0.184
1278	0.192
1323	0.200
1343	0.224

CURVE 10

T	∈
873	0.148
939	0.177
1008	0.182
1063	0.196
1118	0.206
1169	0.216
1220	0.225*
1268	0.233*
1316	0.242*

CURVE 11

T	∈
930	0.132
1010	0.144
1066	0.160
1117	0.172
1160	0.188
1206	0.202
1253	0.212
1296	0.220
1337	0.230

CURVE 12

T	∈
848	0.168
923	0.184
988	0.200
1043	0.212
1093	0.223
1148	0.232
1198	0.238
1248	0.248
1298	0.256

CURVE 13

T	∈
895	0.140
968	0.154
1040	0.167
1091	0.178
1143	0.192
1197	0.188
1250	0.206
1310	0.206
1349	0.212

CURVE 14

T	∈
856	0.160
918	0.188
1000	0.202
1055	0.216
1106	0.220
1163	0.226*
1213	0.232
1260	0.240
1313	0.252

CURVE 15*

T	∈
873	0.156
937	0.172
1006	0.184
1061	0.202
1115	0.208
1170	0.218
1215	0.226
1263	0.236
1313	0.240

CURVE 16*

T	∈
857	0.164
926	0.182
988	0.200
1047	0.212
1100	0.220
1153	0.228
1203	0.236
1248	0.248
1301	0.254

CURVE 17

T	∈
875	0.156*
933	0.174*
990	0.196*
1039	0.208*
1090	0.224*
1133	0.242
1179	0.250
1223	0.264
1266	0.272

CURVE 18

T	∈
733	0.216
788	0.260
843	0.292
895	0.316
943	0.338
998	0.336
1060	0.334
1113	0.334
1157	0.346

CURVE 19

T	∈
923	0.176*
980	0.197*
1033	0.216
1076	0.230
1123	0.244
1169	0.260
1210	0.270
1250	0.284

CURVE 20

T	∈
743	0.236
813	0.254
868	0.284
923	0.294
973	0.316
1037	0.316
1103	0.306
1156	0.314
1196	0.324

CURVE 21

T	∈
1008	0.128
1104	0.141*
1200	0.155
1300	0.163
1404	0.165

CURVE 22

T	∈
1012	0.113
1132	0.129
1232	0.142
1324	0.153
1412	0.164*
1484	0.176
1564	0.183
1656	0.193
1764	0.204
1864	0.217
1956	0.229
2046	0.241
2150	0.243
2244	0.247
2324	0.252
2404	0.259

CURVE 23

T	∈
1005	0.858
1059	0.791
1061	0.786
1067	0.772
1070	0.764
1074	0.755
1078	0.746

CURVE 24

T	∈
1051	0.746
1070	0.740
1090	0.732
1090	0.732*
1091	0.729
1104	0.727
1119	0.724

CURVE 25

T	∈
1055	0.738

CURVE 25 (cont.)

T	∈
1091	0.723
1091	0.723*
1119	0.718
1134	0.713
1148	0.711
1150	0.708*
1152	0.704
1170	0.701
1190	0.699
1203	0.697
1221	0.693
1240	0.690
1260	0.690
1264	0.688

CURVE 26

T	∈
1145	0.681
1143	0.686*
1143	0.684*
1140	0.686
1140	0.686*
1140	0.684*
1140	0.681*
1138	0.682

CURVE 27

T	∈
1044	0.744*
1090	0.697
1122	0.688
1144	0.679*
1172	0.674
1204	0.671
1229	0.671
1257	0.668

CURVE 28*

T	∈
1146	0.678
1145	0.676
1142	0.678

CURVE 29

T	∈
1279	0.638
1255	0.639
1213	0.641

CURVE 29 (cont.)

T	∈
1170	0.643
1128	0.651
1084	0.668
1046	0.706

CURVE 30

T	∈
1258	0.666
1257	0.670*
1266	0.619

CURVE 31

T	∈
1266	0.619*
1225	0.619
1182	0.620
1141	0.623
1101	0.634
1068	0.660
1049	0.676

CURVE 32

T	∈
1266	0.617*
1276	0.617*
1289	0.616
1303	0.619
1315	0.618
1327	0.619
1344	0.619
1356	0.640
1368	0.620
1368	0.623

CURVE 33*

T	∈
1271	0.614
1271	0.613
1266	0.626
1146	0.616
1058	0.662

CURVE 34

T	∈
1045	0.670
1058	0.662
1088	0.633
1145	0.617

*Not shown on plot

DATA TABLE NO. 143 (continued)

T	∈
CURVE 34 (cont.)	
1205	0.614
1255	0.608
1315	0.613*
1363	0.619*
1366	0.615*
1372	0.608
1379	0.599
CURVE 35	
1376	0.584
1366	0.599
1354	0.582
CURVE 36	
1405	0.475
1342	0.461
1283	0.450
1223	0.489
1169	0.430
1112	0.423
1055	0.421
CURVE 37	
1089	0.426
1154	0.430
1213	0.434
1271	0.434
1326	0.426
1388	0.423
1463	0.399
1486	0.374
1504	0.365
CURVE 38	
1416	0.351
1376	0.348
1328	0.339
1266	0.327
1213	0.319
1153	0.307
1073	0.294
CURVE 39	
1043	0.275
1094	0.284
CURVE 39 (cont.)	
1166	0.297
1194	0.300
1255	0.308
1314	0.313
1368	0.320
1420	0.325
1481	0.336
CURVE 40	
1479	0.334
1416	0.329
1371	0.317
1321	0.311
1265	0.300
1208	0.292
1178	0.287
1106	0.273
1051	0.268
CURVE 41*	
1478	0.327
1478	0.327
1478	0.325
CURVE 42	
1479	0.324
1421	0.319
1371	0.316*
1320	0.302
1264	0.294
1209	0.284
1147	0.284
1104	0.266
1053	0.260
CURVE 43	
1400	0.156
1600	0.178
1800	0.199
2000	0.217
2200	0.232
2400	0.245
2500	0.250
CURVE 44	
1803	0.195
1955	0.207
2069	0.210
2185	0.214
2251	0.245
2346	0.223
2260	0.217
2210	0.246
2129	0.209
2053	0.207
1966	0.197
1898	0.194
1786	0.187
1788	0.186
1673	0.170
1506	0.164
1342	0.153
1244	0.144
1422	0.160
1613	0.173
1800	0.188
1942	0.197
2063	0.212
2192	0.216
2494	0.240
CURVE 45*	
1200	0.140
1300	0.149
1400	0.158
1500	0.167
1600	0.176
1700	0.184
1800	0.191
1900	0.198
2000	0.204
2100	0.210
2200	0.217
2300	0.223
2400	0.229
2500	0.233

*Not shown on plot

SPECIFICATION TABLE NO. 144 NORMAL TOTAL EMITTANCE OF NIOBIUM

Curve No.	Ref. No.	Year	Temperature Range, K	Geometry θ'	Reported Error, %	Composition (weight percent), Specifications and Remarks
1	85	1963	1000-2440	$\sim 0^\circ$	± 5	Optically smooth, opaque, contamination-free ribbon (0.005 in. thick); measured in vacuum (3×10^{-7} to 1×10^{-5} mm Hg); data extracted from smooth curve.

DATA TABLE NO. 144 NORMAL TOTAL EMITTANCE OF NIOBIUM

[Temperature, T, K; Emittance, ϵ]

T	ϵ
CURVE 1*	
1000	0.086
1200	0.110
1400	0.132
1600	0.153
1800	0.171
2000	0.188
2200	0.203
2400	0.215
2440	0.218

*Not shown on plot

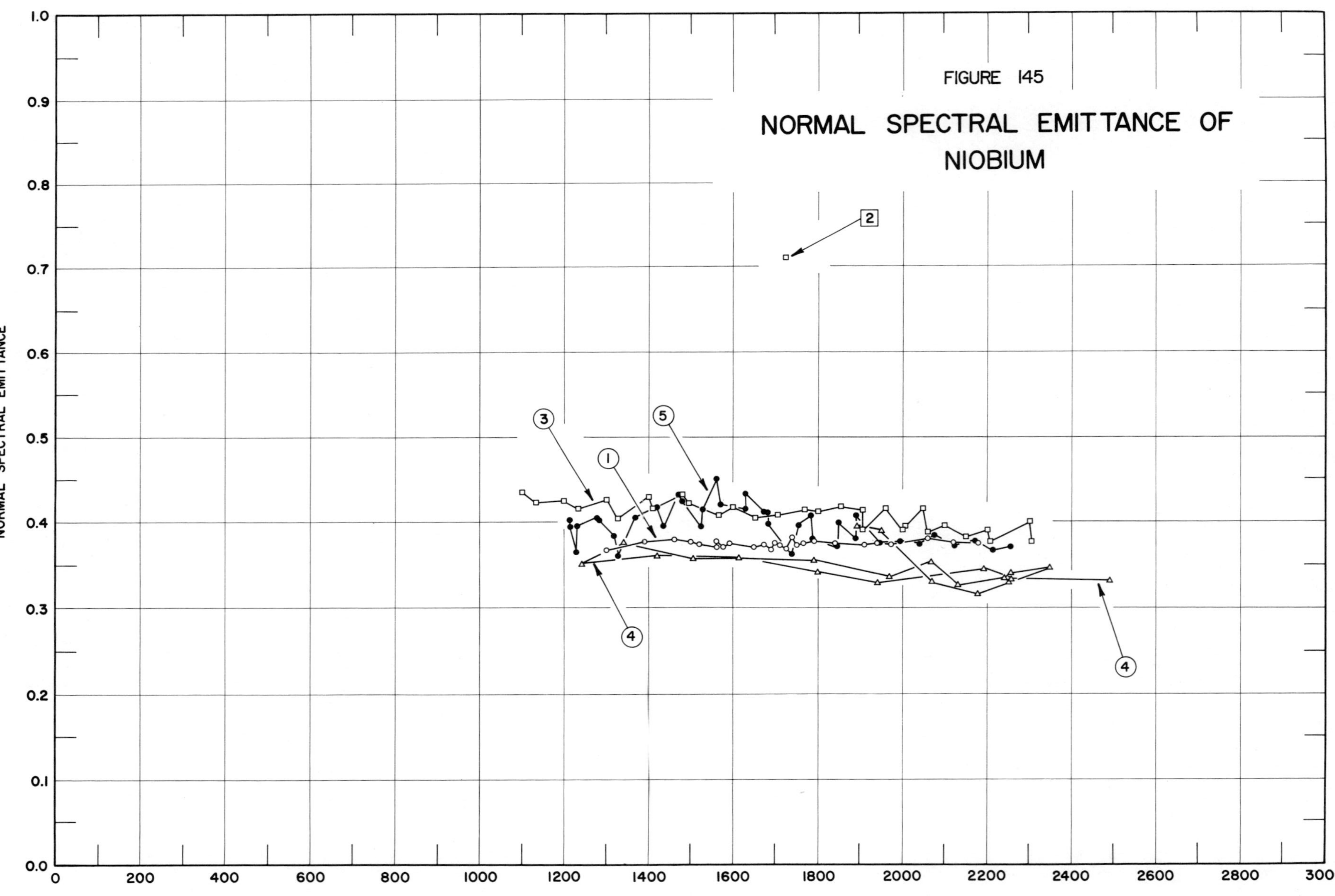
FIGURE 145
NORMAL SPECTRAL EMITTANCE OF
NIOBIUM
NORMAL SPECTRAL EMITTANCE
TEMPERATURE, K
1.0
0.9
0.8
0.7
0.6
0.5
0.4
0.3
0.2
0.1
0.0
0
200
400
600
800
1000
1200
1400
1600
1800
2000
2200
2400
2600
2800
3000
1
2
3
4
5

SPECIFICATION TABLE NO. 145 NORMAL SPECTRAL EMITTANCE OF NIOBIUM

Curve No.	Ref. No.	Year	Wavelength μ	Temperature Range, K	Geometry θ'	Reported Error, %	Composition (weight percent), Specifications and Remarks
1	55	1935	0.667	1300-2180	$\sim 0^\circ$		Heat treated for over 500 hrs; measured in vacuum ($< 10^{-6}$ mm Hg).
2	19	1914	0.65	1723	$\sim 0^\circ$	1	Film; tungsten substrate; melted in hydrogen then oxidized in air by heating; measured in air; Pt reference ($\epsilon = 0.33$ for $\lambda = 0.650\ \mu$ at all temp).
3	85	1963	0.65	1100-2304	$\sim 0^\circ$	± 5	Optically smooth, opaque, contamination - free ribbon (0.005 in. thick); measured in vacuum (3×10^{-7} to 1×10^{-5} mm Hg).
4	291	1963	0.66	1893-2494	$\sim 0^\circ$		0.4-0.6 Ta, 0.01-0.03 Mg, 0.007-0.009 Ti, 0.004-0.006 Fe, 0.001-0.003 Mn, 0.0004-0.0006 Cu, Nb balance; average values.
5	325	1963	0.65	1215-2255	$\sim 0^\circ$		No details given.

DATA TABLE NO. 145 NORMAL SPECTRAL EMITTANCE OF NIOBIUM

[Temperature, T, K; Emittance, ∈; Wavelength, λ, μ]

CURVE 1
$\lambda = 0.667$

T	∈
1300	0.368
1390	0.376
1460	0.380
1500	0.377
1520	0.374
1560	0.370
1560	0.377
1575	0.370
1590	0.374
1650	0.370
1675	0.373
1690	0.368
1700	0.376
1710	0.372
1725	0.369
1740	0.381
1750	0.372
1765	0.374
1790	0.377
1840	0.374
1910	0.371
1940	0.374
1975	0.372
2060	0.380
2120	0.375
2180	0.374

CURVE 2
$\lambda = 0.65$

T	∈
1723	0.71

CURVE 3
$\lambda = 0.65$

T	∈
1100	0.436
1132	0.422
1200	0.425
1232	0.416
1300	0.427
1324	0.403
1400	0.430
1410	0.416
1480	0.432

CURVE 3 (cont.)

T	∈
1496	0.422
1566	0.408
1600	0.418
1652	0.403
1704	0.409
1768	0.414
1800	0.411
1856	0.418
1904	0.414
1906	0.390
1960	0.416
2000	0.390
2004	0.395
2048	0.415
2060	0.387
2100	0.396
2150	0.381
2200	0.390
2204	0.377
2300	0.400
2304	0.376

CURVE 4
$\lambda = 0.66$

T	∈
1893	0.343
1950	0.339
2069	0.330
2179	0.315
2250	0.329
2346	0.347
2258	0.340
2240	0.333
2132	0.327
2066	0.352
1966	0.337
1789	0.353
1506	0.358
1342	0.372
1242	0.351
1422	0.360
1613	0.358
1800	0.341
1942	0.329
2192	0.343
2259	0.332
2494	0.331

CURVE 5
$\lambda = 0.65$

T	∈
1215	0.402
1216	0.396
1229	0.397
1230	0.365
1279	0.405
1285	0.401
1318	0.384
1324	0.360
1367	0.407
1420	0.419
1437	0.396
1477	0.434
1481	0.426
1525	0.393
1527	0.416
1561	0.451
1572	0.421
1628	0.417
1632	0.384
1676	0.412
1686	0.411
1686	0.399
1740	0.362
1756	0.397
1783	0.407
1787	0.380
1845	0.371
1852	0.399
1889	0.381
1891	0.407
1946	0.376
1996	0.378
2040	0.375
2078	0.383
2126	0.372
2177	0.378
2217	0.366
2255	0.370

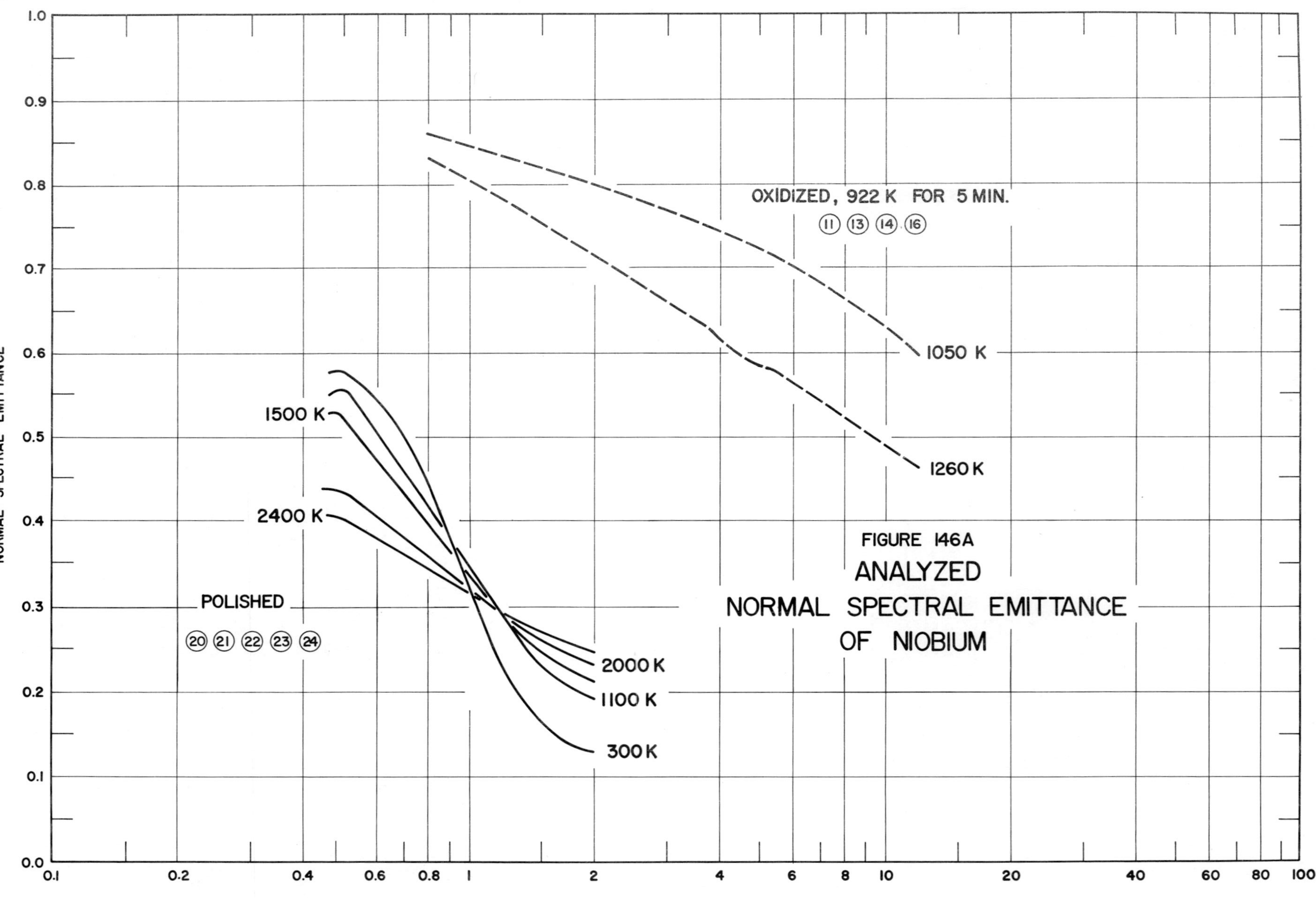

FIGURE 146A ANALYZED NORMAL SPECTRAL EMITTANCE OF NIOBIUM

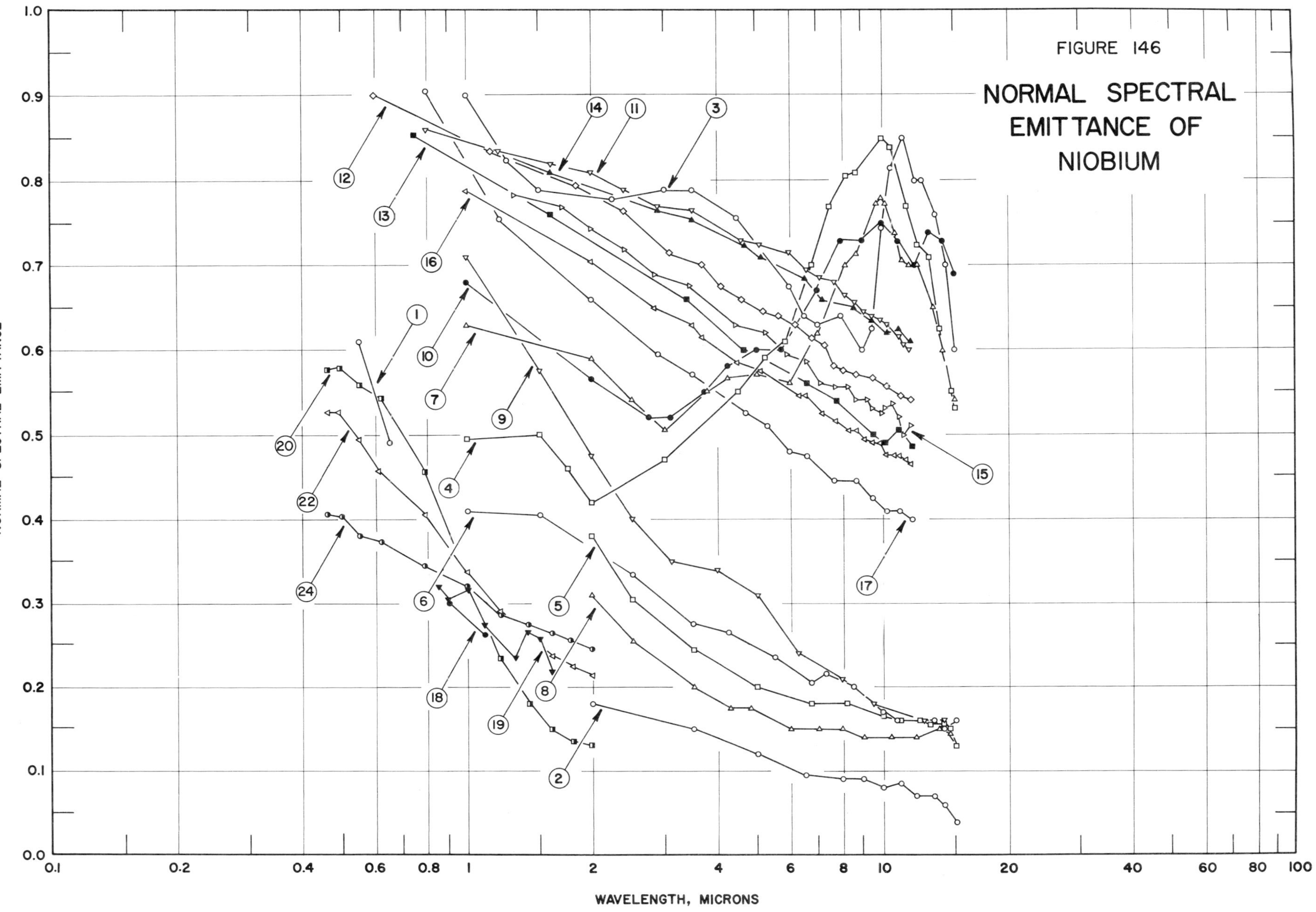

FIGURE 146
NORMAL SPECTRAL EMITTANCE OF NIOBIUM
NORMAL SPECTRAL EMITTANCE
WAVELENGTH, MICRONS
1.0
0.9
0.8
0.7
0.6
0.5
0.4
0.3
0.2
0.1
0.0
0.1
0.2
0.4
0.6
0.8
1
2
4
6
8
10
20
40
60
80
100
1
2
3
4
5
6
7
8
9
10
11
12
13
14
15
16
17
18
19
20
22
24

SPECIFICATION TABLE NO. 146 NORMAL SPECTRAL EMITTANCE OF NIOBIUM

Curve No.	Ref. No.	Year	Temperature K	Wavelength Range, μ	Geometry θ'	Reported Error, %	Composition (weight percent), Specifications and Remarks
1	19	1914	<M.P.	0.55-0.65	~0°	1	Film; tungsten substrate; measured in hydrogen; Pt reference ($\epsilon = 0.33$ for $\lambda = 0.650$ μ and $\epsilon = 0.38$ for $\lambda = 0.547$ μ at all temp).
2	86	1961	523	2.00-15.00	~0°	± 5	As received; data extracted from smooth curve.
3	86	1961	773	1.00-15.00	~0°	± 5	Different sample, same as curve 2 specimen and conditions.
4	86	1961	1023	1.00-15.00	~0°	± 5	Different sample, same as curve 2 specimen and conditions.
5	86	1961	523	2.00-15.00	~0°	± 5	Different sample, same as curve 2 specimen and conditions except heated in argon at 1366 K for 30 min.
6	86	1961	773	1.00-15.00	~0°	± 5	Different sample, same as curve 5 specimen and conditions.
7	86	1961	1023	1.00-15.00	~0°	± 5	Different sample, same as curve 5 specimen and conditions.
8	86	1961	523	2.00-15.00	~0°	± 5	Different sample, same as curve 2 specimen and conditions except heated in vacuum (2.2×10^{-5} mm Hg) at 1366 K for 30 min.
9	86	1961	773	1.00-15.00	~0°	± 5	Different sample, same as curve 8 specimen and conditions.
10	86	1961	1023	1.00-15.00	~0°	± 5	Different sample, same as curve 8 specimen and conditions.
11	12	1962	1050	0.80-11.65	~0°	± 3	Oxidized in air at 922 K for 5 min, then brushed with fine steel wool; measured in vacuum ($< 10^{-7}$ mm Hg); run A.
12	12	1962	1150	0.60-11.70	~0°	± 3	Above specimen and conditions; run B.
13	12	1962	1260	0.75-11.75	~0°	± 3	Above specimen and conditions; run C.
14	12	1962	1050	1.15-11.70	~0°	± 3	Above specimen and conditions; run D.
15	12	1962	1144	1.30-11.70	~0°	± 3	Above specimen and conditions; run E.
16	12	1962	1255	1.00-11.70	~0°	± 3	Above specimen and conditions; run F.
17	12	1962	1366	0.80-11.75	~0°	± 3	Above specimen and conditions; run G.
18	241	1963	1953	0.9-1.1	~0°		Single crystal; oriented so that surface of interest coincided with closed packed plane; optically polished; heated at several hundred degrees above temperature of interest in vacuum; computed from optical constants.
19	241	1963	2003	0.85-1.6	~0°		Above specimen and conditions.
20	323	1966	300	0.46-1.99	0°		Filament (0.25-0.32 mm in dia); baked for 1 hr at 798 K in vacuum, cooled, heated for 5-10 min in vacuum, and cooled; measured in argon (600 mm Hg); data calculated from optical constants; mechanically polished surface.
21	323	1966	1100	0.46-2.00	0°		Above specimen and conditions.
22	323	1966	1500	0.46-2.00	0°		Above specimen and conditions.
23	323	1966	2000	0.45-1.99	0°		Above specimen and conditions.
24	323	1966	2400	0.46-2.00	0°		Above specimen and conditions.

DATA TABLE NO. 146 NORMAL SPECTRAL EMITTANCE OF NIOBIUM

[Wavelength, λ, μ; Emittance, ∈; Temperature, T, K]

CURVE 1
T = <M.P.

λ	∈
0.55	0.61
0.65	0.49

CURVE 2
T = 523

λ	∈
2.00	0.180
3.50	0.150
5.00	0.120
6.50	0.095
8.00	0.090
9.00	0.090
10.00	0.080
11.00	0.085
12.00	0.070
13.25	0.070
14.00	0.060
15.00	0.040

CURVE 3
T = 773

λ	∈
1.00	0.900
1.25	0.825
1.50	0.790
2.25	0.780
3.00	0.790
3.50	0.790
4.50	0.755
6.00	0.675
6.50	0.640
7.00	0.630
8.00	0.640
9.00	0.600
9.50	0.625
10.00	0.745
10.50	0.815
11.20	0.850
12.00	0.800
12.50	0.800
13.50	0.760
14.25	0.700
15.00	0.600

CURVE 4
T = 1023

λ	∈
1.00	0.495
1.50	0.500
1.75	0.460
2.00	0.420
3.00	0.470
4.50	0.550
5.25	0.590
5.85	0.610
6.80	0.700
7.50	0.770
8.25	0.805
8.75	0.810
10.00	0.850
10.50	0.840
11.50	0.770
12.10	0.725
13.00	0.710
13.85	0.625
14.70	0.550
15.00	0.530

CURVE 5
T = 523

λ	∈
2.00	0.380
2.50	0.305
3.50	0.245
5.00	0.200
6.75	0.180
8.20	0.180
10.00	0.165
11.00	0.160
12.25	0.160
13.00	0.155
14.00	0.155
14.50	0.150
15.00	0.130

CURVE 6
T = 773

λ	∈
1.00	0.410
1.50	0.405
2.50	0.335

CURVE 6 (cont.)

λ	∈
3.50	0.275
4.25	0.265
5.50	0.235
6.75	0.205
7.30	0.215
8.50	0.200
10.00	0.170
10.75	0.160
12.25	0.160*
13.25	0.160
14.00	0.150
14.50	0.150*
15.00	0.160

CURVE 7
T = 1023

λ	∈
1.00	0.630
2.00	0.590
2.50	0.540
3.00	0.505
3.80	0.550
4.25	0.565
5.00	0.570
6.00	0.560
7.00	0.620
8.20	0.700
8.75	0.715
9.75	0.775
10.00	0.780
10.25	0.775
10.75	0.740
11.25	0.705
11.60	0.700
12.10	0.700
13.20	0.650
14.00	0.600
15.00	0.540

CURVE 8
T = 523

λ	∈
2.00	0.310
2.50	0.255

CURVE 8 (cont.)

λ	∈
3.50	0.200
4.30	0.175
4.80	0.175
6.00	0.150
7.00	0.150
8.00	0.150
9.00	0.140
10.50	0.140
12.00	0.140
13.75	0.150
14.50	0.145
15.00	0.130*

CURVE 9
T = 773

λ	∈
1.00	0.710
1.50	0.575
2.00	0.475
2.50	0.400
3.10	0.350
4.00	0.340
5.00	0.310
6.25	0.240
8.00	0.210
9.50	0.180
11.00	0.160*
12.50	0.160
14.00	0.160
15.00	0.160*

CURVE 10
T = 1023

λ	∈
1.00	0.680
2.00	0.565
2.75	0.520
3.10	0.520
3.75	0.550
4.25	0.580
5.00	0.600
5.75	0.600
7.00	0.670
8.00	0.730

CURVE 10 (cont.)

λ	∈
9.00	0.730
10.00	0.750
11.00	0.730
12.00	0.700
13.00	0.740
14.00	0.730
15.00	0.690

CURVE 11
T = 1050

λ	∈
0.80	0.860
1.20	0.835
1.60	0.820
2.00	0.810
2.40	0.790
2.90	0.770
3.50	0.765
4.60	0.730
5.10	0.725
6.00	0.715
6.60	0.695
7.15	0.685
7.75	0.680
8.20	0.665
8.65	0.655
9.10	0.645
9.50	0.640
9.95	0.635
10.30	0.630
11.00	0.615
11.30	0.605
11.65	0.600

CURVE 12
T = 1150

λ	∈
0.60	0.900
1.15	0.835
1.85	0.795
2.40	0.765
3.10	0.715
3.70	0.700
4.10	0.675

CURVE 12 (cont.)

λ	∈
4.60	0.660
5.20	0.645
5.65	0.640
6.20	0.630
6.80	0.615
7.30	0.605
7.70	0.580
8.10	0.575
8.70	0.570
9.55	0.565
10.30	0.555
11.05	0.545
11.70	0.540

CURVE 13
T = 1260

λ	∈
0.75	0.855
1.60	0.760
3.40	0.660
4.65	0.600
5.25	0.590*
6.60	0.560
7.80	0.540
9.55	0.500
10.25	0.490
11.00	0.505
11.75	0.485

CURVE 14
T = 1050

λ	∈
1.15	0.835*
1.60	0.810
2.90	0.765
3.50	0.755
4.70	0.725
5.15	0.710
6.55	0.685
7.20	0.660
8.60	0.650
9.50	0.635
10.30	0.620
11.00	0.625
11.70	0.610

CURVE 15
T = 1144

λ	∈
1.30	0.785
1.70	0.770
2.00	0.745
2.40	0.720
2.85	0.690
3.45	0.675
4.45	0.630
5.25	0.620
5.95	0.595
6.60	0.585
7.15	0.560
7.75	0.555
8.30	0.555
8.65	0.540
9.25	0.540
9.55	0.530
10.00	0.525
10.25	0.530
10.65	0.535
11.00	0.520
11.35	0.500
11.70	0.510

CURVE 16
T = 1255

λ	∈
1.00	0.790
2.00	0.705
2.85	0.650
3.50	0.630
3.70	0.615
4.50	0.585
5.15	0.575
6.35	0.545
6.55	0.545
7.20	0.525
7.80	0.515
8.35	0.505
8.70	0.505
9.10	0.495
9.50	0.490
9.95	0.490
10.35	0.475
10.70	0.475
11.00	0.475

CURVE 16 (cont.)

λ	∈
11.40	0.470
11.70	0.465

CURVE 17
T = 1366

λ	∈
0.80	0.905
1.20	0.755
2.00	0.660
2.90	0.595
3.50	0.570
4.70	0.525
5.30	0.510
6.00	0.480
6.60	0.475
7.70	0.445
8.70	0.445
9.50	0.425
10.30	0.410
11.00	0.410
11.75	0.400

CURVE 18
T = 1953

λ	∈
0.9	0.300
1.1	0.262

CURVE 19
T = 2003

λ	∈
0.85	0.320
0.9	0.303
1.0	0.315
1.1	0.274
1.3	0.233
1.4	0.266
1.5	0.256
1.6	0.216

*Not shown on plot

DATA TABLE NO. 146 (continued)

λ	ϵ	λ	ϵ
CURVE 20 T = 300		CURVE 23* T = 2000	
0.46	0.575	0.45	0.437
0.49	0.578	0.50	0.434
0.55	0.558	0.56	0.413
0.61	0.541	0.62	0.389
0.79	0.456	0.79	0.361
1.00	0.318*	1.00	0.321
1.19	0.232	1.19	0.290
1.39	0.181	1.39	0.265
1.59	0.149	1.59	0.253
1.79	0.134	1.80	0.241
1.99	0.130	1.99	0.233
CURVE 21* T = 1100		CURVE 24 T = 2400	
0.46	0.550	0.46	0.406
0.50	0.556	0.50	0.402
0.56	0.525	0.55	0.379
0.61	0.487	0.62	0.372
0.80	0.419	0.79	0.345
1.00	0.346	0.99	0.321
1.20	0.289	1.20	0.287
1.39	0.245	1.40	0.273
1.59	0.219	1.59	0.273
1.79	0.203	1.80	0.255
2.00	0.191	2.00	0.246
CURVE 22 T = 1500			
0.46	0.527		
0.49	0.527		
0.55	0.494		
0.61	0.458		
0.79	0.404		
0.99	0.338		
1.19	0.289		
1.40	0.265*		
1.60	0.236		
1.79	0.223		
2.00	0.213		

*Not shown on plot

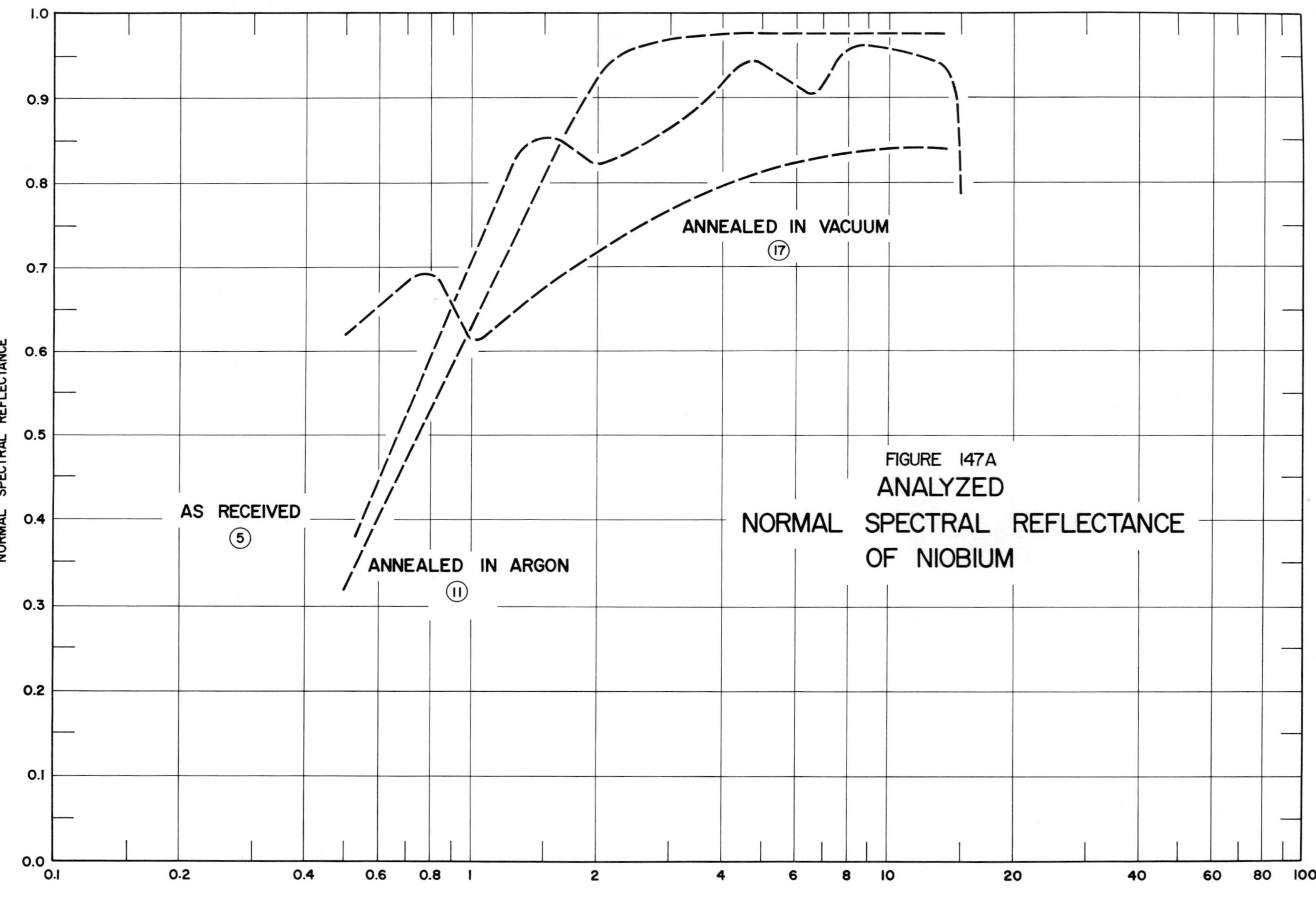

FIGURE 147A
ANALYZED
NORMAL SPECTRAL REFLECTANCE
OF NIOBIUM

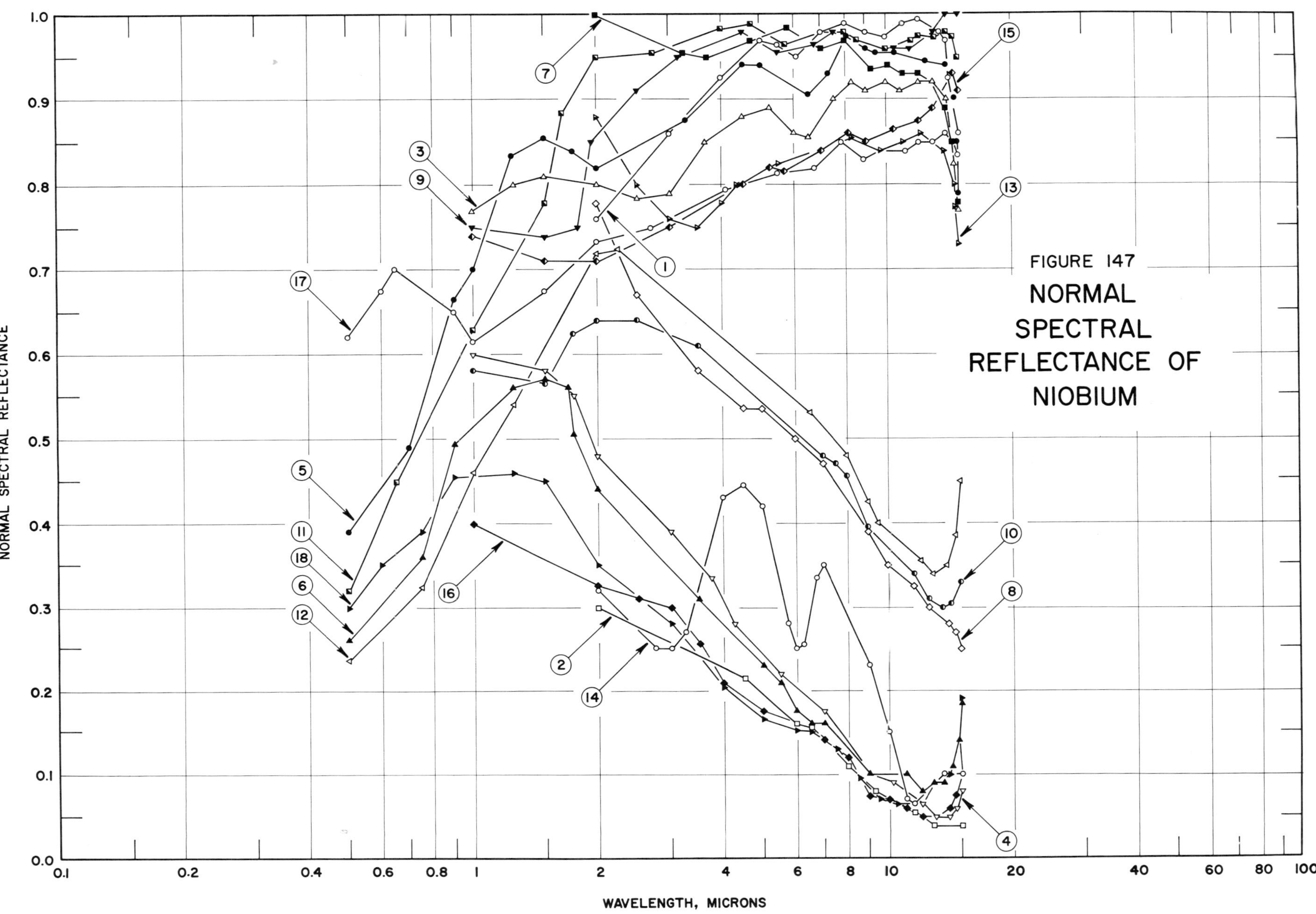
FIGURE 147
NORMAL SPECTRAL REFLECTANCE OF NIOBIUM
NORMAL SPECTRAL REFLECTANCE
WAVELENGTH, MICRONS
0.0
0.1
0.2
0.3
0.4
0.5
0.6
0.7
0.8
0.9
1.0
0.1
0.2
0.4
0.6
0.8
1
2
4
6
8
10
20
40
60
80
100
1
2
3
4
5
6
7
8
9
10
11
12
13
14
15
16
17
18

SPECIFICATION TABLE NO. 147 NORMAL SPECTRAL REFLECTANCE OF NIOBIUM

Curve No.	Ref. No.	Year	Temperature K	Wavelength Range, μ	Geometry θ	θ'	ω'	Reported Error, %	Composition (weight percent), Specifications and Remarks
1	86	1961	~322	2.00-15.00	~0°		2π	< 2	As received; data extracted from smooth curve; hohlraum at 523 K; converted from R (2π, 0°).
2	86	1961	~322	2.00-15.00	~0°		2π	< 2	Above specimen and conditions; diffuse component only.
3	86	1961	~322	1.00-15.00	~0°		2π	< 2	Different sample, same as Curve 1 specimen and conditions except hohlraum at 773 K.
4	86	1961	~322	1.00-15.00	~0°		2π	< 2	Above specimen and conditions; diffuse component only.
5	86	1961	~322	0.50-15.00	~0°		2π	< 2	Different sample, same as Curve 1 specimen and conditions except hohlraum at 1273 K.
6	86	1961	~322	0.50-15.00	~0°		2π	< 2	Above specimen and conditions; diffuse component only.
7	86	1961	~322	2.00-15.00	~0°		2π	< 2	Different sample, same as Curve 1 specimen and conditions except heated in argon at 1366 K for 30 min.
8	86	1961	~322	2.00-15.00	~0°		2π	< 2	Above specimen and conditions; diffuse component only.
9	86	1961	~322	1.00-15.00	~0°		2π	< 2	Different sample, same as Curve 7 specimen and conditions except hohlraum at 773 K.
10	86	1961	~322	1.00-15.00	~0°		2π	< 2	Above specimen and conditions; diffuse component only.
11	86	1961	~322	0.50-15.00	~0°		2π	< 2	Different sample, same as Curve 7 specimen and conditions except hohlraum at 1273 K.
12	86	1961	~322	0.50-15.00	~0°		2π	< 2	Above specimen and conditions; diffuse component only.
13	86	1961	~322	2.00-15.00	~0°		2π	< 2	Different sample, same as Curve 1 specimen and conditions except heated in vacuum (2.2×10^{-5} mm Hg) at 1366 K for 30 min.
14	86	1961	~322	2.00-15.00	~0°		2π	< 2	Above specimen and conditions; diffuse component only.
15	86	1961	~322	1.00-15.00	~0°		2π	< 2	Different sample, same as Curve 13 specimen and conditions except hohlraum at 773 K.
16	86	1961	~322	1.00-15.00	~0°		2π	< 2	Above specimen and conditions; diffuse component only.
17	86	1961	~322	0.50-15.00	~0°		2π	< 2	Different sample, same as Curve 13 specimen and conditions except hohlraum at 1273 K.
18	86	1961	~322	0.50-15.00	~0°		2π	< 2	Above specimen and conditions; diffuse component only.

DATA TABLE NO. 147 NORMAL SPECTRAL REFLECTANCE OF NIOBIUM

[Wavelength, λ, μ; Reflectance, ρ; Temperature, T, K]

CURVE 1
T = ~322

λ	ρ
2.00	0.760
3.00	0.860
4.00	0.925
5.00	0.970
5.50	0.965
6.10	0.950
7.00	0.980
8.00	0.990
9.00	0.980
10.00	0.975
11.00	0.990
12.00	0.995
13.50	0.980
14.00	0.970
14.15	0.925
15.00	0.860

CURVE 2
T = ~322

λ	ρ
2.00	0.300
4.50	0.215
6.00	0.160
6.50	0.155
8.00	0.110
9.25	0.080
11.50	0.055
12.75	0.040
15.00	0.040

CURVE 3
T = ~322

λ	ρ
1.00	0.770
1.25	0.800
1.50	0.810
2.00	0.800
2.50	0.785
3.00	0.790
3.65	0.850
4.50	0.880
5.25	0.890
6.00	0.860

CURVE 3 (cont.)

λ	ρ
7.50	0.900
8.30	0.920
9.00	0.910
10.00	0.920
10.75	0.910
12.00	0.920
13.00	0.920
14.00	0.900
14.65	0.825
15.00	0.770

CURVE 4
T = ~322

λ	ρ
1.00	0.600
1.50	0.580
1.75	0.550
2.00	0.480
3.00	0.390
3.75	0.335
4.25	0.280
5.50	0.220
7.00	0.175
9.00	0.100*
10.25	0.090
12.00	0.065
13.00	0.050
14.00	0.050
14.50	0.060
15.00	0.080

CURVE 5
T = ~322

λ	ρ
0.50	0.390
0.70	0.490
0.90	0.665
1.00	0.700
1.25	0.835
1.50	0.855
1.75	0.840
2.00	0.820
3.30	0.875
4.50	0.940

CURVE 5 (cont.)

λ	ρ
5.00	0.940
6.50	0.905
7.25	0.930
8.10	0.975
9.00	0.960
9.50	0.955
10.50	0.955
12.50	0.945
14.00	0.940
14.55	0.900
14.90	0.850
15.00	0.790

CURVE 6
T = ~322

λ	ρ
0.50	0.260
0.75	0.360
0.90	0.495
1.25	0.560
1.50	0.570
1.70	0.560
1.75	0.505
2.00	0.440
3.50	0.310
5.00	0.230
5.50	0.210
6.00	0.175
6.50	0.160
7.00	0.160
9.00	0.100
11.00	0.100
12.00	0.080
12.75	0.090
13.50	0.090
14.25	0.110
14.75	0.140
15.00	0.185

CURVE 7
T = ~322

λ	ρ
2.00	1.000
3.25	0.955

CURVE 7 (cont.)

λ	ρ
3.70	0.950
4.75	0.970
5.80	0.985
7.00	0.960
8.00	0.970
9.25	0.935
10.10	0.940
11.00	0.930
12.00	0.930
13.00	0.920*
14.00	0.890
14.60	0.850
15.00	0.780

CURVE 8
T = ~322

λ	ρ
2.00	0.780
2.50	0.670
3.50	0.580
4.50	0.535
5.00	0.535
6.00	0.500
7.00	0.470
9.00	0.390
10.00	0.350
11.50	0.325
12.50	0.300
14.00	0.280
14.50	0.270
15.00	0.250

CURVE 9
T = ~322

λ	ρ
1.00	0.750
1.50	0.740
1.80	0.750
1.95	0.850
2.50	0.910
3.15	0.950
4.50	0.980
5.50	0.955
6.75	0.965

CURVE 9 (cont.)

λ	ρ
7.50	0.980
9.00	0.960*
10.50	0.960
11.50	0.960
13.00	0.980
14.00	1.000
15.00	1.000

CURVE 10
T = ~322

λ	ρ
1.00	0.580
1.50	0.565
1.75	0.625
2.00	0.640
2.50	0.640
3.50	0.610
7.00	0.480
7.50	0.470
8.00	0.455
9.00	0.395
11.50	0.340
12.50	0.310
13.50	0.300
14.25	0.305
15.00	0.330

CURVE 11
T = ~322

λ	ρ
0.50	0.320
0.65	0.450
1.00	0.630
1.50	0.780
1.65	0.885
2.00	0.950
2.75	0.955
4.00	0.985
4.75	0.990
5.75	0.965
7.00	0.980*
8.00	0.980
8.55	0.970
10.00	0.970

CURVE 11 (cont.)

λ	ρ
11.50	0.970
12.00	0.975
13.25	0.975
14.00	0.980
14.50	0.975
15.00	0.950

CURVE 12
T = ~322

λ	ρ
0.50	0.235
0.75	0.325
1.00	0.460
1.25	0.540
2.00	0.720
2.25	0.725
6.50	0.530
8.00	0.480
9.00	0.425
9.50	0.400
12.00	0.355
12.75	0.340
13.75	0.350
14.50	0.385
15.00	0.450

CURVE 13
T = ~322

λ	ρ
2.00	0.880
2.50	0.800
3.00	0.760
3.50	0.750
4.00	0.780
4.35	0.800
5.50	0.825
7.00	0.840*
8.25	0.855
9.75	0.840
11.00	0.850
12.15	0.860
13.75	0.840
14.60	0.800
14.85	0.775
15.00	0.730

CURVE 14
T = ~322

λ	ρ
2.00	0.320
2.75	0.250
3.00	0.250
3.25	0.270
4.00	0.430
4.50	0.445
5.00	0.420
5.75	0.280
6.00	0.250
6.25	0.255
6.75	0.335
7.00	0.350
9.00	0.230
10.00	0.150
11.00	0.070
11.50	0.065
13.00	0.090*
13.50	0.100
15.00	0.100

CURVE 15
T = ~322

λ	ρ
1.00	0.740
1.50	0.710
2.00	0.710
3.00	0.750
4.50	0.800
5.25	0.820
5.70	0.815
7.00	0.840
8.15	0.860
9.00	0.850
10.50	0.865
12.00	0.875
13.00	0.890
14.25	0.930*
14.50	0.930
15.00	0.910

CURVE 16
T = ~322

λ	ρ
1.00	0.400
2.00	0.325
2.50	0.310
3.00	0.300
3.50	0.255
4.00	0.210
5.00	0.175
6.00	0.160*
7.00	0.140
8.00	0.120
9.00	0.075
10.00	0.070
11.00	0.060
12.00	0.050
13.00	0.050
14.00	0.060
14.50	0.075
15.00	0.100*

CURVE 17
T = ~322

λ	ρ
0.50	0.620
0.70	0.675
0.75	0.700
0.90	0.650
1.00	0.615
1.50	0.675
2.00	0.735
2.70	0.750
4.10	0.795
5.50	0.815
6.75	0.820
7,85	0.850
8.90	0.830
9.60	0.840*
11.25	0.840
12.00	0.850
13.00	0.850
14.00	0.860
14.75	0.850*
15.00	0.835

*Not shown on plot

DATA TABLE NO. 147 (continued)

λ	ρ
CURVE 18 $T = \sim 322$	
0.50	0.300
0.60	0.350
0.75	0.390
0.90	0.455
1.25	0.460
1.50	0.450
2.00	0.350
3.00	0.280
4.00	0.205
5.00	0.165
6.00	0.150
6.50	0.150
7.50	0.130
8.50	0.095
9.50	0.070
10.50	0.065
12.00	0.065*
13.00	0.090*
14.00	0.100
14.75	0.140*
15.00	0.190

*Not shown on plot

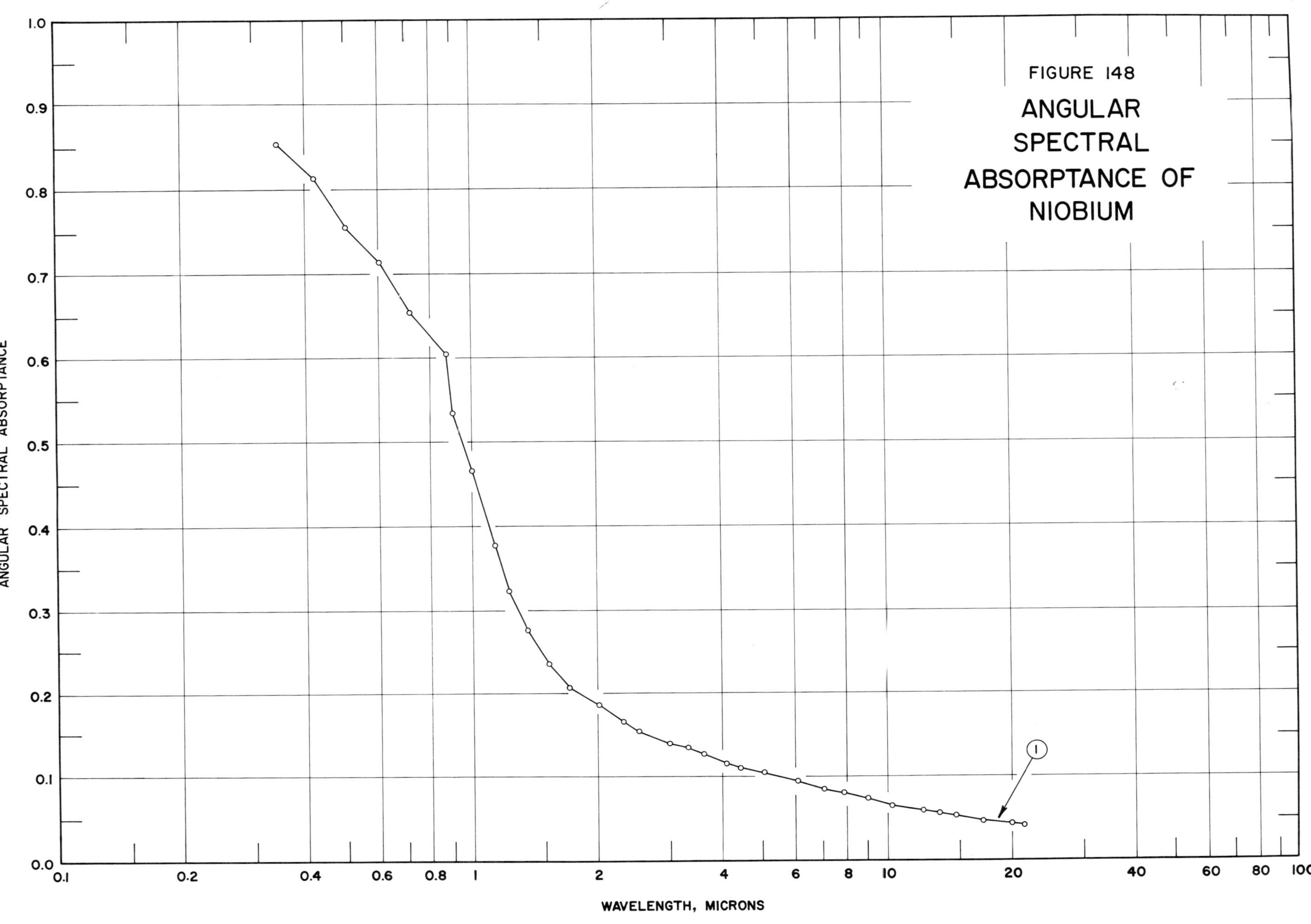
FIGURE 148
ANGULAR SPECTRAL ABSORPTANCE OF NIOBIUM
1
ANGULAR SPECTRAL ABSORPTANCE
1.0
0.9
0.8
0.7
0.6
0.5
0.4
0.3
0.2
0.1
0.0
0.1
0.2
0.4
0.6
0.8
1
2
4
6
8
10
20
40
60
80
100
WAVELENGTH, MICRONS

SPECIFICATION TABLE NO. 148 ANGULAR SPECTRAL ABSORPTANCE OF NIOBIUM

Curve No.	Ref. No.	Year	Temperature K	Wavelength Range, μ	Geometry θ	Reported Error, %	Composition (weight percent), Specifications and Remarks
1	225	1965	306	0.344-21.4	25°		Rolled plate niobium from Belmont Smelting and Refining Works; measured in dry nitrogen; heated cavity at approx 1056 K with platinum reference; authors assumed α = 1-R(2π, 25°).

DATA TABLE NO. 148 ANGULAR SPECTRAL ABSORPTANCE OF NIOBIUM

[Wavelength, λ, μ ; Abosrptance, α ; Temperature, T, K]

λ	α
CURVE 1	
T = 306	
0.344	0.853
0.419	0.813
0.504	0.755
0.607	0.713
0.719	0.655
0.776	0.607
0.902	0.533
1.01	0.466
1.13	0.378
1.23	0.324
1.37	0.277
1.53	0.236
1.71	0.208
2.04	0.186
2.31	0.166
2.52	0.154
2.99	0.139
3.30	0.134
3.61	0.127
4.10	0.115
4.43	0.110
5.04	0.103
6.10	0.093
7.06	0.084
7.93	0.080
9.02	0.074
10.3	0.067
12.2	0.060
13.4	0.057
14.8	0.053
17.1	0.048
20.0	0.043
21.4	0.041

SPECIFICATION TABLE NO. 149 NORMAL SPECTRAL EMITTANCE OF OSMIUM

Curve No.	Ref. No.	Year	Wavelength μ	Temperature Range, K	Geometry θ'	Reported Error, %	Composition (weight percent), Specifications and Remarks
1	135	1961	0.655	1240-2500	$\sim 0^\circ$		99.5 Os from Englehard Industries, Inc.; prepared from powder; heated at 673 K for 1 hr and presintered at 1473 K for 1 hr; measured in vacuum; data extracted from smooth curve.

DATA TABLE NO. 149 NORMAL SPECTRAL EMITTANCE OF OSMIUM

[Temperature, T, K; Emittance, ϵ; Wavelength, λ, μ]

T	ϵ
CURVE 1*	
$\lambda = 0.655$	
1240	0.550
1300	0.506
1400	0.462
1500	0.430
1600	0.409
1700	0.395
1800	0.387
1900	0.382
2000	0.380
2100	0.378
2200	0.380
2300	0.383
2400	0.387
2500	0.392

*Not shown on plot

SPECIFICATION TABLE NO. 150 HEMISPHERICAL TOTAL EMITTANCE OF PALLADIUM

Curve No.	Ref. No.	Year	Temperature Range, K	Reported Error, %	Composition (weight percent), Specifications and Remarks
1	343	1964	1125-1567		Measured in vacuum (10^{-9} mm Hg).

DATA TABLE NO. 150 HEMISPHERICAL TOTAL EMITTANCE OF PALLADIUM

[Temperature, T, K; Emittance, ϵ]

T	ϵ
CURVE 1*	
1125	0.135
1146	0.129
1149	0.135
1155	0.133
1196	0.143
1229	0.139
1234	0.142
1287	0.144
1287	0.150
1300	0.154
1300	0.150
1303	0.146
1350	0.158
1367	0.156
1385	0.156
1411	0.161
1420	0.168
1424	0.160
1473	0.174
1494	0.173
1508	0.173
1537	0.184
1550	0.177
1565	0.179
1567	0.185

*Not shown on plot

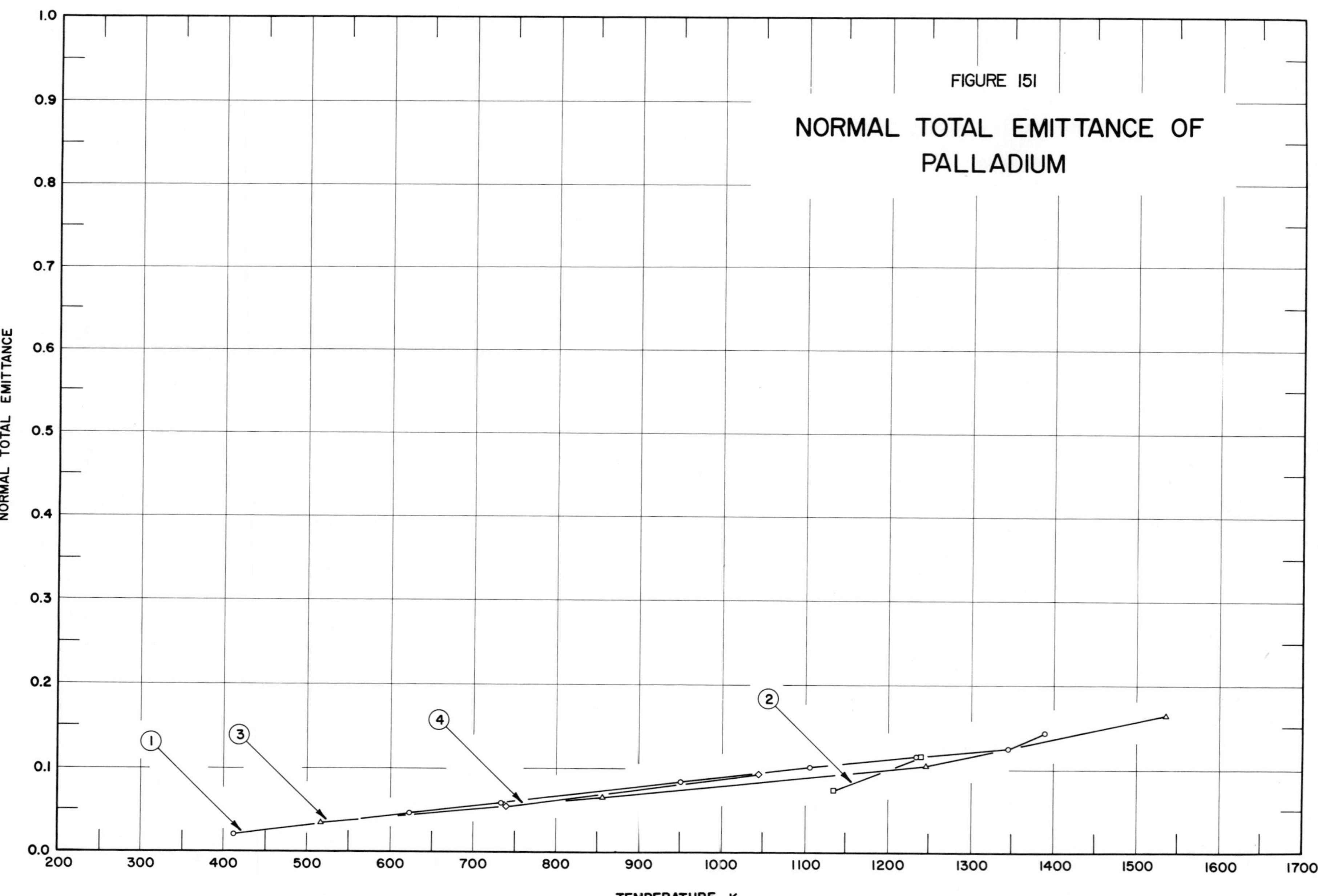
FIGURE 151
NORMAL TOTAL EMITTANCE OF PALLADIUM
NORMAL TOTAL EMITTANCE
TEMPERATURE, K
0.0
0.1
0.2
0.3
0.4
0.5
0.6
0.7
0.8
0.9
1.0
200
300
400
500
600
700
800
900
1000
1100
1200
1300
1400
1500
1600
1700
1
2
3
4

SPECIFICATION TABLE NO. 151 NORMAL TOTAL EMITTANCE OF PALLADIUM

Curve No.	Ref. No.	Year	Temperature Range, K	Geometry θ'	Reported Error, %	Composition (weight percent), Specifications and Remarks
1	34	1957	411-1389	$\sim 0°$	± 10	Pure; strip (0.005 in. thick); same results obtained for 4 different surface treatments (a. as received, b. cleaned with liquid detergent, c. polished, d. oxidized in air at red heat for 30 min.); measured in air (5×10^{-4} mm Hg); increasing temp, Cycle 1.
2	34	1957	1239-1133	$\sim 0°$	± 10	Above specimen and conditions; decreasing temp, Cycle 1.
3	34	1957	516-1533	$\sim 0°$	± 10	Curve 1 specimen and conditions; increasing temp, Cycle 2.
4	34	1957	739-1044	$\sim 0°$	± 10	Curve 1 specimen and conditions; increasing temp, Cycle 3.

DATA TABLE NO. 151 NORMAL TOTAL EMITTANCE OF PALLADIUM

[Temperature, T, K; Emittance, ϵ]

T	ϵ
CURVE 1	
411	0.020
622	0.047
733	0.059
950	0.084
1105	0.101
1233	0.114
1344	0.125
1389	0.144
CURVE 2	
1239	0.115
1133	0.075
CURVE 3	
516	0.035
855	0.066
1244	0.104
1533	0.165
CURVE 4	
739	0.055
1044	0.093

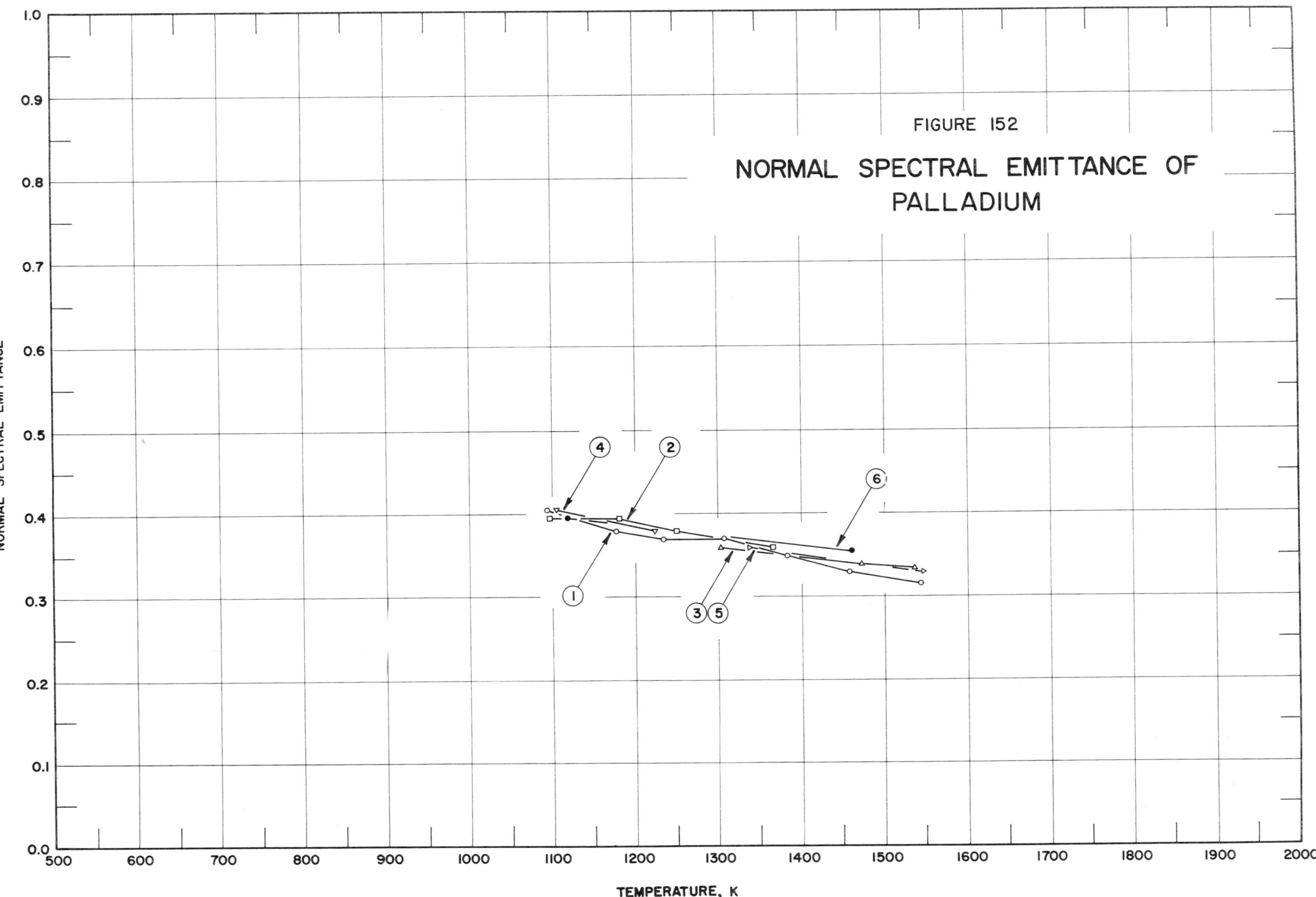
FIGURE 152
NORMAL SPECTRAL EMITTANCE OF PALLADIUM
NORMAL SPECTRAL EMITTANCE
TEMPERATURE, K
0.0
0.1
0.2
0.3
0.4
0.5
0.6
0.7
0.8
0.9
1.0
500
600
700
800
900
1000
1100
1200
1300
1400
1500
1600
1700
1800
1900
2000
1
2
3
4
5
6

SPECIFICATION TABLE NO. 152 NORMAL SPECTRAL EMITTANCE OF PALLADIUM

Curve No.	Ref. No.	Year	Wavelength μ	Temperature Range, K	Geometry θ'	Reported Error, %	Composition (weight percent), Specifications and Remarks
1	34	1957	0.665	1094-1541	$\sim 0°$		Pure metal; same results obtained for 3 different surface treatments (a. as received, b. cleaned with liquid detergent, c. polished); measured in air (5×10^{-4} mm Hg); increasing temp, Cycle 1.
2	34	1957	0.665	1366-1097	$\sim 0°$		Above specimen and conditions; decreasing temp, Cycle 1.
3	34	1957	0.665	1302-1536	$\sim 0°$		Curve 1 specimen and conditions; Cycle 2.
4	34	1957	0.665	1222-1105	$\sim 0°$		Above specimen and conditions; decreasing temp, Cycle 2.
5	34	1957	0.665	1339-1547	$\sim 0°$		Curve 1 specimen and conditions; Cycle 3.
6	34	1957	0.665	1461-1119	$\sim 0°$		Above specimen and conditions; decreasing temp, Cycle 3.

DATA TABLE NO. 152 NORMAL SPECTRAL EMITTANCE OF PALLADIUM

[Temperature, T, K; Emittance, ϵ; Wavelength, λ, μ]

T	ϵ
CURVE 1 $\lambda = 0.665$	
1094	0.405
1177	0.380
1233	0.370
1308	0.370
1383	0.350
1458	0.330
1541	0.315
CURVE 2 $\lambda = 0.665$	
1366	0.360
1250	0.380
1180	0.395
1097	0.395
CURVE 3 $\lambda = 0.665$	
1302	0.360
1472	0.340
1536	0.335
CURVE 4 $\lambda = 0.665$	
1222	0.380
1105	0.405
CURVE 5 $\lambda = 0.665$	
1339	0.36
1547	0.33
CURVE 6 $\lambda = 0.665$	
1461	0.355
1119	0.395

SPECIFICATION TABLE NO. 153 NORMAL SPECTRAL EMITTANCE OF PALLADIUM

Curve No.	Ref. No.	Year	Temperature K	Wavelength Range, μ	Geometry θ'	Reported Error, %	Composition (weight percent), Specifications and Remarks
1	19	1914	1803	0.55-0.65	$\sim 0°$	1	Film; tungsten substrate; measured in hydrogen; Pt reference ($\epsilon = 0.33$ for $\lambda = 0.650\ \mu$ at all temp).

DATA TABLE NO. 153 NORMAL SPECTRAL EMITTANCE OF PALLADIUM

[Wavelength, λ, μ; Emittance, ϵ; Temperature, T, K]

λ	ϵ
CURVE 1*	
T = 1803	
0.55	0.38
0.65	0.33

*Not shown on plot

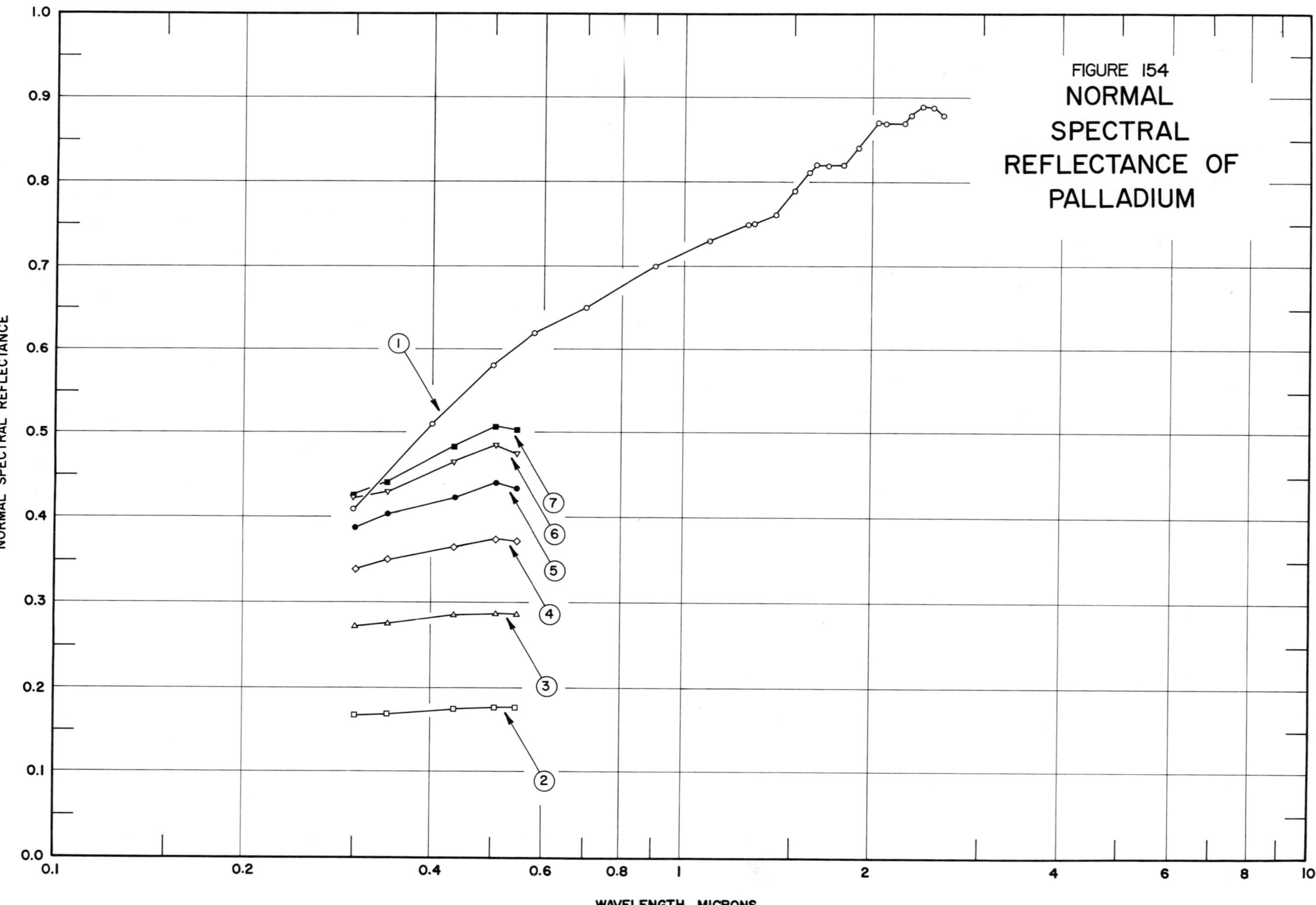
FIGURE 154
NORMAL
SPECTRAL
REFLECTANCE OF
PALLADIUM
NORMAL SPECTRAL REFLECTANCE
1.0
0.9
0.8
0.7
0.6
0.5
0.4
0.3
0.2
0.1
0.0
WAVELENGTH, MICRONS
0.1
0.2
0.4
0.6
0.8
1
2
4
6
8
10
1
2
3
4
5
6
7

SPECIFICATION TABLE NO. 154 NORMAL SPECTRAL REFLECTANCE OF PALLADIUM

Curve No.	Ref. No.	Year	Temperature K	Wavelength Range, μ	Geometry θ	θ'	ω'	Reported Error, %	Composition (weight percent), Specifications and Remarks
1	34	1957	298	0.30-2.60	9°		2π	± 4	Pure metal; as received; data extracted from a smooth curve; magnesium carbonate reference.
2	251	1957	298	0.3021-0.5461	$\sim 0^\circ$	$\sim 0^\circ$			Vacuum deposited film (5.05 m μ thick); measured in air.
3	251	1957	298	0.3021-0.5461	$\sim 0^\circ$	$\sim 0^\circ$			Different sample, same as above specimen and conditions except 10.1 m μ thick.
4	251	1957	298	0.3021-0.5461	$\sim 0^\circ$	$\sim 0^\circ$			Different sample, same as above specimen and conditions except 15.2 m μ thick.
5	251	1957	298	0.3021-0.5461	$\sim 0^\circ$	$\sim 0^\circ$			Different sample, same as above specimen and conditions except 20.2 m μ thick.
6	251	1957	298	0.3021-0.5461	$\sim 0^\circ$	$\sim 0^\circ$			Different sample, same as above specimen and conditions except 25.3 m μ thick.
7	251	1957	298	0.3021-0.5461	$\sim 0^\circ$	$\sim 0^\circ$			Different sample, same as above specimen and conditions except 30.3 m μ thick.

DATA TABLE NO. 154 NORMAL SPECTRAL REFLECTANCE OF PALLADIUM

[Wavelength, λ, μ; Reflectance, ρ; Temperature, T, K]

λ	ρ	λ	ρ
CURVE 1 T = 298		CURVE 4 T = 298	
0.30	0.41	0.3021	0.340
0.40	0.51	0.3404	0.351
0.50	0.58	0.4358	0.367
0.58	0.62	0.5085	0.375
0.70	0.65	0.5461	0.372
0.90	0.70		
1.10	0.73	CURVE 5 T = 298	
1.26	0.75		
1.30	0.75		
1.40	0.76	0.3021	0.388
1.50	0.79	0.3404	0.403
1.58	0.81	0.4358	0.423
1.62	0.82	0.5085	0.441
1.70	0.82	0.5461	0.435
1.80	0.82		
1.90	0.84	CURVE 6 T = 298	
2.02	0.87		
2.10	0.87		
2.24	0.87	0.3021	0.423
2.30	0.88	0.3404	0.431
2.40	0.89	0.4358	0.465
2.50	0.89	0.5085	0.486
2.60	0.88	0.5461	0.477
CURVE 2 T = 298		CURVE 7 T = 298	
0.3021	0.168	0.3021	0.425
0.3404	0.169	0.3404	0.441
0.4358	0.175	0.4358	0.484
0.5085	0.176	0.5085	0.507
0.5461	0.177	0.5461	0.503
CURVE 3 T = 298			
0.3021	0.272		
0.3404	0.276		
0.4358	0.286		
0.5085	0.288		
0.5461	0.288		

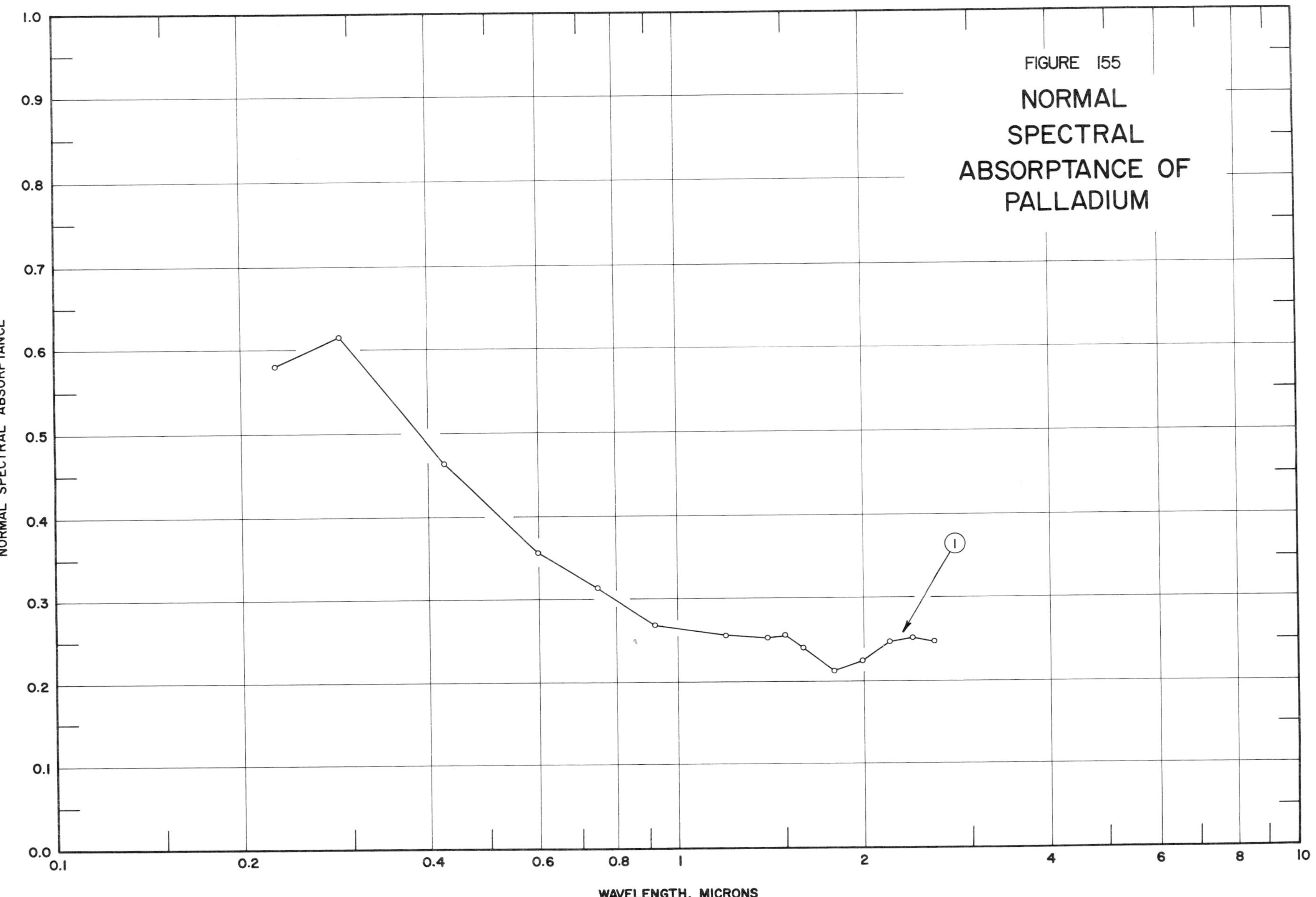

FIGURE 155 NORMAL SPECTRAL ABSORPTANCE OF PALLADIUM

SPECIFICATION TABLE NO. 155 NORMAL SPECTRAL ABSORPTANCE OF PALLADIUM

Curve No.	Ref. No.	Year	Temperature K	Wavelength Range, μ	Geometry θ	Reported Error, %	Composition (weight percent), Specifications and Remarks
1	307	1954	~298	0.224-2.600	~0°		Data extracted from smooth curve.

DATA TABLE NO. 155 NORMAL SPECTRAL ABSORPTANCE OF PALLADIUM

λ	α
CURVE 1 T = ~298	
0.224	0.581
0.290	0.615
0.425	0.464
0.602	0.357
0.745	0.314
0.918	0.269
1.200	0.256
1.402	0.252
1.496	0.256
1.599	0.240
1.795	0.212
1.999	0.224
2.200	0.248
2.400	0.252
2.600	0.248

SPECIFICATION TABLE NO. 156 NORMAL SOLAR ABSORPTANCE OF PALLADIUM

Curve No.	Ref. No.	Year	Temperature Range, K	Geometry θ	Reported Error, %	Composition (weight percent), Specifications and Remarks
1	34	1957	298	9°		Pure metal; as received (bright); computed from spectral reflectance data for sea level conditions.
2	34	1957	298	9°		Above specimen and conditions except computed for above atmosphere conditions.

DATA TABLE NO. 156 NORMAL SOLAR ABSORPTANCE OF PALLADIUM

[Temperature, T, K; Absorptance, α]

T	α
CURVE 1*	
298	0.345
CURVE 2*	
298	0.343

*Not shown on plot

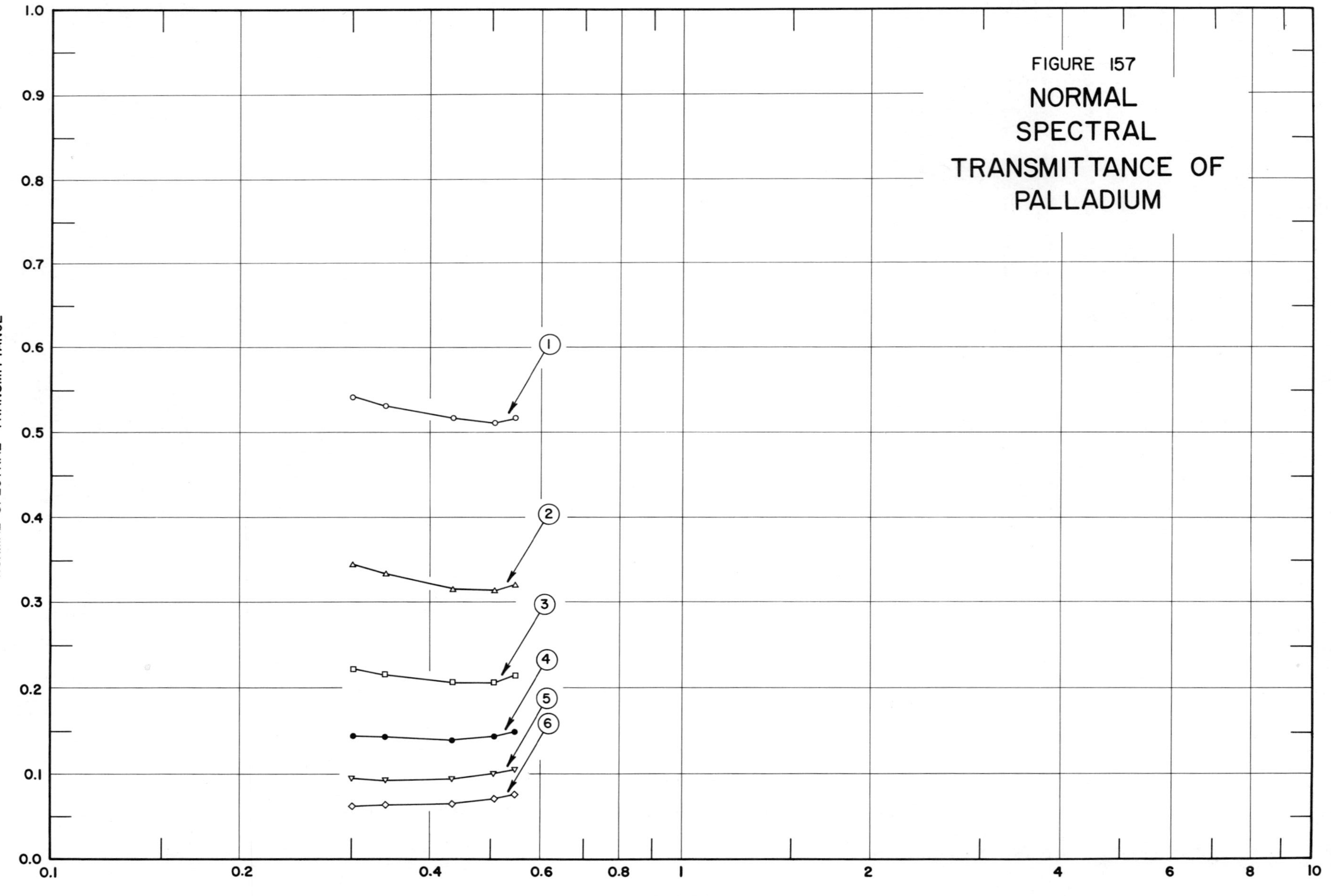

FIGURE 157
NORMAL
SPECTRAL
TRANSMITTANCE OF
PALLADIUM
NORMAL SPECTRAL TRANSMITTANCE
WAVELENGTH, MICRONS
1.0
0.9
0.8
0.7
0.6
0.5
0.4
0.3
0.2
0.1
0.0
0.1
0.2
0.4
0.6
0.8
1
2
4
6
8
10
1
2
3
4
5
6

SPECIFICATION TABLE NO. 157 NORMAL SPECTRAL TRANSMITTANCE OF PALLADIUM

Curve No.	Ref. No.	Year	Temperature K	Wavelength Range, μ	Geometry θ	θ' ω'	Reported Error, %	Composition (weight percent), Specifications and Remarks
1	251	1957	298	0.3021-0.5461	$\sim 0^\circ$	$\sim 0^\circ$		Vacuum deposited film (5.05 m μ thick); measured in air.
2	251	1957	298	0.3021-0.5461	$\sim 0^\circ$	$\sim 0^\circ$		Different sample, same as above specimen and conditions except 10.1 m μ thick.
3	251	1957	298	0.3021-0.5461	$\sim 0^\circ$	$\sim 0^\circ$		Different sample, same as above specimen and conditions except 15.2 m μ thick.
4	251	1957	298	0.3021-0.5461	$\sim 0^\circ$	$\sim 0^\circ$		Different sample, same as above specimen and conditions except 20.2 m μ thick.
5	251	1957	298	0.3021-0.5461	$\sim 0^\circ$	$\sim 0^\circ$		Different sample, same as above specimen and conditions except 25.3 m μ thick.
6	251	1957	298	0.3021-0.5461	$\sim 0^\circ$	$\sim 0^\circ$		Different sample, same as above specimen and conditions except 30.3 m μ thick.

DATA TABLE NO. 157 NORMAL SPECTRAL TRANSMITTANCE OF PALLADIUM

[Wavelength, λ, μ; Transmittance, τ; Temperature, T, K]

λ	τ	λ	τ
CURVE 1 T = 298		CURVE 6 T = 298	
0.3021	0.542	0.3021	0.062
0.3404	0.531	0.3404	0.063
0.4358	0.517	0.4358	0.067
0.5085	0.512	0.5085	0.072
0.5461	0.517	0.5461	0.077
CURVE 2 T = 298			
0.3021	0.344		
0.3404	0.333		
0.4358	0.317		
0.5085	0.314		
0.5461	0.321		
CURVE 3 T = 298			
0.3021	0.222		
0.3404	0.216		
0.4358	0.207		
0.5085	0.207		
0.5461	0.215		
CURVE 4 T = 298			
0.3021	0.144		
0.3404	0.143		
0.4358	0.140		
0.5085	0.143		
0.5461	0.150		
CURVE 5 T = 298			
0.3021	0.096		
0.3404	0.095		
0.4358	0.096		
0.5085	0.101		
0.5461	0.107		

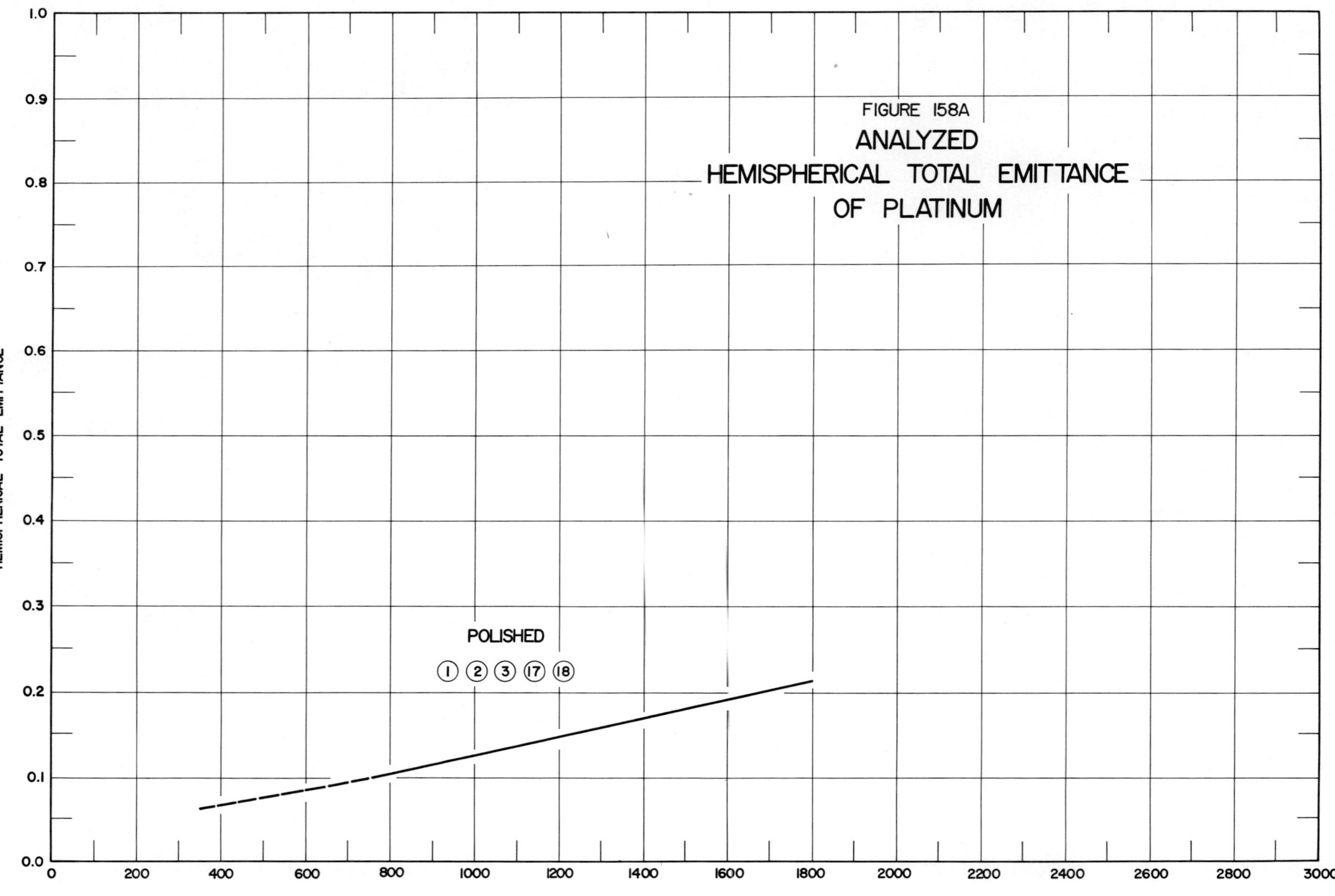

FIGURE 158A ANALYZED HEMISPHERICAL TOTAL EMITTANCE OF PLATINUM

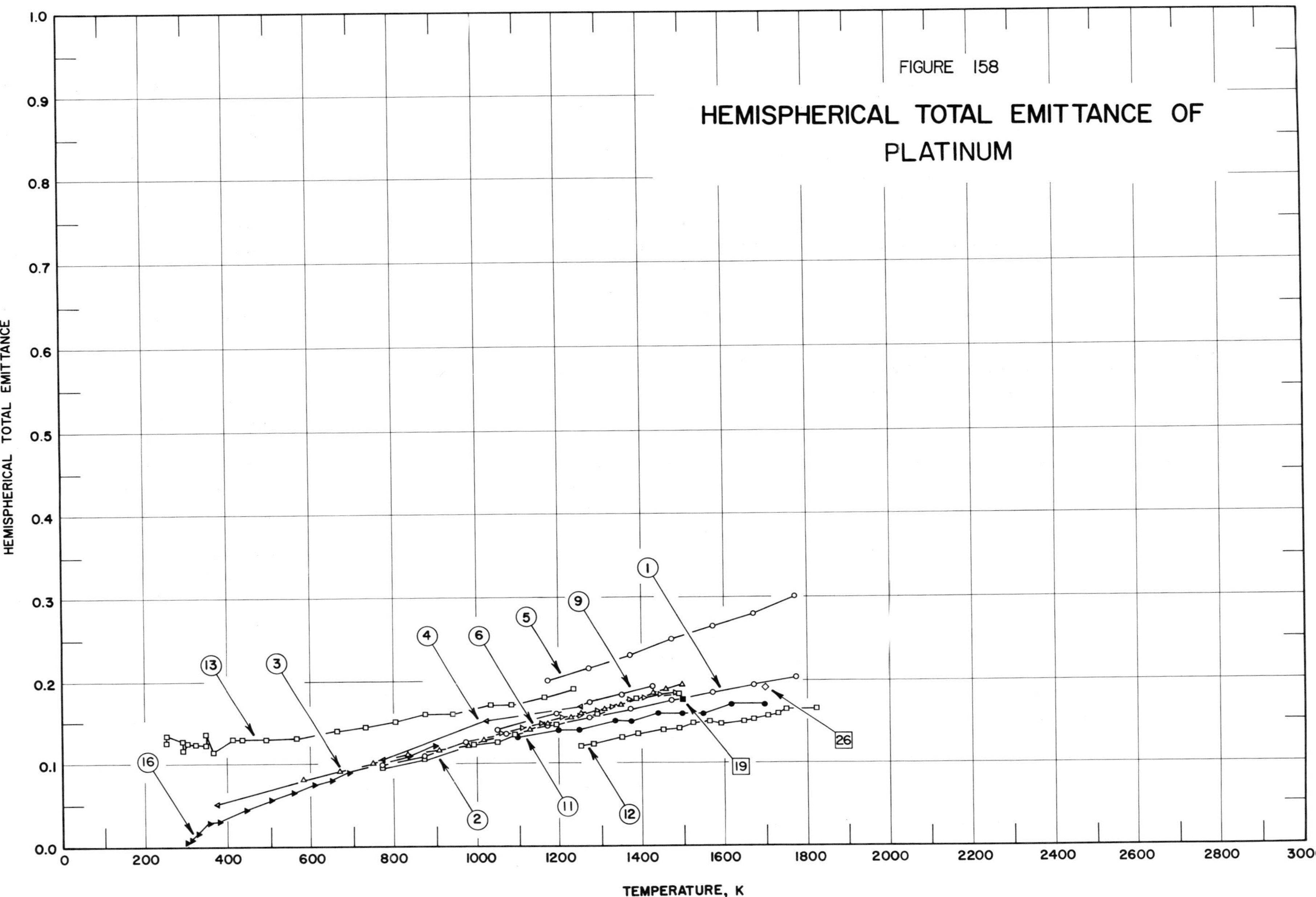
FIGURE 158
HEMISPHERICAL TOTAL EMITTANCE OF PLATINUM
HEMISPHERICAL TOTAL EMITTANCE
TEMPERATURE, K
0.0
0.1
0.2
0.3
0.4
0.5
0.6
0.7
0.8
0.9
1.0
0
200
400
600
800
1000
1200
1400
1600
1800
2000
2200
2400
2600
2800
3000
1
2
3
4
5
6
9
11
12
13
16
19
26

SPECIFICATION TABLE NO. 158 HEMISPHERICAL TOTAL EMITTANCE OF PLATINUM

Curve No.	Ref. No.	Year	Temperature Range, K	Reported Error, %	Composition (weight percent), Specifications and Remarks
1	54	1962	773-1773	± 4	Clean polished surface; measured in vacuum (10^{-4} to 10^{-6} mm Hg).
2	54	1962	773-1493	± 4	Clean rolled surface; measured in vacuum (10^{-4} to 10^{-6} mm Hg).
3	56	1921	580-1500		Measured in vacuum.
4	57	1963	373-1253	2	Wire; special grade MPTU 4292-53 (0.1977 mm dia); measured in vacuum (10^{-5} to 10^{-6} mm Hg).
5	58	1961	1173-1773	± 2.5	Flashed in vacuum at 2023 K; measured in vacuum (< 5 x 10^{-6} mm Hg); data extracted from smooth curve.
6	59	1961	1015-1480	± 5	99.95 pure with traces of gold, palladium, rhodium, and silver; strip (0.003 in. thick); as received (cold rolled); measured in vacuum (10^{-4} mm Hg); cycle 2, increasing temperature.
7	59	1961	1016-1476	± 5	Above specimen and conditions; cycle 3, increasing temperature.
8	59	1961	1014-1467	± 5	Above specimen and conditions; cycle 4, increasing temperature.
9	59	1961	1058-1429	± 5	Above specimen and conditions; cycle 4, decreasing temperature.
10	59	1961	774-1493	± 5	99.95 pure with traces of gold, palladium, rhodium, and silver; strip (0.003 in. thick); as received (cold rolled); measured in vacuum (10^{-4} mm Hg).
11	62	1954	1100-1700		Spectroscopically pure.
12	61	1915	1258-1823	< 8	Thin strip.
13	27	1960	258-1238		Disc; polished on a cloth lap saturated with water and alumina; measured in vacuum (5 x 10^{-5} mm Hg).
14	56	1921	574-1500		Measured in vacuum.
15	60	1924	301.9-1509		Measured in vacuum (10^{-6} mm Hg); [Author's designation: Filament 1].
16	60	1924	301.9-901.8		Measured in vacuum (10^{-6} mm Hg); [Author's designation: Filament 2].
17	67	1962	529-1173	± 2.7	99.5 Pt, 0.2 Ir, 0.05 Pd, 0.03 Fe, 0.02 Cu; rough hand polished with 4-p metalographic paper, polished with A and B alumina, and finally polished with diamond paste; measured in vacuum (5 x 10^{-6} mm Hg).
18	67	1962	363-1273	± 2.7	Different sample, same as curve 17 specimen and conditions.
19	252	1964	1500		Polished; measured in vacuum (10^{-5} mm Hg).
20	252	1964	1500		Above specimen and conditions except heated for 26 min at 1500 K.
21	252	1964	1500		Above specimen and conditions except heated additional 13 min at 1500 K.
22	252	1964	1500		Above specimen and conditions except heated additional 16 min at 1500 K.
23	252	1964	1500		Above specimen and conditions except heated additional 43 min at 1500 K.
24	252	1964	1500		Above specimen and conditions except heated additional 205 min at 1500 K.
25	252	1964	1500		Above specimen and conditions except heated additional 184 min at 1500 K.
26	252	1964	1700		Above specimen and conditions except heated additional 5 min at 1700 K.
27	252	1964	1700		Above specimen and conditions except heated additional 173 min at 1700 K.
28	252	1964	1500		Above specimen and conditions except heated additional 3 min at 1500 K.

SPECIFICATION TABLE NO. 158 (continued)

Curve No.	Ref. No.	Year	Temperature Range, K	Reported Error, %	Composition (weight percent), Specifications and Remarks
29	252	1964	1500		Above specimen and conditions except heated additional 56 min at 1500 K.
30	252	1964	1682		Above specimen and conditions except heated additional 1 min at 1682 K.
31	252	1964	1682		Above specimen and conditions except heated additional 282 min at 1682 K.

DATA TABLE NO. 158 HEMISPHERICAL TOTAL EMITTANCE OF PLATINUM

[Temperature, T, K; Emittance, ∈]

CURVE 1

T	∈
773	0.098
873	0.109
973	0.124
1073	0.135
1173	0.144
1273	0.154
1373	0.165
1473	0.176
1573	0.185
1673	0.194
1773	0.202

CURVE 2

T	∈
773	0.095
873	0.105
993	0.121
1053	0.125
1093	0.133
1098	0.131*
1198	0.146
1293	0.159
1388	0.168
1493	0.184

CURVE 3

T	∈
580	0.080
670	0.090
750	0.100
835	0.110
910	0.115
980	0.120
1020	0.127
1070	0.135*
1130	0.140
1173	0.147
1230	0.155
1265	0.160
1310	0.165
1350	0.170
1385	0.177*
1430	0.185
1460	0.190
1500	0.195

CURVE 4

T	∈
373	0.050
773	0.104
1123	0.150
1253	0.168

CURVE 5

T	∈
1173	0.200
1273	0.215
1373	0.230
1473	0.250
1573	0.265
1673	0.280
1773	0.300

CURVE 6

T	∈
1015	0.1295*
1062	0.1365
1111	0.1428
1159	0.1481
1206	0.1557
1250	0.1586
1291	0.1634
1330	0.1688
1369	0.1790
1406	0.1799
1445	0.1835
1480	0.1865

CURVE 7*

T	∈
1016	0.1294
1064	0.1358
1161	0.1474
1248	0.1589
1330	0.1688
1406	0.1793
1476	0.1868

CURVE 8*

T	∈
1014	0.1300
1061	0.1375
1112	0.1465
1158	0.1498
1203	0.1588
1247	0.1610
1287	0.1676
1325	0.1730
1363	0.1790
1399	0.1830
1435	0.1904
1467	0.1933

CURVE 9

T	∈
1058	0.1400
1198	0.1600
1279	0.1725
1356	0.1830
1429	0.1933

CURVE 10*

T	∈
774	0.095
873	0.105
873	0.108
991	0.120
1054	0.125
1091	0.133
1095	0.131
1193	0.146
1293	0.159
1389	0.168
1493	0.184

CURVE 11

T	∈
1100	0.13
1200	0.14
1250	0.14
1335	0.15
1375	0.15
1440	0.16
1500	0.16
1550	0.16
1620	0.17
1700	0.17

CURVE 12

T	∈
1258	0.120
1283	0.122
1298	0.125*
1353	0.130
1398	0.135
1453	0.140
1493	0.142
1523	0.147
1563	0.150
1598	0.147
1648	0.150
1673	0.152
1703	0.157
1733	0.160
1753	0.165
1823	0.165

CURVE 13

T	∈
258	0.124
258	0.134
283	0.126
283	0.116
303	0.123
323	0.122
348	0.121
348	0.126*
348	0.136
368	0.133
413	0.129
438	0.129
493	0.129
563	0.130
663	0.140
733	0.144
803	0.150
878	0.160
943	0.160
1033	0.170
1083	0.170
1163	0.180
1238	0.190

CURVE 14*

T	∈
574	0.080
674	0.091
748	0.098
836	0.109
909	0.117
978	0.121
1021	0.129
1071	0.136
1125	0.142
1173	0.147
1223	0.154
1263	0.160
1308	0.166
1347	0.172
1382	0.177
1427	0.184
1454	0.189
1500	0.192

CURVE 15*

T	∈
301.9	0.00216
313.9	0.00795
331.8	0.01578
355.7	0.0231
380.8	0.0311
446.1	0.0454
505.6	0.0568
559.1	0.0660
607.1	0.0743
652.0	0.0814
693.2	0.0881
769.1	0.1002
838.2	0.111
901.8	0.120
961.4	0.128
1017.8	0.136
1071	0.143
1080	0.144
1121	0.150
1132	0.151
1181	0.157
1274	0.168
1360	0.178
1443	0.186
1509	0.193

CURVE 16

T	∈
301.9	0.00205
313.9	0.00741
331.8	0.01465
355.7	0.0207
380.8	0.0288
446.1	0.0424
505.6	0.0536
559.1	0.0630
607.1	0.0715
652.0	0.0789
693.2	0.0859
769.1	0.0985*
838.2	0.1094
901.8	0.119

CURVE 17*

T	∈
529	0.082
533	0.088
773	0.099
798	0.100
1028	0.138
1173	0.157

CURVE 18*

T	∈
363	0.080
533	0.082
768	0.100
933	0.127
1098	0.152
1273	0.177

CURVE 19

T	∈
1500	0.1795

CURVE 20*

T	∈
1500	0.1788

CURVE 21*

T	∈
1500	0.1791

CURVE 22*

T	∈
1500	0.1791

CURVE 23*

T	∈
1500	0.1778

CURVE 24*

T	∈
1500	0.1778

CURVE 25*

T	∈
1500	0.1777

CURVE 26

T	∈
1700	0.1957

CURVE 27*

T	∈
1700	0.1979

CURVE 28*

T	∈
1500	0.1813

CURVE 29*

T	∈
1500	0.1819

CURVE 30*

T	∈
1682	0.1967

CURVE 31*

T	∈
1682	0.1998

*Not shown on plot

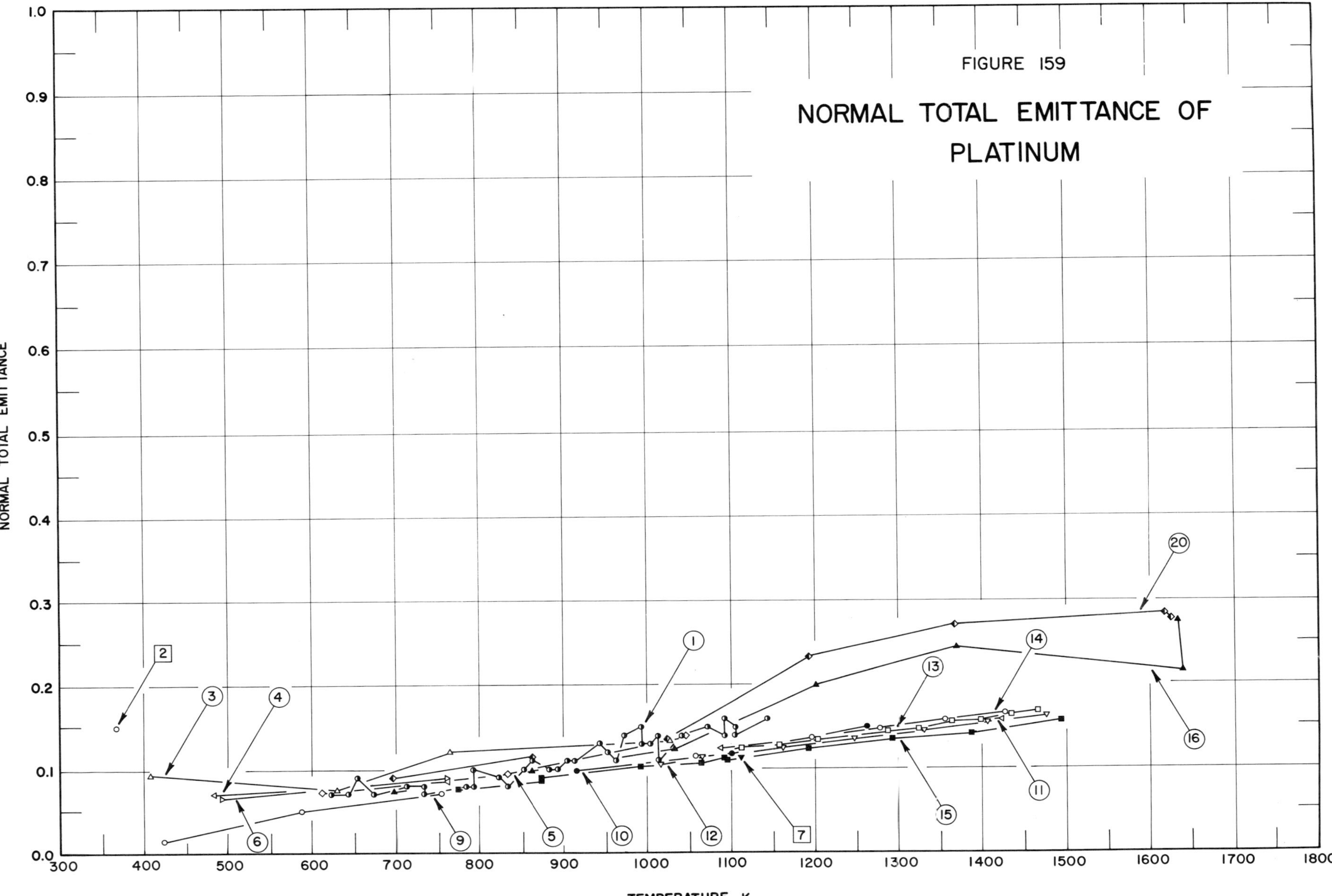
FIGURE 159
NORMAL TOTAL EMITTANCE OF
PLATINUM
NORMAL TOTAL EMITTANCE
1.0
0.9
0.8
0.7
0.6
0.5
0.4
0.3
0.2
0.1
0.0
TEMPERATURE, K
300
400
500
600
700
800
900
1000
1100
1200
1300
1400
1500
1600
1700
1800
1
2
3
4
5
6
7
9
10
11
12
13
14
15
16
20

SPECIFICATION TABLE NO. 159 NORMAL TOTAL EMITTANCE OF PLATINUM

Curve No.	Ref. No.	Year	Temperature Range, K	Geometry θ'	Reported Error, %	Composition (weight percent), Specifications and Remarks
1	68	1962	623-1143	$\sim 0^\circ$		Strip; polished; measured in air.
2	40	1962	367	$\sim 0^\circ$		Platinum clad Carpenter No. 20 stainless steel.
3	40	1962	408-1028	$\sim 0^\circ$		Highly polished; increasing temp, cycle 1.
4	40	1962	1028-483	$\sim 0^\circ$		Above specimen and conditions; decreasing temp, cycle 1.
5	40	1962	611-1047	$\sim 0^\circ$		Above specimen and conditions; increasing temp, cycle 2.
6	40	1962	1047-491	$\sim 0^\circ$		Above specimen and conditions; decreasing temp, cycle 2.
7	30	1963	1111	$\sim 0^\circ$		Heated at 672 K in air for 3 hrs, at 1089 K in vacuum for 3 hrs, and at T > 1217 K for 1 hr; polished with aluminum oxide and cleaned with water; computed from NSE and corrected to 1111 K.
8	32	1959	77. 6	$\sim 0^\circ$	13	Polished on a buffing wheel, cleaned with detergent, rinsed with distilled water; measured in vacuum (10^{-5} mm Hg).
9	34	1957	422-755	$\sim 0^\circ$	± 10	Pure; strip (0.005 in. thick); same results obtained for 4 different surface treatments (a. as received, b. cleaned with liquid detergent, c. polished, d. oxidized in air at red heat for 30 min); measured in air (5×10^{-4} mm Hg); increasing temp, cycle 1.
10	34	1957	916-1261	$\sim 0^\circ$	± 10	Above specimen and conditions; cycle 2.
11	34	1957	1089-1422	$\sim 0^\circ$	± 10	Above specimen and conditions; cycle 3.
12	59	1961	1016-1476	$\sim 0^\circ$	± 5	99.95 pure with traces of gold, palladium, rhodium, and silver; strip (0.003 in. thick); as received (cold rolled); measured in vacuum (10^{-4} mm Hg); cycle 3, increasing temp.
13	59	1961	1014-1467	$\sim 0^\circ$	± 5	Above specimen and conditions; cycle 4, increasing temp.
14	59	1961	1467-1058	$\sim 0^\circ$	± 5	Above specimen and conditions; cycle 4, decreasing temp.
15	59	1961	774-1493	$\sim 0^\circ$	± 5	99.95 pure; traces of gold, palladium, rhodium, and silver; strip (0.003 in. thick); as received (cold rolled); measured in vacuum (10^{-4} mm Hg).
16	227	1964	697-1639	$\sim 0^\circ$		>99.9 Pt; as received; preheated in vacuum at 1300 K for 1 hr; surface roughness after emittance test 4.7 microinches (rms); measured in vacuum (7×10^{-5} mm Hg); $\omega' = 3.4 \times 10^{-4}$ sr; [Author's designation: Sample 1].
17	227	1964	699-1642	$\sim 0^\circ$		Above specimen and conditions except second cycle.
18	227	1964	703-1625	$\sim 0^\circ$		>99.9 Pt; Glas-Shot blasted; preheated in vacuum at 1300 K for 1 hr; surface roughness before emittance test 169 microinches (rms), after emittance test 79 microinches (rms); measured in vacuum ($< 6 \times 10^{-5}$ mm Hg); $\omega' = 3.4 \times 10^{-4}$ sr; [Author's designation: Sample 3].
19	227	1964	1625-986	$\sim 0^\circ$		Above specimen and conditions; cooling cycle.
20	227	1964	697-1625	$\sim 0^\circ$		>99.9 Pt; Glas-Shot blasted; preheated in vacuum at 1300 K for 1 hr; surface roughness before emittance test 91 microinches (rms), after emittance test 38 microinches (rms); measured in vacuum ($< 6 \times 10^{-5}$ mm Hg); $\omega' = 3.4 \times 10^{-4}$ sr; [Author's designation: Sample 6].
21	227	1964	1625-1193	$\sim 0^\circ$		Above specimen and conditions; cooling cycle.

DATA TABLE NO. 159 NORMAL TOTAL EMITTANCE OF PLATINUM

[Temperature, T, K; Emittance, ε]

T	ε
CURVE 1	
623	0.07
643	0.07
653	0.09
673	0.07
713	0.08
733	0.08
733	0.07
773	0.08*
783	0.08
793	0.10
793	0.08
823	0.09
833	0.08
853	0.10
863	0.11
883	0.10
893	0.10
903	0.11
913	0.11
943	0.13
953	0.12
963	0.11
973	0.14
993	0.15
993	0.13
1003	0.13
1013	0.14
1013	0.11
1043	0.14
1073	0.15
1093	0.14
1093	0.16
1103	0.15
1103	0.14
1143	0.16
CURVE 2	
367	0.15
CURVE 3	
408	0.093
630	0.075
765	0.120
1028	0.133
CURVE 4	
1028	0.133*
761	0.085
483	0.070
CURVE 5	
611	0.072
833	0.095
1047	0.140
CURVE 6	
1047	0.140*
761	0.090
491	0.065
CURVE 7	
1111	0.112
CURVE 8*	
76.6	0.03
CURVE 9	
422	0.015
589	0.050
755	0.070
CURVE 10	
916	0.099
1100	0.119
1261	0.150
CURVE 11	
1089	0.124
1422	0.157
CURVE 12	
1016	0.1056
1064	0.1136
1161	0.1249
CURVE 12 (cont.)	
1248	0.1368
1330	0.1456
1406	0.1537
1476	0.1628
CURVE 13	
1014	0.1058*
1061	0.1147*
1112	0.1237
1158	0.1283
1203	0.1336
1247	0.1385*
1287	0.1437
1325	0.1487
1363	0.1540
1399	0.1584
1435	0.1633
1467	0.1686
CURVE 14	
1467	0.1686*
1429	0.1665
1356	0.1588
1279	0.1494
1198	0.1381
1058	0.1161
CURVE 15	
774	0.076
873	0.086
873	0.090
991	0.101
1054	0.106
1091	0.112
1095	0.110
1193	0.123
1293	0.136
1389	0.142
1493	0.157
CURVE 16	
697	0.074
863	0.098
1033	0.122
1202	0.200
1369	0.242
1639	0.212
1633	0.274
CURVE 17*	
699	0.077
871	0.102
1032	0.125
1200	0.213
1362	0.242
1497	0.269
1642	0.258
CURVE 18*	
703	0.105
867	0.122
1034	0.155
1202	0.239
1370	0.279
1618	0.296
1625	0.296
CURVE 19*	
1625	0.296
1370	0.267
1209	0.232
986	0.129
CURVE 20	
697	0.090
863	0.114
1025	0.135
1194	0.232
1369	0.271
1618	0.283
1625	0.278
CURVE 21*	
1625	0.278
1373	0.253
1193	0.216

*Not shown on plot

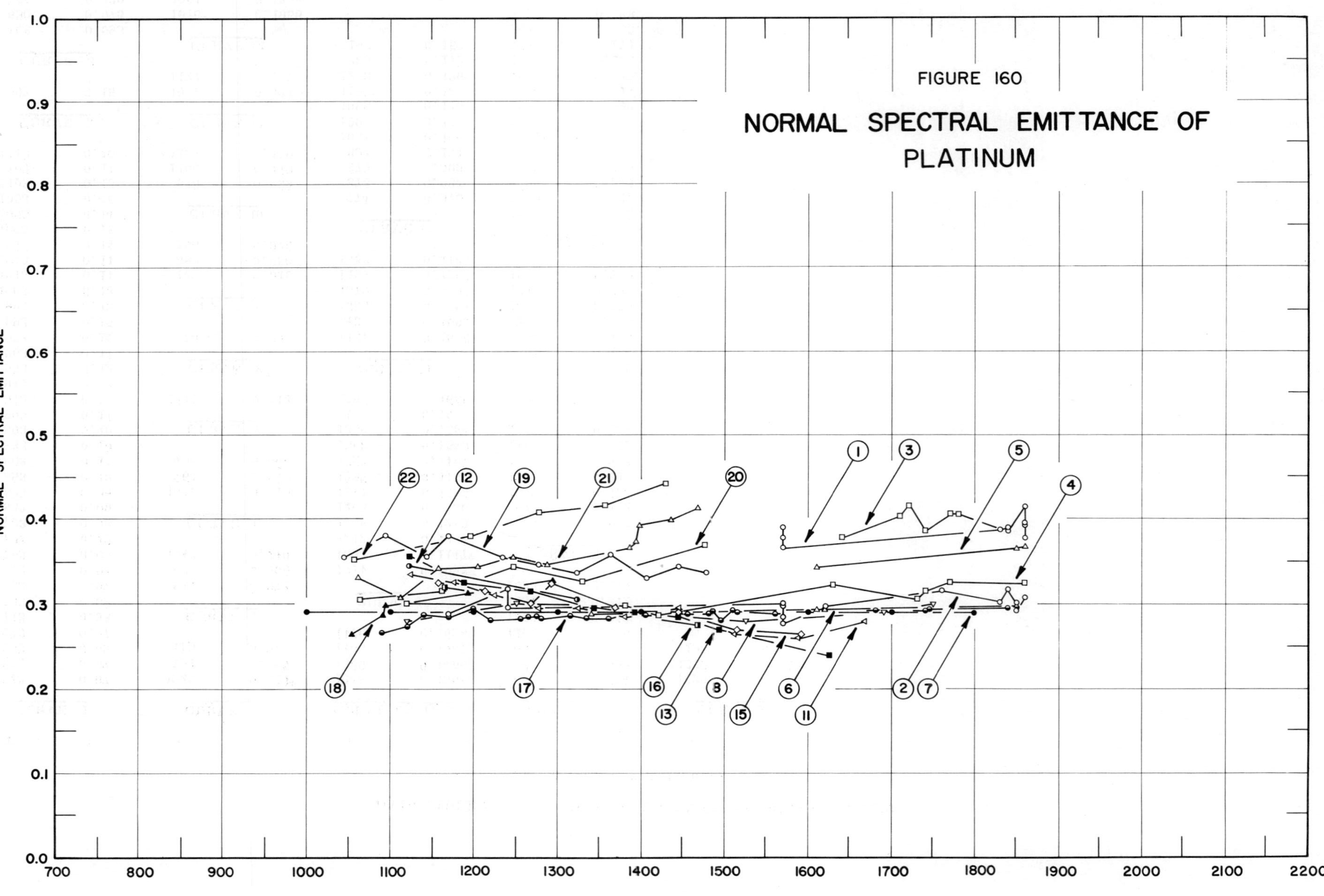
FIGURE 160
NORMAL SPECTRAL EMITTANCE OF PLATINUM
NORMAL SPECTRAL EMITTANCE
TEMPERATURE, K
1.0
0.9
0.8
0.7
0.6
0.5
0.4
0.3
0.2
0.1
0.0
700
800
900
1000
1100
1200
1300
1400
1500
1600
1700
1800
1900
2000
2100
2200
22
12
19
21
20
1
3
5
4
18
17
16
13
8
15
6
11
2
7

SPECIFICATION TABLE NO. 160 NORMAL SPECTRAL EMITTANCE OF PLATINUM

Curve No.	Ref. No.	Year	Wavelength μ	Temperature Range, K	Geometry θ'	Reported Error, %	Composition (weight percent), Specifications and Remarks
1	51	1926	0.463	1570-1860	$\sim 0^\circ$		Measured in argon.
2	51	1926	0.665	1170-1860	$\sim 0^\circ$		Above specimen and conditions.
3	51	1926	0.463	1640-1860	$\sim 0^\circ$		Measured in argon.
4	51	1926	0.665	1120-1860	$\sim 0^\circ$		Above specimen and conditions.
5	51	1926	0.463	1610-1860	$\sim 0^\circ$		Measured in argon.
6	51	1926	0.665	1340-1850	$\sim 0^\circ$		Above specimen and conditions.
7	62	1954	0.655	1000-1800	$\sim 0^\circ$	$< \pm 7$	Spectroscopically pure; wire.
8	62	1954	0.665	1120-1750	$\sim 0^\circ$	$< \pm 7$	Foil.
9	19	1914	0.55	< M. P.	$\sim 0^\circ$	1	Measured in hydrogen.
10	19	1914	0.650	< M. P.	$\sim 0^\circ$	1	Measured in hydrogen.
11	34	1957	0.665	1125-1669	$\sim 0^\circ$		Pure metal; same results obtained for 3 different surface treatments (a. as received, b. cleaned with liquid detergent, c. polished); measured in air (5×10^{-4} mm Hg); increasing temp, cycle 1.
12	34	1957	0.665	1669-1122	$\sim 0^\circ$		Above specimen and conditions; decreasing temp, cycle 1.
13	34	1957	0.665	1122-1625	$\sim 0^\circ$		Above specimen and conditions; cycle 2.
14	34	1957	0.665	1625-1136	$\sim 0^\circ$		Above specimen and conditions; decreasing temp, cycle 2.
15	34	1957	0.665	1158-1591	$\sim 0^\circ$		Above specimen and conditions; cycle 3.
16	34	1957	0.665	1591-1166	$\sim 0^\circ$		Above specimen and conditions; decreasing temp, cycle 3.
17	64	1938	0.66	1090-1840	10°		Polished with rouge.
18	59	1961	0.65	1054-1293	$\sim 0^\circ$	± 5	99.95 pure with traces of gold, palladium, rhodium, and silver; strip (0.003 in. thick); as received (cold rolled); measured in vacuum (10^{-4} mm Hg).
19	59	1961	0.650	1045-1479	$\sim 0^\circ$	± 5	99.95 pure with traces of gold palladium, rhodium, and silver; strip (0.003 in. thick); as received (cold rolled); cycle 1, increasing temp.
20	59	1961	0.650	1064-1476	$\sim 0^\circ$	± 5	Above specimen and conditions; cycle 3, increasing temp.
21	59	1961	0.65	1061-1467	$\sim 0^\circ$	± 5	Above specimen and conditions; cycle 4, increasing temp.
22	59	1961	0.65	1469-1058	$\sim 0^\circ$	± 5	Above specimen and conditions; cycle 4, decreasing temp.

DATA TABLE NO. 160 NORMAL SPECTRAL EMITTANCE OF PLATINUM

[Temperature, T, K; Emittance, ∈; Wavelength, λ, μ]

T	∈
CURVE 1 λ = 0.463	
1570	0.390
1570	0.379
1570	0.366
1830	0.389
1840	0.390
1840	0.388
1860	0.414
1860	0.395
1860	0.393
1860	0.378
CURVE 2 λ = 0.665	
1170	0.289
1240	0.318
1240	0.296
1570	0.300
1570	0.298
1570	0.284
1570	0.278
1620	0.298
1760	0.316
1830	0.301
1840	0.317
1850	0.301
1850	0.291
1860	0.308
CURVE 3 λ = 0.463	
1640	0.379
1710	0.402
1720	0.416
1740	0.386
1770	0.405
1780	0.405
1830	0.389*
1840	0.390*
1840	0.388*
1860	0.414*
1860	0.395*
1860	0.392*
1860	0.378*
CURVE 4 λ = 0.665	
1120	0.300
1380	0.299
1420	0.287
1630	0.321
1730	0.306
1740	0.315
1770	0.326
1770	0.325*
1860	0.325
CURVE 5 λ = 0.463	
1610	0.343
1850	0.366
1860	0.367
CURVE 6 λ = 0.665	
1340	0.288
1610	0.294
1620	0.294
1850	0.299
CURVE 7 λ = 0.655	
1000	0.29
1100	0.29
1200	0.29
1300	0.29
1400	0.29
1600	0.29
1700	0.29
1800	0.29
CURVE 8 λ = 0.665	
1120	0.28
1250	0.30
1400	0.29*
1525	0.28
1600	0.29*
CURVE 8 (cont.)	
1690	0.29
1750	0.30
CURVE 9* λ = 0.55	
<M. P.	0.38
CURVE 10* λ = 0.650	
<M. P.	0.33
CURVE 11 λ = 0.665	
1125	0.335
1177	0.325
1225	0.310
1277	0.295
1325	0.295
1380	0.285
1444	0.295
1511	0.265
1589	0.260
1669	0.280
CURVE 12 λ = 0.665	
1669	0.280*
1322	0.305
1122	0.345
CURVE 13 λ = 0.665	
1122	0.355
1189	0.325
1269	0.315
1344	0.295
1391	0.290
1444	0.285
1491	0.270
1625	0.240
CURVE 14* λ = 0.665	
1625	0.240*
1491	0.280
1269	0.305
1136	0.320
CURVE 15 λ = 0.665	
1158	0.325
1214	0.315
1269	0.30
1291	0.325
1369	0.295
1444	0.29
1514	0.27
1591	0.265
CURVE 16 λ = 0.665	
1591	0.265*
1469	0.275
1166	0.320
CURVE 17 λ = 0.66	
1090	0.267
1120	0.275
1140	0.287
1170	0.285
1200	0.292
1220	0.282
1255	0.283
1265	0.286
1275	0.285
1280	0.283
1300	0.290*
1315	0.287
1335	0.283
1360	0.283
1390	0.285*
1405	0.287
1420	0.288*
1455	0.288
CURVE 17 (cont.)	
1480	0.289
1485	0.291
1495	0.281
1510	0.292
1515	0.290
1515	0.289*
1560	0.289
1680	0.293
1695	0.290*
1740	0.291
1745	0.294
1840	0.296
CURVE 18 λ = 0.65	
1054	0.265
1091	0.287
1095	0.298
1193	0.312
1293	0.329
CURVE 19 λ = 0.650	
1045	0.3546
1095	0.3804
1145	0.3546
1170	0.3804
1234	0.3654
1278	0.3477
1321	0.3374
1362	0.3582
1406	0.3307
1443	0.3443
1479	0.3374
CURVE 20 λ = 0.650	
1064	0.3054
1161	0.3177
1248	0.3443
1330	0.3274
1476	0.3692
CURVE 21 λ = 0.65	
1061	0.3307
1112	0.3086
1158	0.3410
1203	0.3443
1247	0.3547
1287	0.3477
1385	0.3654
1393	0.3729
1399	0.3921
1435	0.3999
1467	0.4121
CURVE 22 λ = 0.65	
1469	0.4121*
1429	0.4421
1356	0.4162
1279	0.4079
1198	0.3804
1058	0.3512

*Not shown on plot

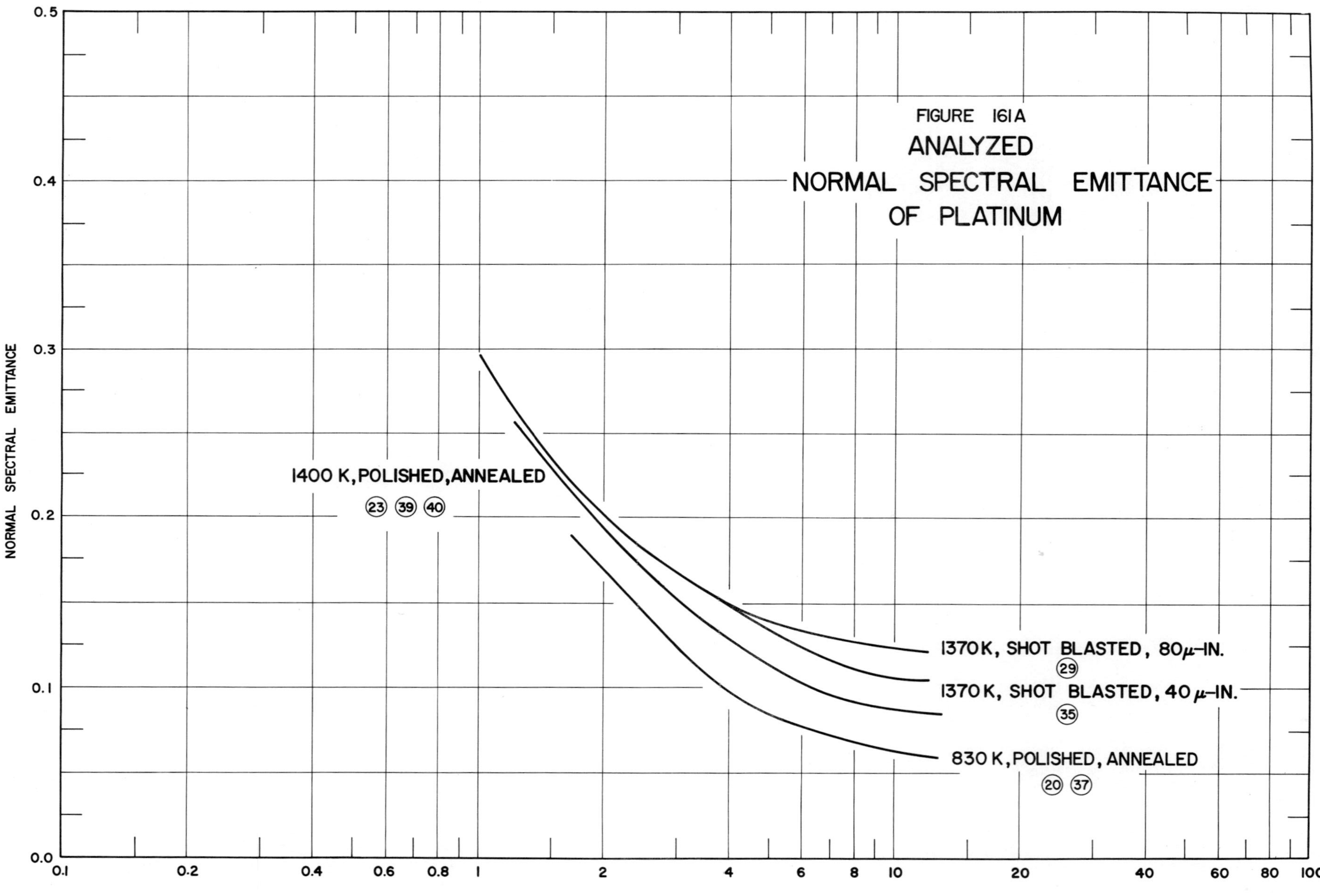

FIGURE 161A
ANALYZED
NORMAL SPECTRAL EMITTANCE
OF PLATINUM
1400 K, POLISHED, ANNEALED
(23) (39) (40)
1370 K, SHOT BLASTED, 80 μ-IN.
(29)
1370 K, SHOT BLASTED, 40 μ-IN.
(35)
830 K, POLISHED, ANNEALED
(20) (37)
NORMAL SPECTRAL EMITTANCE
0.0
0.1
0.2
0.3
0.4
0.5
WAVELENGTH, MICRONS
0.1
0.2
0.4
0.6
0.8
1
2
4
6
8
10
20
40
60
80
100

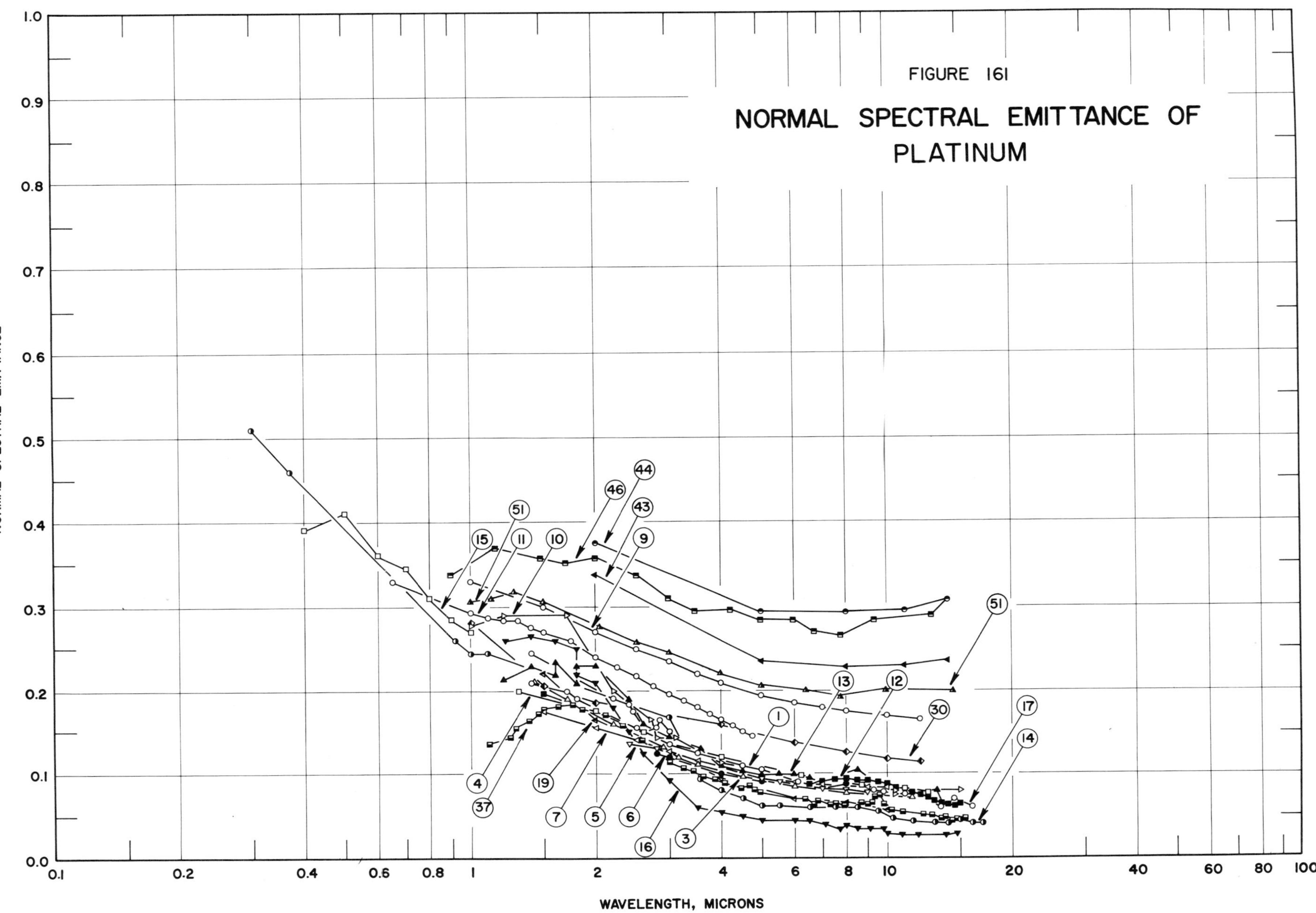

FIGURE 161
NORMAL SPECTRAL EMITTANCE OF PLATINUM

SPECIFICATION TABLE NO. 161 NORMAL SPECTRAL EMITTANCE OF PLATINUM

Curve No.	Ref. No.	Year	Temperature K	Wavelength Range, μ	Geometry θ'	Reported Error, %	Composition (weight percent), Specifications and Remarks
1	30	1963	1088	1.4–6.1	~0°		NBS platinum; heated in air for 1 hr at 1523 K, cooled for 12 hrs, and heated at T > 1088 K for 1 hr; measured in argon; data extracted from smooth curve.
2	30	1963	1108	1.5–5.6	~0°		Above specimen and conditions except heated for 23 additional hrs at T > 1088 K.
3	30	1963	1088	1.42–11.50	~0°		Above specimen and conditions except heated at T > 1083 K for 11 additional hrs; measured after 6 mos.
4	30	1963	1328	1.30–12.50	~0°		Above specimen and conditions except heated at T > 1277 K for 1 additional hr.
5	30	1963	1216	2.4–9.0	~0°		Mechanically polished with aluminum oxide and cleaned with water (surface roughness 0.025 μ peak to peak and 10 μ lateral); heated at 672 K for 3 hrs, at 1088 K for 6 hrs, and at T > 1216 K for 1 hr; measured in vacuum (5×10^{-6} mm Hg); data extracted from smooth curve; [Author's designation: Sample 3].
6	30	1963	1180	2.8–8.8	~0°		Above specimen and conditions except heated at T > 1111 K for 48 additional hrs; measured in argon.
7	30	1963	1383	1.5–10.0	~0°		Above specimen and conditions except heated at T > 1222 K for 4 additional hrs; measured in argon.
8	30	1963	1383	1.5–10.0	~0°		Above specimen and conditions except heated for 30 additional hrs at 1460 K.
9	30	1963	1444	1.0–12.0	~0°		Above specimen and conditions except heated for 58 additional hrs at 1477 K; measured in vacuum (5×10^{-6} mm Hg).
10	63	1962	973	1.00–15.00	~0°	< 8.9	Foil; as received.
11	37	1947	1398	0.65–4.75	~0°	4	99.8 pure; polished and washed in alcohol; measured in vacuum.
12	66	1960	1400	1.5–15.0	~0°	5	Strip; annealed at 1525 K; surface became crystalline after annealing.
13	65	1962	1311	1.20–14.00	~0°		Measured in air; calculated from $\epsilon = 1 - \rho$.
14	65	1962	294	0.30–17.00	~0°		Measured in air; calculated from $\epsilon = 1 - \rho$.
15	65	1962	294	0.40–1.00	~0°		Measured in air; calculated from $\epsilon = 1 - \rho$.
16	65	1962	294	1.21–14.80	~0°		Measured in air; calculated from $\epsilon = 1 - \rho$.
17	65	1962	1289	1.40–15.00	~0°		Measured in air.
18	245	1965	1424	1.19–14.1	~0°	5	Measured in argon.
19	227	1964	697	1.5–10	~0°		> 99.9 Pt; as received; preheated in vacuum at 1300 K for 1 hr; surface roughness after emittance test 4.7 microinches (rms); measured in vacuum (7×10^{-5} mm Hg); $\omega' = 3.4 \times 10^{-4}$ sr; [Author's designation: Sample 1].
20	227	1964	863	1.5–10	~0°		Above specimen and conditions.
21	227	1964	1033	1.5–10	~0°		Above specimen and conditions.
22	227	1964	1202	1–10	~0°		Above specimen and conditions.
23	227	1964	1369	1–12	~0°		Above specimen and conditions.

SPECIFICATION TABLE NO. 161 (continued)

Curve No.	Ref. No.	Year	Temperature K	Wavelength Range, μ	Geometry θ'	Reported Error, %	Composition (weight percent), Specifications and Remarks
24	227	1964	1633	0.65-12	~0°		Above specimen and conditions.
25	227	1964	703	1.5-8	~0°		99.9 Pt; Glas-Shot blasted; preheated in vacuum at 1300 K for 1 hr; surface roughness before emittance test 169 microinches (rms), after emittance test 79 microinches (rms); measured in vacuum ($<6 \times 10^{-5}$ mm Hg); $\omega' = 3.4 \times 10^{-4}$ sr; [Author's designation: Sample 3].
26	227	1964	867	1.5-10	~0°		Above specimen and conditions.
27	227	1964	1034	1-12	~0°		Above specimen and conditions.
28	227	1964	1202	1-10	~0°		Above specimen and conditions.
29	227	1964	1370	1-12	~0°		Above specimen and conditions.
30	227	1964	1618	1-12	~0°		Above specimen and conditions.
31	227	1964	697	1.5-10	~0°		>99.9 Pt; Glas-Shot blasted; preheated in vacuum at 1300 K for 1 hr; surface roughness before emittance test 91 microinches (rms), after emittance test 38 microinches (rms); measured in vacuum ($<6 \times 10^{-5}$ mm Hg); $\omega' = 3.4 \times 10^{-4}$ sr; [Author's designation: Sample 6].
32	227	1964	863	1.5-10	~0°		Above specimen and conditions.
33	227	1964	1025	1.5-12	~0°		Above specimen and conditions.
34	227	1964	1194	1-12	~0°		Above specimen and conditions.
35	227	1964	1369	1-12	~0°		Above specimen and conditions.
36	227	1964	1618	1-10	~0°		Above specimen and conditions.
37	254	1963	800	1.10-15.13	~0°		Smooth surface; washed in hot tap water with detergent, rinsed in hot tap water, then in distilled water, and finally in ethyl alcohol; annealed in an electrically heated, silicone carbide furnace, then heated at 1523 K for 1 hr and cooled; mean of 3 determinations on each of 6 specimens.
38	254	1963	1100	1.10-15.26	~0°		Above specimen and conditions.
39	254	1963	1400	1.08-15.21	~0°		Above specimen and conditions.
40	172	1961	1400	2-15	~0°		Measured in nitrogen.
41	292	1965	1400	2-14	0°		Polished.
42	292	1965	1400	2-14	0°		Polished.
43	292	1965	1400	2-14	0°		Grit-blasted.
44	292	1965	1400	2-14	0°		Grit-blasted.
45	317	1962	1455	1.000-14.06	0°		Sanded to a 4/0 finish with emery paper and buffed to mirror finish with aluminum oxide; aged 50 hrs above 1366 K and 18 hrs at 1255 K; measured in vacuum; computed from reflectance.
46	317	1962	1472	0.8995-14.09	0°		Sanded to a 4/0 finish with emery paper and buffed to mirror finish with aluminum oxide; aged 46 hrs above 1366 K and 18 hrs at 1255 K; measured in vacuum; computed from reflectance.

SPECIFICATION TABLE NO. 161 (continued)

Curve No.	Ref. No.	Year	Temperature K	Wavelength Range, μ	Geometry θ'	Reported Error, %	Composition (weight percent), Specifications and Remarks
47	317	1962	1473	1.000-14.22	0°		Sanded to a 4/0 finish with emery paper and buffed to mirror finish with aluminum oxide; aged 22 hrs above 1366 K; measured in vacuum; computed from reflectance.
48	317	1962	1474	1.000-13.93	0°		Sanded to a 4/0 finish with emery paper and buffed to mirror finish with aluminum oxide; aged 16 hrs above 1366 K; measured in vacuum; computed from reflectance.
49	317	1962	1479	1.000-14.13	0°		Sanded to a 4/0 finish with emery paper and buffed to mirror finish with aluminum oxide; aged 10 hrs above 1366 K and 6 hrs at 1255 K; measured in vacuum; reflectance data.
50	317	1962	1255	1.000-14.06	0°		Sanded to a 4/0 finish with emery paper and buffed to mirror finish with aluminum oxide; aged 6 hrs at 1255 K; measured in argon; computed from reflectance.
51	317	1962	294	1.000-14.32	~0°		Sanded to a 4/0 finish with emery paper and buffed to mirror finish with aluminum oxide; annealed after 20 hrs at 1255 K and 80 hrs at 1422 K; computed from reflectance data.
52	317	1962	294	1.000-14.06	~0°		Sanded to a 4/0 finish with emery paper and buffed to mirror finish with aluminum oxide; annealed 3 hrs at 672 K and 3 hrs at 922 K; computed from reflectance data.
53	317	1962	294	1.000-14.06	~0°		Sanded to a 4/0 finish with emery paper and buffed to mirror finish with aluminum oxide; measured immediately after polishing; computed from reflectance data.
54	317	1962	294	1.000-14.42	~0°		Sanded to a 4/0 finish with emery paper and buffed to mirror finish with aluminum oxide; preceded by runs above 1366 K; aged 6-96 hrs; surface roughness 2.5 μ; computed from reflectance data.

DATA TABLE NO. 161 NORMAL SPECTRAL EMITTANCE OF PLATINUM

[Wavelength, λ, μ; Emittance, ∈; Temperature, T, K]

CURVE 1 (T = 1088)

λ	∈
1.4	0.210
1.7	0.200
1.8	0.190
2.5	0.155
3.0	0.135
3.5	0.125
4.5	0.110
5.0	0.105
6.1	0.090

CURVE 2* (T = 1108)

λ	∈
1.5	0.180
2.0	0.160
2.3	0.150
2.6	0.140
3.0	0.125
3.5	0.115
3.8	0.110
4.6	0.100
5.2	0.096
5.6	0.094

CURVE 3 (T = 1088)

λ	∈
1.42	0.212
1.70	0.190
2.20	0.160
2.90	0.132
3.15	0.120
3.50	0.112
4.50	0.097
6.00	0.085
8.00	0.077
11.50	0.071

CURVE 4 (T = 1328)

λ	∈
1.30	0.200
2.00	0.175
2.60	0.150
3.00	0.136*
4.00	0.120
6.20	0.098
8.70	0.086
11.00	0.081
12.50	0.077

CURVE 5 (T = 1216)

λ	∈
2.4	0.135
2.8	0.130
3.5	0.110*
5.0	0.094
5.5	0.090
6.0	0.086*
7.0	0.082
8.0	0.080
9.0	0.078

CURVE 6 (T = 1180)

λ	∈
2.8	0.125
3.0	0.120
4.0	0.100
5.0	0.090
6.0	0.084*
7.0	0.080*
8.0	0.087
8.8	0.086*

CURVE 7 (T = 1383)

λ	∈
1.5	0.175
2.0	0.155
2.5	0.140
3.0	0.125
3.5	0.115
5.0	0.096*
6.0	0.090
7.0	0.086
8.0	0.084*
9.0	0.083
10.0	0.083

CURVE 8* (T = 1383)

λ	∈
1.5	0.175
2.0	0.155
2.5	0.140
3.0	0.125
3.5	0.115
5.0	0.096
6.0	0.090
7.0	0.086
8.0	0.084
9.0	0.083
10.0	0.083

CURVE 9 (T = 1444)

λ	∈
1.0	0.330
1.5	0.300
2.0	0.270
2.5	0.250
3.0	0.235
3.5	0.220
4.0	0.210
5.0	0.195
6.0	0.185
7.0	0.180
8.0	0.175
10.0	0.170
12.0	0.165

CURVE 10 (T = 973)

λ	∈
1.00	0.277
1.20	0.290
1.70	0.290
2.20	0.200
2.70	0.165
2.80	0.142
3.10	0.142
4.00	0.115
5.00	0.102*
6.00	0.090*
8.00	0.085*
10.40	0.075
10.40	0.080
15.00	0.080

CURVE 11 (T = 1398)

λ	∈
0.65	0.330
1.0	0.293
1.1	0.287
1.2	0.284
1.3	0.284
1.4	0.276
1.5	0.270
1.75	0.255
2.0	0.240
2.25	0.228
2.50	0.218
2.75	0.206
3.0	0.196
3.25	0.188
3.5	0.180
3.75	0.172
4.0	0.165
4.25	0.157
4.5	0.150
4.75	0.145

CURVE 12 (T = 1400)

λ	∈
1.5	0.197
2.0	0.179*
2.5	0.154*
3.0	0.139*
3.5	0.128*
4.0	0.117*
4.5	0.108*
5.0	0.100*
5.5	0.094*
6.0	0.090*
6.5	0.088
7.0	0.090
7.5	0.092
8.0	0.093
8.5	0.091
9.0	0.091
9.5	0.090
10.0	0.087
10.5	0.083
11.0	0.080*
11.5	0.078
12.0	0.075
12.5	0.072
13.0	0.068
13.5	0.066
14.0	0.064
14.5	0.063
15.0	0.065

CURVE 13 (T = 1311)

λ	∈
1.20	0.215
1.40	0.230
1.60	0.220
1.60	0.235
1.80	0.210
1.80	0.230
2.00	0.230
2.20	0.200*
2.40	0.190
2.60	0.160
3.00	0.145
3.55	0.130
4.00	0.110
4.50	0.104
5.00	0.100
5.50	0.100
6.00	0.100
6.50	0.095
7.00	0.090*
8.50	0.104
9.00	0.090*
10.00	0.085*
11.50	0.070*
12.00	0.079*
12.50	0.075*
13.20	0.080
14.00	0.065*

CURVE 14 (T = 294)

λ	∈
0.30	0.510
0.37	0.460
0.92	0.260
1.00	0.245
1.10	0.245
3.00	0.118
3.55	0.092
4.00	0.080
4.50	0.070
5.00	0.061
5.50	0.061
6.50	0.060
7.50	0.060
8.50	0.060
9.50	0.055
10.20	0.048
11.50	0.044
13.00	0.042
14.00	0.041
15.00	0.045
16.00	0.041
17.00	0.041

CURVE 15 (T = 294)

λ	∈
0.40	0.390
0.50	0.410
0.60	0.360
0.70	0.345
0.80	0.310
0.90	0.285
1.00	0.270

CURVE 16 (T = 294)

λ	∈
1.21	0.260
1.40	0.265
1.60	0.260
1.80	0.250
1.80	0.220
2.00	0.210
2.20	0.180
2.40	0.150
2.60	0.124
3.00	0.092
3.50	0.060
4.00	0.055
4.50	0.050
5.00	0.045
6.00	0.045
6.50	0.044
7.10	0.040
7.70	0.035
8.00	0.039
8.50	0.035
9.10	0.035
9.80	0.035
10.00	0.030
10.90	0.029
11.90	0.029
13.00	0.029
14.80	0.030

CURVE 17 (T = 1289)

λ	∈
1.40	0.245
1.50	0.200*
1.70	0.200*
2.20	0.190
2.40	0.184
2.45	0.176
2.55	0.154*
2.80	0.155
2.85	0.164
3.00	0.150
3.50	0.120*
5.60	0.090*
6.00	0.090*
6.50	0.087*
7.75	0.092*
8.20	0.085*
9.30	0.080
9.80	0.079
10.40	0.075*
11.00	0.075
12.60	0.070*
13.40	0.060
14.60	0.070
15.00	0.060

CURVE 18* (T = 1424)

λ	∈
1.19	0.199
1.39	0.189
1.58	0.180
1.79	0.176
1.99	0.168
2.17	0.163
2.38	0.158
2.58	0.148
2.77	0.141
2.99	0.136
3.44	0.126
3.98	0.117
4.47	0.113
5.01	0.107
5.94	0.0993
7.08	0.0946
8.02	0.0898
9.10	0.0841
10.0	0.0832
11.0	0.0802
12.0	0.0838
14.1	0.0793

CURVE 19 (T = 697)

λ	∈
1.5	0.220
2	0.163
4	0.093
6	0.070
8	0.066
10	0.056

CURVE 20* (T = 863)

λ	∈
1.5	0.220
2	0.163
4	0.105
6	0.084
8	0.077
10	0.068

*Not shown on plot

DATA TABLE NO. 161 (continued)

λ	ϵ
CURVE 21*	
T = 1033	
1.5	0.225
2	0.178
4	0.112
6	0.095
8	0.083
10	0.076
CURVE 22*	
T = 1202	
1	0.296
1.5	0.230
2	0.187
4	0.125
6	0.098
8	0.081
10	0.079
CURVE 23*	
T = 1369	
1	0.281
1.5	0.225
2	0.193
4	0.141
6	0.111
8	0.093
10	0.084
12	0.090
CURVE 24*	
T = 1633	
0.65	0.29
1	0.245
1.5	0.167
2	0.156
4	0.141
6	0.128
8	0.112
10	0.104
12	0.109

λ	ϵ
CURVE 25*	
T = 703	
1.5	0.300
2	0.214
4	0.121
6	0.101
8	0.100
CURVE 26*	
T = 867	
1.5	0.269
2	0.203
4	0.133
6	0.118
8	0.107
10	0.097
CURVE 27*	
T = 1034	
1	0.284
1.5	0.269
2	0.196
4	0.137
6	0.117
8	0.115
10	0.113
12	0.100
CURVE 28*	
T = 1202	
1	0.300
1.5	0.248
2	0.204
4	0.150
6	0.136
8	0.124
10	0.121
CURVE 29*	
T = 1370	
1	0.283
1.5	0.224
2	0.189

λ	ϵ
CURVE 29 (cont.)*	
4	0.149
6	0.135
8	0.129
10	0.127
12	0.114
CURVE 30	
T = 1618	
1	0.280
1.5	0.206
2	0.186
4	0.159
6	0.138
8	0.126
10	0.118
12	0.115
CURVE 31*	
T = 697	
1.5	0.275
2	0.197
4	0.107
6	0.089
8	0.081
10	0.071
CURVE 32*	
T = 863	
1.5	0.264
2	0.190
4	0.118
6	0.099
8	0.089
10	0.084
CURVE 33*	
T = 1025	
1.5	0.250
2	0.198
4	0.132
6	0.107
8	0.094
10	0.093
12	0.084

λ	ϵ
CURVE 34*	
T = 1194	
1	0.334
1.5	0.252
2	0.201
4	0.145
6	0.115
8	0.107
10	0.091
12	0.091
CURVE 35*	
T = 1369	
1	0.288
1.5	0.233
2	0.197
4	0.149
6	0.129
8	0.111
10	0.106
12	0.105
CURVE 36*	
T = 1618	
1	0.201
1.5	0.175
2	0.149
4	0.133
6	0.117
8	0.108
10	0.107
CURVE 37	
T = 800	
1.10	0.136
1.22	0.142
1.27	0.153
1.38	0.162
1.46	0.170
1.51	0.177
1.62	0.181
1.75	0.184
1.86	0.178
2.11	0.170

λ	ϵ
CURVE 37 (cont.)	
2.32	0.156
2.56	0.140
3.01	0.116
3.23	0.107
3.41	0.102
3.60	0.095
3.83	0.092
4.06	0.089
4.44	0.082
4.62	0.084
4.79	0.080
4.93	0.078
6.02	0.070*
6.28	0.070
6.65	0.064
6.75	0.066
7.27	0.063
7.52	0.064
7.93	0.062
8.67	0.066
9.00	0.065
9.18	0.066
9.36	0.075
9.54	0.076
9.83	0.064
10.13	0.058
10.93	0.055
12.15	0.051
12.89	0.051
13.51	0.049
13.89	0.049
14.30	0.047
14.70	0.048
14.86	0.048*
15.13	0.048
CURVE 38*	
T = 1100	
1.10	0.187
1.16	0.201
1.21	0.209
1.42	0.219
1.60	0.215
1.79	0.205
2.39	0.169

λ	ϵ
CURVE 38 (cont.)*	
2.85	0.143
3.06	0.133
3.68	0.114
4.12	0.109
4.37	0.111
4.55	0.100
4.72	0.102
5.30	0.094
5.62	0.093
6.04	0.090
6.16	0.087
6.41	0.087
6.64	0.083
6.76	0.084
7.42	0.082
7.57	0.081
8.05	0.080
8.41	0.080
8.80	0.083
9.30	0.082
9.64	0.090
10.12	0.078
10.56	0.072
11.31	0.070
11.99	0.068
12.50	0.064
12.89	0.065
13.24	0.064
13.70	0.064
13.82	0.063
14.09	0.064
14.17	0.063
15.11	0.072
15.26	0.071
CURVE 39*	
T = 1400	
1.08	0.250
1.17	0.249
1.45	0.238
1.64	0.224
2.39	0.183
3.07	0.152
3.68	0.131
3.90	0.129

λ	ϵ
CURVE 39 (cont.)*	
4.32	0.123
4.49	0.117
4.65	0.119
4.98	0.115
5.70	0.106
6.23	0.101
6.46	0.101
6.72	0.097
6.89	0.098
7.03	0.095
7.65	0.095
8.15	0.092
8.90	0.096
9.25	0.094
9.74	0.102
10.54	0.090
10.91	0.088
11.41	0.086
12.19	0.084
12.92	0.084
13.50	0.087
14.00	0.088
14.20	0.091
14.67	0.096
14.89	0.100
15.08	0.107
15.21	0.106
CURVE 40*	
T = 1400	
2	0.179
3	0.139
4	0.117
5	0.100
6	0.090
7	0.090
8	0.093
9	0.091
10	0.087
11	0.080
12	0.075
13	0.068
14	0.064
15	0.065

λ	ϵ
CURVE 41*	
T = 1400	
2	0.175
5	0.098
8	0.080
11	0.070
14	0.065
CURVE 42*	
T = 1400	
2	0.177
5	0.100
8	0.080
11	0.070
14	0.065
CURVE 43	
T = 1400	
2	0.338
5	0.233
8	0.224
11	0.229
14	0.237
CURVE 44	
T = 1400	
2	0.373
5	0.296
8	0.294
11	0.297
14	0.308
CURVE 45*	
T = 1455	
1.000	0.2918
1.140	0.2918
1.390	0.2780
1.687	0.2553
2.094	0.2393
2.523	0.2183
3.428	0.1795
3.767	0.1754
4.487	0.1622
4.932	0.1611

*Not shown on plot

DATA TABLE NO. 161 (continued)

λ	∈
CURVE 45 (cont.)*	
5.741	0.1462
7.464	0.1365
9.795	0.1377
12.02	0.1396
14.06	0.1337
CURVE 46 T = 1472	
0.8995	0.3389
1.143	0.3690
1.496	0.3540
1.698	0.3516
2.000	0.3573
2.495	0.3365
2.999	0.3098
3.475	0.2938
4.236	0.2979
5.023	0.2825
5.970	0.2825
6.699	0.2717
7.780	0.2679
9.354	0.2845
12.82	0.2904
14.09	0.3091*
CURVE 47* T = 1473	
1.000	0.2193
1.148	0.2786
1.253	0.2924
1.521	0.2845
1.762	0.2754
2.489	0.2388
2.992	0.2218
4.027	0.1959
4.989	0.1888
5.929	0.1762
6.950	0.1734
7.943	0.1750
9.462	0.1726
10.96	0.1787
12.47	0.1875
14.22	0.1858

λ	∈
CURVE 48* T = 1474	
1.000	0.2582
1.222	0.2588
1.496	0.2421
1.799	0.2265
2.291	0.2085
2.992	0.1778
3.882	0.1556
4.560	0.1449
5.212	0.1406
5.902	0.1413
7.962	0.1312
9.376	0.1262
10.16	0.1277
11.97	0.1191
13.00	0.1222
13.93	0.1225
CURVE 49* T = 1479	
1.000	0.2163
1.297	0.2047
1.698	0.1888
2.148	0.1730
2.673	0.1507
3.491	0.1309
3.999	0.1236
4.966	0.1162
5.984	0.1069
7.031	0.1021
7.962	0.09529
9.162	0.09226
10.96	0.08852
12.19	0.08954
13.06	0.08690
14.13	0.08913
CURVE 50* T = 1255	
1.000	0.1702
1.151	0.1746
1.403	0.1706
1.714	0.1607

λ	∈
CURVE 50 (cont.)*	
2.109	0.1429
3.020	0.1135
4.018	0.09863
4.498	0.09529
5.495	0.08690
6.442	0.08395
7.603	0.07781
10.00	0.07499
12.11	0.07414
13.21	0.07113
14.06	0.06638
CURVE 51 T = 294	
1.000	0.3069
1.109	0.3098
1.256	0.3155
1.489	0.3083
1.991	0.2774
2.495	0.2571
2.992	0.2466
3.999	0.2213
4.989	0.2070
6.368	0.2014
7.727	0.1928
9.931	0.2014
14.32	0.2066
CURVE 52* T = 294	
1.000	0.1403
1.199	0.1726
1.496	0.1879
1.795	0.1807
2.118	0.1600
2.355	0.1300
2.559	0.1097
2.871	0.0908
3.162	0.0706
3.516	0.0610
3.990	0.0544
4.519	0.0498
5.309	0.0419

λ	∈
CURVE 52 (cont.)*	
5.808	0.0398
6.637	0.0381
7.709	0.0320
8.551	0.0324
10.47	0.0284
12.33	0.0297
14.06	0.0337
CURVE 53* T = 294	
1.000	0.1406
1.102	0.1618
1.253	0.1783
1.500	0.1871
1.722	0.1837
2.000	0.1694
2.208	0.1500
2.489	0.1167
2.799	0.09485
3.365	0.07228
3.999	0.05916
4.977	0.04776
5.728	0.04325
6.546	0.04266
7.161	0.04000
8.035	0.03540
9.661	0.03199
10.86	0.02864
12.97	0.03076
14.06	0.03365
CURVE 54* T = 294	
1.000	0.2735
1.094	0.2812
1.202	0.2845
1.403	0.2767
1.592	0.2667
1.799	0.2618
1.972	0.2489
2.506	0.2254
2.992	0.2042
4.018	0.1702

λ	∈
CURVE 54 (cont.)*	
4.977	0.1510
5.998	0.1409
6.950	0.1334
7.980	0.1285
8.995	0.1277
10.00	0.1245
11.02	0.1222
13.00	0.1216
14.03	0.1225
14.42	0.1277

* Not shown on plot

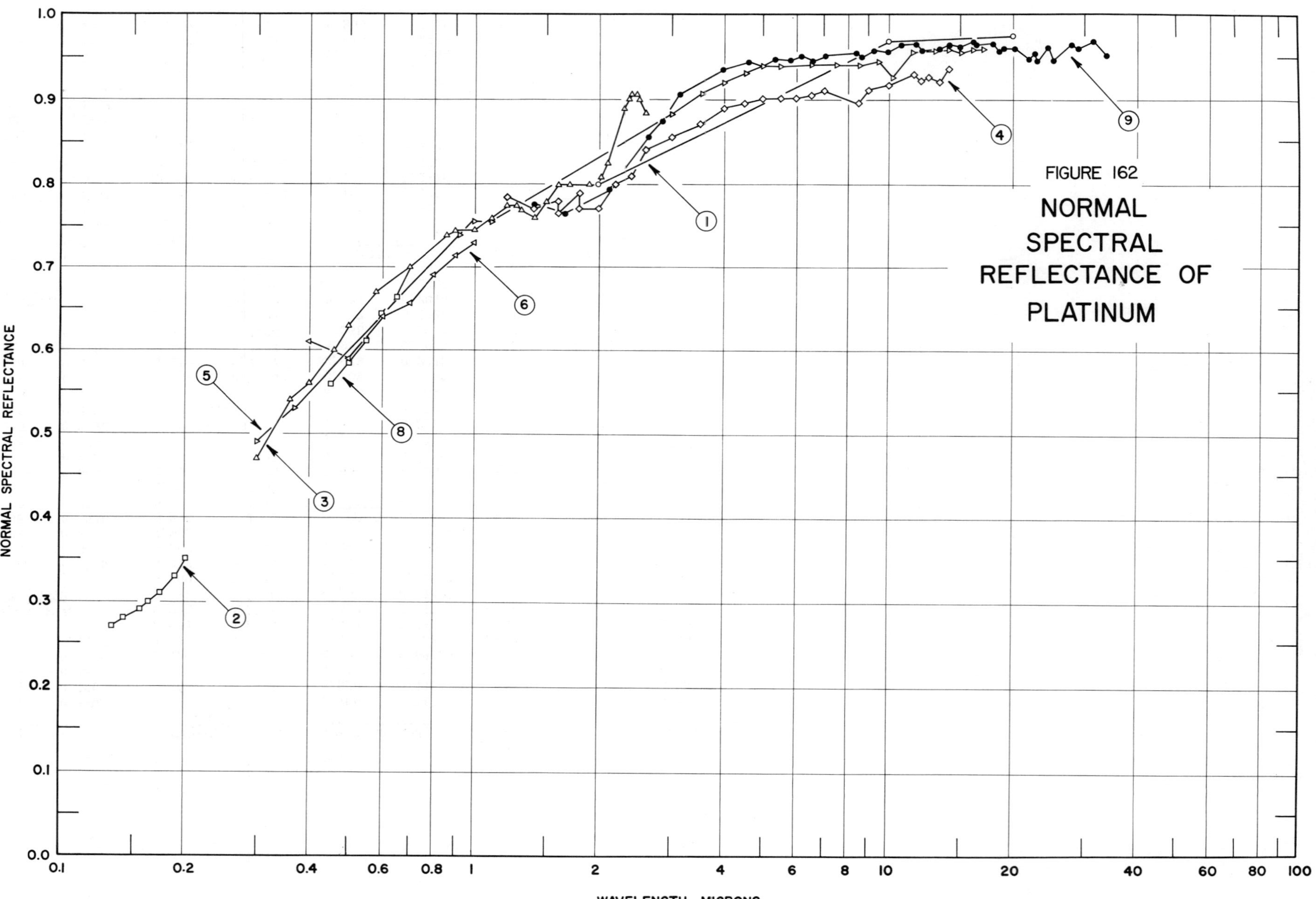
FIGURE 162
NORMAL SPECTRAL REFLECTANCE OF PLATINUM
NORMAL SPECTRAL REFLECTANCE
1.0
0.9
0.8
0.7
0.6
0.5
0.4
0.3
0.2
0.1
0.0
0.1
0.2
0.4
0.6
0.8
1
2
4
6
8
10
20
40
60
80
100
WAVELENGTH, MICRONS
1
2
3
4
5
6
8
9

SPECIFICATION TABLE NO. 162 NORMAL SPECTRAL REFLECTANCE OF PLATINUM

Curve No.	Ref. No.	Year	Temperature K	Wavelength Range, μ	Geometry θ	θ'	ω'	Reported Error, %	Composition (weight percent), Specifications and Remarks
1	128	1962	298	2.00-20.00	$\sim 0^{\circ}$		2π		Plate; as received; $\omega' = 0.034$ sr; converted from $R(2\pi, 0^{\circ})$.
2	124	1941	298	0.1347-0.2026	$\sim 0^{\circ}$	$\sim 0^{\circ}$			Opaque film on glass; deposited by cathodic sputtering; measured in vacuum (10^{-3}mm Hg).
3	34	1957	298	0.30-2.60	9°		2π	± 4	Pure metal; as received; data extracted from a smooth curve; magnesium carbonate reference.
4	165	1962	1311	1.20-14.00	$\sim 0^{\circ}$		2π		Measured in air.
5	65	1962	294	0.30-17.00	$\sim 0^{\circ}$		2π		Measured in air.
6	65	1962	294	0.40-1.00	$\sim 0^{\circ}$		2π		Measured in air.
7	65	1962	294	1.21-14.80	$\sim 0^{\circ}$		2π		Measured in air.
8	228	1900	298	0.45-0.70	$\sim 0^{\circ}$	$\sim 0^{\circ}$			Pure; mirror like surface; platinum filament lamp source.
9	244	1963	311	1.40-33.7	$\sim 0^{\circ}$		2π		Foil (0.005 in. thick); converted from $R(2\pi, 0^{\circ})$.

DATA TABLE NO. 162 NORMAL SPECTRAL REFLECTANCE OF PLATINUM

[Wavelength, λ, μ; Reflectance, ρ; Temperature, T, K]

λ	ρ
CURVE 1 T = 298	
2.00	0.800
10.00	0.970
20.00	0.975
CURVE 2 T = 298	
0.1347	0.27
0.1438	0.28
0.1570	0.29
0.1640	0.30
0.1757	0.31
0.1901	0.33
0.2026	0.35
CURVE 3 T = 298	
0.30	0.470
0.36	0.540
0.40	0.560
0.46	0.600
0.50	0.630
0.58	0.670
0.70	0.700
0.86	0.740
0.90	0.745
1.00	0.745
1.10	0.760
1.20	0.775
1.26	0.775
1.30	0.770
1.40	0.760
1.50	0.780
1.60	0.800
1.70	0.800
1.90	0.800
1.98	0.800*
2.02	0.810
2.10	0.825
2.30	0.890
2.38	0.900
CURVE 3 (cont.)	
2.40	0.905
2.46	0.905
2.50	0.900
2.60	0.885
CURVE 4 T = 1311	
1.20	0.785
1.40	0.770
1.60	0.780
1.60	0.765
1.80	0.790
1.80	0.770
2.00	0.770
2.20	0.800
2.40	0.810
2.60	0.840
3.00	0.855
3.55	0.870
4.00	0.890
4.50	0.896
5.00	0.900
5.50	0.900
6.00	0.900
6.50	0.905
7.00	0.910
8.50	0.896
9.00	0.910
10.00	0.915
11.50	0.930
12.00	0.921
12.50	0.925
13.20	0.920
14.00	0.935
CURVE 5 T = 294	
0.30	0.490
0.37	0.530
0.92	0.740
1.00	0.755
CURVE 5 (cont.)	
1.10	0.755
3.00	0.882
3.55	0.908
4.00	0.920
4.50	0.930
5.00	0.939
5.50	0.939
6.50	0.940
7.50	0.940
8.50	0.940
9.50	0.945
10.20	0.952
11.50	0.956
13.00	0.958
14.00	0.959
15.00	0.955
16.00	0.959
17.00	0.959
CURVE 6 T = 294	
0.40	0.610
0.50	0.590
0.60	0.640
0.70	0.655
0.80	0.690
0.90	0.715
1.00	0.730
CURVE 7* T = 294	
1.21	0.740
1.40	0.735
1.60	0.740
1.80	0.750
1.80	0.780
2.00	0.790
2.20	0.820
2.40	0.850
2.60	0.876
3.00	0.908
CURVE 7* (cont.)	
3.50	0.940
4.00	0.945
4.50	0.950
5.00	0.955
6.00	0.955
6.50	0.956
7.10	0.960
7.70	0.965
8.00	0.961
8.50	0.965
9.10	0.965
9.80	0.965
10.00	0.970
10.90	0.971
11.90	0.971
13.00	0.971
14.80	0.970
CURVE 8 T = 298	
0.45	0.558
0.50	0.584
0.55	0.611
0.60	0.642
0.65	0.663
0.70	0.701*
CURVE 9 T = 311	
1.40	0.777
1.66	0.764
2.14	0.794
2.63	0.857
2.87	0.875
3.15	0.905
3.99	0.934
4.60	0.942
4.95	0.938*
5.37	0.946
5.80	0.944
6.15	0.951
CURVE 9 (cont.)	
6.59	0.945
7.03	0.951
8.41	0.955
8.65	0.950
9.24	0.958
10.1	0.956
10.7	0.963
11.6	0.963
12.1	0.958
13.3	0.960
14.1	0.966
14.9	0.962
16.0	0.970
16.2	0.966
17.9	0.967
18.5	0.958
19.1	0.961
20.3	0.961
21.9	0.948
22.7	0.953
23.0	0.946
24.4	0.961
25.1	0.947
27.8	0.964
28.9	0.960
31.4	0.970
33.7	0.951

*Not shown on plot

SPECIFICATION TABLE NO. 163 ANGULAR SPECTRAL REFLECTANCE OF PLATINUM

Curve No.	Ref. No.	Year	Temperature K	Wavelength Range, μ	Geometry θ	θ'	ω'	Reported Error, %	Composition (weight percent), Specifications and Remarks
1	306	1959	~298	0.0590-0.2201	~18°	~18°			Pt film (300 Å thick) evaporated from multistranded helical tungsten filaments in 10 to 20 sec at 5 x 10^{-5} mm Hg.

DATA TABLE NO. 163 ANGULAR SPECTRAL REFLECTANCE OF PLATINUM

[Wavelength, λ, μ; Reflectance, ρ; Temperature, T, K]

λ	ρ
CURVE 1* T = ~298	
0.0590	0.209
0.0730	0.136
0.0902	0.159
0.1000	0.183
0.1101	0.207
0.1217	0.233
0.1328	0.248
0.1486	0.260
0.1615	0.265
0.1798	0.270
0.2005	0.303
0.2201	0.328

*Not shown on plot

SPECIFICATION TABLE NO. 164 ANGULAR SPECTRAL REFLECTANCE OF PLATINUM

Curve No.	Ref. No.	Year	Temperature K	Wavelength, μ	Angular Range°	Geometry θ	Geometry θ'	Geometry ω'	Reported Error, %	Composition (weight percent), Specifications and Remarks
1	148	1961	298	2	10-80	θ		2π	<2	No details given.

DATA TABLE NO. 164 ANGULAR SPECTRAL REFLECTANCE OF PLATINUM

[Angle, θ, °; Reflectance, ρ; Temperature, T, K; Wavelength, λ, μ]

θ	ρ
CURVE 1*	
T = 298	
$\lambda = 2$	
10	0.810
20	0.810
30	0.805
35	0.800
40	0.790
45	0.780
50	0.780
60	0.765
65	0.735
70	0.705
75	0.665
80	0.640

*Not shown on plot

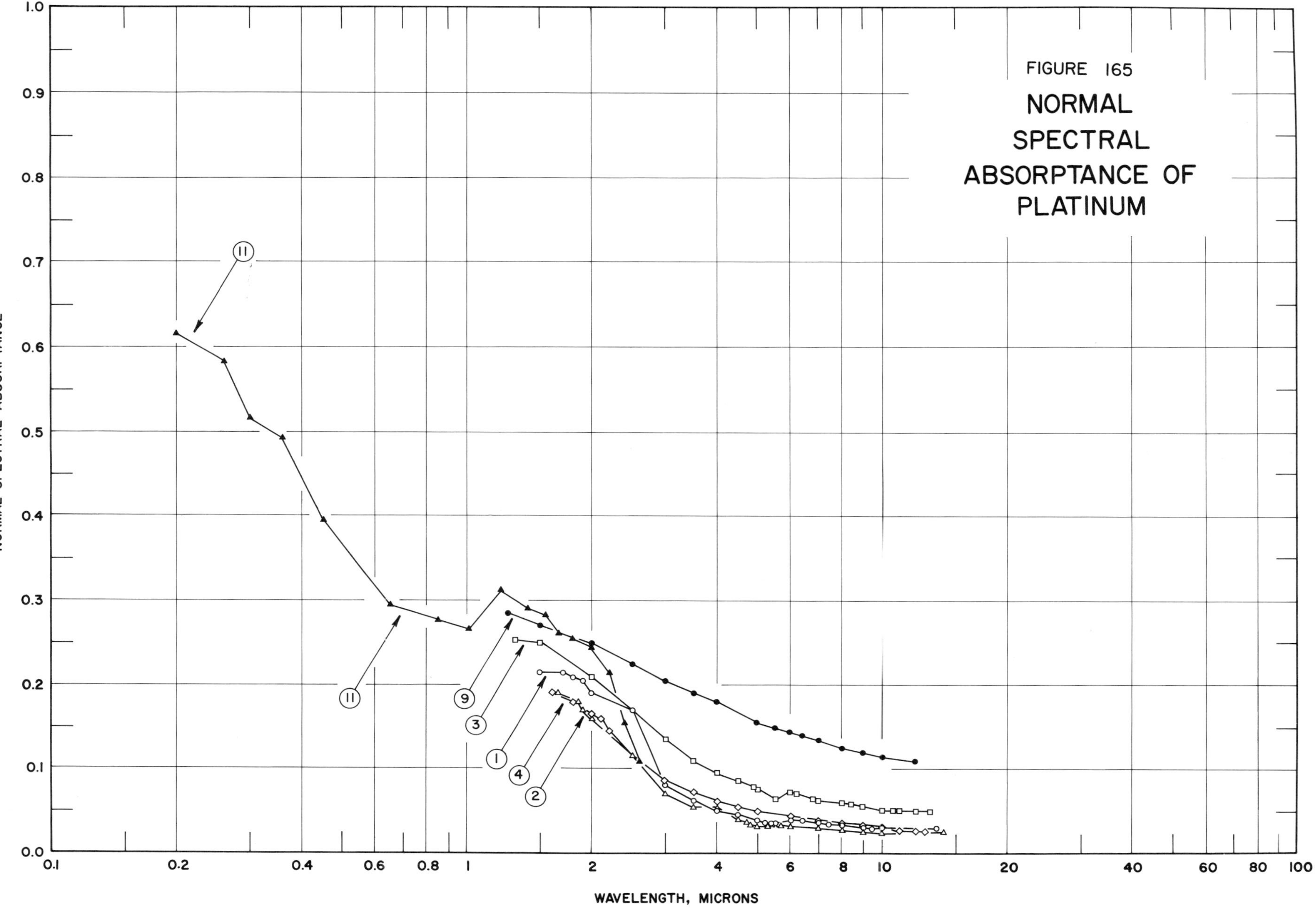
FIGURE 165
NORMAL SPECTRAL ABSORPTANCE OF PLATINUM
NORMAL SPECTRAL ABSORPTANCE
1.0
0.9
0.8
0.7
0.6
0.5
0.4
0.3
0.2
0.1
0.0
0.1
0.2
0.4
0.6
0.8
1
2
4
6
8
10
20
40
60
80
100
WAVELENGTH, MICRONS
11
11
9
3
1
4
2

SPECIFICATION TABLE NO. 165 NORMAL SPECTRAL ABSORPTANCE OF PLATINUM

Curve No.	Ref. No.	Year	Temperature K	Wavelength Range, μ	Geometry θ	Reported Error, %	Composition (weight percent), Specifications and Remarks
1	30	1963	294	1.50–13.50	~0°		NBS platinum; heated in air for 1 hr at 1523 K and cooled for 12 hrs; measured in air; data extracted from smooth curve.
2	30	1963	294	1.65–14.00	~0°		NBS platinum; heated in air for 1 hr at 1523 K, cooled for 12 hrs, heated at T > 1088 K for 24 hrs and at 1523 K for 1 hr; measured in air; data extracted from smooth curve.
3	30	1963	294	1.30–13.00	~0°		Above specimen and conditions except after 6 months heated at T > 1083 K for 11 hrs and at T > 1277 K for 1 hr; measured after a total of 8 months.
4	30	1963	294	1.60–12.50	~0°		Mechanically polished with aluminum oxide and cleaned with water (surface roughness 0.025 μ peak to peak and 10 μ lateral); data extracted from smooth curve; 3 different samples gave the same results; [Author's designation: Sample 3].
5	30	1963	294	1.60–12.50	~0°		Above specimen and conditions except heated at 672 K for 3 hrs; measured in air.
6	30	1963	294	1.60–12.50	~0°		Above specimen and conditions except heated at 1088 K for 3 hrs; measured in vacuum (5×10^{-6} mm Hg).
7	30	1963	294	1.65–12.50	~0°		Above specimen and conditions except heated at 1088 K for 3 additional hrs; measured in air.
8	30	1963	294	1.50–12.50	~0°		Mechanically polished with aluminum oxide and cleaned with water (surface roughness 0.025 μ peak to peak and 25 μ lateral); heated at 672 K for 3 hrs, at 1088 K for 6 hrs, at T > 1111 K for 53 hrs, and at 1460 K for 30 hrs; measured in air; data extracted from smooth curve; [Author's designation: Sample 3].
9	30	1963	294	1.25–12.00	~0°		Above specimen and conditions except heated for 58 additional hrs at 1477 K; measured in air.
10	245	1965	294	1.42–12.0	~0°	< 5	Measured in hohlraum.
11	307	1954	~294	0.202–2.600	~0°		Data extracted from smooth curve.

DATA TABLE NO. 165 NORMAL SPECTRAL ABSORPTANCE OF PLATINUM

[Wavelength, λ, μ; Absorptance, α; Temperature, T, K]

CURVE 1
T = 294

λ	α
1.50	0.215
1.70	0.215
1.80	0.210
1.90	0.205
2.00	0.190
2.50	0.170
3.00	0.080
3.50	0.062
4.00	0.050
4.50	0.046
5.00	0.039
5.20	0.037
5.40	0.037
5.50	0.037
6.00	0.039
6.40	0.039
7.00	0.037
7.40	0.035
8.00	0.033
9.00	0.031
9.40	0.030
10.00	0.030
13.50	0.030

CURVE 2
T = 294

λ	α
1.65	0.190
1.85	0.180
1.90	0.170
2.00	0.160
2.50	0.115
3.00	0.070
3.50	0.054
4.00	0.053
4.50	0.039
4.70	0.036
4.80	0.033
5.00	0.032
5.30	0.032
5.70	0.033
6.00	0.032
7.00	0.030
8.00	0.028
9.00	0.027

CURVE 2 (cont.)

λ	α
10.00	0.026
14.00	0.026

CURVE 3
T = 294

λ	α
1.30	0.255
1.50	0.250
2.00	0.210
2.50	0.170*
3.00	0.135
3.50	0.110
4.00	0.096
4.50	0.086
4.90	0.078
5.00	0.076
5.50	0.064
6.00	0.072
6.20	0.070
6.80	0.064
7.00	0.063
8.00	0.060
8.40	0.058
9.00	0.055
10.00	0.051
10.50	0.051
11.00	0.051
12.00	0.050
13.00	0.050

CURVE 4
T = 294

λ	α
1.60	0.190
1.80	0.180
2.00	0.165
2.10	0.160
2.20	0.145
2.50	0.115*
3.00	0.088
3.50	0.072
4.00	0.062
4.50	0.055
5.00	0.050
6.00	0.044

CURVE 4 (cont.)

λ	α
7.00	0.039
8.00	0.036
9.00	0.034
10.00	0.032
11.00	0.028
12.00	0.027
12.50	0.027

CURVE 5*
T = 294

λ	α
1.60	0.190
1.80	0.180
2.00	0.165
2.10	0.160
2.20	0.145
2.50	0.115
3.00	0.088
3.50	0.072
4.00	0.062
4.50	0.055
5.00	0.050
6.00	0.044
7.00	0.039
8.00	0.036
9.00	0.034
10.00	0.032
11.00	0.028
12.00	0.027
12.50	0.027

CURVE 6*
T = 294

λ	α
1.60	0.190
1.80	0.180
2.00	0.165
2.10	0.160
2.20	0.145
2.50	0.115
3.00	0.088
3.50	0.072
4.00	0.062
4.50	0.055
5.00	0.050

CURVE 6 (cont.)*

λ	α
6.00	0.044
7.00	0.039
8.00	0.036
9.00	0.034
10.00	0.032
11.00	0.028
12.00	0.027
12.50	0.027

CURVE 7*
T = 294

λ	α
1.65	0.195
1.80	0.190
2.00	0.170
2.20	0.152
2.50	0.115
3.00	0.080
3.50	0.062
4.00	0.052
5.00	0.042
6.00	0.037
7.00	0.034
8.00	0.032
9.00	0.031
10.00	0.029
11.00	0.028
12.00	0.027
12.50	0.027

CURVE 8*
T = 294

λ	α
1.50	0.240
1.80	0.220
2.00	0.200
2.20	0.175
2.40	0.150
3.00	0.092
3.50	0.072
4.00	0.058
5.00	0.047
6.00	0.043
7.00	0.041
7.60	0.040

CURVE 8 (cont.)*

λ	α
8.00	0.040
9.00	0.041
9.60	0.040
10.00	0.039
11.00	0.032
12.00	0.031
12.50	0.030

CURVE 9
T = 294

λ	α
1.25	0.285
1.50	0.270
2.00	0.250
2.50	0.225
3.00	0.205
3.50	0.190
4.00	0.180
5.00	0.155
5.50	0.150
6.00	0.145
6.40	0.140
7.00	0.135
8.00	0.125
9.00	0.120
10.00	0.115
12.00	0.110

CURVE 10*
T = 294

λ	α
1.42	0.205
1.61	0.205
1.81	0.204
1.99	0.199
2.17	0.174
2.43	0.140
2.66	0.107
2.86	0.0951
2.98	0.0811
3.45	0.0621
4.05	0.0527
5.02	0.0415
6.00	0.0383
7.00	0.0349

CURVE 10 (cont.)*

λ	α
8.02	0.0331
8.97	0.0310
10.0	0.0296
10.9	0.0291
12.0	0.0282

CURVE 11
T = ~298

λ	α
0.202	0.617
0.260	0.582
0.304	0.518
0.361	0.492
0.448	0.397
0.651	0.293
0.851	0.276
1.015	0.264
1.207	0.312
1.404	0.290
1.536	0.282
1.649	0.261
1.799	0.255
2.000	0.244
2.200	0.214
2.399	0.153
2.600	0.109

*Not shown on plot

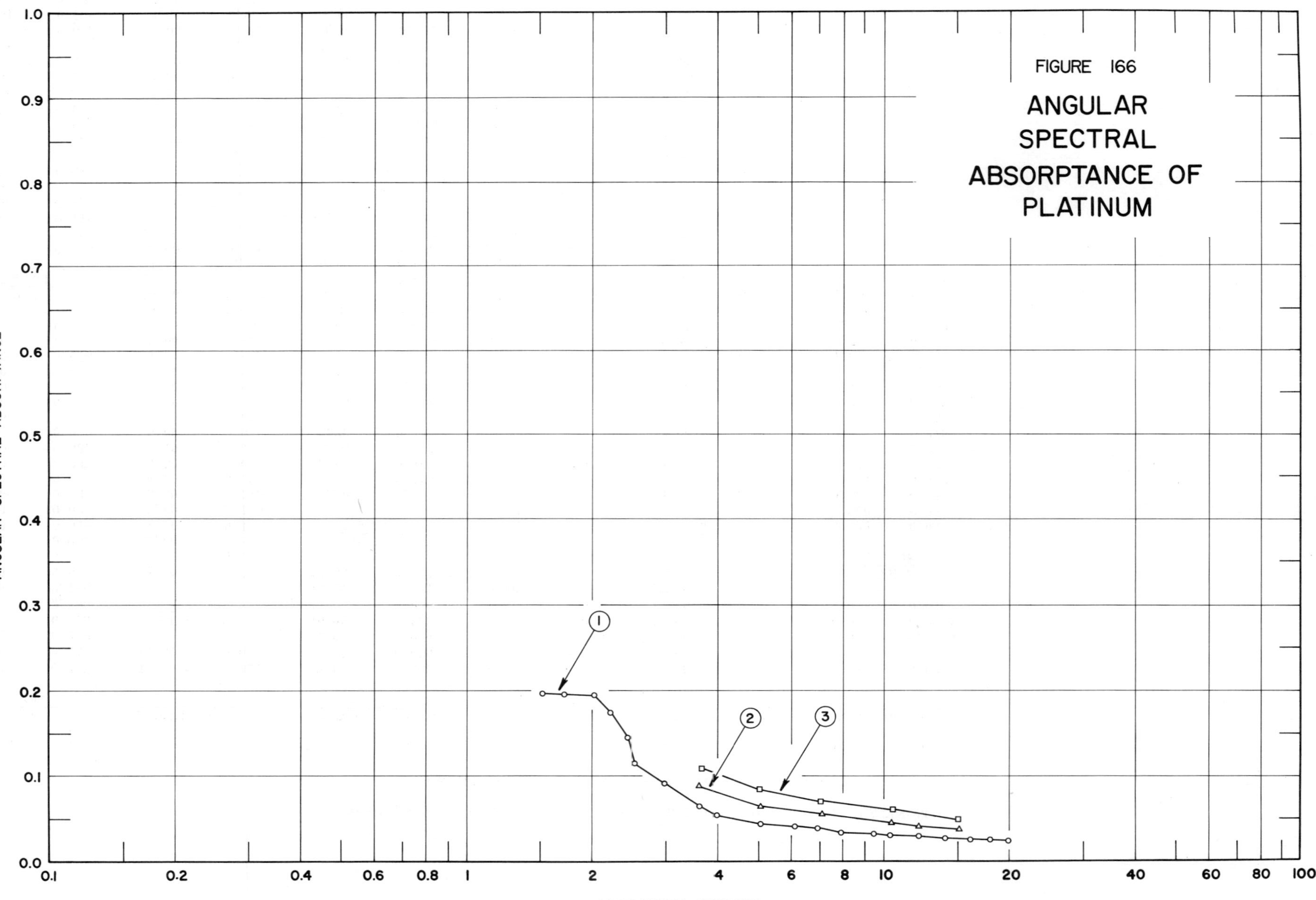

FIGURE 166
ANGULAR SPECTRAL ABSORPTANCE OF PLATINUM
ANGULAR SPECTRAL ABSORPTANCE
WAVELENGTH, MICRONS
0.0
0.1
0.2
0.3
0.4
0.5
0.6
0.7
0.8
0.9
1.0
0.1
0.2
0.4
0.6
0.8
1
2
4
6
8
10
20
40
60
80
100
1
2
3

SPECIFICATION TABLE NO. 166 ANGULAR SPECTRAL ABSORPTANCE OF PLATINUM

Curve No.	Ref. No.	Year	Temperature K	Wavelength Range, μ	Geometry θ	Reported Error, %	Composition (weight percent), Specifications and Remarks
1	225	1965	294	1.52-19.9	25°		99.99 Pt (cold-rolled plate) from Wilkinson Co.; measured in dry nitrogen; heated cavity at approx 1056 K with platinum reference; authors assumed $\alpha = 1-R(2\pi, 25^\circ)$.
2	225	1965	556	3.60-15.1	25°		Above specimen and conditions.
3	225	1965	833	3.65-15.1	25°		Above specimen and conditions.

DATA TABLE NO. 166 ANGULAR SPECTRAL ABSORPTANCE OF PLATINUM

[Wavelength, λ, μ; Absorptance, α; Temperature, T, K]

λ	α
CURVE 1 T = 294	
1.52	0.198
1.71	0.197
2.02	0.195
2.21	0.173
2.42	0.146
2.52	0.116
2.99	0.091
3.61	0.064
3.99	0.053
5.04	0.043
6.10	0.041
6.95	0.039
7.87	0.036
9.44	0.033
10.3	0.031
12.1	0.031
14.0	0.028
16.1	0.027
17.9	0.027
19.9	0.026
CURVE 2 T = 556	
3.60	0.088
5.06	0.064
7.05	0.055
10.3	0.046
12.1	0.041
15.1	0.038
CURVE 3 T = 833	
3.65	0.109
5.05	0.084
7.02	0.070
10.5	0.061
15.1	0.049

SPECIFICATION TABLE NO. 167 NORMAL SOLAR ABSORPTANCE OF PLATINUM

Curve No.	Ref. No.	Year	Temperature Range, K	Geometry θ	Reported Error, %	Composition (weight percent), Specifications and Remarks
1	34	1957	298	9°		Pure metal; as received (bright); computed from spectral reflectance data for sea level conditions.
2	34	1957	298	9°		Above specimen and conditions except computed for above atmosphere conditions.
3	80	1956	298	9°		Platinum; as received (bright); computed from spectral reflectivity for sea level conditions.
4	80	1956	298	9°		Above specimen and conditions except computed for above atmosphere conditions.

DATA TABLE NO. 167 NORMAL SOLAR ABSORPTANCE OF PLATINUM

[Temperature, T, K; Absorptance, α]

T	α
CURVE 1*	
298	0.307
CURVE 2*	
298	0.308
CURVE 3*	
298	0.307
CURVE 4*	
298	0.308

*Not shown on plot

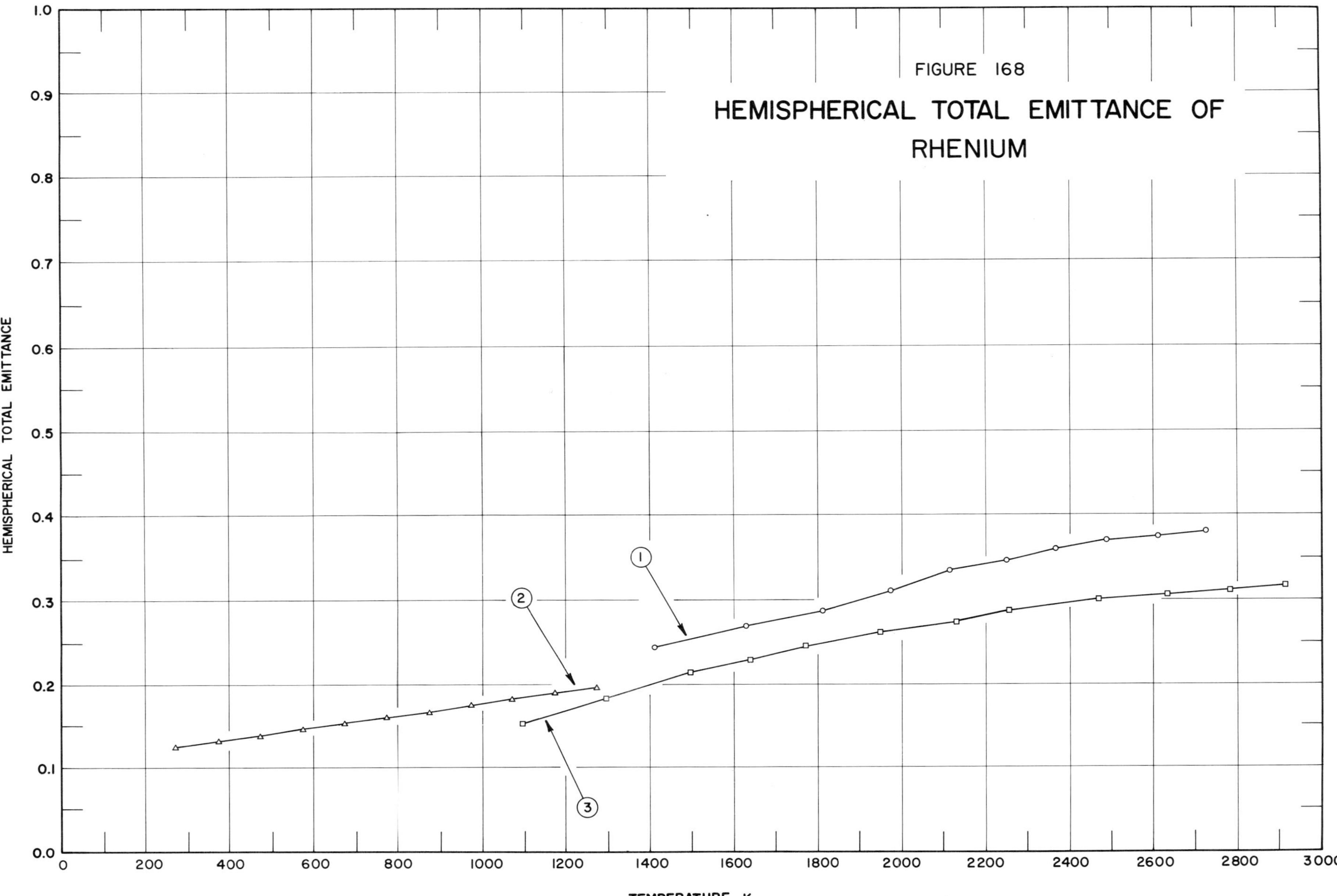
FIGURE 168
HEMISPHERICAL TOTAL EMITTANCE OF RHENIUM
HEMISPHERICAL TOTAL EMITTANCE
TEMPERATURE, K
1.0
0.9
0.8
0.7
0.6
0.5
0.4
0.3
0.2
0.1
0.0
0
200
400
600
800
1000
1200
1400
1600
1800
2000
2200
2400
2600
2800
3000
1
2
3

SPECIFICATION TABLE NO. 168 HEMISPHERICAL TOTAL EMITTANCE OF RHENIUM

Curve No.	Ref. No.	Year	Temperature Range, K	Reported Error, %	Composition (weight percent), Specifications and Remarks
1	71	1962	1410-2725	± 10	Measured in vacuum ($< 10^{-5}$ mm Hg).
2	255	1964	273-1273		Measured in vacuum (5×10^{-7} mm Hg).
3	343	1964	1096-2918		Measured in vacuum (10^{-9} mm Hg).

DATA TABLE NO. 168 HEMISPHERICAL TOTAL EMITTANCE OF RHENIUM

[Temperature, T, K; Emittance, ϵ]

T	ϵ
CURVE 1	
1410	0.245
1630	0.270
1810	0.288
1975	0.310
2115	0.335
2250	0.347
2370	0.360
2490	0.370
2610	0.373
2725	0.381
CURVE 2	
273	0.123
373	0.130
473	0.138
573	0.145
673	0.152
773	0.160
873	0.167
973	0.175
1073	0.182
1173	0.189
1273	0.197
CURVE 3	
1096	0.152
1297	0.183
1499	0.213
1640	0.229
1772	0.247
1947	0.262
2130	0.274
2258	0.288
2469	0.301
2633	0.307
2782	0.311
2918	0.317

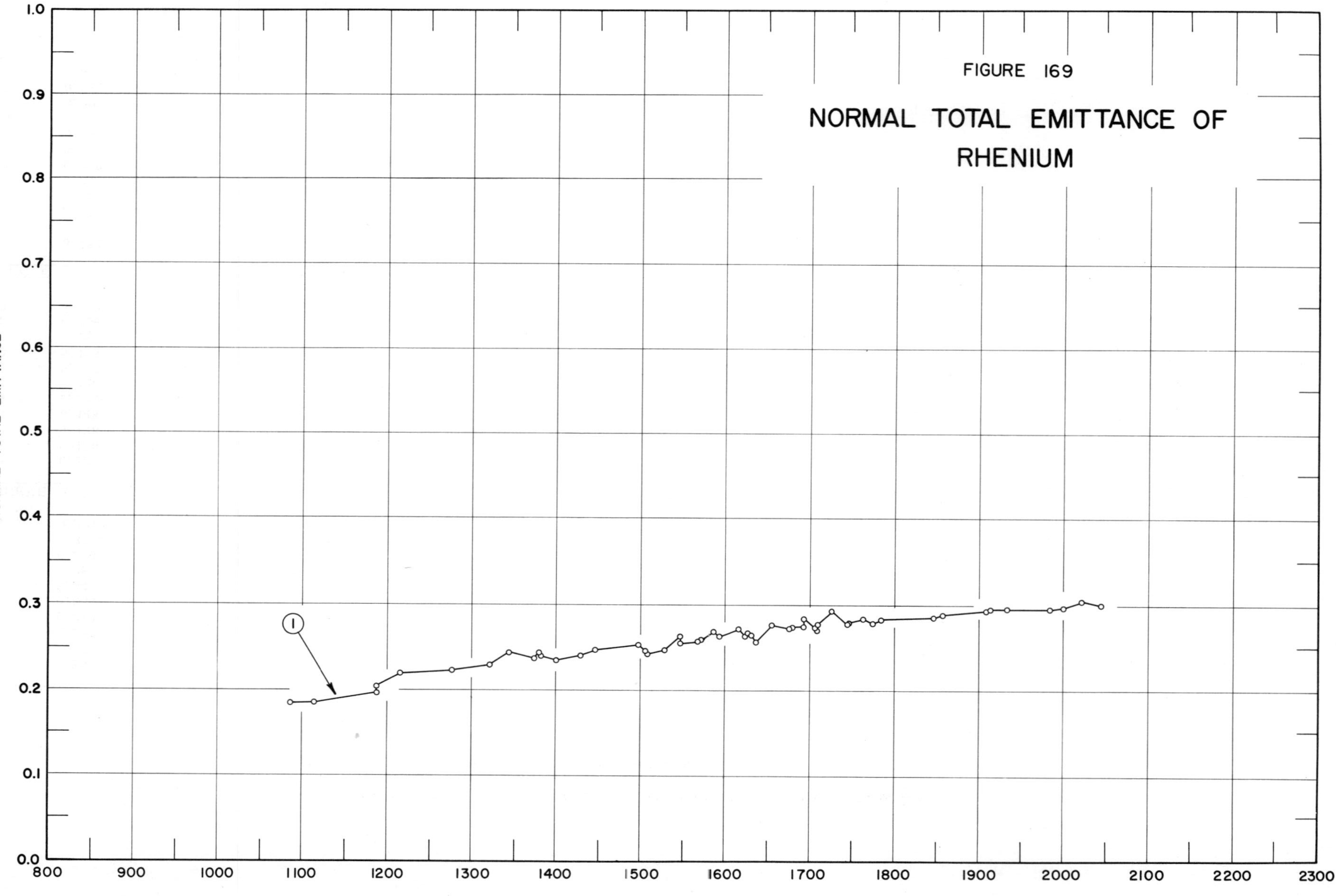

FIGURE 169
NORMAL TOTAL EMITTANCE OF RHENIUM

SPECIFICATION TABLE NO. 169 NORMAL TOTAL EMITTANCE OF RHENIUM

Curve No.	Ref. No.	Year	Temperature Range, K	Geometry θ'	Reported Error, %	Composition (weight percent), Specifications and Remarks
1	256	1963	1085-2045	$\sim 0°$		Heated several times to temperatures of the order of 2000 K; measured in vacuum (10^{-6} to 10^{-8} mm Hg).

DATA TABLE NO. 169 NORMAL TOTAL EMITTANCE OF RHENIUM

[Temperature, T, K; Emittance, ϵ]

T	ϵ	T	ϵ
CURVE 1		**CURVE 1 (cont.)**	
1085	0.185	1858	0.289
1114	0.185	1909	0.294
1188	0.197	1913	0.296
1188	0.202	1933	0.297
1215	0.219	1984	0.297
1276	0.222	2000	0.299
1320	0.230	2022	0.305
1342	0.246	2045	0.301
1373	0.238		
1379	0.246		
1381	0.240		
1400	0.237		
1429	0.241		
1429	0.242*		
1446	0.249		
1497	0.254		
1504	0.248		
1507	0.246		
1528	0.249		
1547	0.263		
1548	0.256		
1568	0.258		
1571	0.260		
1586	0.270		
1592	0.263		
1616	0.272		
1623	0.263		
1626	0.269		
1631	0.267		
1637	0.258		
1655	0.278		
1675	0.273		
1680	0.275		
1692	0.277		
1692	0.284		
1706	0.275		
1709	0.271		
1709	0.279		
1726	0.293		
1745	0.279		
1748	0.280		
1764	0.285		
1774	0.280		
1783	0.283		
1847	0.287		

*Not shown on plot

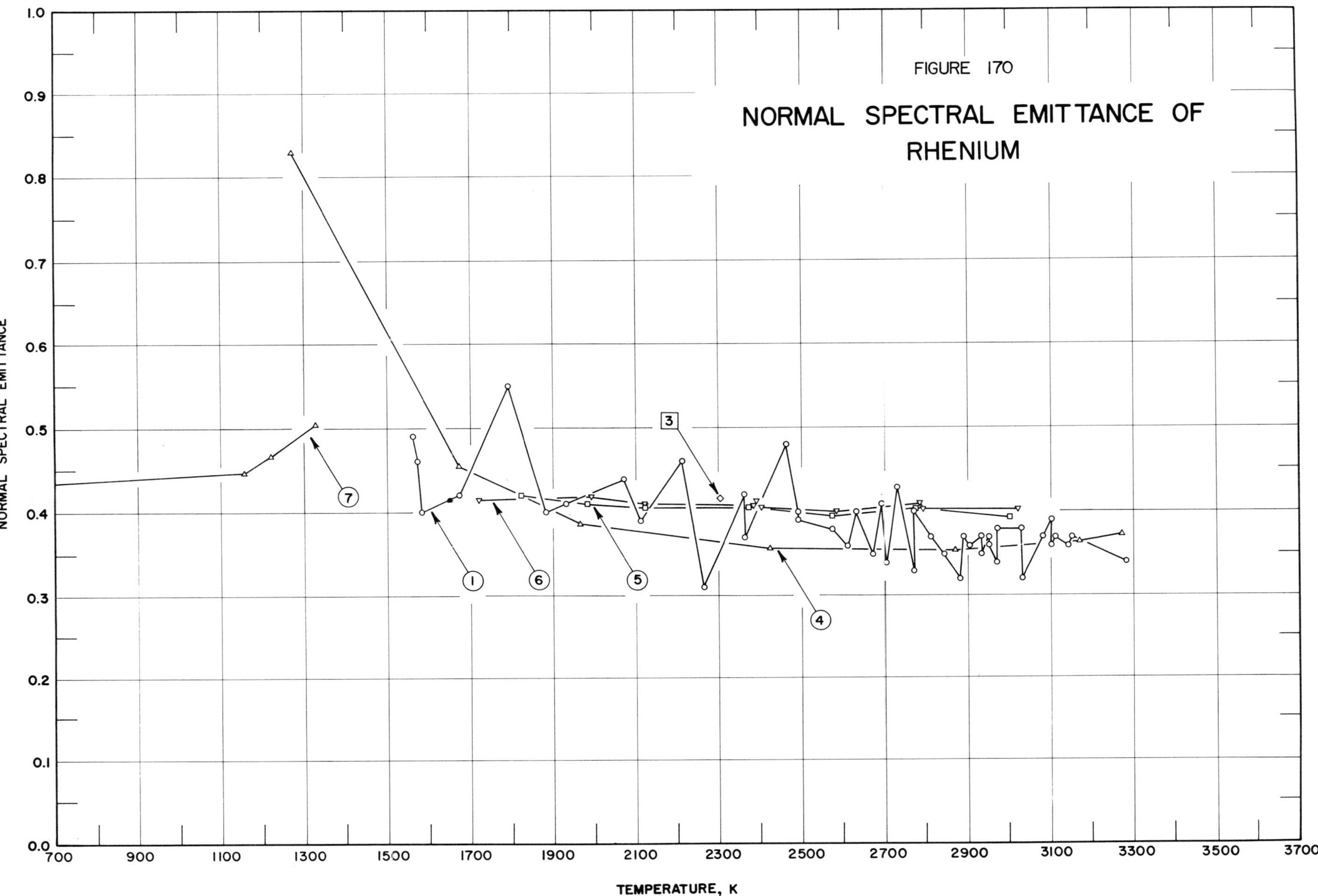
FIGURE 170
NORMAL SPECTRAL EMITTANCE OF RHENIUM
NORMAL SPECTRAL EMITTANCE
TEMPERATURE, K
0.0
0.1
0.2
0.3
0.4
0.5
0.6
0.7
0.8
0.9
1.0
700
900
1100
1300
1500
1700
1900
2100
2300
2500
2700
2900
3100
3300
3500
3700
7
1
6
5
3
4

SPECIFICATION TABLE NO. 170 NORMAL SPECTRAL EMITTANCE OF RHENIUM

Curve No.	Ref. No.	Year	Wavelength μ	Temperature Range, K	Geometry θ'	Reported Error, %	Composition (weight percent), Specifications and Remarks
1	91	1955	0.655	1563-3283	$\sim 0°$		Prepared by methods of powder metallurgy.
2	90	1954	0.655	298-1156	$\sim 0°$	± 2	Crystal-bar; polished; measured in vacuum; authors assumed $\epsilon = 1 - \rho$.
3	90	1954	0.655	2303	$\sim 0°$		Wire (0.015 - mil.); measured in vacuum.
4	89	1953	0.655	1273-3273	$\sim 0°$		Pressed powdered rhenium bar with a matte surface; author reports: unreliable emissivity values between temps 1273 K to 2073 K.
5	88	1955	0.65	1823-3003	$\sim 0°$		Tube (0.020 in. thick); measured in vacuum.
6	88	1955	0.65	1723-3023	$\sim 0°$		Tube (0.020 in. thick); measured in vacuum.
7	120	1953	0.655	293-1327	$\sim 0°$		Solid rod ground to form a hollow cylinder; polished; measured in vacuum; at each temperature emissivity is the average of several runs.

DATA TABLE NO. 170 NORMAL SPECTRAL EMITTANCE OF RHENIUM

[Temperature, T,K; Emittance, ϵ; Wavelength, λ,μ]

T	ϵ	T	ϵ	T	ϵ
CURVE 1		CURVE 1 (cont.)		CURVE 6 (cont.)	
λ = 0.655		3113	0.37	1993	0.419
		3143	0.36	1993	0.416*
1563	0.49	3153	0.37	2123	0.410
1573	0.46	3283	0.34	2383	0.408
1583	0.40			2393	0.411
1673	0.42	CURVE 2*		2403	0.405
1793	0.55	λ = 0.655		2583	0.400
1883	0.40			2783	0.410
1933	0.41	298	0.425	2793	0.402
2073	0.44	1156	0.448	3023	0.401
2113	0.39				
2213	0.46	CURVE 3		CURVE 7	
2263	0.31	λ = 0.655		λ = 0.655	
2363	0.42				
2363	0.37	2303	0.416	293	0.425*
2463	0.48			1150	0.448
2493	0.40	CURVE 4		1220	0.467
2493	0.39	λ = 0.655		1327	0.505
2573	0.38				
2633	0.40	1273	0.830		
2613	0.36	1673	0.455		
2673	0.35	1968	0.388		
2693	0.41	2423	0.358		
2703	0.34	2873	0.355		
2733	0.43	3173	0.365		
2773	0.33	3273	0.371		
2773	0.40				
2813	0.37	CURVE 5			
2843	0.35	λ = 0.65			
2883	0.32				
2893	0.37	1823	0.420		
2903	0.36	1983	0.410		
2933	0.37	2123	0.405		
2933	0.35	2373	0.405		
2953	0.37	2573	0.395		
2953	0.36	2773	0.401		
2973	0.34	3003	0.392		
2973	0.38				
3033	0.38	CURVE 6			
3033	0.32	λ = 0.65			
3083	0.37				
3103	0.39				
3103	0.36	1723	0.415		

* Not shown on plot

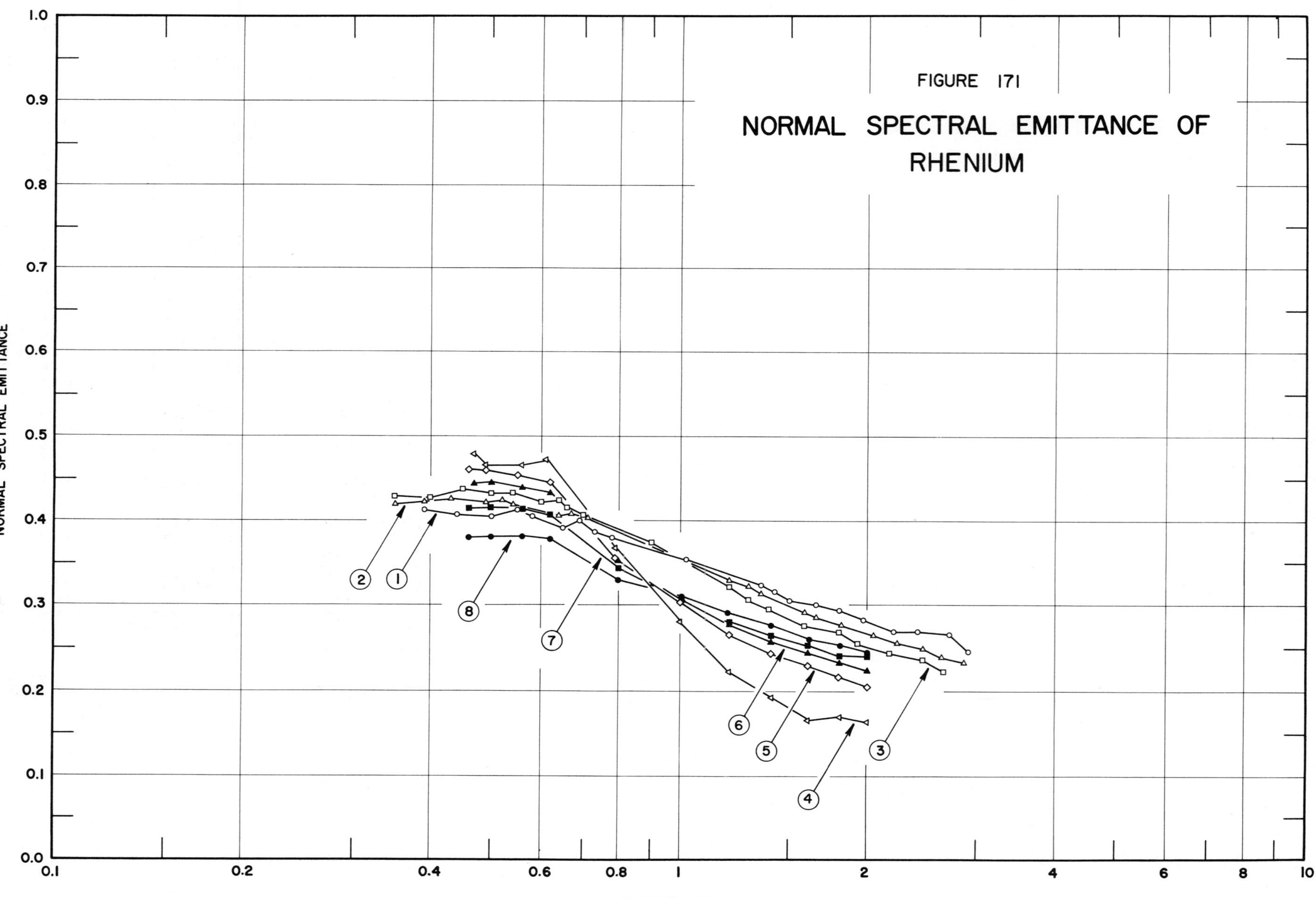

FIGURE 171
NORMAL SPECTRAL EMITTANCE OF RHENIUM

SPECIFICATION TABLE NO. 171 NORMAL SPECTRAL EMITTANCE OF RHENIUM

Curve No.	Ref. No.	Year	Temperature K	Wavelength Range, μ	Geometry θ'	Reported Error, %	Composition (weight percent), Specifications and Remarks
1	88	1955	3045	0.39-2.90	~0°	± 6	Tube (0.020 in. thick); heated at 673 K and at 2473 K for 15 min in vacuum; measured in argon (50 cm Hg).
2	88	1955	2388	0.35-2.85	~0°	± 6	Tube (0.020 in. thick); heated at 673 K in vacuum for 1 hr, then cooled and reheated to 2573 K; measured in vacuum (5×10^{-6} mm Hg).
3	88	1955	1810	0.35-2.65	~0°	± 7	Tube (0.020 in. thick); heated at 673 K in vacuum for 1 hr, then cooled and reheated to 2573 K; measured in vacuum (5×10^{-6} mm Hg).
4	323	1966	300	0.47-2.00	0°		Filament (0.25-0.32 mm in diameter); baked for 1 hr at 798 K in vacuum, cooled, heated for 5-10 min in vacuum, and cooled; measured in argon (600 mm Hg); data calculated from optical constants.
5	323	1966	1100	0.46-2.00	0°		Above specimen and conditions.
6	323	1966	1500	0.47-2.00	0°		Above specimen and conditions.
7	323	1966	2000	0.46-2.00	0°		Above specimen and conditions.
8	323	1966	2500	0.46-2.00	0°		Above specimen and conditions.

DATA TABLE NO. 171 NORMAL SPECTRAL EMITTANCE OF RHENIUM

[Wavelength, λ, μ; Emittance, ϵ; Temperature, T, K]

λ	ϵ
CURVE 1 T = 3045	
0.39	0.412
0.44	0.408
0.50	0.405
0.55	0.412
0.58	0.405
0.65	0.391
0.69	0.400
0.73	0.388
0.78	0.380
1.02	0.355
1.35	0.323
1.42	0.317
1.50	0.305
1.65	0.300
1.80	0.295
1.97	0.283
2.20	0.270
2.40	0.270
2.70	0.257
2.90	0.248
CURVE 2 T = 2388	
0.35	0.420
0.39	0.422
0.43	0.427
0.49	0.422
0.52	0.425
0.54	0.420
0.64	0.408
0.67	0.410
0.71	0.403
1.20	0.330
1.28	0.322
1.35	0.313
1.57	0.291
1.65	0.287
1.82	0.278
2.05	0.267
2.22	0.257
2.45	0.250
2.62	0.240
2.85	0.233
CURVE 3 T = 1810	
0.35	0.430
0.40	0.429
0.45	0.437
0.50	0.433
0.54	0.433
0.60	0.423
0.64	0.424
0.66	0.417
0.70	0.408
0.90	0.375
1.20	0.321
1.28	0.307
1.38	0.297
1.57	0.277
1.80	0.270
1.93	0.256
2.17	0.244
2.45	0.238
2.65	0.223
CURVE 4 T = 300	
0.47	0.480
0.49	0.467
0.56	0.467
0.61	0.472
0.79	0.368
1.00	0.280
1.20	0.222
1.41	0.192
1.61	0.176
1.81	0.170
2.00	0.164
CURVE 5 T = 1100	
0.46	0.461
0.49	0.460
0.55	0.453
0.62	0.447
0.79	0.357
1.00	0.302
CURVE 5 (cont.)	
1.20	0.267
1.40	0.245
1.60	0.229
1.81	0.217
2.00	0.205
CURVE 6 T = 1500	
0.47	0.445
0.50	0.447
0.56	0.440
0.62	0.434
0.80	0.352
1.01	0.305
1.20	0.278
1.40	0.258
1.60	0.244
1.80	0.232
2.00	0.223
CURVE 7 T = 2000	
0.46	0.416
0.50	0.418
0.56	0.416
0.62	0.409
0.80	0.343
1.00	0.303*
1.20	0.280
1.41	0.264
1.60	0.254
1.80	0.241
2.00	0.239
CURVE 8 T = 2500	
0.46	0.380
0.50	0.381
0.56	0.381
0.62	0.378
0.80	0.330
CURVE 8 (cont.)	
1.01	0.310
1.19	0.291
1.40	0.277
1.61	0.260
1.81	0.252
2.00	0.246

* Not shown on plot

SPECIFICATION TABLE NO. 172 HEMISPHERICAL TOTAL EMITTANCE OF RHODIUM

Curve No.	Ref. No.	Year	Temperature Range, K	Reported Error, %	Composition (weight percent), Specifications and Remarks
1	343	1964	1142-2124		Measured in vacuum (10^{-9} mm Hg).

DATA TABLE NO. 172 HEMISPHERICAL TOTAL EMITTANCE OF RHODIUM

[Temperature, T, K; Emittance, ϵ]

T	ϵ
CURVE 1*	
1142	0.101
1278	0.120
1416	0.135
1553	0.150
1750	0.168
1962	0.180
2124	0.189

*Not shown on plot

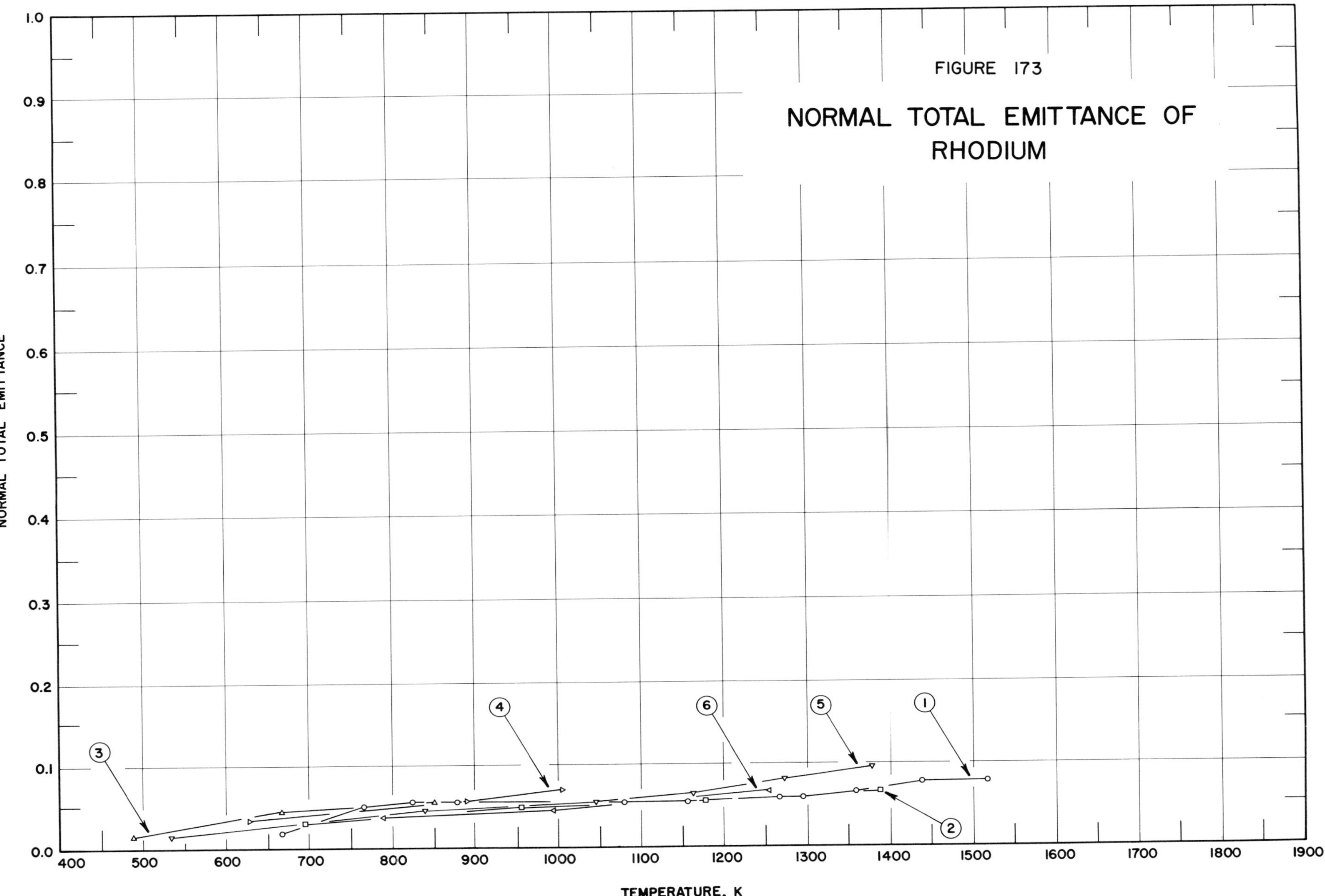

FIGURE 173
NORMAL TOTAL EMITTANCE OF RHODIUM
NORMAL TOTAL EMITTANCE
TEMPERATURE, K
0.0
0.1
0.2
0.3
0.4
0.5
0.6
0.7
0.8
0.9
1.0
400
500
600
700
800
900
1000
1100
1200
1300
1400
1500
1600
1700
1800
1900
1
2
3
4
5
6

SPECIFICATION TABLE NO. 173 NORMAL TOTAL EMITTANCE OF RHODIUM

Curve No.	Ref. No.	Year	Temperature Range, K	Geometry θ'	Reported Error, %	Composition (weight percent), Specifications and Remarks
1	34	1957	666-1516	$\sim 0°$	± 10	Pure; strip (0.005 in. thick); oxidized in air at red heat for 30 min.; measured in air (5×10^{-4} mm Hg); increasing temp, Cycle 1.
2	34	1957	1389-694	$\sim 0°$	± 10	Above specimen and conditions; decreasing temp, Cycle 1.
3	34	1957	489-850	$\sim 0°$	± 10	Pure; strip (0.005 in. thick); same results obtained for 3 different surface treatments (a. as received, b. cleaned with liquid detergent, c. polished); measured in air (5×10^{-4} mm Hg); increasing temp, Cycle 1.
4	34	1957	1003-628	$\sim 0°$	± 10	Above specimen and conditions; decreasing temp, Cycle 1.
5	34	1957	533-1378	$\sim 0°$	± 10	Above specimen and conditions; Cycle 2.
6	34	1957	789-1255	$\sim 0°$	± 10	Above specimen and conditions; Cycle 3.

DATA TABLE NO. 173 NORMAL TOTAL EMITTANCE OF RHODIUM

[Temperature, T, K; Emittance, ϵ]

T	ϵ
CURVE 1	
666	0.019
766	0.050
822	0.055
878	0.054
1078	0.053
1155	0.054
1266	0.059
1294	0.059
1358	0.064
1439	0.077
1516	0.078
CURVE 2	
1389	0.065
1178	0.056
955	0.049
694	0.030
CURVE 3	
489	0.017
666	0.044
850	0.055
CURVE 4	
1003	0.068
889	0.055
628	0.034
CURVE 5	
533	0.016
839	0.044
1044	0.054
1161	0.063
1272	0.080
1378	0.094
CURVE 6	
789	0.037
994	0.044
1255	0.066

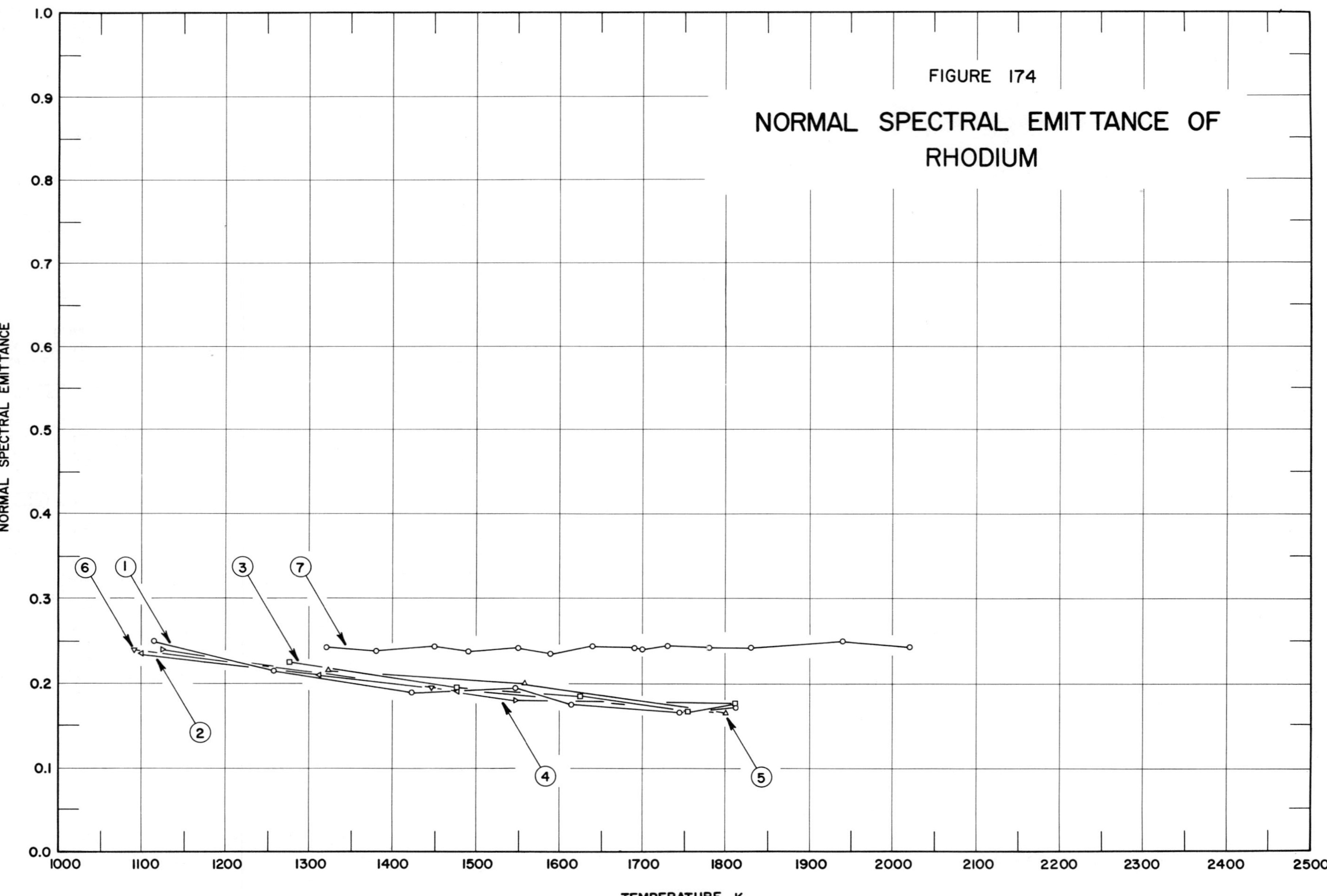
FIGURE 174
NORMAL SPECTRAL EMITTANCE OF RHODIUM
NORMAL SPECTRAL EMITTANCE
TEMPERATURE, K
1.0
0.9
0.8
0.7
0.6
0.5
0.4
0.3
0.2
0.1
0.0
1000
1100
1200
1300
1400
1500
1600
1700
1800
1900
2000
2100
2200
2300
2400
2500
6
1
3
7
2
4
5

SPECIFICATION TABLE NO. 174 NORMAL SPECTRAL EMITTANCE OF RHODIUM

Curve No.	Ref. No.	Year	Wavelength μ	Temperature Range, K	Geometry θ'	Reported Error, %	Composition (weight percent), Specifications and Remarks
1	34	1957	0.665	1114-1811	$\sim 0°$		Pure metal; same results obtained for 3 different surface treatments (a. as received, b. cleaned with liquid detergent, c. polished); measured in air (5×10^{-4} mm Hg); increasing temp , Cycle 1.
2	34	1957	0.665	1477-1100	$\sim 0°$		Above specimen and conditions; decreasing temp , Cycle 1.
3	34	1957	0.665	1277-1811	$\sim 0°$		Above specimen and conditions; Cycle 2.
4	34	1957	0.665	1811-1125	$\sim 0°$		Above specimen and conditions; decreasing temp , Cycle 2.
5	34	1957	0.665	1322-1800	$\sim 0°$		Above specimen and conditions; Cycle 3.
6	34	1957	0.665	1800-1091	$\sim 0°$		Above specimen and conditions; decreasing temp , Cycle 3.
7	55	1935	0.667	1320-2020	$\sim 0°$		Heat treated for over 700 hrs; measured in vacuum (4×10^{-8} mm Hg).

DATA TABLE NO. 174 NORMAL SPECTRAL EMITTANCE OF RHODIUM

[Temperature, T, K; Emittance, ϵ; Wavelength, λ, μ]

T	ϵ	T	ϵ
CURVE 1 $\lambda = 0.665$		CURVE 7 $\lambda = 0.667$	
1114	0.25	1320	0.243
1258	0.215	1380	0.239
1422	0.19	1450	0.244
1547	0.195	1490	0.238
1614	0.175	1550	0.241
1744	0.165	1590	0.235
1811	0.170	1640	0.244
		1690	0.242
CURVE 2 $\lambda = 0.665$		1700	0.240
		1730	0.744
		1780	0.241
1477	0.190	1830	0.241
1311	0.210	1940	0.250
1100	0.235	2020	0.242
CURVE 3 $\lambda = 0.665$			
1277	0.225		
1477	0.195		
1625	0.185		
1755	0.165		
1811	0.175		
CURVE 4 $\lambda = 0.665$			
1811	0.175*		
1547	0.180		
1125	0.240		
CURVE 5 $\lambda = 0.665$			
1322	0.215		
1558	0.200		
1800	0.165		
CURVE 6 $\lambda = 0.665$			
1800	0.165*		
1447	0.195		
1091	0.240		

*Not shown on plot

SPECIFICATION TABLE NO. 175 NORMAL SPECTRAL EMITTANCE OF RHODIUM

Curve No.	Ref. No.	Year	Temperature K	Wavelength Range, μ	Geometry θ '	Reported Error, %	Composition (weight percent), Specifications and Remarks
1	19	1914	<M. P.	0.55-0.65	~0°	1	Film; tungsten substrate; measured in hydrogen; Pt reference ($\epsilon = 0.33$ for $\lambda = 0.650\ \mu$ and $\epsilon = 0.38$ for $\lambda = 0.547\ \mu$ at all temp).

DATA TABLE NO. 175 NORMAL SPECTRAL EMITTANCE OF RHODIUM

[Wavelength, λ, μ; Emittance, ϵ; Temperature, T, K]

λ	ϵ
CURVE 1*	
T = < M. P.	
0.55	0.29
0.65	0.29

*Not shown on plot

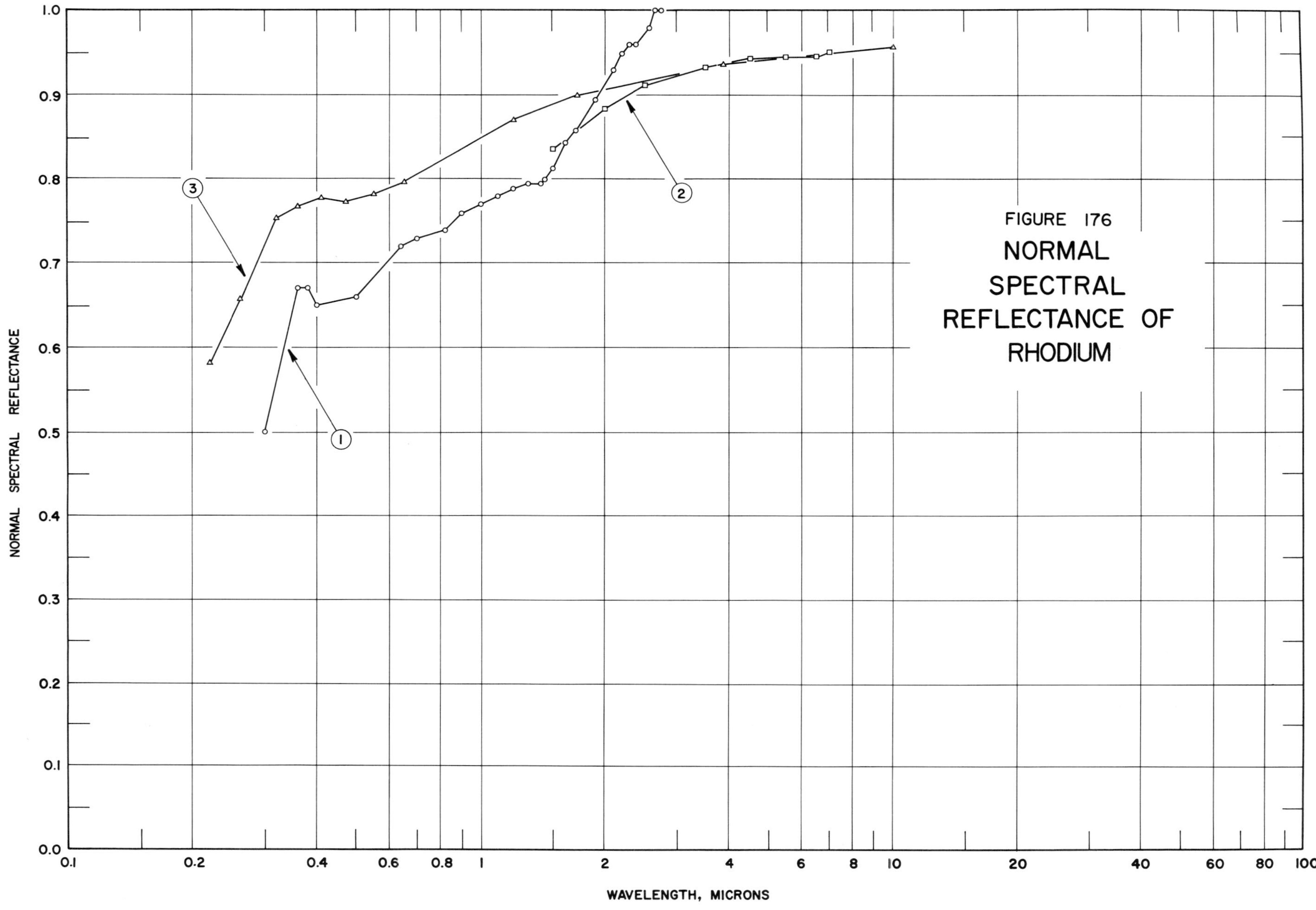

FIGURE 176 NORMAL SPECTRAL REFLECTANCE OF RHODIUM

SPECIFICATION TABLE NO. 176 NORMAL SPECTRAL REFLECTANCE OF RHODIUM

Curve No.	Ref. No.	Year	Temperature K	Wavelength Range, μ	Geometry θ	θ'	ω'	Reported Error, %	Composition (weight percent), Specifications and Remarks
1	34	1957	298	0.30-2.70	9°		2π	± 4	Pure metal; as received; data extracted from a smooth curve; magnesium carbonate reference.
2	362	1965	298	1.5-7.0	7°		2π		Evaporated film; average values.
3	371	1955	298	0.22-10.00	~0°	~0°			Freshly deposited film; measured directly after preparation; data extracted from smooth curve.

DATA TABLE NO. 176   NORMAL SPECTRAL REFLECTANCE OF RHODIUM

[Wavelength, λ, μ; Reflectance, ρ; Temperature, T, K]

λ	ρ	λ	ρ
CURVE 1 T = 298		CURVE 3 T = 298	
0.30	0.500	0.22	0.582
0.36	0.670	0.26	0.658
0.38	0.670	0.32	0.753
0.40	0.650	0.36	0.769
0.50	0.660	0.41	0.779
0.64	0.720	0.47	0.774
0.70	0.730	0.55	0.783
0.82	0.740	0.65	0.797
0.90	0.760	1.20	0.871
1.00	0.770	1.71	0.900
1.10	0.780	3.86	0.937
1.20	0.790	10.00	0.958
1.30	0.795		
1.40	0.795		
1.42	0.800		
1.50	0.815		
1.60	0.845		
1.70	0.860		
1.90	0.895		
2.10	0.930		
2.20	0.950		
2.30	0.960		
2.38	0.960		
2.56	0.980		
2.64	1.000		
2.70	1.000		
CURVE 2 T = 298			
1.5	0.8383		
2.0	0.8850		
2.5	0.9104		
3.5	0.9339		
4.5	0.9428		
5.5	0.9470		
6.5	0.9474		
7.0	0.9510		

* Not shown on plot

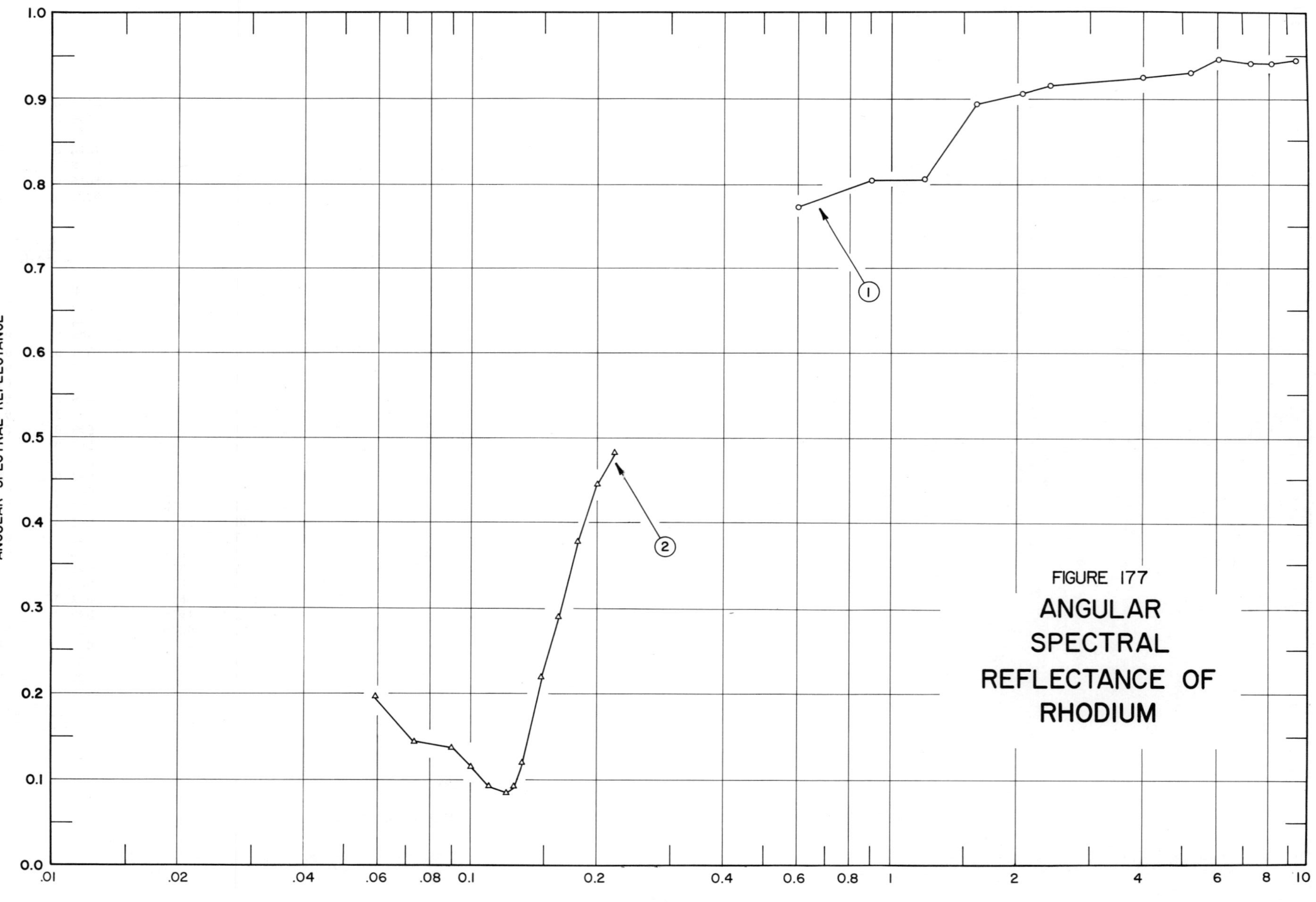

FIGURE 177
ANGULAR SPECTRAL REFLECTANCE OF RHODIUM

SPECIFICATION TABLE NO. 177 ANGULAR SPECTRAL REFLECTANCE OF RHODIUM

Curve No.	Ref. No.	Year	Temperature K	Wavelength Range, μ	Geometry θ	Geometry θ'	Geometry ω'	Reported Error, %	Composition (weight percent), Specifications and Remarks
1	132	1911	298	0.60-9.30	15°	15°		≤ 3	Silvered glass mirror reference.
2	306	1959	~298	0.0589-0.2200	~18°	~18°			Rh film (300 Å thick); evaporated from multistranded helical tungsten filaments in 10 to 20 sec at 5 x 10^{-5} mm Hg .

DATA TABLE NO. 177 ANGULAR SPECTRAL REFLECTANCE OF RHODIUM

[Wavelength, λ, μ; Reflectance, ρ; Temperature, T, K]

λ	ρ
CURVE 1	
T = 298	
0.60	0.775
0.90	0.805
1.20	0.807
1.60	0.895
2.05	0.905
2.40	0.915
4.05	0.925
5.20	0.930
6.07	0.945
7.20	0.940
8.05	0.940
9.30	0.945
CURVE 2	
T = ~298	
0.0589	0.198
0.0730	0.144
0.0900	0.138
0.1000	0.116
0.1101	0.093
0.1215	0.083
0.1269	0.092
0.1324	0.120
0.1485	0.221
0.1615	0.289
0.1798	0.378
0.2003	0.445
0.2200	0.482

SPECIFICATION TABLE NO. 178 ANGULAR SPECTRAL ABSORPTANCE OF RHODIUM

Curve No.	Ref. No.	Year	Temperature K	Wavelength Range, μ	Geometry θ	Reported Error, %	Composition (weight percent), Specifications and Remarks
1	225	1965	306	1.98-21.0	25°		Rhodium electroplated on polished nickel; polished with optical grade rouge with cotton moistened in ethyl alcohol; measured in dry nitrogen; heated cavity at approx 1056 K with platinum reference; authors assumed $\alpha = 1-R(2\pi, 25^\circ)$.

DATA TABLE NO. 178 ANGULAR SPECTRAL ABSORPTANCE OF RHODIUM

[Wavelength, λ, μ; Absorptance, α; Temperature, T, K]

λ	α
CURVE 1*	
T = 306	
1.98	0.154
2.63	0.133
4.18	0.105
9.08	0.071
12.1	0.057
15.0	0.048
17.8	0.044
20.0	0.044
21.0	0.043

*Not shown on plot

SPECIFICATION TABLE NO. 179 NORMAL SOLAR ABSORPTANCE OF RHODIUM

Curve No.	Ref. No.	Year	Temperature Range, K	Geometry θ	Reported Error, %	Composition (weight percent), Specifications and Remarks
1	34	1957	298	9°		Pure metal; as received (bright); computed from spectral reflectance data for sea level conditions.
2	34	1957	298	9°		Above specimen and conditions except computed for above atmosphere conditions.

DATA TABLE NO. 179 NORMAL SOLAR ABSORPTANCE OF RHODIUM

[Temperature, T, K; Absorptance, α]

T	α
CURVE 1*	
298	0.269
CURVE 2*	
298	0.263

*Not shown on plot

SPECIFICATION TABLE NO. 180 NORMAL SPECTRAL EMITTANCE OF RUTHENIUM

Curve No.	Ref. No.	Year	Wavelength μ	Temperature Range, K	Geometry θ'	Reported Error, %	Composition (weight percent), Specifications and Remarks
1	135	1961	0.655	1200-2500	• $\sim 0^\circ$		Commercial purity (< 1.0 impurities); prepared from powder; heated at 673 K for 1 hr and presintered at 1473 K for 1 hr; measured in vacuum; data extracted from smooth curve.

DATA TABLE NO. 180 NORMAL SPECTRAL EMITTANCE OF RUTHENIUM

[Temperature, T, K; Emittance, ϵ; Wavelength, λ, μ]

T	ϵ
CURVE 1*	
$\lambda = 0.655$	
1200	0.450
1300	0.403
1400	0.367
1500	0.343
1600	0.322
1700	0.316
1800	0.311
1900	0.308
2000	0.307
2100	0.307
2200	0.310
2300	0.314
2400	0.319
2500	0.324

*Not shown on plot

SPECIFICATION TABLE NO. 181 NORMAL TOTAL EMITTANCE OF SILICON

Curve No.	Ref. No.	Year	Temperature Range, K	Geometry θ'	Reported Error, %	Composition (weight percent), Specifications and Remarks
1	357	1961	100-251	~0°		IRC solar cell specimen; data extracted from smooth curve.

DATA TABLE NO. 181 NORMAL TOTAL EMITTANCE OF SILICON

[Temperature, T, K; Emittance, ϵ]

T	ϵ
CURVE 1	
100	0.187
137	0.192
178	0.208
212	0.222
251	0.235

SPECIFICATION TABLE NO. 182 NORMAL SPECTRAL EMITTANCE OF SILICON

Curve No.	Ref. No.	Year	Wavelength μ	Temperature Range, K	Geometry θ'	Reported Error, %	Composition (weight percent), Specifications and Remarks
1	20	1957	0.65	1000-1688	$\sim 0^\circ$	± 10	Etched; measured in vacuum (10^{-7}-10^{-8} mm Hg).

DATA TABLE NO. 182 NORMAL SPECTRAL EMITTANCE OF SILICON

[Temperature, T, K; Emittance, ∈; Wavelength, λ, μ]

T	∈
CURVE 1 $\lambda = 0.65$	
1000	0.64
1100	0.62
1200	0.60
1300	0.57
1400	0.54
1500	0.50
1600	0.48
1688	0.46

FIGURE 183 A

ANALYZED NORMAL SPECTRAL EMITTANCE OF SILICON

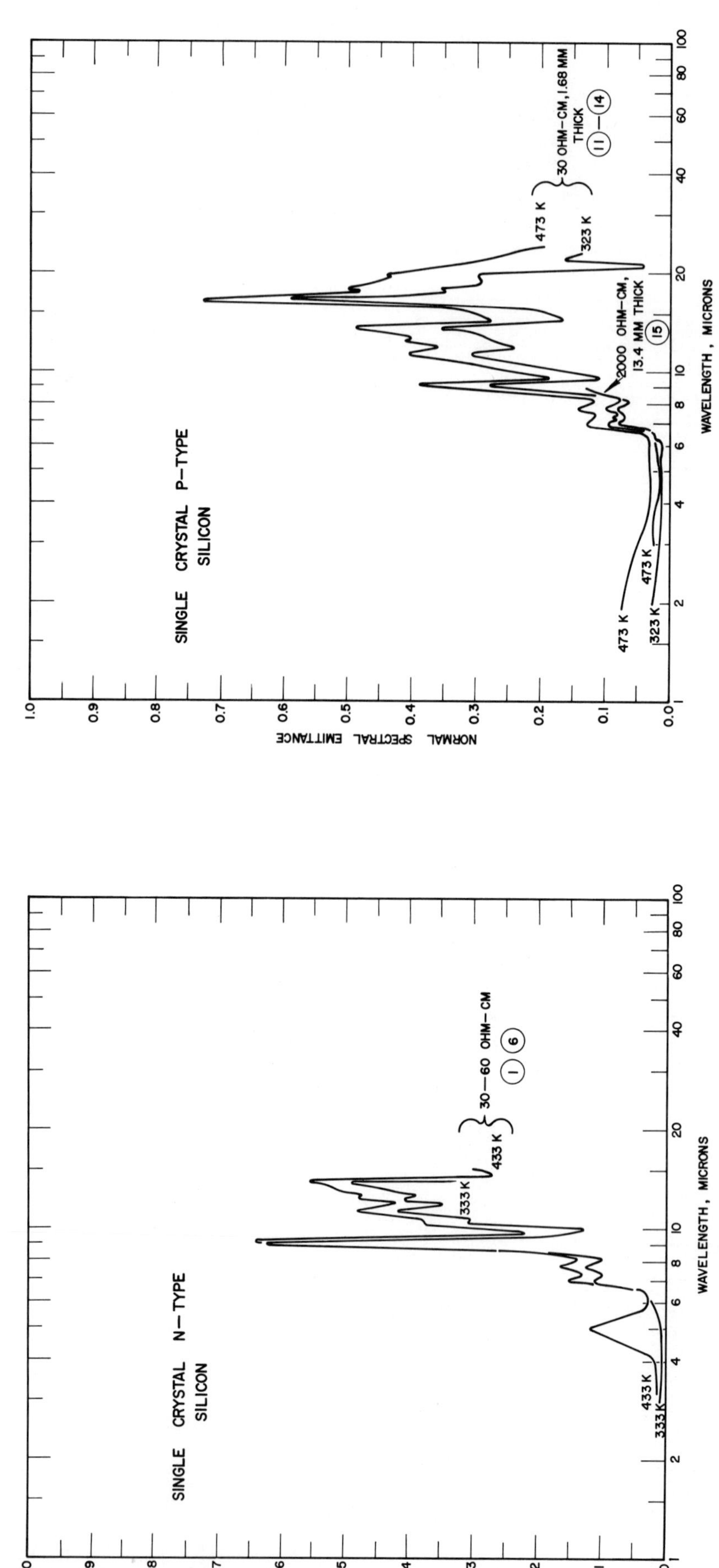

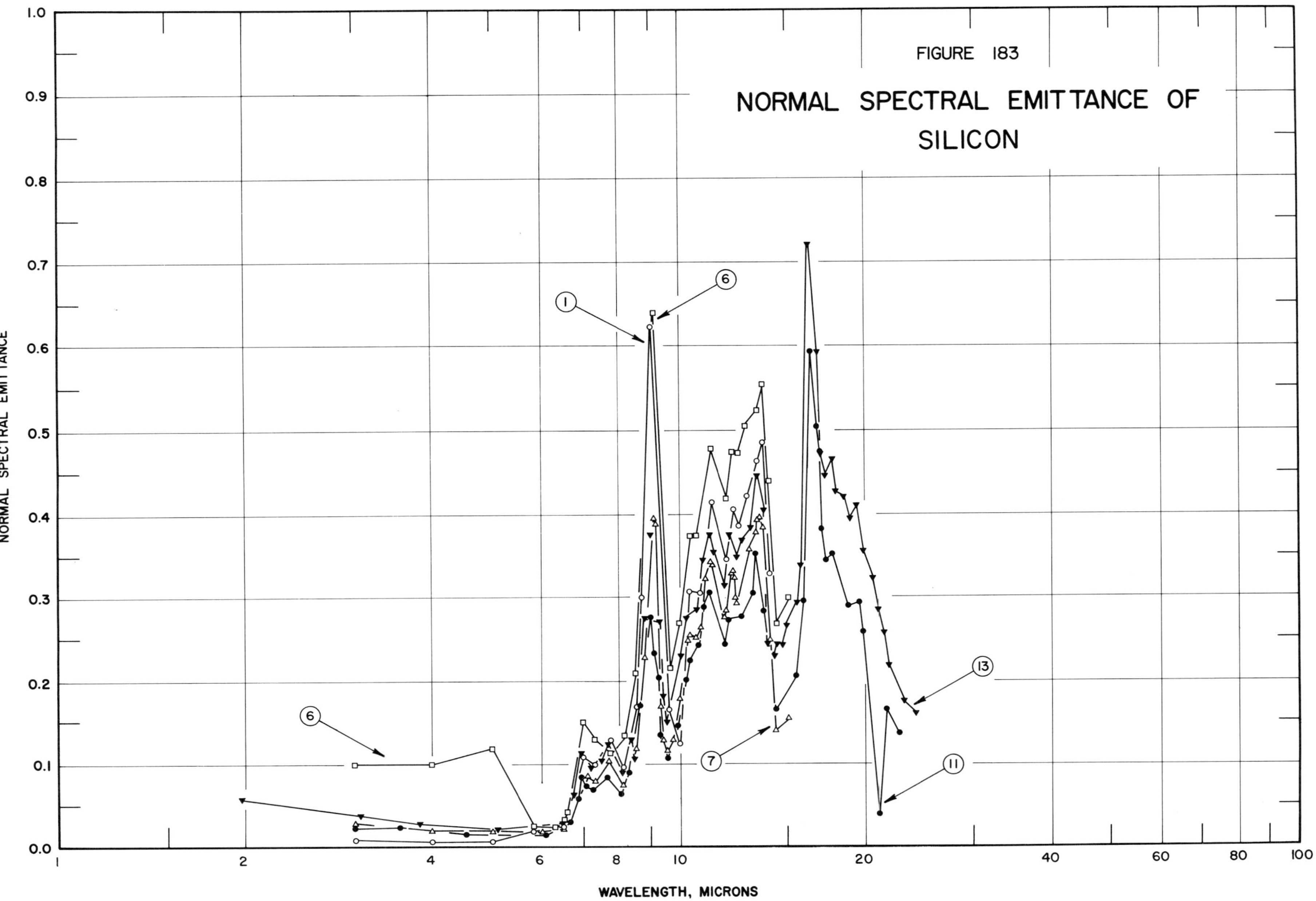
FIGURE 183
NORMAL SPECTRAL EMITTANCE OF
SILICON
NORMAL SPECTRAL EMITTANCE
WAVELENGTH, MICRONS
0.0
0.1
0.2
0.3
0.4
0.5
0.6
0.7
0.8
0.9
1.0
1
2
4
6
8
10
20
40
60
80
100
1
6
6
7
11
13

SPECIFICATION TABLE NO. 183 NORMAL SPECTRAL EMITTANCE OF SILICON

Curve No.	Ref. No.	Year	Temperature K	Wavelength Range, λ	Geometry θ'	Reported Error, %	Composition (weight percent), Specifications and Remarks
1	21	1961	333	3.0-14.0	~0°		n-type single crystal silicon; resistivity 30-60 ohm cm; ground and polished; measured in vacuum; data extracted from smooth curve.
2	21	1961	353	3.0-15.0	~0°		Above specimen and conditions.
3	21	1961	375	3.0-14.9	~0°		Above specimen and conditions.
4	21	1961	393	3.0-15.0	~0°		Above specimen and conditions.
5	21	1961	413	3.0-15.0	~0°		Above specimen and conditions.
6	21	1961	433	3.0-15.0	~0°		Above specimen and conditions.
7	21	1961	313	3.0-15.0	~0°		p-type single crystal silicon; ground and polished; measured in vacuum; data extracted from smooth curve.
8	21	1961	353	3.0-15.0	~0°		Above specimen and conditions.
9	21	1961	393	3.0-15.0	~0°		Above specimen and conditions.
10	21	1961	433	3.0-15.0	~0°		Above specimen and conditions.
11	214	1962	323	3.00-22.69	~0°		Single crystal p-type silicon (1.68 mm thick); resistivity 30 ohm cm; data extracted from smooth curve.
12	214	1962	373	2.29-24.00	~0°		Above specimen and conditions.
13	214	1962	423	1.94-24.01	~0°		Above specimen and conditions.
14	214	1962	473	1.94-23.99	~0°		Above specimen and conditions.
15	214	1962	473	2.00-8.94	~0°		Different sample (13.4 mm thick); same as above specimen and conditions; resistivity 2000 ohm cm.
16	357	1961	298	1.18-15.97	~0°		IRC solar cell specimen; data extracted from smooth curve.

DATA TABLE NO. 183 NORMAL SPECTRAL EMITTANCE OF SILICON

[Wavelength, λ, μ; Emittance, ∈; Temperature, T, K]

CURVE 1 T = 333

λ	μ
3.0	0.010
4.0	0.008
5.0	0.008
5.8	0.020
6.5	0.026
7.0	0.110
7.3	0.100
7.75	0.129
8.1	0.097
8.5	0.170
8.7	0.300
9.0	0.625
9.6	0.166
10.0	0.125
10.4	0.308
10.6	0.305
11.3	0.416
11.9	0.348
12.25	0.408
12.5	0.388
12.8	0.422
13.4	0.465
13.6	0.488
14.0	0.330

CURVE 2* T = 353

λ	μ
3.0	0.010
4.0	0.009
5.0	0.010
5.8	0.021
6.5	0.028
7.0	0.115
7.3	0.102
7.75	0.131
8.1	0.102
8.5	0.170
8.7	0.300
9.1	0.625
9.6	0.162
10.0	0.130
10.4	0.315
10.6	0.310
11.3	0.425

CURVE 2 (cont.)*

λ	μ
11.9	0.360
12.25	0.427
12.5	0.400
12.8	0.436
13.3	0.460
13.6	0.488
14.0	0.350
14.2	0.250
14.5	0.230
15.0	0.235

CURVE 3* T = 375

λ	μ
3.0	0.010
4.0	0.009
5.0	0.010
5.8	0.022
6.5	0.029
7.0	0.125
7.3	0.110
7.75	0.141
8.15	0.113
8.5	0.175
8.7	0.300
9.1	0.626
9.6	0.175
10.0	0.240
10.4	0.332
10.6	0.330
11.3	0.442
11.9	0.375
12.25	0.434
12.45	0.410
12.8	0.450
13.4	0.475
13.6	0.505
14.0	0.355
14.35	0.215
14.9	0.254

CURVE 4* T = 393

λ	μ
3.0	0.010
4.0	0.010
5.0	0.010
5.8	0.024
6.5	0.030
7.0	0.130
7.3	0.113
7.75	0.148
8.15	0.117
8.5	0.180
8.7	0.300
9.1	0.625
9.6	0.190
10.0	0.250
10.4	0.345
10.6	0.338
11.3	0.452
11.9	0.390
12.25	0.450
12.45	0.425
12.9	0.478
13.4	0.482
13.6	0.510
14.0	0.360
14.2	0.270
14.5	0.254
15.0	0.264

CURVE 5* T = 413

λ	μ
3.0	0.010
4.0	0.010
5.0	0.012
5.8	0.025
6.5	0.030
7.0	0.140
7.3	0.122
7.75	0.155
8.2	0.125
8.5	0.200
8.7	0.300
9.14	0.630
9.65	0.205
10.0	0.265

CURVE 5 (cont.)*

λ	μ
10.4	0.365
10.7	0.365
11.35	0.475
11.9	0.415
12.2	0.472
12.5	0.465
12.8	0.500
13.1	0.505
13.4	0.522
13.6	0.548
14.0	0.410
14.2	0.280
14.4	0.261
15.0	0.280

CURVE 6 T = 433

λ	μ
3.0	0.010*
4.0	0.010*
5.0	0.120
5.8	0.027
6.3	0.025
6.5	0.034
6.6	0.043
7.0	0.150
7.3	0.130
7.75	0.165
8.15	0.133
8.5	0.210
8.7	0.300*
9.1	0.640
9.65	0.217
10.0	0.270
10.4	0.375
10.65	0.375
11.35	0.480
11.9	0.420
12.2	0.476
12.50	0.475
12.8	0.505
13.4	0.524
13.6	0.555
14.0	0.440
14.4	0.269
15.0	0.300

CURVE 7 T = 313

λ	μ
3.0	0.030
4.0	0.021
5.0	0.020
6.0	0.020
6.5	0.024
7.1	0.088
7.3	0.080
7.7	0.103
8.1	0.076
8.5	0.120
8.8	0.230
9.1	0.398
9.15	0.390
9.3	0.170
9.4	0.130
9.55	0.118
9.7	0.130
10.0	0.180
10.3	0.250
10.4	0.255
10.6	0.252
10.7	0.265
11.0	0.323
11.2	0.344
11.3	0.340
11.8	0.278
11.95	0.286
12.2	0.330
12.25	0.333
12.3	0.325
13.35	0.300
12.4	0.294
13.0	0.360
13.3	0.380
13.4	0.395
13.5	0.398
13.65	0.385
14.0	0.250
14.3	0.140
15.0	0.155

CURVE 8* T = 353

λ	μ
3.0	0.033
4.0	0.022
5.0	0.021
6.0	0.022
6.5	0.025
7.05	0.108
7.3	0.090
7.7	0.120
8.1	0.086
8.5	0.150
8.8	0.240
9.1	0.405
9.15	0.390
9.35	0.200
9.5	0.150
9.6	0.140
9.75	0.150
10.0	0.186
10.3	0.270
10.4	0.282
10.55	0.273
10.70	0.287
11.3	0.285
11.8	0.315
11.9	0.310
12.0	0.316
12.3	0.370
12.4	0.355
12.5	0.350
12.75	0.375
13.0	0.390
13.25	0.404
13.35	0.404
13.4	0.405
13.6	0.433
14.0	0.265
14.25	0.171
14.4	0.175
14.6	0.171
15.0	0.185

CURVE 9* T = 393

λ	μ
3.0	0.030
4.0	0.026
5.0	0.023
6.0	0.025
6.5	0.032
7.05	0.115
7.30	0.100
7.75	0.128
8.10	0.099
8.5	0.160
9.0	0.400
9.10	0.412
9.2	0.400
9.4	0.200
9.5	0.170
9.65	0.152
9.8	0.162
10.0	0.190
10.4	0.295
10.5	0.300
10.65	0.298
10.75	0.305
11.0	0.375
11.2	0.400
11.35	0.408
11.50	0.395
11.8	0.350
11.9	0.345
12.0	0.350
12.3	0.404
12.4	0.395
12.5	0.383
12.65	0.385
12.8	0.420
13.0	0.425
13.2	0.420
13.6	0.480
14.0	0.365
14.4	0.185
14.6	0.205
15.0	0.222

CURVE 10* T = 433

λ	μ
3.0	0.040
4.0	0.027
5.0	0.023
6.0	0.027
6.5	0.038
7.05	0.133
7.35	0.110
7.75	0.142
8.15	0.110
8.8	0.280
9.0	0.400
9.1	0.412
9.2	0.400
9.4	0.240
9.5	0.200
9.65	0.179
9.8	0.190
10.4	0.320
10.5	0.322
10.65	0.320
10.8	0.340
11.0	0.390
11.2	0.420
11.3	0.425
11.5	0.420
11.8	0.380
11.95	0.362
12.0	0.370
12.2	0.410
12.35	0.418
12.5	0.408
12.6	0.404
12.65	0.408
12.85	0.450
13.0	0.460
13.2	0.460
13.3	0.458
13.6	0.488
14.0	0.375
14.25	0.240
14.5	0.210
14.8	0.190
15.0	0.194

DATA TABLE NO. 183 (continued)

CURVE 11
T = 323

λ	μ
3.00	0.024
3.53	0.025
4.52	0.016
6.08	0.016
6.66	0.030
6.87	0.060
6.93	0.085
7.06	0.074
7.24	0.070
7.64	0.085
8.05	0.065
8.27	0.090
8.63	0.171
9.01	0.278
9.08	0.233
9.21	0.203
9.29	0.135
9.54	0.108
9.90	0.146
10.20	0.201
10.40	0.226
10.72	0.242
10.99	0.290
11.20	0.307
11.80	0.243
12.11	0.273
12.59	0.278
13.02	0.305
13.36	0.354
13.65	0.285
14.31	0.166
15.45	0.206
15.82	0.297
16.27	0.594
16.66	0.505
16.83	0.477
17.01	0.382
17.30	0.347
17.66	0.353
18.89	0.291
19.52	0.296
19.95	0.259
21.03	0.041
21.72	0.166
22.69	0.138

CURVE 12*
T = 373

λ	μ
2.29	0.027
2.78	0.031
3.66	0,028
4.70	0.021
6.14	0.019
6.67	0.037
6.94	0.089
7.15	0.078
7.46	0.083
7.74	0.098
8.06	0.076
8.31	0.101
9.00	0.328
9.32	0.148
9.62	0.125
10.25	0.227
10.50	0.240
11.14	0.323
11.80	0.292
12.08	0.311
12.33	0.298
12.68	0.324
13.07	0.334
13.43	0.385
13.68	0.292
13.94	0.231
14.25	0.189
14.54	0.201
14.72	0.198
15.03	0.212
15.46	0.236
15.79	0.296
15.98	0.513
16.31	0.650
16.67	0.567
16.91	0.427
17.38	0.380
17.75	0.396
18.54	0.340
19.11	0.327
19.42	0.342
20.12	0.291
20.56	0.269
21.75	0.191
22.63	0.152
23.59	0.131
24.00	0.123

CURVE 13
T = 423

λ	μ
1.94	0.059
3.06	0.039
3.82	0.029
5.10	0.022
6.45	0.030
6.72	0.063
6.96	0.114
7.19	0.096
7.48	0.104
7.68	0.123
8.07	0.090
8.36	0.130
8.84	0.277
9.00	0.377
9.25	0.271
9.41	0.181
9.55	0.150
10.10	0.230
10.39	0.275
10.70	0.286
10.97	0.347
11.22	0.376
11.47	0.355
11.85	0.315
12.15	0.377
12.45	0.350
12.63	0.370
13.11	0.386
13.48	0.448
13.70	0.406
13.90	0.244
14.27	0.230
14.48	0.245
14.71	0.242
14.95	0.266
15.51	0.296
15.69	0.339
16.29	0.720
16.75	0.592
16.93	0.472
17.25	0.447
17.63	0.467
17.94	0.429
18.50	0.421
18.96	0.395
19.45	0.411
19.98	0.356

CURVE 13 (cont.)

λ	μ
20.57	0.325
21.00	0.286
21.45	0.259
21.86	0.219
23.12	0.176
24.01	0.160

CURVE 14*
T = 473

λ	μ
1.94	0.073
2.70	0.058
3.35	0.040
4.44	0.029
6.48	0.034
6.97	0.128
7.44	0.115
7.77	0.144
8.08	0.117
8.36	0.140
8.65	0.240
9.07	0.389
9.33	0.290
9.56	0.185
10.03	0.252
10.16	0.289
10.72	0.319
10.87	0.343
11.23	0.408
11.90	0.360
12.22	0.411
12.39	0.405
12.68	0.426
12.94	0.431
13.48	0.489
13.92	0.375
14.31	0.273
15.44	0.313
15.68	0.347
15.88	0.524
16.20	0.729
16.75	0.637
17.30	0.480
17.65	0.506
18.57	0.450
19.19	0.431
19.51	0.438
21.44	0.302

CURVE 14 (cont.)*

λ	μ
23.08	0.235
23.99	0.200

CURVE 15*
T = 473

λ	μ
2.00	0.024
2.87	0.016
4.15	0.012
5.13	0.011
5.34	0.013
5.63	0.014
5.83	0.015
6.07	0.017
6.23	0.021
6.51	0.024
6.58	0.032
6.89	0.095
7.11	0.083
7.25	0.085
7.40	0.082
7.71	0.098
7.87	0.092
8.18	0.079
8.46	0.106
8.94	0.128

CURVE 16*
T = 298

λ	μ
1.18	0.962
1.79	0.968
2.75	0.952
3.61	0.849
4.27	0.755
4.50	0.706
4.95	0.606
5.87	0.512
8.55	0.341
8.96	0.328
9.41	0.341
9.86	0.358
11.82	0.278
12.86	0.295
14.08	0.280
15.02	0.263
15.97	0.255

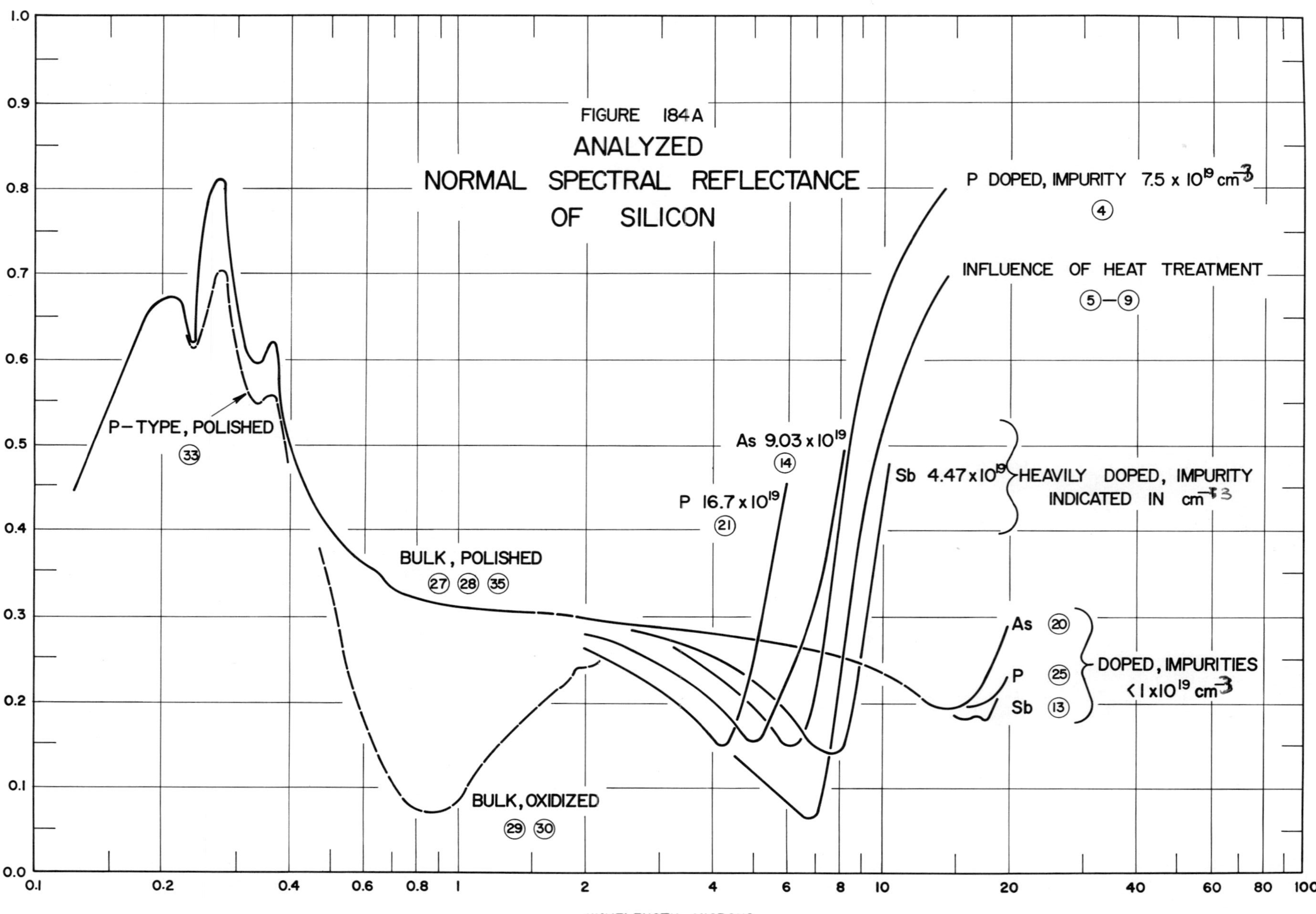
FIGURE 184A
ANALYZED
NORMAL SPECTRAL REFLECTANCE
OF SILICON
P DOPED, IMPURITY 7.5 x 10^19 cm^-3
(4)
INFLUENCE OF HEAT TREATMENT
(5)—(9)
Sb 4.47 x 10^19
HEAVILY DOPED, IMPURITY
INDICATED IN cm^-3
As 9.03 x 10^19
(14)
P 16.7 x 10^19
(21)
As (20)
P (25)
Sb (13)
DOPED, IMPURITIES
< 1 x 10^19 cm^-3
BULK, POLISHED
(27) (28) (35)
BULK, OXIDIZED
(29) (30)
P-TYPE, POLISHED
(33)
1.0
0.9
0.8
0.7
0.6
0.5
0.4
0.3
0.2
0.1
0.0
0.1
0.2
0.4
0.6
0.8
1
2
4
6
8
10
20
40
60
80
100
WAVELENGTH, MICRONS

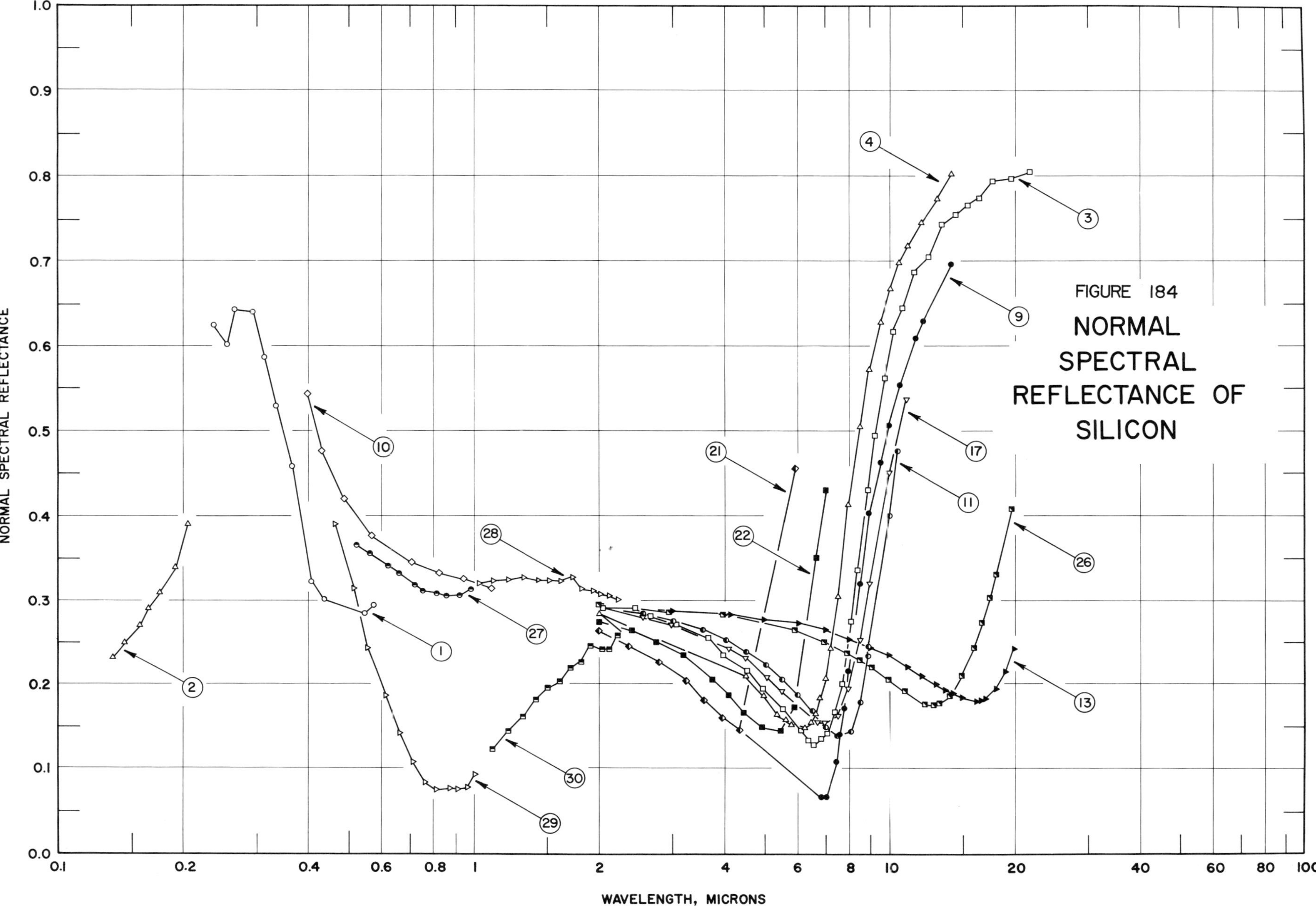
FIGURE 184
NORMAL SPECTRAL REFLECTANCE OF SILICON
NORMAL SPECTRAL REFLECTANCE
WAVELENGTH, MICRONS
0.0
0.1
0.2
0.3
0.4
0.5
0.6
0.7
0.8
0.9
1.0
0.1
0.2
0.4
0.6
0.8
1
2
4
6
8
10
20
40
60
80
100
1
2
3
4
9
10
11
13
17
21
22
26
27
28
29
30

SPECIFICATION TABLE NO. 184 NORMAL SPECTRAL REFLECTANCE OF SILICON

Curve No.	Ref. No.	Year	Temperature K	Wavelength Range, λ	Geometry θ	θ'	ω'	Reported Error, %	Composition (weight percent), Specifications and Remarks
1	133	1934	298	0.2350-0.5780	~0°	~0°			Cold worked; annealed in an inert gas for 6-24 hrs; polished; stored in dilute solutions of NaOH, NaOH + NaF, and HNO_3.
2	124	1941	298	0.1347-0.2026	~0°	~0°			Polished with tin oxide; surface somewhat pitted and scratched; measured in vacuum (0.001 mm Hg).
3	259	1966	298	2.03-21.80	~0°	~0°		<4	n-type silicon.
4	231	1964	298	2.00-14.03	~0°	~0°			Phosphorous doped, ($N = 7.5 \times 10^{19}$ cm^{-3}); polished; measured in vacuum ($<10^{-7}$ mm Hg).
5	231	1964	298	2.00-14.05	~0°	~0°			Above specimen and conditions except heat treated at 1310 K for 30 min.
6	231	1964	298	2.00-13.00	~0°	~0°			Above specimen and conditions except heat treated for 1 hr at 1310 K.
7	231	1964	298	2.00-14.03	~0°	~0°			Above specimen and conditions except heat treated at 1310 K for 1.5 hrs.
8	231	1964	298	1.99-14.04	~0°	~0°			Above specimen and conditions except heat treated at 1310 K for 2 hrs.
9	231	1964	298	1.99-14.04	~0°	~0°			Above specimen and conditions except heat treated at 1310 K for 3.5 hrs.
10	260	1966	298	0.399-1.101	~0°	~0°			n-type silicon photocell (concentration 4×10^{20} atoms cm^{-3}); data extracted from smooth curve; silicon photocell and aluminized mirror reference standards.
11	261	1962	298	2.05-10.47	~0°	~0°			n-type silicon doped with antimony; carrier concentration $N = 4.47 \times 10^{19}$ cm^{-3}; aluminum mirror reference standard.
12	261	1962	298	2.03-14.46	~0°	~0°			Different sample, same as above specimen and conditions except carrier concentration $N = 1.66 \times 10^{19}$ cm^{-3}.
13	261	1962	298	2.03-19.99	~0°	~0°			Different sample, same as above specimen and conditions except carrier concentration $N = 0.832 \times 10^{19}$ cm^{-3}.
14	261	1962	298	2.00-8.06	~0°	~0°			Different sample, same as above specimen and conditions except doped with arsenic; carrier concentration $N = 9.03 \times 10^{19}$ cm^{-3}.
15	261	1962	298	2.00-8.01	~0°	~0°			Different sample, same as above specimen and conditions except carrier concentration $N = 7.92 \times 10^{19}$ cm^{-3}.
16	261	1962	298	2.01-10.03	~0°	~0°			Different sample, same as above specimen and conditions except carrier concentration $N = 6.37 \times 10^{19}$ cm^{-3}.
17	261	1962	298	2.02-11.01	~0°	~0°			Different sample, same as above specimen and conditions except carrier concentration $N = 5.05 \times 10^{19}$ cm^{-3}.
18	261	1962	298	2.01-13.99	~0°	~0°			Different sample, same as above specimen and conditions except carrier concentration $N = 3.48 \times 10^{19}$ cm^{-3}.
19	261	1962	298	2.01-13.99	~0°	~0°			Different sample, same as above specimen and conditions except carrier concentration $N = 2.84 \times 10^{19}$ cm^{-3}.
20	261	1962	298	1.99-20.00	~0°	~0°			Different sample, same as above specimen and conditions except carrier concentration $N = 0.877 \times 10^{19}$ cm^{-3}.

SPECIFICATION TABLE NO. 184 (continued)

Curve No.	Ref. No.	Year	Temperature K	Wavelength Range, λ	Geometry θ	θ'	ω'	Reported Error, %	Composition (weight percent), Specifications and Remarks
21	261	1962	298	2.00-5.94	~0°	~0°			Different sample, same as above specimen and conditions except doped with phosphorous; carrier concentration $N = 16.7 \times 10^{19}$ cm^{-3}.
22	261	1962	298	2.01-7.02	~0°	~0°			Different sample, same as above specimen and conditions except carrier concentration $N = 10.22 \times 10^{19}$ cm^{-3}.
23	261	1962	298	2.01-11.96	~0°	~0°			Different sample, same as above specimen and conditions except carrier concentration $N = 4.38 \times 10^{19}$ cm^{-3}.
24	261	1962	298	2.03-14.94	~0°	~0°			Different sample, same as above specimen and conditions except carrier concentration $N = 2.05 \times 10^{19}$ cm^{-3}.
25	261	1962	298	2.01-20.00	~0°	~0°			Different sample, same as above specimen and conditions except carrier concentration $N = 0.74 \times 10^{19}$ cm^{-3}.
26	261	1962	298	2.01-19.67	~0°	~0°			Different sample, same as above specimen and conditions except carrier concentration $N = 1.27 \times 10^{19}$ cm^{-3}.
27	327	1960	298	0.52-0.98	~0°		2π	<3	MgO reference.
28	327	1960	298	1.02-2.21	~0°	~0°			Above specimen; different apparatus used for measurement.
29	327	1960	298	0.46-1.00	~0°		2π	<3	Oxidized silicon; MgO reference.
30	327	1960	298	1.11-2.21	~0°	~0°			Above specimen; different apparatus used for measurement.
31	361	1962	298	0.38-25.01	~0°	~0°			IRC silicon solar cell; data extracted from smooth curve.
32	357	1961	298	1.17-15.98	~0°	~0°			IRC solar cell specimen; data extracted from smooth curve.
33	363	1963	298	0.394-0.228	~0°	~0°			p-type silicon; sliced to 1-1.5 cm thickness; ground and polished with diamond paste, then final polished with Al_2O_3 on flannel; etched and cleaned; impurity 10^{20} B atoms cm^{-3}; effect of B impurity levels on peaks given in reference.
34	364	1968	~298	5.00-11.01	10.5°	10.5°			n-type silicon wafers; diffused with antimony using deposit and slumping method; ambient atmospheres, nitrogen for deposition, and dry oxygen for slumping.
35	365	1968	~298	0.124-1.823	~0°	~0°			Bulk silicon.

DATA TABLE NO. 184 NORMAL SPECTRAL REFLECTANCE OF SILICON

[Wavelength, λ, μ; Reflectance, ρ; Temperature T, K]

λ	ρ
CURVE 1	
T = 298	
0.2350	0.623
0.2537	0.601
0.2650	0.642
0.2930	0.640
0.3125	0.587
0.3340	0.528
0.3660	0.456
0.4060	0.322
0.4355	0.301
0.5460	0.283
0.5780	0.294
CURVE 2	
T = 298	
0.1347	0.23
0.1438	0.25
0.1570	0.27
0.1640	0.29
0.1757	0.31
0.1901	0.34
0.2026	0.39
CURVE 3	
T = 298	
2.03	0.289
2.43	0.289
2.65	0.281
3.05	0.271
3.67	0.255
3.99	0.234
4.55	0.216
4.96	0.195
5.56	0.170
6.10	0.146
6.36	0.132
6.56	0.128
6.85	0.134
7.08	0.141
7.39	0.167
7.64	0.200
8.08	0.275
8.40	0.337
8.88	0.430
9.22	0.494
CURVE 3 (cont.)	
9.78	0.560
10.21	0.618
10.83	0.646
11.55	0.688
12.43	0.704
13.34	0.743
14.36	0.755
15.32	0.767
16.32	0.774
17.83	0.794
19.80	0.798
21.80	0.804
CURVE 4	
T = 298	
2.00	0.284
4.50	0.210
4.99	0.186
5.36	0.163
5.62	0.157
5.80	0.151
6.21	0.148
6.44	0.154
6.60	0.165
6.79	0.184
7.00	0.208
7.19	0.242
7.49	0.305
7.95	0.412
8.47	0.504
8.98	0.571
9.50	0.629
10.00	0.668
10.47	0.696
11.02	0.719
11.98	0.748
13.00	0.776
14.03	0.802
CURVE 5*	
T = 298	
2.00	0.284
5.82	0.124
6.01	0.111
6.18	0.105
CURVE 5 (cont.)*	
6.60	0.105
6.79	0.109
7.02	0.123
7.22	0.150
7.40	0.179
7.60	0.220
7.99	0.306
8.50	0.407
9.00	0.486
9.52	0.545
9.99	0.585
10.53	0.621
11.05	0.666
11.53	0.676
12.02	0.697
13.05	0.727
14.05	0.765
CURVE 6*	
T = 298	
2.00	0.285
6.43	0.089
7.01	0.103
7.23	0.124
7.40	0.148
7.58	0.181
7.77	0.220
7.98	0.263
8.48	0.361
9.01	0.443
9.48	0.499
10.00	0.548
11.52	0.636
12.02	0.655
13.00	0.692
CURVE 7*	
T = 298	
2.00	0.285
6.82	0.069
7.21	0.092
7.39	0.122
7.78	0.186
8.48	0.336
8.97	0.422
CURVE 7 (cont.)	
9.49	0.480
10.00	0.531
10.51	0.567
11.03	0.600
11.51	0.620
12.02	0.640
14.03	0.721
CURVE 8*	
T = 298	
1.99	0.284
6.82	0.067
7.02	0.067
7.41	0.108
7.58	0.140
7.78	0.171
7.95	0.216
8.47	0.320
8.96	0.402
9.48	0.462
9.98	0.507
10.50	0.552
11.52	0.609
12.03	0.630
14.04	0.696
CURVE 9	
T = 298	
1.99	0.284*
6.82	0.067
7.02	0.067
7.41	0.108
7.58	0.140
7.78	0.171
7.95	0.216
8.47	0.320
8.96	0.402
9.48	0.462
9.98	0.507
10.50	0.552
11.52	0.609
12.03	0.630
14.04	0.696
CURVE 10	
T = 298	
0.399	0.542
0.432	0.475
0.488	0.419
0.571	0.374
0.706	0.344
0.825	0.332
0.955	0.325
1.101	0.313
CURVE 11	
T = 298	
2.05	0.292*
2.53	0.283
3.01	0.276
3.52	0.265
4.03	0.252
4.50	0.238
5.04	0.222
5.52	0.206
5.99	0.186
6.52	0.168
7.00	0.149
7.48	0.139
8.03	0.143
8.52	0.178
8.97	0.234
8.97	0.244
10.02	0.400
10.47	0.475
CURVE 12*	
T = 298	
2.03	0.291
2.94	0.283
3.52	0.275
4.01	0.269
4.53	0.267
5.02	0.264
5.47	0.258
6.00	0.253
6.52	0.243
6.97	0.236
7.50	0.228
7.96	0.218
CURVE 12 (cont.)*	
8.48	0.210
9.01	0.203
9.50	0.194
9.98	0.183
10.51	0.175
10.99	0.164
11.50	0.160
12.00	0.160
12.49	0.163
13.02	0.175
13.52	0.193
13.97	0.214
14.46	0.237
CURVE 13	
T = 298	
2.03	0.291*
2.98	0.287
4.06	0.283
4.99	0.277
6.02	0.272
7.00	0.265
8.00	0.254
9.01	0.244
9.98	0.234
10.99	0.221
11.98	0.210
12.98	0.200
13.45	0.193
14.02	0.189
14.95	0.183
16.02	0.180
16.50	0.180
16.96	0.182
18.00	0.195
18.99	0.216
19.99	0.242
CURVE 14*	
T = 298	
2.00	0.279
2.50	0.262
2.98	0.243
3.52	0.221
4.01	0.200
CURVE 14 (cont.)*	
4.48	0.178
5.03	0.157
7.04	0.327
8.06	0.491
CURVE 15*	
T = 298	
2.00	0.287
2.55	0.275
2.98	0.257
3.52	0.236
4.02	0.213
4.48	0.192
5.03	0.173
5.56	0.157
6.50	0.184
7.00	0.258
8.01	0.432
CURVE 16*	
T = 298	
2.01	0.290
2.50	0.276
2.94	0.267
3.27	0.259
4.04	0.237
4.48	0.215
4.96	0.193
5.49	0.175
6.72	0.163
7.06	0.173
7.51	0.216
8.03	0.288
8.99	0.442
10.03	0.547
CURVE 17	
T = 298	
2.02	0.289*
2.53	0.279
2.98	0.271
4.04	0.242
4.52	0.231
5.03	0.208
CURVE 17 (cont.)	
5.51	0.190
6.70	0.154
6.99	0.154
7.48	0.162
7.99	0.194
8.53	0.252
9.00	0.320
10.06	0.450
11.01	0.537
CURVE 18*	
T = 298	
2.01	0.288
2.48	0.280
2.95	0.277
3.51	0.270
4.04	0.266
4.47	0.248
4.98	0.240
5.49	0.231
5.95	0.216
6.43	0.200
7.00	0.187
7.50	0.174
7.96	0.164
8.45	0.163
8.98	0.164
10.03	0.235
11.05	0.331
12.02	0.427
13.99	0.551
CURVE 19*	
T = 298	
2.01	0.289
2.52	0.279
3.01	0.275
3.50	0.276
4.02	0.266
4.46	0.259
4.98	0.250
5.52	0.239
5.96	0.230
6.47	0.220
6.98	0.208

DATA TABLE NO. 184 (continued)

λ	ρ
CURVE 19 (cont.)*	
7.52	0.196
8.52	0.177
8.97	0.171
9.52	0.170
10.01	0.170
10.54	0.185
11.03	0.201
11.50	0.230
12.02	0.268
12.52	0.307
13.01	0.348
13.99	0.419
CURVE 20*	
T = 298	
1.99	0.291
2.98	0.286
4.04	0.281
5.04	0.273
5.99	0.268
7.01	0.261
7.99	0.249
8.52	0.245
9.00	0.239
10.00	0.226
11.05	0.213
12.03	0.203
13.01	0.194
13.50	0.193
13.99	0.191
14.50	0.191
14.98	0.193
15.54	0.193
15.99	0.197
17.00	0.213
18.03	0.235
20.00	0.292
CURVE 21	
T = 298	
2.00	0.264
2.35	0.247
2.79	0.227
3.22	0.203

λ	ρ
CURVE 21 (ccnt.)	
3.57	0.181
3.95	0.160
4.37	0.146
5.94	0.456
CURVE 22	
T = 298	
2.01	0.274
2.40	0.263
2.73	0.250
3.19	0.234
3.74	0.205
4.11	0.187
4.45	0.166
4.94	0.149
5.45	0.145
5.87	0.172
6.65	0.350
7.02	0.430
CURVE 23*	
T = 298	
2.01	0.294
2.97	0.282
3.44	0.269
3.97	0.257
4.54	0.242
5.36	0.221
5.79	0.209
6.36	0.192
6.93	0.172
7.50	0.157
8.21	0.153
8.80	0.168
10.00	0.275
10.54	0.339
10.98	0.400
11.49	0.451
11.96	0.496
CURVE 24*	
T = 298	
2.03	0.294

λ	ρ
CURVE 24 (cont.)	
3.01	0.281
3.97	0.275
4.45	0.267
5.17	0.260
5.60	0.252
6.21	0.243
6.73	0.234
7.34	0.221
8.00	0.206
8.59	0.193
9.15	0.182
9.77	0.173
10.34	0.170
10.92	0.170
11.39	0.175
11.98	0.191
12.91	0.242
13.50	0.284
14.01	0.318
14.94	0.397
CURVE 25*	
T = 298	
2.01	0.294
2.97	0.289
3.93	0.282
4.96	0.279
5.95	0.273
6.96	0.266
7.97	0.260
8.98	0.254
9.96	0.241
11.03	0.236
11.47	0.229
11.96	0.224
12.46	0.218
13.00	0.211
13.98	0.200
14.94	0.198
15.76	0.194
16.14	0.194
16.75	0.195
17.50	0.201
17.99	0.206
18.49	0.210

λ	ρ
CURVE 25 (cont.)*	
18.96	0.220
20.00	0.237
CURVE 26	
T = 298	
2.01	0.295
2.96	0.286
3.98	0.282
5.91	0.264
6.99	0.250
7.96	0.237
8.45	0.228
9.04	0.220
9.96	0.204
10.96	0.191
12.13	0.177
12.77	0.176
13.19	0.177
13.99	0.185
14.94	0.211
15.98	0.243
16.58	0.273
17.25	0.302
18.00	0.331
19.67	0.408
CURVE 27	
T = 298	
0.52	0.363
0.56	0.354
0.62	0.341
0.66	0.332
0.72	0.319
0.75	0.311
0.81	0.308
0.86	0.304
0.92	0.305
0.98	0.312
CURVE 28	
T = 298	
1.02	0.320
1.11	0.323

λ	ρ
CURVE 28 (cont.)	
1.21	0.325
1.31	0.328
1.42	0.325
1.52	0.324
1.62	0.323
1.72	0.328
1.81	0.315
1.92	0.311
2.01	0.308
2.11	0.305
2.21	0.301
CURVE 29	
T = 298	
0.46	0.389
0.51	0.314
0.55	0.242
0.61	0.183
0.66	0.141
0.71	0.106
0.76	0.082
0.81	0.074
0.87	0.076
0.91	0.076
0.96	0.077
1.00	0.091
CURVE 30	
T = 298	
1.11	0.121
1.21	0.142
1.31	0.160
1.41	0.180
1.51	0.194
1.61	0.201
1.70	0.218
1.81	0.225
1.91	0.246
2.02	0.241
2.10	0.241
2.21	0.258

λ	ρ
CURVE 31*	
T = 298	
0.38	0.091
0.53	0.053
0.75	0.033
0.96	0.033
1.16	0.063
1.56	0.077
2.55	0.076
3.24	0.058
3.99	0.119
4.75	0.281
5.98	0.480
6.41	0.532
7.00	0.558
7.55	0.596
8.03	0.629
8.96	0.623
9.33	0.596
10.27	0.658
10.84	0.680
11.91	0.711
13.81	0.743
16.49	0.761
20.81	0.773
21.88	0.763
25.01	0.764
CURVE 32*	
T = 298	
1.17	0.042
1.76	0.036
2.74	0.052
3.45	0.135
4.23	0.248
4.46	0.298
4.80	0.375
5.24	0.436
6.29	0.526
8.59	0.665
8.97	0.676
9.41	0.660
9.88	0.645
11.94	0.724
12.87	0.709
13.97	0.723

λ	ρ
CURVE 32 (cont.)*	
14.93	0.740
15.98	0.749
CURVE 33*	
T = 298	
0.394	0.479
0.392	0.489
0.389	0.502
0.383	0.517
0.380	0.530
0.376	0.540
0.372	0.548
0.367	0.557
0.360	0.559
0.356	0.558
0.349	0.549
0.339	0.548
0.330	0.548
0.324	0.552
0.318	0.560
0.313	0.570
0.310	0.577
0.306	0.587
0.303	0.599
0.300	0.611
0.297	0.625
0.295	0.635
0.292	0.647
0.291	0.659
0.290	0.671
0.286	0.683
0.283	0.694
0.277	0.701
0.272	0.703
0.266	0.697
0.263	0.687
0.259	0.676
0.256	0.667
0.253	0.658
0.251	0.648
0.248	0.639
0.245	0.630
0.243	0.622
0.238	0.613
0.235	0.613

λ	ρ
CURVE 33 (cont.)*	
0.231	0.621
0.228	0.629
CURVE 34*	
T = ~298	
5.00	0.258
6.00	0.221
7.01	0.196
7.60	0.175
8.00	0.184
9.01	0.219
9.99	0.279
11.01	0.335
CURVE 35*	
T = ~298	
0.124	0.449
0.138	0.500
0.155	0.560
0.177	0.639
0.191	0.668
0.206	0.670
0.225	0.670
0.229	0.650
0.234	0.620
0.239	0.671
0.245	0.719
0.249	0.770
0.254	0.782
0.258	0.796
0.263	0.804
0.268	0.808
0.273	0.810
0.278	0.808
0.283	0.780
0.289	0.758
0.293	0.719
0.298	0.682
0.306	0.641
0.317	0.610
0.327	0.600
0.336	0.598
0.346	0.598
0.356	0.615

* Not shown on plot

DATA TABLE NO. 184 (continued)

λ	ρ
CURVE 35 (cont.)*	
0.366	0.624
0.371	0.610
0.373	0.682
0.383	0.545
0.391	0.511
0.403	0.498
0.413	0.475
0.420	0.462
0.430	0.452
0.449	0.435
0.469	0.418
0.494	0.403
0.516	0.389
0.532	0.379
0.558	0.371
0.587	0.365
0.635	0.350
0.681	0.338
0.700	0.334
0.729	0.331
0.770	0.328
0.810	0.322
0.861	0.319
0.911	0.316
0.961	0.314
1.050	0.311
1.117	0.309
1.240	0.308
1.319	0.305
1.425	0.305
1.610	0.301
1.823	0.301

* Not shown on plot

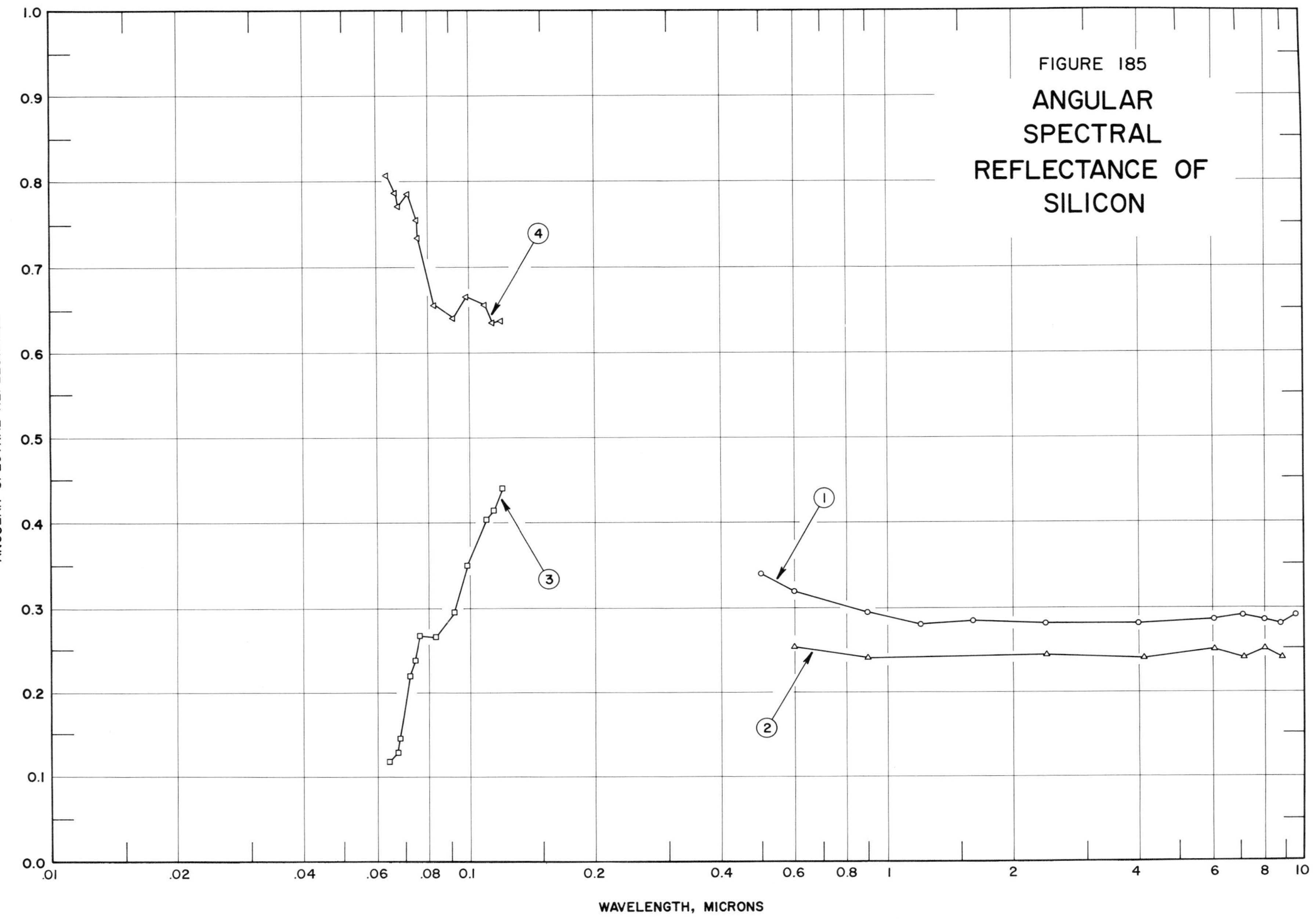

FIGURE 185
ANGULAR SPECTRAL REFLECTANCE OF SILICON
ANGULAR SPECTRAL REFLECTANCE
WAVELENGTH, MICRONS
1.0
0.9
0.8
0.7
0.6
0.5
0.4
0.3
0.2
0.1
0.0
.01
.02
.04
.06
.08
0.1
0.2
0.4
0.6
0.8
1
2
4
6
8
10
1
2
3
4

SPECIFICATION TABLE NO. 185 ANGULAR SPECTRAL REFLECTANCE OF SILICON

Curve No.	Ref. No.	Year	Temperature K	Wavelength Range, λ	Geometry θ	θ'	ω'	Reported Error, %	Composition (weight percent), Specifications and Remarks
1	132	1911	298	0.50-9.50	15°	15°		≤ 3	Obtained from Kahlbaum; polished with a fine grade of emery paper covered with a mixture of tin oxide and a little graphite; free from scratches; silvered glass mirror reference.
2	132	1911	298	0.60-8.80	15°	15°		≤ 3	Obtained from Carborundum Co.; different sample, same as curve 1 specimen and conditions except took a poor polish.
3	326	1962	298	0.064-0.120	20°	20°		<10	n-type single crystal cut along the (111) plane; polished mechanically; soaked with HF for 90 sec, rinsed several times with acetone; mounted immediately after rinsing, evacuated.
4	326	1962	298	0.064-0.120	70°	70°		< 5	Above specimen and conditions.
5	366	1963	~298	2.0-50.0	30°	30°			Specimen 1 cm thick; data extracted from smooth curve; aluminum mirror standard.

DATA TABLE NO. 185 ANGULAR SPECTRAL REFLECTANCE OF SILICON

[Wavelength, λ, μ; Reflectance, ρ; Temperature, T, K]

λ	ρ	λ	ρ
CURVE 1		CURVE 4	
T = 298		T = 298	
0.50	0.340	0.064	0.808
0.60	0.320	0.067	0.738
0.90	0.295	0.068	0.736
1.20	0.280	0.072	0.736
1.60	0.283	0.075	0.756
2.40	0.280	0.076	0.735
4.00	0.280	0.083	0.654
6.05	0.285	0.092	0.641
7.10	0.290	0.099	0.667
8.00	0.283	0.109	0.657
8.75	0.280	0.114	0.636
9.50	0.290	0.120	0.638
CURVE 2		CURVE 5*	
T = 298		T = 298	
0.60	0.253	2.0	0.342
0.90	0.240	5.5	0.382
2.40	0.245	7.1	0.310
4.10	0.240	8.0	0.327
6.03	0.240	8.6	0.299
7.10	0.240	9.6	0.319
8.00	0.250	11.4	0.284
8.80	0.240	13.0	0.303
		17.5	0.353
CURVE 3		21.8	0.355
T = 298		25.1	0.366
		31.9	0.341
0.064	0.118	36.0	0.336
0.067	0.128	42.3	0.349
0.068	0.144	45.4	0.364
0.072	0.219	48.5	0.426
0.074	0.238	50.0	0.399
0.076	0.267		
0.083	0.264		
0.092	0.294		
0.099	0.350		
0.109	0.402		
0.114	0.414		
0.120	0.440		

SPECIFICATION TABLE NO. 186 NORMAL SPECTRAL ABSORPTANCE OF SILICON

Curve No.	Ref. No.	Year	Temperature K	Wavelength Range, μ	Geometry θ	Reported Error, %	Composition (weight percent), Specifications and Remarks
1	367	1963	77	16.0-44.0	0°		n-type silicon; 2 mm thick; measured in vacuum; data extracted from smooth curve.

DATA TABLE NO. 186 NORMAL SPECTRAL ABSORPTANCE OF SILICON

[Wavelength, λ, μ; Absorptance, α; Temperature, T, K]

λ	α
CURVE 1*	
T = 77	
16.0	0.34
16.4	0.41
16.6	0.48
17.0	0.38
17.7	0.23
18.4	0.25
18.4	0.21
19.1	0.17
19.6	0.18
21.0	0.09
22.3	0.05
23.9	0.04
26.0	0.03
33.9	0.07
37.7	0.13
44.0	0.19

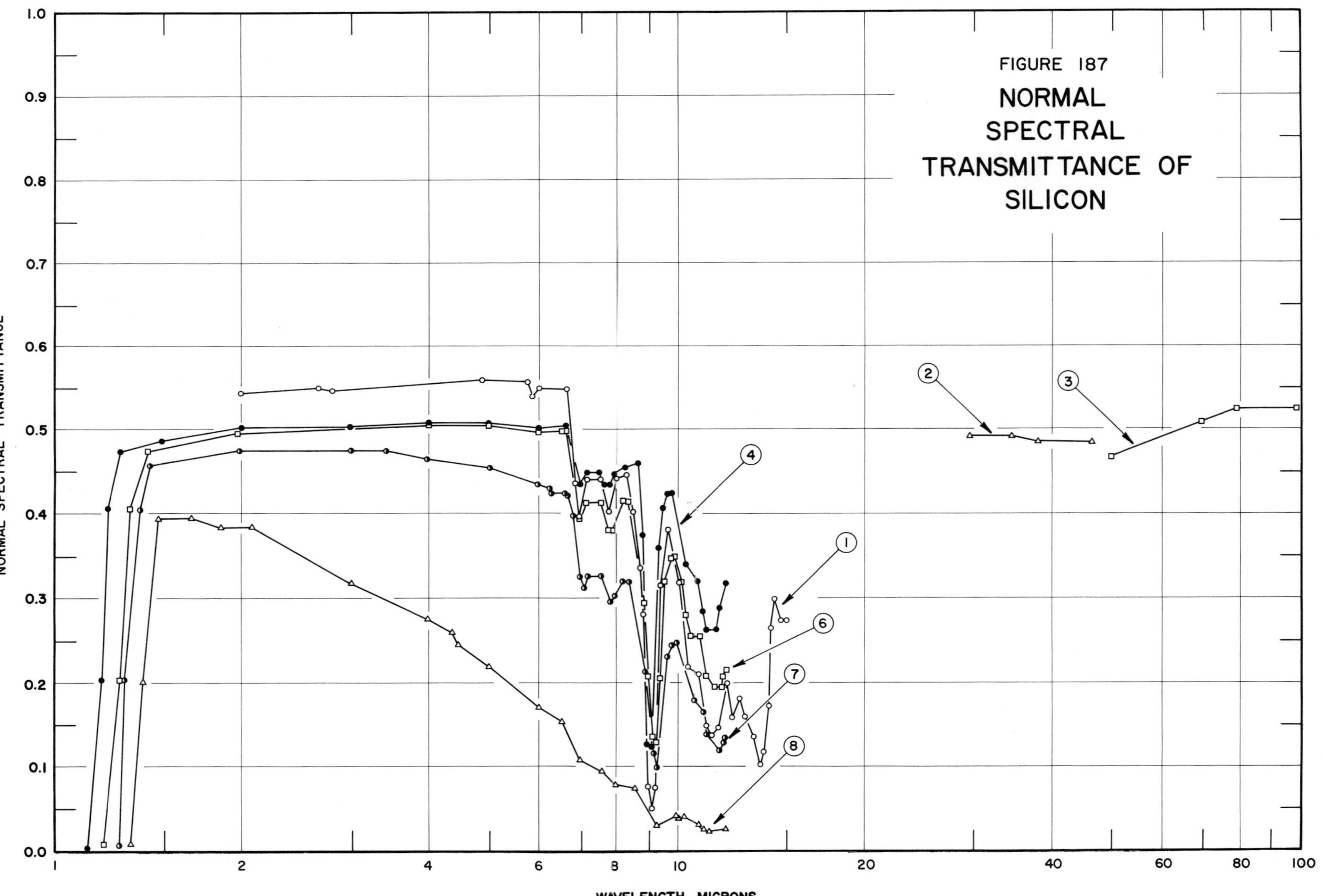

FIGURE 187
NORMAL SPECTRAL TRANSMITTANCE OF SILICON
NORMAL SPECTRAL TRANSMITTANCE
WAVELENGTH, MICRONS
1.0
0.9
0.8
0.7
0.6
0.5
0.4
0.3
0.2
0.1
0.0
1
2
4
6
8
10
20
40
60
80
100
1
2
3
4
6
7
8

SPECIFICATION TABLE NO. 187 NORMAL SPECTRAL TRANSMITTANCE OF SILICON

Curve No.	Ref. No.	Year	Temperature K	Wavelength Range, μ	Geometry θ	θ'	ω'	Reported Error, %	Composition (weight percent), Specifications and Remarks
1	226	1963	298	2-15	~0°	~0°			6 ppb boron and approx 20 ppb phosphorus; from Knapic Electro-physics, Inc; both surfaces polished optical flat to within 5 green mercury fringes; refractive index 3.453; data extracted from smooth curve.
2	262	1965	298	29.5-46.2	~0°	~0°			Electrical resistivity 15 ohm cm.
3	262	1965	298	49.8-98.7	~0°	~0°			Electrical resistivity 200 ohm cm.
4	288	1964	298	1.14-11.98	~0°	~0°			n-type silicon; 6 ppb B and 20 ppb P; single crystal; electrical resistivity 5 ohm cm; polished; nernst glower source; data extracted from smooth curve.
5	288	1964	373	1.16-11.98	~0°	~0°			Above specimen and conditions.
6	288	1964	473	1.21-12.00	~0°	~0°			Above specimen and conditions.
7	288	1964	573	1.27-11.98	~0°	~0°			Above specimen and conditions.
8	288	1964	673	1.32-12.00	~0°	~0°			Above specimen and conditions.
9	366	1963	~298	2.0-50.0	0°	0°			Specimen 1 cm thick; data extracted from smooth curve.
10	368	1963	298	40.9-98.5	0°	0°			p-type silicon; doped with In; thickness 0.28 mm; resistivity 0.9 ohm cm.
11	368	1963	298	40.8-98.4	0°	0°			Different sample, same as above specimen and conditions except thickness 0.48 mm.
12	368	1963	298	40.8-98.6	0°	0°			Different sample, same as above specimen and conditions except thickness 1.1 mm.
13	369	1951	298	12.5-38.4	~0°	~0°			Optically polished; thickness 2 mm; uncorrected for reflection losses.
14	370	1965	300	18.1-225.4	~0°	~0°			Thickness 2 mm; data extracted from smooth curve.

DATA TABLE NO. 187 NORMAL SPECTRAL TRANSMITTANCE OF SILICON

[Wavelength, λ, μ; Transmittance, τ; Temperature, T, K]

CURVE 1 T = 298

λ	τ
2.00	0.542
2.67	0.550
2.80	0.548
4.88	0.560
5.76	0.558
5.85	0.539
6.03	0.550
6.66	0.549
6.97	0.396
7.18	0.440
7.53	0.440
7.78	0.402
8.01	0.442
8.29	0.447
8.48	0.401
8.73	0.337
8.82	0.281
8.98	0.078
9.08	0.050
9.18	0.077
9.37	0.317
9.67	0.381
10.02	0.319
10.36	0.219
10.76	0.211
11.10	0.149
11.36	0.136
11.66	0.148
12.06	0.199
12.26	0.158
12.39	0.158*
12.58	0.181
12.76	0.159
13.41	0.134
13.64	0.101
13.88	0.118
14.04	0.173
14.19	0.265
14.40	0.299
14.69	0.273
15.00	0.273

CURVE 2 T = 298

λ	τ
29.5	0.492
34.5	0.492
38.0	0.488
46.2	0.487

CURVE 3 T = 298

λ	τ
49.8	0.468
69.2	0.509
79.1	0.524
98.7	0.525

CURVE 4 T = 298

λ	τ
1.14	0.002
1.19	0.202
1.24	0.404
1.28	0.474
1.49	0.487
2.00	0.501
2.99	0.503
4.00	0.508
5.01	0.507
6.02	0.501
6.63	0.504
6.88	0.437*
6.99	0.436
7.16	0.449
7.49	0.449
7.64	0.434
7.81	0.434
7.97	0.448
8.25	0.455
8.65	0.410
8.79	0.374
8.99	0.127
9.09	0.123
9.33	0.360
9.49	0.408
9.63	0.424
9.78	0.424
10.28	0.340

CURVE 4 (cont.)

λ	τ
10.73	0.321
10.99	0.284
11.16	0.263
11.60	0.263
11.70	0.289
11.98	0.318

CURVE 5* T = 373

λ	τ
1.16	0.007
1.20	0.203
1.29	0.459
1.34	0.469
1.99	0.498
2.99	0.502
3.98	0.508
5.02	0.508
6.01	0.500
6.64	0.501
6.91	0.412
6.99	0.414
7.10	0.431
7.49	0.433
7.71	0.414
7.84	0.414
7.99	0.428
8.27	0.437
8.80	0.348
8.85	0.310
8.99	0.208
9.10	0.134
9.35	0.315
9.49	0.373
9.71	0.394
10.00	0.373
10.24	0.325
10.50	0.297
10.74	0.297
11.00	0.259
11.12	0.237
11.23	0.226
11.60	0.229
11.76	0.251
11.98	0.266

CURVE 6 T = 473

λ	τ
1.21	0.008
1.27	0.202
1.33	0.405
1.42	0.473
1.97	0.495
4.01	0.505
5.00	0.505
6.00	0.498
6.52	0.498
6.61	0.498
6.97	0.395
7.13	0.412
7.56	0.412
7.77	0.380
7.88	0.380
8.18	0.416
8.34	0.416
8.85	0.294
8.97	0.208
9.10	0.136
9.20	0.129
9.35	0.206
9.52	0.320
9.76	0.348
9.87	0.350
10.11	0.319
10.32	0.279
10.50	0.257
10.80	0.257
11.18	0.209
11.53	0.196
11.75	0.196
11.79	0.208
12.00	0.217

CURVE 7 T = 573

λ	τ
1.27	0.007
1.30	0.203
1.37	0.405
1.44	0.457
1.98	0.475
3.00	0.475

CURVE 7 (cont.)

λ	τ
3.42	0.475
3.98	0.463
5.02	0.453
5.99	0.434
6.24	0.431
6.28	0.425
6.61	0.424
6.65	0.422
6.80	0.398
6.96	0.326
7.07	0.312
7.17	0.327
7.56	0.327
7.83	0.297
7.98	0.302
8.16	0.320
8.37	0.320
8.89	0.212
9.12	0.117
9.24	0.099
9.61	0.231
9.76	0.245
9.98	0.248
10.63	0.179
11.00	0.167
11.19	0.138
11.71	0.119
11.88	0.128
11.98	0.134

CURVE 8 T = 673

λ	τ
1.32	0.008
1.39	0.201
1.47	0.394
1.66	0.395
1.86	0.383
2.08	0.383
3.00	0.319
3.99	0.276
4.35	0.260
4.46	0.247
5.01	0.220
6.00	0.171

CURVE 8 (cont.)

λ	τ
6.53	0.153
6.99	0.108
7.57	0.095
7.98	0.078
8.56	0.075
9.22	0.030
9.98	0.042
10.08	0.038
10.24	0.040
10.85	0.031
11.01	0.025
11.75	0.023
12.00	0.026

CURVE 9* T = 298

λ	τ
2.0	0.432
4.9	0.434
6.6	0.410
6.8	0.360
7.4	0.378
7.8	0.273
8.4	0.362
8.4	0.002
9.1	0.002
9.5	0.277
11.0	0.047
11.9	0.090
12.4	0.042
13.0	0.061
13.5	0.052
13.8	0.142
14.6	0.173
15.2	0.129
16.4	0.026
17.1	0.047
17.7	0.047
18.4	0.068
18.9	0.044
20.4	0.096
23.6	0.199
27.8	0.223
31.2	0.223
37.5	0.194

CURVE 9 (cont.)*

λ	τ
45.2	0.094
50.0	0.054

CURVE 10* T = 298

λ	τ
40.9	0.280
42.4	0.292
45.0	0.283
48.6	0.292
51.4	0.274
51.4	0.285
54.5	0.285
57.4	0.273
60.4	0.273
60.4	0.243
63.4	0.266
63.4	0.293
67.4	0.276
71.9	0.295
79.2	0.210
82.0	0.214
85.1	0.205
89.1	0.197
91.9	0.204
98.5	0.224

CURVE 11* T = 298

λ	τ
40.8	0.201
42.3	0.201
45.0	0.163
48.7	0.163
51.4	0.172
54.5	0.172
54.5	0.154
57.4	0.154
57.4	0.162
60.4	0.155
63.5	0.173
63.5	0.182
67.4	0.134
71.8	0.113
79.2	0.105

CURVE 11 (cont.)*

λ	τ
82.1	0.105
85.0	0.103
89.1	0.103
91.9	0.104
98.4	0.142

CURVE 12* T = 298

λ	τ
40.8	0.079
42.4	0.063
45.0	0.070
48.7	0.034
51.4	0.042
51.4	0.062
54.6	0.062
54.6	0.042
57.4	0.042
57.4	0.053
60.4	0.053
60.4	0.042
63.5	0.062
63.4	0.074
67.4	0.025
79.2	0.025
82.2	0.000
85.2	0.022
89.2	0.024
98.6	0.010

CURVE 13* T = 298

λ	τ
12.5	0.25
12.8	0.22
13.1	0.19
13.5	0.18
13.5	0.18
13.8	0.25
14.2	0.36
14.7	0.32
15.1	0.30
15.6	0.27
16.1	0.05
16.3	0.03

DATA TABLE NO. 187 (continued)

λ	τ
CURVE 13 (cont.)*	
16.6	0.05
17.2	0.18
17.8	0.17
18.5	0.20
19.2	0.20
19.3	0.14
20.0	0.25
20.8	0.28
21.7	0.37
22.7	0.36
23.8	0.36
25.0	0.36
26.3	0.37
27.7	0.37
29.4	0.37
31.2	0.37
33.3	0.38
34.4	0.39
35.7	0.37
37.0	0.37
38.4	0.36
CURVE 14*	
T = 300	
18.1	0.223
18.8	0.271
19.5	0.241
21.7	0.394
23.5	0.442
25.0	0.442
27.2	0.423
29.3	0.486
30.4	0.494
33.4	0.457
42.2	0.480
44.7	0.468
51.1	0.407
59.8	0.426
186.6	0.433
225.4	0.446

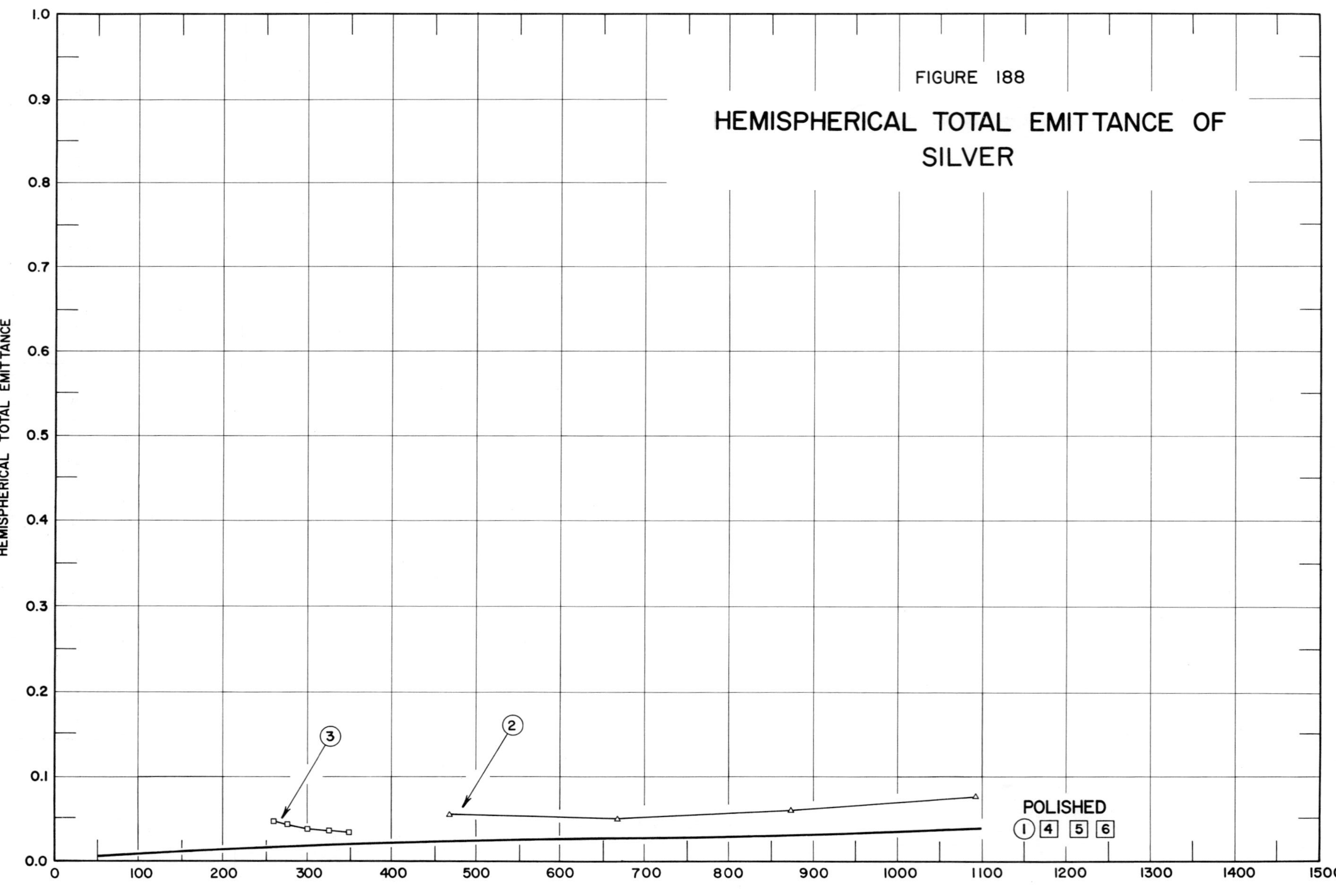
FIGURE 188
HEMISPHERICAL TOTAL EMITTANCE OF
SILVER
HEMISPHERICAL TOTAL EMITTANCE
1.0
0.9
0.8
0.7
0.6
0.5
0.4
0.3
0.2
0.1
0.0
3
2
POLISHED
1 4 5 6
0
100
200
300
400
500
600
700
800
900
1000
1100
1200
1300
1400
1500
TEMPERATURE, K

SPECIFICATION TABLE NO. 188 HEMISPHERICAL TOTAL EMITTANCE OF SILVER

Curve No.	Ref. No.	Year	Temperature Range, K	Reported Error, %	Composition (weight percent), Specifications and Remarks
1	47	1961	668–1093	< 10	Commercial rolled plate; 99.9 pure; ground with 600 grit carborundum and polished on a wet cloth lap with unlevigated jewelers rouge; measured in vacuum (10^{-5} mm Hg).
2	47	1961	468–1093	< 10	Above specimen and conditions after thermal etching.
3	26	1962	260–350	3	Commercial sheet; cleaned with both sodium dichromate and dilute nitric acid solutions; buffed on a felt buffing wheel and cleaned with CCl_4 and acetone; measured in vacuum (1 x 10^{-6} mm Hg); data extracted from smooth curve.
4	23	1955	300	±20	Silver plating (0.0003 in. thick); matte surface; measured in vacuum (<3 x 10^{-6} mm Hg).
5	23	1955	300	±20	Silver plating (0.0003 in. thick); lume surface; measured in vacuum (<3 x 10^{-6} mm Hg).
6	3	1955	76	5	Solvent cleaned; measured in vacuum (10^{-6} to 10^{-7} mm Hg); emittance for 300 K black body incident radiation; authors assumed $\alpha = \epsilon$.

DATA TABLE NO. 188 HEMISPHERICAL TOTAL EMITTANCE OF SILVER

[Temperature, T, K; Emittance, ϵ]

T	ϵ
CURVE 1	
668	0.03
873	0.03
1093	0.04
CURVE 2	
468	0.055
668	0.050
873	0.060
1093	0.075
CURVE 3	
260	0.047
275	0.043
300	0.038
325	0.035
350	0.033
CURVE 4	
300	0.020
CURVE 5	
300	0.017
CURVE 6	
76	0.008

SPECIFICATION TABLE NO. 189 NORMAL TOTAL EMITTANCE OF SILVER

Curve No.	Ref. No.	Year	Temperature Range, K	Geometry θ'	Reported Error, %	Composition (weight percent), Specifications and Remarks
1	15	1947	373	$\sim 0^\circ$		Polished.
2	355	1929	310-643	$\sim 0^\circ$		Polished.
3	356	1967	383-1174	$\sim 0^\circ$		Electrolytic silver, 99.99 Ag; mechanically polished with felt; washed with distilled alcohol; heated in air for 2 hrs at constant temperature before the temperature was raised to the next step.
4	356	1967	1041-404	$\sim 0^\circ$		Above specimen and conditions except cooling.

DATA TABLE NO. 189 NORMAL TOTAL EMITTANCE OF SILVER

[Temperature, T, K; Emittance, ∈]

T	∈
CURVE 1*	
373	0.052
CURVE 2*	
310	0.0221
366	0.0252
421	0.0292
477	0.0315
532	0.0295
588	0.0308
643	0.0312
CURVE 3*	
383	0.021
392	0.022
432	0.018
508	0.017
521	0.019
575	0.018
606	0.019
658	0.018
743	0.020
758	0.021
843	0.024
860	0.028
948	0.029
1023	0.032
1142	0.035
1174	0.036
CURVE 4*	
1041	0.034
890	0.033
807	0.032
679	0.031
544	0.030
404	0.029

*Not shown on plot

SPECIFICATION TABLE NO. 190 NORMAL SPECTRAL EMITTANCE OF SILVER

Curve No.	Ref. No.	Year	Wavelength μ	Temperature Range, K	Geometry θ'	Reported Error, %	Composition (weight percent), Specifications and Remarks
1	35	1914	0.66	975-2095	$\sim 0°$		Measured in burning hydrogen.

DATA TABLE NO. 190 NORMAL SPECTRAL EMITTANCE OF SILVER

[Temperature, T, K; Emittance, ϵ; Wavelength, λ, μ]

T	ϵ
CURVE 1*	
$\lambda = 0.66$	
975	0.65
1050	0.60
1260	0.50
1400	0.50
1540	0.60
1680	0.60
1820	0.60
1960	0.60
2095	0.65

*Not shown on plot

SPECIFICATION TABLE NO. 191 NORMAL SPECTRAL EMITTANCE OF SILVER

Curve No.	Ref. No.	Year	Temperature K	Wavelength Range, μ	Geometry θ'	Reported Error, %	Composition (weight percent), Specifications and Remarks
1	19	1914	1213	0.55-0.65	$\sim 0°$	1	Film; tungsten substrate; measured in hydrogen; Pt reference ($\epsilon = 0.33$ for $\lambda = 0.650\ \mu$ and $\epsilon = 0.38$ for $\lambda = 0.547\ \mu$ at all temp).

DATA TABLE NO. 191 NORMAL SPECTRAL EMITTANCE OF SILVER

[Wavelength, λ, μ; Emittance, ϵ; Temperature, T, K]

λ	ϵ
CURVE 1*	
T = 1213	
0.55	<0.35
0.65	0.044

*Not shown on plot

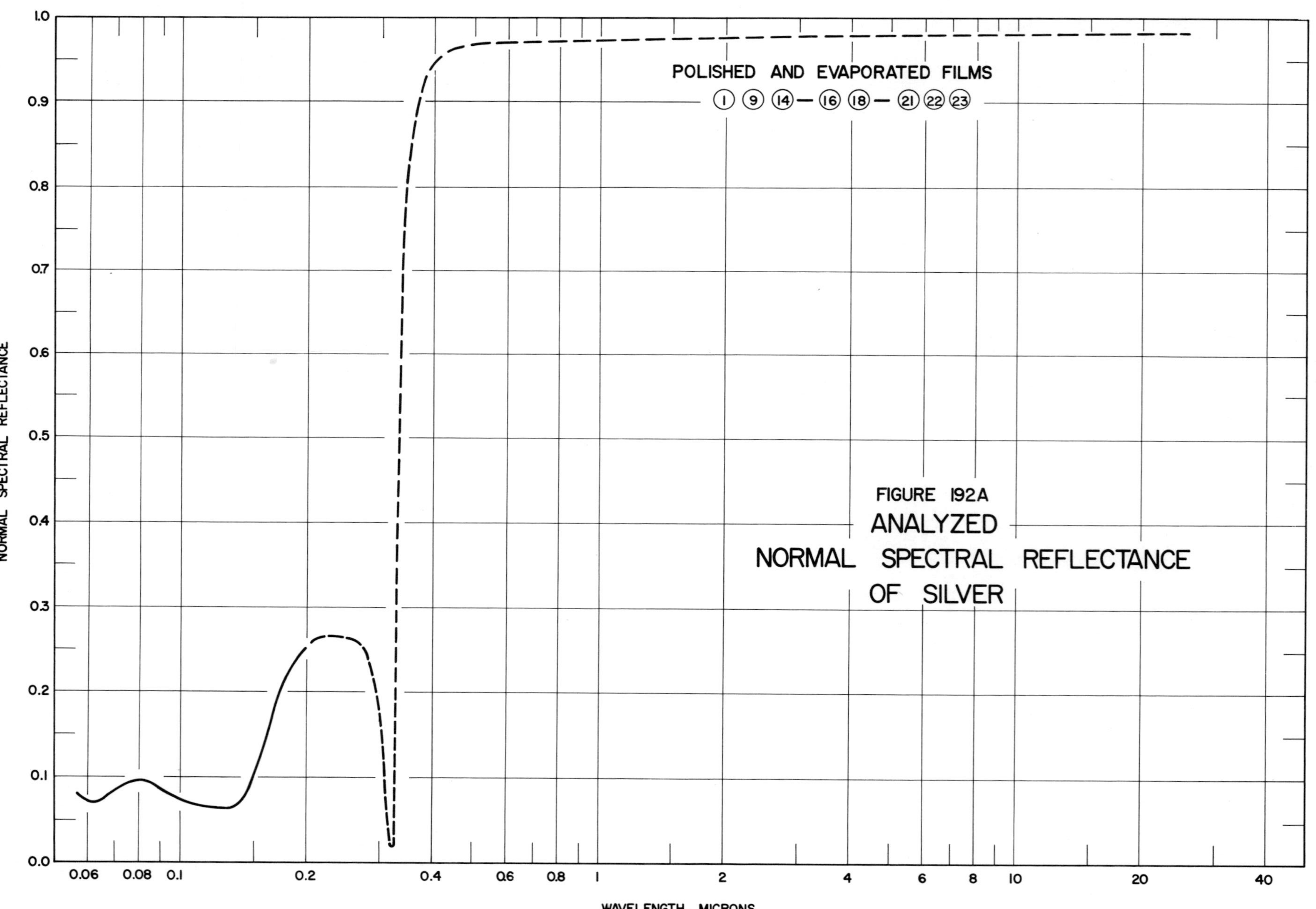

FIGURE 192A
ANALYZED
NORMAL SPECTRAL REFLECTANCE
OF SILVER

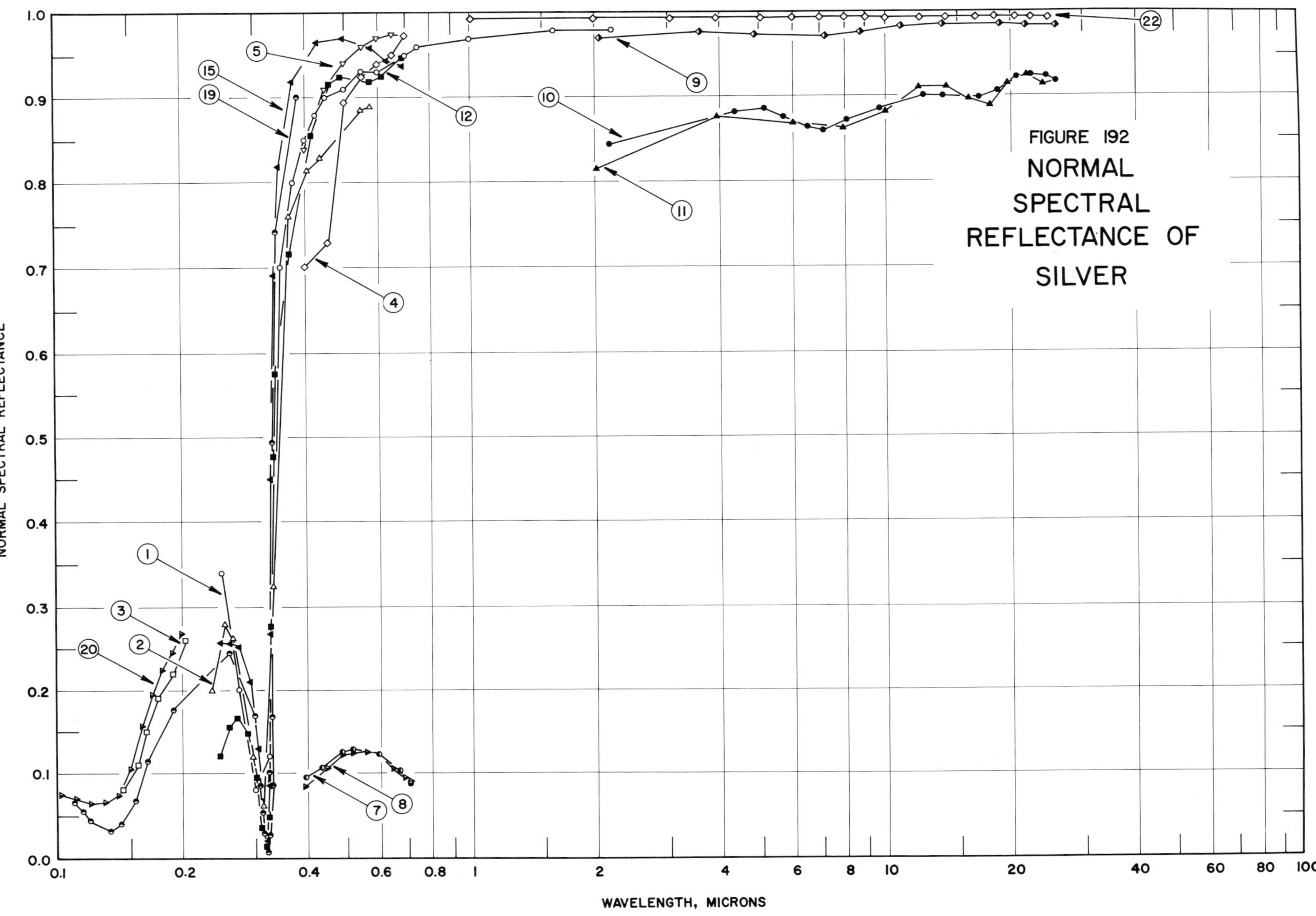
FIGURE 192
NORMAL
SPECTRAL
REFLECTANCE OF
SILVER
NORMAL SPECTRAL REFLECTANCE
WAVELENGTH, MICRONS
0.0
0.1
0.2
0.3
0.4
0.5
0.6
0.7
0.8
0.9
1.0
0.1
0.2
0.4
0.6
0.8
1
2
4
6
8
10
20
40
60
80
100
1
2
3
4
5
7
8
9
10
11
12
15
19
20
22

SPECIFICATION TABLE NO. 192 NORMAL SPECTRAL REFLECTANCE OF SILVER

Curve No.	Ref. No.	Year	Temperature K	Wavelength Range, μ	Geometry θ	θ'	ω'	Reported Error, %	Composition (weight percent), Specifications and Remarks
1	130	1934	298	0.250-2.200	~0°	~0°			Deposited on a mirror by evaporation.
2	133	1934	298	0.2350-0.5780	~0°	~0°			Cold worked; annealed in an inert gas for 6-24 hrs; polished; stored in dilute solutions of NaOH, NaOH + NaF, and HNO_3.
3	124	1941	298	0.1347-0.2026	~0°	~0°			An opaque film on glass; deposited chemically; polished with rouge and cotton wool; measured in vacuum (0.001 mm Hg).
4	360	1942	298	0.1000-0.2130	~0°	~0°			Vacuum evaporated film; measured in vacuum (10^{-5} mm Hg); data extracted from smooth curve.
5	353	1963	298	0.0304-0.1671	10°	10°		<3	>99.95 pure opaque film (~2000 Å thick); deposition rate ~1000 Å sec^{-1}; measured in situ (10^{-5} - 8 x 10^{-6} mm Hg).
6	228	1900	298	0.45-0.70	~0°	~0°			Pure; mirror-like surface; platinum filament lamp source.
7	263	1958	298	0.3974-0.7070	~0°	~0°		0.5	Thin film of silver on quartz; vacuum deposited; measured in air.
8	263	1958	298	0.3969-0.7098	~0°	~0°			Above specimen and conditions except measured in vacuum (10^{-4} - 5 x 10^{-5} mm Hg).
9	223	1962	298	2.06-25.89	~0°		2π		Polished; converted from R (2π, 0°); data extracted from smooth curve.
10	223	1962	298	2.17-26.00	~0°		2π		Above specimen and conditions except after damage by particle impact.
11	223	1962	77	2.02-25.92	~0°		2π		Above specimen and conditions.
12	297	1963	298	0.248-0.689	~10°	~10°			Hand lapped with silicon carbide papers, lapped on a metallographic polishing wheel, and electrolytically slide polished; tungsten filament source for 0.365 to 0.689 μ and a hydrogen lamp source at lower wavelengths.
13	297	1963	298	0.248-0.689	~10°	~10°			Different sample, same as above specimen and conditions except has heavier electrolytic polish.
14	297	1963	298	0.248-0.689	~10°	~10°			Hand lapped with silicon carbide papers, lapped on a metallographic polishing wheel, and electrolytically slide polished; tungsten filament source for 0.365 to 0.689 μ and a hydrogen lamp source at lower wavelengths.
15	297	1963	298	0.248-0.689	~10°	~10°			Hand lapped with silicon carbide papers, lapped on a metallographic polishing wheel, and electrolytically slide polished; tungsten filament source for 0.365 to 0.689 μ and a hydrogen lamp source at lower wavelengths.
16	297	1963	298	0.248-0.689	~10°	~10°			Different sample, same as above specimen and conditions except exposed to room atmosphere for one day.
17	297	1963	298	0.248-0.689	~10°	~10°			Different sample, same as curve 12 specimen and conditions except exposed to room atmosphere for four days.
18	313	1962	300	0.055-0.124	~0°	~0°			Electrolytically polished; data extracted from smooth curve.
19	313	1961	300	0.110-0.384	~0°	~0°			Electrolytically polished; data extracted from smooth curve.
20	315	1965	295	0.1029-0.1998	~4°	~4°			>99.99 Ag film (approx 1200 Å thick); evaporated in vacuum (10^{-6} mm Hg) on glass; measured in vacuum within 10-20 sec of completion of deposition.

SPECIFICATION TABLE NO. 192 (continued)

Curve No.	Ref. No.	Year	Temperature K	Wavelength Range, μ	Geometry θ	θ'	ω'	Reported Error, %	Composition (weight percent), Specifications and Remarks
21	315	1965	295	0.1028-0.1999	~4°	~4°			>99.99 Ag film (approx 1200 Å thick); evaporated in vacuum (10^{-6} mm Hg) on glass; measured after 5 weeks exposure to air.
22	354	1968	298	1.00-24.60	~0°	~0°			99.999 pure opaque film (650-1000 Å thick); ultrahigh vacuum (10^{-9} mm Hg) evaporated (25 Å sec^{-1}) onto fused quartz optical flats with supersmooth finish; surface roughness 7 Å rms; average values of several specimens.
23	358	1962	298	0.39-2.1	0°		2π		No details given.
24	359	1965	298	3.0-30.1	~5°	~5°			>99.999 Ag; opaque ultrahigh vacuum film; vacuum evaporated (10^{-9} mm Hg) on supersmooth fused quartz optical flats; measured in nitrogen.

DATA TABLE NO. 192 NORMAL SPECTRAL REFLECTANCE OF SILVER

[Wavelength, λ, μ; Reflectance, ρ; Temperature, T, K]

CURVE 1*
T = 298

λ	ρ
0.250	0.34
0.275	0.20
0.300	0.08
0.325	0.12
0.350	0.70
0.375	0.80
0.375	0.80
0.400	0.85
0.425	0.88
0.450	0.90
0.500	0.91
0.550	0.93
0.600	0.93
0.700	0.95
0.750	0.96
1.000	0.97
1.600	0.98
2.200	0.98

CURVE 2
T = 298

λ	ρ
0.2350	0.200
0.2537	0.279
0.2650	0.262
0.2970	0.119
0.3125	0.061
0.3340	0.325
0.3660	0.760
0.4060	0.819
0.4355	0.830
0.5460	0.887
0.5780	0.890

CURVE 3
T = 298

λ	ρ
0.1347	0.05
0.1438	0.08
0.1570	0.11
0.1640	0.15
0.1757	0.19
0.1901	0.22
0.2026	0.26

CURVE 4*
T = 298

λ	ρ
0.1000	0.050
0.1205	0.050
0.1349	0.054
0.1454	0.069
0.1567	0.096
0.1643	0.120
0.1741	0.153
0.1881	0.198
0.1972	0.229
0.2042	0.247
0.2130	0.263

CURVE 5*
T = 298

λ	ρ
0.0304	0.045
0.0406	0.017
0.0461	0.056
0.0584	0.087
0.0735	0.092
0.0932	0.102
0.1048	0.069
0.1216	0.060
0.1311	0.057
0.1470	0.049
0.1671	0.126

CURVE 6*
T = 298

λ	ρ
0.45	0.906
0.50	0.918
0.55	0.925
0.60	0.930
0.65	0.936
0.70	0.946

CURVE 7
T = 298

λ	ρ
0.3974	0.0947
0.4360	0.1069
0.4846	0.1241
0.5180	0.1282

CURVE 7 (cont.)

λ	ρ
0.5955	0.1202
0.6629	0.1014
0.7070	0.0873

CURVE 8
T = 298

λ	ρ
0.3969	0.0884
0.4483	0.1049
0.4905	0.1202
0.5153	0.1252
0.5598	0.1262
0.5978	0.1226*
0.6425	0.1103
0.6874	0.0966
0.7098	0.0890

CURVE 9*
T = 298

λ	ρ
2.06	0.971
3.60	0.978
4.87	0.976
7.20	0.972
8.76	0.978
10.96	0.984
13.70	0.987
18.95	0.986
21.79	0.985
25.89	0.983

CURVE 10
T = 298

λ	ρ
2.17	0.846
4.35	0.882
5.13	0.887
5.73	0.878
6.52	0.865
7.10	0.861
8.11	0.872
9.75	0.888
12.42	0.900
13.93	0.900
16.98	0.899

CURVE 10 (cont.)

λ	ρ
18.72	0.907
20.65	0.922
22.46	0.926
24.65	0.923
26.00	0.918

CURVE 11
T = 77

λ	ρ
2.02	0.817
3.94	0.878
6.01	0.870
7.97	0.862
10.00	0.882
12.01	0.910
14.08	0.911
15.98	0.897
17.92	0.890
19.82	0.916
21.91	0.924
24.04	0.913
25.92	0.917*

CURVE 12
T = 298

λ	ρ
0.248	0.121
0.260	0.157
0.272	0.166
0.288	0.148
0.302	0.094
0.312	0.037
0.319	0.013
0.324	0.049
0.329	0.277
0.334	0.477
0.339	0.573
0.366	0.718
0.416	0.856
0.459	0.916
0.488	0.924
0.574	0.919
0.617	0.925
0.689	0.948

CURVE 13*
T = 298

λ	ρ
0.248	0.212
0.259	0.229
0.266	0.229
0.287	0.196
0.304	0.124
0.312	0.056
0.319	0.012
0.323	0.032
0.328	0.220
0.333	0.449
0.342	0.651
0.358	0.808
0.376	0.869
0.418	0.900
0.449	0.908
0.481	0.920
0.556	0.922
0.596	0.922
0.689	0.952

CURVE 14*
T = 298

λ	ρ
0.248	0.257
0.260	0.272
0.277	0.255
0.292	0.213
0.306	0.117
0.319	0.021
0.325	0.073
0.327	0.276
0.332	0.537
0.342	0.760
0.362	0.889
0.386	0.944
0.430	0.965
0.477	0.966
0.558	0.957
0.633	0.945
0.689	0.948

CURVE 15
T = 298

λ	ρ
0.248	0.257*
0.262	0.257
0.275	0.252
0.292	0.210
0.305	0.129
0.313	0.060*
0.320	0.020
0.325	0.085
0.327	0.267
0.330	0.450
0.337	0.691
0.347	0.819
0.376	0.919
0.432	0.967
0.498	0.971
0.579	0.959
0.633	0.943
0.689	0.937

CURVE 16*
T = 298

λ	ρ
0.248	0.249
0.265	0.252
0.281	0.237
0.294	0.196
0.303	0.133
0.312	0.059
0.320	0.018
0.325	0.075
0.328	0.276
0.332	0.496
0.338	0.674
0.348	0.780
0.376	0.883
0.420	0.939
0.479	0.963
0.528	0.964
0.582	0.954
0.636	0.941
0.689	0.939

CURVE 17*
T = 298

λ	ρ
0.248	0.229
0.261	0.237
0.273	0.234
0.287	0.210
0.300	0.153
0.310	0.067
0.319	0.018
0.325	0.037
0.327	0.325
0.330	0.346
0.332	0.488
0.339	0.647
0.355	0.780
0.382	0.858
0.434	0.919
0.494	0.947
0.530	0.951
0.605	0.940
0.689	0.938

CURVE 18*
T = 300

λ	ρ
0.055	0.113
0.056	0.096
0.058	0.080
0.062	0.068
0.065	0.072
0.071	0.085
0.079	0.096
0.083	0.095
0.090	0.085
0.102	0.065
0.124	0.053

CURVE 19
T = 300

λ	ρ
0.110	0.066
0.115	0.055
0.120	0.044
0.134	0.032
0.143	0.040
0.154	0.068

*Not shown on plot

DATA TABLE NO. 192 (continued)

λ	ρ
CURVE 19 (cont.)	
0.165	0.113
0.191	0.177
0.261	0.246
0.300	0.169
0.308	0.084
0.312	0.054
0.315	0.029
0.318	0.014*
0.321	0.007
0.322	0.010
0.325	0.027
0.328	0.084
0.329	0.168
0.331	0.492
0.340	0.743
0.384	0.900
CURVE 20*	
T = 295	
0.1029	0.075
0.1115	0.070
0.1214	0.065
0.1300	0.066
0.1400	0.073
0.1500	0.104
0.1604	0.156
0.1700	0.195
0.1799	0.223
0.1902	0.245
0.1998	0.263
CURVE 21*	
T = 295	
0.1028	0.073
0.1215	0.064
0.1400	0.064
0.1607	0.143
0.1800	0.214
0.1999	0.256
CURVE 22	
T = 298	
1.00	0.9937
1.98	0.9941
CURVE 22 (cont.)	
3.01	0.9941
3.99	0.9944
5.00	0.9945
5.97	0.9946
6.98	0.9949
7.96	0.9946
8.99	0.9947
9.98	0.9949
12.00	0.9947
13.93	0.9950
16.26	0.9949
18.11	0.9951
20.28	0.9950
22.13	0.9949
24.60	0.9949
CURVE 23*	
T = 298	
0.39	0.69
0.4	0.792
0.45	0.842
0.5	0.90
0.55	0.90
0.55	0.9175
0.6	0.927
0.65	0.94
0.7	0.955
0.75	0.962
0.8	0.967
0.85	0.97
0.9	0.98
0.95	0.985
1.0	0.97
1.1	0.99
1.2	0.99
1.3	0.99
1.4	0.995
1.6	1.00
1.8	0.98
2.0	0.98
2.1	1.00
CURVE 24*	
T = 298	
3.0	0.9944
4.0	0.9945
5.0	0.9947
6.0	0.9948
7.0	0.9953
8.0	0.9951
9.0	0.9952
10.0	0.9951
11.0	0.9951
12.0	0.9952
13.0	0.9952
14.0	0.9958
16.0	0.9955
18.0	0.9956
20.0	0.9959
22.0	0.9954
24.0	0.9959
26.1	0.9958
28.0	0.9958
30.1	0.9958

* Not shown on plot

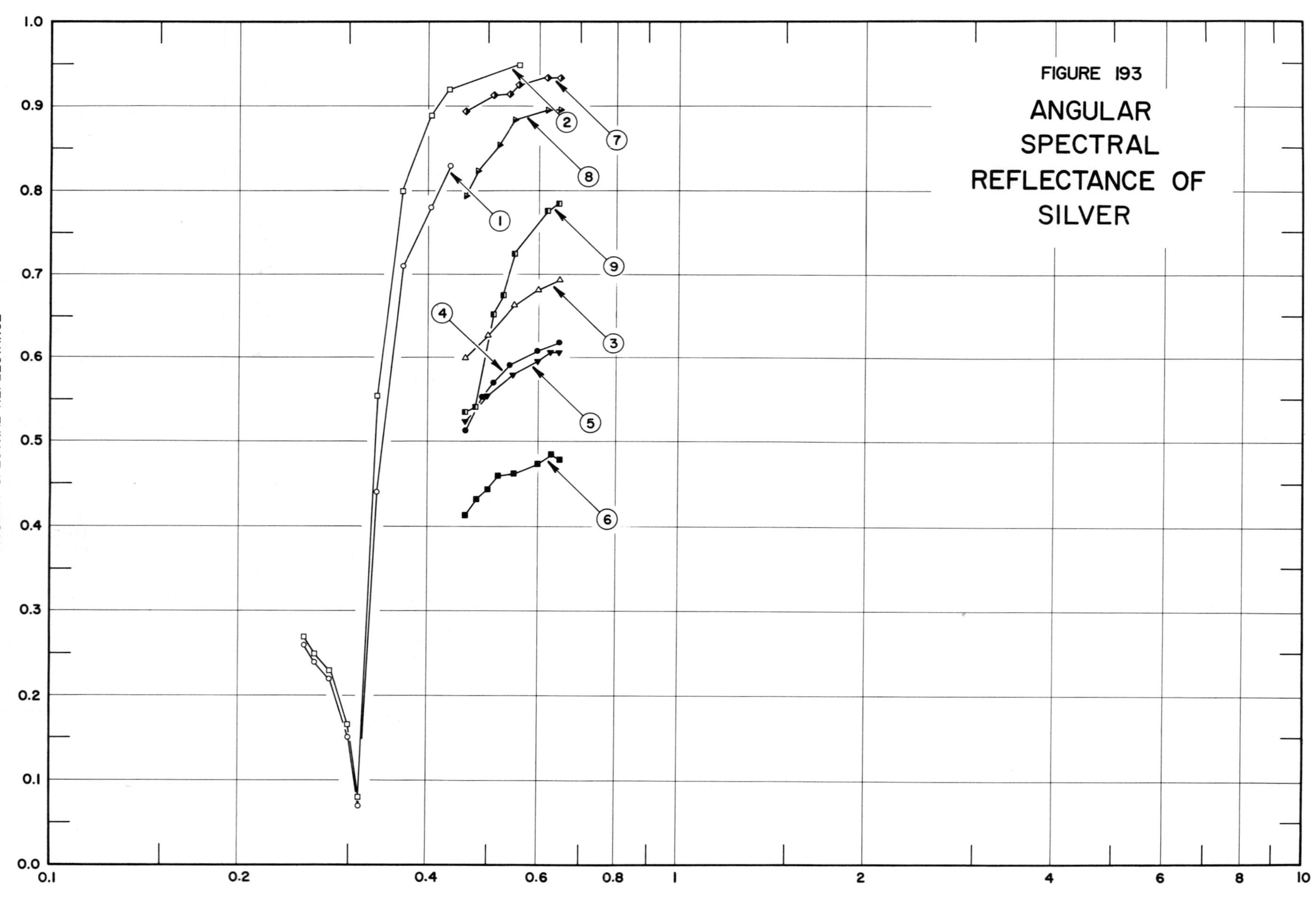

FIGURE 193
ANGULAR SPECTRAL REFLECTANCE OF SILVER
ANGULAR SPECTRAL REFLECTANCE
WAVELENGTH, MICRONS
0.0
0.1
0.2
0.3
0.4
0.5
0.6
0.7
0.8
0.9
1.0
0.1
0.2
0.4
0.6
0.8
1
2
4
6
8
10
1
2
3
4
5
6
7
8
9

SPECIFICATION TABLE NO. 193 ANGULAR SPECTRAL REFLECTANCE OF SILVER

Curve No.	Ref. No.	Year	Temperature K	Wavelength Range, μ	Geometry θ	θ'	ω'	Reported Error, %	Composition (weight percent), Specifications and Remarks
1	143	1928	298	0.255-0.435	45°	45°			Chemically desposited on glass and aged.
2	143	1928	298	0.255-0.560	45°	45°			Chemically deposited on glass; freshly prepared.
3	264	1939	298	0.46-0.65	45°	45°			Pure; rolled; chrome plated or rhodium plated mirrors reference.
4	264	1939	298	0.46-0.65	45°	45°			Above specimen and conditions except annealed at 773 K for 10 min.
5	264	1939	298	0.46-0.65	45°	45°			Above specimen and conditions except annealed at 973 K for 10 min.
6	264	1939	298	0.46-0.65	45°	45°			Above specimen and conditions except annealed at 1173 K for 10 min.
7	264	1939	298	0.46-0.65	45°	45°			Pure; cold rolled; polished; measured immediately after polishing; chrome plated or rhodium plated mirror references.
8	264	1939	298	0.46-0.65	45°	45°			Above specimen and conditions; measured after exposing for 21 days to atmosphere containing little sulphur dioxide and moisture.
9	264	1939	298	0.46-0.65	45°	45°			Above specimen and conditions; except 52 days additional exposure to the same atmosphere.
10	353	1963	298	0.0304-0.1671	30°	30°		<3	>99.95 pure; opaque film (~2000 Å thick); deposition rate ~1000 Å sec^{-1}; measured in situ (10^{-5} - 8 x 10^{-6} mm Hg).
11	353	1963	298	0.0304-0.1671	50°	50°		<3	Above specimen and conditions.
12	353	1963	298	0.0304-0.1671	70°	70°		<3	Above specimen and conditions.

DATA TABLE NO. 193 ANGULAR SPECTRAL REFLECTANCE OF SILVER

[Wavelength, λ, μ; Reflectance, ρ; Temperature, T, K]

λ	ρ
CURVE 1 T = 298	
0.255	0.26
0.265	0.24
0.280	0.22
0.300	0.15
0.312	0.07
0.332	0.44
0.365	0.71
0.405	0.78
0.435	0.83
CURVE 2 T = 298	
0.255	0.270
0.265	0.250
0.280	0.230
0.300	0.165
0.312	0.080
0.333	0.555
0.365	0.800
0.405	0.890
0.433	0.920
0.560	0.950
CURVE 3 T = 298	
0.46	0.600
0.50	0.629
0.55	0.665
0.60	0.681
0.65	0.696
CURVE 4 T = 298	
0.46	0.513
0.49	0.553
0.51	0.571
0.54	0.592
0.60	0.609
0.65	0.620
CURVE 5 T = 298	
0.46	0.522
0.50	0.555
0.55	0.580
0.60	0.596
0.63	0.608
0.65	0.608
CURVE 6 T = 298	
0.46	0.414
0.48	0.432
0.50	0.444
0.52	0.460
0.55	0.462
0.60	0.476
0.63	0.487
0.65	0.480
CURVE 7 T = 298	
0.46	0.895
0.51	0.914
0.54	0.914
0.56	0.926
0.62	0.935
0.65	0.935
CURVE 8 T = 298	
0.46	0.797
0.48	0.823
0.52	0.853
0.55	0.885
0.62	0.898
0.65	0.898
CURVE 9 T = 298	
0.46	0.535
0.48	0.591
0.51	0.652
0.53	0.676
0.55	0.726
0.62	0.777
0.65	0.785
CURVE 10* T = 298	
0.0304	0.046
0.0406	0.025
0.0461	0.056
0.0584	0.094
0.0735	0.104
0.0932	0.123
0.1048	0.089
0.1216	0.075
0.1311	0.068
0.1470	0.056
0.1671	0.161
CURVE 11* T = 298	
0.0304	0.095
0.0406	0.075
0.0461	0.117
0.0584	0.151
0.0735	0.167
0.0932	0.184
0.1048	0.158
0.1216	0.133
0.1311	0.131
0.1470	0.135
0.1671	0.268
CURVE 12* T = 298	
0.0304	0.295
0.0406	0.296
CURVE 12 (cont.)*	
0.0461	0.349
0.0584	0.338
0.0735	0.167
0.0932	0.363
0.1048	0.354
0.1216	0.337
0.1311	0.341
0.1470	0.359
0.1671	0.511

*Not shown on plot

SPECIFICATION TABLE NO. 194 HEMISPHERICAL INTEGRATED ABSORPTANCE OF SILVER

Curve No.	Ref. No.	Year	Temperature Range, K	Reported Error, %	Composition (weight percent), Specifications and Remarks
1	23	1955	77.3	± 10	Plating (0.0003 in. thick); lume surface; measured in vacuum ($<3 \times 10^{-6}$ mm Hg); absorptance for 300 K blackbody incident radiation.
2	23	1955	77.3	± 10	Plating (0.0003 in. thick); matte surface; measured in vacuum ($<3 \times 10^{-6}$ mm Hg); absorptance for 300 K blackbody incident radiation.
3	23	1955	77.3	± 10	Plating (0.0003 in. thick); dry buffed; measured in vacuum ($<3 \times 10^{-6}$ mm Hg); absorptance for 300 K blackbody incident radiation.
4	23	1955	76	5	Solvent cleaned; measured in vacuum (10^{-6} to 10^{-7} mm Hg); absorptance for 300 K blackbody incident radiation.

DATA TABLE NO. 194 HEMISPHERICAL INTEGRATED ABSORPTANCE OF SILVER

[Temperature, T, K; Absorptance, α]

T	α
CURVE 1*	
77.3	0.0083
CURVE 2*	
77.3	0.012
CURVE 3*	
77.3	0.016
CURVE 4*	
76	0.008

*Not shown on plot

SPECIFICATION TABLE NO. 195 NORMAL SPECTRAL ABSORPTANCE OF SILVER

Curve No.	Ref. No.	Year	Wavelength μ	Temperature Range, K	Geometry θ	Reported Error, %	Composition (weight percent), Specifications and Remarks
1	35	1914	0.66	873-2073	$\sim 0^{\circ}$		Measured in burning hydrogen; author assumed $\alpha = \epsilon$.

DATA TABLE NO. 195 NORMAL SPECTRAL ABSORPTANCE OF SILVER

[Temperature, T, K; Absorptance, α; Wavelength, λ, μ]

T	α
CURVE 1*	
$\lambda = 0.66$	
873	0.055
2073	0.055

*Not shown on plot

SPECIFICATION TABLE NO. 196 NORMAL SPECTRAL ABSORPTANCE OF SILVER

Curve No.	Ref. No.	Year	Temperature K	Wavelength Range, μ	Geometry θ	Reported Error, %	Composition (weight percent), Specifications and Remarks
1	307	1954	~298	0.226-2.600	~0°		5 mils thick; data extracted from smooth curve.

DATA TABLE NO. 196 NORMAL SPECTRAL ABSORPTANCE OF SILVER

[Wavelength, λ, μ; Absorptance, α; Temperature, T, K]

λ	α
CURVE 1*	
T = ~298	
0.226	0.844
0.238	0.880
0.265	0.895
0.318	0.930
0.331	0.904
0.327	0.757
0.338	0.625
0.354	0.543
0.480	0.383
0.579	0.270
0.666	0.212
0.821	0.164
0.990	0.147
1.186	0.159
1.402	0.147
1.589	0.139
1.800	0.148
2.000	0.170
2.200	0.199
2.399	0.179
2.600	0.133

*Not shown on plot

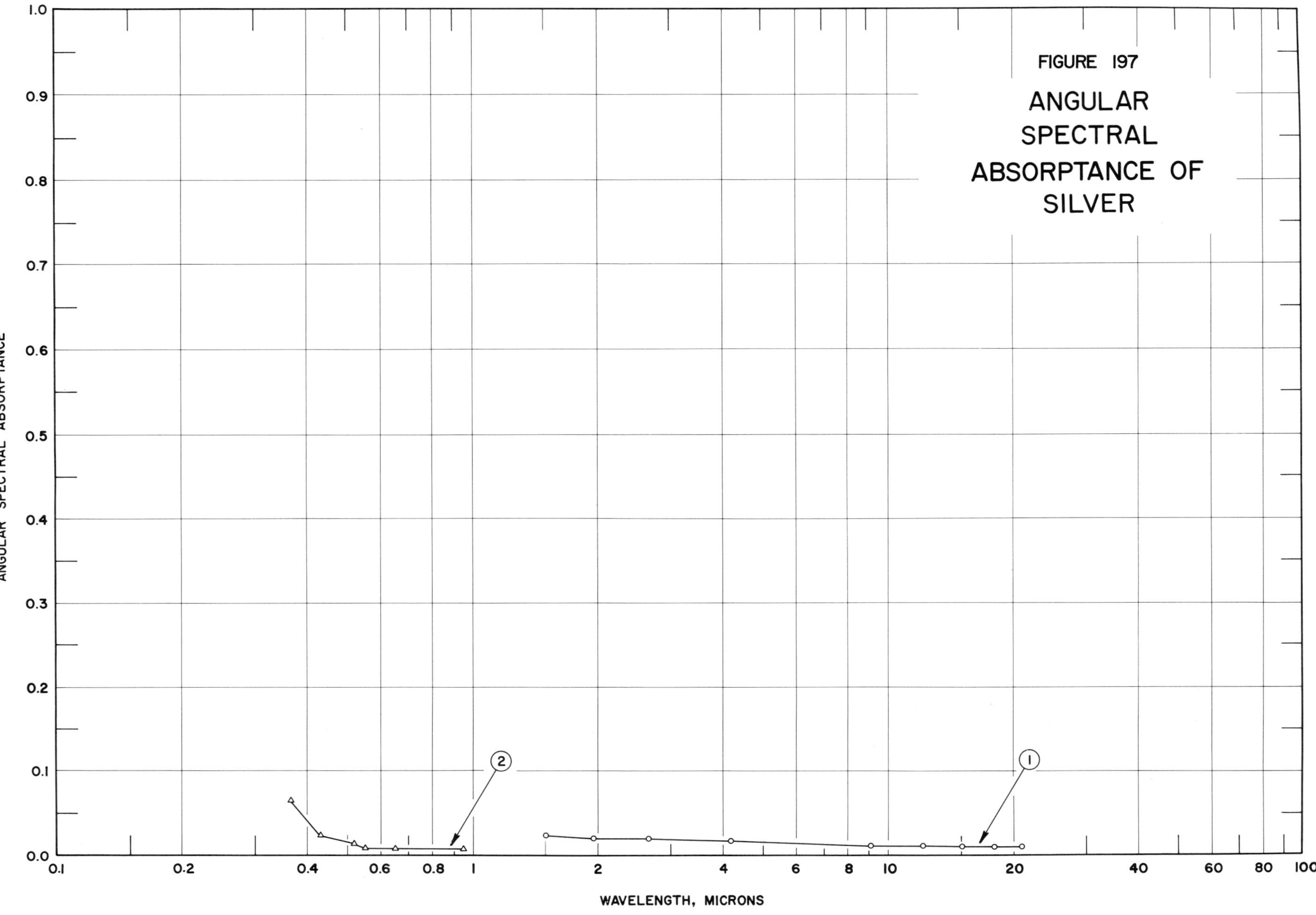
FIGURE 197
ANGULAR
SPECTRAL
ABSORPTANCE OF
SILVER
ANGULAR SPECTRAL ABSORPTANCE
1.0
0.9
0.8
0.7
0.6
0.5
0.4
0.3
0.2
0.1
0.0
2
1
0.1
0.2
0.4
0.6
0.8
1
2
4
6
8
10
20
40
60
80
100
WAVELENGTH, MICRONS

SPECIFICATION TABLE NO. 197 ANGULAR SPECTRAL ABSORPTANCE OF SILVER

Curve No.	Ref. No.	Year	Temperature K	Wavelength Range, μ	Geometry θ	Reported Error, %	Composition (weight percent), Specifications and Remarks
1	225	1965	306	1.51-20.9	25°		Silver electroplated on polished nickel; polished with optical grade rouge with cotton moistened in ethyl alcohol; measured in dry nitrogen; heated cavity at approx 1056 K with platinum reference; authors assumed $\alpha = 1-R(2\pi, 25°)$.
2	287	1956	4.2	0.365-0.949	15°		99.98 pure; machined from vacuum melted ingot, faced with a diamond cutting tool, mechanically polished with metallographic polishing paper, electropolished, annealed for ~10 hrs either in vacuum or in hydrogen at ~303 K below melting point, electropolished, rinsed, and mounted in the apparatus under a helium atmosphere to prevent oxidation; measured in vacuum.

DATA TABLE NO. 197 ANGULAR SPECTRAL ABSORPTANCE OF SILVER

[Wavelength, λ, μ; Absorptance, α; Temperature, T, K]

λ	α
CURVE 1 T = 306	
1.51	0.024
1.96	0.021
2.64	0.020
4.19	0.017
9.12	0.012
12.1	0.011
15.1	0.011
18.0	0.011
20.9	0.011
CURVE 2 T = 4.2	
0.365	0.065
0.430	0.026
0.518	0.014
0.555	0.009
0.651	0.008
0.949	0.008

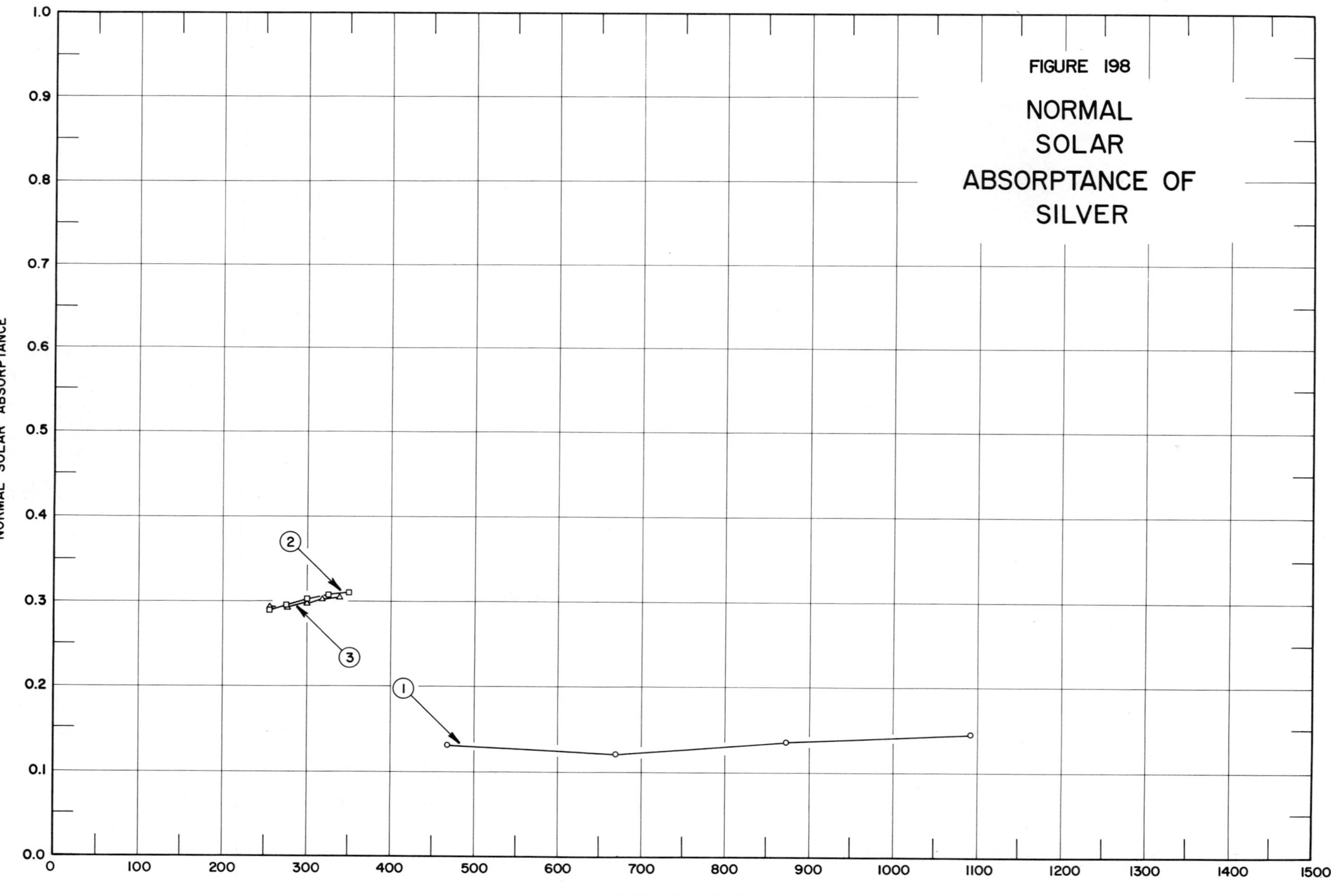
FIGURE 198
NORMAL SOLAR ABSORPTANCE OF SILVER
NORMAL SOLAR ABSORPTANCE
TEMPERATURE, K
0.0
0.1
0.2
0.3
0.4
0.5
0.6
0.7
0.8
0.9
1.0
0
100
200
300
400
500
600
700
800
900
1000
1100
1200
1300
1400
1500
1
2
3

SPECIFICATION TABLE NO. 198 NORMAL SOLAR ABSORPTANCE OF SILVER

Curve No.	Ref. No.	Year	Temperature Range, K	Geometry θ	Reported Error, %	Composition (weight percent), Specifications and Remarks
1	47	1961	468-1093	$\sim 0^\circ$	< 10	Commercial rolled plate; 99.9 pure; ground with 600 grit Carborundum and polished on a wet cloth lap with unlevigated jewelers rouge; measured in vacuum.
2	26	1962	255-350	$\sim 0^\circ$	5	Commercial sheet; cleaned with both sodium dichromate and dilute nitric acid solutions; buffed on a felt buffing wheel and cleaned with $C Cl_4$ and acetone; measured in vacuum (1×10^{-6} mm Hg); data extracted from smooth curve.
3	26	1964	255-339	$\sim 0^\circ$		Commercially pure; cleaned, buffed, cleaned with carbon tetrachloride and acetone; measured in vacuum (1×10^{-6} mm Hg); data extracted from smooth curve.

DATA TABLE NO. 198 NORMAL SOLAR ABSORPTANCE OF SILVER

[Temperature, T, K; Absorptance, α]

T	α
CURVE 1	
468	0.130
668	0.120
873	0.135
1093	0.145
CURVE 2	
255	0.290
275	0.295
300	0.302
325	0.308
350	0.310
CURVE 3	
255	0.292
277	0.296
300	0.299
318	0.302
339	0.306

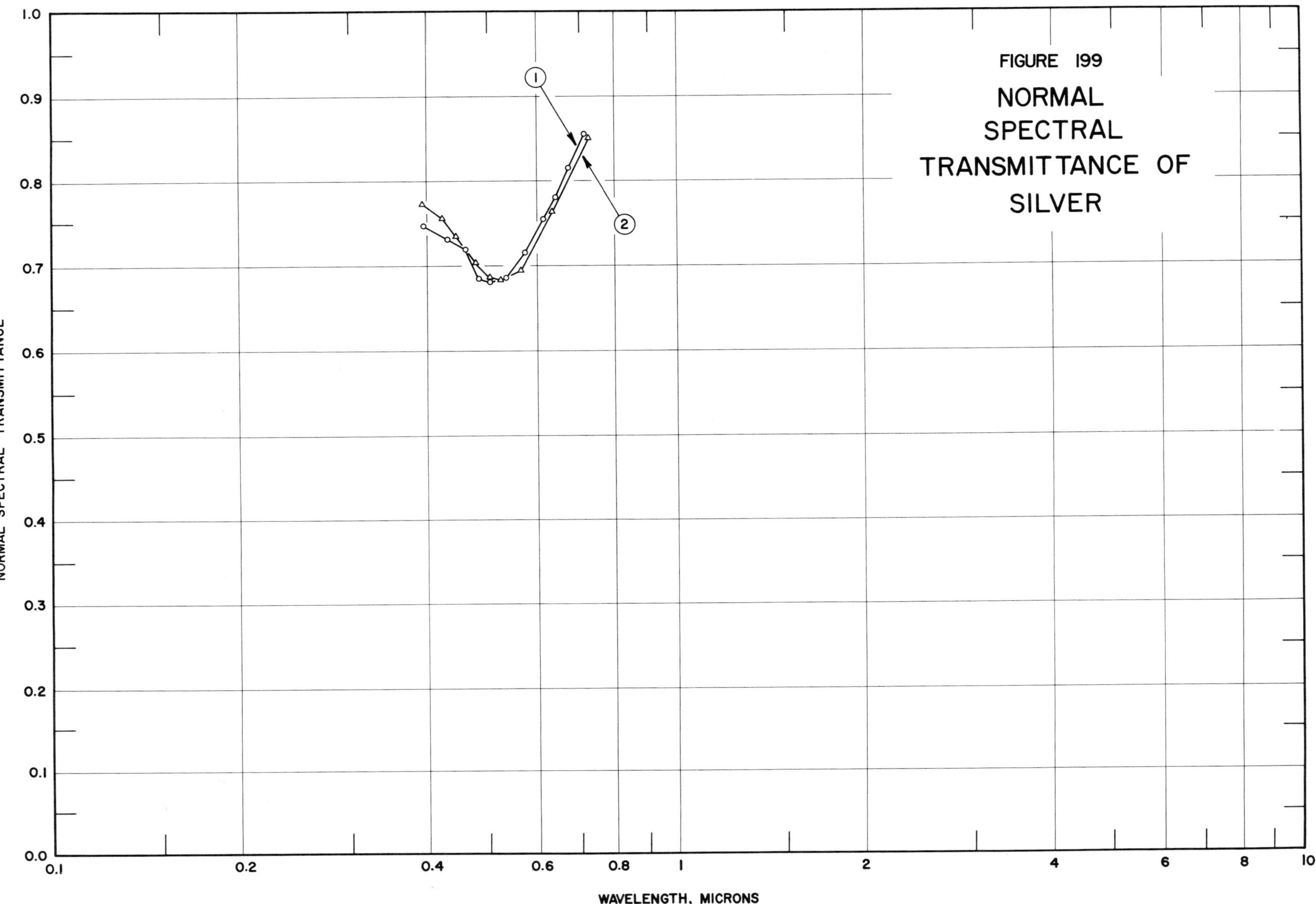
FIGURE 199
NORMAL SPECTRAL TRANSMITTANCE OF SILVER
1
2
NORMAL SPECTRAL TRANSMITTANCE
1.0
0.9
0.8
0.7
0.6
0.5
0.4
0.3
0.2
0.1
0.0
0.1
0.2
0.4
0.6
0.8
1
2
4
6
8
10
WAVELENGTH, MICRONS

SPECIFICATION TABLE NO. 199 NORMAL SPECTRAL TRANSMITTANCE OF SILVER

Curve No.	Ref. No.	Year	Temperature K	Wavelength Range, μ	Geometry θ θ' ω'	Reported Error, %	Composition (weight percent), Specifications and Remarks
1	263	1958	298	0.3950-0.7159	~0° ~0°	0.5	Thin film of silver on quartz; vacuum deposited; measured in air.
2	263	1958	298	0.3948-0.7274	~0° ~0°	0.5	Above specimen and conditions except measured in vacuum (10^{-4} - 5 x 10^{-5} mm Hg).

DATA TABLE NO. 199 NORMAL SPECTRAL TRANSMITTANCE OF SILVER

[Wavelength, λ, μ; Transmittance, τ; Temperature, T, K]

λ	τ
CURVE 1	
T = 298	
0.3950	0.7486
0.4318	0.7319
0.4609	0.7109
0.4854	0.6867
0.5054	0.6810
0.5362	0.6875
0.5765	0.7150
0.6152	0.7541
0.6434	0.7816
0.6771	0.8180
0.7159	0.8578
CURVE 2	
T = 298	
0.3948	0.7724
0.4229	0.7566
0.4477	0.7337
0.4799	0.7028
0.5025	0.6866
0.5268	0.6825
0.5648	0.6973
0.6345	0.7604
0.7274	0.8521

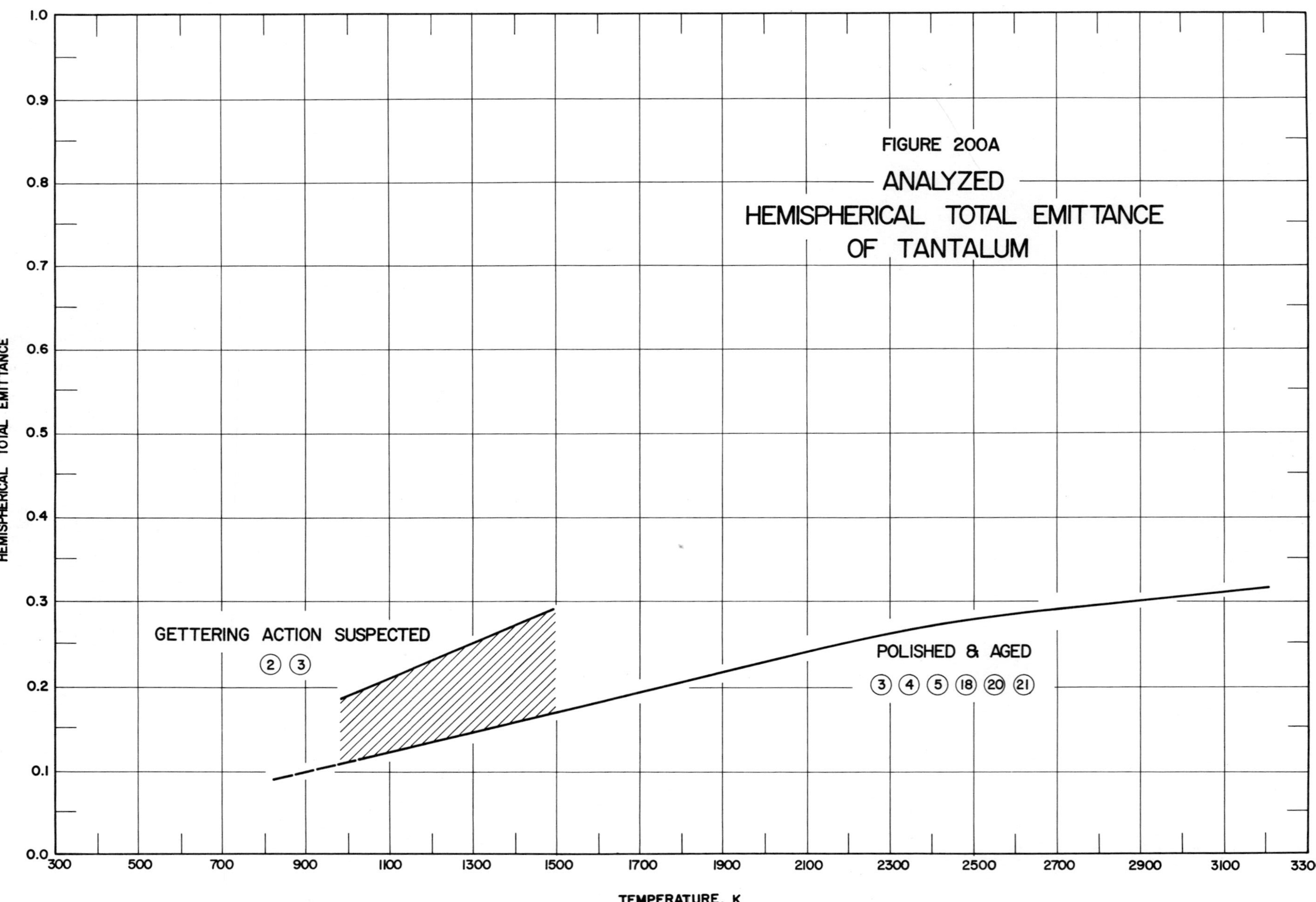
FIGURE 200A
ANALYZED
HEMISPHERICAL TOTAL EMITTANCE
OF TANTALUM
GETTERING ACTION SUSPECTED
2 3
POLISHED & AGED
3 4 5 18 20 21
HEMISPHERICAL TOTAL EMITTANCE
1.0
0.9
0.8
0.7
0.6
0.5
0.4
0.3
0.2
0.1
0.0
300
500
700
900
1100
1300
1500
1700
1900
2100
2300
2500
2700
2900
3100
3300
TEMPERATURE, K

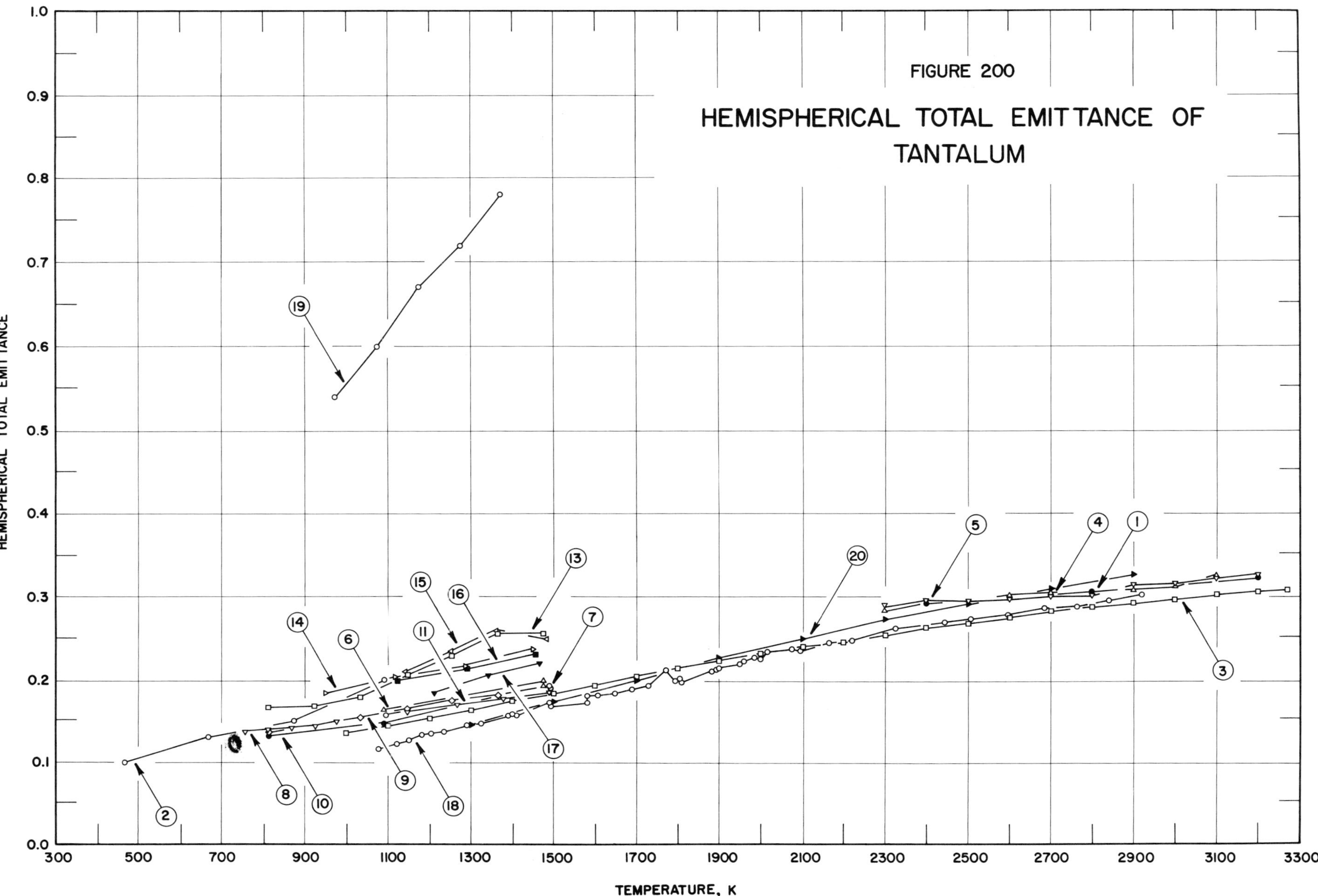
FIGURE 200
HEMISPHERICAL TOTAL EMITTANCE OF TANTALUM
HEMISPHERICAL TOTAL EMITTANCE
TEMPERATURE, K
0.0
0.1
0.2
0.3
0.4
0.5
0.6
0.7
0.8
0.9
1.0
300
500
700
900
1100
1300
1500
1700
1900
2100
2300
2500
2700
2900
3100
3300
1
2
3
4
5
6
7
8
9
10
11
13
14
15
16
17
18
19
20

SPECIFICATION TABLE NO. 200 HEMISPHERICAL TOTAL EMITTANCE OF TANTALUM

Curve No.	Ref. No.	Year	Temperature Range, K	Reported Error, %	Composition (weight percent), Specifications and Remarks
1	84	1962	2400-3200		Polished with successively finer abrasive papers (numbers 1, 0, 00, 000 and 0000); measured in argon.
2	47	1961	468-1093	< 10	Commercially pure; arc cast or sintered; disc (0.04 in. thick); ground with 600 grit carborundum and polished on a wet cloth lap with unlevigated jewelers rouge; measured in vacuum (10^{-5} mm Hg).
3	93	1939	1000-3269		Regular stock from Fansteel Metallurgical Corp.; tantalum was sealed in glass bulb, outgassed, flashed, sealed in clean bulb, and outgassed at 2023 K for 5 hrs with seven, interposed, 30 sec flashes at 2673 K; data calculated from total radiation intensity by means of Stefan-Boltzmann law.
4	70	1960	2300-3100		< 0.02 Si, 0.005 Fe, 0.003 Mo, 0.0008 C, 0.052 others, Ta balance; pressed and sintered metal powder; hot and cold rolled; successively polished with No. 1-, 0-, 00-, 000-, and 0000- abrasive papers; measured in argon (> 760 mm Hg); [Author's designation: Tantalum No. 1].
5	70	1960	2300-3200		0.0035 Nb, 0.0032 O_2, 0.0028 Fe, 0.0016 C, < 0.0010 N_2, 0.0175 others, Ta balance; cast, in vacuum, in water-cooled copper mold using consumable electrode, cold rolled, swaged, and cold drawn; successively polished with No. 1-, 0-, 00-, 000-, and 0000- abrasive papers; measured in argon (> 760 mm Hg); [Author's designation: Tantalum No. 2].
6	96	1964	1088-1478		Measured in vacuum (< 10^{-7} mm Hg); temperature measured with thermocouple; heating cycle; [Author's designation: first specimen].
7	96	1964	1098-1492		Above specimen and conditions except temperature measured with optical pyrometer.
8	96	1964	758-978		Curve 6 specimen and conditions; run 2.
9	96	1964	811-1477		Curve 6 specimen and conditions; run 3, heating cycle.
10	96	1964	1477-811		Above specimen and conditions; cooling cycle.
11	96	1964	1149-1492		Curve 9 specimen and conditions except temperature measured with optical pyrometer.
12	96	1964	1492-1098		Above specimen and conditions; cooling cycle.
13	96	1964	811-1478		Different sample, same as curve 6 specimen and conditions; [Author's designation: second specimen].
14	96	1964	1478-949		Above specimen and conditions; cooling cycle.
15	96	1964	1140-1486		Curve 13 specimen and conditions except temperature measured with optical pyrometer.
16	96	1964	1486-1121		Above specimen and conditions; cooling cycle.
17	85	1963	1216-1464	± 5	Unaged, optically smooth, opaque, contamination-free ribbon (0.005 in. thick); measured in vacuum (3×10^{-7} to 1×10^{-5} mm Hg).
18	85	1963	1075-2920	± 5	Aged, optically smooth, opaque, contamination-free ribbon (0.005 in. thick); measured in vacuum (3×10^{-7} to 1×10^{-5} mm Hg).
19	58	1961	973-1373	± 2.5	Lightly etched in hydrofluoric acid; fine matte finish; measured in vacuum (< 5×10^{-6} mm Hg); data extracted from smooth curve.

SPECIFICATION TABLE NO. 200 (continued)

Curve No.	Ref. No.	Year	Temperature Range, K	Reported Error, %	Composition (weight percent), Specifications and Remarks
20	328	1966	1300-2900	± 10	99.61 Ta, 0.33 Nb, < 0.01 Fe, < 0.01 Ti, < 0.01 Si, ~0.02 Mo, ~0.014 W; produced by electron-beam melting in vacuum; specific gravity at 293 K 16.57 g cm^{-3}; electrical resistivity at 293 K 13.7 x 10^{-6} Ω cm; measured in vacuum (~10^{-5} mm Hg).
21	328	1966	1300-2700	± 10	Above specimen and conditions except corrected for linear expansion.

DATA TABLE NO. 200 HEMISPHERICAL TOTAL EMITTANCE OF TANTALUM

[Temperature, T, K; Emittance, ∈]

T	∈
CURVE 1	
2400	0.292
2800	0.306
3200	0.321
CURVE 2	
468	0.10
668	0.13
873	0.15
1093	0.20
CURVE 3	
1000	0.136
1100	0.144
1200	0.153
1300	0.163
1400	0.174
1500	0.184
1600	0.194
1700	0.205
1800	0.215
1900	0.223
2000	0.232
2100	0.240
2200	0.247
2300	0.254
2400	0.261
2500	0.269
2600	0.276
2700	0.282
2800	0.288
2900	0.293
3000	0.298
3100	0.302
3200	0.306
3269	0.309
CURVE 4	
2300	0.284
2400	0.293*
2600	0.302
2700	0.304
2800	0.306*
2900	0.309
3000	0.312
3100	0.326
CURVE 5	
2300	0.290
2400	0.296
2500	0.296
2600	0.298
2700	0.302
2800	0.302
2900	0.313
3000	0.314
3100	0.322
3200	0.326
CURVE 6	
1088	0.164
1478	0.200
1478	0.197
CURVE 7	
1098	0.158
1489	0.194
1492	0.190
CURVE 8	
758	0.139
810	0.140
866	0.142
923	0.144
978	0.150
CURVE 9	
811	0.139
922	0.145*
1033	0.154
1144	0.165
1256	0.175
1366	0.183
1477	0.195*
CURVE 10	
1477	0.195*
1090	0.147
811	0.133
CURVE 11	
1149	0.162
CURVE 11 (cont.)	
1265	0.170
1378	0.178
1492	0.187
CURVE 12*	
1492	0.187
1098	0.144
CURVE 13	
811	0.167
922	0.169
1033	0.180
1144	0.207
1255	0.230
1366	0.256
1478	0.256
CURVE 14	
1478	0.256*
1450	0.237
1283	0.218
1116	0.203
949	0.185
CURVE 15	
1140	0.210
1249	0.234
1363	0.259
1486	0.250
CURVE 16	
1486	0.250*
1458	0.232
1288	0.215
1121	0.200
CURVE 17	
1216	0.185
1340	0.206
1464	0.220
CURVE 18	
1075	0.116
CURVE 18 (cont.)	
1120	0.122
1148	0.126
1180	0.133
1202	0.135
1236	0.137
1288	0.145
1324	0.147
1388	0.157
1400	0.159
1408	0.158
1490	0.171
1492	0.167
1580	0.171
1580	0.181
1604	0.181
1648	0.185
1690	0.190
1696	0.192*
1728	0.194
1770	0.212
1796	0.200
1804	0.201
1808	0.199
1880	0.211
1892	0.213
1900	0.215
1948	0.220
1960	0.223
1984	0.228
2000	0.227
2018	0.234
2076	0.238
2092	0.236
2104	0.240*
2164	0.246
2196	0.247*
2220	0.249
2296	0.258*
2324	0.261
2400	0.261*
2444	0.270
2504	0.273
2594	0.277
2686	0.286
2762	0.289
2840	0.296
2920	0.301
CURVE 19	
973	0.540
1073	0.600
1173	0.670
1273	0.720
1373	0.780
CURVE 20	
1300	0.145
1500	0.174
1700	0.200
1900	0.225
2100	0.250
2300	0.273
2500	0.292
2700	0.310
2900	0.328
CURVE 21*	
1300	0.143
1500	0.171
1700	0.196
1900	0.220
2100	0.243
2300	0.264
2500	0.282
2700	0.298

* Not shown on plot

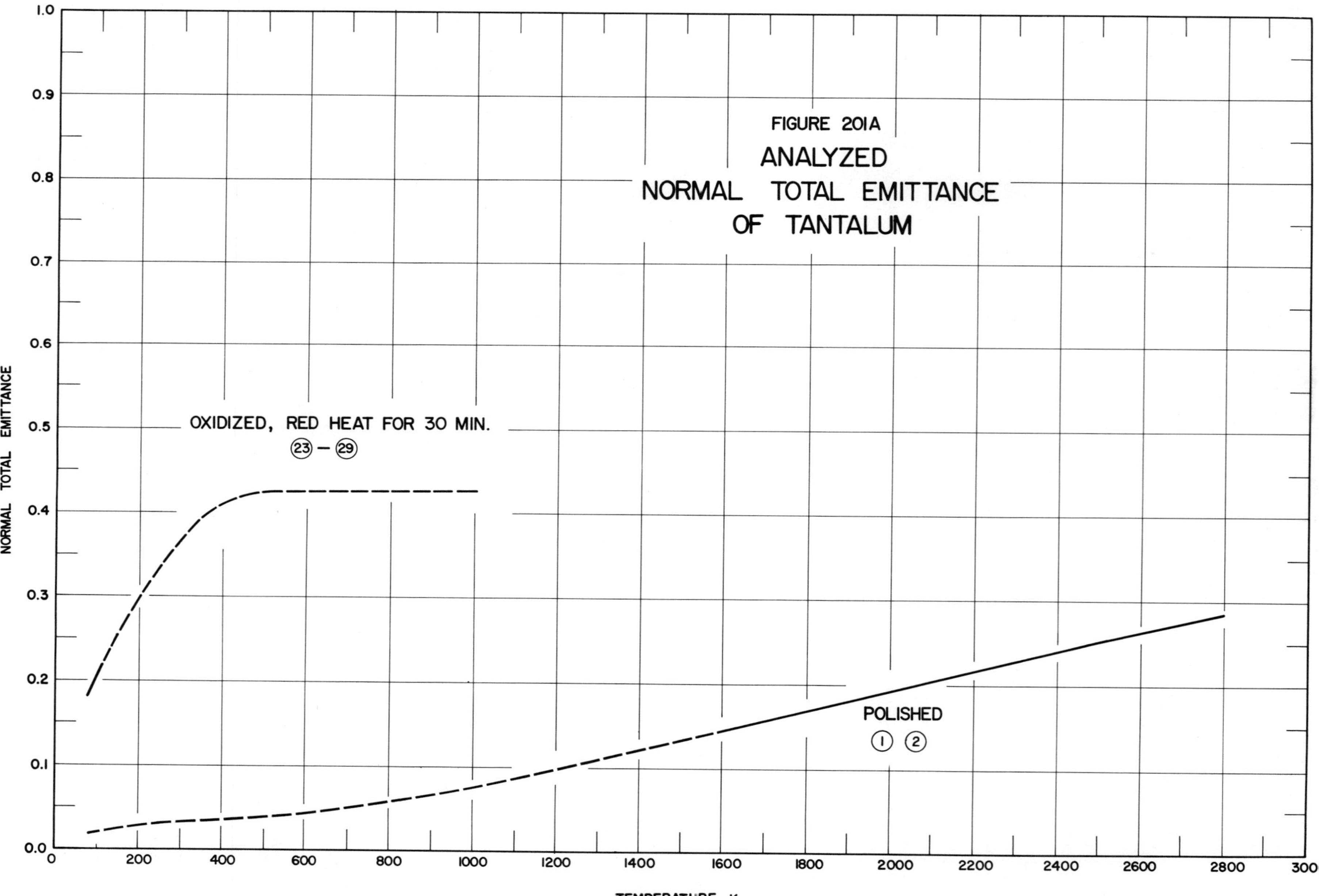

FIGURE 201A
ANALYZED
NORMAL TOTAL EMITTANCE
OF TANTALUM

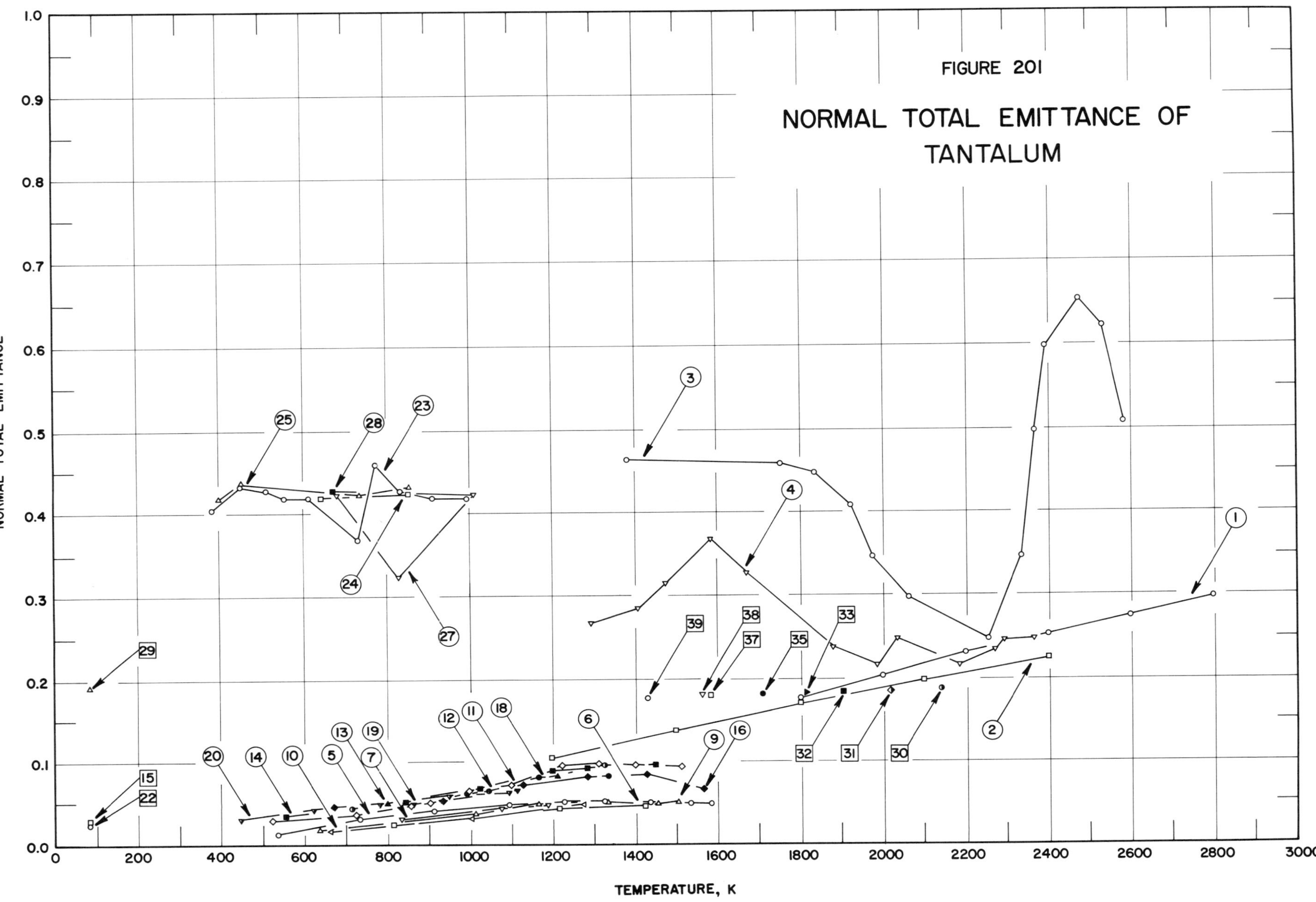
FIGURE 201
NORMAL TOTAL EMITTANCE OF TANTALUM
NORMAL TOTAL EMITTANCE
TEMPERATURE, K
0.0
0.1
0.2
0.3
0.4
0.5
0.6
0.7
0.8
0.9
1.0
0
200
400
600
800
1000
1200
1400
1600
1800
2000
2200
2400
2600
2800
3000

SPECIFICATION TABLE NO. 201 NORMAL TOTAL EMITTANCE OF TANTALUM

Curve No.	Ref. No.	Year	Temperature Range, K	Geometry θ'	Reported Error, %	Composition (weight percent), Specifications and Remarks
1	95	1963	1800-2800	$\sim 0°$		99.9 Ta; from Bram Chemical Co.; polished with carbide paper of 240, 400, and 600 grit, respectively, then with silk cloth and felt cloth; washed in acetone, then alcohol, and dried with dry nitrogen; data calculated from spectral data over all wavelengths and extracted from smooth curve.
2	85	1963	1200-2400	$\sim 0°$	± 5	Optically smooth, opaque, contamination-free ribbon (0.005 in. thick); measured in vacuum (3×10^{-7} to 1×10^{-5} mm Hg); data extracted from smooth curve.
3	75	1962	1383-2589	$\sim 0°$		Ground to a smooth finish; measured in dried argon or helium; data extracted from smooth curve.
4	92	1963	1295-2366	$\sim 0°$		Polished; surface roughness measured with profilometer 0.5 - 1.0 μ (RMS); measured in vacuum (3×10^{-8} to 4×10^{-8} mm Hg).
5	34	1957	539-1583	$\sim 0°$	± 10	Pure (Fansteel); strip (0.005 in. thick); as received; measured in air (5×10^{-4} mm Hg); increasing temp, cycle 1.
6	34	1957	1583-816	$\sim 0°$	± 10	Above specimen and conditions; decreasing temp, cycle 1.
7	34	1957	833-1189	$\sim 0°$	± 10	Above specimen and conditions; increasing temp, cycle 2.
8	34	1957	1189-622	$\sim 0°$	± 10	Above specimen and conditions; decreasing temp, cycle 2.
9	34	1957	633-1505	$\sim 0°$	± 10	Above specimen and conditions; increasing temp, cycle 3.
10	34	1957	1505-661	$\sim 0°$	± 10	Above specimen and conditions; decreasing temp, cycle 3.
11	34	1957	522-1511	$\sim 0°$	± 10	Pure (Fansteel); strip (0.005 in. thick); cleaned with liquid detergent; measured in air (5×10^{-4} mm Hg); increasing temp, cycle 1.
12	34	1957	716-1511	$\sim 0°$	± 10	Above specimen and conditions; cycle 2.
13	34	1957	1211-800	$\sim 0°$	± 10	Above specimen and conditions; decreasing temp, cycle 2.
14	34	1957	555-1450	$\sim 0°$	± 10	Curve 11 specimen and conditions; increasing temp, cycle 3.
15	34	1957	83.2	$\sim 0°$	± 10	Pure (Fansteel); strip (0.005 in. thick); as received or cleaned with liquid detergent; measured in air (5×10^{-4} mm Hg).
16	34	1957	672-1566	$\sim 0°$	± 10	Pure (Fansteel); strip (0.005 in. thick); polished; measured in air (5×10^{-4} mm Hg); increasing temp, cycle 1.
17	34	1957	1566-672	$\sim 0°$	± 10	Above specimen and conditions; decreasing temp, cycle 1.
18	34	1957	1333-994	$\sim 0°$	± 10	Curve 16 specimen and conditions; decreasing temp, cycle 2.
19	34	1957	1116-861	$\sim 0°$	± 10	Above specimen and conditions; decreasing temp, cycle 2.
20	34	1957	450-1094	$\sim 0°$	± 10	Curve 16 specimen and conditions; increasing temp, cycle 3.
21	34	1957	1094-839	$\sim 0°$	± 10	Above specimen and conditions; decreasing temp, cycle 3.
22	34	1957	83.2	$\sim 0°$	± 10	Different sample, same as curve 16 specimen and conditions.

SPECIFICATION TABLE NO. 201 (continued)

Curve No.	Ref. No.	Year	Temperature Range, K	Geometry θ'	Reported Error, %	Composition (weight percent), Specifications and Remarks
23	34	1957	383-997	$\sim 0^\circ$	± 10	Pure (Fansteel); strip (0.005 in. thick); oxidized in air at red heat for 30 min; measured in air (5×10^{-4} mm Hg); increasing temp, cycle 1.
24	34	1957	997-644	$\sim 0^\circ$	± 10	Above specimen and conditions; decreasing temp, cycle 1.
25	34	1957	400-855	$\sim 0^\circ$	± 10	Curve 23 specimen and conditions; increasing temp, cycle 2.
26	34	1957	855-455	$\sim 0^\circ$	± 10	Above specimen and conditions; decreasing temp, cycle 2.
27	34	1957	683-1011	$\sim 0^\circ$	± 10	Curve 23 specimen and conditions; increasing temp, cycle 3.
28	34	1957	1011-687	$\sim 0^\circ$	± 10	Above specimen and conditions; decreasing temp, cycle 3.
29	34	1957	83.2	$\sim 0^\circ$	± 10	Different sample, same as curve 23 specimen and conditions.
30	267	1965	2142	$\sim 0^\circ$		Obtained from Fansteel Metallurgical Corp; thermal conductivity 0.6657 W $cm^{-1} K^{-1}$; specific heat 0.0401 Cal $sec^{-1} cm^{-1} K^{-1}$.
31	267	1965	2020	$\sim 0^\circ$		Obtained from Fansteel Metallurgical Corp; thermal conductivity 0.5808 W $cm^{-1} K^{-1}$; specific heat 0.0397 Cal $sec^{-1} cm^{-1} K^{-1}$.
32	267	1965	1907	$\sim 0^\circ$		Obtained from Fansteel Metallurgical Corp; thermal conductivity 0.6364 W $cm^{-1} K^{-1}$; specific heat 0.0394 Cal $sec^{-1} cm^{-1} K^{-1}$.
33	267	1965	1817	$\sim 0^\circ$		Obtained from Fansteel Metallurgical Corp; thermal conductivity 0.6172 W $cm^{-1} K^{-1}$; specific heat 0.0391 Cal $sec^{-1} cm^{-1} K^{-1}$.
34	267	1965	1811	$\sim 0^\circ$		Obtained from Fansteel Metallurgical Corp; thermal conductivity 0.5929 W $cm^{-1} K^{-1}$; specific heat 0.0391 Cal $sec^{-1} cm^{-1} K^{-1}$.
35	267	1965	1714	$\sim 0^\circ$		Obtained from Fansteel Metallurgical Corp; thermal conductivity 0.6418 W $cm^{-1} K^{-1}$; specific heat 0.0388 Cal $sec^{-1} cm^{-1} K^{-1}$.
36	267	1965	1712	$\sim 0^\circ$		Obtained from Fansteel Metallurgical Corp; thermal conductivity 0.6393 W $cm^{-1} K^{-1}$; specific heat 0.0388 Cal $sec^{-1} cm^{-1} K^{-1}$.
37	267	1965	1585	$\sim 0^\circ$		Obtained from Fansteel Metallurgical Corp; thermal conductivity 0.6192 W $cm^{-1} K^{-1}$; specific heat 0.0384 Cal $sec^{-1} cm^{-1} K^{-1}$.
38	267	1965	1563	$\sim 0^\circ$		Obtained from Fansteel Metallurgical Corp; thermal conductivity 0.5854 W $cm^{-1} K^{-1}$; specific heat 0.0384 Cal $sec^{-1} cm^{-1} K^{-1}$.
39	267	1965	1434	$\sim 0^\circ$		Obtained from Fansteel Metallurgical Corp; thermal conductivity 0.5715 W $cm^{-1} K^{-1}$; specific heat 0.0379 Cal $sec^{-1} cm^{-1} K^{-1}$.

DATA TABLE NO. 201 NORMAL TOTAL EMITTANCE OF TANTALUM

[Temperature, T, K; Emittance, ε]

	T	ε
CURVE 1	1800	0.180
	2000	0.205
	2200	0.231
	2400	0.255
	2600	0.278
	2800	0.300
CURVE 2	1200	0.106
	1500	0.140
	1800	0.172
	2100	0.200
	2400	0.225
CURVE 3	1383	0.465
	1755	0.460
	1839	0.450
	1922	0.410
	1978	0.350
	2061	0.300
	2255	0.250
	2339	0.350
	2367	0.500
	2394	0.600
	2478	0.655
	2533	0.625
	2589	0.510
CURVE 4	1295	0.270
	1408	0.287
	1478	0.317
	1585	0.370
	1672	0.330
	1880	0.240
	1989	0.218
	2033	0.250
	2186	0.218
	2272	0.235
	2291	0.247
	2366	0.250
CURVE 5	539	0.013
	733	0.032
	911	0.041
	1094	0.049
	1228	0.052
	1322	0.052
	1439	0.051
	1533	0.050
	1583	0.050
CURVE 6	1583	0.050*
	1422	0.048
	1216	0.045
	816	0.026
CURVE 7	833	0.031
	1078	0.044
	1189	0.048
CURVE 8*	1189	0.048
	855	0.031
	622	0.018
CURVE 9	633	0.019
	1011	0.037
	1166	0.050
	1333	0.051
	1455	0.050
	1505	0.051
CURVE 10	1505	0.051*
	1278	0.050
	1005	0.032
	661	0.018
CURVE 11	522	0.030
	722	0.037
	855	0.049
	905	0.051
	1000	0.066
	1100	0.072
	1222	0.097
	1311	0.099
	1400	0.098
	1511	0.096
CURVE 12	716	0.045
	1044	0.065
	1322	0.098
	1511	0.097*
CURVE 13	1211	0.050
	800	0.083
CURVE 14	555	0.035
	844	0.051
	1022	0.068
	1200	0.090
	1283	0.092
	1450	0.098
CURVE 15	83.2	0.030
CURVE 16	672	0.047
	933	0.053
	1128	0.072
	1283	0.083
	1428	0.085
	1566	0.067
CURVE 17*	1566	0.067
	1272	0.081
	922	0.053
	672	0.046
CURVE 18	1333	0.082
	1166	0.081
	994	0.063
CURVE 19	1116	0.066
	861	0.050
CURVE 20	450	0.031
	622	0.043
	789	0.050
	950	0.060
	1094	0.062
CURVE 21*	1094	0.062
	839	0.050
CURVE 22	83.2	0.025
CURVE 23	383	0.405
	450	0.435
	511	0.430
	555	0.420
	616	0.420
	733	0.370
	778	0.460
	833	0.430
	911	0.420
	997	0.420
CURVE 24	997	0.420
	855	0.425
	644	0.420
CURVE 25	400	0.420
	455	0.440
	739	0.425
	855	0.435
CURVE 26*	855	0.435
	628	0.425
	455	0.440
CURVE 27	683	0.425
	828	0.325
	1011	0.425
CURVE 28	1011	0.425
	687	0.430
CURVE 29	83.2	0.192
CURVE 30	2142	0.191
CURVE 31	2020	0.189
CURVE 32	1907	0.188
CURVE 33	1817	0.186
CURVE 34*	1811	0.186
CURVE 35	1714	0.185
CURVE 36*	1712	0.185
CURVE 37	1585	0.183
CURVE 38	1563	0.183
CURVE 39	1434	0.180

*Not shown on plot

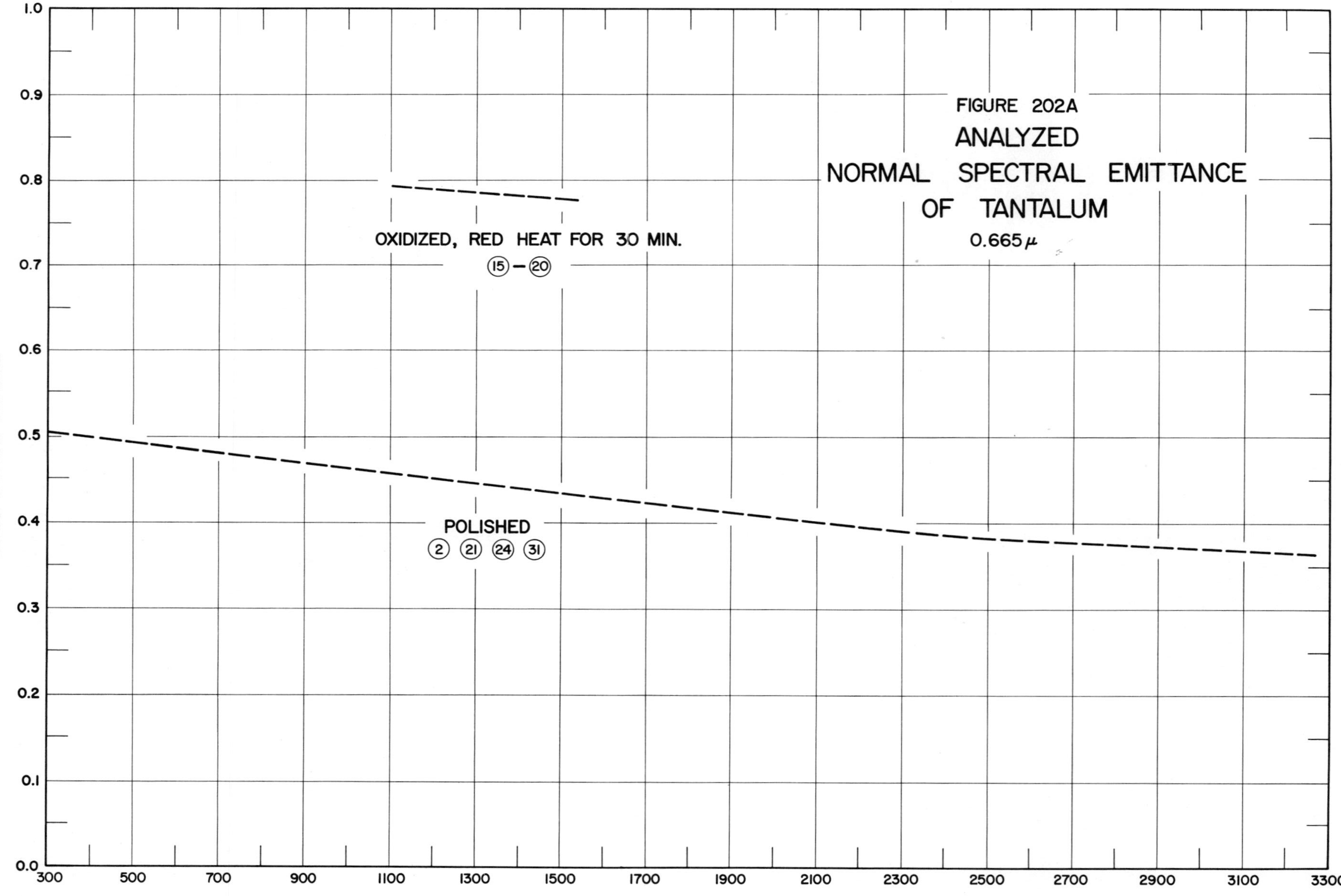

FIGURE 202A ANALYZED NORMAL SPECTRAL EMITTANCE OF TANTALUM 0.665μ

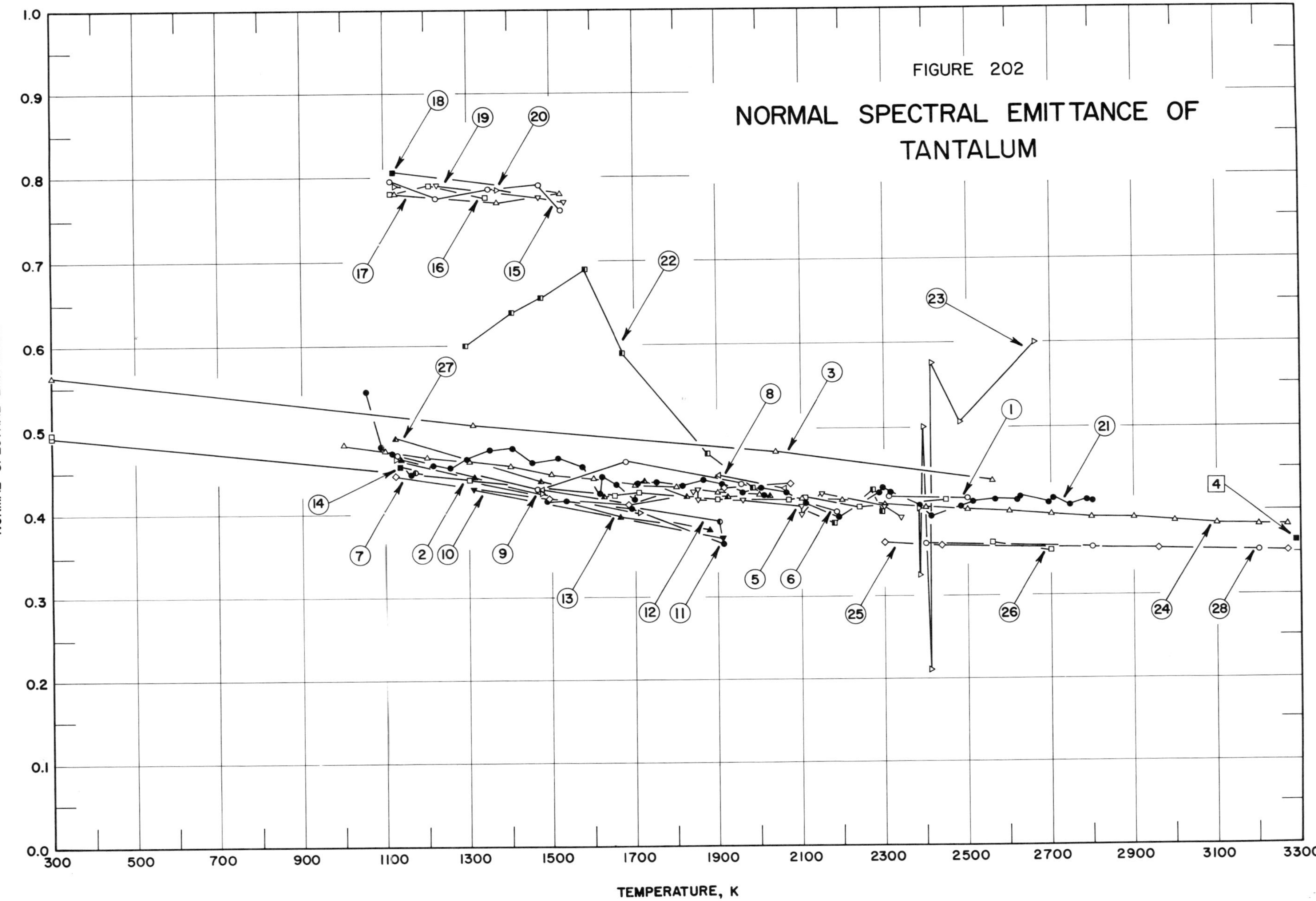

FIGURE 202

NORMAL SPECTRAL EMITTANCE OF TANTALUM

SPECIFICATION TABLE NO. 202 NORMAL SPECTRAL EMITTANCE OF TANTALUM

Curve No.	Ref. No.	Year	Wavelength μ	Temperature Range, K	Geometry θ'	Reported Error, %	Composition (weight percent), Specifications and Remarks
1	51	1926	0.665	2310-2500	$\sim 0^\circ$		Pre-war tantalum.
2	51	1926	0.665	300-2450	$\sim 0^\circ$		Polished; values corrected to account for surface pitting; author assumed $\epsilon = 1 - \rho$.
3	51	1926	0.463	300-2560	$\sim 0^\circ$		Polished; values corrected to account for surface pitting; author assumed $\epsilon = 1 - \rho$.
4	51	1926	0.665	3290	$\sim 0^\circ$		W reference; calculated from brightness temp.
5	51	1926	0.665	1840-2340	$\sim 0^\circ$		Mo reference; calculated from brightness temp.
6	80	1956	0.665	1127-2184	$\sim 0^\circ$	<2.7	As received; measured in vacuum (2.0-7.0 x 10^{-4} mm Hg); first cycle.
7	80	1956	0.665	1123-2074	$\sim 0^\circ$	<1.7	Above specimen and conditions; second cycle.
8	80	1956	0.665	1124-2060	$\sim 0^\circ$	<1.7	Above specimen and conditions; third cycle.
9	34	1957	0.665	1122-1711	$\sim 0^\circ$		Pure metal (Fansteel); same results obtained for 3 different surface treatments (a. as received, b. cleaned with liquid detergent, c. polished); measured in air (5 x 10^{-4} mm Hg); increasing temp, cycle 1.
10	34	1957	0.665	1911-1311	$\sim 0^\circ$		Above specimen and conditions; decreasing temp, cycle 1.
11	34	1957	0.665	1122-1914	$\sim 0^\circ$		Above specimen and conditions; increasing temp, cycle 2.
12	34	1957	0.665	1914-1169	$\sim 0^\circ$		Above specimen and conditions; decreasing temp, cycle 2.
13	34	1957	0.665	1133-1880	$\sim 0^\circ$		Above specimen and conditions; increasing temp, cycle 3.
14	34	1957	0.665	1880-1136	$\sim 0^\circ$		Above specimen and conditions; decreasing temp, cycle 3.
15	34	1957	0.665	1116-1525	$\sim 0^\circ$		Pure metal (Fansteel); oxidized in air at red heat for 30 min; measured in air (5 x 10^{-4} mm Hg); increasing temp, cycle 1.
16	34	1957	0.665	1525-1116	$\sim 0^\circ$		Above specimen and conditions; decreasing temp, cycle 1.
17	34	1957	0.665	1227-1525	$\sim 0^\circ$		Above specimen and conditions; increasing temp, cycle 2.
18	34	1957	0.665	1525-1122	$\sim 0^\circ$		Above specimen and conditions; decreasing temp, cycle 2.
19	34	1957	0.665	1227-1533	$\sim 0^\circ$		Above specimen and conditions; increasing temp, cycle 3.
20	34	1957	0.665	1533-1127	$\sim 0^\circ$		Above specimen and conditions; decreasing temp, cycle 3.
21	85	1963	0.65	1056-2800	$\sim 0^\circ$	±5	Optically smooth, opaque, contamination-free ribbon (0.005 in. thick); measured in vacuum (3 x 10^{-7} to 1 x 10^{-5} mm Hg).
22	92	1963	0.65	1294-2294	$\sim 0^\circ$		Polished; surface roughness measured with profilometer 0.5-1.0 μ (RMS); measured in vacuum (3 x 10^{-8} to 4 x 10^{-8} mm Hg).
23	75	1962	0.69	2383-2661	$\sim 0^\circ$		Ground to a smooth finish; measured in dried argon or helium.

SPECIFICATION TABLE NO. 202 (continued)

Curve No.	Ref. No.	Year	Wavelength μ	Temperature Range, K	Geometry θ'	Reported Error, %	Composition (weight percent), Specifications and Remarks
24	93	1939	0.665	1000-3269	~0°		Regular stock from Fansteel Metallurgical Corp; tantalum was sealed in glass bulb, outgassed, flashed, sealed in clean bulb, and outgassed at 2023 K for 5 hrs with seven, interposed, 30 sec flashes at 2673 K; data extracted from smooth curve.
25	70	1960	0.65	2300-3270	~0°		0.0035 Nb, 0.0032 O_2, 0.0028 Fe, 0.0016 C, <0.0010 N_2, 0.0175 others, Ta balance; cast in vacuum in water-cooled copper mold using consumable electrode, cold rolled, swaged and cold drawn; successively polished with No. 1-, 0 , 00-, 000-, and 0000- abrasive papers; measured in argon (>760 mm Hg); [Author's designation: Tantalum No. 2].
26	70	1960	0.65	2560-2700	~0°		<0.02 Si, 0.005 Fe, 0.003 Mo, 0.0008 C, 0.052 others, Ta balance; pressed and sintered metal powder; hot and cold rolled; successively polished with No. 1-, 0-, 00-, 000-, and 0000- abrasive papers; measured in argon (>760 mm Hg); [Author's designation: Tantalum No. 1].
27	94	1960	0.650	1123-2023	~0°		Pure tantalum sheet (0.1 to 0.2 mm thick).
28	84	1962	0.65	2400-3200	~0°		Polished with successively finer abrasive papers (numbers 1-, 0-, 00-, 000- and 0000-); measured in argon.
29	325	1963	0.65	1190-2545	~0°		Specimen obtained from foil; mean square deviation of roughness 0.65 μ.
30	325	1963	0.65	1187-2275	~0°		Polished surface; mean square deviation of roughness 0.9 μ.
31	325	1963	0.65	1196-2548	~0°		Mean square deviation of roughness 3.2 μ.

DATA TABLE NO. 202 NORMAL SPECTRAL EMITTANCE OF TANTALUM

[Temperature, T, K; Emittance, ∈; Wavelength, λ, μ]

T	∈
CURVE 1 $\lambda = 0.665$	
2310	0.419
2500	0.416
CURVE 2 $\lambda = 0.665$	
300	0.496
300	0.492
1300	0.442
1650	0.422
1710	0.423
1900	0.416
2070	0.415
2110	0.419
2240	0.406
2450	0.413
CURVE 3 $\lambda = 0.463$	
300	0.563
1310	0.506
2040	0.472
2560	0.438
CURVE 4 $\lambda = 0.665$	
3290	0.364
CURVE 5 $\lambda = 0.665$	
1840	0.427
1850	0.429
1850	0.415
1960	0.415
2100	0.408
2100	0.397
2150	0.421
2300	0.405
2340	0.393
CURVE 6 $\lambda = 0.665$	
1127	0.470
1466	0.430
1679	0.460
1933	0.435
2184	0.400
CURVE 7 $\lambda = 0.665$	
1123	0.448
1495	0.418
1681	0.410
1917	0.430
2074	0.435
CURVE 8 $\lambda = 0.665$	
1124	0.47*
1477	0.43
1685	0.41*
1905	0.44
2060	0.43
CURVE 9 $\lambda = 0.665$	
1122	0.465
1477	0.425
1711	0.400
CURVE 10 $\lambda = 0.665$	
1911	0.370
1311	0.430
CURVE 11 $\lambda = 0.665$	
1122	0.445*
1491	0.415
CURVE 11 (cont.)	
1533	0.415
1691	0.405
1914	0.365
CURVE 12 $\lambda = 0.665$	
1914	0.365*
1902	0.390
1169	0.450
CURVE 13 $\lambda = 0.665$	
1133	0.465
1311	0.445
1491	0.415*
1666	0.395
1880	0.380
CURVE 14 $\lambda = 0.665$	
1880	0.380*
1136	0.455
CURVE 15 $\lambda = 0.665$	
1116	0.795
1225	0.775
1352	0.785
1472	0.790
1525	0.760
CURVE 16 $\lambda = 0.665$	
1525	0.760*
1341	0.775
1208	0.790
1116	0.780
CURVE 17 $\lambda = 0.665$	
1227	0.780
1372	0.770
1525	0.780
CURVE 18 $\lambda = 0.665$	
1525	0.780
1122	0.805
CURVE 19 $\lambda = 0.665$	
1227	0.790
1475	0.775
1533	0.770
CURVE 20 $\lambda = 0.665$	
1533	0.770*
1372	0.785
1127	0.790
CURVE 21 $\lambda = 0.65$	
1056	0.545
1088	0.480
1116	0.471
1160	0.449
1216	0.459
1256	0.455
1296	0.464
1352	0.475
1404	0.478
1456	0.460
1516	0.466
1576	0.455
1616	0.423
1620	0.445
1656	0.436
CURVE 21 (cont.)	
1700	0.417
1704	0.437
1752	0.438
1816	0.434
1864	0.440
1912	0.434
1960	0.424
2004	0.430
2012	0.422
2064	0.425
2108	0.410
2112	0.415*
2192	0.394
2288	0.422
2296	0.430
2316	0.424
2384	0.408
2412	0.394
2484	0.404
2516	0.412
2568	0.414
2620	0.413
2626	0.418
2696	0.410
2704	0.416
2746	0.408
2788	0.414
2800	0.411
CURVE 22 $\lambda = 0.65$	
1294	0.600
1405	0.640
1478	0.657
1583	0.690
1672	0.590
1878	0.470
1986	0.430
2180	0.388
2272	0.427
2294	0.400
CURVE 23 $\lambda = 0.69$	
2383	0.405
2389	0.325
2394	0.500
2411	0.210
2417	0.575
2483	0.505
2661	0.600
CURVE 24 $\lambda = 0.665$	
1000	0.481
1100	0.476
1200	0.469
1300	0.462
1400	0.456
1500	0.449
1600	0.442
1700	0.437
1800	0.432
1900	0.426
2000	0.421
2100	0.417
2200	0.413
2300	0.409
2400	0.405
2500	0.402
2600	0.400
2700	0.397
2800	0.394
2900	0.391
3000	0.388
3100	0.386
3200	0.384
3269	0.383
CURVE 25 $\lambda = 0.65$	
2300	0.364
2440	0.360
2960	0.355
3270	0.351
CURVE 26 $\lambda = 0.65$	
2560	0.362
2700	0.353
CURVE 27 $\lambda = 0.650$	
1123	0.49
1473	0.44
1623	0.42
1723	0.44
1823	0.42
1923	0.42
2023	0.42
CURVE 28 $\lambda = 0.65$	
2400	0.361
2800	0.356
3200	0.352
CURVE 29* $\lambda = 0.65$	
1190	0.433
1198	0.445
1206	0.413
1228	0.426
1241	0.386
1285	0.425
1293	0.423
1309	0.389
1311	0.399
1375	0.425
1383	0.417
1394	0.442
1397	0.424
1478	0.417
1481	0.439
1492	0.413
1568	0.428
1573	0.415
1583	0.426
CURVE 29 (cont.)*	
1682	0.416
1690	0.404
1732	0.400
1791	0.411
1824	0.396
1831	0.405
1894	0.428
1910	0.419
1932	0.414
2004	0.415
2018	0.423
2039	0.414
2093	0.414
2115	0.426
2124	0.421
2214	0.417
2309	0.429
2313	0.414
2326	0.423
2332	0.411
2376	0.416
2376	0.428
2398	0.424
2416	0.424
2498	0.408
2499	0.421
2512	0.418
2545	0.420
CURVE 30* $\lambda = 0.65$	
1187	0.453
1211	0.460
1248	0.429
1280	0.451
1310	0.426
1370	0.460
1420	0.457
1467	0.480
1497	0.458
1528	0.490
1557	0.457
1575	0.480

*Not shown on plot

DATA TABLE NO. 202 (continued)

T	ε	T	ε	T	ε
CURVE 30 (cont.)*		CURVE 31 (cont.)*		CURVE 31 (cont.)*	
1589	0.447	1598	0.441	2435	0.408
1617	0.475	1624	0.448	2489	0.410
1685	0.438	1629	0.437	2491	0.403
1697	0.455	1650	0.442	2548	0.403
1706	0.436	1650	0.429		
1737	0.450	1662	0.441		
1768	0.443	1683	0.428		
1819	0.447	1694	0.447		
1869	0.447	1707	0.425		
1939	0.432	1741	0.453		
1979	0.435	1742	0.429		
2027	0.435	1754	0.437		
2077	0.437	1792	0.420		
2109	0.430	1795	0.448		
2173	0.437	1810	0.443		
2198	0.437	1824	0.449		
2275	0.421	1833	0.436		
		1855	0.422		
CURVE 31*		1873	0.425		
$\lambda = 0.65$		1887	0.432		
		1918	0.428		
1196	0.478	1944	0.428		
1229	0.458	1947	0.438		
1231	0.420	1977	0.425		
1245	0.428	1990	0.429		
1273	0.430	2004	0.434		
1273	0.474	2016	0.427		
1301	0.418	2063	0.431		
1312	0.450	2072	0.422		
1321	0.439	2089	0.421		
1344	0.430	2106	0.421		
1344	0.457	2133	0.421		
1361	0.446	2137	0.432		
1395	0.462	2156	0.426		
1412	0.447	2165	0.440		
1441	0.469	2175	0.416		
1447	0.457	2214	0.428		
1464	0.447	2236	0.428		
1484	0.447	2268	0.422		
1486	0.461	2279	0.424		
1499	0.461	2308	0.420		
1522	0.452	2329	0.419		
1535	0.452	2344	0.419		
1549	0.448	2352	0.411		
1564	0.459	2391	0.411		
1564	0.447	2414	0.392		
1576	0.443	2414	0.408		

*Not shown on plot

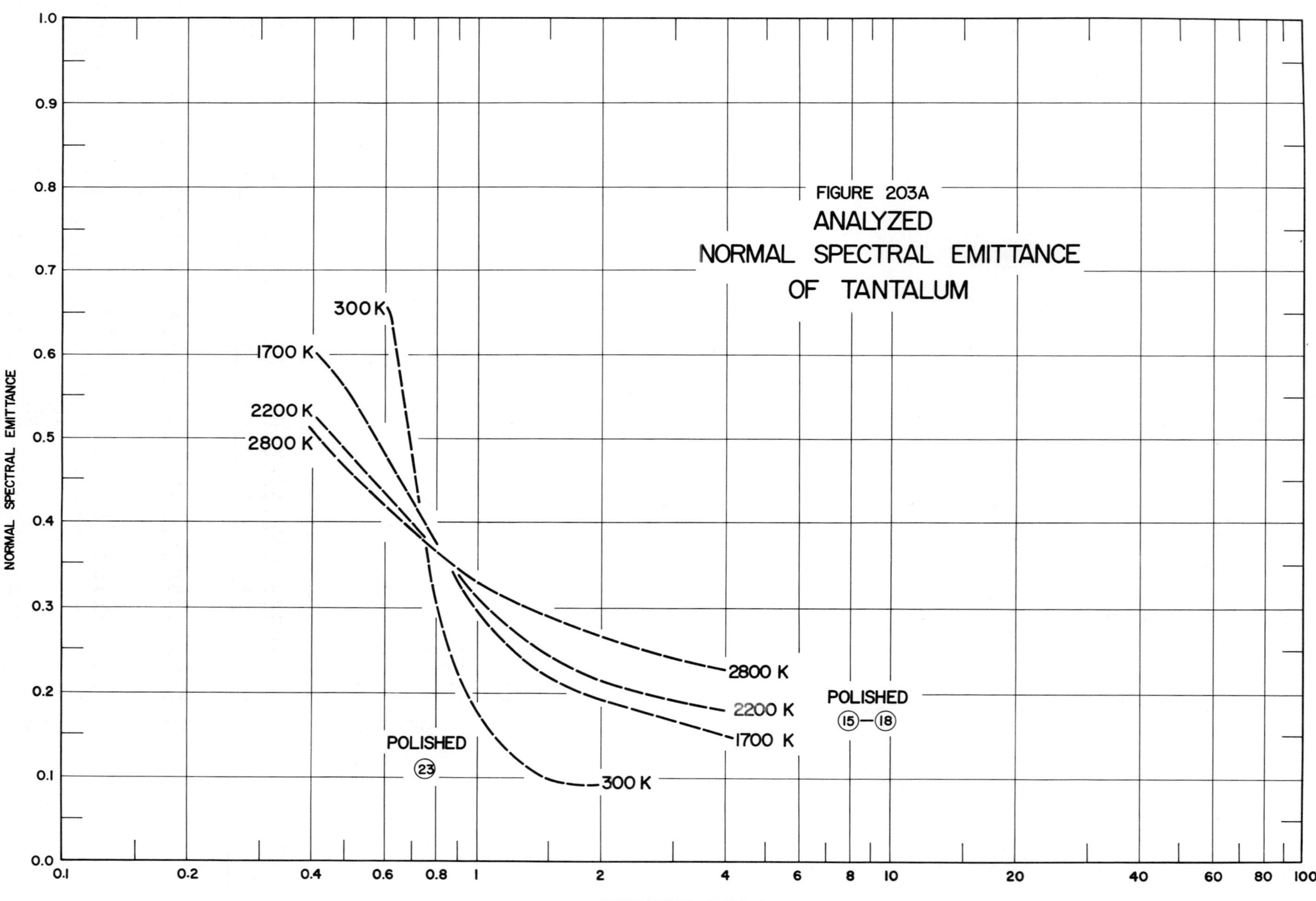

FIGURE 203A
ANALYZED
NORMAL SPECTRAL EMITTANCE
OF TANTALUM
300 K
1700 K
2200 K
2800 K
2800 K
2200 K
1700 K
POLISHED
(15)—(18)
POLISHED
(23)
300 K
NORMAL SPECTRAL EMITTANCE
1.0
0.9
0.8
0.7
0.6
0.5
0.4
0.3
0.2
0.1
0.0
0.1
0.2
0.4
0.6
0.8
1
2
4
6
8
10
20
40
60
80
100
WAVELENGTH, MICRONS

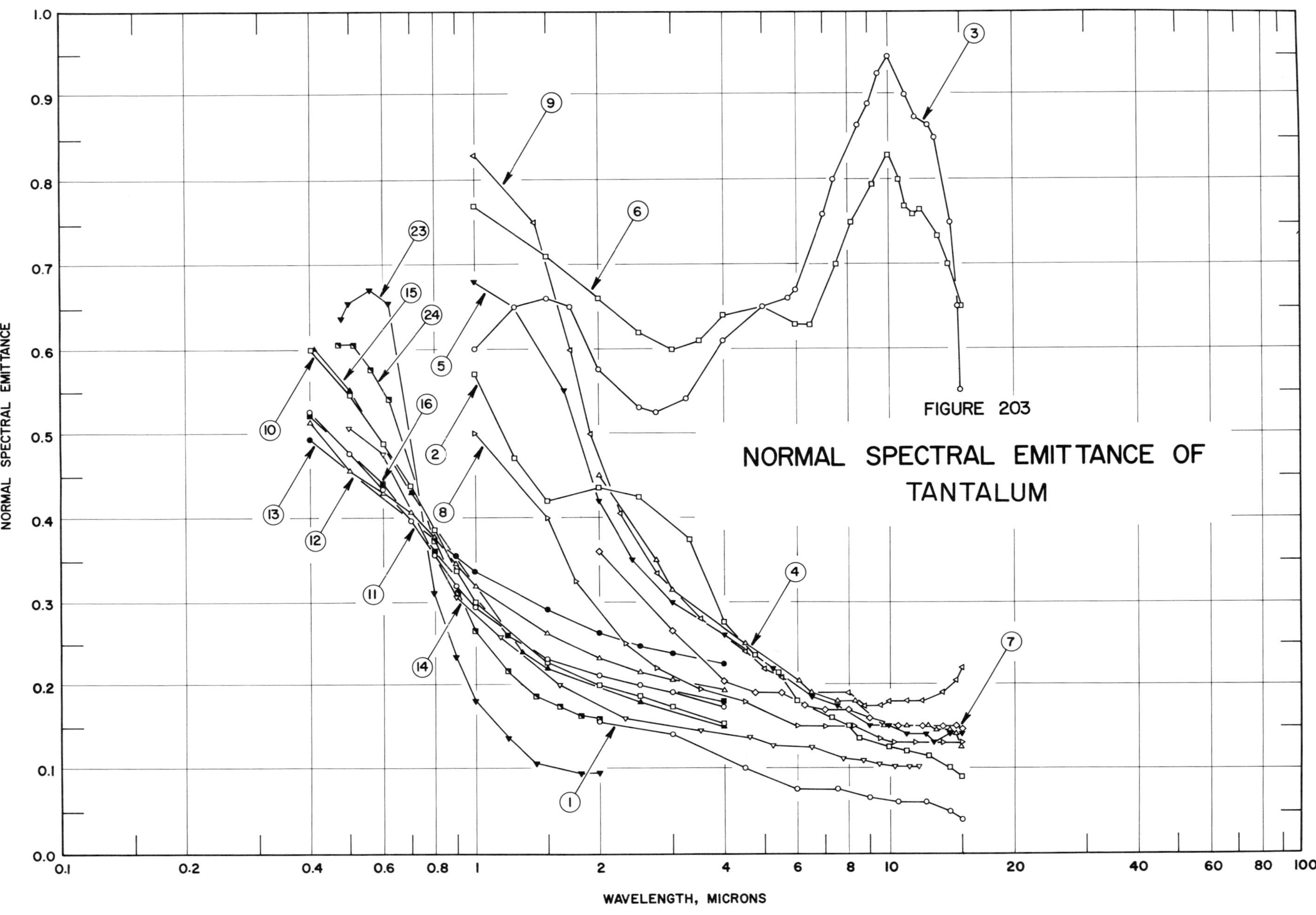

FIGURE 203
NORMAL SPECTRAL EMITTANCE OF TANTALUM

SPECIFICATION TABLE NO. 203 NORMAL SPECTRAL EMITTANCE OF TANTALUM

Curve No.	Ref. No.	Year	Temperature K	Wavelength Range, μ	Geometry θ'	Reported Error, %	Composition (weight percent), Specifications and Remarks
1	86	1961	523	2. 00-15. 00	$\sim0^\circ$	±5	As received; data extracted from smooth curve.
2	86	1961	773	1. 00-15. 00	$\sim0^\circ$	±5	Different sample, same as curve 1 specimen and conditions.
3	86	1961	1023	1. 00-15. 00	$\sim0^\circ$	±5	Different sample, same as curve 1 specimen and conditions.
4	86	1961	523	2. 00-15. 00	$\sim0^\circ$	±5	Different sample, same as curve 1 specimen and conditions except heated in argon at 1366 K for 30 min.
5	86	1961	773	1. 00-15. 00	$\sim0^\circ$	±5	Different sample, same as curve 4 specimen and conditions.
6	86	1961	1023	1. 00-15. 00	$\sim0^\circ$	±5	Different sample, same as curve 4 specimen and conditions.
7	86	1961	523	2. 00-15. 00	$\sim0^\circ$	±5	Different sample, same as curve 1 specimen and conditions except heated in vacuum (2. 2 x 10^{-5} mm Hg) at 1366 K for 30 min.
8	86	1961	773	1. 00-15. 00	$\sim0^\circ$	±5	Different sample, same as curve 7 specimen and conditions.
9	86	1961	1023	1. 00-15. 00	$\sim0^\circ$	±5	Different sample, same as curve 7 specimen and conditions.
10	95	1963	1700	0. 40-4. 00	$\sim0^\circ$		99. 9 Ta from Bram Chemical Co. ; polished with carbide paper of 240, 400, and 600 grit, respectively, then with silk cloth and felt cloth; washed in acetone, then alcohol, and dried with dry nitrogen; data extracted from smooth curve.
11	95	1963	2200	0. 40-4. 00	$\sim0^\circ$		Above specimen and conditions except heat treated at 2400 K.
12	95	1963	2400	0. 40-4. 00	$\sim0^\circ$		Above specimen and conditions.
13	95	1963	2800	0. 40-4. 00	$\sim0^\circ$		Above specimen and conditions.
14	12	1962	1478	0. 40-11. 80	$\sim0^\circ$		Measured in vacuum ($<10^{-7}$ mm Hg).
15	76	1962	1700	0. 410-4. 000	$\sim0^\circ$		Prepared from micronized powder; hot pressed at >2273 K; sintered, polished, etched, then degassed by heating to 2973 K; measured in argon; data extracted from smooth curve.
16	76	1962	2200	0. 400-4. 000	$\sim0^\circ$		Above specimen and conditions.
17	76	1962	2400	0. 400-4. 000	$\sim0^\circ$		Above specimen and conditions.
18	76	1962	2800	0. 400-4. 000	$\sim0^\circ$		Above specimen and conditions.
19	113	1961	1723	0. 402-4. 97	$\sim0^\circ$		Very thin and tenacious surface film (possibly tantalum nitride) detected; measured in vacuum; data extracted from smooth curve.
20	113	1961	2023	0. 405-4. 91	$\sim0^\circ$		Above specimen and conditions.
21	113	1961	2323	0. 417-4. 97	$\sim0^\circ$		Above specimen and conditions.
22	113	1961	2673	0. 410-4. 95	$\sim0^\circ$		Above specimen and conditions.

SPECIFICATION TABLE NO. 203 (continued)

Curve No.	Ref. No.	Year	Temperature K	Wavelength Range, μ	Geometry θ'	Reported Error, %	Composition (weight percent), Specifications and Remarks
23	323	1966	300	0.48-2.00	0°		Filament (0.25-0.32 mm in dia); baked for 1 hr at 798 K in vacuum; cooled, heated for 5-10 min in vacuum, and cooled; measured in argon (600 mm Hg); data calculated from optical constants.
24	323	1966	1100	0.47-2.00	0°		Above specimen and conditions.
25	323	1966	1500	0.47-2.00	0°		Above specimen and conditions.
26	323	1966	2000	0.47-2.01	0°		Above specimen and conditions.
27	323	1966	2500	0.48-2.00	0°		Above specimen and conditions.

DATA TABLE NO. 203 NORMAL SPECTRAL EMITTANCE OF TANTALUM

[Wavelength, λ, μ; Emittance, ∈; Temperature, T, K]

λ	∈
CURVE 1 T = 523	
2.00	0.155
3.00	0.140
4.50	0.100
6.00	0.075
7.50	0.075
9.00	0.065
10.50	0.060
12.25	0.060
14.00	0.050
15.00	0.040
CURVE 2 T = 773	
1.00	0.570
1.25	0.470
1.50	0.420
2.00	0.435
2.50	0.425
3.30	0.375
4.00	0.275
4.75	0.235
5.40	0.215
6.00	0.180
7.25	0.160
8.00	0.150
8.50	0.135
10.00	0.125
11.00	0.120
12.50	0.115
14.00	0.100
15.00	0.090
CURVE 3 T = 1023	
1.00	0.600
1.25	0.650
1.50	0.660
1.70	0.650
2.00	0.575
2.50	0.530
CURVE 3 (cont.)	
2.75	0.525
3.25	0.540
4.00	0.610
5.00	0.650
5.75	0.660
6.00	0.670
7.00	0.760
7.40	0.800
8.50	0.865
9.00	0.890
9.50	0.925
10.00	0.945
11.00	0.900
11.60	0.875
12.50	0.865
13.00	0.850
14.10	0.750
14.75	0.650
15.00	0.550
CURVE 4 T = 523	
2.00	0.450
2.75	0.350
3.00	0.315
4.50	0.250
6.05	0.205
6.50	0.190
7.50	0.180
8.25	0.180
9.70	0.150
11.00	0.150
12.40	0.150
13.00	0.145
14.00	0.145
14.50	0.140
15.00	0.125
CURVE 5 T = 773	
1.00	0.680
CURVE 5 (cont.)	
1.25	0.650*
1.65	0.550
2.00	0.420
2.40	0.350
3.00	0.300
4.00	0.260
5.25	0.220
6.50	0.185
7.50	0.175
9.00	0.150
10.00	0.150
11.00	0.140
12.25	0.140
12.75	0.130
14.00	0.140
15.00	0.140
CURVE 6 T = 1023	
1.00	0.770
1.50	0.710
2.00	0.660
2.50	0.620
3.00	0.600
3.50	0.610
4.00	0.640
5.00	0.650*
6.00	0.630
6.50	0.630
7.50	0.700
8.20	0.750
9.20	0.795
10.00	0.830
10.65	0.800
11.00	0.770
11.50	0.760
12.00	0.765
13.25	0.735
14.40	0.700
15.00	0.650
CURVE 7 T = 523	
2.00	0.360
3.00	0.265
4.00	0.205
4.75	0.190
5.50	0.190
6.25	0.175
7.00	0.170
8.00	0.170
9.00	0.160
10.50	0.150
12.00	0.150
13.50	0.150
14.50	0.150
15.00	0.145
CURVE 8 T = 773	
1.00	0.500
1.50	0.400
1.75	0.325
2.30	0.250
2.75	0.220
3.50	0.195
4.50	0.180
6.00	0.150
7.00	0.150
8.20	0.150
9.50	0.135
10.25	0.130
11.50	0.130
13.50	0.130
15.00	0.130
CURVE 9 T = 1023	
1.00	0.830
1.40	0.750
1.70	0.600
1.90	0.500
2.25	0.405
CURVE 9 (cont.)	
2.75	0.335
3.50	0.280
4.50	0.240
5.50	0.210
6.50	0.190*
8.00	0.190
8.75	0.175
9.50	0.175
10.00	0.180
11.00	0.180
12.00	0.180
13.50	0.190
14.50	0.205
15.00	0.220
CURVE 10 T = 1700	
0.40	0.600
0.50	0.545
0.60	0.487
0.70	0.437
0.80	0.386
0.90	0.337
1.00	0.300
1.50	0.228
2.00	0.200
2.50	0.186
3.00	0.172
4.00	0.152
CURVE 11 T = 2200	
0.40	0.524
0.50	0.476
0.60	0.432
0.70	0.396
0.80	0.356
0.90	0.320
1.00	0.294
1.50	0.230
2.00	0.212
CURVE 11 (cont.)	
2.50	0.200
3.00	0.190
4.00	0.173
CURVE 12 T = 2400	
0.40	0.513
0.50	0.457
0.60	0.430
0.70	0.407
0.80	0.378
0.90	0.346
1.00	0.320
1.50	0.262
2.00	0.232
2.50	0.216
3.00	0.206
4.00	0.193
CURVE 13 T = 2800	
0.40	0.494
0.50	0.457*
0.60	0.430*
0.70	0.407*
0.80	0.378*
0.90	0.355
1.00	0.338
1.50	0.290
2.00	0.263
2.50	0.249
3.00	0.239
4.00	0.226
CURVE 14 T = 1478	
0.40	0.515*
0.60	0.505
0.70	0.475
0.90	0.305
CURVE 14 (cont.)	
1.15	0.257
1.60	0.200
2.30	0.160
3.50	0.145
4.60	0.137
5.25	0.128
6.50	0.125
7.75	0.112
8.70	0.110
9.50	0.105
10.30	0.102
11.10	0.102
11.80	0.102
CURVE 15 T = 1700	
0.410	0.600
0.500	0.550
0.700	0.430
1.000	0.300*
1.300	0.240
1.500	0.220
2.500	0.180
4.000	0.150
CURVE 16 T = 2200	
0.400	0.520
0.600	0.440
0.800	0.360
1.000	0.300*
1.200	0.260
1.500	0.230*
2.000	0.210*
4.000	0.180
CURVE 17* T = 2400	
0.400	0.490
0.500	0.460
CURVE 17 (cont.)*	
0.700	0.400
0.800	0.380
1.000	0.310
1.500	0.260
2.000	0.228
2.500	0.210
4.000	0.190
CURVE 18* T = 2800	
0.400	0.510
0.500	0.460
0.700	0.400
0.800	0.380
1.000	0.330
1.500	0.290
2.000	0.267
3.000	0.240
4.000	0.230
CURVE 19* T = 1723	
0.402	0.563
0.466	0.529
0.550	0.487
0.635	0.446
0.693	0.418
0.760	0.388
0.837	0.355
0.950	0.312
1.10	0.274
1.34	0.231
1.59	0.203
1.85	0.188
2.15	0.180
2.60	0.178
3.25	0.168
3.72	0.160
4.32	0.158
4.97	0.154

* Not shown on plot

DATA TABLE NO. 203 (continued)

λ	∈
CURVE 20*	**T = 2023**
0.405	0.538
0.492	0.499
0.562	0.469
0.653	0.433
0.699	0.411
0.851	0.370
0.972	0.333
1.05	0.308
1.18	0.278
1.31	0.254
1.48	0.235
1.73	0.220
2.04	0.209
2.55	0.201
3.08	0.193
3.68	0.185
4.11	0.182
4.91	0.179
CURVE 21*	**T = 2323**
0.417	0.521
0.503	0.485
0.559	0.461
0.634	0.437
0.706	0.411
0.831	0.375
0.952	0.339
1.09	0.305
1.21	0.286
1.47	0.254
1.71	0.234
1.97	0.220
2.24	0.214
2.39	0.214
2.86	0.212
3.69	0.206
4.36	0.199
4.97	0.189
CURVE 22*	**T = 2673**
0.410	0.513
0.486	0.484
0.550	0.462
0.621	0.440
0.698	0.412
0.765	0.394
0.859	0.368
0.984	0.337
1.16	0.304
1.31	0.285
1.64	0.264
2.23	0.241
3.02	0.222
3.84	0.212
4.95	0.206
CURVE 23	**T = 300**
0.48	0.638
0.50	0.653
0.56	0.670
0.62	0.654
0.80	0.310
0.90	0.232
1.00	0.180
1.20	0.134
1.41	0.107
1.80	0.093
2.00	0.093
CURVE 24	**T = 1100**
0.47	0.606
0.51	0.604
0.56	0.573
0.62	0.540
0.80	0.378
0.91	0.312
CURVE 24 (cont.)	
1.00	0.265
1.20	0.217
1.40	0.187
1.60	0.173
1.81	0.162
2.00	0.157
CURVE 25*	**T = 1500**
0.47	0.593
0.49	0.583
0.69	0.560
0.61	0.512
0.80	0.379
0.90	0.335
1.00	0.301
1.20	0.258
1.40	0.226
1.61	0.203
1.81	0.196
2.00	0.189
CURVE 26*	**T = 2000**
0.47	0.570
0.50	0.539
0.56	0.509
0.63	0.486
0.80	0.381
0.90	0.340
1.00	0.315
1.20	0.278
1.41	0.253
1.60	0.236
1.81	0.224
2.01	0.214
CURVE 27*	**T = 2500**
0.48	0.521
0.51	0.511
0.56	0.493
0.62	0.463
0.80	0.379
0.90	0.348
1.00	0.324
1.20	0.289
1.40	0.276
1.60	0.258
1.80	0.244
2.00	0.232

* Not shown on plot

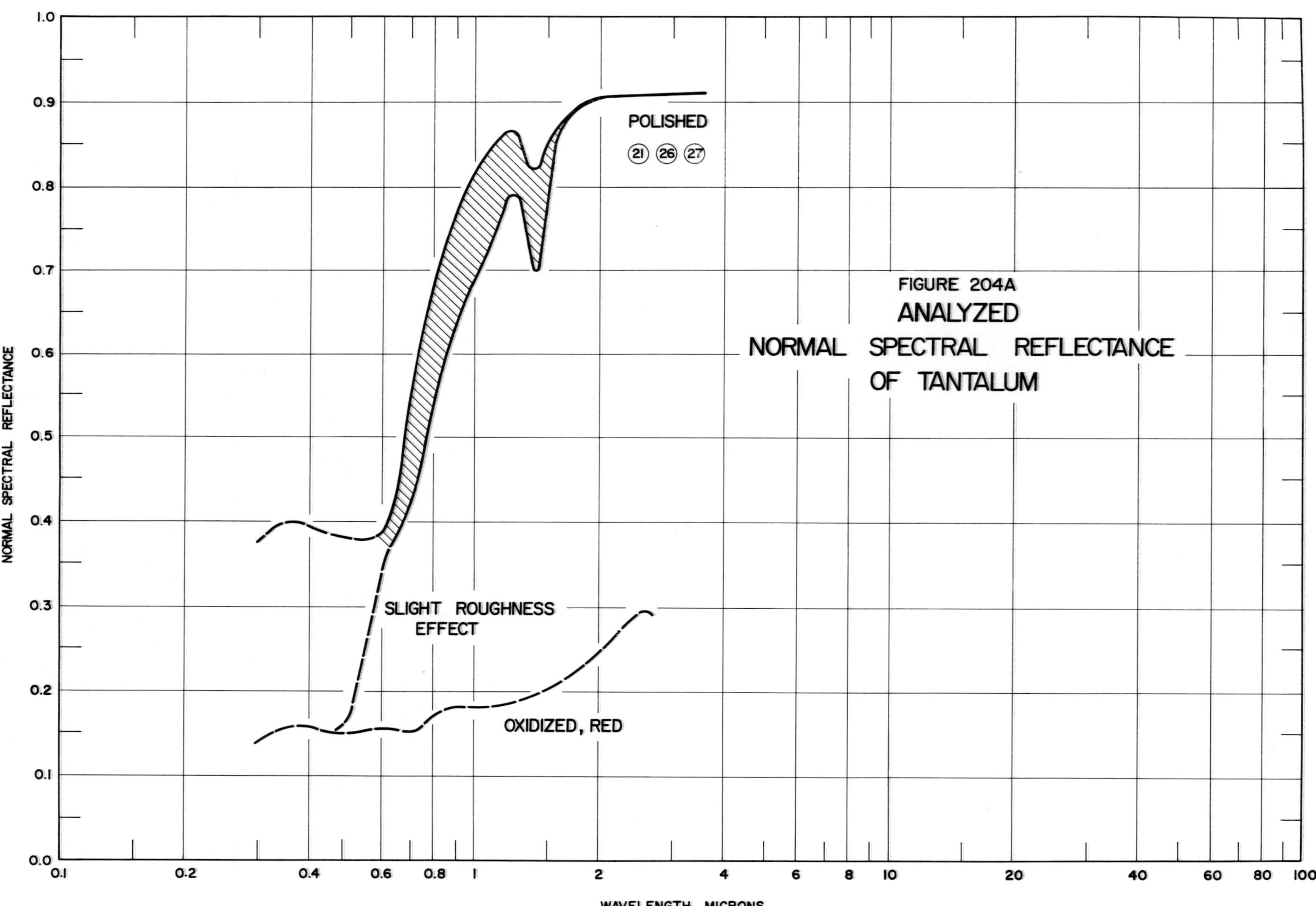

FIGURE 204A
ANALYZED
NORMAL SPECTRAL REFLECTANCE
OF TANTALUM

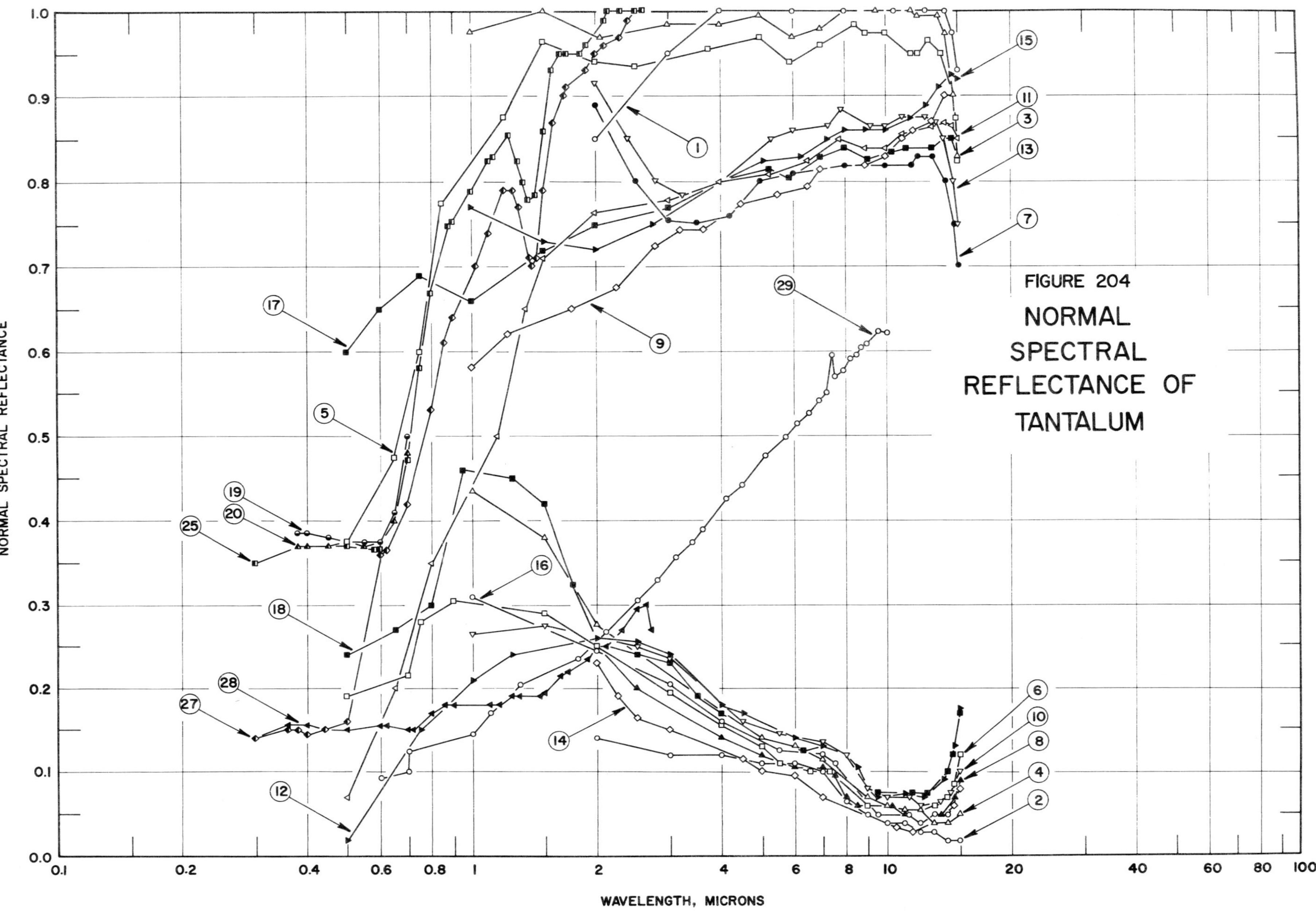
FIGURE 204
NORMAL SPECTRAL REFLECTANCE OF TANTALUM
NORMAL SPECTRAL REFLECTANCE
WAVELENGTH, MICRONS
0.0
0.1
0.2
0.3
0.4
0.5
0.6
0.7
0.8
0.9
1.0
0.1
0.2
0.4
0.6
0.8
1
2
4
6
8
10
20
40
60
80
100
1
2
3
4
5
6
7
8
9
10
11
12
13
14
15
16
17
18
19
20
25
27
28
29

SPECIFICATION TABLE NO. 204 NORMAL SPECTRAL REFLECTANCE OF TANTALUM

Curve No.	Ref. No.	Year	Temperature K	Wavelength Range, μ	Geometry θ	θ'	ω'	Reported Error, %	Composition (weight percent), Specifications and Remarks
1	86	1961	~322	2.00-15.00	~0°		2π	< 2	As received; data extracted from smooth curve; hohlraum at 523 K; converted from R (2π, 0°).
2	86	1961	~322	2.00-15.00	~0°		2π	< 2	Above specimen and conditions; diffuse component only.
3	86	1961	~322	1.00-15.00	~0°		2π	< 2	Different sample,same as Curve 1 specimen and conditions except hohlraum at 773 K.
4	86	1961	~322	1.00-15.00	~0°		2π	< 2	Above specimen and conditions; diffuse component only.
5	86	1961	~322	0.50-15.00	~0°		2π	< 2	Different sample, same as Curve 1 specimen and conditions except hohlraum at 1273 K.
6	86	1961	~322	0.50-15.00	~0°		2π	< 2	Above specimen and conditions; diffuse component only.
7	86	1961	~322	2.00-15.00	~0°		2π	< 2	Different sample, same as Curve 1 specimen and conditions except heated in argon at 1366 K for 30 min.
8	86	1961	~322	2.00-15.00	~0°		2π	< 2	Above specimen and conditions; diffuse component only.
9	86	1961	~322	1.00-15.00	~0°		2π	< 2	Different sample, same as Curve 7 specimen and conditions except hohlraum at 773 K.
10	86	1961	~322	1.00-15.00	~0°		2π	< 2	Above specimen and conditions; diffuse component only.
11	86	1961	~322	0.50-15.00	~0°		2π	< 2	Different sample, same as Curve 7 specimen and conditions except hohlraum at 1273 K.
12	86	1961	~322	0.50-15.00	~0°		2π	< 2	Above specimen and conditons; diffuse component only.
13	86	1961	~322	2.00-15.00	~0°		2π	< 2	Different sample, same as Curve 1 specimen and conditions except heated in vacuum (2.2×10^{-5} mm Hg) at 1366 K for 30 min.
14	86	1961	~322	2.00-15.00	~0°		2π	< 2	Above specimen and conditions; diffuse component only.
15	86	1961	~322	1.00-15.00	~0°		2π	< 2	Different sample, same as Curve 13 specimen and conditions except hohlraum at 773 K.
16	86	1961	~322	1.00-15.00	~0°		2π	< 2	Above specimen and conditions; diffuse component only.
17	86	1961	~322	0.50-15.00	~0°		2π	< 2	Different sample, same as Curve 13 specimen and conditions except hohlraum at 1273 K.
18	86	1961	~322	0.50-15.00	~0°		2π	< 2	Above specimen and conditions; diffuse component only.
19	29	1956	298	0.38-0.70	9°		2π		As received; data extracted from smooth curve; MgO reference.
20	29	1956	298	0.38-0.70	9°		2π		Above specimen and conditions except rotated 90° in its own plane.
21	29	1956	298	0.38-0.70	9°		2π		Smoothed and cleaned; data extracted from smooth curve; MgO reference.
22	29	1956	298	0.38-0.70	9°		2π		Above specimen and conditions except rotated 90° in its own plane.

SPECIFICATION TABLE NO. 204 (continued)

Curve No.	Ref. No.	Year	Temperature K	Wavelength Range, μ	Geometry θ	θ'	ω'	Reported Error, %	Composition (weight percent), Specifications and Remarks
23	29	1956	298	0.38-0.70	9°		2π		Polished; data extracted from smooth curve; MgO reference.
24	29	1956	298	0.38-0.70	9°		2π		Above specimen and conditions except rotated 90° in its own plane.
25	34	1957	298	0.30-2.60	9°		2π		Pure metal (Fansteel); as received; data extracted from a smooth curve; magnesium carbonate reference.
26	34	1957	298	0.30-2.60	9°		2π	±4	Pure metal (Fansteel); cleaned with liquid detergent; data extracted from a smooth curve; magnesium carbonate reference.
27	34	1957	298	0.30-2.60	9°		2π	±4	Pure metal (Fansteel); polished; data extracted from a smooth curve; magnesium carbonate reference.
28	34	1957	298	0.30-2.70	9°		2π	±4	Pure metal (Fansteel); oxidized in air at red heat for 30 min; data extracted from a smooth curve; magnesium carbonate reference.
29	340	1910	298	0.6-10.1	$<15°$	$<15°$			Pure; polished.

DATA TABLE NO. 204 NORMAL SPECTRAL REFLECTANCE OF TANTALUM

[Wavelength, λ, μ; Reflectance, ρ; Temperature, T, K]

λ	ρ
CURVE 1 T = ~322	
2.00	0.850
3.00	0.950
4.00	1.000
6.00	1.000
8.00	1.000
10.50	1.000
12.50	1.000
14.00	1.000
14.65	0.975
15.00	0.930
CURVE 2 T = ~322	
2.00	0.140
3.00	0.120
4.00	0.120
5.00	0.110
6.00	0.110
7.00	0.100
8.00	0.065
9.00	0.050
10.00	0.040
11.00	0.040
12.00	0.030
13.00	0.030
14.00	0.020
15.00	0.020
CURVE 3 T = ~322	
1.00	0.975
1.50	1.000
2.05	0.970
3.00	0.985
4.00	0.985
5.00	0.995
6.00	0.970
7.00	0.980
8.00	1.000*
9.50	1.000
11.50	1.000

λ	ρ
CURVE 3 (cont.)	
12.00	0.995
13.50	0.995
14.00	0.975
14.60	0.900
15.00	0.830
CURVE 4 T = ~322	
1.00	0.435
1.50	0.380
2.00	0.275
5.00	0.140
6.00	0.130
7.00	0.115
9.00	0.070
10.00	0.060
11.00	0.055
12.00	0.055
13.00	0.040
14.00	0.040
15.00	0.050
CURVE 5 T = ~322	
0.50	0.375
0.65	0.475
0.75	0.600
0.85	0.775
1.20	0.875
1.50	0.965
2.00	0.940
2.50	0.935
3.75	0.955
5.00	0.970
5.90	0.940
7.00	0.960
8.25	0.985
9.00	0.975
10.00	0.975
11.50	0.950
12.00	0.950
12.75	0.965

λ	ρ
CURVE 5 (cont.)	
13.75	0.950
14.70	0.900*
14.90	0.875
15.00	0.825
CURVE 6 T = ~322	
0.50	0.190
0.70	0.215
0.75	0.280
0.90	0.305
1.50	0.290
2.00	0.250
3.00	0.195
4.00	0.155
5.00	0.130
5.50	0.110
6.50	0.100
7.25	0.100
9.00	0.060
10.00	0.060*
11.00	0.055*
12.00	0.055*
13.00	0.060
14.00	0.070
14.50	0.085
15.00	0.120
CURVE 7 T = ~322	
2.00	0.890
2.50	0.800
3.00	0.755
3.50	0.752
4.20	0.760
5.00	0.800
6.00	0.810
8.00	0.820
10.00	0.820
11.50	0.820
12.00	0.830
13.00	0.830

λ	ρ
CURVE 7 (cont.)	
14.00	0.800
14.75	0.750
15.00	0.700
CURVE 8 T = ~322	
2.00	0.250*
2.50	0.200
4.00	0.140
5.00	0.120
6.00	0.105
7.00	0.105
7.50	0.095
8.00	0.070
8.50	0.060
10.25	0.060
11.00	0.050
13.50	0.050
14.50	0.070
15.00	0.090
CURVE 9 T = ~322	
1.00	0.580
1.25	0.625
1.75	0.650
2.25	0.675
2.80	0.725
3.20	0.745
3.65	0.745
4.50	0.775
5.50	0.785
6.50	0.795
7.00	0.815
8.00	0.820*
9.00	0.820
10.00	0.830
11.00	0.850
11.75	0.860
13.00	0.870
14.00	0.900
15.00	0.900*

λ	ρ
CURVE 10 T = ~322	
1.00	0.265
1.50	0.275
2.50	0.250
3.00	0.235
4.50	0.160
5.50	0.145
7.00	0.135
8.00	0.120
9.00	0.080
10.00	0.070
11.25	0.070
12.00	0.060
13.50	0.065
14.25	0.075
15.00	0.100
CURVE 11 T = ~322	
0.50	0.070
0.65	0.200
0.80	0.350
1.15	0.500
1.35	0.650
1.50	0.710
2.00	0.765
3.00	0.780
4.00	0.800
5.25	0.810
6.50	0.825
7.80	0.850
9.00	0.840
10.00	0.840
11.00	0.855
13.00	0.865
14.00	0.870
14.50	0.865
15.00	0.850
CURVE 12 T = ~322	
0.50	0.020

λ	ρ
CURVE 12 (cont.)	
0.75	0.150
1.00	0.210
1.25	0.240
2.00	0.260
2.50	0.255
3.00	0.240
4.00	0.180
4.50	0.170
6.00	0.140
7.00	0.130
8.00	0.120*
8.50	0.105
9.00	0.080*
9.50	0.070
11.00	0.075
12.25	0.070
13.75	0.090
14.50	0.130
15.00	0.175
CURVE 13 T = ~322	
2.00	0.915
2.40	0.850
2.80	0.800
3.25	0.785
4.00	0.800*
5.30	0.850
6.00	0.860
7.25	0.865
7.85	0.885
9.25	0.865
10.00	0.865
11.00	0.875
12.50	0.875
13.25	0.870
13.90	0.850
14.65	0.800
15.00	0.750

λ	ρ
CURVE 14 T = ~322	
2.00	0.230
2.25	0.190
2.50	0.165
3.00	0.150
4.50	0.115
5.00	0.100
6.00	0.095
7.00	0.070
10.50	0.035
11.50	0.030
13.00	0.040*
14.50	0.060
15.00	0.080
CURVE 15 T = ~322	
1.00	0.770
1.50	0.730
2.00	0.720
2.75	0.750
4.00	0.800*
5.10	0.825
6.25	0.830
7.25	0.850
8.00	0.860
9.00	0.860
10.00	0.860
11.50	0.875
12.50	0.890
13.50	0.910
14.50	0.925
15.00	0.920
CURVE 16 T = ~322	
1.00	0.310
2.00	0.245
3.00	0.205
4.00	0.160
5.50	0.125
7.00	0.120

λ	ρ
CURVE 16 (cont.)	
9.50	0.050
11.25	0.050
12.00	0.040
13.00	0.050
14.00	0.050
15.00	0.080*
CURVE 17 T = ~322	
0.50	0.600
0.60	0.650
0.75	0.690
1.00	0.660
1.50	0.720
2.00	0.750
3.00	0.770
4.00	0.800
5.25	0.815
5.90	0.805
7.00	0.830
8.00	0.840
9.10	0.825
10.35	0.835
11.25	0.840
13.00	0.840
14.00	0.850*
14.50	0.850
15.00	0.830*
CURVE 18 T = ~322	
0.50	0.240
0.65	0.270
0.80	0.300
0.95	0.460
1.25	0.450
1.50	0.420
1.75	0.325
2.00	0.255*
2.50	0.240
3.00	0.230
3.50	0.190

*Not shown on plot

DATA TABLE NO. 204 (continued)

CURVE 18 (cont.)

λ	ρ
4.00	0.170
6.25	0.125
7.00	0.135*
8.00	0.120*
9.00	0.080*
9.50	0.075
11.50	0.075
12.50	0.075
14.00	0.100
14.50	0.120
15.00	0.170

CURVE 19
T = 298

λ	ρ
0.38	0.385
0.40	0.385
0.45	0.380
0.50	0.375*
0.55	0.375
0.60	0.375
0.65	0.410
0.70	0.500

CURVE 20
T = 298

λ	ρ
0.38	0.370
0.40	0.370
0.45	0.370
0.50	0.375*
0.55	0.370
0.60	0.375*
0.65	0.400
0.70	0.480

CURVE 21*
T = 298

λ	ρ
0.38	0.390
0.40	0.395
0.45	0.395
0.50	0.390
0.55	0.380
0.60	0.380
0.65	0.420
0.70	0.500

CURVE 22*
T = 298

λ	ρ
0.38	0.380
0.40	0.390
0.45	0.390
0.50	0.380
0.55	0.370
0.60	0.370
0.65	0.415
0.70	0.480

CURVE 23*
T = 298

λ	ρ
0.38	0.360
0.40	0.365
0.45	0.370
0.50	0.370
0.55	0.370
0.60	0.370
0.65	0.390
0.70	0.430

CURVE 24*
T = 298

λ	ρ
0.38	0.350
0.40	0.360
0.45	0.370
0.50	0.370
0.55	0.360
0.60	0.360
0.65	0.380
0.70	0.420

CURVE 25
T = 298

λ	ρ
0.30	0.350
0.38	0.370*
0.50	0.370
0.58	0.365
0.60	0.365
0.70	0.470
0.75	0.580
0.80	0.670
0.88	0.750
0.90	0.755

CURVE 25 (cont.)

λ	ρ
1.00	0.790
1.10	0.825
1.12	0.830
1.22	0.855
1.30	0.825
1.34	0.800
1.38	0.780
1.42	0.785
1.50	0.860
1.58	0.930
1.64	0.950
1.70	0.950
1.82	0.950
1.90	0.960
2.10	0.990
2.14	1.000
2.30	1.000
2.50	1.000
2.60	1.000

CURVE 26*
T = 298

λ	ρ
0.30	0.370
0.36	0.405
0.40	0.390
0.50	0.380
0.58	0.375
0.60	0.380
0.70	0.490
0.75	0.580
0.80	0.670
0.88	0.750
0.90	0.755
1.00	0.790
1.10	0.825
1.12	0.830
1.20	0.865
1.26	0.850
1.30	0.830
1.34	0.800
1.38	0.790
1.42	0.800
1.50	0.860
1.52	0.880
1.56	0.930
1.60	0.950

CURVE 26*(cont.)

λ	ρ
1.64	0.955
1.70	0.960
1.80	0.960
1.90	0.970
2.10	0.990
2.14	1.000
2.30	1.000
2.50	1.000
2.60	1.000

CURVE 27
T = 298

λ	ρ
0.30	0.140
0.36	0.150
0.38	0.150
0.40	0.145
0.44	0.150
0.50	0.160
0.60	0.360
0.62	0.365
0.70	0.420
0.80	0.530
0.86	0.610
0.90	0.640
1.02	0.700
1.10	0.740
1.20	0.790
1.26	0.790
1.30	0.770
1.38	0.710
1.40	0.700
1.42	0.710
1.50	0.790
1.58	0.870
1.68	0.900
1.70	0.910
1.90	0.930
2.00	0.950
2.10	0.960
2.30	0.970
2.40	0.990
2.50	1.000*
2.60	1.000*

CURVE 28
T = 298

λ	ρ
0.30	0.140*
0.36	0.155
0.40	0.155
0.44	0.150*
0.50	0.150
0.60	0.155
0.62	0.155
0.70	0.150
0.72	0.150
0.80	0.170
0.86	0.180
0.90	0.180
1.10	0.180
1.16	0.180
1.24	0.190
1.30	0.190
1.46	0.190
1.50	0.195
1.62	0.215
1.70	0.220
1.90	0.235
2.02	0.250*
2.10	0.250
2.30	0.270
2.50	0.295
2.62	0.300
2.70	0.270

CURVE 29
T = 298

λ	ρ
0.6	0.092
0.7	0.100
0.7	0.124
1.0	0.146
1.1	0.170
1.3	0.203
1.8	0.236
2.1	0.267
2.5	0.305
2.8	0.330
3.1	0.357
3.4	0.373
3.6	0.389
4.1	0.427
4.5	0.441

CURVE 29(cont.)

λ	ρ
5.1	0.476
5.7	0.499
6.1	0.513
6.5	0.527
6.9	0.541
7.2	0.551
7.4	0.596
7.5	0.569
7.9	0.576
8.2	0.591
8.5	0.595
8.7	0.604
9.0	0.609
9.6	0.625
10.1	0.621

* Not shown on plot

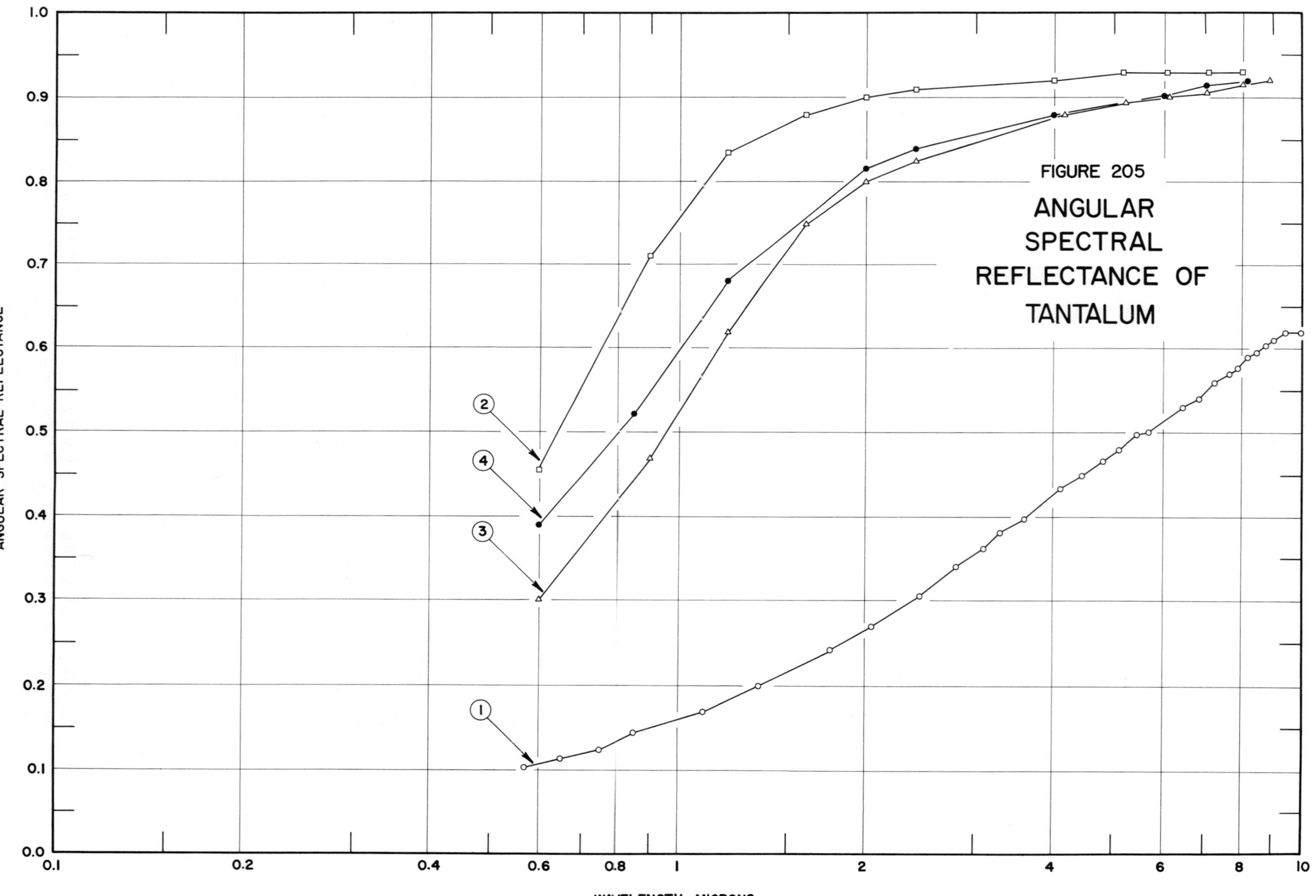

FIGURE 205
ANGULAR SPECTRAL REFLECTANCE OF TANTALUM
ANGULAR SPECTRAL REFLECTANCE
WAVELENGTH, MICRONS
1.0
0.9
0.8
0.7
0.6
0.5
0.4
0.3
0.2
0.1
0.0
0.1
0.2
0.4
0.6
0.8
1
2
4
6
8
10
2
4
3
1

SPECIFICATION TABLE NO. 205 ANGULAR SPECTRAL REFLECTANCE OF TANTALUM

Curve No.	Ref. No.	Year	Temperature K	Wavelength Range, μ	Geometry θ	ϑ'	ω'	Reported Error, %	Composition (weight percent), Specifications and Remarks
1	132	1911	298	0.57-9.95	15°	15°		≤3	Polished; silvered glass mirror reference; three series of observations.
2	132	1911	298	0.60-8.00	15°	15°		≤3	Excellent polish; silvered glass mirror reference.
3	132	1911	298	0.60-8.85	15°	15°		≤3	Polished with fine dry emery paper; silvered glass mirror reference.
4	132	1911	298	0.60-8.15	15°	15°		≤3	Different sample, same as curve 3 specimen and conditions except also polished with graphite and alcohol.

DATA TABLE NO. 205 ANGULAR SPECTRAL REFLECTANCE OF TANTALUM

[Wavelength, λ, μ; Reflectance, ρ; Temperature, T, K]

λ	ρ
CURVE 1 T = 298	
0.57	0.103
0.65	0.115
0.75	0.125
0.85	0.145
1.10	0.170
1.35	0.200
1.77	0.243
2.05	0.270
2.45	0.305
2.80	0.340
3.10	0.363
3.30	0.380
3.60	0.397
4.13	0.435
4.45	0.450
4.80	0.467
5.10	0.480
5.45	0.497
5.70	0.500
6.45	0.530
6.85	0.540
7.25	0.560
7.65	0.570
7.90	0.577
8.20	0.590
8.45	0.595
8.77	0.603
9.00	0.610
9.40	0.620
9.95	0.620
CURVE 2 T = 298	
0.60	0.455
0.90	0.710
1.20	0.835
1.60	0.880
2.00	0.900
2.40	0.910
4.00	0.920
5.15	0.930
6.05	0.930

λ	ρ
CURVE 2 (cont.)	
7.05	0.930
8.00	0.930
CURVE 3 T = 298	
0.60	0.300
0.90	0.470
1.20	0.620
1.60	0.750
2.00	0.800
2.40	0.825
4.15	0.880
5.20	0.895
6.10	0.900
7.00	0.905
8.00	0.915
8.85	0.920
CURVE 4 T = 298	
0.60	0.390
0.85	0.523
1.20	0.680
2.00	0.817
2.40	0.840
4.00	0.880
6.00	0.902
7.00	0.915
8.15	0.920

*Not shown on plot

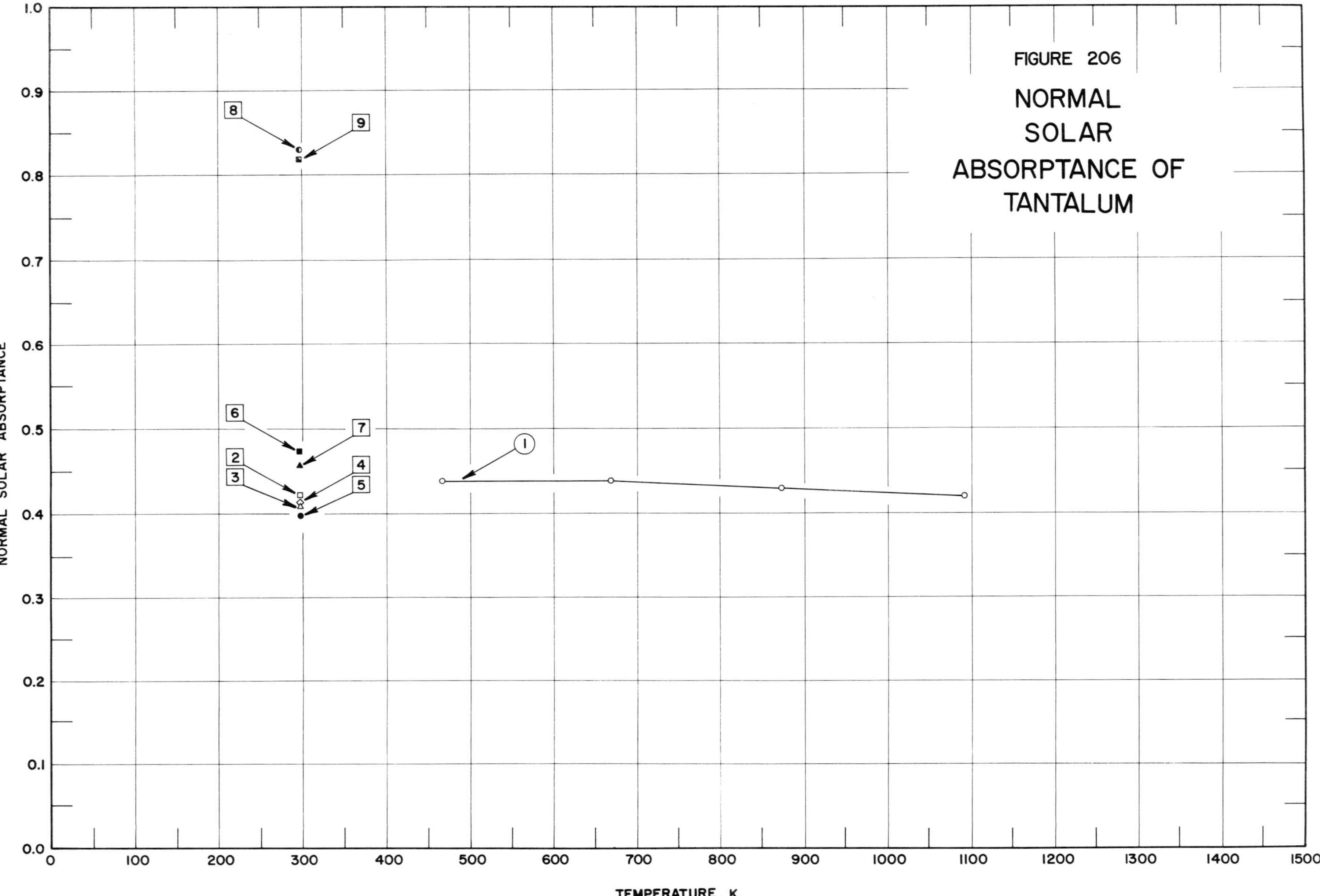
FIGURE 206
NORMAL
SOLAR
ABSORPTANCE OF
TANTALUM
NORMAL SOLAR ABSORPTANCE
TEMPERATURE, K
0.0
0.1
0.2
0.3
0.4
0.5
0.6
0.7
0.8
0.9
1.0
0
100
200
300
400
500
600
700
800
900
1000
1100
1200
1300
1400
1500
1
2
3
4
5
6
7
8
9

SPECIFICATION TABLE NO. 206 NORMAL SOLAR ABSORPTANCE OF TANTALUM

Curve No.	Ref. No.	Year	Temperature Range, K	Geometry θ	Reported Error, %	Composition (weight percent), Specifications and Remarks
1	47	1961	468-1093	$\sim 0^\circ$		Commercially pure; arc cast or sintered; disc (0.04 in. thick); ground with 600 grit carborundum and polished on a wet cloth lap with unlevigated jewelers rouge; measured in vacuum (10^{-5} mm Hg).
2	34	1957	298	9°		Pure metal (Fansteel); as received; computed from spectral reflectance data for sea level conditions.
3	34	1957	298	9°		Above specimen and conditions except computed for above atmosphere conditions.
4	34	1957	298	9°		Different sample, same as curve 2 specimen and conditions except cleaned with liquid detergent.
5	34	1957	298	9°		Above specimen and conditions except computed for above atmosphere conditions.
6	34	1957	298	9°		Different sample, same as curve 2 specimen and conditions except polished.
7	34	1957	298	9°		Above specimen and conditions except computed for above atmosphere conditions.
8	34	1957	298	9°		Different sample, same as curve 2 specimen and conditions except oxidized in air at red heat for 30 min.
9	34	1957	298	9°		Above specimen and conditons except computed for above atmosphere conditions.

DATA TABLE NO. 206 NORMAL SOLAR ABSORPTANCE OF TANTALUM

[Temperature, T, K; Absorptance, α]

T	α
CURVE 1	
468	0.44
668	0.44
873	0.43
1093	0.42
CURVE 2	
298	0.422
CURVE 3	
298	0.409
CURVE 4	
298	0.413
CURVE 5	
298	0.399
CURVE 6	
298	0.474
CURVE 7	
298	0.457
CURVE 8	
298	0.831
CURVE 9	
298	0.820

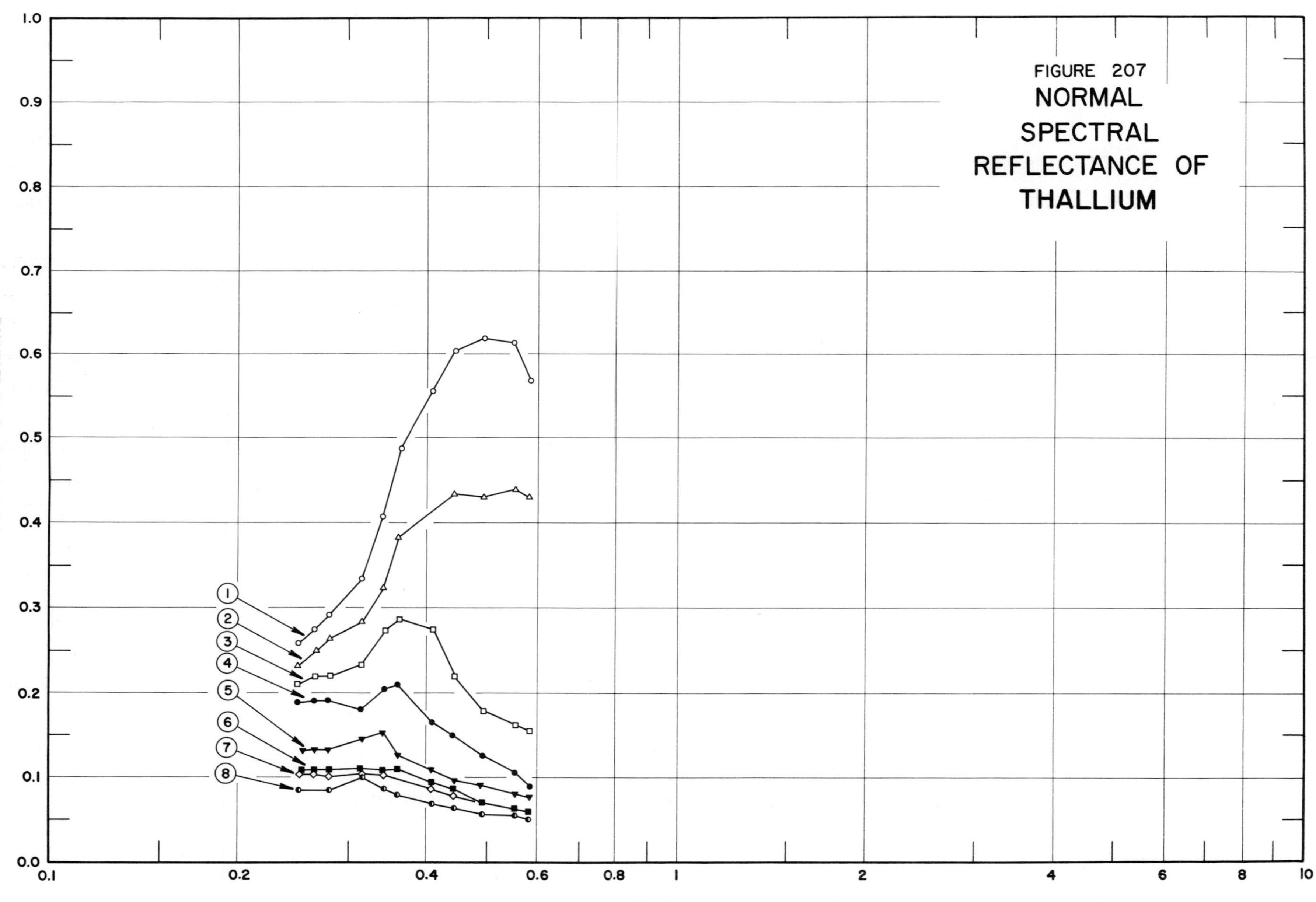

FIGURE 207
NORMAL
SPECTRAL
REFLECTANCE OF
THALLIUM
NORMAL SPECTRAL REFLECTANCE
WAVELENGTH, MICRONS
1.0
0.9
0.8
0.7
0.6
0.5
0.4
0.3
0.2
0.1
0.0
0.1
0.2
0.4
0.6
0.8
1
2
4
6
8
10
1
2
3
4
5
6
7
8

SPECIFICATION TABLE NO. 207 NORMAL SPECTRAL REFLECTANCE OF THALLIUM

Curve No.	Ref. No.	Year	Temperature K	Wavelength Range, μ	Geometry θ	θ'	ω'	Reported Error, %	Composition (weight percent), Specifications and Remarks
1	266	1964	298	0.251-0.583	$\sim$0°	$\sim$0°			Thin film (30 m μ thick) vacuum deposited on quartz; measured in vacuum.
2	266	1964	298	0.249-0.581	$\sim$0°	$\sim$0°			Different sample, same as above specimen and conditions except 20 m μ thick.
3	266	1964	298	0.249-0.582	$\sim$0°	$\sim$0°			Different sample, same as above specimen and conditions except 15 m μ thick.
4	266	1964	298	0.249-0.582	$\sim$0°	$\sim$0°			Different sample, same as above specimen and conditions except 10.4 m μ thick.
5	266	1964	298	0.254-0.583	$\sim$0°	$\sim$0°			Different sample, same as above specimen and conditions except 10 m μ thick.
6	266	1964	298	0.253-0.583	$\sim$0°	$\sim$0°			Different sample, same as above specimen and conditions except 5.2 m μ thick.
7	266	1964	298	0.252-0.582	$\sim$0°	$\sim$0°			Different sample, same as above specimen and conditions except 5 m μ thick.
8	266	1964	298	0.252-0.581	$\sim$0°	$\sim$0°			Different sample, same as above specimen and conditions except 2.6 m μ thick.

DATA TABLE NO. 207 NORMAL SPECTRAL REFLECTANCE OF THALLIUM

[Wavelength, λ, μ; Reflectance, ρ; Temperature, T, K]

CURVE 1 (T = 298)

λ	ρ
0.251	0.259
0.265	0.275
0.280	0.292
0.315	0.334
0.340	0.407
0.363	0.488
0.409	0.557
0.444	0.602
0.494	0.620
0.553	0.610
0.583	0.570

CURVE 2 (T = 298)

λ	ρ
0.249	0.232
0.266	0.250
0.280	0.264
0.315	0.285
0.340	0.324
0.361	0.382
0.442	0.433
0.493	0.431
0.554	0.439
0.581	0.431

CURVE 3 (T = 298)

λ	ρ
0.249	0.210
0.265	0.220
0.281	0.220
0.315	0.233
0.342	0.273
0.362	0.288
0.410	0.276
0.444	0.220
0.494	0.179
0.555	0.162
0.582	0.156

CURVE 4 (T = 298)

λ	ρ
0.249	0.189
0.265	0.191
0.279	0.191
0.314	0.181
0.342	0.204
0.360	0.210
0.410	0.165
0.441	0.150
0.493	0.126
0.553	0.105
0.582	0.089

CURVE 5 (T = 298)

λ	ρ
0.254	0.131
0.266	0.132
0.280	0.132
0.316	0.146
0.341	0.152
0.361	0.126
0.409	0.109
0.443	0.096
0.490	0.091
0.552	0.080
0.583	0.077

CURVE 6 (T = 298)

λ	ρ
0.253	0.109
0.266	0.109
0.281	0.109
0.315	0.111
0.341	0.109
0.361	0.110
0.410	0.095
0.443	0.087
0.493	0.071
0.553	0.064
0.583	0.061

CURVE 7 (T = 298)

λ	ρ
0.252	0.102
0.265	0.102
0.280	0.101
0.316	0.103
0.342	0.102
0.409	0.087
0.442	0.078
0.493	0.070*
0.553	0.063*
0.582	0.061*

CURVE 8 (T = 298)

λ	ρ
0.252	0.085
0.281	0.083
0.317	0.100
0.342	0.086
0.360	0.079
0.410	0.069
0.443	0.064
0.493	0.057
0.554	0.054
0.581	0.050

* Not shown on plot

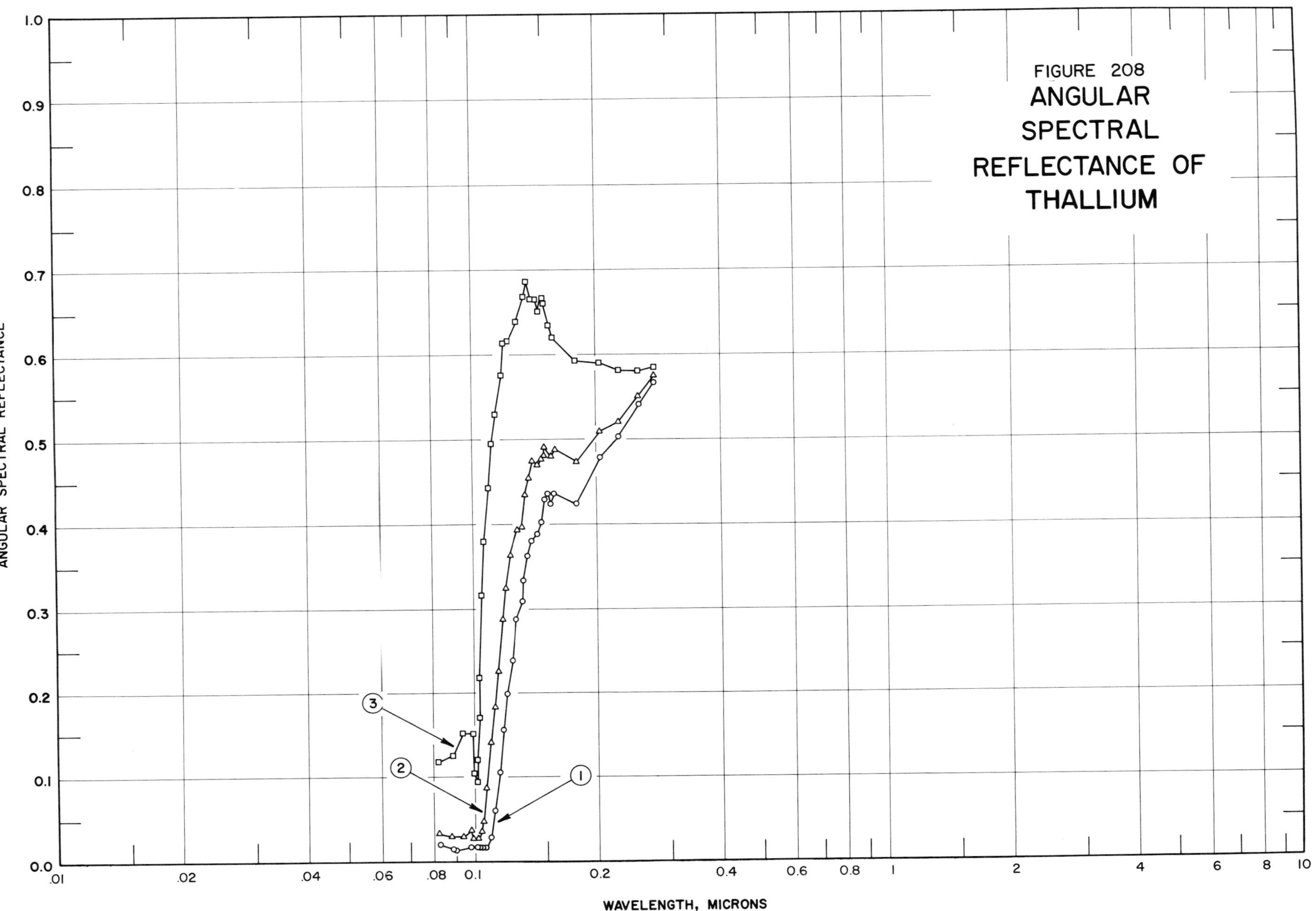
FIGURE 208
ANGULAR SPECTRAL REFLECTANCE OF THALLIUM
ANGULAR SPECTRAL REFLECTANCE
WAVELENGTH, MICRONS
1.0
0.9
0.8
0.7
0.6
0.5
0.4
0.3
0.2
0.1
0.0
.01
.02
.04
.06
.08
0.1
0.2
0.4
0.6
0.8
1
2
4
6
8
10
1
2
3

SPECIFICATION TABLE NO. 208 ANGULAR SPECTRAL REFLECTANCE OF THALLIUM

Curve No.	Ref. No.	Year	Temperature K	Wavelength Range, μ	Geometry θ	θ'	ω'	Reported Error, %	Composition (weight percent), Specifications and Remarks
1	310	1966	298	0.083-0.278	20°	20°			99.9 Tl film (2400 Å thick); evaporated at 10^{-6} mm Hg on glass slide at room temperature with source-to-substrate distance of 7 cm; not exposed to air.
2	310	1966	298	0.082-0.278	45°	45°			Above specimen and conditions.
3	310	1966	298	0.082-0.278	70°	70°			Above specimen and conditions.

DATA TABLE NO. 208 ANGULAR SPECTRAL REFLECTANCE OF THALLIUM

[Wavelength, λ, μ; Reflectance, ρ; Temperature, T, K]

λ	ρ
CURVE 1	
T = 298	
0.083	0.021
0.089	0.016
0.094	0.014
0.098	0.018
0.102	0.018
0.103	0.018
0.105	0.018
0.107	0.018
0.109	0.029
0.112	0.060
0.115	0.104
0.118	0.155
0.121	0.199
0.124	0.238
0.127	0.288
0.132	0.309
0.133	0.333
0.136	0.362
0.139	0.380
0.143	0.389
0.147	0.402
0.150	0.429
0.152	0.437
0.155	0.423
0.158	0.437
0.179	0.424
0.204	0.479
0.228	0.502
0.253	0.541
0.278	0.566
CURVE 2	
T = 298	
0.082	0.034
0.088	0.031
0.094	0.031
0.098	0.038
0.099	0.028
0.102	0.028
0.104	0.037
0.105	0.049
0.107	0.087
0.110	0.140
0.112	0.182
CURVE 2 (cont.)	
0.115	0.226
0.118	0.288
0.121	0.326
0.124	0.363
0.128	0.393
0.132	0.398
0.134	0.436
0.137	0.454
0.140	0.474
0.144	0.470
0.147	0.478
0.150	0.481
0.151	0.491
0.155	0.481
0.159	0.488
0.179	0.473
0.204	0.509
0.228	0.521
0.253	0.550
0.278	0.574
CURVE 3	
T = 298	
0.082	0.118
0.089	0.126
0.094	0.151
0.098	0.151
0.099	0.105
0.102	0.094
0.102	0.120
0.103	0.169
0.103	0.218
0.105	0.316
0.107	0.380
0.109	0.442
0.112	0.495
0.114	0.529
0.118	0.573
0.121	0.613
0.123	0.616
0.128	0.640
0.134	0.669
0.137	0.688
0.140	0.666
0.143	0.666
CURVE 3 (cont.)	
0.147	0.651
0.150	0.668
0.151	0.661
0.155	0.634
0.158	0.621
0.179	0.592
0.203	0.590
0.228	0.581
0.253	0.580
0.278	0.586

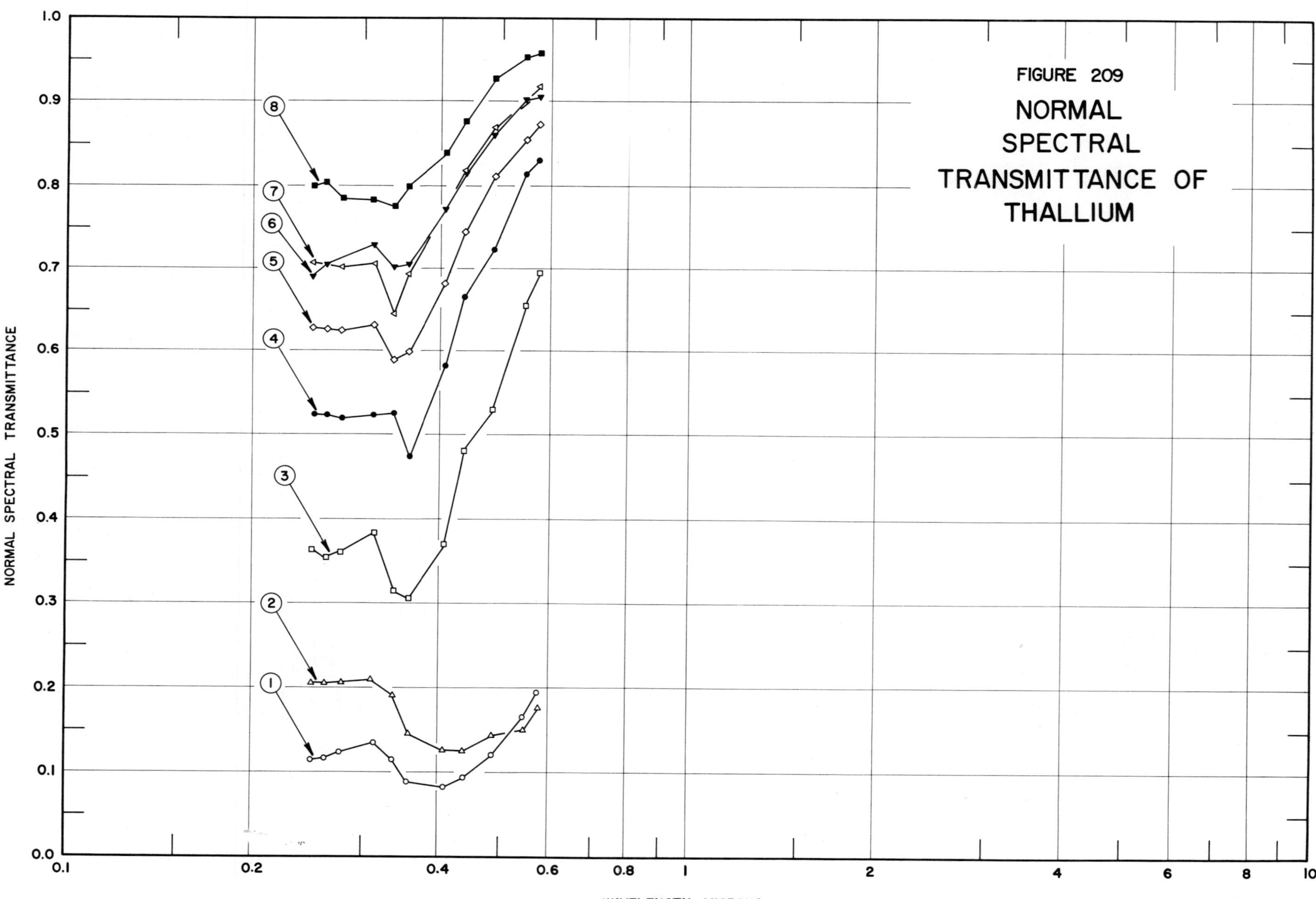
FIGURE 209
NORMAL SPECTRAL TRANSMITTANCE OF THALLIUM
NORMAL SPECTRAL TRANSMITTANCE
WAVELENGTH, MICRONS
1.0
0.9
0.8
0.7
0.6
0.5
0.4
0.3
0.2
0.1
0.0
0.1
0.2
0.4
0.6
0.8
1
2
4
6
8
10
8
7
6
5
4
3
2
1

SPECIFICATION TABLE NO. 209 NORMAL SPECTRAL TRANSMITTANCE OF THALLIUM

Curve No.	Ref. No.	Year	Temperature K	Wavelength Range, μ	Geometry θ	θ'	ω'	Reported Error, %	Composition (weight percent), Specifications and Remarks
1	266	1964	298	0.251-0.577	~0°	~0°			Film (30 m μ thick); vacuum deposited; measured in vacuum.
2	266	1964	298	0.251-0.579	~0°	~0°			Different sample, same as above specimen and conditions except 20 m μ thick.
3	266	1964	298	0.250-0.578	~0°	~0°			Different sample, same as above specimen and conditions except 15 m μ thick.
4	266	1964	298	0.252-0.576	~0°	~0°			Different sample, same as above specimen and conditions except 10.4 m μ thick.
5	266	1964	298	0.250-0.577	~0°	~0°			Different sample, same as above specimen and conditions except 10 m μ thick.
6	266	1964	298	0.250-0.577	~0°	~0°			Different sample, same as above specimen and conditions except 5.2 m μ thick.
7	266	1964	298	0.250-0.577	~0°	~0°			Different sample, same as above specimen and conditions except 5 m μ thick.
8	266	1964	298	0.250-0.577	~0°	~0°			Different sample, same as above specimen and conditions except 2.6 m μ thick.

DATA TABLE NO. 209 NORMAL SPECTRAL TRANSMITTANCE OF THALLIUM

[Wavelength, λ, μ; Transmittance, τ; Temperature, T, K]

λ	τ	λ	τ	λ	τ
CURVE 1 T = 298		CURVE 4 T = 298		CURVE 7 T = 298	
0.251	0.113	0.252	0.522	0.250	0.708
0.263	0.117	0.264	0.522	0.263	0.705*
0.278	0.123	0.279	0.519	0.278	0.701
0.315	0.134	0.313	0.522	0.314	0.705
0.338	0.116	0.338	0.524	0.338	0.647
0.358	0.089	0.359	0.475	0.356	0.694
0.410	0.082	0.409	0.582	0.439	0.819
0.440	0.094	0.438	0.667	0.489	0.870
0.489	0.121	0.489	0.722	0.577	0.919
0.548	0.167	0.549	0.815		
0.577	0.196	0.576	0.830	CURVE 8 T = 298	
CURVE 2 T = 298		CURVE 5 T = 298		0.250	0.800
				0.262	0.803
0.251	0.205	0.250	0.629	0.279	0.786
0.263	0.204	0.264	0.628	0.312	0.784
0.279	0.206	0.278	0.627	0.338	0.775
0.312	0.209	0.313	0.632	0.356	0.800
0.338	0.191	0.338	0.590	0.409	0.839
0.359	0.145	0.357	0.599	0.439	0.878
0.408	0.126	0.407	0.681	0.490	0.929
0.440	0.125	0.439	0.744	0.548	0.953
0.489	0.143	0.490	0.811	0.577	0.958
0.549	0.150	0.549	0.854		
0.579	0.177	0.577	0.873		
CURVE 3 T = 298		CURVE 6 T = 298			
0.250	0.364	0.250	0.689		
0.264	0.359	0.264	0.704		
0.279	0.361	0.313	0.729		
0.315	0.384	0.338	0.701		
0.338	0.315	0.356	0.704		
0.358	0.306	0.409	0.771		
0.408	0.371	0.440	0.813		
0.438	0.482	0.489	0.860		
0.488	0.530	0.548	0.901		
0.549	0.656	0.577	0.908		
0.578	0.697				

*Not shown on plot

SPECIFICATION TABLE NO. 210 NORMAL SPECTRAL EMITTANCE OF THORIUM

Curve No.	Ref. No.	Year	Wavelength μ	Temperature Range, K	Geometry θ'	Reported Error, %	Composition (weight percent), Specifications and Remarks
1	55	1935	0.667	1300-1655	$\sim 0°$		Heat treated for 600 hrs.

DATA TABLE NO. 210 NORMAL SPECTRAL EMITTANCE OF THORIUM

[Temperature, T, K; Emittance, ϵ; Wavelength, λ, μ]

T	ϵ
CURVE 1*	
$\lambda = 0.667$	
1300	0.384
1350	0.381
1380	0.386
1430	0.372
1435	0.379
1530	0.377
1550	0.374
1590	0.381
1655	0.382

*Not shown on plot

SPECIFICATION TABLE NO. 211 NORMAL SPECTRAL EMITTANCE OF THORIUM

Curve No.	Ref. No.	Year	Temperature K	Wavelength Range, μ	Geometry θ'	Reported Error, %	Composition (weight percent), Specifications and Remarks
1	19	1914	1623	0.55-0.65	$\sim 0°$	1	Film; tungsten substrate; melted in hydrogen then oxidized in air by heating; measured in air; Pt reference (ϵ = 0.33 for λ = 0.650 μ and ϵ = 0.38 for λ = 0.547 μ at all temp).
2	19	1914	<M. P.	0.55-0.65	$\sim 0°$	1	Film; tungsten substrate; measured in hydrogen; Pt reference (ϵ = 0.33 for λ = 0.650 μ and ϵ = 0.38 for λ = 0.547 μ at all temp).

DATA TABLE NO. 211 NORMAL SPECTRAL EMITTANCE OF THORIUM

[Wavelength, λ, μ; Emittance, ϵ; Temperature, T, K]

λ	ϵ
CURVE 1* T = 1623	
0.55	0.69
0.65	0.57
CURVE 2* T = < M. P.	
0.55	0.36
0.65	0.36

*Not shown on plot

SPECIFICATION TABLE NO. 212 HEMISPHERICAL TOTAL EMITTANCE OF TIN

Curve No.	Ref. No.	Year	Temperature Range, K	Reported Error, %	Composition (weight percent), Specifications and Remarks
1	3	1955	76	5	Foil (0.001 in. thick); cleaned; measured in vacuum (10^{-6} to 10^{-7} mm Hg); emittance for 300 K black body incident radiation; authors assumed $\alpha = \epsilon$.
2	4	1953	76		Mill foil (0.001 in. thick); measured in vacuum ($< 10^{-6}$ mm Hg); emittance for 294 K black body radiation; authors assumed $\alpha = \epsilon$.
3	4	1953	76		Mill foil (0.001 in. thick); sample partially covered with frost from a condensible gas; measured in vacuum ($< 10^{-6}$ mm Hg); emittance for 294 K black body radiation; authors assumed $\alpha = \epsilon$.

DATA TABLE NO. 212 HEMISPHERICAL TOTAL EMITTANCE OF TIN

[Temperature, T, K; Emittance, ∈]

T	∈
CURVE 1*	
76	0.013
CURVE 2*	
76	0.0136
CURVE 3*	
76	0.105

*Not shown on plot

SPECIFICATION TABLE NO. 213 NORMAL TOTAL EMITTANCE OF TIN

Curve No.	Ref. No.	Year	Temperature Range, K	Geometry θ'	Reported Error, %	Composition (weight percent), Specifications and Remarks
1	16	1937	367	$\sim 0^\circ$	± 1.1	Polished sheet.
2	15	1947	373	$\sim 0^\circ$		Commercial plating on sheet iron.
3	15	1947	373	$\sim 0^\circ$		Polished.

DATA TABLE NO. 213 NORMAL TOTAL EMITTANCE OF TIN

[Temperature, T, K; Emittance, ϵ]

T	ϵ
CURVE 1*	
367	0.05
CURVE 2*	
373	0.084
CURVE 3*	
373	0.069

*Not shown on plot

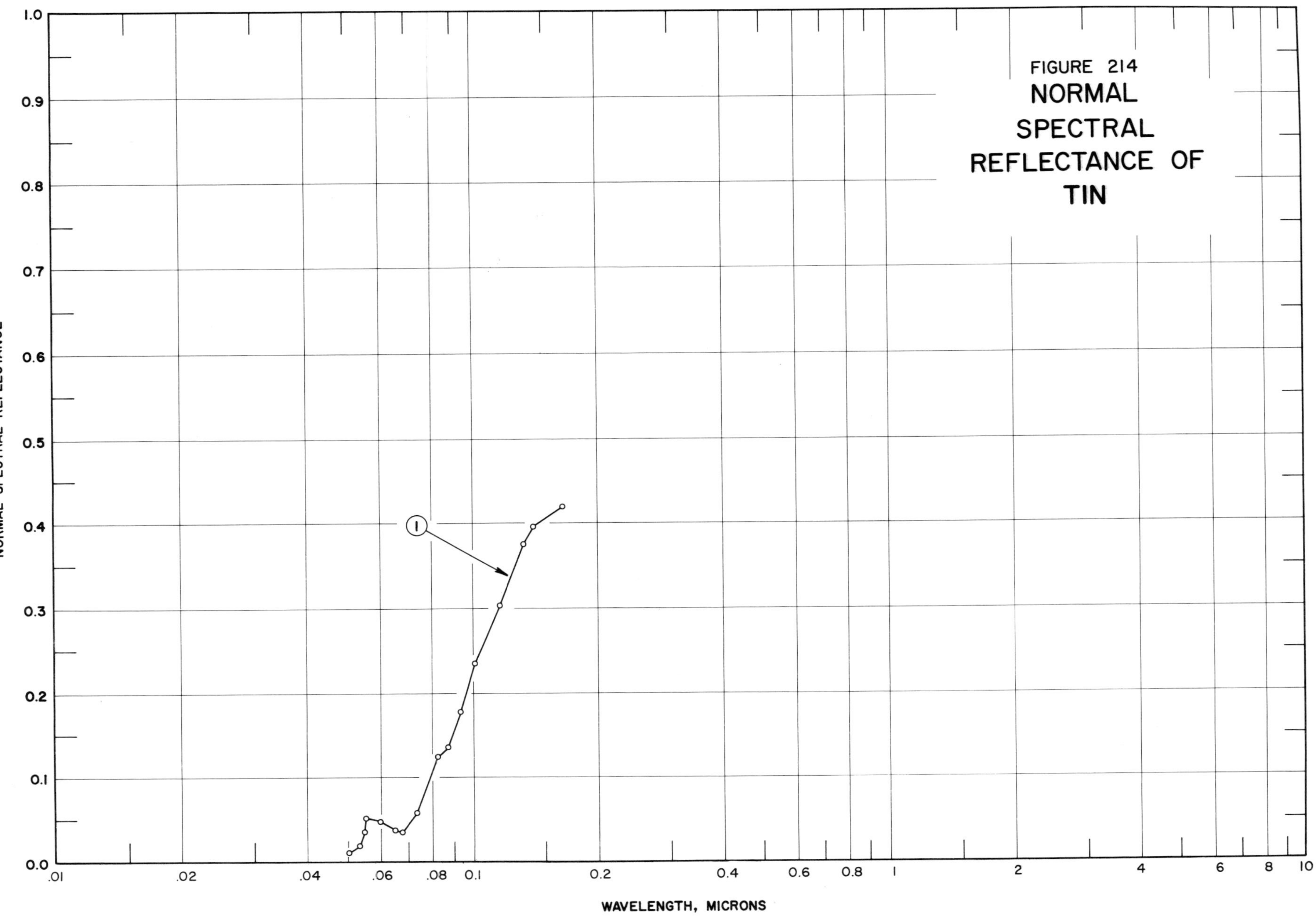

FIGURE 214 NORMAL SPECTRAL REFLECTANCE OF TIN

SPECIFICATION TABLE NO. 214 NORMAL SPECTRAL REFLECTANCE OF TIN

Curve No.	Ref. No.	Year	Temperature K	Wavelength Range, μ	Geometry θ	θ'	ω'	Reported Error, %	Composition (weight percent), Specifications and Remarks
1	349	1965	298	0.0504 - 0.1653	~0°	~0°			Evaporated film; glass substrate; data extracted from smooth curve.

DATA TABLE NO. 214 NORMAL SPECTRAL REFLECTANCE OF TIN

[Wavelength, λ, μ; Reflectance, ρ; Temperature, T, K]

λ	ρ
CURVE 1	
T = 298	
0.0504	0.011
0.0530	0.019
0.0546	0.035
0.0571	0.052
0.0596	0.048
0.0646	0.037
0.0670	0.035
0.0729	0.058
0.0821	0.124
0.0867	0.134
0.0932	0.178
0.1016	0.234
0.1170	0.302
0.1333	0.374
0.1409	0.397
0.1653	0.420

SPECIFICATION TABLE NO. 215 HEMISPHERICAL INTEGRATED ABSORPTANCE OF TIN

Curve No.	Ref. No.	Year	Temperature Range, K	Reported Error, %	Composition (weight percent), Specifications and Remarks
1	3	1955	76	5	Foil (0.001 in. thick) ; cleaned; measured in vacuum (10^{-6} to 10^{-7} mm Hg) ; absorptance for 300 K black body incident radiation.
2	4	1953	76		Mill foil (0.001 in. thick) ; measured in vacuum ($< 10^{-6}$ mm Hg) ; absorptance for 294 K black body incident radiation.
3	4	1953	76		Mill foil (0.001 in. thick) ; sample partially covered with frost from a condensible gas ; measured in vacuum ($< 10^{-6}$ mm Hg) ; absorptance for 294 K black body incident radiation.

DATA TABLE NO. 215 HEMISPHERICAL INTEGRATED ABSORPTANCE OF TIN

[Temperature, T, K; Absorptance, α]

T	α
CURVE 1*	
76	0.013
CURVE 2*	
76	0.0136
CURVE 3*	
76	0.105

*Not shown on plot

SPECIFICATION TABLE NO. 216 NORMAL INTEGRATED ABSORPTANCE OF TIN

Curve No.	Ref. No.	Year	Temperature Range, K	Geometry θ	Reported Error, %	Composition (weight percent), Specifications and Remarks
1	134	1952	2	$\sim 0^{\circ}$	1	Electropolished; absorptance for 298 K black body incident radiation.

DATA TABLE NO. 216 NORMAL INTEGRATED ABSORPTANCE OF TIN

[Temperature, T, K; Absorptance, α]

T	α
CURVE 1*	
2	0.0124

*Not shown on plot

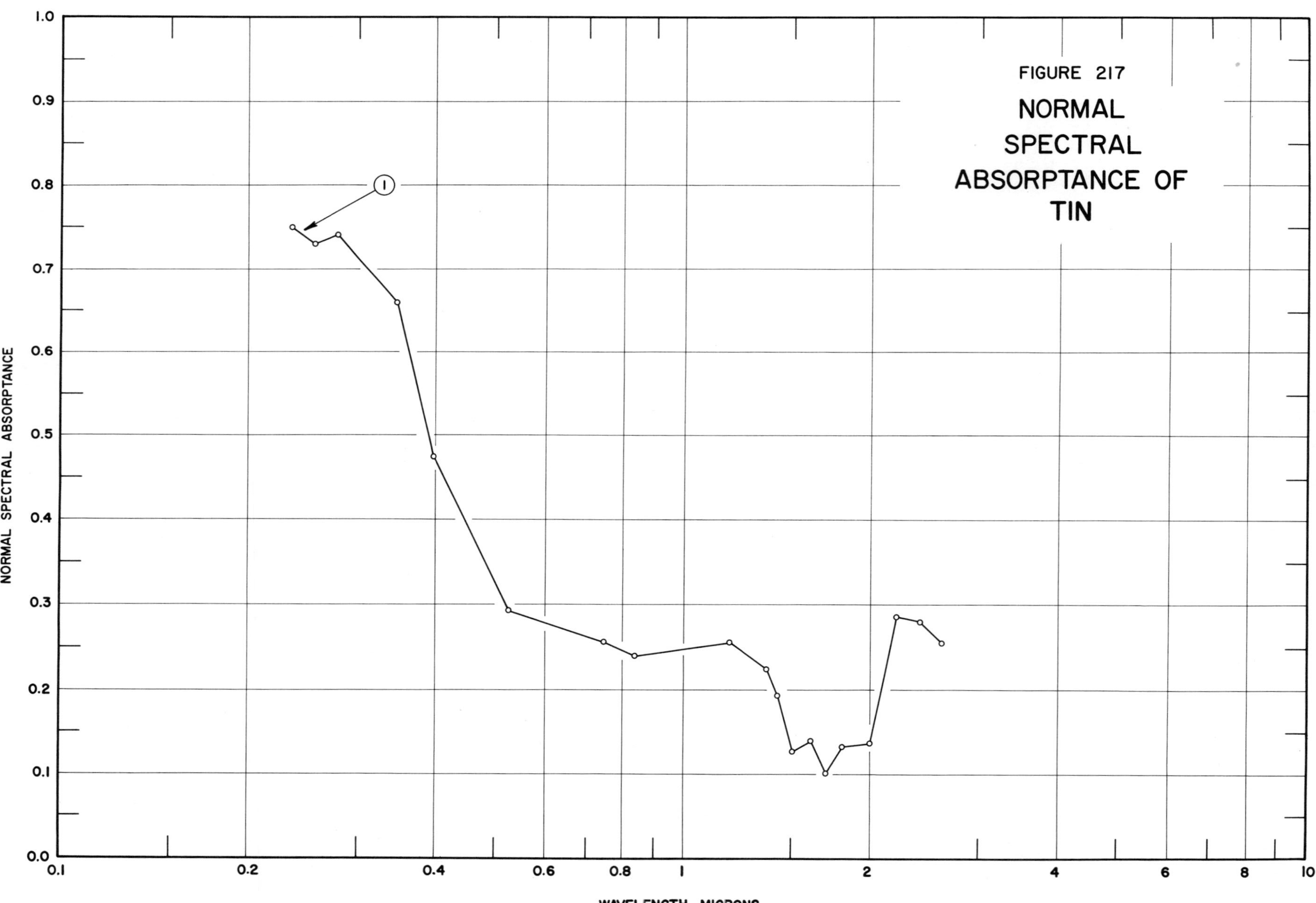
FIGURE 217
NORMAL
SPECTRAL
ABSORPTANCE OF
TIN
1
NORMAL SPECTRAL ABSORPTANCE
1.0
0.9
0.8
0.7
0.6
0.5
0.4
0.3
0.2
0.1
0.0
0.1
0.2
0.4
0.6
0.8
1
2
4
6
8
10
WAVELENGTH, MICRONS

SPECIFICATION TABLE NO. 217 NORMAL SPECTRAL ABSORPTANCE OF TIN

Curve No.	Ref. No.	Year	Temperature K	Wavelength Range, μ	Geometry θ	Reported Error, %	Composition (weight percent), Specifications and Remarks
1	307	1954	~298	0.235-2.600	~0°		Tin, 5 mils thick; data extracted from smooth curve.

DATA TABLE NO. 217 NORMAL SPECTRAL ABSORPTANCE OF TIN

[Wavelength, λ, μ; Absorptance, α; Temperature, T, K]

λ	α
CURVE 1 T = ~298	
0.235	0.750
0.255	0.731
0.278	0.740
0.346	0.660
0.397	0.475
0.525	0.293
0.745	0.256
0.835	0.240
1.185	0.257
1.364	0.224
1.419	0.193
1.502	0.128
1.601	0.139
1.697	0.100
1.803	0.132
2.000	0.187
2.200	0.286
2.400	0.280
2.600	0.255

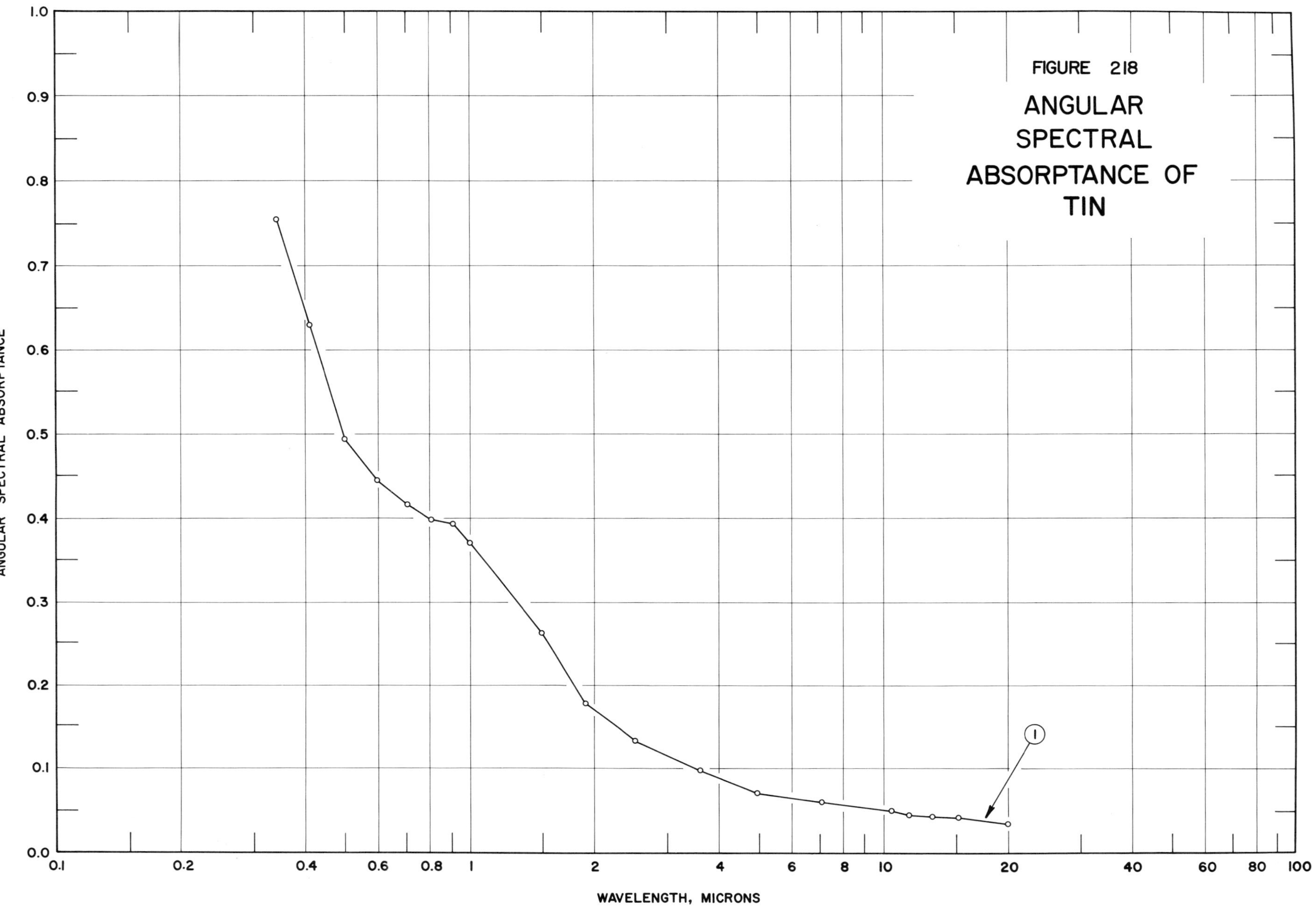
FIGURE 218
ANGULAR SPECTRAL ABSORPTANCE OF TIN
1
ANGULAR SPECTRAL ABSORPTANCE
0.0
0.1
0.2
0.3
0.4
0.5
0.6
0.7
0.8
0.9
1.0
0.1
0.2
0.4
0.6
0.8
1
2
4
6
8
10
20
40
60
80
100
WAVELENGTH, MICRONS

SPECIFICATION TABLE NO. 218 ANGULAR SPECTRAL ABSORPTANCE OF TIN

Curve No.	Ref. No.	Year	Temperature K	Wavelength Range, μ	Geometry θ	Reported Error, %	Composition (weight percent), Specifications and Remarks
1	225	1965	306	0.343-20.0	25°		Tin from Belmont Smelting and Refining Works; measured in dry nitrogen; heated cavity at approx 1056 K with platinum reference; authors assumed $\alpha = 1 - R(2\pi, 25°)$.

DATA TABLE NO. 218 ANGULAR SPECTRAL ABSORPTANCE OF TIN

[Wavelength, λ, μ; Absorptance, α; Temperature, T, K]

λ	α
CURVE 1 T = 306	
0.343	0.755
0.413	0.631
0.500	0.494
0.597	0.446
0.705	0.417
0.813	0.399
0.910	0.393
1.00	0.371
1.49	0.262
1.91	0.178
2.52	0.132
3.60	0.098
4.98	0.070
7.10	0.061
10.4	0.051
11.5	0.047
13.2	0.045
15.2	0.042
20.0	0.036

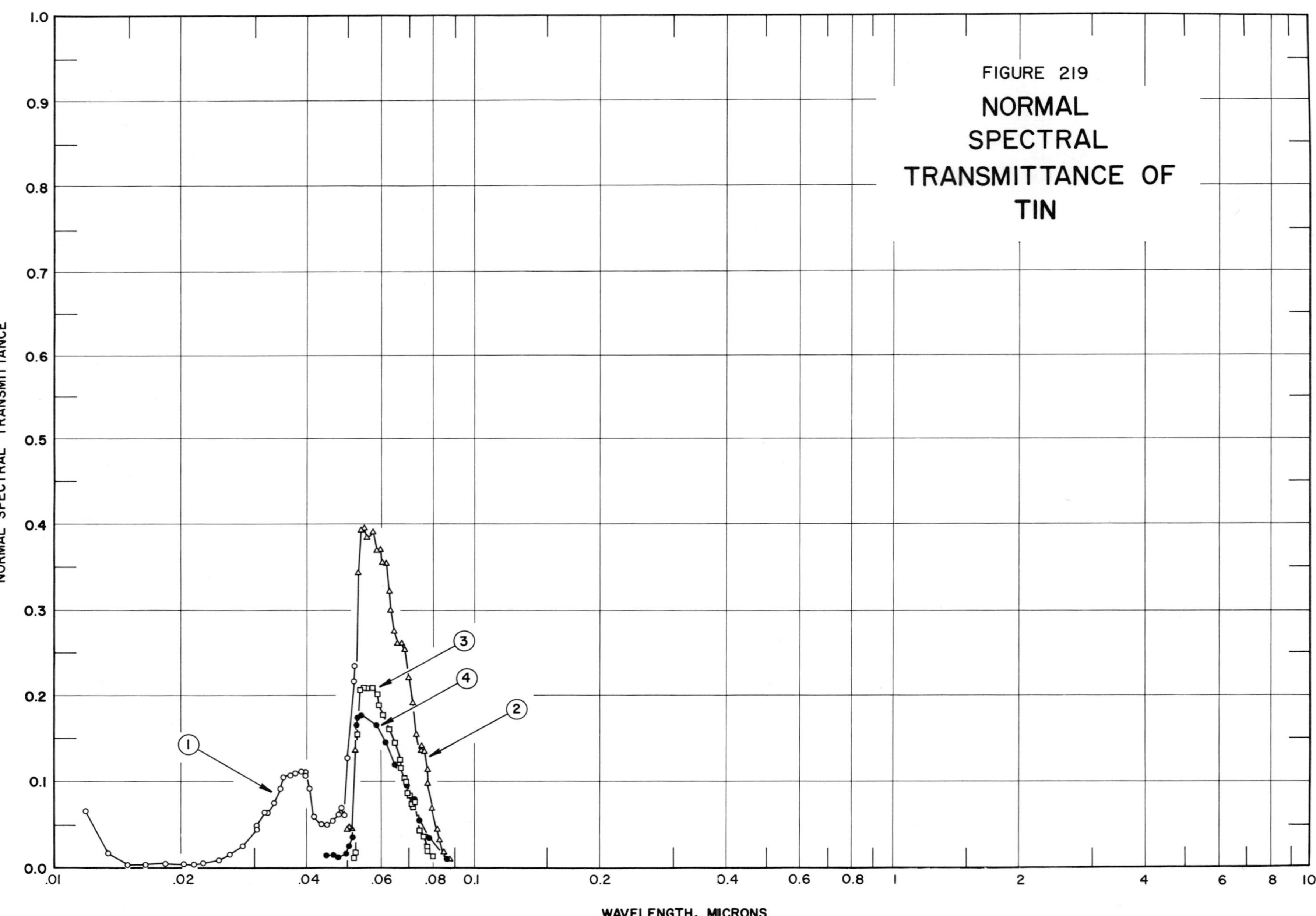

FIGURE 219 NORMAL SPECTRAL TRANSMITTANCE OF TIN

SPECIFICATION TABLE NO. 219 NORMAL SPECTRAL TRANSMITTANCE OF TIN

Curve No.	Ref. No.	Year	Temperature K	Wavelength Range, μ	Geometry θ	θ'	ω'	Reported Error, %	Composition (weight percent), Specifications and Remarks
1	329	1966	298	0.0118-0.0519	0°	0°			Unbacked tin film (680 Å thick); 99.95 pure; evaporated with rate of 3000 Å sec^{-1} at 10^{-5} mm Hg.
2	329	1966	298	0.0501-0.0879	0°	0°			Unbacked tin film (860 Å thick); 99.95 pure; evaporated with rate of 3000 Å sec^{-1} at 10^{-5} mm Hg.
3	329	1966	298	0.0519-0.0801	0°	0°			Unbacked tin film (1690 Å thick); 99.95 pure; evaporated with rate of 3000 Å sec^{-1} at 10^{-5} mm Hg.
4	349	1965	298	0.0454-0.0873	~0°	~0°			Unbacked evaporated film (1020 Å thick); data extracted from smooth curve.

DATA TABLE NO. 219 NORMAL SPECTRAL TRANSMITTANCE OF TIN

[Wavelength, λ, μ; Transmittance, τ; Temperature, T, K

CURVE 1
T = 298

λ	τ
0.0118	0.0670
0.0133	0.0183
0.0149	0.0026
0.0164	0.0023
0.0184	0.0027
0.0202	0.0025
0.0215	0.0022
0.0225	0.0030
0.0247	0.0098
0.0261	0.0142
0.0281	0.0252
0.0304	0.0443
0.0304	0.0497
0.0317	0.0625
0.0322	0.0625
0.0335	0.0748
0.0346	0.0908
0.0351	0.1038
0.0365	0.1076
0.0373	0.1096
0.0388	0.1117
0.0396	0.1072
0.0398	0.1107
0.0408	0.0912
0.0418	0.0586
0.0434	0.0501
0.0452	0.0506
0.0462	0.0532
0.0477	0.0603
0.0485	0.0695
0.0494	0.0708
0.0500	0.1271
0.0518	0.2158
0.0519	0.2333

CURVE 2
T = 298

λ	τ
0.0501	0.0445
0.0505	0.0485
0.0510	0.0466
0.0520	0.1368
0.0526	0.3436
0.0535	0.3945
0.0546	0.3963
0.0553	0.3837
0.0573	0.3908
0.0583	0.3698
0.0598	0.3698
0.0606	0.3548
0.0615	0.3548
0.0627	0.3221
0.0634	0.3006
0.0646	0.2754
0.0657	0.2618
0.0670	0.2618
0.0684	0.2535
0.0700	0.2218
0.0716	0.1905
0.0728	0.1549
0.0744	0.1349
0.0748	0.1406
0.0761	0.1337
0.0774	0.1143
0.0785	0.0982
0.0801	0.0692
0.0820	0.0455
0.0833	0.0325
0.0851	0.0195
0.0879	0.0102

CURVE 3
T = 298

λ	τ
0.0519	0.0100
0.0523	0.0186
0.0526	0.1521
0.0535	0.2070
0.0546	0.2099
0.0553	0.2099
0.0574	0.2099
0.0586	0.2014
0.0596	0.1879
0.0605	0.1762
0.0628	0.1592
0.0646	0.1432
0.0661	0.1253
0.0671	0.1159
0.0681	0.1028
0.0685	0.1005
0.0699	0.0863
0.0702	0.0824
0.0716	0.0724
0.0719	0.0692
0.0727	0.0625
0.0744	0.0425
0.0761	0.0368
0.0773	0.0264
0.0788	0.0193
0.0801	0.0123

CURVE 4
T = 298

λ	τ
0.0454	0.012
0.0463	0.012
0.0477	0.011
0.0498	0.017
0.0508	0.026
0.0515	0.035
0.0521	0.163
0.0525	0.172
0.0537	0.175
0.0582	0.163
0.0614	0.145
0.0643	0.119
0.0689	0.096
0.0721	0.078
0.0743	0.053
0.0785	0.034
0.0873	0.010

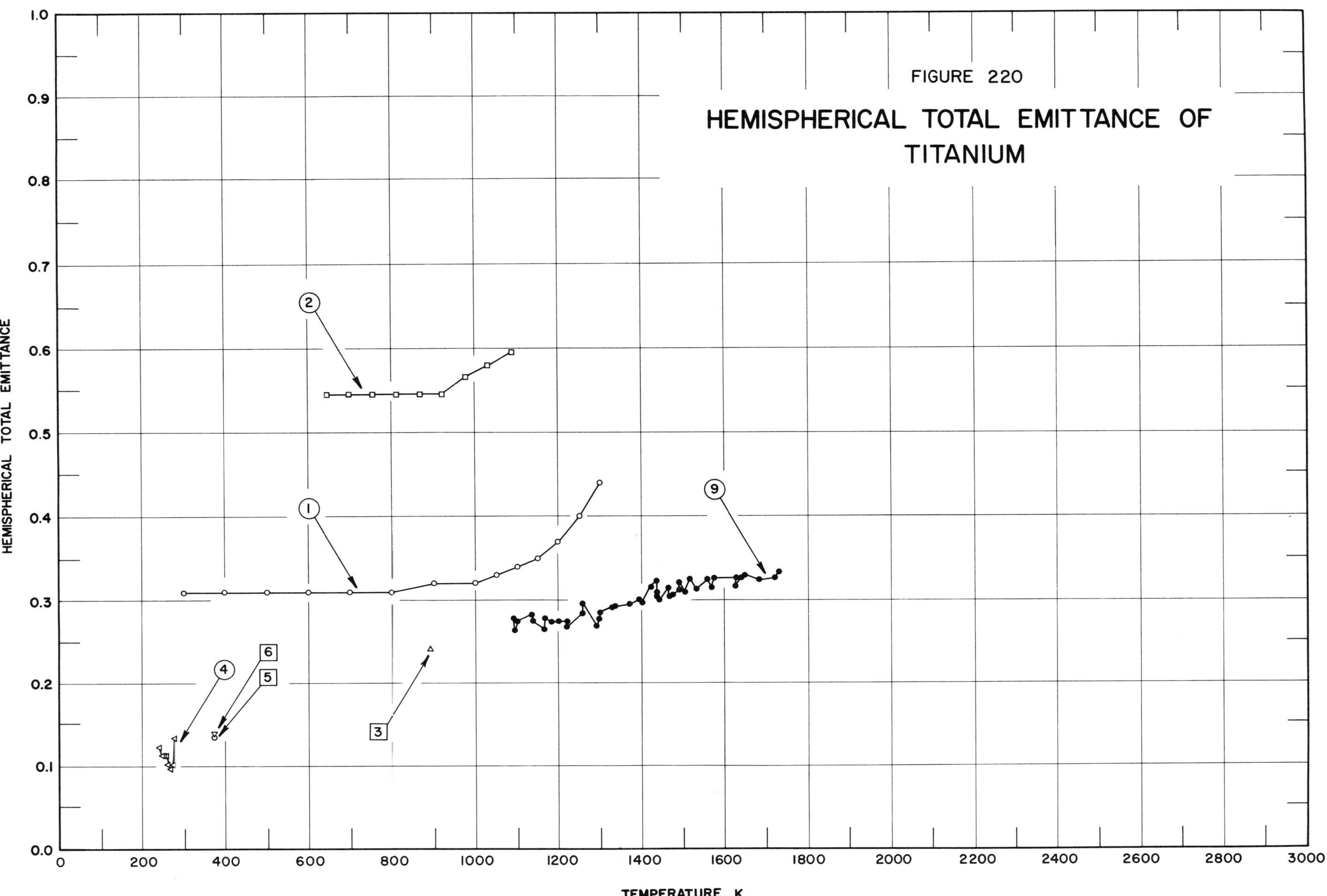

FIGURE 220

HEMISPHERICAL TOTAL EMITTANCE OF TITANIUM

SPECIFICATION TABLE NO. 220 HEMISPHERICAL TOTAL EMITTANCE OF TITANIUM

Curve No.	Ref. No.	Year	Temperature Range, K	Reported Error, %	Composition (weight percent), Specifications and Remarks
1	97	1949	300-1300	±10	Commercial wire (0.0424 cm diameter); measured in vacuum ($< 10^{-5}$ mm Hg).
2	22	1958	644-1089	≤ 2	TMCA Ti - 75 A; nominal composition: 0.3 max Fe, 0.08 max C, 0.05 max N_2, 0.015 max H_2, Ti balance; stably oxidized in quiescent air at 1089 K.
3	98	1960	891	± 8	Polished; measured in vacuum (10^{-5} mm Hg).
4	6	1963	241-279	± 3	Pure; polished; measured in vacuum (10^{-3} mm Hg).
5	8	1963	373		Prefinished with 600 grit aluminum oxide powder on felt and electropolished; measured in vacuum (10^{-5} mm Hg).
6	8	1963	373		Above specimen and conditions except bombarded with hydrogen ions (3.25×10^{20} ions cm^{-2}).
7	8	1963	373		Above specimen and conditions except bombarded with hydrogen ions (6.5×10^{20} ions cm^{-2}).
8	8	1963	373		Above specimen and conditions except bombarded with hydrogen ions (7.9×10^{20} ions cm^{-2}).
9	343	1964	1089-1733		Measured in vacuum (10^{-9} mm Hg).

DATA TABLE NO. 220 HEMISPHERICAL TOTAL EMITTANCE OF TITANIUM

[Temperature, T, K; Emittance, ∈]

T	∈
CURVE 1	
300	0.31
400	0.31
500	0.31
600	0.31
700	0.31
800	0.31
900	0.32
1000	0.32
1050	0.33
1100	0.34
1150	0.35
1200	0.37
1250	0.40
1300	0.44
CURVE 2	
644	0.545
700	0.545
755	0.545
811	0.545
866	0.545
922	0.545
978	0.565
1033	0.580
1089	0.595
CURVE 3	
891	0.241
CURVE 4	
241	0.120
247	0.112
252	0.112
256	0.112
262	0.100
267	0.096
274	0.100
279	0.132
CURVE 5	
373	0.134
CURVE 6	
373	0.138
CURVE 7*	
373	0.138
CURVE 8*	
373	0.138
CURVE 9	
1089	0.278
1095	0.262
1100	0.276
1135	0.282
1140	0.275
1163	0.263
1166	0.278
1186	0.273
1200	0.276
1223	0.276
1222	0.269
1259	0.283
1259	0.295
1290	0.269
1298	0.278
1300	0.284
1329	0.291
1335	0.292
1370	0.293
1396	0.300
1403	0.297
1426	0.315
1435	0.322
1435	0.309
1435	0.303
1444	0.299
1463	0.314
1466	0.303
1474	0.307
1490	0.312
1490	0.321
1508	0.309
1514	0.325
1532	0.312
CURVE 9 (cont.)	
1559	0.326
1568	0.313
1577	0.327
1627	0.328
1626	0.317
1640	0.328
1648	0.330
1685	0.323
1722	0.328
1733	0.334

* Not shown on plot

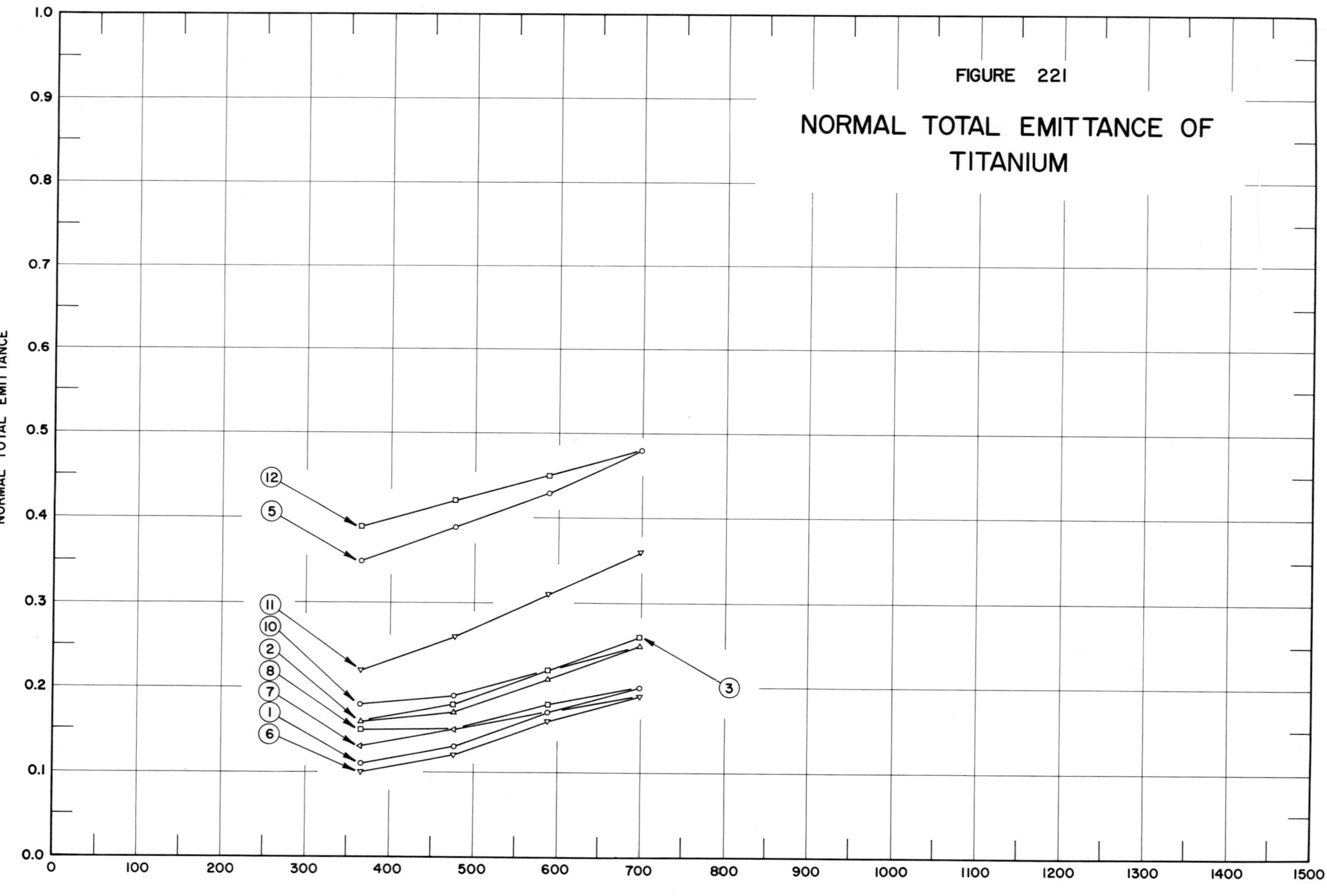
FIGURE 221
NORMAL TOTAL EMITTANCE OF
TITANIUM
NORMAL TOTAL EMITTANCE
0.0
0.1
0.2
0.3
0.4
0.5
0.6
0.7
0.8
0.9
1.0
TEMPERATURE, K
0
100
200
300
400
500
600
700
800
900
1000
1100
1200
1300
1400
1500
12
5
11
10
2
8
7
1
6
3

SPECIFICATION TABLE NO. 221 NORMAL TOTAL EMITTANCE OF TITANIUM

Curve No.	Ref. No.	Year	Temperature Range, K	Geometry θ'	Reported Error, %	Composition (weight percent), Specifications and Remarks
1	99	1958	367-700	~ 0°	< 27	Ti-75A; AMS 4901; nominal composition: 0.1 Fe, 0.02 N, 0.04 max C, 0.08 max W, Ti balance; heated at 580.5 K for 306 hrs; measured in air.
2	99	1958	367-700	~ 0°	< 19	Different sample, same as curve 1 specimen and conditions except heated at 705.5 K for 100 hrs.
3	99	1958	367-700	~ 0°	< 19	Different sample, same as curve 1 specimen and conditions except heated at 711 K for 306 hrs.
4	99	1958	367-700	~ 0°	< 19	Different sample, same as curve 1 specimen and conditions except heated at 739.5 K for 303 hrs.
5	99	1958	367-700	~ 0°	< 8	Different sample, same as curve 1 specimen and conditions except heated at 813 K for 303 hrs.
6	99	1958	367-700	~ 0°	< 30	Different sample, same as curve 1 specimen and conditions except no heat treatment.
7	99	1958	367-700	~ 0°	< 30	Ti-75A; AMS 4901; nominal composition: 0.1 Fe, 0.02 N_2, 0.04 max C, 0.08 max W, Ti balance; calculated from spectral reflectance factor data.
8	99	1958	367-700	~ 0°	< 27	Different sample, same as curve 7 specimen and conditions except heated at 580.5 K for 306 hrs.
9	99	1958	367-700	~ 0°	12.5	Different sample, same as curve 7 specimen and conditions except heated at 705.5 K for 100 hrs.
10	99	1958	367-700	~ 0°	< 19	Different sample, same as curve 7 specimen and conditions except heated at 711 K for 306 hrs.
11	99	1958	367-700	~ 0°	< 19	Different sample, same as curve 7 specimen and conditions except heated at 739 K for 303 hrs.
12	99	1958	367-700	~ 0°	< 8	Different sample, same as curve 7 specimen and conditions except heated at 813 K for 303 hrs.

DATA TABLE NO. 221 NORMAL TOTAL EMITTANCE OF TITANIUM

[Temperature, T, K; Emittance ϵ]

T	ϵ	T	ϵ
CURVE 1		CURVE 7	
367	0.11	367	0.13
478	0.13	478	0.15
589	0.17	589	0.17*
700	0.20	700	0.19*
CURVE 2		CURVE 8	
367	0.16	367	0.15
478	0.17	478	0.15*
589	0.21	589	0.18
700	0.25	700	0.20*
CURVE 3		CURVE 9 *	
367	0.16*	367	0.16
478	0.18	478	0.18
589	0.22	589	0.21
700	0.26	700	0.23
CURVE 4 *		CURVE 10	
367	0.16	367	0.18
478	0.18	478	0.19
589	0.22	589	0.22*
700	0.25	700	0.25*
CURVE 5		CURVE 11	
367	0.35	367	0.22
478	0.39	478	0.26
589	0.43	589	0.31
700	0.48	700	0.36
CURVE 6		CURVE 12	
367	0.10	367	0.39
478	0.12	478	0.42
589	0.16	589	0.45
700	0.19	700	0.48

* Not shown on plot

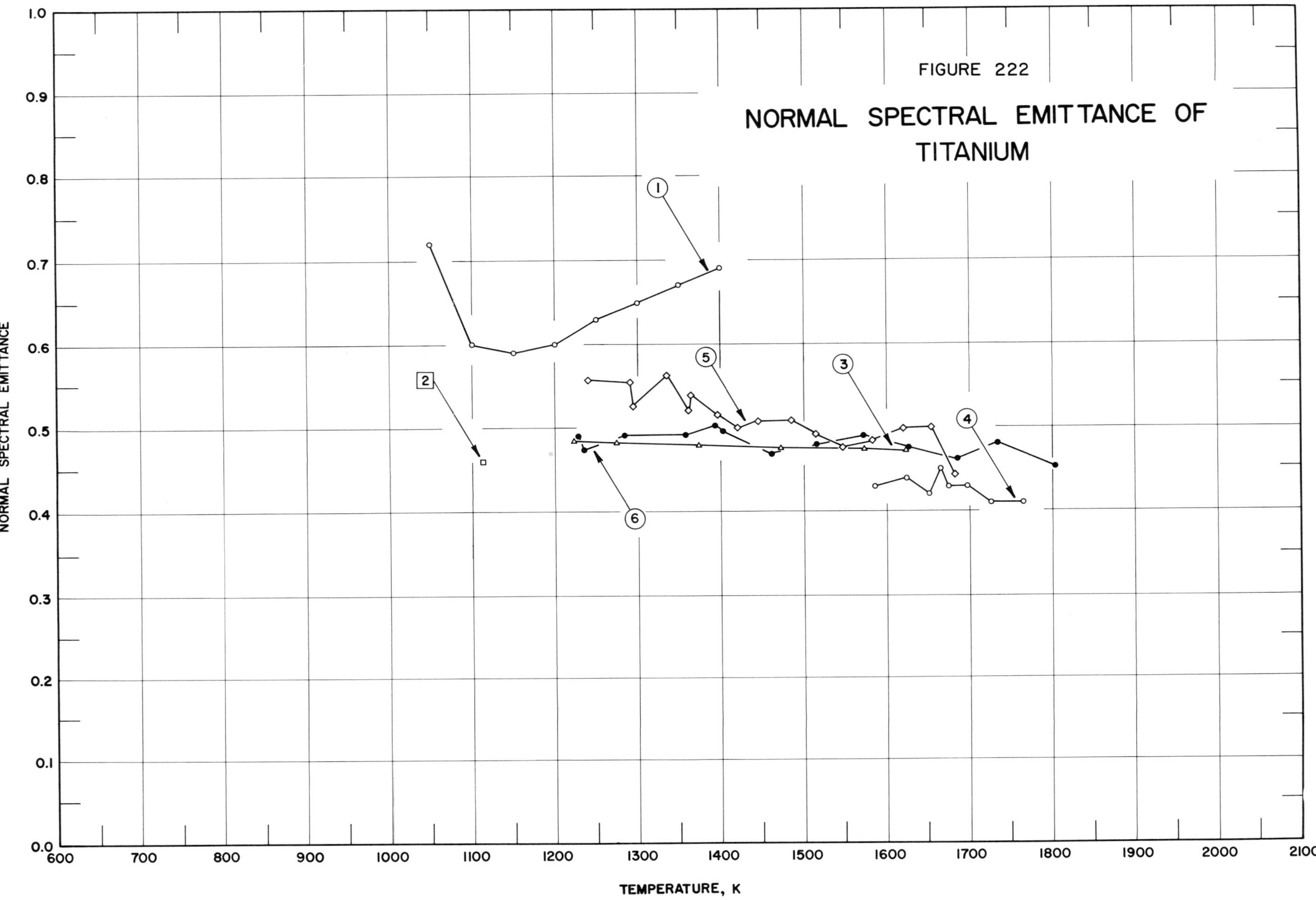

FIGURE 222
NORMAL SPECTRAL EMITTANCE OF TITANIUM
NORMAL SPECTRAL EMITTANCE
TEMPERATURE, K
0.0
0.1
0.2
0.3
0.4
0.5
0.6
0.7
0.8
0.9
1.0
600
700
800
900
1000
1100
1200
1300
1400
1500
1600
1700
1800
1900
2000
2100
1
2
3
4
5
6

SPECIFICATION TABLE NO. 222 NORMAL SPECTRAL EMITTANCE OF TITANIUM

Curve No.	Ref. No.	Year	Wavelength μ	Temperature Range, K	Geometry θ '	Reported Error, %	Composition (weight percent), Specifications and Remarks
1	97	1949	0.655	1050-1400	~0°	± 3	Rod (0.25 in. diameter); measured in vacuum ($< 10^{-5}$ mm Hg).
2	100	1950	0.652	1113	~0°	2.2	0.19 N, 0.096 Fe, 0.073 O, 0.059 C, 0.056 Si, <0.002 Ni; obtained in powder form, sintered in vacuo, worked into fully dense bars, and turned into cylinder; hexagonal crystal structure; polished by normal metallographic methods using magnesia in final stages; measured in vacuum (2×10^{-5} to 8×10^{-6} mm Hg); mean value of three runs.
3	100	1950	0.652	1223-1623	~0°	2.2	Different sample, same as above specimen and conditions except body centered cubic structure.
4	101	1953	0.65	1587-1764	~0°		99.884 Pure; ϵ values are average of three readings at the corresponding temperatures.
5	268	1967	0.65	1241-1683	~0°		99.5 Ti; measured in vacuum (10^{-9} mm Hg); temperature measured by thermocouple.
6	268	1967	0.65	1229-1804	~0°		Different sample, same as above specimen and conditions except temperature measured by pyrometer.

DATA TABLE NO. 222 NORMAL SPECTRAL EMITTANCE OF TITANIUM

[Temperature, T, K; Emittance, ϵ; Wavelength, λ, μ]

T	ϵ	T	ϵ
CURVE 1		**CURVE 5**	
$\lambda = 0.655$		$\lambda = 0.65$	
1050	0.72	1241	0.558
1100	0.60	1292	0.553
1150	0.59	1295	0.527
1200	0.60	1336	0.562
1250	0.63	1362	0.521
1300	0.65	1366	0.538
1350	0.67	1398	0.517
1400	0.69	1422	0.500
		1446	0.508
CURVE 2		1486	0.508
$\lambda = 0.652$		1515	0.491
		1549	0.477
1113	0.459	1584	0.483
		1621	0.499
CURVE 3		1657	0.500
$\lambda = 0.652$		1683	0.443
1223	0.484	**CURVE 6**	
1273	0.482	$\lambda = 0.65$	
1373	0.479		
1473	0.476	1229	0.489
1573	0.473	1236	0.472
1623	0.471	1284	0.490
		1359	0.491
CURVE 4		1394	0.502
$\lambda = 0.65$		1404	0.496
		1462	0.468
1587	0.43	1517	0.479
1624	0.44	1572	0.489
1651	0.42	1628	0.473
1667	0.45	1688	0.460
1675	0.43	1734	0.480
1698	0.43	1804	0.451
1725	0.41		
1764	0.41		

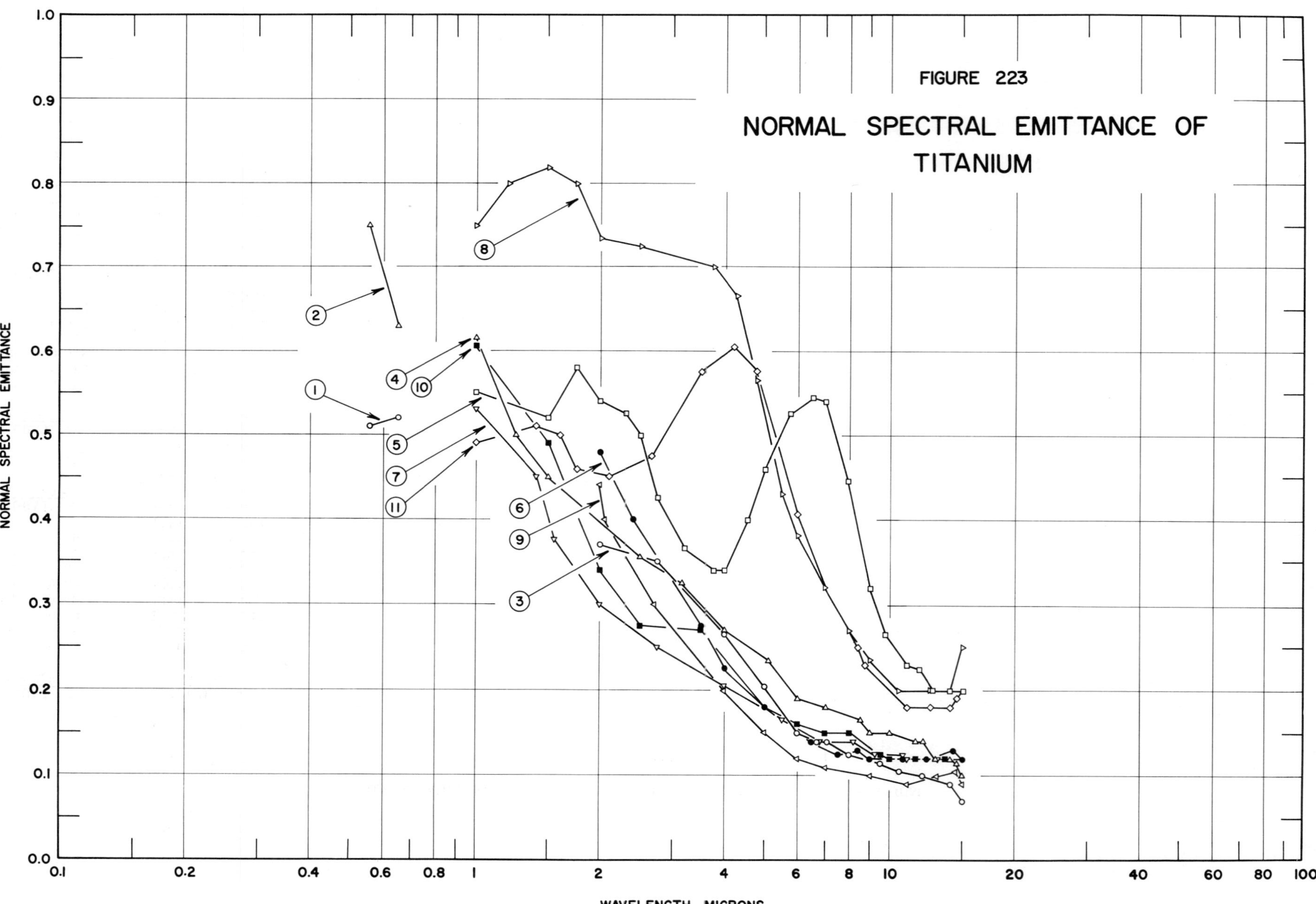

FIGURE 223
NORMAL SPECTRAL EMITTANCE OF TITANIUM

SPECIFICATION TABLE NO. 223 NORMAL SPECTRAL EMITTANCE OF TITANIUM

Curve No.	Ref. No.	Year	Temperature K	Wavelength Range, μ	Geometry θ'	Reported Error, %	Composition (weight percent), Specifications and Remarks
1	19	1914	1723	0.55-0.65	$\sim 0°$	1	Film; tungsten substrate; melted in hydrogen then oxidized in air by heating; measured in air; Pt reference ($\epsilon = 0.33$ for $\lambda = 0.650\ \mu$ and $\epsilon = 0.38$ for $\lambda = 0.547\ \mu$ at all temp).
2	19	1914	1823	0.55-0.65	$\sim 0°$	1	Film; tungsten substrate; measured in hydrogen; Pt reference ($\epsilon = 0.33$ for $\lambda = 0.650\ \mu$ and $\epsilon = 0.38$ for $\lambda = 0.547\ \mu$ at all temp).
3	86	1961	523	2.00-15.00	$\sim 0°$	± 5	As received; data extracted from smooth curve.
4	86	1961	773	1.00-15.00	$\sim 0°$	± 5	Different sample, same as curve 3 specimen and conditions.
5	86	1961	1023	1.00-15.00	$\sim 0°$	± 5	Different sample, same as curve 3 specimen and conditions.
6	86	1961	523	2.00-15.00	$\sim 0°$	± 5	Different sample, same as curve 3 specimen and conditions except heated in air at 1073 K for 30 min.
7	86	1961	773	1.00-15.00	$\sim 0°$	± 5	Different sample, same as curve 6 specimen and conditions.
8	86	1961	1023	1.00-15.00	$\sim 0°$	± 5	Different sample, same as curve 6 specimen and conditions.
9	86	1961	523	2.00-15.00	$\sim 0°$	± 5	Different sample, same as curve 3 specimen and conditions except heated in vacuum (2.8×10^{-5} mm Hg) at 1073 K for 30 min.
10	86	1961	773	1.00-15.00	$\sim 0°$	± 5	Different sample, same as curve 9 specimen and conditions.
11	86	1961	1023	1.00-15.00	$\sim 0°$	± 5	Different sample, same as curve 9 specimen and conditions.

DATA TABLE NO. 223 NORMAL SPECTRAL EMITTANCE OF TITANIUM

[Wavelength, λ, μ; Emittance, ϵ; Temperature, T, K]

λ	ϵ
CURVE 1 T = 1723	
0.55	0.51
0.65	0.52
CURVE 2 T = 1823	
0.55	0.75
0.65	0.63
CURVE 3 T = 523	
2.00	0.370
2.75	0.350
4.00	0.265
5.00	0.205
6.00	0.150
6.70	0.140
7.05	0.140
8.00	0.125
9.50	0.115
10.50	0.105
12.00	0.100
14.00	0.090
15.00	0.070
CURVE 4 T = 773	
1.00	0.615
1.25	0.500
1.50	0.450
2.50	0.355
3.15	0.325
4.00	0.270
5.10	0.235
6.00	0.190
7.00	0.180
8.50	0.165
9.00	0.150
10.00	0.150
11.50	0.140
12.00	0.140
CURVE 4 (cont.)	
12.90	0.120
14.00	0.120
14.50	0.115
15.00	0.100
CURVE 5 T = 1023	
1.00	0.550
1.50	0.520
1.75	0.580
2.00	0.540
2.30	0.525
2.50	0.500
2.75	0.425
3.20	0.365
3.75	0.340
4.00	0.340
4.55	0.400
5.00	0.460
5.75	0.525
6.50	0.545
7.00	0.540
7.95	0.445
9.00	0.320
9.75	0.265
11.00	0.230
11.75	0.225
12.75	0.200
14.00	0.200
15.00	0.200
CURVE 6 T = 523	
2.00	0.480
2.40	0.400
3.50	0.275
4.00	0.225
5.00	0.180
6.25	0.140
7.50	0.125
8.20	0.130
9.00	0.120
CURVE 6 (cont.)	
10.75	0.120
12.25	0.120
14.25	0.130
15.00	0.120
CURVE 7 T = 773	
1.00	0.530
1.40	0.450
1.55	0.375
2.00	0.300
2.75	0.250
4.00	0.205
5.50	0.165
6.75	0.140
8.20	0.140
9.25	0.125
10.75	0.125
11.00	0.120
13.00	0.120
15.00	0.120*
CURVE 8 T = 1023	
1.00	0.750
1.20	0.800
1.50	0.820
1.75	0.800
2.00	0.735
2.50	0.725
3.75	0.700
4.25	0.665
4.75	0.565
5.50	0.430
6.00	0.380
7.00	0.320
8.00	0.270
9.00	0.235
10.50	0.200
12.50	0.200
14.00	0.200*
15.00	0.250
CURVE 9 T = 523	
2.00	0.440
2.05	0.400
2.70	0.300
4.00	0.200
5.00	0.150
6.00	0.120
7.00	0.110
9.00	0.100
11.00	0.090
13.00	0.100
14.40	0.105
15.00	0.090
CURVE 10 T = 773	
1.00	0.605
1.50	0.490
2.00	0.340
2.50	0.275
3.50	0.220
5.00	0.180*
6.00	0.160
7.00	0.150
8.00	0.150
9.50	0.125
10.00	0.120
11.50	0.120
13.50	0.120
15.00	0.120*
CURVE 11 T = 1023	
1.00	0.490
1.40	0.510
1.60	0.500
1.75	0.460
2.10	0.450
2.65	0.475
3.50	0.575
4.10	0.605
4.75	0.575
CURVE 11 (cont.)	
6.00	0.405
7.00	0.320*
8.40	0.250
8.75	0.230
11.00	0.180
12.50	0.180
14.00	0.180
14.50	0.190
15.00	0.200*

* Not shown on plot

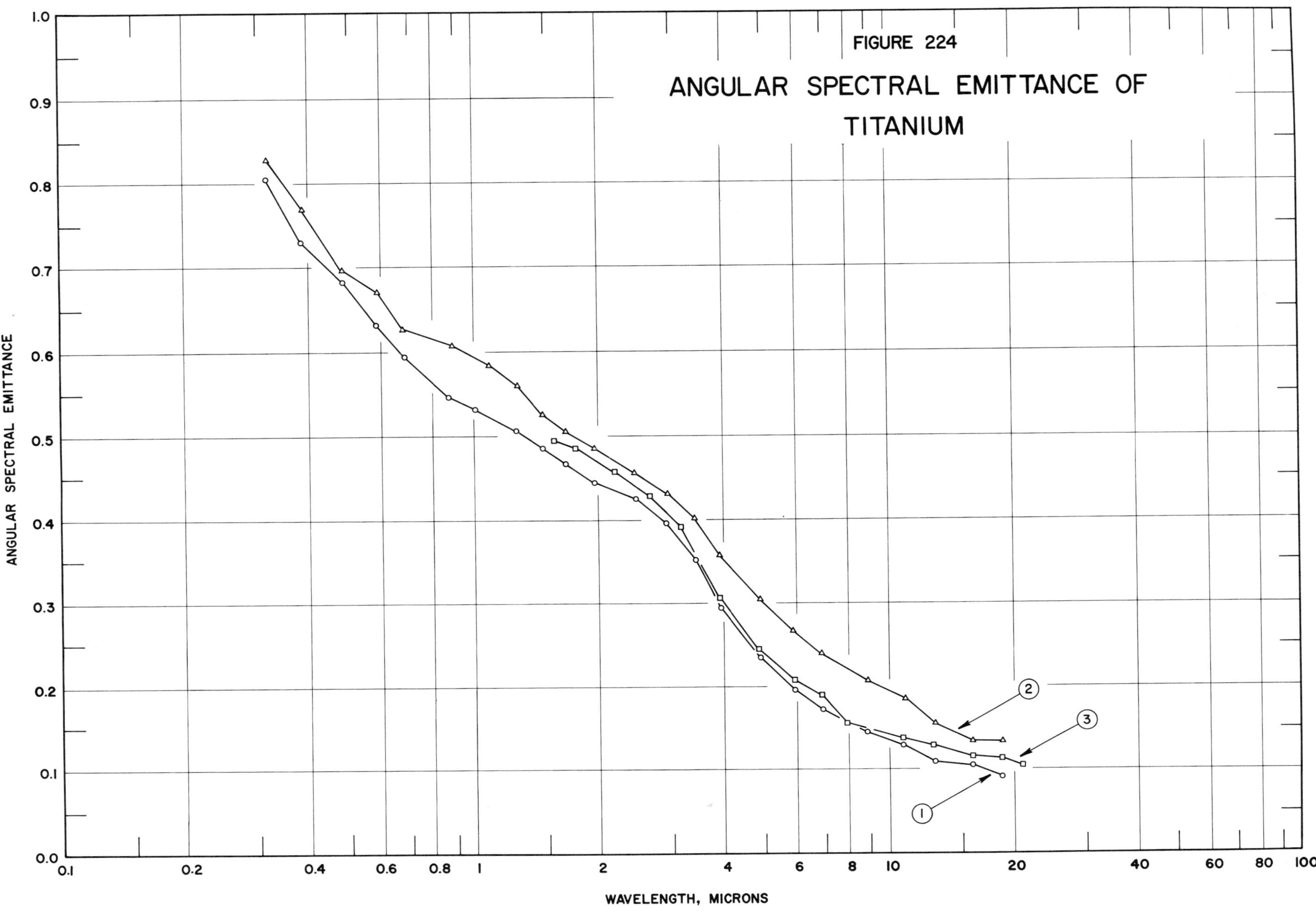

FIGURE 224
ANGULAR SPECTRAL EMITTANCE OF TITANIUM

SPECIFICATION TABLE NO. 224 ANGULAR SPECTRAL EMITTANCE OF TITANIUM

Curve No.	Ref. No.	Year	Temperature K	Wavelength Range, μ	Geometry θ'	Reported Error, %	Composition (weight percent), Specifications and Remarks
1	219	1965	306	0.319-18.8	25°		99 Ti plate from Crucible Steel Co.; lapped; RMS surface roughness 2 microinches; authors assumed $\epsilon = \alpha = 1 - \rho(25°, 2\pi)$.
2	219	1965	306	0.319-18.8	25°		99 Ti plate from Crucible Steel Co.; honed; RMS surface roughness 4 microinches; authors assumed $\epsilon = \alpha = 1 - \rho(25°, 2\pi)$.
3	219	1965	306	1.56-21.0	25°		99 Ti plate from Crucible Steel Co.; ground; RMS surface roughness 16 microinches; authors assumed $\epsilon = \alpha = 1 - \rho(25°, 2\pi)$.

DATA TABLE NO. 224 ANGULAR SPECTRAL EMITTANCE OF TITANIUM

[Wavelength, λ, μ; Emittance, ϵ; Temperature, T, K]

λ	ϵ
CURVE 1	
T = 306	
0.319	0.804
0.385	0.731
0.482	0.682
0.581	0.632
0.681	0.594
0.869	0.547
1.07	0.531
1.26	0.507
1.47	0.486
1.66	0.467
1.96	0.444
2.46	0.425
2.91	0.396
3.42	0.351
3.93	0.293
4.90	0.233
5.96	0.197
6.92	0.172
8.83	0.146
10.8	0.129
12.9	0.110
15.9	0.105
18.8	0.091
CURVE 2	
T = 306	
0.319	0.830
0.387	0.771
0.483	0.698
0.583	0.671
0.678	0.628
0.883	0.608
1.08	0.583
1.27	0.560
1.47	0.526
1.66	0.505
1.95	0.486
2.44	0.457
2.93	0.431
3.41	0.401
3.91	0.357
4.91	0.303
5.92	0.267
CURVE 2 (cont.)	
6.87	0.239
8.93	0.208
10.9	0.183
12.9	0.154
15.9	0.133
18.8	0.133
CURVE 3	
T = 306	
1.56	0.494
1.77	0.486
2.20	0.457
2.67	0.428
3.16	0.390
3.90	0.306
4.86	0.244
5.92	0.208
6.90	0.189
8.89	0.156
10.8	0.138
12.8	0.129
15.9	0.116
18.8	0.114
21.0	0.105

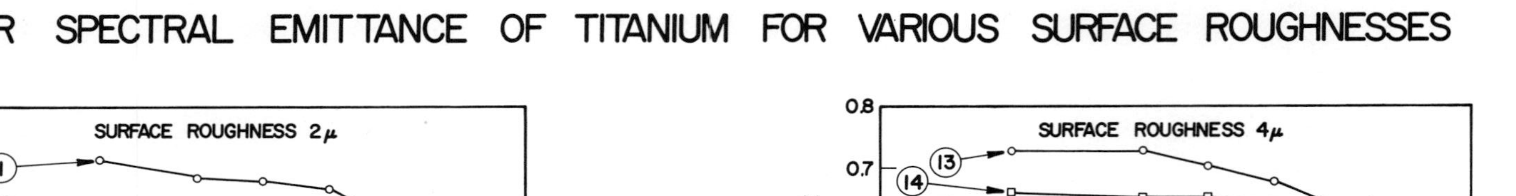

FIGURE 225

ANGULAR SPECTRAL EMITTANCE OF TITANIUM FOR VARIOUS SURFACE ROUGHNESSES

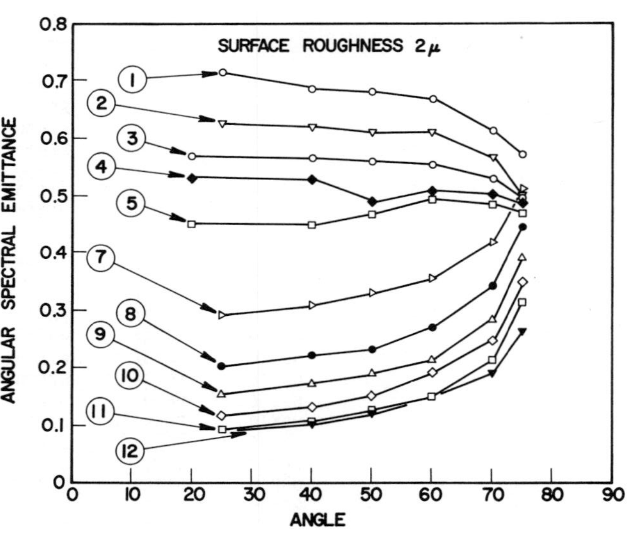

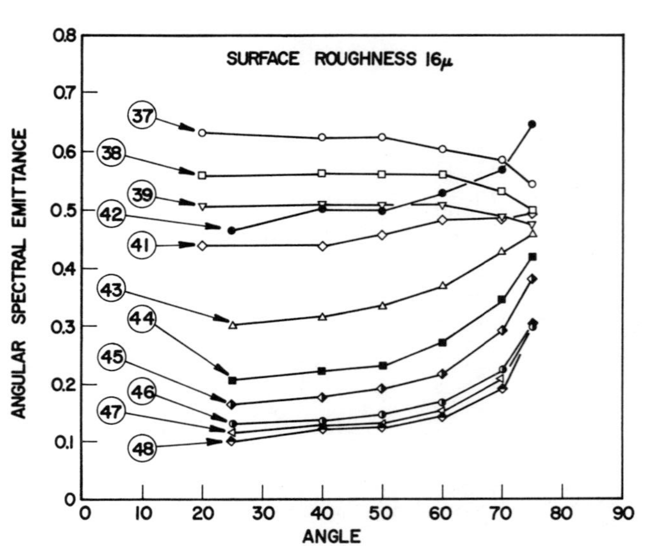

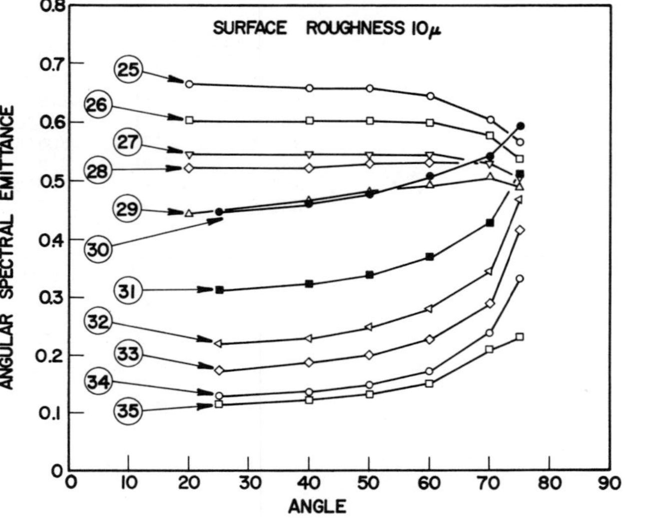

SURFACE ROUGHNESS 16μ
ANGULAR SPECTRAL EMITTANCE
ANGLE
37
38
39
42
41
43
44
45
46
47
48

SPECIFICATION TABLE NO. 225 ANGULAR SPECTRAL EMITTANCE OF TITANIUM

Curve No.	Ref. No.	Year	Temperature K	Wavelength, μ	Angular Range,°	Geometry θ'	Reported Error, %	Composition (weight percent), Specifications and Remarks
1	219	1965	306	0.43	25-75	θ'		Titanium; lapped; RMS surface roughness 2 microinches; authors assumed $\epsilon = \alpha = 1 - \rho(\theta', 2\pi)$.
2	219	1965	306	0.56	25-75	θ'		Above specimen and conditions.
3	219	1965	306	0.72	20-75	θ'		Above specimen and conditions.
4	219	1965	306	1	20-75	θ'		Above specimen and conditions.
5	219	1965	306	1.67	20-75	θ'		Above specimen and conditions.
6	219	1965	306	2	25-75	θ'		Above specimen and conditions.
7	219	1965	306	4	25-75	θ'		Above specimen and conditions.
8	219	1965	306	6	25-75	θ'		Above specimen and conditions.
9	219	1965	306	8.2	25-75	θ'		Above specimen and conditions.
10	219	1965	306	12	25-75	θ'		Above specimen and conditions.
11	219	1965	306	16	25-75	θ'		Above specimen and conditions.
12	219	1965	306	20	25-75	θ'		Above specimen and conditions.
13	219	1965	306	0.43	20-75	θ'		Titanium; honed; RMS surface roughness 4 microinches; authors assumed $\epsilon = \alpha = 1 - \rho(\theta', 2\pi)$.
14	219	1965	306	0.56	20-75	θ'		Above specimen and conditions.
15	219	1965	306	0.72	20-75	θ'		Above specimen and conditions.
16	219	1965	306	1	20-75	θ'		Above specimen and conditions.
17	219	1965	306	1.67	20-75	θ'		Above specimen and conditions.
18	219	1965	306	2	25-75	θ'		Above specimen and conditions.
19	219	1965	306	4	25-75	θ'		Above specimen and conditions.
20	219	1965	306	6	25-75	θ'		Above specimen and conditions.
21	219	1965	306	8.2	25-75	θ'		Above specimen and conditions.
22	219	1965	306	12	25-75	θ'		Above specimen and conditions.
23	219	1965	306	16	25-75	θ'		Above specimen and conditions.
24	219	1965	306	20	25-75	θ'		Above specimen and conditions.
25	219	1965	306	0.43	20-75	θ'		Titanium; ground; RMS surface roughness 10 microinches; authors assumed $\epsilon = \alpha = 1 - \rho(\theta', 2\pi)$.
26	219	1965	306	0.56	20-75	θ'		Above specimen and conditions.
27	219	1965	306	0.72	20-75	θ'		Above specimen and conditions.

SPECIFICATION TABLE NO. 225 (continued)

Curve No.	Ref. No.	Year	Temperature K	Wavelength, μ	Angular Range,°	Geometry θ'	Reported Error, %	Composition (weight percent), Specifications and Remarks
28	219	1965	306	1	20-75	θ'		Above specimen and conditions.
29	219	1965	306	1.67	20-75	θ'		Above specimen and conditions.
30	219	1965	306	2	25-75	θ'		Above specimen and conditions.
31	219	1965	306	4	25-75	θ'		Above specimen and conditions.
32	219	1965	306	6	25-75	θ'		Above specimen and conditions.
33	219	1965	306	8.2	25-75	θ'		Above specimen and conditions.
34	219	1965	306	12	25-75	θ'		Above specimen and conditions.
35	219	1965	306	16	25-75	θ'		Above specimen and conditions.
36	219	1965	306	20	25-75	θ'		Above specimen and conditions.
37	219	1965	306	0.43	20-75	θ'		Titanium; ground; RMS surface roughness 16 microinches; authors assumed $\epsilon = \alpha = 1 - \rho(\theta', 2\pi)$.
38	219	1965	306	0.56	20-75	θ'		Above specimen and conditions.
39	219	1965	306	0.72	20-75	θ'		Above specimen and conditions.
40	219	1965	306	1	20-75	θ'		Above specimen and conditions.
41	219	1965	306	1.67	20-75	θ'		Above specimen and conditions.
42	219	1965	306	2	25-75	θ'		Above specimen and conditions.
43	219	1965	306	4	25-75	θ'		Above specimen and conditions.
44	219	1965	306	6	25-75	θ'		Above specimen and conditions.
45	219	1965	306	8.2	25-75	θ'		Above specimen and conditions.
46	219	1965	306	12	25-75	θ'		Above specimen and conditions.
47	219	1965	306	16	25-75	θ'		Above specimen and conditions.
48	219	1965	306	20	25-75	θ'		Above specimen and conditions.

DATA TABLE NO. 225 ANGULAR SPECTRAL EMITTANCE OF TITANIUM

[Angle, θ',°; Emittance, ∈; Temperature, T,K; Wavelength, λ, μ]

θ'	∈	θ'	∈	θ'	∈	θ'	∈	θ'	∈	θ'	∈	θ'	∈	θ'	∈	θ'	∈
CURVE 1*		CURVE 5*		CURVE 9*		CURVE 13*		CURVE 17*		CURVE 21*		CURVE 25*		CURVE 29*		CURVE 33*	
T = 306		T = 306		T = 306		T = 306		T = 306		T = 306		T = 306		T = 306		T = 306	
λ = 0.43		λ = 1.67		λ = 8.2		λ = 0.43		λ = 1.67		λ = 8.2		λ = 0.43		λ = 1.67		λ = 8.2	
25	0.714	20	0.450	25	0.156	20	0.728	20	0.527	25	0.217	20	0.665	20	0.442	25	0.172
40	0.685	40	0.450	40	0.172	40	0.728	40	0.527	40	0.231	40	0.658	40	0.465	40	0.187
50	0.681	50	0.468	50	0.189	50	0.704	50	0.532	50	0.243	50	0.658	50	0.481	50	0.201
60	0.668	60	0.494	60	0.213	60	0.680	60	0.541	60	0.263	60	0.645	60	0.492	60	0.228
70	0.615	70	0.485	70	0.286	70	0.637	70	0.530	70	0.340	70	0.606	70	0.506	70	0.290
75	0.572	75	0.471	75	0.389	75	0.580	75	0.495	75	0.472	75	0.567	75	0.489	75	0.417
CURVE 2*		CURVE 6*		CURVE 10*		CURVE 14*		CURVE 18*		CURVE 22*		CURVE 26*		CURVE 30*		CURVE 34*	
T = 306		T = 306		T = 306		T = 306		T = 306		T = 306		T = 306		T = 306		T = 306	
λ = 0.56		λ = 2		λ = 12		λ = 0.56		λ = 2		λ = 12		λ = 0.56		λ = 2		λ = 12	
25	0.626	25	0.449	25	0.117	20	0.661	25	0.471	25	0.167	20	0.603	25	0.447	25	0.130
40	0.621	40	0.457	40	0.133	40	0.654	40	0.484	40	0.176	40	0.603	40	0.464	40	0.138
50	0.610	50	0.472	50	0.152	50	0.654	50	0.501	50	0.190	50	0.603	50	0.478	50	0.150
60	0.610	60	0.501	60	0.192	60	0.641	60	0.519	60	0.214	60	0.600	60	0.507	60	0.173
70	0.567	70	0.527	70	0.248	70	0.603	70	0.563	70	0.282	70	0.579	70	0.543	70	0.240
75	0.500	75	0.613	75	0.347	75	0.533	75	0.641	75	0.398	75	0.539	75	0.595	75	0.333
CURVE 3*		CURVE 7*		CURVE 11*		CURVE 15*		CURVE 19*		CURVE 23*		CURVE 27*		CURVE 31*		CURVE 35*	
T = 306		T = 306		T = 306		T = 306		T = 306		T = 306		T = 306		T = 306		T = 306	
λ = 0.72		λ = 4		λ = 16		λ = 0.72		λ = 4		λ = 16		λ = 0.72		λ = 4		λ = 16	
20	0.569	25	0.293	25	0.092	20	0.624	25	0.344	25	0.126	20	0.546	25	0.312	25	0.115
40	0.566	40	0.311	40	0.107	40	0.618	40	0.368	40	0.138	40	0.546	40	0.324	40	0.123
50	0.559	50	0.330	50	0.127	50	0.618	50	0.387	50	0.149	50	0.546	50	0.339	50	0.133
60	0.554	60	0.355	60	0.151	60	0.582	60	0.409	60	0.172	60	0.546	60	0.370	60	0.151
70	0.529	70	0.417	70	0.214	70	0.565	70	0.469	70	0.230	70	0.530	70	0.429	70	0.210
75	0.497	75	0.513	75	0.315	75	0.512	75	0.555	75	0.334	75	0.507	75	0.512	75	0.232
CURVE 4*		CURVE 8*		CURVE 12*		CURVE 16*		CURVE 20*		CURVE 24*		CURVE 28*		CURVE 32*		CURVE 36*	
T = 306		T = 306		T = 306		T = 306		T = 306		T = 306		T = 306		T = 306		T = 306	
λ = 1		λ = 6		λ = 20		λ = 1		λ = 6		λ = 20		λ = 1		λ = 6		λ = 20	
20	0.531	25	0.202	25	0.093	20	0.590	25	0.267	25	0.126	20	0.523	25	0.220	25	0.109
40	0.527	40	0.222	40	0.103	40	0.589	40	0.267	40	0.132	40	0.523	40	0.231	40	0.117
50	0.490	50	0.233	50	0.122	50	0.574	50	0.289	50	0.144	50	0.529	50	0.250	50	0.130
60	0.509	60	0.273	60	0.151	60	0.548	60	0.310	60	0.170	60	0.533	60	0.282	60	0.150
70	0.503	70	0.343	70	0.191	70	0.532	70	0.387	70	0.233	70	0.530	70	0.346	70	0.205
75	0.488	75	0.446	75	0.265	75	0.506	75	0.509	75	0.349	75	0.506	75	0.469	75	0.313

* Not shown on plot

DATA TABLE NO. 225 (continued)

θ'	∈	θ'	∈	θ'	∈
CURVE 37*		CURVE 41*		CURVE 45*	
T = 306		T = 306		T = 306	
λ = 0.43		λ = 1.67		λ = 8.2	
20	0.631	20	0.437	25	0.165
40	0.623	40	0.437	40	0.177
50	0.622	50	0.456	50	0.191
60	0.602	60	0.481	60	0.217
70	0.584	70	0.487	70	0.291
75	0.543	75	0.491	75	0.381
CURVE 38*		CURVE 42*		CURVE 46*	
T = 306		T = 306		T = 306	
λ = 0.56		λ = 2		λ = 12	
20	0.557	25	0.464	25	0.131
40	0.561	40	0.507	40	0.136
50	0.560	50	0.498	50	0.147
60	0.560	60	0.527	60	0.168
70	0.531	70	0.568	70	0.224
75	0.498	75	0.644	75	0.299
CURVE 39*		CURVE 43*		CURVE 47*	
T = 306		T = 306		T = 306	
λ = 0.72		λ = 4		λ = 16	
20	0.506	25	0.300	25	0.117
40	0.508	40	0.315	40	0.128
50	0.508	50	0.344	50	0.130
60	0.508	60	0.368	60	0.151
70	0.487	70	0.425	70	0.209
75	0.474	75	0.507	75	0.299
CURVE 40*		CURVE 44*		CURVE 48*	
T = 306		T = 306		T = 306	
λ = 1		λ = 6		λ = 20	
20	0.501	25	0.208	25	0.104
40	0.501	40	0.222	40	0.126
50	0.505	50	0.231	50	0.126
60	0.521	60	0.271	60	0.144
70	0.521	70	0.344	70	0.190
75	0.500	75	0.418	75	0.300

* Not shown on plot

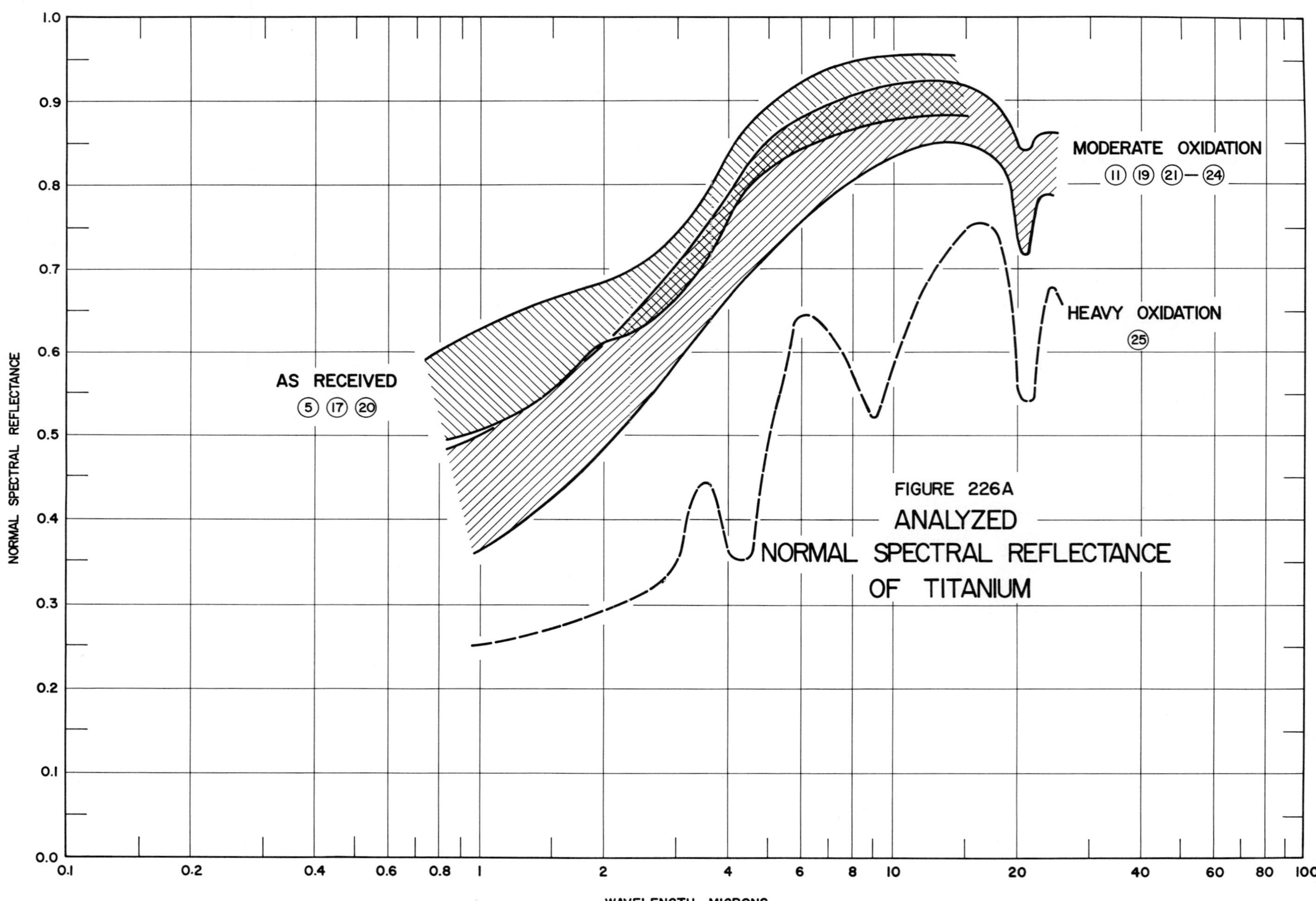

MODERATE OXIDATION
⑪ ⑲ ㉑ — ㉔
HEAVY OXIDATION
㉕
AS RECEIVED
⑤ ⑰ ⑳
FIGURE 226A
ANALYZED
NORMAL SPECTRAL REFLECTANCE
OF TITANIUM
NORMAL SPECTRAL REFLECTANCE
1.0
0.9
0.8
0.7
0.6
0.5
0.4
0.3
0.2
0.1
0.0
0.1
0.2
0.4
0.6
0.8
1
2
4
6
8
10
20
40
60
80
100
WAVELENGTH, MICRONS

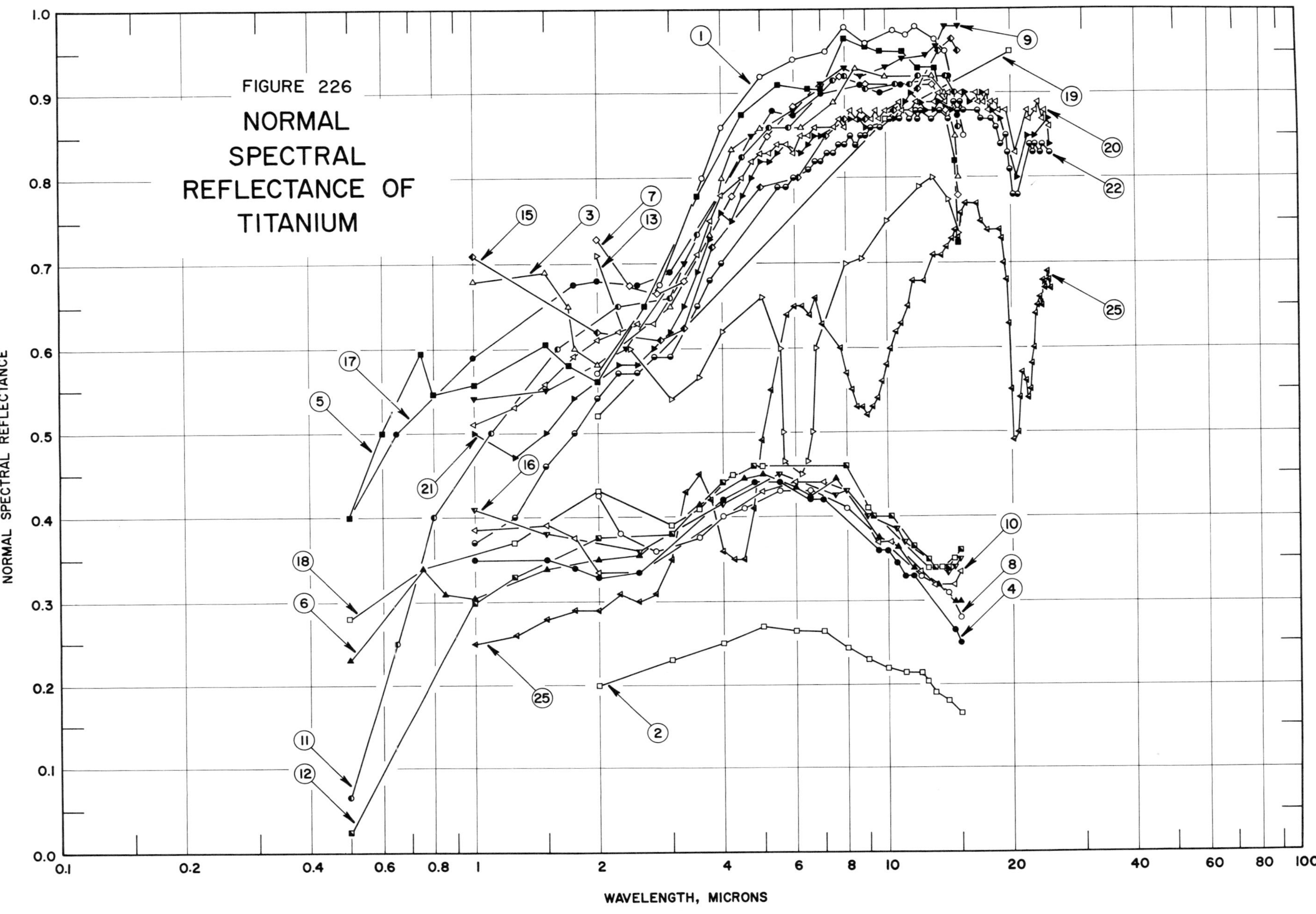

FIGURE 226
NORMAL SPECTRAL REFLECTANCE OF TITANIUM

SPECIFICATION TABLE NO. 226 NORMAL SPECTRAL REFLECTANCE OF TITANIUM

Curve No.	Ref. No.	Year	Temperature K	Wavelength Range, μ	Geometry θ θ'	Geometry ω'	Reported Error, %	Composition (weight percent), Specifications and Remarks
1	86	1961	~322	2.00-15.50	~0°	2π	<2	As received; data extracted from smooth curve; hohlraum at 523 K; converted from R $(2\pi, 0)$.
2	86	1961	~322	2.00-15.00	~0°	2π	<2	Above specimen and conditions; diffuse component only.
3	86	1961	~322	1.00-15.00	~0°	2π	<2	Different sample, same as curve 1 specimen and conditions except hohlraum at 773 K.
4	86	1961	~322	1.00-15.00	~0°	2π	<2	Above specimen and conditions; diffuse component only.
5	86	1961	~322	0.50-15.00	~0°	2π	<2	Different sample, same as curve 1 specimen and conditions except hohlraum at 1023 K.
6	86	1961	~322	0.50-15.00	~0°	2π	<2	Above specimen and conditions; diffuse component only.
7	86	1961	~322	2.00-15.00	~0°	2π	<2	Different sample, same as curve 1 specimen and conditions except heated in air at 1073 K for 30 min.
8	86	1961	~322	2.00-15.00	~0°	2π	<2	Above specimen and conditions; diffuse component only.
9	86	1961	~322	1.00-15.00	~0°	2π	<2	Different sample, same as curve 7 specimen and conditions except hohlraum at 773 K.
10	86	1961	~322	1.00-15.00	~0°	2π	<2	Above specimen and conditions; diffuse component only.
11	86	1961	~322	0.50-15.00	~0°	2π	<2	Different sample, same as curve 7 specimen and conditions except hohlraum at 1273 K.
12	86	1961	~322	0.50-15.00	~0°	2π	<2	Above specimen and conditions; diffuse component only.
13	86	1961	~322	2.00-15.00	~0°	2π	<2	Different sample, same as curve 1 specimen and conditions except heated in vacuum $(2.8 \times 10^{-5}$ mm Hg) at 1073 K for 30 min.
14	86	1961	~322	2.00-15.00	~0°	2π	<2	Above specimen and conditions; diffuse component only.
15	86	1961	~322	1.00-15.00	~0°	2π	<2	Different sample, same as curve 13 specimen and conditions except hohlraum at 773 K.
16	86	1961	~322	1.00-15.00	~0°	2π	<2	Above specimen and conditions; diffuse component only.
17	86	1961	~322	0.50-15.00	~0°	2π	<2	Different sample, same as curve 13 specimen and conditions except hohlraum at 1273 K.
18	86	1961	~322	0.50-15.00	~0°	2π	<2	Above specimen and conditions; diffuse component only.
19	128	1962	298	2.0-20.0	~0°	2π		Film (0.15 μ thick); evaporated on aluminum foil; oxidized at 672 K for 3 hrs in air; cavity at 1033 K; $\omega = 0.034$ Sr.; converted from $R(2\pi, 0)$.
20	99	1958	311	1.00-25.00	~0°	2π		Ti-75A; AMS 4901; nominal composition: 0.1 Fe, 0.02 N_2, 0.04 max C, 0.08 max W, Ti balance; hohlraum at 1089 K; converted from $R(2\pi, 0)$.
21	99	1958	311	1.00-25.00	~0°	2π		Different sample, same as curve 20 specimen and conditions except heated at 580.5 K for 303 hrs.
22	99	1958	311	1.00-25.00	~0°	2π		Different sample, same as curve 20 specimen and conditions except heated at 705.5 K for 100 hrs.
23	99	1958	311	1.00-25.00	~0°	2π		Different sample, same as curve 20 specimen and conditions except heated at 711 K for 306 hrs.

SPECIFICATION TABLE NO. 226 (continued)

Curve No.	Ref. No.	Year	Temperature K	Wavelength Range, μ	Geometry θ	θ'	ω'	Reported Error, %	Composition (weight percent), Specifications and Remarks
24	99	1958	311	1.00-25.00	$\sim 0^\circ$		2π		Different sample, same as curve 20 specimen and conditions except heated at 739.5 K for 303 hrs.
25	99	1950	311	1.00-25.00	$\sim 0^\circ$		2π		Different sample, same as curve 20 specimen and conditions except heated at 813 K for 303 hrs.

DATA TABLE NO. 226 NORMAL SPECTRAL REFLECTANCE OF TITANIUM

[Wavelength, λ, μ; Reflectance, ρ; Temperature, T, K]

λ	ρ
CURVE 1 T = ~322	
2.00	0.570
2.85	0.675
3.60	0.800
4.00	0.860
5.00	0.920
6.00	0.940
7.20	0.950
8.00	0.980
9.00	0.960
10.50	0.975
11.20	0.970
11.80	0.980
13.25	0.965
14.00	0.950
15.50	0.850
CURVE 2 T = ~322	
2.00	0.200
3.00	0.230
4.00	0.250
5.00	0.270
6.00	0.265
7.00	0.265
8.00	0.245
9.00	0.230
10.00	0.220
11.00	0.215
12.00	0.215
12.50	0.205*
13.00	0.190
14.00	0.180
15.00	0.165
CURVE 3 T = ~322	
1.00	0.680
1.50	0.690
1.70	0.650
1.75	0.600
2.00	0.580
CURVE 3 (cont.)	
3.00	0.650
3.75	0.735
4.00	0.800
4.25	0.835
5.00	0.860
6.25	0.860
7.50	0.890
8.50	0.930
10.00	0.920
11.50	0.920*
13.00	0.920
14.00	0.900
14.80	0.850
15.00	0.800
CURVE 4 T = ~322	
1.00	0.350
1.50	0.350
1.75	0.340
2.00	0.330
2.50	0.335
4.00	0.420
4.75	0.440
5.50	0.440
6.50	0.420
7.00	0.420
9.50	0.360
10.00	0.360
10.50	0.345
11.00	0.330
11.50	0.330
14.50	0.265
15.00	0.250
CURVE 5 T = ~322	
0.50	0.400
0.60	0.500
0.75	0.595
0.80	0.545
1.00	0.555
CURVE 5 (cont.)	
1.50	0.605
1.70	0.580
2.00	0.560
2.60	0.650
3.50	0.780
4.50	0.875
5.50	0.910
6.50	0.905
7.00	0.905
8.00	0.965
9.00	0.955
9.75	0.950
11.00	0.950
12.00	0.930
13.25	0.930
14.00	0.900*
14.70	0.820
15.00	0.725
CURVE 6 T = ~322	
0.50	0.230
0.75	0.340
0.85	0.310
1.00	0.305
1.50	0.340
2.00	0.350
2.50	0.355
3.50	0.415
4.50	0.445
5.00	0.450
6.00	0.435
6.50	0.425
7.50	0.445
9.50	0.375
10.00	0.370*
11.50	0.340
13.25	0.320
14.50	0.300
15.00	0.300
CURVE 7 T = ~322	
2.00	0.730
2.40	0.675
2.80	0.665
3.25	0.680
4.25	0.780
5.20	0.850
6.00	0.885
7.00	0.900*
7.80	0.920
9.00	0.910
11.00	0.910*
13.00	0.910
14.00	0.880
14.50	0.850*
15.00	0.780
CURVE 8 T = ~322	
2.00	0.425
2.25	0.380
2.75	0.360
3.50	0.375
4.00	0.400
4.50	0.410
5.50	0.430
6.50	0.430
8.00	0.410
9.50	0.375*
10.50	0.365*
12.00	0.330
14.00	0.310
14.50	0.300*
15.00	0.280
CURVE 9 T = ~322	
1.00	0.540
1.50	0.550
2.35	0.600
3.25	0.700
4.00	0.800*
CURVE 9 (cont.)	
4.75	0.850
6.00	0.880
7.00	0.910
8.00	0.930
8.75	0.920
10.00	0.930
11.00	0.940
12.50	0.945
13.35	0.955
14.00	0.980
15.00	0.980
CURVE 10 T = ~322	
1.00	0.385
1.50	0.390
1.75	0.375
2.00	0.335
2.50	0.335*
4.00	0.400*
5.00	0.430
6.00	0.440
7.00	0.440
8.00	0.430*
9.50	0.370
10.25	0.370
12.00	0.335
13.00	0.320
14.50	0.320
15.00	0.335
CURVE 11 T = ~322	
0.50	0.065
0.65	0.250
0.80	0.400
1.10	0.500
1.60	0.600
2.25	0.650
3.00	0.660
3.50	0.735
4.50	0.825
CURVE 11 (cont.)	
5.25	0.860
5.85	0.860
7.50	0.915
8.00	0.920
9.00	0.905
10.50	0.910
11.50	0.910
12.00	0.920
13.00	0.920*
14.00	0.920
14.70	0.900
15.00	0.860
CURVE 12 T = ~322	
0.50	0.025
1.00	0.300
1.25	0.330
2.00	0.375
3.00	0.380
4.00	0.440
4.75	0.460
8.00	0.460
9.25	0.400
10.25	0.400
11.50	0.365
12.50	0.350
13.00	0.340
14.00	0.340
15.00	0.360
CURVE 13 T = ~322	
2.00	0.710
2.40	0.600
3.00	0.540
3.50	0.565
4.00	0.620
5.00	0.660
5.55	0.600
5.60	0.500
5.65	0.465
CURVE 13 (cont.)	
6.10	0.450
6.40	0.465
6.60	0.500
6.75	0.600
8.00	0.700
8.75	0.705
10.00	0.750
12.00	0.790
13.00	0.800
14.20	0.775
15.00	0.735
CURVE 14* T = ~322	
2.00	0.445
3.00	0.375
4.00	0.405
5.00	0.435
6.00	0.440
7.00	0.440
9.00	0.390
13.00	0.320
14.00	0.310
15.00	0.280
CURVE 15 T = ~322	
1.00	0.710
2.00	0.620
2.85	0.610
3.25	0.625
3.80	0.720
5.00	0.790
6.15	0.800
7.20	0.850
8.00	0.870
9.00	0.870
10.50	0.880
12.00	0.905
13.65	0.950
14.50	0.965
15.00	0.950
CURVE 16 T = ~322	
1.00	0.410
1.50	0.380
2.50	0.360
4.00	0.415
5.50	0.450
6.25	0.450*
7.50	0.425
8.00	0.430
9.00	0.400
10.50	0.385
11.00	0.370
13.00	0.340*
14.00	0.335
14.50	0.340
15.00	0.350
CURVE 17 T = ~322	
0.50	0.400*
0.65	0.500
1.00	0.590
1.75	0.675
2.00	0.680
2.50	0.675
3.00	0.690
4.00	0.800*
5.35	0.880
6.00	0.875
7.00	0.900
8.00	0.920*
8.75	0.910
9.75	0.900
10.90	0.910
12.00	0.920*
12.90	0.910*
14.25	0.920
14.75	0.900*
15.00	0.875

* Not shown on plot

DATA TABLE NO. 226 (continued)

λ	ρ
CURVE 18	**T = ~322**
0.50	0.280
0.75	0.340*
1.25	0.370
1.50	0.380*
2.00	0.430
3.00	0.390
3.50	0.410
4.25	0.450
5.00	0.460
8.00	0.460*
9.00	0.410
12.50	0.340
13.50	0.340
14.50	0.350
15.00	0.350*
CURVE 19	**T = 298**
2.0	0.52
10.0	0.87
20.0	0.95
CURVE 20	**T = 311**
1.00	0.51
1.25	0.53
1.50	0.55
1.75	0.59
2.00	0.61
2.25	0.62
2.50	0.63
2.75	0.63
3.00	0.65*
3.25	0.68*
3.50	0.71
3.75	0.75
4.00	0.78
4.25	0.78*
4.50	0.80
4.75	0.82
5.00	0.83
5.25	0.83
5.50	0.84
CURVE 20 (cont.)	
5.75	0.84
6.00	0.83
6.25	0.85
6.50	0.85
6.75	0.86
7.00	0.86*
7.25	0.86*
7.50	0.86
7.75	0.87
8.00	0.86
8.25	0.88
8.50	0.87
8.75	0.88
9.00	0.87*
9.25	0.87
9.50	0.88
9.75	0.87
10.00	0.88
10.25	0.88*
10.50	0.88*
10.75	0.89
11.00	0.88
11.25	0.88*
11.50	0.88
11.75	0.89
12.00	0.89*
12.25	0.89*
12.50	0.89*
12.75	0.89*
13.00	0.89*
13.25	0.89
13.50	0.90
13.75	0.89*
14.00	0.89
14.25	0.90
14.50	0.90
14.75	0.90
15.00	0.89
15.25	0.90
15.50	0.90*
15.75	0.90*
16.00	0.90*
16.25	0.90*
16.50	0.90
16.75	0.89
17.00	0.89*
CURVE 20 (cont.)	
17.25	0.89
17.50	0.90
17.75	0.88
18.00	0.89
18.25	0.89*
18.50	0.89
18.75	0.88
19.00	0.88*
19.25	0.88*
19.50	0.88
19.75	0.86*
20.00	0.85*
20.25	0.84*
20.50	0.83
20.75	0.84*
21.00	0.85*
21.25	0.87*
21.50	0.88*
21.75	0.88*
22.00	0.88
22.25	0.87
22.50	0.87*
22.75	0.87*
23.00	0.88
23.25	0.89
23.50	0.89*
23.75	0.87*
24.00	0.87
24.25	0.88
24.50	0.87*
24.75	0.87*
25.00	0.86
CURVE 21	**T = 311**
1.00	0.50
1.25	0.47
1.50	0.50
1.75	0.54
2.00	0.56*
2.25	0.58
2.50	0.58
2.75	0.60
3.00	0.62
3.25	0.65
CURVE 21 (cont.)	
3.50	0.69
3.75	0.73
4.00	0.76
4.25	0.75
4.50	0.78
4.75	0.80
5.00	0.82
5.25	0.82
5.50	0.83
5.75	0.84*
6.00	0.83*
6.25	0.83
6.50	0.84
6.75	0.85
7.00	0.85
7.25	0.85*
7.50	0.86*
7.75	0.87*
8.00	0.86*
8.25	0.87
8.50	0.87*
8.75	0.87
9.00	0.86
9.25	0.87*
9.50	0.87*
9.75	0.87*
10.00	0.88*
10.25	0.87*
10.50	0.87
10.75	0.88*
11.00	0.89*
11.25	0.89
11.50	0.90
11.75	0.89*
12.00	0.89*
12.25	0.89
12.50	0.88
12.75	0.88*
13.00	0.88*
13.25	0.88
13.50	0.89
13.75	0.89*
14.00	0.89*
14.25	0.89
14.50	0.88
14.75	0.89*
CURVE 21 (cont.)	
15.00	0.89*
15.25	0.90*
15.50	0.90
15.75	0.89*
16.00	0.89
16.25	0.89*
16.50	0.90*
16.75	0.90*
17.00	0.89*
17.25	0.90
17.50	0.89*
17.75	0.89*
18.00	0.88*
18.25	0.88
18.50	0.88*
18.75	0.88*
19.00	0.87
19.25	0.87*
19.50	0.86*
19.75	0.85*
20.00	0.83*
20.25	0.80*
20.50	0.80*
20.75	0.80
21.00	0.82*
21.25	0.83*
21.50	0.84*
21.75	0.84*
22.00	0.85
22.25	0.85*
22.50	0.85*
22.75	0.85*
23.00	0.85
23.25	0.86*
23.50	0.86*
23.75	0.85*
24.00	0.86*
24.25	0.85*
24.50	0.86*
24.75	0.87
25.00	0.84
CURVE 22	**T = 311**
1.00	0.37
CURVE 22 (cont.)	
1.25	0.40
1.50	0.46
1.75	0.50
2.00	0.54
2.25	0.57
2.50	0.57
2.75	0.59
3.00	0.59
3.25	0.62*
3.50	0.65
3.75	0.68
4.00	0.70
4.25	0.71*
4.50	0.73*
4.75	0.75*
5.00	0.76*
5.25	0.77*
5.50	0.79
5.75	0.79
6.00	0.80
6.25	0.80*
6.50	0.81
6.75	0.82
7.00	0.82
7.25	0.83
7.50	0.83
7.75	0.84
8.00	0.84
8.25	0.85
8.50	0.84
8.75	0.85
9.00	0.85
9.25	0.86
9.50	0.86*
9.75	0.86
10.00	0.87*
10.25	0.87
10.50	0.87*
10.75	0.87
11.00	0.88*
11.25	0.87*
11.50	0.87
11.75	0.88
12.00	0.87
12.25	0.88
12.50	0.88*
CURVE 22 (cont.)	
12.75	0.87*
13.00	0.87
13.25	0.88*
13.50	0.88
13.75	0.88*
14.00	0.88*
14.25	0.87
14.50	0.88*
14.75	0.89
15.00	0.88
15.25	0.89
15.50	0.88
15.75	0.88*
16.00	0.88*
16.25	0.88*
16.50	0.88*
16.75	0.88
17.00	0.87
17.25	0.87*
17.50	0.87*
17.75	0.87*
18.00	0.87
18.25	0.86*
18.50	0.86
18.75	0.85*
19.00	0.84
19.25	0.85*
19.50	0.85
19.75	0.83*
20.00	0.81*
20.25	0.78
20.50	0.78*
20.75	0.78
21.00	0.79*
21.25	0.81*
21.50	0.82*
21.75	0.83*
22.00	0.83*
22.25	0.84
22.50	0.83*
22.75	0.83
23.00	0.84
23.25	0.83*
23.50	0.83*
23.75	0.83
24.00	0.84
CURVE 22 (cont.)	
24.25	0.83
24.50	0.83*
24.75	0.83*
25.00	0.83
CURVE 23*	**T = 311**
1.00	0.37
1.25	0.35
1.50	0.42
1.75	0.47
2.00	0.51
2.25	0.55
2.50	0.56
2.75	0.57
3.00	0.59
3.25	0.61
3.50	0.63
3.75	0.66
4.00	0.69
4.25	0.70
4.50	0.71
4.75	0.73
5.00	0.75
5.25	0.76
5.50	0.77
5.75	0.77
6.00	0.78
6.25	0.78
6.50	0.79
6.75	0.79
7.00	0.79
7.25	0.80
7.50	0.81
7.75	0.81
8.00	0.81
8.25	0.82
8.50	0.82
8.75	0.81
9.00	0.82
9.25	0.83
9.50	0.83
9.75	0.84
10.00	0.84
10.25	0.85

* Not shown on plot

DATA TABLE NO. 226 (continued)

λ	ρ	λ	ρ	λ	ρ	λ	ρ	λ	ρ	λ	ρ
CURVE 23 (cont.)*		CURVE 23 (cont.)*		CURVE 24 (cont.)*		CURVE 24 (cont.)*		CURVE 25 (cont.)		CURVE 25 (cont.)	
10.50	0.84	22.00	0.83	8.25	0.78	19.75	0.77	6.00	0.65	17.50	0.74
10.75	0.85	22.25	0.82	8.50	0.78	20.00	0.73	6.25	0.65	17.75	0.74*
11.00	0.85	22.50	0.83	8.75	0.79	20.25	0.69	6.50	0.64	18.00	0.74*
11.25	0.86	22.75	0.84	9.00	0.80	20.50	0.68	6.75	0.66	18.25	0.74*
11.50	0.86	23.00	0.85	9.25	0.79	20.75	0.70	7.00	0.63	18.50	0.74*
11.75	0.85	23.25	0.84	9.50	0.79	21.00	0.71	7.25	0.62*	18.75	0.74
12.00	0.86	23.50	0.83	9.75	0.80	21.25	0.74	7.50	0.61*	19.00	0.73
12.25	0.86	23.75	0.83	10.00	0.81	21.50	0.76	7.75	0.60	19.25	0.70
12.50	0.85	24.00	0.84	10.25	0.81	21.75	0.76	8.00	0.57	19.50	0.68
12.75	0.85	24.25	0.85	10.50	0.82	22.00	0.77	8.25	0.55	19.75	0.63
13.00	0.86	24.50	0.83	10.75	0.82	22.25	0.78	8.50	0.53	20.00	0.55
13.25	0.86	24.75	0.82	11.00	0.81	22.50	0.77	8.75	0.53	20.25	0.49
13.50	0.86	25.00	0.80	11.25	0.82	22.75	0.78	9.00	0.52	20.50	0.49*
13.75	0.86			11.50	0.82	23.00	0.79	9.25	0.53	20.75	0.50
14.00	0.86	CURVE 24*		11.75	0.83	23.25	0.78	9.50	0.54	21.00	0.54
14.25	0.86	T = 311		12.00	0.81	23.50	0.79	9.75	0.56	21.25	0.57
14.50	0.86			12.25	0.83	23.75	0.78	10.00	0.58	21.50	0.57*
14.75	0.86	1.00	0.31	12.50	0.83	24.00	0.78	10.25	0.60	21.75	0.56
15.00	0.86	1.25	0.28	12.75	0.83	24.25	0.80	10.50	0.62	22.00	0.54
15.25	0.87	1.50	0.20	13.00	0.83	24.50	0.78	10.75	0.63	22.25	0.55
15.50	0.87	1.75	0.16	13.25	0.83	24.75	0.78	11.00	0.65*	22.50	0.58
15.75	0.87	2.00	0.17	13.50	0.84	25.00	0.75	11.25	0.65	22.75	0.60
16.00	0.88	2.25	0.22	13.75	0.84			11.50	0.68	23.00	0.64
16.25	0.88	2.50	0.26	14.00	0.84	CURVE 25		11.75	0.68*	23.25	0.65
16.50	0.87	2.75	0.30	14.25	0.81	T = 311		12.00	0.68*	23.50	0.66
16.75	0.87	3.00	0.33	14.50	0.82			12.25	0.68	23.75	0.65
17.00	0.87	3.25	0.38	14.75	0.84	1.00	0.25	12.50	0.69*	24.00	0.68
17.25	0.87	3.50	0.43	15.00	0.83	1.25	0.26	12.75	0.70*	24.25	0.67
17.50	0.87	3.75	0.49	15.25	0.84	1.50	0.28	13.00	0.71	24.50	0.69
17.75	0.86	4.00	0.53	15.50	0.84	1.75	0.29	13.25	0.71*	24.75	0.68
18.00	0.86	4.25	0.56	15.75	0.85	2.00	0.29	13.50	0.71	25.00	0.67
18.25	0.86	4.50	0.58	16.00	0.85	2.25	0.31	13.75	0.72*		
18.50	0.86	4.75	0.62	16.25	0.85	2.50	0.30	14.00	0.72		
18.75	0.86	5.00	0.64	16.50	0.84	2.75	0.31	14.25	0.73*		
19.00	0.85	5.25	0.66	16.75	0.85	3.00	0.35	14.50	0.73		
19.25	0.84	5.50	0.67	17.00	0.84	3.25	0.43	14.75	0.74		
19.50	0.84	5.75	0.68	17.25	0.84	3.50	0.45	15.00	0.73		
19.75	0.83	6.00	0.70	17.50	0.83	3.75	0.42	15.25	0.76		
20.00	0.80	6.25	0.71	17.75	0.84	4.00	0.36	15.50	0.77		
20.25	0.78	6.50	0.73	18.00	0.83	4.25	0.35	15.75	0.77*		
20.50	0.77	6.75	0.74	18.25	0.83	4.50	0.35	16.00	0.77*		
20.75	0.78	7.00	0.73	18.50	0.82	4.75	0.41	16.25	0.77*		
21.00	0.79	7.25	0.75	18.75	0.82	5.00	0.49	16.50	0.77		
21.25	0.80	7.50	0.76	19.00	0.81	5.25	0.55	16.75	0.76*		
21.50	0.82	7.75	0.77	19.25	0.81	5.50	0.60	17.00	0.75		
21.75	0.82	8.00	0.78	19.50	0.79	5.75	0.64	17.25	0.75*		

* Not shown on plot

FIGURE 227A(1)

ANGULAR SPECTRAL REFLECTANCE OF TITANIUM

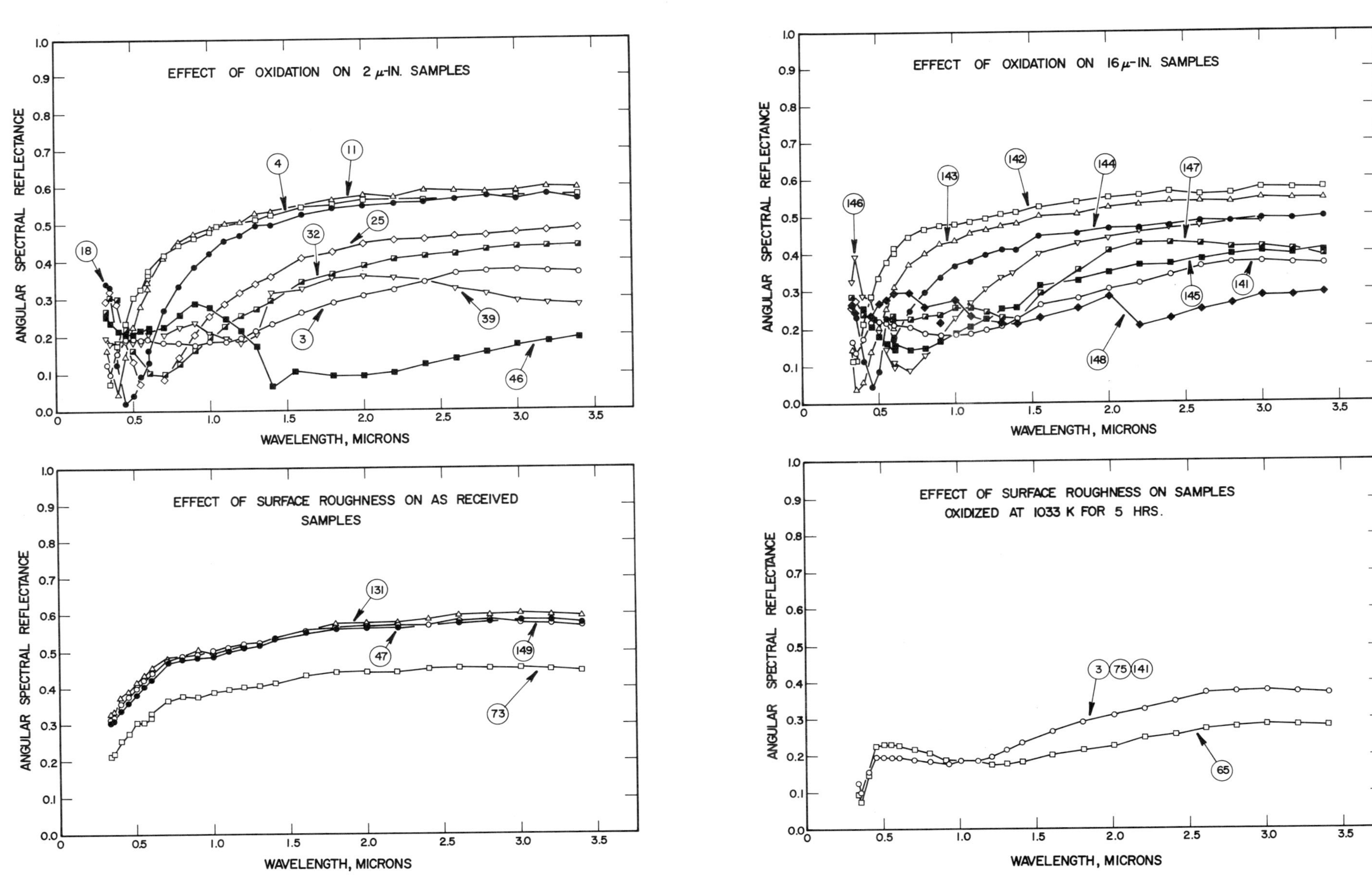

FIGURE 227A(2)

ANGULAR SPECTRAL REFLECTANCE OF TITANIUM

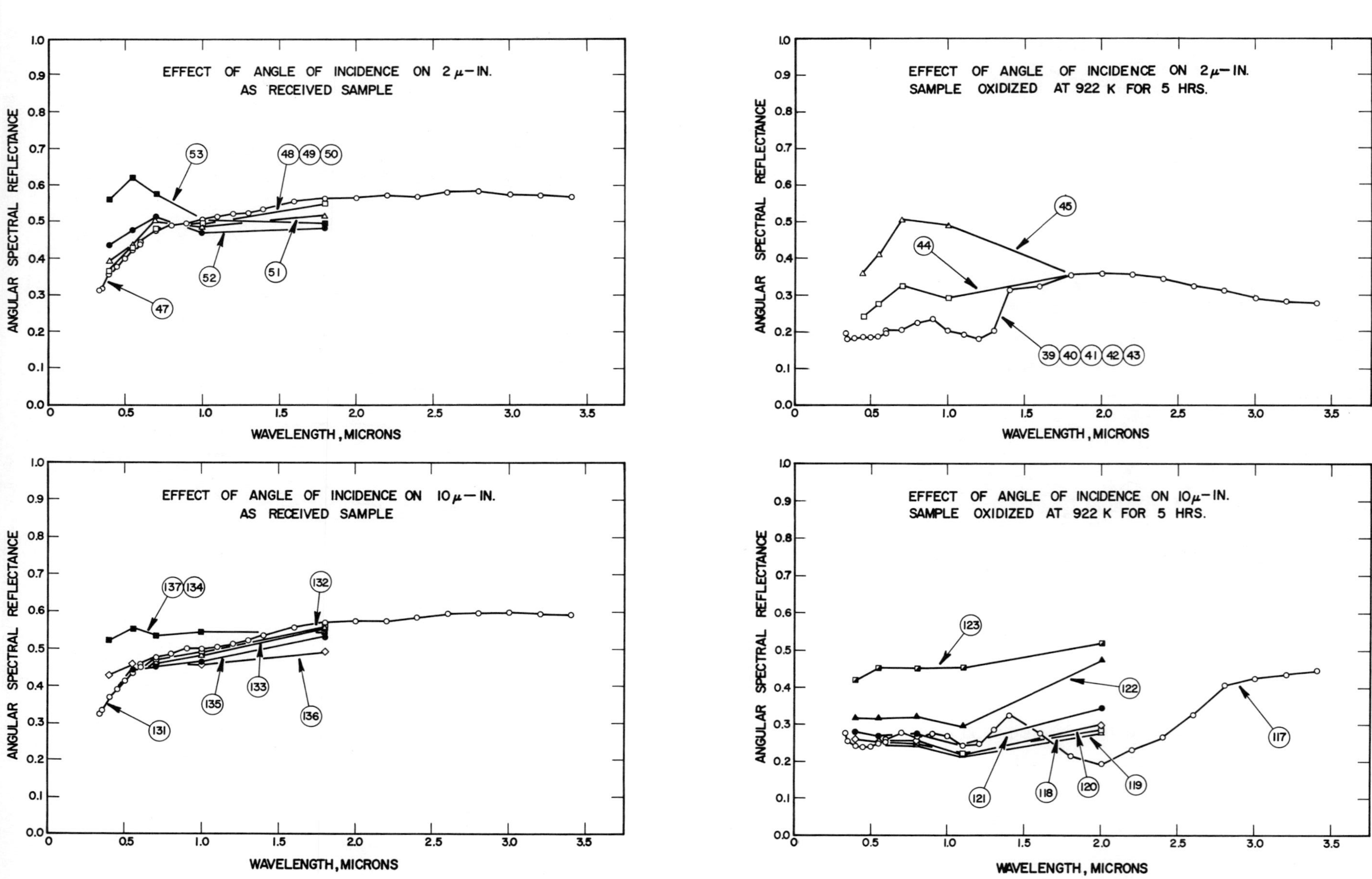

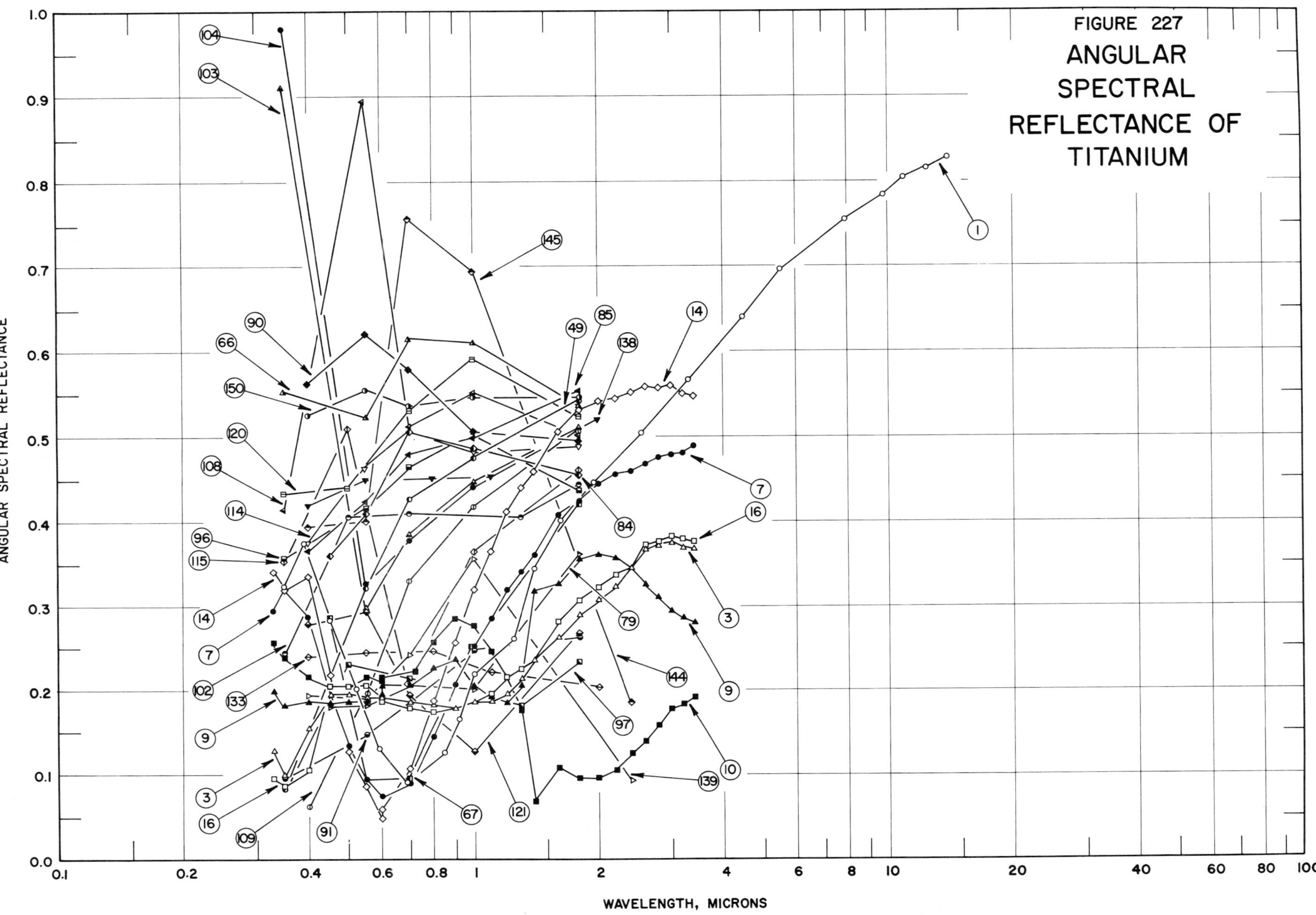

FIGURE 227 ANGULAR SPECTRAL REFLECTANCE OF TITANIUM

SPECIFICATION TABLE NO. 227 ANGULAR SPECTRAL REFLECTANCE OF TITANIUM

Curve No.	Ref. No.	Year	Temperature K	Wavelength Range, μ	Geometry θ	θ'	ω'	Reported Error, %	Composition (weight percent), Specifications and Remarks
1	219	1965	306	0.35-14.00	25°		2π		Titanium; oxidized for 5 hrs at 811 K; thickness of oxide layer 0.05μ (measured by weight increase); surface roughness 16 microinches; data extracted from smooth curve.
2	219	1965	306	0.35-14.00	25°		2π		Titanium; oxidized for 5 hrs at 755 K; thickness of oxide layer 0.07μ (measured by weight increase); surface roughness 16 microinches; data extracted from smooth curve.
3	270	1962	298	0.33-3.4	20°		2π		Commercially pure from Crucible Steel Co.; honed and lapped; surface roughness before oxidation 2μ in. RMS; oxidized by heating in air at 1033 K for 5 hrs; oxide thickness by weight difference 3.62μ; gray color after oxidation; MgO reference; [Author's designation: Specimen No. 1-1].
4	270	1962	298	0.33-3.4	20°		2π		Commercially pure from Crucible Steel Co.; honed and lapped; surface roughness before oxidation 2μ in. RMS; oxidized by heating in air at 589 K for 5 hrs; oxide thickness by weight difference 0.02 μ; light gold color after oxidation; MgO reference; [Author's designation: Specimen No. 1-2]
5	270	1962	298	0.35-1.8	30°		2π		Above specimen and conditions.
6	270	1962	298	0.35-1.8	40°		2π		Above specimen and conditions.
7	270	1962	298	0.35-1.8	50°		2π		Above specimen and conditions.
8	270	1962	298	0.35-1.8	60°		2π		Above specimen and conditions.
9	270	1962	298	0.35-1.8	70°		2π		Above specimen and conditions.
10	270	1962	298	0.35-1.8	80°		2π		Above specimen and conditions.
11	270	1962	298	0.33-3.4	20°		2π		Commercially pure from Crucible Steel Co.; honed and lapped; surface roughness before oxidation 2μ in. RMS; oxidized by heating in air at 644 K for 5 hrs; light yellow color after oxidation; MgO reference; [Author's designation: Specimen No. 1-3]
12	270	1962	298	0.35-1.8	30°		2π		Above specimen and conditions.
13	270	1962	298	0.35-1.8	40°		2π		Above specimen and conditions.
14	270	1962	298	0.35-1.8	50°		2π		Above specimen and conditions.
15	270	1962	298	0.35-1.8	60°		2π		Above specimen and conditions.
16	270	1962	298	0.35-1.8	70°		2π		Above specimen and conditions.
17	270	1962	298	0.35-1.8	80°		2π		Above specimen and conditions.
18	270	1962	298	0.33-3.4	20°		2π		Commercially pure from Crucible Steel Co.; honed and lapped; surface roughness before oxidation 2μ in. RMS; oxidized by heating in air at 700 K for 5 hrs; oxide thickness by weight difference 0.02μ; gold-purple color after oxidation; MgO reference; [Author's designation: Specimen No. 1-4]
19	270	1962	298	0.35-1.8	30°		2π		Above specimen and conditions.
20	270	1962	298	0.35-1.8	40°		2π		Above specimen and conditions.

SPECIFICATION TABLE NO. 227 (continued)

Curve No.	Ref. No.	Year	Temperature K	Wavelength Range, μ	Geometry θ	θ'	ω	Reported Error, %	Composition (weight percent), Specifications and Remarks
21	270	1962	298	0.35-1.8	50°		2π		Above specimen and conditions.
22	270	1962	298	0.35-1.8	60°		2π		Above specimen and conditions.
23	270	1962	298	0.35-1.8	70°		2π		Above specimen and conditions.
24	270	1962	298	0.35-1.8	80°		2π		Above specimen and conditions.
25	270	1962	298	0.33-3.4	20°		2π		Commercially pure from Crucible Steel Co.; honed and lapped; surface roughness before oxidation 2μ in. RMS; oxidized by heating in air at 755 K for 5 hrs; oxide thickness by weight difference 0.03 μ; dark blue color after oxidation; MgO reference; [Author's designation: Specimen No. 1-5].
26	270	1962	298	0.45-1.8	30°		2π		Above specimen and conditions.
27	270	1962	298	0.45-1.8	40°		2π		Above specimen and conditions.
28	270	1962	298	0.45-1.8	50°		2π		Above specimen and conditions.
29	270	1962	298	0.45-1.8	60°		2π		Above specimen and conditions.
30	270	1962	298	0.45-1.8	70°		2π		Above specimen and conditions.
31	270	1962	298	0.45-1.8	80°		2π		Above specimen and conditions.
32	270	1962	298	0.33-3.4	20°		2π		Commercially pure from Crucible Steel Co.; honed and lapped; surface roughness before oxidation 2μ in. RMS; oxidized by heating in air at 811 K for 5 hrs; oxide thickness by weight difference 0.12μ; medium blue color after oxidation; MgO reference; [Author's designation: Specimen No. 1-6].
33	270	1962	298	0.35-1.8	30°		2π		Above specimen and conditions.
34	270	1962	298	0.35-1.8	40°		2π		Above specimen and conditions.
35	270	1962	298	0.35-1.8	50°		2π		Above specimen and conditions.
36	270	1962	298	0.35-1.8	60°		2π		Above specimen and conditions.
37	270	1962	298	0.35-1.8	70°		2π		Above specimen and conditions.
38	270	1962	298	0.35-1.8	80°		2π		Above specimen and conditions.
39	270	1962	298	0.33-3.4	20°		2π		Commercially pure from Crucible Steel Co.; honed and lapped; surface roughness before oxidation 2μ in. RMS; oxidized by heating in air at 922 K for 5 hrs; oxide thickness by weight difference 0.53μ; gray color after oxidation; MgO reference; [Author's designation: Specimen No. 1-7].
40	270	1962	298	0.45-1.8	30°		2π		Above specimen and conditions.
41	270	1962	298	0.45-1.8	40°		2π		Above specimen and conditions.
42	270	1962	298	0.45-1.8	50°		2π		Above specimen and conditions.
43	270	1962	298	0.45-1.8	60°		2π		Above specimen and conditions.
44	270	1962	298	0.45-1.8	70°		2π		Above specimen and conditions.

SPECIFICATION TABLE NO. 227 (continued)

Curve No.	Ref. No.	Year	Temperature K	Wavelength Range, μ	Geometry θ	θ' ω'	Reported Error, %	Composition (weight percent), Specifications and Remarks
45	270	1962	298	0.45-1.8	80°	2π		Above specimen and conditions.
46	270	1962	298	0.33-3.4	20°	2π		Commercially pure from Crucible Steel Co.; honed and lapped; surface roughness before oxidation 2μ in. RMS; oxidized by heating in air at 866 K for 5 hrs; oxide thickness by weight difference 0.20μ; gray color after oxidation; MgO reference; [Author's designation: Specimen No. 1-8].
47	270	1962	298	0.33-3.4	20°	2π		Commercially pure from Crucible Steel Co.; honed and lapped; surface roughness 2μ in. RMS; silver color; MgO reference; [Author's designation: Specimen No. 1-9].
48	270	1962	298	0.4-1.8	30°	2π		Above specimen and conditions.
49	270	1962	298	0.4-1.8	40°	2π		Above specimen and conditions.
50	270	1962	298	0.4-1.8	50°	2π		Above specimen and conditions.
51	270	1962	298	0.4-1.8	60°	2π		Above specimen and conditions.
52	270	1962	298	0.4-1.8	70°	2π		Above specimen and conditions.
53	270	1962	298	0.4-1.8	80°	2π		Above specimen and conditions.
54	270	1962	298	0.33-3.4	20°	2π		Commercially pure from Crucible Steel Co.; honed and lapped; surface roughness before oxidation 2μ in. RMS; oxidized by heating in air at 978 K for 5 hrs; oxide thickness by weight difference 1.37μ; gray color after oxidation; MgO reference; [Author's designation: Specimen No. 1-10].
55	270	1962	298	0.35-1.8	30°	2π		Above specimen and conditions.
56	270	1962	298	0.35-1.8	40°	2π		Above specimen and conditions.
57	270	1962	298	0.35-1.8	50°	2π		Above specimen and conditions.
58	270	1962	298	0.35-1.8	60°	2π		Above specimen and conditions.
59	270	1962	298	0.35-1.8	70°	2π		Above specimen and conditions.
60	270	1962	298	0.35-1.8	80°	2π		Above specimen and conditions.
61	270	1962	298	0.33-3.4	20°	2π		Commercially pure from Crucible Steel Co.; honed and lapped; surface roughness before oxidation 2 μ in. RMS; oxidized by heating in air at 700 K for 5 hrs; gold-purple color after oxidation; MgO reference; [Author's designation: Specimen No. 1-11].
62	270	1962	298	0.33-3.4	20°	2π		Commercially pure from Crucible Steel Co.; honed and lapped; surface roughness before oxidation 2μ in. RMS; oxidized by heating in air at 755 K for 5 hrs; oxide thickness by weight difference 0.04μ; dark blue color after oxidation; MgO reference; [Author's designation: Specimen No. 1-12].
63	270	1962	298	0.33-3.4	20°	2π		Commercially pure from Crucible Steel Co.; honed and lapped; surface roughness before oxidation 2 μ in. RMS; oxidized by heating in air at 811 K for 5.15 hrs; oxide thickness by weight difference 0.13μ; light blue color after oxidation; MgO reference; [Author's designation: Specimen No. 1-13].

SPECIFICATION TABLE NO. 227 (continued)

Curve No.	Ref. No.	Year	Temperature K	Wavelength Range, μ	Geometry θ	θ' ω'	Reported Error, %	Composition (weight percent), Specifications and Remarks
64	270	1962	298	0.33-3.4	20°	2π		Commercially pure from Crucible Steel Co.; honed and lapped; surface roughness before oxidation 2μ in. RMS; oxidized by heating in air at 1033 K for 5 hrs; oxide thickness by weight difference 3.48μ; gray color after oxidation; MgO reference; [Author's designation: Specimen No. 1-19].
65	270	1962	298	0.33-3.4	20°	2π		Commercially pure from Crucible Steel Co.; honed and lapped; surface roughness before oxidation 4μ in. RMS; oxidized by heating in air at 1033 K for 5 hrs; oxide thickness by weight difference 4.65μ; gray color after oxidation; MgO reference; [Author's designation: Specimen No. 2-1].
66	270	1962	298	0.33-3.4	20°	2π		Commercially pure from Crucible Steel Co.; honed and lapped; surface roughness before oxidation 4μ in. RMS; oxidized by heating in air at 589 K for 5 hrs; oxide thickness by weight difference 0.03μ; gold color after oxidation; MgO reference; [Author's designation: Specimen No. 2-2].
67	270	1962	298	0.33-3.4	20°	2π		Commercially pure from Crucible Steel Co.; honed and lapped; surface roughness before oxidation 4μ in. RMS; oxidized by heating in air at 644 K for 0.03 hrs; oxide thickness by weight difference 0.03μ; deep gold color after oxidation; MgO reference; [Author's designation: Specimen No. 2-3].
68	270	1962	298	0.33-3.4	20°	2π		Commercially pure from Crucible Steel Co.; honed and lapped; surface roughness before oxidation 4μ in. RMS; oxidized by heating in air at 700 K for 5 hrs; oxide thickness by weight difference 0.07μ; purple color after oxidation; MgO reference; [Author's designation: Specimen No. 2-4].
69	270	1962	298	0.33-3.4	20°	2π		Commercially pure from Crucible Steel Co.; honed and lapped; surface roughness before oxidation 4μ in. RMS; oxidized by heating in air at 755 K for 5 hrs; oxide thickness by weight difference 0.05μ; medium blue color after oxidation; MgO reference; [Author's designation: Specimen No. 2-5].
70	270	1962	298	0.33-3.4	20°	2π		Commercially pure from Crucible Steel Co.; honed and lapped; surface roughness before oxidation 4μ in. RMS; oxidized by heating in air at 811 K for 5 hrs; oxide thickness by weight difference 0.14μ; light blue color after oxidation; MgO reference; [Author's designation: Specimen No. 2-6].
71	270	1962	298	0.33-3.4	20°	2π		Commercially pure from Crucible Steel Co.; honed and lapped; surface roughness before oxidation 4μ in. RMS; oxidized by heating in air at 922 K for 5 hrs; oxide thickness by weight difference 0.73μ; gray color after oxidation; MgO reference; [Author's designation: Specimen No. 2-7].
72	270	1962	298	0.33-3.4	20°	2π		Commercially pure from Crucible Steel Co.; honed and lapped; surface roughness before oxidation 4μ in. RMS; oxidized by heating in air at 866 K for 5 hrs; oxide thickness by weight difference 0.28μ; gray color after oxidation; MgO reference; [Author's designation: Specimen No. 2-8].
73	270	1962	298	0.33-3.4	20°	2π		Commercially pure from Crucible Steel Co.; honed and lapped; surface roughness 4μ in. RMS; silver color; MgO reference; [Author's designation: Specimen No. 2-9].

SPECIFICATION TABLE NO. 227 (continued)

Curve No.	Ref. No.	Year	Temperature K	Wavelength Range, μ	Geometry θ	θ'	ω'	Reported Error, %	Composition (weight percent), Specifications and Remarks
74	270	1962	298	0.33-3.4	20°		2π		Commercially pure from Crucible Steel Co.; honed and lapped; surface roughness before oxidation 4 μ in. RMS; oxidized by heating in air at 978 K for 5 hrs; oxide thickness by weight difference 1.86 μ; gray color after oxidation; MgO reference; [Author's designation: Specimen No. 2-10].
75	270	1962	298	0.33-3.4	20°		2π		Commercially pure from Crucible Steel Co.; honed and lapped; surface roughness before oxidation 10 μ in. RMS; oxidized by heating in air at 1033 K for 5 hrs; oxide thickness by weight difference 4.33 μ; gray color after oxidation; MgO reference; [Author's designation: Specimen No. 3-1].
76	270	1962	298	0.35-1.8	30°		2π		Above specimen and conditions.
77	270	1962	298	0.35-1.8	40°		2π		Above specimen and conditions.
78	270	1962	298	0.35-1.8	50°		2π		Above specimen and conditions.
79	270	1962	298	0.35-1.8	60°		2π		Above specimen and conditions.
80	270	1962	298	0.35-1.8	70°		2π		Above specimen and conditions.
81	270	1962	298	0.35-1.8	80°		2π		Above specimen and conditions.
82	270	1962	298	0.33-3.4	20°		2π		Commercially pure from Crucible Steel Co.; honed and lapped; surface roughness before oxidation 10 μ in. RMS; oxidized by heating in air at 589 K for 5 hrs; oxide thickness by weight difference 0.01 μ; gold color after oxidation; MgO reference; [Author's designation: Specimen No. 3-2].
83	270	1962	298	0.35-1.8	30°		2π		Above specimen and conditions.
84	270	1962	298	0.35-1.8	40°		2π		Above specimen and conditions.
85	270	1962	298	0.35-1.8	50°		2π		Above specimen and conditions.
86	270	1962	298	0.35-1.8	60°		2π		Above specimen and conditions.
87	270	1962	298	0.35-1.8	70°		2π		Above specimen and conditions.
88	270	1962	298	0.35-1.8	80°		2π		Above specimen and conditions.
89	270	1962	298	0.33-3.4	20°		2π		Commercially pure from Crucible Steel Co.; honed and lapped; surface roughness before oxidation 10 μ in. RMS; oxidized by heating in air at 644 K for 5 hrs; oxide thickness by weight difference 0.02 μ; deep gold color after oxidation; MgO reference; [Author's designation: Specimen No. 3-3].
90	270	1962	298	0.4-1.8	30°		2π		Above specimen and conditions.
91	270	1962	298	0.4-1.8	40°		2π		Above specimen and conditions.
92	270	1962	298	0.4-1.8	50°		2π		Above specimen and conditions.
93	270	1962	298	0.4-1.8	60°		2π		Above specimen and conditions.
94	270	1962	298	0.4-1.8	70°		2π		Above specimen and conditions.
95	270	1962	298	0.4-1.8	80°		2π		Above specimen and conditions.

SPECIFICATION TABLE NO. 227 (continued)

Curve No.	Ref. No.	Year	Temperature K	Wavelength Range, μ	Geometry θ	θ'	ω'	Reported Error, %	Composition (weight percent), Specifications and Remarks
96	270	1962	298	0.33-3.4	20°		2π		Commercially pure from Crucible Steel Co.; honed and lapped; surface roughness before oxidation 10 μ in. RMS; oxidized by heating in air at 700 K for 5 hrs; oxide thickness by weight difference 0.06 μ; purple color after oxidation; MgO reference; [Author's designation: Specimen No. 3-4].
97	270	1962	298	0.35-1.8	30°		2π		Above specimen and conditions.
98	270	1962	298	0.35-1.8	40°		2π		Above specimen and conditions.
99	270	1962	298	0.35-1.8	50°		2π		Above specimen and conditions.
100	270	1962	298	0.35-1.8	60°		2π		Above specimen and conditions.
101	270	1962	298	0.35-1.8	70°		2π		Above specimen and conditions.
102	270	1962	298	0.35-1.8	80°		2π		Above specimen and conditions.
103	270	1962	298	0.33-3.4	20°		2π		Commercially pure from Crucible Steel Co.; honed and lapped; surface roughness before oxidation 10 μ in. RMS; oxidized by heating in air at 755 K for 5 hrs; oxide thickness by weight difference 0.02 μ; light blue color after oxidation; MgO reference; [Author's designation: Specimen No. 3-5].
104	270	1962	298	0.4-1.8	30°		2π		Above specimen and conditions.
105	270	1962	298	0.4-1.8	40°		2π		Above specimen and conditions.
106	270	1962	298	0.4-1.8	50°		2π		Above specimen and conditions.
107	270	1962	298	0.4-1.8	60°		2π		Above specimen and conditions.
108	270	1962	298	0.4-1.8	70°		2π		Above specimen and conditions.
109	270	1962	298	0.4-1.8	80°		2π		Above specimen and conditions.
110	270	1962	298	0.33-3.4	20°		2π		Commercially pure from Crucible Steel Co.; honed and lapped; surface roughness before oxidation 10 μ in. RMS; oxidized by heating in air at 811 K for 5 hrs; oxide thickness by weight difference 0.12 μ; light blue color after oxidation; MgO reference; [Author's designation: Specimen No. 3-6].
111	270	1962	298	0.35-1.8	30°		2π		Above specimen and conditions.
112	270	1962	298	0.35-1.8	40°		2π		Above specimen and conditions.
113	270	1962	298	0.35-1.8	50°		2π		Above specimen and conditions.
114	270	1962	298	0.35-1.8	60°		2π		Above specimen and conditions.
115	270	1962	298	0.35-1.8	70°		2π		Above specimen and conditions.
116	270	1962	298	0.35-1.8	80°		2π		Above specimen and conditions.
117	270	1962	298	0.33-3.4	20°		2π		Commercially pure from Crucible Steel Co.; honed and lapped; surface roughness before oxidation 10 μ in. RMS; oxidized by heating in air at 922 K for 5 hrs; oxide thickness by weight difference 0.72 μ; gray color after oxidation; MgO reference; [Author's designation: Specimen No. 3-7].

SPECIFICATION TABLE NO. 227 (continued)

Curve No.	Ref. No.	Year	Temperature K	Wavelength Range, μ	Geometry θ	θ'	ω'	Reported Error, %	Composition (weight percent), Specifications and Remarks
118	270	1962	298	0.4-2.0	30°		2π		Above specimen and conditions.
119	270	1962	298	0.4-2.0	40°		2π		Above specimen and conditions.
120	270	1962	298	0.4-2.0	50°		2π		Above specimen and conditions.
121	270	1962	298	0.4-2.0	60°		2π		Above specimen and conditions.
122	270	1962	298	0.4-2.0	70°		2π		Above specimen and conditions.
123	270	1962	298	0.4-2.0	80°		2π		Above specimen and conditions.
124	270	1962	298	0.33-3.4	20°		2π		Commercially pure from Crucible Steel Co.; honed and lapped; surface roughness before oxidation 10 μ in. RMS; oxidized by heating in air at 866 K for 5 hrs; oxide thickness by weight difference 0.96 μ; gray color after oxidation; MgO reference; [Author's designation: Specimen No. 3-8]
125	270	1962	298	0.4-2.4	30°		2π		Above specimen and conditions.
126	270	1962	298	0.4-2.4	40°		2π		Above specimen and conditions.
127	270	1962	298	0.4-2.4	50°		2π		Above specimen and conditions.
128	270	1962	298	0.4-2.4	60°		2π		Above specimen and conditions.
129	270	1962	298	0.4-2.4	70°		2π		Above specimen and conditions.
130	270	1962	298	0.4-2.4	80°		2π		Above specimen and conditions.
131	270	1962	298	0.33-3.4	20°		2π		Commercially pure from Crucible Steel Co.; honed and lapped; surface roughness 10μ in. RMS; silver color; MgO reference; [Author's designation: Specimen No. 3-9].
132	270	1962	298	0.4-1.8	30°		2π		Above specimen and conditions.
133	270	1962	298	0.4-1.8	40°		2π		Above specimen and conditions.
134	270	1962	298	0.4-1.8	50°		2π		Above specimen and conditions.
135	270	1962	298	0.4-1.8	60°		2π		Above specimen and conditions.
136	270	1962	298	0.4-1.8	70°		2π		Above specimen and conditions.
137	270	1962	298	0.4-1.8	80°		2π		Above specimen and conditions.
138	270	1962	298	0.33-3.4	20°		2π		Commercially pure from Crucible Steel Co.; honed and lapped; surface roughness before oxidation 10 μ in. RMS; oxidized by heating in air at 978 K for 5 hrs; oxide thickness by weight difference 1.73 μ; gray color after oxidation; MgO reference; [Author's designation: Specimen No. 3-10].
139	270	1962	298	0.33-3.4	20°		2π		Commercially pure from Crucible Steel Co.; honed and lapped; surface roughness before oxidation 10 μ in. RMS; oxidized by heating in air at 700 K for 5 hrs; purple color after oxidation; MgO reference;[Author's designation: Specimen No. 3-11].

SPECIFICATION TABLE NO. 227 (continued)

Curve No.	Ref. No.	Year	Temperature K	Wavelength Range, μ	Geometry θ	θ'	ω'	Reported Error, %	Composition (weight percent), Specifications and Remarks
140	270	1962	298	0.33-3.4	20°		2π		Commercially pure from Crucible Steel Co.; honed and lapped; surface roughness before oxidation 10 μ in. RMS; oxidized by heating in air at 1033 K for 5 hrs; oxide thickness by weight difference 4.21 μ; gray color after oxidation; MgO reference; [Author's designation: Specimen No. 3-19].
141	270	1962	298	0.33-3.4	20°		2π		Commercially pure from Crucible Steel Co.; honed and lapped; surface roughness before oxidation 16 μ in. RMS; oxidized by heating in air at 1033 K for 5 hrs; oxide thickness by weight difference 2.90 μ; gray color after oxidation; MgO reference; [Author's designation: Specimen No. 4-1].
142	270	1962	298	0.33-3.4	20°		2π		Commercially pure from Crucible Steel Co.; honed and lapped; surface roughness before oxidation 16 μ in. RMS; oxidized by heating in air at 589 K for 5 hrs; oxide thickness by weight difference 0.02 μ; light gold color after oxidation; MgO reference; [Author's designation: Specimen No. 4-2].
143	270	1962	298	0.33-3.4	20°		2π		Commercially pure from Crucible Steel Co.; honed and lapped; surface roughness before oxidation 16 μ in. RMS; oxidized by heating in air at 644 K for 5 hrs; oxide thickness by weight difference 0.01 μ; light yellow color after oxidation; MgO reference; [Author's designation: Specimen No. 4-3].
144	270	1962	298	0.33-3.4	20°		2π		Commercially pure from Crucible Steel Co.; honed and lapped; surface roughness before oxidation 16 μ in. RMS; oxidized by heating in air at 700 K for 5 hrs; oxide thickness by weight difference 0.03 μ; gold-purple color after oxidation; MgO reference; [Author's designation: Specimen No. 4-4].
145	270	1962	298	0.33-3.4	20°		2π		Commercially pure from Crucible Steel Co.; honed and lapped; surface roughness before oxidation 16 μ in. RMS; oxidized by heating in air at 755 K for 5 hrs; dark blue color after oxidation; MgO reference; [Author's designation: Specimen No. 4-5].
146	270	1962	298	0.33-3.4	20°		2π		Commercially pure from Crucible Steel Co.; honed and lapped; surface roughness before oxidation 16 μ in. RMS; oxidized by heating in air at 810.9 K for 5 hrs; oxide thickness by weight difference 0.05 μ; medium blue color after oxidation; MgO reference; [Author's designation: Specimen No. 4-6].
147	270	1962	298	0.33-3.4	20°		2π		Commercially pure from Crucible Steel Co.; honed and lapped; surface roughness before oxidation 16 μ in. RMS; oxidized by heating in air at 922 K for 5 hrs; oxide thickness by wieght difference 0.49 μ; gray color after oxidation; MgO reference; [Author's designation: Specimen No. 4-7].
148	270	1962	298	0.33-3.4	20°		2π		Commercially pure from Crucible Steel Co.; honed and lapped; surface roughness before oxidation 16 μ in. RMS; oxidized by heating in air at 866.5 K for 5 hrs; oxide thickness by weight difference 0.18 μ; gray color after oxidation; MgO reference; [Author's designation: Specimen No. 4-8].
149	270	1962	298	0.33-3.4	20°		2π		Commercially pure from Crucible Steel Co.; honed and lapped; surface roughness 16 μ in. RMS; silver color; MgO reference; [Author's designation: Specimen No. 4-9].
150	270	1962	298	0.33-3.4	20°		2π		Commercially pure from Crucible Steel Co.; honed and lapped; surface roughness before oxidation 16 μ in. RMS; oxidized by heating in air at 977.6 K for 5 hrs; oxide thickness by weight difference 1.38 μ; gray color after oxidation; MgO reference; [Author's designation: Specimen No. 4-10].

DATA TABLE NO. 227 ANGULAR SPECTRAL REFLECTANCE OF TITANIUM

[Wavelength, λ, μ; Reflectance, ρ; Temperature, T, K]

λ	ρ
CURVE 1 T = 306	
0.35	0.324
0.39	0.375
0.45	0.282
0.52	0.201
0.59	0.130
0.69	0.090
0.85	0.126
0.92	0.163
1.00	0.219
1.25	0.260
1.40	0.345
1.62	0.401
1.95	0.447
2.54	0.503
3.31	0.567
4.46	0.641
5.52	0.698
6.97	0.755
9.08	0.785
10.97	0.805
12.44	0.818
14.00	0.829
CURVE 2 T = 306	
0.35	0.290
0.38	0.300
0.41	0.256
0.43	0.187
0.47	0.130
0.52	0.091
0.60	0.074
0.71	0.121
0.81	0.198
0.90	0.248
1.00	0.289
1.17	0.337
1.44	0.417
2.05	0.486
3.34	0.570
4.81	0.664
5.97	0.722
6.99	0.756
8.99	0.794
CURVE 2 (cont.)	
11.03	0.824
12.58	0.835
14.00	0.837
CURVE 3 T = 298	
0.33	0.128
0.35	0.100
0.40	0.154
0.45	0.194
0.50	0.195
0.55	0.191
0.6	0.192
0.6	0.192
0.7	0.187
0.8	0.182
0.9	0.178
1.0	0.187
1.1	0.187
1.2	0.196
1.3	0.215
1.4	0.234
1.6	0.262
1.8	0.289
2.0	0.308
2.2	0.324
2.4	0.346
2.6	0.368
2.8	0.372
3.0	0.373
3.2	0.370
3.4	0.368
CURVE 4 T = 298	
0.35	0.076
0.40	0.173
0.45	0.234
0.50	0.304
0.55	0.328
0.6	0.368
0.6	0.378
0.7	0.419
0.8	0.445
CURVE 4 (cont.)	
0.9	0.464
1.0	0.479
1.1	0.491
1.3	0.510
1.4	0.525
1.6	0.545
1.8	0.553
2.0	0.565
2.4	0.565
3.0	0.575
3.4	0.579
CURVE 5 T = 298	
0.35	0.081
0.55	0.321
0.7	0.426
1.0	0.474
1.8	0.541
CURVE 6 T = 298	
0.35	0.092
0.55	0.321
0.7	0.431
1.0	0.474
1.8	0.541
CURVE 7 T = 298	
0.35	0.113
0.55	0.335
0.7	0.444
1.0	0.469
1.8	0.536
CURVE 8 T = 298	
0.35	0.159
0.55	0.356
0.7	0.470
1.0	0.459
1.8	0.528
CURVE 9 T = 298	
0.35	0.249
0.55	0.417
0.7	0.502
1.0	0.440
1.8	0.497
CURVE 10 T = 298	
0.35	0.444
0.55	0.500
0.7	0.732
1.0	0.474
1.8	0.536
CURVE 11 T = 298	
0.33	0.161
0.35	0.072
0.40	0.048
0.45	0.146
0.50	0.225
0.55	0.280
0.6	0.329
0.6	0.345
0.7	0.414
0.8	0.449
0.9	0.474
1.0	0.488
1.1	0.501
1.2	0.506
1.3	0.529
1.4	0.528
1.6	0.559
1.8	0.563
2.0	0.574
2.2	0.571
2.4	0.590
2.6	0.586
2.8	0.583
3.0	0.583
3.2	0.599
3.4	0.596
CURVE 12 T = 298	
0.35	0.072
0.55	0.273
0.7	0.418
1.0	0.493
1.8	0.563
CURVE 13 T = 298	
0.35	0.081
0.55	0.298
0.7	0.421
1.0	0.493
1.8	0.556
CURVE 14 T = 298	
0.35	0.092
0.55	0.310
0.7	0.439
1.0	0.498
1.8	0.545
CURVE 15 T = 298	
0.35	0.138
0.55	0.350
0.7	0.460
1.0	0.498
1.8	0.535
CURVE 16 T = 298	
0.35	0.254
0.55	0.424
0.7	0.505
1.0	0.507
1.8	0.506
CURVE 17 T = 298	
0.35	0.442
0.55	0.540
0.7	0.604
1.0	0.605
1.8	0.529
CURVE 18 T = 298	
0.33	0.341
0.35	0.333
0.40	0.125
0.45	0.024
0.50	0.044
0.55	0.098
0.6	0.133
0.6	0.162
0.7	0.271
0.8	0.335
0.9	0.389
1.0	0.424
1.1	0.457
1.2	0.471
1.3	0.495
1.4	0.499
1.6	0.525
1.8	0.544
2.0	0.550
2.2	0.555
2.4	0.561
2.6	0.568
2.8	0.575
3.0	0.578
3.2	0.580
3.4	0.579
CURVE 19 T = 298	
0.35	0.333
0.55	0.106
0.7	0.276
1.0	0.424
1.8	0.538
CURVE 20 T = 298	
0.35	0.333
0.55	0.123
0.7	0.290
1.0	0.432
1.8	0.526
CURVE 21 T = 298	
0.35	0.342
0.55	0.154
0.7	0.312
1.0	0.440
1.8	0.526
CURVE 22 T = 298	
0.35	0.352
0.55	0.214
0.7	0.350
1.0	0.454
1.8	0.510
CURVE 23 T = 298	
0.35	0.429
0.55	0.343
0.7	0.425
1.0	0.487
1.8	0.490
CURVE 24 T = 298	
0.35	0.552
0.55	0.552
0.7	0.615
1.0	0.610
1.8	0.532
CURVE 25 T = 298	
0.33	0.294
0.35	0.323
0.40	0.288
0.45	0.194
0.50	0.132
0.55	0.093
0.6	0.074
0.6	0.074
0.7	0.089
0.8	0.142
0.9	0.207
1.0	0.251
1.1	0.285
1.2	0.319
1.3	0.340
1.4	0.360
1.6	0.408
1.8	0.424
2.0	0.445
2.2	0.455
2.4	0.459
2.6	0.468
2.8	0.471
3.0	0.479
3.2	0.481
3.4	0.489
CURVE 26 T = 298	
0.45	0.284
0.55	0.093
0.7	0.094
1.0	0.251
1.8	0.424
CURVE 27 T = 298	
0.45	0.279
0.55	0.096
0.7	0.105
1.0	0.261
1.8	0.424

DATA TABLE NO. 227 (continued)

λ	ρ
CURVE 28	
T = 298	
0.45	0.284
0.55	0.112
0.7	0.125
1.0	0.274
1.8	0.424
CURVE 29	
T = 298	
0.45	0.282
0.55	0.158
0.7	0.167
1.0	0.304
1.8	0.424
CURVE 30	
T = 298	
0.45	0.332
0.55	0.266
0.7	0.267
1.0	0.372
1.8	0.452
CURVE 31	
T = 298	
0.45	0.460
0.55	0.487
0.7	0.454
1.0	0.555
1.8	0.550
CURVE 32	
T = 298	
0.33	0.267
0.35	0.304
0.40	0.298
0.45	0.223
0.50	0.161
0.55	0.127
0.6	0.103
0.6	0.103
0.7	0.099
0.8	0.128
CURVE 32 (cont.)	
0.9	0.163
1.0	0.197
1.1	0.226
1.2	0.256
1.3	0.270
1.4	0.293
1.6	0.344
1.8	0.366
2.0	0.388
2.2	0.407
2.4	0.412
2.6	0.419
2.8	0.430
3.0	0.438
3.2	0.439
3.4	0.439
CURVE 33	
T = 298	
0.35	0.304
0.55	0.124
0.7	0.100
1.0	0.197
1.8	0.366
CURVE 34	
T = 298	
0.35	0.304
0.55	0.124
0.7	0.109
1.0	0.203
1.8	0.366
CURVE 35	
T = 298	
0.35	0.197
0.55	0.136
0.7	0.129
1.0	0.220
1.8	0.366
CURVE 36	
T = 298	
0.35	0.191
0.55	0.170
0.7	0.173
1.0	0.252
1.8	0.377
CURVE 37	
T = 298	
0.35	0.209
0.55	0.248
0.7	0.278
1.0	0.331
1.8	0.410
CURVE 38	
T = 298	
0.35	0.292
0.55	0.450
0.7	0.465
1.0	0.525
1.8	0.509
CURVE 39	
T = 298	
0.33	0.199
0.35	0.181
0.40	0.183
0.45	0.184
0.50	0.185
0.55	0.186
0.6	0.197
0.6	0.207
0.7	0.207
0.8	0.227
0.9	0.236
1.0	0.207
1.1	0.192
1.2	0.182
1.3	0.206
1.4	0.318
1.6	0.326
1.8	0.356
2.0	0.360
CURVE 39 (cont.)	
2.2	0.356
2.4	0.346
2.6	0.326
2.8	0.311
3.0	0.292
3.2	0.284
3.4	0.280
CURVE 40	
T = 298	
0.45	0.182
0.55	0.181
0.7	0.207
1.0	0.203
1.8	0.360
CURVE 41	
T = 298	
0.45	0.171
0.55	0.186
0.7	0.211
1.0	0.201
1.8	0.356
CURVE 42	
T = 298	
0.45	0.175
0.55	0.196
0.7	0.228
1.0	0.203
1.8	0.356
CURVE 43	
T = 298	
0.45	0.188
0.55	0.224
0.7	0.259
1.0	0.228
1.8	0.356
CURVE 44	
T = 298	
0.45	0.241
0.55	0.276
0.7	0.323
1.0	0.294
1.8	0.356
CURVE 45	
T = 298	
0.45	0.360
0.55	0.410
0.7	0.505
1.0	0.486
1.8	0.455
CURVE 46	
T = 298	
0.33	0.256
0.35	0.238
0.40	0.216
0.45	0.204
0.50	0.205
0.55	0.216
0.6	0.211
0.6	0.216
0.7	0.222
0.8	0.256
0.9	0.286
1.0	0.276
1.1	0.246
1.2	0.216
1.3	0.176
1.4	0.068
1.6	0.107
1.8	0.096
2.0	0.095
2.2	0.103
2.4	0.124
2.6	0.139
2.8	0.156
3.0	0.178
3.2	0.182
3.4	0.190
CURVE 47	
T = 298	
0.33	0.312
0.35	0.319
0.40	0.356
0.45	0.378
0.50	0.400
0.55	0.421
0.6	0.437
0.6	0.442
0.7	0.474
0.8	0.489
0.9	0.494
1.0	0.506
1.1	0.511
1.2	0.520
1.3	0.524
1.4	0.537
1.6	0.558
1.8	0.564
2.0	0.568
2.2	0.572
2.4	0.570
2.6	0.581
2.8	0.584
3.0	0.577
3.2	0.573
3.4	0.570
CURVE 48	
T = 298	
0.4	0.366
0.55	0.424
0.7	0.479
1.0	0.499
1.8	0.552
CURVE 49	
T = 298	
0.4	0.366
0.55	0.421
0.7	0.479
1.0	0.486
1.8	0.546
CURVE 50	
T = 298	
0.4	0.372
0.55	0.425
0.7	0.484
1.0	0.486
1.8	0.541
CURVE 51	
T = 298	
0.4	0.392
0.55	0.436
0.7	0.500
1.0	0.484
1.8	0.518
CURVE 52	
T = 298	
0.4	0.438
0.55	0.475
0.7	0.513
1.0	0.470
1.8	0.484
CURVE 53	
T = 298	
0.4	0.561
0.55	0.621
0.7	0.579
1.0	0.506
1.8	0.495
CURVE 54	
T = 298	
0.33	0.109
0.35	0.095
0.40	0.106
0.45	0.116
0.50	0.127
0.55	0.137
0.6	0.138
0.6	0.157
0.7	0.182
0.8	0.197
CURVE 54 (cont.)	
0.9	0.222
1.0	0.241
1.1	0.251
1.2	0.245
1.3	0.254
1.4	0.274
1.6	0.252
1.8	0.260
2.0	0.331
2.2	0.366
2.4	0.374
2.6	0.353
2.8	0.317
3.0	0.300
3.2	0.284
3.4	0.274
CURVE 55	
T = 298	
0.35	0.096
0.55	0.147
0.7	0.180
1.0	0.248
1.8	0.262
CURVE 56	
T = 298	
0.35	0.103
0.55	0.147
0.7	0.178
1.0	0.236
1.8	0.265
CURVE 57	
T = 298	
0.35	0.112
0.55	0.152
0.7	0.187
1.0	0.241
1.8	0.278

DATA TABLE NO. 227 (continued)

λ	ρ
CURVE 58	
T = 298	
0.35	0.142
0.55	0.188
0.7	0.213
1.0	0.263
1.8	0.296
CURVE 59	
T = 298	
0.35	0.193
0.55	0.254
0.7	0.276
1.0	0.325
1.8	0.332
CURVE 60	
T = 298	
0.35	0.358
0.55	0.415
0.7	0.464
1.0	0.500
1.8	0.436
CURVE 61	
T = 298	
0.33	0.199
0.35	0.228
0.40	0.077
0.45	0.058
0.50	0.098
0.55	0.138
0.6	0.157
0.6	0.206
0.7	0.286
0.8	0.336
0.9	0.375
1.0	0.405
1.1	0.422
1.2	0.441
1.3	0.455
1.4	0.465
1.6	0.485
1.8	0.490
CURVE 61 (cont.)	
2.0	0.501
2.2	0.511
2.4	0.510
2.6	0.511
2.8	0.525
3.0	0.519
3.2	0.515
3.4	0.520
CURVE 62	
T = 298	
0.33	0.341
0.35	0.371
0.40	0.336
0.45	0.218
0.50	0.127
0.55	0.084
0.6	0.059
0.6	0.049
0.7	0.108
0.8	0.187
0.9	0.256
1.0	0.320
1.1	0.364
1.2	0.412
1.3	0.440
1.4	0.459
1.6	0.505
1.8	0.530
2.0	0.540
2.2	0.544
2.4	0.551
2.6	0.558
2.8	0.556
3.0	0.560
3.2	0.550
3.4	0.548
CURVE 63	
T = 298	
0.33	0.312
0.35	0.286
0.40	0.269
0.45	0.310
0.50	0.341
CURVE 63 (cont.)	
0.55	0.334
0.6	0.334
0.6	0.324
0.7	0.236
0.8	0.168
0.9	0.108
1.0	0.089
1.1	0.098
1.2	0.118
1.3	0.157
1.4	0.205
1.6	0.281
1.8	0.337
2.0	0.402
2.2	0.403
2.4	0.411
2.6	0.428
2.8	0.469
3.0	0.464
3.2	0.469
3.4	0.466
CURVE 64	
T = 298	
0.33	0.095
0.35	0.086
0.40	0.154
0.45	0.203
0.50	0.204
0.55	0.207
0.6	0.192
0.6	0.187
0.7	0.178
0.8	0.173
0.9	0.178
1.0	0.187
1.1	0.196
1.2	0.216
1.3	0.225
1.4	0.234
1.6	0.281
1.8	0.308
2.0	0.322
2.2	0.338
2.4	0.346
CURVE 64 (cont.)	
2.6	0.372
2.8	0.377
3.0	0.382
3.2	0.379
3.4	0.377
CURVE 65	
T = 298	
0.33	0.095
0.35	0.071
0.40	0.144
0.45	0.223
0.50	0.229
0.55	0.230
0.6	0.226
0.6	0.226
0.7	0.217
0.8	0.207
0.9	0.187
1.0	0.187
1.1	0.187
1.2	0.177
1.3	0.176
1.4	0.181
1.6	0.199
1.8	0.212
2.0	0.222
2.2	0.244
2.4	0.252
2.6	0.270
2.8	0.276
3.0	0.282
3.2	0.280
3.4	0.278
CURVE 66	
T = 298	
0.33	0.151
0.35	0.095
0.40	0.076
0.45	0.116
0.50	0.156
0.55	0.196
0.6	0.221
CURVE 66 (cont.)	
0.6	0.256
0.7	0.306
0.8	0.345
0.9	0.380
1.0	0.394
1.1	0.403
1.2	0.417
1.3	0.430
1.4	0.445
1.6	0.460
1.8	0.476
2.0	0.488
2.2	0.492
2.4	0.505
2.6	0.502
2.8	0.515
3.0	0.510
3.2	0.510
3.4	0.507
CURVE 67	
T = 298	
0.33	0.232
0.35	0.204
0.40	0.067
0.45	0.058
0.50	0.078
0.55	0.108
0.6	0.138
0.6	0.162
0.7	0.222
0.8	0.266
0.9	0.306
1.0	0.320
1.1	0.334
1.2	0.353
1.3	0.362
1.4	0.392
1.6	0.407
1.8	0.414
2.0	0.421
2.2	0.436
2.4	0.444
2.6	0.446
2.8	0.551
CURVE 67 (cont.)	
3.0	0.450
3.2	0.451
3.4	0.499
CURVE 68	
T = 298	
0.33	0.312
0.35	0.352
0.40	0.197
0.45	0.087
0.50	0.044
0.55	0.044
0.6	0.059
0.6	0.079
0.7	0.148
0.8	0.202
0.9	0.256
1.0	0.295
1.1	0.314
1.2	0.338
1.3	0.352
1.4	0.371
1.6	0.398
1.8	0.414
2.0	0.445
2.2	0.441
2.4	0.449
2.6	0.446
2.8	0.460
3.0	0.455
3.2	0.455
3.4	0.458
CURVE 69	
T = 298	
0.33	0.265
0.35	0.305
0.40	0.306
0.45	0.252
0.50	0.195
0.55	0.157
0.6	0.118
0.6	0.113
0.7	0.084
CURVE 69 (cont.)	
0.8	0.089
0.9	0.118
1.0	0.152
1.1	0.177
1.2	0.206
1.3	0.230
1.4	0.254
1.6	0.286
1.8	0.308
2.0	0.340
2.2	0.347
2.4	0.355
2.6	0.367
2.8	0.377
3.0	0.382
3.2	0.379
3.4	0.377
CURVE 70	
T = 298	
0.33	0.265
0.35	0.238
0.40	0.248
0.45	0.252
0.50	0.268
0.55	0.270
0.6	0.265
0.6	0.270
0.7	0.217
0.8	0.168
0.9	0.128
1.0	0.108
1.1	0.098
1.2	0.098
1.3	0.112
1.4	0.127
1.6	0.160
1.8	0.188
2.0	0.213
2.2	0.244
2.4	0.252
2.6	0.270
2.8	0.290
3.0	0.300
3.2	0.298
3.4	0.301
CURVE 71	
T = 298	
0.33	0.227
0.35	0.209
0.40	0.202
0.45	0.198
0.50	0.204
0.55	0.206
0.6	0.212
0.6	0.226
0.7	0.242
0.8	0.227
0.9	0.247
1.0	0.236
1.1	0.221
1.2	0.236
1.3	0.274
1.4	0.288
1.6	0.242
1.8	0.207
2.0	0.218
2.2	0.162
2.4	0.294
2.6	0.348
2.8	0.396
3.0	0.419
3.2	0.420
3.4	0.422
CURVE 72	
T = 298	
0.33	0.237
0.35	0.224
0.40	0.202
0.45	0.184
0.50	0.175
0.55	0.176
0.6	0.177
0.6	0.192
0.7	0.198
0.8	0.198
0.9	0.212
1.0	0.256
1.1	0.275
1.2	0.294
1.3	0.298
1.4	0.263

DATA TABLE NO. 227 (continued)

λ	ρ
CURVE 72 (cont.)	
1.6	0.247
1.8	0.222
2.0	0.180
2.2	0.150
2.4	0.131
2.6	0.112
2.8	0.097
3.0	0.095
3.2	0.090
3.4	0.085
CURVE 73	
T = 298	
0.33	0.213
0.35	0.219
0.40	0.255
0.45	0.272
0.50	0.302
0.55	0.304
0.6	0.319
0.6	0.329
0.7	0.365
0.8	0.375
0.9	0.375
1.0	0.388
1.1	0.398
1.2	0.402
1.3	0.406
1.4	0.415
1.6	0.436
1.8	0.443
2.0	0.445
2.2	0.445
2.4	0.454
2.6	0.455
2.8	0.455
3.0	0.455
3.2	0.451
3.4	0.449
CURVE 74	
T = 298	
0.33	0.118
0.35	0.105

λ	ρ
CURVE 74 (cont.)	
0.40	0.135
0.45	0.136
0.50	0.136
0.55	0.137
0.6	0.137
0.6	0.142
0.7	0.157
0.8	0.168
0.9	0.173
1.0	0.196
1.1	0.206
1.2	0.226
1.3	0.240
1.4	0.254
1.6	0.272
1.8	0.279
2.0	0.284
2.2	0.291
2.4	0.304
2.6	0.298
2.8	0.294
3.0	0.296
3.2	0.294
3.4	0.287
CURVE 75	
T = 298	
0.33	0.114
0.35	0.085
0.40	0.154
0.45	0.223
0.50	0.234
0.55	0.230
0.6	0.226
0.6	0.226
0.7	0.212
0.8	0.197
0.9	0.182
1.0	0.182
1.1	0.182
1.2	0.181
1.3	0.181
1.4	0.195
1.6	0.218
1.8	0.240

λ	ρ
CURVE 75 (cont.)	
2.0	0.256
2.2	0.272
2.4	0.285
2.6	0.307
2.8	0.313
3.0	0.318
3.2	0.320
3.4	0.322
CURVE 76	
T = 298	
0.35	0.085
0.5	0.230
0.7	0.215
1.3	0.181
1.8	0.232
CURVE 77	
T = 298	
0.35	0.098
0.5	0.234
0.7	0.226
1.3	0.181
1.8	0.228
CURVE 78	
T = 298	
0.35	0.108
0.5	0.248
0.7	0.236
1.3	0.188
1.8	0.226
CURVE 79	
T = 298	
0.35	0.134
0.5	0.276
0.7	0.262
1.3	0.212
1.8	0.240

λ	ρ
CURVE 80	
T = 298	
0.35	0.166
0.5	0.319
0.7	0.316
1.3	0.263
1.8	0.286
CURVE 81	
T = 298	
0.35	0.242
0.5	0.407
0.7	0.410
1.3	0.405
1.8	0.442
CURVE 82	
T = 298	
0.33	0.118
0.35	0.086
0.40	0.154
0.45	0.213
0.50	0.263
0.55	0.304
0.6	0.334
0.6	0.349
0.7	0.405
0.8	0.429
0.9	0.449
1.0	0.452
1.1	0.461
1.2	0.471
1.3	0.485
1.4	0.498
1.6	0.514
1.8	0.520
2.0	0.535
2.2	0.535
2.4	0.546
2.6	0.545
2.8	0.551
3.0	0.555
3.2	0.555
3.4	0.551

λ	ρ
CURVE 83	
T = 298	
0.35	0.091
0.55	0.298
0.7	0.386
1.0	0.448
1.8	0.510
CURVE 84	
T = 298	
0.35	0.098
0.55	0.327
0.7	0.378
1.0	0.444
1.8	0.505
CURVE 85	
T = 298	
0.35	0.125
0.55	0.378
0.7	0.386
1.0	0.444
1.8	0.500
CURVE 86	
T = 298	
0.35	0.158
0.55	0.445
0.7	0.392
1.0	0.456
1.8	0.489
CURVE 87	
T = 298	
0.35	0.253
0.55	0.608
0.7	0.419
1.0	0.454
1.8	0.468

λ	ρ
CURVE 88	
T = 298	
0.35	0.414
0.55	0.896
0.7	0.511
1.0	0.555
1.8	0.500
CURVE 89	
T = 298	
0.33	0.180
0.35	0.114
0.40	0.058
0.45	0.102
0.50	0.156
0.55	0.196
0.6	0.236
0.6	0.256
0.7	0.336
0.8	0.375
0.9	0.415
1.0	0.423
1.1	0.437
1.2	0.451
1.3	0.465
1.4	0.474
1.6	0.500
1.8	0.510
2.0	0.516
2.2	0.534
2.4	0.533
2.6	0.535
2.8	0.543
3.0	0.550
3.2	0.546
3.4	0.543
CURVE 90	
T = 298	
0.4	0.061
0.55	0.196
0.7	0.330
1.0	0.418
1.8	0.507

λ	ρ
CURVE 91	
T = 298	
0.4	0.067
0.55	0.201
0.7	0.311
1.0	0.415
1.8	0.501
CURVE 92	
T = 298	
0.4	0.086
0.55	0.264
0.7	0.315
1.0	0.418
1.8	0.496
CURVE 93	
T = 298	
0.4	0.119
0.55	0.248
0.7	0.328
1.0	0.418
1.8	0.480
CURVE 94	
T = 298	
0.4	0.206
0.55	0.318
0.7	0.356
1.0	0.418
1.8	0.445
CURVE 95	
T = 298	
0.4	0.374
0.55	0.462
0.7	0.510
1.0	0.486
1.8	0.480

λ	ρ
CURVE 96	
T = 298	
0.33	0.332
0.35	0.352
0.40	0.159
0.45	0.058
0.50	0.049
0.55	0.069
0.6	0.089
0.6	0.113
0.7	0.212
0.8	0.266
0.9	0.316
1.0	0.359
1.1	0.394
1.2	0.397
1.3	0.436
1.4	0.435
1.6	0.456
1.8	0.466
2.0	0.473
2.2	0.496
2.4	0.491
2.6	0.502
2.8	0.506
3.0	0.510
3.2	0.511
3.4	0.511
CURVE 97	
T = 298	
0.35	0.352
0.5	0.051
0.7	0.218
1.0	0.364
1.8	0.460
CURVE 98	
T = 298	
0.35	0.352
0.5	0.058
0.7	0.232
1.0	0.364
1.8	0.435

DATA TABLE NO. 227 (continued)

λ	ρ
CURVE 99 T = 298	
0.35	0.368
0.5	0.076
0.7	0.254
1.0	0.374
1.8	0.430
CURVE 100 T = 298	
0.35	0.368
0.5	0.106
0.7	0.292
1.0	0.394
1.8	0.425
CURVE 101 T = 298	
0.35	0.368
0.5	0.218
0.7	0.384
1.0	0.454
1.8	0.426
CURVE 102 T = 298	
0.35	0.432
0.5	0.439
0.7	0.530
1.0	0.590
1.8	0.523
CURVE 103 T = 298	
0.33	0.284
0.35	0.276
0.40	0.288
0.45	0.301
0.50	0.302
0.55	0.294
0.6	0.280
0.6	0.285
0.7	0.207
CURVE 103 (cont.)	
0.8	0.158
0.9	0.128
1.0	0.126
1.1	0.133
1.2	0.142
1.3	0.166
1.4	0.190
1.6	0.233
1.8	0.265
2.0	0.308
2.2	0.328
2.4	0.336
2.6	0.362
2.8	0.377
3.0	0.382
3.2	0.379
3.4	0.377
CURVE 104 T = 298	
0.4	0.278
0.55	0.294
0.7	0.194
1.0	0.128
1.8	0.267
CURVE 105 T = 298	
0.4	0.274
0.55	0.289
0.7	0.180
1.0	0.132
1.8	0.272
CURVE 106 T = 298	
0.4	0.274
0.55	0.289
0.7	0.185
1.0	0.145
1.8	0.284
CURVE 107 T = 298	
0.4	0.278
0.55	0.299
0.7	0.292
1.0	0.180
1.8	0.310
CURVE 108 T = 298	
0.4	0.307
0.55	0.334
0.7	0.273
1.0	0.272
1.8	0.364
CURVE 109 T = 298	
0.4	0.398
0.55	0.449
0.7	0.440
1.0	0.525
1.8	0.536
CURVE 110 T = 298	
0.33	0.274
0.35	0.255
0.40	0.274
0.45	0.315
0.50	0.302
0.55	0.284
0.6	0.256
0.6	0.260
0.7	0.178
0.8	0.128
0.9	0.118
1.0	0.123
1.1	0.128
1.2	0.152
1.3	0.171
1.4	0.195
1.6	0.233
1.8	0.265
2.0	0.303
CURVE 110 (cont.)	
2.2	0.323
2.4	0.327
2.6	0.340
2.8	0.372
3.0	0.368
3.2	0.370
3.4	0.368
CURVE 111 T = 298	
0.35	0.239
0.55	0.280
0.7	0.168
0.9	0.118
1.8	0.267
CURVE 112 T = 298	
0.35	0.239
0.55	0.274
0.7	0.168
0.9	0.124
1.8	0.272
CURVE 113 T = 298	
0.35	0.239
0.55	0.274
0.7	0.173
0.9	0.143
1.8	0.284
CURVE 114 T = 298	
0.35	0.271
0.55	0.294
0.7	0.204
0.9	0.184
1.8	0.308
CURVE 115 T = 298	
0.35	0.303
0.55	0.333
0.7	0.294
0.9	0.278
1.8	0.370
CURVE 116 T = 298	
0.35	0.475
0.55	0.450
0.7	0.466
0.9	0.479
1.8	0.554
CURVE 117 T = 298	
0.33	0.274
0.35	0.252
0.40	0.240
0.45	0.238
0.50	0.239
0.55	0.245
0.6	0.246
0.6	0.260
0.7	0.276
0.8	0.262
0.9	0.271
1.0	0.266
1.1	0.241
1.2	0.247
1.3	0.284
1.4	0.322
1.6	0.276
1.8	0.217
2.0	0.194
2.2	0.230
2.4	0.262
2.6	0.325
2.8	0.405
3.0	0.423
3.2	0.434
3.4	0.440
CURVE 118 T = 298	
0.4	0.240
0.55	0.245
0.8	0.246
1.1	0.221
2.0	0.282
CURVE 119 T = 298	
0.4	0.248
0.55	0.245
0.8	0.244
1.1	0.216
2.0	0.278
CURVE 120 T = 298	
0.4	0.256
0.55	0.249
0.8	0.250
1.1	0.218
2.0	0.299
CURVE 121 T = 298	
0.4	0.279
0.55	0.267
0.8	0.271
1.1	0.241
2.0	0.345
CURVE 122 T = 298	
0.4	0.318
0.55	0.317
0.8	0.320
1.1	0.298
2.0	0.471
CURVE 123 T = 298	
0.4	0.419
0.55	0.450
0.8	0.450
1.1	0.452
2.0	0.519
CURVE 124 T = 298	
0.33	0.246
0.35	0.228
0.40	0.207
0.45	0.223
0.50	0.200
0.55	0.196
0.6	0.197
0.6	0.216
0.7	0.246
0.8	0.286
0.9	0.340
1.0	0.364
1.1	0.349
1.2	0.324
1.3	0.284
1.4	0.244
1.6	0.179
1.8	0.130
2.0	0.104
2.2	0.094
2.4	0.089
2.6	0.093
2.8	0.106
3.0	0.109
3.2	0.108
3.4	0.112
CURVE 125 T = 298	
0.4	0.183
0.55	0.189
0.7	0.240
1.0	0.356
2.4	0.090
CURVE 126 T = 298	
0.4	0.192
0.55	0.186
0.7	0.246
1.0	0.342
2.4	0.094
CURVE 127 T = 298	
0.4	0.198
0.55	0.194
0.7	0.252
1.0	0.334
2.4	0.101
CURVE 128 T = 298	
0.4	0.216
0.55	0.218
0.7	0.466
1.0	0.356
2.4	0.111
CURVE 129 T = 298	
0.4	0.258
0.55	0.276
0.7	0.315
1.0	0.440
2.4	0.128
CURVE 130 T = 298	
0.4	0.354
0.55	0.401
0.7	0.758
1.0	0.697
2.4	0.185

DATA TABLE NO. 227 (continued)

CURVE 131
T = 298

λ	ρ
0.33	0.326
0.35	0.333
0.40	0.370
0.45	0.390
0.50	0.415
0.55	0.436
0.6	0.452
0.6	0.457
0.7	0.479
0.8	0.489
0.9	0.503
1.0	0.501
1.1	0.506
1.2	0.515
1.3	0.524
1.4	0.537
1.6	0.558
1.8	0.572
2.0	0.573
2.2	0.576
2.4	0.585
2.6	0.595
2.8	0.598
3.0	0.600
3.2	0.595
3.4	0.592

CURVE 132
T = 298

λ	ρ
0.55	0.436
0.4	0.370
0.7	0.475
1.0	0.494
1.8	0.566

CURVE 133
T = 298

λ	ρ
0.55	0.436
0.4	0.370
0.7	0.465
1.0	0.486
1.8	0.560

CURVE 134
T = 298

λ	ρ
0.55	0.436
0.4	0.370
0.7	0.461
1.0	0.476
1.8	0.550

CURVE 135
T = 298

λ	ρ
0.55	0.440
0.4	0.370
0.7	0.456
1.0	0.466
1.8	0.532

CURVE 136
T = 298

λ	ρ
0.55	0.460
0.4	0.430
0.7	0.465
1.0	0.461
1.8	0.498

CURVE 137
T = 298

λ	ρ
0.55	0.554
0.4	0.525
0.7	0.535
1.0	0.547
1.8	0.544

CURVE 138
T = 298

λ	ρ
0.33	0.256
0.35	0.233
0.40	0.216
0.45	0.208
0.50	0.205
0.55	0.206
0.6	0.206
0.6	0.211

CURVE 138 (cont.)

λ	ρ
0.7	0.222
0.8	0.227
0.9	0.246
1.0	0.270
1.1	0.275
1.2	0.285
1.3	0.298
1.4	0.312
1.6	0.320
1.8	0.337
2.0	0.364
2.2	0.356
2.4	0.346
2.6	0.326
2.8	0.331
3.0	0.342
3.2	0.334
3.4	0.341

CURVE 139
T = 298

λ	ρ
0.33	0.313
0.35	0.304
0.40	0.144
0.45	0.067
0.50	0.059
0.55	0.079
0.6	0.108
0.6	0.118
0.7	0.217
0.8	0.286
0.9	0.340
1.0	0.384
1.1	0.403
1.2	0.422
1.3	0.445
1.4	0.465
1.6	0.485
1.8	0.495
2.0	0.506
2.2	0.511
2.4	0.510
2.6	0.515
2.8	0.525
3.0	0.528
3.2	0.524
3.4	0.520

CURVE 140
T = 298

λ	ρ
0.33	0.123
0.35	0.095
0.40	0.163
0.45	0.222
0.50	0.234
0.55	0.226
0.6	0.226
0.6	0.216
0.7	0.207
0.8	0.197
0.9	0.187
1.0	0.187
1.1	0.187
1.2	0.191
1.3	0.195
1.4	0.215
1.6	0.233
1.8	0.260
2.0	0.275
2.2	0.300
2.4	0.308
2.6	0.340
2.8	0.350
3.0	0.346
3.2	0.343
3.4	0.350

CURVE 141
T = 298

λ	ρ
0.33	0.166
0.35	0.138
0.40	0.173
0.45	0.218
0.50	0.219
0.55	0.216
0.6	0.211
0.6	0.211
0.7	0.202
0.8	0.187
0.9	0.182
1.0	0.182
1.1	0.187
1.2	0.196
1.3	0.207

CURVE 141 (cont.)

λ	ρ
1.4	0.227
1.6	0.262
1.8	0.279
2.0	0.303
2.2	0.319
2.4	0.341
2.6	0.363
2.8	0.372
3.0	0.373
3.2	0.370
3.4	0.368

CURVE 142
T = 298

λ	ρ
0.33	0.114
0.35	0.115
0.40	0.213
0.45	0.288
0.50	0.332
0.55	0.373
0.6	0.403
0.6	0.414
0.7	0.444
0.8	0.464
0.9	0.474
1.0	0.479
1.1	0.486
1.2	0.491
1.3	0.505
1.4	0.508
1.6	0.525
1.8	0.539
2.0	0.550
2.2	0.553
2.4	0.562
2.6	0.558
2.8	0.560
3.0	0.575
3.2	0.571
3.4	0.575

CURVE 143
T = 298

λ	ρ
0.33	0.146
0.35	0.039
0.40	0.059
0.45	0.137
0.50	0.206
0.55	0.251
0.6	0.206
0.6	0.310
0.7	0.369
0.8	0.400
0.9	0.425
1.0	0.434
1.1	0.452
1.2	0.462
1.3	0.471
1.4	0.479
1.6	0.500
1.8	0.505
2.0	0.521
2.2	0.529
2.4	0.534
2.6	0.536
2.8	0.537
3.0	0.548
3.2	0.545
3.4	0.543

CURVE 144
T = 298

λ	ρ
0.33	0.260
0.35	0.233
0.40	0.062
0.45	0.044
0.50	0.083
0.55	0.127
0.6	0.177
0.6	0.172
0.7	0.246
0.8	0.296
0.9	0.335
1.0	0.365
1.1	0.379
1.2	0.398
1.3	0.411

CURVE 144 (cont.)

λ	ρ
1.4	0.410
1.6	0.447
1.8	0.452
2.0	0.464
2.2	0.466
2.4	0.473
2.6	0.482
2.8	0.486
3.0	0.494
3.2	0.490
3.4	0.497

CURVE 145
T = 298

λ	ρ
0.33	0.256
0.35	0.276
0.40	0.255
0.45	0.208
0.50	0.180
0.55	0.162
0.6	0.147
0.6	0.153
0.7	0.143
0.8	0.148
0.9	0.168
1.0	0.187
1.1	0.206
1.2	0.226
1.3	0.250
1.4	0.254
1.6	0.311
1.8	0.327
2.0	0.346
2.2	0.366
2.4	0.370
2.6	0.381
2.8	0.395
3.0	0.403
3.2	0.399
3.4	0.403

CURVE 146
T = 298

λ	ρ
0.33	0.322
0.35	0.390
0.40	0.284
0.45	0.281
0.50	0.195
0.55	0.142
0.6	0.108
0.6	0.094
0.7	0.089
0.8	0.128
0.9	0.178
1.0	0.226
1.1	0.266
1.2	0.304
1.3	0.334
1.4	0.342
1.6	0.398
1.8	0.424
2.0	0.440
2.2	0.455
2.4	0.464
2.6	0.470
2.8	0.475
3.0	0.488
3.2	0.490
3.4	0.493

CURVE 147
T = 298

λ	ρ
0.33	0.284
0.35	0.266
0.40	0.240
0.45	0.223
0.50	0.214
0.55	0.216
0.6	0.221
0.6	0.227
0.7	0.227
0.8	0.232
0.9	0.237
1.0	0.256
1.1	0.256
1.2	0.246
1.3	0.230

DATA TABLE NO. 227 (continued)

λ	ρ
CURVE 147 (cont.)	
1.4	0.229
1.6	0.296
1.8	0.356
2.0	0.407
2.2	0.426
2.4	0.426
2.6	0.423
2.8	0.415
3.0	0.411
3.2	0.402
3.4	0.398
CURVE 148	
T = 298	
0.33	0.265
0.35	0.247
0.40	0.212
0.45	0.233
0.50	0.263
0.55	0.279
0.6	0.297
0.6	0.300
0.7	0.296
0.8	0.256
0.9	0.217
1.0	0.172
1.1	0.133
1.2	0.128
1.3	0.113
1.4	0.112
1.6	0.126
1.8	0.154
2.0	0.185
2.2	0.202
2.4	0.225
2.6	0.246
2.8	0.266
3.0	0.283
3.2	0.285
3.4	0.290
CURVE 149	
T = 298	
0.33	0.303
0.35	0.309
0.40	0.336
0.45	0.359
0.50	0.380
0.55	0.406
0.6	0.423
0.6	0.437
0.7	0.469
0.8	0.479
0.9	0.484
1.0	0.486
1.1	0.501
1.2	0.510
1.3	0.519
1.4	0.532
1.6	0.553
1.8	0.563
2.0	0.564
2.2	0.567
2.4	0.570
2.6	0.576
2.8	0.580
3.0	0.582
3.2	0.578
3.4	0.575
CURVE 150	
T = 298	
0.33	0.237
0.35	0.219
0.40	0.202
0.45	0.204
0.50	0.205
0.55	0.216
0.6	0.226
0.6	0.231
0.7	0.256
0.8	0.281
0.9	0.281
1.0	0.300
1.1	0.304
1.2	0.314
CURVE 150 (cont.)	
1.3	0.323
1.4	0.322
1.6	0.344
1.8	0.360
2.0	0.346
2.2	0.319
2.4	0.327
2.6	0.344
2.8	0.377
3.0	0.392
3.2	0.392
3.4	0.400

SPECIFICATION TABLE NO. 228 NORMAL SOLAR REFLECTANCE OF TITANIUM

Curve No.	Ref. No.	Year	Temperature Range, K	Geometry θ	θ'	ω'	Reported Error, %	Composition (weight percent), Specifications and Remarks
1	269	1960	298	$\sim 0°$		2π		Calculated from spectral data.
2	269	1960	298	$\sim 0°$		2π		No details given.

DATA TABLE NO. 228 NORMAL SOLAR REFLECTANCE OF TITANIUM

[Temperature, T, K; Reflectance, ρ]

T	ρ
CURVE 1*	
298	0.36
CURVE 2*	
298	0.37

* Not shown on plot

SPECIFICATION TABLE NO. 229 ANGULAR SPECTRAL ABSORPTANCE OF TITANIUM

Curve No.	Ref. No.	Year	Temperature K	Wavelength Range, μ	Geometry θ	Reported Error, %	Composition (weight percent), Specifications and Remarks
1	225	1965	294	1.48-23.2	25°		99 Ti; lapped plate; surface roughness 2 μ RMS; measured in dry nitrogen; heated cavity at approx 1056 K with platinum reference; authors assumed $\alpha = 1 - R(2\pi, 25°)$.

DATA TABLE NO. 229 ANGULAR SPECTRAL ABSORPTANCE OF TITANIUM

[Wavelength, λ, μ; Absorptance, α; Temperature, T, K]

λ	α
CURVE 1*	
T = 294	
1.48	0.450
1.78	0.433
1.97	0.420
2.47	0.402
3.03	0.370
3.48	0.327
3.95	0.290
4.94	0.233
5.96	0.197
6.92	0.174
8.63	0.160
9.86	0.145
11.0	0.133
13.0	0.105
15.0	0.096
17.8	0.090
20.0	0.081
21.1	0.088
23.2	0.076

* Not shown on plot

SPECIFICATION TABLE NO. 230 NORMAL SPECTRAL TRANSMITTANCE OF TITANIUM

Curve No.	Ref. No.	Year	Temperature K	Wavelength Range, μ	Geometry θ θ' ω'	Reported Error, %	Composition (weight percent), Specifications and Remarks
1	308	1965	298	0.0325-0.0697	$\sim 0°$ $\sim 0°$		Unbacked Ti film (525 ± 50 Å thick); 99.9 pure; evaporated at 2 x 10^{-5} mm Hg at room temperature; error ~10% for $\lambda > 0.035\mu$ and 25% for $\lambda < 0.035\mu$.
2	168	1961	298	0.0383-0.0688	$\sim 0°$ $\sim 0°$		Unbacked Ti film (1020 Å thick); exposed to air; measured in vacuum.

DATA TABLE NO. 230 NORMAL SPECTRAL TRANSMITTANCE OF TITANIUM

[Wavelength, λ, μ; Transmittance, τ; Temperature, T, K]

λ	τ	λ	τ
CURVE 1*		CURVE 2*	
T = 298		T = 298	
0.0325	0.007	0.0383	0.0519
0.0337	0.024	0.0399	0.0478
0.0348	0.051	0.0411	0.0417
0.0355	0.079	0.0449	0.0253
0.0363	0.092	0.0472	0.0208
0.0368	0.088	0.0480	0.0189
0.0378	0.127	0.0488	0.0140
0.0392	0.235	0.0500	0.0117
0.0400	0.220	0.0508	0.0095
0.0405	0.218	0.0522	0.0075
0.0423	0.195	0.0528	0.0060
0.0431	0.175	0.0553	0.0046
0.0437	0.169	0.0564	0.0026
0.0451	0.144	0.0573	0.0017
0.0463	0.125	0.0600	0.0017
0.0470	0.121	0.0637	0.0007
0.0479	0.118	0.0688	0.0000
0.0484	0.101		
0.0494	0.079		
0.0504	0.087		
0.0525	0.075		
0.0528	0.071		
0.0532	0.065		
0.0537	0.069		
0.0549	0.054		
0.0569	0.047		
0.0579	0.042		
0.0596	0.030		
0.0608	0.026		
0.0614	0.022		
0.0626	0.016		
0.0642	0.010		
0.0660	0.008		
0.0667	0.005		
0.0681	0.003		
0.0697	0.000		

* Not shown on plot

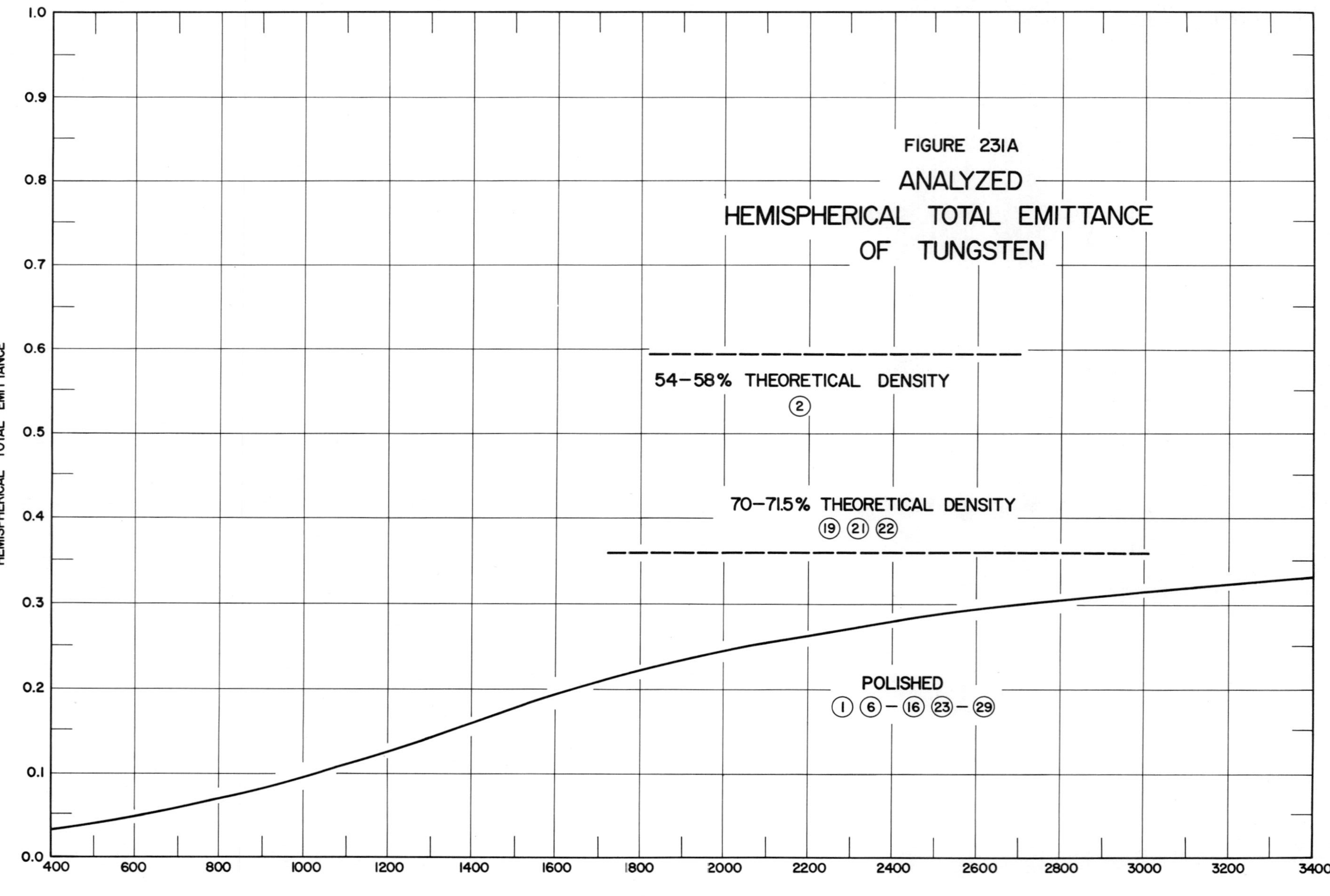
FIGURE 231A
ANALYZED
HEMISPHERICAL TOTAL EMITTANCE
OF TUNGSTEN
54–58% THEORETICAL DENSITY
(2)
70–71.5% THEORETICAL DENSITY
(19) (21) (22)
POLISHED
(1) (6) – (16) (23) – (29)
HEMISPHERICAL TOTAL EMITTANCE
1.0
0.9
0.8
0.7
0.6
0.5
0.4
0.3
0.2
0.1
0.0
400
600
800
1000
1200
1400
1600
1800
2000
2200
2400
2600
2800
3000
3200
3400
TEMPERATURE, K

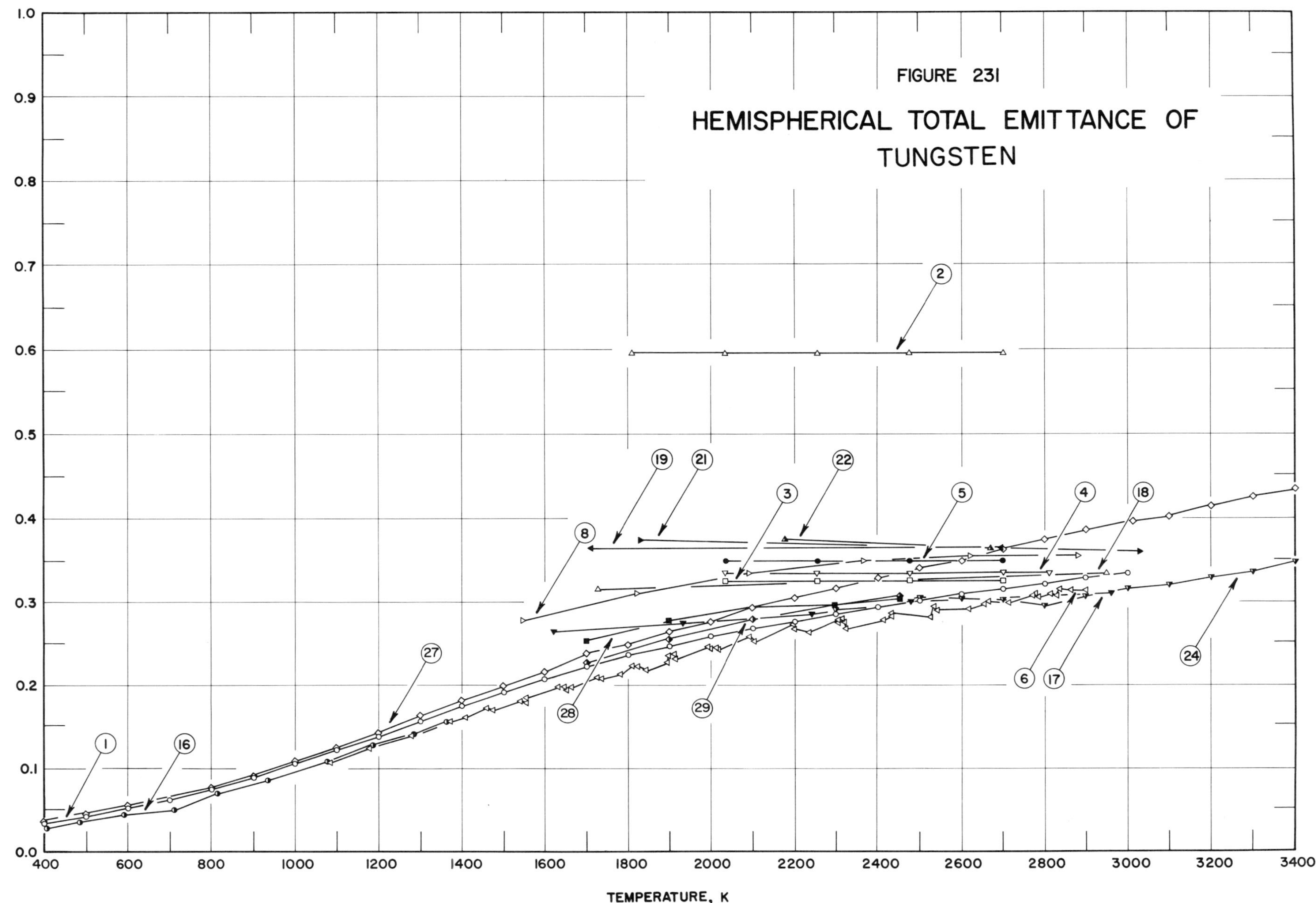

FIGURE 231
HEMISPHERICAL TOTAL EMITTANCE OF TUNGSTEN

SPECIFICATION TABLE NO. 231 HEMISPHERICAL TOTAL EMITTANCE OF TUNGSTEN

Curve No.	Ref. No.	Year	Temperature Range, K	Reported Error, %	Composition (weight percent), Specifications and Remarks
1	102	1933	273-3000		High purity; highly polished; surface contained many grooves; data extracted from smooth curve.
2	103	1966	1811-2700		99. 8 pure; porosity 42-46%; fabricated by gravity sintering tungsten particles 0. 006 to 0. 01 in. in size; fired for a long time at T >2478 K; measured in vacuum; data extracted from smooth curve.
3	103	1966	2033-2700		99. 8 pure; etched; polished; measured in vacuum; data extracted from smooth curve.
4	103	1966	2033-2811		99. 8 pure; etched; polished; measured in vacuum; data extracted from smooth curve.
5	103	1966	2033-2700		99. 8 pure; etched; polished; measured in vacuum; data extracted from smooth curve.
6	85	1963	1084-2900	± 5	Optically smooth, opaque, contamination-free ribbon (0. 005 in. thick); aged at 2400 K for approximately 30 min; measured in vacuum (3×10^{-7} to 1×10^{-5} mm Hg); temperature determined with optical pyrometer.
7	85	1963	1084-2976	± 5	Different sample, same as curve 6 specimen; same conditions except temperature determined with thermocouple.
8	104	1960	1545-2880	± 10	Spectrographically pure; measured in vacuum ($<10^{-5}$ mm Hg).
9	12	1962	515-734	± 2. 7	Measured in vacuum ($<10^{-4}$ mm Hg); Run No. 1A.
10	12	1962	808-629	± 2. 7	Above specimen and conditions; Run No. 1B.
11	12	1962	409-935	± 2. 7	Above specimen and conditions; Run No. 2A.
12	12	1962	1086-596	± 2. 7	Above specimen and conditions; Run No. 2B.
13	12	1962	400-1081	± 2. 7	Above specimen and conditions; Run No. 3.
14	12	1962	418-1287	± 2. 7	Above specimen and conditions; Run No. 4A.
15	12	1962	1363-1264	± 2. 7	Above specimen and conditions; Run No. 4B.
16	12	1962	404-1359	± 2. 7	Above specimen and conditions; Run No. 5.
17	105	1959	1620-2960		Spectrographically pure; measured in vacuum ($<10^{-4}$ mm Hg).
18	84	1962	1723-2948		Porous rod; 90% of theoretical density; measured in argon; [Author's designation: Specimen A].
19	84	1962	1713-2688		Porous rod; 71. 5% of theoretical density; measured in argon; [Author's designation: Specimen B].
20	84	1962	1718-2798		Porous rod; 90% of theoretical density; measured in argon; [Author's designation: Specimen C].
21	84	1962	1828-3028		Porous rod; 70% of theoretical density; measured in argon; [Author's designation: Specimen D].
22	84	1962	2178-2668		Porous rod; 70% of theoretical density; measured in argon; [Author's designation: Specimen E].
23	84	1962	2298-3000		Density >99% of theoretical value; polished with successively finer abrasive papers (numbers 1, 0, 00, 000, and 0000); measured in argon.
24	70	1960	2300-3400		0. 04 Mo, 0. 002 Cu, 0. 008 others, 99. 95 W; pressed and sintered metal powder; hot-worked; successively polished with No. 1-, 0-, 00-, 000-, and 0000- abrasive papers; measured in argon (>760 mm Hg); [Author's designation: Tungsten No. 2].
25	70	1960	2300-3400		0. 04 Mo, 0. 006 O_2, 0. 005 Ni, 0. 005 Ti, 0. 004 Fe, 0. 027 others, W balance; pressed and sintered metal powder; hot-worked; successively polished with No. 1-, 0-, 00-, 000-, and 0000- abrasive papers; measured in argon (>760 mm Hg); [Author's designation: Tungsten No. 1].

SPECIFICATION TABLE NO. 231 (continued)

Curve No.	Ref. No.	Year	Temperature Range, K	Reported Error, %	Composition (weight percent), Specifications and Remarks
26	271	1921	1000-3500		Tungsten filament.
27	272	1916	300-3540		Tungsten filaments aged 24 hrs at 2400 K; measured in vacuum; smoothed values.
28	345	1966	1700-2450		99.9 pure; foil (20μ thick); heated in vacuum of around 5 x 10^{-6} mm Hg at 2200 K for 1 hr; each data point is the mean of ten measurements.
29	345	1966	1699-2450		Above specimen and conditions except after measurement of emittance at 2450 K and data reported is only one reading per temperature setting; above 1900 K data was taken from smooth curve.

DATA TABLE NO. 231 HEMISPHERICAL TOTAL EMITTANCE OF TUNGSTEN

[Temperature, T, K; Emittance, ϵ]

T	ϵ
CURVE 1	
273	0.022*
293	0.023*
300	0.024*
400	0.034
500	0.042
600	0.052
700	0.062
800	0.074
900	0.089
1000	0.105
1100	0.121
1200	0.138
1300	0.156
1400	0.174
1500	0.192
1600	0.207
1700	0.222
1800	0.236
1900	0.248
2000	0.259
2100	0.269
2200	0.278
2300	0.286
2400	0.294
2500	0.301
2600	0.309
2700	0.315
2800	0.321
2900	0.329
3000	0.334
CURVE 2	
1811	0.595
2033	0.595
2255	0.595
2478	0.595
2700	0.595
CURVE 3	
2033	0.325
2255	0.325
CURVE 3 (cont.)	
2478	0.325
2700	0.325
CURVE 4	
2033	0.335
2255	0.335
2478	0.335
2700	0.335
2811	0.335
CURVE 5	
2033	0.350
2255	0.350
2478	0.350
2700	0.350
CURVE 6	
1084	0.107
1178	0.123
1278	0.139
1370	0.155
1408	0.161
1460	0.172
1474	0.169
1540	0.181
1556	0.179
1556	0.185
1630	0.198
1646	0.198
1656	0.195
1662	0.198
1724	0.210
1736	0.208
1780	0.213
1810	0.223
1822	0.222
1842	0.218
1894	0.227
1898	0.235
1910	0.237
CURVE 6 (cont.)	
1914	0.231
1992	0.246
2000	0.243
2014	0.244
2018	0.242
2090	0.258
2104	0.252
2196	0.274
2200	0.268
2234	0.263
2304	0.279
2304	0.276
2316	0.280
2320	0.276
2324	0.268
2416	0.279
2432	0.283
2434	0.286
2526	0.282
2536	0.295
2544	0.290
2618	0.291
2654	0.297
2664	0.300
2712	0.299
2776	0.307
2780	0.309
2784	0.305
2816	0.309
2830	0.307
2838	0.314
2862	0.313
2900	0.313
CURVE 7*	
1084	0.110
1100	0.107
1164	0.124
1176	0.121
1260	0.139
1284	0.136
1360	0.157
CURVE 7 (cont.)*	
1376	0.152
1412	0.160*
1452	0.170
1460	0.168
1472	0.171
1540	0.179
1544	0.186
1556	0.183
1560	0.180
1640	0.194
1644	0.195
1672	0.194
1720	0.205
1724	0.207
1732	0.205
1792	0.211
1824	0.220
1824	0.217
1844	0.216
1904	0.227
1920	0.299
1932	0.227
2004	0.239
2024	0.238
2100	0.249
2112	0.251
2116	0.248
2216	0.266
2228	0.258
2240	0.261
2336	0.265
2340	0.270
2356	0.262
2436	0.273
2450	0.279
2464	0.278
2464	0.275
2532	0.278
2564	0.289
2564	0.284
2580	0.280
2636	0.283
2676	0.293
CURVE 7 (cont.)*	
2680	0.290
2684	0.294
2728	0.293
2796	0.303
2800	0.300
2836	0.307
2856	0.303
2872	0.311
2884	0.305
2912	0.314
2924	0.311
2976	0.313
CURVE 8	
1545	0.278
1820	0.310
2090	0.335
2365	0.350
2620	0.355
2880	0.355
CURVE 9*	
515	0.043
601	0.052
734	0.066
CURVE 10*	
808	0.077
629	0.055
CURVE 11*	
409	0.033
524	0.043
634	0.055
739	0.067
837	0.078
935	0.094
CURVE 12*	
1086	0.119
839	0.077
596	0.048
CURVE 13*	
400	0.035
583	0.050
686	0.056
785	0.071
885	0.083
1000	0.102
1081	0.115
CURVE 14*	
418	0.036
570	0.048
690	0.061
780	0.068
879	0.081
990	0.099
1075	0.114
1147	0.126
1287	0.144
CURVE 15*	
1363	0.161
1264	0.139
CURVE 16	
404	0.028
483	0.036
591	0.045
711	0.050
817	0.069
936	0.085
1078	0.108
1185	0.127
1281	0.141
1359	0.155
CURVE 17	
1620	0.265
1930	0.275
2240	0.285
2480	0.300
2960	0.310
CURVE 18	
1723	0.315
2948	0.335
CURVE 19	
1713	0.365
2688	0.365
CURVE 20*	
1718	0.365
2798	0.345
CURVE 21	
1828	0.375
3028	0.360
CURVE 22	
2178	0.375
2668	0.365
CURVE 23*	
2298	0.268
3000	0.31
CURVE 24	
2300	0.290
2400	0.294*
2500	0.304
2600	0.304
2700	0.302
CURVE 24 (cont.)	
2800	0.295
2900	0.305
3000	0.315
3100	0.320
3200	0.330
3300	0.336
3400	0.348
CURVE 25*	
2300	0.273
2400	0.274
2500	0.284
2600	0.289
2700	0.290
2800	0.298
2900	0.302
3000	0.310
3100	0.318
3200	0.325
3300	0.328
3400	0.336
CURVE 26*	
1000	0.100
1500	0.194
2000	0.264
2500	0.313
3000	0.347
3500	0.369
CURVE 27*	
300	0.0237
400	0.0330
500	0.0421
600	0.0514
700	0.0626
800	0.0755
900	0.0904
1000	0.1069
1100	0.1247

* Not shown on plot

DATA TABLE NO. 231 (continued)

T	ϵ
CURVE 27 (cont.)*	
1200	0.1435
1300	0.1625
1400	0.1813
1500	0.1995
1600	0,2170
1700	0.2338
1800	0.2494
1900	0.2643
2000	0.2785
2100	0.2920
2200	0.3046
2300	0.3174
2400	0.3290
2500	0.3407
2600	0.3517
2700	0.3625
2800	0.3733
2900	0.3839
3000	0.3941
3100	0.4041
3200	0.4142
3300	0.4239
3400	0.4334
3500	0.4428
3540	0.4467
CURVE 28	
1700	0.252
1901	0.276
2100	0.290*
2299	0.296
2450	0.302
CURVE 29*	
1699	0.227
1899	0.255
2101	0.279
2300	0.296
2450	0.302

* Not shown on plot

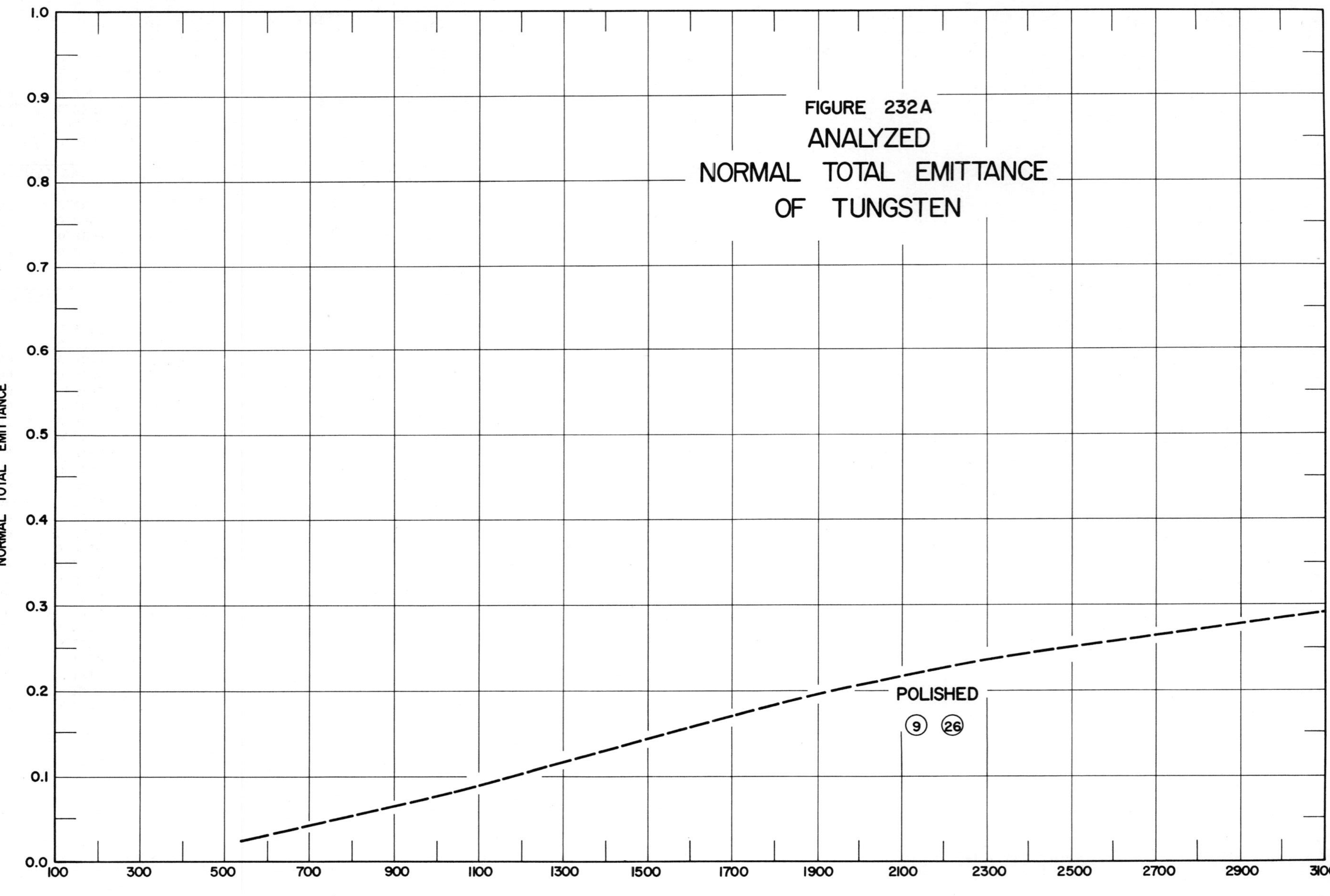
FIGURE 232A
ANALYZED
NORMAL TOTAL EMITTANCE
OF TUNGSTEN
POLISHED
9 26
NORMAL TOTAL EMITTANCE
1.0
0.9
0.8
0.7
0.6
0.5
0.4
0.3
0.2
0.1
0.0
100
300
500
700
900
1100
1300
1500
1700
1900
2100
2300
2500
2700
2900
3100
TEMPERATURE, K

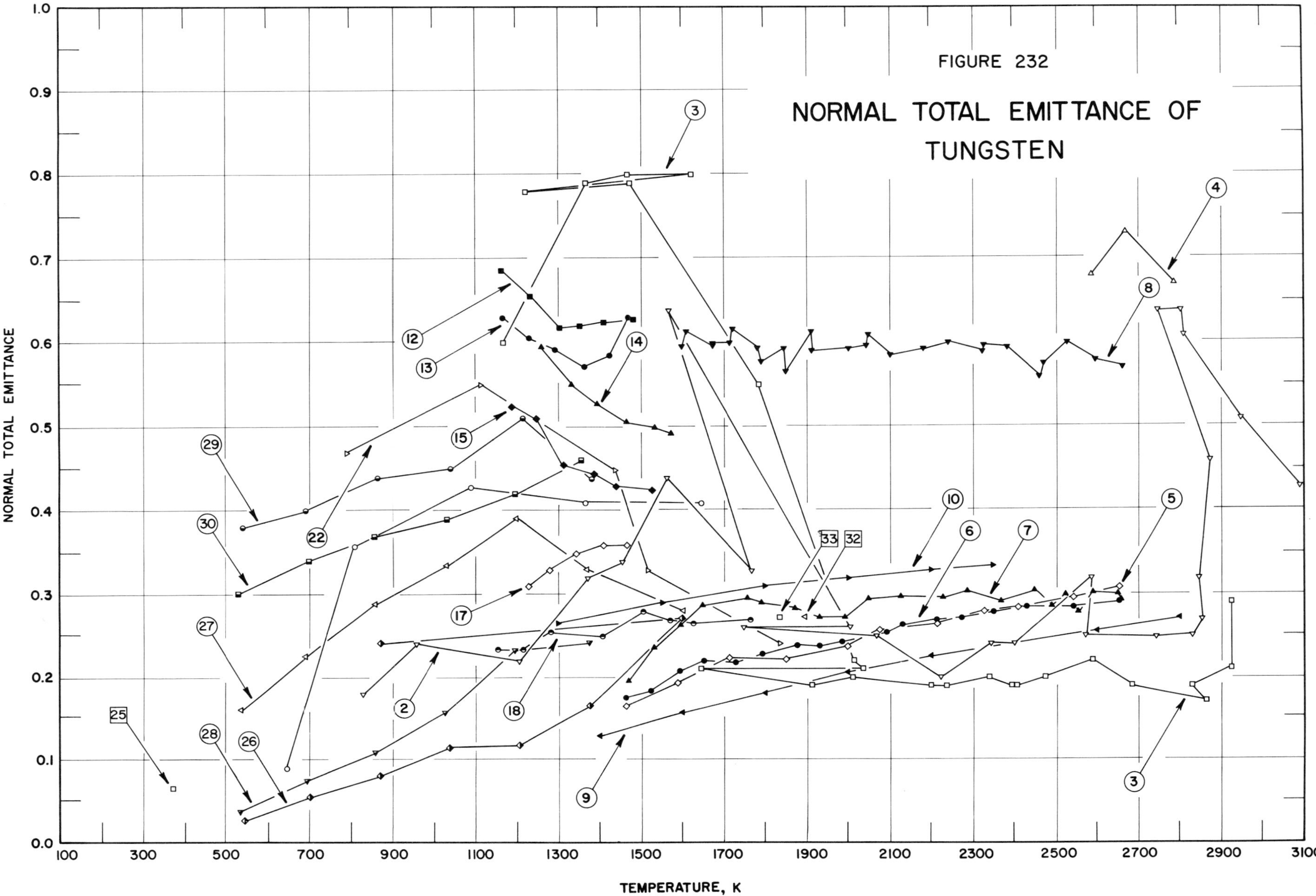
FIGURE 232
NORMAL TOTAL EMITTANCE OF TUNGSTEN
NORMAL TOTAL EMITTANCE
TEMPERATURE, K
0.0
0.1
0.2
0.3
0.4
0.5
0.6
0.7
0.8
0.9
1.0
100
300
500
700
900
1100
1300
1500
1700
1900
2100
2300
2500
2700
2900
3100
2
3
4
5
6
7
8
9
10
12
13
14
15
17
18
22
25
26
27
28
29
30
32
33

SPECIFICATION TABLE NO. 232 NORMAL TOTAL EMITTANCE OF TUNGSTEN

Curve No.	Ref. No.	Year	Temperature Range, K	Geometry θ '	Reported Error, %	Composition (weight percent), Specifications and Remarks
1	107	1960	644-1644	~0°		Powdered tungsten compacted and sintered into ingot, then machined to size and polished; slightly oxidized during experimentation; measured in demoisturized helium gas.
2	106	1962	828-3089	~0°	10-20	Arc cast; whisker growth around edges appeared at 1372 K; at 1567 K spots appeared on surface and then disappeared; measured in argon.
3	106	1962	1172-2922	~0°	10-20	Hot pressed; spots appeared on the specimen surface and specimen started smoking at 1783 K; at 2017 K surface cleared and smoking stopped; measured in argon; run 1, increasing temp; author reports data unreliable for $T < 1645$ K.
4	106	1962	2589-2789	~0°	10-20	Above specimen and conditions; run 1, decreasing temp; author reports data as unreliable.
5	103	1966	1461-2656	~0°		99.8 pure; etched; polished; measured in vacuum; [Author's designation: Specimen S3].
6	103	1966	1461-2656	~0°		99.8 pure; etched; polished; measured in vacuum; [Author's designation: Specimen S4].
7	103	1966	1467-2656	~0°		99.8 pure; etched; polished; measured in vacuum; [Author's designation: Specimen S5].
8	103	1966	1600-2661	~0°		99.8 pure; porosity 42-46%; fabricated by gravity sintering tungsten particles 0.006 to 0.001 in. in size; fired for a long time at $T > 2478$ K; measured in vacuum.
9	85	1963	1400-2800	~0°	± 5	Optically smooth, opaque, contamination-free ribbon (0.005 in. thick); aged at 2400 K for approximately 30 min; measured in vacuum (3×10^{-7} to 1×10^{-5} mm Hg); data extracted from smooth curve.
10	108	1929	1300-2350	~0°		Filament (0.01 in. thick); 5% of filament converted to W_2C; data extracted from smooth curve.
11	108	1929	1350-2450	~0°		Filament; data extracted from smooth curve.
12	109	1961	1165-1482	~0°		Tungsten powder with grain diameter 20 μ compressed into billets; sintered for 20 hrs; measured in vacuum (4×10^{-6} to 2×10^{-5} mm Hg); run 1.
13	109	1961	1168-1470	~0°		Above specimen and conditions; run 2.
14	109	1961	1260-1575	~0°		Different sample, same as curve 12 specimen and conditions except grain diameter 10 μ.
15	109	1961	1190-1528	~0°		Tungsten powder with grain diameter 1 μ compressed into billets; sintered for 20 hrs; measured in vacuum (4×10^{-6} to 2×10^{-5} mm Hg); run 1.
16	109	1961	1172-1435	~0°		Above specimen and conditions; run 2.
17	109	1961	1228-1465	~0°		Tungsten powder with grain diameter 1 μ compressed into billets; measured in vacuum (4×10^{-6} to 2×10^{-5} mm Hg).
18	109	1961	1155-1765	~0°		Different sample, same as curve 17 specimen and conditions.
19	109	1961	1255-1670	~0°		Above specimen and conditions.
20	95	1963	1600-3000	~0°		99.9 W from Carbide Specialty Co.; polished with carbide paper of 240, 400, and 600 grit, respectively, and then with silk cloth and felt cloth; washed in acetone, then alcohol, and dried with dry nitrogen; data calculated from spectral data and extracted from smooth curve.

SPECIFICATION TABLE NO. 232 (continued)

Curve No.	Ref. No.	Year	Temperature Range, K	Geometry θ'	Reported Error, %	Composition (weight percent), Specifications and Remarks
21	75	1962	1561-2922	~0°		Ground to a smooth finish; measured in dried argon or helium; data extracted from smooth curve.
22	75	1962	789-1839	~0°		Air tarnished; N_2 purge.
23	76	1962	1600-3000	~0°		Prepared from micronized powder; hot pressed at >2273 K; sintered, polished, etched, then degassed by heating to ~973 K; measured in argon; data extracted from smooth curve; computed from spectral data.
24	76	1962	1800-2800	~0°		Prepared from micronized powder; hot pressed at >2273 K; sintered, polished, etched, then degassed by heating to ~973 K; measured in argon; data extracted from smooth curve; computed from spectral data.
25	15	1947	373	~0°		Polished coating.
26	227	1964	544-1603	~0°		Impurities <40 ppm; as received; surface roughness 1.5 microinches rms before emittance test, 2.4 microinches rms after emittance test; preheated in vacuum at 1000 K for 0.5 hr; measured in vacuum (8 x 10^{-5} mm Hg); [Author's designation: Sample 1].
27	227	1964	536-1603	~0°		Impurities <40 ppm; grit blasted; surface roughness 17 microinches rms, 21 microinches rms after emittance test; preheated in vacuum at 1000 K for 0.5 hr; measured in vacuum (8 x 10^{-5} mm Hg); [Author's designation: Sample 2].
28	227	1964	536-1378	~0°		Above specimen and conditions except second temperature cycle.
29	227	1964	542-1383	~0°		Impurities <40 ppm; grit blasted; surface roughness 110 microinches rms before emittance test, 38 microinches rms after emittance test; preheated in vacuum at 1000 K for 0.5 hr; measured in vacuum (5 x 10^{-6} mm Hg); $\omega' = 3.4 \times 10^{-4}$ sr; [Author's designation: Sample 3].
30	227	1964	529-1358	~0°		Above specimen and conditions except second cycle.
31	267	1965	1930	~0°		Obtained from Fansteel Metallurgical Corp; thermal conductivity 0.950 W cm^{-1} K^{-1}; specific heat 0.0390 Cal sec^{-1} cm^{-1} K^{-1}.
32	267	1965	1898	~0°		Obtained from Fansteel Metallurgical Corp; thermal conductivity 1.029 W cm^{-1} K^{-1}; specific heat 0.0389 Cal sec^{-1} cm^{-1} K^{-1}.
33	267	1965	1835	~0°		Obtained from Fansteel Metallurgical Corp; thermal conductivity 1.029 W cm^{-1} K^{-1}; specific heat 0.0386 Cal sec^{-1} cm^{-1} K^{-1}.
34	267	1965	1748	~0°		Obtained from Fansteel Metallurgical Corp; thermal conductivity 1.063 W cm^{-1} K^{-1}; specific heat 0.0382 Cal sec^{-1} cm^{-1} K^{-1}.
35	267	1965	1678	~0°		Obtained from Fansteel Metallurgical Corp; thermal conductivity 1.021 W cm^{-1} K^{-1}; specific heat 0.0379 Cal sec^{-1} cm^{-1} K^{-1}.
36	267	1965	1578	~0°		Obtained from Fansteel Metallurgical Corp; thermal conductivity 1.113 W cm^{-1} K^{-1}; specific heat 0.0375 Cal sec^{-1} cm^{-1} K^{-1}.
37	267	1965	1536	~0°		Obtained from Fansteel Metallurgical Corp; thermal conductivity 1.146 W cm^{-1} K^{-1}; specific heat 0.0373 Cal sec^{-1} cm^{-1} K^{-1}.

SPECIFICATION TABLE NO. 232 (continued)

Curve No.	Ref. No.	Year	Temperature Range, K	Geometry θ'	Reported Error, %	Composition (weight percent), Specifications and Remarks
38	267	1965	1513	~0°		Obtained from Fansteel Metallurgical Corp; thermal conductivity 1.130 W cm^{-1} K^{-1}; specific heat 0.0372 Cal sec^{-1} cm^{-1} K^{-1}.
39	267	1965	1373	~0°		Obtained from Fansteel Metallurgical Corp; thermal conductivity 1.021 W cm^{-1} K^{-1}; specific heat 0.0366 Cal sec^{-1} cm^{-1} K^{-1}.

DATA TABLE NO. 232 NORMAL TOTAL EMITTANCE OF TUNGSTEN

[Temperature, T,K; Emittance, ∈]

T	∈
CURVE 1	
644	0.09
811	0.36
1089	0.43
1367	0.41
1644	0.41
CURVE 2	
828	0.18
958	0.24
1206	0.22
1372	0.32
1456	0.34
1561	0.44
1767	0.33
1567	0.64
2006	0.26
1750	0.26
2067	0.25
2222	0.20
2345	0.24
2400	0.24
2589	0.32
2572	0.25
2744	0.25
2833	0.25
2855	0.27
2850	0.32
2878	0.46
2750	0.64
2805	0.64
2811	0.61
2950	0.51
2950	0.51*
3089	0.43
CURVE 3	
1172	0.60
1317	0.79
1471	0.80
1622	0.80
1222	0.78
1478	0.79
1783	0.55
CURVE 3 (cont.)	
2017	0.22
2039	0.21
1645	0.21
1911	0.19
2011	0.20
2200	0.19
2239	0.19
2339	0.20
2395	0.19
2406	0.19
2478	0.20
2589	0.22
2683	0.19
2861	0.17
2828	0.19
2922	0.21
2922	0.21*
2922	0.29
CURVE 4	
2589	0.68
2672	0.73
2789	0.67
CURVE 5	
1461	0.165
1589	0.193
1717	0.224
1850	0.222
2000	0.237
2078	0.257
2217	0.265
2327	0.280
2408	0.285
2545	0.297
2656	0.308
CURVE 6	
1461	0.175
1522	0.184
1597	0.207
1650	0.220
CURVE 6 (cont.)	
1728	0.218
1794	0.228
1875	0.239
1930	0.238
1992	0.242
2094	0.254
2133	0.264
2214	0.270
2278	0.272
2350	0.279
2430	0.285
2545	0.285
2656	0.291
CURVE 7	
1467	0.196
1533	0.236
1597	0.263
1644	0.287
1755	0.295
1789	0.290
1869	0.283
1928	0.274
1994	0.272
2050	0.295
2122	0.298
2228	0.297
2283	0.303
2366	0.292
2450	0.305
2492	0.287
2522	0.300
2558	0.279
2589	0.302
2650	0.301
2656	0.293
CURVE 8	
1600	0.595
1611	0.614
1666	0.596
1666	0.600
1719	0.600
CURVE 8 (cont.)	
1722	0.617
1783	0.593
1789	0.577
1847	0.593
1850	0.565
1911	0.612
1914	0.590
2003	0.592
2044	0.598
2050	0.609
2103	0.585
2183	0.592
2244	0.600
2325	0.590
2328	0.597
2385	0.595
2461	0.560
2472	0.575
2533	0.600
2600	0.580
2661	0.570
CURVE 9	
1400	0.128
1600	0.156
1800	0.181
2000	0.205
2200	0.224
2400	0.240*
2600	0.256
2800	0.271
CURVE 10	
1300	0.265
1550	0.290
1800	0.310
2000	0.320
2200	0.330
2350	0.335
CURVE 11*	
1350	0.160
CURVE 11* (cont.)	
1550	0.200
1750	0.230
2000	0.260
2200	0.280
2400	0.300
2450	0.303*
CURVE 12	
1165	0.685
1235	0.655
1305	0.619
1355	0.621
1412	0.625
1482	0.628
CURVE 13	
1168	0.630
1232	0.605
1295	0.592
1365	0.571
1425	0.585
1470	0.630
CURVE 14	
1260	0.595
1335	0.550
1395	0.528
1465	0.505
1532	0.500
1575	0.492
CURVE 15	
1190	0.525
1250	0.510
1315	0.455
1390	0.445
1440	0.430
1528	0.425
CURVE 16*	
1172	0.535
CURVE 16* (cont.)	
1245	0.490
1325	0.440
1385	0.410
1435	0.420
CURVE 17	
1228	0.310
1280	0.330
1345	0.350
1410	0.360
1465	0.360
CURVE 18	
1155	0.235
1215	0.235
1285	0.255
1412	0.250
1502	0.280
1572	0.270
1625	0.265
1765	0.270
CURVE 19*	
1255	0.280
1340	0.260
1470	0.230
1535	0.230
1630	0.230
1670	0.240
CURVE 20*	
1600	0.197
1800	0.218
2000	0.240
2200	0.261
2400	0.279
2600	0.296
2800	0.313
3000	0.328
CURVE 21*	
1561	0.24
1922	0.24
2478	0.24
2922	0.24
CURVE 22	
789	0.470
1117	0.550
1439	0.450
1519	0.330
1839	0.240
CURVE 23*	
1600	0.200
2000	0.245
2400	0.280
2800	0.312
3000	0.330
CURVE 24*	
1800	0.185
2200	0.232
2600	0.280
2800	0.300
CURVE 25	
373	0.066
CURVE 26	
544	0.025
702	0.054
870	0.080
1039	0.114
1208	0.117
1378	0.164
1603	0.27
CURVE 27	
536	0.159
CURVE 27 (cont.)	
693	0.224
860	0.288
1033	0.334
1200	0.390
1373	0.330
1603	0.280
CURVE 28	
536	0.037
698	0.073
860	0.108
1029	0.156
1196	0.230
1378	0.240
CURVE 29	
542	0.38
692	0.40
863	0.44
1040	0.45
1214	0.51
1383	0.44
CURVE 30	
529	0.30
700	0.34
858	0.37
1033	0.39
1198	0.42
1358	0.46
CURVE 31*	
1930	0.275
CURVE 32	
1898	0.274
CURVE 33	
1835	0.272

* Not shown on plot

DATA TABLE NO. 232 (continued)

T	ϵ
CURVE 34*	
1748	0.269
CURVE 35*	
1678	0.267
CURVE 36*	
1578	0.264
CURVE 37*	
1536	0.263
CURVE 38*	
1513	0.262
CURVE 39*	
1373	0.257

* Not shown on plot

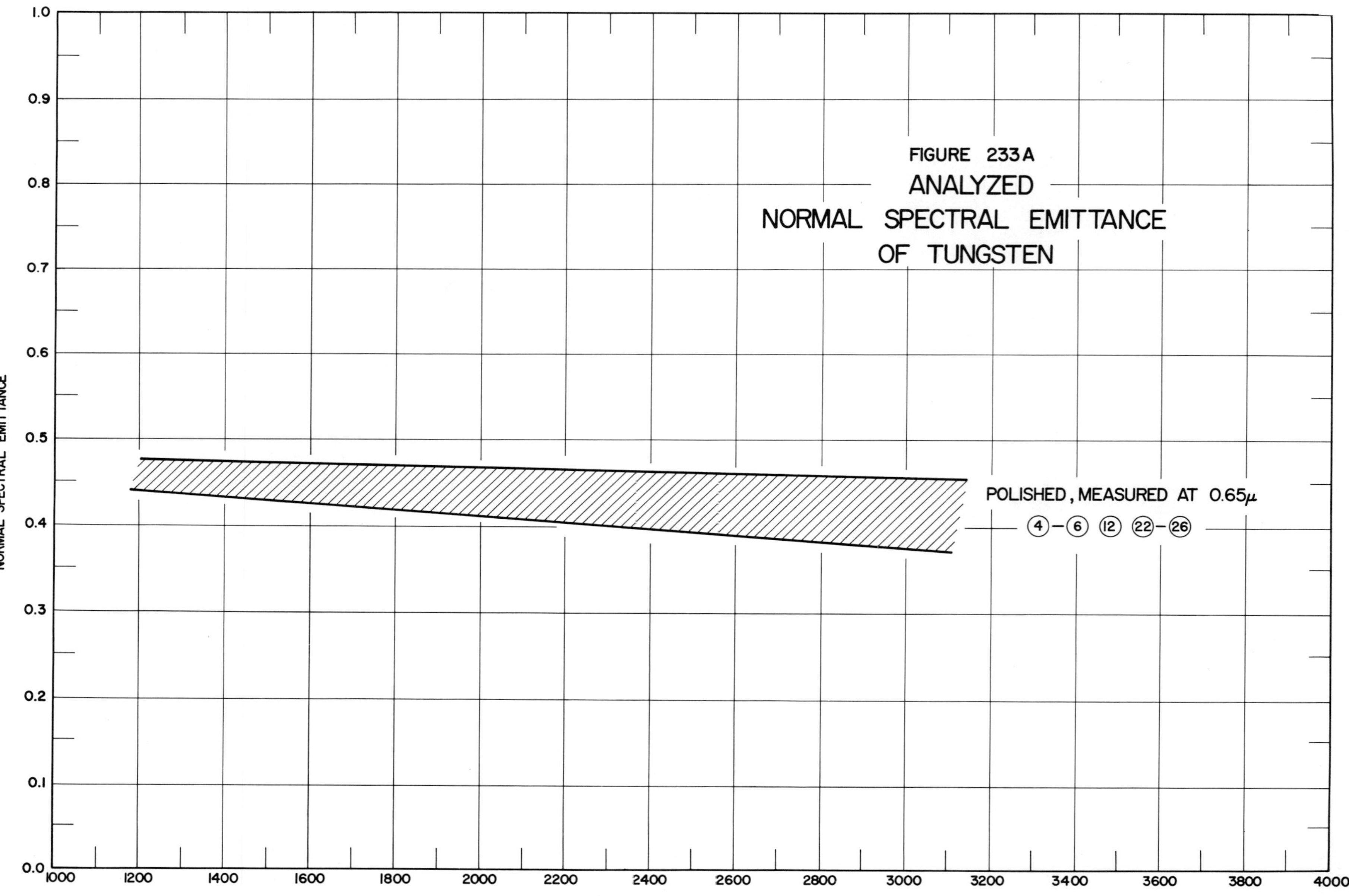
FIGURE 233A
ANALYZED
NORMAL SPECTRAL EMITTANCE
OF TUNGSTEN
POLISHED, MEASURED AT 0.65μ
④–⑥ ⑫ ㉒–㉖
NORMAL SPECTRAL EMITTANCE
TEMPERATURE, K
1.0
0.9
0.8
0.7
0.6
0.5
0.4
0.3
0.2
0.1
0.0
1000
1200
1400
1600
1800
2000
2200
2400
2600
2800
3000
3200
3400
3600
3800
4000

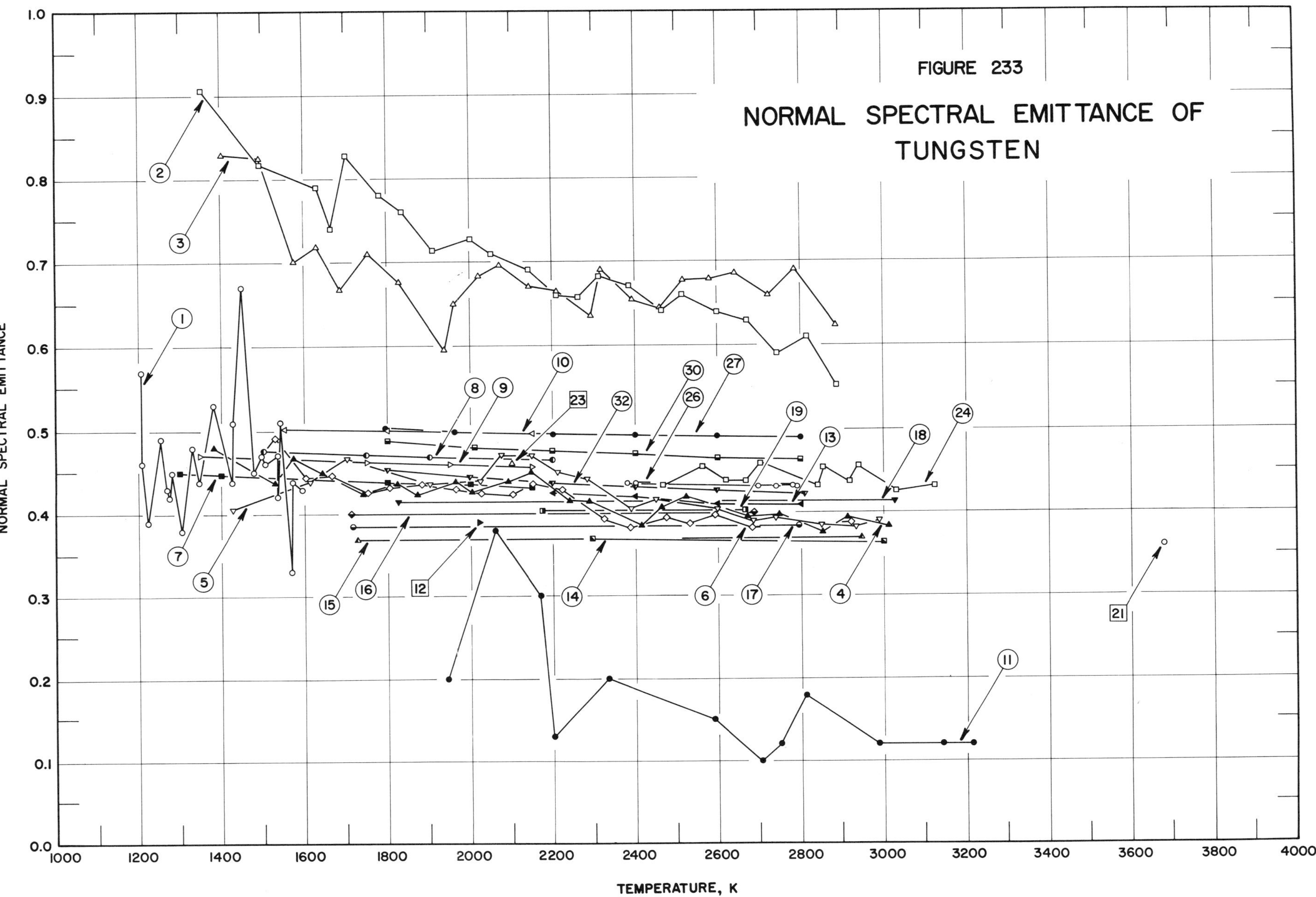

FIGURE 233
NORMAL SPECTRAL EMITTANCE OF TUNGSTEN
NORMAL SPECTRAL EMITTANCE
TEMPERATURE, K
0.0
0.1
0.2
0.3
0.4
0.5
0.6
0.7
0.8
0.9
1.0
1000
1200
1400
1600
1800
2000
2200
2400
2600
2800
3000
3200
3400
3600
3800
4000
1
2
3
4
5
6
7
8
9
10
11
12
13
14
15
16
17
18
19
21
23
24
26
27
30
32

SPECIFICATION TABLE NO. 233 NORMAL SPECTRAL EMITTANCE OF TUNGSTEN

Curve No.	Ref. No.	Year	Wavelength μ	Temperature Range, K	Geometry θ'	Reported Error, %	Composition (weight percent), Specifications and Remarks
1	110	1961	0.65	1206-1596	~0°		Tungsten powder (1 micron in diameter) compressed at 25000 psi; sintered for 20 hrs at 1780 K.
2	103	1966	0.65	1355-2889	~0°		99.8 pure; porosity 42-46%; fabricated by gravity sintering 0.008 to 0.010 in. tungsten particles; fired for a long time at T >2478 K; measured in vacuum.
3	103	1966	0.65	1405-2889	~0°		99.8 pure; porosity 42-46%; fabricated by gravity sintering 0.008 to 0.010 in. tungsten particles; fired for a long time at T >2478 K; measured in hydrogen.
4	103	1966	0.65	1383-3011	~0°		99.8 pure; polished; measured in vacuum.
5	103	1966	0.65	1428-2994	~0°		99.8 pure; etched; measured in vacuum.
6	103	1966	0.65	1472-2922	~0°		99.8 pure; polished; measured in hydrogen.
7	108	1929	0.660	1300-2150	~0°		Tubular pressed filament; data extracted from smooth curve.
8	108	1929	0.472	1500-2200	~0°		Above specimen and conditions.
9	108	1929	0.660	1350-2150	~0°		Tubular pressed tungsten filament; partially carbonized to W_2C by heating in acetylene vapor at 1950 to 2150 K; data extracted from smooth curve.
10	108	1929	0.472	1550-2150	~0°		Above specimen and conditions.
11	75	1962	0.69	1944-3217	~0°		Ground to a smooth finish; measured in dried argon or helium.
12	19	1914	0.650	2023	~0°	1	Film; tungsten substrate; measured in hydrogen; Pt reference (ϵ = 0.33 for λ = 0.650 μ at all temp).
13	77	1936	0.66	1300-2800	~0°	±2	Extruded from paste; sintered in hydrogen; polished electrolytically in KOH bath; polished with 00, 000, and 0000 polishing papers; measured in vacuum (10^{-5} mm Hg).
14	84	1962	0.65	2298-3000	~0°		Density >99% of theoretical value; polished with successively finer abrasive papers (numbers 1, 0, 00, 000 and 0000); measured in argon.
15	84	1962	0.65	1723-2948	~0°		Porous rod; 90% of theoretical density; measured in argon; [Author's designation: Specimen A].
16	84	1962	0.65	1713-2688	~0°		Porous rod; 71.5% of theoretical density; measured in argon; [Author's designation: Specimen B].
17	84	1962	0.65	1718-2798	~0°		Porous rod; 90% of theoretical density; measured in argon; [Author's designation: Specimen C].
18	84	1962	0.65	1828-3028	~0°		Porous rod; 70% of theoretical density; measured in argon; [Author's designation: Specimen D].
19	84	1962	0.65	2178-2668	~0°		Porous rod; 70% of theoretical density; measured in argon; [Author's designation: Specimen E].
20	70	1960	0.65	2300-3100	~0°		0.04 Mo, 0.006 O_2, 0.005 Ni, 0.005 Ti, 0.004 Fe, 0.027 others, W balance; pressed and sintered metal powder; hot worked; successively polished with No. 1-, 0-, 00-, 000-, and 0000- abrasive papers; measured in argon (>760 mm Hg); [Author's designation: Tungsten No. 1].

SPECIFICATION TABLE NO. 233 (continued)

Curve No.	Ref. No.	Year	Wavelength μ	Temperature Range, K	Geometry θ'	Reported Error, %	Composition (weight percent), Specifications and Remarks
21	70	1960	0.65	3680	~0°		0.04 Mo, 0.002 Cu, 0.008 others, 99.95 W; pressed and sintered metal powder; hot worked; successively polished with No. 1-, 0-, 00-, 000-, and 0000- abrasive papers; measured in argon (>760 mm Hg); [Author's designation: Tungsten No. 2].
22	111	1936	0.669	1200-2100	~0°		Heated in vacuum (10^{-7} mm Hg) at 2000 K for 700 hrs; measured in vacuum.
23	111	1936	0.669	2100	~0°		Above specimen and conditions except heat treated at 2100 K for 2 more hrs.
24	274	1964	0.65	2466-3125	~0°		Final polished with No. 600 abrasive cloth and a diamond lap; measured in vacuum (10^{-5} to 10^{-6} mm Hg).
25	177	1933	0.660	1300-2000	~0°		Electrolytically polished in KOH and 00, 000, 0000 polishing paper; measured in vacuum.
26	333	1962	0.650	2380-2790	~0°		Tungsten ribbon.
27	331	1917	0.3400	1800-2800	~0°		Measured in nitrogen; data is mean of three or more measurements; data extracted from smooth curve.
28	331	1917	0.3800	1800-2800	~0°		Above specimen and conditions.
29	331	1917	0.4200	1800-2800	~0°		Above specimen and conditions.
30	331	1917	0.4800	1800-2800	~0°		Above specimen and conditions.
31	331	1917	0.5000	1800-2800	~0°		Above specimen and conditions.
32	331	1917	0.5400	1800-2800	~0°		Above specimen and conditions.

DATA TABLE NO. 233 NORMAL SPECTRAL EMITTANCE OF TUNGSTEN

[Temperature, T, K; Emittance, ∈; Wavelength, λ, u]

T	∈
CURVE 1	
λ = 0.65	
1206	0.57
1208	0.46
1221	0.39
1256	0.49
1269	0.43
1274	0.42
1280	0.45
1302	0.38
1333	0.48
1346	0.44
1383	0.53
1428	0.44
1429	0.51
1452	0.67
1478	0.45
1495	0.47
1503	0.46
1535	0.47
1535	0.42
1542	0.51
1565	0.33
1570	0.44
1596	0.43
CURVE 2	
λ = 0.65	
1355	0.906
1494	0.819
1630	0.790
1666	0.740
1708	0.828
1783	0.780
1839	0.760
1911	0.714
2003	0.728
2055	0.710
2144	0.690
2211	0.660
2264	0.657
2311	0.682
2394	0.670
2466	0.644
CURVE 2 (cont.)	
λ = 0.65	
2517	0.660
2600	0.640
2672	0.630
2744	0.590
2819	0.610
2889	0.554
CURVE 3	
λ = 0.65	
1405	0.830
1494	0.825
1575	0.700
1633	0.718
1686	0.667
1755	0.710
1828	0.675
1939	9.595
1961	0.650
2022	0.683
2072	0.696
2144	0.670
2211	0.665
2294	0.636
2319	0.690
2394	0.655
2464	0.645
2517	0.676
2581	0.677
2644	0.687
2725	0.660
2786	0.690
2889	0.625
CURVE 4	
λ = 0.65	
1383	0.482
1530	0.439
1572	0.467
1644	0.450
1741	0.426
1825	0.439
CURVE 4 (cont.)	
λ = 0.65	
1872	0.424
1964	0.440
2005	0.427
2091	0.441
2147	0.450
2241	0.415
2289	0.414
2414	0.386
2464	0.408
2522	0.420
2603	0.407*
2672	0.397
2747	0.398
2853	0.378
2914	0.395
3011	0.383
CURVE 5	
λ = 0.65	
1428	0.406
1616	0.441
1653	0.468
1905	0.438
1958	0.439*
2025	0.442
2078	0.471
2150	0.470
2211	0.450
2283	0.443
2389	0.405
2447	0.417
2419	0.408*
2600	0.405
2686	0.390
2742	0.395
2856	0.385
2939	0.383
2994	0.391
CURVE 6	
λ = 0.65	
1472	0.450*
1533	0.492
1605	0.445
1666	0.447
1755	0.425
1805	0.433
1883	0.437
1964	0.430
2025	0.425
2105	0.424
2155	0.437
2222	0.429
2322	0.393
2389	0.384
2478	0.395
2533	0.387
2594	0.397
2683	0.382*
2756	0.400*
2850	0.382*
2922	0.389
CURVE 7	
λ = 0.660	
1300	0.450
1400	0.448
1600	0.444*
1800	0.440
2000	0.436
2150	0.432
CURVE 8	
λ = 0.472	
1500	0.476
1750	0.472
1900	0.468
2200	0.464
CURVE 9	
λ = 0.660	
1350	0.472
1550	0.468*
1750	0.464
1950	0.460
2150	0.456
CURVE 10	
λ = 0.472	
1550	0.502
1800	0.500
2150	0.496
CURVE 11	
λ = 0.69	
1944	0.20
2061	0.38
2172	0.30
2200	0.13
2333	0.20
2589	0.15
2705	0.10
2755	0.12
2811	0.18
2989	0.12
3144	0.12
3217	0.12
CURVE 12	
λ = 0.650	
2023	0.39
CURVE 13	
λ = 0.66	
1300	0.450*
1700	0.440*
2000	0.430*
2200	0.425
2400	0.420
2600	0.410
2800	0.410
CURVE 14	
λ = 0.65	
2298	0.37
3000	0.365
CURVE 15	
λ = 0.65	
1723	0.37
2948	0.37
CURVE 16	
λ = 0.65	
1713	0.40
2688	0.40
CURVE 17	
λ = 0.65	
1718	0.385
2798	0.385
CURVE 18	
λ = 0.65	
1828	0.415
3028	0.415
CURVE 19	
λ = 0.65	
2178	0.404
2668	0.404
CURVE 20*	
λ = 0.65	
2300	0.370
3100	0.365
CURVE 21	
λ = 0.65	
3680	0.360
CURVE 22*	
λ = 0.669	
1200	0.46
2000	0.46
2100	0.43
CURVE 23	
λ = 0.669	
2100	0.46
CURVE 24	
λ = 0.65	
2466	0.436
2562	0.457
2622	0.441
2671	0.441
2702	0.460
2841	0.433
2854	0.455
2919	0.441
2940	0.458
3032	0.428
3125	0.434
CURVE 25*	
λ = 0.660	
1300	0.457
1700	0.441
2000	0.431
CURVE 26	
λ = 0.650	
2380	0.438
2690	0.435
2780	0.433
2400	0.438
2740	0.433
2790	0.432
CURVE 27	
λ = 0.3400	
1800	0.501
1958	0.499
2200	0.496
2400	0.494
2599	0.492
2800	0.491
CURVE 28*	
λ = 0.3800	
1800	0.502
2000	0.497
2300	0.492
2602	0.489
2800	0.487
CURVE 29*	
λ = 0.4200	
1800	0.497
1999	0.492
2201	0.487
2399	0.483
2600	0.480
2800	0.478
CURVE 30	
λ = 0.4800	
1800	0.488
1999	0.481
2200	0.476
2400	0.471
2601	0.468
2800	0.464
CURVE 31*	
λ = 0.5000	
1800	0.470
2001	0.464
2200	0.458
2400	0.453
2600	0.449
2800	0.446

* Not shown on plot

DATA TABLE NO. 233 (continued)

T	ϵ
CURVE 32 $\lambda = 0.5400$	
1800	0.452
1999	0.444
2201	0.438
2400	0.433
2599	0.428
2800	0.425

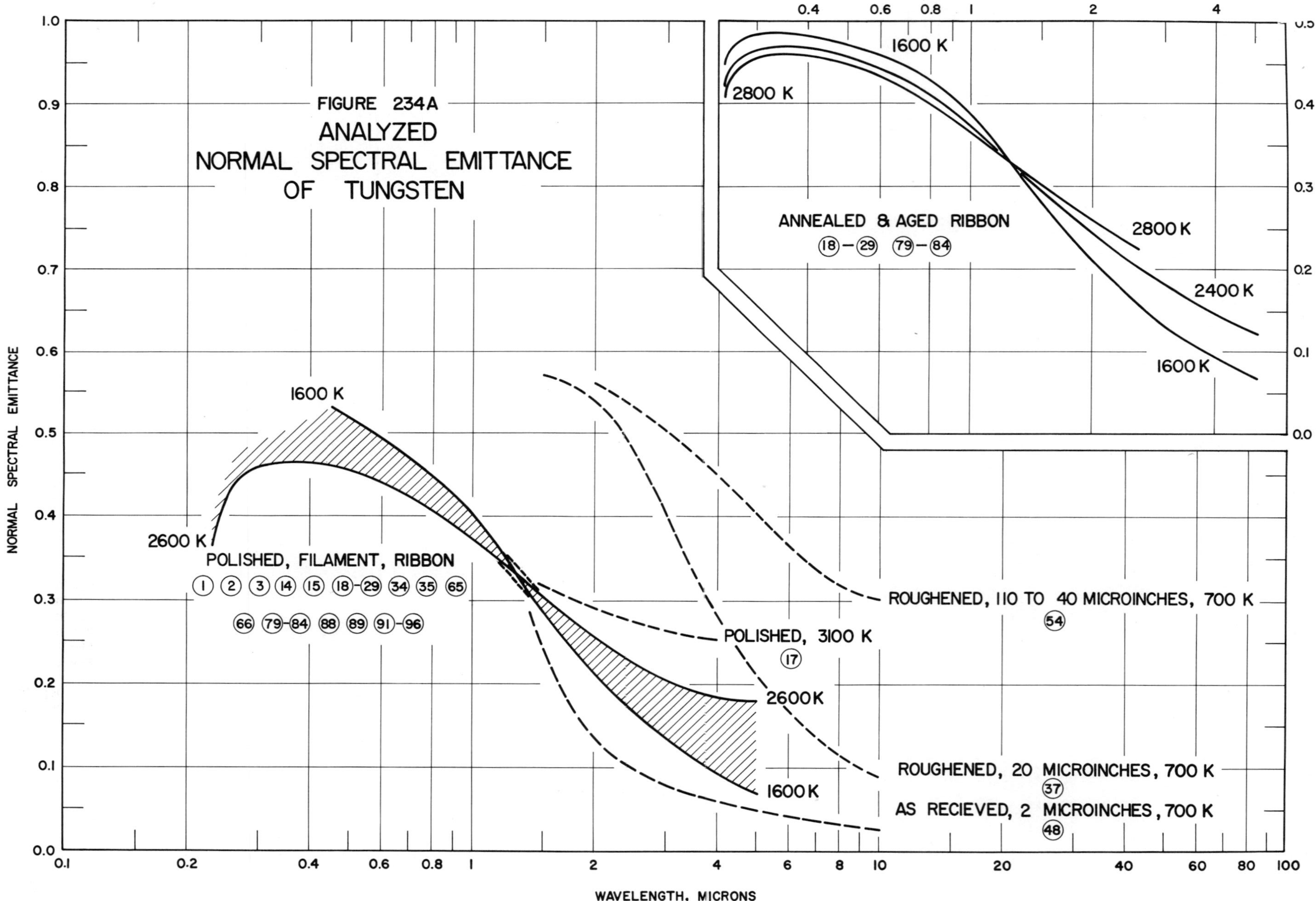
FIGURE 234A
ANALYZED
NORMAL SPECTRAL EMITTANCE
OF TUNGSTEN
NORMAL SPECTRAL EMITTANCE
WAVELENGTH, MICRONS
ANNEALED & AGED RIBBON
(18)-(29) (79)-(84)
1600 K
2800 K
2400 K
POLISHED, FILAMENT, RIBBON
(1) (2) (3) (14) (15) (18)-(29) (34) (35) (65)
(66) (79)-(84) (88) (89) (91)-(96)
2600 K
POLISHED, 3100 K
(17)
ROUGHENED, 110 TO 40 MICROINCHES, 700 K
(54)
ROUGHENED, 20 MICROINCHES, 700 K
(37)
AS RECIEVED, 2 MICROINCHES, 700 K
(48)

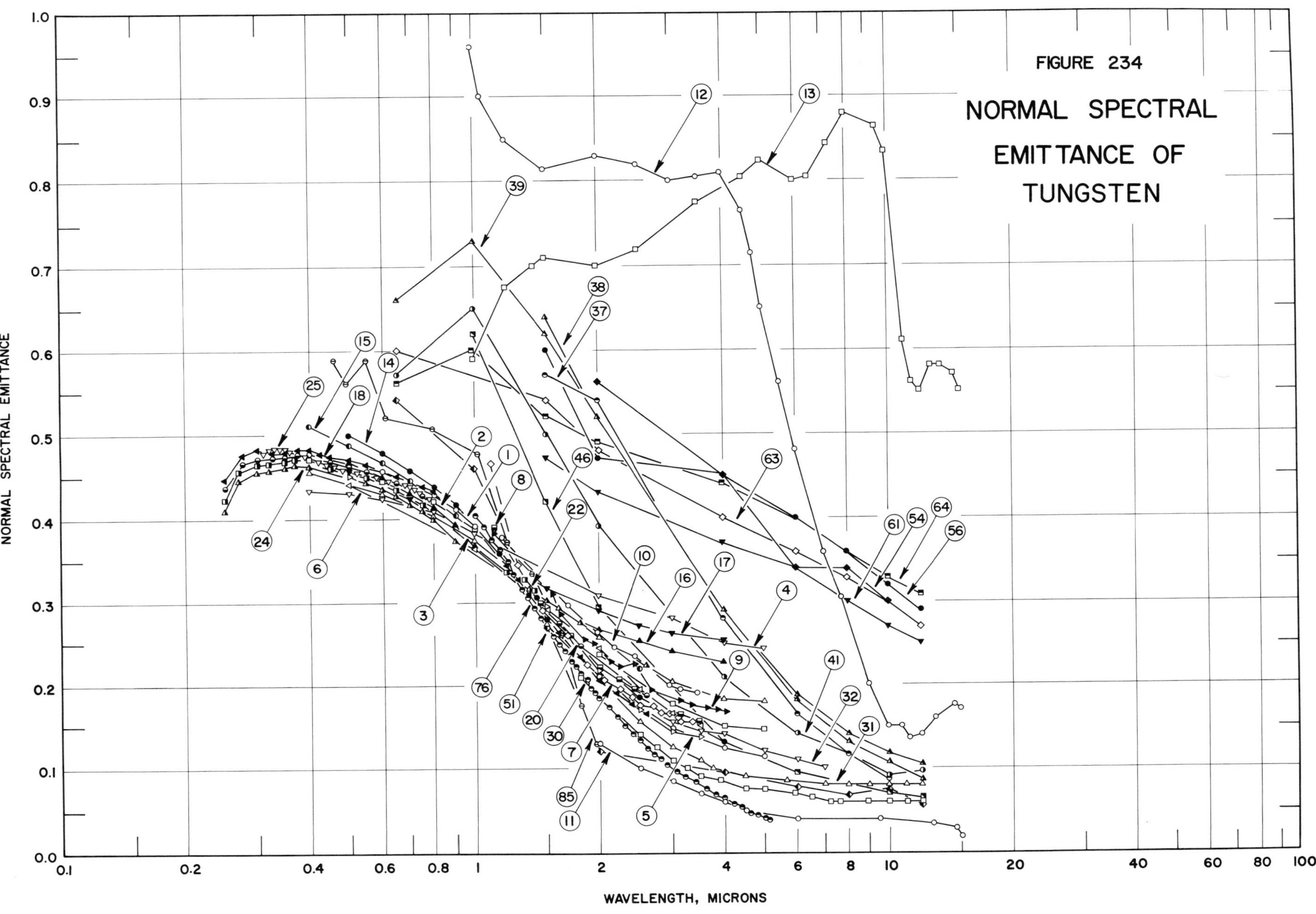

FIGURE 234
NORMAL SPECTRAL EMITTANCE OF TUNGSTEN
NORMAL SPECTRAL EMITTANCE
WAVELENGTH, MICRONS

SPECIFICATION TABLE NO. 234 NORMAL SPECTRAL EMITTANCE OF TUNGSTEN

Curve No.	Ref. No.	Year	Temperature K	Wavelength Range, μ	Geometry θ'	Reported Error, %	Composition (weight percent), Specifications and Remarks
1	112	1963	1605	0.4-5.0	$\sim 0^\circ$		Polished single crystal tungsten; measured in argon (1140 mm Hg) along 002 crystal plane; data extracted from smooth curve.
2	112	1963	2140	0.4-5.0	$\sim 0^\circ$		Above specimen and conditions.
3	112	1963	2639	0.4-5.0	$\sim 0^\circ$		Above specimen and conditions.
4	112	1963	2650	0.4-5.0	$\sim 0^\circ$		Above specimen and conditions.
5	113	1961	1830	0.5-3.5	$\sim 0^\circ$		Measured in argon (760 mm Hg); data extracted from smooth curve.
6	113	1961	2040	0.5-3.5	$\sim 0^\circ$		Above specimen and conditions.
7	113	1961	1316	1.10-3.47	$\sim 0^\circ$		Trace of surface oxidation observed; measured in vacuum.
8	113	1961	1340	1.12-3.48	$\sim 0^\circ$		Above specimen and conditions.
9	113	1961	1382	1.13-4.04	$\sim 0^\circ$		Above specimen and conditions except measured in argon (760 mm Hg).
10	113	1961	1429	1.16-3.48	$\sim 0^\circ$		Above specimen and conditions.
11	86	1961	523	2.00-15.00	$\sim 0^\circ$	±5	As received; data extracted from smooth curve.
12	86	1961	773	1.00-15.00	$\sim 0^\circ$		Different sample, same as curve 11 specimen and conditions.
13	86	1961	1023	1.00-15.00	$\sim 0^\circ$	±5	Different sample, same as curve 11 specimen and conditions.
14	95	1963	1600	0.50-4.00	$\sim 0^\circ$		99.9 W from Carbide Specialty Co.; polished with carbide paper of 240, 400, and 600 grit, respectively, and then with silk cloth and felt cloth; washed in acetone, then alcohol, and dried with dry nitrogen; data extracted from smooth curve.
15	95	1963	2000	0.40-4.00	$\sim 0^\circ$		Above specimen and conditions.
16	95	1963	2800	0.40-4.00	$\sim 0^\circ$		Above specimen and conditions.
17	95	1963	3100	0.40-4.00	$\sim 0^\circ$		Above specimen and conditions.
18	114	1954	1600	0.25-2.60	$\sim 0^\circ$	0.1	0.014 - 0.015 Fe, 0.004 - 0.008 Si, 0.001 - 0.003 Mn, 0.0003 - 0.0006 Mg, W balance; heated at 2400 K, treated in hydrogen, annealed at 2400 K for 100 hrs; measured in vacuum (5×10^{-6} mm Hg); data extracted from smooth curve.
19	114	1954	1800	0.25-2.60	$\sim 0^\circ$	0.1	Above specimen and conditions.
20	114	1954	2000	0.25-2.60	$\sim 0^\circ$	0.1	Above specimen and conditions.
21	114	1954	2200	0.25-2.60	$\sim 0^\circ$	0.1	Above specimen and conditions.
22	114	1954	2400	0.25-2.60	$\sim 0^\circ$	0.1	Above specimen and conditions.
23	114	1954	2600	0.25-2.60	$\sim 0^\circ$	0.1	Above specimen and conditions except measured in argon (500 mm Hg).
24	114	1954	2800	0.25-2.60	$\sim 0^\circ$	0.1	Above specimen and conditions.
25	115	1959	1600	0.310-0.800	$\sim 0^\circ$		Better than 99.99 percent pure; heated at 2750 K for 1/2 hr in vacuum, then annealed at 2500 K for 30 hrs, at 2800 K for 1/2 hr, and at 2500 K for 20 hrs; measured in vacuum (3×10^{-9} to 9.5×10^{-8} mm Hg).

SPECIFICATION TABLE NO. 234 (continued)

Curve No.	Ref. No.	Year	Temperature K	Wavelength Range, μ	Geometry θ'	Reported Error, %	Composition (weight percent), Specifications and Remarks
26	115	1959	1800	0.310-0.800	~0°		Above specimen and conditions.
27	115	1959	2000	0.310-0.800	~0°		Above specimen and conditions.
28	115	1959	2200	0.310-0.800	~0°		Above specimen and conditions.
29	115	1959	2400	0.310-0.800	~0°		Above specimen and conditions.
30	12	1962	1200	0.50-12.00	~0°	±3	Measured in vacuum ($<10^{-7}$ mm Hg).
31	12	1962	1428	0.50-12.00	~0°	±3	Above specimen and conditions.
32	12	1962	1972	0.45-7.00	~0°	±3	Above specimen and conditions.
33	241	1963	1660	1.10-1.70	0	<1	Single crystal; oriented so that surface of interest coincided with closed packed plane; optically polished; heated at several hundred degrees above temperature of interest in 90 Ar + 10 H atm; computed from optical constants.
34	241	1963	1790	0.9-1.7	0°	<1	Above specimen and conditions.
35	241	1963	1950	0.90-1.70	0°	<1	Above specimen and conditions.
36	241	1963	2050	0.90-1.70	0°	<1	Above specimen and conditions.
37	227	1964	693	1.5-10	~0°		Impurities <40 ppm; grit blasted; surface roughness 17 microinches rms, 21 microinches rms after emittance test; preheated in vacuum at 1000 K for 0.5 hr; measured in vacuum (8×10^{-6} mm Hg); [Author's designation: Sample 2].
38	227	1964	860	1.5-12	~0°		Above specimen and conditions.
39	227	1964	1033	0.65-12	~0°		Above specimen and conditions.
40	227	1964	1200	0.65-12	~0°		Above specimen and conditions.
41	227	1964	1373	0.65-12	~0°		Above specimen and conditions.
42	227	1964	1603	0.65-4	~0°		Above specimen and conditions.
43	227	1964	698	1.5-10	~0°		Above specimen and conditions except second temperature cycle.
44	227	1964	860	1.5-10	~0°		Above specimen and conditions.
45	227	1964	1029	1-12	~0°		Above specimen and conditions.
46	227	1964	1196	1-12	~0°		Above specimen and conditions.
47	227	1964	1378	1.5-4	~0°		Above specimen and conditions.
48	227	1964	702	1.5-10	~0°		Impurities <40 ppm; as received; surface roughness 1.5 microinches rms before emittance test, 2.4 microinches rms after emittance; preheated in vacuum for 2 hrs; measured in vacuum (7×10^{-6} mm Hg); [Author's designation: Sample 1].
49	227	1964	870	1.5-10	~0°		Above specimen and conditions.
50	227	1964	1039	1.5-12	~0°		Above specimen and conditions.

SPECIFICATION TABLE NO. 234 (continued)

Curve No.	Ref. No.	Year	Temperature K	Wavelength Range, μ	Geometry θ'	Reported Error, %	Composition (weight percent), Specifications and Remarks
51	227	1964	1208	0.65-12	~0°		Above specimen and conditions.
52	227	1964	1378	0.65-4	~0°		Above specimen and conditions.
53	227	1964	1603	0.65-4	~0°		Above specimen and conditions.
54	227	1962	692	2-10	~0°		Impurities <40 ppm; grit blasted; surface roughness 110 microinches rms before emittance test, 38 microinches after emittance test; preheated in vacuum at 1000 K for 0.5 hr; measured in vacuum (5×10^{-5} mm Hg); $\omega' = 3.4 \times 10^{-4}$ sr.; [Author's designation: Sample 3].
55	227	1964	863	1.5-12	~0°		Above specimen and conditions.
56	227	1964	1040	1.5-12	~0°		Above specimen and conditions.
57	227	1964	1214	0.65-12	~0°		Above specimen and conditions.
58	227	1964	1383	0.65-12	~0°		Above specimen and conditions.
59	227	1964	699	1.5-10	~0°		Above specimen and conditions.
60	227	1964	529	1.5-10	~0°		Above specimen and conditions.
61	227	1964	700	1.5-12	~0°		Above specimen and conditions.
62	227	1964	858	1.5-12	~0°		Above specimen and conditions.
63	227	1964	1198	0.65-12	~0°		Above specimen and conditions.
64	227	1964	1358	0.65-12	~0°		Above specimen and conditions.
65	76	1962	1600	0.50-4.00	~0°		Prepared from micronized powder; hot pressed at >2273 K; sintered, polished, etched, then degassed by heating to ~973 K; measured in argon; data extracted from smooth curve.
66	76	1962	2000	0.40-4.00	~0°		Above specimen and conditions.
67	76	1962	2800	0.40-4.00	~0°		Above specimen and conditions.
68	76	1962	3100	0.40-4.00	~0°		Above specimen and conditions.
69	239	1959	1429	1.157-3.486	~0°	5	Highly polished; measured in argon.
70	239	1959	1382	1.127-4.038	~0°	5	Different sample, same as above specimen and conditions.
71	239	1959	1340	1.120-3.485	~0°	5	Different sample, same as above specimen and conditions.
72	239	1959	1316	1.096-3.459	~0°	5	Different sample, same as above specimen and conditions.
73	237	1965	1800	0.467-0.698	~0°		Chemically pure; measured in vacuum; authors assumed $\epsilon = 1-\rho$ and computed ρ from optical constants.
74	237	1965	2150	0.467-0.698	~0°		Above specimen and conditions.
75	237	1965	2520	0.467-0.698	~0°		Above specimen and conditions.

SPECIFICATION TABLE NO. 234 (continued)

Curve No.	Ref. No.	Year	Temperature K	Wavelength Range, μ	Geometry θ'	Reported Error, %	Composition (weight percent), Specifications and Remarks
76	338	1965	1244	1.00-5.10	0°		Ribbon; black body (at 1336 K) used as reference standard.
77	338	1965	1339	1.00-5.10	0°		Above specimen and conditions.
78	338	1965	1413	1.00-5.10	0°		Above specimen and conditions.
79	338	1965	1629	1.00-5.10	0°		Above specimen and conditions.
80	338	1965	1833	1.00-5.10	0°		Above specimen and conditions.
81	338	1965	2002	1.00-5.10	0°		Above specimen and conditions.
82	338	1965	2160	1.00-5.10	0°		Above specimen and conditions.
83	338	1965	2327	1.00-5.10	0°		Above specimen and conditions.
84	338	1965	2441	1.00-5.10	0°		Above specimen and conditions.
85	323	1966	300	0.46-2.00	0°		Filament (0.25-0.32 mm in dia); baked for 1 hr at 798 K in vacuum, cooled, heated for 5-10 min in vacuum, and cooled; measured in argon (600 mm Hg); data calculated from optical constants.
86	323	1966	1100	0.45-2.00	0°		Above specimen and conditions.
87	323	1966	1500	0.46-2.00	0°		Above specimen and conditions.
88	323	1966	2000	0.46-2.00	0°		Above specimen and conditions.
89	323	1966	2500	0.46-2.00	0°		Above specimen and conditions.
90	331	1917	2143	0.3478-0.5641	~0°	5	Measured in nitrogen; data is mean of three or more measurements.
91	333	1962	1600	0.230-0.269	~0°		Tungsten ribbon; data extracted from smooth curve
92	333	1962	1800	0.229-0.270	~0°		Above specimen and conditions.
93	333	1962	2000	0.229-0.271	~0°		Above specimen and conditions.
94	333	1962	2200	0.231-0.268	~0°		Above specimen and conditions.
95	333	1962	2400	0.231-0.268	~0°		Above specimen and conditions.
96	333	1962	2600	0.231-0.269	~0°		Above specimen and conditions.
97	333	1962	2800	0.229-0.268	~0°		Above specimen and conditions.

DATA TABLE NO. 284 NORMAL SPECTRAL EMITTANCE OF TUNGSTEN

[Wavelength, λ, μ; Emittance, ∈; Temperature, T, K]

λ	∈
CURVE 1	**T = 1605**
0.4	0.478
0.6	0.455
0.8	0.427
0.9	0.410
1.0	0.390
2.0	0.210
3.0	0.147
4.0	0.123
5.0	0.113
CURVE 2	**T = 2140**
0.4	0.470
0.6	0.445
0.8	0.422
1.0	0.387
2.0	0.237
3.0	0.177
4.0	0.150
5.0	0.145
CURVE 3	**T = 2639**
0.4	0.455
0.6	0.430
0.8	0.403
1.0	0.375
2.0	0.258
3.0	0.205
4.0	0.183
5.0	0.180
CURVE 4	**T = 2650**
0.4	0.433
0.5	0.430
0.6	0.425
0.8	0.405
1.0	0.383
CURVE 4 (cont.)	
2.0	0.308
3.0	0.270
4.0	0.250
5.0	0.243
CURVE 5	**T = 1830**
0.5	0.450
1.0	0.380*
1.5	0.295
2.0	0.218
2.5	0.170
3.0	0.146
3.5	0.138
CURVE 6	**T = 2040**
0.5	0.440
1.0	0.365
1.5	0.293
2.0	0.244
2.5	0.195
3.0	0.165
3.5	0.150
CURVE 7	**T = 1316**
1.10	0.465
1.27	0.345
1.47	0.290
1.67	0.260
1.86	0.225
2.05	0.208
2.24	0.195
2.40	0.185
2.50	0.175
2.70	0.174
2.82	0.167
2.97	0.165
3.10	0.160
CURVE 7 (cont.)	
3.12	0.156
3.35	0.155
3.47	0.150*
CURVE 8	**T = 1340**
1.12	0.390
1.45	0.300
1.70	0.260
2.00	0.220
2.22	0.208
2.42	0.195
2.92	0.165*
3.13	0.164
3.48	0.156
CURVE 9	**T = 1382**
1.13	0.382
1.55	0.310
1.62	0.285
1.70	0.263*
1.85	0.255
1.94	0.250
2.14	0.228
2.25	0.221
2.43	0.225
2.68	0.193
2.92	0.185
3.12	0.183
3.33	0.175
3.62	0.172
3.84	0.170
4.04	0.169
CURVE 10	**T = 1429**
1.16	0.379
1.68	0.299
1.99	0.263
CURVE 10 (cont.)	
2.17	0.246
2.43	0.235
2.93	0.200
3.12	0.195
3.48	0.190
CURVE 11	**T = 523**
2.00	0.130
2.50	0.100
3.00	0.085
3.50	0.070
4.00	0.060
4.50	0.050
6.00	0.040
9.50	0.040
12.75	0.035
14.50	0.030
15.00	0.020
CURVE 12	**T = 773**
1.00	0.960
1.05	0.900
1.20	0.850
1.50	0.815
2.00	0.830
2.50	0.820
3.00	0.800
3.50	0.805
4.00	0.810
4.50	0.765
4.75	0.715
5.00	0.650
5.50	0.560
6.00	0.480
7.00	0.360
7.75	0.305
9.00	0.200
10.00	0.150
10.75	0.150
CURVE 12 (cont.)	
11.25	0.135
12.00	0.140
13.00	0.160
14.50	0.175
15.00	0.170
CURVE 13	**T = 1023**
1.00	0.590
1.20	0.675
1.40	0.700
1.50	0.710
2.00	0.700
2.50	0.720
3.50	0.775
4.50	0.805
5.00	0.825
6.00	0.800
6.50	0.805
7.25	0.845
8.00	0.880
9.50	0.865
10.00	0.835
11.00	0.610
11.50	0.560
12.00	0.550
12.75	0.580
13.50	0.580
14.50	0.570
15.00	0.550
CURVE 14	**T = 1600**
0.50	0.500
0.60	0.478
0.70	0.457
0.80	0.437
0.90	0.415
1.00	0.390*
1.15	0.360
1.42	0.306
CURVE 14 (cont.)	
1.50	0.280
2.00	0.208*
2.50	0.184
3.00	0.163*
4.00	0.130
CURVE 15	**T = 2000**
0.40	0.510
0.50	0.488
0.60	0.466
0.70	0.445
0.80	0.425*
0.90	0.404
1.00	0.381*
1.42	0.306*
1.50	0.282*
2.00	0.236*
2.50	0.220
3.00	0.204*
4.00	0.180*
CURVE 16	**T = 2800**
0.40	0.478*
0.50	0.467
0.60	0.450
0.70	0.430
0.80	0.412
0.90	0.393
1.00	0.367
1.42	0.306*
1.50	0.294*
2.00	0.267
2.50	0.252
3.00	0.240
4.00	0.228
CURVE 17	**T = 3100**
0.40	0.472*
0.50	0.458
0.60	0.442*
0.70	0.426
0.80	0.412*
0.90	0.396*
1.00	0.380*
1.50	0.318
2.00	0.290
2.50	0.272
3.00	0.262
4.00	0.252
CURVE 18	**T = 1600**
0.25	0.448
0.27	0.476
0.30	0.482
0.32	0.478
0.35	0.479
0.37	0.482
0.40	0.481
0.42	0.478
0.45	0.474
0.50	0.469
0.55	0.464
0.60	0.456*
0.65	0.450
0.70	0.444*
0.75	0.438
0.80	0.432
0.90	0.412*
1.00	0.390*
1.20	0.344
1.27	0.328
1.35	0.310
1.50	0.280*
1.60	0.263
1.80	0.234
2.00	0.210*
2.20	0.190
CURVE 18 (cont.)	
2.40	0.176
2.60	0.164
CURVE 19*	**T = 1800**
0.25	0.442
0.27	0.471
0.30	0.478
0.32	0.476
0.35	0.476
0.37	0.479
0.40	0.478
0.45	0.470
0.50	0.466
0.55	0.460
0.60	0.452
0.65	0.446
0.70	0.440
0.75	0.434
0.80	0.426
0.90	0.406
1.00	0.386
1.20	0.345
1.27	0.328
1.35	0.312
1.50	0.284
1.60	0.269
1.80	0.242
2.40	0.187
2.60	0.175
CURVE 20	**T = 2000**
0.25	0.436
0.27	0.466
0.30	0.470
0.32	0.472
0.35	0.473
0.37	0.476
0.40	0.474*
0.45	0.467

* Not shown on plot

DATA TABLE NO. 234 (continued)

λ	ε
CURVE 20 (cont.) T = 2000	
0.50	0.462
0.55	0.456
0.60	0.448*
0.65	0.442
0.75	0.428
0.80	0.420*
0.90	0.400*
1.00	0.382*
1.20	0.342*
1.27	0.328*
1.35	0.313*
1.50	0.288*
1.60	0.273
1.80	0.247
2.40	0.196*
2.60	0.187
CURVE 21* T = 2200	
0.25	0.430
0.27	0.460
0.30	0.470
0.32	0.468
0.35	0.470
0.37	0.473
0.40	0.470
0.45	0.464
0.50	0.458
0.55	0.453
0.60	0.444
0.65	0.438
0.70	0.432
0.75	0.423
0.80	0.414
0.90	0.396
1.00	0.377
1.20	0.340
1.32	0.328
1.50	0.284
1.60	0.278
1.80	0.255
2.40	0.205
2.60	0.195

λ	ε
CURVE 22 T = 2400	
0.25	0.422
0.27	0.456
0.30	0.465
0.32	0.465
0.35	0.467
0.37	0.470
0.40	0.468*
0.45	0.460
0.50	0.455*
0.55	0.450
0.60	0.440*
0.65	0.434
0.70	0.428*
0.75	0.418
0.80	0.409*
0.90	0.396*
1.00	0.373*
1.20	0.339
1.32	0.328
1.40	0.313
1.50	0.288*
1.60	0.273*
1.80	0.247*
2.40	0.196*
2.60	0.185*
CURVE 23* T = 2600	
0.25	0.416
0.27	0.450
0.30	0.460
0.33	0.461
0.35	0.466
0.40	0.464
0.45	0.457
0.50	0.451
0.55	0.446
0.60	0.438
0.65	0.430
0.70	0.423
0.75	0.414
0.80	0.404
0.90	0.388

λ	ε
CURVE 23 (cont.)	
1.00	0.360
1.20	0.339
1.32	0.328
1.40	0.317
1.50	0.299
1.60	0.288
1.80	0.268
2.40	0.224
2.60	0.214
CURVE 24 T = 2800	
0.25	0.410
0.27	0.445
0.30	0.456
0.32	0.457
0.35	0.461
0.37	0.463
0.40	0.461
0.45	0.454*
0.50	0.448*
0.55	0.443
0.60	0.434
0.65	0.427
0.70	0.419
0.75	0.410
0.80	0.400
0.90	0.373
1.00	0.367*
1.20	0.337*
1.32	0.328*
1.35	0.318
1.50	0.302
1.60	0.292
1.80	0.274
2.40	0.233*
2.60	0.224
CURVE 25 T = 1600	
0.310	0.479
0.320	0.482
0.330	0.482

λ	ε
CURVE 25 (cont.)	
0.340	0.481
0.360	0.480
0.370	0.479*
0.380	0.477*
0.390	0.475
0.400	0.473*
0.420	0.469
0.440	0.465
0.460	0.462
0.480	0.459
0.500	0.457*
0.520	0.455
0.540	0.453*
0.560	0.452
0.580	0.450
0.600	0.447
0.620	0.445
0.640	0.442*
0.660	0.441
0.680	0.440*
0.700	0.437
0.720	0.434
0.740	0.430
0.760	0.427*
0.780	0.424
0.800	0.422*
CURVE 26* T = 1800	
0.310	0.476
0.320	0.479
0.330	0.480
0.340	0.479
0.350	0.479
0.360	0.478
0.370	0.476
0.380	0.475
0.390	0.473
0.400	0.471
0.420	0.467
0.440	0.463
0.460	0.460
0.480	0.457
0.500	0.455

λ	ε
CURVE 26 (cont.)	
0.520	0.453
0.540	0.451
0.460	0.449
0.580	0.447
0.600	0.444
0.620	0.441
0.640	0.438
0.660	0.436
0.680	0.435
0.700	0.433
0.720	0.429
0.740	0.426
0.760	0.423
0.780	0.421
0.800	0.419
CURVE 27* T = 2000	
0.310	0.474
0.320	0.476
0.330	0.477
0.340	0.477
0.350	0.476
0.360	0.475
0.370	0.474
0.380	0.473
0.390	0.471
0.400	0.469
0.420	0.466
0.440	0.462
0.460	0.459
0.480	0.456
0.500	0.453
0.520	0.450
0.540	0.448
0.560	0.446
0.580	0.443
0.600	0.440
0.620	0.437
0.640	0.434
0.660	0.432
0.680	0.430
0.700	0.428
0.720	0.425

λ	ε
CURVE 27 (cont.)	
0.740	0.422
0.760	0.420
0.780	0.418
0.800	0.416
CURVE 28* T = 2200	
0.310	0.471
0.320	0.473
0.330	0.474
0.340	0.474
0.350	0.474
0.360	0.473
0.370	0.472
0.380	0.471
0.390	0.469
0.400	0.468
0.420	0.464
0.440	0.461
0.460	0.457
0.480	0.454
0.500	0.451
0.520	0.448
0.540	0.446
0.560	0.443
0.580	0.440
0.600	0.437
0.620	0.433
0.640	0.430
0.660	0.428
0.680	0.426
0.700	0.424
0.720	0.421
0.740	0.419
0.760	0.416
0.780	0.415
0.800	0.413
CURVE 29* T = 2400	
0.310	0.468
0.320	0.471
0.330	0.472

λ	ε
CURVE 29 (cont.)	
0.340	0.472
0.350	0.472
0.360	0.471
0.370	0.470
0.380	0.469
0.390	0.467
0.400	0.466
0.420	0.463
0.440	0.459
0.460	0.456
0.480	0.452
0.500	0.449
0.520	0.446
0.540	0.443
0.560	0.441
0.580	0.437
0.600	0.434
0.620	0.430
0.640	0.426
0.660	0.424
0.680	0.421
0.700	0.419
0.720	0.417
0.740	0.415
0.760	0.413
0.780	0.412
0.800	0.411
CURVE 30 T = 1200	
0.50	0.465*
0.66	0.450*
0.70	0.445*
1.00	0.380*
1.80	0.210
2.50	0.140
2.75	0.125
3.00	0.110
3.25	0.100
3.50	0.090
3.90	0.085
4.50	0.075
5.00	0.075
5.90	0.070

λ	ε
CURVE 30 (cont.)	
6.45	0.065
7.20	0.060
7.60	0.060
8.90	0.060
10.00	0.060
11.00	0.060
12.00	0.060
CURVE 31 T = 1428	
0.50	0.456*
0.60	0.445*
0.70	0.428*
1.00	0.380*
1.50	0.275
2.00	0.225*
2.50	0.155
3.00	0.125
3.50	0.110
3.75	0.100
4.50	0.090
5.65	0.085
7.00	0.080
8.00	0.080
9.00	0.080
10.00	0.080
11.00	0.080
12.00	0.080
CURVE 32 T = 1972	
0.45	0.455*
0.50	0.446*
0.55	0.444*
0.60	0.440*
0.65	0.433*
0.70	0.423*
0.80	0.410*
1.50	0.275*
2.00	0.225*
2.50	0.175*
3.00	0.155
3.50	0.145

* Not shown on plot

DATA TABLE NO. 234 (continued)

λ	ϵ
CURVE 32 (cont.) T = 1972	
4.00	0.140
5.00	0.120
6.00	0.110
7.00	0.100
CURVE 33* T = 1660	
1.10	0.340
1.20	0.332
1.30	0.314
1.40	0.300
1.50	0.276
1.55	0.272
1.60	0.260
1.70	0.243
CURVE 34* T = 1790	
0.9	0.394
1.0	0.370
1.1	0.353
1.2	0.330
1.3	0.312
1.4	0.306
1.5	0.275
1.6	0.258
1.7	0.249
CURVE 35* T = 1950	
0.90	0.403
1.00	0.374
1.10	0.358
1.20	0.332
1.25	0.320
1.30	0.316
1.40	0.304
1.50	0.282
1.55	0.277
1.60	0.267
1.70	0.257
CURVE 36* T = 2050	
0.90	0.384
0.95	0.368
1.00	0.365
1.05	0.354
1.10	0.343
1.20	0.324
1.25	0.317
1.30	0.306
1.40	0.296
1.45	0.280
1.50	0.277
1.55	0.270
1.60	0.262
1.70	0.250
CURVE 37 T = 693	
1.5	0.57
2	0.54
4	0.28
6	0.165
8	0.118
10	0.088
CURVE 38 T = 860	
1.5	0.64
2	0.52
4	0.29
6	0.181
8	0.130
10	0.107
12	0.085
CURVE 39 T = 1033	
0.65	0.66
1	0.73
1.5	0.62
2	0.52
4	0.29
CURVE 39 (cont.)	
6	0.188
8	0.140
10	0.118
12	0.104
CURVE 40* T = 1200	
0.65	0.67
1	0.73
1.5	0.61
2	0.52
4	0.29
6	0.185
8	0.143
10	0.119
12	0.103
CURVE 41 T = 1373	
0.65	0.57
1	0.65
1.5	0.50
2	0.39
4	0.21
6	0.141
8	0.119*
10	0.090
12	0.096
CURVE 42* T = 1603	
0.65	0.58
1.5	0.38
2	0.28
4	0.18
CURVE 43* T = 698	
1.5	0.36
2	0.21
4	0.082
CURVE 43 (cont.)*	
6	0.057
8	0.043
10	0.034
CURVE 44* T = 860	
1.5	0.41
2	0.25
4	0.10
6	0.071
8	0.059
10	0.050
CURVE 45* T = 1029	
1	0.55
1.5	0.40
2	0.25
4	0.110
6	0.081
8	0.067
10	0.058
12	0.051
CURVE 46 T = 1196	
1	0.62
1.5	0.42
2	0.29
4	0.129*
6	0.095
8	0.080
10	0.070
12	0.063
CURVE 47* T = 1378	
1.5	0.32
2	0.25
4	0.16
CURVE 48* T = 702	
1.5	0.24
2	0.13
4	0.060
6	0.046
8	0.035
10	0.026
CURVE 49* T = 870	
1.5	0.26
2	0.15
4	0.074
6	0.056
8	0.048
10	0.044
CURVE 50* T = 1039	
1.5	0.32
2	0.18
4	0.088
6	0.069
8	0.057
10	0.055
12	0.050
CURVE 51 T = 1208	
0.65	0.54
1	0.46
1.5	0.27
2	0.17
4	0.093
6	0.077
8	0.068
10	0.072
12	0.058
CURVE 52* T = 1378	
0.65	0.46
1.5	0.22
2	0.14
4	0.11
CURVE 53* T = 1603	
0.65	0.52
1.5	0.29
2	0.21
4	0.18
CURVE 54 T = 692	
2	0.56
4	0.45
6	0.34
8	0.34
10	0.30
CURVE 55* T = 863	
1.5	0.61
2	0.54
4	0.45
6	0.40
8	0.35
10	0.28
12	0.28
CURVE 56 T = 1040	
1.5	0.60
2	0.47
4	0.45*
6	0.40
8	0.36
10	0.32
12	0.29
CURVE 57* T = 1214	
0.65	0.59
1	0.70
1.5	0.61
2	0.55
4	0.46
6	0.41
8	0.37
10	0.34
12	0.30
CURVE 58* T = 1383	
0.65	0.52
1	0.56
1.5	0.48
2	0.45
4	0.40
6	0.36
8	0.32
10	0.30
12	0.27
CURVE 59* T = 699	
1.5	0.46
2	0.41
4	0.36
6	0.32
8	0.29
10	0.26
CURVE 60* T = 529	
1.5	0.49
2	0.44
4	0.37
6	0.33
8	0.30
10	0.26
CURVE 61 T = 700	
1.5	0.47
2	0.43
4	0.37
6	0.34*
8	0.30
10	0.27
12	0.25
CURVE 62* T = 858	
1.5	0.53
2	0.45
4	0.39
6	0.35
8	0.31
10	0.28
12	0.27
CURVE 63 T = 1198	
0.65	0.60
1.5	0.54
2	0.48
4	0.40
6	0.36
8	0.33
10	0.30*
12	0.27
CURVE 64 T = 1358	
0.65	0.56
1	0.60
1.5	0.52
2	0.49
4	0.44
6	0.40*
8	0.36*
10	0.33
12	0.31

* Not shown on plot

DATA TABLE NO. 234 (continued)

λ	∈
CURVE 65*	
T = 1600	
0.50	0.495
0.70	0.450
0.90	0.410
1.15	0.360
1.50	0.290
1.72	0.240
2.00	0.210
2.10	0.200
2.50	0.185
4.00	0.128
CURVE 66*	
T = 2000	
0.40	0.505
0.50	0.480
0.70	0.440
0.90	0.400
1.15	0.350
1.50	0.290
1.75	0.260
2.00	0.240
2.50	0.220
4.00	0.180
CURVE 67*	
T = 2800	
0.40	0.480
0.50	0.460
0.70	0.425
0.90	0.390
1.21	0.330
1.50	0.295
2.00	0.260
3.00	0.231
4.00	0.220
CURVE 68*	
T = 3100	
0.40	0.470
0.50	0.451
0.70	0.420
0.90	0.390
CURVE 68(cont.)*	
T = 3100	
1.50	0.320
2.00	0.291
3.00	0.260
4.00	0.240
CURVE 69*	
T = 1429	
1.157	0.378
1.676	0.298
1.984	0.261
2.164	0.246
2.416	0.235
2.929	0.199
3.122	0.194
3.486	0.189
CURVE 70*	
T = 1382	
1.127	0.382
1.551	0.302
1.615	0.286
1.693	0.272
1.846	0.255
1.934	0.252
2.135	0.229
2.239	0.223
2.422	0.224
2.685	0.191
2.919	0.184
3.128	0.181
3.335	0.173
3.627	0.171
3.843	0.170
4.038	0.169
CURVE 71*	
T = 1340	
1.120	0.386
1.449	0.295
1.696	0.262
2.002	0.223
2.204	0.208
CURVE 71(cont.)*	
T = 1340	
2.420	0.193
2.919	0.165
3.121	0.164
3.485	0.159
CURVE 72*	
T = 1316	
1.096	0.464
1.264	0.345
1.457	0.290
1.655	0.259
1.860	0.224
2.049	0.206
2.231	0.194
2.389	0.182
2.539	0.175
2.694	0.171
2.823	0.167
2.970	0.165
3.092	0.158
3.220	0.154
3.340	0.154
3.459	0.150
CURVE 73*	
T = 1800	
0.467	0.470
0.499	0.466
0.548	0.460
0.578	0.456
0.654	0.445
0.698	0.441
CURVE 74*	
T = 2150	
0.467	0.463
0.499	0.460
0.548	0.453
0.578	0.449
0.654	0.437
0.698	0.431
CURVE 75*	
T = 2520	
0.467	0.454
0.499	0.452
0.548	0.446
0.578	0.441
0.654	0.429
0.698	0.423
CURVE 76	
T = 1244	
1.00	0.402
1.05	0.389
1.10	0.373
1.15	0.360
1.20	0.345
1.24	0.333
1.30	0.318
1.35	0.305
1.40	0.292
1.45	0.280
1.50	0.269*
1.55	0.259
1.60	0.249
1.65	0.240
1.70	0.231
1.75	0.222
1.80	0.214
1.85	0.206
1.90	0.198
1.95	0.191
2.00	0.184
2.10	0.172
2.20	0.161
2.30	0.150
2.40	0.141
2.50	0.133
2.60	0.124
2.70	0.117
2.80	0.110
2.90	0.104
3.00	0.099
3.20	0.089
3.40	0.082
3.60	0.075
3.80	0.068
CURVE 76(cont.)	
T = 1244	
4.00	0.063
4.20	0.058
4.40	0.053
4.60	0.049
4.80	0.045
5.00	0.041
5.10	0.039
CURVE 77*	
T = 1339	
1.00	0.398
1.05	0.385
1.10	0.369
1.15	0.358
1.20	0.344
1.24	0.333
1.30	0.319
1.35	0.306
1.40	0.294
1.45	0.283
1.50	0.272
1.55	0.262
1.60	0.253
1.65	0.244
1.70	0.236
1.75	0.227
1.80	0.219
1.85	0.212
1.90	0.204
1.95	0.197
2.00	0.194
2.10	0.180
2.20	0.169
2.30	0.158
2.40	0.149
2.50	0.141
2.60	0.132
2.70	0.125
2.80	0.118
2.90	0.112
3.00	0.106
3.20	0.097
3.40	0.090
3.60	0.082
CURVE 77(cont.)*	
T = 1339	
3.80	0.076
4.00	0.071
4.20	0.066
4.40	0.061
4.60	0.057
4.80	0.053
5.00	0.048
5.10	0.047
CURVE 78*	
T = 1413	
1.00	0.396
1.05	0.382
1.10	0.368
1.15	0.357
1.20	0.343
1.24	0.333
1.30	0.320
1.35	0.308
1.40	0.296
1.45	0.286
1.50	0.275
1.55	0.265
1.60	0.256
1.65	0.248
1.70	0.240
1.75	0.232
1.80	0.224
1.85	0.217
1.90	0.209
1.95	0.203
2.00	0.197
2.10	0.186
2.20	0.175
2.30	0.165
2.40	0.156
2.50	0.147
2.60	0.139
2.70	0.132
2.80	0.125
2.90	0.119
3.00	0.114
3.20	0.104
3.40	0.096
CURVE 78(cont.)*	
T = 1413	
3.60	0.089
3.80	0.083
4.00	0.077
4.20	0.072
4.40	0.067
4.60	0.062
4.80	0.058
5.00	0.054
5.10	0.052
CURVE 79*	
T = 1629	
1.00	0.388
1.05	0.377
1.10	0.364
1.15	0.354
1.20	0.342
1.24	0.333
1.30	0.321
1.35	0.311
1.40	0.301
1.45	0.292
1.50	0.282
1.55	0.274
1.60	0.265
1.65	0.257
1.70	0.250
1.75	0.242
1.80	0.236
1.85	0.229
1.90	0.222
1.95	0.217
2.00	0.210
2.10	0.199
2.20	0.189
2.30	0.180
2.40	0.171
2.50	0.163
2.60	0.156
2.70	0.150
2.80	0.143
2.90	0.137
3.00	0.132
3.20	0.122
CURVE 79(cont.)*	
T = 1629	
3.40	0.114
3.60	0.107
3.80	0.100
4.00	0.094
4.20	0.088
4.40	0.083
4.60	0.078
4.80	0.074
5.00	0.070
5.10	0.068
CURVE 80*	
T = 1833	
1.00	0.382
1.05	0.372
1.10	0.361
1.15	0.352
1.20	0.342
1.24	0.333
1.30	0.322
1.35	0.313
1.40	0.304
1.45	0.296
1.50	0.288
1.55	0.280
1.60	0.273
1.65	0.266
1.70	0.259
1.75	0.252
1.80	0.245
1.85	0.239
1.90	0.233
1.95	0.227
2.00	0.221
2.10	0.211
2.20	0.201
2.30	0.192
2.40	0.184
2.50	0.176
2.60	0.169
2.70	0.163
2.80	0.157
2.90	0.154
3.00	0.146
3.20	0.136

* Not shown on plot

DATA TABLE NO. 234 (continued)

λ	∈
CURVE 80 (cont.)*	
T = 1833	
3.40	0.128
3.60	0.121
3.80	0.114
4.00	0.108
4.20	0.102
4.40	0.097
4.60	0.092
4.80	0.088
5.00	0.084
5.10	0.082
CURVE 81*	
T = 2002	
1.00	0.378
1.05	0.369
1.10	0.359
1.15	0.351
1.20	0.341
1.24	0.333
1.30	0.323
1.35	0.314
1.40	0.306
1.45	0.298
1.50	0.291
1.55	0.284
1.60	0.277
1.65	0.271
1.70	0.264
1.75	0.257
1.80	0.251
1.85	0.245
1.90	0.239
1.95	0.234
2.00	0.229
2.10	0.220
2.20	0.211
2.30	0.202
2.40	0.194
2.50	0.187
2.60	0.180
2.70	0.173
2.80	0.167
2.90	0.162
3.00	0.157
CURVE 81 (cont.)*	
T = 2002	
3.20	0.147
3.40	0.139
3.60	0.132
3.80	0.125
4.00	0.119
4.20	0.113
4.40	0.108
4.60	0.103
4.80	0.099
5.00	0.095
5.10	0.093
CURVE 82*	
T = 2160	
1.00	0.374
1.05	0.366
1.10	0.357
1.15	0.349
1.20	0.341
1.24	0.333
1.30	0.324
1.35	0.316
1.40	0.308
1.45	0.301
1.50	0.294
1.55	0.287
1.60	0.280
1.65	0.274
1.70	0.268
1.75	0.262
1.80	0.251
1.85	0.251
1.90	0.245
1.95	0.241
2.00	0.236
2.10	0.227
2.20	0.218
2.30	0.210
2.40	0.202
2.50	0.195
2.60	0.189
2.70	0.183
2.80	0.177
2.90	0.172
CURVE 82 (cont.)*	
T = 2160	
3.00	0.167
3.20	0.157
3.40	0.149
3.60	0.142
3.80	0.135
4.00	0.129
4.20	0.124
4.40	0.119
4.60	0.114
4.80	0.110
5.00	0.106
5.10	0.104
CURVE 83*	
T = 2327	
1.00	0.371
1.05	0.364
1.10	0.356
1.15	0.348
1.20	0.340
1.24	0.333
1.30	0.324
1.35	0.317
1.40	0.310
1.45	0.303
1.50	0.297
1.55	0.291
1.60	0.285
1.65	0.279
1.70	0.273
1.75	0.268
1.80	0.262
1.85	0.257
1.90	0.252
1.95	0.247
2.00	0.243
2.10	0.234
2.20	0.225
2.30	0.217
2.40	0.209
2.50	0.202
2.60	0.195
2.70	0.189
2.80	0.184
CURVE 83 (cont.)*	
T = 2327	
2.90	0.179
3.00	0.174
3.20	0.166
3.40	0.157
3.60	0.151
3.80	0.144
4.00	0.139
4.20	0.134
4.40	0.128
4.60	0.124
4.80	0.120
5.00	0.116
5.10	0.114
CURVE 84*	
T = 2441	
1.00	0.369
1.05	0.362
1.10	0.354
1.15	0.347
1.20	0.339
1.24	0.333
1.30	0.325
1.35	0.318
1.40	0.312
1.45	0.305
1.50	0.299
1.55	0.294
1.60	0.288
1.65	0.282
1.70	0.277
1.75	0.272
1.80	0.266
1.85	0.262
1.90	0.257
1.95	0.252
2.00	0.248
2.10	0.239
2.20	0.231
2.30	0.223
2.40	0.216
2.50	0.209
2.60	0.202
2.70	0.196
CURVE 84 (cont.)*	
T = 2441	
2.80	0.191
2.90	0.186
3.00	0.181
3.20	0.172
3.40	0.164
3.60	0.158
3.80	0.151
4.00	0.146
4.20	0.140
4.40	0.136
4.60	0.132
4.80	0.127
5.00	0.124
5.10	0.122
CURVE 85	
T = 300	
0.46	0.583
0.49	0.560
0.55	0.537
0.61	0.519
0.79	0.506
0.99	0.428
1.19	0.371
1.40	0.332
1.60	0.262
1.80	0.175
2.00	0.128
CURVE 86*	
T = 1100	
0.45	0.531
0.49	0.526
0.54	0.517
0.61	0.505
0.80	0.467
0.99	0.414
1.19	0.359
1.41	0.315
1.60	0.271
1.79	0.217
2.00	0.185
CURVE 87*	
T = 1500	
0.46	0.529
0.49	0.519
0.55	0.507
0.61	0.494
0.79	0.450
1.00	0.405
1.19	0.357
1.40	0.315
1.61	0.268
1.79	0.240
2.00	0.208
CURVE 88*	
T = 2000	
0.46	0.506
0.49	0.503
0.55	0.489
0.61	0.482
0.79	0.438
0.99	0.398
1.20	0.360
1.40	0.312
1.60	0.270
1.79	0.250
2.00	0.221
CURVE 89*	
T = 2500	
0.46	0.485
0.49	0.484
0.55	0.480
0.61	0.467
0.80	0.423
0.99	0.385
1.20	0.352
1.40	0.303
1.60	0.282
1.80	0.262
2.00	0.237
CURVE 90*	
T = 2143	
0.3478	0.485
0.3717	0.495
0.3956	0.493
0.4196	0.503
0.4435	0.470
0.4677	0.475
0.4916	0.463
0.5158	0.453
0.5400	0.438
0.5641	0.415
CURVE 91*	
T = 1600	
0.230	0.393
0.244	0.438
0.264	0.469
0.280	0.478
0.299	0.481
0.335	0.477
0.356	0.479
0.382	0.481
0.452	0.473
0.561	0.462
0.625	0.451
0.753	0.439
0.853	0.424
0.963	0.400
1.13	0.360
1.27	0.328
1.50	0.282
1.79	0.234
2.14	0.196
2.40	0.176
2.69	0.160
CURVE 92*	
T = 1800	
0.229	0.379
0.241	0.424
0.258	0.456
0.274	0.471
0.299	0.477
0.327	0.474
CURVE 92 (cont.)*	
T = 1800	
0.365	0.477
0.383	0.478
0.450	0.470
0.545	0.461
0.612	0.450
0.731	0.437
0.828	0.423
0.972	0.393
1.10	0.363
1.27	0.329
1.48	0.289
1.70	0.254
1.97	0.222
2.34	0.191
2.70	0.170
CURVE 93*	
T = 2000	
0.229	0.374
0.240	0.415
0.259	0.451
0.279	0.468
0.302	0.473
0.323	0.471
0.343	0.471
0.380	0.475
0.450	0.466
0.553	0.456
0.604	0.448
0.706	0.436
0.807	0.420
0.939	0.394
1.06	0.368
1.28	0.328
1.56	0.281
1.84	0.243
2.09	0.220
2.40	0.198
2.71	0.180

* Not shown on plot

DATA TABLE NO. 234 (continued)

λ	ε	λ	ε
CURVE 94* T = 2200		CURVE 96* T = 2600	
0.231	0.364	0.231	0.363
0.247	0.427	0.241	0.400
0.264	0.452	0.260	0.437
0.285	0.466	0.283	0.454
0.306	0.469	0.301	0.460
0.331	0.467	0.340	0.462
0.375	0.471	0.377	0.465
0.451	0.463	0.419	0.460
0.553	0.453	0.502	0.450
0.604	0.444	0.567	0.443
0.704	0.432	0.654	0.430
0.807	0.414	0.756	0.413
0.918	0.392	0.883	0.390
1.27	0.329	1.04	0.364
1.60	0.279	1.27	0.328
2.00	0.236	1.60	0.289
2.40	0.206	2.00	0.250
2.68	0.191	2.41	0.224
		2.69	0.210
CURVE 95* T = 2400		CURVE 97* T = 2800	
0.231	0.363	0.229	0.350
0.238	0.397	0.238	0.386
0.247	0.419	0.248	0.412
0.258	0.439	0.268	0.440
0.273	0.455	0.293	0.455
0.293	0.463	0.332	0.457
0.314	0.465	0.364	0.462
0.331	0.464	0.384	0.462
0.373	0.469	0.450	0.453
0.400	0.466	0.554	0.443
0.451	0.459	0.604	0.433
0.526	0.452	0.704	0.420
0.562	0.448	0.822	0.397
0.608	0.440	0.979	0.371
0.706	0.428	1.11	0.350
0.820	0.407	1.27	0.328
0.942	0.384	1.51	0.302
1.14	0.350	1.85	0.270
1.27	0.329	2.21	0.245
1.55	0.290	2.68	0.220
1.83	0.259		
2.21	0.228		
2.68	0.201		

* Not shown on plot

SPECIFICATION TABLE NO. 235 ANGULAR SPECTRAL EMITTANCE OF TUNGSTEN

Curve No.	Ref. No.	Year	Temperature K	Wavelength Range, μ	Geometry θ'	Reported Error, %	Composition (weight percent), Specifications and Remarks
1	351	1962	2200	0.73-4.04	70°	±2	Emittance value was determined by measuring normalized emittance over a range of angles from 60° to 80°.

DATA TABLE NO. 235 ANGULAR SPECTRAL EMITTANCE OF TUNGSTEN

[Wavelength, λ, μ; Emittance, $\in$; Temperature, T, K]

λ	$\in$
CURVE 1*	
T = 2200	
0.73	0.450
2.05	0.213
3.01	0.171
4.04	0.150

* Not shown on plot

SPECIFICATION TABLE NO. 236 ANGULAR SPECTRAL EMITTANCE OF TUNGSTEN

Curve No.	Ref. No.	Year	Temperature K	Wavelength, μ	Angular Range, °	Geometry θ'	Reported Error, %	Composition (weight percent), Specifications and Remarks
1	147	1960	1740	0.65	7.5-81.0	θ'		Lamp filament.

DATA TABLE NO. 236 ANGULAR SPECTRAL EMITTANCE OF TUNGSTEN

[Angle, θ', °; Emittance, ϵ; Temperature, T, K; Wavelength, λ, μ]

θ'	ϵ
CURVE 1*	
T = 1740	
λ = 0.65	
7.5	0.460
22.0	0.470
41.0	0.485
50.5	0.500
63.5	0.535
71.0	0.560
76.5	0.560
81.0	0.530

* Not shown on plot

SPECIFICATION TABLE NO. 237 ANGULAR INTEGRATED REFLECTANCE OF TUNGSTEN

Curve No.	Ref. No.	Year	Temperature K	Angular Range, °	Geometry θ	θ'	ω'	Reported Error, %	Composition (weight percent), Specifications and Remarks
1	275	1926	298	76-82	θ	θ'			Polished by rolling; incident radiation polarized perpendicular to the plane of incidence; visible range of spectrum; $\theta = \theta'$.
2	275	1962	298	76-82	θ	θ'			Above specimen and conditions except incident radiation polarized parallel to the plane of incidence.

DATA TABLE NO. 237 ANGULAR INTEGRATED REFLECTANCE OF TUNGSTEN

[Angle, θ, °; Reflectance, ρ; Temperature, T, K]

θ	ρ
CURVE 1*	
T = 298	
76	0.121
77	0.115
78	0.110
78	0.110
79	0.111
80	0.116
81	0.124
82	0.142
CURVE 2*	
T = 298	
76	0.812
76	0.825
77	0.830
78	0.840
78	0.854
79	0.845
80	0.856
81	0.866
82	0.884

* Not shown on plot

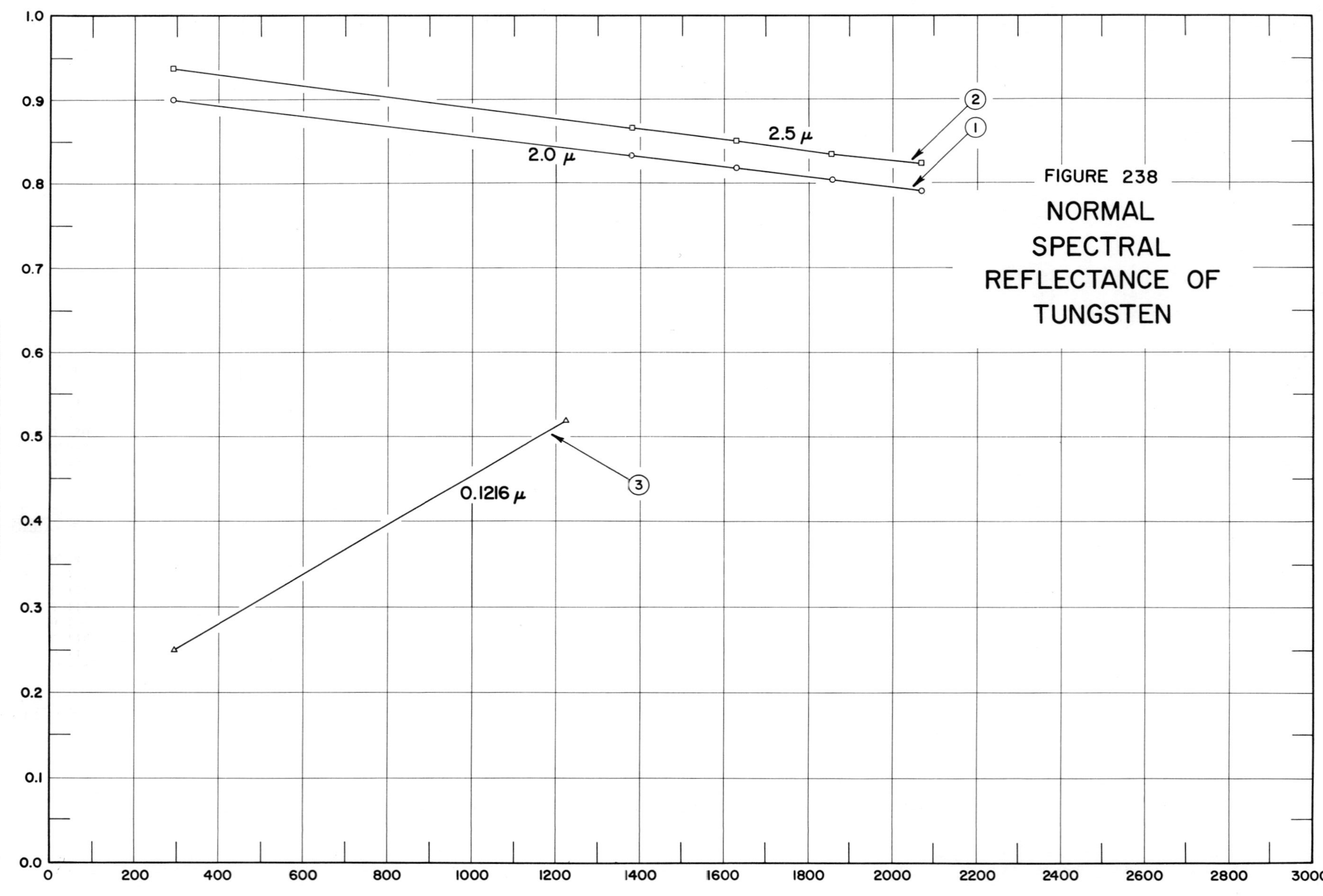
FIGURE 238
NORMAL SPECTRAL REFLECTANCE OF TUNGSTEN
NORMAL SPECTRAL REFLECTANCE
TEMPERATURE, K
2.5 μ
2.0 μ
0.1216 μ
1
2
3
1.0
0.9
0.8
0.7
0.6
0.5
0.4
0.3
0.2
0.1
0.0
0
200
400
600
800
1000
1200
1400
1600
1800
2000
2200
2400
2600
2800
3000

SPECIFICATION TABLE NO. 238 NORMAL SPECTRAL REFLECTANCE OF TUNGSTEN

Curve No.	Ref. No.	Year	Wavelength μ	Temperature Range, K	Geometry θ θ' ω'	Reported Error, %	Composition (weight percent), Specifications and Remarks
1	150	1919	2.0	293-2067	10° 10°		Heated in vacuum at T > 2067 K; rough polished and then finished on a pitch surface with fine rouge; measured in vacuum.
2	150	1919	2.5	293-2067	10° 10°		Above specimen and conditions.
3	332	1966	0.1216	298-1223	$\sim 0^\circ$ $\sim 0^\circ$		0.025-mm foil.

DATA TABLE NO. 238 NORMAL SPECTRAL REFLECTANCE OF TUNGSTEN

[Temperature, T, K; Reflectance, ρ; Wavelength, λ, μ]

T	ρ
CURVE 1	
$\lambda = 2.0$	
293	0.900
1380	0.833
1632	0.818
1859	0.803
2067	0.790
CURVE 2	
$\lambda = 2.5$	
293	0.938
1380	0.866
1632	0.850
1859	0.835
2067	0.823
CURVE 3	
$\lambda = 0.1216$	
298	0.25
1223	0.52

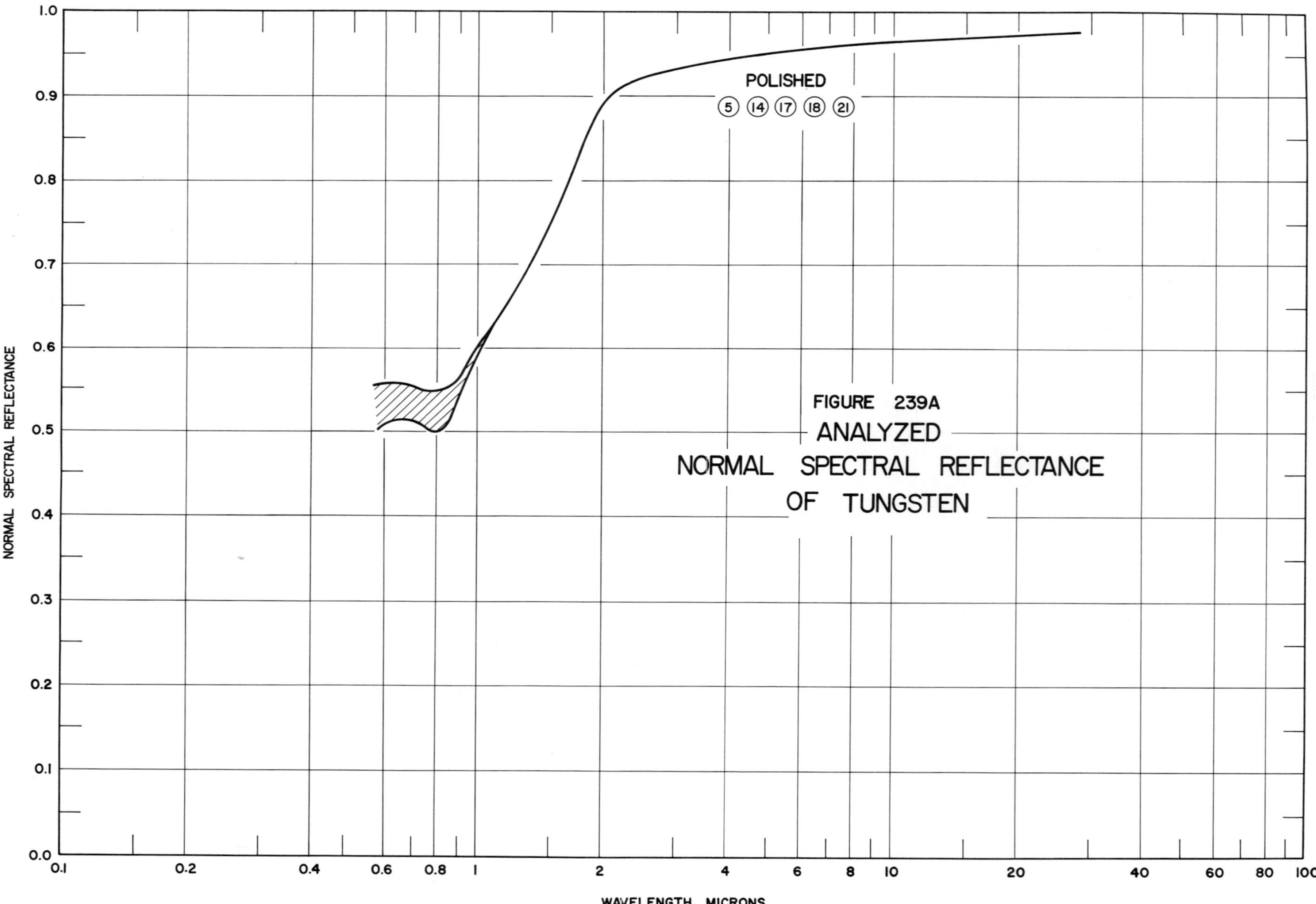

FIGURE 239A
ANALYZED
NORMAL SPECTRAL REFLECTANCE
OF TUNGSTEN

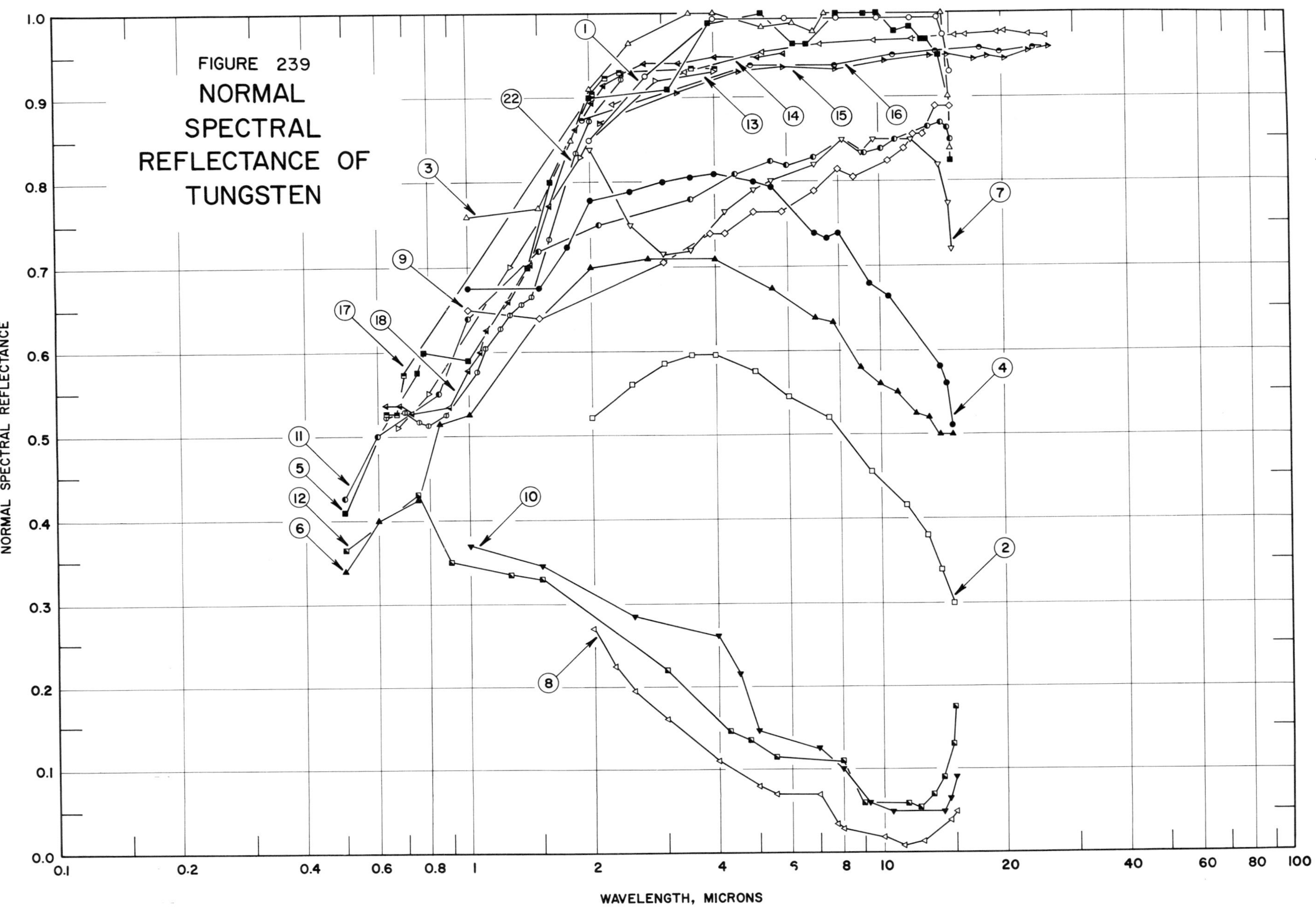

FIGURE 239 NORMAL SPECTRAL REFLECTANCE OF TUNGSTEN

SPECIFICATION TABLE NO. 239 NORMAL SPECTRAL REFLECTANCE OF TUNGSTEN

Curve No.	Ref. No.	Year	Temperature K	Wavelength Range, μ	Geometry θ	θ'	ω'	Reported Error, %	Composition (weight percent), Specifications and Remarks
1	86	1962	~322	2.00-15.00	~0°		2π	<2	As received; data extracted from smooth curve; hohlraum at 523 K; converted from R(2π,0°).
2	86	1962	~322	2.00-15.00	~0°		2π	<2	Above specimen and conditions; diffuse component only.
3	86	1962	~322	1.00-15.00	~0°		2π	<2	Different sample, same as curve 1 specimen and conditions except hohlraum at 773 K.
4	86	1962	~322	1.00-15.00	~0°		2π	<2	Above specimen and conditions; diffuse component only.
5	86	1962	~322	0.50-15.00	~0°		2π	<2	Different sample, same as curve 1 specimen and conditions except hohlraum at 1273 K.
6	86	1962	~322	0.50-15.00	~0°		2π	<2	Above specimen and conditions; diffuse component only.
7	86	1962	~322	2.00-15.00	~0°		2π	<2	Heated in argon at 1366 K for 30 min; data extracted from smooth curve; hohlraum at 523 K; converted from R(2π, 0°).
8	86	1962	~322	2.00-15.00	~0°		2π	<2	Above specimen and conditions; diffuse component only.
9	86	1962	~322	1.00-15.00	~0°		2π	<2	Different sample, same as curve 7 specimen and conditions except hohlraum at 773 K.
10	86	1962	~322	1.00-15.00	~0°		2π	<2	Above specimen and conditions; diffuse component only.
11	86	1962	~322	0.50-15.00	~0°		2π	<2	Different sample, same as curve 7 specimen and conditions except hohlraum at 1273 K.
12	86	1962	~322	0.50-15.00	~0°		2π	<2	Above specimen and conditions; diffuse component only.
13	149	1917	298	0.67-4.00	<15°	<15°			Polished.
14	223	1962	298	2.29-25.99	~0°		2π		Polished; converted from R(2π, 0°).
15	223	1962	298	2.12-26.00	~0°		2π		Above specimen and conditions except after particle impact.
16	223	1962	77	1.92-26.00	~0°		2π		Above specimen and conditions.
17	330	1917	298	0.63-4.06	~0°	~0°			X-ray target, exceptionally pure; highly polished; silvered glass mirror reference standard; observed data were corrected for absorption by the silver mirror.
18	330	1917	298	0.63-5.99	~0°	~0°			X-ray target, exceptionally pure; highly polished; silvered glass mirror reference standard; observed data were corrected for absorption by the silver mirror
19	330	1917	298	0.60-1.06	~0°	~0°			X-ray target, exceptionally pure; highly polished; right-angled glass prism reference standard; observed data were corrected for losses by reflection at the surface of the reference standard.
20	330	1917	298	0.59-2.40	~0°	~0°			Pure; highly polished but showed pits of microscopic size on surface; silver glass mirror reference standard; observed data were corrected for absorption by the silver mirror.
21	330	1917	298	0.58-4.09	~0°	~0°			Lamp filament material; surface ground and polished; surface appeared highly polished and free from minute pits.
22	330	1917	298	0.63-2.40	~0°	~0°			Heated in hydrogen nearly to its melting point thus producing the purest material; surface had a fairly high polish but not entirely free from fine pits; glass prism and silver mirror were used as reference standards over the wavelength ranges 0.6-1.5 μ and 1-2.5 μ, respectively; data corrected for absorption and reflection of the reference standards.

DATA TABLE NO. 239 NORMAL SPECTRAL REFLECTANCE OF TUNGSTEN

[Wavelength, λ, μ; Reflectance, ρ; Temperature, T, K]

λ	ρ
CURVE 1 T = ~322	
2.00	0.850
2.75	0.925
4.00	0.995
6.00	0.995
8.00	0.995
10.00	0.995
12.00	0.995
14.00	0.995
14.50	0.975
15.00	0.930
CURVE 2 T = ~322	
2.00	0.520
2.50	0.560
3.00	0.585
3.50	0.595
4.00	0.595
5.00	0.575
6.00	0.545
7.50	0.520
9.50	0.455
11.50	0.415
13.00	0.380
14.00	0.340
15.00	0.300
CURVE 3 T = ~322	
1.00	0.760
1.50	0.770
1.80	0.850
2.00	0.910
2.50	0.965
3.50	1.000
4.00	1.000
5.25	0.985
6.25	0.990
7.00	0.980
7.45	1.000
CURVE 3 (cont.)	
14.40	1.000
14.75	0.900
15.00	0.840
CURVE 4 T = ~322	
1.00	0.675
1.50	0.675
1.75	0.725
2.00	0.780
2.50	0.790
3.00	0.800
3.50	0.805
4.00	0.810
5.00	0.800
5.50	0.795
7.00	0.740
7.50	0.735
8.00	0.740
9.50	0.680
10.50	0.665
14.00	0.580
14.50	0.560
15.00	0.510
CURVE 5 T = ~322	
0.50	0.410
0.60	0.500*
0.75	0.575
0.78	0.600
1.00	0.590
1.40	0.700
1.60	0.800
2.00	0.900
3.10	0.910
3.90	0.990
5.25	1.000
6.25	0.965
6.75	0.965
8.00	1.000
CURVE 5 (cont.)	
9.25	1.000
10.00	1.000
11.00	0.980
12.00	0.985
12.75	0.970
13.25	0.970
14.10	0.950
14.70	0.900*
15.00	0.825
CURVE 6 T = ~322	
0.50	0.340
0.60	0.400
0.75	0.425
0.85	0.515
1.00	0.525
1.50	0.640*
2.00	0.700
2.75	0.710
4.00	0.710
5.50	0.675
7.00	0.640
7.75	0.635
9.00	0.580
10.00	0.560
11.00	0.550
12.25	0.525
13.25	0.520
14.00	0.500
15.00	0.500
CURVE 7 T = ~322	
2.00	0.840
2.50	0.750
3.00	0.715
3.50	0.720
4.25	0.765
5.00	0.790
5.50	0.800
CURVE 7 (cont.)	
7.00	0.820
8.20	0.850
9.10	0.835
9.75	0.850
11.00	0.850*
12.00	0.850
14.00	0.820
14.75	0.775
15.00	0.720
CURVE 8 T = ~322	
2.00	0.270
2.25	0.225
2.50	0.195
3.00	0.160
4.00	0.110
5.00	0.080
5.50	0.070
7.00	0.070
7.75	0.035
8.00	0.030
10.00	0.020
11.25	0.010
12.50	0.015
14.50	0.040
15.00	0.050
CURVE 9 T = ~322	
1.00	0.650
1.50	0.640
3.00	0.705
3.90	0.740
4.25	0.740
5.00	0.765
5.85	0.765
7.00	0.790
8.00	0.815
8.75	0.805
10.50	0.825
CURVE 9 (cont.)	
11.50	0.840
12.25	0.855
12.80	0.855
13.85	0.890
15.00	0.890
CURVE 10 T = ~322	
1.00	0.370
1.50	0.345
2.50	0.285
4.00	0.260
4.50	0.215
5.00	0.145
7.00	0.125
8.00	0.100
9.25	0.060
10.50	0.050
14.00	0.050
14.50	0.065
15.00	0.090
CURVE 11 T = ~322	
0.50	0.425
0.60	0.500
0.85	0.550
1.00	0.640
1.50	0.720
2.10	0.750
3.50	0.780
4.50	0.810
5.50	0.825
6.00	0.820
7.00	0.830
8.25	0.850*
9.25	0.835
10.25	0.840
11.00	0.850
12.00	0.860*
13.25	0.865
CURVE 11 (cont.)	
14.25	0.870
14.75	0.865
15.00	0.850
CURVE 12 T = ~322	
0.50	0.365
0.75	0.430
0.90	0.350
1.25	0.335
1.50	0.330
3.00	0.220
4.25	0.145
4.75	0.135
5.50	0.115
8.00	0.110
9.00	0.060
11.50	0.060
12.25	0.055
13.25	0.070
14.00	0.090
14.75	0.130
15.00	0.175
CURVE 13 T = 298	
0.67	0.51
0.80	0.55
1.27	0.70
1.90	0.83
2.00	0.85*
2.90	0.92
4.00	0.93
CURVE 14 T = 298	
2.29	0.892
3.45	0.931
5.30	0.954
7.33	0.964
9.92	0.969
CURVE 14 (cont.)	
12.25	0.971
14.46	0.976
16.44	0.976
19.49	0.979
20.54	0.980
23.23	0.977
25.99	0.974
CURVE 15 T = 298	
2.12	0.870
3.27	0.906
4.60	0.930
5.88	0.935
7.97	0.933
10.44	0.942
13.41	0.950
14.60	0.950
17.19	0.947
18.54	0.949
20.15	0.947
23.16	0.954
26.00	0.960
CURVE 16 T = 77	
1.92	0.874
4.96	0.939
7.93	0.938
10.96	0.951
13.92	0.953
17.97	0.959
19.96	0.954
23.94	0.959
26.00	0.959*

* Not shown on plot

DATA TABLE NO. 239 (continued)

CURVE 17
T = 298

λ	ρ
0.63	0.527
0.67	0.527
0.99	0.572
2.03	0.903
2.19	0.924
2.41	0.933
3.56	0.935
4.06	0.935

CURVE 18
T = 298

λ	ρ
0.63	0.537
0.68	0.537
0.73	0.528
0.90	0.533
1.00	0.578
1.06	0.600
1.12	0.626
1.26	0.660
1.43	0.702
1.60	0.772
1.86	0.863
2.03	0.896
2.18	0.916
2.41	0.929
2.73	0.940
3.32	0.941
4.06	0.949
5.18	0.949
5.99	0.952

CURVE 19*
T = 298

λ	ρ
0.60	0.526
0.61	0.533
0.66	0.535
0.72	0.529
0.76	0.517
0.80	0.517
0.89	0.526
1.06	0.607

CURVE 20*
T = 298

λ	ρ
0.59	0.505
0.62	0.513
0.67	0.516
0.72	0.505
0.79	0.494
0.87	0.515
0.99	0.561
1.12	0.613
1.27	0.641
1.43	0.678
1.59	0.756
1.85	0.847
2.02	0.884
2.18	0.907
2.40	0.920

CURVE 21*
T = 298

λ	ρ
0.58	0.553
0.63	0.556
0.67	0.554
0.73	0.548
0.79	0.544
0.87	0.546
0.88	0.557
0.99	0.613
1.04	0.642
1.12	0.694
1.25	0.748
1.43	0.802
1.59	0.836
1.84	0.875
2.04	0.899
2.18	0.908
2.40	0.913
4.09	0.940

CURVE 22
T = 298

λ	ρ
0.63	0.526
0.70	0.529
0.76	0.517
0.80	0.512
0.88	0.523
1.05	0.576
1.12	0.604
1.20	0.629
1.27	0.644
1.35	0.656
1.43	0.666
1.59	0.734
1.85	0.836
2.01	0.874
2.40	0.922

* Not shown on plot

SPECIFICATION TABLE NO. 240 ANGULAR SPECTRAL REFLECTANCE OF TUNGSTEN

Curve No.	Ref. No.	Year	Temperature K	Wavelength Range, μ	Geometry θ θ' ω'	Reported Error, %	Composition (weight percent), Specifications and Remarks
1	132	1911	298	0.45-10.20	15° 15°	≤ 3	Polished; silvered glass mirror reference.

DATA TABLE NO. 240 ANGULAR SPECTRAL REFLECTANCE OF TUNGSTEN

[Wavelength, λ, μ; Reflectance, ρ; Temperature, T, K]

λ	ρ
CURVE 1*	
T = 298	
0.45	0.465
0.50	0.480
0.55	0.495
0.65	0.510
0.75	0.535
0.90	0.575
1.05	0.625
1.25	0.680
1.65	0.780
2.25	0.865
2.45	0.890
3.05	0.905
3.65	0.915
4.10	0.930
4.70	0.935
5.70	0.940
6.10	0.940
6.85	0.950
7.55	0.950
7.85	0.950
8.80	0.955
9.35	0.950
10.20	0.955

* Not shown on plot

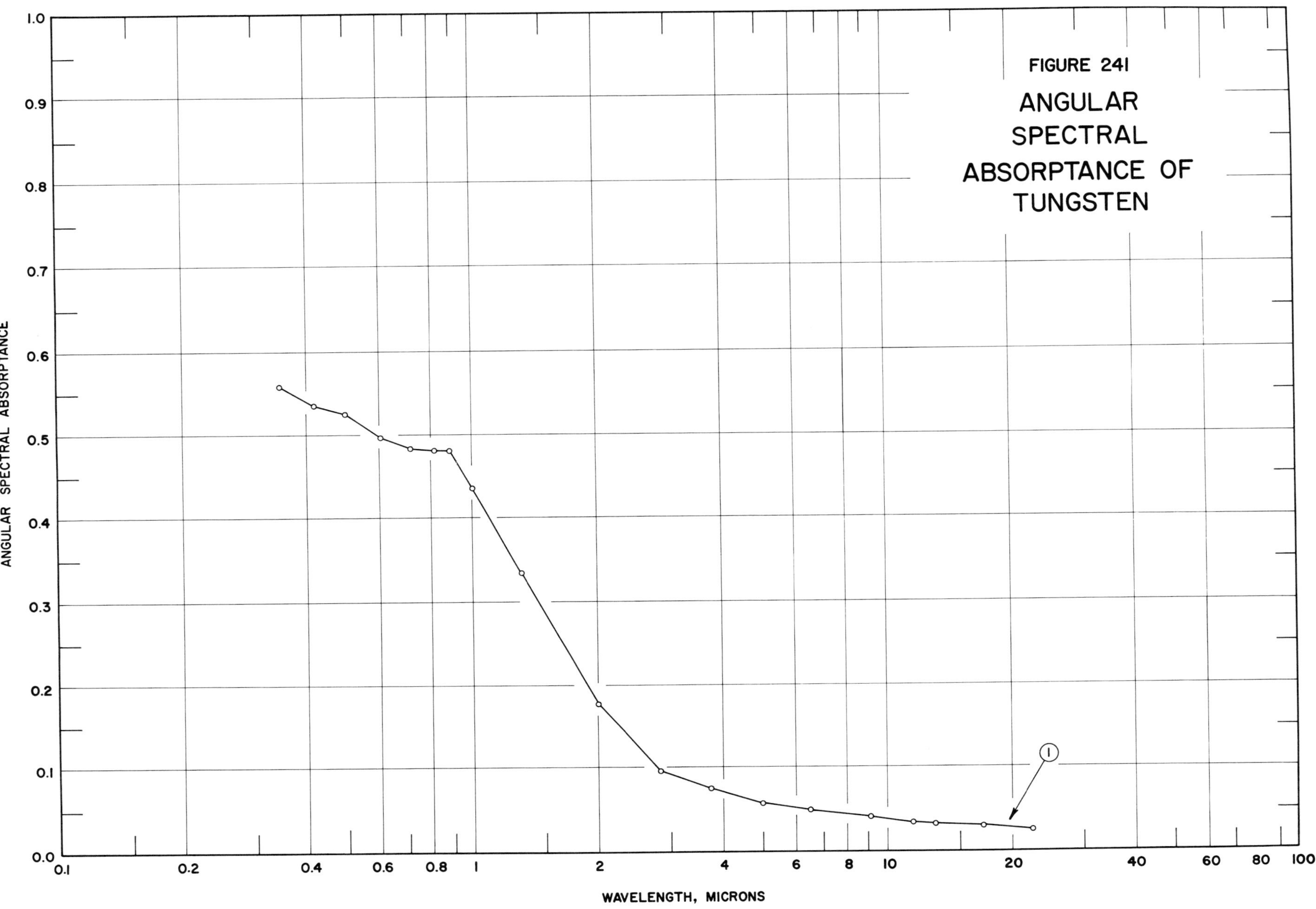
FIGURE 241
ANGULAR SPECTRAL ABSORPTANCE OF TUNGSTEN
ANGULAR SPECTRAL ABSORPTANCE
WAVELENGTH, MICRONS
1
1.0
0.9
0.8
0.7
0.6
0.5
0.4
0.3
0.2
0.1
0.0
0.1
0.2
0.4
0.6
0.8
1
2
4
6
8
10
20
40
60
80
100

SPECIFICATION TABLE NO. 241 ANGULAR SPECTRAL ABSORPTANCE OF TUNGSTEN

Curve No.	Ref. No.	Year	Temperature K	Wavelength Range, μ	Geometry θ	Reported Error, %	Composition (weight percent), Specifications and Remarks
1	225	1965	306	0.345-22.5	25°		Tungsten from Belmont Smelting and Refining Works; lapped; measured in dry nitrogen; heated cavity at approx 1056 K with platinum reference; authors assumed $\alpha = 1-R(2\pi, 25°)$.

DATA TABLE NO. 241 ANGULAR SPECTRAL ABSORPTANCE OF TUNGSTEN

[Wavelength, λ,μ; Absorptance, α; Temperature, T, K]

λ	α
CURVE 1 T = 306	
0.345	0.557
0.418	0.535
0.500	0.525
0.604	0.498
0.715	0.485
0.819	0.481
0.887	0.481
1.01	0.435
1.32	0.333
2.01	0.178
2.83	0.097
3.75	0.074
5.00	0.057
6.53	0.049
9.10	0.041
11.5	0.034
13.1	0.032
17.1	0.030
22.5	0.026

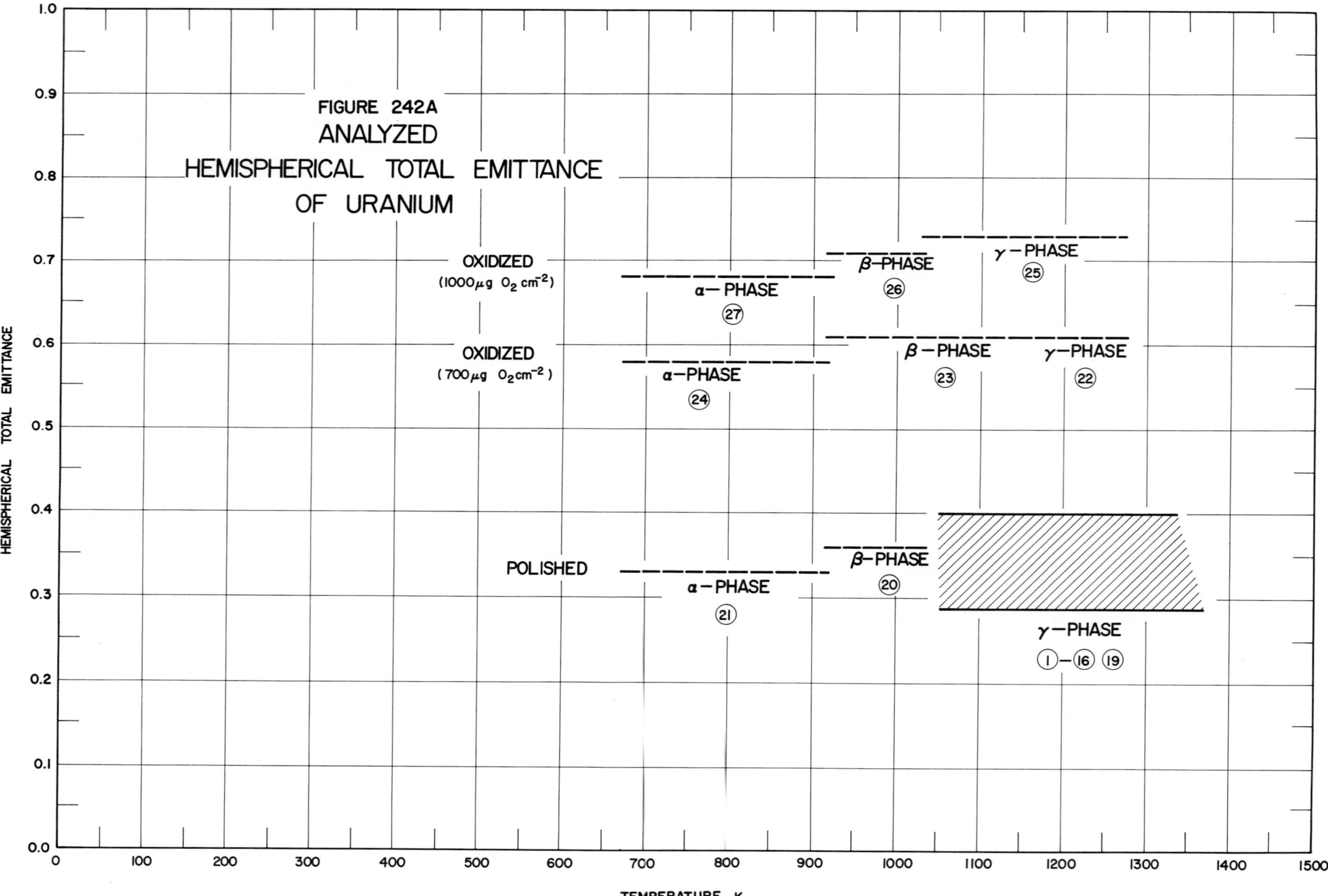

FIGURE 242A ANALYZED HEMISPHERICAL TOTAL EMITTANCE OF URANIUM

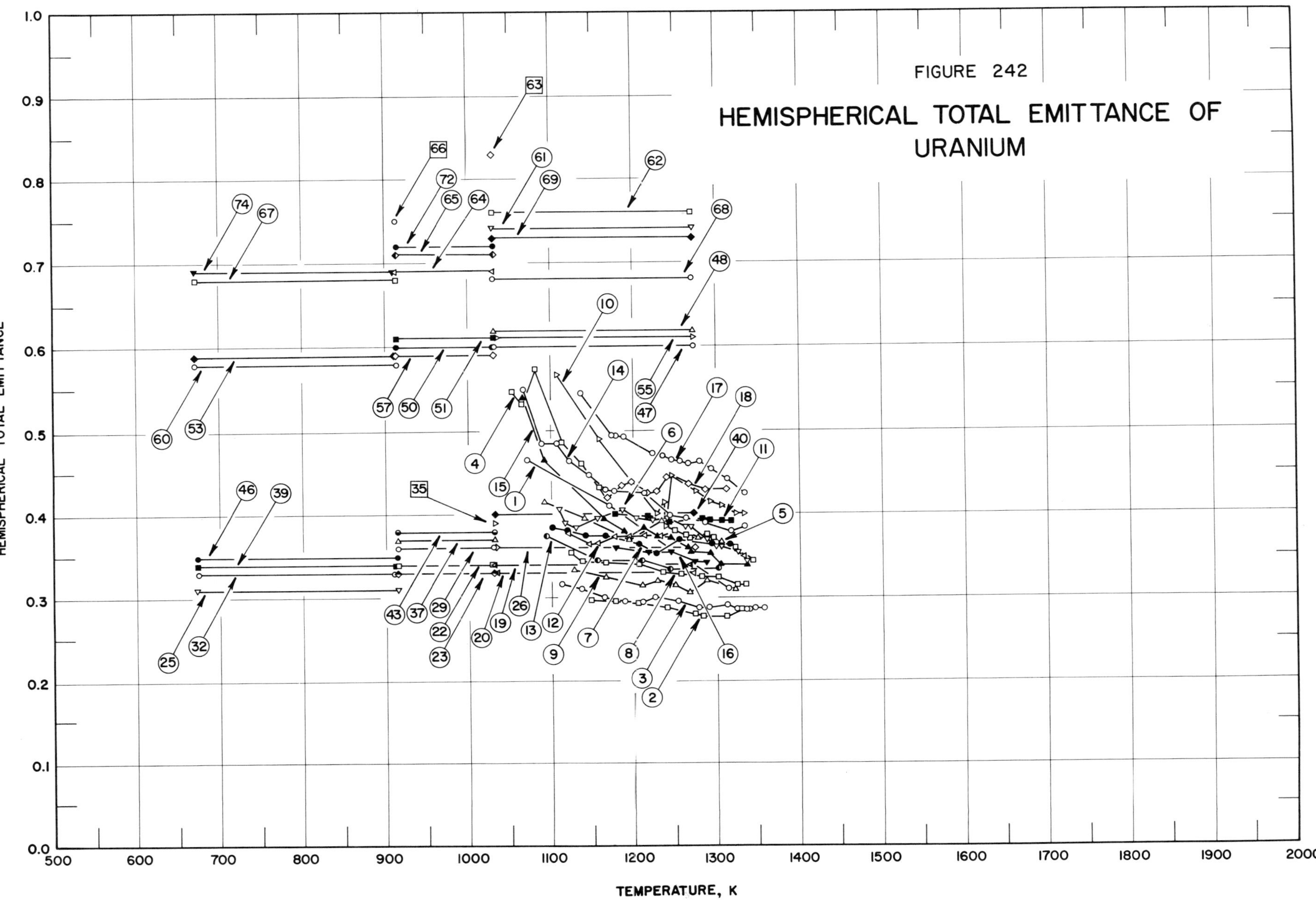

FIGURE 242
HEMISPHERICAL TOTAL EMITTANCE OF URANIUM

SPECIFICATION TABLE NO. 242 HEMISPHERICAL TOTAL EMITTANCE OF URANIUM

Curve No.	Ref. No.	Year	Temperature Range, K	Reported Error, %	Composition (weight percent), Specifications and Remarks
1	116	1957	1073-1314		Tube formed from strip; hot rolled, annealed and quenched, cold rolled to 5 mils, and heated in vacuum at 873 K for 4 hrs; measured in vacuum; run 1, heating cycle; [Author's designation; Specimen 1].
2	116	1957	1148-1330		Above specimen and conditions; cooling cycle.
3	116	1957	1113-1357		Above specimen and conditions; run 2, heating cycle.
4	116	1957	1052-1345		Different sample, same as curve 1 specimen and conditions; run 1, heating cycle; [Author's designation: Specimen 2].
5	116	1957	1092-1306		Above specimen and conditions; cooling cycle.
6	116	1957	1110-1338		Above specimen and conditions; run 2, heating cycle.
7	116	1957	1101-1316		Above specimen and conditions; cooling cycle.
8	116	1957	1123-1335		Above specimen and conditions; run 3, heating cycle.
9	116	1957	1127-1321		Above specimen and conditions; cooling cycle.
10	116	1957	1109-1333		Different sample, same as curve 1 specimen; same conditions except heated in vacuum at 875 K for 4 hrs; run 1, heating cycle; [Author's designation : Specimen 3].
11	116	1957	1178-1319		Above specimen and conditions; cooling cycle.
12	116	1957	1120-1333		Above specimen and conditions; run 2, heating cycle.
13	116	1957	1095-1301		Above specimen and conditions; cooling cycle.
14	116	1957	1068-1334		Different sample, same as curve 10 specimen and conditions; [Author's designation: Specimen 4].
15	116	1957	1068-1337		Different sample, same as curve 14 specimen and conditions; run 1, heating cycle; [Author's designation: Specimen 5].
16	116	1957	1178-1287		Above specimen and conditions; cooling cycle.
17	116	1957	1137-1334		Above specimen and conditions except 2.5 cm^3 of oxygen at atmospheric pressure was added to the vacuum chamber while specimen was at 1173 K (vacuum chamber pressure 2 x 10^{-5} mm Hg); oxygen was then evacuated; run 1, heating cycle.
18	116	1957	1166-1311		Above specimen and conditions; cooling cycle.
19	277	1963	1033-1273		High purity uranium (~250 ppm impurities); γ phase; polished; surface roughness 1 μ; measured in vacuum (3 x 10^{-7} mm Hg); measured using transient technique; data is average of several runs.
20	277	1963	916-1033		Above specimen and conditions except β phase.
21	277	1963	673-916		Above specimen and conditions except α phase.
22	277	1963	1033-1273		Different sample, same as curve 19 specimen and conditions except oxidized (~700 $\mu g\ O_2\ cm^{-2}$).
23	277	1963	916-1033		Above specimen and conditions except β phase.

SPECIFICATION TABLE NO. 242 (continued)

Curve No.	Ref. No.	Year	Temperature Range, K	Reported Error, %	Composition (weight percent), Specifications and Remarks
24	277	1963	673-916		Above specimen and conditions except α phase.
25	277	1963	1033-1273		Different sample, same as curve 19 specimen and conditions except oxidized (~1000 μg O_2 cm^{-2}).
26	277	1963	916-1033		Above specimen and conditions except β phase.
27	277	1963	673-916		Above specimen and conditions except α phase.

DATA TABLE NO. 242 HEMISPHERICAL TOTAL EMITTANCE OF URANIUM

[Temperature, T, K; Emittance, ∈]

T	∈
CURVE 1	
1073	0.465
1170	0.410
1314	0.310
CURVE 2	
1148	0.297
1176	0.295
1204	0.292
1240	0.288
1273	0.280
1283	0.278
1311	0.278
1330	0.285
CURVE 3	
1113	0.316
1136	0.310
1161	0.300
1188	0.296
1210	0.294
1225	0.300
1253	0.295
1278	0.288
1290	0.288
1312	0.290
1325	0.285
1335	0.286
1340	0.286
1346	0.288
1357	0.288
CURVE 4	
1052	0.548
1066	0.531
1081	0.573
1113	0.486
1137	0.460
1159	0.431
1221	0.354*
1239	0.365*
1240	0.341*
CURVE 4 (cont.)	
1250	0.380
1262	0.375
1268	0.372
1288	0.376
1296	0.372
1323	0.360
1345	0.345
CURVE 5	
1092	0.415
1140	0.395
1194	0.370
1232	0.391
1245	0.388*
1276	0.372
1306	0.368
CURVE 6	
1110	0.405
1116	0.390
1130	0.384
1155	0.395
1185	0.405
1201	0.396
1221	0.392
1240	0.395
1261	0.385
1268	0.385
1289	0.370
1300	0.362
1305	0.362
1326	0.355
1338	0.345
CURVE 7	
1101	0.385
1120	0.381
1141	0.375
1165	0.375
1206	0.365
1227	0.353
CURVE 7 (cont.)	
1255	0.370
1295	0.364
1316	0.363
CURVE 8	
1123	0.354
1138	0.344
1165	0.341
1206	0.340
1235	0.330
1256	0.329
1281	0.325
1304	0.323
1325	0.315
1335	0.315
CURVE 9	
1127	0.334
1165	0.325
1210	0.314
1229	0.320
1250	0.315
1267	0.305
1286	0.320
1321	0.310
CURVE 10	
1109	0.567
1159	0.490
1229	0.400
1236	0.414
1238	0.385
1245	0.449
1274	0.429
1291	0.415
1306	0.410
1323	0.400
1333	0.400
CURVE 11	
1178	0.400
1218	0.398
1243	0.390
1281	0.394
1291	0.393
1308	0.392
1319	0.392
CURVE 12	
1120	0.379
1144	0.365
1155	0.366
1177	0.373
1194	0.373
1213	0.374
1231	0.373
1257	0.370*
1267	0.380*
1283	0.373*
1306	0.365*
1333	0.350
CURVE 13	
1095	0.375
1153	0.346
1210	0.345
1244	0.334
1301	0.335
CURVE 14	
1068	0.550
1090	0.485
1108	0.485
1123	0.463
1146	0.449
1165	0.430
1176	0.429
1213	0.425
1243	0.398
1263	0.395
1286	0.390
CURVE 14 (cont.)	
1318	0.380
1334	0.385
CURVE 15*	
1068	0.540
1093	0.466
1163	0.395
1188	0.380
1190	0.375*
1211	0.384
1228	0.375
1244	0.372
1265	0.360
1268	0.355
1291	0.354
1303	0.340
1337	0.340
CURVE 16	
1178	0.360
1218	0.355
1243	0.355
1271	0.345
1287	0.342
CURVE 17*	
1137	0.545
1173	0.495
1178	0.495
1188	0.493
1222	0.472
1235	0.470
1246	0.465
1256	0.463
1266	0.460
1280	0.463
1294	0.454
1312	0.442
1334	0.425
CURVE 18*	
1166	0.420
1185	0.436
1198	0.440
1216	0.425
1228	0.428
1238	0.445
1266	0.438
1286	0.430
1311	0.430
CURVE 19	
1033	0.37
1273	0.37
CURVE 20	
916	0.36
1033	0.36
CURVE 21	
673	0.33
916	0.33
CURVE 22	
1033	0.61
1273	0.61
CURVE 23	
916	0.61
1033	0.61
CURVE 24	
673	0.58
916	0.58
CURVE 25	
1033	0.73
1273	0.73
CURVE 26	
916	0.71
1033	0.71
CURVE 27	
673	0.68
916	0.68

*Not shown on plot

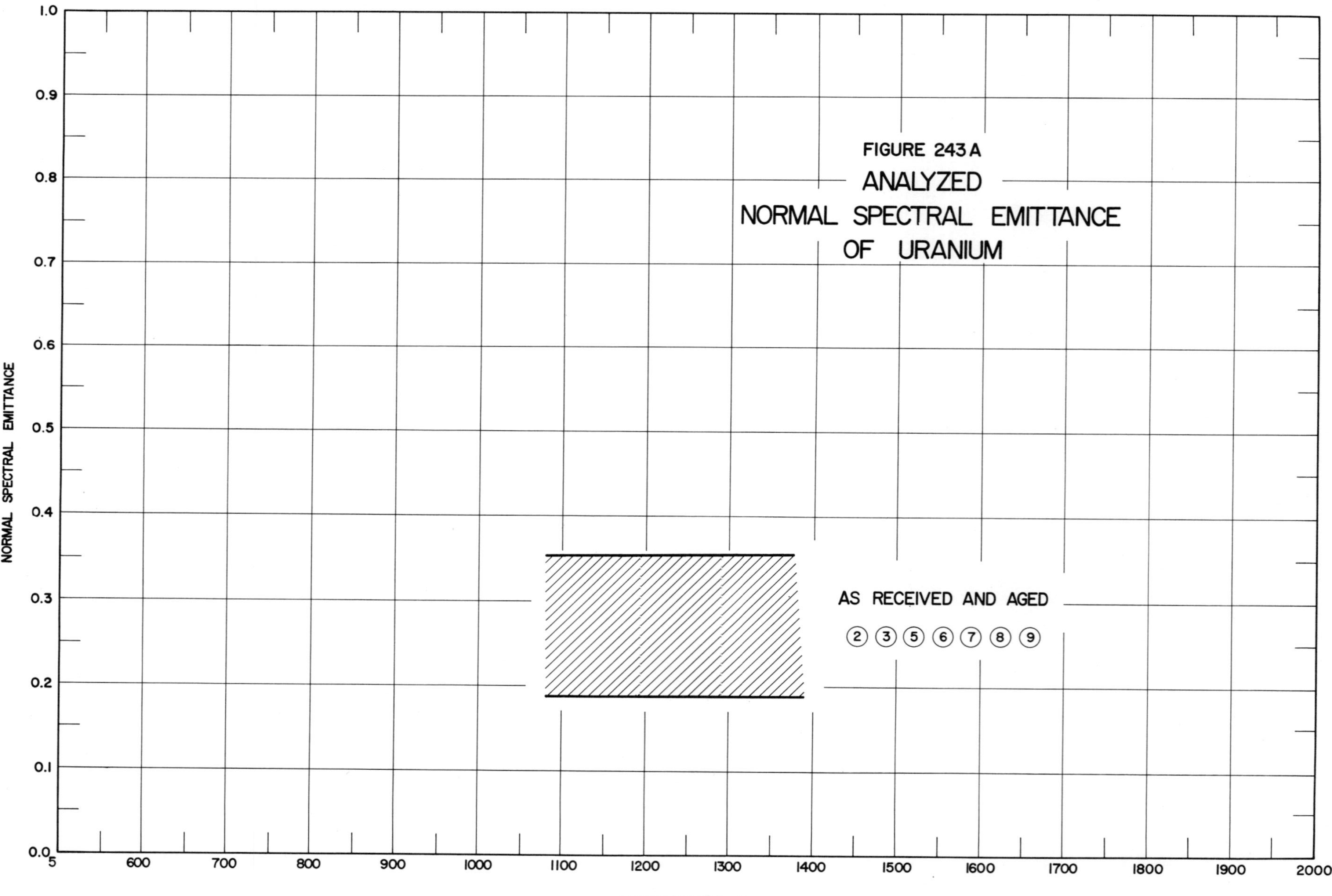

FIGURE 243A
ANALYZED
NORMAL SPECTRAL EMITTANCE
OF URANIUM

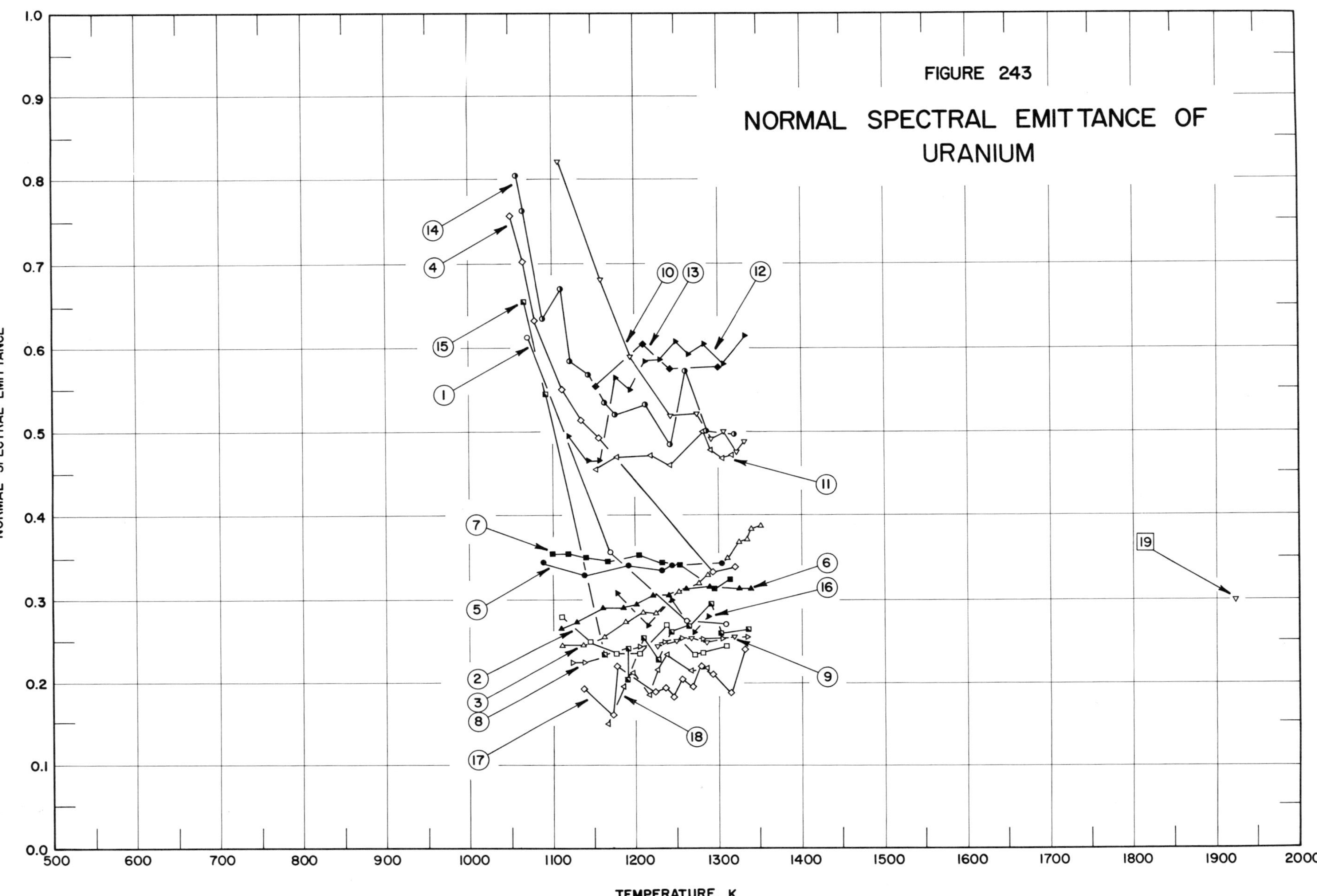
FIGURE 243
NORMAL SPECTRAL EMITTANCE OF
URANIUM
NORMAL SPECTRAL EMITTANCE
TEMPERATURE, K
1.0
0.9
0.8
0.7
0.6
0.5
0.4
0.3
0.2
0.1
0.0
500
600
700
800
900
1000
1100
1200
1300
1400
1500
1600
1700
1800
1900
2000
14
4
15
1
10
13
12
11
7
5
6
16
2
3
8
9
18
17
19

SPECIFICATION TABLE NO. 243 NORMAL SPECTRAL EMITTANCE OF URANIUM

Curve No.	Ref. No.	Year	Wavelength μ	Temperature Range, K	Geometry θ'	Reported Error, %	Composition (weight percent), Specifications and Remarks
1	116	1957	0.65	1071-1312	~0°		Tube formed from strip; hot rolled, annealed and quenched, cold rolled to 5 mils, and heated in vacuum at 873 K for 4 hrs; measured in vacuum; run 1, heating cycle; [Author's designation: Specimen 1].
2	116	1957	0.65	1111-1311	~0°		Above specimen and conditions; cooling cycle.
3	116	1957	0.65	1113-1353	~0°		Above specimen and conditions; run 2, heating cycle.
4	116	1957	0.65	1052-1321	~0°		Different sample, same as curve 1 specimen and conditions; run 1, heating cycle; [Author's designation: Specimen 2].
5	116	1957	0.65	1091-1305	~0°		Above specimen and conditions; cooling cycle.
6	116	1957	0.65	1111-1340	~0°		Above specimen and conditions; run 2, heating cycle.
7	116	1957	0.65	1101-1316	~0°		Above specimen and conditions; cooling cycle.
8	116	1957	0.65	1123-1336	~0°		Above specimen and conditions; run 3, heating cycle.
9	116	1957	0.65	1126-1320	~0°		Above specimen and conditions; cooling cycle.
10	116	1957	0.65	1110-1333	~0°		Different sample, same as curve 1 specimen and conditions; run 1, heating cycle; [Author's designation: Specimen 3].
11	116	1957	0.65	1153-1319	~0°		Above specimen and conditions; cooling cycle.
12	116	1957	0.65	1120-1334	~0°		Above specimen and conditions; run 2, heating cycle.
13	116	1957	0.65	1153-1301	~0°		Above specimen and conditions; cooling cycle.
14	116	1957	0.65	1058-1320	~0°		Different sample, same as curve 10 specimen and conditions; [Author's designation: Specimen 4].
15	116	1957	0.65	1067-1338	~0°		Different sample, same as curve 14 specimen and conditions; run 1, heating cycle; [Author's designation: Specimen 5].
16	116	1957	0.65	1178-1288	~0°		Above specimen and conditions; cooling cycle.
17	116	1957	0.65	1138-1333	~0°		Above specimen and conditions except 2.5 cm^3 of oxygen at atmospheric pressure was added to the vacuum chamber while specimen was at 1173 K (vacuum chamber pressure 2 x 10^{-5} mm Hg); oxygen was then evacuated; run 1, heating cycle.
18	116	1957	0.65	1166-1286	~0°		Above specimen and conditions; cooling cycle.
19	19	1914	0.65	1923	~0°	1	Film; tungsten substrate; melted in hydrogen then oxidized in air by heating; measured in air; Pt reference (ϵ= 0.33 for λ = 0.650 at all temp).
20	117	1939	0.67	< M.P.	~0°		Annealed in vacuum; surface coated with oxide; measured in vacuum (10^{-8} mm Hg).

DATA TABLE NO. 243 NORMAL SPECTRAL EMITTANCE OF URANIUM

[Temperature, T, K; Emittance, ϵ; Wavelength, λ, μ]

T	ϵ
CURVE 1	
λ = 0.65	
1071	0.614
1170	0.357
1263	0.275
1312	0.273
CURVE 2	
λ = 0.65	
1111	0.280
1146	0.250
1176	0.235
1205	0.235
1239	0.270
1273	0.233
1283	0.235
1311	0.243
CURVE 3	
λ = 0.65	
1113	0.245
1136	0.245
1163	0.255
1188	0.274
1210	0.285
1225	0.284
1253	0.310
1277	0.320
1288	0.330
1313	0.350
1326	0.370
1336	0.372
1341	0.385
1346	0.388*
1353	0.388
CURVE 4	
λ = 0.65	
1052	0.757
1067	0.703
1080	0.634
1114	0.550
CURVE 4 (cont.)	
1136	0.515
1157	0.492
1296	0.337
1321	0.340
CURVE 5	
λ = 0.65	
1091	0.346
1139	0.330
1193	0.342
1233	0.336
1245	0.342
1305	0.345
CURVE 6	
λ = 0.65	
1111	0.266
1117	0.273*
1130	0.274
1161	0.290
1168	0.290*
1185	0.291
1201	0.295
1222	0.305
1240	0.305
1262	0.314
1268	0.313*
1290	0.316
1300	0.314*
1305	0.318*
1327	0.314
1340	0.314
CURVE 7	
λ = 0.65	
1101	0.355
1120	0.355
1142	0.351
1166	0.348
1206	0.353
1233	0.345
CURVE 7 (cont.)	
1253	0.344
1295	0.315
1316	0.325
CURVE 8	
λ = 0.65	
1123	0.225
1138	0.225
1164	0.233*
1206	0.244
1235	0.250
1256	0.255
1281	0.254
1305	0.254
1326	0.255*
1336	0.256
CURVE 9	
λ = 0.65	
1126	0.225*
1211	0.244*
1228	0.245
1251	0.250
1268	0.255
1286	0.249
1320	0.255
CURVE 10	
λ = 0.65	
1110	0.820
1160	0.680
1195	0.590
1243	0.520
1275	0.522
1292	0.490
1308	0.499
1323	0.476
1333	0.488
CURVE 11	
λ = 0.65	
1153	0.455
1179	0.470
1220	0.472
1243	0.460
1282	0.500
1291	0.478
1307	0.468
1319	0.472
CURVE 12	
λ = 0.65	
1120	0.495
1144	0.466
1156	0.466
1178	0.565
1194	0.550
1213	0.585
1230	0.586
1251	0.608
1266	0.594
1284	0.604
1308	0.581
1334	0.616
CURVE 13	
λ = 0.65	
1153	0.554
1211	0.605
1244	0.576
1301	0.578
CURVE 14	
λ = 0.65	
1058	0.805
1066	0.763
1090	0.636
1111	0.670
1123	0.585
1145	0.568
1163	0.535
CURVE 14 (cont.)	
1176	0.522
1213	0.532
1243	0.485
1262	0.573
1286	0.501
1320	0.498
CURVE 15	
λ = 0.65	
1067	0.656
1094	0.545
1163	0.235
1191	0.242
1191	0.204
1211	0.255
1228	0.228
1245	0.262
1265	0.270
1268	0.275*
1292	0.296
1303	0.260
1338	0.265
CURVE 16	
λ = 0.65	
1178	0.308
1216	0.270
1243	0.300
1271	0.262
1288	0.280
CURVE 17	
λ = 0.65	
1138	0.194
1173	0.160
1178	0.220
1187	0.214*
1223	0.189
1236	0.193
1247	0.183
1256	0.204
CURVE 17 (cont.)	
1269	0.195
1279	0.220
1293	0.210
1316	0.188
1333	0.240
CURVE 18	
λ = 0.65	
1166	0.149
1186	0.196
1197	0.212
1216	0.185
1228	0.216
1238	0.234
1268	0.215
1286	0.218
CURVE 19	
λ = 0.65	
1923	0.30
CURVE 20*	
λ = 0.67	
<M.P.	0.51

* Not shown on plot

SPECIFICATION TABLE NO. 244 NORMAL SPECTRAL EMITTANCE OF URANIUM

Curve No.	Ref. No.	Year	Temperature K	Wavelength Range, μ	Geometry θ'	Reported Error, %	Composition (weight percent), Specifications and Remarks
1	19	1914	< M.P.	0.55-0.65	$\sim 0^\circ$	1	Film; tungsten substrate; measured in hydrogen; Pt reference (ϵ = 0.33 for λ = 0.650 μ and ϵ = 0.38 for λ = 0.547 μ at all temp).

DATA TABLE NO. 244 NORMAL SPECTRAL EMITTANCE OF URANIUM

[Wavelength, λ, μ; Emittance, ϵ; Temperature, T, K]

λ	ϵ
CURVE 1*	
T = <M.P.	
0.55	0.77
0.65	0.55

* Not shown on plot

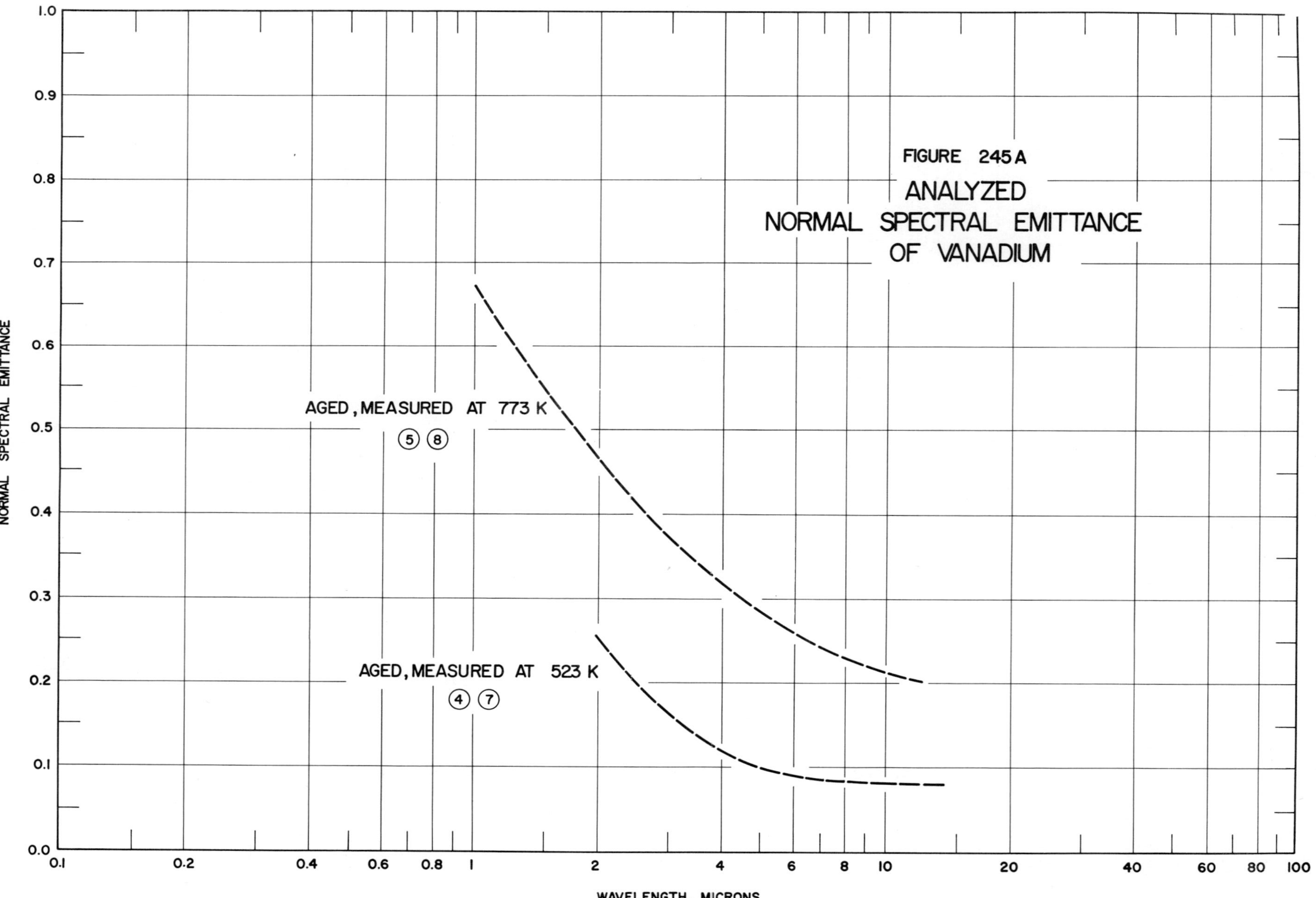
FIGURE 245 A
ANALYZED
NORMAL SPECTRAL EMITTANCE
OF VANADIUM
AGED, MEASURED AT 773 K
5 8
AGED, MEASURED AT 523 K
4 7
NORMAL SPECTRAL EMITTANCE
1.0
0.9
0.8
0.7
0.6
0.5
0.4
0.3
0.2
0.1
0.0
0.1
0.2
0.4
0.6
0.8
1
2
4
6
8
10
20
40
60
80
100
WAVELENGTH, MICRONS

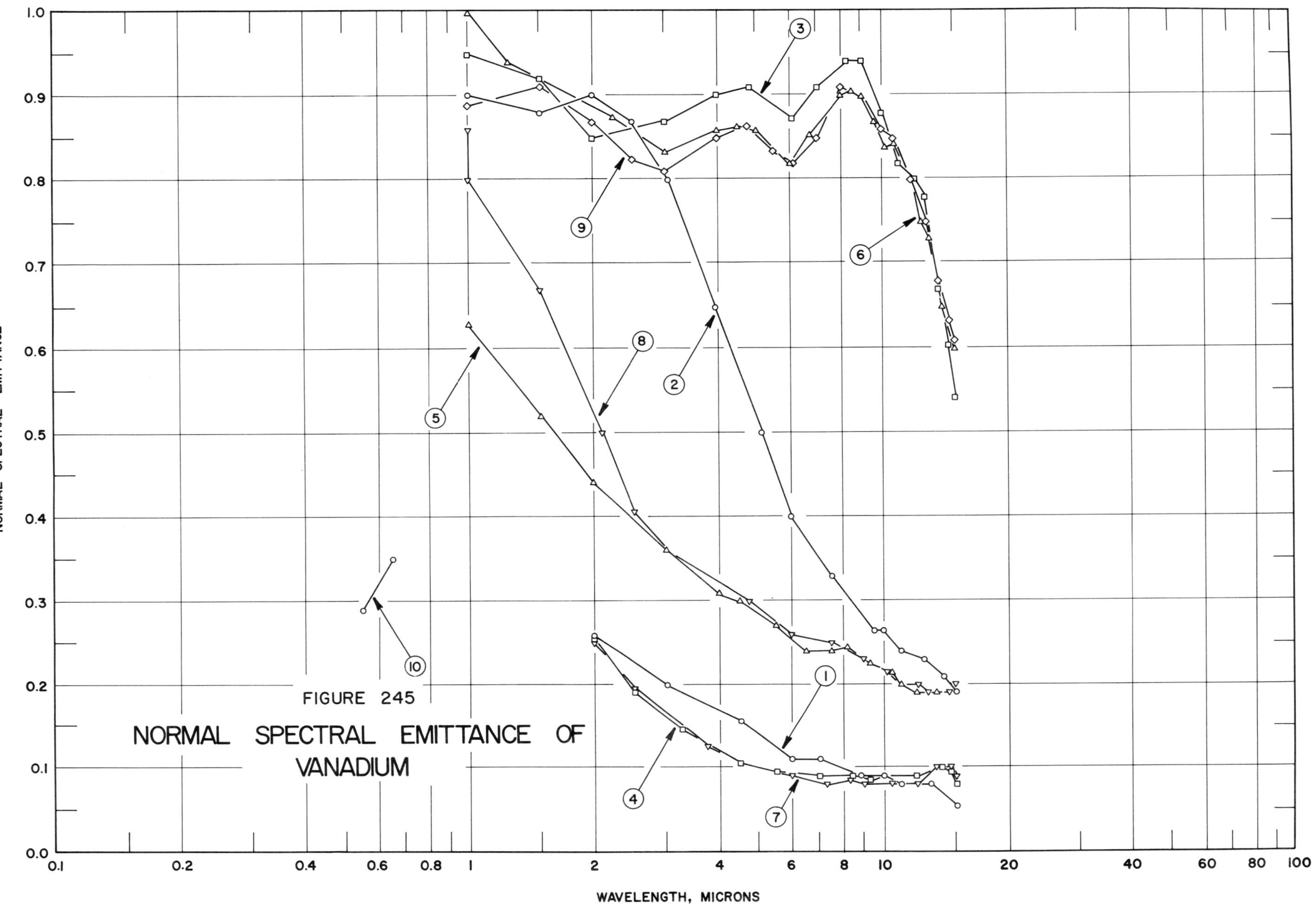

FIGURE 245

NORMAL SPECTRAL EMITTANCE OF VANADIUM

SPECIFICATION TABLE NO. 245 NORMAL SPECTRAL EMITTANCE OF VANADIUM

Curve No.	Ref. No.	Year	Temperature K	Wavelength Range, μ	Geometry θ'	Reported Error, %	Composition (weight percent), Specifications and Remarks
1	86	1961	523	2.00-15.00	$\sim 0°$	± 5	As received; data extracted from smooth curve.
2	86	1961	773	1.00-15.00	$\sim 0°$	± 5	Different sample, same as curve 1 specimen and conditions.
3	86	1961	1023	1.00-15.00	$\sim 0°$	± 5	Different sample, same as curve 1 specimen and conditions.
4	86	1961	523	2.00-15.00	$\sim 0°$	± 5	Different sample, same as curve 1 specimen and conditions except heated in argon at 589 K for 30 min.
5	86	1961	773	1.00-15.00	$\sim 0°$	± 5	Different sample, same as curve 4 specimen and conditions.
6	86	1961	1023	1.00-15.00	$\sim 0°$	± 5	Different sample, same as curve 4 specimen and conditions.
7	86	1961	523	2.00-15.00	$\sim 0°$	± 5	Different sample, same as curve 1 specimen and conditions except heated in vacuum (2.1×10^{-5} mm Hg) at 589 K for 30 min.
8	86	1961	773	1.00-15.00	$\sim 0°$	± 5	Different sample, same as curve 7 specimen and conditions.
9	86	1961	1023	1.00-15.00	$\sim 0°$	± 5	Different sample, same as curve 7 specimen and conditions.
10	19	1914	1843	0.55-0.65	$\sim 0°$	1	Film; tungsten substrate; measured in hydrogen; Pt reference ($\epsilon = 0.33$ for $\lambda = 0.650\ \mu$ and $\epsilon = 0.38$ for $\lambda = 0.547\ \mu$ at all temp).

DATA TABLE NO. 245 NORMAL SPECTRAL EMITTANCE OF VANADIUM

[Wavelength, λ, μ; Emittance, ϵ; Temperature, T, K]

λ	ϵ
CURVE 1	
T = 523	
2.00	0.260
3.00	0.200
4.50	0.155
6.00	0.110
7.00	0.110
8.80	0.090
10.00	0.090
11.00	0.080
13.00	0.080
15.00	0.055
CURVE 2	
T = 773	
1.00	0.900
1.50	0.880
2.00	0.900
2.50	0.870
3.05	0.800
3.95	0.650
5.10	0.500
6.00	0.400
7.50	0.330
9.50	0.265
10.00	0.265
11.00	0.240
12.50	0.230
14.00	0.210
15.00	0.190
CURVE 3	
T = 1023	
1.00	0.950
1.50	0.920
2.00	0.850
3.00	0.870
4.00	0.900
4.80	0.910
6.10	0.875
7.00	0.910
8.25	0.940
CURVE 3 (cont.)	
9.00	0.940
10.00	0.880
11.00	0.820
12.00	0.800
12.75	0.780
13.75	0.670
14.50	0.605
15.00	0.540
CURVE 4	
T = 523	
2.00	0.255
2.50	0.190
3.25	0.145
4.50	0.105
5.50	0.095
7.00	0.090
8.40	0.090
9.25	0.085
10.00	0.090*
12.00	0.090
13.75	0.100
14.50	0.095
15.00	0.080
CURVE 5	
T = 773	
1.00	0.630
1.50	0.520
2.00	0.440
3.00	0.360
4.00	0.310
4.50	0.300
5.50	0.270
6.50	0.240
7.50	0.240
8.20	0.245
9.25	0.225
10.50	0.215
11.00	0.200
12.00	0.190
13.50	0.190
15.00	0.190*
CURVE 6	
T = 1023	
1.00	1.000
1.25	0.940
2.25	0.875
3.00	0.835
4.00	0.860
4.50	0.865
5.00	0.860
6.00	0.820
6.75	0.855
8.00	0.900
8.50	0.905
9.00	0.900
9.60	0.870
10.25	0.840
10.75	0.845
11.70	0.800*
12.50	0.750
13.00	0.730
14.00	0.650
15.00	0.600
CURVE 7	
T = 523	
2.00	0.250
2.50	0.195
3.75	0.125
4.50	0.105*
6.00	0.090
7.25	0.080
8.25	0.085
9.00	0.080
10.50	0.080
12.00	0.080
13.50	0.100
14.50	0.100
15.00	0.090
CURVE 8	
T = 773	
1.00	0.860
1.00	0.800
CURVE 8 (cont.)	
1.50	0.670
2.10	0.500
2.50	0.405
3.00	0.360*
4.75	0.300
6.00	0.260
7.50	0.250
9.00	0.230
10.25	0.215
11.00	0.200*
12.25	0.200
12.75	0.190
14.50	0.190
15.00	0.200
CURVE 9	
T = 1023	
1.00	0.890
1.50	0.910
2.00	0.870
2.50	0.825
3.00	0.810
4.00	0.850
4.75	0.865
5.50	0.835
6.15	0.820
7.00	0.850
8.00	0.910
9.00	0.900*
10.00	0.860
10.70	0.850
11.75	0.800
12.80	0.750
13.75	0.680
14.50	0.635
15.00	0.610
CURVE 10	
T = 1843	
0.55	0.29
0.65	0.35

* Not shown on plot

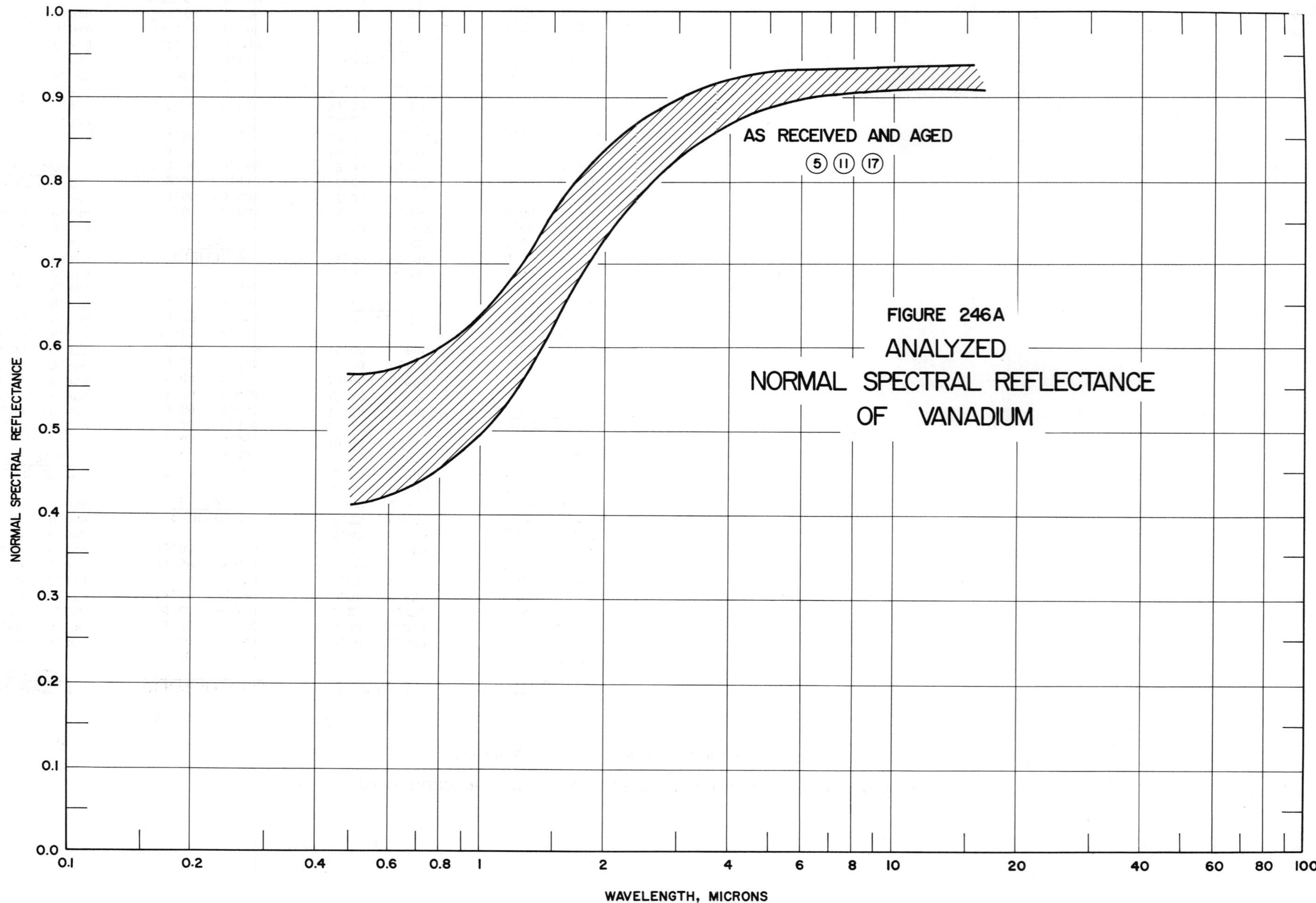

AS RECEIVED AND AGED
5 11 17
FIGURE 246A
ANALYZED
NORMAL SPECTRAL REFLECTANCE
OF VANADIUM
NORMAL SPECTRAL REFLECTANCE
1.0
0.9
0.8
0.7
0.6
0.5
0.4
0.3
0.2
0.1
0.0
0.1
0.2
0.4
0.6
0.8
1
2
4
6
8
10
20
40
60
80
100
WAVELENGTH, MICRONS

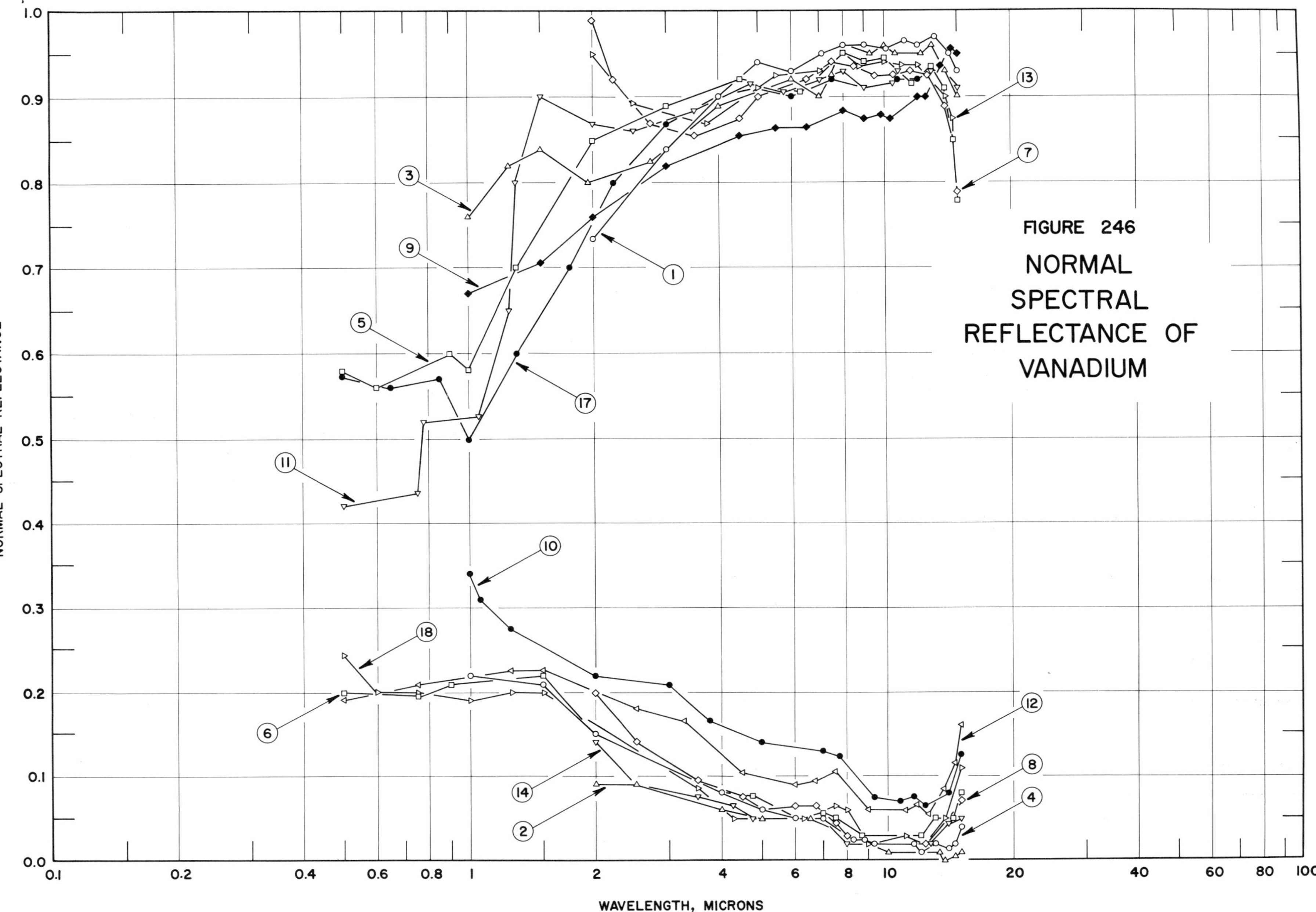

FIGURE 246 NORMAL SPECTRAL REFLECTANCE OF VANADIUM

SPECIFICATION TABLE NO. 246 NORMAL SPECTRAL REFLECTANCE OF VANADIUM

Curve No.	Ref. No.	Year	Temperature K	Wavelength Range, μ	Geometry θ	θ'	ω'	Reported Error, %	Composition (weight percent), Specifications and Remarks
1	86	1961	~ 322	2.00-15.00	~0°		2π	< 2	As received; data extracted from a smooth curve; hohlraum at 523 K; converted from $R(2\pi, 0)$.
2	86	1961	~ 322	2.00-15.00	~0°		2π	< 2	Above specimen and conditions; diffuse component only.
3	86	1961	~ 322	1.00-15.00	~0°		2π	< 2	Different sample, same as Curve 1 specimen and conditions except hohlraum at 773 K.
4	86	1961	~ 322	1.00-15.00	~0°		2π	< 2	Above specimen and conditions; diffuse component only.
5	86	1961	~ 322	0.50-15.00	~0°		2π	< 2	Different sample, same as Curve 1 specimen and conditions except hohlraum at 1273 K.
6	86	1961	~ 322	0.50-15.00	~0°		2π	< 2	Above specimen and conditions; diffuse component only.
7	86	1961	~ 322	2.00-15.00	~0°		2π	< 2	Different sample, same as Curve 1 specimen and conditions except heated in argon at 589 K for 30 min.
8	86	1961	~ 322	2.00-15.00	~0°		2π	< 2	Above specimen and conditions; diffuse component only.
9	86	1961	~ 322	1.00-15.00	~0°		2π	< 2	Different sample, same as Curve 7 specimen and conditions except hohlraum at 773 K.
10	86	1961	~ 322	1.00-15.00	~0°		2π	< 2	Above specimen and conditions; diffuse component only.
11	86	1961	~ 322	0.50-15.00	~0°		2π	< 2	Different sample, same as Curve 7 specimen and conditions except hohlraum at 1273 K.
12	86	1961	~ 322	0.50-15.00	~0°		2π	< 2	Above specimen and conditions; diffuse component only.
13	86	1961	~ 322	2.00-15.00	~0°		2π	< 2	Different sample, same as Curve 1 specimen and conditions except heated in vacuum (2.1×10^{-5} mm Hg) at 589 K for 30 min.
14	86	1961	~ 322	2.00-15.00	~0°		2π	< 2	Above specimen and conditions; diffuse component only.
15	86	1961	~ 322	1.00-15.00	~0°		2π	< 2	Different sample, same as Curve 13 specimen and conditions except hohlraum at 773 K.
16	86	1961	~ 322	0.90-15.00	~0°		2π	< 2	Above specimen and conditions; diffuse component only.
17	86	1961	~ 322	0.50-15.00	~0°		2π	< 2	Different sample, same as Curve 13 specimen and conditions except hohlraum at 1273 K.
18	86	1961	~ 322	0.50-15.00	~0°		2π	< 2	Above specimen and conditions; diffuse component only.

DATA TABLE NO. 246 NORMAL SPECTRAL REFLECTANCE OF VANADIUM

[Wavelength, λ, μ; Reflectance, ρ; Temperature, T, K]

λ	ρ
CURVE 1	**T = ~322**
2.00	0.735
3.00	0.840
4.00	0.900
5.00	0.940
6.00	0.930
7.10	0.950
8.00	0.960
9.00	0.960
10.10	0.955
11.15	0.965
12.00	0.960
13.10	0.970
14.20	0.950
15.00	0.930
CURVE 2	**T = ~322**
2.00	0.090
2.50	0.090
4.00	0.060
5.00	0.050
6.50	0.050
10.00	0.010
13.25	0.010
13.75	0.000
14.50	0.005
15.00	0.010
CURVE 3	**T = ~322**
1.00	0.760
1.25	0.820
1.50	0.840
1.95	0.800
2.75	0.825
4.00	0.890
6.00	0.920
7.00	0.900
7.50	0.925
8.00	0.960*
9.25	0.950
CURVE 3 (cont.)	
10.00	0.960
10.60	0.950
12.25	0.950
13.00	0.960
14.00	0.930
15.00	0.900
CURVE 4	**T = ~322**
1.00	0.220
1.50	0.210
2.00	0.150
4.00	0.080
5.00	0.060
6.00	0.050
7.00	0.050
8.25	0.025
8.80	0.025
9.25	0.020
11.50	0.020
12.00	0.010
13.00	0.020
14.00	0.015
14.50	0.020
15.00	0.040
CURVE 5	**T = ~322**
0.50	0.580
0.60	0.560
0.90	0.600
1.00	0.580
1.30	0.700
2.00	0.850
3.00	0.890
4.50	0.920
6.30	0.905
8.00	0.950
9.00	0.940
10.00	0.945
11.70	0.915
13.00	0.935
CURVE 5 (cont.)	
14.00	0.910
14.75	0.850
15.00	0.780
CURVE 6	**T = ~322**
0.50	0.200
0.75	0.195
0.90	0.210
1.50	0.220
2.00	0.150*
4.00	0.080*
4.75	0.075
6.00	0.050*
7.00	0.055
7.50	0.050
8.50	0.030
12.00	0.030
13.00	0.050
14.25	0.050
15.00	0.080
CURVE 7	**T = ~322**
2.00	0.990
2.25	0.920
2.75	0.870
3.50	0.855
4.50	0.875
5.00	0.900
6.50	0.920
7.50	0.940
8.50	0.935
9.50	0.925
10.50	0.925
11.50	0.930
12.75	0.925
14.00	0.890
14.60	0.850*
15.00	0.790
CURVE 8	**T = ~322**
2.00	0.200
2.50	0.140
3.50	0.095
4.50	0.075
5.00	0.060*
6.00	0.065
6.75	0.065
7.50	0.045
8.00	0.030
9.25	0.020*
12.25	0.020
15.00	0.070
CURVE 9	**T = ~322**
1.00	0.670
1.50	0.705
2.00	0.760
3.00	0.820
4.50	0.855
5.50	0.865
6.50	0.865
8.00	0.885
9.00	0.875
9.85	0.880
10.30	0.875
12.00	0.900
12.50	0.900
13.75	0.935
14.50	0.955
15.00	0.950
CURVE 10	**T = ~322**
1.00	0.340
1.05	0.310
1.25	0.275
2.00	0.220
3.00	0.210
3.75	0.165
5.00	0.140
CURVE 10 (cont.)	
7.00	0.130
7.75	0.125
9.25	0.075
10.75	0.070
11.50	0.075
12.25	0.065
14.00	0.080
15.00	0.125
CURVE 11	**T = ~322**
0.50	0.420
0.75	0.435
0.78	0.520
1.05	0.525
1.25	0.650
1.30	0.800
1.50	0.900
2.00	0.870
2.50	0.860
3.50	0.885
4.80	0.915
5.75	0.905
7.00	0.920
8.00	0.930
9.00	0.910
10.50	0.915
10.80	0.930
13.00	0.930
14.00	0.930*
15.00	0.910
CURVE 12	**T = ~322**
0.50	0.190
0.75	0.210
1.25	0.225
1.50	0.225
2.50	0.180
3.25	0.165
4.50	0.105
6.00	0.090
CURVE 12 (cont.)	
6.75	0.095
7.50	0.105
9.00	0.060
11.00	0.060
11.65	0.065
12.15	0.055
13.75	0.085
14.50	0.115
15.00	0.160
CURVE 13	**T = ~322**
2.00	0.950
2.50	0.895
3.75	0.870
5.00	0.910
5.50	0.925
7.00	0.930
8.00	0.950*
8.75	0.935
10.00	0.940
11.00	0.935
12.00	0.935
14.00	0.900
14.50	0.875
14.80	0.850*
15.00	0.790*
CURVE 14	**T = ~322**
2.00	0.140
2.50	0.090*
3.50	0.075
4.25	0.065
4.75	0.050
7.00	0.050*
8.00	0.020
12.50	0.020
14.00	0.045
15.00	0.050
CURVE 15*	**T = ~322**
1.00	0.680
2.00	0.790
2.75	0.850
4.00	0.900
5.00	0.910
6.00	0.910
7.50	0.925
8.00	0.940
8.80	0.920
10.50	0.935
11.75	0.950
13.00	0.950
14.00	0.990
14.50	0.990
15.00	0.980
CURVE 16*	**T = ~322**
0.90	0.210
1.25	0.215
1.75	0.195
2.25	0.160
3.00	0.140
4.00	0.080
6.00	0.080
7.00	0.090
8.00	0.080
9.00	0.040
9.50	0.030
11.25	0.040
12.25	0.035
13.00	0.040
14.00	0.045
15.00	0.075
CURVE 17	**T = ~322**
0.50	0.575
0.65	0.560
0.85	0.570
1.00	0.500
CURVE 17 (cont.)	
1.30	0.600
1.75	0.700
2.25	0.800
3.00	0.870
4.00	0.900*
5.00	0.910*
6.00	0.900
7.50	0.920
9.00	0.910*
10.75	0.920
12.00	0.920
13.00	0.930*
14.00	0.930*
15.00	0.910*
CURVE 18	**T = ~322**
0.50	0.245
0.60	0.200
0.75	0.200
1.00	0.190
1.25	0.200
1.50	0.200
3.50	0.085
4.25	0.050
6.25	0.050
7.50	0.065
8.00	0.060
9.00	0.020
11.00	0.030
12.00	0.020*
13.75	0.050
15.00	0.110

* Not shown on plot

SPECIFICATION TABLE NO. 247 ANGULAR SPECTRAL REFLECTANCE OF VANADIUM

Curve No.	Ref. No.	Year	Temperature K	Wavelength Range, μ	Geometry θ θ' ω'	Reported Error, %	Composition (weight percent), Specifications and Remarks
1	132	1911	298	0.57-8.80	15° 15°	≤ 3	Highly polished; silvered glass mirror reference.

DATA TABLE NO. 247 ANGULAR SPECTRAL REFLECTANCE OF VANADIUM

[Wavelength, λ, μ; Reflectance, ρ; Temperature, T, K]

λ	ρ
CURVE 1*	
T = 298	
0.57	0.577
0.90	0.600
1.20	0.620
1.60	0.665
2.40	0.710
4.05	0.785
6.10	0.850
7.10	0.880
8.00	0.900
8.80	0.905

* Not shown on plot

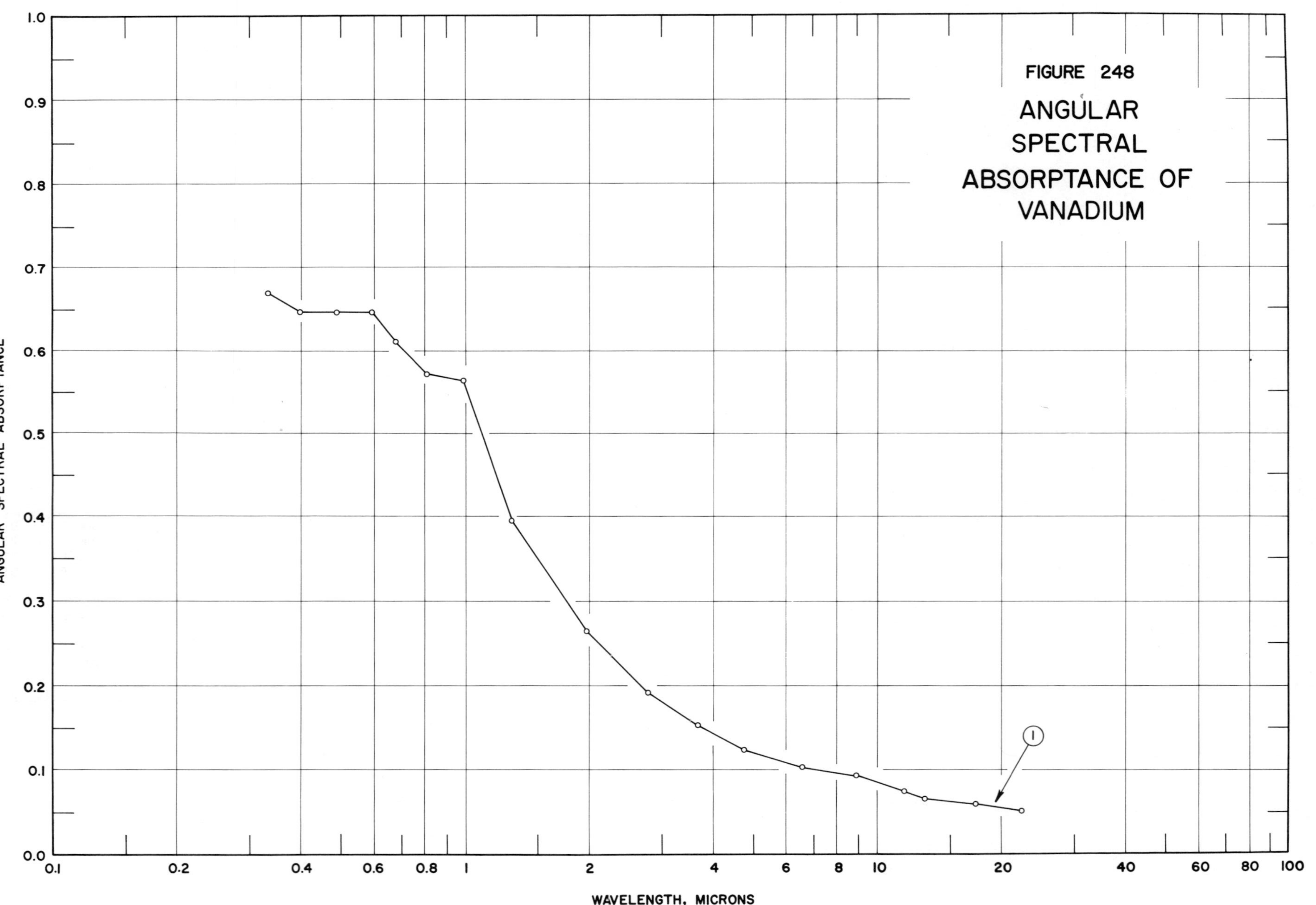

FIGURE 248 ANGULAR SPECTRAL ABSORPTANCE OF VANADIUM

SPECIFICATION TABLE NO. 248 ANGULAR SPECTRAL ABSORPTANCE OF VANADIUM

Curve No.	Ref. No.	Year	Temperature K	Wavelength Range, μ	Geometry θ	Reported Error, %	Composition (weight percent), Specifications and Remarks
1	225	1965	306	0.332-22.3	25°		Vanadium (rolled plate) from Belmont Smelting and Refining Works; measured in dry nitrogen; heated cavity at approx 1056 K with platinum reference; authors assumed $\alpha = 1-R(2\pi, 25°)$.

DATA TABLE NO. 248 ANGULAR SPECTRAL ABSORPTANCE OF VANADIUM

[Wavelength, λ, μ; Absorptance, α; Temperature, T, K]

λ	α
CURVE 1 T = 306	
0.332	0.668
0.398	0.646
0.488	0.646
0.590	0.646
0.679	0.611
0.805	0.570
0.982	0.562
1.29	0.395
1.97	0.265
2.79	0.191
3.66	0.152
4.76	0.123
6.59	0.102
8.95	0.093
11.5	0.073
13.0	0.066
17.2	0.060
22.3	0.052

SPECIFICATION TABLE NO. 249 NORMAL SPECTRAL EMITTANCE OF YTTRIUM

Curve No.	Ref. No.	Year	Wavelength μ	Temperature Range, K	Geometry θ'	Reported Error, %	Composition (weight percent), Specifications and Remarks
1	19	1914	0.65	< M.P.	$\sim 0^\circ$	1	Film; tungsten substrate; measured in hydrogen; Pt reference ($\epsilon = 0.33$ for $\lambda = 0.650\ \mu$ at all temp).

DATA TABLE NO. 249 NORMAL SPECTRAL EMITTANCE OF YTTRIUM

[Temperature, T, K; Emittance, ϵ; Wavelength, λ, μ]

T	ϵ
CURVE 1*	
$\lambda = 0.65$	
<M.P.	0.35

* Not shown on plot

SPECIFICATION TABLE NO. 250 HEMISPHERICAL TOTAL EMITTANCE OF ZINC

Curve No.	Ref. No.	Year	Temperature Range, K	Reported Error, %	Composition (weight percent), Specifications and Remarks
1	3	1955	76	5	Foil (0.0065 in. thick); solvent cleaned; measured in vacuum (10^{-6} to 10^{-7} mm Hg); emittance for 300 K black body incident radiation; authors assumed $\alpha = \epsilon$.

DATA TABLE NO. 250 HEMISPHERICAL TOTAL EMITTANCE OF ZINC

[Temperature, T, K; Emittance, ϵ]

T	ϵ
CURVE 1*	
76	0.02

* Not shown on plot

SPECIFICATION TABLE NO. 251 NORMAL TOTAL EMITTANCE OF ZINC

Curve No.	Ref. No.	Year	Temperature Range, K	Geometry θ'	Reported Error, %	Composition (weight percent), Specifications and Remarks
1	14	1913	673	$\sim 0^\circ$		Cleaned, polished, and oxidized.
2	15	1947	373	$\sim 0^\circ$		Galvanized sheet iron.
3	16	1937	367	$\sim 0^\circ$	± 1.1	Galvanized iron.

DATA TABLE NO. 251 NORMAL TOTAL EMITTANCE OF ZINC

[Temperature, T, K; Emittance, ϵ]

T	ϵ
CURVE 1*	
673	0.110
CURVE 2*	
373	0.21
CURVE 3*	
367	0.07

* Not shown on plot

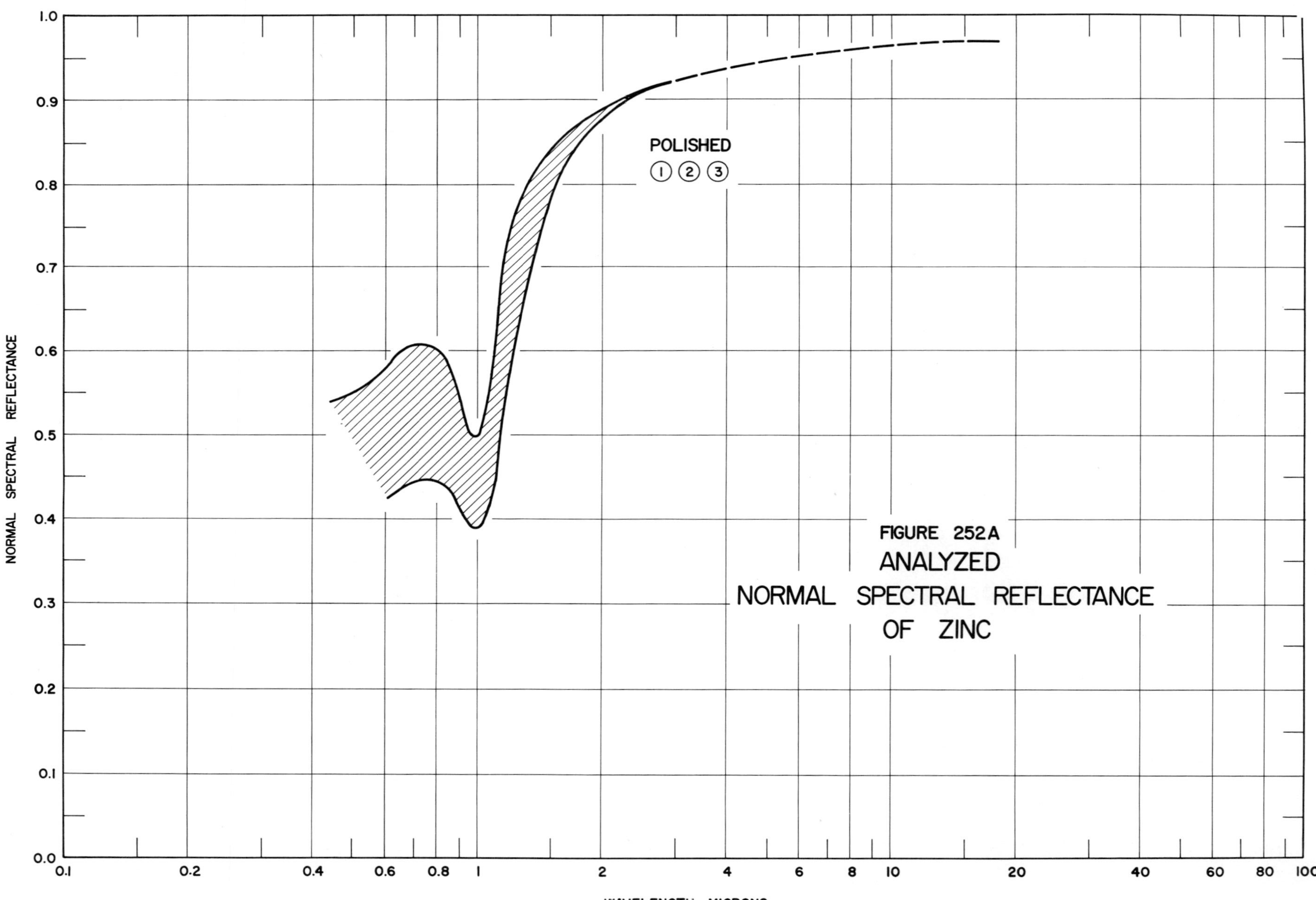

FIGURE 252A
ANALYZED
NORMAL SPECTRAL REFLECTANCE
OF ZINC

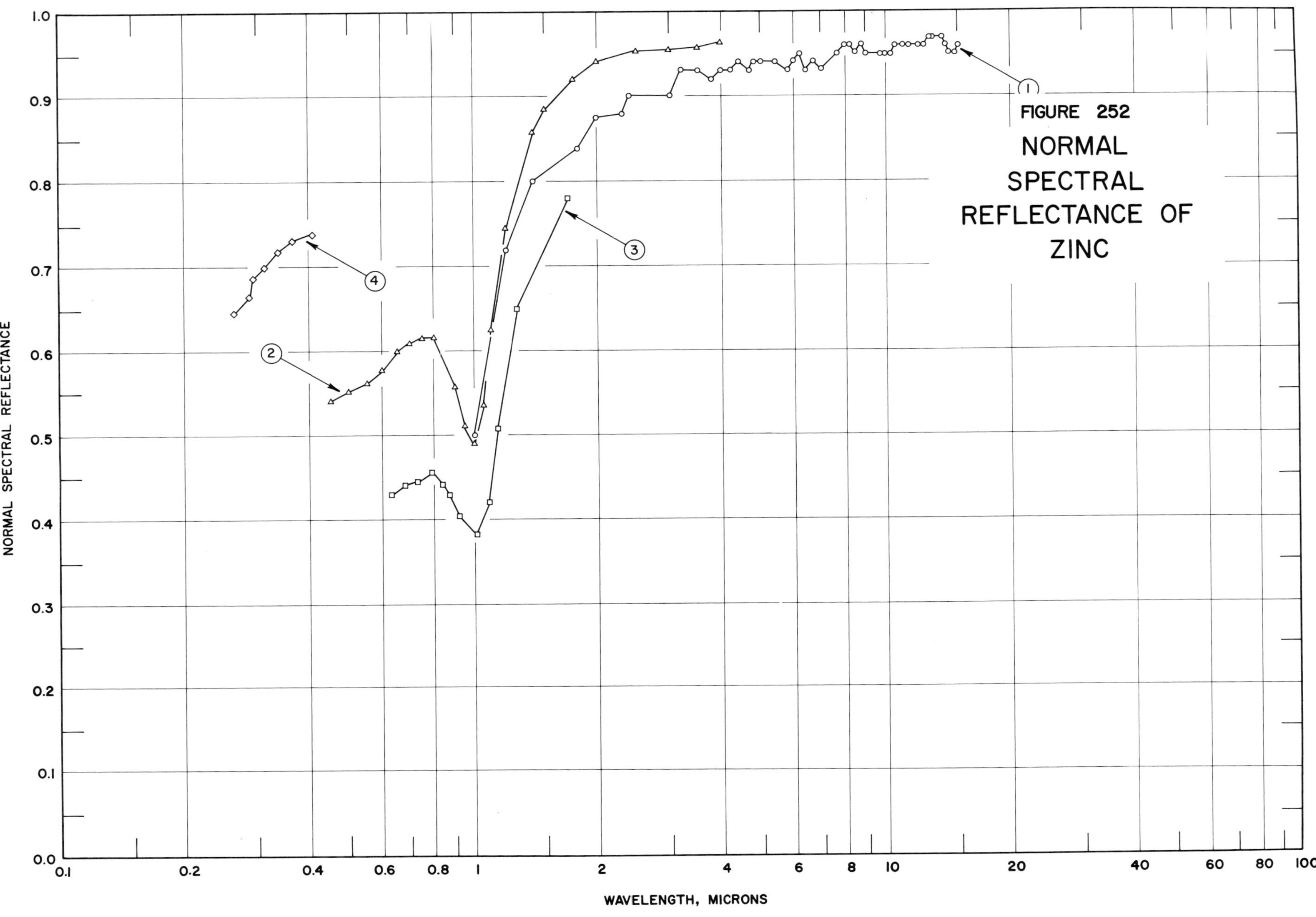
FIGURE 252
NORMAL
SPECTRAL
REFLECTANCE OF
ZINC
1
2
3
4
NORMAL SPECTRAL REFLECTANCE
1.0
0.9
0.8
0.7
0.6
0.5
0.4
0.3
0.2
0.1
0.0
0.1
0.2
0.4
0.6
0.8
1
2
4
6
8
10
20
40
60
80
100
WAVELENGTH, MICRONS

SPECIFICATION TABLE NO. 252 NORMAL SPECTRAL REFLECTANCE OF ZINC

Curve No.	Ref. No.	Year	Temperature K	Wavelength Range, μ	Geometry θ θ' ω'	Reported Error, %	Composition (weight percent), Specifications and Remarks
1	126	1953	298	1.0-15.0	5° 2π	± 2	Polished; data extracted from smooth curve; converted from R(2π, 5°).
2	151	1920	298	0.45-4.00	15° 15°		Polished; silvered mirror reference.
3	152	1920	298	0.63-1.70	15° 15°		Silver mirror reference.
4	224	1931	298	0.2653-0.4045	~5° 2π		Acid-etched.

DATA TABLE NO. 252 NORMAL SPECTRAL REFLECTANCE OF ZINC

[Wavelength, λ, μ; Reflectance, ρ; Temperature, T, K]

λ	ρ
CURVE 1	
T = 298	
1.0	0.50
1.2	0.72
1.4	0.80
1.8	0.84
2.0	0.87
2.3	0.88
2.4	0.90
3.0	0.90
3.2	0.93
3.5	0.93
3.8	0.92
4.0	0.93
4.2	0.93
4.4	0.94
4.7	0.93
4.8	0.94
5.0	0.94
5.4	0.94
5.8	0.93
6.0	0.94
6.2	0.95
6.4	0.93
6.7	0.94
7.0	0.93
7.7	0.95
8.0	0.96
8.2	0.96
8.5	0.95
8.8	0.96
9.0	0.95
9.8	0.95
10.0	0.95
10.2	0.95
10.4	0.96
11.0	0.96
11.4	0.96
12.0	0.96
12.4	0.96
12.7	0.97
13.0	0.97
13.8	0.97
14.0	0.96
CURVE 1 (cont.)	
14.2	0.95
14.8	0.95
15.0	0.96
CURVE 2	
T = 298	
0.45	0.540
0.50	0.550
0.55	0.560
0.60	0.575
0.65	0.600
0.70	0.610
0.75	0.615
0.80	0.615
0.90	0.555
0.95	0.510
1.00	0.490
1.05	0.535
1.10	0.625
1.20	0.747
1.40	0.858
1.50	0.884
1.75	0.920
2.00	0.940
2.50	0.953
3.00	0.955
3.50	0.958
4.00	0.962
CURVE 3	
T = 298	
0.63	0.430
0.68	0.440
0.73	0.445
0.79	0.455
0.84	0.441
0.87	0.430
0.92	0.404
1.01	0.382
1.09	0.421
1.14	0.508
CURVE 3 (cont.)	
1.28	0.650
1.70	0.780
CURVE 4	
T = 298	
0.2653	0.645
0.2891	0.665
0.2964	0.688
0.3130	0.700
0.3338	0.719
0.3662	0.732
0.4045	0.739

SPECIFICATION TABLE NO. 253 HEMISPHERICAL INTEGRATED ABSORPTANCE OF ZINC

Curve No.	Ref. No.	Year	Temperature Range, K	Reported Error, %	Composition (weight percent), Specifications and Remarks
1	3	1955	76	5	Foil (0.0065 in. thick); solvent cleaned; measured in vacuum (10^{-6} to 10^{-7} mm Hg); absorptance for 300 K black body incident radiation.

DATA TABLE NO. 253 HEMISPHERICAL INTEGRATED ABSORPTANCE OF ZINC

[Temperature, T, K; Absorptance, α]

T	α
CURVE 1*	
76	0.02

* Not shown on plot

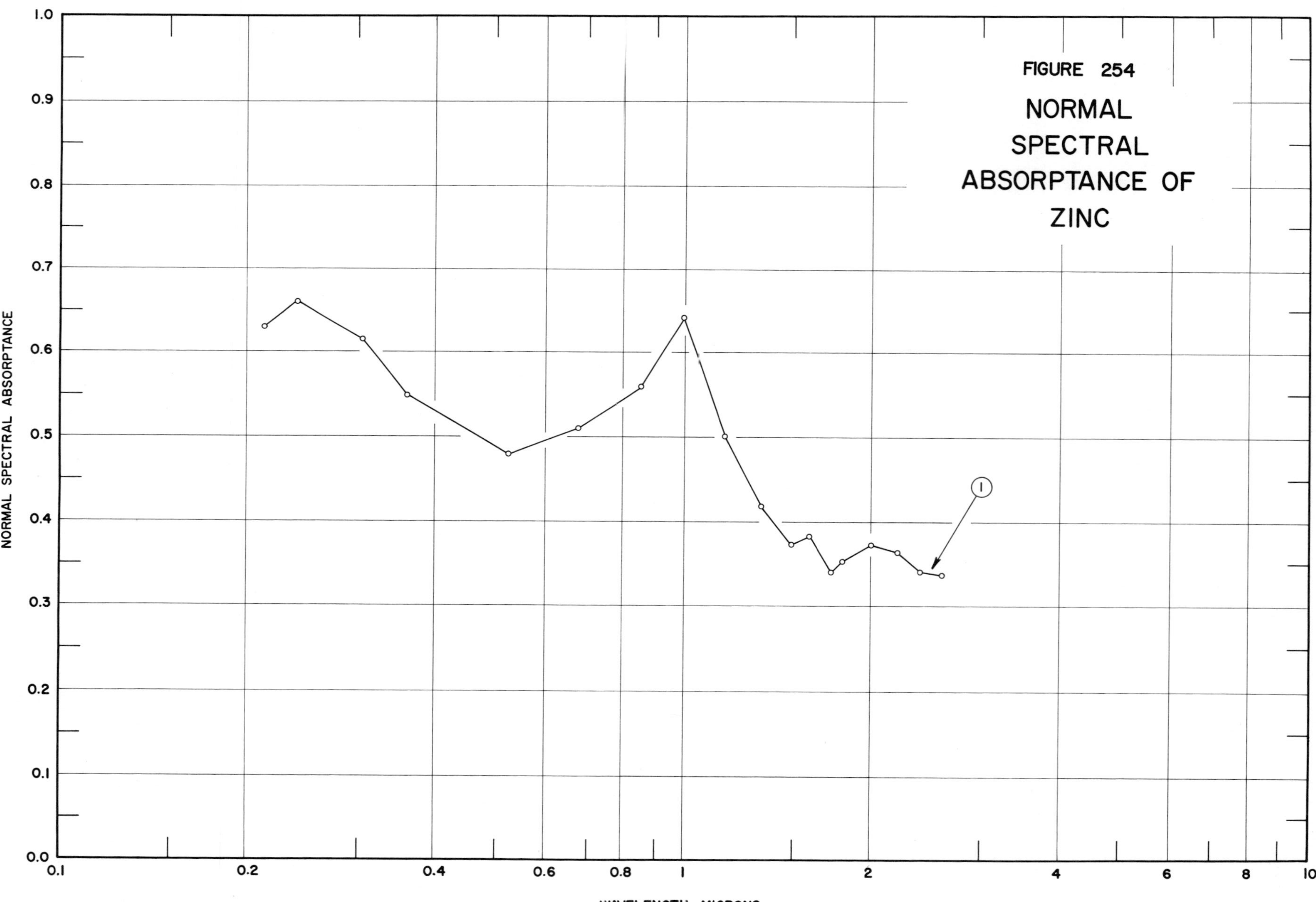
FIGURE 254
NORMAL
SPECTRAL
ABSORPTANCE OF
ZINC
1
NORMAL SPECTRAL ABSORPTANCE
1.0
0.9
0.8
0.7
0.6
0.5
0.4
0.3
0.2
0.1
0.0
0.1
0.2
0.4
0.6
0.8
1
2
4
6
8
10
WAVELENGTH, MICRONS

SPECIFICATION TABLE NO. 254 NORMAL SPECTRAL ABSORPTANCE OF ZINC

Curve No.	Ref. No.	Year	Temperature K	Wavelength Range, μ	Geometry θ	Reported Error, %	Composition (weight percent), Specifications and Remarks
1	307	1954	~298	0.213-2.600	~0°		Zinc, 5 mils thick; data extracted from smooth curve.

DATA TABLE NO. 254 NORMAL SPECTRAL ABSORPTANCE OF ZINC

[Wavelength, λ, μ; Absorptance, α; Temperature, T, K]

λ	α
CURVE 1 T = ~298	
0.213	0.630
0.241	0.661
0.305	0.617
0.361	0.550
0.525	0.481
0.678	0.511
0.851	0.562
0.999	0.643
1.169	0.501
1.332	0.419
1.498	0.373
1.589	0.382
1.725	0.341
1.799	0.353
2.000	0.372
2.200	0.365
2.400	0.341
2.600	0.338

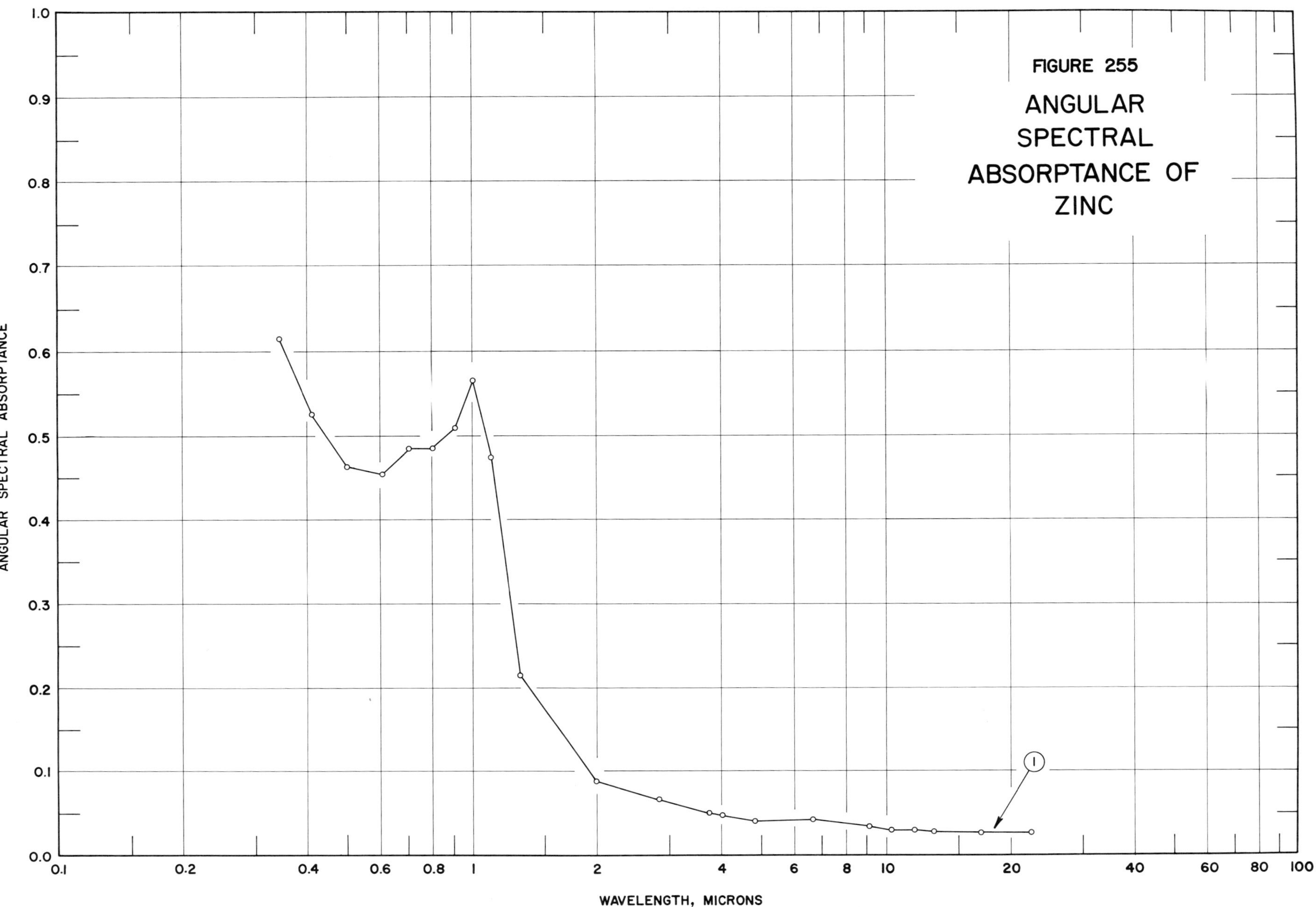

FIGURE 255 ANGULAR SPECTRAL ABSORPTANCE OF ZINC

SPECIFICATION TABLE NO. 255 ANGULAR SPECTRAL ABSORPTANCE OF ZINC

Curve No.	Ref. No.	Year	Temperature K	Wavelength Range, μ	Geometry θ	Reported Error, %	Composition (weight percent), Specifications and Remarks
1	225	1965	306	0.344-22.5	25°		99.9 Zn from Belmont Smelting and Refining Works; measured in dry nitrogen; heated cavity at approx. 1056 K with platinum reference; authors assumed $\alpha = 1 - R(2\pi, 25°)$.

DATA TABLE NO. 255 ANGULAR SPECTRAL ABSORPTANCE OF ZINC

[Wavelength, λ, μ; Absorptance, α; Temperature, T, K]

λ	α
CURVE 1 T = 306	
0.344	0.615
0.412	0.526
0.500	0.462
0.605	0.453
0.705	0.486
0.802	0.486
0.910	0.509
1.01	0.565
1.13	0.472
1.31	0.214
2.00	0.088
2.84	0.066
3.72	0.050
4.03	0.048
4.81	0.040
6.64	0.042
9.16	0.034
10.3	0.030
11.7	0.030
13.1	0.028
17.0	0.027
22.5	0.027

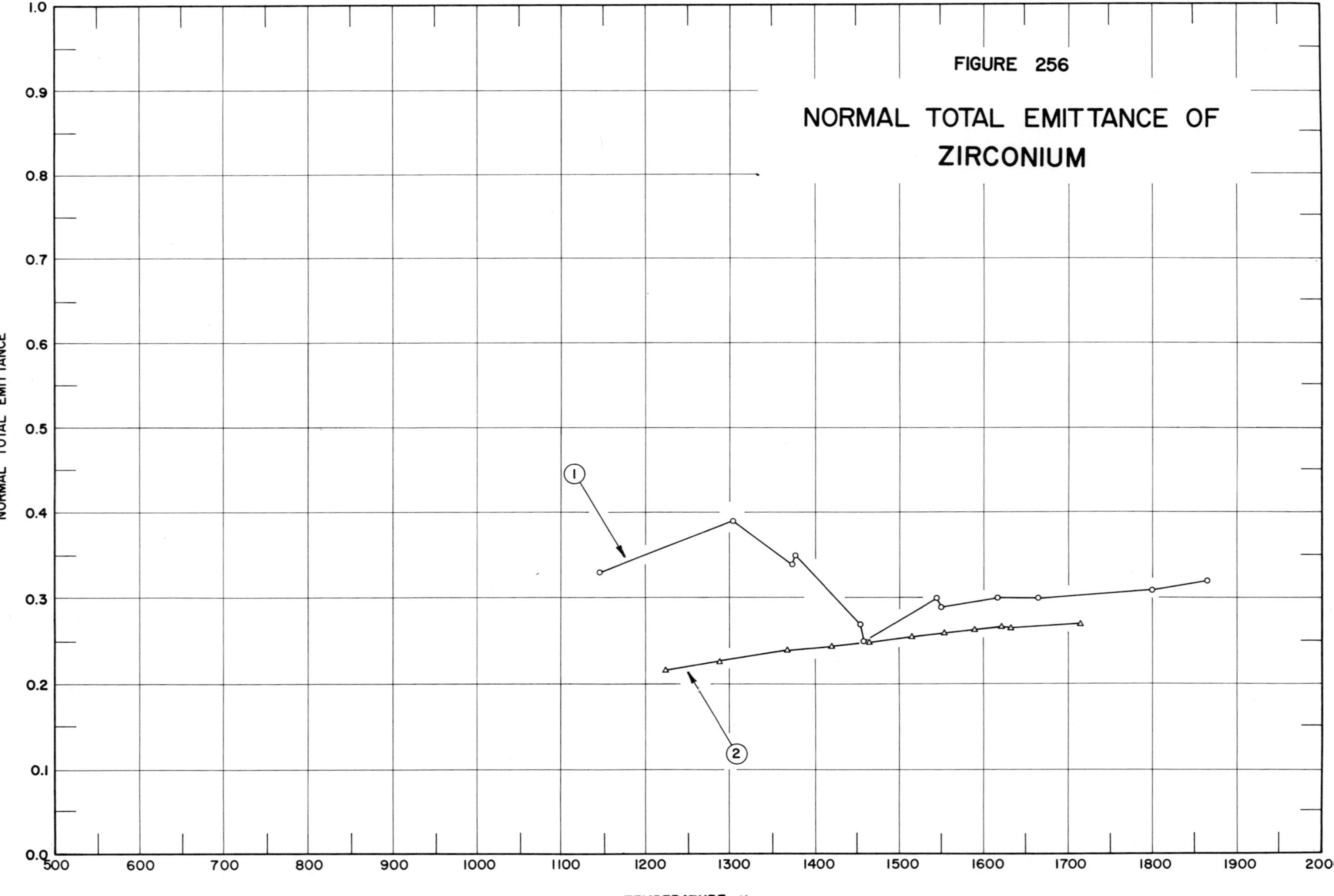
FIGURE 256
NORMAL TOTAL EMITTANCE OF
ZIRCONIUM
NORMAL TOTAL EMITTANCE
1.0
0.9
0.8
0.7
0.6
0.5
0.4
0.3
0.2
0.1
0.0
500
600
700
800
900
1000
1100
1200
1300
1400
1500
1600
1700
1800
1900
2000
TEMPERATURE, K
1
2

SPECIFICATION TABLE NO. 256 NORMAL TOTAL EMITTANCE OF ZIRCONIUM

Curve No.	Ref. No.	Year	Temperature Range, K	Geometry θ'	Reported Error, %	Composition (weight percent), Specifications and Remarks
1	18	1963	1144–1866	$\sim 0°$		As received.
2	334	1965	1222–1715	$\sim 0°$		$>$ 99.5 Zr; iodide zirconium; vacuum annealed; density 6.45 g cm^{-3}.

DATA TABLE NO. 256 NORMAL TOTAL EMITTANCE OF ZIRCONIUM

[Temperature, T, K; Emittance, ϵ]

T	ϵ
CURVE 1	
1144	0.33
1303	0.39
1372	0.34
1378	0.35
1453	0.27
1458	0.25
1544	0.30
1550	0.29
1616	0.30
1664	0.30
1800	0.31
1866	0.32
CURVE 2	
1222	0.217
1288	0.226
1368	0.239
1419	0.246
1464	0.250
1514	0.256
1553	0.260
1590	0.263
1620	0.267
1632	0.265
1715	0.271

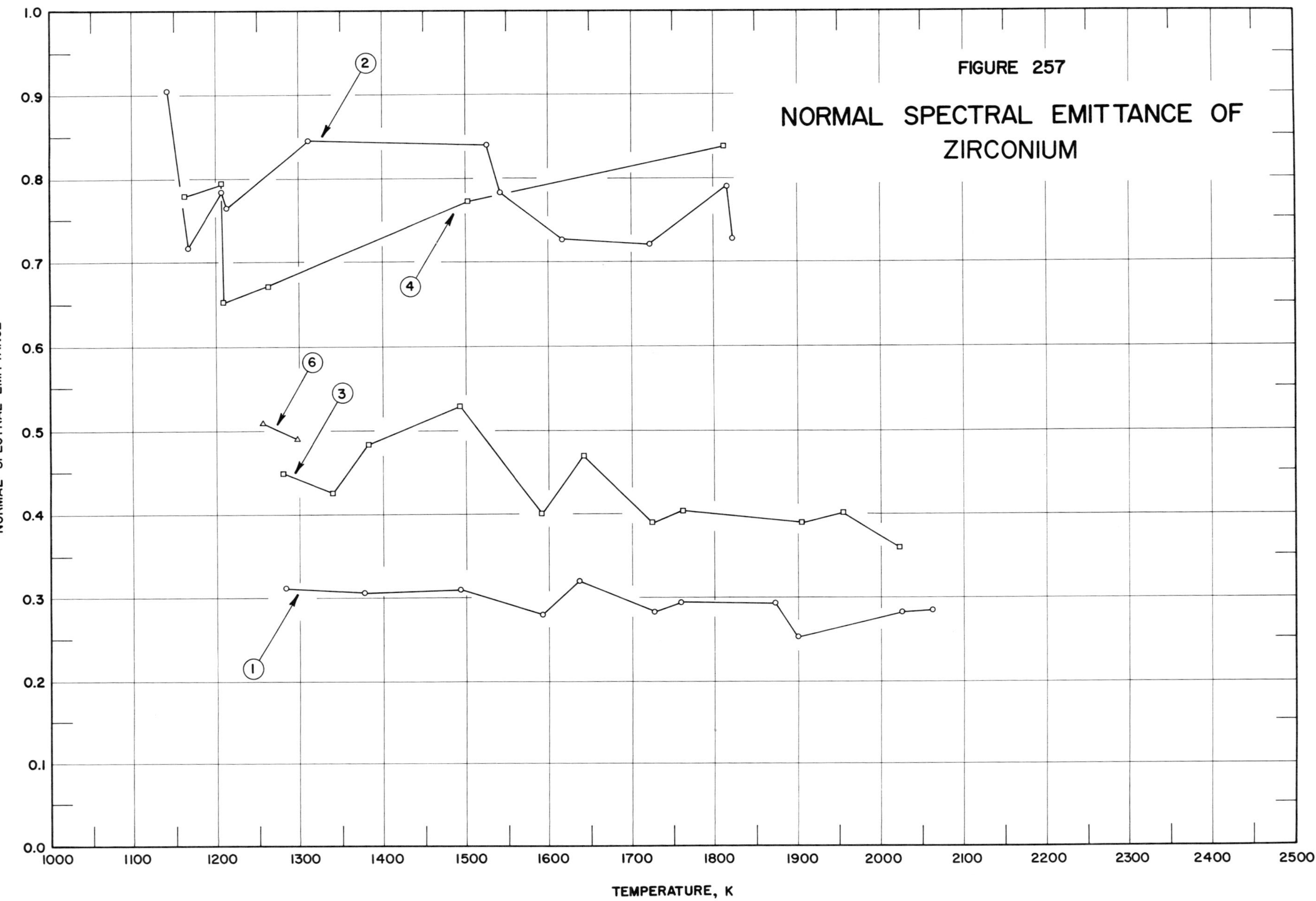
FIGURE 257
NORMAL SPECTRAL EMITTANCE OF
ZIRCONIUM
NORMAL SPECTRAL EMITTANCE
1.0
0.9
0.8
0.7
0.6
0.5
0.4
0.3
0.2
0.1
0.0
1000
1100
1200
1300
1400
1500
1600
1700
1800
1900
2000
2100
2200
2300
2400
2500
TEMPERATURE, K
2
4
6
3
1

SPECIFICATION TABLE NO. 257 NORMAL SPECTRAL EMITTANCE OF ZIRCONIUM

Curve No.	Ref. No.	Year	Wavelength μ	Temperature Range, K	Geometry θ'	Reported Error, %	Composition (weight percent), Specifications and Remarks
1	118	1960	2.3	1283-2063	$\sim 0^\circ$		Measured in dry argon.
2	118	1960	2.3	1143-1823	$\sim 0^\circ$		Oxidized; measured in argon with low moisture content.
3	118	1960	0.65	1280-2023	$\sim 0^\circ$		Measured in dry argon.
4	118	1960	0.65	1163-1813	$\sim 0^\circ$		Oxidized; measured in argon with low moisture content.
5	19	1914	0.65	< M.P.	$\sim 0^\circ$	1	Film; tungsten substrate; measured in hydrogen; Pt reference ($\epsilon = 0.33$ for $\lambda = 0.650\ \mu$ at all temp).
6	276	1951	0.65	1255-1298	$\sim 0^\circ$		Pure Zr; author assumed $\epsilon = 1 - \rho$.

DATA TABLE NO. 257 NORMAL SPECTRAL EMITTANCE OF ZIRCONIUM

[Temperature, T, K; Emittance, ϵ; Wavelength, λ, μ]

T	ϵ	T	ϵ
CURVE 1		CURVE 4	
$\lambda = 2.3$		$\lambda = 0.65$	
1283	0.312	1163	0.780
1378	0.308	1208	0.794
1493	0.310	1210	0.652
1593	0.280	1263	0.671
1638	0.320	1503	0.772
1728	0.283	1813	0.839
1760	0.294		
1873	0.292	CURVE 5*	
1900	0.252	$\lambda = 0.65$	
2026	0.282		
2063	0.284	<M.P.	0.32
CURVE 2		CURVE 6	
$\lambda = 2.3$		$\lambda = 0.65$	
1143	0.906	1255	0.51
1168	0.718	1298	0.49
1208	0.784		
1213	0.764		
1311	0.848		
1528	0.840		
1543	0.782		
1618	0.727		
1723	0.721		
1818	0.790		
1823	0.728		
CURVE 3			
$\lambda = 0.65$			
1280	0.450		
1340	0.428		
1383	0.484		
1493	0.530		
1593	0.400		
1643	0.470		
1726	0.390		
1761	0.404		
1905	0.390		
1956	0.401		
2023	0.360		

* Not shown on plot

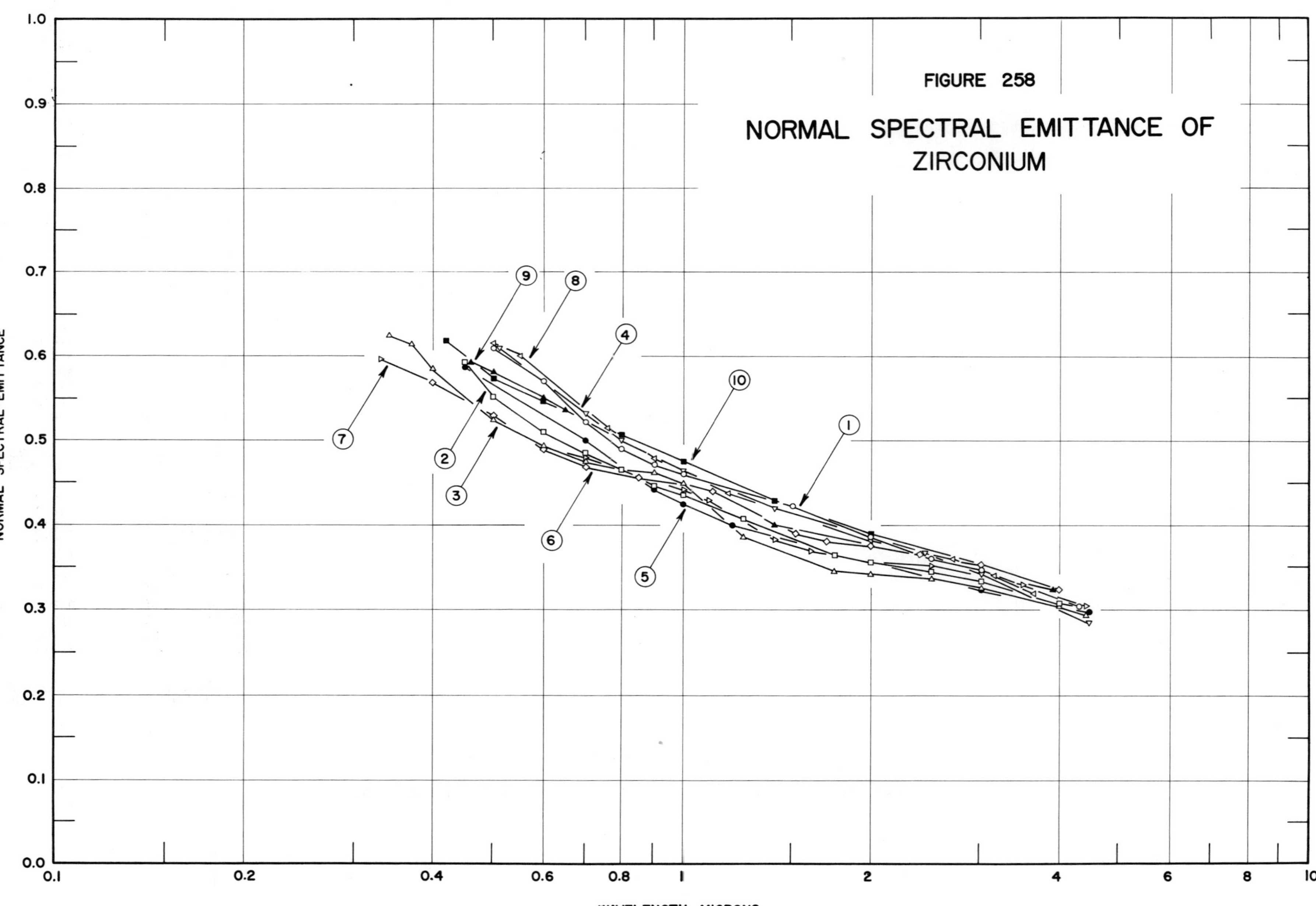

FIGURE 258

NORMAL SPECTRAL EMITTANCE OF ZIRCONIUM

SPECIFICATION TABLE NO. 258 NORMAL SPECTRAL EMITTANCE OF ZIRCONIUM

Curve No.	Ref. No.	Year	Temperature K	Wavelength Range, μ	Geometry θ '	Reported Error, %	Composition (weight percent), Specifications and Remarks
1	95	1963	1400	0.50-4.30	$\sim 0°$		99.9 Zr from Carborundum Metals Corp.; polished with carbide paper of 240 grit, 400 grit, and 600 grit, and then with silk cloth and felt cloth; washed in acetone, then alcohol, and dried with dry nitrogen; measured in vacuum (10^{-6} mm Hg); data extracted from smooth curve; [Author's designation: Specimen No. 2].
2	95	1963	1600	0.45-4.30	$\sim 0°$		Above specimen and conditions.
3	95	1963	2000	0.34-4.40	$\sim 0°$		Above specimen and conditions.
4	95	1963	1400	0.51-4.45	$\sim 0°$		Different sample, same as curve 1 specimen and conditions; [Author's designation: Specimen No. 4].
5	95	1963	1600	0.45-4.45	$\sim 0°$		Above specimen and conditions.
6	95	1963	1800	0.40-4.00	$\sim 0°$		Above specimen and conditions.
7	95	1963	2000	0.33-4.40	$\sim 0°$		Above specimen and conditions.
8	95	1963	1400	0.50-3.65	$\sim 0°$		Different sample, same as curve 1 specimen and conditions; [Author's designation: Specimen No. 5].
9	95	1963	1600	0.46-3.90	$\sim 0°$		Above specimen and conditions.
10	95	1963	1800	0.42-3.90	$\sim 0°$		Above specimen and conditions.

DATA TABLE NO. 258 NORMAL SPECTRAL EMITTANCE OF ZIRCONIUM

[Wavelength, λ, μ; Emittance, ϵ; Temperature, T, K]

λ	ϵ
CURVE 1	**T = 1400**
0.50	0.610
0.60	0.572
0.70	0.522
0.80	0.490
0.90	0.472
1.00	0.460
1.50	0.422
2.00	0.386
2.50	0.360
3.00	0.348
4.30	0.302
CURVE 2	**T = 1600**
0.45	0.594
0.50	0.552
0.60	0.510
0.70	0.486
0.80	0.466
0.90	0.448
1.00	0.436
1.25	0.408
1.75	0.364
2.00	0.356
2.50	0.346
3.00	0.334
4.00	0.308
4.30	0.302*
CURVE 3	**T = 2000**
0.34	0.626
0.37	0.616
0.40	0.586
0.50	0.524
0.60	0.494
0.70	0.476
0.80	0.468*
0.90	0.462
1.00	0.450
1.25	0.386
CURVE 3 (cont.)	
1.75	0.348
2.00	0.342
2.50	0.338
3.00	0.328
4.00	0.304
4.40	0.294
CURVE 4	**T = 1400**
0.51	0.610
0.70	0.531
0.80	0.500
1.00	0.462
1.40	0.420
2.00	0.388*
2.45	0.368
3.00	0.342
4.45	0.285
CURVE 5	**T = 1600**
0.45	0.588
0.70	0.500
0.80	0.467*
0.90	0.443
1.00	0.425
1.20	0.400
2.00	0.355*
3.00	0.326
4.45	0.297
CURVE 6	**T = 1800**
0.40	0.570
0.50	0.530
0.60	0.490
0.70	0.470
0.85	0.457
1.00	0.450*
1.12	0.440
1.52	0.390
CURVE 6 (cont.)	
1.70	0.380
2.00	0.375
2.40	0.367
3.00	0.352
4.00	0.323
CURVE 7	**T = 2000**
0.33	0.597
0.40	0.570*
0.50	0.530*
0.60	0.492*
0.70	0.480
0.85	0.456*
1.00	0.442
1.10	0.430
1.40	0.381
1.60	0.370
2.00	0.357*
2.50	0.351
3.00	0.342*
3.50	0.330
4.40	0.303
CURVE 8	**T = 1400**
0.50	0.615
0.55	0.600
0.76	0.514
0.90	0.480
1.00	0.462*
1.18	0.438
2.00	0.382
2.70	0.360
3.15	0.340
3.65	0.320
CURVE 9	**T = 1600**
0.46	0.595
0.50	0.581
CURVE 9 (cont.)	
0.60	0.552
0.65	0.538
0.90	0.462*
1.00	0.444*
1.40	0.400
2.00	0.368*
3.00	0.343*
3.90	0.325
CURVE 10	**T = 1800**
0.42	0.620
0.50	0.575
0.60	0.549
0.80	0.507
1.00	0.478
1.40	0.430
2.00	0.390
3.00	0.355*
3.90	0.325*

* Not shown on plot

SPECIFICATION TABLE NO. 259 NORMAL SPECTRAL REFLECTANCE OF ZIRCONIUM

Curve No.	Ref. No.	Year	Wavelength μ	Temperature Range, K	Geometry θ	θ'	ω'	Reported Error, %	Composition (weight percent), Specifications and Remarks
1	276	1951	0.65	1255-1298	~0°	~0°			Pure Zr.

DATA TABLE NO. 259 NORMAL SPECTRAL REFLECTANCE OF ZIRCONIUM

[Temperature, T, K; Reflectance, ρ; Wavelength, λ, μ]

T	ρ
CURVE 1*	
$\lambda = 0.65$	
1255	0.49
1298	0.51

* Not shown on plot

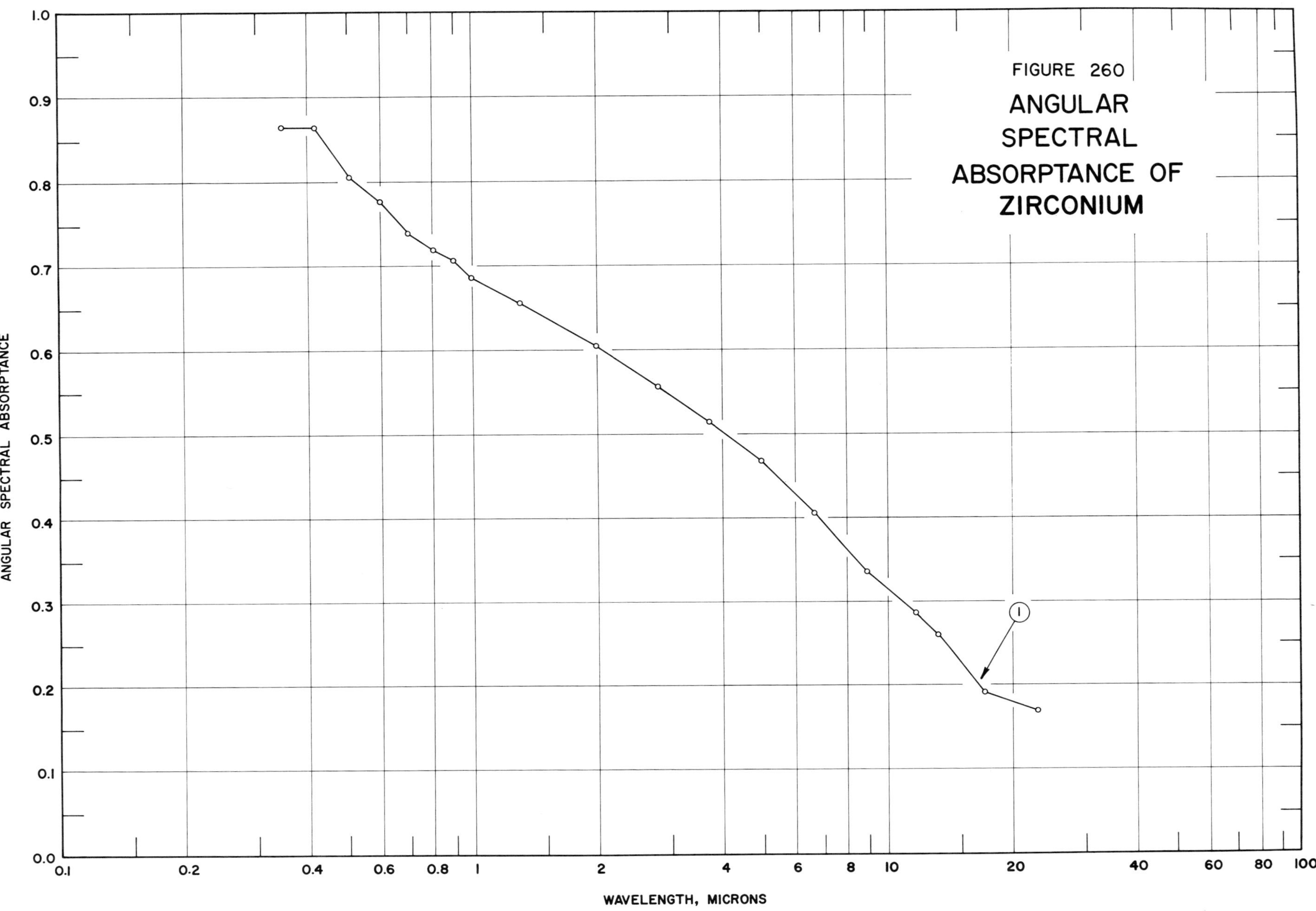

FIGURE 260 ANGULAR SPECTRAL ABSORPTANCE OF ZIRCONIUM

SPECIFICATION TABLE NO. 260 ANGULAR SPECTRAL ABSORPTANCE OF ZIRCONIUM

Curve No.	Ref. No.	Year	Temperature K	Wavelength Range, μ	Geometry θ	Reported Error, %	Composition (weight percent), Specifications and Remarks
1	225	1965	306	0.347-23.0	25°		Zirconium (rolled plate) from Belmont Smelting and Refining Works; measured in dry nitrogen; heated cavity at approx. 1056 K with platinum reference; authors assumed $\alpha = 1 - R(2\pi, 25°)$.

DATA TABLE NO. 260 ANGULAR SPECTRAL ABSORPTANCE OF ZIRCONIUM

[Wavelength, λ, μ; Absorptance, α; Temperature, T, K]

λ	α
CURVE 1 T = 306	
0.347	0.865
0.415	0.840
0.505	0.807
0.600	0.778
0.700	0.740
0.802	0.721
0.900	0.706
0.995	0.686
1.30	0.655
1.99	0.604
2.81	0.556
3.72	0.514
4.97	0.468
6.64	0.407
8.93	0.337
11.6	0.287
13.2	0.260
17.1	0.191
23.0	0.169

2. BINARY ALLOYS

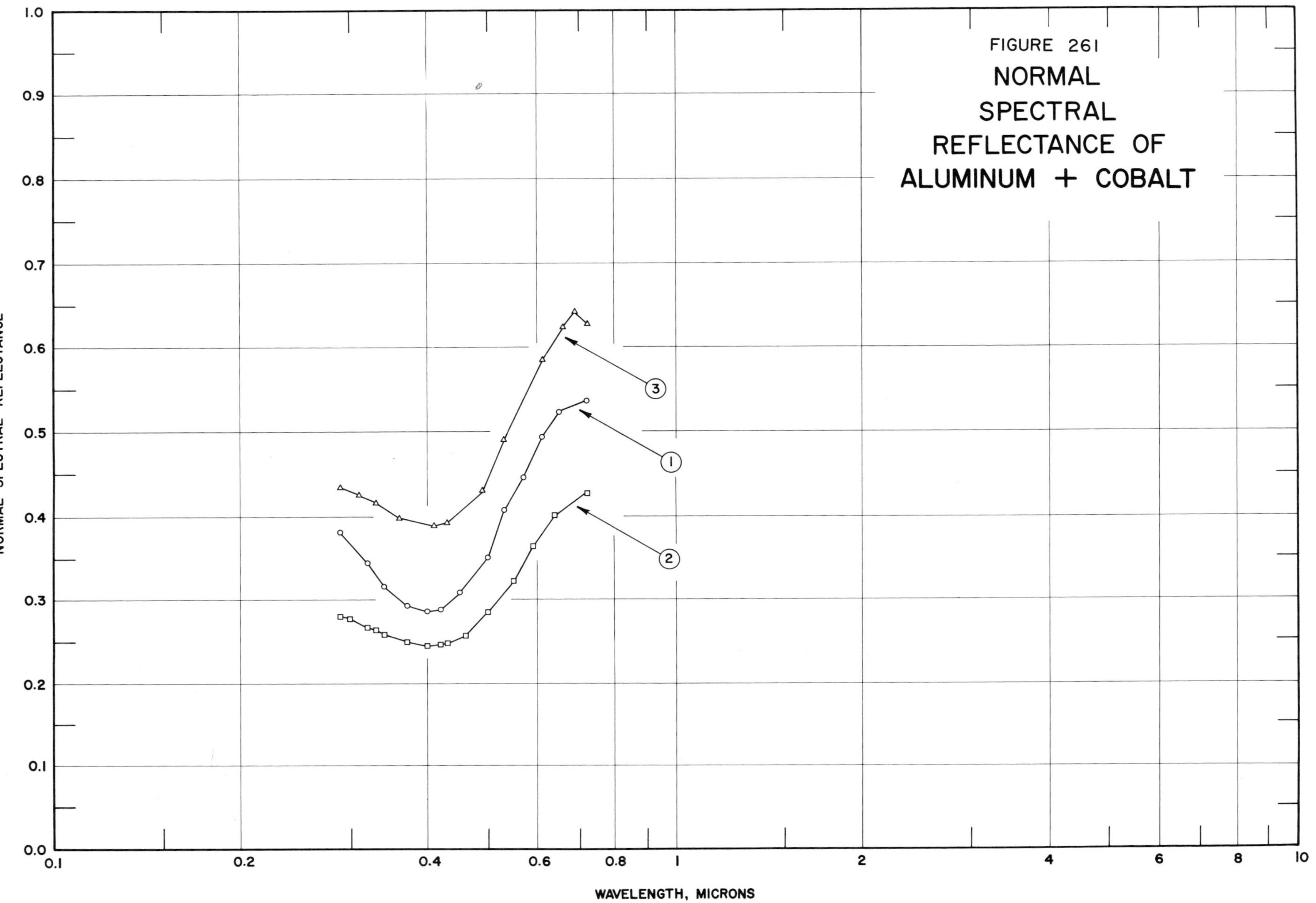

FIGURE 261 NORMAL SPECTRAL REFLECTANCE OF ALUMINUM + COBALT

SPECIFICATION TABLE NO. 261 NORMAL SPECTRAL REFLECTANCE OF [ALUMINUM + COBALT] ALLOYS

Curve No.	Ref. No.	Year	Temperature K	Wavelength Range, μ	Geometry θ	Geometry θ'	Geometry ω'	Reported Error, %	Composition (weight percent), Specifications and Remarks
1	301	1966	298	0.29-0.72	$\sim 0°$		$\sim 2\pi$		50 Al, 50 Co; CsCl structure; polished with silicon carbide paper, alumina powder and diamond paste progressively; annealed in vacuum (3 x 10^{-6} mm Hg) up to 973 K and slowly cooled; data extracted from smooth curve.
2	301	1966	298	0.29-0.72	$\sim 0°$		$\sim 2\pi$		52 Al, 48 Co; CsCl structure; polished with silicon carbide paper, alumina powder and diamond paste progressively; annealed in vacuum (3 x 10^{-6} mm Hg) up to 973 K and slowly cooled; data extracted from smooth curve.
3	301	1966	298	0.29-0.72	$\sim 0°$		$\sim 2\pi$		54 Al, 46 Co; CsCl structure; polished with silicon carbide paper, aluminum powder and diamond paste progressively; annealed in vacuum (3 x 10^{-6} mm Hg) up to 973 K and slowly cooled; data extracted from smooth curve.

DATA TABLE NO. 261 NORMAL SPECTRAL REFLECTANCE OF [ALUMINUM + COBALT] ALLOYS

[Wavelength, λ, μ; Reflectance, ρ; Temperature, T, K]

λ	ρ
CURVE 1 T = 298	
0.29	0.380
0.32	0.346
0.34	0.317
0.37	0.293
0.40	0.287
0.42	0.289
0.45	0.309
0.50	0.351
0.53	0.408
0.57	0.447
0.61	0.494
0.65	0.522
0.72	0.537
CURVE 2 T = 298	
0.29	0.281
0.30	0.278
0.32	0.268
0.33	0.264
0.34	0.259
0.37	0.250
0.40	0.245
0.42	0.248
0.43	0.249
0.46	0.258
0.50	0.286
0.55	0.322
0.59	0.364
0.64	0.400
0.72	0.428
CURVE 3 T = 298	
0.29	0.434
0.31	0.426
0.33	0.416
0.36	0.399
0.41	0.389
0.43	0.392
0.49	0.431
0.53	0.491
0.61	0.587
CURVE 3 (cont.)	
0.66	0.626
0.69	0.642
0.72	0.629

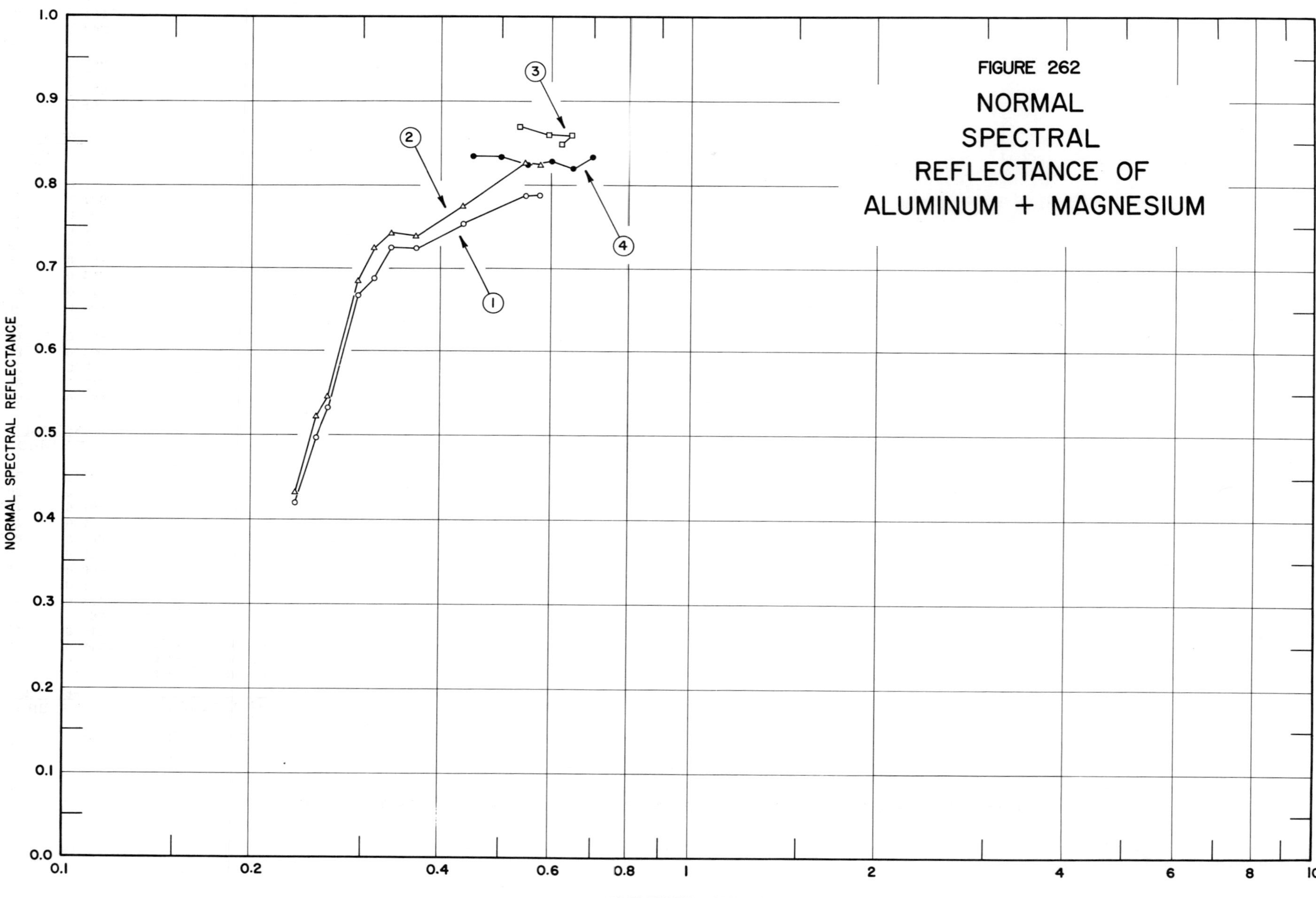
FIGURE 262
NORMAL
SPECTRAL
REFLECTANCE OF
ALUMINUM + MAGNESIUM
NORMAL SPECTRAL REFLECTANCE
WAVELENGTH, MICRONS
1.0
0.9
0.8
0.7
0.6
0.5
0.4
0.3
0.2
0.1
0.0
0.1
0.2
0.4
0.6
0.8
1
2
4
6
8
10
1
2
3
4

SPECIFICATION TABLE NO. 262 NORMAL SPECTRAL REFLECTANCE OF [ALUMINUM + MAGNESIUM] ALLOYS

Curve No.	Ref. No.	Year	Temperature K	Wavelength Range, μ	Geometry θ	θ'	ω'	Reported Error, %	Composition (weight percent), Specifications and Remarks
1	133	1934	298	0.2350-0.5780	~0°	~0°			5 Mg, Al balance; cold worked; annealed in an inert gas for 6-24 hrs; polished; stored in dilute solutions of NaOH, NaOH + NaF, and HNO_3.
2	133	1934	298	0.2350-0.5780	~0°	~0°			20 Mg, Al balance; specimen preparation same as curve 1.
3	152	1920	298	0.533-0.624	~0°	~0°			69 Al, 31 Mg; polished by using alumina on broadcloth moistened with alcohol; silver mirror reference; converted from R (0°, 0°); [Author's designation: A29].
4	228	1900	298	0.45-0.70	~0°	~0°			66 Al, 34 Mg; mirror like surface; platinum filament lamp source.

DATA TABLE NO. 262 NORMAL SPECTRAL REFLECTANCE OF [ALUMINUM + MAGNESIUM] ALLOYS

[Wavelength, λ, μ; Reflectance, ρ; Temperature, T, K]

λ	ρ
CURVE 1 T = 298	
0.2350	0.420
0.2537	0.497
0.2650	0.532
0.2970	0.668
0.3125	0.689
0.3340	0.727
0.3660	0.725
0.4355	0.754
0.5460	0.788
0.5780	0.789
CURVE 2 T = 298	
0.2350	0.433
0.2537	0.521
0.2650	0.547
0.2970	0.686
0.3125	0.727
0.3340	0.743
0.3660	0.740
0.4355	0.776
0.5460	0.827
0.5780	0.825
CURVE 3 T = 298	
0.533	0.870
0.595	0.861
0.649	0.860
0.624	0.850
CURVE 4 T = 298	
0.45	0.834
0.50	0.833
0.55	0.827
0.60	0.830
0.65	0.821
0.70	0.833

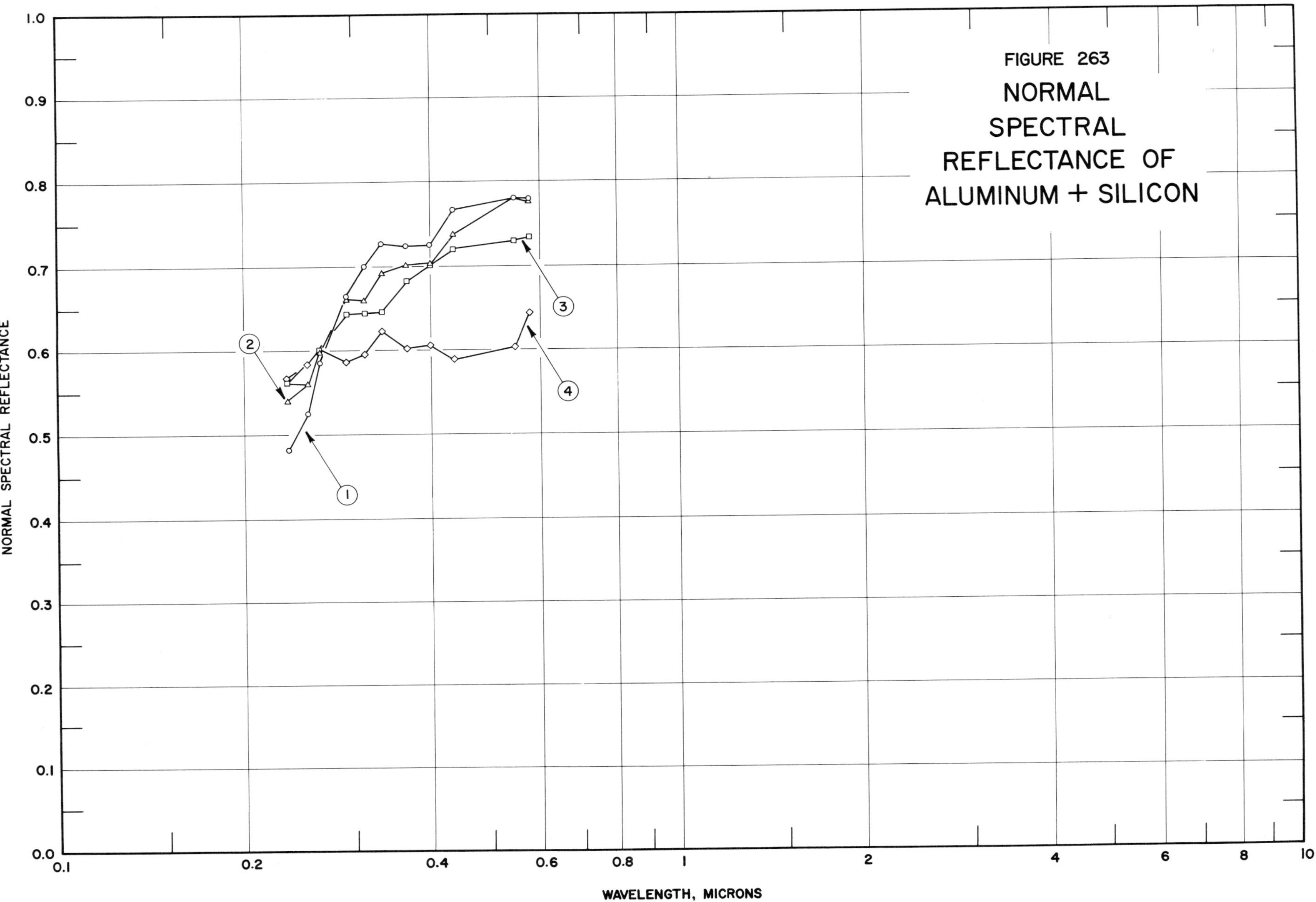
FIGURE 263
NORMAL
SPECTRAL
REFLECTANCE OF
ALUMINUM + SILICON
NORMAL SPECTRAL REFLECTANCE
1.0
0.9
0.8
0.7
0.6
0.5
0.4
0.3
0.2
0.1
0.0
0.1
0.2
0.4
0.6
0.8
1
2
4
6
8
10
WAVELENGTH, MICRONS
1
2
3
4

SPECIFICATION TABLE NO. 263 NORMAL SPECTRAL REFLECTANCE OF [ALUMINUM + SILICON] ALLOYS

Curve No.	Ref. No.	Year	Temperature K	Wavelength Range, μ	Geometry θ	θ'	ω'	Reported Error, %	Composition (weight percent), Specifications and Remarks
1	133	1934	298	0.2350-0.5780	~0°	~0°			98 Al, Si balance; cold worked; annealed in an inert gas for 6-24 hrs; polished; stored in dilute solutions of NaOH, NaOH + NaF, and HNO_3.
2	133	1934	298	0.2350-0.5780	~0°	~0°			94 Al, Si balance; cold worked; annealed in an inert gas for 6-24 hrs; polished; stored in dilute solutions of NaOH, NaOH + NaF, and HNO_3.
3	133	1934	298	0.2350-0.5780	~0°	~0°			88 Al, Si balance; cold worked; annealed in an inert gas for 6-24 hrs; polished; stored in dilute solutions of NaOH, NaOH + NaF, and HNO_3.
4	133	1934	298	0.2350-0.5780	~0°	~0°			80 Al, Si balance; cold worked; annealed in an inert gas for 6-24 hrs; polished; stored in dilute solutions of NaOH, NaOH + NaF, and HNO_3.

DATA TABLE NO. 263 NORMAL SPECTRAL REFLECTANCE OF [ALUMINUM + SILICON] ALLOYS

[Wavelength, λ, μ; Reflectance, ρ; Temperature, T, K]

λ	ρ	λ	ρ
CURVE 1 T = 298		CURVE 4 T = 298	
0.2350	0.482	0.2350	0.568
0.2537	0.524	0.2537	0.584
0.2650	0.586	0.2650	0.589*
0.2930	0.665	0.2930	0.587
0.3125	0.700	0.3125	0.596
0.3340	0.727	0.3340	0.623
0.3660	0.723	0.3660	0.601
0.4060	0.724	0.4060	0.605
0.4355	0.767	0.4355	0.590
0.5460	0.780	0.5460	0.602
0.5780	0.779	0.5780	0.644
CURVE 2 T = 298			
0.2350	0.541		
0.2537	0.560		
0.2650	0.598		
0.2930	0.662		
0.3125	0.660		
0.3340	0.693		
0.3660	0.702		
0.4060	0.703		
0.4355	0.738		
0.5460	0.780*		
0.5780	0.778		
CURVE 3 T = 298			
0.2350	0.562		
0.2537	0.561*		
0.2650	0.600		
0.2930	0.643		
0.3125	0.645		
0.3340	0.647		
0.3660	0.682		
0.4060	0.701		
0.4355	0.720		
0.5460	0.730		
0.5780	0.734		

* Not shown on plot

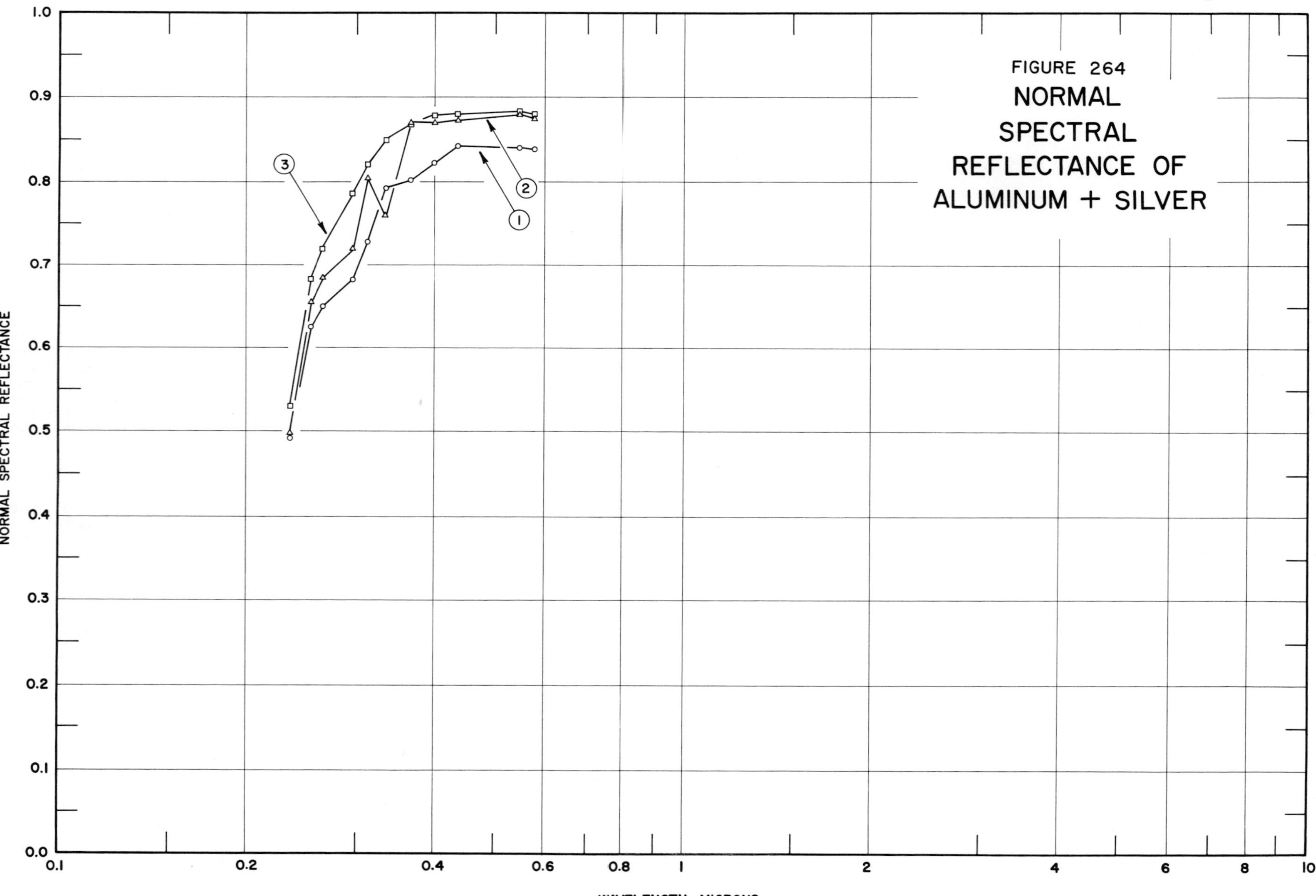
FIGURE 264
NORMAL
SPECTRAL
REFLECTANCE OF
ALUMINUM + SILVER
NORMAL SPECTRAL REFLECTANCE
WAVELENGTH, MICRONS
1.0
0.9
0.8
0.7
0.6
0.5
0.4
0.3
0.2
0.1
0.0
0.1
0.2
0.4
0.6
0.8
1
2
4
6
8
10
1
2
3

SPECIFICATION TABLE NO. 264 NORMAL SPECTRAL REFLECTANCE OF [ALUMINUM + SILVER] ALLOYS

Curve No.	Ref. No.	Year	Temperature K	Wavelength Range, μ	Geometry θ	θ'	ω'	Reported Error, %	Composition (weight percent), Specifications and Remarks
1	133	1934	298	0.2350-0.5780	~0°	~0°			94 Al, Ag balance; cold worked; annealed in an inert gas for 6-24 hrs; polished; stored in dilute solutions of NaOH, NaOH + NaF, and HNO_3.
2	133	1934	298	0.2350-0.5780	~0°	~0°			90 Al, Ag balance; cold worked; annealed in an inert gas for 6-24 hrs; polished; stored in dilute solutions of NaOH, NaOH + NaF, and HNO_3.
3	133	1934	298	0.2350-0.5780	~0°	~0°			72 Al, Ag balance; cold worked; annealed in an inert gas for 6-24 hrs; polished; stored in dilute solutions of NaOH, NaOH + NaF, and HNO_3.

DATA TABLE NO. 264 NORMAL SPECTRAL REFLECTANCE OF [ALUMINUM + SILVER] ALLOYS

[Wavelength, λ, μ; Reflectance, ρ; Temperature, T, K]

λ	ρ
CURVE 1	
T = 298	
0.2350	0.492
0.2537	0.627
0.2650	0.650
0.2970	0.681
0.3125	0.728
0.3340	0.792
0.3660	0.801
0.4060	0.823
0.4355	0.842
0.5460	0.840
0.5780	0.838
CURVE 2	
T = 298	
0.2350	0.498
0.2537	0.654
0.2650	0.683
0.2970	0.720
0.3125	0.803
0.3340	0.760
0.3660	0.871
0.4060	0.870
0.4355	0.873
0.5460	0.880
0.5780	0.875
CURVE 3	
T = 298	
0.2350	0.530
0.2537	0.682
0.2650	0.719
0.2970	0.786
0.3125	0.821
0.3340	0.850
0.3660	0.869
0.4060	0.878
0.4355	0.880
0.5460	0.882
0.5780	0.880

SPECIFICATION TABLE NO. 265 NORMAL INTEGRATED ABSORPTANCE OF [BISMUTH + TIN] ALLOYS

Curve No.	Ref. No.	Year	Temperature Range, K	Geometry θ	Reported Error, %	Composition (weight percent), Specifications and Remarks
1	134	1952	2	~0°		Eutectic; electropolished; absorptance for 298 K black body incident radiation.

DATA TABLE NO. 265 NORMAL INTEGRATED ABSORPTANCE OF [BISMUTH + TIN] ALLOYS

[Temperature, T, K; Absorptance, α]

T	α
CURVE 1*	
2	0.0763

* Not shown on plot

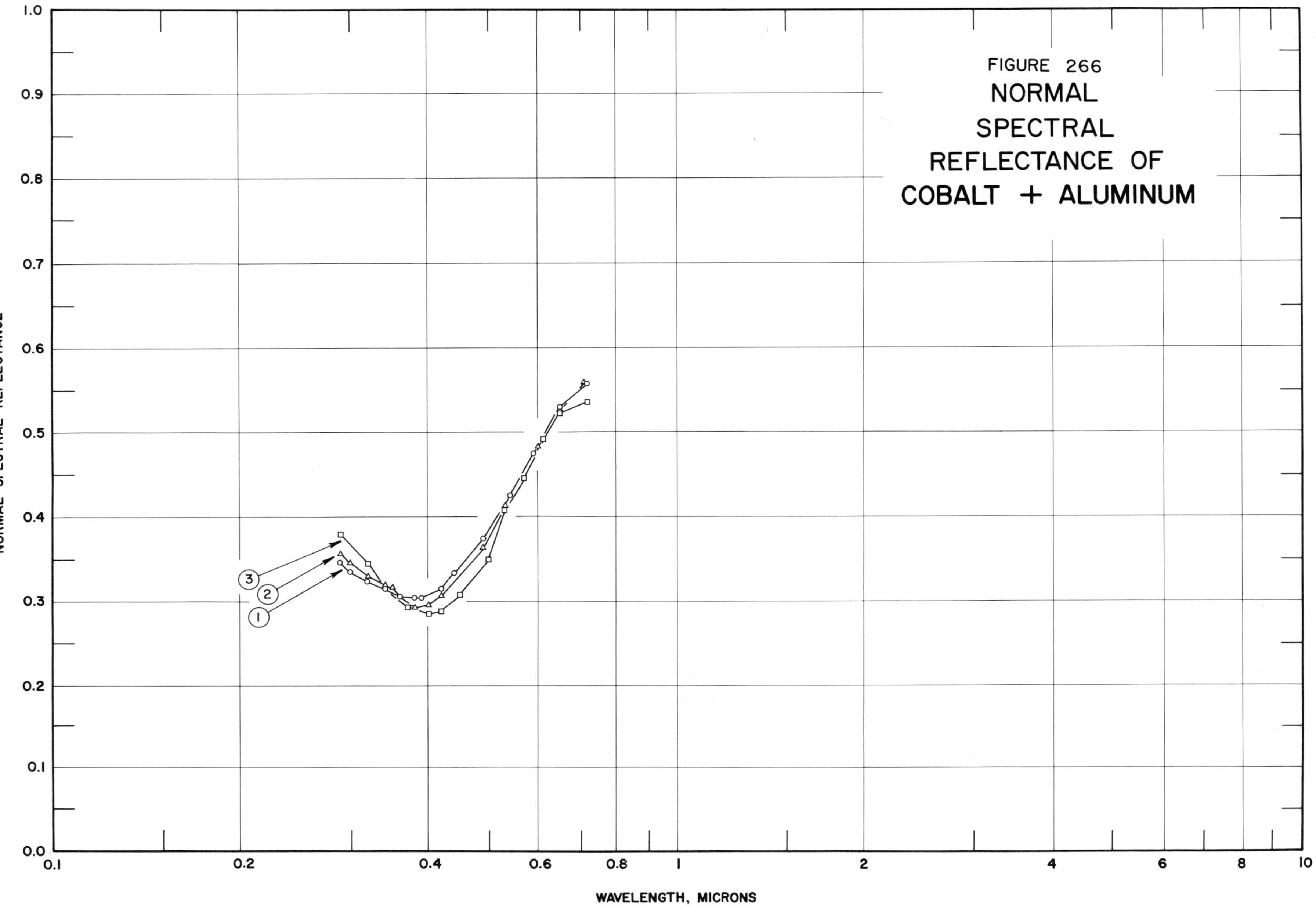
FIGURE 266
NORMAL
SPECTRAL
REFLECTANCE OF
COBALT + ALUMINUM
NORMAL SPECTRAL REFLECTANCE
1.0
0.9
0.8
0.7
0.6
0.5
0.4
0.3
0.2
0.1
0.0
0.1
0.2
0.4
0.6
0.8
1
2
4
6
8
10
WAVELENGTH, MICRONS
3
2
1

SPECIFICATION TABLE NO. 266 NORMAL SPECTRAL REFLECTANCE OF [COBALT + ALUMINUM] ALLOYS

Curve No.	Ref. No.	Year	Temperature K	Wavelength Range, μ	Geometry θ	θ'	ω'	Reported Error, %	Composition (weight percent), Specifications and Remarks
1	301	1966	298	0.29-0.72	$\sim 0°$		$\sim 2\pi$		54 Co, 46 Al; CsCl structure; polished with silicon carbide paper, alumina powder and diamond paste progressively; annealed in vacuum (3×10^{-6} mm Hg) up to 973 K and slowly cooled; data extracted from smooth curve.
2	301	1966	298	0.29-0.71	$\sim 0°$		$\sim 2\pi$		52 Co, 48 Al; CsCl structure; polished with silicon carbide paper, alumina powder and diamond paste progressively; annealed in vacuum (3×10^{-6} mm Hg) up to 973 K and slowly cooled; data extracted from smooth curve.
3	301	1966	298	0.29-0.72	$\sim 0°$		$\sim 2\pi$		50 Co, 50 Al; CsCl structure; polished with silicon carbide paper, alumina powder and diamond paste progressively; annealed in vacuum (3×10^{-6} mm Hg) up to 973 K and slowly cooled; data extracted from smooth curve.

DATA TABLE NO. 266 NORMAL SPECTRAL REFLECTANCE OF [COBALT + ALUMINUM] ALLOYS

[Wavelength, λ, μ; Reflectance, ρ; Temperature, T, K]

λ	ρ
CURVE 1 T = 298	
0.29	0.348
0.30	0.336
0.32	0.324
0.34	0.316
0.36	0.306
0.38	0.304
0.39	0.304
0.42	0.316
0.44	0.334
0.49	0.375
0.54	0.427
0.59	0.477
0.65	0.531
0.72	0.559
CURVE 2 T = 298	
0.29	0.358
0.30	0.348
0.32	0.331
0.33	0.320
0.34	0.318
0.36	0.305*
0.38	0.293
0.40	0.297
0.42	0.308
0.49	0.364
0.53	0.414
0.60	0.485
0.65	0.528
0.71	0.561
CURVE 3 T = 298	
0.29	0.380
0.32	0.346
0.34	0.317*
0.37	0.293
0.40	0.287
0.42	0.289
0.45	0.309
0.50	0.351
0.53	0.408
CURVE 3 (cont.)	
0.57	0.447
0.61	0.494
0.65	0.522
0.72	0.537

* Not shown on plot

SPECIFICATION TABLE NO. 267 NORMAL SPECTRAL EMITTANCE OF [COBALT + IRON] ALLOYS

Curve No.	Ref. No.	Year	Wavelength μ	Temperature Range, K	Geometry θ'	Reported Error, %	Composition (weight percent), Specifications and Remarks
1	174	1948	0.667	975-1540	$\sim 0°$		60 Co, 40 Fe; measured in hydrogen.

DATA TABLE NO. 267 NORMAL SPECTRAL EMITTANCE OF [COBALT + IRON] ALLOYS

[Temperature, T, K; Emittance, ϵ; Wavelength, λ, μ]

T	ϵ
CURVE 1*	
$\lambda = 0.667$	
975	0.405
1025	0.343
1025	0.332
1035	0.318
1038	0.315
1045	0.305
1048	0.310
1050	0.315
1060	0.335
1065	0.330
1070	0.342
1075	0.333
1080	0.323
1085	0.320
1085	0.322
1100	0.322
1105	0.318
1120	0.315
1130	0.313
1135	0.318
1155	0.315
1160	0.318
1170	0.317
1205	0.318
1210	0.312
1285	0.287
1335	0.290
1345	0.288
1360	0.290
1380	0.292
1400	0.295
1430	0.295
1445	0.290
1460	0.295
1490	0.297
1515	0.290
1525	0.292
1540	0.290

* Not shown on plot

SPECIFICATION TABLE NO. 268 NORMAL SPECTRAL EMITTANCE OF [COBALT + NICKEL] ALLOYS

Curve No.	Ref. No.	Year	Wavelength μ	Temperature Range, K	Geometry θ'	Reported Error, %	Composition (weight percent), Specifications and Remarks
1	44	1948	0.667	1210-1366	~0°		65 Co, 35 Ni.

DATA TABLE NO. 268 NORMAL SPECTRAL EMITTANCE [COBALT + NICKEL] ALLOYS

[Temperature, T, K; Emittance, ϵ; Wavelength, λ, μ]

T	ϵ
CURVE 1*	
$\lambda = 0.667$	
1210	0.346
1210	0.342
1223	0.336
1224	0.335
1228	0.329
1228	0.334
1232	0.327
1232	0.329
1234	0.328
1236	0.324
1240	0.324
1242	0.321
1244	0.326
1248	0.331
1248	0.337
1251	0.331
1251	0.333
1252	0.339
1253	0.345
1255	0.337
1256	0.334
1262	0.331
1272	0.329
1276	0.329
1280	0.327
1284	0.328
1248	0.329
1316	0.327
1320	0.326
1322	0.330
1366	0.330

* Not shown on plot

SPECIFICATION TABLE NO. 269 NORMAL TOTAL EMITTANCE OF [COPPER + NICKEL] ALLOYS

Curve No.	Ref. No.	Year	Temperature Range, K	Geometry θ'	Reported Error, %	Composition (weight percent), Specifications and Remarks
1	15	1947	373	~0°		Polished.

DATA TABLE NO. 269 NORMAL TOTAL EMITTANCE OF [COPPER + NICKEL] ALLOYS

[Temperature, T, K; Emittance, ϵ]

T	ϵ
CURVE 1*	
373	0.059

* Not shown on plot

SPECIFICATION TABLE NO. 270 NORMAL SPECTRAL REFLECTANCE OF [COPPER + TIN] ALLOYS

Curve No.	Ref. No.	Year	Temperature K	Wavelength Range, μ	Geometry θ	θ' ω'	Reported Error, %	Composition (weight percent), Specifications and Remarks
1	228	1900	298	0.45-0.70	$\sim 0^\circ$	$\sim 0^\circ$		68.2 Cu, 31.8 Sn; mirror like surface; platinum filament lamp source.
2	228	1900	298	0.45-0.70	$\sim 0^\circ$	$\sim 0^\circ$		Different sample, same as above specimen and conditions.

DATA TABLE NO. 270 NORMAL SPECTRAL REFLECTANCE OF [COPPER + TIN] ALLOYS

[Wavelength, λ, μ; Reflectance, ρ; Temperature, T, K]

λ	ρ
CURVE 1*	
T = 298	
0.45	0.629
0.50	0.632
0.55	0.640
0.60	0.643
0.65	0.656
0.70	0.673
CURVE 2*	
T = 298	
0.45	0.619
0.50	0.633
0.55	0.640
0.60	0.644
0.65	0.654
0.70	0.685

* Not shown on plot

SPECIFICATION TABLE NO. 271 HEMISPHERICAL TOTAL EMITTANCE OF [COPPER + ZINC] ALLOYS

Curve No.	Ref. No.	Year	Temperature Range, K	Reported Error, %	Composition (weight percent), Specifications and Remarks
1	3	1955	76	5	Yellow brass; 65 Cu, 35 Zn; shim stock (0.001 in. thick); measured in vacuum (10^{-6} to 10^{-7} mm Hg); emittance for 300 K black body incident radiation; authors assumed $\alpha = \epsilon$.
2	9	1960	77		Brass; nominal composition: 65 Cu, 35 Zn; clean; hand polished, some scratches; cleaned with acetone; measured in vacuum ($\sim 1 \times 10^{-5}$ mm Hg).
3	9	1960	77		Brass; nominal composition: 65 Cu, 35 Zn; tarnished; partly oxidized with gas flame; cleaned with acetone; measured in vacuum ($\sim 1 \times 10^{-5}$ mm Hg).

DATA TABLE NO. 271 HEMISPHERICAL TOTAL EMITTANCE OF [COPPER + ZINC] ALLOYS

[Temperature, T, K; Emittance, ϵ]

T	ϵ
CURVE 1*	
76	0.029
CURVE 2*	
77	0.10
CURVE 3*	
77	0.11

* Not shown on plot

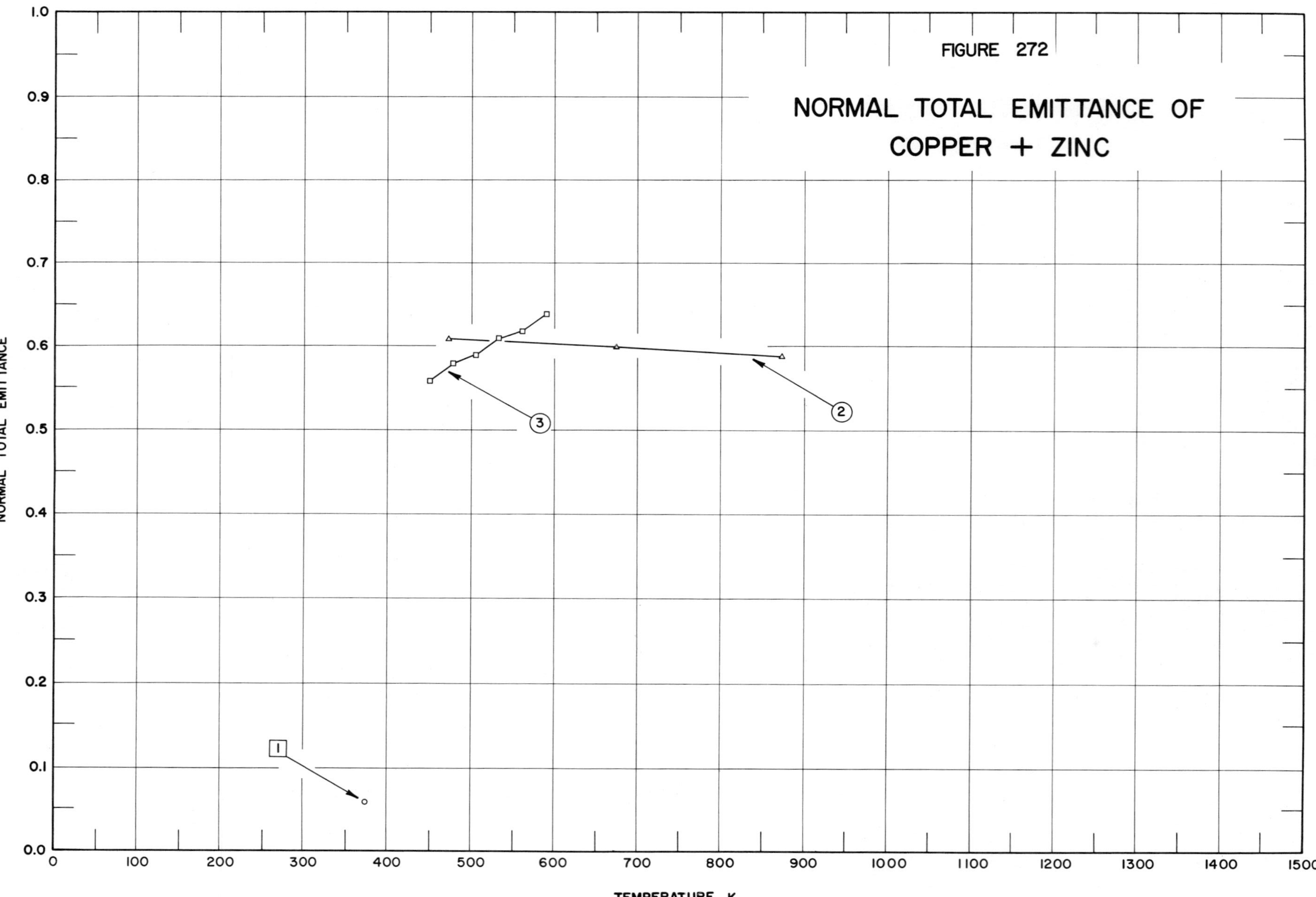

FIGURE 272
NORMAL TOTAL EMITTANCE OF
COPPER + ZINC
1
2
3
NORMAL TOTAL EMITTANCE
1.0
0.9
0.8
0.7
0.6
0.5
0.4
0.3
0.2
0.1
0.0
0
100
200
300
400
500
600
700
800
900
1000
1100
1200
1300
1400
1500
TEMPERATURE, K

SPECIFICATION TABLE NO. 272 NORMAL TOTAL EMITTANCE OF [COPPER + ZINC] ALLOYS

Curve No.	Ref. No.	Year	Temperature Range, K	Geometry θ'	Reported Error, %	Composition (weight percent), Specifications and Remarks
1	15	1947	373	$\sim 0°$		Brass; nominal composition: 65 Cu, 35 Zn; polished.
2	14	1913	473-873	$\sim 0°$		Brass; nominal composition: 65 Cu, 35 Zn; cleaned, polished, and oxidized.
3	302	1958	450-589	$\sim 0°$		Oxidized brass.

DATA TABLE NO. 272 NORMAL TOTAL EMITTANCE OF [COPPER + ZINC] ALLOYS

[Temperature, T, K; Emittance, ∈]

T	∈
CURVE 1	
373	0.059
CURVE 2	
473	0.610
673	0.600
873	0.589
CURVE 3	
450	0.56
478	0.58
505	0.59
533	0.61
561	0.62
589	0.64

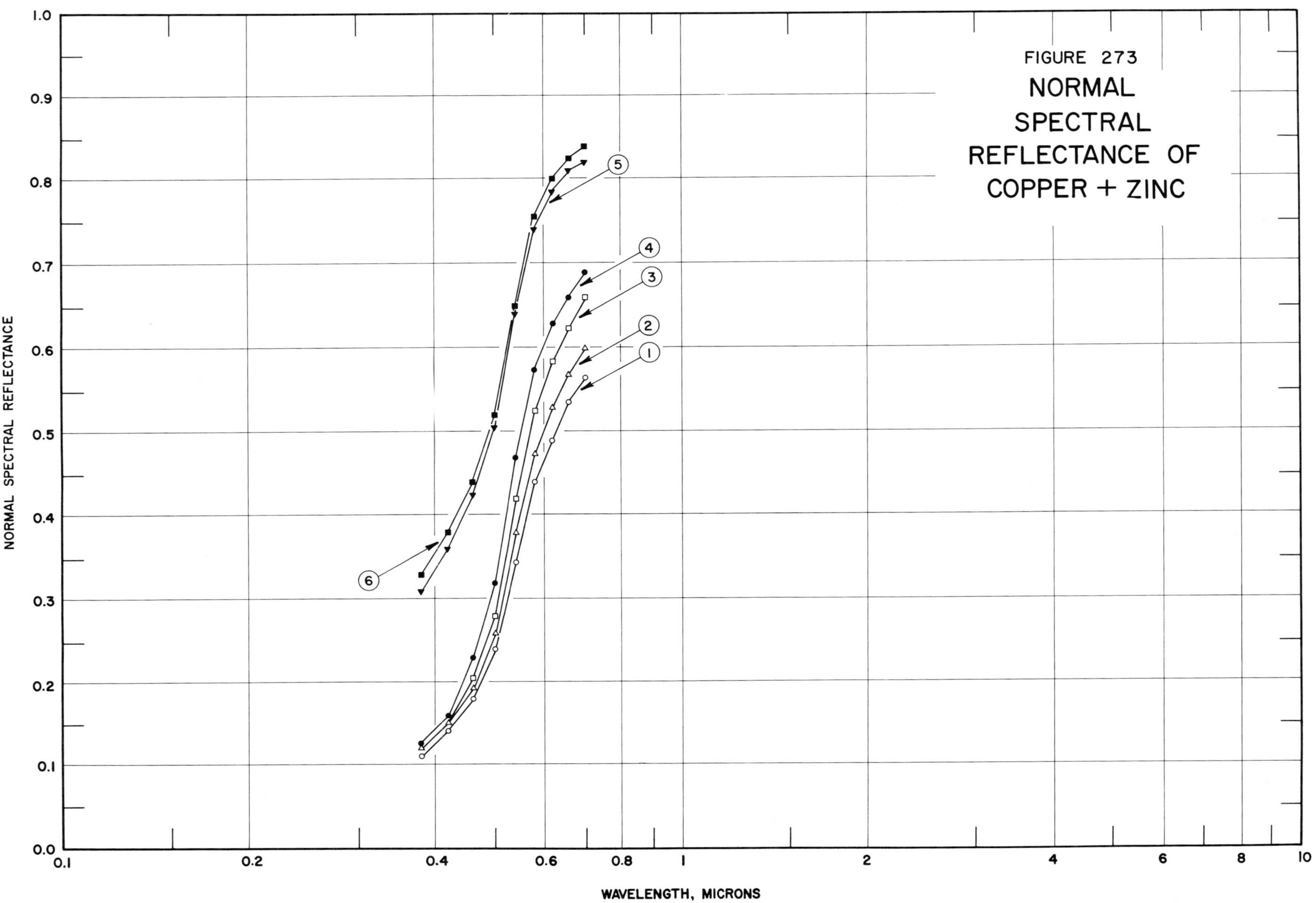
FIGURE 273
NORMAL
SPECTRAL
REFLECTANCE OF
COPPER + ZINC
NORMAL SPECTRAL REFLECTANCE
WAVELENGTH, MICRONS
1.0
0.9
0.8
0.7
0.6
0.5
0.4
0.3
0.2
0.1
0.0
0.1
0.2
0.4
0.6
0.8
1
2
4
6
8
10
1
2
3
4
5
6

SPECIFICATION TABLE NO. 273 NORMAL SPECTRAL REFLECTANCE OF [COPPER + ZINC] ALLOYS

Curve No.	Ref. No.	Year	Temperature K	Wavelength Range, μ	Geometry θ	θ'	ω'	Reported Error, %	Composition (weight percent), Specifications and Remarks
1	29	1956	298	0.38-0.70	9°		2π		Bronze; federal specs. QQ-B-667; nominal composition: 65 Cu, 35 Zn; as received; data extracted from smooth curve; MgO reference.
2	29	1956	298	0.38-0.70	9°		2π		Above specimen and conditions except rotated 90° in its own plane.
3	29	1956	298	0.38-0.70	9°		2π		Different sample, same as curve 1 specimen and conditions except smooth and clean.
4	29	1956	298	0.38-0.70	9°		2π		Above specimen and conditions except rotated 90° in its own plane.
5	29	1956	298	0.38-0.70	9°		2π		Different sample, same as curve 1 specimen and conditions except polished.
6	29	1956	298	0.38-0.70	9°		2π		Above specimen and conditions except rotated 90° in its own plane.

DATA TABLE NO. 273 NORMAL SPECTRAL REFLECTANCE OF [COPPER + ZINC] ALLOYS

[Wavelength, λ, μ; Reflectance, ρ; Temperature, T, K]

λ	ρ
CURVE 1 T = 298	
0.38	0.110
0.42	0.140
0.46	0.180
0.50	0.240
0.54	0.345
0.58	0.440
0.62	0.490
0.66	0.535
0.70	0.565
CURVE 2 T = 298	
0.38	0.120
0.42	0.150
0.46	0.195
0.50	0.260
0.54	0.380
0.58	0.475
0.62	0.530
0.66	0.570
0.70	0.600
CURVE 3 T = 298	
0.38	0.120*
0.42	0.150*
0.46	0.205
0.50	0.280
0.54	0.420
0.58	0.525
0.62	0.585
0.66	0.625
0.70	0.660
CURVE 4 T = 298	
0.38	0.125
0.42	0.160
0.46	0.230
0.50	0.320
CURVE 4 (cont.)	
0.54	0.470
0.58	0.575
0.62	0.630
0.66	0.660
0.70	0.690
CURVE 5 T = 298	
0.38	0.310
0.42	0.360
0.46	0.425
0.50	0.505
0.54	0.640
0.58	0.740
0.62	0.785
0.66	0.810
0.70	0.820
CURVE 6 T = 298	
0.38	0.330
0.42	0.370
0.46	0.440
0.50	0.520
0.54	0.650
0.58	0.755
0.62	0.800
0.66	0.825
0.70	0.840

* Not shown on plot

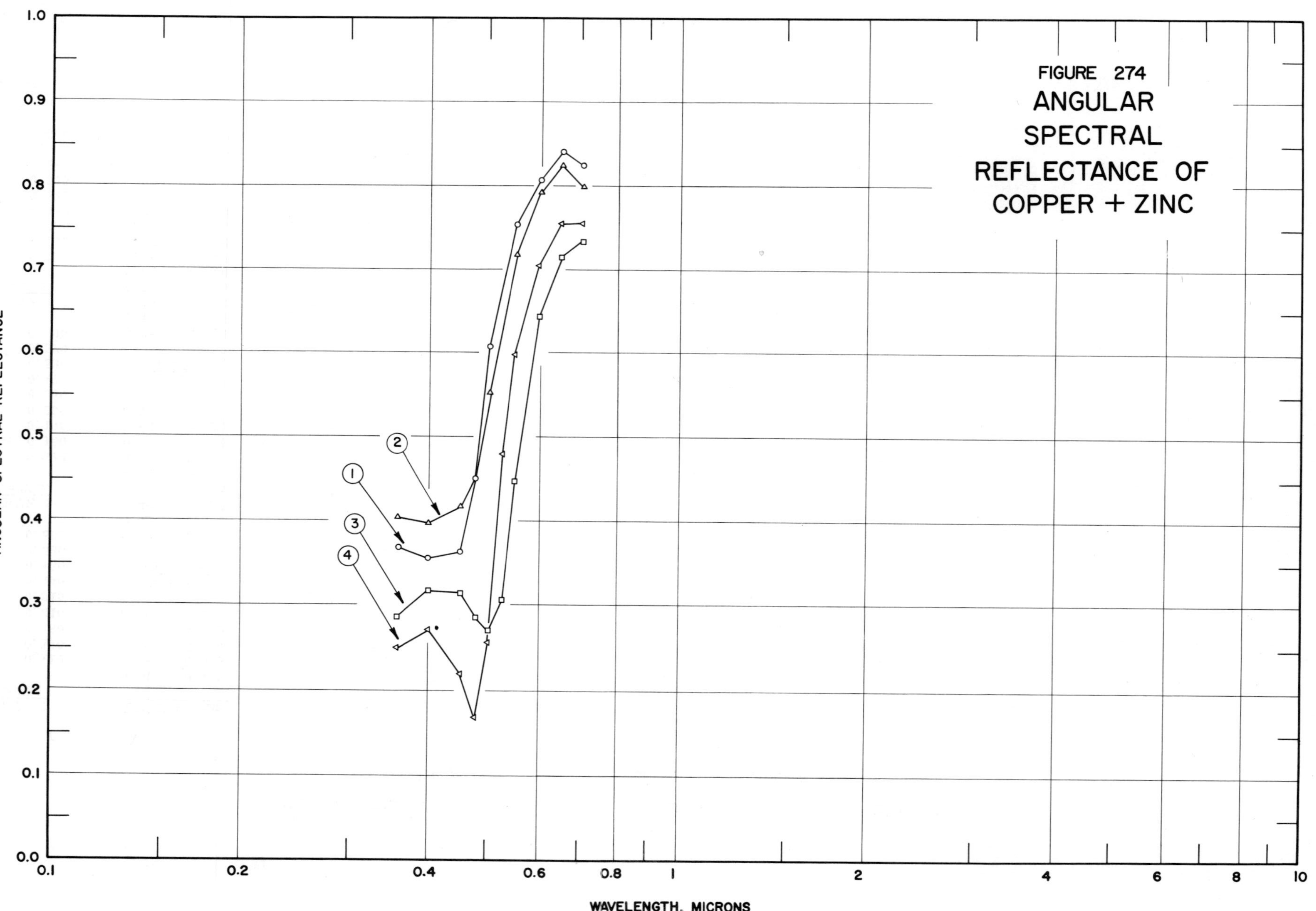

FIGURE 274 ANGULAR SPECTRAL REFLECTANCE OF COPPER + ZINC

SPECIFICATION TABLE NO. 274 ANGULAR SPECTRAL REFLECTANCE OF [COPPER + ZINC] ALLOYS

Curve No.	Ref. No.	Year	Temperature K	Wavelength Range, μ	Geometry θ	θ'	ω'	Reported Error, %	Composition (weight percent), Specifications and Remarks
1	175	1957	300	0.358-0.700	18°		2π		α Brass; nominal composition: 65 Cu, 35 Zn; polished.
2	175	1957	524	0.358-0.700	18°		2π		Different sample, same as above specimen and conditions except measured in dry nitrogen.
3	175	1957	548	0.355-0.700	18°		2π		R brass; nominal composition: 65 Cu, 35 Zn; polished; measured in dry nitrogen.
4	175	1957	300	0.356-0.700	18°		2π		R brass; nominal composition: 65 Cu, 35 Zn; polished.

DATA TABLE NO. 274 ANGULAR SPECTRAL REFLECTANCE OF [COPPER + ZINC] ALLOYS

[Wavelength, λ, μ; Reflectance, ρ; Temperature, T, K]

λ	ρ
CURVE 1 T = 300	
0.358	0.369
0.400	0.355
0.450	0.364
0.475	0.451
0.500	0.607
0.550	0.753
0.600	0.807
0.650	0.841
0.700	0.824
CURVE 2 T = 524	
0.358	0.405
0.400	0.399
0.450	0.418
0.475	0.451*
0.500	0.553
0.550	0.718
0.600	0.793
0.650	0.826
0.700	0.800
CURVE 3 T = 548	
0.355	0.285
0.400	0.318
0.450	0.314
0.474	0.286
0.499	0.270
0.522	0.306
0.548	0.448
0.599	0.643
0.649	0.716
0.700	0.734
CURVE 4 T = 300	
0.356	0.250
0.400	0.271
0.450	0.219
0.474	0.168

λ	ρ
CURVE 4 (cont.)	
0.499	0.256
0.523	0.480
0.548	0.598
0.599	0.703
0.649	0.755
0.700	0.756

* Not shown on plot

SPECIFICATION TABLE NO. 275 HEMISPHERICAL INTEGRATED ABSORPTANCE OF [COPPER + ZINC] ALLOYS

Curve No.	Ref. No.	Year	Temperature Range, K	Reported Error, %	Composition (weight percent), Specifications and Remarks
1	3	1955	76	5	Yellow brass; 65 Cu, 35 Zn; shim stock (0.001 in. thick); measured in vacuum (10^{-6} to 10^{-7} mm Hg); absorptance for 300 K black body incident radiation.

DATA TABLE NO. 275 HEMISPHERICAL INTEGRATED ABSORPTANCE OF [COPPER + ZINC] ALLOYS

[Temperature, T, K; Absorptance, α]

T	α
CURVE 1*	
76	0.029

* Not shown on plot

SPECIFICATION TABLE NO. 276 NORMAL INTEGRATED ABSORPTANCE OF [COPPER + ZINC] ALLOYS

Curve No.	Ref. No.	Year	Temperature Range, K	Geometry θ	Reported Error, %	Composition (weight percent), Specifications and Remarks
1	134	1952	2	~0°	1	Brass; nominal composition: 65 Cu, 35 Zn; electropolished; absorptance for 298 K black body incident radiation.

DATA TABLE NO. 276 NORMAL INTEGRATED ABSORPTANCE OF [COPPER + ZINC] ALLOYS

[Temperature, T, K; Absorptance, α]

T	α
CURVE 1*	
2	0.0178

* Not shown on plot

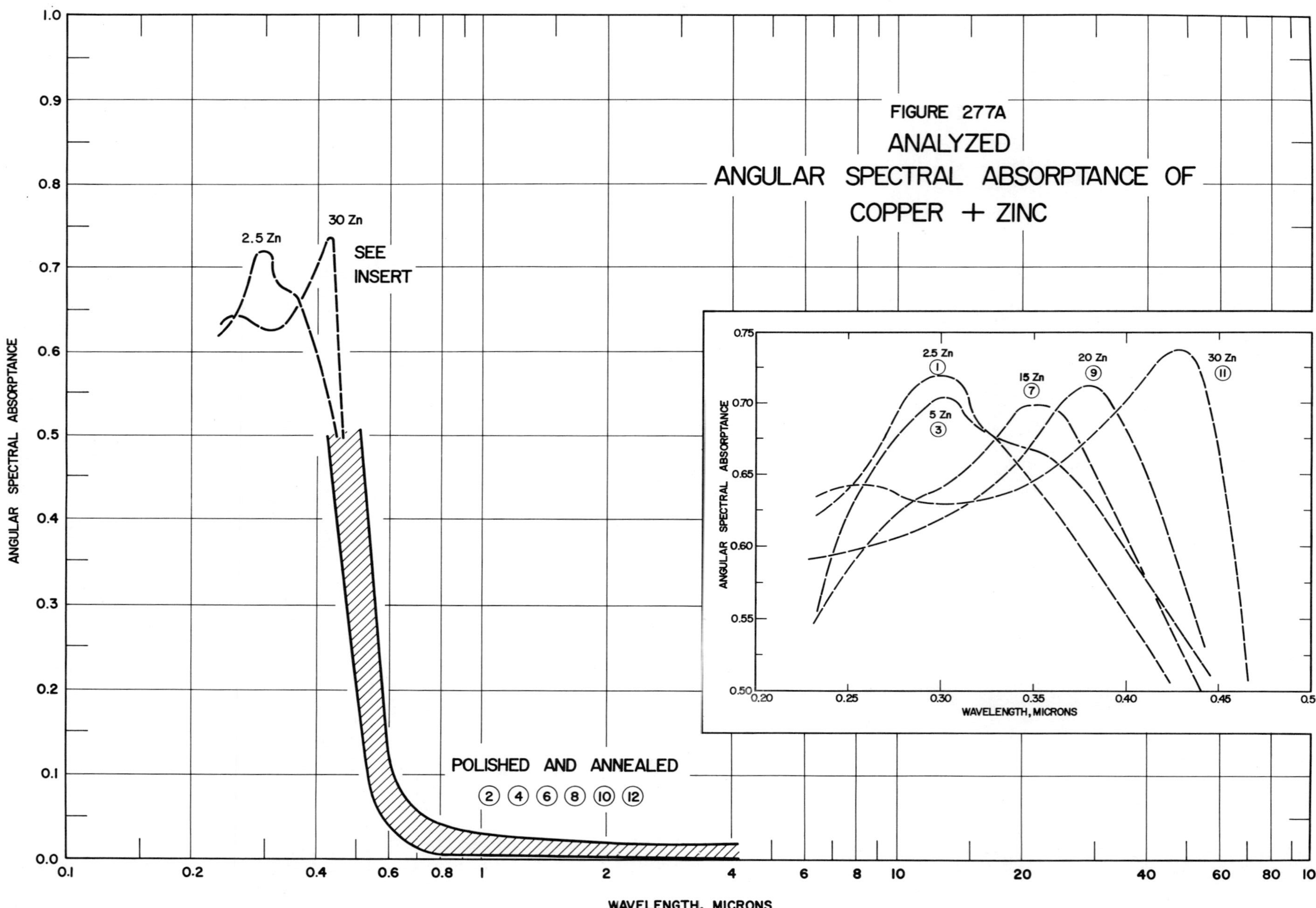

FIGURE 277A
ANALYZED
ANGULAR SPECTRAL ABSORPTANCE OF
COPPER + ZINC

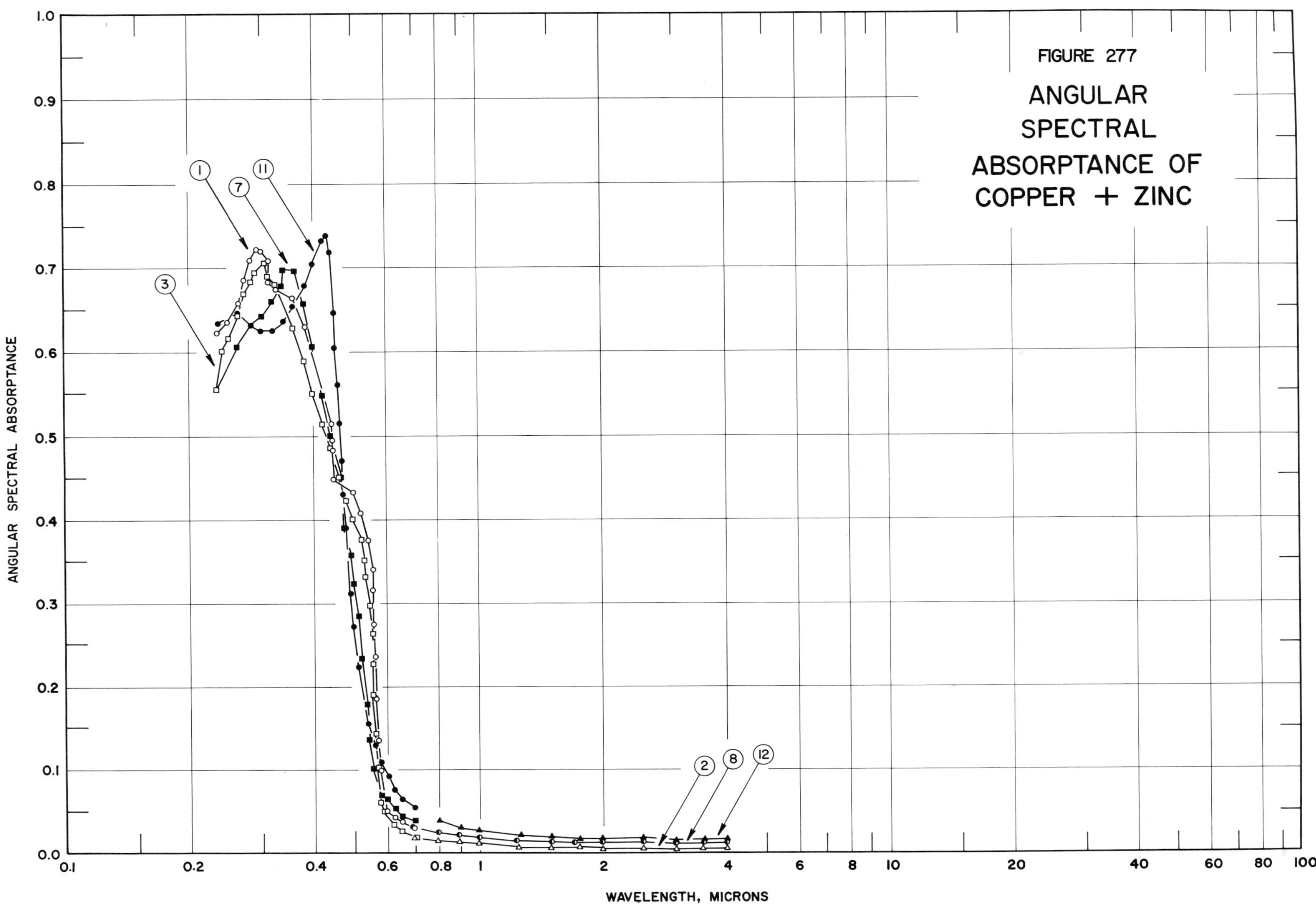

FIGURE 277
ANGULAR SPECTRAL ABSORPTANCE OF COPPER + ZINC

SPECIFICATION TABLE NO. 277 ANGULAR SPECTRAL ABSORPTANCE OF [COPPER + ZINC] ALLOYS

Curve No.	Ref. No.	Year	Temperature K	Wavelength Range, μ	Geometry θ	Reported Error, %	Composition (weight percent), Specifications and Remarks
1	303	1959	4.2	0.2346–0.6992	15°		2.5 Zn; cold worked, maintained at 1073 K for 24 hrs, and annealed in helium; electro-polished and washed; measured in vacuum.
2	303	1959	4.2	0.697–4.00	15°		Above specimen and conditions.
3	303	1959	4.2	0.2331–0.7000	15°		5 Zn; cold worked, maintained at 1073 K for 24 hrs, and annealed in helium; electro-polished and washed; measured in vacuum.
4	303	1959	4.2	0.798–4.00	15°		Above specimen and conditions.
5	303	1959	4.2	0.2323–0.6997	15°		10 Zn; cold worked, maintained at 1073 K for 24 hrs, and annealed in helium; electro-polished and washed; measured in vacuum.
6	303	1959	4.2	0.697–4.00	15°		Above specimen and conditions.
7	303	1959	4.2	0.2334–0.7014	15°		15 Zn; cold worked, maintained at 1073 K for 24 hrs, and annealed in helium; electro-polished and washed; measured in vacuum.
8	303	1959	4.2	0.691–4.01	15°		Above specimen and conditions.
9	303	1959	4.2	0.2333–0.6993	15°		20 Zn; cold worked, maintained at 1073 K for 24 hrs, and annealed in helium; electro-polished and washed; measured in vacuum.
10	303	1959	4.2	0.704–4.01	15°		Above specimen and conditions.
11	303	1959	4.2	0.2352–0.6997	15°		30 Zn; cold worked, maintained at 1073 K for 24 hrs, and annealed in helium; electro-polished and washed; measured in vacuum.
12	303	1959	4.2	0.801–4.00	15°		Above specimen and conditions.

DATA TABLE NO. 277 ANGULAR SPECTRAL ABSORPTANCE OF [COPPER + ZINC] ALLOYS

[Wavelength, λ, μ; Absorptance, α; Temperature, T, K]

CURVE 1
T = 4.2

λ	α
0.2346	0.622
0.2494	0.636
0.2631	0.658
0.2723	0.685
0.2826	0.709
0.2930	0.722
0.3024	0.721
0.3128	0.709
0.3183	0.683
0.3291	0.675
0.3599	0.663
0.3822	0.631
0.4431	0.514
0.4560	0.495
0.4616	0.482
0.4806	0.449
0.5015	0.432
0.5211	0.408
0.5414	0.375
0.5507	0.341
0.5555	0.317
0.5589	0.274
0.5623	0.234
0.5673	0.186
0.5716	0.134
0.5755	0.099
0.5991	0.051
0.6220	0.042
0.6514	0.038
0.6992	0.030

CURVE 2
T = 4.2

λ	α
0.697	0.0194
0.794	0.0162
0.894	0.0135
0.985	0.0119
1.24	0.0089
1.50	0.0079
1.76	0.0080
2.00	0.0076
2.50	0.0075
3.01	0.0068

CURVE 2 (cont.)

λ	α
3.51	0.0066
4.00	0.0070

CURVE 3
T = 4.2

λ	α
0.2331	0.556
0.2416	0.601
0.2519	0.618
0.2630	0.643
0.2727	0.670
0.2835	0.683
0.2910	0.696
0.3008	0.707
0.3108	0.690
0.3255	0.680
0.3599	0.629
0.3805	0.589
0.4004	0.550
0.4200	0.513
0.4403	0.484
0.4609	0.450
0.4806	0.422
0.5004	0.401
0.5213	0.377
0.5325	0.351
0.5384	0.332
0.5442	0.298
0.5518	0.263
0.5558	0.227
0.5598	0.189
0.5643	0.142
0.5689	0.102
0.5793	0.061
0.5900	0.051
0.6210	0.034
0.6496	0.027
0.7000	0.019

CURVE 4*
T = 4.2

λ	α
0.798	0.0178
0.894	0.0147
0.996	0.0136

CURVE 4 (cont.)*

λ	α
1.24	0.0107
1.51	0.0099
1.77	0.0088
1.99	0.0079
2.52	0.0089
2.99	0.0079
3.50	0.0080
4.00	0.0086

CURVE 5*
T = 4.2

λ	α
0.2323	0.556
0.2478	0.591
0.2617	0.616
0.2815	0.645
0.3016	0.672
0.3213	0.692
0.3393	0.691
0.3582	0.657
0.3801	0.618
0.4014	0.572
0.4214	0.535
0.4423	0.493
0.4604	0.460
0.4806	0.406
0.4988	0.361
0.5159	0.317
0.5236	0.285
0.5318	0.250
0.5355	0.221
0.5389	0.191
0.5449	0.154
0.5521	0.127
0.5598	0.095
0.5682	0.073
0.5989	0.056
0.6204	0.042
0.6509	0.036
0.6997	0.029

CURVE 6*
T = 4.2

λ	α
0.697	0.0223

CURVE 6 (cont.)*

λ	α
0.787	0.0205
0.891	0.0178
0.993	0.0161
1.24	0.0127
1.50	0.0127
1.76	0.0112
1.99	0.0112
2.50	0.0114
3.01	0.0104
3.51	0.0104
4.00	0.0101

CURVE 7
T = 4.2

λ	α
0.2334	0.552*
0.2628	0.606
0.2827	0.632*
0.3025	0.642
0.3200	0.660
0.3381	0.679
0.3409	0.698
0.3620	0.697
0.3811	0.658
0.4020	0.605
0.4217	0.548
0.4408	0.500
0.4615	0.450
0.4817	0.391
0.4923	0.358
0.5016	0.323
0.5104	0.284
0.5208	0.232
0.5322	0.178
0.5409	0.136
0.5606	0.101
0.5835	0.069
0.6004	0.065
0.6262	0.054
0.6515	0.045
0.7014	0.039

CURVE 8
T = 4.2

λ	α
0.691	0.0319*
0.798	0.0254
0.888	0.0221
0.993	0.0191
1.24	0.0162
1.50	0.0150
1.76	0.0137
1.99	0.0138
2.50	0.0142
3.01	0.0129
3.50	0.0129
4.01	0.0137

CURVE 9*
T = 4.2

λ	α
0.2333	0.592
0.2483	0.597
0.2617	0.600
0.2824	0.608
0.3215	0.630
0.3428	0.661
0.3614	0.695
0.3816	0.713
0.4025	0.673
0.4129	0.645
0.4232	0.605
0.4323	0.569
0.4418	0.537
0.4604	0.453
0.4732	0.418
0.4825	0.378
0.4898	0.332
0.4994	0.283
0.5094	0.235
0.5204	0.197
0.5302	0.155
0.5404	0.128
0.5594	0.099
0.5801	0.083
0.5992	0.056
0.6232	0.046
0.6513	0.039
0.6993	0.033

CURVE 10*
T = 4.2

λ	α
0.704	0.0387
0.795	0.0294
0.910	0.0253
0.998	0.0237
1.24	0.0194
1.49	0.0173
1.74	0.0162
2.00	0.0162
2.51	0.0162
3.01	0.0149
3.51	0.0149
4.01	0.0149

CURVE 11
T = 4.2

λ	α
0.2352	0.636
0.2652	0.643
0.2824	0.632
0.3028	0.626
0.3217	0.627
0.3408	0.638
0.3614	0.654
0.3826	0.679
0.4017	0.704
0.4217	0.734
0.4319	0.740
0.4413	0.719
0.4514	0.647
0.4563	0.603
0.4610	0.560
0.4657	0.514
0.4716	0.470
0.4761	0.431
0.4807	0.390
0.4912	0.311
0.5015	0.271
0.5107	0.222
0.5400	0.155
0.5601	0.128
0.5788	0.109
0.6003	0.091
0.6261	0.076
0.6511	0.063
0.6997	0.053

CURVE 12
T = 4.2

λ	α
0.801	0.0399
0.893	0.0305
0.998	0.0273
1.25	0.0222
1.50	0.0205
1.76	0.0189
1.99	0.0186
2.50	0.0194
3.01	0.0179
3.51	0.0174
4.00	0.0183

* Not shown on plot

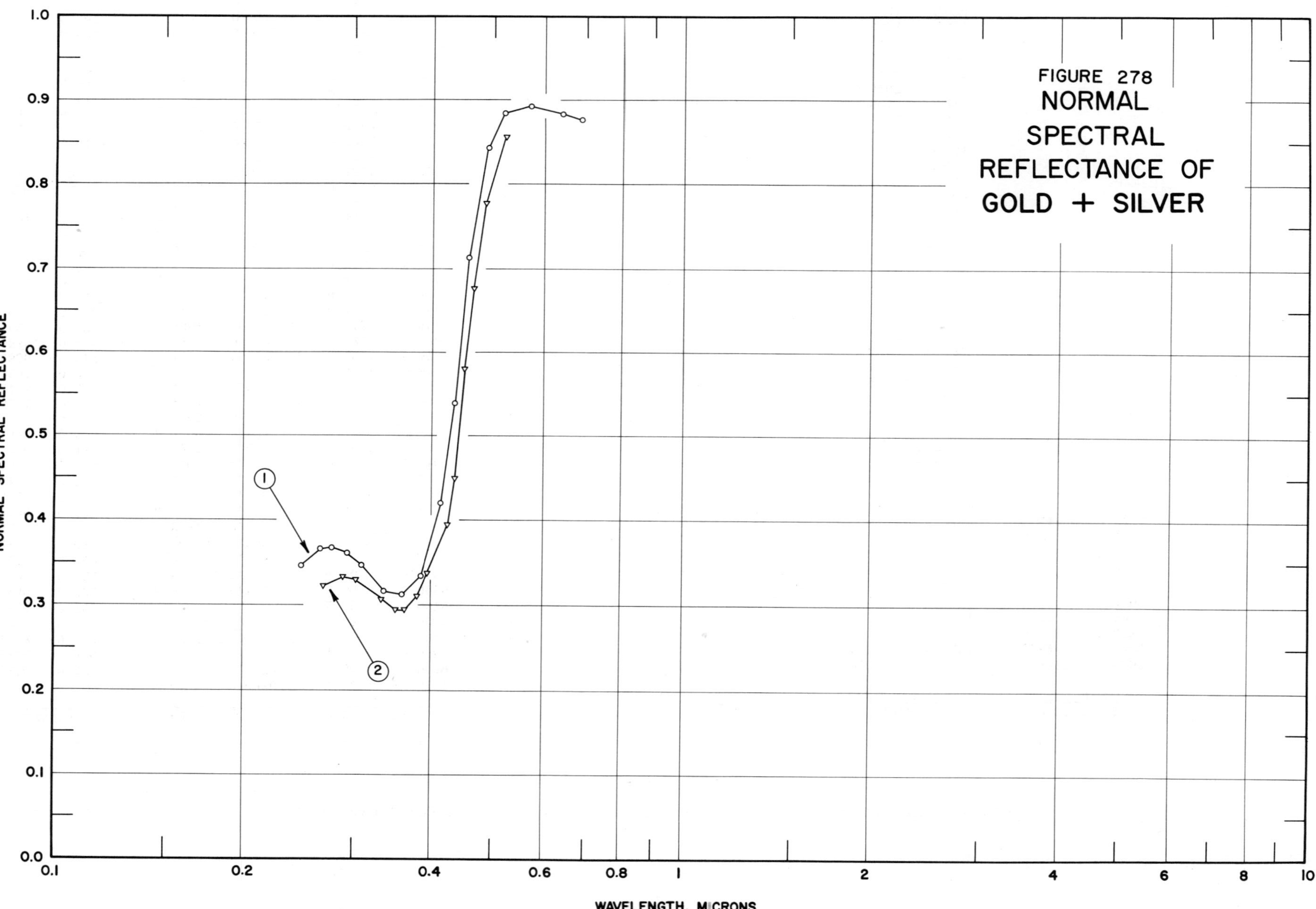

FIGURE 278
NORMAL
SPECTRAL
REFLECTANCE OF
GOLD + SILVER
NORMAL SPECTRAL REFLECTANCE
WAVELENGTH, MICRONS
1.0
0.9
0.8
0.7
0.6
0.5
0.4
0.3
0.2
0.1
0.0
0.1
0.2
0.4
0.6
0.8
1
2
4
6
8
10
1
2

SPECIFICATION TABLE NO. 278 NORMAL SPECTRAL REFLECTANCE OF [GOLD + SILVER] ALLOYS

Curve No.	Ref. No.	Year	Temperature K	Wavelength Range, μ	Geometry θ	θ'	ω'	Reported Error, %	Composition (weight percent), Specifications and Remarks
1	297	1963	298	0.248-0.689	~10°	~10°			50 Au and 50 Ag; hand lapped with silicon carbide papers, lapped on a metallographic polishing wheel, and electrolytically slide polished; tungsten filament source for 0.365 to 0.689 μ and a hydrogen lamp source at lower wavelengths; sample slightly etched during electrolytic polishing; data extracted from smooth curve.
2	350	1965	298	0.269-0.521	~0°	~0°			50 Au and 50 Ag film (~2000 Å thick); evaporated on a quartz plate at 423 K in vacuum (< 3 x 10^{-6} mm Hg); annealed at 673 K for 1.5 hr; data extracted from smooth curve.

DATA TABLE NO. 278 NORMAL SPECTRAL REFLECTANCE OF [GOLD + SILVER] ALLOYS

[Wavelength, λ, μ; Reflectance, ρ; Temperature, T, K]

λ	ρ
CURVE 1	
T = 298	
0.248	0.348
0.266	0.365
0.277	0.367
0.294	0.361
0.310	0.348
0.336	0.317
0.359	0.312
0.384	0.334
0.413	0.420
0.434	0.539
0.456	0.712
0.490	0.845
0.519	0.887
0.571	0.894
0.642	0.883
0.689	0.878
CURVE 2	
T = 298	
0.269	0.322
0.290	0.332
0.302	0.330
0.333	0.306
0.351	0.295
0.363	0.295
0.380	0.310
0.394	0.338
0.423	0.393
0.435	0.449
0.449	0.580
0.464	0.676
0.484	0.778
0.521	0.858

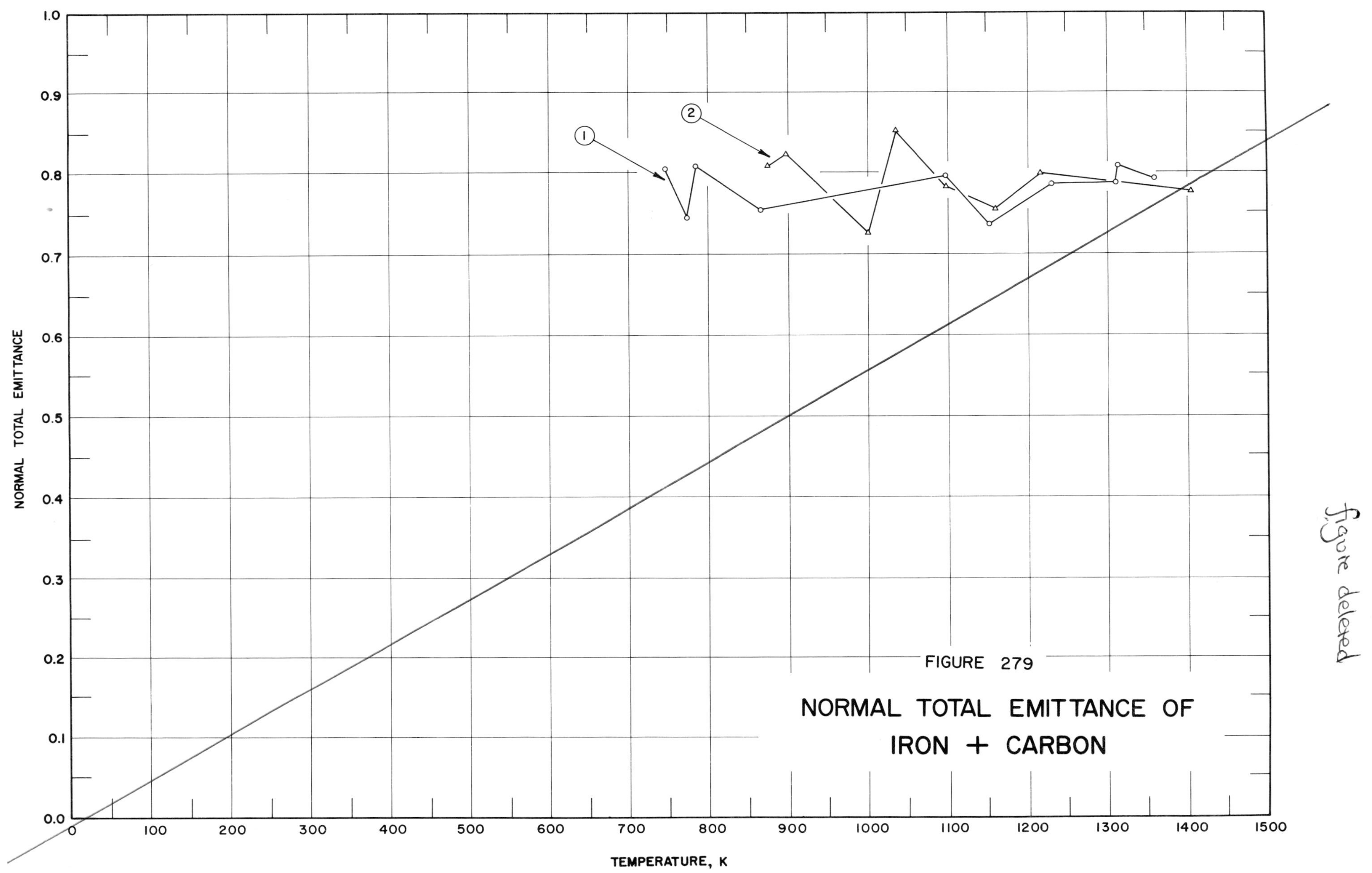

FIGURE 279

NORMAL TOTAL EMITTANCE OF IRON + CARBON

SPECIFICATION TABLE NO. 279 NORMAL TOTAL EMITTANCE OF [IRON + CARBON] ALLOYS

Curve No.	Ref. No.	Year	Temperature Range, K	Geometry θ'	Reported Error, %	Composition (weight percent), Specifications and Remarks
1	273	1962	747-1360	$\sim0^\circ$		Universal Cyclops Unitemp No. 41; nominal composition: 0.65-1.35 C, Fe balance; annealed in air at 1366 K for 1 hr; measure in air; [Author's designation: Specimen No. 3].
2	273	1962	875-1405	$\sim0^\circ$		Different sample,same as above specimen and conditions; [Author's designation: Sample No. 5].

deleted

deleted

DATA TABLE NO. 279 NORMAL TOTAL EMITTANCE OF [IRON + CARBON] ALLOYS

[Temperature, T, K; Emittance, ∈]

T	∈
CURVE 1	
747	0.803
773	0.743
784	0.808
867	0.754
1098	0.795
1152	0.735
1231	0.785
1311	0.736
1313	0.758
1360	0.791
CURVE 2	
875	0.809
897	0.822
1000	0.725
1037	0.852
1099	0.782
1160	0.754
1216	0.799
1405	0.776

SPECIFICATION TABLE NO. 280 NORMAL SPECTRAL REFLECTANCE OF [IRON + CHROMIUM] ALLOYS

Curve No.	Ref. No.	Year	Temperature K	Wavelength Range, μ	Geometry θ	θ'	ω'	Reported Error, %	Composition (weight percent), Specifications and Remarks
1	173	1947	298	0.55-0.65	~0°	~0°			Stainless steel SF11; 13 Cr, Fe balance.

DATA TABLE NO. 280 NORMAL SPECTRAL REFLECTANCE OF [IRON + CHROMIUM] ALLOYS

[Wavelength, λ, μ; Reflectance, ρ; Temperature, T, K]

λ	ρ
CURVE 1*	
T = 298	
0.55	0.501
0.65	0.648

* Not shown on plot

SPECIFICATION TABLE NO. 281 ANGULAR SPECTRAL REFELCTANCE OF [IRON + CHROMIUM] ALLOYS

Curve No.	Ref. No.	Year	Temperature K	Wavelength Range, μ	Geometry θ	θ'	ω'	Reported Error, %	Composition (weight percent), Specifications and Remarks
1	143	1928	298	0.255-0.560	45°	45°			Stainless steel; 13 Cr, Fe balance; highly polished.

DATA TABLE NO. 281 ANGULAR SPECTRAL REFLECTANCE OF [IRON + CHROMIUM] ALLOYS

[Wavelength, λ, μ; Reflectance, ρ; Temperature, T, K]

λ	ρ
CURVE 1*	
T = 298	
0.255	0.41
0.265	0.41
0.280	0.42
0.300	0.46
0.312	0.48
0.335	0.50
0.365	0.54
0.405	0.56
0.435	0.58
0.560	0.60

* Not shown on plot

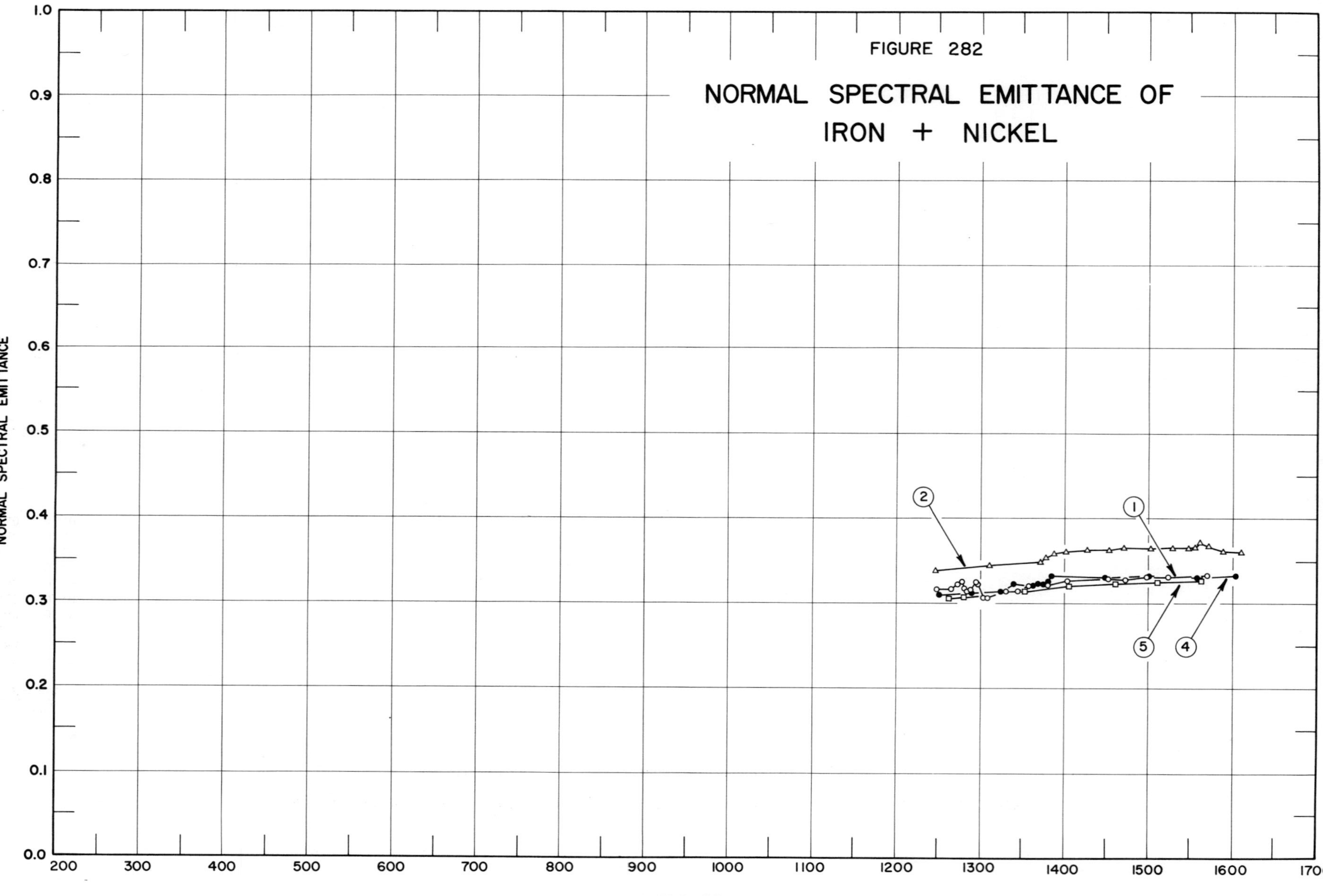
FIGURE 282
NORMAL SPECTRAL EMITTANCE OF
IRON + NICKEL
NORMAL SPECTRAL EMITTANCE
1.0
0.9
0.8
0.7
0.6
0.5
0.4
0.3
0.2
0.1
0.0
200
300
400
500
600
700
800
900
1000
1100
1200
1300
1400
1500
1600
1700
TEMPERATURE, K
2
1
5
4

SPECIFICATION TABLE NO. 282 NORMAL SPECTRAL EMITTANCE OF [IRON + NICKEL] ALLOYS

Curve No.	Ref. No.	Year	Wavelength μ	Temperature Range, K	Geometry θ'	Reported Error, %	Composition (weight percent), Specifications and Remarks
1	176	1951	0.67	1249-1571	~0°		50 Fe, 50 Ni; powders were mixed in desired proportions, compressed at 70,000 psi at 298 K, heated at 1373 K in flowing hydrogen; cold rolled then heated in hydrogen at 1273 K.
2	176	1951	0.67	1248-1611	~0°		56 Fe, 44 Ni; specimen preparation same as curve 1.
3	176	1951	0.67	1271-1597	~0°		58 Fe, 42 Ni; specimen preparation same as curve 1.
4	176	1951	0.67	1252-1604	~0°		62 Fe, 38 Ni; specimen preparation same as curve 1.
5	176	1951	0.67	1262-1563	~0°		75 Fe, 25 Ni; specimen preparation same as curve 1.

DATA TABLE NO. 282 NORMAL SPECTRAL EMITTANCE OF [IRON + NICKEL] ALLOYS

[Temperature, T, K; Emittance, ∈; Wavelength, λ, μ]

T	∈	T	∈	T	∈
CURVE 1		CURVE 3*		CURVE 5 (cont.)	
λ = 0.67		λ = 0.67			
1249	0.317	1271	0.321	1406	0.320
1266	0.317	1297	0.328	1461	0.322
1274	0.322	1302	0.327	1511	0.325
1279	0.325	1308	0.328	1563	0.327
1281	0.317	1326	0.332		
1285	0.313	1332	0.333		
1287	0.314*	1347	0.339		
1290	0.316	1350	0.340		
1296	0.325	1371	0.371		
1297	0.322	1377	0.340		
1303	0.306	1428	0.341		
1309	0.306	1441	0.341		
1331	0.314	1459	0.355		
1345	0.314	1487	0.354		
1358	0.320	1539	0.353		
1381	0.321	1565	0.350		
1404	0.327	1597	0.350		
1452	0.329				
1473	0.328	CURVE 4			
1499	0.331	λ = 0.67			
1524	0.331				
1571	0.332	1252	0.309		
		1291	0.312		
CURVE 2		1324	0.313		
λ = 0.67		1339	0.322		
		1356	0.320*		
1248	0.339	1362	0.321		
1311	0.344	1362	0.322*		
1372	0.349	1369	0.323		
1378	0.353	1375	0.322		
1389	0.360	1381	0.325		
1401	0.361	1386	0.332		
1428	0.362	1450	0.330		
1453	0.362	1500	0.332		
1471	0.365	1558	0.331		
1503	0.364	1604	0.332		
1529	0.365				
1549	0.365	CURVE 5			
1556	0.366	λ = 0.67			
1561	0.371				
1572	0.367	1262	0.304		
1589	0.361	1280	0.307		
1611	0.360	1353	0.313		

* Not shown on plot

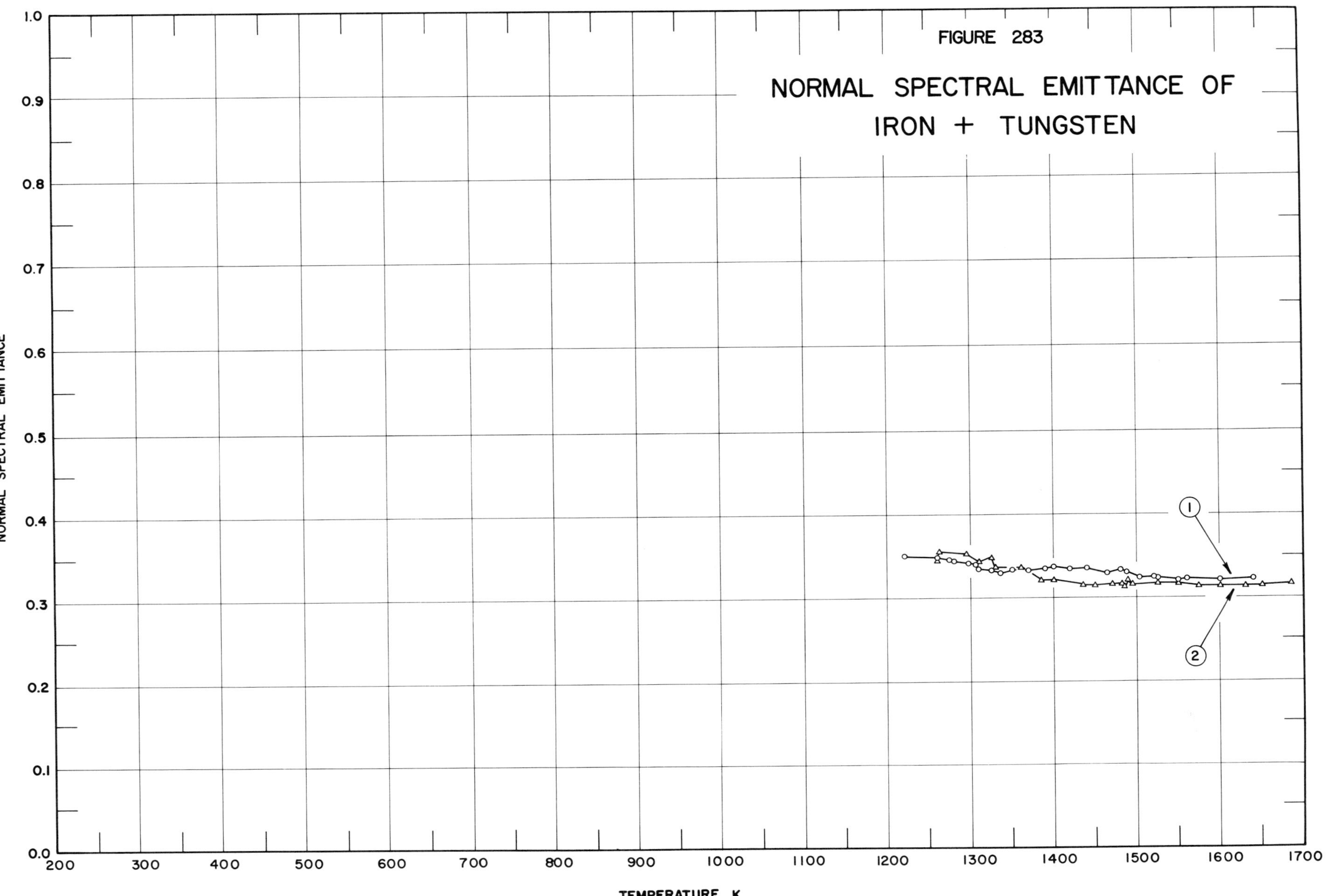
FIGURE 283
NORMAL SPECTRAL EMITTANCE OF
IRON + TUNGSTEN
NORMAL SPECTRAL EMITTANCE
1.0
0.9
0.8
0.7
0.6
0.5
0.4
0.3
0.2
0.1
0.0
200
300
400
500
600
700
800
900
1000
1100
1200
1300
1400
1500
1600
1700
TEMPERATURE, K
1
2

SPECIFICATION TABLE NO. 283 NORMAL SPECTRAL EMITTANCE OF [IRON + TUNGSTEN] ALLOYS

Curve No.	Ref. No.	Year	Wavelength μ	Temperature Range, K	Geometry θ'	Reported Error, %	Composition (weight percent), Specifications and Remarks
1	174	1948	0.667	1220-1640	$\sim 0^\circ$		82 Fe, 18 W; heated to 1375 K; measured in hydrogen; increasing temp.
2	174	1948	0.667	1685-1260	$\sim 0^\circ$		82 Fe, 18 W; heated to 1680 K; measured in hydrogen; decreasing temp.

DATA TABLE NO. 283 NORMAL SPECTRAL EMITTANCE OF [IRON + TUNGSTEN] ALLOYS

[Temperature, T, K; Emittance, ϵ; Wavelength, λ, μ]

T	ϵ
CURVE 1 $\lambda = 0.667$	
1220	0.350
1260	0.349
1275	0.347
1280	0.346
1298	0.342
1305	0.341
1310	0.336
1325	0.334
1335	0.331
1350	0.334
1370	0.333
1390	0.336
1400	0.338
1420	0.335
1440	0.335
1465	0.330
1480	0.333
1488	0.331
1503	0.324
1520	0.326
1525	0.324
1550	0.321
1560	0.322
1600	0.321
1640	0.322
CURVE 2 $\lambda = 0.667$	
1685	0.317
1650	0.314
1630	0.313
1600	0.313
1575	0.313
1550	0.317
1525	0.317
1495	0.315
1490	0.320
1485	0.312
1482	0.316
1470	0.316
1450	0.314

T	ϵ
CURVE 2 (cont.)	
1435	0.315
1400	0.321
1385	0.321
1360	0.336
1330	0.337
1325	0.348
1310	0.344
1295	0.352
1262	0.355
1260	0.347

SPECIFICATION TABLE NO. 284 HEMISPHERICAL TOTAL EMITTANCE OF [LEAD + TIN] ALLOYS

Curve No.	Ref. No.	Year	Temperature Range, K	Reported Error, %	Composition (weight percent), Specifications and Remarks
1	9	1960	77		40 Sn-60 Pb solder; surface applied with air-gas torch; cleaned with acetone; measured in vacuum (~1 x 10^{-5} mm Hg).

DATA TABLE NO. 284 HEMISPHERICAL TOTAL EMITTANCE OF [LEAD + TIN] ALLOYS

[Temperature, T, K; Emittance, ϵ]

T	ϵ
CURVE 1*	
77	0.047

* Not shown on plot

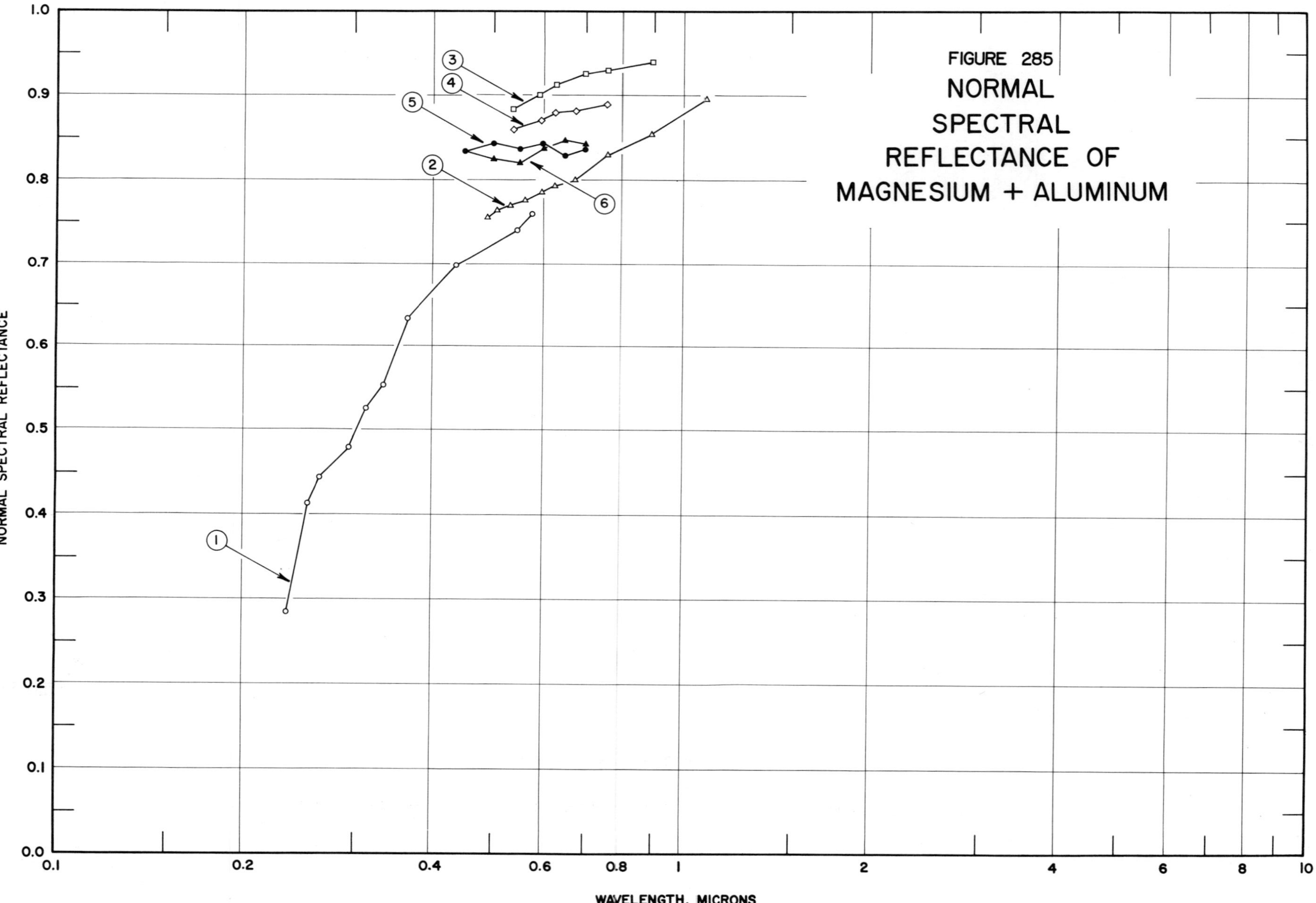
FIGURE 285
NORMAL
SPECTRAL
REFLECTANCE OF
MAGNESIUM + ALUMINUM
NORMAL SPECTRAL REFLECTANCE
WAVELENGTH, MICRONS
1.0
0.9
0.8
0.7
0.6
0.5
0.4
0.3
0.2
0.1
0.0
0.1
0.2
0.4
0.6
0.8
1
2
4
6
8
10
3
4
5
2
6
1

SPECIFICATION TABLE NO. 285 NORMAL SPECTRAL REFLECTANCE OF [MAGNESIUM + ALUMINUM] ALLOYS

Curve No.	Ref. No.	Year	Temperature K	Wavelength Range, μ	Geometry θ	θ'	ω'	Reported Error, %	Composition (weight percent), Specifications and Remarks
1	133	1934	298	0.2350-0.5780	~0°	~0°			70 Mg, Al balance; cold worked; annealed in an inert gas for 6 to 24 hrs; polished; stored in dilute solutions of NaOH, NaOH + NaF, and HNO_3.
2	152	1920	298	0.490-1.190	~0°	~0°			54 Mg, 46 Al; polished with a pitch polisher using rouge; silver mirror reference; converted from R (0°,0°); [Author's designation: R1A].
3	152	1920	298	0.535-0.895	~0°	~0°			Different sample, same as above specimen and conditions except polished with alumina on broadcloth moistened with alcohol; [Author's designation: $R1A_1$].
4	152	1920	298	0.538-0.761	~0°	~0°			Different sample, same as curve 2 specimen and conditions; [Author's designation: R1].
5	228	1900	298	0.45-0.70	~0°	~0°			73 Mg, 27 Al; mirror like surface; platinum filament lamp source.
6	228	1900	298	0.45-0.70	~0°	~0°			60 Mg, 40 Al; mirror like surface; platinum filament lamp source.

DATA TABLE NO. 285 NORMAL SPECTRAL REFLECTANCE OF [MAGNESIUM + ALUMINUM] ALLOYS

[Wavelength, λ, μ; Reflectance, ρ; Temperature, T, K]

λ	ρ	λ	ρ
CURVE 1 (T = 298)		CURVE 5 (T = 298)	
0.2350	0.286	0.45	0.834
0.2537	0.413	0.50	0.845
0.2650	0.444	0.55	0.838
0.2970	0.480	0.60	0.845
0.3125	0.525	0.65	0.83
0.3340	0.553	0.70	0.838
0.3660	0.635		
0.4355	0.699	CURVE 6 (T = 298)	
0.5460	0.740		
0.5780	0.766	0.45	0.834*
CURVE 2 (T = 298)		0.50	0.825
		0.55	0.821
		0.60	0.838
0.490	0.755	0.65	0.849
0.506	0.763	0.70	0.844
0.531	0.770		
0.560	0.775		
0.591	0.786		
0.625	0.794		
0.674	0.800		
0.760	0.831		
0.894	0.854		
1.190	0.897		
CURVE 3 (T = 298)			
0.535	0.885		
0.591	0.900		
0.629	0.912		
0.700	0.927		
0.760	0.930		
0.895	0.940		
CURVE 4 (T = 298)			
0.538	0.860		
0.596	0.871		
0.626	0.880		
0.678	0.882		
0.761	0.890		

* Not shown on plot

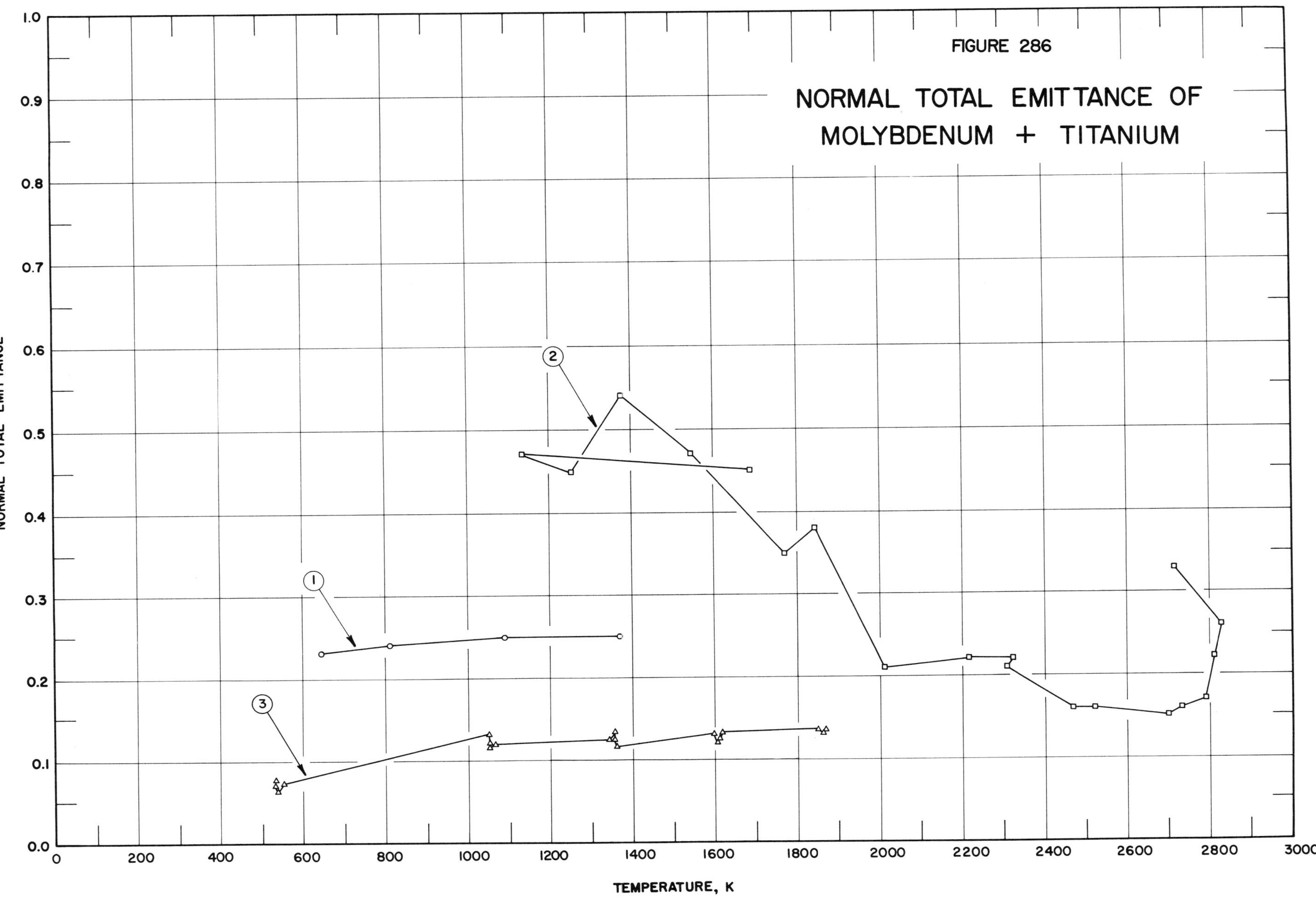

FIGURE 286
NORMAL TOTAL EMITTANCE OF
MOLYBDENUM + TITANIUM
NORMAL TOTAL EMITTANCE
TEMPERATURE, K
1.0
0.9
0.8
0.7
0.6
0.5
0.4
0.3
0.2
0.1
0.0
0
200
400
600
800
1000
1200
1400
1600
1800
2000
2200
2400
2600
2800
3000
1
2
3

SPECIFICATION TABLE NO. 286 NORMAL TOTAL EMITTANCE OF [MOLYBDENUM + TITANIUM] ALLOYS

Curve No.	Ref. No.	Year	Temperature Range, K	Geometry θ'	Reported Error, %	Composition (weight percent), Specifications and Remarks
1	107	1960	644-1367	~0°	±20	Molybdenum alloy;0.5 Ti, Mo balance; arc casted; 1.62 in. diameter billet recrystallized at 1867 K and rolled; measured in demoisturized helium gas.
2	106	1962	1689-2717	~0°	10-20	0.5 Ti, Mo balance; hot pressed; measured in argon.
3	304	1960	533-1864	0°	± 5	Mo + 0.5 Ti from Universal Cyclops Steel Corp; 0.024-0.031 C, 0.43-0.48 Ti, and 0.06 others; recrystallized for 35 min at 1866 K; measured in 90 Ar + 10 H_2 at 78 cm Hg.

DATA TABLE NO. 286 NORMAL TOTAL EMITTANCE OF [MOLYBDENUM + TITANIUM] ALLOYS

[Temperature, T, K; Emittance, ∈]

T	∈	T	∈
CURVE 1		CURVE 3	
644	0.23	533	0.070
811	0.24	533	0.078
1089	0.25	539	0.064
1367	0.25	550	0.072
		1049	0.130
CURVE 2		1052	0.122
		1052	0.114
1689	0.45	1065	0.129
1139	0.47	1342	0.124
1255	0.45	1357	0.133
1378	0.54	1357	0.125
1544	0.47	1360	0.116
1772	0.35	1599	0.130
1844	0.38	1604	0.120
2011	0.21	1610	0.127
2217	0.22	1619	0.132
2322	0.22	1849	0.137
2311	0.21	1861	0.131
2467	0.16	1864	0.135
2522	0.16		
2700	0.15		
2733	0.16		
2794	0.17		
2811	0.22		
2828	0.26		
2717	0.33		

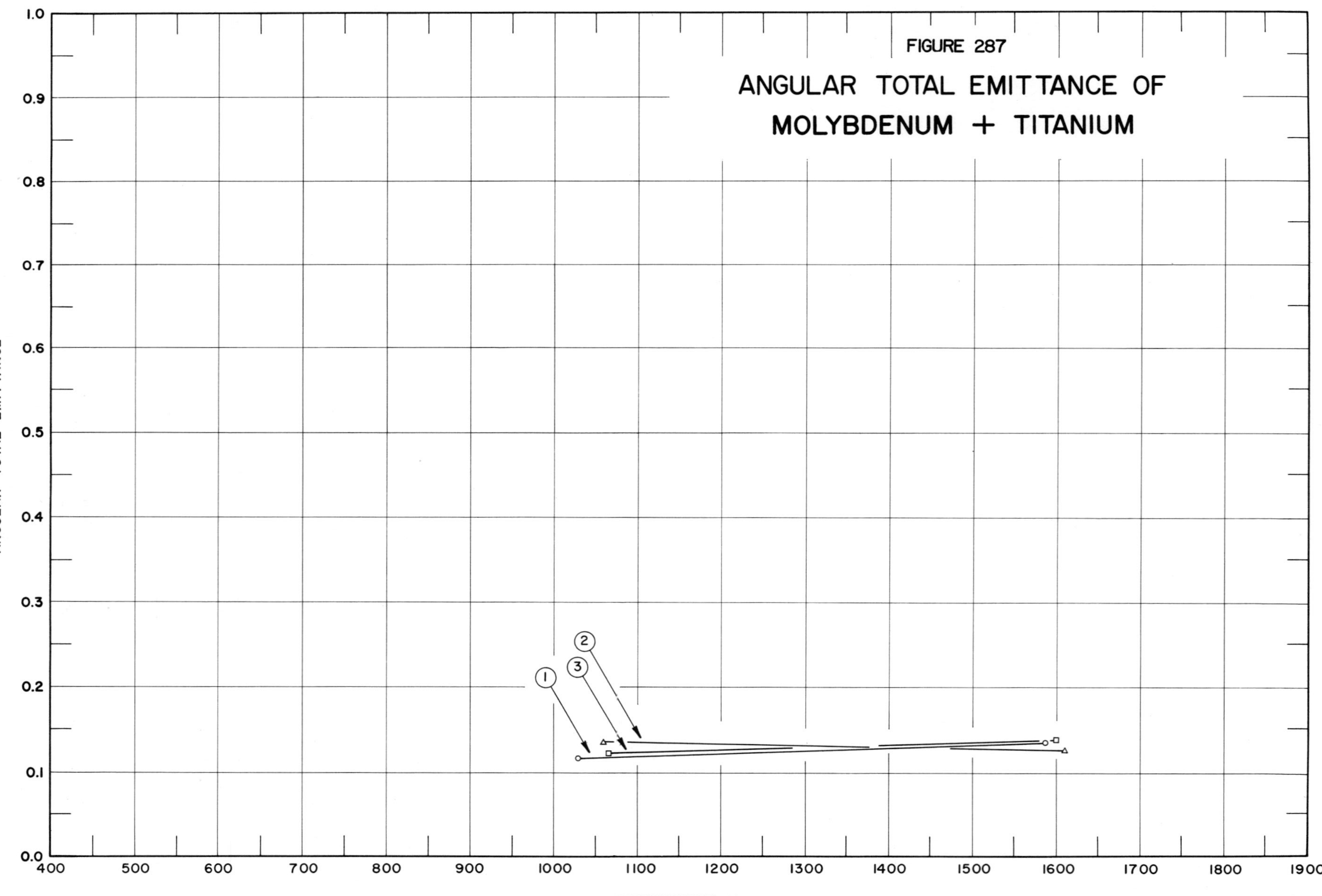
FIGURE 287
ANGULAR TOTAL EMITTANCE OF
MOLYBDENUM + TITANIUM
ANGULAR TOTAL EMITTANCE
1.0
0.9
0.8
0.7
0.6
0.5
0.4
0.3
0.2
0.1
0.0
400
500
600
700
800
900
1000
1100
1200
1300
1400
1500
1600
1700
1800
1900
TEMPERATURE, K
1
2
3

SPECIFICATION TABLE NO. 287 ANGULAR TOTAL EMITTANCE OF [MOLYBDENUM + TITANIUM] ALLOYS

Curve No.	Ref. No.	Year	Temperature Range, K	Geometry θ'	Reported Error, %	Composition (weight percent), Specifications and Remarks
1	304	1960	1030-1588	30°	±5	Mo + 0.5 Ti from Universal Cyclops Steel Corp; 0.024-0.031 C, 0.43-0.48 Ti, and 0.06 others; recrystallized for 35 min at 1866 K; measured in 90 Ar + 10 H_2 at 78 cm Hg.
2	304	1960	1058-1611	45°	±5	Above specimen and conditions.
3	304	1960	1065-1600	60°	±5	Above specimen and conditions.

DATA TABLE NO. 287 ANGULAR TOTAL EMITTANCE OF [MOLYBDENUM + TITANIUM] ALLOYS

[Temperature, T, K; Emittance, ∈]

T	∈
CURVE 1	
1030	0.117
1588	0.134
CURVE 2	
1058	0.134
1611	0.126
CURVE 3	
1065	0.122
1600	0.139

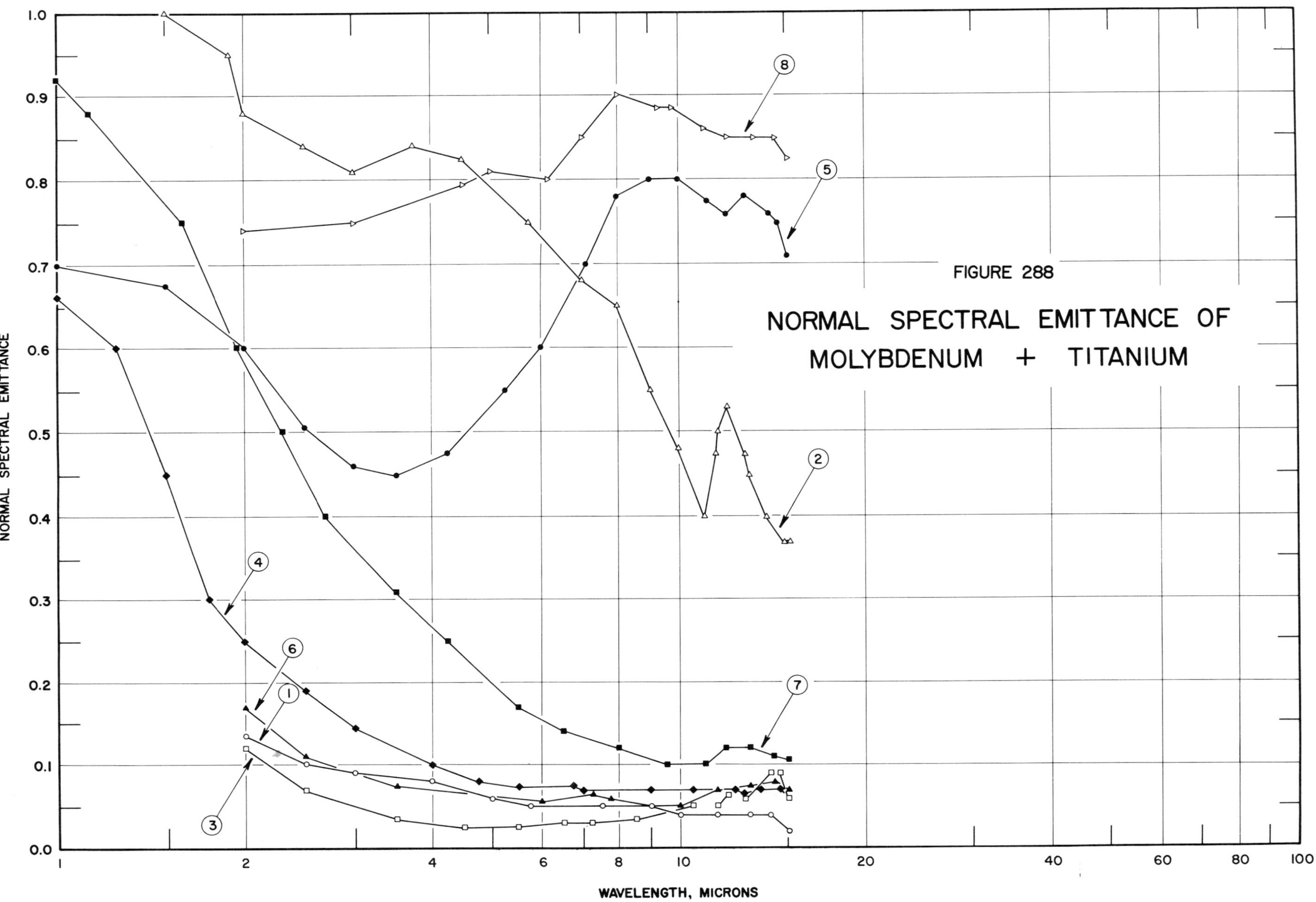
FIGURE 288
NORMAL SPECTRAL EMITTANCE OF
MOLYBDENUM + TITANIUM
NORMAL SPECTRAL EMITTANCE
WAVELENGTH, MICRONS
1.0
0.9
0.8
0.7
0.6
0.5
0.4
0.3
0.2
0.1
0.0
1
2
4
6
8
10
20
40
60
80
100
1
2
3
4
5
6
7
8

SPECIFICATION TABLE NO. 288 NORMAL SPECTRAL EMITTANCE OF [MOLYBDENUM + TITANIUM] ALLOYS

Curve No.	Ref. No.	Year	Temperature K	Wavelength Range, μ	Geometry θ'	Reported Error, %	Composition (weight percent), Specifications and Remarks
1	86	1961	523	2.00-15.00	$\sim 0°$	± 5	99.5 Mo, 0.5 Ti; as received; data extracted from smooth curve.
2	86	1961	773	1.00-15.00	$\sim 0°$	± 5	Different sample, same as curve 1 specimen and conditions.
3	86	1961	523	2.00-15.00	$\sim 0°$	± 5	Different sample, same as curve 1 specimen and conditions except heated in argon at 1367 K for 30 min.
4	86	1961	773	1.00-15.00	$\sim 0°$	± 5	Different sample, same as curve 3 specimen and conditions.
5	86	1961	1023	1.00-15.00	$\sim 0°$	± 5	Different sample, same as curve 3 specimen and conditions.
6	86	1961	523	2.00-15.00	$\sim 0°$	± 5	Different sample, same as curve 1 specimen and conditions except heated in vacuum (2.2×10^{-4} mm Hg) at 1367 K for 30 min.
7	86	1961	773	1.00-15.00	$\sim 0°$	± 5	Different sample, same as curve 6 specimen and conditions.
8	86	1961	1023	2.00-15.00	$\sim 0°$	± 5	Different sample, same as curve 6 specimen and conditions.

DATA TABLE NO. 288 NORMAL SPECTRAL EMITTANCE OF [MOLYBDENUM + TITANIUM] ALLOYS

[Wavelength, λ,μ; Emittance, ∈; Temperature, T, K]

λ	∈
CURVE 1 T = 523	
2.00	0.135
2.50	0.100
3.00	0.090
4.00	0.080
5.00	0.060
5.75	0.050
7.50	0.050
9.00	0.050
10.00	0.040
11.50	0.040
13.00	0.040
14.00	0.040
15.00	0.020
CURVE 2 T = 773	
1.00	1.000
1.50	1.000
1.90	0.950
2.00	0.880
2.50	0.840
3.00	0.810
3.75	0.840
4.50	0.825
5.75	0.750
7.00	0.680
8.00	0.650
9.00	0.550
10.00	0.480
11.00	0.400
11.50	0.475
11.55	0.500
12.00	0.530
12.75	0.475
13.00	0.450
13.80	0.400
14.75	0.370
15.00	0.370
CURVE 3 T = 523	
2.00	0.120
2.50	0.070
3.50	0.035
4.50	0.025
5.50	0.025
6.50	0.030
7.20	0.030
8.50	0.035
10.50	0.050
11.50	0.050
12.00	0.065
12.80	0.060
14.00	0.090
14.50	0.090
15.00	0.060
CURVE 4 T = 773	
1.00	0.660
1.25	0.600
1.50	0.450
1.75	0.300
2.00	0.250
2.50	0.190
3.00	0.145
4.00	0.100
4.75	0.080
5.50	0.075
6.75	0.075
7.00	0.070
9.00	0.070
10.50	0.070
12.25	0.070
12.60	0.065
13.50	0.070
14.50	0.070
15.00	0.060*
CURVE 5 T = 1023	
1.00	0.700
CURVE 5 (cont.)	
1.50	0.675
2.00	0.600
2.50	0.505
3.00	0.460
3.50	0.450
4.25	0.475
5.25	0.550
6.00	0.600
7.10	0.700
8.00	0.780
9.00	0.800
10.00	0.800
11.25	0.775
12.00	0.760
12.80	0.780
14.00	0.760
14.50	0.750
15.00	0.710
CURVE 6 T = 523	
2.00	0.170
2.50	0.110
3.50	0.075
5.00	0.060*
6.00	0.055
7.25	0.065
7.75	0.050
9.00	0.050*
10.00	0.050
11.50	0.070
13.00	0.075
14.25	0.080
15.00	0.070
CURVE 7 T = 773	
1.00	0.920
1.25	0.880
1.60	0.750
1.95	0.600
2.30	0.500
CURVE 7 (cont.)	
2.70	0.400
3.50	0.310
4.25	0.250
5.50	0.170
6.50	0.140
8.00	0.120
9.50	0.100
11.00	0.100
11.80	0.120
13.00	0.120
14.10	0.110
15.00	0.105
CURVE 8 T = 1023	
2.00	0.740
3.00	0.750
4.50	0.795
5.00	0.810
6.15	0.800
7.00	0.850
8.00	0.900
9.25	0.885
9.75	0.885
11.00	0.860
12.00	0.850
13.25	0.850
14.35	0.850
15.00	0.825

* Not shown on plot

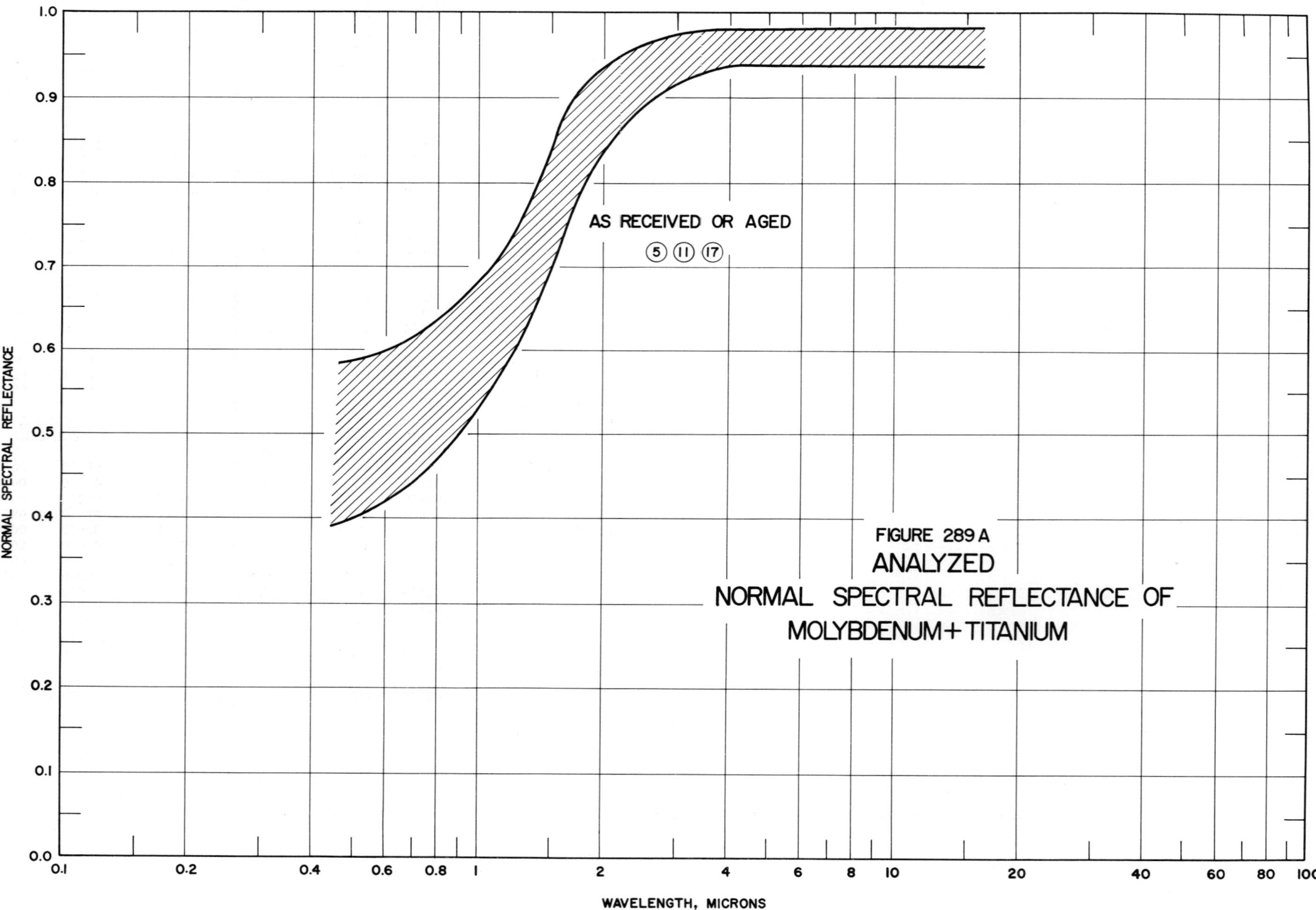

FIGURE 289 A
ANALYZED
NORMAL SPECTRAL REFLECTANCE OF
MOLYBDENUM + TITANIUM

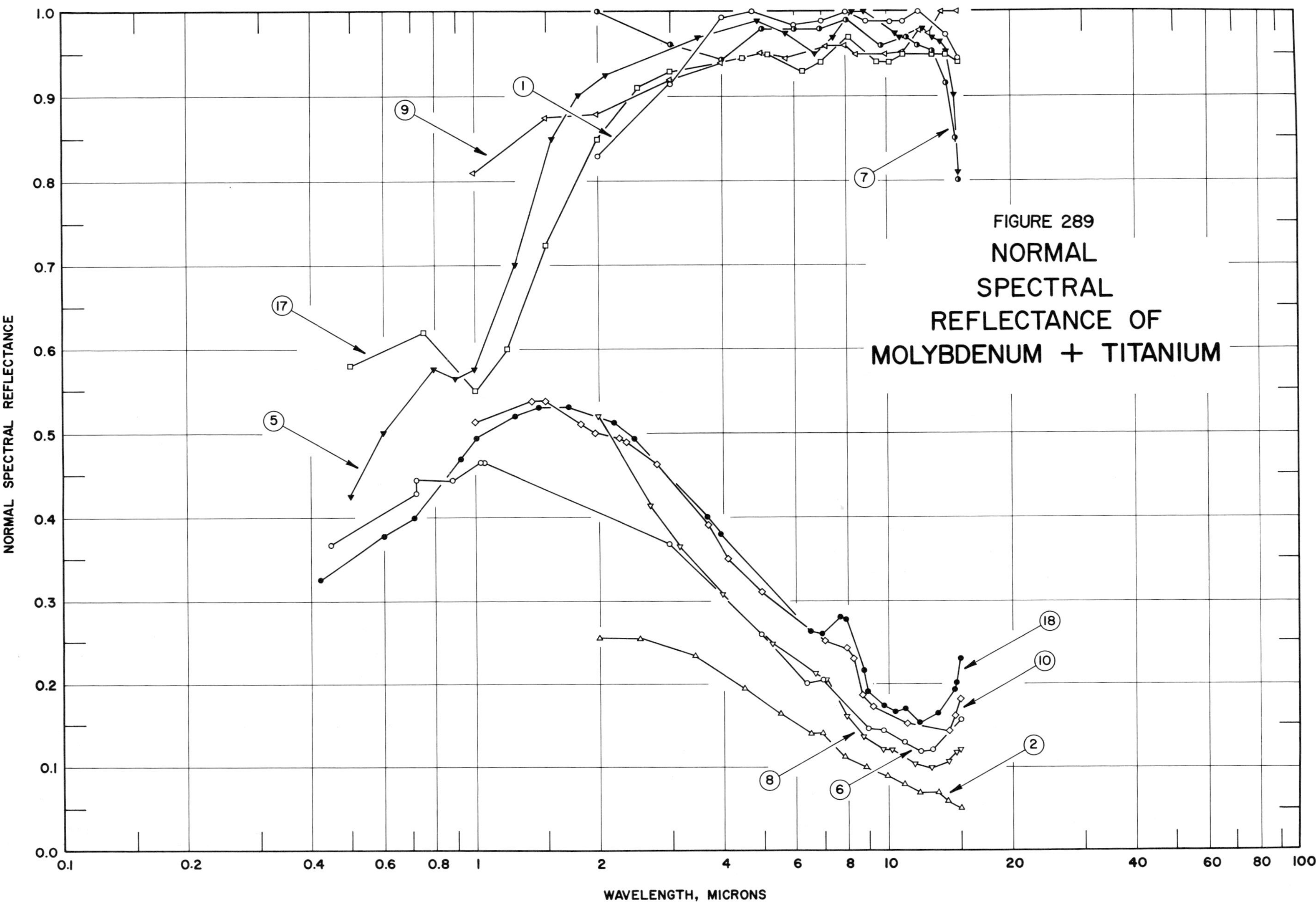
FIGURE 289
NORMAL
SPECTRAL
REFLECTANCE OF
MOLYBDENUM + TITANIUM
NORMAL SPECTRAL REFLECTANCE
WAVELENGTH, MICRONS
0.0
0.1
0.2
0.3
0.4
0.5
0.6
0.7
0.8
0.9
1.0
0.1
0.2
0.4
0.6
0.8
1
2
4
6
8
10
20
40
60
80
100
1
2
5
6
7
8
9
10
17
18

SPECIFICATION TABLE NO. 289 NORMAL SPECTRAL REFLECTANCE OF [MOLYBDENUM + TITANIUM] ALLOYS

Curve No.	Ref. No.	Year	Temperature K	Wavelength Range, μ	Geometry θ	θ'	ω'	Reported Error, %	Composition (weight percent), Specifications and Remarks
1	86	1961	~ 322	2.00-15.00	~0°		2π	< 2	99.5 Mo, 0.5 Ti; as received; data extracted from smooth curve; hohlraum at 523 K.
2	86	1961	~ 322	2.00-14.99	~0°		2π	< 2	Above specimen and conditions; diffuse component only.
3	86	1961	~ 322	1.00-15.00	~0°		2π	<2	Different sample, same as curve 1 specimen and conditions except hohlraum at 773 K.
4	86	1961	~ 322	1.00-15.02	~0°		2π	< 2	Above specimen and conditions; diffuse component only.
5	86	1961	~ 322	0.50-15.00	~0°		2π	< 2	Different sample, same as curve 1 specimen and conditions except hohlraum at 1273 K.
6	86	1961	~ 322	0.44-15.02	~0°		2π	<2	Above specimen and conditions; diffuse component only.
7	86	1961	~ 322	2.00-15.00	~0°		2π	< 2	Different sample, same as curve 1 specimen and conditions except heated in argon at 1366 K for 30 min.
8	86	1961	~ 322	2.00-15.00	~0°		2π	< 2	Above specimen and conditions; diffuse component only.
9	86	1961	~ 322	1.00-15.00	~0°		2π	< 2	Different sample, same as curve 7 specimen and conditions except hohlraum at 773 K.
10	86	1961	~ 322	1.00-15.00	~0°		2π	< 2	Above specimen and conditions; diffuse component only.
11	86	1961	~ 322	0.50-15.00	~0°		2π	< 2	Different sample, same as curve 7 specimen and conditions except hohlraum at 1273 K.
12	86	1961	~ 322	0.47-15.00	~0°		2π	< 2	Above specimen and conditions; diffuse component only.
13	86	1961	~ 322	2.00-15.00	~0°		2π	< 2	Different sample, same as curve 1 specimen and conditions except heated in vacuum (2.2×10^{-4} mm Hg) at 1366 K for 30 min.
14	86	1961	~ 322	1.97-15.02	~0°		2π	< 2	Above specimen and conditions; diffuse component only.
15	86	1961	~ 322	1.00-14.00	~0°		2π	< 2	Different sample, same as curve 13 specimen and conditions except hohlraum at 773 K.
16	86	1961	~ 322	1.00-14.98	~0°		2π	< 2	Above specimen and conditions; diffuse component only.
17	86	1961	~ 322	0.50-15.00	~0°		2π	< 2	Different sample, same as curve 13 specimen and conditions except hohlraum at 1273 K.
18	86	1961	~ 322	0.42-15.00	~0°		2π	< 2	Above specimen and conditions; diffuse component only.

DATA TABLE NO. 289 NORMAL SPECTRAL REFLECTANCE OF [MOLYBDENUM + TITANIUM] ALLOYS

[Wavelength, λ, μ; Reflectance, ρ; Temperature, T, K]

λ	ρ
CURVE 1, T = ~322	
2.00	0.830
3.00	0.915
4.00	0.995
4.75	1.000
6.00	0.985
7.00	0.990
8.00	1.000
9.00	0.990
10.25	0.990
11.00	0.990
12.00	1.000
14.00	0.975
15.00	0.945
CURVE 2, T = ~322	
2.00	0.256
2.49	0.254
3.40	0.232
4.48	0.194
5.46	0.165
6.47	0.140
6.97	0.140
7.83	0.113
8.96	0.100
9.98	0.089
10.98	0.079
11.97	0.070
13.23	0.070
13.99	0.061
14.99	0.051
CURVE 3*, T = ~322	
1.00	0.800
1.35	0.850
1.75	0.875
3.10	0.900
4.00	0.940
4.85	0.930
5.85	0.930
7.00	0.925
CURVE 3 (cont.)*	
7.90	0.970
8.80	0.965
10.00	0.970
10.60	0.960
12.00	0.960
13.75	0.950
14.25	0.925
14.55	0.900
15.00	0.800
CURVE 4*, T = ~322	
1.00	0.449
1.24	0.475
1.48	0.484
1.70	0.474
1.85	0.420
1.99	0.401
2.99	0.345
4.98	0.252
6.24	0.195
6.98	0.186
7.48	0.171
8.98	0.142
10.50	0.110
11.32	0.102
11.83	0.100
12.51	0.091
13.00	0.086
14.60	0.086
15.02	0.091
CURVE 5, T = ~322	
0.50	0.425
0.60	0.500
0.80	0.575
0.90	0.565
1.00	0.575
1.25	0.700
1.55	0.850
1.80	0.900
2.10	0.925
CURVE 5 (cont.)	
3.50	0.970
4.90	0.990
5.75	0.975
6.75	0.950
7.50	0.970
8.25	1.000
8.90	1.000
10.50	0.975
10.80	0.970
12.15	0.980
13.00	0.970
13.50	0.965
14.00	0.950
14.65	0.900
15.00	0.810
CURVE 6, T = ~322	
0.44	0.368
0.72	0.429
0.72	0.443
0.88	0.443
1.34	0.464
1.52	0.464
2.95	0.368
4.95	0.259
6.31	0.200
6.96	0.205
8.97	0.146
9.73	0.144
10.97	0.130
11.98	0.119
12.76	0.121
15.02	0.157
CURVE 7, T = ~322	
2.00	1.000
3.00	0.960
4.00	0.940
5.00	0.980
6.00	0.980
6.90	0.980
CURVE 7 (cont.)	
8.00	0.990
9.75	0.960
11.20	0.970
11.90	0.960
13.00	0.950
14.00	0.915
14.75	0.850
15.00	0.800
CURVE 8, T = ~322	
2.00	0.520
2.67	0.414
3.12	0.366
4.00	0.308
5.25	0.248
6.67	0.212
7.04	0.205
7.99	0.160
8.76	0.136
9.73	0.121
10.22	0.121
11.66	0.103
12.75	0.099
14.01	0.108
14.76	0.118
15.00	0.120
CURVE 9, T = ~322	
1.00	0.810
1.50	0.875
2.00	0.880
3.00	0.920
4.00	0.940
5.00	0.950
5.75	0.945
7.20	0.960
8.00	0.960
8.50	0.950
10.00	0.950
11.00	0.950
12.15	0.980
CURVE 9 (cont.)	
12.70	0.975
13.80	1.000
15.00	1.000
CURVE 10, T = ~322	
1.00	0.512
1.37	0.538
1.48	0.538
1.81	0.511
1.96	0.501
2.24	0.493
2.33	0.490
2.77	0.464
3.66	0.391
4.08	0.351
4.98	0.311
7.01	0.251
7.99	0.242
8.22	0.230
8.66	0.187
9.19	0.172
11.02	0.151
14.01	0.142
14.69	0.161
15.00	0.181
CURVE 11*, T = ~322	
0.50	0.400
0.75	0.450
1.00	0.570
1.25	0.650
1.50	0.750
1.75	0.850
2.15	0.925
2.75	0.975
3.10	0.980
4.25	0.980
5.00	0.990
6.00	0.975
6.75	0.970
7.75	0.980
CURVE 11 (cont.)*	
8.75	0.970
10.00	0.970
11.50	0.970
13.00	0.970
14.00	0.980
14.50	0.980
15.00	0.965
CURVE 12*, T = ~322	
0.47	0.361
0.97	0.418
1.22	0.500
1.47	0.527
1.70	0.539
1.96	0.539
2.40	0.524
3.29	0.455
4.66	0.351
4.98	0.331
5.70	0.303
7.25	0.255
7.58	0.249
8.00	0.240
8.58	0.207
8.98	0.185
9.55	0.180
10.00	0.171
11.25	0.158
12.88	0.149
13.51	0.160
14.00	0.179
14.46	0.208
14.84	0.250
15.00	0.290
CURVE 13*, T = ~322	
2.00	1.000
2.50	0.925
3.00	0.890
3.50	0.885
4.50	0.910
CURVE 13 (cont.)*	
5.50	0.940
6.50	0.950
7.50	0.960
8.00	0.970
9.00	0.950
10.00	0.960
11.00	0.950
12.00	0.960
13.00	0.940
13.75	0.930
14.65	0.900
14.80	0.850
15.00	0.800
CURVE 14*, T = ~322	
1.97	0.533
2.21	0.487
2.60	0.438
3.24	0.388
3.72	0.362
4.21	0.339
6.98	0.248
8.00	0.207
9.00	0.177
11.00	0.143
12.01	0.131
13.28	0.133
14.04	0.137
14.69	0.143
15.02	0.151
CURVE 15*, T = ~322	
1.00	0.860
1.60	0.840
2.50	0.875
3.25	0.925
4.50	0.955
5.00	0.960
5.75	0.950
7.25	0.960
8.20	0.970
CURVE 15 (cont.)*	
9.00	0.950
10.50	0.965
12.00	0.985
13.00	1.000
14.00	1.000
CURVE 16*, T = ~322	
1.00	0.548
1.47	0.529
1.93	0.498
2.32	0.486
2.73	0.479
2.95	0.470
3.71	0.416
4.19	0.388
4.95	0.349
5.97	0.320
6.94	0.290
7.97	0.255
8.97	0.207
10.95	0.171
11.65	0.151
13.38	0.151
13.98	0.160
14.26	0.166
14.64	0.180
14.98	0.200
CURVE 17, T = ~322	
0.50	0.580
0.75	0.620
1.00	0.550
1.20	0.600
1.50	0.725
2.00	0.850
2.50	0.910
3.00	0.930
4.50	0.945
5.20	0.950
6.25	0.930
7.00	0.940

* Not shown on plot

DATA TABLE NO. 289 (continued)

λ	ρ
CURVE 17 (cont.)	
8.15	0.970
9.50	0.940
10.25	0.940
11.00	0.950
13.00	0.950
14.00	0.950
15.00	0.940
CURVE 18 T = ~322	
0.42	0.327
0.60	0.378
0.71	0.400
0.92	0.469
1.01	0.495
1.25	0.520
1.43	0.530
1.69	0.530
2.18	0.512
2.43	0.493
3.67	0.400
3.94	0.380
6.44	0.262
6.94	0.261
7.68	0.280
7.97	0.278
8.57	0.217
8.96	0.190
9.71	0.173
10.23	0.167
10.98	0.170
11.87	0.152
13.27	0.163
14.48	0.192
14.67	0.201
15.00	0.230

SPECIFICATION TABLE NO. 290 HEMISPHERICAL TOTAL EMITTANCE OF [MOLYBDENUM + TUNGSTEN] ALLOYS

Curve No.	Ref. No.	Year	Temperature Range, K	Reported Error, %	Composition (weight percent), Specifications and Remarks
1	58	1961	973-1473	±2.5	50 Mo, 50W; measured in vacuum ($<5 \times 10^{-6}$ mm Hg); data extracted from smooth curve.

DATA TABLE NO. 290 HEMISPHERICAL TOTAL EMITTANCE OF [MOLYBDENUM + TUNGSTEN] ALLOYS

[Temperature, T, K; Emittance, ∈]

T	∈
CURVE 1*	
973	0.315
1073	0.350
1173	0.380
1273	0.400
1373	0.430
1473	0.460

* Not shown on plot

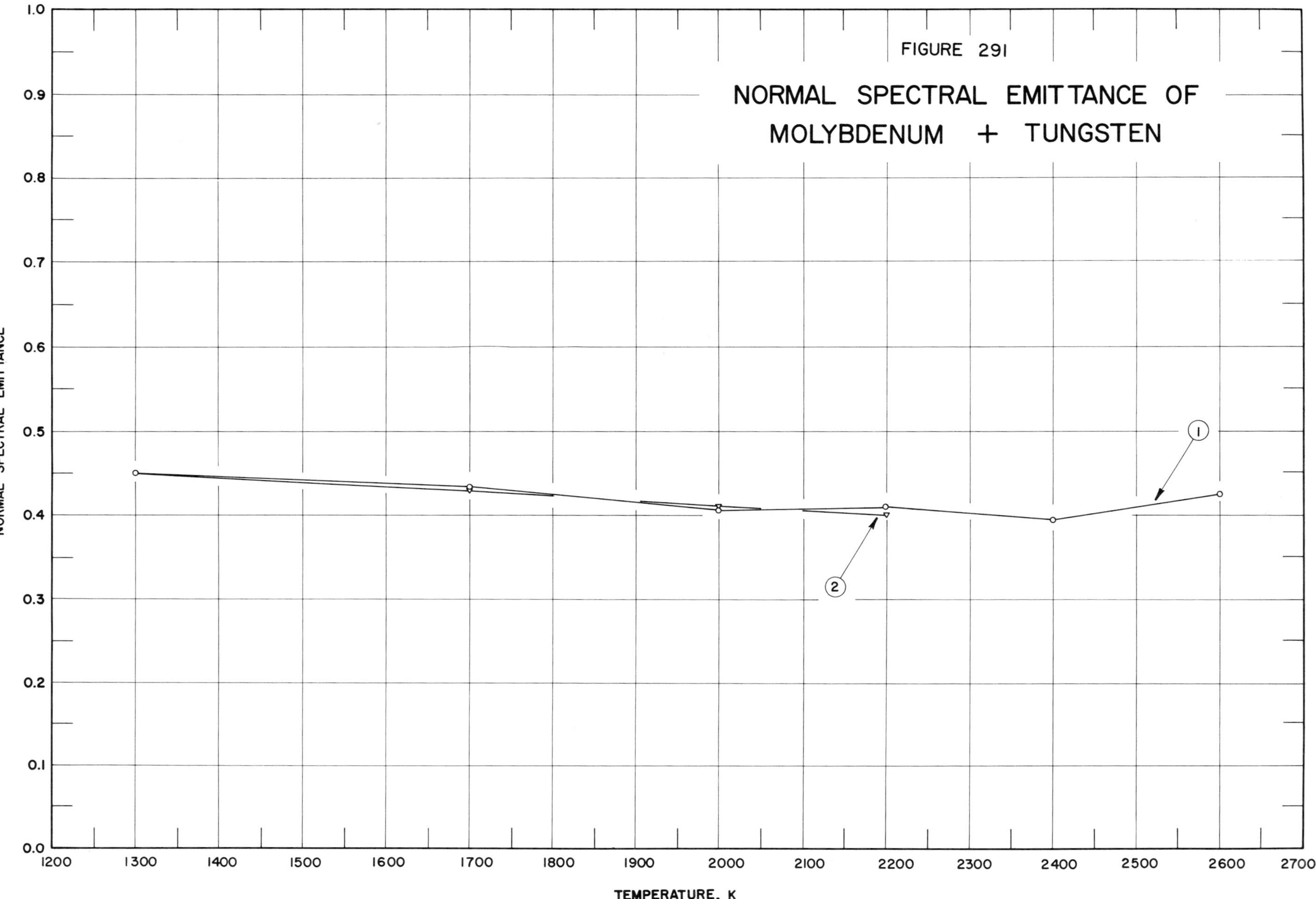
FIGURE 291
NORMAL SPECTRAL EMITTANCE OF
MOLYBDENUM + TUNGSTEN
NORMAL SPECTRAL EMITTANCE
TEMPERATURE, K
1.0
0.9
0.8
0.7
0.6
0.5
0.4
0.3
0.2
0.1
0.0
1200
1300
1400
1500
1600
1700
1800
1900
2000
2100
2200
2300
2400
2500
2600
2700
1
2

SPECIFICATION TABLE NO. 291 NORMAL SPECTRAL EMITTANCE OF [MOLYBDENUM + TUNGSTEN] ALLOYS

Curve No.	Ref. No.	Year	Wavelength μ	Temperature Range, K	Geometry θ'	Reported Error, %	Composition (weight percent), Specifications and Remarks
1	77	1936	0.66	1300-2600	~0°	±2	25.0 W, Mo balance; extruded from paste; sintered in hydrogen; polished electrolytically in KOH bath; polished with 00, 000, 0000 polishing papers; measured in vacuum (10^{-5} mm Hg).
2	177	1933	0.660	1300-2200	~0°	±2	75 Mo, 25 W; sintered at 1873 K for 1 hr; electrolytically polished in KOH and then by using 00, 000, 0000 polishing papers; measured in vacuum.

DATA TABLE NO. 291 NORMAL SPECTRAL EMITTANCE OF [MOLYBDENUM + TUNGSTEN] ALLOYS

[Temperature, T, K; Emittance, ∈; Wavelength, λ, μ]

T	∈
CURVE 1	
$\lambda = 0.66$	
1300	0.450
1700	0.435
2000	0.405
2200	0.410
2400	0.395
2600	0.425
CURVE 2	
$\lambda = 0.660$	
1300	0.450*
1700	0.430
2000	0.410
2200	0.400

* Not shown on plot

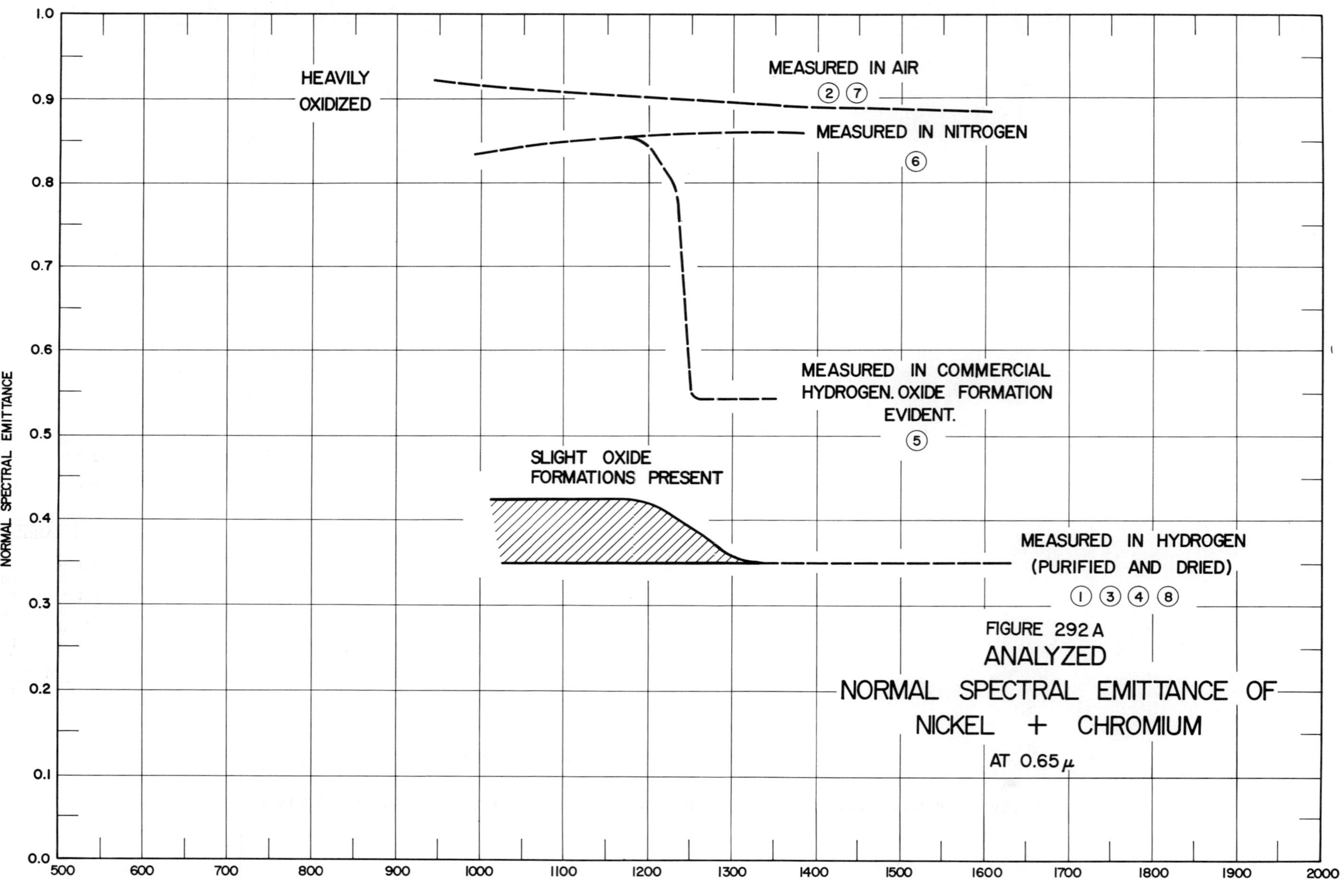

FIGURE 292 A
ANALYZED
NORMAL SPECTRAL EMITTANCE OF
NICKEL + CHROMIUM
AT 0.65 μ

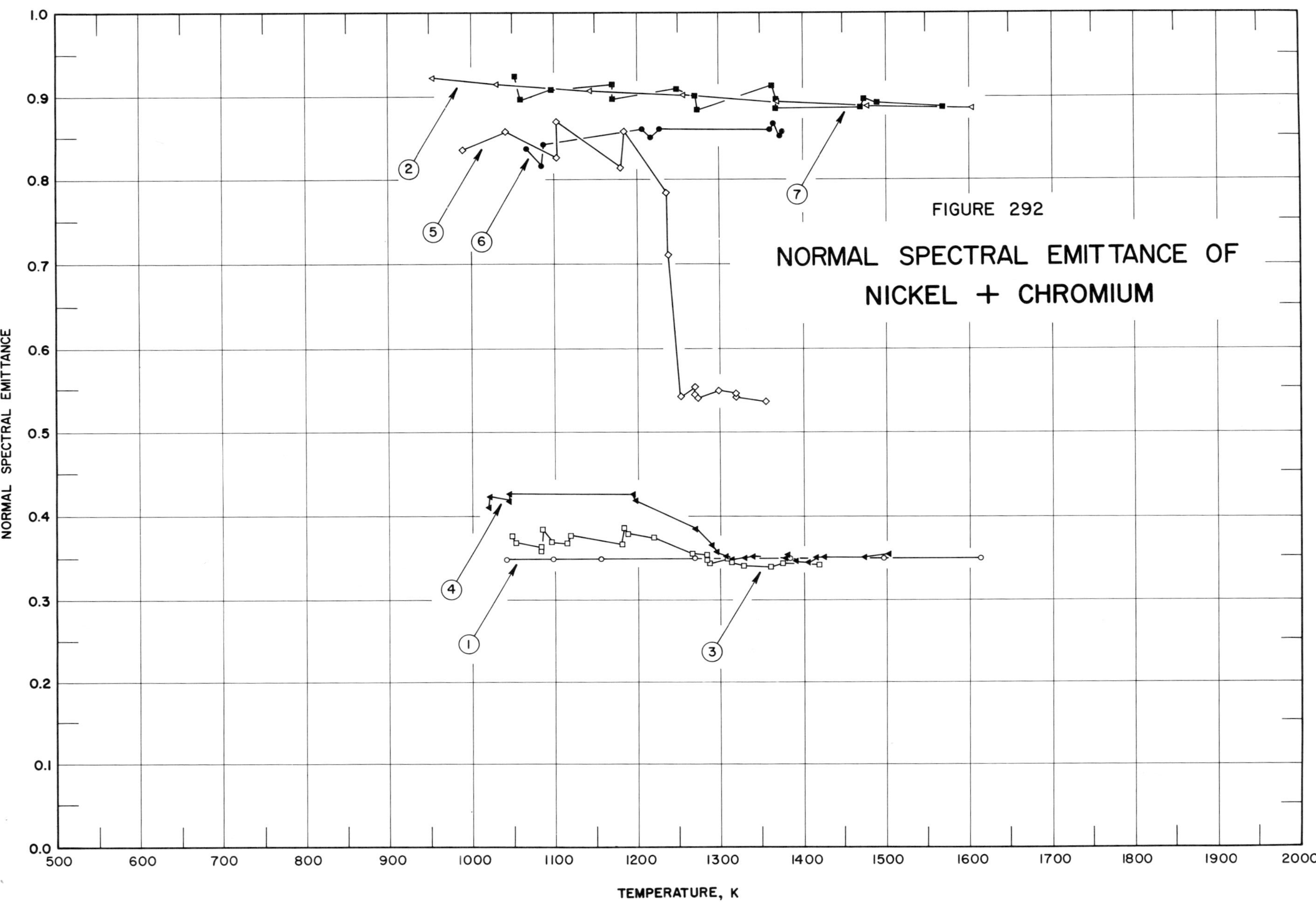
FIGURE 292
NORMAL SPECTRAL EMITTANCE OF
NICKEL + CHROMIUM
NORMAL SPECTRAL EMITTANCE
TEMPERATURE, K
1.0
0.9
0.8
0.7
0.6
0.5
0.4
0.3
0.2
0.1
0.0
500
600
700
800
900
1000
1100
1200
1300
1400
1500
1600
1700
1800
1900
2000
1
2
3
4
5
6
7

SPECIFICATION TABLE NO. 292 NORMAL SPECTRAL EMITTANCE OF [NICKEL + CHROMIUM] ALLOYS

Curve No.	Ref. No.	Year	Wavelength μ	Temperature Range, K	Geometry θ'	Reported Error, %	Composition (weight percent), Specifications and Remarks
1	163	1939	0.65	1041-1612	$\sim 0^\circ$		80 Ni, 20 Cr; polished; heated at 1478 K for several hrs in hydrogen; measured in hydrogen; data extracted from smooth curve.
2	163	1939	0.65	954-1604	$\sim 0^\circ$		80 Ni, 20 Cr; polished; heated at 1478 K for several hrs in air; measured in air; data extracted from smooth curve.
3	163	1939	0.65	1048-1418	$\sim 0^\circ$		80 Ni, 20 Cr; polished with rouge paper; heated at 1478 K in hydrogen; traces of oxides present; measured in purified and dried hydrogen; data extracted from smooth curve.
4	163	1939	0.65	970-1502	$\sim 0^\circ$		80 Ni, 20 Cr; polished with rouge paper; heated at 1478 K in hydrogen; traces of oxides present; measured in purified and dried hydrogen; data extracted from smooth curve.
5	163	1939	0.65	989-1355	$\sim 0^\circ$		80 Ni, 20 Cr; polished with rouge paper; heated at 1478 K; oxides present; measured in commercial hydrogen; data extracted from smooth curve.
6	163	1939	0.65	1067-1374	$\sim 0^\circ$		80 Ni, 20 Cr; polished with rouge paper; heated at 1478 K; oxides present; measured in commercial nitrogen; data extracted from smooth curve.
7	163	1939	0.65	1052-1570	$\sim 0^\circ$		80 Ni, 20 Cr; polished with rouge paper; heated in air at 1478 K for several hrs; measured in air; data extracted from smooth curve.
8	163	1939	0.65	1087-1590	$\sim 0^\circ$		80 Ni, 20 Cr; polished with rouge paper; heated at 1478 K in purified hydrogen; measured in purified and dried hydrogen.

DATA TABLE NO. 292 NORMAL SPECTRAL EMITTANCE OF [NICKEL + CHROMIUM] ALLOYS

[Temperature, T, K; Emittance, ∈; Wavelength, λ, μ]

T	∈
CURVE 1 $\lambda = 0.65$	
1041	0.35
1098	0.35
1154	0.35
1268	0.35
1382	0.35
1497	0.35
1612	0.35
CURVE 2 $\lambda = 0.65$	
954	0.922
1032	0.915
1145	0.907
1257	0.901
1370	0.894
1479	0.888
1604	0.886
CURVE 3 $\lambda = 0.65$	
1048	0.377
1052	0.369
1082	0.363
1083	0.359
1084	0.384
1087	0.370
1114	0.369
1119	0.378
1180	0.367
1182	0.380
1188	0.380
1219	0.374
1265	0.355
1282	0.353
1282	0.348
1286	0.344
1308	0.349
1313	0.346
1328	0.341
1360	0.340
1375	0.344
1418	0.342

T	∈
CURVE 4 $\lambda = 0.65$	
970	0.411
971	0.422
1044	0.417
1044	0.425
1195	0.425
1198	0.418
1269	0.384
1289	0.365
1295	0.357
1308	0.350
1314	0.347
1330	0.350
1340	0.352
1380	0.350
1382	0.353
1391	0.347
1405	0.346
1405	0.351
1424	0.351
1473	0.351
1502	0.353
CURVE 5 $\lambda = 0.65$	
989	0.836
1042	0.808
1102	0.827
1102	0.869
1180	0.812
1184	0.858
1235	0.784
1238	0.711
1254	0.542
1269	0.553
1269	0.545
1272	0.540
1299	0.549
1319	0.547
1319	0.544
1355	0.537

T	∈
CURVE 6 $\lambda = 0.65$	
1067	0.834
1084	0.817
1088	0.841
1205	0.860
1218	0.850
1227	0.860
1361	0.859
1365	0.867
1372	0.851
1374	0.858
CURVE 7 $\lambda = 0.65$	
1052	0.923
1059	0.898
1097	0.909
1170	0.913
1170	0.898
1248	0.909
1270	0.900
1272	0.882
1363	0.912
1369	0.897
1369	0.886
1470	0.886
1474	0.897
1490	0.891
1570	0.887
CURVE 8* $\lambda = 0.65$	
1087	0.350
1087	0.363
1087	0.360
1149	0.365
1149	0.365
1210	0.365
1273	0.350
1273	0.350
1273	0.348
1335	0.345
1335	0.350

T	∈
CURVE 8 (cont.)*	
1335	0.348
1399	0.340
1399	0.350
1399	0.351
1399	0.340
1462	0.345
1462	0.352
1462	0.348
1462	0.345
1462	0.350
1462	0.340
1526	0.350
1526	0.350
1590	0.352
1590	0.351

* Not shown on plot

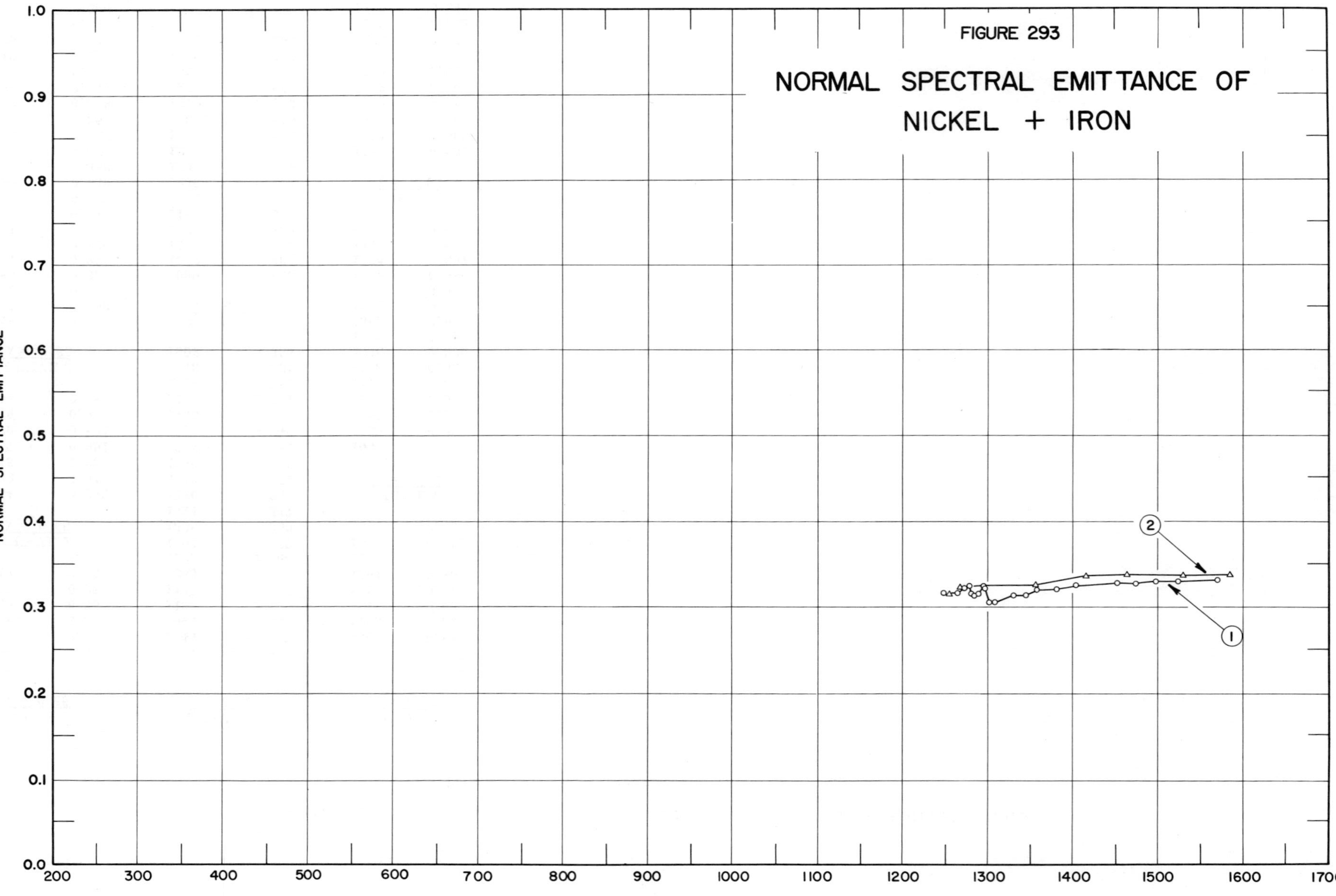
FIGURE 293
NORMAL SPECTRAL EMITTANCE OF
NICKEL + IRON
NORMAL SPECTRAL EMITTANCE
TEMPERATURE, K
1.0
0.9
0.8
0.7
0.6
0.5
0.4
0.3
0.2
0.1
0.0
200
300
400
500
600
700
800
900
1000
1100
1200
1300
1400
1500
1600
1700
2
1

SPECIFICATION TABLE NO. 293 NORMAL SPECTRAL EMITTANCE OF [NICKEL + IRON] ALLOYS

Curve No.	Ref. No.	Year	Wavelength μ	Temperature Range, K	Geometry θ'	Reported Error, %	Composition (weight percent), Specifications and Remarks
1	176	1951	0.67	1249-1571	$\sim 0°$		50 Fe, 50 Ni; powders were mixed in desired proportions, compressed at 70,000 psi at 298 K, heated at 1373 K in flowing hydrogen; cold rolled then heated in hydrogen at 1273 K.
2	176	1951	0.67	1254-1584	$\sim 0°$		75 Ni, 25 Fe; specimen preparation same as curve 1.

DATA TABLE NO. 293 NORMAL SPECTRAL EMITTANCE OF [NICKEL + IRON] ALLOYS

[Temperature, T, K; Emittance, ∈; Wavelength, λ, μ]

T	∈
CURVE 1	
λ = 0.67	
1249	0.317
1266	0.317
1274	0.322
1279	0.325
1281	0.317
1285	0.313
1287	0.314*
1290	0.316
1296	0.325
1297	0.322
1303	0.306
1309	0.306
1331	0.314
1345	0.314
1358	0.320
1381	0.321
1404	0.327
1452	0.329
1473	0.328
1499	0.331
1524	0.331
1571	0.332
CURVE 2	
λ = 0.67	
1254	0.316
1269	0.323
1296	0.324*
1356	0.327
1416	0.338
1463	0.339
1530	0.338
1584	0.339

* Not shown on plot

deleted

SPECIFICATION TABLE NO. 294 NORMAL TOTAL EMITTANCE OF [NICKEL + SILVER] ALLOYS

Curve No.	Ref. No.	Year	Temperature Range, K	Geometry θ'	Reported Error, %	Composition (weight percent), Specifications and Remarks
1	15	1947	373	~0°		Polished.

deleted

DATA TABLE NO. 294 NORMAL TOTAL EMITTANCE OF [NICKEL + SILVER] ALLOYS

[Temperature, T, K; Emittance, ∈]

T	∈
CURVE 1*	
373	0.135

* Not shown on plot

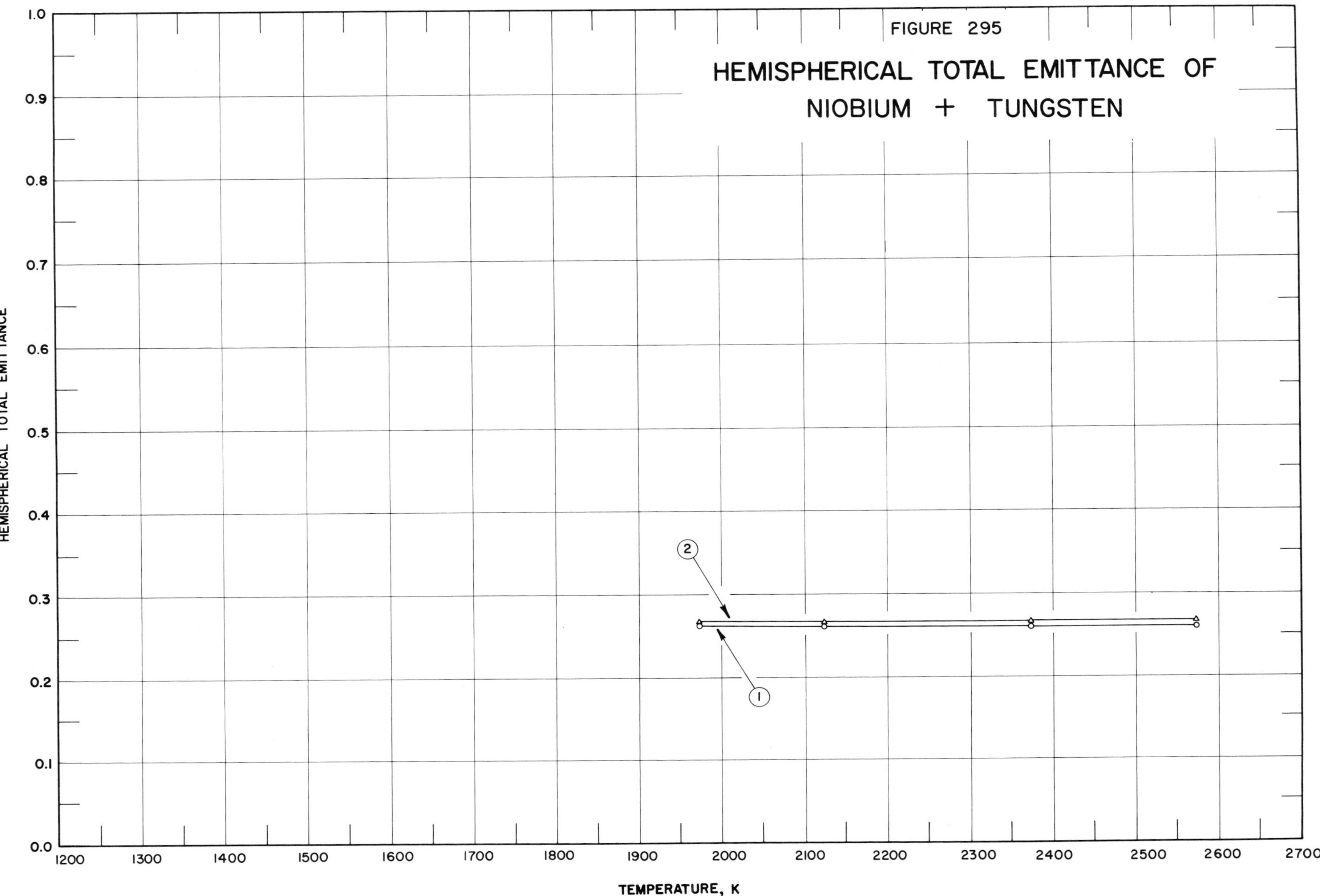

FIGURE 295
HEMISPHERICAL TOTAL EMITTANCE OF
NIOBIUM + TUNGSTEN

SPECIFICATION TABLE NO. 295 HEMISPHERICAL TOTAL EMITTANCE OF [NIOBIUM + TUNGSTEN] ALLOYS

Curve No.	Ref. No.	Year	Temperature Range, K	Reported Error, %	Composition (weight percent), Specifications and Remarks
1	84	1962	1973-2573		85 Nb, 15 W; polished with successively finer abrasive papers (numbers 1, 0, 00, 000, and 0000); measured in argon.
2	84	1962	1973-2573		90 Nb, 10 W; specimen preparation same as curve 1.

DATA TABLE NO. 295 HEMISPHERICAL TOTAL EMITTANCE OF [NIOBIUM + TUNGSTEN] ALLOYS

[Temperature, T, K; Emittance, ϵ]

T	ϵ
CURVE 1	
1973	0.262
2173	0.261
2373	0.260
2573	0.259
CURVE 2	
1973	0.267
2173	0.267
2373	0.267
2573	0.267

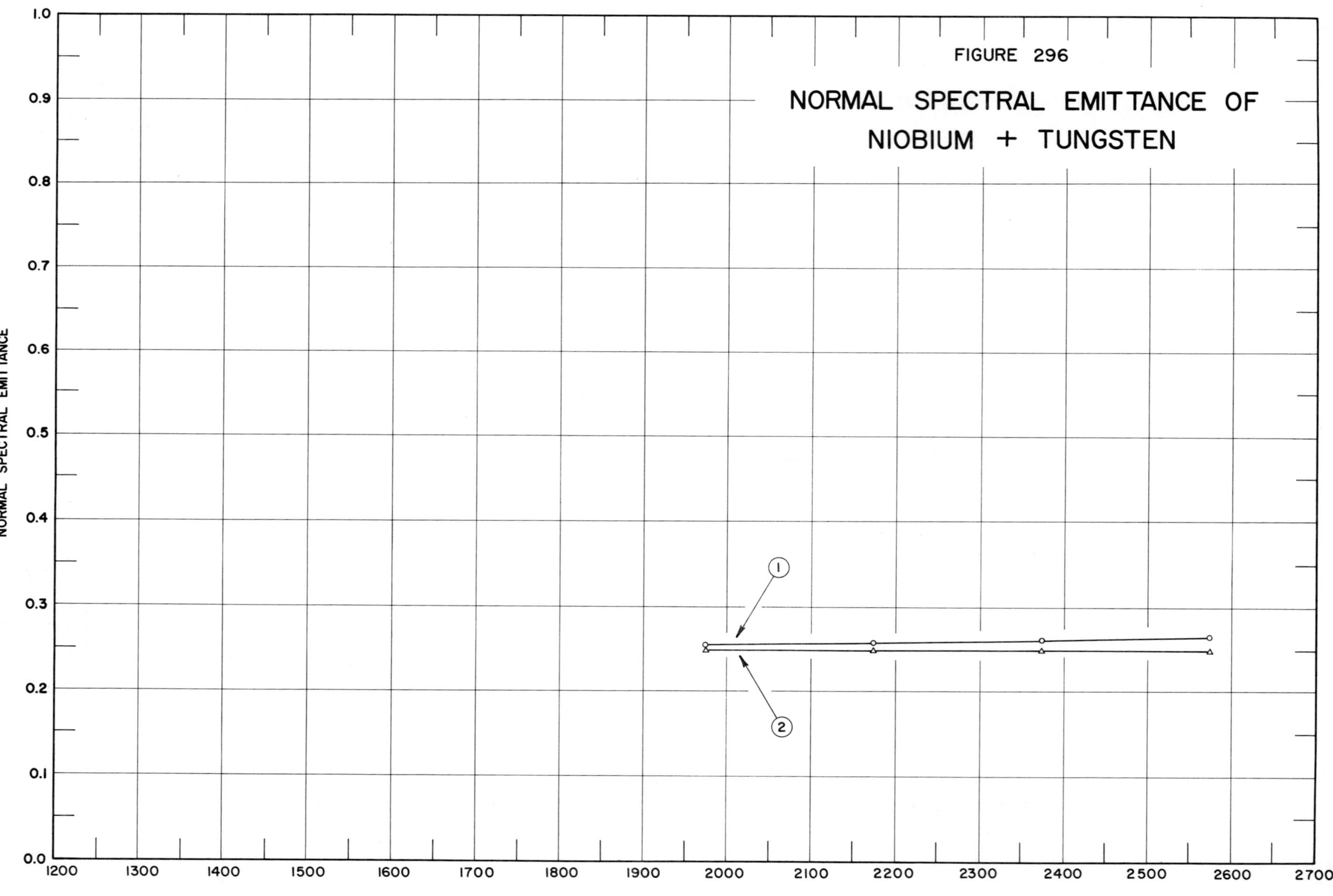
FIGURE 296
NORMAL SPECTRAL EMITTANCE OF
NIOBIUM + TUNGSTEN
1
2
NORMAL SPECTRAL EMITTANCE
1.0
0.9
0.8
0.7
0.6
0.5
0.4
0.3
0.2
0.1
0.0
1200
1300
1400
1500
1600
1700
1800
1900
2000
2100
2200
2300
2400
2500
2600
2700
TEMPERATURE, K

SPECIFICATION TABLE NO. 296 NORMAL SPECTRAL EMITTANCE OF [NIOBIUM + TUNGSTEN] ALLOYS

Curve No.	Ref. No.	Year	Wavelength μ	Temperature Range, K	Geometry θ'	Reported Error, %	Composition (weight percent), Specifications and Remarks
1	84	1962	0.65	1973-2573	$\sim 0^\circ$		85 Nb, 15W; polished with successively finer abrasive papers (numbers 1, 0, 00, 000, and 0000); measured in argon.
2	84	1962	0.65	1973-2573	$\sim 0^\circ$		90 Nb, 10 W; specimen preparation same as curve 1.

DATA TABLE NO. 296 NORMAL SPECTRAL EMITTANCE OF [NIOBIUM + TUNGSTEN] ALLOYS

[Temperature, T, K; Emittance, ∈; Wavelength, λ, μ]

T	∈
CURVE 1 λ = 0.65	
1973	0.255
2173	0.258
2373	0.261
2573	0.265
CURVE 2 λ = 0.65	
1973	0.250
2173	0.250
2373	0.250
2573	0.250

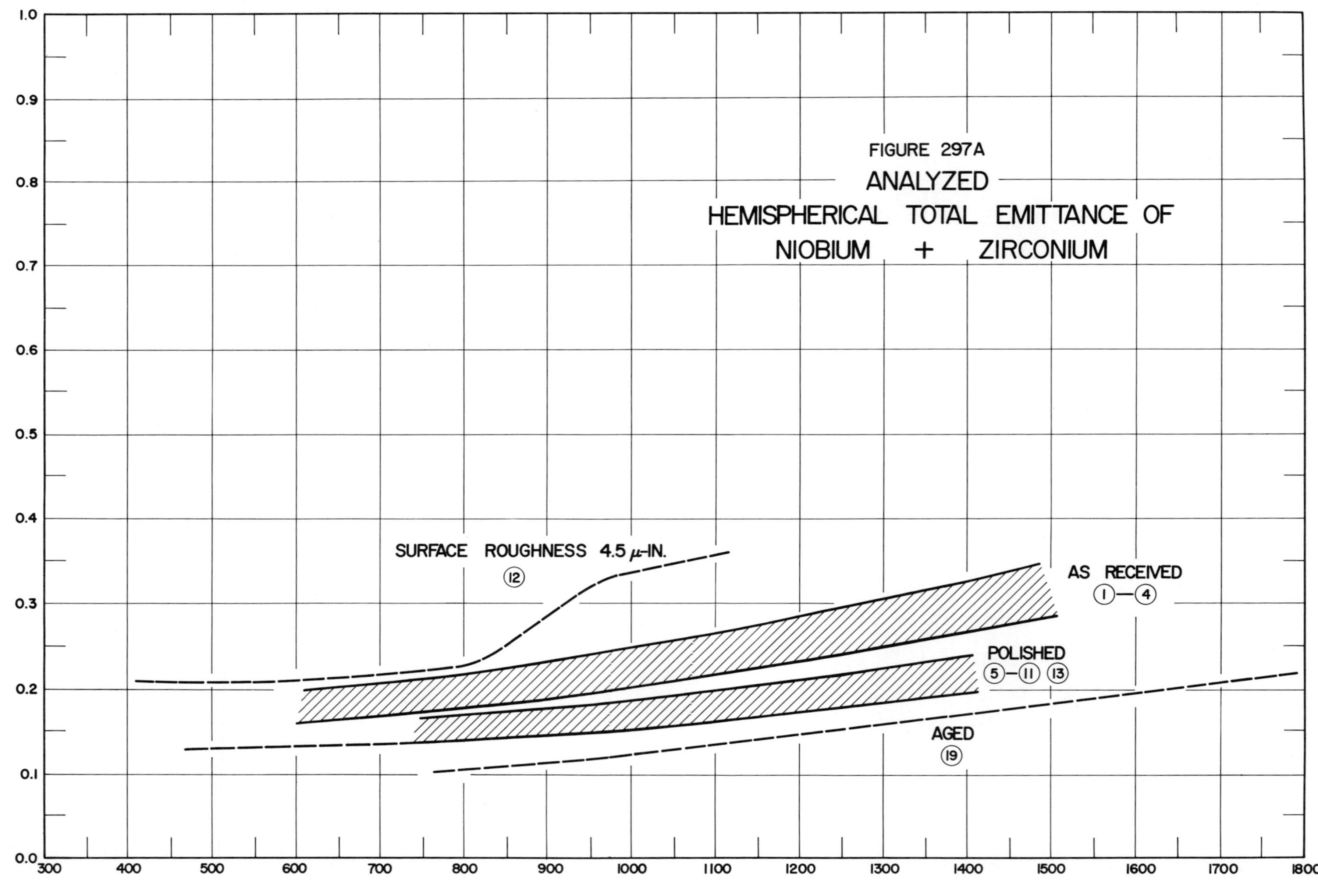

FIGURE 297A
ANALYZED
HEMISPHERICAL TOTAL EMITTANCE OF
NIOBIUM + ZIRCONIUM

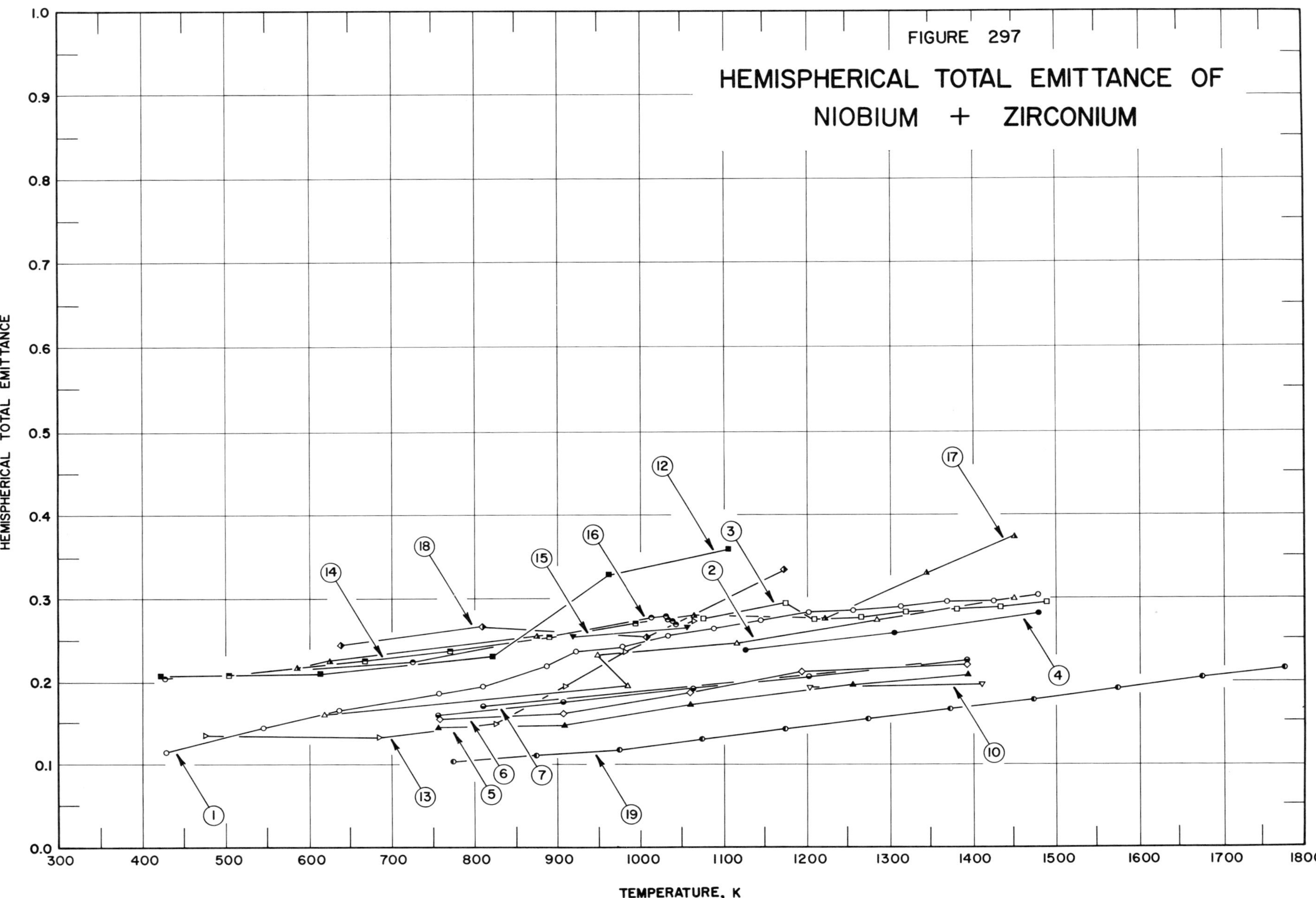
FIGURE 297
HEMISPHERICAL TOTAL EMITTANCE OF
NIOBIUM + ZIRCONIUM
HEMISPHERICAL TOTAL EMITTANCE
TEMPERATURE, K
0.0
0.1
0.2
0.3
0.4
0.5
0.6
0.7
0.8
0.9
1.0
300
400
500
600
700
800
900
1000
1100
1200
1300
1400
1500
1600
1700
1800
1
2
3
4
5
6
7
10
12
13
14
15
16
17
18
19

SPECIFICATION TABLE NO. 297 HEMISPHERICAL TOTAL EMITTANCE OF [NIOBIUM + ZIRCONIUM] ALLOYS

Curve No.	Ref. No.	Year	Temperature Range, K	Reported Error, %	Composition (weight percent), Specifications and Remarks
1	12	1962	428-1479	±2.7	99 Nb, 1 Zr; as received; measured in vacuum (<2.4 x 10^{-6} mm Hg); run no. 1, heating.
2	12	1962	1450-618	±2.7	Above specimen and conditions; run no. 1, cooling.
3	12	1962	1075-1489	±2.3	Above specimen and conditions; heating.
4	12	1962	1479-1127	±2.3	Above specimen and conditions; cooling.
5	12	1962	755-1394	±2.7	99 Nb, 1 Zr; polished; measured in vacuum (<3.1 x 10^{-6} mm Hg); run no. 1.
6	12	1962	757-1393	±2.7	Above specimen and conditions; run no. 2.
7	12	1962	755-1393	±2.7	Above specimen and conditions; run no. 3.
8	12	1962	757-1393	±2.7	Above specimen and conditions; run no. 4.
9	12	1962	757-1391	±2.7	Above specimen and conditions; run no. 5.
10	12	1962	1202-1411	±2.3	Above specimen and conditions; run no. 1a.
11	12	1962	1163-1403	±2.3	Above specimen and conditions; run no. 2a.
12	67	1963	421-1107		Nb-1 Zr alloy; roughened to a 45 microinch surface finish; measured in vacuum (5 x 10^{-6} mm Hg); [Author's designation: Specimen 3].
13	67	1963	474-1063		Nb-1 Zr alloy; hand polished; heated in steps to 1063 K; measured in vacuum (5 x 10^{-6} mm Hg); heating; [Author's designation: Specimen 1].
14	67	1963	993-503		Above specimen and conditions except cooling.
15	67	1963	919-1057		Above specimen and conditions except annealed at 1063 K for 15 hrs.
16	67	1963	1042-426		Nb-1 Zr alloy; hand polished; heated to 1043 K; measured in vacuum (5 x 10^{-6} mm Hg); cooling; [Author's designation: Specimen 2].
17	67	1963	585-1451		Above specimen and conditions except heating.
18	67	1963	1172-638		Above specimen and conditions except cooling.
19	255	1964	773-1773		Nb-1 Zr; measured in vacuum (5 x 10^{-7} mm Hg); last run of several runs.

DATA TABLE NO. 297 HEMISPHERICAL TOTAL EMITTANCE OF [NIOBIUM + ZIRCONIUM] ALLOYS

[Temperature, T, K; Emittance, ∈]

T	∈
CURVE 1	
428	0.114
545	0.145
646	0.166
757	0.187
810	0.196
887	0.219
922	0.236
978	0.242
1033	0.255
1089	0.264
1145	0.274
1201	0.283
1256	0.286
1313	0.291
1369	0.298
1425	0.298
1479	0.305
CURVE 2	
1450	0.300
1284	0.275
1116	0.247
949	0.232
983	0.195
618	0.161
CURVE 3	
1075	0.277
1175	0.295
1209	0.275
1265	0.278
1320	0.285
1381	0.288
1434	0.290
1489	0.297
CURVE 4	
1479	0.282
1305	0.258
1127	0.238
CURVE 5	
755	0.144
909	0.147
1060	0.171
1254	0.197
1394	0.208
CURVE 6	
757	0.154
907	0.162
1059	0.188
1193	0.212
1393	0.219
CURVE 7	
755	0.160
908	0.175
1063	0.192
1201	0.205
1393	0.224
810	0.170
CURVE 8*	
757	0.171
853	0.177
1060	0.194
1197	0.212
1393	0.195
820	0.180
CURVE 9*	
757	0.179
909	0.184
1060	0.202
1196	0.220
1391	0.241
CURVE 10	
1202	0.194
1411	0.198
CURVE 11*	
1163	0.213
1403	0.213
CURVE 12	
421	0.207
614	0.209
821	0.229
962	0.329
1107	0.359
CURVE 13*	
474	0.135
683	0.132
824	0.149
910	0.195
981	0.237
1063	0.274
CURVE 14	
993	0.271
890	0.253
770	0.236
668	0.225
503	0.209
CURVE 15	
919	0.253
1057	0.267
CURVE 16	
1042	0.269
1039	0.273
1033	0.276
1033	0.277*
1031	0.279
1012	0.278
725	0.223
426	0.203
CURVE 17	
585	0.217
625	0.224
875	0.254
1064	0.280
1221	0.276
1346	0.331
1451	0.376
CURVE 18	
1172	0.334
1007	0.303
809	0.268
638	0.246
CURVE 19	
773	0.103
873	0.110
973	0.117
1073	0.130
1173	0.142
1273	0.154
1373	0.167
1473	0.179
1573	0.192
1673	0.204*
1773	0.215*

* Not shown on plot

SPECIFICATION TABLE NO. 298 NORMAL TOTAL EMITTANCE OF [NIOBIUM + ZIRCONIUM] ALLOYS

Curve No.	Ref. No.	Year	Temperature Range, K	Geometry θ'	Reported Error, %	Composition (weight percent), Specifications and Remarks
1	106	1962	822-2744	$\sim 0°$	10	0.5 Zr, Nb balance; hot pressed; measured in argon.

DATA TABLE NO. 298 NORMAL TOTAL EMITTANCE OF [NIOBIUM + ZIRCONIUM] ALLOYS

[Temperature, T, K; Emittance, ϵ]

T	ϵ
CURVE 1*	
822	0.36
967	0.58
1111	0.59
1222	0.53
1339	0.51
1489	0.48
1617	0.48
1733	0.47
1739	0.50
1600	0.47
1939	0.47
2172	0.31
2255	0.31
2467	0.30
2539	0.29
2589	0.31
2744	0.25

* Not shown on plot

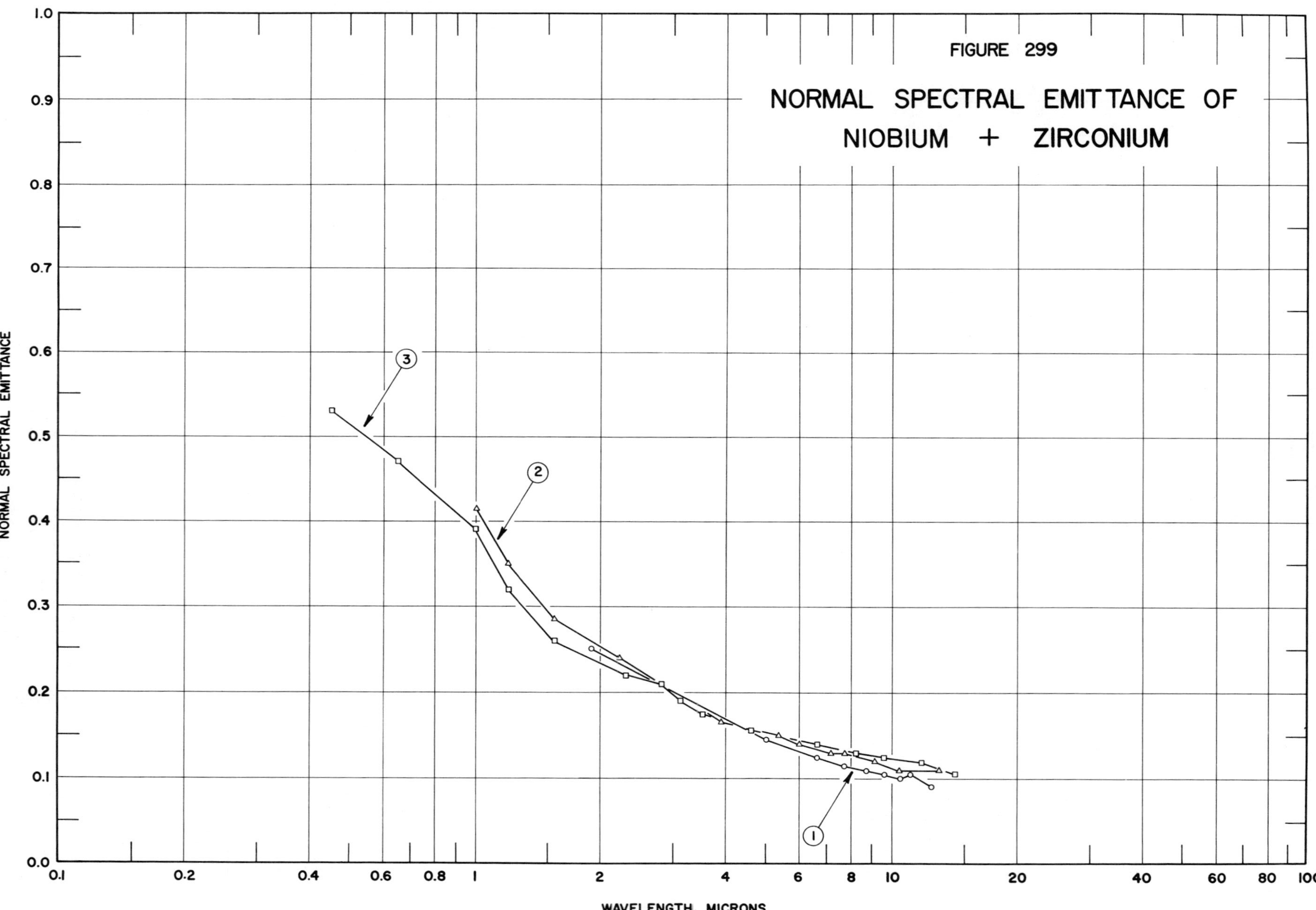
FIGURE 299
NORMAL SPECTRAL EMITTANCE OF
NIOBIUM + ZIRCONIUM
NORMAL SPECTRAL EMITTANCE
1.0
0.9
0.8
0.7
0.6
0.5
0.4
0.3
0.2
0.1
0.0
0.1
0.2
0.4
0.6
0.8
1
2
4
6
8
10
20
40
60
80
100
WAVELENGTH, MICRONS
1
2
3

SPECIFICATION TABLE NO. 299 NORMAL SPECTRAL EMITTANCE OF [NIOBIUM + ZIRCONIUM] ALLOYS

Curve No.	Ref. No.	Year	Temperature K	Wavelength Range, μ	Geometry θ'	Reported Error, %	Composition (weight percent), Specifications and Remarks
1	12	1962	755	1.90-12.40	$\sim 0^\circ$	± 3	99 Nb, 1 Zr; polished; measured in vacuum ($<10^{-7}$ mm Hg).
2	12	1962	1061	1.00-13.00	$\sim 0^\circ$	± 3	Above specimen and conditions.
3	12	1962	1366	0.45-14.10	$\sim 0^\circ$	± 3	Above specimen and conditions.

DATA TABLE NO. 299 NORMAL SPECTRAL EMITTANCE OF [NIOBIUM + ZIRCONIUM] ALLOYS

[Wavelength, λ, μ; Emittance, ϵ; Temperature, T, K]

λ	ϵ
CURVE 1	
T = 755	
1.90	0.250
5.00	0.145
6.60	0.125
7.70	0.115
8.70	0.110
9.60	0.105
10.40	0.100
11.05	0.105
12.40	0.090
CURVE 2	
T = 1061	
1.00	0.415
1.20	0.350
1.55	0.285
2.25	0.240
3.90	0.165
5.35	0.150
6.00	0.140
7.15	0.130
7.70	0.130
9.10	0.120
10.40	0.110
13.00	0.110
CURVE 3	
T = 1366	
0.45	0.530
0.65	0.470
1.00	0.390
1.20	0.320
1.55	0.260
2.30	0.220
2.80	0.210
3.10	0.190
3.50	0.175
4.60	0.155
6.60	0.140
8.20	0.130
9.55	0.125
11.80	0.120
14.10	0.105

SPECIFICATION TABLE NO. 300 HEMISPHERICAL TOTAL EMITTANCE OF [PLATINUM + RHODIUM] ALLOYS

Curve No.	Ref. No.	Year	Temperature Range, K	Reported Error, %	Composition (weight percent), Specifications and Remarks
1	305	1961	873–1723	±1.5	Pt – 10 Rh wires (0.00057 in. dia) from Johnson Matthey; measured in vacuum (<10^{-4} mm Hg).

DATA TABLE NO. 300 HEMISPHERICAL TOTAL EMITTANCE OF [PLATINUM + RHODIUM] ALLOYS

[Temperature, T, K; Emittance, ϵ]

T	ϵ
CURVE 1*	
873	0.1287
973	0.1450
1073	0.1599
1173	0.1723
1273	0.1837
1373	0.1937
1473	0.2032
1573	0.2122
1673	0.2206
1723	0.2243

* Not shown on plot

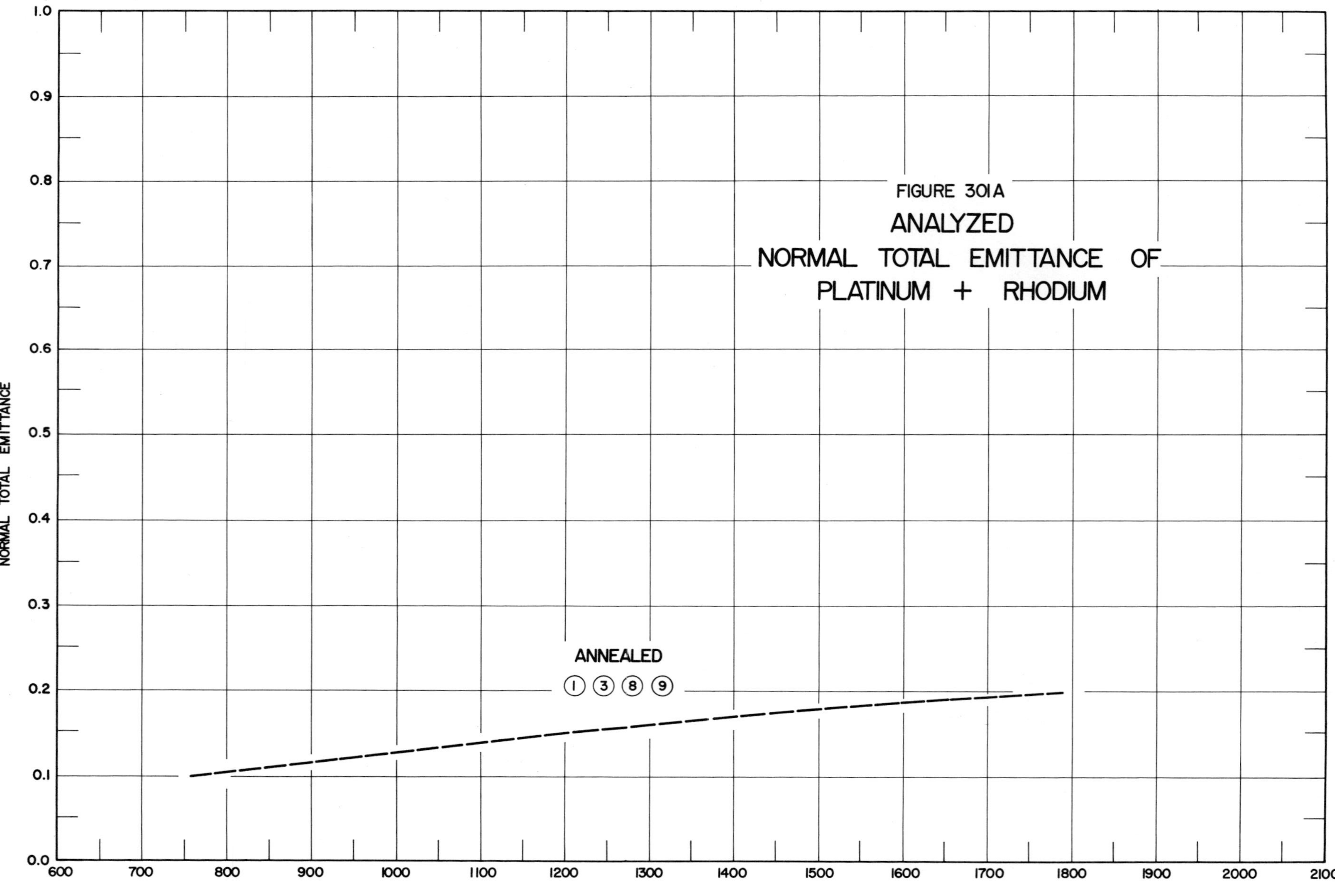
FIGURE 301A
ANALYZED
NORMAL TOTAL EMITTANCE OF
PLATINUM + RHODIUM
ANNEALED
1 3 8 9
NORMAL TOTAL EMITTANCE
TEMPERATURE, K
0.0
0.1
0.2
0.3
0.4
0.5
0.6
0.7
0.8
0.9
1.0
600
700
800
900
1000
1100
1200
1300
1400
1500
1600
1700
1800
1900
2000
2100

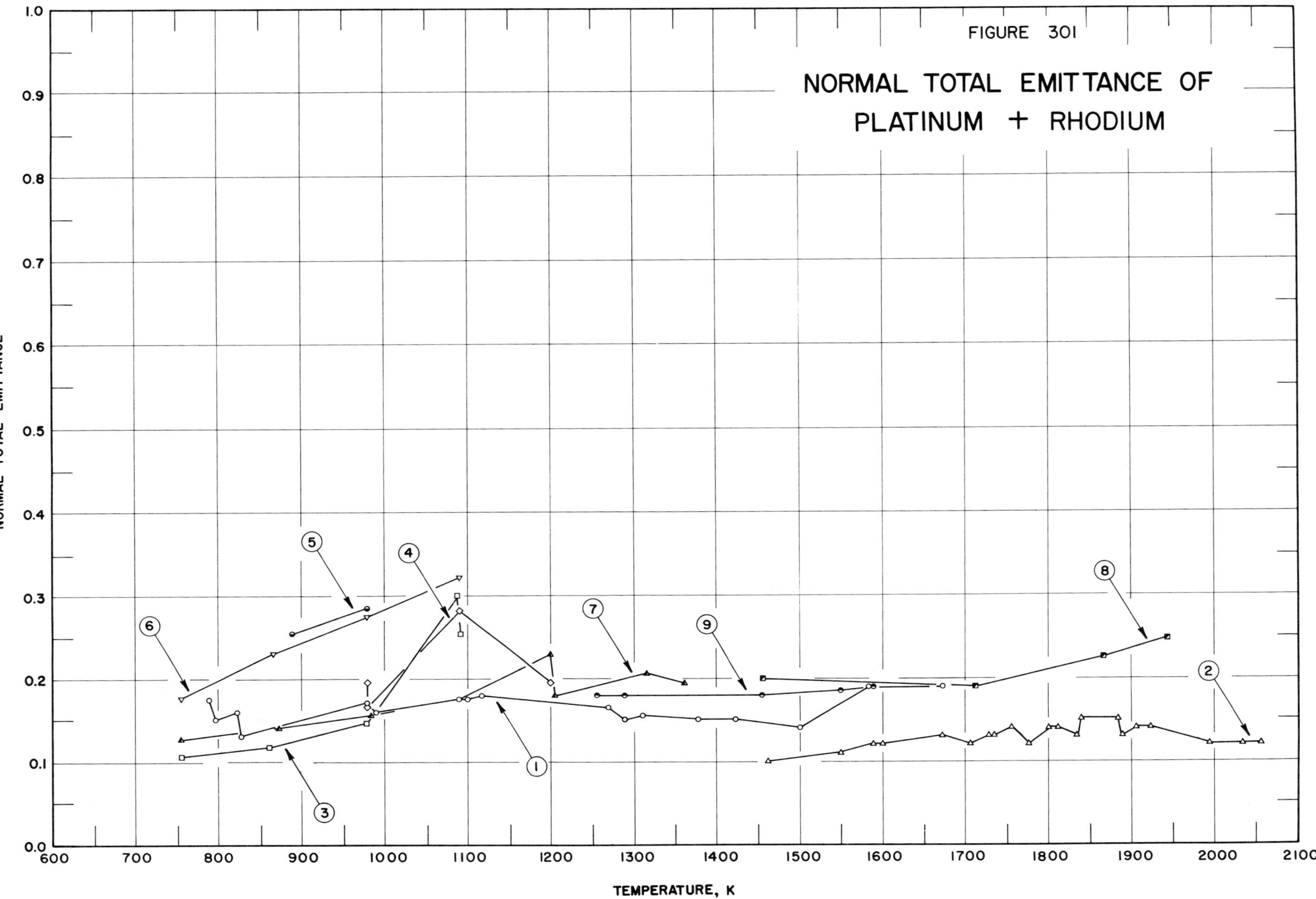

FIGURE 301

NORMAL TOTAL EMITTANCE OF PLATINUM + RHODIUM

SPECIFICATION TABLE NO. 301 NORMAL TOTAL EMITTANCE OF [PLATINUM + RHODIUM] ALLOYS

Curve No.	Ref. No.	Year	Temperature Range, K	Geometry θ'	Reported Error, %	Composition (weight percent), Specifications and Remarks
1	75	1962	789-1672	~0°		87 Pt, 13 Rh; annealed in air at 1866 K for 1 hr; measured in nitrogen.
2	171	1962	1461-2055	~0°	±10	13 Rh, Pt balance; thin sheet.
3	65	1962	755-1091	~0°		13 Rh, Pt balance; run 1.
4	65	1962	978-1200	~0°		Above specimen and conditions; run 2.
5	65	1962	883-978	~0°		Above specimen and conditions; run 3.
6	65	1962	755-1089	~0°		Above specimen and conditions; run 4.
7	65	1962	755-1361	~0°		Above specimen and conditions; run 5.
8	65	1962	1455-1944	~0°		Above specimen and conditions; run 6.
9	65	1962	1255-1589	~0°		Above specimen and conditions; run 7.

DATA TABLE NO. 301 NORMAL TOTAL EMITTANCE OF [PLATINUM + RHODIUM] ALLOYS

[Temperature, T, K; Emittance, ε]

CURVE 1

T	ε
789	0.175
797	0.150
822	0.160
828	0.130
978	0.170
989	0.160
1089	0.175
1100	0.175
1117	0.180
1269	0.165
1289	0.150
1311	0.155
1378	0.150
1422	0.150
1500	0.140
1583	0.190
1672	0.190

CURVE 2

T	ε
1461	0.10
1550	0.11
1589	0.12
1600	0.12
1672	0.13
1705	0.12
1728	0.13
1733	0.13
1755	0.14
1775	0.12
1800	0.14
1811	0.14
1833	0.13
1839	0.15
1883	0.15
1889	0.13
1905	0.14
1922	0.14
1994	0.12
2033	0.12
2055	0.12

CURVE 3

T	ε
755	0.105
861	0.117
978	0.147
1086	0.300
1091	0.253

CURVE 4

T	ε
978	0.196
978	0.167
1089	0.282
1200	0.196

CURVE 5

T	ε
883	0.255
978	0.285

CURVE 6

T	ε
755	0.177
866	0.230
978	0.275
1089	0.322

CURVE 7

T	ε
755	0.125
872	0.140
983	0.155
1089	0.175*
1200	0.229
1205	0.180
1316	0.205
1361	0.194

CURVE 8

T	ε
1455	0.200
1711	0.190
1866	0.224
1944	0.248

CURVE 9

T	ε
1255	0.180
1289	0.180
1455	0.180
1550	0.185
1589	0.190

* Not shown on plot

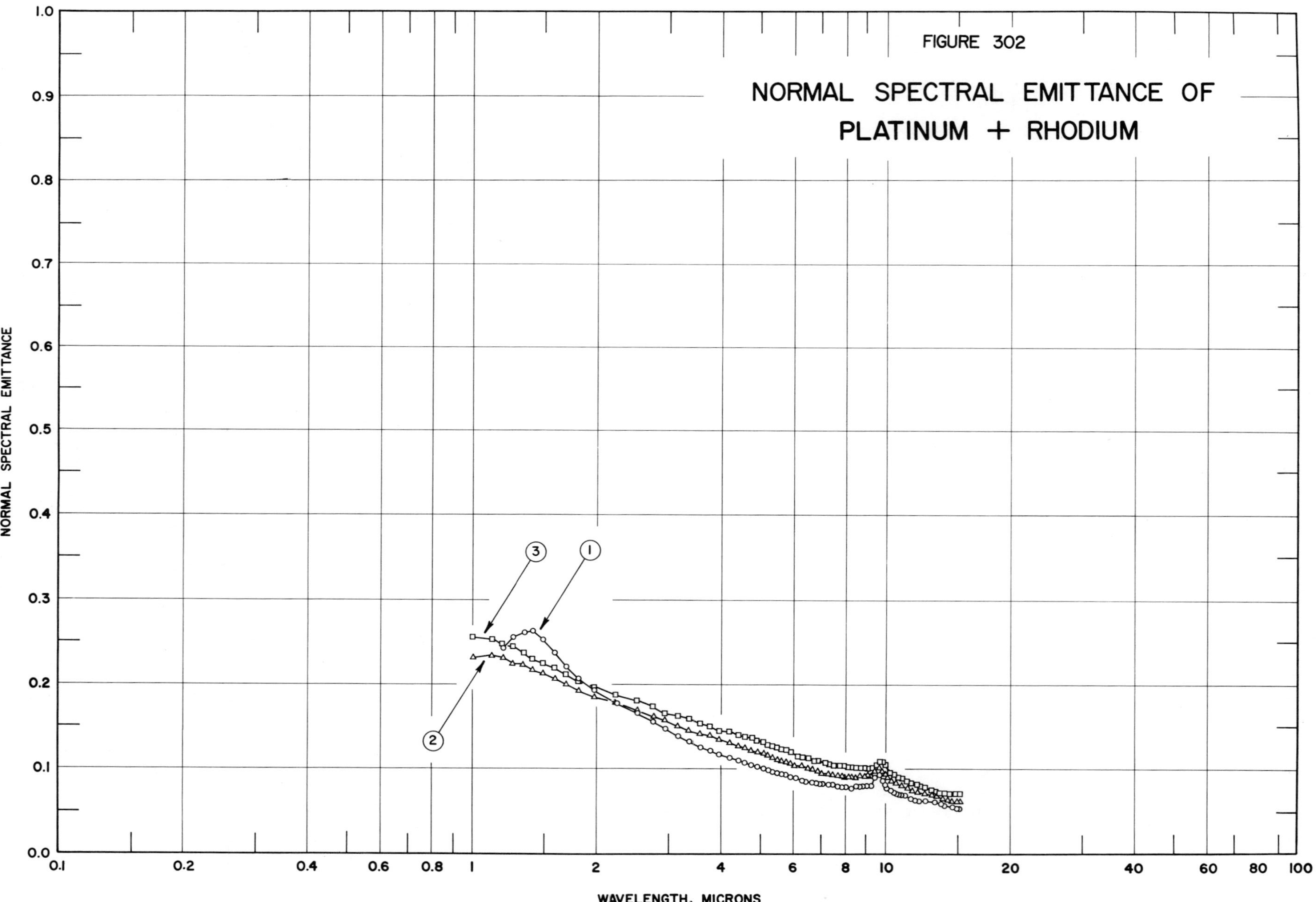
FIGURE 302
NORMAL SPECTRAL EMITTANCE OF
PLATINUM + RHODIUM
NORMAL SPECTRAL EMITTANCE
WAVELENGTH, MICRONS
1.0
0.9
0.8
0.7
0.6
0.5
0.4
0.3
0.2
0.1
0.0
0.1
0.2
0.4
0.6
0.8
1
2
4
6
8
10
20
40
60
80
100
1
2
3

SPECIFICATION TABLE NO. 302 NORMAL SPECTRAL EMITTANCE OF [PLATINUM + RHODIUM] ALLOYS

Curve No.	Ref. No.	Year	Temperature K	Wavelength Range, μ	Geometry θ'	Reported Error, %	Composition (weight percent), Specifications and Remarks
1	298	1964	800	1.18-15.11	$\sim 0°$		Platinum - 13% rhodium alloy; washed and rinsed; annealed at 1523 K for 1 hr; measured in nitrogen; computed from $\rho(0°, 2\pi)$.
2	298	1964	1100	1.07-15.11	$\sim 0°$		Different sample, same as above specimen and conditions.
3	298	1964	1300	1.07-15.11	$\sim 0°$		Different sample, same as curve 1 specimen and conditions.

DATA TABLE NO. 302 NORMAL SPECTRAL EMITTANCE OF [PLATINUM + RHODIUM] ALLOYS

[Wavelength, λ, μ; Emittance, ∈; Temperature, T, K]

λ	∈
CURVE 1	**T = 800**
1.18	0.242
1.25	0.255
1.33	0.261
1.40	0.262
1.48	0.252
1.58	0.237
1.68	0.221
1.81	0.206
1.98	0.192
2.23	0.177
2.51	0.165
2.73	0.154
2.92	0.147
3.14	0.138
3.35	0.132
3.56	0.126
3.77	0.122
3.98	0.117
4.20	0.113
4.41	0.110
4.58	0.108
4.76	0.105
4.91	0.102
5.07	0.101
5.21	0.099
5.34	0.097
5.47	0.097
5.60	0.095
5.76	0.093
5.90	0.091
6.04	0.089
6.16	0.087*
6.28	0.087
6.41	0.086
6.53	0.086*
6.64	0.085
6.76	0.084*
6.87	0.084
6.97	0.084
7.07	0.084
7.19	0.083*
7.32	0.082
7.44	0.082
7.56	0.081*
CURVE 1 (cont.)	
7.66	0.080
7.77	0.080*
7.88	0.079
7.98	0.079*
8.08	0.079
8.18	0.078*
8.28	0.078
8.38	0.079*
8.47	0.080*
8.57	0.080
8.66	0.080
8.75	0.081*
8.84	0.081
8.93	0.081*
9.01	0.080
9.10	0.080*
9.18	0.081*
9.26	0.081
9.35	0.086*
9.43	0.091
9.51	0.094*
9.67	0.093
9.83	0.086
10.00	0.081
10.15	0.078
10.30	0.076
10.46	0.073
10.61	0.072
10.76	0.071
10.91	0.070
11.05	0.070
11.19	0.069*
11.32	0.068*
11.45	0.068*
11.58	0.066
11.72	0.066*
11.84	0.065
11.97	0.066*
12.09	0.064
12.22	0.065*
12.35	0.064*
12.48	0.064
12.60	0.062*
12.72	0.062*
12.85	0.062*
CURVE 1 (cont.)	
12.98	0.062*
13.10	0.062*
13.22	0.062
13.34	0.060*
13.46	0.060*
13.58	0.060
13.69	0.059*
13.81	0.059*
13.93	0.058
14.03	0.058*
14.14	0.058*
14.25	0.057*
14.36	0.057*
14.47	0.057
14.58	0.056*
14.68	0.055*
14.79	0.056*
14.90	0.056
15.01	0.057*
15.11	0.056
CURVE 2	**T = 1100**
1.07	0.231
1.12	0.233
1.18	0.231
1.25	0.225
1.33	0.223
1.40	0.218
1.48	0.213
1.58	0.207
1.68	0.200
1.81	0.191
1.98	0.185
2.23	0.177
2.51	0.168
2.73	0.161
2.92	0.157
3.14	0.150
3.35	0.146
3.56	0.142
3.77	0.139
3.98	0.135
4.20	0.131
4.41	0.127
CURVE 2 (cont.)	
4.58	0.125
4.76	0.122
4.91	0.120
4.58	0.125
4.76	0.122
4.91	0.120
5.07	0.118
5.21	0.116
5.34	0.114
5.47	0.112
5.60	0.110
5.76	0.109
5.90	0.107
6.04	0.105
6.16	0.105*
6.28	0.104
6.41	0.103*
6.53	0.101
6.64	0.100
6.76	0.098*
6.87	0.098
6.97	0.096*
7.07	0.096
7.19	0.095*
7.32	0.095
7.44	0.094
7.56	0.094*
7.66	0.093
7.77	0.092*
7.88	0.092
7.98	0.091*
8.08	0.090
8.18	0.091
8.28	0.091
8.38	0.090*
8.47	0.091
8.57	0.092*
8.66	0.092
8.75	0.093*
8.84	0.092*
8.93	0.092
9.01	0.093*
9.10	0.093
9.18	0.092*
9.26	0.092
CURVE 2 (cont.)	
9.35	0.093*
9.43	0.097
9.51	0.100
9.67	0.102
9.83	0.098
10.00	0.094
10.15	0.090
10.30	0.086
10.46	0.084*
10.61	0.084
10.76	0.082*
10.91	0.081
11.05	0.080*
11.19	0.078*
11.32	0.078
11.45	0.077*
11.58	0.076
11.72	0.075*
11.84	0.074*
11.97	0.074
12.09	0.073*
12.22	0.073*
12.35	0.072*
12.48	0.072
12.60	0.072*
12.72	0.071*
12.85	0.070*
12.98	0.070
13.10	0.069*
13.22	0.069
13.34	0.069*
13.46	0.069*
13.58	0.068*
13.69	0.068
13.81	0.067*
13.93	0.068*
14.03	0.066*
14.14	0.066*
14.25	0.065
14.36	0.065
14.47	0.064*
14.58	0.065*
14.68	0.065*
14.79	0.065*
14.90	0.064
CURVE 2 (cont.)	
15.01	0.064*
15.11	0.064
CURVE 3	**T = 1300**
1.07	0.254
1.12	0.252
1.18	0.247
1.25	0.243
1.33	0.237
1.40	0.231
1.48	0.226
1.58	0.220
1.68	0.212
1.81	0.204
1.98	0.197
2.23	0.187
2.51	0.180
2.73	0.173
2.92	0.166
3.14	0.162
3.35	0.159
3.56	0.153
3.77	0.150
3.98	0.146
4.20	0.144
4.41	0.140
4.58	0.138
4.76	0.137
4.91	0.133
5.07	0.132
5.21	0.129
5.34	0.128
5.47	0.127
5.60	0.125
5.76	0.123
5.90	0.121
6.04	0.119*
6.16	0.116
6.28	0.116
6.41	0.116*
6.53	0.113
6.64	0.112*
6.76	0.111
CURVE 3 (cont.)	
6.87	0.110
6.97	0.109*
7.07	0.108*
7.19	0.108
7.32	0.107
7.44	0.107*
7.56	0.106
7.66	0.104*
7.77	0.104
7.88	0.104*
7.98	0.104
8.08	0.104*
8.18	0.103
8.28	0.103
8.38	0.102*
8.47	0.102*
8.57	0.102
8.66	0.102*
8.75	0.102
8.84	0.103*
8.93	0.102
9.01	0.103*
9.10	0.102
9.18	0.102*
9.26	0.102*
9.35	0.102
9.43	0.102*
9.51	0.105
9.67	0.109
9.83	0.110
10.00	0.105
10.15	0.100*
10.30	0.097
10.46	0.095
10.61	0.092*
10.76	0.091
10.91	0.090
11.05	0.089*
11.19	0.088*
11.32	0.087
11.45	0.085*
11.58	0.085
11.72	0.084*
11.84	0.083*
11.97	0.082
CURVE 3 (cont.)	
12.09	0.080*
12.22	0.081
12.35	0.080*
12.48	0.079
12.60	0.078*
12.72	0.078*
12.85	0.077*
12.98	0.076
13.10	0.075*
13.22	0.075
13.34	0.075*
13.46	0.074*
13.58	0.073
13.69	0.072*
13.81	0.073*
13.93	0.073
14.03	0.072*
14.14	0.072*
14.25	0.072
14.36	0.071*
14.47	0.072*
14.58	0.070*
14.68	0.072
14.79	0.071*
14.90	0.070*
15.01	0.072*
15.11	0.072

* Not shown on plot

SPECIFICATION TABLE NO. 303 NORMAL SPECTRAL REFLECTANCE OF [SILVER + ALUMINUM] ALLOYS

Curve No.	Ref. No.	Year	Temperature K	Wavelength Range, μ	Geometry θ	θ'	ω'	Reported Error, %	Composition (weight percent), Specifications and Remarks
1	133	1934	298	0.2350-0.5780	~0°	~0°			32 Al, Ag balance; cold worked; annealed in an inert gas for 6-24 hrs; polished; stored in dilute solutions of NaOH, NaOH + NaF, and HNO_3.

DATA TABLE NO. 303 NORMAL SPECTRAL REFLECTANCE OF [SILVER + ALUMINUM] ALLOYS

[Wavelength, λ, μ; Reflectance, ρ; Temperature, T, K]

λ	ρ
CURVE 1*	
T = 298	
0.2350	0.418
0.2537	0.482
0.2650	0.521
0.2970	0.549
0.3125	0.578
0.3340	0.601
0.3660	0.680
0.4060	0.702
0.4355	0.740
0.5460	0.759
0.5780	0.760

* Not shown on plot

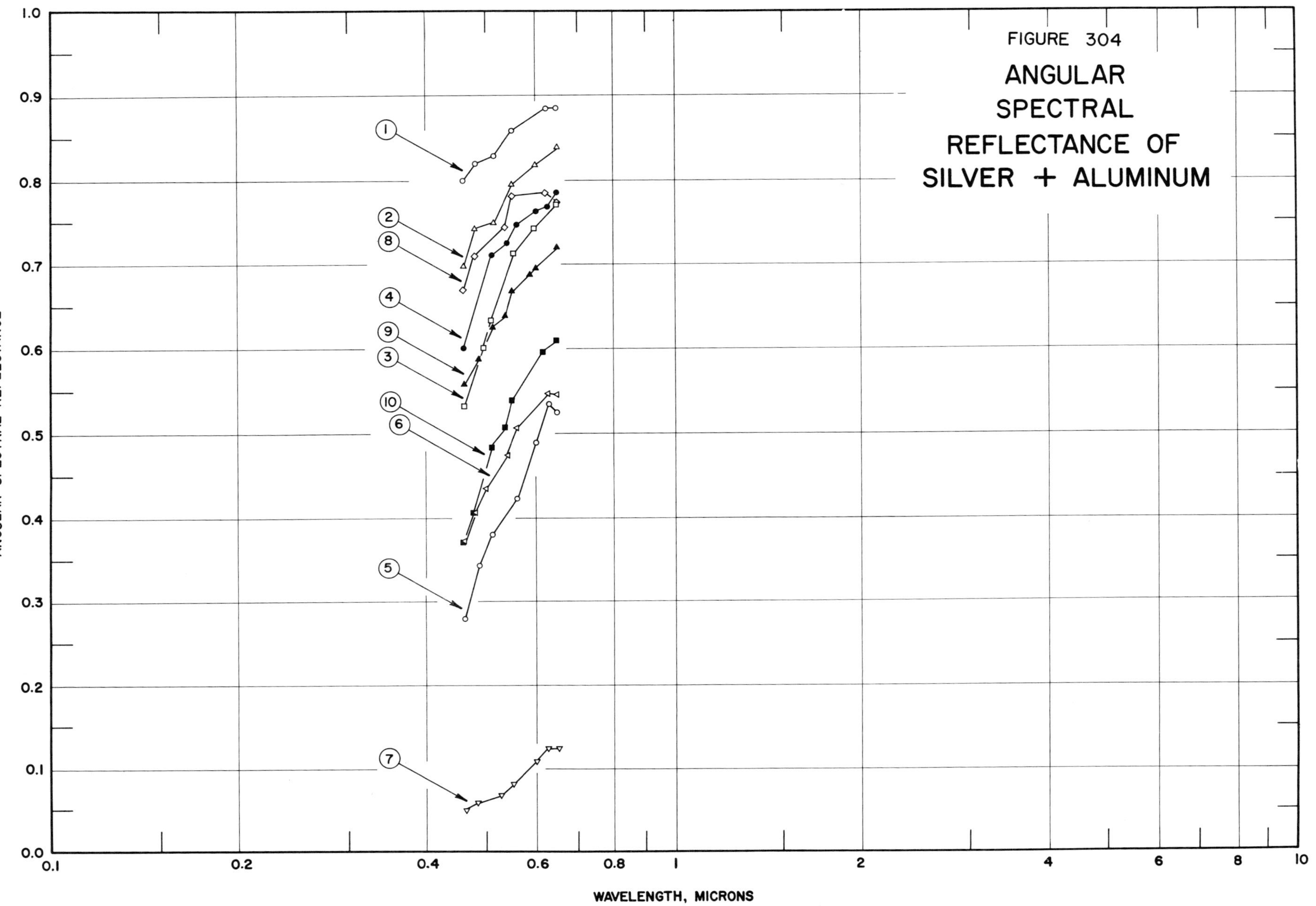
FIGURE 304
ANGULAR
SPECTRAL
REFLECTANCE OF
SILVER + ALUMINUM
ANGULAR SPECTRAL REFLECTANCE
1.0
0.9
0.8
0.7
0.6
0.5
0.4
0.3
0.2
0.1
0.0
0.1
0.2
0.4
0.6
0.8
1
2
4
6
8
10
WAVELENGTH, MICRONS
1
2
8
4
9
3
10
6
5
7

SPECIFICATION TABLE NO. 304 ANGULAR SPECTRAL REFLECTANCE OF [SILVER + ALUMINUM] ALLOYS

Curve No.	Ref. No.	Year	Temperature K	Wavelength Range, μ	Geometry θ	θ'	ω'	Reported Error, %	Composition (weight percent), Specifications and Remarks
1	264	1939	298	0.4611-0.6499	45°	45°			98.77 Ag, 1.23 Al; cold rolled; polished; measured immediately after polishing; chrome plated or rhodium plated mirror references; data extracted from smooth curve.
2	264	1939	298	0.4603-0.6500	45°	45°			Above specimen and condition except measured after exposing for 21 days to an atmosphere containing little sulphur dioxide and moisture.
3	264	1939	298	0.4602-0.6500	45°	45°			Above specimen and conditions except 52 days additional exposure to the same atmosphere.
4	264	1939	298	0.4600-0.6501	45°	45°			98.77 Ag, 1.23 Al; ground and degreased; chrome plated or rhodium plated mirror references; data extracted from smooth curve.
5	264	1939	298	0.4600-0.6499	45°	45°			Above specimen and conditions except oxidized.
6	264	1939	298	0.4611-0.6499	45°	45°			95.8 Ag, 4.2 Al; ground; degreased; chrome plated or rhodium plated mirror references; data extracted from smooth curve.
7	264	1939	298	0.4622-0.6500	45°	45°			Above specimen and conditions except oxidized.
8	264	1939	298	0.4608-0.6500	45°	45°			95.8 Ag, 4.2 Al; cold rolled; polished; measured immediately after polishing; chrome plated or rhodium plated mirror references; data extracted from smooth curve.
9	264	1939	298	0.4604-0.6500	45°	45°			Above specimen and conditions except measured after exposing for 21 days to an atmosphere containing little sulphur dioxide and moisture.
10	264	1939	298	0.4609-0.6499	45°	45°			Above specimen and conditions except 52 days additional exposure to the same atmosphere.

DATA TABLE NO. 304 ANGULAR SPECTRAL REFLECTANCE OF [SILVER + ALUMINUM] ALLOYS

[Wavelength, λ, μ; Reflectance, ρ; Temperature, T, K]

λ	ρ
CURVE 1 T = 298	
0.4611	0.800
0.4816	0.821
0.5130	0.829
0.5500	0.860
0.6236	0.887
0.6499	0.887
CURVE 2 T = 298	
0.4603	0.699
0.4812	0.742
0.5142	0.750
0.5502	0.796
0.6002	0.819
0.6500	0.839
CURVE 3 T = 298	
0.4602	0.532
0.4952	0.601
0.5094	0.635
0.5502	0.712
0.5999	0.742
0.6500	0.771
CURVE 4 T = 298	
0.4600	0.600
0.5105	0.710
0.5405	0.726
0.5614	0.748
0.6014	0.762
0.6287	0.769
0.6501	0.787
CURVE 5 T = 298	
0.4600	0.280
0.4885	0.343
0.5104	0.380
CURVE 5 (cont.)	
0.5606	0.424
0.6005	0.491
0.6304	0.537
0.6499	0.527
CURVE 6 T = 298	
0.4611	0.372
0.4808	0.408
0.5060	0.436
0.5411	0.476
0.5610	0.508
0.6284	0.549
0.6499	0.548
CURVE 7 T = 298	
0.4622	0.050
0.4835	0.051
0.5260	0.068
0.5512	0.081
0.6006	0.109
0.6264	0.124
0.6500	0.124
CURVE 8 T = 298	
0.4608	0.671
0.4814	0.710
0.5358	0.745
0.5511	0.781
0.6203	0.786
0.6500	0.774
CURVE 9 T = 298	
0.4604	0.560
0.4869	0.589
0.5113	0.627
0.5365	0.640
0.5504	0.669
CURVE 9 (cont.)	
0.5879	0.689
0.6004	0.698
0.6500	0.721
CURVE 10 T = 298	
0.4609	0.372
0.4773	0.408
0.5114	0.485
0.5348	0.509
0.5498	0.541
0.6165	0.597
0.6499	0.611

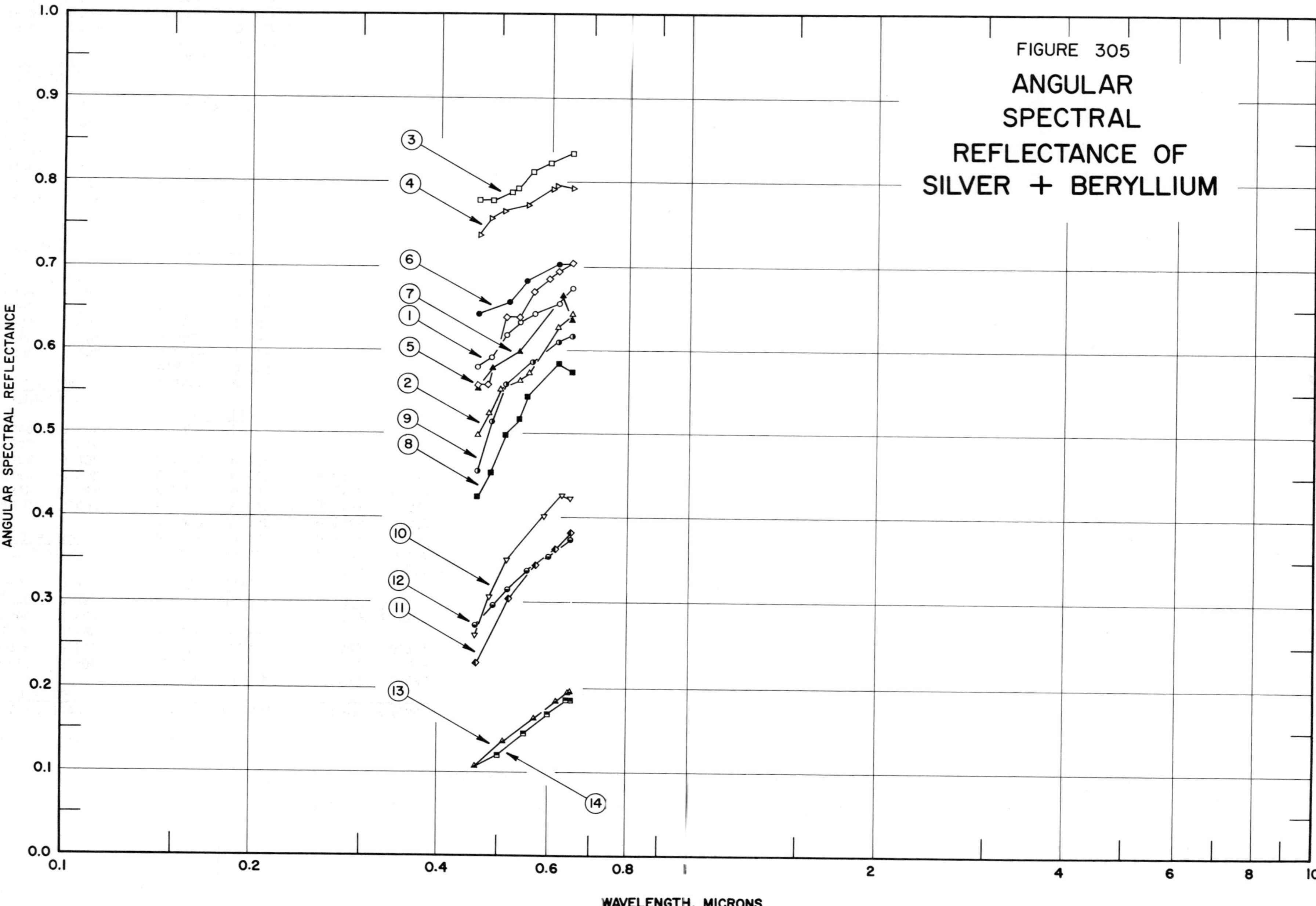

FIGURE 305
ANGULAR SPECTRAL REFLECTANCE OF SILVER + BERYLLIUM

SPECIFICATION TABLE NO. 305 ANGULAR SPECTRAL REFLECTANCE OF [SILVER + BERYLLIUM] ALLOYS

Curve No.	Ref. No.	Year	Temperature K	Wavelength Range, μ	Geometry θ	θ'	ω'	Reported Error, %	Composition (weight percent), Specifications and Remarks
1	264	1939	298	0.4617-0.6500	45°	45°			99.32 Ag, 0.68 Be; cold rolled; polished; chrome plated or rhodium plated mirror references; data extracted from smooth curve.
2	264	1939	298	0.4608-0.6500	45°	45°			Above specimen and conditions except oxidized.
3	264	1939	298	0.4604-0.6500	45°	45°			99.32 Ag, 0.68 Be; cold rolled; polished; measured immediately after polishing; chrome plated or rhodium plated mirror references; data extracted from smooth curve.
4	264	1939	298	0.4605-0.6500	45°	45°			Above specimen and conditions except measured after exposing for 21 days to an atmosphre containing little sulphur dioxide and moisture.
5	264	1939	298	0.4605-0.6500	45°	45°			Above specimen and conditions except 52 days additional exposure to the same atmosphere.
6	264	1939	298	0.4601-0.6500	45°	45°			99.32 Ag, 0.68 Be; cold rolled; polished; measured immediately after polishing; chrome plated or rhodium plated mirror references; data extracted from smooth curve.
7	264	1939	298	0.4599-0.6500	45°	45°			Above specimen and conditions except measured after exposing for 21 days to an atmosphere containing little sulphur dioxide and moisture.
8	264	1939	298	0.4601-0.6500	45°	45°			Above specimen and conditions except 52 days additional exposure to the same atmosphere.
9	264	1939	298	0.4600-0.6501	45°	45°			99.32 Ag, 0.68 Be; ground and degreased; chrome plated or rhodium plated mirror references; data extracted from smooth curve.
10	264	1939	298	0.4595-0.6500	45°	45°			Above specimen and conditions except oxidized during the annealing process.
11	264	1939	298	0.4606-0.6501	45°	45°			Above specimen and conditions except stored in moist atmosphere for 42 days.
12	264	1939	298	0.4604-0.6500	45°	45°			98.4 Ag, 1.6 Be; ground and degreased; chrome plated or rhodium plated mirror references; data extracted from smooth curve.
13	264	1939	298	0.4599-0.6500	45°	45°			Above specimen and conditions except oxidized during the annealing process.
14	264	1939	298	0.4599-0.6500	45°	45°			Above specimen and conditions except stored in moist atmosphere for 30 days.

DATA TABLE NO. 305 ANGULAR SPECTRAL REFLECTANCE OF [SILVER + BERYLLIUM] ALLOYS

[Wavelength, λ, μ; Reflectance, ρ; Temperature, T, K]

λ	ρ
CURVE 1	
T = 298	
0.4617	0.581
0.4827	0.592
0.5105	0.620
0.5367	0.637
0.5676	0.647
0.6194	0.658
0.6500	0.675
CURVE 2	
T = 298	
0.4608	0.499
0.4816	0.527
0.5092	0.554
0.5372	0.565
0.5555	0.573
0.6177	0.630
0.6500	0.645
CURVE 3	
T = 298	
0.4604	0.779
0.4848	0.779
0.5197	0.790
0.5316	0.793
0.5626	0.813
0.5998	0.824
0.6500	0.836
CURVE 4	
T = 298	
0.4605	0.738
0.4816	0.758
0.5096	0.756
0.5507	0.772
0.6054	0.792
0.6192	0.797
0.6500	0.794
CURVE 5	
T = 298	
0.4605	0.560
0.4797	0.560
0.5103	0.640
0.5369	0.640
0.5650	0.671
0.5997	0.687
0.6186	0.697
0.6500	0.705
CURVE 6	
T = 298	
0.4601	0.645
0.5141	0.658
0.5514	0.682
0.6184	0.703
0.6500	0.703*
CURVE 7	
T = 298	
0.4599	0.556
0.4861	0.579
0.5338	0.599
0.6281	0.664
0.6500	0.638
CURVE 8	
T = 298	
0.4601	0.427
0.4835	0.453
0.5100	0.499
0.5378	0.518
0.5521	0.544
0.6190	0.584
0.6500	0.575
CURVE 9	
T = 298	
0.4600	0.456
0.4853	0.516
0.5104	0.560
CURVE 9 (cont.)	
0.5620	0.588
0.6185	0.611
0.6501	0.619
CURVE 10	
T = 298	
0.4595	0.262
0.4830	0.307
0.5161	0.350
0.5907	0.402
0.6296	0.428
0.6500	0.422
CURVE 11	
T = 298	
0.4606	0.230
0.5193	0.305
0.5721	0.345
0.6134	0.362
0.6501	0.382
CURVE 12	
T = 298	
0.4604	0.274
0.4901	0.298
0.5177	0.317
0.5542	0.338
0.6013	0.354
0.6500	0.375
CURVE 13	
T = 298	
0.4599	0.108
0.5105	0.137
0.5707	0.163
0.6173	0.185
0.6442	0.197
0.6500	0.197
CURVE 14	
T = 298	
0.4599	0.106*
0.5005	0.120
0.5501	0.147
0.6004	0.169
0.6410	0.187
0.6500	0.187

* Not shown on plot

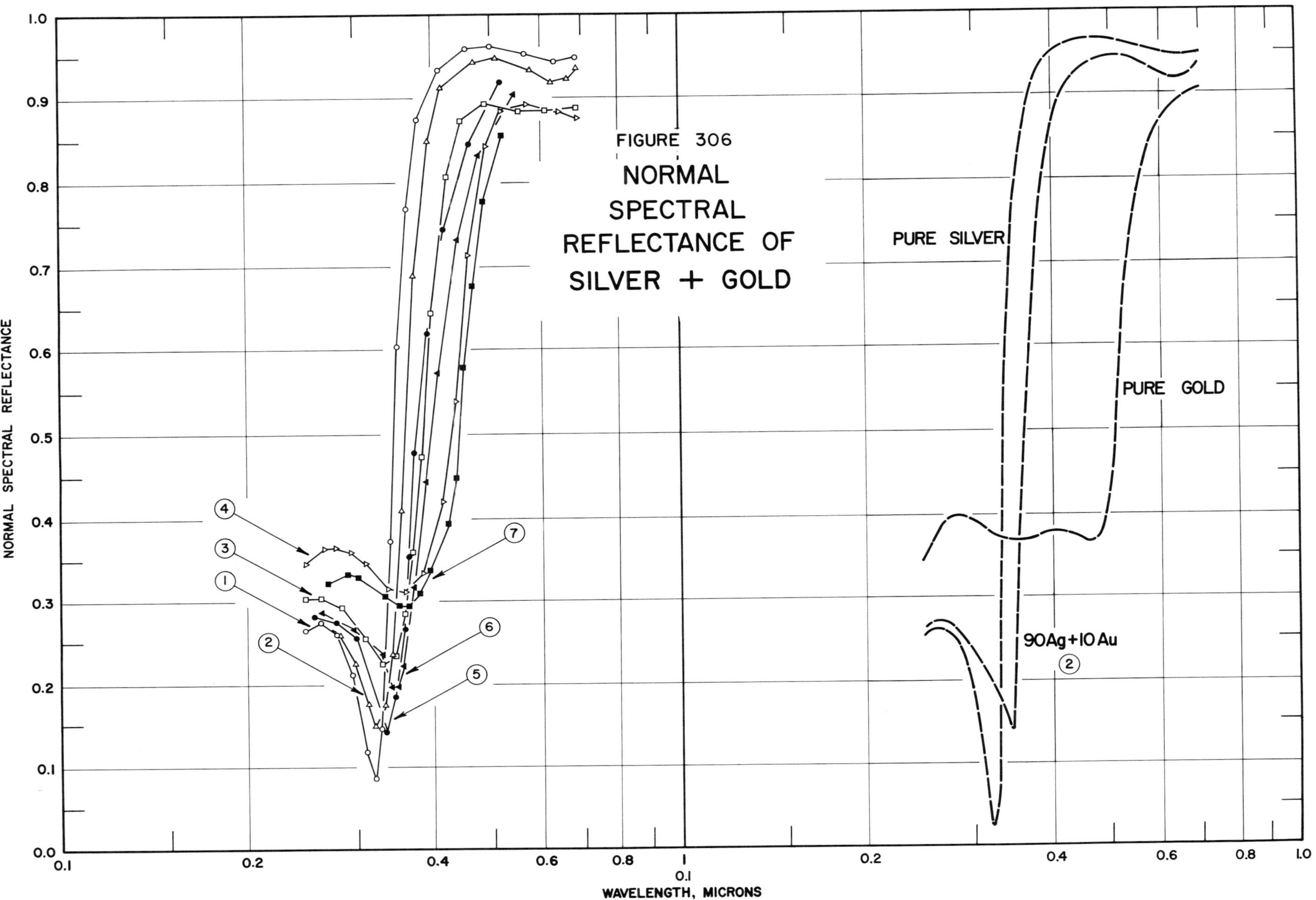

FIGURE 306
NORMAL SPECTRAL REFLECTANCE OF SILVER + GOLD

SPECIFICATION TABLE NO. 306 NORMAL SPECTRAL REFLECTANCE OF [SILVER + GOLD] ALLOYS

Curve No.	Ref. No.	Year	Temperature K	Wavelength Range, μ	Geometry θ	θ'	ω'	Reported Error, %	Composition (weight percent), Specifications and Remarks
1	297	1963	298	0.248-0.689	~10°	~10°			95 Ag and 5 Au; hand lapped with silicon carbide papers, lapped on a metallographic polishing wheel, and electrolytically slide polished; tungsten filament source for 0.365 to 0.689 μ and a hydrogen lamp source at lower wavelengths; data extracted from smooth curve.
2	297	1963	298	0.248-0.689	~10°	~10°			90 Ag and 10 Au; hand lapped with silicon carbide papers, lapped on a metallographic polishing wheel, and electrolytically slide polished; tungsten filament source for 0.365 to 0.689 μ and a hydrogen lamp source at lower wavelengths; data extracted from smooth curve.
3	297	1963	298	0.248-0.689	~10°	~10°			80 Ag and 20 Au; hand lapped with silicon carbide papers, lapped on a metallographic polishing wheel, and electrolytically slide polished; tungsten filament source for 0.365 to 0.689 μ and a hydrogen lamp source at lower wavelengths; sample slightly etched during electrolytic polishing; data extracted from smooth curve.
4	297	1963	298	0.248-0.689	~10°	~10°			50 Ag and 50 Au; hand lapped with silicon carbide papers, lapped on a metallographic polishing wheel, and electrolytically slide polished; tungsten filament source for 0.365 to 0.689 μ and a hydrogen lamp source at lower wavelengths; sample slightly etched during electrolytic polishing; data extracted from smooth curve.
5	350	1965	298	0.257-0.519	~0°	~0°			95 Ag and 5 Au film (~ 2000 Å thick); evaporated on a quartz plate at 423 K in vacuum ($< 3 \times 10^{-6}$ mm Hg); annealed at 673 K for 1.5 hr; data extracted from smooth curve.
6	350	1965	298	0.264-0.544	~0°	~0°			90 Ag and 10 Au film (~2000 Å thick); evaporated on a quartz plate at 423 K in vacuum ($< 3 \times 10^{-6}$ mm Hg); annealed at 673 K for 1.5 hr; data extracted from smooth curve.
7	350	1965	298	0.269-0.521	~0°	~0°			50 Ag and 50 Au film (~2000 Å thick); evaporated on a quartz plate at 423 K in vacuum ($< 3 \times 10^{-6}$ mm Hg); annealed at 673 K for 1.5 hr; data extracted from smooth curve.

DATA TABLE NO. 306 NORMAL SPECTRAL REFLECTANCE OF [SILVER + GOLD] ALLOYS

[Wavelength, λ, μ; Reflectance, ρ; Temperature, T, K]

CURVE 1
T = 298

λ	ρ
0.248	0.266
0.262	0.274
0.278	0.260
0.295	0.211
0.311	0.118
0.320	0.084
0.328	0.145
0.339	0.372
0.351	0.603
0.363	0.769
0.380	0.876
0.412	0.937
0.456	0.960
0.506	0.962
0.569	0.954
0.633	0.945
0.689	0.950

CURVE 2
T = 298

λ	ρ
0.248	0.266*
0.262	0.277*
0.281	0.259
0.298	0.225
0.312	0.175
0.320	0.149
0.332	0.173
0.341	0.237
0.353	0.411
0.372	0.690
0.395	0.851
0.416	0.913
0.470	0.945
0.510	0.949
0.579	0.934
0.626	0.920
0.663	0.923
0.689	0.937

CURVE 3
T = 298

λ	ρ
0.248	0.302
0.262	0.303
0.283	0.293
0.309	0.256
0.329	0.224
0.346	0.235
0.359	0.285
0.370	0.360
0.383	0.473
0.399	0.645
0.423	0.808
0.448	0.873
0.490	0.893
0.554	0.886
0.611	0.751
0.689	0.889

CURVE 4
T = 298

λ	ρ
0.248	0.348
0.266	0.365
0.277	0.367
0.294	0.361
0.310	0.348
0.336	0.317
0.359	0.312
0.384	0.334
0.413	0.420
0.434	0.539
0.456	0.712
0.490	0.845
0.519	0.887
0.571	0.894
0.642	0.883
0.689	0.878

CURVE 5
T = 298

λ	ρ
0.257	0.282
0.276	0.277
0.299	0.256
0.334	0.141
0.345	0.183
0.358	0.267
0.364	0.355
0.372	0.479
0.392	0.621
0.419	0.745
0.461	0.848
0.519	0.921

CURVE 6
T = 298

λ	ρ
0.264	0.288
0.297	0.267
0.330	0.235
0.341	0.198
0.350	0.198
0.356	0.223
0.372	0.318
0.389	0.444
0.409	0.574
0.441	0.732
0.479	0.833
0.544	0.907

CURVE 7
T = 298

λ	ρ
0.269	0.322
0.290	0.332
0.302	0.330
0.333	0.306
0.351	0.295
0.363	0.295
0.380	0.310

CURVE 7 (cont.)

λ	ρ
0.394	0.338
0.423	0.393
0.435	0.449
0.449	0.580
0.464	0.676
0.484	0.778
0.521	0.858

* Not shown on plot

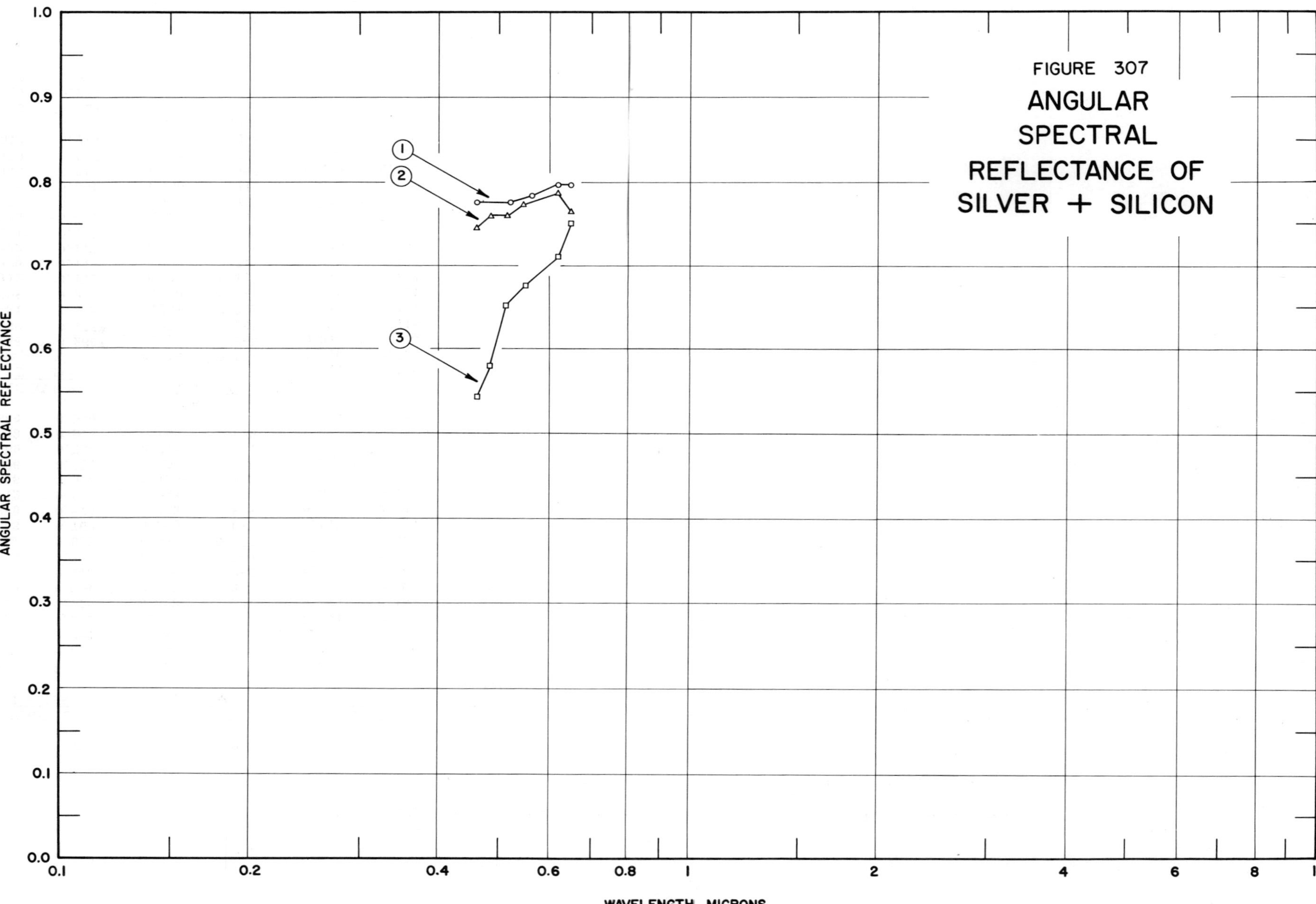
FIGURE 307
ANGULAR
SPECTRAL
REFLECTANCE OF
SILVER + SILICON
ANGULAR SPECTRAL REFLECTANCE
1.0
0.9
0.8
0.7
0.6
0.5
0.4
0.3
0.2
0.1
0.0
0.1
0.2
0.4
0.6
0.8
1
2
4
6
8
10
WAVELENGTH, MICRONS
1
2
3

SPECIFICATION TABLE NO. 307 ANGULAR SPECTRAL REFLECTANCE OF [SILVER + SILICON] ALLOYS

Curve No.	Ref. No.	Year	Temperature K	Wavelength Range, μ	Geometry θ	Geometry θ'	Geometry ω'	Reported Error, %	Composition (weight percent), Specifications and Remarks
1	264	1939	298	0.4600–0.6500	45°	45°			99.4 Ag, 0.6 Si; cold rolled; polished; measured immediately after polishing; chrome plated or rhodium plated mirror references; data extracted from smooth curve.
2	264	1939	298	0.4600–0.6500	45°	45°			Above specimen and conditions except measured after exposing for 21 days to an atmosphere containing little sulphur dioxide and moisture.
3	264	1939	298	0.4601–0.6501	45°	45°			Above specimen and conditions except 52 days additional exposure to the same atmosphere.

DATA TABLE NO. 307 ANGULAR SPECTRAL REFLECTANCE OF [SILVER + SILICON] ALLOYS

[Wavelength, λ, μ; Reflectance, ρ; Temperature, T, K]

λ	ρ
CURVE 1	
T = 298	
0.4600	0.777
0.5201	0.777
0.5645	0.785
0.6195	0.798
0.6500	0.798
CURVE 2	
T = 298	
0.4600	0.747
0.4841	0.760
0.5130	0.760
0.5456	0.773
0.6180	0.788
0.6500	0.765
CURVE 3	
T = 298	
0.4601	0.546
0.4821	0.581
0.5115	0.652
0.5510	0.677
0.6190	0.711
0.6501	0.751

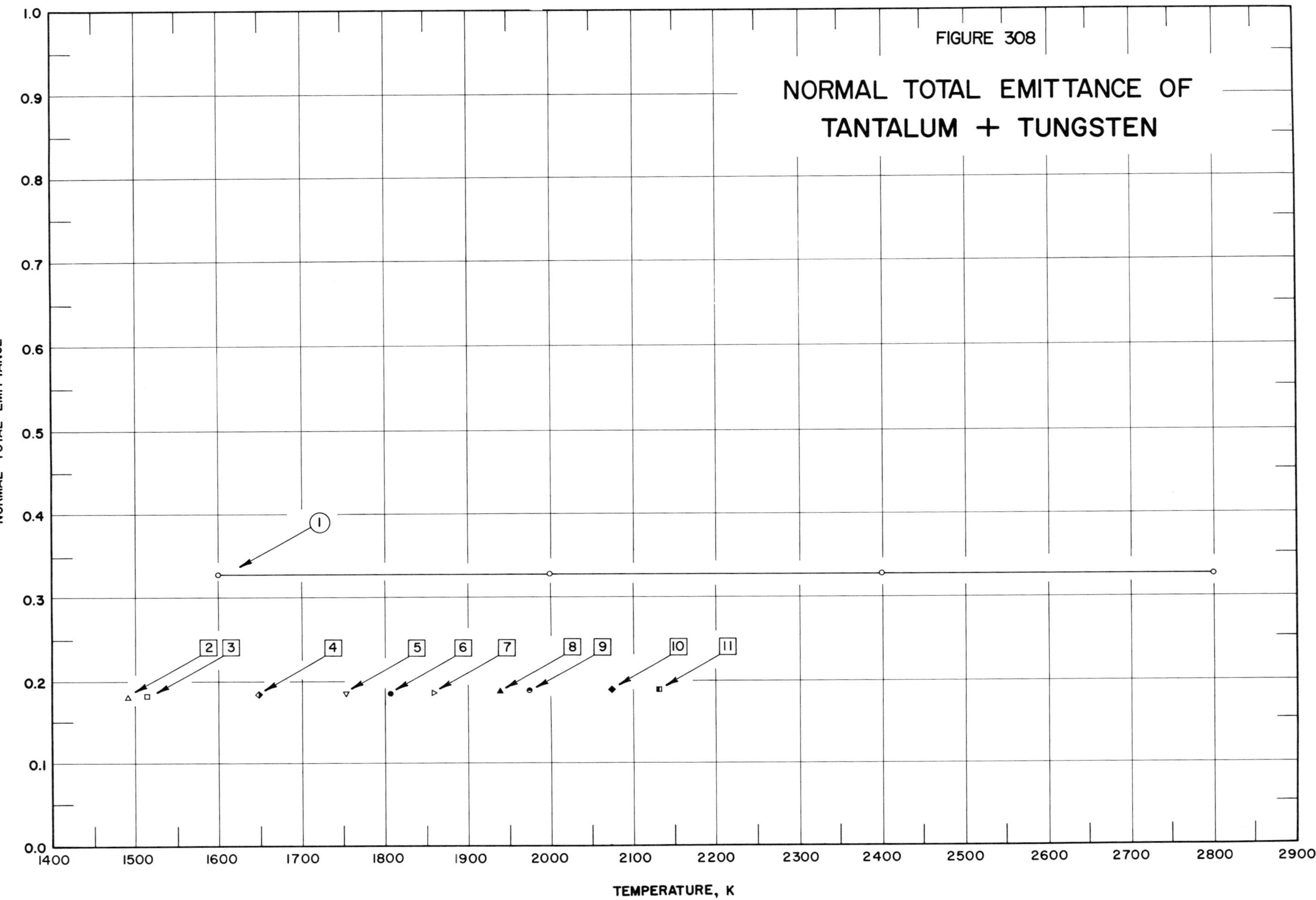
FIGURE 308
NORMAL TOTAL EMITTANCE OF
TANTALUM + TUNGSTEN
NORMAL TOTAL EMITTANCE
TEMPERATURE, K
1.0
0.9
0.8
0.7
0.6
0.5
0.4
0.3
0.2
0.1
0.0
1400
1500
1600
1700
1800
1900
2000
2100
2200
2300
2400
2500
2600
2700
2800
2900
1
2
3
4
5
6
7
8
9
10
11

SPECIFICATION TABLE NO. 308 NORMAL TOTAL EMITTANCE OF [TANTALUM + TUNGSTEN] ALLOYS

Curve No.	Ref. No.	Year	Temperature Range, K	Geometry θ'	Reported Error, %	Composition (weight percent), Specifications and Remarks
1	84	1962	1600-2800	~0°		90 Ta, 10 W; polished with successively finer abrasive papers (numbers 1, 0, 00, 000, and 0000); measured in argon.
2	267	1965	1491	~0°		Obtained from Thermophysics Branch, AFML, RTD; thermal conductivity 0.5247 w $cm^{-1} K^{-1}$; specific heat 0.0381 cal $sec^{-1} cm^{-1} K^{-1}$.
3	267	1965	1514	~0°		Obtained from Thermophysics Branch, AFML, RTD; thermal conductivity 0.5297 w $cm^{-1} K^{-1}$; specific heat 0.0382 cal $sec^{-1} cm^{-1} K^{-1}$.
4	267	1965	1648	~0°		Obtained from Thermophysics Branch, AFML, RTD; thermal conductivity 0.5640 w $cm^{-1} K^{-1}$; specific heat 0.0386 cal $sec^{-1} cm^{-1} K^{-1}$.
5	267	1965	1753	~0°		Obtained from Thermophysics Branch, AFML, RTD; thermal conductivity 0.5280 w $cm^{-1} K^{-1}$; specific heat 0.0389 cal $sec^{-1} cm^{-1} K^{-1}$.
6	267	1965	1807	~0°		Obtained from Thermophysics Branch, AFML, RTD; thermal conductivity 0.4941 w $cm^{-1} K^{-1}$; specific heat 0.0391 cal $sec^{-1} cm^{-1} K^{-1}$.
7	267	1965	1857	~0°		Obtained from Thermophysics Branch, AFML, RTD; thermal conductivity 0.4929 w $cm^{-1} K^{-1}$; specific heat 0.0392 cal $sec^{-1} cm^{-1} K^{-1}$.
8	267	1965	1938	~0°		Obtained from Thermophysics Branch, AFML, RTD; thermal conductivity 0.4724 w $cm^{-1} K^{-1}$; specific heat 0.0395 cal $sec^{-1} cm^{-1} K^{-1}$.
9	267	1965	1973	~0°		Obtained from Thermophysics Branch, AFML, RTD; thermal conductivity 0.5008 w $cm^{-1} K^{-1}$; specific heat 0.0396 cal $sec^{-1} cm^{-1} K^{-1}$.
10	267	1965	2072	~0°		Obtained from Thermophysics Branch, AFML, RTD; thermal conductivity 0.5544 w $cm^{-1} K^{-1}$; specific heat 0.0399 cal $sec^{-1} cm^{-1} K^{-1}$.
11	267	1965	2130	~0°		Obtained from Thermophysics Branch, AFML, RTD; thermal conductivity 0.5130 w $cm^{-1} K^{-1}$; specific heat 0.0400 cal $sec^{-1} cm^{-1} K^{-1}$.

DATA TABLE NO. 308 NORMAL TOTAL EMITTANCE OF [TANTALUM + TUNGSTEN] ALLOYS

[Temperature, T, K; Emittance, ϵ]

T	ϵ
CURVE 1	
1600	0.328
2000	0.328
2400	0.328
2800	0.328
CURVE 2	
1491	0.181
CURVE 3	
1514	0.182
CURVE 4	
1648	0.184
CURVE 5	
1753	0.185
CURVE 6	
1807	0.186
CURVE 7	
1857	0.187
CURVE 8	
1938	0.188
CURVE 9	
1973	0.189
CURVE 10	
2072	0.190
CURVE 11	
2130	0.190

*Not shown on plot

SPECIFICATION TABLE NO. 309 NORMAL SPECTRAL EMITTANCE OF [TANTALUM + TUNGSTEN] ALLOYS

Curve No.	Ref. No.	Year	Wavelength μ	Temperature Range, K	Geometry θ'	Reported Error, %	Composition (weight percent), Specifications and Remarks
1	84	1962	0.65	1600-2800	$\sim 0^\circ$		90 Ta, 10 W; polished with successively finer abrasive papers (numbers 1, 0, 00, 000, and 0000); measured in argon.

DATA TABLE NO. 309 NORMAL SPECTRAL EMITTANCE OF [TANTALUM + TUNGSTEN] ALLOYS

[Temperature, T, K; Emittance, ϵ; Wavelength, λ, μ]

T	ϵ
CURVE 1*	
$\lambda = 0.65$	
1600	0.347
2000	0.347
2400	0.347
2800	0.347

*Not shown on plot

SPECIFICATION TABLE NO. 310 NORMAL INTEGRATED ABSORPTANCE OF [TIN + INDIUM] ALLOYS

Curve No.	Ref. No.	Year	Temperature Range, K	Geometry θ'	Reported Error, %	Composition (weight percent), Specifications and Remarks
1	134	1952	2	$\sim 0^{\circ}$	1	94.6 Sn, 5.4 In; electropolished; absorptance for 298 K black body incident radiation.
2	134	1952	2	$\sim 0^{\circ}$	1	99 Sn, 1 In; electropolished; absorptance for 298 K black body incident radiation.

DATA TABLE NO. 310 NORMAL INTEGRATED ABSORPTANCE OF [TIN + INDIUM] ALLOYS

[Temperature, T, K; Emittance, ϵ]

T	ϵ
CURVE 1*	
2	0.0174
CURVE 2*	
2	0.0125

* Not shown on plot

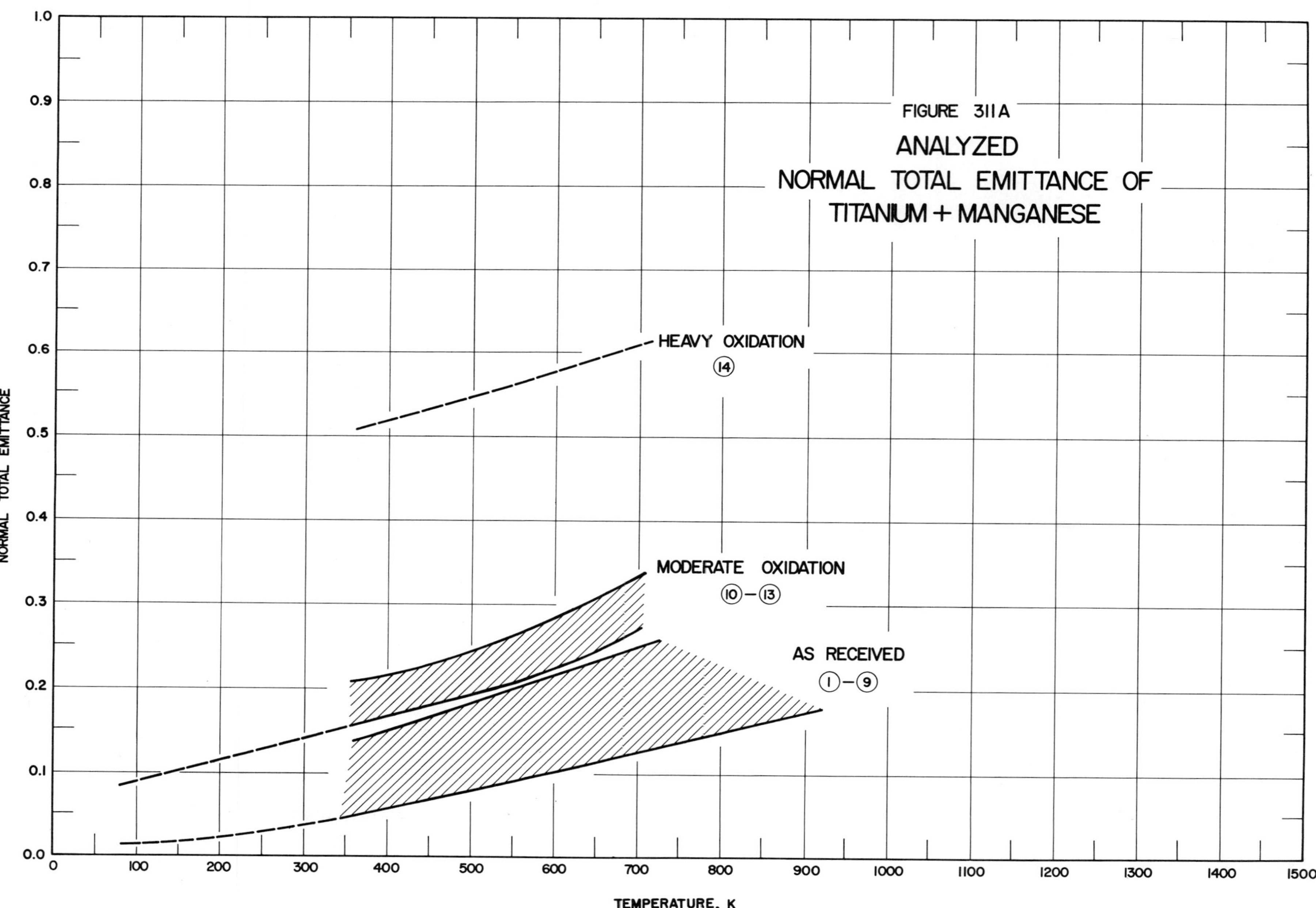
FIGURE 311 A
ANALYZED
NORMAL TOTAL EMITTANCE OF
TITANIUM + MANGANESE
HEAVY OXIDATION
14
MODERATE OXIDATION
10 – 13
AS RECEIVED
1 – 9
NORMAL TOTAL EMITTANCE
0.0
0.1
0.2
0.3
0.4
0.5
0.6
0.7
0.8
0.9
1.0
TEMPERATURE, K
0
100
200
300
400
500
600
700
800
900
1000
1100
1200
1300
1400
1500

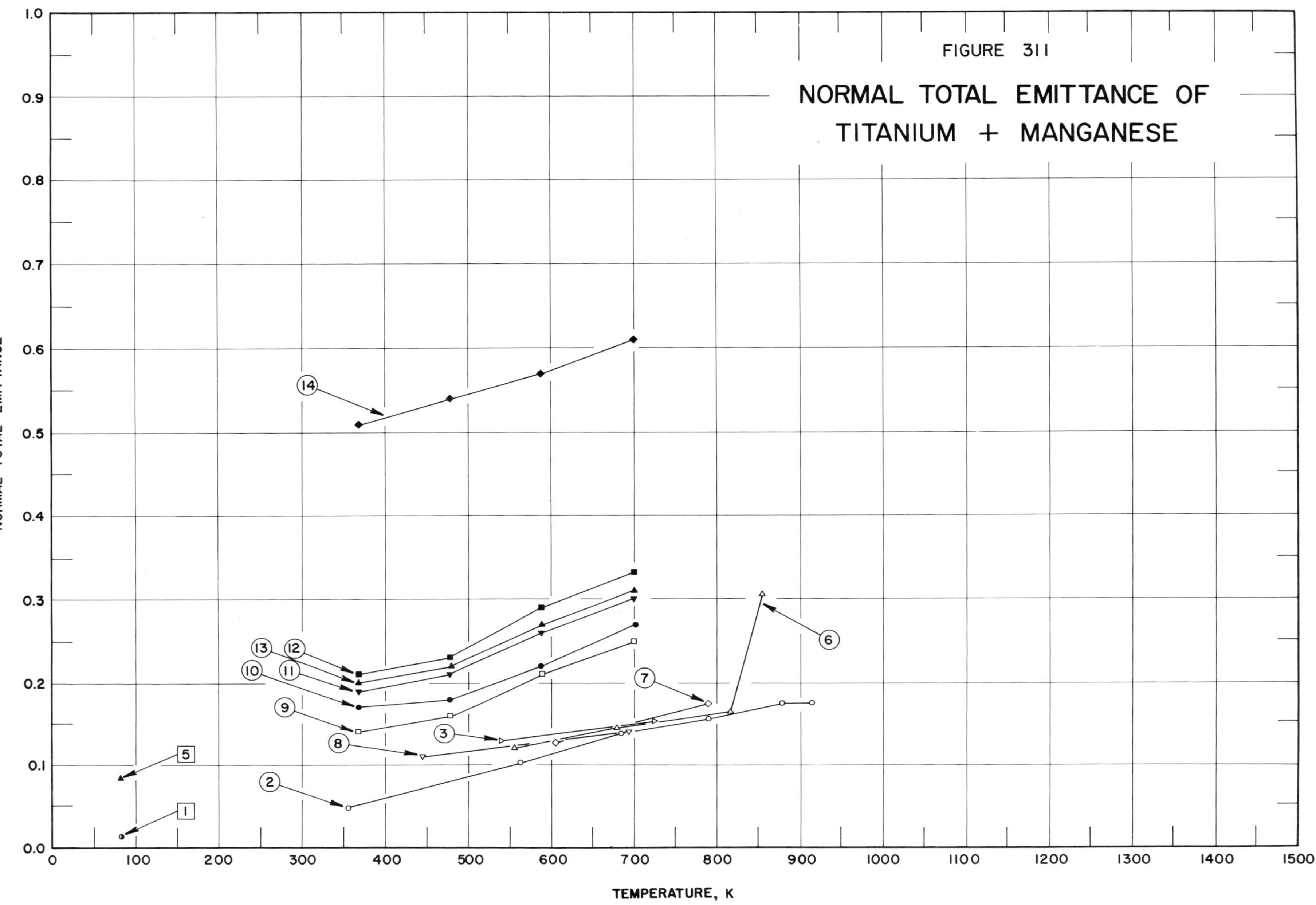

FIGURE 311

NORMAL TOTAL EMITTANCE OF TITANIUM + MANGANESE

SPECIFICATION TABLE NO. 311 NORMAL TOTAL EMITTANCE OF [TITANIUM + MANGANESE] ALLOYS

Curve No.	Ref. No.	Year	Temperature Range, K	Geometry θ'	Reported Error, %	Composition (weight percent), Specifications and Remarks
1	34	1957	83.2	$\sim 0^{\circ}$	±10	Titanium alloy C-110 M; nominal composition: 8 Mn, Ti balance; same results obtained for 3 different surface treatments; as received, cleaned with liquid detergent, and polished; measured in vacuum (5×10^{-4} mm Hg).
2	34	1957	355-916	$\sim 0^{\circ}$	±10	Different sample, same as curve 1 specimen and conditions; increasing temp, cycle 1.
3	34	1957	539-722	$\sim 0^{\circ}$	±10	Above specimen and conditions; cycle 2.
4	34	1957	789	$\sim 0^{\circ}$	±10	Above specimen and conditions; cycle 3.
5	34	1957	83.2	$\sim 0^{\circ}$	±10	Different sample, same as curve 1 specimen and conditions except oxidized in air at red heat for 30 min.
6	34	1957	555-855	$\sim 0^{\circ}$	±10	Different sample, same as curve 5 specimen and conditions; increasing temp, cycle 1.
7	34	1957	605-789	$\sim 0^{\circ}$	±10	Above specimen and conditions; cycle 2.
8	34	1957	444-694	$\sim 0^{\circ}$	±10	Above specimen and conditions; cycle 3.
9	99	1958	367-700	$\sim 0^{\circ}$	<21	Titanium alloy C 110 M; AMS 4908; nominal composition: 8 Mn, Ti balance; measured in air.
10	99	1958	367-700	$\sim 0^{\circ}$	<18	Different sample, same as curve 9 specimen and conditions except heated at 581 K for 306 hrs.
11	99	1958	367-700	$\sim 0^{\circ}$	<16	Different sample, same as curve 9 specimen and conditions except heated at 706 K for 100 hrs.
12	99	1958	367-700	$\sim 0^{\circ}$	<14	Different sample, same as curve 9 specimen and conditions except heated at 711 K for 306 hrs.
13	99	1958	367-700	$\sim 0^{\circ}$	<15	Different sample, same as curve 9 specimen and conditions except heated at 739 K for 303 hrs.
14	99	1958	367-700	$\sim 0^{\circ}$	< 6	Different sample, same as curve 9 specimen and conditions except heated at 813 K for 303 hrs.

DATA TABLE NO. 311 NORMAL TOTAL EMITTANCE OF [TITANIUM + MANGANESE] ALLOYS

[Temperature, T, K; Emittance, ϵ]

T	ϵ	T	ϵ
CURVE 1		CURVE 10	
83.2	0.014	367	0.17
		478	0.18
CURVE 2		589	0.22
		700	0.27
355	0.049		
561	0.101	CURVE 11	
683	0.137		
789	0.153	367	0.19
878	0.174	478	0.21
916	0.174	589	0.26
		700	0.30
CURVE 3			
		CURVE 12	
539	0.129		
722	0.152	367	0.21
		478	0.23
CURVE 4*		589	0.29
		700	0.33
789	0.153		
		CURVE 13	
CURVE 5			
		367	0.20
83.2	0.083	478	0.22
		589	0.27
CURVE 6		700	0.31
555	0.120	CURVE 14	
678	0.145		
816	0.165	367	0.51
855	0.305	478	0.54
		589	0.57
CURVE 7		700	0.61
605	0.125		
789	0.175		
CURVE 8			
444	0.110		
694	0.140		
CURVE 9			
367	0.14		
478	0.16		
589	0.21		
700	0.25		

*Not shown on plot

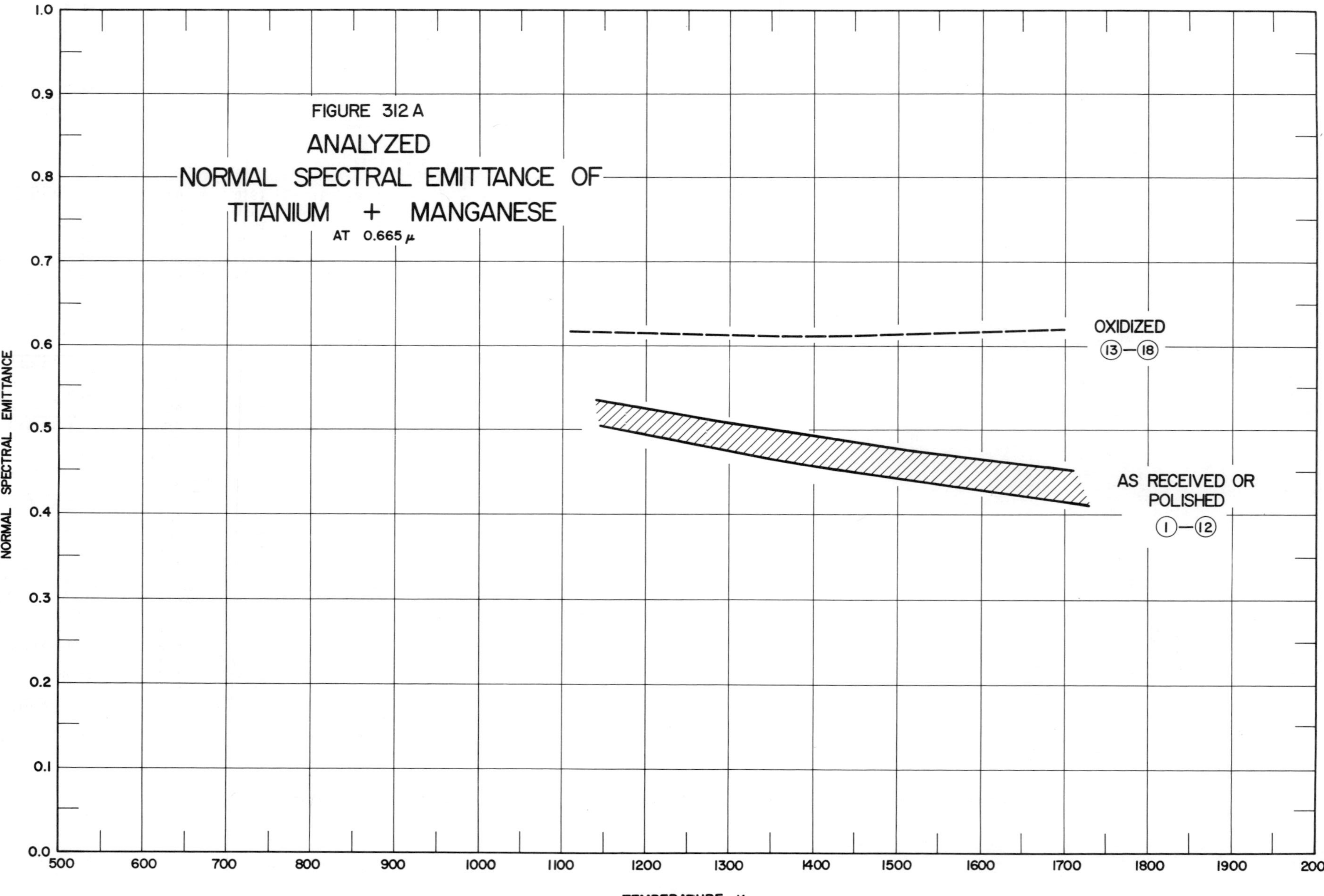
FIGURE 312 A
ANALYZED
NORMAL SPECTRAL EMITTANCE OF
TITANIUM + MANGANESE
AT 0.665 μ
OXIDIZED
(13)—(18)
AS RECEIVED OR
POLISHED
(1)—(12)
NORMAL SPECTRAL EMITTANCE
1.0
0.9
0.8
0.7
0.6
0.5
0.4
0.3
0.2
0.1
0.0
500
600
700
800
900
1000
1100
1200
1300
1400
1500
1600
1700
1800
1900
2000
TEMPERATURE, K

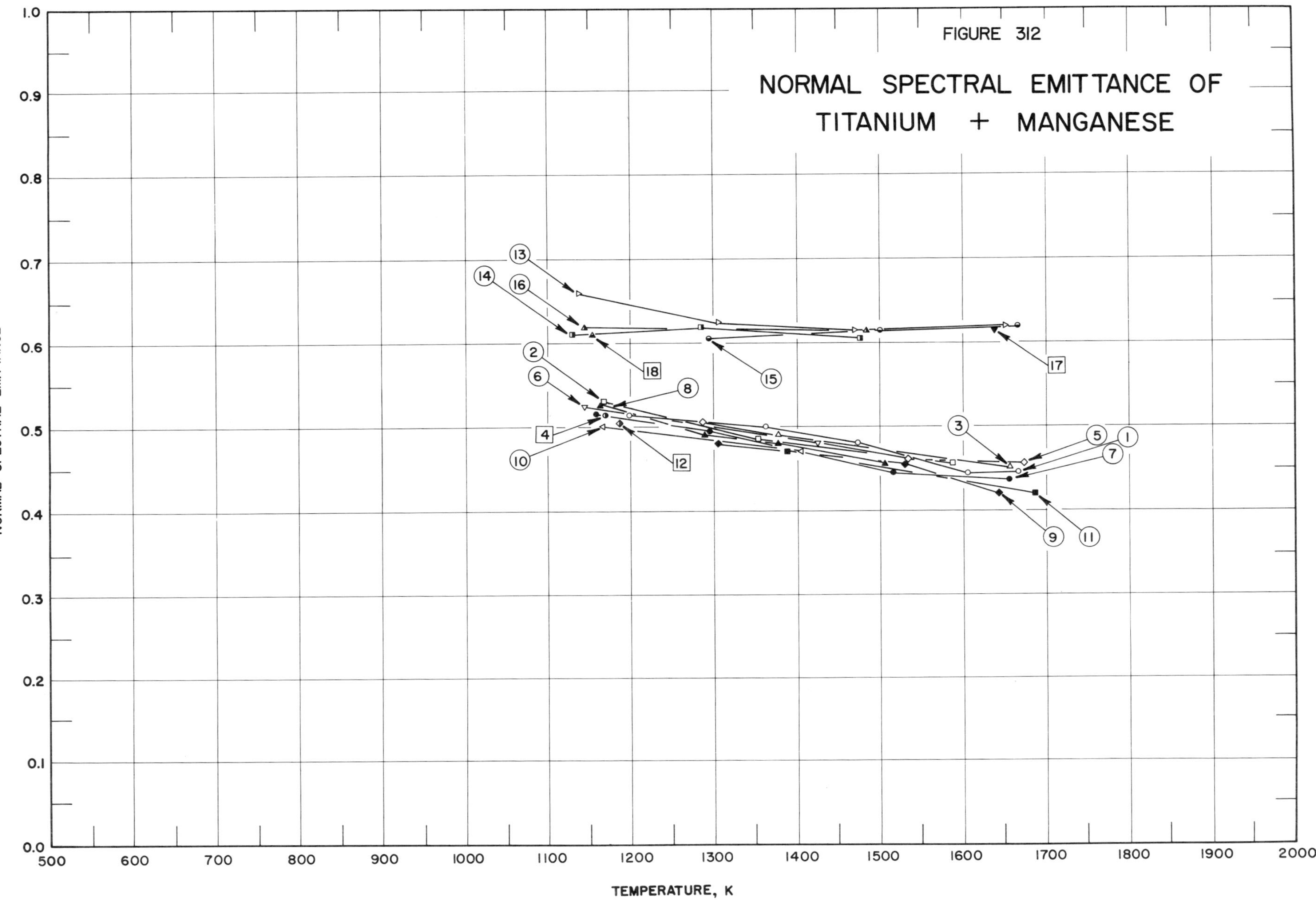
FIGURE 312
NORMAL SPECTRAL EMITTANCE OF
TITANIUM + MANGANESE
NORMAL SPECTRAL EMITTANCE
TEMPERATURE, K
0.0
0.1
0.2
0.3
0.4
0.5
0.6
0.7
0.8
0.9
1.0
500
600
700
800
900
1000
1100
1200
1300
1400
1500
1600
1700
1800
1900
2000
1
2
3
4
5
6
7
8
9
10
11
12
13
14
15
16
17
18

SPECIFICATION TABLE NO. 312 NORMAL SPECTRAL EMITTANCE OF [TITANIUM + MANGANESE] ALLOYS

Curve No.	Ref. No.	Year	Wavelength μ	Temperature Range, K	Geometry θ'	Reported Error, %	Composition (weight percent), Specifications and Remarks
1	34	1957	0.665	1194-1666	~0°		Titanium alloy C 110 M; nominal composition: 8 Mn, Ti balance; same results obtained for 2 different surface treatments; as received, cleaned with liquid detergent; measured in vacuum (5 x 10^{-4} mm Hg); increasing temp, cycle 1.
2	34	1957	0.665	1589-1166	~0°		Above specimen and conditions; decreasing temp, cycle 1.
3	34	1957	0.665	1377-1655	~0°		Above specimen and conditions; cycle 2.
4	34	1957	0.665	1169	~0°		Above specimen and conditions; decreasing temp, cycle 2.
5	34	1957	0.665	1286-1672	~0°		Above specimen and conditions; cycle 3.
6	34	1957	0.665	1422-1144	~0°		Above specimen and conditions; decreasing temp, cycle 3.
7	34	1957	0.665	1158-1655	~0°		Different sample, same as curve 1 specimen and conditions except polished.
8	34	1957	0.665	1505-1161	~0°		Above specimen and conditions; decreasing temp, cycle 1.
9	34	1957	0.665	1305-1644	~0°		Above specimen and conditions; cycle 2.
10	34	1957	0.665	1402-1164	~0°		Above specimen and conditions; decreasing temp, cycle 2.
11	34	1957	0.665	1389-1686	~0°		Above specimen and conditions; cycle 3.
12	34	1957	0.665	1186	~0°		Above specimen and conditions; decreasing temp, cycle 3.
13	34	1957	0.665	1136-1650	~0°		Different sample, same as curve 1 specimen and conditions except oxidized in air at red heat for 30 min.
14	34	1957	0.665	1477-1130	~0°		Above specimen and conditions; decreasing temp, cycle 1.
15	34	1957	0.665	1294-1666	~0°		Above specimen and conditions; cycle 2.
16	34	1957	0.665	1483-1144	~0°		Above specimen and conditions; decreasing temp, cycle 2.
17	34	1957	0.665	1639	~0°		Above specimen and conditions; cycle 3.
18	34	1957	0.665	1152	~0°		Above specimen and conditions; decreasing temp, cycle 3.

DATA TABLE NO. 312 NORMAL SPECTRAL EMITTANCE OF [TITANIUM + MANGANESE] ALLOYS

[Temperature, T, K; Emittance, ∈; Wavelength, λ, μ]

T	∈
CURVE 1	
λ = 0.665	
1194	0.515
1361	0.500
1472	0.480
1605	0.445
1666	0.445
CURVE 2	
λ = 0.665	
1589	0.455
1352	0.485
1166	0.530
CURVE 3	
λ = 0.665	
1377	0.490
1655	0.450
CURVE 4	
λ = 0.665	
1169	0.515
CURVE 5	
λ = 0.665	
1286	0.505
1533	0.460
1672	0.455
CURVE 6	
λ = 0.665	
1422	0.480
1144	0.525
CURVE 7	
λ = 0.665	
1158	0.515
1294	0.495
1516	0.445
1655	0.435
CURVE 8	
λ = 0.665	
1505	0.455
1377	0.480
1289	0.490
1161	0.525
CURVE 9	
λ = 0.665	
1305	0.480
1530	0.455
1644	0.420
CURVE 10	
λ = 0.665	
1402	0.47
1164	0.50
CURVE 11	
λ = 0.665	
1389	0.470
1686	0.420
CURVE 12	
λ = 0.665	
1186	0.505
CURVE 13	
λ = 0.665	
1136	0.660
1305	0.625
1469	0.615
1650	0.620
CURVE 14	
λ = 0.665	
1477	0.605
1283	0.620
1130	0.610
CURVE 15	
λ = 0.665	
1294	0.605
1500	0.615
1666	0.620
CURVE 16	
λ = 0.665	
1483	0.615
1144	0.620
CURVE 17	
λ = 0.665	
1639	0.615
CURVE 18	
λ = 0.665	
1152	0.610

* Not shown on plot

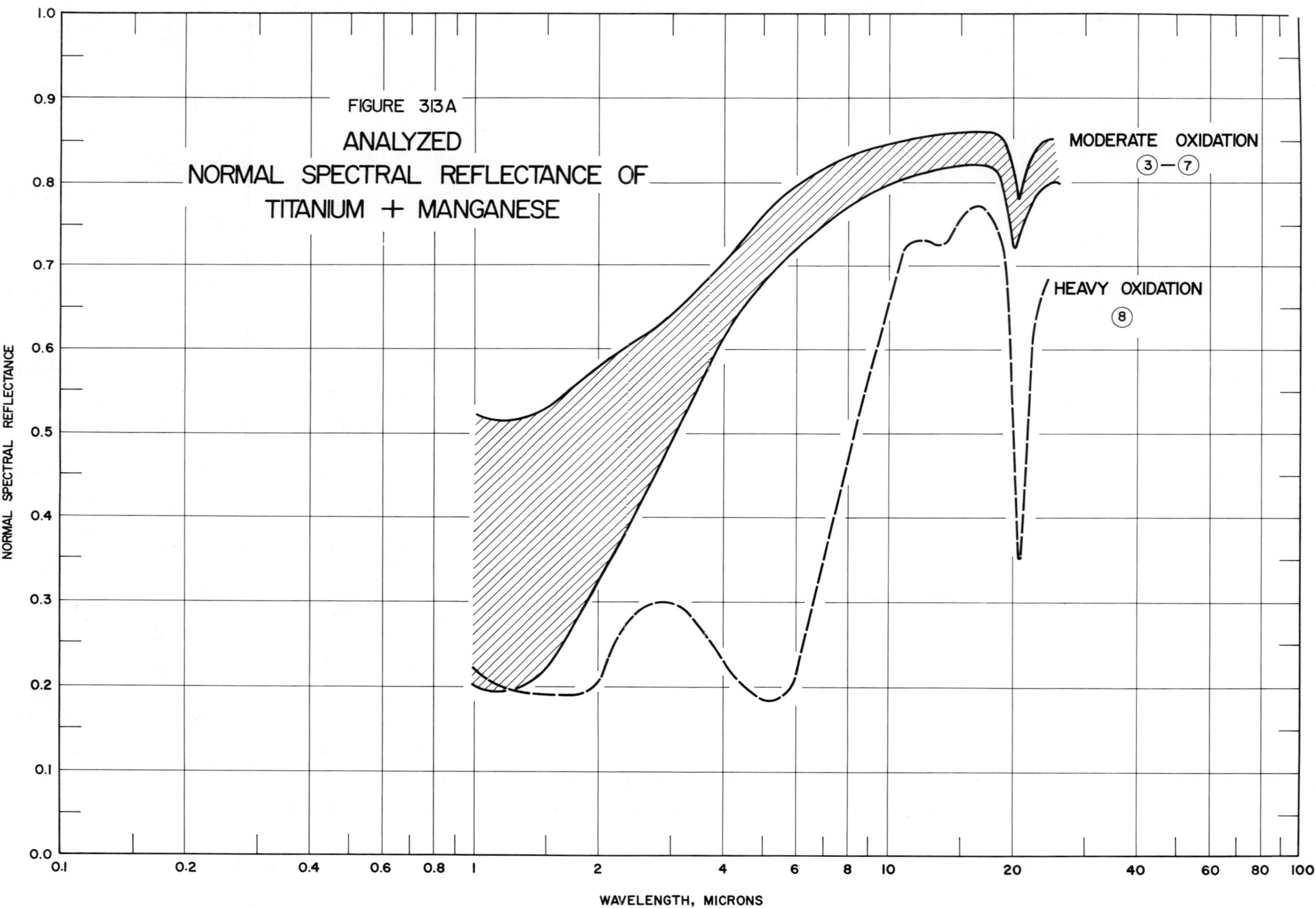

FIGURE 313A
ANALYZED
NORMAL SPECTRAL REFLECTANCE OF
TITANIUM + MANGANESE

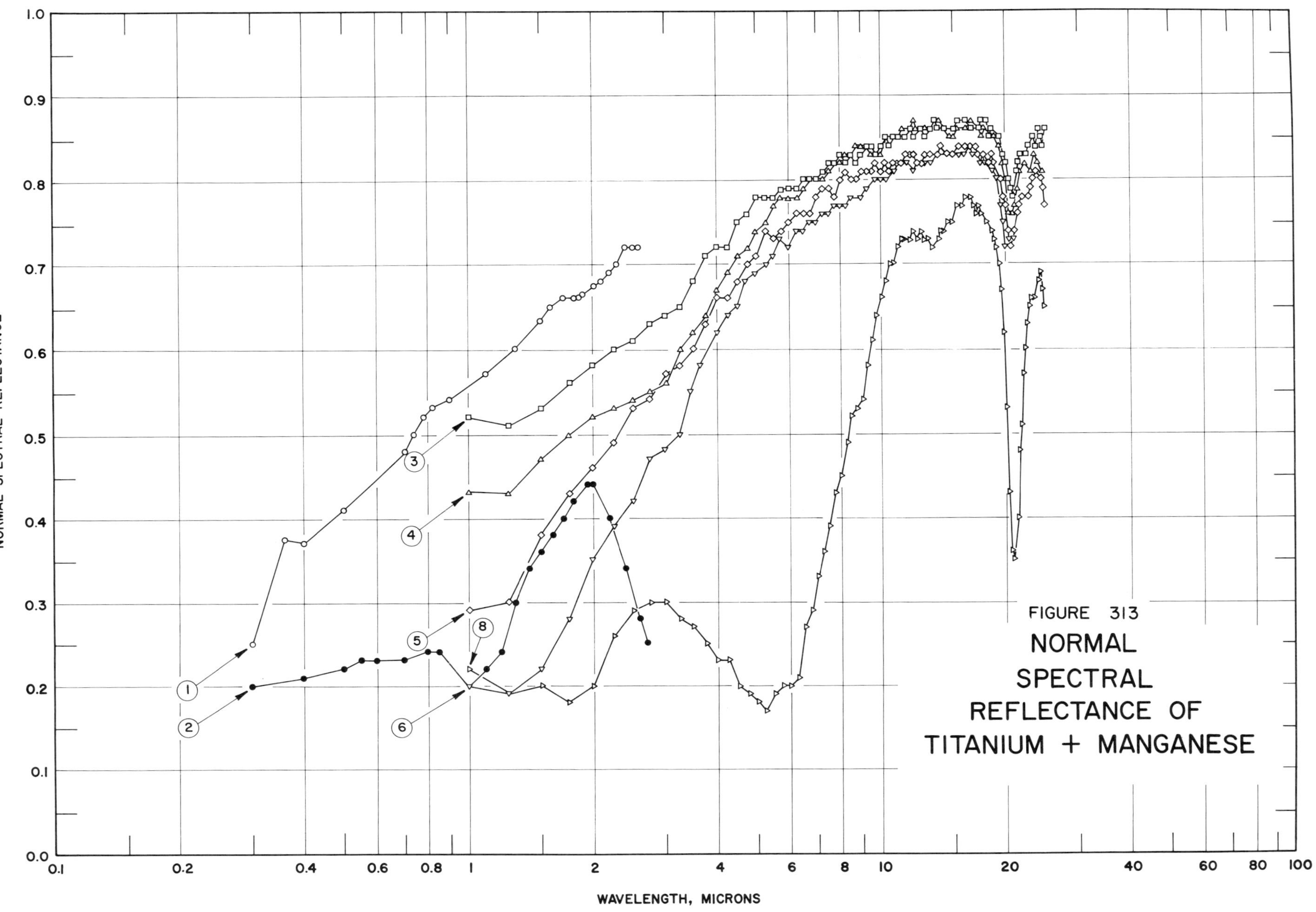
FIGURE 313
NORMAL
SPECTRAL
REFLECTANCE OF
TITANIUM + MANGANESE
NORMAL SPECTRAL REFLECTANCE
WAVELENGTH, MICRONS
0.0
0.1
0.2
0.3
0.4
0.5
0.6
0.7
0.8
0.9
1.0
0.1
0.2
0.4
0.6
0.8
1
2
4
6
8
10
20
40
60
80
100
1
2
3
4
5
6
8

SPECIFICATION TABLE NO. 313 NORMAL SPECTRAL REFLECTANCE OF [TITANIUM + MANGANESE] ALLOYS

Curve No.	Ref. No.	Year	Temperature K	Wavelength Range, μ	Geometry θ	θ'	ω'	Reported Error, %	Composition (weight percent), Specifications and Remarks
1	34	1957	298	0.30-2.60	9°		2π	±4	Titanium alloy C 110 M; nominal composition: 8 Mn, Ti balance; as received; data extracted from smooth curve.
2	146	1958	298	0.30-2.70	9°		2π		Titanium alloy C 110 M; nominal composition: 8 Mn, Ti balance; oxidized at 922 K for 30 min.
3	99	1958	311	1.0-25.00	~0°		2π		Titanium alloy C 110 M; AMS 4908; nominal composition: 8 Mn, Ti balance; hohlraum at 1089 K; converted from $R(2\pi, 0°)$.
4	99	1958	311	1.0-25.00	~0°		2π		Different sample, same as curve 3 specimen and conditions except heated at 581 K for 306 hrs.
5	99	1958	311	1.0-25.00	~0°		2π		Different sample, same as curve 3 specimen and conditions except heated at 706 K for 100 hrs.
6	99	1958	311	1.0-25.00	~0°		2π		Different sample, same as curve 3 specimen and conditions except heated at 711 K for 306 hrs.
7	99	1958	311	1.0-25.00	~0°		2π		Different sample, same as curve 3 specimen and conditions except heated at 739 K for 303 hrs.
8	99	1958	311	1.0-25.00	~0°		2π		Different sample, same as curve 3 specimen and conditions except heated at 813 K for 303 hrs.

DATA TABLE NO. 313 NORMAL SPECTRAL REFLECTANCE OF [TITANIUM + MANGANESE] ALLOYS

[Wavelength, λ, μ; Reflectance, ρ; Temperature, T, K]

λ	ρ
CURVE 1	**T = 298**
0.30	0.250
0.36	0.375
0.40	0.370
0.50	0.410
0.70	0.480
0.74	0.500
0.78	0.520
0.82	0.530
0.90	0.540
1.10	0.570
1.30	0.600
1.50	0.635
1.58	0.650
1.70	0.660
1.82	0.660
1.86	0.660
1.90	0.665
2.02	0.675
2.10	0.680
2.20	0.690
2.30	0.700
2.40	0.720
2.50	0.720
2.60	0.720
CURVE 2	**T = 298**
0.30	0.20
0.40	0.21
0.50	0.22
0.55	0.23
0.60	0.23
0.70	0.23
0.80	0.24
0.85	0.24
1.00	0.20*
1.10	0.22
1.20	0.24
1.30	0.30
1.40	0.34
1.50	0.36
1.60	0.38
1.70	0.40
CURVE 2 (cont.)	
1.80	0.42
1.95	0.44
2.00	0.44
2.20	0.40
2.40	0.34
2.60	0.28
2.70	0.25
CURVE 3	**T = 311**
1.0	0.52
1.25	0.51
1.50	0.53
1.75	0.56
2.00	0.58
2.25	0.60
2.50	0.61
2.75	0.63
3.00	0.64
3.25	0.65
3.50	0.68
3.75	0.71
4.00	0.72
4.25	0.72
4.50	0.75
4.75	0.76
5.00	0.78
5.25	0.78
5.50	0.78
5.75	0.79
6.00	0.79
6.25	0.79
6.50	0.80
6.75	0.80
7.00	0.80
7.25	0.81
7.50	0.82
7.75	0.82
8.00	0.83
8.25	0.82
8.50	0.83
8.75	0.82
9.00	0.83
9.25	0.84
CURVE 3 (cont.)	
9.50	0.84
9.75	0.83
10.00	0.84
10.25	0.85
10.50	0.84
10.75	0.85
11.00	0.85
11.25	0.85*
11.50	0.85
11.75	0.86
12.00	0.85
12.25	0.85*
12.50	0.86
12.75	0.85
13.00	0.85*
13.25	0.86
13.50	0.87
13.75	0.86
14.00	0.86*
14.25	0.86
14.50	0.86*
14.75	0.86*
15.00	0.86
15.25	0.87
15.50	0.87*
15.75	0.87*
16.00	0.87
16.25	0.87*
16.50	0.86
16.75	0.86*
17.00	0.86*
17.25	0.86
17.50	0.87
17.75	0.87*
18.00	0.87
18.25	0.85
18.50	0.86
18.75	0.86*
19.00	0.86*
19.25	0.85
19.50	0.85*
19.75	0.83
20.00	0.82
20.25	0.80
20.50	0.79
CURVE 3 (cont.)	
20.75	0.78
21.00	0.80*
21.25	0.81
21.50	0.82
21.75	0.83
22.00	0.83*
22.25	0.83*
22.50	0.83
22.75	0.84
23.00	0.84*
23.25	0.85
23.50	0.85*
23.75	0.84
24.00	0.86
24.25	0.85
24.50	0.84
24.75	0.85*
25.00	0.86
CURVE 4	**T = 311**
1.0	0.43
1.25	0.43
1.50	0.47
1.75	0.50
2.00	0.52
2.25	0.53
2.50	0.54
2.75	0.55
3.00	0.56
3.25	0.60
3.50	0.62
3.75	0.64
4.00	0.67
4.25	0.69
4.50	0.71
4.75	0.72
5.00	0.74
5.25	0.75
5.50	0.77
5.75	0.78
6.00	0.78
6.25	0.78
6.50	0.79
CURVE 4 (cont.)	
6.75	0.80*
7.00	0.80*
7.25	0.80
7.50	0.81
7.75	0.82*
8.00	0.82
8.25	0.83
8.50	0.83*
8.75	0.84
9.00	0.84
9.25	0.84*
9.50	0.83
9.75	0.83*
10.00	0.83
10.25	0.85*
10.50	0.85*
10.75	0.85*
11.00	0.85*
11.25	0.86
11.50	0.86*
11.75	0.86*
12.00	0.87
12.25	0.86
12.50	0.86*
12.75	0.86
13.00	0.86*
13.25	0.86*
13.50	0.87*
13.75	0.87
14.00	0.87*
14.25	0.86*
14.50	0.86*
14.75	0.85
15.00	0.85
15.25	0.86
15.50	0.87*
15.75	0.87*
16.00	0.86
16.25	0.87*
16.50	0.87
16.75	0.86*
17.00	0.86*
17.25	0.87*
17.50	0.85
17.75	0.86
CURVE 4 (cont.)	
18.00	0.85*
18.25	0.85*
18.50	0.85*
18.75	0.85
19.00	0.85*
19.25	0.84
19.50	0.84*
19.75	0.82
20.00	0.80
20.25	0.76
20.50	0.76*
20.75	0.76
21.00	0.77
21.25	0.79
21.50	0.81
21.75	0.81*
22.00	0.81*
22.25	0.82
22.50	0.82*
22.75	0.83*
23.00	0.81
23.25	0.84*
23.50	0.83
23.75	0.84*
24.00	0.82
24.25	0.82*
24.50	0.82*
24.75	0.81
25.00	0.81*
CURVE 5	**T = 311**
1.0	0.29
1.25	0.30
1.50	0.38
1.75	0.43
2.00	0.46
2.25	0.49
2.50	0.53
2.75	0.54
3.00	0.57
3.25	0.58
3.50	0.60
3.75	0.63
CURVE 5 (cont.)	
4.00	0.66
4.25	0.66
4.50	0.68
4.75	0.70
5.00	0.71
5.25	0.74
5.50	0.73
5.75	0.74
6.00	0.75
6.25	0.76
6.50	0.76
6.75	0.76
7.00	0.78
7.25	0.79
7.50	0.79
7.75	0.78
8.00	0.80
8.25	0.81
8.50	0.80
8.75	0.80
9.00	0.81
9.25	0.81
9.50	0.81
9.75	0.82
10.00	0.81
10.25	0.82
10.50	0.81
10.75	0.82
11.00	0.82
11.25	0.82
11.50	0.83
11.75	0.83
12.00	0.83
12.25	0.82
12.50	0.82
12.75	0.82
13.00	0.83*
13.25	0.83
13.50	0.83*
13.75	0.83
14.00	0.84
14.25	0.84*
14.50	0.83
14.75	0.83
15.00	0.83
15.25	0.83*
CURVE 5 (cont.)	
15.50	0.84
15.75	0.84*
16.00	0.84
16.25	0.84*
16.50	0.84
16.75	0.84*
17.00	0.84
17.25	0.83
17.50	0.83*
17.75	0.83
18.00	0.83*
18.25	0.82
18.50	0.83
18.75	0.82
19.00	0.82*
19.25	0.82*
19.50	0.80
19.75	0.78
20.00	0.77
20.25	0.74
20.50	0.72
20.75	0.74
21.00	0.74
21.25	0.76
21.50	0.77
21.75	0.77*
22.00	0.78
22.25	0.78*
22.50	0.78*
22.75	0.78
23.00	0.79
23.25	0.80
23.50	0.80*
23.75	0.81
24.00	0.81*
24.25	0.81*
24.50	0.80
24.75	0.79
25.00	0.77

* Not shown on plot

DATA TABLE NO. 313 (continued)

λ	ρ
CURVE 6	
T = 311	
1.0	0.20
1.25	0.19
1.50	0.22
1.75	0.28
2.00	0.35
2.25	0.39
2.50	0.42
2.75	0.47
3.00	0.48
3.25	0.50
3.50	0.55
3.75	0.58
4.00	0.62
4.25	0.64
4.50	0.65
4.75	0.68
5.00	0.69
5.25	0.70
5.50	0.71
5.75	0.73
6.00	0.72
6.25	0.74
6.50	0.74
6.75	0.75
7.00	0.75
7.25	0.76
7.50	0.76
7.75	0.77
8.00	0.77
8.25	0.77
8.50	0.78
8.75	0.78*
9.00	0.78
9.25	0.79
9.50	0.79*
9.75	0.80
10.00	0.80
10.25	0.80
10.50	0.81*
10.75	0.81
11.00	0.82*
11.25	0.82*
11.50	0.82
11.75	0.82*
12.00	0.81

λ	ρ
CURVE 6 (cont.)	
12.25	0.82*
12.50	0.82*
12.75	0.82
13.00	0.82*
13.25	0.82
13.50	0.82*
13.75	0.83*
14.00	0.83*
14.25	0.83*
14.50	0.83*
14.75	0.83*
15.00	0.83*
15.25	0.83
15.50	0.83
15.75	0.83*
16.00	0.83
16.25	0.84*
16.50	0.84*
16.75	0.83
17.00	0.83*
17.25	0.83*
17.50	0.82
17.75	0.83*
18.00	0.82
18.25	0.82
18.50	0.82*
18.75	0.81
19.00	0.81*
19.25	0.80
19.50	0.77
19.75	0.75
20.00	0.67
20.25	0.67*
20.50	0.68
20.75	0.72*
21.00	0.73
21.25	0.74*
21.50	0.76*
21.75	0.77*
22.00	0.78*
22.25	0.78*
22.50	0.78*
22.75	0.78*
23.00	0.77*
23.25	0.78*
23.50	0.76*

λ	ρ
CURVE 6 (cont.)	
23.75	0.77*
24.00	0.76*
24.25	0.77*
24.50	0.76*
24.75	0.78*
25.00	0.76*
CURVE 7*	
T = 311	
1.0	0.23
1.25	0.22
1.50	0.20
1.75	0.18
2.00	0.17
2.25	0.17
2.50	0.19
2.75	0.22
3.00	0.24
3.25	0.26
3.50	0.30
3.75	0.34
4.00	0.37
4.25	0.42
4.50	0.44
4.75	0.47
5.00	0.52
5.25	0.54
5.50	0.57
5.75	0.59
6.00	0.63
6.25	0.63
6.50	0.64
6.75	0.67
7.00	0.68
7.25	0.69
7.50	0.71
7.75	0.73
8.00	0.74
8.25	0.73
8.50	0.74
8.75	0.75
9.00	0.75
9.25	0.77
9.50	0.77
9.75	0.77

λ	ρ
CURVE 7 (cont.)*	
10.00	0.78
10.25	0.80
10.50	0.80
10.75	0.80
11.00	0.81
11.25	0.80
11.50	0.82
11.75	0.82
12.00	0.81
12.25	0.80
12.50	0.80
12.75	0.80
13.00	0.81
13.25	0.80
13.50	0.82
13.75	0.80
14.00	0.81
14.25	0.81
14.50	0.82
14.75	0.82
15.00	0.82
15.25	0.82
15.50	0.81
15.75	0.82
16.00	0.82
16.25	0.81
16.50	0.81
16.75	0.81
17.00	0.81
17.25	0.80
17.50	0.79
17.75	0.80
18.00	0.79
18.25	0.79
18.50	0.78
18.75	0.78
19.00	0.71
19.25	0.76
19.50	0.75
19.75	0.70
20.00	0.63
20.25	0.58
20.50	0.55
20.75	0.56
21.00	0.60
21.25	0.64

λ	ρ
CURVE 7 (cont.)*	
21.50	0.68
21.75	0.70
22.00	0.70
22.25	0.72
22.50	0.72
22.75	0.73
23.00	0.74
23.25	0.73
23.50	0.74
23.75	0.74
24.00	0.74
24.25	0.75
24.50	0.73
24.75	0.72
25.00	0.71
CURVE 8	
T = 311	
1.0	0.22
1.25	0.19*
1.50	0.20
1.75	0.18
2.00	0.20
2.25	0.26
2.50	0.29
2.75	0.30
3.00	0.30
3.25	0.28
3.50	0.27
3.75	0.25
4.00	0.23
4.25	0.23
4.50	0.20
4.75	0.19
5.00	0.18
5.25	0.17
5.50	0.19
5.75	0.20
6.00	0.20
6.25	0.21
6.50	0.27
6.75	0.29
7.00	0.33
7.25	0.36
7.50	0.39

λ	ρ
CURVE 8 (cont.)	
7.75	0.43
8.00	0.45
8.25	0.49
8.50	0.52
8.75	0.53
9.00	0.54
9.25	0.58
9.50	0.61
9.75	0.64
10.00	0.66
10.25	0.68
10.50	0.70
10.75	0.70
11.00	0.72
11.25	0.73
11.50	0.73
11.75	0.73
12.00	0.74
12.25	0.73
12.50	0.74
12.75	0.73
13.00	0.73
13.25	0.72
13.50	0.72*
13.75	0.73
14.00	0.74
14.25	0.74
14.50	0.75
14.75	0.75*
15.00	0.75
15.25	0.77
15.50	0.77*
15.75	0.77
16.00	0.78
16.25	0.78*
16.50	0.78
16.75	0.77
17.00	0.76
17.25	0.77
17.50	0.76
17.75	0.76*
18.00	0.75
18.25	0.75*
18.50	0.74
18.75	0.73
19.00	0.72

λ	ρ
CURVE 8 (cont.)	
19.25	0.70
19.50	0.67
19.75	0.62
20.00	0.53
20.25	0.43
20.50	0.36
20.75	0.35
21.00	0.35*
21.25	0.40
21.50	0.48
21.75	0.51
22.00	0.57
22.25	0.60
22.50	0.63
22.75	0.65
23.00	0.66
23.25	0.66*
23.50	0.66
23.75	0.66*
24.00	0.68
24.25	0.69
24.50	0.67
24.75	0.65
25.00	0.65*

* Not shown on plot

SPECIFICATION TABLE NO. 314 NORMAL SOLAR ABSORPTANCE OF [TITANIUM + MANGANESE] ALLOYS

Curve No.	Ref. No.	Year	Temperature Range, K	Geometry θ	Reported Error, %	Composition (weight percent), Specifications and Remarks
1	34	1957	298	9°		Titanium alloy C 110 M; nominal composition: 8 Mn, Ti balance; as received (bright); computed from spectral reflectance data for sea level conditions.
2	34	1957	298	9°		Above specimen and conditions except computed for above atmosphere conditions.
3	146	1958	298	9°		Titanium alloy C 110 M; nominal composition: 8 Mn, Ti balance; oxidized; computed from spectral reflectance data for sea level conditions.
4	146	1958	298	9°		Above specimen and conditions except computed for above atmosphere conditions.

DATA TABLE NO. 314 NORMAL SOLAR ABSORPTANCE OF [TITANIUM + MANGANESE] ALLOYS

[Temperature, T, K; Absorptance, α]

T	α
CURVE 1*	
298	0.506
CURVE 2*	
298	0.504
CURVE 3*	
298	0.763
CURVE 4*	
298	0.762

* Not shown on plot

SPECIFICATION TABLE NO. 315 HEMISPHERICAL TOTAL EMITTANCE OF [TUNGSTEN + MOLYBDENUM] ALLOYS

Curve No.	Ref. No.	Year	Temperature Range, K	Reported Error, %	Composition (weight percent), Specifications and Remarks
1	84	1962	2400-3200		80 W, 20 Mo; polished with successively finer abrasive papers (numbers 1, 0, 00, 000, and 0000); measured in argon.

DATA TABLE NO. 315 HEMISPHERICAL TOTAL EMITTANCE OF [TUNGSTEN + MOLYBDENUM] ALLOYS

[Temperature, T, K; Emittance, ϵ]

T	ϵ
CURVE 1*	
2400	0.30
2600	0.30
2800	0.30
3000	0.30
3200	0.30

* Not shown on plot

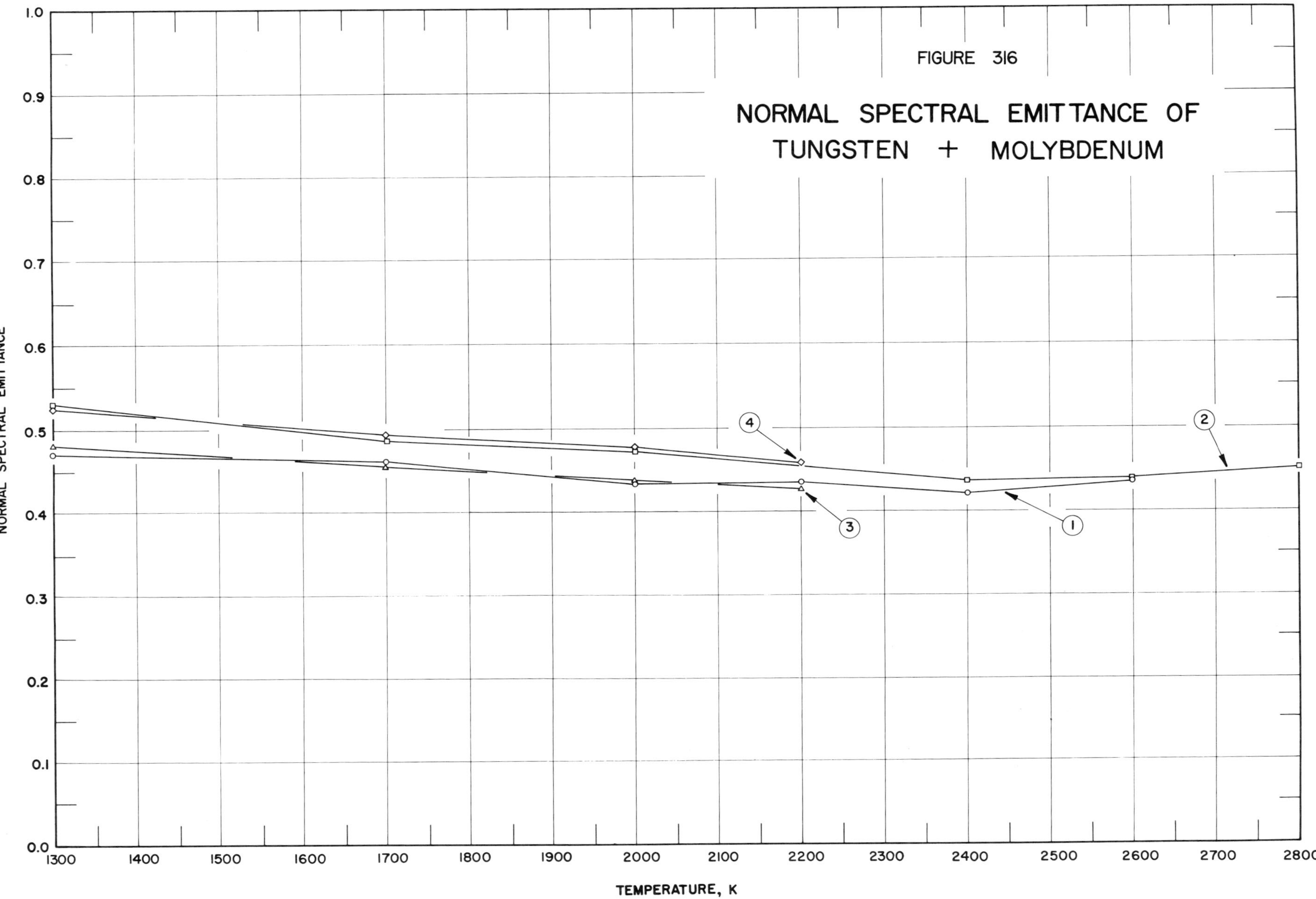
FIGURE 316
NORMAL SPECTRAL EMITTANCE OF
TUNGSTEN + MOLYBDENUM
NORMAL SPECTRAL EMITTANCE
TEMPERATURE, K
1.0
0.9
0.8
0.7
0.6
0.5
0.4
0.3
0.2
0.1
0.0
1300
1400
1500
1600
1700
1800
1900
2000
2100
2200
2300
2400
2500
2600
2700
2800
4
2
3
1

SPECIFICATION TABLE NO. 316 NORMAL SPECTRAL EMITTANCE OF [TUNGSTEN + MOLYBDENUM] ALLOYS

Curve No.	Ref. No.	Year	Wavelength μ	Temperature Range, K	Geometry θ'	Reported Error, %	Composition (weight percent), Specifications and Remarks
1	77	1936	0.660	1300-2600	~0°	±2	62.5 W, 37.5 Mo; extruded from paste; sintered in hydrogen; polished electrolytically in KOH bath; polished with 00, 000, and 0000 polishing papers; measured in vacuum (10^{-5} mm Hg).
2	77	1936	0.660	1300-2800	~0°	±2	87.5 W, 12.5 Mo; specimen preparation same as curve 1.
3	177	1933	0.660	1300-2200	~0°	±2	62.5 W, 37.5 Mo; sintered at 1873 K for 1 hr; electrolytically polished in KOH and then by using 00, 000, 0000 polishing papers; measured in vacuum.
4	177	1933	0.660	1300-2200	~0°	±2	87.5 W, 12.5 Mo; specimen preparation same as curve 1.
5	84	1962	0.65	2400-3200	~0°		80 W, 20 Mo; polished with successively finer abrasive papers (numbers 1, 0, 00, 000, and 0000); measured in argon.

DATA TABLE NO. 316 NORMAL SPECTRAL EMITTANCE OF [TUNGSTEN + MOLYBDENUM] ALLOYS

[Temperature, T, K; Emittance, ∈; Wavelength, λ, μ]

T	∈
CURVE 1 $\lambda = 0.660$	
1300	0.470
1700	0.460
2000	0.433
2200	0.435
2400	0.420
2600	0.435
CURVE 2 $\lambda = 0.660$	
1300	0.530
1700	0.485
2000	0.470
2400	0.435
2600	0.440
2800	0.450
CURVE 3 $\lambda = 0.660$	
1300	0.480
1700	0.456
2000	0.438
2200	0.427
CURVE 4 $\lambda = 0.660$	
1300	0.526
1700	0.491
2000	0.475
2200	0.457
CURVE 5* $\lambda = 0.65$	
2400	0.34
2600	0.34
2800	0.34
3000	0.34
3200	0.34

* Not shown on plot

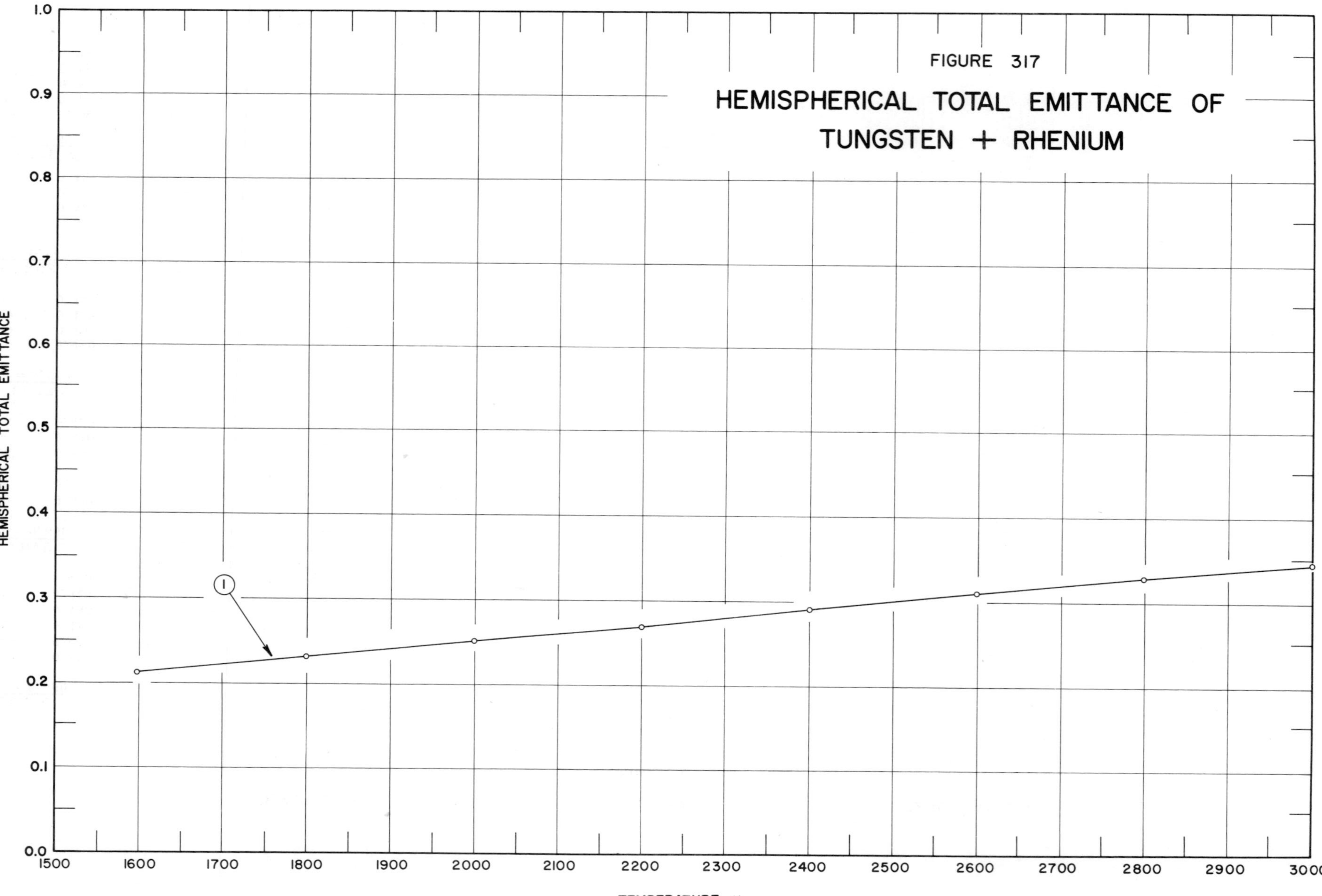
FIGURE 317
HEMISPHERICAL TOTAL EMITTANCE OF
TUNGSTEN + RHENIUM
HEMISPHERICAL TOTAL EMITTANCE
TEMPERATURE, K
1.0
0.9
0.8
0.7
0.6
0.5
0.4
0.3
0.2
0.1
0.0
1500
1600
1700
1800
1900
2000
2100
2200
2300
2400
2500
2600
2700
2800
2900
3000
1

SPECIFICATION TABLE NO. 317 HEMISPHERICAL TOTAL EMITTANCE OF [TUNGSTEN + RHENIUM] ALLOYS

Curve No.	Ref. No.	Year	Temperature Range, K	Reported Error, %	Composition (weight percent), Specifications and Remarks
1	335	1966	1600-3000	± 10	VR-27-VT; 27 Re; produced by arc melting in vacuum; specimen made from ingot subjected to rotary swaging; surface ground to a class 8 finish (Russian standard); annealed 2 hrs at 2200 K in vacuum; measured in vacuum (1-5 x 10^{-5} mm Hg).

DATA TABLE NO. 317 HEMISPHERICAL TOTAL EMITTANCE OF [TUNGSTEN + RHENIUM] ALLOYS

[Temperature, T, K; Emittance, ϵ]

T	ϵ
CURVE 1	
1600	0.211
1800	0.231
2000	0.251
2200	0.270
2400	0.290
2600	0.310
2800	0.329
3000	0.347

SPECIFICATION TABLE NO. 318 NORMAL SPECTRAL EMITTANCE OF [TUNGSTEN + RHENIUM] ALLOYS

Curve No.	Ref. No.	Year	Wavelength μ	Temperature Range, K	Geometry θ'	Reported Error, %	Composition (weight percent), Specifications and Remarks
1	289	1966	0.65	1229–1753	~0°		24.0 Re, 0.026 Si, 0.017 Fe, 0.0033 Ca, 0.00083 Cu, 0.00092 Mg, and balance W; highly polished.

DATA TABLE NO. 318 NORMAL SPECTRAL EMITTANCE OF [TUNGSTEN + RHENIUM] ALLOYS

[Temperature, T, K; Emittance, ϵ; Wavelength, λ, μ]

T	ϵ
CURVE 1*	
$\lambda = 0.65$	
1229	0.500
1328	0.489
1414	0.456
1487	0.445
1559	0.425
1594	0.419
1615	0.420
1691	0.416
1753	0.410

* Not shown on plot

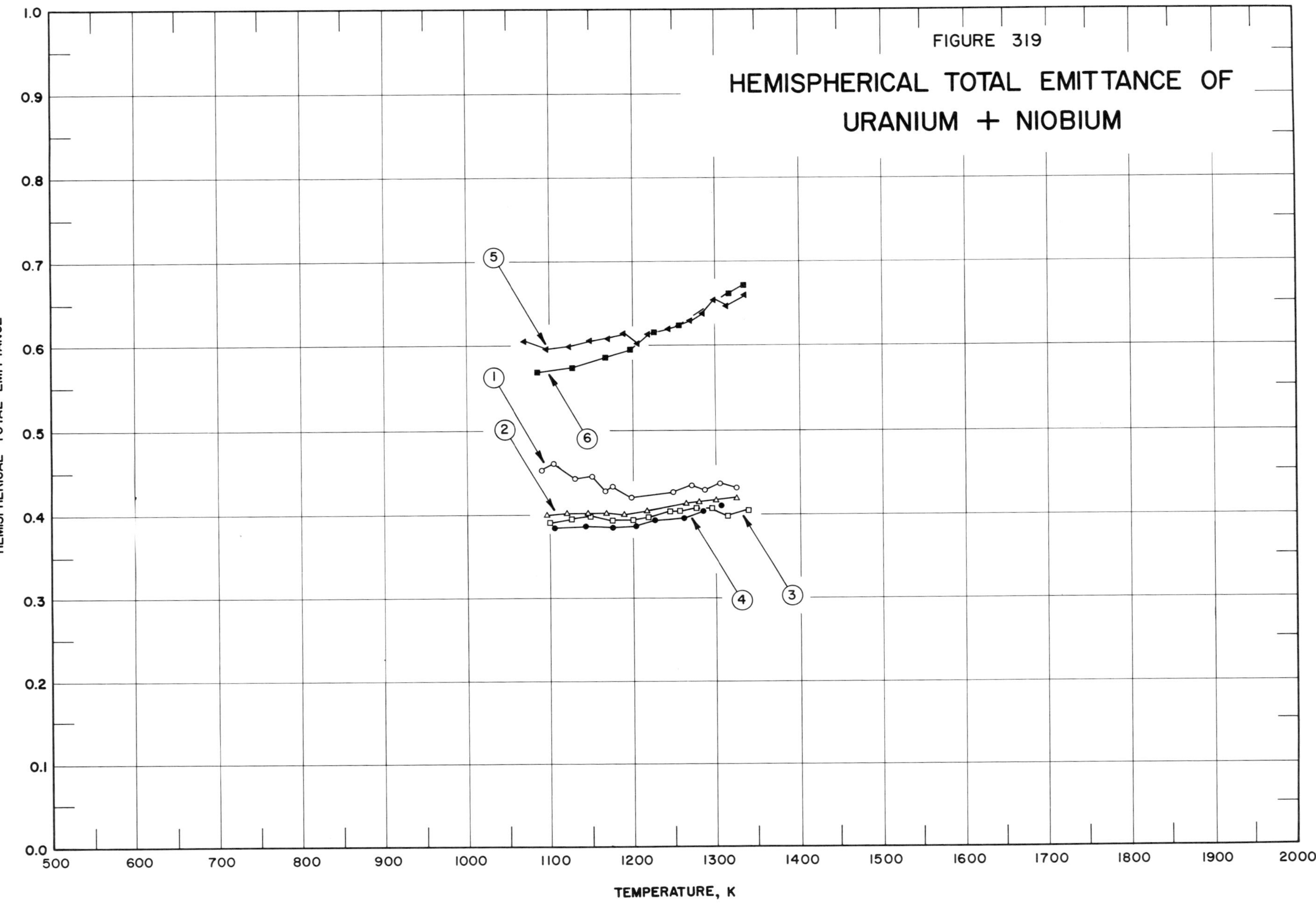
FIGURE 319
HEMISPHERICAL TOTAL EMITTANCE OF
URANIUM + NIOBIUM
HEMISPHERICAL TOTAL EMITTANCE
1.0
0.9
0.8
0.7
0.6
0.5
0.4
0.3
0.2
0.1
0.0
TEMPERATURE, K
500
600
700
800
900
1000
1100
1200
1300
1400
1500
1600
1700
1800
1900
2000
1
2
3
4
5
6

SPECIFICATION TABLE NO. 319 HEMISPHERICAL TOTAL EMITTANCE OF [URANIUM + NIOBIUM] ALLOYS

Curve No.	Ref. No.	Year	Temperature Range, K	Reported Error, %	Composition (weight percent), Specifications and Remarks
1	116	1957	1090-1323		90 U, 10 Nb; annealed in vacuum for 1 hr at 1008 K and water quenched, then heated in vacuum for 4 hrs at 873 K; run 1, increasing temp; [Author's designation: Specimen 12].
2	116	1957	1323-1095		Above specimen and conditions; run 1, decreasing temp.
3	116	1957	1100-1339		Above specimen and conditions; run 2, increasing temp.
4	116	1957	1307-1104		Above specimen and conditions; run 2, decreasing temp.
5	116	1957	1070-1336		Above specimen and conditions except oxygen was added to the specimen by adding 15 individual charges of oxygen (2.5 cm^3 at atmospheric pressure per charge) to the vacuum chamber while the specimen was at 1073 K; chamber re-evacuated; run 1, increasing temp.
6	116	1957	1334-1083		Above specimen and conditions; run 1, decreasing temp.

DATA TABLE NO. 319 HEMISPHERICAL TOTAL EMITTANCE OF [URANIUM + NIOBIUM] ALLOYS

[Temperature, T, K; Emittance, ∈]

T	∈	T	∈
CURVE 1		**CURVE 4**	
1090	0.452	1307	0.410
1103	0.461	1283	0.402
1130	0.445	1261	0.395
1150	0.437	1225	0.392
1165	0.429	1202	0.387
1164	0.434	1174	0.385
1198	0.420	1142	0.387
1248	0.427	1104	0.383
1270	0.436		
1286	0.430	**CURVE 5**	
1303	0.438		
1323	0.432	1070	0.604
		1096	0.596
CURVE 2		1123	0.599
		1149	0.606
1323	0.420	1170	0.609
1300	0.418	1189	0.615
1279	0.415	1206	0.602
1263	0.413	1219	0.615
1216	0.403	1243	0.621
1189	0.400	1270	0.630
1166	0.401	1284	0.640
1144	0.401	1299	0.653
1119	0.401	1314	0.648
1095	0.399	1336	0.660
CURVE 3		**CURVE 6**	
1100	0.391	1334	0.671
1123	0.395	1317	0.662
1148	0.400	1255	0.625
1173	0.393	1227	0.617
1200	0.393	1198	0.596
1218	0.398	1166	0.587
1243	0.403	1127	0.573
1257	0.403	1083	0.569
1275	0.408		
1293	0.408		
1314	0.399		
1339	0.405		

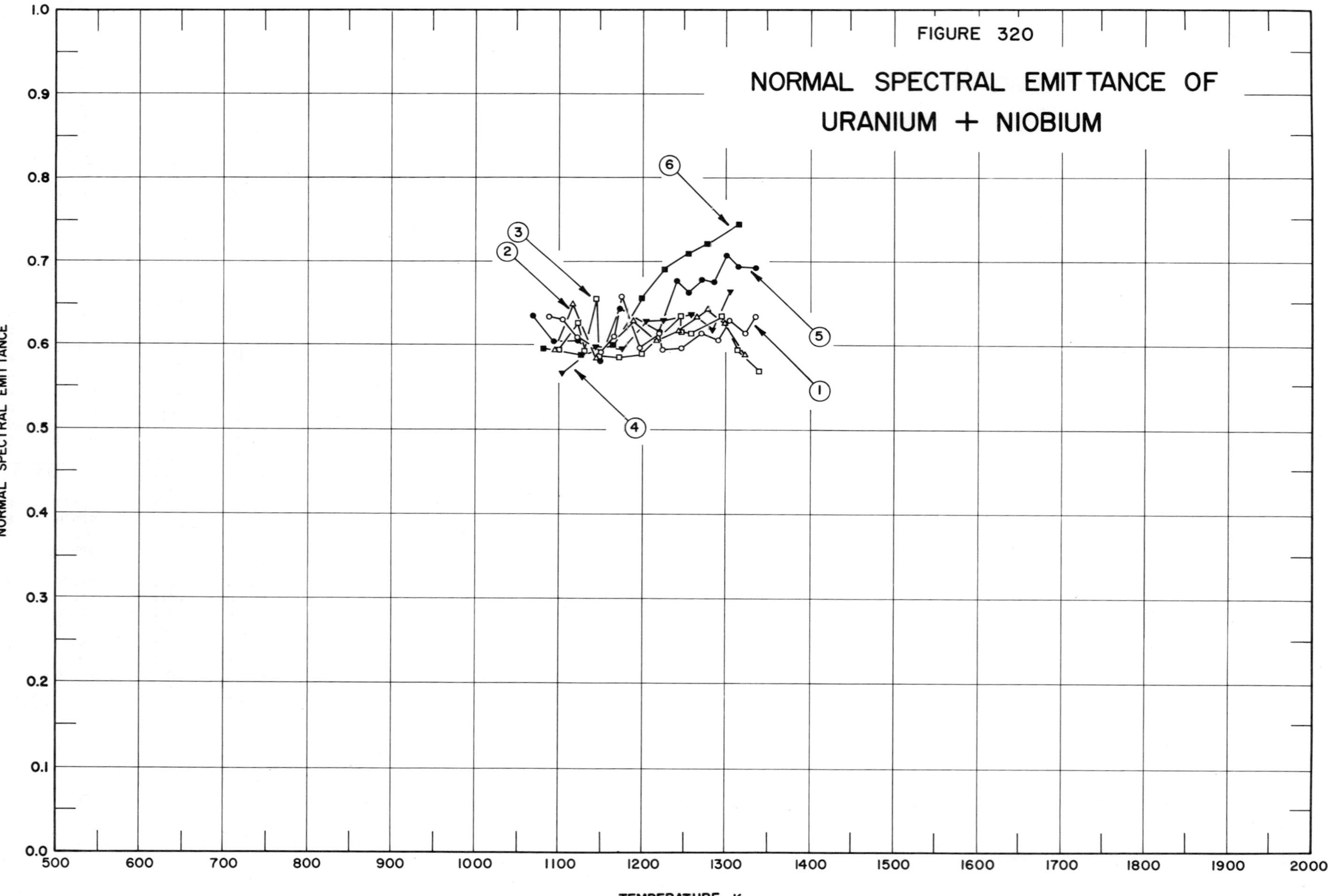
FIGURE 320
NORMAL SPECTRAL EMITTANCE OF
URANIUM + NIOBIUM
NORMAL SPECTRAL EMITTANCE
1.0
0.9
0.8
0.7
0.6
0.5
0.4
0.3
0.2
0.1
0.0
500
600
700
800
900
1000
1100
1200
1300
1400
1500
1600
1700
1800
1900
2000
TEMPERATURE, K
1
2
3
4
5
6

SPECIFICATION TABLE NO. 320 NORMAL SPECTRAL EMITTANCE OF [URANIUM + NIOBIUM] ALLOYS

Curve No.	Ref. No.	Year	Wavelength μ	Temperature Range, K	Geometry θ'	Reported Error, %	Composition (weight percent), Specifications and Remarks
1	116	1957	0.65	1088-1336	~0°		90 U, 10 Nb; annealed in vacuum for 1 hr at 1008 K and water quenched, then heated in vacuum for 4 hrs at 873 K; run 1, increasing temp; [Author's designation: Specimen 12].
2	116	1957	0.65	1322-1094	~0°		Above specimen and conditions; run 1, decreasing temp.
3	116	1957	0.65	1100-1340	~0°		Above specimen and conditions; run 2, increasing temp.
4	116	1957	0.65	1306-1103	~0°		Above specimen and conditions; run 2, decreasing temp.
5	116	1957	0.65	1069-1337	~0°		Above specimen and conditions except oxygen was added to the specimen by adding 15 individual charges of oxygen (2.5 cm^3 at atmospheric pressure per charge) to the vacuum chamber while the specimen was at 1073 K; chamber re-evacuated; run 1, increasing temp.
6	116	1957	0.65	1315-1081	~0°		Above specimen and conditions; run 1, decreasing temp.

DATA TABLE NO. 320 NORMAL SPECTRAL EMITTANCE OF [URANIUM + NIOBIUM] ALLOYS

[Temperature, T, K; Emittance, ∈; Wavelength, λ, μ]

CURVE 1 ($\lambda = 0.65$)

T	∈
1088	0.635
1104	0.681
1122	0.608
1150	0.591
1165	0.610
1173	0.658
1197	0.597
1220	0.613
1223	0.593
1246	0.597
1269	0.615
1290	0.605
1303	0.630
1322	0.614
1336	0.636

CURVE 2 ($\lambda = 0.65$)

T	∈
1322	0.589
1299	0.628
1278	0.646
1264	0.637
1242	0.619
1217	0.605
1188	0.630
1144	0.585
1116	0.650
1094	0.593

CURVE 3 ($\lambda = 0.65$)

T	∈
1100	0.593
1122	0.628
1131	0.592
1145	0.655
1171	0.584
1199	0.589
1246	0.636
1246	0.618
1258	0.614
1294	0.636
1314	0.594
1340	0.568

CURVE 4 ($\lambda = 0.65$)

T	∈
1306	0.663
1283	0.620
1259	0.638
1225	0.629
1203	0.629
1176	0.583
1143	0.598
1103	0.567

CURVE 5 ($\lambda = 0.65$)

T	∈
1069	0.634
1094	0.603
1122	0.607
1149	0.580
1172	0.644
1220	0.616
1241	0.677
1256	0.662
1270	0.678
1285	0.674
1300	0.708
1314	0.692
1337	0.692

CURVE 6 ($\lambda = 0.65$)

T	∈
1315	0.743
1278	0.722
1255	0.710
1226	0.690
1199	0.657
1163	0.600
1126	0.588
1081	0.597

SPECIFICATION TABLE NO. 321 NORMAL SPECTRAL REFLECTANCE OF [ZINC + ALUMINUM] ALLOYS

Curve No.	Ref. No.	Year	Temperature K	Wavelength Range, μ	Geometry θ	θ'	ω'	Reported Error, %	Composition (weight percent), Specifications and Remarks
1	152	1920	298	0.53-2.39	~0°	~0°			78 Zn, 22 Al; polished with a pitch polisher using rouge; silver mirror reference; converted from R(0°,0°); [Author's designation: Z12].

DATA TABLE NO. 321 NORMAL SPECTRAL REFLECTANCE OF [ZINC + ALUMINUM] ALLOYS

[Wavelength, λ, μ; Reflectance, ρ; Temperature, T, K]

λ	ρ
CURVE 1*	
T = 298	
0.53	0.539
0.59	0.538
0.66	0.505
0.72	0.484
0.79	0.440
0.87	0.428
1.02	0.460
1.20	0.565
1.61	0.650
2.39	0.700

* Not shown on plot

3. MULTIPLE ALLOYS

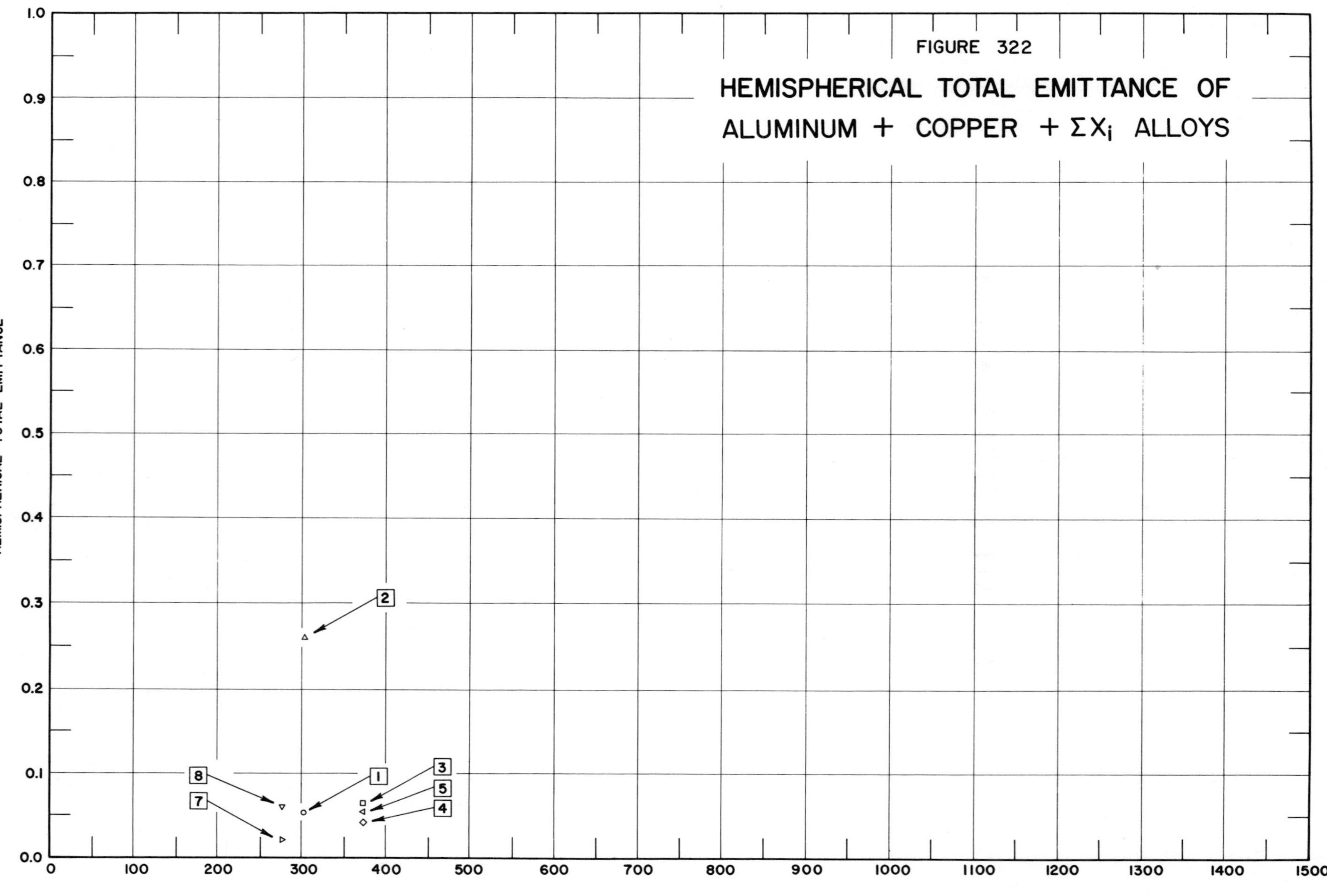
FIGURE 322
HEMISPHERICAL TOTAL EMITTANCE OF
ALUMINUM + COPPER + ΣXi ALLOYS
HEMISPHERICAL TOTAL EMITTANCE
1.0
0.9
0.8
0.7
0.6
0.5
0.4
0.3
0.2
0.1
0.0
0
100
200
300
400
500
600
700
800
900
1000
1100
1200
1300
1400
1500
TEMPERATURE, K
2
8
7
1
3
5
4

SPECIFICATION TABLE NO. 322 HEMISPHERICAL TOTAL EMITTANCE OF [ALUMINUM + COPPER + ΣX_i] ALLOYS

Curve No.	Ref. No.	Year	Temperature Range, K	Reported Error, %	Composition (weight percent), Specifications and Remarks
1	24	1948	303		Aluminum alloy 24-ST; nominal composition: 4.5 Cu, 1.5 Mg, 0.6 Mn, Al balance; measured in air.
2	24	1948	303		Different sample, same as above specimen and conditions except weathered.
3	8	1963	373		Aluminum alloy 2024; nominal composition: 4.5 Cu, 1.5 Mg, 0.6 Mn, Al balance; prefinished with 600 grit aluminum oxide powder on felt; electropolished; measured in vacuum (10^{-6} mm Hg).
4	8	1963	373		Above specimen and conditions except ion bombarded (3.2×10^{20} ions/cm^2).
5	8	1963	373		Above specimen and conditions except ion bombarded (6.5×10^{20} ions/cm^2).
6	8	1963	373		Above specimen and conditions except ion bombarded (9.8×10^{20} ions/cm^2).
7	45	1961	278		Aluminum alloy 2024; nominal composition: 4.5 Cu, 1.5 Mg, 0.6 Mn, Al balance; as received.
8	45	1961	278		Different sample, same as above specimen and conditions except machine polished and degreased.

DATA TABLE NO. 322 HEMISPHERICAL TOTAL EMITTANCE OF [ALUMINUM + COPPER + ΣX_i] ALLOYS

[Temperature, T, K; Emittance, ϵ]

T	ϵ
CURVE 1	
303	0.052
CURVE 2	
303	0.26
CURVE 3	
373	0.063
CURVE 4	
373	0.041
CURVE 5	
373	0.053
CURVE 6*	
373	0.062
CURVE 7	
278	0.02
CURVE 8	
278	0.06

* Not shown on plot

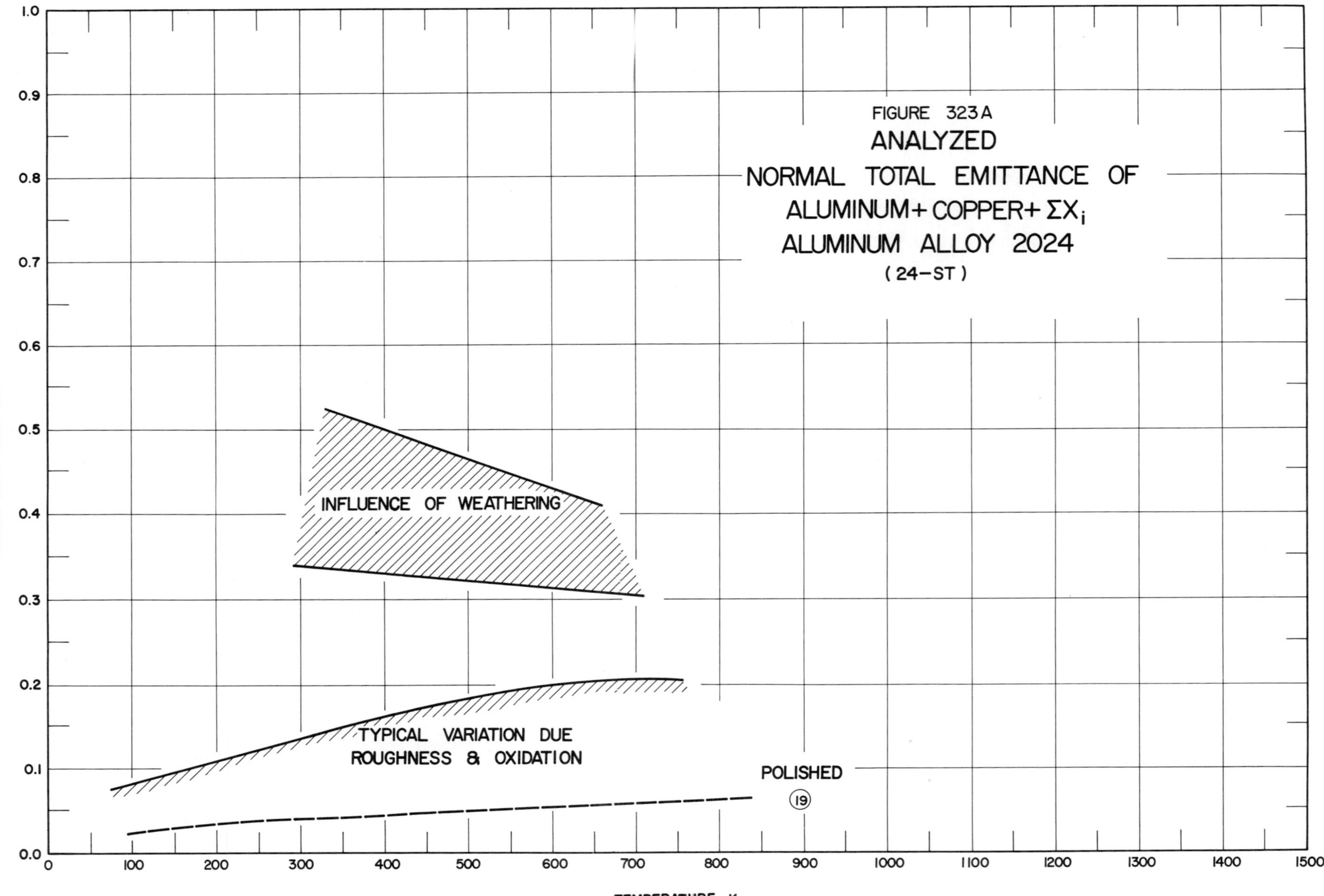

FIGURE 323A
ANALYZED
NORMAL TOTAL EMITTANCE OF
ALUMINUM + COPPER + ΣX_i
ALUMINUM ALLOY 2024
(24–ST)

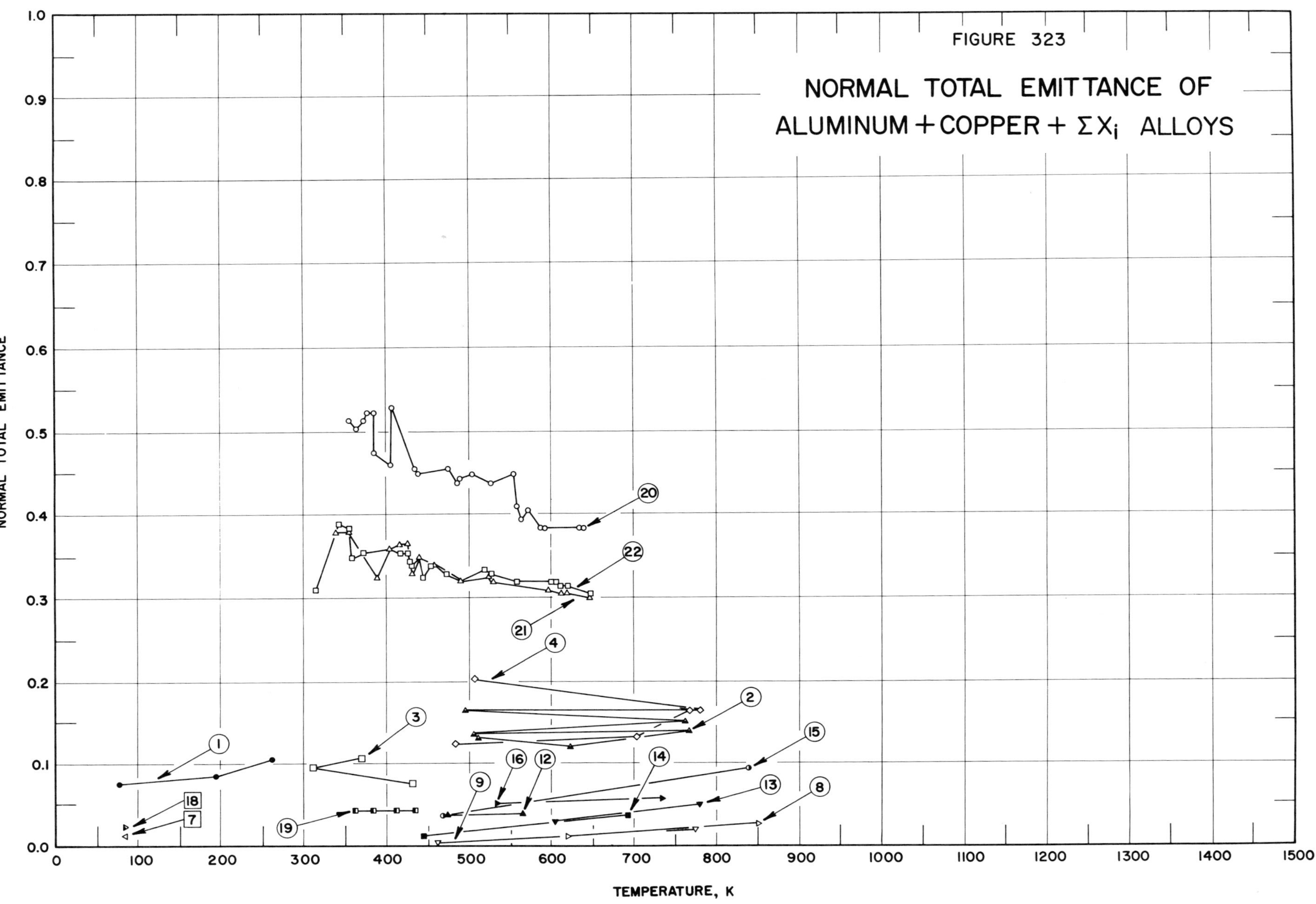
FIGURE 323
NORMAL TOTAL EMITTANCE OF
ALUMINUM + COPPER + ΣXi ALLOYS
NORMAL TOTAL EMITTANCE
TEMPERATURE, K
1.0
0.9
0.8
0.7
0.6
0.5
0.4
0.3
0.2
0.1
0.0
0
100
200
300
400
500
600
700
800
900
1000
1100
1200
1300
1400
1500
1
2
3
4
7
8
9
12
13
14
15
16
18
19
20
21
22

SPECIFICATION TABLE NO. 323 NORMAL TOTAL EMITTANCE OF [ALUMINUM + COPPER + ΣX_i] ALLOYS

Curve No.	Ref. No.	Year	Temperature Range, K	Geometry θ'	Reported Error, %	Composition (weight percent), Specifications and Remarks
1	160	1954	77.8-261	~0°		Aluminum alloy 24 ST; nominal composition: 4.5 Cu, 1.5 Mg, 0.6 Mn, Al balance; cleaned with methyl alcohol; measured in vacuum (10^{-3} mm Hg); [Author's designation: Sample 2].
2	160	1954	494-767	~0°		Different sample, same as above specimen and conditions except measured in argon (10^{-3} mm Hg); heating and cooling; [Author's designation: Sample 6].
3	160	1954	312-430	~0°		Aluminum alloy 24 ST; nominal composition: 4.5 Cu, 1.5 Mg, 0.6 Mn, Al balance; scrubbed with Bon Ami on a wet cloth; washed and dried; wiped with toluene and alcohol; measured in vacuum (10^{-3} mm Hg); [Author's designation: Sample 7].
4	160	1954	482-780	~0°		Different sample, same as above specimen and conditions; [Author's designation: Sample 8].
5	160	1954	311-432	~0°		Different sample, same as above specimen and conditions except polished; finished with a wool buff and rouge and washed; surface free from scratches; [Author's designation: Sample 4].
6	160	1954	489-754	~0°		Different sample, same as above specimen and conditions; [Author's designation: Sample 9].
7	34	1957	83.2	~0°	±10	Aluminum alloy 24 ST (Alclad); nominal composition: 4.5 Cu, 1.5 Mg, 0.6 Mn, Al balance; as received; measured in vacuum (5 x 10^{-4} mm Hg).
8	34	1957	619-850	~0°	±10	Different sample, same as above specimen and conditions except increasing temp, cycle 1.
9	34	1957	461-773	~0°	±10	Above specimen and conditions; cycle 2.
10	34	1957	475-850	~0°	±10	Above specimen and conditions; cycle 3.
11	34	1957	466-822	~0°	±10	Different sample, same as curve 7 specimen and conditions except same results obtained for 2 different surface treatments: cleaned with liquid detergent and polished.
12	34	1957	564-472	~0°	±10	Above specimen and conditions; decreasing temp, cycle 1.
13	34	1957	603-779	~0°	±10	Above specimen and conditions; cycle 2.
14	34	1957	444-691	~0°	±10	Above specimen and conditions; cycle 3.
15	34	1957	469-839	~0°	±10	Different sample, same as curve 7 specimen and conditions except oxidized in air at red heat for 30 min.
16	34	1957	533-733	~0°	±10	Above specimen and conditons; cycle 2.
17	34	1957	461-783	~0°	±10	Above specimen and conditions; cycle 3.
18	34	1957	83.2	~0°	±10	Different sample, same as curve 7 specimen and conditions except same results obtained for 3 different surface treatments: polished, cleaned with liquid detergent, and oxidized in air at red heat for 30 min.

SPECIFICATION TABLE NO. 323 (continued)

Curve No.	Ref. No.	Year	Temperature Range, K	Geometry θ'	Reported Error, %	Composition (weight percent), Specifications and Remarks
19	157	1944	361-433	$\sim 0^\circ$	< 10	Aluminum alloy 24-ST (Alclad); nominal composition: 4.5 Cu, 1.5 Mg, 0.6 Mn, Al balance; measured in air; data extracted from smooth curve.
20	155	1948	355-639	$\sim 0^\circ$		Aluminum alloy 24-ST; nominal composition: 4.5 Cu, 1.5 Mg, 0.6 Mn, Al balance; weathered.
21	155	1948	339-647	$\sim 0^\circ$		Above specimen and conditions.
22	155	1948	314-647	$\sim 0^\circ$		Different sample, same as above specimen and conditions.

DATA TABLE NO. 323 NORMAL TOTAL EMITTANCE OF [ALUMINUM + COPPER + ΣX_i] ALLOYS

[Temperature, T, K; Emittance, ϵ]

T	ϵ
CURVE 1	
77.8	0.075
194	0.085
261	0.105
CURVE 2	
511	0.130
622	0.120
767	0.140
506	0.135
761	0.150
494	0.165
CURVE 3	
430	0.075
312	0.094
370	0.106
CURVE 4	
482	0.122
702	0.132
767	0.165
497	0.163*
780	0.165
508	0.204
CURVE 5*	
432	0.071
311	0.090
373	0.099
CURVE 6*	
522	0.114
754	0.134
498	0.208
750	0.145
544	0.145
732	0.134
489	0.160
CURVE 7	
83.2	0.011
CURVE 8	
619	0.0110
850	0.0261
CURVE 9	
461	0.0025
773	0.0200
CURVE 10*	
475	0.0600
725	0.0170
850	0.0261
CURVE 11*	
466	0.003
625	0.011
822	0.053
CURVE 12	
564	0.039
472	0.038
CURVE 13	
603	0.030
779	0.050
CURVE 14	
444	0.012
691	0.038
CURVE 15	
469	0.037
839	0.094
CURVE 16	
533	0.051
733	0.057
CURVE 17*	
461	0.049
605	0.051
783	0.056
CURVE 18	
83.2	0.022
CURVE 19	
361	0.042
383	0.042
411	0.042
433	0.042
CURVE 20	
355	0.515
364	0.505
372	0.515
378	0.525
386	0.525
386	0.475
405	0.460
408	0.530
433	0.455
439	0.450
475	0.455
486	0.440
489	0.445
503	0.450
528	0.440
553	0.450
558	0.410
564	0.395
572	0.405
588	0.385
592	0.385
633	0.385
639	0.385
CURVE 21	
339	0.380
355	0.380
389	0.325
403	0.360
417	0.365
425	0.365
430	0.330
439	0.350
458	0.340
489	0.320
522	0.325
528	0.320
597	0.310
611	0.305
619	0.305
647	0.300
CURVE 22	
314	0.310
342	0.390
355	0.385
358	0.350
372	0.355
403	0.360*
417	0.355
425	0.355
428	0.345
430	0.340
439	0.350*
444	0.325
453	0.340
458	0.340*
472	0.330
486	0.320*
519	0.335
528	0.330
558	0.320
600	0.320
605	0.320
611	0.315
619	0.315
647	0.305

* Not shown on plot

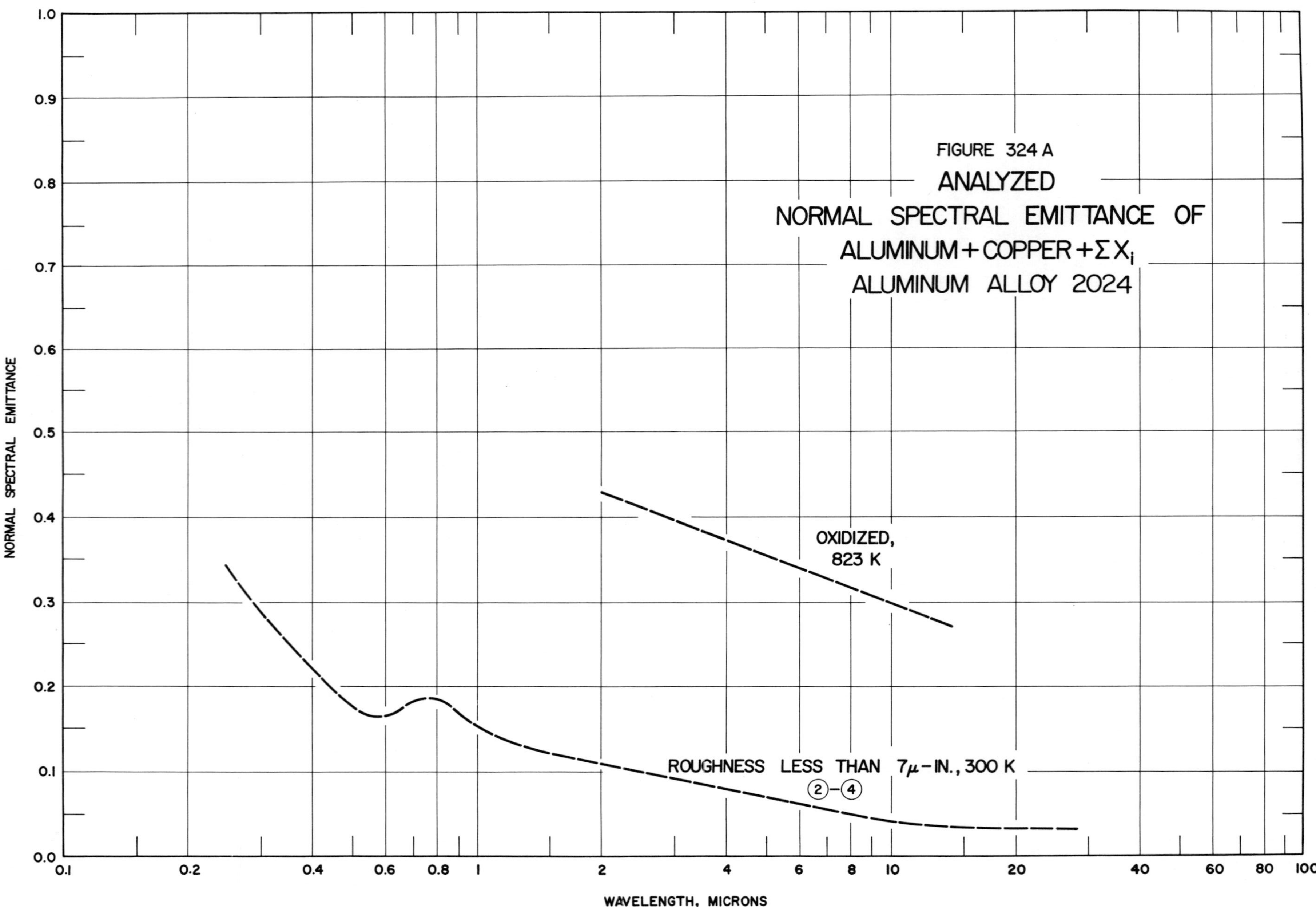

FIGURE 324 A
ANALYZED
NORMAL SPECTRAL EMITTANCE OF
ALUMINUM + COPPER + ΣX_i
ALUMINUM ALLOY 2024

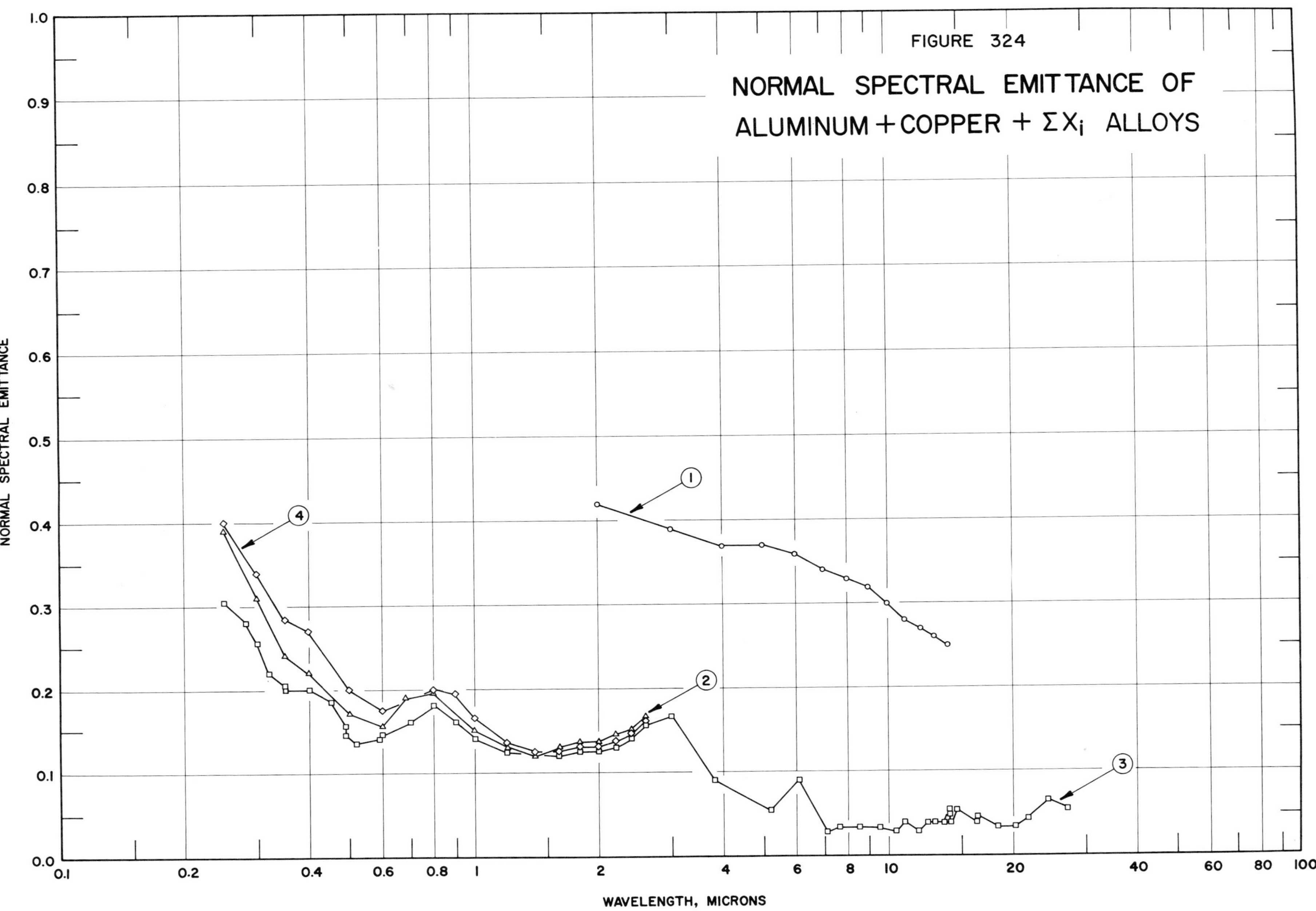

FIGURE 324
NORMAL SPECTRAL EMITTANCE OF
ALUMINUM + COPPER + ΣXi ALLOYS
NORMAL SPECTRAL EMITTANCE
WAVELENGTH, MICRONS
1.0
0.9
0.8
0.7
0.6
0.5
0.4
0.3
0.2
0.1
0.0
0.1
0.2
0.4
0.6
0.8
1
2
4
6
8
10
20
40
60
80
100
1
2
3
4

SPECIFICATION TABLE NO. 324 NORMAL SPECTRAL EMITTANCE OF [ALUMINUM + COPPER + ΣX_i] ALLOYS

Curve No.	Ref. No.	Year	Temperature K	Wavelength Range, μ	Geometry θ'	Reported Error, %	Composition (weight percent), Specifications and Remarks
1	156	1960	823	2.0-14.0	$\sim 0^\circ$		Aluminum alloy 2024; nominal composition: 4.5 Cu, 1.5 Mg, 0.6 Mn, Al balance; oxidized in air for 2 hrs; measured in air.
2	159	1963	323	0.25-2.60	5°		Aluminum alloy 2024; nominal composition: 4.5 Cu, 1.5 Mg, 0.6 Mn, Al balance; surface roughness 1.5-1.8 microinches (center line average); measured in nitrogen; computed by $\epsilon = 1-R\ (2\pi, 5^\circ)$; [Author's designation: Specimen 1].
3	159	1963	323	0.25-27	5°		Different sample, same as above specimen and conditions except surface roughness 6.0-7.0 microinches (center line average); [Author's designation: Specimen 3].
4	159	1963	323	0.25-2.60	5°		Different sample, same as above specimen and conditions except surface roughness 2.8-3.9 microinches (center line average); [Author's designation: Specimen 4].

DATA TABLE NO. 324 NORMAL SPECTRAL EMITTANCE OF [ALUMINUM + COPPER + ΣX_i] ALLOYS

[Wavelength, λ, μ; Emittance, ϵ; Temperature, T, K]

λ	ϵ
CURVE 1 T = 823	
2.0	0.42
3.0	0.39
4.0	0.37
5.0	0.37
6.0	0.36
7.0	0.34
8.0	0.33
9.0	0.32
10.0	0.30
11.0	0.28
12.0	0.27
13.0	0.26
14.0	0.25
CURVE 2 T = 323	
0.25	0.390
0.30	0.310
0.35	0.240
0.40	0.220
0.50	0.170
0.60	0.155
0.68	0.190
0.80	0.195
1.00	0.150
1.20	0.130
1.40	0.120
1.60	0.130
1.80	0.135
2.00	0.135
2.20	0.145
2.40	0.150
2.60	0.165
CURVE 3 T = 323	
0.25	0.305
0.28	0.280
0.30	0.255
0.32	0.220
0.35	0.205
0.35	0.200
CURVE 3 (cont.)	
0.40	0.200
0.45	0.185
0.49	0.155
0.49	0.145
0.52	0.135
0.59	0.140
0.60	0.145
0.70	0.160
0.80	0.180
0.90	0.160
1.00	0.140
1.20	0.125
1.60	0.120
1.80	0.125
2.00	0.125
2.20	0.130
2.40	0.140
2.60	0.155
3.00	0.165
3.80	0.090
5.20	0.055
6.10	0.090
7.10	0.030
7.60	0.035
8.50	0.035
9.50	0.035
10.30	0.030
10.90	0.040
11.70	0.030
12.30	0.040
12.90	0.040
13.60	0.040
14.00	0.055
14.05	0.050
14.10	0.040
14.60	0.055
16.25	0.040
16.25	0.048
18.25	0.035
20.10	0.035
21.70	0.045
24.30	0.065
27.00	0.055
CURVE 4 T = 323	
0.25	0.400
0.30	0.340
0.35	0.285
0.40	0.270
0.50	0.200
0.60	0.175
0.80	0.200
0.90	0.195
1.00	0.165
1.20	0.135
1.40	0.125
1.60	0.125
1.80	0.130
2.00	0.130
2.20	0.135
2.40	0.145
2.60	0.160

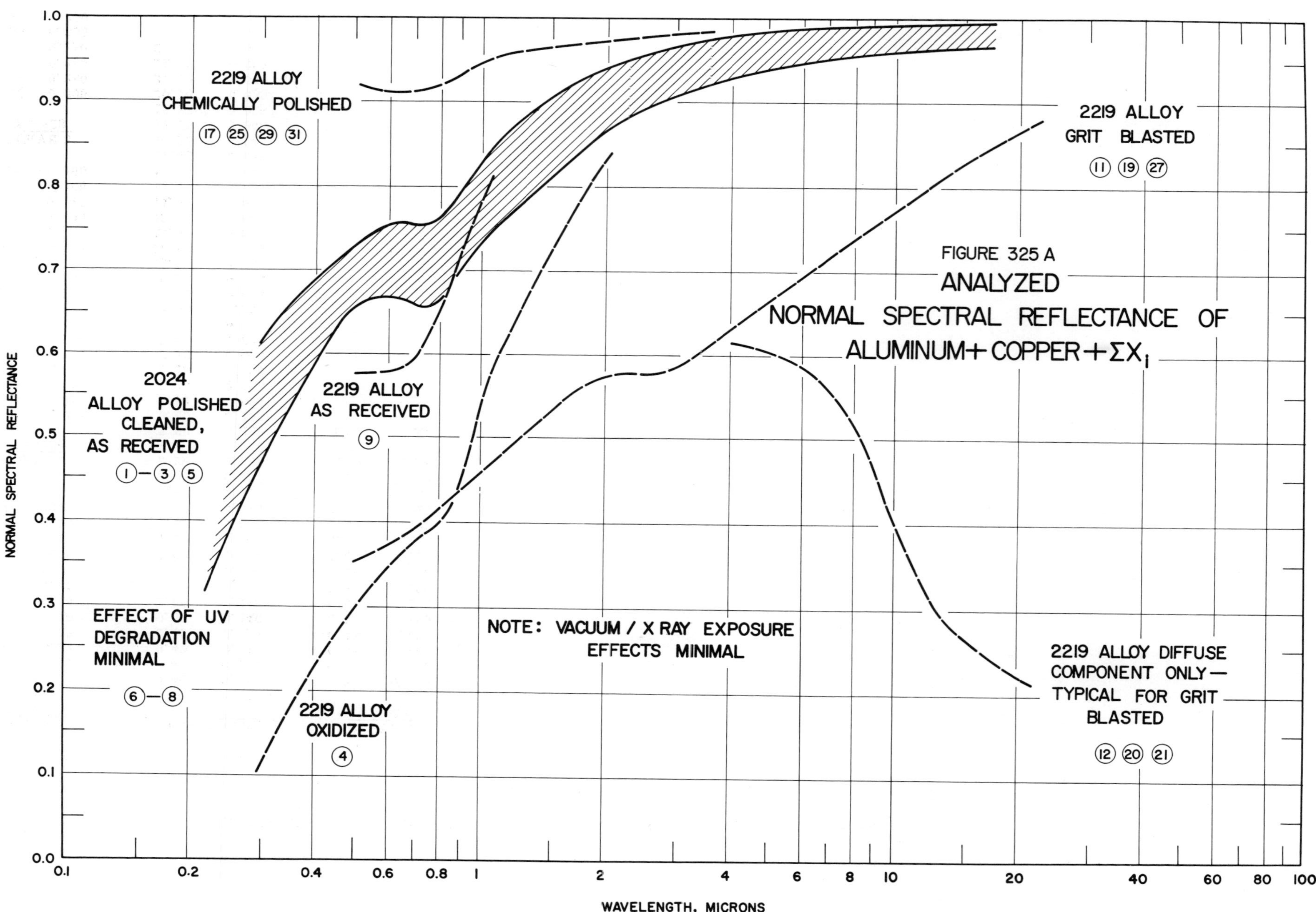

FIGURE 325 A
ANALYZED
NORMAL SPECTRAL REFLECTANCE OF
ALUMINUM + COPPER + ΣX_i

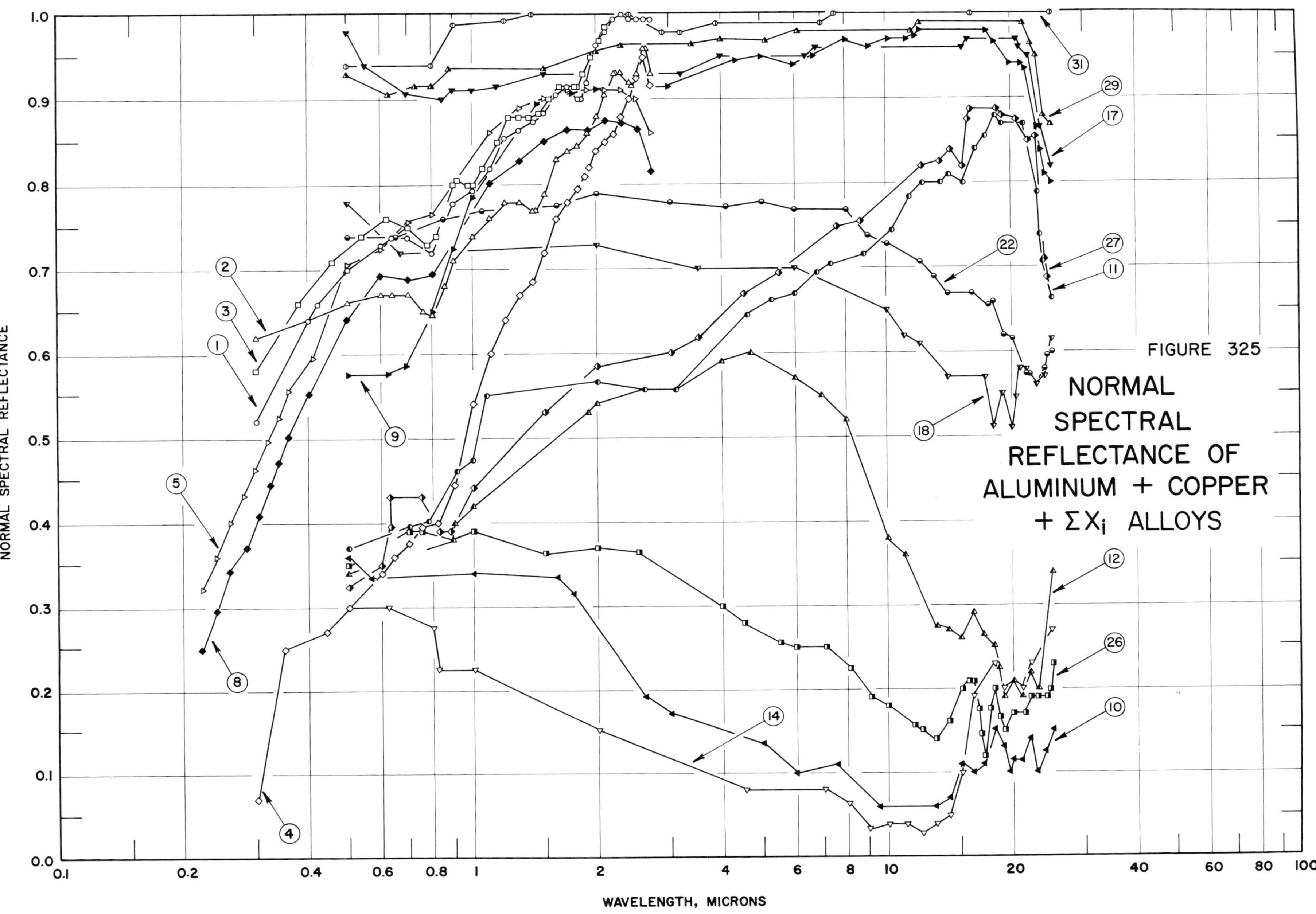

FIGURE 325
NORMAL SPECTRAL REFLECTANCE OF ALUMINUM + COPPER + ΣX_i ALLOYS

SPECIFICATION TABLE NO. 325 NORMAL SPECTRAL REFLECTANCE OF [ALUMINUM + COPPER + ΣX_i] ALLOYS

Curve No.	Ref. No.	Year	Temperature K	Wavelength Range, μ	Geometry θ	θ'	ω'	Reported Error, %	Composition (weight percent), Specifications and Remarks
1	34	1957	298	0.30-2.70	9°		2π	± 4	Aluminum alloy 24 ST; nominal composition: 4.5 Cu, 1.5 Mg, 0.6 Mn, Al balance; as received; data extracted from smooth curve.
2	34	1957	298	0.30-2.70	9°		2π	± 4	Different sample, same as curve 1 specimen and conditions except polished.
3	34	1957	298	0.30-2.70	9°		2π	± 4	Different sample, same as curve 1 specimen and conditions except cleaned with liquid detergent.
4	34	1957	298	0.30-2.70	9°		2π	± 4	Different sample, same as curve 1 specimen and conditions except oxidized in air at red heat for 30 min.
5	278	1960	378	0.221-2.715	~0°		2π		Aluminum alloy 24 ST; nominal composition: 4.5 Cu, 1.5 Mg, 0.6 Mn, Al balance; cleaned; measured in vacuum (10^{-5} mm Hg); MgO reference standard.
6	278	1960	378	0.221-2.713	~0°		2π		Above specimen and conditions except exposed in UV radiation for 20 hrs; G. E. Type UA-3 lamp source.
7	278	1960	378	0.220-2.748	~0°		2π		Above specimen and conditions except exposed in UV radiation for 60 hrs; G. E. Type UA-3 lamp source.
8	278	1960	378	0.221-2.715	~0°		2π		Above specimen and conditions except exposed in UV radiation for 100 hrs; G. E. Type UA-3 lamp source.
9	153	1962	~322	0.50-25.00	~0°		2π	< 2	Aluminum alloy 2219, Mil-A-8920; nominal composition: 6.3 Cu, 0.30 Mn, 0.18 Zr, 0.10 V, 0.06 Ti, Al balance; as received; cleaned; data extracted from smooth curve; hohlraum at 1273 K; converted from R(2π, 0°).
10	153	1962	~322	0.50-25.00	~0°		2π	< 2	Above specimen and conditions; diffuse component only.
11	153	1962	~322	0.50-25.00	~0°		2π	< 2	Different sample, same as curve 9 specimen and conditions except grit blasted using 60 grit silicon carbide with air pressure of 110 to 120 psi for 30 to 45 sec.
12	153	1962	~322	0.50-25.00	~0°		2π	< 2	Above specimen and conditions; diffuse component only.
13	153	1962	~322	0.50-25.00	~0°		2π	< 2	Different sample, same as curve 9 specimen and conditions except exposed to vacuum ($< 4 \times 10^{-8}$ mm Hg) for 24 hrs.
14	153	1962	~322	0.50-25.00	~0°		2π	< 2	Above specimen and conditions; diffuse component only.
15	153	1962	~322	0.50-25.00	~0°		2π	< 2	Different sample, same as curve 9 specimen and conditions except x-ray exposed in vacuum ($< 4 \times 10^{-8}$ mm Hg) for 24 hrs.
16	153	1962	~322	0.50-25.00	~0°		2π	< 2	Above specimen and conditions; diffuse component only.
17	153	1962	~322	0.50-25.00	~0°		2π	< 2	Different sample, same as curve 9 specimen and conditions except chem-milled using the Turco 9H process for 3 min.
18	153	1962	~322	0.50-25.00	~0°		2π	< 2	Above specimen and conditions; diffuse component only.
19	153	1962	~322	0.50-25.00	~0°		2π	< 2	Different sample, same as curve 9 specimen and conditions except chem-milled using the Turco 9H process for 3 min; grit blasted using 60 grit silicon carbide with air pressure of 110 to 120 psi for 30 to 45 sec.
20	153	1962	~322	0.50-25.00	~0°		2π	< 2	Above specimen and conditions; diffuse component only.

SPECIFICATION TABLE NO. 325 (continued)

Curve No.	Ref. No.	Year	Temperature K	Wavelength Range, μ	Geometry θ θ'	ω'	Reported Error, %	Composition (weight percent), Specifications and Remarks
21	153	1962	~322	0.50-25.00	~0°	2π	< 2	Different sample, same as curve 9 specimen and conditions except chem-milled using the Turco 9H process for 3 min; exposed to vacuum ($<4 \times 10^{-8}$ mm Hg) for 24 hrs.
22	153	1962	~322	0.50-25.00	~0°	2π	< 2	Above specimen and conditions; diffuse component only.
23	153	1962	~322	0.50-25.00	~0°	2π	< 2	Different sample, same as curve 9 specimen and conditions except chem-milled using the Turco 9H process for 3 min; x-ray exposed in vacuum (4×10^{-8} mm Hg) for 24 hrs.
24	153	1962	~322	0.50-25.00	~0°	2π	< 2	Above specimen and conditions; diffuse component only.
25	153	1962	~322	0.50-25.00	~0°	2π	< 2	Different sample, same as curve 9 specimen and conditions except chemically polished using the Alcoa process with a 2 min immersion at 358 K.
26	153	1962	~322	0.50-25.00	~0°	2π	< 2	Above specimen and conditions; diffuse component only.
27	153	1962	~322	0.50-25.00	~0°	2π	< 2	Different sample, same as curve 9 specimen and conditions except chemically polished using the Alcoa process with a 2 min immersion at 358 K; grit blasted using 60 grit silicon carbide with air pressure of 110 to 120 psi for 30 to 45 sec.
28	153	1962	~322	0.50-25.00	~0°	2π	< 2	Above specimen and conditions; diffuse component only.
29	153	1962	~322	0.50-25.00	~0°	2π	< 2	Different sample, same as curve 9 specimen and conditions except chemically polished using the Alcoa process with a 2 min immersion at 358 K; exposed to vacuum ($<4 \times 10^{-8}$ mm Hg) for 24 hrs.
30	153	1962	~322	0.50-25.00	~0°	2π	< 2	Above specimen and conditions; diffuse component only.
31	153	1962	~322	0.50-25.00	~0°	2π	< 2	Different sample, same as curve 9 specimen and conditions except chemically polished using the Alcoa process with a 2 min immersion at 358 K; x-ray exposed in vacuum (4×10^{-8} mm Hg) for 24 hrs.
32	153	1962	~322	0.50-25.00	~0°	2π	< 2	Above specimen and conditions; diffuse component only.

DATA TABLE NO. 325 NORMAL SPECTRAL REFLECTANCE OF [ALUMINUM + COPPER + ΣX_i] ALLOYS

[Wavelength, λ, μ; Reflectance, ρ; Temperature, T, K]

λ	ρ
CURVE 1 T = 298	
0.30	0.520
0.40	0.640
0.42	0.660
0.50	0.700
0.60	0.730
0.64	0.740
0.70	0.740
0.72	0.740*
0.80	0.720
0.90	0.780
1.00	0.795
1.10	0.820
1.20	0.855
1.30	0.865
1.40	0.875
1.50	0.885
1.60	0.905
1.70	0.915
1.80	0.900
1.82	0.900
1.90	0.920
2.00	0.965
2.10	0.980
2.20	0.995
2.30	1.000
2.40	0.995
2.50	0.995
2.60	0.995
2.70	0.995
CURVE 2 T = 298	
0.30	0.620
0.40	0.640*
0.50	0.660
0.60	0.670
0.64	0.670
0.70	0.670
0.76	0.650
0.80	0.645
0.86	0.680
0.90	0.710
1.00	0.740
CURVE 2 (cont.)	
1.10	0.760
1.20	0.780
1.30	0.780
1.40	0.770
1.42	0.770
1.50	0.790
1.60	0.830
1.70	0.840
1.80	0.845
1.90	0.860
2.00	0.880
2.10	0.905
2.20	0.930
2.24	0.930
2.30	0.930
2.40	0.920
2.44	0.915
2.50	0.930
2.60	0.960
2.62	0.960
2.70	0.930
CURVE 3 T = 298	
0.30	0.580
0.38	0.660
0.46	0.710
0.54	0.740
0.62	0.760
0.70	0.750
0.78	0.730
0.82	0.740
0.90	0.800
0.92	0.805
0.98	0.800
1.00	0.800
1.06	0.820
1.14	0.850
1.22	0.880
1.30	0.880
1.38	0.880
1.46	0.885
1.54	0.900
1.62	0.915
CURVE 3 (cont.)	
1.70	0.915*
1.78	0.915
1.80	0.915
1.86	0.930
1.94	0.950
2.02	0.970
2.10	0.985
2.18	0.995*
2.26	1.000*
2.34	1.000*
2.42	0.995*
2.50	0.995*
2.58	0.995*
2.66	0.995*
2.70	0.995*
CURVE 4 T = 298	
0.30	0.070
0.35	0.250
0.44	0.270
0.50	0.300
0.60	0.340
0.64	0.360
0.70	0.375
0.75	0.395
0.82	0.400
0.90	0.445
1.00	0.540
1.10	0.600
1.20	0.640
1.30	0.670
1.40	0.685
1.50	0.720
1.60	0.760
1.70	0.780
1.80	0.795
1.88	0.810
1.92	0.820
2.00	0.840
2.10	0.850
2.20	0.860
2.30	0.880
2.40	0.900
CURVE 4 (cont.)	
2.50	0.925
2.60	0.950
2.70	0.915
CURVE 5 T = 378	
0.221	0.321
0.240	0.359
0.260	0.400
0.280	0.431
0.299	0.462
0.320	0.496
0.340	0.523
0.360	0.553
0.410	0.596
0.501	0.706
0.601	0.727
0.702	0.755
0.801	0.764
1.115	0.862
1.312	0.893
1.513	0.904
1.715	0.909
1.911	0.909
2.116	0.905
2.314	0.908
2.512	0.898
2.715	0.857
CURVE 6* T = 378	
0.221	0.291
0.240	0.333
0.260	0.371
0.279	0.401
0.300	0.434
0.319	0.474
0.340	0.510
0.359	0.533
0.409	0.574
0.501	0.676
0.601	0.714
0.700	0.708
CURVE 6 (cont.)	
0.801	0.717
1.114	0.822
1.314	0.852
1.517	0.871
1.717	0.880
1.918	0.883
2.111	0.886
2.315	0.893
2.514	0.879
2.713	0.825
CURVE 7* T = 378	
0.220	0.266
0.240	0.312
0.260	0.355
0.280	0.372
0.300	0.412
0.321	0.445
0.341	0.486
0.360	0.512
0.409	0.562
0.501	0.661
0.602	0.702
0.706	0.705
0.803	0.707
1.112	0.810
1.319	0.839
1.517	0.860
1.715	0.871
1.918	0.873
2.138	0.888
2.317	0.883
2.546	0.876
2.748	0.823
CURVE 8 T = 378	
0.221	0.249
0.240	0.293
0.259	0.342
0.284	0.371
0.304	0.408
CURVE 8 (cont.)	
0.325	0.445
0.340	0.470
0.360	0.501
0.409	0.550
0.502	0.640
0.601	0.692
0.702	0.688
0.803	0.693
1.117	0.801
1.310	0.828
1.517	0.850
1.718	0.863
1.920	0.863
2.116	0.876
2.317	0.872
2.520	0.863
2.715	0.815
CURVE 9 T = ~ 322	
0.50	0.575
0.62	0.575
0.69	0.585
0.80	0.650
0.90	0.725
1.00	0.785
1.45	0.895
1.75	0.905
2.00	0.910
3.00	0.912
4.35	0.945
5.00	0.950
6.00	0.940
6.70	0.950
8.00	0.970
9.10	0.960
10.10	0.970
11.20	0.970
11.70	0.975
12.00	0.980
17.25	0.980
18.10	0.968
19.90	0.940
21.00	0.940
CURVE 9 (cont.)	
21.50	0.935
23.00	0.865
23.50	0.840
24.10	0.810
25.00	0.800
CURVE 10 T = ~322	
0.50	0.360
0.57	0.335
1.00	0.340
1.60	0.335
1.75	0.315
2.60	0.190
3.00	0.170
5.00	0.135
6.00	0.100
7.50	0.110
9.50	0.060
13.00	0.060
14.00	0.070
15.00	0.110
16.00	0.100
17.00	0.110
18.00	0.150
19.00	0.080
19.60	0.100
20.00	0.115
21.00	0.115
22.00	0.140
23.00	0.100
24.00	0.125
25.00	0.150
CURVE 11 T = ~322	
0.50	0.370*
0.70	0.397
0.78	0.402
0.91	0.460
1.00	0.472
1.75	0.550
2.00	0.565
CURVE 11 (cont.)	
2.60	0.555
3.10	0.555
4.60	0.645
5.30	0.662
6.00	0.670
6.80	0.695
7.35	0.705
8.85	0.717
10.30	0.745
11.30	0.785
12.15	0.800
13.45	0.800
14.10	0.810
15.30	0.800
16.30	0.840
17.20	0.855
18.20	0.880
19.00	0.870
21.30	0.870
23.10	0.790
23.50	0.740
23.85	0.705
25.00	0.662
CURVE 12 T = ~322	
0.50	0.340
0.87	0.380
0.90	0.400
1.00	0.420
1.90	0.530
2.00	0.540
2.60	0.555*
3.10	0.555*
4.00	0.590
4.70	0.600
6.00	0.570
7.00	0.550
8.00	0.520
10.00	0.380
11.00	0.360
13.10	0.275
14.00	0.270
15.00	0.260

* Not shown on plot

DATA TABLE NO. 325 (continued)

λ	ρ
CURVE 12 (cont.)	
16.00	0.290
17.00	0.265
18.00	0.250
18.50	0.225
19.00	0.190
20.00	0.210
21.00	0.190
22.00	0.220
23.00	0.200
25.00	0.340
CURVE 13*	**T = ~322**
0.50	0.565
0.60	0.610
0.65	0.635
0.78	0.685
0.93	0.735
1.00	0.785
1.25	0.845
1.65	0.910
2.00	0.937
2.60	0.950
4.40	0.950
5.00	0.960
6.00	0.960
6.50	0.965
7.60	0.980
11.50	0.980
12.10	0.970
13.00	0.980
21.20	0.980
21.70	0.958
22.10	0.950
23.20	0.950
23.60	0.925
23.90	0.870
25.00	0.870
CURVE 14	**T = ~322**
0.50	0.300*
0.62	0.300
0.80	0.275
0.82	0.225

λ	ρ
CURVE 14 (cont.)	
1.00	0.225
2.00	0.150
4.50	0.080
7.00	0.080
8.00	0.065
9.00	0.035
10.00	0.040
11.00	0.040
12.00	0.030
13.00	0.040
14.00	0.050
15.00	0.100
16.00	0.190
18.00	0.230
19.00	0.200
20.20	0.210*
21.00	0.200
22.20	0.230
25.00	0.270
CURVE 15*	**T = ~322**
0.50	0.450
0.65	0.560
0.78	0.585
0.83	0.610
0.88	0.700
1.00	0.750
1.20	0.865
1.45	0.895
1.75	0.910
2.40	0.910
2.90	0.920
3.80	0.950
4.70	0.970
7.20	0.970
7.70	0.990
16.00	0.990
25.00	0.990
CURVE 16*	**T = ~322**
0.50	0.280
0.70	0.330

λ	ρ
CURVE 16 (cont.)*	
0.75	0.330
0.82	0.340
0.90	0.330
2.00	0.310
4.50	0.225
6.00	0.200
8.00	0.200
8.90	0.160
9.50	0.150
11.30	0.150
12.00	0.140
13.00	0.150
14.00	0.180
15.00	0.200
16.00	0.160
17.00	0.160
18.00	0.200
19.00	0.150
20.00	0.180
21.00	0.160
22.00	0.200
23.00	0.170
25.00	0.240
CURVE 17	**T = ~322**
0.50	0.980
0.55	0.940
0.70	0.907
0.85	0.900
0.90	0.910
1.00	0.910
1.15	0.915
1.50	0.930
3.20	0.930
4.00	0.950
6.40	0.950
6.80	0.960
15.20	0.960
15.80	0.970
20.60	0.970
21.00	0.960
22.00	0.950
23.70	0.865
25.00	0.820

λ	ρ
CURVE 18	**T = ~322**
0.50	0.780
0.67	0.720
2.00	0.730
3.50	0.700
6.00	0.700
10.00	0.650
11.00	0.620
12.00	0.610
14.00	0.570
17.20	0.570
18.00	0.510
19.00	0.550
20.00	0.510
20.50	0.545
21.00	0.580
21.70	0.580
23.00	0.560
24.00	0.570
25.00	0.615
CURVE 19*	**T = ~322**
0.50	0.320
0.58	0.345
0.72	0.425
0.78	0.425
0.83	0.400
0.88	0.400
1.00	0.430
1.50	0.490
2.00	0.580
3.00	0.580
3.60	0.615
4.40	0.670
5.25	0.693
6.40	0.705
8.10	0.750
9.30	0.750
11.30	0.815
11.90	0.830
14.20	0.830
14.70	0.825
15.00	0.820
15.50	0.840

λ	ρ
CURVE 19 (cont.)*	
16.00	0.895
16.70	0.900
18.70	0.900
19.60	0.882
21.00	0.880
21.60	0.870
23.00	0.810
23.50	0.775
23.90	0.718
25.00	0.680
CURVE 20*	**T = ~322**
0.50	0.290
0.62	0.325
0.75	0.350
0.80	0.400
1.00	0.450
1.50	0.510
2.00	0.580
4.00	0.650
5.10	0.670
6.00	0.660
7.50	0.660
8.00	0.660
9.00	0.620
10.00	0.600
13.00	0.550
14.00	0.540
15.00	0.550
16.20	0.530
17.00	0.530
18.00	0.500
19.00	0.450
20.00	0.440
21.10	0.420
22.00	0.440
23.00	0.410
24.00	0.410
25.00	0.450
CURVE 21*	**T = ~322**
0.50	0.890
0.60	0.870

λ	ρ
CURVE 21 (cont.)*	
0.73	0.880
0.93	0.885
1.00	0.890
1.70	0.890
2.00	0.970
2.80	0.918
4.50	0.920
5.00	0.930
6.80	0.930
7.50	0.948
8.50	0.950
9.30	0.940
9.90	0.940
10.60	0.948
11.00	0.950
17.50	0.950
18.10	0.960
20.40	0.960
21.10	0.970
22.30	0.935
23.00	0.925
23.50	0.900
23.75	0.870
24.10	0.860
25.00	0.860
CURVE 22	**T = ~322**
0.50	0.740
0.65	0.740
0.85	0.760
1.05	0.770
1.60	0.775
2.00	0.790
3.00	0.780
4.10	0.775
5.00	0.780
6.00	0.770
8.00	0.770
9.00	0.740
10.00	0.730
11.20	0.705
13.00	0.690
14.00	0.670
16.00	0.670
17.50	0.655

λ	ρ
CURVE 22 (cont.)	
18.00	0.660
19.00	0.620
20.00	0.615
21.00	0.580*
21.50	0.575
22.00	0.575
23.00	0.560*
24.00	0.580
24.40	0.595
25.00	0.600
CURVE 23*	**T = ~322**
0.50	0.880
0.78	0.880
0.90	0.912
1.00	0.920
1.50	0.940
2.00	0.940
2.70	0.390
3.25	0.930
3.90	0.950
7.20	0.950
7.60	0.960
15.20	0.960
15.70	0.990
20.00	0.990
25.00	0.990
CURVE 24*	**T = ~322**
0.50	0.700
0.70	0.720
0.80	0.725
0.85	0.740
1.00	0.745
1.30	0.780
1.60	0.795
1.90	0.800
3.00	0.800
4.00	0.800
5.50	0.800
6.10	0.790
7.00	0.800
8.00	0.780

λ	ρ
CURVE 24 (cont.)*	
9.00	0.760
10.00	0.740
11.00	0.730
12.00	0.720
13.00	0.710
14.50	0.700
17.00	0.700
20.00	0.700
23.00	0.700
25.00	0.700
CURVE 25*	**T = ~322**
0.50	0.900
0.70	0.880
0.80	0.890
0.93	0.985
1.00	1.000
2.00	0.980
2.90	0.970
3.80	0.970
4.90	0.980
5.50	0.975
6.70	0.960
7.00	0.960
8.10	0.990
10.00	0.970
11.00	0.990
18.30	0.990
19.70	0.973
20.90	0.970
21.80	0.930
23.00	0.900
24.00	0.850
25.00	0.833
CURVE 26	**T = ~322**
0.50	0.350
0.70	0.390
0.75	0.390
0 90	0.380
1.00	0.390
1.50	0.365
2.00	0.370

* Not shown on plot

DATA TABLE NO. 325 (continued)

λ	ρ
CURVE 26 (cont.)	
2.50	0.365
4.00	0.300
4.50	0.280
5.50	0.255
6.00	0.250
7.10	0.250
8.10	0.225
9.10	0.190
10.00	0.180
11.50	0.155
12.00	0.150
13.00	0.140
14.00	0.160
15.00	0.200
15.50	0.210
16.00	0.210
16.50	0.175
16.60	0.145
17.00	0.120
17.60	0.175
18.00	0.200
18.50	0.165
19.00	0.150
20.00	0.170
21.20	0.170
22.00	0.190
23.00	0.190
24.10	0.190
24.60	0.200
25.00	0.230
CURVE 27 T = ~322	
0.50	0.325
0.60	0.350
0.63	0.395
0.63	0.430
0.75	0.430
0.83	0.390
0.88	0.390
1.00	0.440
1.50	0.530
2.00	0.585
3.00	0.600
3.50	0.620
CURVE 27 (cont.)	
4.50	0.670
5.50	0.695
7.60	0.750
8.70	0.755
12.10	0.820
13.35	0.825
14.20	0.840
15.20	0.820
15.60	0.875
15.90	0.888
18.40	0.888
19.00	0.880
20.60	0.875
22.00	0.850
23.00	0.805
24.00	0.710
24.30	0.688
25.00	0.663*
CURVE 28* T = ~322	
0.50	0.415
0.70	0.430
0.76	0.440
0.95	0.450
1.30	0.500
1.50	0.530
2.40	0.585
2.80	0.590
3.50	0.600
4.20	0.625
5.00	0.640
5.50	0.625
6.10	0.600
6.50	0.590
7.00	0.590
8.00	0.550
9.00	0.480
10.00	0.440
11.50	0.370
12.80	0.325
14.00	0.300
14.50	0.325
15.00	0.350
CURVE 28 (cont.)*	
16.20	0.325
17.20	0.285
18.00	0.280
19.00	0.240
20.00	0.250
21.00	0.230
22.00	0.260
23.00	0.255
24.00	0.300
25.00	0.350
CURVE 29 T = ~322	
0.50	0.930
0.63	0.905
0.73	0.915
0.80	0.915
0.88	0.935
1.50	0.935
2.00	0.958
2.30	0.963
3.40	0.963
4.00	0.970
5.15	0.970
6.15	0.980
11.50	0.980
12.00	0.990
21.20	0.990
22.20	0.965
23.00	0.950
24.00	0.880
25.00	0.870
CURVE 30* T = ~322	
0.50	0.420
0.65	0.430
0.80	0.440
0.85	0.375
1.00	0.390
1.65	0.375
2.20	0.365
3.00	0.340
4.00	0.300
CURVE 30 (cont.)*	
5.00	0.270
6.00	0.250
7.00	0.240
8.00	0.235
9.00	0.180
10.00	0.170
11.50	0.150
12.00	0.150
13.00	0.140
14.00	0.150
15.00	0.150
16.00	0.170
17.10	0.170
18.00	0.190
18.50	0.175
19.00	0.140
19.60	0.165
20.00	0.185
21.00	0.160
22.00	0.180
23.00	0.160
24.00	0.200
25.00	0.225
CURVE 31 T = ~322	
0.50	0.940
0.80	0.940
0.90	0.990
1.20	0.993
1.40	1.000
2.40	1.000
2.90	0.980
3.20	0.980
3.90	0.990
7.30	0.990
7.50	1.000
16.00	1.000
25.00	1.000
CURVE 32* T = ~322	
0.50	0.410
0.65	0.450
CURVE 32 (cont.)*	
0.85	0.460
1.00	0.425
1.50	0.410
2.00	0.400
3.20	0.375
4.40	0.330
4.80	0.325
6.20	0.280
7.50	0.275
8.50	0.255
9.20	0.225
10.50	0.215
11.50	0.200
13.00	0.200
14.00	0.200
14.50	0.215
15.00	0.250
16.00	0.250
17.10	0.230
18.00	0.250
18.60	0.225
19.10	0.200
20.00	0.240
21.00	0.215
21.50	0.240
22.00	0.265
23.00	0.255
24.00	0.280
25.00	0.300

* Not shown on plot

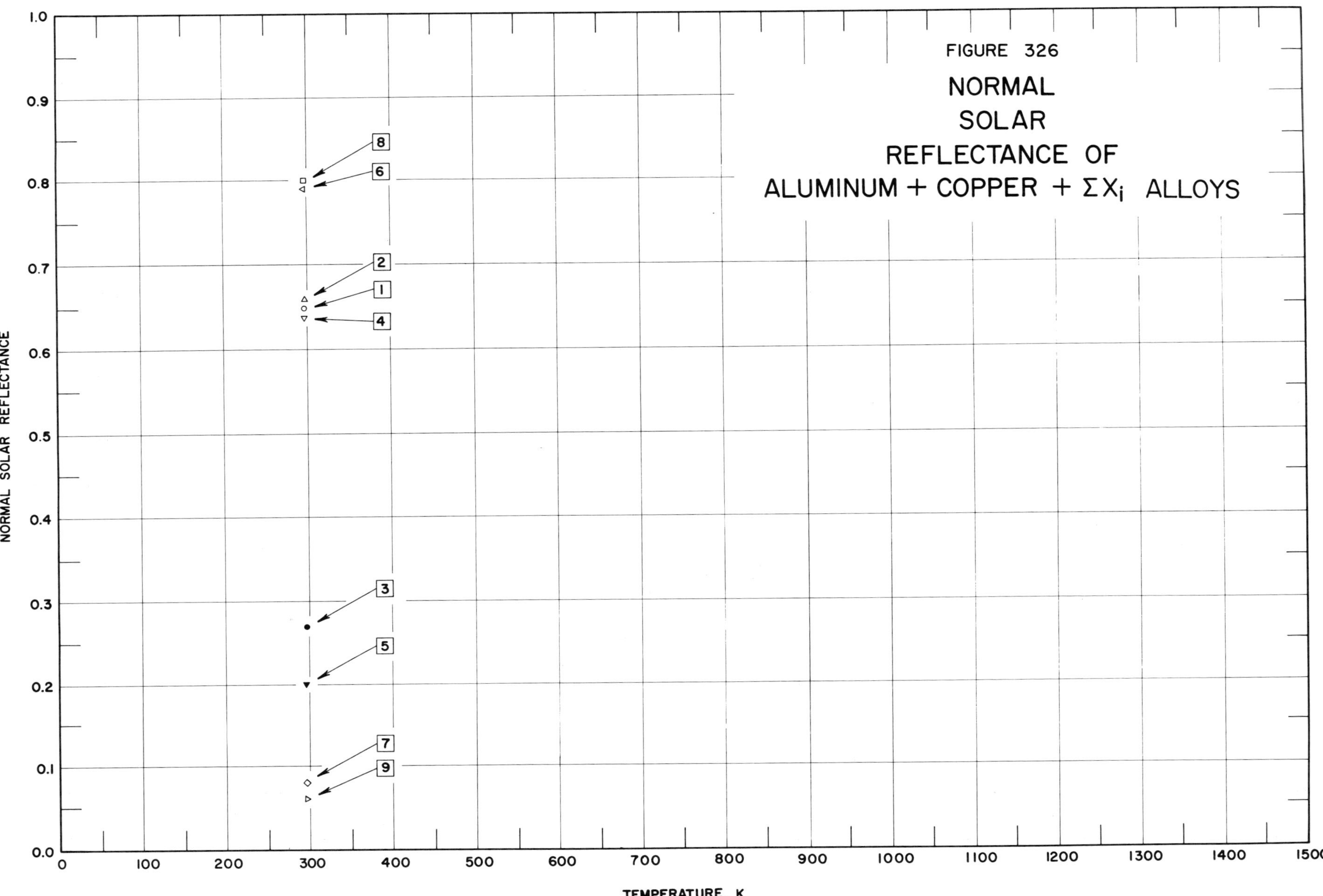

FIGURE 326 NORMAL SOLAR REFLECTANCE OF ALUMINUM + COPPER + ΣX_i ALLOYS

SPECIFICATION TABLE NO. 326 NORMAL SOLAR REFLECTANCE OF [ALUMINUM + COPPER + ΣX_i] ALLOYS

Curve No.	Ref. No.	Year	Temperature Range, K	Geometry θ	θ'	ω	Reported Error, %	Composition (weight percent), Specifications and Remarks
1	40	1962	298	~0°		2π		Aluminum 24SO Alclad; nominal composition: 4.5 Cu, 1.5 Mg, 0.6 Mn, Al balance; polished; measured in air.
2	131	1964	298	~0°		2π		L 72 (British Aircraft Material Specification); nominal composition: 3.8-4.8 Cu, 1.0 Fe, 0.4-1.2 Mn, 0.6-0.9 Si, 0.55-0.85 Mg, 0.3 Ti, 0.2 Ni, 0.2 Zn, 0.05 Pb, 0.05 Sn, 0.3 Ti and/or Cr, Al balance; as received; freshly prepared; MgO reference; computed from spectral data.
3	131	1964	298	~0°		2π		Above specimen and conditions; diffuse component only.
4	131	1964	298	~0°		2π		Different sample, same as curve 2 specimen and conditions.
5	131	1964	298	~0°		2π		Above specimen and conditions; diffuse component only.
6	131	1964	298	~0°		2π		Different sample, same as curve 2 specimen and conditions except buffed.
7	131	1964	298	~0°		2π		Above specimen and conditions; diffuse component only.
8	131	1964	298	~0°		2π		Different sample, same as curve 6 specimen and conditions.
9	131	1964	298	~0°		2π		Above specimen and conditions; diffuse component only.
10	160	1954	311	~0°		2π		Aluminum alloy 24-ST; nominal composition: 4.5 Cu, 1.5 Mg, 0.6 Mn, Al balance; heated to 136 K; clean and smooth surface; measured in air; calculated from solar absorptance.
11	160	1954	311	~0°		2π		Above specimen and conditions except reheated to 563 K.
12	160	1954	311	~0°		2π		Above specimen and conditions except reheated to 755 K.
13	160	1954	311	~0°		2π		Different sample, same as curve 10 specimen and conditions except heated to 317 K; polished; surface free from scratches.
14	160	1954	311	~0°		2π		Above specimen and conditions except reheated to 559 K.
15	160	1954	311	~0°		2π		Above specimen and conditions except reheated to 745 K.
16	160	1954	311	~0°		2π		Different sample, same as curve 10 specimen and conditions except heated to 325 K; cleaned with methyl alcohol.
17	160	1954	311	~0°		2π		Above specimen and conditions except reheated to 603 K.
18	160	1954	311	~0°		2π		Above specimen and conditions except reheated to 755 K.

DATA TABLE NO. 326 NORMAL SOLAR REFLECTANCE OF [ALUMINUM + COPPER + ΣX_i] ALLOYS

[Temperature, T, K; Reflectance, ρ]

T	ρ	T	ρ
CURVE 1		CURVE 13*	
298	0.65	311	0.752
CURVE 2		CURVE 14*	
298	0.66	311	0.823
CURVE 3		CURVE 15*	
298	0.27	311	0.570
CURVE 4		CURVE 16*	
298	0.64	311	0.512
CURVE 5		CURVE 17*	
298	0.20	311	0.541
CURVE 6		CURVE 18*	
298	0.79	311	0.270
CURVE 7			
298	0.08		
CURVE 8			
298	0.80		
CURVE 9			
298	0.06		
CURVE 10*			
311	0.549		
CURVE 11*			
311	0.420		
CURVE 12*			
311	0.329		

*Not shown on plot

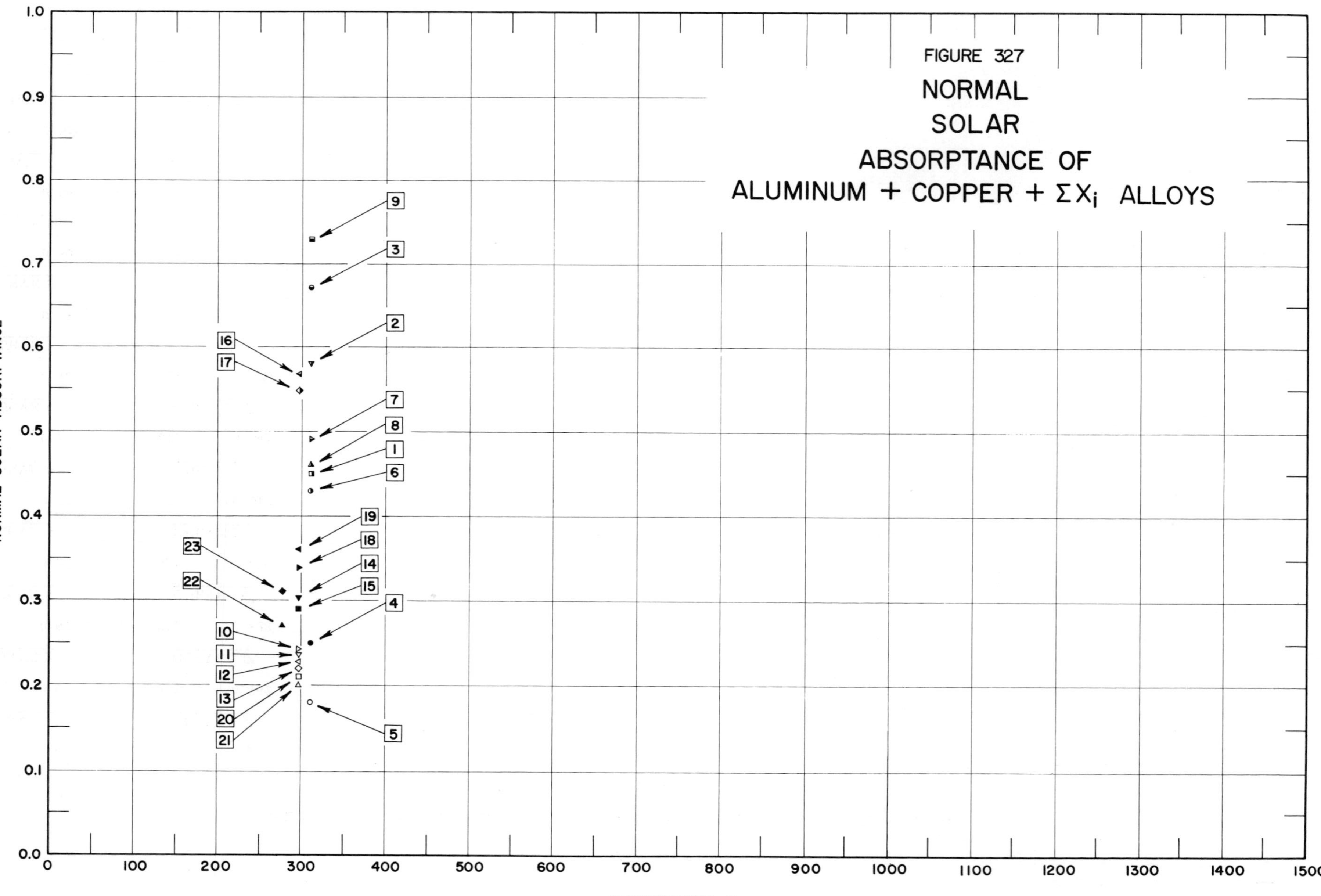
FIGURE 327
NORMAL
SOLAR
ABSORPTANCE OF
ALUMINUM + COPPER + ΣXi ALLOYS
NORMAL SOLAR ABSORPTANCE
TEMPERATURE, K
1.0
0.9
0.8
0.7
0.6
0.5
0.4
0.3
0.2
0.1
0.0
0
100
200
300
400
500
600
700
800
900
1000
1100
1200
1300
1400
1500
9
3
2
16
17
7
8
1
6
19
18
14
15
4
23
22
10
11
12
13
20
21
5

SPECIFICATION TABLE NO. 327 NORMAL SOLAR ABSORPTANCE OF [ALUMINUM + COPPER + ΣX_i] ALLOYS

Curve No.	Ref. No.	Year	Temperature Range, K	Geometry θ	Reported Error, %	Composition (weight percent), Specifications and Remarks
1	160	1954	311	$\sim 0^\circ$		Aluminum alloy 24-ST; nominal composition: 4.5 Cu, 1.5 Mg, 0.6 Mn, Al balance; heated to 310 K; clean and smooth surface; measured in air at sea level.
2	160	1954	311	$\sim 0^\circ$		Above specimen and conditions except reheated to 560 K.
3	160	1954	311	$\sim 0^\circ$		Above specimen and conditions except reheated to 755 K.
4	160	1954	311	$\sim 0^\circ$		Different sample, same as curve 1 specimen and conditions except heated to 307 K; polished; surface free from scratches.
5	160	1954	311	$\sim 0^\circ$		Above specimen and conditions except reheated to 560 K.
6	160	1954	311	$\sim 0^\circ$		Above specimen and conditions except reheated to 755 K.
7	160	1954	311	$\sim 0^\circ$		Different sample, same as curve 1 specimen and conditions except heated to 324 K; cleaned with methyl alcohol.
8	160	1954	311	$\sim 0^\circ$		Above specimen and conditions except reheated to 603 K.
9	160	1954	311	$\sim 0^\circ$		Above specimen and conditions except reheated to 755 K.
10	34	1957	298	9°		Aluminum alloy 24-ST; nominal composition: 4.5 Cu, 1.5 Mg, 0.6 Mn, Al balance; as received; computed from spectral reflectance data for sea level conditions.
11	34	1957	298	9°		Above specimen and conditions except computed for above atmosphere conditions.
12	34	1957	298	9°		Different sample, same as curve 10 specimen and conditions except cleaned with liquid detergent.
13	34	1957	298	9°		Above specimen and conditions except computed for above atmosphere conditions.
14	34	1957	298	9°		Different sample, same as curve 10 specimen and conditions except polished.
15	34	1957	298	9°		Above specimen and conditions except computed for above atmosphere conditions.
16	34	1957	298	9°		Different sample, same as curve 10 specimen and conditions except oxidized in air at red heat for 30 min.
17	34	1957	298	9°		Above specimen and conditions except computed for above atmosphere conditions.
18	131	1964	298	$\sim 0^\circ$		L 72 (British Aircraft Material Specification); nominal composition: 3.8-4.8 Cu, 1.0 Fe, 0.4-1.2 Mn, 0.6-0.9 Si, 0.55-0.85 Mg, 0.3 Ti, 0.2 Ni, 0.2 Zn, 0.05 Pb, 0.05 Sn, 0.3 Ti and/or Cr, Al balance; as received; freshly prepared; calculated from $(1-\rho)$.
19	131	1964	298	$\sim 0^\circ$		Different sample, same as above specimen and conditions.
20	131	1964	298	$\sim 0^\circ$		Different sample, same as above specimen and conditions except buffed.
21	131	1964	298	$\sim 0^\circ$		Different sample, same as above specimen and conditions.
22	45	1961	278	$\sim 0^\circ$	10	Aluminum alloy 2024; nominal composition: 4.5 Cu, 1.5 Mg, 0.6 Mn, Al balance; as received; extraterrestrial.
23	45	1961	278	$\sim 0^\circ$	10	Different sample, same as above specimen and conditions except machine polished and degreased.

DATA TABLE NO. 327 NORMAL SOLAR ABSORPTANCE OF [ALUMINUM + COPPER + ΣX_i] ALLOYS

[Temperature, T, K; Absorptance, α]

T	α	T	α
CURVE 1		CURVE 13	
311	0.45	298	0.220
CURVE 2		CURVE 14	
311	0.58	298	0.302
CURVE 3		CURVE 15	
311	0.67	298	0.290
CURVE 4		CURVE 16	
311	0.25	298	0.568
CURVE 5		CURVE 17	
311	0.18	298	0.548
CURVE 6		CURVE 18	
311	0.43	298	0.34
CURVE 7		CURVE 19	
311	0.49	298	0.36
CURVE 8		CURVE 20	
311	0.46	298	0.21
CURVE 9		CURVE 21	
311	0.73	298	0.20
CURVE 10		CURVE 22	
298	0.242	278	0.27
CURVE 11		CURVE 23	
298	0.236	278	0.31
CURVE 12			
298	0.228		

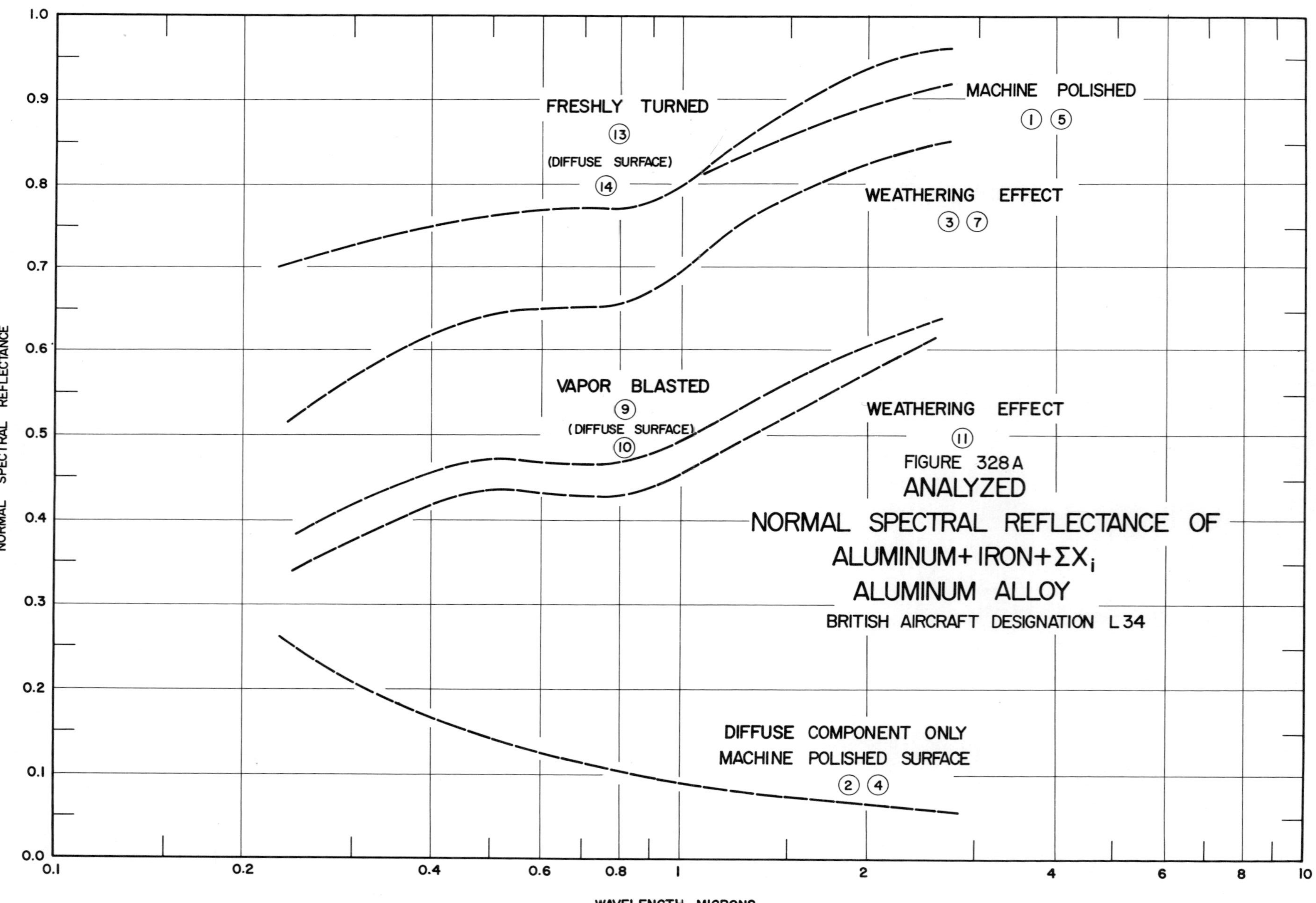
FRESHLY TURNED
13
(DIFFUSE SURFACE)
14
MACHINE POLISHED
1 5
WEATHERING EFFECT
3 7
VAPOR BLASTED
9
(DIFFUSE SURFACE)
10
WEATHERING EFFECT
11
FIGURE 328A
ANALYZED
NORMAL SPECTRAL REFLECTANCE OF
ALUMINUM+IRON+ΣX_i
ALUMINUM ALLOY
BRITISH AIRCRAFT DESIGNATION L 34
DIFFUSE COMPONENT ONLY
MACHINE POLISHED SURFACE
2 4
NORMAL SPECTRAL REFLECTANCE
1.0
0.9
0.8
0.7
0.6
0.5
0.4
0.3
0.2
0.1
0.0
0.1
0.2
0.4
0.6
0.8
1
2
4
6
8
10
WAVELENGTH, MICRONS

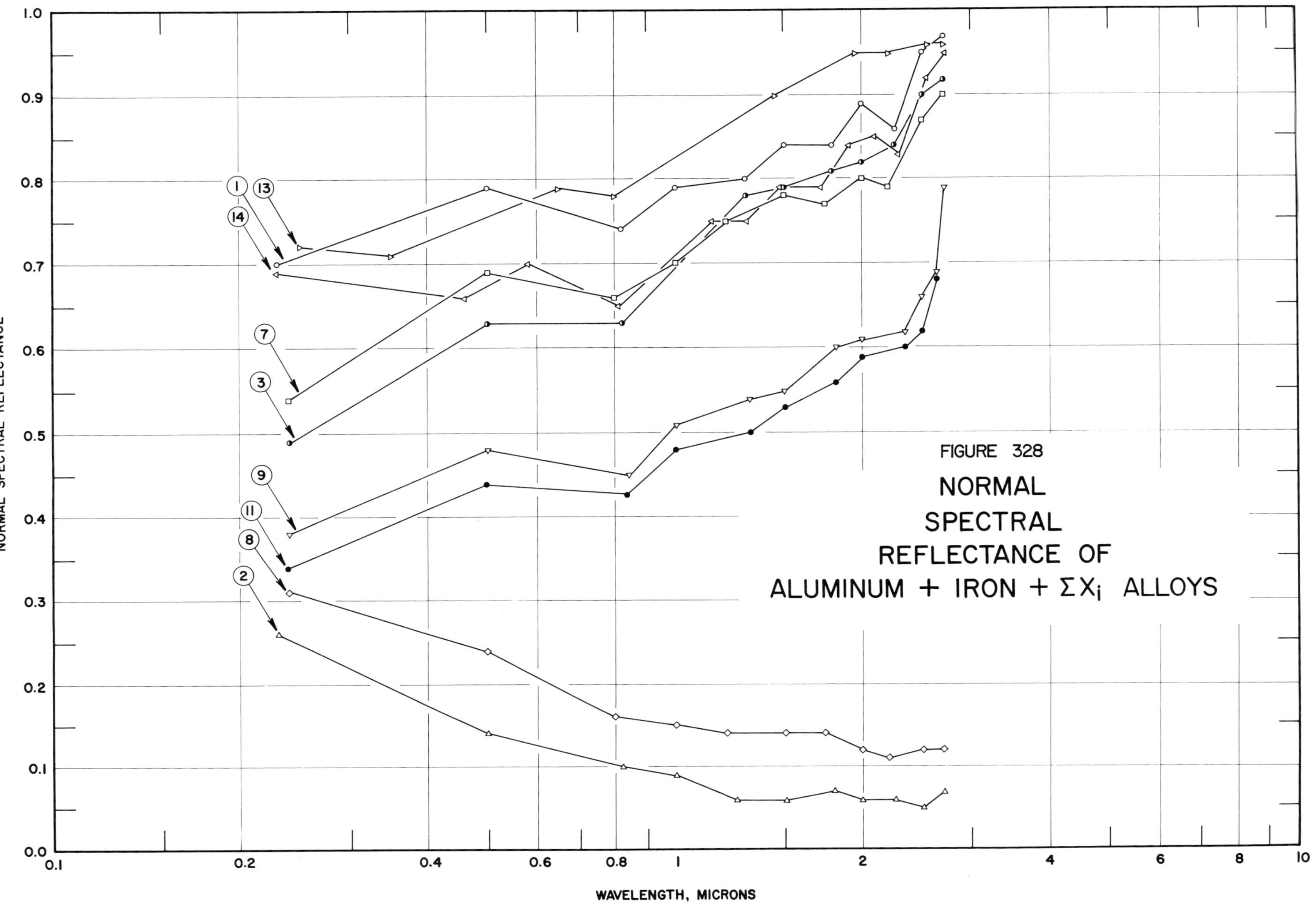

FIGURE 328
NORMAL
SPECTRAL
REFLECTANCE OF
ALUMINUM + IRON + ΣXi ALLOYS
NORMAL SPECTRAL REFLECTANCE
WAVELENGTH, MICRONS
1.0
0.9
0.8
0.7
0.6
0.5
0.4
0.3
0.2
0.1
0.0
0.1
0.2
0.4
0.6
0.8
1
2
4
6
8
10
1
13
14
7
3
9
11
8
2

SPECIFICATION TABLE NO. 328 NORMAL SPECTRAL REFLECTANCE OF [ALUMINUM + IRON + ΣX_i] ALLOYS

Curve No.	Ref. No.	Year	Temperature K	Wavelength Range, μ	Geometry θ	θ'	ω'	Reported Error, %	Composition (weight percent), Specifications and Remarks
1	131	1964	298	0.230-2.700	$\sim 0°$		2π		L 34 (British Aircraft Material Specification); nominal composition: 99.0 min Al, 0.7 Fe, 0.5 Si, 0.15 Ti, 0.10 Cu, 0.1 Mg, 0.1 Mn, 0.1 Ni, 0.10 Zn, 0.05 Sn, 0.05 Pb; machine polished; freshly prepared; data extracted from smooth curve; MgO reference.
2	131	1964	298	0.230-2.700	$\sim 0°$		2π		Above specimen and conditions; diffuse component only.
3	131	1964	298	0.230-2.700	$\sim 0°$		2π		Curve 1 specimen and conditions except specimen exposed to outside atmosphere for 1 month.
4	131	1964	298	0.230-2.700	$\sim 0°$		2π		Above specimen and conditions; diffuse component only.
5	131	1964	298	0.240-2.700	$\sim 0°$		2π		Different sample, same as curve 1 specimen and conditions.
6	131	1964	298	0.240-2.700	$\sim 0°$		2π		Above specimen and conditions; diffuse component only.
7	131	1964	298	0.240-2.700	$\sim 0°$		2π		Curve 1 specimen and conditions except specimen exposed within clean laboratory area for 1 month.
8	131	1964	298	0.240-2.700	$\sim 0°$		2π		Above specimen and conditions; diffuse component only.
9	131	1964	298	0.240-2.710	$\sim 0°$		2π		Different sample, same as curve 1 specimen and conditions except vapor blasted; freshly prepared.
10	131	1964	298	0.240-2.630	$\sim 0°$		2π		Above specimen and conditions; diffuse component only.
11	131	1964	298	0.240-2.630	$\sim 0°$		2π		Curve 9 specimen and conditions except specimen exposed within clean laboratory area for 1 month.
12	131	1964	298	0.240-2.630	$\sim 0°$		2π		Above specimen and conditions; diffuse component only.
13	131	1964	298	0.250-2.700	$\sim 0°$		2π		Different sample, same as curve 1 specimen and conditions except turned; freshly prepared.
14	131	1964	298	0.230-2.710	$\sim 0°$		2π		Above specimen and conditions; diffuse component only.
15	131	1964	298	0.250-2.700	$\sim 0°$		2π		Curve 13 specimen and conditions except specimen exposed within clean laboratory area for 1 month.
16	131	1964	298	0.240-2.700	$\sim 0°$		2π		Above specimen and conditions; diffuse component only.

DATA TABLE NO. 328 NORMAL SPECTRAL REFLECTANCE OF [ALUMINUM + IRON + ΣX_i] ALLOYS

[Wavelength, λ, μ; Reflectance, ρ; Temperature, T, K]

CURVE 1, T = 298

λ	ρ
0.230	0.70
0.500	0.79
0.820	0.74
1.000	0.79
1.300	0.80
1.500	0.84
1.790	0.84
2.000	0.89
2.250	0.86
2.500	0.95
2.700	0.97

CURVE 2, T = 298

λ	ρ
0.230	0.26
0.500	0.14
0.820	0.10
1.000	0.09
1.300	0.06
1.500	0.06
1.790	0.07
2.000	0.06
2.250	0.06
2.500	0.05
2.700	0.07

CURVE 3, T = 298

λ	ρ
0.230	0.49
0.500	0.63
0.820	0.63
1.000	0.70*
1.300	0.78
1.500	0.79
1.790	0.81
2.000	0.82
2.250	0.84
2.500	0.90
2.700	0.92

CURVE 4*, T = 298

λ	ρ
0.230	0.27
0.500	0.15
0.820	0.10
1.000	0.09
1.300	0.06
1.500	0.06
1.790	0.06
2.000	0.05
2.250	0.05
2.500	0.05
2.700	0.06

CURVE 5*, T = 298

λ	ρ
0.240	0.70
0.500	0.79
0.800	0.75
1.000	0.78
1.210	0.80
1.500	0.82
1.740	0.83
2.000	0.88
2.200	0.87
2.500	0.96
2.700	0.97

CURVE 6*, T = 298

λ	ρ
0.240	0.25
0.500	0.14
0.800	0.09
1.000	0.09
1.210	0.08
1.500	0.08
1.740	0.06
2.000	0.05
2.200	0.05
2.500	0.04
2.700	0.04

CURVE 7, T = 298

λ	ρ
0.240	0.54
0.500	0.69
0.800	0.66
1.000	0.70
1.210	0.75
1.500	0.78
1.740	0.77
2.000	0.80
2.200	0.79
2.500	0.87
2.700	0.90

CURVE 8, T = 298

λ	ρ
0.240	0.31
0.500	0.24
0.800	0.16
1.000	0.15
1.210	0.14
1.500	0.14
1.740	0.14
2.000	0.12
2.200	0.11
2.500	0.12
2.700	0.12

CURVE 9, T = 298

λ	ρ
0.240	0.38
0.500	0.48
0.840	0.45
1.000	0.51
1.320	0.54
1.500	0.55
1.810	0.60
2.000	0.61
2.330	0.62
2.500	0.66
2.630	0.69
2.710	0.79

CURVE 10*, T = 298

λ	ρ
0.240	0.38
0.500	0.45
0.840	0.45
1.000	0.51
1.320	0.54
1.500	0.56
1.810	0.60
2.000	0.61
2.330	0.62
2.500	0.63
2.630	0.68

CURVE 11, T = 298

λ	ρ
0.240	0.34
0.500	0.44
0.840	0.43
1.000	0.48
1.320	0.50
1.500	0.53
1.810	0.56
2.000	0.59
2.330	0.60
2.500	0.62
2.630	0.68

CURVE 12*, T = 298

λ	ρ
0.240	0.34
0.500	0.44
0.840	0.43
1.000	0.48
1.320	0.50
1.500	0.53
1.810	0.56
2.000	0.59
2.330	0.60
2.500	0.62
2.630	0.68

CURVE 13, T = 298

λ	ρ
0.250	0.72
0.350	0.71
0.650	0.79
0.800	0.78
1.450	0.90
1.950	0.95
2.200	0.95
2.550	0.96
2.700	0.96

CURVE 14, T = 298

λ	ρ
0.230	0.69
0.460	0.66
0.580	0.70
0.810	0.65
1.150	0.75
1.310	0.75
1.480	0.79
1.720	0.79
1.910	0.84
2.100	0.85
2.290	0.83
2.540	0.92
2.710	0.95

CURVE 15*, T = 298

λ	ρ
0.250	0.76
0.450	0.75
0.690	0.80
0.810	0.79
1.200	0.89
1.600	0.96
1.830	0.96
2.000	0.99
2.210	0.96
2.430	0.97
2.700	0.97

CURVE 16*, T = 298

λ	ρ
0.240	0.70
0.360	0.69
0.580	0.74
0.800	0.69
1.000	0.78
1.400	0.82
1.800	0.85
2.030	0.90
2.430	0.94
2.700	0.98

* Not shown on plot

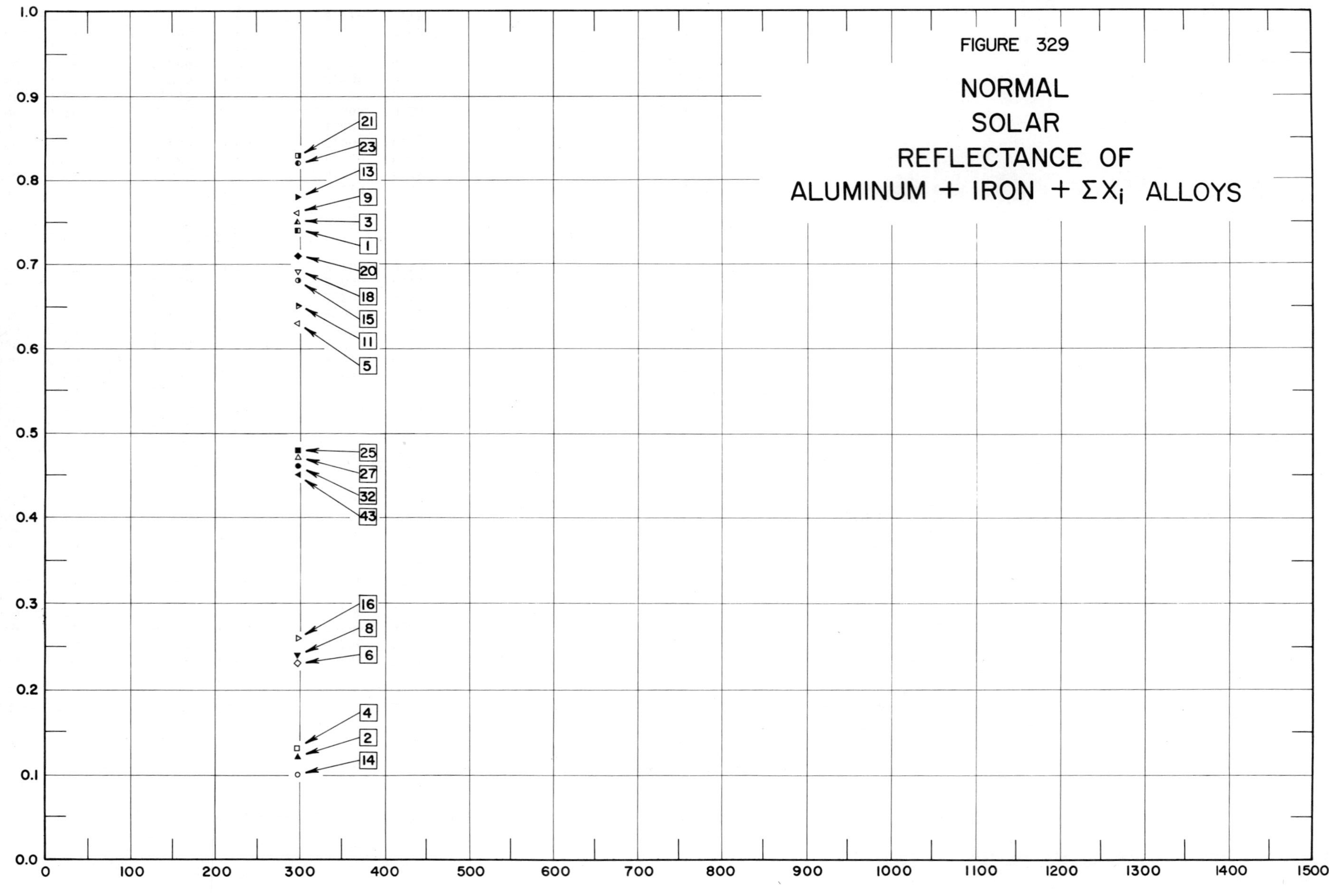
FIGURE 329
NORMAL
SOLAR
REFLECTANCE OF
ALUMINUM + IRON + ΣXi ALLOYS
NORMAL SOLAR REFLECTANCE
TEMPERATURE, K
1.0
0.9
0.8
0.7
0.6
0.5
0.4
0.3
0.2
0.1
0.0
0
100
200
300
400
500
600
700
800
900
1000
1100
1200
1300
1400
1500
21
23
13
9
3
1
20
18
15
11
5
25
27
32
43
16
8
6
4
2
14

SPECIFICATION TABLE NO. 329 NORMAL SOLAR REFLECTANCE OF [ALUMINUM + IRON + ΣX_i] ALLOYS

Curve No.	Ref. No.	Year	Temperature Range, K	Geometry θ	θ'	ω'	Reported Error, %	Composition (weight percent), Specifications and Remarks
1	131	1964	298	~0°		2π		L 34 (British Aircraft Material Specification); nominal composition: 99.0 min Al, 0.7 Fe, 0.5 Si, 0.15 Ti, 0.10 Cu, 0.1 Mg, 0.1 Mn, 0.1 Ni, 0.10 Zn, 0.05 Sn, 0.05 Pb; machine polished; freshly prepared; MgO reference; computed from spectral data.
2	131	1964	298	~0°		2π		Above specimen and conditions; diffuse component only.
3	131	1964	298	~0°		2π		Curve 1 specimen and conditions except exposed within clean laboratory area for 1 month.
4	131	1964	298	~0°		2π		Above specimen and conditions; diffuse component only.
5	131	1964	298	~0°		2π		Different sample, same as curve 1 specimen and conditions.
6	131	1964	298	~0°		2π		Above specimen and conditions; diffuse component only.
7	131	1964	298	~0°		2π		Curve 5 specimen and conditions except exposed within clean laboratory area for 1 month.
8	131	1964	298	~0°		2π		Above specimen and conditions; diffuse component only.
9	131	1964	298	~0°		2π		Different sample, same as curve 1 specimen and conditions.
10	131	1964	298	~0°		2π		Above specimen and conditions; diffuse component only.
11	131	1964	298	~0°		2π		Curve 9 specimen and conditions except exposed to outside atmosphere for 1 month.
12	131	1964	298	~0°		2π		Above specimen and conditions; diffuse component only.
13	131	1964	298	~0°		2π		Different sample, same as curve 1 specimen and conditions.
14	131	1964	298	~0°		2π		Above specimen and conditions; diffuse component only.
15	131	1964	298	~0°		2π		Curve 13 specimen and conditions except exposed to outside atmosphere for 1 month.
16	131	1964	298	~0°		2π		Above specimen and conditions; diffuse component only.
17	131	1964	298	~0°		2π		Different sample, same as curve 1 specimen and conditions except turned; freshly prepared.
18	131	1964	298	~0°		2π		Above specimen and conditions; diffuse component only.
19	131	1964	298	~0°		2π		Curve 17 specimen and conditions except exposed within clean laboratory area for 1 month.
20	131	1964	298	~0°		2π		Above specimen and conditions; diffuse component only.
21	131	1964	298	~0°		2π		Different sample, same as curve 17 specimen and conditions.
22	131	1964	298	~0°		2π		Above specimen and conditions; diffuse component only.
23	131	1964	298	~0°		2π		Curve 21 specimen and conditions except exposed within clean laboratory area for 1 month.
24	131	1964	298	~0°		2π		Above specimen and conditions; diffuse component only.
25	131	1964	298	~0°		2π		Different sample, same as curve 1 specimen and conditions except vapor blasted with 400 mesh alumina; 70 psi nozzle pressure; freshly prepared.
26	131	1964	298	~0°		2π		Above specimen and conditions; diffuse component only.
27	131	1964	298	~0°		2π		Curve 25 specimen and conditions except exposed within clean laboratory area for 1 month.

SPECIFICATION TABLE NO. 329 (continued)

Curve No.	Ref. No.	Year	Temperature Range, K	Geometry θ	θ'	ω'	Reported Error, %	Composition (weight percent), Specifications and Remarks
28	131	1964	298	~0°		2π		Above specimen and conditions; diffuse component only.
29	131	1964	298	~0°		2π		Different sample, same as curve 25 specimen and conditions except 50 psi nozzle pressure.
30	131	1964	298	~0°		2π		Above specimen and conditions; diffuse component only.
31	131	1964	298	~0°		2π		Different sample, same as curve 25 specimen and conditions except 30 psi nozzle pressure.
32	131	1964	298	~0°		2π		Above specimen and conditions; diffuse component only.
33	131	1964	298	~0°		2π		Different sample, same as curve 25 specimen and conditions.
34	131	1964	298	~0°		2π		Above specimen and conditions; diffuse component only.
35	131	1964	298	~0°		2π		Different sample, same as curve 25 specimen and conditions except 240 mesh alumina.
36	131	1964	298	~0°		2π		Above specimen and conditions; diffuse component only.
37	131	1964	298	~0°		2π		Different sample, same as curve 25 specimen and conditions except 120 mesh alumina.
38	131	1964	298	~0°		2π		Above specimen and conditions; diffuse component only.

DATA TABLE NO. 329 NORMAL SOLAR REFLECTANCE OF [ALUMINUM + IRON + ΣX_i] ALLOYS

[Temperature, T, K; Reflectance, ρ]

T	ρ	T	ρ	T	ρ	T	ρ
CURVE 1		CURVE 13		CURVE 25		CURVE 37*	
298	0.74	298	0.78	298	0.48	298	0.48
CURVE 2		CURVE 14		CURVE 26*		CURVE 38*	
298	0.12	298	0.10	298	0.48	298	0.48
CURVE 3		CURVE 15		CURVE 27			
298	0.75	298	0.68	298	0.47		
CURVE 4		CURVE 16		CURVE 28*			
298	0.13	298	0.26	298	0.47		
CURVE 5		CURVE 17*		CURVE 29*			
298	0.63	298	0.78	298	0.48		
CURVE 6		CURVE 18		CURVE 30*			
298	0.23	298	0.69	298	0.48		
CURVE 7*		CURVE 19*		CURVE 31*			
298	0.63	298	0.78	298	0.48		
CURVE 8		CURVE 20		CURVE 32*			
298	0.24	298	0.71	298	0.48		
CURVE 9		CURVE 21		CURVE 33*			
298	0.76	298	0.83	298	0.47		
CURVE 10*		CURVE 22*		CURVE 34*			
298	0.12	298	0.71	298	0.48		
CURVE 11		CURVE 23		CURVE 35*			
298	0.65	298	0.82	298	0.45		
CURVE 12*		CURVE 24*		CURVE 36*			
298	0.24	298	0.75	298	0.45		

* Not shown on plot

SPECIFICATION TABLE NO. 330 HEMISPHERICAL TOTAL EMITTANCE OF [ALUMINUM + MAGNESIUM + ΣX_i] ALLOYS

Curve No.	Ref. No.	Year	Temperature Range, K	Reported Error, %	Composition (weight percent), Specifications and Remarks
1	45	1961	278		Aluminum alloy 6061; nominal composition: 1.0 Mg, 0.6 Si, 0.25 Cr, 0.25 Cu, Al balance; as received.
2	45	1961	278		Different sample, same as above specimen and conditions except machine polished and degreased.
3	45	1961	278	10	Different sample, same as above specimen and conditions except sandblasted (120 size grit).
4	48	1964	302	< 7.7	Aluminum alloy 6061-T6; nominal composition: 1.0 Mg, 0.6 Si, 0.25 Cr, 0.25 Cu, Al balance; buffed; measured in vacuum (2×10^{-7} mm Hg).
5	48	1964	303	< 7.7	Above specimen and conditions; second trial.
6	48	1964	302	< 7.7	Above specimen and conditions; third trial.
7	48	1964	302	< 7.7	Curve 4 specimen and conditions; different analysis of observed data.
8	48	1964	302	< 7.7	Above specimen and conditions; second trial.

DATA TABLE NO. 330 HEMISPHERICAL TOTAL EMITTANCE OF [ALUMINUM + MAGNESIUM + ΣX_i] ALLOYS

[Temperature, T, K; Emittance, ϵ]

T	ϵ
CURVE 1*	
278	0.04
CURVE 2*	
278	0.04
CURVE 3*	
278	0.41
CURVE 4*	
302	0.0506
CURVE 5*	
303	0.0500
CURVE 6*	
302	0.0506
CURVE 7*	
302	0.0490
CURVE 8*	
302	0.0501

*Not shown on plot

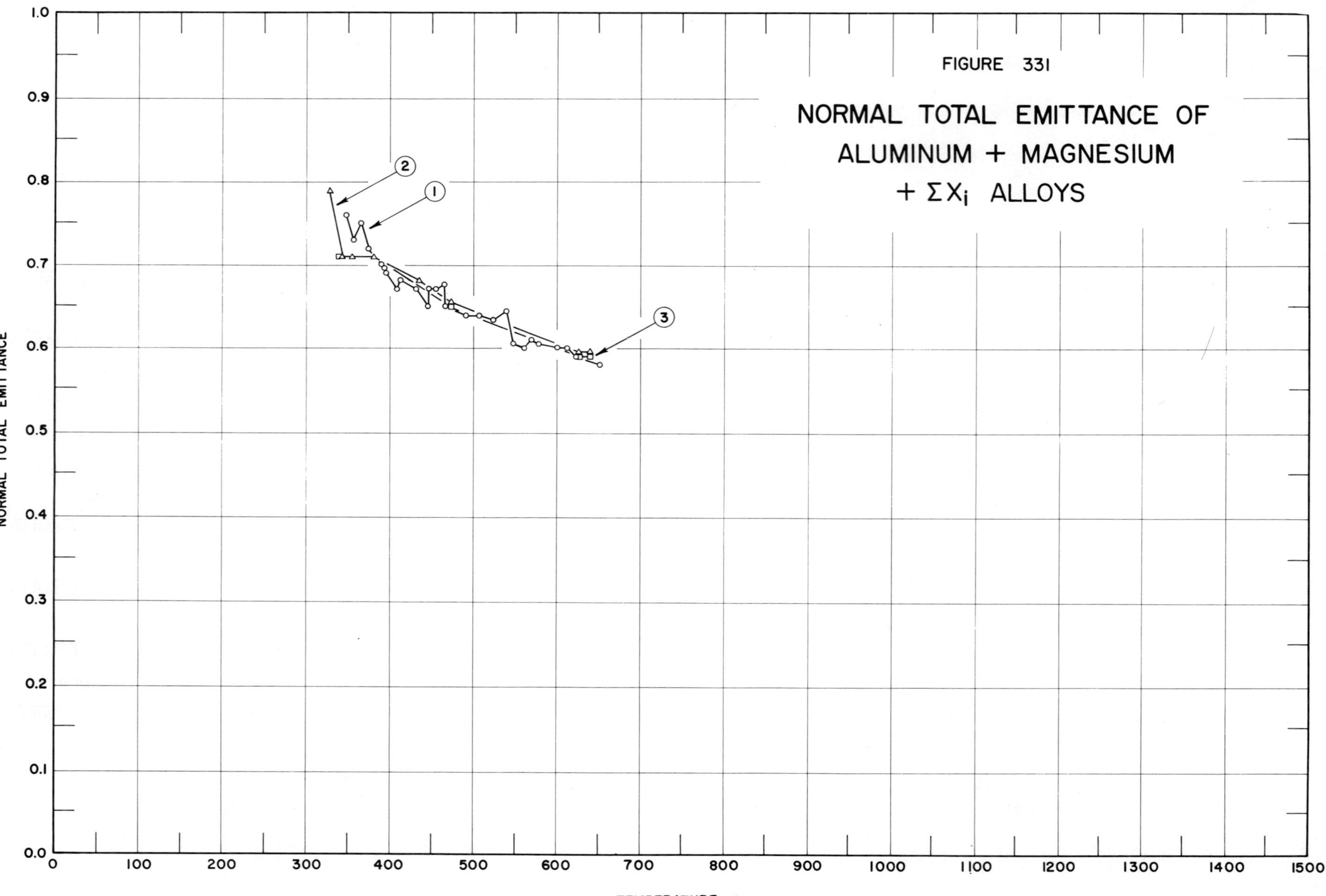
FIGURE 331
NORMAL TOTAL EMITTANCE OF
ALUMINUM + MAGNESIUM
+ ΣXi ALLOYS
NORMAL TOTAL EMITTANCE
TEMPERATURE, K
1.0
0.9
0.8
0.7
0.6
0.5
0.4
0.3
0.2
0.1
0.0
0
100
200
300
400
500
600
700
800
900
1000
1100
1200
1300
1400
1500
2
1
3

SPECIFICATION TABLE NO. 331 NORMAL TOTAL EMITTANCE OF [ALUMINUM + MAGNESIUM + ΣX_i] ALLOYS

Curve No.	Ref. No.	Year	Temperature Range, K	Geometry θ'	Reported Error, %	Composition (weight percent), Specifications and Remarks
1	155	1948	347-650	~0°		Aluminum alloy 53-SO; nominal composition: 1.3 Mg, 0.7 Si, 0.25 Cr, Al balance; weathered.
2	155	1948	328-639	~0°		Above specimen and conditions.
3	155	1948	339-639	~0°		Different sample, same as above specimen and conditions.

DATA TABLE NO. 331 NORMAL TOTAL EMITTANCE OF [ALUMINUM + MAGNESIUM + ΣX_i] ALLOYS

[Temperature, T, K; Emittance, ϵ]

T	ϵ
CURVE 1	
347	0.760
355	0.730
364	0.750
372	0.720
389	0.700
392	0.695
394	0.690
408	0.670
411	0.680
430	0.670
444	0.650
447	0.670
453	0.670
463	0.675
464	0.650
489	0.640
505	0.640
522	0.635
539	0.645
547	0.605
560	0.600
569	0.610
578	0.605
600	0.600
611	0.600
622	0.590
650	0.580
CURVE 2	
328	0.790
342	0.710
353	0.710
380	0.710
433	0.680
472	0.655
625	0.595
639	0.595
CURVE 3	
339	0.710
353	0.710*
380	0.710*
CURVE 3 (cont.)	
433	0.680*
472	0.650
625	0.590
639	0.590

* Not shown on plot

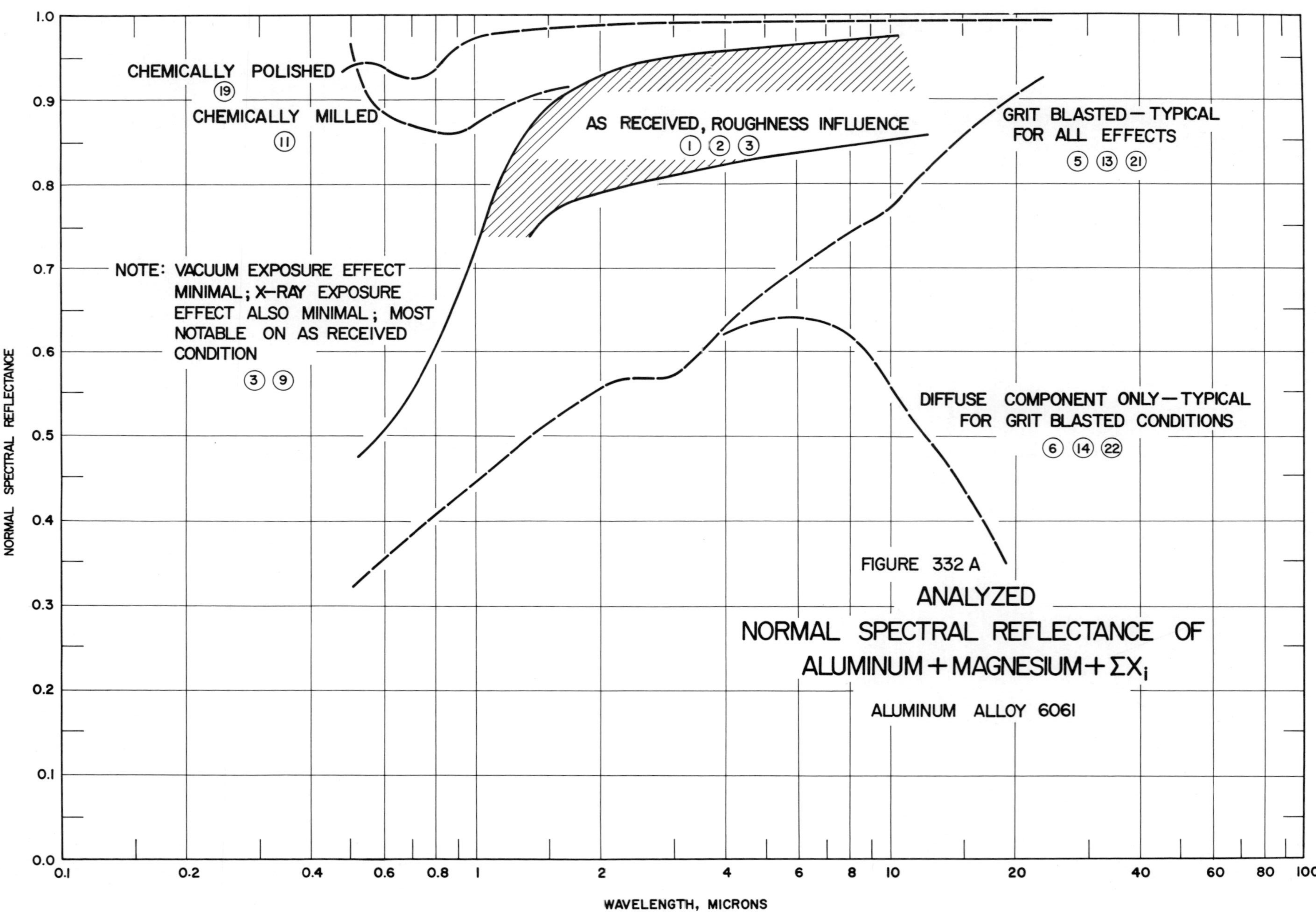

FIGURE 332 A
ANALYZED
NORMAL SPECTRAL REFLECTANCE OF ALUMINUM + MAGNESIUM + ΣX_i
ALUMINUM ALLOY 6061

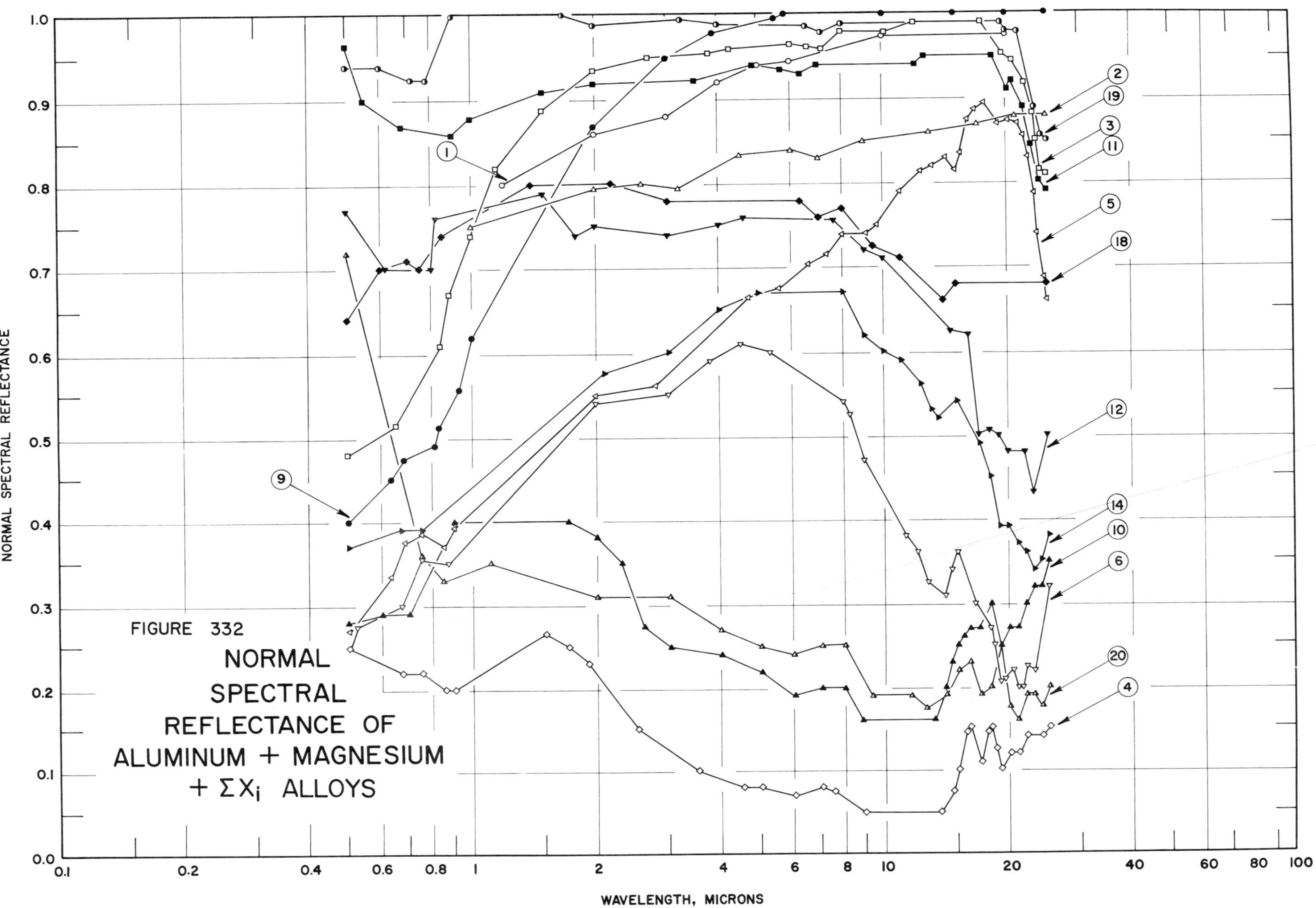

FIGURE 332
NORMAL SPECTRAL REFLECTANCE OF ALUMINUM + MAGNESIUM + ΣX_i ALLOYS

SPECIFICATION TABLE NO. 332 NORMAL SPECTRAL REFLECTANCE OF [ALUMINUM + MAGNESIUM + ΣX_i] ALLOYS

Curve No.	Ref. No.	Year	Temperature K	Wavelength Range, μ	Geometry θ θ'	ω'	Reported Error, %	Composition (weight percent), Specifications and Remarks
1	128	1962	298	1.20-20.00	~0°	2π		Aluminum alloy 6061-0; nominal composition: 1.0 Mg, 0.6 Si, 0.25 Cr, 0.25 Cu, Al balance; commercial finish; surface roughness 1.5μ rms; converted from $R(2\pi, 0°)$; $\omega' = 0.034$ Sr.
2	128	1962	298	1.00-25.00	~0°	2π		Different sample, same as above specimen and conditions except surface roughness 12.5μ rms.
3	153	1962	~322	0.50-25.00	~0°	2π	< 2	Aluminum alloy 6061-T6, QQ-A-327; nominal composition: 1.0 Mg, 0.6 Si, 0.25 Cr, 0.25 Cu, Al balance; as received; cleaned; data extracted from smooth curve; hohlraum at 1273 K; converted from $R(2\pi, 0°)$.
4	153	1962	~322	0.50-25.00	~0°	2π	< 2	Above specimen and conditions; diffuse component only.
5	153	1962	~322	0.50-25.00	~0°	2π	< 2	Different sample, same as Curve 3 specimen and conditions except grit blasted using 60 grit silicon carbide with air pressure of 110 to 120 psi for 30 to 45 sec.
6	153	1962	~322	0.50-25.00	~0°	2π	< 2	Above specimen and conditions; diffuse component only.
7	153	1962	~322	0.50-25.00	~0°	2π	< 2	Different sample, same as Curve 3 specimen and conditions except exposed to vacuum (< 4 x 10^{-8} mm Hg) for 24 hrs.
8	153	1962	~322	0.50-25.00	~0°	2π	< 2	Above specimen and conditions; diffuse component only.
9	153	1962	~322	0.50-25.00	~0°	2π	< 2	Different sample, same as Curve 3 specimen and conditions except X-Ray exposed in vacuum (4 x 10^{-8} mm Hg) for 24 hrs.
10	153	1962	~322	0.50-25.00	~0°	2π	< 2	Above specimen and conditions; diffuse component only.
11	153	1962	~322	0.50-25.00	~0°	2π	< 2	Different sample, same as Curve 3 specimen and conditions except chem-milled using the Turco 9H process for 3 min.
12	153	1962	~322	0.50-25.00	~0°	2π	< 2	Above specimen and conditions; diffuse component only.
13	153	1962	~322	0.50-25.00	~0°	2π	< 2	Different sample, same as Curve 3 specimen and conditions except chem-milled using the Turco 9H process for 3 min; grit blasted using 60 grit silicon carbide with air pressure of 110 to 120 psi for 30 to 45 sec.
14	153	1962	~322	0.50-25.00	~0°	2π	< 2	Above specimen and conditions; diffuse component only.
15	153	1962	~322	0.50-25.00	~0°	2π	< 2	Different sample, same as Curve 3 specimen and conditions except chem-milled using the Turco 9H process for 3 min; exposed to vacuum (< 4 x 10^{-8} mm Hg) for 24 hrs.
16	153	1962	~322	0.50-25.00	~0°	2π	< 2	Above specimen and conditions; diffuse component only.
17	153	1962	~322	0.50-25.00	~0°	2π	< 2	Different sample, same as Curve 3 specimen and conditions except chem-milled using the Turco 9H process for 3 min; X-Ray exposed in vacuum (4 x 10^{-8} mm Hg) for 24 hrs.
18	153	1962	~322	0.50-25.00	~0°	2π	< 2	Above specimen and conditions; diffuse component only.
19	153	1962	~322	0.50-25.00	~0°	2π	< 2	Different sample, same as Curve 3 specimen and conditions except chemically polished using the Alcoa process with a 2 min immersion at 358 K.

SPECIFICATION TABLE NO. 332 (continued)

Curve No.	Ref. No.	Year	Temperature K	Wavelength Range, μ	Geometry θ	θ'	ω'	Reported Error, %	Composition (weight percent), Specifications and Remarks
20	153	1962	~322	0.50-25.00	~0°		2π	< 2	Above specimen and conditions; diffuse component only.
21	153	1962	~322	0.50-25.00	~0°		2π	< 2	Different sample, same as Curve 3 specimen and conditions except chemically polished using the Alcoa process with a 2 min immersion at 358 K; grit blasted using 60 grit silicon carbide with air pressure of 110 to 120 psi for 30 to 45 sec.
22	153	1962	~322	0.50-25.00	~0°		2π	< 2	Above specimen and conditions; diffuse component only.
23	153	1962	~322	0.50-25.00	~0°		2π	< 2	Different sample, same as Curve 3 specimen and conditions except chemically polished using the Alcoa process with a 2 min immersion at 358 K; exposed to vacuum ($< 4 \times 10^{-8}$ mm Hg) for 24 hrs.
24	153	1962	~322	0.50-25.00	~0°		2π	< 2	Above specimen and conditions; diffuse component only.
25	153	1962	~322	0.50-25.00	~0°		2π	< 2	Different sample, same as Curve 3 specimen and conditions except chemically polished using the Alcoa process with a 2 min immersion at 358 K; X-Ray exposed in vacuum (4×10^{-8} mm Hg) for 24 hrs.
26	153	1962	~322	0.50-25.00	~0°		2π	< 2	Above specimen and conditions; diffuse component only.

DATA TABLE NO. 332 NORMAL SPECTRAL REFLECTANCE OF [ALUMINUM + MAGNESIUM + ΣX_i] ALLOYS

[Wavelength, λ, μ; Reflectance, ρ; Temperature, T, K]

λ	ρ
CURVE 1	**T = 298**
1.20	0.800
2.00	0.860
3.00	0.880
4.00	0.920
5.00	0.940
6.00	0.945
10.00	0.975
20.00	0.975
CURVE 2	**T = 298**
1.00	0.750
2.00	0.795
2.60	0.800
3.20	0.795
4.25	0.835
6.00	0.840
7.00	0.830
9.00	0.850
13.00	0.860
17.00	0.870
21.00	0.880
25.00	0.880
CURVE 3	**T = ~322**
0.50	0.480
0.65	0.515
0.84	0.610
0.88	0.670
1.00	0.740
1.15	0.820
1.50	0.890
2.00	0.935
2.70	0.950
3.80	0.956
5.25	0.960
6.00	0.967
6.60	0.962
7.15	0.960
8.00	0.980
CURVE 3 (cont.)	
10.20	0.980
12.00	0.990
17.40	0.990
19.65	0.952
20.70	0.945
22.10	0.918
23.20	0.882
23.70	0.850
24.10	0.815
25.00	0.810
CURVE 4	**T = ~322**
0.50	0.250
0.67	0.220
0.75	0.220
0.85	0.200
0.90	0.200
1.50	0.265
1.70	0.250
1.90	0.230
2.50	0.150
3.50	0.100
4.50	0.080
5.00	0.080
6.00	0.070
7.00	0.080
7.50	0.075
8.90	0.050
13.50	0.050
14.50	0.075
15.00	0.100
15.70	0.145
16.00	0.150
17.00	0.110
17.70	0.145
18.00	0.150
18.50	0.125
19.00	0.100
20.00	0.120
21.00	0.120
22.00	0.140
24.00	0.140
25.00	0.150
CURVE 5	**T = ~322**
0.50	0.270
0.63	0.335
0.68	0.375
0.75	0.385
0.85	0.370
0.90	0.392
2.00	0.550
2.80	0.561
4.70	0.665
5.60	0.675
6.60	0.702
7.30	0.715
8.00	0.740
9.10	0.740
9.70	0.750
11.00	0.790
12.40	0.813
13.20	0.820
14.20	0.830
15.00	0.813
15.50	0.835
16.10	0.875
16.80	0.888
17.70	0.893
19.10	0.870
20.20	0.873
21.20	0.870
22.00	0.855
22.50	0.830
23.25	0.788
23.75	0.740
24.75	0.687
25.00	0.660
CURVE 6	**T = ~322**
0.50	0.250*
0.52	0.275
0.67	0.300
0.75	0.355
0.87	0.350
2.00	0.540
CURVE 6 (cont.)	
3.00	0.550
3.80	0.590
4.50	0.610
5.30	0.600
8.00	0.540
8.30	0.525
9.00	0.470
11.20	0.380
12.00	0.360
12.70	0.325
14.00	0.310
14.50	0.340
15.00	0.360
16.50	0.300
18.00	0.270
18.30	0.250
19.00	0.205
19.50	0.210
20.30	0.220
21.00	0.200
21.30	0.200
22.00	0.225
23.00	0.220
25.00	0.320
CURVE 7*	**T = ~322**
0.50	0.440
0.63	0.500
0.78	0.520
1.00	0.650
2.00	0.887
2.70	0.905
4.20	0.952
5.30	0.952
6.75	0.970
8.50	0.973
9.10	0.980
11.40	0.980
12.10	0.990
21.20	0.900
22.40	0.955
23.20	0.945
CURVE 7*(cont.)	
24.00	0.862
25.00	0.850
CURVE 8*	**T = ~322**
0.50	0.250
0.67	0.300
0.85	0.280
1.00	0.320
1.35	0.315
1.75	0.285
2.00	0.260
2.50	0.215
3.50	0.140
4.00	0.120
5.00	0.100
6.00	0.080
8.00	0.100
9.00	0.060
9.50	0.055
10.00	0.060
11.00	0.070
11.05	0.065
12.00	0.055
13.00	0.070
13.50	0.070
14.00	0.090
14.50	0.150
15.00	0.190
15.50	0.215
16.00	0.230
17.00	0.240
18.00	0.250
18.20	0.250
19.00	0.220
19.50	0.230
20.00	0.250
21.00	0.240
22.00	0.270
23.00	0.280
24.00	0.280
25.00	0.310
CURVE 9	**T = ~322**
0.50	0.400
0.63	0.450
0.68	0.473
0.81	0.490
0.83	0.512
0.93	0.557
1.00	0.620
2.00	0.870
3.00	0.950
3.90	0.980
5.50	0.995
5.80	1.000
10.00	1.000
15.00	1.000
20.00	1.000
25.00	1.000
CURVE 10	**T = ~322**
0.50	0.280
0.60	0.290
0.70	0.290
0.90	0.400
1.70	0.400
2.00	0.380
2.30	0.350
2.60	0.275
3.00	0.250
4.00	0.240
5.00	0.220
6.00	0.190
7.00	0.200
8.00	0.200
8.80	0.160
13.10	0.160
14.00	0.200
14.50	0.230
15.00	0.250
15.50	0.260
16.00	0.270
17.00	0.270
18.00	0.300
CURVE 10 (cont.)	
19.00	0.250
20.00	0.270
21.00	0.270
22.00	0.300
23.00	0.320
24.00	0.320
25.00	0.350
CURVE 11	**T = ~322**
0.50	0.965
0.55	0.900
0.68	0.870
0.90	0.860
1.00	0.880
1.50	0.910
2.00	0.920
3.50	0.922
4.90	0.940
5.70	0.935
6.30	0.930
7.00	0.940
12.00	0.940
12.75	0.950
18.55	0.950
20.10	0.910
20.85	0.920
22.00	0.890
23.30	0.845
24.00	0.800
25.00	0.790
CURVE 12	**T = ~322**
0.50	0.770
0.62	0.700
0.80	0.700
0.82	0.760
1.50	0.790
1.80	0.740
2.00	0.750
3.00	0.740
CURVE 12 (cont.)	
4.00	0.750
4.60	0.760
7.60	0.755
9.00	0.720
10.00	0.710
14.50	0.625
16.00	0.620
17.00	0.500
18.00	0.505
19.00	0.500
20.00	0.480
22.00	0.480
23.00	0.430
25.00	0.500
CURVE 13*	**T = ~322**
0.50	0.320
0.63	0.350
0.68	0.390
0.79	0.400
0.88	0.425
1.00	0.450
2.00	0.558
2.25	0.565
2.70	0.565
3.15	0.575
4.15	0.645
5.10	0.670
6.30	0.707
7.40	0.730
8.00	0.750
9.25	0.750
12.00	0.820
14.00	0.825
14.60	0.835
15.20	0.858
16.00	0.900
17.70	0.900
18.50	0.882
21.25	0.880
23.00	0.820
23.40	0.788

* Not shown on plot

DATA TABLE NO. 332 (continued)

λ	ρ
CURVE 13*(cont.)	
23.70	0.723
25.00	0.672
CURVE 14	
T = ~322	
0.50	0.370
0.67	0.390
0.75	0.390
2.10	0.575
3.00	0.600
4.00	0.650
5.00	0.670
8.00	0.670
9.00	0.620
10.00	0.600
11.00	0.590
12.30	0.560
13.00	0.530
13.50	0.520
15.00	0.540
17.00	0.490
18.00	0.450
19.00	0.390
20.00	0.390
21.00	0.370
22.00	0.360
23.00	0.340
24.00	0.340
25.00	0.380
CURVE 15*	
T = ~322	
0.50	0.920
0.60	0.890
0.73	0.885
0.85	0.870
1.00	0.900
1.60	0.910
2.00	0.923
2.60	0.932
3.50	0.932
4.30	0.930
5.65	0.950

λ	ρ
CURVE 15*(cont.)	
8.50	0.950
9.80	0.938
12.10	0.960
14.35	0.960
15.10	0.950
20.50	0.950
21.10	0.960
22.20	0.925
22.90	0.920
23.35	0.910
23.80	0.860
24.15	0.840
25.00	0.850
CURVE 16*	
T = ~322	
0.50	0.730
0.80	0.710
1.00	0.730
1.50	0.740
2.00	0.760
4.00	0.750
4.50	0.750
5.00	0.760
7.00	0.760
11.00	0.790
13.00	0.640
15.00	0.620
17.00	0.620
18.00	0.610
19.00	0.580
20.00	0.580
21.00	0.550
23.00	0.550
25.00	0.590
CURVE 17*	
T = ~322	
0.50	0.830
0.73	0.875
0.93	0.880
1.00	0.910
1.50	0.930

λ	ρ
CURVE 17*(cont.)	
2.00	0.958
3.00	0.928
3.90	0.950
6.20	0.950
6.80	0.960
8.30	0.960
8.85	0.950
10.45	0.950
11.00	0.960
18.00	0.960
25.00	0.960
CURVE 18	
T = ~322	
0.50	0.640
0.60	0.700
0.70	0.710
0.75	0.700
0.85	0.740
1.40	0.800
2.20	0.800
3.00	0.780
6.30	0.780
7.00	0.760
8.00	0.770
9.50	0.725
11.00	0.710
14.00	0.660
15.00	0.680
25.00	0.680
CURVE 19	
T = ~322	
0.50	0.940
0.60	0.940
0.72	0.925
0.78	0.925
0.90	1.000
1.68	1.000
2.00	0.990
3.25	0.995
4.00	0.990
6.55	0.987

λ	ρ
CURVE 19 (cont.)	
7.15	0.980
8.00	0.990
19.50	0.990
20.00	0.980
21.20	0.980
23.30	0.890
24.20	0.855
25.00	0.850
CURVE 20	
T = ~322	
0.50	0.720
0.75	0.360
0.85	0.330
1.10	0.350
2.00	0.310
3.00	0.310
4.00	0.270
5.00	0.250
6.00	0.240
7.00	0.250
8.00	0.250
9.30	0.190
11.50	0.190
12.50	0.175
14.00	0.190
15.00	0.220
16.00	0.230
17.00	0.190
18.00	0.200
19.00	0.250*
20.00	0.175
21.00	0.160
22.00	0.190
23.00	0.190
24.00	0.175
25.00	0.200
CURVE 21*	
T = ~322	
0.50	0.322
0.68	0.350
0.78	0.350

λ	ρ
CURVE 21*(cont.)	
0.80	0.400
1.00	0.457
1.35	0.495
2.00	0.548
3.40	0.580
5.30	0.645
5.70	0.655
6.30	0.667
7.70	0.725
8.20	0.733
9.00	0.730
9.40	0.735
10.55	0.765
11.30	0.775
12.50	0.810
13.20	0.810
14.00	0.820
14.40	0.818
15.00	0.800
15.25	0.810
15.40	0.855
16.00	0.900
17.50	0.900
18.90	0.870
21.20	0.870
22.60	0.820
23.20	0.795
23.70	0.722
24.00	0.695
25.00	0.653
CURVE 22*	
T = ~322	
0.50	0.290
0.57	0.310
0.75	0.310
0.85	0.395
2.00	0.540
3.40	0.580
5.00	0.600
6.00	0.580
7.00	0.580
8.00	0.560
9.00	0.490

λ	ρ
CURVE 22*(cont.)	
10.00	0.450
12.00	0.400
14.00	0.340
15.00	0.360
16.50	0.320
18.00	0.300
19.00	0.240
20.00	0.260
21.00	0.240
22.00	0.260
23.00	0.240
25.00	0.330
CURVE 23*	
T = ~322	
0.50	0.940
0.60	0.968
0.68	0.950
0.90	0.950
1.00	0.970
1.75	0.970
2.00	0.980
3.00	0.980
4.00	0.970
5.60	0.970
6.40	0.960
7.00	0.960
8.10	0.980
10.30	0.980
11.00	0.970
11.50	0.970
12.40	0.988
13.10	0.990
14.10	0.970
15.00	0.980
23.00	0.980
23.40	0.960
23.70	0.915
24.20	0.898
25.00	0.890

λ	ρ
CURVE 24*	
T = ~322	
0.50	0.400
0.65	0.415
0.85	0.310
1.00	0.325
1.65	0.300
2.60	0.300
4.00	0.250
6.00	0.220
8.00	0.220
9.00	0.175
12.00	0.150
15.00	0.200
16.00	0.250
19.00	0.250
20.00	0.280
21.00	0.260
22.00	0.300
23.00	0.290
25.00	0.350
CURVE 25*	
T = ~322	
0.50	0.980
0.85	0.940
0.93	0.970
1.00	0.980
1.25	0.995
1.40	1.000
9.00	1.000
18.00	1.000
25.00	1.000
CURVE 26*	
T = ~322	
0.50	0.475
0.62	0.515
0.75	0.525
0.90	0.555
4.20	0.390
6.00	0.340
7.20	0.340
8.00	0.325

λ	ρ
CURVE 26*(cont.)	
9.00	0.285
10.00	0.270
12.00	0.270
13.00	0.280
14.00	0.310
14.50	0.325
14.80	0.330
15.00	0.330
16.00	0.300
17.00	0.275
18.00	0.300
19.00	0.240
20.00	0.270
21.20	0.260
22.00	0.320
23.50	0.320
24.00	0.330
25.00	0.280

* Not shown on plot

SPECIFICATION TABLE NO. 333 NORMAL SOLAR ABSORPTANCE OF [ALUMINUM + MAGNESIUM + ΣX_i] ALLOYS

Curve No.	Ref. No.	Year	Temperature Range, K	Geometry θ	Reported Error, %	Composition (weight percent), Specifications and Remarks
1	45	1961	278	~0°	10	Alumirum alloy 6061; nominal composition: 1.0 Mg, 0.6 Si, 0.25 Cr, 0.25 Cu, Al balance; as received; extraterrestrial.
2	45	1961	278	~0°	10	Different sample, same as above specimen and conditions except machine polished and degreased.
3	45	1961	278	~0°	10	Different sample, same as above specimen and conditions except sandblasted (120 size grit).

DATA TABLE NO. 333 NORMAL SOLAR ABSORPTANCE OF [ALUMINUM + MAGNESIUM + ΣX_i] ALLOYS

[Temperature, T, K; Absorptance, α]

T	α
CURVE 1*	
278	0.41
CURVE 2*	
278	0.35
CURVE 3*	
278	0.60

* Not shown on plot

SPECIFICATION TABLE NO. 334 HEMISPHERICAL TOTAL EMITTANCE OF [ALUMINUM + MANGANESE + ΣX_i] ALLOYS

Curve No.	Ref. No.	Year	Temperature Range, K	Reported Error, %	Composition (weight percent), Specifications and Remarks
1	22	1958	589-811	≤ 2	Aluminum alloy AA 3003; nominal composition: 1.0-1.5 Mn, 0.7 max Fe, 0.6 max Si, 0.2 max Cu, 0.1 max Zn, Al balance; stably oxidized at 811 K.

DATA TABLE NO. 334 HEMISPHERICAL TOTAL EMITTANCE OF [ALUMINUM + MANGANESE + ΣX_i] ALLOYS

[Temperature, T, K; Emittance, ϵ]

T	ϵ
CURVE 1*	
589	0.400
644	0.400
700	0.395
755	0.400
811	0.400

* Not shown on plot

SPECIFICATION TABLE NO. 335 NORMAL TOTAL EMITTANCE OF [ALUMINUM + MANGANESE + ΣX_i] ALLOYS

Curve No.	Ref. No.	Year	Temperature Range, K	Geometry θ'	Reported Error, %	Composition (weight percent), Specifications and Remarks
1	22	1958	589-811	~0°	≤ 2	Aluminum alloy AA 3003; nominal composition: 1.0-1.5 Mn, 0.7 max Fe, 0.6 max Si, 0.2 max Cu, 0.1 max Zn, Al balance; stably oxidized at 811 K.

DATA TABLE NO. 335 NORMAL TOTAL EMITTANCE OF [ALUMINUM + MANGANESE + ΣX_i] ALLOYS

[Temperature, T, K; Emittance, ϵ]

T	ϵ
CURVE 1*	
589	0.400
644	0.400
700	0.405
755	0.410
811	0.415

* Not shown on plot

SPECIFICATION TABLE NO. 336 HEMISPHERICAL TOTAL EMITTANCE OF [ALUMINUM + ZINC + ΣX_i] ALLOYS

Curve No.	Ref. No.	Year	Temperature Range, K	Reported Error, %	Composition (weight percent), Specifications and Remarks
1	24	1948	303		Aluminum alloy 75-ST (alclad); nominal composition: 1.0 Zn; measured in air.
2	24	1948	301		Different sample, same as above specimen and conditions except aerobright polished.

DATA TABLE NO. 336 HEMISPHERICAL TOTAL EMITTANCE OF [ALUMINUM + ZINC + ΣX_i] ALLOYS

[Temperature, T, K; Emittance, ϵ]

T	ϵ
CURVE 1*	
303	0.020
CURVE 2*	
301	0.07

* Not shown on plot

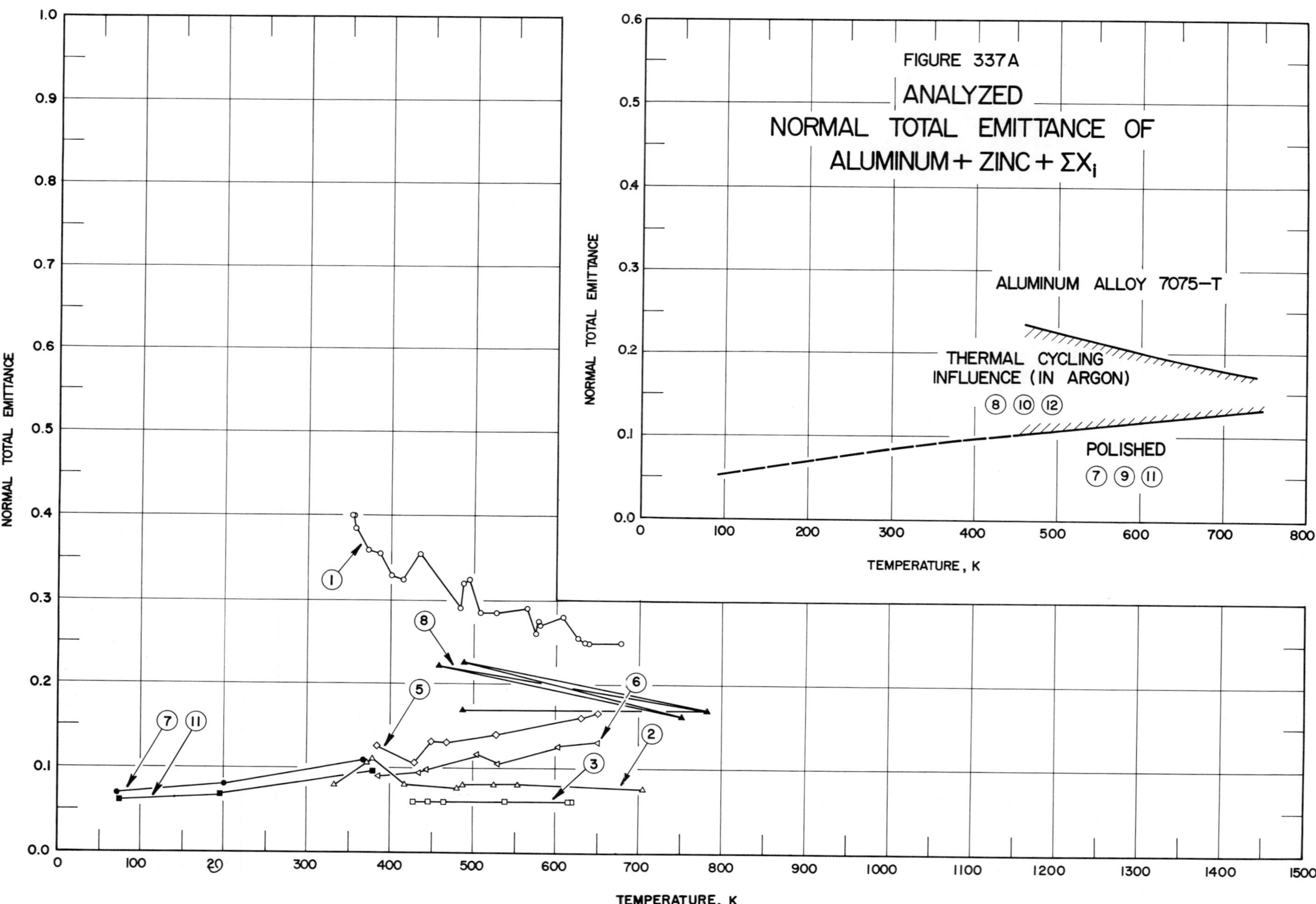
FIGURE 337A
ANALYZED
NORMAL TOTAL EMITTANCE OF
ALUMINUM + ZINC + ΣXi
ALUMINUM ALLOY 7075-T
THERMAL CYCLING
INFLUENCE (IN ARGON)
8 10 12
POLISHED
7 9 11
NORMAL TOTAL EMITTANCE
TEMPERATURE, K
1
2
3
5
6
8
7 11

SPECIFICATION TABLE NO. 337 NORMAL TOTAL EMITTANCE OF [ALUMINUM + ZINC + ΣX_i] ALLOYS

Curve No.	Ref. No.	Year	Temperature Range, K	Geometry θ'	Reported Error, %	Composition (weight percent), Specifications and Remarks
1	155	1948	353-678	~0°		Aluminum alloy 75-ST (alclad); nominal composition: 5.6 Zn, 2.5 Mg, 1.6 Cu, 0.30 Cr, Al balance; effect of viewing configuration change apparent.
2	155	1948	333-705	~0°		Above specimen and conditions; another change in viewing configuration made.
3	155	1948	428-619	~0°		Above specimen and conditions; another change in viewing configuration made.
4	155	1948	428-619	~0°		Different sample, same as above specimen and conditions; unpolished.
5	155	1948	383-650	~0°		Different sample, same as above specimen and conditions except polished with aerobright and Bon Ami.
6	155	1948	386-650	~0°		Above specimen and conditions; repeat measurement one day later.
7	160	1954	74.8-369	~0°		Aluminum alloy 75 ST; nominal composition: 5.6 Zn, 2.5 Mg, 1.6 Cu, 0.30 Cr, Al balance; cleaned with methyl alcohol; heating; measured in air (10^{-3} mm Hg); [Author's designation: Sample 1].
8	160	1954	458-782	~0°		Different sample, same as curve 7 specimen and conditions except measured in argon (10^{-3} mm Hg); [Author's designation: Sample 5].
9	160	1954	80.4-388	~0°		Different sample, same as curve 7 specimen and conditions except scrubbed with Bon Ami on a wet cloth, washed and dried, wiped with toluene and alcohol; [Author's designation: Sample 7].
10	160	1954	482-799	~0°		Different sample, same as curve 9 specimen and conditions except measured in argon (10^{-3} mm Hg); [Author's designation: Sample 8].
11	160	1954	77.1-380	~0°		Different sample, same as curve 7 specimen and conditions except polished and then finished with a wool buff and rouge and washed; surface free from scratches; [Author's designation: Sample 6].
12	160	1954	490-802	~0°		Different sample, same as curve 11 specimen and conditions except measured in argon (10^{-3} mm Hg); [Author's designation: Sample 9].

DATA TABLE NO. 337 NORMAL TOTAL EMITTANCE OF [ALUMINUM + ZINC + ΣX_i] ALLOYS

[Temperature, T, K; Emittance, ϵ]

T	ϵ	T	ϵ	T	ϵ
CURVE 1		CURVE 4*		CURVE 9*	
353	0.400	428	0.065	80.4	0.065
355	0.400	433	0.060	198	0.081
358	0.385	447	0.065	388	0.117
372	0.360	464	0.055		
386	0.355	514	0.055	CURVE 10*	
400	0.330	539	0.055		
414	0.325	547	0.055	515	0.121
433	0.355	617	0.060	705	0.129
483	0.290	619	0.060	799	0.159
486	0.320			557	0.188
494	0.325	CURVE 5		699	0.182
508	0.285			482	0.203
528	0.285	383	0.125		
564	0.290	428	0.105	CURVE 11	
575	0.260	450	0.130		
578	0.275	469	0.130	77.1	0.061
580	0.270	528	0.140	196	0.067
608	0.280	630	0.160	380	0.095
625	0.255	650	0.165		
633	0.250			CURVE 12*	
639	0.250	CURVE 6			
678	0.250			540	0.104
		386	0.090	802	0.123
CURVE 2		433	0.095	778	0.147
		444	0.098	490	0.223
333	0.080	505	0.115	769	0.150
372	0.105	530	0.105	519	0.208
378	0.110	603	0.125		
417	0.080	650	0.130		
480	0.075				
486	0.080	CURVE 7			
525	0.080				
553	0.080	74.8	0.069		
705	0.075	202	0.080		
		369	0.109		
CURVE 3					
		CURVE 8			
428	0.060				
447	0.060	488	0.168		
464	0.060	782	0.169		
539	0.060	458	0.222		
614	0.060	753	0.162		
619	0.060	489	0.226		

* Not shown on plot

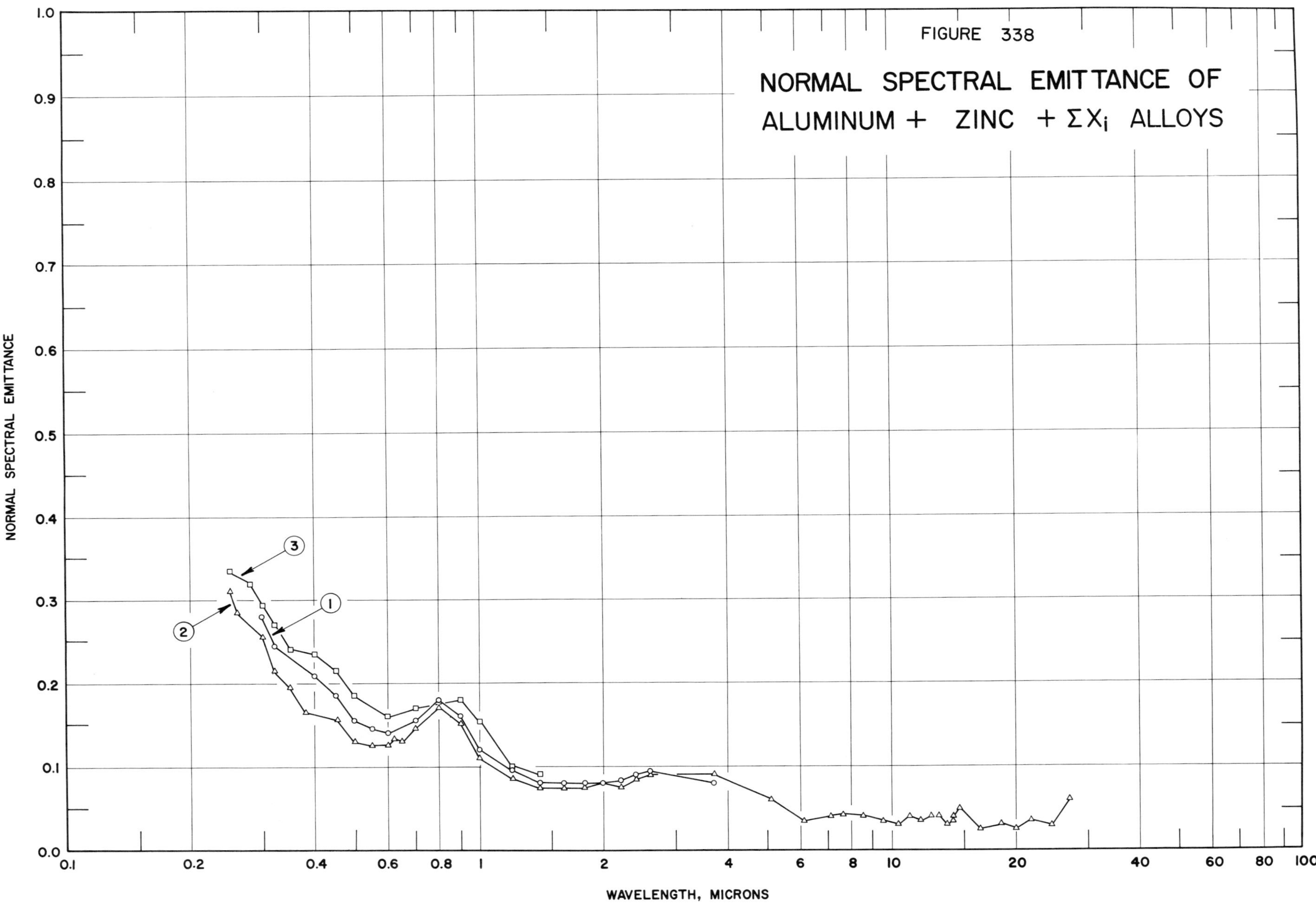
FIGURE 338
NORMAL SPECTRAL EMITTANCE OF
ALUMINUM + ZINC + ΣXi ALLOYS
NORMAL SPECTRAL EMITTANCE
1.0
0.9
0.8
0.7
0.6
0.5
0.4
0.3
0.2
0.1
0.0
WAVELENGTH, MICRONS
0.1
0.2
0.4
0.6
0.8
1
2
4
6
8
10
20
40
60
80
100
1
2
3

SPECIFICATION TABLE NO. 338 NORMAL SPECTRAL EMITTANCE OF [ALUMINUM + ZINC + ΣX_i] ALLOYS

Curve No.	Ref. No.	Year	Temperature K	Wavelength Range, μ	Geometry θ'	Reported Error, %	Composition (weight percent), Specifications and Remarks
1	159	1963	323	0.30-3.70	5°		Aluminum alloy 7075; nominal composition: 5.6 Zn, 2.5 Mg, 1.6 Cu, 0.30 Cr, Al balance; surface roughness 3.2-4.4 microinches (center line average); measured in nitrogen; computed by $\epsilon = 1 - R(2\pi, 5^\circ)$; [Author's designation: Specimen 1].
2	159	1963	323	0.25-27.00	5°		Different sample, same as above specimen and conditions except surface roughness 1.9-2.5 microinches (center line average); [Author's designation: Specimen 3].
3	159	1963	323	0.25-1.40	5°		Different sample, same as above specimen and conditions except surface roughness 3.2-4.5 microinches (center line average); [Author's designation: Specimen 4].

DATA TABLE NO. 338 NORMAL SPECTRAL EMITTANCE OF [ALUMINUM + ZINC + ΣX_i] ALLOYS

[Wavelength, λ, μ; Emittance, ϵ; Temperature, T K]

λ	ϵ	λ	ϵ
CURVE 1 T = 323		**CURVE 2 (cont.)**	
		1.80	0.075
0.30	0.280	2.00	0.080*
0.32	0.245	2.20	0.075
0.40	0.210	2.40	0.085
0.45	0.185	2.60	0.090
0.50	0.155	3.70	0.090
0.55	0.145	5.10	0.060
0.60	0.140	6.10	0.035
0.70	0.155	7.10	0.040
0.80	0.180	7.60	0.042
0.90	0.160	8.50	0.040
1.00	0.120	9.50	0.035
1.20	0.095	10.30	0.030
1.40	0.080	11.00	0.040
1.60	0.080	11.70	0.035
1.80	0.080	12.40	0.040
2.00	0.080	13.00	0.040
2.20	0.082	13.60	0.030
2.40	0.090	14.05	0.035
2.60	0.095	14.10	0.040
3.70	0.080	14.70	0.050
		16.30	0.025
CURVE 2 T = 323		18.40	0.030
		20.00	0.025
		21.80	0.035
0.25	0.310	24.40	0.030
0.26	0.285	27.00	0.060
0.30	0.255		
0.32	0.215	**CURVE 3** T = 323	
0.35	0.195		
0.38	0.165		
0.45	0.155	0.25	0.335
0.50	0.130	0.28	0.320
0.55	0.125	0.30	0.295
0.60	0.125	0.32	0.270
0.62	0.132	0.35	0.240
0.65	0.130	0.40	0.235
0.70	0.145	0.45	0.215
0.80	0.170	0.50	0.185
0.90	0.150	0.60	0.160
1.00	0.110	0.70	0.170
1.20	0.085	0.90	0.180
1.40	0.075	1.00	0.152
1.60	0.075	1.20	0.100
		1.40	0.090

* Not shown on plot

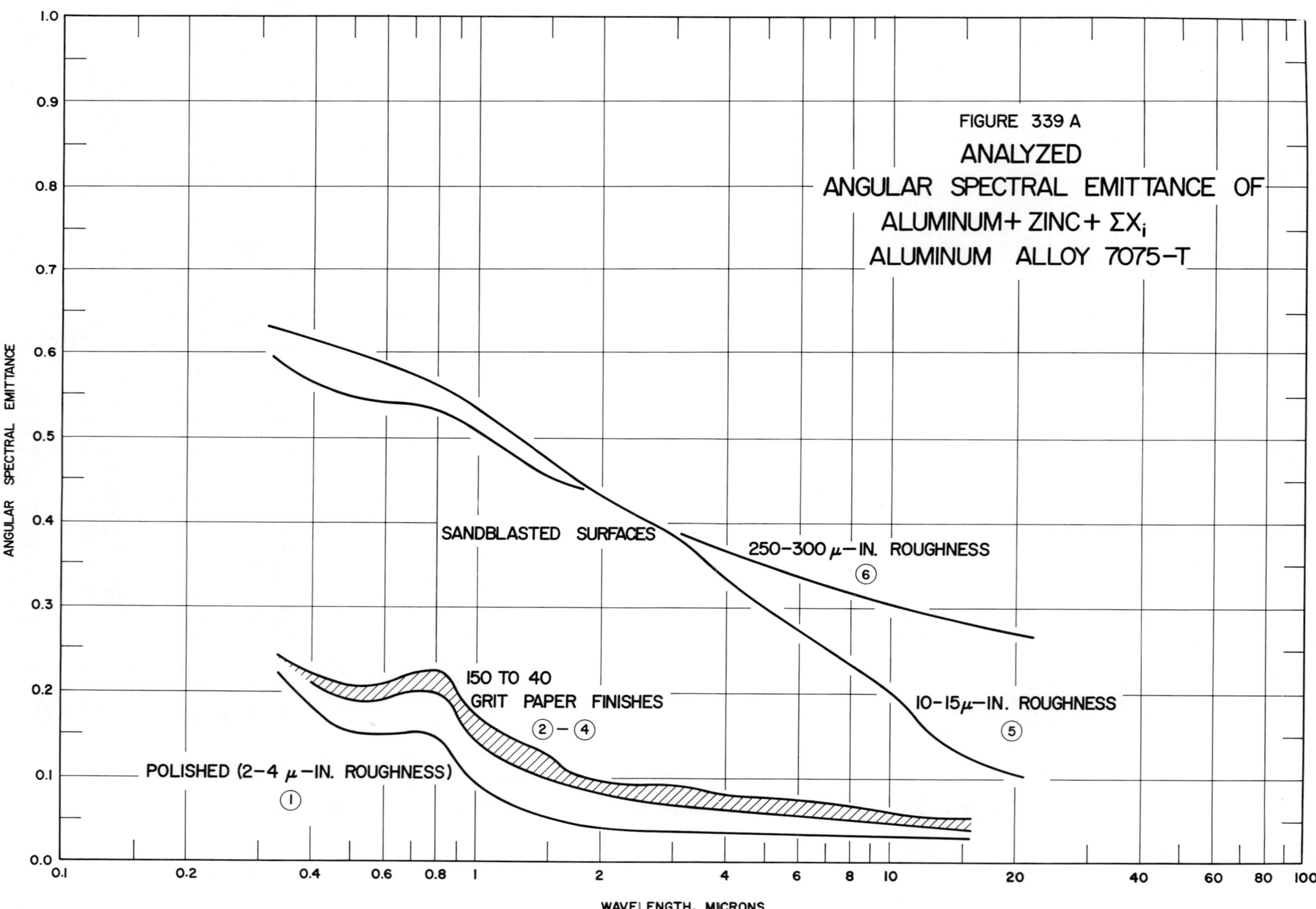
FIGURE 339 A
ANALYZED
ANGULAR SPECTRAL EMITTANCE OF
ALUMINUM+ ZINC+ ΣX_i
ALUMINUM ALLOY 7075-T
SANDBLASTED SURFACES
250-300 μ-IN. ROUGHNESS
⑥
150 TO 40
GRIT PAPER FINISHES
② - ④
10-15μ-IN. ROUGHNESS
⑤
POLISHED (2-4 μ-IN. ROUGHNESS)
①
ANGULAR SPECTRAL EMITTANCE
1.0
0.9
0.8
0.7
0.6
0.5
0.4
0.3
0.2
0.1
0.0
0.1
0.2
0.4
0.6
0.8
1
2
4
6
8
10
20
40
60
80
100
WAVELENGTH, MICRONS

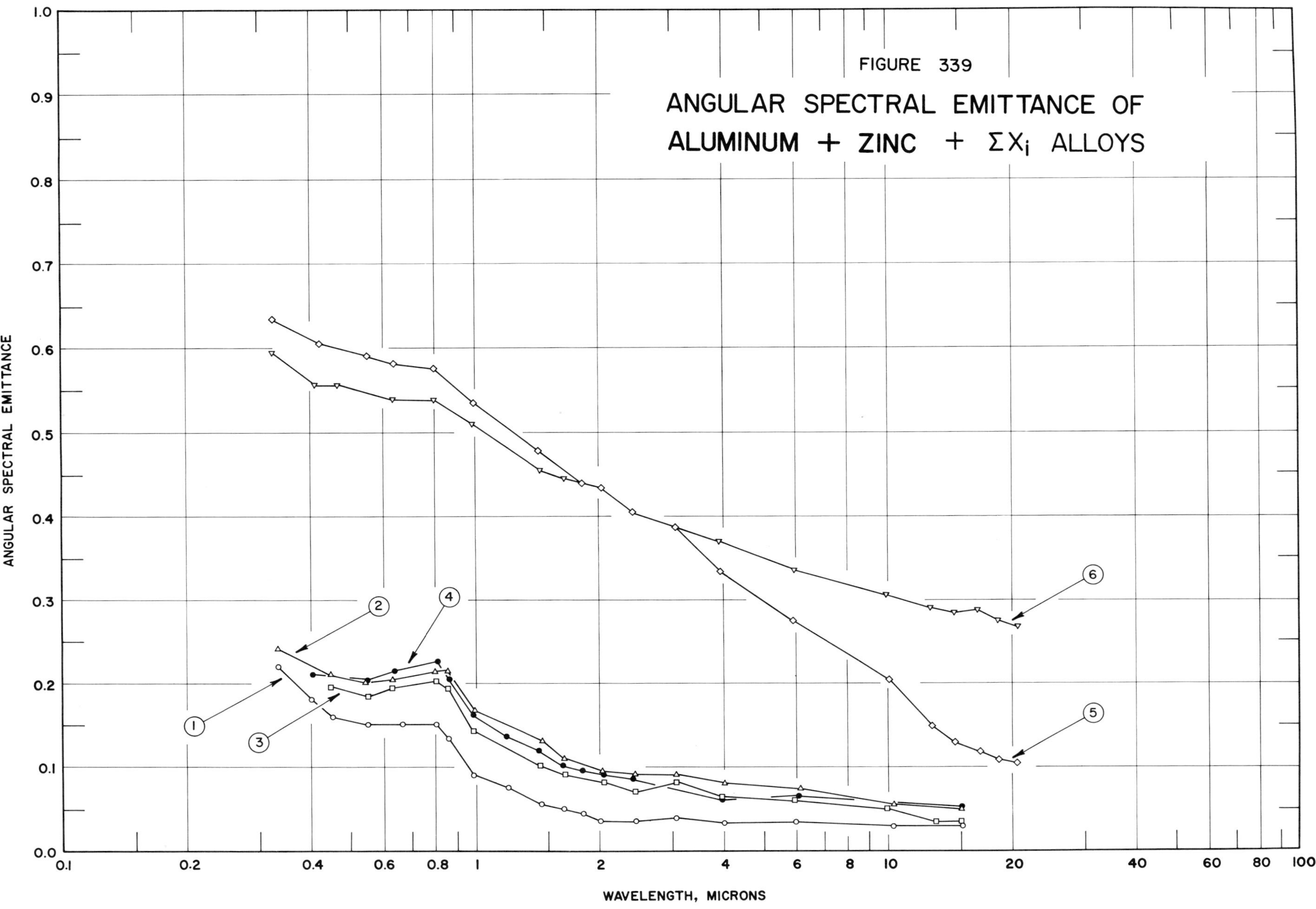

FIGURE 339
ANGULAR SPECTRAL EMITTANCE OF
ALUMINUM + ZINC + ΣX_i ALLOYS
ANGULAR SPECTRAL EMITTANCE
1.0
0.9
0.8
0.7
0.6
0.5
0.4
0.3
0.2
0.1
0.0
0.1
0.2
0.4
0.6
0.8
1
2
4
6
8
10
20
40
60
80
100
WAVELENGTH, MICRONS
1
2
3
4
5
6

SPECIFICATION TABLE NO. 339 ANGULAR SPECTRAL EMITTANCE OF [ALUMINUM + ZINC + ΣX_i] ALLOYS

Curve No.	Ref. No.	Year	Temperature K	Wavelength Range, μ	Geometry θ'	Reported Error, %	Composition (weight percent), Specifications and Remarks
1	219	1965	306	0.334-15.1	25°		Aluminum 7075-T6; nominal composition: 5.6 Zn, 2.5 Mg, 1.6 Cu, 0.30 Cr, and Al balance; polished; surface roughness 2-4 microinches (RMS); authors assumed $\epsilon = \alpha = 1-\rho(25°, 2\pi)$.
2	219	1965	306	0.333-15.0	25°		Aluminum 7075-T6; nominal composition: 5.6 Zn, 2.5 Mg, 1.6 Cu, 0.30 Cr, and Al balance; sanded with 150 grit paper (grit sieve opening 104 μ); RMS surface roughness in microinches: in line 10-15, across 70-90; authors assumed $\epsilon = \alpha = 1-\rho(25°, 2\pi)$.
3	219	1965	306	0.456-15.0	25°		Aluminum 7075-T6; nominal composition: 5.6 Zn, 2.5 Mg, 1.6 Cu, 0.30 Cr, and Al balance; sanded with 80 grit paper (grit sieve opening 175 μ); RMS surface roughness in microinches: in line 20-60, across 150-170; authors assumed $\epsilon = \alpha = 1-\rho(25°, 2\pi)$.
4	219	1965	306	0.405-15.2	25°		Aluminum 7075-T6; nominal composition: 5.6 Zn, 2.5 Mg, 1.6 Cu, 0.30 Cr, and Al balance; sanded with 40 grit paper (grit sieve opening 42 μ); RMS surface roughness in microinches: in line 50-100, across 270-300; authors assumed $\epsilon = \alpha = 1-\rho(25°, 2\pi)$.
5	219	1965	306	0.327-20.5	25°		Aluminum 7075-T6; nominal composition: 5.6 Zn, 2.5 Mg, 1.6 Cu, 0.30 Cr, and Al balance; sandblasted with 250 mesh silicon carbide (mesh opening 60 μ); RMS surface roughness 10-15 microinches; authors assumed $\epsilon = \alpha = 1-\rho(25°, 2\pi)$.
6	219	1965	306	0.327-20.7	25°		Aluminum 7075-T6; nominal composition: 5.6 Zn, 2.5 Mg, 1.6 Cu, 0.30 Cr, and Al balance; sandblasted with 60 mesh silicon carbide (mesh opening 250 μ); RMS surface roughness 250-300 microinches; authors assumed $\epsilon = \alpha = 1-\rho(25°, 2\pi)$.

DATA TABLE NO. 339 ANGULAR SPECTRAL EMITTANCE OF [ALUMINUM + ZINC + ΣX_i] ALLOYS

[Wavelength, λ, μ; Emittance, ϵ; Temperature, T, K]

CURVE 1, T = 306

λ	ϵ
0.334	0.220
0.402	0.180
0.454	0.159
0.547	0.150
0.637	0.150
0.804	0.150
0.861	0.134
0.989	0.090
1.20	0.076
1.44	0.056
1.63	0.050
1.82	0.045
2.02	0.037
2.45	0.037
3.08	0.039
4.00	0.034
5.98	0.034
10.3	0.030
15.1	0.030

CURVE 2, T = 306

λ	ϵ
0.333	0.240
0.454	0.211
0.541	0.200
0.631	0.203
0.804	0.214
0.859	0.214
0.993	0.168
1.45	0.131
1.64	0.111
2.03	0.096
2.42	0.091
3.07	0.091
4.00	0.080
6.07	0.075
10.3	0.056
15.0	0.050

CURVE 3, T = 306

λ	ϵ
0.456	0.196
0.547	0.183
0.629	0.193
0.803	0.202
0.859	0.193
0.988	0.142
1.44	0.101
1.65	0.091
2.04	0.081
2.43	0.070
3.08	0.081
3.96	0.066
5.97	0.060
9.95	0.050
13.0	0.035
15.0	0.035

CURVE 4, T = 306

λ	ϵ
0.405	0.211
0.545	0.203
0.638	0.216
0.813	0.227
0.863	0.204
0.986	0.161
1.19	0.136
1.43	0.119
1.63	0.101
1.82	0.096
2.04	0.091
2.40	0.086
3.97	0.071
6.03	0.066
15.2	0.051

CURVE 5, T = 306

λ	ϵ
0.327	0.634
0.411	0.604
0.550	0.590
0.640	0.581
0.800	0.575
0.993	0.533

CURVE 5 (cont.), T = 306

λ	ϵ
1.43	0.478
1.82	0.439
2.02	0.433
2.42	0.405
3.07	0.387
3.93	0.334
5.90	0.275
10.1	0.203
12.8	0.149
14.6	0.129
16.8	0.118
18.6	0.109
20.5	0.104

CURVE 6, T = 306

λ	ϵ
0.327	0.594
0.411	0.556
0.463	0.556
0.637	0.538
0.800	0.538
0.986	0.509
1.44	0.454
1.64	0.446
2.02	0.432*
2.43	0.404*
3.06	0.386*
3.91	0.370
5.94	0.337
9.95	0.306
12.76	0.290
14.6	0.284
16.6	0.288
18.6	0.275
20.7	0.268

* Not shown on plot

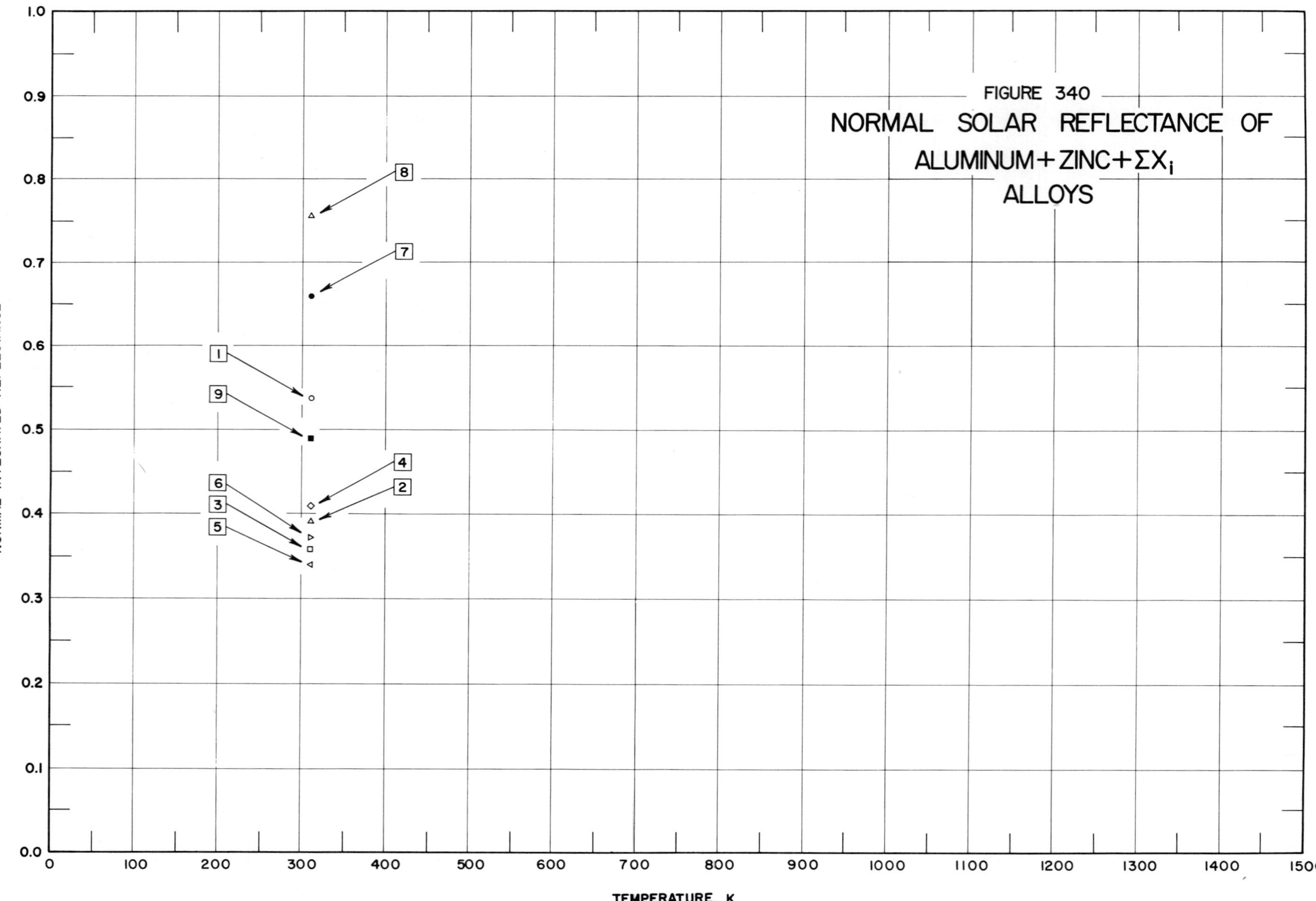
FIGURE 340
NORMAL SOLAR REFLECTANCE OF
ALUMINUM + ZINC + ΣXi
ALLOYS
NORMAL INTEGRATED REFLECTANCE
TEMPERATURE, K
1.0
0.9
0.8
0.7
0.6
0.5
0.4
0.3
0.2
0.1
0.0
0
100
200
300
400
500
600
700
800
900
1000
1100
1200
1300
1400
1500
8
7
1
9
4
2
6
3
5

SPECIFICATION TABLE NO. 340 NORMAL SOLAR REFLECTANCE OF [ALUMINUM + ZINC + ΣX_i] ALLOYS

Curve No.	Ref. No.	Year	Temperature Range, K	Geometry θ	Geometry θ'	Geometry ω'	Reported Error, %	Composition (weight percent), Specifications and Remarks
1	160	1954	311	~0°		2π		Aluminum alloy 75 ST; nominal composition: 5.6 Zn, 2.5 Mg, 1.6 Cu, 0.30 Cr, Al balance; heated to 322 K; clean and smooth surface; measured in air; calculated from solar absorptance.
2	160	1954	311	~0°		2π		Above specimen and conditions except reheated to 558 K.
3	160	1954	311	~0°		2π		Above specimen and conditions except reheated to 797 K.
4	160	1954	311	~0°		2π		Different sample, same as curve 1 specimen and conditions except heated to 319 K; cleaned with methyl alcohol.
5	160	1954	311	~0°		2π		Above specimen and conditions except reheated to 617 K.
6	160	1954	311	~0°		2π		Above specimen and conditions except reheated to 755 K.
7	160	1954	311	~0°		2π		Different sample, same as curve 1 specimen and conditions except heated to 308 K; polished; surface free from scratches.
8	160	1954	311	~0°		2π		Above specimen and conditions except reheated to 568 K.
9	160	1954	311	~0°		2π		Above specimen and conditions except reheated to 784 K.
10	40	1962	298	~0°		2π		Aluminum alloy 75 ST (alclad); nominal composition: 5.6 Zn, 2.5 Mg, 1.6 Cu, 0.30 Cr, Al balance; measured in air.

DATA TABLE NO. 340 NORMAL SOLAR REFLECTANCE OF [ALUMINUM + ZINC + ΣX_i] ALLOYS

[Temperature, T, K; Reflectance, ρ]

T	ρ
CURVE 1	
311	0.538
CURVE 2	
311	0.391
CURVE 3	
311	0.358
CURVE 4	
311	0.409
CURVE 5	
311	0.341
CURVE 6	
311	0.371
CURVE 7	
311	0.659
CURVE 8	
311	0.756
CURVE 9	
311	0.489
CURVE 10*	
298	0.67

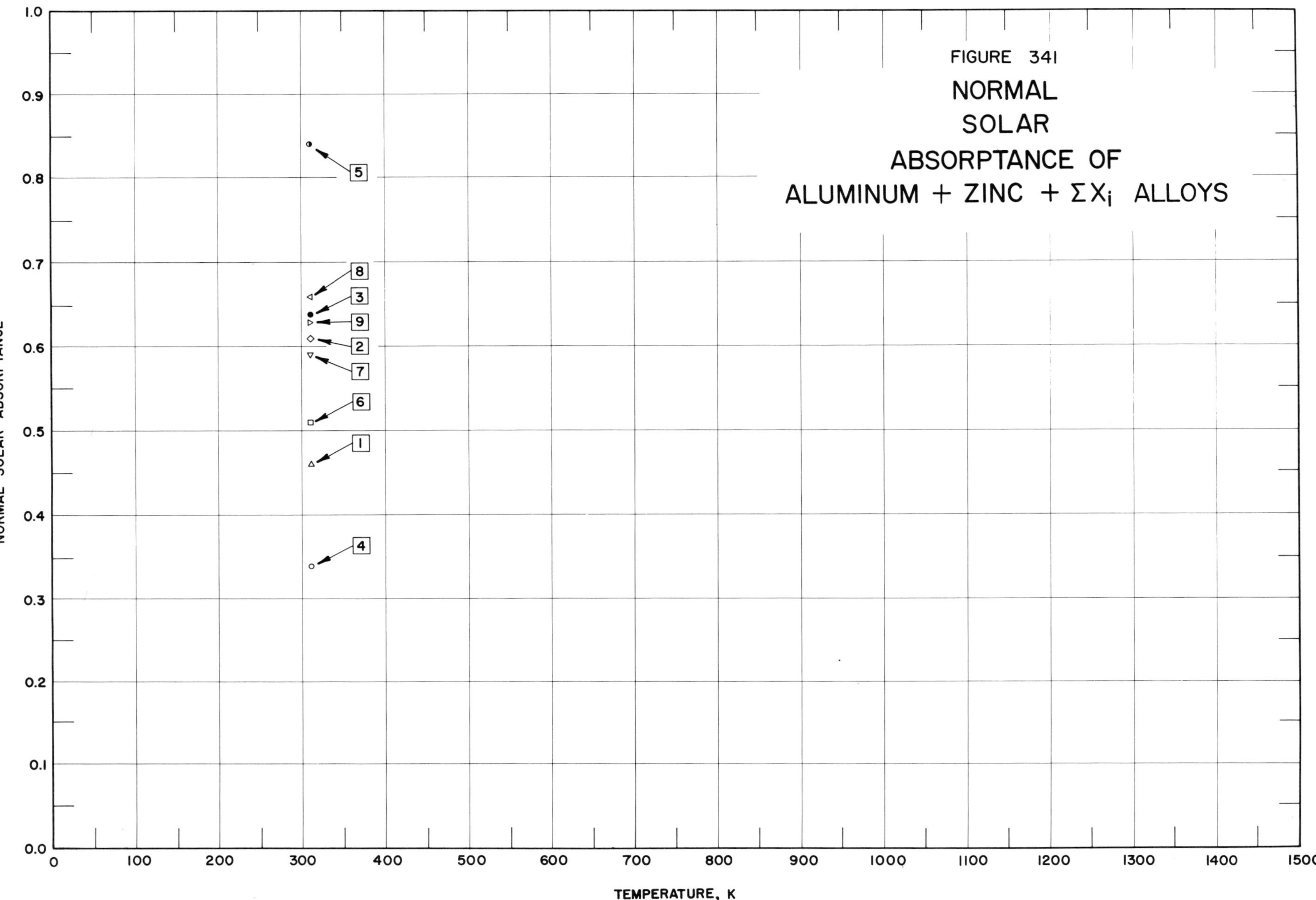
FIGURE 341
NORMAL
SOLAR
ABSORPTANCE OF
ALUMINUM + ZINC + ΣXi ALLOYS
NORMAL SOLAR ABSORPTANCE
TEMPERATURE, K
1.0
0.9
0.8
0.7
0.6
0.5
0.4
0.3
0.2
0.1
0.0
0
100
200
300
400
500
600
700
800
900
1000
1100
1200
1300
1400
1500
5
8
3
9
2
7
6
1
4

SPECIFICATION TABLE NO. 341 NORMAL SOLAR ABSORPTANCE OF [ALUMINUM + ZINC + ΣX_i] ALLOYS

Curve No.	Ref. No.	Year	Temperature Range, K	Geometry θ	Reported Error, %	Composition (weight percent), Specifications and Remarks
1	160	1954	311	~0°		Aluminum alloy 75-ST; nominal composition: 5.6 Zn, 2.5 Mg, 1.6 Cu, 0.30 Cr, Al balance; heated to 326 K; clean and smooth surface; measured in air at sea level.
2	160	1954	311	~0°		Above specimen and conditions except reheated to 559 K.
3	160	1954	311	~0°		Above specimen and conditions except reheated to 794 K.
4	160	1954	311	~0°		Different sample, same as Curve 1 specimen and conditions except heated to 320 K; polished; surface free from scratches.
5	160	1954	311	~0°		Above specimen and conditions except reheated to 567 K.
6	160	1954	311	~0°		Above specimen and conditions except reheated to 783 K.
7	160	1954	311	~0°		Different sample, same as Curve 1 specimen and conditions except heated to 319 K; cleaned with methyl alcohol.
8	160	1954	311	~0°		Above specimen and conditions except reheated to 617 K.
9	160	1954	311	~0°		Above specimen and conditions except reheated to 755 K.

DATA TABLE NO. 341 NORMAL SOLAR ABSORPTANCE OF [ALUMINUM + ZINC + ΣX_i] ALLOYS

[Temperature, T, K; Absorptance, α]

T	α
CURVE 1	
311	0.46
CURVE 2	
311	0.61
CURVE 3	
311	0.64
CURVE 4	
311	0.34
CURVE 5	
311	0.84
CURVE 6	
311	0.51
CURVE 7	
311	0.59
CURVE 8	
311	0.66
CURVE 9	
311	0.63

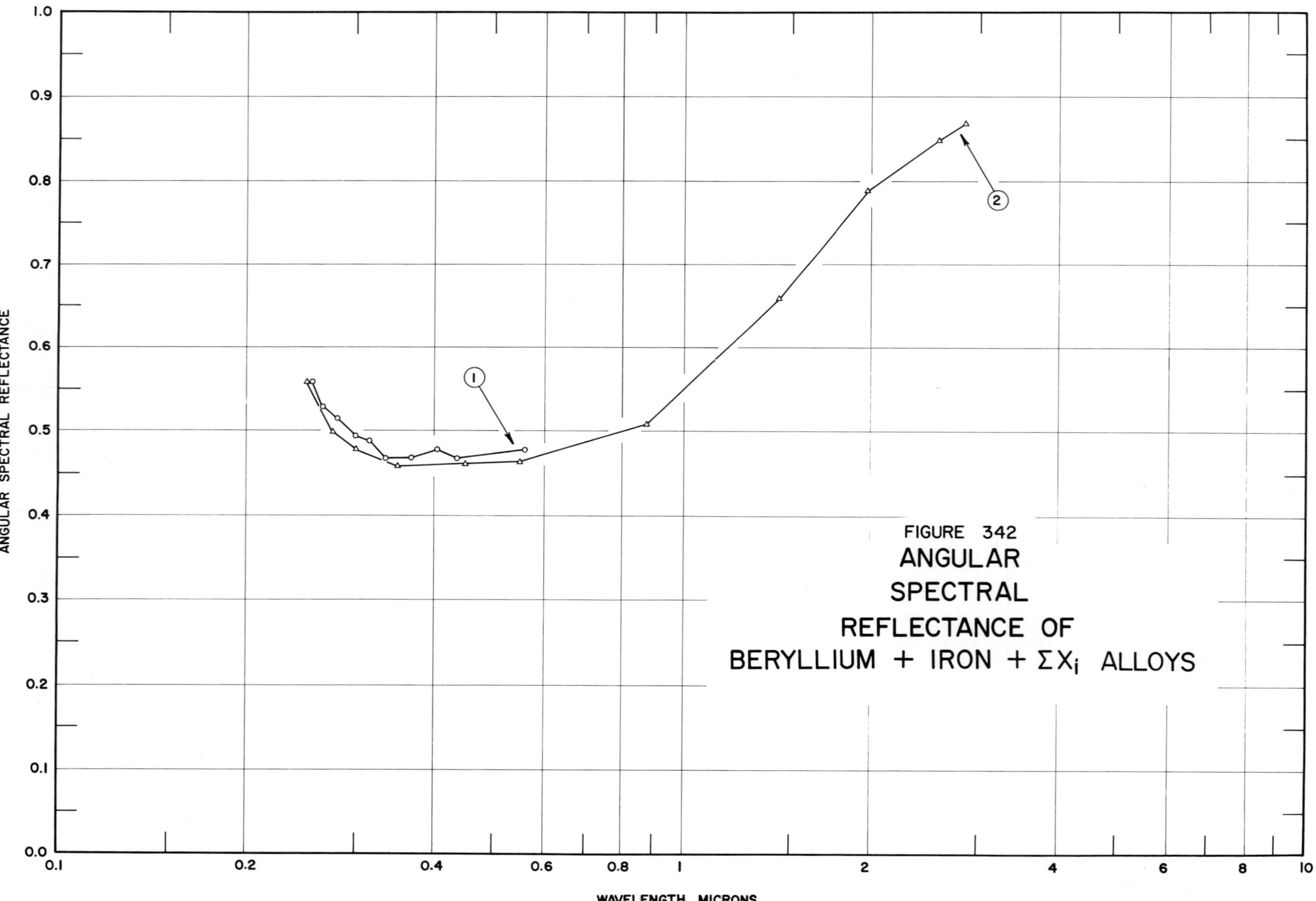

FIGURE 342
ANGULAR
SPECTRAL
REFLECTANCE OF
BERYLLIUM + IRON + ΣX_i ALLOYS
ANGULAR SPECTRAL REFLECTANCE
WAVELENGTH, MICRONS
1.0
0.9
0.8
0.7
0.6
0.5
0.4
0.3
0.2
0.1
0.0
0.1
0.2
0.4
0.6
0.8
1
2
4
6
8
10
1
2

SPECIFICATION TABLE NO. 342 ANGULAR SPECTRAL REFLECTANCE OF [BERYLLIUM + IRON + ΣX_i] ALLOYS

Curve No.	Ref. No.	Year	Temperature K	Wavelength Range, μ	Geometry θ	θ'	ω'	Reported Error, %	Composition (weight percent), Specifications and Remarks
1	143	1928	298	0.255-0.560	45°	45°			98.7 Be, 1 Fe, 0.2 Ba, trace of Si; ground and polished; surface contained a few scratches and cracks.
2	143	1928	298	0.250-2.825	45°	45°			Different sample, same as above specimen and conditions.

DATA TABLE NO. 342 ANGULAR SPECTRAL REFLECTANCE OF [BERYLLIUM + IRON + ΣX_i] ALLOYS

[Wavelength, λ, μ; Reflectance, ρ; Temperature, T, K]

λ	ρ
CURVE 1 T = 298	
0.255	0.560
0.265	0.530
0.280	0.515
0.300	0.495
0.315	0.490
0.335	0.470
0.368	0.470
0.405	0.480
0.435	0.470
0.560	0.480
CURVE 2 T = 298	
0.250	0.560
0.275	0.500
0.300	0.480
0.350	0.460
0.450	0.462
0.550	0.465
0.875	0.510
1.425	0.660
1.975	0.790
2.575	0.850
2.825	0.870

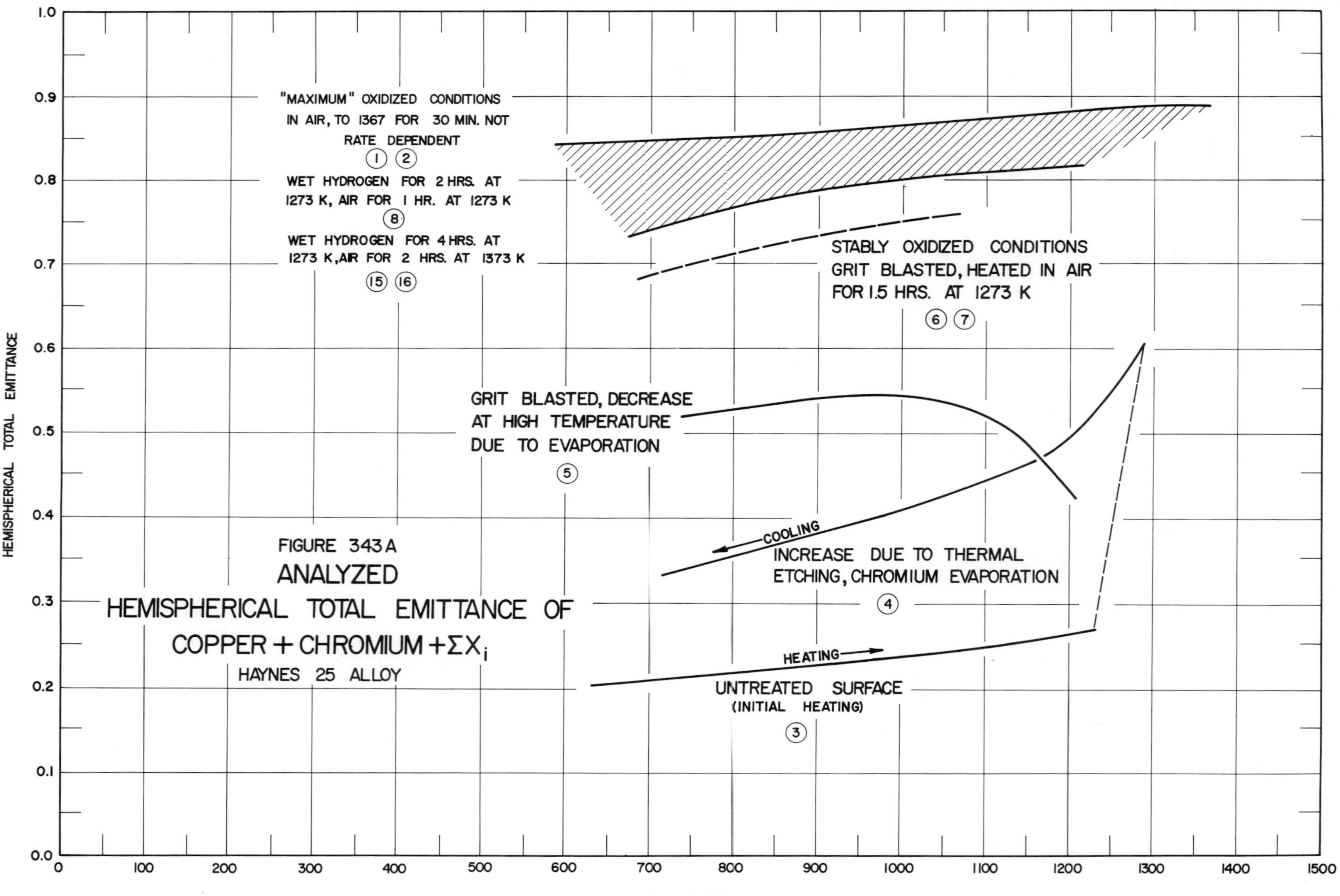

FIGURE 343A
ANALYZED
HEMISPHERICAL TOTAL EMITTANCE OF
COPPER + CHROMIUM + ΣX_i
HAYNES 25 ALLOY

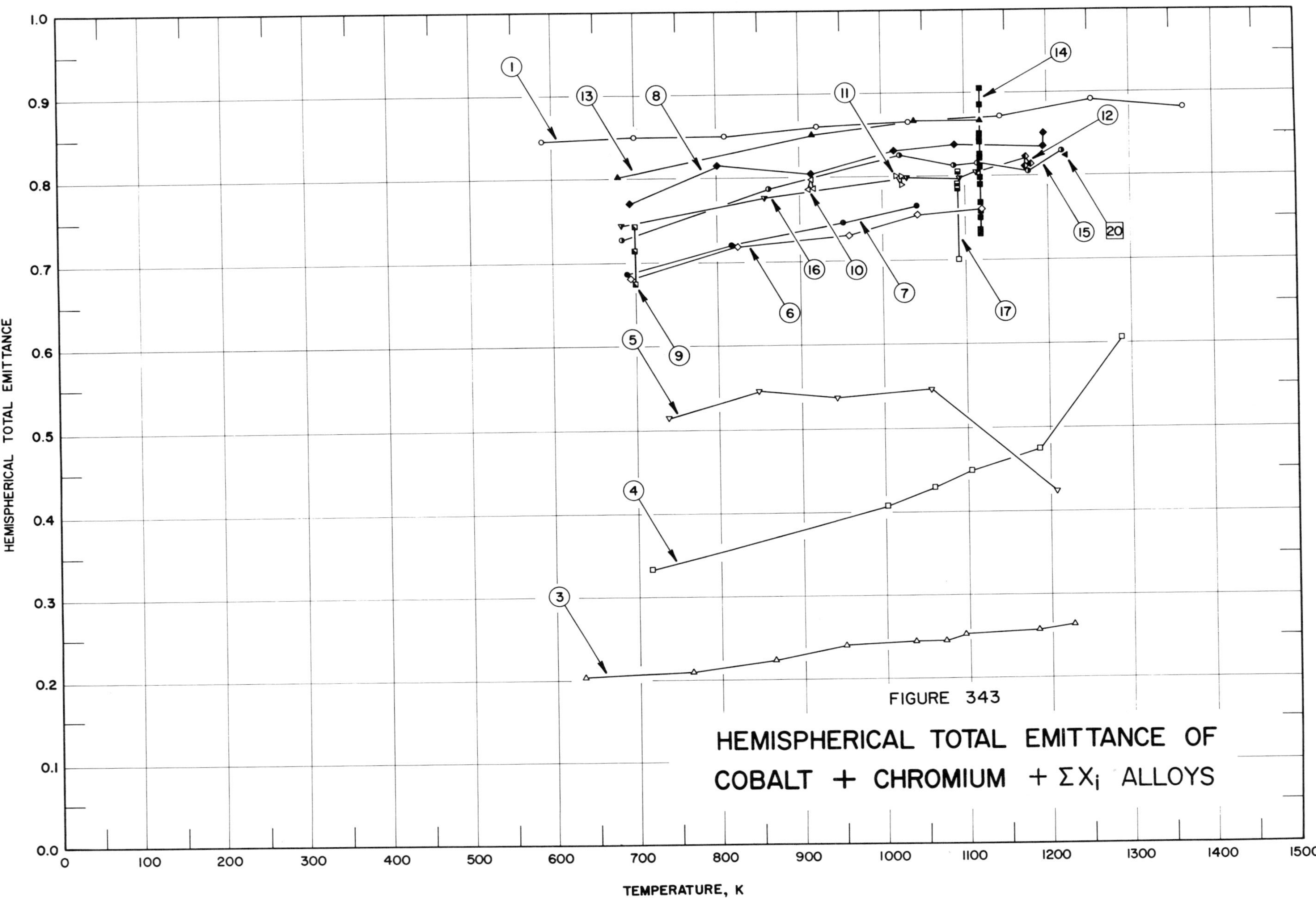

FIGURE 343

HEMISPHERICAL TOTAL EMITTANCE OF COBALT + CHROMIUM + ΣX$_i$ ALLOYS

SPECIFICATION TABLE NO. 343 HEMISPHERICAL TOTAL EMITTANCE OF [COBALT + CHROMIUM + ΣX_i] ALLOYS

Curve No.	Ref. No.	Year	Temperature Range, K	Reported Error, %	Composition (weight percent), Specifications and Remarks
1	161	1959	589-1367		Haynes alloy 25; nominal composition: 19-21 Cr, 14-16 W, 9-11 Ni, 2 max Fe, 0.15 max C, Co balance; mechanically polished; washed in alcohol and distilled water; dried by a hot-air blast; oxidized in air at 1367 K for 30 min; author assumed normal total emittance = hemispherical total emittance; diffuse surface.
2	161	1959	589-1367		Different sample, same as above specimen and conditions except gradually oxidized in still air at temperatures between 0 to 1367 K.
3	279	1964	633-1228		Haynes 25; 19.00-21.00 Cr, 14.00-16.00 W, 9.00-11.00 Ni, 3.00 max Fe, 1.0-2.0 Mn, 1.00 max Si, 0.05-0.15 C, Co balance; cleaned untreated surface; measured in vacuum ($<10^{-6}$ mm Hg); increasing temperature.
4	279	1964	1289-717		Above specimen and conditions except decreasing temperature; thermal etching occurred.
5	279	1964	739-1209		Different sample, same as curve 3 specimen and conditions except cleaned and silicon carbide blasted; [Author's designation: Sample B].
6	279	1964	694-1121		Different sample, same as curve 3 specimen and conditions except cleaned, silicon carbide blasted, oxidized in air for 1.5 hrs at 1273 K; [Author's designation: Sample A].
7	279	1964	1044-690		Above specimen and conditions except decreasing temperature.
8	279	1964	694-1197		Different sample, same as curve 3 specimen and conditions except cleaned; silicon carbide blasted, and oxidized in a wet hydrogen atmosphere for 2 hrs at 1273 K followed by oxidation in still air for 1 hr; [Author's designation: Sample No. 1].
9	279	1964	700		Above specimen and conditions; 6 hr stability trial.
10	279	1964	912-918		Above specimen and conditions; 3 hr stability trial.
11	279	1964	1018-1023		Above specimen and conditions; 2 hr stability trial.
12	279	1964	1176-1181		Above specimen and conditions; 3.5 hr stability trial.
13	279	1964	680-1120		Different sample, same as curve 3 specimen and conditions except cleaned; silicon carbide blasted and oxidized in a wet hydrogen atmosphere for 2 hrs at 1273 K followed by oxidation in still air for 1.5 hrs; [Author's designation: Sample No. 3].
14	279	1964	1120		Above specimen and conditions; 103 hr stability trial.
15	279	1964	684-1219		Different sample, same as curve 3 specimen and conditions except cleaned; silicon carbide blasted and oxidized in wet hydrogen atmosphere for 4 hrs at 1273 K and followed by oxidation in still air for 2 hrs at 1373 K; [Author's designation: Sample No. 5].
16	279	1964	1177-684		Above specimen and conditions; decreasing temperature.
17	279	1964	1093		Different sample, same as above specimen and conditions; 100 hr stability trial; [Author's designation: Sample No. 4].
18	279	1964	1123		Above specimen and conditions; 25 hr stability trial.
19	279	1964	1173		Above specimen and conditions; 25 hr stability trial.
20	279	1964	1223		Above specimen and conditions; 5 hr stability trial.

DATA TABLE NO. 343 HEMISPHERICAL TOTAL EMITTANCE OF [COBALT + CHROMIUM + ΣX_i] ALLOYS

[Temperature, T, K; Emittance, ϵ]

T	ϵ
CURVE 1	
589	0.845
700	0.850
811	0.850
922	0.860
1033	0.865
1144	0.870
1255	0.890
1367	0.880
CURVE 2*	
589	0.845
700	0.850
811	0.850
922	0.860
1033	0.875
1144	0.875
1255	0.890
1367	0.890
CURVE 3	
633	0.203
765	0.209
864	0.223
951	0.240
1035	0.243
1072	0.247
1097	0.252
1183	0.258
1228	0.263
CURVE 4	
1289	0.606
1189	0.473
1105	0.449
1060	0.430
1002	0.408
717	0.334
CURVE 5	
739	0.514
849	0.546
943	0.536
1058	0.544
1209	0.423
CURVE 6	
694	0.680
824	0.719
961	0.731
1044	0.754
1121	0.761
CURVE 7	
1044	0.765
954	0.744
818	0.721
690	0.686
CURVE 8	
694	0.771
801	0.816
914	0.804
1016	0.831
1090	0.838
1197	0.836
1197	0.851
CURVE 9	
700	0.673
700	0.713
700	0.742
CURVE 10	
912	0.786
916	0.798
918	0.787
CURVE 11	
1018	0.800
1023	0.789
1023	0.800
CURVE 12	
1176	0.811
1177	0.822
1181	0.812
CURVE 13	
680	0.802
917	0.851
1040	0.866
1120	0.866
CURVE 14	
1120	0.903
1120	0.885
1120	0.844
1120	0.850
1120	0.841
1120	0.841*
1120	0.821
1120	0.826
1120	0.810
1120	0.824*
1120	0.809*
1120	0.815*
1120	0.812*
1120	0.799
1120	0.789
1120	0.771
1120	0.759
1120	0.750
1120	0.764*
1120	0.753*
1120	0.752*
1120	0.736
1120	0.731
CURVE 15	
684	0.728
863	0.787
1022	0.826
1089	0.812
1117	0.814
1176	0.806
1219	0.830
CURVE 16	
1177	0.819*
1116	0.804
1095	0.798
1031	0.799
859	0.777
684	0.742
CURVE 17	
1093	0.700
1093	0.786
1093	0.790
1093	0.799*
1093	0.803*
1093	0.804*
1093	0.804*
1093	0.805*
1093	0.805
CURVE 18*	
1123	0.805
1123	0.805
1123	0.805
1123	0.805
1123	0.805
CURVE 19*	
1173	0.815
1173	0.816
1173	0.819
CURVE 20	
1223	0.823
1223	0.823*

* Not shown on plot

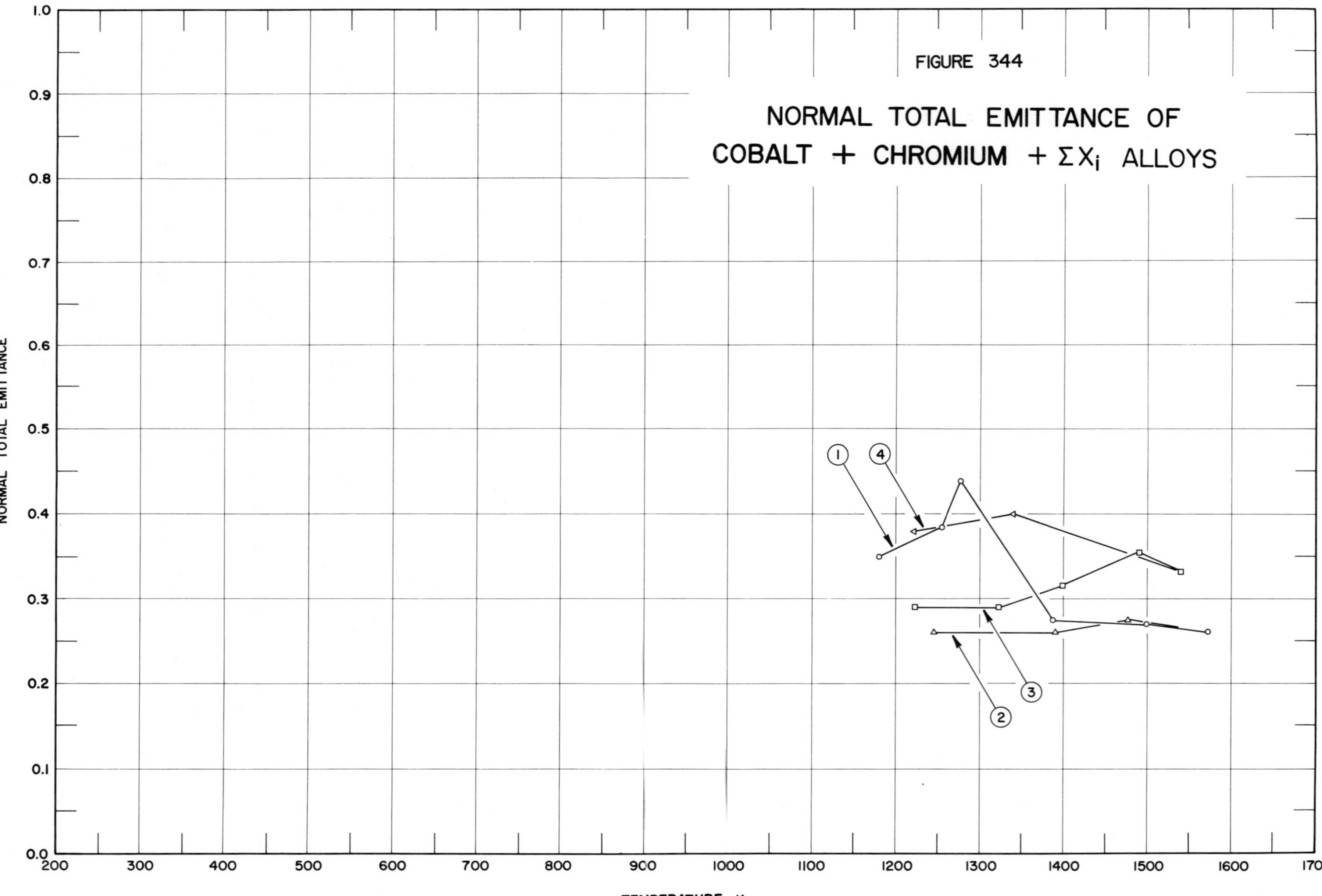
FIGURE 344
NORMAL TOTAL EMITTANCE OF
COBALT + CHROMIUM + ΣX_i ALLOYS
1
4
3
2
NORMAL TOTAL EMITTANCE
1.0
0.9
0.8
0.7
0.6
0.5
0.4
0.3
0.2
0.1
0.0
200
300
400
500
600
700
800
900
1000
1100
1200
1300
1400
1500
1600
1700
TEMPERATURE, K

SPECIFICATION TABLE NO. 344 NORMAL TOTAL EMITTANCE OF [COBALT + CHROMIUM + ΣX_i] ALLOYS

Curve No.	Ref. No.	Year	Temperature Range, K	Geometry θ'	Reported Error, %	Composition (weight percent), Specifications and Remarks
1	92	1963	1180-1572	$\sim 0°$		Haynes Alloy 25 (L-605); 19-21 Cr, 14-16 W, 9-11 Ni, 1-2 Mn, 1.0 Si, 3 Fe, 0.05-0.15 C, Co balance; polished; surface roughness 0.7 to 1 μ (RMS) measured with profilometer; measured in vacuum (3 to 4 x 10^{-4} mm Hg).
2	92	1963	1572-1247	$\sim 0°$		Above specimen and conditions; descending temperature.
3	92	1963	1222-1539	$\sim 0°$		Different sample, same as curve 1 specimen and conditions except sandblasted; surface roughness 70 to 90 μ (RMS).
4	92	1963	1539-1222	$\sim 0°$		Above specimen and conditions; descending temperature.

DATA TABLE NO. 344 NORMAL TOTAL EMITTANCE OF [COBALT + CHROMIUM + ΣX_i] ALLOYS

[Temperature, T, K; Emittance, ϵ]

T	ϵ
CURVE 1	
1180	0.350
1255	0.385
1278	0.440
1389	0.275
1500	0.270
1572	0.260
CURVE 2	
1572	0.260*
1478	0.275
1391	0.260
1247	0.260
CURVE 3	
1222	0.290
1322	0.290
1400	0.315
1491	0.355
1539	0.332
CURVE 4	
1539	0.332*
1339	0.400
1222	0.380

* Not shown on plot

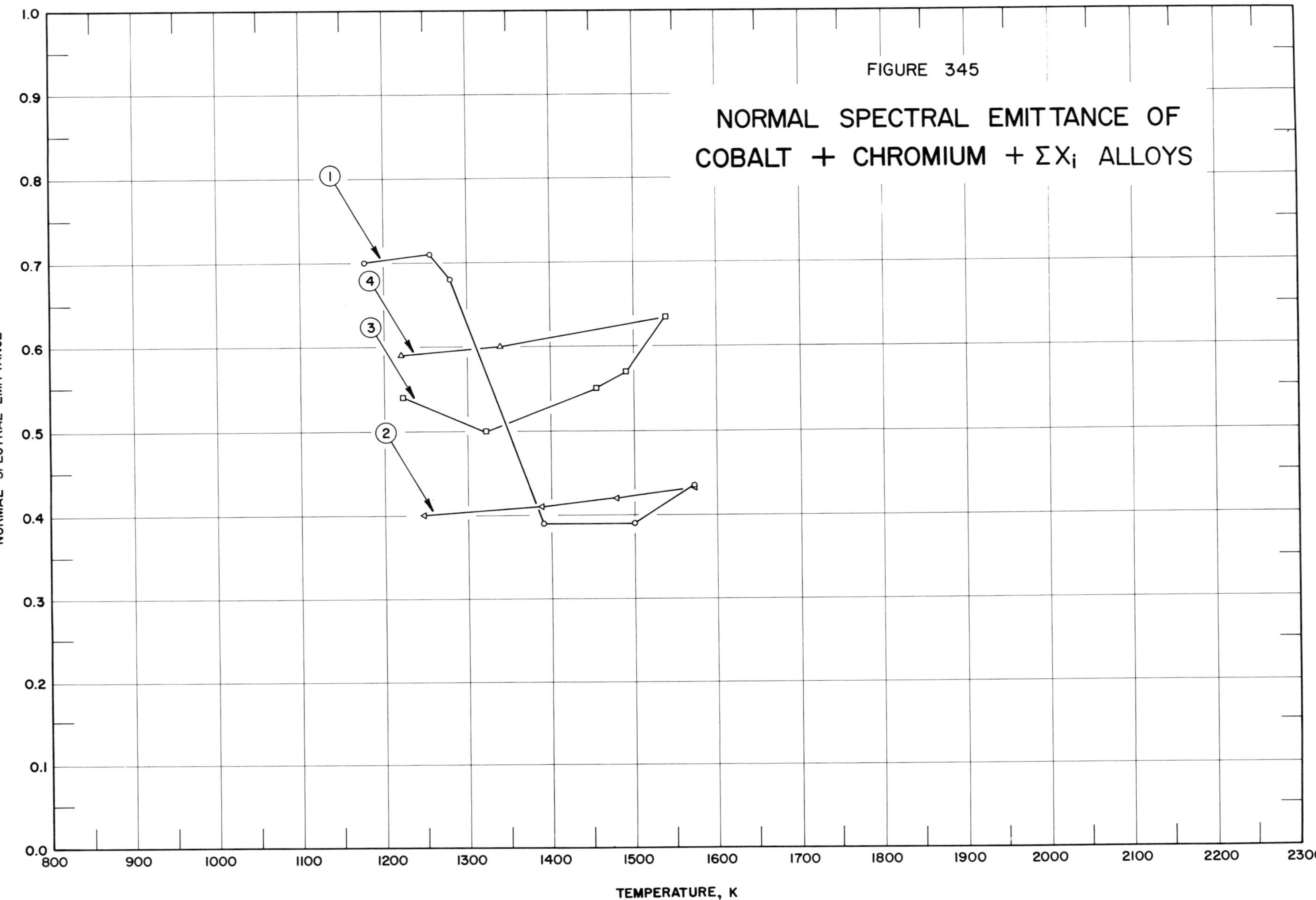

FIGURE 345
NORMAL SPECTRAL EMITTANCE OF
COBALT + CHROMIUM + ΣXi ALLOYS
NORMAL SPECTRAL EMITTANCE
TEMPERATURE, K
1.0
0.9
0.8
0.7
0.6
0.5
0.4
0.3
0.2
0.1
0.0
800
900
1000
1100
1200
1300
1400
1500
1600
1700
1800
1900
2000
2100
2200
2300
1
4
3
2

SPECIFICATION TABLE NO. 345 NORMAL SPECTRAL EMITTANCE OF [COBALT + CHROMIUM + ΣX_i] ALLOYS

Curve No.	Ref. No.	Year	Wavelength μ	Temperature Range, K	Geometry θ'	Reported Error, %	Composition (weight percent), Specifications and Remarks
1	92	1963	0.65	1178-1572	$\sim 0°$		Haynes Alloy 25 (L-605); 19-21 Cr, 14-16 W, 9-11 Ni, 1-2 Mn, 1.0 Si, 3 Fe, 0.05-0.15 C, Co balance; polished; surface roughness 0.7 to 1 μ (RMS) measured with profilometer; measured in vacuum (3 to 4 x 10^{-4} mm Hg); authors assumed specimen was a gray body.
2	92	1963	0.65	1572-1247	$\sim 0°$		Above specimen and conditions; descending temperature.
3	92	1963	0.65	1222-1539	$\sim 0°$		Different sample, same as curve 1 specimen and conditions except sand blasted; surface roughness 70 to 90 μ (RMS).
4	92	1963	0.65	1539-1220	$\sim 0°$		Above specimen and conditions; descending temperature.

DATA TABLE NO. 345 NORMAL SPECTRAL EMITTANCE OF [COBALT + CHROMIUM + ΣX_i] ALLOYS

[Temperature, T, K; Emittance, ϵ; Wavelength, λ, μ]

T	ϵ
CURVE 1 $\lambda = 0.65$	
1178	0.700
1255	0.710
1280	0.680
1391	0.390
1500	0.390
1572	0.435
CURVE 2 $\lambda = 0.65$	
1572	0.435
1478	0.420
1389	0.410
1247	0.400
CURVE 3 $\lambda = 0.65$	
1222	0.540
1322	0.500
1455	0.550
1491	0.570
1539	0.635
CURVE 4 $\lambda = 0.65$	
1539	0.635*
1339	0.600
1220	0.590

*Not shown on plot

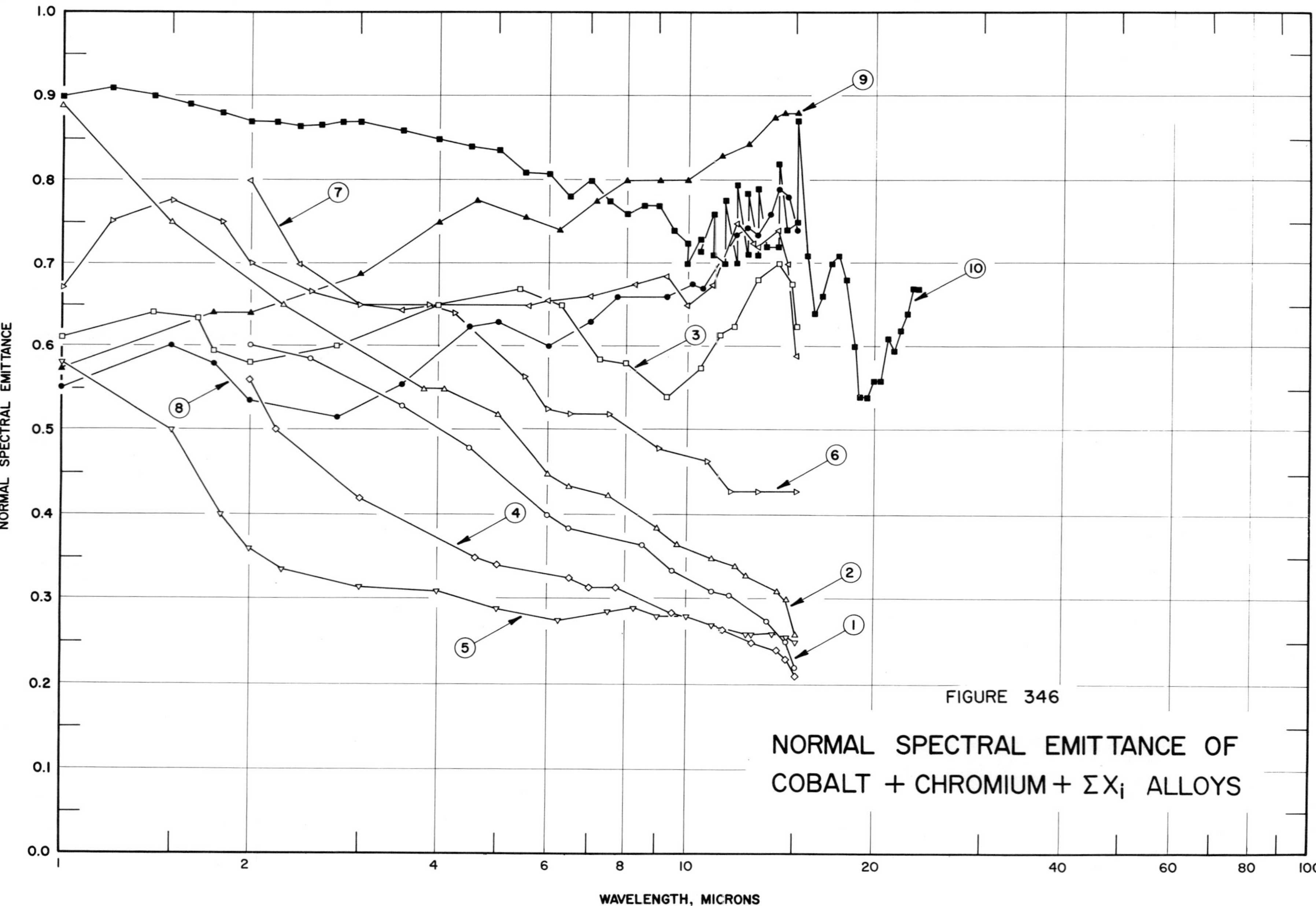

FIGURE 346

NORMAL SPECTRAL EMITTANCE OF COBALT + CHROMIUM + ΣX_i ALLOYS

SPECIFICATION TABLE NO. 346 NORMAL SPECTRAL EMITTANCE OF [COBALT + CHROMIUM + ΣX_i] ALLOYS

Curve No.	Ref. No.	Year	Temperature K	Wavelength Range, μ	Geometry θ'	Reported Error, %	Composition (weight percent), Specifications and Remarks
1	86	1961	523	2.00-15.00	$\sim 0°$	±5	Haynes Stellite 25; nominal composition: 19-21 Cr, 14-16 W, 9-11 Ni, 2 max Fe, 0.15 max C, Co balance; as received; data extracted from smooth curve.
2	86	1961	773	1.00-15.00	$\sim 0°$	±5	Different sample, same as above specimen and conditions.
3	86	1961	1023	1.00-15.00	$\sim 0°$	±5	Different sample, same as above specimen and conditions.
4	86	1961	523	2.00-15.00	$\sim 0°$	±5	Different sample, same as above specimen and conditions except heated in vacuum (7.6×10^{-5} mm Hg) at 1089 K for 30 min.
5	86	1961	773	1.00-15.00	$\sim 0°$	±5	Different sample, same as above specimen and conditions.
6	86	1961	1023	1.00-15.00	$\sim 0°$	±5	Different sample, same as above specimen and conditions.
7	86	1961	523	2.00-15.00	$\sim 0°$	±5	Different sample, same as above specimen and conditions except heated in air at 1089 K for 30 min.
8	86	1961	773	1.00-15.00	$\sim 0°$	±5	Different sample, same as above specimen and conditions.
9	86	1961	1023	1.00-15.00	$\sim 0°$	±5	Different sample, same as above specimen and conditions.
10	65	1962	1028	1.00-23.50	$\sim 0°$		Haynes Stellite 25; nominal composition: 19-21 Cr, 14-16 W, 9-11 Ni, 2 max Fe, 0.15 max C, Co balance; oxidized; measured in air.

DATA TABLE NO. 346 NORMAL SPECTRAL EMITTANCE OF [COBALT + CHROMIUM + ΣX_i] ALLOYS

[Wavelength, λ, μ; Emittance, ϵ; Temperature, T, K]

CURVE 1 T = 523

λ	∈
2.00	0.600
2.50	0.585
3.50	0.530
4.50	0.480
6.00	0.400
6.50	0.385
8.50	0.365
9.50	0.335
11.00	0.310
11.75	0.305
13.50	0.275
14.50	0.250
15.00	0.220

CURVE 2 T = 773

λ	∈
1.00	0.890
1.00	0.860
1.50	0.750
2.25	0.650
3.80	0.550
4.10	0.550
5.00	0.520
6.00	0.450
6.50	0.435
7.50	0.425
9.00	0.385
9.65	0.365
11.00	0.350
12.00	0.340
12.50	0.330
14.00	0.310
14.50	0.300
15.00	0.260

CURVE 3 T = 1023

λ	∈
1.00	0.610
1.40	0.640
1.65	0.635
1.75	0.595
2.00	0.580
2.75	0.600
4.00	0.650
5.40	0.670
6.30	0.650
7.25	0.585
8.00	0.580
9.25	0.540
10.50	0.575
11.30	0.615
11.90	0.625
13.00	0.680
14.00	0.700
14.75	0.675
15.00	0.625

CURVE 4 T = 523

λ	∈
2.00	0.560
2.20	0.500
3.00	0.420
4.60	0.350
5.00	0.340
6.50	0.325
7.00	0.315
7.75	0.315
9.50	0.285
11.50	0.265
12.75	0.250
14.00	0.240
14.50	0.230
15.00	0.210

CURVE 5 T = 773

λ	∈
1.00	0.580
1.50	0.500
1.80	0.400
2.00	0.360
2.25	0.335
3.00	0.315
4.00	0.310
5.00	0.290
6.25	0.275
7.50	0.285
8.25	0.290
9.00	0.280
10.00	0.280
11.00	0.270
12.50	0.260
12.75	0.260
13.75	0.260
14.50	0.257
15.00	0.250

CURVE 6 T = 1023

λ	∈
1.00	0.670
1.20	0.750
1.50	0.775
1.80	0.750
2.00	0.700
2.50	0.665
3.00	0.650
3.85	0.650
4.25	0.640
5.50	0.565
6.00	0.525
6.50	0.520
7.50	0.520
9.00	0.480
10.75	0.465
11.75	0.430
13.00	0.430
15.00	0.430

CURVE 7 T = 523

λ	∈
2.00	0.800
2.40	0.700
3.00	0.650*
3.50	0.645
4.00	0.650*
5.60	0.650
6.00	0.655
7.00	0.660
8.25	0.675
9.25	0.685
10.00	0.650
11.00	0.675
12.00	0.750
12.75	0.725
13.00	0.720
13.90	0.740
14.55	0.700
15.00	0.590

CURVE 8 T = 773

λ	∈
1.00	0.550
1.50	0.600
1.75	0.580
2.00	0.535
2.75	0.515
3.50	0.555
4.50	0.625
5.00	0.630
6.00	0.600
7.00	0.630
7.75	0.660
9.25	0.660
10.20	0.675
10.60	0.670
12.00	0.735
12.50	0.745
13.00	0.735
13.60	0.760
14.00	0.790
14.50	0.780
15.00	0.740

CURVE 9 T = 1023

λ	∈
1.00	0.575
1.75	0.640
2.00	0.640
3.00	0.685
4.00	0.750
4.60	0.775
5.50	0.755
6.25	0.740
7.15	0.775
8.00	0.800
9.00	0.800
10.00	0.800
11.30	0.830
12.50	0.845
13.75	0.875
14.25	0.880
15.00	0.880

CURVE 10 T = 1028

λ	∈
1.00	0.900
1.20	0.910
1.40	0.900
1.60	0.890
1.80	0.880
2.00	0.870
2.20	0.870
2.40	0.865
2.60	0.865
2.80	0.870
3.00	0.870
3.50	0.860
4.00	0.850
4.50	0.840
5.00	0.836
5.50	0.810
6.00	0.808
6.50	0.780
7.00	0.800
7.50	0.775
8.00	0.760
8.50	0.770
9.00	0.770
9.50	0.740
10.00	0.725
10.00	0.700
10.50	0.730
10.50	0.715
11.00	0.760
11.00	0.710
11.50	0.700
11.50	0.775
12.00	0.700
12.00	0.795
12.50	0.710
12.50	0.785
13.00	0.710
13.00	0.790
13.40	0.720
14.00	0.720
14.00	0.820
14.40	0.740
15.00	0.750
15.00	0.870
15.60	0.710
16.00	0.640
16.50	0.660
17.00	0.700
17.50	0.710
18.00	0.680
18.60	0.600
19.00	0.540
19.50	0.540
20.00	0.560
20.40	0.560
21.00	0.610
21.50	0.595
22.00	0.620
22.50	0.640
23.00	0.670
23.50	0.670

* Not shown on plot

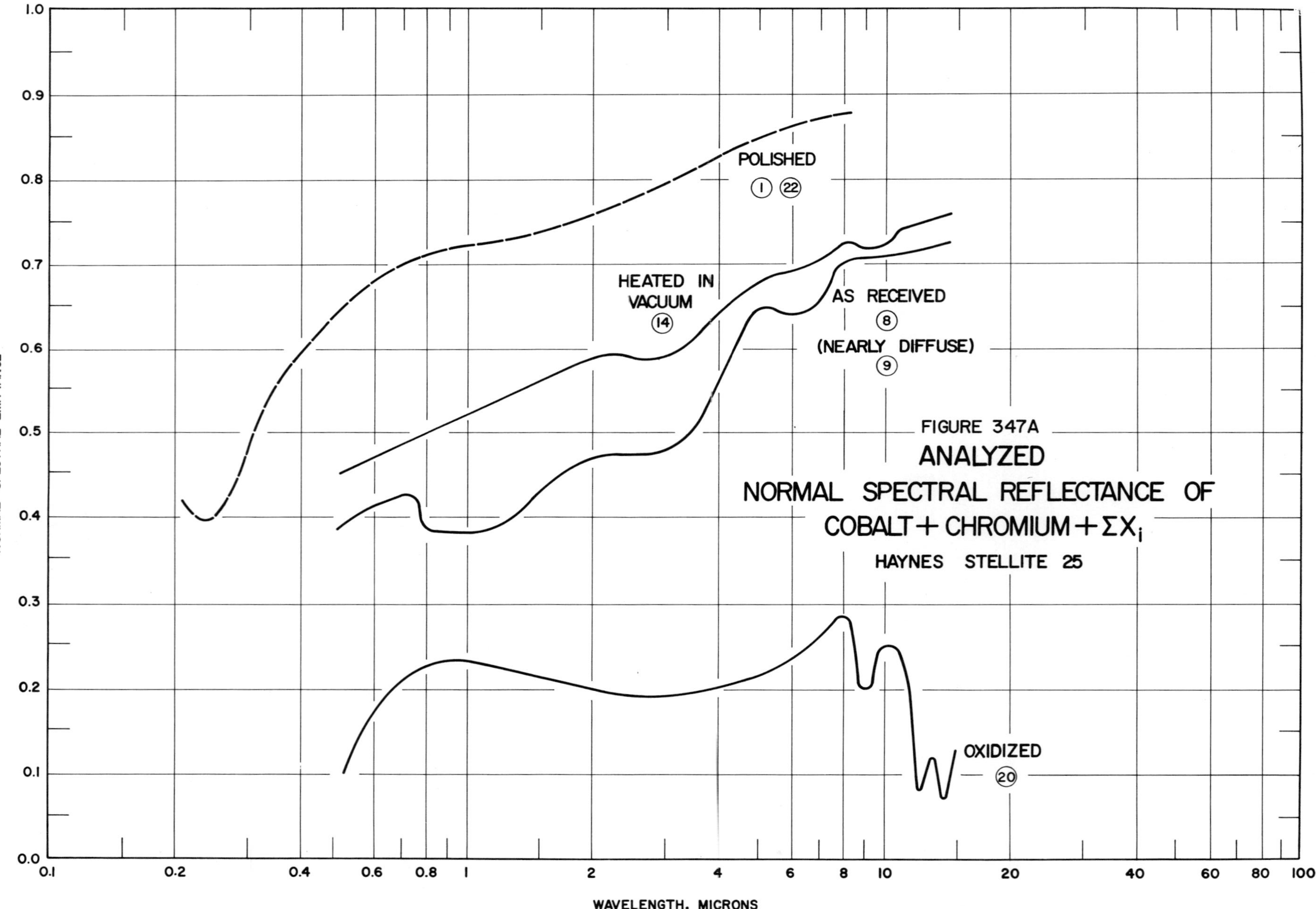

FIGURE 347A
ANALYZED
NORMAL SPECTRAL REFLECTANCE OF
COBALT + CHROMIUM + ΣXi
HAYNES STELLITE 25
POLISHED
1 22
HEATED IN VACUUM
14
AS RECEIVED
8
(NEARLY DIFFUSE)
9
OXIDIZED
20
NORMAL SPECTRAL EMITTANCE
0.0
0.1
0.2
0.3
0.4
0.5
0.6
0.7
0.8
0.9
1.0
WAVELENGTH, MICRONS
0.1
0.2
0.4
0.6
0.8
1
2
4
6
8
10
20
40
60
80
100

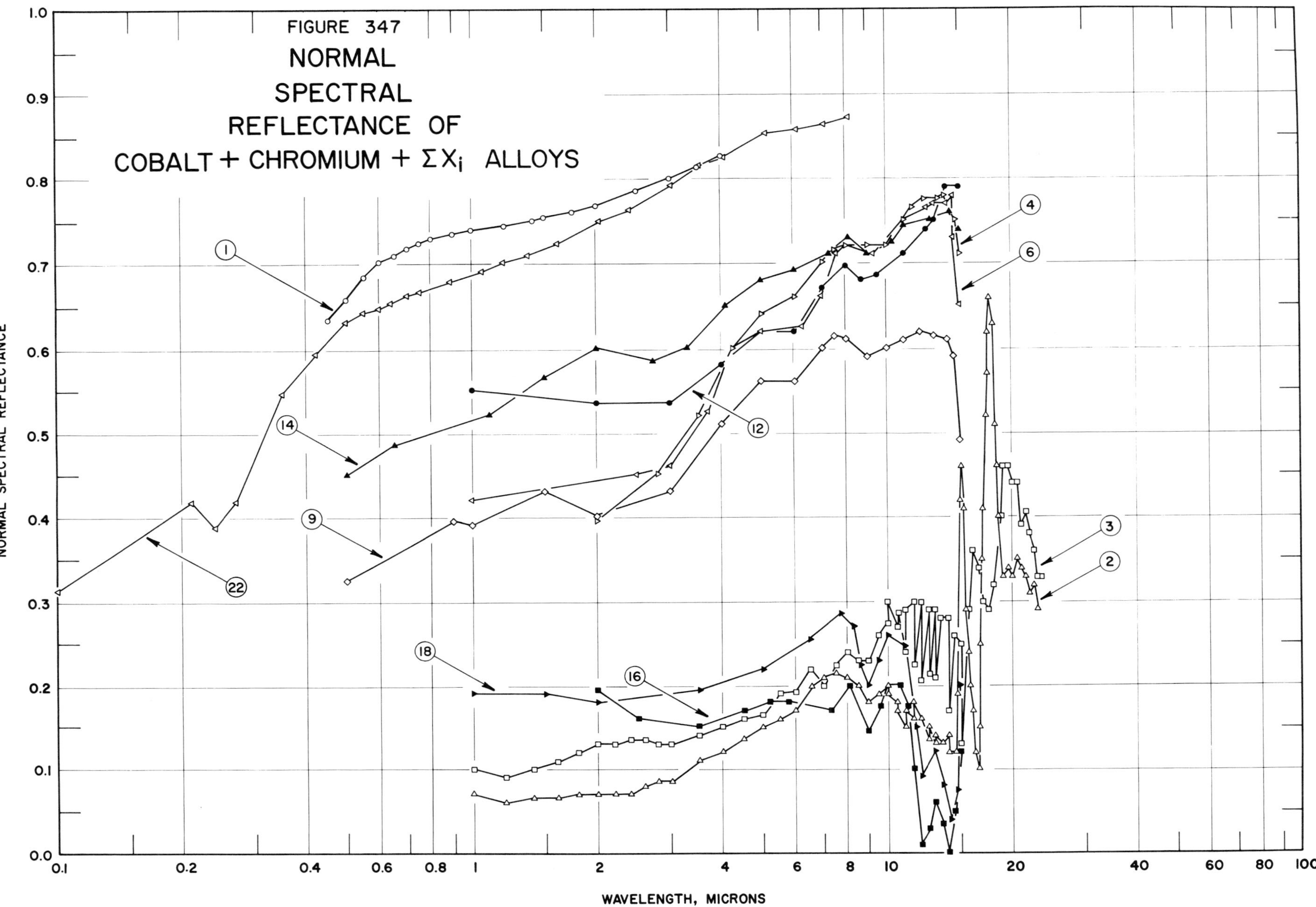
FIGURE 347
NORMAL
SPECTRAL
REFLECTANCE OF
COBALT + CHROMIUM + ΣX_i ALLOYS
NORMAL SPECTRAL REFLECTANCE
WAVELENGTH, MICRONS
1.0
0.9
0.8
0.7
0.6
0.5
0.4
0.3
0.2
0.1
0.0
0.1
0.2
0.4
0.6
0.8
1
2
4
6
8
10
20
40
60
80
100
1
2
3
4
6
9
12
14
16
18
22

SPECIFICATION TABLE NO. 347 NORMAL SPECTRAL REFLECTANCE OF [COBALT + CHROMIUM + ΣX_i] ALLOYS

Curve No.	Ref. No.	Year	Temperature K	Wavelength Range, μ	Geometry θ	θ'	ω'	Reported Error, %	Composition (weight percent), Specifications and Remarks
1	151	1920	298	0.45-4.00	~0°	~0°			Stellite from Haynes Stellite Co.; nominal composition: 65 Co, 25 Cr, and 10 Mo; silvered mirror reference.
2	65	1962	294	1.00-23.00	7°		2π		Haynes Stellite 25; nominal composition: 19-21 Cr, 14-16 W, 9-11 Ni, 2 max Fe, 0.15 max C, Co balance; oxidized; measured in air.
3	65	1962	1028	1.00-23.50	~0°		2π		Different sample, same as above specimen and conditions except calculated from (1-∈).
4	86	1961	~322	2.00-15.00	~0°		2π	< 2	Haynes Stellite 25; nominal composition: 19-21 Cr, 14-16 W, 9-11 Ni, 2 max Fe, 1.00-2.00 Mn, 1.00 Si, 0.15 max C, 0.04 P, 0.03 S, Co balance; as received; data extracted from smooth curve; hohlraum at 523 K; converted from R(2π, 0°).
5	86	1961	~322	2.00-15.00	~0°		2π	< 2	Above specimen and conditions; diffuse component only.
6	86	1961	~322	1.00-15.00	~0°		2π	< 2	Different sample, same as curve 4 specimen and conditions except hohlraum at 773 K.
7	86	1961	~322	1.00-15.00	~0°		2π	< 2	Above specimen and conditions; diffuse component only.
8	86	1961	~322	0.50-15.00	~0°		2π	< 2	Different sample, same as curve 4 specimen and conditions except hohlraum at 1273 K.
9	86	1961	~322	0.50-15.00	~0°		2π	< 2	Above specimen and conditions; diffuse component only.
10	86	1961	~322	2.00-15.00	~0°		2π	< 2	Different sample, same as curve 4 specimen and conditions except heated in vacuum (7.6 x 10^{-5}mm Hg) at 1255 K for 30 min.
11	86	1961	~322	2.00-15.00	~0°		2π	< 2	Above specimen and conditions; diffuse component only.
12	86	1961	~322	1.00-15.00	~0°		2π	< 2	Different sample, same as curve 10 specimen and conditions except hohlraum at 773 K.
13	86	1961	~322	1.00-15.00	~0°		2π	< 2	Above specimen and conditions; diffuse component only.
14	86	1961	~322	0.50-15.00	~0°		2π	< 2	Different sample, same as curve 10 specimen and conditions except hohlraum at 1273 K.
15	86	1961	~322	0.55-15.00	~0°		2π	< 2	Above specimen and conditions; diffuse component only.
16	86	1961	~322	2.00-15.00	~0°		2π	< 2	Different sample, same as curve 4 specimen and conditions except heated in air at 1255 K for 30 min.
17	86	1961	~322	2.00-15.00	~0°		2π	< 2	Above specimen and conditions; diffuse component only.
18	86	1961	~322	1.00-15.00	~0°		2π	< 2	Different sample, same as curve 16 specimen and conditions except hohlraum at 773 K.
19	86	1961	~322	1.00-15.00	~0°		2π	< 2	Above specimen and conditions; diffuse component only.
20	86	1961	~322	0.50-15.00	~0°		2π	< 2	Different sample, same as curve 16 specimen and conditions except hohlraum at 1273 K.
21	86	1961	~322	0.50-15.00	~0°		2π	< 2	Above specimen and conditions; diffuse component only.
22	330	1917	298	0.10-8.19	~0°	~0°			Stellite; nominal composition: 65 Co, 25 Cr, 10 Mo; highly polished.

DATA TABLE NO. 347 NORMAL SPECTRAL REFLECTANCE OF [COBALT + CHROMIUM + ΣX_i] ALLOYS

[Wavelength, λ, μ; Reflectance, ρ; Temperature, T, K]

CURVE 1, T = 298

λ	ρ
0.45	0.635
0.50	0.658
0.55	0.683
0.60	0.701
0.65	0.710
0.70	0.718
0.75	0.724
0.80	0.730
0.90	0.735
1.00	0.740
1.20	0.745
1.40	0.750
1.50	0.753
1.75	0.760
2.00	0.768
2.50	0.786
3.00	0.800
3.50	0.814
4.00	0.828

CURVE 2, T = 294

λ	ρ
1.00	0.070
1.20	0.060
1.40	0.065
1.60	0.065
1.80	0.070
2.00	0.070
2.20	0.070
2.40	0.070
2.60	0.080
2.80	0.085
3.00	0.085
3.50	0.110
4.00	0.120
4.50	0.135
5.00	0.150
5.50	0.160
6.00	0.170
6.50	0.200
7.00	0.210
7.50	0.215
8.00	0.210
8.50	0.200
9.00	0.180
9.50	0.190
10.00	0.200
10.00	0.190
10.50	0.180
10.50	0.170
11.00	0.150
11.00	0.170
11.50	0.160
11.50	0.180
12.00	0.160
12.50	0.135
12.50	0.140
13.00	0.130
13.00	0.140
13.50	0.130
14.00	0.140
14.00	0.120
14.50	0.120
14.70	0.190
15.00	0.420
15.10	0.460
15.30	0.410
15.40	0.290
15.60	0.240
15.80	0.200
16.00	0.170
16.20	0.120
16.50	0.100
16.60	0.150
16.70	0.250
16.90	0.350
17.00	0.410
17.30	0.520
17.40	0.570
17.50	0.620
17.70	0.660
18.00	0.630
18.20	0.510
18.30	0.460
18.50	0.400
19.00	0.330
19.50	0.340
20.00	0.330
20.50	0.350
21.00	0.340
21.50	0.330
22.00	0.310
22.50	0.320
23.00	0.290

CURVE 3, T = 1028

λ	ρ
1.00	0.100
1.20	0.090
1.40	0.100
1.60	0.110
1.80	0.120
2.00	0.130
2.20	0.130
2.40	0.135
2.60	0.135
2.80	0.130
3.00	0.130
3.50	0.140
4.00	0.150
4.50	0.160
5.00	0.164
5.50	0.190
6.00	0.192
6.50	0.220
7.00	0.200
7.50	0.225
8.00	0.240
8.50	0.230
9.00	0.230
9.50	0.260
10.00	0.275
10.00	0.300
10.50	0.270
10.50	0.285
11.00	0.240
11.00	0.290
11.50	0.300
11.50	0.225
12.00	0.300
12.00	0.205
12.50	0.290
12.50	0.215
13.00	0.290
13.00	0.210
13.40	0.280
14.00	0.280
14.00	0.170
14.40	0.260
15.00	0.250
15.00	0.130
15.60	0.290
16.00	0.360
16.50	0.340
17.00	0.300
17.50	0.290
18.00	0.320
18.60	0.400
19.00	0.460
19.50	0.460
20.00	0.440
20.40	0.440
21.00	0.390
21.50	0.405
22.00	0.380
22.50	0.360
23.00	0.330
23.50	0.330

CURVE 4, T = ~322

λ	ρ
2.00	0.395
2.80	0.450
3.50	0.520
4.25	0.600
5.00	0.640
6.00	0.660
7.00	0.700
7.50	0.715
8.00	0.720
9.00	0.720
10.00	0.720
11.00	0.750
11.50	0.765
12.25	0.775
13.25	0.775
13.75	0.780
14.75	0.750
15.00	0.710

CURVE 5*, T = ~322

λ	ρ
2.00	0.420
3.00	0.480
4.00	0.560
5.00	0.610
5.50	0.615
6.00	0.610
7.00	0.630
7.75	0.650
8.50	0.640
9.50	0.620
11.00	0.650
12.00	0.650
13.00	0.645
14.00	0.615
15.00	0.590

CURVE 6, T = ~322

λ	ρ
1.00	0.420
2.50	0.450
3.00	0.460
3.70	0.525
4.25	0.600*
5.00	0.620
6.25	0.625
7.00	0.660
7.60	0.710
8.00	0.720*
9.30	0.710
11.00	0.750*
12.50	0.765
13.00	0.770
14.00	0.770
14.50	0.730
15.00	0.650

CURVE 7*, T = ~322

λ	ρ
1.00	0.440
1.50	0.440
2.00	0.460
2.50	0.475
3.00	0.493
4.00	0.585
5.00	0.620
6.00	0.650
7.00	0.680
8.00	0.705
8.75	0.700
9.00	0.690
10.00	0.700
11.00	0.710
12.00	0.720
13.00	0.720
14.00	0.710
15.00	0.610

CURVE 8*, T = ~322

λ	ρ
0.50	0.385
0.55	0.410
0.75	0.425
0.80	0.385
1.20	0.385
1.50	0.430
1.80	0.460
2.00	0.470
3.00	0.470
3.75	0.525
4.00	0.565
5.00	0.650
6.00	0.640
7.00	0.660
7.75	0.700
8.50	0.710
10.00	0.710
11.50	0.750
13.00	0.750
14.00	0.725
14.60	0.675
14.90	0.625
15.00	0.575

CURVE 9, T = ~322

λ	ρ
0.50	0.325
0.90	0.395
1.00	0.390
1.50	0.430
2.00	0.400
3.00	0.430
4.00	0.510
5.00	0.560
6.00	0.560
7.00	0.600
7.50	0.615
8.00	0.610
9.00	0.590
10.00	0.600
11.00	0.610
12.00	0.620
13.00	0.615
14.00	0.610
14.50	0.590
15.00	0.490

CURVE 10*, T = ~322

λ	ρ
2.00	0.620
2.50	0.535
3.00	0.495
3.75	0.525
4.20	0.575
5.00	0.600
6.00	0.590
6.75	0.625
7.50	0.660
8.25	0.665
9.00	0.655
10.50	0.675
12.00	0.700
13.00	0.710
13.75	0.715
14.50	0.695
15.00	0.640

CURVE 11*, T = ~322

λ	ρ
2.00	0.650
3.00	0.512
3.10	0.510
4.00	0.570
5.00	0.610
6.00	0.580
7.00	0.630
8.00	0.630
9.00	0.620
10.00	0.610
11.00	0.615
12.00	0.610
13.00	0.620
14.00	0.610
15.00	0.550

CURVE 12, T = ~ 322

λ	ρ
1.00	0.550
2.00	0.535
3.00	0.535
4.00	0.580
5.00	0.620*
6.00	0.620
7.00	0.670
8.00	0.695
8.75	0.680
9.50	0.685
11.00	0.710
12.50	0.740

* Not shown on plot

DATA TABLE NO. 347 (continued)

λ	ρ
CURVE 12 (cont.)	
13.00	0.750
14.00	0.790
15.00	0.790
CURVE 13*	
T = ~322	
1.00	0.470
1.50	0.445
2.00	0.545
2.50	0.555
3.00	0.570
4.00	0.620
5.00	0.660
6.00	0.670
7.00	0.690
8.00	0.700
9.00	0.670
9.50	0.665
10.00	0.680
11.00	0.690
12.00	0.700
13.00	0.710
14.00	0.730
15.00	0.720
CURVE 14	
T = ~322	
0.50	0.450
0.65	0.485
1.10	0.520
1.50	0.565
2.00	0.600
2.75	0.585
3.30	0.600
4.10	0.650
5.00	0.680
6.00	0.690
7.25	0.710
8.10	0.730
9.00	0.710
10.30	0.725
11.00	0.745
12.75	0.750

λ	ρ
CURVE 14 (cont.)	
14.15	0.760
14.75	0.750*
15.00	0.740
CURVE 15*	
T = ~322	
0.55	0.405
1.00	0.490
2.00	0.580
3.00	0.600
4.00	0.660
5.00	0.695
5.50	0.705
6.00	0.700
6.50	0.700
7.00	0.710
8.00	0.720
9.00	0.700
10.00	0.690
11.00	0.700
12.00	0.710
13.00	0.710
14.00	0.710
15.00	0.690
CURVE 16	
T = ~322	
2.00	0.195
2.50	0.160
3.50	0.150
4.50	0.170
5.20	0.180
5.75	0.180
7.30	0.220
8.15	0.200
9.00	0.145
9.60	0.175
10.00	0.200*
10.60	0.200
11.15	0.175
11.50	0.100
12.00	0.010
12.50	0.030

λ	ρ
CURVE 16 (cont.)	
13.00	0.060
13.50	0.035
14.00	0.000
14.65	0.050
15.00	0.120
CURVE 17*	
T = ~322	
2.00	0.210
3.00	0.160
3.50	0.155
4.00	0.170
5.00	0.190
6.00	0.200
7.00	0.230
8.00	0.210
9.00	0.150
10.00	0.210
11.00	0.167
12.00	0.020
13.00	0.070
14.00	0.005
15.00	0.120
CURVE 18	
T = ~322	
1.00	0.190
1.50	0.190
2.00	0.180
3.50	0.195
5.00	0.220
6.50	0.255
7.75	0.285
8.25	0.270
8.60	0.225
9.00	0.200
9.50	0.230
10.00	0.260
11.00	0.245
11.60	0.150
12.00	0.090
12.90	0.120
13.50	0.080

λ	ρ
CURVE 18 (cont.)	
14.10	0.040
14.65	0.075
15.00	0.200
CURVE 19*	
T = ~322	
1.00	0.160
2.00	0.180
3.00	0.175
4.00	0.190
5.00	0.215
6.00	0.227
7.00	0.250
8.00	0.260
9.15	0.177
10.00	0.212
10.50	0.220
11.00	0.220
12.00	0.070
13.00	0.100
14.00	0.040
15.00	0.175
CURVE 20*	
T = ~322	
0.50	0.100
0.60	0.175
0.75	0.225
1.00	0.235
2.00	0.200
3.00	0.190
4.50	0.210
5.25	0.225
5.75	0.225
6.75	0.250
8.00	0.285
8.50	0.240
9.00	0.200
9.75	0.250
10.50	0.250
11.25	0.225
11.60	0.150
12.00	0.080

λ	ρ
CURVE 20* (cont.)	
12.50	0.100
13.00	0.120
13.50	0.090
14.00	0.070
14.70	0.125
15.00	0.210
CURVE 21*	
T = ~322	
0.50	0.180
0.60	0.205
0.70	0.205
1.00	0.225
2.00	0.195
3.00	0.190
4.00	0.200
5.00	0.230
6.00	0.240
7.00	0.260
8.00	0.280
9.00	0.190
10.00	0.260
11.00	0.240
11.50	0.180
12.00	0.080
13.00	0.110
14.00	0.070
15.00	0.210
CURVE 22	
T = 298	
0.10	0.317
0.21	0.419
0.24	0.389
0.27	0.420
0.35	0.548
0.42	0.597
0.50	0.633
0.55	0.644
0.60	0.649
0.64	0.656
0.70	0.665
0.75	0.669

λ	ρ
CURVE 22 (cont.)	
0.89	0.681
1.06	0.692
1.20	0.702
1.37	0.711
1.62	0.726
2.04	0.751
2.41	0.766
3.04	0.794
3.57	0.818
4.05	0.828
5.17	0.856
6.11	0.860
7.10	0.867
8.19	0.875

* Not shown on plot

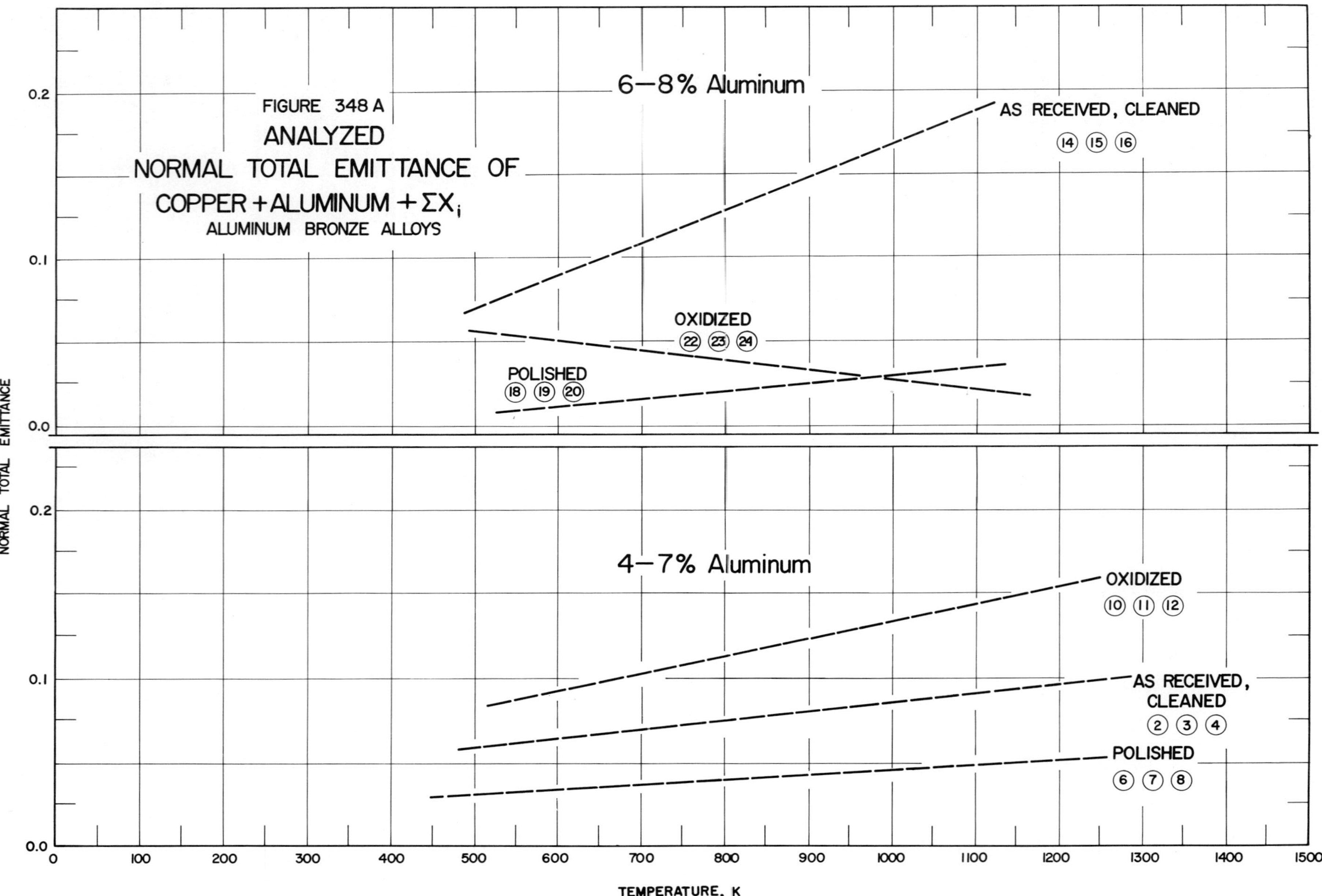
FIGURE 348 A
ANALYZED
NORMAL TOTAL EMITTANCE OF
COPPER + ALUMINUM + ΣX_i
ALUMINUM BRONZE ALLOYS
6–8% Aluminum
AS RECEIVED, CLEANED
14 15 16
OXIDIZED
22 23 24
POLISHED
18 19 20
4–7% Aluminum
OXIDIZED
10 11 12
AS RECEIVED, CLEANED
2 3 4
POLISHED
6 7 8
NORMAL TOTAL EMITTANCE
0.0
0.1
0.2
TEMPERATURE, K
0
100
200
300
400
500
600
700
800
900
1000
1100
1200
1300
1400
1500

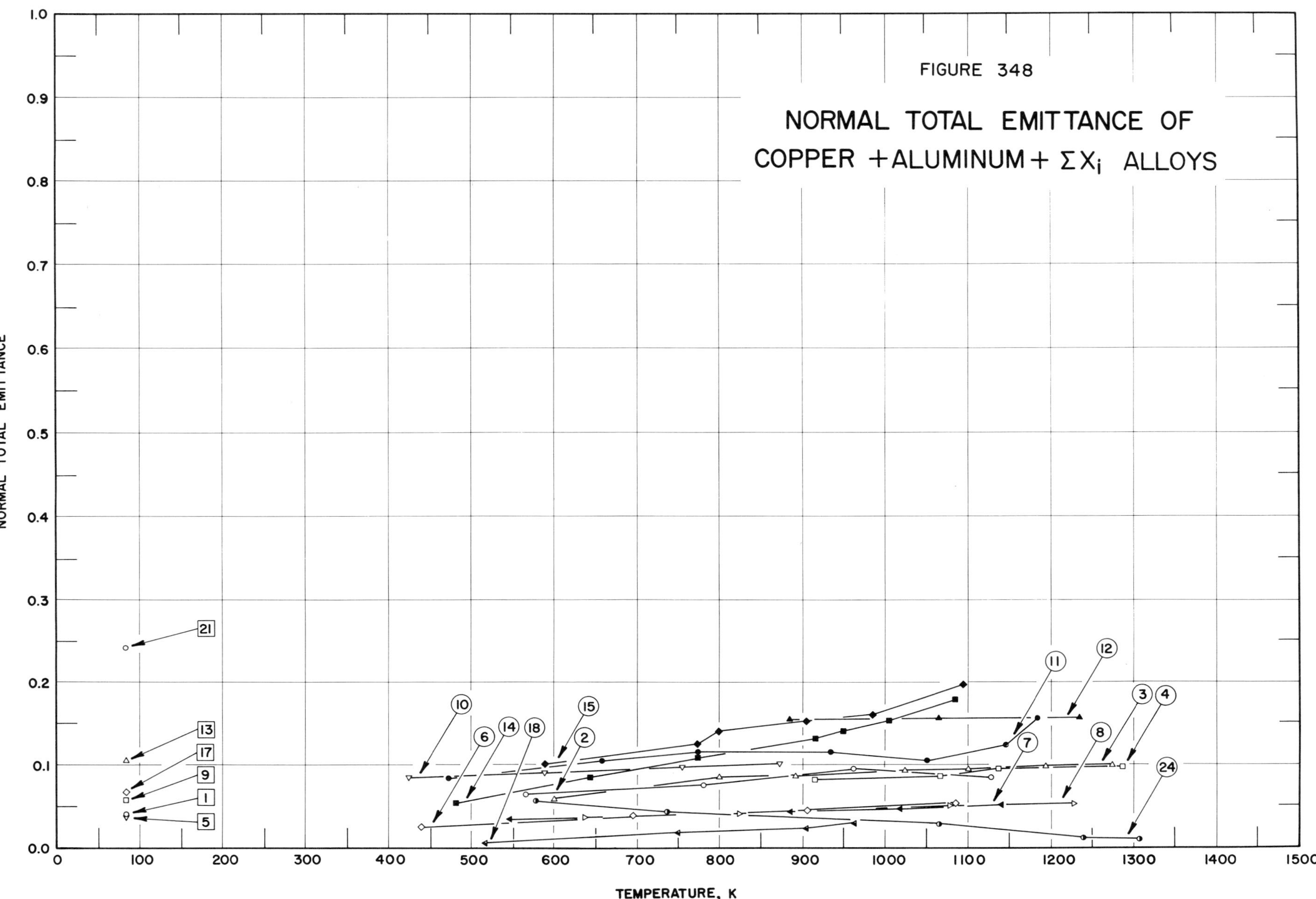
FIGURE 348
NORMAL TOTAL EMITTANCE OF
COPPER + ALUMINUM + ΣX_i ALLOYS
NORMAL TOTAL EMITTANCE
TEMPERATURE, K
0.0
0.1
0.2
0.3
0.4
0.5
0.6
0.7
0.8
0.9
1.0
0
100
200
300
400
500
600
700
800
900
1000
1100
1200
1300
1400
1500
21
13
17
9
1
5
10
6
14
18
15
2
11
12
7
8
3
4
24

SPECIFICATION TABLE NO. 348 NORMAL TOTAL EMITTANCE OF [COPPER + ALUMINUM + ΣX_i] ALLOYS

Curve No.	Ref. No.	Year	Temperature Range, K	Geometry θ'	Reported Error, %	Composition (weight percent), Specifications and Remarks
1	34	1957	83.2	$\sim 0^\circ$	±10	Aluminum bronze; nominal composition: 92.0-96.0 Cu, 4.0-7.0 Al, 0.50 max Fe; as received or cleaned with liquid detergent; measured in vacuum (5×10^{-4} mm Hg).
2	34	1957	566-1128	$\sim 0^\circ$	±10	Different sample, same as above specimen and conditions; increasing temp, cycle 1.
3	34	1957	600-1272	$\sim 0^\circ$	±10	Above specimen and conditions; cycle 2.
4	34	1957	914-1286	$\sim 0^\circ$	±10	Above specimen and conditions; cycle 3.
5	34	1957	83.2	$\sim 0^\circ$	±10	Different sample, same as curve 1 specimen and conditions except polished.
6	34	1957	439-1083	$\sim 0^\circ$	±10	Different sample, same as above specimen and conditions; increasing temp, cycle 1.
7	34	1957	547-1139	$\sim 0^\circ$	±10	Above specimen and conditions; cycle 2.
8	34	1957	636-1228	$\sim 0^\circ$	±10	Above specimen and conditions; cycle 3.
9	34	1957	83.2	$\sim 0^\circ$	±10	Different sample, same as curve 1 specimen and conditions except oxidized in air at red heat for 30 min.
10	34	1957	422-872	$\sim 0^\circ$	±10	Different sample, same as above specimen and conditions; increasing temp, cycle 1.
11	34	1957	472-1183	$\sim 0^\circ$	±10	Above specimen and conditions; cycle 2.
12	34	1957	883-1233	$\sim 0^\circ$	±10	Above specimen and conditions; cycle 3.
13	34	1957	83.2	$\sim 0^\circ$	±10	Aluminum bronze; nominal composition: 88.0-92.5 Cu, 6.0-8.0 Al, 1.5-3.5 Fe; as received or cleaned with liquid detergent; measured in vacuum (5×10^{-4} mm Hg).
14	34	1957	480-1083	$\sim 0^\circ$	±10	Different sample, same as above specimen and conditions; increasing temp, cycle 1.
15	34	1957	589-1094	$\sim 0^\circ$	±10	Above specimen and conditions; cycle 2.
16	34	1957	755-1300	$\sim 0^\circ$	±10	Above specimen and conditions; cycle 3.
17	34	1957	83.2	$\sim 0^\circ$	±10	Different sample, same as curve 13 specimen and conditions except polished.
18	34	1957	516-961	$\sim 0^\circ$	±10	Different sample, same as above specimen and conditions; increasing temp, cycle 1.
19	34	1957	661-1039	$\sim 0^\circ$	±10	Above specimen and conditions; cycle 2.
20	34	1957	578-1205	$\sim 0^\circ$	±10	Above specimen and conditions; cycle 3.
21	34	1957	83.2	$\sim 0^\circ$	±10	Different sample, same as curve 13 specimen and conditions except oxidized in air at red heat for 30 min.
22	34	1957	480-1230	$\sim 0^\circ$	±10	Different sample, same as above specimen and conditions; increasing temp, cycle 1.
23	34	1957	628-1150	$\sim 0^\circ$	±10	Above specimen and conditions; cycle 2.
24	34	1957	578-1308	$\sim 0^\circ$	±10	Above specimen and conditions; cycle 3.

DATA TABLE NO. 348 NORMAL TOTAL EMITTANCE OF [COPPER + ALUMINUM + ΣX_i] ALLOYS

[Temperature, T, K; Emittance, ∈]

T	∈
CURVE 1	
83.2	0.041
CURVE 2	
566	0.063
780	0.076
961	0.095
1128	0.087
CURVE 3	
600	0.059
800	0.085
891	0.086
1022	0.092
1100	0.094
1194	0.099
1272	0.100
CURVE 4	
914	0.081
1066	0.086
1136	0.095
1286	0.098
CURVE 5	
83.2	0.038
CURVE 6	
439	0.023
694	0.039
905	0.046
1083	0.053
CURVE 7	
547	0.034
883	0.042
1016	0.047
1139	0.051
CURVE 8	
636	0.037
822	0.041
1078	0.050
1228	0.052
CURVE 9	
83.2	0.058
CURVE 10	
422	0.084
589	0.090
755	0.098
872	0.101
CURVE 11	
472	0.083
658	0.104
772	0.115
933	0.114
1050	0.104
1144	0.122
1183	0.157
CURVE 12	
883	0.153
1064	0.157
1233	0.157
CURVE 13	
83.2	0.104
CURVE 14	
480	0.053
644	0.085
772	0.108
916	0.131
950	0.141
CURVE 14 (cont.)	
1005	0.152
1083	0.179
CURVE 15	
589	0.101
772	0.126
800	0.141
905	0.152
986	0.161
1094	0.198
CURVE 16*	
755	0.140
900	0.156
1014	0.176
1122	0.198
1216	0.157
1300	0.157
CURVE 17	
83.2	0.067
CURVE 18	
516	0.005
750	0.019
903	0.022
961	0.029
CURVE 19*	
661	0.019
866	0.027
1039	0.040
CURVE 20*	
578	0.020
803	0.025
947	0.027
CURVE 20 (cont.)*	
1044	0.040
1125	0.042
1205	0.043
CURVE 21	
83.2	0.241
CURVE 22*	
480	0.055
627	0.046
861	0.035
1027	0.025
1150	0.017
1230	0.014
CURVE 23*	
628	0.052
922	0.036
1150	0.022
CURVE 24	
578	0.057
736	0.043
1066	0.029
1239	0.011
1308	0.010

* Not shown on plot

FIGURE 349

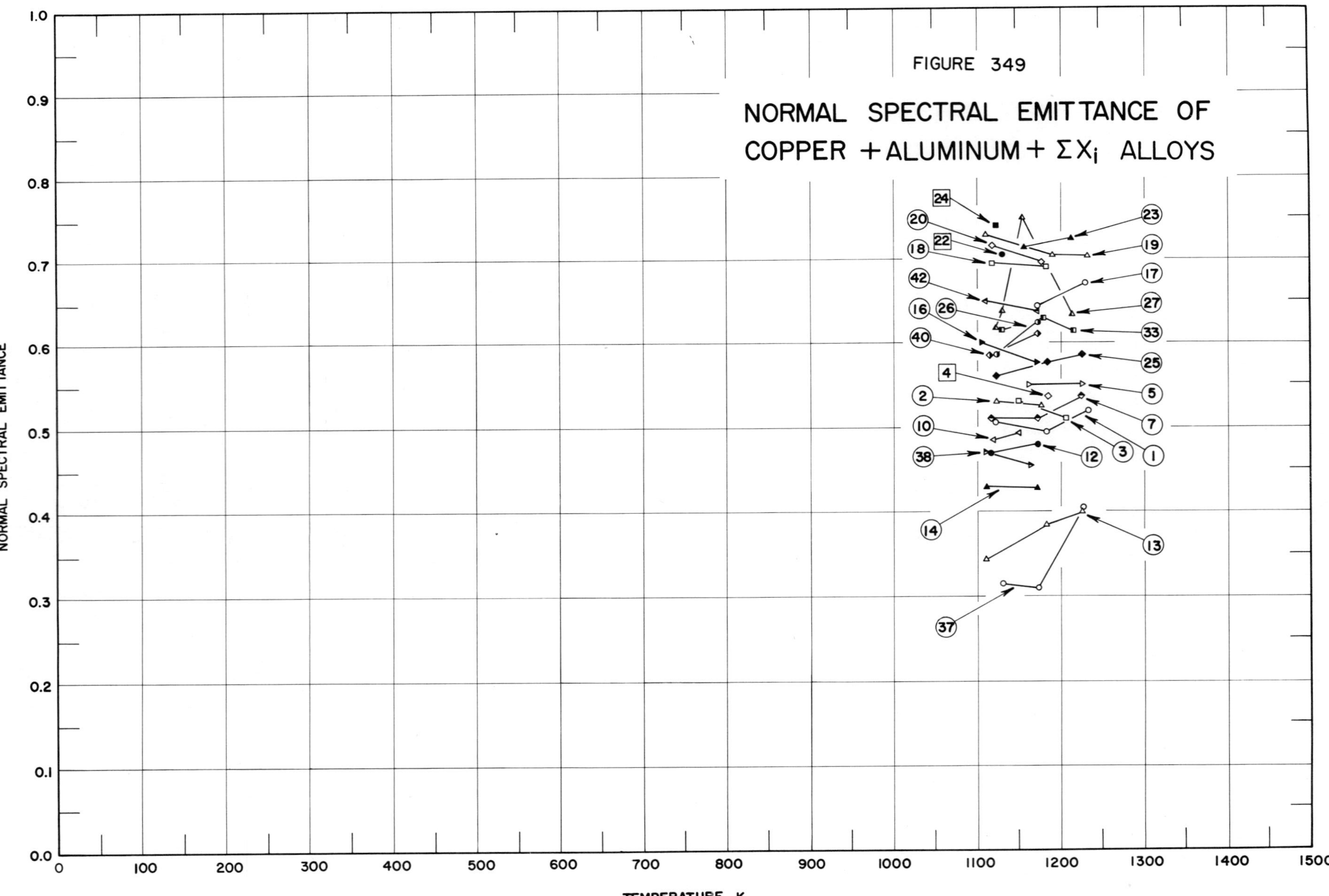
NORMAL SPECTRAL EMITTANCE OF
COPPER + ALUMINUM + ΣXi ALLOYS
NORMAL SPECTRAL EMITTANCE
TEMPERATURE, K
0.0 0.1 0.2 0.3 0.4 0.5 0.6 0.7 0.8 0.9 1.0
0 100 200 300 400 500 600 700 800 900 1000 1100 1200 1300 1400 1500
24
20
22
18
42
16
26
40
4
2
10
38
14
37
23
19
17
27
33
25
5
7
12
3
1
13

SPECIFICATION TABLE NO. 349 NORMAL SPECTRAL EMITTANCE OF [COPPER + ALUMINUM + ΣX_i] ALLOYS

Curve No.	Ref. No.	Year	Wavelength μ	Temperature Range, K	Geometry θ'	Reported Error, %	Composition (weight percent), Specifications and Remarks
1	34	1957	0.665	1122-1233	~0°		Bronze, 6-8 Al; as received; measured in vacuum (5×10^{-4} mm Hg); increasing temp, cycle 1.
2	34	1957	0.665	1177-1122	~0°		Above specimen and conditions; decreasing temp, cycle 1.
3	34	1957	0.665	1150-1208	~0°		Above specimen and conditions; cycle 2.
4	34	1957	0.665	1186	~0°		Above specimen and conditions; decreasing temp, cycle 2.
5	34	1957	0.665	1161-1227	~0°		Above specimen and conditions; cycle 3.
6	34	1957	0.665	1172-1127	~0°		Above specimen and conditions; decreasing temp, cycle 3.
7	34	1957	0.665	1116-1227	~0°		Different sample, same as curve 1 specimen and conditions except cleaned with a liquid detergent.
8	34	1957	0.665	1158-1116	~0°		Above specimen and conditions; decreasing temp, cycle 1.
9	34	1957	0.665	1166-1226	~0°		Above specimen and conditions; cycle 2.
10	34	1957	0.665	1150-1119	~0°		Above specimen and conditions; decreasing temp, cycle 2.
11	34	1957	0.665	1189-1227	~0°		Above specimen and conditions; cycle 3.
12	34	1957	0.665	1172-1116	~0°		Above specimen and conditions; decreasing temp, cycle 3.
13	34	1957	0.665	1111-1226	~0°		Different sample, same as curve 1 specimen and conditions except polished.
14	34	1957	0.665	1172-1111	~0°		Above specimen and conditions; decreasing temp, cycle 1.
15	34	1957	0.665	1183-1222	~0°		Above specimen and conditions; cycle 2.
16	34	1957	0.665	1172-1105	~0°		Above specimen and conditions; decreasing temp, cycle 2.
17	34	1957	0.665	1172-1230	~0°		Above specimen and conditions; cycle 3.
18	34	1957	0.665	1183-1119	~0°		Above specimen and conditions; decreasing temp, cycle 3.
19	34	1957	0.665	1111-1233	~0°		Different sample, same as curve 1 specimen and conditions except oxidized in air at red heat for 30 min.
20	34	1957	0.665	1177-1119	~0°		Above specimen and conditions; decreasing temp, cycle 1.
21	34	1957	0.665	1186-1226	~0°		Above specimen and conditions; cycle 2.
22	34	1957	0.665	1130	~0°		Above specimen and conditions; decreasing temp, cycle 2.
23	34	1957	0.665	1158-1214	~0°		Above specimen and conditions; cycle 3.
24	34	1957	0.665	1122	~0°		Above specimen and conditions; decreasing temp, cycle 3.
25	34	1957	0.665	1122-1228	~0°		Bronze, 4-7 Al; as received; measured in vacuum (5×10^{-4} mm Hg); increasing temp, cycle 1.
26	34	1957	0.665	1172-1122	~0°		Above specimen and conditions; decreasing temp, cycle 1.
27	34	1957	0.665	1122-1216	~0°		Above specimen and conditions; cycle 2.

SPECIFICATION TABLE NO. 349 (continued)

Curve No.	Ref. No.	Year	Wavelength μ	Temperature Range, K	Geometry θ'	Reported Error, %	Composition (weight percent), Specifications and Remarks
28	34	1957	0.665	1172	$\sim 0^{\circ}$		Above specimen and conditions; decreasing temp, cycle 2.
29	34	1957	0.665	1172-1228	$\sim 0^{\circ}$		Above specimen and conditions; cycle 3.
30	34	1957	0.665	1154-1100	$\sim 0^{\circ}$		Above specimen and conditions; decreasing temp, cycle 3.
31	34	1957	0.665	1130-1228	$\sim 0^{\circ}$		Different sample, same as curve 25 specimen and conditions except cleaned in liquid detergent.
32	34	1957	0.665	1166-1122	$\sim 0^{\circ}$		Above specimen and conditions; decreasing temp, cycle 1.
33	34	1957	0.665	1130-1216	$\sim 0^{\circ}$		Above specimen and conditions; cycle 2.
34	34	1957	0.665	1178-1122	$\sim 0^{\circ}$		Above specimen and conditions; decreasing temp, cycle 2.
35	34	1957	0.665	1178-1228	$\sim 0^{\circ}$		Above specimen and conditions; cycle 3.
36	34	1957	0.665	1161-1116	$\sim 0^{\circ}$		Above specimen and conditions; decreasing temp, cycle 3.
37	34	1957	0.665	1130-1228	$\sim 0^{\circ}$		Different sample, same as curve 25 specimen and conditions except polished.
38	34	1957	0.665	1164-1111	$\sim 0^{\circ}$		Above specimen and conditions; decreasing temp, cycle 1.
39	34	1957	0.665	1178-1228	$\sim 0^{\circ}$		Above specimen and conditions; cycle 2.
40	34	1957	0.665	1172-1116	$\sim 0^{\circ}$		Above specimen and conditions; decreasing temp, cycle 2.
41	34	1957	0.665	1172-1222	$\sim 0^{\circ}$		Above specimen and conditions; cycle 3.
42	34	1957	0.665	1172-1111	$\sim 0^{\circ}$		Above specimen and conditions; decreasing temp, cycle 3.

DATA TABLE NO. 349 NORMAL SPECTRAL EMITTANCE OF [COPPER + ALUMINUM + ΣX_i] ALLOYS

[Temperature, T, K; Emittance, ∈; Wavelength, λ, μ]

T	∈
CURVE 1	**λ = 0.665**
1122	0.505
1183	0.495
1233	0.520
CURVE 2	**λ = 0.665**
1177	0.525
1122	0.530
CURVE 3	**λ = 0.665**
1150	0.530
1208	0.510
CURVE 4	**λ = 0.665**
1186	0.535
CURVE 5	**λ = 0.665**
1161	0.550
1227	0.550
CURVE 6*	**λ = 0.665**
1172	0.535
1127	0.540
CURVE 7	**λ = 0.665**
1116	0.510
1172	0.510
1227	0.535
CURVE 8*	**λ = 0.665**
1158	0.515
1116	0.520
CURVE 9*	**λ = 0.665**
1166	0.505
1226	0.505
CURVE 10	**λ = 0.665**
1150	0.495
1119	0.485
CURVE 11*	**λ = 0.665**
1189	0.498
1227	0.516
CURVE 12	**λ = 0.665**
1172	0.480
1116	0.470
CURVE 13	**λ = 0.665**
1111	0.345
1183	0.385
1226	0.400
CURVE 14	**λ = 0.665**
1172	0.429
1111	0.430
CURVE 15*	**λ = 0.665**
1183	0.495
1222	0.535
CURVE 16	**λ = 0.665**
1172	0.575
1105	0.600
CURVE 17	**λ = 0.665**
1172	0.645
1230	0.670
CURVE 18	**λ = 0.665**
1183	0.690
1119	0.695
CURVE 19	**λ = 0.665**
1111	0.730
1191	0.705
1233	0.703
CURVE 20	**λ = 0.665**
1177	0.695
1119	0.718
CURVE 21*	**λ = 0.665**
1186	0.710
1226	0.708
CURVE 22	**λ = 0.665**
1130	0.706
CURVE 23	**λ = 0.665**
1158	0.715
1214	0.725
CURVE 24	**λ = 0.665**
1122	0.740
CURVE 25	**λ = 0.665**
1122	0.560
1186	0.575
1228	0.585
CURVE 26	**λ = 0.665**
1172	0.625
1122	0.585
CURVE 27	**λ = 0.665**
1122	0.620
1130	0.640
1155	0.750
1216	0.635
CURVE 28*	**λ = 0.665**
1172	0.618
CURVE 29*	**λ = 0.665**
1172	0.610
1228	0.598
CURVE 30*	**λ = 0.665**
1154	0.610
1100	0.575
CURVE 31*	**λ = 0.665**
1130	0.545
1178	0.565
1228	0.585
CURVE 32*	**λ = 0.665**
1166	0.600
1122	0.590
CURVE 33	**λ = 0.665**
1130	0.615
1180	0.630
1216	0.615
CURVE 34*	**λ = 0.665**
1178	0.620
1122	0.630
CURVE 35*	**λ = 0.665**
1178	0.635
1228	0.630
CURVE 36*	**λ = 0.665**
1161	0.630
1116	0.635
CURVE 37	**λ = 0.665**
1130	0.315
1172	0.310
1228	0.405
CURVE 38	**λ = 0.665**
1164	0.455
1111	0.470
CURVE 39*	**λ = 0.665**
1178	0.490
1228	0.550
CURVE 40	**λ = 0.665**
1172	0.610
1116	0.585
CURVE 41*	**λ = 0.665**
1172	0.590
1222	0.630
CURVE 42	**λ = 0.665**
1172	0.640
1111	0.650

* Not shown on plot

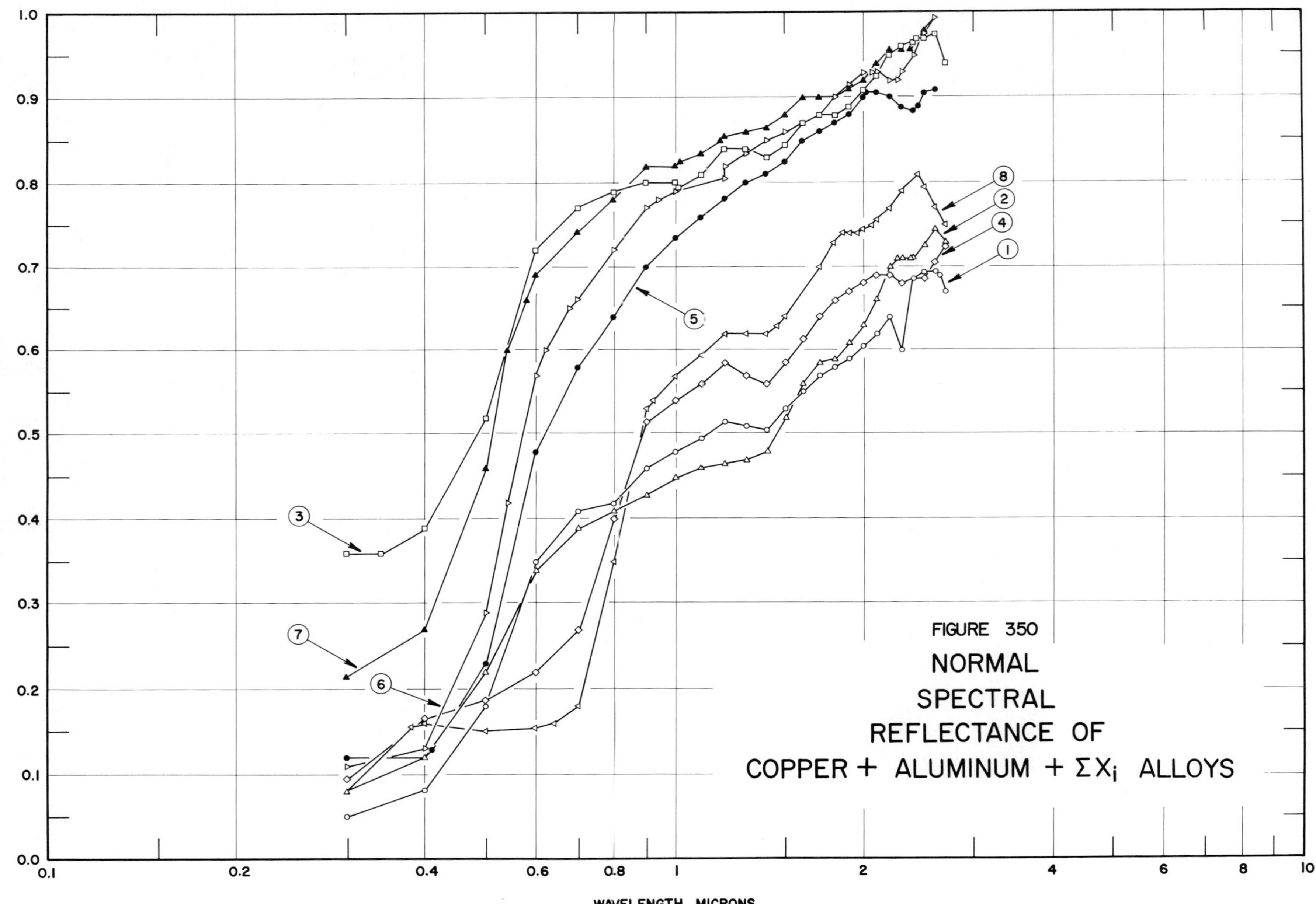

FIGURE 350 NORMAL SPECTRAL REFLECTANCE OF COPPER + ALUMINUM + ΣX_i ALLOYS

SPECIFICATION TABLE NO. 350 NORMAL SPECTRAL REFLECTANCE OF [COPPER + ALUMINUM + ΣX_i] ALLOYS

Curve No.	Ref. No.	Year	Temperature K	Wavelength Range, μ	Geometry θ	θ'	ω'	Reported Error, %	Composition (weight percent), Specifications and Remarks
1	34	1957	298	0.30-2.70	9°		2π	±4	Bronze, 6-8 Al; as received; data extracted from smooth curve.
2	34	1957	298	0.30-2.70	9°		2π	±4	Different sample, same as above specimen and conditions except cleaned with liquid detergent.
3	34	1957	298	0.30-2.70	9°		2π	±4	Different sample, same as above specimen and conditions except polished.
4	34	1957	298	0.30-2.70	9°		2π	±4	Different sample, same as above specimen and conditions except oxidized in air at red heat for 30 min.
5	34	1957	298	0.30-2.60	9°		2π	±4	Bronze, 4-7 Al; as received; data extracted from smooth curve.
6	34	1957	298	0.30-2.60	9°		2π	±4	Different sample, same as above specimen and conditions except cleaned with liquid detergent.
7	34	1957	298	0.30-2.60	9°		2π	±4	Different sample, same as above specimen and conditions except polished.
8	34	1957	298	0.30-2.70	9°		2π	±4	Different sample, same as above specimen and conditions except oxidized in air at red heat for 30 min.

DATA TABLE NO. 350 NORMAL SPECTRAL REFLECTANCE OF [COPPER + ALUMINUM + ΣX_i] ALLOYS

[Wavelength, λ, μ; Reflectance, ρ; Temperature, T, K]

CURVE 1
T = 298

λ	ρ
0.30	0.050
0.40	0.080
0.50	0.180
0.60	0.350
0.70	0.410
0.80	0.420
0.90	0.460
1.00	0.480
1.10	0.495
1.20	0.515
1.30	0.510
1.40	0.505
1.50	0.530
1.60	0.550
1.70	0.570
1.80	0.580
1.90	0.590
2.00	0.605
2.10	0.620
2.20	0.640
2.30	0.660
2.40	0.685
2.50	0.695
2.60	0.695
2.64	0.690
2.70	0.670

CURVE 2
T = 298

λ	ρ
0.30	0.080
0.40	0.120
0.50	0.220
0.60	0.340
0.70	0.390
0.80	0.410
0.90	0.430
1.00	0.450
1.10	0.460
1.20	0.465
1.30	0.470
1.40	0.480
1.50	0.520
1.60	0.560

CURVE 2 (cont.)

λ	ρ
1.70	0.585
1.80	0.590
1.90	0.610
2.00	0.630
2.10	0.660
2.20	0.700
2.26	0.710
2.30	0.710
2.38	0.710
2.40	0.710
2.50	0.725
2.60	0.745
2.70	0.730

CURVE 3
T = 298

λ	ρ
0.30	0.360
0.34	0.360
0.40	0.390
0.50	0.520
0.60	0.720
0.70	0.770
0.80	0.790
0.90	0.800
1.00	0.800
1.02	0.795
1.10	0.810
1.20	0.840
1.30	0.840
1.40	0.830
1.50	0.845
1.60	0.870
1.70	0.880
1.80	0.880
1.90	0.890
2.00	0.910
2.10	0.925
2.20	0.950
2.30	0.960
2.40	0.965
2.42	0.970
2.50	0.970
2.60	0.975
2.70	0.940

CURVE 4
T = 298

λ	ρ
0.30	0.095
0.40	0.165
0.50	0.185
0.60	0.220
0.70	0.270
0.80	0.400
0.90	0.515
1.00	0.540
1.10	0.560
1.20	0.585
1.30	0.570
1.40	0.560
1.50	0.585
1.60	0.615
1.70	0.640
1.80	0.660
1.90	0.670
2.00	0.680
2.10	0.690
2.20	0.690
2.30	0.680
2.40	0.685*
2.50	0.685
2.60	0.705
2.70	0.725

CURVE 5
T = 298

λ	ρ
0.30	0.120
0.40	0.120*
0.42	0.130
0.50	0.230
0.60	0.480
0.70	0.580
0.80	0.640
0.90	0.700
1.00	0.735
1.10	0.760
1.20	0.780
1.30	0.800
1.40	0.810
1.50	0.825
1.60	0.850

CURVE 5 (cont.)

λ	ρ
1.70	0.860
1.80	0.870
1.90	0.880
2.00	0.900
2.02	0.905
2.10	0.905
2.20	0.900
2.30	0.890
2.40	0.885
2.44	0.890
2.50	0.905
2.60	0.960

CURVE 6
T = 298

λ	ρ
0.30	0.110
0.40	0.130
0.50	0.290
0.54	0.420
0.60	0.570
0.62	0.600
0.68	0.650
0.70	0.660
0.80	0.720
0.90	0.770
0.94	0.780
1.00	0.790
1.20	0.805
1.20	0.820
1.30	0.835
1.40	0.850
1.50	0.860
1.60	0.870*
1.70	0.880*
1.80	0.900
1.90	0.915
2.00	0.930
2.06	0.930
2.10	0.930
2.20	0.920
2.26	0.920
2.30	0.930
2.40	0.950
2.50	0.975
2.60	0.995

CURVE 7
T = 298

λ	ρ
0.30	0.215
0.40	0.270
0.50	0.460
0.54	0.600
0.58	0.630
0.60	0.690
0.70	0.740
0.80	0.730
0.90	0.820
1.00	0.820
1.04	0.825
1.10	0.835
1.18	0.850
1.20	0.855
1.30	0.860
1.40	0.865
1.50	0.880
1.60	0.900
1.70	0.900
1.80	0.900*
1.90	0.910
2.00	0.920
2.10	0.940
2.18	0.950*
2.20	0.955
2.30	0.955
2.36	0.955
2.40	0.960
2.50	0.980
2.60	0.995*

CURVE 8
T = 298

λ	ρ
0.30	0.080*
0.38	0.155
0.40	0.160
0.50	0.150
0.60	0.155
0.64	0.160
0.70	0.180
0.80	0.350
0.90	0.530
0.92	0.540

CURVE 8 (cont.)

λ	ρ
1.00	0.570
1.10	0.595
1.20	0.620
1.30	0.620
1.40	0.620
1.46	0.630
1.50	0.640
1.70	0.700
1.80	0.730
1.86	0.740
1.90	0.740
1.96	0.740
2.00	0.745
2.06	0.750
2.10	0.755
2.20	0.770
2.30	0.790
2.40	0.810
2.44	0.810
2.50	0.795
2.60	0.770
2.70	0.750

* Not shown on plot

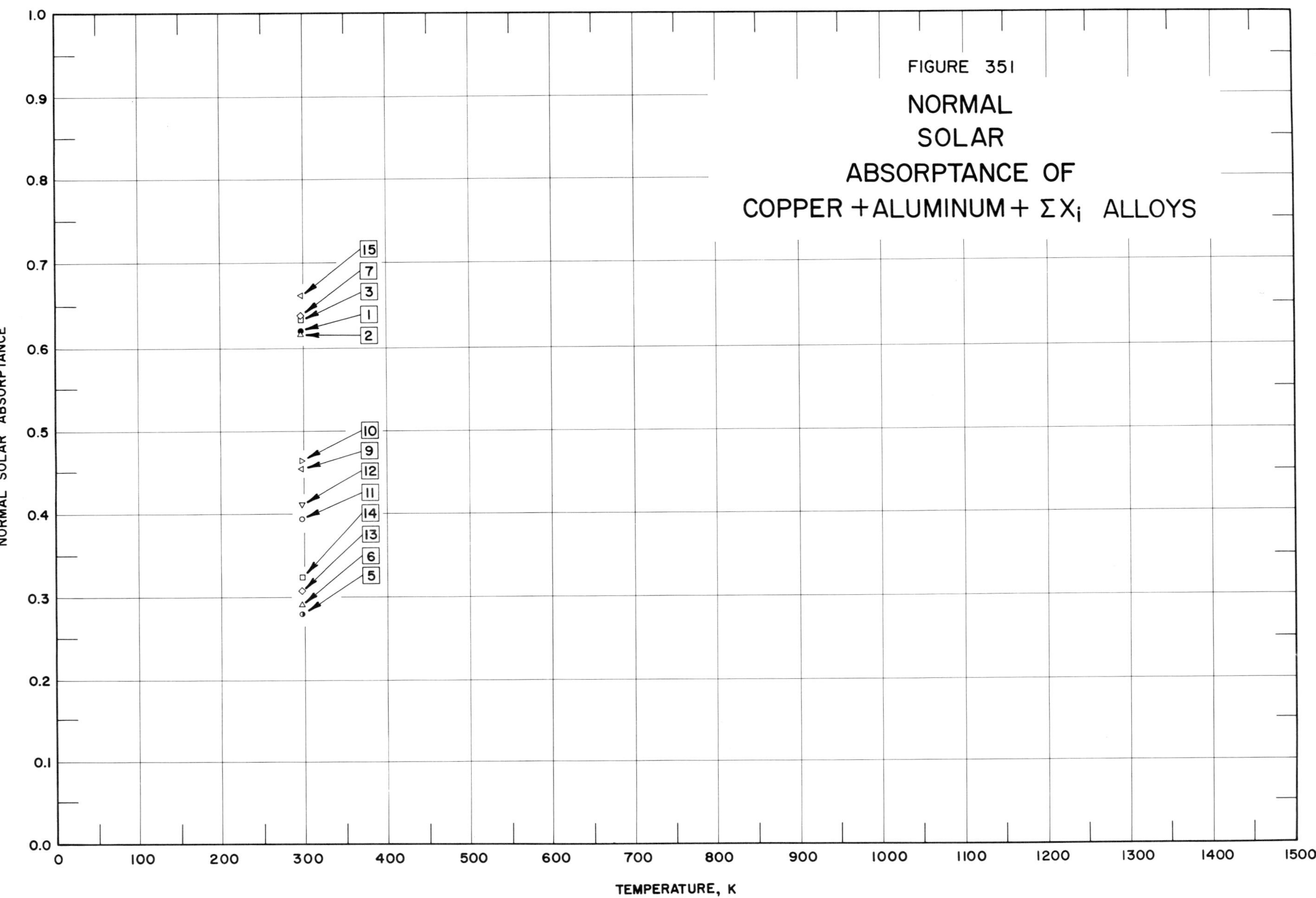
FIGURE 351
NORMAL
SOLAR
ABSORPTANCE OF
COPPER + ALUMINUM + ΣXi ALLOYS
NORMAL SOLAR ABSORPTANCE
TEMPERATURE, K
1.0
0.9
0.8
0.7
0.6
0.5
0.4
0.3
0.2
0.1
0.0
0
100
200
300
400
500
600
700
800
900
1000
1100
1200
1300
1400
1500
15
7
3
1
2
10
9
12
11
14
13
6
5

SPECIFICATION TABLE NO. 351 NORMAL SOLAR ABSORPTANCE OF [COPPER + ALUMINUM + ΣX_i] ALLOYS

Curve No.	Ref. No.	Year	Temperature Range, K	Geometry θ	Reported Error, %	Composition (weight percent), Specifications and Remarks
1	34	1957	298	9°		Bronze, 6-8 Al; as received; computed from spectral reflectance data for sea level conditions.
2	34	1957	298	9°		Above specimen and conditions except computed for above atmosphere conditions.
3	34	1957	298	9°		Different sample, same as curve 1 specimen and conditions except cleaned with liquid detergent.
4	34	1957	298	9°		Above specimen and conditions except computed for above atmosphere conditions.
5	34	1957	298	9°		Different sample, same as curve 1 specimen and conditions except polished.
6	34	1957	298	9°		Above specimen and conditions except computed for above atmosphere conditions.
7	34	1957	298	9°		Different sample, same as curve 1 specimen and conditions except oxidized in air at red heat for 30 min.
8	34	1957	298	9°		Above specimen and conditions except computed for above atmosphere conditions.
9	34	1957	298	9°		Bronze, 4-7 Al; as received; computed from spectral reflectance data for sea level conditions.
10	34	1957	298	9°		Above specimen and conditions except computed for above atmosphere conditions.
11	34	1957	298	9°		Different sample, same as curve 9 specimen and conditions except cleaned with liquid detergent.
12	34	1957	298	9°		Above specimen and conditions except computed for above atmosphere conditions.
13	34	1957	298	9°		Different sample, same as curve 9 specimen and conditions except polished.
14	34	1957	298	9°		Above specimen and conditions except computed for above atmosphere conditions.
15	34	1957	298	9°		Different sample, same as curve 9 specimen and conditions except oxidized in air at red heat for 30 min.
16	34	1957	298	9°		Above specimen and conditions except computed for above atmosphere conditions.

DATA TABLE NO. 351 NORMAL SOLAR ABSORPTANCE OF [COPPER + ALUMINUM + ΣX_i] ALLOYS

[Temperature, T, K; Absorptance, α]

T	α	T	α
CURVE 1		CURVE 13	
298	0.621	298	0.308
CURVE 2		CURVE 14	
298	0.618	298	0.325
CURVE 3		CURVE 15	
298	0.634	298	0.662
CURVE 4*		CURVE 16*	
298	0.631	298	0.642
CURVE 5			
298	0.281		
CURVE 6			
298	0.292		
CURVE 7			
298	0.640		
CURVE 8*			
298	0.632		
CURVE 9			
298	0.455		
CURVE 10			
298	0.464		
CURVE 11			
298	0.395		
CURVE 12			
298	0.411		

* Not shown on plot

SPECIFICATION TABLE NO. 352 NORMAL SPECTRAL REFLECTANCE OF [COPPER + NICKEL + ΣX_i] ALLOYS

Curve No.	Ref. No.	Year	Temperature K	Wavelength Range, μ	Geometry ϑ θ' ω'	Reported Error, %	Composition (weight percent), Specifications and Remarks
1	228	1900	298	0.45–0.70	$\sim 0^\circ$ $\sim 0^\circ$		41 Cu, 26 Ni, 24 Sn, 8 Fe, 1 Sb; mirror-like surface.

DATA TABLE NO. 352 NORMAL SPECTRAL REFLECTANCE OF [COPPER + NICKEL + ΣX_i] ALLOYS

[Wavelength, λ, μ; Reflectance, ρ; Temperature, T, K]

λ	ρ
CURVE 1*	
T = 298	
0.45	0.491
0.50	0.493
0.55	0.483
0.60	0.475
0.65	0.497
0.70	0.549

* Not shown on plot

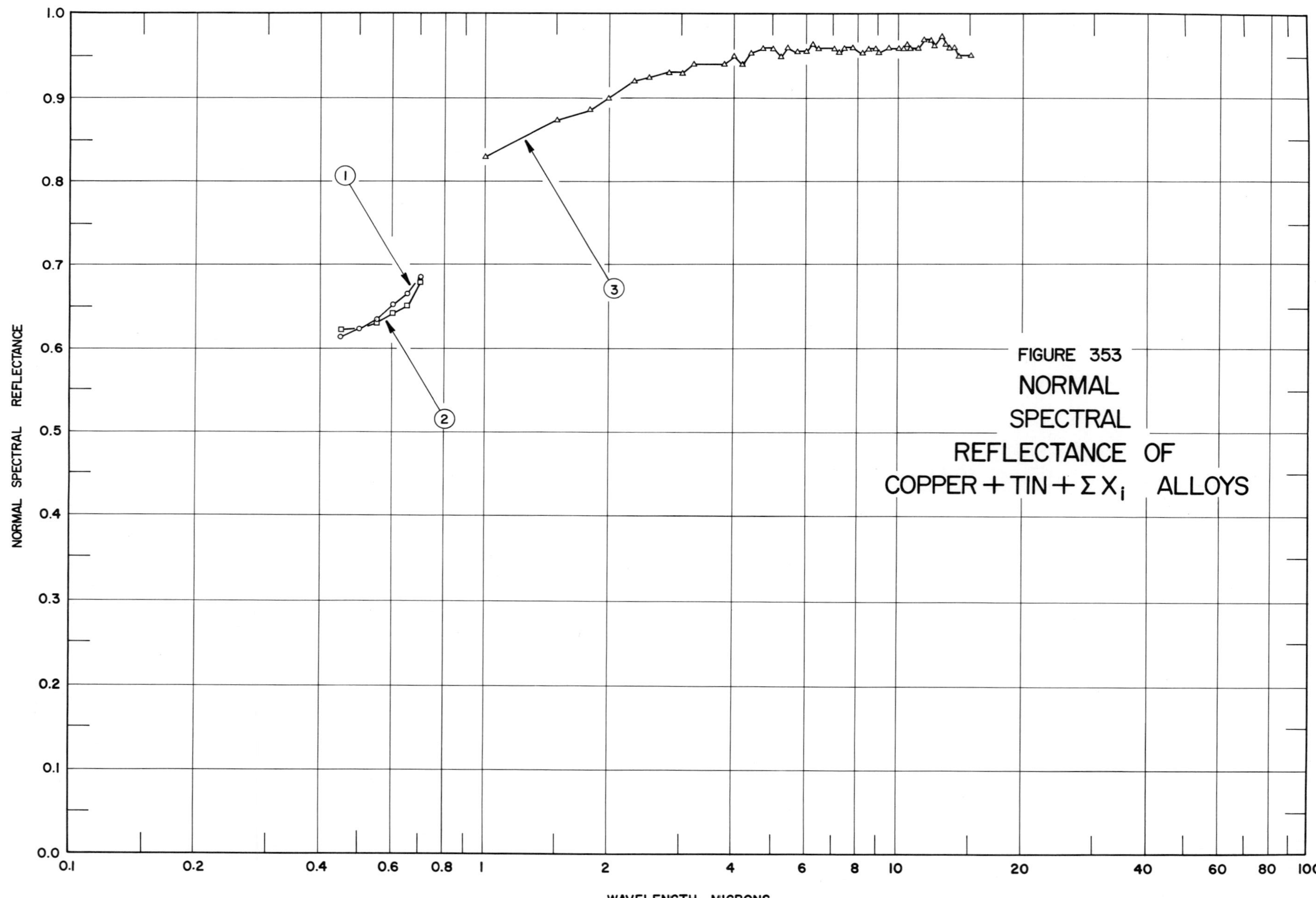

FIGURE 353
NORMAL
SPECTRAL
REFLECTANCE OF
COPPER + TIN + Σ X_i ALLOYS
NORMAL SPECTRAL REFLECTANCE
WAVELENGTH, MICRONS
1.0
0.9
0.8
0.7
0.6
0.5
0.4
0.3
0.2
0.1
0.0
0.1
0.2
0.4
0.6
0.8
1
2
4
6
8
10
20
40
60
80
100
1
2
3

SPECIFICATION TABLE NO. 353 NORMAL SPECTRAL REFLECTANCE OF [COPPER + TIN + ΣX_i] ALLOYS

Curve No.	Ref. No.	Year	Temperature K	Wavelength Range, μ	Geometry, θ	θ'	ω'	Reported Error, %	Composition (weight percent), Specifications and Remarks
1	228	1900	298	0.45-0.70	$\sim 0^\circ$	$\sim 0^\circ$			60 Cu, 30 Sn, 10 Ag; mirror like surface; platinum filament lamp source.
2	228	1900	298	0.45-0.70	$\sim 0^\circ$	$\sim 0^\circ$			66 Cu, 22 Sn, 12 Zn; mirror like surface; platinum filament lamp source.
3	126	1953	298	1.0-15.0	$\sim 5^\circ$		2π		Phosphor bronze; polished; data extracted from smooth curve; converted from $R(2\pi, 5^\circ)$.

DATA TABLE NO. 353 NORMAL SPECTRAL REFLECTANCE OF [COPPER + TIN + ΣX_i] ALLOYS

[Wavelength, λ, μ; Reflectance, ρ; Temperature, T, K]

λ	ρ
CURVE 1	
T = 298	
0.45	0.615
0.50	0.625
0.55	0.636
0.60	0.652
0.65	0.666
0.70	0.686
CURVE 2	
T = 298	
0.45	0.624
0.50	0.625
0.55	0.634
0.60	0.642
0.65	0.651
0.70	0.680
CURVE 3	
T = 298	
1.0	0.830
1.5	0.875
1.8	0.880
2.0	0.900
2.3	0.920
2.5	0.925
2.8	0.930
3.0	0.930
3.2	0.940
3.8	0.940
4.0	0.950
4.2	0.940
4.4	0.955
4.7	0.960
5.0	0.960
5.2	0.950
5.4	0.960
5.7	0.955
6.0	0.955
6.2	0.965
6.4	0.960
7.0	0.960
7.2	0.955
7.4	0.960
7.8	0.960
CURVE 3 (cont.)	
8.2	0.950
8.5	0.960
8.8	0.960
9.0	0.955
9.5	0.960
10.0	0.960
10.2	0.960
10.5	0.965
10.7	0.960
11.2	0.960
11.5	0.970
12.0	0.970
12.2	0.965
12.8	0.975
13.0	0.965
13.2	0.960
13.6	0.960
14.0	0.950
15.0	0.950

* Not shown on plot

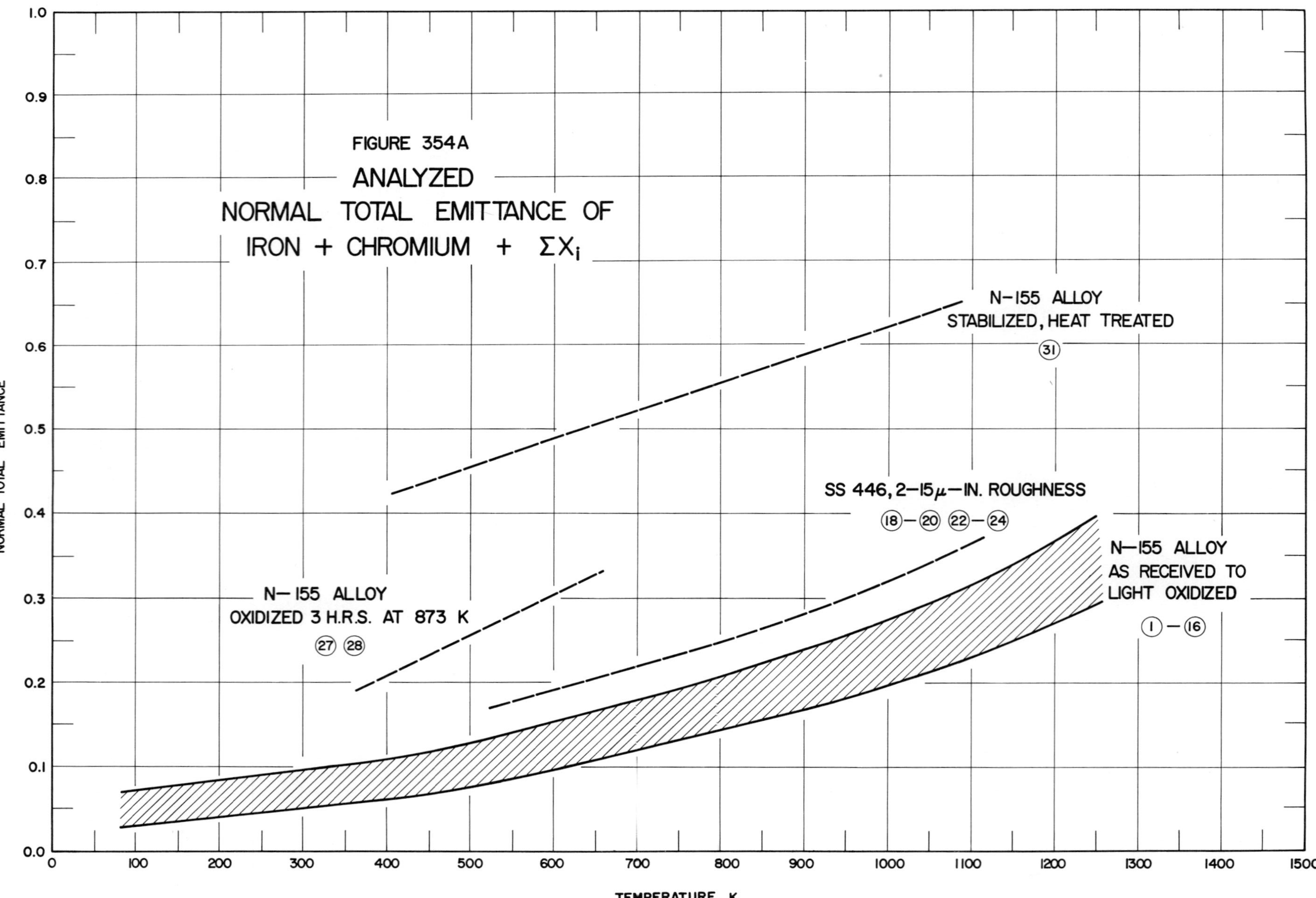
FIGURE 354A
ANALYZED
NORMAL TOTAL EMITTANCE OF
IRON + CHROMIUM + ΣXi
N-155 ALLOY
STABILIZED, HEAT TREATED
(31)
SS 446, 2-15μ-IN. ROUGHNESS
(18)-(20) (22)-(24)
N-155 ALLOY
AS RECEIVED TO
LIGHT OXIDIZED
(1)-(16)
N-155 ALLOY
OXIDIZED 3 H.R.S. AT 873 K
(27) (28)
NORMAL TOTAL EMITTANCE
TEMPERATURE, K
0.0
0.1
0.2
0.3
0.4
0.5
0.6
0.7
0.8
0.9
1.0
0
100
200
300
400
500
600
700
800
900
1000
1100
1200
1300
1400
1500

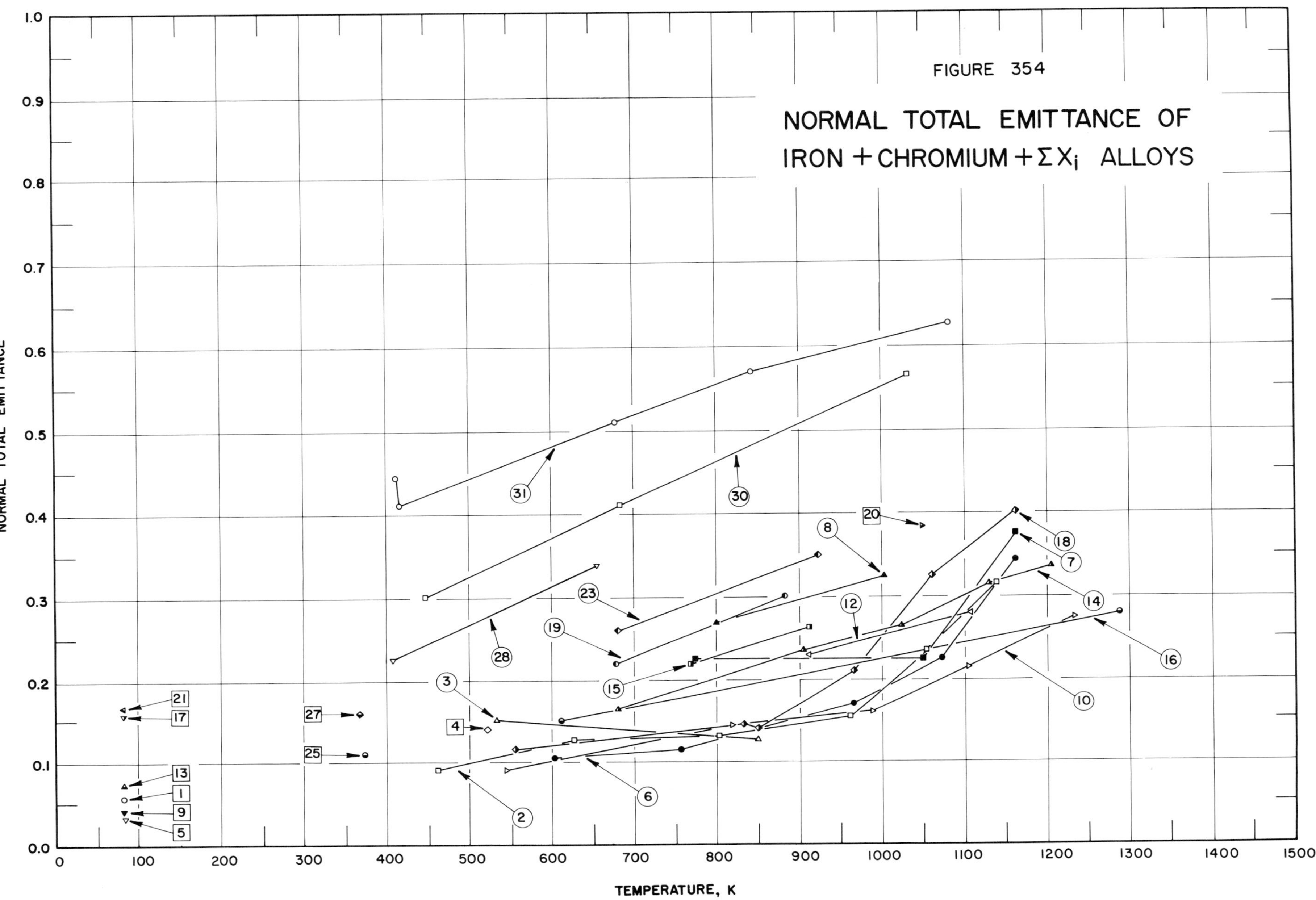
FIGURE 354
NORMAL TOTAL EMITTANCE OF
IRON + CHROMIUM + Σ Xi ALLOYS
NORMAL TOTAL EMITTANCE
TEMPERATURE, K
0.0
0.1
0.2
0.3
0.4
0.5
0.6
0.7
0.8
0.9
1.0
0
100
200
300
400
500
600
700
800
900
1000
1100
1200
1300
1400
1500
1
2
3
4
5
6
7
8
9
10
12
13
14
15
16
17
18
19
20
21
23
25
27
28
30
31

SPECIFICATION TABLE NO. 354 NORMAL TOTAL EMITTANCE OF [IRON + CHROMIUM + ΣX_i] ALLOYS

Curve No.	Ref. No.	Year	Temperature Range, K	Geometry θ'	Reported Error, %	Composition (weight percent), Specifications and Remarks
1	34	1957	83.2	~0°	±10	Cobalt alloy N-155; nominal composition: 21 Cr, 20 Co, 20 Ni, 3 Mo, 3 W, 1.5 Mn, 1 Nb, 0.5 Si, 0.15 C, 0.15 N, Fe balance; as received; measured in vacuum (5 x 10^{-4} mm Hg).
2	34	1957	461-1139	~0°	±10	Different sample, same as above specimen and conditions; increasing temp, cycle 1.
3	34	1957	533-850	~0°	±10	Above specimen and conditions; cycle 2.
4	34	1957	522	~0°	±10	Above specimen and conditions; cycle 3.
5	34	1957	83.2	~0°	±10	Different sample, same as curve 1 specimen and conditions except cleaned with liquid detergent.
6	34	1957	603-1161	~0°	±10	Different sample, same as above specimen and conditions; increasing temp, cycle 1.
7	34	1957	772-1161	~0°	±10	Above specimen and conditions; cycle 2.
8	34	1957	800-1003	~0°	±10	Above specimen and conditions; cycle 3.
9	34	1957	83.2	~0°	±10	Different sample, same as curve 1 specimen and conditions except polished.
10	34	1957	544-1233	~0°	±10	Different sample, same as above specimen and conditions; increasing temp, cycle 1.
11	34	1957	811-989	~0°	±10	Above specimen and conditions; cycle 2.
12	34	1957	911-1108	~0°	±10	Above specimen and conditions; cycle 3.
13	34	1957	83.2	~0°	±10	Different sample, same as curve 1 specimen and conditions except oxidized in air at red heat for 30 min.
14	34	1957	680-1205	~0°	±10	Different sample, same as above specimen and conditions; increasing temp, cycle 1.
15	34	1957	769-911	~0°	±10	Above specimen and conditions; cycle 2.
16	34	1957	611-1289	~0°	±10	Above specimen and conditions; cycle 3.
17	34	1957	83.2	~0°	±10	Stainless steel type 446; nominal composition: 23.00-27.00 Cr, 1.5 max Mn, 1.00 max Si, 0.25 max N, 0.20 max C, Fe balance; surface roughness ~2 microinches rms; measured in vacuum (5 x 10^{-4} mm Hg).
18	34	1957	555-1161	~0°	±10	Different sample, same as above specimen and conditions; increasing temp, cycle 1.
19	34	1957	678-883	~0°	±10	Above specimen and conditions; cycle 2.
20	34	1957	1050	~0°	±10	Above specimen and conditions; cycle 3.
21	34	1957	83.2	~0°	±10	Different sample, same as curve 17 specimen and conditions except surface roughness ~15 microinches rms.
22	34	1957	461-750	~0°	±10	Different sample, same as above specimen and conditions; increasing temp, cycle 1.
23	34	1957	680-922	~0°	±10	Above specimen and conditions; cycle 2.
24	34	1957	680-1155	~0°	±10	Above specimen and conditions; cycle 3.
25	15	1947	373	~0°		Alleghany alloy No. 66; nominal composition: 16-18 Cr, 0.12 max C, Fe balance; polished.

SPECIFICATION TABLE NO. 354 (continued)

Curve No.	Ref. No.	Year	Temperature Range, K	Geometry θ'	Reported Error, %	Composition (weight percent), Specifications and Remarks
26	40	1962	408-1061	~0°		Cobalt alloy N-155 (surface N-1); nominal composition: 21 Cr, 20 Co, 20 Ni, 3 Mo, 3 W, 1.5 Mn, 1 Nb, 0.5 Si, 0.15 C, 0.15 N, Fe balance; as received; increasing temp.
27	40	1962	367	~0°		Different sample, same as above specimen and conditions except highly polished; mirror finish; oxide formation at 873 K for 3 hrs.
28	40	1962	408-655	~0°		Above specimen and conditions; decreasing temp.
29	40	1962	447-1061	~0°		Different sample, same as curve 26 specimen and conditions except surface N-2; increasing temp.
30	40	1962	1033-447	~0°		Above specimen and conditions; decreasing temp.
31	40	1962	411-1084	~0°		Different sample, same as above specimen and conditions except heat treated; same results for increasing and decreasing temp.

DATA TABLE NO. 354 NORMAL TOTAL EMITTANCE OF [IRON + CHROMIUM + ΣX_i] ALLOYS

[Temperature, T, K; Emittance, ϵ]

T	ϵ
CURVE 1	
83.2	0.058
CURVE 2	
461	0.090
628	0.125
803	0.130
961	0.155
1055	0.235
1139	0.315
CURVE 3	
533	0.150
850	0.125
CURVE 4	
522	0.140
CURVE 5	
83.2	0.033
CURVE 6	
603	0.105
758	0.115
966	0.170
1072	0.225
1161	0.345
CURVE 7	
772	0.225
1050	0.225
1161	0.375
CURVE 8	
800	0.270
1003	0.325
CURVE 9	
83.2	0.041
CURVE 10	
544	0.090
819	0.145
989	0.160
1105	0.215
1233	0.275
CURVE 11*	
811	0.190
989	0.240
CURVE 12	
911	0.230
1108	0.280
CURVE 13	
83.2	0.072
CURVE 14	
680	0.165
905	0.235
1025	0.265
1130	0.315
1205	0.335
CURVE 15	
769	0.220
911	0.265
CURVE 16	
611	0.150
1289	0.280
CURVE 17	
83.2	0.158
CURVE 18	
555	0.115
833	0.145
850	0.140
966	0.210
1061	0.325
1161	0.400
CURVE 19	
678	0.220
883	0.300
CURVE 20	
1050	0.385
CURVE 21	
83.2	0.167
CURVE 22*	
461	0.165
750	0.185
CURVE 23	
680	0.260
922	0.350
CURVE 24*	
680	0.175
836	0.215
994	0.220
1155	0.420
CURVE 25	
373	0.11
CURVE 26*	
408	0.165
444	0.150
CURVE 26 (cont.)*	
683	0.175
891	0.155
1061	0.550
CURVE 27	
367	0.16
CURVE 28	
408	0.225
655	0.338
CURVE 29*	
447	0.190
722	0.220
855	0.290
916	0.325
964	0.400
1061	0.600
CURVE 30	
1033	0.566
683	0.410
447	0.300
CURVE 31	
411	0.445
416	0.410
677	0.510
844	0.570
1084	0.628

* Not shown on plot

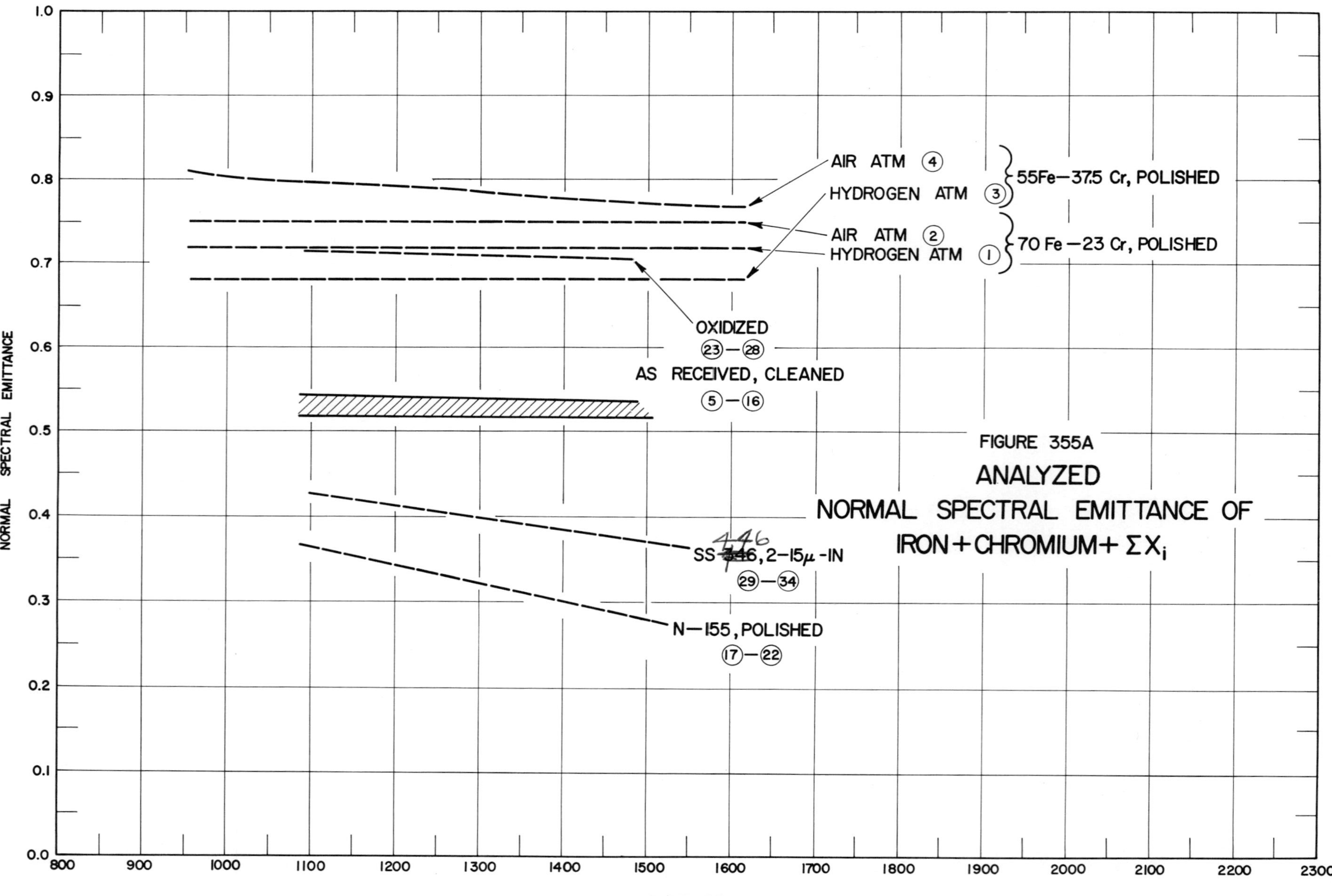

AIR ATM (4)
HYDROGEN ATM (3)
55Fe—37.5 Cr, POLISHED
AIR ATM (2)
HYDROGEN ATM (1)
70 Fe—23 Cr, POLISHED
OXIDIZED
(23)—(28)
AS RECEIVED, CLEANED
(5)—(16)
FIGURE 355A
ANALYZED
NORMAL SPECTRAL EMITTANCE OF
IRON + CHROMIUM + ΣX_i
SS 446, 2—15μ-IN
(29)—(34)
N—155, POLISHED
(17)—(22)
NORMAL SPECTRAL EMITTANCE
1.0
0.9
0.8
0.7
0.6
0.5
0.4
0.3
0.2
0.1
0.0
800
900
1000
1100
1200
1300
1400
1500
1600
1700
1800
1900
2000
2100
2200
2300
TEMPERATURE, K

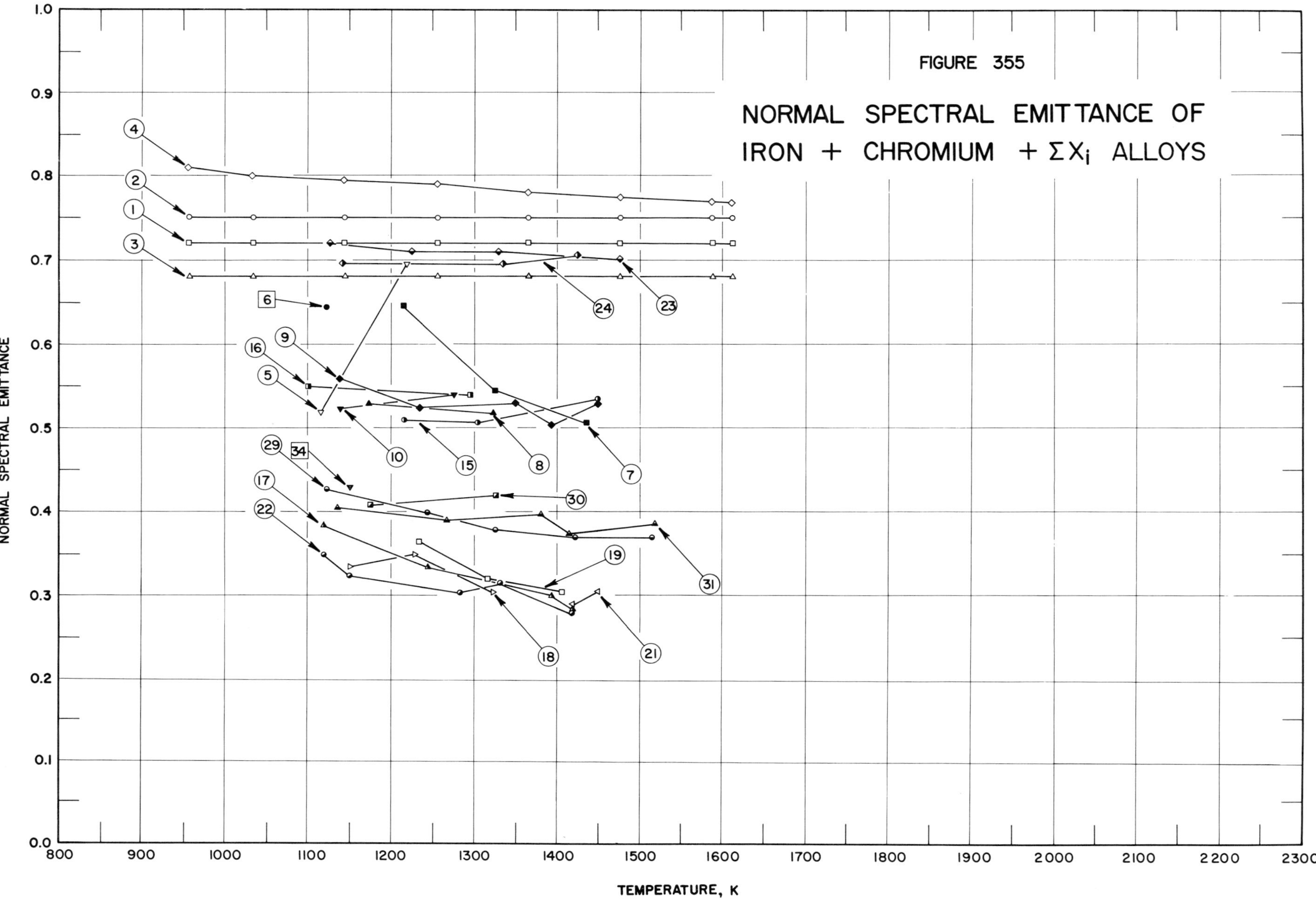
FIGURE 355
NORMAL SPECTRAL EMITTANCE OF
IRON + CHROMIUM + ΣXi ALLOYS
NORMAL SPECTRAL EMITTANCE
TEMPERATURE, K
1.0
0.9
0.8
0.7
0.6
0.5
0.4
0.3
0.2
0.1
0.0
800
900
1000
1100
1200
1300
1400
1500
1600
1700
1800
1900
2000
2100
2200
2300
4
2
1
3
6
24
23
9
16
5
10
15
8
7
29
34
30
17
22
19
31
18
21

SPECIFICATION TABLE NO. 355 NORMAL SPECTRAL EMITTANCE OF [IRON + CHROMIUM + ΣX_i] ALLOYS

Curve No.	Ref. No.	Year	Wavelength μ	Temperature Range, K	Geometry θ'	Reported Error, %	Composition (weight percent), Specifications and Remarks
1	163	1939	0.65	956-1611	~0°		70 Fe, 23 Cr, 5 Al, 2 Co; polished with rouge paper; heated in hydrogen at 1478 K for several hrs; measured in hydrogen; data extracted from smooth curve.
2	163	1939	0.65	956-1611	~0°		Different sample, same as above specimen and conditions except heated in air at 1478 K for several hrs; measured in air.
3	163	1939	0.65	956-1611	~0°		55 Fe, 37.5 Cr, 7.5 Al; polished with rouge paper; heated at 1478 K; measured in hydrogen; data extracted from smooth curve.
4	163	1939	0.65	956-1611	~0°		Different sample, same as above specimen and conditions except oxidized in air at 1478 K; measured in air.
5	34	1957	0.665	1116-1219	~0°		Cobalt alloy N-155; nominal composition: 21 Cr, 20 Co, 20 Ni, 3 Mo, 3 W, 1.5 Mn, 1 Nb, 0.5 Si, 0.15 C, 0.15 N, Fe balance; as received; measured in vacuum (5×10^{-4} mm Hg); increasing temp, cycle 1.
6	34	1957	0.665	1122	~0°		Above specimen and conditions; decreasing temp, cycle 1.
7	34	1957	0.665	1216-1436	~0°		Above specimen and conditions; cycle 2.
8	34	1957	0.665	1322-1172	~0°		Above specimen and conditions; decreasing temp, cycle 2.
9	34	1957	0.665	1139-1450	~0°		Above specimen and conditions; cycle 3.
10	34	1957	0.665	1277-1139	~0°		Above specimen and conditions; decreasing temp, cycle 3.
11	34	1957	0.665	1116-1483	~0°		Different sample, same as curve 5 specimen and conditions except cleaned in liquid detergent; increasing temp, cycle 1.
12	34	1957	0.665	1344-1105	~0°		Above specimen and conditions; decreasing temp, cycle 1.
13	34	1957	0.665	1227-1477	~0°		Above specimen and conditions; cycle 2.
14	34	1957	0.665	1372-1164	~0°		Above specimen and conditions; decreasing temp, cycle 2.
15	34	1957	0.665	1216-1450	~0°		Above specimen and conditions; cycle 3.
16	34	1957	0.665	1294-1100	~0°		Above specimen and conditions; decreasing temp, cycle 3.
17	34	1957	0.665	1119-1419	~0°		Different sample, same as curve 5 specimen and conditions except polished; increasing temp, cycle 1.
18	34	1957	0.665	1322-1150	~0°		Above specimen and conditions; decreasing temp, cycle 1.
19	34	1957	0.665	1233-1405	~0°		Above specimen and conditions; cycle 2.
20	34	1957	0.665	1325-1119	~0°		Above specimen and conditions; decreasing temp, cycle 2.
21	34	1957	0.665	1419-1450	~0°		Above specimen and conditions; cycle 3.
22	34	1957	0.665	1119-1419	~0°		Above specimen and conditions; decreasing temp, cycle 3.
23	34	1957	0.665	1127-1477	~0°		Different sample, same as curve 5 specimen and conditions except oxidized in air at red heat for 30 min; increasing temp, cycle 1.

SPECIFICATION TABLE NO. 355 (continued)

Curve No.	Ref. No.	Year	Wavelength μ	Temperature Range, K	Geometry θ'	Reported Error, %	Composition (weight percent), Specifications and Remarks
24	34	1957	0.665	1141-1425	~0°		Above specimen and conditions; decreasing temp, cycle 1.
25	34	1957	0.665	1152-1475	~0°		Above specimen and conditions; cycle 2.
26	34	1957	0.665	1219-1452	~0°		Above specimen and conditions; decreasing temp, cycle 2.
27	34	1957	0.665	1311	~0°		Above specimen and conditions; cycle 3.
28	34	1957	0.665	1143-1289	~0°		Above specimen and conditions; decreasing temp, cycle 3.
29	34	1957	0.665	1122-1516	~0°		Stainless steel type 446; nominal composition: 23.00-27.00 Cr, 1.5 max Mn, 1.00 max Si, 0.25 max N, 0.20 max C, Fe balance; same results obtained for two samples with surface roughnesses 2 and 15 microinches rms; measured in vacuum (5×10^{-4} mm Hg); increasing temp, cycle 1.
30	34	1957	0.665	1327-1175	~0°		Above specimen and conditions; decreasing temp, cycle 1.
31	34	1957	0.665	1136-1519	~0°		Above specimen and conditions; cycle 2.
32	34	1957	0.665	1283-1139	~0°		Above specimen and conditions; decreasing temp, cycle 2.
33	34	1957	0.665	1216-1505	~0°		Above specimen and conditions; cycle 3.
34	34	1957	0.665	1150	~0°		Above specimen and conditions; decreasing temp, cycle 3.

DATA TABLE NO. 355 NORMAL SPECTRAL EMITTANCE OF [IRON + CHROMIUM + ΣX_i] ALLOYS

[Temperature, T, K; Emittance, ∈; Wavelength, λ, μ]

T	∈
CURVE 1 λ = 0.65	
956	0.72
1033	0.72
1144	0.72
1256	0.72
1367	0.72
1478	0.72
1589	0.72
1611	0.72
CURVE 2 λ = 0.65	
956	0.75
1033	0.75
1144	0.75
1256	0.75
1367	0.75
1478	0.75
1589	0.75
1611	0.75
CURVE 3 λ = 0.65	
956	0.68
1033	0.68
1144	0.68
1256	0.68
1367	0.68
1478	0.68
1589	0.68
1611	0.68
CURVE 4 λ = 0.65	
956	0.810
1033	0.800
1144	0.795
1256	0.790
1367	0.780
1478	0.775
1589	0.770
1611	0.769
CURVE 5 λ = 0.665	
1116	0.520
1219	0.695
CURVE 6 λ = 0.665	
1122	0.645
CURVE 7 λ = 0.665	
1216	0.645
1327	0.545
1436	0.508
CURVE 8 λ = 0.665	
1322	0.520
1172	0.530
CURVE 9 λ = 0.665	
1139	0.560
1233	0.525
1350	0.530
1394	0.505
1450	0.530
CURVE 10 λ = 0.665	
1277	0.540
1139	0.525
CURVE 11* λ = 0.665	
1116	0.540
1275	0.550
1377	0.505
1483	0.540
CURVE 12* λ = 0.665	
1344	0.520
1205	0.540
1105	0.515
CURVE 13* λ = 0.665	
1227	0.535
1361	0.510
1477	0.515
CURVE 14* λ = 0.665	
1372	0.540
1164	0.550
CURVE 15 λ = 0.665	
1216	0.510
1305	0.508
1450	0.535
CURVE 16 λ = 0.665	
1294	0.540
1100	0.550
CURVE 17 λ = 0.665	
1119	0.385
1244	0.335
1394	0.300
1419	0.285
CURVE 18 λ = 0.665	
1322	0.303
1227	0.350
1150	0.335
CURVE 19 λ = 0.665	
1233	0.365
1316	0.320
1405	0.305
CURVE 20* λ = 0.665	
1325	0.315
1119	0.365
CURVE 21 λ = 0.665	
1419	0.290
1450	0.305
CURVE 22 λ = 0.665	
1119	0.350
1150	0.325
1283	0.303
1333	0.315
1419	0.280
CURVE 23 λ = 0.665	
1127	0.720
1225	0.707
1330	0.710
1477	0.700
CURVE 24 λ = 0.665	
1141	0.696
1336	0.695
1425	0.706
CURVE 25* λ = 0.665	
1152	0.725
CURVE 25 (cont.)*	
1327	0.706
1427	0.713
1455	0.695
1475	0.715
CURVE 26* λ = 0.665	
1219	0.705
1308	0.720
1452	0.725
CURVE 27* λ = 0.665	
1311	0.693
CURVE 28* λ = 0.665	
1143	0.712
1289	0.710
CURVE 29 λ = 0.665	
1122	0.428
1244	0.399
1327	0.379
1422	0.370
1516	0.370
CURVE 30 λ = 0.665	
1327	0.420
1175	0.408
CURVE 31 λ = 0.665	
1136	0.405
1266	0.390
1380	0.398
1416	0.375
1519	0.386
CURVE 32* λ = 0.665	
1283	0.410
1139	0.415
CURVE 33* λ = 0.665	
1216	0.420
1505	0.355
CURVE 34 λ = 0.665	
1150	0.430

* Not shown on plot

FIGURE 356 A

NORMAL SPECTRAL EMITTANCE OF IRON+CHROMIUM+ΣX_i

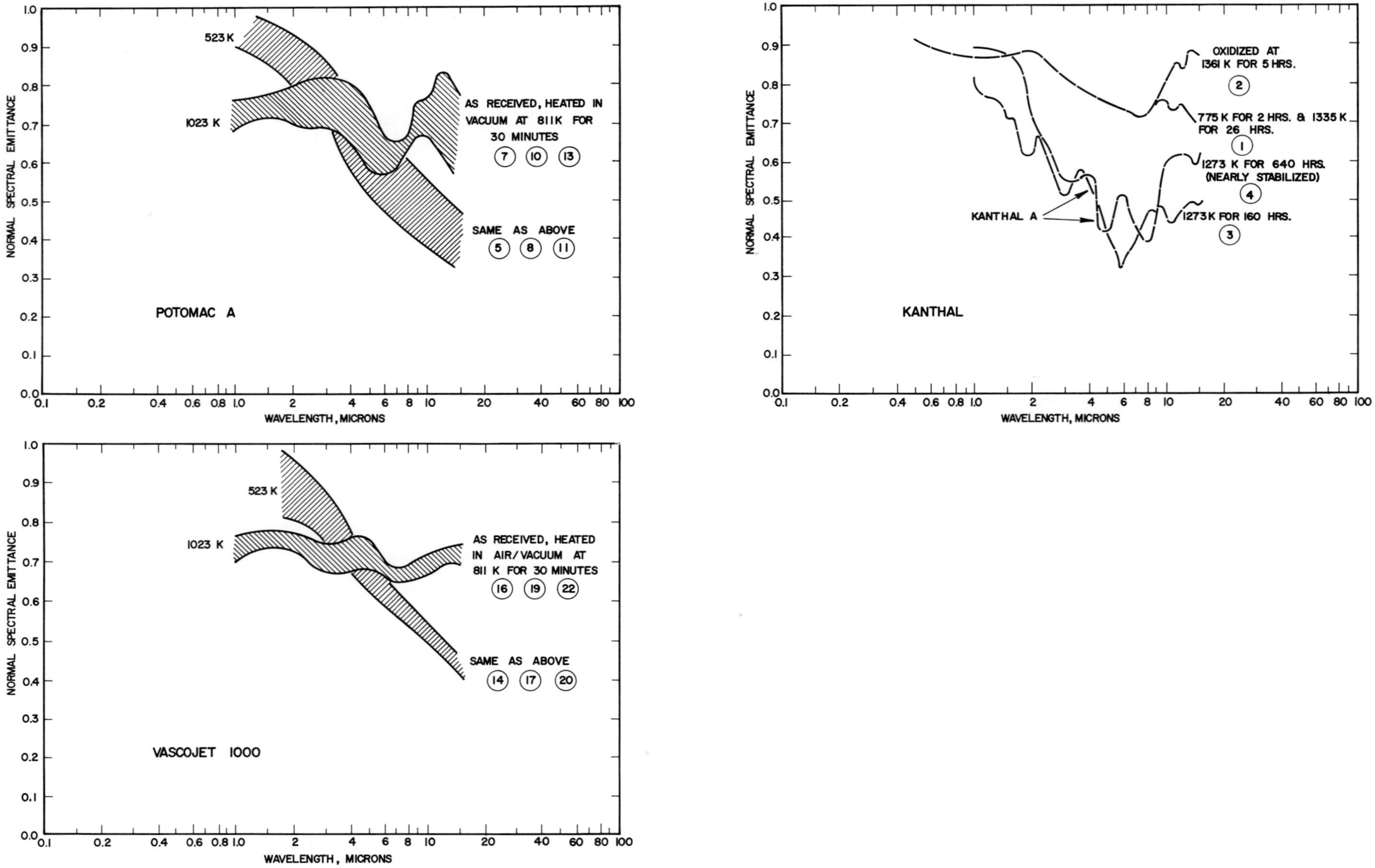

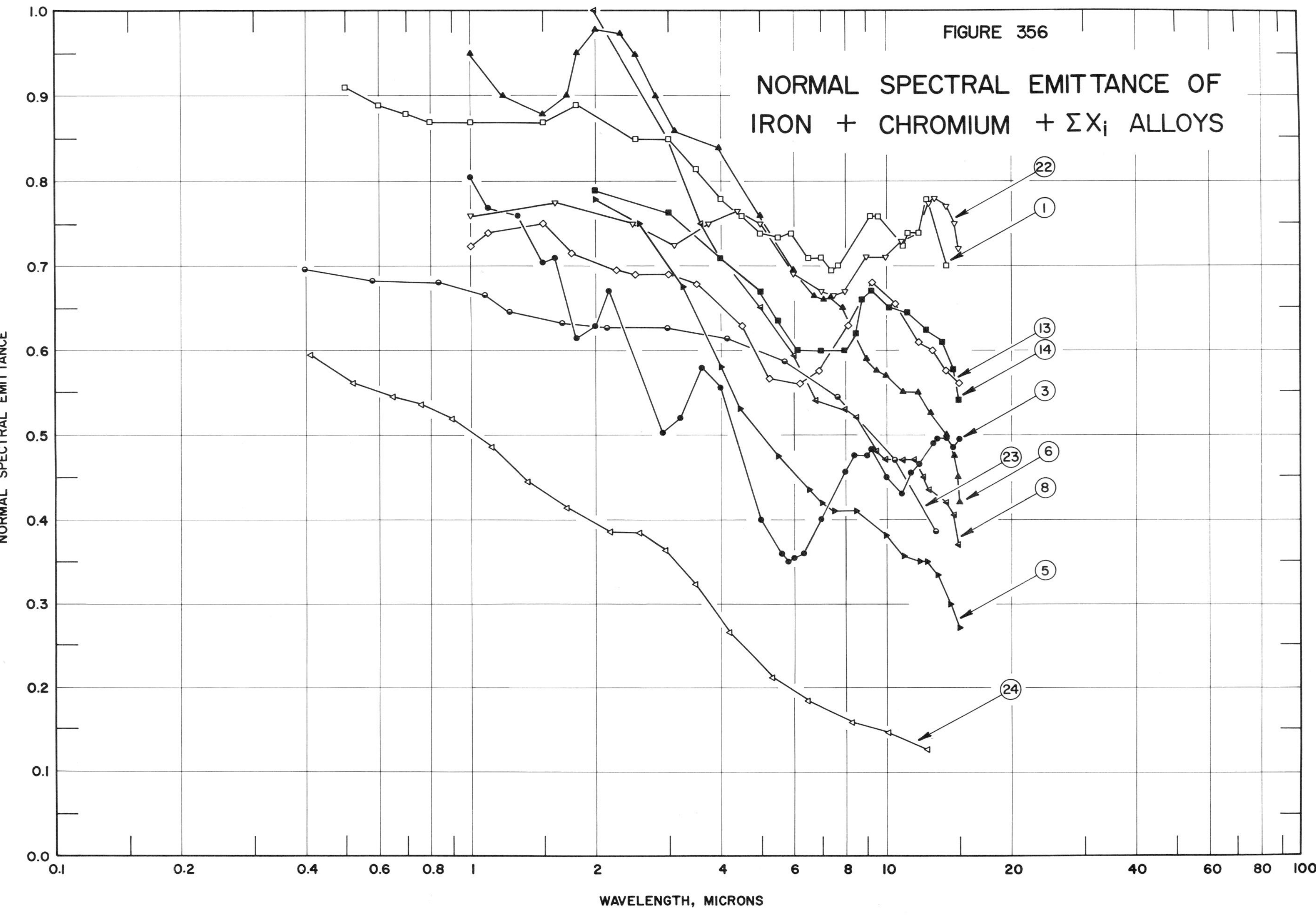
FIGURE 356
NORMAL SPECTRAL EMITTANCE OF
IRON + CHROMIUM + Σ X_i ALLOYS
22
1
13
14
3
23
6
8
5
24
NORMAL SPECTRAL EMITTANCE
1.0
0.9
0.8
0.7
0.6
0.5
0.4
0.3
0.2
0.1
0.0
0.1
0.2
0.4
0.6
0.8
1
2
4
6
8
10
20
40
60
80
100
WAVELENGTH, MICRONS

SPECIFICATION TABLE NO. 356 NORMAL SPECTRAL EMITTANCE OF [IRON + CHROMIUM + ΣX_i] ALLOYS

Curve No.	Ref. No.	Year	Temperature K	Wavelength Range, μ	Geometry θ'	Reported Error, %	Composition (weight percent), Specifications and Remarks
1	30	1963	1339	0.5-14.0	~0°		Kanthal; oxidized; heated at 775 K for 2 hrs and at 1335 K for 26 hrs; measured in argon; data extracted from smooth curve.
2	30	1963	1363	1.4-14.0	~0°		Above specimen and conditions; heated at 1361 K for 5 hrs.
3	172	1961	1400	1.00-15.00	~0°		Kanthal A; nominal composition: 23.4 Cr, 6.2 Al, 1.9 Co, 0.06 C, Fe balance; mechanically polished; oxidized in air for 160 hrs at 1273 K; measured in nitrogen; data extracted from smooth curve.
4	172	1961	1400	1.00-15.00	~0°		Different sample, same as above specimen and conditions except oxidized in air for 640 hrs at 1273 K.
5	86	1961	523	2.00-15.00	~0°	±5	Potomac A; nominal composition: 5.0 Cr, 1.3 Mo, 0.90 Si, 0.5 V, 0.40 C, 0.30 Mn, Fe balance; as received; data extracted from smooth curve.
6	86	1961	773	1.00-15.00	~0°	±5	Different sample, same as curve 5 specimen and conditions.
7	86	1961	1023	1.00-15.00	~0°	±5	Different sample, same as curve 5 specimen and conditions.
8	86	1961	523	2.00-15.00	~0°	±5	Different sample, same as curve 5 specimen and conditions.
9	86	1961	773	1.00-15.00	~0°	±5	Different sample, same as curve 8 specimen and conditions.
10	86	1961	1023	1.00-15.00	~0°	±5	Different sample, same as curve 8 specimen and conditions.
11	86	1961	523	2.00-15.00	~0°	±5	Different sample, same as curve 5 specimen and conditions except heated in vacuum (3.6×10^{-5} mm Hg) at 811 K for 30 min.
12	86	1961	773	1.00-15.00	~0°	±5	Different sample, same as curve 11 specimen and conditions.
13	86	1961	1023	1.00-15.00	~0°	±5	Different sample, same as curve 11 specimen and conditions.
14	86	1961	523	2.00-15.00	~0°	±5	Vascojet 1000; nominal composition: 5 Cr, 1.3 Mo, 0.5 V, 0.40 C, Fe balance; as received; data extracted from smooth curve.
15	86	1961	773	1.00-15.00	~0°	±5	Different sample, same as curve 14 specimen and conditions.
16	86	1961	1023	1.00-15.00	~0°	±5	Different sample, same as curve 14 specimen and conditions.
17	86	1961	523	2.00-15.00	~0°	±5	Different sample, same as curve 14 specimen and conditions except heated in air at 811 K for 30 min.
18	86	1961	773	1.00-15.00	~0°	±5	Different sample, same as curve 17 specimen and conditions.
19	86	1961	1023	1.00-15.00	~0°	±5	Different sample, same as curve 17 specimen and conditions.
20	86	1961	523	2.00-15.00	~0°	±5	Different sample, same as curve 14 specimen and conditions except heated in vacuum (3.6×10^{-5} mm Hg) at 811 K for 30 min.
21	86	1961	773	1.00-15.00	~0°	±5	Different sample, same as curve 20 specimen and conditions.
22	86	1961	1023	1.00-15.00	~0°	±5	Different sample, same as curve 20 specimen and conditions.

SPECIFICATION TABLE NO. 356 (continued)

Curve No.	Ref. No.	Year	Temperature K	Wavelength Range, μ	Geometry θ'	Reported Error, %	Composition (weight percent), Specifications and Remarks
23	280	1959	~298	0.40-13.20	~0°		Stainless Steel AISI 430; nominal composition: 14.00-18.00 Cr, 1.00 max Mn, 1.00 max Si, 0.12 max C, Fe balance; sand blasted; data extracted from smooth curve.
24	280	1959	~298	0.41-12.74	~0°		Above specimen and conditions except plain surface.

DATA TABLE NO. 356 NORMAL SPECTRAL EMITTANCE OF [IRON + CHROMIUM + ΣX_i] ALLOYS

[Wavelength, λ, μ; Emittance, ϵ; Temperature, T, K]

CURVE 1, T = 1339

λ	ϵ
0.5	0.910
0.6	0.890
0.7	0.880
0.8	0.870
1.0	0.870
1.5	0.870
1.8	0.890
2.5	0.850
3.0	0.850
3.5	0.815
4.0	0.780
4.5	0.760
5.0	0.740
5.5	0.735
5.9	0.740
6.5	0.710
7.0	0.710
7.4	0.695
7.7	0.700
9.2	0.760
9.6	0.760
11.0	0.725
11.2	0.740
12.0	0.740
12.6	0.780
14.0	0.700

CURVE 2*, T = 1363

λ	ϵ
1.4	0.850
1.7	0.850
2.0	0.870
2.5	0.840
3.0	0.790
3.5	0.760
4.0	0.760
5.0	0.740
5.5	0.750
6.0	0.750
6.4	0.730
7.0	0.730
7.8	0.730

CURVE 2 (cont.)*

λ	ϵ
9.0	0.760
10.0	0.820
11.0	0.860
11.8	0.840
12.2	0.840
13.0	0.890
14.0	0.850

CURVE 3, T = 1400

λ	ϵ
1.00	0.805
1.10	0.770
1.30	0.760
1.50	0.703
1.60	0.710
1.80	0.615
2.00	0.630
2.15	0.670
2.90	0.502
3.20	0.520
3.60	0.580
4.00	0.555
5.00	0.400
5.60	0.360
5.80	0.350
6.00	0.353
6.30	0.360
7.00	0.400
8.00	0.455
8.40	0.475
9.00	0.475
9.20	0.482
10.00	0.450
10.90	0.430
11.40	0.453
12.00	0.465
13.00	0.490
13.20	0.495
14.00	0.495
14.50	0.485
15.00	0.495

CURVE 4*, T = 1400

λ	ϵ
1.00	0.892
1.50	0.765
1.70	0.750
1.80	0.710
2.00	0.700
2.05	0.695
2.40	0.643
2.65	0.660
3.00	0.550
3.25	0.530
3.40	0.550
3.70	0.610
4.00	0.560
4.40	0.450
4.50	0.420
4.80	0.413
5.00	0.418
5.30	0.440
5.70	0.500
6.00	0.514
6.20	0.510
6.80	0.430
7.10	0.400
7.40	0.390
7.80	0.388
8.20	0.406
8.40	0.408
8.60	0.430
9.00	0.470
10.00	0.590
11.00	0.610
12.00	0.620
13.00	0.620
14.00	0.595
15.00	0.620

CURVE 5, T = 523

λ	ϵ
2.00	0.780
2.55	0.750
3.25	0.675
4.00	0.580

CURVE 5 (cont.)

λ	ϵ
4.45	0.530
5.50	0.475
6.50	0.435
7.00	0.420
7.50	0.410
8.50	0.410
10.00	0.380
11.00	0.357
12.00	0.350
12.50	0.350
13.25	0.335
14.25	0.300
15.00	0.270

CURVE 6, T = 773

λ	ϵ
1.00	0.950
1.20	0.900
1.50	0.880
1.70	0.900
1.80	0.950
2.00	0.980
2.30	0.975
2.50	0.950
2.80	0.900
3.10	0.860
3.95	0.840
5.00	0.760
6.00	0.693
6.75	0.665
7.10	0.660
7.40	0.665
7.90	0.650
9.00	0.590
9.50	0.575
10.00	0.570
11.00	0.550
12.00	0.550
12.80	0.525
14.00	0.500
14.60	0.475
14.90	0.450
15.00	0.420

CURVE 7*, T = 1023

λ	ϵ
1.00	0.640
1.15	0.700
1.50	0.740
1.75	0.730
2.00	0.700
3.00	0.740
3.90	0.700
5.00	0.610
5.50	0.575
6.00	0.560
7.20	0.600
8.00	0.640
8.50	0.695
9.00	0.730
9.50	0.720
10.35	0.720
10.80	0.735
11.50	0.725
12.50	0.690
13.40	0.650
14.30	0.600
15.00	0.545

CURVE 8, T = 523

λ	ϵ
2.00	1.000
3.00	0.850*
3.60	0.750
4.00	0.710*
5.00	0.650
6.00	0.595
6.80	0.540
8.00	0.530
8.50	0.520
9.50	0.480
10.00	0.470
11.00	0.470
11.60	0.470
12.25	0.450
12.75	0.435
14.00	0.420
14.50	0.405
15.00	0.370

CURVE 9*, T = 773

λ	ϵ
1.00	0.785
2.00	0.720
3.00	0.695
3.50	0.695
4.25	0.700
4.75	0.685
5.50	0.602
6.00	0.565
6.50	0.530
7.50	0.490
8.50	0.445
8.85	0.425
10.00	0.390
11.40	0.350
12.10	0.350
12.70	0.330
14.50	0.330
15.00	0.340

CURVE 10*, T = 1023

λ	ϵ
1.00	0.760
1.50	0.765
2.25	0.800
2.85	0.820
4.00	0.810
5.00	0.750
5.75	0.695
6.25	0.665
7.15	0.650
7.80	0.655
8.20	0.680
8.75	0.740
9.00	0.760
10.00	0.760
10.50	0.770
11.50	0.825
12.25	0.840
13.35	0.820
15.00	0.780

CURVE 11*, T = 523

λ	ϵ
2.00	0.770
2.25	0.700
2.55	0.650
3.45	0.550
4.00	0.500
4.50	0.475
4.85	0.475
5.[illegible]0	0.440
6.00	0.430
7.00	0.420
8.00	0.400
9.00	0.380
9.90	0.360
11.00	0.350
11.50	0.335
13.00	0.320
14.00	0.310
14.50	0.300
15.00	0.260

CURVE 12*, T = 773

λ	ϵ
1.00	0.990
1.40	0.900
1.70	0.800
2.00	0.705
2.30	0.650
2.75	0.610
3.25	0.585
4.00	0.585
5.00	0.555
6.00	0.510
6.50	0.505
7.10	0.510
8.50	0.495
9.00	0.490
9.50	0.475
10.00	0.450
11.00	0.450
12.10	0.450
12.60	0.440
14.00	0.440
15.00	0.440

* Not shown on plot

DATA TABLE NO. 356 (continued)

λ	ε
CURVE 13 T = 1023	
1.00	0.725
1.10	0.740
1.50	0.750
1.75	0.715
2.25	0.695
2.50	0.690
3.00	0.690
3.50	0.680
4.50	0.630
5.25	0.565
6.20	0.560
6.90	0.575
8.15	0.630
9.30	0.680
10.50	0.655
12.00	0.610
13.00	0.600
14.00	0.575
15.00	0.560
CURVE 14 T = 523	
2.00	0.790
3.00	0.765
4.00	0.710
5.00	0.670
5.50	0.635
6.10	0.600
7.00	0.600
8.00	0.600
8.50	0.620
8.80	0.660
9.25	0.670
10.25	0.650
11.25	0.645
12.50	0.625
13.75	0.610
14.50	0.577
15.00	0.540
CURVE 15* T = 773	
1.00	1.000
1.20	0.925
1.60	0.890
2.00	0.910
2.50	0.910
3.50	0.885
4.10	0.860
4.85	0.780
5.75	0.715
6.75	0.655
7.00	0.650
8.00	0.650
8.70	0.615
9.00	0.590
10.00	0.580
11.10	0.545
12.00	0.540
12.55	0.520
13.75	0.505
14.40	0.490
14.80	0.460
15.00	0.425
CURVE 16* T = 1023	
1.00	0.690
1.35	0.750
1.60	0.760
2.10	0.735
2.70	0.735
3.00	0.750
4.25	0.732
5.40	0.705
6.00	0.670
7.25	0.670
8.15	0.660
9.00	0.675
9.50	0.675
10.10	0.670
11.00	0.700
12.00	0.765
13.00	0.800
13.50	0.805
14.00	0.800
CURVE 16 (cont.)*	
14.65	0.750
14.90	0.700
15.00	0.660
CURVE 17* T = 523	
2.00	1.000
3.00	0.890
4.00	0.775
5.00	0.715
5.50	0.680
6.00	0.630
6.60	0.600
7.75	0.575
9.00	0.550
10.00	0.500
10.75	0.500
12.00	0.470
13.25	0.430
14.00	0.430
14.50	0.425
15.00	0.400
CURVE 18* T = 773	
1.00	0.810
1.75	0.700
2.25	0.640
2.65	0.630
3.10	0.650
3.50	0.705
4.00	0.740
4.50	0.745
5.40	0.725
6.10	0.675
6.75	0.660
7.50	0.650
8.50	0.615
9.50	0.590
10.00	0.565
11.00	0.550
12.50	0.535
13.25	0.530
14.50	0.530
15.00	0.520
CURVE 19* T = 1023	
1.00	0.745
1.25	0.750
2.00	0.725
3.00	0.660
4.00	0.670
5.00	0.690
5.50	0.690
6.75	0.650
7.75	0.600
8.10	0.590
9.25	0.605
9.90	0.605
11.00	0.675
12.00	0.740
13.00	0.770
13.30	0.770
14.00	0.750
15.00	0.700
CURVE 20* T = 523	
2.00	0.950
2.50	0.815
3.00	0.720
3.25	0.690
3.50	0.675
4.10	0.655
4.60	0.625
5.50	0.590
6.00	0.580
7.00	0.570
8.00	0.565
9.00	0.550
10.00	0.530
10.50	0.515
11.50	0.500
13.00	0.470
14.00	0.450
14.60	0.430
15.00	0.400
CURVE 21* T = 773	
1.00	0.890
1.50	0.850
2.00	0.750
2.25	0.700
2.75	0.660
3.55	0.665
5.00	0.680
5.50	0.665
6.00	0.660
7.25	0.700
8.25	0.735
8.75	0.728
9.50	0.703
10.50	0.720
11.50	0.700
12.10	0.690
13.25	0.635
14.25	0.590
14.75	0.580
15.00	0.580
CURVE 22 T = 1023	
1.00	0.760
1.60	0.775
2.45	0.750
3.10	0.725
3.75	0.750
4.40	0.765
5.00	0.750
6.00	0.690
7.00	0.670
7.50	0.665
8.00	0.670
9.00	0.710
10.00	0.710
10.90	0.730
12.10	0.740*
12.75	0.775
13.15	0.780
14.00	0.770
14.60	0.750
15.00	0.720
CURVE 23* T = ~298	
0.40	0.697
0.58	0.682
0.84	0.681
1.08	0.667
1.24	0.647
1.66	0.632
2.12	0.628
2.99	0.628
4.16	0.613
5.74	0.587
7.68	0.544
10.45	0.469
13.20	0.386
CURVE 24* T = ~298	
0.41	0.595
0.52	0.560
0.65	0.543
0.76	0.536
0.90	0.519
1.12	0.486
1.37	0.444
1.70	0.413
2.16	0.387
2.55	0.383
2.93	0.364
3.46	0.323
4.19	0.266
5.30	0.212
6.43	0.183
8.28	0.158
10.12	0.147
12.74	0.127

* Not shown on plot

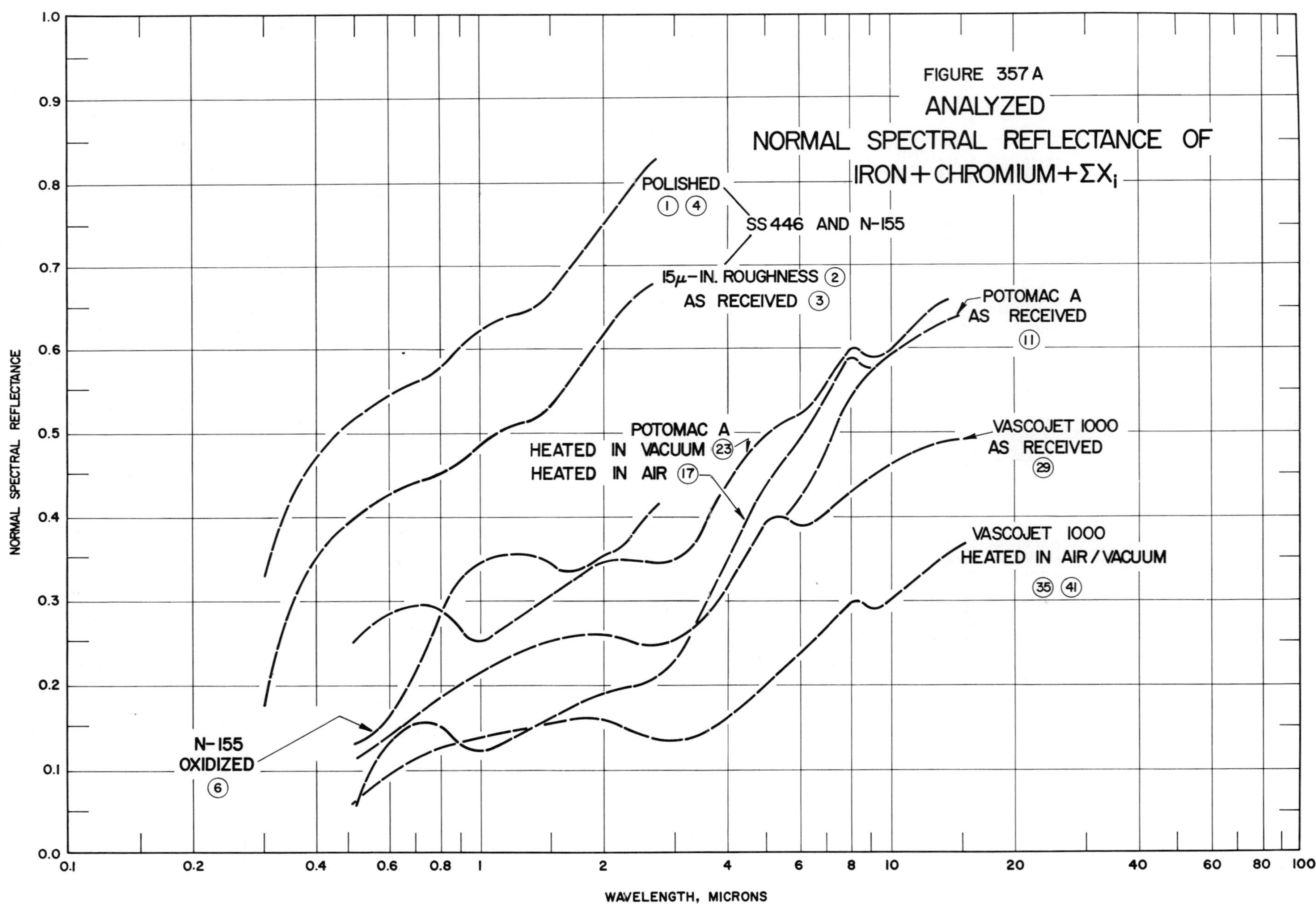

FIGURE 357 A
ANALYZED
NORMAL SPECTRAL REFLECTANCE OF
IRON + CHROMIUM + ΣXi
POLISHED
1 4
SS 446 AND N-155
15μ-IN. ROUGHNESS 2
AS RECEIVED 3
POTOMAC A
AS RECEIVED
11
POTOMAC A
HEATED IN VACUUM 23
HEATED IN AIR 17
VASCOJET 1000
AS RECEIVED
29
VASCOJET 1000
HEATED IN AIR / VACUUM
35 41
N-155
OXIDIZED
6
NORMAL SPECTRAL REFLECTANCE
1.0
0.9
0.8
0.7
0.6
0.5
0.4
0.3
0.2
0.1
0.0
0.1
0.2
0.4
0.6
0.8
1
2
4
6
8
10
20
40
60
80
100
WAVELENGTH, MICRONS

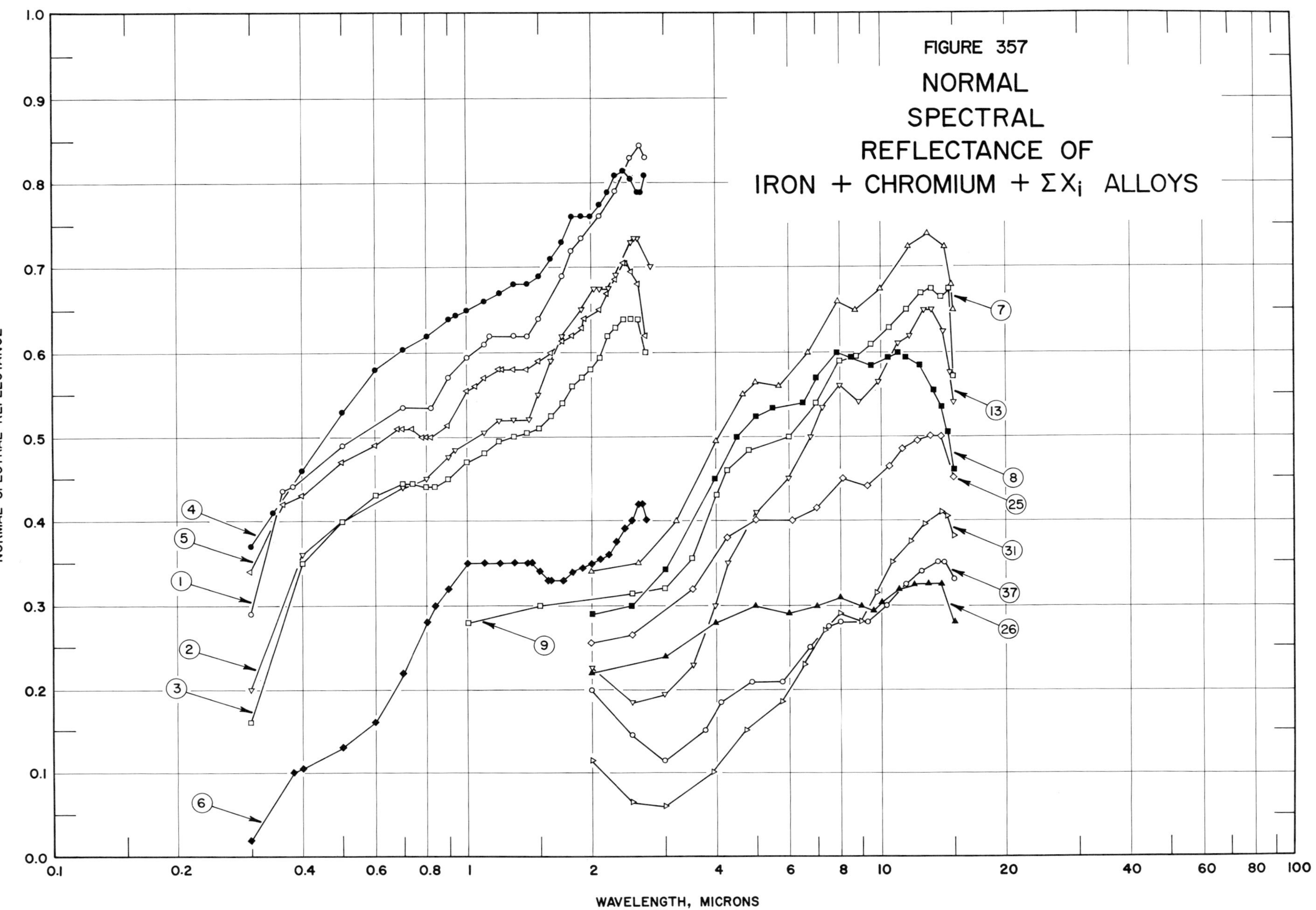
FIGURE 357
NORMAL
SPECTRAL
REFLECTANCE OF
IRON + CHROMIUM + ΣXi ALLOYS
NORMAL SPECTRAL REFLECTANCE
WAVELENGTH, MICRONS
0.0
0.1
0.2
0.3
0.4
0.5
0.6
0.7
0.8
0.9
1.0
0.1
0.2
0.4
0.6
0.8
1
2
4
6
8
10
20
40
60
80
100
1
2
3
4
5
6
7
8
9
13
25
26
31
37

SPECIFICATION TABLE NO. 357 NORMAL SPECTRAL REFLECTANCE OF [IRON + CHROMIUM + ΣX_i] ALLOYS

Curve No.	Ref. No.	Year	Temperature K	Wavelength Range, μ	Geometry θ	θ'	ω'	Reported Error, %	Composition (weight percent), Specifications and Remarks
1	34	1957	298	0.30-2.70	9°		2π	±4	Stainless steel type 446; nominal composition: 23.00-27.00 Cr, 1.5 max Mn, 1.00 max Si, 0.25 max N, 0.20 max C, Fe balance; surface finish ~2 microinches rms; data extracted from a smooth curve.
2	34	1957	298	0.30-2.70	9°		2π	±4	Different sample, same as above specimen and conditions; surface roughness ~15 microinches rms.
3	34	1957	298	0.30-2.70	9°		2π	±4	Cobalt alloy N-155; nominal composition: 21 Cr, 20 Co, 20 Ni, 3 Mo, 3 W, 1.5 Mn, 1 Nb, 0.5 Si, 0.15 C, 0.15 N, Fe balance; as received; data extracted from smooth curve.
4	34	1957	298	0.30-2.70	9°		2π	±4	Different sample, same as curve 3 specimen and conditions except polished.
5	34	1957	298	0.30-2.70	9°		2π	±4	Different sample, same as curve 3 specimen and conditions except cleaned with liquid detergent.
6	34	1957	298	0.30-2.70	9°		2π	±4	Different sample, same as curve 3 specimen and conditions except oxidized in air at red heat for 30 min.
7	86	1961	~322	2.00-15.00	~0°		2π	<2	Potomac A; nominal composition: 5.0 Cr, 1.3 Mo, 0.90 Si, 0.5 V, 0.40 C, 0.30 Mn, Fe balance; as received; data extracted from smooth curve; hohlraum at 523 K; converted from R $(2\pi, 0^\circ)$.
8	86	1961	~322	2.00-15.00	~0°		2π	<2	Above specimen and conditions; diffuse component only.
9	86	1961	~322	1.00-15.00	~0°		2π	<2	Different sample, same as curve 7 specimen and conditions except hohlraum at 773 K.
10	86	1961	~322	1.00-15.00	~0°		2π	<2	Above specimen and conditions; diffuse component only.
11	86	1961	~322	0.50-15.00	~0°		2π	<2	Different sample, same as curve 7 specimen and conditions except hohlraum at 1273 K.
12	86	1961	~322	0.75-15.00	~0°		2π	<2	Above specimen and conditions; diffuse component only.
13	86	1961	~322	2.00-15.00	~0°		2π	<2	Different sample, same as curve 7 specimen and conditions except heated in air at 811 K for 30 min.
14	86	1961	~322	2.00-15.00	~0°		2π	<2	Above specimen and conditions; diffuse component only.
15	86	1961	~322	1.00-15.00	~0°		2π	<2	Different sample, same as curve 13 specimen and conditions except hohlraum at 773 K.
16	86	1961	~322	1.00-15.00	~0°		2π	<2	Above specimen and conditions; diffuse component only.
17	86	1961	~322	0.50-15.00	~0°		2π	<2	Different sample, same as curve 13 specimen and conditions except hohlraum at 1273 K.
18	86	1961	~322	0.50-15.00	~0°		2π	<2	Above specimen and conditions; diffuse component only.
19	86	1961	~322	2.00-15.00	~0°		2π	<2	Different sample, same as curve 7 specimen and conditions except heated in vacuum (3.6×10^{-6} mm Hg) at 811 K for 30 min.
20	86	1961	~322	2.00-15.00	~0°		2π	<2	Above specimen and conditions; diffuse component only.
21	86	1961	~322	1.00-15.00	~0°		2π	<2	Different sample, same as curve 19 specimen and conditions except hohlraum at 773 K.
22	86	1961	~322	1.00-15.00	~0°		2π	<2	Above specimen and conditions; diffuse component only.

SPECIFICATION TABLE NO. 357 (continued)

Curve No.	Ref. No.	Year	Temperature K	Wavelength Range, μ	Geometry θ	θ'	ω'	Reported Error, %	Composition (weight percent), Specifications and Remarks
23	86	1961	~322	0.50-15.00	~0°		2π	<2	Different sample, same as curve 19 specimen and conditions except hohlraum 1273 K.
24	86	1961	~322	0.50-15.00	~0°		2π	<2	Above specimen and conditions; diffuse component only.
25	86	1961	~322	2.00-15.00	~0°		2π	<2	Vascojet 1000; nominal composition: 5 Cr, 1.3 Mo, 0.5 V, 0.40 C, Fe balance; as received; data extracted from smooth curve; hohlraum at 523 K; converted from R (2π, 0°).
26	86	1961	~322	2.00-15.00	~0°		2π	<2	Above specimen and conditions; diffuse component only.
27	86	1961	~322	1.00-15.00	~0°		2π	<2	Different sample, same as curve 25 specimen and conditions except hohlraum at 773 K.
28	86	1961	~322	1.00-15.00	~0°		2π	<2	Above specimen and conditions; diffuse component only.
29	86	1961	~322	0.50-15.00	~0°		2π	<2	Different sample, same as curve 25 specimen and conditions except hohlraum at 1273 K.
30	86	1961	~322	0.50-15.00	~0°		2π	<2	Above specimen and conditions; diffuse component only.
31	86	1961	~322	2.00-15.00	~0°		2π	<2	Different sample, same as curve 25 specimen and conditions except heated in air at 811 K for 30 min.
32	86	1961	~322	2.00-15.00	~0°		2π	<2	Above specimen and conditions; diffuse component only.
33	86	1961	~322	1.00-15.00	~0°		2π	<2	Different sample, same as curve 31 specimen and conditions except hohlraum at 773 K.
34	86	1961	~322	1.00-15.00	~0°		2π	<2	Above specimen and conditions; diffuse component only.
35	86	1961	~322	0.50-15.00	~0°		2π	<2	Different sample, same as curve 31 specimen and conditions except hohlraum at 1273 K.
36	86	1961	~322	0.50-15.00	~0°		2π	<2	Above specimen and conditions; diffuse component only.
37	86	1961	~322	2.00-15.00	~0°		2π	<2	Different sample, same as curve 25 specimen and conditions except heated in vacuum (3.6×10^{-5} mm Hg) at 811 K for 30 min.
38	86	1961	~322	2.00-15.00	~0°		2π	<2	Above specimen and conditions; diffuse component only.
39	86	1961	~322	1.00-15.00	~0°		2π	<2	Different sample, same as curve 37 specimen and conditions except hohlraum at 773 K.
40	86	1961	~322	1.00-15.00	~0°		2π	<2	Above specimen and conditions; diffuse component only.
41	86	1961	~322	0.50-15.00	~0°		2π	<2	Different sample, same as curve 37 specimen and conditions except hohlraum at 1273 K.
42	86	1961	~322	0.50-15.00	~0°		2π	<2	Above specimen and conditions; diffuse component only.

DATA TABLE NO. 357 NORMAL SPECTRAL REFLECTANCE OF [IRON + CHROMIUM + ΣX_i] ALLOYS

[Wavelength, λ, μ; Reflectance, ρ; Temperature, T, K]

λ	ρ
CURVE 1	
T = 298	
0.30	0.290
0.36	0.435
0.38	0.440
0.50	0.490
0.70	0.535
0.82	0.535
0.90	0.570
1.00	0.595
1.10	0.610
1.14	0.620
1.30	0.620
1.40	0.620
1.50	0.640
1.70	0.690
1.80	0.720
1.90	0.735
2.10	0.760
2.30	0.790
2.50	0.830
2.62	0.845
2.70	0.830
CURVE 2	
T = 298	
0.30	0.200
0.40	0.360
0.50	0.400*
0.70	0.440
0.80	0.450
0.90	0.475
0.94	0.485
1.10	0.505
1.20	0.520
1.30	0.520
1.42	0.520
1.50	0.550
1.60	0.590
1.70	0.620
1.90	0.650
2.04	0.675
2.10	0.675
2.22	0.675
2.30	0.690
2.50	0.730
CURVE 2 (cont.)	
2.54	0.735
2.60	0.735
2.70	0.700
CURVE 3	
T = 298	
0.30	0.160
0.40	0.350
0.50	0.400
0.60	0.430
0.70	0.445
0.74	0.445
0.80	0.440
0.84	0.440
0.90	0.450
1.00	0.470
1.10	0.480
1.20	0.495
1.30	0.500
1.40	0.505
1.50	0.510
1.60	0.525
1.70	0.540
1.80	0.560
1.90	0.570
2.00	0.580
2.10	0.595
2.20	0.620
2.30	0.630
2.40	0.640
2.50	0.640
2.60	0.640
2.70	0.600
CURVE 4	
T = 298	
0.30	0.370
0.34	0.410
0.40	0.460
0.50	0.530
0.60	0.580
0.70	0.605
0.80	0.620
CURVE 4 (cont.)	
0.90	0.640
0.94	0.645
1.00	0.650
1.10	0.660
1.20	0.670
1.30	0.680
1.40	0.680
1.50	0.690
1.60	0.710
1.70	0.730
1.80	0.760
1.90	0.760
2.00	0.760
2.10	0.775
2.20	0.790
2.30	0.810
2.40	0.815
2.50	0.805
2.60	0.790
2.64	0.790
2.70	0.810
CURVE 5	
T = 298	
0.30	0.340
0.36	0.420
0.40	0.430
0.50	0.470
0.60	0.490
0.68	0.510
0.70	0.510
0.74	0.510
0.78	0.500
0.80	0.500
0.82	0.500
0.90	0.515
1.00	0.550
1.04	0.560
1.10	0.570
1.20	0.580
1.22	0.580
1.30	0.580
1.40	0.580
1.50	0.590
CURVE 5 (cont.)	
1.60	0.600
1.70	0.615
1.80	0.620
1.90	0.630
1.91	0.640
2.10	0.650
2.20	0.670
2.30	0.685
2.40	0.705
2.44	0.705
2.50	0.695
2.60	0.680
2.70	0.620
CURVE 6	
T = 298	
0.30	0.020
0.38	0.100
0.40	0.105
0.50	0.130
0.60	0.160
0.70	0.220
0.80	0.280
0.84	0.300
0.90	0.320
1.00	0.350
1.10	0.350
1.20	0.350
1.30	0.350
1.40	0.350
1.44	0.350
1.50	0.340
1.58	0.330
1.60	0.330
1.70	0.330
1.80	0.340
1.90	0.345
2.00	0.350
2.10	0.355
2.20	0.360
2.30	0.375
2.40	0.390
2.50	0.400
2.60	0.420
2.64	0.420
2.70	0.400
CURVE 7	
T = ~322	
2.00	0.340
2.60	0.350
3.20	0.400
4.00	0.495
4.65	0.550
5.00	0.565
5.75	0.560
6.75	0.600
7.90	0.660
8.75	0.650
10.00	0.675
11.75	0.725
13.00	0.740
14.25	0.725
14.90	0.680
15.00	0.650
CURVE 8	
T = ~322	
2.00	0.290
2.50	0.300
3.00	0.342
4.00	0.450
4.50	0.500
5.00	0.525
5.50	0.535
6.00	0.530
6.50	0.540
7.00	0.570
7.85	0.600
8.50	0.595
9.50	0.585
10.50	0.595
11.00	0.600
11.50	0.595
12.50	0.585
13.50	0.555
14.00	0.535
14.50	0.505
15.00	0.460
CURVE 9	
T = ~322	
1.00	0.280
1.50	0.300
2.50	0.315
3.00	0.320
3.50	0.355
4.00	0.430
4.25	0.460
4.80	0.485
6.00	0.500
7.00	0.540
8.00	0.590
8.75	0.595
9.50	0.610
10.50	0.630
11.50	0.650
12.50	0.670
13.25	0.675
14.00	0.665
14.75	0.625
15.00	0.570
CURVE 10*	
T = ~322	
1.00	0.250
1.50	0.265
2.00	0.300
2.50	0.330
3.00	0.350
3.50	0.385
4.00	0.460
4.50	0.495
5.00	0.515
5.50	0.520
6.00	0.520
6.50	0.530
7.00	0.560
7.50	0.590
8.00	0.605
8.50	0.605
9.00	0.600
9.50	0.595
10.00	0.590
10.50	0.595
CURVE 10 (cont.)*	
11.00	0.605
11.50	0.597
12.50	0.580
13.50	0.560
14.00	0.535
14.50	0.500
15.00	0.450
CURVE 11*	
T = ~322	
0.50	0.100
0.75	0.150
0.90	0.230
1.00	0.230
1.50	0.280
2.25	0.245
3.00	0.225
3.50	0.250
3.80	0.300
4.00	0.350
5.10	0.425
5.75	0.415
6.60	0.450
7.60	0.525
8.05	0.550
9.25	0.575
10.50	0.600
11.50	0.605
13.00	0.640
14.00	0.600
14.70	0.550
15.05	0.500
15.10	0.475
15.00	0.450
CURVE 12*	
T = ~322	
0.75	0.150
0.85	0.260
1.00	0.260
1.25	0.300
1.50	0.325
1.75	0.300
CURVE 12 (cont.)*	
1.95	0.200
2.25	0.170
2.50	0.165
2.75	0.165
3.35	0.200
3.75	0.250
4.00	0.305
4.50	0.340
5.00	0.367
5.50	0.355
5.90	0.350
6.50	0.375
7.00	0.420
7.50	0.455
8.00	0.470
9.00	0.470
10.00	0.470
11.00	0.470
12.00	0.470
12.50	0.470
13.00	0.460
14.00	0.420
14.50	0.395
15.00	0.315
CURVE 13	
T = ~322	
2.00	0.225
2.50	0.185
3.00	0.195
3.50	0.230
3.95	0.300
4.25	0.350
5.00	0.410
6.00	0.450
6.80	0.500
7.25	0.535
8.00	0.560
8.90	0.540
9.90	0.565
11.00	0.610
11.75	0.620
12.80	0.650
13.25	0.650

* Not shown on plot

DATA TABLE NO. 357 (continued)

λ	ρ
CURVE 13 (cont.)	
14.15	0.625
14.75	0.575
15.00	0.540
CURVE 14*	
T = ~322	
2.00	0.250
2.50	0.210
2.75	0.205
3.00	0.210
3.50	0.245
4.00	0.340
4.25	0.385
4.50	0.410
5.00	0.435
5.50	0.460
6.00	0.475
6.50	0.500
7.00	0.540
7.50	0.570
8.00	0.580
8.50	0.570
9.00	0.550
9.50	0.535
10.00	0.560
10.50	0.575
11.00	0.580
12.00	0.580
12.50	0.580
13.00	0.570
14.00	0.540
14.50	0.515
15.00	0.470
CURVE 15*	
T = ~322	
1.00	0.100
1.25	0.100
1.70	0.125
2.20	0.190
3.00	0.225
3.55	0.300
4.10	0.375
4.50	0.425

λ	ρ
CURVE 15 (cont.)*	
5.25	0.475
6.00	0.500
6.90	0.550
7.60	0.600
8.00	0.610
8.80	0.600
9.50	0.600
10.10	0.620
10.70	0.660
11.50	0.680
12.65	0.700
13.75	0.735
14.20	0.740
14.75	0.725
15.00	0.710
CURVE 16*	
T = ~322	
1.00	0.110
1.50	0.130
2.00	0.175
2.50	0.195
3.00	0.200
3.50	0.255
3.75	0.335
4.00	0.375
4.50	0.405
5.00	0.415
6.00	0.460
7.00	0.510
7.50	0.540
8.00	0.560
8.50	0.545
9.00	0.535
9.50	0.540
10.50	0.565
11.00	0.580
12.00	0.580
13.00	0.580
14.00	0.580
14.50	0.575
15.00	0.550

λ	ρ
CURVE 17*	
T = ~322	
0.50	0.060
0.70	0.160
1.00	0.120
1.50	0.140
2.00	0.190
2.55	0.200
3.20	0.250
3.90	0.340
4.75	0.425
5.50	0.465
6.50	0.510
7.25	0.555
8.00	0.590
8.75	0.575
9.25	0.570
10.00	0.580
11.00	0.625
12.00	0.645
13.00	0.660
13.50	0.655
14.30	0.625
15.00	0.570
CURVE 18*	
T = ~322	
0.50	0.060
0.70	0.160
1.00	0.130
1.50	0.125
2.00	0.170
2.50	0.190
2.75	0.200
3.50	0.287
4.25	0.380
5.00	0.445
6.00	0.490
7.00	0.547
7.50	0.580
8.00	0.590
9.00	0.570
9.50	0.560
10.00	0.570
10.50	0.590
11.00	0.600

λ	ρ
CURVE 18 (cont.)*	
12.00	0.600
12.50	0.595
13.50	0.580
14.00	0.562
14.50	0.530
15.00	0.480
CURVE 19*	
T = ~322	
2.00	0.425
2.50	0.385
3.00	0.360
3.25	0.365
3.65	0.415
4.50	0.475
5.00	0.500
5.40	0.500
6.50	0.545
7.50	0.570
8.00	0.570
9.00	0.540
9.50	0.520
10.50	0.500
11.25	0.450
12.00	0.390
12.50	0.355
13.75	0.300
15.00	0.250
CURVE 20*	
T = ~322	
2.00	0.430
2.50	0.390
3.00	0.370
3.50	0.395
4.00	0.440
5.00	0.500
5.50	0.510
6.00	0.532
7.00	0.565
7.50	0.570
8.00	0.560
8.50	0.560
9.50	0.560

λ	ρ
CURVE 20 (cont.)*	
10.00	0.560
10.50	0.570
11.00	0.575
12.00	0.570
13.00	0.560
14.00	0.540
14.50	0.525
15.00	0.470
CURVE 21*	
T = ~322	
1.00	0.370
1.50	0.385
2.25	0.375
3.00	0.370
3.25	0.385
4.00	0.475
4.60	0.525
5.20	0.550
5.40	0.550
6.50	0.575
7.50	0.610
8.15	0.630
9.70	0.610
10.50	0.640
11.00	0.670
12.25	0.700
13.25	0.715
14.25	0.730
15.00	0.720
CURVE 22*	
T = ~322	
1.00	0.370
1.50	0.380
2.00	0.345
2.50	0.325
2.75	0.320
3.00	0.330
3.50	0.365
4.00	0.435
4.50	0.485
5.00	0.507
5.50	0.515

λ	ρ
CURVE 22 (cont.)*	
6.00	0.520
6.50	0.525
7.00	0.540
8.00	0.580
8.50	0.580
9.00	0.570
9.50	0.570
10.00	0.580
11.00	0.610
12.00	0.610
13.00	0.610
14.00	0.610
14.50	0.605
15.00	0.590
CURVE 23*	
T = ~322	
0.50	0.250
0.75	0.300
1.00	0.250
1.40	0.300
2.00	0.345
2.80	0.345
3.25	0.355
3.60	0.400
4.50	0.475
5.00	0.500
6.00	0.520
6.80	0.550
8.00	0.600
9.25	0.585
10.50	0.610
11.50	0.635
12.50	0.645
13.50	0.650
14.00	0.650
14.70	0.635
15.00	0.610
CURVE 24*	
T = ~322	
0.50	0.170
0.75	0.230
1.00	0.315

λ	ρ
CURVE 24 (cont.)*	
1.25	0.365
1.50	0.385
1.75	0.385
2.00	0.370
2.50	0.350
2.80	0.345
3.00	0.350
3.50	0.385
4.00	0.432
5.00	0.500
6.00	0.530
7.00	0.557
7.50	0.580
8.00	0.590
8.50	0.585
9.50	0.560
10.00	0.560
10.50	0.565
11.00	0.585
11.50	0.585
12.50	0.575
13.00	0.575
14.00	0.550
14.50	0.550
15.00	0.560
CURVE 25	
T = ~322	
2.00	0.255
2.50	0.265
3.50	0.320
4.25	0.380
5.00	0.400
6.10	0.400
7.00	0.425
8.10	0.450
9.25	0.440
10.50	0.465
11.25	0.485
12.25	0.495
13.25	0.500
14.00	0.500
15.00	0.450

λ	ρ
CURVE 26	
T = ~322	
2.00	0.220
3.00	0.240
4.00	0.280
5.00	0.300
6.00	0.290
7.00	0.300
8.00	0.310
9.00	0.300
9.60	0.295
10.00	0.305
11.00	0.320
12.00	0.325
13.00	0.325
14.00	0.325
15.00	0.280
CURVE 27*	
T = ~322	
1.00	0.250
2.00	0.250
3.00	0.295
4.00	0.355
5.00	0.385
6.25	0.400
7.00	0.410
8.00	0.450
9.35	0.450
11.00	0.480
12.00	0.490
13.25	0.490
14.00	0.485
14.75	0.460
15.00	0.420
CURVE 28*	
T = ~322	
1.00	0.222
1.50	0.240
2.00	0.225
2.50	0.225
3.50	0.270
4.50	0.315
5.00	0.320

* Not shown on plot

DATA TABLE NO. 357 (continued)

λ	ρ
CURVE 28 (cont.)*	
6.00	0.320
7.00	0.340
7.50	0.350
8.50	0.350
9.00	0.345
9.50	0.345
10.00	0.350
11.00	0.365
12.00	0.370
13.00	0.380
14.00	0.380
15.00	0.330
CURVE 29*	
T = ~322	
0.50	0.130
0.75	0.195
0.90	0.190
1.50	0.240
2.00	0.260
2.50	0.255
3.35	0.275
4.00	0.300
5.00	0.375
5.35	0.380
6.00	0.370
7.00	0.395
8.25	0.445
8.75	0.440
9.75	0.450
11.00	0.470
12.00	0.480
13.00	0.485
13.75	0.485
14.50	0.450
14.85	0.400
15.00	0.360
CURVE 30*	
T = ~322	
0.50	0.100
0.75	0.165
0.85	0.160
0.90	0.165
1.00	0.165
CURVE 30 (cont.)*	
1.50	0.190
2.00	0.230
3.00	0.245
4.00	0.300
4.50	0.320
5.00	0.325
6.00	0.310
7.00	0.335
8.00	0.350
9.00	0.335
10.00	0.340
11.00	0.360
12.00	0.360
13.00	0.370
14.00	0.360
15.00	0.310
CURVE 31	
T = ~322	
2.00	0.115
2.50	0.065
3.00	0.060
3.90	0.100
4.70	0.150
5.75	0.185
6.50	0.230
7.25	0.270
8.00	0.290
9.00	0.280
9.75	0.315
10.60	0.350
11.75	0.375
12.75	0.395
14.00	0.410
14.50	0.405
15.00	0.380
CURVE 32*	
T = ~322	
2.00	0.130
2.50	0.062
3.00	0.065
4.00	0.100
5.00	0.125
5.50	0.130
CURVE 32 (cont.)*	
6.00	0.150
7.00	0.200
7.85	0.220
8.85	0.200
10.00	0.240
11.00	0.270
12.00	0.275
13.00	0.305
14.00	0.330
14.50	0.330
15.00	0.310
CURVE 33*	
T = ~322	
1.00	0.170
1.50	0.170
2.00	0.115
2.25	0.105
2.75	0.100
3.00	0.105
4.25	0.130
5.25	0.175
6.50	0.225
7.25	0.265
8.00	0.290
8.90	0.270
10.00	0.310
11.00	0.340
11.75	0.345
12.25	0.360
14.00	0.410
15.00	0.410
CURVE 34*	
T = ~322	
1.00	0.130
1.50	0.140
2.00	0.117
3.00	0.090
3.50	0.095
4.00	0.115
5.00	0.150
6.00	0.192
7.00	0.210
8.00	0.240
CURVE 34 (cont.)*	
9.00	0.220
9.50	0.230
10.00	0.250
11.00	0.280
12.00	0.290
12.50	0.290
13.00	0.310
14.00	0.340
15.00	0.350
CURVE 35*	
T = ~322	
0.50	0.025
0.75	0.085
1.00	0.140
1.50	0.160
2.50	0.120
3.15	0.110
4.00	0.130
5.20	0.190
5.70	0.200
6.75	0.260
7.60	0.300
8.15	0.315
8.90	0.295
10.00	0.330
11.50	0.370
12.25	0.365
13.00	0.390
13.75	0.400
14.45	0.400
15.00	0.370
CURVE 36*	
T = ~322	
0.50	0.025
0.75	0.085
1.00	0.135
1.50	0.145
2.00	0.135
3.00	0.100
3.50	0.100
4.00	0.115
5.00	0.150
6.00	0.180
CURVE 36 (cont.)*	
7.00	0.220
8.00	0.260
8.50	0.255
9.00	0.230
10.00	0.260
11.00	0.290
12.00	0.290
13.00	0.310
14.00	0.330
14.50	0.335
15.00	0.320
CURVE 37	
T = ~322	
2.00	0.200
2.50	0.145
3.00	0.115
3.75	0.150
4.10	0.185
4.85	0.210
5.75	0.210
6.75	0.250
7.50	0.275
8.00	0.280
9.25	0.280
10.25	0.300
11.50	0.325
12.50	0.340
13.75	0.350
14.25	0.350
15.00	0.330
CURVE 38*	
T = ~322	
2.00	0.185
3.00	0.110
3.50	0.120
4.00	0.170
4.50	0.190
5.00	0.190
5.50	0.190
6.00	0.195
7.00	0.220
7.50	0.230
8.00	0.230
CURVE 38 (cont.)*	
9.00	0.230
9.50	0.225
10.00	0.230
11.00	0.250
11.50	0.260
12.00	0.260
12.50	0.265
13.00	0.270
14.00	0.270
15.00	0.260
CURVE 39*	
T = ~322	
1.00	0.225
1.50	0.220
2.50	0.175
3.10	0.150
4.00	0.200
4.75	0.235
5.50	0.240
6.50	0.260
7.50	0.280
8.10	0.290
8.50	0.285
9.50	0.295
10.75	0.310
12.00	0.320
12.50	0.325
13.75	0.370
14.50	0.390
15.00	0.390
CURVE 40*	
T = ~322	
1.00	0.185
1.50	0.200
2.00	0.170
3.00	0.130
3.50	0.140
4.00	0.170
5.00	0.190
6.00	0.200
7.00	0.220
8.00	0.230
9.00	0.225
CURVE 40 (cont.)*	
10.00	0.230
11.00	0.250
12.00	0.260
13.00	0.270
14.00	0.290
15.00	0.300
CURVE 41*	
T = ~322	
0.50	0.090
0.75	0.150
1.50	0.185
1.75	0.190
2.50	0.165
3.00	0.140
3.75	0.180
4.25	0.210
5.00	0.230
6.00	0.240
7.00	0.260
8.00	0.295
9.00	0.280
9.50	0.275
10.50	0.290
11.75	0.310
13.00	0.325
14.40	0.340
15.00	0.300
CURVE 42*	
T = ~322	
0.50	0.085
1.00	0.145
1.50	0.180
1.75	0.185
2.00	0.180
2.50	0.155
3.00	0.130
3.15	0.130
4.00	0.170
5.00	0.190
6.00	0.200
7.00	0.220
7.50	0.230
8.00	0.235
CURVE 42 (cont.)*	
9.00	0.220
9.50	0.215
10.00	0.220
11.00	0.240
11.50	0.250
12.00	0.250
13.00	0.260
14.00	0.275
14.50	0.275
15.00	0.260

* Not shown on plot

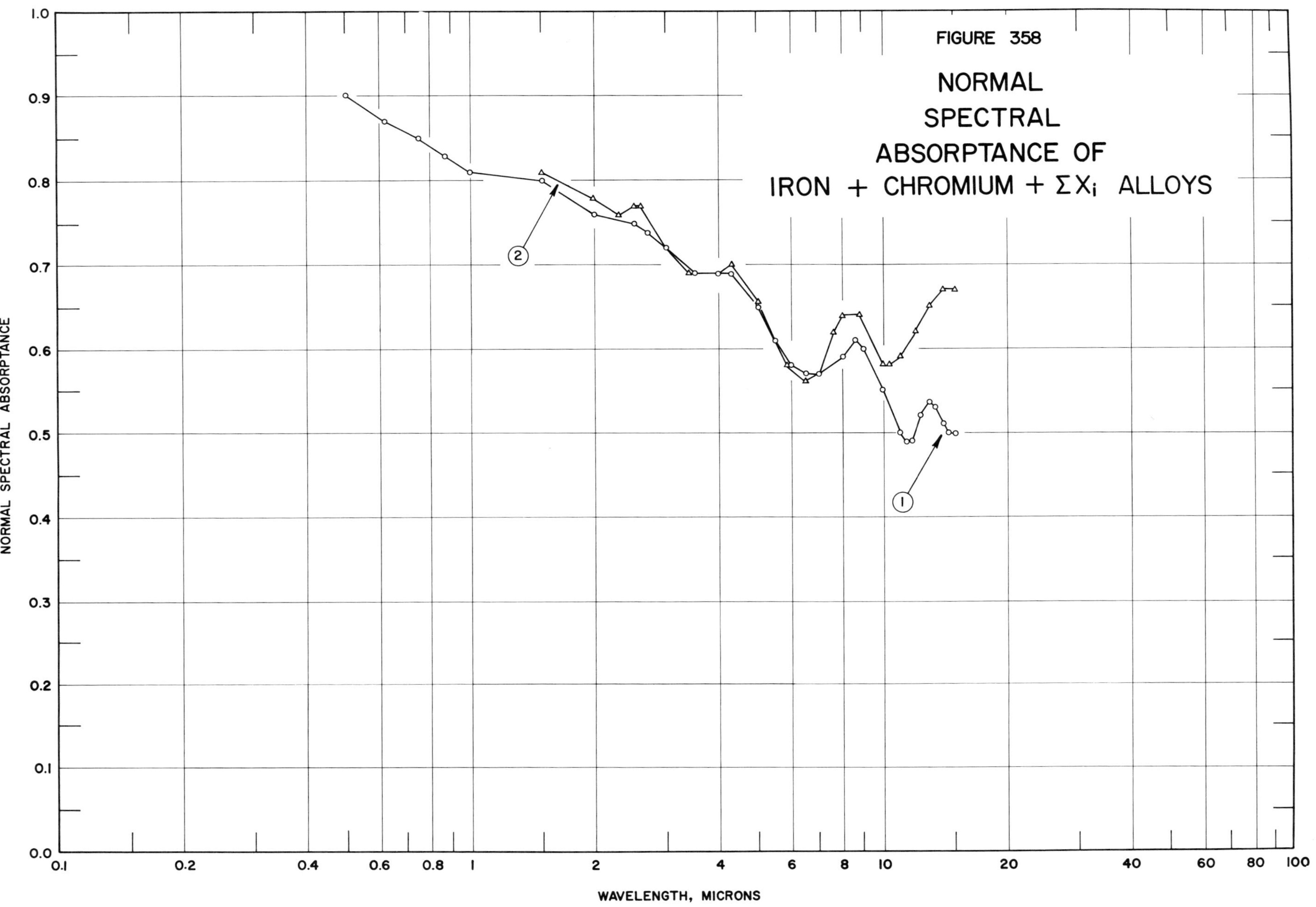
FIGURE 358
NORMAL
SPECTRAL
ABSORPTANCE OF
IRON + CHROMIUM + ΣXi ALLOYS
2
1
NORMAL SPECTRAL ABSORPTANCE
1.0
0.9
0.8
0.7
0.6
0.5
0.4
0.3
0.2
0.1
0.0
0.1
0.2
0.4
0.6
0.8
1
2
4
6
8
10
20
40
60
80
100
WAVELENGTH, MICRONS

SPECIFICATION TABLE NO. 358 NORMAL SPECTRAL ABSORPTANCE OF [IRON + CHROMIUM + ΣX_i] ALLOYS

Curve No.	Ref. No.	Year	Temperature K	Wavelength Range, μ	Geometry θ	Reported Error, %	Composition (weight percent), Specifications and Remarks
1	30	1963	294	0.50-15.00	~0°		Kanthal; oxidized; measured in air; data extracted from smooth curve.
2	30	1963	775	1.5-15.0	~0°		Above specimen and conditions; heated at 775 K for 2 hrs.

DATA TABLE NO. 358 NORMAL SPECTRAL ABSORPTANCE OF [IRON + CHROMIUM + ΣX_i] ALLOYS

[Wavelength, λ, μ; Absorptance, α; Temperature, T, K]

λ	α
CURVE 1	**T = 294**
0.50	0.900
0.62	0.870
0.75	0.850
0.87	0.830
1.00	0.810
1.50	0.800
2.00	0.760
2.50	0.750
2.70	0.740
3.00	0.720
3.50	0.690
4.00	0.690
4.30	0.690
5.00	0.650
5.50	0.610
6.00	0.580
6.50	0.570
7.00	0.570
8.00	0.590
8.60	0.610
9.00	0.600
10.00	0.550
11.00	0.500
11.40	0.490
11.80	0.490
12.40	0.520
13.00	0.535
13.40	0.530
14.00	0.510
14.40	0.500
15.00	0.500
CURVE 2	**T = 775**
1.5	0.810
2.0	0.780
2.3	0.760
2.5	0.770
2.6	0.770
3.0	0.720*
3.4	0.690
4.0	0.690*
4.3	0.700
CURVE 2 (cont.)	
5.0	0.655
5.5	0.610*
6.0	0.580
6.5	0.560
7.0	0.570*
7.6	0.620
8.0	0.640
8.8	0.640
10.0	0.580
10.4	0.580
11.0	0.590
12.0	0.602
13.0	0.650
14.0	0.670
15.0	0.670

* Not shown on plot

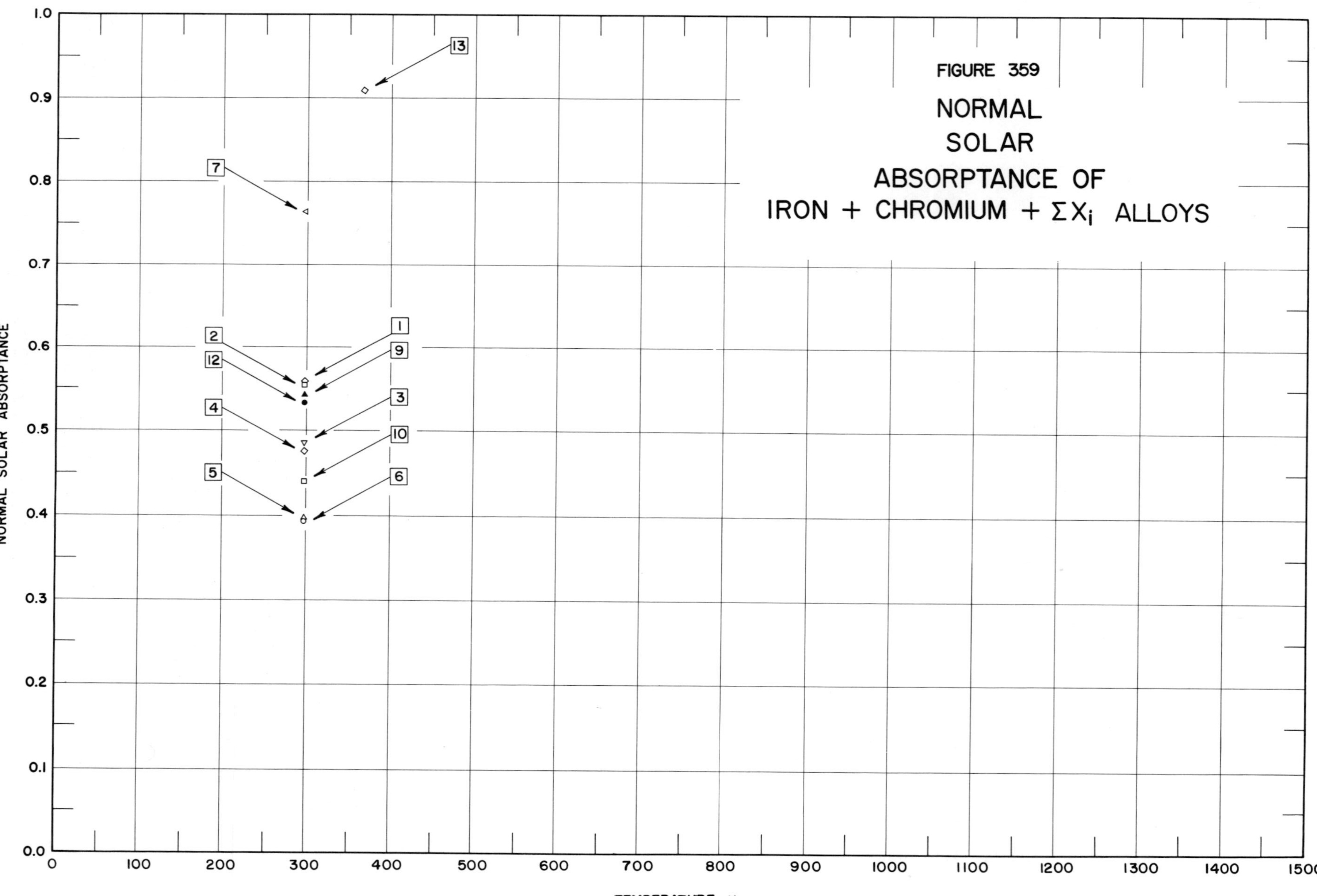
FIGURE 359
NORMAL
SOLAR
ABSORPTANCE OF
IRON + CHROMIUM + ΣXi ALLOYS
NORMAL SOLAR ABSORPTANCE
TEMPERATURE, K
0.0
0.1
0.2
0.3
0.4
0.5
0.6
0.7
0.8
0.9
1.0
0
100
200
300
400
500
600
700
800
900
1000
1100
1200
1300
1400
1500
1
2
3
4
5
6
7
9
10
12
13

SPECIFICATION TABLE NO. 359 NORMAL SOLAR ABSORPTANCE OF [IRON + CHROMIUM + ΣX_i] ALLOYS

Curve No.	Ref. No.	Year	Temperature Range, K	Geometry θ	Reported Error, %	Composition (weight percent), Specifications and Remarks
1	34	1957	298	9°		Cobalt alloy N-155; nominal composition: 21 Cr, 20 Co, 20 Ni, 3 Mo, 3 W, 1.5 Mn, 1 Nb, 0.5 Si, 0.15 C, 0.15 N, Fe balance; as received; computed from spectral reflectance data for sea level conditions.
2	34	1957	298	9°		Above specimen and conditions except computed for above atmosphere conditions.
3	34	1957	298	9°		Different sample, same as curve 1 specimen and conditions except cleaned with liquid detergent.
4	34	1957	298	9°		Above specimen and conditions except computed for above atmosphere conditions.
5	34	1957	298	9°		Different sample, same as curve 1 specimen and conditions except polished.
6	34	1957	298	9°		Above specimen and conditions except computed for above atmosphere conditions.
7	34	1957	298	9°		Different sample, same as curve 1 specimen and conditions except oxidized in air at red heat for 30 min.
8	34	1957	298	9°		Above specimen and conditions except computed for above atmosphere conditions.
9	34	1957	298	9°		Stainless steel type 446; nominal composition: 23.00-27.00 Cr, 1.5 max Mn, 1.00 max Si, 0.25 max N, 0.20 max C, Fe balance; surface roughness ~2 microinches rms; computed from spectral reflectance data for sea level conditions.
10	34	1957	298	9°		Above specimen and conditions except computed for above atmosphere conditions.
11	34	1957	298	9°		Different sample, same as curve 9 specimen and conditions; surface roughness ~15 microinches rms.
12	34	1957	298	9°		Above specimen and conditions except computed for above atmosphere conditions.
13	40	1962	367	~0°	±0.01	Cobalt alloy N-155 (surface N-1); nominal composition: 21 Cr, 20 Co, 20 Ni, 3 Mo, 3 W, 1.5 Mn, 1 Nb, 0.5 Si, 0.15 C, 0.15 N, Fe balance; highly polished; mirror finish; oxide formation at 873 K for 3 hrs.

DATA TABLE NO. 359 NORMAL SOLAR ABSORPTANCE OF [IRON + CHROMIUM + ΣX_i] ALLOYS

[Temperature, T, K; Absorptance, α]

T	α	T	α
CURVE 1		CURVE 13	
298	0.559	367	0.91
CURVE 2			
298	0.555		
CURVE 3			
298	0.486		
CURVE 4			
298	0.477		
CURVE 5			
298	0.398		
CURVE 6			
298	0.394		
CURVE 7			
298	0.764		
CURVE 8*			
298	0.763		
CURVE 9			
298	0.543		
CURVE 10			
298	0.442		
CURVE 11*			
298	0.544		
CURVE 12			
298	0.533		

* Not shown on plot

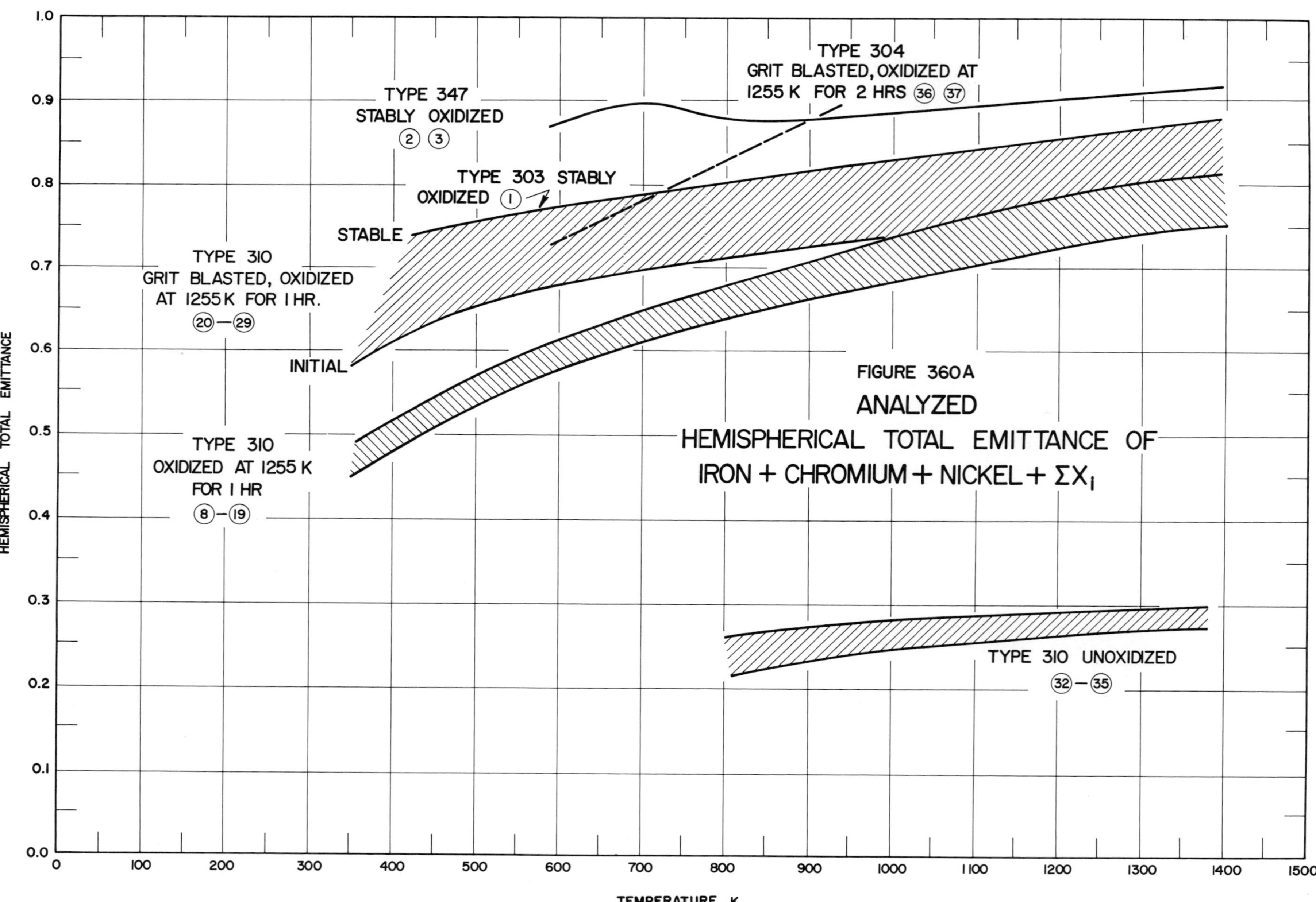
TYPE 304
GRIT BLASTED, OXIDIZED AT
1255 K FOR 2 HRS (36) (37)
TYPE 347
STABLY OXIDIZED
(2) (3)
TYPE 303 STABLY
OXIDIZED (1)
STABLE
TYPE 310
GRIT BLASTED, OXIDIZED
AT 1255K FOR 1 HR.
(20)—(29)
INITIAL
FIGURE 360A
ANALYZED
HEMISPHERICAL TOTAL EMITTANCE OF
IRON + CHROMIUM + NICKEL + ΣX_i
TYPE 310
OXIDIZED AT 1255 K
FOR 1 HR
(8)—(19)
TYPE 310 UNOXIDIZED
(32)—(35)
HEMISPHERICAL TOTAL EMITTANCE
1.0
0.9
0.8
0.7
0.6
0.5
0.4
0.3
0.2
0.1
0.0
0
100
200
300
400
500
600
700
800
900
1000
1100
1200
1300
1400
1500
TEMPERATURE, K

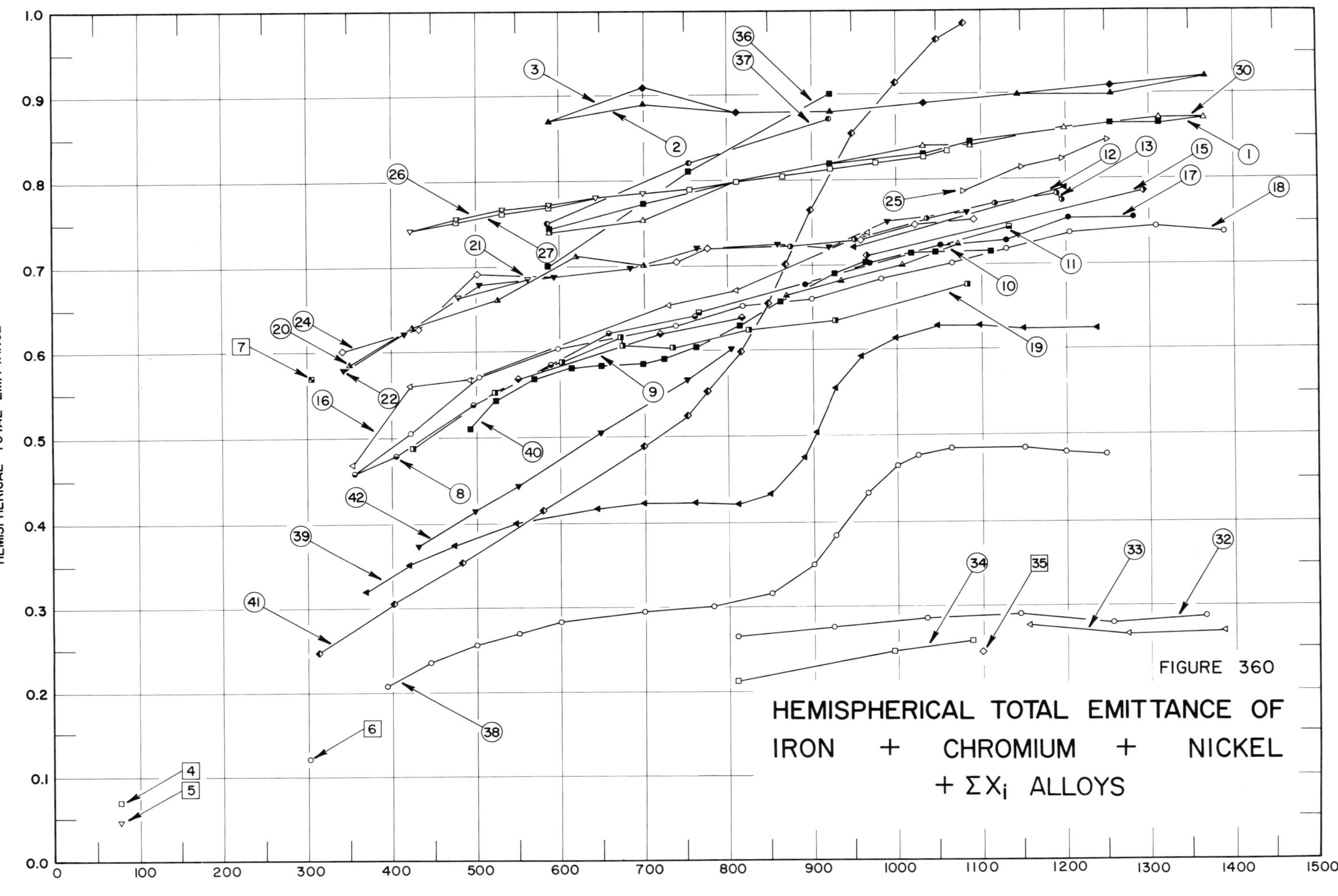
FIGURE 360
HEMISPHERICAL TOTAL EMITTANCE OF
IRON + CHROMIUM + NICKEL
+ ΣX_i ALLOYS
HEMISPHERICAL TOTAL EMITTANCE
TEMPERATURE, K

SPECIFICATION TABLE NO. 360 HEMISPHERICAL TOTAL EMITTANCE OF [IRON + CHROMIUM + NICKEL + ΣX_i] ALLOYS

Curve No.	Ref. No.	Year	Temperature Range, K	Reported Error, %	Composition (weight percent), Specifications and Remarks
1	22	1958	589-1367	≤ 2	Stainless steel AISI 303; nominal composition: 17.00-19.00 Cr, 8.00-10.00 Ni, 2.00 max Mn, 1.00 max Si, 0.15 min S, 0.15 max C, Fe balance; stably oxidized at 1367 K in quiescent air.
2	161	1959	589-1367		Stainless Steel 347; nominal composition: 68-73 Fe, 17-19 Cr, 9-12 Ni, 0.80 max Nb, 0.08 max C; mechanically polished, washed in alcohol and distilled water, dried by a hot-air blast; oxidized for 30 min. in air at 1367 K; author assumed normal total emittance = hemispherical total emittance.
3	161	1959	589-1367		Different sample, same as above specimen and conditions except gradually oxidized in still air at temperatures from 300 to 1367 K.
4	3	1955	76		Commercial ball type stainless steel 302; nominal composition: 17.00-19.00 Cr, 8.00-10.00 Ni, 2.00 max Mn, 1.00 max Si, 0.15 max C, Fe balance; solvent cleaned; measured in vacuum (10^{-6} to 10^{-7} mm Hg); emittance for 300 K black body incident radiation; authors assumed $\alpha = \epsilon$.
5	3	1955	76		Different sample, same as above specimen and conditions.
6	24	1948	301		Stainless steel 18-8; nominal composition: 18.45 Cr, 8.79 Ni, 0.50 Mn, 0.10 C, Fe balance; aerobright polish; measured in air.
7	24	1958	305		Different sample, same as above specimen and conditions except chromic and sulfuric acid treated; weathered.
8	12	1962	355.9-761.2	± 2.7	Stainless steel AISI 310; nominal composition: 24.00-26.00 Cr, 19.00-22.00 Ni, 2.00 max Mn, 1.50 max Si, 0.25 max C, Fe balance; oxidized at 1255 K for 1 hr; measured in vacuum (< 10^{-4} mm Hg); run no. 1A.
9	12	1962	817.9-549.5	± 2.7	Above specimen and conditions; run no. 1B.
10	12	1962	870.4-1074.8	± 2.7	Above specimen and conditions; run no. 1C.
11	12	1962	765.4-1134.3	± 2.7	Above specimen and conditions; run no. 1D.
12	12	1962	1200.4-950.1	± 2.7	Above specimen and conditions; run no. 1E.
13	12	1962	874.0-1197.0	± 2.7	Above specimen and conditions; run no. 2A.
14	12	1962	1121.8-1198.7	± 2.7	Above specimen and conditions; run no. 2B.
15	12	1962	1294.3-965.1	± 2.7	Above specimen and conditions; run no. 2C.
16	12	1962	354.3-966.5	± 2.7	Above specimen and conditions; run no. 3A.
17	12	1962	892.6-1281.2	± 2.7	Above specimen and conditions; run no. 3B.
18	12	1962	355.9-1389.0	± 2.7	Above specimen and conditions; run no. 4.
19	12	1962	424.5-1083.7	± 2.7	Above specimen and conditions; run no. 5.

SPECIFICATION TABLE NO. 360 (continued)

Curve No.	Ref. No.	Year	Temperature Range, K	Reported Error, %	Composition (weight percent), Specifications and Remarks
20	12	1962	350.4-699.8	± 2.7	Stainless steel AISI 310; nominal composition: 24.00-26.00 Cr, 19.00-22.00 Ni, 2.00 max Mn, 1.50 max Si, 0.25 max C, Fe balance; grit blasted with aluminum oxide No. 90 (PMC 304A), then oxidized in air at 1255 K for 1 hr; measured in vacuum ($<8 \times 10^{-4}$ mm Hg); run no. 1A.
21	12	1962	562.9-479.5	± 2.7	Above specimen and conditions; run no. 1B.
22	12	1962	344.3-1083.2	± 2.7	Above specimen and conditions; run no. 2A.
23	12	1962	997.9-630.4	± 2.7	Above specimen and conditions; run no. 2B.
24	12	1962	341.5-1092.6	± 2.7	Above specimen and conditions; run no. 3.
25	12	1962	1079.3-1250.4	± 2.7	Above specimen and conditions; run no. 4.
26	12	1962	422-700	± 2.7	Different sample, same as curve 20 specimen and conditions; run no. 1.
27	12	1962	422-1061	± 2.7	Above specimen and conditions; run no. 2.
28	12	1962	422-1061	± 2.7	Above specimen and conditions; run no. 3.
29	12	1962	1067-1063	± 2.7	Above specimen and conditions; measurements taken at intervals during 330.3 hr endurance run.
30	164	1961	589-1366	± 2.5	Stainless steel type 303; nominal composition: 17.00-19.00 Cr, 8.00-10.00 Ni, 2.00 max Mn, 0.15 max C, Fe balance; mechanically polished and cleaned; oxidized in quiescent air at 1366 K for 30 min; measured in air.
31	4	1953	76		Stainless steel 302; nominal composition: 17.00-19.00 Cr, 8.00-10.00 Ni, 2.00 max Mn, 1.00 max Si, 0.15 max C, Fe balance; foil (0.005 in. thick); solvent cleaned; measured in vacuum ($< 10^{-6}$ mm Hg); emittance for 294 K black body radiation; authors assumed $\alpha = \epsilon$.
32	96	1963	811-1366		AISI-310 Stainless Steel; nominal composition: 24.00-26.00 Cr, 19.00-22.00 Ni, 2.00 max Mn, 1.50 max Si, 0.25 max C, Fe balance; measured in vacuum (4.4×10^{-7} to 2.0×10^{-6} mm Hg); temperature measured with thermocouple; Run No. 1, heating cycle.
33	96	1963	1156-1388		Above specimen and conditions except temperature measured with optical pyrometer.
34	96	1963	810-1089		Curve 32 specimen and conditions; Run No. 2.
35	96	1963	1100		Curve 33 specimen and conditions; Run No. 2.
36	295	1963	588-922		Stainless steel type 304 ELC; nominal composition: 18.00-20.00 Cr, 8.00-12.00 Ni, 2.00 max Mn, 1.00 max Si, 0.08 max C, Fe balance; sandblasted; oxidized at 1255 K for 2 hrs; measured in vacuum (10^{-5} mm Hg).
37	295	1963	588-922		Stainless steel type 304; nominal composition: 18.00-20.00 Cr, 8.00-12.00 Ni, 2.00 max Mn, 1.00 max Si, 0.08 max C, Fe balance; sandblasted; oxidized at 1255 K for 2 hrs; measured in vacuum (10^{-5} mm Hg).

SPECIFICATION TABLE NO. 360 (continued)

Curve No.	Ref. No.	Year	Temperature Range, K	Reported Error, %	Composition (weight percent), Specifications and Remarks
38	299	1958	392-1249	> 6	Multimet; nominal composition: 21 Cr, 20 Co, 20 Ni, 3 Mo, 3 W, 1.5 Mn, 1 Nb, 0.5 Si, 0.15 C, 0.15 M, Fe balance; polished surface.
39	299	1958	369-1238	> 6	Above specimen and conditions except rough surface.
40	299	1958	492-1113	> 6	Above specimen and conditions except surface preheated in air to 1100 K.
41	299	1958	312-1081	> 6	Stainless steel 18-8; nominal composition: 18.45 Cr, 8.79 Ni, 0.50 Mn, 0.10 C, Fe balance; polished.
42	299	1958	431-805	> 6	Above specimen and conditions except preheated at 1000 K in air.

DATA TABLE NO. 360 HEMISPHERICAL TOTAL EMITTANCE OF [IRON + CHROMIUM + NICKEL + ΣX_i] ALLOYS

[Temperature, T, K; Emittance, ϵ]

T	ϵ
CURVE 1	
589	0.745
700	0.775
811	0.800*
922	0.820
1033	0.830
1089	0.845
1200	0.860*
1255	0.865
1311	0.865
1367	0.870*
CURVE 2	
589	0.87
700	0.89
811	0.88*
922	0.88
1033	0.89*
1144	0.90
1255	0.90
1367	0.92
CURVE 3	
589	0.87*
700	0.91
811	0.88
922	0.88*
1033	0.89
1144	0.90*
1255	0.91
1367	0.92*
CURVE 4	
76	0.07
CURVE 5	
76	0.048
CURVE 6	
301	0.12
CURVE 7	
305	0.57
CURVE 8	
355.9	0.459
405.4	0.480
497.6	0.539
589.5	0.586
658.4	0.623
761.2	0.642
CURVE 9	
817.9	0.641
719.3	0.623
549.5	0.569
CURVE 10	
870.4	0.666
934.0	0.682
1006.5	0.700
1074.8	0.725
CURVE 11	
765.4	0.644
1134.3	0.743
CURVE 12	
1200.4	0.790
950.1	0.722
CURVE 13	
874.0	0.723
951.2	0.730
1037.3	0.754
1117.9	0.773
1191.8	0.784
1197.0	0.777
CURVE 14*	
1121.8	0.774
1194.0	0.781
1198.7	0.770
CURVE 15	
1294.3	0.787
965.1	0.711
CURVE 16	
354.3	0.469
421.2	0.561
491.5	0.569
729.3	0.655
810.4	0.671
966.5	0.738
CURVE 17	
892.6	0.678
967.9	0.703
1052.9	0.712
1130.7	0.729
1204.5	0.754
1281.2	0.755
CURVE 18	
355.9	0.458*
422.6	0.506
513.7	0.571
597.0	0.603
737.3	0.631
817.9	0.653
899.3	0.661
982.6	0.684
1067.3	0.702
1130.7	0.719
1205.4	0.736
1308.4	0.744
1389.0	0.737
CURVE 19	
424.5	0.488
521.5	0.552
602.6	0.589
671.5	0.618
673.2	0.608
734.3	0.604
825.1	0.626
927.0	0.636
1083.7	0.677
CURVE 20	
350.4	0.586
424.8	0.630
526.8	0.662
619.3	0.712
699.8	0.702
CURVE 21	
562.9	0.684
479.5	0.665
CURVE 22	
344.3	0.580
415.4	0.623
504.8	0.678
594.5	0.687
684.3	0.698
763.4	0.722
860.1	0.726
920.9	0.722
989.8	0.751
1083.2	0.761
CURVE 23*	
997.9	0.741
997.3	0.741
931.2	0.726
870.7	0.739
809.0	0.731
753.4	0.727
CURVE 23*(cont.)	
691.2	0.710
630.4	0.708
CURVE 24	
341.5	0.601
433.7	0.628
501.5	0.692
595.4	0.689*
739.3	0.706
777.0	0.722
872.6	0.726*
958.2	0.730
1023.2	0.748
1092.6	0.752
CURVE 25	
1079.3	0.786
1148.7	0.814
1196.5	0.823
1250.4	0.845
CURVE 26	
422	0.742
478	0.756
533	0.768
589	0.772
644	0.781
700	0.786
CURVE 27	
422	0.741*
478	0.752
533	0.762
589	0.770
644	0.779*
700	0.786*
755	0.790
811	0.798
866	0.805
922	0.812
CURVE 27 (cont.)	
978	0.820
1033	0.828
1061	0.833
CURVE 28*	
422	0.738
478	0.755
533	0.763
589	0.773
644	0.784
700	0.790
755	0.797
811	0.805
866	0.811
922	0.818
978	0.826
1061	0.833
CURVE 29*	
1067	0.817
1065	0.825
1064	0.828
1064	0.828
1064	0.828
1064	0.828
1064	0.830
1064	0.830
1064	0.828
1064	0.828
1064	0.830
1064	0.830
1063	0.831
1063	0.832
1064	0.830
1064	0.830
1064	0.830
1064	0.830
1064	0.830
1063	0.832
1063	0.832
1063	0.832
1062	0.835
CURVE 29*(cont.)	
1063	0.834
1063	0.833
1063	0.832
1063	0.832
1063	0.833
1063	0.835
CURVE 30	
589	0.740
700	0.755
811	0.800*
922	0.820*
1033	0.840
1089	0.840
1200	0.860
1255	0.865*
1311	0.870
1366	0.870
CURVE 31*	
76	0.0485
CURVE 32	
811	0.264
923	0.276
1034	0.285
1144	0.289
1255	0.279
1366	0.287
CURVE 33	
1156	0.277
1273	0.264
1388	0.269
CURVE 34	
810	0.211
996	0.246
1089	0.258
CURVE 35	
1100	0.244
CURVE 36	
588	0.70
755	0.81
922	0.90
CURVE 37	
588	0.75
755	0.82
922	0.87
CURVE 38	
392	0.207
443	0.235
499	0.256
549	0.270
600	0.282
699	0.294
781	0.300
851	0.316
900	0.350
927	0.385
965	0.434
1000	0.466
1025	0.478
1064	0.484
1151	0.483
1201	0.480
1249	0.478
CURVE 39	
369	0.321
420	0.352
474	0.376
548	0.400
644	0.416
699	0.423
761	0.423
812	0.421

* Not shown on plot

DATA TABLE NO. 360 (continued)

T	ϵ
CURVE 39 (cont.)	
850	0.432
891	0.477
904	0.503
928	0.557
959	0.592
999	0.615
1050	0.628
1099	0.630
1151	0.627
1238	0.625
CURVE 40	
492	0.509
524	0.542
570	0.568
614	0.580
650	0.585
699	0.586
724	0.590
762	0.603
814	0.629
864	0.658
928	0.689
965	0.702
1018	0.713
1048	0.715
1113	0.716
CURVE 41	
312	0.248
401	0.304
482	0.354
579	0.416
700	0.491
752	0.527
776	0.552
816	0.599
849	0.654
869	0.701
898	0.765
949	0.853
1000	0.916
1049	0.964
1081	0.985

T	ϵ
CURVE 42	
431	0.372
499	0.413
550	0.444
649	0.505
752	0.567
805	0.602

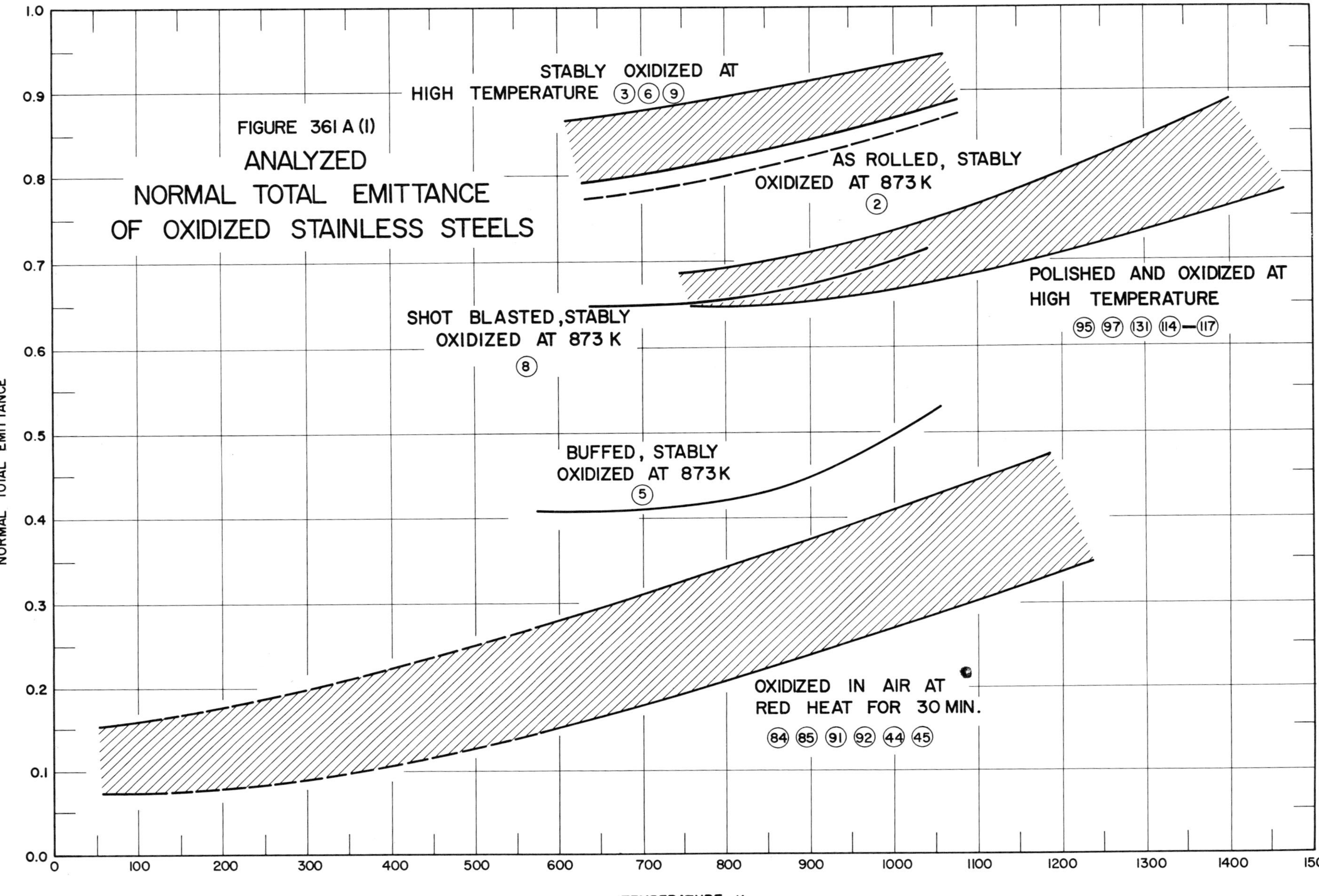

FIGURE 361 A (I)
ANALYZED
NORMAL TOTAL EMITTANCE
OF OXIDIZED STAINLESS STEELS
STABLY OXIDIZED AT
HIGH TEMPERATURE (3) (6) (9)
AS ROLLED, STABLY
OXIDIZED AT 873 K
(2)
POLISHED AND OXIDIZED AT
HIGH TEMPERATURE
(95) (97) (131) (114)—(117)
SHOT BLASTED, STABLY
OXIDIZED AT 873 K
(8)
BUFFED, STABLY
OXIDIZED AT 873 K
(5)
OXIDIZED IN AIR AT
RED HEAT FOR 30 MIN.
(84) (85) (91) (92) (44) (45)
NORMAL TOTAL EMITTANCE
1.0
0.9
0.8
0.7
0.6
0.5
0.4
0.3
0.2
0.1
0.0
TEMPERATURE, K
0
100
200
300
400
500
600
700
800
900
1000
1100
1200
1300
1400
1500

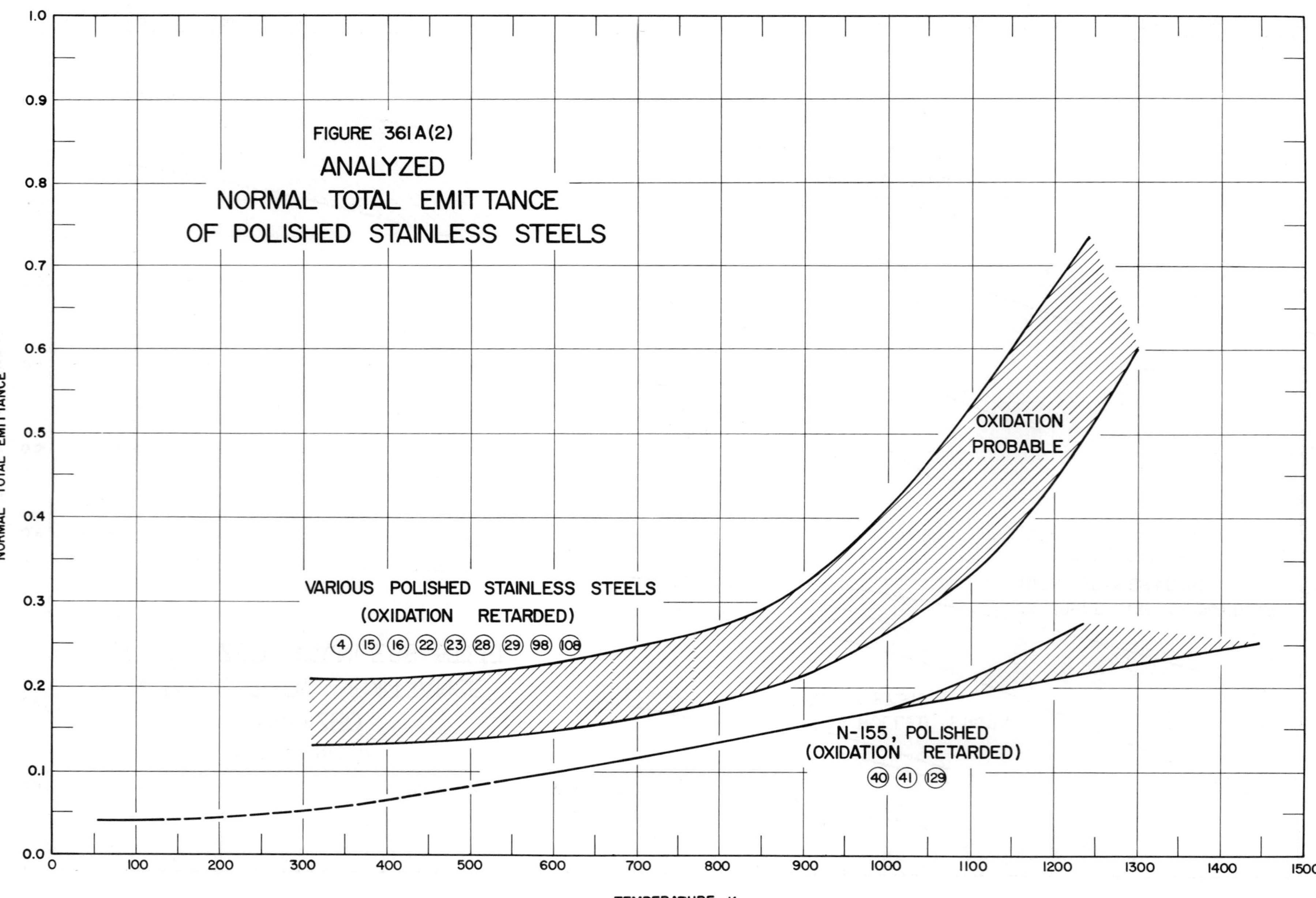
FIGURE 361A(2)
ANALYZED
NORMAL TOTAL EMITTANCE
OF POLISHED STAINLESS STEELS
OXIDATION
PROBABLE
VARIOUS POLISHED STAINLESS STEELS
(OXIDATION RETARDED)
4 15 16 22 23 28 29 98 108
N-155, POLISHED
(OXIDATION RETARDED)
40 41 129
NORMAL TOTAL EMITTANCE
TEMPERATURE, K
0.0
0.1
0.2
0.3
0.4
0.5
0.6
0.7
0.8
0.9
1.0
0
100
200
300
400
500
600
700
800
900
1000
1100
1200
1300
1400
1500

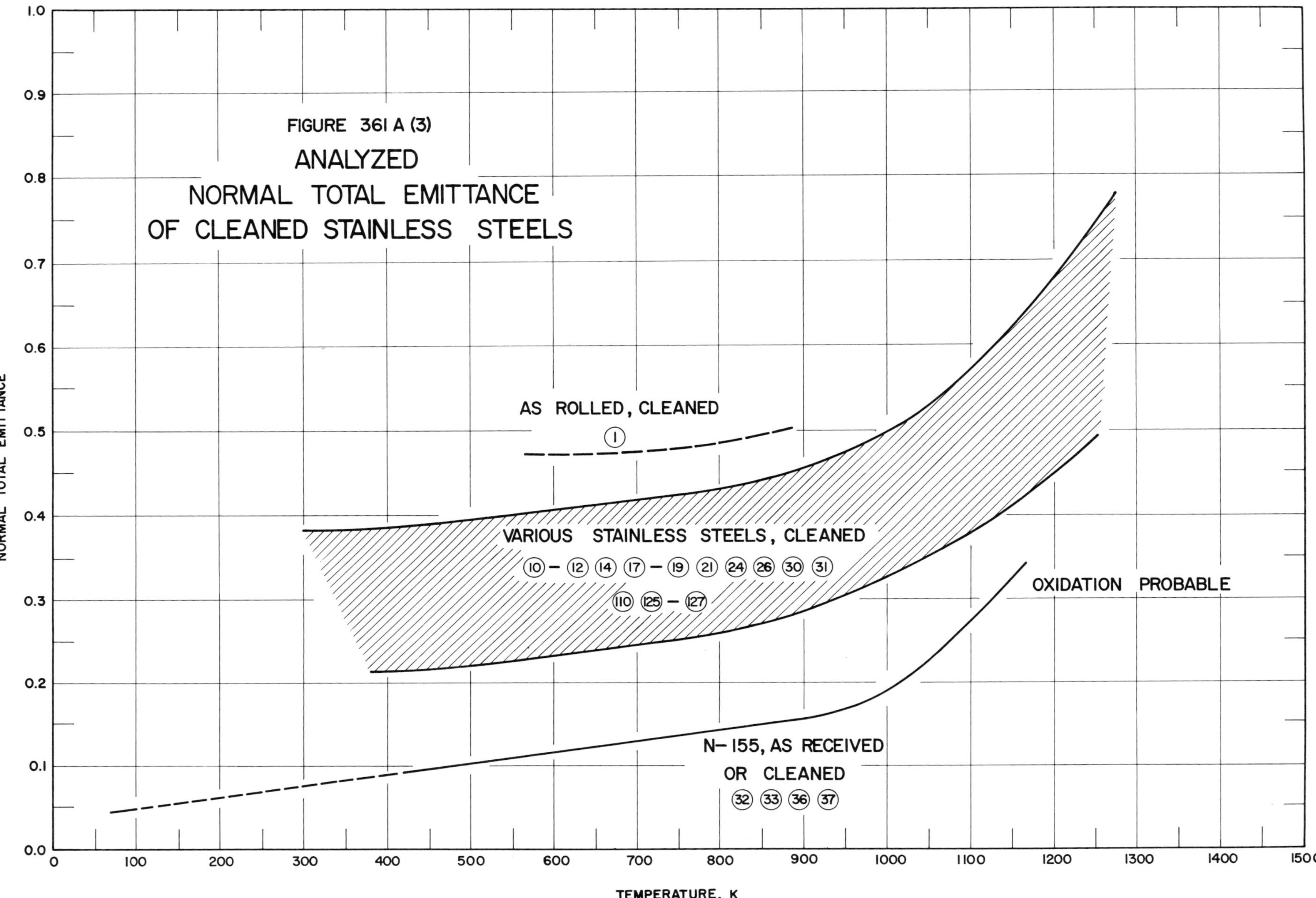
FIGURE 361 A (3)
ANALYZED
NORMAL TOTAL EMITTANCE
OF CLEANED STAINLESS STEELS
AS ROLLED, CLEANED
1
VARIOUS STAINLESS STEELS, CLEANED
10 – 12 14 17 – 19 21 24 26 30 31
110 125 – 127
OXIDATION PROBABLE
N– 155, AS RECEIVED
OR CLEANED
32 33 36 37
NORMAL TOTAL EMITTANCE
1.0
0.9
0.8
0.7
0.6
0.5
0.4
0.3
0.2
0.1
0.0
0
100
200
300
400
500
600
700
800
900
1000
1100
1200
1300
1400
1500
TEMPERATURE, K

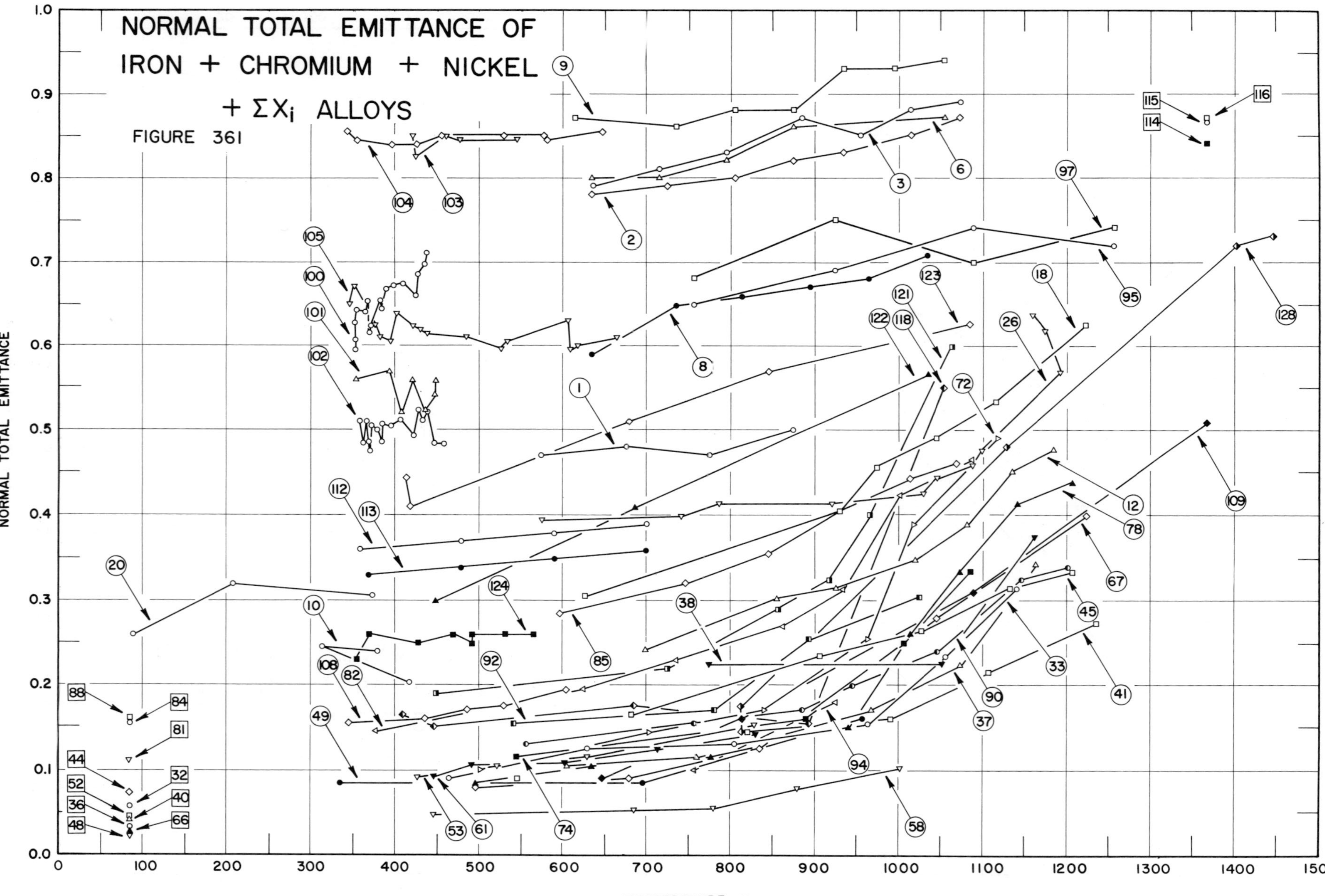
NORMAL TOTAL EMITTANCE OF
IRON + CHROMIUM + NICKEL
+ ΣXi ALLOYS
NORMAL TOTAL EMITTANCE
TEMPERATURE, K
0.0
0.1
0.2
0.3
0.4
0.5
0.6
0.7
0.8
0.9
1.0
0
100
200
300
400
500
600
700
800
900
1000
1100
1200
1300
1400
1500

FIGURE 361

SPECIFICATION TABLE NO. 361 NORMAL TOTAL EMITTANCE OF [IRON + CHROMIUM + NICKEL + ΣX_i] ALLOYS

Curve No.	Ref. No.	Year	Temperature Range, K	Geometry θ'	Reported Error, %	Composition (weight percent), Specifications and Remarks
1	68	1952	573-873	~0°		Stainless steel Vickers F.D.P.; nominal composition: 18 Cr, 8 Ni, Fe balance; as rolled; cleaned with CCl_4; measured in air.
2	68	1952	633-1073	~0°		Different sample, same as curve 1 specimen and conditions; oxidized at 873 K until steady state reached.
3	68	1952	633-1073	~0°		Different sample, same as curve 1 specimen and conditions except oxidized at 1173 K until steady state reached.
4	68	1952	663-893	~0°		Different sample, same as curve 1 specimen and conditions except buffed.
5	68	1952	573-1053	~0°		Different sample, same as curve 4 specimen and conditions; oxidized at 873 K until steady state reached.
6	68	1952	633-1053	~0°		Different sample, same as curve 4 specimen and conditions except oxidized at 1173 K until steady state reached.
7	68	1952	583-873	~0°		Different sample, same as curve 1 specimen and conditions except shot blasted with fused alumina.
8	68	1952	633-1033	~0°		Different sample, same as curve 7 specimen and conditions; oxidized at 873 K until steady state reached.
9	68	1952	613-1053	~0°		Different sample, same as curve 7 specimen and conditions except oxidized at 1173 K until steady state reached.
10	160	1954	377-414	~0°		Stainless steel 301; nominal composition: 16.00-18.00 Cr, 6.00-8.00 Ni, 2.00 max Mn, 1.00 max Si, 0.15 max C, Fe balance; cleaned with methyl alcohol; measured in vacuum (10^{-3} mm Hg); [Author's designation: Sample 6].
11	160	1954	505-1280	~0°		Different sample, same as above specimen and conditions except measured in argon (10^{-3} mm Hg); [Author's designation: Sample 11].
12	160	1954	697-1183	~0°		Above specimen and conditions.
13	160	1954	88.9-364	~0°		Different sample, same as curve 10 specimen and conditions except scrubbed with Bon Ami on a wet cloth, washed and dried, wiped with tolvene and alcohol; [Author's designation: Sample 2].
14	160	1954	439-1347	~0°		Different sample, same as above specimen and conditions except measured in argon (10^{-3} mm Hg); heating and cooling; [Author's designation: Sample 12].
15	160	1954	305-412	~0°		Different sample, same as curve 10 specimen and conditions except polished, then finished with a wool buff and rouge and washed; surface free from scratches; [Author's designation: Sample 4].
16	160	1954	469-1258	~0°		Different sample, same as above specimen and conditions except measured in argon (10^{-3} mm Hg); [Author's designation: Sample 13].
17	160	1954	310-423	~0°		Stainless steel 316; nominal composition: 16.00-18.00 Cr, 10.00-14.00 Ni, 2.00-3.00 Mo, 2.00 max Mn, 1.00 max Si, 0.08 max C; cleaned with methyl alcohol; measured in vacuum (10^{-3} mm Hg); heating; [Author's designation: Sample 1].

SPECIFICATION TABLE NO. 361 (continued)

Curve No.	Ref. No.	Year	Temperature Range, K	Geometry θ'	Reported Error, %	Composition (weight percent), Specifications and Remarks
18	160	1954	623-1221	~0°		Different sample, same as above specimen and conditions except measured in argon (10^{-3} mm Hg); [Author's designation: Sample 10].
19	160	1954	496-1219	~0°		Above specimen and conditions.
20	160	1954	88.9-372	~0°		Different sample, same as curve 17 specimen and conditions except scrubbed with Bon Ami on a wet cloth, washed and dried, wiped with toluene and alcohol; [Author's designation: Sample 2].
21	160	1954	517-1402	~0°		Different sample, same as above specimen and conditions except measured in argon (10^{-3} mm Hg); [Author's designation: Sample 11].
22	160	1954	314-432	~0°		Different sample, same as curve 17 specimen and conditions except polished and then finished with a wool buff and rouge and washed; surface free from scratches; [Author's designation: Sample 3].
23	160	1954	281-1417	~0°		Different sample, same as above specimen and conditions except measured in argon (10^{-3} mm Hg); [Author's designation: Sample 12].
24	160	1954	318-423	~0°		Stainless steel 347; nominal composition: 17.00-19.00 Cr, 9.00-13.00 Ni, 2.00 max Mn, 1.00 max Si, 0.08 max C, 10 x C min Nb-Ta, Fe balance; cleaned with methyl alcohol; measured in vacuum (10^{-3} mm Hg); heating; [Author's designation: Sample 7].
25	160	1954	319-423	~0°		Different sample, same as curve 24 specimen and conditions; [Author's designation: Sample 13].
26	160	1954	574-1191	~0°		Different sample, same as curve 24 specimen and conditions except measured in argon (10^{-3} mm Hg); [Author's designation: Sample 12].
27	160	1954	474-1192	~0°		Above specimen and conditions except heating and cooling.
28	160	1954	303-426	~0°		Different sample, same as curve 24 specimen and conditions except polished and then finished with a wool buff and rouge and washed; surface free from scratches; [Author's designation: Sample 6].
29	160	1954	488-1294	~0°		Different sample, same as curve 28 specimen and conditions except measured in argon (10^{-3} mm Hg); [Author's designation: Sample 15].
30	160	1954	303-434	~0°		Different sample, same as curve 24 specimen and conditions except scrubbed with Bon Ami on a wet cloth, washed and dried, wiped with toluene and alcohol; [Author's designation: Sample 5].
31	160	1954	493-1267	~0°		Different sample, same as curve 30 specimen and conditions except measured in argon (10^{-3} mm Hg); [Author's designation: Sample 14].
32	34	1957	83.2	~0°	± 10	Cobalt alloy N-155; nominal composition: 21 Cr, 20 Co, 20 Ni, 3 Mo, 3 W, 1.5 Mn, 1 Nb, 0.5 Si, 0.15 C, 0.15 N, Fe balance; as received; measured in vacuum (5 x 10^{-4} mm Hg).
33	34	1957	461-1139	~0°	± 10	Different sample, same as above specimen and conditions; increasing temp, cycle 1.
34	34	1957	533-850	~0°	± 10	Above specimen and conditions; cycle 2.

SPECIFICATION TABLE NO. 361 (continued)

Curve No.	Ref. No.	Year	Temperature Range, K	Geometry θ'	Reported Error, %	Composition (weight percent), Specifications and Remarks
35	34	1957	522	~0°	± 10	Above specimen and conditions; cycle 3.
36	34	1957	83.2	~0°	± 10	Different sample, same as curve 32 specimen and conditions except cleaned with liquid detergent.
37	34	1957	603-1161	~0°	± 10	Different sample, same as above specimen and conditions; increasing temp, cycle 1.
38	34	1957	772-1161	~0°	± 10	Above specimen and conditions; cycle 2.
39	34	1957	800-1003	~0°	± 10	Above specimen and conditions; cycle 3.
40	34	1957	83.2	~0°	± 10	Different sample, same as curve 32 specimen and conditions except polished.
41	34	1957	544-1233	~0°	± 10	Different sample, same as above specimen and conditions; increasing temp, cycle 1.
42	34	1957	811-989	~0°	± 10	Above specimen and conditions; cycle 2.
43	34	1957	911-1108	~0°	± 10	Above specimen and conditions; cycle 3.
44	34	1957	83.2	~0°	± 10	Different sample, same as curve 32 specimen and conditions except oxidized in air at red heat for 30 min.
45	34	1957	680-1205	~0°	± 10	Different sample, same as above specimen and conditions; increasing temp, cycle 1.
46	34	1957	769-911	~0°	± 10	Above specimen and conditions; cycle 2.
47	34	1957	611-1289	~0°	± 10	Above specimen and conditions; cycle 3.
48	34	1957	83.2	~0°	± 10	Stainless steel type PH 15-7 Mo; nominal composition: 15 Cr, 7 Ni, 2.25 Mo, 1.15 Al, 0.70 Mn, 0.40 Si, 0.07 C, Fe balance; surface roughness ~2 microinches rms; measured in vacuum (5 x 10^{-4} mm Hg).
49	34	1957	333-955	~0°	± 10	Different sample, same as above specimen and conditions; increasing temp, cycle 1.
50	34	1957	661-866	~0°	± 10	Above specimen and conditions; cycle 2.
51	34	1957	511-950	~0°	± 10	Above specimen and conditions; cycle 3.
52	34	1957	83.2	~0°	± 10	Different sample, same as curve 48 specimen and conditions; surface roughness ~15 microinches.
53	34	1957	425-828	~0°	± 10	Different sample, same as above specimen and conditions; increasing temp, cycle 1.
54	34	1957	483-878	~0°	± 10	Above specimen and conditions; cycle 2.
55	34	1957	625-844	~0°	± 10	Above specimen and conditions; cycle 3.
56	34	1957	83.2	~0°	± 10	Stainless steel type 17-7 PH; nominal composition: 17 Cr, 7 Ni, 1.15 Al, 0.70 Mn, 0.40 Si, 0.07 C, Fe balance; surface roughness ~2 microinches rms; measured in vacuum (5 x 10^{-4} mm Hg).
57	34	1957	472-755	~0°	± 10	Different sample, same as above specimen and conditions except increasing temp, cycle 1.
58	34	1957	444-1000	~0°	± 10	Above specimen and conditions; cycle 2.
59	34	1957	650-933	~0°	± 10	Above specimen and conditions; cycle 3.

SPECIFICATION TABLE NO. 361 (continued)

Curve No.	Ref. No.	Year	Temperature Range, K	Geometry θ'	Reported Error, %	Composition (weight percent), Specifications and Remarks
60	34	1957	83.2	$\sim 0°$	± 10	Different sample, same as curve 56 specimen and conditions; surface roughness ~15 microinches rms.
61	34	1957	444-828	$\sim 0°$	± 10	Different sample, same as above specimen and conditions except increasing temp, cycle 1.
62	34	1957	661-1053	$\sim 0°$	± 10	Above specimen and conditions; cycle 2.
63	34	1957	605	$\sim 0°$	± 10	Above specimen and conditions; decreasing temp, cycle 2.
64	34	1957	478-722	$\sim 0°$	± 10	Above specimen and conditions; cycle 3.
65	34	1957	616	$\sim 0°$	± 10	Above specimen and conditions; decreasing temp, cycle 3.
66	34	1957	83.2	$\sim 0°$	± 10	Stainless steel type 316; nominal composition: 16.00-18.00 Cr, 10.00-14.00 Ni, 2.00-3.00 Mo, 2.00 max Mn, 1.00 max Si, 0.08 max C, Fe balance; surface roughness ~2 microinches rms; measured in vacuum (5 x 10^{-4} mm Hg).
67	34	1957	494-1222	$\sim 0°$	± 10	Different sample, same as above specimen and conditions except increasing temp, cycle 1.
68	34	1957	505-1039	$\sim 0°$	± 10	Above specimen and conditions; cycle 2.
69	34	1957	83.2	$\sim 0°$	± 10	Different sample, same as curve 66 specimen and conditions; surface roughness ~15 microinches rms.
70	34	1957	444-628	$\sim 0°$	± 10	Different sample, same as above specimen and conditions except increasing temp, cycle 1.
71	34	1957	466-855	$\sim 0°$	± 10	Above specimen and conditions; cycle 2.
72	34	1957	500-1116	$\sim 0°$	± 10	Above specimen and conditions; cycle 3.
73	34	1957	83.2	$\sim 0°$	± 10	Stainless steel type 321; nominal composition: 17.00-19.00 Cr, 9.00-12.00 Ni, 2.00 max Mn, 1.00 max Si, 0.08 max C, 5 x C min Ti, Fe balance; bright finish; measured in vacuum (5 x 10^{-4} mm Hg).
74	34	1957	544-1083	$\sim 0°$	± 10	Different sample, same as above specimen and conditions except increasing temp, cycle 1.
75	34	1957	622-894	$\sim 0°$	± 10	Above specimen and conditions; cycle 2.
76	34	1957	872-1122	$\sim 0°$	± 10	Above specimen and conditions; cycle 3.
77	34	1957	83.2	$\sim 0°$	± 10	Different sample, same as curve 73 specimen and conditions; surface roughness ~2 microinches rms.
78	34	1957	494-1205	$\sim 0°$	± 10	Different sample, same as above specimen and conditions except increasing temp, cycle 1.
79	34	1957	633-994	$\sim 0°$	± 10	Above specimen and conditions; cycle 2.
80	34	1957	728-1061	$\sim 0°$	± 10	Above specimen and conditions; cycle 3.
81	34	1957	83.2	$\sim 0°$	± 10	Different sample, same as curve 73 specimen and conditions except dull finish; surface roughness ~6 microinches rms.
82	34	1957	375-1089	$\sim 0°$	± 10	Different sample, same as above specimen and conditions.
83	34	1957	561-761	$\sim 0°$	± 10	Above specimen and conditions; cycle 2.

SPECIFICATION TABLE NO. 361 (continued)

Curve No.	Ref. No.	Year	Temperature Range, K	Geometry θ'	Reported Error, %	Composition (weight percent), Specifications and Remarks
84	34	1957	83.2	~0°	± 10	Different sample, same as curve 73 specimen and conditions except oxidized in air at red heat for 30 min; surface roughness ~6 microinches rms.
85	34	1957	594-1069	~0°	± 10	Different sample, same as above specimen and conditions except increasing temp, cycle 1.
86	34	1957	544-789	~0°	± 10	Above specimen and conditions; cycle 2.
87	34	1957	950	~0°	± 10	Above specimen and conditions; cycle 3.
88	34	1957	83.2	~0°	± 10	Stainless steel type AM 350; nominal composition: 16.50 Cr, 4.25 Ni, 2.75 Mo, 0.75 Mn, 0.35 Si, 0.10 C, 0.10 N, Fe balance; surface roughness ~2 microinches rms; measured in vacuum (5 x 10^{-4} mm Hg).
89	34	1957	422-605	~0°	± 10	Different sample, same as above specimen and conditions except increasing temp, cycle 1.
90	34	1957	553-1200	~0°	± 10	Above specimen and conditions; cycle 2.
91	34	1957	83.2	~0°	± 10	Different sample, same as curve 88 specimen and conditions except oxidized in air at red heat for 30 min.
92	34	1957	539-1022	~0°	± 10	Different sample, same as above specimen and conditions except increasing temp, cycle 1.
93	34	1957	900-1161	~0°	± 10	Above specimen and conditions; cycle 2.
94	170	1959	755-922	~0°		Stainless steel 321; titanium stabilized 18 Cr, 8 Ni austenitic stainless steel; electropolished; computed from spectral data.
95	170	1959	755-1255	~0°		Different sample, same as curve 94 specimen and conditions except also oxidized in air at 1255 K for 1/2 hr.
96	170	1959	755-922	~0°		Different sample, same as curve 94 specimen and conditions except sandblasted by 40-mesh glass sand and air at 40 psi.
97	170	1959	755-1255	~0°		Different sample, same as curve 94 specimen and conditions except also oxidized in air at 1255 K for 1/2 hr.
98	15	1947	373	~0°		Alleghany metal; nominal composition: 17-20 Cr, 7-10 Ni, 0.50 max Mn, 0.20 C, Fe balance; No. 4 polish.
99	157	1944	356-441	~0°		Stainless steel 18-8; nominal composition: 18.45 Cr, 8.79 Ni, 0.50 Mn, 0.10 C, Fe balance; oxidized at 811 K; measured in air.
100	157	1944	350-435	~0°		Different sample, same as curve 99 specimen and conditions except oxidized at 1089 K.
101	157	1944	351-446	~0°		Different sample, same as curve 99 specimen and conditions except chromic and sulfuric blackened.
102	157	1944	355-456	~0°		Different sample, same as curve 99 specimen and conditions except sand blasted.
103	155	1948	419-594	~0°		Stainless steel 18-8; nominal composition: 18.45 Cr, 8.79 Ni, 0.50 Mn, 0.10 C, Fe balance; sand blasted and weathered.
104	155	1948	342-646	~0°		Different sample, same as curve 103 specimen and conditions except oxidized at 1089 K and weathered.

SPECIFICATION TABLE NO. 361 (continued)

Curve No.	Ref. No.	Year	Temperature Range, K	Geometry θ'	Reported Error, %	Composition (weight percent), Specifications and Remarks
105	155	1948	344-661	~0°		Different sample, same as curve 103 specimen and conditions except chromic and sulphuric acid treated and weathered.
106	155	1948	455-650	~0°		Different sample, same as curve 103 specimen and conditions except unpolished.
107	155	1948	353-655	~0°		Different sample, same as curve 103 specimen and conditions.
108	155	1948	344-603	~0°		Above specimen and conditions except polished with Aerobright and Bon Ami.
109	107	1960	644-1644	~0°	± 20	Stainless steel 301; nominal composition: 16.00-18.00 Cr, 6.00-8.00 Ni, 2.00 Mn, 1.00 max Si, 0.15 max C, Fe balance; measured in demoisturized helium gas.
110	99	1958	366-699	~0°	< 9	Type 321 corrosion-resistant steel; MIL-S-6721; nominal composition: 17.00-19.00 Cr, 9.00-12.00 Ni, 2.00 max Mn, 1.00 max Si, 0.08 max C, 5 x C min Ti, Fe balance; measured in air.
111	99	1958	366-699	~0°	< 9	Different sample, same as curve 110 specimen and conditions except heated at 647 K for 1000 hrs.
112	99	1958	366-699	~0°		Different sample, same as curve 110 specimen and conditions except calculated from spectral R $(2\pi, 0°)$.
113	99	1958	366-699	~0°		Different sample, same as curve 112 specimen and conditions except heated at 647 K for 1000 hrs.
114	164	1961	1366	~0°	± 2.5	Stainless steel type 303; nominal composition: 17.00-19.00 Cr, 8.00-10.00 Ni, 2.00 max Mn, 1.00 max Si, 0.15 min S, 0.15 max C, Fe balance; mechanically polished and cleaned; oxidized in quiescent air at 1366 K for 10 min; measured in air.
115	164	1961	1366	~0°	± 2.5	Above specimen and conditions except oxidized in quiescent air at 1366 K for 25 min.
116	164	1961	1366	~0°	± 2.5	Above specimen and conditions except oxidized in quiescent air at 1366 K for 40 min.
117	164	1961	1366	~0°	± 2.5	Above specimen and conditions except oxidized in quiescent air at 1366 K for 70 min.
118	40	1962	408-1061	~0°		Cobalt alloy N-155 (surface N-1); nominal composition: 21 Cr, 20 Co, 20 Ni, 3 Mo, 3 W, 1.5 Mn, 1 Nb, 0.5 Si, 0.15 C, 0.15 N, Fe balance; as received; increasing temp.
119	40	1962	408-655	~0°		Above specimen and conditions; decreasing temp.
120	40	1962	367	~0°		Different sample, same as curve 118 specimen and conditions except highly polished, mirror finish; oxide formation at 873 K for 3 hrs.
121	40	1962	447-1061	~0°		Different sample, same as curve 118 specimen and conditions except surface N-2; increasing temp.
122	40	1962	1033-447	~0°		Above specimen and conditions; decreasing temp.
123	40	1962	411-1084	~0°		Different sample, same as curve 121 specimen and conditions except heat treated; same results for increasing and decreasing temp.
124	40	1962	352-564	~0°		Poroloy (18-8 stainless steel); nominal composition: 18.45 Cr, 8.79 Ni, 0.50 Mn, 0.10 C, Fe balance; porosity between 28 and 31%.

SPECIFICATION TABLE NO. 361 (continued)

Curve No.	Ref. No.	Year	Temperature Range, K	Geometry θ'	Reported Error, %	Composition (weight percent), Specifications and Remarks
125	40	1962	367	~0°		Different sample, same as above specimen and conditions except porosity 28%.
126	40	1962	367	~0°		Different sample, same as above specimen and conditions except porosity 31%.
127	40	1962	367	~0°		Different sample, same as above specimen and conditions except porosity 43%.
128	75	1962	811-1444	~0°		Stainless steel 304; nominal composition: 18.00-20.00 Cr, 8.00-12.00 Ni, 2.00 max Mn, 1.00 max Si, 0.08 max C, Fe balance; machine finished; helium purge.
129	92	1963	1328-1466	~0°		Haynes Alloy N-155 (Multimet); 23.98-36.15 Fe, 19-21 Ni, 18.5-21 Co, 20-22.5 Cr, 2-3 W, 0.75-1.25 Nb and Ta, 2.5-3.5 Mo, 1.0-2.0 Mn, 0.5 max Cu, 1.0 max Si, 0.03 max S, 0.04 max P, 0.1-0.2 N_2, 0.08-0.16 C; polished; surface roughness 1 to 2 μ (RMS) measured with profilometer; measured in vacuum (3 to 4 x 10^{-4} mm Hg); 1st cycle.
130	92	1963	1289-1600	~0°		Above specimen and conditions; 2nd cycle.
131	92	1963	1239-1452	~0°		Curve 129 specimen and conditions except oxidized.
132	273	1962	805-1442	~0°		Stainless steel 304; nominal composition: 18.00-20.00 Cr, 8.00-12.00 Ni, 2.00 max Mn, 1.00 max Si, 0.08 max C, Fe balance; mechanical finish; measured in He gas.

DATA TABLE NO. 361 NORMAL TOTAL EMITTANCE OF [IRON + CHROMIUM + NICKEL + ΣX_i] ALLOYS

[Temperature, T, K; Emittance, ∈]

T	∈
CURVE 1	
573	0.47
673	0.48
773	0.47
873	0.50
CURVE 2	
633	0.77
723	0.79
803	0.80
873	0.82
933	0.83
1013	0.85
1073	0.87
CURVE 3	
633	0.79
713	0.81
793	0.83
883	0.87
953	0.85
1013	0.88
1073	0.89
CURVE 4*	
663	0.22
733	0.24
783	0.26
853	0.28
893	0.28
CURVE 5*	
573	0.42
653	0.40
763	0.42
853	0.42
933	0.46
993	0.50
1053	0.52
CURVE 6	
633	0.85
713	0.85
793	0.87
873	0.86
1053	0.87
CURVE 7*	
583	0.40
683	0.43
733	0.44
783	0.45
833	0.47
873	0.49
CURVE 8	
633	0.59
733	0.65
813	0.66
893	0.67
963	0.68
1033	0.71
CURVE 9	
613	0.87
733	0.86
803	0.88
873	0.88
933	0.93
993	0.93
1053	0.94
CURVE 10	
414	0.202
312	0.246
377	0.240
CURVE 11*	
531	0.223
1242	0.487
1239	0.603
CURVE 11* (cont.)	
1280	0.577
507	0.557
1218	0.547
505	0.573
CURVE 12	
697	0.241
852	0.301
924	0.315
1018	0.344
1080	0.389
1133	0.451
1183	0.477
CURVE 13*	
88.9	0.245
214	0.260
364	0.275
CURVE 14*	
464	0.300
706	0.270
872	0.320
947	0.345
1114	0.460
1197	0.580
1247	0.620
1347	0.370
556	0.230
1206	0.460
1189	0.525
1189	0.580
1172	0.690
439	0.440
488	0.400
1222	0.630
547	0.540
CURVE 15*	
412	0.100
305	0.145
390	0.203
CURVE 16*	
526	0.233
707	0.212
908	0.261
1067	0.466
1128	0.536
1186	0.640
1239	0.729
1258	0.675
469	0.427
1178	0.708
485	0.549
1185	0.699
483	0.502
CURVE 17*	
423	0.247
310	0.280
364	0.285
CURVE 18	
623	0.306
929	0.405
971	0.457
1043	0.490
1114	0.532
1221	0.626
CURVE 19*	
496	0.329
786	0.319
1164	0.585
1219	0.674
510	0.572
1132	0.663
504	0.581
CURVE 20	
88.9	0.260
206	0.320
372	0.305
CURVE 21*	
538	0.314
792	0.312
929	0.330
1051	0.396
1160	0.559
1217	0.663
1282	0.777
1321	0.657
1397	0.431
1402	0.362
523	0.331
1164	0.410
1329	0.552
1350	0.510
533	0.529
1329	0.504
517	0.517
CURVE 22*	
432	0.169
314	0.196
371	0.176
CURVE 23*	
679	0.174
893	0.203
1029	0.283
1127	0.362
1169	0.431
1214	0.594
1280	0.629
1300	0.597
1417	0.320
502	0.226
1147	0.282
1327	0.316
281	0.262
CURVE 24*	
423	0.368
318	0.394
360	0.368
CURVE 25*	
423	0.424
319	0.463
359	0.424
CURVE 26	
574	0.394
739	0.398
785	0.414
919	0.413
1029	0.426
1045	0.444
1088	0.458
1098	0.474
1191	0.568
1173	0.620
1159	0.639
CURVE 27*	
525	0.394
1192	0.495
1159	0.637
482	0.493
1177	0.650
474	0.516
CURVE 28*	
426	0.150
303	0.169
370	0.173
CURVE 29*	
532	0.211
719	0.217
919	0.232
1046	0.281
1170	0.412
1232	0.553
1290	0.624
1294	0.656
488	0.388
1175	0.607
537	0.488
CURVE 29* (cont.)	
1183	0.636
547	0.497
CURVE 30*	
434	0.324
303	0.381
368	0.336
CURVE 31*	
493	0.367
661	0.385
923	0.437
1017	0.475
1111	0.629
1194	0.696
1247	0.717
1267	0.705
525	0.561
548	0.485
1151	0.654
550	0.521
CURVE 32	
83.2	0.058
CURVE 33	
461	0.090
628	0.125
803	0.130
961	0.155
1055	0.235
1139	0.315
CURVE 34*	
533	0.150
850	0.125
CURVE 35*	
522	0.140
CURVE 36	
83.2	0.033
CURVE 37	
603	0.105
758	0.115
966	0.170
1072	0.225
1161	0.345
CURVE 38	
772	0.225
1050	0.225
1161	0.375
CURVE 39*	
800	0.270
1003	0.325
CURVE 40	
83.2	0.041
CURVE 41	
544	0.090
819	0.145
989	0.160
1105	0.215
1233	0.275
CURVE 42*	
811	0.190
989	0.240
CURVE 43*	
911	0.230
1108	0.280
CURVE 44	
83.2	0.072

* Not shown on plot

DATA TABLE NO. 361 (continued)

T	∈
CURVE 45	
680	0.165
905	0.235
1025	0.265
1130	0.315
1205	0.335
CURVE 46*	
769	0.220
911	0.265
CURVE 47*	
611	0.150
1289	0.280
CURVE 48	
83.2	0.022
CURVE 49	
333	0.085
694	0.085
955	0.160
CURVE 50*	
661	0.090
866	0.115
CURVE 51*	
511	0.080
678	0.100
950	0.140
CURVE 52	
83.2	0.044
CURVE 53	
425	0.091
519	0.103
628	0.116
828	0.152
CURVE 54*	
483	0.082
616	0.103
733	0.132
816	0.153
878	0.172
CURVE 55*	
625	0.131
844	0.196
CURVE 56*	
83.2	0.022
CURVE 57*	
472	0.046
666	0.053
755	0.057
CURVE 58	
444	0.049
683	0.053
778	0.056
878	0.079
1000	0.101
CURVE 59*	
650	0.052
933	0.086
CURVE 60*	
83.2	0.044
CURVE 61	
444	0.093
489	0.105
600	0.106
711	0.123
828	0.143
CURVE 62*	
661	0.106
939	0.154
1053	0.147
CURVE 63*	
605	0.103
CURVE 64*	
478	0.103
722	0.126
CURVE 65*	
616	0.101
CURVE 66	
83.2	0.027
CURVE 67	
494	0.080
678	0.090
833	0.125
1044	0.280
1222	0.400
CURVE 68*	
505	0.090
666	0.100
772	0.095
1039	0.300
CURVE 69*	
83.2	0.045
CURVE 70*	
444	0.100
628	0.120
CURVE 71*	
466	0.100
661	0.120
761	0.145
855	0.170
CURVE 72*	
500	0.100
700	0.145
839	0.170
961	0.255
1016	0.390
1116	0.490
CURVE 73*	
83.2	0.044
CURVE 74	
544	0.115
889	0.160
1005	0.250
1083	0.335
CURVE 75*	
622	0.165
894	0.275
CURVE 76*	
872	0.275
1044	0.350
1122	0.420
CURVE 77*	
83.2	0.036
CURVE 78	
494	0.085
633	0.105
775	0.115
939	0.150
1011	0.260
1072	0.335
1139	0.415
1205	0.440
CURVE 79*	
633	0.280
844	0.345
955	0.360
994	0.390
CURVE 80*	
728	0.320
1061	0.425
CURVE 81	
83.2	0.111
CURVE 82	
375	0.145
622	0.195
733	0.230
861	0.270
933	0.315
1000	0.423
1089	0.465
CURVE 83*	
561	0.260
761	0.345
CURVE 84	
83.2	0.155
CURVE 85	
594	0.285
744	0.320
844	0.355
930	0.405*
1011	0.445
1069	0.460
CURVE 86*	
544	0.300
789	0.360
CURVE 87*	
950	0.420
CURVE 88	
83.2	0.161
CURVE 89*	
422	0.110
605	0.125
CURVE 90	
553	0.130
755	0.155
883	0.170
941	0.200
1044	0.240
1144	0.325
1200	0.340
CURVE 91*	
83.2	0.111
CURVE 92	
539	0.155
778	0.170
891	0.255
1022	0.305
CURVE 93*	
900	0.290
1005	0.345
1161	0.390
CURVE 94	
755	0.10
922	0.18
CURVE 95	
755	0.65
922	0.69
1089	0.74
1255	0.72
CURVE 96*	
755	0.34
922	0.38
CURVE 97	
755	0.68
922	0.75
1089	0.70
1255	0.74
CURVE 98*	
373	0.13
CURVE 99*	
356	0.312
357	0.328
359	0.314
CURVE 99*(cont.)	
394	0.340
395	0.348
396	0.354
405	0.347
406	0.348
407	0.358
421	0.346
422	0.341
423	0.358
436	0.348
436	0.356
438	0.364
440	0.366
441	0.366
CURVE 100	
350	0.628
351	0.606
351	0.594
352	0.644
353	0.644*
364	0.642
365	0.654
365	0.656*
367	0.616
369	0.624
372	0.624*
373	0.628*
380	0.654
382	0.645
386	0.668
396	0.672
398	0.670*
406	0.674
410	0.676*
422	0.660
423	0.654*
424	0.686
432	0.698
435	0.712

* Not shown on plot

DATA TABLE NO. 361 (continued)

T	ε
CURVE 101	
351	0.560
391	0.570
405	0.522
418	0.560
433	0.524
445	0.542
446	0.560
CURVE 102	
355	0.510
360	0.484
364	0.510
366	0.486
369	0.476
370	0.504
377	0.500
383	0.486
383	0.508
393	0.504
395	0.506*
404	0.511
419	0.492
425	0.524
429	0.510
435	0.522
438	0.518*
444	0.484
456	0.484
CURVE 103	
419	0.850
422	0.825
458	0.850
475	0.845
594	0.845
CURVE 104	
342	0.855
353	0.845
394	0.840
425	0.840
453	0.850

T	ε
CURVE 104 (cont.)	
528	0.850
577	0.850
579	0.845
646	0.855
CURVE 105	
344	0.650
350	0.670
375	0.625
380	0.610
392	0.605
397	0.610*
400	0.640
419	0.625
428	0.620
433	0.620*
436	0.615
442	0.615*
483	0.610
525	0.595
533	0.605
605	0.630
608	0.595
617	0.600
661	0.610
CURVE 106*	
455	0.205
472	0.170
542	0.225
550	0.204
583	0.210
594	0.230
632	0.220
650	0.225
CURVE 107*	
353	0.180
380	0.170
414	0.180
503	0.190
539	0.195

T	ε
CURVE 107* (cont.)	
563	0.203
592	0.200
611	0.210
655	0.205
CURVE 108	
344	0.155
433	0.160
483	0.170
528	0.175
603	0.195
CURVE 109	
644	0.09
811	0.16
1089	0.31
1367	0.51
1644	0.72*
CURVE 110*	
366	0.31
477	0.31
588	0.31
699	0.33
CURVE 111*	
366	0.31
477	0.33
588	0.31
699	0.33
CURVE 112	
366	0.36
477	0.37
588	0.38
699	0.39

T	ε
CURVE 113	
366	0.33
477	0.34
588	0.35
699	0.36
CURVE 114	
1366	0.841
CURVE 115	
1366	0.867
CURVE 116	
1366	0.870
CURVE 117*	
1366	0.874
CURVE 118	
408	0.165
444	0.150
683	0.175
891	0.155
1061	0.550
CURVE 119*	
408	0.225
655	0.338
CURVE 120*	
367	0.16
CURVE 121	
447	0.190
722	0.220
855	0.290
916	0.325
964	0.400
1061	0.600

T	ε
CURVE 122	
1033	0.566
683	0.410
447	0.300
CURVE 123	
411	0.445
416	0.410
677	0.510
844	0.570
1084	0.628
CURVE 124	
352	0.23
358	0.26
425	0.25
425	0.26*
469	0.26
489	0.25
490	0.26
491	0.26*
530	0.26
564	0.26
CURVE 125*	
367	0.23
CURVE 126*	
367	0.23
CURVE 127*	
367	0.32
CURVE 128	
811	0.145
811	0.175
1128	0.480
1400	0.720
1444	0.730

T	ε
CURVE 129*	
1328	0.230
1369	0.250
1466	0.245
CURVE 130*	
1289	0.205
1354	0.220
1392	0.220
1490	0.240
1549	0.250*
1575	0.270*
1600	0.290*
CURVE 131*	
1239	0.720
1255	0.730
1311	0.760
1355	0.780
1452	0.700
CURVE 132*	
805	0.135
808	0.161
1128	0.477
1402	0.711
1442	0.722

* Not shown on plot

SPECIFICATION TABLE NO. 362 ANGULAR TOTAL EMITTANCE OF [IRON + CHROMIUM + NICKEL + ΣX_i] ALLOYS

Curve No.	Ref. No.	Year	Temperature K	Angular Range, °	Geometry θ'	Reported Error, %	Composition (weight percent), Specifications and Remarks
1	281	1966	705	0-70	θ'	< 6	1 Cr 18 Ni 9 Ti; nominal composition: 17.0-20.0 Cr, 8.0-11.0 Ni, 2.0 Mn, 0.8 Si, 0.8 max Ti, 0.12 C, 0.035 P, 0.030 S, Fe balance; rolled; mechanically polished; surface roughness 10 to 18.7 μ; data extracted from smooth curve.
2	281	1966	863	0-70	θ'	< 6	Above specimen and conditions.
3	281	1966	971	0-70	θ'	< 6	Above specimen and conditions.
4	281	1966	640	0-70	θ'	< 6	Above specimen and conditions.
5	281	1966	1045	0-70	θ'	< 6	Above specimen and conditions.
6	281	1966	729	0-70	θ'	< 6	Different sample, same as curve 1 specimen and conditions except surface roughness 3.2 - 6.3 μ.
7	281	1966	848	0-70	θ'	< 6	Above specimen and conditions.
8	281	1966	948	0-70	θ'	< 6	Above specimen and conditions.
9	281	1966	730	0-70	θ'	< 6	Above specimen and conditions.
10	281	1966	956	0-70	θ'	< 6	Above specimen and conditions.

DATA TABLE NO. 362 ANGULAR TOTAL EMITTANCE OF [IRON + CHROMIUM + NICKEL + ΣX_i] ALLOYS

[Angle, θ', °; Emittance, ∈; Temperature, T, K]

θ'	∈	θ'	∈	θ'	∈
CURVE 1* T = 705		CURVE 5* T = 1045		CURVE 9* T = 730	
0	0.35	0	0.70	0	0.30
10	0.36	10	0.70	10	0.30
20	0.35	20	0.71	20	0.30
30	0.31	30	0.72	30	0.30
40	0.33	40	0.72	40	0.30
45	0.40	50	0.72	50	0.30
50	0.45	60	0.71	60	0.30
60	0.45	70	0.66	70	0.30
70	0.46				
CURVE 2* T = 863		CURVE 6* T = 729		CURVE 10* T = 956	
0	0.48	0	0.29	0	0.39
10	0.49	10	0.28	10	0.39
20	0.49	20	0.28	20	0.39
30	0.50	30	0.28	30	0.39
40	0.50	40	0.28	40	0.39
50	0.51	50	0.28	50	0.40
60	0.52	60	0.28	60	0.39
70	0.53	70	0.29	70	0.39
CURVE 3* T = 971		CURVE 7* T = 848			
0	0.60	0	0.30		
10	0.60	10	0.31		
20	0.60	20	0.31		
30	0.61	30	0.31		
40	0.68	40	0.32		
50	0.68	50	0.30		
60	0.67	60	0.31		
70	0.60	70	0.32		
CURVE 4* T = 640		CURVE 8* T = 948			
0	0.58	0	0.39		
10	0.57	10	0.40		
20	0.56	20	0.40		
30	0.56	30	0.41		
40	0.65	40	0.40		
50	0.65	50	0.42		
60	0.64	60	0.41		
70	0.56	70	0.41		

* Not shown on plot

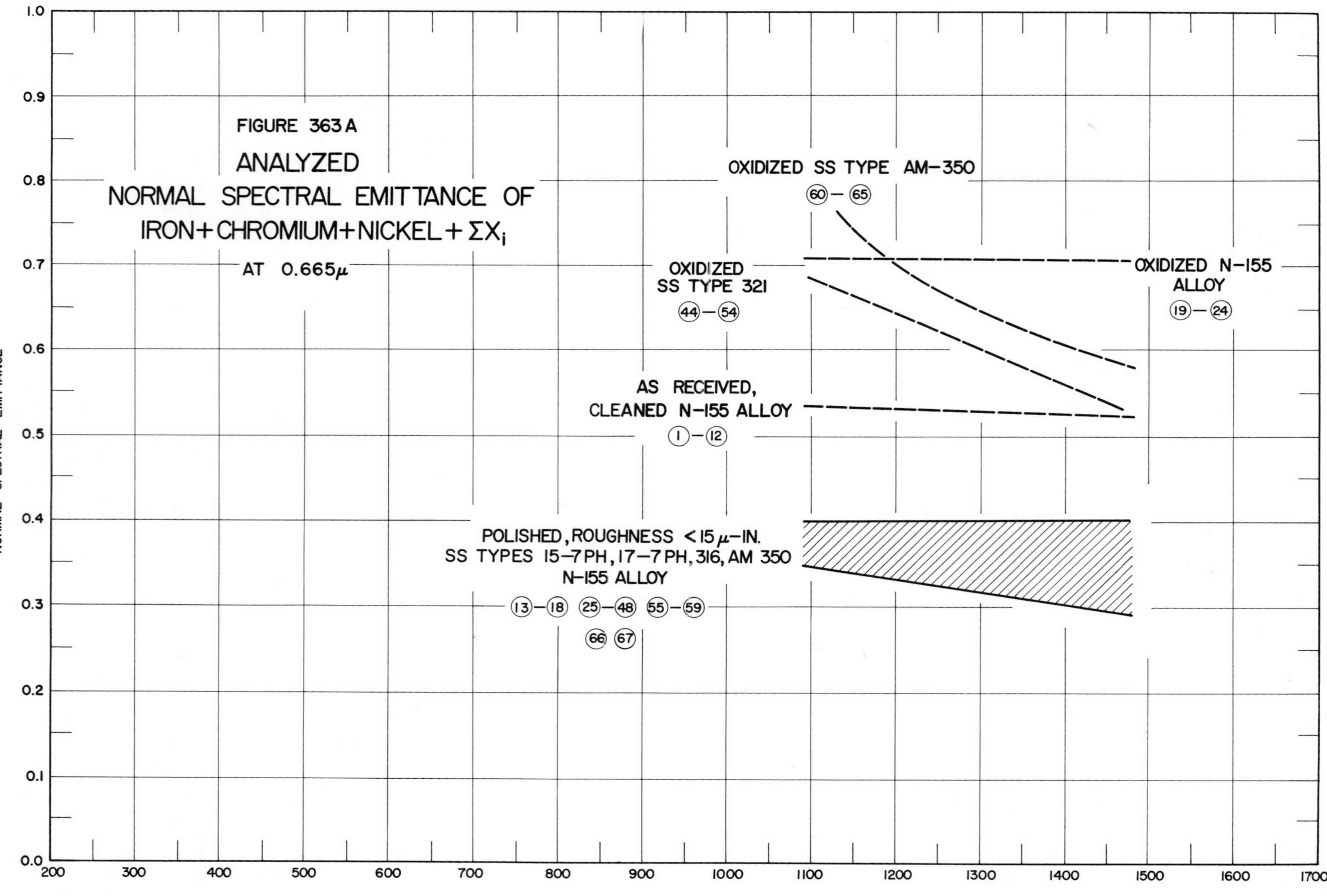
FIGURE 363A
ANALYZED
NORMAL SPECTRAL EMITTANCE OF
IRON+CHROMIUM+NICKEL+ΣX_i
AT 0.665μ
OXIDIZED SS TYPE AM-350
60 — 65
OXIDIZED
SS TYPE 321
44 — 54
OXIDIZED N-155
ALLOY
19 — 24
AS RECEIVED,
CLEANED N-155 ALLOY
1 — 12
POLISHED, ROUGHNESS <15 μ-IN.
SS TYPES 15-7PH, 17-7PH, 316, AM 350
N-155 ALLOY
13 — 18 25 — 48 55 — 59
66 67
NORMAL SPECTRAL EMITTANCE
1.0
0.9
0.8
0.7
0.6
0.5
0.4
0.3
0.2
0.1
0.0
TEMPERATURE, K
200
300
400
500
600
700
800
900
1000
1100
1200
1300
1400
1500
1600
1700

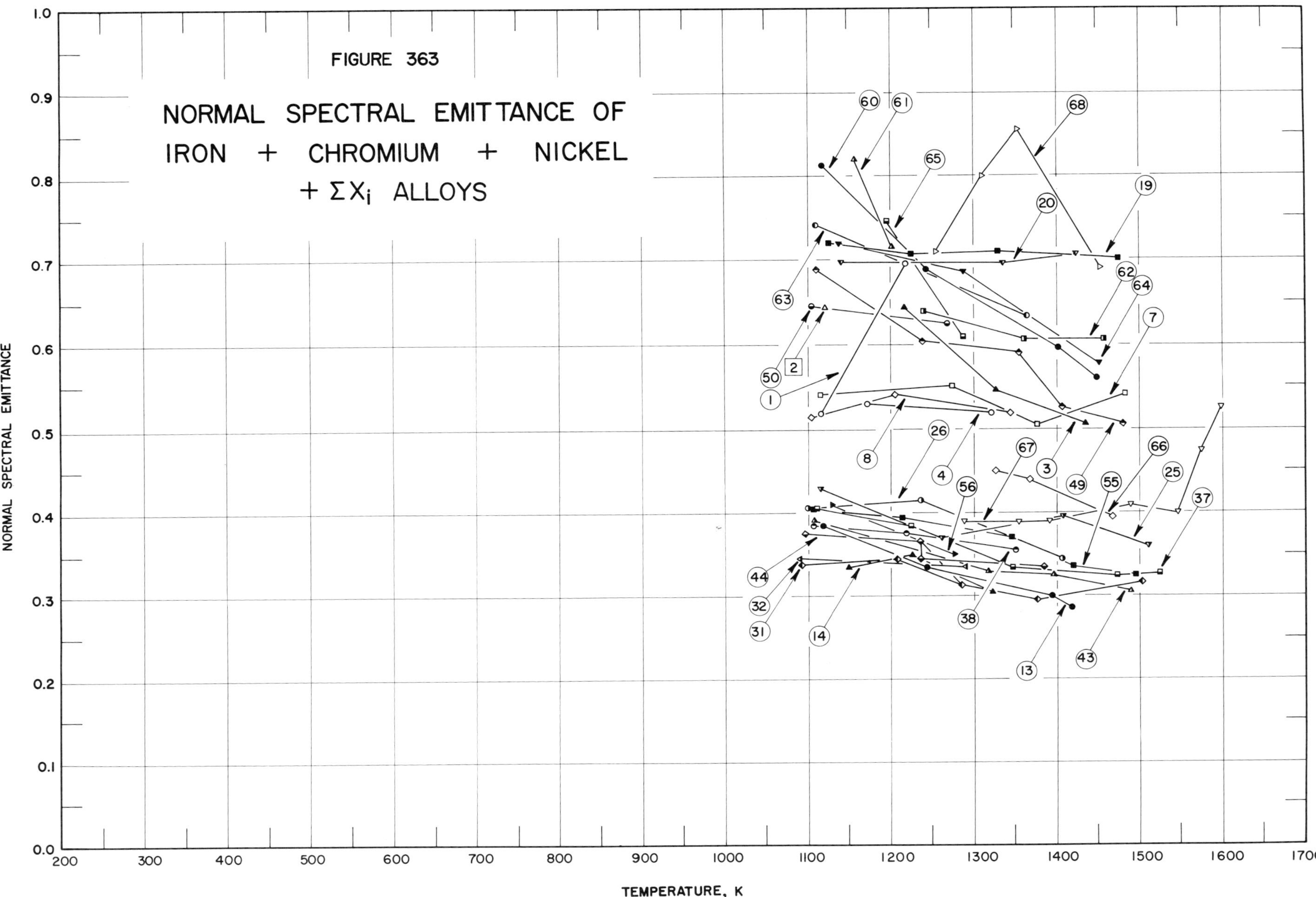
FIGURE 363
NORMAL SPECTRAL EMITTANCE OF
IRON + CHROMIUM + NICKEL
+ ΣX_i ALLOYS
NORMAL SPECTRAL EMITTANCE
TEMPERATURE, K
0.0
0.1
0.2
0.3
0.4
0.5
0.6
0.7
0.8
0.9
1.0
200
300
400
500
600
700
800
900
1000
1100
1200
1300
1400
1500
1600
1700
60
61
68
65
19
20
62
64
7
63
2
50
1
8
26
4
67
3
49
55
66
25
37
56
44
32
31
14
38
13
43

SPECIFICATION TABLE NO. 363 NORMAL SPECTRAL EMITTANCE OF [IRON + CHROMIUM + NICKEL + ΣX_i] ALLOYS

Curve No.	Ref. No.	Year	Wavelength μ	Temperature Range, K	Geometry θ'	Reported Error, %	Composition (weight percent), Specifications and Remarks
1	34	1957	0.665	1116-1219	~0°		Cobalt alloy N-155; nominal composition: 21 Cr, 20 Co, 20 Ni, 3 Mo, 3 W, 1.5 Mn, 1. Nb, 0.5 Si, 0.15 C, 0.15 N, Fe balance; as received; measured in vacuum (5×10^{-4} mm Hg); increasing temp, cycle 1.
2	34	1957	0.665	1122	~0°		Above specimen and conditions; decreasing temp, cycle 1.
3	34	1957	0.665	1216-1436	~0°		Above specimen and conditions; cycle 2.
4	34	1957	0.665	1322-1172	~0°		Above specimen and conditions; decreasing temp, cycle 2.
5	34	1957	0.665	1139-1450	~0°		Above specimen and conditions; cycle 3.
6	34	1957	0.665	1277-1139	~0°		Above specimen and conditions; decreasing temp, cycle 3.
7	34	1957	0.665	1116-1483	~0°		Different sample, same as curve 1 specimen and conditions except cleaned in liquid detergent; increasing temp, cycle 1.
8	34	1957	0.665	1344-1105	~0°		Above specimen and conditions; decreasing temp, cycle 1.
9	34	1957	0.665	1227-1477	~0°		Above specimen and conditions; cycle 2.
10	34	1957	0.665	1372-1164	~0°		Above specimen and conditions; decreasing temp, cycle 2.
11	34	1957	0.665	1216-1450	~0°		Above specimen and conditions; cycle 3.
12	34	1957	0.665	1294-1100	~0°		Above specimen and conditions; decreasing temp, cycle 3.
13	34	1957	0.665	1119-1419	~0°		Different sample, same as curve 1 specimen and conditions except polished; increasing temp, cycle 1.
14	34	1957	0.665	1322-1150	~0°		Above specimen and conditions; decreasing temp, cycle 1.
15	34	1957	0.665	1233-1405	~0°		Above specimen and conditions; cycle 2.
16	34	1957	0.665	1325-1119	~0°		Above specimen and conditions; decreasing temp, cycle 2.
17	34	1957	0.665	1419-1450	~0°		Above specimen and conditions; cycle 3.
18	34	1957	0.665	1119-1419	~0°		Above specimen and conditions; decreasing temp, cycle 3.
19	34	1957	0.665	1127-1477	~0°		Different sample, same as curve 1 specimen and conditions except oxidized in air at red heat for 30 min; increasing temp, cycle 1.
20	34	1957	0.665	1141-1425	~0°		Above specimen and conditions; decreasing temp, cycle 1.
21	34	1957	0.665	1152-1475	~0°		Above specimen and conditions; cycle 2.
22	34	1957	0.665	1219-1452	~0°		Above specimen and conditions; decreasing temp, cycle 2.
23	34	1957	0.665	1311	~0°		Above specimen and conditions; cycle 3.
24	34	1957	0.665	1143-1289	~0°		Above specimen and conditions; decreasing temp, cycle 3.

SPECIFICATION TABLE NO. 363 (continued)

Curve No.	Ref. No.	Year	Wavelength μ	Temperature Range, K	Geometry θ'	Reported Error, %	Composition (weight percent), Specifications and Remarks
25	34	1957	0.665	1114-1511	$\sim 0°$		Stainless steel type PH 15-7 Mo; nominal composition: 15 Cr, 7 Ni, 2.25 Mo, 1.15 Al, 0.70 Mn, 0.40 Si, 0.07 C, Fe balance; same results obtained for two different surface conditions: ~2 and 15 microinches rms; measured in vacuum (5×10^{-4} mm Hg); increasing temp, cycle 1.
26	34	1957	0.665	1408-1100	$\sim 0°$		Above specimen and conditions; decreasing temp, cycle 1.
27	34	1957	0.665	1244-1511	$\sim 0°$		Above specimen and conditions; cycle 2.
28	34	1957	0.665	1372-1102	$\sim 0°$		Above specimen and conditions; decreasing temp, cycle 2.
29	34	1957	0.665	1236-1516	$\sim 0°$		Above specimen and conditions; cycle 3.
30	34	1957	0.665	1116	$\sim 0°$		Above specimen and conditions; decreasing temp, cycle 3.
31	34	1957	0.665	1092-1502	$\sim 0°$		Stainless steel type 17-7 PH; nominal composition : 17 Cr, 7 Ni, 1.15 Al, 0.70 Mn, 0.40 Si, 0.07 C, Fe balance; same results obtained for two different surface conditions: ~2 and 15 microinches rms; measured in vacuum (5×10^{-4} mm Hg); increasing temp, cycle 1.
32	34	1957	0.665	1291-1091	$\sim 0°$		Above specimen and conditions; decreasing temp, cycle 1.
33	34	1957	0.665	1202-1500	$\sim 0°$		Above specimen and conditions; cycle 2.
34	34	1957	0.665	1100	$\sim 0°$		Above specimen and conditions; decreasing temp, cycle 2.
35	34	1957	0.665	1269-1491	$\sim 0°$		Above specimen and conditions; cycle 3.
36	34	1957	0.665	1433-1105	$\sim 0°$		Above specimen and conditions; decreasing temp, cycle 3.
37	34	1957	0.665	1111-1525	$\sim 0°$		Stainless steel type 316; nominal composition : 16.00-18.00 Cr, 10.00-14.00 Ni, 2.00-3.00 Mo, 2.00 max Mn, 1.00 max Si, 0.08 max C, Fe balance; same results obtained for two different surface conditions: ~2 and 15 microinches rms; measured in vacuum (5×10^{-4} mm Hg); increasing temp, cycle 1.
38	34	1957	0.665	1351-1108	$\sim 0°$		Above specimen and conditions; decreasing temp, cycle 1.
39	34	1957	0.665	1194-1508	$\sim 0°$		Above specimen and conditions; cycle 2.
40	34	1957	0.665	1430-1227	$\sim 0°$		Above specimen and conditions; decreasing temp, cycle 2.
41	34	1957	0.665	1097-1497	$\sim 0°$		Above specimen and conditions; cycle 3.
42	34	1957	0.665	1325-1097	$\sim 0°$		Above specimen and conditions; decreasing temp, cycle 3.
43	34	1957	0.665	1108-1490	$\sim 0°$		Stainless steel type 321; nominal composition : 17.00-19.00 Cr, 9.00-12.00 Ni, 2.00 max Mn, 1.00 max Si, 0.08 max C, 5 x C min Ti, Fe balance; same results obtained for 3 different surface conditions : 1. surface roughness ~2 microinches, 2. dull finish with surface roughness ~6 microinches, 3. bright finish; measured in vacuum (5×10^{-4} mm Hg); increasing temp, cycle 1.
44	34	1957	0.665	1386-1097	$\sim 0°$		Above specimen and conditions; decreasing temp, cycle 1.
45	34	1957	0.665	1246-1461	$\sim 0°$		Above specimen and conditions; cycle 2.
46	34	1957	0.665	1302-1086	$\sim 0°$		Above specimen and conditions; decreasing temp, cycle 2.

SPECIFICATION TABLE NO. 363 (continued)

Curve No.	Ref. No.	Year	Wavelength μ	Temperature Range, K	Geometry θ'	Reported Error, %	Composition (weight percent), Specifications and Remarks
47	34	1957	0.665	1250-1475	~0°		Above specimen and conditions; cycle 3.
48	34	1957	0.665	1300-1114	~0°		Above specimen and conditions; decreasing temp, cycle 3.
49	34	1957	0.665	1111-1480	~0°		Stainless steel type 321; nominal composition : 17.00-19.00 Cr, 9.00-12.00 Ni, 2.00 max Mn, 1.00 max Si, 0.08 max C, 5 x C min Ti, Fe balance; dull finish with surface roughness ~6 microinches rms; oxidized in air at red heat for 30 min; measured in vacuum (5 x 10^{-4} mm Hg); increasing temp, cycle 1.
50	34	1957	0.665	1269-1105	~0°		Above specimen and conditions; decreasing temp, cycle 1.
51	34	1957	0.665	1239-1486	~0°		Above specimen and conditions; cycle 2.
52	34	1957	0.665	1372-1125	~0°		Above specimen and conditions; decreasing temp, cycle 2.
53	34	1957	0.665	1269-1461	~0°		Above specimen and conditions; cycle 3.
54	34	1957	0.665	1419-1186	~0°		Above specimen and conditions; decreasing temp, cycle 3.
55	34	1957	0.665	1108-1497	~0°		Stainless steel type AM 350; nominal composition : 16.50 Cr, 4.25 Ni, 2.75 Mo, 0.75 Mn, 0.10 N, 0.10 C, Fe balance; same results obtained for two different surface conditions: ~2 and 15 microinches rms; measured in vacuum (5 x 10^{-4} mm Hg); increasing temp, cycle 1.
56	34	1957	0.665	1277-1130	~0°		Above specimen and conditions; decreasing temp, cycle 1.
57	34	1957	0.665	1111-1472	~0°		Above specimen and conditions; cycle 2.
58	34	1957	0.665	1289	~0°		Above specimen and conditions; decreasing temp, cycle 2.
59	34	1957	0.665	1139-1475	~0°		Above specimen and conditions; cycle 3.
60	34	1957	0.665	1119-1450	~0°		Stainless steel type AM 350; nominal composition : 16.50 Cr, 4.25 Ni, 2.75 Mo, 0.75 Mn, 0.10 N, 0.10 C, Fe balance; surface roughness ~2 microinches rms; oxidized in air at red heat for 30 min; measured in vacuum (5 x 10^{-4} mm Hg); increasing temp, cycle 1.
61	34	1957	0.665	1202-1158	~0°		Above specimen and conditions; decreasing temp, cycle 1.
62	34	1957	0.665	1241-1458	~0°		Above specimen and conditions; cycle 2.
63	34	1957	0.665	1366-1111	~0°		Above specimen and conditions; decreasing temp, cycle 2.
64	34	1957	0.665	1139-1452	~0°		Above specimen and conditions; cycle 3.
65	34	1957	0.665	1289-1197	~0°		Above specimen and conditions; decreasing temp, cycle 3.
66	92	1963	0.65	1328-1468	~0°		Haynes Alloy N-155 (Multimet); 23.98-36.15 Fe, 19-21 Ni, 18.5-21 Co, 20-22.5 Cr, 2-3 W, 0.75-1.25 Nb and Ta, 2.5-3.5 Mo, 1.0-2.0 Mn, 0.5 max Cu, 1.0 max Si, 0.03 max S, 0.04 max P, 0.1-0.2 N_2, 0.08-0.16 C; polished; surface roughness 1-2 μ (RMS) measured with profilometer; measured in vacuum (3 to 4 x 10^{-4} mm Hg); authors assumed specimen was a grey body; 1st cycle.
67	92	1963	0.65	1290-1600	~0°		Above specimen and conditions; 2nd cycle.

SPECIFICATION TABLE NO. 363 (continued)

Curve No.	Ref. No.	Year	Wavelength μ	Temperature Range, K	Geometry θ'	Reported Error, %	Composition (weight percent), Specifications and Remarks
68	92	1963	0.65	1255-1453	$\sim 0°$		Curve 66 specimen and conditions except oxidized.

DATA TABLE NO. 363 NORMAL SPECTRAL EMITTANCE OF [IRON + CHROMIUM + NICKEL + ΣX_i] ALLOYS

[Temperature, T, K; Emittance, ∈; Wavelength, λ, μ]

T	∈
CURVE 1 λ = 0.665	
1116	0.520
1219	0.695
CURVE 2 λ = 0.665	
1122	0.645
CURVE 3 λ = 0.665	
1216	0.645
1327	0.545
1436	0.508
CURVE 4 λ = 0.665	
1322	0.520
1172	0.530
CURVE 5* λ = 0.665	
1139	0.560
1233	0.525
1350	0.530
1394	0.505
1450	0.530
CURVE 6* λ = 0.665	
1277	0.540
1139	0.525
CURVE 7 λ = 0.665	
1116	0.540
1275	0.550
1377	0.505
1483	0.540

T	∈
CURVE 8 λ = 0.665	
1344	0.520
1205	0.540
1105	0.515
CURVE 9* λ = 0.665	
1227	0.535
1361	0.510
1477	0.515
CURVE 10* λ = 0.665	
1372	0.540
1164	0.550
CURVE 11* λ = 0.665	
1216	0.510
1305	0.508
1450	0.535
CURVE 12* λ = 0.665	
1294	0.540
1100	0.550
CURVE 13 λ = 0.665	
1119	0.385
1244	0.335
1394	0.300
1419	0.285
CURVE 14 λ = 0.665	
1322	0.303
1227	0.350
1150	0.335

T	∈
CURVE 15* λ = 0.665	
1233	0.365
1316	0.320
1405	0.305
CURVE 16* λ = 0.665	
1325	0.315
1119	0.365
CURVE 17* λ = 0.665	
1419	0.290
1450	0.305
CURVE 18* λ = 0.665	
1119	0.350
1150	0.325
1283	0.303
1333	0.315
1419	0.280
CURVE 19 λ = 0.665	
1127	0.720
1225	0.707
1330	0.710
1477	0.700
CURVE 20 λ = 0.665	
1141	0.696
1336	0.695
1425	0.706

T	∈
CURVE 21* λ = 0.665	
1152	0.725
1327	0.706
1427	0.713
1455	0.695
1475	0.715
CURVE 22* λ = 0.665	
1219	0.705
1308	0.720
1452	0.725
CURVE 23* λ = 0.665	
1311	0.693
CURVE 24* λ = 0.665	
1143	0.712
1289	0.710
CURVE 25 λ = 0.665	
1114	0.430
1261	0.370
1408	0.395
1511	0.360
CURVE 26 λ = 0.665	
1408	0.345
1236	0.415
1100	0.405
CURVE 27* λ = 0.665	
1244	0.390

T	∈
CURVE 27 (cont.)*	
1405	0.370
1511	0.340
CURVE 28* λ = 0.665	
1372	0.370
1102	0.375
CURVE 29* λ = 0.665	
1236	0.365
1314	0.365
1461	0.375
1516	0.375
CURVE 30* λ = 0.665	
1116	0.415
CURVE 31 λ = 0.665	
1092	0.338
1208	0.345
1286	0.312
1377	0.295
1502	0.317
CURVE 32 λ = 0.665	
1291	0.335
1091	0.345
CURVE 33* λ = 0.665	
1202	0.350
1214	0.324
1380	0.310
1500	0.277

T	∈
CURVE 34* λ = 0.665	
1100	0.360
CURVE 35* λ = 0.665	
1269	0.330
1491	0.289
CURVE 36* λ = 0.665	
1433	0.290
1252	0.295
1105	0.330
CURVE 37 λ = 0.665	
1111	0.405
1225	0.385
1347	0.335
1472	0.325
1525	0.328
CURVE 38 λ = 0.665	
1351	0.355
1219	0.375
1108	0.385
CURVE 39* λ = 0.665	
1194	0.375
1325	0.340
1430	0.325
1508	0.315

T	∈
CURVE 40* λ = 0.665	
1430	0.355
1227	0.355
CURVE 41* λ = 0.665	
1097	0.398
1252	0.376
1325	0.358
1497	0.344
CURVE 42* λ = 0.665	
1325	0.385
1097	0.370
CURVE 43 λ = 0.665	
1108	0.392
1316	0.330
1397	0.326
1490	0.305
CURVE 44 λ = 0.665	
1386	0.336
1236	0.365
1236	0.345
1097	0.375
CURVE 45* λ = 0.665	
1246	0.335
1322	0.355
1386	0.315
1461	0.286

T	∈
CURVE 46* λ = 0.665	
1302	0.335
1086	0.405
CURVE 47* λ = 0.665	
1250	0.362
1336	0.333
1475	0.310
CURVE 48* λ = 0.665	
1300	0.355
1114	0.384
CURVE 49 λ = 0.665	
1111	0.688
1239	0.602
1355	0.589
1408	0.525
1480	0.505
CURVE 50 λ = 0.665	
1269	0.625
1105	0.646
CURVE 51* λ = 0.665	
1239	0.635
1364	0.545
1486	0.535
CURVE 52* λ = 0.665	
1372	0.575
1125	0.685

* Not shown on plot

DATA TABLE NO. 363 (continued)

T	∈
CURVE 53*	**λ = 0.665**
1269	0.582
1402	0.565
1461	0.518
CURVE 54*	**λ = 0.665**
1419	0.540
1186	0.660
CURVE 55	**λ = 0.665**
1108	0.405
1214	0.395
1347	0.370
1421	0.335
1497	0.325
CURVE 56	**λ = 0.665**
1277	0.350
1130	0.410
CURVE 57*	**λ = 0.665**
1111	0.383
1194	0.375
1355	0.355
1472	0.348
CURVE 58*	**λ = 0.665**
1289	0.370
CURVE 59*	**λ = 0.665**
1139	0.395
1475	0.320
CURVE 60	**λ = 0.665**
1119	0.812
1244	0.688
1402	0.595
1450	0.560
CURVE 61	**λ = 0.665**
1202	0.715
1158	0.820
CURVE 62	**λ = 0.665**
1241	0.640
1361	0.605
1458	0.605
CURVE 63	**λ = 0.665**
1366	0.633
1111	0.740
CURVE 64	**λ = 0.665**
1139	0.720
1289	0.685
1452	0.578
CURVE 65	**λ = 0.665**
1289	0.608
1197	0.745
CURVE 66	**λ = 0.65**
1328	0.450
1368	0.440
1468	0.395
CURVE 67	**λ = 0.65**
1290	0.390
1355	0.390
1392	0.390
1490	0.410
1549	0.400
1576	0.475
1600	0.525
CURVE 68	**λ = 0.65**
1255	0.710
1311	0.800
1355	0.855
1453	0.690

* Not shown on plot

FIGURE 364 A

NORMAL SPECTRAL EMITTANCE OF IRON+CHROMIUM+NICKEL+ΣX_i

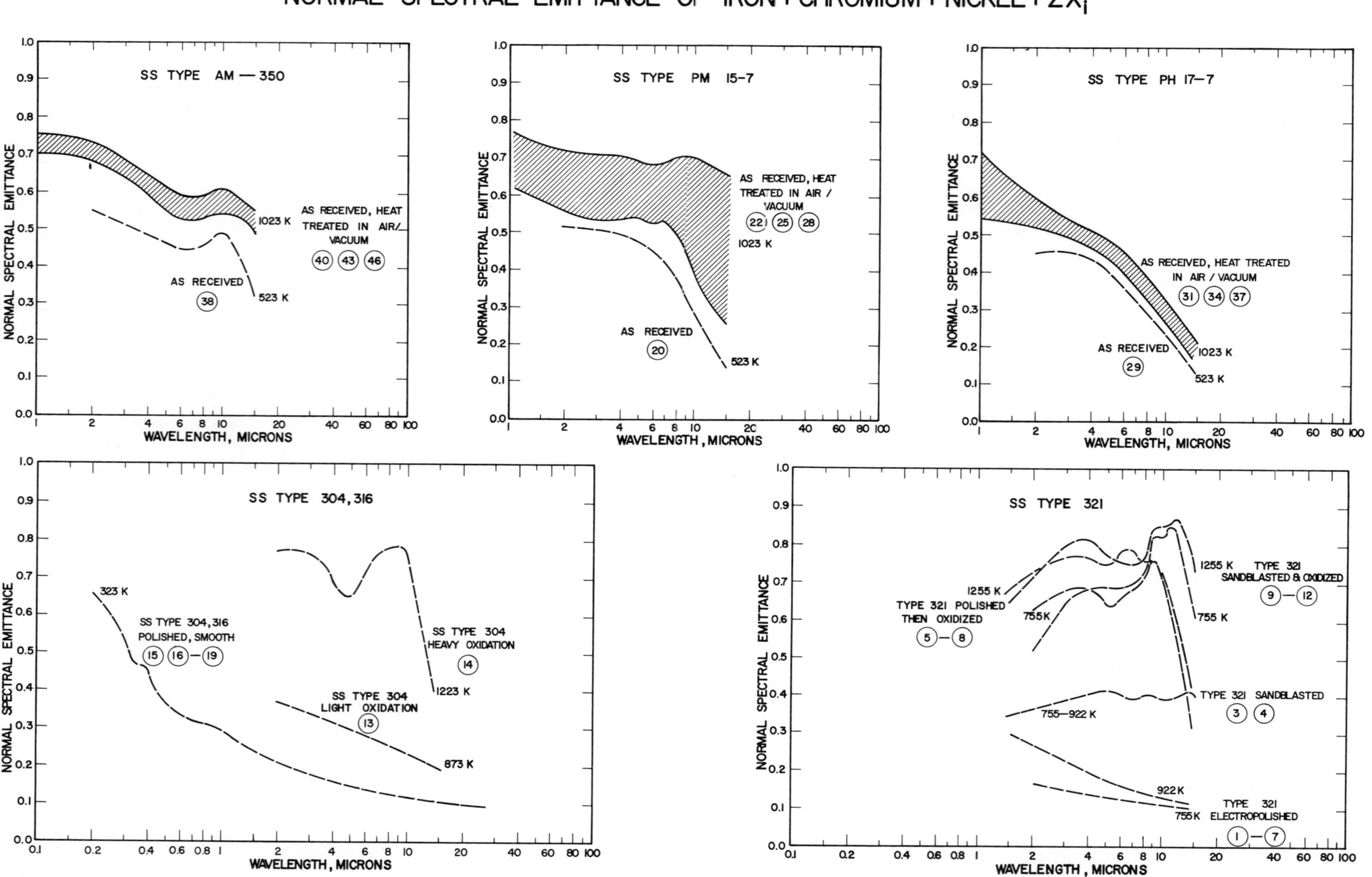

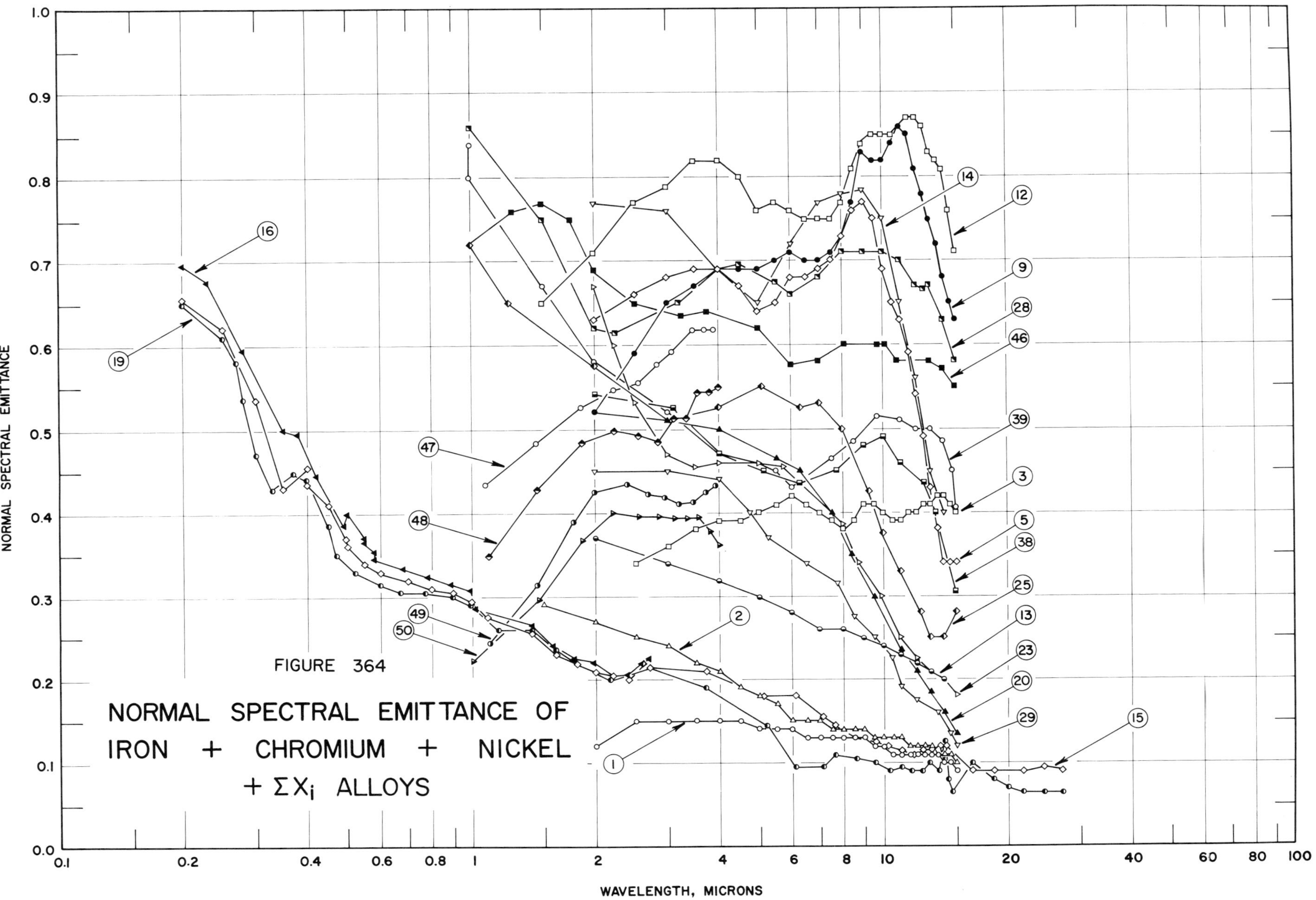

FIGURE 364

NORMAL SPECTRAL EMITTANCE OF IRON + CHROMIUM + NICKEL + ΣX_i ALLOYS

SPECIFICATION TABLE NO. 364 NORMAL SPECTRAL EMITTANCE OF [IRON + CHROMIUM + NICKEL + ΣX_i] ALLOYS

Curve No.	Ref. No.	Year	Temperature K	Wavelength Range, μ	Geometry θ'	Reported Error, %	Composition (weight percent), Specifications and Remarks
1	170	1959	755	2.0-15.0	$\sim 0°$	±4	Stainless steel 321; titanium-stabilized 18 Cr, 8 Ni austenitic stainless steel; electropolished.
2	170	1959	922	1.5-15.0	$\sim 0°$	±4	Above specimen and conditions.
3	170	1959	755	2.5-15.0	$\sim 0°$	±4	Different sample, same as curve 1 specimen and conditions; also sandblasted by 40-mesh glass sand and air at 40 psi.
4	170	1959	922	1.5-15.0	$\sim 0°$	±4	Above specimen and conditions.
5	170	1959	755	2.0-15.0	$\sim 0°$	±4	Different sample, same as curve 1 specimen and conditions except oxidized in air at 1255 K for 0.50 hr.
6	170	1959	922	2.0-15.0	$\sim 0°$	±4	Above specimen and conditions.
7	170	1959	1089	1.5-15.0	$\sim 0°$	±4	Above specimen and conditions.
8	170	1959	1255	1.5-15.0	$\sim 0°$	±4	Above specimen and conditions.
9	170	1959	755	2.0-15.0	$\sim 0°$	±4	Different sample, same as curve 1 specimen and conditions; also sandblasted by 40-mesh glass sand and air at 40 psi; oxidized in air at 1255 K for 0.50 hr.
10	170	1959	922	1.5-15.0	$\sim 0°$	±4	Above specimen and conditions.
11	170	1959	1089	1.5-15.0	$\sim 0°$	±4	Above specimen and conditions.
12	170	1959	1255	1.5-15.0	$\sim 0°$	±4	Above specimen and conditions.
13	156	1960	873	2.0-14.0	$\sim 0°$	±4	Stainless steel 304; nominal composition : 18.00-20.00 Cr, 8.00-12.00 Ni, 2.00 max Mn, 1.00 max Si, 0.08 max C, Fe balance; oxidized in air for 3 hrs at 873 K; measured in air.
14	156	1960	1273	2.0-14.0	$\sim 0°$	±4	Different sample, same as above specimen and conditions except oxidized in air for 6 hrs at 1273 K.
15	159	1963	323	0.20-27.00	5°		Stainless steel 304; nominal composition : 18.00-20.00 Cr, 8.00-12.00 Ni, 2.00 max Mn, 1.00 max Si, 0.08 max C, Fe balance; surface roughness 0.75 microinches (center line average); measured in nitrogen; computed from $\epsilon = 1 - R(2\pi, 5°)$; [Author's designation : specimen 1].
16	159	1963	323	0.20-2.68	5°		316 stainless steel; nominal composition : 16.00-18.00 Cr, 10.00-14.00 Ni, 2.00-3.00 Mo, 2.00 max Mn, 1.00 max Si, 0.08 max C, Fe balance; surface roughness 0.75-1.3 microinches (center line average); measured in nitrogen; computed from $\epsilon = 1 - R(2\pi, 5°)$; [Author's designation: specimen 1]
17	159	1963	323	0.22-27.10	5°		Different sample, same as above specimen and conditions; surface roughness 1.75-2.00 microinches (center line average); [Author's designation : specimen 3].
18	159	1963	323	0.22-27.00	5°		Different sample, same as above specimen and conditions; surface roughness 4.5-7.0 microinches (center line average); [Author's designation : specimen 4].

SPECIFICATION TABLE NO. 364 (continued)

Curve No.	Ref. No.	Year	Temperature K	Wavelength Range, μ	Geometry θ'	Reported Error, %	Composition (weight percent), Specifications and Remarks
19	159	1963	323	0.20-27.10	5°		Different sample, same as above specimen and conditions except roughened and electropolished; surface roughness 0.6 microinch (center line average); [Author's designation : Specimen 2].
20	86	1961	523	2.00-15.00	$\sim 0^{\circ}$	± 5	Steel PH 15-7 Mo, NAA LB0160-130; nominal composition : 15.00 Cr, 7.00 Ni, 2.25 Mo, 1.15 Al, 0.70 Mn, 0.40 Si, 0.07 C, Fe balance; as received; data extracted from smooth curve.
21	86	1961	773	1.00-15.00	$\sim 0^{\circ}$	± 5	Different sample, same as curve 20 specimen and conditions.
22	86	1961	1023	1.00-15.00	$\sim 0^{\circ}$	± 5	Different sample, same as curve 20 specimen and conditions.
23	86	1961	523	2.00-15.00	$\sim 0^{\circ}$	± 5	Different sample, same as curve 20 specimen and conditions except heated in air at 755 K for 30 min.
24	86	1961	773	1.00-15.00	$\sim 0^{\circ}$	± 5	Different sample, same as curve 23 specimen and conditions.
25	86	1961	1023	1.00-15.00	$\sim 0^{\circ}$	± 5	Different sample, same as curve 23 specimen and conditions.
26	86	1961	523	2.00-15.00	$\sim 0^{\circ}$	± 5	Different sample, same as curve 20 specimen and conditions except heated in vacuum (4.4×10^{-5} mm Hg) at 755 K for 30 min.
27	86	1961	773	1.00-15.00	$\sim 0^{\circ}$	± 5	Different sample, same as curve 26 specimen and conditions.
28	86	1961	1023	1.00-15.00	$\sim 0^{\circ}$	± 5	Different sample, same as curve 26 specimen and conditions.
29	86	1961	523	2.00-15.00	$\sim 0^{\circ}$	± 5	Steel, 17-7 PH, MIL-S-25043; nominal composition : 17.00 Cr, 7.00 Ni, 1.15 Al, 0.70 Mn, 0.40 Si, 0.07 C, Fe balance; as received; data extracted from smooth curve.
30	86	1961	773	1.00-15.00	$\sim 0^{\circ}$	± 5	Different sample, same as curve 29 specimen and conditions.
31	86	1961	1023	1.00-15.00	$\sim 0^{\circ}$	± 5	Different sample, same as curve 29 specimen and conditions.
32	86	1961	523	2.00-15.00	$\sim 0^{\circ}$	± 5	Different sample, same as curve 29 specimen and conditions except heated in air at 755 K for 30 min.
33	86	1961	773	1.00-15.00	$\sim 0^{\circ}$	± 5	Different sample, same as curve 32 specimen and conditions.
34	86	1961	1023	1.00-15.00	$\sim 0^{\circ}$	± 5	Different sample, same as curve 32 specimen and conditions.
35	86	1961	523	2.00-15.00	$\sim 0^{\circ}$	± 5	Different sample, same as curve 29 specimen and conditions except heated in vacuum (4.4×10^{-5} mm Hg) at 755 K for 30 min.
36	86	1961	773	1.00-15.00	$\sim 0^{\circ}$	± 5	Different sample, same as curve 35 specimen and conditions.
37	86	1961	1023	1.00-15.00	$\sim 0^{\circ}$	± 5	Different sample, same as curve 35 specimen and conditions.
38	86	1961	523	2.00-15.00	$\sim 0^{\circ}$	± 5	Steel AM 350; nominal composition : 16.50 Cr, 4.25 Ni, 2.75 Mo, 0.75 Mn, 0.35 Si, 0.10 C, 0.10 N, Fe balance; as received; data extracted from smooth curve.
39	86	1961	773	1.00-15.00	$\sim 0^{\circ}$	± 5	Different sample, same as curve 38 specimen and conditions.
40	86	1961	1023	1.00-15.00	$\sim 0^{\circ}$	± 5	Different sample, same as curve 38 specimen and conditions.

SPECIFICATION TABLE NO. 364 (continued)

Curve No.	Ref. No.	Year	Temperature K	Wavelength Range, μ	Geometry θ'	Reported Error, %	Composition (weight percent), Specifications and Remarks
41	86	1961	523	2.00-15.00	~0°	± 5	Different sample, same as curve 38 specimen and conditions except heated in air at 755 K for 30 min.
42	86	1961	773	1.00-15.00	~0°	± 5	Different sample, same as curve 41 specimen and conditions.
43	86	1961	1023	1.00-15.00	~0°	± 5	Different sample, same as curve 41 specimen and conditions.
44	86	1961	523	2.00-15.00	~0°	± 5	Different sample, same as curve 38 specimen and conditions except heated in vacuum (4.4×10^{-5} mm Hg) at 755 K for 30 min.
45	86	1961	773	1.00-15.00	~0°	± 5	Different sample, same as curve 44 specimen and conditions.
46	86	1961	1023	1.00-15.00	~0°	± 5	Different sample, same as curve 44 specimen and conditions.
47	239	1959	1371	1.085-3.908	~0°	5	Stainless steel 321; 17.79 Cr, 10.54 Ni, 1.24 Mn, 0.545 Si, 0.46 Ti, 0.22 Mo, 0.048 C, 0.025 P, 0.020 S, Fe balance; polished; measured in argon; oxidation suspected.
48	239	1959	1346	1.104-4.013	~0°	5	Different sample, same as above specimen and conditions.
49	239	1959	1282	1.100-3.915	~0°	5	Different sample, same as above specimen and conditions.
50	239	1959	1269	1.096-4.000	~0°	5	Different sample, same as above specimen and conditions.
51	239	1959	1310	1.096-4.002	~0°	5	Different sample, same as above specimen and conditions.

DATA TABLE NO. 364 NORMAL SPECTRAL EMITTANCE OF [IRON + CHROMIUM + NICKEL + ΣX_i] ALLOYS

[Wavelength, λ μ; Emittance, ∈; Temperature, T, K]

λ	∈
CURVE 1 T = 755	
2.0	0.12
2.5	0.15
3.0	0.15
3.5	0.15
4.0	0.15
4.5	0.15
5.0	0.14
5.5	0.14
6.0	0.14
6.5	0.13
7.0	0.13
7.5	0.13
8.0	0.13
8.5	0.13
9.0	0.13
9.5	0.12
10.0	0.12
10.5	0.11
11.0	0.11
11.5	0.11
12.0	0.11
12.5	0.11
13.0	0.11
13.5	0.11
14.0	0.10
14.5	0.10
15.0	0.09
CURVE 2 T = 922	
1.5	0.29
2.0	0.27
2.5	0.25
3.0	0.24
3.5	0.22
4.0	0.21
4.5	0.19
5.0	0.18
5.5	0.17
6.0	0.15
6.5	0.15
7.0	0.15
CURVE 2 (cont.)	
7.5	0.14
8.0	0.14
8.5	0.14
9.0	0.14
9.5	0.13
10.0	0.13
10.5	0.13
11.0	0.13
11.5	0.12
12.0	0.12
12.5	0.12
13.0	0.12
13.5	0.11*
14.0	0.11
14.5	0.11
15.0	0.10
CURVE 3 T = 755	
2.5	0.34
3.0	0.36
3.5	0.38
4.0	0.39
4.5	0.39
5.0	0.40
5.5	0.41
6.0	0.42
6.5	0.41
7.0	0.40
7.5	0.39
8.0	0.38
8.5	0.39
9.0	0.41
9.5	0.41
10.0	0.40
10.5	0.39
11.0	0.39
11.5	0.40
12.0	0.40
12.5	0.41
13.0	0.41
13.5	0.42
14.0	0.42
CURVE 3 (cont.)	
14.5	0.41
15.0	0.40
CURVE 4* T = 922	
1.5	0.34
2.0	0.36
2.5	0.37
3.0	0.38
3.5	0.41
4.0	0.41
4.5	0.41
5.0	0.40
5.5	0.41
6.0	0.40
6.5	0.38
7.0	0.38
7.5	0.38
8.0	0.39
8.5	0.40
9.0	0.41
9.5	0.41
10.0	0.41
10.5	0.42
11.0	0.41
11.5	0.41
12.0	0.40
12.5	0.42
13.0	0.44
13.5	0.44
14.0	0.44
14.5	0.43
15.0	0.42
CURVE 5 T = 755	
2.0	0.63
2.5	0.66
3.0	0.68
3.5	0.69
4.0	0.69
4.5	0.67
CURVE 5 (cont.)	
5.0	0.64
5.5	0.65
6.0	0.68
6.5	0.68
7.0	0.69
7.5	0.70
8.0	0.73
8.5	0.76
9.0	0.77
9.5	0.75
10.0	0.69
10.5	0.65
11.0	0.63
11.5	0.59
12.0	0.54
12.5	0.49
13.0	0.43
13.5	0.38
14.0	0.34
14.5	0.34
15.0	0.34
CURVE 6* T = 922	
2.0	0.60
2.5	0.66
3.0	0.69
3.5	0.70
4.0	0.69
4.5	0.68
5.0	0.65
5.5	0.68
6.0	0.69
6.5	0.71
7.0	0.72
7.5	0.73
8.0	0.73
8.5	0.74
9.0	0.75
9.5	0.74
10.0	0.73
10.5	0.70
11.0	0.67
CURVE 6 (cont.)	
11.5	0.64
12.0	0.59
12.5	0.52
13.0	0.46
13.5	0.41
14.0	0.37
14.5	0.36
15.0	0.36
CURVE 7* T = 1089	
1.5	0.70
2.0	0.76
2.5	0.78
3.0	0.79
3.5	0.78
4.0	0.76
4.5	0.74
5.0	0.68
5.5	0.73
6.0	0.74
6.5	0.76
6.7	0.77
7.5	0.78
8.0	0.78
8.5	0.79
9.0	0.80
9.5	0.78
10.0	0.76
10.5	0.74
11.0	0.71
11.5	0.68
12.0	0.63
12.5	0.58
13.0	0.51
13.5	0.46
14.0	0.42
14.5	0.40
15.0	0.39
CURVE 8* T = 1255	
1.5	0.67
2.0	0.73
2.5	0.75
3.0	0.77
3.5	0.77
4.0	0.77
4.5	0.76
5.0	0.70
5.5	0.77
6.0	0.78
6.5	0.80
7.0	0.78
7.5	0.77
8.0	0.76
8.5	0.76
9.0	0.77
9.5	0.75
10.0	0.72
10.5	0.69
11.0	0.66
11.5	0.64
12.0	0.61
12.5	0.57
13.0	0.54
13.5	0.49
14.0	0.44
14.5	0.42
15.0	0.40
CURVE 9 T = 755	
2.0	0.52
2.5	0.59
3.0	0.65
3.5	0.67
4.0	0.69*
4.5	0.69
5.0	0.69
5.5	0.70
6.0	0.71
6.5	0.70
7.0	0.70
CURVE 9 (cont.)	
7.5	0.71
8.0	0.73*
8.5	0.77
9.0	0.83
9.5	0.82
10.0	0.82
10.5	0.84
11.0	0.86
11.5	0.85
12.0	0.81
12.5	0.78
13.0	0.75
13.5	0.72
14.0	0.68
14.5	0.65
15.0	0.63
CURVE 10* T = 922	
1.5	0.65
2.0	0.65
2.5	0.68
3.0	0.72
3.5	0.73
4.0	0.74
4.5	0.75
5.0	0.75
5.5	0.74
6.0	0.74
6.5	0.74
7.0	0.73
7.5	0.75
8.0	0.76
8.5	0.78
9.0	0.83
9.5	0.84
10.0	0.84
10.5	0.85
11.0	0.86
11.5	0.86
12.0	0.86
12.5	0.83
13.0	0.80

* Not shown on plot

DATA TABLE NO. 364 (continued)

λ	∈
CURVE 10 (cont.)*	
13.5	0.77
14.0	0.75
14.5	0.72
15.0	0.69
CURVE 11*	
T = 1089	
1.5	0.65
2.0	0.65
2.5	0.68
3.0	0.72
3.5	0.73
4.0	0.74
4.5	0.75
5.0	0.76
5.5	0.76
6.0	0.76
6.5	0.75
7.0	0.75
7.5	0.75
8.0	0.76
8.5	0.78
9.0	0.83
9.5	0.84
10.0	0.84
10.5	0.84
11.0	0.84
11.5	0.84
12.0	0.84
12.5	0.81
13.0	0.78
13.5	0.76
14.0	0.73
14.5	0.70
15.0	0.67
CURVE 12	
T = 1255	
1.5	0.65
2.0	0.71
2.5	0.77
3.0	0.79
3.5	0.82
4.0	0.82

λ	∈
CURVE 12 (cont.)	
4.5	0.80
5.0	0.76
5.5	0.77
6.0	0.76
6.5	0.75
7.0	0.75
7.5	0.75
8.0	0.77
8.5	0.81
9.0	0.84
9.5	0.85
10.0	0.85
10.5	0.85
11.0	0.86*
11.5	0.87
12.0	0.87
12.5	0.86
13.0	0.83
13.5	0.82
14.0	0.81
14.5	0.76
15.0	0.71
CURVE 13	
T = 873	
2.0	0.370
3.0	0.340
4.0	0.320
5.0	0.300
6.0	0.280
7.0	0.260
8.0	0.260
9.0	0.250
10.0	0.240
11.0	0.230
12.0	0.220
13.0	0.210
14.0	0.200
CURVE 14	
T = 1273	
2.0	0.770
3.0	0.760
4.0	0.690*

λ	∈
CURVE 14 (cont.)	
5.0	0.650
6.0	0.720
7.0	0.770
8.0	0.780
9.0	0.785
10.0	0.750
11.0	0.650
12.0	0.560
13.0	0.450
14.0	0.400
CURVE 15	
T = 323	
0.20	0.655
0.25	0.620
0.30	0.535
0.33	0.430
0.40	0.455
0.40	0.435
0.45	0.410
0.49	0.370
0.50	0.360
0.55	0.340
0.60	0.330
0.70	0.320
0.80	0.310
0.90	0.305
1.00	0.295
1.18	0.275
1.40	0.255
1.60	0.230
1.80	0.220
2.00	0.210
2.20	0.205
2.40	0.200
2.60	0.220
2.70	0.215
3.70	0.210
5.10	0.180
6.10	0.180
7.10	0.155
7.60	0.145
8.50	0.130*
9.55	0.125
10.20	0.120

λ	∈
CURVE 15 (cont.)	
11.00	0.115
11.70	0.110
12.40	0.115
13.00	0.116
13.60	0.120
14.00	0.100*
14.10	0.120
16.30	0.090
18.30	0.090
21.70	0.090
24.40	0.095
27.00	0.090
CURVE 16	
T = 323	
0.20	0.695
0.23	0.675
0.28	0.595
0.35	0.500
0.38	0.495
0.42	0.445
0.49	0.385
0.50	0.400
0.55	0.370
0.55	0.365
0.58	0.355
0.58	0.345
0.68	0.334
0.78	0.325
0.89	0.315
0.99	0.308
1.18	0.285
1.40	0.265
1.58	0.240
1.78	0.225
1.98	0.220
2.18	0.205*
2.38	0.205
2.58	0.220
2.68	0.225

λ	∈
CURVE 17*	
T = 323	
0.22	0.689
0.24	0.634
0.25	0.600
0.30	0.550
0.31	0.495
0.32	0.455
0.37	0.465
0.40	0.456
0.40	0.445
0.45	0.400
0.48	0.365
0.50	0.360
0.55	0.335
0.58	0.320
0.68	0.310
0.80	0.310
0.90	0.305
0.98	0.290
1.18	0.275
1.40	0.260
1.60	0.235
1.80	0.215
1.98	0.210
2.20	0.200
2.40	0.200
2.58	0.210
2.68	0.215
2.98	0.220
3.70	0.180
5.10	0.160
6.10	0.140
7.10	0.130
7.60	0.140
8.60	0.130
9.50	0.120
10.30	0.105
11.00	0.120
11.70	0.105
12.30	0.105
13.00	0.107
14.10	0.090
14.50	0.100
15.10	0.108
15.60	0.105
16.20	0.085

λ	∈
CURVE 17 (cont.)*	
18.30	0.075
20.05	0.062
21.80	0.060
24.42	0.070
27.10	0.060
CURVE 18*	
T = 323	
0.22	0.640
0.25	0.585
0.27	0.560
0.30	0.510
0.32	0.445
0.35	0.400
0.37	0.425
0.40	0.420
0.45	0.390
0.50	0.355
0.50	0.345
0.55	0.340
0.55	0.335
0.60	0.325
0.60	0.320
0.70	0.320
0.80	0.320
0.90	0.315
1.00	0.310
1.20	0.285
1.40	0.265
1.60	0.250
1.80	0.235
2.00	0.225
2.20	0.215
2.40	0.215
2.60	0.226
2.70	0.220
3.80	0.204
5.10	0.170
6.10	0.145
7.20	0.140
7.65	0.140
8.60	0.140
9.50	0.125
10.30	0.120
11.00	0.125

λ	∈
CURVE 18 (cont.)*	
11.80	0.115
12.40	0.120
13.00	0.120
13.70	0.125
14.10	0.125
14.10	0.090
14.70	0.120
16.30	0.080
18.30	0.080
21.80	0.090
24.40	0.095
27.00	0.092
CURVE 19	
T = 323	
0.20	0.650
0.25	0.610
0.27	0.580
0.28	0.535
0.30	0.470
0.33	0.428
0.37	0.449
0.40	0.440
0.45	0.385
0.47	0.350
0.52	0.330
0.60	0.315
0.67	0.305
0.77	0.305
0.90	0.300
0.99	0.290
1.16	0.270
1.40	0.260
1.60	0.235
1.80	0.220*
1.98	0.210*
2.18	0.200
2.40	0.205*
2.60	0.220*
2.70	0.215*
3.70	0.190
5.20	0.145
6.10	0.095
7.10	0.095
7.60	0.110

* Not shown on plot

DATA TABLE NO. 364 (continued)

λ	∈
CURVE 19 (cont.)	
8.60	0.105
9.50	0.100
10.30	0.090
11.00	0.095
11.60	0.090
12.30	0.090
12.90	0.100
13.60	0.090
14.05	0.125
14.18	0.080
14.60	0.065
16.30	0.100
18.30	0.080
20.00	0.070
21.80	0.065
24.40	0.065
27.10	0.065
CURVE 20	**T = 523**
2.00	0.520*
3.00	0.510
4.00	0.500
5.50	0.465
6.25	0.450
7.50	0.400
8.40	0.350
9.50	0.300
11.00	0.235
12.00	0.210
13.00	0.185
14.00	0.160
15.00	0.135
CURVE 21*	**T = 773**
1.00	0.690
1.50	0.560
2.00	0.500
3.00	0.450
3.50	0.450
4.10	0.475
4.60	0.475
5.75	0.440

λ	∈
CURVE 21 (cont.)*	
6.25	0.430
6.75	0.430
8.00	0.380
9.00	0.320
10.00	0.280
11.00	0.240
12.00	0.210
13.00	0.190
14.00	0.170
15.00	0.145
CURVE 22*	**T = 1023**
1.00	0.635
1.10	0.700
1.50	0.750
2.00	0.700
2.50	0.675
2.95	0.675
3.60	0.695
4.75	0.675
6.00	0.640
6.50	0.635
7.25	0.655
7.75	0.655
8.75	0.620
9.50	0.590
10.00	0.580
11.00	0.530
12.35	0.450
13.50	0.380
14.50	0.330
15.00	0.330
CURVE 23	**T = 523**
2.00	0.670
2.25	0.600
2.50	0.530
3.00	0.470
3.50	0.455
4.00	0.460
5.00	0.460
5.75	0.455

λ	∈
CURVE 23 (cont.)	
7.50	0.400*
8.00	0.385
8.75	0.340
9.90	0.300
11.00	0.250
12.00	0.225
13.00	0.210*
14.00	0.200*
15.00	0.180
CURVE 24*	**T = 773**
1.00	0.600
1.35	0.550
1.50	0.500
1.60	0.400
2.00	0.325
2.50	0.320
3.50	0.340
4.50	0.370
6.00	0.360
7.50	0.325
9.00	0.270
10.40	0.215
11.40	0.180
12.25	0.170
13.50	0.170
14.50	0.170
15.00	0.160
CURVE 25	**T = 1023**
1.00	0.720
1.25	0.650
2.00	0.575*
3.00	0.510*
4.00	0.525
5.10	0.550
6.30	0.525
7.00	0.530
8.00	0.500
9.25	0.425
10.00	0.375
11.00	0.330

λ	∈
CURVE 25 (cont.)	
12.25	0.280
13.00	0.250
14.00	0.250
15.00	0.280
CURVE 26*	**T = 523**
2.00	0.555
3.00	0.465
4.00	0.415
4.50	0.430
5.00	0.440
6.00	0.400
6.75	0.390
8.00	0.360
9.00	0.300
10.00	0.270
11.00	0.225
12.00	0.200
12.50	0.190
13.10	0.190
13.80	0.175
14.50	0.155
15.00	0.150
CURVE 27*	**T = 773**
1.00	0.725
1.50	0.600
2.00	0.450
2.50	0.365
3.00	0.340
3.75	0.350
4.80	0.390
6.00	0.370
8.00	0.350
9.00	0.300
10.20	0.250
11.50	0.210
13.00	0.190
14.00	0.180
15.00	0.180

λ	∈
CURVE 28	**T = 1023**
1.00	0.860
1.50	0.750
2.00	0.620
2.25	0.615
3.20	0.650
4.00	0.690
4.50	0.695
5.50	0.675
6.00	0.660
7.00	0.680
8.00	0.710
9.00	0.710
10.00	0.710
11.00	0.700
12.00	0.670
12.50	0.665
13.00	0.670
14.00	0.630
15.00	0.580
CURVE 29	**T = 523**
2.00	0.450
3.00	0.450
4.00	0.440
4.75	0.400
5.25	0.370
6.50	0.340
7.75	0.315
8.50	0.275
9.50	0.250
10.50	0.225
11.00	0.190
12.00	0.175
13.50	0.160
14.50	0.135
15.00	0.120
CURVE 30*	**T = 773**
1.00	0.630
1.25	0.550
1.50	0.490

λ	∈
CURVE 30 (cont.)*	
2.50	0.435
3.25	0.420
3.75	0.430
4.75	0.400
6.00	0.370
7.00	0.350
8.00	0.310
9.00	0.270
10.50	0.225
11.00	0.205
12.00	0.190
13.25	0.165
15.00	0.140
CURVE 31*	**T = 1023**
1.00	0.540
2.50	0.505
3.50	0.480
4.00	0.470
5.00	0.430
6.00	0.400
7.50	0.355
8.80	0.300
9.50	0.265
10.50	0.240
11.50	0.220
12.50	0.200
13.50	0.180
14.25	0.180
15.00	0.195
CURVE 32*	**T = 523**
2.00	0.610
2.25	0.550
2.75	0.500
3.50	0.450
4.50	0.415
5.25	0.400
6.50	0.350
7.25	0.315
8.00	0.310
9.00	0.270

λ	∈
CURVE 32 (cont.)*	
10.00	0.240
11.00	0.220
12.00	0.200
13.00	0.190
14.00	0.190
14.50	0.185
15.00	0.170
CURVE 33*	**T = 773**
1.00	0.560
1.50	0.490
2.00	0.400
2.50	0.350
3.25	0.325
4.50	0.330
5.00	0.330
6.00	0.300
7.25	0.280
8.25	0.260
9.00	0.230
10.50	0.200
11.50	0.175
12.50	0.160
13.00	0.160
14.00	0.160
15.00	0.160
CURVE 34*	**T = 1023**
1.00	0.740
1.50	0.650
1.80	0.575
2.00	0.550
2.50	0.520
3.50	0.500
4.50	0.475
6.00	0.420
8.00	0.375
9.00	0.320
10.00	0.275
11.30	0.250
11.75	0.235
12.50	0.225

λ	∈
CURVE 34 (cont.)*	
13.50	0.220
14.35	0.225
15.00	0.250
CURVE 35*	**T = 523**
2.00	0.515
3.00	0.410
4.00	0.360
5.50	0.330
7.00	0.280
7.50	0.265
8.05	0.265
9.00	0.230
10.50	0.200
12.00	0.180
13.00	0.170
14.00	0.170
14.50	0.165
15.00	0.150
CURVE 36*	**T = 773**
1.00	0.650
1.40	0.600
1.75	0.500
2.00	0.425
2.25	0.375
2.50	0.350
3.00	0.340
4.50	0.345
5.50	0.325
6.50	0.295
8.00	0.280
9.00	0.240
9.75	0.225
10.50	0.200
12.00	0.180
13.00	0.170
13.50	0.160
15.00	0.160

* Not shown on plot

DATA TABLE NO. 364 (continued)

λ	∈
CURVE 37* T = 1023	
1.00	0.720
1.50	0.700
2.00	0.600
2.20	0.550
2.50	0.525
3.00	0.500
4.00	0.485
5.00	0.470
5.50	0.440
6.00	0.410
7.00	0.385
8.00	0.360
9.00	0.320
9.75	0.290
11.00	0.270
12.00	0.240
12.75	0.230
14.00	0.230
14.70	0.250
15.00	0.280
CURVE 38 T = 523	
2.00	0.540
3.10	0.525
4.00	0.470
5.15	0.450
6.25	0.435
7.75	0.450
9.00	0.480
10.00	0.490
11.00	0.460
12.50	0.435
13.40	0.400
15.00	0.305
CURVE 39 T = 773	
1.00	0.840
1.00	0.800
1.50	0.670
2.00	0.580
3.00	0.520
CURVE 39 (cont.)	
4.00	0.470*
5.00	0.460*
5.50	0.450
6.00	0.430
7.50	0.465
8.50	0.485
9.75	0.515
11.00	0.510
12.00	0.500
13.00	0.500
14.00	0.485
14.70	0.450
15.00	0.405
CURVE 40* T = 1023	
1.00	0.700
1.50	0.710
2.00	0.700
3.00	0.640
4.00	0.600
5.25	0.550
6.25	0.515
7.50	0.525
9.00	0.540
9.80	0.540
11.00	0.530
11.75	0.540
13.50	0.525
14.00	0.510
15.00	0.485
CURVE 41* T = 523	
2.00	0.690
2.10	0.650
2.50	0.575
3.50	0.500
5.00	0.450
6.00	0.425
7.50	0.440
9.00	0.475
10.00	0.490
11.25	0.465
CURVE 41 (cont.)*	
11.75	0.460
13.00	0.460
13.75	0.450
14.50	0.420
15.00	0.380
CURVE 42* T = 773	
1.00	0.570
1.30	0.550
1.75	0.475
1.85	0.435
2.50	0.385
3.25	0.365
4.50	0.350
6.00	0.350
6.75	0.355
7.50	0.375
8.50	0.400
9.85	0.420
11.00	0.395
12.00	0.390
13.00	0.370
14.00	0.360
15.00	0.350
CURVE 43* T = 1023	
1.00	0.750
1.35	0.750
2.50	0.695
3.50	0.655
4.50	0.620
5.50	0.570
6.15	0.550
7.50	0.575
8.75	0.600
10.00	0.610
11.00	0.600
12.25	0.575
13.00	0.560
13.75	0.550
14.55	0.555
15.00	0.570
CURVE 44* T = 523	
2.00	0.625
2.25	0.555
3.00	0.475
4.00	0.420
5.00	0.390
6.00	0.380
7.50	0.395
9.00	0.420
9.75	0.430
10.75	0.410
12.25	0.410
14.00	0.380
14.60	0.360
15.00	0.320
CURVE 45* T = 773	
1.00	0.685
1.75	0.550
2.50	0.425
3.25	0.370
4.00	0.375
4.75	0.380
5.50	0.370
6.25	0.360
7.25	0.375
8.00	0.400
9.50	0.410
10.00	0.420
11.50	0.420
13.00	0.420
14.25	0.435
14.75	0.425
15.00	0.410
CURVE 46 T = 1023	
1.00	0.720*
1.25	0.760
1.50	0.770
1.75	0.750
2.00	0.690
2.50	0.650
CURVE 46 (cont.)	
3.25	0.635
3.75	0.640
5.00	0.620
6.00	0.575
7.00	0.580
8.10	0.600
9.75	0.600
11.10	0.600
11.80	0.580
13.00	0.580
14.00	0.570
15.00	0.550
CURVE 47 T = 1371	
1.085	0.436
1.442	0.486
1.853	0.528
2.221	0.548
2.552	0.553
2.840	0.578
3.097	0.592
3.467	0.619
3.700	0.619
3.908	0.619
CURVE 48 T = 1346	
1.104	0.350
1.456	0.430
1.857	0.487
2.223	0.500
2.566	0.493
2.840	0.487
3.111	0.514
3.342	0.514
3.582	0.544
3.803	0.544
4.013	0.551
CURVE 49 T = 1282	
1.100	0.245
CURVE 49 (cont.)	
1.457	0.316
1.761	0.389
2.042	0.427
2.393	0.434
2.694	0.422
2.979	0.419
3.215	0.412
3.463	0.413
3.706	0.426
3.915	0.432
CURVE 50 T = 1269	
1.096	0.224
1.457	0.298
1.860	0.368
2.217	0.402
2.552	0.398
2.851	0.398
3.107	0.396
3.343	0.395
3.588	0.396
3.806	0.379
4.000	0.362
CURVE 51* T = 1310	
1.096	0.293
1.259	0.356
1.463	0.378
1.857	0.437
2.216	0.476
2.553	0.474
2.843	0.458
3.107	0.446
3.340	0.421
3.584	0.411
3.800	0.399
4.002	0.383

* Not shown on plot

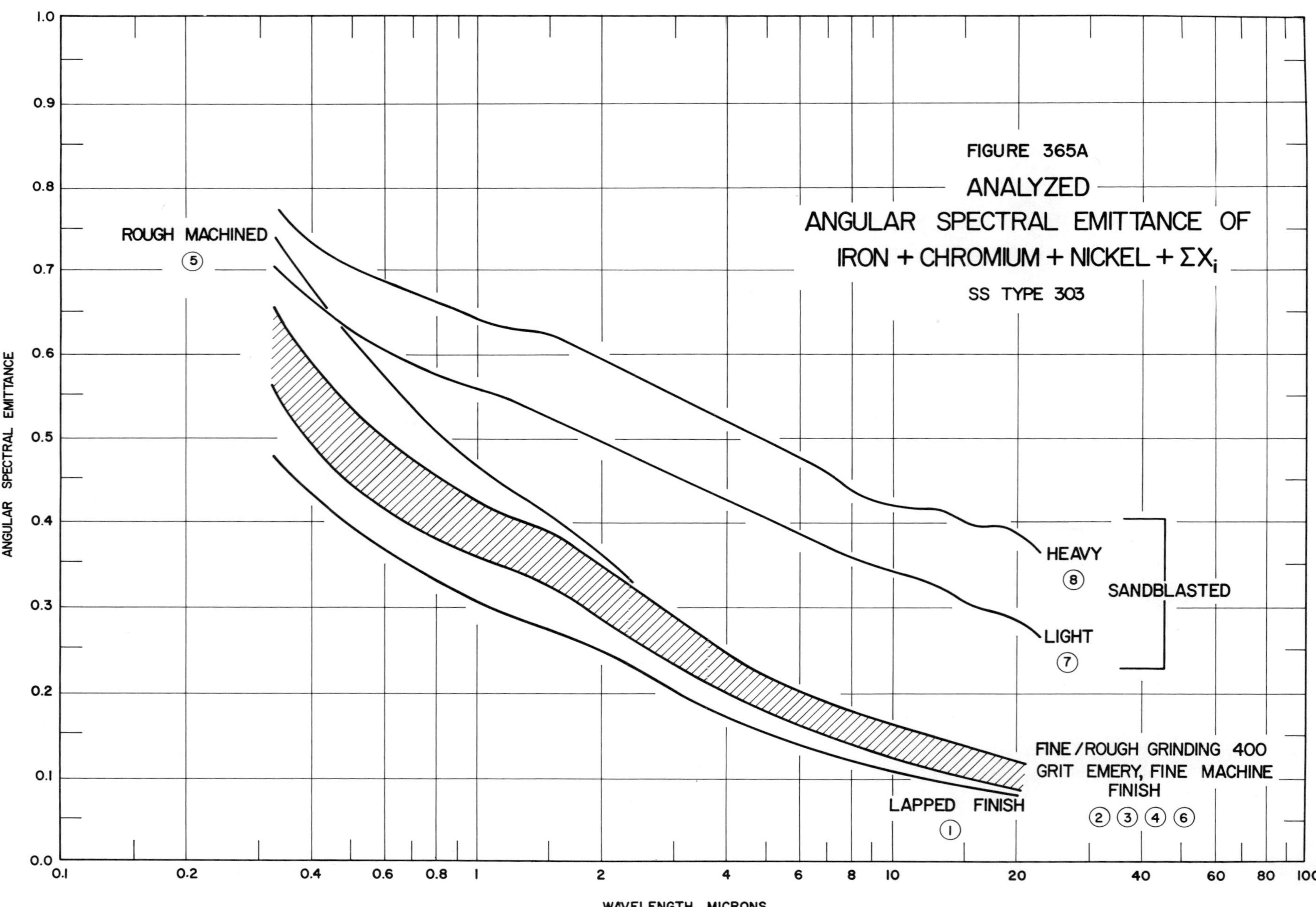
FIGURE 365A
ANALYZED
ANGULAR SPECTRAL EMITTANCE OF
IRON + CHROMIUM + NICKEL + ΣX_i
SS TYPE 303
ROUGH MACHINED
⑤
HEAVY
⑧
SANDBLASTED
LIGHT
⑦
FINE / ROUGH GRINDING 400
GRIT EMERY, FINE MACHINE
FINISH
② ③ ④ ⑥
LAPPED FINISH
①
ANGULAR SPECTRAL EMITTANCE
1.0
0.9
0.8
0.7
0.6
0.5
0.4
0.3
0.2
0.1
0.0
0.1
0.2
0.4
0.6
0.8
1
2
4
6
8
10
20
40
60
80
100
WAVELENGTH, MICRONS

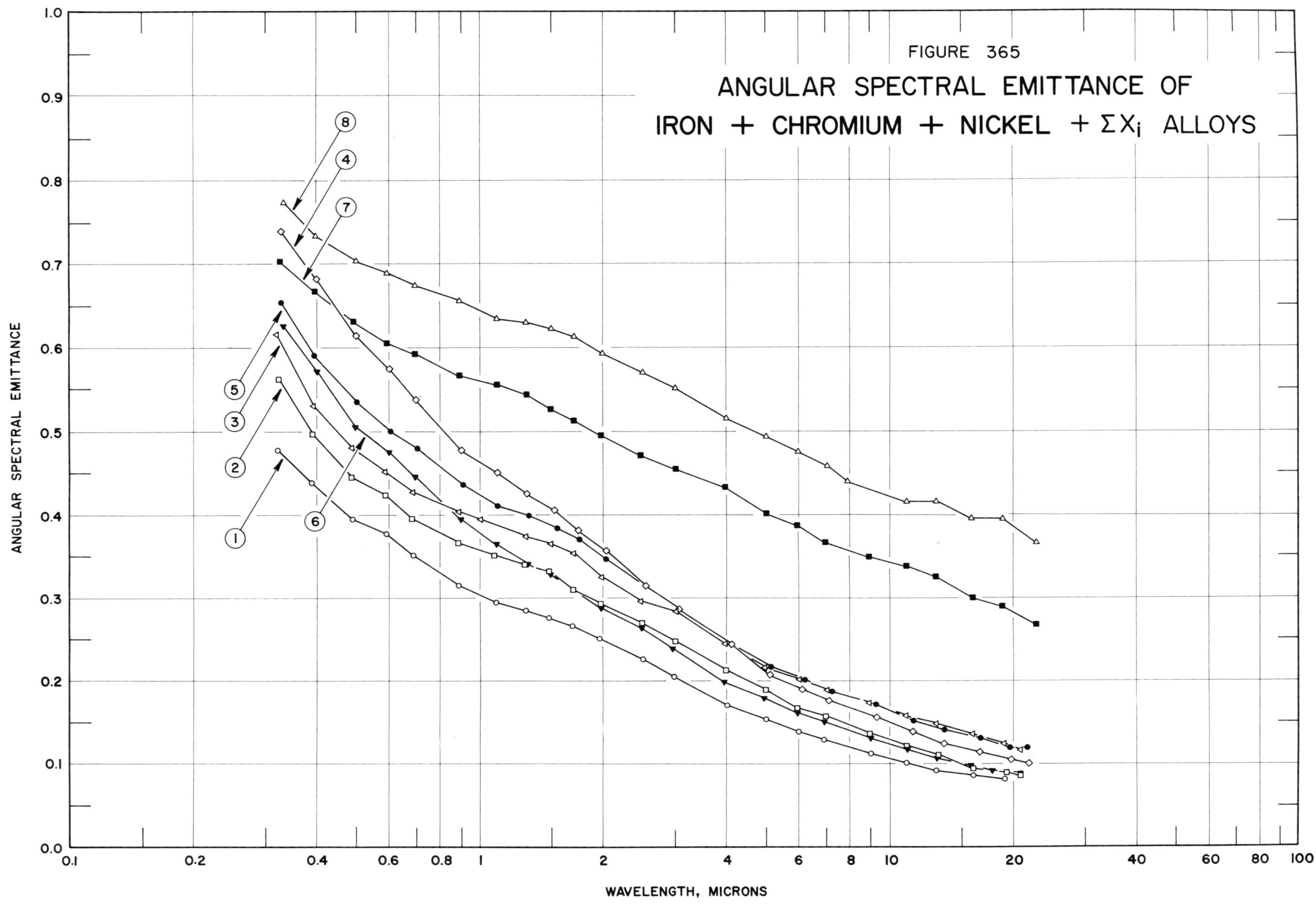

FIGURE 365

ANGULAR SPECTRAL EMITTANCE OF IRON + CHROMIUM + NICKEL + ΣX_i ALLOYS

SPECIFICATION TABLE NO. 365 ANGULAR SPECTRAL EMITTANCE OF [IRON + CHROMIUM + NICKEL + ΣX_i] ALLOYS

Curve No.	Ref. No.	Year	Temperature K	Wavelength Range, μ	Geometry θ'	Reported Error, %	Composition (weight percent), Specifications and Remarks
1	219	1965	306	0.322-19.1	25°		Stainless Steel 303; 17-19 Cr, 8-10 Ni, 2 Mg, 1 Si, 0.6 Mo, 0.6 Zr, 0.15 C, 0.15 Se, and Fe balance; cloth lapped with No. 2 polishing alumina (0.5 μ); authors assumed $\epsilon = \alpha = 1 - \rho(25°, 2\pi)$.
2	219	1965	306	0.323-20.9	25°		Stainless Steel 303; 17-19 Cr, 8-10 Ni, 2 Mg, 1 Si, 0.6 Mo, 0.6 Zr, 0.15 C, 0.15 Se, and Fe balance; ground fine on Grit A 60 Grade T Bond 100 grinding wheel; authors assumed $\epsilon = \alpha = 1 - \rho(25°, 2\pi)$.
3	219	1965	306	0.321-20.9	25°		Different sample, same as above specimen and conditions except ground rough.
4	219	1965	306	0.330-21.9	25°		Stainless Steel 303; 17-19 Cr, 8-10 Ni, 2 Mg, 1 Si, 0.6 Mo, 0.6 Zr, 0.15 C, 0.15 Se, and Fe balance; machined fine; authors assumed $\epsilon = \alpha = 1 - \rho(25°, 2\pi)$.
5	219	1965	306	0.330-21.8	25°		Different sample, same as above specimen and conditions except machined rough.
6	219	1965	306	0.335-20.8	25°		Stainless Steel 303; 17-19 Cr, 8-10 Ni, 2 Mg, 1 Si, 0.6 Mo, 0.6 Zr, 0.15 C, 0.15 Se, and Fe balance; sanded with 400 grit emery paper; authors assumed $\epsilon = \alpha = 1 - \rho(25°, 2\pi)$.
7	219	1965	306	0.329-22.9	25°		Stainless Steel 303; 17-19 Cr, 8-10 Ni, 2 Mg, 1 Si, 0.6 Mo, 0.6 Zr, 0.15 C, 0.15 Se, and Fe balance; lightly sandblasted from normal direction with No. 80 Al_2O_3 grit (mesh opening 175 μ); authors assumed $\epsilon = \alpha = 1 - \rho(25°, 2\pi)$.
8	219	1965	306	0.333-22.9	25°		Different sample, same as above specimen and conditions except heavily sandblasted.

DATA TABLE NO. 365 ANGULAR SPECTRAL EMITTANCE OF [IRON + CHROMIUM + NICKEL + ΣX_i] ALLOYS

[Wavelength, λ, μ; Emittance, ϵ; Temperature, T, K]

λ	ϵ
CURVE 1	
T = 306	
0.322	0.478
0.389	0.438
0.490	0.394
0.592	0.377
0.692	0.351
0.887	0.315
1.09	0.294
1.29	0.285
1.48	0.276
1.68	0.266
1.97	0.250
2.51	0.225
2.99	0.204
4.01	0.169
5.02	0.152
6.04	0.138
6.98	0.128
9.04	0.112
11.0	0.100
13.0	0.091
16.0	0.086
19.1	0.081
CURVE 2	
T = 306	
0.323	0.561
0.391	0.498
0.488	0.446
0.590	0.423
0.682	0.395
0.883	0.366
1.08	0.351
1.28	0.340
1.48	0.332
1.69	0.310
1.98	0.292
2.49	0.269
3.01	0.248
3.99	0.213
5.00	0.189
5.98	0.167
7.02	0.157
9.00	0.135
CURVE 2 (cont.)	
11.0	0.121
13.2	0.111
16.0	0.094
19.3	0.089
20.9	0.085
CURVE 3	
T = 306	
0.321	0.617
0.392	0.530
0.492	0.480
0.588	0.451
0.686	0.428
0.891	0.403
1.08	0.395
1.29	0.372
1.49	0.363
1.69	0.352
1.99	0.327
2.48	0.297
3.01	0.282
3.99	0.244
4.99	0.217
6.04	0.201
7.02	0.188
8.99	0.172
11.0	0.157
13.1	0.148
16.0	0.135
19.1	0.123
20.9	0.117
CURVE 4	
T = 306	
0.330	0.740
0.401	0.682
0.501	0.614
0.604	0.573
0.700	0.538
0.904	0.478
1.11	0.451
1.31	0.426
1.53	0.406
CURVE 4 (cont.)	
1.75	0.381
2.04	0.353
2.55	0.315
3.08	0.286
4.14	0.243
5.12	0.208
6.15	0.189
7.18	0.176
9.38	0.155
11.4	0.138
13.6	0.124
16.7	0.115
19.9	0.106
21.9	0.100
CURVE 5	
T = 306	
0.330	0.653
0.395	0.590
0.502	0.533
0.604	0.500
0.702	0.480
0.910	0.436
1.11	0.411
1.32	0.399
1.55	0.382
1.76	0.370
2.04	0.347
2.56	0.312*
3.07	0.286*
4.17	0.242*
5.15	0.217
6.21	0.201
7.23	0.187
9.33	0.171
11.4	0.150
13.6	0.140
16.7	0.130
19.8	0.119
21.8	0.119
CURVE 6	
T = 306	
0.335	0.625
0.402	0.570
0.501	0.504
0.601	0.475
0.697	0.446
0.900	0.394
1.10	0.363
1.30	0.340
1.49	0.328
1.69	0.310*
1.99	0.287
2.48	0.262
2.98	0.238
3.98	0.199
4.99	0.178
6.00	0.159
6.98	0.149
9.02	0.129
11.1	0.119
13.0	0.108
15.8	0.098
18.9	0.090
20.8	0.088
CURVE 7	
T = 306	
0.329	0.702
0.398	0.667
0.497	0.630
0.597	0.603
0.695	0.592
0.897	0.566
1.10	0.553
1.30	0.543
1.50	0.527
1.70	0.512
1.98	0.496
2.48	0.471
3.02	0.455
3.99	0.432
5.02	0.400
5.97	0.387
6.98	0.366
CURVE 7 (cont.)	
8.97	0.349
11.0	0.338
13.1	0.324
15.9	0.300
18.9	0.290
22.9	0.267
CURVE 8	
T = 306	
0.333	0.773
0.399	0.733
0.500	0.702
0.597	0.689
0.697	0.673
0.897	0.656
1.10	0.635
1.30	0.630
1.50	0.622
1.70	0.612
2.00	0.592
2.51	0.570
3.01	0.551
4.01	0.516
5.01	0.494
6.00	0.476
7.05	0.459
8.97	0.439
11.0	0.416
13.0	0.416
15.8	0.395
18.9	0.395
22.9	0.366

* Not shown on plot

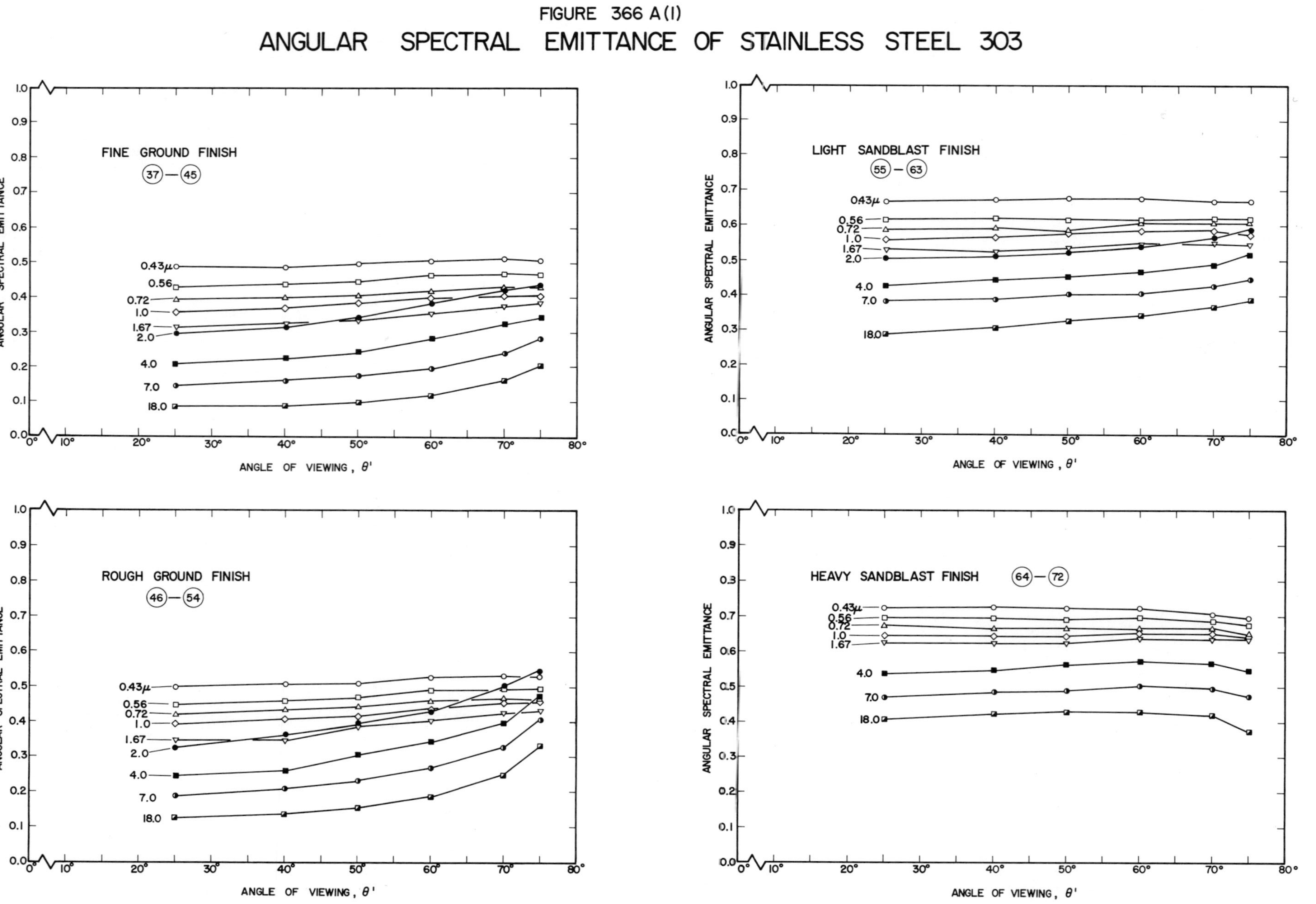

FIGURE 366 A (1)

ANGULAR SPECTRAL EMITTANCE OF STAINLESS STEEL 303

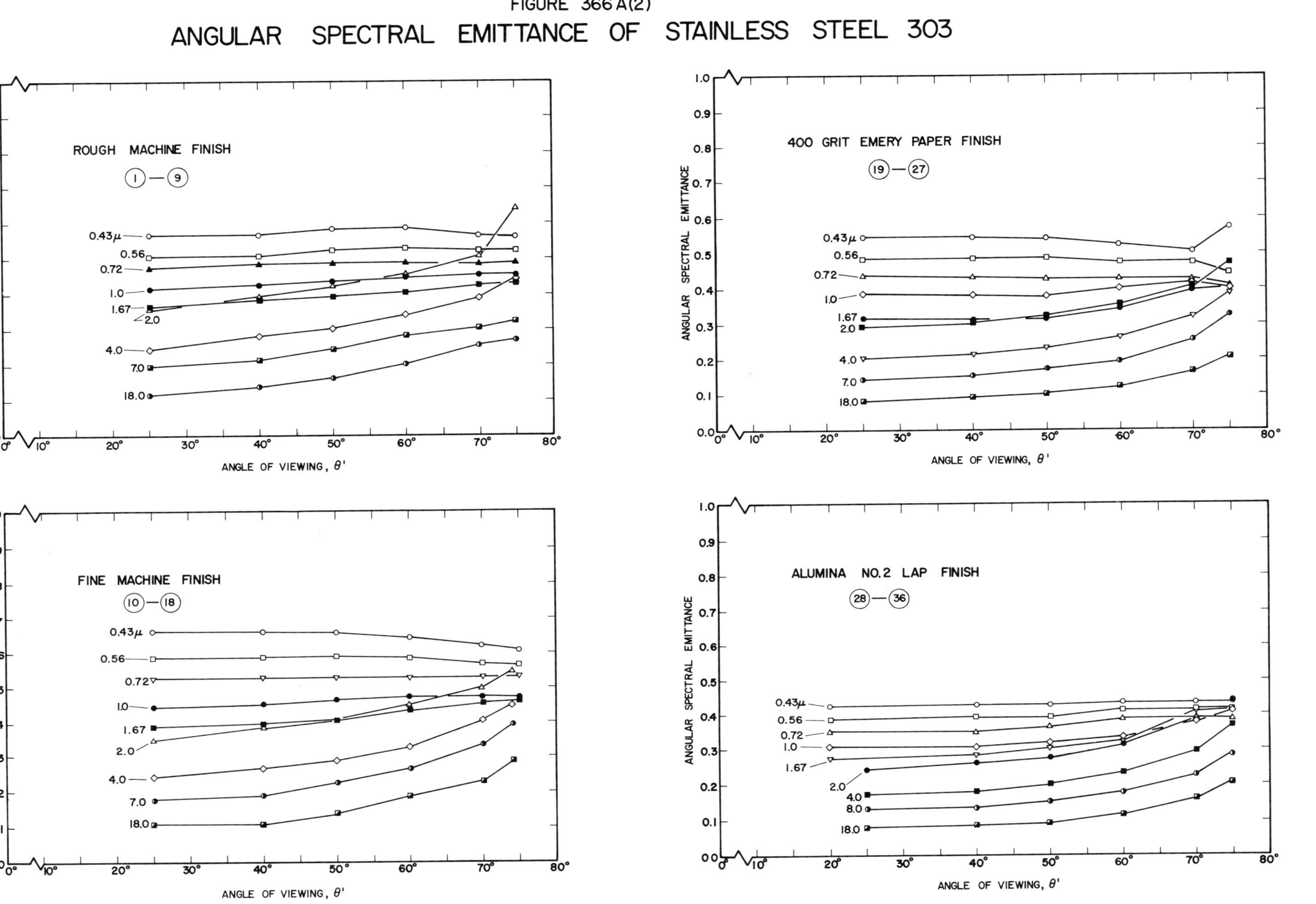
FIGURE 366 A(2)
ANGULAR SPECTRAL EMITTANCE OF STAINLESS STEEL 303
ROUGH MACHINE FINISH
1 — 9
400 GRIT EMERY PAPER FINISH
19 — 27
FINE MACHINE FINISH
10 — 18
ALUMINA NO.2 LAP FINISH
28 — 36
0.43μ
0.56
0.72
1.0
1.67
2.0
4.0
7.0
8.0
18.0
ANGULAR SPECTRAL EMITTANCE
ANGLE OF VIEWING, θ'

SPECIFICATION TABLE NO. 366 ANGULAR SPECTRAL EMITTANCE OF [IRON + CHROMIUM + NICKEL + ΣX_i] ALLOYS

Curve No.	Ref. No.	Year	Temperature K	Wavelength, μ	Angular Range, °	Geometry θ'	Reported Error, %	Composition (weight percent), Specifications and Remarks
1	219	1965	306	0.43	25-75	θ'		Stainless Steel 303; 17-19 Cr, 8-10 Ni, 2 Mg, 1 Si, 0.6 Mo, 0.6 Zr, 0.15 C, 0.15 Se, and Fe balance; machined rough; authors assumed $\epsilon = \alpha = 1 - \rho(\theta', 2\pi)$.
2	219	1965	306	0.56	25-75	θ'		Above specimen and conditions.
3	219	1965	306	0.72	25-75	θ'		Above specimen and conditions.
4	219	1965	306	1	25-75	θ'		Above specimen and conditions.
5	219	1965	306	1.67	25-75	θ'		Above specimen and conditions.
6	219	1965	306	2	25-75	θ'		Above specimen and conditions.
7	219	1965	306	4	25-75	θ'		Above specimen and conditions.
8	219	1965	306	7	25-75	θ'		Above specimen and conditions.
9	219	1965	306	18	25-75	θ'		Above specimen and conditions.
10	219	1965	306	0.43	25-75	θ'		Stainless Steel 303; 17-19 Cr, 8-10 Ni, 2 Mg, 1 Si, 0.6 Mo, 0.6 Zr, 0.15 C, 0.15 Se, and Fe balance; machined fine; authors assumed $\epsilon = \alpha = 1 - \rho(\theta', 2\pi)$.
11	219	1965	306	0.56	25-75	θ'		Above specimen and conditions.
12	219	1965	306	0.72	25-75	θ'		Above specimen and conditions.
13	219	1965	306	1	25-75	θ'		Above specimen and conditions.
14	219	1965	306	1.67	25-75	θ'		Above specimen and conditions.
15	219	1965	306	2	25-74	θ'		Above specimen and conditions.
16	219	1965	306	4	25-74	θ'		Above specimen and conditions.
17	219	1965	306	7	25-74	θ'		Above specimen and conditions.
18	219	1965	306	18	25-74	θ'		Above specimen and conditions.
19	219	1965	306	0.43	25-75	θ'		Stainless Steel 303; 17-19 Cr, 8-10 Ni, 2 Mg, 1 Si, 0.6 Mo, 0.6 Zr, 0.15 C, 0.15 Se, and Fe balance; sanded with 400 grit emery paper; authors assumed $\epsilon = \alpha = 1 - \rho(\theta', 2\pi)$.
20	219	1965	306	0.56	25-75	θ'		Above specimen and conditions.
21	219	1965	306	0.72	25-75	θ'		Above specimen and conditions.
22	219	1965	306	1	25-75	θ'		Above specimen and conditions.
23	219	1965	306	1.67	25-75	θ'		Above specimen and conditions.
24	219	1965	306	2	25-75	θ'		Above specimen and conditions.
25	219	1965	306	4	25-75	θ'		Above specimen and conditions.
26	219	1965	306	7	25-75	θ'		Above specimen and conditions.
27	219	1965	306	18	25-75	θ'		Above specimen and conditions.

SPECIFICATION TABLE NO. 366 (continued)

Curve No.	Ref. No.	Year	Temperature K	Wavelength, μ	Angular Range, °	Geometry θ'	Reported Error, %	Composition (weight percent), Specifications and Remarks
28	219	1965	306	0.43	20–75	θ'		Stainless Steel 303; 17–19 Cr, 8–10 Ni, 2 Mg, 1 Si, 0.6 Mo, 0.6 Zr, 0.15 C, 0.15 Se, and Fe balance; cloth lapped with No. 2 polishing alumina (0.5 μ); authors assumed $\epsilon = \alpha = 1 - \rho(\theta', 2\pi)$.
29	219	1965	306	0.56	20–75	θ'		Above specimen and conditions.
30	219	1965	306	0.72	20–75	θ'		Above specimen and conditions.
31	219	1965	306	1	20–75	θ'		Above specimen and conditions.
32	219	1965	306	1.67	20–75	θ'		Above specimen and conditions.
33	219	1965	306	2	25–75	θ'		Above specimen and conditions.
34	219	1965	306	4	25–75	θ'		Above specimen and conditions.
35	219	1965	306	7	25–75	θ'		Above specimen and conditions.
36	219	1965	306	18	25–75	θ'		Above specimen and conditions.
37	219	1965	306	0.43	25–75	θ'		Stainless Steel 303; 17–19 Cr, 8–10 Ni, 2 Mg, 1 Si, 0.6 Mo, 0.6 Zr, 0.15 C, 0.15 Se, and Fe balance; ground fine; authors assumed $\epsilon = \alpha = 1 - \rho(\theta', 2\pi)$.
38	219	1965	306	0.56	25–75	θ'		Above specimen and conditions.
39	219	1965	306	0.72	25–75	θ'		Above specimen and conditions.
40	219	1965	306	1	25–75	θ'		Above specimen and conditions.
41	219	1965	306	1.67	25–75	θ'		Above specimen and conditions.
42	219	1965	306	2	25–75	θ'		Above specimen and conditions.
43	219	1965	306	4	25–75	θ'		Above specimen and conditions.
44	219	1965	306	7	25–75	θ'		Above specimen and conditions.
45	219	1965	306	18	25–75	θ'		Above specimen and conditions.
46	219	1965	306	0.43	25–75	θ'		Stainless Steel 303; 17–19 Cr, 8–10 Ni, 2 Mg, 1 Si, 0.6 Mo, 0.6 Zr, 0.15 C, 0.15 Se, and Fe balance; ground rough; authors assumed $\epsilon = \alpha = 1 - \rho(\theta', 2\pi)$.
47	219	1965	306	0.56	25–75	θ'		Above specimen and conditions.
48	219	1965	306	0.72	25–75	θ'		Above specimen and conditions.
49	219	1965	306	1	25–75	θ'		Above specimen and conditions.
50	219	1965	306	1.67	25–75	θ'		Above specimen and conditions.
51	219	1965	306	2	25–74	θ'		Above specimen and conditions.
52	219	1965	306	4	25–74	θ'		Above specimen and conditions.
53	219	1965	306	7	25–74	θ'		Above specimen and conditions.
54	219	1965	306	18	25–74	θ'		Above specimen and conditions.

SPECIFICATION TABLE NO. 366 (continued)

Curve No.	Ref. No.	Year	Temperature K	Wavelength, μ	Angular Range, °	Geometry θ'	Reported Error, %	Composition (weight percent), Specifications and Remarks
55	219	1965	306	0.43	25-75	θ'		Stainless Steel 303; 17-19 Cr, 8-10 Ni, 2 Mg, 1 Si, 0.6 Mo, 0.6 Zr, 0.15 C, 0.15 Se, Fe balance; sandblasted lightly with No. 80 Al_2O_3 grit; authors assumed $\epsilon = \alpha = 1 - \rho(\theta', 2\pi)$.
56	219	1965	306	0.56	25-75	θ'		Above specimen and conditions.
57	219	1965	306	0.72	25-75	θ'		Above specimen and conditions.
58	219	1965	306	1	25-75	θ'		Above specimen and conditions.
59	219	1965	306	1.67	25-75	θ'		Above specimen and conditions.
60	219	1965	306	2	25-75	θ'		Above specimen and conditions.
61	219	1965	306	4	25-75	θ'		Above specimen and conditions.
62	219	1965	306	7	25-75	θ'		Above specimen and conditions.
63	219	1965	306	18	25-75	θ'		Above specimen and conditions.
64	219	1965	306	0.43	25-75	θ'		Stainless Steel 303; 17-19 Cr, 8-10 Ni, 2 Mg, 1 Si, 0.6 Mo, 0.6 Zr, 0.15 C, 0.15 Se, and Fe balance; sandblasted heavily with No. 80 Al_2O_3 grit; authors assumed $\epsilon = \alpha = 1 - \rho(\theta', 2\pi)$.
65	219	1965	306	0.56	25-75	θ'		Above specimen and conditions.
66	219	1965	306	0.72	25-75	θ'		Above specimen and conditions.
67	219	1965	306	1	25-75	θ'		Above specimen and conditions.
68	219	1965	306	1.67	25-75	θ'		Above specimen and conditions.
69	219	1965	306	2	25-74	θ'		Above specimen and conditions.
70	219	1965	306	4	25-74	θ'		Above specimen and conditions.
71	219	1965	306	7	25-74	θ'		Above specimen and conditions.
72	219	1965	306	18	25-74	θ'		Above specimen and conditions.

DATA TABLE NO. 366 ANGULAR SPECTRAL EMITTANCE OF [IRON + CHROMIUM + NICKEL + ΣX_i] ALLOYS

[Angle, θ', °; Emittance, ∈; Temperature, T, K; Wavelength, λ, μ]

θ'	∈	θ'	∈	θ'	∈	θ'	∈	θ'	∈	θ'	∈	θ'	∈	θ'	∈
CURVE 1*		CURVE 5*		CURVE 9*		CURVE 13*		CURVE 17*		CURVE 21*		CURVE 25*		CURVE 29*	
T = 306		T = 306		T = 306		T = 306		T = 306		T = 306		T = 306		T = 306	
λ = 0.43		λ = 1.67		λ = 18		λ = 1		λ = 7		λ = 0.72		λ = 4		λ = 0.56	
25	0.568	25	0.368	25	0.119	25	0.446	25	0.173	25	0.439	25	0.201	20	0.388
40	0.568	40	0.388	40	0.138	40	0.452	40	0.189	40	0.432	40	0.212	40	0.391
50	0.582	50	0.396	50	0.162	50	0.464	50	0.225	50	0.429	50	0.231	50	0.391
60	0.584	60	0.406	60	0.203	60	0.472	60	0.263	60	0.429	60	0.262	60	0.411
70	0.565	70	0.426	70	0.258	70	0.472	70	0.332	70	0.429	70	0.321	70	0.413
75	0.561	75	0.431	75	0.270	75	0.472	74	0.391	75	0.407	75	0.386	75	0.413
CURVE 2*		CURVE 6*		CURVE 10*		CURVE 14*		CURVE 18*		CURVE 22*		CURVE 26*		CURVE 30*	
T = 306		T = 306		T = 306		T = 306		T = 306		T = 306		T = 306		T = 306	
λ = 0.56		λ = 2		λ = 0.43		λ = 1.67		λ = 18		λ = 1		λ = 7		λ = 0.72	
25	0.508	25	0.362	25	0.661	25	0.385	25	0.108	25	0.386	25	0.143	20	0.350
40	0.508	40	0.389	40	0.661	40	0.397	40	0.108	40	0.381	40	0.157	40	0.350
50	0.525	50	0.422	50	0.660	50	0.407	50	0.137	50	0.381	50	0.172	50	0.365
60	0.530	60	0.451	60	0.644	60	0.435	60	0.187	60	0.400	60	0.197	60	0.383
70	0.525	70	0.510	70	0.620	70	0.455	70	0.229	70	0.418	70	0.252	70	0.389
75	0.521	75	0.641	75	0.609	75	0.460	74	0.286	75	0.406	75	0.323	75	0.386
CURVE 3*		CURVE 7*		CURVE 11*		CURVE 15*		CURVE 19*		CURVE 23*		CURVE 27*		CURVE 31*	
T = 306		T = 306		T = 306		T = 306		T = 306		T = 306		T = 306		T = 306	
λ = 0.72		λ = 4		λ = 0.56		λ = 2		λ = 0.43		λ = 1.67		λ = 18		λ = 1	
25	0.472	25	0.245	25	0.587	25	0.348	25	0.547	25	0.318	25	0.087	20	0.309
40	0.487	40	0.283	40	0.587	40	0.384	40	0.547	40	0.316	40	0.093	40	0.309
50	0.487	50	0.305	50	0.589	50	0.407	50	0.541	50	0.316	50	0.103	50	0.321
60	0.490	60	0.344	60	0.584	60	0.448	60	0.526	60	0.345	60	0.122	60	0.336
70	0.488	70	0.390	70	0.568	70	0.497	70	0.506	70	0.398	70	0.166	70	0.382
75	0.488	75	0.445	75	0.564	74	0.547	75	0.471	75	0.405	75	0.209	75	0.412
CURVE 4*		CURVE 8*		CURVE 12*		CURVE 16*		CURVE 20*		CURVE 24*		CURVE 28*		CURVE 32*	
T = 306		T = 306		T = 306		T = 306		T = 306		T = 306		T = 306		T = 306	
λ = 1		λ = 7		λ = 0.72		λ = 4		λ = 0.56		λ = 2		λ = 0.43		λ = 1.67	
25	0.419	25	0.198	25	0.525	25	0.242	25	0.487	25	0.290	20	0.426	20	0.272
40	0.429	40	0.215	40	0.528	40	0.266	40	0.485	40	0.302	40	0.429	40	0.284
50	0.436	50	0.246	50	0.528	50	0.288	50	0.486	50	0.325	50	0.429	50	0.304
60	0.447	60	0.286	60	0.528	60	0.326	60	0.476	60	0.356	60	0.433	60	0.322
70	0.454	70	0.308	70	0.530	70	0.406	70	0.473	70	0.408	70	0.433	70	0.410
75	0.455	75	0.325	75	0.531	74	0.450	75	0.446	75	0.471	75	0.433	75	0.414

* Not shown on plot

DATA TABLE NO. 366 (continued)

θ'	ε	θ'	ε	θ'	ε	θ'	ε	θ'	ε	θ'	ε	θ'	ε	θ'	ε
CURVE 33*		CURVE 37*		CURVE 41*		CURVE 45*		CURVE 49*		CURVE 53*		CURVE 57*		CURVE 61*	
T = 306		T = 306		T = 306		T = 306		T = 306		T = 306		T = 306		T = 306	
λ = 2		λ = 0.43		λ = 1.67		λ = 18		λ = 1		λ = 7		λ = 0.72		λ = 4	
25	0.244	25	0.487	25	0.311	25	0.085	25	0.392	25	0.189	25	0.588	25	0.428
40	0.261	40	0.487	40	0.321	40	0.089	40	0.407	40	0.209	40	0.591	40	0.445
50	0.279	50	0.499	50	0.336	50	0.098	50	0.411	50	0.230	50	0.588	50	0.454
60	0.316	60	0.507	60	0.354	60	0.118	60	0.434	60	0.268	60	0.603	60	0.469
70	0.385	70	0.513	70	0.374	70	0.161	70	0.455	70	0.327	70	0.604	70	0.487
75	0.438	75	0.509	75	0.383	75	0.203	75	0.459	74	0.405	75	0.606	75	0.518
CURVE 34*		CURVE 38*		CURVE 42*		CURVE 46*		CURVE 50*		CURVE 54*		CURVE 58*		CURVE 62*	
T = 306		T = 306		T = 306		T = 306		T = 306		T = 306		T = 306		T = 306	
λ = 4		λ = 0.56		λ = 2		λ = 0.43		λ = 1.67		λ = 18		λ = 1		λ = 7	
25	0.172	25	0.428	25	0.295	25	0.498	25	0.347	25	0.127	25	0.559	25	0.381
40	0.183	40	0.438	40	0.318	40	0.507	40	0.347	40	0.133	40	0.565	40	0.388
50	0.202	50	0.444	50	0.341	50	0.509	50	0.384	50	0.152	50	0.573	50	0.403
60	0.233	60	0.463	60	0.381	60	0.528	60	0.401	60	0.186	60	0.585	60	0.408
70	0.296	70	0.469	70	0.420	70	0.530	70	0.422	70	0.248	70	0.585	70	0.429
75	0.368	75	0.469	75	0.439	75	0.530	75	0.431	74	0.330	75	0.575	75	0.450
CURVE 35*		CURVE 39*		CURVE 43*		CURVE 47*		CURVE 51*		CURVE 55*		CURVE 59*		CURVE 63*	
T = 306		T = 306		T = 306		T = 306		T = 306		T = 306		T = 306		T = 306	
λ = 7		λ = 0.72		λ = 4		λ = 0.56		λ = 2		λ = 0.43		λ = 1.67		λ = 18	
25	0.131	25	0.392	25	0.208	25	0.448	25	0.323	25	0.668	25	0.528	25	0.287
40	0.139	40	0.400	40	0.223	40	0.460	40	0.363	40	0.671	40	0.524	40	0.306
50	0.154	50	0.405	50	0.243	50	0.470	50	0.394	50	0.675	50	0.532	50	0.326
60	0.180	70	0.417	60	0.281	60	0.490	60	0.431	60	0.674	60	0.548	60	0.343
70	0.230	70	0.429	70	0.325	70	0.497	70	0.505	70	0.669	70	0.550	70	0.367
75	0.285	75	0.429	75	0.354	75	0.497	74	0.548	75	0.668	75	0.547	75	0.389
CURVE 36*		CURVE 40*		CURVE 44*		CURVE 48*		CURVE 52*		CURVE 56*		CURVE 60*		CURVE 64*	
T = 306		T = 306		T = 306		T = 306		T = 306		T = 306		T = 306		T = 306	
λ = 18		λ = 1		λ = 7		λ = 0.72		λ = 4		λ = 0.56		λ = 2		λ = 0.43	
25	0.080	25	0.357	25	0.147	25	0.418	25	0.243	25	0.616	25	0.504	25	0.721
40	0.088	40	0.369	40	0.160	40	0.432	40	0.260	40	0.620	40	0.510	40	0.724
50	0.093	50	0.383	50	0.174	50	0.442	50	0.302	50	0.617	50	0.522	50	0.724
60	0.119	60	0.399	60	0.197	60	0.459	60	0.341	60	0.617	60	0.542	60	0.721
70	0.161	70	0.407	70	0.241	70	0.467	70	0.398	70	0.620	70	0.564	70	0.705
75	0.207	75	0.404	75	0.281	75	0.467	74	0.468	75	0.619	75	0.589	75	0.697

* Not shown on plot

DATA TABLE NO. 366 (continued)

θ'	∈	θ'	∈
CURVE 65*		**CURVE 69***	
T = 306		T = 306	
λ = 0.56		λ = 2	
25	0.697	25	0.607
40	0.695	40	0.621
50	0.695	50	0.628
60	0.694	60	0.640
70	0.689	70	0.640
75	0.679	74	0.638
CURVE 66*		**CURVE 70***	
T = 306		T = 306	
λ = 0.72		λ = 4	
25	0.671	25	0.537
40	0.666	40	0.550
50	0.666	50	0.565
60	0.666	60	0.571
70	0.666	70	0.568
75	0.655	74	0.548
CURVE 67*		**CURVE 71***	
T = 306		T = 306	
λ = 1		λ = 7	
25	0.644	25	0.469
40	0.644	40	0.486
50	0.644	50	0.489
60	0.652	60	0.503
70	0.652	70	0.492
75	0.644	74	0.473
CURVE 68*		**CURVE 72***	
T = 306		T = 306	
λ = 1.67		λ = 18	
25	0.623	25	0.407
40	0.623	40	0.421
50	0.623	50	0.429
60	0.637	60	0.429
70	0.637	70	0.420
75	0.637	74	0.371

* Not shown on plot

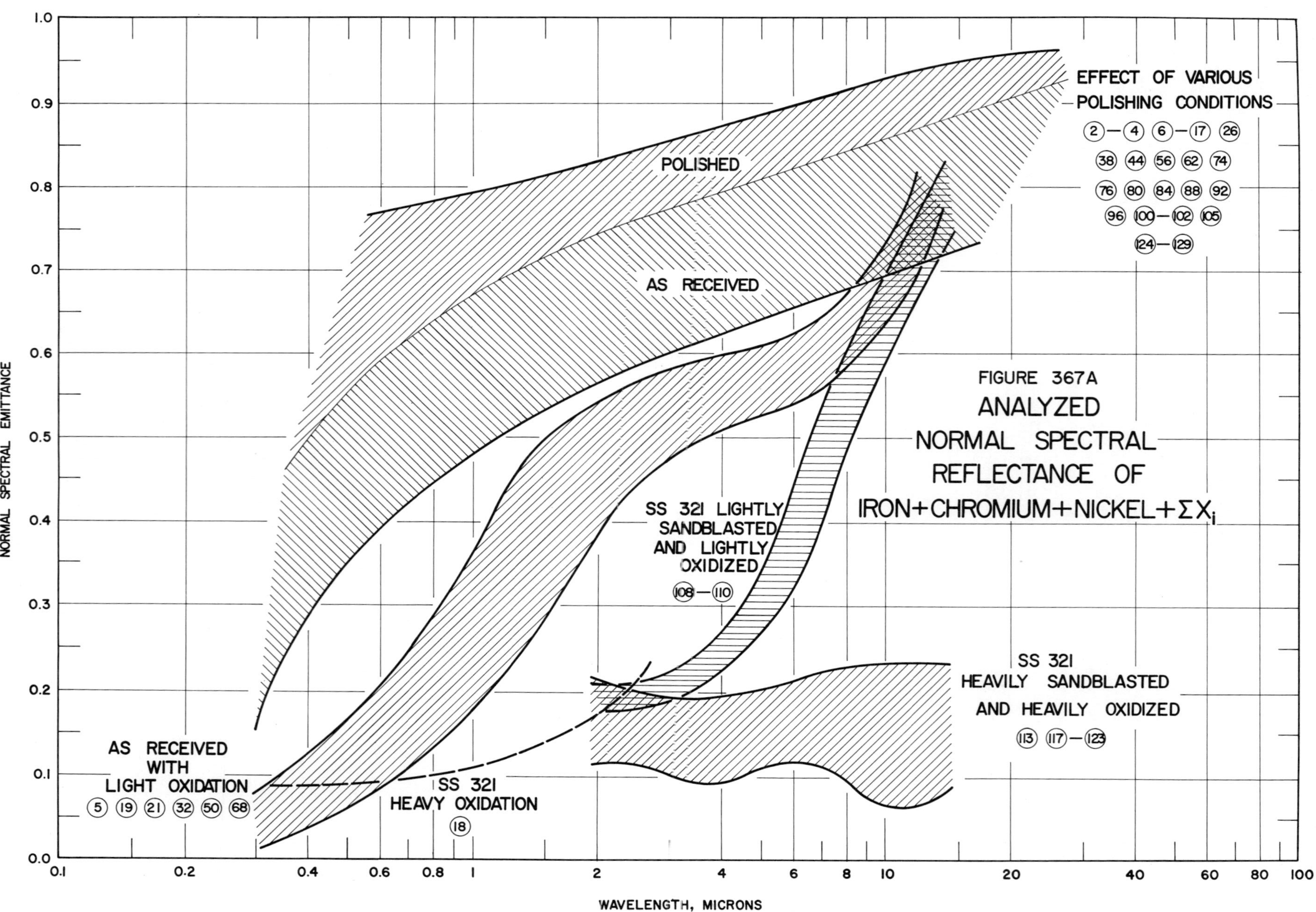
FIGURE 367A
ANALYZED
NORMAL SPECTRAL
REFLECTANCE OF
IRON+CHROMIUM+NICKEL+ΣX_i
EFFECT OF VARIOUS
POLISHING CONDITIONS
2 — 4 6 — 17 26
38 44 56 62 74
76 80 84 88 92
96 100 — 102 105
124 — 129
POLISHED
AS RECEIVED
SS 321 LIGHTLY
SANDBLASTED
AND LIGHTLY
OXIDIZED
108 — 110
SS 321
HEAVILY SANDBLASTED
AND HEAVILY OXIDIZED
113 117 — 123
AS RECEIVED
WITH
LIGHT OXIDATION
5 19 21 32 50 68
SS 321
HEAVY OXIDATION
18
NORMAL SPECTRAL EMITTANCE
1.0
0.9
0.8
0.7
0.6
0.5
0.4
0.3
0.2
0.1
0.0
0.1
0.2
0.4
0.6
0.8
1
2
4
6
8
10
20
40
60
80
100
WAVELENGTH, MICRONS

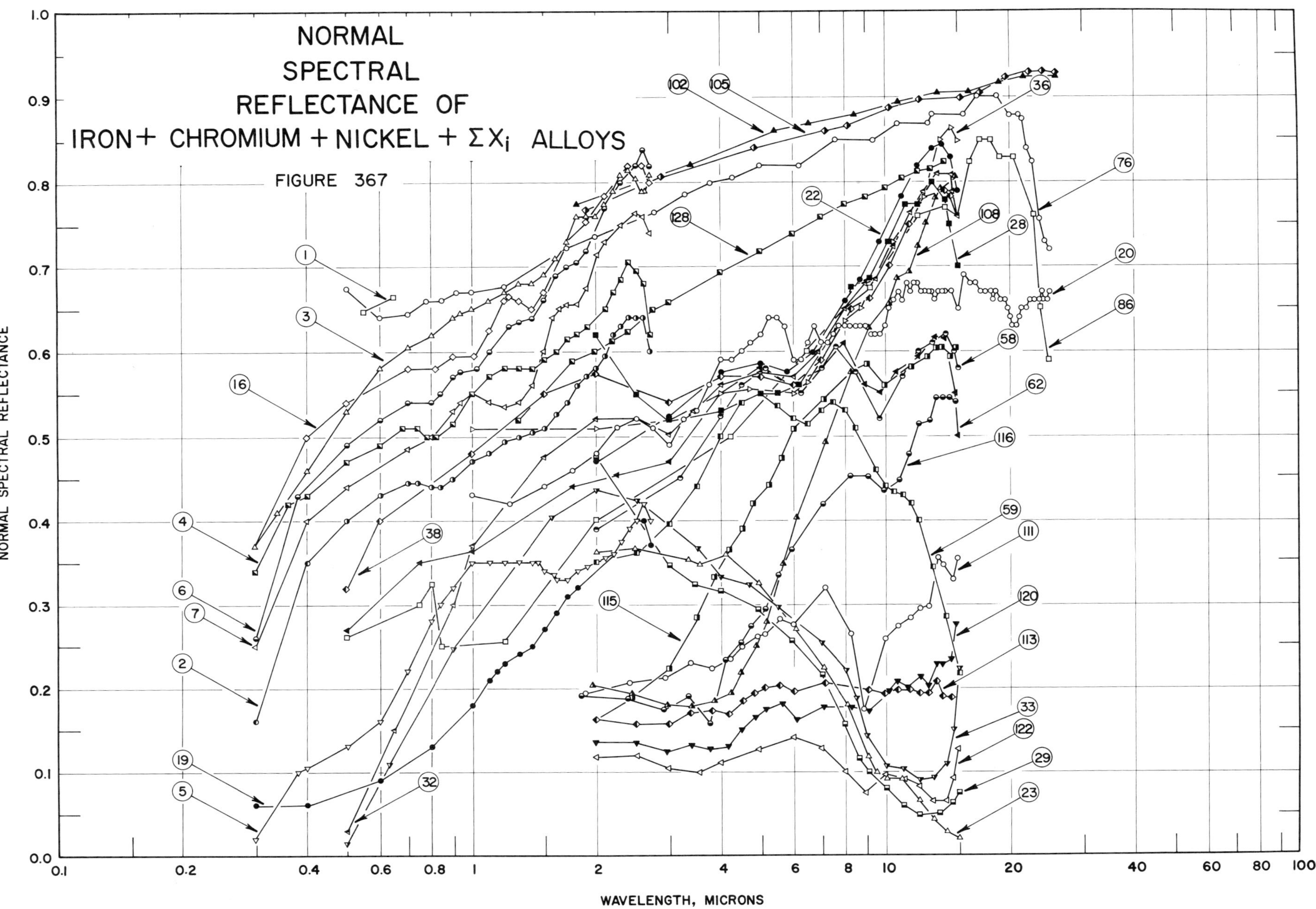
NORMAL
SPECTRAL
REFLECTANCE OF
IRON + CHROMIUM + NICKEL + ΣXi ALLOYS
FIGURE 367
NORMAL SPECTRAL REFLECTANCE
WAVELENGTH, MICRONS
0.0
0.1
0.2
0.3
0.4
0.5
0.6
0.7
0.8
0.9
1.0
0.1
0.2
0.4
0.6
0.8
1
2
4
6
8
10
20
40
60
80
100
1
2
3
4
5
6
7
16
19
20
22
23
28
29
32
33
36
38
58
59
62
76
86
102
105
108
111
113
115
116
120
122
128

SPECIFICATION TABLE NO. 367 NORMAL SPECTRAL REFLECTANCE OF [IRON + CHROMIUM + NICKEL + ΣX_i] ALLOYS

Curve No.	Ref. No.	Year	Temperature K	Wavelength Range, μ	Geometry θ	θ'	ω'	Reported Error, %	Composition (weight percent), Specifications and Remarks
1	173	1947	298	0.55-0.65	~0°	~0°			Stainless steel SF 20; 18 Cr, 8 Ni, Fe balance.
2	34	1957	298	0.30-2.70	9°		2π	± 4	Cobalt alloy N-155; nominal composition: 21 Cr, 20 Co, 20 Ni, 3 Mo, 3 W, 1.5 Mn, 1 Cb, 0.5 Si, 0.15 C, 0.15 N, Fe balance; as received; data extracted from smooth curve.
3	34	1957	298	0.30-2.70	9°		2π	± 4	Different sample, same as curve 2 specimen and conditions except polished.
4	34	1957	298	0.30-2.70	9°		2π	± 4	Different sample, same as curve 2 specimen and conditions except cleaned with liquid detergent.
5	34	1957	298	0.30-2.70	9°		2π	± 4	Different sample, same as curve 2 specimen and conditions except oxidized in air at red heat for 30 min.
6	34	1957	298	0.30-2.70	9°		2π	± 4	Stainless steel type PH 15-7 Mo; nominal composition: 15.00 Cr, 7.00 Ni, 2.25 Mo, 1.15 Al, 0.70 Mn, 0.40 Si, 0.07 C, Fe balance; surface roughness ~2 microinches rms; data extracted from a smooth curve.
7	34	1957	298	0.30-2.70	9°		2π	± 4	Different sample, same as above specimen and conditions except surface roughness ~15 microinches rms.
8	34	1957	298	0.30-2.70	9°		2π	± 4	Stainless steel type 17-7 PH; nominal composition: 17.00 Cr, 7.00 Ni, 1.15 Al, 0.70 Mn, 0.40 Si, 0.07 C, Fe balance; surface roughness ~2 microinches rms; data extracted from a smooth curve.
9	34	1957	298	0.30-2.70	9°		2π	± 4	Different sample, same as above specimen and conditions except surface roughness ~15 microinches rms.
10	34	1957	298	0.30-2.70	9°		2π	± 4	Stainless steel type 316; nominal composition, 16.00-18.00 Cr, 10.00-14.00 Ni, 2.00-3.00 Mo, 2.00 max Mn, 1.00 max Si, 0.08 max C, Fe balance; surface roughness 2 microinches rms; data extracted from a smooth curve.
11	34	1957	298	0.30-2.70	9°		2π	± 4	Different sample, same as above specimen and conditions except surface roughness ~15 microinches rms.
12	34	1957	298	0.30-2.70	9°		2π	± 4	Stainless steel type 321; nominal composition: 17.00-19.00 Cr, 9.00-12.00 Ni, 2.00 max Mn, 1.00 max Si, 0.08 max C, 5 x C min Ti, Fe balance; surface roughness ~2 microinches rms; data extracted from a smooth curve.
13	34	1957	298	0.30-2.70	9°		2π	± 4	Different sample, same as curve 12 specimen and conditions except surface roughness ~6 microinches rms; dull finish.
14	34	1957	298	0.30-2.70	9°		2π	± 4	Different sample, same as curve 12 specimen and conditions except surface roughness ~15 microinches rms.
15	34	1957	298	0.36-2.72	9°		2π	± 4	Different sample, same as curve 12 specimen and conditions except bright finish.
16	34	1957	298	0.30-2.70	9°		2π	± 4	Stainless steel type AM350; nominal composition: 16.50 Cr, 4.25 Ni, 2.75 Mo, 0.75 Mn, 0.35 Si, 0.10 C, 0.10 N, Fe balance; surface roughness ~2 microinches rms; data extracted from a smooth curve.
17	34	1957	298	0.30-2.70	9°		2π	± 4	Different sample, same as above specimen and conditions except surface roughness ~15 microinches rms.

SPECIFICATION TABLE NO. 367 (continued)

Curve No.	Ref. No.	Year	Temperature K	Wavelength Range, μ	Geometry θ	θ'	ω'	Reported Error, %	Composition (weight percent), Specifications and Remarks
18	146	1958	298	0.30-2.70	~0°		2π		Stainless steel type 321; nominal composition: 17.00-19.00 Cr, 9.00-12.00 Ni, 2.00 max Mn, 1.00 max Si, 0.08 max C, 5 x C min Ti, Fe balance; oxidized at 922 K for 30 min.
19	146	1958	298	0.30-2.70	~0°		2π		Stainless steel type AM 350; nominal composition: 16.50 Cr, 4.25 Ni, 2.75 Mo, 0.75 Mn, 0.35 Si, 0.10 C, 0.10 N, Fe balance; oxidized at 922 K for 30 min.
20	99	1958	311	1.0-25.00	~0°		2π		Type 321 corrosion resistant steel; MIL-S-6721; nominal composition: 17.00-19.00 Cr, 9.00-12.00 Ni, 2.00 max Mn, 1.00 max Si, 0.08 max C, 5 x C min Ti, Fe balance; hohlraum at 1089 K; converted from R (2π, 0°).
21	99	1958	311	1.0-25.00	~0°		2π		Different sample, same as above specimen and conditions except heated at 647 K for 1000 hrs.
22	86	1961	~322	2.00-15.00	~0°		2π	< 2	Steel PH 15-7 Mo, NAA LBO 160-130; nominal composition: 15.00 Cr, 7.00 Ni, 2.25 Mo, 1.15 Al, 0.70 Mn, 0.40 Si, 0.07 C, Fe balance; as received; data extracted from smooth curve; hohlraum at 523 K; converted from R (2π, 0°).
23	86	1961	~322	2.00-15.00	~0°		2π	< 2	Above specimen and conditions; diffuse component only.
24	86	1961	~322	1.00-15.00	~0°		2π	< 2	Different sample, same as curve 22 specimen and conditions except hohlraum at 773 K.
25	86	1961	~322	1.00-15.00	~0°		2π	< 2	Above specimen and conditions; diffuse component only.
26	86	1961	~322	0.50-15.00	~0°		2π	< 2	Different sample, same as curve 22 specimen and conditions except hohlraum at 1273 K.
27	86	1961	~322	0.50-15.00	~0°		2π	< 2	Above specimen and conditions; diffuse component only.
28	86	1961	~322	2.00-15.00	~0°		2π	< 2	Different sample, same as curve 22 specimen and conditions except heated in air at 755 K for 30 min.
29	86	1961	~322	2.00-15.00	~0°		2π	< 2	Above specimen and conditions; diffuse component only.
30	86	1961	~322	1.00-15.00	~0°		2π	< 2	Different sample, same as curve 28 specimen and conditions except hohlraum at 773 K.
31	86	1961	~322	1.00-15.00	~0°		2π	< 2	Above specimen and conditions; diffuse component only.
32	86	1961	~322	0.50-15.00	~0°		2π	< 2	Different sample, same as curve 28 specimen and conditions except hohlraum at 1273 K.
33	86	1961	~322	0.50-15.00	~0°		2π	< 2	Above specimen and conditions; diffuse component only.
34	86	1961	~322	2.00-15.00	~0°		2π	< 2	Different sample, same as curve 22 specimen and conditions except heated in vacuum (4.4 x 10^{-5} mm Hg) at 755 K for 30 min.
35	86	1961	~322	2.00-15.00	~0°		2π	< 2	Above specimen and conditions; diffuse component only.
36	86	1961	~322	1.00-15.00	~0°		2π	< 2	Different sample, same as curve 34 specimen and conditions except hohlraum at 773 K.
37	86	1961	~322	1.00-15.00	~0°		2π	< 2	Above specimen and conditions; diffuse component only.
38	86	1961	~322	0.50-15.00	~0°		2π	< 2	Different sample, same as curve 34 specimen and conditions except hohlraum at 1273 K.
39	86	1961	~322	0.50-15.00	~0°		2π	< 2	Above specimen and conditions; diffuse component only.

SPECIFICATION TABLE NO. 367 (continued)

Curve No.	Ref. No.	Year	Temperature K	Wavelength Range, μ	Geometry θ	θ'	ω'	Reported Error, %	Composition (weight percent), Specifications and Remarks
40	86	1961	~322	2.00-15.00	~0°		2π	< 2	17-7 PH steel, MIL-S-25043; nominal composition: 17.00 Cr, 7.00 Ni, 1.15 Al, 0.70 Mn, 0.40 Si, 0.07 C, Fe balance; as received; data extracted from smooth curve; hohlraum at 523 K; converted from R (2π, 0°).
41	86	1961	~322	2.00-15.00	~0°		2π	< 2	Above specimen and conditions; diffuse component only.
42	86	1961	~322	1.00-15.50	~0°		2π	< 2	Different sample, same as curve 40 specimen and conditions except hohlraum at 773 K.
43	86	1961	~322	1.00-15.00	~0°		2π	< 2	Above specimen and conditions; diffuse component only.
44	86	1961	~322	0.50-15.00	~0°		2π	< 2	Different sample, same as curve 40 specimen and conditions except hohlraum at 1273 K.
45	86	1961	~322	0.50-15.00	~0°		2π	< 2	Above specimen and conditions; diffuse component only.
46	86	1961	~322	2.00-15.00	~0°		2π	< 2	Different sample, same as curve 40 specimen and conditions except heated in air at 755 K for 30 min.
47	86	1961	~322	2.00-15.00	~0°		2π	< 2	Above specimen and conditions; diffuse component only.
48	86	1961	~322	1.00-15.00	~0°		2π	< 2	Different sample, same as curve 46 specimen and conditions except hohlraum at 773 K.
49	86	1961	~322	1.00-15.00	~0°		2π	< 2	Above specimen and conditions; diffuse component only.
50	86	1961	~322	0.50-15.00	~0°		2π	< 2	Different sample, same as curve 46 specimen and conditions except hohlraum at 1273 K.
51	86	1961	~322	0.50-15.00	~0°		2π	< 2	Above specimen and conditions; diffuse component only.
52	86	1961	~322	2.00-15.00	~0°		2π	< 2	Different sample, same as curve 40 specimen and conditions except heated in vacuum (4.4×10^{-5} mm Hg) at 755 K for 30 min.
53	86	1961	~322	2.00-15.00	~0°		2π	< 2	Above specimen and conditions; diffuse component only.
54	86	1961	~322	1.00-15.00	~0°		2π	< 2	Different sample, same as curve 52 specimen and conditions except hohlraum at 773 K.
55	86	1961	~322	1.00-15.00	~0°		2π	< 2	Above specimen and conditions; diffuse component only.
56	86	1961	~322	0.50-15.00	~0°		2π	< 2	Different sample, same as curve 52 specimen and conditions except hohlraum at 1273 K.
57	86	1961	~322	0.50-15.00	~0°		2π	< 2	Above specimen and conditions; diffuse component only.
58	86	1961	~322	2.00-15.00	~0°		2π	< 2	Steel AM 350; nominal composition: 16.50 Cr, 4.25 Ni, 2.75 Mo, 0.75 Mn, 0.35 Si, 0.10 C, 0.10 N, Fe balance; as received; data extracted from smooth curve; hohlraum at 523 K; converted from R (2π, 0°).
59	86	1961	~322	2.00-15.00	~0°		2π	< 2	Above specimen and conditions; diffuse component only.
60	86	1961	~322	1.00-15.00	~0°		2π	< 2	Different sample, same as curve 58 specimen and conditions except hohlraum at 773 K.
61	86	1961	~322	1.00-15.00	~0°		2π	< 2	Above specimen and conditions; diffuse component only.
62	86	1961	~322	0.50-15.00	~0°		2π	< 2	Different sample, same as curve 58 specimen and conditions except hohlraum at 1273 K.
63	86	1961	~322	0.50-15.00	~0°		2π	< 2	Above specimen and conditions; diffuse component only.
64	86	1961	~322	2.00-15.00	~0°		2π	< 2	Different sample, same as curve 58 specimen and conditions except heated in air at 755 K for 30 min.

SPECIFICATION TABLE NO. 367 (continued)

Curve No.	Ref. No.	Year	Temperature K	Wavelength Range, μ	Geometry θ	θ'	ω'	Reported Error, %	Composition (weight percent), Specifications and Remarks
65	86	1961	~322	2.00-15.00	~0°		2π	< 2	Above specimen and conditions; diffuse component only.
66	86	1961	~322	1.00-15.00	~0°		2π	< 2	Different sample, same as curve 64 specimen and conditions except hohlraum at 773 K.
67	86	1961	~322	1.00-15.00	~0°		2π	< 2	Above specimen and conditions; diffuse component only.
68	86	1961	~322	0.50-15.00	~0°		2π	< 2	Different sample, same as curve 64 specimen and conditions except hohlraum at 1273 K.
69	86	1961	~322	0.50-15.00	~0°		2π	< 2	Above specimen and conditions; diffuse component only.
70	86	1961	~322	2.00-15.00	~0°		2π	< 2	Different sample, same as curve 58 specimen and conditions except heated in vacuum (4.4×10^{-5} mm Hg) at 755 K for 30 min.
71	86	1961	~322	2.00-15.00	~0°		2π	< 2	Above specimen and conditions; diffuse component only.
72	86	1961	~322	1.00-15.00	~0°		2π	< 2	Different sample, same as curve 70 specimen and conditions except hohlraum at 773 K.
73	86	1961	~322	1.00-15.00	~0°		2π	< 2	Above specimen and conditions; diffuse component only.
74	86	1961	~322	0.50-15.00	~0°		2π	< 2	Different sample, same as curve 70 specimen and conditions except hohlraum at 1273 K.
75	86	1961	~322	0.50-15.00	~0°		2π	< 2	Above specimen and conditions; diffuse component only.
76	153	1962	~322	0.50-25.00	~0°		2π	< 2	Stainless steel 301, Mil-S-5059; nominal composition: 16.00-18.00 Cr, 6.00-8.00 Ni, 2.00 max Mn, 1.00 max Si, 0.15 C, Fe balance; as received; cleaned; data extracted from smooth curve; hohlraum at 1273 K; converted from R (2π, 0°).
77	153	1962	~322	0.50-25.00	~0°		2π	< 2	Above specimen and conditions; diffuse component only.
78	153	1962	~322	0.50-25.00	~0°		2π	< 2	Different sample, same as curve 76 specimen and conditions except grit blasted using 60 grit silicon carbide with air pressure of 110 to 120 psi for 30 to 45 sec.
79	153	1962	~322	0.50-25.00	~0°		2π	< 2	Above specimen and conditions; diffuse component only.
80	153	1962	~322	0.50-25.00	~0°		2π	< 2	Different sample, same as curve 76 specimen and conditions except exposed to vacuum ($<4 \times 10^{-8}$ mm Hg) for 24 hrs.
81	153	1962	~322	0.50-25.00	~0°		2π	< 2	Above specimen and conditions; diffuse component only.
82	153	1962	~322	0.50-25.00	~0°		2π	< 2	Different sample, same as curve 76 specimen and conditions except X-Ray exposed in vacuum (4×10^{-8} mm Hg) for 24 hrs.
83	153	1962	~322	0.50-25.00	~0°		2π	< 2	Above specimen and conditions; diffuse component only.
84	153	1962	~322	0.50-25.00	~0°		2π	< 2	Different sample, same as curve 76 specimen and conditions except chem-milled at 298 K for 10 min.
85	153	1962	~322	0.50-25.00	~0°		2π	< 2	Above specimen and conditions; diffuse component only.
86	153	1962	~322	0.50-25.00	~0°		2π	< 2	Different sample, same as curve 76 specimen and conditions except chem-milled at 298 K for 10 min; grit blasted using 60 grit silicon carbide with air pressure of 110 to 120 psi for 30 to 45 sec.
87	153	1962	~322	0.50-25.00	~0°		2π	< 2	Above specimen and conditions; diffuse component only.

SPECIFICATION TABLE NO. 367 (continued)

Curve No.	Ref. No.	Year	Temperature K	Wavelength Range, μ	Geometry θ	θ'	ω'	Reported Error, %	Composition (weight percent), Specifications and Remarks
88	153	1962	~322	0.50-25.00	~0°		2π	< 2	Different sample, same as curve 76 specimen and conditions except chem-milled at 298 K for 10 min; exposed to vacuum (<4 x 10^{-8} mm Hg) for 24 hrs.
89	153	1962	~322	0.50-25.00	~0°		2π	< 2	Above specimen and conditions; diffuse component only.
90	153	1962	~322	0.50-25.00	~0°		2π	< 2	Different sample, same as curve 76 specimen and conditions except chem-milled at 298 K for 10 min; X-Ray exposed in vacuum (4 x 10^{-8} mm Hg) for 24 hrs.
91	153	1962	~322	0.50-25.00	~0°		2π	< 2	Above specimen and conditions; diffuse component only.
92	153	1962	~322	0.50-25.00	~0°		2π	< 2	Different sample, same as curve 76 specimen and conditions except electropolished for 5 min.
93	153	1962	~322	0.50-25.00	~0°		2π	< 2	Above specimen and conditions; diffuse component only.
94	153	1962	~322	0.50-25.00	~0°		2π	< 2	Different sample, same as curve 76 specimen and conditions except electropolished for 5 min; grit blasted using 60 grit silicon carbide with air pressure of 110 to 120 psi for 30 to 45 sec.
95	153	1962	~322	0.50-25.00	~0°		2π	< 2	Above specimen and conditions; diffuse component only.
96	153	1962	~322	0.50-25.00	~0°		2π	< 2	Different sample, same as curve 76 specimen and conditions except electropolished for 5 min; exposed to vacuum (<4 x 10^{-8} mm Hg) for 24 hrs.
97	153	1962	~322	0.50-25.00	~0°		2π	< 2	Above specimen and conditions; diffuse component only.
98	153	1962	~322	0.50-25.00	~0°		2π	< 2	Different sample, same as curve 76 specimen and conditions except electropolished for 5 min; X-Ray exposed in vacuum (4 x 10^{-8} mm Hg) for 24 hrs.
99	153	1962	~322	0.50-25.00	~0°		2π	< 2	Above specimen and conditions; diffuse component only.
100	65	1962	294	0.40-14.00	~0°		2π		Stainless steel 321; nominal composition: 17.00-19.00 Cr, 9.00-12.00 Ni, 2.00 max Mn, 1.00 max Si, 0.40 min Ti, 0.08 max C, Fe balance; mechanically polished.
101	65	1962	294	0.60-14.10	~0°		2π		Above specimen and conditions except roughened with sandpaper; surface roughness 1.25μ.
102	223	1962	298	1.81-26.01	~0°		2π		Stainless steel 304; nominal composition, 18.00-20.00 Cr, 8.00-12.00 Ni, 2.00 max Mn, 1.00 max Si, 0.08 max C, Fe balance; polished; converted from R(2π, 0°); data extracted from smooth curve.
103	223	1962	298	1.94-26.01	~0°		2π		Above specimen and conditions except damaged by particle impact.
104	223	1962	77	1.90-25.99	~0°		2π		Above specimen and conditions.
105	223	1962	298	1.90-25.91	~0°		2π		Stainless steel 316; nominal composition, 16.00-18.00 Cr, 10.00-14.00 Ni, 2.00-3.00 Mo, 2.00 max Mn, 1.00 max Si, 0.08 max C, Fe balance; polished; converted from R (2π, 0°); data extracted from smooth curve.
106	223	1962	298	1.97-26.00	~0°		2π		Above specimen and conditions except damaged by partical impact.
107	223	1962	77	1.94-26.00	~0°		2π		Above specimen and conditions.

SPECIFICATION TABLE NO. 367 (continued)

Curve No.	Ref. No.	Year	Temperature K	Wavelength Range, μ	Geometry θ	θ'	ω'	Reported Error, %	Composition (weight percent), Specifications and Remarks
108	283	1966	~300	1.95-14.82	~0°		2π	< 2	Stainless Steel 316; nominal composition: 16.00-18.00 Cr, 10.00-14.00 Ni, 2.00-3.00 Mo, 2.00 max Mn, 1.00 max Si, 0.08 max C, Fe balance; sandblasted (325 grit); average surface roughness 22 to 25 microinches; oxidized at 922 K for 1 hr in air; converted from R (2π, 0°); [Authors' designation: Specimen No. 1].
109	283	1966	~300	1.86-14.93	~0°		2π	< 2	Different sample, same as curve 108 specimen and conditions except oxidized at 922 K for 4 hrs in air; [Authors' designation: Specimen No. 3].
110	283	1966	~300	1.89-14.84	~0°		2π	< 2	Different sample, same as curve 108 specimen and conditions except oxidized at 922 K for 8 hrs in air; [Authors' designation: Specimen No. 5].
111	283	1966	~300	1.87-14.93	~0°		2π	< 2	Different sample, same as curve 108 specimen and conditions except oxidized at 922 K for 16 hrs in air; [Authors' designation: Specimen No. 7].
112	283	1966	~300	2.00-14.95	~0°		2π	< 2	Different sample, same as curve 108 specimen and conditions except oxidized at 922 K for 24 hrs in air; [Authors' designation: Specimen No. 9].
113	283	1966	~300	2.00-14.43	~0°		2π	< 2	Different sample, same as curve 108 specimen and conditions except oxidized at 1255 K for 2 hrs in air; [Authors' designation: Specimen No. 31].
114	283	1966	~300	2.01-14.96	~0°		2π	< 2	Different sample, same as curve 108 specimen and conditions except oxidized at 1255 K for 6 hrs in air; [Authors' designation: Specimen No. 33].
115	283	1966	~300	1.99-14.86	~0°		2π	< 2	Different sample, same as curve 108 specimen and conditions except sandblasted (80 grit); average surface roughness 60 to 62 microinches; [Authors' designation: Specimen No. 2].
116	283	1966	~300	1.84-14.89	~0°		2π	< 2	Different sample, same as curve 115 specimen and conditions except oxidized at 922 K for 4 hrs in air; [Authors' designation: Specimen No. 4].
117	283	1966	~300	1.90-14.95	~0°		2π	< 2	Different sample, same as curve 115 specimen and conditions except oxidized at 922 K for 8 hrs in air; [Authors' designation: Specimen No. 6].
118	283	1966	~300	1.86-14.92	~0°		2π	< 2	Different sample, same as curve 115 specimen and conditions except oxidized at 922 K for 16 hrs in air; [Authors' designation: Specimen No. 8].
119	283	1966	~300	1.96-14.91	~0°		2π	< 2	Different sample, same as curve 115 specimen and conditions except oxidized at 922 K for 24 hrs in air; [Authors' designation: Specimen No. 10].
120	283	1966	~300	2.00-14.89	~0°		2π	< 2	Different sample, same as curve 115 specimen and conditions except oxidized at 1255 K for 2 hrs in air; [Authors' designation: Specimen No. 32].
121	283	1966	~300	2.01-14.97	~0°		2π	< 2	Different sample, same as curve 115 specimen and conditions except oxidized at 1255 K for 6 hrs in air; [Authors' designation: Specimen No. 34].
122	283	1966	~300	2.00-15.00	~0°		2π	< 2	Different sample, same as curve 108 specimen and conditions except sandblasted (40 grit); average surface roughness 220 to 250 microinches; oxidized at 1255 K for 16 hrs in air; [Authors' designation: Specimen No. 35].
123	283	1966	~300	2.00-15.05	~0°		2π	< 2	Different sample, same as curve 108 specimen and conditions except oxidized at 1255 K for 24 hrs in air; [Authors' designation: Specimen No. 36].

SPECIFICATION TABLE NO. 367 (continued)

Curve No.	Ref. No.	Year	Temperature K	Wavelength Range, μ	Geometry θ	θ'	ω'	Reported Error, %	Composition (weight percent), Specifications and Remarks
124	318	1961	298	1.20–14.00	0°		2π		Stainless Steel 321; nominal composition: 17.00–19.00 Cr, 9.00–12.00 Ni, 2.00 max Mn, 1.00 max Si, 0.08 max C, 5 x C min Ti, Fe balance; sanded with increasingly finer grades of emery paper; buffed to a high polish with a buffing wheel using relevigated alumina as the abrasive compound; rinsed in water and alcohol successively and immediately dried; fresh sample; [Authors' designation: S0].
125	318	1961	298	0.40–2.40	0°		2π		Above specimen and conditions except MgO reference.
126	318	1961	298	1.20–14.00	0°		2π		Stainless Steel 321; nominal composition: 17.00–19.00 Cr, 9.00–12.00 Ni, 2.00 max Mn, 1.00 max Si, 0.08 max C, 5 x C min Ti, Fe balance; sanded with 0000 emery paper; sanded in one direction to produce a fine-grooved appearance; peak to peak height of asperity as measured by profilometer 0.06 μ; mean groove spacing 5μ; mean groove angle 2°; fresh sample; [Authors' designation: S1].
127	318	1961	298	0.40–2.29	0°		2π		Above specimen and conditions except MgO reference.
128	318	1961	298	1.29–14.00	0°		2π		Stainless Steel 321; nominal composition: 17.00–19.00 Cr, 9.00–12.00 Ni, 2.00 max Mn, 1.00 max Si, 0.08 max C, 5 x C min Ti, Fe balance; sanded with grade 1 emery paper; sanded in one direction to produce a fine-grooved appearance; peak to peak height of asperity as measured by profilometer 1.25μ; mean groove spaceing 9μ; mean groove angle 15°; fresh sample; [Authors' designation: S3].
129	318	1961	298	0.60–2.20	0°		2π		Above specimen and conditions except MgO reference.

DATA TABLE NO. 367 NORMAL SPECTRAL REFLECTANCE OF [IRON + CHROMIUM + NICKEL + ΣX_i] ALLOYS

[Wavelength, λ, μ; Reflectance, ρ; Temperature, T, K]

λ	ρ
CURVE 1 T = 298	
0.55	0.648
0.65	0.664
CURVE 2 T = 298	
0.30	0.160
0.40	0.350
0.50	0.400
0.60	0.430
0.70	0.445
0.74	0.445
0.80	0.440
0.84	0.440
0.90	0.450
1.00	0.470
1.10	0.480
1.20	0.495
1.30	0.500
1.40	0.505
1.50	0.510
1.60	0.525
1.70	0.540
1.80	0.560
1.90	0.570
2.00	0.580
2.10	0.595
2.20	0.620
2.30	0.630
2.40	0.640
2.50	0.640
2.60	0.640
2.70	0.600
CURVE 3 T = 298	
0.30	0.370
0.34	0.410
0.40	0.460
0.50	0.530
0.60	0.580
0.70	0.605

λ	ρ
CURVE 3 (cont.)	
0.80	0.620
0.90	0.640
0.94	0.645
1.00	0.650
1.10	0.660
1.20	0.670
1.30	0.680
1.40	0.680
1.50	0.690
1.60	0.710
1.70	0.730
1.80	0.760
1.90	0.760
2.00	0.760
2.10	0.775
2.20	0.790
2.30	0.810
2.40	0.815
2.50	0.805
2.60	0.790
2.64	0.790
2.70	0.810
CURVE 4 T = 298	
0.30	0.340
0.36	0.420
0.40	0.430
0.50	0.470
0.60	0.490
0.68	0.510
0.70	0.510*
0.74	0.510
0.78	0.500
0.80	0.500*
0.82	0.500
0.90	0.515
1.00	0.550
1.04	0.560*
1.10	0.570
1.20	0.580
1.22	0.580*
1.30	0.580

λ	ρ
CURVE 4 (cont.)	
1.40	0.580
1.50	0.590
1.60	0.600
1.70	0.615
1.80	0.620
1.90	0.630
1.91	0.640*
2.10	0.650
2.20	0.670
2.30	0.685
2.40	0.705
2.44	0.705
2.50	0.695
2.60	0.680
2.70	0.620
CURVE 5 T = 298	
0.30	0.020
0.38	0.100
0.40	0.105
0.50	0.130
0.60	0.160
0.70	0.220
0.80	0.280
0.84	0.300
0.90	0.320
1.00	0.350
1.10	0.350
1.20	0.350
1.30	0.350
1.40	0.350
1.44	0.350
1.50	0.340
1.58	0.330
1.60	0.330
1.70	0.330
1.80	0.340
1.90	0.345
2.00	0.350*
2.10	0.355
2.20	0.360
2.30	0.375

λ	ρ
CURVE 5 (cont.)	
2.40	0.390
2.50	0.400
2.60	0.420
2.64	0.420*
2.70	0.400
CURVE 6 T = 298	
0.30	0.260
0.38	0.430
0.50	0.490
0.60	0.520
0.70	0.540
0.80	0.540
0.84	0.550
0.90	0.570
0.94	0.575
1.02	0.580
1.10	0.600
1.22	0.630
1.30	0.635
1.40	0.640
1.50	0.660
1.60	0.690
1.70	0.700
1.80	0.705
1.90	0.720
2.10	0.770
2.30	0.800
2.50	0.820
2.60	0.840
2.64	0.840*
2.70	0.820
CURVE 7 T = 298	
0.30	0.250
0.40	0.400
0.50	0.440
0.70	0.485
0.80	0.500
0.90	0.530

λ	ρ
CURVE 7 (cont.)	
0.94	0.540
1.06	0.550*
1.10	0.540
1.20	0.535
1.30	0.540
1.40	0.560
1.50	0.600
1.58	0.640
1.62	0.650
1.70	0.655
1.80	0.655
1.90	0.675
2.02	0.715
2.10	0.730
2.30	0.750
2.46	0.765*
2.50	0.765
2.54	0.765*
2.60	0.760
2.70	0.740
CURVE 8* T = 298	
0.30	0.340
0.40	0.470
0.50	0.520
0.70	0.570
0.80	0.580
0.90	0.600
1.10	0.630
1.20	0.650
1.30	0.660
1.34	0.660
1.50	0.690
1.60	0.720
1.62	0.725
1.70	0.725
1.74	0.720
1.86	0.725
1.90	0.730
2.10	0.765
2.22	0.785
2.30	0.790

λ	ρ
CURVE 8* (cont.)	
2.50	0.790
2.62	0.795
2.70	0.775
CURVE 9* T = 298	
0.30	0.340
0.40	0.410
0.50	0.440
0.70	0.470
0.80	0.500
0.90	0.530
0.94	0.540
1.10	0.550
1.20	0.560
1.30	0.550
1.40	0.540
1.50	0.580
1.60	0.640
1.70	0.650
1.80	0.660
1.90	0.670
2.02	0.700
2.10	0.710
2.30	0.720
2.50	0.730
2.58	0.730
2.70	0.700
CURVE 10* T = 298	
0.30	0.330
0.40	0.480
0.50	0.550
0.60	0.590
0.70	0.610
0.78	0.605
0.80	0.605
0.90	0.640
1.00	0.635
1.10	0.660
1.20	0.680

λ	ρ
CURVE 10* (cont.)	
1.24	0.680
1.30	0.670
1.40	0.660
1.50	0.680
1.60	0.700
1.70	0.720
1.80	0.740
1.90	0.760
2.00	0.770
2.10	0.785
2.20	0.795
2.30	0.805
2.40	0.820
2.50	0.825
2.60	0.830
2.70	0.810
CURVE 11* T = 298	
0.30	0.260
0.36	0.370
0.42	0.410
0.50	0.440
0.66	0.490
0.70	0.490
0.80	0.490
0.90	0.515
1.10	0.540
1.18	0.550
1.30	0.560
1.42	0.560
1.50	0.585
1.66	0.630
1.70	0.635
1.90	0.660
2.10	0.700
2.30	0.720
2.50	0.730
2.58	0.740
2.70	0.700

λ	ρ
CURVE 12* T = 298	
0.30	0.320
0.42	0.480
0.50	0.520
0.60	0.550
0.70	0.570
0.76	0.570
0.82	0.580
0.90	0.605
1.00	0.610
1.10	0.620
1.20	0.630
1.30	0.640
1.40	0.650
1.50	0.675
1.60	0.705
1.70	0.720
1.80	0.730
1.90	0.740
2.00	0.755
2.10	0.765
2.20	0.775
2.30	0.790
2.38	0.810
2.42	0.810
2.50	0.795
2.60	0.770
2.70	0.790
CURVE 13* T = 298	
0.30	0.240
0.40	0.370
0.42	0.400
0.50	0.430
0.60	0.460
0.66	0.470
0.70	0.470
0.80	0.470
0.90	0.480
1.00	0.500
1.10	0.510
1.20	0.510

* Not shown on plot

DATA TABLE NO. 367 (continued)

CURVE 13* (cont.)

λ	ρ
1.30	0.520
1.38	0.530
1.40	0.530
1.50	0.530
1.58	0.525
1.60	0.530
1.70	0.575
1.80	0.610
1.90	0.610
2.00	0.610
2.10	0.615
2.20	0.625
2.30	0.635
2.40	0.650
2.50	0.650
2.60	0.640
2.70	0.635

CURVE 14*
T = 298

λ	ρ
0.30	0.250
0.40	0.420
0.50	0.460
0.60	0.480
0.70	0.500
0.76	0.500
0.80	0.495
0.82	0.500
0.84	0.520
0.90	0.540
1.00	0.540
1.10	0.550
1.20	0.560
1.30	0.565
1.40	0.570
1.50	0.610
1.60	0.670
1.62	0.680
1.70	0.680
1.80	0.680
1.90	0.680
2.00	0.690
2.10	0.710
2.20	0.730

CURVE 14* (cont.)

λ	ρ
2.30	0.745
2.40	0.750
2.50	0.760
2.60	0.760
2.70	0.710

CURVE 15*
T = 298

λ	ρ
0.36	0.410
0.40	0.430
0.50	0.445
0.60	0.465
0.70	0.480
0.80	0.505
0.90	0.530
1.00	0.560
1.10	0.570
1.20	0.580
1.30	0.580
1.40	0.580
1.50	0.590
1.60	0.610
1.62	0.680
1.70	0.640
1.80	0.665
1.90	0.690
2.00	0.710
2.10	0.710
2.20	0.710
2.30	0.730
2.40	0.770
2.50	0.785
2.54	0.790
2.64	0.790
2.72	0.780

CURVE 16
T = 298

λ	ρ
0.30	0.370*
0.40	0.500
0.50	0.540
0.70	0.580
0.82	0.580

CURVE 16 (cont.)

λ	ρ
0.90	0.595
1.02	0.595
1.10	0.625
1.20	0.660
1.22	0.665
1.30	0.660
1.40	0.650
1.50	0.670
1.70	0.730*
1.90	0.755
2.10	0.785
2.30	0.805*
2.42	0.820
2.50	0.820*
2.60	0.820
2.70	0.800

CURVE 17*
T = 298

λ	ρ
0.30	0.240
0.40	0.400
0.50	0.440
0.70	0.475
0.72	0.480
0.80	0.480
0.90	0.500
1.10	0.530
1.22	0.550
1.30	0.555
1.40	0.555
1.50	0.580
1.70	0.640
1.90	0.670
2.10	0.700
2.30	0.720
2.46	0.740
2.50	0.740
2.60	0.740
2.70	0.710

CURVE 18*
T = 298

λ	ρ
0.30	0.08
0.40	0.10

CURVE 18* (cont.)

λ	ρ
0.50	0.11
0.60	0.12
0.70	0.12
0.95	0.10
1.00	0.11
1.20	0.12
1.40	0.13
1.70	0.13
1.75	0.13
1.80	0.14
2.00	0.17
2.15	0.20
2.20	0.21
2.40	0.21
2.50	0.22
2.55	0.24
2.60	0.24
2.70	0.20

CURVE 19
T = 298

λ	ρ
0.30	0.06
0.40	0.06
0.60	0.09
0.80	0.13
1.00	0.18
1.10	0.21
1.15	0.22
1.20	0.23
1.30	0.24
1.40	0.25
1.50	0.27
1.60	0.29
1.70	0.31
1.80	0.32
2.20	0.36*
2.50	0.40*
2.60	0.40
2.70	0.37

CURVE 20
T = 311

λ	ρ
1.0	0.43
1.25	0.42
1.50	0.44
1.75	0.46
2.00	0.48
2.25	0.51
2.50	0.52
2.75	0.51
3.00	0.49
3.25	0.52
3.50	0.53
3.75	0.56
4.00	0.59
4.25	0.59
4.50	0.60
4.75	0.61
5.00	0.62
5.25	0.64
5.50	0.64
5.75	0.63
6.00	0.59
6.25	0.59
6.50	0.61
6.75	0.63
7.00	0.61
7.25	0.61
7.50	0.62
7.75	0.63
8.00	0.63
8.25	0.63
8.50	0.63
8.75	0.63
9.00	0.63
9.25	0.62
9.50	0.62
9.75	0.62
10.00	0.63
10.25	0.65
10.50	0.66
10.75	0.67
11.00	0.66
11.25	0.68
11.50	0.67
11.75	0.68

CURVE 20 (cont.)

λ	ρ
12.00	0.68
12.25	0.67
12.50	0.67
12.75	0.67
13.00	0.67
13.25	0.66
13.50	0.67
13.75	0.67
14.00	0.67
14.25	0.67*
14.50	0.67
14.75	0.65*
15.00	0.65
15.25	0.69*
15.50	0.69
15.75	0.69*
16.00	0.68
16.25	0.68*
16.50	0.68
16.75	0.67*
17.00	0.67
17.25	0.68*
17.50	0.67
17.75	0.68*
18.00	0.67
18.25	0.66
18.50	0.67
18.75	0.66*
19.00	0.66
19.25	0.66*
19.50	0.66
19.75	0.65
20.00	0.64
20.25	0.63
20.50	0.63*
20.75	0.63
21.00	0.64
21.25	0.65
21.50	0.65*
21.75	0.65*
22.00	0.65
22.25	0.65*
22.50	0.66
22.75	0.66*
23.00	0.66

CURVE 20 (cont.)

λ	ρ
23.25	0.66*
23.50	0.66*
23.75	0.66
24.00	0.67
24.25	0.66
24.50	0.66*
24.75	0.66
25.00	0.67

CURVE 21*
T = 311

λ	ρ
1.0	0.32
1.25	0.33
1.50	0.39
1.75	0.43
2.00	0.47
2.25	0.50
2.50	0.52
2.75	0.54
3.00	0.54
3.25	0.56
3.50	0.57
3.75	0.59
4.00	0.61
4.25	0.61
4.50	0.64
4.75	0.65
5.00	0.65
5.25	0.66
5.50	0.67
5.75	0.68
6.00	0.67
6.25	0.67
6.50	0.67
6.75	0.69
7.00	0.69
7.25	0.69
7.50	0.69
7.75	0.70
8.00	0.70
8.25	0.69
8.50	0.67
8.75	0.67
9.00	0.67

CURVE 21 (cont.)

λ	ρ
9.25	0.66
9.50	0.67
9.75	0.68
10.00	0.68
10.25	0.69
10.50	0.69
10.75	0.69
11.00	0.70
11.25	0.70
11.50	0.70
11.75	0.71
12.00	0.70
12.25	0.70
12.50	0.70
12.75	0.70
13.00	0.70
13.25	0.70
13.50	0.69
13.75	0.68
14.00	0.68
14.25	0.66
14.50	0.66
14.75	0.66
15.00	0.65
15.25	0.65
15.50	0.64
15.75	0.64
16.00	0.64
16.25	0.64
16.50	0.63
16.75	0.63
17.00	0.63
17.25	0.63
17.50	0.63
17.75	0.63
18.00	0.64
18.25	0.64
18.50	0.64
18.75	0.63
19.00	0.62
19.25	0.62
19.50	0.63
19.75	0.62
20.00	0.61
20.25	0.62

* Not shown on plot

DATA TABLE NO. 367 (continued)

λ	ρ
CURVE 21* (cont.)	
20.50	0.62
20.75	0.63
21.00	0.64
21.25	0.65
21.50	0.65
21.75	0.65
22.00	0.64
22.25	0.64
22.50	0.64
22.75	0.65
23.00	0.65
23.25	0.64
23.50	0.64
23.75	0.65
24.00	0.64
25.25	0.64
24.50	0.65
24.75	0.62
25.00	0.62
CURVE 22 T =~322	
2.00	0.470
3.00	0.525
4.00	0.575
5.00	0.585
5.80	0.575
6.75	0.600
8.00	0.660
8.75	0.685
9.75	0.730
11.00	0.785
12.00	0.820
13.00	0.840
13.75	0.845
14.50	0.830
15.00	0.790
CURVE 23 T =~322	
2.00	0.361
2.49	0.368
3.33	0.353
CURVE 23 (cont.)	
3.63	0.346
4.14	0.351
4.98	0.324
6.01	0.270
7.01	0.225
8.01	0.180
9.02	0.119
9.52	0.100
10.00	0.091
10.99	0.091
12.01	0.066
13.01	0.043
14.01	0.030
15.00	0.021
CURVE 24* T =~322	
1.00	0.550
2.00	0.510
2.50	0.500
3.25	0.530
4.00	0.600
5.00	0.630
6.00	0.650
7.00	0.675
8.00	0.740
9.00	0.765
10.25	0.800
11.50	0.815
12.00	0.820
13.00	0.840
13.50	0.840
14.10	0.825
14.70	0.775
15.00	0.710
CURVE 25* T =~322	
1.00	0.471
1.50	0.482
2.49	0.541
3.50	0.559
4.01	0.566
CURVE 25* (cont.)	
4.49	0.550
5.00	0.502
5.49	0.467
6.23	0.432
7.00	0.371
7.99	0.310
8.74	0.250
10.01	0.190
11.01	0.151
12.91	0.126
13.01	0.105
13.98	0.094
14.49	0.094
14.76	0.093
15.00	0.100
CURVE 26* T =~322	
0.50	0.410
0.75	0.475
0.85	0.420
1.00	0.430
1.60	0.490
2.65	0.550
3.75	0.600
4.50	0.640
5.40	0.660
6.25	0.665
7.00	0.690
8.25	0.760
8.75	0.760
10.00	0.790
11.00	0.810
12.25	0.810
13.00	0.835
13.75	0.820
14.50	0.755
15.00	0.650
CURVE 27* T =~322	
0.50	0.300
0.68	0.347
CURVE 27* (cont.)	
0.74	0.370
0.83	0.365
0.88	0.376
0.99	0.376
1.56	0.447
1.98	0.498
2.75	0.523
3.99	0.539
4.98	0.507
5.48	0.476
6.00	0.421
6.51	0.391
7.00	0.372
8.00	0.301
8.99	0.226
10.00	0.192
11.01	0.159
12.01	0.140
14.02	0.140
14.48	0.149
15.00	0.175
CURVE 28 T =~322	
2.00	0.620
2.50	0.550
3.00	0.520
4.00	0.530
5.00	0.550
5.50	0.550
6.20	0.560
7.50	0.620*
8.35	0.675
9.10	0.685
10.10	0.730
11.10	0.775
12.00	0.775
13.00	0.800
14.00	0.780
14.55	0.750
15.00	0.700
CURVE 29 T =~322	
2.00	0.475
3.00	0.347
3.44	0.327
4.00	0.318
4.98	0.293
5.98	0.253
7.00	0.217
7.98	0.157
8.60	0.117
9.17	0.100
10.00	0.080
11.02	0.060
12.00	0.050
13.52	0.050
14.50	0.063
15.00	0.075
CURVE 30* T =~322	
1.00	0.380
1.50	0.425
2.25	0.480
3.00	0.535
3.50	0.555
4.00	0.570
5.50	0.570
6.50	0.585
7.50	0.625
8.15	0.665
9.00	0.685
9.25	0.685
10.20	0.725
11.10	0.775
12.00	0.800
13.00	0.840
14.00	0.870
15.00	0.880
CURVE 31* T =~322	
1.00	0.350
1.51	0.363
CURVE 31* (cont.)	
2.00	0.360
2.99	0.340
3.77	0.302
4.49	0.276
5.99	0.233
7.01	0.211
8.01	0.191
8.50	0.166
9.01	0.124
9.53	0.101
9.99	0.091
10.51	0.091
11.00	0.081
11.49	0.064
12.00	0.059
13.00	0.061
14.01	0.061
14.52	0.061
15.00	0.050
CURVE 32 T =~322	
0.50	0.030
0.65	0.150
0.90	0.300
1.00	0.370
1.50	0.475
2.00	0.520
2.50	0.520*
3.00	0.505
3.50	0.530
4.00	0.560
5.00	0.575
6.00	0.570
6.85	0.600
8.00	0.650
9.00	0.680
9.50	0.685
10.50	0.725
11.50	0.765
12.50	0.790
13.50	0.810
14.50	0.810
14.65	0.790
15.00	0.760
CURVE 33 T =~322	
0.50	0.015
0.63	0.101
0.90	0.248
1.57	0.402
2.00	0.434
2.50	0.424
3.50	0.367
4.00	0.331
4.68	0.324
5.49	0.299
7.00	0.252
8.01	0.222
8.51	0.187
9.01	0.143
10.00	0.108
11.01	0.102
12.01	0.090
13.00	0.094
14.00	0.111
14.52	0.150
15.00	0.225
CURVE 34* T =~322	
2.00	0.630
2.50	0.560
3.00	0.510
3.50	0.505
4.50	0.530
5.50	0.550
6.75	0.565
8.00	0.635
9.50	0.685
10.50	0.725
11.50	0.765
12.50	0.785
13.25	0.790
14.25	0.775
15.00	0.750
CURVE 35* T =~322	
2.00	0.500
2.03	0.450
2.49	0.365
2.99	0.334
3.98	0.308
4.98	0.278
6.00	0.228
7.01	0.221
7.52	0.202
8.01	0.163
9.01	0.111
10.00	0.082
11.00	0.073
12.01	0.057
13.02	0.052
13.52	0.054
14.01	0.062
14.51	0.069
15.00	0.072
CURVE 36 T =~322	
1.00	0.510
2.00	0.510
3.00	0.520*
4.00	0.550
4.75	0.555
6.00	0.550
6.50	0.565
8.00	0.635
8.80	0.650
10.50	0.730
12.15	0.785
13.75	0.850
14.50	0.865
15.00	0.850
CURVE 37* T =~322	
1.00	0.470
2.00	0.469

* Not shown on plot

DATA TABLE NO. 367 (continued)

CURVE 37*(cont.)

λ	ρ
3.00	0.407
3.49	0.370
4.47	0.336
4.97	0.318
5.98	0.283
7.02	0.249
8.01	0.209
8.51	0.183
9.01	0.142
10.01	0.112
11.00	0.094
11.99	0.072
13.00	0.072
14.00	0.077
14.51	0.085
15.00	0.100

CURVE 38
T = ~322

λ	ρ
0.50	0.320
0.60	0.400
1.00	0.480
1.50	0.550
2.00	0.575
3.00	0.540
4.00	0.570
5.00	0.570
6.00	0.560
7.00	0.590
8.35	0.650
9.25	0.665
10.25	0.700
11.25	0.750
12.50	0.775*
14.00	0.790
14.50	0.785
15.00	0.760*

CURVE 39*
T = ~322

λ	ρ
0.50	0.272
0.79	0.352
1.00	0.426

CURVE 39*(cont.)

λ	ρ
1.97	0.501
2.48	0.483
2.96	0.416
4.00	0.352
4.98	0.325
5.99	0.282
6.99	0.253
8.01	0.214
9.01	0.131
10.00	0.103
11.00	0.095
12.00	0.073
13.02	0.083
13.99	0.110
14.51	0.144
15.00	0.212

CURVE 40*
T = ~322

λ	ρ
2.00	0.470
3.00	0.550
4.00	0.610
5.00	0.660
5.75	0.665
6.20	0.680
6.75	0.700
7.75	0.750
8.00	0.760
9.00	0.770
9.75	0.800
10.50	0.805
12.00	0.850
13.00	0.860
14.00	0.860
14.75	0.825
15.00	0.790

CURVE 41*
T = ~322

λ	ρ
2.00	0.440
2.49	0.475
3.02	0.499
3.47	0.516

CURVE 41*(cont.)

λ	ρ
4.00	0.925
4.50	0.507
6.01	0.411
7.00	0.341
8.00	0.277
9.02	0.208
9.52	0.185
10.01	0.168
11.01	0.134
11.51	0.119
12.00	0.108
12.49	0.090
12.99	0.080
14.00	0.069
14.50	0.063
15.00	0.054

CURVE 42*
T = ~322

λ	ρ
1.00	0.560
1.50	0.575
1.75	0.510
2.00	0.490
2.50	0.480
3.00	0.500
3.50	0.530
4.00	0.550
5.00	0.540
5.50	0.525
6.40	0.540
7.00	0.565
8.00	0.635
9.00	0.680
10.50	0.735
12.00	0.790
13.00	0.800
14.00	0.800
14.90	0.750
15.50	0.690

CURVE 43*
T = ~322

λ	ρ
1.00	0.508
1.34	0.522
1.61	0.519
2.01	0.506
2.49	0.515
2.99	0.537
3.50	0.545
3.99	0.543
4.48	0.525
4.98	0.490
5.49	0.462
6.00	0.422
6.48	0.386
7.02	0.364
8.00	0.307
8.50	0.265
9.00	0.233
10.01	0.184
11.01	0.151
12.00	0.124
13.00	0.101
13.51	0.092
13.99	0.087
14.50	0.088
15.00	0.100

CURVE 44*
T = ~322

λ	ρ
0.50	0.410
0.70	0.490
0.80	0.450
1.10	0.455
1.50	0.490
1.75	0.425
2.00	0.425
2.40	0.450
2.50	0.500
3.00	0.540
4.00	0.560
5.70	0.520
6.00	0.550
7.00	0.590
8.00	0.660
9.50	0.700

CURVE 44*(cont.)

λ	ρ
11.00	0.750
12.00	0.780
13.00	0.795
14.00	0.780
14.70	0.750
15.00	0.720

CURVE 45*
T = ~322

λ	ρ
0.50	0.410
0.76	0.470
0.88	0.443
1.16	0.451
1.49	0.476
1.71	0.460
1.87	0.416
1.99	0.401
2.49	0.368
3.00	0.350
3.99	0.321
4.50	0.311
4.81	0.311
5.48	0.277
5.99	0.245
6.48	0.226
7.00	0.222
7.49	0.201
8.24	0.174
9.00	0.124
9.99	0.103
11.03	0.083
12.00	0.071
12.52	0.071
13.49	0.075
14.01	0.082
14.52	0.101
15.00	0.160

CURVE 46*
T = ~322

λ	ρ
2.00	0.610
2.50	0.550
3.00	0.525
3.75	0.550

CURVE 46*(cont.)

λ	ρ
5.00	0.630
6.50	0.680
6.80	0.700
8.00	0.750
9.00	0.775
10.00	0.785
10.25	0.785
11.00	0.820
11.50	0.825
13.00	0.840
13.50	0.840
14.40	0.800
14.80	0.750
15.00	0.720

CURVE 47*
T = ~322

λ	ρ
2.00	0.600
2.16	0.556
2.64	0.522
3.68	0.510
4.45	0.498
4.96	0.478
5.97	0.409
6.49	0.385
7.16	0.350
7.79	0.302
8.73	0.231
9.51	0.193
10.01	0.176
11.52	0.135
12.01	0.122
13.01	0.113
13.51	0.111
14.01	0.115
14.52	0.122
15.00	0.133

CURVE 48*
T = ~322

λ	ρ
1.00	0.460
1.50	0.480
2.00	0.490
2.60	0.485

CURVE 48*(cont.)

λ	ρ
3.50	0.500
4.50	0.540
5.25	0.580
6.00	0.585
6.80	0.650
7.75	0.700
8.60	0.720
9.15	0.720
10.35	0.750
12.00	0.800
13.50	0.830
14.75	0.850
15.00	0.850

CURVE 49*
T = ~322

λ	ρ
1.00	0.330
1.50	0.352
1.99	0.403
2.49	0.453
3.17	0.503
3.98	0.522
4.49	0.512
5.00	0.493
6.02	0.434
7.01	0.373
7.75	0.329
8.51	0.276
9.04	0.243
10.04	0.190
10.54	0.171
11.00	0.159
11.98	0.141
13.00	0.120
13.49	0.117
13.99	0.123
14.49	0.134
15.00	0.150

CURVE 50*
T = ~322

λ	ρ
0.50	0.065
0.60	0.200
0.85	0.350

CURVE 50*(cont.)

λ	ρ
1.15	0.450
1.50	0.500
1.65	0.550
2.10	0.495
2.75	0.530
3.50	0.580
5.00	0.640
6.50	0.690
8.00	0.750
8.50	0.750
10.00	0.780
11.00	0.810
12.50	0.825
13.50	0.835
14.25	0.825
15.00	0.780

CURVE 51*
T = ~322

λ	ρ
0.50	0.065
0.54	0.153
0.69	0.250
0.87	0.348
1.04	0.398
1.52	0.472
1.95	0.506
2.46	0.516
2.97	0.503
3.48	0.537
3.97	0.557
4.95	0.517
5.99	0.449
6.68	0.423
7.48	0.375
7.98	0.345
8.47	0.307
8.99	0.262
9.50	0.229
9.99	0.216
10.99	0.194
11.51	0.178
11.99	0.163
12.51	0.155
12.99	0.155
13.51	0.164

* Not shown on plot

DATA TABLE NO. 367 (continued)

λ	ρ
CURVE 51*(cont.)	
14.00	0.177
14.51	0.208
14.76	0.239
15.00	0.275
CURVE 52* T = ~322	
2.00	0.590
2.25	0.550
2.75	0.530
3.65	0.550
4.50	0.600
5.50	0.630
6.50	0.660
7.50	0.705
8.25	0.735
8.80	0.735
10.00	0.775
11.00	0.800
12.00	0.820
13.00	0.825
13.50	0.825
14.25	0.800
14.75	0.750
15.00	0.710
CURVE 53* T = ~322	
2.00	0.591
2.52	0.534
3.00	0.524
4.02	0.528
4.49	0.513
5.01	0.496
6.00	0.429
7.01	0.361
7.50	0.326
8.01	0.278
8.51	0.227
9.01	0.198
10.00	0.160
11.00	0.128
11.98	0.109

λ	ρ
CURVE 53*(cont.)	
13.00	0.098
13.50	0.098
13.99	0.098
14.50	0.104
15.00	0.116
CURVE 54* T = ~322	
1.00	0.515
1.50	0.520
2.50	0.500
3.50	0.510
4.15	0.525
5.20	0.585
5.85	0.585
6.75	0.650
7.50	0.690
8.50	0.725
9.50	0.740
10.50	0.780
11.50	0.815
12.10	0.830
12.75	0.840
14.25	0.880
15.00	0.870
CURVE 55* T = ~322	
1.00	0.410
1.34	0.390
1.80	0.412
2.16	0.452
2.65	0.499
2.96	0.525
3.46	0.552
3.96	0.550
4.61	0.526
5.01	0.499
5.63	0.449
6.48	0.405
7.50	0.348
8.00	0.327
8.48	0.280
8.99	0.219

λ	ρ
CURVE 55*(cont.)	
10.00	0.178
10.99	0.151
11.99	0.129
13.00	0.111
13.49	0.102
14.00	0.100
14.52	0.101
15.00	0.125
CURVE 56* T = ~322	
0.50	0.425
0.95	0.500
1.50	0.560
1.75	0.565
3.00	0.565
3.75	0.600
4.25	0.630
5.10	0.650
6.10	0.670
7.30	0.725
8.00	0.740
9.25	0.755
10.50	0.780
11.50	0.800
12.50	0.810
14.00	0.830
14.65	0.815
15.00	0.790
CURVE 57* T = ~322	
0.50	0.360
0.77	0.429
0.86	0.399
1.01	0.360
1.51	0.426
1.79	0.501
2.01	0.551
2.46	0.551
2.96	0.544
3.48	0.563
3.98	0.570

λ	ρ
CURVE 57*(cont.)	
4.53	0.552
5.48	0.503
6.16	0.452
6.99	0.401
8.02	0.337
8.51	0.290
9.00	0.245
10.00	0.197
11.02	0.160
11.50	0.155
12.99	0.155
14.00	0.306
14.50	0.177
15.00	0.200
CURVE 58 T = ~322	
2.00	0.390
3.20	0.450
4.00	0.525
4.50	0.560
5.15	0.570
6.25	0.550
7.00	0.580
7.60	0.605
8.50	0.575
9.65	0.520
11.00	0.570
12.00	0.600
13.00	0.610
14.00	0.620
14.75	0.600
15.00	0.580
CURVE 59 T = ~322	
2.00	0.350
2.50	0.365
3.00	0.395
3.50	0.440
4.00	0.500
4.50	0.540
5.00	0.550*

λ	ρ
CURVE 59 (cont.)	
5.50	0.535
6.00	0.520
6.50	0.515
7.00	0.530
7.50	0.540
8.00	0.530
8.50	0.510
9.50	0.460
10.00	0.440
10.50	0.435
11.00	0.430
11.50	0.420
12.00	0.400
13.00	0.345
14.00	0.285
15.00	0.220
CURVE 60* T = ~322	
1.00	0.415
1.50	0.410
2.25	0.450
2.75	0.445
3.60	0.515
4.50	0.565
5.00	0.580
6.00	0.550
7.00	0.570
7.75	0.590
8.75	0.575
9.50	0.560
10.50	0.570
11.25	0.590
12.50	0.605
13.75	0.620
14.00	0.615
14.65	0.575
15.00	0.540
CURVE 61* T = ~322	
1.00	0.490
1.25	0.445

λ	ρ
CURVE 61*(cont.)	
1.50	0.430
2.00	0.440
2.50	0.460
2.75	0.460
3.00	0.475
3.50	0.515
4.00	0.560
4.50	0.585
4.75	0.585
5.00	0.585
5.50	0.565
6.00	0.500
7.00	0.570
8.00	0.570
8.50	0.555
9.00	0.530
10.00	0.480
11.00	0.480
11.50	0.450
13.00	0.370
14.00	0.310
14.50	0.275
15.00	0.225
CURVE 62 T = ~322	
0.50	0.270
0.75	0.350
1.00	0.365
1.75	0.440
2.25	0.455
3.00	0.470
4.00	0.550*
5.00	0.580
5.85	0.560*
7.00	0.580*
8.00	0.610
9.00	0.560
9.75	0.550
10.75	0.575
12.00	0.595
13.25	0.610
14.00	0.610
14.70	0.550*
15.00	0.500

λ	ρ
CURVE 63* T = ~322	
0.50	0.270
0.55	0.310
0.75	0.330
0.85	0.385
1.00	0.380
1.50	0.425
2.00	0.450
2.50	0.455
3.00	0.455
3.25	0.465
3.50	0.485
4.00	0.545
4.50	0.570
5.00	0.580
5.50	0.560
6.00	0.540
6.50	0.550
7.00	0.555
7.50	0.555
8.00	0.550
8.25	0.540
8.50	0.520
8.75	0.495
9.00	0.485
10.00	0.465
11.00	0.450
12.00	0.420
13.00	0.380
14.00	0.330
14.50	0.315
15.00	0.310
CURVE 64* T = ~322	
2.00	0.475
2.35	0.400
3.00	0.355
3.50	0.405
4.00	0.475
4.75	0.530
5.25	0.540
6.00	0.535
7.00	0.575
8.10	0.590

λ	ρ
CURVE 64*(cont.)	
8.70	0.550
9.00	0.525
9.50	0.515
10.00	0.520
10.80	0.550
12.00	0.550
13.00	0.550
14.00	0.535
15.00	0.480
CURVE 65* T = ~322	
2.00	0.470
2.50	0.400
3.00	0.370
3.25	0.370
3.50	0.390
4.00	0.470
4.50	0.510
5.00	0.520
5.50	0.510
6.00	0.500
6.50	0.510
7.00	0.535
7.50	0.545
8.00	0.530
8.50	0.500
9.00	0.460
9.50	0.445
10.00	0.440
10.50	0.445
11.00	0.460
11.25	0.460
12.00	0.440
14.00	0.340
14.50	0.315
15.00	0.280
CURVE 66* T = ~322	
1.00	0.280
2.00	0.360
2.70	0.400
3.50	0.470

* Not shown on plot

DATA TABLE NO. 367 (continued)

λ	ρ
CURVE 66*(cont.)	
4.30	0.550
5.00	0.585
5.50	0.590
6.25	0.585
7.60	0.620
8.50	0.590
9.70	0.545
11.00	0.580
11.75	0.590
12.75	0.590
13.50	0.590
14.00	0.600
15.00	0.600
CURVE 67* T =∼322	
1.00	0.220
1.50	0.290
2.00	0.350
2.50	0.375
3.00	0.400
4.00	0.485
4.50	0.510
5.00	0.530
5.25	0.535
5.50	0.530
6.00	0.510
6.50	0.525
7.00	0.540
7.50	0.545
8.00	0.540
8.50	0.515
9.00	0.480
9.50	0.460
10.00	0.455
10.50	0.475
11.00	0.470
11.50	0.470
12.00	0.460
13.00	0.425
13.50	0.400
14.00	0.375
14.50	0.355
15.00	0.340
CURVE 68* T =∼322	
0.50	0.050
0.70	0.150
1.00	0.220
1.25	0.300
1.50	0.380
2.25	0.400
3.50	0.475
4.50	0.550
5.50	0.580
6.00	0.580
6.50	0.590
8.00	0.620
9.00	0.550
10.00	0.530
11.00	0.570
12.00	0.580
12.50	0.570
14.00	0.570
15.00	0.550
CURVE 69* T =∼322	
0.50	0.005
0.75	0.150
1.00	0.250
1.25	0.290
1.50	0.350
2.00	0.410
2.50	0.435
3.00	0.442
3.50	0.480
4.50	0.550
5.00	0.570
5.50	0.580
6.00	0.570
6.50	0.570
8.00	0.570
8.50	0.550
9.00	0.500
9.50	0.480
10.00	0.470
10.50	0.485
11.25	0.500
CURVE 69*(cont.)	
12.00	0.485
13.00	0.450
14.00	0.400
14.50	0.385
15.00	0.380
CURVE 70* T =∼322	
2.00	0.500
2.50	0.425
3.00	0.390
3.75	0.455
4.50	0.505
5.25	0.520
6.20	0.520
7.50	0.575
8.00	0.570
8.75	0.525
9.50	0.505
11.00	0.550
12.00	0.555
13.50	0.575
14.00	0.575
14.75	0.550
15.00	0.500
CURVE 71* T =∼322	
2.00	0.495
2.50	0.460
3.00	0.435
3.50	0.465
4.00	0.520
4.50	0.560
5.00	0.570
5.50	0.565
6.50	0.560
7.00	0.570
8.00	0.570
8.50	0.520
9.00	0.510
9.50	0.495
10.00	0.485
CURVE 71*(cont.)	
10.50	0.490
11.00	0.500
11.50	0.487
12.00	0.470
14.00	0.370
15.00	0.310
CURVE 72* T =∼322	
1.00	0.490
1.90	0.460
2.85	0.475
3.80	0.525
4.50	0.580
5.50	0.600
6.35	0.600
8.00	0.620
8.90	0.575
9.75	0.560
10.35	0.575
10.80	0.600
12.00	0.600
13.00	0.615
14.00	0.650
14.70	0.665
15.00	0.660
CURVE 73* T =∼322	
1.00	0.465
1.50	0.475
2.00	0.470
3.00	0.450
3.50	0.485
4.00	0.542
4.50	0.565
5.00	0.585
5.50	0.585
6.00	0.580
6.50	0.585
8.00	0.590
8.50	0.565
9.00	0.530
CURVE 73*(cont.)	
9.50	0.500
10.00	0.500
10.50	0.510
11.00	0.530
11.50	0.530
12.00	0.510
14.50	0.405
15.00	0.390
CURVE 74* T =∼322	
0.50	0.305
0.70	0.350
1.00	0.420
1.50	0.470
2.50	0.465
3.00	0.475
4.00	0.555
5.00	0.600
6.25	0.600
7.25	0.610
8.00	0.620
8.60	0.585
9.00	0.560
9.80	0.560
10.75	0.580
11.50	0.605
12.50	0.605
13.50	0.605
14.50	0.610
15.00	0.600
CURVE 75* T =∼322	
0.50	0.170
0.75	0.240
0.75	0.290
1.00	0.320
1.25	0.375
1.50	0.450
1.75	0.480
2.25	0.500
3.00	0.495
CURVE 75*(cont.)	
3.50	0.520
4.50	0.580
5.00	0.590
5.50	0.590
6.00	0.585
6.50	0.575
7.00	0.580
7.50	0.585
8.00	0.590
8.25	0.585
8.50	0.570
9.00	0.510
9.50	0.490
10.00	0.480
10.50	0.500
11.00	0.510
11.50	0.500
12.50	0.470
13.50	0.425
14.00	0.410
15.00	0.420
CURVE 76 T =∼322	
0.50	0.675
0.60	0.640
0.70	0.645
0.78	0.660
0.85	0.660
0.92	0.670
1.00	0.670
1.20	0.675
1.75	0.725
2.00	0.737
2.80	0.765
3.30	0.787
3.80	0.800
4.30	0.805
5.00	0.820
6.20	0.820
7.60	0.850
9.40	0.850
10.80	0.870
12.60	0.870
CURVE 76 (cont.)	
13.00	0.880
15.30	0.880
16.80	0.900
18.40	0.900
20.00	0.880
21.00	0.880
21.40	0.875
22.00	0.840
22.90	0.825
23.70	0.755
24.40	0.730
25.00	0.720
CURVE 77* T =∼322	
0.50	0.240
0.60	0.220
0.75	0.225
0.90	0.245
1.10	0.235
1.50	0.230
2.00	0.180
3.00	0.150
4.00	0.125
5.00	0.110
6.00	0.100
7.00	0.100
8.20	0.075
8.70	0.075
10.00	0.050
11.00	0.060
12.00	0.040
12.70	0.050
13.30	0.050
14.00	0.060
15.00	0.090
17.00	0.090
18.00	0.110
19.00	0.080
20.00	0.100
21.00	0.090
22.00	0.110
23.00	0.090
24.00	0.100
25.00	0.120
CURVE 78* T =∼322	
0.50	0.250
0.73	0.300
0.88	0.315
0.90	0.340
1.00	0.345
2.00	0.400
2.50	0.418
4.65	0.520
5.50	0.555
6.30	0.580
8.00	0.650
8.70	0.662
12.10	0.760
13.10	0.750
14.10	0.770
15.00	0.750
15.70	0.775
16.00	0.795
16.90	0.810
21.80	0.810
22.60	0.780
23.20	0.740
24.00	0.645
25.00	0.590
CURVE 79* T =∼322	
0.50	0.160
0.75	0.200
0.85	0.210
1.00	0.220
1.35	0.300
1.50	0.335
2.00	0.375
3.20	0.415
5.00	0.370
6.00	0.320
6.50	0.295
8.00	0.250
8.50	0.225
9.00	0.185
10.00	0.150
11.50	0.120
13.00	0.100

* Not shown on plot

DATA TABLE NO. 367 (continued)

CURVE 79* (cont.)

λ	ρ
14.00	0.105
15.00	0.150
15.50	0.130
15.80	0.100
16.50	0.085
17.15	0.100
18.00	0.130
18.50	0.110
19.00	0.090
20.00	0.120
21.00	0.110
22.00	0.130
23.00	0.120
23.50	0.135
24.00	0.160
25.00	0.200

CURVE 80*
T =~322

λ	ρ
0.50	0.500
0.63	0.540
0.68	0.585
0.78	0.595
0.85	0.620
1.10	0.635
1.35	0.650
2.00	0.725
3.30	0.790
4.50	0.805
6.30	0.855
9.10	0.880
10.20	0.880
11.70	0.900
15.20	0.900
16.30	0.925
18.80	0.945
20.10	0.925
21.20	0.920
22.20	0.885
23.00	0.890
24.00	0.850
25.00	0.850

CURVE 81*
T =~322

λ	ρ
0.50	0.260
0.65	0.300
0.75	0.310
0.85	0.280
0.95	0.285
1.50	0.245
2.50	0.215
3.00	0.200
4.20	0.150
5.50	0.120
6.20	0.110
7.00	0.125
7.50	0.125
8.00	0.120
9.00	0.080
10.00	0.080
11.00	0.070
12.20	0.070
13.10	0.070
14.20	0.095
15.00	0.100
15.60	0.150
16.00	0.190
17.00	0.190
18.10	0.210
19.00	0.170
20.00	0.180
21.00	0.200
22.10	0.220
23.20	0.210
24.00	0.230
25.00	0.265

CURVE 82*
T =~322

λ	ρ
0.50	0.315
0.63	0.370
0.75	0.535
0.78	0.600
0.87	0.680
0.90	0.690
1.10	0.690

CURVE 82* (cont.)

λ	ρ
1.50	0.700
3.00	0.800
3.60	0.805
5.20	0.835
7.00	0.850
8.50	0.876
9.40	0.880
10.10	0.895
11.15	0.895
11.80	0.900
17.00	0.900
21.00	0.900
25.00	0.900

CURVE 83*
T =~322

λ	ρ
0.50	0.160
0.55	0.200
0.75	0.250
1.25	0.250
2.00	0.250
2.30	0.250
3.00	0.270
4.00	0.250
5.00	0.240
6.00	0.220
8.20	0.220
9.00	0.190
10.50	0.180
12.00	0.170
13.00	0.180
14.00	0.200
15.00	0.220
16.00	0.200
17.10	0.200
19.00	0.170
20.00	0.200
21.10	0.200
22.00	0.230
22.80	0.230
24.00	0.250
25.00	0.270

CURVE 84*
T =~322

λ	ρ
0.50	0.675
0.60	0.645
0.68	0.645
0.75	0.680
1.00	0.680
1.50	0.720
3.00	0.780
4.80	0.830
6.00	0.840
7.30	0.860
8.30	0.855
10.00	0.880
11.00	0.880
12.90	0.900
16.00	0.900
17.00	0.910
18.00	0.900
19.20	0.900
21.00	0.880
22.00	0.850
23.00	0.820
23.70	0.770
24.50	0.740
25.00	0.730

CURVE 85*
T =~322

λ	ρ
0.50	0.260
0.65	0.220
0.75	0.225
0.85	0.250
1.20	0.230
1.40	0.225
2.00	0.180
3.50	0.135
4.60	0.110
5.50	0.100
7.10	0.100
8.00	0.080
9.00	0.060
10.00	0.060
11.00	0.060

CURVE 85* (cont.)

λ	ρ
12.00	0.050
12.60	0.060
14.00	0.060
15.00	0.100
16.00	0.100
17.10	0.100
18.00	0.130
19.00	0.080
20.00	0.100
21.20	0.100
22.00	0.120
23.10	0.110
24.20	0.125
25.00	0.145

CURVE 86
T =~322

λ	ρ
0.50	0.260
0.75	0.300
0.80	0.325
0.82	0.250
1.20	0.255
2.00	0.400
4.20	0.500
6.50	0.600
8.00	0.650*
9.20	0.675
12.00	0.760
14.00	0.770
15.00	0.760*
16.00	0.825
17.00	0.850
18.00	0.850
19.00	0.830
20.50	0.830
23.00	0.760
23.90	0.650
25.00	0.590

CURVE 87*
T =~322

λ	ρ
0.50	0.175
0.52	0.225
0.65	0.275
0.95	0.290
1.70	0.350
3.00	0.420
4.00	0.400
5.20	0.350
6.20	0.300
8.00	0.230
9.00	0.170
10.00	0.140
11.00	0.120
12.00	0.100
13.20	0.090
14.00	0.100
14.50	0.130
15.00	0.155
16.00	0.140
16.50	0.140
18.00	0.170
18.40	0.155
18.70	0.125
19.00	0.110
19.50	0.150
21.00	0.130
21.50	0.150
22.00	0.170
23.00	0.160
24.00	0.200
25.00	0.260

CURVE 88*
T =~322

λ	ρ
0.50	0.590
0.75	0.650
0.90	0.670
1.05	0.660
1.50	0.670
2.40	0.750
3.10	0.790
4.00	0.800

CURVE 88* (cont.)

λ	ρ
6.00	0.840
7.00	0.850
8.00	0.870
9.40	0.870
10.10	0.880
11.50	0.880
13.00	0.900
15.50	0.900
18.00	0.940
20.00	0.920
21.10	0.920
22.50	0.895
23.00	0.890
24.00	0.850
25.00	0.850

CURVE 89*
T =~322

λ	ρ
0.50	0.190
0.70	0.220
0.85	0.200
1.05	0.200
1.60	0.175
2.50	0.140
3.20	0.125
4.50	0.085
5.00	0.080
6.50	0.080
7.50	0.080
9.00	0.040
10.00	0.040
11.50	0.040
12.00	0.030
13.40	0.030
14.00	0.040
14.60	0.100
15.30	0.175
16.00	0.230
17.00	0.230
18.00	0.240
19.00	0.200
20.00	0.230
21.00	0.230

CURVE 89* (cont.)

λ	ρ
22.00	0.240
23.00	0.250
24.00	0.260
25.00	0.290

CURVE 90*
T =~322

λ	ρ
0.50	0.490
0.80	0.600
0.85	0.670
1.05	0.670
2.00	0.745
3.20	0.810
4.50	0.825
5.50	0.850
7.20	0.850
7.80	0.870
11.40	0.870
12.80	0.890
14.70	0.890
15.90	0.940
20.00	0.940
25.00	0.940

CURVE 91*
T =~322

λ	ρ
0.50	0.220
0.75	0.295
0.80	0.300
0.90	0.290
1.00	0.350
1.50	0.400
1.60	0.350
1.75	0.250
2.00	0.200
3.00	0.200
4.00	0.170
5.60	0.150
8.00	0.150
9.00	0.110
9.60	0.100
11.50	0.100

* Not shown on plot

DATA TABLE NO. 367 (continued)

λ	ρ
CURVE 91*(cont.)	
12.50	0.100
14.00	0.100
15.00	0.150
16.00	0.160
17.00	0.160
18.00	0.200
19.00	0.140
20.00	0.180
21.00	0.160
22.00	0.200
23.20	0.190
25.00	0.230
CURVE 92*	
T =∼322	
0.50	0.800
0.63	0.770
0.75	0.800
0.85	0.790
0.90	0.800
1.00	0.770
1.25	0.765
1.50	0.770
3.00	0.850
5.50	0.885
8.60	0.920
12.00	0.920
15.00	0.920
15.40	0.950
17.10	0.950
18.00	0.970
20.00	0.940
22.00	0.900
23.00	0.870
25.00	0.830
CURVE 93*	
T =∼322	
0.50	0.080
0.65	0.050
0.85	0.080
1.10	0.080
1.50	0.100
CURVE 93*(cont.)	
1.85	0.070
2.00	0.050
3.00	0.040
4.00	0.030
5.00	0.030
6.00	0.040
8.00	0.040
9.00	0.020
10.00	0.030
11.00	0.030
12.50	0.030
13.50	0.030
15.00	0.050
15.50	0.100
16.00	0.150
17.00	0.120
18.00	0.150
19.00	0.100
19.80	0.150
20.50	0.170
21.80	0.150
23.00	0.100
24.00	0.100
25.00	0.120
CURVE 94*	
T =∼322	
0.50	0.280
0.75	0.320
0.90	0.325
1.00	0.350
1.50	0.350
3.00	0.440
4.10	0.495
5.10	0.525
6.20	0.570
8.90	0.630
12.00	0.720
14.00	0.730
15.00	0.720
16.00	0.785
18.00	0.810
19.40	0.810
CURVE 94*(cont.)	
20.00	0.800
21.00	0.800
22.50	0.750
23.10	0.735
24.00	0.625
25.00	0.580
CURVE 95*	
T =∼322	
0.50	0.260
0.75	0.320
1.00	0.320
1.50	0.350
2.20	0.400
3.00	0.425
3.90	0.435
5.00	0.420
7.00	0.350
8.20	0.300
9.00	0.290
9.50	0.250
9.80	0.220
10.50	0.200
12.00	0.160
13.50	0.140
14.00	0.140
15.00	0.200
16.00	0.200
17.00	0.190
18.00	0.200
18.50	0.175
19.00	0.150
20.00	0.170
21.00	0.160
22.00	0.190
23.00	0.165
24.00	0.230
25.00	0.290
CURVE 96*	
T =∼322	
0.50	0.790
0.63	0.780
CURVE 96*(cont.)	
0.75	0.800
1.25	0.775
2.00	0.810
3.50	0.850
4.20	0.855
5.60	0.880
7.00	0.900
9.20	0.900
10.20	0.920
11.20	0.915
12.00	0.930
14.00	0.930
15.20	0.900
16.50	0.950
18.00	0.980
19.70	0.950
21.10	0.950
23.00	0.900
24.00	0.830
25.00	0.830
CURVE 97*	
T =∼322	
0.50	0.065
0.60	0.075
0.75	0.120
0.80	0.110
0.85	0.030
1.50	0.030
2.50	0.030
3.00	0.040
4.00	0.030
5.00	0.040
6.20	0.040
7.00	0.050
8.00	0.050
9.00	0.030
10.00	0.030
11.50	0.030
13.00	0.040
14.00	0.050
14.70	0.100
15.50	0.150
16.40	0.175
CURVE 97*(cont.)	
17.00	0.180
18.00	0.200
19.00	0.160
20.00	0.180
21.00	0.170
22.00	0.200
23.10	0.190
24.20	0.190
25.00	0.210
CURVE 98*	
T =∼322	
0.50	0.670
0.63	0.675
0.70	0.720
0.80	0.730
0.83	0.770
1.50	0.770
2.50	0.850
5.00	0.890
8.00	0.930
10.00	0.930
11.50	0.935
12.00	0.940
13.50	0.940
15.00	0.940
16.00	1.000
20.00	1.000
25.00	1.000
CURVE 99*	
T =∼322	
0.50	0.120
0.60	0.185
0.85	0.225
1.50	0.225
1.90	0.175
2.00	0.130
3.00	0.160
3.50	0.150
4.30	0.150
5.00	0.160
6.00	0.150
CURVE 99*(cont.)	
7.40	0.180
8.20	0.180
9.00	0.150
11.00	0.150
12.70	0.150
14.00	0.200
15.00	0.250
16.00	0.200
16.50	0.200
17.30	0.225
18.00	0.250
19.00	0.200
20.00	0.250
21.00	0.250
22.00	0.250
24.00	0.230
25.00	0.280
CURVE 100*	
T = 294	
0.40	0.887
0.60	0.895
0.80	0.902
1.00	0.913
1.20	0.918
1.60	0.928
2.00	0.933
2.50	0.943
3.00	0.948
4.00	0.956
5.00	0.960
6.00	0.954
7.00	0.965
8.00	0.959
9.00	0.960
10.00	0.973
11.00	0.975
13.00	0.976
14.00	0.976
CURVE 101*	
T = 294	
0.60	0.872
0.80	0.875
1.00	0.880
1.20	0.885
1.40	0.892
1.60	0.897
1.80	0.901
2.00	0.904
2.40	0.908
2.80	0.914
3.00	0.916
4.00	0.924
5.00	0.930
6.00	0.937
7.00	0.940
8.00	0.945
9.00	0.944
10.00	0.948
11.00	0.951
12.00	0.954
13.00	0.955
14.10	0.957
CURVE 102	
T = 298	
1.81	0.777
3.40	0.822
5.42	0.862
6.55	0.871
8.44	0.881
10.77	0.897
13.42	0.907
15.90	0.908
18.98	0.919
21.65	0.923
26.01	0.924
CURVE 103*	
T = 298	
1.94	0.739
3.39	0.798
5.92	0.852
CURVE 103(cont.)*	
7.77	0.866
10.02	0.887
12.09	0.896
14.02	0.898
16.65	0.902
18.70	0.898
20.67	0.890
22.32	0.895
24.41	0.899
26.01	0.899
CURVE 104*	
T = 77	
1.90	0.736
5.01	0.827
7.99	0.866
11.02	0.883
13.95	0.904
15.81	0.903
18.02	0.888
22.01	0.889
23.98	0.895
25.99	0.899
CURVE 105	
T = 298	
1.90	0.770
2.89	0.809
4.83	0.842
7.20	0.862
8.18	0.869
10.27	0.890
12.17	0.899
15.22	0.900
17.01	0.906
19.85	0.923
22.20	0.930
24.02	0.931
25.91	0.929

* Not shown on plot

DATA TABLE NO. 367 (continued)

CURVE 106* T = 298

λ	ρ
1.97	0.742
3.53	0.792
5.32	0.831
7.38	0.854
9.38	0.866
12.75	0.884
14.18	0.887
16.47	0.880
19.41	0.876
21.40	0.879
23.51	0.878
26.00	0.879

CURVE 107* T = 77

λ	ρ
1.94	0.742
3.88	0.808
5.87	0.845
7.91	0.864
9.90	0.875
11.90	0.890
13.89	0.900
15.87	0.885
17.81	0.871
19.86	0.873
21.86	0.882
23.85	0.878
26.00	0.881

CURVE 108 T = ~300

λ	ρ
1.95	0.206
2.42	0.195
2.98	0.181
3.39	0.179
3.82	0.185
4.21	0.197
4.52	0.222
4.85	0.251
5.16	0.280
5.62	0.349
6.10	0.404
7.18	0.495
8.29	0.577

CURVE 108 (cont.)

λ	ρ
9.17	0.629
10.02	0.657
10.69	0.688
11.45	0.696
12.07	0.728
12.68	0.753,
13.32	0.784
13.71	0.794
14.44	0.793
14.82	0.809

CURVE 109* T = ~300

λ	ρ
1.86	0.202
2.38	0.202
2.94	0.189
3.38	0.201
3.76	0.224
4.09	0.247
4.46	0.279
4.86	0.308
5.14	0.338
5.56	0.391
6.01	0.445
7.15	0.528
8.23	0.589
9.18	0.621
9.99	0.651
10.70	0.691
11.43	0.734
12.08	0.754
12.77	0.766
13.30	0.798
13.83	0.789
14.38	0.756
14.93	0.751

CURVE 110* T = ~300

λ	ρ
1.89	0.192
2.44	0.201
3.02	0.203
3.43	0.212
3.79	0.218
4.20	0.234

CURVE 110 (cont.)*

λ	ρ
4.52	0.252
4.85	0.273
5.16	0.290
5.61	0.327
6.12	0.365
7.18	0.441
8.27	0.501
9.15	0.534
10.02	0.568
10.70	0.609
11.39	0.637
12.15	0.681
12.66	0.700
13.28	0.762
13.87	0.742
14.39	0.717
14.84	0.705

CURVE 111 T = ~300

λ	ρ
1.87	0.195
2.40	0.209
2.94	0.214
3.39	0.231
3.80	0.224
4.20	0.236
4.49	0.250
4.92	0.262
5.13	0.266
5.58	0.282
6.01	0.277
7.15	0.321
8.26	0.264
9.10	0.174
10.01	0.259
10.75	0.274
11.43	0.284
12.17	0.297
12.73	0.299
13.34	0.356
13.80	0.348
14.49	0.331
14.93	0.353

CURVE 112* T = ~300

λ	ρ
2.00	0.204
2.52	0.215
3.08	0.221
3.47	0.231
3.83	0.233
4.30	0.247
4.60	0.256
4.91	0.264
5.15	0.275
5.66	0.275
6.16	0.305
7.20	0.371
8.33	0.350
9.18	0.272
10.05	0.331
10.79	0.336
11.49	0.343
12.19	0.371
12.75	0.384
13.31	0.468
13.89	0.471
14.48	0.444
14.95	0.426

CURVE 113 T = ~300

λ	ρ
2.00	0.163
2.49	0.158
3.00	0.157
3.37	0.171
3.80	0.173
4.17	0.169
4.54	0.186
4.81	0.195
5.12	0.202
5.57	0.205
6.05	0.198
7.14	0.208
9.14	0.147
9.99	0.193
10.70	0.199
11.39	0.199
12.16	0.195
12.71	0.195
13.26	0.210
13.72	0.190
14.43	0.189

CURVE 114* T = ~300

λ	ρ
2.01	0.185
2.46	0.178
2.98	0.167
3.41	0.177
4.19	0.177
4.89	0.197
5.61	0.202
6.11	0.208
7.15	0.218
8.22	0.182
9.11	0.150
10.08	0.206
10.74	0.218
11.48	0.228
12.15	0.221
12.73	0.225
13.38	0.261
13.83	0.236
14.46	0.247
14.96	0.293

CURVE 115 T = ~300

λ	ρ
1.99	0.164*
2.42	0.190
3.00	0.223
3.49	0.285
3.83	0.333
4.19	0.367
4.53	0.391
4.89	0.421
5.24	0.442
5.62	0.477
6.02	0.510
7.12	0.544
8.26	0.577*
9.08	0.587
9.98	0.561
10.76	0.575
11.50	0.582
12.03	0.599*
12.66	0.596
13.32	0.606
13.71	0.606
14.35	0.597
14.86	0.604

CURVE 116 T = ~300

λ	ρ
1.84	0.191
2.39	0.188
2.91	0.177
3.34	0.190
3.78	0.208
4.10	0.233
4.47	0.255
4.76	0.275
5.14	0.295
5.51	0.336
5.95	0.366
7.08	0.420
8.24	0.452
9.16	0.452
9.99	0.437
10.70	0.449
11.40	0.480
12.03	0.516
12.79	0.518
13.26	0.545
13.82	0.548
14.40	0.548
14.89	0.540

CURVE 117* T = ~300

λ	ρ
1.90	0.207
2.45	0.217
2.98	0.224
3.34	0.237
3.80	0.241
4.16	0.258
4.52	0.273
4.88	0.282
5.20	0.294
5.64	0.299
6.06	0.299
7.12	0.317
8.29	0.284
9.17	0.216
10.01	0.239
10.70	0.238
11.39	0.240
12.14	0.276
12.71	0.276
13.28	0.315
13.88	0.315

CURVE 117 (cont.)*

λ	ρ
14.38	0.295
14.95	0.302

CURVE 118* T = ~300

λ	ρ
1.86	0.179
2.50	0.195
2.98	0.207
3.39	0.221
3.74	0.228
4.15	0.246
4.51	0.264
4.87	0.284
5.17	0.300
5.58	0.314
6.06	0.328
7.16	0.364
8.24	0.313
9.15	0.222
9.96	0.234
10.66	0.249
11.40	0.259
12.13	0.305
12.73	0.330
13.34	0.388
13.79	0.391
14.44	0.348
14.92	0.353

CURVE 119* T = ~300

λ	ρ
1.96	0.178
2.57	0.182
3.05	0.183
3.47	0.188
3.86	0.194
4.22	0.194
4.49	0.217
4.89	0.231
5.22	0.244
5.66	0.288
6.09	0.294
7.18	0.338
8.20	0.326
9.16	0.269
10.03	0.282
10.74	0.270

CURVE 119 (cont.)*

λ	ρ
11.48	0.253
12.13	0.242
12.73	0.236
13.33	0.249
13.92	0.251
14.53	0.276
14.91	0.269

CURVE 120 T = ~300

λ	ρ
2.00	0.136
2.49	0.136
2.98	0.123
3.41	0.132
3.78	0.128
4.18	0.131
4.54	0.150
4.85	0.165
5.15	0.176
5.62	0.181
6.09	0.162
7.13	0.178
8.24	0.178
9.17	0.172
10.04	0.197
10.72	0.209
11.41	0.201
12.20	0.215
12.70	0.203
13.36	0.229
13.83	0.229
14.42	0.234
14.89	0.277

CURVE 121* T = ~300

λ	ρ
2.01	0.164
2.46	0.162
3.03	0.152
3.47	0.160
4.18	0.158
4.95	0.182
5.57	0.189
6.09	0.193
7.22	0.195
8.24	0.158
9.14	0.143

* Not shown on plot

DATA TABLE NO. 367 (continued)

λ	ρ
CURVE 121 (cont.)*	
10.01	0.180
10.72	0.191
11.42	0.195
12.24	0.194
12.76	0.194
13.29	0.230
13.89	0.213
14.48	0.213
14.97	0.272
CURVE 122	
T = ~300	
2.00	0.119
2.51	0.120
2.99	0.106
3.54	0.100
4.00	0.112
4.98	0.128
6.01	0.141
7.00	0.129
8.01	0.101
9.00	0.077
10.02	0.098
11.00	0.091
12.03	0.084
13.04	0.067
14.03	0.067
14.59	0.093
15.00	0.129
CURVE 123*	
T = ~300	
2.00	0.144
2.52	0.153
2.96	0.160
3.99	0.177
5.01	0.189
5.99	0.186
6.98	0.177
8.04	0.144
8.98	0.114
10.04	0.120
11.01	0.112
12.00	0.107
13.00	0.092
CURVE 123 (cont.)*	
14.03	0.079
14.57	0.101
15.05	0.145
CURVE 124*	
T = 298	
1.20	0.669
1.40	0.688
1.60	0.708
1.80	0.724
2.00	0.734
2.20	0.741
2.40	0.760
2.60	0.774
3.00	0.790
4.00	0.826
5.00	0.841
6.00	0.856
7.00	0.861
8.00	0.876
9.00	0.881
11.00	0.897
13.00	0.907
14.00	0.914
CURVE 125*	
T = 298	
0.40	0.543
0.60	0.569
0.80	0.613
1.00	0.647
1.20	0.669
1.40	0.689
1.60	0.705
1.80	0.714
2.00	0.738
2.20	0.763
2.40	0.770
CURVE 126*	
T = 298	
1.20	0.637
1.40	0.649
2.00	0.697
2.20	0.708
2.40	0.719
2.60	0.730
3.00	0.761
3.50	0.781
4.00	0.800
5.00	0.822
6.00	0.836
7.00	0.851
8.00	0.865
10.00	0.880
12.00	0.889
14.00	0.900
CURVE 127*	
T = 298	
0.40	0.506
0.60	0.538
0.80	0.571
1.00	0.608
1.20	0.629
1.40	0.651
1.60	0.663
1.80	0.679
2.00	0.702
2.29	0.713
CURVE 128	
T = 298	
1.29	0.520
1.75	0.589
2.00	0.600
2.20	0.614
2.40	0.624
2.55	0.640*
2.75	0.650
3.00	0.659
4.00	0.694
5.00	0.720
6.00	0.740
7.00	0.760
CURVE 128 (cont.)	
8.00	0.776
9.00	0.784
10.00	0.796
11.00	0.804
12.00	0.816
13.01	0.818
14.00	0.825
CURVE 129*	
T = 298	
0.60	0.480
0.80	0.495
1.00	0.521
1.40	0.559
1.60	0.580
1.80	0.596
2.00	0.610
2.20	0.628

* Not shown on plot

SPECIFICATION TABLE NO. 368 ANGULAR SPECTRAL REFLECTANCE OF [IRON + CHROMIUM + NICKEL + ΣX_i] ALLOYS

Curve No.	Ref. No.	Year	Temperature K	Wavelength Range, μ	Geometry θ	θ'	ω'	Reported Error, %	Composition (weight percent), Specifications and Remarks
1	284	1929	298	0.326-0.560	45°	45°		10	Uranus 10; nominal composition: 17-19 Cr, 8-10 Ni, 2 max Mn, 0.08-0.20 C, Fe balance; polished mirror surface; measured in air.
2	284	1929	298	0.326-0.560	45°	45°		10	Above specimen and conditions except exposed to air for 1 month.

DATA TABLE NO. 368 ANGULAR SPECTRAL REFLECTANCE OF [IRON + CHROMIUM + NICKEL + ΣX_i] ALLOYS

[Wavelength, λ, μ; Reflectance, ρ; Temperature, T, K]

λ	ρ
CURVE 1*	
T = 298	
0.326	0.56
0.341	0.62
0.361	0.65
0.405	0.64
0.467	0.66
0.507	0.67
0.560	0.67
CURVE 2*	
T = 298	
0.326	0.55
0.341	0.55
0.361	0.57
0.405	0.61
0.467	0.64
0.507	0.64
0.560	0.67

* Not shown on plot

SPECIFICATION TABLE NO. 369 ANGULAR SPECTRAL REFLECTANCE OF [IRON + CHROMIUM + NICKEL + ΣX_i] ALLOYS

Curve No.	Ref. No.	Year	Temperature K	Wavelength, μ	Angular Range	Geometry θ	Geometry θ'	Geometry ω'	Reported Error, %	Composition (weight percent), Specifications and Remarks
1	318	1961	298	0.65	10-75	θ		2π		Stainless Steel 321; nominal composition: 17.00-19.00 Cr, 9.00-12.00 Ni, 2.00 max Mn, 1.00 max Si, 0.08 max C, 5 x C min Ti, Fe balance; sanded with increasingly finer grades of emery paper; buffed to a high polish with a buffing wheel using relevigated alumina as the abrasive compound; rinsed in water and alcohol successively and immediately dried; fresh sample; MgO reference; incident beam polarized in a direction perpendicular to the incident plane of the sample; [Author's designation: S0].
2	318	1961	298	1.80	10-75	θ		2π		Above specimen and conditions.
3	318	1961	298	0.65	10-75	θ		2π		Stainless Steel 321; nominal composition: 17.00-19.00 Cr, 9.00-12.00 Ni, 2.00 max Mn, 1.00 max Si, 0.08 max C, 5 x C min Ti, Fe balance; sanded with 0000 emery paper; sanded in one direction to produce a fine-grooved appearance; peak to peak height of asperity as measured by profilometer 0.06μ; mean groove spacing 5μ; mean groove angle 2°; fresh sample; MgO reference; incident beam polarized in a direction perpendicular to the incident plane of the sample; [Author's designation: S1].
4	318	1961	298	1.80	10-75	θ		2π		Above specimen and conditions.
5	318	1961	298	0.65	10-75	θ		2π		Stainless Steel 321; nominal composition: 17.00-19.00 Cr, 9.00-12.00 Ni, 2.00 max Mn, 1.00 max Si, 0.08 max C, 5 x C min Ti, Fe balance; sanded with 00 emery paper; sanded in one direction to produce a fine-grooved appearance; peak to peak height of asperity as measured by profilometer 0.5μ; mean groove spacing 9μ; mean groove angle 6°; fresh sample; MgO reference; incident beam polarized in a direction perpendicular to the incident plane of the sample; [Author's designation: S2].
6	318	1961	298	1.80	10-75	θ		2π		Above specimen and conditions.
7	318	1961	298	0.65	10-75	θ		2π		Stainless Steel 321; nominal composition: 17.00-19.00 Cr, 9.00-12.00 Ni, 2.00 max Mn, 1.00 max Si, 0.08 max C, 5 x C min Ti, Fe balance; sanded with grade 1 emery paper; sanded in one direction to produce a fine-grooved appearance; peak to peak height of asperity as measured by profilometer 1.25μ; mean groove spacing 9μ; mean groove angle 15°; fresh sample; MgO reference; incident beam polarized in a direction perpendicular to the incident plane of the sample; [Author's designation: S3].
8	318	1961	298	1.80	10-75	θ		2π		Above specimen and conditions.

DATA TABLE NO. 369 ANGULAR SPECTRAL REFLECTANCE CF [IRON + CHROMIUM + NICKEL + ΣX_i] ALLOYS

[Angle, θ, °; Reflectance, ρ; Temperature, T, K; Wavelength, λ, μ]

θ	ρ	θ	ρ	θ	ρ
CURVE 1*		CURVE 4 (cont.)*		CURVE 8*	
T = 298				T = 298	
λ = 0.65		40	0.689	λ = 1.80	
		50	0.693		
10	0.607	60	0.649	10	0.600
20	0.607	70	0.624	20	0.605
30	0.607	75	0.605	30	0.597
40	0.618			40	0.597
50	0.617	CURVE 5*		50	0.597
60	0.607	T = 298		60	0.590
70	0.607	λ = 0.65		70	0.578
75	0.611			75	0.591
		10	0.502		
CURVE 2*		20	0.502		
T = 298		30	0.502		
λ = 1.80		40	0.511		
		50	0.512		
10	0.714	60	0.513		
20	0.714	70	0.532		
30	0.714	75	0.590		
40	0.728				
50	0.728	CURVE 6*			
60	0.704	T = 298			
70	0.670	λ = 1.80			
75	0.660				
		10	0.622		
CURVE 3*		20	0.628		
T = 298		30	0.632		
λ = 0.65		40	0.639		
		50	0.643		
10	0.547	60	0.629		
20	0.545	70	0.615		
30	0.545	75	0.615		
40	0.545				
50	0.545	CURVE 7*			
60	0.545	T = 298			
70	0.557	λ = 0.65			
75	0.598				
		10	0.485		
CURVE 4*		20	0.481		
T = 298		30	0.486		
λ = 1.80		40	0.491		
		50	0.483		
10	0.678	60	0.502		
20	0.674	70	0.503		
30	0.674	75	0.492		

* Not shown on plot

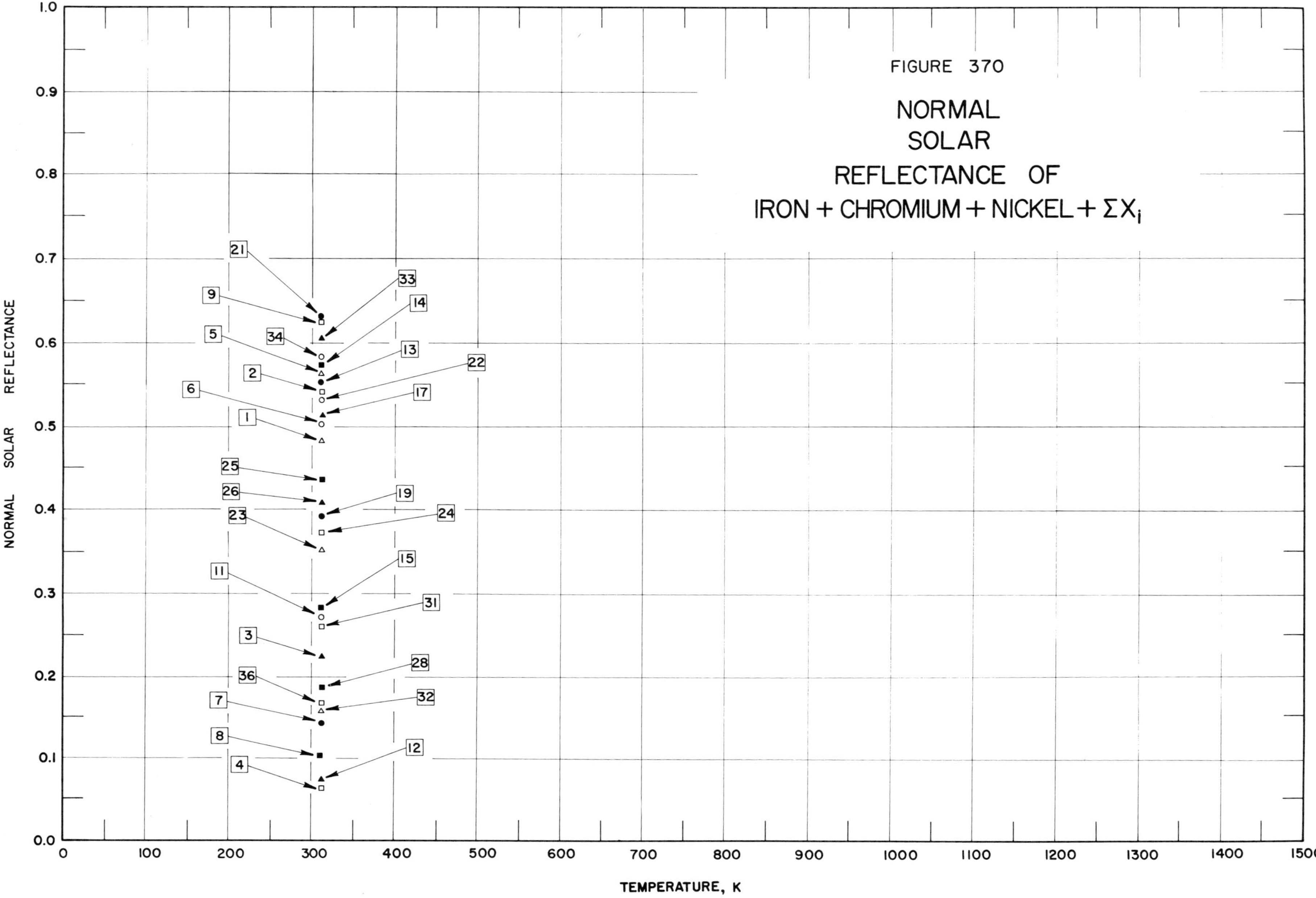
FIGURE 370
NORMAL
SOLAR
REFLECTANCE OF
IRON + CHROMIUM + NICKEL + ΣX_i
NORMAL SOLAR REFLECTANCE
TEMPERATURE, K
1.0 0.9 0.8 0.7 0.6 0.5 0.4 0.3 0.2 0.1 0.0
0 100 200 300 400 500 600 700 800 900 1000 1100 1200 1300 1400 1500
21 9 5 34 2 6 1 25 26 23 11 3 36 7 8 4
33 14 13 22 17 19 24 15 31 28 32 12

SPECIFICATION TABLE NO. 370 NORMAL SOLAR REFLECTANCE OF [IRON + CHROMIUM + NICKEL + ΣX_i] ALLOYS

Curve No.	Ref. No.	Year	Temperature Range, K	Geometry θ	θ'	ω'	Reported Error, %	Composition (weight percent), Specifications and Remarks
1	160	1954	311	~0°		2π		Stainless steel 301; nominal composition: 16.00-18.00 Cr, 6.00-8.00 Ni, 2.00 max Mn, 1.00 max Si, 0.15 max C, Fe balance; heated to 308 K; clean and smooth surface; measured in air; calculated from solar absorptance.
2	160	1954	311	~0°		2π		Above specimen and conditions except heated to 628 K.
3	160	1954	311	~0°		2π		Above specimen and conditions except heated to 944 K.
4	160	1954	311	~0°		2π		Above specimen and conditions except heated to 1220 K.
5	160	1954	311	~0°		2π		Different sample, same as curve 1 specimen and conditions except heated to 311 K; cleaned with methyl alcohol.
6	160	1954	311	~0°		2π		Above specimen and conditions except heated to 608 K.
7	160	1954	311	~0°		2π		Above specimen and conditions except heated to 928 K.
8	160	1954	311	~0°		2π		Above specimen and conditions except heated to 1206 K.
9	160	1954	311	~0°		2π		Different sample, same as curve 1 specimen and conditions except heated to 323 K; polished; surface free from scratches.
10	160	1954	311	~0°		2π		Above specimen and conditions except heated to 598 K.
11	160	1954	311	~0°		2π		Above specimen and conditions except heated to 935 K.
12	160	1954	311	~0°		2π		Above specimen and conditions except heated to 1179 K.
13	160	1954	311	~0°		2π		Stainless steel 316; nominal composition: 16.00-18.00 Cr, 10.00-14.00 Ni, 2.00-3.00 Mo, 2.00 max Mn, 1.00 max Si, 0.08 max C, Fe balance; heated to 305 K; clean and smooth surface; measured in air; calculated from solar absorptance.
14	160	1954	311	~0°		2π		Above specimen and conditions except heated to 613 K.
15	160	1954	311	~0°		2π		Above specimen and conditions except heated to 942 K.
16	160	1954	311	~0°		2π		Above specimen and conditions except heated to 1325 K.
17	160	1954	311	~0°		2π		Different sample, same as curve 13 specimen and conditions except heated to 312 K; cleaned with methyl alcohol.
18	160	1954	311	~0°		2π		Above specimen and conditions except heated to 623 K.
19	160	1954	311	~0°		2π		Above specimen and conditions except heated to 915 K.
20	160	1954	311	~0°		2π		Above specimen and conditions except heated to 1131 K.
21	160	1954	311	~0°		2π		Different sample, same as curve 13 specimen and conditions except heated to 312 K; polished; surface free from scratches.
22	160	1954	311	~0°		2π		Above specimen and conditions except heated to 600 K.
23	160	1954	311	~0°		2π		Above specimen and conditions except heated to 934 K.
24	160	1954	311	~0°		2π		Above specimen and conditions except heated to 1323 K.

SPECIFICATION TABLE NO. 370 (continued)

Curve No.	Ref. No.	Year	Temperature Range, K	Geometry θ	θ'	ω'	Reported Error, %	Composition (weight percent), Specifications and Remarks
25	160	1954	311	$\sim 0°$		2π		Stainless steel 347; nominal composition: 17.00-19.00 Cr, 9.00-13.00 Ni, 2.00 max Mn, 1.00 max Si, 0.08 max C, 10 x C min Nb-Ta, Fe balance; heated to 315 K; clean and smooth surface; measured in air; calculated from solar absorptance.
26	160	1954	311	$\sim 0°$		2π		Above specimen and conditions except heated to 612 K.
27	160	1954	311	$\sim 0°$		2π		Above specimen and conditions except heated to 937 K.
28	160	1954	311	$\sim 0°$		2π		Above specimen and conditions except heated to 1145 K.
29	160	1954	311	$\sim 0°$		2π		Different sample, same as curve 25 specimen and conditions except heated to 325 K; cleaned with methyl alcohol.
30	160	1954	311	$\sim 0°$		2π		Above specimen and conditions except heated to 622 K.
31	160	1954	311	$\sim 0°$		2π		Above specimen and conditions except heated to 930 K.
32	160	1954	311	$\sim 0°$		2π		Above specimen and conditions except heated to 1175 K.
33	160	1954	311	$\sim 0°$		2π		Different sample, same as curve 25 specimen and conditions except heated to 320 K; polished; surface free from scratches.
34	160	1954	311	$\sim 0°$		2π		Above specimen and conditions except heated to 680 K.
35	160	1954	311	$\sim 0°$		2π		Above specimen and conditions except heated to 950 K.
36	160	1954	311	$\sim 0°$		2π		Above specimen and conditions except heated to 1172 K.

DATA TABLE NO. 370 NORMAL SOLAR REFLECTANCE OF [IRON + CHROMIUM + NICKEL + ΣX_i] ALLOYS

[Temperature, T, K; Reflectance, ρ]

T	ρ	T	ρ	T	ρ
CURVE 1		CURVE 13		CURVE 25	
311	0.481	311	0.552	311	0.438
CURVE 2		CURVE 14		CURVE 26	
311	0.541	311	0.571	311	0.408
CURVE 3		CURVE 15		CURVE 27*	
311	0.223	311	0.282	311	0.109
CURVE 4		CURVE 16*		CURVE 28	
311	0.061	311	0.270	311	0.188
CURVE 5		CURVE 17		CURVE 29*	
311	0.563	311	0.513	311	0.389
CURVE 6		CURVE 18*		CURVE 30*	
311	0.502	311	0.483	311	0.389
CURVE 7		CURVE 19		CURVE 31	
311	0.143	311	0.391	311	0.260
CURVE 8		CURVE 20*		CURVE 32	
311	0.101	311	0.111	311	0.159
CURVE 9		CURVE 21		CURVE 33	
311	0.624	311	0.632	311	0.606
CURVE 10*		CURVE 22		CURVE 34	
311	0.534	311	0.532	311	0.583
CURVE 11		CURVE 23		CURVE 35*	
311	0.271	311	0.352	311	0.229
CURVE 12		CURVE 24		CURVE 36	
311	0.072	311	0.373	311	0.168

* Not shown on plot

SPECIFICATION TABLE NO. 371 HEMISPHERICAL INTEGRATED ABSORPTANCE OF [IRON + CHROMIUM + NICKEL + ΣX_i] ALLOYS

Curve No.	Ref. No.	Year	Temperature Range, K	Reported Error, %	Composition (weight percent), Specifications and Remarks
1	3	1955	76		Commercial ball type stainless steel 302; nominal composition: 17.00-19.00 Cr, 8.00-10.00 Ni, 2.00 Mn, 1.00 Si, 0.15 C, 0.045 P, 0.03 S, and Fe balance; solvent cleaned; measured in vacuum (10^{-6} to 10^{-7} mm Hg); absorptance for 300 K black body incident radiation.
2	3	1955	76		Different sample, same as above specimen and conditions except sheet (0.005 in. thick).
3	4	1953	76		Stainless Steel 302; nominal composition: 17.00-19.00 Cr, 8.00-10.00 Ni, 2.00 max Mn, 1.00 max Si, 0.15 max C, Fe balance; foil (0.005 in. thick); solvent cleaned; measured in vacuum ($< 10^{-6}$ mm Hg); absorptance for 294 K black body incident radiation.

DATA TABLE NO. 371 HEMISPHERICAL INTEGRATED ABSORPTANCE OF [IRON + CHROMIUM + NICKEL + ΣX_i] ALLOYS

[Temperature, T, K; Absorptance, α]

T	α
CURVE 1*	
76	0.07
CURVE 2*	
76	0.048
CURVE 3*	
76	0.0485

* Not shown on plot

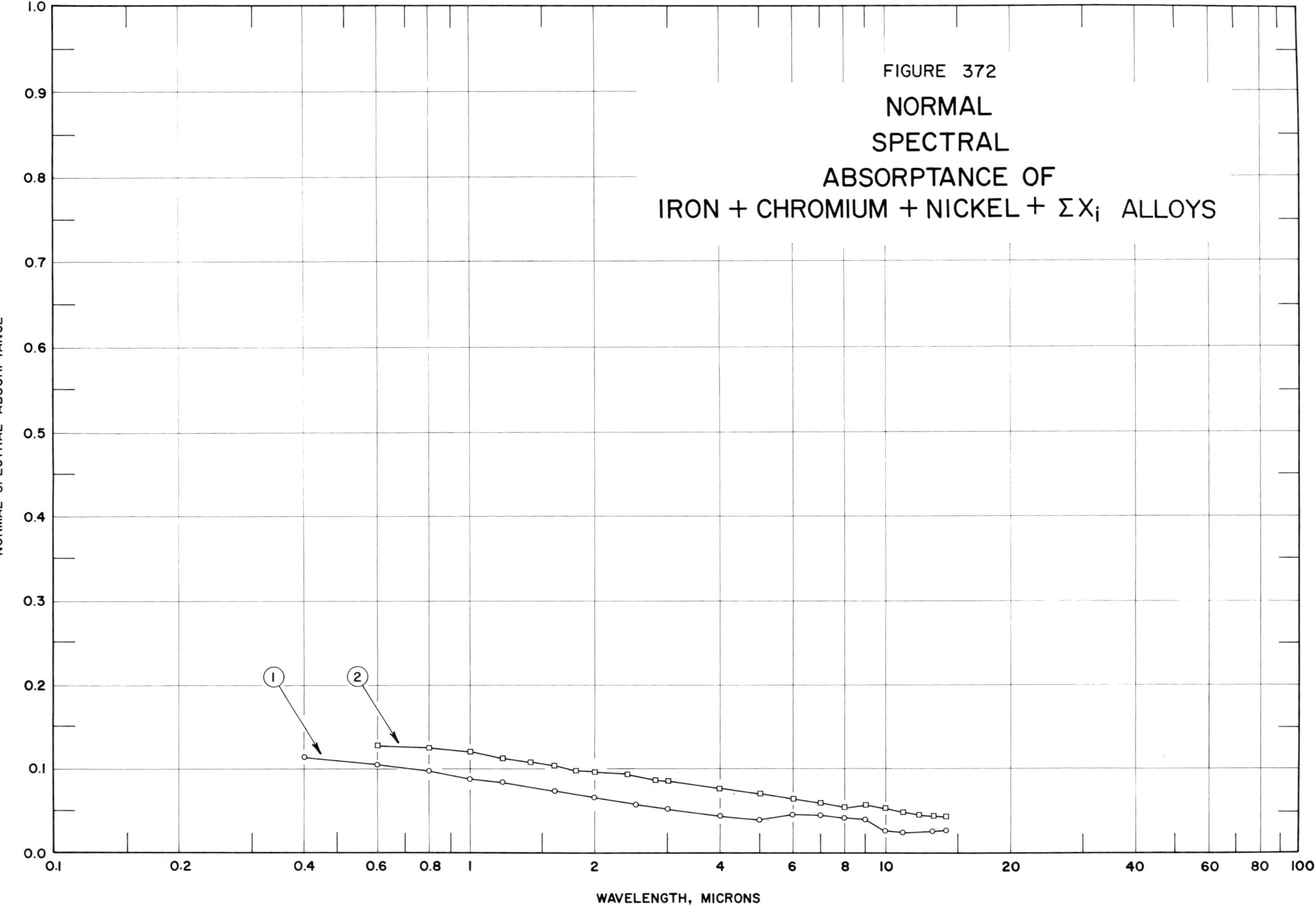

FIGURE 372
NORMAL
SPECTRAL
ABSORPTANCE OF
IRON + CHROMIUM + NICKEL + ΣX_i ALLOYS

SPECIFICATION TABLE NO. 372 NORMAL SPECTRAL ABSORPTANCE OF [IRON + CHROMIUM + NICKEL + ΣX_i] ALLOYS

Curve No.	Ref. No.	Year	Temperature K	Wavelength Range, μ	Geometry θ	Reported Error, %	Composition (weight percent), Specifications and Remarks
1	65	1962	294	0.40-14.00	$\sim 0°$		Stainless steel 321; nominal composition: 17.00-19.00 Cr, 9.00-12.00 Ni, 2.00 max Mn, 1.00 max Si, 0.40 min Ti, 0.08 max C, Fe balance; mechanically polished; calculated from $(1 - \rho)$.
2	65	1962	294	0.60-14.10	$\sim 0°$		Above specimen and conditions except roughened with sandpaper; surface roughness 1.25μ.

DATA TABLE NO. 372 NORMAL SPECTRAL ABSORPTANCE OF [IRON + CHROMIUM + NICKEL + ΣX_i] ALLOYS

[Wavelength, λ, μ; Absorptance, α; Temperature, T, K]

λ	α
CURVE 1	
T = 294	
0.40	0.113
0.60	0.105
0.80	0.098
1.00	0.087
1.20	0.082
1.60	0.072
2.00	0.067
2.50	0.057
3.00	0.052
4.00	0.044
5.00	0.040
6.00	0.046
7.00	0.045
8.00	0.041
9.00	0.040
10.00	0.027
11.00	0.025
13.00	0.024
14.00	0.024
CURVE 2	
T = 294	
0.60	0.128
0.80	0.125
1.00	0.120
1.20	0.115
1.40	0.108
1.60	0.103
1.80	0.099
2.00	0.096
2.40	0.092
2.80	0.086
3.00	0.084
4.00	0.076
5.00	0.070
6.00	0.063
7.00	0.060
8.00	0.055
9.00	0.056
10.00	0.052
11.00	0.049
12.00	0.046
13.00	0.045
14.10	0.043

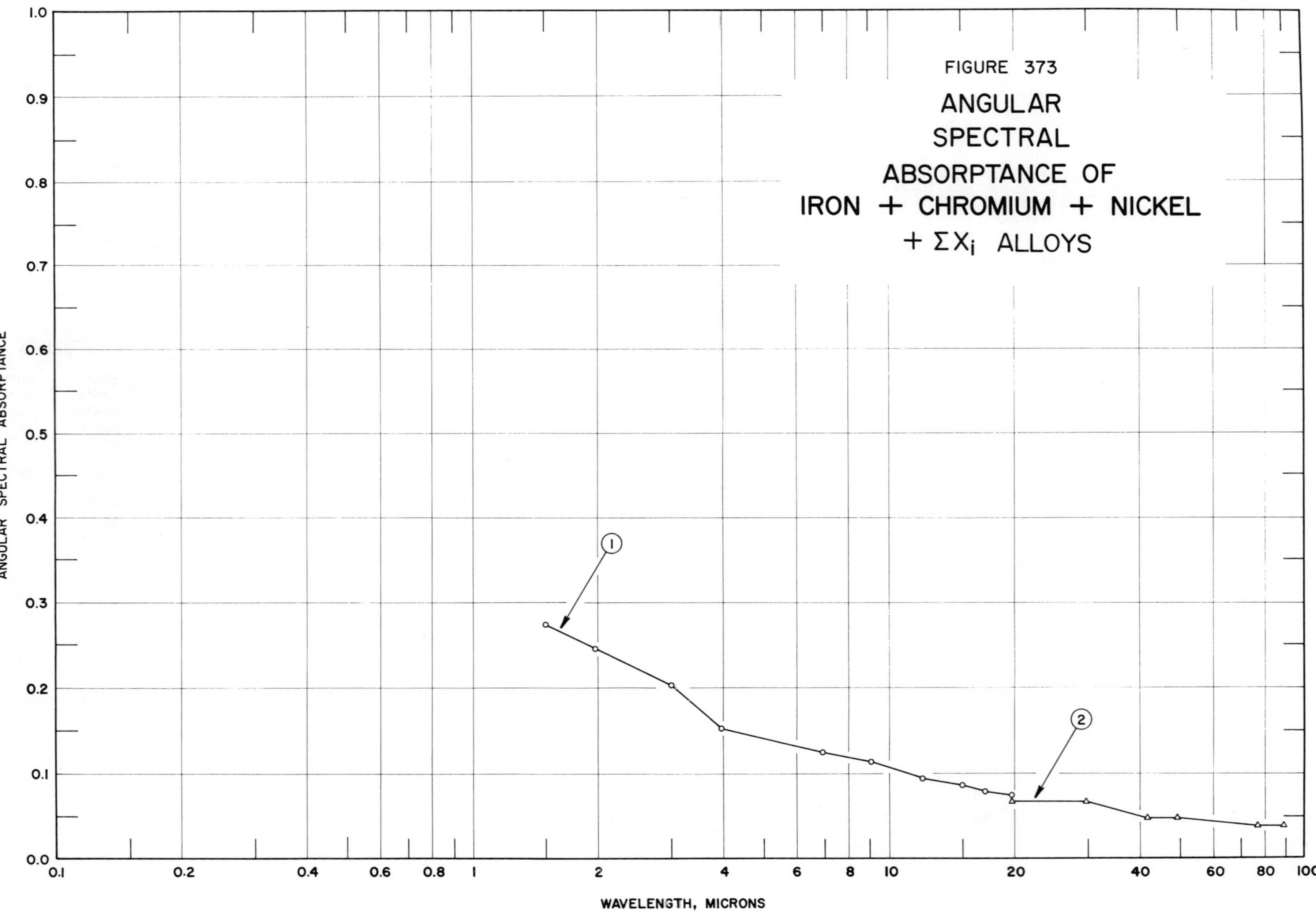

FIGURE 373
ANGULAR SPECTRAL ABSORPTANCE OF IRON + CHROMIUM + NICKEL + ΣX_i ALLOYS

SPECIFICATION TABLE NO. 373 ANGULAR SPECTRAL ABSORPTANCE OF [IRON + CHROMIUM + NICKEL + ΣX_i] ALLOYS

Curve No.	Ref. No.	Year	Temperature K	Wavelength Range, μ	Geometry θ	Reported Error, %	Composition (weight percent), Specifications and Remarks
1	225	1965	306	1.50–19.8	25°		Stainless Steel 303; nominal composition: 17.00–19.00 Cr, 8.00–10.00 Ni, 2.00 max Mn, 1.00 max Si, 0.15 min S, 0.15 max C, and Fe balance; lapped; measured in dry nitrogen; heated cavity at approx 1056 K with platinum reference; authors assumed α = 1-R(2π, 25°).
2	225	1965	306	19.8–86.3	25°		Stainless Steel 303; nominal composition: 17.00–19.00 Cr, 8.00–10.00 Ni, 2.00 max Mn, 1.00 max Si, 0.15 min S, 0.15 max C, and Fe balance; lapped.

DATA TABLE NO. 373 ANGULAR SPECTRAL ABSORPTANCE OF [IRON + CHROMIUM + NICKEL + ΣX_i] ALLOYS

[Wavelength, λ, μ; Absorptance, α; Temperature, T, K]

λ	α
CURVE 1	
T = 306	
1.50	0.272
1.98	0.246
3.00	0.202
4.97	0.152
6.97	0.125
9.04	0.114
12.0	0.093
15.0	0.087
17.1	0.079
19.8	0.075
CURVE 2	
T = 306	
19.8	0.068
29.7	0.068
41.5	0.049
49.1	0.049
76.6	0.040
86.3	0.040

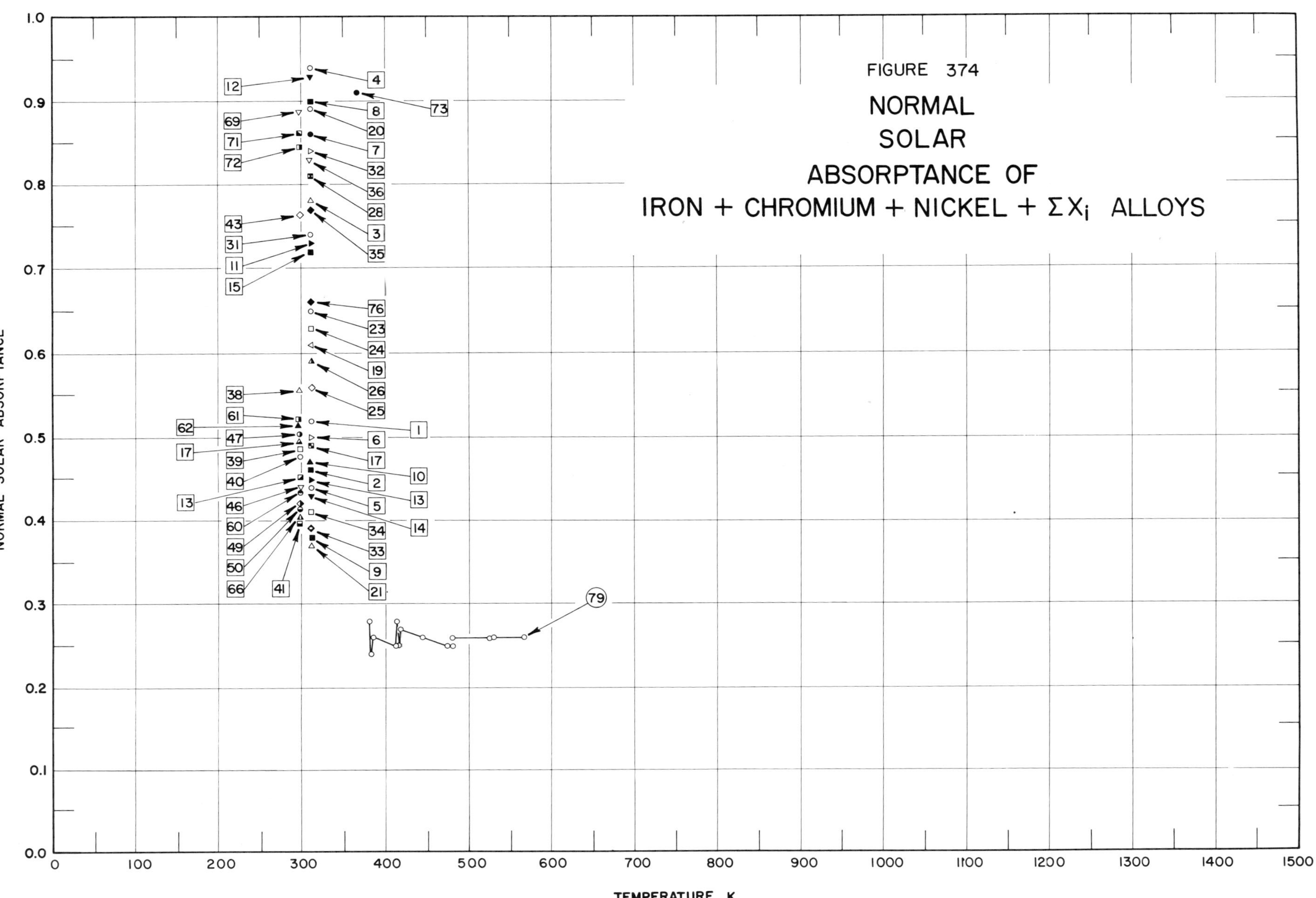

FIGURE 374
NORMAL
SOLAR
ABSORPTANCE OF
IRON + CHROMIUM + NICKEL + ΣX_i ALLOYS
NORMAL SOLAR ABSORPTANCE
TEMPERATURE, K

SPECIFICATION TABLE NO. 374 NORMAL SOLAR ABSORPTANCE OF [IRON + CHROMIUM + NICKEL + ΣX_i] ALLOYS

Curve No.	Ref. No.	Year	Temperature Range, K	Geometry θ	Reported Error, %	Composition (weight percent), Specifications and Remarks
1	160	1954	311	~0°		Stainless steel 301; nominal composition : 16.00-18.00 Cr, 6.00-8.00 Ni, 2.00 max Mn, 1.00 max Si, 0.15 max C, Fe balance;heated to 308 K; clean and smooth surface; measured in air at sea level.
2	160	1954	311	~0°		Above specimen and conditions except reheated to 632 K.
3	160	1954	311	~0°		Above specimen and conditions except reheated to 947 K.
4	160	1954	311	~0°		Above specimen and conditions except reheated to 1219 K.
5	160	1954	311	~0°		Different sample, same as curve 1 specimen and conditions except heated to 309 K; cleaned with methyl alcohol.
6	160	1954	311	~0°		Above specimen and conditions except reheated to 610 K.
7	160	1954	311	~0°		Above specimen and conditions except reheated to 930 K.
8	160	1954	311	~0°		Above specimen and conditions except reheated to 1205 K.
9	160	1954	311	~0°		Different sample, same as curve 1 specimen and conditions except heated to 322 K; polished; surface free from scratches.
10	160	1954	311	~0°		Above specimen and conditions except reheated to 602 K.
11	160	1954	311	~0°		Above specimen and conditions except reheated to 938 K.
12	160	1954	311	~0°		Above specimen and conditions except reheated to 1180 K.
13	160	1954	311	~0°		Stainless steel 316; nominal composition : 16.00-18.00 Cr, 10.00-14.00 Ni, 2.00-3.00 Mo, 2.00 max Mn, 1.00 max Si, 0.08 max C, Fe balance; heated to 303 K; clean and smooth surface; measured in air at sea level.
14	160	1954	311	~0°		Above specimen and conditions except reheated to 613 K.
15	160	1954	311	~0°		Above specimen and conditions except reheated to 943 K.
16	160	1954	311	~0°		Above specimen and conditions except reheated to 1322 K.
17	160	1954	311	~0°		Different sample, same as curve 13 specimen and conditions except heated to 310 K; cleaned with methyl alcohol.
18	160	1954	311	~0°		Above specimen and conditions except reheated to 625 K.
19	160	1954	311	~0°		Above specimen and conditions except reheated to 918 K.
20	160	1954	311	~0°		Above specimen and conditions except reheated to 1130 K.
21	160	1954	311	~0°		Different sample,same as curve 13 specimen and conditions except heated to 309 K; polished; surface free from scratches.
22	160	1954	311	~0°		Above specimen and conditions except reheated to 601 K.
23	160	1954	311	~0°		Above specimen and conditions except reheated to 929 K.
24	160	1954	311	~0°		Above specimen and conditions except reheated to 1321 K.

SPECIFICATION TABLE NO. 374 (continued)

Curve No.	Ref. No.	Year	Temperature Range, K	Geometry θ	Reported Error, %	Composition (weight percent), Specifications and Remarks
25	160	1954	311	~0°		Stainless steel 347; nominal composition: 17.00-19.00 Cr, 9.00-13.00 Ni, 2.00 max Mn, 1.00 max Si, 0.08 max C, 10 x C min Nb-Ta, Fe balance; heated to 314 K; clean and smooth surface; measured in air at sea level.
26	160	1954	311	~0°		Above specimen and conditions except reheated to 611 K.
27	160	1954	311	~0°		Above specimen and conditions except reheated to 938 K.
28	160	1954	311	~0°		Above specimen and conditions except reheated to 1144 K.
29	160	1954	311	~0°		Different sample, same as curve 25 specimen and conditions except heated to 327 K; cleaned with methyl alcohol.
30	160	1954	311	~0°		Above specimen and conditions except reheated to 620 K.
31	160	1954	311	~0°		Above specimen and conditions except reheated to 674 K.
32	160	1954	311	~0°		Above specimen and conditions except reheated to 1172 K.
33	160	1954	311	~0°		Different sample, same as curve 25 specimen and conditions except heated to 320 K; polished; surface free from scratches.
34	160	1954	311	~0°		Above specimen and conditions except reheated to 680 K.
35	160	1954	311	~0°		Above specimen and conditions except reheated to 949 K.
36	160	1954	311	~0°		Above specimen and conditions except reheated to 1172 K.
37	34	1957	298	9°		Cobalt alloy N-155; nominal composition: 21 Cr, 20 Co, 20 Ni, 3 Mo, 3 W, 1.5 Mn, 1 Nb, 0.5 Si, 0.15 C, 0.15 N, Fe balance; as received; compute from spectral reflectance data for sea level conditions.
38	34	1957	298	9°		Above specimen and conditions except computed for above atmosphere conditions.
39	34	1957	298	9°		Different sample, same as curve 37 specimen and conditions except cleaned with liquid detergent.
40	34	1957	298	9°		Above specimen and conditions except computed for above atmosphere conditions.
41	34	1957	298	9°		Different sample, same as curve 37 specimen and conditions except polished.
42	34	1957	298	9°		Above specimen and conditions except computed for above atmosphere conditions.
43	34	1957	298	9°		Different sample, same as curve 37 specimen and conditions except oxidized in air at red heat for 30 min.
44	34	1957	298	9°		Above specimen and conditions except computed for above atmosphere conditions.
45	34	1957	298	9°		Stainless steel type PH 15-7 Mo; nominal composition: 15.00 Cr, 7.00 Ni, 2.25 Mo, 1.15 Al, 0.70 Mn, 0.40 Si, 0.07 C, Fe balance; surface roughness ~2 micro-inches rms; computed from spectral reflectance data for sea level conditions.
46	34	1957	298	9°		Above specimen and conditions except computed for above atmosphere conditions.

SPECIFICATION TABLE NO. 374 (continued)

Curve No.	Ref. No.	Year	Temperature Range, K	Geometry θ	Reported Error, %	Composition (weight percent), Specifications and Remarks
47	34	1957	298	9°		Different sample, same as curve 45 specimen and conditions except surface roughness ~15 microinches rms.
48	34	1957	298	9°		Above specimen and conditions except computed for above atmosphere conditions.
49	34	1957	298	9°		Stainless steel type 17-7 Ph; nominal composition: 17.00 Cr, 7.00 Ni, 1.15 Al, 0.70 Mn, 0.40 Si, 0.07 C, Fe balance; surface roughness ~2 microinches rms; computed from spectral reflectance data for sea level conditions.
50	34	1957	298	9°		Above specimen and conditions except computed for above atmosphere conditions.
51	34	1957	298	9°		Different sample, same as curve 49 specimen and conditions except surface roughness ~15 microinches rms.
52	34	1957	298	9°		Above specimen and conditions except computed for above atmosphere conditions.
53	34	1957	298	9°		Stainless steel type 316; nominal composition: 16.00-18.00 Cr, 10.00-14.00 Ni, 2.00-3.00 Mo, 2.00 max Mn, 1.00 max Si, 0.08 max C, Fe balance; surface roughness ~2 microinches rms; computed from spectral reflectance data for sea level conditions.
54	34	1957	298	9°		Above specimen and conditions except computed for above atmosphere conditions.
55	34	1957	298	9°		Different sample, same as curve 53 specimen and conditions except surface roughness ~15 microinches rms.
56	34	1957	298	9°		Above specimen and conditions except computed for above atmosphere conditions.
57	34	1957	298	9°		Stainless steel type 321; nominal composition: 17.00-19.00 Cr, 9.00-12.00 Ni, 2.0 max Mn, 1.0 max Si, 0.08 max C, 5 x C min Ti, Fe balance; surface roughness ~2 microinches rms; computed from spectral reflectance data for sea level conditions.
58	34	1957	298	9°		Above specimen and conditions except computed for above atmosphere conditions.
59	34	1957	298	9°		Different sample, same as curve 57 specimen and conditions except surface roughness ~15 microinches rms.
60	34	1957	298	9°		Above specimen and conditions except computed for above atmosphere conditions.
61	34	1957	298	9°		Different sample, same as curve 57 specimen and conditions except dull finish; surface roughness ~6 microinches rms.
62	34	1957	298	9°		Above specimen and conditions except computed for above atmosphere conditions.
63	34	1957	298	9°		Different sample, same as curve 57 specimen and conditions except bright finish.
64	34	1957	298	9°		Above specimen and conditions except computed for above atmosphere conditions.
65	34	1957	298	9°		Stainless steel type AM-350; nominal composition: 16.50 Cr, 4.25 Ni, 2.75 Mo, 0.75 Mn, 0.10 N, 0.10 C, Fe balance; surface roughness ~2 microinches rms; computed from spectral reflectance data for sea level conditions.
66	34	1957	298	9°		Above specimen and conditions except computed for above atmosphere conditions.

SPECIFICATION TABLE NO. 374 (continued)

Curve No.	Ref. No.	Year	Temperature Range, K	Geometry θ	Reported Error, %	Composition (weight percent), Specifications and Remarks
67	34	1957	298	9°		Different sample, same as curve 65 specimen and conditions except surface roughness ~15 microinches rms.
68	34	1957	298	9°		Above specimen and conditions except computed for above atmosphere conditions.
69	146	1958	298	9°		Stainless steel type 321; nominal composition; 17.00-19.00 Cr, 9.00-12.00 Ni, 2.00 max Mn, 1.00 max Si, 0.08 max C, 5 x C min Ti, Fe balance; oxidized; computed from spectral reflectance data for sea level conditions.
70	146	1958	298	9°		Above specimen and conditions except computed for above atmosphere conditions.
71	146	1958	298	9°		Stainless steel type AM-350; nominal composition: 16.50 Cr, 4.25 Ni, 2.75 Mo, 0.75 Mn, 0.35 Si, 0.10 C, 0.10 N, Fe balance; oxidized; computed from spectral reflectance data for sea leavel conditions.
72	146	1958	298	9°		Above specimen and conditions except computed for above atmosphere conditions.
73	40	1962	367	~0°		Cobalt alloy N-155 (surface N-1); nominal composition: 21 Cr, 20 Co, 20 Ni, 3 Mo, 3 W, 1.5 Mn, 1 Nb, 0.5 Si, 0.15 C, 0.15 N, Fe balance; highly polished, mirror finish; oxide formation at 873 K for 3 hrs.
74	40	1962	311	~0°		Poroloy (18-8 stainless steel); nominal composition: 18.45 Cr, 8.79 Ni, 0.50 Mn, 0.10 C, Fe balance; porosity 28%.
75	40	1962	311	~0°		Different sample, same as above specimen and conditions except porosity 31%.
76	40	1962	311	~0°		Different sample, same as above specimen and conditions except porosity 43%.
77	40	1962	311	~0°		Tyler 20 x 200 mesh (18-8 stainless steel); nominal composition: 18.45 Cr, 8.79 Ni, 0.50 Mn, 0.10 C, Fe balance.
78	40	1962	311	~0°		Different sample, same as above specimen and conditions except 20 x 350 mesh.
79	40	1962	380-569	~0°		Different sample, same as curve 77 specimen and conditions.

DATA TABLE NO. 374 NORMAL SOLAR ABSORPTANCE OF [IRON + CHROMIUM + NICKEL + ΣX_i]

[Temperature, T, K; Absorptance, α]

T	α	T	α	T	α	T	α	T	α	T	α	T	α	T	α
CURVE 1		CURVE 12		CURVE 23		CURVE 34		CURVE 45*		CURVE 56*		CURVE 67*		CURVE 78*	
311	0.52	311	0.93	311	0.65	311	0.41	298	0.451	298	0.499	298	0.512	311	0.73
CURVE 2		CURVE 13		CURVE 24		CURVE 35		CURVE 46		CURVE 57*		CURVE 68*		CURVE 79	
311	0.46	311	0.45	311	0.63	311	0.77	298	0.442	298	0.424	298	0.500	380	0.28
CURVE 3		CURVE 14		CURVE 25		CURVE 36		CURVE 47		CURVE 58*		CURVE 69		381	0.24
311	0.78	311	0.43	311	0.56	311	0.83	298	0.504	298	0.422	298	0.889	383	0.26
CURVE 4		CURVE 15		CURVE 26		CURVE 37*		CURVE 48*		CURVE 59*		CURVE 70*		414	0.25
311	0.94	311	0.72	311	0.59	298	0.559	298	0.496	298	0.492	298	0.886	414	0.28
CURVE 5		CURVE 16*		CURVE 27*		CURVE 38		CURVE 49		CURVE 60*		CURVE 71		416	0.25
311	0.44	311	0.73	311	0.89	298	0.555	298	0.420	298	0.486	298	0.862	417	0.27
CURVE 6		CURVE 17		CURVE 28		CURVE 39		CURVE 50		CURVE 61		CURVE 72		444	0.26
311	0.50	311	0.49	311	0.81	298	0.486	298	0.416	298	0.522	298	0.846	475	0.25
CURVE 7		CURVE 18*		CURVE 29*		CURVE 40		CURVE 51*		CURVE 62		CURVE 73		480	0.25
311	0.86	311	0.52	311	0.61	298	0.477	298	0.504	298	0.516	367	0.91	480	0.26
CURVE 8		CURVE 19		CURVE 30*		CURVE 41		CURVE 52*		CURVE 63*		CURVE 74*		527	0.26
311	0.90	311	0.61	311	0.61	298	0.398	298	0.497	298	0.503	311	0.63	530	0.26
CURVE 9		CURVE 20		CURVE 31		CURVE 42*		CURVE 53*		CURVE 64*		CURVE 75*		569	0.26
311	0.38	311	0.89	311	0.74	298	0.394	298	0.395	298	0.490	311	0.63		
CURVE 10		CURVE 21		CURVE 32		CURVE 43		CURVE 54*		CURVE 65*		CURVE 76			
311	0.47	311	0.37	311	0.84	298	0.764	298	0.393	298	0.412	311	0.66		
CURVE 11		CURVE 22*		CURVE 33		CURVE 44*		CURVE 55*		CURVE 66		CURVE 77*			
311	0.73	311	0.47	311	0.39	298	0.763	298	0.506	298	0.402	311	0.77		

* Not shown on plot

SPECIFICATION TABLE NO. 375 HEMISPHERICAL TOTAL EMITTANCE OF [IRON + MANGANESE + ΣX_i] ALLOYS

Curve No.	Ref. No.	Year	Temperature Range, K	Reported Error, %	Composition (weight percent), Specifications and Remarks
1	22	1958	589-1089	≤ 2	Mild Steel, AISI C1020; nominal composition: 0.30-0.60 Mn, 0.18-0.23 C, 0.050 max S, 0.040 max P, Fe balance; stably oxidized at 1089 K in quiescent air.

DATA TABLE NO. 375 HEMISPHERICAL TOTAL EMITTANCE OF [IRON + MANGANESE + ΣX_i] ALLOYS

[Temperature, T, K; Emittance, ϵ]

T	ϵ
CURVE 1*	
589	0.830
644	0.855
700	0.895
756	0.900
811	0.905
867	0.925
922	0.930
978	0.925
1033	0.905
1089	0.885

* Not shown on plot

SPECIFICATION TABLE NO. 376 NORMAL TOTAL EMITTANCE OF [IRON + MANGANESE + ΣX_i] ALLOYS

Curve No.	Ref. No.	Year	Temperature Range, K	Geometry θ'	Reported Error, %	Composition (weight percent), Specifications and Remarks
1	22	1958	589-1089	~0°	≤ 2	Mild Steel, AISI C1020; nominal composition: 0.30-0.60 Mn, 0.18-0.23 C, 0.050 max S, 0.040 max P, Fe balance; stably oxidized at 1089 K in quiescent air.

DATA TABLE NO. 376 NORMAL TOTAL EMITTANCE OF [IRON + MANGANESE + ΣX_i] ALLOYS

[Temperature, T, K; Emittance, ϵ]

T	ϵ
CURVE 1*	
589	0.850
644	0.875
700	0.905
756	0.910
811	0.920
867	0.940
922	0.945
978	0.940
1033	0.920
1089	0.890

* Not shown on plot

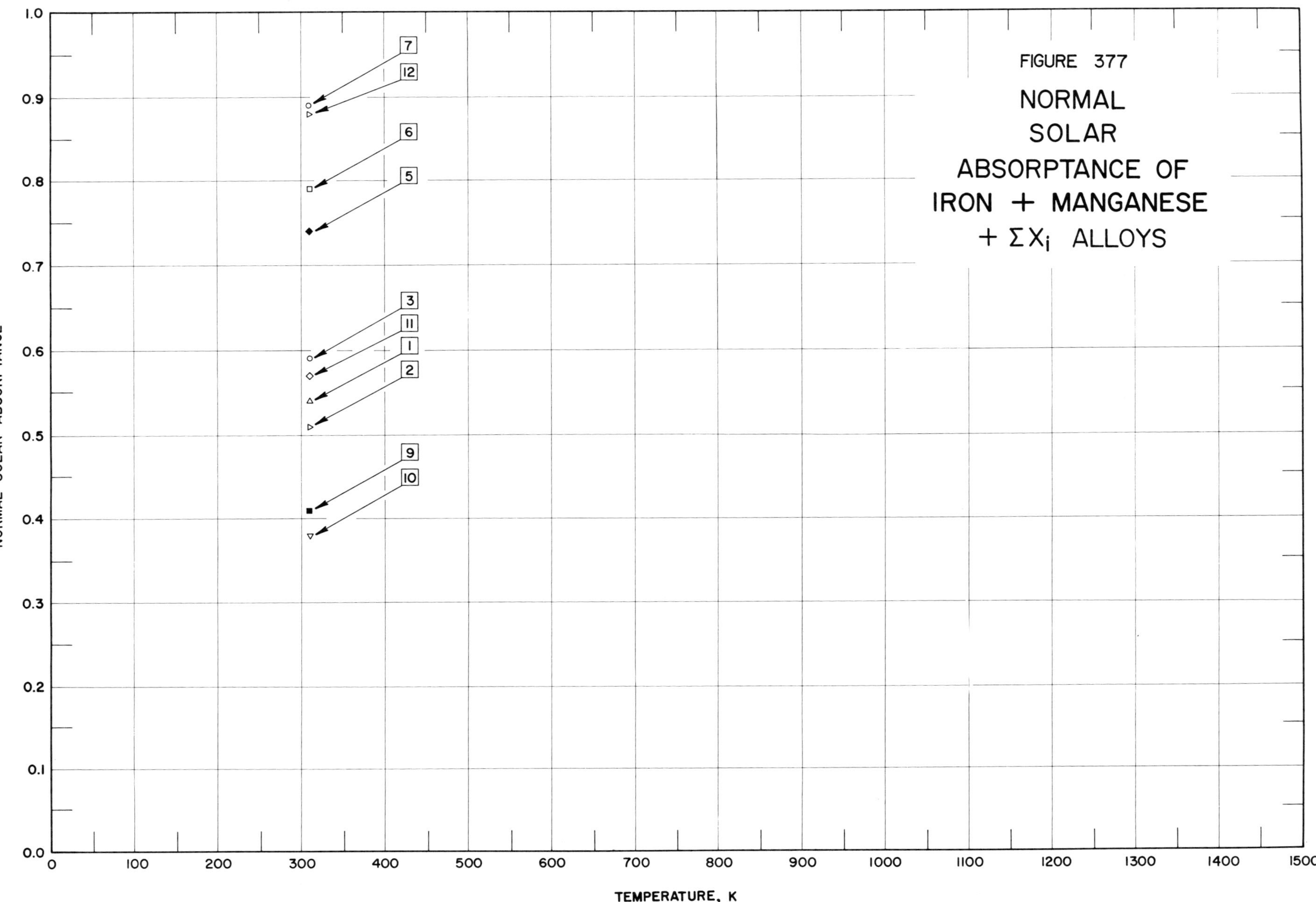
FIGURE 377
NORMAL
SOLAR
ABSORPTANCE OF
IRON + MANGANESE
+ ΣXi ALLOYS
NORMAL SOLAR ABSORPTANCE
TEMPERATURE, K
1.0
0.9
0.8
0.7
0.6
0.5
0.4
0.3
0.2
0.1
0.0
0
100
200
300
400
500
600
700
800
900
1000
1100
1200
1300
1400
1500
7
12
6
5
3
11
1
2
9
10

SPECIFICATION TABLE NO. 377 NORMAL SOLAR ABSORPTANCE OF [IRON + MANGANESE + ΣX_i] ALLOYS

Curve No.	Ref. No.	Year	Temperature Range, K	Geometry θ	Reported Error, %	Composition (weight percent), Specifications and Remarks
1	160	1954	311	$\sim 0°$		Mild Steel; nominal composition: 0.25-0.50 Mn, 0.05-0.10 C, ≤ 0.05 S, ≤ 0.04 P, Fe balance; heated to 316 K; clean and smooth surface; measured in air at sea level.
2	160	1954	311	$\sim 0°$		Above specimen and conditions except reheated to 608 K.
3	160	1954	311	$\sim 0°$		Above specimen and conditions except reheated to 954 K.
4	160	1954	311	$\sim 0°$		Above specimen and conditions except reheated to 1264 K.
5	160	1954	311	$\sim 0°$		Mild Steel; nominal composition: 0.25-0.50 Mn, 0.05-0.10 C, ≤ 0.05 S, ≤ 0.04 P, Fe balance; heated to 322 K; cleaned with methyl alcohol; measured in air at sea level.
6	160	1954	311	$\sim 0°$		Above specimen and conditions except reheated to 625 K.
7	160	1954	311	$\sim 0°$		Above specimen and conditions except reheated to 954 K.
8	160	1954	311	$\sim 0°$		Above specimen and conditions except reheated to 1339 K.
9	160	1954	311	$\sim 0°$		Mild Steel; nominal composition: 0.25-0.50 Mn, 0.05-0.10 C, ≤ 0.05 S, ≤ 0.04 P, Fe balance; heated to 313 K; polished; surface free from scratches; measured in air at sea level.
10	160	1954	311	$\sim 0°$		Above specimen and conditions except reheated to 610 K.
11	160	1954	311	$\sim 0°$		Above specimen and conditions except reheated to 936 K.
12	160	1954	311	$\sim 0°$		Above specimen and conditions except reheated to 1264 K.

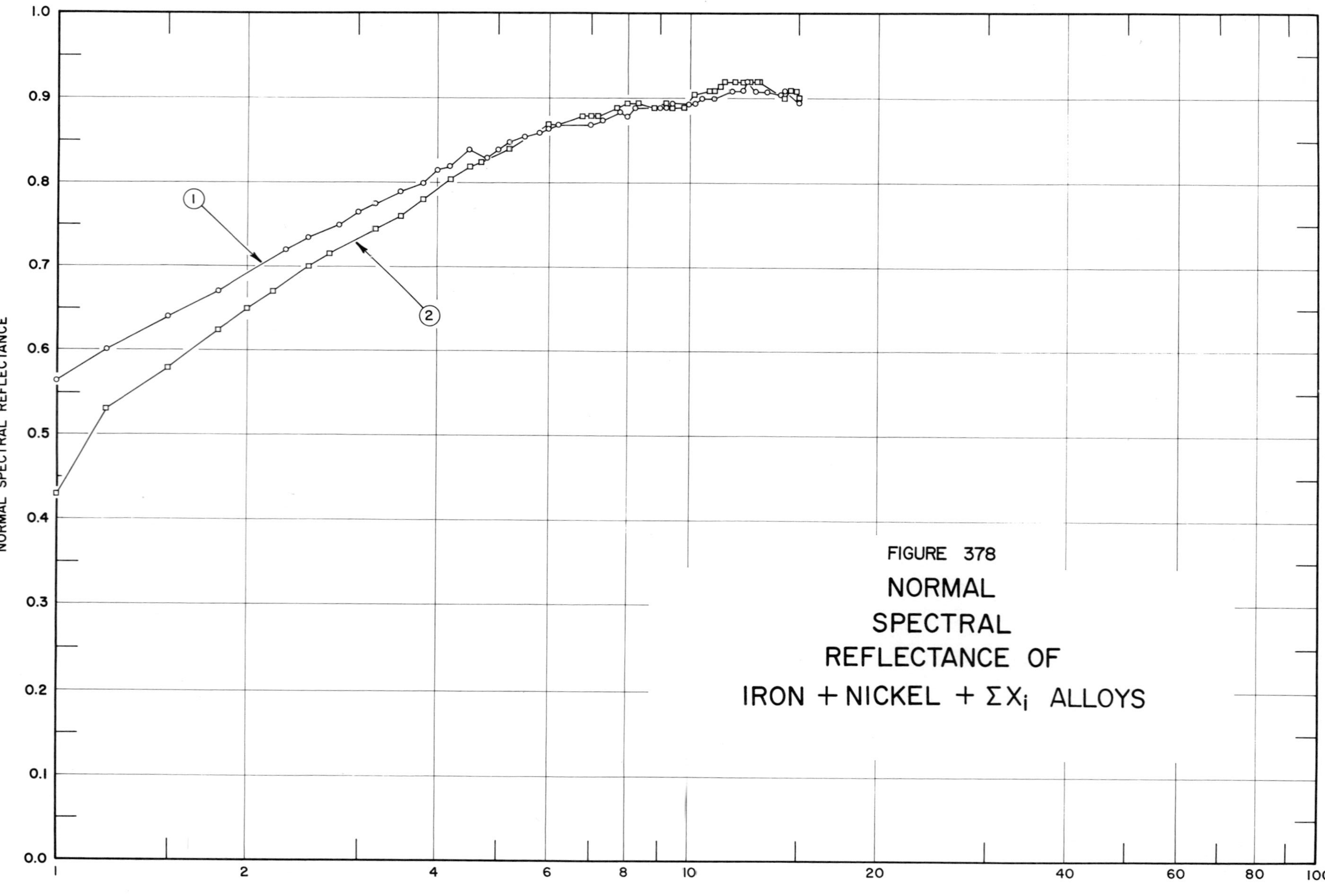

FIGURE 378 NORMAL SPECTRAL REFLECTANCE OF IRON + NICKEL + ΣX_i ALLOYS

SPECIFICATION TABLE NO. 378 NORMAL SPECTRAL REFLECTANCE OF [IRON + NICKEL + ΣX_i] ALLOYS

Curve No.	Ref. No.	Year	Temperature K	Wavelength Range, μ	Geometry θ	θ'	ω	Reported Error, %	Composition (weight percent), Specifications and Remarks
1	126	1953	298	1.0-15.0	5°		2π	± 2	Kovar; 29 Ni, 17 Co, 0.2 Mn, Fe balance; buffed and polished; data extracted from smooth curve; converted from R(2π, 5°).
2	126	1953	298	1.0-15.0	5°		2π	± 2	Different sample, same as above specimen and conditions.

DATA TABLE NO. 378 NORMAL SPECTRAL REFLECTANCE OF [IRON + NICKEL + ΣX_i] ALLOYS

[Wavelength, λ, μ; Reflectance, ρ; Temperature, T, K]

λ	ρ	λ	ρ
CURVE 1 T = 298		CURVE 2 T = 298	
1.0	0.565	1.0	0.430
1.2	0.600	1.2	0.530
1.5	0.640	1.5	0.580
1.8	0.670	1.8	0.625
2.3	0.720	2.0	0.650
2.5	0.735	2.2	0.670
2.8	0.750	2.5	0.700
3.0	0.765	2.7	0.715
3.2	0.775	3.2	0.745
3.5	0.790	3.5	0.760
3.8	0.800	3.8	0.780
4.0	0.815	4.2	0.805
4.2	0.820	4.5	0.820
4.5	0.840	4.7	0.825
4.8	0.830	5.2	0.840
5.0	0.840	6.0	0.870
5.2	0.850	6.2	0.870*
5.5	0.855	6.8	0.880
5.8	0.860	7.0	0.880
6.0	0.865	7.2	0.880
6.2	0.870	7.7	0.890
7.0	0.870	8.0	0.895
7.3	0.875	8.3	0.895
7.8	0.885	8.8	0.890
8.0	0.880	9.2	0.895
8.2	0.890	9.4	0.890
9.0	0.890	9.8	0.890
9.2	0.890	10.2	0.905
9.4	0.895	10.8	0.910
10.0	0.895	11.0	0.910
10.2	0.895	11.2	0.915
10.5	0.900	11.4	0.920
11.0	0.900	11.8	0.920
11.7	0.910	12.2	0.920
12.2	0.910	12.5	0.920
12.4	0.920	12.8	0.920
12.8	0.910	13.0	0.920
13.4	0.910	14.2	0.900
14.0	0.905	14.5	0.910
14.2	0.910	14.8	0.910
15.0	0.895	15.0	0.900

* Not shown on plot

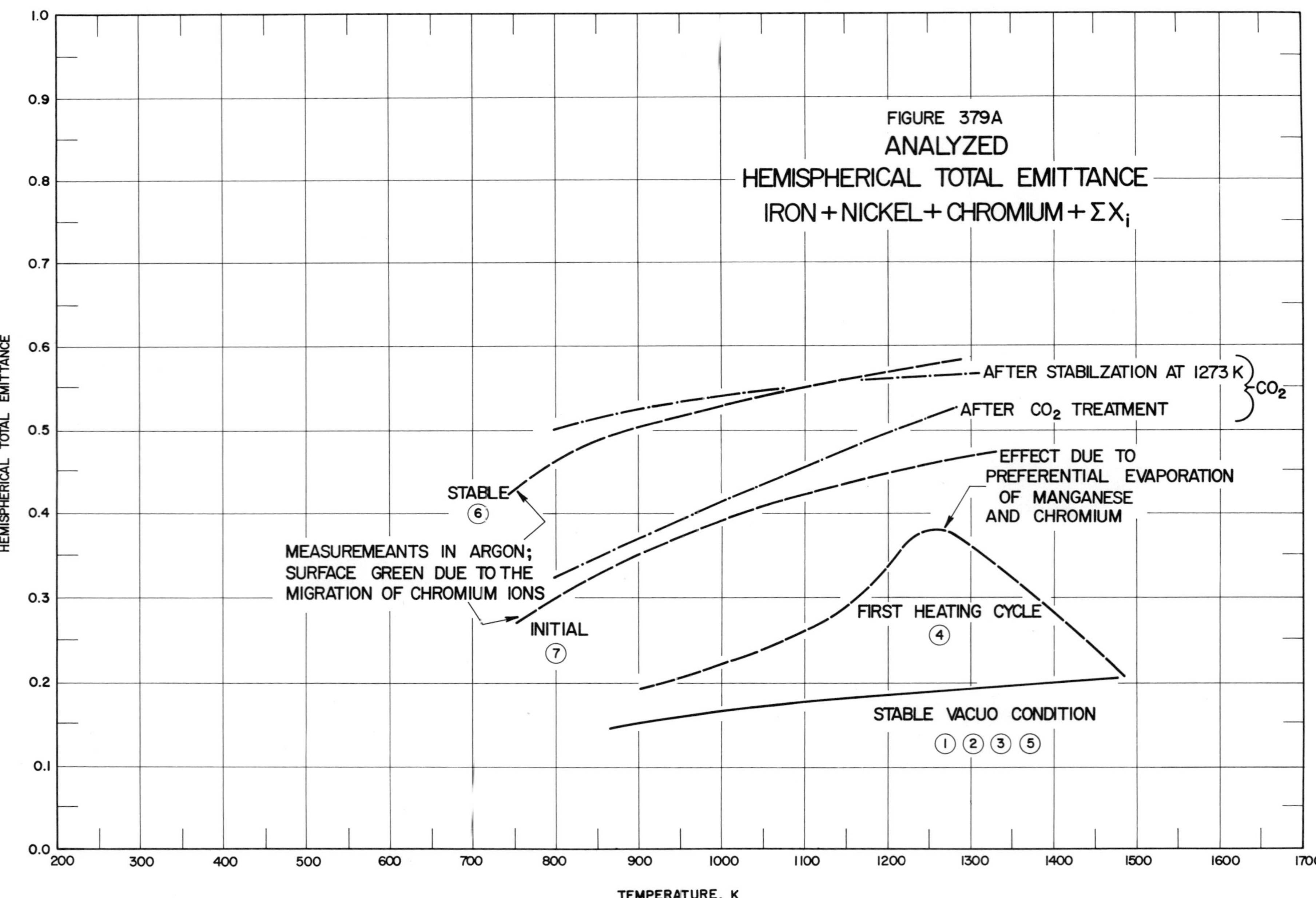
FIGURE 379A
ANALYZED
HEMISPHERICAL TOTAL EMITTANCE
IRON + NICKEL + CHROMIUM + ΣX_i
AFTER STABILZATION AT 1273 K
CO_2
AFTER CO_2 TREATMENT
EFFECT DUE TO PREFERENTIAL EVAPORATION OF MANGANESE AND CHROMIUM
STABLE
6
MEASUREMEANTS IN ARGON; SURFACE GREEN DUE TO THE MIGRATION OF CHROMIUM IONS
INITIAL
7
FIRST HEATING CYCLE
4
STABLE VACUO CONDITION
1 2 3 5
HEMISPHERICAL TOTAL EMITTANCE
0.0
0.1
0.2
0.3
0.4
0.5
0.6
0.7
0.8
0.9
1.0
TEMPERATURE, K
200
300
400
500
600
700
800
900
1000
1100
1200
1300
1400
1500
1600
1700

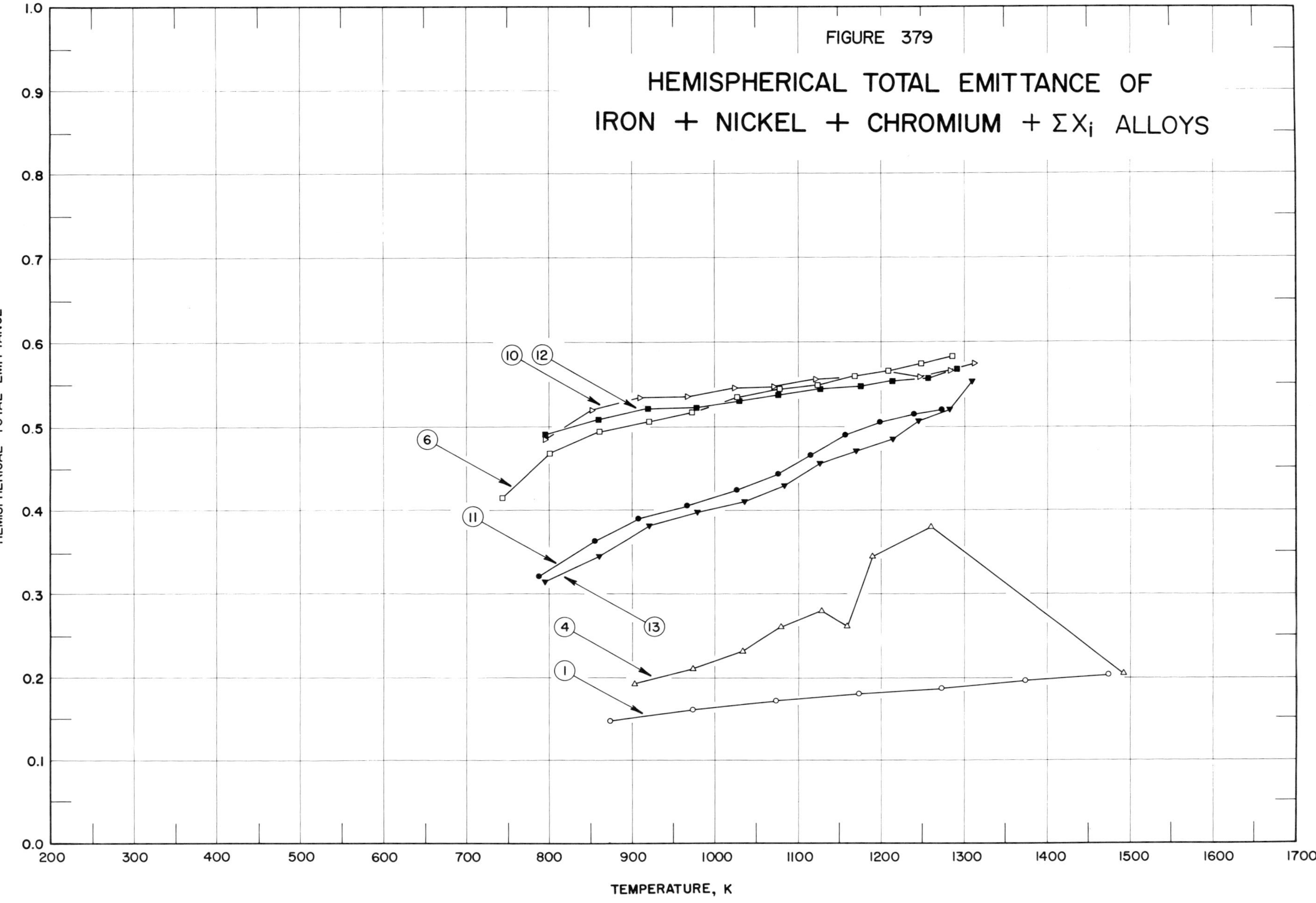

FIGURE 379

HEMISPHERICAL TOTAL EMITTANCE OF IRON + NICKEL + CHROMIUM + ΣX_i ALLOYS

SPECIFICATION TABLE NO. 379 HEMISPHERICAL TOTAL EMITTANCE OF [IRON + NICKEL + CHROMIUM + ΣX_i] ALLOYS

Curve No.	Ref. No.	Year	Temperature Range, K	Reported Error, %	Composition (weight percent), Specifications and Remarks
1	294	1964	873-1473	5	Nb-stabilized steel; 25 Ni, 20 Cr, 0.7 Mn, traces of Cu and Nb, Fe balance; annealed in hydrogen, cleaned in trichlorethylene; measured in vacuum (10^{-4} mm Hg); data taken after initial cycle; [Authors' designation: Sample A].
2	294	1964	873-1473	5	Above specimen and conditions; [Authors' designation: Sample B].
3	294	1964	873-1473	5	Above specimen and conditions; [Authors' designation: Sample C].
4	294	1964	902-1492	5	Curve 1 specimen and conditions; first cycle.
5	294	1964	946-1490	5	Curve 1 specimen and conditions.
6	294	1964	743-1287	5	Curve 1 specimen and conditions except measured in argon (0.01 atm pressure); stable condition; [Authors' designation: Sample D].
7	294	1964	753-1306	5	Above specimen and conditions; second run.
8	294	1964	1003	5	Above specimen and conditions except measured in CO_2 (760 mm Hg); measurements taken within first 22 hrs of heating the sample; [Authors' designation: Sample F].
9	294	1964	873	5	Above specimen and conditions; [Authors' designation: Sample E].
10	294	1964	795-1312	5	Above specimen and conditions except measured in CO_2 (760 mm Hg); [Authors' designation: Sample F]
11	294	1964	788-1275	5	Curve 10 specimen and conditions except measured after stabilization.
12	294	1964	796-1292	5	Different sample, same as curve 10 specimen and conditions; [Authors' designation: Sample E].
13	294	1964	795-1311	5	Curve 12 specimen and conditions except measured after stabilization.

DATA TABLE NO. 379 HEMISPHERICAL TOTAL EMITTANCE OF [IRON + NICKEL + CHROMIUM + ΣX_i] ALLOYS

[Temperature, T, K; Emittance, ∈]

T	∈
CURVE 1	
873	0.148
973	0.161
1073	0.171
1173	0.180
1273	0.188
1373	0.196
1473	0.202
CURVE 2*	
873	0.146
973	0.160
1073	0.170
1173	0.180
1273	0.189
1373	0.197
1473	0.204
CURVE 3*	
873	0.143
973	0.157
1073	0.170
1173	0.181
1273	0.192
1373	0.200
1473	0.208
CURVE 4	
902	0.192
973	0.211
1033	0.231
1079	0.260
1128	0.281
1159	0.261
1190	0.347
1260	0.381
1492	0.204
CURVE 5*	
946	0.160
1032	0.168
1110	0.176
1184	0.181
1252	0.188
1320	0.192
1381	0.196
1438	0.197
1490	0.201
CURVE 6	
743	0.414
801	0.468
861	0.494
921	0.507
973	0.518
1027	0.535
1078	0.543
1124	0.549
1169	0.560
1209	0.566
1249	0.573
1287	0.582
CURVE 7*	
753	0.267
855	0.332
979	0.375
1088	0.416
1182	0.440
1263	0.461
1306	0.476
CURVE 8*	
1003	0.271
1003	0.414
1003	0.441
1003	0.430
1003	0.433
1003	0.451
1003	0.451
1003	0.451
CURVE 9*	
873	0.243
873	0.298
873	0.331
873	0.353
873	0.375
873	0.381
873	0.399
873	0.395
CURVE 10	
795	0.485
851	0.521
910	0.534
967	0.537
1023	0.547
1071	0.548
1121	0.556
1166	0.559*
1210	0.566*
1248	0.559
1285	0.567
1312	0.574
CURVE 11	
788	0.321
854	0.363
908	0.390
967	0.404
1027	0.423
1076	0.444
1114	0.467
1158	0.490
1199	0.505
1241	0.514
1275	0.521
CURVE 12	
796	0.490
859	0.509
919	0.522
978	0.522
1030	0.530
CURVE 12 (cont.)	
1077	0.538
1127	0.546
1175	0.548
1214	0.553
1258	0.558
1292	0.568
CURVE 13	
795	0.312
860	0.345
921	0.381
979	0.398
1037	0.410
1083	0.430
1128	0.454
1171	0.470
1215	0.483
1246	0.507
1282	0.520
1311	0.552

* Not shown on plot

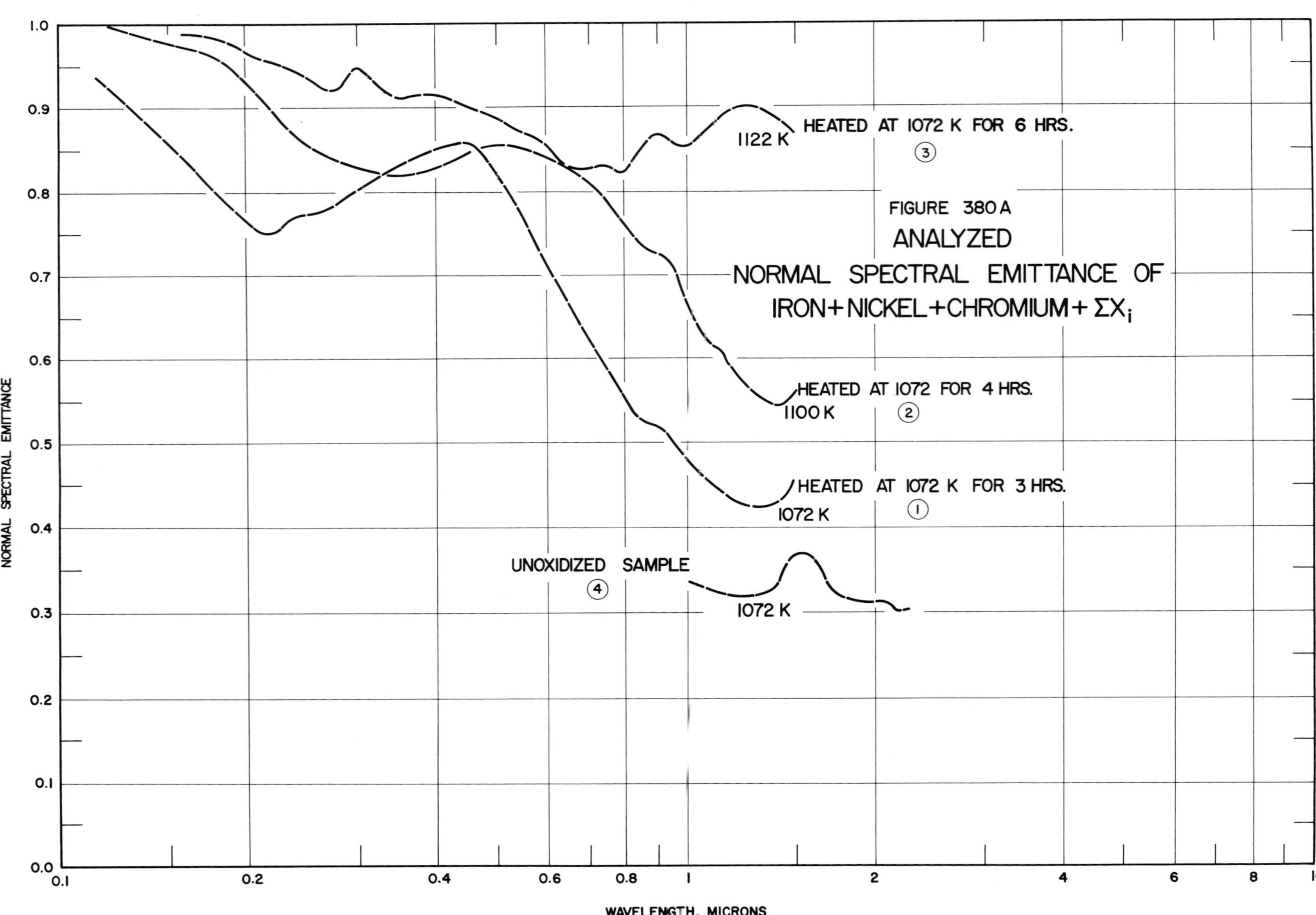

FIGURE 380A
ANALYZED
NORMAL SPECTRAL EMITTANCE OF IRON+NICKEL+CHROMIUM+ΣX_i

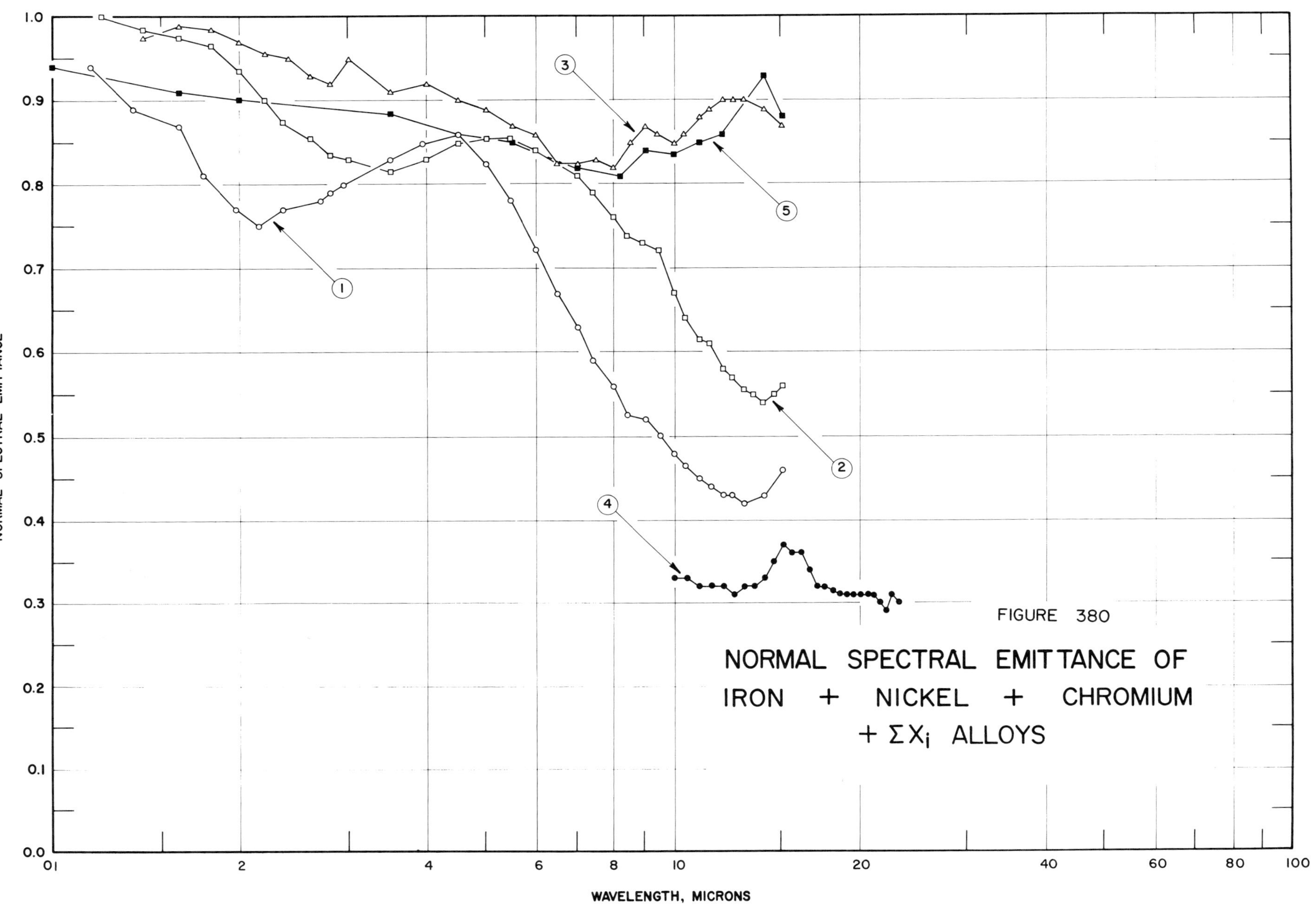

FIGURE 380

NORMAL SPECTRAL EMITTANCE OF IRON + NICKEL + CHROMIUM + ΣX_i ALLOYS

SPECIFICATION TABLE NO. 380 NORMAL SPECTRAL EMITTANCE OF [IRON + NICKEL + CHROMIUM + ΣX_i] ALLOYS

Curve No.	Ref. No.	Year	Temperature K	Wavelength Range, μ	Geometry θ'	Reported Error, %	Composition (weight percent), Specifications and Remarks
1	65	1962	1072	1.15-15.00	~0°		Stainless steel A286; nominal composition: 24.0-27.0 Ni, 13.5-16.0 Cr, 1.75-2.25 Ti, 1.0-2.0 Mn, 1.0-1.5 Mo, 0.1-0.5 V, 0.4-1.0 Si, 0.35 Al, 0.08 C, 0.04 P, 0.03 S, 0.001-0.01 B, Fe balance; heated at 1072 K for 2 hrs; measured in air.
2	65	1962	1100	1.20-15.00	~0°		Different sample, same as curve 1 specimen and conditions except heated at 1072 K for 4 hrs.
3	65	1962	1122	1.40-15.00	~0°		Different sample, same as curve 1 specimen and conditions except heated at 1072 K for 6 hrs.
4	65	1962	1072	10.00-23.00	~0°		Different sample, same as curve 1 specimen and conditions except unoxidized.
5	65	1962	1122	1.00-15.00	~0°		Different sample, same as curve 1 specimen and conditions except oxidized.

DATA TABLE NO. 380 NORMAL SPECTRAL EMITTANCE OF [IRON + NICKEL + CHROMIUM + ΣX_i] ALLOYS

[Wavelength, λ, μ; Emittance, ϵ; Temperature, T, K]

λ	ϵ
CURVE 1	**T = 1072**
1.15	0.940
1.35	0.890
1.60	0.870
1.75	0.810
1.97	0.770
2.15	0.750
2.35	0.770
2.60	0.780
2.80	0.790
2.95	0.800
3.50	0.830
3.95	0.850
4.50	0.860
5.00	0.825
5.45	0.780
6.00	0.720
6.50	0.670
7.00	0.630
7.40	0.590
8.00	0.560
8.40	0.525
9.00	0.520
9.50	0.500
10.00	0.480
10.40	0.465
11.00	0.450
11.50	0.440
12.00	0.430
12.40	0.430
13.00	0.420
14.00	0.430
15.00	0.460
CURVE 2	**T = 1100**
1.20	1.000
1.40	0.985
1.60	0.975
1.80	0.965
2.00	0.935
2.20	0.900
2.35	0.875
CURVE 2 (cont.)	
2.60	0.855
2.80	0.835
3.00	0.830
3.50	0.815
4.00	0.830
4.50	0.850
5.00	0.855
5.45	0.855
6.00	0.840
6.50	0.840
7.00	0.810
7.40	0.790
8.00	0.760
8.40	0.738
8.90	0.730
9.45	0.720
10.00	0.670
10.40	0.640
11.00	0.615
11.40	0.610
12.00	0.580
12.45	0.570
13.00	0.555
13.45	0.550
13.90	0.540
14.50	0.550
15.00	0.560
CURVE 3	**T = 1122**
1.40	0.975
1.60	0.990
1.80	0.985
2.00	0.970
2.20	0.955
2.40	0.950
2.60	0.930
2.80	0.920
3.00	0.950
3.50	0.910
4.00	0.920
4.50	0.900
5.00	0.890
CURVE 3 (cont.)	
5.50	0.870
6.00	0.860
6.50	0.825
7.00	0.825
7.50	0.830
8.00	0.820
8.50	0.850
9.00	0.870
9.40	0.860
10.00	0.850
10.40	0.860
11.00	0.880
11.40	0.890
12.00	0.900
12.50	0.900
13.00	0.900
14.00	0.890
15.00	0.870
CURVE 4	**T = 1072**
10.00	0.330
10.50	0.330
11.00	0.320
11.50	0.320
12.00	0.320
12.50	0.310
13.00	0.320
13.50	0.320
14.00	0.330
14.50	0.350
15.00	0.370
15.50	0.360
16.00	0.360
16.50	0.340
17.00	0.320
17.50	0.320
18.00	0.315
18.50	0.310
19.00	0.310
19.50	0.310
20.00	0.310
20.50	0.310
CURVE 4 (cont.)	
21.00	0.310
21.50	0.300
22.00	0.290
22.40	0.310
23.00	0.300
CURVE 5	**T = 1122**
1.00	0.940
1.60	0.910
2.00	0.900
3.50	0.885
4.50	0.860*
5.50	0.850
7.00	0.820
8.20	0.810
9.00	0.840
10.00	0.836
11.00	0.850
12.00	0.860
14.00	0.930
15.00	0.880

* Not shown on plot

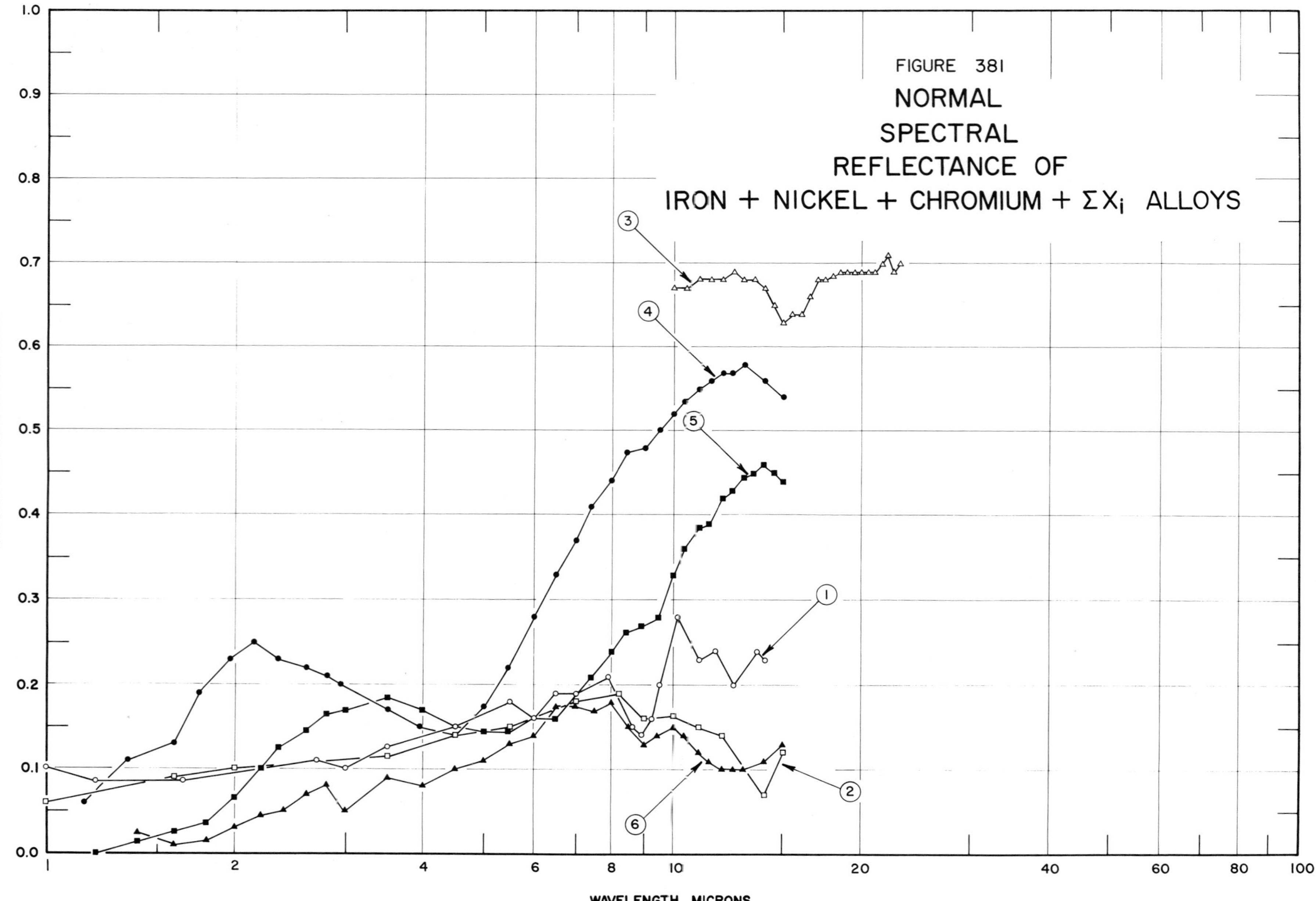
FIGURE 381
NORMAL
SPECTRAL
REFLECTANCE OF
IRON + NICKEL + CHROMIUM + ΣXi ALLOYS
NORMAL SPECTRAL REFLECTANCE
WAVELENGTH, MICRONS
0.0
0.1
0.2
0.3
0.4
0.5
0.6
0.7
0.8
0.9
1.0
1
2
4
6
8
10
20
40
60
80
100
1
2
3
4
5
6

SPECIFICATION TABLE NO. 381 NORMAL SPECTRAL REFLECTANCE OF [IRON + NICKEL + CHROMIUM + ΣX_i] ALLOYS

Curve No.	Ref. No.	Year	Temperature K	Wavelength Range, μ	Geometry θ	Geometry θ'	Geometry ω'	Reported Error, %	Composition (weight percent), Specifications and Remarks
1	65	1962	294	1.00-14.00	~0°		2π		Stainless steel A 286; nominal composition: 24.0-27.0 Ni, 13.5-16.0 Cr, 1.75-2.25 Ti, 1.0-2.0 Mn, 1.0-1.5 Mo, 0.1-0.5 V, 0.4-1.0 Si, 0.35 Al, 0.08 C, 0.04 P, 0.03 S, 0.001-0.01 B, Fe balance; oxidized; measured in air.
2	65	1962	1122	1.00-15.00	~0°		2π		Different sample, same as curve 1 specimen and conditions except calculated from $(1 - \epsilon)$.
3	65	1962	1072	10.00-23.00	~0°		2π		Different sample, same as curve 1 specimen and conditions except unoxidized.
4	65	1962	1072	1.15-15.00	~0°		2π		Different sample, same as curve 1 specimen and conditions except heated at 1072 K for 2 hrs; measured in air; calculated from $(1 - \epsilon)$.
5	65	1962	1100	1.20-15.00	~0°		2π		Different sample, same as curve 4 specimen and conditions except heated at 1072 K for 4 hrs.
6	65	1962	1122	1.40-15.00	~0°		2π		Different sample, same as curve 4 specimen and conditions except heated at 1072 K for 6 hrs.

DATA TABLE NO. 381 NORMAL SPECTRAL REFLECTANCE OF [IRON +NICKEL + CHROMIUM + ΣX_i] ALLOYS

[Wavelength, λ, μ; Reflectance, ρ; Temperature, T, K]

λ	ρ
CURVE 1	**T = 294**
1.00	0.100
1.20	0.086
1.65	0.086
2.70	0.110
3.00	0.100
3.50	0.126
4.50	0.150
5.50	0.180
6.00	0.160
6.50	0.190
7.00	0.190
7.90	0.210
8.60	0.150
8.90	0.140
9.20	0.160
9.50	0.200
10.20	0.280
11.00	0.230
11.70	0.240
12.50	0.200
13.60	0.240
14.00	0.230
CURVE 2	**T = 1122**
1.00	0.060
1.60	0.090
2.00	0.100
3.50	0.115
4.50	0.140
5.50	0.150
7.00	0.180
8.20	0.190
9.00	0.160
10.00	0.164
11.00	0.150
12.00	0.140
14.00	0.070
15.00	0.120
CURVE 3	**T = 1072**
10.00	0.670
10.50	0.670
11.00	0.680
11.50	0.680
12.00	0.680
12.50	0.690
13.00	0.680
13.50	0.680
14.00	0.670
14.50	0.650
15.00	0.630
15.50	0.640
16.00	0.640
16.50	0.660
17.00	0.680
17.50	0.680
18.00	0.685
18.50	0.690
19.00	0.690
19.50	0.690
20.00	0.690
20.50	0.690
21.00	0.690
21.50	0.700
22.00	0.710
22.40	0.690
23.00	0.700
CURVE 4	**T = 1072**
1.15	0.060
1.35	0.110
1.60	0.130
1.75	0.190
1.97	0.230
2.15	0.250
2.35	0.230
2.60	0.220
2.80	0.210
2.95	0.200
3.50	0.170
3.95	0.150
4.50	0.140*
CURVE 4 (cont.)	
5.00	0.175
5.45	0.220
6.00	0.280
6.50	0.330
7.00	0.370
7.40	0.410
8.00	0.440
8.40	0.475
9.00	0.480
9.50	0.500
10.00	0.520
10.40	0.535
11.00	0.550
11.50	0.560
12.00	0.570
12.40	0.570
13.00	0.580
14.00	0.560
15.00	0.540
CURVE 5	**T = 1100**
1.20	0.000
1.40	0.015
1.60	0.025
1.80	0.035
2.00	0.065
2.20	0.100
2.35	0.125
2.60	0.145
2.30	0.165
3.00	0.170
3.50	0.185
4.00	0.170
4.50	0.150*
5.00	0.145
5.45	0.145
6.00	0.160*
6.50	0.160
7.00	0.190*
7.40	0.210
8.00	0.240
8.40	0.262
8.90	0.270
CURVE 5 (cont.)	
9.45	0.280
10.00	0.330
10.40	0.360
11.00	0.385
11.40	0.390
12.00	0.420
12.45	0.430
13.00	0.445
13.45	0.450
13.90	0.460
14.50	0.450
15.00	0.440
CURVE 6	**T = 1122**
1.40	0.025
1.60	0.010
1.80	0.015
2.00	0.030
2.20	0.045
2.40	0.050
2.60	0.070
2.80	0.080
3.00	0.050
3.50	0.090
4.00	0.080
4.50	0.100
5.00	0.110
5.50	0.130
6.00	0.140
6.50	0.175
7.00	0.175
7.50	0.170
8.00	0.180
8.50	0.150
9.00	0.130
9.40	0.140
10.00	0.150
10.40	0.140
11.00	0.120
11.40	0.110
12.00	0.100
12.50	0.100
13.00	0.100
CURVE 6 (cont.)	
14.00	0.110
15.00	0.130

* Not shown on plot

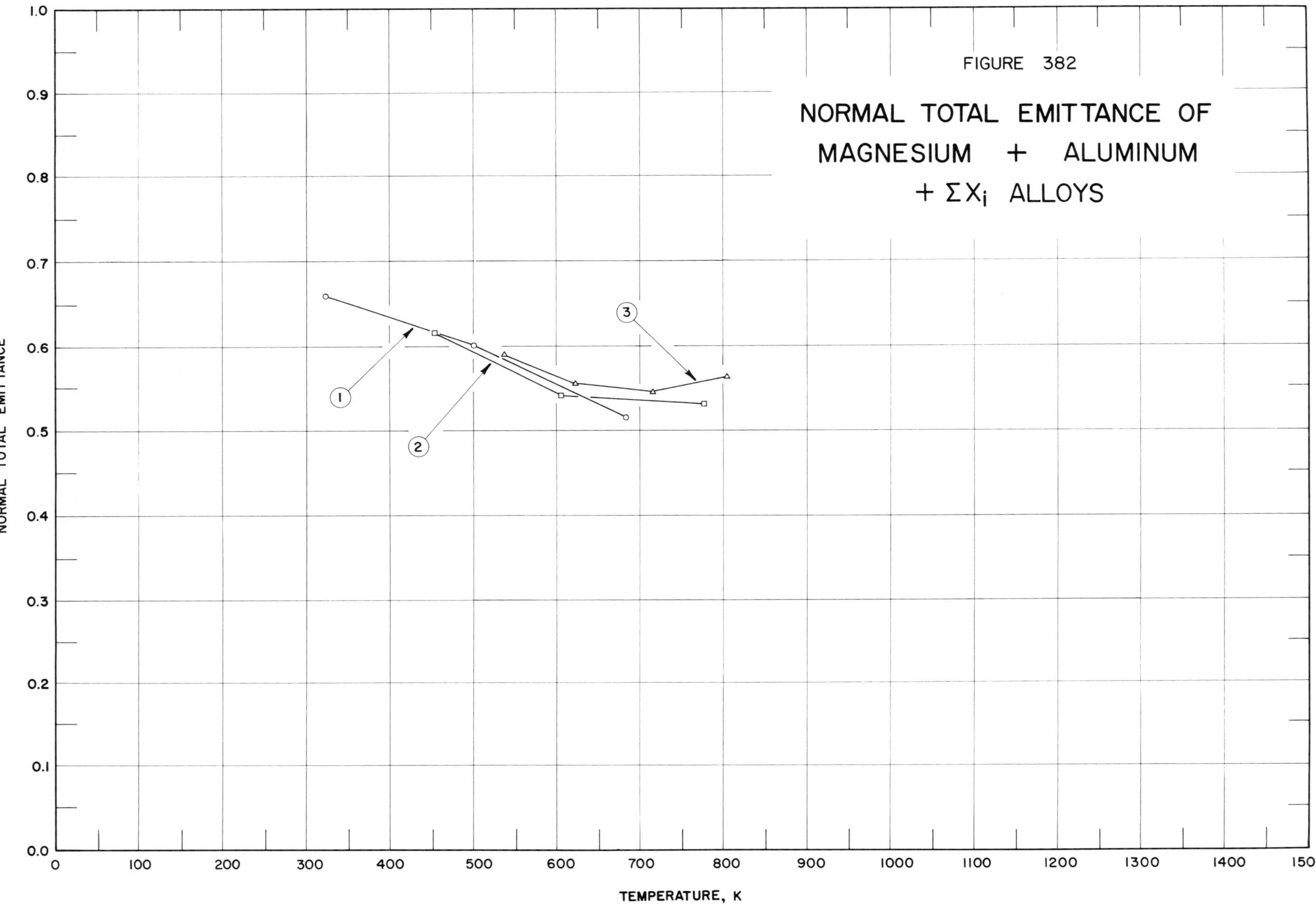
FIGURE 382
NORMAL TOTAL EMITTANCE OF
MAGNESIUM + ALUMINUM
+ ΣXi ALLOYS
NORMAL TOTAL EMITTANCE
TEMPERATURE, K
1.0
0.9
0.8
0.7
0.6
0.5
0.4
0.3
0.2
0.1
0.0
0
100
200
300
400
500
600
700
800
900
1000
1100
1200
1300
1400
1500
1
2
3

SPECIFICATION TABLE NO. 382 NORMAL TOTAL EMITTANCE OF [MAGNESIUM + ALUMINUM + ΣX_i] ALLOYS

Curve No.	Ref. No.	Year	Temperature Range, K	Geometry θ'	Reported Error, %	Composition (weight percent), Specifications and Remarks
1	34	1957	322–683	~0°	± 10	Magnesium alloy AZ-31; nominal composition: 3 Al, 1 Zn, 0.2 Mn, Mg balance; strip (0.005 in. thick); as received; measured in vacuum (5 x 10^{-4} mm Hg); increasing temp, cycle 1.
2	34	1957	455–778	~0°	± 10	Above specimen and conditions; cycle 2.
3	34	1957	539–805	~0°	± 10	Above specimen and conditions; cycle 3.

DATA TABLE NO. 382 NORMAL TOTAL EMITTANCE OF [MAGNESIUM + ALUMINUM + ΣX_i] ALLOYS

[Temperature, T, K; Emittance, ϵ]

T	ϵ
CURVE 1	
322	0.660
500	0.600
683	0.515
CURVE 2	
455	0.615
605	0.540
778	0.530
CURVE 3	
539	0.590
622	0.555
716	0.545
805	0.565

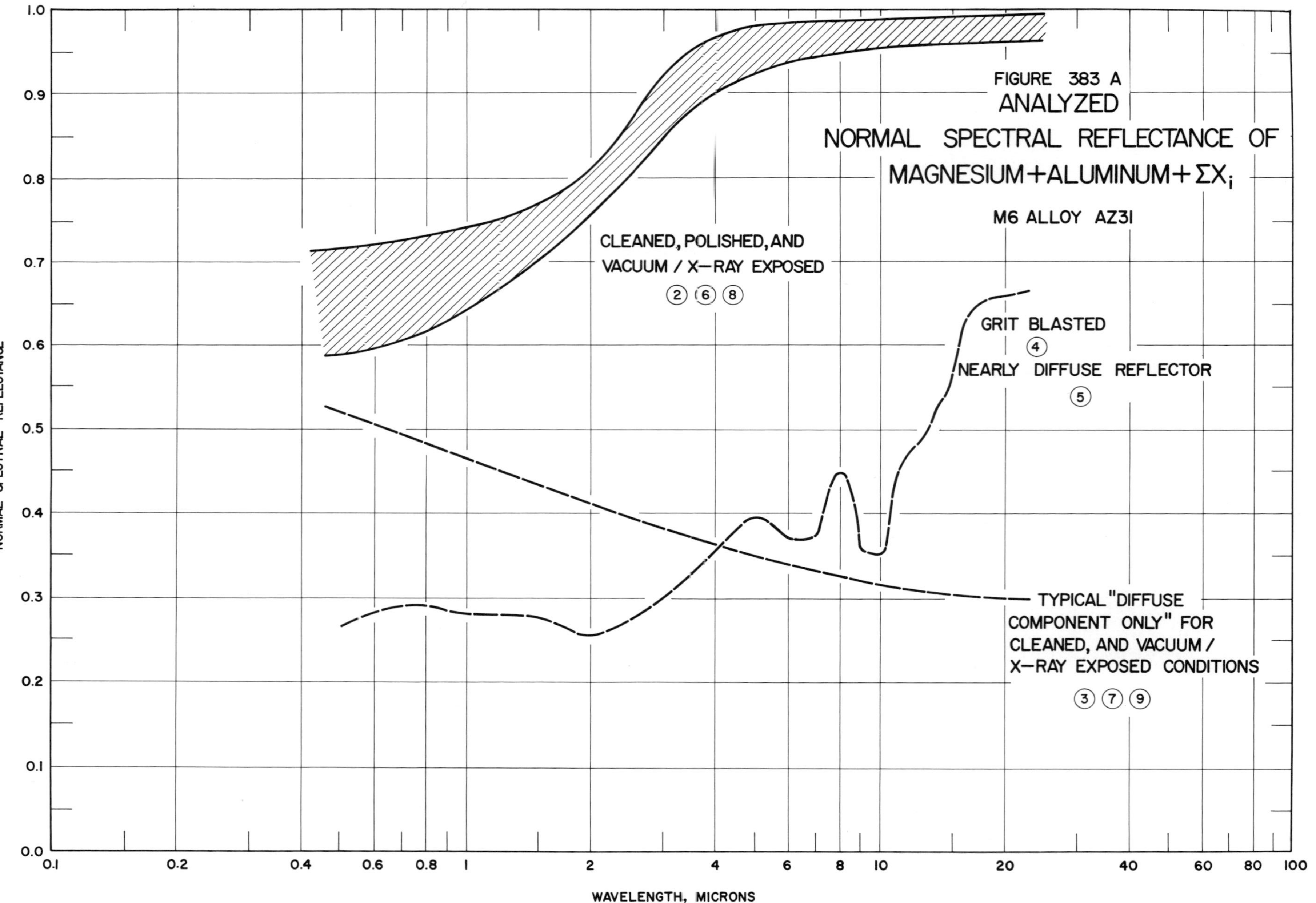
FIGURE 383 A
ANALYZED
NORMAL SPECTRAL REFLECTANCE OF
MAGNESIUM+ALUMINUM+ΣX$_i$
M6 ALLOY AZ31
CLEANED, POLISHED, AND
VACUUM / X-RAY EXPOSED
② ⑥ ⑧
GRIT BLASTED
④
NEARLY DIFFUSE REFLECTOR
⑤
TYPICAL "DIFFUSE
COMPONENT ONLY" FOR
CLEANED, AND VACUUM /
X-RAY EXPOSED CONDITIONS
③ ⑦ ⑨
NORMAL SPECTRAL REFLECTANCE
WAVELENGTH, MICRONS
1.0
0.9
0.8
0.7
0.6
0.5
0.4
0.3
0.2
0.1
0.0
0.1
0.2
0.4
0.6
0.8
1
2
4
6
8
10
20
40
60
80
100

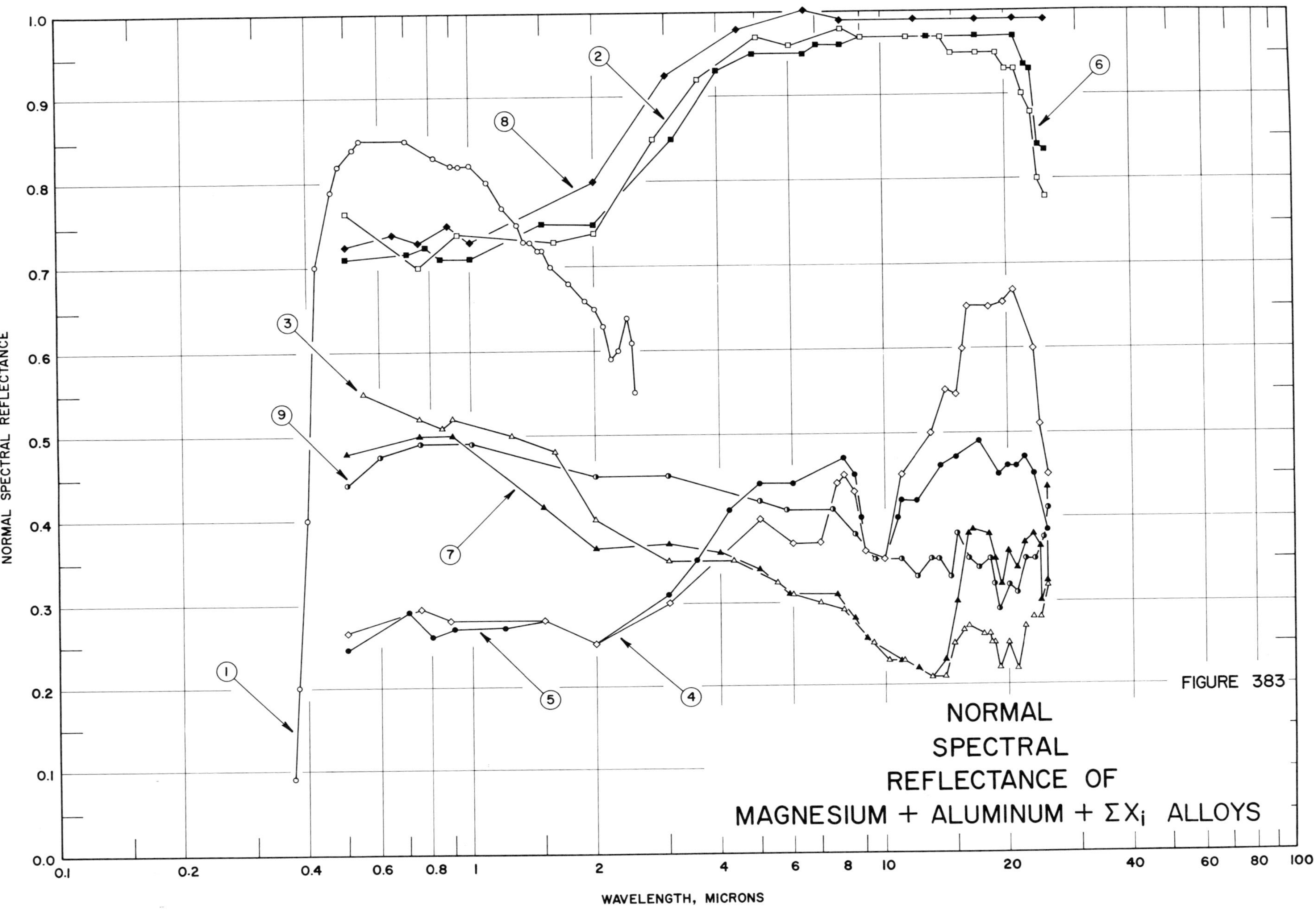
1.0
0.9
0.8
0.7
0.6
0.5
0.4
0.3
0.2
0.1
0.0
NORMAL SPECTRAL REFLECTANCE
0.1
0.2
0.4
0.6
0.8
1
2
4
6
8
10
20
40
60
80
100
WAVELENGTH, MICRONS
1
2
3
4
5
6
7
8
9
FIGURE 383
NORMAL
SPECTRAL
REFLECTANCE OF
MAGNESIUM + ALUMINUM + ΣXi ALLOYS

SPECIFICATION TABLE NO. 383 NORMAL SPECTRAL REFLECTANCE OF [MAGNESIUM + ALUMINUM + ΣX_i] ALLOYS

Curve No.	Ref. No.	Year	Temperature K	Wavelength Range, μ	Geometry θ	θ'	ω	Reported Error, %	Composition (weight percent), Specifications and Remarks
1	34	1957	298	0.37-2.50	9°		2π	±4	Magnesium alloy AZ 31; nominal composition: 3 Al, 1 Zn, 0.2 Mn, Mg balance; as received; data extracted from smooth curve; magnesium carbonate reference.
2	153	1962	~322	0.50-25.00	~0°		2π	< 2	Magnesium alloy AZ 31, QQ-M-44; nominal composition: 3 Al, 1 Zn, 0.2 Mn, Mg balance; as received; cleaned; data extracted from smooth curve; hohlraum at 1273 K; converted from R (2π, 0°).
3	153	1962	~322	0.55-25.00	~0°		2π	< 2	Above specimen and conditions; diffuse component only.
4	153	1962	~322	0.50-25.00	~0°		2π	< 2	Different sample, same as curve 2 specimen and conditions except grit blasted using 60 grit silicon carbide with air pressure of 110 to 120 psi for 30 to 45 sec.
5	153	1962	~322	0.50-25.00	~0°		2π	< 2	Above specimen and conditions; diffuse component only.
6	153	1962	~322	0.50-25.00	~0°		2π	< 2	Different sample, same as curve 2 specimen and conditions except exposed to vacuum (< 4×10^{-8} mm Hg) for 24 hrs.
7	153	1962	~322	0.50-25.00	~0°		2π	< 2	Above specimen and conditions; diffuse component only.
8	153	1962	~322	0.50-25.00	~0°		2π	< 2	Different sample, same as curve 2 specimen and conditions except X-ray exposed in vacuum (4×10^{-8} mm Hg) for 24 hrs.
9	153	1962	~322	0.50-25.00	~0°		2π	< 2	Above specimen and conditions; diffuse component only.
10	122	1961	298	0.4-20	~0°		2π		Magnesium AZ 31 B; nominal composition: 3 Al, 1 Zn, 0.2 Mn, Mg balance; mechanically polished; measured in vacuum (~10^{-6} mm Hg); [Author's designation: Specimen No. 156].

DATA TABLE NO. 383 NORMAL SPECTRAL REFLECTANCE OF [MAGNESIUM + ALUMINUM + ΣX_i] ALLOYS

[Wavelength, λ, μ; Reflectance, ρ; Temperature, T, K]

CURVE 1, T = 298

λ	ρ
0.37	0.09
0.38	0.20
0.40	0.40
0.42	0.70
0.46	0.79
0.48	0.82
0.52	0.84
0.54	0.85
0.70	0.85
0.82	0.83
0.90	0.82
0.94	0.82
1.00	0.82
1.10	0.80
1.20	0.77
1.30	0.75
1.35	0.73
1.40	0.73
1.47	0.72
1.50	0.72
1.58	0.70
1.72	0.68
1.90	0.66
2.00	0.65
2.10	0.63
2.20	0.59
2.30	0.60
2.40	0.64
2.46	0.61
2.50	0.55

CURVE 2, T = ~322

λ	ρ
0.50	0.765
0.75	0.700
0.93	0.740
1.60	0.730
2.00	0.740
2.80	0.850
3.60	0.920
5.00	0.970
6.00	0.960
8.00	0.980
9.00	0.970
11.50	0.970
14.00	0.970
14.90	0.950
17.00	0.950
19.00	0.950
20.00	0.930
21.00	0.930
22.00	0.900
23.00	0.880
24.00	0.800
25.00	0.780

CURVE 3, T = ~322

λ	ρ
0.55	0.550
0.75	0.520
0.85	0.510
0.90	0.520
1.25	0.500
1.60	0.480
2.00	0.400
3.00	0.350
4.30	0.350
5.50	0.325
6.00	0.310
7.00	0.300
8.00	0.290
9.20	0.250
10.20	0.230
11.10	0.230
13.00	0.210
14.00	0.210
14.80	0.250
15.50	0.265
16.00	0.270
17.40	0.260
18.00	0.260
18.50	0.250
19.00	0.220
20.00	0.250
21.00	0.220
22.00	0.270
23.00	0.280
24.00	0.280
25.00	0.320

CURVE 4, T = ~322

λ	ρ
0.50	0.265
0.75	0.290
0.88	0.280
1.50	0.280
2.00	0.250
3.00	0.300
5.00	0.400
6.00	0.370
7.00	0.370
7.70	0.440
8.00	0.450
8.50	0.430
9.00	0.360
10.00	0.350
11.00	0.450
13.00	0.500
14.10	0.550
15.00	0.545
15.50	0.600
16.00	0.650
18.00	0.650
19.50	0.655
20.80	0.670
23.10	0.600
24.00	0.510
25.00	0.450

CURVE 5, T = ~322

λ	ρ
0.50	0.245
0.70	0.290
0.80	0.260
0.90	0.270
1.20	0.270
1.50	0.280*
2.00	0.250*
3.00	0.310
3.50	0.350
4.20	0.410
5.00	0.440
6.00	0.440
8.00	0.470
8.50	0.450
8.80	0.400
9.00	0.360*
10.00	0.350*
10.80	0.400
11.00	0.420
12.00	0.420
13.80	0.460
15.00	0.470
17.00	0.490
17.20	0.490*
19.00	0.450
20.00	0.460
21.00	0.480
22.00	0.470
23.20	0.450
25.00	0.385

CURVE 6, T = ~322

λ	ρ
0.50	0.710
0.70	0.715
0.78	0.725
0.85	0.710
1.00	0.710
1.50	0.750
2.00	0.750
3.10	0.850
4.00	0.930
4.90	0.950
6.50	0.950
7.00	0.960
8.00	0.960
9.00	0.970*
13.00	0.970
17.00	0.970
21.00	0.970
22.20	0.935
23.00	0.930
24.00	0.840
25.00	0.835

CURVE 7, T = ~322

λ	ρ
0.50	0.480
0.75	0.500
0.90	0.500
1.50	0.415
2.00	0.365
3.00	0.370
4.00	0.360
5.00	0.340
5.90	0.310
7.70	0.310
8.50	0.280
9.10	0.255
11.00	0.230
12.00	0.220
13.00	0.210*
14.00	0.230
15.00	0.300
16.00	0.380
16.40	0.385
18.00	0.380
18.60	0.350
19.10	0.320
20.00	0.360
21.00	0.340
22.00	0.370
23.00	0.380
24.00	0.365
25.00	0.435

CURVE 8, T = ~322

λ	ρ
0.50	0.725
0.65	0.740
0.75	0.730
0.88	0.750
1.00	0.730
2.00	0.800
3.00	0.925
4.50	0.980
6.50	1.000
8.00	0.990
12.00	0.990
17.00	0.990
21.00	0.990
25.00	0.990

CURVE 9, T = ~322

λ	ρ
0.50	0.440
0.60	0.475
0.75	0.490
1.50	0.490
2.00	0.450
3.00	0.450
5.00	0.420
5.80	0.410
7.50	0.410
8.50	0.380
9.50	0.350
11.00	0.350
12.00	0.330
13.00	0.350
13.50	0.350
15.00	0.380
16.00	0.350
17.00	0.340
18.00	0.350
18.50	0.320
19.00	0.290
20.00	0.320
21.00	0.310
22.00	0.350
23.40	0.350
24.40	0.375
25.00	0.410

CURVE 10, T = 298

λ	ρ
0.4	0.586
0.45	0.592
0.5	0.593
0.6	0.623
0.7	0.638
0.8	0.647
0.9	0.709
1.0	0.696
1.2	0.736
1.4	0.751
1.6	0.762
1.8	0.761
2.0	0.767
2.5	0.814
3.0	0.814
3.5	0.912
4	0.922
5	0.943
6	0.953
7	0.956
8	0.958
9	0.960
10	0.956
11	0.975
12	0.975
13	0.979
14	0.984
15	0.980
16	0.979
17	0.978
18	0.987
19	0.993
20	0.981

* Not shown on plot

SPECIFICATION TABLE NO. 384 NORMAL SOLAR ABSORPTANCE OF [MAGNESIUM + ALUMINUM + ΣX_i] ALLOYS

Curve No.	Ref. No.	Year	Temperature Range, K	Geometry θ	Reported Error, %	Composition (weight percent), Specifications and Remarks
1	34	1957	298	9°		Magnesium alloy AZ 31; nominal composition: 3 Al, 1 Zn, 0.2 Mn, Mg balance; computed from spectral reflectance data for sea level conditions.
2	34	1957	298	9°		Above specimen and conditions except computed for above atmosphere conditions.

DATA TABLE NO. 384 NORMAL SOLAR ABSORPTANCE OF [MAGNESIUM + ALUMINUM + ΣX_i] ALLOYS

[Temperature, T, K; Absorptance, α]

T	α
CURVE 1*	
298	0.210
CURVE 2*	
298	0.268

* Not shown on plot

SPECIFICATION TABLE NO. 385 NORMAL TOTAL EMITTANCE OF [MAGNESIUM + THORIUM + ΣX_i] ALLOYS

Curve No.	Ref. No.	Year	Temperature Range, K	Geometry θ'	Reported Error, %	Composition (weight percent), Specifications and Remarks
1	34	1957	450-619	$\sim 0^\circ$	±10	Magnesium alloy HK-31; nominal composition: 3.25 Th, 0.7 Zr, Mg balance; as received; measured in vacuum (5×10^{-4} mm Hg); increasing temp, cycle 1.
2	34	1957	394-850	$\sim 0^\circ$	±10	Above specimen and conditions; cycle 2.

DATA TABLE NO. 385 NORMAL TOTAL EMITTANCE OF [MAGNESIUM + THORIUM + ΣX_i] ALLOYS

[Temperature, T, K; Emittance, ϵ]

T	ϵ
CURVE 1*	
450	0.610
619	0.530
CURVE 2*	
394	0.635
561	0.600
700	0.535
850	0.540

* Not shown on plot

SPECIFICATION TABLE NO. 386 NORMAL SPECTRAL REFLECTANCE OF [MAGNESIUM + THORIUM + ΣX_i] ALLOYS

Curve No.	Ref. No.	Year	Temperature K	Wavelength Range, μ	Geometry θ	θ'	ω'	Reported Error, %	Composition (weight percent), Specifications and Remarks
1	34	1957	298	0.37-2.50	9^c		2π	±4	Magnesium alloy HK-31; nominal composition: 3.25 Th, 0.7 Zr, Mg balance; data extracted from smooth curve; magnesium carbonate reference.

DATA TABLE NO. 386 NORMAL SPECTRAL REFLECTANCE OF [MAGNESIUM + THORIUM + ΣX_i] ALLOYS

[Wavelength, λ, μ; Reflectance ρ; Temperature, T, K]

λ	ρ
CURVE 1*	
T = 298	
0.37	0.09
0.38	0.20
0.40	0.40
0.42	0.70
0.44	0.78
0.46	0.83
0.48	0.86
0.54	0.88
0.64	0.89
0.70	0.88
0.82	0.86
0.94	0.87
1.00	0.87
1.10	0.84
1.30	0.80
1.40	0.77
1.50	0.75
1.62	0.73
1.74	0.72
1.80	0.72
1.92	0.69
2.02	0.64
2.10	0.59
2.20	0.51
2.24	0.50
2.40	0.54
2.50	0.50

* Not shown on plot

SPECIFICATION TABLE NO. 387 NORMAL SOLAR ABSORPTANCE OF [MAGNESIUM + THORIUM + ΣX_i] ALLOYS

Curve No.	Ref. No.	Year	Temperature Range, K	Geometry θ	Reported Error, %	Composition (weight percent), Specifications and Remarks
1	34	1957	298	9°		Magnesium alloy HK-31; nominal composition: 3.25 Th, 0.7 Zr, Mg balance; computed from spectral reflectance data for sea level conditions.
2	34	1957	298	9°		Above specimen and conditions except computed for above atmosphere conditions.

DATA TABLE NO. 387 NORMAL SOLAR ABSORPTANCE OF [MAGNESIUM + THORIUM + ΣX_i] ALLOYS

[Temperature, T, K; Absorptance, α]

T	α
CURVE 1*	
298	0.175
CURVE 2*	
298	0.251

* Not shown on plot

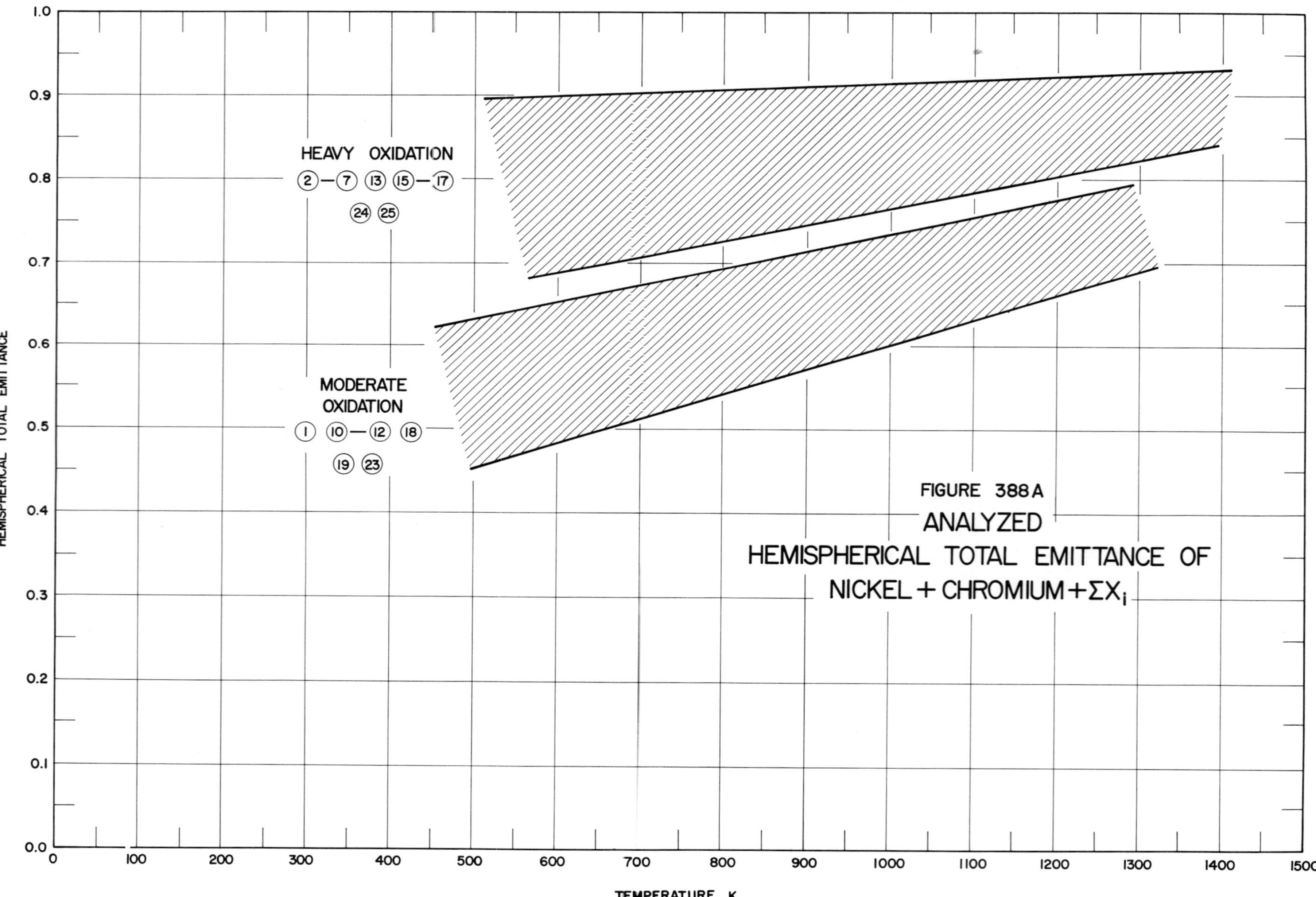

FIGURE 388A
ANALYZED
HEMISPHERICAL TOTAL EMITTANCE OF
NICKEL + CHROMIUM + ΣX_i

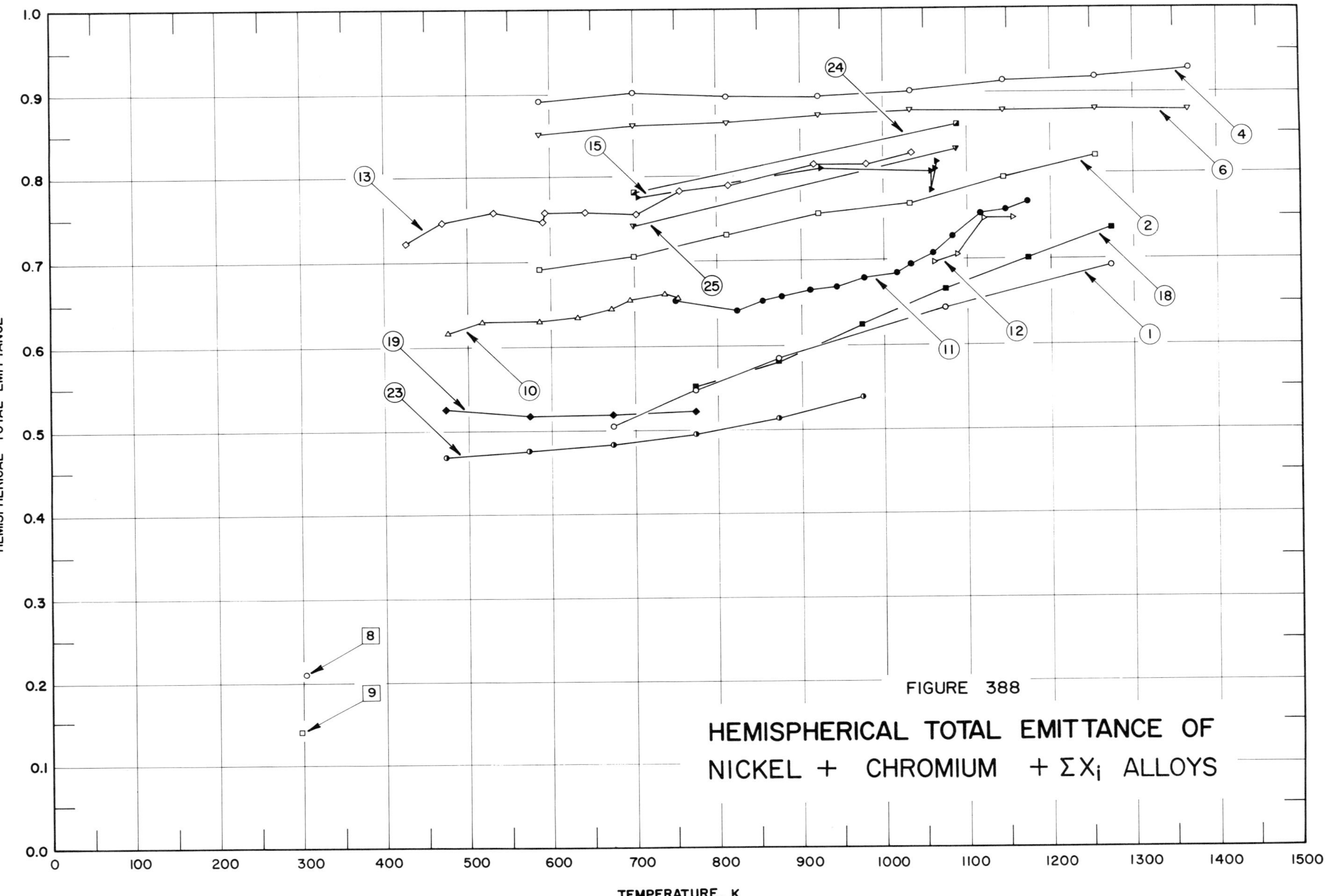

FIGURE 388

HEMISPHERICAL TOTAL EMITTANCE OF NICKEL + CHROMIUM + ΣX_i ALLOYS

SPECIFICATION TABLE NO. 338 HEMISPHERICAL TOTAL EMITTANCE OF [NICKEL + CHROMIUM + ΣX_i] ALLOYS

Curve No.	Ref. No.	Year	Temperature Range, K	Reported Error, %	Composition (weight percent), Specifications and Remarks
1	1	1955	673-1273		Inconel; nominal composition: 78 Ni, 15 Cr, 7 Fe, 0.35 Mn, 0.20 Si, 0.04 C; computed from normal spectral emittance.
2	169	1957	589-1255	< 5	Inconel; 77.0 Ni, 15 Cr, 7.0 Fe, 0.25 Mn, 0.25 Si, 0.2 Cu, 0.08 C, 0.007 S; etched and polished; stably oxidized at 1367 K for 9 min; authors assumed hemispherical total emittance = normal total emittance.
3	169	1957	589-1255	< 5	Different sample, same as above specimen and conditions except stably oxidized at 1367 K for 13 min.
4	169	1957	589-1367	< 5	Inconel X; 70.00 min Ni, 14.00-16.00 Cr, 5.00-9.00 Fe, 2.25-2.75 Ti, 0.7-1.20 Nb, 0.4-1.00 Al, 0.30-1.00 Mn, 0.50 max Si, 0.20 max Cu, 0.08 max C, 0.01 max S; etched and polished; stably oxidized at 1367 K for 30 min; authors assumed hemispherical total emittance = normal total emittance.
5	169	1957	700-1367	< 5	Different sample, same as above specimen and conditions except stably oxidized by gradual heating to a maximum temp of 1367 K in air.
6	161	1959	589-1367		Haynes alloy X; 42-52 Ni, 20.5-23 Cr, 17-20 Fe, 8-10 Mo, 0.5-2.5 Co, 1.00 max Mn, 1.00 max Si, 0.2-1.00 W, 0.05-0.15 C; mechanically polished; oxidized in air at 1367 K for 30 min; authors assumed hemispherical total emittance = normal total emittance.
7	161	1959	589-1367		Different sample, same as above specimen and conditions except gradually oxidized in still air at temperatures from 589 to 1367 K.
8	24	1948	303		Inconel; nominal composition: 78 Ni, 15 Cr, 7 Fe, 0.35 Mn, 0.20 Si, 0.04 C; dull finish; measured in air.
9	24	1948	299		Different sample, same as above specimen and conditions except cleaned; smooth surface.
10	12	1962	477.1-753.5	± 2.7	Nichrome; oxidized in air at 1255 K for 2 hrs; measured in vacuum (<2.0×10^{-5} mm Hg); Run No. 1.
11	12	1962	750.8-1173.6	± 2.7	Above specimen and conditions; Run No. 2A.
12	12	1962	1156.9-1060.3	± 2.7	Above specimen and conditions; Run No. 2B.
13	12	1962	426-1035	± 2.7	Nichrome; oxidized in oxygen at 1255 K for 15 min; measured in vacuum (<2.5×10^{-5} mm Hg); Run No. 1.
14	12	1962	1043-1059	± 2.7	Above specimen and conditions; measurements taken at intervals during 168.8 hr endurance run.
15	12	1962	705-1069	± 2.7	Above specimen and conditions; measurements taken at intervals during 280.4 hr endurance run; Run No. 2.
16	164	1961	589-1255	± 2.5	Inconel; nominal composition: 77.0 Ni, 15.0 Cr, 7.0 Fe, 0.25 Si, 0.25 Mn, 0.2 Cu, 0.08 C, 0.007 S; polished and etched; oxidized in quiescent air at 1366 K for 9 min; measured in air.

SPECIFICATION TABLE NO. 388 (continued)

Curve No.	Ref. No.	Year	Temperature Range, K	Reported Error, %	Composition (weight percent), Specifications and Remarks
17	164	1961	589-1366	± 2.5	Inconel X; nominal composition: 70.00 min Ni, 14.00-16.00 Cr, 5.00-9.00 Fe, 2.25-2.75 Ti, 0.7-1.20 Nb, 0.4-1.00 Al, 0.30-1.00 Mn, 0.50 max Si, 0.20 max Cu, 0.08 max C, 0.01 max S; polished and etched; oxidized in quiescent air at 1366 K for 30 min; measured in air.
18	58	1961	773-1273	± 2.5	Nimonic 75; nominal composition: 20 Cr, 5 max Fe, 1 max Mn, 1 max Si, 0.40 Ti, 0.12 C, Ni balance; measured in vacuum ($<5 \times 10^{-6}$ mm Hg); data extracted from smooth curve.
19	296	1960	473-773		Inconel; 15.0 Cr, 7.0 Fe, 0.25 Mn, 0.25 Si, 0.2 Cu, 0.08 C, 0.007 S, Ni balance; sandblasted; measured in vacuum (4×10^{-5} mm Hg); measurements taken at 10 min intervals; first cycle.
20	296	1960	473-848		Above specimen and conditions except heated at 848 K for 1.5 hrs; cycle 2.
21	296	1960	473-893		Above specimen and conditions except heated at 893 K for 1 hr; cycle 3.
22	296	1960	473-1068		Above specimen and conditions except heated at 1068 K for 2.5 hrs; cycle 4.
23	296	1960	473-973		Above specimen and conditions; cycle 5.
24	295	1963	700-1089		Inconel X; nominal composition: 73 Ni, 15 Cr, 7 Fe, 2.5 Ti, 1 Nb, 0.9 Al, 0.70 Mn, 0.30 Si, 0.04 C; sandblasted and oxidized at 1353 K for 0.5 hr; measured in vacuum (10^{-5} mm Hg).
25	295	1963	700-1089		Inconel 702; nominal composition: 78 Ni, 16 Cr, 3.5 Al, 0.5 Ti, 0.35 Fe, 0.1 Mn, 0.03 C; sandblasted and oxidized at 1353 K for 0.5 hr; measured in vacuum (10^{-5} mm Hg).

DATA TABLE NO. 388 HEMISPHERICAL TOTAL EMITTANCE OF [NICKEL + CHROMIUM + ΣX_i] ALLOYS

[Temperature, T, K; Emittance, ∈]

T	∈
CURVE 1	
673	0.504
773	0.547
773	0.547
873	0.584
873	0.584
873	0.584
1073	0.643
1073	0.643
1073	0.643
1273	0.690
1273	0.690
1273	0.690
CURVE 2	
589	0.690
700	0.705
811	0.730
922	0.755
1033	0.765
1144	0.795
1255	0.820
CURVE 3*	
589	0.690
700	0.705
811	0.730
922	0.760
1033	0.770
1144	0.800
1255	0.820
CURVE 4	
589	0.890
700	0.900
811	0.895
922	0.895
1033	0.900
1144	0.912
1255	0.916
1367	0.925
CURVE 5*	
700	0.890
811	0.900
922	0.900
1033	0.906
1144	0.912
1255	0.916
1367	0.925
CURVE 6	
589	0.850
700	0.860
811	0.862
922	0.870
1033	0.876
1144	0.875
1255	0.875
1367	0.875
CURVE 7*	
589	0.850
700	0.860
811	0.862
922	0.870
1033	0.880
1144	0.875
1255	0.875
1367	0.875
CURVE 8	
303	0.21
CURVE 9	
299	0.14
CURVE 10	
477.1	0.618
518.7	0.630
587.1	0.630
632.9	0.636
CURVE 10 (cont.)	
674.1	0.646
696.9	0.655
739.0	0.661
753.5	0.657
CURVE 11	
750.8	0.653
824.7	0.642
853.2	0.651
879.6	0.658
911.5	0.664
944.8	0.667
978.0	0.679
1017.1	0.682
1033.5	0.692
1060.3	0.709
1082.5	0.727
1117.9	0.752
1149.8	0.757
1173.6	0.768
CURVE 12	
1156.9	0.758
1120.4	0.749
1089.6	0.754
1060.3	0.747
CURVE 13	
426	0.721
470	0.746
533	0.753
591	0.747
594	0.757
641	0.753
702	0.754
756	0.781
814	0.783
918	0.812
980	0.811
1035	0.825
CURVE 14*	
1059	0.820
1059	0.827
1049	0.807
1050	0.835
1050	0.828
1049	0.828
1043	0.845
1047	0.845
1045	0.831
1049	0.828
1048	0.833
1049	0.830
1049	0.830
1051	0.833
1049	0.838
1048	0.832
CURVE 15	
705	0.775
924	0.808
1061	0.806
1058	0.804
1058	0.780
1056	0.807*
1057	0.803*
1056	0.804*
1055	0.803*
1059	0.798*
1055	0.808*
1055	0.805*
1055	0.799*
1055	0.796*
1056	0.806*
1055	0.810*
1065	0.814
1065	0.810*
1063	0.814*
1065	0.812*
1062	0.816*
1065	0.812*
1067	0.816*
1069	0.811*
1069	0.811*
CURVE 15 (cont.)	
1060	0.811*
1068	0.809*
1068	0.814*
1065	0.808*
CURVE 16*	
589	0.690
700	0.710
811	0.730
922	0.755
1033	0.780
1144	0.800
1255	0.820
CURVE 17*	
589	0.895
700	0.895
811	0.900
922	0.900
1033	0.905
1144	0.910
1255	0.920
1366	0.925
CURVE 18	
773	0.550
873	0.580
973	0.625
1073	0.665
1173	0.700
1273	0.735
CURVE 19	
473	0.527
573	0.518
673	0.519
773	0.522
CURVE 20*	
473	0.493
848	0.538
848	0.543
848	0.570
848	0.579
848	0.586
848	0.592
848	0.596
848	0.603
848	0.606
848	0.609
848	0.612
848	0.615
848	0.612
CURVE 21*	
473	0.527
893	0.658
893	0.670
893	0.693
893	0.696
893	0.701
893	0.704
893	0.708
CURVE 22*	
473	0.568
1068	0.736
1068	0.703
1068	0.653
1068	0.643
1068	0.635
1068	0.626
1068	0.617
1068	0.610
1068	0.606
1068	0.599
1068	0.586
1068	0.579
1068	0.576
1068	0.573
1068	0.571
1068	0.566
1068	0.469
CURVE 23	
473	0.468
573	0.474
673	0.482
773	0.494
873	0.512
973	0.538
CURVE 24	
700	0.78
1089	0.86
CURVE 25	
700	0.74
1089	0.83

* Not shown on plot

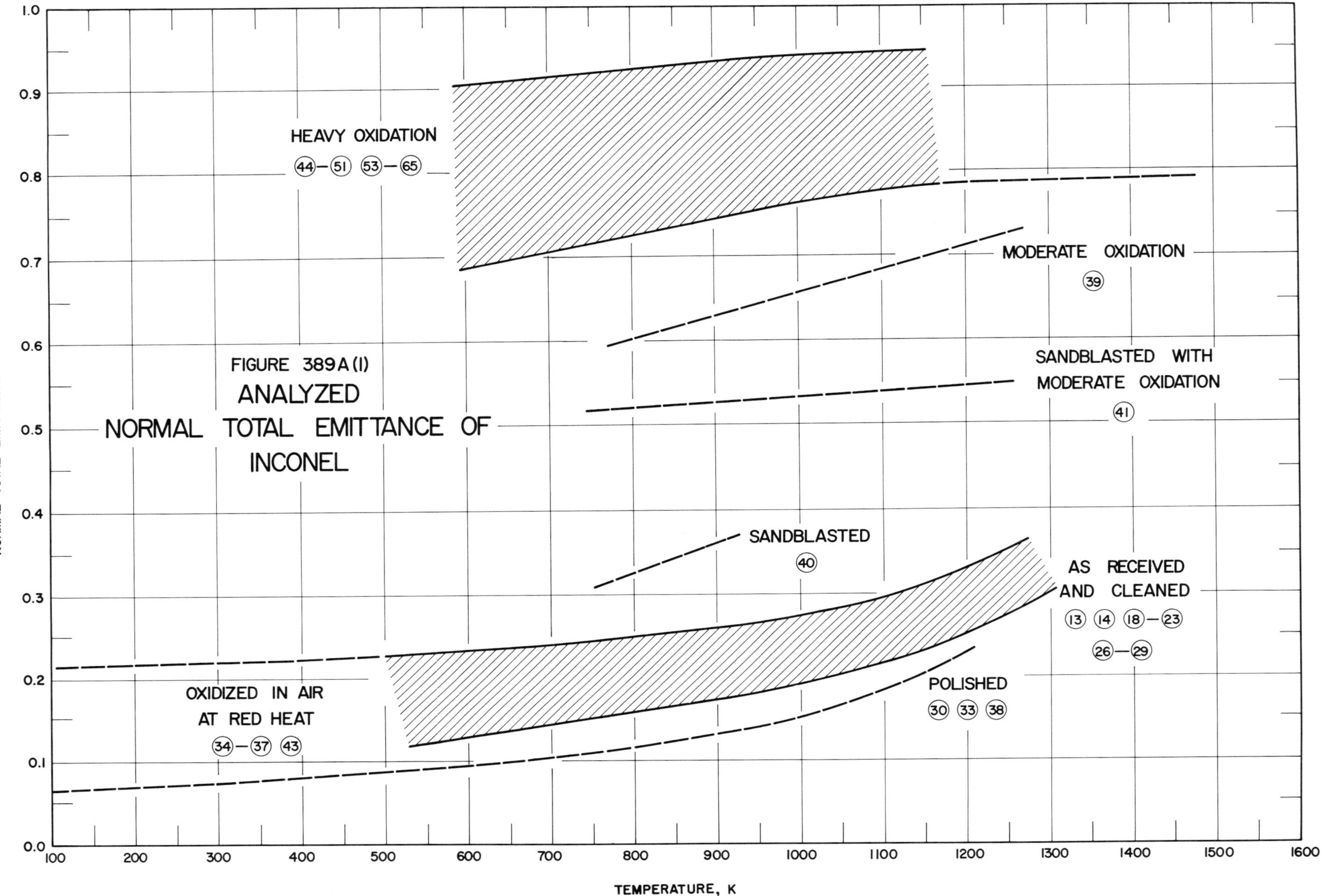

FIGURE 389A (I) ANALYZED NORMAL TOTAL EMITTANCE OF INCONEL

FIGURE 389A (2)

NORMAL TOTAL EMITTANCE OF NICKEL + CHROMIUM + ΣX_i

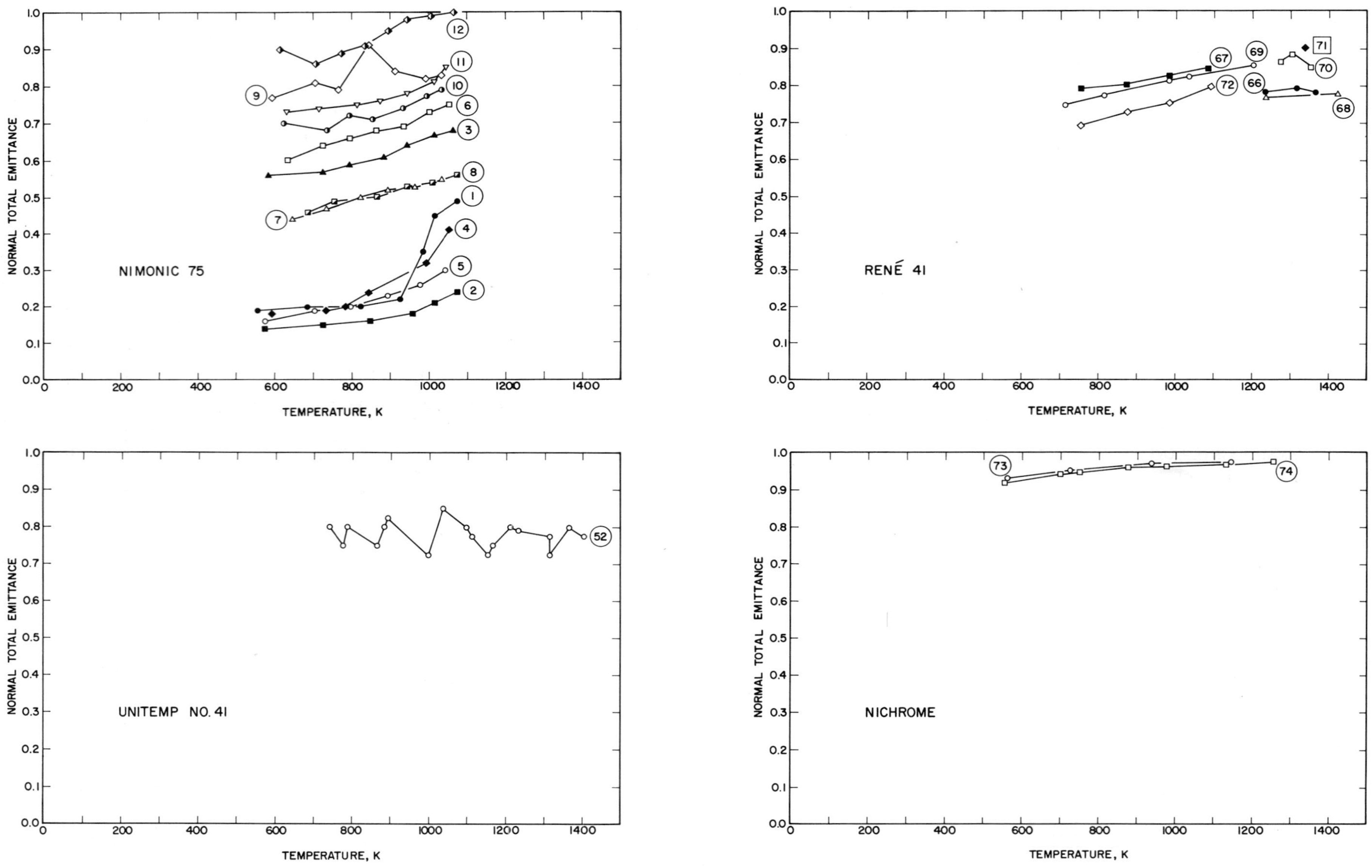
NIMONIC 75
RENÉ 41
UNITEMP NO. 41
NICHROME
NORMAL TOTAL EMITTANCE
TEMPERATURE, K
1
2
3
4
5
6
7
8
9
10
11
12
66
67
68
69
70
71
72
73
74
52

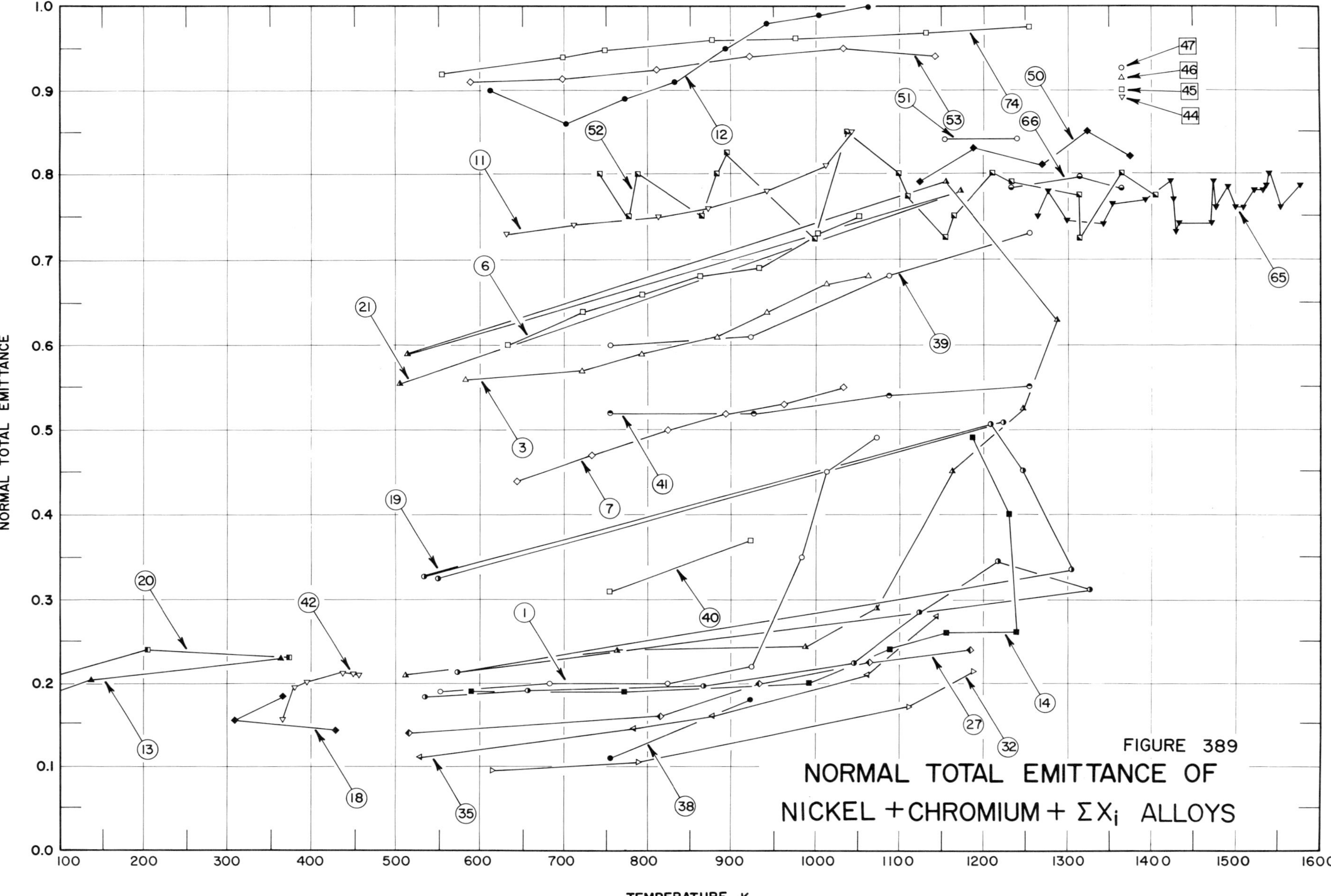

FIGURE 389
NORMAL TOTAL EMITTANCE OF NICKEL + CHROMIUM + ΣX_i ALLOYS

SPECIFICATION TABLE NO. 389 NORMAL TOTAL EMITTANCE OF [NICKEL + CHROMIUM + ΣX_i] ALLOYS

Curve No.	Ref. No.	Year	Temperature Range, K	Geometry θ'	Reported Error, %	Composition (weight percent), Specifications and Remarks
1	68	1952	553-1073	$\sim 0°$		Nimonic 75; nominal composition: 20 Cr, 5 max Fe, 1 max Mn, 1 max Si, 0.40 Ti, 0.12 C, Ni balance; as rolled; cleaned with CCl_4; measured in air.
2	68	1952	573-1073	$\sim 0°$		Different sample, same as curve 1 specimen and conditions except buffed.
3	68	1952	583-1063	$\sim 0°$		Different sample, same as curve 1 specimen and conditions except shot blasted with fused alumina.
4	68	1952	593-1053	$\sim 0°$		Different sample, same as curve 1 specimen and conditions; oxidized at 873 K until steady state reached.
5	68	1952	573-1043	$\sim 0°$		Different sample, same as curve 4 specimen and conditions except buffed.
6	68	1952	633-1053	$\sim 0°$		Different sample, same as curve 4 specimen and conditions except shot blasted with fused alumina.
7	68	1952	643-1033	$\sim 0°$		Different sample, same as curve 1 specimen and conditions; oxidized at 1173 K until steady state reached.
8	68	1952	683-1073	$\sim 0°$		Different sample, same as curve 7 specimen and conditions except buffed.
9	68	1952	593-1033	$\sim 0°$		Different sample, same as curve 7 specimen and conditions except shot blasted with fused alumina.
10	68	1952	623-1033	$\sim 0°$		Different sample, same as curve 1 specimen and conditions; oxidized at 1473 K until steady state reached.
11	68	1952	633-1043	$\sim 0°$		Different sample, same as curve 10 specimen and conditions except buffed.
12	68	1952	613-1063	$\sim 0°$		Different sample, same as curve 10 specimen and conditions except shot blasted with fused alumina.
13	160	1954	88.9-364	$\sim 0°$		Inconel B; nominal composition: 72 min Ni, 16-18 Cr, 9.5 max Fe, 1 max Mn, 0.15 C; cleaned with methyl alcohol; measured in vacuum (10^{-3}mm Hg); [Author's designation: Sample 1].
14	160	1954	589-1239	$\sim 0°$		Different sample, same as above specimen and conditions except measured in argon (10^{-3}mm Hg); [Author's designation: Sample 4].
15	160	1954	497-1306	$\sim 0°$		Above specimen and conditions except heating and cooling.
16	160	1954	307-425	$\sim 0°$		Different sample, same as curve 13 specimen and conditions except polished and then finished with a wool buff and rouge; washed; surface free from scratches; [Author's designation: Sample 3].
17	160	1954	515-1293	$\sim 0°$		Different sample, same as above specimen and conditions except measured in argon (10^{-3}mm Hg); [Author's designation: Sample 6].
18	160	1954	309-428	$\sim 0°$		Different sample, same as curve 13 specimen and conditions except scrubbed with Bon Ami on a wet cloth, washed and dried, wiped with toluene and alcohol; [Author's designation: Sample 2].

SPECIFICATION TABLE NO. 389 (continued)

Curve No.	Ref. No.	Year	Temperature Range, K	Geometry θ'	Reported Error, %	Composition (weight percent), Specifications and Remarks
19	160	1954	532-1328	$\sim 0°$		Different sample, same as above specimen and conditions except measured in argon (10^{-3}mm Hg); [Author's designation: Sample 5].
20	160	1954	97.2-372	$\sim 0°$		Inconel X; nominal composition: 73 Ni, 15 Cr, 7 Fe, 2.5 Ti, 1 Nb, 0.9 Al, 0.70 Mn, 0.30 Si, 0.04 C; cleaned with methyl alcohol; measured in vacuum (10^{-3}mm Hg); heating; [Author's designation: Sample 1].
21	160	1954	506-1289	$\sim 0°$		Different sample, same as above specimen and conditions except measured in argon (10^{-3}mm Hg); [Author's designation: Sample 2].
22	160	1954	316-439	$\sim 0°$		Different sample, same as curve 20 specimen and conditions except scrubbed with Bon Ami on a wet cloth, washed and dried, wiped with toluene and alcohol; [Author's designation: Sample 3].
23	160	1954	513-1279	$\sim 0°$		Different sample, same as above specimen and conditions except measured in argon (10^{-3}mm Hg); [Author's designation: Sample 4].
24	160	1954	303-424	$\sim 0°$		Different sample, same as curve 20 specimen and conditions except polished and then finished with a wool buff and rouge; washed; surface free from scratches; [Author's designation: Sample 5].
25	160	1954	516-1263	$\sim 0°$		Different sample, same as above specimen and conditions except measured in argon (10^{-3}mm Hg); [Author's designation: Sample 6].
26	34	1957	83.2	$\sim 0°$	± 10	Inconel X; nominal composition: 73 Ni, 15 Cr, 7 Fe, 2.5 Ti, 1 Nb, 0.9 Al, 0.70 Mn, 0.30 Si, 0.04 C; strip (0.005 in. thick); as received or cleaned with liquid detergent; measured in vacuum (5 x 10^{-4}mm Hg).
27	34	1957	516-1183	$\sim 0°$	± 10	Different sample, same as above specimen and conditions except increasing temp, cycle 1.
28	34	1957	739-983	$\sim 0°$	± 10	Above specimen and conditions; cycle 2.
29	34	1957	961	$\sim 0°$	± 10	Above specimen and conditions; cycle 3.
30	34	1957	83.2	$\sim 0°$	± 10	Inconel X; nominal composition: 73 Ni, 15 Cr, 7 Fe, 2.5 Ti, 1 Nb, 0.9 Al, 0.70 Mn, 0.30 Si, 0.04 C; strip (0.005 in. thick); polished; measured in vacuum (5 x 10^{-4}mm Hg).
31	34	1957	550-778	$\sim 0°$	± 10	Different sample, same as above specimen and conditions except increasing temp, cycle 1.
32	34	1957	614-1189	$\sim 0°$	± 10	Above specimen and conditions; cycle 2.
33	34	1957	722-1022	$\sim 0°$	± 10	Above specimen and conditions; cycle 3.
34	34	1957	83.2	$\sim 0°$	± 10	Inconel X; nominal composition: 73 Ni, 15 Cr, 7 Fe, 2.5 Ti, 1 Nb, 0.9 Al, 0.70 Mn, 0.30 Si, 0.04 C; strip (0.005 in. thick); oxidized in air at red heat for 30 min; measured in vacuum (5 x 10^{-4}mm Hg).
35	34	1957	528-1144	$\sim 0°$	± 10	Different sample, same as above specimen and conditions except increasing temp, cycle 1.
36	34	1957	589-1116	$\sim 0°$	± 10	Above specimen and conditions; cycle 2.
37	34	1957	600-889	$\sim 0°$	± 10	Above specimen and conditions; cycle 3.
38	170	1959	755-922	$\sim 0°$		Inconel; nominal composition: 80 Ni, 14 Cr, 6 Fe; electropolished.

SPECIFICATION TABLE NO. 389 (continued)

Curve No.	Ref. No.	Year	Temperature Range, K	Geometry θ'	Reported Error, %	Composition (weight percent), Specifications and Remarks
39	170	1959	755-1255	~0°		Different sample, same as above specimen and conditions; oxidized in air at 1255 K for 1/2 hr.
40	170	1959	755-922	~0°		Different sample, same as above specimen and conditions except sandblasted with 40-mesh glass sand and air at 40 psi.
41	170	1959	755-1255	~0°		Different sample, same as above specimen and conditions; oxidized in air at 1255 K for 1/2 hr.
42	157	1944	366-455	~0°		Inconel; nominal composition: 78 Ni, 15 Cr, 7 Fe, 0.35 Mn, 0.20 Si, 0.04 C; measured in air.
43	157	1944	359-443	~0°		Different sample, same as above specimen and conditions; oxidized at 1033 K to a brassy color.
44	164	1961	1366	~0°	± 2.5	Inconel X; nominal composition: 73 Ni, 15 Cr, 7 Fe, 2.5 Ti, 1 Nb, 0.9 Al, 0.70 Mn, 0.30 Si; polished and etched; oxidized in quiescent air at 1366 K for 1 mi[illegible]; measured in air.
45	164	1961	1366	~0°	± 2.5	Above specimen and conditions except oxidized in quiescent air at 1366 K for 10 min.
46	164	1961	1366	~0°	± 2.5	Above specimen and conditions except oxidized in quiescent air at 1366 K for 20 min.
47	164	1961	1366	~0°	± 2.5	Above specimen and conditions except oxidized in quiescent air at 1366 K for 30 min.
48	164	1961	1366	~0°	± 2.5	Above specimen and conditions except oxidized in quiescent air at 1366 K for 45 min.
49	164	1961	1366	~0°	± 2.5	Above specimen and conditions except oxidized in quiescent air at 1366 K for 60 min.
50	109	1961	1125-1375	~0°		Inconel; nominal composition: 78 Ni, 15 Cr, 7 Fe, 0.35 Mn, 0.20 Si, 0.04 C; oxidized; highly rough surface; measured in vacuum (4×10^{-6} mm Hg); run 1.
51	109	1961	1155-1240	~0°		Above specimen and conditions; run 2.
52	75	1962	744-1405	~0°		Unitemp No. 41; nominal composition: 19 Cr, 11 Co, 10 Mo, 3 Fe, 3 Ti, 1.5 Al, 0.1 C, trace B, Ni balance; preoxidized in air at 1366 K; measured in air.
53	165	1962	589-1144	~0°		Inconel; nominal composition: 78 Ni, 15 Cr, 7 Fe, 0.35 Mn, 0.20 Si, 0.04 C; oxidized by spraying with sodium dichromate; held at 700 K for 30 min; initial run; [Author's designation: Specimen 1].
54	165	1962	589-1144	~0°		Above specimen and conditions except heated at 1144 K for 30 min.
55	165	1962	589-1255	~0°		Different sample, same as curve 53 specimen and conditions; [Author's designation: Specimen 2].
56	165	1962	589-1255	~0°		Above specimen and conditions except heated at 1255 K for 30 min.
57	165	1962	1144	~0°		Different sample, same as curve 53 specimen and conditions; emittance constant at 1144 K for 30 min.
58	165	1962	1255	~0°		Different sample, same as curve 53 specimen and conditions.
59	165	1962	1255	~0°		Above specimen and conditions except heated at 1255 K for 5 min in air.
60	165	1962	1255	~0°		Curve 53 specimen and conditions except heated at 1255 K for 10 min in air.
61	165	1962	1255	~0°		Curve 53 specimen and conditions except heated at 1255 K for 15 min in air.

SPECIFICATION TABLE NO. 389 (continued)

Curve No.	Ref. No.	Year	Temperature Range, K	Geometry θ'	Reported Error, %	Composition (weight percent), Specifications and Remarks
62	165	1962	1255	~ 0°		Curve 53 specimen and conditions except heated at 1255 K for 20 min in air.
63	165	1962	1255	~ 0°		Curve 53 specimen and conditions except heated at 1255 K for 25 min in air.
64	165	1962	1255	~ 0°		Curve 53 specimen and conditions except heated at 1255 K for 30 min in air.
65	171	1962	1266-1578	~ 0°	± 10	Inconel X; nominal composition: 73 Ni, 15 Cr, 7 Fe, 2.5 Ti, 1 Nb, 0.9 Al, 0.70 Mn, 0.30 Si, 0.04 C; oxidized.
66	65	1962	1233-1366	~ 0°		Rene' 41; nominal composition: 19 Cr, 11 Co, 10 Mo, 3 Fe, 3 Ti, 1.5 Al, 0.1 C, trace B, Ni balance; run 1: [Author's designation: Sample No. 1].
67	65	1962	755-1089	~ 0°		Above specimen and conditions; run 2.
68	65	1962	1239-1422	~ 0°		Above specimen and conditions; run 3.
69	65	1962	711-1205	~ 0°		Above specimen and conditions; run 4.
70	65	1962	1278-1355	~ 0°		Above specimen and conditions; run 5.
71	65	1962	1339	~ 0°		Above specimen and conditions; run 6.
72	65	1962	755-1094	~ 0°		Different sample, same as curve 66 specimen and conditions; run 1: [Author's designation: sample no. 2].
73	28	1958	561-1150	~ 0°		Nichrome.
74	28	1958	555-1255	~ 0°		Nichrome.

DATA TABLE NO. 389 NORMAL TOTAL EMITTANCE OF [NICKEL + CHROMIUM + ΣX_i] ALLOYS

[Temperature, T, K; Emittance, ϵ]

CURVE 1

T	ϵ
553	0.19
683	0.20
823	0.20
923	0.22
983	0.35
1013	0.45
1073	0.49

CURVE 2*

T	ϵ
573	0.14
723	0.15
848	0.16
958	0.18
1013	0.21
1073	0.24

CURVE 3

T	ϵ
583	0.56
722	0.57
793	0.59
883	0.61
943	0.64
1013	0.67
1063	0.68

CURVE 4*

T	ϵ
593	0.18
733	0.19
783	0.20
843	0.24
993	0.32
1053	0.41

CURVE 5*

T	ϵ
573	0.16
703	0.19
798	0.20
893	0.23
978	0.26
1043	0.30

CURVE 6

T	ϵ
633	0.60
723	0.64
793	0.66
863	0.68
933	0.69
1003	0.73
1053	0.75

CURVE 7

T	ϵ
643	0.44
733	0.47
823	0.50
893	0.52
963	0.53
1033	0.55

CURVE 8*

T	ϵ
683	0.46
753	0.49
863	0.50
943	0.53
1008	0.54
1073	0.56

CURVE 9*

T	ϵ
593	0.77
703	0.81
763	0.79
843	0.91
913	0.84
993	0.82
1033	0.83

CURVE 10*

T	ϵ
623	0.70
733	0.68
793	0.72
853	0.71
933	0.74
993	0.77
1033	0.79

CURVE 11

T	ϵ
633	0.73
713	0.74
813	0.75
873	0.76
943	0.78
1013	0.81
1043	0.85

CURVE 12

T	ϵ
613	0.90
703	0.86
773	0.89
833	0.91
893	0.95
943	0.98
1003	0.99
1063	1.00

CURVE 13

T	ϵ
88.9	0.180*
139	0.205
364	0.230

CURVE 14

T	ϵ
589	0.190
772	0.190
992	0.200
1089	0.240
1156	0.260
1239	0.260
1231	0.400
1189	0.490

CURVE 15*

T	ϵ
501	0.190
1306	0.280
1264	0.225
1172	0.590
497	0.325
1156	0.555
1172	0.510

CURVE 15*(cont.)

T	ϵ
939	0.495
756	0.430
497	0.350

CURVE 16*

T	ϵ
425	0.175
307	0.200
370	0.178

CURVE 17*

T	ϵ
519	0.223
846	0.235
952	0.247
1063	0.308
1172	0.391
1235	0.464
515	0.399
1262	0.395
532	0.440
1293	0.399
542	0.323
1279	0.395
530	0.355

CURVE 18

T	ϵ
428	0.142
309	0.155
367	0.185

CURVE 19

T	ϵ
534	0.184
657	0.191
867	0.197
1047	0.225
1123	0.286
1217	0.346
1328	0.311
573	0.213
1304	0.335
1248	0.451
1209	0.506

CURVE 19 (cont.)

T	ϵ
532	0.328
1223	0.509
550	0.326

CURVE 20

T	ϵ
97.2	0.200*
205	0.240
372	0.230

CURVE 21

T	ϵ
511	0.210
764	0.240
989	0.245
1072	0.290
1164	0.450
1247	0.525
1289	0.630
1156	0.790
514	0.590
1172	0.780
506	0.555

CURVE 22*

T	ϵ
439	0.174
316	0.217
318	0.191

CURVE 23*

T	ϵ
527	0.200
629	0.201
774	0.197
945	0.208
1052	0.242
1147	0.389
1213	0.462
1279	0.551
1233	0.661
529	0.575
1133	0.753
513	0.600

CURVE 24*

T	ϵ
424	0.171
303	0.199
377	0.207

CURVE 25*

T	ϵ
573	0.217
728	0.195
940	0.205
1061	0.463
1153	0.655
1228	0.705
1263	0.597
1250	0.603
530	0.386
1170	0.733
516	0.583
1147	0.731
517	0.614

CURVE 26*

T	ϵ
83.2	0.055

CURVE 27

T	ϵ
516	0.140
816	0.160
933	0.200
1064	0.225
1183	0.240

CURVE 28*

T	ϵ
739	0.165
983	0.195

CURVE 29*

T	ϵ
961	0.200

CURVE 30*

T	ϵ
83.2	0.067

CURVE 31*

T	ϵ
550	0.105
778	0.110

CURVE 32

T	ϵ
614	0.095
789	0.115
1111	0.170
1189	0.215

CURVE 33*

T	ϵ
722	0.130
911	0.170
1022	0.185

CURVE 34*

T	ϵ
83.2	0.067

CURVE 35

T	ϵ
528	0.110
783	0.145
878	0.160
1061	0.210
1144	0.280

CURVE 36*

T	ϵ
589	0.140
922	0.210
1116	0.325

CURVE 37*

T	ϵ
600	0.165
889	0.245

CURVE 38

T	ϵ
755	0.11
922	0.18

CURVE 39

T	ϵ
755	0.60
922	0.61
1089	0.68
1255	0.73

CURVE 40

T	ϵ
755	0.31
922	0.37

CURVE 41

T	ϵ
755	0.52
927	0.52
1089	0.54
1255	0.55

CURVE 42

T	ϵ
366	0.156
380	0.196
394	0.201
436	0.212
450	0.211
455	0.210

CURVE 43*

T	ϵ
359	0.184
376	0.188
377	0.182
378	0.182
393	0.180
393	0.182
406	0.186
407	0.185
408	0.185
419	0.180
420	0.182
421	0.184
426	0.188
429	0.190
439	0.183
440	0.182

* Not shown on plot

DATA TABLE NO. 389 (continued)

T	∈
CURVE 43*(cont.)	
441	0.180
443	0.190
443	0.178
CURVE 44	
1366	0.890
CURVE 45	
1366	0.900
CURVE 46	
1366	0.914
CURVE 47	
1366	0.926
CURVE 48*	
1366	0.926
CURVE 49*	
1366	0.926
CURVE 50	
1125	0.79
1190	0.83
1270	0.81
1325	0.85
1375	0.82
CURVE 51	
1155	0.84
1240	0.84
CURVE 52	
744	0.800
778	0.750
CURVE 52 (cont.)	
789	0.800
866	0.750
883	0.800
894	0.825
1000	0.725
1039	0.850
1100	0.800
1111	0.775
1155	0.725
1166	0.750
1211	0.800
1233	0.790
1316	0.775
1316	0.725
1366	0.800
1405	0.775
CURVE 53	
589	0.910
700	0.912
811	0.925
922	0.940
1033	0.950
1144	0.941
CURVE 54*	
589	0.910
700	0.912
811	0.925
922	0.933
1033	0.950
1144	0.941
CURVE 55*	
589	0.890
700	0.912
811	0.940
922	0.947
1033	0.970
1144	0.980
1255	0.980
CURVE 56*	
589	0.890
644	0.830
700	0.790
811	0.770
922	0.780
1033	0.790
1144	0.810
1255	0.820
CURVE 57*	
1144	0.94
CURVE 58*	
1255	0.93
CURVE 59*	
1255	0.89
CURVE 60*	
1255	0.86
CURVE 61*	
1255	0.85
CURVE 62*	
1255	0.83
CURVE 63*	
1255	0.83
CURVE 64*	
1255	0.82
CURVE 65	
1266	0.750
1278	0.780
CURVE 65 (cont.)	
1300	0.745
1344	0.740
1355	0.765
1394	0.770
1422	0.790
1428	0.770
1430	0.731
1433	0.740
1472	0.740
1475	0.790
1478	0.760
1492	0.785
1500	0.760
1511	0.760
1522	0.780
1533	0.780
1539	0.785
1541	0.800
1555	0.760
1578	0.785
CURVE 66	
1233	0.785
1316	0.795
1366	0.782
CURVE 67*	
755	0.794
872	0.805
983	0.829
1089	0.848
CURVE 68*	
1239	0.770
1422	0.780
CURVE 69*	
711	0.750
816	0.775
983	0.815
1039	0.825
1205	0.855
CURVE 70*	
1278	0.866
1305	0.887
1355	0.855
CURVE 71*	
1339	0.905
CURVE 72*	
755	0.695
878	0.730
983	0.757
1094	0.800
CURVE 73*	
561	0.930
728	0.951
939	0.970
1150	0.975
CURVE 74	
555	0.920
700	0.940
750	0.949
878	0.960
978	0.962
1133	0.969
1255	0.975

* Not shown on plot

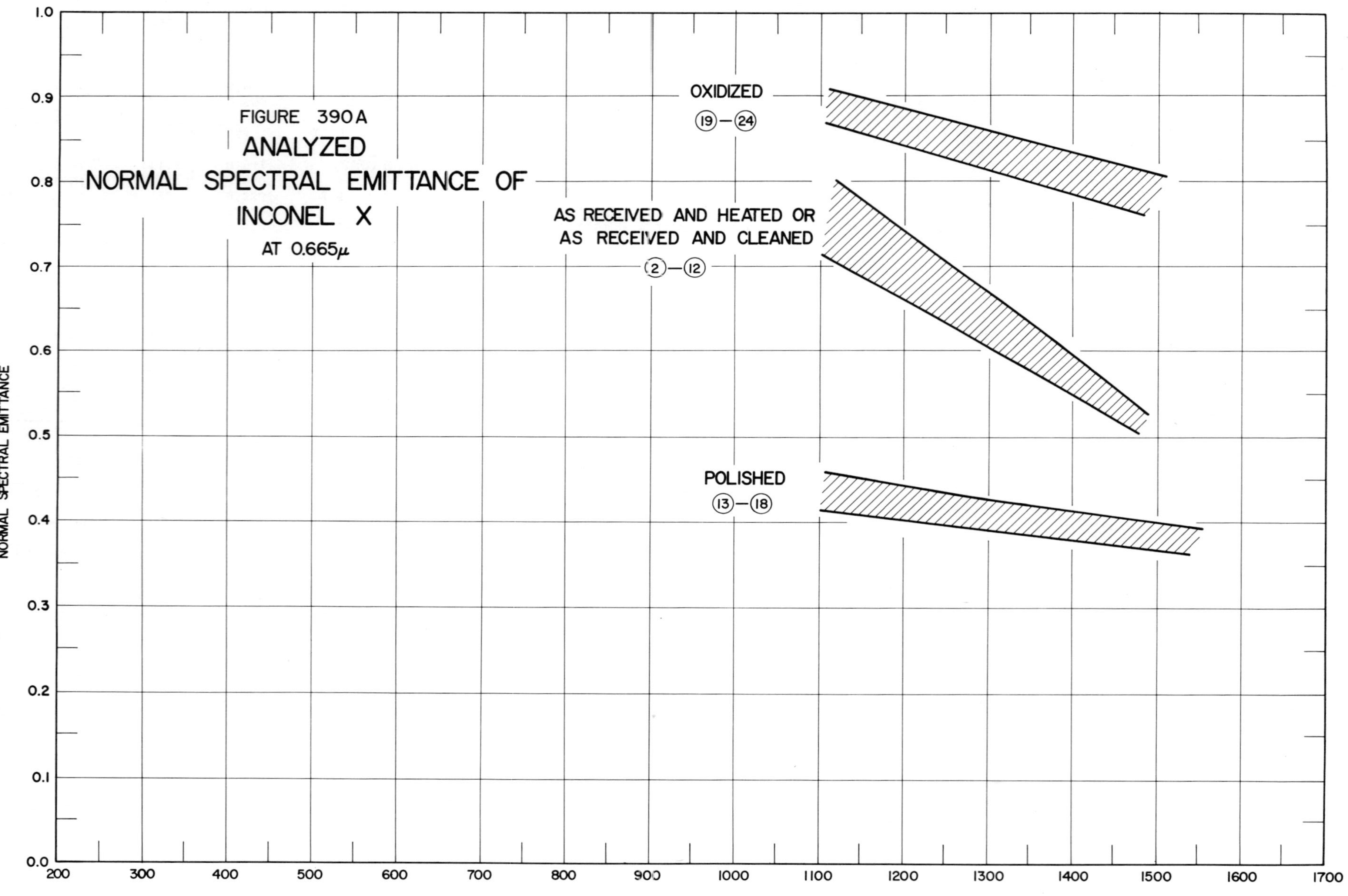
FIGURE 390A
ANALYZED
NORMAL SPECTRAL EMITTANCE OF
INCONEL X
AT 0.665μ
OXIDIZED
19 — 24
AS RECEIVED AND HEATED OR
AS RECEIVED AND CLEANED
2 — 12
POLISHED
13 — 18
NORMAL SPECTRAL EMITTANCE
1.0
0.9
0.8
0.7
0.6
0.5
0.4
0.3
0.2
0.1
0.0
200
300
400
500
600
700
800
900
1000
1100
1200
1300
1400
1500
1600
1700
TEMPERATURE, K

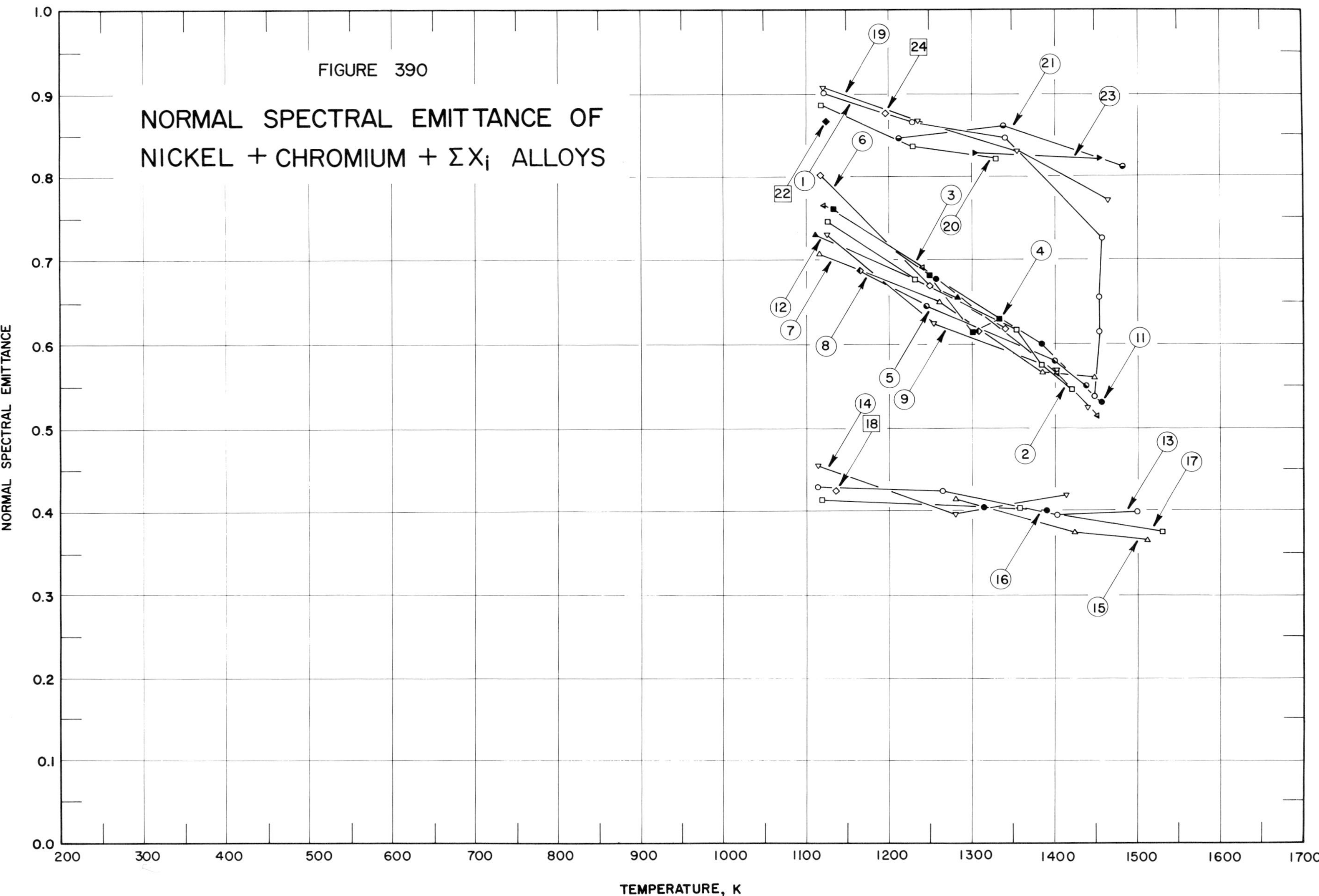
FIGURE 390
NORMAL SPECTRAL EMITTANCE OF
NICKEL + CHROMIUM + ΣX_i ALLOYS
NORMAL SPECTRAL EMITTANCE
TEMPERATURE, K
1.0
0.9
0.8
0.7
0.6
0.5
0.4
0.3
0.2
0.1
0.0
200
300
400
500
600
700
800
900
1000
1100
1200
1300
1400
1500
1600
1700
19
24
21
23
6
1
22
3
20
4
12
7
8
5
9
11
14
18
2
13
17
16
15

SPECIFICATION TABLE NO. 390 NORMAL SPECTRAL EMITTANCE OF [NICKEL + CHROMIUM + ΣX_i] ALLOYS

Curve No.	Ref. No.	Year	Wavelength μ	Temperature Range, K	Geometry θ'	Reported Error, %	Composition (weight percent), Specifications and Remarks
1	34	1957	0.665	1122-1450	~0°		Inconel X; nominal composition: 73 Ni, 15 Cr, 7 Fe, 2.5 Ti, 1 Nb, 0.9 Al, 0.7 Mn, 0.3 Si, 0.04 C; as received; measured in vacuum (5×10^{-4} mm Hg); increasing temp, cycle 1.
2	34	1957	0.665	1422-1127	~0°		Above specimen and conditions; decreasing temp, cycle 1.
3	34	1957	0.665	1122-1452	~0°		Above specimen and conditions; cycle 2.
4	34	1957	0.665	1333-1133	~0°		Above specimen and conditions; decreasing temp, cycle 2.
5	34	1957	0.665	1247-1439	~0°		Above specimen and conditions; cycle 3.
6	34	1957	0.665	1341-1119	~0°		Above specimen and conditions; decreasing temp, cycle 3.
7	34	1957	0.665	1116-1450	~0°		Different sample, same as curve 1 specimen and conditions except cleaned with liquid detergent.
8	34	1957	0.665	1308-1166	~0°		Above specimen and conditions; decreasing temp, cycle 1.
9	34	1957	0.665	1127-1441	~0°		Above specimen and conditions; cycle 2.
10	34	1957	0.665	1264-1125	~0°		Above specimen and conditions; decreasing temp, cycle 2.
11	34	1957	0.665	1258-1458	~0°		Above specimen and conditions; cycle 3.
12	34	1957	0.665	1283-1111	~0°		Above specimen and conditions; decreasing temp, cycle 3.
13	34	1957	0.665	1114-1500	~0°		Different sample, same as curve 1 specimen and conditions except polished.
14	34	1957	0.665	1414-1114	~0°		Above specimen and conditions; decreasing temp, cycle 1.
15	34	1957	0.665	1280-1511	~0°		Above specimen and conditions; cycle 2.
16	34	1957	0.665	1391-1264	~0°		Above specimen and conditions; decreasing temp, cycle 2.
17	34	1957	0.665	1119-1530	~0°		Above specimen and conditions; cycle 3.
18	34	1957	0.665	1136	~0°		Above specimen and conditions; decreasing temp, cycle 3.
19	34	1957	0.665	1122-1466	~0°		Different sample, same as curve 1 specimen and conditions except oxidized in air at red heat for 30 min.
20	34	1957	0.665	1330-1119	~0°		Above specimen and conditions; decreasing temp, cycle 1.
21	34	1957	0.665	1214-1483	~0°		Above specimen and conditions; cycle 2.
22	34	1957	0.665	1125	~0°		Above specimen and conditions; decreasing temp, cycle 2.
23	34	1957	0.665	1302-1455	~0°		Above specimen and conditions; cycle 3.
24	34	1957	0.665	1197	~0°		Above specimen and conditions; decreasing temp, cycle 3.

DATA TABLE NO. 390 NORMAL SPECTRAL EMITTANCE OF [NICKEL + CHROMIUM + ΣX_i] ALLOYS

[Temperature, T, K; Emittance, ϵ; wavelength, λ, μ]

T	ϵ
CURVE 1	
$\lambda = 0.665$	
1122	0.900
1230	0.865
1341	0.845
1458	0.725
1455	0.655
1455	0.615
1450	0.537
1450	0.535*
CURVE 2	
$\lambda = 0.665$	
1422	0.545
1386	0.575
1355	0.617
1233	0.675
1127	0.745
CURVE 3	
$\lambda = 0.665$	
1122	0.765
1241	0.690
1402	0.565
1452	0.515
CURVE 4	
$\lambda = 0.665$	
1333	0.630
1302	0.615
1250	0.680
1133	0.760
CURVE 5	
$\lambda = 0.665$	
1247	0.645
1400	0.580
1439	0.550
CURVE 6	
$\lambda = 0.665$	
1341	0.620
1250	0.668
1119	0.800
CURVE 7	
$\lambda = 0.665$	
1116	0.707
1261	0.650
1386	0.565
1450	0.560
CURVE 8	
$\lambda = 0.665$	
1308	0.615
1166	0.685
CURVE 9	
$\lambda = 0.665$	
1127	0.730
1255	0.625
1402	0.567*
1441	0.525
CURVE 10*	
$\lambda = 0.665$	
1264	0.640
1125	0.745
CURVE 11	
$\lambda = 0.665$	
1258	0.675
1386	0.60
1458	0.53
CURVE 12	
$\lambda = 0.665$	
1283	0.655
1111	0.730
CURVE 13	
$\lambda = 0.665$	
1114	0.430
1266	0.425
1402	0.397
1500	0.400
CURVE 14	
$\lambda = 0.665$	
1414	0.420
1280	0.398
1114	0.455
CURVE 15	
$\lambda = 0.665$	
1280	0.415
1425	0.375
1511	0.365
CURVE 16	
$\lambda = 0.665$	
1391	0.400
1264	0.405
CURVE 17	
$\lambda = 0.665$	
1119	0.413
1358	0.403
1530	0.375
CURVE 18	
$\lambda = 0.665$	
1136	0.425
CURVE 19	
$\lambda = 0.665$	
1122	0.905
1236	0.865
1355	0.830
1466	0.770
CURVE 20	
$\lambda = 0.665$	
1330	0.820
1230	0.835
1119	0.885
CURVE 21	
$\lambda = 0.665$	
1214	0.845
1339	0.860
1483	0.810
CURVE 22	
$\lambda = 0.665$	
1125	0.865
CURVE 23	
$\lambda = 0.665$	
1302	0.825
1455	0.820
CURVE 24	
$\lambda = 0.665$	
1197	0.875

* Not shown on plot

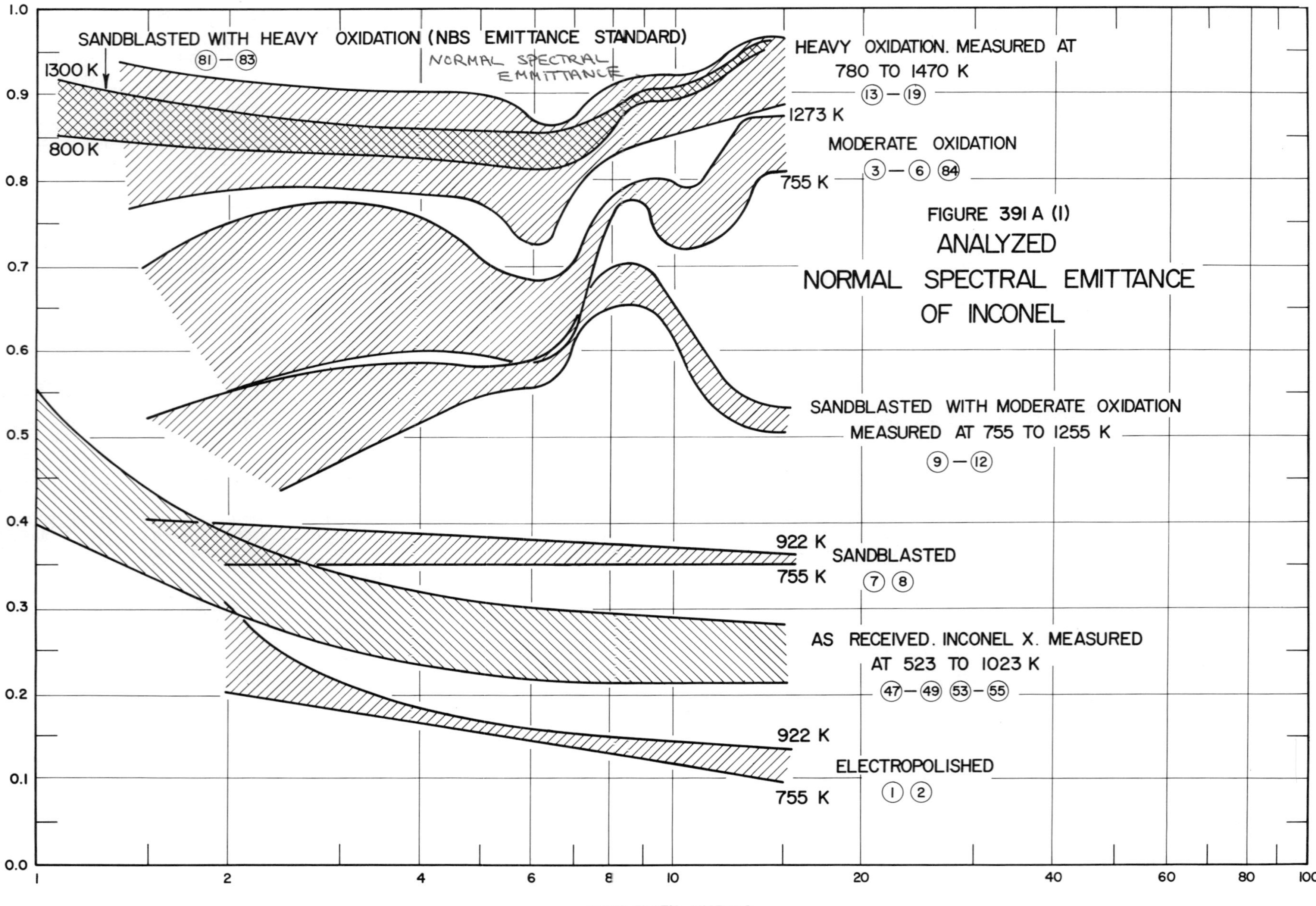

FIGURE 391 A (I) ANALYZED NORMAL SPECTRAL EMITTANCE OF INCONEL

FIGURE 391 A(2)

NORMAL SPECTRAL EMITTANCE OF NICKEL + CHROMIUM + ΣX_i

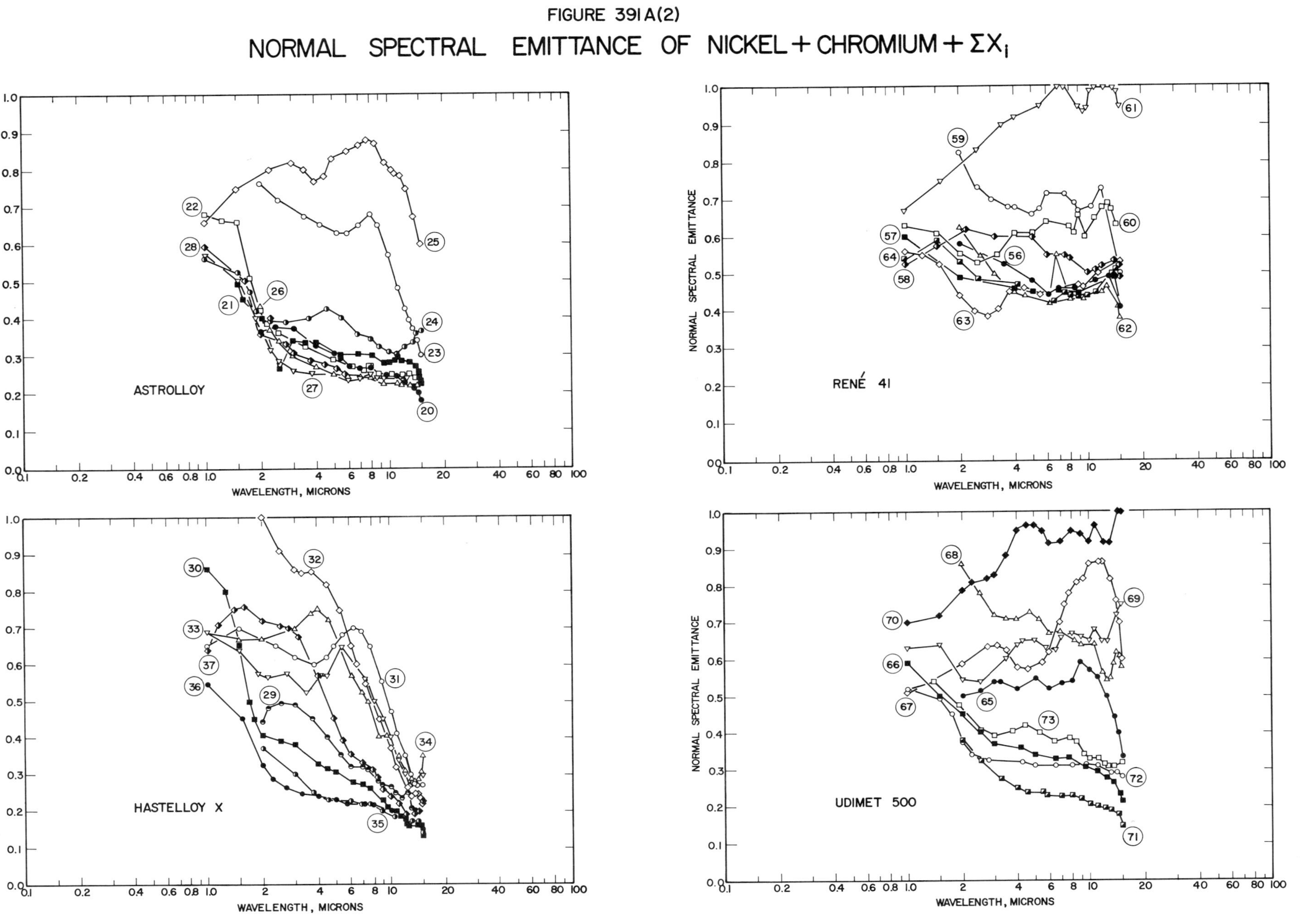

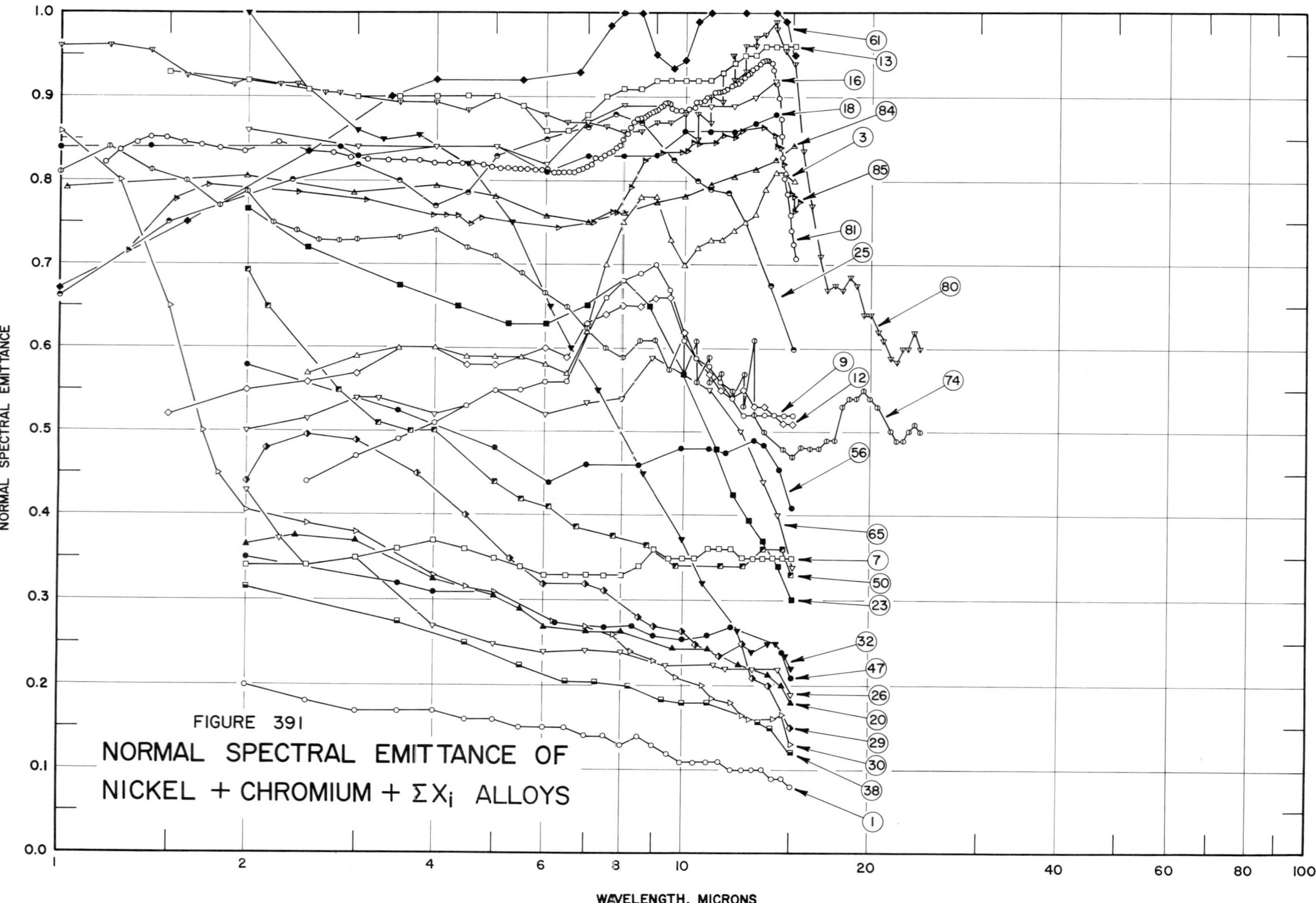

FIGURE 391
NORMAL SPECTRAL EMITTANCE OF NICKEL + CHROMIUM + ΣX_i ALLOYS

SPECIFICATION TABLE NO. 391 NORMAL SPECTRAL EMITTANCE OF [NICKEL + CHROMIUM + ΣX_i] ALLOYS

Curve No.	Ref. No.	Year	Temperature K	Wavelength Range, μ	Geometry θ'	Reported Error, %	Composition (weight percent), Specifications and Remarks
1	170	1959	755	2.0-15.0	~0°	± 4	Inconel; nominal composition: 80 Ni, 14 Cr, 6 Fe; electropolished.
2	170	1959	922	2.0-15.0	~0°	± 4	Above specimen and conditions.
3	170	1959	755	2.5-15.0	~0°	± 4	Different sample, same as curve 1 specimen and conditions; oxidized in air at 1255 K for 1/2 hr.
4	170	1959	922	1.5-15.0	~0°	± 4	Above specimen and conditions.
5	170	1959	1089	1.5-15.0	~0°	± 4	Above specimen and conditions.
6	170	1959	1255	1.5-15.0	~0°	± 4	Above specimen and conditions.
7	170	1959	755	2.0-15.0	~0°	± 4	Different sample, same as curve 1 specimen and conditions except sandblasted with 40-mesh glass sand and air at 40 psi.
8	170	1959	922	1.5-15.0	~0°	± 4	Above specimen and conditions.
9	170	1959	755	2.5-15.0	~0°	± 4	Different sample, same as curve 1 specimen and conditions; oxidized in air at 1255 K for 1/2 hr.
10	170	1959	922	2.0-15.0	~0°	± 4	Above specimen and conditions.
11	170	1959	1089	2.0-15.0	~0°	± 4	Above specimen and conditions.
12	170	1959	1255	1.5-15.0	~0°	± 4	Above specimen and conditions.
13	66	1959	1400	1.5-15.0	~0°	± 3	Inconel; nominal composition: 78 Ni, 15 Cr, 7 Fe, 0.35 Mn, 0.20 Si, 0.04 C; cleaned by sandblasting; heated in air for 6 hrs at high temp to form an oxide coating; average value of several runs.
14	66	1959	1050	1.5-15.0	~0°	± 3	Different sample, same as above specimen and conditions.
15	66	1959	780	1.5-15.0	~0°	± 3	Different sample, same as above specimen and conditions.
16	156	1960	873	2.0-14.0	~0°	± 4	Inconel; nominal composition: 78 Ni, 15 Cr, 7 Fe, 0.35 Mn, 0.20 Si, 0.04 C; oxidized in air; measured in air.
17	156	1960	1273	1.0-14.0	~0°	± 4	Above specimen and conditions.
18	30	1963	1253	1.0-14.0	~0°		Inconel; nominal composition: 78 Ni, 15 Cr, 7 Fe, 0.35 Mn, 0.20 Si, 0.04 C; oxidized; heated at 1250 K for 2 hrs; measured in argon.
19	30	1963	1470	1.0-14.0	~0°		Different sample, same as above specimen and conditions except heated at 1467 K for 2 hrs.
20	86	1961	523	2.00-15.00	~0°	± 5	Astrolloy; nominal composition: 15 Cr, 15 Co, 5 Mo, 4.3 Al, 3.5 Ti, 0.05 C, 0.03 B, Ni balance; as received; data extracted from smooth curve.
21	86	1961	773	1.50-15.00	~0°	± 5	Different sample, same as curve 20 specimen and conditions.
22	86	1961	1023	1.00-15.00	~0°	± 5	Different sample, same as curve 20 specimen and conditions.
23	86	1961	523	2.00-15.00	~0°	± 5	Different sample, same as curve 20 specimen and conditions except heated in air at 1255 K for 30 min.

SPECIFICATION TABLE NO. 391 (continued)

Curve No.	Ref. No.	Year	Temperature K	Wavelength Range, μ	Geometry θ'	Reported Error, %	Composition (weight percent), Specifications and Remarks
24	86	1961	773	1.00-15.00	$\sim 0°$	± 5	Different sample, same as curve 23 specimen and conditions.
25	86	1961	1023	1.00-15.00	$\sim 0°$	± 5	Different sample, same as curve 23 specimen and conditions.
26	86	1961	523	2.00-15.00	$\sim 0°$	± 5	Different sample, same as curve 20 specimen and conditions except heated in vacuum (7.6 x 10^{-5} mm Hg) at 1255 K for 30 min.
27	86	1961	773	1.00-15.00	$\sim 0°$	± 5	Different sample, same as curve 26 specimen and conditions.
28	86	1961	1023	1.00-15.00	$\sim 0°$	± 5	Different sample, same as curve 26 specimen and conditions.
29	86	1961	523	2.00-15.00	$\sim 0°$	± 5	Hastelloy X, AMS 5536 C; nominal composition: 22.0 Cr, 20 Fe, 9 Mo, 0.15 C, Ni balance; as received; data extracted from smooth curve.
30	86	1961	773	1.00-15.00	$\sim 0°$	± 5	Different sample, same as curve 29 specimen and conditions.
31	86	1961	1023	1.00-15.00	$\sim 0°$	± 5	Different sample, same as curve 29 specimen and conditions.
32	86	1961	523	2.00-15.00	$\sim 0°$	± 5	Different sample, same as curve 29 specimen and conditions except heated in argon at 1366 K for 30 min.
33	86	1961	773	1.00-15.00	$\sim 0°$	± 5	Different sample, same as curve 32 specimen and conditions.
34	86	1961	1023	1.00-15.00	$\sim 0°$	± 5	Different sample, same as curve 32 specimen and conditions.
35	86	1961	523	2.00-15.00	$\sim 0°$	± 5	Different sample, same as curve 29 specimen and conditions except heated in vacuum (2.2 x 10^{-5} mm Hg) at 1366 K for 30 min.
36	86	1961	773	1.00-15.00	$\sim 0°$	± 5	Different sample, same as curve 35 specimen and conditions.
37	86	1961	1023	1.00-15.00	$\sim 0°$	± 5	Different sample, same as curve 35 specimen and conditions.
38	86	1961	523	2.00-15.00	$\sim 0°$	± 5	Inconel 702; nominal composition: 78 Ni, 16 Cr, 3.5 Al, 0.5 Ti, 0.35 Fe, 0.1 Mn, 0.03 C; as received; data extracted from smooth curve.
39	86	1961	773	1.50-15.00	$\sim 0°$	± 5	Different sample, same as curve 38 specimen and conditions.
40	86	1961	1023	1.00-15.00	$\sim 0°$	± 5	Different sample, same as curve 38 specimen and conditions.
41	86	1961	523	2.00-15.00	$\sim 0°$	± 5	Different sample, same as curve 38 specimen and conditions except heated in air at 2073 K for 30 min.
42	86	1961	773	1.00-15.00	$\sim 0°$	± 5	Different sample, same as curve 41 specimen and conditions.
43	86	1961	1023	1.00-15.00	$\sim 0°$	± 5	Different sample, same as curve 41 specimen and conditions.
44	86	1961	523	2.00-15.00	$\sim 0°$	± 5	Different sample, same as curve 38 specimen and conditions except heated in vacuum (7.6 x 10^{-5} mm Hg) at 2073 K for 30 min.
45	86	1961	773	1.00-15.00	$\sim 0°$	± 5	Different sample, same as curve 44 specimen and conditions.
46	86	1961	1023	1.00-15.00	$\sim 0°$	± 5	Different sample, same as curve 44 specimen and conditions.

SPECIFICATION TABLE NO. 391 (continued)

Curve No.	Ref. No.	Year	Temperature K	Wavelength Range, μ	Geometry θ'	Reported Error, %	Composition (weight percent), Specifications and Remarks
47	86	1961	523	2.00-15.00	~0°	± 5	Inconel X, AMS 5542; nominal composition: 73 Ni, 15 Cr, 7 Fe, 2.5 Ti, 1 Nb, 0.9 Al, 0.70 Mn, 0.30 Si, 0.04 C; as received; data extracted from smooth curve.
48	86	1961	773	1.00-15.00	~0°	± 5	Different sample, same as curve 47 specimen and conditions.
49	86	1961	1023	1.00-15.00	~0°	± 5	Different sample, same as curve 47 specimen and conditions.
50	86	1961	523	2.00-15.00	~0°	± 5	Different sample, same as curve 47 specimen and conditions except heated in air at 1089 K for 30 min.
51	86	1961	773	1.00-15.00	~0°	± 5	Different sample, same as curve 50 specimen and conditions.
52	86	1961	1023	1.00-15.00	~0°	± 5	Different sample, same as curve 50 specimen and conditions.
53	86	1961	523	2.00-15.00	~0°	± 5	Different sample, same as curve 47 specimen and conditions except heated in vacuum (6.8×10^{-6} mm Hg) at 1089 K for 30 min.
54	86	1961	773	1.00-15.00	~0°	± 5	Different sample, same as curve 53 specimen and conditions.
55	86	1961	1023	1.00-15.00	~0°	± 5	Different sample, same as curve 53 specimen and conditions.
56	86	1961	523	2.00-15.00	~0°	± 5	Rene' 41; nominal composition: 19 Cr, 11 Co, 10 Mo, 3 Fe, 3 Ti, 1.5 Al, 0.1 C, trace B, Ni balance; as received; data extracted from smooth curve.
57	86	1961	773	1.00-15.00	~0°	± 5	Different sample, same as curve 56 specimen and conditions.
58	86	1961	1023	1.00-15.00	~0°	± 5	Different sample, same as curve 56 specimen and conditions.
59	86	1961	523	2.00-15.00	~0°	± 5	Different sample, same as curve 56 specimen and conditions except heated in air at 1089 K for 30 min.
60	86	1961	773	1.00-15.00	~0°	± 5	Different sample, same as curve 59 specimen and conditions.
61	86	1961	1023	1.00-15.00	~0°	± 5	Different sample, same as curve 59 specimen and conditions.
62	86	1961	523	2.00-15.00	~0°	± 5	Different sample, same as curve 56 specimen and conditions except heated in vacuum (7.6×10^{-6} mm Hg) at 1089 K for 30 min.
63	86	1961	773	1.00-15.00	~0°	± 5	Different sample, same as curve 62 specimen and conditions.
64	86	1961	1023	1.00-15.00	~0°	± 5	Different sample, same as curve 62 specimen and conditions.
65	86	1961	523	2.00-15.00	~0°	± 5	Udimet 500; nominal composition: 19 Cr, 19 Co, 4 Mo, 4 max Fe, 3 Ti, 2.9 Al, 0.1 C, trace B, Ni balance; as received; data extracted from smooth curve.
66	86	1961	773	1.00-15.00	~0°	± 5	Different sample, same as curve 65 specimen and conditions.
67	86	1961	1023	1.00-15.00	~0°	± 5	Different sample, same as curve 65 specimen and conditions.
68	86	1961	523	2.00-15.00	~0°	± 5	Different sample, same as curve 65 specimen and conditions except heated in air at 1255 K for 30 min.
69	86	1961	773	1.00-15.00	~0°	± 5	Different sample, same as curve 68 specimen and conditions.
70	86	1961	1023	1.00-15.00	~0°	± 5	Different sample, same as curve 68 specimen and conditions.

SPECIFICATION TABLE NO. 391 (continued)

Curve No.	Ref. No.	Year	Temperature K	Wavelength Range, μ	Geometry θ'	Reported Error, %	Composition (weight percent), Specifications and Remarks
71	86	1961	523	2.00-15.00	~0°	± 5	Different sample, same as curve 65 specimen and conditions except heated in vacuum (7.6 x 10^{-5} mm Hg) at 1255 K for 30 min.
72	86	1961	773	1.00-15.00	~0°	± 5	Different sample, same as curve 71 specimen and conditions.
73	86	1961	1023	1.00-15.00	~0°	± 5	Different sample, same as curve 71 specimen and conditions.
74	65	1962	1033	1.0-24.0	~0°		Inconel; nominal composition: 78 Ni, 15 Cr, 7 Fe, 0.35 Mn, 0.20 Si, 0.04 C; oxidized; measured in air.
75	65	1962	1061	1.60-14.00	~0°		Different sample, same as curve 74 specimen and conditions except measured in argon.
76	65	1962	1089	1.25-14.00	~0°		Different sample, same as curve 74 specimen and conditions.
77	65	1962	1089	1.60-14.00	~0°		Different sample, same as curve 74 specimen and conditions.
78	65	1962	1255	1.20-14.00	~0°		Different sample, same as curve 74 specimen and conditions except measured in vacuum.
79	65	1962	1011	1.00-24.00	~0°		M 252; nominal composition: 19 Cr, 10 Co, 10 Mo, 2.50 Ti, 2.0 Fe, 1.00 Al, 0.5 Mn, 0.5 Si, 0.15 C, 0.005 B, Ni balance; oxidized; measured in air.
80	65	1962	1041	1.00-24.00	~0°		Rene' 41; nominal composition: 19 Cr, 11 Co, 10 Mo, 3 Fe, 3 Ti, 1.5 Al, 0.1 C, Ni balance; oxidized, measured in air.
81	298	1964	800	1.18-15.11	~0°		Inconel; nominal composition: 78 Ni, 15 Cr, 7 Fe, 0.35 Mn, 0.20 Si, 0.04 C; sandblasted; cleaned and rinsed; oxidized at 1340 K for 24 hrs, then at 1100 K for 24 hrs; measured in nitrogen; computed from $\rho(0°, 2\pi)$.
82	298	1964	1100	1.07-15.11	~0°		Different sample, same as above specimen and conditions.
83	298	1964	1300	1.07-15.11	~0°		Different sample, same as curve 81 specimen and conditions.
84	201	1963	1273	1.02-14.98	~0°	5	Inconel; nominal composition: 78 Ni, 15 Cr, 7 Fe, 0.35 Mn, 0.20 Si, 0.04 C; oxidized in air at 1273 K for 20 min.
85	254	1963	800	1.28-15.26	~0°		Inconel; nominal composition: 78 Ni, 15 Cr, 7 Fe, 0.35 Mn, 0.20 Si, 0.04 C; machined from 0.053 in. Inconel sheet; cleaned with acetone; sandblasted with 60-mesh fused alumina grit at 70 psi; cleaned ultrasonically in acetone, passivated for one min in 10% nitric acid at 316 K, rinsed in distilled water and distilled acetone; heated at 1340 for 24 hrs, then cooled; mean of 3 determinations on each of 6 specimens.
86	254	1963	1100	1.10-15.19	~0°		Above specimen and conditions.
87	254	1963	1300	1.09-15.17	~0°		Above specimen and conditions.

DATA TABLE NO. 391 NORMAL SPECTRAL EMITTANCE OF [NICKEL + CHROMIUM + ΣX_i] ALLOYS

[Wavelength, λ, μ; Emittance, ϵ; Temperature, T, K]

CURVE 1 (T = 755)

λ	ϵ
2.0	0.20
2.5	0.18
3.0	0.17
3.5	0.17
4.0	0.17
4.5	0.16
5.0	0.16
5.5	0.15
6.0	0.15
6.5	0.15
7.0	0.14
7.5	0.14
8.0	0.13
8.5	0.14
9.0	0.13
9.5	0.12
10.0	0.11
10.5	0.11
11.0	0.11
11.5	0.11
12.0	0.10
12.5	0.10
13.0	0.10
13.5	0.10
14.0	0.09
14.5	0.09
15.0	0.08

CURVE 2* (T = 922)

λ	ϵ
2.0	0.30
2.5	0.24
3.0	0.21
3.5	0.19
4.0	0.18
4.5	0.17
5.0	0.16
5.5	0.17
6.0	0.16
6.5	0.15
7.0	0.15
7.5	0.15
8.0	0.15
8.5	0.15
9.0	0.15
9.5	0.15
10.0	0.14
10.5	0.14
11.0	0.14
11.5	0.14
12.0	0.14
12.5	0.13
13.0	0.13
13.5	0.13
14.0	0.13
14.5	0.12
15.0	0.12

CURVE 3 (T = 755)

λ	ϵ
2.5	0.57
3.0	0.59
3.5	0.60
4.0	0.60
4.5	0.59
5.0	0.59
5.5	0.59
6.0	0.58
6.5	0.57
7.0	0.62
7.5	0.70
8.0	0.75
8.5	0.78
9.0	0.78
9.5	0.73
10.0	0.70
10.5	0.72
11.0	0.73
11.5	0.73
12.0	0.74
12.5	0.75
13.0	0.76
13.5	0.79
14.0	0.81
14.5	0.81
15.0	0.80

CURVE 4* (T = 922)

λ	ϵ
1.5	0.58
2.0	0.59
2.5	0.60
3.0	0.61
3.5	0.61
4.0	0.63
4.5	0.62
5.0	0.61
5.5	0.60
6.0	0.59
6.5	0.59
7.0	0.64
7.5	0.71
8.0	0.75
8.5	0.78
9.0	0.79
9.5	0.79
10.0	0.80
10.5	0.78
11.0	0.77
11.5	0.77
12.0	0.77
12.5	0.78
13.0	0.79
13.5	0.81
14.0	0.84
14.5	0.83
15.0	0.82

CURVE 5* (T = 1089)

λ	ϵ
1.5	0.68
2.0	0.71
2.5	0.72
3.0	0.69
3.5	0.71
4.0	0.69
4.5	0.66
5.0	0.65
5.5	0.65
6.0	0.64
6.5	0.66
7.0	0.68
7.5	0.74
8.0	0.78
8.5	0.78
9.0	0.79
9.5	0.78
10.0	0.78
10.5	0.77
11.0	0.76
11.5	0.76
12.0	0.78
12.5	0.79
13.0	0.80
13.5	0.81
14.0	0.82
14.5	0.83
15.0	0.83

CURVE 6* (T = 1255)

λ	ϵ
1.5	0.70
2.0	0.75
2.5	0.76
3.0	0.75
3.5	0.79
4.0	0.76
4.5	0.72
5.0	0.71
5.5	0.69
6.0	0.68
6.5	0.68
7.0	0.70
7.5	0.74
8.0	0.77
8.5	0.79
9.0	0.80
9.5	0.80
10.0	0.78
10.5	0.78
11.0	0.79
11.5	0.81
12.0	0.83
12.5	0.85
13.0	0.87
13.5	0.87
14.0	0.87
14.5	0.85
15.0	0.84

CURVE 7 (T = 755)

λ	ϵ
2.0	0.34
2.5	0.34
3.0	0.35
3.5	0.36
4.0	0.37
4.5	0.36
5.0	0.35
5.5	0.34
6.0	0.33
6.5	0.33
7.0	0.33
7.5	0.33
8.0	0.33
8.5	0.34
9.0	0.36
9.5	0.35
10.0	0.35
10.5	0.35
11.0	0.36
11.5	0.36
12.0	0.36
12.5	0.35
13.0	0.35
13.5	0.35
14.0	0.35
14.5	0.35
15.0	0.35

CURVE 8* (T = 922)

λ	ϵ
1.5	0.41
2.0	0.38
2.5	0.41
3.0	0.41
3.5	0.40
4.0	0.39
4.5	0.39
5.0	0.38
5.5	0.37
6.0	0.36
6.5	0.36
7.0	0.36
7.5	0.35
8.0	0.35
8.5	0.36
9.0	0.37
9.5	0.36
10.0	0.36
10.5	0.36
11.0	0.36
11.5	0.37
12.0	0.37
12.5	0.37
13.0	0.37
13.5	0.39
14.0	0.39
14.5	0.39
15.0	0.38

CURVE 9 (T = 755)

λ	ϵ
2.5	0.44
3.0	0.47
3.5	0.49
4.0	0.51
4.5	0.53
5.0	0.55
5.5	0.55
6.0	0.56
6.5	0.56
7.0	0.62*
7.5	0.66
8.0	0.68
8.5	0.69
9.0	0.70
9.5	0.67
10.0	0.61
10.5	0.59
11.0	0.58
11.5	0.56
12.0	0.54
12.5	0.52
13.0	0.52
13.5	0.52
14.0	0.52
14.5	0.52
15.0	0.52

CURVE 10* (T = 922)

λ	ϵ
2.0	0.50
2.5	0.51
3.0	0.51
3.5	0.53
4.0	0.54
4.5	0.55
5.0	0.57
5.5	0.56
6.0	0.56
6.5	0.56
7.0	0.63
7.5	0.66
8.0	0.67
8.5	0.68
9.0	0.67
9.5	0.75
10.0	0.61
10.5	0.60
11.0	0.58
11.5	0.57
12.0	0.53
12.5	0.52
13.0	0.51
13.5	0.52
14.0	0.53
14.5	0.53
15.0	0.53

CURVE 11* (T = 1089)

λ	ϵ
2.0	0.49
2.5	0.51
3.0	0.53
3.5	0.54
4.0	0.56
4.5	0.57
5.0	0.57
5.5	0.58
6.0	0.58
6.5	0.59
7.0	0.65
7.5	0.67
8.0	0.69
8.5	0.70
9.0	0.70
9.5	0.61
10.0	0.65
10.5	0.63
11.0	0.61
11.5	0.58
12.0	0.56
12.5	0.54
13.0	0.53
13.5	0.53
14.0	0.53
14.5	0.53
15.0	0.53

CURVE 12 (T = 1255)

λ	ϵ
1.5	0.52
2.0	0.55
2.5	0.56
3.0	0.57
3.5	0.60*
4.0	0.60*
4.5	0.58
5.0	0.58
5.5	0.59*
6.0	0.60
6.5	0.59
7.0	0.63

* Not shown on plot

DATA TABLE NO. 391 (continued)

λ	∈
CURVE 12 (cont.)	
7.5	0.64
8.0	0.65
8.5	0.65
9.0	0.66
9.5	0.66
10.0	0.62
10.5	0.59*
11.0	0.57
11.5	0.55
12.0	0.54*
12.5	0.55
13.0	0.53
13.5	0.53
14.0	0.52*
14.5	0.51
15.0	0.51
CURVE 13 T = 1400	
1.5	0.93
2.0	0.92
2.5	0.91
3.0	0.90
3.5	0.90
4.0	0.90
4.5	0.90
5.0	0.90
5.5	0.89
6.0	0.86
6.5	0.86
7.0	0.88
7.5	0.90
8.0	0.91
8.5	0.91
9.0	0.92
9.5	0.92
10.0	0.92
10.5	0.92
11.0	0.92
11.5	0.93
12.0	0.94
12.5	0.95
13.0	0.95
13.5	0.96
CURVE 13 (cont.)	
14.0	0.96
14.5	0.96
15.0	0.96
CURVE 14* T = 1050	
1.5	0.88
2.0	0.88
2.5	0.86
3.0	0.86
3.5	0.85
4.0	0.85
4.5	0.84
5.0	0.84
5.5	0.83
6.0	0.80
6.5	0.80
7.0	0.84
7.5	0.87
8.0	0.90
8.5	0.90
9.0	0.90
9.5	0.90
10.0	0.89
10.5	0.90
11.0	0.90
11.5	0.91
12.0	0.92
12.5	0.93
13.0	0.94
13.5	0.94
14.0	0.94
14.5	0.94
15.0	0.94
CURVE 15* T = 780	
1.5	0.77
2.0	0.79
2.5	0.78
3.0	0.77
3.5	0.78
4.0	0.78
CURVE 15* (cont.)	
4.5	0.78
5.0	0.77
5.5	0.77
6.0	0.72
6.5	0.74
7.0	0.78
7.5	0.82
8.0	0.85
8.5	0.87
9.0	0.87
9.5	0.86
10.0	0.86
10.5	0.87
11.0	0.88
11.5	0.88
12.0	0.89
12.5	0.91
13.0	0.92
13.5	0.93
14.0	0.90
14.5	0.89
15.0	0.87
CURVE 16 T = 873	
2.0	0.86
3.0	0.84
5.0	0.84
6.0	0.82
7.0	0.87
8.0	0.89
11.0	0.89
12.0	0.89
13.0	0.90
14.0	0.92
CURVE 17* T = 1273	
1.0	0.90
2.0	0.88
3.0	0.88
4.0	0.89
CURVE 17* (cont.)	
5.0	0.88
6.0	0.87
7.0	0.90
9.0	0.91
11.0	0.91
12.0	0.92
13.0	0.92
14.0	0.93
CURVE 18 T = 1253	
1.0	0.84
1.4	0.84
2.8	0.84
3.0	0.83
4.0	0.84*
5.0	0.84*
6.0	0.81
7.0	0.83
8.0	0.83
9.0	0.83
10.0	0.86
11.0	0.86
12.0	0.86
13.0	0.87
14.0	0.88
CURVE 19* T = 1470	
1.0	0.915
1.4	0.875
1.8	0.880
2.0	0.870
2.8	0.870
3.0	0.870
4.0	0.860
5.0	0.850
6.0	0.840
7.0	0.855
8.0	0.845
9.0	0.880
10.0	0.900
CURVE 19* (cont.)	
11.0	0.910
12.0	0.910
13.0	0.900
14.0	0.920
CURVE 20 T = 523	
2.00	0.365
2.40	0.375
3.00	0.370
4.00	0.325
5.00	0.305
5.50	0.290
6.00	0.270
7.00	0.265
8.00	0.265
9.65	0.245
11.00	0.245
12.35	0.225
13.75	0.215
14.50	0.200
15.00	0.180
CURVE 21* T = 773	
1.50	0.490
1.60	0.450
2.00	0.400
2.50	0.365
3.00	0.340
3.50	0.335
4.00	0.335
5.50	0.300
6.75	0.300
8.10	0.300
9.50	0.280
10.00	0.280
11.20	0.290
11.95	0.280
13.00	0.280
14.00	0.270
14.50	0.255
14.75	0.240
15.00	0.220
CURVE 22* T = 1023	
1.00	0.680
1.25	0.650
1.50	0.600
1.75	0.510
2.00	0.420
2.15	0.385
2.50	0.360
3.50	0.325
5.00	0.285
6.25	0.270
7.25	0.260
7.75	0.270
8.00	0.270
8.75	0.250
10.50	0.250
12.00	0.250
13.00	0.250
14.00	0.240
15.00	0.230
CURVE 23 T = 523	
2.00	0.765
2.50	0.720
3.50	0.675
4.35	0.650
5.25	0.630
6.00	0.630
7.00	0.650
8.00	0.680
8.80	0.650
10.00	0.570
11.25	0.480
12.10	0.425
12.75	0.395
13.50	0.370
14.25	0.340
15.00	0.300
CURVE 24* T = 773	
1.00	0.560
1.50	0.520
CURVE 24* (cont.)	
1.75	0.470
1.90	0.420
2.25	0.392
2.75	0.390
3.65	0.400
4.60	0.425
5.50	0.400
6.75	0.355
8.00	0.345
8.75	0.325
10.00	0.310
11.00	0.305
12.10	0.325
13.50	0.350
14.00	0.360
15.00	0.365
CURVE 25 T = 1023	
1.00	0.660
1.50	0.750
2.35	0.800
3.00	0.820
3.50	0.800
4.00	0.770
4.50	0.785
5.00	0.830
6.00	0.850
7.00	0.865
7.75	0.880
8.50	0.870
9.60	0.825
10.50	0.800
11.00	0.790
11.75	0.786
12.50	0.750
13.75	0.675
15.00	0.600
CURVE 26 T = 523	
2.00	0.430
2.25	0.370
2.50	0.340*
CURVE 26 (cont.)	
3.00	0.300*
4.00	0.270
5.00	0.250
6.00	0.240
7.00	0.242
8.00	0.240
9.40	0.225
11.25	0.225
11.75	0.220
13.00	0.220
14.25	0.220
15.00	0.190
CURVE 27* T = 773	
1.00	0.570
1.50	0.500
1.85	0.400
2.00	0.360
2.25	0.315
2.50	0.285
3.00	0.260
3.75	0.250
5.00	0.250
6.00	0.232
7.00	0.240
8.00	0.250
8.60	0.240
9.50	0.240
11.00	0.240
12.50	0.240
14.00	0.240
15.00	0.230
CURVE 28* T = 1023	
1.00	0.595
1.65	0.500
2.25	0.400
2.75	0.330
3.00	0.305
3.75	0.285
4.50	0.275
5.20	0.265

* Not shown on plot

DATA TABLE NO. 391 (continued)

λ	∈
CURVE 28* (cont.)	
5.75	0.250
7.00	0.260
8.00	0.260
9.00	0.245
9.75	0.240
11.00	0.240
12.50	0.240
14.00	0.250
15.00	0.260
CURVE 29 T = 523	
2.00	0.440
2.15	0.480
2.50	0.497
3.00	0.490
3.75	0.450
4.50	0.400
5.30	0.350
6.00	0.320
7.00	0.320
7.50	0.312
8.50	0.280
9.00	0.270
10.00	0.265
10.50	0.250
11.50	0.235
12.50	0.250
13.00	0.210
13.75	0.200
15.00	0.150
CURVE 30 T = 773	
1.00	0.860
1.25	0.800
1.50	0.650
1.70	0.500
1.80	0.450
2.00	0.405
2.50	0.390
3.00	0.380
4.00	0.330
4.50	0.315
CURVE 30 (cont.)	
5.00	0.310
6.20	0.275
7.00	0.270
7.75	0.260
8.25	0.240
9.00	0.230
9.75	0.210
10.75	0.200
11.20	0.185
12.00	0.180
12.50	0.165
12.75	0.160
14.00	0.160
14.50	0.165
15.00	0.130
CURVE 31* T = 1023	
1.00	0.650
1.50	0.700
2.35	0.650
3.00	0.620
3.85	0.600
4.50	0.620
4.90	0.650
5.50	0.682
6.30	0.700
7.00	0.690
7.80	0.650
9.00	0.550
10.00	0.470
10.75	0.410
12.00	0.350
13.00	0.300
13.50	0.275
14.00	0.270
15.00	0.270
CURVE 32 T = 523	
2.00	1.000
2.50	0.910*
3.00	0.860
3.30	0.850

λ	∈
CURVE 32 (cont.)	
3.75	0.855
4.50	0.820
5.30	0.750
6.10	0.650
6.60	0.600
7.30	0.550
8.65	0.450
10.00	0.370
10.75	0.320
12.25	0.265
12.50	0.250*
13.00	0.240
13.75	0.250
14.10	0.250
14.65	0.235
15.00	0.220
CURVE 33* T = 773	
1.00	0.690
1.50	0.640
1.90	0.575
2.15	0.565
2.70	0.575
3.50	0.625
4.15	0.675
4.50	0.677
5.40	0.650
6.50	0.590
7.15	0.550
8.20	0.500
9.00	0.450
10.00	0.400
11.00	0.340
12.00	0.300
12.50	0.275
13.00	0.260
14.00	0.270
15.00	0.300
CURVE 34* T = 1023	
1.00	0.690
1.50	0.669
CURVE 34* (cont.)	
2.00	0.670
3.00	0.700
3.65	0.740
4.00	0.750
4.60	0.725
5.40	0.650
6.05	0.575
7.00	0.525
7.55	0.500
8.35	0.450
9.50	0.405
11.00	0.350
12.00	0.310
13.00	0.285
13.50	0.280
14.10	0.285
14.60	0.310
15.00	0.350
CURVE 35* T = 523	
2.00	0.370
3.00	0.300
3.75	0.250
4.50	0.230
6.00	0.225
7.15	0.220
7.70	0.220
9.00	0.200
10.50	0.185
12.00	0.170
13.00	0.170
14.00	0.160
14.50	0.155
15.00	0.140
CURVE 36* T = 773	
1.00	0.545
1.55	0.450
2.00	0.325
2.25	0.285
2.65	0.262
3.25	0.245

λ	∈
CURVE 36* (cont.)	
4.00	0.240
5.00	0.230
5.75	0.220
7.00	0.220
8.00	0.220
10.00	0.200
11.00	0.190
12.75	0.170
13.75	0.174
15.00	0.160
CURVE 37* T = 1023	
1.00	0.640
1.15	0.710
1.40	0.750
1.60	0.760
2.00	0.720
2.50	0.705
2.80	0.700
3.15	0.675
4.00	0.570
4.90	0.450
5.50	0.390
6.00	0.355
7.00	0.330
8.00	0.310
8.50	0.290
9.00	0.260
10.00	0.240
11.00	0.220
11.75	0.195
12.00	0.190
13.25	0.190
14.40	0.200
15.00	0.230
CURVE 38 T = 523	
2.00	0.315
3.50	0.275
4.50	0.250
5.50	0.225
6.50	0.205
CURVE 38 (cont.)	
7.25	0.205
8.20	0.200
9.25	0.185
10.00	0.180
11.00	0.180
12.50	0.165*
13.25	0.157
13.95	0.150
15.00	0.120
CURVE 39* T = 773	
1.50	0.290
2.00	0.290
2.95	0.250
3.50	0.245
4.00	0.245
5.25	0.225
6.15	0.210
6.50	0.210
7.00	0.245
7.85	0.245
9.00	0.200
10.00	0.200
11.50	0.190
12.50	0.185
13.00	0.180
13.85	0.150
14.30	0.145
15.00	0.140
CURVE 40* T = 1023	
1.00	0.470
2.00	0.570
2.50	0.635
2.75	0.675
3.05	0.700
3.45	0.705
4.00	0.690
5.00	0.590
5.35	0.555
6.00	0.505
6.25	0.475

λ	∈
CURVE 40* (cont.)	
6.70	0.450
7.90	0.425
9.00	0.380
9.50	0.362
10.50	0.355
11.50	0.325
12.50	0.310
14.00	0.275
15.00	0.260
CURVE 41* T = 523	
2.00	0.580
2.75	0.500
3.25	0.465
4.00	0.455
4.50	0.430
5.00	0.390
6.50	0.335
8.00	0.295
9.00	0.270*
10.00	0.240
10.50	0.230
12.00	0.235
13.50	0.240
14.25	0.237
15.00	0.210
CURVE 42* T = 773	
1.00	0.590
1.40	0.550
1.60	0.500
2.00	0.400
2.50	0.360
3.50	0.325
4.50	0.310
5.50	0.305
6.50	0.300
7.15	0.300
7.55	0.310
8.20	0.330
9.00	0.320
10.00	0.320
CURVE 42* (cont.)	
10.25	0.315
10.70	0.320
11.00	0.330
11.50	0.390
12.45	0.500
13.00	0.565
13.50	0.610
14.00	0.625
14.50	0.635
15.00	0.670
CURVE 43* T = 1023	
1.00	0.750
1.75	0.770
2.25	0.770
3.00	0.780
4.30	0.825
5.00	0.855
5.50	0.855
5.85	0.850
7.25	0.860
7.75	0.870
8.00	0.870
9.00	0.810
10.00	0.725
10.80	0.650
12.00	0.600
13.00	0.580
13.50	0.580
14.10	0.600
14.50	0.615
15.00	0.620
CURVE 44* T = 523	
2.00	0.560
2.25	0.500
2.50	0.460
3.00	0.420
3.50	0.395
4.00	0.365
5.25	0.355
6.00	0.350

* Not shown on plot

DATA TABLE NO. 391 (continued)

CURVE 44*(cont.)

λ	ε
7.50	0.350
9.00	0.350
10.50	0.335
11.50	0.325
12.50	0.310
13.50	0.295
14.50	0.275
15.00	0.260

CURVE 45*
T = 773

λ	ε
1.00	0.470
1.75	0.350
2.00	0.290
2.25	0.250
2.75	0.215
3.25	0.200
4.25	0.200
5.25	0.270
6.00	0.190
8.00	0.200
9.00	0.190
10.00	0.190
11.15	0.180
12.30	0.180
12.75	0.170
14.00	0.170
15.00	0.170

CURVE 46*
T = 1023

λ	ε
1.00	0.550
1.25	0.475
1.75	0.400
2.25	0.340
3.00	0.270
3.60	0.250
4.25	0.240
5.25	0.240
6.00	0.220
6.75	0.220
7.70	0.225
8.00	0.230
8.75	0.225

CURVE 46*(cont.)

λ	ε
9.50	0.210
11.00	0.210
12.00	0.210
12.75	0.200
15.00	0.210

CURVE 47
T = 523

λ	ε
2.00	0.350
3.50	0.320
4.00	0.310
5.00	0.310*
6.25	0.275
7.50	0.270
8.30	0.270
9.00	0.260
10.00	0.255
11.00	0.260
12.00	0.270
14.00	0.250*
14.50	0.240
15.00	0.210

CURVE 48*
T = 773

λ	ε
1.00	0.545
1.10	0.500
1.50	0.440
2.00	0.385
2.50	0.350
3.50	0.310
4.50	0.290
5.00	0.280
6.50	0.280
7.25	0.275
8.00	0.280
10.00	0.280
11.00	0.270
12.25	0.270
12.90	0.280
14.00	0.255
15.00	0.220

CURVE 49*
T = 1023

λ	ε
1.00	0.400
1.50	0.490
1.75	0.550
2.00	0.580
2.50	0.580
3.00	0.570
4.00	0.510
4.75	0.485
6.00	0.405
7.00	0.375
8.00	0.350
9.00	0.320
10.00	0.310
11.00	0.310
12.00	0.300
13.50	0.300
14.50	0.300
15.00	0.290

CURVE 50
T = 523

λ	ε
2.00	0.695
2.15	0.650
2.80	0.550
3.25	0.510
3.65	0.500
4.00	0.500
5.00	0.440
5.50	0.420
6.10	0.410
6.75	0.385
7.75	0.375
8.75	0.365
9.75	0.340
11.50	0.340
12.50	0.340
13.50	0.360
14.50	0.360
15.00	0.330

CURVE 51*
T = 773

λ	ε
1.00	0.560
1.50	0.480

CURVE 51*(cont.)

λ	ε
1.75	0.400
1.90	0.365
2.25	0.360
2.75	0.365
4.00	0.400
5.50	0.350
6.50	0.335
7.25	0.330
7.90	0.330
9.00	0.320
9.90	0.320
11.10	0.310
12.50	0.330
13.75	0.355
15.00	0.380

CURVE 52*
T = 1023

λ	ε
1.00	0.600
1.50	0.620
2.00	0.600
2.65	0.575
3.50	0.595
4.25	0.615
5.00	0.600
5.80	0.560
6.15	0.560
7.35	0.590
8.25	0.588
9.00	0.575
10.00	0.570
11.10	0.550
12.25	0.525
13.00	0.500
14.50	0.490
15.00	0.490

CURVE 53*
T = 523

λ	ε
2.00	0.400
2.50	0.325
3.40	0.275
4.00	0.265
5.50	0.265

CURVE 53*(cont.)

λ	ε
6.00	0.260
7.65	0.260
8.00	0.265
8.75	0.260
10.00	0.260
10.75	0.250
12.00	0.250
14.00	0.240
14.50	0.230
15.00	0.210

CURVE 54*
T = 773

λ	ε
1.00	0.440
1.50	0.380
2.00	0.280
2.50	0.230
3.00	0.210
3.40	0.205
4.25	0.225
4.50	0.225
5.25	0.210
6.00	0.210
7.50	0.225
8.00	0.230
8.75	0.220
9.75	0.227
11.00	0.240
12.75	0.240
14.50	0.240
15.00	0.230

CURVE 55*
T = 1023

λ	ε
1.00	0.510
1.50	0.450
2.00	0.375
2.50	0.295
3.00	0.260
3.75	0.250
5.00	0.250
5.90	0.230
7.00	0.240
8.25	0.250

CURVE 55*(cont.)

λ	ε
9.00	0.240
9.50	0.240
11.00	0.250
12.00	0.250
13.00	0.250
14.50	0.265
15.00	0.280

CURVE 56
T = 523

λ	ε
2.00	0.580
3.50	0.525
5.00	0.480
6.10	0.440
7.00	0.460
8.50	0.460
9.90	0.480
11.00	0.480
11.70	0.475
13.00	0.490
13.50	0.485
14.25	0.455
15.00	0.410

CURVE 57*
T = 773

λ	ε
1.00	0.600
1.20	0.550
1.50	0.510
2.00	0.490
2.30	0.490
2.75	0.475
4.00	0.460
5.00	0.450
5.90	0.430
7.00	0.450
7.75	0.460
9.00	0.450
9.50	0.450
11.00	0.480
11.50	0.485
12.90	0.485
14.00	0.490
14.60	0.457
15.00	0.410

CURVE 58*
T = 1023

λ	ε
1.00	0.525
1.50	0.575
1.70	0.605
2.15	0.620
2.75	0.610
3.10	0.600
4.10	0.610
5.00	0.600
5.50	0.570
6.00	0.550
7.50	0.550
8.00	0.540
8.60	0.515
9.50	0.503
10.00	0.500
11.00	0.510
12.00	0.520
12.75	0.515
14.00	0.535
14.50	0.530
14.80	0.520
15.00	0.490

CURVE 59*
T = 523

λ	ε
2.00	0.830
2.20	0.775
2.50	0.735
3.00	0.700
3.70	0.680
4.00	0.680
4.70	0.660
5.00	0.660
5.50	0.675
6.00	0.715
6.40	0.725
7.50	0.715
8.50	0.690
9.00	0.670
9.50	0.665
10.50	0.680
11.25	0.700
12.00	0.730
12.25	0.730

CURVE 59*(cont.)

λ	ε
12.75	0.720
13.85	0.650
14.40	0.600
14.75	0.550
15.00	0.500

CURVE 60*
T = 773

λ	ε
1.00	0.630
1.50	0.610
2.00	0.555
2.50	0.530
2.80	0.530
3.20	0.550
3.75	0.600
4.00	0.610
5.00	0.610
6.00	0.640
6.70	0.650
8.00	0.630
8.50	0.610
9.00	0.600
9.75	0.600
11.00	0.650
12.00	0.680
13.00	0.690
13.50	0.675
14.25	0.635
15.00	0.570

CURVE 61
T = 1023

λ	ε
1.00	0.670
1.60	0.750
2.50	0.835
3.40	0.900
4.00	0.920
5.50	0.920
5.80	0.930
6.60	0.985
7.00	1.000
7.50	1.000
9.00	0.950
9.60	0.935

* Not shown on plot

DATA TABLE NO. 391 (continued)

λ	∈
CURVE 61 (cont.)	
10.00	0.945
10.50	0.990
11.00	1.000
12.50	1.000
14.00	1.000
14.50	0.990
15.00	0.950
CURVE 62* T = 523	
2.00	0.625
2.60	0.550
3.10	0.500
4.00	0.450
4.50	0.440
5.15	0.440
5.75	0.425
6.20	0.420
6.75	0.425
8.00	0.430
9.50	0.430
10.15	0.430
10.90	0.450
12.00	0.450
12.75	0.465
13.50	0.460
14.25	0.435
14.70	0.410
15.00	0.380
CURVE 63* T = 773	
1.00	0.560
1.25	0.550
1.55	0.525
2.00	0.440
2.40	0.400
2.85	0.385
3.25	0.405
3.75	0.450
4.00	0.460
4.50	0.460
5.00	0.450
5.50	0.440

λ	∈
CURVE 63* (cont.)	
6.50	0.445
7.50	0.455
8.50	0.465
9.00	0.470
9.50	0.465
10.50	0.480
11.50	0.500
12.75	0.525
13.75	0.540
14.50	0.542
15.00	0.530
CURVE 64* T = 1023	
1.00	0.540
1.25	0.580
1.50	0.590
1.80	0.575
2.00	0.530
2.50	0.485
2.90	0.470
4.15	0.470
5.00	0.450
6.50	0.425
7.50	0.445
8.00	0.440
9.00	0.440
10.00	0.440
11.00	0.450
11.75	0.450
13.50	0.500
14.25	0.515
15.00	0.500
CURVE 65 T = 523	
2.00	0.500
2.50	0.515
3.00	0.540
3.25	0.540
4.00	0.520
4.50	0.530*
5.00	0.550*
6.00	0.520

λ	∈
CURVE 65 (cont.)	
7.00	0.535
8.00	0.540
8.90	0.590
10.00	0.570*
11.00	0.550
12.40	0.500
13.50	0.440
14.20	0.400
15.00	0.335
CURVE 66* T = 773	
1.00	0.590
1.10	0.550
1.50	0.500
2.00	0.450
2.50	0.400
3.00	0.370
3.50	0.360
4.15	0.360
5.00	0.340
5.57	0.330
6.50	0.330
7.75	0.331
8.25	0.325
9.50	0.305
11.00	0.290
12.15	0.275
13.50	0.265
14.25	0.255
14.75	0.235
15.00	0.215
CURVE 67* T = 1023	
1.00	0.510
2.00	0.590
2.75	0.635
3.10	0.640
3.50	0.625
4.00	0.580
4.25	0.575
4.60	0.575
5.50	0.590

λ	∈
CURVE 67* (cont.)	
6.00	0.620
6.85	0.700
7.30	0.750
7.80	0.780
8.50	0.810
9.25	0.820
10.00	0.860
11.30	0.868
11.75	0.865
13.00	0.820
14.00	0.760
14.60	0.700
14.85	0.650
15.00	0.600
CURVE 68* T = 523	
2.00	0.860
2.50	0.780
3.00	0.720
3.50	0.700
4.00	0.710
4.65	0.730
5.25	0.710
6.00	0.670
6.25	0.670
6.90	0.675
8.25	0.650
9.00	0.640
9.80	0.650
10.50	0.640
11.30	0.600
11.90	0.565
12.50	0.545
12.75	0.545
13.00	0.550
13.50	0.590
14.00	0.620
14.50	0.610
15.00	0.580
CURVE 69* T = 773	
1.00	0.630
1.25	0.640

λ	∈
CURVE 69* (cont.)	
1.50	0.640
1.75	0.600
2.00	0.545
2.25	0.540
2.50	0.540
2.85	0.550
3.50	0.600
4.00	0.640
4.25	0.650
5.00	0.650
5.75	0.630
6.10	0.620
6.50	0.625
7.00	0.660
8.00	0.670
9.00	0.660
10.00	0.650
10.80	0.680
11.05	0.680
11.75	0.650
12.50	0.630
12.85	0.635
14.00	0.720
14.70	0.750
15.00	0.750
CURVE 70* T = 1023	
1.00	0.700
1.50	0.720
2.00	0.790
2.25	0.810
2.75	0.820
3.00	0.830
3.50	0.885
4.00	0.950
4.50	0.965
5.00	0.965
5.50	0.950
6.00	0.915
6.25	0.910
7.00	0.920
8.00	0.950
9.00	0.940
10.00	0.920

λ	∈
CURVE 70* (cont.)	
10.25	0.925
10.85	0.965
12.00	0.920
13.00	0.920
13.75	0.960
14.50	1.000
15.00	1.000
CURVE 71* T = 523	
2.00	0.380
2.50	0.325
3.25	0.275
4.00	0.250
4.50	0.240
5.50	0.240
5.75	0.230
7.00	0.230
8.00	0.230
9.00	0.220
10.00	0.205
11.00	0.200
12.00	0.195
13.00	0.190
14.25	0.180
15.00	0.150
CURVE 72* T = 773	
1.00	0.520
1.50	0.495
1.75	0.450
2.00	0.375
2.25	0.340
2.75	0.325
3.25	0.320
4.25	0.320
5.00	0.310
6.50	0.310
8.00	0.310
9.50	0.310
11.00	0.310
11.80	0.290
13.00	0.290

λ	∈
CURVE 72* (cont.)	
14.00	0.290
15.00	0.280
CURVE 73* T = 1023	
1.00	0.520
1.40	0.540
1.75	0.520
1.90	0.475
2.00	0.450
2.50	0.405
3.00	0.390
3.75	0.405
4.40	0.420
5.30	0.400
6.40	0.375
7.50	0.385
8.20	0.375
9.25	0.340
10.00	0.330
11.10	0.330
12.00	0.315
12.75	0.310
13.50	0.310
15.00	0.320
CURVE 74 T = 1033	
1.0	0.810
1.2	0.840
1.4	0.812
1.6	0.800
1.8	0.770
2.0	0.787
2.2	0.750
2.4	0.740
2.6	0.730
2.8	0.730
3.0	0.730
3.5	0.732
4.0	0.740
4.5	0.720
5.0	0.710
5.5	0.690

λ	∈
CURVE 74 (cont.)	
6.0	0.666
6.5	0.650
7.0	0.625
7.5	0.600
8.0	0.590
8.5	0.610
9.0	0.610
9.5	0.575
10.0	0.620*
10.0	0.570*
10.5	0.610
10.5	0.560
11.0	0.590
11.0	0.560
11.5	0.570
11.5	0.560*
12.0	0.550
12.0	0.540*
12.5	0.570
12.5	0.530
13.0	0.610
13.0	0.520*
13.5	0.500
14.5	0.480
15.0	0.470
15.5	0.482
16.0	0.480
16.5	0.480
17.0	0.490
17.5	0.490
18.0	0.530
18.5	0.540
19.0	0.540
19.5	0.550
20.0	0.540
20.5	0.530
21.5	0.500
22.0	0.490
22.5	0.490
23.0	0.500
23.5	0.510
24.0	0.500

* Not shown on plot

DATA TABLE NO. 391 (continued)

λ	ϵ
CURVE 75*	**T = 1061**
1.60	0.710
2.00	0.700
2.60	0.710
3.00	0.700
4.00	0.690
5.00	0.665
6.00	0.655
7.00	0.665
8.00	0.670
9.00	0.660
10.00	0.650
11.00	0.650
12.00	0.670
13.00	0.715
14.00	0.740
CURVE 76*	**T = 1089**
1.25	0.750
1.40	0.780
1.65	0.740
1.85	0.765
2.00	0.760
2.20	0.760
2.40	0.745
2.60	0.735
2.80	0.748
3.00	0.735
3.50	0.705
4.00	0.700
4.50	0.700
5.00	0.680
5.50	0.655
6.00	0.650
6.50	0.660
7.00	0.685
7.50	0.670
8.00	0.645
8.50	0.670
9.00	0.650
9.50	0.660
10.00	0.650
CURVE 76*(cont.)	
10.55	0.665
11.00	0.650
11.50	0.675
12.00	0.670
12.50	0.675
13.00	0.700
13.50	0.730
14.00	0.735
CURVE 77*	**T = 1089**
1.60	0.720
2.00	0.715
3.00	0.715
4.00	0.700
5.00	0.680
6.00	0.660
7.00	0.670
8.00	0.675
9.00	0.675
10.00	0.650
11.00	0.650
12.00	0.680
13.00	0.715
14.00	0.715
CURVE 78*	**T = 1255**
1.20	0.780
1.40	0.708
1.80	0.740
2.00	0.760
2.20	0.765
2.40	0.760
2.60	0.765
2.75	0.745
3.00	0.750
3.50	0.720
4.00	0.700
4.50	0.680
5.00	0.680
5.50	0.670
CURVE 78*(cont.)	
6.00	0.670
6.50	0.670
7.00	0.680
7.50	0.670
8.00	0.675
8.50	0.670
9.00	0.640
9.50	0.630
10.00	0.650
10.50	0.660
11.00	0.670
11.50	0.690
12.00	0.700
12.50	0.710
13.00	0.690
13.50	0.725
14.00	0.750
CURVE 79*	**T = 1011**
1.00	0.82
1.20	0.83
1.40	0.84
1.60	0.84
1.80	0.83
2.00	0.84
2.20	0.83
2.40	0.84
2.60	0.85
2.80	0.84
3.00	0.84
3.50	0.84
4.00	0.84
4.50	0.84
5.00	0.82
5.50	0.82
6.00	0.81
6.50	0.81
7.00	0.80
7.50	0.81
8.00	0.83
8.45	0.86
9.00	0.87
CURVE 79*(cont.)	
9.45	0.86
10.00	0.80
10.00	0.79
10.40	0.76
11.00	0.73
11.00	0.72
11.50	0.72
12.00	0.73
12.00	0.75
12.40	0.73
12.45	0.71
13.00	0.89
13.00	0.86
13.50	0.92
14.00	0.87
14.50	0.83
15.00	0.76
15.50	0.70
16.00	0.61
16.45	0.56
17.00	0.58
17.50	0.57
18.00	0.54
18.50	0.52
19.00	0.53
19.40	0.53
20.00	0.54
20.40	0.55
21.00	0.55
21.40	0.54
22.00	0.55
22.50	0.58
23.00	0.59
23.40	0.59
24.00	0.59
CURVE 80	**T = 1041**
1.00	0.960
1.20	0.960
1.40	0.955
1.60	0.925
1.80	0.915
CURVE 80 (cont.)	
2.00	0.920*
2.25	0.915
2.40	0.915
2.65	0.905
2.80	0.903
3.00	0.900*
3.50	0.895
4.00	0.895
4.50	0.885
5.00	0.900*
5.50	0.890*
6.00	0.880
6.50	0.870
7.00	0.870*
7.50	0.865
8.00	0.860
8.50	0.860
9.00	0.870
9.50	0.870
10.00	0.880
10.00	0.860*
10.50	0.850
10.50	0.880
11.00	0.870
11.00	0.900
11.50	0.895
11.50	0.930*
11.90	0.950
12.00	0.920
12.50	0.930
12.50	0.960
13.00	0.960*
13.00	0.970
13.50	0.975
14.00	0.990
14.00	0.980
14.50	0.955
15.00	0.890
15.00	0.910*
15.50	0.835
16.00	0.770
16.50	0.710
17.00	0.670
17.50	0.675
CURVE 80 (cont.)	
18.00	0.670
18.50	0.685
19.00	0.675
19.50	0.640
20.00	0.640
20.50	0.620
21.00	0.610
21.50	0.590
22.00	0.585
22.50	0.600
23.00	0.600
23.50	0.620
24.00	0.600
CURVE 81	**T = 800**
1.18	0.820
1.25	0.835
1.33	0.845
1.40	0.851
1.48	0.851
1.58	0.847
1.68	0.842
1.81	0.839
1.98	0.835
2.23	0.845
2.51	0.833*
2.73	0.832
2.92	0.828
3.14	0.827
3.35	0.826
3.56	0.825
3.77	0.824
3.98	0.821
4.20	0.821
4.41	0.821
4.58	0.821
4.76	0.819
4.91	0.817
5.07	0.816
5.21	0.817
5.34	0.816
5.47	0.816
CURVE 81 (cont.)	
5.60	0.815
5.76	0.813
5.90	0.813
6.04	0.813
6.16	0.811
6.28	0.811
6.41	0.811
6.53	0.811
6.64	0.811
6.76	0.813
6.87	0.815
6.97	0.818
7.07	0.821
7.19	0.828
7.32	0.828
7.44	0.831
7.56	0.833
7.66	0.836
7.77	0.839
7.88	0.842
7.98	0.850
8.08	0.859
8.18	0.865
8.28	0.871
8.38	0.875
8.47	0.875
8.57	0.877
8.66	0.879
8.75	0.881
8.84	0.882
8.93	0.884
9.01	0.884
9.10	0.889
9.18	0.891
9.26	0.893
9.35	0.894
9.43	0.892
9.51	0.888
9.67	0.884
9.83	0.883
10.00	0.883
10.15	0.886
10.30	0.889
10.46	0.894
CURVE 81 (cont.)	
10.61	0.895
10.76	0.897
10.91	0.900
11.05	0.902
11.19	0.905
11.32	0.906
11.45	0.907
11.58	0.908
11.72	0.911
11.84	0.913
11.97	0.915
12.09	0.916
12.22	0.918
12.35	0.920
12.48	0.924
12.60	0.929
12.72	0.930
12.85	0.933
12.98	0.934
13.10	0.938
13.22	0.941
13.34	0.942
13.46	0.943
13.58	0.944
13.69	0.943
13.81	0.940
13.93	0.933
14.03	0.917
14.14	0.898
14.25	0.876
14.36	0.852
14.47	0.828
14.58	0.805
14.68	0.782
14.79	0.759
14.90	0.739
15.01	0.723
15.11	0.708

* Not shown on plot

DATA TABLE NO. 391 (continued)

CURVE 82* T = 1100

λ	∈
1.07	0.894
1.12	0.895
1.18	0.894
1.25	0.893
1.33	0.891
1.40	0.889
1.48	0.886
1.58	0.881
1.68	0.877
1.81	0.874
1.98	0.869
2.23	0.867
2.51	0.863
2.73	0.857
2.93	0.855
3.14	0.854
3.35	0.852
3.56	0.850
3.77	0.849
3.98	0.848
4.20	0.846
4.41	0.845
4.58	0.844
4.76	0.844
4.91	0.842
5.07	0.843
5.21	0.844
5.34	0.845
5.47	0.846
5.60	0.845
5.76	0.840
5.90	0.839
6.04	0.840
6.16	0.838
6.28	0.839
6.41	0.840
6.53	0.839
6.64	0.840
6.76	0.842
6.87	0.846
6.97	0.848
7.07	0.850
7.19	0.852
7.32	0.853
7.44	0.854
7.56	0.856

CURVE 82 (cont.)*

λ	∈
7.66	0.857
7.77	0.860
7.88	0.862
7.98	0.867
8.08	0.871
8.18	0.876
8.28	0.880
8.38	0.884
8.47	0.886
8.57	0.887
8.66	0.889
8.75	0.889
8.84	0.890
8.93	0.893
9.01	0.894
9.10	0.895
9.18	0.897
9.26	0.898
9.35	0.899
9.43	0.900
9.51	0.900
9.67	0.898
9.83	0.899
10.00	0.899
10.15	0.900
10.30	0.901
10.46	0.902
10.61	0.904
10.76	0.907
10.91	0.909
11.05	0.912
11.19	0.914
11.32	0.916
11.45	0.916
11.58	0.919
11.72	0.921
11.84	0.922
11.97	0.923
12.09	0.925
12.22	0.927
12.35	0.930
12.48	0.932
12.60	0.935
12.72	0.937
12.85	0.938
12.98	0.940
13.10	0.944
13.22	0.949
13.34	0.951
13.46	0.953
13.58	0.954
13.69	0.954
13.81	0.954
13.93	0.954
14.03	0.953
14.14	0.948
14.25	0.938
14.36	0.924
14.47	0.904
14.58	0.891
14.68	0.873
14.79	0.854
14.90	0.833
15.01	0.813
15.11	0.803

CURVE 83* T = 1300

λ	∈
1.07	0.917
1.12	0.914
1.18	0.912
1.25	0.910
1.33	0.908
1.40	0.904
1.48	0.898
1.58	0.894
1.68	0.888
1.81	0.883
1.98	0.879
2.23	0.874
2.51	0.869
2.73	0.865
2.92	0.863
3.14	0.861
3.35	0.860
3.56	0.860
3.77	0.860
3.98	0.860
4.20	0.858
4.41	0.857
4.58	0.857
4.76	0.856
4.91	0.856

CURVE 83 (cont.)*

λ	∈
5.07	0.856
5.21	0.856
5.34	0.856
5.47	0.856
5.60	0.856
5.76	0.855
5.90	0.854
6.04	0.855
6.16	0.855
6.28	0.856
6.41	0.857
6.53	0.857
6.64	0.857
6.76	0.858
6.87	0.860
6.97	0.861
7.07	0.861
7.19	0.863
7.32	0.865
7.44	0.867
7.56	0.869
7.66	0.870
7.77	0.873
7.88	0.875
7.98	0.877
8.08	0.880
8.18	0.882
8.28	0.884
8.38	0.888
8.47	0.894
8.57	0.894
8.66	0.895
8.75	0.896
8.84	0.897
8.93	0.898
9.01	0.900
9.10	0.901
9.18	0.899
9.26	0.902
9.35	0.903
9.43	0.903
9.51	0.904
9.67	0.906
9.83	0.904
10.00	0.903
10.15	0.904
10.30	0.905
10.46	0.908
10.61	0.909
10.76	0.910
10.91	0.912
11.05	0.914
11.19	0.917
11.32	0.920
11.45	0.921
11.58	0.922
11.72	0.923
11.84	0.926
11.97	0.929
12.09	0.930
12.22	0.932
12.35	0.933
12.48	0.936
12.60	0.939
12.72	0.943
12.85	0.944
12.98	0.947
13.10	0.949
13.22	0.951
13.34	0.955
13.46	0.956
13.58	0.958
13.69	0.958
13.81	0.960
13.93	0.960
14.03	0.958
14.14	0.955
14.25	0.949
14.36	0.940
14.47	0.931
14.58	0.919
14.68	0.901
14.79	0.884
14.90	0.864
15.01	0.845
15.11	0.827

CURVE 84 T = 1273

λ	∈
1.02	0.790
1.99	0.803
2.96	0.785
4.01	0.792

CURVE 84 (cont.)

λ	∈
4.98	0.781
6.01	0.758
7.00	0.750
8.04	0.761
9.02	0.773
10.02	0.781
11.04	0.793
12.01	0.804
12.99	0.812
13.99	0.827
14.98	0.841

CURVE 85 T = 800

λ	∈
1.28	0.715
1.53	0.777
1.72	0.794
2.41	0.785
3.09	0.777
3.94	0.758
4.14	0.758
4.32	0.758
4.53	0.749
4.73	0.756
5.31	0.750*
6.25	0.743
7.16	0.750
7.65	0.762
7.79	0.762
8.18	0.792
8.60	0.823
9.14	0.844
9.81	0.844
10.14	0.837
10.58	0.846
11.20	0.846
11.55	0.855
12.04	0.852
12.26	0.861
13.43	0.864
13.95	0.854
14.15	0.840
14.25	0.840
14.38	0.820
14.79	0.784
14.93	0.782
15.04	0.762
15.26	0.776

CURVE 86* T = 1100

λ	∈
1.10	0.835
1.38	0.864
1.74	0.844
2.38	0.815
2.64	0.813
2.86	0.805
3.27	0.796
4.12	0.785
5.02	0.785
5.45	0.784
5.85	0.778
6.11	0.784
6.37	0.779
6.61	0.781
6.95	0.774
7.14	0.783
7.91	0.800
8.68	0.845
9.74	0.866
10.18	0.856
11.39	0.874
13.27	0.887
14.01	0.883
14.65	0.837
15.09	0.788
15.19	0.787

CURVE 87* T = 1300

λ	∈
1.09	0.888
1.19	0.883
1.37	0.886
1.91	0.849
2.64	0.828
3.66	0.810
4.32	0.805
4.52	0.800
4.85	0.807
6.97	0.802
7.97	0.821
8.75	0.857
9.60	0.868
10.54	0.878
10.84	0.872
11.28	0.884
11.68	0.882
12.40	0.886
13.02	0.897
13.87	0.898
14.40	0.878
15.06	0.814
15.17	0.818

* Not shown on plot

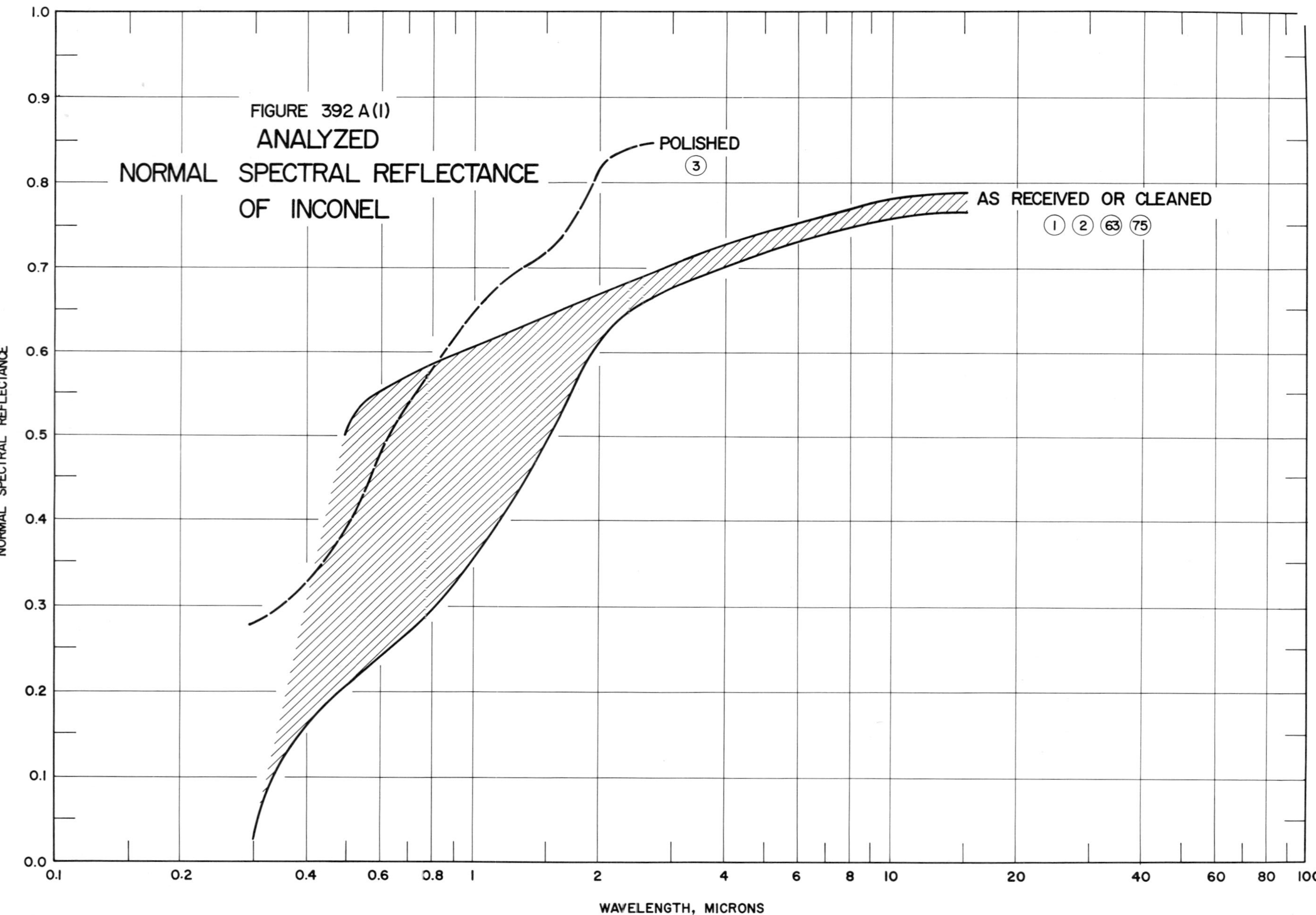

FIGURE 392 A(I)
ANALYZED
NORMAL SPECTRAL REFLECTANCE
OF INCONEL

FIGURE 392 A (2)

NORMAL SPECTRAL REFLECTANCE OF NICKEL + CHROMIUM + ΣX_i

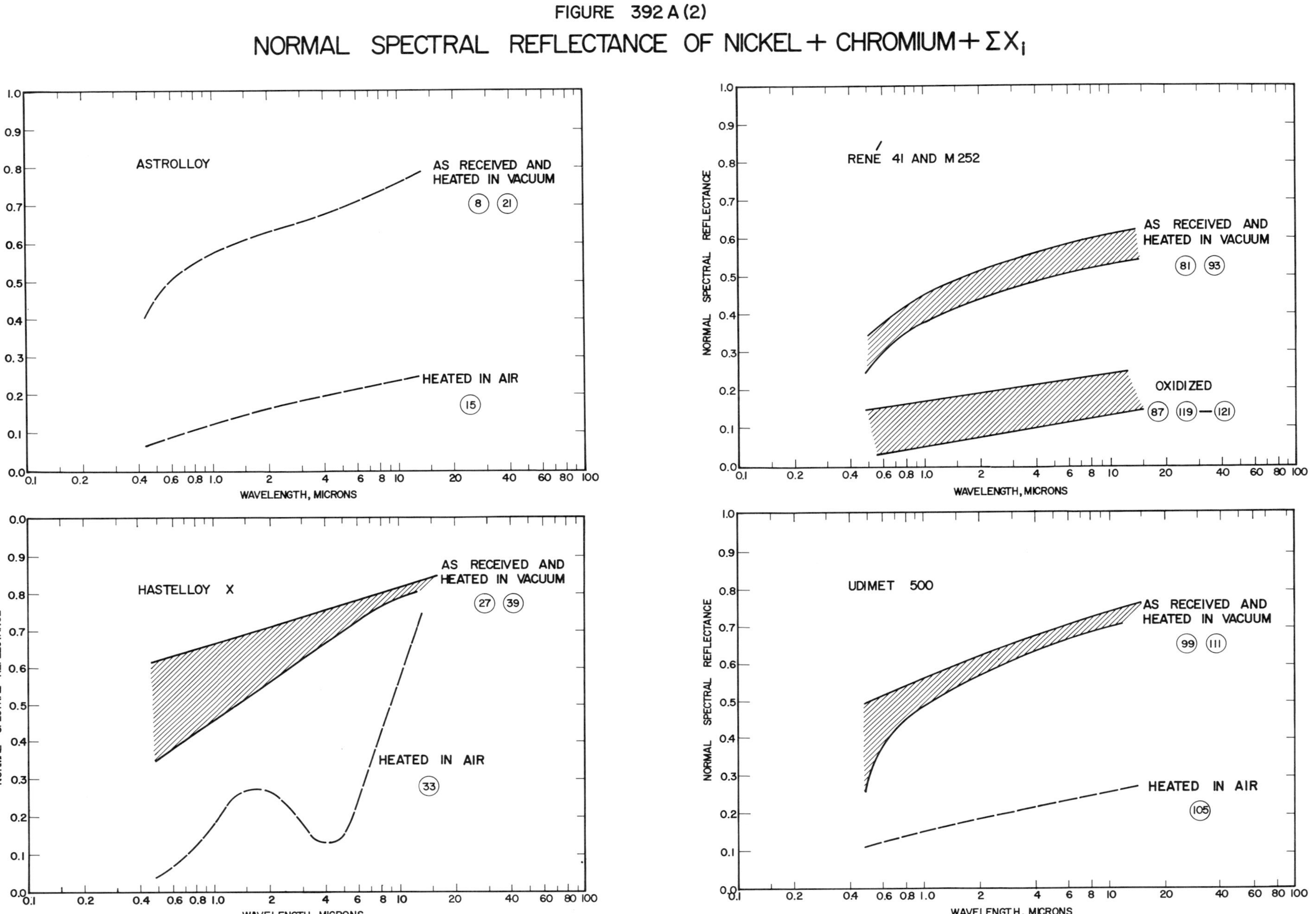

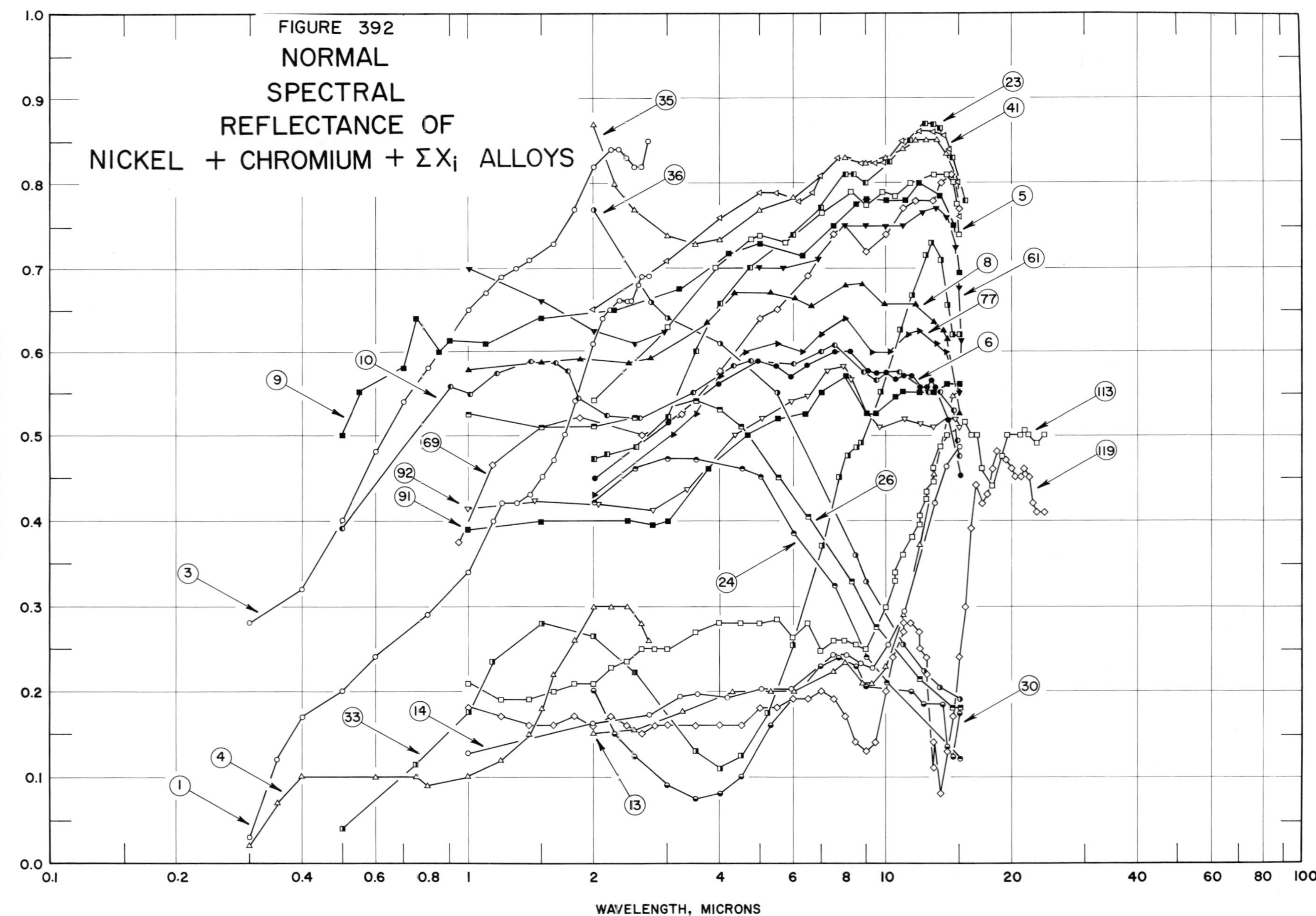

FIGURE 392 NORMAL SPECTRAL REFLECTANCE OF NICKEL + CHROMIUM + ΣX_i ALLOYS

SPECIFICATION TABLE NO. 392 NORMAL SPECTRAL REFLECTANCE OF [NICKEL + CHROMIUM + ΣX_i] ALLOYS

Curve No.	Ref. No.	Year	Temperature K	Wavelength Range, μ	Geometry θ	θ'	ω'	Reported Error, %	Composition (weight percent), Specifications and Remarks
1	146	1958	298	0.30-2.70	9°		2π		Inconel X; nominal composition: 73 Ni, 15 Cr, 7 Fe, 2.5 Ti, 1 Nb, 0.9 Al, 0.70 Mn, 0.30 Si, 0.04 C; as received.
2	146	1958	298	0.30-2.70	9°		2π		Different sample, same as above specimen and conditions except cleaned.
3	146	1958	298	0.30-2.70	~9°		2π		Different sample, same as curve 1 specimen and conditions except polished.
4	146	1958	298	0.30-2.70	~9°		2π		Different sample, same as curve 1 specimen and conditions except oxidized at 922 K for 30 min.
5	86	1961	~322	2.00-15.00	~0°		2π	<2	Astrolloy; nominal composition: 15 Cr, 15 Co, 5 Mo, 4.3 Al, 3.5 Ti, 0.05 C, 0.03 B, Ni balance; as received; data extracted from smooth curve; hohlraum at 523 K; converted from $R(2\pi, 0^\circ)$.
6	86	1961	~322	2.00-15.04	~0°		2π	<2	Above specimen and conditions; diffuse component only.
7	86	1961	~322	1.00-15.00	~0°		2π	<2	Different sample, same as curve 5 specimen and conditions except hohlraum at 773 K.
8	86	1961	~322	1.00-15.00	~0°		2π	<2	Above specimen and conditions; diffuse component only.
9	86	1961	~322	0.50-15.00	~0°		2π	<2	Different sample, same as curve 5 specimen and conditions except hohlraum at 1273 K.
10	86	1961	~322	0.50-14.99	~0°		2π	<2	Above specimen and conditions; diffuse component only.
11	86	1961	~322	2.00-15.00	~0°		2π	<2	Different sample, same as curve 5 specimen and conditions except heated in air at 1255 K for 30 min.
12	86	1961	~322	2.00-14.99	~0°		2π	<2	Above specimen and conditions; diffuse component only.
13	86	1961	~322	2.00-15.00	~0°		2π	<2	Different sample, same as curve 11 specimen and conditions except hohlraum at 773 K.
14	86	1961	~322	1.00-15.00	~0°		2π	<2	Above specimen and conditions; diffuse component only.
15	86	1961	~322	0.50-15.00	~0°		2π	<2	Different sample, same as curve 11 specimen and conditions except hohlraum at 1273 K.
16	86	1961	~322	0.55-15.00	~0°		2π	<2	Above specimen and conditions; diffuse component only.
17	86	1961	~322	2.00-15.00	~0°		2π	<2	Different sample, same as curve 5 specimen and conditions except heated in vacuum (7.6×10^{-5} mm Hg) at 1255 K for 30 min.
18	86	1961	~322	2.00-15.00	~0°		2π	<2	Above specimen and conditions; diffuse component only.
19	86	1961	~322	1.00-15.00	~0°		2π	<2	Different sample, same as curve 17 specimen and conditions except hohlraum at 773 K.
20	86	1961	~322	1.00-15.00	~0°		2π	<2	Above specimen and conditions; diffuse component only.
21	86	1961	~322	0.50-15.00	~0°		2π	<2	Different sample, same as curve 17 specimen and conditions except hohlraum at 1273 K.

SPECIFICATION TABLE NO. 392 (continued)

Curve No.	Ref. No.	Year	Temperature K	Wavelength Range, μ	Geometry θ	θ'	ω'	Reported Error, %	Composition (weight percent), Specifications and Remarks
22	86	1961	~322	0.50-14.99	~0°		2π	<2	Above specimen and conditions; diffuse component only.
23	86	1961	~322	2.00-15.50	~0°		2π	<2	Hastelloy X, AMS5536C; nominal composition: 22.0 Cr, 20 Fe, 9 Mo, 0.15 C, Ni balance; as received; data extracted from smooth curve; hohlraum at 523 K; converted from $R(2\pi, \mathring{0})$.
24	86	1961	~322	2.00-15.00	~0°		2π	<2	Above specimen and conditions; diffuse component only.
25	86	1961	~322	1.00-15.00	~0°		2π	<2	Different sample, same as curve 23 specimen and conditions except hohlraum at 773 K.
26	86	1961	~322	1.00-15.00	~0°		2π	<2	Above specimen and conditions; diffuse component only.
27	86	1961	~322	0.50-15.00	~0°		2π	<2	Different sample, same as curve 23 specimen and conditions except hohlraum at 1023 K.
28	86	1961	~322	0.50-15.00	~0°		2π	<2	Above specimen and conditions; diffuse component only.
29	86	1961	~322	2.00-15.00	~0°		2π	<2	Different sample, same as curve 23 specimen and conditions except heated in argon at 1366 K for 30 min.
30	86	1961	~322	2.00-15.00	~0°		2π	<2	Above specimen and conditions; diffuse component only.
31	86	1961	~322	1.00-15.00	~0°		2π	<2	Different sample, same as curve 29 specimen and conditions except hohlraum at 773 K.
32	86	1961	~322	1.00-15.00	~0°		2π	<2	Above specimen and conditions; diffuse component only.
33	86	1961	~322	0.50-15.00	~0°		2π	<2	Different sample, same as curve 29 specimen and conditions except hohlraum at 1273 K.
34	86	1961	~322	0.50-15.00	~0°		2π	<2	Above specimen and conditions; diffuse component only.
35	86	1961	~322	2.00-15.00	~0°		2π	<2	Different sample, same as curve 23 specimen and conditions except heated in vacuum (2.2×10^{-5} mm Hg) at 1366 K for 30 min.
36	86	1961	~322	2.00-15.00	~0°		2π	<2	Above specimen and conditions; diffuse component only.
37	86	1961	~322	1.00-15.00	~0°		2π	<2	Different sample, same as curve 35 specimen and conditions except hohlraum at 773 K.
38	86	1961	~322	1.00-15.00	~0°		2π	<2	Above specimen and conditions; diffuse component only.
39	86	1961	~322	0.50-15.00	~0°		2π	<2	Different sample, same as curve 35 specimen and conditions except hohlraum at 1023 K.
40	86	1961	~322	0.50-15.00	~0°		2π	<2	Above specimen and conditions; diffuse component only.
41	86	1961	~322	2.00-15.00	~0°		2π	<2	Inconel 702; nominal composition: 78 Ni, 16 Cr, 3.5 Al, 0.5 Ti, 0.35 Fe, 0.1 Mn, 0.03 C; as received; data extracted from smooth curve; hohlraum at 523 K; converted from $R(2\pi, \mathring{0})$
42	86	1961	~322	2.00-15.00	~0°		2π	<2	Above specimen and conditions; diffuse component only.

SPECIFICATION TABLE NO. 392 (continued)

Curve No.	Ref. No.	Year	Temperature K	Wavelength Range, μ	Geometry θ	θ'	ω'	Reported Error, %	Composition (weight percent), Specifications and Remarks
43	86	1961	~322	1.00-15.00	~0°		2π	<2	Different sample, same as curve 41 specimen and conditions except hohlraum at 773 K.
44	86	1961	~322	1.00-15.00	~0°		2π	<2	Above specimen and conditions; diffuse component only.
45	86	1961	~322	0.50-15.00	~0°		2π	<2	Different sample, same as curve 41 specimen and conditions except hohlraum at 1273 K.
46	86	1961	~322	0.55-15.00	~0°		2π	<2	Above specimen and conditions; diffuse component only.
47	86	1961	~322	2.00-15.00	~0°		2π	<2	Different sample, same as curve 41 specimen and conditions except heated in air at 1255 K for 30 min.
48	86	1961	~322	2.00-15.00	~0°		2π	<2	Above specimen and conditions; diffuse component only.
49	86	1961	~322	1.00-15.00	~0°		2π	<2	Different sample, same as curve 47 specimen and conditions except hohlraum at 773 K.
50	86	1961	~322	1.00-15.00	~0°		2π	<2	Above specimen and conditions; diffuse component only.
51	86	1961	~322	0.50-15.00	~0°		2π	<2	Different sample, same as curve 47 specimen and conditions except hohlraum at 1273 K.
52	86	1961	~322	0.50-15.00	~0°		2π	<2	Above specimen and conditions; diffuse component only.
53	86	1961	~322	2.00-15.00	~0°		2π	<2	Different sample, same as curve 41 specimen and conditions except heated in vacuum (7.6 x 10^{-5} mm Hg) at 1255 K for 30 min.
54	86	1961	~322	2.00-15.00	~0°		2π	<2	Above specimen and conditions; diffuse component only.
55	86	1961	~322	1.00-15.00	~0°		2π	<2	Different sample, same as curve 53 specimen and conditions except hohlraum at 773 K.
56	86	1961	~322	1.00-15.00	~0°		2π	<2	Above specimen and conditions; diffuse component only.
57	86	1961	~322	0.50-15.00	~0°		2π	<2	Different sample, same as curve 53 specimen and conditions except hohlraum at 1273 K.
58	86	1961	~322	0.50-15.00	~0°		2π	<2	Above specimen and conditions; diffuse component only.
59	86	1961	~322	2.00-15.00	~0°		2π	<2	Inconel X, AMS5542; nominal composition: 73 Ni, 15 Cr, 7 Fe, 2.5 T, 1 Nb, 0.9 Al, 0.70 Mn, 0.30 Si, 0.04 C; as received; data extracted from smooth curve; hohlraum at 523 K; converted from R(2π, 0̂).
60	86	1961	~322	2.00-15.00	~0°		2π	<2	Above specimen and conditions; diffuse component only.
61	86	1961	~322	1.00-15.00	~0°		2π	<2	Different sample, same as curve 59 specimen and conditions except hohlraum at 773 K.
62	86	1961	~322	1.00-15.00	~0°		2π	<2	Above specimen and conditions; diffuse component only.
63	86	1961	~322	0.50-15.00	~0°		2π	<2	Different sample, same as curve 59 specimen and conditions except hohlraum at 1273 K.

SPECIFICATION TABLE NO. 392 (continued)

Curve No.	Ref. No.	Year	Temperature K	Wavelength Range, μ	Geometry θ	θ'	ω'	Reported Error, %	Composition (weight percent), Specifications and Remarks
64	86	1961	~322	0.50-15.00	~0°		2π	<2	Above specimen and conditions; diffuse component only.
65	86	1961	~322	2.00-15.00	~0°		2π	<2	Different sample, same as curve 59 specimen and conditions except heated in air at 1089 K for 30 min.
66	86	1961	~322	2.00-15.00	~0°		2π	<2	Above specimen and conditions; diffuse component only.
67	86	1961	~322	1.00-15.00	~0°		2π	<2	Different sample, same as curve 65 specimen and conditions except hohlraum at 1023 K.
68	86	1961	~322	1.00-15.00	~0°		2π	<2	Above specimen and conditions; diffuse component only.
69	86	1961	~322	0.95-15.00	~0°		2π	<2	Different sample, same as curve 65 specimen and conditions except hohlraum at 1273 K.
70	86	1961	~322	0.88-14.99	~0°		2π	<2	Above specimen and conditions; diffuse component only.
71	86	1961	~322	2.00-15.00	~0°		2π	<2	Different sample, same as curve 59 specimen and conditions except heated in vacuum (6.8×10^{-5} mm Hg) at 1089 K for 30 min.
72	86	1961	~322	2.00-15.00	~0°		2π	<2	Above specimen and conditions; diffuse component only.
73	86	1961	~322	1.00-15.00	~0°		2π	<2	Different sample, same as curve 71 specimen and conditions except hohlraum at 773 K.
74	86	1961	~322	1.00-15.00	~0°		2π	<2	Above specimen and conditions; diffuse component only.
75	86	1961	~322	0.50-15.00	~0°		2π	<2	Different sample, same as curve 71 specimen and conditions except hohlraum at 1273 K.
76	86	1961	~322	0.50-15.00	~0°		2π	<2	Above specimen and conditions; diffuse component only.
77	86	1961	~322	2.00-15.00	~0°		2π	<2	René 41; nominal composition: 19 Cr, 11 Co, 10 Mo, 3 Fe, 3 Ti, 1.5 Al, 0.1 C, trace B, Ni balance; as received; data extracted from smooth curve; hohlraum at 523 K; converted from R(2π, θ̂).
78	86	1961	~322	2.00-15.00	~0°		2π	<2	Above specimen and conditions; diffuse component only.
79	86	1961	~322	1.00-15.00	~0°		2π	<2	Different sample, same as curve 77 specimen and conditions except hohlraum at 773 K.
80	86	1961	~322	1.50-15.00	~0°		2π	<2	Above specimen and conditions; diffuse component only.
81	86	1961	~322	0.50-15.00	~0°		2π	<2	Different sample, same as curve 77 specimen and conditions except hohlraum at 1273 K.
82	86	1961	~322	0.50-15.00	~0°		2π	<2	Above specimen and conditions; diffuse component only.
83	86	1961	~322	2.00-15.00	~0°		2π	<2	Different sample, same as curve 77 specimen and conditions except heated in air at 1255 K for 30 min.
84	86	1961	~322	2.00-15.00	~0°		2π	<2	Above specimen and conditions; diffuse component only.

SPECIFICATION TABLE NO. 392 (continued)

Curve No.	Ref. No.	Year	Temperature K	Wavelength Range, μ	Geometry θ	θ'	ω'	Reported Error, %	Composition (weight percent), Specifications and Remarks
85	86	1961	~322	1.00-15.00	~0°		2π	<2	Different sample, same as curve 83 specimen and conditions except hohlraum at 773 K.
86	86	1961	~322	1.00-15.00	~0°		2π	<2	Above specimen and conditions; diffuse component only.
87	86	1961	~322	0.50-15.00	~0°		2π	<2	Different sample, same as curve 83 specimen and conditions except hohlraum at 1273 K.
88	86	1961	~322	0.50-15.00	~0°		2π	<2	Above specimen and conditions; diffuse component only.
89	86	1961	~322	2.00-15.00	~0°		2π	<2	Different sample, same as curve 77 specimen and conditions except heated in vacuum (7.6×10^{-5} mm Hg) at 1255 K for 30 min.
90	86	1961	~322	2.00-15.00	~0°		2π	<2	Above specimen and conditions; diffuse component only.
91	86	1961	~322	1.00-15.00	~0°		2π	<2	Different sample, same as curve 89 specimen and conditions except hohlraum at 773 K.
92	86	1961	~322	1.00-14.99	~0°		2π	<2	Above specimen and conditions; diffuse component only.
93	86	1961	~322	0.50-15.00	~0°		2π	<2	Different sample, same as curve 89 specimen and conditions except hohlraum at 1273 K.
94	86	1961	~322	0.50-14.99	~0°		2π	<2	Above specimen and conditions; diffuse component only.
95	86	1961	~322	2.00-15.00	~0°		2π	<2	Udimet 500; nominal composition: 19 Cr, 19 Co, 4 Mo, 4 max Fe, 3 Ti, 2.9 Al, 0.1 C, trace B, Ni balance; as received; data extracted from smooth curve; hohlraum at 523 K.
96	86	1961	~322	2.00-15.00	~0°		2π	<2	Above specimen and conditions; diffuse component only.
97	86	1961	~322	1.00-15.00	~0°		2π	<2	Different sample, same as curve 95 specimen and conditions except hohlraum at 773 K.
98	86	1961	~322	1.00-15.00	~0°		2π	<2	Above specimen and conditions; diffuse component only.
99	86	1961	~322	0.50-15.00	~0°		2π	<2	Different sample, same as curve 95 specimen and conditions except hohlraum at 1273 K.
100	86	1961	~322	0.50-14.99	~0°		2π	<2	Above specimen and conditions; diffuse component only.
101	86	1961	~322	2.00-15.00	~0°		2π	<2	Different sample, same as curve 95 specimen and conditions except heated in air at 1255 K for 30 min.
102	86	1961	~322	2.00-15.00	~0°		2π	<2	Above specimen and conditions; diffuse component only.
103	86	1961	~322	1.00-15.00	~0°		2π	<2	Different sample, same as curve 101 specimen and conditions except hohlraum at 773 K.
104	86	1961	~322	1.00-15.00	~0°		2π	<2	Above specimen and conditions; diffuse component only.
105	86	1961	~322	0.50-15.00	~0°		2π	<2	Different sample, same as curve 101 specimen and conditions except hohlraum at 1273 K.

SPECIFICATION TABLE NO. 392 (continued)

Curve No.	Ref. No.	Year	Temperature K	Wavelength Range, μ	Geometry θ	θ'	ω'	Reported Error, %	Composition (weight percent), Specifications and Remarks
106	86	1961	~322	0.50-15.00	~0°		2π	<2	Above specimen and conditions; diffuse component only.
107	86	1961	~322	2.00-15.00	~0°		2π	<2	Different sample, same as curve 95 specimen and conditions except heated in vacuum (7.6 x 10^{-5} mm Hg) at 1255 K for 30 min.
108	86	1961	~322	2.00-15.00	~0°		2π	<2	Above specimen and conditions; diffuse component only.
109	86	1961	~322	1.00-15.00	~0°		2π	<2	Different sample, same as curve 107 specimen and conditions except hohlraum at 773 K.
110	86	1961	~322	1.00-15.00	~0°		2π	<2	Above specimen and conditions; diffuse component only.
111	86	1961	~322	0.50-15.00	~0°		2π	<2	Different sample, same as curve 107 specimen and conditions except hohlraum at 1273 K.
112	86	1961	~322	0.50-15.00	~0°		2π	<2	Above specimen and conditions; diffuse component only.
113	65	1962	294	1.0-24.0	~0°		2π		Inconel; nominal composition: 78 Ni, 15 Cr, 7 Fe, 0.35 Mn, 0.20 Si, 0.04 C; oxidized; measured in air.
114	65	1962	1033	1.0-24.0	~0°		2π		Different sample, same as above specimen and conditions except calculated from (1 - ∈).
115	65	1962	1061	1.60-14.00	~0°		2π		Different sample, same as curve 113 specimen and conditions except measured in argon.
116	65	1962	1089	1.60-14.00	~0°		2π		Different sample, same as curve 113 specimen and conditions.
117	65	1962	1089	1.25-14.00	~0°		2π		Different sample, same as curve 113 specimen and conditions.
118	65	1962	1255	1.20-14.00	~0°		2π		Different sample, same as curve 113 specimen and conditions except measured in vacuum.
119	65	1962	1011	1.00-24.00	~0°		2π		M 252; nominal composition: 19 Cr, 10 Co, 10 Mo, 2.50 Ti, 2.0 Fe, 1.00 Al, 0.5 Mn, 0.5 Si, 0.15 C, 0.005 B, Ni balance; oxidized; measured in air; calculated from (1 - ∈).
120	65	1962	294	1.00-23.00	~0°		2π		René 41; nominal composition: 19 Cr, 11 Co, 10 Mo, 3 Fe, 3 Ti, 1.5 Al, 0.1 C, B, Ni balance; oxidized; measured in air.
121	65	1962	1041	1.00-24.00	~0°		2π		Different sample, same as above specimen and conditions except calculated from (1 - ∈).

DATA TABLE NO. 392 NORMAL SPECTRAL REFLECTANCE OF [NICKEL + CHROMIUM + ΣX_i] ALLOYS

[Wavelength, λ, μ; Reflectance, ρ; Temperature, T, K]

CURVE 1 T = 298

λ	ρ
0.30	0.03
0.35	0.12
0.40	0.17
0.50	0.20
0.60	0.24
0.80	0.29
1.00	0.34
1.15	0.40
1.20	0.42
1.30	0.42
1.40	0.43
1.50	0.45
1.60	0.47
1.70	0.50
1.80	0.54
2.00	0.61
2.10	0.64
2.20	0.65
2.30	0.66
2.40	0.66
2.45	0.66
2.55	0.68
2.60	0.69
2.70	0.69

CURVE 2* T = 298

λ	ρ
0.30	0.03
0.35	0.12
0.40	0.16
0.50	0.18
0.60	0.20
0.70	0.26
0.80	0.29
1.00	0.36
1.10	0.40
1.20	0.43
1.30	0.43
1.40	0.44
1.45	0.45
1.50	0.46
1.60	0.50
1.80	0.56

CURVE 2 (cont.)*

λ	ρ
1.90	0.60
2.00	0.62
2.10	0.65
2.20	0.66
2.30	0.67
2.40	0.68
2.60	0.68
2.70	0.68

CURVE 3 T = 298

λ	ρ
0.30	0.28
0.40	0.32
0.50	0.40
0.60	0.48
0.70	0.54
0.80	0.58
1.00	0.65
1.10	0.67
1.20	0.69
1.30	0.70
1.40	0.71
1.60	0.73
1.80	0.77
2.00	0.82
2.20	0.84
2.30	0.84
2.40	0.83
2.50	0.82
2.60	0.82
2.70	0.85

CURVE 4 T = 298

λ	ρ
0.30	0.02
0.35	0.07
0.40	0.10
0.60	0.10
0.75	0.10
0.80	0.09
1.00	0.10
1.20	0.12

CURVE 4 (cont.)

λ	ρ
1.40	0.15
1.50	0.18
1.60	0.22
1.80	0.26
2.00	0.30
2.20	0.30
2.40	0.30
2.60	0.28
2.70	0.26

CURVE 5 T = ~322

λ	ρ
2.00	0.540
3.00	0.630
3.90	0.700
4.75	0.735
5.00	0.740
5.70	0.730
7.00	0.765
8.25	0.790
9.00	0.775
9.85	0.790
10.50	0.785
11.50	0.800
13.00	0.810
14.00	0.810
14.50	0.800
14.90	0.775
15.00	0.740

CURVE 6 T = ~322

λ	ρ
2.00	0.449
3.00	0.513
3.98	0.560
4.96	0.588
5.46	0.581
5.92	0.568
6.49	0.581
7.55	0.600
8.20	0.600
9.19	0.577
9.51	0.573

CURVE 6 (cont.)

λ	ρ
10.02	0.573
10.61	0.565
11.01	0.570
11.50	0.570
12.02	0.556
12.50	0.556
12.78	0.562
13.11	0.553
14.04	0.516
15.04	0.451

CURVE 7* T = ~322

λ	ρ
1.00	0.690
1.50	0.660
2.00	0.615
2.50	0.625
3.50	0.680
4.25	0.705
5.00	0.730
6.10	0.710
7.25	0.735
8.50	0.790
9.25	0.790
10.10	0.775
11.00	0.790
12.00	0.790
13.00	0.800
14.00	0.785
15.00	0.780

CURVE 8 T = ~322

λ	ρ
1.00	0.579
1.52	0.589
1.86	0.590
2.41	0.584
2.72	0.592
3.71	0.636
4.35	0.669
5.29	0.669
6.03	0.662
6.63	0.651

CURVE 8 (cont.)

λ	ρ
8.01	0.681
8.80	0.681
9.98	0.655
11.89	0.655
13.00	0.635
13.63	0.627
14.01	0.615
15.00	0.525

CURVE 9 T = ~322

λ	ρ
0.50	0.500
0.55	0.550
0.70	0.580
0.75	0.640
0.85	0.600
0.90	0.615
1.10	0.610
1.50	0.640
2.25	0.650
3.20	0.675
4.20	0.715
5.00	0.730
6.30	0.715
7.50	0.750
8.50	0.775
9.00	0.780
10.00	0.780
11.15	0.780
12.00	0.800
13.50	0.785
14.50	0.750
15.00	0.695

CURVE 10 T = ~322

λ	ρ
0.50	0.390
0.90	0.558
1.02	0.549
1.17	0.574
1.42	0.588
1.61	0.587
1.73	0.575

CURVE 10 (cont.)

λ	ρ
1.82	0.542
2.15	0.523
2.59	0.520
3.46	0.550
4.35	0.581
4.79	0.589
6.07	0.585
7.01	0.600
7.52	0.607
8.97	0.572
9.49	0.563
10.07	0.573*
10.87	0.573
12.61	0.550
13.50	0.550
14.47	0.529
14.77	0.512*
14.93	0.492
14.99	0.473

CURVE 11* T = ~322

λ	ρ
2.00	0.190
2.30	0.150
2.75	0.140
3.20	0.150
3.75	0.185
4.25	0.190
5.50	0.170
5.50	0.170
6.25	0.180
7.00	0.200
7.75	0.210
8.75	0.195
9.50	0.220
10.50	0.265
11.50	0.320
12.50	0.400
13.00	0.450
13.35	0.500
14.25	0.520
14.80	0.500
15.00	0.480

CURVE 12* T = ~322

λ	ρ
2.00	0.200
2.31	0.165
2.61	0.151
3.23	0.176
3.98	0.200
4.99	0.187
5.47	0.177
6.26	0.184
6.78	0.191
8.60	0.191
10.50	0.267
11.66	0.336
11.99	0.376
12.49	0.406
13.28	0.442
14.02	0.453
14.50	0.453
14.99	0.431

CURVE 13 T = ~322

λ	ρ
2.00	0.150
2.50	0.155
3.25	0.175
4.30	0.200
5.30	0.200
6.00	0.200
7.50	0.225
8.00	0.235
8.80	0.210
9.25	0.210
10.00	0.230
11.00	0.290
12.00	0.370
13.00	0.455
14.00	0.520*
14.50	0.545
15.00	0.550*

CURVE 14 T = ~322

λ	ρ
1.00	0.128
1.99	0.163
2.71	0.172
3.24	0.192
3.55	0.197
4.19	0.192
5.06	0.204
5.98	0.204
7.48	0.243
8.10	0.243
8.77	0.232
9.26	0.228
10.26	0.255
11.17	0.294
13.01	0.420
13.99	0.462
15.00	0.485

CURVE 15* T = ~322

λ	ρ
0.50	0.075
0.75	0.100
1.00	0.140
1.40	0.140
2.00	0.165
2.50	0.175
3.15	0.175
4.75	0.200
6.15	0.200
6.50	0.210
7.20	0.240
7.85	0.250
8.75	0.230
9.25	0.230
10.10	0.250
11.50	0.335
12.50	0.405
13.50	0.470
14.25	0.490
14.75	0.485
15.00	0.470

* Not shown on plot

DATA TABLE NO. 392 (continued)

λ	ρ
CURVE 16*	**T = ~322**
0.55	0.066
0.65	0.085
0.77	0.085
1.06	0.147
1.54	0.151
2.04	0.181
2.39	0.183
3.24	0.183
5.15	0.215
5.79	0.205
6.50	0.205
7.37	0.237
7.83	0.237
8.77	0.213
9.37	0.216
9.98	0.232
11.29	0.310
11.78	0.331
12.17	0.351
13.30	0.414
14.04	0.434
14.63	0.432
15.00	0.400
CURVE 17*	**T = ~322**
2.00	0.740
2.50	0.680
3.10	0.650
4.00	0.675
5.15	0.725
6.50	0.750
7.50	0.770
8.00	0.770
9.25	0.760
10.00	0.760
11.50	0.770
13.00	0.785
13.75	0.775
14.50	0.750
14.80	0.725
15.00	0.680
CURVE 18*	**T = ~322**
2.00	0.649
2.99	0.559
3.27	0.553
3.60	0.560
3.98	0.578
4.79	0.601
6.50	0.601
7.01	0.610
8.01	0.611
9.26	0.582
10.15	0.576
11.51	0.551
12.10	0.552
13.42	0.535
14.30	0.518
14.81	0.493
15.00	0.450
CURVE 19*	**T = ~322**
1.00	0.570
2.00	0.565
2.50	0.575
3.50	0.630
4.50	0.700
5.00	0.710
6.00	0.710
7.00	0.730
8.15	0.760
8.75	0.750
10.00	0.750
11.50	0.775
13.00	0.790
13.50	0.815
14.00	0.820
15.00	0.820
CURVE 20*	**T = ~322**
1.00	0.548
2.67	0.549
4.19	0.607
4.99	0.618
CURVE 20 (cont.)*	
5.64	0.619
7.37	0.642
7.98	0.637
9.77	0.598
12.04	0.598
13.04	0.590
13.76	0.600
14.65	0.600
15.00	0.587
CURVE 21*	**T = ~322**
0.50	0.390
0.60	0.475
0.75	0.540
1.00	0.515
1.20	0.565
2.00	0.610
2.50	0.615
2.75	0.620
3.50	0.650
4.50	0.690
5.50	0.705
6.50	0.715
8.00	0.745
8.85	0.730
9.80	0.730
11.50	0.760
12.50	0.760
13.50	0.770
14.00	0.770
14.60	0.760
15.00	0.740
CURVE 22*	**T = ~322**
0.50	0.183
0.50	0.275
0.97	0.453
1.30	0.500
1.82	0.552
2.31	0.580
2.83	0.591
CURVE 22 (cont.)*	
3.76	0.628
4.16	0.637
4.97	0.652
6.32	0.652
7.12	0.665
7.99	0.672
8.31	0.665
8.69	0.641
9.75	0.623
11.22	0.623
11.71	0.605
13.69	0.605
14.38	0.585
14.99	0.573
CURVE 23	**T = ~322**
2.00	0.470
2.15	0.475
2.50	0.485
3.00	0.520
3.50	0.600
4.00	0.655
4.75	0.700
6.00	0.740
7.00	0.770
8.00	0.810
8.40	0.810
9.00	0.800
10.20	0.825
11.40	0.850
12.50	0.870
13.00	0.870
13.50	0.865
14.50	0.830
15.50	0.780
CURVE 24	**T = ~322**
2.00	0.420
2.50	0.460
3.00	0.470
3.50	0.470
4.50	0.460
CURVE 24 (cont.)	
5.00	0.450
6.00	0.385
7.50	0.325
9.00	0.240
10.00	0.210
15.00	0.120
CURVE 25*	**T = ~322**
1.00	0.530
1.50	0.515
2.50	0.565
3.50	0.625
4.50	0.670
5.75	0.675
6.50	0.700
7.50	0.725
8.50	0.770
9.00	0.780
9.70	0.775
10.15	0.785
10.75	0.800
12.00	0.800
12.75	0.820
13.50	0.800
14.25	0.800
15.00	0.790
CURVE 26	**T = ~322**
1.00	0.525
1.50	0.510
2.00	0.510
2.50	0.520
3.50	0.540
4.00	0.530
4.50	0.510
5.50	0.450
6.50	0.405
8.25	0.330
9.50	0.275
12.00	0.215
14.50	0.180
15.00	0.180
CURVE 27*	**T = ~322**
0.50	0.350
0.60	0.400
0.90	0.460
1.05	0.460
1.60	0.520
2.00	0.527
3.00	0.615
4.00	0.670
5.00	0.710
5.50	0.710
6.10	0.700
6.30	0.700
6.75	0.725
7.50	0.770
8.00	0.790
8.80	0.800
9.25	0.800
10.00	0.810
11.00	0.820
12.00	0.820
12.50	0.825
12.90	0.830
13.50	0.825
14.20	0.800
14.70	0.750
14.90	0.700
15.00	0.650
CURVE 28*	**T = ~322**
0.50	0.380
0.60	0.420
0.75	0.450
0.85	0.420
0.90	0.430
1.00	0.435
15.0	0.470
2.00	0.530
2.50	0.545
3.00	0.550
4.00	0.540
5.00	0.510
5.50	0.480
6.00	0.440
CURVE 28 (cont.)*	
6.50	0.410
7.50	0.380
8.00	0.370
9.00	0.300
9.50	0.285
11.00	0.260
11.50	0.245
12.00	0.240
14.00	0.240
14.50	0.260
15.00	0.300
CURVE 29*	**T = ~322**
2.00	0.150
2.25	0.100
2.50	0.075
2.95	0.055
3.25	0.060
3.80	0.100
4.45	0.175
5.10	0.250
5.85	0.325
6.60	0.400
7.10	0.450
7.60	0.490
8.50	0.530
9.50	0.565
10.00	0.590
10.50	0.630
11.25	0.670
12.00	0.700
12.85	0.730
13.20	0.725
13.55	0.700
13.95	0.650
14.10	0.635
14.50	0.615
15.00	0.600
CURVE 30	**T = ~322**
2.00	0.200
2.25	0.150
CURVE 30 (cont.)	
2.50	0.125
3.00	0.090
3.50	0.075
4.00	0.080
4.50	0.100
5.25	0.160
7.00	0.230
7.75	0.240
8.50	0.230
9.00	0.205
11.50	0.200
12.25	0.185
13.75	0.185
14.00	0.135
14.50	0.125
14.75	0.130*
15.00	0.175
CURVE 31*	**T = ~322**
1.00	0.180
1.50	0.255
2.00	0.225
2.25	0.195
2.50	0.150
2.80	0.120
3.10	0.120
3.75	0.160
4.50	0.205
5.00	0.240
5.65	0.300
6.40	0.375
7.10	0.450
7.45	0.490
8.00	0.515
9.00	0.545
9.75	0.570
10.25	0.600
10.80	0.650
11.50	0.690
12.00	0.710
12.50	0.740
12.90	0.760
13.50	0.725
14.00	0.710
14.50	0.720
15.00	0.745

* Not shown on plot

DATA TABLE NO. 392 (continued)

λ	ρ
CURVE 32*	
T = ~322	
1.00	0.160
1.50	0.200
2.00	0.225
2.25	0.220
3.00	0.130
3.50	0.100
4.00	0.090
4.75	0.165
6.50	0.270
7.50	0.310
8.00	0.300
9.00	0.270
9.75	0.250
11.00	0.250
12.50	0.235
13.00	0.230
13.50	0.190
14.00	0.170
14.50	0.190
15.00	0.220
CURVE 33	
T = ~322	
0.50	0.040
0.75	0.115
1.00	0.175
1.15	0.235
1.50	0.280
2.00	0.265
2.50	0.225
3.50	0.130
4.00	0.110
4.50	0.125
5.20	0.175
6.00	0.255
7.00	0.370
7.75	0.450
8.10	0.475
8.50	0.485
8.75	0.490
9.70	0.550
10.75	0.625
11.50	0.665
12.50	0.715
12.95	0.730

λ	ρ
CURVE 33 (cont.)	
13.50	0.710
14.00	0.655
14.50	0.620
15.00	0.620
CURVE 34*	
T = ~322	
0.50	0.040
0.75	0.115
1.00	0.140
1.50	0.250
1.75	0.255
2.00	0.250
2.50	0.210
3.50	0.130
4.00	0.115
4.50	0.135
5.00	0.180
5.50	0.210
8.00	0.300
8.25	0.300
9.00	0.270
12.00	0.250
12.50	0.240
14.00	0.190
14.50	0.195
15.00	0.225
CURVE 35	
T = ~322	
2.00	0.870
2.25	0.800
2.50	0.770
3.00	0.740
3.50	0.730
4.00	0.735
5.00	0.770
6.00	0.785
7.00	0.810*
8.00	0.830
9.00	0.825
10.00	0.830
11.00	0.840
11.70	0.850

λ	ρ
CURVE 35 (cont.)	
12.50	0.850
13.25	0.850
14.00	0.835
14.70	0.800*
15.00	0.750*
CURVE 36	
T = ~322	
2.00	0.770
2.75	0.660
3.00	0.640
4.00	0.610
5.50	0.550
8.50	0.360
9.00	0.330
11.00	0.255
12.25	0.225
13.50	0.205
15.00	0.190
CURVE 37*	
T = ~322	
1.00	0.760
1.65	0.700
2.00	0.660
2.50	0.665
3.50	0.715
4.50	0.755
5.25	0.770
5.75	0.770
6.25	0.760
7.00	0.770
8.00	0.800
8.70	0.800
9.25	0.800
10.50	0.815
11.50	0.830
12.20	0.840
12.70	0.840
13.75	0.875
14.50	0.885
15.00	0.875

λ	ρ
CURVE 38*	
T = ~322	
1.00	0.650
1.25	0.690
1.50	0.695
2.00	0.665
3.50	0.660
4.50	0.630
6.00	0.540
7.50	0.460
9.50	0.325
10.25	0.295
11.00	0.220
12.00	0.185
13.50	0.160
14.00	0.160
14.50	0.170
15.00	0.190
CURVE 39*	
T = ~322	
0.50	0.615
0.75	0.685
1.00	0.665
1.50	0.710
2.00	0.710
2.50	0.710
3.50	0.725
4.50	0.750
4.90	0.760
6.00	0.760
6.75	0.775
8.00	0.800
9.10	0.800
10.25	0.800
11.00	0.820
11.70	0.820
12.00	0.830
13.25	0.845
14.00	0.850
14.50	0.850
15.00	0.830

λ	ρ
CURVE 40*	
T = ~322	
0.50	0.515
0.75	0.585
1.00	0.640
1.50	0.690
2.00	0.700
2.50	0.700
4.00	0.670
5.00	0.625
6.50	0.545
7.25	0.515
9.50	0.340
12.00	0.265
13.50	0.260
14.25	0.290
14.75	0.335
15.00	0.400
CURVE 41	
T = ~322	
2.00	0.650
3.00	0.710
4.00	0.760
5.00	0.790
5.50	0.790
6.20	0.780
6.70	0.790
7.00	0.810
7.75	0.830
8.80	0.825
9.50	0.825
9.90	0.825
11.00	0.850
12.00	0.860
13.00	0.860
13.75	0.855
14.25	0.840
14.70	0.800
15.00	0.760
CURVE 42*	
T = ~322	
2.00	0.470
2.50	0.480

λ	ρ
CURVE 42 (cont.)*	
3.00	0.520
3.50	0.555
4.00	0.560
5.00	0.540
5.50	0.510
6.00	0.480
6.50	0.470
8.00	0.400
9.25	0.315
10.75	0.245
13.00	0.175
14.00	0.130
14.50	0.120
15.00	0.120
CURVE 43*	
T = ~322	
1.00	0.750
1.30	0.750
1.60	0.735
2.00	0.700
2.40	0.695
3.00	0.710
4.00	0.760
5.00	0.790
5.50	0.785
6.00	0.780
7.00	0.800
8.00	0.840
8.80	0.840
9.80	0.850
10.50	0.850
11.00	0.850
11.75	0.860
12.75	0.860
13.50	0.860
14.20	0.840
14.70	0.800
15.00	0.730
CURVE 44*	
T = ~322	
1.00	0.710
1.75	0.625

λ	ρ
CURVE 44 (cont.)*	
2.00	0.610
2.50	0.615
3.50	0.645
4.00	0.650
4.75	0.635
5.50	0.605
8.00	0.450
8.50	0.410
9.00	0.365
11.00	0.270
12.50	0.210
13.50	0.190
14.50	0.170
15.00	0.170
CURVE 45*	
T = ~322	
0.50	0.420
0.60	0.500
0.80	0.550
0.80	0.675
0.90	0.700
1.00	0.690
1.50	0.730
2.05	0.750
3.10	0.760
4.00	0.790
4.45	0.795
4.85	0.810
6.10	0.810
7.00	0.830
7.60	0.835
8.00	0.850
9.00	0.840
10.35	0.850
11.60	0.870
12.75	0.870
13.75	0.870
14.45	0.850
14.95	0.800
15.00	0.750

λ	ρ
CURVE 46*	
T = ~322	
0.55	0.530
0.75	0.620
0.90	0.590
0.95	0.610
1.00	0.600
1.50	0.625
2.50	0.655
3.25	0.670
4.25	0.670
5.00	0.640
6.50	0.540
7.00	0.520
8.00	0.460
9.00	0.380
12.00	0.260
14.00	0.240
14.50	0.245
14.75	0.260
15.00	0.300
CURVE 47*	
T = ~322	
2.00	0.210
2.25	0.175
2.50	0.160
3.00	0.150
3.50	0.155
4.50	0.180
5.00	0.200
5.70	0.210
6.25	0.200
7.25	0.160
8.25	0.130
9.00	0.115
9.50	0.130
10.00	0.160
10.75	0.230
11.30	0.300
11.90	0.400
12.45	0.500
12.90	0.575
13.75	0.600
14.25	0.600
14.75	0.575
15.00	0.550

* Not shown on plot

DATA TABLE NC. 392 (continued)

λ	ρ
CURVE 48*	
T = ~322	
2.00	0.090
2.50	0.115
3.00	0.125
3.50	0.130
4.00	0.130
4.50	0.140
5.00	0.150
6.00	0.140
7.00	0.120
8.00	0.080
10.00	0.080
11.00	0.100
12.50	0.120
13.00	0.130
14.00	0.115
14.50	0.110
15.00	0.120
CURVE 49*	
T = ~322	
1.00	0.110
1.50	0.150
2.25	0.160
3.00	0.180
4.00	0.180
4.50	0.195
4.90	0.225
5.50	0.225
7.00	0.200
8.00	0.170
8.75	0.150
9.50	0.165
10.50	0.235
11.25	0.330
12.00	0.450
12.50	0.530
13.00	0.590
14.00	0.640
15.00	0.670
CURVE 50*	
T = ~322	
1.00	0.110

λ	ρ
CURVE 50 (cont.)*	
2.00	0.140
2.50	0.220
3.00	0.250
4.25	0.255
5.00	0.270
7.25	0.270
8.00	0.250
9.25	0.205
11.00	0.190
12.00	0.170
12.50	0.160
13.00	0.160
13.50	0.145
14.25	0.145
15.00	0.170
CURVE 51*	
T = ~322	
0.50	0.030
0.65	0.070
0.90	0.100
1.00	0.150
1.50	0.190
2.25	0.200
3.20	0.225
4.00	0.225
5.00	0.260
6.00	0.245
7.00	0.230
8.00	0.200
9.00	0.180
9.50	0.200
10.50	0.290
11.45	0.400
12.20	0.500
12.60	0.560
13.10	0.595
13.75	0.605
14.50	0.590
15.00	0.565
CURVE 52*	
T = ~322	
0.50	0.060

λ	ρ
CURVE 52 (cont.)*	
0.75	0.100
1.00	0.155
1.50	0.200
2.00	0.210
3.00	0.200
3.75	0.190
4.25	0.195
5.00	0.210
8.25	0.175
9.75	0.150
11.25	0.160
12.00	0.155
13.00	0.165
14.00	0.160
14.50	0.170
15.00	0.210
CURVE 53*	
T = ~322	
2.00	0.780
2.50	0.700
3.00	0.670
4.00	0.700
5.40	0.750
6.00	0.755
7.30	0.800
8.00	0.810
9.00	0.805
9.50	0.805
10.00	0.820
10.80	0.820
11.75	0.840
13.00	0.840
14.00	0.820
14.50	0.800
15.00	0.750
CURVE 54*	
T = ~322	
2.00	0.730
2.50	0.625
3.00	0.580
3.25	0.570
3.75	0.595

λ	ρ
CURVE 54 (cont.)*	
4.50	0.625
5.00	0.620
5.75	0.585
6.00	0.560
6.50	0.555
7.00	0.550
9.50	0.385
10.50	0.335
11.50	0.285
12.50	0.250
13.50	0.230
15.00	0.200
CURVE 55*	
T = ~322	
1.00	0.720
1.50	0.690
2.00	0.690
3.00	0.720
4.00	0.750
5.00	0.770
6.10	0.770
7.00	0.800
8.00	0.800
9.00	0.800
10.50	0.815
12.00	0.830
13.00	0.850
14.00	0.880
14.50	0.885
15.00	0.880
CURVE 56*	
T = ~322	
1.00	0.640
1.50	0.615
2.25	0.625
2.75	0.625
4.00	0.640
4.50	0.635
5.50	0.595
6.50	0.565
8.00	0.490
9.00	0.410

λ	ρ
CURVE 56 (cont.)*	
11.00	0.320
12.00	0.272
14.00	0.230
14.50	0.230
15.00	0.250
CURVE 57*	
T = ~322	
0.50	0.520
0.65	0.600
0.90	0.660
1.50	0.725
2.00	0.745
2.80	0.725
3.75	0.745
5.00	0.770
5.75	0.765
6.75	0.785
7.70	0.810
8.25	0.810
9.00	0.800
10.00	0.810
11.00	0.810
11.50	0.815
12.25	0.825
13.00	0.840
13.75	0.845
14.50	0.835
15.00	0.820
CURVE 58*	
T = ~322	
0.50	0.390
0.75	0.510
1.00	0.625
1.25	0.660
2.00	0.690
3.00	0.670
4.00	0.680
5.00	0.660
6.50	0.590
7.00	0.580
8.00	0.540
9.50	0.425
11.00	0.355

λ	ρ
CURVE 58 (cont.)*	
12.50	0.310
13.00	0.300
14.00	0.300
14.50	0.325
15.00	0.400
CURVE 59*	
T = ~322	
2.00	0.570
3.00	0.650
4.00	0.720
5.00	0.760
6.00	0.740
6.75	0.760
7.50	0.785
8.00	0.780
9.00	0.770
9.50	0.765
10.50	0.780
11.50	0.785
12.50	0.790
13.10	0.800
14.00	0.790
15.00	0.840
CURVE 60*	
T = ~322	
2.00	0.520
3.12	0.588
4.13	0.649
5.18	0.683
5.52	0.679
6.16	0.664
7.00	0.682
7.96	0.684
8.64	0.655
9.03	0.629
10.24	0.593
11.74	0.552
12.96	0.497
14.04	0.429
15.00	0.350

λ	ρ
CURVE 61	
T = ~322	
1.00	0.700
1.50	0.660
2.00	0.625
2.50	0.610
2.95	0.625
3.90	0.700*
5.00	0.700
5.65	0.700
6.90	0.710
8.00	0.750
9.00	0.750
10.00	0.750
11.00	0.750
12.25	0.765
13.25	0.770
14.00	0.760
14.65	0.725
15.00	0.675
15.10	0.615
15.00	0.550
CURVE 62*	
T = ~322	
1.00	0.679
1.50	0.663
2.46	0.632
2.85	0.652
3.61	0.702
4.05	0.712
4.64	0.712
5.01	0.721
6.82	0.725
7.26	0.739
7.87	0.737
9.01	0.702
9.50	0.678
10.78	0.623
11.05	0.622
11.74	0.589
12.81	0.534
14.00	0.479
14.82	0.424
15.00	0.375

λ	ρ
CURVE 63*	
T = ~322	
0.50	0.500
0.75	0.575
1.00	0.595
2.00	0.660
2.25	0.660
2.75	0.665
4.00	0.730
5.00	0.750
6.25	0.735
7.25	0.760
8.00	0.690
9.00	0.770
10.00	0.780
11.00	0.780
12.75	0.780
14.00	0.765
14.70	0.725
14.90	0.700
15.00	0.650
CURVE 64*	
T = ~322	
0.50	0.575
0.76	0.630
0.87	0.585
0.96	0.612
1.02	0.603
1.50	0.631
1.82	0.621
2.19	0.610
2.72	0.617
3.25	0.643
4.31	0.696
5.00	0.712
5.60	0.707
6.00	0.701
6.61	0.702
7.25	0.713
7.79	0.710
8.35	0.692
8.79	0.654
9.70	0.620
10.29	0.611
11.15	0.577

* Not shown on plot

DATA TABLE NO. 392 (continued)

λ	ρ
CURVE 64 (cont.)*	
12.03	0.542
12.75	0.522
14.37	0.464
15.00	0.450
CURVE 65* T = ~322	
2.00	0.500
2.10	0.450
2.50	0.415
2.90	0.410
3.25	0.425
3.70	0.475
4.25	0.550
4.80	0.600
5.80	0.640
6.50	0.675
7.75	0.740
9.10	0.725
10.25	0.750
11.00	0.770
11.75	0.770
13.00	0.760
14.00	0.720
14.60	0.675
15.00	0.620
CURVE 66* T = ~322	
2.00	0.460
2.17	0.424
2.63	0.394
3.07	0.394
3.72	0.426
4.71	0.481
5.85	0.520
7.46	0.591
7.99	0.591
8.71	0.563
10.02	0.541
11.21	0.521
12.16	0.489
12.77	0.452
13.23	0.419
CURVE 66 (cont.)*	
13.89	0.383
14.63	0.341
15.00	0.300
CURVE 67* T = ~322	
1.00	0.350
1.50	0.330
1.75	0.350
1.80	0.425
2.00	0.460
2.65	0.475
3.75	0.530
4.50	0.575
5.20	0.605
5.80	0.610
6.75	0.655
8.00	0.720
9.00	0.710
10.00	0.735
11.00	0.775
11.85	0.800
13.00	0.800
14.00	0.800
15.00	0.800
CURVE 68* T = ~322	
1.00	0.340
1.38	0.346
1.68	0.367
2.07	0.408
2.67	0.408
3.03	0.426
3.54	0.470
3.97	0.506
4.96	0.548
6.69	0.617
7.37	0.640
7.77	0.644
9.00	0.601
9.82	0.576
11.08	0.526
12.07	0.505
CURVE 68 (cont.)*	
12.73	0.463
13.35	0.400
13.77	0.356
14.34	0.316
14.68	0.309
15.00	0.325
CURVE 69 T = ~322	
0.95	0.375
1.15	0.465
1.50	0.510*
1.85	0.520
2.60	0.500
3.25	0.525
4.00	0.575
5.00	0.640
5.50	0.650
6.50	0.690
7.50	0.740
8.00	0.750*
9.00	0.720
10.00	0.740
11.00	0.770
11.75	0.780
13.00	0.780
13.60	0.800
14.25	0.810
14.70	0.800*
15.00	0.770
CURVE 70* T = ~322	
0.88	0.300
1.05	0.375
1.52	0.444
1.97	0.472
2.34	0.476
2.74	0.492
4.15	0.590
4.57	0.611
5.25	0.630
7.68	0.681
8.20	0.676
CURVE 70 (cont.)*	
8.69	0.643
9.13	0.616
9.80	0.608
10.75	0.564
11.80	0.528
12.82	0.474
14.21	0.405
14.70	0.401
14.99	0.425
CURVE 71* T = ~322	
2.00	0.750
2.50	0.665
3.10	0.635
3.65	0.650
4.45	0.690
5.25	0.705
6.40	0.725
7.40	0.750
7.90	0.750
9.00	0.740
9.50	0.735
10.50	0.745
11.25	0.750
12.50	0.750
13.30	0.750
14.00	0.740
14.75	0.700
15.00	0.670
CURVE 72* T = ~322	
2.00	0.683
2.34	0.623
2.80	0.576
3.07	0.569
3.74	0.589
4.62	0.627
5.99	0.627
6.49	0.640
7.35	0.683
7.99	0.698
8.69	0.693
CURVE 72 (cont.)*	
10.01	0.666
11.26	0.639
12.75	0.597
14.00	0.544
15.00	0.479
CURVE 73* T = ~322	
1.00	0.530
1.50	0.550
2.00	0.625
2.25	0.640
2.75	0.650
3.50	0.680
4.25	0.710
5.00	0.730
6.00	0.740
6.50	0.745
7.55	0.775
8.00	0.780
9.25	0.765
10.50	0.775
11.50	0.785
12.25	0.790
12.80	0.795
13.75	0.215
14.50	0.825
15.00	0.820
CURVE 74* T = ~322	
1.00	0.621
1.53	0.605
1.90	0.585
2.26	0.576
2.63	0.580
4.25	0.641
5.25	0.647
5.79	0.645
6.34	0.664
8.02	0.735
8.69	0.727
10.02	0.714
11.00	0.693
12.68	0.649
CURVE 74 (cont.)*	
13.04	0.632
14.01	0.613
15.00	0.583
CURVE 75* T = ~322	
0.50	0.500
0.60	0.550
0.75	0.585
1.00	0.600
1.35	0.650
1.70	0.665
2.60	0.650
3.40	0.670
4.25	0.700
5.10	0.720
5.60	0.710
6.50	0.735
7.30	0.760
8.00	0.770
9.00	0.750
10.10	0.750
11.25	0.760
12.00	0.770
13.00	0.770
14.20	0.780
14.75	0.775
15.00	0.760
CURVE 76* T = ~322	
0.50	0.275
0.68	0.375
0.68	0.440
0.90	0.524
1.23	0.562
1.56	0.569
1.98	0.564
2.27	0.573
2.83	0.613
3.98	0.650
5.24	0.676
5.94	0.669
6.33	0.683
8.22	0.743
CURVE 76 (cont.)*	
8.75	0.736
9.68	0.722
11.10	0.708
12.08	0.668
13.00	0.641
14.59	0.598
15.00	0.600
CURVE 77 T = ~322	
2.00	0.430
3.10	0.500
3.50	0.525
4.00	0.570
4.60	0.600
5.50	0.610
6.30	0.600
7.00	0.620
8.00	0.640
9.30	0.600
10.20	0.600
11.40	0.620
12.00	0.625
13.10	0.610
14.00	0.600
15.00	0.550*
CURVE 78* T = ~322	
2.00	0.440
3.70	0.550
4.20	0.582
5.00	0.602
5.51	0.602
6.24	0.606
7.03	0.630
7.49	0.639
8.17	0.630
8.81	0.605
9.42	0.581
10.02	0.579
10.97	0.591
12.02	0.589
13.30	0.579
CURVE 78 (cont.)	
13.95	0.560
14.59	0.500
15.00	0.459
CURVE 79* T = ~322	
1.00	0.500
1.50	0.485
2.25	0.515
2.75	0.515
3.50	0.550
4.50	0.600
5.00	0.610
5.80	0.600
7.00	0.620
8.20	0.640
8.75	0.620
9.50	0.620
10.70	0.620
12.00	0.620
13.50	0.610
14.00	0.600
14.50	0.585
15.00	0.560
CURVE 80* T = ~322	
1.50	0.500
1.80	0.492
2.31	0.495
2.96	0.520
3.52	0.550
4.45	0.592
4.97	0.600
5.97	0.601
7.84	0.648
8.08	0.650
8.66	0.627
9.05	0.600
12.72	0.598
14.20	0.569
14.74	0.534
14.91	0.512
15.00	0.475

* Not shown on plot

DATA TABLE NO. 392 (continued)

CURVE 81*
T = ~322

λ	ρ
0.50	0.350
0.65	0.390
0.80	0.460
1.00	0.450
1.75	0.480
2.50	0.495
3.00	0.500
3.60	0.550
4.15	0.580
5.00	0.600
6.25	0.585
7.00	0.600
8.00	0.630
9.00	0.610
10.00	0.610
11.00	0.610
12.00	0.600
13.00	0.600
14.00	0.580
14.60	0.550
15.00	0.490

CURVE 82*
T = ~322

λ	ρ
0.50	0.350
0.57	0.405
0.92	0.459
1.01	0.457
1.63	0.494
2.24	0.499
2.86	0.515
3.44	0.548
4.14	0.598
4.84	0.618
5.43	0.614
6.10	0.588
6.58	0.593
7.95	0.628
8.40	0.621
8.66	0.617
9.27	0.589
9.94	0.579

CURVE 82 (cont.)*

λ	ρ
10.94	0.599
12.96	0.599
13.99	0.582
14.60	0.551
15.00	0.491

CURVE 83*
T = ~322

λ	ρ
2.00	0.170
2.50	0.125
3.00	0.120
4.00	0.135
4.75	0.175
5.25	0.180
5.50	0.170
6.00	0.125
6.50	0.105
7.00	0.100
7.50	0.120
8.50	0.175
9.00	0.185
10.00	0.170
11.00	0.145
12.00	0.110
13.00	0.140
14.00	0.200
14.75	0.235
15.00	0.240

CURVE 84*
T = ~322

λ	ρ
2.00	0.145
2.60	0.138
3.36	0.144
3.99	0.155
5.00	0.201
5.45	0.190
6.02	0.131
6.80	0.109
7.36	0.118
8.17	0.157

CURVE 84 (cont.)*

λ	ρ
9.00	0.180
9.83	0.176
10.57	0.154
11.13	0.143
12.00	0.111
13.02	0.142
13.35	0.140
14.40	0.130
15.00	0.142

CURVE 85*
T = ~322

λ	ρ
1.00	0.110
1.40	0.150
2.50	0.170
3.75	0.175
4.25	0.195
4.55	0.225
5.00	0.240
6.00	0.200
6.70	0.150
7.10	0.140
7.75	0.180
8.50	0.230
9.00	0.250
9.50	0.255
10.50	0.240
12.15	0.180
13.00	0.190
14.00	0.235
15.00	0.310

CURVE 86*
T = ~322

λ	ρ
1.00	0.170
1.47	0.180
1.88	0.173
2.35	0.158
2.73	0.160
3.24	0.165
3.76	0.165

CURVE 86 (cont.)*

λ	ρ
4.17	0.177
4.80	0.212
5.11	0.218
5.80	0.197
6.31	0.175
6.93	0.139
7.32	0.141
8.09	0.173
9.03	0.200
9.74	0.208
10.73	0.194
12.01	0.159
12.42	0.150
13.01	0.168
13.60	0.159
14.15	0.138
15.00	0.172

CURVE 87*
T = ~322

λ	ρ
0.50	0.100
0.70	0.130
0.90	0.175
1.00	0.190
2.00	0.195
3.00	0.180
3.75	0.175
4.50	0.210
5.00	0.240
5.50	0.230
6.50	0.180
7.00	0.160
7.50	0.175
8.50	0.225
9.30	0.250
9.90	0.250
10.20	0.200
10.50	0.120
10.75	0.200
11.00	0.230
13.30	0.160
12.00	0.190

CURVE 87 (cont.)*

λ	ρ
12.25	0.120
12.60	0.150
13.00	0.200
14.00	0.225
15.00	0.270

CURVE 88*
T = ~322

λ	ρ
0.50	0.100
0.57	0.126
0.86	0.144
1.05	0.191
1.51	0.179
2.08	0.200
3.22	0.177
3.89	0.176
4.27	0.201
4.97	0.252
5.33	0.251
6.00	0.217
6.90	0.151
7.24	0.151
8.32	0.203
9.03	0.234
9.67	0.247
10.03	0.241
10.18	0.203
10.50	0.125
10.78	0.152
10.98	0.211
11.26	0.152
11.60	0.159
12.00	0.171
12.22	0.122
12.59	0.139
12.93	0.177
13.20	0.193
14.00	0.202
14.51	0.211
15.00	0.250

CURVE 89*
T = ~322

λ	ρ
2.00	0.325
2.40	0.250
3.00	0.200
3.50	0.225
3.90	0.300
4.00	0.330
4.50	0.360
5.00	0.370
6.00	0.310
6.25	0.315
6.75	0.400
7.50	0.460
8.00	0.480
9.00	0.465
10.00	0.470
10.75	0.480
11.50	0.475
12.50	0.470
13.50	0.470
14.50	0.460
15.00	0.420

CURVE 90*
T = ~322

λ	ρ
2.00	0.406
2.35	0.338
2.96	0.284
3.35	0.292
3.83	0.329
4.24	0.354
4.98	0.360
5.41	0.330
6.09	0.270
6.43	0.293
7.17	0.344
7.75	0.345
8.81	0.323
10.19	0.308
11.34	0.272
12.41	0.238
13.27	0.222
14.02	0.218
15.00	0.200

CURVE 91
T = ~322

λ	ρ
1.00	0.390
1.50	0.400
2.40	0.400
2.75	0.395
3.00	0.400
3.75	0.460
4.65	0.500
5.50	0.520
6.40	0.525
7.00	0.550
8.00	0.570
9.00	0.525
9.50	0.525
10.50	0.545
11.00	0.550
12.00	0.550
13.00	0.550
14.00	0.560
15.00	0.560

CURVE 92
T = ~322

λ	ρ
1.00	0.414
1.44	0.422
2.05	0.419
2.76	0.412
3.34	0.436
4.35	0.500
5.04	0.519
5.98	0.539
6.51	0.545
7.24	0.575
7.93	0.581
8.35	0.566
9.13	0.521*
9.65	0.509
11.02	0.518
12.11	0.512
12.99	0.509
14.25	0.520*
14.64	0.518
14.99	0.509

CURVE 93*
T = ~322

λ	ρ
0.50	0.250
0.65	0.325
1.00	0.375
1.75	0.430
2.00	0.440
2.75	0.420
3.00	0.410
3.65	0.450
4.25	0.500
5.00	0.530
6.00	0.540
6.75	0.550
7.75	0.575
8.15	0.580
8.75	0.550
9.25	0.515
9.75	0.515
10.50	0.535
11.25	0.555
12.00	0.550
12.75	0.550
13.75	0.540
14.50	0.535
15.00	0.520

CURVE 94*
T = ~322

λ	ρ
0.50	0.200
0.90	0.327
1.19	0.378
1.77	0.422
2.35	0.436
2.95	0.418
3.32	0.430
3.76	0.475
4.29	0.520
5.00	0.539
6.06	0.542
6.80	0.561
8.02	0.590
8.43	0.585
8.90	0.550
9.65	0.514
10.24	0.506

* Not shown on plot

DATA TABLE NO. 392 (continued)

CURVE 94 (cont.)*

λ	ρ
11.02	0.519
12.25	0.521
12.99	0.511
14.29	0.519
14.99	0.500

CURVE 95*
T = ~322

λ	ρ
2.00	0.510
2.75	0.550
3.50	0.615
4.25	0.680
5.00	0.705
6.00	0.690
6.60	0.700
7.50	0.740
8.00	0.740
9.00	0.720
10.00	0.730
11.00	0.760
12.00	0.785
13.00	0.790
14.00	0.790
14.55	0.775
15.00	0.730

CURVE 96*
T = ~322

λ	ρ
2.00	0.425
2.97	0.392
4.52	0.359
6.04	0.339
6.46	0.329
8.00	0.329
8.99	0.319
9.99	0.319
11.01	0.301
12.01	0.291
13.01	0.279
14.03	0.251
15.00	0.221

CURVE 97*
T = ~322

λ	ρ
1.00	0.620
1.50	0.595
2.00	0.600
2.60	0.595
3.25	0.625
4.00	0.675
5.00	0.700
6.00	0.680
6.75	0.690
7.40	0.715
8.00	0.740
8.50	0.740
9.00	0.735
10.00	0.745
11.00	0.765
12.00	0.775
12.75	0.770
13.50	0.770
14.00	0.770
14.60	0.730
15.00	0.660

CURVE 98*
T = ~322

λ	ρ
1.00	0.650
1.99	0.599
2.83	0.581
3.15	0.586
3.76	0.650
4.26	0.677
4.97	0.681
5.99	0.668
6.66	0.678
7.80	0.699
8.25	0.693
8.67	0.668
9.01	0.650
9.99	0.627
11.02	0.602
12.01	0.572
13.00	0.524
14.01	0.476
14.76	0.432
15.00	0.390

CURVE 99*
T = ~322

λ	ρ
0.50	0.480
0.75	0.565
1.00	0.565
1.50	0.600
2.00	0.590
2.50	0.580
3.15	0.600
3.50	0.625
4.00	0.650
4.75	0.680
5.00	0.680
5.75	0.660
6.60	0.675
7.50	0.710
8.00	0.730
9.00	0.720
9.75	0.725
11.00	0.750
11.75	0.750
12.60	0.750
13.20	0.765
14.00	0.750
14.60	0.700
14.85	0.650
15.00	0.600

CURVE 100*
T = ~322

λ	ρ
0.50	0.475
0.62	0.504
0.85	0.552
0.91	0.581
1.01	0.571
1.51	0.611
1.70	0.593
1.73	0.550
1.72	0.454
2.00	0.402
2.50	0.453
2.77	0.485
3.24	0.522
4.49	0.596
4.97	0.610
5.47	0.597

CURVE 100 (cont.)*

λ	ρ
5.84	0.579
6.99	0.602
8.01	0.611
8.51	0.595
9.21	0.552
10.00	0.533
11.00	0.521
12.00	0.494
12.98	0.461
13.98	0.436
14.99	0.394

CURVE 101*
T = ~322

λ	ρ
2.00	0.170
2.75	0.120
3.50	0.105
4.50	0.110
6.00	0.150
7.25	0.185
8.25	0.205
9.00	0.205
10.25	0.165
10.75	0.180
11.50	0.230
12.25	0.305
13.00	0.360
13.70	0.250
14.00	0.180
14.60	0.150
15.00	0.140

CURVE 102*
T = ~322

λ	ρ
2.00	0.195
2.71	0.152
3.49	0.136
4.25	0.144
5.25	0.167
6.16	0.200
7.98	0.231
8.53	0.212
9.11	0.190
10.00	0.202

CURVE 102 (cont.)*

λ	ρ
11.60	0.216
12.17	0.200
12.86	0.212
13.26	0.201
13.98	0.147
14.52	0.115
15.00	0.100

CURVE 103*
T = ~322

λ	ρ
1.00	0.160
2.00	0.155
3.00	0.150
3.25	0.150
5.00	0.170
6.00	0.205
6.50	0.220
7.00	0.230
7.50	0.250
8.25	0.265
9.50	0.250
10.50	0.235
10.75	0.235
11.25	0.270
12.00	0.340
13.00	0.410
13.25	0.400
13.65	0.300
14.00	0.225
14.25	0.210
15.00	0.200

CURVE 104*
T = ~322

λ	ρ
1.00	0.163
1.51	0.175
2.00	0.169
2.71	0.165
3.07	0.171
3.97	0.160
4.73	0.166
5.47	0.186
6.22	0.219
6.99	0.230

CURVE 104 (cont.)*

λ	ρ
8.01	0.250
8.64	0.237
8.99	0.220
9.99	0.229
11.16	0.243
12.01	0.232
12.99	0.228
13.73	0.184
14.34	0.153
15.00	0.153

CURVE 105*
T = ~322

λ	ρ
0.50	0.100
0.55	0.150
0.75	0.160
1.00	0.140
1.50	0.130
1.75	0.175
2.00	0.200
2.75	0.185
3.50	0.175
4.25	0.175
5.50	0.210
6.25	0.235
7.25	0.250
7.75	0.260
8.00	0.285
8.50	0.290
9.50	0.270
10.10	0.250
10.50	0.175
10.70	0.160
10.80	0.175
11.00	0.260
11.35	0.240
11.65	0.275
12.00	0.350
12.75	0.310
12.85	0.375
13.00	0.440
13.15	0.375
13.85	0.275
14.50	0.235
15.00	0.220

CURVE 106*
T = ~322

λ	ρ
0.50	0.035
0.50	0.069
0.61	0.085
0.68	0.138
0.78	0.139
0.94	0.193
1.08	0.194
1.50	0.177
2.00	0.198
2.36	0.193
2.99	0.179
4.37	0.169
5.22	0.199
5.98	0.217
6.99	0.249
7.50	0.253
8.08	0.275
8.49	0.260
9.00	0.228
9.64	0.232
10.00	0.239
10.43	0.174
10.73	0.150
10.90	0.175
11.00	0.248
11.41	0.210
12.00	0.241
12.25	0.235
12.63	0.201
12.87	0.226
13.01	0.249
13.70	0.201
14.00	0.170
14.51	0.170
15.00	0.180

CURVE 107*
T = ~322

λ	ρ
2.00	0.630
2.50	0.550
3.00	0.520
3.50	0.550
3.75	0.575
4.50	0.610

CURVE 107 (cont.)*

λ	ρ
5.50	0.625
6.25	0.630
7.50	0.665
8.00	0.680
9.00	0.650
9.25	0.650
10.50	0.680
12.00	0.710
13.25	0.710
14.40	0.700
14.85	0.675
15.00	0.650

CURVE 108*
T = ~322

λ	ρ
2.00	0.625
2.48	0.576
3.00	0.541
3.73	0.576
4.73	0.625
5.48	0.639
6.25	0.642
7.01	0.661
8.01	0.671
9.02	0.647
10.52	0.635
12.03	0.611
13.00	0.583
13.98	0.543
14.78	0.484
15.00	0.450

CURVE 109*
T = ~322

λ	ρ
1.00	0.470
1.50	0.540
2.00	0.530
2.75	0.550
3.80	0.600
4.50	0.630
5.75	0.650
7.00	0.660
7.90	0.685
9.00	0.660

* Not shown on plot

DATA TABLE NO. 392 (continued)

λ	ρ
CURVE 109 (cont.)*	
9.60	0.650
10.40	0.675
11.00	0.700
12.00	0.720
13.00	0.730
13.75	0.755
14.40	0.765
15.00	0.760
CURVE 110*	
T = ~322	
1.00	0.492
2.10	0.552
2.80	0.566
3.33	0.587
3.97	0.634
4.96	0.664
6.15	0.672
6.99	0.701
7.98	0.710
8.99	0.691
9.60	0.673
11.00	0.671
11.99	0.660
13.00	0.631
14.00	0.611
15.00	0.592
CURVE 111*	
T = ~322	
0.50	0.250
0.60	0.375
0.75	0.450
1.25	0.525
2.00	0.595
2.50	0.580
3.00	0.560
3.70	0.600
4.00	0.630
5.00	0.660
5.50	0.665
6.25	0.660
7.25	0.675
8.25	0.695
9.25	0.685

λ	ρ
CURVE 111 (cont.)*	
10.00	0.675
11.00	0.710
12.00	0.720
13.25	0.735
14.00	0.740
14.55	0.725
14.90	0.690
15.00	0.660
CURVE 112*	
T = ~ 322	
0.50	0.288
0.63	0.367
0.98	0.464
1.54	0.534
1.98	0.574
2.29	0.589
2.97	0.589
3.98	0.643
4.99	0.669
6.10	0.669
7.00	0.681
8.00	0.699
8.50	0.684
8.99	0.657
9.74	0.642
11.48	0.650
12.49	0.626
13.50	0.593
14.50	0.568
15.00	0.561
CURVE 113	
T = 294	
1.0	0.210
1.2	0.190
1.4	0.190
1.6	0.200
1.8	0.210
2.0	0.210
2.2	0.228
2.4	0.236
2.6	0.250
2.8	0.250
3.0	0.250

λ	ρ
CURVE 113 (cont.)	
3.5	0.270
4.0	0.280
4.5	0.280
5.0	0.280
5.5	0.285
6.0	0.263
6.5	0.280
7.0	0.248
7.5	0.260
8.0	0.260
8.5	0.255
9.0	0.250
9.5	0.278*
10.0	0.300
10.5	0.330
10.5	0.340
11.0	0.360
11.5	0.380
12.0	0.395
12.0	0.405
12.5	0.425
12.5	0.433
13.0	0.445
13.0	0.460
13.5	0.485
14.0	0.500
14.0	0.525*
14.5	0.500*
15.0	0.510*
15.5	0.515
16.0	0.500
16.5	0.500
17.0	0.460
17.5	0.460*
18.0	0.440
18.5	0.440*
19.0	0.475
19.5	0.500
20.0	0.500*
20.5	0.500*
21.0	0.500
21.5	0.505
22.0	0.500
22.5	0.500*
23.0	0.490
23.5	0.500*
24.0	0.500

λ	ρ
CURVE 114*	
T = 1033	
1.0	0.190
1.2	0.160
1.4	0.188
1.6	0.200
1.8	0.230
2.0	0.213
2.2	0.250
2.4	0.260
2.6	0.270
2.8	0.270
3.0	0.270
3.5	0.268
4.0	0.260
4.5	0.280
5.0	0.290
5.5	0.310
6.0	0.334
6.5	0.350
7.0	0.375
7.5	0.400
8.0	0.410
8.5	0.390
9.0	0.390
9.5	0.425
10.0	0.380
10.0	0.430
10.5	0.390
10.5	0.440
11.0	0.410
11.0	0.440
11.5	0.430
11.5	0.440
12.0	0.450
12.0	0.460
12.5	0.430
12.5	0.470
13.0	0.390
13.0	0.480
13.5	0.500
14.5	0.520
15.0	0.530
15.5	0.518
16.0	0.520
16.5	0.520
17.0	0.510
17.5	0.510

λ	ρ
CURVE 114 (cont.)*	
18.0	0.470
18.5	0.460
19.0	0.460
19.5	0.450
20.0	0.460
20.5	0.470
21.5	0.500
22.0	0.510
22.5	0.510
23.0	0.500
23.5	0.490
24.0	0.500
CURVE 115*	
T = 1061	
1.60	0.290
2.00	0.300
2.60	0.290
3.00	0.300
4.00	0.310
5.00	0.335
6.00	0.345
7.00	0.335
8.00	0.330
9.00	0.340
10.00	0.350
11.00	0.350
12.00	0.330
13.00	0.285
14.00	0.260
CURVE 116*	
T = 1089	
1.60	0.280
2.00	0.285
3.00	0.285
4.00	0.300
5.00	0.320
6.00	0.340
7.00	0.330
8.00	0.325
9.00	0.325
10.00	0.350
11.00	0.350
12.00	0.320

λ	ρ
CURVE 116 (cont.)*	
13.00	0.285
14.00	0.285
CURVE 117*	
T = 1089	
1.25	0.250
1.40	0.220
1.65	0.260
1.85	0.235
2.00	0.240
2.20	0.240
2.40	0.255
2.60	0.265
2.80	0.252
3.00	0.265
3.50	0.295
4.00	0.300
4.50	0.300
5.00	0.320
5.50	0.345
6.00	0.350
6.50	0.340
7.00	0.315
7.50	0.330
8.00	0.355
8.50	0.330
9.00	0.350
9.50	0.340
10.00	0.350
10.55	0.335
11.00	0.350
11.50	0.325
12.00	0.330
12.50	0.325
13.00	0.300
13.50	0.270
14.00	0.265
CURVE 118*	
T = 1255	
1.20	0.220
1.40	0.292
1.80	0.260
2.00	0.240
2.20	0.235

λ	ρ
CURVE 118 (cont.)*	
2.40	0.240
2.60	0.235
2.75	0.255
3.00	0.250
3.50	0.280
4.00	0.300
4.50	0.320
5.00	0.320
5.50	0.330
6.00	0.330
6.50	0.330
7.00	0.320
7.50	0.330
8.00	0.325
8.50	0.330
9.00	0.360
9.50	0.370
10.00	0.350
10.50	0.340
11.00	0.330
11.50	0.310
12.00	0.300
12.50	0.290
13.00	0.310
13.50	0.275
14.00	0.250
CURVE 119	
T - 1011	
1.00	0.18
1.20	0.17
1.40	0.16
1.60	0.16
1.80	0.17
2.00	0.16
2.20	0.17
2.40	0.16
2.60	0.15
2.80	0.16
3.00	0.16
3.50	0.16
4.00	0.16
4.50	0.16
5.00	0.18
5.50	0.18
6.00	0.19

λ	ρ
CURVE 119 (cont.)	
6.50	0.19
7.00	0.20
7.50	0.19
8.00	0.17
8.45	0.14
9.00	0.13
9.45	0.14
10.00	0.20
10.00	0.21*
10.40	0.24
11.00	0.27
11.00	0.28
11.50	0.28
12.00	0.27
12.00	0.25
12.40	0.17
12.45	0.19
13.00	0.11
13.00	0.14
13.50	0.08
14.00	0.13
14.50	0.17
15.00	0.24
15.50	0.30
16.00	0.39
16.45	0.44
14.00	0.42
14.50	0.43
18.00	0.46
18.50	0.48
19.00	0.47*
19.40	0.47
20.00	0.46
20.40	0.45
21.00	0.45
21.40	0.46
22.00	0.45
22.50	0.42
23.00	0.41
23.40	0.41*
24.00	0.41
CURVE 120*	
T = 294	
1.00	0.115
1.20	0.100
1.40	0.090

* Not shown on plot

DATA TABLE NO. 392 (continued)

CURVE 120 (cont.)*

λ	ρ
1.60	0.075
1.80	0.085
2.00	0.085
2.25	0.095
2.40	0.095
2.65	0.100
2.80	0.105
3.00	0.115
3.50	0.120
4.00	0.130
4.50	0.140
5.00	0.150
5.50	0.165
6.00	0.180
6.50	0.190
7.00	0.195
7.50	0.196
8.00	0.196
8.40	0.196
9.00	0.200
9.50	0.200
10.00	0.210
10.00	0.170
10.40	0.240
10.50	0.180
11.00	0.270
11.00	0.200
11.40	0.280
11.50	0.240
12.00	0.265
12.00	0.240
12.50	0.210
12.60	0.260
13.00	0.220
13.00	0.290
13.50	0.245
13.50	0.330
14.00	0.280
14.00	0.330
14.50	0.380
15.00	0.420
15.50	0.460
16.00	0.485
16.50	0.400
17.00	0.425
17.50	0.460

CURVE 120 (cont.)*

λ	ρ
18.00	0.550
18.50	0.530
19.00	0.480
19.50	0.465
20.00	0.445
20.50	0.425
21.00	0.405
21.50	0.410
22.50	0.410
23.00	0.410

CURVE 121*
T = 1041

λ	ρ
1.00	0.060
1.20	0.060
1.40	0.065
1.60	0.075
1.80	0.085
2.00	0.080
2.25	0.085
2.40	0.085
2.65	0.095
2.80	0.097
3.00	0.100
3.50	0.105
4.00	0.105
4.50	0.115
5.00	0.100
5.50	0.110
6.00	0.120
6.50	0.130
7.00	0.130
7.50	0.135
8.00	0.140
8.50	0.140
9.00	0.130
9.50	0.130
10.00	0.120
10.00	0.140
10.50	0.145
10.50	0.120
11.00	0.130
11.00	0.100
11.50	0.105
11.50	0.070

CURVE 121 (cont.)*

λ	ρ
11.90	0.050
12.00	0.080
12.50	0.070
12.50	0.040
13.00	0.040
13.00	0.030
13.50	0.025
14.00	0.010
14.00	0.020
14.50	0.045
15.00	0.110
15.00	0.090
15.50	0.165
16.00	0.230
16.50	0.290
17.00	0.330
17.50	0.325
18.00	0.330
18.50	0.315
19.00	0.325
19.50	0.360
20.00	0.360
20.50	0.380
21.00	0.390
21.50	0.410
22.00	0.415
22.50	0.400
23.00	0.400
23.50	0.380
24.00	0.400

* Not shown on plot

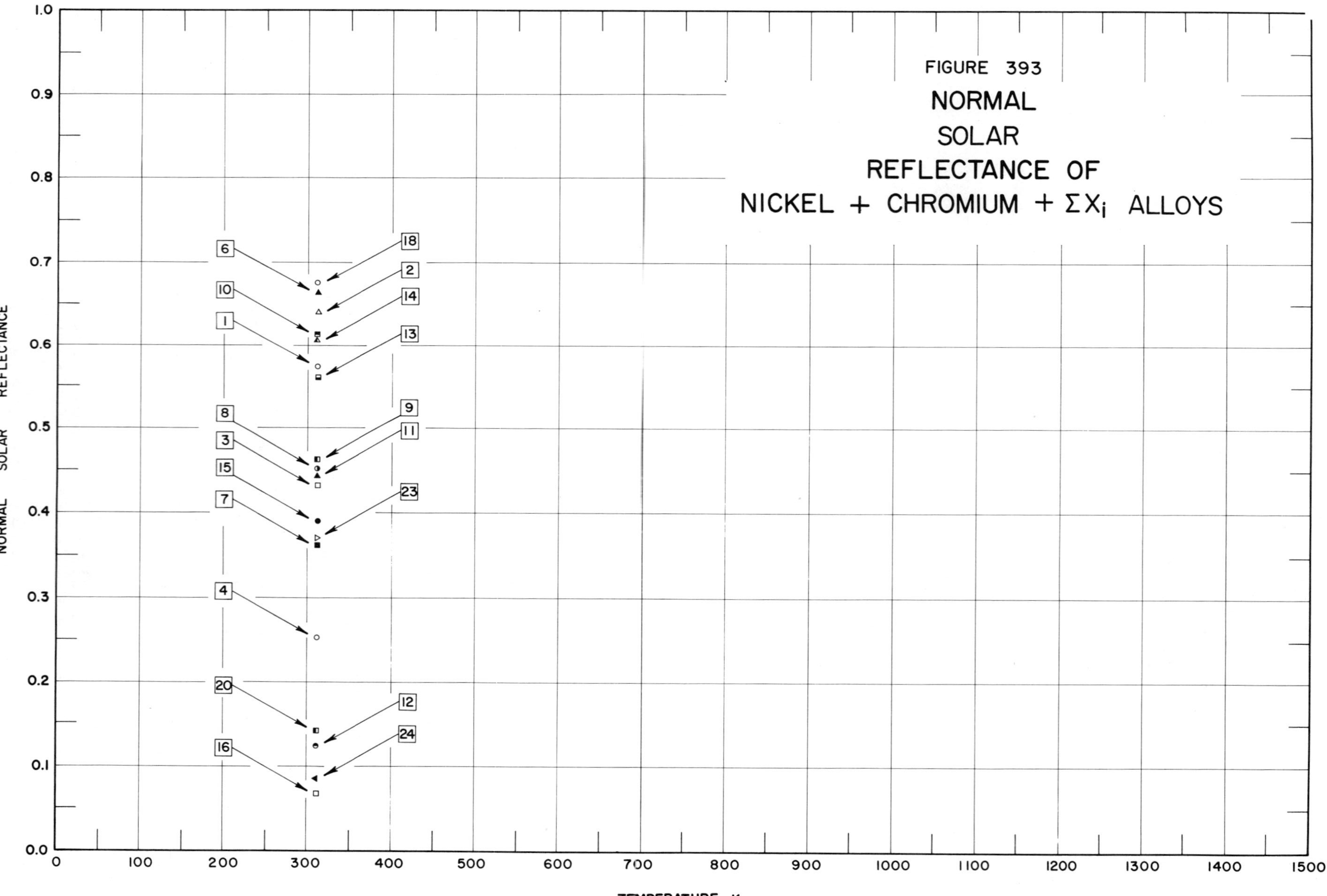

FIGURE 393
NORMAL
SOLAR
REFLECTANCE OF
NICKEL + CHROMIUM + ΣX_i ALLOYS
NORMAL SOLAR REFLECTANCE
TEMPERATURE, K
0.0
0.1
0.2
0.3
0.4
0.5
0.6
0.7
0.8
0.9
1.0
0
100
200
300
400
500
600
700
800
900
1000
1100
1200
1300
1400
1500
6
10
1
8
3
15
7
4
20
16
18
2
14
13
9
11
23
12
24

SPECIFICATION TABLE NO. 393 NORMAL SOLAR REFLECTANCE OF [NICKEL + CHROMIUM + ΣX_i] ALLOYS

Curve No.	Ref. No.	Year	Temperature Range, K	Geometry θ θ' ω'		Reported Error, %	Composition (weight percent), Specifications and Remarks
1	160	1954	311	~0°	2π		Inconel B; nominal composition: 72 min Ni, 16-18 Cr, 9.5 max Fe, 1 max Mn, 0.15 C; heated to 301 K; clean and smooth surface; measured in air; calculated from solar absorptance.
2	160	1954	311	~0°	2π		Above specimen and conditions except heated to 616 K.
3	160	1954	311	~0°	2π		Above specimen and conditions except heated to 921 K.
4	160	1954	311	~0°	2π		Above specimen and conditions except heated to 1313 K.
5	160	1954	311	~0°	2π		Different sample, same as curve 1 specimen and conditions except heated to 320 K; polished.
6	160	1954	311	~0°	2π		Above specimen and conditions except heated to 677 K.
7	160	1954	311	~0°	2π		Above specimen and conditions except heated to 977 K.
8	160	1954	311	~0°	2π		Above specimen and conditions except heated to 1254 K.
9	160	1954	311	~0°	2π		Different sample, same as curve 1 specimen and conditions except heated to 324 K; cleaned with methyl alcohol.
10	160	1954	311	~0°	2π		Above specimen and conditions except heated to 611 K.
11	160	1954	311	~0°	2π		Above specimen and conditions except heated to 933 K.
12	160	1954	311	~0°	2π		Above specimen and conditions except heated to 1135 K.
13	160	1954	311	~0°	2π		Inconel X; nominal composition: 73 Ni, 15 Cr, 7 Fe, 2.5 Ti, 1 Nb, 0.9 Al, 0.70 Mn, 0.30 Si, 0.04 C; heated to 315 K; clean and smooth surface; measured in air; calculated from solar absorptance.
14	160	1954	311	~0°	2π		Above specimen and conditions except heated to 611 K.
15	160	1954	311	~0°	2π		Above specimen and conditions except heated to 939 K.
16	160	1954	311	~0°	2π		Above specimen and conditions except heated to 1207 K.
17	160	1954	311	~0°	2π		Different sample, same as curve 13 specimen and conditions except heated to 306 K; cleaned with methyl alcohol.
18	160	1954	311	~0°	2π		Above specimen and conditions except heated to 612 K.
19	160	1954	311	~0°	2π		Above specimen and conditions except heated to 947 K.
20	160	1954	311	~0°	2π		Above specimen and conditions except heated to 1228 K.
21	160	1954	311	~0°	2π		Different sample, same as curve 13 specimen and conditions except heated to 307 K; polished.
22	160	1954	311	~0°	2π		Above specimen and conditions except heated to 590 K.
23	160	1954	311	~0°	2π		Above specimen and conditions except heated to 945 K.
24	160	1954	311	~0°	2π		Above specimen and conditions except heated to 1232 K.

DATA TABLE NO. 393 NORMAL SOLAR REFLECTANCE OF [NICKEL + CHROMIUM + ΣX_i] ALLOYS

[Temperature, T, K; Reflectance, ρ]

T	ρ	T	ρ	T	ρ
CURVE 1		CURVE 12		CURVE 23	
311	0.573	311	0.121	311	0.370
CURVE 2		CURVE 13		CURVE 24	
311	0.640	311	0.561	311	0.085
CURVE 3		CURVE 14			
311	0.432	311	0.609		
CURVE 4		CURVE 15			
311	0.252	311	0.390		
CURVE 5*		CURVE 16			
311	0.642	311	0.069		
CURVE 6		CURVE 17*			
311	0.662	311	0.563		
CURVE 7		CURVE 18			
311	0.361	311	0.674		
CURVE 8		CURVE 19*			
311	0.452	311	0.559		
CURVE 9		CURVE 20			
311	0.460	311	0.141		
CURVE 10		CURVE 21*			
311	0.614	311	0.612		
CURVE 11		CURVE 22*			
311	0.443	311	0.669		

* Not shown on plot

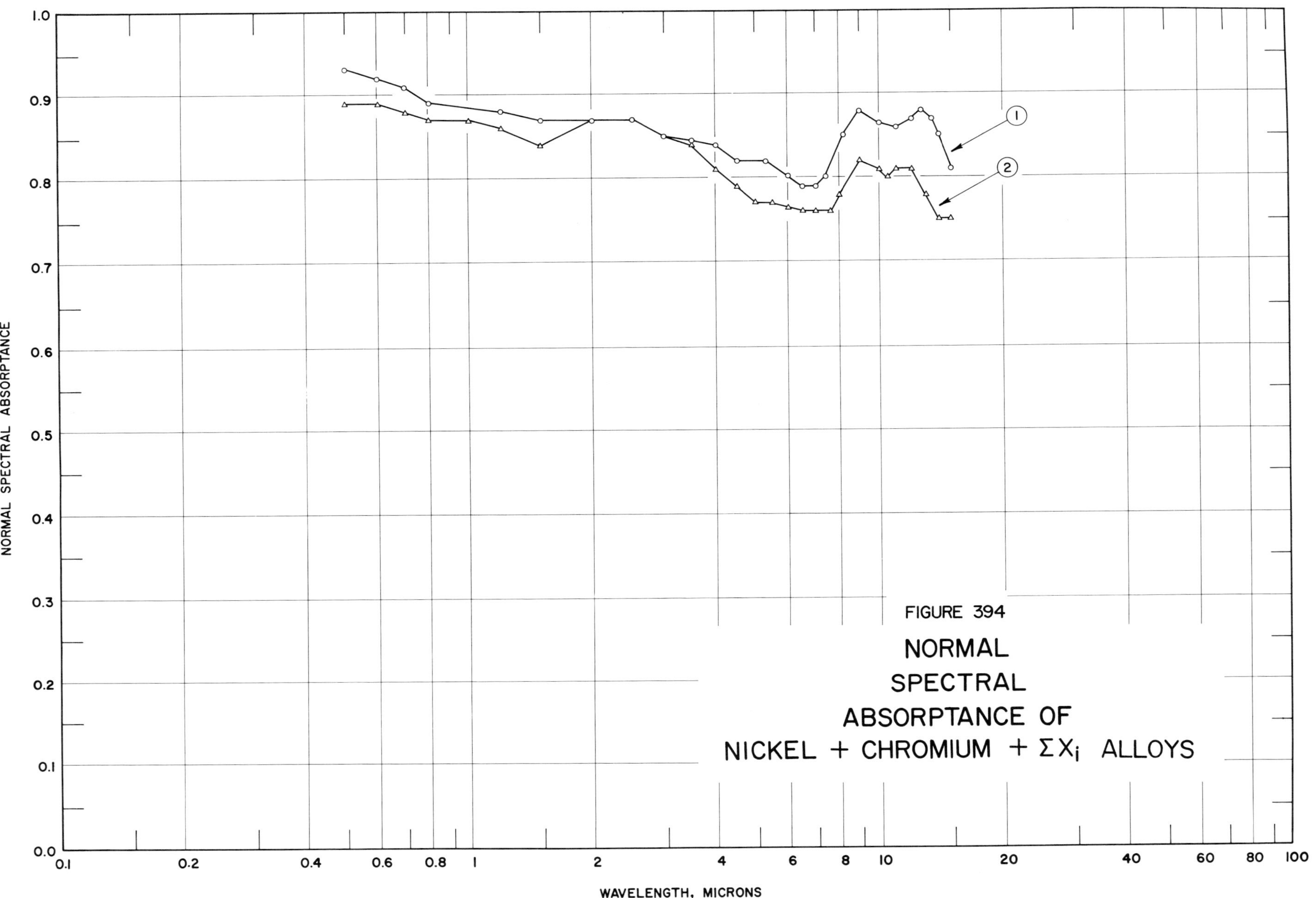
FIGURE 394
NORMAL
SPECTRAL
ABSORPTANCE OF
NICKEL + CHROMIUM + ΣXᵢ ALLOYS
NORMAL SPECTRAL ABSORPTANCE
WAVELENGTH, MICRONS
1.0
0.9
0.8
0.7
0.6
0.5
0.4
0.3
0.2
0.1
0.0
0.1
0.2
0.4
0.6
0.8
1
2
4
6
8
10
20
40
60
80
100
1
2

SPECIFICATION TABLE NO. 394 NORMAL SPECTRAL ABSORPTANCE OF [NICKEL + CHROMIUM + ΣX_i] ALLOYS

Curve No.	Ref. No.	Year	Temperature K	Wavelength Range, μ	Geometry θ	Reported Error, %	Composition (weight percent), Specifications and Remarks
1	30	1963	294	0.5-15.0	$\sim 0°$		Inconel; nominal composition: 78 Ni, 15 Cr, 7 Fe, 0.35 Mn, 0.20 Si, 0.04 C; oxidized; measured in air; data extracted from smooth curve.
2	30	1963	294	0.5-15.0	$\sim 0°$		Above specimen and conditions; heated at 1250 K for 2 hrs and at 1467 K for 2 hrs.

DATA TABLE NO. 394 NORMAL SPECTRAL ABSORPTANCE OF [NICKEL + CHROMIUM + ΣX_i] ALLOYS

[Wavelength, λ, μ; Absorptance, α; Temperature, T, K]

λ	α
CURVE 1	
T = 294	
0.5	0.930
0.6	0.920
0.7	0.910
0.8	0.890
1.2	0.880
1.5	0.870
2.0	0.870
2.5	0.870
3.0	0.850
3.5	0.845
4.0	0.840
4.5	0.820
5.3	0.820
6.0	0.800
6.5	0.790
7.0	0.790
7.4	0.800
8.2	0.850
9.0	0.880
10.0	0.865
11.0	0.860
12.0	0.870
12.6	0.880
13.4	0.870
14.0	0.850
15.0	0.810
CURVE 2	
T = 294	
0.5	0.890
0.6	0.890
0.7	0.880
0.8	0.870
1.0	0.870
1.2	0.860
1.5	0.840
2.0	0.870*
2.5	0.870*
3.0	0.850*
3.5	0.840
4.0	0.810
4.5	0.790

λ	α
CURVE 2 (cont.)	
5.0	0.770
5.5	0.770
6.0	0.765
6.5	0.760
7.0	0.760
7.6	0.760
8.0	0.780
9.0	0.820
10.0	0.810
10.4	0.800
11.0	0.810
12.0	0.810
13.0	0.780
14.0	0.750
15.0	0.750

* Not shown on plot

SPECIFICATION TABLE NO. 395 ANGULAR SPECTRAL ABSORPTANCE OF [NICKEL + CHROMIUM + ΣX_i] ALLOYS

Curve No.	Ref. No.	Year	Temperature K	Wavelength Range, μ	Geometry θ	Reported Error, %	Composition (weight percent), Specifications and Remarks
1	225	1965	306	0.371-17.0	25°		Inconel X (rolled plate); nominal composition: 73 Ni, 15 Cr, 7 Fe, 2.5 Ti, 1 Nb, 0.9 Al, 0.70 Mn, 0.30 Si, and 0.04 C; measured in dry nitrogen; heated cavity at approx 1056 K with platinum reference; authors assumed $\alpha = 1 - R(2\pi, 25°)$.

DATA TABLE NO. 395 ANGULAR SPECTRAL ABSORPTANCE OF [NICKEL + CHROMIUM + ΣX_i] ALLOYS

[Wavelength, λ, μ; Absorptance, α; Temperature, T, K]

λ	α
CURVE 1*	
T = 306	
0.371	0.652
0.410	0.614
0.496	0.562
0.596	0.525
0.689	0.501
0.794	0.481
0.891	0.468
0.995	0.447
1.28	0.406
1.48	0.394
1.95	0.374
2.76	0.333
3.74	0.299
4.99	0.255
6.50	0.210
8.89	0.200
11.4	0.185
13.0	0.185
17.0	0.127

* Not shown on plot

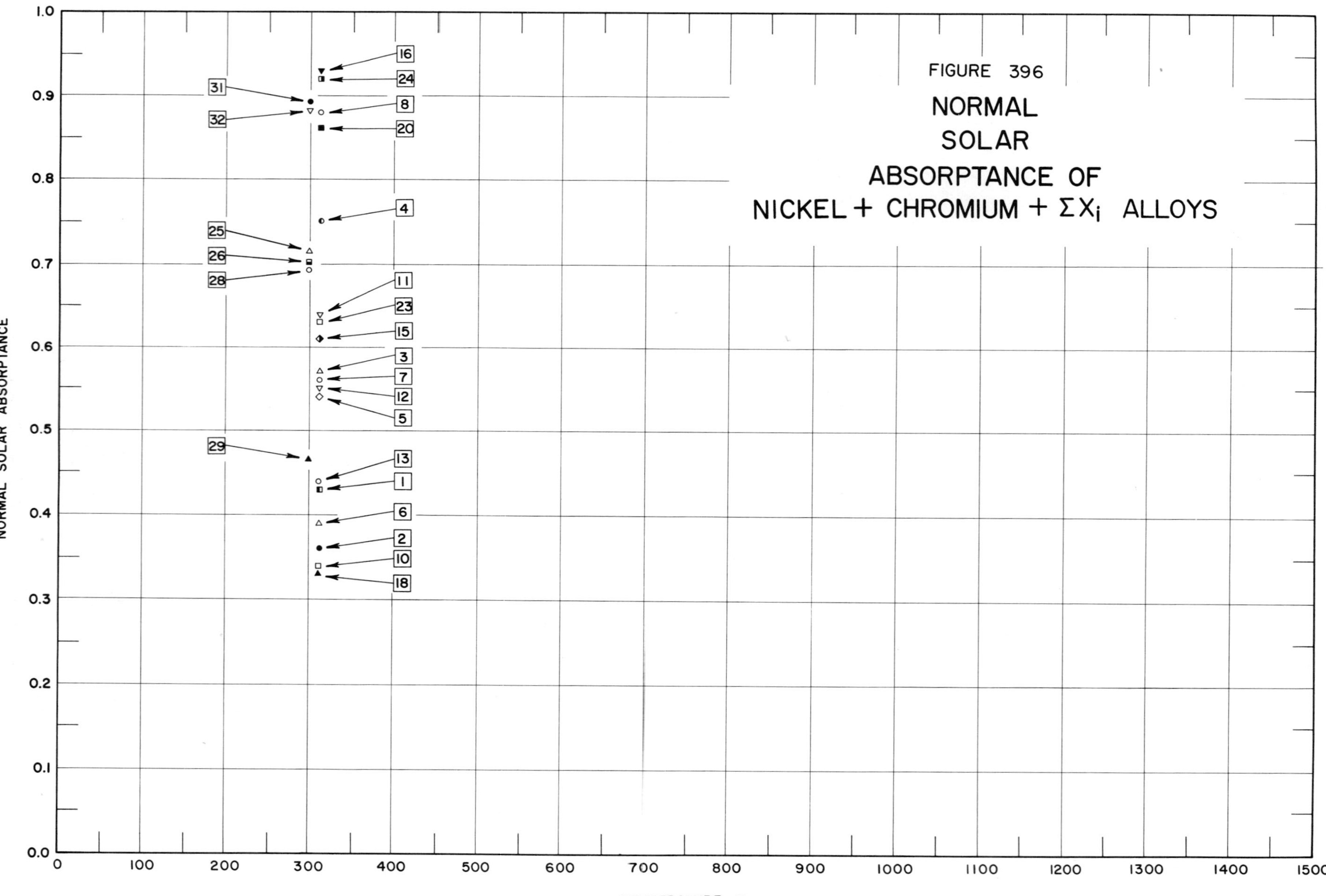

FIGURE 396 NORMAL SOLAR ABSORPTANCE OF NICKEL + CHROMIUM + ΣX_i ALLOYS

SPECIFICATION TABLE NO. 396 NORMAL SOLAR ABSORPTANCE OF [NICKEL + CHROMIUM + ΣX_i] ALLOYS

Curve No.	Ref. No.	Year	Temperature Range, K	Geometry θ	Reported Error, %	Composition (weight percent), Specifications and Remarks
1	160	1954	311	~0°		Inconel B; nominal composition: 72 min Ni, 16-18 Cr, 9.5 max Fe, 1 max Mn, 0.15 C; heated to 305 K; clean and smooth surface; measured in air at sea level.
2	160	1954	311	~0°		Above specimen and conditions except reheated to 617 K.
3	160	1954	311	~0°		Above specimen and conditions except reheated to 923 K.
4	160	1954	311	~0°		Above specimen and conditions except reheated to 1311 K.
5	160	1954	311	~0°		Different sample, same as curve 1 specimen and conditions except heated to 325 K; cleaned with methyl alcohol.
6	160	1954	311	~0°		Above specimen and conditions except reheated to 613 K.
7	160	1954	311	~0°		Above specimen and conditions except reheated to 935 K.
8	160	1954	311	~0°		Above specimen and conditions except reheated to 1133 K.
9	160	1954	311	~0°		Different sample, same as curve 1 specimen and conditions except heated to 320 K; polished.
10	160	1954	311	~0°		Above specimen and conditions except reheated to 680 K.
11	160	1954	311	~0°		Above specimen and conditions except reheated to 945 K.
12	160	1954	311	~0°		Above specimen and conditions except reheated to 1255 K.
13	160	1954	311	~0°		Inconel X; nominal composition: 73 Ni, 15 Cr, 7 Fe, 2.5 Ti, 1 Nb, 0.9 Al, 0.70 Mn, 0.30 Si, 0.04 C; heated to 314 K; clean and smooth surface; measured in air at sea level.
14	160	1954	311	~0°		Above specimen and conditions except reheated to 614 K.
15	160	1954	311	~0°		Above specimen and conditions except reheated to 938 K.
16	160	1954	311	~0°		Above specimen and conditions except reheated to 1200 K.
17	160	1954	311	~0°		Different sample, same as curve 13 specimen and conditions except heated to 309 K; cleaned with methyl alcohol.
18	160	1954	311	~0°		Above specimen and conditions except reheated to 614 K.
19	160	1954	311	~0°		Above specimen and conditions except reheated to 948 K.
20	160	1954	311	~0°		Above specimen and conditions except reheated to 1228 K.
21	160	1954	311	~0°		Different sample, same as curve 13 specimen and conditions except heated to 310 K; polished.
22	160	1954	311	~0°		Above specimen and conditions except reheated to 596 K.
23	160	1954	311	~0°		Above specimen and conditions except reheated to 947 K.
24	160	1954	311	~0°		Above specimen and conditions except reheated to 1228 K.

SPECIFICATION TABLE NO. 396 (continued)

Curve No.	Ref. No.	Year	Temperature Range, K	Geometry θ	Reported Error, %	Composition (weight percent), Specifications and Remarks
25	146	1958	298	9°		Inconel X; nominal composition: 73 Ni, 15 Cr, 7 Fe, 2.5 Ti, 1 Nb, 0.9 Al, 0.7 Mn, 0.3 Si, 0.04 C; as received; computed from spectral reflectance data for sea level conditions.
26	146	1958	298	9°		Above specimen and conditions except computed for above atmosphere conditions.
27	146	1958	298	9°		Different sample, same as curve 25 specimen and conditions except cleaned.
28	146	1958	298	9°		Above specimen and conditions except computed for above atmosphere conditions.
29	146	1958	298	9°		Different sample, same as curve 25 specimen and conditions except polished.
30	146	1958	298	9°		Above specimen and conditions except computed for above atmosphere conditions.
31	146	1958	298	9°		Different sample, same as curve 25 specimen and conditions except oxidized.
32	146	1958	298	9°		Above specimen and conditions except computed for above atmosphere conditions.

DATA TABLE NO. 396 NORMAL SOLAR ABSORPTANCE OF [NICKEL + CHROMIUM + ΣX_i] ALLOYS

[Temperature, T, K; Absorptance, α]

T	α	T	α	T	α
CURVE 1		CURVE 13		CURVE 25	
311	0.43	311	0.44	298	0.715
CURVE 2		CURVE 14*		CURVE 26	
311	0.36	311	0.39	298	0.701
CURVE 3		CURVE 15		CURVE 27*	
311	0.57	311	0.61	298	0.706
CURVE 4		CURVE 16		CURVE 28	
311	0.75	311	0.93	298	0.691
CURVE 5		CURVE 17*		CURVE 29	
311	0.54	311	0.44	298	0.463
CURVE 6		CURVE 18		CURVE 30*	
311	0.39	311	0.33	298	0.463
CURVE 7		CURVE 19*		CURVE 31	
311	0.56	311	0.44	298	0.891
CURVE 8		CURVE 20		CURVE 32	
311	0.88	311	0.86	298	0.881
CURVE 9*		CURVE 21*			
311	0.36	311	0.39		
CURVE 10		CURVE 22*			
311	0.34	311	0.33		
CURVE 11		CURVE 23			
311	0.64	311	0.63		
CURVE 12		CURVE 24			
311	0.55	311	0.92		

* Not shown on plot

SPECIFICATION TABLE NO. 397 HEMISPHERICAL TOTAL EMITTANCE [NICKEL + COBALT + ΣX_i] ALLOYS

Curve No.	Ref. No.	Year	Temperature Range, K	Reported Error, %	Composition (weight percent), Specifications and Remarks
1	81	1941	973-1213	<1	Konel; nominal composition: 73.07 Ni, 17.16 Co, 8.8 Ti, 0.55 Si, 0.26 Al, 0.16 Mn; filament used in type DL20 radio tube; coated with oxide; heated in hydrogen at 1098 K for 15 min.

DATA TABLE NO. 397 HEMISPHERICAL TOTAL EMITTANCE OF [NICKEL + COBALT + ΣX_i] ALLOYS

[Temperature, T, K; Emittance, ϵ]

T	ϵ
CURVE 1*	
973	0.452
1213	0.412

* Not shown on plot

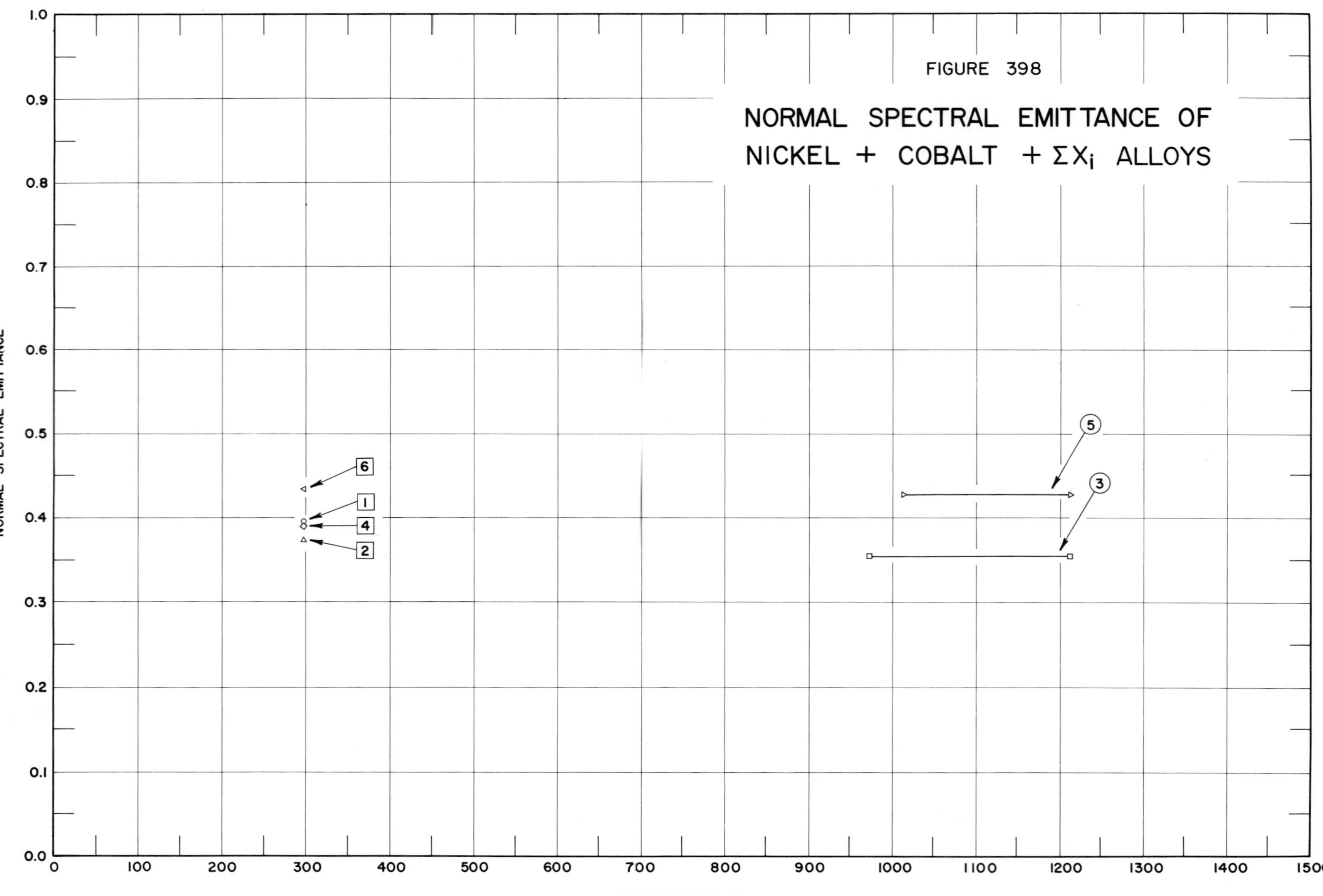
FIGURE 398
NORMAL SPECTRAL EMITTANCE OF
NICKEL + COBALT + ΣXi ALLOYS
NORMAL SPECTRAL EMITTANCE
1.0
0.9
0.8
0.7
0.6
0.5
0.4
0.3
0.2
0.1
0.0
0
100
200
300
400
500
600
700
800
900
1000
1100
1200
1300
1400
1500
TEMPERATURE, K
6
1
4
2
5
3

SPECIFICATION TABLE NO. 398 NORMAL SPECTRAL EMITTANCE OF [NICKEL + COBALT + ΣX_i] ALLOYS

Curve No.	Ref. No.	Year	Wavelength μ	Temperature Range, K	Geometry θ'	Reported Error, %	Composition (weight percent), Specifications and Remarks
1	81	1941	0.65	298	$\sim 0^{\circ}$		Konel; nominal composition: 73.07 Ni, 17.16 Co, 8.8 Ti, 0.55 Si, 0.26 Al, 0.16 Mn; filament used in type DL20 radio tube; coated with oxide; heated in hydrogen at 1098 K for 15 min.
2	81	1941	0.65	298	$\sim 0^{\circ}$		Different sample, same as above specimen and conditions except authors assumed $\alpha = \epsilon$.
3	81	1941	0.65	973-1213	$\sim 0^{\circ}$	< 1	Different sample, same as above specimen and conditions.
4	81	1941	0.65	298	$\sim 0^{\circ}$		Different sample, same as curve 1 specimen and conditions except filament used in type DL21 radio tube.
5	81	1941	0.65	1013-1213	$\sim 0^{\circ}$	< 1	Different sample, same as above specimen and conditions except authors assumed $\alpha = \epsilon$.
6	81	1941	0.65	298	$\sim 0^{\circ}$		Different sample, same as above specimen and conditions.

DATA TABLE NO. 398 NORMAL SPECTRAL EMITTANCE OF [NICKEL + COBALT + ΣX_i] ALLOYS

[Temperature, T, K; Emittance, ϵ; Wavelength, λ, μ]

T	ϵ
CURVE 1 $\lambda = 0.65$	
298	0.397
CURVE 2 $\lambda = 0.65$	
298	0.374
CURVE 3 $\lambda = 0.65$	
973	0.355
1213	0.355
CURVE 4 $\lambda = 0.65$	
298	0.391
CURVE 5 $\lambda = 0.65$	
1013	0.429
1213	0.429
CURVE 6 $\lambda = 0.65$	
298	0.436

* Not shown on plot

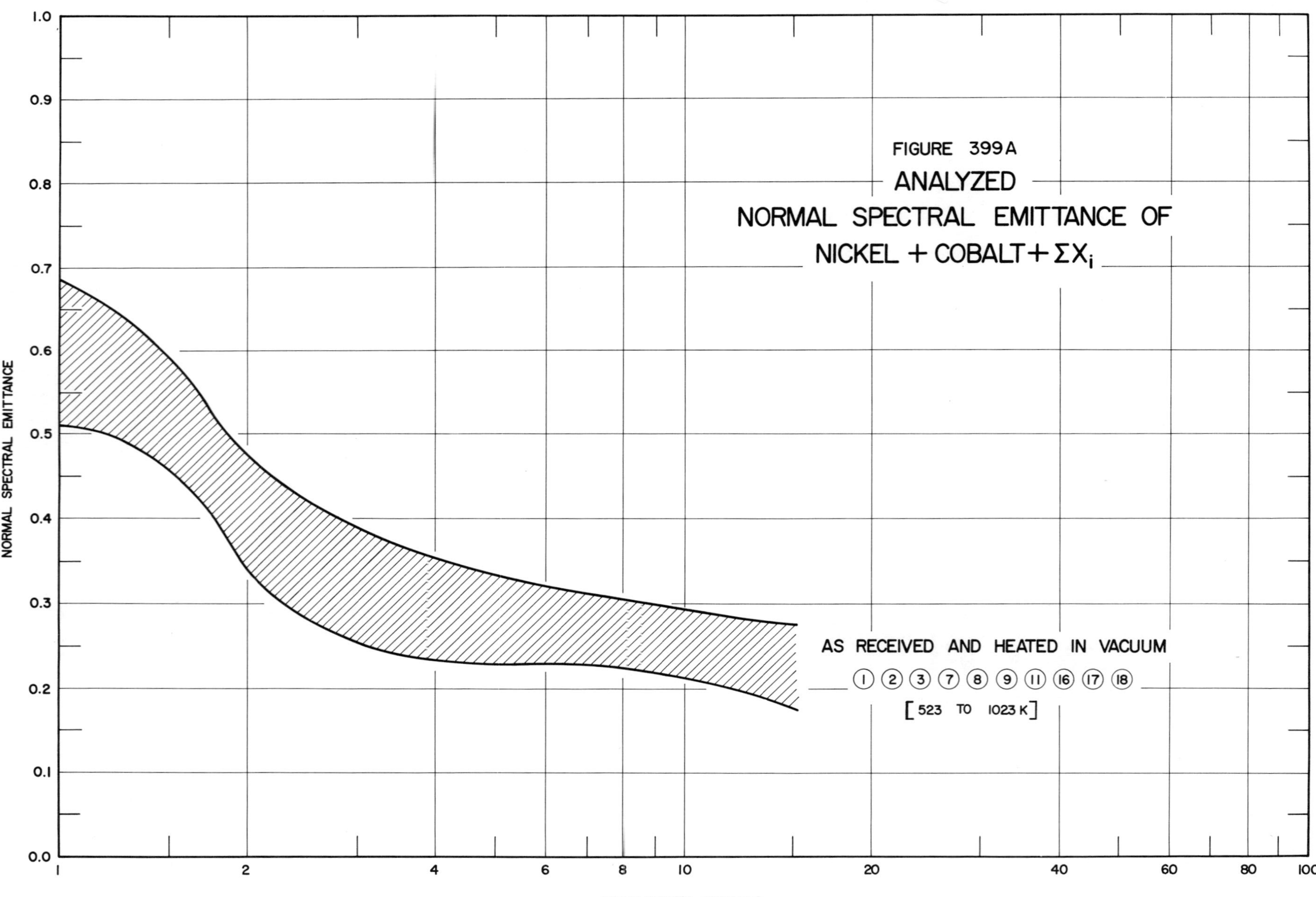
FIGURE 399A
ANALYZED
NORMAL SPECTRAL EMITTANCE OF
NICKEL + COBALT + ΣX_i
AS RECEIVED AND HEATED IN VACUUM
1 2 3 7 8 9 11 16 17 18
[523 TO 1023 K]
NORMAL SPECTRAL EMITTANCE
1.0
0.9
0.8
0.7
0.6
0.5
0.4
0.3
0.2
0.1
0.0
1
2
4
6
8
10
20
40
60
80
100
WAVELENGTH, MICRONS

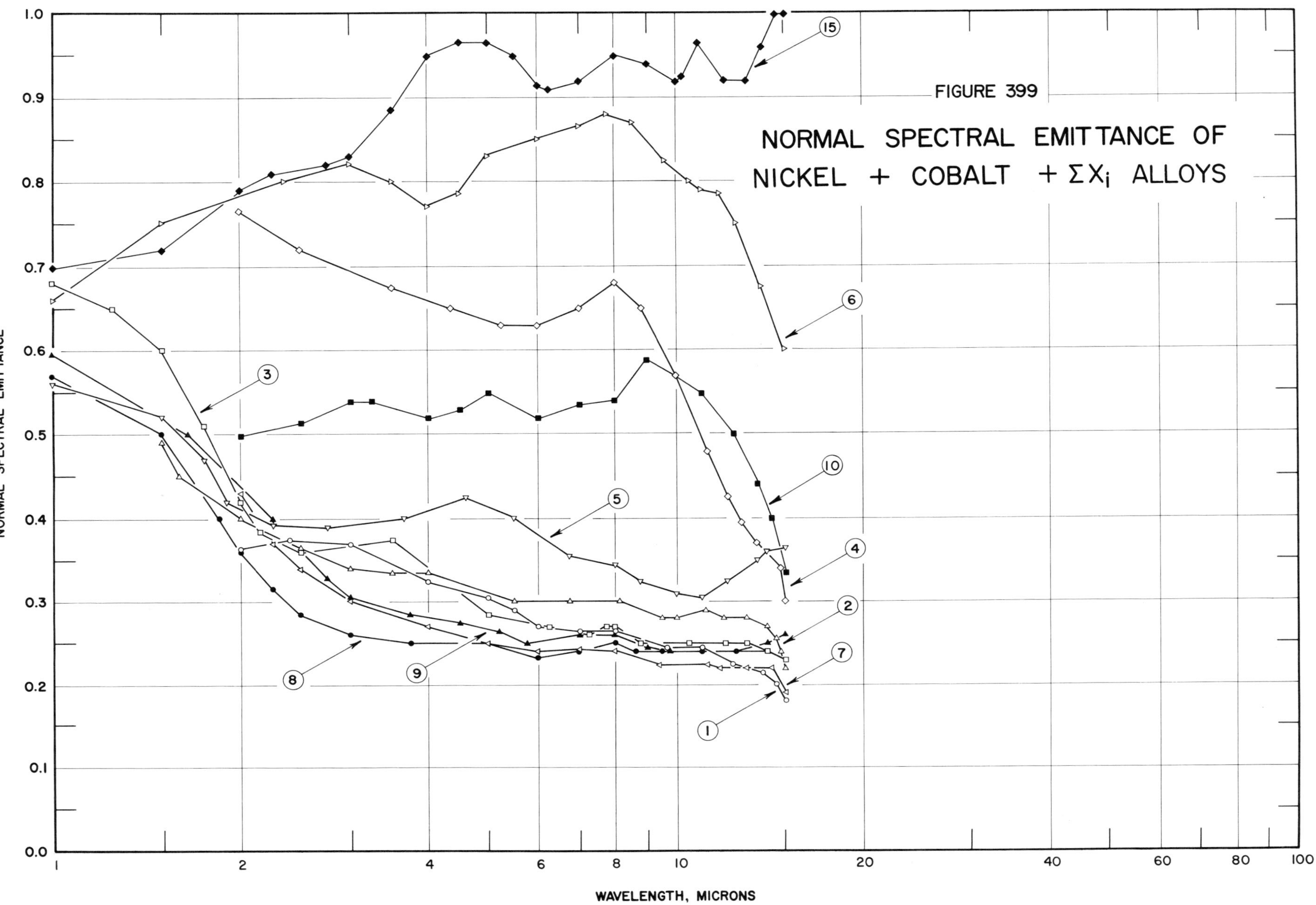
FIGURE 399
NORMAL SPECTRAL EMITTANCE OF
NICKEL + COBALT + ΣXi ALLOYS
NORMAL SPECTRAL EMITTANCE
WAVELENGTH, MICRONS
1.0
0.9
0.8
0.7
0.6
0.5
0.4
0.3
0.2
0.1
0.0
1
2
4
6
8
10
20
40
60
80
100
15
6
3
10
5
4
2
7
8
9
1

SPECIFICATION TABLE NO. 399 NORMAL SPECTRAL EMITTANCE OF [NICKEL + COBALT + ΣX_i] ALLOYS

Curve No.	Ref. No.	Year	Temperature K	Wavelength Range, μ	Geometry θ'	Reported Error, %	Composition (weight percent), Specifications and Remarks
1	86	1961	523	2.00-15.00	$\sim 0^\circ$	± 5	Astrolloy; nominal composition: 15 Co, 15 Cr, 5 Mo, 4.3 Al, 3.5 Ti, 0.05 C, 0.03 B, Ni balance; as received; data extracted from smooth curve.
2	86	1961	773	1.50-15.00	$\sim 0^\circ$	± 5	Different sample, same as curve 1 specimen and conditions.
3	86	1961	1023	1.00-15.00	$\sim 0^\circ$	± 5	Different sample, same as curve 1 specimen and conditions.
4	86	1961	523	2.00-15.00	$\sim 0^\circ$	± 5	Different sample, same as curve 1 specimen and conditions except heated in air at 1255 K for 30 min.
5	86	1961	773	1.00-15.00	$\sim 0^\circ$	± 5	Different sample, same as curve 4 specimen and conditions.
6	86	1961	1023	1.00-15.00	$\sim 0^\circ$	± 5	Different sample, same as curve 4 specimen and conditions.
7	86	1961	523	2.00-15.00	$\sim 0^\circ$	± 5	Different sample, same as curve 1 specimen and conditions except heated in vacuum (7.6×10^{-5} mm Hg) at 1255 K for 30 min.
8	86	1961	773	1.00-15.00	$\sim 0^\circ$	± 5	Different sample, same as curve 7 specimen and conditions.
9	86	1961	1023	1.00-15.00	$\sim 0^\circ$	± 5	Different sample, same as curve 7 specimen and conditions.
10	86	1961	523	2.00-15.00	$\sim 0^\circ$	± 5	Udimet 500; nominal composition: 19 Co, 19 Cr, 4 Mo, 4 max Fe, 3 Ti, 2.9 Al, 0.1 C, trace B, Ni balance; as received; data extracted from smooth curve.
11	86	1961	773	1.00-15.00	$\sim 0^\circ$	± 5	Different sample, same as curve 10 specimen and conditions.
12	86	1961	1023	1.00-15.00	$\sim 0^\circ$	± 5	Different sample, same as curve 10 specimen and conditions.
13	86	1961	523	2.00-15.00	$\sim 0^\circ$	± 5	Different sample, same as curve 10 specimen and conditions except heated in air at 1255 K for 30 min.
14	86	1961	773	1.00-15.00	$\sim 0^\circ$	± 5	Different sample, same as curve 13 specimen and conditions.
15	86	1961	1023	1.00-15.00	$\sim 0^\circ$	± 5	Different sample, same as curve 13 specimen and conditions.
16	86	1961	523	2.00-15.00	$\sim 0^\circ$	± 5	Different sample, same as curve 10 specimen and conditions except heated in vacuum (7.6×10^{-5} mm Hg) at 1255 K for 30 min.
17	86	1961	773	1.00-15.00	$\sim 0^\circ$	± 5	Different sample, same as curve 16 specimen and conditions.
18	86	1961	1023	1.00-15.00	$\sim 0^\circ$	± 5	Different sample, same as curve 16 specimen and conditions.

DATA TABLE NO. 399 NORMAL SPECTRAL EMITTANCE OF [NICKEL + COBALT + ΣX_i] ALLOYS

[Wavelength, λ, μ; Emittance, ∈; Temperature, T, K]

λ	∈
CURVE 1 T = 523	
2.00	0.365
2.40	0.375
3.00	0.370
4.00	0.325
5.00	0.305
5.50	0.290
6.00	0.270
7.00	0.265
8.00	0.265
9.65	0.245
11.00	0.245
12.35	0.225
13.75	0.215
14.50	0.200
15.00	0.180
CURVE 2 T = 773	
1.50	0.490
1.60	0.450
2.00	0.400
2.50	0.365
3.00	0.340
3.50	0.335
4.00	0.335
5.50	0.300
6.75	0.300
8.10	0.300
9.50	0.280
10.00	0.280
11.20	0.290
11.95	0.280
13.00	0.280
14.00	0.270
14.50	0.255
14.75	0.240
15.00	0.220
CURVE 3 T = 1023	
1.00	0.680
1.25	0.650
1.50	0.600
1.75	0.510

λ	∈
CURVE 3 (cont.)	
2.00	0.420
2.15	0.385
2.50	0.360
3.50	0.325
5.00	0.285
6.25	0.270
7.25	0.260
7.75	0.270
8.00	0.270
8.75	0.250
10.50	0.250
12.00	0.250
13.00	0.250
14.00	0.240
15.00	0.230
CURVE 4 T = 523	
2.00	0.765
2.50	0.720
3.50	0.675
4.35	0.650
5.25	0.630
6.00	0.630
7.00	0.650
8.00	0.680
8.80	0.650
10.00	0.570
11.25	0.480
12.10	0.425
12.75	0.395
13.50	0.370
14.75	0.340
15.00	0.300
CURVE 5 T = 773	
1.00	0.560
1.50	0.520
1.75	0.470
1.90	0.420
2.25	0.392
2.75	0.390
3.65	0.400
4.60	0.425

λ	∈
CURVE 5 (cont.)	
5.50	0.400
6.75	0.355
8.00	0.345
8.75	0.325
10.00	0.310
11.00	0.305
12.10	0.325
13.50	0.350
14.00	0.360
15.00	0.365
CURVE 6 T = 1023	
1.00	0.660
1.50	0.750
2.35	0.800
3.00	0.820
3.50	0.800
4.00	0.770
4.50	0.785
5.00	0.830
6.00	0.850
7.00	0.865
7.75	0.880
8.50	0.870
9.60	0.825
10.50	0.800
11.00	0.790
11.75	0.786
12.50	0.750
13.75	0.675
15.00	0.600
CURVE 7 T = 523	
2.00	0.430
2.25	0.370
2.50	0.340
3.00	0.300
4.00	0.270
5.00	0.250
6.00	0.240
7.00	0.242
8.00	0.240
9.40	0.225

λ	∈
CURVE 7 (cont.)	
11.25	0.225
11.75	0.220
13.00	0.220
14.25	0.220
15.00	0.190
CURVE 8 T = 773	
1.00	0.570
1.50	0.500
1.85	0.400
2.00	0.360
2.25	0.315
2.50	0.285
3.00	0.260
3.75	0.250
5.00	0.250*
6.00	0.232
7.00	0.240
8.00	0.250
8.60	0.240
9.50	0.240
11.00	0.240
12.50	0.240
14.00	0.240*
15.00	0.230*
CURVE 9 T = 1023	
1.00	0.595
1.65	0.500
2.25	0.400
2.75	0.330
3.00	0.305
3.75	0.285
4.50	0.275
5.20	0.265
5.75	0.250
7.00	0.260
8.00	0.260
9.00	0.245
9.75	0.240
11.00	0.240*
12.50	0.240*
14.00	0.250
15.00	0.260

λ	∈
CURVE 10 T = 523	
2.00	0.500
2.50	0.515
3.00	0.540
3.25	0.540
4.00	0.520
4.50	0.530
5.00	0.550
6.00	0.520
7.00	0.535
8.00	0.540
8.90	0.590
10.00	0.570*
11.00	0.550
12.40	0.500
13.50	0.440
14.20	0.400
15.00	0.335
CURVE 11* T = 773	
1.00	0.590
1.10	0.550
1.50	0.500
2.00	0.450
2.50	0.400
3.00	0.370
3.50	0.360
4.15	0.360
5.00	0.340
5.75	0.330
6.50	0.330
7.75	0.331
8.25	0.325
9.50	0.305
11.00	0.290
12.15	0.275
13.50	0.265
14.25	0.255
14.75	0.235
15.00	0.215
CURVE 12* T = 1023	
1.00	0.510
2.00	0.590

λ	∈
CURVE 12* (cont.)	
2.75	0.635
3.10	0.640
3.50	0.625
4.00	0.580
4.25	0.575
4.60	0.575
5.50	0.590
6.00	0.620
6.85	0.700
7.30	0.750
7.80	0.780
8.50	0.810
9.25	0.820
10.00	0.860
11.30	0.868
11.75	0.865
13.00	0.820
14.00	0.760
14.60	0.700
14.85	0.650
15.00	0.600
CURVE 13* T = 523	
2.00	0.860
2.50	0.780
3.00	0.720
3.50	0.700
4.00	0.710
4.65	0.730
5.25	0.710
6.00	0.670
6.25	0.670
6.90	0.675
8.25	0.650
9.00	0.640
9.80	0.650
10.50	0.640
11.30	0.600
11.90	0.565
12.50	0.545
12.75	0.545
13.00	0.550
13.50	0.590
14.00	0.620
14.50	0.610
15.00	0.580

λ	∈
CURVE 14* T = 773	
1.00	0.630
1.25	0.640
1.50	0.640
1.75	0.600
2.00	0.545
2.25	0.540
2.50	0.540
2.85	0.550
3.50	0.600
4.00	0.640
4.25	0.650
5.00	0.650
5.75	0.630
6.10	0.620
6.50	0.625
7.00	0.660
8.00	0.670
9.00	0.660
10.00	0.650
10.80	0.680
11.05	0.680
11.75	0.650
12.50	0.630
12.85	0.635
14.00	0.720
14.70	0.750
15.00	0.750
CURVE 15 T = 1023	
1.00	0.700
1.50	0.720
2.00	0.790
2.25	0.810
2.75	0.820
3.00	0.830
3.50	0.885
4.00	0.950
4.50	0.965
5.00	0.965
5.50	0.950
6.00	0.915
6.25	0.910
7.00	0.920
8.00	0.950
9.00	0.940

λ	ρ
CURVE 15 (cont.)	
10.00	0.920
10.25	0.925
10.85	0.965
12.00	0.920
13.00	0.920
13.75	0.960
14.50	1.000
15.00	1.000
CURVE 16* T = 523	
2.00	0.380
2.50	0.325
3.25	0.275
4.00	0.250
4.50	0.240
5.50	0.240
5.75	0.230
7.00	0.230
8.00	0.230
9.00	0.220
10.00	0.205
11.00	0.200
12.00	0.195
13.00	0.190
14.25	0.180
15.00	0.150
CURVE 17* T = 773	
1.00	0.520
1.50	0.495
1.75	0.450
2.00	0.375
2.25	0.340
2.75	0.325
3.25	0.320
4.25	0.320
5.00	0.310
6.50	0.310
8.00	0.310
9.50	0.310
11.00	0.310
11.80	0.290
13.00	0.290
14.00	0.290
15.00	0.280

* Not shown on plot

DATA TABLE NO. 399 (continued)

λ	ρ
CURVE 18*	
T = 1023	
1.00	0.520
1.40	0.540
1.75	0.520
1.90	0.475
2.00	0.450
2.50	0.405
3.00	0.390
3.75	0.405
4.40	0.420
5.30	0.400
6.40	0.375
7.50	0.385
8.20	0.375
9.25	0.340
10.00	0.330
11.10	0.330
12.00	0.315
12.75	0.310
13.50	0.310
15.00	0.320

* Not shown on plot

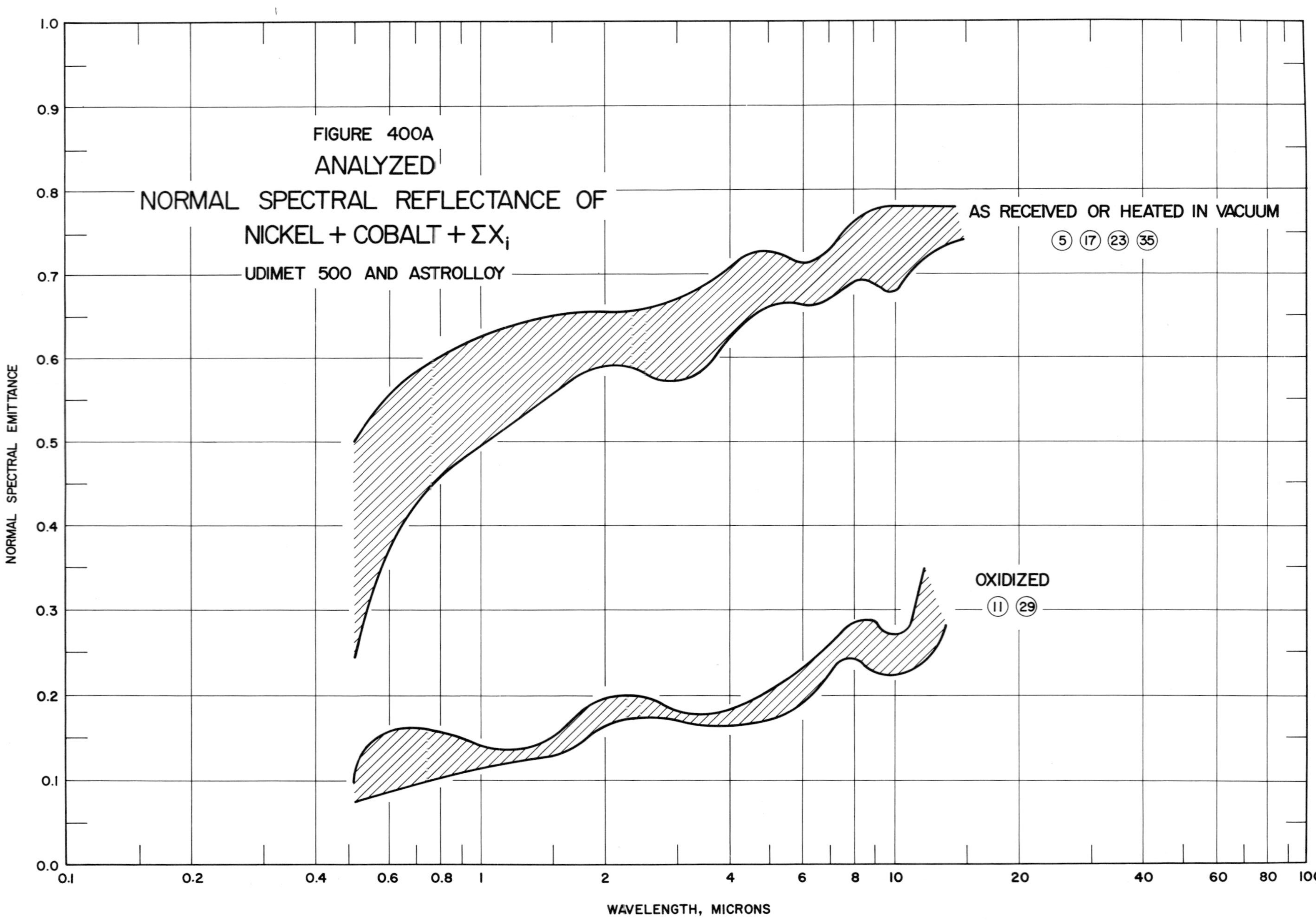
FIGURE 400A
ANALYZED
NORMAL SPECTRAL REFLECTANCE OF
NICKEL + COBALT + ΣX_i
UDIMET 500 AND ASTROLLOY
AS RECEIVED OR HEATED IN VACUUM
5 17 23 35
OXIDIZED
11 29
NORMAL SPECTRAL EMITTANCE
1.0
0.9
0.8
0.7
0.6
0.5
0.4
0.3
0.2
0.1
0.0
0.1
0.2
0.4
0.6
0.8
1
2
4
6
8
10
20
40
60
80
100
WAVELENGTH, MICRONS

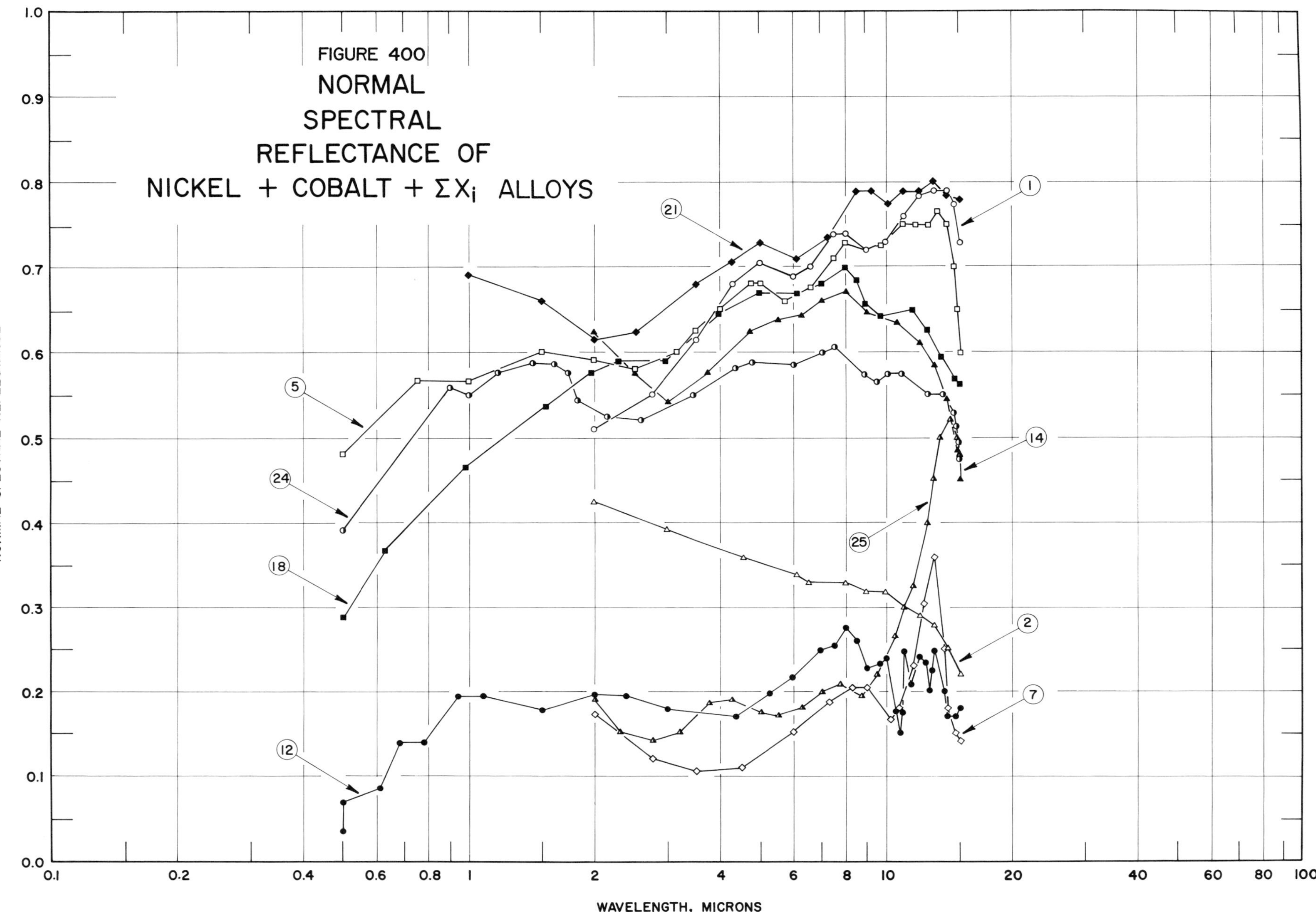
FIGURE 400
NORMAL
SPECTRAL
REFLECTANCE OF
NICKEL + COBALT + ΣX_i ALLOYS
NORMAL SPECTRAL REFLECTANCE
WAVELENGTH, MICRONS
1.0
0.9
0.8
0.7
0.6
0.5
0.4
0.3
0.2
0.1
0.0
0.1
0.2
0.4
0.6
0.8
1
2
4
6
8
10
20
40
60
80
100
1
21
5
24
18
14
25
2
7
12

SPECIFICATION TABLE NO. 400 NORMAL SPECTRAL REFLECTANCE OF [NICKEL + COBALT + ΣX_i] ALLOYS

Curve No.	Ref. No.	Year	Temperature K	Wavelength Range, μ	Geometry θ	θ'	ω'	Reported Error, %	Composition (weight percent), Specifications and Remarks
1	86	1961	~ 322	2.00-15.00	~0°		2π	< 2	Udimet 500; nominal composition: 19 Co, 19 Cr, 4 Mo, 4 max Fe, 3 Ti, 2.9 Al, 0.1 C, trace B, Ni balance; as received; data extracted from smooth curve; hohlraum at 523 K; converted from $R(2\pi, 0^\circ)$.
2	86	1961	~ 322	2.00-15.00	~0°		2π	< 2	Above specimen and conditions; diffuse component only.
3	86	1961	~ 322	1.00-15.00	~0°		2π	< 2	Different sample, same as curve 1 specimen and conditions except hohlraum at 773 K.
4	86	1961	~ 322	1.00-15.00	~0°		2π	< 2	Above specimen and conditions; diffuse component only.
5	86	1961	~ 322	0.50-15.00	~0°		2π	< 2	Different sample, same as curve 1 specimen and conditions except hohlraum at 1273 K.
6	86	1961	~ 322	0.50-14.99	~0°		2π	< 2	Above specimen and conditions; diffuse component only.
7	86	1961	~ 322	2.00-15.00	~0°		2π	< 2	Different sample, same as curve 1 specimen and conditions except heated in air at 1255 K for 30 min.
8	86	1961	~ 322	2.00-15.00	~0°		2π	< 2	Above specimen and conditions; diffuse component only.
9	86	1961	~ 322	1.00-15.00	~0°		2π	< 2	Different sample, same as curve 7 specimen and conditions except hohlraum at 773 K.
10	86	1961	~ 322	1.00-15.00	~0°		2π	< 2	Above specimen and conditions; diffuse component only.
11	86	1961	~ 322	0.50-15.00	~0°		2π	< 2	Different sample, same as curve 7 specimen and conditions except hohlraum at 1273 K.
12	86	1961	~ 322	0.50-15.00	~0°		2π	< 2	Above specimen and conditions; diffuse component only.
13	86	1961	~ 322	2.00-15.00	~0°		2π	< 2	Different sample, same as curve 1 specimen and conditions except heated in vacuum (7.6×10^{-5} mm Hg) at 1255 K for 30 min.
14	86	1961	~ 322	2.00-15.00	~0°		2π	< 2	Above specimen and conditions; diffuse component only.
15	86	1961	~ 322	1.00-15.00	~0°		2π	< 2	Different sample, same as curve 13 specimen and conditions except hohlraum at 773 K.
16	86	1961	~ 322	1.00-15.00	~0°		2π	< 2	Above specimen and conditions; diffuse component only.
17	86	1961	~ 322	0.50-15.00	~0°		2π	< 2	Different sample, same as curve 13 specimen and conditions except hohlraum at 1273 K.
18	86	1961	~ 322	0.50-15.00	~0°		2π	< 2	Above specimen and conditions; diffuse component only.
19	86	1961	~ 322	2.00-15.00	~0°		2π	< 2	Astrolloy; nominal composition: 15 Co, 15 Cr, 5 Mo, 4.3 Al, 3.5 Ti, 0.05 C, 0.03 B, Ni balance; as received; data extracted from smooth curve; hohlraum at 523 K; converted from $R(2\pi, 0^\circ)$.
20	86	1961	~ 322	2.00-15.04	~0°		2π	< 2	Above specimen and conditions; diffuse component only.
21	86	1961	~ 322	1.00-15.00	~0°		2π	< 2	Different sample, same as curve 19 specimen and conditions except hohlraum at 773 K.
22	86	1961	~ 322	1.00-15.00	~0°		2π	< 2	Above specimen and conditions; diffuse component only.
23	86	1961	~ 322	0.50-15.00	~0°		2π	< 2	Different sample, same as curve 19 specimen and conditions except hohlraum at 1273 K.
24	86	1961	~ 322	0.50-14.99	~0°		2π	< 2	Above specimen and conditions; diffuse component only.
25	86	1961	~ 322	2.00-15.00	~0°		2π	< 2	Different sample, same as curve 19 specimen and conditions except heated in air at 1255 K for 30 min.

SPECIFICATION TABLE NO. 400 (continued)

Curve No.	Ref. No.	Year	Temperature K	Wavelength Range, μ	Geometry θ	θ'	ω'	Reported Error, %	Composition (weight percent), Specifications and Remarks
26	86	1961	~ 322	2.00-14.99	~0°		2π	< 2	Above specimen and conditions; diffuse component only.
27	86	1961	~ 322	2.00-15.00	~0°		2π	< 2	Different sample, same as curve 25 specimen and conditions except hohlraum at 773 K.
28	86	1961	~ 322	1.00-15.00	~0°		2π	< 2	Above specimen and conditions; diffuse component only.
29	86	1961	~ 322	0.50-15.00	~0°		2π	< 2	Different sample, same as curve 25 specimen and conditions except hohlraum at 1273 K.
30	86	1961	~ 322	0.55-15.00	~0°		2π	< 2	Above specimen and conditions; diffuse component only.
31	86	1961	~ 322	2.00-15.00	~0°		2π	< 2	Different sample, same as curve 19 specimen and conditions except heated in vacuum (7.6 x 10^{-5}mm Hg) at 1255 K for 30 min.
32	86	1961	~ 322	2.00-15.00	~0°		2π	< 2	Above specimen and conditions; diffuse component only.
33	86	1961	~ 322	1.00-15.00	~0°		2π	< 2	Different sample, same as curve 31 specimen and conditions except hohlraum at 773 K.
34	86	1961	~ 322	1.00-15.00	~0°		2π	< 2	Above specimen and conditions; diffuse component only.
35	86	1961	~ 322	0.50-15.00	~0°		2π	< 2	Different sample, same as curve 31 specimen and conditions except hohlraum at 1273 K.
36	86	1961	~ 322	0.50-14.99	~0°		2π	< 2	Above specimen and conditions; diffuse component only.

DATA TABLE NO. 400 NORMAL SPECTRAL REFLECTANCE OF [NICKEL + COBALT + ΣX_i] ALLOYS

[Wavelength, λ, μ; Reflectance, ρ; Temperature, T, K]

λ	ρ
CURVE 1	**T = ~ 322**
2.00	0.510
2.75	0.550
3.50	0.615
4.25	0.680
5.00	0.705
6.00	0.690
6.60	0.700
7.50	0.740
8.00	0.740
9.00	0.720
10.00	0.730
11.00	0.760
12.00	0.785
13.00	0.790
14.00	0.790
14.50	0.775
15.00	0.730
CURVE 2	**T = ~ 322**
2.00	0.425
2.97	0.392
4.52	0.359
6.04	0.339
6.46	0.329
8.00	0.329
8.99	0.319
9.99	0.319
11.01	0.301
12.01	0.291
13.01	0.279
14.03	0.251
15.00	0.221
CURVE 3*	**T = ~ 322**
1.00	0.620
1.50	0.595
2.00	0.600
2.60	0.595
3.25	0.625
CURVE 3* (cont.)	
4.00	0.675
5.00	0.700
6.00	0.680
6.75	0.690
7.40	0.715
8.00	0.740
8.50	0.740
9.00	0.735
10.00	0.745
11.00	0.765
12.00	0.775
12.75	0.770
13.50	0.770
14.00	0.770
14.60	0.730
15.00	0.660
CURVE 4*	**T = ~ 322**
1.00	0.650
1.99	0.599
2.83	0.581
3.15	0.586
3.76	0.650
4.26	0.677
4.97	0.681
5.99	0.668
6.66	0.678
7.80	0.699
8.25	0.693
8.67	0.668
9.01	0.650
9.99	0.627
11.02	0.602
12.01	0.572
13.00	0.524
14.01	0.476
14.76	0.432
15.00	0.390
CURVE 5	**T = ~ 322**
0.50	0.480
0.75	0.565
1.00	0.565
1.50	0.600
2.00	0.590
2.50	0.580
3.15	0.600
3.50	0.625
4.00	0.650
4.75	0.680
5.00	0.680
5.75	0.660
6.60	0.675
7.50	0.710
8.00	0.730
9.00	0.720*
9.75	0.725
11.00	0.750
11.75	0.750
12.60	0.750
13.20	0.765
14.00	0.750
14.60	0.700
14.85	0.650
15.00	0.600
CURVE 6*	**T = ~ 322**
0.50	0.475
0.62	0.504
0.85	0.552
0.91	0.581
1.01	0.571
1.51	0.611
1.70	0.593
1.73	0.550
1.72	0.454
2.00	0.402
2.50	0.453
2.77	0.485
3.24	0.522
4.49	0.596
CURVE 6* (cont.)	
4.97	0.610
5.47	0.597
5.84	0.579
6.99	0.602
8.01	0.611
8.51	0.595
9.21	0.552
10.00	0.533
11.00	0.521
12.00	0.494
12.98	0.461
13.98	0.436
14.99	0.394
CURVE 7	**T = ~ 322**
2.00	0.170
2.75	0.120
3.50	0.105
4.50	0.110
6.00	0.150
7.25	0.185
8.25	0.205
9.00	0.205
10.25	0.165
10.75	0.180
11.50	0.230
12.25	0.305
13.00	0.360
13.70	0.250
14.00	0.180
14.60	0.150
15.00	0.140
CURVE 8*	**T = ~ 322**
2.00	0.195
2.71	0.152
3.49	0.136
4.25	0.144
5.25	0.167
6.16	0.200
CURVE 8* (cont.)	
7.98	0.231
8.53	0.212
9.11	0.190
10.00	0.202
11.00	0.216
11.60	0.216
12.17	0.200
12.86	0.212
13.26	0.201
13.98	0.147
14.52	0.115
15.00	0.100
CURVE 9*	**T = ~ 322**
1.00	0.160
2.00	0.155
3.00	0.150
3.25	0.150
5.00	0.170
6.00	0.205
6.50	0.220
7.00	0.230
7.50	0.250
8.25	0.265
9.50	0.250
10.50	0.235
10.75	0.235
11.25	0.270
12.00	0.340
13.00	0.410
13.25	0.400
13.65	0.300
14.00	0.225
14.25	0.210
15.00	0.200
CURVE 10*	**T = ~ 322**
1.00	0.163
1.51	0.175
2.00	0.169
CURVE 10* (cont.)	
2.71	0.165
3.07	0.171
3.97	0.160
4.73	0.166
5.47	0.186
6.22	0.219
6.99	0.230
8.01	0.250
8.64	0.237
8.99	0.220
9.99	0.229
11.16	0.243
12.01	0.232
12.99	0.228
13.73	0.184
14.34	0.153
15.00	0.153
CURVE 11*	**T = ~ 322**
0.50	0.100
0.55	0.150
0.75	0.160
1.00	0.140
1.50	0.130
1.75	0.175
2.00	0.200
2.75	0.185
3.50	0.175
4.25	0.175
5.50	0.210
6.25	0.235
7.25	0.250
7.75	0.260
8.00	0.285
8.50	0.290
9.50	0.270
10.10	0.250
10.50	0.175
10.70	0.160
10.80	0.175
11.00	0.260
11.35	0.240
CURVE 11* (cont.)	
11.65	0.275
12.00	0.350
12.75	0.310
12.85	0.375
13.00	0.440
13.15	0.375
13.85	0.275
14.50	0.235
15.00	0.220
CURVE 12	**T = ~322**
0.50	0.035
0.50	0.069
0.61	0.085
0.68	0.138
0.78	0.139
0.94	0.193
1.08	0.194
1.50	0.177
2.00	0.198
2.36	0.193
2.99	0.179
4.37	0.169
5.22	0.199
5.98	0.217
6.99	0.249
7.50	0.253
8.08	0.275
8.49	0.260
9.00	0.228
9.64	0.232
10.00	0.239
10.43	0.174
10.73	0.150
10.90	0.175
11.00	0.248
11.41	0.210
12.00	0.241
12.25	0.235
12.63	0.201
12.87	0.226
13.01	0.249
CURVE 12 (cont.)	
13.70	0.201
14.00	0.170
14.51	0.170
15.00	0.180
CURVE 13*	**T = ~ 322**
2.00	0.630
2.50	0.550
3.00	0.520
3.50	0.550
3.75	0.575
4.50	0.610
5.50	0.625
6.25	0.630
7.50	0.665
8.00	0.680
9.00	0.650
9.25	0.650
10.50	0.680
12.00	0.710
13.25	0.710
14.40	0.700
14.85	0.675
15.00	0.650
CURVE 14	**T = ~ 322**
2.00	0.625
2.48	0.576
3.00	0.541
3.73	0.576
4.73	0.625
5.48	0.639
6.25	0.642
7.01	0.661
8.01	0.671
9.02	0.647
10.52	0.635
12.03	0.611
13.00	0.583
13.98	0.543
14.78	0.484
15.00	0.450

* Not shown on plot

DATA TABLE NO. 400 (continued)

λ	ρ
CURVE 15*	
T = ~ 322	
1.00	0.470
1.50	0.540
2.00	0.530
2.75	0.550
3.80	0.600
4.50	0.630
5.75	0.650
7.00	0.660
7.90	0.685
9.00	0.660
9.60	0.650
10.40	0.675
11.00	0.700
12.00	0.720
13.00	0.730
13.75	0.755
14.40	0.765
15.00	0.760
CURVE 16*	
T = ~ 322	
1.00	0.492
2.10	0.552
2.80	0.566
3.33	0.587
3.97	0.634
4.96	0.664
6.15	0.672
6.99	0.701
7.98	0.710
8.99	0.691
9.60	0.673
11.00	0.671
11.99	0.660
13.00	0.631
14.00	0.611
15.00	0.592
CURVE 17*	
T = ~ 322	
0.50	0.250
0.60	0.375
CURVE 17* (cont.)	
0.75	0.450
1.25	0.525
2.00	0.595
2.50	0.580
3.00	0.560
3.70	0.600
4.00	0.630
5.00	0.660
5.50	0.665
6.25	0.660
7.25	0.675
8.25	0.695
9.25	0.685
10.00	0.675
11.00	0.710
12.00	0.720
13.25	0.735
14.00	0.740
14.55	0.725
14.90	0.690
15.00	0.660
CURVE 18	
T = ~ 322	
0.50	0.288
0.63	0.367
0.98	0.464
1.54	0.534
1.98	0.574
2.29	0.589
2.97	0.589
3.98	0.643
4.99	0.669
6.10	0.669
7.00	0.681
8.00	0.699
8.50	0.684
8.99	0.657
9.74	0.642
11.48	0.650
12.49	0.626
13.50	0.593
14.50	0.568
15.00	0.561
CURVE 19*	
T = ~ 322	
2.00	0.540
3.00	0.630
3.90	0.700
4.75	0.735
5.00	0.740
5.70	0.730
7.00	0.765
8.25	0.790
9.00	0.775
9.85	0.790
10.50	0.785
11.50	0.800
13.00	0.810
14.00	0.810
14.50	0.800
14.90	0.775
15.00	0.740
CURVE 20*	
T = ~ 322	
2.00	0.449
3.00	0.513
3.98	0.560
4.96	0.588
5.46	0.581
5.92	0.568
6.49	0.581
7.55	0.600
8.20	0.600
9.19	0.577
9.51	0.573
10.02	0.573
10.61	0.565
11.01	0.570
11.50	0.570
12.02	0.556
12.50	0.556
12.78	0.562
13.11	0.553
14.04	0.516
15.04	0.451
CURVE 21	
T = ~ 322	
1.00	0.690
1.50	0.660
2.00	0.615
2.50	0.625
3.50	0.680
4.25	0.705
5.00	0.730
6.10	0.710
7.25	0.735
8.50	0.790
9.25	0.790
10.10	0.775
11.00	0.790
12.00	0.790
13.00	0.800
14.00	0.785
15.00	0.780
CURVE 22*	
T = ~ 322	
1.00	0.579
1.52	0.589
1.86	0.590
2.41	0.584
2.72	0.592
3.71	0.636
4.35	0.669
5.29	0.669
6.03	0.662
6.63	0.651
8.01	0.681
8.80	0.681
9.98	0.655
11.89	0.655
13.00	0.635
13.63	0.627
14.01	0.615
15.00	0.525
CURVE 23*	
T = ~ 322	
0.50	0.500
0.55	0.550
CURVE 23* (cont.)	
0.70	0.580
0.75	0.640
0.85	0.600
0.90	0.615
1.10	0.610
1.50	0.640
2.25	0.650
3.20	0.675
4.20	0.715
5.00	0.730
6.30	0.715
7.50	0.750
8.50	0.775
9.00	0.780
10.00	0.780
11.15	0.780
12.00	0.800
13.50	0.785
14.50	0.750
15.00	0.695
CURVE 24	
T = ~ 322	
0.50	0.390
0.90	0.558
1.02	0.549
1.17	0.574
1.42	0.588
1.61	0.587
1.73	0.575
1.82	0.542
2.15	0.523
2.59	0.520
3.46	0.550
4.35	0.581
4.79	0.589
6.07	0.585
7.01	0.600
7.52	0.607
8.97	0.572
9.49	0.563
10.07	0.573
10.87	0.573
12.61	0.550
CURVE 24 (cont.)	
13.50	0.550
14.47	0.529
14.77	0.512
14.93	0.492
14.99	0.473
CURVE 25	
T = ~ 322	
2.00	0.190
2.30	0.150
2.75	0.140
3.20	0.150
3.75	0.185
4.25	0.190
5.00	0.175
5.50	0.170
6.25	0.180
7.00	0.200
7.75	0.210
8.75	0.195
9.50	0.220
10.50	0.265
11.50	0.325
12.50	0.400
13.00	0.450
13.35	0.500
14.25	0.520
14.80	0.500
15.00	0.480
CURVE 26*	
T = ~ 322	
2.00	0.200
2.31	0.165
2.61	0.151
3.23	0.176
3.98	0.200
4.99	0.187
5.47	0.177
6.26	0.184
6.78	0.191
8.60	0.191
10.50	0.267
CURVE 26* (cont.)	
11.66	0.336
11.99	0.376
12.49	0.406
13.28	0.442
14.02	0.453
14.50	0.453
14.99	0.431
CURVE 27*	
T = ~322	
2.00	0.150
2.50	0.155
3.25	0.175
4.30	0.200
5.30	0.200
6.00	0.200
7.50	0.225
8.00	0.235
8.80	0.210
9.25	0.210
10.00	0.230
11.00	0.290
12.00	0.370
13.00	0.455
14.00	0.520
14.50	0.545
15.00	0.550
CURVE 28*	
T = ~322	
1.00	0.128
1.99	0.163
2.71	0.172
3.24	0.192
3.55	0.197
4.19	0.192
5.06	0.204
5.98	0.204
7.48	0.243
8.10	0.243
8.77	0.232
9.26	0.228
10.26	0.255
CURVE 28* (cont.)	
11.17	0.294
13.01	0.420
13.99	0.462
15.00	0.485
CURVE 29*	
T = ~ 322	
0.50	0.075
0.75	0.100
1.00	0.140
1.40	0.140
2.00	0.165
2.50	0.175
3.15	0.175
4.75	0.200
6.15	0.200
6.50	0.210
7.20	0.240
7.85	0.250
8.75	0.230
9.25	0.230
10.10	0.250
11.50	0.335
12.50	0.405
13.50	0.470
14.25	0.490
14.75	0.485
15.00	0.470
CURVE 30*	
T = ~ 322	
0.55	0.066
0.65	0.085
0.77	0.085
1.06	0.147
1.54	0.151
2.04	0.181
2.39	0.183
3.24	0.183
5.15	0.215
5.79	0.205
6.50	0.205
7.37	0.237

* Not shown on plot

DATA TABLE NO. 400 (continued)

λ	ρ
CURVE 30* (cont.)	
7.83	0.237
8.77	0.213
9.37	0.216
9.98	0.232
11.29	0.310
11.78	0.331
12.17	0.351
13.30	0.414
14.04	0.434
14.63	0.432
15.00	0.400
CURVE 31*	
T = ~ 322	
2.00	0.740
2.50	0.680
3.10	0.650
4.00	0.675
5.15	0.725
6.50	0.750
7.50	0.770
8.00	0.770
9.25	0.760
10.00	0.760
11.50	0.770
13.00	0.785
13.75	0.775
14.50	0.750
14.80	0.725
15.00	0.680
CURVE 32*	
T = ~ 322	
2.00	0.649
2.99	0.559
3.27	0.553
3.60	0.560
3.98	0.578
4.79	0.601
6.50	0.601
7.01	0.610
8.01	0.611
9.26	0.582
10.15	0.576

λ	ρ
CURVE 32* (cont.)	
11.51	0.551
12.10	0.552
13.42	0.535
14.30	0.518
14.81	0.493
15.00	0.450
CURVE 33*	
T = ~ 322	
1.00	0.570
2.00	0.565
2.50	0.575
3.50	0.630
4.50	0.700
5.00	0.710
6.00	0.710
7.00	0.730
8.15	0.760
8.75	0.750
10.00	0.750
11.50	0.775
13.00	0.790
13.50	0.815
14.00	0.820
15.00	0.820
CURVE 34*	
T = ~ 322	
1.00	0.548
2.67	0.549
4.19	0.607
4.99	0.618
5.64	0.619
7.37	0.642
7.98	0.637
9.77	0.598
12.04	0.598
13.04	0.590
13.76	0.600
14.65	0.600
15.00	0.587

λ	ρ
CURVE 35*	
T = ~ 322	
0.50	0.390
0.60	0.475
0.75	0.540
1.00	0.515
1.20	0.565
2.00	0.610
2.50	0.615
2.75	0.620
3.50	0.650
4.50	0.690
5.50	0.705
6.50	0.715
8.00	0.745
8.85	0.730
9.80	0.730
11.50	0.760
12.50	0.760
13.50	0.770
14.00	0.770
14.60	0.760
15.00	0.740
CURVE 36*	
T = ~ 322	
0.50	0.183
0.50	0.275
0.97	0.453
1.30	0.500
1.82	0.552
2.31	0.580
2.83	0.591
3.76	0.628
4.16	0.637
4.97	0.652
6.32	0.652
7.12	0.665
7.99	0.672
8.31	0.665
8.69	0.641
9.75	0.623
11.22	0.623
11.71	0.605
13.69	0.605
14.38	0.585
14.99	0.573

* Not shown on plot

SPECIFICATION TABLE NO. 401 HEMISPHERICAL TOTAL EMITTANCE OF [NICKEL + COPPER + ΣX_i] ALLOYS

Curve No.	Ref. No.	Year	Temperature Range, K	Reported Error, %	Composition (weight percent), Specifications and Remarks
1	24	1948	304		Monel; nominal composition: 63 Ni, 30 Cu, 1.5 Si; cleaned; measured in air.
2	9	1960	77		Monel; nominal composition: 63 Ni, 30 Cu, 1.5 Si; hand polished; cleaned with acetone; measured in vacuum ($\sim 1 \times 10^{-5}$ mm Hg).

DATA TABLE NO. 401 HEMISPHERICAL TOTAL EMITTANCE OF [NICKEL + COPPER + ΣX_i] ALLOYS

[Temperature, T, K; Emittance, ϵ]

T	ϵ
CURVE 1*	
304	0.12
CURVE 2*	
77	0.11

* Not shown on plot

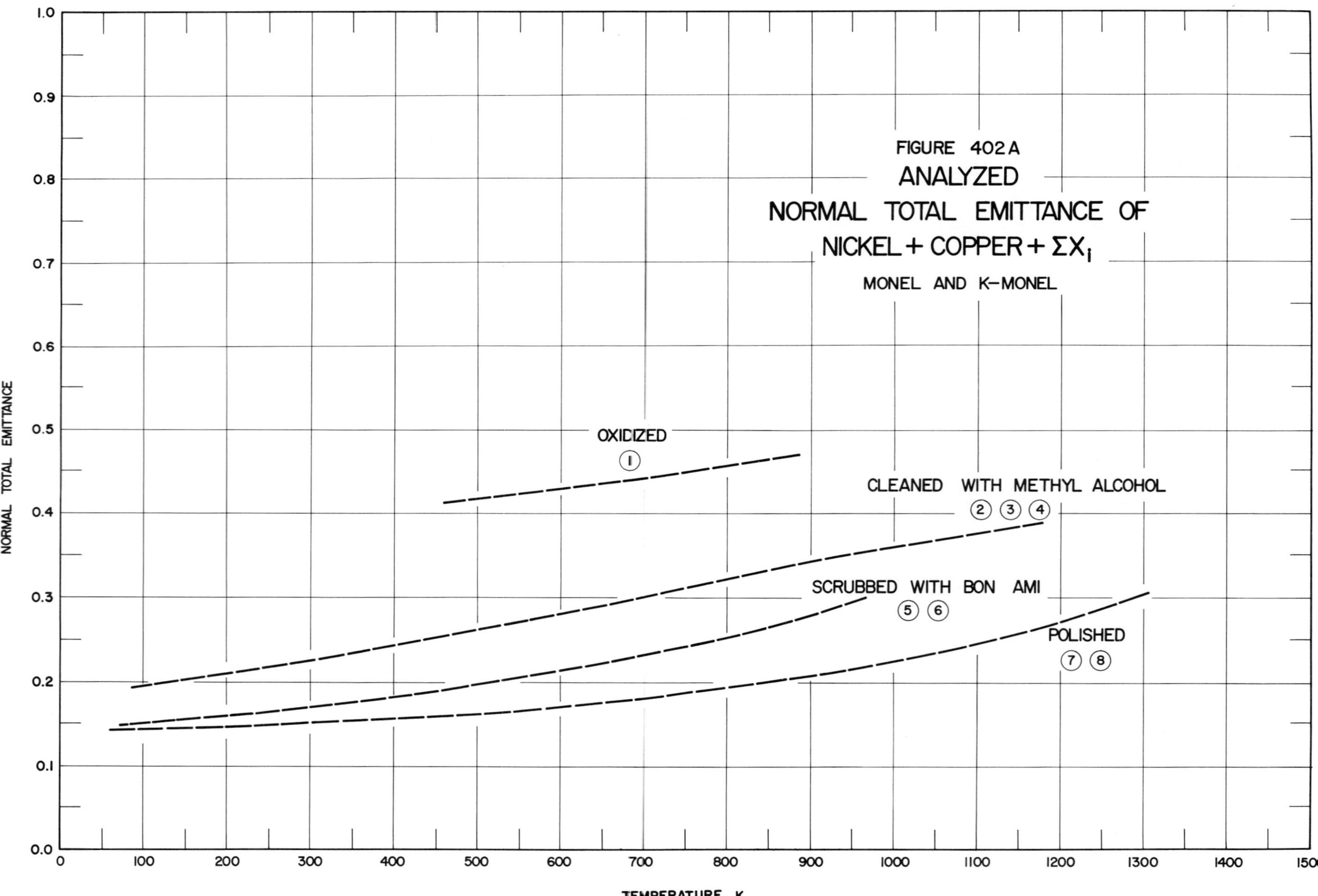
FIGURE 402A
ANALYZED
NORMAL TOTAL EMITTANCE OF
NICKEL + COPPER + ΣXi
MONEL AND K-MONEL
OXIDIZED
1
CLEANED WITH METHYL ALCOHOL
2 3 4
SCRUBBED WITH BON AMI
5 6
POLISHED
7 8
NORMAL TOTAL EMITTANCE
1.0
0.9
0.8
0.7
0.6
0.5
0.4
0.3
0.2
0.1
0.0
0
100
200
300
400
500
600
700
800
900
1000
1100
1200
1300
1400
1500
TEMPERATURE, K

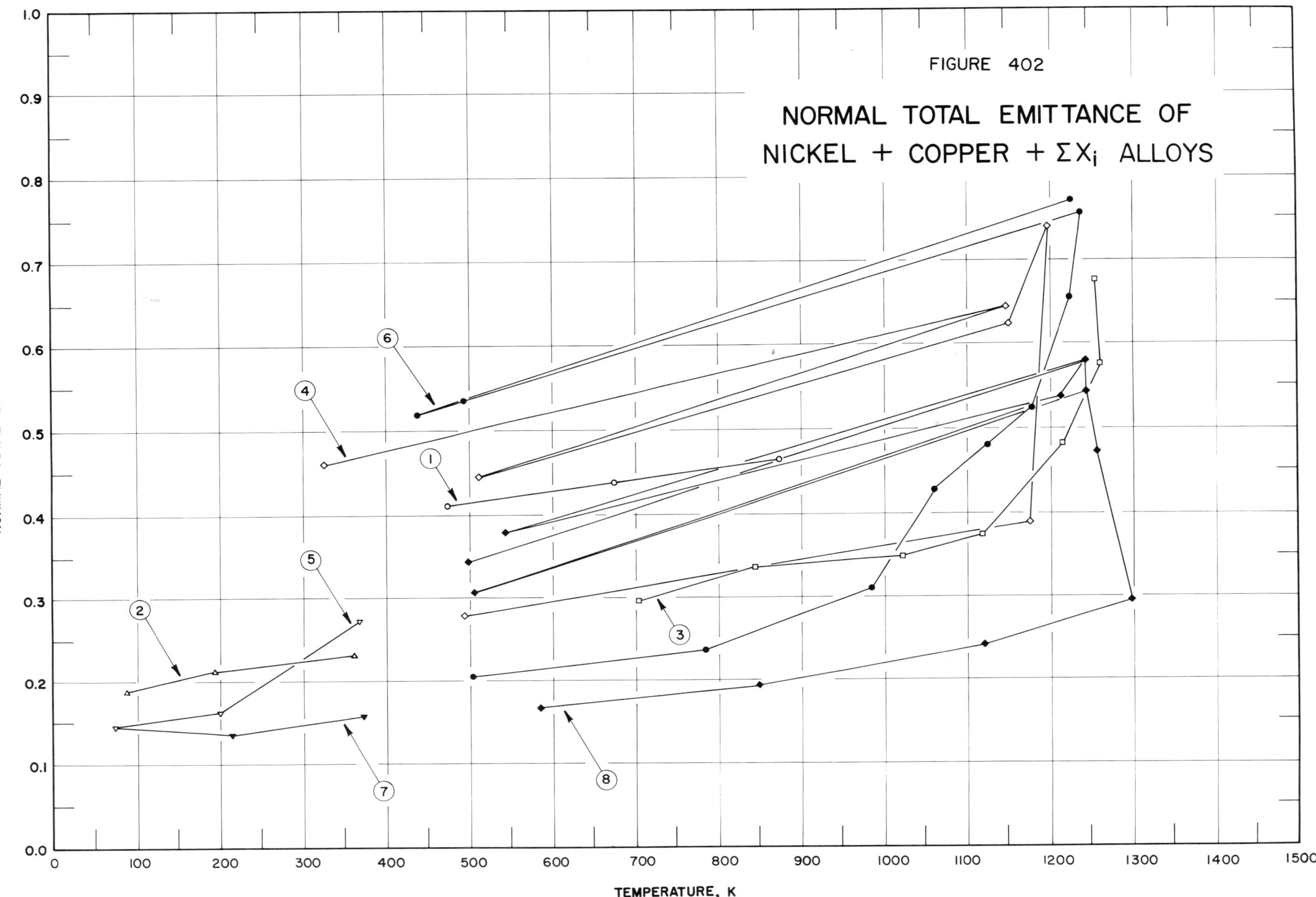
FIGURE 402
NORMAL TOTAL EMITTANCE OF
NICKEL + COPPER + ΣX_i ALLOYS
NORMAL TOTAL EMITTANCE
TEMPERATURE, K
0.0
0.1
0.2
0.3
0.4
0.5
0.6
0.7
0.8
0.9
1.0
0
100
200
300
400
500
600
700
800
900
1000
1100
1200
1300
1400
1500
1
2
3
4
5
6
7
8

SPECIFICATION TABLE NO. 402 NORMAL TOTAL EMITTANCE OF [NICKEL + COPPER + ΣX_i] ALLOYS

Curve No.	Ref. No.	Year	Temperature Range, K	Geometry θ'	Reported Error, %	Composition (weight percent), Specifications and Remarks
1	14	1913	473-873	~0°		Monel; nominal composition: 63 Ni, 30 Cu, 1.5 Si; cleaned, polished, and oxidized.
2	160	1954	88.7-360	~0°		K-Monel 5700; cleaned with methyl alcohol; measured in vacuum (10^{-3}mm Hg); [Author's designation: Sample 2].
3	160	1954	702-1255	~0°		Different sample, same as above specimen and conditions except measured in argon (10^{-3}mm Hg); [Author's designation: Sample 2].
4	160	1954	325-1198	~0°		Above specimen and conditions.
5	160	1954	72.6-366	~0°		Different sample, same as curve 2 specimen and conditions except scrubbed with Bon Ami on a wet cloth, washed and dried, wiped with toluene and alcohol; [Author's designation: Sample 6].
6	160	1954	438-1237	~0°		Different sample, same as above specimen and conditions except measured in argon (10^{-3}mm Hg); [Author's designation: Sample 10].
7	160	1954	74.8-372	~0°		Different sample, same as curve 2 specimen and conditions except polished and then finished with a wool buff and rouge and washed; surface free from scratches; [Author's designation: Sample 7].
8	160	1954	499-1298	~0°		Different sample, same as above specimen and conditions except measured in argon (10^{-3}mm Hg); [Author's designation: Sample 11].

DATA TABLE NO. 402 NORMAL TOTAL EMITTANCE OF [NICKEL + COPPER + ΣX_i] ALLOYS

[Temperature, T, K; Emittance, ϵ]

T	ϵ
CURVE 1	
473	0.411
673	0.439
873	0.463
CURVE 2	
88.7	0.188
194	0.210
360	0.231
CURVE 3	
702	0.296
843	0.338
1021	0.349
1118	0.373
1216	0.482
1260	0.573
1255	0.674
CURVE 4	
492	0.279
1175	0.390
1198	0.738
1150	0.625
512	0.445
1149	0.646
325	0.459
CURVE 5	
72.6	0.143
200	0.161
366	0.272
CURVE 6	
502	0.203
784	0.237
985	0.310
1060	0.429
1125	0.481
1179	0.523

T	ϵ
CURVE 6 (cont.)	
1224	0.654
1237	0.756
493	0.535
1228	0.770
438	0.519
CURVE 7	
74.8	0.143*
214	0.134
372	0.157
CURVE 8	
585	0.166
849	0.193
1119	0.241
1298	0.296
1258	0.472
1244	0.542
505	0.307
1215	0.537
543	0.380
1243	0.530
499	0.343

* Not shown on plot

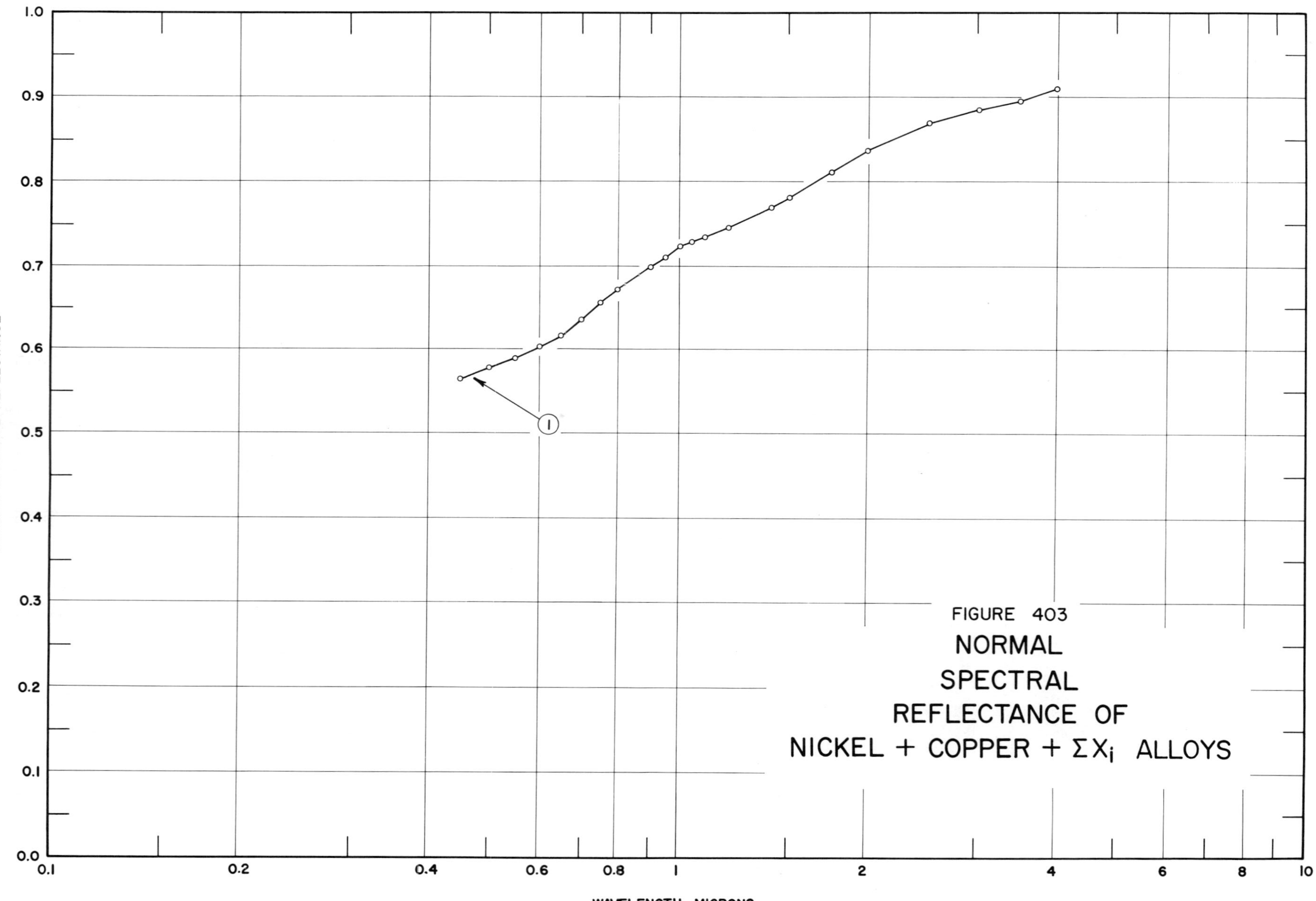

FIGURE 403
NORMAL
SPECTRAL
REFLECTANCE OF
NICKEL + COPPER + ΣX_i ALLOYS

SPECIFICATION TABLE NO. 403 NORMAL SPECTRAL REFLECTANCE OF [NICKEL + COPPER + ΣX_i] ALLOYS

Curve No.	Ref. No.	Year	Temperature K	Wavelength Range, μ	Geometry θ	θ'	ω'	Reported Error, %	Composition (weight percent), Specifications and Remarks
1	151	1920	298	0.45-4.00	<15°	<15°			Monel from the International Nickel Co; 68-70 Ni, 1.5 Fe, Cu balance; optically flat and highly polished surface; silvered mirror reference.

DATA TABLE NO. 403 NORMAL SPECTRAL REFLECTANCE OF [NICKEL + COPPER + ΣX_i] ALLOYS

[Wavelength, λ, μ; Reflectance, ρ; Temperature, T, K]

λ	ρ
CURVE 1	
T = 298	
0.45	0.565
0.50	0.578
0.55	0.590
0.60	0.602
0.65	0.618
0.70	0.637
0.75	0.656
0.80	0.672
0.90	0.700
0.95	0.711
1.00	0.723
1.05	0.730
1.10	0.736
1.20	0.748
1.40	0.770
1.50	0.782
1.75	0.812
2.00	0.838
2.50	0.870
3.00	0.887
3.50	0.895
4.00	0.910

* Not shown on plot

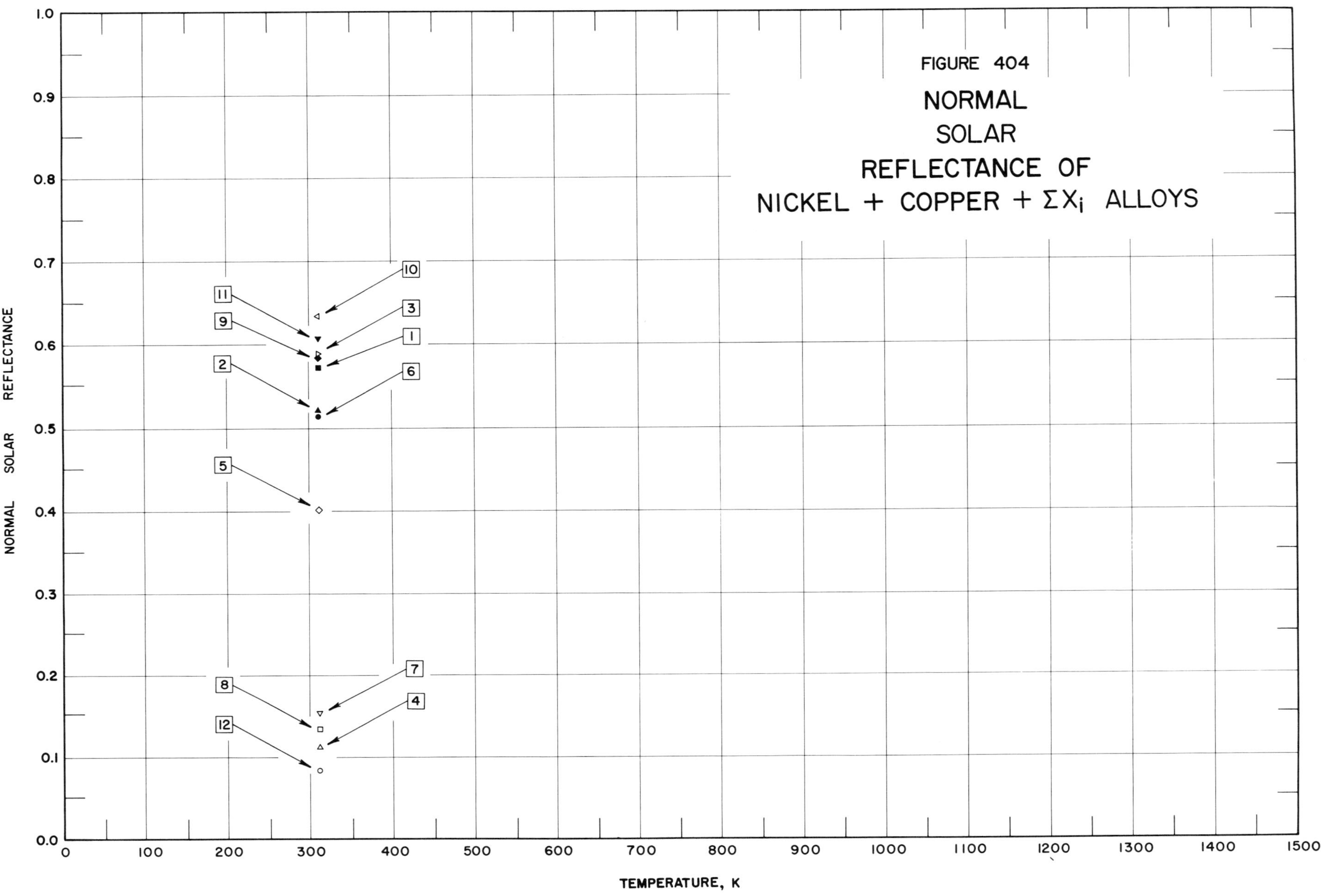
FIGURE 404
NORMAL
SOLAR
REFLECTANCE OF
NICKEL + COPPER + ΣX_i ALLOYS
NORMAL SOLAR REFLECTANCE
TEMPERATURE, K
0
100
200
300
400
500
600
700
800
900
1000
1100
1200
1300
1400
1500
0.0
0.1
0.2
0.3
0.4
0.5
0.6
0.7
0.8
0.9
1.0
1
2
3
4
5
6
7
8
9
10
11
12

SPECIFICATION TABLE NO. 404 NORMAL SOLAR REFLECTANCE OF [NICKEL + COPPER + ΣX_i] ALLOYS

Curve No.	Ref. No.	Year	Temperature Range, K	Geometry θ	θ'	ω'	Reported Error, %	Composition (weight percent), Specifications and Remarks
1	160	1954	311	~0°		2π		K-Monel 5700; heated to 308 K; clean and smooth surface; measured in air; calculated from solar absorptance.
2	160	1954	311	~0°		2π		Above specimen and conditions except heated to 615 K.
3	160	1954	311	~0°		2π		Above specimen and conditions except heated to 928 K.
4	160	1954	311	~0°		2π		Above specimen and conditions except reheated to 1225 K.
5	160	1954	311	~0°		2π		Different sample, same as curve 1 specimen and conditions except heated to 325 K; cleaned with methyl alcohol.
6	160	1954	311	~0°		2π		Above specimen and conditions except heated to 632 K.
7	160	1954	311	~0°		2π		Above specimen and conditions except heated to 935 K.
8	160	1954	311	~0°		2π		Above specimen and conditions except heated to 1225 K.
9	160	1954	311	~0°		2π		Different sample, same as curve 1 specimen and conditions except heated to 319 K; polished; surface free from scratches.
10	160	1954	311	~0°		2π		Above specimen and conditions except reheated to 592 K.
11	160	1954	311	~0°		2π		Above specimen and conditions except reheated to 938 K.
12	160	1954	311	~0°		2π		Above specimen and conditions except reheated to 1255 K.

DATA TABLE NO. 404 NORMAL SOLAR REFLECTANCE OF [NICKEL + COPPER + ΣX_i] ALLOYS

[Temperature, T, K; Reflectance, ρ]

T	ρ
CURVE 1	
311	0.571
CURVE 2	
311	0.521
CURVE 3	
311	0.589
CURVE 4	
311	0.111
CURVE 5	
311	0.401
CURVE 6	
311	0.513
CURVE 7	
311	0.151
CURVE 8	
311	0.132
CURVE 9	
311	0.583
CURVE 10	
311	0.634
CURVE 11	
311	0.608
CURVE 12	
311	0.082

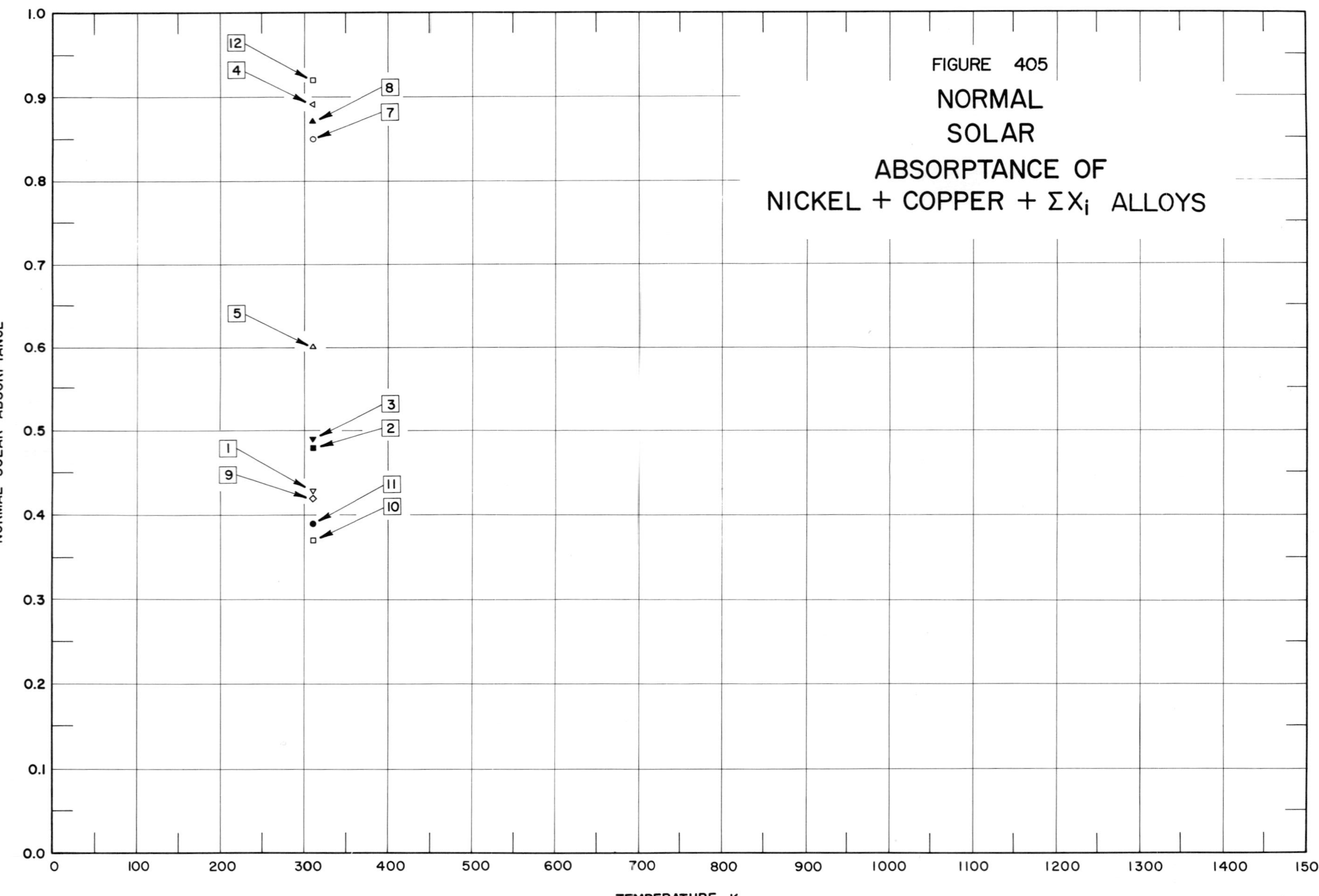

FIGURE 405
NORMAL
SOLAR
ABSORPTANCE OF
NICKEL + COPPER + ΣXi ALLOYS
NORMAL SOLAR ABSORPTANCE
TEMPERATURE, K
1.0
0.9
0.8
0.7
0.6
0.5
0.4
0.3
0.2
0.1
0.0
0
100
200
300
400
500
600
700
800
900
1000
1100
1200
1300
1400
1500
12
4
8
7
5
3
2
1
9
11
10

SPECIFICATION TABLE NO. 405 NORMAL SOLAR ABSORPTANCE OF [NICKEL + COPPER + ΣX_i] ALLOYS

Curve No.	Ref. No.	Year	Temperature Range, K	Geometry θ'	Reported Error, %	Composition (weight percent), Specifications and Remarks
1	160	1954	311	~0°		K-Monel 5700; heated to 309 K; clean and smooth surface; measured in air at sea level.
2	160	1954	311	~0°		Above specimen and conditions except reheated to 616 K.
3	160	1954	311	~0°		Above specimen and conditions except reheated to 933 K.
4	160	1954	311	~0°		Above specimen and conditions except reheated to 1228 K.
5	160	1954	311	~0°		Different sample, same as curve 1 specimen and conditions except heated to 323 K; cleaned with methyl alcohol.
6	160	1954	311	~0°		Above specimen and conditions except reheated to 637 K.
7	160	1954	311	~0°		Above specimen and conditions except reheated to 937 K.
8	160	1954	311	~0°		Above specimen and conditions except reheated to 1228 K.
9	160	1954	311	~0°		Different sample, same as curve 1 specimen and conditions except heated to 316 K; polished; surface free from scratches.
10	160	1954	311	~0°		Above specimen and conditions except reheated to 593 K.
11	160	1954	311	~0°		Above specimen and conditions except reheated to 941 K.
12	160	1954	311	~0°		Above specimen and conditions except reheated to 1255 K.

DATA TABLE NO. 405 NORMAL SOLAR ABSORPTANCE OF [NICKEL + COPPER + ΣX_i] ALLOYS

[Temperature, T, K; Absorptance, α]

T	α
CURVE 1	
311	0.43
CURVE 2	
311	0.48
CURVE 3	
311	0.49
CURVE 4	
311	0.89
CURVE 5	
311	0.60
CURVE 6*	
311	0.49
CURVE 7	
311	0.85
CURVE 8	
311	0.87
CURVE 9	
311	0.42
CURVE 10	
311	0.37
CURVE 11	
311	0.39
CURVE 12	
311	0.92

* Not shown on plot

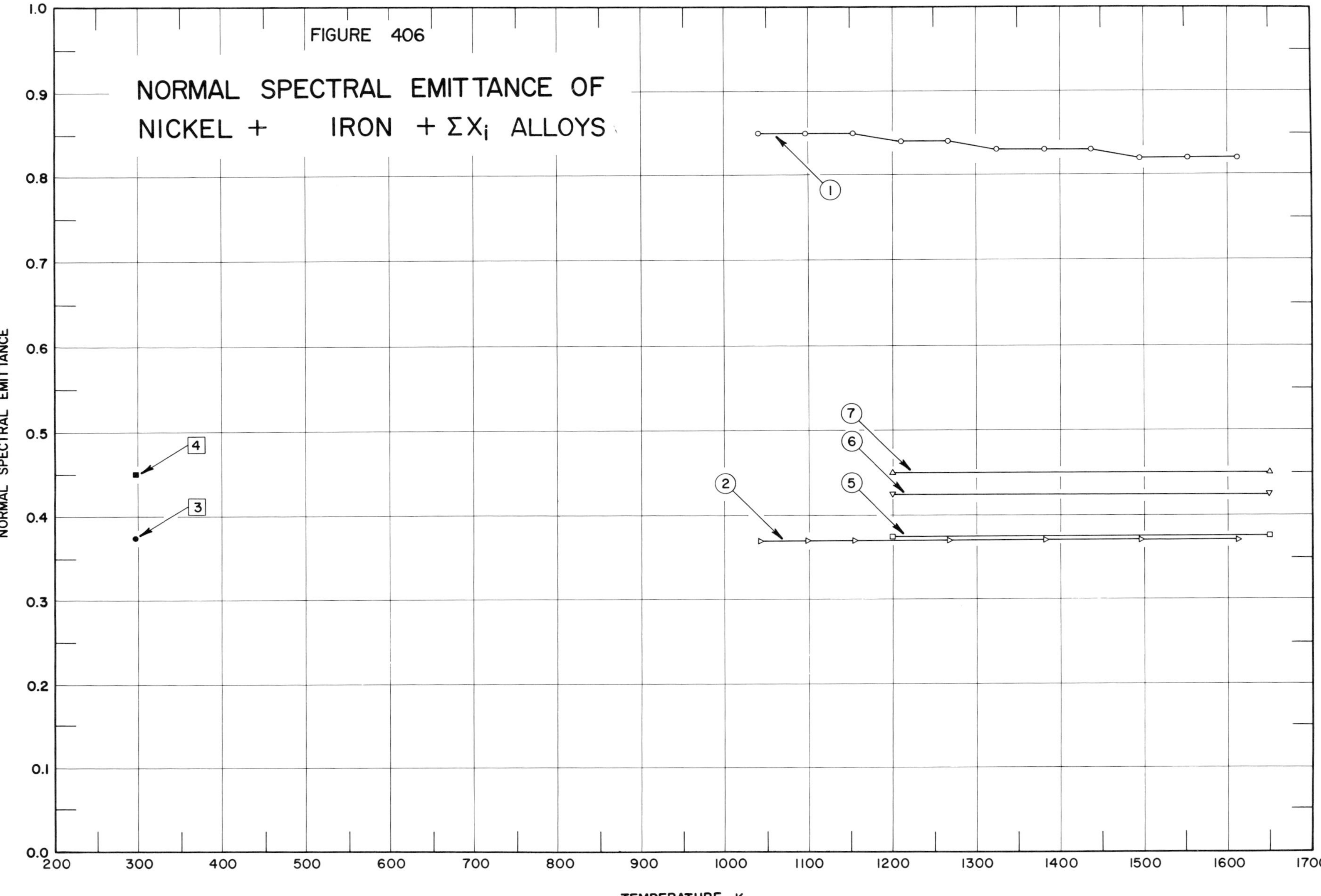
FIGURE 406
NORMAL SPECTRAL EMITTANCE OF
NICKEL + IRON + ΣXi ALLOYS
NORMAL SPECTRAL EMITTANCE
TEMPERATURE, K
0.0
0.1
0.2
0.3
0.4
0.5
0.6
0.7
0.8
0.9
1.0
200
300
400
500
600
700
800
900
1000
1100
1200
1300
1400
1500
1600
1700
1
2
3
4
5
6
7

SPECIFICATION TABLE NO. 406 NORMAL SPECTRAL EMITTANCE OF [NICKEL + IRON + ΣX_i] ALLOYS

Curve No.	Ref. No.	Year	Wavelength μ	Temperature Range, K	Geometry θ'	Reported Error, %	Composition (weight percent), Specifications and Remarks
1	163	1939	0.65	1041-1612	~0°		60 Ni, 24 Fe, 16 Cr; polished with rough paper; heated in air at 1478 K for several hours; measured in air; data extracted from smooth curve.
2	163	1939	0.65	1041-1612	~0°		Different sample, same as above specimen and conditions except heated at 1478 K for several hours in hydrogen; measured in hydrogen.
3	51	1926	0.665	298	~0°		98.8 Ni, 0.75 Fe, 0.15 Cu, 0.15 Mn, 0.15 Si, trace C, trace P, trace S; measured in argon.
4	51	1926	0.460	298	~0°		Above specimen and conditions.
5	51	1926	0.665	1200-1650	~0°		Above specimen and conditions.
6	51	1926	0.535	1200-1650	~0°		Above specimen and conditions.
7	51	1926	0.460	1200-1650	~0°		Above specimen and conditions.

DATA TABLE NO. 406 NORMAL SPECTRAL EMITTANCE OF [NICKEL + IRON + ΣX_i] ALLOYS

[Temperature, T, K; Emittance, ϵ; Wavelength, λ, μ]

T	ϵ	T	ϵ
CURVE 1 $\lambda = 0.65$		CURVE 7 $\lambda = 0.460$	
1041	0.85	1200	0.450
1098	0.85	1650	0.450
1154	0.85		
1211	0.84		
1268	0.84		
1325	0.83		
1382	0.83		
1439	0.83		
1497	0.82		
1554	0.82		
1612	0.82		
CURVE 2 $\lambda = 0.65$			
1041	0.37		
1078	0.37		
1154	0.37		
1268	0.37		
1382	0.37		
1497	0.37		
1612	0.37		
CURVE 3 $\lambda = 0.665$			
298	0.375		
CURVE 4 $\lambda = 0.460$			
298	0.450		
CURVE 5 $\lambda = 0.665$			
1200	0.375		
1650	0.375		
CURVE 6 $\lambda = 0.535$			
1200	0.425		
1650	0.425		

* Not shown on plot

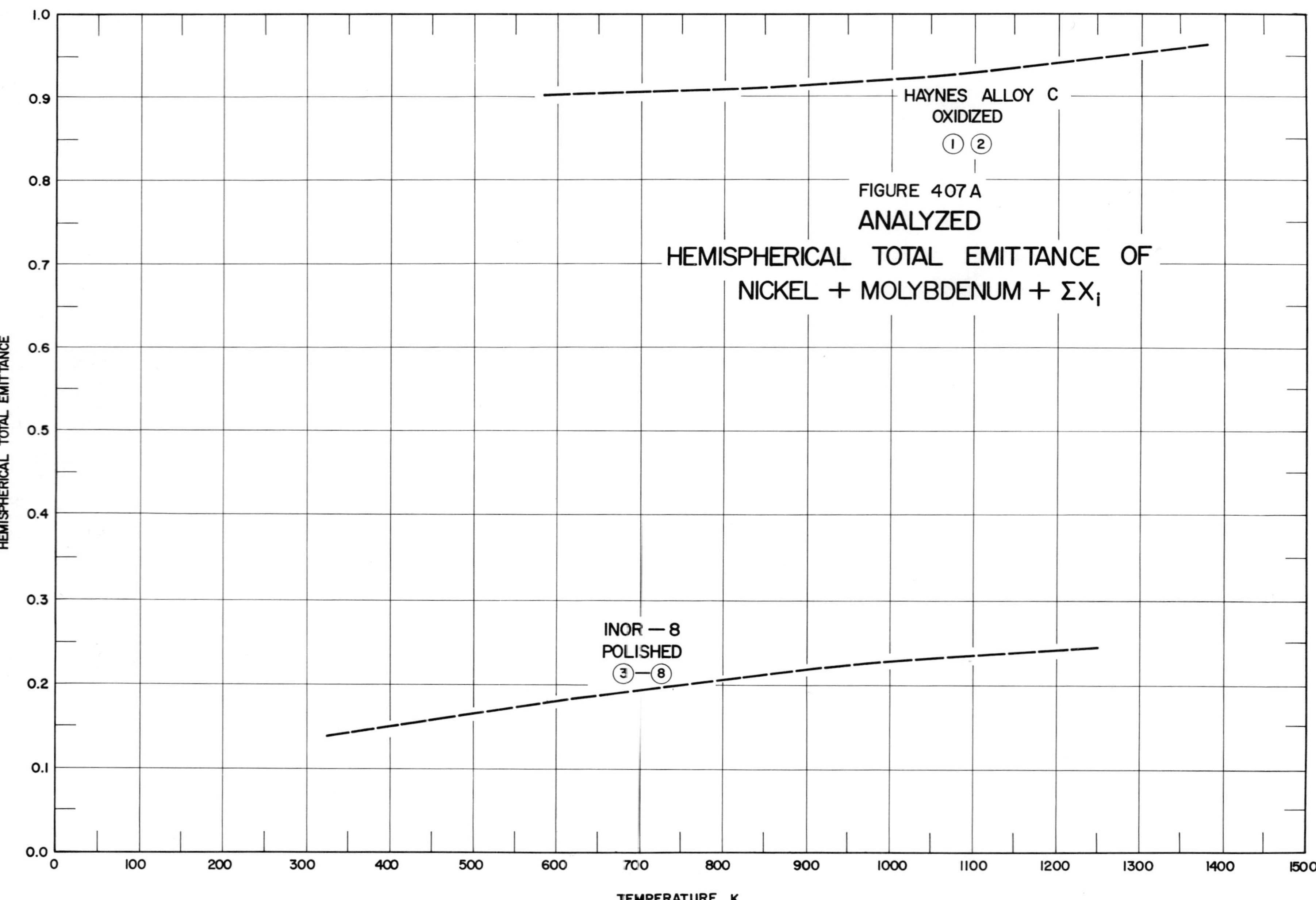

HAYNES ALLOY C
OXIDIZED
1 2
FIGURE 407 A
ANALYZED
HEMISPHERICAL TOTAL EMITTANCE OF
NICKEL + MOLYBDENUM + ΣX_i
INOR — 8
POLISHED
3 — 8
HEMISPHERICAL TOTAL EMITTANCE
1.0
0.9
0.8
0.7
0.6
0.5
0.4
0.3
0.2
0.1
0.0
0
100
200
300
400
500
600
700
800
900
1000
1100
1200
1300
1400
1500
TEMPERATURE, K

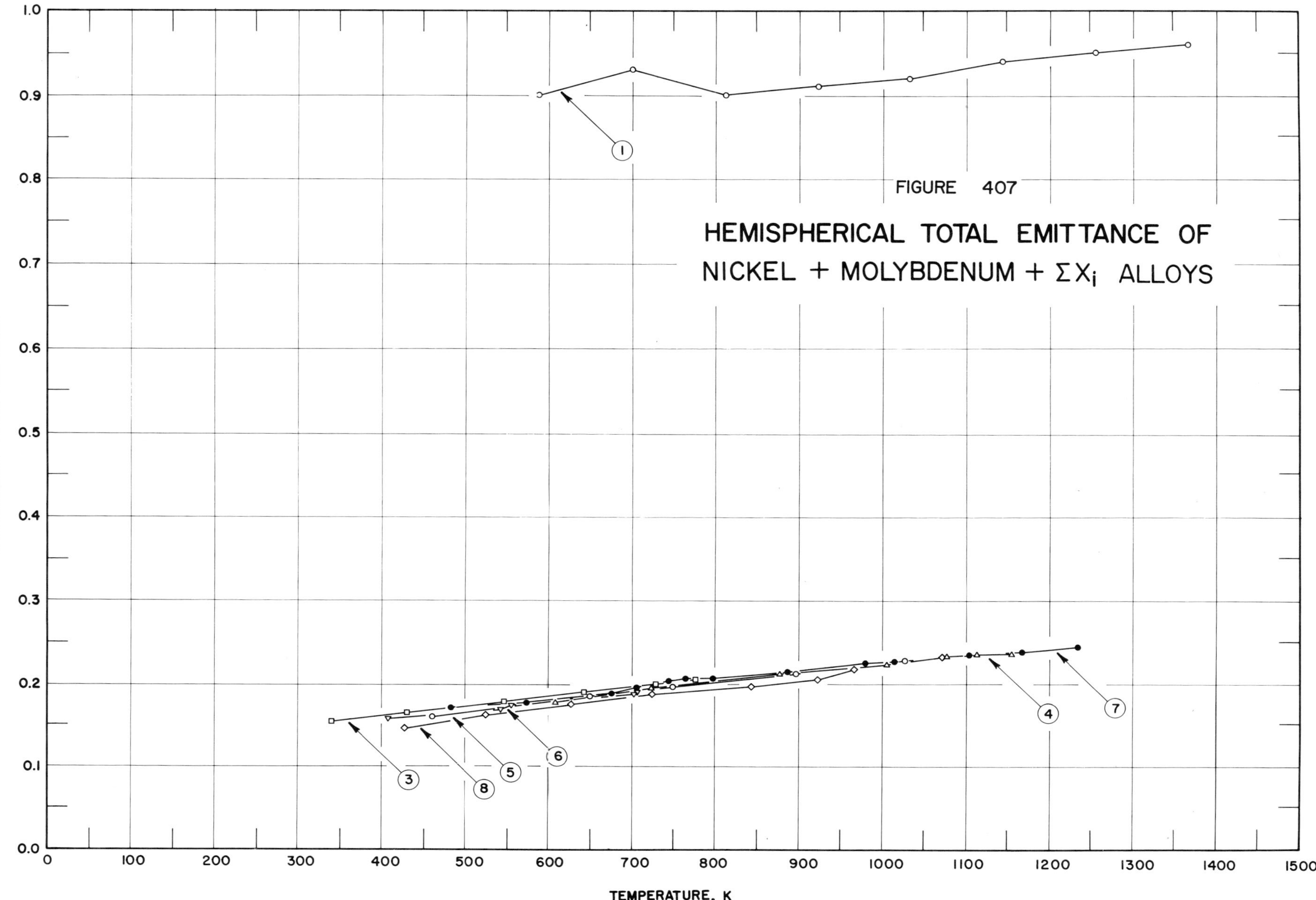
FIGURE 407
HEMISPHERICAL TOTAL EMITTANCE OF
NICKEL + MOLYBDENUM + ΣX_i ALLOYS
HEMISPHERICAL TOTAL EMITTANCE
TEMPERATURE, K
0.0
0.1
0.2
0.3
0.4
0.5
0.6
0.7
0.8
0.9
1.0
0
100
200
300
400
500
600
700
800
900
1000
1100
1200
1300
1400
1500
1
3
4
5
6
7
8

SPECIFICATION TABLE NO. 407 HEMISPHERICAL TOTAL EMITTANCE OF [NICKEL + MOLYBDENUM + ΣX_i] ALLOYS

Curve No.	Ref. No.	Year	Temperature Range, K	Reported Error, %	Composition (weight percent), Specifications and Remarks
1	161	1959	589-1367		Haynes Alloy C; 52-60 Ni, 16-18 Mo, 15.5-17.5 Cr, 4.5-7.0 Fe, 3.75-5.25 W, 0.15 max C; mechanically polished; oxidized in air at 1367 K for 30 min; author assumed normal total emittance = hemispherical total emittance.
2	161	1959	589-1367		Different sample, same as above specimen and conditions except gradually oxidized in still air at temperatures between 0 and 1367 K.
3	67	1963	340-776	± 2.7	Inor-8; nominal composition: 17 Mo, 7 Cr, 5 max Fe, 0.8 max Mn, 0.5 max Al + Ti, 0.06 C, Ni balance; manually polished; measured in vacuum (5×10^{-6} mm Hg).
4	67	1963	609-1155	± 2.7	Different sample, same as above specimen and conditions; heating; [Authors' designation: Specimen 4].
5	67	1963	1027-460	± 2.7	Above specimen and conditions except cooling.
6	67	1963	409-707	± 2.7	Different sample, same as Curve 3 specimen and conditions except surface roughness <1 microinch; [Authors' designation: Specimen 5].
7	67	1963	481-1234	± 2.7	Different sample, same as above specimen and conditions; [Authors' designation: Specimen 6].
8	67	1963	1074-428	± 2.7	Above specimen and conditions except cooling.

DATA TABLE NO. 407 HEMISPHERICAL TOTAL EMITTANCE OF [NICKEL + MOLYBDENUM + ΣX_i] ALLOYS

[Temperature, T, K; Emittance, ϵ]

T	ϵ	T	ϵ
CURVE 1		CURVE 5 (cont.)	
589	0.90	650	0.185
700	0.93	460	0.160
811	0.90		
922	0.91	CURVE 6	
1033	0.92		
1144	0.94	409	0.157
1255	0.95	412	0.156*
1367	0.96	542	0.169
		548	0.171*
CURVE 2*		554	0.173
		701	0.189
589	0.90	705	0.191*
700	0.93	707	0.191
811	0.90		
922	0.91	CURVE 7	
1033	0.92		
1144	0.94	481	0.170
1255	0.95	574	0.178
1367	0.96	676	0.189
		705	0.195
CURVE 3		744	0.203
		764	0.207
340	0.153	799	0.207
430	0.164	886	0.215
546	0.179	980	0.225
642	0.190	1014	0.227
729	0.200	1104	0.234
776	0.205	1169	0.239
		1234	0.244
CURVE 4			
		CURVE 8	
609	0.177		
723	0.196	1074	0.229*
878	0.211	1072	0.231
1006	0.223	966	0.219
1078	0.232	921	0.205
1114	0.235	843	0.199
1155	0.236	723	0.187
		627	0.175
CURVE 5		524	0.161
		428	0.147
1027	0.228		
896	0.212		
750	0.197		

* Not shown on plot

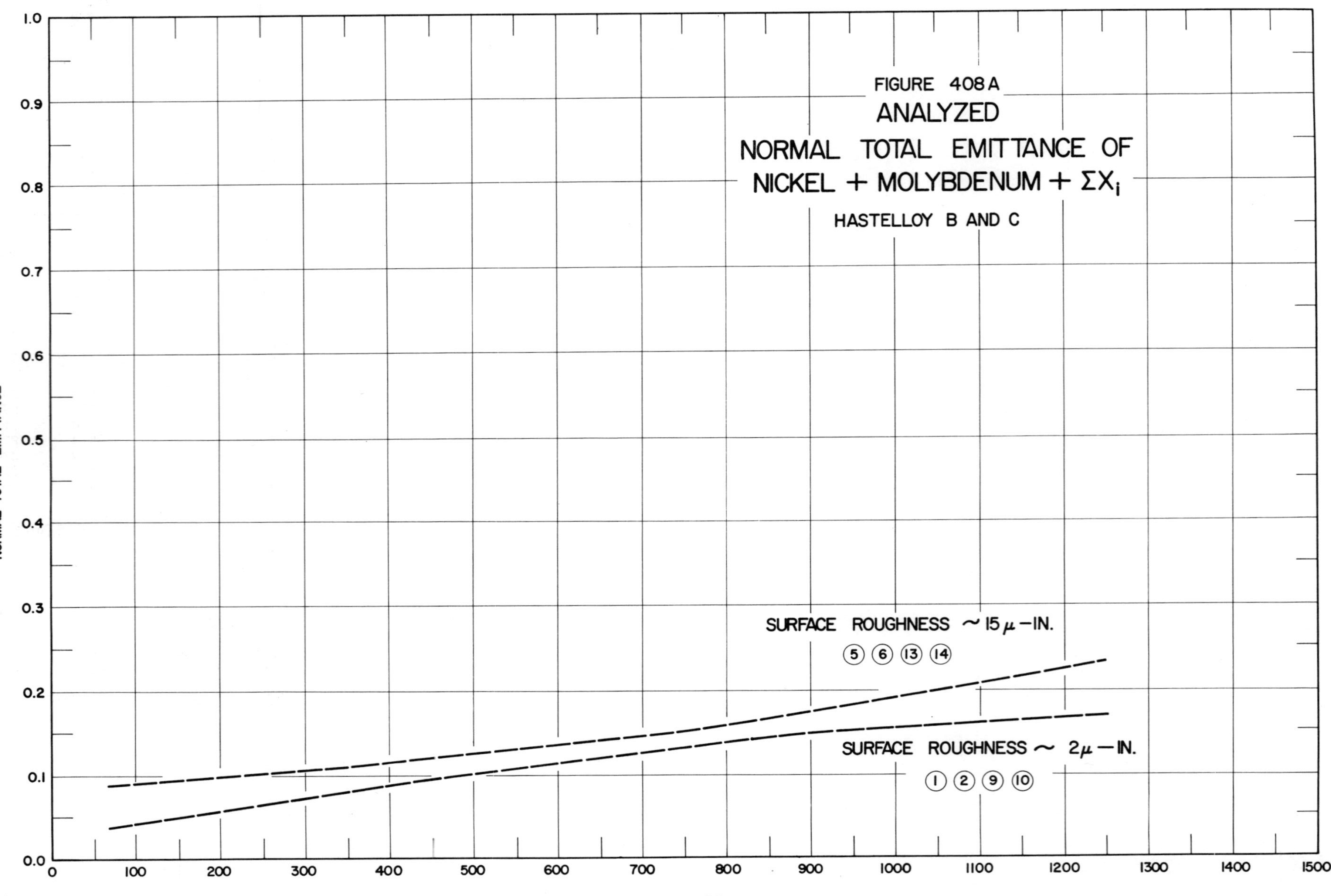
FIGURE 408A
ANALYZED
NORMAL TOTAL EMITTANCE OF
NICKEL + MOLYBDENUM + ΣX$_i$
HASTELLOY B AND C
SURFACE ROUGHNESS ~ 15μ-IN.
5 6 13 14
SURFACE ROUGHNESS ~ 2μ-IN.
1 2 9 10
NORMAL TOTAL EMITTANCE
TEMPERATURE, K
0.0
0.1
0.2
0.3
0.4
0.5
0.6
0.7
0.8
0.9
1.0
0
100
200
300
400
500
600
700
800
900
1000
1100
1200
1300
1400
1500

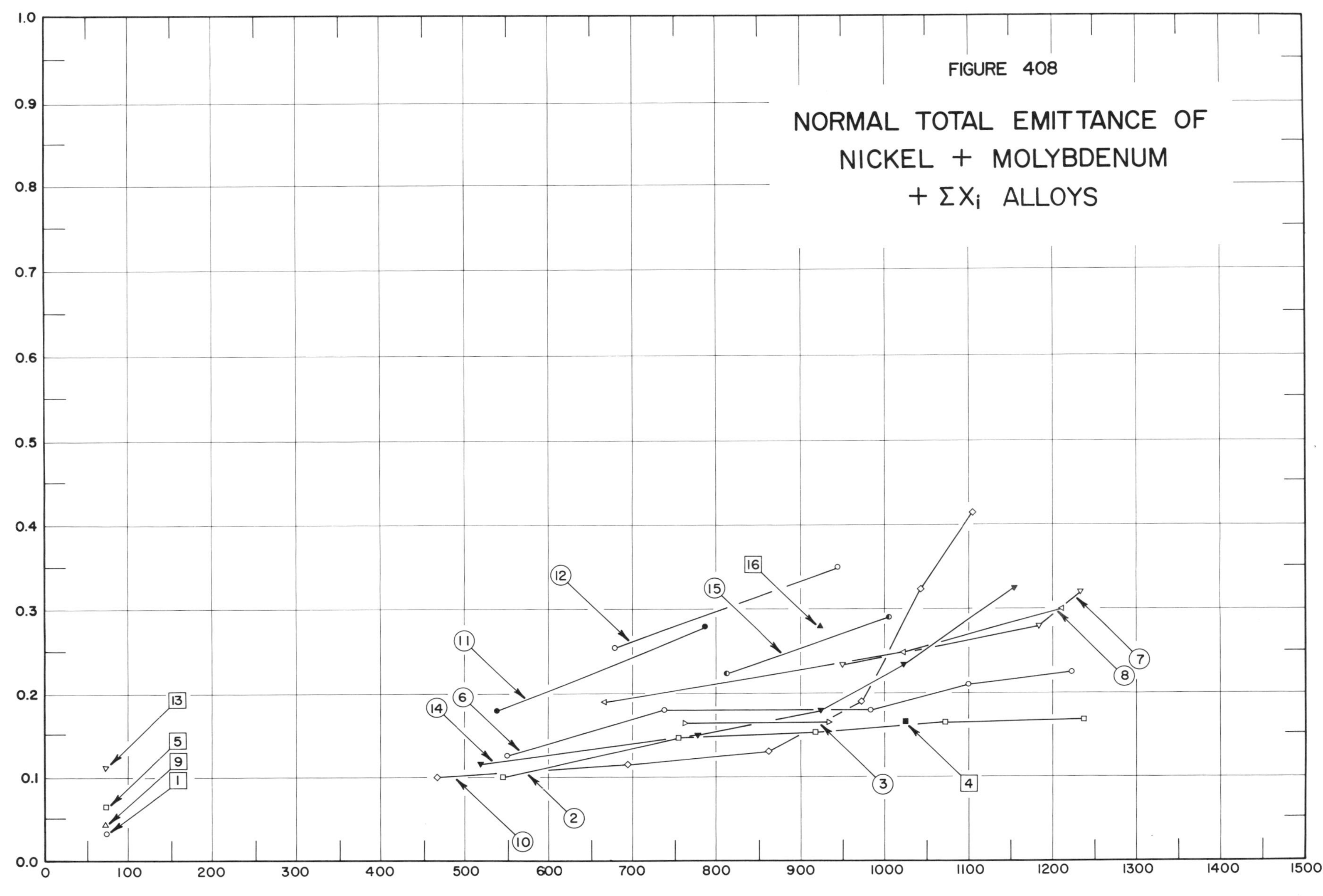
FIGURE 408
NORMAL TOTAL EMITTANCE OF
NICKEL + MOLYBDENUM
+ ΣXi ALLOYS
NORMAL TOTAL EMITTANCE
TEMPERATURE, K
0.0 0.1 0.2 0.3 0.4 0.5 0.6 0.7 0.8 0.9 1.0
0 100 200 300 400 500 600 700 800 900 1000 1100 1200 1300 1400 1500
1 2 3 4 5 6 7 8 9 10 11 12 13 14 15 16

SPECIFICATION TABLE NO. 408 NORMAL TOTAL EMITTANCE OF [NICKEL + MOLYBDENUM + ΣX_i] ALLOYS

Curve No.	Ref. No.	Year	Temperature Range, K	Geometry θ'	Reported Error, %	Composition (weight percent), Specifications and Remarks
1	34	1957	83.2	~0°	±10	Hastelloy B; nominal composition: 62 Ni, 28 Mo, 5 Fe; surface roughness ~2 microinches rms; measured in vacuum (5×10^{-4} mmHg).
2	34	1957	544-1239	~0°	± 10	Different sample, same as above specimen and conditions except increasing temp, cycle 1.
3	34	1957	761-933	~0°	± 10	Above specimen and conditions; cycle 2.
4	34	1957	1025	~0°	± 10	Above specimen and conditions; cycle 3.
5	34	1957	83.2	~0°	± 10	Different sample, same as curve 1 specimen and conditions; surface roughness ~15 microinches rms.
6	34	1957	550-1222	~0°	± 10	Different sample, same as above specimen and conditions except increasing temp, cycle 1.
7	34	1957	950-1233	~0°	± 10	Above specimen and conditions; cycle 2.
8	34	1957	666-1211	~0°	± 10	Above specimen and conditions; cycle 3.
9	34	1957	83.2	~0°	± 10	Hastelloy C, grade AMS 5530 C; nominal composition: 54 Ni, 17 Mo, 15 Cr, 5 Fe, 4 W; annealed; surface roughness ~2 microinches rms; measured in vacuum (5×10^{-4}mm Hg).
10	34	1957	466-1105	~0°	± 10	Different sample, same as above specimen and conditions except increasing temp, cycle 1.
11	34	1957	539-786	~0°	± 10	Above specimen and conditions; cycle 2.
12	34	1957	678-944	~0°	± 10	Above specimen and conditions; cycle 3.
13	34	1957	83.2	~0°	± 10	Different sample, same as curve 9 specimen and conditions; surface roughness ~15 microinches rms.
14	34	1957	519-1155	~0°	± 10	Different sample, same as above specimen and conditions except increasing temp, cycle 1.
15	34	1957	811-1005	~0°	± 10	Above specimen and conditions; cycle 2.
16	34	1957	922	~0°	± 10	Above specimen and conditions; cycle 3.

DATA TABLE NO. 408 NORMAL TOTAL EMITTANCE OF [NICKEL + MOLYBDENUM + ΣX_i] ALLOYS

[Temperature, T, K; Emittance, ϵ]

T	ϵ
CURVE 1	
83.2	0.032
CURVE 2	
544	0.100
755	0.147
919	0.151
1072	0.165
1239	0.169
CURVE 3	
761	0.164
933	0.164
CURVE 4	
1025	0.166
CURVE 5	
83.2	0.065
CURVE 6	
550	0.125
739	0.180
983	0.180
1100	0.210
1222	0.225
CURVE 7	
950	0.235
1183	0.280
1233	0.320
CURVE 8	
666	0.190
1022	0.250
1211	0.300
CURVE 9	
83.2	0.043
CURVE 10	
466	0.100
694	0.115
861	0.130
972	0.190
1044	0.325
1105	0.415
CURVE 11	
539	0.180
786	0.280
CURVE 12	
678	0.255
944	0.350
CURVE 13	
83.2	0.111
CURVE 14	
519	0.115
778	0.150
922	0.180
1022	0.235
1155	0.325
CURVE 15	
811	0.225
1005	0.290
CURVE 16	
922	0.280

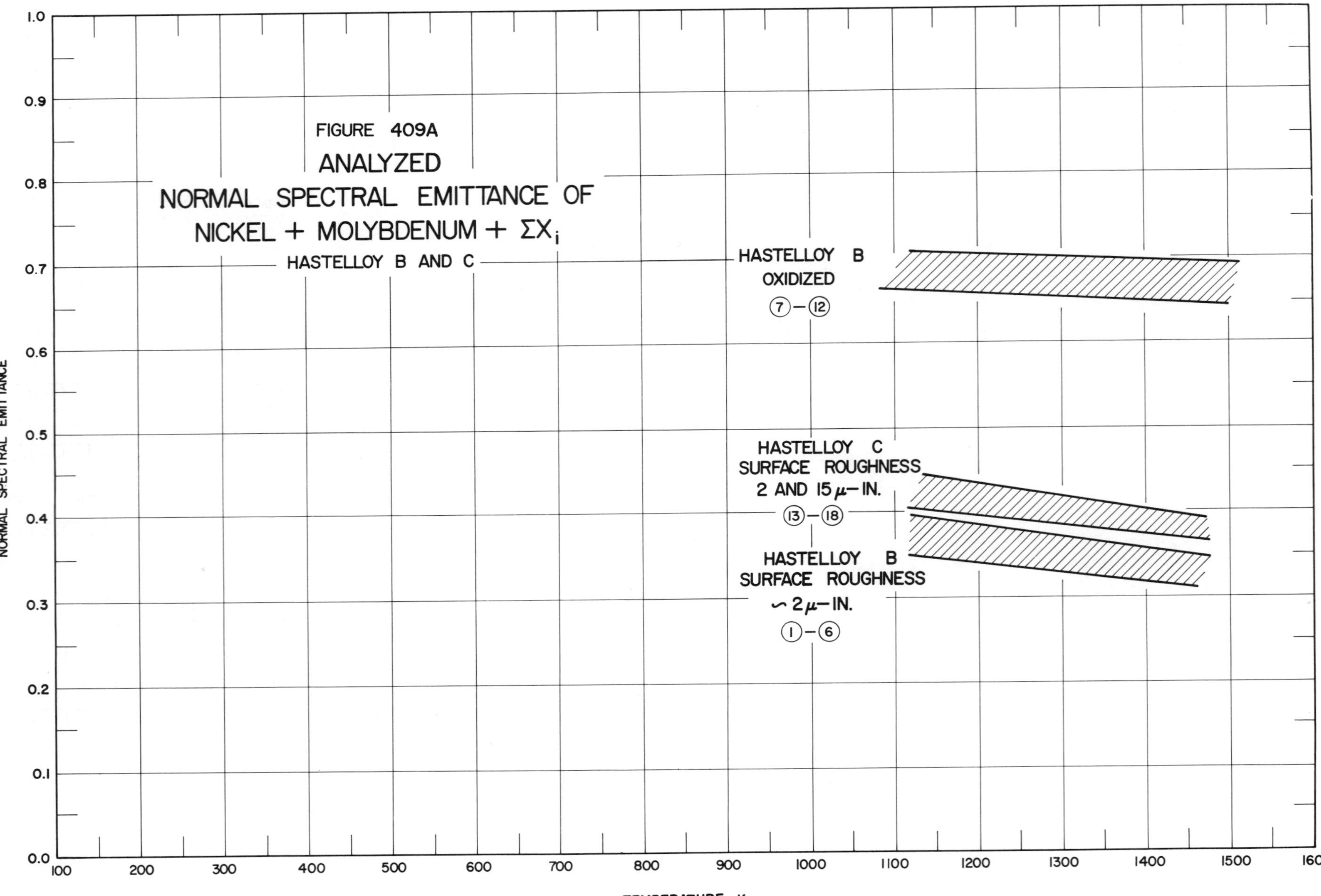
FIGURE 409A
ANALYZED
NORMAL SPECTRAL EMITTANCE OF
NICKEL + MOLYBDENUM + ΣX_i
HASTELLOY B AND C
HASTELLOY B
OXIDIZED
(7)—(12)
HASTELLOY C
SURFACE ROUGHNESS
2 AND 15 μ–IN.
(13)—(18)
HASTELLOY B
SURFACE ROUGHNESS
~2 μ–IN.
(1)—(6)
NORMAL SPECTRAL EMITTANCE
1.0
0.9
0.8
0.7
0.6
0.5
0.4
0.3
0.2
0.1
0.0
100
200
300
400
500
600
700
800
900
1000
1100
1200
1300
1400
1500
1600
TEMPERATURE, K

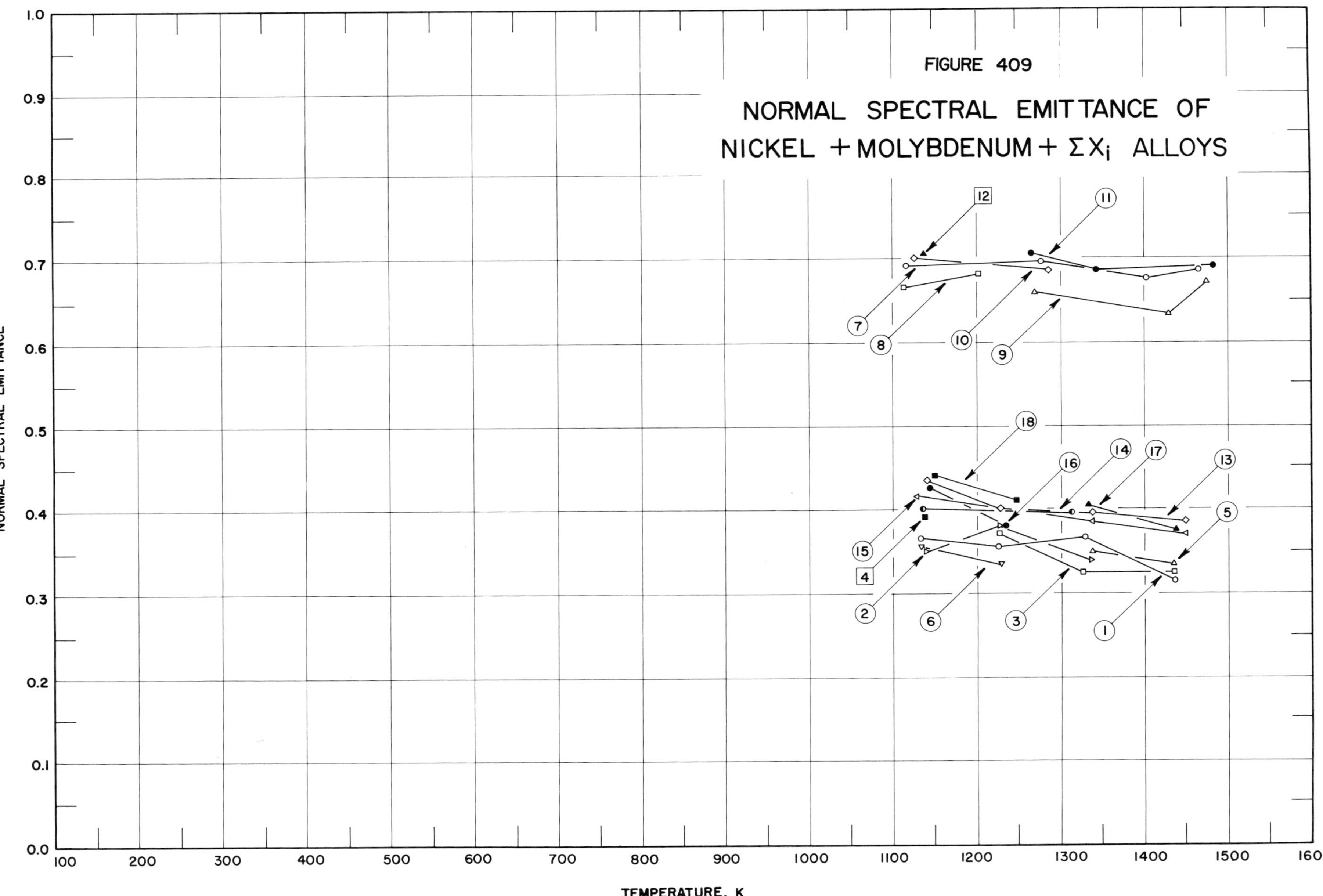

FIGURE 409
NORMAL SPECTRAL EMITTANCE OF
NICKEL + MOLYBDENUM + ΣX_i ALLOYS
NORMAL SPECTRAL EMITTANCE
TEMPERATURE, K
1.0
0.9
0.8
0.7
0.6
0.5
0.4
0.3
0.2
0.1
0.0
100
200
300
400
500
600
700
800
900
1000
1100
1200
1300
1400
1500
1600
12
11
7
8
10
9
18
16
14
17
13
5
15
4
2
6
3
1

SPECIFICATION TABLE NO. 409 NORMAL SPECTRAL EMITTANCE OF [NICKEL + MOLYBDENUM + ΣX_i] ALLOYS

Curve No.	Ref. No.	Year	Wavelength μ	Temperature Range, K	Geometry θ'	Reported Error, %	Composition (weight percent), Specifications and Remarks
1	34	1957	0.665	1133-1436	~0°		Hastelloy B; nominal composition: 62 Ni, 28 Mo, 5 Fe; surface roughness ~2 microinches (rms); measured in vacuum (5×10^{-4} mm Hg); increasing temp, cycle 1.
2	34	1957	0.665	1336-1139	~0°		Above specimen and conditions; decreasing temp, cycle 1.
3	34	1957	0.665	1227-1436	~0°		Above specimen and conditions; cycle 2.
4	34	1957	0.665	1139	~0°		Above specimen and conditions; decreasing temp, cycle 2.
5	34	1957	0.665	1339-1439	~0°		Above specimen and conditions; cycle 3.
6	34	1957	0.665	1230-1133	~0°		Above specimen and conditions; decreasing temp, cycle 3.
7	34	1957	0.665	1116-1466	~0°		Different sample, same as curve 1 specimen and conditions; oxidized in air at red heat for 30 min.
8	34	1957	0.665	1202-1114	~0°		Above specimen and conditions; decreasing temp, cycle 1.
9	34	1957	0.665	1270-1475	~0°		Above specimen and conditions; cycle 2.
10	34	1957	0.665	1286-1127	~0°		Above specimen and conditions; decreasing temp, cycle 2.
11	34	1957	0.665	1266-1483	~0°		Above specimen and conditions; cycle 3.
12	34	1957	0.665	1139	~0°		Above specimen and conditions; decreasing temp, cycle 3.
13	34	1957	0.665	1141-1450	~0°		Hastelloy C; nominal composition: 54 Ni, 17 Mo, 15 Cr, 5 Fe, 4 W; surface roughness ~2 and 15 microinches (rms); measured in vacuum (5×10^{-4} mm Hg); increasing temp, cycle 1.
14	34	1957	0.665	1314-1136	~0°		Above specimen and conditions; decreasing temp, cycle 1.
15	34	1957	0.665	1230-1450	~0°		Above specimen and conditions; cycle 2.
16	34	1957	0.665	1233-1144	~0°		Above specimen and conditions; decreasing temp, cycle 2.
17	34	1957	0.665	1333-1439	~0°		Above specimen and conditions; cycle 3.
18	34	1957	0.665	1247-1150	~0°		Above specimen and conditions; decreasing temp, cycle 3.

DATA TABLE NO. 409 NORMAL SPECTRAL EMITTANCE OF [NICKEL + MOLYBDENUM + ΣX_i] ALLOYS

[Temperature, T, K; Emittance, ϵ; Wavelength, λ, μ]

T	ϵ	T	ϵ	T	ϵ
CURVE 1	$\lambda = 0.665$	CURVE 8	$\lambda = 0.665$	CURVE 15	$\lambda = 0.665$
1133	0.365	1202	0.680	1230	0.415
1225	0.355	1114	0.665	1339	0.385
1330	0.365			1450	0.370
1436	0.315	CURVE 9	$\lambda = 0.665$		
		1270	0.660	CURVE 16	$\lambda = 0.665$
CURVE 2	$\lambda = 0.665$	1430	0.635	1233	0.380
1336	0.340	1475	0.670	1144	0.425
1227	0.380				
1139	0.350	CURVE 10	$\lambda = 0.665$	CURVE 17	$\lambda = 0.665$
		1286	0.685	1333	0.405
CURVE 3	$\lambda = 0.665$	1127	0.700	1439	0.375
1227	0.370				
1327	0.325	CURVE 11	$\lambda = 0.665$	CURVE 18	$\lambda = 0.665$
1436	0.325	1266	0.705	1247	0.410
		1344	0.685	1150	0.440
CURVE 4	$\lambda = 0.665$	1483	0.690		
1139	0.390				
		CURVE 12	$\lambda = 0.665$		
CURVE 5	$\lambda = 0.665$	1139	0.705		
1339	0.350				
1439	0.335	CURVE 13	$\lambda = 0.665$		
		1141	0.435		
CURVE 6	$\lambda = 0.665$	1227	0.400		
1230	0.335	1339	0.395		
1133	0.355	1450	0.385		
CURVE 7	$\lambda = 0.665$	CURVE 14	$\lambda = 0.665$		
1116	0.690	1314	0.395		
1277	0.695	1136	0.400		
1402	0.675				
1466	0.685				

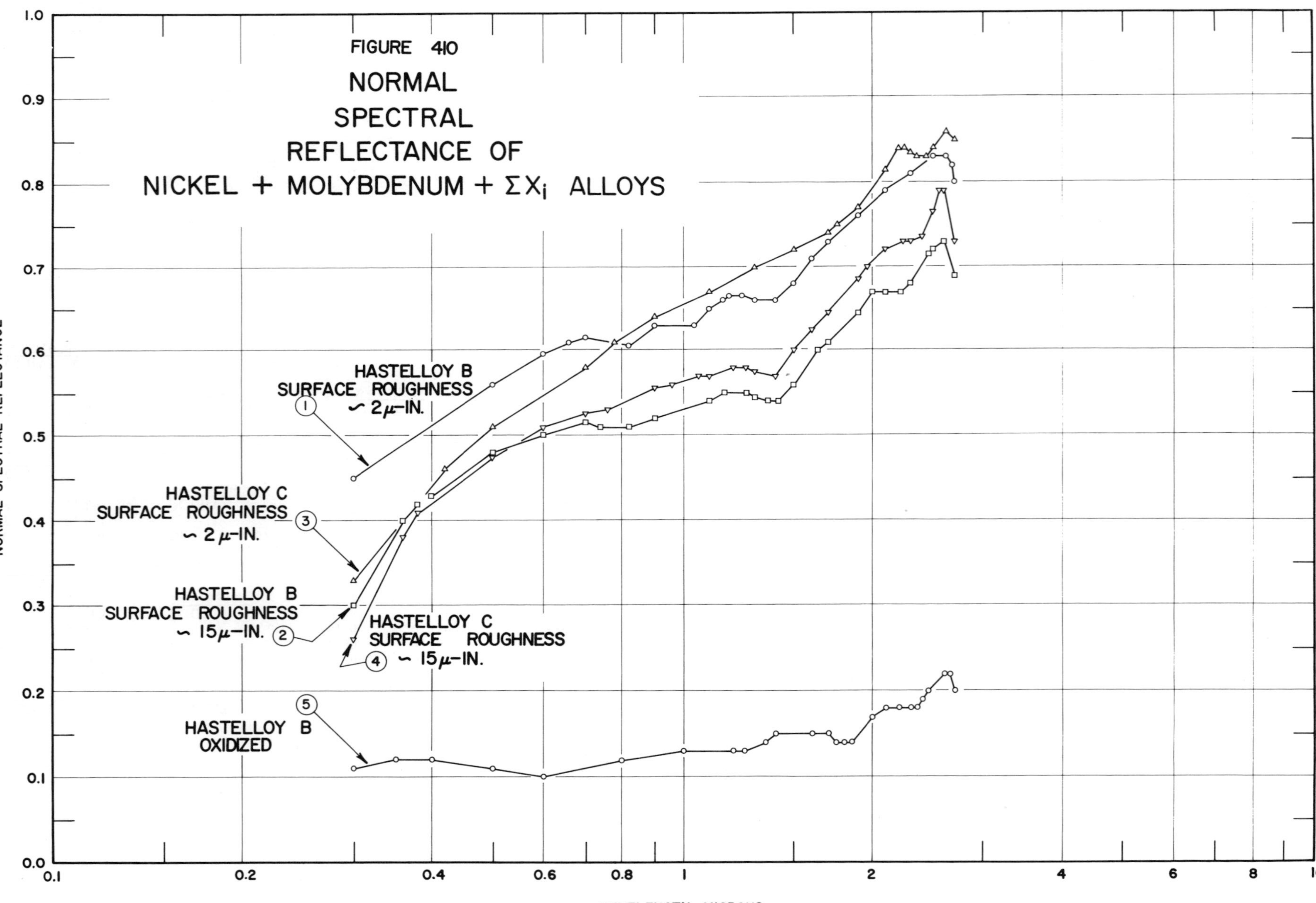
FIGURE 410
NORMAL
SPECTRAL
REFLECTANCE OF
NICKEL + MOLYBDENUM + ΣX_i ALLOYS
① HASTELLOY B SURFACE ROUGHNESS ~ 2μ-IN.
③ HASTELLOY C SURFACE ROUGHNESS ~ 2μ-IN.
② HASTELLOY B SURFACE ROUGHNESS ~ 15μ-IN.
④ HASTELLOY C SURFACE ROUGHNESS ~ 15μ-IN.
⑤ HASTELLOY B OXIDIZED
NORMAL SPECTRAL REFLECTANCE
WAVELENGTH, MICRONS

SPECIFICATION TABLE NO. 410 NORMAL SPECTRAL REFLECTANCE OF [NICKEL + MOLYBDENUM + ΣX_i] ALLOYS

Curve No.	Ref. No.	Year	Temperature K	Wavelength Range, μ	Geometry θ	Geometry θ'	Geometry ω'	Reported Error, %	Composition (weight percent), Specifications and Remarks
1	34	1957	298	0.30-2.70	9°		2π	± 4	Hastelloy B; nominal composition: 62 Ni, 28 Mo, 5 Fe; surface roughness ~2 microinches rms; data extracted from smooth curve.
2	34	1957	298	0.30-2.70	9°		2π	± 4	Different sample, same as above specimen and conditions; surface roughness ~15 microinches rms.
3	34	1957	298	0.30-2.70	9°		2π	± 4	Hastelloy C; nominal composition: 54 Ni, 17 Mo, 15 Cr, 5 Fe, 4 W; surface roughness ~2 microinches rms; data extracted from smooth curve.
4	34	1957	298	0.30-2.70	9°		2π	± 4	Different sample, same as above specimen and conditions; surface roughness ~15 microinches rms.
5	146	1958	298	0.30-2.70	~0°		2π		Hastelloy B; nominal composition: 62 Ni, 28 Mo, 5 Fe; oxidized at 922 K for 30 min.

DATA TABLE NO. 410 NORMAL SPECTRAL REFLECTANCE OF [NICKEL + MOLYBDENUM + ΣX_i] ALLOYS

[Wavelength, λ, μ; Reflectance, ρ; Temperature, T, K]

CURVE 1
T = 298

λ	ρ
0.30	0.450
0.50	0.560
0.60	0.595
0.66	0.610
0.70	0.615
0.82	0.605
0.90	0.630
1.04	0.630
1.10	0.650
1.16	0.660
1.18	0.665
1.24	0.665
1.30	0.660
1.40	0.660
1.50	0.680
1.60	0.710
1.70	0.730
1.90	0.760
2.10	0.790
2.30	0.810
2.50	0.830
2.62	0.830
2.68	0.820
2.70	0.800

CURVE 2
T = 298

λ	ρ
0.30	0.300
0.36	0.400
0.38	0.420
0.40	0.430
0.50	0.480
0.60	0.500
0.70	0.515
0.74	0.510
0.82	0.510
0.90	0.520
1.10	0.540
1.16	0.550
1.26	0.550
1.30	0.545
1.36	0.540

CURVE 2 (cont.)

λ	ρ
1.42	0.540
1.50	0.560
1.64	0.600
1.70	0.610
1.90	0.645
2.00	0.670
2.10	0.670
2.22	0.670
2.30	0.680
2.46	0.715
2.50	0.720
2.60	0.730
2.70	0.690

CURVE 3
T = 298

λ	ρ
0.30	0.330
0.42	0.460
0.50	0.510
0.70	0.580
0.78	0.610
0.90	0.640
1.10	0.670
1.30	0.700
1.50	0.720
1.70	0.740
1.76	0.750
1.90	0.770
2.10	0.815
2.20	0.840
2.24	0.840
2.30	0.835
2.34	0.830
2.44	0.830
2.50	0.840
2.62	0.860
2.70	0.850

CURVE 4
T = 298

λ	ρ
0.30	0.260
0.36	0.380

CURVE 4 (cont.)

λ	ρ
0.38	0.410
0.40	0.430*
0.50	0.475
0.60	0.510
0.70	0.525
0.76	0.530
0.90	0.555
0.96	0.560
1.06	0.570
1.10	0.570
1.20	0.580
1.26	0.580
1.30	0.575
1.40	0.570
1.50	0.600
1.60	0.625
1.70	0.645
1.90	0.685
1.96	0.700
2.10	0.720
2.24	0.730
2.30	0.730
2.40	0.745
2.50	0.765
2.56	0.790
2.60	0.790
2.70	0.730

CURVE 5
T = 298

λ	ρ
0.30	0.11
0.35	0.12
0.40	0.12
0.50	0.11
0.60	0.10
0.80	0.12
1.00	0.13
1.20	0.13
1.25	0.13
1.35	0.14
1.40	0.15
1.60	0.15
1.70	0.15

CURVE 5 (cont.)

λ	ρ
1.75	0.14
1.80	0.14
1.85	0.14
2.00	0.17
2.10	0.18
2.20	0.18
2.30	0.18
2.35	0.18
2.40	0.19
2.45	0.20
2.60	0.22
2.65	0.22
2.70	0.20

* Not shown on plot

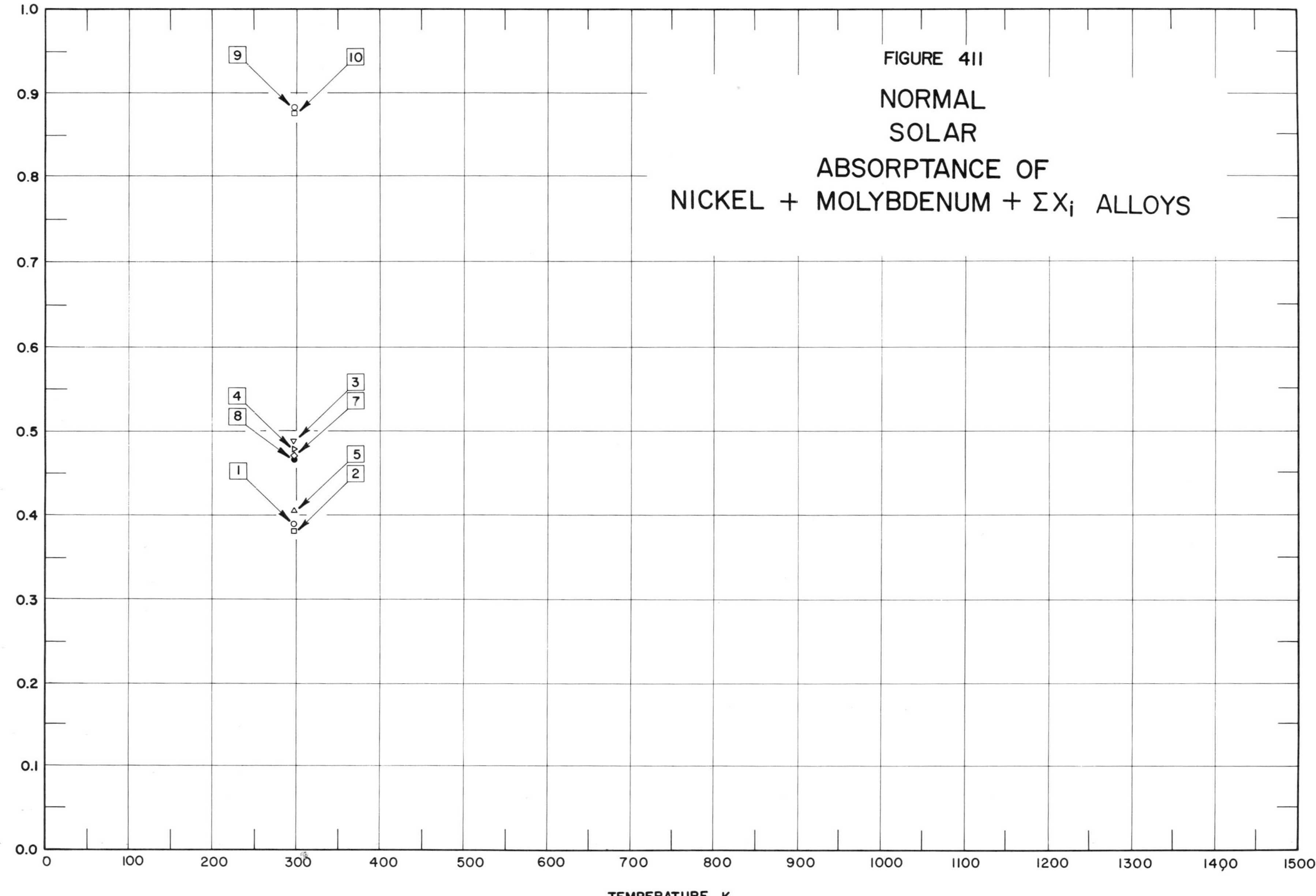
FIGURE 411
NORMAL
SOLAR
ABSORPTANCE OF
NICKEL + MOLYBDENUM + ΣXi ALLOYS
NORMAL SOLAR ABSORPTANCE
1.0
0.9
0.8
0.7
0.6
0.5
0.4
0.3
0.2
0.1
0.0
0
100
200
300
400
500
600
700
800
900
1000
1100
1200
1300
1400
1500
TEMPERATURE, K
9
10
4
8
3
7
1
5
2

SPECIFICATION TABLE NO. 411 NORMAL SOLAR ABSORPTANCE OF [NICKEL + MOLYBDENUM + ΣX_i] ALLOYS

Curve No.	Ref. No.	Year	Temperature Range, K	Geometry θ	Reported Error, %	Composition (weight percent), Specifications and Remarks
1	34	1957	298	9°		Hastelloy B; nominal composition: 62 Ni, 28 Mo, 5 Fe; surface roughness ~2 microinches rms; computed from spectral reflectance data for sea level conditions.
2	34	1957	298	9°		Above specimen and conditions except computed for above atmosphere conditions.
3	34	1957	298	9°		Different sample, same as curve 1 specimen and conditions; surface roughness ~15 microinches rms.
4	34	1957	298	9°		Above specimen and conditions except computed for above atmosphere conditions.
5	34	1957	298	9°		Hastelloy C; nominal composition: 54 Ni, 17 Mo, 15 Cr, 5 Fe, 4 W; surface roughness ~2 microinches rms; computed from spectral reflectance data for sea level conditions.
6	34	1957	298	9°		Above specimen and conditions except computed for above atmosphere conditions.
7	34	1957	298	9°		Different sample, same as curve 5 specimen and conditions; surface roughness ~15 microinches rms.
8	34	1957	298	9°		Above specimen and conditions except computed for above atmosphere conditions.
9	146	1958	298	9°		Hastelloy B; nominal composition: 62 Ni, 28 Mo, 5 Fe; oxidized; computed from spectral reflectance data for sea level conditions.
10	146	1958	298	9°		Above specimen and conditions except computed for above atmosphere conditions.

DATA TABLE NO. 411 NORMAL SOLAR ABSORPTANCE OF [NICKEL + MOLYBDENUM + ΣX_i] ALLOYS

[Temperature, T, K; Absorptance, α]

T	α
CURVE 1	
298	0.390
CURVE 2	
298	0.381
CURVE 3	
298	0.487
CURVE 4	
298	0.478
CURVE 5	
298	0.405
CURVE 6*	
298	0.403
CURVE 7	
298	0.470
CURVE 8	
298	0.466
CURVE 9	
298	0.882
CURVE 10	
298	0.875

* Not shown on plot

SPECIFICATION TABLE NO. 412 NORMAL TOTAL EMITTANCE OF [NIOBIUM + MOLYBDENUM + ΣX_i] ALLOYS

Curve No.	Ref. No.	Year	Temperature Range, K	Geometry θ'	Reported Error, %	Composition (weight percent), Specifications and Remarks
1	107	1960	644-1644	$\sim 0^{\circ}$	±20	Niobium alloy; 10 Mo, 10 Ti, Nb balance; surface finish 63; measured in demoisturized helium gas.

DATA TABLE NO. 412 NORMAL TOTAL EMITTANCE OF [NIOBIUM + MOLYBDENUM + ΣX_i] ALLOYS

[Temperature, T, K; Emittance, ϵ]

T	ϵ
CURVE 1*	
644	0.24
811	0.36
1089	0.53
1366	0.69
1644	0.80

* Not shown on plot

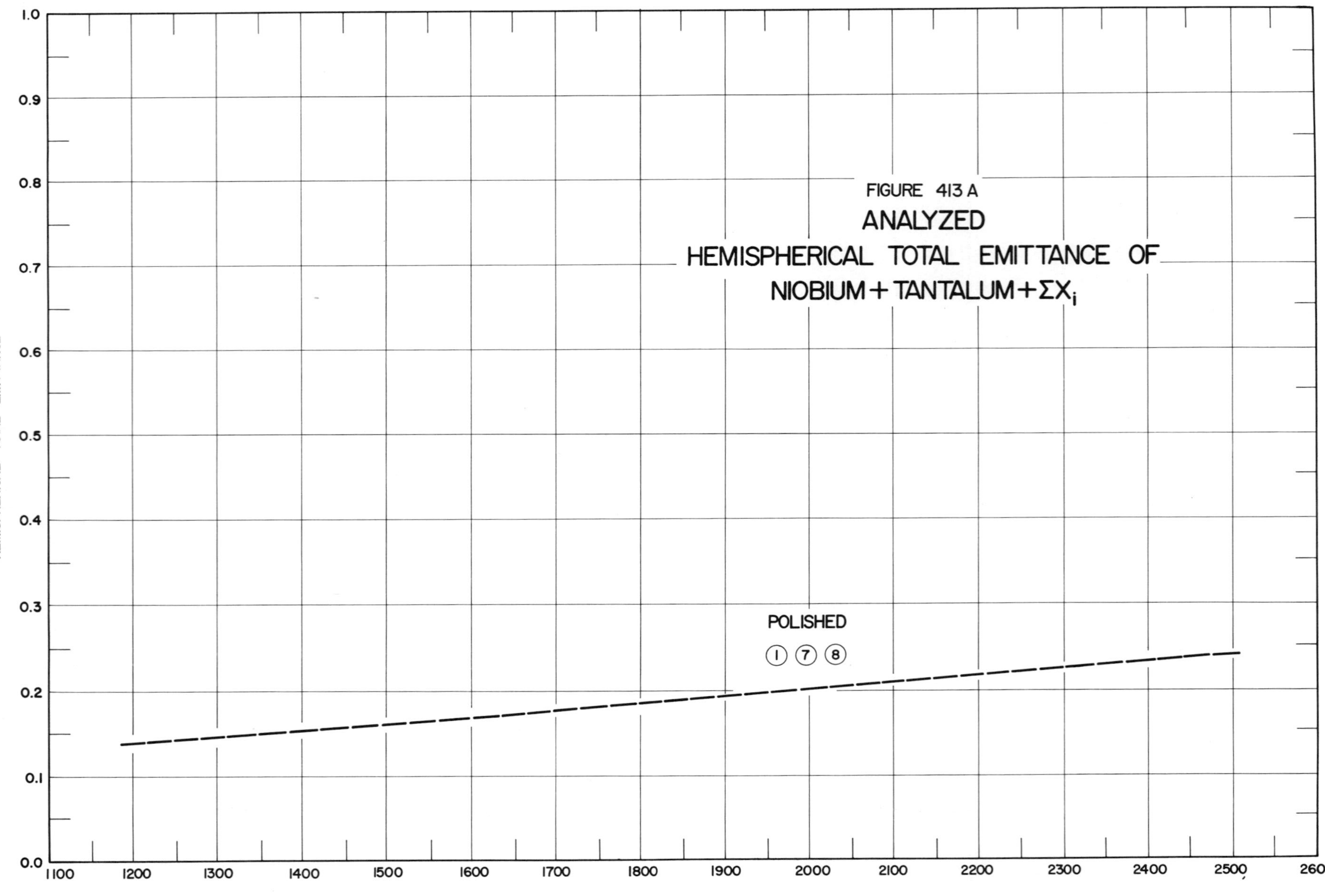
FIGURE 413 A
ANALYZED
HEMISPHERICAL TOTAL EMITTANCE OF
NIOBIUM + TANTALUM + ΣX_i
POLISHED
1 7 8
HEMISPHERICAL TOTAL EMITTANCE
1.0
0.9
0.8
0.7
0.6
0.5
0.4
0.3
0.2
0.1
0.0
TEMPERATURE, K
1100
1200
1300
1400
1500
1600
1700
1800
1900
2000
2100
2200
2300
2400
2500
2600

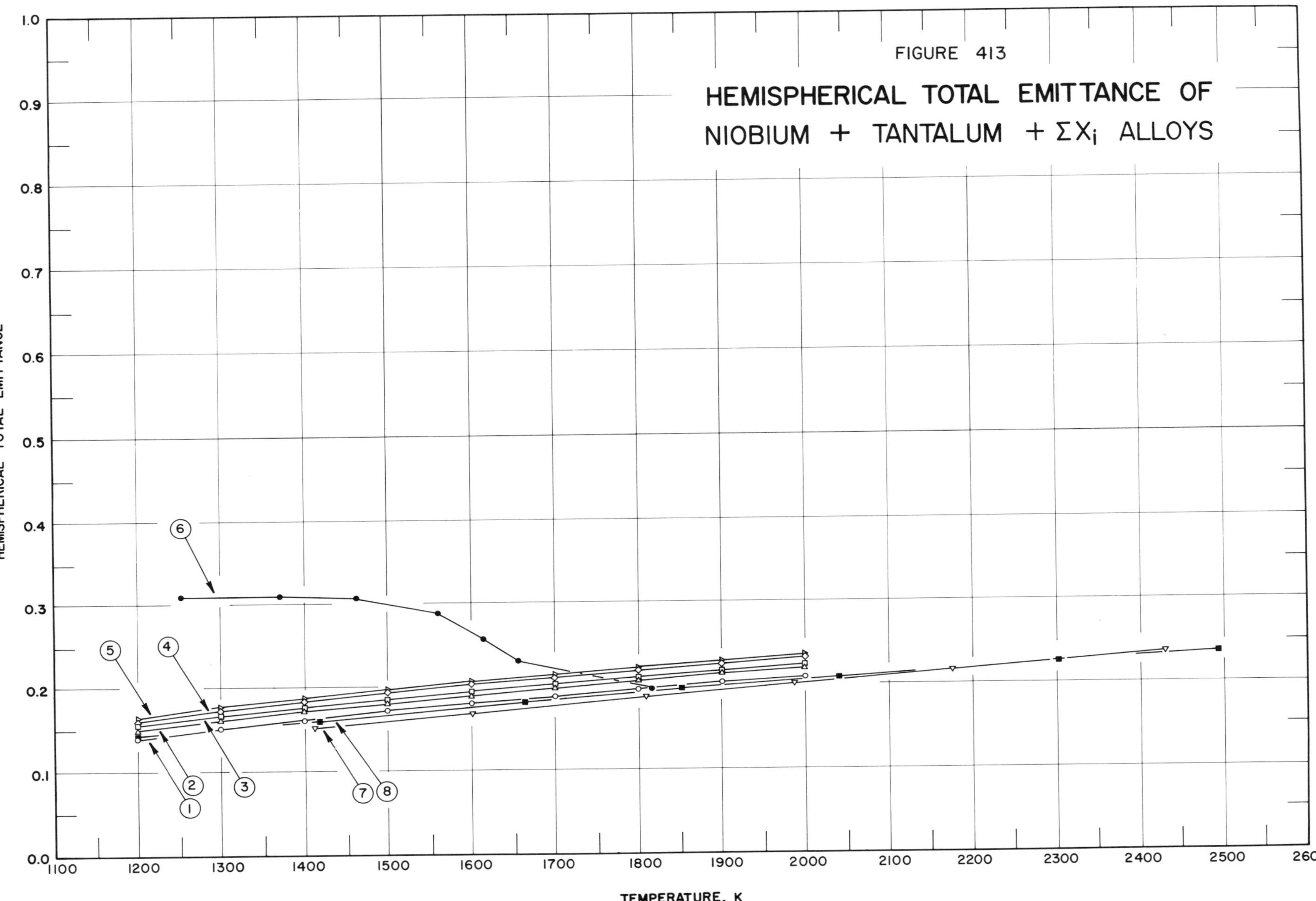

FIGURE 413

HEMISPHERICAL TOTAL EMITTANCE OF NIOBIUM + TANTALUM + ΣX_i ALLOYS

SPECIFICATION TABLE NO. 413 HEMISPHERICAL TOTAL EMITTANCE OF [NIOBIUM + TANTALUM + ΣX_i] ALLOYS

Curve No.	Ref. No.	Year	Temperature Range, K	Reported Error, %	Composition (weight percent), Specifications and Remarks
1	162	1964	1200-2000	< 1.9	99.26 Nb, 0.5 Ta, 0.06 Fe, 0.03 Si, 0.026 Ti; mean arithmetical profile 0.02-0.025 μ; measured in vacuum.
2	162	1964	1200-2000	< 1.9	Different sample, same as above specimen and conditions except mean arithmetical profile 0.4-0.5 μ.
3	162	1964	1200-2000	< 1.9	Different sample, same as above specimen and conditions except mean arithmetical profile 0.5-0.63 μ.
4	162	1964	1200-2000	< 1.9	Different sample, same as above specimen and conditions except mean arithmetical profile 0.8-1.0 μ.
5	162	1964	1200-2000	< 1.9	Different sample, same as above specimen and conditions except mean arithmetical profile 1.0-1.25 μ.
6	291	1963	1252-1815	6	0.4-0.6 Ta, 0.01-0.03 Mg, 0.007-0.009 Ti, 0.004-0.006 Fe, 0.001-0.003 Mn, 0.0004-0.0006 Cu, Nb balance; data extracted from smooth curve.
7	291	1963	1413-2430		Different sample, same as above specimen and conditions.
8	291	1963	1200-2494		Different sample, same as above specimen and conditions.

DATA TABLE NO. 413 HEMISPHERICAL TOTAL EMITTANCE OF [NIOBIUM + TANTALUM + ΣX_i] ALLOYS

[Temperature, T, K; Emittance, ∈]

T	∈	T	∈
CURVE 1		CURVE 5	
1200	0.140	1200	0.163
1300	0.151	1300	0.175
1400	0.161	1400	0.186
1500	0.171	1500	0.196
1600	0.180	1600	0.205
1700	0.188	1700	0.213
1800	0.196	1800	0.221
1900	0.203	1900	0.228
2000	0.209	2000	0.236
CURVE 2		CURVE 6	
1200	0.150	1252	0.310
1300	0.161	1371	0.310
1400	0.171	1463	0.307
1500	0.180	1560	0.288
1600	0.189	1614	0.257
1700	0.198	1657	0.231
1800	0.206	1815	0.196
1900	0.213		
2000	0.220	CURVE 7	
CURVE 3		1413	0.151
		1601	0.167
1200	0.154	1810	0.186
1300	0.165	1988	0.201
1400	0.175	2178	0.217
1500	0.185	2430	0.237
1600	0.194		
1700	0.202	CURVE 8	
1800	0.210		
1900	0.217	1200	0.142
2000	0.224	1417	0.160
		1664	0.181
CURVE 4		1852	0.197
		2041	0.208
1200	0.160	2302	0.225
1300	0.172	2494	0.237
1400	0.183		
1500	0.193		
1600	0.202		
1700	0.210		
1800	0.218		
1900	0.225		
2000	0.232		

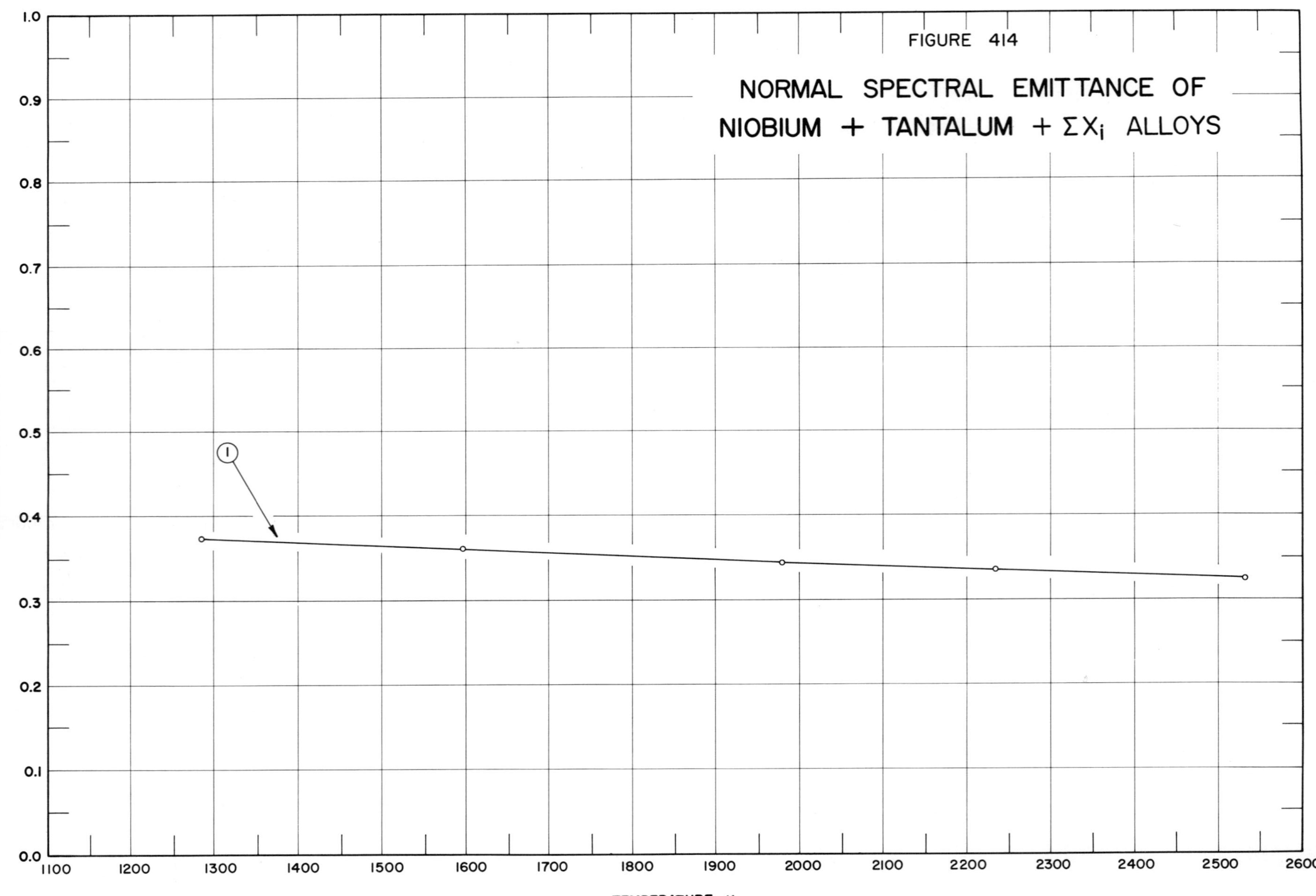
FIGURE 414
NORMAL SPECTRAL EMITTANCE OF
NIOBIUM + TANTALUM + ΣXi ALLOYS
NORMAL SPECTRAL EMITTANCE
TEMPERATURE, K
1.0
0.9
0.8
0.7
0.6
0.5
0.4
0.3
0.2
0.1
0.0
1100
1200
1300
1400
1500
1600
1700
1800
1900
2000
2100
2200
2300
2400
2500
2600
1

SPECIFICATION TABLE NO. 414 NORMAL SPECTRAL EMITTANCE OF [NIOBIUM + TANTALUM + ΣX_i] ALLOYS

Curve No.	Ref. No.	Year	Wavelength μ	Temperature Range, K	Geometry θ'	Reported Error, %	Composition (weight percent), Specifications and Remarks
1	291	1963	0.66	1285-2532	~0°		0.4-0.6 Ta, 0.01-0.03 Mg, 0.007-0.009 Ti, 0.004-0.006 Fe, 0.001-0.003 Mn, 0.0004-0.0006 Cu, Nb balance; data extracted from smooth curve.

DATA TABLE NO. 414 NORMAL SPECTRAL EMITTANCE OF [NIOBIUM + TANTALUM + ΣX_i] ALLOYS

[Temperature, T, K; Emittance, ϵ; Wavelength, λ, μ]

T	ϵ
CURVE 1	
$\lambda = 0.66$	
1285	0.373
1597	0.360
1979	0.345
2234	0.335
2532	0.323

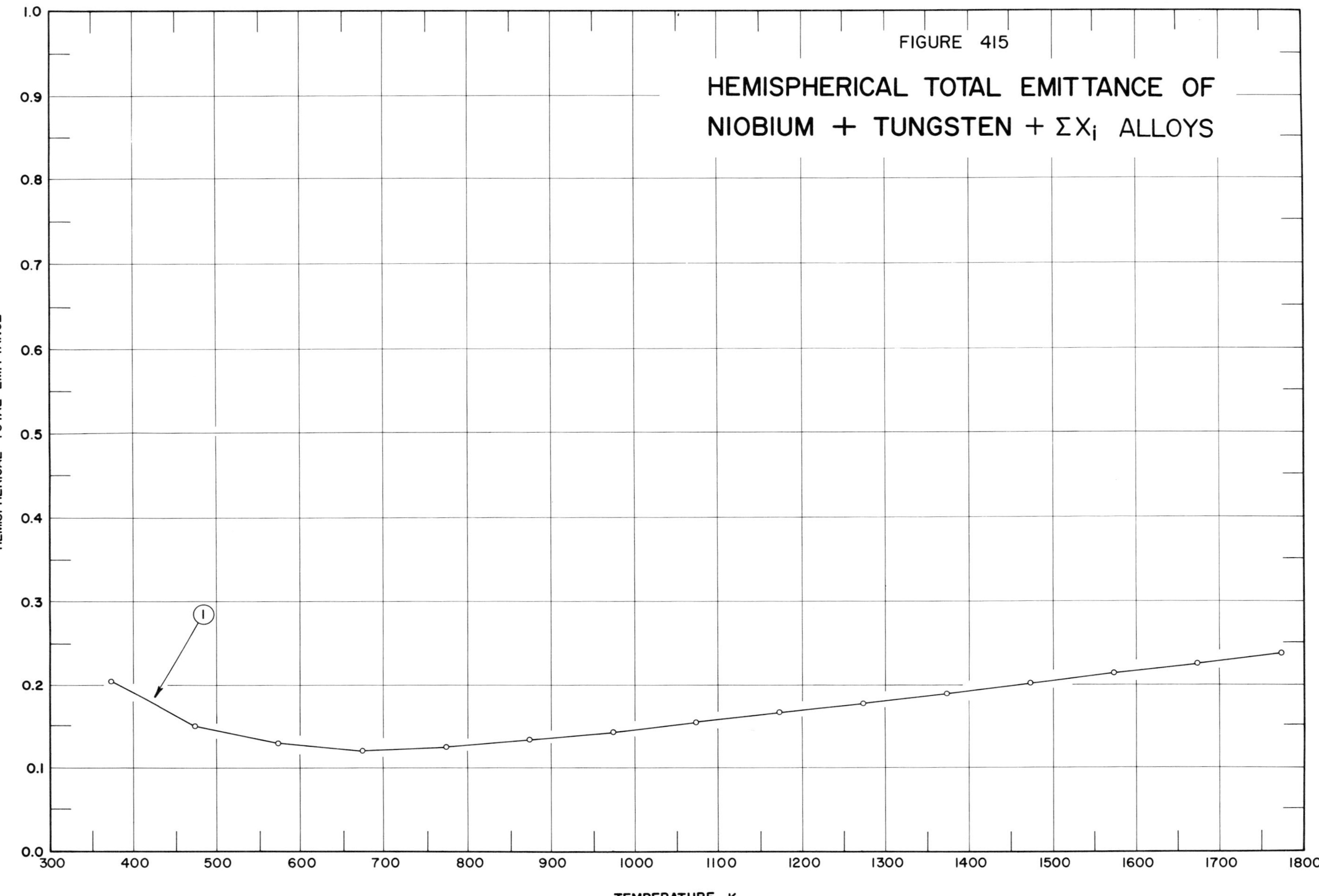

FIGURE 415
HEMISPHERICAL TOTAL EMITTANCE OF
NIOBIUM + TUNGSTEN + ΣXi ALLOYS
1
HEMISPHERICAL TOTAL EMITTANCE
1.0
0.9
0.8
0.7
0.6
0.5
0.4
0.3
0.2
0.1
0.0
300
400
500
600
700
800
900
1000
1100
1200
1300
1400
1500
1600
1700
1800
TEMPERATURE, K

SPECIFICATION TABLE NO. 415 HEMISPHERICAL TOTAL EMITTANCE OF [NIOBIUM + TUNGSTEN + ΣX_i] ALLOYS

Curve No.	Ref. No.	Year	Temperature Range, K	Reported Error, %	Composition (weight percent), Specifications and Remarks
1	255	1964	373-1773		D-43; 10 W, 1 Zr, 1 C, Nb balance; measured in vacuum (5×10^{-7} mm Hg) ; authors stated that data below 773 K is not valid.

DATA TABLE NO. 415 HEMISPHERICAL TOTAL EMITTANCE OF [NIOBIUM + TUNGSTEN + ΣX_i] ALLOYS

[Temperature, T, K; Emittance, ϵ]

T	ϵ
CURVE 1	
373	0.205
473	0.150
573	0.129
673	0.120
773	0.124
873	0.133
973	0.143
1073	0.154
1173	0.167
1273	0.178
1373	0.190
1473	0.202
1573	0.214
1673	0.226
1773	0.238

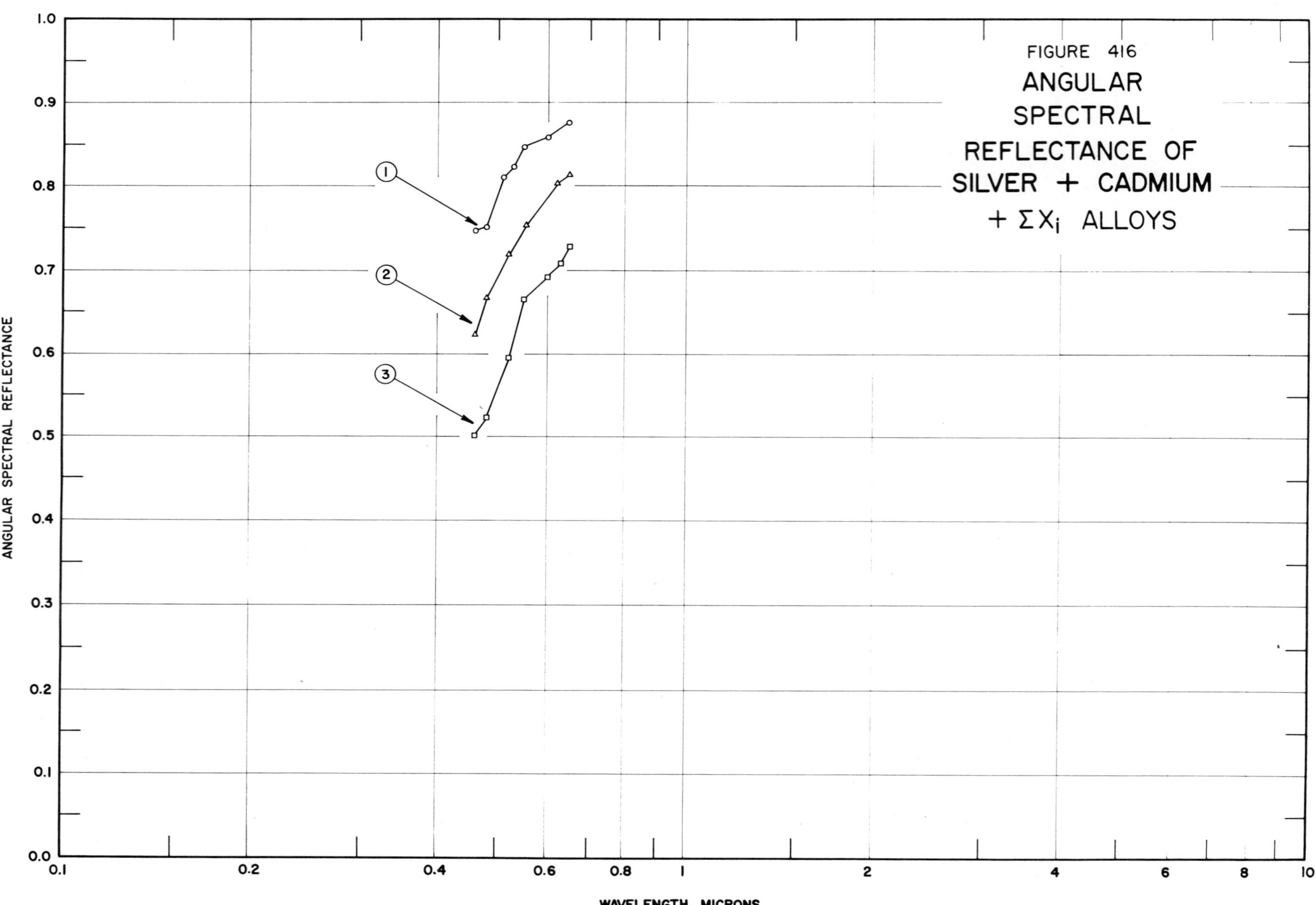
FIGURE 416
ANGULAR
SPECTRAL
REFLECTANCE OF
SILVER + CADMIUM
+ ΣXi ALLOYS
ANGULAR SPECTRAL REFLECTANCE
WAVELENGTH, MICRONS
1.0
0.9
0.8
0.7
0.6
0.5
0.4
0.3
0.2
0.1
0.0
0.1
0.2
0.4
0.6
0.8
1
2
4
6
8
10
1
2
3

SPECIFICATION TABLE NO. 416 ANGULAR SPECTRAL REFLECTANCE OF [SILVER + CADMIUM + ΣX_i] ALLOYS

Curve No.	Ref. No.	Year	Temperature K	Wavelength Range, μ	Geometry θ	θ'	ω'	Reported Error, %	Composition (weight percent), Specifications and Remarks
1	264	1939	298	0.46-0.65	45°	45°			91.97 Ag, 6.70 Cd, 1.33 Al; cold rolled; polished; measured immediately after polishing; chrome plated or rhodium plated mirror reference.
2	264	1939	298	0.46-0.65	45°	45°			Above specimen and conditions except measured after exposing for 21 days to an atmosphere containing little sulphur dioxide and moisture.
3	264	1939	298	0.46-0.65	45°	45°			Above specimen and conditions except 52 days additional exposure to the same atmosphere.

DATA TABLE NO. 416 ANGULAR SPECTRAL REFLECTANCE OF [SILVER + CADMIUM + ΣXi] ALLOYS

[Wavelength, λ, μ; Reflectance, ρ; Temperature, T, K]

λ	ρ
CURVE 1 T = 298	
0.46	0.747
0.48	0.751
0.51	0.810
0.53	0.823
0.55	0.849
0.60	0.860
0.65	0.877
CURVE 2 T = 298	
0.46	0.622
0.48	0.668
0.52	0.719
0.55	0.753
0.62	0.803
0.65	0.814
CURVE 3 T = 298	
0.46	0.501
0.48	0.522
0.52	0.596
0.55	0.667
0.60	0.692
0.63	0.709
0.65	0.729

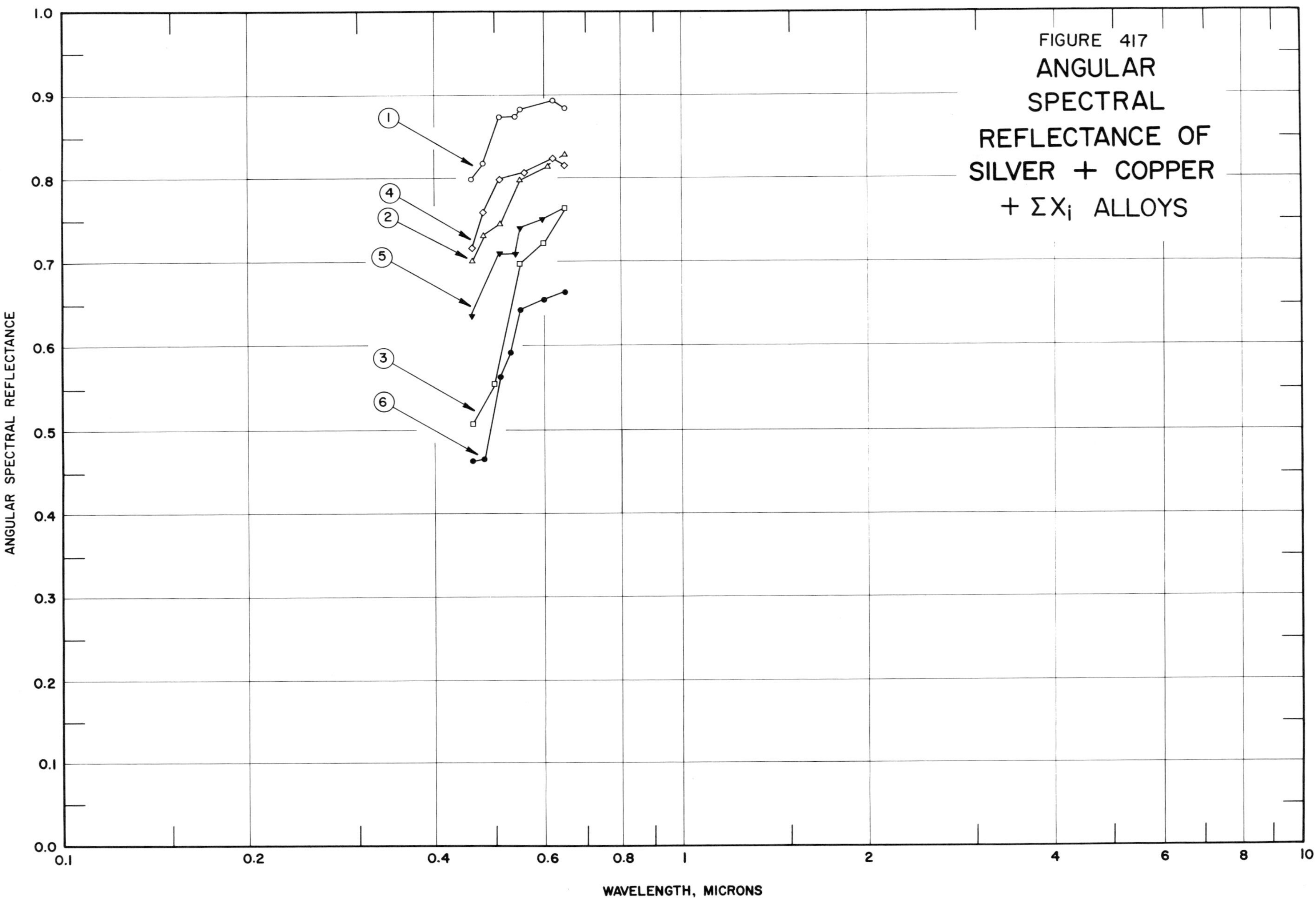

FIGURE 417 ANGULAR SPECTRAL REFLECTANCE OF SILVER + COPPER + ΣX_i ALLOYS

SPECIFICATION TABLE NO. 417 ANGULAR SPECTRAL REFLECTANCE OF [SILVER + COPPER + ΣX_i] ALLOYS

Curve No.	Ref. No.	Year	Temperature K	Wavelength Range, μ	Geometry θ	Geometry θ'	Geometry ω'	Reported Error, %	Composition (weight percent), Specifications and Remarks
1	264	1939	298	0.46-0.65	45°	45°			92.77 Ag, 5.86 Cu, 1.37 Al; cold rolled; polished; measured immediately after polishing; chrome plated or rhodium plated mirror reference.
2	264	1939	298	0.46-0.65	45°	45°			Above specimen and conditions except measured after exposing for 21 days to an atmosphere containing little sulphur dioxide and moisture.
3	264	1939	298	0.46-0.65	45°	45°			Above specimen and conditions except 52 days additional exposure to the same atmosphere.
4	264	1939	298	0.46-0.65	45°	45°			92.69 Ag, 6.51 Cu, 0.80 Be; cold rolled; polished; measured immediately after polishing; chrome plated or rhodium plated mirror reference.
5	264	1939	298	0.46-0.65	45°	45°			Above specimen and conditions except measured after exposing for 21 days to an atmosphere containing little sulphur dioxide and moisture.
6	264	1939	298	0.46-0.65	45°	45°			Above specimen and conditions except 52 days additional exposure to the same atmosphere.

DATA TABLE NO. 417 ANGULAR SPECTRAL REFLECTANCE OF [SILVER + COPPER + ΣX_i] ALLOYS

[Wavelength, λ, μ; Reflectance, ρ; Temperature, T, K]

λ	ρ	λ	ρ
CURVE 1 T = 298		**CURVE 5** T = 298	
0.46	0.800	0.46	0.637
0.48	0.819	0.51	0.709
0.51	0.873	0.54	0.709
0.54	0.875	0.55	0.739
0.55	0.883	0.60	0.750
0.62	0.894	0.65	0.764*
0.65	0.887		
CURVE 2 T = 298		**CURVE 6** T = 298	
0.46	0.701	0.46	0.462
0.48	0.732	0.48	0.465
0.51	0.747	0.51	0.563
0.55	0.799	0.53	0.592
0.61	0.815	0.55	0.642
0.65	0.830	0.60	0.655
		0.65	0.665
CURVE 3 T = 298			
0.46	0.509		
0.50	0.557		
0.55	0.699		
0.60	0.722		
0.65	0.764		
CURVE 4 T = 298			
0.46	0.716		
0.48	0.760		
0.51	0.799		
0.56	0.808		
0.62	0.825		
0.65	0.816		

* Not shown on plot

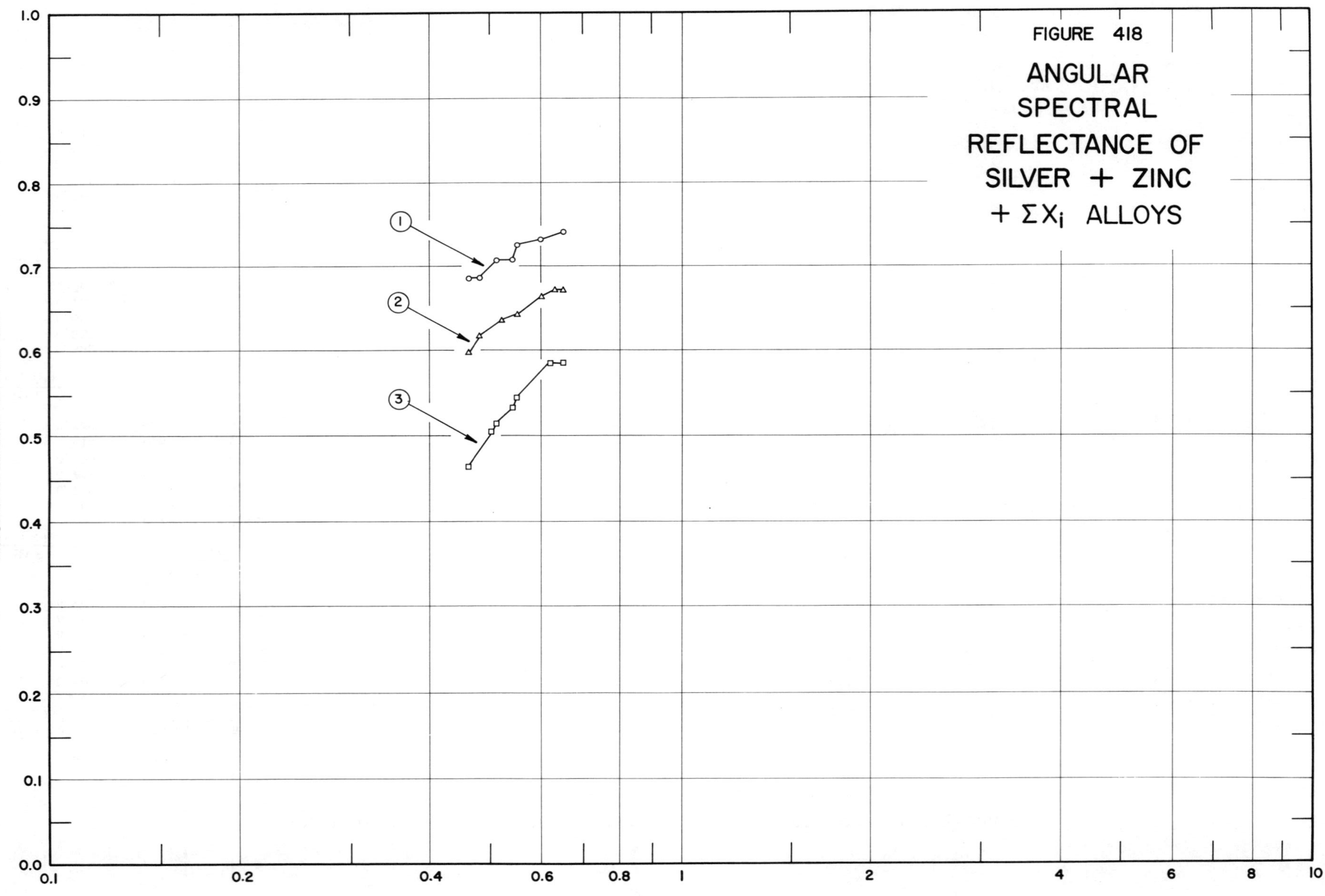
FIGURE 418
ANGULAR SPECTRAL REFLECTANCE OF SILVER + ZINC + ΣX_i ALLOYS
ANGULAR SPECTRAL REFLECTANCE
1.0
0.9
0.8
0.7
0.6
0.5
0.4
0.3
0.2
0.1
0.0
WAVELENGTH, MICRONS
0.1
0.2
0.4
0.6
0.8
1
2
4
6
8
10
1
2
3

SPECIFICATION TABLE NO. 418 ANGULAR SPECTRAL REFLECTANCE OF [SILVER + ZINC + ΣX_i] ALLOYS

Curve No.	Ref. No.	Year	Temperature K	Wavelength Range, μ	Geometry θ	Geometry θ'	Geometry ω'	Reported Error, %	Composition (weight percent), Specifications and Remarks
1	264	1939	298	0.46-0.65	45°	45°			92.57 Ag, 6.98 Zn, 0.45 Be; cold rolled; polished; measured immediately after polishing; chrome plated or rhodium plated mirror reference.
2	264	1939	298	0.46-0.65	45°	45°			Above specimen and conditions except measured after exposing for 21 days to an atmosphere containing little sulphur dioxide and moisture.
3	264	1939	298	0.46-0.65	45°	45°			Above specimen and conditions except 52 days additional exposure to the same atmosphere.

DATA TABLE NO. 418 ANGULAR SPECTRAL REFLECTANCE OF [SILVER + ZINC + ΣX_i] ALLOYS

[Wavelength, λ, μ; Reflectance, ρ; Temperature, T, K]

λ	ρ
CURVE 1	
T = 298	
0.46	0.686
0.48	0.686
0.51	0.707
0.54	0.708
0.55	0.726
0.60	0.732
0.65	0.741
CURVE 2	
T = 298	
0.46	0.599
0.48	0.619
0.52	0.638
0.55	0.645
0.60	0.664
0.63	0.672
0.65	0.672
CURVE 3	
T = 298	
0.46	0.466
0.50	0.507
0.51	0.516
0.54	0.533
0.55	0.547
0.62	0.587
0.65	0.587

SPECIFICATION TABLE NO. 419 HEMISPHERICAL TOTAL EMITTANCE OF [TANTALUM + TUNGSTEN + ΣX_i] ALLOYS

Curve No.	Ref. No.	Year	Temperature Range, K	Reported Error, %	Composition (weight percent), Specifications and Remarks
1	255	1964	773-1773		T-111; 8 W, 2 Hf, Ta balance; measured in vacuum (5×10^{-7} mm Hg).

DATA TABLE NO. 419 HEMISPHERICAL TOTAL EMITTANCE OF [TANTALUM + TUNGSTEN + ΣX_i] ALLOYS

[Temperature, T, K; Emittance, ∈]

T	∈
CURVE 1*	
773	0.081
873	0.096
973	0.111
1073	0.126
1173	0.141
1273	0.156
1373	0.170
1473	0.184
1573	0.199
1673	0.213
1773	0.227

* Not shown on plot

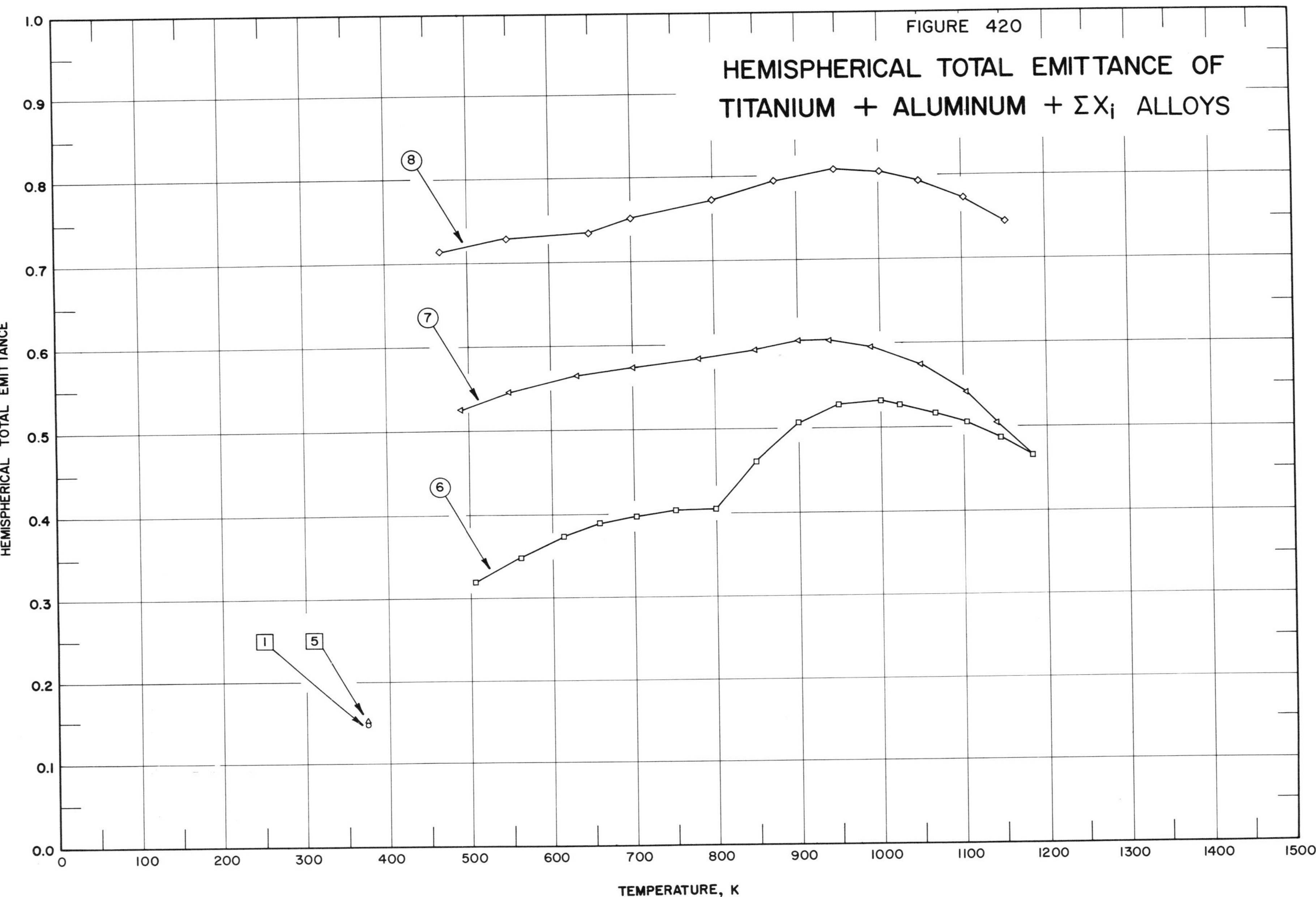

FIGURE 420

HEMISPHERICAL TOTAL EMITTANCE OF TITANIUM + ALUMINUM + ΣX_i ALLOYS

SPECIFICATION TABLE NO. 420 HEMISPHERICAL TOTAL EMITTANCE OF [TITANIUM + ALUMINUM + ΣX_i] ALLOYS

Curve No.	Ref. No.	Year	Temperature Range, K	Reported Error, %	Composition (weight percent), Specifications and Remarks
1	8	1963	373		6 Al, 4 V, Ti balance; prefinished with 600 grit silicon carbide paper; electropolished; measured in vacuum (10^{-5} mm Hg).
2	8	1963	373		Above specimen and conditions except ion bombarded (1.6×10^{20} ions/cm^2).
3	352	1963	373		Titanium alloy; 6 Al, 4 V, Ti balance; measured in vacuum (1×10^{-5} mm Hg).
4	352	1963	373		Above specimen and conditions except hydrogen ion bombarded (1.59×10^{20} ions-cm^{-2}).
5	352	1963	373		Above specimen and conditions except hydrogen ion bombarded (9.88×10^{20} ions-cm^{-2}).
6	299	1958	506-1183	> 6	TA 6 V; nominal composition: 5.5-7.0 Al, 3.5-4.5 V, 0.25 max Fe, 0.08 max C, Ti balance; rough surface; as received; data extracted from smooth curve.
7	299	1958	490-1182	> 6	Above specimen and conditions except preheated at 1150 K in air.
8	299	1958	468-1153	> 6	TA 5 E; rough surface; as received; data extracted from smooth curve.

DATA TABLE NO. 420 HEMISPHERICAL TOTAL EMITTANCE OF [TITANIUM + ALUMINUM + ΣX_i] ALLOYS

[Temperature, T, K; Emittance, ϵ]

T	ϵ
CURVE 1	
373	0.148
CURVE 2*	
373	0.147
CURVE 3*	
373	0.148
CURVE 4*	
373	0.147
CURVE 5	
373	0.151
CURVE 6	
506	0.320
561	0.350
613	0.373
658	0.389
701	0.398
750	0.404
799	0.404
849	0.461
900	0.507
950	0.528
1000	0.531
1023	0.527
1068	0.518
1102	0.506
1145	0.487
1183	0.466
CURVE 7	
490	0.527
549	0.546
632	0.564
700	0.574
781	0.584
CURVE 7 (cont.)	
849	0.596
901	0.605
939	0.605
989	0.597
1050	0.573
1102	0.540
1140	0.505
1182	0.466*
CURVE 8	
468	0.717
548	0.730
598	0.737
699	0.752
799	0.773
873	0.797
947	0.809
1001	0.806
1049	0.795
1102	0.773
1153	0.744

* Not shown on plot

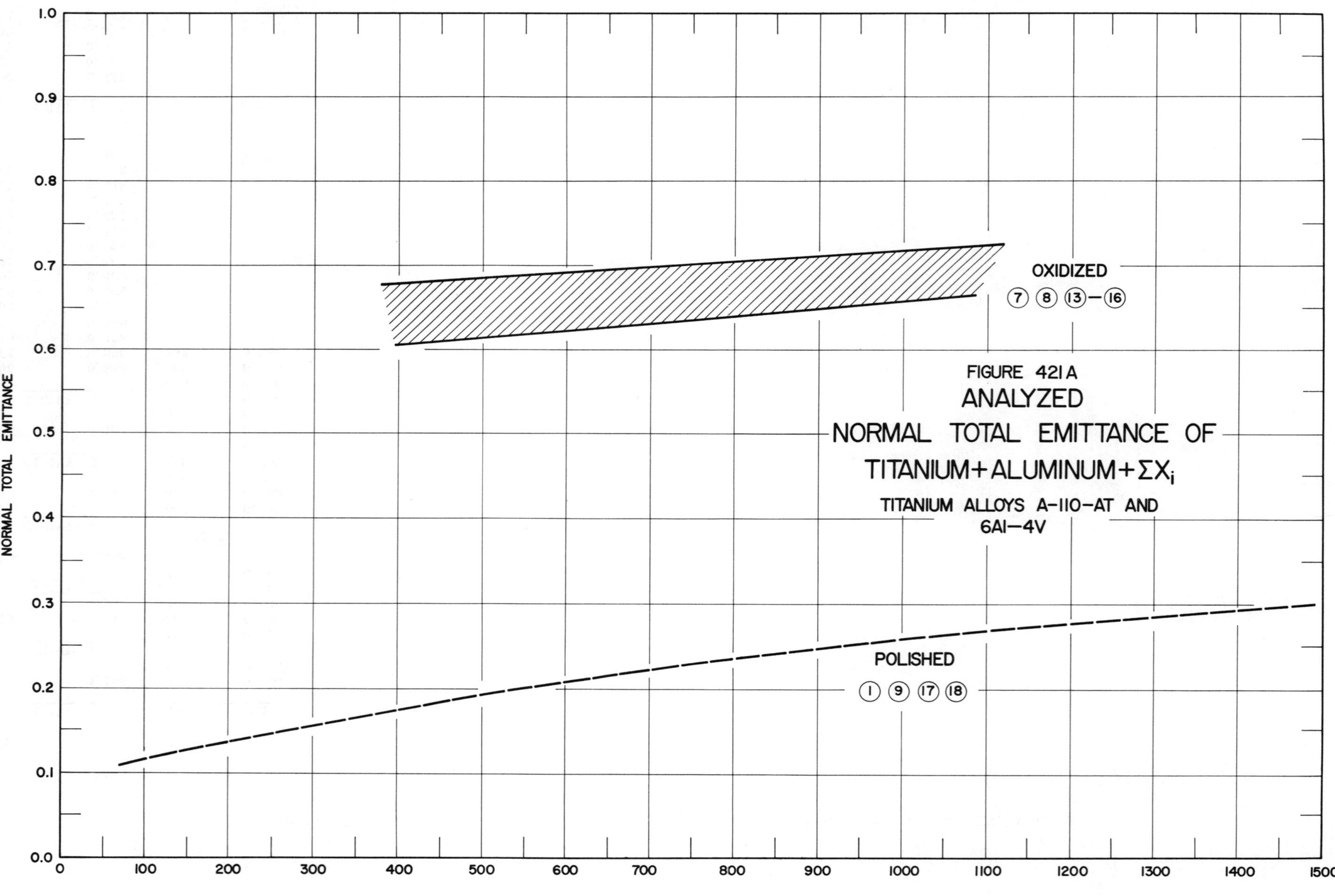

FIGURE 421A
ANALYZED
NORMAL TOTAL EMITTANCE OF
TITANIUM+ALUMINUM+ΣX_i
TITANIUM ALLOYS A-110-AT AND
6Al-4V

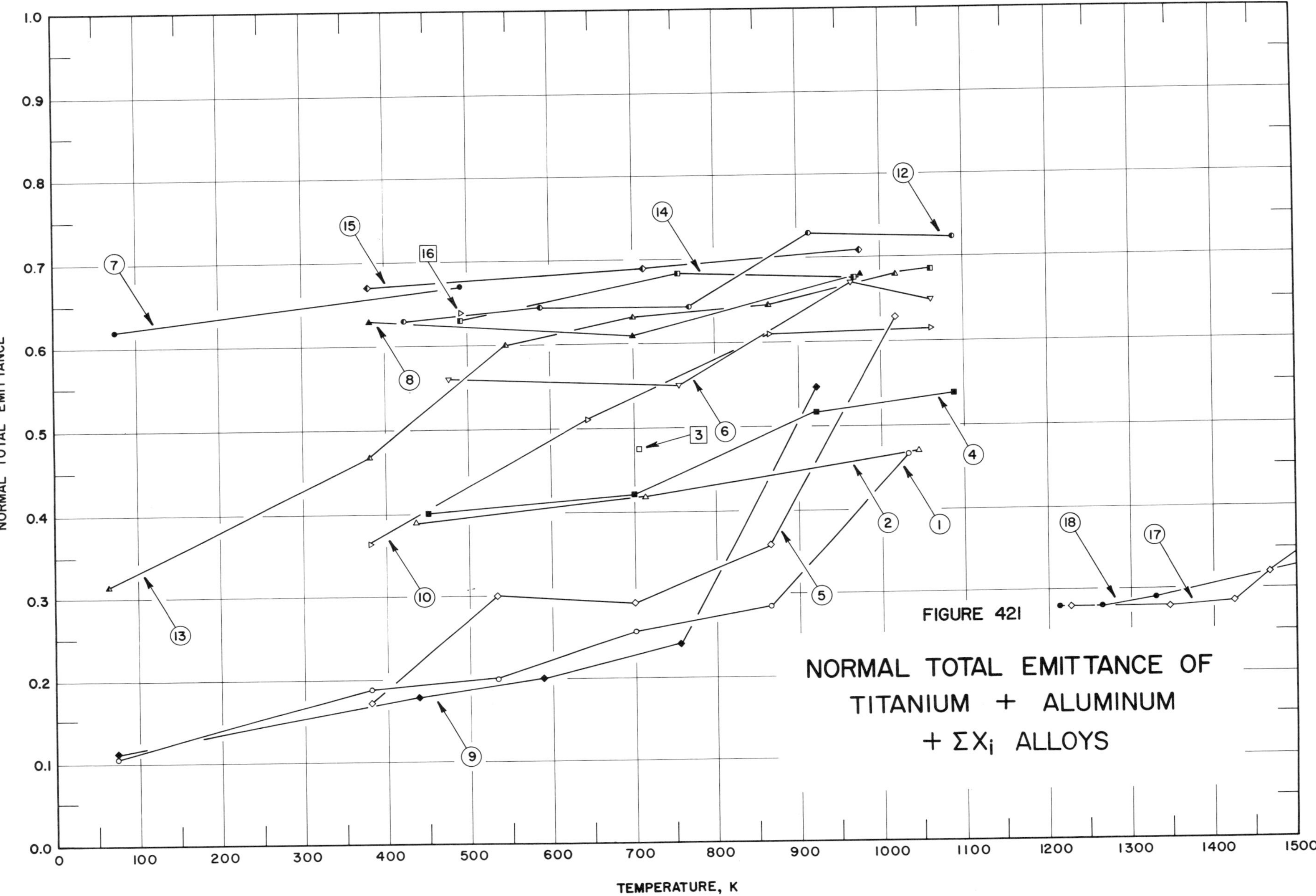

FIGURE 421

NORMAL TOTAL EMITTANCE OF TITANIUM + ALUMINUM + ΣX_i ALLOYS

SPECIFICATION TABLE NO. 421 NORMAL TOTAL EMITTANCE OF [TITANIUM + ALUMINUM + ΣX_i] ALLOYS

Curve No.	Ref. No.	Year	Temperature Range, K	Geometry θ'	Reported Error, %	Composition (weight percent), Specifications and Remarks
1	146	1958	74.8-1033	~0°		Titanium alloy A-110-AT; nominal composition: 5 Al, 2.5 Sn, Ti balance; polished; measured in air; increasing temp, cycle 1.
2	146	1958	436-1047	~0°		Above specimen and conditions; cycle 2.
3	146	1958	707	~0°		Above specimen and conditions; decreasing temp, cycle 2.
4	146	1958	450-1089	~0°		Above specimen and conditions; cycle 3.
5	146	1958	380-1019	~0°		Different sample, same as curve 1 specimen and conditions except oxidized at 922 K for 30 min.
6	146	1958	477-1061	~0°		Above specimen and conditions; cycle 2.
7	146	1958	491-74.8	~0°		Above specimen and conditions; decreasing temp, cycle 2.
8	146	1958	380-977	~0°		Above specimen and conditions; cycle 3.
9	146	1958	74.8-922	~0°		Titanium alloy 6 Al - 4 V; nominal composition: 6 Al, 4 V, Ti balance; polished; measured in air; increasing temp, cycle 1.
10	146	1958	380-1061	~0°		Above specimen and conditions; cycle 2.
11	146	1958	1061-380	~0°		Above specimen and conditions; decreasing temp, cycle 2.
12	146	1958	422-1089	~0°		Above specimen and conditions; cycle 3.
13	146	1958	74.8-1019	~0°		Different sample, same as curve 9 specimen and conditions except oxidized at 922 K for 30 min.
14	146	1958	491-1061	~0°		Above specimen and conditions; cycle 2.
15	146	1958	380-977	~0°		Above specimen and conditions; cycle 3.
16	146	1958	491	~0°		Above specimen and conditions; decreasing temp, cycle 3.
17	92	1963	1229-1566	~0°		Titanium Alloy 6 Al-4 V; 5.5-6.5 Al, 3.5-4.5 V, 0.1 max C, 0.3 max Fe, 0.05 max N_2, 0.0125 max H_2, 0.15 max O_2, Ti balance; polished; surface roughness 2 to 3 μ (RMS) measured with profilometer; measured in vacuum (3 to 4 x 10^{-4} mm Hg).
18	92	1963	1566-1215	~0°		Above specimen and conditions; descending temperature.

DATA TABLE NO. 421 NORMAL TOTAL EMITTANCE OF [TITANIUM + ALUMINUM + ΣX_i] ALLOYS

[Temperature, T, K; Emittance, ϵ]

T	ϵ	T	ϵ	T	ϵ
CURVE 1		CURVE 8		CURVE 14	
74.8	0.104	380	0.63	491	0.630
380	0.188	700	0.61	755	0.682
533	0.200	977	0.68	969	0.676
700	0.255			1061	0.684
866	0.284	CURVE 9			
1033	0.464			CURVE 15	
		74.8	0.110		
CURVE 2		436	0.178	380	0.670
		589	0.200	714	0.690
436	0.390	755	0.240	977	0.710
714	0.418	922	0.545		
1047	0.470			CURVE 16	
		CURVE 10			
CURVE 3				491	0.640
		380	0.364		
707	0.474	644	0.510	CURVE 17	
		866	0.610		
CURVE 4		1061	0.616	1229	0.280
				1349	0.280
450	0.400	CURVE 11*		1426	0.285
700	0.421			1469	0.320
922	0.518	1061	0.616	1513	0.330*
1089	0.538	380	0.632	1566	0.320*
CURVE 5		CURVE 12		CURVE 18	
380	0.17	422	0.630	1566	0.320*
533	0.30	589	0.645	1331	0.290
700	0.29	769	0.643	1266	0.280
866	0.36	914	0.730	1215	0.280
1019	0.63	1089	0.724		
CURVE 6		CURVE 13			
477	0.56	74.8	0.314		
755	0.55	380	0.468		
964	0.67	547	0.600		
1061	0.65	700	0.632		
		866	0.645		
CURVE 7		1019	0.680		
491	0.67				
74.8	0.62				

* Not shown on plot

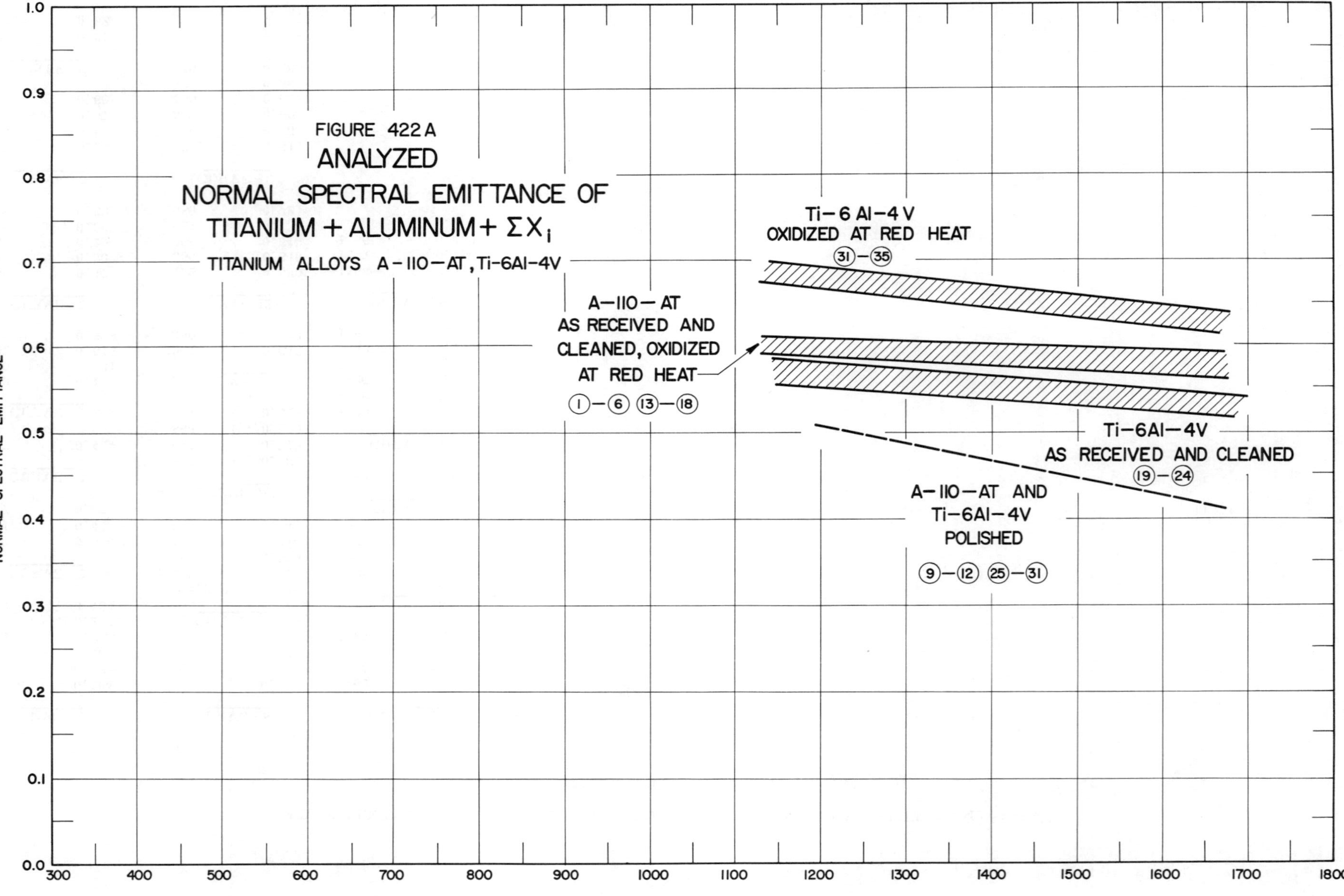

FIGURE 422 A
ANALYZED
NORMAL SPECTRAL EMITTANCE OF
TITANIUM + ALUMINUM + ΣX_i
TITANIUM ALLOYS A-110-AT, Ti-6Al-4V
Ti-6 Al-4 V
OXIDIZED AT RED HEAT
31-35
A-110-AT
AS RECEIVED AND
CLEANED, OXIDIZED
AT RED HEAT
1-6 13-18
Ti-6Al-4V
AS RECEIVED AND CLEANED
19-24
A-110-AT AND
Ti-6Al-4V
POLISHED
9-12 25-31
NORMAL SPECTRAL EMITTANCE
0.0
0.1
0.2
0.3
0.4
0.5
0.6
0.7
0.8
0.9
1.0
TEMPERATURE, K
300
400
500
600
700
800
900
1000
1100
1200
1300
1400
1500
1600
1700
1800

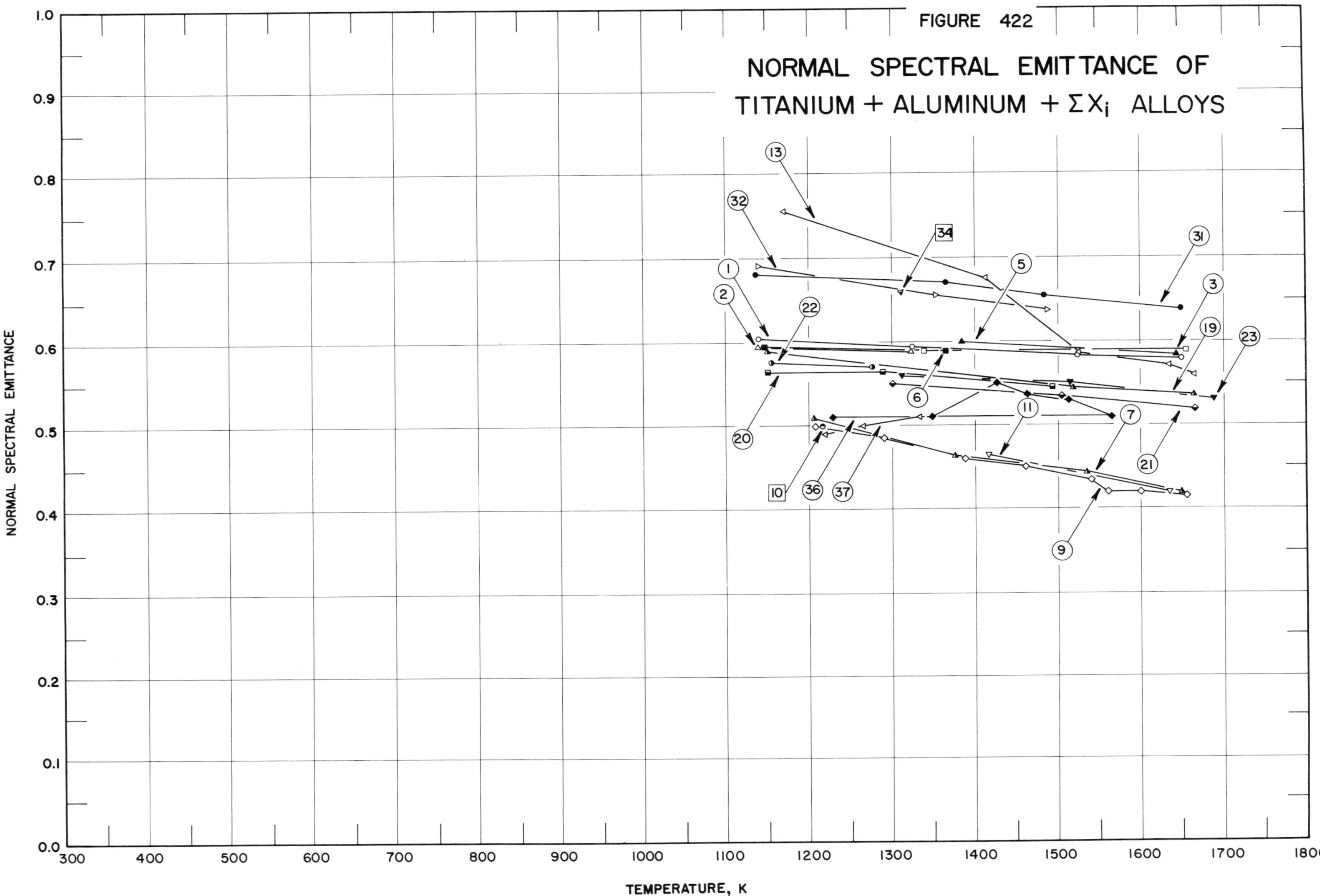
FIGURE 422
NORMAL SPECTRAL EMITTANCE OF
TITANIUM + ALUMINUM + ΣXi ALLOYS
NORMAL SPECTRAL EMITTANCE
TEMPERATURE, K
1.0
0.9
0.8
0.7
0.6
0.5
0.4
0.3
0.2
0.1
0.0
300
400
500
600
700
800
900
1000
1100
1200
1300
1400
1500
1600
1700
1800
13
32
34
31
1
5
3
2
22
19
23
6
11
7
20
21
10
36
37
9

SPECIFICATION TABLE NO. 422 NORMAL SPECTRAL EMITTANCE OF [TITANIUM + ALUMINUM + ΣX_i] ALLOYS

Curve No.	Ref. No.	Year	Wavelength μ	Temperature Range, K	Geometry θ'	Reported Error, %	Composition (weight percent), Specifications and Remarks
1	34	1957	0.665	1139-1650	~0°		Titanium alloy A-110-AT; nominal composition: 5 Al, 2.5 Sn, Ti balance; as received and cleaned with liquid detergent; measured in vacuum (5 x 10^{-4} mm Hg); increasing temp, cycle 1.
2	34	1957	0.665	1322-1139	~0°		Above specimen and conditions; decreasing temp, cycle 1.
3	34	1957	0.665	1339-1655	~0°		Above specimen and conditions; cycle 2.
4	34	1957	0.665	1533-1133	~0°		Above specimen and conditions; decreasing temp, cycle 2.
5	34	1957	0.665	1383-1644	~0°		Above specimen and conditions; cycle 3.
6	34	1957	0.665	1366-1147	~0°		Above specimen and conditions; decreasing temp, cycle 3.
7	34	1957	0.665	1205-1650	~0°		Different sample, same as curve 1 specimen and conditions except polished.
8	34	1957	0.665	1216	~0°		Above specimen and conditions; decreasing temp, cycle 1.
9	34	1957	0.665	1208-1655	~0°		Above specimen and conditions; cycle 2.
10	34	1957	0.665	1216	~0°		Above specimen and conditions; decreasing temp, cycle 2.
11	34	1957	0.665	1416-1633	~0°		Above specimen and conditions; cycle 3.
12	34	1957	0.665	1452-1236	~0°		Above specimen and conditions; decreasing temp, cycle 3.
13	34	1957	0.665	1171-1666	~0°		Different sample, same as curve 1 specimen and conditions except oxidized in air at red heat for 30 min.
14	34	1957	0.665	1325-1133	~0°		Above specimen and conditions; decreasing temp, cycle 1.
15	34	1957	0.665	1325-1633	~0°		Above specimen and conditions; cycle 2.
16	34	1957	0.665	1347-1133	~0°		Above specimen and conditions; decreasing temp, cycle 2.
17	34	1957	0.665	1333-1655	~0°		Above specimen and conditions; cycle 3.
18	34	1957	0.665	1514-1155	~0°		Above specimen and conditions; decreasing temp, cycle 3.
19	34	1957	0.665	1150-1666	~0°		Titanium alloy Ti-6Al-4V; nominal composition: 6 Al, 4 V, Ti balance; as received and cleaned with liquid detergent; measured in vacuum (5 x 10^{-4} mm Hg); increasing temp, cycle 1.
20	34	1957	0.665	1494-1150	~0°		Above specimen and conditions; decreasing temp, cycle 1.
21	34	1957	0.665	1300-1666	~0°		Above specimen and conditions; cycle 2.
22	34	1957	0.665	1277-1155	~0°		Above specimen and conditions; decreasing temp, cycle 2.
23	34	1957	0.665	1311-1689	~0°		Above specimen and conditions; cycle 3.
24	34	1957	0.665	1480-1150	~0°		Above specimen and conditions; decreasing temp, cycle 3.
25	34	1957	0.665	1166-1694	~0°		Different sample, same as curve 19 specimen and conditions except polished.
26	34	1957	0.665	1439-1166	~0°		Above specimen and conditions; decreasing temp, cycle 1.

SPECIFICATION TABLE NO. 422 (continued)

Curve No.	Ref. No.	Year	Wavelength μ	Temperature Range, K	Geometry θ'	Reported Error, %	Composition (weight percent), Specifications and Remarks
27	34	1957	0.665	1333-1689	~0°		Above specimen and conditions; cycle 2.
28	34	1957	0.665	1500-1177	~0°		Above specimen and conditions; decreasing temp, cycle 2.
29	34	1957	0.665	1677	~0°		Above specimen and conditions; cycle 3.
30	34	1957	0.665	1194	~0°		Above specimen and conditions; decreasing temp, cycle 3.
31	34	1957	0.665	1136-1650	~0°		Different sample, same as curve 19 specimen and conditions except oxidized in air at red heat for 30 min.
32	34	1957	0.665	1489-1139	~0°		Above specimen and conditions; decreasing temp, cycle 1.
33	34	1957	0.665	1294-1650	~0°		Above specimen and conditions; cycle 2.
34	34	1957	0.665	1311	~0°		Above specimen and conditions; decreasing temp, cycle 2.
35	34	1957	0.665	1600-1158	~0°		Above specimen and conditions; decreasing temp, cycle 3.
36	92	1963	0.65	1229-1566	~0°		Titanium Alloy 6 Al - 4 V; 5.5-6.5 Al, 3.5-4.5 V, 0.1 max C, 0.3 max Fe, 0.05 max N_2, 0.0125 max H_2, 0.15 max O_2, Ti balance; polished; surface roughness 2 to 3 μ (RMS) measured with profilometer; measured in vacuum (3 to 4 x 10^{-4} mm Hg); authors assumed specimen was a grey body.
37	92	1963	0.65	1566-1218	~0°		Above specimen and conditions; descending temperature.

DATA TABLE NO. 422 NORMAL SPECTRAL EMITTANCE OF [TITANIUM + ALUMINUM + ΣX_i] ALLOYS

[Temperature, T, K; Emittance, ϵ; Wavelength, λ, μ]

T	ϵ
CURVE 1 $\lambda = 0.665$	
1139	0.605
1325	0.595
1525	0.585
1650	0.580
CURVE 2 $\lambda = 0.665$	
1322	0.590
1139	0.595
CURVE 3 $\lambda = 0.665$	
1339	0.590
1655	0.590
CURVE 4* $\lambda = 0.665$	
1533	0.590
1133	0.585
CURVE 5 $\lambda = 0.665$	
1383	0.600
1644	0.585
CURVE 6 $\lambda = 0.665$	
1366	0.590
1147	0.595
CURVE 7 $\lambda = 0.665$	
1205	0.510
1377	0.465
1536	0.445
1650	0.420
CURVE 8* $\lambda = 0.665$	
1216	0.515
CURVE 9 $\lambda = 0.665$	
1208	0.500
1291	0.485
1389	0.460
1461	0.450
1539	0.435
1561	0.420
1600	0.420
1655	0.415
CURVE 10 $\lambda = 0.665$	
1216	0.500
CURVE 11 $\lambda = 0.665$	
1416	0.465
1633	0.420
CURVE 12* $\lambda = 0.655$	
1452	0.455
1236	0.510
CURVE 13 $\lambda = 0.665$	
1171	0.755
1414	0.675
1527	0.585
1636	0.570
1666	0.560
CURVE 14* $\lambda = 0.665$	
1325	0.600
1133	0.595
CURVE 15* $\lambda = 0.665$	
1325	0.585
1455	0.580
1633	0.575
CURVE 16* $\lambda = 0.665$	
1347	0.590
1133	0.605
CURVE 17* $\lambda = 0.665$	
1333	0.590
1533	0.595
1655	0.570
CURVE 18* $\lambda = 0.665$	
1514	0.580
1155	0.600
CURVE 19 $\lambda = 0.665$	
1150	0.590
1519	0.545
1666	0.535
CURVE 20 $\lambda = 0.665$	
1494	0.545
1289	0.565
1150	0.565
CURVE 21 $\lambda = 0.665$	
1300	0.550
1505	0.535
1666	0.520
CURVE 22 $\lambda = 0.665$	
1277	0.570
1155	0.575
CURVE 23 $\lambda = 0.665$	
1311	0.560
1516	0.550
1689	0.530
CURVE 24* $\lambda = 0.665$	
1480	0.565
1150	0.550
CURVE 25* $\lambda = 0.665$	
1166	0.500
1311	0.485
1511	0.445
1658	0.425
1694	0.405
CURVE 26* $\lambda = 0.665$	
1439	0.455
1244	0.500
1166	0.510
CURVE 27* $\lambda = 0.665$	
1333	0.485
1491	0.460
1689	0.420
CURVE 28* $\lambda = 0.665$	
1500	0.440
1177	0.500
CURVE 29* $\lambda = 0.665$	
1677	0.425
CURVE 30* $\lambda = 0.665$	
1194	0.510
CURVE 31 $\lambda = 0.665$	
1136	0.680
1366	0.670
1483	0.655
1650	0.640
CURVE 32 $\lambda = 0.665$	
1489	0.640
1352	0.655
1139	0.690
CURVE 33* $\lambda = 0.665$	
1294	0.670
1500	0.640
1650	0.635
CURVE 34 $\lambda = 0.665$	
1311	0.660
CURVE 35* $\lambda = 0.665$	
1600	0.620
1158	0.670
CURVE 36 $\lambda = 0.65$	
1229	0.510
1349	0.510
1426	0.550
1464	0.535
1513	0.530
1566	0.510
CURVE 37 $\lambda = 0.65$	
1566	0.510*
1333	0.510
1266	0.500
1218	0.490

* Not shown on plot

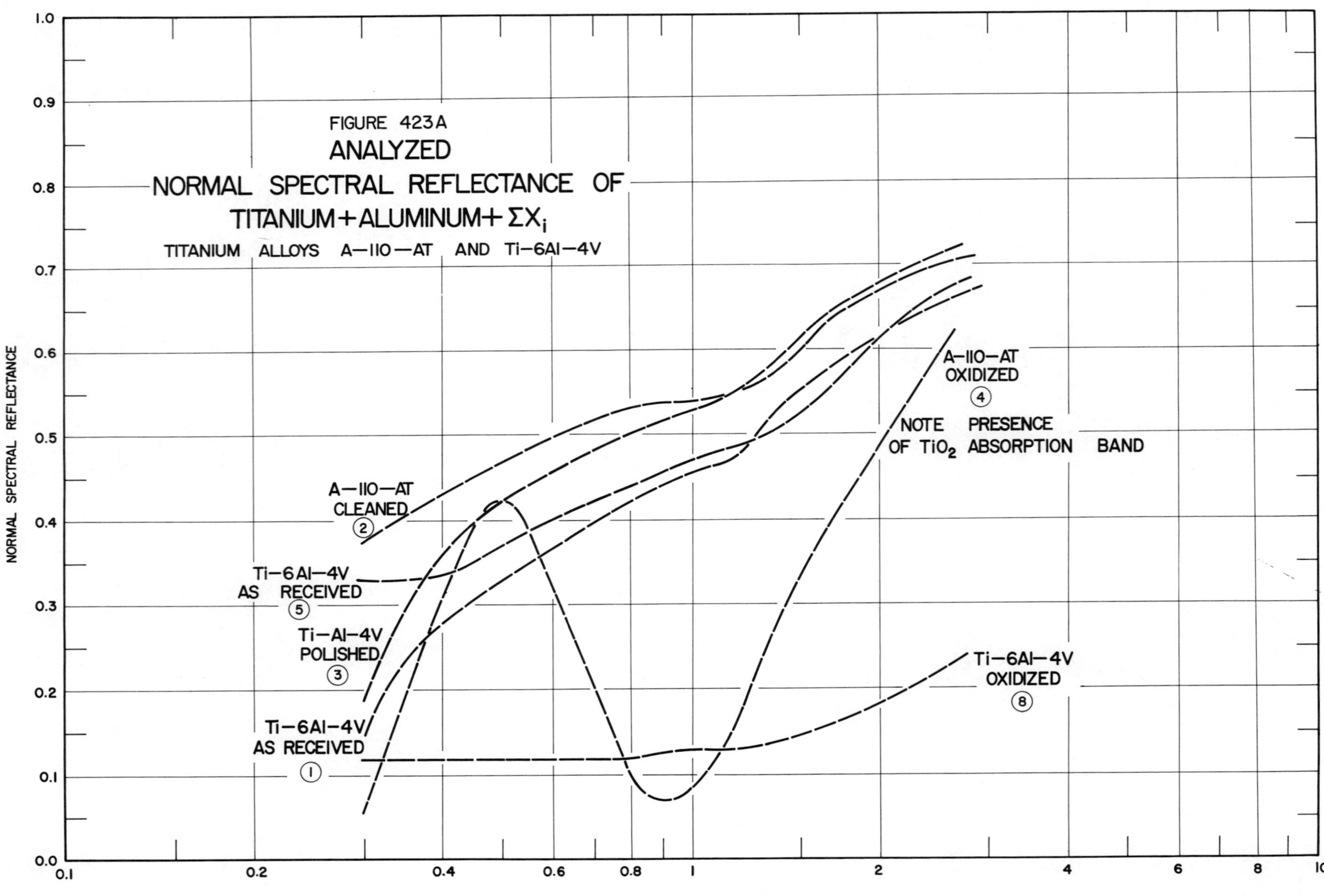
FIGURE 423A
ANALYZED
NORMAL SPECTRAL REFLECTANCE OF
TITANIUM+ALUMINUM+ΣX_i
TITANIUM ALLOYS A-110-AT AND Ti-6Al-4V
A-110-AT CLEANED (2)
Ti-6Al-4V AS RECEIVED (5)
Ti-Al-4V POLISHED (3)
Ti-6Al-4V AS RECEIVED (1)
A-110-AT OXIDIZED (4)
NOTE PRESENCE OF TiO_2 ABSORPTION BAND
Ti-6Al-4V OXIDIZED (8)
NORMAL SPECTRAL REFLECTANCE
WAVELENGTH, MICRONS
0.0 0.1 0.2 0.3 0.4 0.5 0.6 0.7 0.8 0.9 1.0
0.1 0.2 0.4 0.6 0.8 1 2 4 6 8 10

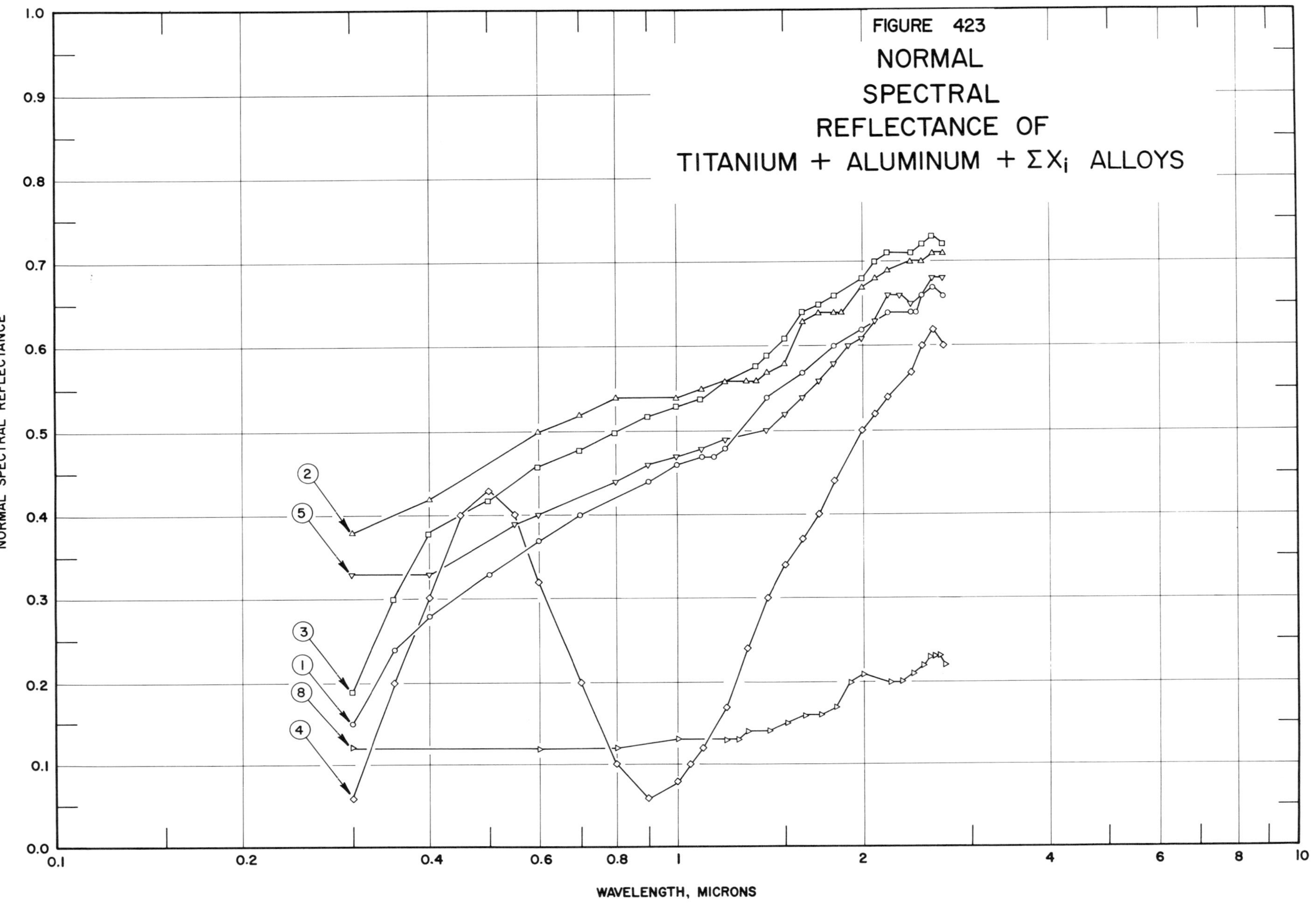
FIGURE 423
NORMAL
SPECTRAL
REFLECTANCE OF
TITANIUM + ALUMINUM + ΣX$_i$ ALLOYS
NORMAL SPECTRAL REFLECTANCE
1.0
0.9
0.8
0.7
0.6
0.5
0.4
0.3
0.2
0.1
0.0
0.1
0.2
0.4
0.6
0.8
1
2
4
6
8
10
WAVELENGTH, MICRONS
2
5
3
1
8
4

SPECIFICATION TABLE NO. 423 NORMAL SPECTRAL REFLECTANCE OF [TITANIUM + ALUMINUM + ΣX_i] ALLOYS

Curve No.	Ref. No.	Year	Temperature K	Wavelength Range, μ	Geometry θ	Geometry θ' ω'	Reported Error, %	Composition (weight percent), Specifications and Remarks
1	146	1958	298	0.30-2.70	9°	2π		Titanium alloy A-110-AT; nominal composition: 5 Al, 2.5 Sn, Ti balance; as received.
2	146	1958	298	0.30-2.70	9°	2π		Different sample, same as curve 1 specimen and conditions except cleaned.
3	146	1958	298	0.30-2.70	9°	2π		Different sample, same as curve 1 specimen and conditions except polished.
4	146	1958	298	0.30-2.70	9°	2π		Different sample, same as curve 1 specimen and conditions except oxidized at 922 K for 30 min.
5	146	1958	298	0.30-2.70	9°	2π		Titanium alloy Ti-6Al-4V; nominal composition: 6 Al, 4 V, Ti balance; as received.
6	146	1958	298	0.30-2.70	9°	2π		Different sample, same as curve 5 specimen and conditions except cleaned.
7	146	1958	298	0.30-2.75	9°	2π		Different sample, same as curve 5 specimen and conditions except polished.
8	146	1958	298	0.30-2.70	9°	2π		Different sample, same as curve 5 specimen and conditions except oxidized at 922 K for 30 min.

DATA TABLE NO. 423 NORMAL SPECTRAL REFLECTANCE OF [TITANIUM + ALUMINUM + ΣX_i] ALLOYS

[Wavelength, λ, μ; Reflectance, ρ; Temperature, T, K]

λ	ρ
CURVE 1	**T = 298**
0.30	0.15
0.35	0.24
0.40	0.28
0.50	0.33
0.60	0.37
0.70	0.40
0.90	0.44
1.00	0.46
1.10	0.47
1.15	0.47
1.20	0.48
1.40	0.54
1.60	0.57
1.80	0.60
2.00	0.62
2.20	0.64
2.40	0.64
2.45	0.64
2.50	0.66
2.60	0.67
2.70	0.66
CURVE 2	**T = 298**
0.30	0.38
0.40	0.42
0.60	0.50
0.70	0.52
0.80	0.54
1.00	0.54
1.10	0.55
1.20	0.56
1.30	0.56
1.35	0.56
1.40	0.57
1.50	0.58
1.60	0.63
1.70	0.64
1.80	0.64
1.85	0.64
2.00	0.67
CURVE 2 (cont.)	
2.10	0.68
2.20	0.69
2.40	0.70
2.50	0.70
2.60	0.71
2.70	0.71
CURVE 3	**T = 298**
0.30	0.19
0.35	0.30
0.40	0.38
0.50	0.42
0.60	0.46
0.70	0.48
0.80	0.50
0.90	0.52
1.00	0.53
1.10	0.54
1.20	0.56*
1.35	0.58
1.40	0.59
1.50	0.61
1.60	0.64
1.70	0.65
1.80	0.66
2.00	0.68
2.10	0.70
2.20	0.71
2.40	0.71
2.50	0.72
2.60	0.73
2.70	0.72
CURVE 4	**T = 298**
0.30	0.06
0.35	0.20
0.40	0.30
0.45	0.40
0.50	0.43
CURVE 4 (cont.)	
0.55	0.40
0.60	0.32
0.70	0.20
0.80	0.10
0.90	0.06
1.00	0.08
1.05	0.10
1.10	0.12
1.20	0.17
1.30	0.24
1.40	0.30
1.50	0.34
1.60	0.37
1.70	0.40
1.80	0.44
2.00	0.50
2.10	0.52
2.20	0.54
2.40	0.57
2.50	0.60
2.60	0.62
2.70	0.60
CURVE 5	**T = 298**
0.30	0.33
0.40	0.33
0.55	0.39
0.60	0.40
0.80	0.44
0.90	0.46
1.00	0.47
1.10	0.48
1.20	0.49
1.40	0.50
1.50	0.52
1.60	0.54
1.70	0.56
1.80	0.58
1.90	0.60
2.00	0.61
2.10	0.63
CURVE 5 (cont.)	
2.20	0.66
2.30	0.66
2.40	0.65
2.50	0.66*
2.60	0.68
2.70	0.68
CURVE 6*	**T = 298**
0.30	0.38
0.40	0.42
0.50	0.45
0.60	0.47
0.70	0.49
0.80	0.50
0.90	0.51
1.00	0.52
1.20	0.54
1.40	0.55
1.50	0.56
1.60	0.59
1.65	0.60
1.70	0.61
1.80	0.62
2.00	0.64
2.10	0.66
2.20	0.67
2.30	0.67
2.40	0.67
2.50	0.68
2.60	0.70
2.70	0.69
CURVE 7*	**T = 298**
0.30	0.34
0.40	0.36
0.50	0.40
0.70	0.48
0.80	0.52
0.90	0.53
CURVE 7* (cont.)	
1.00	0.54
1.20	0.56
1.40	0.58
1.50	0.60
1.60	0.62
1.70	0.63
1.80	0.64
1.90	0.66
2.00	0.67
2.10	0.68
2.20	0.70
2.30	0.70
2.40	0.71
2.50	0.72
2.60	0.73
2.70	0.72
2.75	0.72
CURVE 8	**T = 298**
0.30	0.12
0.60	0.12
0.80	0.12
1.00	0.13
1.20	0.13
1.25	0.13
1.30	0.14
1.40	0.14
1.50	0.15
1.60	0.16
1.70	0.16
1.80	0.17
1.90	0.20
2.00	0.21
2.20	0.20
2.30	0.20
2.40	0.21
2.50	0.22
2.55	0.23
2.60	0.23
2.65	0.23
2.70	0.22

* Not shown on plot

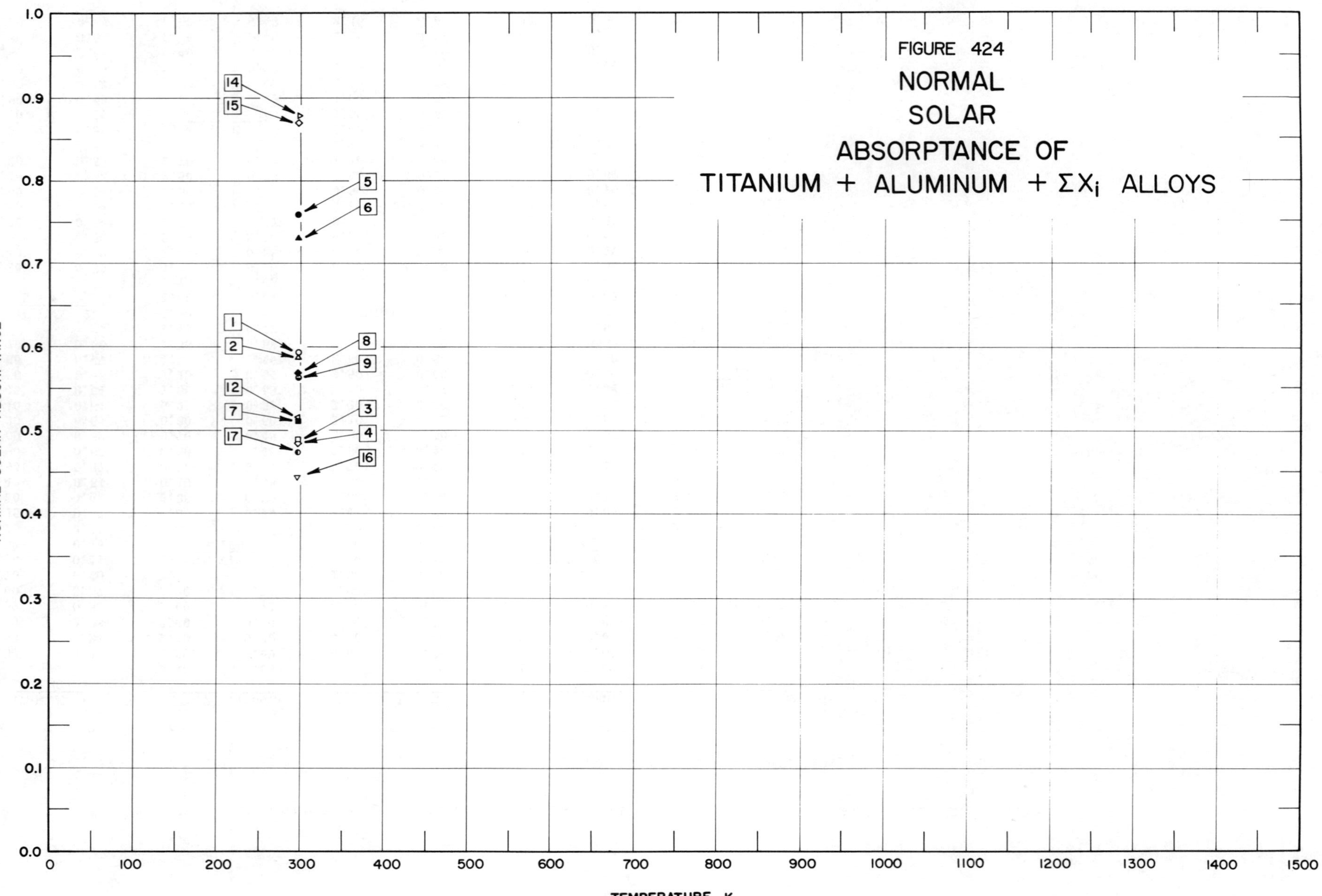
FIGURE 424
NORMAL
SOLAR
ABSORPTANCE OF
TITANIUM + ALUMINUM + ΣXi ALLOYS
NORMAL SOLAR ABSORPTANCE
TEMPERATURE, K
1.0
0.9
0.8
0.7
0.6
0.5
0.4
0.3
0.2
0.1
0.0
0
100
200
300
400
500
600
700
800
900
1000
1100
1200
1300
1400
1500
14
15
5
6
1
2
8
9
12
7
3
4
17
16

SPECIFICATION TABLE NO. 424 NORMAL SOLAR ABSORPTANCE OF [TITANIUM + ALUMINUM + ΣX_i] ALLOYS

Curve No.	Ref. No.	Year	Temperature Range, K	Geometry θ	Reported Error, %	Composition (weight percent), Specifications and Remarks
1	146	1958	298	9°		Titanium alloy A-110-AT; nominal composition: 5 Al, 2.5 Sn, Ti balance; as received; computed for spectral reflectance data for sea level conditions.
2	146	1958	298	9°		Above specimen and conditions except computed for above atmosphere conditions.
3	146	1958	298	9°		Different sample, same as curve 1 specimen and conditions except cleaned.
4	146	1958	298	9°		Above specimen and conditions except computed for above atmosphere conditions.
5	146	1958	298	9°		Different sample, same as curve 1 specimen and conditions except oxidized.
6	146	1958	298	9°		Above specimen and conditions except computed for above atmosphere conditions.
7	146	1958	298	9°		Different sample, same as curve 5 specimen and conditions except polished.
8	146	1958	298	9°		Titanium alloy Ti-6Al-4V; nominal composition: 6 Al, 4 V, Ti balance; as received; computed from spectral reflectance data for sea level conditions.
9	146	1958	298	9°		Above specimen and conditions except computed for above atmosphere conditions.
10	146	1958	298	9°		Different sample, same as curve 8 specimen and conditions except cleaned.
11	146	1958	298	9°		Above specimen and conditions except computed for above atmosphere conditions.
12	146	1958	298	9°		Different sample, same as curve 8 specimen and conditions except polished.
13	146	1958	298	9°		Above specimen and conditions except computed for above atmosphere conditions.
14	146	1958	298	9°		Different sample, same as curve 8 specimen and conditions except oxidized.
15	146	1958	298	9°		Above specimen and conditions except computed for above atmosphere conditions.
16	352	1963	298	~0°		Titanium alloy; 6Al, 4 V, Ti balance; computed from spectral reflectance.
17	352	1963	298	~0°		Above specimen and conditions except hydrogen ion bombarded (0.11 x 10^{20} ions-cm^{-2}).
18	352	1963	298	~0°		Above specimen and conditions except hydrogen ion bombarded (0.47 x 10^{20} ions-cm^{-2}).
19	352	1963	298	~0°		Above specimen and conditions except hydrogen ion bombarded (1.27 x 10^{20} ions-cm^{-2}).
20	352	1963	298	~0°		Above specimen and conditions except hydrogen ion bombarded (3.07 x 10^{20} ions-cm^{-2}).
21	352	1963	298	~0°		Above specimen and conditions except hydrogen ion bombarded (7.70 x 10^{20} ions-cm^{-2}).
22	352	1963	298	~0°		Above specimen and conditions except hydrogen ion bombarded (9.87 x 10^{20} ions-cm^{-2}).

DATA TABLE NO. 424 NORMAL SOLAR ABSORPTANCE OF [TITANIUM + ALUMINUM + ΣX_i] ALLOYS

[Temperature, T, K; Absorptance, α]

T	α	T	α
CURVE 1		CURVE 13*	
298	0.592	298	0.511
CURVE 2		CURVE 14	
298	0.588	298	0.878
CURVE 3		CURVE 15	
298	0.489	298	0.869
CURVE 4		CURVE 16	
298	0.486	298	0.444
CURVE 5		CURVE 17	
298	0.759	298	0.474
CURVE 6		CURVE 18*	
298	0.730	298	0.507
CURVE 7		CURVE 19*	
298	0.511	298	0.515
CURVE 8		CURVE 20*	
298	0.568	298	0.512
CURVE 9		CURVE 21*	
298	0.563	298	0.512
CURVE 10*		CURVE 22*	
298	0.509	298	0.511
CURVE 11*			
298	0.508		
CURVE 12			
298	0.515		

* Not shown on plot

SPECIFICATION TABLE NO. 425 HEMISPHERICAL TOTAL EMITTANCE OF [TITANIUM + MANGANESE + ΣX_i] ALLOYS

Curve No.	Ref. No.	Year	Temperature Range, K	Reported Error, %	Composition (weight percent), Specifications and Remarks
1	164	1961	589-1089	± 2.5	Titanium alloy RS-120; nominal composition: 94.170 Ti, 5.700 Mn, 0.106 C, 0.022 Ni; polished and cleaned; oxidized in quiescent air at 1089 K for 80 min; measured in air.

DATA TABLE NO. 425 HEMISPHERICAL TOTAL EMITTANCE OF [TITANIUM + MANGANESE + ΣX_i] ALLOYS

[Temperature, T, K; Emittance, ϵ]

T	ϵ
CURVE 1*	
589	0.660
644	0.670
700	0.680
755	0.691
811	0.694
866	0.693
922	0.700
978	0.710
1033	0.712
1089	0.715

* Not shown on plot

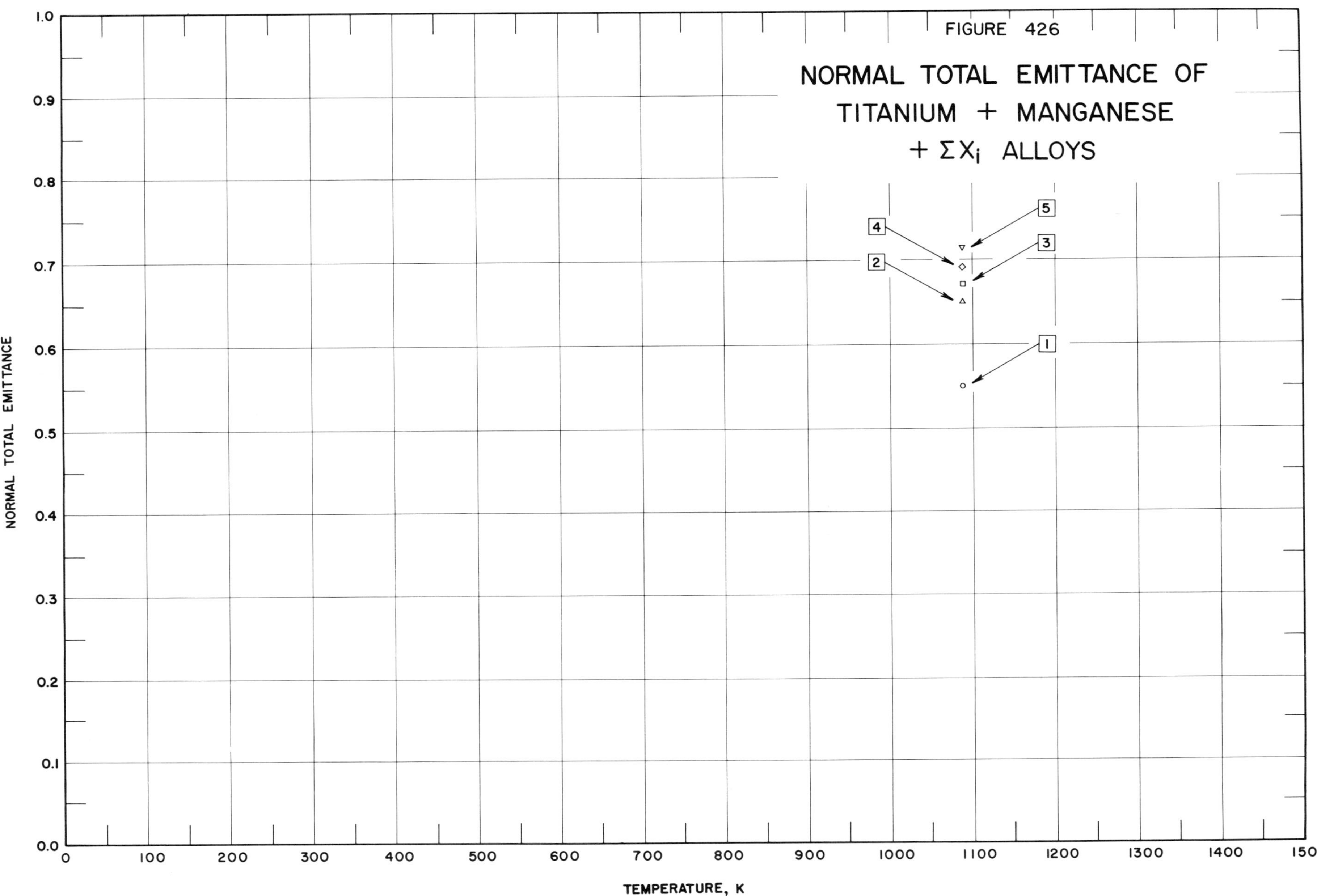

FIGURE 426
NORMAL TOTAL EMITTANCE OF
TITANIUM + MANGANESE
+ ΣX_i ALLOYS
5
4
3
2
1
NORMAL TOTAL EMITTANCE
1.0
0.9
0.8
0.7
0.6
0.5
0.4
0.3
0.2
0.1
0.0
TEMPERATURE, K
0
100
200
300
400
500
600
700
800
900
1000
1100
1200
1300
1400
1500

SPECIFICATION TABLE NO. 426 NORMAL TOTAL EMITTANCE OF [TITANIUM + MANGANESE + ΣX_i] ALLOYS

Curve No.	Ref. No.	Year	Temperature Range, K	Geometry θ'	Reported Error, %	Composition (weight percent), Specifications and Remarks
1	164	1961	1089	~0°	± 2.5	Titanium alloy RS-120; nominal composition: 94.170 Ti, 5.700 Mn, 0.106 C, 0.022 Ni; polished and cleaned; oxidized in quiescent air at 1089 K for 5 min; measured in air.
2	164	1961	1089	~0°	± 2.5	Above specimen and conditions except oxidized in quiescent air at 1089 K for 20 min.
3	164	1961	1089	~0°	± 2.5	Above specimen and conditions except oxidized in quiescent air at 1089 K for 35 min.
4	164	1961	1089	~0°	± 2.5	Above specimen and conditions except oxidized in quiescent air at 1089 K for 50 min.
5	164	1961	1089	~0°	± 2.5	Above specimen and conditions except oxidized in quiescent air at 1089 K for 65 min.
6	164	1961	1089	~0°	± 2.5	Above specimen and conditions except oxidized in quiescent air at 1089 K for 80 min.

DATA TABLE NO. 426 NORMAL TOTAL EMITTANCE OF [TITANIUM + MANGANESE + ΣX_i] ALLOYS

[Temperature, T, K; Emittance, ϵ]

T	ϵ
CURVE 1	
1089	0.550
CURVE 2	
1089	0.650
CURVE 3	
1089	0.670
CURVE 4	
1089	0.690
CURVE 5	
1089	0.715
CURVE 6*	
1089	0.716

* Not shown on plot

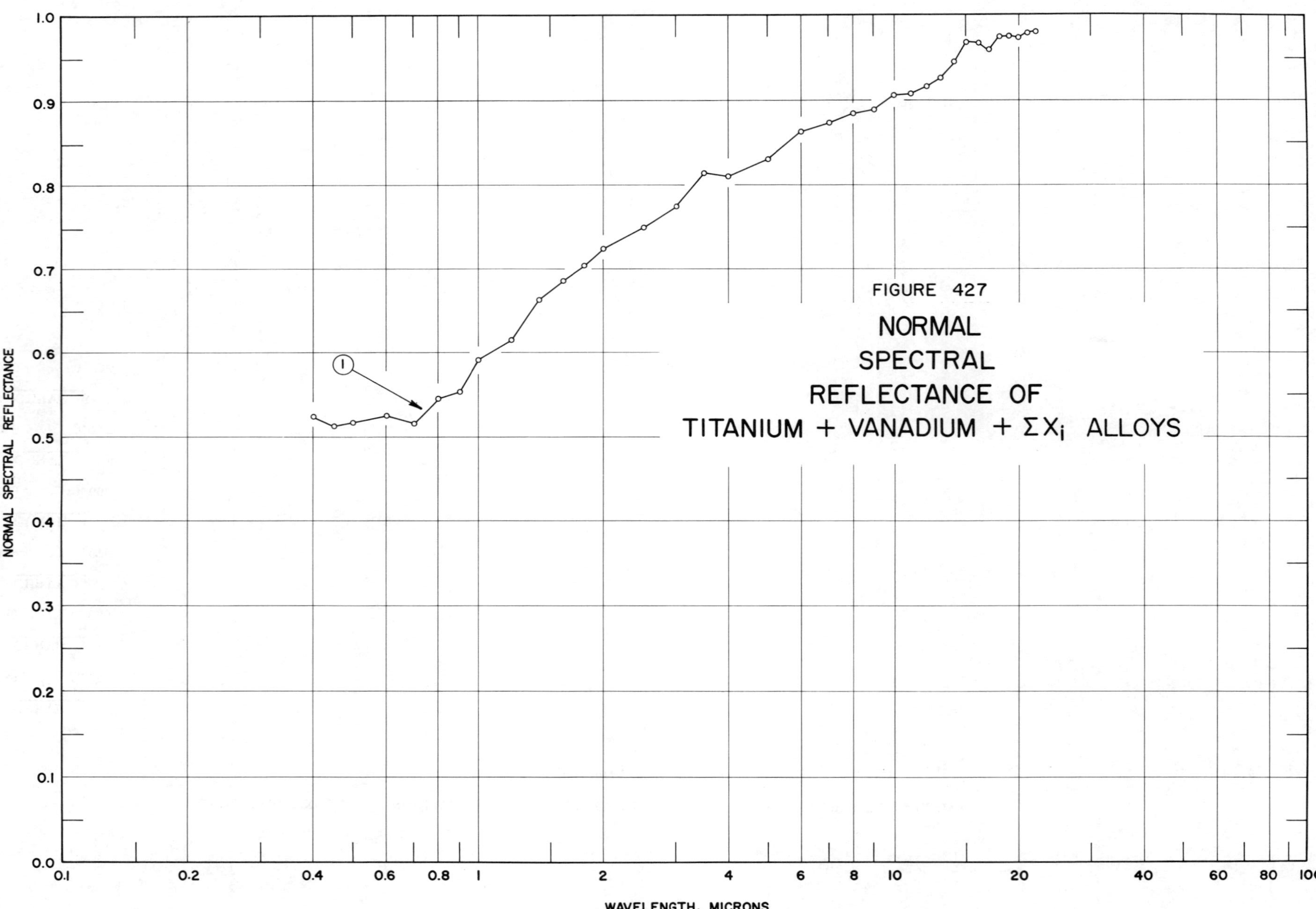

FIGURE 427
NORMAL SPECTRAL REFLECTANCE OF TITANIUM + VANADIUM + ΣX_i ALLOYS

SPECIFICATION TABLE NO. 427 NORMAL SPECTRAL REFLECTANCE OF [TITANIUM + VANADIUM + ΣX_i] ALLOYS

Curve No.	Ref. No.	Year	Temperature K	Wavelength Range, μ	Geometry θ	Geometry θ'	Geometry ω'	Reported Error, %	Composition (weight percent), Specifications and Remarks
1	122	1961	298	0.4-22	~0°		2π		Titanium B120 VAC; nominal composition: 73 Ti, 13 V, 11 Cr, and 3 Al; mechanically polished and electropolished; measured in vacuum (~10^{-6} mm Hg); [Author's designation: Specimen No. 202].

DATA TABLE NO. 427 NORMAL SPECTRAL REFLECTANCE OF [TITANIUM + VANADIUM + ΣX_i] ALLOYS

[Temperature, T, K; Reflectance, ρ; Wavelength, λ, μ]

λ	ρ
CURVE 1	
T = 298	
0.4	0.522
0.45	0.512
0.5	0.517
0.6	0.524
0.7	0.515
0.8	0.543
0.9	0.552
1.0	0.591
1.2	0.616
1.4	0.662
1.6	0.686
1.8	0.703
2.0	0.726
2.5	0.750
3.0	0.777
3.5	0.815
4	0.811
5	0.831
6	0.863
7	0.873
8	0.886
9	0.890
10	0.903
11	0.907
12	0.916
13	0.927
14	0.946
15	0.969
16	0.968
17	0.959
18	0.974
19	0.977
20	0.976
21	0.980
22	0.981

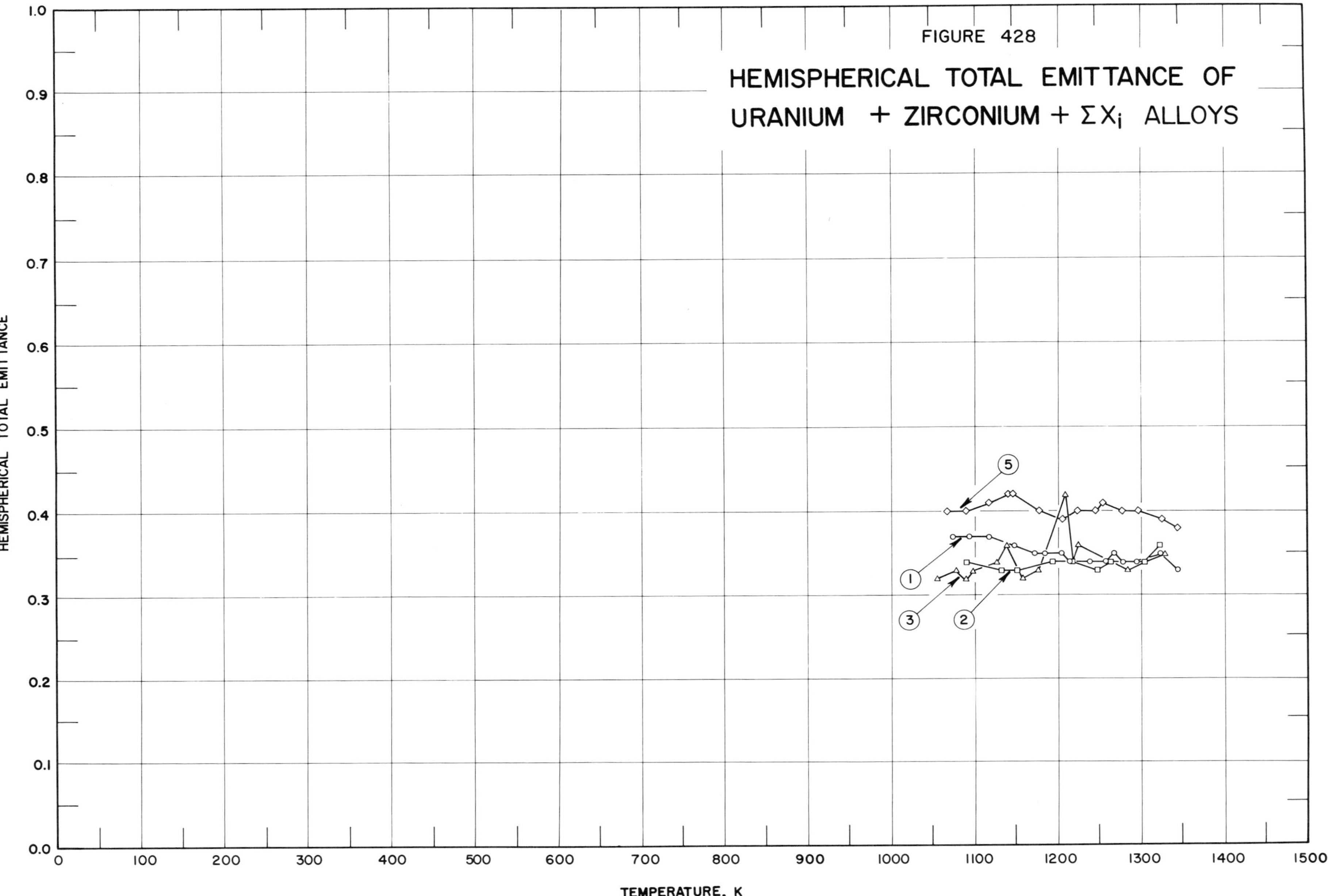
FIGURE 428
HEMISPHERICAL TOTAL EMITTANCE OF
URANIUM + ZIRCONIUM + ΣX_i ALLOYS
HEMISPHERICAL TOTAL EMITTANCE
TEMPERATURE, K
1.0
0.9
0.8
0.7
0.6
0.5
0.4
0.3
0.2
0.1
0.0
0
100
200
300
400
500
600
700
800
900
1000
1100
1200
1300
1400
1500
5
1
3
2

SPECIFICATION TABLE NO. 428 HEMISPHERICAL TOTAL EMITTANCE OF [URANIUM + ZIRCONIUM + ΣX_i] ALLOYS

Curve No.	Ref. No.	Year	Temperature Range, K	Reported Error, %	Composition (weight percent), Specifications and Remarks
1	116	1957	1073-1343		67.5 U, 24.55 Zr, 7.5 Nb, 0.375 Sn, 0.0375 Fe, 0.025 Cr, 0.0125 Ni; annealed in vacuum for 1 hr at 1008 K and water quenched then heated in vacuum for 4 hrs at 873 K; run 1, increasing temp; [Author's designation: Specimen 9].
2	116	1957	1090-1321		Above specimen and conditions; run 1, decreasing temp.
3	116	1957	1053-1328		Above specimen and conditions; run 2, increasing temp.
4	116	1957	1098-1275		Above specimen and conditions; run 2, decreasing temp.
5	116	1957	1068-1343		Above specimen and conditions except oxygen added to the specimen by adding 6 individual charges of oxygen (2.5 cm^3 at atmospheric pressure per charge) to the vacuum chamber while the specimen was at 1073 K; chamber re-evacuated; run 1, increasing temp.
6	116	1957	1090-1313		Above specimen and conditions; run 1, decreasing temp.

DATA TABLE NO. 428 HEMISPHERICAL TOTAL EMITTANCE OF [URANIUM + ZIRCONIUM + ΣX_i] ALLOYS

[Temperature, T, K; Emittance, ϵ]

T	ϵ	T	ϵ
CURVE 1		CURVE 4*	
1073	0.37	1098	0.33
1093	0.37	1125	0.34
1118	0.37	1163	0.33
1148	0.36	1199	0.33
1171	0.35	1243	0.33
1185	0.35	1275	0.34
1203	0.35		
1211	0.34	CURVE 5	
1238	0.34		
1258	0.34	1068	0.40
1268	0.35	1090	0.40
1278	0.34	1118	0.41
1295	0.34	1140	0.42
1321	0.35	1146	0.42
1343	0.33	1178	0.40
		1203	0.39
CURVE 2		1223	0.40
		1240	0.40
1090	0.34	1253	0.41
1131	0.33	1278	0.40
1150	0.33	1298	0.40
1193	0.34	1324	0.39
1219	0.34	1343	0.38
1249	0.33		
1263	0.34	CURVE 6*	
1303	0.34		
1321	0.36	1090	0.34
		1124	0.35
CURVE 3		1163	0.36
		1203	0.36
1053	0.32	1234	0.36
1078	0.33	1264	0.37
1090	0.32	1293	0.36
1098	0.33	1313	0.36
1127	0.34		
1138	0.36		
1158	0.32		
1178	0.33		
1210	0.42		
1219	0.34*		
1225	0.36		
1283	0.33		
1303	0.34*		
1328	0.35		

* Not shown on plot

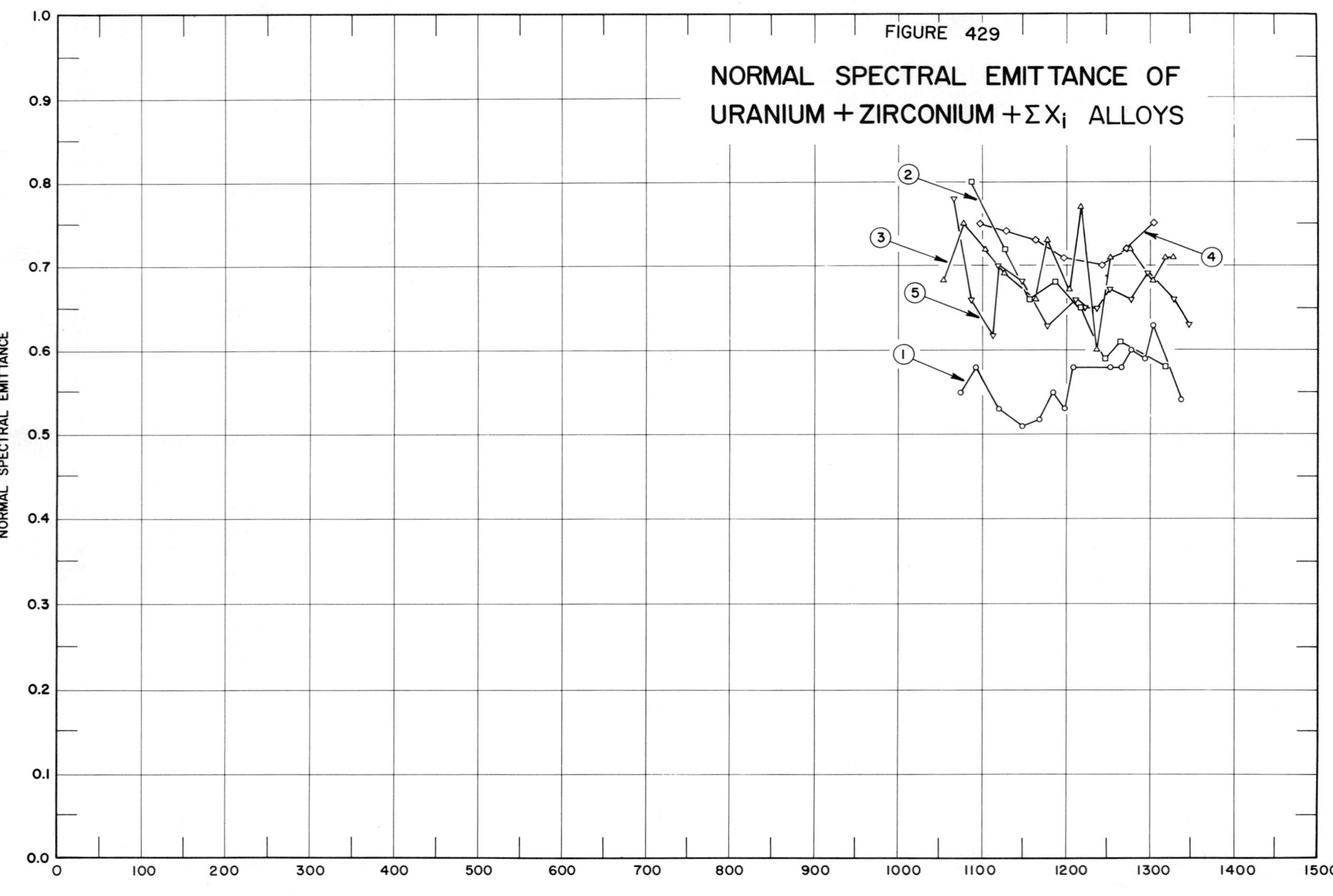
FIGURE 429
NORMAL SPECTRAL EMITTANCE OF
URANIUM + ZIRCONIUM + Σ Xi ALLOYS
1
2
3
4
5
NORMAL SPECTRAL EMITTANCE
1.0
0.9
0.8
0.7
0.6
0.5
0.4
0.3
0.2
0.1
0.0
0
100
200
300
400
500
600
700
800
900
1000
1100
1200
1300
1400
1500
TEMPERATURE, K

SPECIFICATION TABLE NO. 429 NORMAL SPECTRAL EMITTANCE OF [URANIUM + ZIRCONIUM + ΣX_i] ALLOYS

Curve No.	Ref. No.	Year	Wavelength μ	Temperature Range, K	Geometry θ'	Reported Error, %	Composition (weight percent), Specifications and Remarks
1	116	1957	0.65	1073-1338	$\sim 0^\circ$		67.5 U, 24.55 Zr, 7.5 Nb, 0.375 Sn, 0.0375 Fe, 0.025 Cr, 0.0125 Ni; annealed in vacuum for 1 hr at 1008 K and water quenched, then heated in vacuum for 4 hrs at 873 K; run 1, increasing temp; [Author's designation: Specimen 9].
2	116	1957	0.65	1088-1319	$\sim 0^\circ$		Above specimen and conditions; run 1, decreasing temp.
3	116	1957	0.65	1053-1329	$\sim 0^\circ$		Above specimen and conditions; run 2, increasing temp.
4	116	1957	0.65	1098-1303	$\sim 0^\circ$		Above specimen and conditions; run 2, decreasing temp.
5	116	1957	0.65	1068-1348	$\sim 0^\circ$		Above specimen and conditions except oxygen added to the specimen by adding 6 individual charges of oxygen (2.5 cm^3 at atmospheric pressure per charge) to the vacuum chamber while the specimen was at 1073 K; chamber re-evacuated; run 1, increasing temp.
6	116	1957	0.65	1098-1313	$\sim 0^\circ$		Above specimen and conditions; run 1, decreasing temp.

DATA TABLE NO. 429 NORMAL SPECTRAL EMITTANCE OF [URANIUM + ZIRCONIUM + ΣX_i] ALLOYS

[Temperature, T, K; Emittance, ∈; Wavelength, λ, μ]

T	∈	T	∈
CURVE 1		CURVE 4	
λ = 0.65		λ = 0.65	
1073	0.55	1098	0.75
1093	0.58	1129	0.74
1120	0.53	1163	0.73
1147	0.51	1198	0.71
1168	0.52	1243	0.70
1183	0.55	1273	0.72
1198	0.53	1303	0.75
1209	0.58		
1253	0.58	CURVE 5	
1266	0.58	λ = 0.65	
1278	0.60		
1293	0.59	1068	0.78
1303	0.63	1088	0.66
1338	0.54	1113	0.62
		1118	0.70
CURVE 2		1148	0.68
λ = 0.65		1178	0.63
		1211	0.66
1088	0.80	1223	0.65
1128	0.72	1238	0.65
1158	0.66	1253	0.67
1188	0.68	1278	0.66
1218	0.65	1298	0.69
1249	0.59	1328	0.66
1264	0.61	1348	0.63
1319	0.58		
		CURVE 6*	
CURVE 3		λ = 0.65	
λ = 0.65			
		1098	0.69
1053	0.68	1126	0.64
1078	0.75	1163	0.68
1103	0.72	1198	0.64
1127	0.69	1240	0.61
1161	0.66	1264	0.68
1178	0.73	1288	0.62
1203	0.67	1313	0.65
1218	0.77		
1236	0.60		
1253	0.71		
1278	0.72		
1303	0.68		
1318	0.71		
1329	0.71		

* Not shown on plot

SPECIFICATION TABLE NO. 430 NORMAL SPECTRAL EMITTANCE OF [ZIRCONIUM + HAFNIUM + ΣX_i] ALLOYS

Curve No.	Ref. No.	Year	Wavelength μ	Temperature Range, K	Geometry θ'	Reported Error, %	Composition (weight percent), Specifications and Remarks
1	100	1950	0.652	1098	$\sim 0^\circ$	2.8	3.0 Hf, 0.36 Fe, 0.17 N, 0.13 Si, 0.075 O, 0.048 C, 0.02 Ni, Zr balance; obtained in powder form, sintered in vacuo, worked into fully dense bars, and turned into cylinder; hexagonal crystal structure; polished by normal metallographic methods using magnesia in final stages; measured in vacuum (2×10^{-5} to 8×10^{-6} mm Hg); mean value of two runs.
2	100	1950	0.652	1581	$\sim 0^\circ$	2.8	Above specimen and conditions except body centered cubic structure.

DATA TABLE NO. 430 NORMAL SPECTRAL EMITTANCE OF [ZIRCONIUM + HAFNIUM + ΣX_i] ALLOYS

[Temperature, T, K; Emittance, ϵ; Wavelength, λ, μ]

T	ϵ
CURVE 1*	
$\lambda = 0.652$	
1098	0.436
CURVE 2*	
$\lambda = 0.652$	
1581	0.426

* Not shown on plot

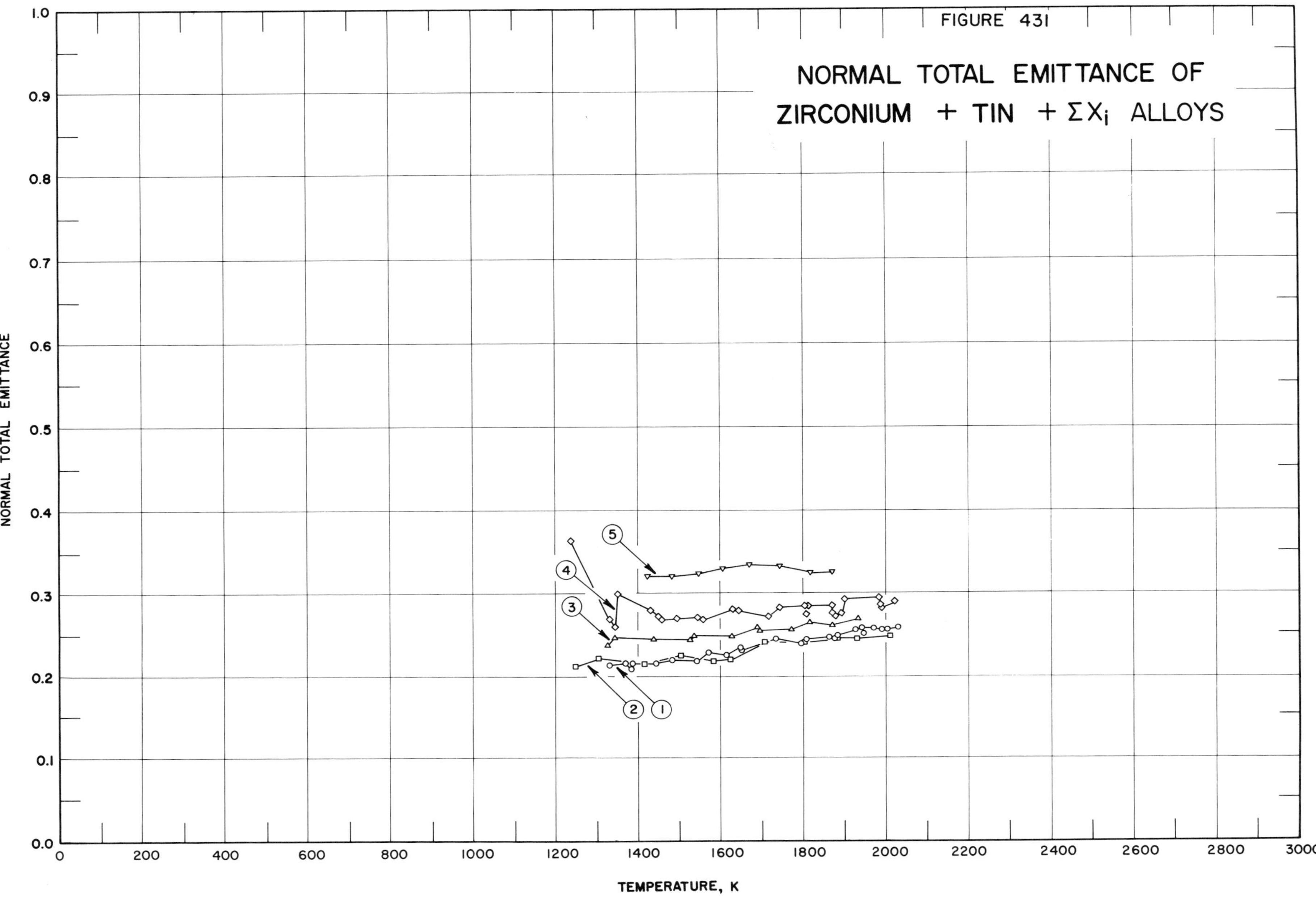
FIGURE 431
NORMAL TOTAL EMITTANCE OF
ZIRCONIUM + TIN + ΣXi ALLOYS
NORMAL TOTAL EMITTANCE
1.0
0.9
0.8
0.7
0.6
0.5
0.4
0.3
0.2
0.1
0.0
0
200
400
600
800
1000
1200
1400
1600
1800
2000
2200
2400
2600
2800
3000
TEMPERATURE, K
5
4
3
2
1

SPECIFICATION TABLE NO. 431 NORMAL TOTAL EMITTANCE OF [ZIRCONIUM + TIN + ΣX_i] ALLOYS

Curve No.	Ref. No.	Year	Temperature Range, K	Geometry θ'	Reported Error, %	Composition (weight percent), Specifications and Remarks
1	166	1957	1333-2038	$\sim0^\circ$	±3	Zircaloy 2; nominal composition: 1.5 Sn, 0.12 Fe, 0.10 Cr, 0.05 Ni, Zr balance; as received.
2	166	1957	1248-2013	$\sim0^\circ$	±3	Above specimen and conditions.
3	166	1957	1328-1938	$\sim0^\circ$	±3	Different sample, same as above specimen and conditions.
4	166	1957	1238-2023	$\sim0^\circ$	±3	Above specimen and conditions except oxygen content raised to 1.8 percent.
5	166	1957	1423-1873	$\sim0^\circ$	±3	Above specimen and conditions except oxygen content raised to 3.6 percent.
6	178	1956	1273-1973	$\sim0^\circ$		Zircaloy 2; nominal composition: 1.5 Sn, 0.12 Fe, 0.10 Cr, 0.05 Ni, Zr balance; as received.
7	178	1956	1273-1973	$\sim0^\circ$		Above specimen and conditions except oxidized (0.014 percent oxygen content).
8	178	1956	1273-1973	$\sim0^\circ$		Above specimen and conditions except further oxidized (0.028 percent oxygen content).
9	178	1956	1473-1873	$\sim0^\circ$		Above specimen and conditions except further oxidized (0.057 percent oxygen content).
10	300	1956	1473-2023	$\sim0^\circ$		Zircaloy 2; nominal composition: 1.5 Sn, 0.12 Fe, 0.10 Cr, 0.05 Ni, Zr balance; as received.
11	300	1956	1473-2023	$\sim0^\circ$		Different sample, same as above specimen and conditions.

DATA TABLE NO. 431 NORMAL TOTAL EMITTANCE OF [ZIRCONIUM + TIN + ΣX_i] ALLOYS

[Temperature, T, K; Emittance, ε]

CURVE 1

T	ε
1333	0.212
1378	0.215
1383	0.208
1388	0.214
1443	0.214
1483	0.219
1543	0.217
1573	0.227
1578	0.225*
1618	0.224
1649	0.233
1653	0.230
1737	0.244
1798	0.238
1808	0.243
1863	0.246
1878	0.243
1883	0.248
1933	0.253
1941	0.257
1943	0.250
1973	0.257
1993	0.254
2003	0.255
2038	0.258

CURVE 2

T	ε
1248	0.210
1303	0.220
1418	0.214
1505	0.223
1581	0.217
1623	0.219
1708	0.240
1803	0.240
1883	0.245
1933	0.242
2013	0.248

CURVE 3

T	ε
1328	0.236
1347	0.246
1438	0.244
1528	0.242

CURVE 3 (cont.)

T	ε
1623	0.248
1693	0.258
1698	0.254
1778	0.256
1818	0.264
1873	0.260
1938	0.269

CURVE 4

T	ε
1238	0.363
1330	0.270
1343	0.259
1353	0.298
1433	0.279
1453	0.272
1458	0.268
1498	0.270
1543	0.270
1558	0.268
1633	0.280
1643	0.278
1718	0.271
1743	0.283
1805	0.284
1807	0.274
1815	0.284
1878	0.283
1879	0.276
1880	0.273
1895	0.276
1901	0.291
1983	0.294
1991	0.286
1993	0.282
2023	0.288

CURVE 5

T	ε
1423	0.320
1483	0.320
1548	0.333
1608	0.330
1678	0.336
1743	0.328
1818	0.325
1873	0.327

CURVE 6*

T	ε
1273	0.21
1373	0.21
1473	0.22
1573	0.23
1673	0.23
1773	0.24
1873	0.25
1973	0.25

CURVE 7*

T	ε
1273	0.24
1373	0.24
1473	0.24
1573	0.25
1673	0.25
1773	0.26
1873	0.26
1973	0.27

CURVE 8*

T	ε
1273	0.26
1373	0.26
1473	0.26
1573	0.27
1673	0.27
1773	0.28
1873	0.28
1973	0.29

CURVE 9*

T	ε
1473	0.32
1573	0.33
1673	0.33
1773	0.33
1873	0.33

CURVE 10*

T	ε
1473	0.21
1573	0.22
1673	0.23
1773	0.24
1873	0.25
1973	0.26
2023	0.26

CURVE 11*

T	ε
1473	0.22
1573	0.22
1673	0.23
1773	0.24
1873	0.25
1973	0.25
2023	0.26

* Not shown on plot

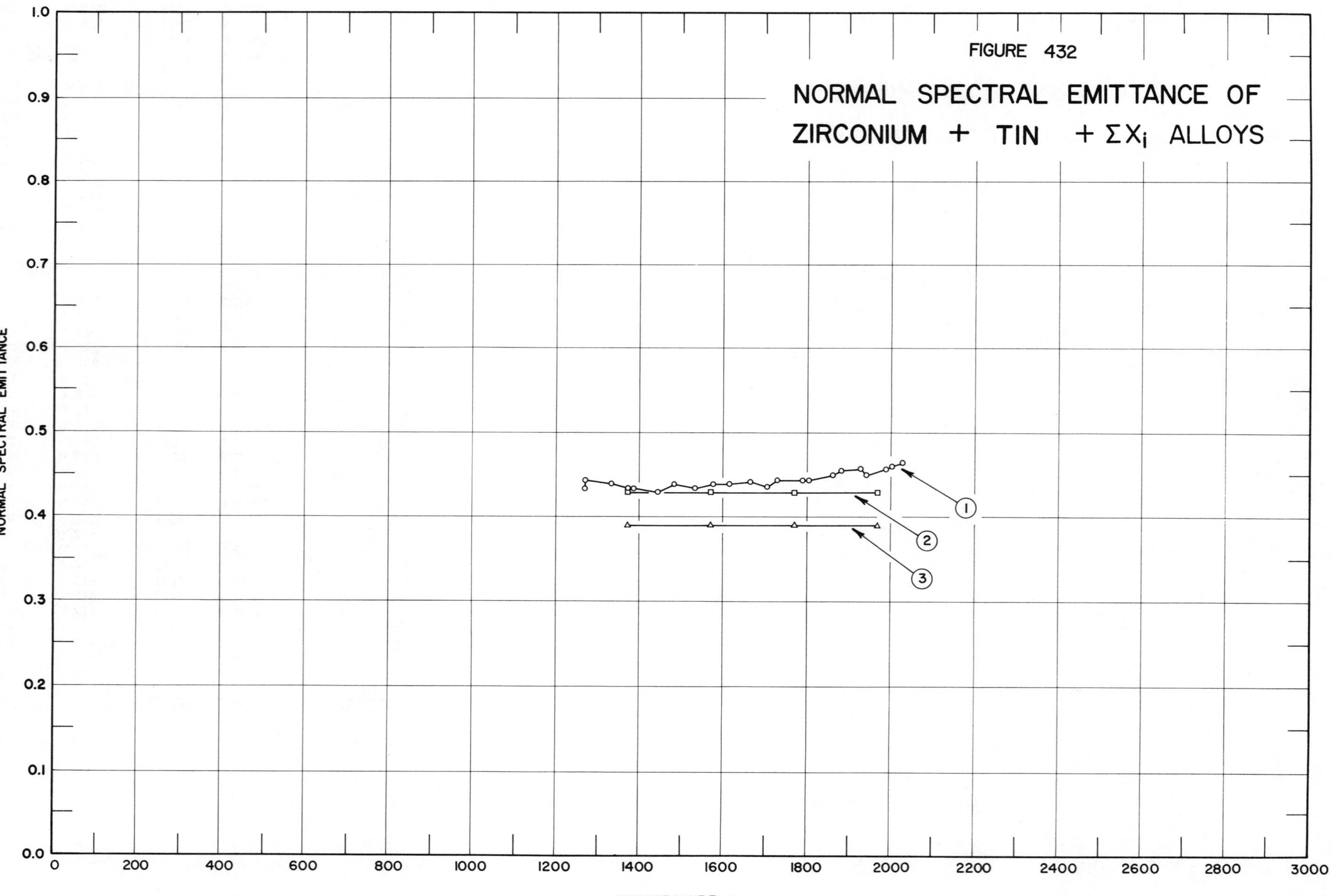
FIGURE 432
NORMAL SPECTRAL EMITTANCE OF
ZIRCONIUM + TIN + ΣXi ALLOYS
1
2
3
NORMAL SPECTRAL EMITTANCE
1.0
0.9
0.8
0.7
0.6
0.5
0.4
0.3
0.2
0.1
0.0
TEMPERATURE, K
0
200
400
600
800
1000
1200
1400
1600
1800
2000
2200
2400
2600
2800
3000

SPECIFICATION TABLE NO. 432 NORMAL SPECTRAL EMITTANCE OF [ZIRCONIUM + TIN + ΣX_i] ALLOYS

Curve No.	Ref. No.	Year	Wavelength μ	Temperature Range, K	Geometry θ'	Reported Error, %	Composition (weight percent), Specifications and Remarks
1	166	1957	0.65	1273-2036	$\sim 0^{\circ}$	±15	Zircaloy 2; nominal composition: 1.5 Sn, 0.12 Fe, 0.10 Cr, 0.05 Ni, Zr balance; as received.
2	166	1957	0.65	1373-1973	$\sim 0^{\circ}$	±15	Above specimen and conditions except oxygen content raised to 0.9 percent.
3	166	1957	0.65	1373-1973	$\sim 0^{\circ}$	±15	Above specimen and conditions except oxygen content raised to 1.6 percent.

DATA TABLE NO. 432 NORMAL SPECTRAL EMITTANCE OF [ZIRCONIUM + TIN + ΣX_i]

[Temperature, T, K; Emittance, ϵ; Wavelength, λ, μ]

T	ϵ
CURVE 1	
$\lambda = 0.65$	
1273	0.435
1276	0.445
1333	0.440
1378	0.435
1388	0.435
1443	0.430
1483	0.440
1537	0.435
1578	0.440
1618	0.440
1663	0.442
1707	0.437
1734	0.444
1792	0.445
1803	0.445
1863	0.450
1883	0.454
1933	0.454
1943	0.450
1993	0.457
2003	0.460
2036	0.465
CURVE 2	
$\lambda = 0.65$	
1373	0.43
1573	0.43
1773	0.43
1973	0.43
CURVE 3	
$\lambda = 0.65$	
1373	0.39
1573	0.39
1773	0.39
1973	0.39

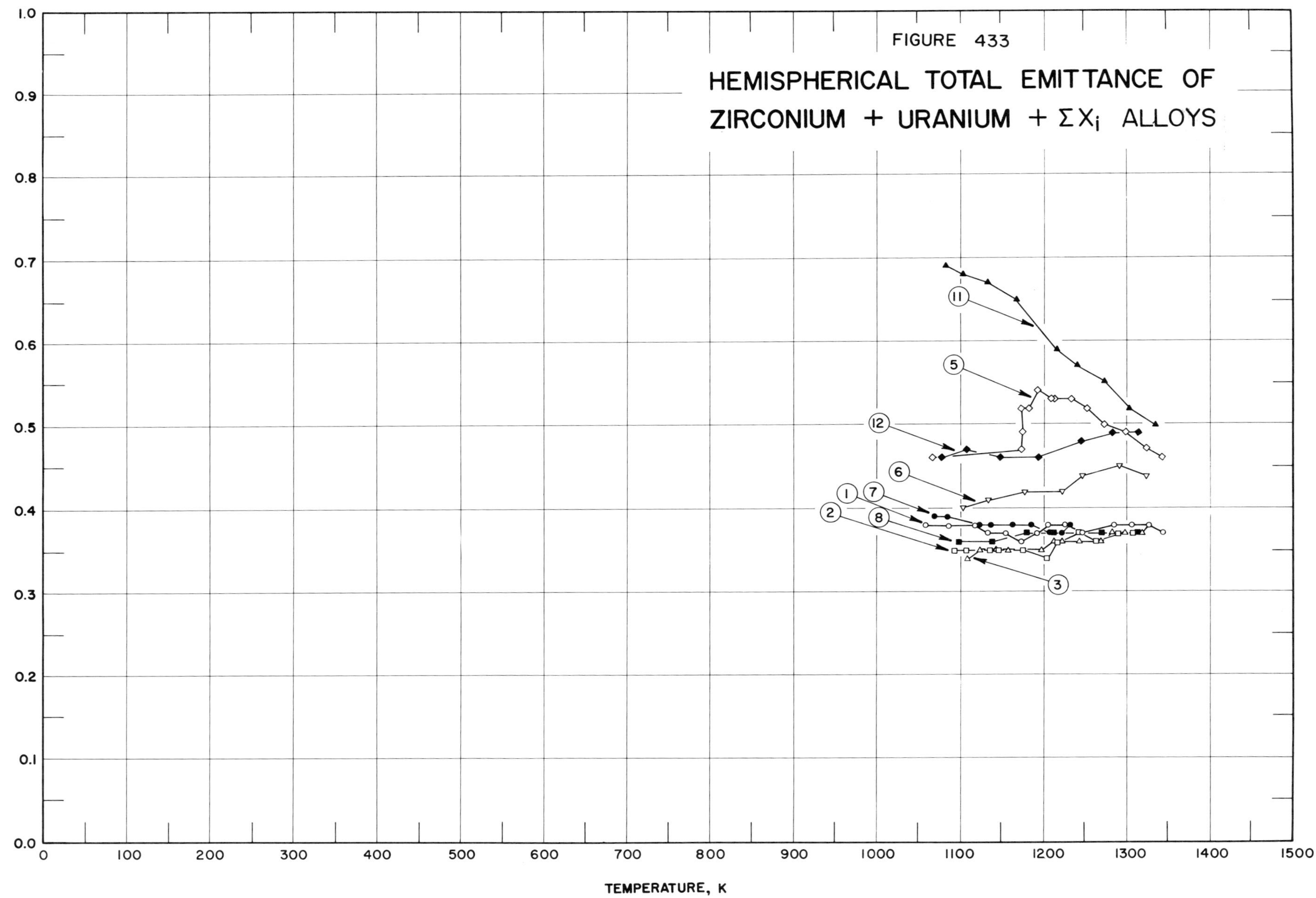

FIGURE 433

HEMISPHERICAL TOTAL EMITTANCE OF ZIRCONIUM + URANIUM + ΣX_i ALLOYS

SPECIFICATION TABLE NO. 433 HEMISPHERICAL TOTAL EMITTANCE OF [ZIRCONIUM + URANIUM + ΣX_i] ALLOYS

Curve No.	Ref. No.	Year	Temperature Range, K	Reported Error, %	Composition (weight percent), Specifications and Remarks
1	116	1957	1058-1343		49.1 Zr, 45 U, 5 Nb, 0.75 Sn, 0.075 Fe, 0.05 Cr, 0.025 Ni; annealed in vacuum for 1 hr at 1008 K and water quenched, then heated in vacuum for 4 hrs at 873 K; run 1, increasing temp; [Author's designation: Specimen 8].
2	116	1957	1308-1093		Above specimen and conditions; run 1, decreasing temp.
3	116	1957	1108-1329		Above specimen and conditions; run 2, increasing temp.
4	116	1957	1293-1093		Above specimen and conditions; run 2, decreasing temp.
5	116	1957	1068-1343		Above specimen and conditions except oxygen added to the specimen by adding 20 individual charges of oxygen (2.5 cm^3 at atmospheric pressure per charge) to the vacuum chamber while the specimen was at 1078 K; chamber re-evacuated; run 1, increasing temp.
6	116	1957	1323-1103		Above specimen and conditions; run 1, decreasing temp.
7	116	1957	1070-1233		73.67 Zr, 22.5 U, 2.5 Nb, 1.13 Sn, 0.09 Fe, 0.075 Cr, 0.038 Ni; annealed in vacuum for 1 hr at 1008 K and water quenched,then heated in vacuum at 873 K for 4 hrs; run 1, increasing temp; [Author's designation: Specimen 10].
8	116	1957	1313-1098		Above specimen and conditions; run 1, decreasing temp.
9	116	1957	1123-1329		Above specimen and conditions; run 2, increasing temp.
10	116	1957	1303-1098		Above specimen and conditions; run 2, decreasing temp.
11	116	1957	1083-1335		Above specimen and conditions except heated at 1073 K in oxygen; run 1, increasing temp.
12	116	1957	1313-1078		Above specimen and conditions; run 1, decreasing temp.

DATA TABLE NO. 433 HEMISPHERICAL TOTAL EMITTANCE OF [ZIRCONIUM + URANIUM + ΣX_i] ALLOYS

[Temperature, T, K; Emittance, ∈]

T	∈
CURVE 1	
1058	0.38
1089	0.38
1118	0.38
1133	0.37
1153	0.37
1173	0.36
1193	0.37
1206	0.38
1228	0.38
1249	0.37
1285	0.38
1308	0.38
1328	0.37
1343	0.36
CURVE 2	
1308	0.37
1290	0.37
1263	0.36
1243	0.37
1218	0.36
1205	0.34
1176	0.35
1146	0.35
1135	0.35
1108	0.35
1093	0.35
CURVE 3	
1108	0.34
1123	0.35
1143	0.35
1185	0.35
1198	0.35
1213	0.36
1223	0.36
1243	0.36
1268	0.36
1283	0.37
1298	0.37
1314	0.37
1329	0.38*
CURVE 4*	
1293	0.36
1268	0.37
1248	0.36
1203	0.36
1148	0.35
1123	0.35
1093	0.35
CURVE 5	
1068	0.46
1173	0.47
1175	0.49
1173	0.52
1181	0.52
1193	0.54
1210	0.53
1213	0.53
1233	0.53
1253	0.52
1273	0.50
1298	0.49
1323	0.47
1343	0.46
CURVE 6	
1323	0.44
1293	0.45
1248	0.44
1223	0.42
1178	0.42
1133	0.41
1103	0.40
CURVE 7	
1070	0.39
1086	0.39
1123	0.38
1138	0.38
1163	0.38
1188	0.38
1208	0.37
CURVE 7 (cont.)	
1221	0.37
1233	0.38
CURVE 8	
1313	0.37
1285	0.37*
1270	0.37
1213	0.37
1180	0.37
1138	0.36
1098	0.36
CURVE 9*	
1123	0.37
1148	0.36
1203	0.36
1225	0.37
1249	0.38
1264	0.38
1288	0.38
1313	0.38
1329	0.39
CURVE 10*	
1303	0.38
1267	0.38
1231	0.38
1199	0.37
1178	0.36
1158	0.37
1125	0.36
1098	0.36
CURVE 11	
1083	0.69
1103	0.68
1133	0.67
1168	0.65
1218	0.59
1240	0.57
CURVE 11 (cont.)	
1273	0.55
1303	0.52
1335	0.50
CURVE 12	
1313	0.49
1283	0.49
1246	0.48
1193	0.46
1148	0.46
1108	0.47
1078	0.46

* Not shown on plot

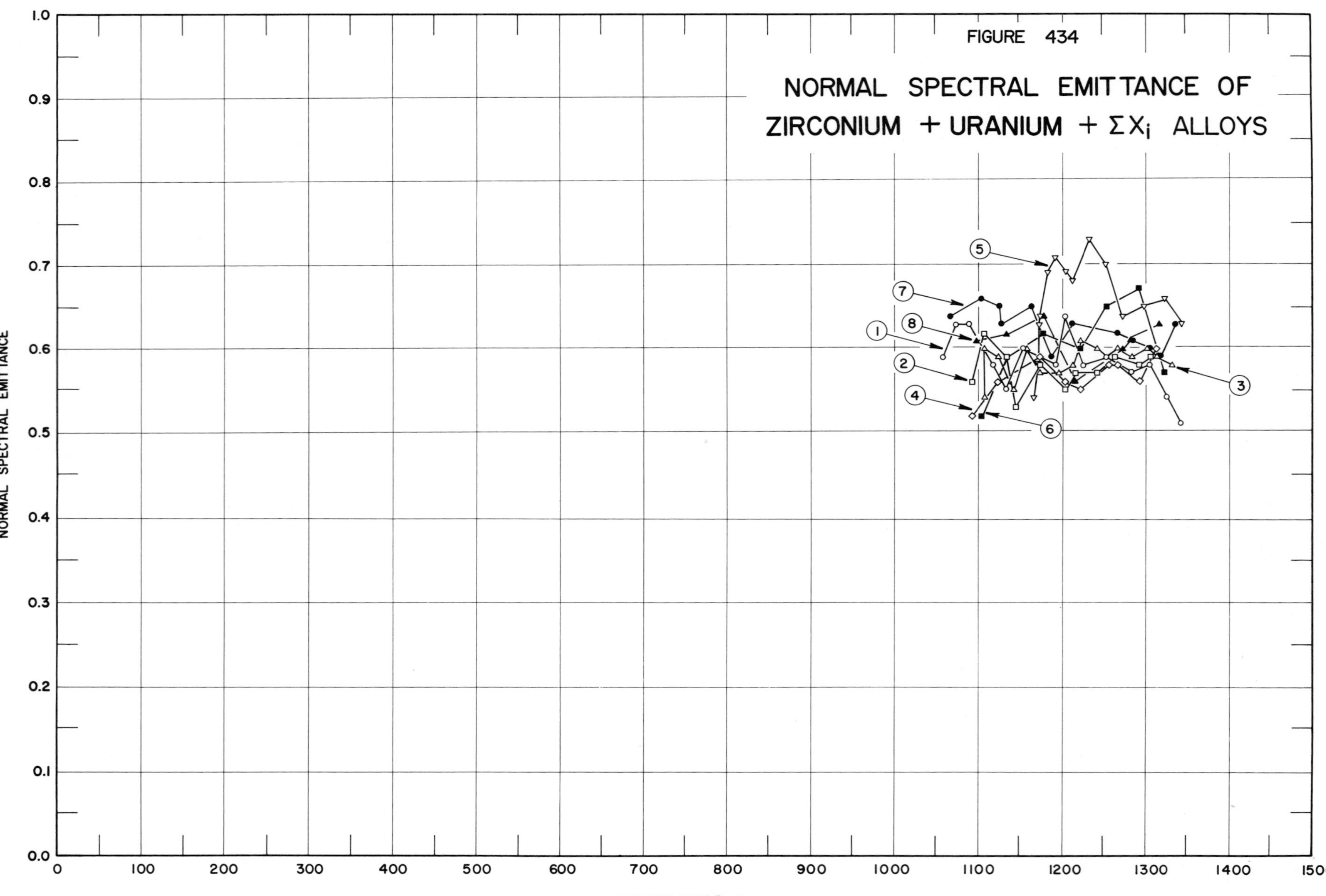
FIGURE 434
NORMAL SPECTRAL EMITTANCE OF
ZIRCONIUM + URANIUM + ΣX$_i$ ALLOYS
NORMAL SPECTRAL EMITTANCE
TEMPERATURE, K
1.0
0.9
0.8
0.7
0.6
0.5
0.4
0.3
0.2
0.1
0.0
0
100
200
300
400
500
600
700
800
900
1000
1100
1200
1300
1400
1500
5
7
8
1
2
4
6
3

SPECIFICATION TABLE NO. 434 NORMAL SPECTRAL EMITTANCE OF [ZIRCONIUM + URANIUM + ΣX_i] ALLOYS

Curve No.	Ref. No.	Year	Wavelength μ	Temperature Range, K	Geometry θ'	Reported Error, %	Composition (weight percent), Specifications and Remarks
1	116	1957	0.65	1058-1343	~0°		49.1 Zr, 45 U, 5 Nb, 0.75 Sn, 0.075 Fe, 0.05 Cr, 0.025 Ni; annealed in vacuum for 1 hr at 1008 K and water quenched,then heated in vacuum for 4 hrs at 873 K; run 1, increasing temp; [Author's designation: Specimen 8].
2	116	1957	0.65	1308-1093	~0°		Above specimen and conditions; run 1, decreasing temp.
3	116	1957	0.65	1108-1331	~0°		Above specimen and conditions; run 2, increasing temp.
4	116	1957	0.65	1313-1093	~0°		Above specimen and conditions; run 2, decreasing temp.
5	116	1957	0.65	1168-1343	~0°		Above specimen and conditions except oxygen added to the specimen by adding 20 individual charges of oxygen (2.5 cm^3 at atmospheric pressure per charge) to the vacuum chamber while the specimen was at 1078 K; chamber re-evacuated; run 1, increasing temp.
6	116	1957	0.65	1323-1103	~0°		Above specimen and conditions; run 1, decreasing temp.
7	116	1957	0.65	1068-1335	~0°		73.67 Zr, 22.5 U, 2.5 Nb, 1.13 Sn, 0.09 Fe, 0.075 Cr, 0.038 Ni; annealed in vacuum for 1 hr at 1008 K and water quenched then,heated in vacuum at 873 K for 4 hrs; run 1, increasing temp; [Author's designation: Specimen 10].
8	116	1957	0.65	1316-1098	~0°		Above specimen and conditions; run 1, decreasing temp.
9	116	1957	0.65	1096-1329	~0°		Above specimen and conditions; run 2, increasing temp.
10	116	1957	0.65	1303-1128	~0°		Above specimen and conditions; run 2, decreasing temp.

DATA TABLE NO. 434 NORMAL SPECTRAL EMITTANCE OF [ZIRCONIUM + URANIUM + ΣX_i] ALLOYS

[Temperature, T, K; Emittance, ∈; Wavelength, λ, μ]

T	∈
CURVE 1	
λ = 0.65	
1058	0.59
1073	0.63
1090	0.63
1118	0.58
1133	0.55
1153	0.60
1193	0.58
1203	0.64
1228	0.58
1253	0.59
1283	0.57
1305	0.58
1325	0.54
1343	0.51
CURVE 2	
λ = 0.65	
1308	0.59
1293	0.58
1265	0.59
1243	0.57
1218	0.57
1204	0.55
1175	0.58
1145	0.53
1134	0.59
1108	0.62
1093	0.56
CURVE 3	
λ = 0.65	
1108	0.54
1108	0.60
1123	0.59
1143	0.55
1158	0.60
1173	0.57
1198	0.57
1213	0.58
1223	0.61
1243	0.60
1253	0.59*
CURVE 3 (cont.)	
1268	0.60
1285	0.59
1301	0.60
1313	0.59
1331	0.58
CURVE 4	
λ = 0.65	
1313	0.60
1293	0.56
1268	0.58
1258	0.58
1223	0.55
1203	0.56
1173	0.59
1149	0.53*
1123	0.56
1093	0.52
CURVE 5	
λ = 0.65	
1168	0.54
1173	0.58
1173	0.63
1173	0.64
1183	0.69
1193	0.71
1205	0.69
1213	0.68
1233	0.73
1253	0.70
1273	0.64
1298	0.65
1323	0.66
1343	0.63
CURVE 6	
λ = 0.65	
1323	0.57
1293	0.67
1253	0.65
1223	0.60
CURVE 6 (cont.)	
1178	0.62
1133	0.59*
1103	0.52
CURVE 7	
λ = 0.65	
1068	0.64
1103	0.66
1126	0.65
1128	0.63
1163	0.65
1188	0.59
1211	0.63
1268	0.62
1285	0.61
1303	0.60
1319	0.59
1335	0.63
CURVE 8	
λ = 0.65	
1316	0.63
1283	0.61
1273	0.60
1213	0.56
1179	0.64
1135	0.62
1098	0.61
CURVE 9*	
λ = 0.65	
1096	0.61
1123	0.62
1148	0.65
1178	0.61
1203	0.63
1225	0.66
1253	0.62
1268	0.67
1293	0.65
1313	0.66
1329	0.64
CURVE 10*	
λ = 0.65	
1303	0.62
1265	0.66
1233	0.60
1203	0.65
1163	0.63
1128	0.61

* Not shown on plot

REFERENCES TO DATA SOURCES

Ref. No.	TPRC No.	
1	6639	Goodman, S., NBS Rept. 4239, 1-13, 1955. [AD 71409]
2	11723	Reynolds, P.M., Brit. J. Appl. Physics, 12 (3), 111-4, 1961.
3	10185	Fulk, M.M., Reynolds, M.M., and Park, O.E., NBS Rept. 3517, 151-7, 1955. [AD 125047]
4	19293	Reynolds, M.M., Fulk, M.M., Weitzel, D.H., and Park, O.E., Nat'l. Bur. Stds. Cryogenic Eng. Lab., Boulder, Colo., NBS Rept. 2484, 1-10, 1953.
5	25038	Jenkins, R.J., Butler, C.P., and Parker, W.J., USNRDL-TR-663, SSD-TDR-62-189, 1-57, 1963. [AD 419067]
6	27325	Carpenter, W.G.D. and Sewell, J.H., Royal Aircraft Establishment, RAE-TN-CPM-20, 1-9, 1963. [AD 418612]
7	26738	Haury, G.L., ASD-TDR-63-146, 1-16, 1963. [AD 411140]
8	22816	Anderson, D.L. and Nothwang, G.J., NASA-TN-D-1646, 1-36, 1963.
9	27333	Ziegler, W.T. and Cheung, H., Advan. Cryog. Eng., 2, 100-3, 1960.
10	4981	Best, G., J. Opt. Soc. Amer., 39 (12), 1009-11, 1949.
11	25811	Carpenter, W.G.D. and Sewell, J.H., Royal Aircraft Establishment (G. Brit.), RAE Rept. CHem-538, 1-6, 1962. [AD 295648]
12	20641	Askwyth, W.H., Yahes, R.J., House, R.D., and Mikk, G., NASA-CR-56496, 56497, 56498, 1-277, 1962.
13	6459	Taylor, C.S. and Edwards, J.D., Heating, Piping, and Air-Conditioning, 11 (1), 59-63, 1939.
14	16426	Randolf, C.P. and Overholzer, M.J., Phys. Rev., 2, 144-52, 1913.
15	9353	Barnes, T.T., Forsythe, W.E., and Adams, E.Q., J. Opt. Soc. Am., 37 (10), 804-7, 1947.
16	6353	McDermott, P.F., Rev. Sci. Instruments, 8 (6), 185-92, 1937.
17	6348	Taylor, C.S., Willey, L.A., Smith, D.W., and Edwards, J.D., Metals and Alloys, 7 (8), 189-92, 1938.
18	30642	Konopken, S. and Klemm, R., NASA-SP-31, 505-13, 1963.
19	22522	Burgess, G.K. and Waltenberg, R.G., Nat'l. Bur. Stds. Bull., 11, 591-605, 1915.
20	8677	Allen, F.G., J. Appl. Phys., 28 (12), 1510-11, 1957.
21	16961	Stierwalt, D.L., Naval Ordnance Lab., NAVWEPS Rept. 7160, NOLC Rept. 537, 1-34, 1961. [AD 250530]
22	6965	Wade, W.R., Langley Aeronaut. Lab., NASA, 1-43, 1958. [AD 153191]
23	5125	Zimmermann, F.J., J. Appl. Phys., 26, 1483-8, 1955.
24	17044	Dunkle, R.V. and Gier, J.T., Univ. Calif., ONR, 1-24, 1948. [ATI 91560]
25	24863	Gaumer, R.E., McKellar, L.A., Streed, E.R., Frame, K.L., and Grammer, J.R., ASME Second Symposium on Thermophysical Properties, Princeton, N.J., 575-87, Jan. 24-6, 1962.
26	35270	Thaler, W.J., Finn, E.J., Treado, P.A., and Nakhleh, J., J. Appl. Opt., 3 (12), 1411-5, 1964.
27	24830	Butler, C.P. and Inn, E.C.Y., First Symposium on Surface Effects on Spacecraft Materials, John Wiley and Sons, Inc., 195-211, 1960.
28	9060	Aref, M.N., ASME-AIChE Joint Heat Transfer Conf., Chicago, Ill., Paper 58-HT-18, 1-15, 1958.
29	8144	Betz, H.T., Olson, O.H., Schurin, B.D., and Morris, J.C., WADC-TR-56-222 (Pt 1), 1-43, 1956. [AD 110458]
30	27592	Seban, R.A., WADD-TR-60-370 (Pt 3), 1-68, 1963. [AD 419028]
31	31741	Southern Research Inst., Birmingham, Ala., NASA-CR-55073, 1-77, 1963.
32	6469	Morris, J.C., Schurin, B., and Olson, O.H., ASME Symp. on Thermal Properties, Purdue Univ., 400-4, 1959.
33	23267	Rudnaya, A.I. and Bostrem, Z.D., Trudy Vsesoyuz Nauch-Issledouatel. Inst. Metrol, (35), 95-107, 1958.
34	6979	Betz, H.T., Olson, O.H., Schurin, B.D., and Morris, J.C., WADC-TR-56-222 (Pt 2), 1-184, 1957. [AD 202493]
35	16427	Bidwell, C.C., Phys. Rev., 3, 439-49, 1914.

Ref. No.	TPRC No.	
36	9910	Stubbs, C.M., Proc. Roy. Soc. (London), A88, 195-205, 1913.
37	5060	Price, D.J., Proc. Phys. Soc. (London), 59 (1), 118-31, 1947.
38	16329	Hurst, C., Proc. Roy. Soc. (London), A142, 466-90, 1933.
39	4783	Wahlin, H.B., Phys. Rev., 73, 1458-9, 1948.
40	24862	Birkebak, R.C., Hartnett, J.P., and Eckert, R.G., ASME Second Symposium on Thermophysical Properties, Princeton, N.J., 563-74, 1962.
41	4113	Ward, L., Proc. Phys. Soc. (London), B69, 339-43, 1956.
42	4830	Wahlin, H.B. and Wright, R., J. Appl. Phys., 13, 40-2, 1942.
43	9899	Lund. H. and Ward, L., Proc. Phys. Soc. (London), B65, 535-40, 1952.
44	7488	Wahlin, H.B. and Knop, H.W., Jr., Phys. Rev., 74, 687-9, 1948.
45	31646	Gaumer, R.E., Clauss, F.J., Sibert, M.E., and Shaw, C.C., WADD-TR-60-773, 117-36, 1961. [AD 267 310]
46	23922	Cline, D. and Kropschot, R.H., Advan. in Cryog. Eng., 7, 534-8, 1962.
47	31650	Butler, C.P., Jenkins, R.J., Rudkin, R.L., and Laughridge, F.I., WADD-TR-60-773, 229-52, 1961. [AD 267 310]
48	31790	Hall, W.M., NASA-CR-56037, 1-8, 1964.
49	9909	Stubbs, C.M. and Prideaux, E.B.R., Proc. Roy. Soc. (London), A87, 451-65, 1912.
50	16284	Worthing, A.G., J. Franklin Inst., 192, 112, 1921.
51	16436	Worthing, A.G., Phys. Rev., 28, 174-89, 1926.
52	4003	Taylor, J.E., J. Opt. Soc. Am., 42, 33-6, 1952.
53	16425	Bidwell, C.C., Phys. Rev., 1, 482-3, 1913.
54	29589	Abbott, G.L., Alvares, N.J., and Parker, W.J., WADD-TR-61-94 (Pt 2), 1-31, 1962. [AD 297 865]
55	16444	Whitney, L.V., Phys. Rev., 48, 458-61, 1935.
56	22542	Davisson, C. and Weeks, J.R., Phys. Rev., 17, 261-3, 1921.
57	41915	Agababov, S.G. and Komarek, A., NLL, RTS-2762, 1-5, 1966.
58	25564	Cade, C.M., IRE Transactions on Electron Devices, ED-8, 56-69, 1961.
59	21110	Abbott, G.L., Alvares, N.J., and Parker, W.J., WADD-TR-61-94, 1-48, 1961. [AD 270 470]
60	16296	Davisson, C. and Weeks, J.R., Jr., J. Opt. Soc. Am., 8, 581-605, 1924.
61	16400	Foote, P.D., J. Wash. Acad. Sci., 5, 1-7, 1915.
62	7189	Krishnan, K.S. and Jain, S.C., Brit. J. Appl. Phys., 5 (12), 1-7, 1915.
63	25127	Shatz, E.A. and McCandless, L.C., ASD-TR-62-443, 1-92, 1962. [AD 281 821]
64	6441	Stephens, R.E., J. Opt. Soc. Am., 29 (4), 158-61, 1939.
65	33603	Seban, R.A., WADD-TR-60-370 (Pt 2), 1-72, 1962. [AD 286 863]
66	10876	Maki, A.G., Stair, R., and Johnston, R.G., J. Res. Nat'l. Bur. Std., 64C (2), 99-102, 1960.
67	30630	McElroy, D.L. and Kollie, T.G., NASA-SP-31, 365-79, 1963.
68	6523	Sully, A.H., Brandes, E.A., and Waterhouse, R.B., Brit. J. Appl. Phys., 3 (3), 97-101, 1952.
69	16635	Rudkin, R.L., USNRDL-TR-433, 1-19, 1960. [AD 244 211]
70	16668	Allen, R.D., Glasier, L.F., Jr., and Jordan, P.L., J. Appl. Phys., 31, 1382-7, 1960.
71	31192	Rudkin, R.L., Parker, W.J., and Jenkins, R.J., Temp. Meas. Control Sci. Ind., 3 (Pt 2), 523-34, 1962.
72	26244	Petrov, V.A., Chekhovskoi, V.Ya., and Sheindlin, A.E., High Temperature, 1 (1), 19-23, 1963.
73	30821	Gordon, A.R. and Muchnik, G.F., High Temperature (USSR), 2 (2), 258-60, 1964.
74	6634	Baldwin, G.J., Shilts, J.L., and Coomes, E.A., Notre Dame Physical Electronic Group Res., ONR, 1-8, 1955. [AD 78 005]
75	24864	Pears, C.D., ASME Second Symposium on Thermophysical Properties, Princeton, N.J., 588-98, 1962.
76	30647	Riethof, T.R. and DeSantis, V.J., NASA-SP-31, 565-84, 1963.
77	6349	Bossart, P.N., Physica, 7 (2), 50-4, 1936.
78	9291	Zwikker, C., Physica, 7, 71-4, 1927.
79	16434	Worthing, A.G., Phys. Rev., 25, 846-57, 1925.

Ref. No.	TPRC No.	
80	10555	Ill. Inst. of Tech., Determination of Emissivity and Reflectivity Data on Aircraft Structural Materials, Progress Rept. No. 9, 1-8, 1956. [AD 122711]
81	11083	Moore, G.E. and Allison, H.W., J. Appl. Phys., 12, 431-5, 1941.
82	16439	Barnes, B.T., Phys. Rev., 34, 1026-30, 1929.
83	22521	Burgess, G.K. and Foote, P.D., Nat'l. Bur. Std. Bull. 11, 41-64, 1915.
84	29504	Allen, R.D., ARS Journal, 32 (6), 965-7, 1962.
85	28741	Abbott, G.L., WADD-TR-61-94 (Pt 3), 1-30, 1963. [AD 435825], [AD 436887]
86	29405	Adams, J.G., Northrop Corp., Novair Dov., 1-259, 1962. [AD 274558]
87	13553	Howl, C.D.A. and Davis, A.F., Brit. J. Appl. Phys., 13 (5), 223-6, 1962.
88	9109	Marple, D.T.F., J. Opt. Soc. Am., 46 (7), 490-4, 1956.
89	10162	Rosenbaum, D.M., Sherwood, E.M., Campbell, I.E., Jaffee, R.I., Sims, C.T., Craighead, C.M., Wyler, E.N., and Todd, F.C., BMI, Investigations of Rhenium, Quarterly Progress Rept. No. 3, 1-40, 1953. [AD 14174]
90	10514	Sims, C.T., Craighead, C.M., Jaffee, R.I., Gideon, D.N., Wayler, E.N., Todd, F.C., Rosenbaum, D.M., Sherwood, E.M., and Campbell, I.E., WADC-TR-54-371, 1-138, 1954. [AD 48279]
91	6460	Sims, C.T., Craighead, C.M., and Jaffee, R.I., AIME Trans., 203, 168-78, 1955.
92	23145	Sklarew, S. and Rabensteine, A.S., Marquardt Corp., Rept. No. PR 281-3Q-1, 1-37, 1963. [AD 299417]
93	11752	Malter, L. and Langmuir, D.B., Phys. Rev., 55, 743-7, 1939.
94	14404	Serebryakova, T.I., Paderno, Yu.B., and Samsonov, G.V., Optics and Spectroscopy (USSR), 8 (3), 212-3, 1960.
95	23126	Coffman, J.A., Kibler, G.M., Lyon, T.F., and Acchione, B.D., WADD-TR-60-646 (Pt 2), 1-183, 1963. [AD 297946]
96	31743	Pratt and Whitney Aircraft, PWA-2309, NASA-CR-58054, 1-83, 1964.
97	5062	Michels, W.C. and Wilford, S., J. Appl. Phys., 20, 1223-6, 1949.
98	18591	Cairns, J.H., J. Sci. Instrum., 37 (3), 84-7, 1960.
99	19724	Bevans, J.T., Gier, J.T., and Dunkle, R.V., Trans. ASME, 80, 1405-16, 1958.
100	5025	Bradshaw, F.J., Proc. Phys. Soc. (London), B63, 573-7, 1950.
101	31016	Edwards, J.W., Johnston, H.L., and Ditmars, W.E., J. Am. Chem. Soc., 75, 2467-70, 1953.
102	20450	Forsythe, W.E. and Watson, E.M., J. Opt. Soc. Am., 24, 114-8, 1934.
103	34926	Isreal, S.L., Hawkins, T.D., and Hyman, S.C., NASA-CR-402, 1-46, 1966.
104	28865	Rudkin, R.L., Parker, W.J., and Westover, R.W., USNRDL-TR-419, 1-27, 1960. [AD 240185-L]
105	10768	Jenkins, R.J., Parker, W.J., and Butler, C.P., USNRDL-TR-348, 1-24, 1959. [AD 226896]
106	26008	Pears, C.D., ASD-TDR-62-765, 1-420, 1963. [AD 298061]
107	16727	Pears, C.D., WADC-TR-59-744, Vol. 3, 99-116, 1960. [AD 247110-L]
108	21878	Barnes, B.T., J. Phys. Chem., 33, 688-91, 1929.
109	22616	Reynolds, T.W. and Kreps, L.W., NASA-TN-D-871, 1-43, 1961. [AD 262047]
110	20863	Lockwood, D.L. and Cybulski, R.J., NASA-TN-D-766, 1-53, 1961. [AD 253876]
111	16445	Wahlin, H.B. and Whitney, L.V., Phys. Rev., 50, 735-8, 1936.
112	26981	Kibler, G.M., Lyon, T.F., Linevsky, M.J., and DeSantis, V.J., Gen. Elec. Co., Refractory Materials Research, Quart. Progress Rept. No. 10, 1-34, 1963. [AD 403529], [AD 405520]
113	16963	Riethof, Gen. Elec. Co., Space Sci. Lab., 1-34, 1961. [AD 250274]
114	16423	DeVos, J.C., Physica, 20, 690-714, 1954.
115	9978	Larrabee, R.D., J. Opt. Soc. Am., 49 (6), 619-25, 1959.
116	9364	Lemmon, A.W., BMI-1192, 1-74, 1957.
117	4440	Hole, W.L. and Wright, R.W., Phys. Rev., 56, 785-7, 1939.
118	10931	Furman, S.C. and McManus, P.A., USAEC, GEAP-3338, 1-46, 1960.
119	27371	Maki, A.G. and Plyler, E.K., J. Res. Nat'l. Bur. Std., C66 (3), 283-7, 1962.
120	10168	Rosenbaum, D.M., Sherwood, E.M., Campbell, I.E., Sims, C.T., Craighead, C.M., Jaffee, R.I., Wyler, E.N., and Todd, F.C., BMI, Investigations of Rhenium, Quart. Progress Rept. No. 4, 1-40, 1953. [AD 17543]

Ref. No.	TPRC No.	
121	30772	Herrington, K., Ofc. Dig., Federation Soc. Paint Technol. 34, 1061-4, 1962.
122	29316	Janssen, J. E., Torborg, R. H., Luck, J. R., and Schmidt, R. N., ASD-TR-61-147, 1-269, 1961. [AD 270 453]
123	27253	Walin, D. R., Gen. Dynamics Convair, Study of Thermal Insulating Materials, Final Rept., 1-200, 1961. [AD 606 107]
124	31731	Johnson, B. K., Proc. Phys. Soc. (London), 53, 258-64, 1941.
125	27424	Bennett, H. E., Bennett, J. M., and Ashley, E. J., J. Opt. Soc. Am., 52, 1245-50, 1962. [AD 404 995]
126	28940	Dunkle, R. V. and Gier, J. T., Inst. of Eng. Res., Calif. Univ., Berkeley, Progress Rept., 1-73, 1953. [AD 16 830]
127	25806	Holland, L. and Williams, B. J., J. Sci. Instr., 32, 287, 1955.
128	24861	Dunkle, R. V., Edwards, D. K., Gier, J. T., Nelson, K. E., and Roddick, R. D., ASME Second Symposium on Thermophysical Properties, Princeton, N. J., 541-62, 1962.
129	36483	Cox, J. T., Hass, G., and Ramsey, J. B., Proc. 1964 Army Sci. Conf., West Point, N. Y., 193-205, 1964. [AD 612 134]
130	31933	Pettit, E., Publ. Astron. Soc. Pacific, 46, 27-31, 1934.
131	25561	Porter, J. and Butler, E. A. W., Royal Aircraft Establishment, RAE-TN-S-64, 1-24, 1964. [AD 612 164]
132	23741	Coblentz, W. W., Bull. Nat'l. Bur. Std., 7 (2), 197-225, 1911.
133	7159	Wulff, J., J. Opt. Soc. Am., 24, 223-6, 1934.
134	7538	Romanathan, K. G., Nat'l. Bur. Std. Circ., 519, 257-9, 1952.
135	20294	Douglass, R. W. and Adkins, E. F., Trans. Met. Soc. AIME, 221 (2), 248-9, 1961.
136	25436	Gier, J. T., Dunkle, R. V., and Bevans, J. T., J. Opt. Soc. Am., 44 (7), 558-62, 1954.
137	22526	Lowery, H. and Moore, R. L., Phil. Mag., 13, 938-52, 1932.
138	19294	Seban, R. A. and Rolling, R. E., WADD-TR-60-370, 1-110, 1960.
139	23290	Rayne, J. A., Phys. Rev. Letters, 3 (11), 512-4, 1959.
140	7464	Biondi, M. A., Phys. Rev., 96 (2), 534-5, 1954.
141	4687	Bueche, F., J. Opt. Soc. Am., 38, 806-10, 1948.
142	896	Weil, R., Proc. Phys. Soc. (London), 59, 781-91, 1947.
143	22516	Coblentz, W. W. and Stair, R. J., Research, Nat'l. Bur. Std., 2, 343-54, 1929.
144	21573	Crowell, C. R., Spitzer, W. G., Howarth, L. E., and LaBate, E. E., Phys. Rev., 127, 2006-15, 1962.
145	20474	Harris, L. and Fowler, P., J. Opt. Soc. Am., 51 (2), 164-7, 1961.
146	10017	Olson, O. H. and Morris, J. C., WADC-TR-56-222 (Pt 2 Suppl. 1), 1-31, 1958. [AD 202494]
147	10461	Blau, H. H., Jr., Chaffee, E., Jasperse, J. R., and Martin, W. S., AFCRC-TN-60-165, 1-71, 1960. [AD 236 394]
148	29595	Edwards, D. K., Gier, J. T., Nelson, K. E., and Roddick, R. D., J. Opt. Soc. Am., 51 (11), 1279-88, 1961.
149	22500	Weniger, W. and Pfund, A. H., J. Franklin Inst., 183, 354-5, 1917.
150	22539	Weniger, W. and Pfund, A. H., Phys. Rev., 14, 427-433, 1919.
151	22523	Coblenz, W. W., Nat'l. Bur. Std. Bull., 16, 249-52, 1920.
152	25234	Waltenburg, R. G. and Coblenz, W. W., Sci. Papers, Nat'l. Bur. Std., 15, 653-7, 1920.
153	29572	Adams, J. G., Northrop Space Labs., Hawthorne, Calif., NSL-62-198, 1-101, 1962.
154	23571	Bevans, J. T., LeVantine, A. D., and Luedke, E. E., Space Tech. Labs. Inc., STL/TR-60-0000-19423, AFBMD-TR-61-12, 1-32, 1960. [AD 255 968]
155	16503	Snyder, N. W., Gier, J. T., Dunkle, R. V., and Possner, L., Univ. Calif., Berkeley, 1-16, 1948. [ATI 90 576] [PB 142 256]
156	16606	Blau, H. H., Jr., March, J. B., Martin, W. S., Jasperse, J. R., and Chaffee, E., AFCRL-TR-60-416, 1-78, 1960. [AD 248 276]
157	12280	Boelter, L. M. K., Bromberg, B. R., and Gier, J. T., Univ. Calif., Berkeley, NACA, ARR-4 A21, 1-13, 1944.
158	24808	Cowling, J. E., Alexander, A. L., and Noonan, F. M., WADD-TR-60-773, 17-37, 1961. [AD 267 310]

Ref. No.	TPRC No.	
159	29202	Research Projects Div., G. C. Marshall Space Flight Center, Huntsville, Ala., NASA-TN-D-1523, NASA-N63-14272, 1-253, 1963.
160	6596	Wilkes, G.B., WACD-TR-54-42, 1-94, 1954. [AD 88066]
161	8135	Wade, W.R., NASA MEMO 1-20-59 L, 1-30, 1959. [AD 209192]
162	29579	Gordon, A.R. and Muchnik, G.F., High Temperature, 2 (4), 505-8, 1964.
163	4310	Roeser, W.F., Proc. Am. Soc. Testing Materials, 39, 780-7, 1939.
164	20772	O'Sullivan, W.J. and Wade, W.R., NASA-TR-T-90, 1-24, 1961.
165	26089	Wade, W.R. and Slemp, W.S., NASA-TN-D-998, 1-35, 1962. [AD 272614]
166	6740	Lemmon, A.W., Jr., USAEC, BMI-1154, 114, 1957.
167	26268	Komarek, A. and Strigin, B.K., High Temperature, 1 (1), 24-6, 1963.
168	40793	Rustigi, O.P., Walker, W.C., and Weissler, G.L., J. Opt. Soc. Am., 51 (12), 1357-9, 1961.
169	6932	O'Sullivan, W.J., Jr. and Wade, W.R., NACA-TN-4121, 1-48, 1957.
170	8277	Richmond, J.C. and Stewart, J.E., NASA Memo 4-9-59 W, 1-30, 1959.
171	30639	Evans, R.J., Clayton, W.A., and Fries, M., NASA-SP-31, 483-88, 1963.
172	28946	Harrison, W.N., Richmond, J.C., Plyler, E.K., Stair, R., and Skramstad, H.K., WADC-TR-59-510 (Pt 2), 1-21, 1961. [AD 259326]
173	3735	Weil, R., Nature, 159, 305, 1947.
174	759	Knop, H.W., Jr., Phys. Rev., 74, 1413-6, 1948.
175	23134	Muldawer, L., AFOSR-TN-57-667, 1-33, 1957. [AD 136656]
176	4799	Wahlin, H.B., Zentner, R., and Martin, J., J. Appl. Physics, 23 (1), 107-8, 1952.
177	6444	Bossart, P.N., Univ. Pittsburgh Bull. No. 30, 59-64, 1933.
178	31929	Lucks, C.F., Gaines, G.B., Deem, H.W., Wood, W.D., and Nexsen, W.E., Jr., USAEC, BMI-1094 (Del.), 84-6, 1956.
179	12198	Shibata, K., J. Opt. Soc. Am., 47 (2), 172-5, 1957.
180	24537	Gannon, R.E. and Linder, B., J. Am. Ceram. Soc., 47 (11), 592-3, 1964.
181	10060	Olson, O.H. and Morris, J.C., WADC-TR-56-222 (Pt 3), 1-96, 1959. [AD 239302]
182	22272	Schatz, E.A., Goldberg, D.M., Pearson, E.G., and Burks, T.L., ASD-TDR-63-657 (Pt 1), 1-181, 1963. [AD 423743]
183	4200	Patterson, J.R., Trans. Brit. Ceram. Soc., 54, 698-705, 1955.
184	25074	Childers, H.M. and Cerceo, J.M., WADD-TR-60-190, 1-66, 1961. [AD 272691]
185	36236	Laszlo, T.S., Gannon, R.E., and Sheehan, P.J., Solar Energy (USA), 8 (4), 105-11, 1964. [AD 611945]
186	10101	Kingery, W.D. and Norton, F.H., M.I.T., USAEC, NYO-6447, 1-16, 1955.
187	29570	Folweiler, R.C., ASD-TDR-62-719, 1-115, 1964. [AD 600370]
188	3340	Seifert, R.L., Phys. Rev., 73, 1181-7, 1948.
189	28431	Durig, J.R., Lord, R.C., Gardner, W.J., and Johnston, L.H., J. Opt. Soc. Am., 52, 1078, 1962.
190	22518	Burgess, G.K., Nat'l. Bur. Std. Bull., 6, 111-9, 1909.
191	24705	McMahon, W.R. and Wilder, D.R., Ames Lab., Iowa State, USAEC, IS-578, 1-48, 1963.
192	16013	Bailey, P.C. and Goldman, A., AFCRC-TR-60-147, 1-75, 1960. [AD 240236]
193	26852	Kaiser, W., Spitzer, W.G., Kaiser, R.H., and Howarth, L.E., Phys. Rev., 127 (6), 1950-4, 1962.
194	4784	Benford, F., Lloyd, G.P., and Schwarz, S., J. Opt. Soc. Am., 38, 445-7, 1948.
195	31733	Avgustinik, A.I., FTD-TT-63-265, 1-7, 1963. [AD 409701]
196	10967	Mergerian, D., Olson, O.H., and Weigandt, A., ARF-1159, Second Quarterly Progress Rept., 1-8, 1960.
197	12026	Seifert, H.S. and Randall, H.M., Rev. Sci. Instruments, 11, 365-8, 1940.
198	8271	Ehlert, T.C. and Margrave, J.L., J. Am. Ceram. Soc., 41 (8), 330, 1958.
199	24582	Wickersheim, K.A. and Lefever, R.A., J. Opt. Soc. Am., 51, 1147-8, 1961.
200	4837	Mollwo, E., Zangew. Phys., 6, 257-60, 1954.
201	21923	Slemp, W.S. and Wade, W.R., NASA-SP-31, 433-9, 1963.
202	27141	Bogdan, L., NASA-CR-27, NAS 8-823, 1-39, 1964.

Ref. No.	TPRC No.	
203	18429	Gottlieb, M., J. Opt. Soc. Am., 50, 350-1, 1960.
204	33682	Kibler, G.M., Lyon, T.F., Linevsky, M.J., and Desantis, V.J., WADD-TR-60-646 (Pt 4), 1-141, 1964. [AD 606 836]
205	15869	Lowrie, R., Crist, R.H., and Schomaker, V., Union Carbide Res. Inst., Research in Physical and Chemical Principles Affecting High Temperature Materials for Rocket Nozzles, 1-58, 1960. [AD 239 305] [PB 159 273]
206	30773	Samsonov, G.V., Fomenko, V.S., and Paderno, Yu.B., Oyneupory, 27 (1), 40-2, 1962.
207	758	Silverman, S., J. Opt. Soc. Am., 38, 989, 1948.
208	25673	Mitchell, C.A., J. Opt. Soc. Am., 52 (3), 341-2, 1962.
209	20761	Shaffer, P.T.B., Development of Ultra Refractory Materials, Progress Rept. No. 25, 1-4, 1961. [AD 261461]
210	26949	Grossman, L.N., Hoyt, E.W., Ingold, H.H., Kaznoff, A., and Sanderson, M.J., Gen. Elec. Co., Pleasanton, Calif., GEST-2009, 1-208, 1962. [AD 296 577]
211	18819	Los Alamos Scientific Lab., Quarterly Status Report on the LASL Plasma Thermocouple Development Program for Period Ending March 20, 1961, USAEC, LAMS-2544, 1-15, 1961.
212	23653	Haas, C. and Corbey, M.M.G., J. Phys. Chem. Solids, 20 (3/4), 197-203, 1961.
213	39680	Schleiger, E.R., AFML-TR-66-148, 1-41, 1966. [AD 801322-L]
214	32537	Stierwalt, D.L. and Potter, R.F., Proc. Intern. Conf. Phys. Semicond., Exeter, Engl., 513-20, 1962.
215	33491	Maki, A.G., Proc. Conf. on Radiative Transfer from Solid Materials, Boston, Mass., 135-41, 1962.
216	29648	Gier, J.T., Possner, L., Test, A.J., Dunkle, R.V., and Bevans, J.T., Calif. Univ., Dept. Eng., USN, NR-05-202, 1-4, 1949. [ATI 59 635]
217	33206	Schmidt, E., Hauzeitschr, V.A.W., and Erftwerk, A.G., Aluminum, 3, 91-6, 1930.
218	30622	Zerlaut, G.A., NASA-SP-31, 275-85, 1963.
219	38391	Edwards, D.K. and Catton, I., From Advances in Thermophysical Properties at Extreme Temperatures and Pressures, 3rd ASME Symp. on Thermophysical Properties, Laf. Ind., 189-99, 1965.
220	36320	Davies, J.M. and Zagieboylo, W., Appl. Opt., 4 (2), 167-74, 1965.
221	37355	Keating, G.M. and Mullings, J.A., NASA-TN-D-2388, 1-45, 1964.
222	24473	Bennett, H.E. and Koehler, W.F., J. Opt. Soc. Am., 50, 1-6, 1960.
223	33512	Leigh, C.H., RAD-TR-62-33, NASA-CR-53235, N64-17590, 1-68, 1962.
224	33287	Taylor, A.H., J. Opt. Soc. Am., 21, 776-84, 1931.
225	38390	Edwards, D.K. and deVolo, N.B., From Advances in Thermophysical Properties at Extreme Temperatures and Pressures, 3rd ASME Symp. on Thermophysical Properties, Purdue Univ., 174-88, 1965.
226	27345	LaBaw, K.B., Olsen, A.L., and Nichols, L.W., Naval Ordnance Test Station, NavWeps 8086, 1-32, 1963. [AD 403 988]
227	35833	Rolling, R.E., Funai, A.I., and Grammer, J.R., USAF, ML-TR-64-363, 1-161, 1964. [AD 466 662]
228	20392	Hagen, E. and Rubens, H., Ann. Physik, 1, 352-75, 1900.
229	33080	Wesolowska, C. and Richard, J., Compt. Rend., 258 (9), 2533-6, 1964.
230	38173	Brekhovskikh, V.F., Inzhen.-Fiz. Zh. (USSR), 7 (5), 66-9, 1964.
231	32234	Spitzer, W.G., Gobeli, G.W., and Trumbore, F.A., J. Appl. Phys., 35 (1), 206-11, 1964.
232	38629	Richard, J., Compt. Rend., 256 (5), 1093-5, 1963.
233	18520	Philip, R., J. Phys. Radium, 20, 535-40, 1959.
234	15016	Philip, R., Compt. Rend., 247 (25), 2322-4, 1958.
235	35518	Hughes, R.S. and Lawson, A.W., Phys. Letters, 25A (6), 473-4, 1967.
236	34758	Shaw, M.L., J. Appl. Phys., 37 (2), 919-20, 1966.
237	36161	Tingwaldt, C., Schley, U., Verch, J., and Takata, S., Optik, 22 (1), 48-59, 1965.
238	30957	Naumann, V.O., Univ. of Wisc., PhD. Thesis, 1-23, 1956.
239	30838	Metzger, J.W., Drexel Inst. of Tech., M.S. Thesis, 1-115, 1959.
240	35988	Davey, J.E. and Pankey, T., J. Appl. Phys., 36 (8), 2571-6, 1965.
241	23414	Martin, W.S., Duchane, E.M., and Blau, H.H., Jr., AFCRL-63-547, 1026, 1963. [AD 428932]
242	26461	Loferski, J.J., Univ. of Penn., Semi-Conductor Res., Quarterly Rept. No. 3, 1-59, 1953. [AD 14 711] [PB 160 835]

Ref. No.	TPRC No.	
243	36689	Rabinovich, K., Canfield, L.R., and Madden, R.P., Appl. Opt., 4 (8), 1005-10, 1965.
244	30611	Hembach, R.J., Hemmerdinger, L., and Katz, A.J., NASA-SP-31, 153-67, 1963.
245	36942	Seban, R.A., J. Heat Transfer ASME, C87 (2), 173-6, 1965.
246	40949	Jain, S.C., Goel, T.C., and Chandra, I., Phys. Letters, 24A (6), 320-1, 1967.
247	31791	Birkebak, R.C., Sparrow, E.M., Eckert, E.R.G., and Ramsey, J.W., Trans. ASME, 86C (2), 193-99, 1964.
248	32639	Lapina, E.A. and Chudnovskii, F.A., High Temperature (USSR), 3 (5), 639-42, 1965.
249	30472	Birkebak, R.C., M.S. Thesis, Univ. Minnesota, 1-27, 1956.
250	40866	Voskressenskii, V.Yu., Peletskii, V.E., and Timrot, D.L., High Temperature, 4 (1), 39-42, 1966.
251	12687	Malé, D. and Trompette, J., J. Phys. Radium, 18 (2), 128-30, 1957.
252	36503	Alvares, N.J., NASA-SP-55, 183-7, 1965.
253	30874	Ehrenreich, H., Philipp, H.R., and Olechna, D.J., Phys. Rev., 131 (6), 2469-77, 1963.
254	21284	Harrison, W.N., Richmond, J.C., Shorten, F.J., and Joseph, H.M., WADC-TR-59-510 (Pt 4), 1-90, 1963. [AD 426 846]
255	30353	Kollie, T.G. and McElroy, D.L., Oak Ridge Nat'l. Lab., USAEC, ORNL-3670, 109-11, 1964.
256	30379	Gaines, G.B. and Sims, C.T., J. Appl. Phys., 34 (9), 2922, 1963.
257	35334	Vasko, A., Czechoslov. J. Physics, 15 (3), 170-7, 1965.
258	32456	Eckart, F. and Henrion, W., Monatsber. Deutschen Akad. Wiss., Berlin, 4 (7), 440-51, 1962.
259	32340	Kuharskii, A.A. and Subashiev, V.K., Soviet Physics-Solid State, 8 (3), 603-6, 1966.
260	40955	Koltun, M.M. and Golovner, T.M., Opt. Spectr'y, 21 (5), 347-50, 1966.
261	29605	Howarth, L.E. and Gilbert, J.F., J. Appl. Phys., 34, 236-7, 1963.
262	9874	Lark-Horovitz, K. and Meissner, K.W., Phys. Rev., 76, 1530, 1949.
263	8386	Trompette, J., Comptes Rendus, 246, 753-6, 1958.
264	31619	Raub, E. and Engel, M., Z. Metallk., 31 (11), 339-44, 1939.
265	38766	Valeev, A.S. and Gisin, M.A., Opt. Spectr'y, 19 (1), 62-5, 1965.
266	24388	Wesolowska, C., Compt. Rend., 258 (21), 5191-4, 1964.
267	38987	Jun, C.K. and Hoch, M., AFML-TR-65-191, 1-11, 1965. [AD 477 224]
268	41008	Seemuller, H. and Stark, D., Z. Physik, 198 (2), 201-4, 1967.
269	29596	Dunkle, R.V., Edwards, D.K., Gier, J.T., and Bevans, J.T., Solar Energy, 4 (2), 27-39, 1960.
270	30750	Malek, G.J., Univ. of Calif., M.S. Thesis, 1-145, 1962.
271	16433	Worthing, A.G. and Forsythe, W.E., Phys. Rev., 18, 144-7, 1921.
272	32539	Langmuir, I., Phys. Rev., 7 (3), 302-30, 1916.
273	29638	Engelke, W.T. and Pears, C.D., Southern Research Inst., SRI-5465-1419-I, 1-10, 1962. [AD 459 521]
274	37801	Deadmore, D., J. Am. Ceram. Soc., 47, 649-50, 1964.
275	16298	Worthing, A.G., J. Opt. Soc. Am., 13, 635-49, 1926.
276	31931	Cubicciotti, D., J. Am. Chem. Soc., 73, 2032-5, 1951.
277	30766	Baker, L., Jr., Mouradian, E.M., and Bingle, J.D., Nucl. Sci. Eng., 15, 218-20, 1963.
278	16009	Noonan, F.M., Alexander, A.L., and Cowling, J.E., NRL Rept. 5503, 1-39, 1960. [AD 240 141]
279	35821	DeSantis, V.J., Gen. Elec. Co., Missile and Space Div., Rept. No. R64 SD60, 1-30, 1964. [AD 466 356]
280	24826	McDonough, R., First Symposium on Surface Effects on Spacecraft Materials, John Wiley and Sons, Inc., 141-51, 1960.
281	41003	Mitor, V.V. and Konopel'Ko, I.N., Thermal Engineering, 13 (7), 92-7, 1966.
282	34454	Brandenberg, W.M., Clausen, O.W., and McKeown, D., J. Opt. Soc. Am., 56 (1), 80-6, 1966.
283	37595	Minura, T., Anagnostou, E., and Colarusso, P.E., NASA-TN-D-3234, 1-56, 1966.
284	33623	Clavier, J., Rev. Optique, 8, 379-91, 1929.
285	33241	Bennett, H.E., Proc. Conf. on Radiative Transfer from Solid Materials, MacMillan Co., N.Y., 166-80, 1962.
286	32538	Biondi, M.A. and Rayne, J.A., Phys. Rev., 115 (6), 1522-30, 1959.

Ref. No.	TPRC No.	
287	44279	Biondi, M.A., Phys. Rev., 102 (4), 964-67, 1956.
288	20810	Gillespie, G.T., Olsen, A.L., and Nichols, L.W., NAVWEPS Rept. 8558, NOTS-TP-3586, 1-28, 1964. [AD 609 036]
289	40152	Husmann, O.K., J. Appl. Phys., 37 (13), 4662-70, 1966.
290	44300	Hunter, W.R., Optical Properties and Electronic Structure of Metals and Alloys, North-Holland Pub. Co., Amsterdam, 136-46, 1966.
291	38325	Petrov, V.A., Checkovskoi, V.Ys., and Sheindlin, A.E., High Temperature (USSR), 1 (3), 416-18, 1963.
292	38726	Clark, H.E., Appl. Opt., 4 (10), 1356-7, 1965.
293	33882	Muller, W.E., Appl. Opt., 5 (5), 876-7, 1966.
294	37539	Howl, D.A. and Davis, A.F., Iron and Steel Institute, J., 202, 523-6, 1964.
295	30625	Funai, A.I., NASA-SP-31, 317-27, 1963.
296	18661	Richmond, J.C. and Harrison, W.N., Am. Ceramic Soc. Bull., 39, 668-73, 1960.
297	32493	Wessel, P.R., Phys. Rev., 132 (5), 2062-4, 1963.
298	26751	Richmond, J.C., DeWitt, D.P., and Hayes, W.D., Jr., NBS-TN-252-ML-TDR-64-257, 55, 1964. [AD 612 812]
299	31957	de L'Estoile, H. and Rosenthal, L., Advisory Group for Aeronautical Research and Development, Paris, France, AGARD-211, N63-21549, 1-82, 1958.
300	31794	Lucks, C.F., Gaines, G.B., Deem, H.W., Wood, W.D., and Nexsen, W.E., Jr., USAEC, BMI-1088, 66-7, 1956.
301	39182	Sambongi, T., Hagiwara, R., and Yamadaya, T., J. Phys. Soc. Japan, 21 (5), 923-5, 1966.
302	10649	Grenis, J.A. and Wong, K., Watertown Arsenal, WAL TR 397.1/1, 11, 1958. [AD 201488]
303	44279	Biondi, A., Phys. Rev., 102 (4), 964-67, 1956.
304	16590	Fieldhouse, I.B., Lang, J.I., and Blau, H.H., Jr., WADC-TR-59-744, 4, 1-78, 1960. [AD 249 166]
305	25754	Bradley, D. and Entwistle, A.G., British J. Appl. Phys., 12 (12), 708-11, 1961.
306	32363	Hass, G. and Tousey, R., J. Opt. Soc. Am., 49 (6), 593-602, 1959.
307	32388	Byrne, R.F. and Mancinelli, L.N., Material Lab., N.Y. Naval Shipyard, 39, 1954. [PB 159 155]
308	36346	Rustgi, Om.P., J. Opt. Soc. Am., 55 (6), 630-4, 1965.
309	39977	Fisher, E.I., Fujita, I., and Weissler, G.L., J. Opt. Soc. Am., 56 (11), 1560-4, 1966.
310	34548	Jelinek, T.M., Hamm, R.N., Arakawa, E.T., and Huebner, R.H., J. Opt. Soc. Am., 56 (2), 185-6, 1966.
311	39976	Henderson, G. and Weaver, C., J. Opt. Soc. Am., 56 (11), 1551-9, 1966.
312	28197	Lowery, H., Wilkinson, H., and Smare, D.L., Phila. Mag., 22, 769-90, 1936.
313	33099	Ehrenreich, H. and Philipp, H.R., Phys. Rev., 128 (4), 1622-9, 1962.
314	33896	Torrance, K.E. and Sparrow, E.M., Trans. ASME, J. Heat Transfer, 88C (2), 223-30, 1966.
315	38254	Canfield, L.R. and Hass, G., J. Opt. Soc. Am., 55 (1), 61-4, 1965.
316	44307	Schuler, C.Chr., Optical Properties of Electronic Structure of Metals and Alloys, North-Holland Pub. Co., Amsterdam, 221-36, 1966.
317	32673	Anderson, A.E., Univ. of Calif., M.S. Thesis, 1-57, 1962.
318	32880	Russell, A.D., Univ. of Calif., M.S. Thesis, 1-47, 1960.
319	24665	Philipp, H.R. and Taft, E.A., Phys. Rev., 113 (4), 1002-5, 1959.
320	39215	Rideout, V.L. and Wemple, S.H., J. Opt. Soc. Am., 56 (6), 749-51, 1966.
321	28015	Braner, A.A. and Chen, R., J. Phys. Chem. Solids, 24, 135-9, 1963.
322	40221	Hass, G., Jacobus, G.F., and Hunter, W.R., J. Opt. Soc. Am., 57 (6), 758-62, 1967.
323	39975	Barnes, T.B., J. Opt. Soc. Am., 56 (11), 1546-50, 1966.
324	17329	Brandt, J.A., Irvine, T.F., Jr., and Eckert, E.R.G., Proc. Heat Trans. Fluid Mech. Inst., Stanford Univ., 220-7, 1960.
325	26230	Krzhizhanovskii, B.A., Kolchenogova, I.P., and Rakov, A.M., High Temperature, 1 (1), 13-18, 1963.
326	26832	Sasaki, T. and Ishiguro, K., Phys. Rev., 127, 1091-2, 1962.
327	26044	Vavilov, V.S. and Galkin, G.N., Fiz. Tverdogo Tela, 1, 1201-4, 1959.

Ref. No.	TPRC No.	
328	42006	Peletskii, V. E. and Voskresenskii, V. Yu., High Temperature, 4 (3), 329-33, 1966.
329	34549	Codling, K., Madden, R. P., Hunter, W. R., and Angel, D. W., J. Opt. Soc. Am., 56 (2), 189-92, 1966.
330	25167	Coblentz, W. W., Bull. Nat'l. Bur. Std., 14, 307-16, 1917.
331	27822	Hulburt, E. O., Astrophys. J., 45, 149-63, 1917.
332	39978	Cairns, R. B. and Samson, J. A. R., J. Opt. Soc. Am., 56 (11), 1568-73, 1966.
333	25454	Good, R. C., Jr., Space Sciences Lab., General Electric Co., AFOSR-5096, 1-80, 1963. [AD 413 974]
334	38942	Timrot, D. L. and Peletskii, V. E., High Temperature, 3 (2), 199-202, 1965.
335	41741	Peletskii, V. E. and Voskrenskii, V. Yu., High Temperature, 4 (2), 293-4, 1966.
336	40413	Schocken, K. and Fountain, J. A., Proc. of Conf. on Spacecraft Coatings Development, NASA-TM-X-56167, 20, 1964.
337	35223	Cairns, R. B. and Samson, J. A. R., J. Opt. Soc. Am., 57 (3), 433-4, 1967.
338	39369	Dmitriev, V. D. and Kholopov, G. K., Zh. Prikl. Spektrosk., Akad Nauk. Belorussk, SSR, 2 (6), 481-8, 1965.
339	39439	Grant, P. M. and Paul, W., J. Appl. Phys., 37 (8), 3110-20, 1966.
340	28621	Coblentz, W. W., Phys. Rev., 30 (5), 645-7, 1910.
341	42894	Vehse, R. C., Arakawa, E. T., and Stanford, J. L., J. Opt. Soc. Am., 57 (4), 551-2, 1967.
342	31937	Gordon, G. D., Rev. Sci. Instr., 31 (11), 1204-8, 1960.
343	37418	Landensperger, W. and Stark, D., Z. Physik, 180 (2), 178-83, 1964.
344	32449	Bennett, H. E., J. Opt. Soc. Am., 53 (12), 1389-94, 1963.
345	43899	Kunyrina, L. I. and Titkov, A. S., High Temperature, 4 (3), 394-8, 1966.
346	44061	Timrot, D. L., Peletskii, V. E., and Voskresenskii, V. Yu., High Temperature (USSR), 4 (6), 808-9, 1966.
347	44308	Blodgett, A. J., Jr., Spicer, W. E., and Yu, A. Y. C., Optical Properties and Electronic Structure of Metals and Alloys, North-Holland Pub. Co., Amsterdam, 246-56, 1966.
348	44310	Sasaki, T. and Ejiri, A., Optical Properties and Electronic Structure of Metals and Alloys, North-Holland Pub. Co., Amsterdam, 417-27, 1966.
349	36342	Rustgi, Om. P. and Weissler, G. L., J. Opt. Soc. Am., 55 (4), 456, 1965.
350	44314	Fukutani, H. and Sueoka, O., Optical Properties and Electronic Structure of Metals and Alloys, North-Holland Pub. Co., Amsterdam, 565-73, 1966.
351	33490	Martin, W. S., Proc. Conf. on Radiative Transfer from Solid Materials, MacMillan Co., 123-32, 1962.
352	35744	Anderson, D. L., Trans. Nat'l. Vacuum Syp., 10, 37-41, 1963.
353	28736	Gournay, L. S., Lab. for Atmospheric and Space Physics, U. of Colorado, Boulder, 1-104, 1963. [AD 409 809]
354	47821	Bennett, H. E., Bennett, J. M., Ashley, E. J., and Motyka, R. J., Phys. Rev., 165 (3), 755-64, 1968.
355	32861	Heiman, R. H., Mechanical Engineering, 51, 355-9, 1929.
356	46982	Zhorov, G. A., High Temperature, 5 (3), 403-9, 1967.
357	29641	Technical Report No. 1, Bausch and Lomb, Military Products Div., Rochester, N. Y., 1-11, 1961.
358	45667	de la Perrelle, E. T. and Herbert, H., Aeronautical Research Council (Gt. Brit.), ARC-CP-601, 1-22, 1962.
359	44304	Bennett, H. E. and Bennett, J. M., Optical Properties and Electronic Structure of Metals, North-Holland Pub. Co., Amsterdam, 175-88, 1966.
360	32603	Banning, M., J. Opt. Soc. Am., 32, 98-102, 1942.
361	29594	Edwards, D. K., Gier, J. T., Nelson, K. E., and Roddick, R. D., Solar Energy, 6 (1), 1-8, 1962.
362	35323	Dunn, S. T., Ph. D. Thesis, Oklahoma State Univ., Stillwater, 1-185, 1965.
363	32600	Bramer, B. R., Vertogen, G., and Penning, P., Solid State Column, 1 (6), 138-43, 1963.
364	48547	Abe, T. and Nishi, Y., Japan J. Appl. Phys., 7 (4), 397-403, 1968.
365	49482	Verleur, H. W., J. Opt. Soc. Am., 58 (10), 1356-64, 1968.
366	30100	McCarthy, D. E., Appl. Opt., 2 (6), 591-603, 1963.
367	45698	Stierwalt, D. L., Bernstein, J. B., and Kirk, D. D., Appl. Opt., 2, 1169-73, 1963.

Ref. No.	TPRC No.	
368	32051	Walles, S., Arkiv Fysik, 25 (4), 33-47, 1963.
369	33154	Lord, R.C., Phys. Rev., 85, 140-1, 1952.
370	42590	Mitsuishi, A., Yoshinaga, H., Yata, K., and Manabe, A., Japan J. Appl. Phys., Suppl., 4 (1), 581-7, 1964, Publ. 1965.
371	40651	Hass, G., J. Opt. Soc. Am., 45 (11), 945-52, 1955.

Material Index

MATERIAL INDEX

Material Name	EMITTANCE						REFLECTANCE									ABSORPTANCE									TRANSMITTANCE								
	Total			Spectral			Integrated			Spectral			Solar			Integrated			Spectral			Solar			Integrated			Spectral			Solar		
	Hemispherical	Normal	Angular	Hemispherical	Normal	Angular	Hemispherical	Normal	Angular	Hemispherical	Normal	Angular	Hemispherical	Normal	Angular	Hemispherical	Normal	Angular	Hemispherical	Normal	Angular	Hemispherical	Normal	Angular	Hemispherical	Normal	Angular	Hemispherical	Normal	Angular	Hemispherical	Normal	Angular
Alcoa No. 2 reflector plate . . .	4 5	-	-	-	-	-	-	-	-	-	-	-	-	-	-	42 43	-	-	-	-	-	-	-	-	-	-	-	-	-	-	-	-	-
Alleghany Alloy No. 66	-	1180	-	-	-	-	-	-	-	-	-	-	-	-	-	-	-	-	-	-	-	-	-	-	-	-	-	-	-	-	-	-	-
Alleghany Metal	-	1225	-	-	-	-	-	-	-	-	-	-	-	-	-	-	-	-	-	-	-	-	-	-	-	-	-	-	-	-	-	-	-
Aluminum	2	8	-	-	12	15	-	-	18	20	24	34 38	-	40	-	42	45	-	-	47	50	-	52	55	-	-	-	-	57	60	-	-	-
Aluminum 1075	-	-	-	-	-	-	-	-	-	-	28	-	-	-	-	-	-	-	-	-	-	-	-	-	-	-	-	-	-	-	-	-	-
Aluminum 1100	-	10	-	-	-	16	-	-	-	-	-	-	-	-	-	-	-	-	-	-	-	-	-	-	-	-	-	-	-	-	-	-	-
Aluminum foil	5 6	9	-	-	-	-	-	-	-	-	26 27	-	-	40	-	43	-	-	-	-	50	-	-	55	-	-	-	-	-	60	-	-	-
Aluminum alloys:																																	
24-ST	1063	1068 1069	-	-	-	-	-	-	-	-	1078	-	-	1084	-	-	-	-	-	-	-	-	1087	-	-	-	-	-	-	-	-	-	-
53-SO	-	1101	-	-	-	-	-	-	-	-	-	-	-	-	-	-	-	-	-	-	-	-	-	-	-	-	-	-	-	-	-	-	-
75-ST	-	1119	-	-	-	-	-	-	-	-	-	-	-	1129	-	-	-	-	-	-	-	-	1132	-	-	-	-	-	-	-	-	-	-
2024	1063	-	-	-	1074	-	-	-	-	-	-	-	-	-	-	-	-	-	-	-	-	-	1087	-	-	-	-	-	-	-	-	-	-
2024-T . . . . (see 24-ST)																																	
2219	-	-	-	-	-	-	-	-	-	-	1078	-	-	-	-	-	-	-	-	-	-	-	-	-	-	-	-	-	-	-	-	-	-

Material Name	EMITTANCE Total Hemispherical	EMITTANCE Total Normal	EMITTANCE Total Angular	EMITTANCE Spectral Hemispherical	EMITTANCE Spectral Normal	EMITTANCE Spectral Angular
Aluminum alloys (continued)						
3003	1112	1114	-	-	-	-
5053-O (see 53-SO)						
6061	1098	-	-	-	-	-
7075	-	-	-	-	1122	-
7075-T (see 75-ST)						
7075-T6	-	-	-	-	-	1126
Alclad 24-SO	-	-	-	-	-	-
Alclad 24-ST	-	1068 1069	-	-	-	-
Alclad 75-ST	1116	1119	-	-	-	-
Alclad 2024-O (see Alclad 24-SO)						
Alclad 2024-T (see Alclad 24-ST)						
Alclad 7075-T (see Alclad 75-ST)						
L34	-	-	-	-	-	-
L72	-	-	-	-	-	-
Aluminum bronze	-	1160	-	-	-	-
Aluminum + Cobalt	-	-	-	-	-	-
Aluminum + Copper + ΣX_i	1062	1066	-	-	1072	-
Aluminum + Iron + ΣX_i	-	-	-	-	-	-

Material Name	REFLECTANCE Integrated Hemispherical	REFLECTANCE Integrated Normal	REFLECTANCE Integrated Angular	REFLECTANCE Spectral Hemispherical	REFLECTANCE Spectral Normal	REFLECTANCE Spectral Angular	REFLECTANCE Solar Hemispherical	REFLECTANCE Solar Normal	REFLECTANCE Solar Angular
Aluminum alloys (continued)									
3003	-	-	-	-	-	-	-	-	-
5053-O (see 53-SO)									
6061	-	-	-	-	1106	-	-	-	-
7075	-	-	-	-	-	-	-	-	-
7075-T (see 75-ST)									
7075-T6	-	-	-	-	-	-	-	-	-
Alclad 24-SO	-	-	-	-	-	-	-	1084	-
Alclad 24-ST	-	-	-	-	-	-	-	-	-
Alclad 75-ST	-	-	-	-	-	-	-	1129	-
Alclad 2024-O (see Alclad 24-SO)									
Alclad 2024-T (see Alclad 24-ST)									
Alclad 7075-T (see Alclad 75-ST)									
L34	-	-	-	-	1092	-	-	1095	-
L72	-	-	-	-	-	-	-	1084	-
Aluminum bronze	-	-	-	-	-	-	-	-	-
Aluminum + Cobalt	-	-	-	-	887	-	-	-	-
Aluminum + Copper + ΣX_i	-	-	-	-	1076	-	-	1083	-
Aluminum + Iron + ΣX_i	-	-	-	-	1090	-	-	1094	-

Material Name	ABSORPTANCE Integrated Hemispherical	ABSORPTANCE Integrated Normal	ABSORPTANCE Integrated Angular	ABSORPTANCE Spectral Hemispherical	ABSORPTANCE Spectral Normal	ABSORPTANCE Spectral Angular	ABSORPTANCE Solar Hemispherical	ABSORPTANCE Solar Normal	ABSORPTANCE Solar Angular
Aluminum alloys (continued)									
3003	-	-	-	-	-	-	-	-	-
5053-O (see 53-SO)									
6061	-	-	-	-	-	-	-	1110	-
7075	-	-	-	-	-	-	-	-	-
7075-T (see 75-ST)									
7075-T6	-	-	-	-	-	-	-	-	-
Alclad 24-SO	-	-	-	-	-	-	-	-	-
Alclad 24-ST	-	-	-	-	-	-	-	-	-
Alclad 75-ST	-	-	-	-	-	-	-	-	-
Alclad 2024-O (see Alclad 24-SO)									
Alclad 2024-T (see Alclad 24-ST)									
Alclad 7075-T (see Alclad 75-ST)									
L34	-	-	-	-	-	-	-	-	-
L72	-	-	-	-	-	-	-	1087	-
Aluminum bronze	-	-	-	-	-	-	-	-	-
Aluminum + Cobalt	-	-	-	-	-	-	-	-	-
Aluminum + Copper + ΣX_i	-	-	-	-	-	-	-	1086	-
Aluminum + Iron + ΣX_i	-	-	-	-	-	-	-	-	-

Material Name	TRANSMITTANCE Integrated Hemispherical	TRANSMITTANCE Integrated Normal	TRANSMITTANCE Integrated Angular	TRANSMITTANCE Spectral Hemispherical	TRANSMITTANCE Spectral Normal	TRANSMITTANCE Spectral Angular	TRANSMITTANCE Solar Hemispherical	TRANSMITTANCE Solar Normal	TRANSMITTANCE Solar Angular
Aluminum alloys (continued)									
3003	-	-	-	-	-	-	-	-	-
5053-O (see 53-SO)									
6061	-	-	-	-	-	-	-	-	-
7075	-	-	-	-	-	-	-	-	-
7075-T (see 75-ST)									
7075-T6	-	-	-	-	-	-	-	-	-
Alclad 24-SO	-	-	-	-	-	-	-	-	-
Alclad 24-ST	-	-	-	-	-	-	-	-	-
Alclad 75-ST	-	-	-	-	-	-	-	-	-
Alclad 2024-O (see Alclad 24-SO)									
Alclad 2024-T (see Alclad 24-ST)									
Alclad 7075-T (see Alclad 75-ST)									
L34	-	-	-	-	-	-	-	-	-
L72	-	-	-	-	-	-	-	-	-
Aluminum bronze	-	-	-	-	-	-	-	-	-
Aluminum + Cobalt	-	-	-	-	-	-	-	-	-
Aluminum + Copper + ΣX_i	-	-	-	-	-	-	-	-	-
Aluminum + Iron + ΣX_i	-	-	-	-	-	-	-	-	-

Material Name	EMITTANCE Total Hemispherical	EMITTANCE Total Normal	EMITTANCE Total Angular	EMITTANCE Spectral Hemispherical	EMITTANCE Spectral Normal	EMITTANCE Spectral Angular	REFLECTANCE Integrated Hemispherical	REFLECTANCE Integrated Normal	REFLECTANCE Integrated Angular	REFLECTANCE Spectral Hemispherical	REFLECTANCE Spectral Normal	REFLECTANCE Spectral Angular	REFLECTANCE Solar Hemispherical	REFLECTANCE Solar Normal	REFLECTANCE Solar Angular	ABSORPTANCE Integrated Hemispherical	ABSORPTANCE Integrated Normal	ABSORPTANCE Integrated Angular	ABSORPTANCE Spectral Hemispherical	ABSORPTANCE Spectral Normal	ABSORPTANCE Spectral Angular	ABSORPTANCE Solar Hemispherical	ABSORPTANCE Solar Normal	ABSORPTANCE Solar Angular	TRANSMITTANCE Integrated Hemispherical	TRANSMITTANCE Integrated Normal	TRANSMITTANCE Integrated Angular	TRANSMITTANCE Spectral Hemispherical	TRANSMITTANCE Spectral Normal	TRANSMITTANCE Spectral Angular	TRANSMITTANCE Solar Hemispherical	TRANSMITTANCE Solar Normal	TRANSMITTANCE Solar Angular
Aluminum + Magnesium	-	-	-	-	-	-	-	-	-	-	890	-	-	-	-	-	-	-	-	-	-	-	-	-	-	-	-	-	-	-	-	-	-
Aluminum + Magnesium + ΣX_i	1098	1100	-	-	-	-	-	-	-	-	1105	-	-	-	-	-	-	-	-	-	-	-	1110	-	-	-	-	-	-	-	-	-	-
Aluminum + Manganese + ΣX_i	1112	1114	-	-	-	-	-	-	-	-	-	-	-	-	-	-	-	-	-	-	-	-	-	-	-	-	-	-	-	-	-	-	-
Aluminum + Silicon	-	-	-	-	-	-	-	-	-	-	893	-	-	-	-	-	-	-	-	-	-	-	-	-	-	-	-	-	-	-	-	-	-
Aluminum + Silver	-	-	-	-	-	-	-	-	-	-	896	-	-	-	-	-	-	-	-	-	-	-	-	-	-	-	-	-	-	-	-	-	-
Aluminum + Zinc + ΣX_i	1116	1118	-	-	1121	1124	-	-	-	-	-	-	-	1128	-	-	-	-	-	-	-	-	1131	-	-	-	-	-	-	-	-	-	-
Antimony	-	-	-	-	-	-	-	-	-	-	-	63	-	-	-	-	-	-	-	-	-	-	-	-	-	-	-	-	66	-	-	-	-
Armco Iron	303	308	-	-	-	-	-	-	-	-	322	-	-	-	-	-	-	-	-	-	-	-	332	-	-	-	-	-	-	-	-	-	-
Astrolloy	-	-	-	-	1363 1412	-	-	-	-	-	1377 1418	-	-	-	-	-	-	-	-	-	-	-	-	-	-	-	-	-	-	-	-	-	-
AZ-31	-	1328	-	-	-	-	-	-	-	-	1332	-	-	-	-	-	-	-	-	-	-	-	1334	-	-	-	-	-	-	-	-	-	-
AZ-31B	-	-	-	-	-	-	-	-	-	-	1332	-	-	-	-	-	-	-	-	-	-	-	-	-	-	-	-	-	-	-	-	-	-
Barium	-	-	-	-	-	-	-	-	-	-	-	68	-	-	-	-	-	-	-	-	-	-	-	-	-	-	-	-	-	-	-	-	-
Beryllium	-	71	-	-	74 76	-	-	-	-	-	78	-	-	-	-	-	-	-	-	-	-	-	-	-	-	-	-	-	82	-	-	-	-
Beryllium extrusion #30	-	-	-	-	-	-	-	-	-	-	79	-	-	-	-	-	-	-	-	-	-	-	-	-	-	-	-	-	-	-	-	-	-
Beryllium + Iron + ΣX_i	-	-	-	-	-	-	-	-	-	-	-	1134	-	-	-	-	-	-	-	-	-	-	-	-	-	-	-	-	-	-	-	-	-
Bismuth	-	-	-	-	-	-	-	-	-	-	-	-	-	-	-	-	-	-	-	-	-	-	-	-	-	-	-	-	85	88	-	-	-
Bismuth + Tin	-	-	-	-	-	-	-	-	-	-	-	-	-	-	-	-	899	-	-	-	-	-	-	-	-	-	-	-	-	-	-	-	-
Brass	912	915	-	-	-	-	-	-	-	-	-	-	-	-	-	-	925	-	-	-	-	-	-	-	-	-	-	-	-	-	-	-	-
Brass, α	-	-	-	-	-	-	-	-	-	-	-	921	-	-	-	-	-	-	-	-	-	-	-	-	-	-	-	-	-	-	-	-	-
Brass, β	-	-	-	-	-	-	-	-	-	-	-	921	-	-	-	-	-	-	-	-	-	-	-	-	-	-	-	-	-	-	-	-	-
Brass, yellow	912	-	-	-	-	-	-	-	-	-	-	-	-	-	-	923	-	-	-	-	-	-	-	-	-	-	-	-	-	-	-	-	-
Bronze	-	-	-	-	1163	-	-	-	-	-	918 1167	-	-	-	-	-	-	-	-	-	-	-	1170	-	-	-	-	-	-	-	-	-	-

Material Name	EMITTANCE						REFLECTANCE									ABSORPTANCE									TRANSMITTANCE								
	Total			Spectral			Integrated			Spectral			Solar			Integrated			Spectral			Solar			Integrated			Spectral			Solar		
	Hemispherical	Normal	Angular	Hemispherical	Normal	Angular	Hemispherical	Normal	Angular	Hemispherical	Normal	Angular	Hemispherical	Normal	Angular	Hemispherical	Normal	Angular	Hemispherical	Normal	Angular	Hemispherical	Normal	Angular	Hemispherical	Normal	Angular	Hemispherical	Normal	Angular	Hemispherical	Normal	Angular
B. S. 1433 Copper	-	-	-	-	-	-	-	-	-	-	-	-	-	173	-	-	-	-	-	-	-	-	194	-	-	-	-	-	-	-	-	-	-
Cadmium	91	-	-	-	-	-	-	-	-	-	-	93	-	-	-	96	-	-	-	-	98	-	-	-	-	-	-	-	-	-	-	-	-
Chromium	101	103	-	-	106	-	-	-	-	-	110	113	-	-	-	-	-	-	-	115	-	-	118	-	-	-	-	-	120	-	-	-	-
Cobalt	-	123	-	-	126	-	-	-	-	-	-	132	-	-	-	-	-	-	-	-	-	-	-	-	-	-	-	-	-	-	-	-	-
Cobalt alloy N-155	-	1180 1181 1222 1226	-	-	1186 1236	-	-	-	-	-	1198 1266	-	-	-	-	-	-	-	-	-	-	-	1207 1301 1303	-	-	-	-	-	-	-	-	-	-
Cobalt + Aluminum	-	-	-	-	-	-	-	-	-	-	901	-	-	-	-	-	-	-	-	-	-	-	-	-	-	-	-	-	-	-	-	-	-
Cobalt + Chromium + ΣX_i	1138	1142	-	-	1145 1148	-	-	-	-	-	1152	-	-	-	-	-	-	-	-	-	-	-	-	-	-	-	-	-	-	-	-	-	-
Cobalt + Iron	-	-	-	-	904	-	-	-	-	-	-	-	-	-	-	-	-	-	-	-	-	-	-	-	-	-	-	-	-	-	-	-	-
Cobalt + Nickel	-	-	-	-	906	-	-	-	-	-	-	-	-	-	-	-	-	-	-	-	-	-	-	-	-	-	-	-	-	-	-	-	-
Cockron home foil	4 5	-	-	-	-	-	-	-	-	-	-	-	-	-	-	42 43	-	-	-	-	-	-	-	-	-	-	-	-	-	-	-	-	-
Copper	136	142	-	-	149 152	-	-	-	-	-	158	165 169	-	172	-	177	179	-	-	181 184	188	191	193	-	-	-	-	-	199	-	-	-	-
Copper, B. S. 1433	-	-	-	-	-	-	-	-	-	-	-	-	-	173 174	-	-	-	-	-	-	-	-	194	-	-	-	-	-	-	-	-	-	-
Copper, OFHC	138	-	-	-	-	-	-	-	-	-	-	-	-	-	-	-	-	-	-	-	189	-	-	-	-	-	-	-	-	-	-	-	-
Copper + Aluminum + ΣX_i	-	1159	-	-	1162	-	-	-	-	-	1166	-	-	-	-	-	-	-	-	-	-	-	1169	-	-	-	-	-	-	-	-	-	-
Copper + Nickel	-	908	-	-	-	-	-	-	-	-	-	-	-	-	-	-	-	-	-	-	-	-	-	-	-	-	-	-	-	-	-	-	-
Copper + Nickel + ΣX_i	-	-	-	-	-	-	-	-	-	-	1172	-	-	-	-	-	-	-	-	-	-	-	-	-	-	-	-	-	-	-	-	-	-
Copper + Tin	-	-	-	-	-	-	-	-	-	-	910	-	-	-	-	-	-	-	-	-	-	-	-	-	-	-	-	-	-	-	-	-	-
Copper + Tin + ΣX_i	-	-	-	-	-	-	-	-	-	-	1174	-	-	-	-	-	-	-	-	-	-	-	-	-	-	-	-	-	-	-	-	-	-
Copper + Zinc	912	914	-	-	-	-	-	-	-	-	917	920	-	-	-	923	925	-	-	-	928	-	-	-	-	-	-	-	-	-	-	-	-

Material Name	EMITTANCE						REFLECTANCE									ABSORPTANCE									TRANSMITTANCE								
	Total			Spectral			Integrated			Spectral			Solar			Integrated			Spectral			Solar			Integrated			Spectral			Solar		
	Hemispherical	Normal	Angular	Hemispherical	Normal	Angular	Hemispherical	Normal	Angular	Hemispherical	Normal	Angular	Hemispherical	Normal	Angular	Hemispherical	Normal	Angular	Hemispherical	Normal	Angular	Hemispherical	Normal	Angular	Hemispherical	Normal	Angular	Hemispherical	Normal	Angular	Hemispherical	Normal	Angular
D-43	1470	-	-	-	-	-	-	-	-	-	-	-	-	-	-	-	-	-	-	-	-	-	-	-	-	-	-	-	-	-	-	-	-
Erbium	-	-	-	-	202	-	-	-	-	-	-	-	-	-	-	-	-	-	-	-	-	-	-	-	-	-	-	-	-	-	-	-	-
Gadolinium	-	-	-	-	-	-	-	-	-	-	204	-	-	-	-	-	-	-	-	-	-	-	-	-	-	-	-	-	207	-	-	-	-
Gallium	-	-	-	-	-	-	-	-	-	-	210	-	-	-	-	-	-	-	-	213	-	-	-	-	-	-	-	-	216	-	-	-	-
Germanium	-	219	-	-	222 224	-	-	-	-	-	231	-	-	-	-	-	-	-	-	-	-	-	-	-	-	-	-	-	236	240	-	-	-
Gold	244	248	-	-	250 254	-	-	-	-	-	258	264	-	267	-	269	-	-	-	271 273	275	-	277	-	-	-	-	-	-	-	-	-	-
Gold + Silver	-	-	-	-	-	-	-	-	-	-	932	-	-	-	-	-	-	-	-	-	-	-	-	-	-	-	-	-	-	-	-	-	-
Hafnium	280	-	-	-	282 284	-	-	-	-	-	-	-	-	-	-	-	-	-	-	-	-	-	-	-	-	-	-	-	-	-	-	-	-
Hastelloy X	-	-	-	-	1364	-	-	-	-	-	1378	-	-	-	-	-	-	-	-	-	-	-	-	-	-	-	-	-	-	-	-	-	-
Haynes Alloy C	1444	-	-	-	-	-	-	-	-	-	-	-	-	-	-	-	-	-	-	-	-	-	-	-	-	-	-	-	-	-	-	-	-
Haynes Alloy No. 25	1140	1143	-	-	1146 1149	-	-	-	-	-	1154	-	-	-	-	-	-	-	-	-	-	-	-	-	-	-	-	-	-	-	-	-	-
Haynes Alloy N-155	-	1227	-	-	1238	-	-	-	-	-	-	-	-	-	-	-	-	-	-	-	-	-	-	-	-	-	-	-	-	-	-	-	-
Haynes Alloy X	1344	-	-	-	-	-	-	-	-	-	-	-	-	-	-	-	-	-	-	-	-	-	-	-	-	-	-	-	-	-	-	-	-
Hastelloy B	-	1448	-	-	1452	-	-	-	-	-	1455	-	-	-	-	-	-	-	-	-	-	-	1458	-	-	-	-	-	-	-	-	-	-
Hastelloy C	-	1448	-	-	1452	-	-	-	-	-	1455	-	-	-	-	-	-	-	-	-	-	-	1458	-	-	-	-	-	-	-	-	-	-
HK-31	-	1336	-	-	-	-	-	-	-	-	1338	-	-	-	-	-	-	-	-	-	-	-	1340	-	-	-	-	-	-	-	-	-	-
Hurwich home foil	4 5	-	-	-	-	-	-	-	-	-	-	-	-	-	-	42 43	-	-	-	-	-	-	-	-	-	-	-	-	-	-	-	-	-
Inconel	1344 1345	1351 1352	-	-	1363 1366	-	-	-	-	-	1382	-	-	-	-	-	-	-	-	1396	-	-	-	-	-	-	-	-	-	-	-	-	-

Material Name	EMITTANCE						REFLECTANCE									ABSORPTANCE									TRANSMITTANCE								
	Total			Spectral			Integrated			Spectral			Solar			Integrated			Spectral			Solar			Integrated			Spectral			Solar		
	Hemispherical	Normal	Angular	Hemispherical	Normal	Angular	Hemispherical	Normal	Angular	Hemispherical	Normal	Angular	Hemispherical	Normal	Angular	Hemispherical	Normal	Angular	Hemispherical	Normal	Angular	Hemispherical	Normal	Angular	Hemispherical	Normal	Angular	Hemispherical	Normal	Angular	Hemispherical	Normal	Angular
Inconel 600 (see Inconel)																																	
Incone 702	1345	–	–	–	1364	–	–	–	–	–	1378	–	–	–	–	–	–	–	–	–	–	–	–	–	–	–	–	–	–	–	–	–	–
Inconel B	–	1350	–	–	–	–	–	–	–	–	–	–	–	1393	–	–	–	–	–	–	–	–	1401	–	–	–	–	–	–	–	–	–	–
Inconel X	1344 1345	1351 1352 1353	–	–	1358 1365	–	–	–	–	–	1377 1379	–	–	1393	–	–	–	–	–	–	1398	–	1401 1402	–	–	–	–	–	–	–	–	–	–
Inconel X-750 (see Inconel X)																																	
Indium	–	–	–	–	–	–	–	–	–	–	–	–	–	–	–	–	–	–	–	–	286	–	–	–	–	–	–	–	–	–	–	–	–
Inor-8	1444	–	–	–	–	–	–	–	–	–	–	–	–	–	–	–	–	–	–	–	–	–	–	–	–	–	–	–	–	–	–	–	–
Iodide Hafnium	280	–	–	–	–	–	–	–	–	–	–	–	–	–	–	–	–	–	–	–	–	–	–	–	–	–	–	–	–	–	–	–	–
Iridium	–	–	–	–	289 291	–	–	–	–	–	294	297 299	–	–	–	–	–	–	–	–	–	–	–	–	–	–	–	–	–	–	–	–	–
Iron	302	306	–	–	310 316	–	–	–	–	–	319 321	324	–	–	–	–	–	–	–	327 329	–	–	332	–	–	–	–	–	–	–	–	–	–
~~Iron + Carbon~~		~~935~~			–	–	–	–	–	–	–	–	–	–	–	–	–	–	–	–	–	–	–	–	–	–	–	–	–	–	–	–	–
Iron + Chromium	–	–	–	–	–	–	–	–	–	–	938	940	–	–	–	–	–	–	–	–	–	–	–	–	–	–	–	–	–	–	–	–	–
Iron + Chromium + ΣX_i	–	1178	–	–	1184 1190	–	–	–	–	–	1196	–	–	–	–	–	–	–	–	1203	–	–	1206	–	–	–	–	–	–	–	–	–	–
Iron + Chromium + Nickel + ΣX_i	1210	1217	1231	–	1235 1242	1253 1256	–	–	–	–	1264	1283 1285	–	1287	–	1291	–	–	–	1293	1296	–	1299	–	–	–	–	–	–	–	–	–	–
Iron + Manganese + ΣX_i	1305	1307	–	–	–	–	–	–	–	–	–	–	–	–	–	–	–	–	–	–	–	–	1309	–	–	–	–	–	–	–	–	–	–
Iron + Nickel	–	–	–	–	942	–	–	–	–	–	–	–	–	–	–	–	–	–	–	–	–	–	–	–	–	–	–	–	–	–	–	–	–
Iron + Nickel + ΣX_i	–	–	–	–	–	–	–	–	–	–	1312	–	–	–	–	–	–	–	–	–	–	–	–	–	–	–	–	–	–	–	–	–	–
Iron + Nickel + Chromium + ΣX_i	1317	–	–	–	1320	–	–	–	–	–	1324	–	–	–	–	–	–	–	–	–	–	–	–	–	–	–	–	–	–	–	–	–	–

Material Name	EMITTANCE Total Hemispherical	Total Normal	Total Angular	Spectral Hemispherical	Spectral Normal	Spectral Angular
Iron + Tungsten	-	-	-	-	945	-
K Monel	-	1428	-	-	-	-
Kaiser foil	45	-	-	-	-	-
Kanthal	-	-	-	-	1192	-
Kanthal A	-	-	-	-	1192	-
Konel	1404	-	-	-	1407	-
Kovar	-	-	-	-	-	-
L34	-	-	-	-	-	-
L72	-	-	-	-	-	-
L120 Magnesium	-	-	-	-	-	-
Lead	335	337	-	-	-	-
Lead + Tin	948	-	-	-	-	-
Lutetium	-	-	-	-	-	-
M252	-	-	-	-	1366	-
Magnesium	353	-	-	-	-	-
Magnesium, L120	-	-	-	-	-	-
Magnesium alloys:						
AZ-31	-	1328	-	-	-	-
AZ-31B	-	-	-	-	-	-
HK-31	-	1336	-	-	-	-

Material Name	REFLECTANCE Integrated Hemispherical	Integrated Normal	Integrated Angular	Spectral Hemispherical	Spectral Normal	Spectral Angular	Solar Hemispherical	Solar Normal	Solar Angular
Iron + Tungsten	-	-	-	-	-	-	-	-	-
K Monel	-	-	-	-	-	-	-	1434	-
Kaiser foil	-	-	-	-	-	-	-	-	-
Kanthal	-	-	-	-	-	-	-	-	-
Kanthal A	-	-	-	-	-	-	-	-	-
Konel	-	-	-	-	-	-	-	-	-
Kovar	-	-	-	-	1313	-	-	-	-
L34	-	-	-	-	1092	-	-	1095	-
L72	-	-	-	-	-	-	-	1084	-
L120 Magnesium	-	-	-	-	-	-	-	361 362	-
Lead	-	-	-	-	-	-	-	-	-
Lead + Tin	-	-	-	-	-	-	-	-	-
Lutetium	-	-	-	-	347	-	-	-	-
M252	-	-	-	-	1382	-	-	-	-
Magnesium	-	-	-	-	356	358	-	360	-
Magnesium, L120	-	-	-	-	-	-	-	361 362	-
Magnesium alloys:									
AZ-31	-	-	-	-	1332	-	-	-	-
AZ-31B	-	-	-	-	1332	-	-	-	-
HK-31	-	-	-	-	1338	-	-	-	-

Material Name	ABSORPTANCE Integrated Hemispherical	Integrated Normal	Integrated Angular	Spectral Hemispherical	Spectral Normal	Spectral Angular	Solar Hemispherical	Solar Normal	Solar Angular
Iron + Tungsten	-	-	-	-	-	-	-	-	-
K Monel	-	-	-	-	-	-	-	1437	-
Kaiser foil	42	-	-	-	-	-	-	-	-
Kanthal	-	-	-	-	1204	-	-	-	-
Kanthal A	-	-	-	-	-	-	-	-	-
Konel	-	-	-	-	-	-	-	-	-
Kovar	-	-	-	-	-	-	-	-	-
L34	-	-	-	-	-	-	-	-	-
L72	-	-	-	-	-	-	-	1087	-
L120 Magnesium	-	-	-	-	-	-	-	365	-
Lead	339	341	-	-	343	345	-	-	-
Lead + Tin	-	-	-	-	-	-	-	-	-
Lutetium	-	-	-	-	-	-	-	-	-
M252	-	-	-	-	-	-	-	-	-
Magnesium	-	-	-	-	-	-	-	364	-
Magnesium, L120	-	-	-	-	-	-	-	365	-
Magnesium alloys:									
AZ-31	-	-	-	-	-	-	-	1334	-
AZ-31B	-	-	-	-	-	-	-	-	-
HK-31	-	-	-	-	-	-	-	1340	-

Material Name	TRANSMITTANCE Integrated Hemispherical	Integrated Normal	Integrated Angular	Spectral Hemispherical	Spectral Normal	Spectral Angular	Solar Hemispherical	Solar Normal	Solar Angular
Iron + Tungsten	-	-	-	-	-	-	-	-	-
K Monel	-	-	-	-	-	-	-	-	-
Kaiser foil	-	-	-	-	-	-	-	-	-
Kanthal	-	-	-	-	-	-	-	-	-
Kanthal A	-	-	-	-	-	-	-	-	-
Konel	-	-	-	-	-	-	-	-	-
Kovar	-	-	-	-	-	-	-	-	-
L34	-	-	-	-	-	-	-	-	-
L72	-	-	-	-	-	-	-	-	-
L120 Magnesium	-	-	-	-	-	-	-	-	-
Lead	-	-	-	-	-	-	-	-	-
Lead + Tin	-	-	-	-	-	-	-	-	-
Lutetium	-	-	-	-	350	-	-	-	-
M252	-	-	-	-	-	-	-	-	-
Magnesium	-	-	-	-	367	-	-	-	-
Magnesium, L120	-	-	-	-	-	-	-	-	-
Magnesium alloys:									
AZ-31	-	-	-	-	-	-	-	-	-
AZ-31B	-	-	-	-	-	-	-	-	-
HK-31	-	-	-	-	-	-	-	-	-

Material Name	EMITTANCE						REFLECTANCE									ABSORPTANCE									TRANSMITTANCE								
	Total			Spectral			Integrated			Spectral			Solar			Integrated			Spectral			Solar			Integrated			Spectral			Solar		
	Hemispherical	Normal	Angular	Hemispherical	Normal	Angular	Hemispherical	Normal	Angular	Hemispherical	Normal	Angular	Hemispherical	Normal	Angular	Hemispherical	Normal	Angular	Hemispherical	Normal	Angular	Hemispherical	Normal	Angular	Hemispherical	Normal	Angular	Hemispherical	Normal	Angular	Hemispherical	Normal	Angular
Magnesium + Aluminum . . .	-	-	-	-	-	-	-	-	-	-	950	-	-	-	-	-	-	-	-	-	-	-	-	-	-	-	-	-	-	-	-	-	-
Magnesium + Aluminum + ΣX_i	-	1327	-	-	-	-	-	-	-	-	1330	-	-	-	-	-	-	-	-	-	-	-	1334	-	-	-	-	-	-	-	-	-	-
Magnesium + Thorium + ΣX_i	-	1336	-	-	-	-	-	-	-	-	1338	-	-	-	-	-	-	-	-	-	-	-	1340	-	-	-	-	-	-	-	-	-	-
Manganese	-	-	-	-	369	-	-	-	-	-	-	371	-	-	-	-	-	-	-	-	374	-	-	-	-	-	-	-	-	-	-	-	-
Mild steel	1305	1307	-	-	-	-	-	-	-	-	-	-	-	-	-	-	-	-	-	-	-	-	1310	-	-	-	-	-	-	-	-	-	-
Mo + 0.5 Ti alloy	-	954	957	-	-	-	-	-	-	-	-	-	-	-	-	-	-	-	-	-	-	-	-	-	-	-	-	-	-	-	-	-	-
Molybdenum	376	383	-	-	387 392	-	-	-	-	-	398	402	-	-	-	-	-	-	-	404	407	-	410	-	-	-	-	-	-	-	-	-	-
Molybdenum + Titanium . . .	-	953	956	-	959	-	-	-	-	-	962	-	-	-	-	-	-	-	-	-	-	-	-	-	-	-	-	-	-	-	-	-	-
Molybdenum + Tungsten . . .	967	-	-	-	969	-	-	-	-	-	-	-	-	-	-	-	-	-	-	-	-	-	-	-	-	-	-	-	-	-	-	-	-
Monel	1423	1428	-	-	-	-	-	-	-	-	1431	-	-	-	-	-	-	-	-	-	-	-	-	-	-	-	-	-	-	-	-	-	-
Monel 400 (see Monel)																																	
Monel K-500 (see K Monel)																																	
Monel 501 (see KR Monel)																																	
Multimet	1214	1227	-	-	1238	-	-	-	-	-	-	-	-	-	-	-	-	-	-	-	-	-	-	-	-	-	-	-	-	-	-	-	-
Nb - 1 Zr alloy	990	-	-	-	-	-	-	-	-	-	-	-	-	-	-	-	-	-	-	-	-	-	-	-	-	-	-	-	-	-	-	-	-
Nichrome	1344	1353	-	-	-	-	-	-	-	-	-	-	-	-	-	-	-	-	-	-	-	-	-	-	-	-	-	-	-	-	-	-	-
Nickel	413	416	-	-	424 434	-	-	440	446	-	454	457	-	-	-	460	-	-	-	462	465	468	470	-	-	-	-	-	-	-	-	-	-
Nickel, Grade A	-	418 419 420	-	-	426 427 428 429	-	-	-	-	-	455	-	-	-	-	-	-	-	-	-	-	-	471	-	-	-	-	-	-	-	-	-	-

Material Name	EMITTANCE						REFLECTANCE									ABSORPTANCE									TRANSMITTANCE								
	Total			Spectral			Integrated			Spectral			Solar			Integrated			Spectral			Solar			Integrated			Spectral			Solar		
	Hemispherical	Normal	Angular	Hemispherical	Normal	Angular	Hemispherical	Normal	Angular	Hemispherical	Normal	Angular	Hemispherical	Normal	Angular	Hemispherical	Normal	Angular	Hemispherical	Normal	Angular	Hemispherical	Normal	Angular	Hemispherical	Normal	Angular	Hemispherical	Normal	Angular	Hemispherical	Normal	Angular
Nickel, Grade NP-3	-	-	-	-	435	-	-	-	-	-	-	-	-	-	-	-	-	-	-	-	-	-	-	-	-	-	-	-	-	-	-	-	-
Nickel + Chromium	-	-	-	-	972	-	-	-	-	-	-	-	-	-	-	-	-	-	-	-	-	-	-	-	-	-	-	-	-	-	-	-	-
Nickel + Chromium + ΣX_i	1342	1347	-	-	1356 1360	-	-	-	-	-	1374	-	-	1392	-	-	-	-	-	1395	1398	-	1400	-	-	-	-	-	-	-	-	-	-
Nickel + Cobalt + ΣX_i	1404	-	-	-	1406 1410	-	-	-	-	-	1416	-	-	-	-	-	-	-	-	-	-	-	-	-	-	-	-	-	-	-	-	-	-
Nickel + Copper + ΣX_i	1423	1426	-	-	-	-	-	-	-	-	1430	-	-	1433	-	-	-	-	-	-	-	-	1436	-	-	-	-	-	-	-	-	-	-
Nickel + Iron	-	-	-	-	976	-	-	-	-	-	-	-	-	-	-	-	-	-	-	-	-	-	-	-	-	-	-	-	-	-	-	-	-
Nickel + Iron + ΣX_i	-	-	-	-	1439	-	-	-	-	-	-	-	-	-	-	-	-	-	-	-	-	-	-	-	-	-	-	-	-	-	-	-	-
Nickel + Molybdenum + ΣX_i	1442	1446	-	-	1450	-	-	-	-	-	1454	-	-	-	-	-	-	-	-	-	-	-	1457	-	-	-	-	-	-	-	-	-	-
~~Nickel + Silver~~	-	979	-	-	-	-	-	-	-	-	-	-	-	-	-	-	-	-	-	-	-	-	-	-	-	-	-	-	-	-	-	-	-
Nimonic 75	1345	1350	-	-	-	-	-	-	-	-	-	-	-	-	-	-	-	-	-	-	-	-	-	-	-	-	-	-	-	-	-	-	-
Niobium	474	480	-	-	482 486	-	-	-	-	-	492	-	-	-	-	-	-	-	-	-	497	-	-	-	-	-	-	-	-	-	-	-	-
Niobium + Molybdenum + ΣX_i	-	1460	-	-	-	-	-	-	-	-	-	-	-	-	-	-	-	-	-	-	-	-	-	-	-	-	-	-	-	-	-	-	-
Niobium + Tantalum + ΣX_i	1463	-	-	-	1466	-	-	-	-	-	-	-	-	-	-	-	-	-	-	-	-	-	-	-	-	-	-	-	-	-	-	-	-
Niobium + Tungsten	981	-	-	-	984	-	-	-	-	-	-	-	-	-	-	-	-	-	-	-	-	-	-	-	-	-	-	-	-	-	-	-	-
Niobium + Tungsten + ΣX_i	1469	-	-	-	-	-	-	-	-	-	-	-	-	-	-	-	-	-	-	-	-	-	-	-	-	-	-	-	-	-	-	-	-
Niobium + Zirconium	988	992	-	-	994	-	-	-	-	-	-	-	-	-	-	-	-	-	-	-	-	-	-	-	-	-	-	-	-	-	-	-	-
OFHC Copper	138	-	-	-	-	-	-	-	-	-	-	-	-	-	-	-	-	-	-	-	189	-	-	-	-	-	-	-	-	-	-	-	-
Osmium	-	-	-	-	500	-	-	-	-	-	-	-	-	-	-	-	-	-	-	-	-	-	-	-	-	-	-	-	-	-	-	-	-
Palladium	502	504	-	-	507 510	-	-	-	-	-	512	-	-	-	-	-	-	-	-	515	-	-	518	-	-	-	-	-	520	-	-	-	-

Material Name	EMITTANCE						REFLECTANCE									ABSORPTANCE									TRANSMITTANCE								
	Total			Spectral			Integrated			Spectral			Solar			Integrated			Spectral			Solar			Integrated			Spectral			Solar		
	Hemispherical	Normal	Angular	Hemispherical	Normal	Angular	Hemispherical	Normal	Angular	Hemispherical	Normal	Angular	Hemispherical	Normal	Angular	Hemispherical	Normal	Angular	Hemispherical	Normal	Angular	Hemispherical	Normal	Angular	Hemispherical	Normal	Angular	Hemispherical	Normal	Angular	Hemispherical	Normal	Angular
Phosphor bronze	-	-	-	-	-	-	-	-	-	-	1175	-	-	-	-	-	-	-	-	-	-	-	-	-	-	-	-	-	-	-	-	-	-
Platinum	524	529	-	-	532 536	-	-	-	-	-	544	547 549	-	-	-	-	-	-	-	551	554	-	557	-	-	-	-	-	-	-	-	-	-
Platinum, NBS	-	-	-	-	538	-	-	-	-	-	-	-	-	-	-	-	-	-	-	552	-	-	-	-	-	-	-	-	-	-	-	-	-
Platinum wire, grade MPTU 4292-53	526	-	-	-	-	-	-	-	-	-	-	-	-	-	-	-	-	-	-	-	-	-	-	-	-	-	-	-	-	-	-	-	-
Platinum + Rhodium	997	1000	-	-	1004	-	-	-	-	-	-	-	-	-	-	-	-	-	-	-	-	-	-	-	-	-	-	-	-	-	-	-	-
Poroloy	-	1226	-	-	-	-	-	-	-	-	-	-	-	-	-	-	-	-	-	-	-	-	1303	-	-	-	-	-	-	-	-	-	-
Potomac A	-	-	-	-	1192	-	-	-	-	-	1198	-	-	-	-	-	-	-	-	-	-	-	-	-	-	-	-	-	-	-	-	-	-
Pt-10 Rh alloy	997	-	-	-	-	-	-	-	-	-	-	-	-	-	-	-	-	-	-	-	-	-	-	-	-	-	-	-	-	-	-	-	-
Pt-13 Rh alloy	-	1002	-	-	1005	-	-	-	-	-	-	-	-	-	-	-	-	-	-	-	-	-	-	-	-	-	-	-	-	-	-	-	-
Rene' 41	-	1353	-	-	1365 1366	-	-	-	-	-	1380 1382	-	-	-	-	-	-	-	-	-	-	-	-	-	-	-	-	-	-	-	-	-	-
Reynolds Wrap	-	-	-	-	-	-	-	-	-	-	-	-	-	40	-	-	-	-	-	-	-	-	-	55	-	-	-	-	-	-	-	-	-
Rhenium	559	562	-	-	565 568	-	-	-	-	-	-	-	-	-	-	-	-	-	-	-	-	-	-	-	-	-	-	-	-	-	-	-	-
Rhodium	571	573	-	-	576 579	-	-	-	-	-	581	584	-	-	-	-	-	-	-	-	587	-	589	-	-	-	-	-	-	-	-	-	-
Ruthenium	-	-	-	-	591	-	-	-	-	-	-	-	-	-	-	-	-	-	-	-	-	-	-	-	-	-	-	-	-	-	-	-	-
Silicon	-	593	-	-	595 598	-	-	-	-	-	604	611	-	-	-	-	-	-	-	614	-	-	-	-	-	-	-	-	616	-	-	-	-
Silver	620	623	-	-	625 627	-	-	-	-	-	630	636	-	-	-	639	-	-	-	641 643	645	-	648	-	-	-	-	-	651	-	-	-	-
Silver + Aluminum	-	-	-	-	-	-	-	-	-	-	1007	1009	-	-	-	-	-	-	-	-	-	-	-	-	-	-	-	-	-	-	-	-	-
Silver + Beryllium	-	-	-	-	-	-	-	-	-	-	-	1012	-	-	-	-	-	-	-	-	-	-	-	-	-	-	-	-	-	-	-	-	-
Silver + Cadmium + ΣX_i	-	-	-	-	-	-	-	-	-	-	-	1472	-	-	-	-	-	-	-	-	-	-	-	-	-	-	-	-	-	-	-	-	-

Material Name	EMITTANCE						REFLECTANCE									ABSORPTANCE									TRANSMITTANCE								
	Total			Spectral			Integrated			Spectral			Solar			Integrated			Spectral			Solar			Integrated			Spectral			Solar		
	Hemispherical	Normal	Angular	Hemispherical	Normal	Angular	Hemispherical	Normal	Angular	Hemispherical	Normal	Angular	Hemispherical	Normal	Angular	Hemispherical	Normal	Angular	Hemispherical	Normal	Angular	Hemispherical	Normal	Angular	Hemispherical	Normal	Angular	Hemispherical	Normal	Angular	Hemispherical	Normal	Angular
Silver + Copper + ΣX_i	-	-	-	-	-	-	-	-	-	-	-	1475	-	-	-	-	-	-	-	-	-	-	-	-	-	-	-	-	-	-	-	-	-
Silver + Gold	-	-	-	-	-	-	-	-	-	-	1015	-	-	-	-	-	-	-	-	-	-	-	-	-	-	-	-	-	-	-	-	-	-
Silver + Silicon	-	-	-	-	-	-	-	-	-	-	-	1018	-	-	-	-	-	-	-	-	-	-	-	-	-	-	-	-	-	-	-	-	-
Silver + Zinc + ΣX_i	-	-	-	-	-	-	-	-	-	-	-	1478	-	-	-	-	-	-	-	-	-	-	-	-	-	-	-	-	-	-	-	-	-
Solder	948	-	-	-	-	-	-	-	-	-	-	-	-	-	-	-	-	-	-	-	-	-	-	-	-	-	-	-	-	-	-	-	-
Stainless Steels:																																	
17-7 PH	-	1223	-	-	1237 1245	-	-	-	-	-	1266 1268	-	-	-	-	-	-	-	-	-	-	-	1302	-	-	-	-	-	-	-	-	-	-
18-8	1212 1214	1225 1226	-	-	-	-	-	-	-	-	-	-	-	-	-	-	-	-	-	-	-	-	1303	-	-	-	-	-	-	-	-	-	-
301	-	1221 1226	-	-	-	-	-	-	-	-	1269	-	-	1288	-	-	-	-	-	-	-	-	1300	-	-	-	-	-	-	-	-	-	-
302	1212 1213	-	-	-	-	-	-	-	-	-	-	-	-	-	-	1291	-	-	-	-	-	-	-	-	-	-	-	-	-	-	-	-	-
303	1212 1213	1226	-	-	-	1254 1258 1259 1260	-	-	-	-	-	-	-	-	-	-	-	-	-	-	1297	-	-	-	-	-	-	-	-	-	-	-	-
304	1213	1227	-	-	1244	-	-	-	-	-	1270	-	-	-	-	-	-	-	-	-	-	-	-	-	-	-	-	-	-	-	-	-	-
304 ELC	1213	-	-	-	-	-	-	-	-	-	-	-	-	-	-	-	-	-	-	-	-	-	-	-	-	-	-	-	-	-	-	-	-
310	1212 1213	-	-	-	-	-	-	-	-	-	-	-	-	-	-	-	-	-	-	-	-	-	-	-	-	-	-	-	-	-	-	-	-
316	-	1221 1224	-	-	1237 1244	-	-	-	-	-	1266 1270 1271	-	-	1288	-	-	-	-	-	-	-	-	1300 1302	-	-	-	-	-	-	-	-	-	-
321	-	1224 1225 1226	-	-	1237 1238 1244 1246	-	-	-	-	-	1266 1267 1270 1272	1285	-	-	-	-	-	-	-	1294	-	-	1302 1303	-	-	-	-	-	-	-	-	-	-
347	1212	1222	-	-	-	-	-	-	-	-	-	-	-	1288	-	-	-	-	-	-	-	-	1301	-	-	-	-	-	-	-	-	-	-

Material Name	EMITTANCE						REFLECTANCE									ABSORPTANCE									TRANSMITTANCE								
	Total			Spectral			Integrated			Spectral			Solar			Integrated			Spectral			Solar			Integrated			Spectral			Solar		
	Hemispherical	Normal	Angular	Hemispherical	Normal	Angular	Hemispherical	Normal	Angular	Hemispherical	Normal	Angular	Hemispherical	Normal	Angular	Hemispherical	Normal	Angular	Hemispherical	Normal	Angular	Hemispherical	Normal	Angular	Hemispherical	Normal	Angular	Hemispherical	Normal	Angular	Hemispherical	Normal	Angular
Stainless Steels (continued)																																				
430	-	-	-	-	1193	-	-	-	-	-	-	-	-	-	-	-	-	-	-	-	-	-	-	-	-	-	-	-	-	-	-	-	-	-	-	-
446	-	1180	-	-	1187	-	-	-	-	-	1198	-	-	-	-	-	-	-	-	-	-	-	1207	-	-	-	-	-	-	-	-	-	-	-	-	-
A286	-	-	-	-	1322	-	-	-	-	-	1325	-	-	-	-	-	-	-	-	-	-	-	-	-	-	-	-	-	-	-	-	-	-	-	-	-
AM 350	-	1225	-	-	1238 1245	-	-	-	-	-	1266 1267 1268	-	-	-	-	-	-	-	-	-	-	-	1302 1303	-	-	-	-	-	-	-	-	-	-	-	-	-
PH 15-7 Mo	-	1223	-	-	1237 1245	-	-	-	-	-	1266 1267	-	-	-	-	-	-	-	-	-	-	-	1301	-	-	-	-	-	-	-	-	-	-	-	-	-
SF 11	-	-	-	-	-	-	-	-	-	-	938	-	-	-	-	-	-	-	-	-	-	-	-	-	-	-	-	-	-	-	-	-	-	-	-	-
SF 20	-	-	-	-	-	-	-	-	-	-	1266	-	-	-	-	-	-	-	-	-	-	-	-	-	-	-	-	-	-	-	-	-	-	-	-	-
Stellite	-	-	-	-	-	-	-	-	-	-	1154	-	-	-	-	-	-	-	-	-	-	-	-	-	-	-	-	-	-	-	-	-	-	-	-	-
T-111	1481	-	-	-	-	-	-	-	-	-	-	-	-	-	-	-	-	-	-	-	-	-	-	-	-	-	-	-	-	-	-	-	-	-	-	-
Tantalum	654	661	-	-	666 672	-	-	-	-	-	678	684	-	-	-	-	-	-	-	-	-	-	687	-	-	-	-	-	-	-	-	-	-	-	-	-
Tantalum + Tungsten	-	1021	-	-	1024	-	-	-	-	-	-	-	-	-	-	-	-	-	-	-	-	-	-	-	-	-	-	-	-	-	-	-	-	-	-	-
Tantalum + Tungsten + ΣX_i . .	1481	-	-	-	-	-	-	-	-	-	-	-	-	-	-	-	-	-	-	-	-	-	-	-	-	-	-	-	-	-	-	-	-	-	-	-
Thallium.	-	-	-	-	-	-	-	-	-	-	690	693	-	-	-	-	-	-	-	-	-	-	-	-	-	-	-	-	696	-	-	-	-	-	-	-
Thorium	-	-	-	-	699 701	-	-	-	-	-	-	-	-	-	-	-	-	-	-	-	-	-	-	-	-	-	-	-	-	-	-	-	-	-	-	-
Tin	703	705	-	-	-	-	-	-	-	-	707	-	-	-	-	710	712	-	-	714	717	-	-	-	-	-	-	-	720	-	-	-	-	-	-	-
Tin + Indium	-	-	-	-	-	-	-	-	-	-	-	-	-	-	-	-	1026	-	-	-	-	-	-	-	-	-	-	-	-	-	-	-	-	-	-	-
Titanium.	723	726	-	-	729 732	735 738	-	-	-	-	744	751	-	769	-	-	-	-	-	-	771	-	-	-	-	-	-	-	773	-	-	-	-	-	-	-
Titanium, AMS 4901	-	727	-	-	-	-	-	-	-	-	746	-	-	-	-	-	-	-	-	-	-	-	-	-	-	-	-	-	-	-	-	-	-	-	-	-
Titanium, Ti-75 A	724	727	-	-	-	-	-	-	-	-	746	-	-	-	-	-	-	-	-	-	-	-	-	-	-	-	-	-	-	-	-	-	-	-	-	-

	EMITTANCE						REFLECTANCE									ABSORPTANCE									TRANSMITTANCE								
	Total			Spectral			Integrated			Spectral			Solar			Integrated			Spectral			Solar			Integrated			Spectral			Solar		
Material Name	Hemispherical	Normal	Angular	Hemispherical	Normal	Angular	Hemispherical	Normal	Angular	Hemispherical	Normal	Angular	Hemispherical	Normal	Angular	Hemispherical	Normal	Angular	Hemispherical	Normal	Angular	Hemispherical	Normal	Angular	Hemispherical	Normal	Angular	Hemispherical	Normal	Angular	Hemispherical	Normal	Angular
Titanium alloys:																																	
6Al-4V	1484	1488	-	-	1492 1493	-	-	-	-	-	1498	-	-	-	-	-	-	-	-	-	-	-	1501	-	-	-	-	-	-	-	-	-	-
A-110-AT	-	1488	-	-	1492	-	-	-	-	-	1498	-	-	-	-	-	-	-	-	-	-	-	1501	-	-	-	-	-	-	-	-	-	-
AMS 4908	-	1030	-	-	-	-	-	-	-	-	1038	-	-	-	-	-	-	-	-	-	-	-	-	-	-	-	-	-	-	-	-	-	-
B120 VAC	-	-	-	-	-	-	-	-	-	-	1509	-	-	-	-	-	-	-	-	-	-	-	-	-	-	-	-	-	-	-	-	-	-
C-110 M	-	1030	-	-	1034	-	-	-	-	-	1038	-	-	-	-	-	-	-	-	-	-	-	1041	-	-	-	-	-	-	-	-	-	-
RS-120	1503	1506	-	-	-	-	-	-	-	-	-	-	-	-	-	-	-	-	-	-	-	-	-	-	-	-	-	-	-	-	-	-	-
TA5E	1484	-	-	-	-	-	-	-	-	-	-	-	-	-	-	-	-	-	-	-	-	-	-	-	-	-	-	-	-	-	-	-	-
TA6V	1484	-	-	-	-	-	-	-	-	-	-	-	-	-	-	-	-	-	-	-	-	-	-	-	-	-	-	-	-	-	-	-	-
Titanium + Aluminum + ΣX_i . .	1483	1486	-	-	1490	-	-	-	-	-	1497	-	-	-	-	-	-	-	-	-	-	-	1500	-	-	-	-	-	-	-	-	-	-
Titanium + Manganese	-	1028	-	-	1032	-	-	-	-	-	1037	-	-	-	-	-	-	-	-	-	-	-	1041	-	-	-	-	-	-	-	-	-	-
Titanium + Manganese + ΣX_i . .	1503	1505	-	-	-	-	-	-	-	-	-	-	-	-	-	-	-	-	-	-	-	-	-	-	-	-	-	-	-	-	-	-	-
Titanium + Vanadium + ΣX_i . .	-	-	-	-	-	-	-	-	-	-	1508	-	-	-	-	-	-	-	-	-	-	-	-	-	-	-	-	-	-	-	-	-	-
Tungsten	776	782	-	-	790 796	808 810	-	-	812	-	814 819	823	-	-	-	-	-	-	-	-	825	-	-	-	-	-	-	-	-	-	-	-	-
Tungsten alloys:																																	
VR-27-VT	1049	-	-	-	-	-	-	-	-	-	-	-	-	-	-	-	-	-	-	-	-	-	-	-	-	-	-	-	-	-	-	-	-
Tungsten + Molybdenum . . .	1043	-	-	-	1045	-	-	-	-	-	-	-	-	-	-	-	-	-	-	-	-	-	-	-	-	-	-	-	-	-	-	-	-
Tungsten + Rhenium	1048	-	-	-	1051	-	-	-	-	-	-	-	-	-	-	-	-	-	-	-	-	-	-	-	-	-	-	-	-	-	-	-	-
Udimet 500	-	-	-	-	1365 1412	-	-	-	-	-	1381 1418	-	-	-	-	-	-	-	-	-	-	-	-	-	-	-	-	-	-	-	-	-	-
Unitemp No. 41	-	936 1352	-	-	-	-	-	-	-	-	-	-	-	-	-	-	-	-	-	-	-	-	-	-	-	-	-	-	-	-	-	-	-
Uranium	828	-	-	-	834 838	-	-	-	-	-	-	-	-	-	-	-	-	-	-	-	-	-	-	-	-	-	-	-	-	-	-	-	-

Material Name	EMITTANCE						REFLECTANCE									ABSORPTANCE									TRANSMITTANCE								
	Total			Spectral			Integrated			Spectral			Solar			Integrated			Spectral			Solar			Integrated			Spectral			Solar		
	Hemispherical	Normal	Angular	Hemispherical	Normal	Angular	Hemispherical	Normal	Angular	Hemispherical	Normal	Angular	Hemispherical	Normal	Angular	Hemispherical	Normal	Angular	Hemispherical	Normal	Angular	Hemispherical	Normal	Angular	Hemispherical	Normal	Angular	Hemispherical	Normal	Angular	Hemispherical	Normal	Angular
Uranium + Niobium	1053	-	-	-	1056	-	-	-	-	-	-	-	-	-	-	-	-	-	-	-	-	-	-	-	-	-	-	-	-	-	-	-	-
Uranium + Zirconium + ΣX_i . .	1511	-	-	-	1514	-	-	-	-	-	-	-	-	-	-	-	-	-	-	-	-	-	-	-	-	-	-	-	-	-	-	-	-
Uranus 10	-	-	-	-	-	-	-	-	-	-	-	1283	-	-	-	-	-	-	-	-	-	-	-	-	-	-	-	-	-	-	-	-	-
Vanadium	-	-	-	-	840	-	-	-	-	-	844	848	-	-	-	-	-	-	-	-	850	-	-	-	-	-	-	-	-	-	-	-	-
Vascojet 1000	-	-	-	-	1192	-	-	-	-	-	1199	-	-	-	-	-	-	-	-	-	-	-	-	-	-	-	-	-	-	-	-	-	-
Vickers F. D. P.	-	1221	-	-	-	-	-	-	-	-	-	-	-	-	-	-	-	-	-	-	-	-	-	-	-	-	-	-	-	-	-	-	-
VR-27-VT	1049	-	-	-	-	-	-	-	-	-	-	-	-	-	-	-	-	-	-	-	-	-	-	-	-	-	-	-	-	-	-	-	-
Yellow brass	912	-	-	-	-	-	-	-	-	-	-	-	-	-	-	923	-	-	-	-	-	-	-	-	-	-	-	-	-	-	-	-	-
Yttrium	-	-	-	-	853	-	-	-	-	-	-	-	-	-	-	-	-	-	-	-	-	-	-	-	-	-	-	-	-	-	-	-	-
Zinc	855	857	-	-	-	-	-	-	-	-	860	-	-	-	-	864	-	-	-	866	869	-	-	-	-	-	-	-	-	-	-	-	-
Zinc + Aluminum	-	-	-	-	-	-	-	-	-	-	1059	-	-	-	-	-	-	-	-	-	-	-	-	-	-	-	-	-	-	-	-	-	-
Zircalloy 2	-	1520	-	-	1523	-	-	-	-	-	-	-	-	-	-	-	-	-	-	-	-	-	-	-	-	-	-	-	-	-	-	-	-
Zirconium	-	872	-	-	875 878	-	-	-	-	-	881	-	-	-	-	-	-	-	-	-	883	-	-	-	-	-	-	-	-	-	-	-	-
Zirconium + Hafnium + ΣX_i . .	-	-	-	-	1517	-	-	-	-	-	-	-	-	-	-	-	-	-	-	-	-	-	-	-	-	-	-	-	-	-	-	-	-
Zirconium + Tin + ΣX_i	-	1519	-	-	1522	-	-	-	-	-	-	-	-	-	-	-	-	-	-	-	-	-	-	-	-	-	-	-	-	-	-	-	-
Zirconium + Uranium + ΣX_i . .	1525	-	-	-	1528	-	-	-	-	-	-	-	-	-	-	-	-	-	-	-	-	-	-	-	-	-	-	-	-	-	-	-	-